IMPORTANT:

HERE IS YOUR REGISTRATION CODE TO ACCESS
YOUR PREMIUM McGRAW-HILL ONLINE RESOURCES.

For key premium online resources you need THIS CODE to gain access. Once the code is entered, you will be able to use the Web resources for the length of your course.

If your course is using **WebCT** or **Blackboard**, you'll be able to use this code to access the McGraw-Hill content within your instructor's online course.

Access is provided if you have purchased a new book. If the registration code is missing from this book, the registration screen on our Website, and within your WebCT or Blackboard course, will tell you how to obtain your new code.

Registering for McGraw-Hill Online Resources

TO gain access to your McGraw-Hill web resources simply follow the steps below:

1. USE YOUR WEB BROWSER TO GO TO: **http://www.mhhe.com/shier10**

2. CLICK ON **FIRST TIME USER**.

3. ENTER THE REGISTRATION CODE* PRINTED ON THE TEAR-OFF BOOKMARK ON THE RIGHT.

4. AFTER YOU HAVE ENTERED YOUR REGISTRATION CODE, CLICK **REGISTER**.

5. FOLLOW THE INSTRUCTIONS TO SET-UP YOUR PERSONAL UserID AND PASSWORD.

6. WRITE YOUR UserID AND PASSWORD DOWN FOR FUTURE REFERENCE.
KEEP IT IN A SAFE PLACE.

TO GAIN ACCESS to the McGraw-Hill content in your instructor's **WebCT** or **Blackboard** course simply log in to the course with the UserID and Password provided by your instructor. Enter the registration code exactly as it appears in the box to the right when prompted by the system. You will only need to use the code the first time you click on McGraw-Hill content.

Thank you, and welcome to your McGraw-Hill online Resources!

*YOUR REGISTRATION CODE CAN BE USED ONLY ONCE TO ESTABLISH ACCESS. IT IS NOT TRANSFERABLE.

0-07-291931-0 SHIER: HOLE'S HUMAN ANATOMY AND PHYSIOLOGY, 10E

REGISTRATION CODE

8MV0-AQAY-LJUP-MK7A-L8D2

Hole's Human Anatomy & Physiology

TENTH EDITION

DAVID SHIER
Washtenaw Community College

JACKIE BUTLER
Grayson County College

RICKI LEWIS
Contributing Editor to The Scientist

Boston   Burr Ridge, IL   Dubuque, IA   Madison, WI   New York   San Francisco   St. Louis
Bangkok   Bogotá   Caracas   Kuala Lumpur   Lisbon   London   Madrid   Mexico City
Milan   Montreal   New Delhi   Santiago   Seoul   Singapore   Sydney   Taipei   Toronto

HOLE'S HUMAN ANATOMY & PHYSIOLOGY, TENTH EDITION

Published by McGraw-Hill, a business unit of The McGraw-Hill Companies, Inc., 1221 Avenue of the Americas, New York, NY 10020. Copyright © 2004, 2002, 1999, 1996 by The McGraw-Hill Companies, Inc. All rights reserved. No part of this publication may be reproduced or distributed in any form or by any means, or stored in a database or retrieval system, without the prior written consent of The McGraw-Hill Companies, Inc., including, but not limited to, in any network or other electronic storage or transmission, or broadcast for distance learning.

Some ancillaries, including electronic and print components, may not be available to customers outside the United States.

This book is printed on acid-free paper.

4 5 6 7 8 9 0 QPD/QPD 0 9 8 7 6 5

ISBN 0–07–291932–9

Publisher: *Martin J. Lange*
Sponsoring editor: *Michelle Watnick*
Senior developmental editor: *Patricia Hesse*
Director of development: *Kristine Tibbetts*
Marketing manager: *James F. Connely*
Senior project manager: *Jayne Klein*
Lead production supervisor: *Sandy Ludovissy*
Media project manager: *Sandra M. Schnee*
Senior media technology producer: *Barbara R. Block*
Designer: *K. Wayne Harms*
Cover/interior designer: *Christopher Reese*
Cover image: *Hoby Finn/Gettyimages*
Senior photo research coordinator: *John Leland*
Photo research: *Billie Porter*
Supplement producer: *Brenda A. Ernzen*
Compositor: *Precision Graphics*
Typeface: *10/12 Melior*
Printer: *Quebecor World Dubuque*

The credits section for this book begins on page 1007 and is considered an extension of the copyright page.

Library of Congress Cataloging-in-Publication Data

Shier, David.
   Hole's human anatomy & physiology / David Shier, Jackie Butler,
Ricki Lewis. — 10th ed.
      p.   cm.
Includes index.
ISBN 0–07–291932–9 (hard copy : alk. paper)
 1. Human physiology. 2. Human anatomy. I. Title: Hole's human
anatomy and physiology. II. Title: Human anatomy & physiology.
III. Title.

QP34.5 .H63   2004
612—dc21   2002015462
                                      CIP

www.mhhe.com

Brief Contents

UNIT ONE

LEVELS OF ORGANIZATION   1

1   Introduction to Human Anatomy and Physiology   1
2   Chemical Basis of Life   37
3   Cells   61
4   Cellular Metabolism   103
5   Tissues   131

UNIT TWO

SUPPORT AND MOVEMENT   157

6   Skin and the Integumentary System   157
7   Skeletal System   181
8   Joints of the Skeletal System   253
9   Muscular System   277

UNIT THREE

INTEGRATION AND COORDINATION   337

10   Nervous System I:
       Basic Structure and Function   337
11   Nervous System II:
       Divisions of the Nervous System   365
12   Somatic and Special Senses   421
13   Endocrine System   467

UNIT FOUR

TRANSPORT   509

14   Blood   509
15   Cardiovascular System   541
16   Lymphatic System and Immunity   607

UNIT FIVE

ABSORPTION AND EXCRETION   643

17   Digestive System   643
18   Nutrition and Metabolism   693
19   Respiratory System   731
20   Urinary System   771
21   Water, Electrolyte, and Acid-Base Balance   807

UNIT SIX

THE HUMAN LIFE CYCLE   829

22   Reproductive Systems   829
23   Pregnancy, Growth, and Development   875
24   Genetics and Genomics   919

Contents

Clinical Connections xi
About the Authors xiii
Preface xiv
The Evolution of a Classic xxiv

UNIT ONE — LEVELS OF ORGANIZATION

CHAPTER 1

Introduction to Human Anatomy and Physiology

Anatomy and Physiology   3
Levels of Organization   4
Characteristics of Life   6
Maintenance of Life   7
Organization of the Human Body   12
Life-Span Changes   19
Anatomical Terminology   21
Some Medical and Applied Sciences   25
CHAPTER SUMMARY   26
CRITICAL THINKING QUESTIONS   27
REVIEW EXERCISES   28
WEB CONNECTIONS   28

CLINICAL APPLICATIONS

1.1   ULTRASONOGRAPHY AND MAGNETIC RESONANCE IMAGING: A TALE OF TWO PATIENTS   6

REFERENCE PLATES

The Human Organism   29

CHAPTER 2

Chemical Basis of Life

Structure of Matter   38
Chemical Constituents of Cells   47
CHAPTER SUMMARY   59
CRITICAL THINKING QUESTIONS   60
REVIEW EXERCISES   60
WEB CONNECTIONS   60

CLINICAL APPLICATIONS

2.1   RADIOACTIVE ISOTOPES REVEAL PHYSIOLOGY   41
2.2   IONIZING RADIATION: FROM THE COLD WAR TO YUCCA MOUNTAIN   45
2.3   CT SCANNING AND PET IMAGING   56

CHAPTER 3

Cells

A Composite Cell   62
Movements Into and Out of the Cell   80
The Cell Cycle   90
Control of Cell Division   92
Stem and Progenitor Cells   94
CHAPTER SUMMARY   99
CRITICAL THINKING QUESTIONS   101
REVIEW EXERCISES   101
WEB CONNECTIONS   102

CLINICAL APPLICATIONS

3.1   FAULTY ION CHANNELS CAUSE DISEASE   68
3.2   THE BLOOD-BRAIN BARRIER   69
3.3   DISEASE AT THE ORGANELLE LEVEL   75

FROM SCIENCE TO TECHNOLOGY

3.1   CLONING TO PRODUCE THERAPEUTIC STEM CELLS   98

C H A P T E R 4

Cellular Metabolism

Metabolic Processes   104
Control of Metabolic Reactions   106
Energy for Metabolic Reactions   107
Cellular Respiration   108
Nucleic Acids and Protein Synthesis   115
Changes in Genetic Information   123

CHAPTER SUMMARY   128
CRITICAL THINKING QUESTIONS   129
REVIEW EXERCISES   130
WEB CONNECTIONS   130

CLINICAL APPLICATIONS

4.1   OVERRIDING A BLOCK IN GLYCOLYSIS   110
4.2   PHENYLKETONURIA   127

FROM SCIENCE TO TECHNOLOGY

4.1   DNA MAKES HISTORY   118
4.2   GENE AMPLIFICATION   124

C H A P T E R 5

Tissues

Epithelial Tissues   133
Connective Tissues   141
Muscle Tissues   148
Nervous Tissues   151
Types of Membranes   152

CHAPTER SUMMARY   153
CRITICAL THINKING QUESTIONS   155
REVIEW EXERCISES   155
WEB CONNECTIONS   156

CLINICAL APPLICATIONS

5.1   ABNORMALITIES OF COLLAGEN   143

FROM SCIENCE TO TECHNOLOGY

5.1   TISSUE ENGINEERING   153

UNIT TWO   SUPPORT AND MOVEMENT

C H A P T E R 6

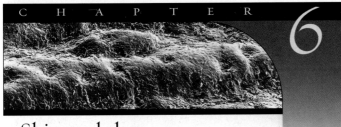

Skin and the Integumentary System

Skin and Its Tissues   158
Accessory Organs of the Skin   165
Regulation of Body Temperature   169
Skin Color   172
Healing of Wounds and Burns   173
Life-Span Changes   175
Common Skin Disorders   176

CHAPTER SUMMARY   178
CRITICAL THINKING QUESTIONS   179
REVIEW EXERCISES   179
WEB CONNECTIONS   180

CLINICAL APPLICATIONS

6.1   SKIN CANCER   163
6.2   HAIR LOSS   166
6.3   ACNE   169
6.4   ELEVATED BODY TEMPERATURE   171

C H A P T E R 7

Skeletal System

Bone Structure   182
Bone Development and Growth   186
Bone Function   191
Skeletal Organization   196
Skull   199
Vertebral Column   209
Thoracic Cage   216
Pectoral Girdle   218
Upper Limb   220
Pelvic Girdle   224
Lower Limb   227
Life-Span Changes   231
Clinical Terms Related to the Skeletal System   232

CHAPTER SUMMARY   234
CRITICAL THINKING QUESTIONS   236
REVIEW EXERCISES   237
WEB CONNECTIONS   237

CLINICAL APPLICATIONS

7.1   FRACTURES   192
7.2   OSTEOPOROSIS   195
7.3   DISORDERS OF THE VERTEBRAL COLUMN   216

REFERENCE PLATES

Human Skull   238

CHAPTER 8

Joints of the Skeletal System

Classification of Joints   254
General Structure of a Synovial Joint   257
Types of Synovial Joints   259
Types of Joint Movements   261
Examples of Synovial Joints   264
Life-Span Changes   271
Clinical Terms Related to Joints   274

CHAPTER SUMMARY   274
CRITICAL THINKING QUESTIONS   276
REVIEW EXERCISES   276
WEB CONNECTIONS   276

CLINICAL APPLICATIONS

8.1   REPLACING JOINTS   269
8.2   JOINT DISORDERS   272

CHAPTER 9

Muscular System

Structure of a Skeletal Muscle   278
Skeletal Muscle Contraction   282
Muscular Responses   290
Smooth Muscles   293
Cardiac Muscle   294
Skeletal Muscle Actions   296
Major Skeletal Muscles   297
Life-Span Changes   325
Clinical Terms Related to the Muscular System   327

CHAPTER SUMMARY   327
CRITICAL THINKING QUESTIONS   330
REVIEW EXERCISES   330
WEB CONNECTIONS   331

CLINICAL APPLICATIONS

9.1   MYASTHENIA GRAVIS   284
9.2   USE AND DISUSE OF SKELETAL MUSCLES   293
9.3   TMJ SYNDROME   300

REFERENCE PLATES

Surface Anatomy   332

UNIT THREE   INTEGRATION AND COORDINATION

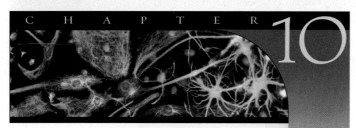

CHAPTER 10

Nervous System I: Basic Structure and Function

General Functions of the Nervous System   338
Classification of Neurons and Neuroglial Cells   343
Cell Membrane Potential   348
The Synapse   354
Impulse Processing   358

CHAPTER SUMMARY   363
CRITICAL THINKING QUESTIONS   364
REVIEW EXERCISES   364
WEB CONNECTIONS   364

CLINICAL APPLICATIONS

10.1   MIGRAINE   340
10.2   MULTIPLE SCLEROSIS   343
10.3   FACTORS AFFECTING IMPULSE CONDUCTION   355
10.4   OPIATES IN THE HUMAN BODY   359
10.5   DRUG ADDICTION   360

CHAPTER 11

Nervous System II: Divisions of the Nervous System

Meninges   367
Ventricles and Cerebrospinal Fluid   368
Spinal Cord   372
Brain   381
Peripheral Nervous System   395
Autonomic Nervous System   407
Life-Span Changes   416
Clinical Terms Related to the Nervous System   416

CHAPTER SUMMARY   417
CRITICAL THINKING QUESTIONS   419
REVIEW EXERCISES   419
WEB CONNECTIONS   420

CLINICAL APPLICATIONS

11.1   CEREBROSPINAL FLUID PRESSURE   371
11.2   USES OF REFLEXES   377
11.3   SPINAL CORD INJURIES   380
11.4   CEREBRAL INJURIES AND ABNORMALITIES   388
11.5   PARKINSON DISEASE   390
11.6   BRAIN WAVES   396
11.7   SPINAL NERVE INJURIES   408

CHAPTER 12

Somatic and Special Senses

Receptors and Sensations   422

Somatic Senses   424

Special Senses   430

Life-Span Changes   462

Clinical Terms Related to the Senses   462

CHAPTER SUMMARY   463

CRITICAL THINKING QUESTIONS   465

REVIEW EXERCISES   466

WEB CONNECTIONS   466

CLINICAL APPLICATIONS

12.1   CANCER PAIN AND CHRONIC PAIN   428

12.2   MIXED-UP SENSES—SYNESTHESIA   430

12.3   SMELL AND TASTE DISORDERS   435

12.4   HEARING LOSS   443

12.5   REFRACTION DISORDERS   457

CHAPTER 13

Endocrine System

General Characteristics of the Endocrine System   468

Hormone Action   469

Control of Hormonal Secretions   477

Pituitary Gland   478

Thyroid Gland   485

Parathyroid Glands   489

Adrenal Glands   490

Pancreas   496

Other Endocrine Glands   498

Stress and Its Effects   500

Life-Span Changes   502

Clinical Terms Related to the Endocrine System   504

CHAPTER SUMMARY   504

CRITICAL THINKING QUESTIONS   506

REVIEW EXERCISES   507

WEB CONNECTIONS   507

CLINICAL APPLICATIONS

13.1   USING HORMONES TO IMPROVE ATHLETIC PERFORMANCE   474

13.2   GROWTH HORMONE UPS AND DOWNS   482

13.3   DISORDERS OF THE ADRENAL CORTEX   496

13.4   DIABETES MELLITUS   499

UNIT FOUR TRANSPORT

CHAPTER 14

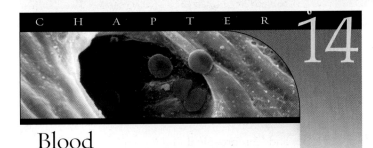

Blood

Blood and Blood Cells   510

Blood Plasma   523

Hemostasis   526

Blood Groups and Transfusions   531

Clinical Terms Related to the Blood   537

CHAPTER SUMMARY   537

CRITICAL THINKING QUESTIONS   539

REVIEW EXERCISES   540

WEB CONNECTIONS   540

CLINICAL APPLICATIONS

14.1   KING GEORGE III AND PORPHYRIA VARIEGATA   517

14.2   LEUKEMIA   522

14.3   THE RETURN OF THE MEDICINAL LEECH   532

14.4   LIVING WITH HEMOPHILIA   533

CHAPTER 15

Cardiovascular System

Structure of the Heart   542

Heart Actions   551

Blood Vessels   562

Blood Pressure   570

Paths of Circulation   580

Arterial System   582

Venous System   591

Life-Span Changes   598
Clinical Terms Related to the Cardiovascular System   601

CHAPTER SUMMARY   603
CRITICAL THINKING QUESTIONS   605
REVIEW EXERCISES   605
WEB CONNECTIONS   606

CLINICAL APPLICATIONS

15.1   ARRHYTHMIAS   562
15.2   BLOOD VESSEL DISORDERS   571
15.3   MEASUREMENT OF ARTERIAL BLOOD
         PRESSURE   572
15.4   SPACE MEDICINE   574
15.5   HYPERTENSION   578
15.6   EXERCISE AND THE CARDIOVASCULAR
         SYSTEM   580
15.7   MOLECULAR CAUSES OF CARDIOVASCULAR
         DISEASE   598
15.8   CORONARY ARTERY DISEASE   600

FROM SCIENCE TO TECHNOLOGY

15.1   REPLACING THE HEART   556
15.2   ALTERING ANGIOGENESIS   564

Lymphatic System and Immunity

Lymphatic Pathways   608
Tissue Fluid and Lymph   611
Lymph Movement   612
Lymph Nodes   612
Thymus and Spleen   614
Body Defenses Against Infection   616
Innate (Nonspecific) Defenses   617
Adaptive (Specific) Defenses or Immunity   619
Life-Span Changes   634
Clinical Terms Related to the Lymphatic System
         and Immunity   637

CHAPTER SUMMARY   637
CRITICAL THINKING QUESTIONS   640
REVIEW EXERCISES   640
WEB CONNECTIONS   641

CLINICAL APPLICATIONS

16.1   IMMUNITY BREAKDOWN: AIDS   635

FROM SCIENCE TO TECHNOLOGY

16.1   IMMUNOTHERAPY   624

UNIT FIVE   ABSORPTION AND EXCRETION

Digestive System

General Characteristics of the Alimentary Canal   644
Mouth   648
Salivary Glands   652
Pharynx and Esophagus   655
Stomach   658
Pancreas   664
Liver   666

Small Intestine   673
Large Intestine   680
Life-Span Changes   686
Clinical Terms Related to the Digestive System   686

CHAPTER SUMMARY   688
CRITICAL THINKING QUESTIONS   690
REVIEW EXERCISES   691
WEB CONNECTIONS   691

CLINICAL APPLICATIONS

17.1   DENTAL CARIES   653
17.2   OH, MY ACHING STOMACH!   663
17.3   HEPATITIS   670
17.4   GALLBLADDER DISEASE   672
17.5   DISORDERS OF THE LARGE INTESTINE   684

FROM SCIENCE TO TECHNOLOGY

17.1   REPLACING THE LIVER   669

CHAPTER 18

Nutrition and Metabolism

Why We Eat   694
Carbohydrates   695
Lipids   697
Proteins   699
Energy Expenditures   702
Vitamins   705
Minerals   712
Healthy Eating   718
Life-Span Changes   724
Clinical Terms Related to Nutrition and Metabolism   725

CHAPTER SUMMARY   725
CRITICAL THINKING QUESTIONS   728
REVIEW EXERCISES   729
WEB CONNECTIONS   729

CLINICAL APPLICATIONS

18.1   OBESITY   704
18.2   DIETARY SUPPLEMENTS—PROCEED WITH CAUTION   720
18.3   NUTRITION AND THE ATHLETE   722

CHAPTER 19

Respiratory System

Why We Breathe   732
Organs of the Respiratory System   733
Breathing Mechanism   745
Control of Breathing   753
Alveolar Gas Exchanges   757
Gas Transport   760
Life-Span Changes   764
Clinical Terms Related to the Respiratory System   764

CHAPTER SUMMARY   766
CRITICAL THINKING QUESTIONS   768
REVIEW EXERCISES   768
WEB CONNECTIONS   769

CLINICAL APPLICATIONS

19.1   THE EFFECTS OF CIGARETTE SMOKING ON THE RESPIRATORY SYSTEM   735
19.2   LUNG IRRITANTS   745
19.3   RESPIRATORY DISORDERS THAT DECREASE VENTILATION: BRONCHIAL ASTHMA AND EMPHYSEMA   752
19.4   EXERCISE AND BREATHING   756
19.5   DISORDERS THAT IMPAIR GAS EXCHANGE: PNEUMONIA, TUBERCULOSIS, AND ADULT RESPIRATORY DISTRESS SYNDROME   759
19.6   EFFECTS OF HIGH ALTITUDE   760

CHAPTER 20

Urinary System

Kidneys   772
Urine Formation   782
Elimination of Urine   795
Life-Span Changes   800
Clinical Terms Related to the Urinary System   802

CHAPTER SUMMARY   802
CRITICAL THINKING QUESTIONS   804
REVIEW EXERCISES   804
WEB CONNECTIONS   805

CLINICAL APPLICATIONS

20.1   CHRONIC KIDNEY FAILURE   776
20.2   GLOMERULONEPHRITIS   780
20.3   THE NEPHROTIC SYNDROME   789
20.4   URINALYSIS: CLUES TO HEALTH   799

CHAPTER 21

Water, Electrolyte, and Acid-Base Balance

Distribution of Body Fluids   808
Water Balance   811
Electrolyte Balance   813
Acid-Base Balance   817
Clinical Terms Related to Water, Electrolyte, and Acid-Base Balance   824

CHAPTER SUMMARY   825
CRITICAL THINKING QUESTIONS   826
REVIEW EXERCISES   826
WEB CONNECTIONS   827

CLINICAL APPLICATIONS

21.1   WATER BALANCE DISORDERS   814
21.2   SODIUM AND POTASSIUM IMBALANCES   818
21.3   ACID-BASE IMBALANCES   822

CONTENTS

CHAPTER 22

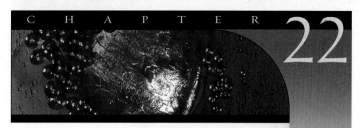

Reproductive Systems

Organs of the Male Reproductive System   830
Testes   830
Male Internal Accessory Organs   838
Male External Reproductive Organs   841
Hormonal Control of Male Reproductive Functions   845
Organs of the Female Reproductive System   846
Ovaries   848
Female Internal Accessory Organs   852
Female External Reproductive Organs   855
Hormonal Control of Female Reproductive
    Functions   857
Mammary Glands   860
Birth Control   862
Sexually Transmitted Diseases   867
Clinical Terms Related to the Reproductive Systems   868

CHAPTER SUMMARY   870
CRITICAL THINKING QUESTIONS   873
REVIEW EXERCISES   873
WEB CONNECTIONS   874

CLINICAL APPLICATIONS
22.1   PROSTATE ENLARGEMENT   840
22.2   MALE INFERTILITY   842
22.3   FEMALE INFERTILITY   861
22.4   TREATING BREAST CANCER   864

CHAPTER 23

Pregnancy, Growth, and Development

Pregnancy   876
Prenatal Period   880
Postnatal Period   905
Aging   912
Clinical Terms Related to Human Growth
    and Development   914

CHAPTER SUMMARY   914
CRITICAL THINKING QUESTIONS   917
REVIEW EXERCISES   917
WEB CONNECTIONS   917

CLINICAL APPLICATIONS
23.1   SOME CAUSES OF BIRTH DEFECTS   894
23.2   JOINED FOR LIFE   902
23.3   HUMAN MILK-THE PERFECT FOOD FOR HUMAN
    BABIES   907
23.4   OLD BEFORE THEIR TIME   913

FROM SCIENCE TO TECHNOLOGY
23.1   ASSISTED REPRODUCTIVE TECHNOLOGIES   878
23.2   PREIMPLANTATION GENETIC DIAGNOSIS   882

CHAPTER 24

Genetics and Genomics

The Emerging Role of Genetics and Genomics
    in Medicine   921
Modes of Inheritance   924
Gene Expression   928
Complex Traits   929
Matters of Sex   931
Chromosome Disorders   934
Gene Therapy   937

CHAPTER SUMMARY   943
CRITICAL THINKING QUESTIONS   944
REVIEW EXERCISES   945
WEB CONNECTIONS   945

CLINICAL APPLICATIONS
24.1   IT'S ALL IN THE GENES   924
24.2   DOWN SYNDROME   936
24.3   GENE THERAPY SUCCESSES AND SETBACKS   940

REFERENCE PLATES
Human Cadavers   947

APPENDIX A
    Periodic Table of Elements   969

APPENDIX B
    Laboratory Tests of Clinical Importance   970

APPENDIX C
    A Closer Look at Cellular Reproduction   974

APPENDIX D
    A Closer Look at DNA Structures   978

Glossary   981
Credits   1007
Index   1011

Clinical Connections

Clinical Applications and From Science to Technology

CHAPTER 1
1.1: *Ultrasonography and Magnetic Resonance Imaging: A Tale of Two Patients* 6

CHAPTER 2
2.1: *Radioactive Isotopes Reveal Physiology* 41
2.2: *Ionizing Radiation: From the Cold War to Yucca Mountain* 45
2.3: *CT Scanning and PET Imaging* 56

CHAPTER 3
3.1: *Faulty Ion Channels Cause Disease* 68
3.2: *The Blood-Brain Barrier* 69
3.3: *Disease at the Organelle Level* 75

3.1: *Cloning to Produce Therapeutic Stem Cells* 98

CHAPTER 4
4.1: *Overriding a Block in Glycolysis* 110
4.2: *Phenylketonuria* 127

4.1: *DNA Makes History* 118
4.2: *Gene Amplification* 124

CHAPTER 5
5.1: *Abnormalities of Collagen* 143

5.1: *Tissue Engineering* 153

CHAPTER 6
6.1: *Skin Cancer* 163
6.2: *Hair Loss* 166
6.3: *Acne* 169
6.4: *Elevated Body Temperature* 171

CHAPTER 7
7.1: *Fractures* 192
7.2: *Osteoporosis* 195
7.3: *Disorders of the Vertebral Column* 216

CHAPTER 8
8.1: *Replacing Joints* 269
8.2: *Joint Disorders* 272

CHAPTER 9
9.1: *Myasthenia Gravis* 284
9.2: *Use and Disuse of Skeletal Muscles* 293
9.3: *TMJ Syndrome* 300

CHAPTER 10
10.1: *Migraine* 340
10.2: *Multiple Sclerosis* 343
10.3: *Factors Affecting Impulse Conduction* 355
10.4: *Opiates in the Human Body* 359
10.5: *Drug Addiction* 360

CHAPTER 11
11.1: *Cerebrospinal Fluid Pressure* 371
11.2: *Uses of Reflexes* 377
11.3: *Spinal Cord Injuries* 380
11.4: *Cerebral Injuries and Abnormalities* 388
11.5: *Parkinson Disease* 390
11.6: *Brain Waves* 396
11.7: *Spinal Nerve Injuries* 408

CHAPTER 12
12.1: *Cancer Pain and Chronic Pain* 428
12.2: *Mixed-up Senses—Synesthesia* 430
12.3: *Smell and Taste Disorders* 435
12.4: *Hearing Loss* 443
12.5: *Refraction Disorders* 457

CHAPTER 13
13.1: *Using Hormones to Improve Athletic Performance* 474
13.2: *Growth Hormone Ups and Downs* 482
13.3: *Disorders of the Adrenal Cortex* 496
13.4: *Diabetes Mellitus* 499

CHAPTER 14
14.1: *King George III and Porphyria Variegata* 517
14.2: *Leukemia* 522
14.3: *The Return of the Medicinal Leech* 532
14.4: *Living with Hemophilia* 533

CHAPTER 15
15.1: *Arrhythmias* 562
15.2: *Blood Vessel Disorders* 571
15.3: *Measurement of Arterial Blood Pressure* 572
15.4: *Space Medicine* 574
15.5: *Hypertension* 578
15.6: *Exercise and the Cardiovascular System* 580
15.7: *Molecular Causes of Cardiovascular Disease* 598
15.8: *Coronary Artery Disease* 600

15.1: *Replacing the Heart* 556
15.2: *Angiogenesis* 564

CHAPTER 16
16.1: *Immunity Breakdown: AIDS* 635

16.1: *Immunotherapy* 624

CHAPTER 17
17.1: *Dental Caries* 653
17.2: *Oh, My Aching Stomach!* 663
17.3: *Hepatitis* 670
17.4: *Gallbladder Disease* 672
17.5: *Disorders of the Large Intestine* 684

17.1: *Replacing the Liver* 669

CHAPTER 18
18.1: *Obesity* 704
18.2: *Dietary Supplements—Proceed with Caution* 720
18.3: *Nutrition and the Athlete* 722

CHAPTER 19
19.1: *The Effects of Cigarette Smoking on the Respiratory System* 735
19.2: *Lung Irritants* 745
19.3: *Respiratory Disorders that Decrease Ventilation: Bronchial Asthma and Emphysema* 752
19.4: *Exercise and Breathing* 756
19.5: *Disorders That Impair Gas Exchange: Pneumonia, Tuberculosis, and Adult Respiratory Distress Syndrome* 759
19.6: *Effects of High Altitude* 760

CHAPTER 20
    20.1: *Chronic Kidney Failure*   *776*
    20.2: *Glomerulonephritis*   *780*
    20.3: *The Nephrotic Syndrome*   *789*
    20.4: *Urinalysis: Clues to Health*   *799*

CHAPTER 21
    21.1: *Water Balance Disorders*   *814*
    21.2: *Sodium and Potassium Imbalances*   *818*
    21.3: *Acid-Base Imbalances*   *822*

CHAPTER 22
    22.1: *Prostate Enlargement*   *840*
    22.2: *Male Infertility*   *842*
    22.3: *Female Infertility*   *861*
    22.4: *Treating Breast Cancer*   *864*

CHAPTER 23
    23.1: *Some Causes of Birth Defects*   *894*
    23.2: *Joined for Life*   *902*
    23.3: *Human Milk-The Perfect Food for Human Babies*   *907*
    23.4: *Old Before Their Time*   *913*

    23.1: *Assisted Reproductive Technologies*   *878*
    23.2: *Preimplantation Genetic Diagnosis*   *882*

CHAPTER 24
    24.1: *It's All in the Genes*   *924*
    24.2: *Down Syndrome*   *936*
    24.3: *Gene Therapy Successes and Setbacks*   *940*

Life-Span Changes

Aging Process   *19*
Aging-Related Changes in the Skin   *175*
Aging-Related Changes in the Skeletal System   *231*
Joint Stiffness   *271*
Signs of Aging in the Muscular System   *325*
Physical and Functional Signs of an Aging Nervous System   *416*
Aging and Diminished Senses   *462*
Changes in the Glands of the Endocrine System   *502*
Aging-Related Changes to the Cardiovascular System   *598*
Aging of the Immune System   *634*
Changes to the Digestive System   *686*
Aging and Changing Nutrition   *724*
Aging-Related Changes in the Respiratory System   *764*
Changes in Structure and Function of the Urinary System   *800*

Clinical Terminology

Clinical Terms Related to the Skeletal System   *232*
Clinical Terms Related to Joints   *274*
Clinical Terms Related to the Muscular System   *327*
Clinical Terms Related to the Nervous System   *416*
Clinical Terms Related to the Senses   *462*
Clinical Terms Related to the Endocrine System   *504*
Clinical Terms Related to the Blood   *537*
Clinical Terms Related to the Cardiovascular System   *601*
Clinical Terms Related to the Lymphatic System and Immunity   *637*
Clinical Terms Related to the Digestive System   *686*
Clinical Terms Related to Nutrition and Metabolism   *725*
Clinical Terms Related to the Respiratory System   *764*
Clinical Terms Related to the Urinary System   *802*
Clinical Terms Related to Water, Electrolyte, and Acid-Base Balance   *824*
Clinical Terms Related to the Reproductive Systems   *868*
Clinical Terms Related to Human Growth and Development   *914*

CLINICAL CONNECTIONS

About the Authors

David Shier

David Shier has accumulated twenty-seven years of experience teaching anatomy and physiology, primarily to premedical, nursing, dental, and allied health students. He has effectively incorporated his extensive teaching experience into another student-friendly revision of *Hole's Human Anatomy & Physiology* and *Hole's Essentials of Human Anatomy and Physiology.* David has published numerous papers and abstracts in the areas of renal and cardiovascular physiology, the endocrinology of fluid and electrolyte balance, and hypertension. A faculty member in the Life Science Department at Washtenaw Community College, he is actively involved in a number of projects dealing with assessment, articulation, and the incorporation of technology into instructional design. David holds a Ph.D. in physiology from the University of Michigan.

Jackie Butler

Jackie Butler's professional background includes work at the University of Texas Health Science Center conducting research about the genetics of bilateral retinoblastoma. She later worked at Houston's M. D. Anderson Hospital conducting research on remission in leukemia patients. Now a popular educator at Grayson County College, Jackie teaches microbiology and human anatomy and physiology for health science majors. Her experience and work with students of various educational backgrounds have contributed significantly to another revision of *Hole's Human Anatomy & Physiology* and *Hole's Essentials of Human Anatomy and Physiology.* Jackie Butler received her B.S. and M.S. degrees from Texas A&M University, focusing on microbiology, including courses in immunology and epidemiology.

Ricki Lewis

Ricki Lewis, author of the McGraw-Hill textbooks *Life* and *Human Genetics,* combines the skills of scientist and journalist. Since earning her Ph.D. in genetics from Indiana University in 1980, she has published more than 3,000 articles in scientific and popular publications. Today Ricki contributes regularly to *The Scientist* and *Biophotonics International,* and has published an essay collection, *Discovery: Windows on the Life Sciences.* She is a genetic counselor for a private medical practice in upstate New York. Ricki brings a molecular, cellular, and genetics perspective, with a journalistic flair, to *Hole's Human Anatomy & Physiology* and *Hole's Essentials of Human Anatomy and Physiology.*

Ricki Lewis, David Shier, Jackie Butler

Preface

The Evolution of a Classic

In biological evolution, a population of organisms changes over time. Molded by natural selection, a successful species becomes the best suited that it can be for a particular environment. In a similar manner, this textbook has evolved over the past quarter century.

From its beginnings as a clear, concise, and exciting grand tour of the human body, John Hole's *Human Anatomy & Physiology* has matured into a modern exploration of the human, from its interacting organ systems to the cellular and molecular underpinnings of the functions of life. In our preface to the seventh edition, when we came on board to continue Dr. Hole's legacy, we termed his work "a classic." That it certainly is, with over one million copies sold worldwide over its 25-year history.

Dr. John W. Hole, Jr.

Dr. Hole tells of his book's origin:

"When I began teaching human anatomy and physiology 35 years ago, the nation was entering an era of increased space exploration, advances in civil rights, and influences of the women's movement. In the 1970s, the floppy disc appeared, rocks were the pets of choice, and *Star Wars* transported us to a galaxy far, far away. Despite the advances made during this era, the available anatomy and physiology textbooks were lacking in some of the features I felt were desirable for my students.

The first edition of Hole's Human Anatomy & Physiology, published in 1978, reflected our efforts to prepare a textbook that would engage students and involve them actively in the learning process. The text included information of particular interest to allied health students and devices to help them relate their classroom knowledge to their future clinical practice. Boxed information illustrated how theory is applied to clinical practice, lists of terms and word parts expanded understanding of technical and medical terminology, and review activities within as well as at the end of each chapter aided the reader in evaluating his or her progress in achieving the chapter objectives.

As I think about the many years of work involved in preparing the first edition of the textbook, I am reminded of how much of it was a team effort, and I will be forever grateful for the help and support from all who were part of the text's development and production. With each edition, the current authors continue to include, expand and improve the features that define this text."

Success came quickly for *Hole's Human Anatomy & Physiology*. One early adopter wrote, "I think it is one of the finest books of its kind I have ever seen. It is an excellent teaching text, the organization is superb, and its explanatory style is highly effective." Such praise is rare indeed for a first edition. By fall 1978, sales confirmed that John Hole's approach had struck a chord, and the publisher declared the textbook "an overwhelming success." Work began on the second edition, and the success exploded. With each revision, the textbook grew. Much of the black-and-white art evolved into full color, and certain chapters underwent a binary fission of sorts, the nervous system expanding into two chapters, and bones and joints given their own turf. New clinical case studies, practical applications, and laboratory applications continued to complement the trademark of clear explanations.

When we took over with the seventh edition a decade ago, space travel had become more common place, pet rocks had vanished, and the Internet was beginning to link us all together. Powerful imaging technologies added new views of anatomy and physiology, as nonstop discoveries in molecular and cellular biology and genetics revealed the mechanisms behind body functions. To embrace new knowledge while at the same time making the material accessible, we introduced a personal touch—compelling vignettes to open chapters and more tales of real people. Coverage of pathology ranged from the tragic to the commonplace to the quirky, usually offset in small boxes or sidebars so as not to interrupt the narrative flow. We delved into historical anecdotes where appropriate for understanding the present, while introducing new biomedical technologies. At the same time, increased coverage of homeostasis and a new feature to end the systems chapters, called InnerConnections, wove the text into a tighter fabric.

Other changes streamlined the learning process. We reorganized the chapter sequence, and placed the clinical case studies, practical applications, and laboratory applications under the umbrella heading of clinical applications. Improvements in art, text, as well as content updating, continued through the eighth edition. The ninth edition introduced a "life-span changes" section at the ends of the systems chapters, and a "reconnect" feature throughout to help the reader integrate the information, and more extensive on-line student resources. The final chapter evolved to become "Genetics and Genomics" to reflect the sequencing of the human genome and the emergence of a new field.

Just as world events helped to inspire the first edition of the book, so too have they influenced this anniversary edition. The vignette for the integumentary system chapter addresses a possible reintroduction of smallpox; that for the respiratory system chapter examines air quality concerns at the World Trade Center site in the months following September 11, 2001. This edition also introduces a developmental backdrop by considering how stem cells contribute to tissues, including two spectacularly redone illustrations, vignettes, a basic section in the Cells chapter, and relevant mentions throughout. Stem cells also star in three of the *From Science to Technology* boxes, which highlight the origins of medical and biotechnologies.

Audience

The tenth edition brings new awareness and reveals a new set of rules. In our evolution as authors we are evolving as teachers. What we and our reviewers do in class is reflected more in this than in previous editions. Students have always come first in our approach to teaching and textbook authoring, but we now feel more excited than ever about the student-oriented, teacher-friendly quality of this text. We have never included detail for its own sake, but we have felt free to include extra detail if the end result is to clarify.

The level of this text is geared toward students in two-semester courses in anatomy and physiology who are pursuing careers in allied health fields and who have minimal background in physical and biological sciences. The first four chapters cover the chemistry and processes. Students who have studied this material previously will view it as a welcomed review, but newcomers will not find it intimidating.

What's New?

Over 25 years have passed, and *Hole's Human Anatomy & Physiology* is still *Hole's Human Anatomy & Physiology*—but with a sharper focus and appearance.

- **Design**—The revitalized text design injects new life into the study of Anatomy and Physiology. Bright, bold, modern colors are used throughout the feature boxes, tables, and chapter openers, making them easy to recognize.
- **Illustrations**—All illustrations have been revised. New art incorporates cutting-edge technology offering vivid depictions of complex processes while maintaining the conceptual base that has established Hole as the most effective "instructional tool" on the market, with a unique focus on the fundamentals. Hole's art focuses on the main concepts by using concise labeling methodology that keeps students from getting bogged down with excessive detail. Difficult concepts are broken down into easy-to-understand illustrations.

- **Chapter Openers**—Chapter opener images give you a closer look inside the wonders of the human body through the technology of scanning electron micrographs, endoscopic photography, and immunofluorescent light micrographs. The authors provide interesting, creative, and thought-provoking vignettes that introduce the chapter topics with readings on such topics as smallpox, heart transplants, and defibrillator implants.

- **From Science to Technology**—The new "From Science to Technology" readings cover topics such as *Cloning to Produce Therapeutic Stem Cells* and *Replacing the Liver.*

- **Clinical Applications**—New topics have been added to the Clinical Application boxes in several chapters. Read updates on Parkinson disease treatment, asthma, and food supplements.

- **Review Exercises and Critical Thinking**—Updated end-of-chapter review exercises help the student check their understanding of the chapter's major ideas. Critical thinking questions encourage the student to apply information to clinical situations.

- **Online Learning Center**—New OLC activities and resources are available for students and instructors.

- **Digital Content Manager**—The *Digital Content Manager*, a multimedia collection of visual resources, allows instructors to utilize artwork from the text in multiple formats to create customized classroom presentations, visually-based tests and quizzes, dynamic course website content, or attractive printed support materials. The digital assets on this cross-platform CD-ROM are grouped by chapter within easy-to-use folders.

Updates and Additions

Chapter 1 reorders topics to provide a more solid foundation for understanding by presenting the internal environment in more detail with unique figures and introducing hierarchy of organization and various organ systems first. New figures on homeostatic mechanisms have also been added.

Chapter 2 features a revised presentation of dissociation of salts in water, a revised presentation of protein structure, and an improved explanation of electron shells and octet rule, and polar bonds. The explanation of saturated/unsaturated fatty acids and fats has also been reordered.

Chapter 3 presents a revised figure on osmosis, which now allows for equilibrium to be reached, thus better illustrating the relationship between intracellular and extracellular fluids. A new section covers stem and progenitor cells.

Chapter 4 offers additional steps shown in translation and a better representation of the relationship between chromosome structure and DNA.

Chapter 5 presents a new vignette on building a blood vessel plus the addition of a *From Science to Technology* reading on tissue engineering. The "types of membranes" topic from chapter 6 has been moved to chapter 5.

Chapter 6 introduces a new, boxed reading on the causes as well as the anatomical and physiological effects of sunburn. A new vignette on smallpox has been added at the beginning of the chapter.

Chapter 7 features revised skeletal figures that present a consistent right side orientation. *Skeletons From the Past* is the new chapter opener vignette.

Chapter 9 presents a clearer relationship between thick and thin filaments, striation pattern, and the explanation of the sliding filament model. A new figure on muscle contraction shows the crossbridge cycle and the relationship to relaxed state. New art for muscles has been added throughout. Terminology is now more consistent with *Terminologia Anatomica*, except when such convention conflicts with current clinical usage.

Chapter 10 introduces a new figure showing the relationship between CNS and PNS, including motor and sensory divisions of PNS and the somatic and autonomic divisions of the motor portion. Unipolar neurons are now shown to have an axon with a central process and a peripheral process.

Chapter 11 features a revised presentation of neuroanatomy distinguishing between gray matter and white matter.

Chapter 12 offers new illustrations of the inner ear.

Chapter 13 provides a new illustration and a new table that compare the nervous and endocrine systems and highlights the importance of target cells.

Chapter 14 introduces the topic of blood with a new vignette on blood substitutes.

Chapter 15 provides added and expanded information on the control of blood pressure; end-diastolic volume, end-systolic volume, and preload. A new vignette on defibrillator implants opens the chapter.

Chapter 16 presents a new section on *Natural Killer Cells (NK)*, includes expanded information on MHC classes, and a new table on the comparison of T cells and B cells. The topic of peanut allergies is featured in the chapter-opening vignette.

Chapter 17 *From Science to Technology 17.1* features a new reading on liver replacement, and the introductory chapter vignette covers a brief history of constipation.

Chapter 18 features an expanded section on appetite control and a new vignette on preventing vitamin D deficiency.

Chapter 20 has improved art pieces presenting kidney anatomy, the countercurrent mechanism, and the mechanism of urine concentration.

Chapter 23 presents a new vignette on multiple births and a new table on the stages and events of early human prenatal development. Pregnancy, the birth process and milk production are now included in this chapter.

Chapter 24 provides an update on human genome sequencing results and chromosomal abnormalities.

Teaching and Learning Supplements

McGraw-Hill offers various tools and teaching products to support the tenth edition of *Hole's Human Anatomy & Physiology*. Students can order supplemental study materials by contacting your local bookstore. Instructors can obtain teaching aids by calling the Customer Service Department at 800-338-3987, visiting our A&P website at www.mhhe.com, or contacting your local McGraw-Hill sales representative.

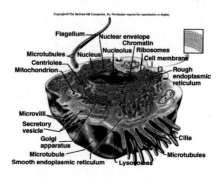

The **Digital Content Manager**, 0-07-243895-9, is a multimedia collection of visual resources that allows instructors to utilize artwork from the text in multiple formats to create customized classroom presentations, visually-based tests and quizzes, dynamic course website content, or attractive printed support materials. The digital assets on this cross-platform CD-ROM are grouped by chapter within the following easy-to-use folders.

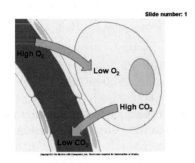

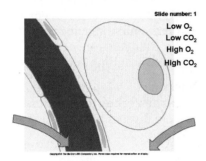

- **Active Art Library** Key Process Figures from the text are saved in manipulable layers that can be isolated and customized to meet the needs of the lecture environment.

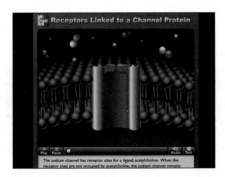

- **Animations Library** Numerous full-color animations of key physiological processes are provided. Harness the visual impact of processes in motion by importing these files into classroom presentations or course websites.

- **Art Libraries** Full-color digital files of all illustrations in the book, plus the same art saved in unlabeled and gray scale versions, can be readily incorporated into lecture presentations, exams, or custom-made classroom materials. These images are also pre-inserted into blank PowerPoint slides for ease of use.

- **Photo Libraries** Digital files of instructionally significant photographs from the text—including cadaver, bone, histology, and surface anatomy images—can be reproduced for multiple classroom uses.

- **PowerPoint Lectures** Ready-made presentations that combine art and lecture notes have been specifically written to cover each of the 24 chapters of the text. Use the PowerPoint lectures as they are, or tailor them to reflect your preferred lecture topics and sequences.

- **Tables Library** Every table that appears in the text is provided in electronic form. You can quickly preview images and incorporate them into PowerPoint or other presentation programs to create your own multimedia presentations. You can also remove and replace labels to suit your own preferences in terminology or level of detail.

Instructor Testing and Resource CD-ROM, 0-07-282738-6, is a cross-platform CD-ROM providing a wealth of resources for the instructor. Supplements featured on this CD-ROM include a computerized test bank utilizing Brownstone Diploma© testing software to quickly create customized exams. This user-friendly program allows instructors to search for questions by topic, format, or difficulty level; edit existing questions or add new ones; and scramble questions and answer keys for multiple versions of the same test.

Other assets on the Instructor's Testing and Resource CD-ROM are grouped within easy-to-use folders. The Instructor's Manual and the Instructor's Manual to accompany the Laboratory Manual are available in both Word and PDF formats. Word files of the test bank are included for those instructors who prefer to work outside of the test generator software.

The ***Instructor's Manual,*** by Michael F. Peters includes supplemental topics and demonstration ideas for your lectures, suggested readings, critical thinking questions, and teaching strategies. The Instructor's Manual is available through the Instructor Resources of the Online Learning Center and the Instructor Testing and Resource CD-ROM.

McGraw-Hill provides **Overhead Transparencies, Labeled** 0-07-243894-0, of all text line art and numerous photos and **Unlabeled** 0-07-284222-9, of key line art and photos.

English/Spanish Glossary for Anatomy and Physiology, 0-07-283118-9, is a complete glossary that includes every key term used in a typical two-semester anatomy and physiology course. Definitions are provided in both English and Spanish. A phonetic guide to pronunciation follows each word in the glossary.

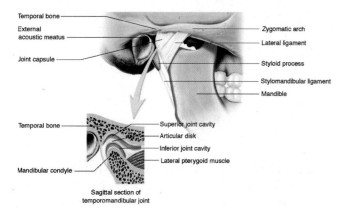

Sagittal section of temporomandibular joint

A Visual Guide for Anatomy and Physiology, 0-07-286378-1, is a visual atlas containing key gross anatomy illustrations that have been enlarged in size to make it easier for students to learn anatomy.

Course Delivery Systems With help from our partners, WebCT, Blackboard, TopClass, eCollege, and other course management systems, professors can take complete control over their course content. These course cartridges also provide online testing and powerful student tracking features. *Hole's Human Anatomy & Physiology* Online Learning Center is available within all of these platforms.

For the Student

MediaPhys 2.0 CD-ROM

This interactive tool offers detailed explanations, high quality illustrations and animations to provide students with a thorough introduction to the world of physiology—giving them a virtual tour of physiological processes. MediaPhys is filled with interactive activities and quizzes to help reinforce physiology concepts that are often difficult to understand.

Online Learning Center (http://www.mhhe.com/shier10)
The OLC offers an extensive array of learning and teaching tools. The site includes quizzes for each chapter, links to websites related to each chapter, clinical applications, interactive activities, art labeling exercises, and case studies. Students can click on a diagram of the human body and get case studies related to the regions they select. Instructor resources at the site include lecture outlines, technology resources, clinical applications, and case studies.

- **Essential Study Partner**
  The ESP contains 120 animations and more than 800 learning activities to help your students grasp complex concepts. Interactive diagrams and quizzes will make learning stimulating and fun for your students. The Essentials Study Partner can be accessed via the Online Learning Center.

- **Live News Feeds**
  The OLC offers course specific real-time news articles to help students stay current with the latest topics in anatomy and physiology.

- **Tutorial Service**
  This free "homework hotline" offers students the opportunity to discuss text questions with our A&P consultant.

- **GetBody Smart.com** is an online examination of human anatomy and physiology.

- **Access Science** is the online version of McGraw-Hill's Encyclopedia of Science & Technology. Link to this site free of charge from the Online Learning Center.

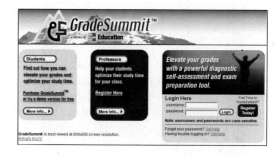

GradeSummit (www.gradesummit.com)
This Internet-based self-assessment service provides students and instructors with diagnostic information about subject strengths and weaknesses. This detailed feedback and direction enables learners and teachers to focus study time on areas where it will be most effective. GradeSummit also enables instructors to measure their students' progress and assess that progress relative to others in their classes and worldwide.

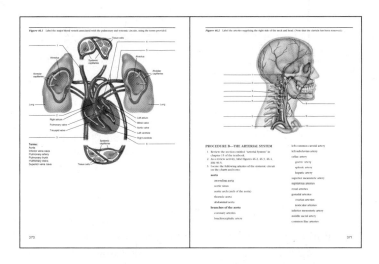

The **Laboratory Manual for Hole's Human Anatomy & Physiology**, Tenth Edition, 0-07-243891-6, by Terry R. Martin, Kishwaukee College
This lab manual is designed to accompany the tenth edition of *Hole's Human Anatomy and Physiology*.

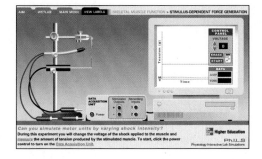

Physiology Interactive Lab Simulations (Ph.I.L.S)
The Ph.I.L.S CD-ROM contains eleven laboratory simulations that allow students to perform experiments without using expensive lab equipment or live animals. This easy-to-use software offers students the flexibility to change the parameters of every lab experiment, with no limit to the amount of times a student can repeat experiments or modify variables. This power to manipulate each experiment reinforces key physiology concepts by helping students to view outcomes, make predictions, and draw conclusions.

Student Study Guide, 0-07-243893-2, by Nancy A. Sickles Corbett contains chapter overviews, chapter objectives, focus questions, mastery tests, study activities, and mastery test answers.

Anatomy and Physiology Laboratory Manual–Fetal Pig, Second Edition, 0-07-243814-2, by Terry R. Martin, provides excellent full-color photos of the dissected fetal pig with corresponding labeled art. It includes World Wide Web activities for many chapters.

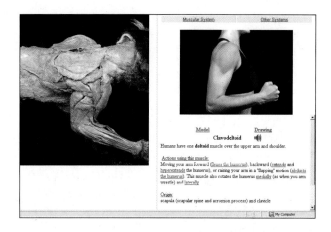

Virtual Anatomy Dissection Review, CD-ROM, 0-07-285621-1, by John Waters, Pennsylvania State University
This multimedia program contains vivid, high quality labeled cat dissection photographs. The program helps students easily identify and review the corresponding structures and functions between the cat and the human body.

Life Science Animation CD-ROM, 0-07-234296-X, contains 125 animations of major biological concepts and processes such as the sliding filament mechanism, active transport, genetic transcription and translation, and other topics that may be difficult for students to visualize.

Laboratory Atlas of Anatomy and Physiology, fourth edition, 0-07-243810-X, by Eder et al., is a full-color atlas containing histology, human skeletal anatomy, human muscular anatomy, dissections, and reference tables.

Acknowledgments

Any textbook is the result of hard work by a large team. Although we directed the revision, many "behind-the-scenes" people at McGraw-Hill were indispensable to the project. We would like to thank our editorial team of Michael Lange, Marty Lange, Kris Tibbetts, Michelle Watnick, and Pat Hesse; our production team, which included Jayne Klein, Sandy Ludovissy, Wayne Harms, John Leland, Sandy Schnee, Barb Block; Joanne Bales, art director, Precision Graphics; and most of all, John Hole, for giving us the opportunity and freedom to continue his classic work. We also thank our wonderfully patient families for their support.

David Shier

Jackie Butler

Ricki Lewis

Reviewers

We would like to acknowledge the valuable contributions of all professors and their students who have provided detailed recommendations for improving chapter content and illustrations throughout the revision process for each edition. Hundreds of professors from the U.S., Canada, and Europe have played a vital role in building a solid foundation for *Hole's Human Anatomy & Physiology.*

First Edition

Edward Barnett, *Kellogg Community College*

Nancy Corbett, *Thomas Jefferson University*

Jesse Dolson, *Delta College*

John Frehn, *Illinois State University*

John Harley, *Eastern Kentucky University*

Theodore Hollis, *Penn State University*

Ann Lesak, *Moraine Community College*

Robert Nabors, *Tarrant County Junior College*

Richard Pflanzer, *Indiana Univerisy*

John Childrey, *Purdue University*

Judy Best, *Purdue University*

Mary Dorhman, *University of Northern Iowa*

Second Edition

Edward M. Barnett, *Kellogg Community College*

William Bednar, *C.S. Mott Community College*

Colin Campbell, *Pima Community College*

Jessie Dolson, *Delta College*

Hester Fassel, *Iowa State University*

Yola Forbes, *Iowa State University*

John Frehn, *Illinois State University*

Cecilia Valle Gonzales, *St. Philip's College*

Terry E. Greathouse, *Cuyahoga Community College*

Joe Harbor, *San Antonio College*

John P. Harley, *Eastern Kentucky University*

Eugene S. Horowitz, *Queensborough Community College*

Anne Lesak, *Moraine Valley Community College*

Robert E. Nabors, *Tarrant County Junior College*

Richard Northrup, *Delta College*

Joseph R. Powell, *Florida Junior College*

Margaret Howarth Przyogda, *Middlesex County College*

Ed Reschke, *Muskegon Community College*

Ethel Sloane, *University of Wisconsin, Milwaukee*

Jane McNamara Bieber, *Virginia Commonwealth University*

Anne Denner, *Indiana State University, Evansville*

Joyce M. Dungan, *University of Evansville*

Jerry O. Erkert, *Santa Fe Community College*

James Ezell, *J. Sargeant Reynolds Community College*

William Garretson, *Valencia Community College*

Mary Etta Hight, *Marshall University*

Edward C. Hurlbut, *Mesa College*

Kenneth L. Jones, *Mt. San Antonio College*

Thomas S. Kaufman, *Montgomery College*

Jack Kildebeck, *Bakersfield College*

Donald S. Kisiel, *Suffolk County Community College*

Mary Linda Lungren, *Community College of Denver*

Margaret May, *Virginia Commonwealth University*

Mary Lou Mulvihill, *William Rainey Harper College*

Patricia M. O'Mahoney, *University of Southern Maine*

Harry S. Reasor, *Miami-Dade Community College*

Jo Ann Robertson, *Western Illinois University*

Curtis Robinson, *Milwaukee Area Technical College*

Maggie Sample, *Valencia Community College*

Elise Schoenfeld, *University of Albuquerque*

Louis Squitieri, *Bronx Community College*

G. Arthur Stephens, *Arapahoe Community College*

Michael J. Timmons, *Moraine Valley Community College*

Kent. M. Van De Graff, *Brigham Young University*

Third Edition

Allan L. Abati, *California State University Long Beach*

Lucille Aulsebrook, *Vanderbilt University*

Shirley Bishel, *Rio Hondo College*

Mary Jane Burge, *Cuyahoga Community College*

Warren Burggren, *University of Massachusetts*

Robert Catlett, *University of Colorado, Colorado Springs*

Philip L. Cooper, *Suffolk County Community College*

Ruthanna Dyer, *Seneca College*

David E. Grosland, *Iowa Central Community College*

William C. Kleinelp, Jr., *Middlesex County College*

Brenda H. Knight, *Catawba Valley Technical College*

Roxine McQuitty, *Milwaukee Area Technical College*

John A. Martin, *Clark College*

T. Pavlovitch, *Pasadena City College*

Frank C. Salter, *Jacksonville State University*

Donald A. Wheeler, *Cuyahoga Community College*

Louis Wigginton, *St. Clair County Community College*

Calvin G. Beams, Jr., *Oklahoma State University*

Donna Edwards, *Olympia Technical Community College*

Steve Hager, *University of Scranton*

Roy Hyle, *Thomas Nelson Community College*

Paula Holloway, *Lewis and Clark Community College*

Mariana Holson, *Olympia Technical Community College*

Mary Lou Mulvihill, *William Rainey Harper College*

Sherry Stair, *Thomas Nelson Community College*

Dave Straley, *University of Dubuque*

Fourth Edition

Thomas S. Kaufman, *Montgomery College*

Robert E. Nabors, *Tarrant County Junior College*

James W. Russell, *Georgia Southwestern College*

Louise Squitieri, *Bronx Community College*

Howard M. Fuld, *Bronx Community College*

Gerald R. Dotson, *Front Range Community College*

Robert D. Morden, *University of Wisconsin–Superior*

Ahmad Kamal, *Olive-Harvey College*

Karen A. Koos, *Rio Hondo College*

Fifth Edition

Richard Anderson, *Modesto Junior College*

Helene Auld, *Northeast Iowa Technical Institute*

Paul Badaracco, *Yuba College*

Phil J. Costa, *Queensborough College of the City University of New York*

Paul R. Holmgren, *Northern Arizona University*

Dennis D. Kalichstein, *Ocean County College*

Anne E. Lesak, *Moraine Valley Community College*

Ronald A. Markle, *Northern Arizona University*

Constance R. Martin, *Hunter College of the City University of New York*

Richard L. Myers, *Southwest Missouri State University*

Fredrick Prince, *Plymouth State College*

Cecelia Thomas, *Hinds Community College*

Carol B. Veil, *Anne Arundel Community College*

Sixth Edition

David Logan, *York University*

Terry R. Martin, *Kishwaukee College*

Aaron E. James, *Gateway Community College*

Dr. Louis A. Giacinti, *Milwaukee Area Technical College*

Clarence C. Wolfe, *Northern Virginia Community College*

Dale A. DesLauriers, *Chaffey College*

Jean S. Helgeson, *Collin County Community College*

Nancy Ann S. Corbett, *Camden College of Arts, & Sciences*

Edwin J. Bessler, *Franciscan University of Steubenville*

Ed Krol, *Henry Ford Community College*

Dwight Kamback, *Northhampton Community College*

Robert Smith, *Forest Park Community College*

John H. Dustman, *Indiana University Northwest*

Seventh Edition

Susan M. Behling, *Concordia University Wisconsin*

Charles H. Bennett, *Kentucky State University*

Barbara A. Bernardi, *Springfield College in Illinois*

Moges Bizuneh, *IVY Tech State College*

Brenda C. Blackwelder, *Central Piedmont Community College*

Stanton Braude, *Washington University, University of Missouri at St. Louis*

Wanda L. Buckland, *Dabney S. Lancaster Community College*

Judith Carpenter, *Columbus State Community College*

Melvin C. Chambliss, *Michigan State University*

F. Jeffrey Chyatte, *University of Maryland*

Karen M. Cianci, *Houghton College*

Rosanne M. Ciccia, *D'Youville College*

Nancy A. Sickles Corbett, *Rutgers The State University of New Jersey*

James E. Cordes, *Louisiana State University at Eunice*

Michael Corral, *Darrow School*

Jean Cremins, *Massachusetts Bay Community College*

Opal H. Dakin, *Hinds Community College*

Patricia R. Daron, *Northern Virginia Community College*

Winifred B. Dickinson, *Franciscan University of Steubenville*

Michael A. Dorset, *Cleveland State Community College*

Victor P. Eroschenko, *University of Idaho*

L. Fleming Fallon, *Jameson Hospital, Columbia University School of Public Health*

Bruce A. Fisher, *Roane State Community College*

Kate Fleury, *Lake Washington Technical College*

Pamela B. Fouché, *Walters State Community College*

Ralph F. Fregosi, *The University of Arizona*

William S. Garlick, *Arizona State University*

Phyllis Gee, *University of Manitoba*

Mike Gehner, *Xavier University*

H. R. Giesman, *North Iowa Area Community College*

Sister Terence Glum, *University of Mary*

Keith R. Graham, *Lutheran College of Health Professions*

Darryl V. Grennell, *Alcorn State University*

Kevin Jon Gyolai, *North Dakota State College of Science*

Ruth L. Hays, *Clemson University*

Jimmie F. Hughey, *St. John's University*

Robert L. Jochen, *Blue Ridge Community College*

Jerry M. Johnson, *Western Baptist College*

Ronald L. Johnson, *Arkansas State University*

Drusilla B. Jolly, *Forsyth Technical Community College*

Joan H. Jones, *Naugatuck Valley Community–Technical College*

Brian E. Jordan, *Lansing Community College*

Kamal I. Kamal, *Valencia Community College–West Campus*

Dwight Kamback, *Northampton Community College*

Judith Kasperek, *Pitt Community College*

Gary Kennedy, *Lethbridge Community College*

Frank G. Kitakis, *Wayne County Community College*

John E. Kovaleski, *Indiana State University*

Jeffrey R. LaDuca, *Canisius College*

Billie S. Lane, *Chattanooga State Technical Community College*

Gina Langley, *Eastern New Mexico University–Ruidoso*

Mary T. Leonard, *University of Dayton*

Mary Katherine Lockwood, *University of New Hampshire*

D. M. Logan, *York University*

Bonita L. Longo, *Community Hospital School of Nursing*

Charmayne Mack, *Rosary College*

Dennis Malek, *Triton College*

Terry R. Martin, *Kishwaukee College*

William J. Mathena, *Kaskaskia College*

Pamela S. McLaughlin, *Madisonville Community College*

Michael C. Meyers, *Montana State University*

Robert D. Muckel, *Doane College*

Shirley Mulcahy, *San Diego Mesa College*

Tara Narayansingh, *University of Manitoba*

J. Felix Palmer, *Tulane University*

Brian K. Paulson, *California University of Pennsylvania*

Carlos F. A. Pinkham, *Norwich University*

Pam Rhyne, *Kennesaw State College*

Kristi Sather-Smith, *Hinds Community College*

Robert A Sharp, *Aquinas College*

Clyde F. Smith, *Odessa College*

Jean E. Smith, *Carroll College*

Shirley N. Smith, *Lansing Community College*

Paulette R. Snyder, *Erie Community College, North*

Janet E. Steele, *University of Nebraska at Kearney*

Stuart S. Sumida, *California State University–San Bernardino*

Donald L. Terpening, *Ulster County Community College*

William R. Tobin, Jr., *Erie Community College South Campus*

Robin Vance, *Union College*

Dianne L. Vermillion, *University of Rochester*

Margaret G. Wade, *Midland College*

Robert C. Wall, *Lake-Sumter Community College*

Garry M. Wallace, *Northwest College*

Leslie Jayne Wallace, *Baker College of Owosso*

Alan R. Wasmoen, *Iowa Central Community College*

Carl F. Wellstead, *West Virginia Institute of Technology*

Philip C. Whitford, *Capital University*

Barbara Wineinger, *Vincennes University Jasper*

Clarence C. Wolfe, *Northern Virginia Community College–Annandale Campus*

Ricky K. Wong, *Los Angeles Trade-Technical College*

Diana L. Wyman, *New Hampshire Technical College*

Special Contributors

Louis A. Giacinti, *Milwaukee Area Technical College*

Charles J. Grossman, *Xavier University, Research Service, Veterans Affairs Medical Center*

Virginia Rivers, *Truckee Meadows Community College*

Kenneth S. Saladin, *Georgia College*

D. M. Van Wynsberghe, *University of Wisconsin–Milwaukee*

Leslie J. Wiemerslage, *Belleville Area College*

Eric A. Wise, *Santa Barbara City College*

Eighth Edition

Janice Asel, *Mitchell Community College*

Beth M. Atkin, *Washington State Community College*

Gordon Atkins, *Andrews University*

Stephanie Sajdak Baiyasi, *Delta College*

Anna Bartosh, *Howard County Junior College*

William R. Belzer, *Clarion University of Pennsylvania–Venango Campus*

Edwin Bessler, *Franciscan University of Steubenville*

E. Beth Bonner, *Delgado Community College*

Ray D. Burkett, *Shelby State Community College*

Rebecca M. Burt, *Southeast Community College–Beatrice Campus*

Jennifer Carr Burtwistle, *Northeast Community College*

Michael S. Capp, *Carlow College*

Holly Carmichael, *Wilson Technical Community College*

Melvin C. Chambliss, *Michigan State University's Veterinary Technology Program*

William H. Chrouser, *Warner Southern College*

Lu Anne Clark, *Lansing Community College*

Barbara J. Cohen, *Delaware County Community College*

Mary Catharine Cox, *Wingate University*

Allen R. Crooker, Jr., *Hartwick College*

Lin Doyle, *Northwest College*

Duane A. Dreyer, *Durham Technical Community College*

Peter I. Ekechukwu, *Horry-Georgetown Technical College*

Barbara F. Ensley, *Haywood Community College*

Gary Estep, *Lubbock Christian University*

Louis A. Giacinti, *Milwaukee Area Technical College*

William A Gibson, *University of New Orleans*

Susan K. Gilmore, *University of Pittsburgh at Bradford Jamestown Community College*

David E. Harris, *Lewiston-Auburn College, University of Southern Maine*

George E. Heath, *University of Maryland Eastern Shore*

Drusilla Beal Jolly, *Forsyth Technical Community College*

Beverly W. Juett, *Midway College*

Kamal I. Kamal, *Valencia Community College, West Campus*

Gary M. Kiebzak, *Miller Orthopaedic Clinic, Charlotte, NC*

Glenn E. Kietzmann, *Wayne State College*

Alan Knowles, *Pensacola Christian College*

Kristin Krause, *Saint Thomas Aquinas College*

Gopal Krishna, *Moberly Area Community College*

Barbara Lax, *Community College of Allegheny County*

Nancy Longlet, *Concordia College*

Lisa Lupini, *Baker College of Flint*

Bradford D. Martin, *La Sierra University*

William J. Mathena, *Kaskaskia College*

Julie A. Medlin, *Northwestern Michigan College*

Jim Miller, *College of the Southwest*

Eli C. Minkoff, *Bates College*

Robert Moldenhauer, *Washtenaw Community College*

James (Jym) C. Moon, *Western Iowa Technical Community College*

David Mork, *Saint Cloud State University*

C. Aubrey Morris, *Pensacola Junior College*

Tony E. Morris, *Fairmont State College*

Steve C. Nunez, *Sauk Valley Community College*

Nicole J. Okazaki, *Southeastern Louisiana University*

Charles M. Page, *El Camino College*

Mark A. Paulissen, *McNeese State University*

Mary S. Rea, *Sage Junior College*

Donald Rodd, *University of Evansville*

Connie E. Rye, *Bevill State Community College*

David A. Sandmire, *University of New England*

Soma Sanyal, *Penn State–Altoona*

Marilyn Shopper, *Johnson County Community College*

Richard Sims, *Jones County Junior College*

Katherine Smalley, *Emporia State University*

Denise L. Smith, *Skidmore College*

Michael E. Smith, *Valdosta State University*

Paul M. Spannbauer, *Hudson Valley Community College*

Marian Spozio, *Jefferson Community College*

Sarah Anne Staples, *Andrew College*

John R. Steele, *Ivy Tech State College*

Dennis M. Sullivan, *Cedarville College*

P. Alleice Summers, *Dyersburg State Community College*

Patricia J. Thomas, *Delgado Community College*

William R. Tobin, *West Valley Central School*

Don Varnado, *Southern Ohio College–Northern Kentucky Campus*

Dianne L. Vermillion, *School of Nursing–University of Rochester*

Garry M. Wallace, *Northwest College*

Norma J. Weekly, *Wilkes Community College*

Christine A. Wilson, *Community College of Allegheny County–Boyce Campus*

Barbara Wineinger, *Vincennes University Jasper*

Clarence C. Wolfe, *Northern Virginia Community College Annandale Campus*

Ninth Edition

Marion Alexander, *University of Manitoba*

Angela J. Andrews, *Redlands Community College*

Martha W. Andrus, *Grambling State University*

Timothy A. Ballard, *University of North Carolina at Wilmington*

Brenda C. Blackwelder, *Central Piedmont Community College*

James Bridger, *Prince George's Community College*

Carolyn Burroughs, *Bossier Parish Community College*

Edward W. Carroll, *Marquette University*

Margaret Chad, *Saskatchewan Institute of Applied Science & Technology*

Lynda B. Collins, *Mississippi College*

Shirley A. Colvin, *Gadsden State Community College*

Wilfrid DuBois, *D'Youville College*

Sondra Dubowsky, *Allen County Community College*

John Erickson, *Ivy Tech State College*

Marilyn Ziegler Franklin, *Grambling State University*

Brent M. Graves, *Northern Michigan University*

Mary Guise, *Mohawk College of Applied Arts & Technology*

Michael J. Harman, *North Harris Montgomery Community College*

Alan G. Heath, *Virginia Polytechnic Institute & State University*

Julie A. Huggins, *Arkansas State University*

Marsha Jones, *Southwestern Community College*

Beverly W. Juett, *Midway College*

Jeffrey S. Kiggins, *Blue Ridge Community College*

Nancy G. Kincaid, *Troy State University Montgomery*

Alan C. Knowles, *Pensacola Christian College*

Donna A. Kreft, *Iowa Central Community College*

Mary Katherine Lockwood, *University of New Hampshire*

Josephine Macias, *West Nebraska Community College*

Qian Frances Moss, *Des Moines Area Community College*

Sheila A. Murray, *Berkshire Community College*

Steve Nunez, *Sauk Valley Community College*

Augustine I. Okonkwo, *Norfolk State University*

Amy Griffin Ouchley, *University of Louisiana at Monroe*

David J. Pierotti, *Northern Arizona University*

John Romanowicz, *International School of Amsterdam*

David K. Saunders, *Emporia State University*

Melvin Schmidt, *McNeese State University*

Brian Shmaefsky, *Kingwood College*

Bharathi P. Sudarsanam, *Labette Community College*

Gary Lee Tieben, *University of Saint Francis*

John M. Wakeman, *Louisiana Tech University*

Murray B. Weinstein, *Erie Community College, City Campus*

Eddie L. Whitson, *Gadsden State Community College*

Tenth Edition

Pegge Alciatore, *University of Louisiana Lafayette*

Vivian T. Anderson, *Oakland Community College–Auburn Hills*

Sharon R. Barnewall, *Columbus State Community College*

Charles J. Biggers, *University of Memphis*

Jennifer Borash, *Horry-Georgetown Technical College*

Karen Borg, *Midlands Technical College*

Sara Brenizer, *Shelton State Community College*

Joseph Cameron, *Hinds Community College*

Kenneth Carpenter, *Southwest Tennessee Community College*

Sandra I. Caudle and students, *Calhoun Community College*

W. Wade Cooper, *Shelton State Community College*

Larry G. DeLay, *Waubonsee Community College*

Nichol Dolby, *Amarillo College*

Ardath Egle, *University of Texas–Pan American*

Mary Catherine Flath and students of Anatomy I and II, *Ashland Community College*

Tom M. Graham, *University of Alabama*

Kathryn Gronlund, *Edison Community College*

Linden C. Haynes, *Hinds Community College*

Jacqueline A. Homan, *South Plains College*

Dale R. Horeth, *Tidewater Community College*

Dianne M. Jedlicka, *The School of the Art Institute of Chicago*

Narinder Kapoor, *Concordia University*

Mary Katherine Lockwood, *University of New Hampshire*

Jane R. Marone, *University of Illinois–Chicago*

William J. Mathena, *Kaskaskia College*

Richard McCloskey, *Boise State University*

W. J. McCracken, *Tallahassee Community College*

Robert C. McReynolds, *San Jacinto College Central*

Stephen H. McReynolds, *Tarleton State University*

Sharon Miles, *Itawamba Community College*

John E. Moore, *Parkland College*

Jesse J. Myers, student, *Oregon State University*

Augustine Okonkwa, *Norfolk State University*

Linda Nichols, *Sante Fe Community College*

Justicia Opoku, *University of Maryland, College Park*

Margaret (Betsy) Ott, *Tyler Junior College*

Julie C. Pilcher, *University of Southern Indiana*

Linda Powell, *Community College of Philadelphia*

Mattie Roig, *Broward Community College*

Melvin Schmidt, *McNeese State University*

Michael Squires, *Columbus State Community College*

Sarah Strong, *Austin Community College*

Mark Wygoda, *McNeese State University*

Canadian Reviewers

Mary T. Guise, *Mohawk College of Applied Arts and Technology*

William (Bill) Magill, *Humber College*

Donna Newhouse, *Lakehead University*

Delia Roberts, *Selkirk College*

Hole's Human Anatomy & Physiology

Over 25 years have passed, and *Hole's Human Anatomy & Physiology* is still *Hole's Human Anatomy & Physiology*—but with a sharper focus and appearance.

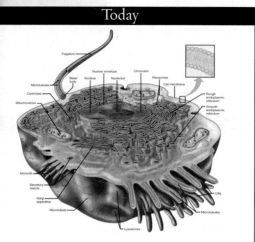

1978

Today

New and Revised Art

incorporates cutting-edge technology offering vivid depictions of complex processes while maintaining the conceptual base that has established Hole as the most effective "instructional tool" on the market with a unique focus on the fundamentals.

Hole's art is focused on the main concepts by using concise labeling methodology that keeps students from getting bogged down with excessive detail.

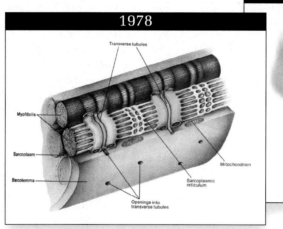

1978

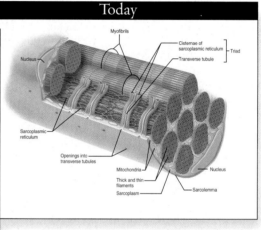

Today

Today

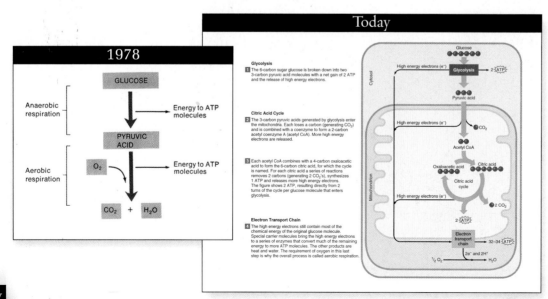

1978

Difficult concepts are illustrated as a step-by-step process.

ART PROGRAM

Correlation of Photomicrographs with Line Art

makes it easier for students to identify specific structures.

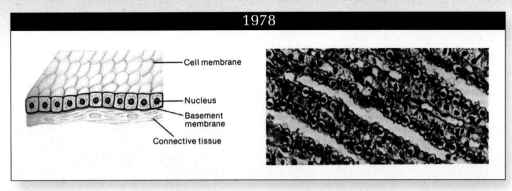

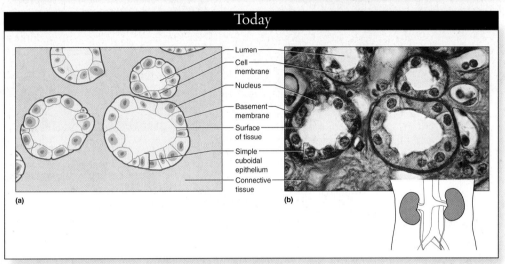

Macroscopic to Microscopic Presentation

makes the connection between gross anatomy and microscopic anatomy.

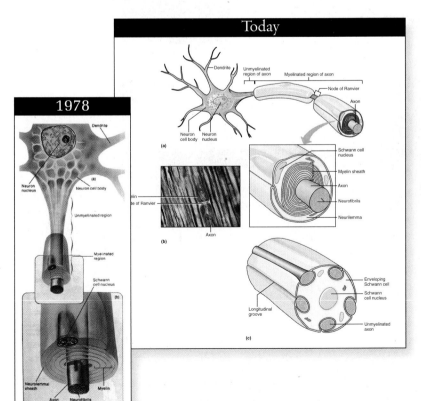

ART PROGRAM

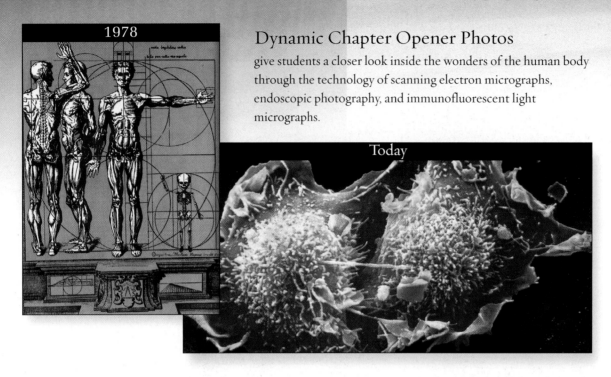

1978

Today

Dynamic Chapter Opener Photos

give students a closer look inside the wonders of the human body through the technology of scanning electron micrographs, endoscopic photography, and immunofluorescent light micrographs.

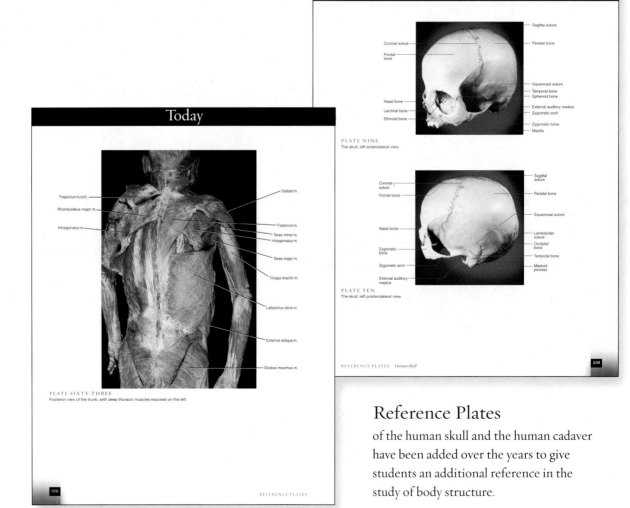

Today

Coronal suture
Frontal bone

Nasal bone
Lacrimal bone
Ethmoid bone

Sagittal suture
Parietal bone

Squamosal suture
Temporal bone
Sphenoid bone
External auditory meatus
Zygomatic arch
Zygomatic bone
Maxilla

PLATE NINE
The skull, left anterolateral view.

Coronal suture
Frontal bone

Nasal bone

Zygomatic bone

Zygomatic arch

External auditory meatus

Sagittal suture
Parietal bone

Squamosal suture

Lambdoidal suture
Occipital bone
Temporal bone
Mastoid process

PLATE TEN
The skull, left posterolateral view.

REFERENCE PLATES   Human Skull

239

Today

Trapezius m.(cut)
Rhomboideus major m.
Infraspinatus m.

Deltoid m.

Trapezius m.
Teres minor m.
Infraspinatus m.

Teres major m.

Triceps brachii m.

Latissimus dorsi m.

External oblique m.

Gluteus maximus m.

PLATE SIXTY-THREE
Posterior view of the trunk, with deep thoracic muscles exposed on the left.

958

REFERENCE PLATES

Reference Plates

of the human skull and the human cadaver have been added over the years to give students an additional reference in the study of body structure.

CLINICAL CONNECTIONS

Textbooks provide a foundation of facts, viewpoints, and overviews. They sequence information and facts to understand issues and create a context for comparing and understanding other sources.

Additional readings engage the students by creating a richer understanding of the concepts presented and provide a real-life connection to anatomy and physiology.

In 1978, John Hole integrated short, boxed readings within the text to help students apply the ideas presented in the narrative to clinical situations. Today, because the author team recognizes the vital role clinical connections play in bridging the gap between facts and real life, they have integrated several engaging formats.

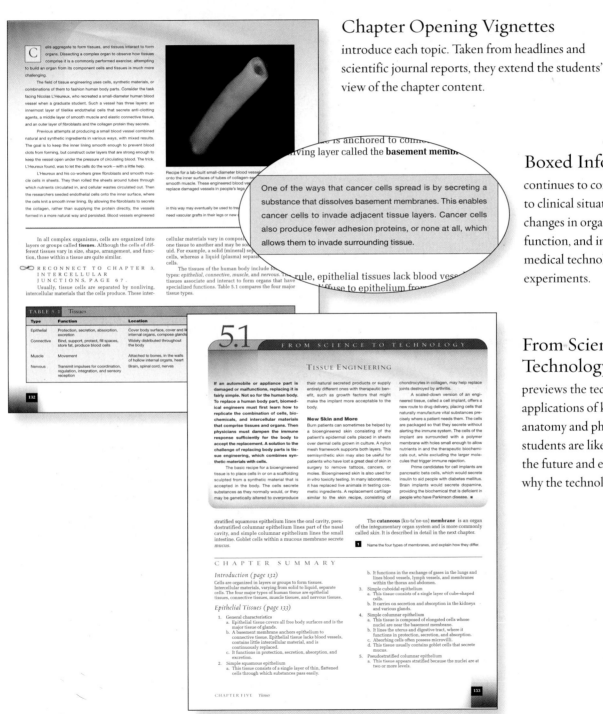

Chapter Opening Vignettes

introduce each topic. Taken from headlines and scientific journal reports, they extend the students' view of the chapter content.

Boxed Information

continues to connect chapter ideas to clinical situations, discusses changes in organ structure and function, and introduces new medical technology or experiments.

From Science to Technology

previews the technological applications of knowledge in anatomy and physiology that students are likely to encounter in the future and explains how and why the technology was developed.

CLINICAL CONNECTIONS

Clinical Applications

encourage students to explore information on related pathology, historical insights, and clinical examples that they are likely to encounter in their careers.

Clinical Terms

expand the students' understanding of medical terminology. It gives students the chance to brush up on phonetic pronunciations and definitions of related terms often used in clinical situations.

Life-Span Changes

There is no escaping the fact that aging is a part of life. Because our organs and organ systems are interrelated, aging-related changes in one influence the functioning of others. These readings chart the changes specific to particular organ systems.

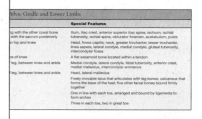

LEARNING SYSTEM

This text evolved because John W. Hole, Jr. had the desire and vision to provide the best possible anatomy and physiology text for his students. The pedagogical elements created were the key to engaging students and involving them actively in the learning process. With each edition, the current authors continue to include, expand, and improve upon the learning system features that define this text.

Understanding Words

includes *root words, stems, prefixes, and suffixes* revealing word meanings and origins. Knowing the roots from these lists help students remember scientific word meanings and understand new terms.

Chapter Objectives

provide a glimpse ahead to important sections of the narrative. They indicate what the student should be able to do after mastering the chapter content.

Key Terms and Pronunciations

anchor the students' understanding of anatomy and physiology. The bold face terms found throughout the narrative are key to building a solid science vocabulary.

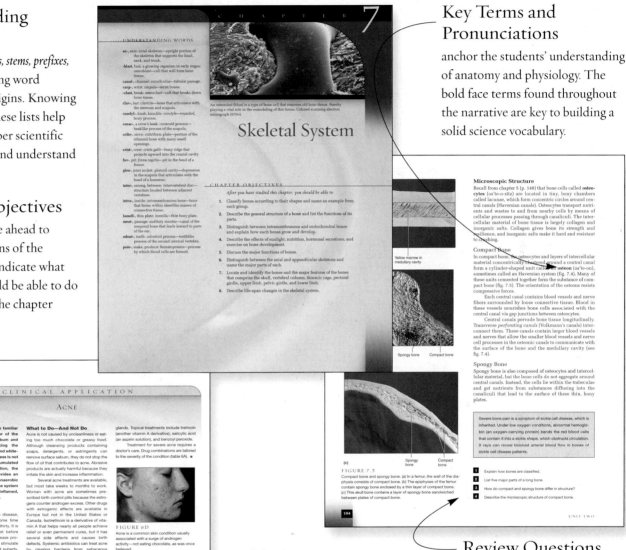

Tables

are designed to organize and summarize sections of the narrative and to present pertinent data.

Reconnect Icon

prompts the student to review key concepts found in previous chapters that will promote their understanding of new information.

Review Questions

occur at the ends of major sections within each chapter. They challenge students to test their mastery of the concepts before moving on to additional topics.

InnerConnections

conceptually link the highlighted body system to every other system. These graphic representations review chapter concepts, make connections, and stress the "big picture" in learning and applying the concepts and facts of anatomy and physiology.

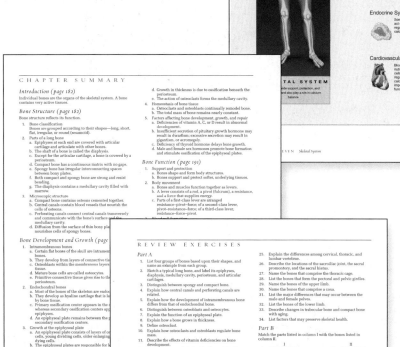

Chapter Summary

provides an outline for the students to use during review of the major ideas presented in the chapter and as a tool in organizing their thoughts.

Review Exercises

check the students' understanding of the major ideas presented in the chapter.

Web Connections

direct students to the Online Learning Center at www.mhhe.com/shier10 where they will find answers to chapter questions, additional quizzes, interactive learning exercises, and other study tools.

Critical Thinking Questions

apply main concepts of the chapter to clinical or research situations and take the student beyond memorization to the utilization of knowledge.

append-, to hang something: *append*icular—pertaining to the upper limbs and lower limbs.

cardi-, heart: peri*cardi*um—membrane that surrounds the heart.

cerebr-, brain: *cerebr*um—largest portion of the brain.

cran-, helmet: *cran*ial—pertaining to the portion of the skull that surrounds the brain.

dors-, back: *dors*al—position toward the back of the body.

homeo-, same: *homeo*stasis—maintenance of a stable internal environment.

-logy, the study of: physio*logy*—study of body functions.

meta-, change: *meta*bolism—chemical changes that occur within the body.

nas-, nose: *nas*al—pertaining to the nose.

orb-, circle: *orb*ital—pertaining to the portion of skull that encircles an eye.

pariet-, wall: *pariet*al membrane—membrane that lines the wall of a cavity.

pelv-, basin: *pelv*ic cavity—basin-shaped cavity enclosed by the pelvic bones.

peri-, around: *peri*cardial membrane—membrane that surrounds the heart.

pleur-, rib: *pleur*al membrane—membrane that encloses the lungs within the rib cage.

-stasis, standing still: homeo*stasis*—maintenance of a stable internal environment.

super-, above: *super*ior—referring to a body part that is located above another.

-tomy, cutting: ana*tomy*—study of structure, which often involves cutting or removing body parts.

A falsely colored scanning electron micrograph shows fat cells (yellow). Almost the entire volume of each cell is occupied by a single lipid droplet containing mostly triglycerides (440×).

Introduction to Human Anatomy and Physiology

CHAPTER OBJECTIVES

After you have studied this chapter, you should be able to

1. Define *anatomy* and *physiology* and explain how they are related.

2. List and describe the major characteristics of life.

3. List and describe the major requirements of organisms.

4. Define *homeostasis* and explain its importance to survival.

5. Describe a homeostatic mechanism.

6. Explain the levels of organization of the human body.

7. Describe the locations of the major body cavities.

8. List the organs located in each major body cavity.

9. Name the membranes associated with the thoracic and abdominopelvic cavities.

10. Name the major organ systems and list the organs associated with each.

11. Describe the general functions of each organ system.

12. Properly use the terms that describe relative positions, body sections, and body regions.

J udith R. had not been wearing a seat belt when the accident occurred because she had to drive only a short distance. She hadn't anticipated the intoxicated driver in the oncoming lane who swerved right in front of her. Thrown several feet, she now lay near her wrecked car as emergency medical technicians immobilized her neck and spine. Terrified, Judith tried to assess her condition. She didn't think she was bleeding, and nothing hurt terribly, but she felt a dull ache in the upper right part of her abdomen.

Minutes later, in the emergency department, a nurse gave Judith a quick exam, checking her blood pressure, pulse and breathing rate, and other vital signs and asking questions. These vital signs reflect underlying metabolic activities necessary for life, and they are important in any medical decision. Because Judith's vital signs were stable, and she was alert, knew who and where she was, and didn't seem to have any obvious life-threatening injuries, transfer to a trauma center was not necessary. However, Judith continued to report abdominal pain. The attending physician ordered abdominal X rays, knowing that about a third of patients with abdominal injuries show no outward sign of a problem. As part of standard procedure, Judith received oxygen and intravenous fluids, and a technician took several tubes of blood for testing.

A young physician approached and smiled at Judith as assistants snipped off her clothing. The doctor carefully looked and listened and gently poked and probed. She was looking for cuts; red areas called hematomas where blood vessels had broken; and treadmarks on the skin. Had Judith been wearing her seat belt, the doctor would have checked for characteristic "seat belt contusions," crushed bones or burst hollow organs caused by the twisting constrictions that can occur at the moment of impact when a person wears a seat belt. Finally, the doctor measured the girth of Judith's abdomen. If her abdomen swelled later on, this could indicate a complication, such as infection or internal bleeding.

On the basis of a hematoma in Judith's upper right abdomen and the continued pain coming from this area, the emergency room physician ordered a computed tomography (CT) scan. The scan revealed a lacerated liver. Judith underwent emergency surgery to remove the small torn portion of this vital organ.

When Judith awoke from surgery, a different physician was scanning her chart, looking up frequently. The doctor was studying her medical history for any notation of a disorder that might impede healing. Judith's history of slow blood clotting, he noted, might slow her recovery from surgery. Next, the physician looked and listened. A bluish discoloration of Judith's side might indicate bleeding from her pancreas, kidney, small intestine, or aorta (the artery leading from the heart). A bluish hue near the navel would also be a bad sign, indicating bleeding from the liver or spleen. Her umbilical area was somewhat discolored.

The doctor gently tapped Judith's abdomen and carefully listened to sounds from her digestive tract. A drumlike resonance could mean that a hollow organ had burst, whereas a dull sound might indicate inter-

The difference between life and death may depend on a health-care professional's understanding of the human body.

nal bleeding. Judith's abdomen produced dull sounds throughout. Plus, her abdomen had swollen, the pain intensifying when the doctor gently pushed on the area. With Judith's heart rate increasing and blood pressure falling, bleeding from the damaged liver was a definite possibility.

Blood tests confirmed the doctor's suspicions. Because blood is a complex mixture of cells and biochemicals, it serves as a barometer of health. Injury or illness disrupts the body's maintenance of specific levels of various biochemicals. This maintenance is called homeostasis. Judith's blood tests revealed that her body had not yet recovered from the accident. Levels of clotting factors produced by her liver were falling, and blood was oozing from her incision, a sign of impaired clotting. Judith's blood glucose level remained elevated, as it had been in the emergency room. Her body was still reacting to the injury.

Based on Judith's blood tests, heart rate, blood pressure, reports of pain, and the physical exam, the doctor sent her back to the operating room. Sure enough, the part of her liver where the injured portion had been removed was still bleeding. When the doctors placed packing material at the wound site, the oozing gradually stopped. Judith returned to the recovery room and, as her condition stabilized, to her room. This time, all went well, and a few days later, she was able to go home. The next time she drove, Judith wore her seat belt!

Imagine yourself as one of the health-care professionals who helped identify Judith R.'s injury and got her on the road back to health. How would you know what to look, listen, and feel for? How would you place the signs and symptoms into a bigger picture that would suggest the appropriate diagnosis? Nurses, doctors, technicians, and other integral members of health-care teams must have a working knowledge of the many intricacies of the human body. How can they begin to understand its astounding complexity? The study of human anatomy and physiology is a daunting, but fascinating and ultimately life-saving, challenge. ■

Our understanding of the human body has a long and interesting history (fig. 1.1). It began with our earliest ancestors, who must have been as curious about how their bodies worked as we are today. At first their interests most likely concerned injuries and illnesses, because healthy bodies demand little attention from their owners. Although they did not have emergency departments to turn to, primitive people certainly suffered from occasional aches and pains, injured themselves, bled, broke bones, developed diseases, and contracted infections.

The change from a hunter-gatherer to an agricultural lifestyle, which occurred from 6,000 to 10,000 years ago in various parts of the world, altered the spectrum of human illnesses. Before agriculture, isolated bands of peoples had little contact with each other, and so infectious diseases did not spread easily, as they do today with our global connections. In addition, these ancient peoples ate wild plants that provided chemicals that combated some parasitic infections.

With agriculture came exposure to pinworms, tapeworms and hookworms in excrement used as fertilizer, and less reliance on the wild plants that offered their protective substances. The rise of urbanization brought even more infectious disease as well as malnutrition, as people became sedentary and altered their diets. Several types of evidence from preserved bones and teeth chronicle these changes. Tooth decay, for example, affected 3% of samples from hunter-gatherers, but 8.7% from farmers, and 17% of samples from city residents. Preserved bones from children reflect increasing malnutrition as people moved from the grasslands to farms to cities. When a child starves or suffers from severe infection, the ends of the long bones stop growing. When health returns, growth resumes, but leaves behind telltale areas of dense bone.

In addition to the changes in health brought about by our own activities, some types of illnesses seem intrinsic to humans. Arthritis, for example, afflicts millions of people today and is also evident in fossils of our ancestors from 3 million years ago, from Neanderthals that lived 100,000 years ago, and from a preserved "ice man" from 5,300 years ago.

The rise of medical science paralleled human prehistory and history. At first, healers relied heavily on superstitions and notions about magic. However, as they tried to help the sick, these early medical workers began to discover useful ways of examining and treating the human body. They observed the effects of injuries, noticed how wounds healed, and examined dead bodies to determine the causes of death. They also found that certain herbs and potions could sometimes be used to treat coughs, headaches, and other common problems. These long-ago physicians began to wonder how these substances, the forerunners of modern drugs, affected body functions in general.

People began asking more questions and seeking answers, setting the stage for the development of modern medical science. Techniques for making accurate observations and performing careful experiments evolved, and knowledge of the human body expanded rapidly.

This new knowledge of the structure and function of the human body required a new, specialized language. Early medical providers devised many terms to name body parts, describe their locations, and explain their functions. These terms, most of which originated from Greek and Latin, formed the basis for the language of anatomy and physiology. (A list of some of the modern medical and applied sciences appears on page 25.)

1 What factors probably stimulated an early interest in the human body?

2 How did human health change as lifestyle changed?

3 What kinds of activities helped promote the development of modern medical science?

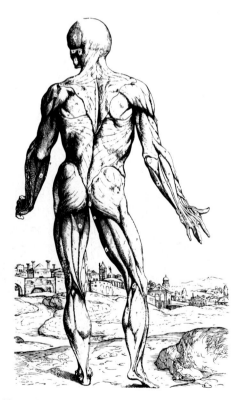

FIGURE 1.1

The study of the human body has a long history, as this illustration from the second book of *De Humani Corporis Fabrica* by Andreas Vesalius, issued in 1543, indicates. Note the similarity to the anatomical position (described on page 21).

Anatomy and Physiology

Two major areas of medical science, **anatomy** (ah-nat'o-me) and **physiology** (fiz"e-ol'o-je) are concerned with how the body maintains life. Anatomy deals with the **structures,** or morphology, of body parts—their forms and organization. Physiology considers the **functions** of these body parts—what they do and how they do it. Although anatomists rely more on examination of the body and physiologists more on experimentation, together their efforts have provided a solid foundation upon which an understanding of how our bodies work is built.

It is difficult to separate the topics of anatomy and physiology because anatomical structures make possible their functions. Parts form a well-organized unit—the **human organism**—and each part plays a role in the operation of the unit as a whole. This functional role depends upon the way the part is constructed. For example, the arrangement of bones and muscles in the human hand, with its long, jointed fingers, makes grasping possible. The heart's powerful muscular walls are structured to contract and propel blood out of the chambers and into blood vessels, and heart valves ensure that blood moves in the proper direction. The shape of the mouth enables it to receive food; teeth are shaped so that they break solid foods into smaller pieces; and the muscular tongue and cheeks are constructed to help mix food particles with saliva and prepare them for swallowing (fig. 1.2).

Anatomy and physiology are ongoing as well as ancient fields. For example, recent research has revealed a previously unknown muscle between two bones in the head, and identified a hormone, ghrelin, that controls fat utilization. The first discovery is anatomical; the second, physiological. Aspects of anatomy and physiology are increasingly being explained at the cellular and molecular levels, especially since researchers sequenced the human genome in 2000—the complete set of genetic instructions for a human body.

1. What are the differences between anatomy and physiology?

2. Why is it difficult to separate the topics of anatomy and physiology?

3. List several examples that illustrate how the structure of a body part makes possible its function.

4. How are anatomy and physiology both old and new fields?

Levels of Organization

Early investigators, limited in their ability to observe small structures, focused their attention on larger body parts. Studies of small structures had to await invention of magnifying lenses and microscopes, which came into use about 400 years ago. These tools revealed that larger body structures were made up of smaller parts, which, in turn, were composed of even smaller ones.

Today, scientists recognize that all materials, including those that comprise the human body, are composed of chemicals. Chemicals consist of tiny particles called **atoms,** which are commonly bound together to form larger particles called **molecules;** small molecules may combine to form larger molecules called **macromolecules.**

Within the human organism, the basic unit of structure and function is a **cell.** Although individual cells vary in size and shape, all share certain characteristics. Human cells contain structures called **organelles** (or″gan-elz′) that carry on specific activities. These organelles are composed of aggregates of large molecules, including proteins, carbohydrates, lipids, and nucleic acids. All cells in a human contain a complete set of genetic instructions, yet use only a subset of them, allowing cells to develop specialized

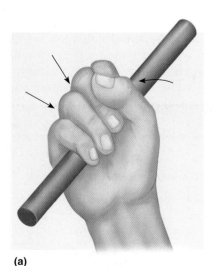
(a)

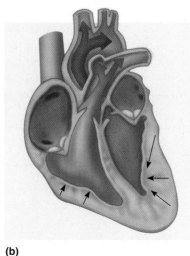
(b)

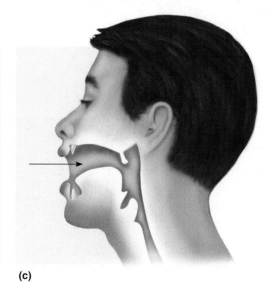
(c)

FIGURE 1.2

The structures of body parts make possible their functions: (a) The hand is adapted for grasping, (b) the heart for pumping blood, and (c) the mouth for receiving food. (Arrows indicate movements associated with these functions.)

functions. All cells share the same characteristics of life and must meet certain requirements to continue living.

Cells are organized into layers or masses that have common functions. Such a group of cells forms a **tissue.** Groups of different tissues form **organs**—complex structures with specialized functions—and groups of organs that function closely together comprise **organ systems.** Interacting organ systems make up an **organism.**

A body part can be described at different levels. The heart, for example, contains muscle, fat, and nervous tissue. These tissues, in turn, are constructed of cells, which contain organelles. All of the structures of life are, ulti-

mately, composed of chemicals (fig. 1.3). Clinical Application 1.1 describes two technologies used to visualize differences based on body chemistry.

Chapters 2–6 discuss these levels of organization in more detail. Chapter 2 describes the atomic and molecular levels; chapter 3 deals with organelles and cellular structures and functions; chapter 4 explores cellular metabolism; chapter 5 describes tissues; and chapter 6 presents the skin and its accessory organs as an example of an organ system. Beginning with chapter 7, the structures and functions of each of the organ systems are described in detail. Table 1.1 lists the levels of organization and some

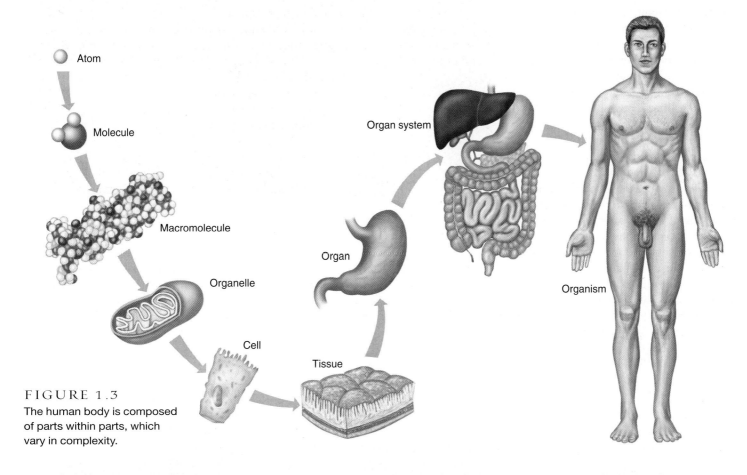

FIGURE 1.3
The human body is composed of parts within parts, which vary in complexity.

TABLE 1.1	Levels of Organization	
Level	**Example**	**Illustration**
Atom	Hydrogen atom, lithium atom	Figure 2.1
Molecule	Water molecule, glucose molecule	Figure 2.7
Macromolecule	Protein molecule, DNA molecule	Figure 2.19
Organelle	Mitochondrion, Golgi apparatus, nucleus	Figure 3.13
Cell	Muscle cell, nerve cell	Figure 5.28
Tissue	Simple squamous epithelium, loose connective tissue	Figure 5.1
Organ	Skin, femur, heart, kidney	Figure 6.2
Organ system	Integumentary system, skeletal system, digestive system	Figure 1.13
Organism	Human	Figure 1.19

ULTRASONOGRAPHY AND MAGNETIC RESONANCE IMAGING: A TALE OF TWO PATIENTS

The two patients enter the hospital medical scanning unit hoping for opposite outcomes. Vanessa Q., who has suffered several pregnancy losses, hopes that an ultrasound exam will reveal that her current pregnancy is continuing to progress normally. Michael P., a sixteen-year-old who has excruciating headaches, is to undergo a magnetic resonance imaging (MRI) scan to assure his physician (and himself!) that the cause of the headache is not a brain tumor.

Both ultrasound and magnetic resonance imaging scans are noninvasive procedures that provide images of soft internal structures. Ultrasonography uses high-frequency sound waves that are beyond the range of human hearing. A technician gently presses a device called a transducer, which emits sound waves, against the skin and moves it slowly over the surface of the area being examined, which in this case is Vanessa's abdomen (fig. 1A).

Prior to the exam, Vanessa drank several glasses of water. Her filled bladder will intensify the contrast between her uterus (and its contents) and nearby organs because as the sound waves from the transducer travel into the body, some of the waves reflect back to the transducer when they reach a border between structures of slightly different densities. Other sound waves continue into deeper tissues, and

FIGURE 1A

Ultrasonography uses reflected sound waves to visualize internal body structures.

some of them are reflected back by still other interfaces. As the reflected sound waves reach the transducer, they are converted into electrical impulses that are amplified and used to create a sectional image of the body's internal structure on a viewing screen. This image is known as a sonogram (fig. 1B).

Glancing at the screen, Vanessa smiles. The image reveals the fetus in her uterus, heart beating and already showing characteristics of a fully developed newborn.

Vanessa's ultrasound exam takes only a few minutes, whereas Michael's MRI scan takes an hour. First, he receives an injection of a dye that provides contrast so that a radiologist examining the scan can distinguish certain brain structures. Then, a nurse wheels the narrow bed on which Michael lies into a chamber surrounded by a powerful magnet and a special radio antenna. The chamber, which looks like a metal doughnut, is the MRI instrument. As Michael set-

corresponding illustrations in this textbook. Table 1.2 summarizes the organ systems, the major organs that comprise them, and their major functions in the order presented in this book. They are discussed in more detail later in this chapter (pages 16–19).

1 How does the human body illustrate levels of organization?

2 What is an organism?

3 How do body parts at different levels of organization vary in complexity?

Characteristics of Life

A scene such as Judith R.'s accident and injury underscores the delicate balance that must be maintained in order to sustain life. In those seconds at the limits of life—the birth of a baby, a trauma scene, or the precise instant of death following a long illness—we often think about just what combination of qualities constitutes this state that we call life. Indeed, although this text addresses the human body, the most fundamental characteristics of life are shared by all organisms. As living organisms, we

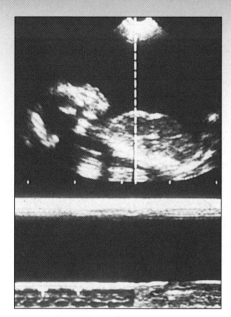

FIGURE 1B
This image resulting from an ultrasonographic procedure reveals the presence of a fetus in the uterus.

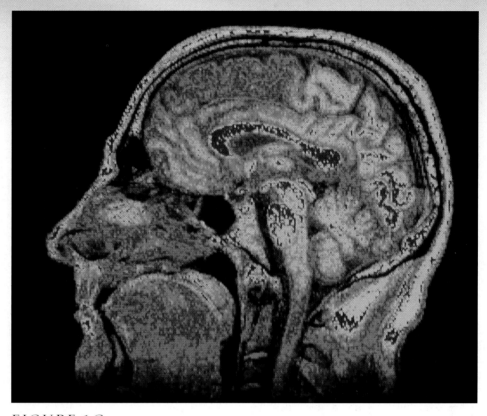

FIGURE 1C
Falsely colored MRI of a human head and brain (sagittal section).

tles back and closes his eyes, a technician activates the device.

The magnet generates a magnetic field that alters the alignment and spin of certain types of atoms within Michael's brain. At the same time, a second rotating magnetic field causes particular types of atoms (such as the hydrogen atoms in body fluids and organic compounds) to release weak radio waves with characteristic frequencies. The nearby antenna receives and amplifies the radio waves, which are then processed by a computer. Within a few minutes, the computer generates a sectional image based on the locations and concentrations of the atoms being studied (fig. 1C). The device continues to produce data, painting portraits of Michael's brain in the transverse, coronal, and sagittal sections (p. 22).

Michael and his parents nervously wait two days for the expert eyes of a radiologist to interpret the MRI scan. Happily, the scan shows normal brain structure. Whatever is causing Michael's headaches, it is not a brain tumor. ■

can move and respond to our surroundings. We start out as small individuals and then grow, eventually able to reproduce. We gain energy by ingesting (taking in), digesting (breaking down), absorbing, and assimilating the nutrients in food. The absorbed substances circulate throughout the internal environment of our bodies. We can then, by the process of respiration, use the energy in these nutrients for such vital functions as growth and repair of body parts. Finally, we excrete wastes from the body. Taken together, these physical and chemical events that release and utilize energy constitute **metabolism** (mĕ-tab′o-liz-m). Table 1.3 summarizes the characteristics of life.

1 What are the characteristics of life?

2 Which physical and chemical events constitute metabolism?

Maintenance of Life

With the exception of an organism's reproductive system, which perpetuates the species, all body structures and functions work in ways that maintain life.

TABLE 1.2   Organ Systems

Organ System	Major Organs	Major Functions
Integumentary	Skin, hair, nails, sweat glands, sebaceous glands	Protect tissues, regulate body temperature, support sensory receptors
Skeletal	Bones, ligaments, cartilages	Provide framework, protect soft tissues, provide attachments for muscles, produce blood cells, store inorganic salts
Muscular	Muscles	Cause movements, maintain posture, produce body heat
Nervous	Brain, spinal cord, nerves, sense organs	Detect changes, receive and interpret sensory information, stimulate muscles and glands
Endocrine	Glands that secrete hormones (pituitary gland, thyroid gland, parathyroid glands, adrenal glands, pancreas, ovaries, testes, pineal gland, and thymus gland)	Control metabolic activities of body structures
Cardiovascular	Heart, arteries, capillaries, veins	Move blood through blood vessels and transport substances throughout body
Lymphatic	Lymphatic vessels, lymph nodes, thymus, spleen	Return tissue fluid to the blood, carry certain absorbed food molecules, defend the body against infection
Digestive	Mouth, tongue, teeth, salivary glands, pharynx, esophagus, stomach, liver, gallbladder, pancreas, small and large intestines	Receive, break down, and absorb food; eliminate unabsorbed material
Respiratory	Nasal cavity, pharynx, larynx, trachea, bronchi, lungs	Intake and output of air, exchange of gases between air and blood
Urinary	Kidneys, ureters, urinary bladder, urethra	Remove wastes from blood, maintain water and electrolyte balance, store and transport urine
Reproductive	Male: scrotum, testes, epididymides, vasa deferentia, seminal vesicles, prostate gland, bulbourethral glands, urethra, penis	Produce and maintain sperm cells, transfer sperm cells into female reproductive tract
	Female: ovaries, uterine tubes, uterus, vagina, clitoris, vulva	Produce and maintain egg cells, receive sperm cells, support development of an embryo and function in birth process

TABLE 1.3   Characteristics of Life

Process	Examples	Process	Examples
Movement	Change in position of the body or of a body part; motion of an internal organ	Digestion	Breakdown of food substances into simpler forms that can be absorbed and used
Responsiveness	Reaction to a change taking place inside or outside the body	Absorption	Passage of substances through membranes and into body fluids
Growth	Increase in body size without change in shape	Circulation	Movement of substances from place to place in body fluids
Reproduction	Production of new organisms and new cells	Assimilation	Changing of absorbed substances into chemically different forms
Respiration	Obtaining oxygen, removing carbon dioxide, and releasing energy from foods (some forms of life do not use oxygen in respiration)	Excretion	Removal of wastes produced by metabolic reactions

Requirements of Organisms

Life depends upon the following environmental factors:

1. **Water** is the most abundant substance in the body. It is required for a variety of metabolic processes, and it provides the environment in which most of them take place. Water also transports substances within organisms and is important in regulating body temperature.

2. **Food** refers to substances that provide organisms with necessary chemicals (nutrients) in addition to water. Nutrients supply energy and raw materials for building new living matter.

3. **Oxygen** is a gas that makes up about one-fifth of the air. It is used in the process of releasing energy from nutrients. The energy, in turn, is used to drive metabolic processes.

4. **Heat** is a form of energy. It is a product of metabolic reactions, and it partly controls the rate at which these reactions occur. Generally, the more heat, the more rapidly chemical reactions take place. *Temperature* is a measure of the amount of heat present.

5. **Pressure** is an application of force on an object or substance. For example, the force acting on the outside of a land organism due to the weight of air above it is called *atmospheric pressure.* In humans, this pressure plays an important role in breathing. Similarly, organisms living under water are subjected to *hydrostatic pressure*—a pressure exerted by a liquid—due to the weight of water above them. In complex organisms, such as humans, heart action produces blood pressure (another form of hydrostatic pressure), which keeps blood flowing through blood vessels.

Although the human organism requires water, food, oxygen, heat, and pressure, these factors alone are not enough to ensure survival. Both the quantities and the qualities of such factors are also important. Table 1.4 summarizes the major requirements of organisms.

Homeostasis

Some organisms exist as single **cells,** the smallest living units. Consider the amoeba, a simple, one-celled organism found in lakes and ponds (fig. 1.4). Despite its simple structure compared to a human, an amoeba has very specific requirements that must be met if it is to survive. As long as the outside world—its **environment**—supports its requirements, an amoeba flourishes. As environmental factors such as temperature, water composition, and food availability become unsatisfactory, the amoeba's survival may be threatened. Although the amoeba has a limited ability to move from one place to another, environmental changes are likely to affect the whole pond, and with no place else to go, the amoeba dies.

In contrast to the amoeba, we humans are composed of about 70 trillion cells that surround themselves with their own environment inside our bodies. Our cells, as parts of organs and organ systems, interact in ways that keep this **internal environment** relatively constant, despite an ever-changing outside environment. Anatomically the internal environment is inside our body, but consists of the fluid that surrounds our cells, *extracellular fluid* (see chapter 21, p. 809). The internal environment protects our cells (and us!) from changes in the outside world that would kill isolated cells such as the amoeba (fig. 1.5). The body's maintenance of a stable internal environment is called **homeostasis,** (ho″me-ō-sta′sis) and it is so important that most of our metabolic energy is spent on it. Many of the tests performed on Judith R. during her

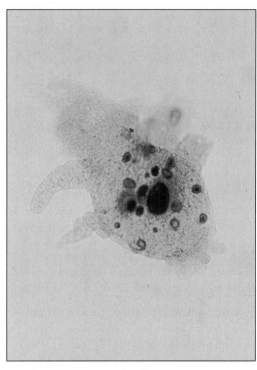

FIGURE 1.4
The amoeba is an organism consisting of a single cell (100×).

TABLE 1.4	Requirements of Organisms				
Factor	**Characteristic**	**Use**	**Factor**	**Characteristic**	**Use**
Water	A chemical substance	For metabolic processes, as a medium for metabolic reactions, to transport substances, and to regulate body temperature	Oxygen	A chemical substance	To help release energy from food substances
			Heat	A form of energy	To help regulate the rates of metabolic reactions
Food	Various chemical substances	To supply energy and raw materials for the production of necessary substances and for the regulation of vital reactions	Pressure	A force	Atmospheric pressure for breathing; hydrostatic pressure to help circulate blood

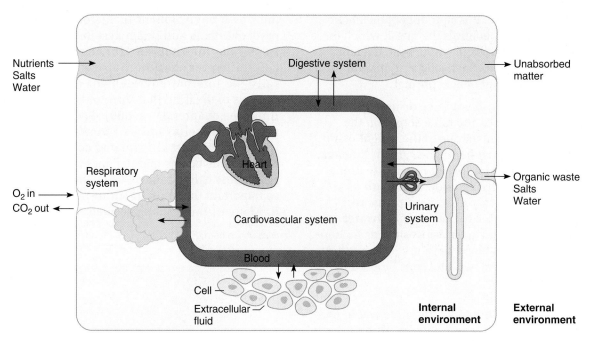

FIGURE 1.5

Our cells lie within an internal environment, which they maintain.

hospitalization (as described in the opening vignette) assessed her body's return to homeostasis.

The body maintains homeostasis through a number of self-regulating control systems, or **homeostatic mechanisms.** These have the following three components in common (fig. 1.6):

1. **Receptors,** which provide information about specific conditions (stimuli) in the internal environment.

2. A **control center,** which includes a **set point,** tells what a particular value should be (such as body temperature at 98.6°F).

3. **Effectors,** such as muscles or glands, which cause responses that alter conditions in the internal environment.

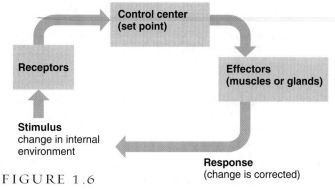

FIGURE 1.6

A homeostatic mechanism monitors an aspect of the internal environment and corrects any changes.

A homeostatic mechanism works as follows. If the receptors measure deviations from the set point, effectors are activated that can return conditions toward normal. As conditions return toward normal, the deviation from the set point progressively lessens, and the effectors are gradually shut down. Such a response is called a **negative feedback** (neg'ah-tiv fēd'bak) mechanism, both because the deviation from the set point is corrected (moves in the opposite or negative direction) and because the correction reduces the action of the effectors. This latter aspect is important because it prevents a correction from going too far.

To better understand this idea of maintaining a stable internal environment, imagine a room equipped with a furnace and an air conditioner. Suppose the room temperature is to remain near 20°C (68°F), so the thermostat is adjusted to a set point of 20°C. Because a thermostat is sensitive to temperature changes, it will signal the furnace to start and the air conditioner to stop whenever the room temperature drops below the set point. If the temperature rises above the set point, the thermostat will cause the furnace to stop and the air conditioner to start. These actions maintain a relatively constant temperature in the room (fig. 1.7).

A similar homeostatic mechanism regulates body temperature in humans (fig. 1.8). The "thermostat" is a temperature-sensitive region in a control center of the brain called the hypothalamus. In healthy persons, the set point of this body thermostat is at or near 37°C (98.6°F).

If a person is exposed to a cold environment and the body temperature begins to drop, the hypothalamus

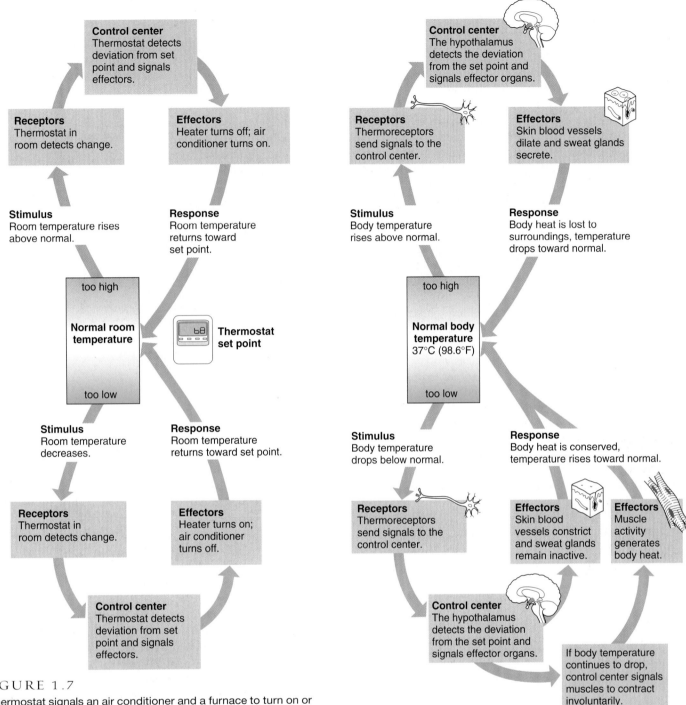

FIGURE 1.7

A thermostat signals an air conditioner and a furnace to turn on or off to maintain a relatively stable room temperature. This system is an example of a homeostatic mechanism.

FIGURE 1.8

The homeostatic mechanism that regulates body temperature.

senses this change and triggers heat-conserving and heat-generating activities. Blood vessels in the skin constrict so that blood flow there is reduced and deeper tissues retain heat. At the same time, small groups of muscle cells may be stimulated to contract involuntarily, an action called shivering. Such muscular contractions produce heat, which helps warm the body.

If a person becomes overheated, the hypothalamus triggers a series of changes that promote loss of body heat.

Sweat glands in the skin secrete watery perspiration. As the water evaporates from the surface, heat is carried away and the skin is cooled. At the same time, blood vessels in the skin dilate. This allows the blood that carries heat from deeper tissues to reach the surface where more heat is lost to the outside. Body temperature regulation is discussed in more detail in chapter 6 (p. 170).

Another homeostatic mechanism regulates the blood pressure in the blood vessels (arteries) leading away from the heart. In this instance, pressure-sensitive areas (sensory receptors) within the walls of these vessels detect changes in blood pressure and signal a pressure control center in the brain. If the blood pressure is above the pressure set point, the brain signals the heart, causing its chambers to contract less rapidly and with less force. Because of decreased heart action, less blood enters the blood vessels, and the pressure inside the vessels decreases. If the blood pressure drops below the set point, the brain center signals the heart to contract more rapidly and with greater force, increasing the pressure in the vessels. Chapter 15 (pp. 572–577) discusses blood pressure regulation in more detail.

A homeostatic mechanism regulates the concentration of the sugar glucose in blood. In this case, cells within an organ called the pancreas determine the set point. If, for example, the concentration of blood glucose increases following a meal, the pancreas detects this change and releases a chemical (insulin) into the blood. Insulin allows glucose to move from the blood into various body cells and to be stored in the liver and muscles. As this occurs, the concentration of blood glucose decreases, and as it reaches the normal set point, the pancreas decreases its release of insulin. If, on the other hand, the blood glucose concentration becomes abnormally low, the pancreas detects this change and releases a different chemical (glucagon) that causes stored glucose to be released into the blood. Chapter 13 (pp. 496–498) discusses regulation of the blood glucose concentration in more detail (see fig. 13.36).

Human physiology offers many other examples of homeostatic mechanisms. A familiar one is the increased respiratory activity that maintains blood levels of oxygen in the internal environment during strenuous exercise. Another is the sensation of thirst the nervous system creates when the internal environment is too concentrated, prompting us to drink. Negative feedback mechanisms also control hormone secretion (see chapter 13, p. 477).

Most feedback mechanisms in the body are negative. Sometimes in physiology, changes stimulate similar changes. Such a process that causes movement away from the normal state is called a *positive feedback mechanism.*

A positive feedback system operates for a short time when a blood clot forms, because the chemicals present in a clot promote further clotting (see chapter 14, pp. 527–528). Positive feedback controls milk production. If a baby suckles with greater force or duration, the mother's mammary glands respond by making more milk. Positive feedback also increases the strength of uterine contractions during childbirth. Because positive feedback mechanisms usually produce unstable conditions, the examples associated with normal health have very specific functions and are relatively short lived.

Homeostatic mechanisms maintain a relatively constant internal environment, yet physiological values may vary slightly in a person from time to time or from one person to the next. Therefore, both normal values for an individual and the idea of a **normal range** for the general population are clinically important. Numerous examples of homeostasis are presented throughout this book, and normal ranges for a number of physiological variables are listed in Appendix B, Laboratory Tests of Clinical Importance, pages 970–973.

1 Which requirements of organisms does the external environment provide?

2 What is the relationship between oxygen use and heat production?

3 Why is homeostasis so important to survival?

4 Describe three homeostatic mechanisms.

Organization of the Human Body

The human organism is a complex structure composed of many parts. The major features of the human body include cavities, various types of membranes, and organ systems.

Body Cavities

The human organism can be divided into an **axial** (ak′se-al) **portion,** which includes the head, neck, and trunk, and an **appendicular** (ap″en-dik′u-lar) **portion,** which includes the upper and lower limbs. Within the axial portion are two major cavities—a **dorsal cavity** and a larger **ventral cavity.** The organs within such a cavity are called **viscera** (vis′er-ah). The dorsal cavity can be subdivided into two parts—the **cranial cavity,** which houses the brain, and the **vertebral canal** (spinal cavity), which contains the spinal cord and is surrounded by sections of the backbone (vertebrae). The ventral cavity consists of a **thoracic** (tho-ras′ik) **cavity** and an **abdominopelvic cavity.** Figure 1.9 shows these major body cavities.

The thoracic cavity is separated from the lower abdominopelvic cavity by a broad, thin muscle called the **diaphragm** (di′ah-fram). When it is at rest, this muscle curves upward into the thorax like a dome. When it contracts during inhalation, it presses down upon the abdominal viscera. The wall of the thoracic cavity is composed of skin, skeletal muscles, and bones. Within the thoracic cavity are the lungs and a region between the lungs, called the **mediastinum** (me″de-as-ti′num). The mediastinum separates the thorax into two compartments that contain the right and left lungs. The remaining thoracic viscera—heart, esophagus, trachea, and thymus gland—are within the mediastinum.

The abdominopelvic cavity, which includes an upper abdominal portion and a lower pelvic portion,

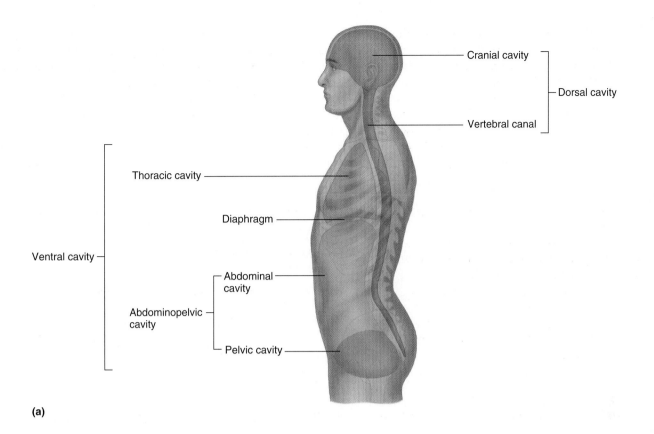

(a)

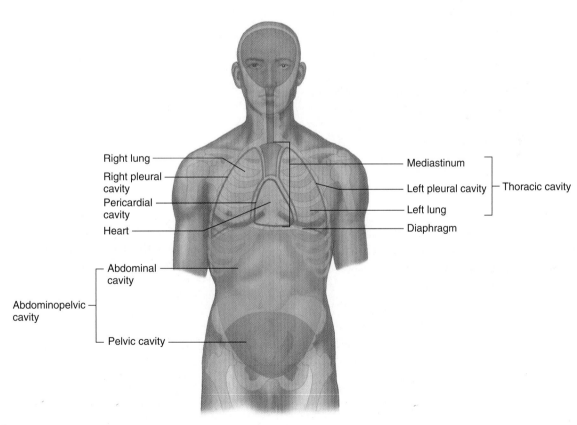

(b)

FIGURE 1.9

Major body cavities. (*a*) Lateral view. (*b*) Anterior view.

extends from the diaphragm to the floor of the pelvis. Its wall primarily consists of skin, skeletal muscles, and bones. The viscera within the **abdominal cavity** include the stomach, liver, spleen, gallbladder, and the small and large intestines.

The **pelvic cavity** is the portion of the abdomino-pelvic cavity enclosed by the pelvic bones. It contains the terminal end of the large intestine, the urinary bladder, and the internal reproductive organs.

Smaller cavities within the head include the following (fig. 1.10):

1. *Oral cavity,* containing the teeth and tongue.

2. *Nasal cavity,* located within the nose and divided into right and left portions by a nasal septum. Several air-filled sinuses are connected to the nasal cavity. These include the sphenoidal and frontal sinuses (see fig. 7.25).

3. *Orbital cavities,* containing the eyes and associated skeletal muscles and nerves.

4. *Middle ear cavities,* containing the middle ear bones.

Thoracic and Abdominopelvic Membranes

Thin **serous membranes** line the walls of the thoracic and abdominal cavities and fold back to cover the organs within these cavities. These membranes secrete a slippery serous fluid that separates the layer lining the wall (parietal layer) from the layer covering the organ (visceral layer). For example, the right and left thoracic compartments, which contain the lungs, are lined with a serous membrane called the *parietal pleura.* This membrane folds back to cover the lungs, forming the *visceral pleura.* A thin film of serous fluid separates the parietal and visceral **pleural** (ploo′ral) **membranes.** Although there is normally no actual space between these two membranes, the potential space between them is called the *pleural cavity.*

The heart, which is located in the broadest portion of the mediastinum, is surrounded by **pericardial** (per″ĭ-kar′de-al) **membranes.** A thin *visceral pericardium* (epicardium) covers the heart's surface and is separated from the *parietal pericardium* by a small amount of serous fluid. The potential space between these membranes is called the *pericardial cavity.* The parietal pericardium is covered by a much thicker third layer, the *fibrous pericardium.* Figure 1.11 shows the membranes associated with the heart and lungs.

In the abdominopelvic cavity, the membranes are called **peritoneal** (per″ĭ-to-ne′al) **membranes.** A *parietal peritoneum* lines the wall, and a *visceral peritoneum* covers each organ in the abdominal cavity. The potential space between these membranes is called the *peritoneal cavity* (fig. 1.12).

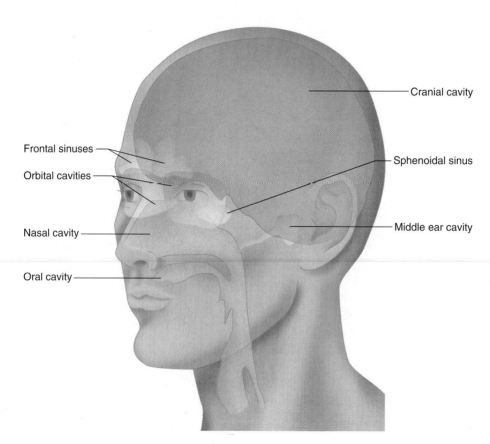

FIGURE 1.10

The cavities within the head include the cranial, oral, nasal, orbital, and middle ear cavities, as well as several sinuses.

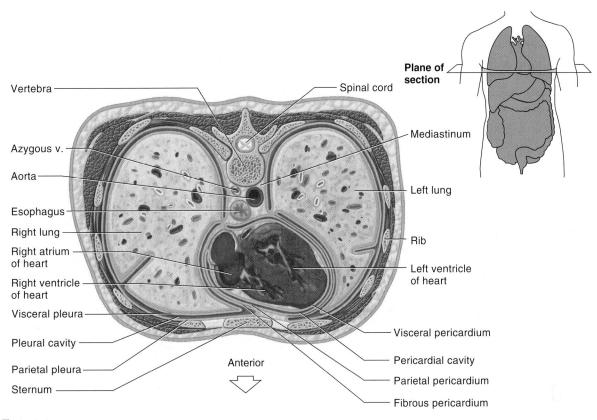

FIGURE 1.11

A transverse section through the thorax reveals the serous membranes associated with the heart and lungs (superior view).

Labels for Figure 1.11:
Vertebra, Spinal cord, Azygous v., Aorta, Mediastinum, Esophagus, Left lung, Right lung, Rib, Right atrium of heart, Left ventricle of heart, Right ventricle of heart, Visceral pleura, Pleural cavity, Visceral pericardium, Parietal pleura, Pericardial cavity, Sternum, Parietal pericardium, Fibrous pericardium, Anterior, Plane of section

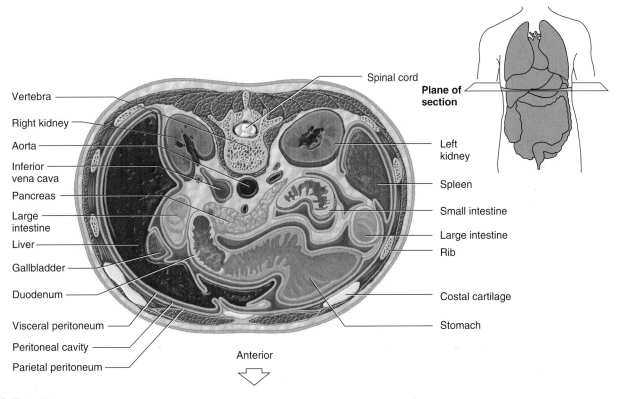

FIGURE 1.12

Transverse section through the abdomen (superior view).

Labels for Figure 1.12:
Vertebra, Spinal cord, Right kidney, Plane of section, Aorta, Left kidney, Inferior vena cava, Pancreas, Spleen, Large intestine, Small intestine, Liver, Large intestine, Gallbladder, Rib, Duodenum, Costal cartilage, Visceral peritoneum, Stomach, Peritoneal cavity, Parietal peritoneum, Anterior

1. What are the viscera?

2. Which organs occupy the dorsal cavity? The ventral cavity?

3. Name the cavities of the head.

4. Describe the membranes associated with the thoracic and abdominopelvic cavities.

5. Distinguish between the parietal and visceral peritoneum.

Organ Systems

The human organism consists of several organ systems, each of which includes a set of interrelated organs that work together to provide specialized functions. The maintenance of homeostasis depends on the coordination of organ systems. A figure called **"InnerConnections"** at the end of certain chapters ties together the ways in which organ systems interact. As you read about each organ system, you may want to consult the illustrations of the human torso in reference plates 1–7 and locate some of the features listed in the descriptions.

Body Covering

The organs of the **integumentary** (in-teg-u-men′tar-e) **system** (fig. 1.13) include the skin and accessory organs such as the hair, nails, sweat glands, and sebaceous glands. These parts protect underlying tissues, help regulate body temperature, house a variety of sensory receptors, and synthesize certain products. Chapter 6 discusses the integumentary system.

Support and Movement

The organs of the skeletal and muscular systems support and move body parts. The **skeletal** (skel′ĕ-tal) **system** (fig. 1.14) consists of the bones as well as the ligaments and cartilages that bind bones together at joints. These parts provide frameworks and protective shields for softer tissues, serve as attachments for muscles, and act together with muscles when body parts move. Tissues within bones also produce blood cells and store inorganic salts.

The muscles are the organs of the **muscular** (mus′ku-lar) **system** (fig. 1.14). By contracting and pulling their ends closer together, they provide the forces that cause body movements. Muscles also help maintain posture and are the primary source of body heat. Chapters 7, 8, and 9 discuss the skeletal and muscular systems.

Integration and Coordination

For the body to act as a unit, its parts must be integrated and coordinated. The nervous and endocrine systems control and adjust various organ functions from time to time, maintaining homeostasis.

The **nervous** (ner′vus) **system** (fig. 1.15) consists of the brain, spinal cord, nerves, and sense organs. Nerve cells within these organs use electrochemical signals called *nerve impulses* (action potentials) to communicate with

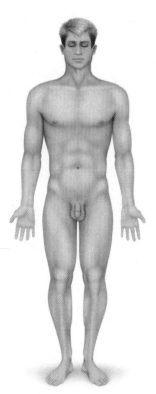

Integumentary system

FIGURE 1.13

The integumentary system covers the body.

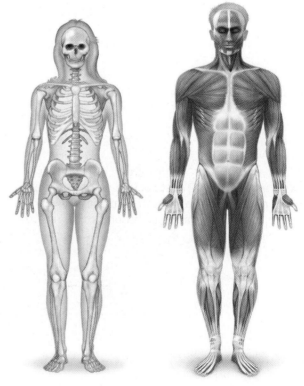

Skeletal system   Muscular system

FIGURE 1.14

The skeletal and muscular organ systems provide support and movement.

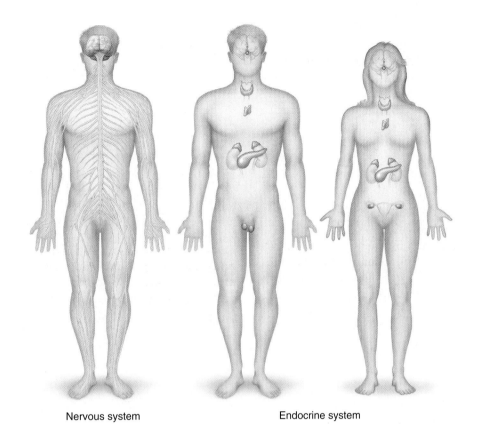

Nervous system   Endocrine system

FIGURE 1.15

The nervous and endocrine organ systems integrate and coordinate body functions.

one another and with muscles and glands. Each impulse produces a relatively short-term effect on its target. Some nerve cells act as specialized sensory receptors that can detect changes occurring inside and outside the body. Other nerve cells receive the impulses transmitted from these sensory units and interpret and act on the information. Still other nerve cells carry impulses from the brain or spinal cord to muscles or glands, stimulating them to contract or to secrete products. Chapters 10 and 11 discuss the nervous system, and chapter 12 discusses sense organs.

The **endocrine** (en′do-krin) **system** (fig. 1.15) includes all the glands that secrete chemical messengers, called *hormones.* Hormones, in turn, travel away from the glands in body fluids such as blood or tissue fluid. Usually a particular hormone affects only a particular group of cells, called its *target cells.* The effect of a hormone is to alter the metabolism of the target cells. Compared to nerve impulses, hormonal effects occur over a relatively long time period.

Organs of the endocrine system include the pituitary, thyroid, parathyroid, and adrenal glands, as well as the pancreas, ovaries, testes, pineal gland, and thymus gland. These are discussed further in chapter 13.

Transport

Two organ systems transport substances throughout the internal environment. The **cardiovascular** (kahr″de-o-vas′ku-lur) **system** (fig. 1.16) includes the heart, arteries,

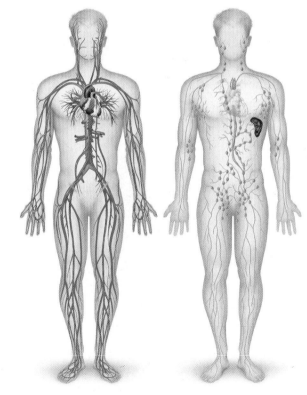

Cardiovascular system   Lymphatic system

FIGURE 1.16

The cardiovascular and lymphatic organ systems transport fluids.

CHAPTER ONE   *Introduction to Human Anatomy and Physiology*

capillaries, veins, and blood. The heart is a muscular pump that helps force blood through the blood vessels. Blood transports gases, nutrients, hormones, and wastes. It carries oxygen from the lungs and nutrients from the digestive organs to all body cells, where these substances are used in metabolic processes. Blood also transports hormones from endocrine glands to their target cells and carries wastes from body cells to the excretory organs, where the wastes are removed from the blood and released to the outside. Blood and the cardiovascular system are discussed in chapters 14 and 15.

The **lymphatic** (lim-fat'ik) **system** (fig. 1.16) is sometimes considered part of the cardiovascular system. It is composed of the lymphatic vessels, lymph fluid, lymph nodes, thymus gland, and spleen. This system transports some of the fluid from the spaces within tissues (tissue fluid) back to the bloodstream and carries certain fatty substances away from the digestive organs. Cells of the lymphatic system, called lymphocytes, defend the body against infections by removing pathogens (disease-causing microorganisms and viruses) from tissue fluid. The lymphatic system is discussed in chapter 16.

Absorption and Excretion

Organs in several systems absorb nutrients and oxygen and excrete wastes. The organs of the **digestive** (di-jest'tiv) **system** (fig. 1.17) receive foods and then break down food molecules into simpler forms that can pass through cell membranes and be absorbed into the internal environment. Materials that are not absorbed are transported outside. Certain digestive organs (chapter 17, pp. 661, 662, 665) also produce hormones and thus function as parts of the endocrine system.

The digestive system includes the mouth, tongue, teeth, salivary glands, pharynx, esophagus, stomach, liver, gallbladder, pancreas, small intestine, and large intestine. Chapter 18 discusses nutrition and metabolism, considering the fate of foods in the body.

The organs of the **respiratory** (re-spi'rah-to"re) **system** (fig. 1.17) take air in and out and exchange gases between the blood and the air. More specifically, oxygen passes from air within the lungs into the blood, and carbon dioxide leaves the blood and enters the air. The nasal cavity, pharynx, larynx, trachea, bronchi, and lungs are parts of this system, which is discussed in chapter 19.

The **urinary** (u'rĭ-ner"e) **system** (fig. 1.17) consists of the kidneys, ureters, urinary bladder, and urethra. The kidneys remove wastes from blood and assist in maintaining the body's water and electrolyte balance. The product of these activities is urine. Other portions of the urinary system store urine and transport it outside the body. Chapter 20 discusses the urinary system. Sometimes the urinary system is called the *excretory system.* However, **excretion** (ek-skre'shun), or waste removal, is also a function of the respiratory system and, to a lesser extent, the digestive and integumentary systems.

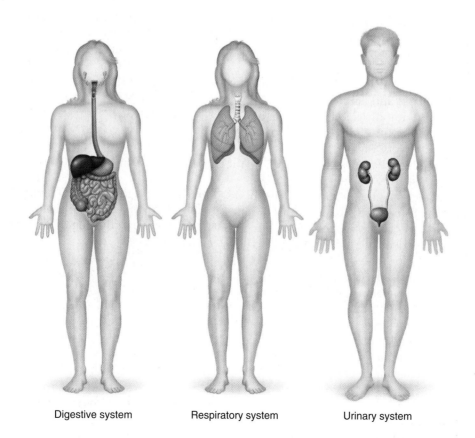

Digestive system   Respiratory system   Urinary system

FIGURE 1.17

The digestive, respiratory, and urinary organ systems absorb nutrients, take in oxygen and release carbon dioxide, and excrete wastes.

Reproduction

Reproduction (re″pro-duk′shun) is the process of producing offspring (progeny). Cells reproduce when they divide and give rise to new cells. The **reproductive** (re″pro-duk′tiv) **system** (fig. 1.18) of an organism, however, produces whole new organisms like itself (see chapter 22).

The male reproductive system includes the scrotum, testes, epididymides, vasa deferentia, seminal vesicles, prostate gland, bulbourethral glands, urethra, and penis. These structures produce and maintain the male sex cells, or sperm cells (spermatozoa). The male reproductive system also transfers these cells from their site of origin into the female reproductive tract.

The female reproductive system consists of the ovaries, uterine tubes, uterus, vagina, clitoris, and vulva. These organs produce and maintain the female sex cells (egg cells or ova), receive the male sex cells (sperm cells), and transport the female sex cells within the female reproductive system. The female reproductive system also supports development of embryos and functions in the birth process.

Figure 1.19 illustrates the organ systems in humans. Also, special looks at various organs and organ systems as a person ages are considered in certain chapters, beginning here.

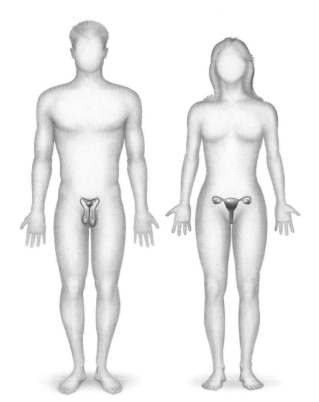

Male reproductive system   Female reproductive system

FIGURE 1.18

The reproductive systems manufacture and transport sex cells. The female reproductive system provides for fetal development and childbirth.

1 Name the major organ systems and list the organs of each system.

2 Describe the general functions of each organ system.

Life-Span Changes

Aging is a part of life. According to the dictionary, aging is the process of becoming mature or old. It is, in essence, the passage of time and the accompanying bodily changes. Because the passage of time is inevitable, so, too, is aging, claims for the anti-aging properties of various diets, cosmetics, pills and skin-care products to the contrary.

Aging occurs from the whole-body level to the microscopic level. Although programmed cell death begins in the fetus, we are usually not very aware of aging until the third decade of life, when a few gray hairs, faint lines etched into facial skin, and minor joint stiffness in the morning remind us that time marches on. A woman over the age of thirty-five attempting to conceive a child might be shocked to learn that she is of "advanced maternal age," because the chances of conceiving an offspring with an abnormal chromosome number increase with the age of the egg. In both sexes, by the fourth or fifth decade, as hair color fades and skin etches become wrinkles, the first signs of adult-onset disorders may appear, such as increased blood pressure that one day may be considered hypertension, and slightly elevated blood glucose that could become diabetes mellitus. A person with a strong family history of heart disease, coupled with unhealthy diet and exercise habits, may be advised to change his or her lifestyle, and perhaps even begin taking a drug to lower serum cholesterol levels. The sixth decade sees grayer or whiter hair, more and deeper skin wrinkles, and a waning immunity that makes vaccinations against influenza and other infectious diseases important. Yet many, if not most, people in their sixties and older have sharp minds and are capable of all sorts of physical activities.

Changes at the tissue, cell and molecular levels explain the familiar signs of aging. Decreased production of the connective tissue proteins collagen and elastin account for the stiffening of skin, and diminished levels of subcutaneous fat are responsible for wrinkling. Proportions of fat to water in the tissues change, with the percentage of fats increasing steadily in women, and increasing until about age sixty in men. These alterations explain why the elderly metabolize certain drugs at different rates than do younger people. As a person ages, tissues atrophy, and as a result, organs shrink.

Cells mark time too, many approaching the end of a limited number of predetermined cell divisions as their chromosome tips whittle down. Such cells reaching the end of their division days may enlarge or die. Some cells may be unable to build the spindle apparatus that pulls apart replicated chromosomes in a cell on the verge of division. Impaired cell division translates into impaired

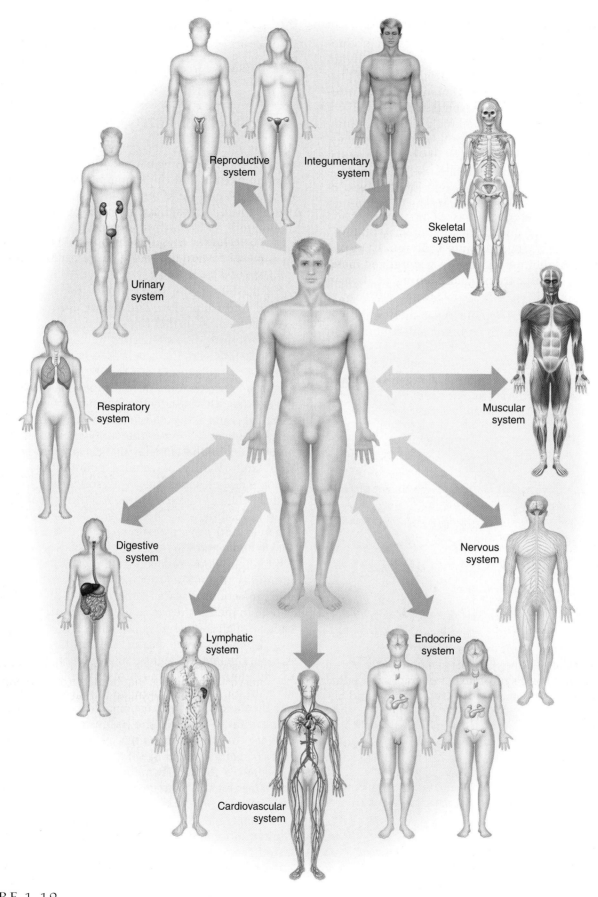

FIGURE 1.19

The organ systems in humans interact to maintain homeostasis.

wound healing, yet at the same time, the inappropriate cell division that underlies cancer becomes more likely. Certain subcellular functions lose efficiency, including the DNA repair that would otherwise patch up mutations, and the transport of substances across cell membranes. Aging cells also have fewer mitochondria, the structures that house the reactions that extract energy from nutrients, and also have fewer lysosomes, the disposal units that break down aged or damaged cell parts.

Just as changes at the tissue level cause organ-level signs of aging, certain biochemical changes fuel cellular aging. Lipofuscin and ceroid pigments accumulate as the cell can no longer prevent formation of damaging oxygen free radicals. A protein called beta amyloid may build up in the brain, contributing, in some individuals, to the development of Alzheimer disease. A generalized metabolic slowdown results from a dampening of thyroid gland function, impairing glucose utilization, the rate of protein synthesis, and production of digestive enzymes. At the whole-body level, we notice slowed metabolism as diminished tolerance to cold, weight gain, and fatigue.

A clearer understanding of the precise steps of the aging process will emerge as researchers identify the roles of each of our genes. For example, many of the molecular and cellular changes of aging may be controlled by the action of one gene, called p21. Its protein product turns on and off about ninety other genes, whose specific actions promote the signs of older age. The p21 gene intervenes when cells are damaged by radiation or toxins, promoting their death, which prevents them from causing disease. It also stimulates production of proteins that are associated with particular disorders seen in aging, including atherosclerosis, Alzheimer disease, and arthritis.

Because our organs and organ systems are interrelated, aging-related changes in one influence the functioning of others. Several chapters in this book conclude with a "Life-Span Changes" section that discusses changes specific to particular organ systems. These changes reflect the natural breakdown of structure and function that accompanies the passage of time, as well as events that are in our genes ("nature") and symptoms or characteristics that might arise as a consequence of lifestyle choices and circumstances ("nurture").

Anatomical Terminology

To communicate effectively with one another, investigators over the ages have developed a set of terms with precise meanings. Some of these terms concern the relative positions of body parts, others refer to imaginary planes along which cuts may be made, and still others describe body regions. When such terms are used, it is assumed that the body is in the **anatomical position;** that is, it is standing erect, the face is forward, and the upper limbs are at the sides, with the palms forward.

Relative Position

Terms of relative position are used to describe the location of one body part with respect to another. They include the following:

1. **Superior** means a part is above another part, or closer to the head. (The thoracic cavity is superior to the abdominopelvic cavity.)

2. **Inferior** means a part is below another part, or toward the feet. (The neck is inferior to the head.)

3. **Anterior** (or ventral) means toward the front. (The eyes are anterior to the brain.)

4. **Posterior** (or dorsal) is the opposite of anterior; it means toward the back. (The pharynx is posterior to the oral cavity.)

5. **Medial** relates to an imaginary midline dividing the body into equal right and left halves. A part is medial if it is closer to this line than another part. (The nose is medial to the eyes.)

6. **Lateral** means toward the side with respect to the imaginary midline. (The ears are lateral to the eyes.) **Ipsilateral** pertains to the same side (the spleen and the descending colon are ipsilateral), whereas **contralateral** refers to the opposite side (the spleen and the gallbladder are contralateral).

7. **Proximal** describes a part that is closer to the trunk of the body or closer to another specified point of reference than another part. (The elbow is proximal to the wrist.)

8. **Distal** is the opposite of proximal. It means a particular body part is farther from the trunk or farther from another specified point of reference than another part. (The fingers are distal to the wrist.)

9. **Superficial** means situated near the surface. (The epidermis is the superficial layer of the skin.) **Peripheral** also means outward or near the surface. It describes the location of certain blood vessels and nerves. (The nerves that branch from the brain and spinal cord are peripheral nerves.)

10. **Deep** describes parts that are more internal. (The dermis is the deep layer of the skin.)

Body Sections

To observe the relative locations and arrangements of internal parts, it is necessary to cut, or section, the body along various planes (figs. 1.20 and 1.21). The following terms describe such planes and sections:

1. **Sagittal** refers to a lengthwise cut that divides the body into right and left portions. If a sagittal section passes along the midline and divides the body into equal parts, it is called median (midsagittal).

CHAPTER ONE *Introduction to Human Anatomy and Physiology*

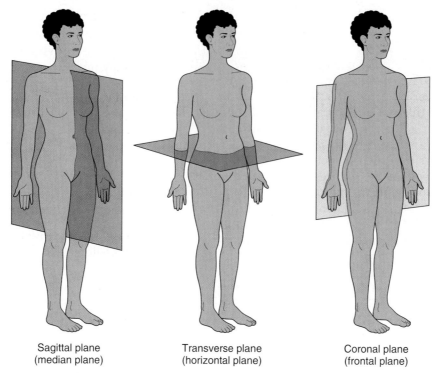

Sagittal plane
(median plane)

Transverse plane
(horizontal plane)

Coronal plane
(frontal plane)

FIGURE 1.20
Observation of internal parts requires sectioning the body along various planes.

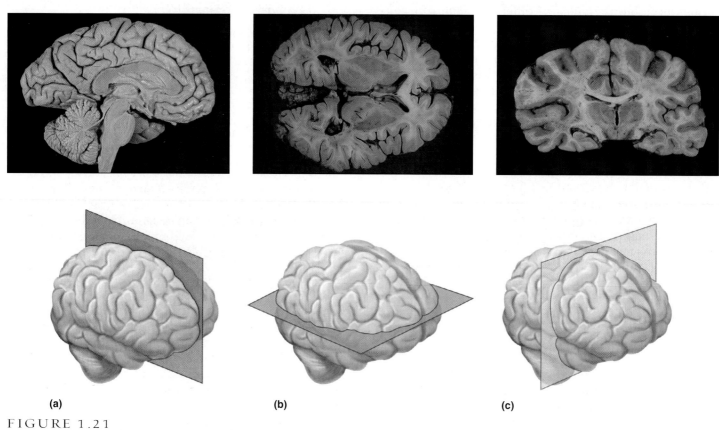

(a)   (b)   (c)

FIGURE 1.21
A human brain sectioned along (a) the sagittal plane, (b) the transverse plane, and (c) the coronal plane.

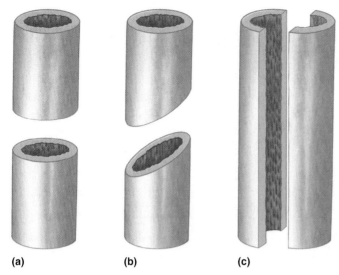

(a)   **(b)**   **(c)**

FIGURE 1.22

Cylindrical parts may be cut in (a) cross section, (b) oblique section, or (c) longitudinal section.

2. **Transverse** (or horizontal) refers to a cut that divides the body into superior and inferior portions.

3. **Coronal** (or frontal) refers to a section that divides the body into anterior and posterior portions.

Sometimes a cylindrical organ such as a blood vessel is sectioned. In this case, a cut across the structure is called a *cross section,* an angular cut is called an *oblique*

section, and a lengthwise cut is called a *longitudinal section* (fig. 1.22).

Body Regions

A number of terms designate body regions. The abdominal area, for example, is subdivided into the following regions, as shown in figure 1.23a:

1. **Epigastric region**   The upper middle portion.

2. **Left** and **right hypochondriac regions**   On each side of the epigastric region.

3. **Umbilical region**   The central portion.

4. **Left** and **right lumbar regions**   On each side of the umbilical region.

5. **Hypogastric region**   The lower middle portion.

6. **Left** and **right iliac (or inguinal) regions**   On each side of the hypogastric region.

The abdominal area may also be subdivided into the following four quadrants, as figure 1.23b illustrates:

1. **Right upper quadrant** (RUQ).

2. **Right lower quadrant** (RLQ).

3. **Left upper quadrant** (LUQ).

4. **Left lower quadrant** (LLQ).

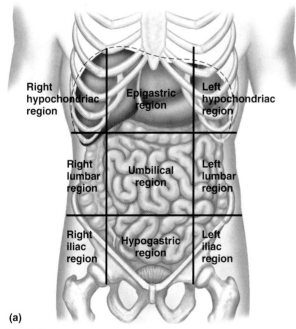

(a)

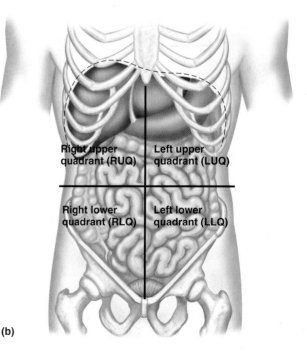

(b)

FIGURE 1.23

There are two common ways to subdivide the abdominal area. (a) The abdominal area is subdivided into nine regions. (b) The abdominal area may also be subdivided into four quadrants.

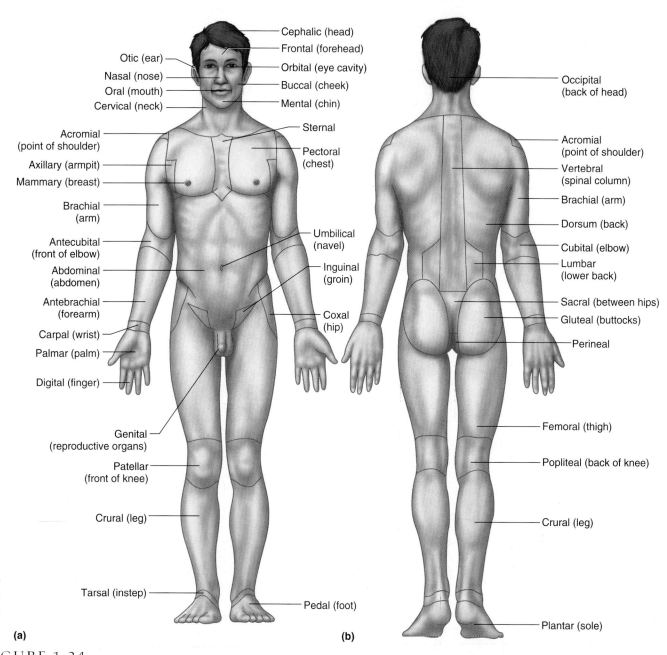

(a)   **(b)**

FIGURE 1.24

Some terms used to describe body regions. (*a*) Anterior regions. (*b*) Posterior regions.

   The following terms are commonly used when referring to various body regions. Figure 1.24 illustrates some of these regions.

abdominal (ab-dom′ĭ-nal) region between the thorax and pelvis
acromial (ah-kro′me-al) point of the shoulder
antebrachial (an″te-bra′ke-al) forearm
antecubital (an″te-ku′bĭ-tal) space in front of the elbow
axillary (ak′sĭ-ler″e) armpit

brachial (bra′ke-al) arm
buccal (buk′al) cheek
carpal (kar′pal) wrist
celiac (se′le-ak) abdomen
cephalic (sĕ-fal′ik) head
cervical (ser′vĭ-kal) neck
costal (kos′tal) ribs
coxal (kok′sal) hip
crural (krōōr′al) leg
cubital (ku′bĭ-tal) elbow
digital (dij′ĭ-tal) finger

dorsum (dor'sum) back
femoral (fem'or-al) thigh
frontal (frun'tal) forehead
genital (jen'i-tal) reproductive organs
gluteal (gloo'te-al) buttocks
inguinal (ing'gwĭ-nal) depressed area of the abdominal wall near the thigh (groin)
lumbar (lum'bar) region of the lower back between the ribs and the pelvis (loin)
mammary (mam'er-e) breast
mental (men'tal) chin
nasal (na'zal) nose
occipital (ok-sip'ĭ-tal) lower posterior region of the head
oral (o'ral) mouth
orbital (or'bi-tal) eye cavity
otic (o'tik) ear
palmar (pahl'mar) palm of the hand
patellar (pah-tel'ar) front of the knee
pectoral (pek'tor-al) chest
pedal (ped'al) foot
pelvic (pel'vik) pelvis
perineal (per"ĭ-ne'al) region between the anus and the external reproductive organs (perineum)
plantar (plan'tar) sole of the foot
popliteal (pop"lĭ-te'al) area behind the knee
sacral (sa'kral) posterior region between the hipbones
sternal (ster'nal) middle of the thorax, anteriorly
tarsal (tahr'sal) instep of the foot (ankle)
umbilical (um-bil'ĭ-kal) navel
vertebral (ver'te-bral) spinal column

1 Describe the anatomical position.

2 Using the appropriate terms, describe the relative positions of several body parts.

3 Describe three types of body sections.

4 Describe the nine regions of the abdomen.

5 Explain how the names of the abdominal regions describe their locations.

Some Medical and Applied Sciences

cardiology (kar"de-ol'o-je) Branch of medical science dealing with the heart and heart diseases.
dermatology (der"mah-tol'o-je) Study of skin and its diseases.
endocrinology (en"do-krĭ-nol'o-je) Study of hormones, hormone-secreting glands, and associated diseases.
epidemiology (ep"ĭ-de"me-ol'o-je) Study of the factors that contribute to determining the distribution and frequency of health-related conditions within a defined human population.
gastroenterology (gas"tro-en"ter-ol'o-je) Study of the stomach and intestines, as well as their diseases.
geriatrics (jer"e-at'riks) Branch of medicine dealing with older individuals and their medical problems.
gerontology (jer"on-tol'o-je) Study of the process of aging and the various problems of older individuals.
gynecology (gi"nĕ-kol'o-je) Study of the female reproductive system and its diseases.
hematology (hem"ah-tol'o-je) Study of blood and blood diseases.
histology (his-tol'o-je) Study of the structure and function of tissues (microscopic anatomy).
immunology (im"u-nol'o-je) Study of the body's resistance to disease.
neonatology (ne"o-na-tol'o-je) Study of newborns and the treatment of their disorders.
nephrology (nĕ-frol'o-je) Study of the structure, function, and diseases of the kidneys.
neurology (nu-rol'o-je) Study of the nervous system in health and disease.
obstetrics (ob-stet'riks) Branch of medicine dealing with pregnancy and childbirth.
oncology (ong-kol'o-je) Study of cancers.
ophthalmology (of"thal-mol'o-je) Study of the eye and eye diseases.
orthopedics (or"tho-pe'diks) Branch of medicine dealing with the muscular and skeletal systems and their problems.
otolaryngology (o"to-lar"in-gol'o-je) Study of the ear, throat, larynx, and their diseases.
pathology (pah-thol'o-je) Study of structural and functional changes within the body associated with disease.
pediatrics (pe"de-at'riks) Branch of medicine dealing with children and their diseases.
pharmacology (fahr"mah-kol'o-je) Study of drugs and their uses in the treatment of diseases.
podiatry (po-di'ah-tre) Study of the care and treatment of the feet.
psychiatry (si-ki'ah-tre) Branch of medicine dealing with the mind and its disorders.
radiology (ra"de-ol'o-je) Study of X rays and radioactive substances, as well as their uses in diagnosing and treating diseases.
toxicology (tok"sĭ-kol'o-je) Study of poisonous substances and their effects on physiology.
urology (u-rol'o-je) Branch of medicine dealing with the urinary and male reproductive systems and their diseases.

Introduction (page 3)

1. Early interest in the human body probably developed as people became concerned about injuries and illnesses. Changes in lifestyle, from hunter-gatherer to farmer to city dweller, were reflected in types of illnesses.
2. Early doctors began to learn how certain herbs and potions affected body functions.
3. The idea that humans could understand forces that caused natural events led to the development of modern science.
4. A set of terms originating from Greek and Latin formed the basis for the language of anatomy and physiology.

Anatomy and Physiology (page 3)

1. Anatomy deals with the form and organization of body parts.
2. Physiology deals with the functions of these parts.
3. The function of a part depends upon the way it is constructed.

Levels of Organization (page 4)

The body is composed of parts that can be considered at different levels of organization.

1. Matter is composed of atoms.
2. Atoms join to form molecules.
3. Organelles consist of aggregates of interacting large molecules.
4. Cells, which are composed of organelles, are the basic units of structure and function of the body.
5. Cells are organized into layers or masses called tissues.
6. Tissues are organized into organs.
7. Organs form organ systems.
8. Organ systems constitute the organism.
9. These parts vary in complexity progressively from one level to the next.

Characteristics of Life (page 6)

Characteristics of life are traits all organisms share.

1. These characteristics include
   a. Movement—changing body position or moving internal parts.
   b. Responsiveness—sensing and reacting to internal or external changes.
   c. Growth—increasing in size without changing in shape.
   d. Reproduction—producing offspring.
   e. Respiration—obtaining oxygen, using oxygen to release energy from foods, and removing gaseous wastes.
   f. Digestion—breaking down food substances into forms that can be absorbed.
   g. Absorption—moving substances through membranes and into body fluids.
   h. Circulation—moving substances through the body in body fluids.
   i. Assimilation—changing substances into chemically different forms.
   j. Excretion—removing body wastes.
2. Metabolism is the acquisition and utilization of energy by an organism.

Maintenance of Life (page 7)

The structures and functions of body parts maintain the life of the organism.

1. Requirements of organisms
   a. Water is used in many metabolic processes, provides the environment for metabolic reactions, and transports substances.
   b. Nutrients supply energy, raw materials for building substances, and chemicals necessary in vital reactions.
   c. Oxygen is used in releasing energy from nutrients; this energy drives metabolic reactions.
   d. Heat is a product of metabolic reactions and helps control rates of these reactions.
   e. Pressure is an application of force; in humans, atmospheric and hydrostatic pressures help breathing and blood movements, respectively.
2. Homeostasis
   a. If an organism is to survive, the conditions within its body fluids must remain relatively stable.
   b. The tendency to maintain a stable internal environment is called homeostasis.
   c. Homeostatic mechanisms involve sensory receptors, a control center with a set point, and effectors.
   d. Homeostatic mechanisms include those that regulate body temperature, blood pressure, and blood glucose concentration.
   e. Homeostatic mechanisms employ negative feedback.

Organization of the Human Body (page 12)

1. Body cavities
   a. The axial portion of the body contains the dorsal and ventral cavities.
      (1) The dorsal cavity includes the cranial cavity and vertebral canal.
      (2) The ventral cavity includes the thoracic and abdominopelvic cavities, which are separated by the diaphragm.
   b. The organs within a body cavity are called viscera.
   c. Other body cavities include the oral, nasal, orbital, and middle ear cavities.
2. Thoracic and abdominopelvic membranes
   Parietal serous membranes line the walls of these cavities; visceral serous membranes cover organs within them. They secrete serous fluid.
   a. Thoracic membranes
      (1) Pleural membranes line the thoracic cavity and cover the lungs.
      (2) Pericardial membranes surround the heart and cover its surface.
      (3) The pleural and pericardial cavities are potential spaces between these membranes.
   b. Abdominopelvic membranes
      (1) Peritoneal membranes line the abdominopelvic cavity and cover the organs inside.
      (2) The peritoneal cavity is a potential space between these membranes.
3. Organ systems
   The human organism consists of several organ systems. Each system includes interrelated organs.
   a. Integumentary system
      (1) The integumentary system covers the body.

(2) It includes the skin, hair, nails, sweat glands, and sebaceous glands.
(3) It protects underlying tissues, regulates body temperature, houses sensory receptors, and synthesizes substances.
b. Skeletal system
(1) The skeletal system is composed of bones and the ligaments and cartilages that bind bones together.
(2) It provides framework, protective shields, and attachments for muscles; it also produces blood cells and stores inorganic salts.
c. Muscular system
(1) The muscular system includes the muscles of the body.
(2) It moves body parts, maintains posture, and produces body heat.
d. Nervous system
(1) The nervous system consists of the brain, spinal cord, nerves, and sense organs.
(2) It receives impulses from sensory parts, interprets these impulses, and acts on them, stimulating muscles or glands to respond.
e. Endocrine system
(1) The endocrine system consists of glands that secrete hormones.
(2) Hormones help regulate metabolism by stimulating target tissues.
(3) It includes the pituitary gland, thyroid gland, parathyroid glands, adrenal glands, pancreas, ovaries, testes, pineal gland, and thymus gland.
f. Digestive system
(1) The digestive system receives foods, breaks down nutrients into forms that can pass through cell membranes, and eliminates materials that are not absorbed.
(2) Some digestive organs produce hormones.
(3) The digestive system includes the mouth, tongue, teeth, salivary glands, pharynx, esophagus, stomach, liver, gallbladder, pancreas, small intestine, and large intestine.
g. Respiratory system
(1) The respiratory system takes in and releases air and exchanges gases between the blood and the air.
(2) It includes the nasal cavity, pharynx, larynx, trachea, bronchi, and lungs.
h. Cardiovascular system
(1) The cardiovascular system includes the heart, which pumps blood, and the blood vessels, which carry blood to and from body parts.
(2) Blood transports oxygen, nutrients, hormones, and wastes.
i. Lymphatic system
(1) The lymphatic system is composed of lymphatic vessels, lymph nodes, thymus, and spleen.

(2) It transports lymph from tissue spaces to the bloodstream and carries certain fatty substances away from the digestive organs. Lymphocytes defend the body against disease-causing agents.
j. Urinary system
(1) The urinary system includes the kidneys, ureters, urinary bladder, and urethra.
(2) It filters wastes from the blood and helps maintain fluid and electrolyte balance.
k. Reproductive systems
(1) The reproductive system enables an organism to produce progeny.
(2) The male reproductive system includes the scrotum, testes, epididymides, vasa deferentia, seminal vesicles, prostate gland, bulbourethral glands, urethra, and penis, which produce, maintain, and transport male sex cells.
(3) The female reproductive system includes the ovaries, uterine tubes, uterus, vagina, clitoris, and vulva, which produce, maintain, and transport female sex cells.

Life-Span Changes (page 19)

Aging occurs from conception on, and has effects at the cell, tissue, organ, and organ system levels.

1. The first signs of aging are noticeable in one's thirties. Female fertility begins to decline during this time.
2. In the forties and fifties, adult-onset disorders may begin.
3. Skin changes reflect less elastin, collagen, and subcutaneous fat.
4. Older people may metabolize certain drugs at different rates than younger people.
5. Cells divide a limited number of times. As DNA repair falters, mutations may accumulate.
6. Oxygen free-radical damage produces certain pigments. Metabolism slows, and beta amyloid protein may build up in the brain.

Anatomical Terminology (page 21)

Investigators use terms with precise meanings to effectively communicate with one another.

1. Relative position
   These terms describe the location of one part with respect to another part.
2. Body sections
   Body sections are planes along which the body may be cut to observe the relative locations and arrangements of internal parts.
3. Body regions
   Special terms designate various body regions.

C R I T I C A L   T H I N K I N G   Q U E S T I O N S

1. In many states, death is defined as "irreversible cessation of total brain function." How is death defined in your state? How is this definition related to the characteristics of life?

2. In health, body parts interact to maintain homeostasis. Illness may threaten homeostasis, requiring treatments. What treatments might be used to help control a patient's

(a) body temperature, (b) blood oxygen concentration, and (c) water content?

3. Suppose two individuals have benign (noncancerous) tumors that produce symptoms because they occupy space and crowd adjacent organs. If one of these persons has a tumor in her ventral cavity and the other has a tumor in his dorsal cavity, which patient would be likely to develop symptoms first? Why?

4. If a patient complained of a stomachache and pointed to the umbilical region as the site of the discomfort, which organs located in this region might be the source of the pain?

5. How could the basic requirements of a human be provided for a patient who is unconscious?

6. What is the advantage of using ultrasonography rather than X rays to visualize a fetus in the uterus, assuming that the same information could be obtained by either method?

REVIEW EXERCISES

Part A

1. Briefly describe the early development of knowledge about the human body.
2. Distinguish between anatomy and physiology.
3. How does a biological structure's form determine its function? Give an example.
4. List and describe ten characteristics of life.
5. Define *metabolism*.
6. List and describe five requirements of organisms.
7. Explain how the idea of homeostasis relates to the five requirements you listed in item 6.
8. Distinguish between heat and temperature.
9. What are two types of pressures that may act upon organisms?
10. How are body temperature, blood pressure, and blood glucose concentration controlled?
11. Describe how homeostatic mechanisms act by negative feedback.
12. How does the human body illustrate the levels of anatomical organization?
13. Distinguish between the axial and appendicular portions of the body.
14. Distinguish between the dorsal and ventral body cavities, and name the smaller cavities within each.
15. What are the viscera?
16. Where is the mediastinum?
17. Describe the locations of the oral, nasal, orbital, and middle ear cavities.
18. How does a parietal membrane differ from a visceral membrane?
19. Name the major organ systems, and describe the general functions of each.
20. List the major organs that comprise each organ system.
21. In what body region did Judith R.'s injury occur?

Part B

1. Name the body cavity housing each of the following organs:
   a. stomach   f. rectum
   b. heart   g. spinal cord
   c. brain   h. esophagus
   d. liver   i. spleen
   e. trachea   j. urinary bladder

2. Write complete sentences using each of the following terms correctly:
   a. superior   h. contralateral
   b. inferior   i. proximal
   c. anterior   j. distal
   d. posterior   k. superficial
   e. medial   l. peripheral
   f. lateral   m. deep
   g. ipsilateral

3. Prepare a sketch of a human body, and use lines to indicate each of the following sections:
   a. sagittal   b. transverse   c. frontal

4. Prepare a sketch of the abdominal area, and indicate the location of each of the following regions:
   a. epigastric   c. hypogastric   e. lumbar
   b. umbilical   d. hypochondriac   f. iliac

5. Prepare a sketch of the abdominal area, and indicate the location of each of the following regions:
   a. right upper quadrant   c. left upper quadrant
   b. right lower quadrant   d. left lower quadrant

6. Provide the common name for the region described by the following terms:
   a. acromial   j. gluteal   s. perineal
   b. antebrachial   k. inguinal   t. plantar
   c. axillary   l. mental   u. popliteal
   d. buccal   m. occipital   v. sacral
   e. celiac   n. orbital   w. sternal
   f. coxal   o. otic   x. tarsal
   g. crural   p. palmar   y. umbilical
   h. femoral   q. pectoral   z. vertebral
   i. genital   r. pedal

WEB CONNECTIONS

The Human Organism

The following series of illustrations show the major organs of the human torso. The first plate illustrates the anterior surface and reveals the superficial muscles on one side. Each subsequent plate exposes deeper organs, including those in the thoracic, abdominal, and pelvic cavities.

Chapters 6–22 of this textbook describe the organ systems of the human organism in detail. As you read them, you may want to refer to these plates to help visualize the locations of organs and the three-dimensional relationships among them. You may also want to study the photographs of human cadavers in the reference plates that follow chapter 24. These photographs illustrate many of the larger organs of the human body.

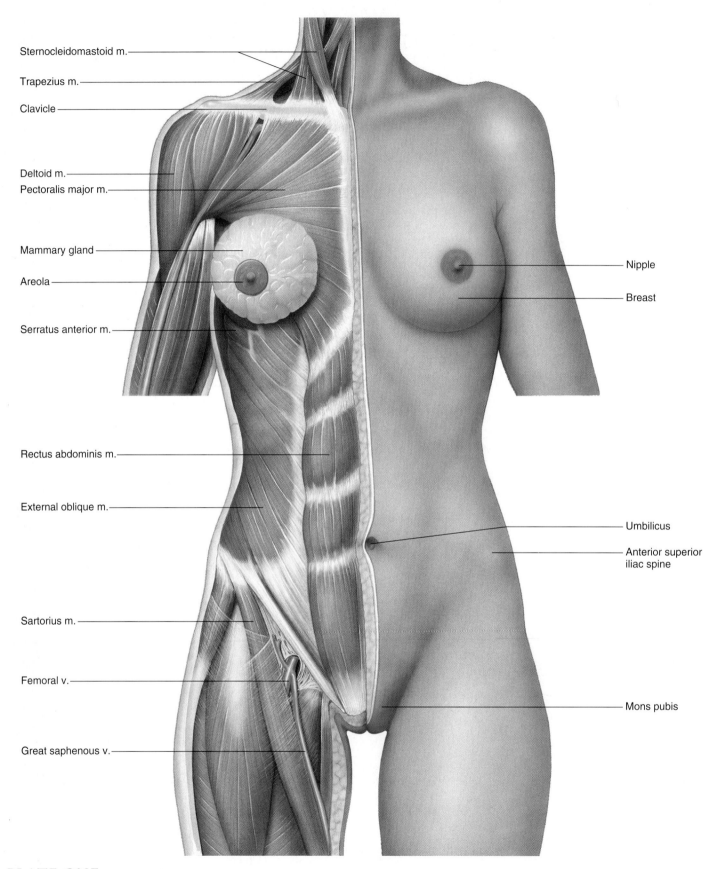

Sternocleidomastoid m.

Trapezius m.

Clavicle

Deltoid m.

Pectoralis major m.

Mammary gland

Areola

Serratus anterior m.

Rectus abdominis m.

External oblique m.

Sartorius m.

Femoral v.

Great saphenous v.

Nipple

Breast

Umbilicus

Anterior superior
iliac spine

Mons pubis

PLATE ONE
Human female torso showing the anterior surface on one side and the superficial muscles exposed on the other side. (*m.* stands for *muscle*, and *v.* stands for *vein.*)

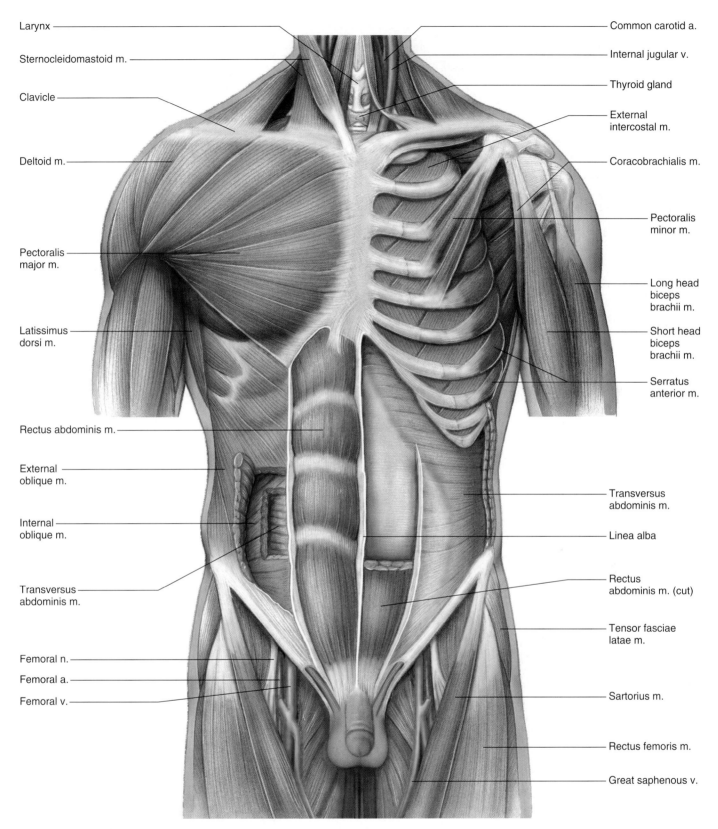

Larynx

Sternocleidomastoid m.

Clavicle

Deltoid m.

Pectoralis major m.

Latissimus dorsi m.

Rectus abdominis m.

External oblique m.

Internal oblique m.

Transversus abdominis m.

Femoral n.

Femoral a.

Femoral v.

Common carotid a.

Internal jugular v.

Thyroid gland

External intercostal m.

Coracobrachialis m.

Pectoralis minor m.

Long head biceps brachii m.

Short head biceps brachii m.

Serratus anterior m.

Transversus abdominis m.

Linea alba

Rectus abdominis m. (cut)

Tensor fasciae latae m.

Sartorius m.

Rectus femoris m.

Great saphenous v.

PLATE TWO

Human male torso with the deeper muscle layers exposed. (*n.* stands for *nerve, a.* stands for *artery,* and *v.* stands for *vein.*)

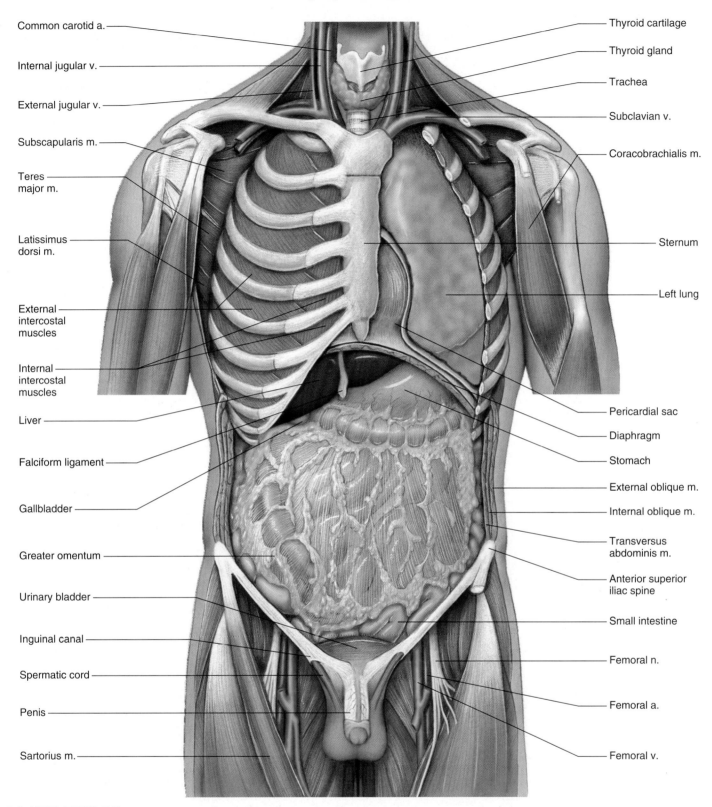

Common carotid a.

Internal jugular v.

External jugular v.

Subscapularis m.

Teres major m.

Latissimus dorsi m.

External intercostal muscles

Internal intercostal muscles

Liver

Falciform ligament

Gallbladder

Greater omentum

Urinary bladder

Inguinal canal

Spermatic cord

Penis

Sartorius m.

Thyroid cartilage

Thyroid gland

Trachea

Subclavian v.

Coracobrachialis m.

Sternum

Left lung

Pericardial sac

Diaphragm

Stomach

External oblique m.

Internal oblique m.

Transversus abdominis m.

Anterior superior iliac spine

Small intestine

Femoral n.

Femoral a.

Femoral v.

PLATE THREE

Human male torso with the deep muscles removed and the abdominal viscera exposed. (*n.* stands for *nerve*, *a.* stands for *artery*, *m.* stands for *muscle*, and *v.* stands for *vein.*)

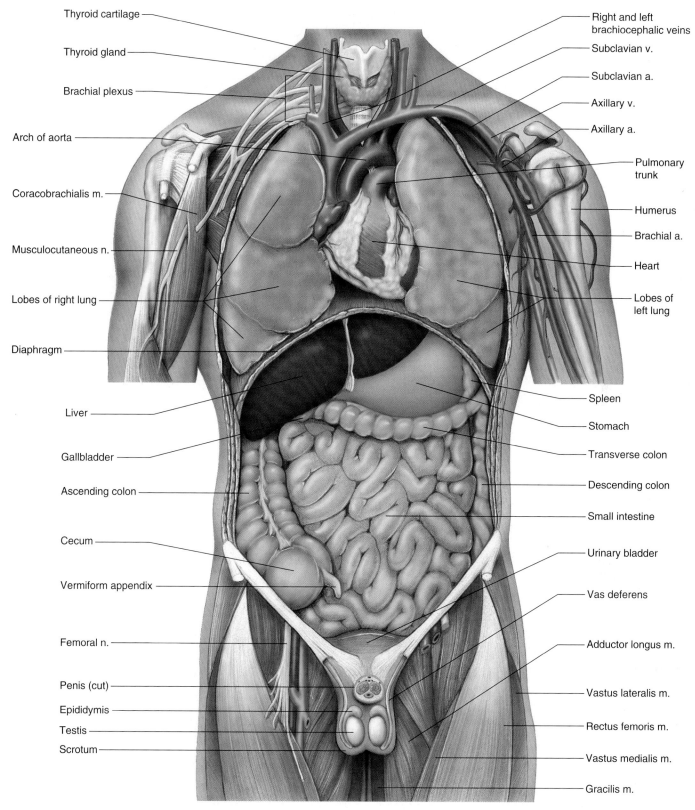

Thyroid cartilage

Thyroid gland

Brachial plexus

Arch of aorta

Coracobrachialis m.

Musculocutaneous n.

Lobes of right lung

Diaphragm

Liver

Gallbladder

Ascending colon

Cecum

Vermiform appendix

Femoral n.

Penis (cut)

Epididymis

Testis

Scrotum

Right and left
brachiocephalic veins

Subclavian v.

Subclavian a.

Axillary v.

Axillary a.

Pulmonary
trunk

Humerus

Brachial a.

Heart

Lobes of
left lung

Spleen

Stomach

Transverse colon

Descending colon

Small intestine

Urinary bladder

Vas deferens

Adductor longus m.

Vastus lateralis m.

Rectus femoris m.

Vastus medialis m.

Gracilis m.

PLATE FOUR

Human male torso with the thoracic and abdominal viscera exposed. (*n.* stands for *nerve, a.* stands for *artery, m.* stands for *muscle,* and *v.* stands for *vein.*)

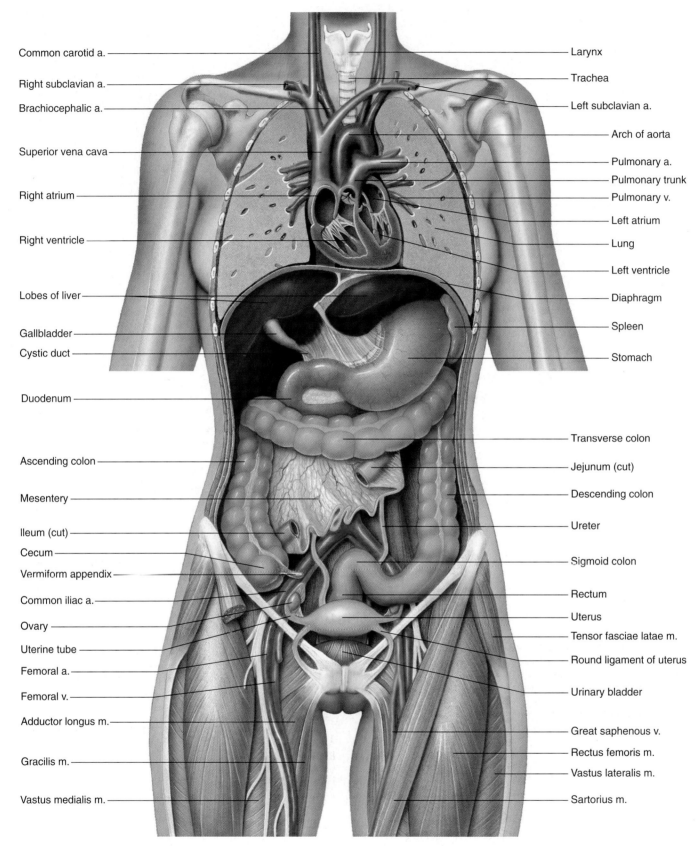

Common carotid a. — Larynx

Right subclavian a. — Trachea

Brachiocephalic a. — Left subclavian a.

Superior vena cava — Arch of aorta

— Pulmonary a.

— Pulmonary trunk

Right atrium — Pulmonary v.

— Left atrium

Right ventricle — Lung

— Left ventricle

Lobes of liver — Diaphragm

Gallbladder — Spleen

Cystic duct — Stomach

Duodenum

— Transverse colon

Ascending colon — Jejunum (cut)

Mesentery — Descending colon

Ileum (cut) — Ureter

Cecum — Sigmoid colon

Vermiform appendix — Rectum

Common iliac a. — Uterus

Ovary — Tensor fasciae latae m.

Uterine tube — Round ligament of uterus

Femoral a. — Urinary bladder

Femoral v.

Adductor longus m. — Great saphenous v.

— Rectus femoris m.

Gracilis m. — Vastus lateralis m.

Vastus medialis m. — Sartorius m.

PLATE FIVE

Human female torso with the lungs, heart, and small intestine sectioned and the liver reflected (lifted back). (*a.* stands for *artery, m.* stands for *muscle,* and *v.* stands for *vein.*)

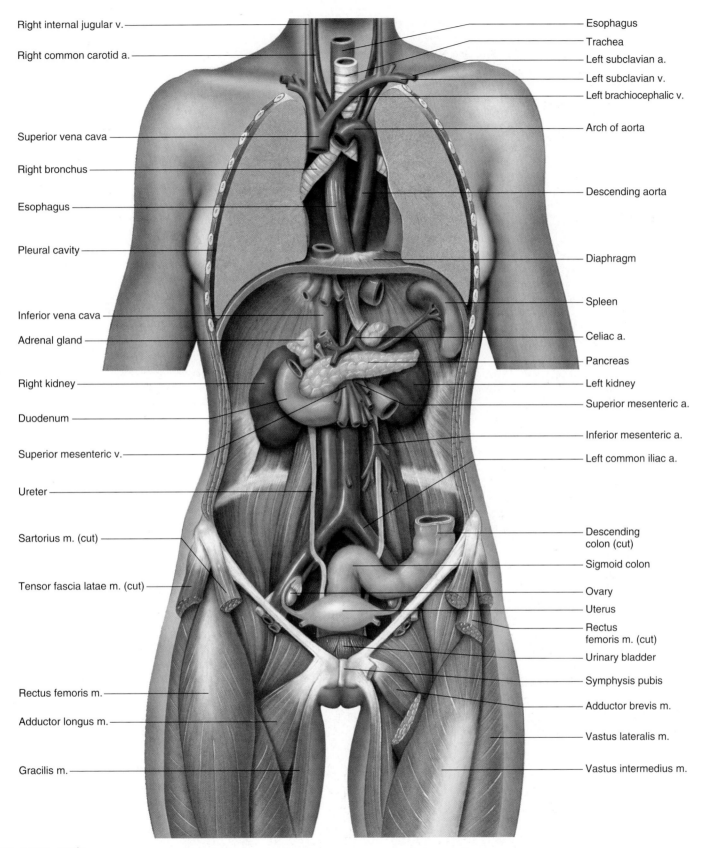

Right internal jugular v.
Right common carotid a.

Superior vena cava

Right bronchus

Esophagus

Pleural cavity

Inferior vena cava

Adrenal gland

Right kidney

Duodenum

Superior mesenteric v.

Ureter

Sartorius m. (cut)

Tensor fascia latae m. (cut)

Rectus femoris m.

Adductor longus m.

Gracilis m.

Esophagus
Trachea
Left subclavian a.
Left subclavian v.
Left brachiocephalic v.

Arch of aorta

Descending aorta

Diaphragm

Spleen

Celiac a.

Pancreas

Left kidney

Superior mesenteric a.

Inferior mesenteric a.

Left common iliac a.

Descending colon (cut)

Sigmoid colon

Ovary
Uterus
Rectus femoris m. (cut)

Urinary bladder

Symphysis pubis

Adductor brevis m.

Vastus lateralis m.

Vastus intermedius m.

PLATE SIX
Human female torso with the heart, stomach, liver, and parts of the intestine and lungs removed. (*a.* stands for *artery, m.* stands for *muscle,* and *v.* stands for *vein.*)

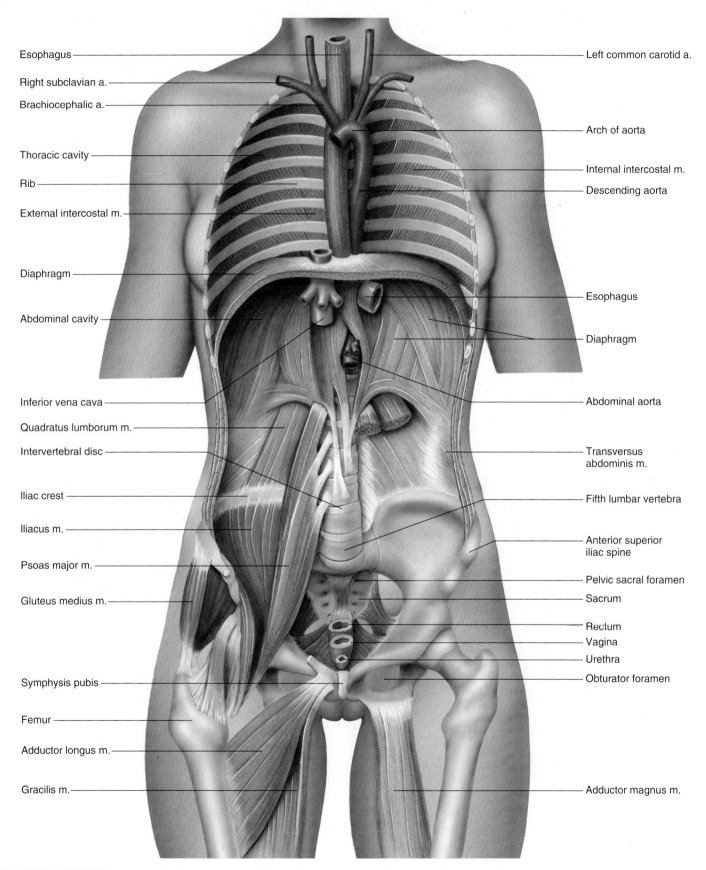

Esophagus

Right subclavian a.

Brachiocephalic a.

Thoracic cavity

Rib

External intercostal m.

Diaphragm

Abdominal cavity

Inferior vena cava

Quadratus lumborum m.

Intervertebral disc

Iliac crest

Iliacus m.

Psoas major m.

Gluteus medius m.

Symphysis pubis

Femur

Adductor longus m.

Gracilis m.

Left common carotid a.

Arch of aorta

Internal intercostal m.

Descending aorta

Esophagus

Diaphragm

Abdominal aorta

Transversus
abdominis m.

Fifth lumbar vertebra

Anterior superior
iliac spine

Pelvic sacral foramen

Sacrum

Rectum

Vagina

Urethra

Obturator foramen

Adductor magnus m.

PLATE SEVEN
Human female torso with the thoracic, abdominal, and pelvic viscera removed. (*a.* stands for *artery* and *m.* stands for *muscle.*)

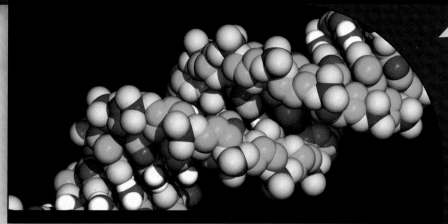

2

The photo shows a portion of a DNA molecule as modeled by a computer. Atoms are shown as color coded balls: carbon (green), oxygen (yellow), nitrogen (blue), phosphorus (red), and hydrogen (white).

Chemical Basis of Life

CHAPTER OBJECTIVES

After you have studied this chapter, you should be able to

1. Explain how the study of living material depends on the study of chemistry.

2. Describe the relationships among matter, atoms, and molecules.

3. Discuss how atomic structure determines how atoms interact.

4. Explain how molecular and structural formulas symbolize the composition of compounds.

5. Describe three types of chemical reactions.

6. Define *pH.*

7. List the major groups of inorganic substances that are common in cells.

8. Describe the general functions of the main classes of organic molecules in cells.

he reunion of the extended Slone family in Kentucky was an unusual event. Not only did ninety relatives gather, but medical researchers also attended, sampling blood from everyone. The reason—the family is very rare in that many members suffer from hereditary pancreatitis, locally known as Slone's disease. In this painful and untreatable condition, the pancreas digests itself. This organ produces digestive enzymes and hormones that regulate the blood glucose level. The researchers were looking for biochemical instructions, in the form of genes, that might explain how the disease arises. This information may also help the many thousands of people who suffer from nonhereditary pancreatitis.

Kevin Slone, who organized the reunion, knew well the ravages of his family's illness. In 1989, as a teenager, he was hospitalized for severe abdominal pain. When he was hospitalized again five years later, three-quarters of his pancreas had become scar tissue. Because many relatives also complained of frequent and severe abdominal pain, Kevin's father, Bobby, began assembling a family tree. Using a computer, he traced more than 700 relatives through nine generations. Although he didn't realize it, Bobby Slone was conducting sophisticated and invaluable genetic research.

David Whitcomb and Garth Ehrlich, geneticists at the University of Pittsburgh, had become interested in hereditary pancreatitis and put the word out that they were looking for a large family in which to hunt for a causative gene. A colleague at a new pancreatitis clinic at the University of Kentucky put them in touch with the Slones and their enormous family tree. Soon after the blood sampling at the family reunion, the researchers identified the biochemical cause of hereditary pancreatitis.

Affected family members have a mutation that blocks normal control of the manufacture of trypsin, a digestive enzyme that breaks down protein. When the powerful enzyme accumulates, it digests the pancreas. A disorder felt painfully at the whole-body level is caused by a problem at the biochemical level. Researchers are using the information provided by the Slone family to develop a diagnostic test and treatments for this debilitating disorder. ■

Chemistry considers the composition of substances and how they change. Although it is possible to study anatomy without much reference to chemistry, it is essential for understanding physiology, because body functions depend on cellular functions that, in turn, result from chemical changes.

As interest in the chemistry of living organisms grew and knowledge of the subject expanded, a field of life science called biological chemistry, or **biochemistry,** emerged. Biochemistry has been important not only in helping explain physiological processes but also in developing many new drugs and methods for treating diseases.

1 Why is a knowledge of chemistry essential to understanding physiology?

2 What is biochemistry?

Structure of Matter

Matter is anything that has weight and takes up space. This includes all the solids, liquids, and gases in our surroundings as well as in our bodies. All matter consists of particles that are organized in specific ways. Table 2.1 lists some particles of matter and their characteristics.

Elements and Atoms

All matter is composed of fundamental substances called **elements** (el'e-mentz). As of early 2002, 114 such elements are known, although naturally occurring matter on earth includes only 92 of them. Among these elements are such common materials as iron, copper, silver, gold, aluminum, carbon, hydrogen, and oxygen. Some elements exist in a pure form, but these and other elements are more commonly parts of chemical combinations called **compounds** (kom'-powndz).

Elements required by the body in large amounts—such as carbon, hydrogen, oxygen, nitrogen, sulfur, and phosphorus—are termed **bulk elements.** These elements make up more than 95% (by weight) of the human body (table 2.2). Elements required in small amounts are called **trace elements.** Many trace elements are important parts of enzymes, which are proteins that regulate the rates of chemical reactions in living organisms. Some elements that are toxic in large amounts, such as arsenic, may actually be vital in very small amounts, and these are called **ultratrace elements.**

Elements are composed of particles called **atoms** (at'omz), which are the smallest complete units of the elements. The atoms that make up each element are chemically identical to one another, but they differ from the atoms that make up other elements. Atoms vary in size, weight, and the way they interact with one another. Some atoms, for instance, can combine either with atoms like themselves or with other kinds of atoms, while other atoms cannot.

Atomic Structure

An atom consists of a central portion called the **nucleus** and one or more **electrons** that constantly move around the nucleus. The nucleus contains one or more relatively large particles, **protons** and usually **neutrons.** Protons and neutrons are about equal in weight, but they are otherwise quite different (fig. 2.1).

Electrons, which are so small that they have almost no weight, carry a single, negative electrical charge (e^-). Each proton carries a single, positive electrical charge

TABLE 2.1	Some Particles of Matter		
Name	**Characteristic**	**Name**	**Characteristic**
Atom	Smallest particle of an element that has the properties of that element	Neutron (n^0)	Particle with about the same weight as a proton; uncharged and thus electrically neutral; found within an atomic nucleus
Electron (e^-)	Extremely small particle with almost no weight; carries a negative electrical charge and is in constant motion around an atomic nucleus	Ion	Particle that is electrically charged because it has gained or lost one or more electrons
Proton (p^+)	Relatively large atomic particle; carries a positive electrical charge and is found within an atomic nucleus	Molecule	Particle formed by the chemical union of two or more atoms

TABLE 2.2	Major Elements in the Human Body (by Weight)	
Major Elements	**Symbol**	**Approximate Percentage of the Human Body**
Oxygen	O	65.0
Carbon	C	18.5
Hydrogen	H	9.5
Nitrogen	N	3.2
Calcium	Ca	1.5
Phosphorus	P	1.0
Potassium	K	0.4
Sulfur	S	0.3
Chlorine	Cl	0.2
Sodium	Na	0.2
Magnesium	Mg	0.1

99.9%

Trace Elements		
Cobalt	Co	
Copper	Cu	
Fluorine	F	
Iodine	I	less than 0.1%
Iron	Fe	
Manganese	Mn	
Zinc	Zn	

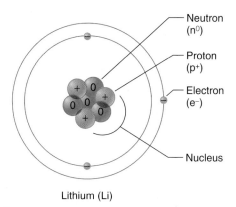

Neutron (n^0)

Proton (p^+)

Electron (e^-)

Nucleus

Lithium (Li)

FIGURE 2.1

In an atom of lithium, three electrons move around a nucleus that consists of three protons and four neutrons.

(p^+). Neutrons are uncharged and thus are electrically neutral (n^0).

Because the nucleus contains protons, this part of an atom is always positively charged. However, the number of electrons outside the nucleus equals the number of protons, so a complete atom is said to have no net charge and is thus electrically neutral.

The atoms of different elements contain different numbers of protons. The number of protons in the atoms of a particular element is called its **atomic number.** Hydrogen, for example, whose atoms contain one proton, has atomic number 1; carbon, whose atoms have six protons, has atomic number 6.

The weight of an atom of an element is primarily due to the protons and neutrons in its nucleus, because the electrons have so little weight. For this reason, the number of protons plus the number of neutrons in each of an element's atoms essentially equals the **atomic weight** of that atom. Thus, the atomic weight of a hydrogen atom, which has only one proton and no neutrons, is approximately 1. The atomic weight of a carbon atom, with six protons and six neutrons, is approximately 12 (table 2.3).

Isotopes

All the atoms of a particular element have the same atomic number because they have the same number of protons and electrons. However, the atoms of an element vary in the number of neutrons in their nuclei; thus, they vary in atomic weight. For example, all oxygen atoms have eight protons in their nuclei. Some, however, have eight neutrons (atomic weight 16), others have nine neutrons (atomic weight 17), and still others have ten neutrons (atomic weight 18). Atoms that have the same atomic numbers but different atomic weights are called **isotopes** (i'so-tōpz) of an element. Because a sample of an element is likely to include more than one isotope, the atomic weight of the element is often considered to be the average weight of the isotopes present. (See Appendix A, Periodic Table of the Elements, p. 969)

The ways atoms interact with one another are due largely to their numbers of electrons. Because the number of electrons in an atom equals its number of protons, all the isotopes of a particular element have the same number

TABLE 2.3 Atomic Structure of Elements 1 Through 12

Element	Symbol	Number	Approximate Atomic Weight	Protons	Neutrons	Electrons in Shells First	Second	Third
Hydrogen	H	1	1	1	0	1		
Helium	He	2	4	2	2	2 (inert)		
Lithium	Li	3	7	3	4	2	1	
Beryllium	Be	4	9	4	5	2	2	
Boron	B	5	11	5	6	2	3	
Carbon	C	6	12	6	6	2	4	
Nitrogen	N	7	14	7	7	2	5	
Oxygen	O	8	16	8	8	2	6	
Fluorine	F	9	19	9	10	2	7	
Neon	Ne	10	20	10	10	2	8 (inert)	
Sodium	Na	11	23	11	12	2	8	1
Magnesium	Mg	12	24	12	12	2	8	2

(For more detail, see Appendix A, Periodic Table of the Elements, p. 969.)

of electrons and chemically react in the same manner. For example, any of the isotopes of oxygen can have the same function in the metabolic reactions of an organism.

Isotopes of an element may be stable, or they may have unstable atomic nuclei that decompose, releasing energy or pieces of themselves until they reach a stable form. Such unstable isotopes are called *radioactive,* and the energy or atomic fragments they emit are called *atomic radiation.* Elements that have radioactive isotopes include oxygen, iodine, iron, phosphorus, and cobalt. Some radioactive isotopes are used to detect and treat disease (Clinical Application 2.1).

Atomic radiation includes three common forms called alpha (α), beta (β), and gamma (γ). Each kind of radioactive isotope produces one or more of these forms of radiation. Alpha radiation consists of particles from atomic nuclei, each of which includes two protons and two neutrons, that move relatively slowly and cannot easily penetrate matter. Beta radiation consists of much smaller particles (electrons) that travel faster and more deeply penetrate matter. Gamma radiation is similar to X-radiation and is the most penetrating of these forms.

1 What is the relationship between matter and elements?

2 Which elements are most common in the human body?

3 How are electrons, protons, and neutrons positioned within an atom?

4 What is an isotope?

5 What is atomic radiation?

Molecules and Compounds

Two or more atoms may combine to form a distinctive kind of particle called a **molecule** (mol'ĕ-kūl). A **molecular formula** is used to depict the numbers and kinds of atoms in a molecule. Such a formula consists of the symbols of the elements in the molecule with numbers as subscripts to indicate how many atoms of each element are present. For example, the molecular formula for water is H_2O, which indicates two atoms of hydrogen and one atom of oxygen in each molecule. The molecular formula for the sugar glucose is $C_6H_{12}O_6$, which means there are six atoms of carbon, twelve atoms of hydrogen, and six atoms of oxygen in a glucose molecule.

If atoms of the same element combine, they produce molecules of that element. Gases of hydrogen (H_2), oxygen (O_2), and nitrogen (N_2) consist of such molecules. If atoms of different elements combine, molecules of substances called **compounds** form. Two atoms of hydrogen, for example, can combine with one atom of oxygen to produce a molecule of the compound water (H_2O), as figure 2.2 shows. Table sugar, baking soda, natural gas, beverage alcohol, and most medical drugs are compounds.

A molecule of a compound always contains definite types and numbers of atoms. A molecule of water (H_2O), for instance, always contains two hydrogen atoms and one oxygen atom. If two hydrogen atoms combine with two oxygen atoms, the compound formed is not water, but hydrogen peroxide (H_2O_2).

Bonding of Atoms

Atoms combine with other atoms by forming **bonds.** When atoms form such bonds, it is the result of interactions involving their electrons.

The electrons of an atom are found in one or more regions of space called **electron shells** around the nucleus. Because electrons have an amount of energy characteristic of the particular shell they are in, the shells are sometimes called *energy shells.* Each electron shell can hold only a limited number of electrons. The maximum number of electrons that each of the first three

RADIOACTIVE ISOTOPES REVEAL PHYSIOLOGY

Vicki L. arrived early at the nuclear medicine department of the health center. As she sat in an isolated cubicle, a doctor in full sterile dress approached with a small metal canister marked with numerous warnings. The doctor carefully unscrewed the top, inserted a straw, and watched as the young woman sipped the fluid within. It tasted like stale water but was actually a solution containing a radioactive isotope, iodine-131.

Vicki's thyroid gland had been removed three months earlier, and this test was to determine whether any active thyroid tissue remained. The thyroid is the only part of the body to metabolize iodine, so if Vicki's body retained any of the radioactive drink, it would mean that some of her cancerous thyroid gland remained. By using a radioactive isotope, her physicians could detect iodine uptake using a scanning device called a scintillation counter (fig. 2A). Figure 2B illustrates iodine-131 uptake in a complete thyroid gland.

The next day, Vicki returned for the scan, which showed that a small amount of thyroid tissue was indeed left and was functioning. This meant another treatment would be necessary. Vicki would drink more of the radioactive iodine, enough to destroy the remaining tissue. This time, she drank the solution while in an isolation room, which

was lined with paper to keep her from contaminating the floor, walls, and furniture. The same physician administered the radioactive iodine. Vicki's physician had this job because his own thyroid had been removed many years earlier, and therefore, the radiation couldn't harm him.

After two days in isolation, Vicki went home with a list of odd instructions. She was to stay away from her children and pets, wash her clothing separately, use disposable utensils and plates, and flush the toilet three times each time she used it. These precautions would minimize her contaminating her family—mom was radioactive!

Iodine-131 is a medically useful radioactive isotope because it has a short *half-life,* a measurement of the time it takes for half of an amount of an isotope to decay to a nonradioactive form. The half-life of iodine-131 is 8.1 days. With the amount of radiation in Vicki's body dissipating by half every 8.1 days, after three months there would be hardly any left. Doctors hoped that the remaining unhealthy thyroid cells would leave her body along with the radioactive iodine.

Isotopes of other elements have different half-lives. The half-life of iron-59 is 45.1 days; that of phosphorus-32 is 14.3 days; that of cobalt-60 is 5.26 years; and that of radium-226 is 1,620 years.

A form of thallium-201 with a half-life of 73.5 hours is commonly used to detect disorders in the blood vessels supplying the heart muscle or to locate regions of damaged heart tissue after a heart attack. Gallium-67, with a half-life of 78 hours, is used to detect and monitor the progress of certain cancers and inflammatory illnesses. These medical procedures inject the isotope into the blood and follow its path using detectors that record images on paper or film.

Radioactive isotopes are also used to assess kidney function, estimate the concentrations of hormones in body fluids, measure blood volume, and study changes in bone density. Cobalt-60 is a radioactive isotope used to treat some cancers. The cobalt emits radiation that damages cancer cells more readily than it does healthy cells. ∎

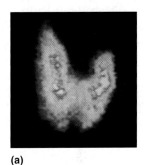

(a)

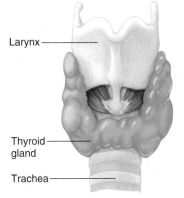

Larynx

Thyroid gland

Trachea

(b)

FIGURE 2B

(*a*) A scan of the thyroid gland twenty-four hours after the patient receives radioactive iodine. Note how closely the scan in (*a*) resembles the shape of the thyroid gland as depicted in (*b*).

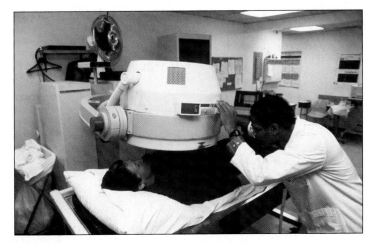

FIGURE 2A

Physicians use scintillation counters such as this to detect radioactive isotopes.

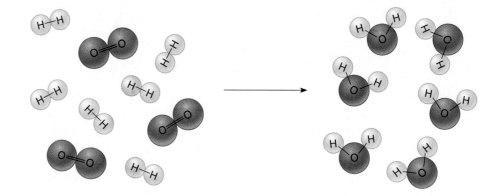

FIGURE 2.2
Under certain conditions, hydrogen molecules can combine with oxygen molecules to form water molecules.

shells can hold for elements of atomic number 18 and under is

First shell (closest to the nucleus)	*2 electrons*
Second shell	*8 electrons*
Third shell	*8 electrons*

More complex atoms may have as many as eighteen electrons in the third shell.

Simplified diagrams such as those in figure 2.3 are used to show electron configuration in atoms. Notice that the single electron of a hydrogen atom is located in the first shell, the two electrons of a helium atom fill its first shell, and the three electrons of a lithium atom occur with two in the first shell and one in the second shell. Lower energy shells, closer to the nucleus, must be filled first.

The number of electrons in the outermost shell of an atom determines whether it will react with another atom. Atoms react in a way that leaves the outermost shell completely filled with electrons, thus achieving a more stable structure. This is sometimes known as the **octet rule,** since, except for the first shell, it takes eight electrons to fill the shells in most of the atoms important in living organisms.

Atoms such as helium, whose outermost electron shells are filled, already have stable structures and are chemically inactive or **inert** (they cannot form chemical bonds). Atoms with incompletely filled outer shells, such

as those of hydrogen or lithium, tend to gain, lose, or share electrons in ways that empty or fill their outer shells. In this way, they achieve stable structures.

Atoms that gain or lose electrons become electrically charged and are called **ions** (i'onz). An atom of sodium, for example, has eleven electrons: two in the first shell, eight in the second shell, and one in the third shell. This atom tends to lose the electron from its outer shell, which leaves the second (now the outermost) shell filled and the new form stable (fig. 2.4a). In the process,

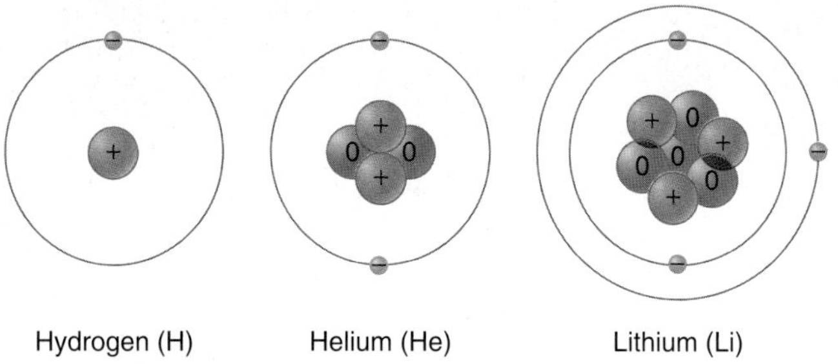

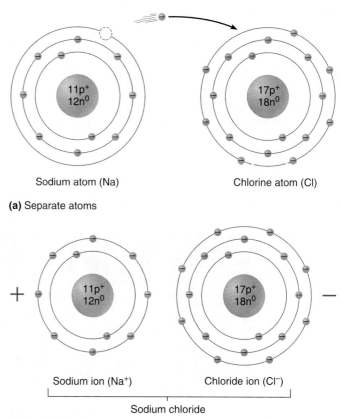

Sodium atom (Na) Chlorine atom (Cl)

(a) Separate atoms

Sodium ion (Na$^+$) Chloride ion (Cl$^-$)

Sodium chloride

(b) Bonded ions

FIGURE 2.4
Formation of an ionic bond. (*a*) If a sodium atom loses an electron to a chlorine atom, the sodium atom becomes a sodium ion (Na$^+$), and the chlorine atom becomes a chloride ion (Cl$^-$). (*b*) These oppositely charged particles attract electrically and join by an ionic bond.

Hydrogen (H) Helium (He) Lithium (Li)

FIGURE 2.3
The single electron of a hydrogen atom is located in its first shell. The two electrons of a helium atom fill its first shell. Two of the three electrons of a lithium atom are in the first shell, and one is in the second shell.

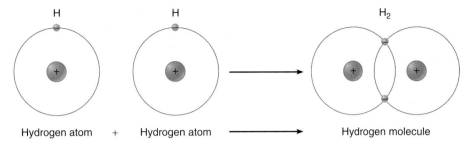

FIGURE 2.5

A hydrogen molecule forms when two hydrogen atoms share a pair of electrons and join by a covalent bond.

sodium is left with eleven protons (11^+) in its nucleus and only ten electrons (10^-). As a result, the atom develops a net electrical charge of 1^+ and is called a sodium ion, symbolized Na^+.

A chlorine atom has seventeen electrons, with two in the first shell, eight in the second shell, and seven in the third shell. An atom of this type tends to accept a single electron, thus filling its outer shell and achieving stability. In the process, the chlorine atom is left with seventeen protons (17^+) in its nucleus and eighteen electrons (18^-). As a result, the atom develops a net electrical charge of 1^- and is called a chloride ion, symbolized Cl^-.

Because oppositely charged ions attract, sodium and chlorine atoms that have formed ions may react together to form a type of chemical bond called an **ionic bond** (electrovalent bond). Sodium ions (Na^+) and chloride ions (Cl^-) uniting in this manner form the compound sodium chloride (NaCl), or table salt (fig. 2.4b). Similarly, a hydrogen atom may lose its single electron and become a hydrogen ion (H^+). Such an ion can bond with a chloride ion (Cl^-) to form hydrogen chloride (HCl, hydrochloric acid).

Atoms may also bond by sharing electrons rather than by gaining or losing them. A hydrogen atom, for example, has one electron in its first shell but requires two electrons to achieve a stable structure. It may fill this shell by combining with another hydrogen atom in such a way that the two atoms share a pair of electrons. As figure 2.5 shows, the two electrons then encircle the nuclei of both atoms, filling the outermost shell, and each atom becomes stable. In this case, the chemical bond between the atoms is called a **covalent bond.**

Usually atoms of each element form a specific number of covalent bonds. Hydrogen atoms form single bonds, oxygen atoms form two bonds, nitrogen atoms form three bonds, and carbon atoms form four bonds. Symbols and lines can be used to represent the bonding capacities of these atoms, as follows:

$$-H \qquad\qquad -O-$$

$$-N- \qquad\qquad -C-$$

Representations such as these can be used to show how atoms bond and are arranged in various molecules. One pair of shared electrons, a single covalent bond, is shown by a single line. Sometimes atoms may share two pairs of electrons (a double covalent bond), or even three pairs (a triple covalent bond), represented by two and three lines, respectively. Illustrations of this type are called **structural formulas** (fig. 2.6). Structural formulas are convenient, but cannot adequately capture the three-dimensional forms of molecules. Figure 2.7 shows a three-dimensional (space-filling) representation of a water molecule.

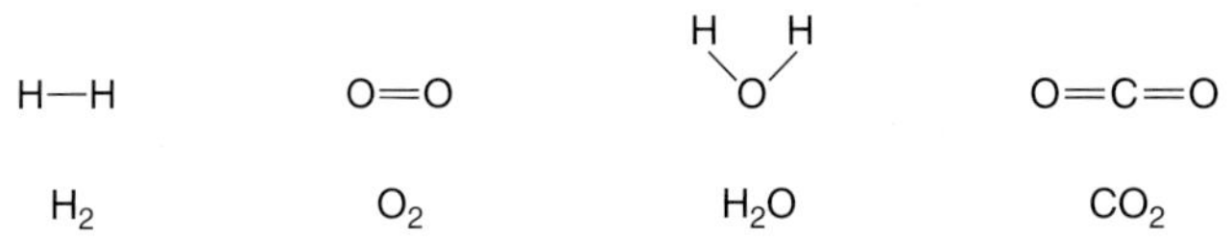

$$H-H \qquad O=O \qquad \overset{\text{H}\quad\text{H}}{\underset{\text{O}}{\diagdown\diagup}} \qquad O=C=O$$

$$H_2 \qquad\quad O_2 \qquad\quad H_2O \qquad\quad CO_2$$

FIGURE 2.6

Structural and molecular formulas for molecules of hydrogen, oxygen, water, and carbon dioxide. Note the double covalent bonds.

FIGURE 2.7

A water molecule (H_2O) can be represented by a three-dimensional model. The white parts represent the hydrogen atoms, and the red part represents oxygen.

Different types of chemical bonds share electrons to different degrees. At one extreme is the ionic bond in which atoms gain or lose electrons. At the other extreme is the covalent bond in which the electrons are shared equally. In between lies the covalent bond in which electrons are not shared equally, resulting in a molecule whose shape gives an uneven distribution of charges. Such a molecule is called **polar.** Unlike an ion, a polar molecule has an equal number of protons and electrons, but one end of the molecule has more than its share of electrons, becoming slightly negative, while the other end of the molecule has less than its share, becoming slightly positive. Typically, polar covalent bonds occur where hydrogen atoms bond to oxygen or nitrogen atoms. Water is an important example of a polar molecule (fig. 2.8a).

The attraction of the positive hydrogen end of a polar molecule to the negative nitrogen or oxygen end of another polar molecule is called a **hydrogen bond.** Hydrogen bonds are weak, particularly at body temperature. For

example, below 0°C, the hydrogen bonds between water molecules shown in figure 2.8b are strong enough to form ice. As the temperature rises, increased molecular movement is sufficient to break the hydrogen bonds, and water becomes a liquid. Even at body temperature, hydrogen bonds are important in protein and nucleic acid structure. In these cases, hydrogen bonds occur between polar regions within different parts of a single, very large molecule. Clinical Application 2.2 examines how radiation that moves electrons can affect human health.

1 Distinguish between a molecule and a compound.

2 What is an ion?

3 Describe two ways that atoms may combine with other atoms.

4 What is a molecular formula? A structural formula?

5 Distinguish between an ion and a polar molecule.

Chemical Reactions

Chemical reactions form or break bonds between atoms, ions, or molecules. Those being changed by the chemical reaction are called **reactants.** Those formed at the reaction's conclusion are called **products.** When two or more atoms, ions, or molecules bond to form a more complex structure, as when hydrogen and oxygen atoms bond to form molecules of water, the reaction is called **synthesis** (sin'thĕ-sis). Such a reaction can be symbolized this way:

$$A + B \rightarrow AB$$

If the bonds of a reactant molecule break to form simpler molecules, atoms, or ions, the reaction is called **decomposition** (de″kom-po-zish′un). For example, molecules of water can decompose to yield the products hydrogen and oxygen. Decomposition is symbolized as follows:

$$AB \rightarrow A + B$$

Synthetic reactions, which build larger molecules from smaller ones, are particularly important in growth of body parts and repair of worn or damaged tissues. Decomposition reactions occur when food substances are digested and they release energy.

A third type of chemical reaction is an **exchange reaction** (replacement reaction). In this reaction, parts of two different kinds of molecules trade positions. The reaction is symbolized as follows:

$$AB + CD \rightarrow AD + CB$$

An example of an exchange reaction is an acid reacting with a base, producing water and a salt. This type of reaction is discussed in the following section.

Many chemical reactions are reversible. This means the product or products can change back to the reactant or

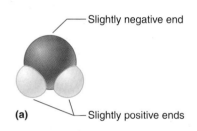

- Slightly negative end
- Slightly positive ends

(a)

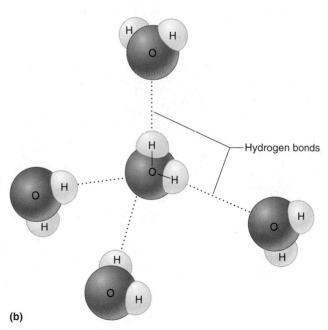

- H H
- O
- Hydrogen bonds
- O H
- H
- H O H
- O H
- H
- O H
- H O H

(b)

FIGURE 2.8

Water is a polar molecule. (a) Water molecules have equal numbers of electrons and protons but are polar because the electrons are shared unequally, creating slightly negative ends and slightly positive ends. (b) Hydrogen bonding between water molecules.

IONIZING RADIATION:
FROM THE COLD WAR TO YUCCA MOUNTAIN

Alpha, beta, and gamma radiation are called ionizing radiation because their energy adds or removes electrons from atoms (fig. 2C). Electrons dislodged by ionizing radiation can affect nearby atoms, disrupting physiology at the chemical level in a variety of ways—causing cancer, clouding the lens of the eye, and interfering with normal growth and development.

In the United States, some people are exposed to very low levels of ionizing radiation, mostly from background radiation, which originates from natural environmental sources (table 2A). This is not true, however, for people who live near sites of atomic weapons manufacture. Epidemiologists are now studying recently uncovered medical records that document illnesses linked to long-term exposure to ionizing radiation in a 1,200-square kilometer area in Germany.

The lake near Oberrothenback, Germany, which appears inviting, contains enough toxins to kill thousands of people, polluted with heavy metals, low-level radioactive chemical waste, and 22,500 tons of arsenic. Radon, a radioactive by-product of uranium, permeates the soil. High death rates affect farm animals and pets that have drunk from the lake. Cancer rates and respiratory disorders among the human residents nearby are well above normal.

The lake in Oberrothenback was once a dump for a factory that produced "yellow cake," a term for processed uranium ore, used to build atomic bombs for the former Soviet Union. In the early 1950s, nearly half a million workers labored here and in surrounding areas in factories and mines. Records released in 1989, after the reunification of Germany, reveal that workers were given perks, such as alcoholic beverages and better wages, to work in the more dangerous areas. The workers paid a heavy price: tens of thousands died of lung ailments.

Today, concern over the health effects of exposure to ionizing radiation centers on the U.S. government's plan to transport tens of thousands of metric tons of high-level nuclear waste from 109 reactors around the country for burial beneath Yucca Mountain, Nevada, by 2010. The waste, currently stored near the reactors, will be buried in impenetrable containers by robots under the mountain. In the reactors, nuclear fuel rods contain uranium oxide, which produces electricity as it decays to plutonium, which gives off gamma rays. Periodically the fuel rods must be replaced, and the spent ones buried. Environmental groups are concerned that the waste could be exposed during transport, and that the facility in the mountain may not adequately contain it. ■

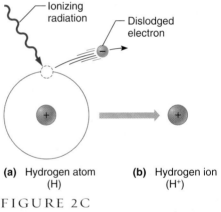

(a) Hydrogen atom (H) **(b)** Hydrogen ion (H⁺)

FIGURE 2C
(*a*) Ionizing radiation may dislodge an electron from an electrically neutral hydrogen atom. (*b*) Without its electron, the hydrogen atom becomes a positively charged hydrogen ion (H^+).

TABLE 2A	Sources of Ionizing Radiation
Background (Natural environmental)	Cosmic rays from space
	Radioactive elements in earth's crust
	Rocks and clay in building materials
	Radioactive elements naturally in the body (potassium-40, carbon-14)
Medical and dental	X rays
	Radioactive substances
Other	Atomic and nuclear weapons
	Mining and processing radioactive minerals
	Radioactive fuels in nuclear power plants
	Radioactive elements in consumer products (luminescent dials, smoke detectors, color TV components)

reactants. A **reversible reaction** is symbolized using a double arrow, as follows:

$$A + B \rightleftarrows AB$$

Whether a reversible reaction proceeds in one direction or another depends on the relative proportions of reactant (or reactants) and product (or products) as well as the amount of energy available. **Catalysts** are molecules that influence the rates (not the direction) of chemical reactions but are not consumed in the reaction.

Acids, Bases, and Salts

The polarity of water causes the ionically bound salts in the internal environment to dissociate from one another. Sodium chloride (NaCl), for example, ionizes into sodium ions (Na^+) and chloride ions (Cl^-) when it dissolves (fig. 2.9). This reaction is represented as

$$NaCl \rightarrow Na^+ + Cl^-$$

Because the resulting solution contains electrically charged particles (ions), it conducts an electric current. Substances that release ions in water are, therefore, called **electrolytes** (e-lek′tro-lītz). Electrolytes that release hydrogen ions (H^+) in water are called **acids.** For example, in water, the compound hydrochloric acid (HCl) releases hydrogen ions (H^+) and chloride ions (Cl^-):

$$HCl \rightarrow H^+ + Cl^-$$

Substances that combine with hydrogen ions are called **bases.** The compound sodium hydroxide (NaOH) in water releases hydroxide ions (OH^-). The hydroxide ions, in turn, can combine with hydrogen ions to form water. Thus, sodium hydroxide is a base:

$$NaOH \rightarrow Na^+ + OH^-$$

(*Note:* Some ions, such as OH^- contain two or more atoms. However, such a group usually behaves like a single atom and remains unchanged during a chemical reaction.)

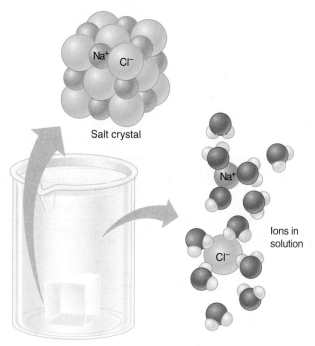

FIGURE 2.9

The polar nature of water molecules causes sodium chloride (NaCl) to dissociate in water, releasing sodium ions (Na^+) and chloride ions (Cl^-).

Acids and bases can react to form water and electrolytes called **salts.** For example, hydrochloric acid and sodium hydroxide react to form water and sodium chloride:

$$HCl + NaOH \rightarrow H_2O + NaCl$$

Table 2.4 summarizes the three types of electrolytes.

Acid and Base Concentrations

Concentrations of acids and bases affect the chemical reactions that constitute many life processes, such as those controlling breathing rate. Thus, the concentrations of these substances in body fluids are of special importance.

Hydrogen ion concentration can be measured in grams of ions per liter of solution. However, because hydrogen ion concentration can cover such a wide range (gastric juice has 0.01 grams H^+/liter; household ammonia has 0.00000000001 grams H^+/liter), a shorthand system called the **pH scale** has been developed. This system tracks the number of decimal places in a hydrogen ion concentration without having to write them out. For example, a solution with a hydrogen ion concentration of 0.1 grams per liter has a pH value of 1.0; a concentration of 0.01 g H^+/L has pH 2.0; 0.001 g H^+/L has pH 3.0; and so forth. Each whole number on the pH scale, which extends from pH 0 to pH 14, represents a tenfold difference in hydrogen ion concentration, and as the hydrogen ion concentration increases, the pH number gets smaller. Thus, a solution with a pH of 6 has ten times the hydrogen ion concentration of a solution with a pH of 7. This means that relatively small changes in pH can reflect large changes in hydrogen ion concentration.

In pure water, which ionizes only slightly, the hydrogen ion concentration is 0.0000001 g/L, and the pH is 7.0. Because water ionizes to release equal numbers of acidic hydrogen ions and basic hydroxide ions, it is *neutral.*

$$H_2O \rightarrow H^+ + OH^-$$

Many bases are present in the body fluids, but because of the way they react in water, the concentration of hydroxide ions is a good estimate of the total base concentration. The concentrations of hydrogen ions and hydroxide ions

TABLE 2.4		Types of Electrolytes
	Characteristic	**Examples**
Acid	Substance that releases hydrogen ions (H^+)	Carbonic acid, hydrochloric acid, acetic acid, phosphoric acid
Base	Substance that releases ions that can combine with hydrogen ions	Sodium hydroxide, potassium hydroxide, magnesium hydroxide, sodium bicarbonate
Salt	Substance formed by the reaction between an acid and a base	Sodium chloride, aluminum chloride, magnesium sulfate

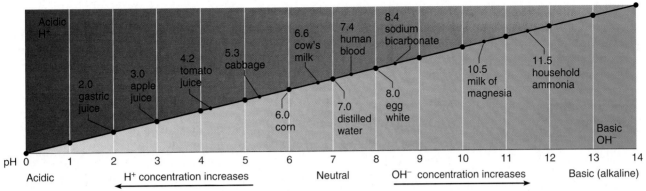

FIGURE 2.10

As the concentration of hydrogen ions (H^+) increases, a solution becomes more acidic, and the pH value decreases. As the concentration of substances that combine with hydrogen ions (such as hydroxide ions) increases, a solution becomes more basic (alkaline), and the pH value increases. The pH values of some common substances are shown.

are always in balance, such that if one increases, the other decreases, and vice versa. Solutions with more hydrogen ions than hydroxide ions are *acidic.* That is, acidic solutions have pH values less than 7.0 (fig. 2.10). Solutions with fewer hydrogen ions than hydroxide ions are *basic* (alkaline); that is, they have pH values greater than 7.0.

Table 2.5 summarizes the relationship between hydrogen ion concentration and pH. Chapter 21 (pp. 819–820) discusses the regulation of hydrogen ion concentrations in the internal environment.

Many fluids in the human body function within a narrow pH range. Illness results when pH changes. The normal pH of blood, for example, is 7.35 to 7.45. Blood pH of 7.5 to 7.8, called **alkalosis,** makes one feel agitated and dizzy. This can be caused by breathing rapidly at high altitudes, taking too many antacids, high fever, anxiety, or mild to moderate vomiting that rids the body of stomach acid. **Acidosis,** in which blood pH falls to 7.0 to 7.3, makes one feel disoriented and fatigued, and breathing may become difficult. This condition can result from severe vomiting that empties the alkaline small intestinal contents, diabetes, brain damage, impaired breathing, and lung and kidney disease. Buffers are chemicals that resist pH change. Buffers are discussed later in chapter 21 (p. 819).

1 Describe three kinds of chemical reactions.

2 Compare the characteristics of an acid, a base, and a salt.

3 What is pH?

Chemical Constituents of Cells

The chemicals that enter into metabolic reactions or are produced by them can be divided into two large groups. Generally, those that contain carbon and hydrogen atoms are called **organic** (or-gan′ik); the rest are called **inorganic** (in″or-gan′ik).

Inorganic substances usually dissolve in water or react with water to release ions; thus, they are *electrolytes.* Many organic compounds also dissolve in water, although as a group, they are more likely to dissolve in organic liquids such as ether or alcohol. Organic compounds that dissolve in water usually do not release ions and are therefore called *nonelectrolytes.*

Inorganic Substances

Common inorganic substances in cells include water, oxygen, carbon dioxide, and inorganic salts.

Water

Water (H_2O) is the most abundant compound in living material and accounts for about two-thirds of the weight of an adult human. It is the major component of blood and other body fluids, including those within cells.

TABLE 2.5	Hydrogen Ion Concentrations and pH	
Grams of H⁺ per Liter	**pH**	
0.00000000000001	14	
0.0000000000001	13	↑
0.000000000001	12	
0.00000000001	11	Increasingly basic
0.0000000001	10	
0.000000001	9	
0.00000001	8	
0.0000001	7	Neutral—neither acidic
0.000001	6	nor basic
0.00001	5	
0.0001	4	
0.001	3	Increasingly acidic
0.01	2	
0.1	1	
1.0	0	↓

When substances dissolve in water, the polar water molecules cause molecules of the substance to separate from each other, or even to break up into ions. These particles are much more likely to take part in chemical reactions. Consequently, most metabolic reactions occur in water.

Water also plays an important role in transporting chemicals within the body. Blood, which is mostly water, carries many vital substances, such as oxygen, sugars, salts, and vitamins, from organs of the digestive and respiratory systems to cells. Blood also carries waste materials, such as carbon dioxide and urea, from these cells to the lungs and kidneys, respectively, which remove them from the blood and release them outside the body.

In addition, water can absorb and transport heat. Blood carries heat released from muscle cells during exercise from deeper parts of the body to the surface. At the same time, water released by skin cells in the form of perspiration can carry heat away by evaporation.

Oxygen

Molecules of oxygen gas (O_2) enter the internal environment through the respiratory organs and are transported throughout the body by the blood, especially by red blood cells. Within cells, organelles use oxygen to release energy from nutrient molecules. The released energy is used to drive the cell's metabolic activities. A continuing supply of oxygen is necessary for cell survival and, ultimately, for the survival of the person.

NO (nitric oxide) and CO (carbon monoxide) are two small chemicals that can harm health, yet are also important to normal physiology. NO is found in smog, cigarettes, and acid rain. CO is a colorless, odorless, lethal gas that is deadly when it leaks from home heating systems or exhaust pipes in closed garages. However, NO and CO are important biological messenger molecules. NO is involved in digestion, memory, immunity, respiration, and circulation. CO functions in the spleen, which recycles old red blood cells, and in the parts of the brain that control memory, smell, and vital functions.

Carbon Dioxide

Carbon dioxide (CO_2) is a simple, carbon-containing inorganic compound. It is produced as a waste product when energy is released during certain metabolic processes. As it moves from cells into surrounding body fluids and blood, most of the carbon dioxide reacts with water to form a weak acid (carbonic acid, H_2CO_3). This acid ionizes, releasing hydrogen ions (H^+) and bicarbonate ions (HCO_3^-), which blood carries to the respiratory organs. There, the chemical reactions reverse, and carbon dioxide gas is produced, eventually to be exhaled.

Inorganic Salts

Inorganic salts are abundant in body fluids. They are the sources of many necessary ions, including ions of sodium (Na^+), chloride (Cl^-), potassium (K^+), calcium (Ca^{+2}), magnesium (Mg^{+2}), phosphate (PO_4^{-2}), carbonate (CO_3^{-2}), bicarbonate (HCO_3^-), and sulfate (SO_4^{-2}). These ions play important roles in metabolic processes, helping to maintain proper water concentrations in body fluids, pH, blood clotting, bone development, energy transfer within cells, and muscle and nerve functions. These electrolytes are regularly gained and lost by the body but must be present in certain concentrations, both inside and outside cells, to maintain homeostasis. Such a condition is called **electrolyte balance.** Disrupted electrolyte balance occurs in certain diseases, and modern medical treatment places considerable emphasis on restoring it. Table 2.6 summarizes the functions of some of the inorganic components of cells.

1 What are the general differences between an organic molecule and an inorganic molecule?

2 What is the difference between an electrolyte and a nonelectrolyte?

3 Define electrolyte balance.

Organic Substances

Important groups of organic substances in cells include carbohydrates, lipids, proteins, and nucleic acids.

Carbohydrates

Carbohydrates (kar″bo-hi′drātz) provide much of the energy that cells require. They also supply materials to build certain cell structures, and they often are stored as reserve energy supplies.

Carbohydrates are water-soluble molecules that contain atoms of carbon, hydrogen, and oxygen. These molecules usually have twice as many hydrogen as oxygen atoms, the same ratio of hydrogen to oxygen as in water molecules (H_2O). This ratio is easy to see in the molecular formulas of the carbohydrates glucose ($C_6H_{12}O_6$) and sucrose ($C_{12}H_{22}O_{11}$).

Carbohydrates are classified by size. Simple carbohydrates, or **sugars,** include the **monosaccharides** (single sugars) and **disaccharides** (double sugars). A monosaccharide may include from three to seven carbon atoms, occurring in a straight chain or a ring (fig. 2.11). Monosaccharides include glucose (dextrose), fructose, and galactose. Disaccharides consist of two 6-carbon units (fig. 2.12b). Sucrose (table sugar) and lactose (milk sugar) are disaccharides.

Complex carbohydrates, also called **polysaccharides,** are built of simple carbohydrates (fig. 2.12c). Cellulose is a polysaccharide made of many glucose molecules, which humans cannot digest. It is important as dietary "fiber." Plant starch is another example. Starch molecules consist of highly branched chains of

TABLE 2.6 — Inorganic Substances Common in Cells

Substance	Symbol or Formula	Functions
I. Inorganic Molecules		
Water	H_2O	Major component of body fluids (chapter 21, p. 808); medium in which most biochemical reactions occur; transports various chemical substances (chapter 14, p. 523); helps regulate body temperature (chapter 6, p. 170)
Oxygen	O_2	Used in release of energy from glucose molecules (chapter 4, p. 111)
Carbon dioxide	CO_2	Waste product that results from metabolism (chapter 4, p. 111); reacts with water to form carbonic acid (chapter 19, p. 762)
II. Inorganic Ions		
Bicarbonate ions	HCO_3^-	Help maintain acid-base balance (chapter 21, p. 819)
Calcium ions	Ca^{+2}	Necessary for bone development (chapter 7, p. 190); muscle contraction (chapter 9, p. 284) and blood clotting (chapter 14, fig. 14.19)
Carbonate ions	CO_3^{-2}	Component of bone tissue (chapter 7, p. 194)
Chloride ions	Cl^-	Help maintain water balance (chapter 21, p. 810)
Hydrogen ions	H^+	pH of the internal environment (chapters 19, p. 754, and 21, p. 817)
Magnesium ions	Mg^{+2}	Component of bone tissue (chapter 7, p. 194); required for certain metabolic processes (chapter 18, p. 715)
Phosphate ions	PO_4^{-3}	Required for synthesis of ATP, nucleic acids, and other vital substances (chapter 4, p. 108); component of bone tissue (chapter 7, p. 194); help maintain polarization of cell membranes (chapter 10, p. 350)
Potassium ions	K^+	Required for polarization of cell membranes (chapter 10, p. 350)
Sodium ions	Na^+	Required for polarization of cell membranes (chapter 10, p. 350); help maintain water balance (chapter 21, p. 810)
Sulfate ions	SO_4^{-2}	Help maintain polarization of cell membranes (chapter 10, p. 350) and acid-base balance (chapter 21, p. 817)

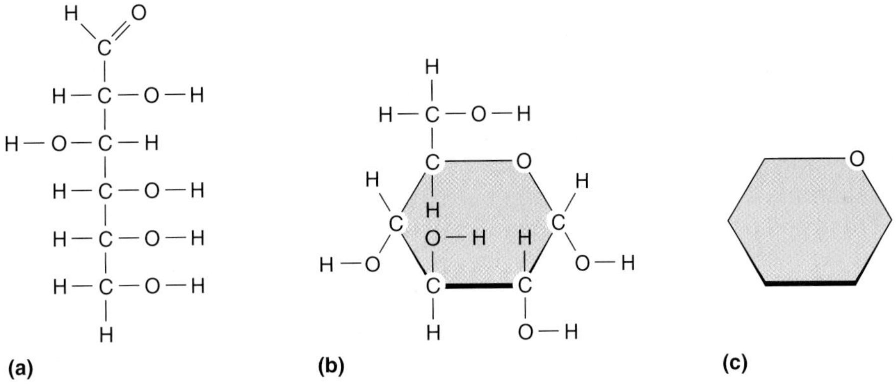

FIGURE 2.11

Structural formulas for glucose. (*a*) Some glucose molecules ($C_6H_{12}O_6$) have a straight chain of carbon atoms. (*b*) More commonly, glucose molecules form a ring structure. (*c*) This shape symbolizes the ring structure of a glucose molecule.

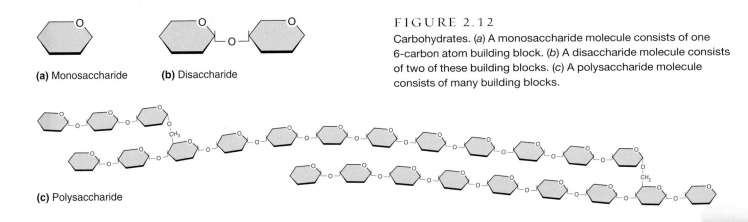

(a) Monosaccharide **(b)** Disaccharide

(c) Polysaccharide

FIGURE 2.12

Carbohydrates. (*a*) A monosaccharide molecule consists of one 6-carbon atom building block. (*b*) A disaccharide molecule consists of two of these building blocks. (*c*) A polysaccharide molecule consists of many building blocks.

glucose molecules connected differently than in cellulose. Humans easily digest starch.

Animals, including humans, synthesize a polysaccharide similar to starch called *glycogen.* Its molecules also are branched chains of sugar units; each branch consists of a dozen or fewer glucose units.

Lipids

Lipids (lip′idz) are a group of organic chemicals that are insoluble in water but soluble in organic solvents, such as ether and chloroform. Lipids include a number of compounds, such as fats, phospholipids, and steroids, that have vital functions in cells and are important constituents of cell membranes (see chapter 3, p. 65). The most common lipids are the *fats,* which are primarily used to supply energy for cellular activities. Fat molecules can supply more energy gram for gram than can carbohydrate molecules. This is why eating a fatty diet leads to weight gain.

Like carbohydrates, fat molecules are composed of carbon, hydrogen, and oxygen atoms. However, fats have a much smaller proportion of oxygen than do carbohydrates. The formula for the fat *tristearin,* $C_{57}H_{110}O_6$, illustrates these characteristic proportions.

The building blocks of fat molecules are **fatty acids** and **glycerol.** Although the glycerol portion of every fat molecule is the same, there are many kinds of fatty acids and, therefore, many kinds of fats. All fatty acid molecules include a carboxyl group (—COOH) at the end of a chain of carbon atoms. Fatty acids differ in the lengths of their carbon atom chains, although such chains usually contain an even number of carbon atoms. The fatty acid chains also may vary in the ways the carbon atoms join. In some cases, the carbon atoms are all linked by single carbon-carbon bonds. This type of fatty acid is called a **saturated fatty acid;** that is, each carbon atom binds as many hydrogen atoms as possible and is thus saturated with hydrogen atoms. Other fatty acid chains have one or more double bonds between carbon atoms. Fatty acids with one double bond are called *monounsaturated fatty*

acids, and those with two or more double bonds are *polyunsaturated fatty acids* (fig. 2.13).

Fatty acids and glycerol are united so that each glycerol molecule combines with three fatty acid molecules. The result is a single fat molecule, or *triglyceride* (fig. 2.14). The fatty acids found in a triglyceride may have different lengths and different degrees of saturation. Therefore, a wide variety of fats exist. Fat molecules that contain only saturated fatty acids are called **saturated fats,** and those that include unsaturated fatty acids are called **unsaturated fats.** Each kind of fat molecule has distinct properties.

A diet rich in saturated fat increases a person's risk of developing atherosclerosis, a serious disease that obstructs blood vessels. Unsaturated, particularly monounsaturated, fats are healthier to eat than saturated fats.

Saturated fats are more abundant in fatty foods that are solids at room temperature, such as butter, lard, and most other animal fats. Unsaturated fats are plentiful in fatty foods that are liquids at room temperature, such as soft margarine and seed oils, including corn oil and soybean oil. Coconut and palm oils, however, are exceptions—they are relatively high in saturated fat.

A *phospholipid* molecule is similar to a fat molecule in that it contains a glycerol portion and fatty acid chains. The phospholipid, however, has only two fatty acid chains and, in place of the third, has a portion containing a phosphate group. This phosphate-containing portion is soluble in water (hydrophilic) and forms the "head" of the molecule, whereas the fatty acid portion is insoluble in water (hydrophobic) and forms a "tail." Figure 2.15 illustrates the molecular structure of cephalin, a phospholipid in blood. Other phospholipids are important in cellular structures.

Steroid molecules are complex structures that include connected rings of carbon atoms (fig. 2.16). Among the more

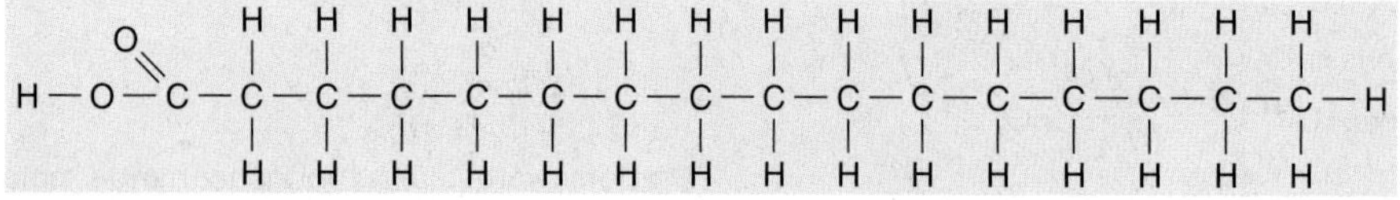

(a) Saturated fatty acid

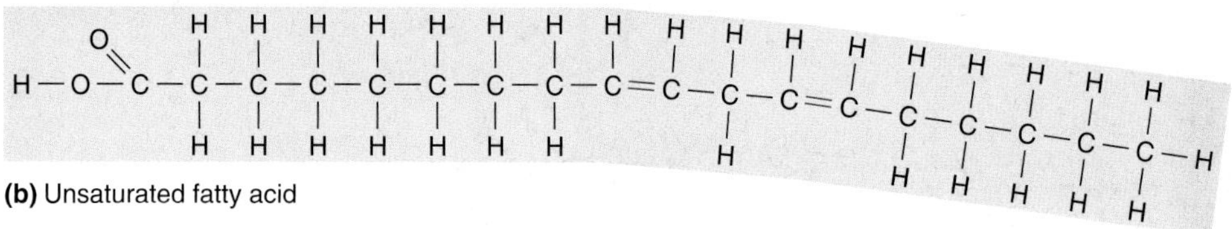

(b) Unsaturated fatty acid

FIGURE 2.13

Fatty acids. (*a*) A molecule of saturated fatty acid and (*b*) a molecule of unsaturated fatty acid. Double bonds between carbon atoms are shown in red. Note that they cause a "kink" in the shape of the molecule.

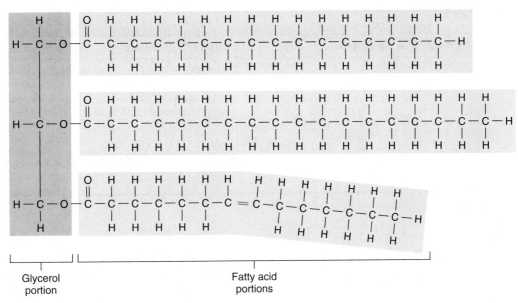

Glycerol portion

Fatty acid portions

FIGURE 2.14

A triglyceride molecule (fat) consists of a glycerol portion and three fatty acid portions. This is an example of an unsaturated fat. The double bond between carbon atoms is shown in red.

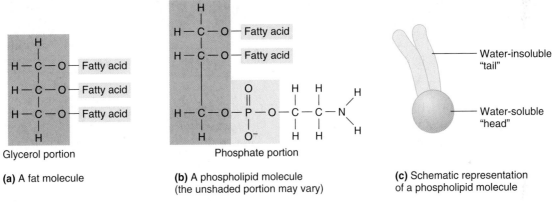

Glycerol portion

(a) A fat molecule

Phosphate portion

(b) A phospholipid molecule
(the unshaded portion may vary)

Water-insoluble
"tail"

Water-soluble
"head"

(c) Schematic representation
of a phospholipid molecule

FIGURE 2.15

Fats and phospholipids. (*a*) A fat molecule (triglyceride) contains a glycerol and three fatty acids. (*b*) In a phospholipid molecule, a phosphate-containing group replaces one fatty acid. (*c*) Schematic representation of a phospholipid.

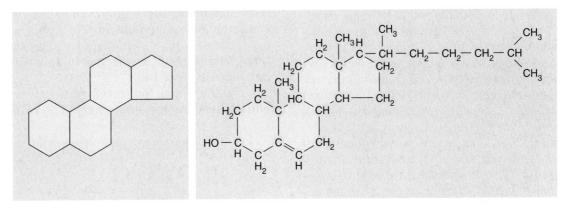

(a) General structure of a steroid

(b) Cholesterol

FIGURE 2.16

Steroid structure. (*a*) The general structure of a steroid. (*b*) The structural formula for cholesterol, a steroid widely distributed in the body.

CHAPTER TWO   *Chemical Basis of Life*

TABLE 2.7 Important Groups of Lipids

Group	Basic Molecular Structure	Characteristics
Triglycerides	Three fatty acid molecules bound to a glycerol molecule	Most common lipid in the body; stored in fat tissue as an energy supply; fat tissue also provides insulation beneath the skin
Phospholipids	Two fatty acid molecules and a phosphate group bound to a glycerol molecule (may also include a nitrogen-containing molecule attached to the phosphate group)	Used as structural components in cell membranes; large amounts are in the liver and parts of the nervous system
Steroids	Four connected rings of carbon atoms	Widely distributed in the body with a variety of functions; includes cholesterol, sex hormones, and certain hormones of the adrenal glands

FIGURE 2.17

Amino acid structure. (*a*) The general structure of an amino acid. Note the amino group and carboxyl group that are common to all amino acid molecules. (*b*) Some representative amino acids and their structural formulas. Each amino acid molecule has a particular shape due to the different R groups.

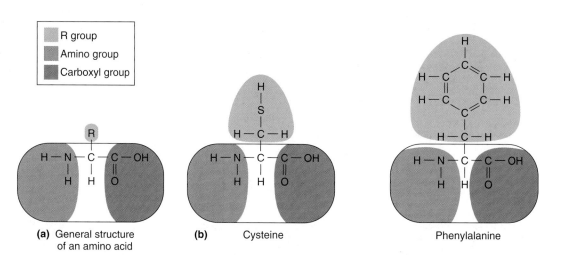

(a) General structure of an amino acid

(b) Cysteine

Phenylalanine

important steroids are cholesterol, which is in all body cells and is used to synthesize other steroids; sex hormones, such as estrogen, progesterone, and testosterone; and several hormones from the adrenal glands. Chapters 13, 14, 20, 21, and 22 discuss these steroids. Table 2.7 summarizes the molecular structures and characteristics of lipids.

Proteins

Proteins (pro′te-inz) have a great variety of functions. Some proteins serve as structural materials, energy sources, or chemical messengers (hormones). Other proteins combine with carbohydrates (glycoproteins) and function as receptors on cell surfaces that bind to particular kinds of molecules. Proteins called *antibodies* recognize and destroy substances that are foreign to the body. Many proteins play vital roles in metabolic processes as **enzymes** (en′zīmz), which are catalysts in living systems. That is, they speed specific chemical reactions without being consumed in the process. (Enzymes are discussed in chapter 4, p. 106.)

Like carbohydrates and lipids, proteins consist of atoms of carbon, hydrogen, and oxygen. In addition, proteins always contain nitrogen atoms and sometimes contain sulfur atoms. The building blocks of proteins are smaller molecules called **amino acids.**

Twenty kinds of amino acids comprise proteins in organisms. Amino acid molecules have an amino group

(—NH$_2$) at one end and a carboxyl group (—COOH) at the other end. Between these groups is a single carbon atom. This central carbon is bonded to a hydrogen atom and to another group of atoms called a *side chain* or *R group* ("R" may be thought of as the "Rest of the molecule"). The composition of the R group distinguishes one type of amino acid from another (fig. 2.17).

Proteins have complex three-dimensional shapes, yet they are assembled from simple chains of amino acids connected by peptide bonds. These are covalent bonds that link the amino end of one amino acid with the carboxyl end of another. Figure 2.18 shows two amino acids connected by a peptide bond. The resulting molecule is a dipeptide. Adding a third amino acid creates a tripeptide. Many amino acids connected in this way constitute a polypeptide (fig. 2.19*a*).

Proteins have four levels of structure: primary, secondary, tertiary and quaternary. The *primary structure*

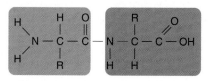

FIGURE 2.18

A peptide bond between two amino acids.

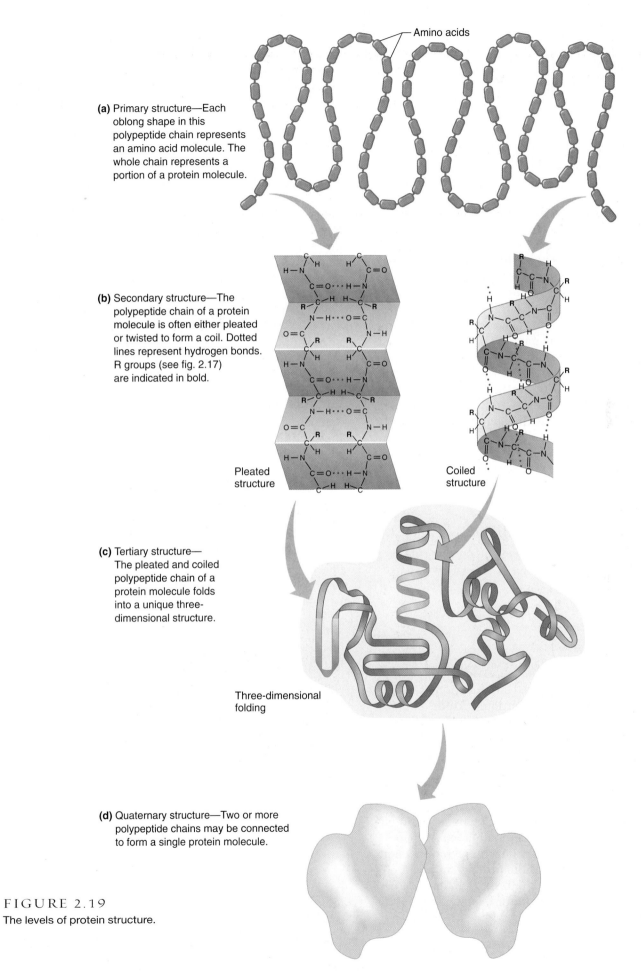

(a) Primary structure—Each oblong shape in this polypeptide chain represents an amino acid molecule. The whole chain represents a portion of a protein molecule.

Amino acids

(b) Secondary structure—The polypeptide chain of a protein molecule is often either pleated or twisted to form a coil. Dotted lines represent hydrogen bonds. R groups (see fig. 2.17) are indicated in bold.

Pleated structure

Coiled structure

(c) Tertiary structure— The pleated and coiled polypeptide chain of a protein molecule folds into a unique three-dimensional structure.

Three-dimensional folding

(d) Quaternary structure—Two or more polypeptide chains may be connected to form a single protein molecule.

FIGURE 2.19

The levels of protein structure.

is the amino acid sequence of the polypeptide chain. Depending on the protein, the primary structure may range from fewer than 100 to more than 5,000 amino acids. The amino acid sequence is characteristic of a particular protein. Thus, the blood protein hemoglobin and the muscle protein myosin have different amino acid sequences.

In the *secondary structure* (fig. 2.19*b*), the polypeptide chain either forms a springlike coil (alpha helix), or it folds back and forth on itself (beta-pleated sheet). Secondary structure is due to hydrogen bonding. Recall that polar molecules result when electrons are not shared evenly in certain covalent bonds. In amino acids, this results in slightly negative oxygen and nitrogen atoms and slightly positive hydrogen atoms. Hydrogen bonding between oxygen and hydrogen atoms in different parts of the molecule determines the secondary structure.

Hydrogen bonding and even covalent bonding between atoms in different parts of a polypeptide can cause yet another level of folding, the *tertiary structure.* As a result, proteins have distinct three-dimensional shapes, or **conformations** (fig. 2.19*c*), which determine function. Some proteins are long and fibrous, such as the keratins that form hair and the threads of fibrin that knit a blood clot. Many proteins are globular. Myoglobin and hemoglobin, which transport oxygen in muscle and blood, respectively, are globular, as are many enzymes.

Protein misfolding can cause disease. Some mutations that cause cystic fibrosis, for example, prevent the encoded protein from assuming its final form and anchoring in the cell membrane, where it normally controls the flow of chloride ions. This dries out certain body fluids, which impairs respiration and digestion. A class of illnesses called transmissible spongiform encephalopathies, which includes "mad cow disease," results when a type of protein called a prion folds into an abnormal form that is infectious—that is, it converts normal prion protein into the pathological form, which riddles the brain with spongy-looking holes. Alzheimer disease results from the cutting of a protein called beta amyloid into pieces of a certain size, which attach and accumulate, forming structures called plaques in parts of the brain controlling memory and cognition.

Various treatments can cause the secondary and tertiary structures of a protein's conformation to fall apart, or *denature.* Because the primary structure (amino acid sequence) remains, sometimes the protein can regain its shape when normal conditions return. High temperature, radiation, pH changes, and certain chemicals (such as urea) can denature proteins.

A familiar example of irreversible protein denaturation is the response of the protein albumin to heat (for example, cooking an egg white). A permanent wave that curls hair also results from protein denaturation. Chemicals first break apart the tertiary structure formed when sulfur-containing amino acids attract each other within keratin molecules. This relaxes the hair. When the chemicals are washed out and the hair set, the sulfur bonds reform, but in different places, changing the appearance of the hair.

Not all proteins are single polypeptide chains. Sometimes several polypeptide chains are connected in a fourth level, or *quaternary structure,* to form a very large protein (fig. 2.19*d*). Hemoglobin is a quaternary protein made up of four separate polypeptide chains.

A protein's conformation determines its function. The amino acid sequence and interactions between the amino acids in a protein determine the conformation. Thus, it is the amino acid sequence of a protein that determines its function in the body. Genes, made of nucleic acid, contain the information for the amino acid sequences of all the body's proteins in a form that the cell can decode.

Nucleic Acids

Nucleic acids (nu-kle′ik as′idz) constitute genes, the instructions that control a cell's activities, and play important roles in protein synthesis. Nucleic acid molecules are very large and complex. They contain atoms of carbon, hydrogen, oxygen, nitrogen, and phosphorus, which form building blocks called **nucleotides.** Each nucleotide consists of a 5-carbon sugar (ribose or deoxyribose), a phosphate group, and one of several organic bases (fig. 2.20). Such nucleotides, linked in a chain, form a **polynucleotide** (fig. 2.21).

There are two major types of nucleic acids. One type is composed of molecules whose nucleotides contain ribose sugar; it is called **RNA** (ribonucleic acid), and it is a single polynucleotide chain. The nucleotides of the second type contain deoxyribose sugar; nucleic acid of this type is called **DNA** (deoxyribonucleic acid), and it is a double polynucleotide chain. Figure 2.22 compares the structure of ribose and deoxyribose, which differ by one oxygen atom. DNA and RNA also differ in the types of bases they contain.

DNA molecules store information in a type of molecular code. Cells use this information to construct specific protein molecules, which have a wide variety of functions. RNA molecules help to synthesize proteins.

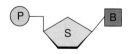

FIGURE 2.20
A nucleotide consists of a 5-carbon sugar (S = sugar), a phosphate group (P = phosphate), and an organic base (B = base).

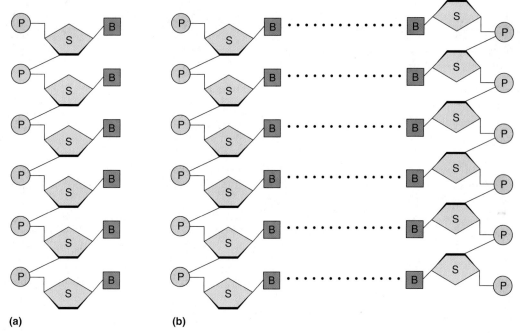

FIGURE 2.21

A schematic representation of nucleic acid structure. A nucleic acid molecule consists of (*a*) one (RNA) or (*b*) two (DNA) polynucleotide chains. DNA chains are held together by hydrogen bonds (dotted lines).

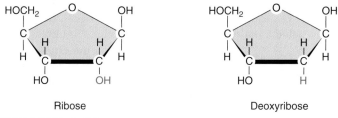

Ribose   Deoxyribose

FIGURE 2.22

The molecules of ribose and deoxyribose differ by a single oxygen atom.

DNA molecules have a unique ability to make copies of, or replicate, themselves. They replicate prior to cell division, and each newly formed cell receives an exact copy of the original cell's DNA molecules. Chapter 4 (p. 116) discusses the storage of information in nucleic acid molecules, use of the information in the manufacture of protein molecules, and how these proteins control metabolic reactions.

Table 2.8 summarizes the four groups of organic compounds. Figure 2.23 shows three-dimensional (space-filling) models of some important molecules, illustrating their shapes. Clinical Application 2.3 describes two techniques used to view human anatomy and physiology.

1 Compare the chemical composition of carbohydrates, lipids, proteins, and nucleic acids.

2 How does an enzyme affect a chemical reaction?

3 What is likely to happen to a protein molecule that is exposed to intense heat or radiation?

4 What are the functions of DNA and RNA?

TABLE 2.8	Organic Compounds in Cells			
Compound	**Elements Present**	**Building Blocks**	**Functions**	**Examples**
Carbohydrates	C,H,O	Simple sugar	Provide energy, cell structure	Glucose, starch
Lipids	C,H,O (often P)	Glycerol, fatty acids, phosphate groups	Provide energy, cell structure	Triglycerides, phospholipids, steroids
Proteins	C,H,O,N (often S)	Amino acids	Provide cell structure, enzymes, energy	Albumins, hemoglobin
Nucleic acids	C,H,O,N,P	Nucleotides	Store information for the synthesis of proteins, control cell activities	RNA, DNA

CT Scanning and PET Imaging

Physicians use two techniques—computerized tomography (CT) scanning and positron emission tomography (PET imaging)—to paint portraits of anatomy and physiology, respectively.

In CT scanning, an X-ray emitting device is positioned around the region of the body being examined. At the same time, an X-ray detector is moved in the opposite direction on the other side of the body. As these parts move, an X-ray beam passes through the body from hundreds of different angles. Because tissues and organs of varying composition absorb X rays differently, the intensity of X rays reaching the detector varies from position to position. A computer records the measurements made by the X-ray detector and combines them mathematically. This creates on a viewing screen a sectional image of the internal body parts (fig. 2D).

Ordinary X-ray techniques produce two-dimensional images known as radiographs, X rays, or films. A CT scan provides three-dimensional information. The CT scan can also clearly differentiate between soft tissues of slightly different densities, such as the liver and kidneys, which cannot be seen in a conventional X-ray image. In this way, a CT scan can detect abnormal tissue, such as a tumor. For example, a CT scan can tell whether a sinus headache that does not respond to antibiotic therapy is caused by a drug-resistant infection or by a tumor.

PET imaging uses radioactive isotopes that naturally emit positrons, which are atypical positively charged electrons, to detect biochemical activity in a specific body part. Useful isotopes in PET imaging include carbon-11, nitrogen-13, oxygen-15, and fluorine-18. When one of these isotopes releases a positron, it interacts with a nearby negatively charged electron. The two particles destroy each other, an event called annihilation. At the moment of destruction, two gamma rays appear and move away from each other in opposite directions. Special equipment detects the gamma radiation.

To produce a PET image of biochemically active tissue, a person is injected with a metabolically active compound that includes a bound positron-emitting isotope. To study the brain, for example, a person is injected with glucose containing fluorine-18. After the brain takes up the isotope-tagged compound, the person rests the head within a circular array of radiation detectors. A device records each time two gamma rays are emitted simultaneously and travel in opposite directions (the result of annihilation). A computer collects and combines the data and generates a cross-sectional image. The image indicates the location and relative concentration of the radioactive isotope in different regions of the brain and can be used to study those parts metabolizing glucose.

PET images reveal the parts of the brain that are affected in such disorders as Huntington disease, Parkinson disease, epilepsy, and Alzheimer disease, and they are used to study blood flow in vessels supplying the brain and heart. The technology is invaluable for detecting the physiological bases of poorly understood behavioral disorders, such as obsessive-compulsive dis-

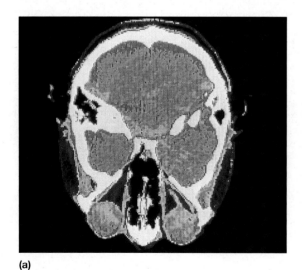

(a)

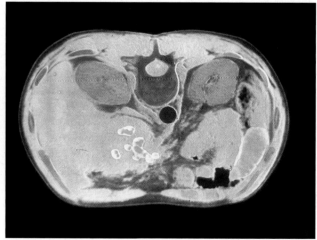

(b)

FIGURE 2D

CT scans of (a) the head and (b) the abdomen.

order. In this condition, a person repeatedly performs a certain behavior, such as washing hands, showering, locking doors, or checking to see that the stove is turned off. PET images of people with this disorder reveal intense activity in two parts of the brain that are quiet in the brains of unaffected individuals. Knowing the site of altered brain activity can help researchers develop more directed drug therapy.

In addition to highlighting biochemical activities behind illness, PET scans allow biologists to track normal brain physiology. Figure 2E shows that different patterns of brain activity are associated with learning and with reviewing something already learned. ∎

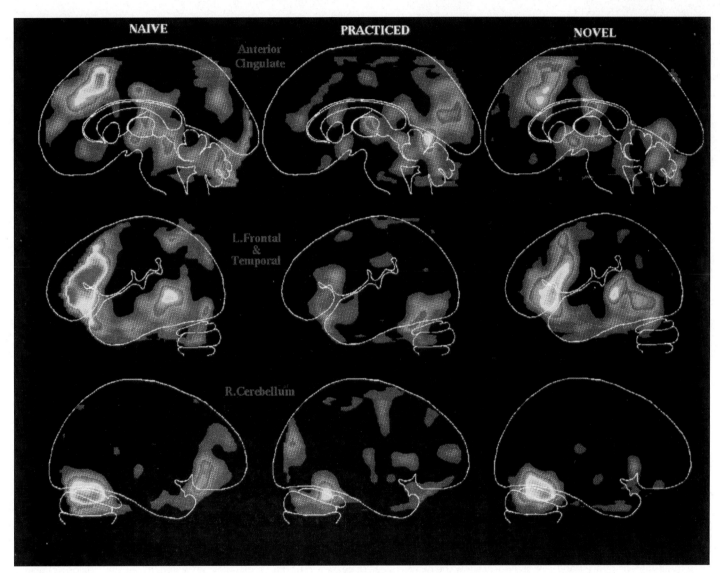

FIGURE 2E

These PET images demonstrate brain changes that accompany learning. The top and bottom views show different parts of the same brain. The "naive" brain on the left has been given a list of nouns and asked to visualize each word. In the middle column, the person has practiced the task, so he can picture the nouns with less brain activity. In the third column, the person receives a new list of nouns. Learning centers in the brain show increased activity.

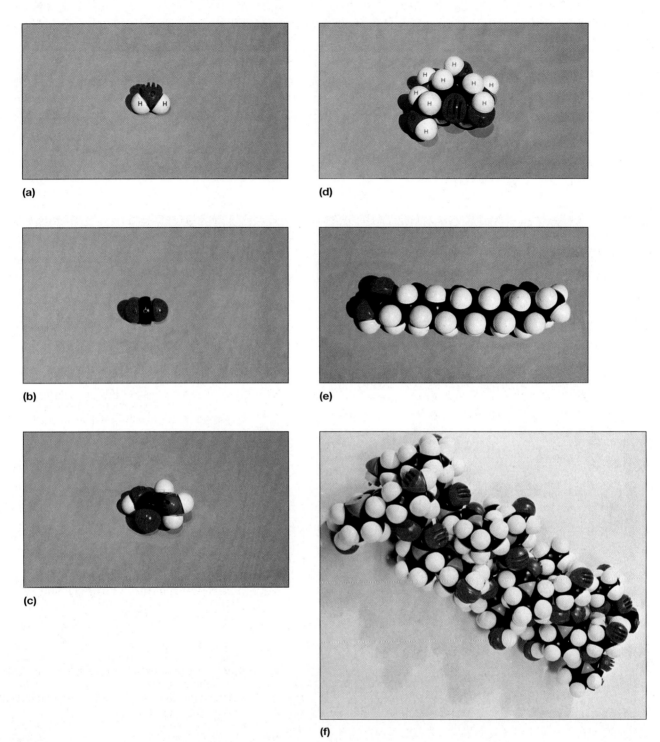

FIGURE 2.23

These three-dimensional (space-filling) models show the relative sizes of several important molecules: (*a*) water, (*b*) carbon dioxide, (*c*) glycine (an amino acid), (*d*) glucose (a monosaccharide), (*e*) a fatty acid, and (*f*) collagen (a protein). White = hydrogen, red = oxygen, blue = nitrogen, black = carbon.

CHAPTER SUMMARY

Introduction (page 38)

Chemistry deals with the composition of substances and changes in their composition. Biochemistry is the chemistry of living organisms.

Structure of Matter (page 38)

Matter is anything that has weight and takes up space.

1. Elements and atoms
   a. Naturally occurring matter on earth is composed of ninety-two elements.
   b. Elements occur most frequently in chemical combinations called compounds.
   c. Elements are composed of atoms.
   d. Atoms of different elements vary in size, weight, and ways of interacting.

2. Atomic structure
   a. An atom consists of electrons surrounding a nucleus, which contains protons and neutrons. The exception is hydrogen, which contains only a proton in its nucleus.
   b. Electrons are negatively charged, protons positively charged, and neutrons uncharged.
   c. A complete atom is electrically neutral.
   d. The atomic number of an element is equal to the number of protons in each atom; the atomic weight is equal to the number of protons plus the number of neutrons in each atom.

3. Isotopes
   a. Isotopes are atoms with the same atomic number but different atomic weights (due to differing numbers of neutrons).
   b. All the isotopes of an element react chemically in the same manner.
   c. Some isotopes are radioactive and release atomic radiation.

4. Molecules and compounds
   a. Two or more atoms may combine to form a molecule.
   b. A molecular formula represents the numbers and kinds of atoms in a molecule.
   c. If atoms of the same element combine, they produce molecules of that element.
   d. If atoms of different elements combine, they form molecules of substances called compounds.

5. Bonding of atoms
   a. When atoms combine, they gain, lose, or share electrons.
   b. Electrons occur in shells around an atomic nucleus.
   c. Atoms with completely filled outer shells are inactive, whereas atoms with incompletely filled outer shells tend to gain, lose, or share electrons and thus achieve stable structures.
   d. Atoms that lose electrons become positively charged; atoms that gain electrons become negatively charged.
   e. Ions with opposite charges attract and join by ionic bonds; atoms that share electrons join by covalent bonds.
   f. A structural formula represents the arrangement of atoms within a molecule.
   g. Polar molecules result from an unequal sharing of electrons.
   h. Hydrogen bonds occur between polar molecules.

6. Chemical reactions
   a. In a chemical reaction, bonds between atoms, ions, or molecules break or form.
   b. Three kinds of chemical reactions are synthesis, in which larger molecules form from smaller particles; decomposition, in which smaller particles form from breakdown of larger molecules; and exchange reactions, in which parts of two different molecules trade positions.
   c. Many reactions are reversible. The direction of a reaction depends upon the proportion of reactants and products, the energy available, and the presence or absence of catalysts.

7. Acids, bases, and salts
   a. Compounds that ionize when they dissolve in water are electrolytes.
   b. Electrolytes that release hydrogen ions are acids, and those that release hydroxide or other ions that react with hydrogen ions are bases.
   c. Acids and bases react to form water and electrolytes called salts.

8. Acid and base concentrations
   a. pH represents the concentration of hydrogen ions (H^+) and hydroxide ions (OH^-) in a solution.
   b. A solution with equal numbers of H^+ and OH^- is neutral and has a pH of 7.0; a solution with more H^+ than OH^- is acidic (pH less than 7.0); a solution with fewer H^+ than OH^- is basic (pH greater than 7.0).
   c. A tenfold difference in hydrogen ion concentration separates each whole number in the pH scale.

Chemical Constituents of Cells (page 47)

Molecules containing carbon and hydrogen atoms are organic and are usually nonelectrolytes; other molecules are inorganic and are usually electrolytes.

1. Inorganic substances
   a. Water is the most abundant compound in cells. Many chemical reactions take place in water. Water transports chemicals and heat and helps release excess body heat.
   b. Oxygen releases energy needed for metabolic activities from glucose and other molecules.
   c. Carbon dioxide is produced when energy is released during metabolic processes.
   d. Inorganic salts provide ions needed in a variety of metabolic processes.
   e. Electrolytes must be present in certain concentrations inside and outside of cells.

2. Organic substances
   a. Carbohydrates provide much of the energy cells require; their building blocks are simple sugar molecules.
   b. Lipids, such as fats, phospholipids, and steroids, supply energy and are used to build cell parts; their building blocks are molecules of glycerol and fatty acids.
   c. Proteins serve as structural materials, energy sources, hormones, cell surface receptors, antibodies, and enzymes which initiate or speed chemical reactions without being consumed.
      (1) The building blocks of proteins are amino acids.
      (2) Proteins vary in the numbers and kinds of amino acids they contain; the sequences of these amino

acids; and their three-dimensional structures, or conformations.

(3) The amino acid sequence determines the protein's conformation.

(4) The protein's conformation determines its function.

(5) Exposure to excessive heat, radiation, electricity, or certain chemicals can denature proteins.

d. Nucleic acids constitute genes, the instructions that control cell activities, and direct protein synthesis.

(1) The two major kinds are RNA and DNA.

(2) Nucleic acid molecules are composed of building blocks called nucleotides.

(3) DNA molecules store information that is used by cell parts to construct specific kinds of protein molecules.

(4) RNA molecules help synthesize proteins.

(5) DNA molecules are replicated and an exact copy of the original cell's DNA is passed to each of the newly formed cells, resulting from cell division.

CRITICAL THINKING QUESTIONS

1. Which acidic and alkaline substances do you encounter daily? What foods do you eat regularly that are acidic? What alkaline foods do you eat?

2. Using the information on page 50 to distinguish between saturated and unsaturated fats, try to list all of the sources of saturated and unsaturated fats you have eaten during the past twenty-four hours.

3. How would you reassure a patient who is about to undergo CT scanning for evaluation of a tumor, and who fears becoming a radiation hazard to family members?

4. Various forms of ionizing radiation, such as that released from X-ray tubes and radioactive substances, are commonly used in the treatment of cancer, yet such exposure can cause adverse effects, including the development of cancers. How would you explain the value of radiation therapy to a cancer patient in light of this seeming contradiction?

5. How would you explain the importance of amino acids and proteins in a diet to a person who is following a diet composed primarily of carbohydrates?

6. Which clinical laboratory tests that you know of are based on chemistry?

7. Explain why the symptoms of many inherited diseases result from abnormal protein function.

REVIEW EXERCISES

1. Distinguish between chemistry and biochemistry.
2. Define *matter*.
3. Explain the relationship between elements and atoms.
4. Define *compound*.
5. List the four most abundant elements in the human body.
6. Describe the major parts of an atom.
7. Distinguish between protons and neutrons.
8. Explain why a complete atom is electrically neutral.
9. Distinguish between atomic number and atomic weight.
10. Define *isotope*.
11. Define *atomic radiation*.
12. Explain the relationship between molecules and compounds.
13. Describe how electrons are arranged within atoms.
14. Explain why some atoms are chemically inert.
15. Distinguish between an ionic bond and a covalent bond.
16. Distinguish between a single covalent bond and a double covalent bond.
17. Distinguish between a molecular formula and a structural formula.

18. Describe three major types of chemical reactions.
19. Define *reversible reaction*.
20. Define *catalyst*.
21. Define *acid, base, salt*, and *electrolyte*.
22. Explain what *pH* measures.
23. Distinguish between organic and inorganic substances.
24. Describe the functions of water and oxygen in the human body.
25. List several ions that cells require, and describe their general functions.
26. Define *electrolyte balance*.
27. Describe the general characteristics of carbohydrates.
28. Distinguish between simple and complex carbohydrates.
29. Describe the general characteristics of lipids.
30. Distinguish between saturated and unsaturated fats.
31. Describe the general characteristics of proteins.
32. Describe the function of an enzyme.
33. Explain how protein molecules may become denatured.
34. Describe the general characteristics of nucleic acids.
35. Explain the general functions of nucleic acids.

WEB CONNECTIONS

Visit the Student Edition of the Online Learning Center at www.mhhe.com/shier10 **for answers to chapter questions, additional quizzes, interactive learning exercises, and other study tools.**

UNDERSTANDING WORDS

cyt-, cell: *cyt*oplasm—fluid between the cell membrane and nuclear envelope.

endo-, within: *endo*plasmic reticulum—complex of membranous structures in the cytoplasm.

hyper-, above: *hyper*tonic—solution that has a greater osmotic pressure than the cytosol.

hypo-, below: *hypo*tonic—solution that has a lesser osmotic pressure than the cytosol.

inter-, between: *inter*phase—stage between mitotic divisions of a cell.

iso-, equal: *iso*tonic—solution that has an osmotic pressure equal to that of the cytosol.

lys-, to break up: *lys*osome—organelle containing enzymes that break down molecules of protein, carbohydrate, or nucleic acid.

mit-, thread: *mit*osis—stage of cell division when chromosomes condense and become visible.

phag-, to eat: *phag*ocytosis—process by which a cell takes in solid particles.

pino-, to drink: *pino*cytosis—process by which a cell takes in tiny droplets of liquid.

pro-, before: *pro*phase—first stage of mitosis.

som, body: ribo*some*—tiny, spherical organelle composed of protein and RNA.

vesic-, bladder: *vesic*le—small, saclike organelle that contains various substances to be transported or secreted.

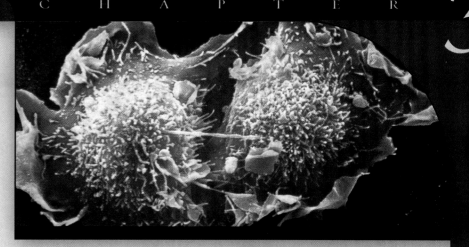

Falsely colored scanning electron micrograph of two daughter cells emerging from the final stage of cell division (3,500×).

Cells

CHAPTER OBJECTIVES

After you have studied this chapter, you should be able to

1. Explain how cells differ from one another.

2. Describe the general characteristics of a composite cell.

3. Explain how the components of a cell's membrane provide its functions.

4. Describe each kind of cytoplasmic organelle and explain its function.

5. Describe the cell nucleus and its parts.

6. Explain how substances move into and out of cells.

7. Describe the cell cycle.

8. Explain how a cell divides.

9. Describe several controls of cell division.

10. Explain how stem cells and progenitor cells make possible growth and repair of tissues.

C ertain people are naturally resistant to HIV, the virus that causes AIDS. One woman received a blood transfusion in 1980 that was later found to be contaminated with HIV, but she never became infected. Some intravenous drug users share needles with people who later develop AIDS and never become ill, and some prostitutes exposed to many HIV-positive men never themselves become infected.

We usually think of avoiding AIDS by avoiding activities that spread the virus, and this is without a doubt the best course. But what protects these people, all of whom have been exposed to HIV? A lucky few individuals cannot contract AIDS because of an abnormality of their cells.

When HIV enters a human body, it approaches certain white blood cells, called CD4 helper T cells, that control the immune system. The virus binds first to receptors called CD4—the receptors are proteins that extend from the cell surface. Once bound, HIV moves down the CD4 receptor and binds another receptor, called CCR5. Only then can the virus enter the cell and start the chain reaction of viral replication that ultimately topples immunity.

Thanks to heredity, 1% of Caucasians in the United States, and far fewer Asians, African Americans, and Native Americans, have cell surfaces that lack the crucial CCR5 HIV docking sites. These lucky few individuals cannot get AIDS, because HIV cannot enter their cells. Another 20% of the Caucasian population (less for others) have half the normal number of CCR5 receptors. These people can become infected, but remain healthy longer than is usual.

Researchers are now applying this knowledge of how AIDS begins at the cellular level to develop vaccines and new treatments. Understanding how HIV interacts with cells, the units of life, has revealed what might finally prove to be HIV's point of vulnerability—a protein portal called CCR5. ■

An adult human body consists of about 75 trillion **cells,** the basic units of an organism. All cells have much in common, yet they come in at least 260 different varieties. Different cell types interact to build tissues, which interact to form organs. Cells with specialized characteristics, such as contractile proteins in muscle cells or the gland cells' ability to secrete substances, are termed **differentiated.** Such specialized cells form from less specialized cells that divide.

Cells vary considerably in size. We measure cell sizes in units called *micrometers* (mi'kro-me"terz). A micrometer equals one thousandth of a millimeter and is symbolized μm. A human egg cell is about 140 μm in diameter and is just barely visible to an unaided eye. This is large when compared to a red blood cell, which is about 7.5 μm in diameter, or the most common types of white blood cells, which vary from 10 to 12 μm in diameter. On the other hand, smooth muscle cells can be between 20 and 500 μm long (fig. 3.1).

Cells vary in shape, and typically their shapes make possible their functions (fig. 3.2). For instance, nerve cells that have long, threadlike extensions many centimeters long transmit nerve impulses from one part of the body to another. Epithelial cells that line the inside of the mouth are thin, flattened, and tightly packed, somewhat like floor tiles. They form a barrier that shields underlying tissue. Muscle cells, which contract and pull structures closer together, are slender and rodlike, with their ends attached to the parts they move. Muscle cells are filled with contractile proteins. An adipose cell is little more than a blob of fat; a B lymphocyte, a type of white blood cell, is an antibody factory.

A Composite Cell

It is not possible to describe a typical cell, because cells vary greatly in size, shape, content, and function. We can, however, consider a hypothetical composite cell that includes many known cell structures (fig. 3.3).

The three major parts of a cell—the **nucleus** (nu'kle-us), the **cytoplasm** (si'to-plazm), and the **cell membrane—** are easily seen under the light microscope if appropriately stained. The nucleus is innermost and is enclosed by a thin membrane called the nuclear envelope. It includes the genetic material (DNA), which directs the cell's functions. The cytoplasm is a mass of fluid that surrounds the nucleus and is itself encircled by the even-thinner cell membrane (also called a plasma membrane). Within the cytoplasm are specialized structures called **cytoplasmic organelles** (or-gan-elz), which can be seen clearly only under the higher magnification of an electron microscope. Organelles, suspended in a liquid called **cytosol,** perform specific functions, in a sense dividing the labor of the cell.

Cells with nuclei, such as those of the human body, are termed *eukaryotic,* meaning "true nucleus." In contrast are the *prokaryotic* ("before nucleus") cells of bacteria. Although bacterial cells lack nuclei and other membrane-bound organelles and are thus simpler than eukaryotic cells, the bacteria are widespread and have existed much longer than eukaryotic cells.

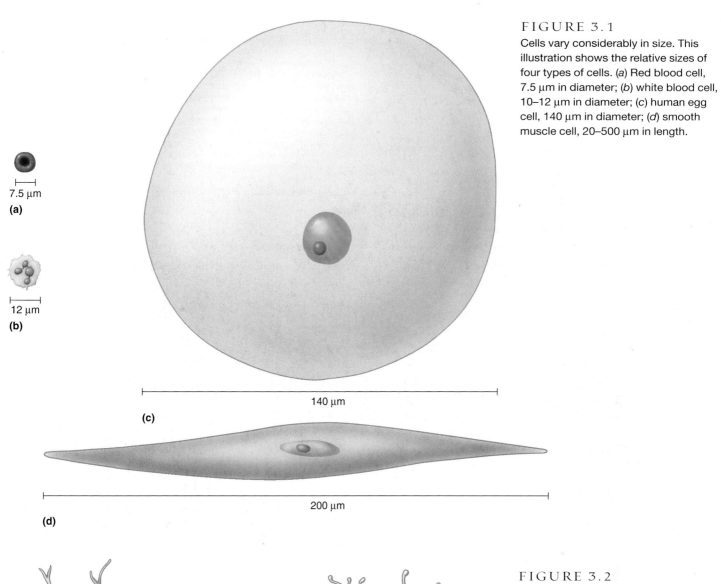

7.5 μm
(a)

12 μm
(b)

140 μm
(c)

200 μm
(d)

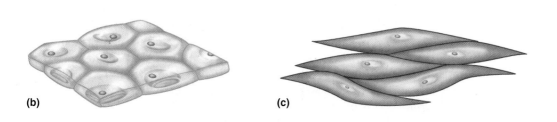

FIGURE 3.2
Cells vary in shape and function. (*a*) A nerve cell transmits impulses from one body part to another. (*b*) Epithelial cells form layers that protect underlying cells. (*c*) Muscle cells contract, pulling structures closer together.

(a)

(b)   (c)

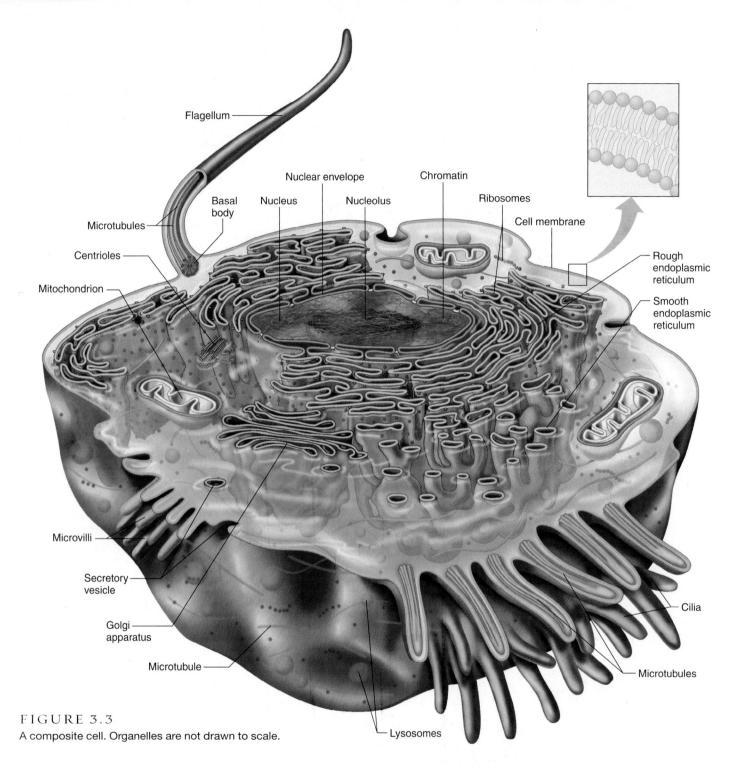

FIGURE 3.3
A composite cell. Organelles are not drawn to scale.

Labels in figure:
Flagellum
Microtubules
Centrioles
Mitochondrion
Basal body
Nucleus
Nuclear envelope
Nucleolus
Chromatin
Ribosomes
Cell membrane
Rough endoplasmic reticulum
Smooth endoplasmic reticulum
Microvilli
Secretory vesicle
Golgi apparatus
Microtubule
Lysosomes
Cilia
Microtubules

1 Define differentiated cell.

2 Name the major parts of a cell.

3 What are the general functions of the cytoplasm and nucleus?

Cell Membrane

The cell membrane is the outermost limit of a cell. Not just a simple boundary, the cell membrane is an actively functioning part of the living material. Many important metabolic reactions take place on its surfaces, and it har-bors molecules that enable cells to communicate and interact.

General Characteristics

The cell membrane is extremely thin—visible only with the aid of an electron microscope (fig. 3.4)—but it is flexible and somewhat elastic. It typically has complex surface features with many outpouchings and infoldings that increase surface area. The cell membrane quickly seals tiny breaks, but if it is extensively damaged, cell contents escape, and the cell dies.

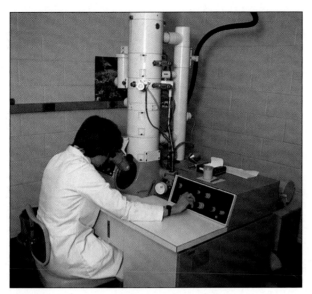

FIGURE 3.4
A transmission electron microscope.

The maximum effective magnification possible using a light microscope is about 1,200×. A confocal microscope is a type of light microscope that passes white or laser light through a pinhole and lens to impinge on the object, which greatly enhances resolution (ability to distinguish fine detail). A transmission electron microscope (TEM) provides an effective magnification of nearly 1,000,000×, whereas a scanning electron microscope (SEM), can provide about 50,000×. Photographs of microscopic objects (micrographs) produced using the light microscope and the transmission electron microscope are typically two-dimensional, but those obtained with the scanning electron microscope have a three-dimensional quality (fig. 3.5). Scanning probe microscopes work differently from light or electron microscopes. They move a probe over a surface and translate the distances into an image.

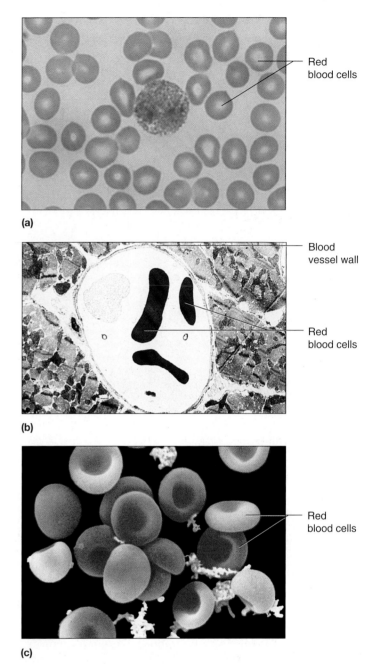

(a)

(b)

(c)

FIGURE 3.5
Human red blood cells as viewed using (a) a light microscope (1,200×), (b) a transmission electron microscope (2,500×), and (c) a scanning electron microscope (1,900×).

In addition to maintaining the integrity of the cell, the membrane controls the entrance and exit of substances, allowing some in while excluding others. A membrane that functions in this manner is *selectively permeable* (per′me-ah-bl). The cell membrane is crucial because it is a conduit between the cell and the extracellular fluids in the body's internal environment. It allows the cell to receive and respond to incoming messages, a process called **signal transduction.** (Signal transduction is described in more detail in chapter 13, p. 472.)

Membrane Structure

The cell membrane is mainly composed of lipids and proteins, with some carbohydrate. Its basic framework is a double layer (bilayer) of phospholipid molecules (see chapter 2 and fig. 2.15) that self-assemble so that their water-soluble (hydrophilic, or "water-loving") "heads," containing phosphate groups, form the surfaces of the membrane, and their water-insoluble (hydrophobic, or "water-fearing") "tails," consisting of fatty acid chains, make up the interior of the membrane (see figs. 3.3 and 3.6). The lipid molecules can move sideways within the plane of the membrane, and collectively they form a thin, but stable fluid film.

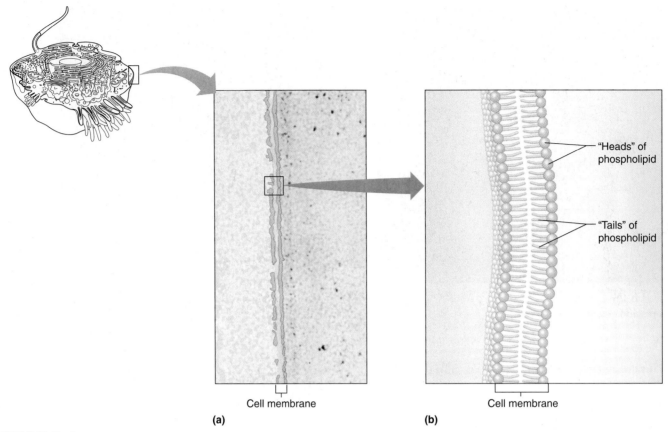

"Heads" of phospholipid

"Tails" of phospholipid

Cell membrane
(a)

Cell membrane
(b)

FIGURE 3.6

The cell membrane is a phospholipid bilayer. (*a*) A transmission electron micrograph of a cell membrane (250,000× micrograph enlarged to 600,000×); (*b*) the framework of the membrane consists of a double layer of phospholipid molecules.

RECONNECT TO CHAPTER 2, LIPIDS, PAGE 50.

Because the interior of the cell membrane consists largely of the fatty acid portions of the phospholipid molecules, it is oily. Molecules that are soluble in lipids, such as oxygen, carbon dioxide, and steroid hormones, can pass through this layer easily; however, the layer is impermeable to water-soluble molecules, such as amino acids, sugars, proteins, nucleic acids, and various ions. Many cholesterol molecules embedded in the interior of the membrane also help make it impermeable to water-soluble substances. In addition, the relatively rigid structure of the cholesterol molecules helps stabilize the cell membrane.

A cell membrane includes only a few types of lipid molecules but many kinds of proteins (fig. 3.7), which provide the specialized functions of the membrane. The membrane proteins can be classified by their shape, locations within the phospholipid bilayer, and function (table 3.1). For example, certain tightly coiled, rodlike molecules span the membrane, extending outward from the cell surface yet also dipping into the cell's interior. These proteins function as *receptors,* specialized to combine with specific kinds of incoming molecules, such as hormones, triggering responses from within the cell (see chapter 13, p. 469).

Members of another group of cell membrane proteins are more compact and globular. Some of these proteins, called integral proteins, are embedded in the interior of the phospholipid bilayer. Typically, they span the membrane and provide routes for small molecules and ions to cross the otherwise impermeable phospholipid bilayer. Some of these integral proteins form "pores" that admit water and others are highly selective and form channels that allow only particular ions to enter. In nerve cells, for example, selective channels control the movements of sodium and potassium ions, which are important in nerve impulse conduction (see chapter 10, p. 349). Clinical Application 3.1 discusses how abnormal ion channels can cause disease.

Peripheral proteins are also globular, and they associate with the surface of the cell membrane. These proteins function as enzymes (see chapter 4, p. 106), and many are part of signal transduction pathways. Other peripheral proteins function as **cellular adhesion molecules** (CAMs) that enable certain cells to touch or bind, discussed at the end of this section. Carbohydrate groups associated with peripheral proteins form glycoproteins that protrude as

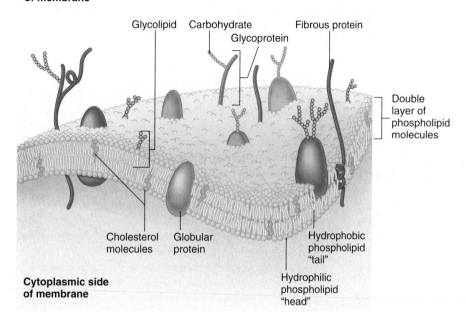

Extracellular side of membrane

Glycolipid Carbohydrate Fibrous protein

Glycoprotein

Double layer of phospholipid molecules

Cholesterol molecules

Globular protein

Hydrophobic phospholipid "tail"

Cytoplasmic side of membrane

Hydrophilic phospholipid "head"

FIGURE 3.7

The cell membrane is composed primarily of phospholipids (and some cholesterol), with proteins scattered throughout the lipid bilayer and associated with its surfaces.

TABLE 3.1	Types of Membrane Proteins
Protein Type	**Function**
Receptor proteins	Receive and transmit messages into a cell
Integral proteins	Form pores, channels, and carriers in cell membrane
Enzymes	Signal transduction
Cellular adhesion molecules	Enable cells to stick to each other
Cell surface proteins	Establish self

branches from a cell's surface, helping cells to recognize and bind to each other. This is important as cells aggregate to form tissues. Cell surface glycoproteins also mark the cells of an individual as "self," and mark cells within the individual as being a particular differentiated cell type. The immune system can distinguish between "self" cell surfaces and "nonself" cell surfaces that may indicate a potential threat, such as the presence of infectious bacteria. When a person's blood or bone marrow is typed for use in a transfusion or transplant, it is the cell surface's protein and glycoprotein topography that is determined and matched with those of potential recipients.

Intercellular Junctions

Some cells, such as blood cells, are separated from each other in fluid-filled spaces or intercellular (in″ter-sel′u-lar) spaces. Many other cell types, however, are tightly packed, with structures called **intercellular junctions** that connect their cell membranes.

In one type of intercellular junction, called a *tight junction,* the membranes of adjacent cells converge and fuse. The area of fusion surrounds the cell like a belt, and the junction closes the space between the cells. Cells that form sheetlike layers, such as those that line the inside of the digestive tract, often are joined by tight junctions. The linings of tiny blood vessels in the brain are extremely tight (Clinical Application 3.2).

Another type of intercellular junction, called a *desmosome,* rivets or "spot welds" adjacent skin cells, so they form a reinforced structural unit. The membranes of certain other cells, such as those in heart muscle and muscle of the digestive tract, are interconnected by tubular channels called *gap junctions.* These channels link the cytoplasm of adjacent cells and allow ions, nutrients (such as sugars, amino acids, and nucleotides), and other small molecules to move between them (fig. 3.8). Table 3.2 summarizes these intercellular junctions.

Cellular Adhesion Molecules

Often cells must interact dynamically and transiently, rather than form permanent attachments. Proteins called cellular adhesion molecules, or CAMs for short, guide cells on the move. Consider a white blood cell moving in the bloodstream to the site of an injury, where it is required to fight infection. Imagine that such a cell must

FAULTY ION CHANNELS CAUSE DISEASE

What do collapsing horses, irregular heartbeats in teenagers, and cystic fibrosis have in common? All result from abnormal ion channels in cell membranes.

Ion channels are tunnels through the lipid bilayer of a biological membrane that consist of protein (see fig. 10.10). These passageways permit electrical signals to pass in and out of membranes in the form of ions. An ion channel functions as a gate, opening or closing to a specific ion in response to certain conditions. Ten million ions can pass through an ion channel in one second. Events that can trigger an ion channel to open or close include a change in voltage across the membrane, binding of a ligand (a molecule that binds specifically to a membrane receptor) to the cell membrane, or receiving biochemical messages from within the cell.

Abundant ion channels include those specific for calcium (Ca^{+2}), chloride (Cl^-), sodium (Na^+), or potassium (K^+). A cell may have a few thousand ion channels specific for each ion.

Many drugs act by affecting ion channels (table 3A). The distribution of specific ion channels on particular cell types explains the symptoms of illnesses that result from abnormal channels. Following are descriptions of three illnesses caused by malfunctioning ion channels.

Hyperkalemic Periodic Paralysis and Sodium Channels

The quarter horse was originally bred to run the quarter mile in the 1600s. Four particularly fast stallions were used to establish much of the current population of nearly 3 million animals. Unfortunately, one of the original stallions had an inherited condition called hyperkalemic periodic paralysis. The horse was indeed a champion, but the disease brought on symptoms undesirable in a racehorse—attacks of weakness and paralysis that caused sudden collapse.

Hyperkalemic periodic paralysis results from abnormal sodium channels in the cell membranes of muscle cells. But the trigger for the temporary paralysis is another ion: potassium. When the blood potassium level rises, as it may following intense exercise, it slightly alters the muscle cell membrane's electrical potential. Normally, this slight change would have no effect. In affected horses, however, the change causes sodium channels to open too widely and admit too much sodium into the cell. The influx of sodium renders the muscle cell unable to respond to nervous stimulation for a short time—but long enough for the racehorse to fall.

Humans can inherit this condition too. In one affected family, several members collapsed after eating bananas! Bananas are very high in potassium, which triggered the symptoms of hyperkalemic periodic paralysis.

Long-QT Syndrome and Potassium Channels

A Norwegian family had four children, all born deaf. Three of the children died at ages four, five, and nine; the fourth so far has been lucky. All of the children inherited from their unaffected "carrier" parents a condition called "long-QT syndrome associated with deafness." They have abnormal potassium channels in the heart muscle and in the inner ear. In the heart, the malfunctioning channels cause fatal arrhythmia. In the inner ear, the abnormal channels alter the concentration of potassium ions in a fluid, impairing hearing.

The inherited form of long-QT syndrome in the Norwegian family is extremely rare, but other forms of the condition are more common, causing 50,000 sudden deaths each year, often in apparently healthy children and young adults.

Diagnosing long-QT syndrome early is essential because the first symptom may be fatal. It is usually diagnosed following a sudden death of a relative or detected on a routine examination of the heart's electrical activity (an electrocardiogram, see fig. 15.22). Drugs, pacemakers, and surgery to remove certain nerves can treat the condition and possibly prevent sudden death.

Cystic Fibrosis and Chloride Channels

A seventeenth-century English saying, "A child that is salty to taste will die shortly after birth," described the consequence of abnormal chloride channels in the inherited illness cystic fibrosis (CF). The disorder affects 1 in 2,500 Caucasians, 1 in 14,000 blacks, and 1 in 90,000 Asians and is inherited from two unaffected parents who are carriers. The major symptoms of impaired breathing, respiratory infections, and a clogged pancreas result from secretion of extremely thick mucus. Severely affected individuals undergo twice-daily exercise sessions to shake free the sticky mucus and take supplemental digestive enzymes to aid pancreatic function. Strong antibiotics are used to combat their frequent lung infections.

The defect that causes CF is abnormal chloride channels in cells lining the lung passageways, ducts in the pancreas, and elsewhere. The primary defect in the chloride channels also impairs sodium channels. The result is salt trapped inside affected cells, which draws moisture in, thickening the surrounding mucus. Experimental gene therapies attempt to correct affected cells' instructions for building chloride channel proteins. ■

TABLE 3A	Drugs That Affect Ion Channels	
Target	**Indication**
Calcium channels	Antihypertensives Antiangina (chest pain)
Sodium channels	Antiarrhythmias, diuretics Local anesthetics
Chloride channels	Anticonvulsants Muscle relaxants
Potassium channels	Antihypertensives, antidiabetics (noninsulin-dependent)

THE BLOOD-BRAIN BARRIER

Perhaps nowhere else in the body are cells attached as firmly and closely as they are in the 400-mile network of capillaries in the brain. The walls of these microscopic blood vessels are but a single cell thick. They form sheets that fold into minute tubules. A century ago, bacteriologist Paul Ehrlich showed the existence of the blood-brain barrier by injecting a dye intravenously. The brain failed to take up the dye, indicating that its blood vessels did not allow the molecules to leave and enter the brain's nervous tissue.

Studies in 1969 using the electron microscope revealed that in the brain, capillary cell membranes overlap to form a barrier of tight junctions. Unlike the cells forming capillary walls elsewhere in the body, which are pocked with vesicles and windowlike portals called clefts, the cells comprising this blood-brain barrier have few vesicles and no clefts. Certain star-shaped brain cells called astrocytes contribute to this barrier as well.

The impenetrable barrier that the capillaries in the brain form shields delicate brain tissue from toxins in the bloodstream and from biochemical fluctuations that could be overwhelming if the brain had to continually respond to them. It also allows selective drug delivery to the periphery—for example some antihistamines do not cause drowsiness because they cannot breach the blood-brain barrier. But all this protection has a limitation—the brain cannot take up many therapeutic drugs that must penetrate to be effective.

By studying the types of molecules embedded in the membranes of the cells forming the barrier, researchers are developing clever ways to sneak drugs into the brain. They can tag drugs to substances that can cross the barrier, design drugs to fit natural receptors in the barrier, or inject substances that temporarily relax the tight junctions forming the barrier. Drugs that can cross the blood-brain barrier could be used to treat Alzheimer disease, Parkinson dis-

ease, brain tumors, and AIDS-related brain infections.

A malfunctioning blood-brain barrier can threaten health. During the Persian Gulf War in 1991, response of the barrier to stress in soldiers caused illness. Many troops were given a drug to protect against the effects of nerve gas on peripheral nerves—those outside the brain and spinal cord. The drug, based on its chemistry, was not expected to cross the blood-brain barrier. However, 213 Israeli soldiers treated with the drug developed brain-based symptoms, including nervousness, insomnia, headaches, drowsiness, and inability to pay attention and to do simple calculations. Further reports from soldiers, and experiments on mice, revealed that under stressful conditions, the blood-brain barrier can temporarily loosen, admitting a drug that it would normally keep out. The blood-brain barrier, then, is not a fixed boundary, but rather a dynamic structure that can change in response to a changing environment. ∎

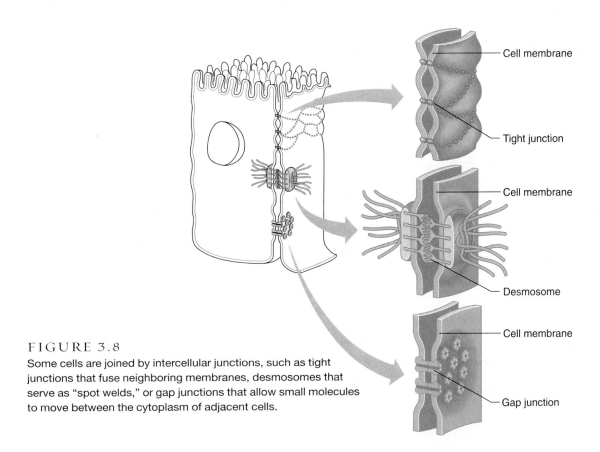

Cell membrane

Tight junction

Cell membrane

Desmosome

Cell membrane

Gap junction

FIGURE 3.8

Some cells are joined by intercellular junctions, such as tight junctions that fuse neighboring membranes, desmosomes that serve as "spot welds," or gap junctions that allow small molecules to move between the cytoplasm of adjacent cells.

TABLE 3.2	Types of Intercellular Junctions	
Type	**Function**	**Location**
Tight junctions	Close space between cells by fusing cell membranes	Cells that line inside of the small intestine
Desmosomes	Bind cells by forming "spot welds" between cell membranes	Cells of the outer skin layer
Gap junctions	Form tubular channels between cells that allow substances to be exchanged	Muscle cells of the heart and digestive tract

reach a woody splinter embedded in a person's palm (fig. 3.9).

Once near the splinter, the white blood cell must slow down in the turbulence of the bloodstream. A type of CAM called a *selectin* does this by coating the white blood cell and providing traction. The white blood cell slows to a roll and binds to carbohydrates on the inner capillary surface. Clotting blood, bacteria, and decaying tissue at the injury site release biochemicals (chemoat-tractants) that attract the white blood cell. Finally, a type of CAM called an *integrin* contacts an adhesion receptor protein protruding into the capillary space near the splinter and pushes up through the capillary cell membrane, grabbing the passing slowed white blood cell and directing it between the tilelike cells of the capillary wall. White blood cells collecting at an injury site produce inflammation and, with the dying bacteria, form pus. (The role of white blood cells in body defense is discussed further in chapter 16, pp. 619–625.)

Cellular adhesion is critical to many functions. CAMs guide cells surrounding an embryo to grow toward maternal cells and form the placenta, the supportive organ linking a pregnant woman to the fetus (see fig. 23.18). Sequences of CAMs help establish the connections between nerve cells that underlie learning and memory.

Abnormal cellular adhesion affects health. Lack of cellular adhesion, for example, eases the journey of cancer cells as they spread from one part of the body to another. Arthritis may occur when white blood cells are reined in by the wrong adhesion molecules and inflame a joint where there isn't an injury.

1 What is a selectively permeable membrane?

2 Describe the chemical structure of a cell membrane.

3 What are some functions of cell membrane proteins?

4 What are the different types of intercellular junctions?

5 What are some of the events of cellular adhesion?

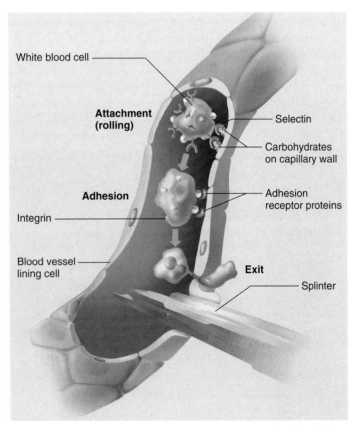

FIGURE 3.9

Cellular adhesion molecules (CAMs) direct white blood cells to injury sites, such as this splinter. Selectin proteins latch onto a rolling white blood cell and bind carbohydrates on the inner blood vessel wall at the same time, slowing the cell from moving at 2,500 micrometers per second to a more leisurely 50 micrometers per second. Chemoattractants are secreted. Then, integrin proteins anchor the white blood cell to the blood vessel wall. Finally, the white blood cell squeezes between lining cells at the injury site and exits the bloodstream.

Cytoplasm

When viewed through a light microscope, cytoplasm usually appears clear with scattered specks. However, a transmission electron microscope (see fig. 3.4), reveals networks of membranes and organelles suspended in a clear liquid, called cytosol. Cytoplasm also contains abundant protein rods and tubules that form a supportive framework called the **cytoskeleton** (si′to-skel-i-tun).

The activities of a cell occur largely in its cytoplasm, where nutrient molecules are received, processed, and used in metabolic reactions. Within the cytoplasm, the following organelles have specific functions:

1. **Endoplasmic reticulum.** The endoplasmic reticulum (en′do-plaz′mik re-tik′u-lum) (ER) is a

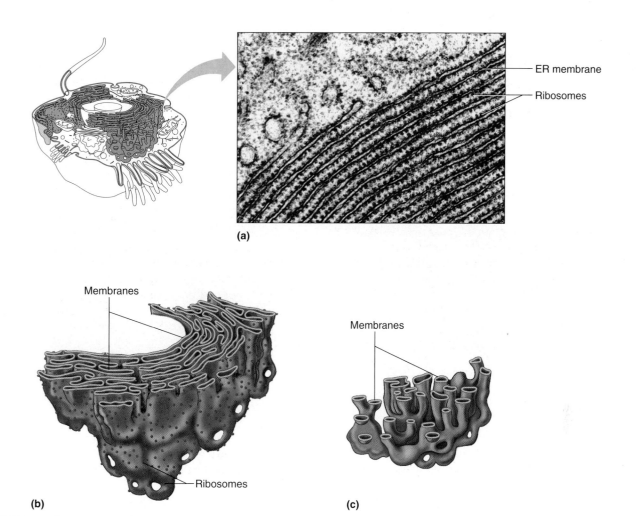

ER membrane

Ribosomes

(a)

Membranes

Ribosomes

(b)

Membranes

(c)

FIGURE 3.10

Endoplasmic reticulum. (*a*) A transmission electron micrograph of rough endoplasmic reticulum (ER) (28,000×). (*b*) Rough ER is dotted with ribosomes, whereas (*c*) smooth ER lacks ribosomes.

complex organelle composed of membrane-bound flattened sacs, elongated canals, and fluid-filled vesicles (fig. 3.10). These membranous parts are interconnected, and they communicate with the cell membrane, the nuclear envelope, and certain cytoplasmic organelles. ER is widely distributed through the cytoplasm, providing a tubular transport system for molecules throughout the cell.

The endoplasmic reticulum participates in the synthesis of protein and lipid molecules. These molecules may leave the cell as secretions or be used within the cell for such functions as producing new ER or cell membrane as the cell grows.

The outer membranous surface of the ER is studded with many tiny, spherical structures called *ribosomes* (ri′bo-sōmz) that give the ER a textured appearance when viewed with an electron microscope. Such endoplasmic reticulum is termed *rough ER*. The ribosomes of rough ER are sites of protein synthesis. The proteins then move through

the tubules of the endoplasmic reticulum to the Golgi apparatus for further processing.

ER that lacks ribosomes is called *smooth ER* (fig. 3.10). It contains enzymes important in lipid synthesis. Smooth ER contains enzymes important in synthesizing lipids, absorbing fats from the digestive tract, and breaking down drugs. In the liver, cells that break down alcohol and drugs have extensive networks of smooth ER.

2. **Ribosomes.** Besides being found on the endoplasmic reticulum, some ribosomes are scattered freely throughout the cytoplasm. All ribosomes are composed of protein and RNA and provide a structural support and enzymatic activity required to link amino acids to form proteins (see chapter 4, p. 119). Ribosomes are the sites of protein synthesis.

3. **Golgi apparatus.** The Golgi apparatus (gol′je ap″ah-ra′tus) is composed of a stack of half a dozen or so flattened, membranous sacs called *cisternae*.

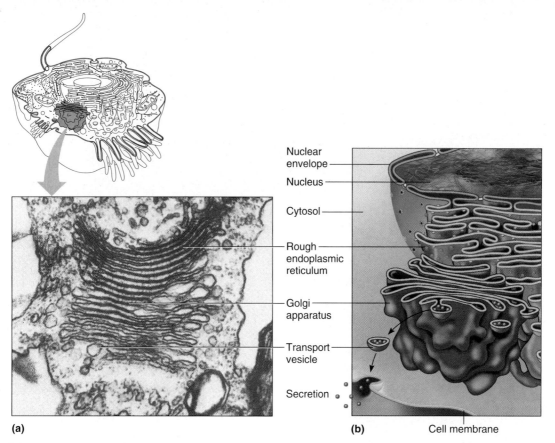

FIGURE 3.11

The Golgi apparatus. (*a*) A transmission electron micrograph of a Golgi apparatus (48,500×). (*b*) The Golgi apparatus consists of membranous sacs that continually receive vesicles from the endoplasmic reticulum and produce vesicles that enclose secretions.

This organelle refines, packages, and delivers proteins synthesized by the ribosomes associated with the ER (fig. 3.11).

Proteins arrive at the Golgi apparatus enclosed in tiny vesicles composed of membrane from the endoplasmic reticulum. These sacs fuse to the membrane at the innermost end of the Golgi apparatus, which is specialized to receive proteins. Previously, in the endoplasmic reticulum these protein molecules were combined with sugar molecules to form glycoproteins.

As the glycoproteins pass from layer to layer through the Golgi stacks, they are modified chemically. For example, sugar molecules may be added or removed from them. When the altered glycoproteins reach the outermost layer, they are packaged in bits of Golgi apparatus membrane that bud off and form transport vesicles. Such a vesicle may then move to the cell membrane, where it fuses and releases its contents to the outside of the cell as a secretion, an example of a process called exocytosis (see page 87). Other vesicles may transport glycoproteins to organelles within the cell (fig. 3.12). Movement of substances within cells by way of vesicles is called *vesicle trafficking.*

Some cells, including certain liver cells and white blood cells (lymphocytes), secrete glycoprotein molecules as rapidly as they are synthesized. However, certain other cells, such as those that manufacture protein hormones, release vesicles containing newly synthesized molecules only when the cells are stimulated. Otherwise, the loaded vesicles remain in the cytoplasm. (Chapter 13, p. 468 discusses hormone secretion.)

Secretory vesicles that originate in the ER not only release substances outside the cell, but also provide new cell membrane. This is especially important during cell growth.

4. **Mitochondria.** Mitochondria (mi″to-kon′dre-ah) are elongated, fluid-filled sacs 2–5 μm long. They often move about slowly in the cytoplasm and can divide. A mitochondrion contains a small amount of DNA that encodes information for making a few kinds of proteins and specialized RNA. However, most proteins used in mitochondrial functions are encoded in the DNA of the nucleus. These proteins are synthesized elsewhere in the cell and then enter the mitochondria.

A mitochondrion (mi″to-kon′dre-on) has two layers—an outer membrane and an inner membrane. The inner membrane is folded extensively in to form shelflike partitions called *cristae.* Small, stalked

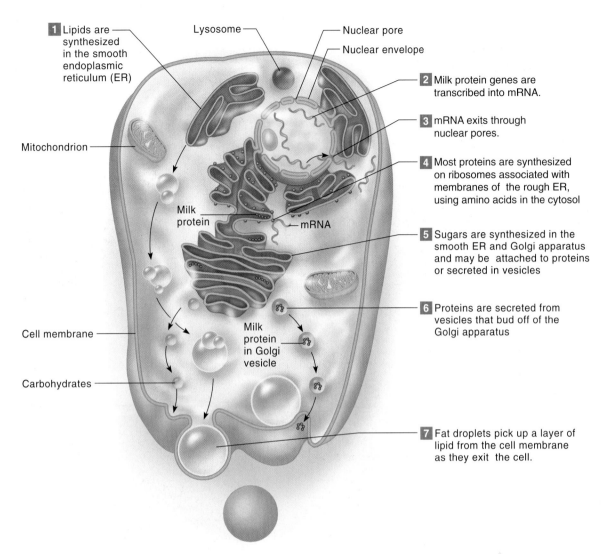

1 Lipids are synthesized in the smooth endoplasmic reticulum (ER)

Lysosome

Nuclear pore

Nuclear envelope

2 Milk protein genes are transcribed into mRNA.

3 mRNA exits through nuclear pores.

Mitochondrion

4 Most proteins are synthesized on ribosomes associated with membranes of the rough ER, using amino acids in the cytosol

Milk protein

mRNA

5 Sugars are synthesized in the smooth ER and Golgi apparatus and may be attached to proteins or secreted in vesicles

6 Proteins are secreted from vesicles that bud off of the Golgi apparatus

Cell membrane

Milk protein in Golgi vesicle

Carbohydrates

7 Fat droplets pick up a layer of lipid from the cell membrane as they exit the cell.

FIGURE 3.12

Milk secretion illustrates how organelles interact to synthesize, transport, store, and export biochemicals (*1–7*). When the baby suckles, he or she receives a chemically complex secretion—milk.

particles that contain enzymes are connected to the cristae. These enzymes and others dissolved in the fluid within the mitochondrion control many of the chemical reactions that release energy from glucose and other organic nutrients. The mitochondrion captures and transfers this newly released energy into special chemical bonds of the molecule **adenosine triphosphate** (ATP), that cells can readily use (fig. 3.13 and chapter 4, p. 108). For this reason, the mitochondrion is sometimes called the "powerhouse" of the cell.

A typical cell has about 1,700 mitochondria, but cells with very high energy requirements, such as muscle, have many thousands of mitochondria. This is why a common symptom of illnesses affecting mitochondria is muscle weakness. Symptoms of these "mitochondrial myopathies" include exercise intolerance and weak and flaccid muscles.

Mitochondria provide glimpses into the past. These organelles are passed to offspring from mothers only, because the mitochondria are excluded from the part of a sperm that enters an egg cell. Evolutionary biologists study the DNA sequences of genes in mitochondria as one way of tracing human origins, back to a long-ago group of ancestors metaphorically called "mitochondrial Eve."

Mitochondria may provide clues to a past far more remote than the beginnings of humankind. According to the widely accepted endosymbiont theory, mitochondria are the remnants of once free-living bacterialike cells that were swallowed by primitive eukaryotic cells. These bacterial passengers remain in our cells today, where they participate in energy reactions. Mitochondria physically resemble bacteria.

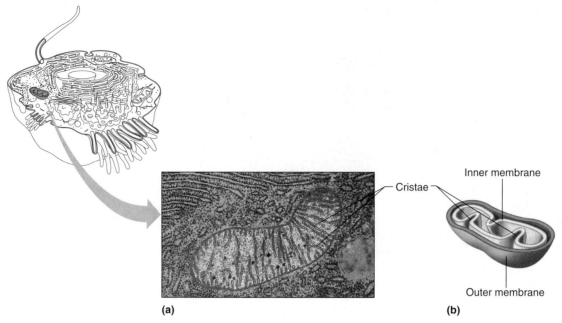

FIGURE 3.13

A mitochondrion. (*a*) A transmission electron micrograph of a mitochondrion (28,000×). (*b*) Cristae partition this saclike organelle.

5. **Lysosomes.** Lysosomes (li'so-sōmz) are the "garbage disposals" of the cell, whose function is to dismantle debris. They are sometimes difficult to identify because their shapes vary so greatly. However, they commonly appear as tiny, membranous sacs (fig. 3.14). These sacs contain powerful enzymes that break down proteins, carbohydrates, and nucleic acids, including foreign particles composed of these substances. Certain white blood cells, for example, engulf bacteria that are then digested by the lysosomal enzymes. This is one way that white blood cells help stop bacterial infections.

   Lysosomes also destroy worn cellular parts. In fact, lysosomes in certain scavenger cells may engulf and digest entire body cells that have been injured. How the lysosomal membrane is able to withstand being digested itself is not well understood, but this organelle sequesters enzymes that can function only under very acidic conditions, preventing them from destroying the cellular contents around them. Human lysosomes contain forty or so different types of enzymes. An abnormality in just one type of lysosomal enzyme can be devastating to health (Clinical Application 3.3).

6. **Peroxisomes** (pĕ-roks'ĭ-somz). Peroxisomes are membranous sacs that resemble lysosomes in size and shape. Although present in all human cells, peroxisomes are most abundant in the liver and kidneys. Peroxisomes contain enzymes, called peroxidases, that catalyze metabolic reactions that

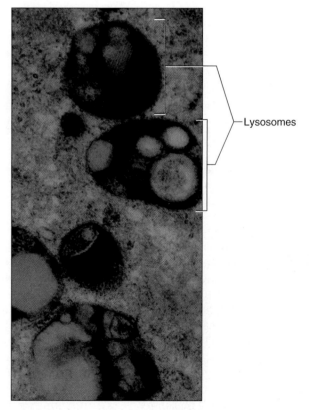

FIGURE 3.14

In this falsely colored transmission electron micrograph, lysosomes appear as membranous sacs (14,100×).

DISEASE AT THE ORGANELLE LEVEL

German physiologist Rudolph Virchow first hypothesized cellular pathology—disease at the cellular level—in the 1850s. Today, new treatments for many disorders are a direct result of understanding a disease process at the cellular level. Here, we examine how three abnormalities—in mitochondria, in peroxisomes, and in lysosomes—cause whole-body symptoms.

MELAS and Mitochondria

Sharon had always been small for her age, easily fatigued, slightly developmentally delayed, and had difficulty with schoolwork. She also had seizures. At age eleven, she suffered a stroke. An astute physician who observed Sharon's mother, Lillian, suspected that the girl's symptoms were all related, and the result of abnormal mitochondria, the organelles that house the biochemical reactions that extract energy from nutrients.

The doctor noticed that Lillian was uncoordinated and had numb hands. When she asked if Lillian ever had migraine headaches, she said that she suffered from them nearly daily, as did her two sisters and one brother. Lillian and her siblings also had diabetes mellitus and muscle weakness. Based on this information, the doctor ordered several blood tests for mother and daughter, which revealed that both had elevated levels of biochemicals (pyruvic acid and lactic acid) that indicated they were unable to extract the maximal energy from nutrients. Muscle biopsies then showed the source of the problem—abnormal mitochondria. Accumulation of these mitochondria in smooth muscle cells in blood vessel walls in the brain caused Sharon's stroke and was probably also causing her seizures.

All of the affected family members were diagnosed with a disorder called MELAS, which stands for the major symptoms—mitochondrial encephalomyopathy, lactic acidosis, and strokelike episodes. Their mitochondria cannot synthesize some of the proteins required to carry out the energy reactions. The responsible gene is part of the DNA in mitochondria, and Lillian's mother transmitted it to all of her children. But because mitochondria are usually inherited only from the mother, Sharon's uncle will not pass MELAS to his children.

Adrenoleukodystrophy (ALD) and Peroxisomes

For young Lorenzo Odone, the first sign of adrenoleukodystrophy was disruptive behavior in school. When he became lethargic, weak, and dizzy, his teachers and parents realized that his problem was not just temper tantrums. His skin darkened, blood sugar levels plummeted, heart rhythm altered, and the levels of electrolytes in his body fluids changed. He lost control over his limbs as his nervous system continued to deteriorate. Lorenzo's parents took him to many doctors. Finally, one of them tested the child's blood for an enzyme normally manufactured in peroxisomes.

Lorenzo's peroxisomes lacked the second most abundant protein in the outer membrane of this organelle. Normally, the missing protein transports an enzyme into the peroxisome. The enzyme controls breakdown of a type of very long chain fatty acid. Without the enzyme, the fatty acid builds up in cells in the brain and spinal cord, eventually stripping these cells of their fatty sheaths, made of a substance called myelin. Without the myelin sheaths, the nerve cells cannot transmit messages fast enough. Death comes in a few years.

For Lorenzo and many other sufferers of ALD, eating a type of triglyceride from canola oil slows the buildup of the very long chain fatty acids for a few years, stalling symptoms. But the treatment eventually impairs blood clotting and other vital functions and fails to halt the progression of the illness.

The disappointment over the failure of "Lorenzo's oil" may be lessened by a drug that activates a different gene, whose protein product can replace the missing or abnormal one in ALD. In cells from children with ALD, the replacement protein stopped the buildup of very long chain fatty acids and also increased the number of peroxisomes.

Tay-Sachs Disease and Lysosomes

Michael was a pleasant, happy infant who seemed to be developing normally until about six months of age. Able to roll over and sit for a few seconds, he suddenly seemed to lose those abilities. Soon, he no longer turned and smiled at his mother's voice, and he did not seem as interested in his mobile. Concerned about Michael's reversals in development, his anxious parents took him to the doctor. It took exams by several specialists to diagnose Michael's Tay-Sachs disease, because, thanks to screening programs in the population groups known to have this inherited illness, fewer than ten new cases appear each year. Michael's parents were not among those ethnic groups and previously had no idea that they both were carriers of the gene that causes this very rare illness.

A neurologist clinched her suspicion of Tay-Sachs by looking into Michael's eyes, where she saw the telltale "cherry red spot" indicating the illness. A look at his cells provided further clues—the lysosomes, tiny enzyme-filled sacs, were swollen to huge proportions. Michael's lysosomes lacked one of the forty types of lysosomal enzymes, resulting in a "lysosomal storage disease" that built up fatty material on his nerve cells. His nervous system would continue to fail, and he would be paralyzed and unable to see or hear by the time he died, before the age of four years.

The cellular and molecular signs of Tay-Sachs disease—the swollen lysosomes and missing enzyme—had been present long before Michael began to lag developmentally. The next time his parents expected a child, they had her tested before birth for the enzyme deficiency. They learned that she would be a carrier like themselves, but not ill. ■

release hydrogen peroxide (H$_2$O$_2$), which is toxic to cells, as a by-product. Peroxisomes also contain an enzyme called catalase, which decomposes hydrogen peroxide.

The outer membrane of a peroxisome contains some forty types of enzymes, which catalyze a variety of biochemical reactions, including

- synthesis of bile acids, which are used in fat digestion
- breakdown of lipids called very long chain fatty acids
- degradation of rare biochemicals
- detoxification of alcohol

Abnormal peroxisomal enzymes can drastically affect health.

7. **Centrosome.** A centrosome (sen′tro-sōm) (central body) is a structure located in the cytoplasm near the nucleus. It is nonmembranous and consists of two hollow cylinders called *centrioles* built of tubelike proteins called microtubules. The centrioles usually lie at right angles to each other. During cell division, the centrioles move away from one another to either side of the nucleus, where they form spindle fibers that pull on and distribute *chromosomes,* (kro′mo-sōmz) which carry DNA information to the newly forming cells (fig. 3.15).

Centrioles also form parts of hairlike cellular projections called cilia and flagella.

8. **Cilia** and **flagella.** Cilia and flagella are motile extensions of certain cells. They are structurally similar and differ mainly in their length and the number present. Both consist of the same number of microtubules organized in a distinct cylindrical pattern.

Cilia are abundant on the free surfaces of some epithelial cells. Each cilium is a tiny, hairlike structure about 10 μm long, which attaches just beneath the cell membrane to a modified centriole called a *basal body.* Cilia occur in precise patterns. They have a "to-and-fro" type of movement that is coordinated so that rows of cilia beat one after the other, producing a wave that sweeps across the ciliated surface. For example, this action propels mucus over the surface of tissues that form the lining of the respiratory tract (fig. 3.16). Chemicals in cigarette smoke destroy cilia, which impairs the respiratory tract's ability to expel bacteria. Infection may result.

A cell usually has only one flagellum, which is much longer than a cilium. A flagellum begins its characteristic undulating, wavelike motion at its

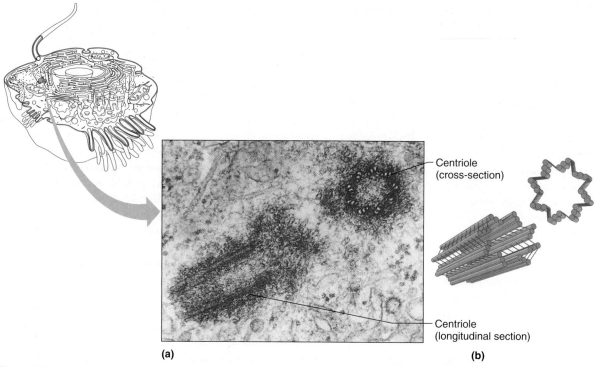

(a) **(b)**

FIGURE 3.15

Centrioles. (*a*) A transmission electron micrograph of the two centrioles in a centrosome (120,000×). (*b*) The centrioles lie at right angles to one another.

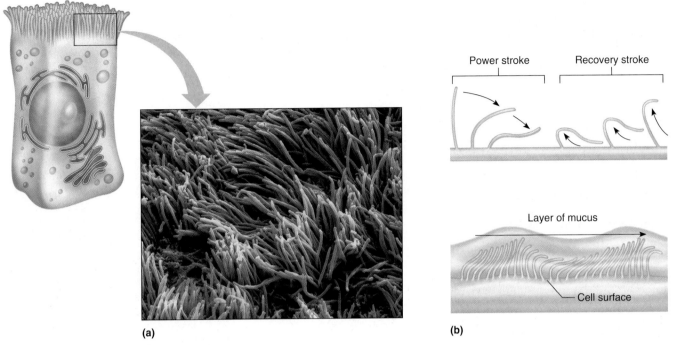

FIGURE 3.16

Cilia are sweeping hairlike extensions. (*a*) Cilia, such as these (arrow), are common on the surfaces of certain cells that form the inner lining of the respiratory tract (5,400×). (*b*) Cilia have a power stroke and a recovery stroke that create a "to-and-fro" movement that sweeps fluids across the tissue surface.

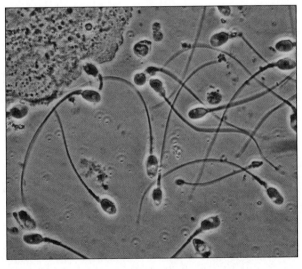

FIGURE 3.17

Flagella form the tails of these human sperm cells (840×).

base. The tail of a sperm cell, for example, is a flagellum that causes the sperm's swimming movements (fig. 3.17 and chapter 22, p. 838). It is the only known flagellum in humans.

9. **Vesicles.** Vesicles (ves'i'k'lz) (vacuoles) are membranous sacs that vary in size and contents. They may form when a portion of the cell membrane folds inward and pinches off. As a result, a tiny, bubblelike vesicle, containing some liquid or solid material that was formerly outside the cell, enters the cytoplasm. The Golgi apparatus and ER also form vesicles. Fleets of vesicles transport many substances into and out of cells in vesicle trafficking.

10. **Microfilaments** and **microtubules.** Two types of threadlike structures in the cytoplasm are microfilaments and microtubules. Microfilaments are tiny rods of the protein actin that typically occur in meshworks or bundles. They cause various kinds of cellular movements. In muscle cells, for example, microfilaments constitute *myofibrils,* which shorten or contract these cells. In other cells, microfilaments associated with the inner surface of the cell membrane aid cell motility (fig. 3.18).

Microtubules are long, slender tubes with diameters two or three times greater than those of microfilaments. They are composed of the globular protein tubulin. Microtubules are usually somewhat rigid and form the cytoskeleton, which helps maintain the shape of the cell (fig. 3.19). In cilia and flagella, microtubules interact to provide movement (see figs. 3.16 and 3.17).

Microtubules also move organelles and structures within the cell. For instance, microtubules are assembled from tubulin subunits in the cytoplasm during cell division and help distribute

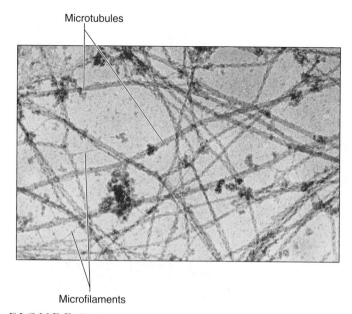

Microtubules

Microfilaments

FIGURE 3.18

A transmission electron micrograph of microfilaments and microtubules within the cytoplasm (35,000×).

chromosomes to the newly forming cells, a process described in more detail later in this chapter. Microtubules also provide conduits for organelles, like the tracks of a roller coaster.

11. **Other structures.** In addition to organelles, cytoplasm contains chemicals called *inclusions*. These usually are in a cell temporarily. Inclusions include stored nutrients such as glycogen and lipids, and pigments such as melanin in the skin.

1 What are the functions of the endoplasmic reticulum?

2 Describe how the Golgi apparatus functions.

3 Why are mitochondria called the "powerhouses" of cells?

4 How do lysosomes function?

5 Describe the functions of microfilaments and microtubules.

6 Distinguish between organelles and inclusions.

Cell Nucleus

A nucleus is a relatively large, usually spherical structure that contains the genetic material (DNA) that directs the activities of the cell. The DNA occurs as extremely long molecules complexed with proteins to form **chromatin fibers,** which are visible with a microscope as **chromosomes** when a cell that is in the process of dividing is stained.

The nucleus is enclosed in a double-layered **nuclear envelope,** which consists of an inner and an outer lipid bilayer membrane. These two membranes have a narrow space between them, but are joined at places that sur-

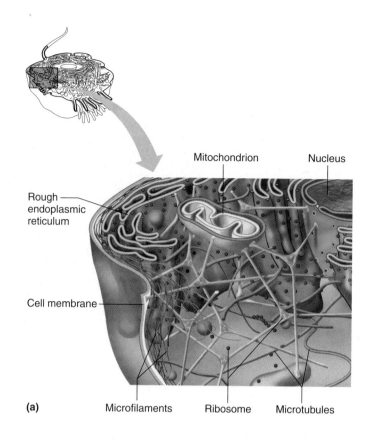

Mitochondrion   Nucleus

Rough endoplasmic reticulum

Cell membrane

(a)   Microfilaments   Ribosome   Microtubules

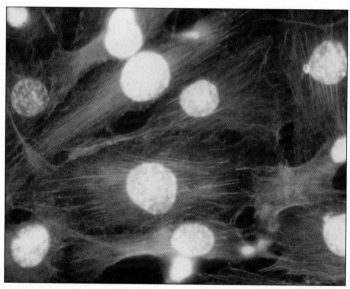

(b)

FIGURE 3.19

Cytoskeleton. (*a*) Microtubules help maintain the shape of a cell by forming an internal "scaffolding," or cytoskeleton, beneath the cell membrane and within the cytoplasm. (*b*) A falsely colored electron micrograph of cells showing the cytoskeleton (750×).

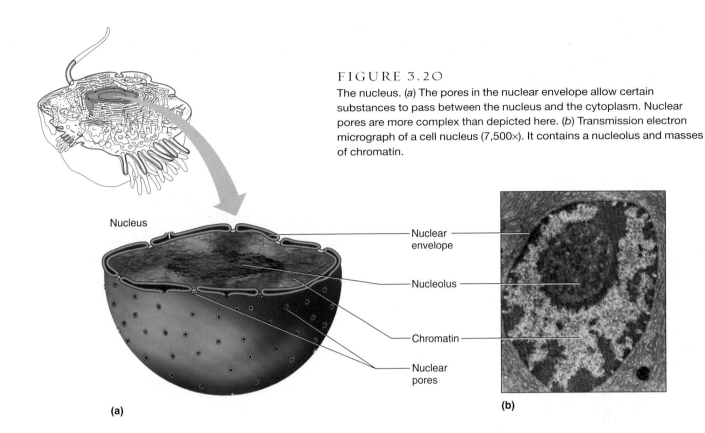

FIGURE 3.20

The nucleus. (*a*) The pores in the nuclear envelope allow certain substances to pass between the nucleus and the cytoplasm. Nuclear pores are more complex than depicted here. (*b*) Transmission electron micrograph of a cell nucleus (7,500×). It contains a nucleolus and masses of chromatin.

Nucleus

Nuclear envelope

Nucleolus

Chromatin

Nuclear pores

(a)

(b)

round openings called **nuclear pores.** These pores are not mere perforations, but channels consisting of more than 100 different types of proteins. Nuclear pores allow certain dissolved substances to move between the nucleus and the cytoplasm (fig. 3.20), most notably molecules of messenger RNA that carry genetic information.

The nucleus contains a fluid (nucleoplasm) in which other structures float. These structures include the following:

1. **Nucleolus.** A nucleolus (nu-kle′o-lus) ("little nucleus") is a small, dense body largely composed of RNA and protein. It has no surrounding membrane and is formed in specialized regions of certain chromosomes. The nucleolus is the site of ribosome production. Once ribosomes form, they migrate through the nuclear pores to the cytoplasm. A cell may have more than one nucleolus. The nuclei of cells that synthesize large amounts of protein, such as those of glands, may contain especially large nucleoli.

2. **Chromatin.** Chromatin consists of loosely coiled fibers in the nuclear fluid. When cell division begins, these fibers more tightly coil to form the rodlike chromosomes. Chromatin fibers are composed of continuous DNA molecules wrapped around clusters of proteins called histones, giving the appearance of beads on a string (see fig. 4.19).

The DNA molecules contain the information for synthesis of proteins. "Chromatin" means colored substance, and "chromosome" means colored body.

Table 3.3 summarizes the structures and functions of organelles.

1 How are the nuclear contents separated from the cytoplasm?

2 What is the function of the nucleolus?

3 What is chromatin?

Cells die in different ways. *Apoptosis* (ap″o-to′sus) is one form of cell death in which the cell manufactures an enzyme that cuts up DNA not protected by histones. Apoptosis is an active process because a new substance is made. Cell contents are packaged up and destroyed by large scavenger cells. Apoptosis is important in shaping the embryo, in maintaining organ form during growth, and in carving the developing immune system and brain from more cells than are actually required. In contrast, *necrosis* is a type of cell death that is a passive response to severe injury. Typically, proteins lose their characteristic shapes, and the cell membrane deteriorates as the cell swells and bursts. Unlike apoptosis, necrosis is not a neat disposal process, but causes great inflammation.

TABLE 3.3 Structures and Functions of Organelles

Organelle	Structure	Function
Cell membrane	Membrane mainly composed of protein and lipid molecules	Maintains integrity of the cell, controls the passage of materials into and out of the cell, and provides for signal transduction
Endoplasmic reticulum	Complex of connected, membrane-bound sacs, canals, and vesicles	Transports materials within the cell, provides attachment for ribosomes, and synthesizes lipids
Ribosomes	Particles composed of protein and RNA molecules	Synthesize proteins
Golgi apparatus	Group of flattened, membranous sacs	Packages and modifies protein molecules for transport and secretion
Mitochondria	Membranous sacs with inner partitions	Release energy from food molecules and transform energy into usable form
Lysosomes	Membranous sacs	Contain enzymes capable of digesting worn cellular parts or substances that enter cells
Peroxisomes	Membranous vesicles	Contain enzymes called peroxidases, important in the breakdown of many organic molecules
Centrosome	Nonmembranous structure composed of two rodlike centrioles	Helps distribute chromosomes to new cells during cell division and initiates formation of cilia
Cilia	Motile projections attached to basal bodies beneath the cell membrane	Propel fluids over cellular surface
Flagella	Motile projections attached to basal bodies beneath the cell membrane	Enable sperm cells to move
Vesicles	Membranous sacs	Contain substances that recently entered the cell and store and transport newly synthesized molecules
Microfilaments and microtubules	Thin rods and tubules	Support cytoplasm and help move substances and organelles within the cytoplasm
Nuclear envelope	Porous double membrane that separates the nuclear contents from the cytoplasm	Maintains the integrity of the nucleus and controls the passage of materials between the nucleus and cytoplasm
Nucleolus	Dense, nonmembranous body composed of protein and RNA molecules	Site of ribosome formation
Chromatin	Fibers composed of protein and DNA molecules	Contains cellular information for synthesizing proteins

Movements Into and Out of the Cell

The cell membrane is a barrier that controls which substances enter and leave the cell. Oxygen and nutrient molecules enter through this membrane, whereas carbon dioxide and other wastes leave through it. These movements involve *physical* (or passive) processes, such as diffusion, facilitated diffusion, osmosis, and filtration, and *physiological* (or active) mechanisms, such as active transport, endocytosis, and exocytosis. The mechanisms by which substances cross the cell membrane are important for understanding many aspects of physiology.

Diffusion

Diffusion (dĭ-fu′zhun) (also called simple diffusion) is the tendency of atoms, molecules, and ions in a liquid or air solution to move from areas of higher concentration to areas of lower concentration, thus becoming more evenly distributed, or more *diffuse.* Diffusion occurs because atoms, molecules, and ions are in constant motion. Each particle travels in a separate path along a straight line until it collides with some other particle and bounces off. Then it moves in its new direction until it collides again and changes direction once more. Because collisions are less likely if there are fewer particles, there is a net movement of particles from an area of higher concentration to an area of lower concentration. This difference in concentrations is called a *concentration gradient,* and atoms, molecules, and ions are said to diffuse down a concentration gradient. With time, the concentration of a given substance becomes uniform throughout a solution. This is the condition of *diffusional equilibrium* (dĭ-fu′zhun-ul e″kwi-lib′re-um). At diffusional equilibrium, although random movements continue, there is no further net movement, and the concentration of a substance is uniform throughout the solution.

Sugar (a solute) put into a glass of water (a solvent), can be used to illustrate diffusion (fig. 3.21). The sugar at first remains in high concentration at the bottom of the glass. As the sugar molecules move about, they may collide with each other or miss each other completely. Since they are less likely to collide with each other in areas where there are fewer sugar molecules, sugar molecules gradually diffuse from areas of high concentration to areas of lower concentration (*down* the concentration gradient), and eventually the sugar molecules become uniformly distributed in the water.

Diffusion of a substance across a membrane can occur only if (1) the cell membrane is permeable to that substance and (2) a concentration gradient exists such that the substance is at a higher concentration on one side of the membrane or the other (fig. 3.22). This principle applies to diffusion of substances across the cell membrane. Consider oxygen and carbon dioxide, two substances to which cell membranes are permeable. In the body, diffusion is the process whereby oxygen enters cells and carbon dioxide leaves cells, but equilibrium is never reached. Intracellular oxygen is always low because oxygen is constantly used up in metabolic reactions. Extracellular oxygen is maintained at a high level by homeostatic mechanisms in the respiratory and cardiovascular systems. Thus, a concentration gradient always allows oxygen to diffuse into the body's cells.

The level of carbon dioxide, produced as a waste product of metabolism, is always high inside cells. Homeostasis maintains a lower extracellular carbon dioxide level, so a concentration gradient always favors carbon dioxide diffusing out of cells (fig. 3.23).

Diffusional equilibrium does not normally occur in organisms. Rather, the term *physiological steady state,* where concentrations of diffusing substances are unequal but stable, is more appropriate.

Some of the previous examples considered imaginary membranes with specific permeabilities. For the cell membrane, permeability is more complex because of its selective nature. Lipid-soluble substances, such as oxygen, carbon dioxide, steroids, and general anesthetics, freely cross the cell membrane by simple diffusion. Small solutes that are not lipid-soluble, such as ions of sodium, potassium and chloride, may diffuse through protein channels in the membrane, described earlier. (Water molecules may also diffuse through similar channels, called pores.) Because this type of movement uses membrane proteins as "helpers," it is considered to be another type of diffusion, called **facilitated diffusion** (fah-sil"i-tat'ed dĭ-fu'zhun). Facilitated diffusion is very important not only for ions, but for larger water-soluble molecules, such as glucose and amino acids.

A number of factors influence the diffusion rate, but those most important in the body are distance, the concentration gradient, and temperature. In general, diffusion is more rapid over shorter distances, larger concentration gradients, and at higher temperatures. Homeostasis maintains all three of these factors at optimum levels.

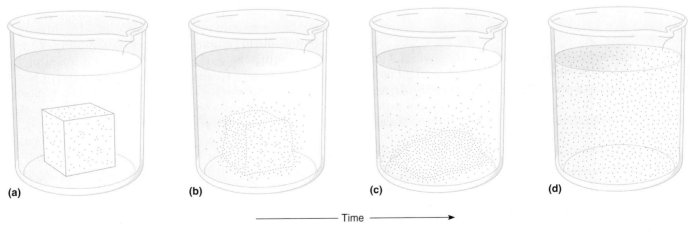

(a) (b) (c) (d)

Time →

FIGURE 3.21

An example of diffusion (a, b, and c). A sugar cube placed in water slowly disappears as the sugar molecules dissolve and then diffuse from regions where they are more concentrated toward regions where they are less concentrated. (d) Eventually, the sugar molecules distribute evenly throughout the water.

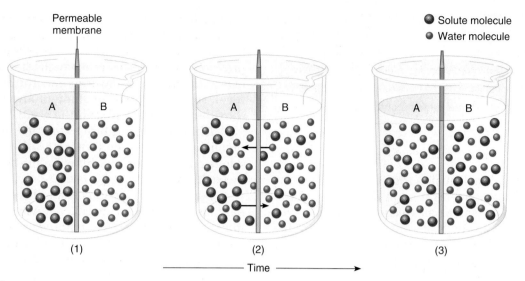

Permeable membrane

○ Solute molecule
○ Water molecule

(1)   (2)   (3)

← Time →

FIGURE 3.22

Diffusion. (*1*) A membrane permeable to water and solute molecules separates a container into two compartments. Compartment *A* contains both types of molecules, while compartment *B* contains only water molecules. (*2*) As a result of molecular motions, solute molecules tend to diffuse from compartment *A* into compartment *B*. Water molecules tend to diffuse from compartment *B* into compartment *A*. (*3*) Eventually, equilibrium is reached.

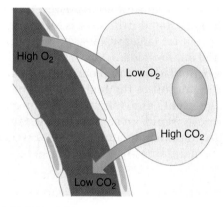

High O_2
Low O_2
High CO_2
Low CO_2

FIGURE 3.23

Oxygen enters cells and carbon dioxide leaves cells by diffusion.

Facilitated Diffusion

Most sugars and amino acids are insoluble in lipids, and they are too large to pass through cell membrane pores. Facilitated diffusion includes not only protein channels, but also certain proteins that function as "carriers" to bring such molecules across the cell membrane. In the facilitated diffusion of glucose, for example, glucose combines with a protein carrier molecule at the surface of the membrane. This union of glucose and carrier molecule changes the shape of the carrier that moves glucose to the inner face of the membrane. The glucose portion is released, and the carrier molecule can return to its original shape to pick up another glucose molecule. The hormone *insulin,* discussed in chapter 13 (p. 496), promotes facilitated diffusion of glucose through the membranes of certain cells.

Facilitated diffusion is similar to simple diffusion in that it can move molecules only from regions of higher concentration toward regions of lower concentration. However, unlike simple diffusion, the number of carrier molecules in the cell membrane limits the rate of facilitated diffusion (fig. 3.24).

Osmosis

Osmosis (oz-mo′sis) is the diffusion of water molecules from a region of higher water concentration to a region of lower water concentration across a selectively permeable membrane, such as a cell membrane. In the following example, assume that the selectively permeable membrane is permeable to water molecules (the solvent) but impermeable to protein molecules (the solute).

In solutions, a higher concentration of solute (protein in this case) means a lower concentration of water; a lower concentration of solute means a higher concentration of water. This is because the solute molecules take up space that water molecules would otherwise occupy.

Just like molecules of other substances, molecules of water will diffuse from areas of higher concentration to areas of lower concentration. In figure 3.25, the greater concentration of protein in compartment *A* means that the water concentration there is less than the concentration of pure water in compartment *B*. Therefore, water diffuses from compartment *B* across the selectively permeable membrane and into compartment *A*. In other words, water moves from compartment *B* into compartment *A* by osmosis. Protein, on the other hand, cannot diffuse out of com-

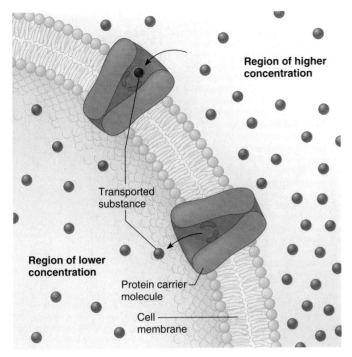

FIGURE 3.24

Facilitated diffusion transports some substances into or out of cells using carrier molecules, from a region of higher concentration to one of lower concentration.

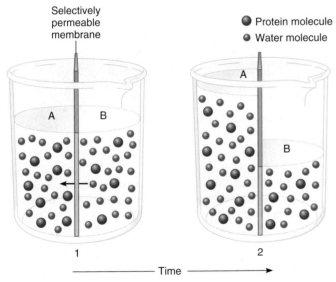

FIGURE 3.25

Osmosis. (1) A selectively permeable membrane separates the container into two compartments. At first, compartment A contains a higher concentration of protein (and a lower concentration of water) than compartment B. As a result of molecular motion, water diffuses by osmosis from compartment B into compartment A. (2) Because the membrane is impermeable to proteins, equilibrium can only be reached by diffusion of water. As water accumulates in compartment A, the water level on that side of the membrane rises.

partment A because the selectively permeable membrane is impermeable to it. Note in figure 3.25 that as osmosis occurs, the level of water on side A rises. This ability of osmosis to generate enough pressure to lift a volume of water is called *osmotic pressure.* Thus equilibrium must be achieved by the osmotic movement of water alone.

The greater the concentration of nonpermeable solute particles (protein in this case) in a solution, the *lower* the water concentration of that solution and the *greater* the osmotic pressure. Water always tends to diffuse toward solutions of greater osmotic pressure.

Since cell membranes are generally permeable to water, water equilibrates by osmosis throughout the body, and the concentration of water and solutes everywhere in the intracellular and extracellular fluids is essentially the same. Therefore, the osmotic pressure of the intracellular and extracellular fluids is the same. Any solution that has the same osmotic pressure as body fluids is called **isotonic.**

Solutions that have a higher osmotic pressure than body fluids are called **hypertonic.** If cells are put into a hypertonic solution, there will be a net movement of water by osmosis out of the cells into the surrounding solution, and the cells shrink. Conversely, cells put into a **hypotonic** solution, which has a lower osmotic pressure than body fluids, tend to gain water by osmosis and swell. Although cell membranes are somewhat elastic, the cells may swell so much that they burst. Figure 3.26 illustrates the effects of the three types of solutions on red blood cells.

It is important to control the concentration of solute in solutions that are infused into body tissues or blood. Otherwise, osmosis may cause cells to swell or shrink, impairing their function. For instance, if red blood cells are placed in distilled water (which is hypotonic to them), water will diffuse into the cells, and they will burst (hemolyze). On the other hand, if red blood cells are exposed to 0.9% NaCl solution (normal saline), the cells will remain unchanged because this solution is isotonic to human cells. Similarly, a 5% solution of glucose is isotonic to human cells. (The lower percentage is needed with NaCl to produce an isotonic solution, in part because NaCl ionizes in solution more completely and produces more solute particles than does glucose.)

Filtration

Molecules move through membranes by diffusion or osmosis because of their random movements. In other instances, molecules are forced through membranes by the process of **filtration** (fil-tra′shun).

Filtration is commonly used to separate solids from water. One method is to pour a mixture of solids and

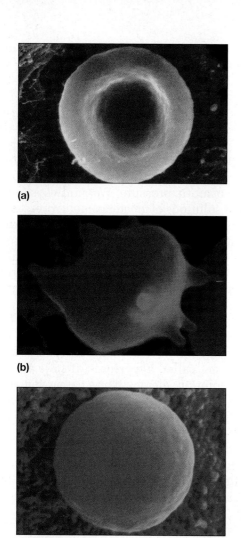

(a)

(b)

(c)

FIGURE 3.26

When red blood cells are placed (a) in an isotonic solution, equal volumes of water enter and leave the cells, and size and shape remain unchanged. (b) In a hypertonic solution, more water leaves than enters, and cells shrink. (c) In a hypotonic solution, more water enters than leaves, and cells swell and may burst (5,000×).

FIGURE 3.27

In filtration of water and solids, gravity forces water through filter paper, while tiny openings in the paper retain the solids. This process is similar to the drip method of preparing coffee.

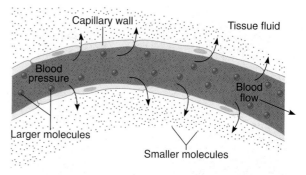

FIGURE 3.28

In filtration in the body, blood pressure forces smaller molecules through tiny openings in the capillary wall. The larger molecules remain inside.

water onto filter paper in a funnel (fig. 3.27). The paper serves as a porous membrane through which the small water molecules can pass, leaving the larger solid particles behind. Hydrostatic pressure, which is created by the weight of water due to gravity, forces the water molecules through to the other side. An example of this is making coffee by the drip method.

In the body, tissue fluid forms when water and dissolved substances are forced out through the thin, porous walls of blood capillaries, but larger particles such as blood protein molecules are left inside (fig. 3.28). The force for this movement comes from blood pressure, generated largely by heart action, which is greater within the vessel than outside it. However, the impermeable proteins tend to hold water in blood vessels by osmosis, thus preventing the formation of excess tissue fluid, a condition called

edema. (Although heart action is an active body process, filtration is still considered passive because it can occur due to the pressure caused by gravity alone.) Filtration is discussed further in chapters 15 (p. 568) and 20 (p. 783).

1 What kinds of substances most readily diffuse through a cell membrane?

2 Explain the differences among diffusion, facilitated diffusion, and osmosis.

3 Distinguish among isotonic, hypertonic, and hypotonic solutions.

4 Explain how filtration occurs in the body.

Active Transport

When molecules or ions pass through cell membranes by diffusion, facilitated diffusion, or osmosis, their net movement is from regions of higher concentration to regions of lower concentration. Sometimes, however, the net movement of particles passing through membranes is in the opposite direction, from a region of lower concentration to one of higher concentration.

Sodium ions, for example, can diffuse slowly through cell membranes. Yet, the concentration of these ions typically remains many times greater outside cells (in the extracellular fluid) than inside cells (in the intracellular fluid). This is because sodium ions are continually moved through the cell membrane from regions of lower concentration (inside) to regions of higher concentration (outside). Movement against a concentration gradient is called **active transport** (ak'tiv trans'port) and requires energy derived from cellular metabolism. Up to 40% of a cell's energy supply may be used for active transport of particles through its membranes.

Active transport is similar to facilitated diffusion in that it uses carrier molecules within cell membranes. As figure 3.29 shows, these carrier molecules are proteins that have binding sites that combine with the specific particles being transported. Such a union triggers release of cellular energy, and this energy alters the shape of the carrier protein. As a result, the "passenger" molecules move through the membrane. Once on the other side, the transported particles are released, and the carrier molecules can accept other passenger molecules at their binding sites. Because they transport substances from regions of low concentration to regions of higher concentration, these carrier proteins are sometimes called "pumps." A sodium/potassium pump, for example, transports sodium ions out of cells and potassium ions into cells.

Particles that are moved across cell membranes by active transport include sugars, amino acids, and sodium, potassium, calcium, and hydrogen ions. Some of these substances are actively transported into cells, and others are transported out. Movements of this type are important to cell survival, particularly maintenance of homeostasis. Some of these movements are described in subsequent chapters as they apply to specific organ systems.

Endocytosis

Two processes use cellular energy to move substances into or out of a cell without actually crossing the cell membrane. In **endocytosis** (en"do-si-to'sis), molecules or other particles that are too large to enter a cell by diffusion or active transport are conveyed within a vesicle that forms from a section of the cell membrane. In **exocytosis** (ex-o-si-to'sis), the reverse process secretes a substance stored in a vesicle from the cell.

The three forms of endocytosis are pinocytosis, phagocytosis, and receptor-mediated endocytosis. In **pinocytosis** (pi"-no-si-to'sis), cells take in tiny droplets of

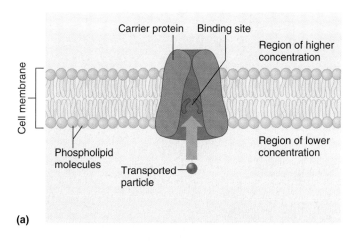

(a)

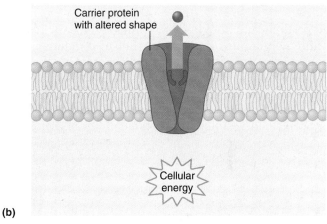

(b)

FIGURE 3.29

Active transport. (a) During active transport, a molecule or an ion combines with a carrier protein, whose shape changes as a result. (b) This process, which requires cellular energy, transports the particle across the cell membrane.

liquid from their surroundings (fig. 3.30). When this happens, a small portion of cell membrane indents (invaginates). The open end of the tubelike part thus formed seals off and produces a small vesicle about 0.1 µm in diameter. This tiny sac detaches from the surface and moves into the cytoplasm. For a time, the vesicular membrane, which was part of the cell membrane, separates its contents from the rest of the cell; however, the membrane eventually breaks down, and the liquid inside becomes part of the cytoplasm. In this way, a cell is able to take in water and the particles dissolved in it, such as proteins, that otherwise might be too large to enter.

Phagocytosis (fag"o-si-to'sis) is similar to pinocytosis, but the cell takes in solids rather than liquid. Certain kinds of cells, including some white blood cells, are called **phagocytes** because they can take in solid particles such as bacteria and cellular debris. When a phagocyte first encounters such a particle, the particle attaches to the cell membrane. This stimulates a portion of the membrane to project outward, surround

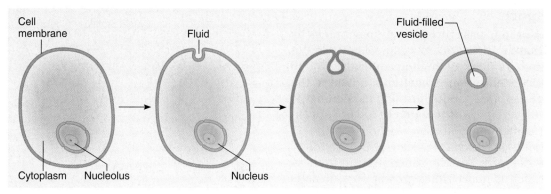

FIGURE 3.30

A cell may take in a tiny droplet of fluid from its surroundings by pinocytosis.

the particle, and slowly draw it inside the cell. The part of the membrane surrounding the solid detaches from the cell's surface, forming a vesicle containing the particle (fig. 3.31). Such a vesicle may be several micrometers in diameter.

Usually, a lysosome soon combines with such a newly formed vesicle, and lysosomal digestive enzymes decompose the contents (fig. 3.32). The products of this decomposition may then diffuse out of the lysosome and into the cytoplasm, where they may be used as raw materials in metabolic processes. Exocytosis may expel any remaining residue. In this way, phagocytic cells dispose of foreign objects, such as dust particles; remove damaged cells or cell parts that are no longer functional;

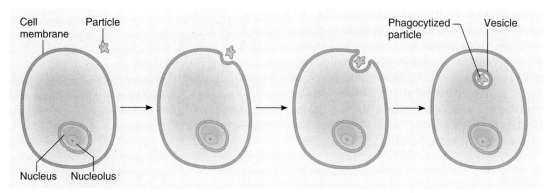

FIGURE 3.31

A cell may take in a solid particle from its surroundings by phagocytosis.

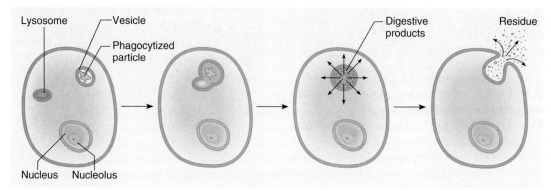

FIGURE 3.32

When a lysosome combines with a vesicle that contains a phagocytized particle, its digestive enzymes may destroy the particle. The products of this intracellular digestion diffuse into the cytoplasm. Any residue may be expelled from the cell by exocytosis.

or destroy disease-causing microorganisms. Phagocytosis is an important line of defense against infection.

Pinocytosis and phagocytosis engulf nonspecifically. In contrast is the more discriminating **receptor-mediated endocytosis,** which moves very specific kinds of particles into the cell. In this mechanism, protein molecules extend through the cell membrane and are exposed on its outer surface. These proteins are receptors to which specific substances from the fluid surroundings of the cell can bind. Molecules that can bind to the receptor sites selectively enter the cell; other kinds of molecules are left outside (fig. 3.33). Molecules that bind specifically to receptors are called *ligands.*

Entry of cholesterol molecules into cells illustrates receptor-mediated endocytosis. Cholesterol molecules synthesized in liver cells are packaged into large spherical particles called *low-density lipoproteins* (LDL). An LDL particle has a coating that contains a binding protein called *apoprotein-B.* The membranes of various body cells have receptors for apoprotein-B. When the liver releases LDL particles into the blood, cells with apoprotein-B receptors can recognize the LDL particles and bind them. Formation of such a receptor-ligand combination stimulates the cell membrane to indent and form a vesicle around the LDL particle. The vesicle carries the LDL particle to a lysosome, where enzymes digest it and release the cholesterol molecules for cellular use.

Receptor-mediated endocytosis is particularly important because it allows cells with the appropriate receptors to remove and process specific kinds of substances from their surroundings, even when these substances are present in very low concentrations. In short, receptor-mediated endocytosis provides specificity.

As a toddler, Stormie Jones already had a blood serum cholesterol level six times normal. Before she died at age ten, she had suffered several heart attacks and had undergone two cardiac bypass surgeries, several heart valve replacements, and finally a heart-liver transplant. The transplant lowered her blood cholesterol to a near-normal level, but she died from the multiple traumas suffered over her short lifetime.

Stormie had the severe form of familial hypercholesterolemia (FH), meaning simply too much cholesterol in the blood. Her liver cells lacked LDL receptors. Blocked from entering cells, cholesterol accumulated in her bloodstream, forming the plaques that caused her heart disease.

Stormie Jones was one in a million. One in 500 people have the milder form of FH, in which liver cells have half the normal number of LDL receptors. These individuals are prone to suffer heart attacks in early adulthood. However, they can delay symptom onset by taking precautions to avoid cholesterol buildup, such as exercising, eating a low-fat diet, not smoking, and taking statin drugs. (These precautions may also benefit individuals not suffering from FH.)

Exocytosis

Exocytosis is essentially the reverse of endocytosis. In exocytosis, substances made within the cell are packaged into a vesicle, which then fuses with the cell membrane, releasing its contents outside the cell. Cells secrete some proteins by this process. Nerve cells use exocytosis to release the neurotransmitter chemicals that signal other nerve cells, muscle cells, or glands (fig. 3.34).

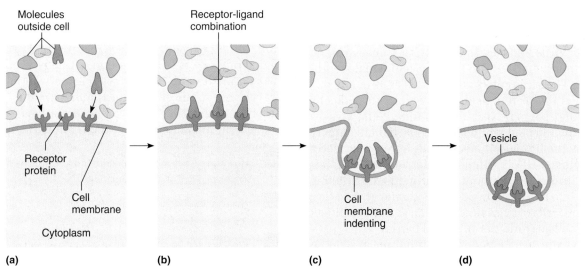

(a) (b) (c) (d)

FIGURE 3.33

Receptor-mediated endocytosis. (*a, b*) A specific substance binds to a receptor site protein. (*c*) The combination of the substance with the receptor-site protein stimulates the cell membrane to indent. (*d*) The resulting vesicle transports the substance into the cytoplasm.

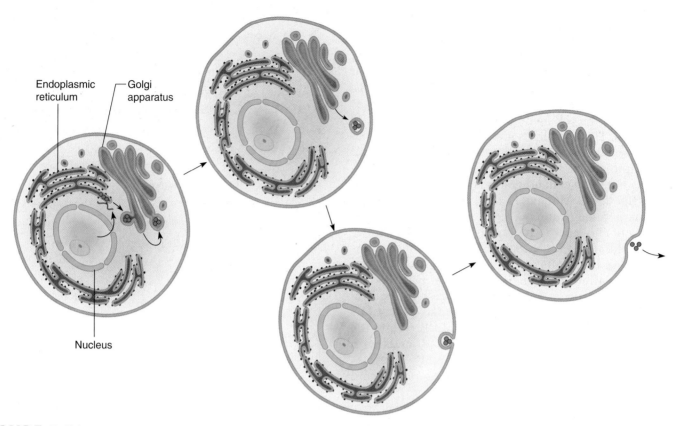

Endoplasmic reticulum

Golgi apparatus

Nucleus

FIGURE 3.34

Exocytosis releases particles, such as newly synthesized proteins, from cells.

Transcytosis

Endocytosis brings a substance into a cell, and exocytosis transports a substance out of a cell. Another process, **transcytosis** (tranz-si-to′-sis), combines endocytosis and exocytosis to selectively and rapidly transport a substance or particle from one end of a cell to the other (fig. 3.35). Transcytosis moves substances across barriers formed by tightly connected cells.

HIV, the virus that causes AIDS, uses transcytosis to cross lining (epithelial) cells in the anus and female reproductive tract. The virus enters white blood cells in mucous secretions, and the secretions then carry the infected cells to an epithelial barrier. Near these lining cells, viruses rapidly exit the infected white blood cells and are quickly enveloped by the lining cell membranes in receptor-mediated endocytosis. HIV particles are ferried, in vesicles, through the lining cell, without actually infecting (taking over) the cell, to exit from the cell membrane at the other end of the cell. After transcytosis, the HIV particles infect white blood cells beyond the epithelial barrier.

Transcytosis also enables the immune system to monitor pathogens in the small intestine, protecting against some forms of food poisoning. Scattered among the small intestinal epithelial cells are rare M cells, so-named because the cell side that faces into the intestine has microfolds that maximize surface area. The other side of the M cell appears punched in, forming a pocket where immune system cells gather. The M cell binds and takes in a bacterium from the intestinal side by endocytosis, then transports it through the cell to the side that faces the immune system cells, where it is released by exocytosis. The immune system sentinels bind parts of the bacteria, and, if they recognize surface features of a pathogen, they signal other cells to mature into antibody-producing cells. The antibodies are then secreted into the bloodstream and travel back to the small intestine, where they destroy the infecting bacteria. Table 3.4 summarizes the types of movement into and out of the cell, inlcuding transcytosis.

1 How does a cell maintain unequal concentrations of ions on opposite sides of a cell membrane?

2 How are facilitated diffusion and active transport similar? How are they different?

3 What is the difference between pinocytosis and phagocytosis?

4 Describe receptor-mediated endocytosis.

5 What does transcytosis accomplish?

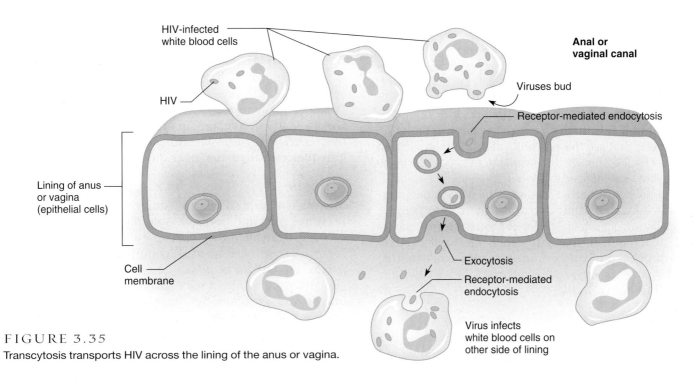

FIGURE 3.35

Transcytosis transports HIV across the lining of the anus or vagina.

TABLE 3.4   Movements Into and Out of the Cell

Process	Characteristics	Source of Energy	Example
I. Passive (Physical) Processes			
A. Simple diffusion	Molecules or ions move from regions of higher concentration toward regions of lower concentration.	Molecular motion	Exchange of oxygen and carbon dioxide in the lungs
B. Facilitated diffusion	Molecules move across the membrane through channels or by carrier molecules from a region of higher concentration to one of lower concentration.	Molecular motion	Movement of glucose through a cell membrane
C. Osmosis	Water molecules move from regions of higher concentration toward regions of lower concentration through a selectively permeable membrane.	Molecular motion	Distilled water entering a cell
D. Filtration	Smaller molecules are forced through porous membranes from regions of higher pressure to regions of lower pressure.	Hydrostatic pressure	Molecules leaving blood capillaries
II. Active (Physiological) Processes			
A. Active transport	Carrier molecules transport molecules or ions through membranes from regions of lower concentration toward regions of higher concentration.	Cellular energy	Movement of various ions and amino acids through membranes
B. Endocytosis			
1. Pinocytosis	Membrane engulfs droplets of liquid from surroundings.	Cellular energy	Membrane-forming vesicles containing large particles dissolved in water
2. Phagocytosis	Membrane engulfs solid particles from surroundings.	Cellular energy	White blood cell membrane engulfing bacterial cell
3. Receptor-mediated endocytosis	Membrane engulfs selected molecules combined with receptor proteins.	Cellular energy	Cell removing cholesterol-containing LDL particles from its surroundings
C. Exocytosis	Vesicles fuse with membrane and release contents outside of the cell.	Cellular energy	Protein secretion, neurotransmitter release
D. Transcytosis	Combines receptor-mediated endocytosis and exocytosis to ferry particles through a cell.	Cellular energy	HIV crossing a cell layer

The Cell Cycle

The series of changes that a cell undergoes, from the time it forms until it divides, is called the **cell cycle** (fig. 3.36). Superficially, this cycle seems rather simple—a newly formed cell grows for a time, and then divides in half to form two new cells, which, in turn, may grow and divide. The two cells that are the products of division are called daughter cells. Yet the specific events of the cycle are quite complex. For ease of study, the cell cycle can be considered to consist of distinct stages, which include interphase, mitosis, cytoplasmic division, and differentiation.

The actions of several types of proteins form "checkpoints" that control the cell cycle. One particularly important checkpoint determines a cell's fate—that is, whether it will continue in the cell cycle and divide, stay specialized yet alive, or die.

Interphase

Once thought to be a time of rest, **interphase** is actually a very active period. During interphase, the cell grows and maintains its routine functions as well as its contributions to the internal environment.

If the cell is developmentally programmed to divide, it must amass important biochemicals and duplicate much of its contents so that two cells can form from one.

For example, the cell must replicate its genetic material and synthesize and assemble the parts of membranes, ribosomes, lysosomes, peroxisomes and mitochondria.

Interphase is divided into phases based on the sequence of activities. DNA is replicated during S phase (S stands for synthesis) and is bracketed by two G phases, G_1 and G_2 (G stands for gap or growth). Structures other than DNA are synthesized during the G phases. Cellular growth occurs then, too (fig. 3.36).

Mitosis

Mitosis is a form of cell division that occurs in somatic (nonsex) cells and produces two daughter cells from an original cell (fig. 3.37). These new cells are genetically identical, each with the full complement of 46 chromosomes. In contrast is *meiosis,* a second form of cell division that occurs only in sex cells (sperm and eggs). Meiosis halves the chromosome number, a mechanism that ensures that when sperm meets egg, the total number of 46 chromosomes is restored. Chapter 22 (pp. 835–836) considers meiosis in detail.

> Mitosis is sometimes called cellular reproduction, because it results in two cells from one—the cell reproduces. This may be confusing, because meiosis is the prelude to human sexual reproduction. Both mitosis and meiosis are forms of cell division, with similar steps but different outcomes, and occurring in different types of cells.

During mitosis, the nuclear contents divide, an event called karyokinesis, and then the cytoplasm is apportioned into the two daughter cells, a process called cytokinesis. Mitosis must be very precise so that each new cell receives a complete copy of the genetic information. Although the chromosomes have already been copied in interphase, it is in mitosis that the chromosome sets are evenly distributed between the two forming cells.

Mitosis is a continuous process, but it is described in stages that indicate the sequence of major events, as follows:

1. **Prophase.** One of the first indications that a cell is going to divide is the condensation of chromatin fibers into tightly coiled rods, the chromosomes. During interphase, when the DNA molecules replicate, each chromosome consists of two identical structures, called chromatids, that are temporarily attached by a region on each called a *centromere.*

   The centrioles of the centrosome replicate just before the onset of mitosis (fig. 3.37*a*), and during prophase, the two newly formed pairs of centrioles move to opposite sides of the cell. Soon the nuclear

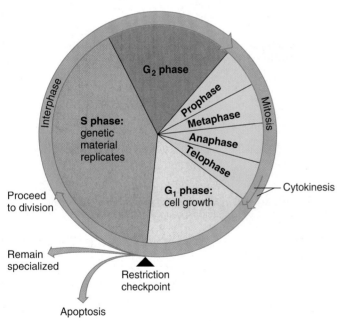

FIGURE 3.36

The cell cycle is divided into interphase, when cellular components duplicate, and cell division (mitosis and cytokinesis), when the cell splits in two, distributing its contents into two cells. Interphase is divided into two gap phases (G_1 and G_2), when specific molecules and structures duplicate, and a synthesis phase (S), when the genetic material replicates. Mitosis can be described as consisting of stages—prophase, metaphase, anaphase, and telophase.

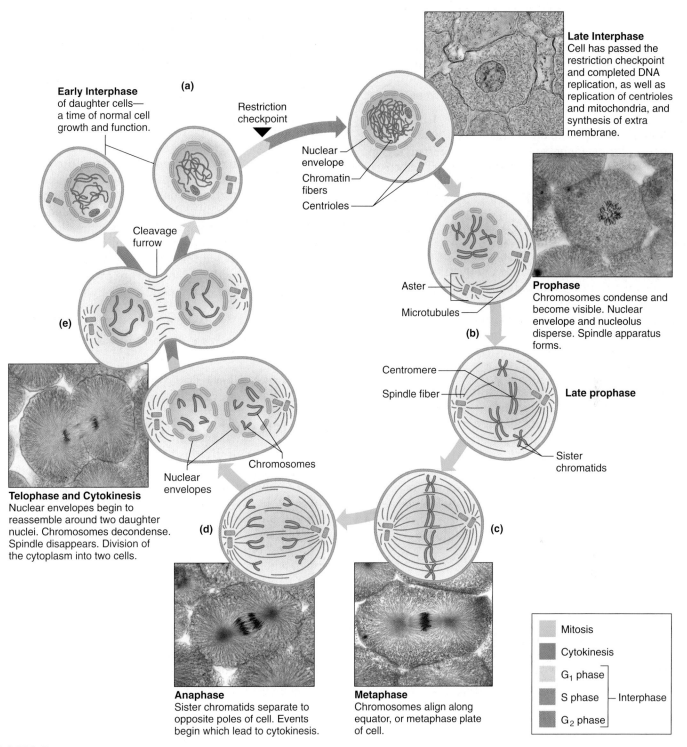

Late Interphase
Cell has passed the restriction checkpoint and completed DNA replication, as well as replication of centrioles and mitochondria, and synthesis of extra membrane.

(a)

Early Interphase
of daughter cells—a time of normal cell growth and function.

Restriction checkpoint

Nuclear envelope
Chromatin fibers
Centrioles

Cleavage furrow

(e)

Aster
Microtubules

Prophase
Chromosomes condense and become visible. Nuclear envelope and nucleolus disperse. Spindle apparatus forms.

(b)

Centromere
Spindle fiber

Late prophase

Sister chromatids

Chromosomes

Nuclear envelopes

Telophase and Cytokinesis
Nuclear envelopes begin to reassemble around two daughter nuclei. Chromosomes decondense. Spindle disappears. Division of the cytoplasm into two cells.

(d)

(c)

Anaphase
Sister chromatids separate to opposite poles of cell. Events begin which lead to cytokinesis.

Metaphase
Chromosomes align along equator, or metaphase plate of cell.

	Mitosis
	Cytokinesis
	G_1 phase
	S phase — Interphase
	G_2 phase

FIGURE 3.37

Mitosis and cytokinesis. (*a*) During interphase, before mitosis, chromosomes are visible only as chromatin fibers. A single pair of centrioles is present, but not visible at this magnification. (*b*) In prophase, as mitosis begins, chromosomes have condensed and are easily visible when stained. The centrioles have replicated, and each pair moves to an opposite end of the cell. The nuclear envelope and nucleolus disappear, and spindle fibers associate with the centrioles and the chromosomes. (*c*) In metaphase, the chromosomes line up midway between the centrioles. (*d*) In anaphase, the centromeres are pulled apart by the spindle fibers, and the chromatids, now individual chromosomes, move in opposite directions. (*e*) In telophase, chromosomes complete their migration and become chromatin, the nuclear envelope reforms, and microtubules disassemble. Cytokinesis, which actually began during anaphase, continues during telophase. Not all chromosomes are shown in these drawings. (Micrographs approximately 360×)

envelope and the nucleolus disperse and are no longer visible. Microtubules are assembled from tubulin proteins in the cytoplasm, and these structures associate with the centrioles and chromosomes. A spindle-shaped array of microtubules (spindle fibers) forms between the centrioles as they move apart (fig. 3.37b).

2. **Metaphase.** Spindle fibers attach to the centromeres of the chromosomes so that a fiber accompanying one chromatid attaches to one centromere and a fiber accompanying the other chromatid attaches to its centromere (fig. 3.37c). The chromosomes move along the spindle fibers and align about midway between the centrioles as a result of microtubule activity.

3. **Anaphase.** Soon the centromeres of the chromatids separate, and these identical chromatids are now considered individual chromosomes. The separated chromosomes move in opposite directions, and once again, the movement results from microtubule activity. The spindle fibers shorten and pull their attached chromosomes toward the centrioles at opposite sides of the cell (fig. 3.37d).

4. **Telophase.** The final stage of mitosis begins when the chromosomes complete their migration toward the centrioles. It is much like the reverse of prophase. As the identical sets of chromosomes approach their respective centrioles, they begin to elongate and unwind from rodlike structures to threadlike structures. A nuclear envelope forms around each chromosome set, and nucleoli become visible within the newly formed nuclei. Finally, the microtubules disassemble into free tubulin molecules (fig. 3.37e).

Table 3.5 summarizes the stages of mitosis.

TABLE 3.5	Major Events in Mitosis and Cytokinesis
Stage	**Major Events**
Prophase	Chromatin condenses into chromosomes; centrioles move to opposite sides of cytoplasm; nuclear membrane and nucleolus disperse; microtubules appear and associate with centrioles and chromatids of chromosomes.
Metaphase	Spindle fibers from the centrioles attach to the centromeres of each chromosome; chromosomes align midway between the centrioles.
Anaphase	Centromeres separate, and chromatids of the chromosomes separate; spindle fibers shorten and pull these new individual chromosomes toward centrioles.
Telophase	Chromosomes elongate and form chromatin threads; nuclear membranes appear around each chromosome set; nucleoli appear; microtubules break down.

Cytoplasmic Division

Cytoplasmic division (cytokinesis) begins during anaphase when the cell membrane starts to constrict around the middle, which it continues to do through telophase. The musclelike contraction of a ring of actin microfilaments pinches off two cells from one. The microfilaments assemble in the cytoplasm and attach to the inner surface of the cell membrane. The contractile ring forms at right angles to the microtubules that pulled the chromosomes to opposite ends of the cell during mitosis. As the ring pinches, it separates the two newly formed nuclei and apportions about half of the organelles into each of the daughter cells. The newly formed cells may differ slightly in size and number of organelles and inclusions, but they have identical chromosomes and thus contain identical DNA information (fig. 3.38). How that DNA is expressed (used to manufacture proteins) determines the specialization of the cell, a point we return to at the chapter's end.

1 Why is precise division of nuclear materials during mitosis important?

2 Describe the events that occur during mitosis.

Control of Cell Division

How often a cell divides is strictly controlled and varies with cell type. Skin cells, blood-forming cells, and cells that line the intestine, for example, divide often and continually. In contrast, cells of the liver divide a specific number of times and then cease—they are alive and specialized, but no longer divide.

Most types of human cells divide from forty to sixty times when grown in the laboratory. Adherence to this limit can be startling. A connective tissue cell from a human fetus divides thirty-five to sixty-three times, the average being about fifty times. However, a similar cell from an adult divides only fourteen to twenty-nine times, as if the cell "knows" how many times it has already divided.

A physical basis for this mitotic clock is the DNA at the tips of chromosomes, called telomeres, where the same six-nucleotide sequence repeats hundreds of times. Each mitosis removes up to 1,200 nucleotides. When the chromosome tips wear down to a certain point, this somehow signals the cell to cease dividing.

Other external and internal factors influence the timing and frequency of mitosis. Within cells, waxing and waning levels of proteins called kinases and cyclins control the cell cycle. Another internal influence is cell size, specifically the ratio between the surface area the cell membrane provides and the cell volume. The larger the cell, the more nutrients it requires to maintain the activities of life. However, a cell's surface area limits the number of nutrient molecules that can enter. Because volume

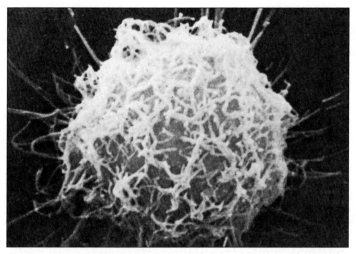

(a)

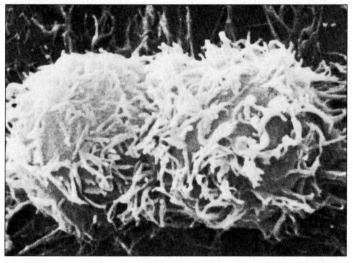

(b)

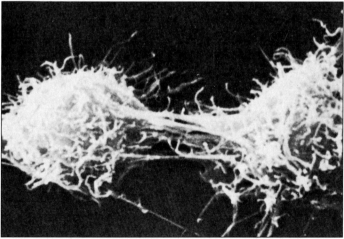

(c)

FIGURE 3.38

Following mitosis, the cytoplasm of a cell divides in two, as seen in these scanning electron micrographs (*a.* 3,750×; *b.* 3,750×; *c.* 3,190×). From *Scanning Electron Microscopy in Biology,* by R. G. Kessel and C. Y. Shih. © 1976 Springer-Verlag.

increases faster than does surface area, a cell can grow too large to efficiently obtain nutrients. Cell division solves this growth problem. The resulting daughter cells are smaller than the original cell and thus have a more favorable surface area-to-volume relationship.

External controls of cell division include hormones and growth factors. Hormones are biochemicals manufactured in a gland and transported in the bloodstream to a site where they exert an effect. Hormones signal mitosis in the lining of a woman's uterus each month, building up the tissue to nurture a possible pregnancy. Similarly, a pregnant woman's hormones stimulate mitosis in her breasts when their function as milk-producing glands will soon be required.

Growth factors are like hormones in function but act closer to their sites of synthesis. Epidermal growth factor, for example, stimulates growth of new skin beneath the scab on a skinned knee. Salivary glands also produce this growth factor. This is why an animal's licking a wound may speed healing.

> Growth factors are used as drugs. Epidermal growth factor (EGF), for example, can hasten healing of a wounded or transplanted cornea, a one-cell-thick layer covering the eye. Normally these cells do not divide. However, cells of a damaged cornea treated with EGF undergo mitosis, restoring a complete cell layer. EGF is also used to help the body accept skin grafts and to stimulate healing of skin ulcers that occur as a complication of diabetes.

Space availability is another external factor that influences the timing and rate of cell division. Healthy cells do not divide if they are surrounded by other cells, a phenomenon called contact (density dependent) inhibition.

Control of cell division is absolutely crucial to health. With too infrequent mitoses, an embryo could not develop, a child could not grow, and wounds would not heal. Too frequent mitoses produce an abnormal growth, or neoplasm, which may form a disorganized mass called a **tumor.**

Tumors are of two types. A *benign* tumor remains in place like a lump, eventually interfering with the function of healthy tissue. A *malignant,* or cancerous, tumor looks quite different—it is invasive, extending into surrounding tissue. A growing malignant tumor may roughly resemble a crab with outreaching claws, which is where the word "cancer" comes from. Cancer cells, if not stopped, eventually reach the circulation and spread, or metastasize, to other sites. Table 3.6 lists characteristics of cancer cells, and figure 3.39 illustrates how cancer cells infiltrate healthy tissue.

TABLE 3.6	Characteristics of Cancer Cells

Loss of cell cycle control

Heritability (a cancer cell divides to form more cancer cells)

Transplantability (a cancer cell implanted into another individual will cause cancer to develop)

Dedifferentiation (loss of specialized characteristics)

Loss of contact inhibition

Ability to induce local blood vessel formation (angiogenesis)

Invasiveness

Ability to metastasize (spread)

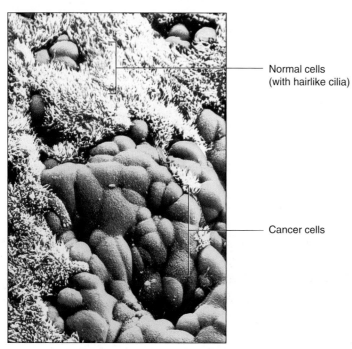

Normal cells (with hairlike cilia)

Cancer cells

FIGURE 3.39

A cancer cell is rounder and less specialized than surrounding healthy cells. It secretes biochemicals that cut through nearby tissue (invasiveness) and other biochemicals that stimulate formation of blood vessels that nurture the tumor's growth (angiogenesis) (2,200×).

Cancer is a collection of disorders distinguished by their site of origin and the affected cell type. Many cancers are treatable with surgery, radiation, chemicals (chemotherapy), or immune system substances used as drugs. A newer approach to treating cancer is to develop molecules that bind to receptors that are unique to, or unusually abundant on, cancer cells, blocking the cells from receiving signals to divide.

Two major types of genes cause cancer. **Oncogenes** activate other genes that increase cell division rate. **Tumor suppressor genes** normally hold mitosis in check. When tumor suppressor genes are removed or otherwise inactivated, this lifts control of the cell cycle, and uncontrolled cell division leading to cancer results (fig. 3.40). Environmental factors, such as exposure to toxic chemicals or radiation, may induce cancer by altering (mutat-

ing) oncogenes and tumor suppressor genes in body (somatic) cells. Cancer may also be the consequence of a failure of normal programmed cell death (apoptosis), resulting in overgrowth.

1. How do cells vary in their rates of division?

2. Which factors control the number of times and the rate at which cells divide?

3. How can too infrequent or too frequent cell division affect health?

4. What is the difference between a benign and a cancerous tumor?

5. What are two ways that genes cause cancer?

Stem and Progenitor Cells

In 1855, German physiologist Rudolph Virchow stated that all cells come from preexisting cells. Until this time, most people thought that life, and cells, sprang from nothingness, or from the nonliving. The fact that cells come from preexisting cells explains how a many-celled organism develops from a single fertilized egg and how body parts grow and injuries heal.

Cells that retain the ability to divide repeatedly allow for this continual growth and renewal (fig. 3.41). A **stem cell** divides mitotically to yield either two daughter cells like itself, or one daughter cell that is a stem cell and one that is partially specialized. The partly specialized cell is intermediate between a stem cell and a fully differentiated cell and is termed a **progenitor cell.** A progenitor is said to be "committed" because its daughter cells can become any of a restricted number of cell types. For example, a neural stem cell divides to give rise to cells that become part of neural tissue (neurons and neuroglial cells), but not part of muscle or bone tissue. All of the 260 or so differentiated cell types in a human body can be traced back through lineages of progenitor and stem cells, although researchers have not identified all of them.

Stem cells and progenitor cells are described in terms of their potential—that is, according to the possible fates of their daughter cells. A fertilized ovum and cells of the very early embryo, when it is just a small ball of cells, are **totipotent,** which means that they can give rise to every cell type (fig. 3.42). In contrast, stem cells that arise later in development as well as progenitor cells are **pluripotent,** which means that their daughter cells can follow any of several pathways, but not all of them.

Researchers are discovering that many, if not all, of the organs in an adult human body harbor very small populations of stem or progenitor cells that are activated when injury or illness occurs. For example, one in 10,000 to 15,000 bone marrow cells is a hematopoietic stem cell, which can give rise to blood and several other cell types. Stem cells in the adult body may have been set aside in the embryo or fetus, as repositories of future healing. Alternatively, or perhaps also, stem cells or progenitor cells may

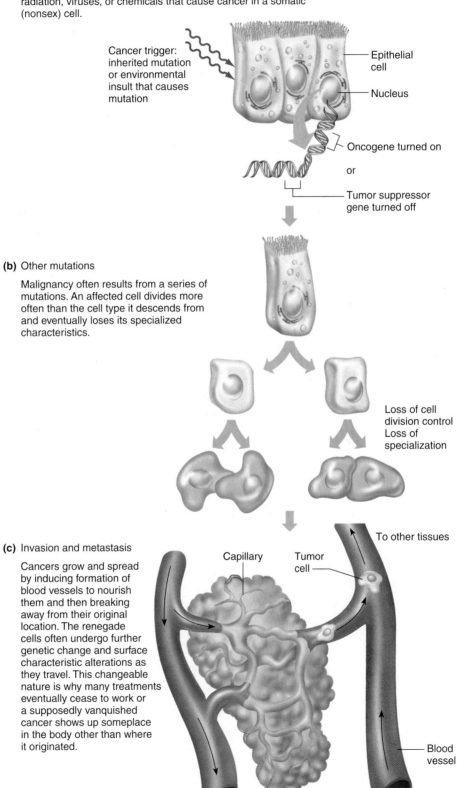

(a) Healthy, specialized cells

In a healthy cell, oncogenes are not overexpressed, and tumor suppressor genes are expressed. As a result, cell division rate is under control. Cancer begins in a single cell when an oncogene is turned on or a tumor suppressor gene is turned off. This initial step may result from an inherited mutation, or from exposure to radiation, viruses, or chemicals that cause cancer in a somatic (nonsex) cell.

Cancer trigger: inherited mutation or environmental insult that causes mutation

Epithelial cell

Nucleus

Oncogene turned on

or

Tumor suppressor gene turned off

(b) Other mutations

Malignancy often results from a series of mutations. An affected cell divides more often than the cell type it descends from and eventually loses its specialized characteristics.

Loss of cell division control
Loss of specialization

(c) Invasion and metastasis

Cancers grow and spread by inducing formation of blood vessels to nourish them and then breaking away from their original location. The renegade cells often undergo further genetic change and surface characteristic alterations as they travel. This changeable nature is why many treatments eventually cease to work or a supposedly vanquished cancer shows up someplace in the body other than where it originated.

To other tissues

Capillary

Tumor cell

Blood vessel

Tumor

FIGURE 3.40
Steps in the development of cancer.

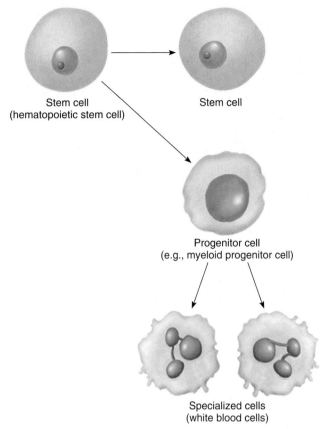

FIGURE 3.41

Stem cells and progenitor cells. A true stem cell divides mitotically to yield two stem cell daughters, or a stem cell and a progenitor cell, which may show the beginnings of differentiation. Progenitor cells give rise to progenitors or more differentiated cells of a restricted lineage.

travel from bone marrow to replace damaged or dead cells in response to certain signals that are sent when injury or disease occurs.

Through development, cells become progressively more specialized. All cells in the human body (except red blood cells, which expel their nuclei), have the same set of genetic instructions, but as cells specialize, they use some genes and ignore others. For example, an immature bone cell (osteoblast) forms from a progenitor cell by manufacturing proteins necessary to bind bone mineral, as well as alkaline phosphatase, an enzyme required for bone formation. An immature muscle cell (myoblast), in contrast, forms from a muscle progenitor cell and accumulates the contractile proteins that define a muscle cell. The term "blast" is often used to describe these fledgling differentiated cells, such as osteoblast and myoblast. The osteoblast does not produce contractile proteins, just as the myoblast does not produce mineral-binding proteins and alkaline phosphatase. The final differentiated cell is like a library. It contains a complete collection of information but accesses only some of it.

From Science to Technology 3.1 describes how stem cell technology is being developed to treat a range of disorders. This technology uses embryonic stem cells, which are derived from the very early embryo and can generate any cell type, as well as rare stem and progenitor cells from adult tissues.

1 Distinguish between a stem cell and a progenitor cell.

2 Distinguish between totipotent and pluripotent.

3 How do cells differentiate?

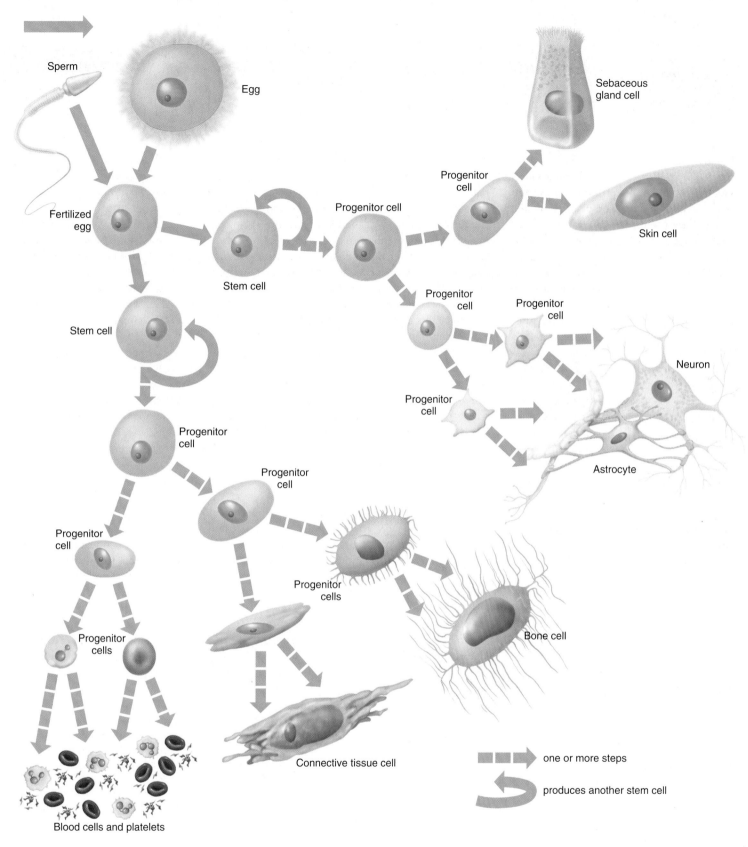

Sperm

Egg

Fertilized egg

Stem cell

Stem cell

Progenitor cell

Progenitor cell

Progenitor cell

Sebaceous gland cell

Skin cell

Progenitor cell

Progenitor cell

Progenitor cell

Progenitor cell

Neuron

Astrocyte

Progenitor cell

Progenitor cells

Progenitor cell

Progenitor cells

Bone cell

Blood cells and platelets

Connective tissue cell

one or more steps

produces another stem cell

FIGURE 3.42

Cell lineages. All cells in the human body ultimately descend from stem cells, through the processes of mitosis and differentiation. This depiction is a simplified view of a few of the pathways that cells follow, grouping the cell types by the closeness of their lineages. Note that the pattern is treelike, or fractal. The differentiated cells on the left are all connective tissues (blood, connective tissue and bone), but the blood cells are more closely related to each other than they are to the other two cell types, and vice versa (cartilage and other connective tissues are omitted). Towards the right, the skin and sebaceous gland cells share a recent progenitor, and both share a more distant progenitor with neurons and astrocytes. It's easy to see the complexity of the lineages of the more than 260 human cell types!

CLONING TO PRODUCE THERAPEUTIC STEM CELLS

In the human body, dividing stem cells and progenitor cells produce the differentiated cell types that enable organs to grow and replace damaged or dead cells. Stem cell technology harnesses this capability so that cell types needed to treat a particular disorder or injury can be cultured in laboratory dishes—some made-to-order.

Stem cells can be derived from embryos or after birth—even from deceased adults. From a biological standpoint, embryonic stem (ES) cells, obtained from the inner cell mass portion of a five-day embryo, or blastocyst, are preferable because they are less likely to provoke rejection by the recipient's immune system. ES cells can also differentiate into more cell types than can the progenitor cells typically found in adult tissues. From an ethical standpoint, however, ES cells are controversial because they must come from embryos.

There are two sources of ES cells. One is to use blastocysts from fertility clinics where couples undergoing *in vitro* fertilization (IVF) freeze extra blastocysts that are not implanted into the woman. This approach could be used to create banks of stored cell types, perhaps stripped of their cell surface molecules so that recipients would not reject them. The other source of ES cells is to use the nucleus from a somatic cell from a particular individual who requires new cells—for example, a person who has suffered a spinal cord injury. In this still hypo-thetical scenario, the nucleus is injected into or fused with a donated egg cell whose nucleus has been removed. The new cell divides in laboratory culture for five days, producing a blastocyst that is a clone—a genetic replica—of the person who donated the somatic cell nucleus. When cells of the inner cell mass are removed from the blastocyst and cultured in dishes to become ES cells, then given particular combinations of growth factors and other biochemicals, they differentiate into what is needed—neurons in the case of a spinal cord injury. The person's body accepts the cells because they are genetically identical to his or her own. This approach is called somatic cell nuclear transfer (fig. 3A). It has been shown to work in cattle and mice but remains highly controversial for humans.

So far, nations are about evenly divided in what they permit—using stem cells from IVF leftovers or somatic cell nuclear transfer, both or neither. Ethical objections focus on intent. Objection to using IVF blastocysts is that they were conceived with the goal of producing a child. Objection to somatic cell nuclear transfer is that the blastocyst is created for the purpose of supplying cells. The United States has considered not only banning use of federal funds for human ES cell research, but of fining and imprisoning researchers who use the cells, even if they obtain them from countries where the work is permitted.

While the debate over ES cell technology continues, investigation into using stem cells found in the adult human body may make the matter moot. In fact, hematopoietic (blood-forming) stem cells from adult bone marrow have been used clinically for half a century, and more recently, they are being obtained from stored umbilical cord blood.

Continuing basic research on stem cells from the adult may have other clinical applications. For example, researchers have identified the stem cells in human breast tissue, small cells tucked between the epithelial and contractile layers of the milk ducts. Learning what stimulates these cells to divide too often may reveal how breast cancer arises. The vignette to chapter 10 describes the discovery of neural stem cells, which hold promise for treating neurodegenerative diseases as well as spinal cord injuries, and From Science to Technology 15.1 explores an experiment that revealed how stem cells in a heart transplant recipient infiltrate the donor heart and then differentiate into precisely what is required to accept the new organ. Stem cell technology may be used to treat less serious conditions, too. The discovery that a single stem cell can give rise to skin cells, hair follicle cells, and sebaceous (oil) gland cells suggests that learning to manipulate them can be used to treat baldness, acne, and removal of unwanted hair! ∎

FIGURE 3A

Somatic cell nuclear transfer, also known as therapeutic cloning, places a nucleus from a somatic cell from a person with a particular illness or injury in an immature egg cell (oocyte) whose nucleus has been removed. Development continues for five days, until the blastocyst stage when cells of the inner cell mass are removed and cultured, forming embryonic stem (ES) cells. When the ES cells are stimulated with the appropriate growth factors, they differentiate as the cell types that the person requires for treatment.

Somatic cell
(skin fibroblast)

Patient

Egg cell
with nucleus
removed

Somatic cell
nuclear transfer

Cell fusion

Cell division

Totipotent
cells

Cell division

Inner cell
mass

Blastocyst

Embryonic stem (ES) cells

Cultured
stem cells

Blood   Muscle   Nervous

Patient

Transplantation

CHAPTER SUMMARY

Introduction (page 62)

Differentiated cells vary considerably in size, shape, and function. The shapes of cells are important in determining their functions. Specialized cells descend from less specialized cells.

A Composite Cell (page 62)

1. A cell includes a nucleus, cytoplasm, and a cell membrane.
2. Cytoplasmic organelles perform specific vital functions, but the nucleus controls the overall activities of the cell.
3. Cell membrane
   a. The cell membrane forms the outermost limit of the living material.
   b. It acts as a selectively permeable passageway that controls the movements of substances between the cell and its surroundings and thus is the site of signal transduction.
   c. It includes protein, lipid, and carbohydrate molecules.
   d. The cell membrane framework mainly consists of a double layer of phospholipid molecules.
   e. Molecules that are soluble in lipids pass through the membrane easily, but water-soluble molecules do not.
   f. Cholesterol molecules help stabilize the membrane.
   g. Proteins provide the special functions of the membrane, as receptors, cell surface markers of self, transporters, enzymes, and cellular adhesion molecules.
   h. Specialized intercellular junctions (tight junctions, desmosomes, and gap junctions) connect cells.
   i. Cell adhesion molecules oversee some cell interactions and movements.
4. Cytoplasm
   a. Cytoplasm contains networks of membranes and organelles suspended in fluid.
   b. Endoplasmic reticulum is composed of connected membranous sacs, canals, and vesicles that provide a tubular communication system and an attachment for ribosomes; it also functions in the synthesis of proteins and lipids.
   c. Ribosomes are particles of protein and RNA that function in protein synthesis.
   d. The Golgi apparatus is a stack of flattened, membranous sacs that package glycoproteins for secretion.
   e. Mitochondria are membranous sacs containing enzymes that catalyze the reactions that release energy from nutrient molecules and transform it into a usable form.
   f. Lysosomes are membranous sacs containing digestive enzymes that destroy debris and worn-out organelles.
   g. Peroxisomes are membranous, enzyme-containing vesicles.
   h. The centrosome is a nonmembranous structure consisting of two centrioles that aid in the distribution of chromosomes during cell division.
   i. Cilia and flagella are motile extensions on some cell surfaces.
      (1) Cilia are numerous tiny, hairlike structures that wave, moving fluids across cell surfaces.
      (2) Flagella are longer extensions such as the tail of a sperm cell.

j. Vesicles are membranous sacs containing substances that recently entered or were produced in the cell.

k. Microfilaments and microtubules are threadlike structures that aid cellular movements and support and stabilize the cytoplasm.

l. Cytoplasm may contain nonliving cellular products, such as nutrients and pigments, called inclusions.

5. Cell nucleus
   a. The nucleus is enclosed in a double-layered nuclear envelope that has nuclear pores that control movement of substances between the nucleus and cytoplasm.
   b. A nucleolus is a dense body of protein and RNA where ribosome synthesis occurs.
   c. Chromatin is composed of loosely coiled fibers of protein and DNA that condense into chromosomes during cell division.

Movements Into and Out of the Cell (page 80)

Movement of substances into and out of the cell may use physical or physiological processes.

1. Diffusion
   a. Diffusion is due to the random movement of molecules in air or liquid solution.
   b. Diffusion is movement of molecules or ions from regions of higher concentration toward regions of lower concentration (down a concentration gradient).
   c. It exchanges oxygen and carbon dioxide in the body.
   d. The most important factors determining the rate of diffusion in the body include distance, the concentration gradient, and temperature.

2. Facilitated diffusion
   a. Facilitated diffusion uses protein channels or carrier molecules in the cell membrane.
   b. This process moves substances such as ions, sugars, and amino acids from regions of higher concentration to regions of lower concentration.

3. Osmosis
   a. Osmosis is a special case of diffusion in which water molecules diffuse from regions of higher water concentration to lower water concentration through a selectively permeable membrane.
   b. Osmotic pressure increases as the number of particles dissolved in a solution increases.
   c. Cells lose water when placed in hypertonic solutions and gain water when placed in hypotonic solutions.
   d. A solution is isotonic when it contains the same concentration of dissolved particles as the cell contents.

4. Filtration
   a. In filtration, molecules move through a membrane from regions of higher hydrostatic pressure toward regions of lower hydrostatic pressure.
   b. Blood pressure filters water and dissolved substances through porous capillary walls.

5. Active transport
   a. Active transport moves molecules or ions from regions of lower concentration to regions of higher concentration.
   b. It requires cellular energy and carrier molecules in the cell membrane.

6. Endocytosis
   a. In pinocytosis, a cell membrane engulfs tiny droplets of liquid.
   b. In phagocytosis, a cell membrane engulfs solid particles.
   c. In receptor-mediated endocytosis, receptor proteins combine with specific molecules in the cell surroundings. The membrane engulfs the combinations.

7. Exocytosis
   a. Exocytosis is the reverse of endocytosis.
   b. In exocytosis, vesicles containing secretions fuse with the cell membrane, releasing the substances to the outside.

8. Transcytosis
   a. Transcytosis combines endocytosis and exocytosis.
   b. In transcytosis, a substance or particle crosses a cell.
   c. Transcytosis is specific.

The Cell Cycle (page 90)

1. The cell cycle includes interphase, mitosis, cytoplasmic division, and differentiation.

2. Interphase
   a. Interphase is the stage when a cell grows, DNA replicates, and new organelles form.
   b. It terminates when the cell begins mitosis.

3. Mitosis
   a. Mitosis is the division and distribution of DNA to daughter cells.
   b. The stages of mitosis include prophase, metaphase, anaphase, and telophase.

4. The cytoplasm divides into two portions with the completion of mitosis.

Control of Cell Division (page 92)

1. Cell division capacities vary greatly among cell types.

2. Chromosome tips that shorten with each mitosis provide a mitotic clock, usually limiting the number of divisions to fifty.

3. Cell division is limited, and controlled by both internal and external factors.

4. As a cell grows, its surface area increases to a lesser degree than its volume, and eventually the area becomes inadequate for the requirements of the living material within the cell. When a cell divides, the daughter cells have more favorable surface area-volume relationships.

5. Growth factors and hormones also stimulate cell division.

6. Cancer is the consequence of a loss of cell cycle control.

Stem and Progenitor Cells (page 94)

1. A stem cell divides to yield another stem cell and a progenitor cell that is partially differentiated.

2. Cells that give rise to any differentiated cell type are totipotent. Cells with more restricted fates are pluripotent.

3. Stem cells may be present in adult organs or migrate from the bone marrow to replace damaged cells—or both.

4. As cells specialize, they express different sets of genes that provide their distinct characteristics.

CRITICAL THINKING QUESTIONS

1. Which process—diffusion, osmosis, or filtration—accounts for the following situations?
   a. Injection of a drug that is hypertonic to the tissues stimulates pain.
   b. A person with extremely low blood pressure stops producing urine.
   c. The concentration of urea in the dialyzing fluid of an artificial kidney is kept low.
2. Which characteristic of cell membranes may explain why fat-soluble substances such as chloroform and ether rapidly affect cells?
3. A person exposed to many X rays may lose white blood cells and become more susceptible to infection. How are these effects related?
4. Exposure to tobacco smoke causes cilia to cease moving and degenerate. Why might this explain why tobacco smokers have an increased incidence of respiratory infections?
5. How would you explain the function of phagocytic cells to a patient with a bacterial infection?
6. How is knowledge of how cell division is controlled important to an understanding of each of the following?
   a. growth
   b. wound healing
   c. cancer
7. Why are enlarged lysosomes a sign of a serious illness?
8. Mutations occur in somatic cells over a person's lifetime. Explain how this might complicate somatic cell nuclear transfer to create cells to be used therapeutically for a particular individual.

REVIEW EXERCISES

1. Use specific examples to illustrate how cells vary in size.
2. Describe how the shapes of nerve, epithelial, and muscle cells are well suited to their functions.
3. Name the major components of a cell, and describe how they interact.
4. Discuss the structure and functions of a cell membrane.
5. How do cilia, flagella, and cell adhesion molecules move cells?
6. Distinguish between organelles and inclusions.
7. Define *selectively permeable*.
8. Describe the chemical structure of a membrane.
9. Explain how the structure of a cell membrane determines which types of substances it admits.
10. Explain the function of membrane proteins.
11. Describe three kinds of intercellular junctions.
12. Describe the structures and functions of each of the following:
    a. endoplasmic reticulum   g. cilium
    b. ribosome   h. flagellum
    c. Golgi apparatus   i. centrosome
    d. mitochondrion   j. vesicle
    e. lysosome   k. microfilament
    f. peroxisome   l. microtubule
13. Describe the structure of the nucleus and the functions of its contents.
14. Distinguish between diffusion and facilitated diffusion.
15. Name three factors that increase the rate of diffusion.
16. Explain how diffusion aids gas exchange within the body.
17. Define *osmosis*.
18. Define *osmotic pressure*.
19. Explain how the number of solute particles in a solution affects its osmotic pressure.
20. Distinguish among solutions that are hypertonic, hypotonic, and isotonic.
21. Define *filtration*.
22. Explain how filtration moves substances through capillary walls.
23. Explain why active transport is called a physiological process, whereas diffusion is called a physical process.
24. Explain the function of carrier molecules in active transport.
25. Distinguish between pinocytosis and phagocytosis.
26. Describe *receptor-mediated endocytosis*. How might it be used to deliver drugs across the blood-brain barrier?
27. Explain how transcytosis includes endocytosis and exocytosis.
28. List the phases in the cell cycle. Why is interphase *not* a time of cellular rest?
29. Name the two processes included in cell division.
30. Describe the major events of mitosis.
31. Explain how the cytoplasm is divided during cell division.
32. Explain what happens during interphase.
33. Define *differentiation*.
34. Explain how differentiation may reflect repression of DNA information.
35. How does loss of genetic control cause cancer?
36. Distinguish between a stem cell and a progenitor cell.
37. Distinguish between a totipotent cell and a pluripotent cell.
38. Explain how differentiated cells can have the same genetic instructions but look and function very differently.

Visit the Student Edition of the Online Learning Center at www.mhhe.com/shier10 for answers to chapter questions, additional quizzes, interactive learning exercises, and other study tools.

UNDERSTANDING WORDS

aer-, air: *aer*obic respiration—respiratory process that requires oxygen.

an-, without: *an*aerobic respiration—respiratory process that does not require oxygen.

ana-, up: *ana*bolism—cellular processes in which smaller molecules are used to build up larger ones.

cata-, down: *cata*bolism—cellular processes in which larger molecules are broken down into smaller ones.

co-, with: *co*enzyme—substance that unites with a protein to complete the structure of an active enzyme molecule.

de-, undoing: *de*amination—process by which nitrogen-containing portions of amino acid molecules are removed.

mut-, change: *mut*ation—change in the genetic information of a cell.

-strat, spread out: sub*strat*e—substance upon which an enzyme acts.

sub-, under: *sub*strate—substance upon which an enzyme acts.

-zym, causing to ferment: en*zym*e—protein that speeds up a chemical reaction without itself being consumed.

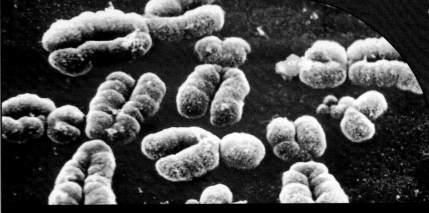

Chromosomes are the structures in the cell nucleus that carry genetic information in the form of DNA molecules. DNA instructs the cell to manufacture the enzymes that are crucial to metabolism (25,000×).

Cellular Metabolism

CHAPTER OBJECTIVES

After you have studied this chapter, you should be able to

1. Distinguish between anabolism and catabolism.

2. Explain how enzymes control metabolic processes.

3. Explain how the reactions of cellular respiration release chemical energy.

4. Describe how cells access energy for their activities.

5. Describe the general metabolic pathways of carbohydrate metabolism.

6. Explain how metabolic pathways are regulated.

7. Describe how DNA molecules store genetic information.

8. Explain how protein synthesis relies on genetic information.

9. Describe how DNA molecules are replicated.

10. Explain how genetic information can be altered and how such a change may affect an organism.

A physician and his seventeen-year-old son ate leftover spaghetti with homemade pesto sauce for several days, each time after it had been unrefrigerated for an hour or two. On the fourth day, the food had a peculiar odor, but the father heated it in a pan anyway. About a half hour after lunch, father and son developed severe abdominal pain. The father recovered, but the son began to behave strangely, becoming listless, then very sleepy. A yellow pallor indicated that his liver was malfunctioning.

Because of the rapid onset of abdominal pain after eating reheated food, food poisoning was likely. Indeed, the boy's body fluids and the pan used to reheat the spaghetti contained *Bacillus cereus,* a type of bacterium that produces a toxin that can cause abdominal pain. In the boy, the toxin took a deadly turn to the liver.

To learn how the bacterial toxin harms the liver, researchers applied toxin from the boy to rat liver cells growing in culture. This experiment revealed that the toxin targets mitochondria, the organelles that house the biochemical reactions that extract energy from nutrients. Specifically, the toxin destroyed the mitochondria's ability to break down fats. Ironically, liver cells have many mitochondria to power the energy-requiring reactions that break down toxins. With his liver mitochondria severely impaired, the boy's liver literally ran out of energy and shut down. He died four days after the spaghetti meal. ■

Metabolic Processes

In every human cell, even in the most sedentary individual, thousands of chemical reactions essential to life occur every second. A special type of protein called an enzyme controls the pace of each reaction. The sum total of chemical reactions within the cell constitutes metabolism.

Many metabolic reactions occur one after the other, with the products of one reaction serving as starting materials of another, forming intricate pathways and cycles that may intersect by sharing intermediate compounds. As a result, metabolism in its entirety may seem enormously complex. However, individual pathways of metabolism are fascinating to study because they reveal how cells function—in essence, how chemistry becomes biology. This chapter explores how metabolic pathways supply a cell with energy and how other biochemical processes enable a cell to produce proteins—including the enzymes that make all of metabolism possible.

Metabolic reactions and pathways are of two types. In **anabolism** (an″ah-bol′liz-m), larger molecules are constructed from smaller ones, requiring input of energy. In **catabolism** (kat″ah-bol-liz-m), larger molecules are broken down into smaller ones, releasing energy.

Anabolism

Anabolism provides all the substances required for cellular growth and repair. For example, an anabolic process called **dehydration synthesis** (de″hi-dra′shun sin′the-sis) joins many simple sugar molecules (monosaccharides) to form larger molecules of glycogen. When a runner consumes pasta the night before a race, digestion breaks down the complex carbohydrates to monosaccharides, which can be absorbed into the bloodstream, which carries these energy-rich molecules to body cells. Here, dehydration synthesis joins the monosaccharides to form glycogen, which stores energy that the runner may not need until later, as the finish line nears. When monosaccharide units join, an —OH (hydroxyl group) from one monosaccharide molecule and an —H (hydrogen atom) from an —OH group of another are removed. As the —H and —OH react to produce a water molecule, the monosaccharides are joined by a shared oxygen atom, as figure 4.1 shows (read from left to right). As the process repeats, the molecular chain extends, forming a polysaccharide.

Similarly, glycerol and fatty acid molecules join by dehydration synthesis in fat (adipose) tissue cells to form fat molecules. In this case, three hydrogen atoms are removed from a glycerol molecule, and an —OH group is removed from each of three fatty acid mole-

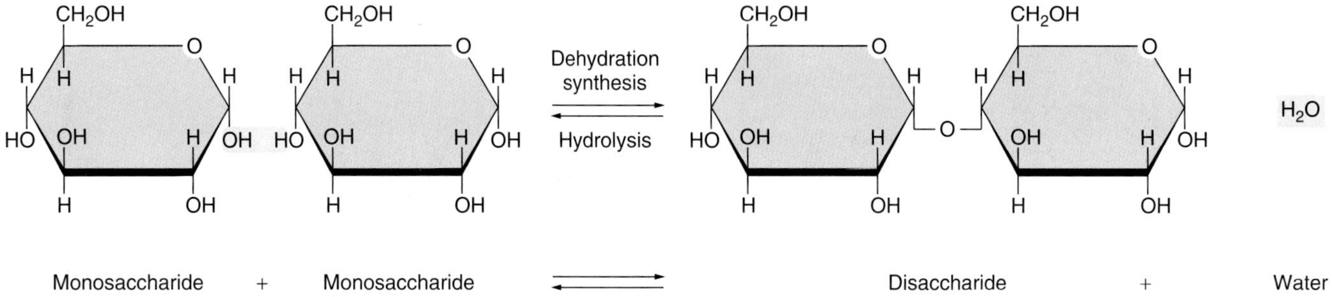

FIGURE 4.1
Two monosaccharides may join by dehydration synthesis to form a disaccharide. In the reverse reaction, the disaccharide is hydrolyzed to two monosaccharides.

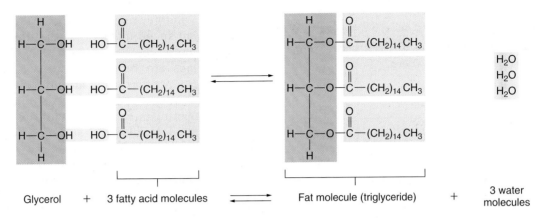

FIGURE 4.2

A glycerol molecule and three fatty acid molecules may join by dehydration synthesis to form a fat molecule (triglyceride). In the reverse reaction, fat is hydrolyzed to three fatty acids and glycerol.

cules, as figure 4.2 shows (read from left to right). The result is three water molecules and a single fat molecule, whose glycerol and fatty acid portions are bound by shared oxygen atoms.

In cells, dehydration synthesis also builds protein molecules by joining amino acid molecules. When two amino acid molecules are united, an —OH from one and an —H from the —NH$_2$ group of another are removed. A water molecule forms, and the amino acid molecules join by a bond between a carbon atom and a nitrogen atom (fig. 4.3; read from left to right). This type of bond, called a *peptide bond,* holds the amino acids together. Two such bound amino acids form a *dipeptide,* and many joined in a chain form a *polypeptide.* Generally, a polypeptide consisting of 100 or more amino acid molecules is called a *protein,* although the boundary between polypeptides and proteins is not precisely defined.

Catabolism

Physiological processes that break down larger molecules into smaller ones constitute catabolism. An example of catabolism is **hydrolysis** (hi-drol'ĭ-sis), which can decompose carbohydrates, lipids, and proteins. A water molecule is used to split these substances into two parts.

The hydrolysis of a disaccharide, for instance, results in two monosaccharide molecules (see fig. 4.1; read from right to left). In this case, the bond between the simple sugars breaks, and the water molecule supplies a hydrogen atom to one sugar molecule and a hydroxyl group to the other. Thus, hydrolysis is the reverse of dehydration synthesis.

Hydrolysis breaks down carbohydrates into monosaccharides; fats into glycerol and fatty acids (see fig. 4.2; read from right to left); proteins into amino acids (see fig. 4.3; read from right to left); and nucleic acids into nucleotides.

Hydrolysis does not occur automatically, even though in the body, water molecules are readily available to provide the necessary —H and —OH. For example, water-soluble substances such as the disaccharide sucrose (table sugar) *dissolve* in a glass of water but do not undergo hydrolysis. Like dehydration synthesis, hydrolysis requires the help of specific enzymes, which are discussed in the next section, Control of Metabolic Reactions.

The reactions of metabolism are often reversible. However, the enzyme that speeds, or catalyzes, an anabolic reaction is often different from that which catalyzes the corresponding catabolic reaction.

Both catabolism and anabolism must be carefully controlled so that the breakdown or energy-releasing reactions occur at rates that are adjusted to the requirements of the building up or energy-utilizing reactions. Any disturbance in this balance is likely to damage or kill cells.

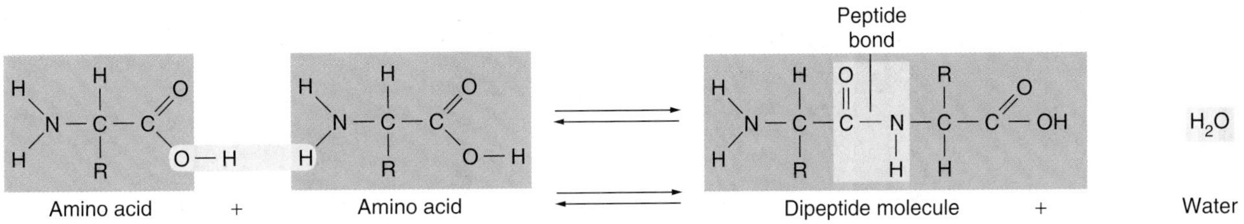

FIGURE 4.3

When two amino acid molecules unite by dehydration synthesis, a peptide bond forms between a carbon atom and a nitrogen atom. In the reverse reaction, a dipeptide is hydrolyzed to two amino acids.

1   What are the general functions of anabolism and catabolism?

2   What substance does the anabolism of monosaccharides form? Of glycerol and fatty acids? Of amino acids?

3   Distinguish between dehydration synthesis and hydrolysis.

Control of Metabolic Reactions

Different kinds of cells may conduct specialized metabolic processes, but all cells perform certain basic reactions, such as the buildup and breakdown of carbohydrates, lipids, proteins, and nucleic acids. These reactions include hundreds of very specific chemical changes that must occur in particular sequences. Enzymes control the rates of these metabolic reactions.

Enzyme Action

Like other chemical reactions, metabolic reactions require energy (activation energy) before they proceed. This is why heat is used to increase the rates of chemical reactions in laboratories. Heat energy increases the rate at which molecules move and the frequency of molecular collisions. These collisions increase the likelihood of interactions among the electrons of the molecules that can form new chemical bonds. The temperature conditions in cells are usually too mild to adequately promote the reactions of life. Enzymes make these reactions possible.

> The antibiotic drug penicillin interferes with enzymes that enable certain bacteria to construct cell walls. As a result, the bacteria die. In this manner, penicillin protects against certain bacterial infections. The drug does not harm human cells because these do not have cell walls.

Enzymes are usually globular proteins that promote specific chemical reactions within cells by lowering the activation energy required to start these reactions. Enzymes can speed metabolic reactions by a factor of a million or more.

Enzymes are required in very small quantities, because as they work, they are not consumed and can, therefore, function repeatedly. Also, each enzyme has *specificity,* acting only on a particular kind of substance, which is called its **substrate** (sub′strāt). For example, the substrate of an enzyme called catalase (found in the peroxisomes of liver and kidney cells) is hydrogen peroxide, a toxic by-product of certain metabolic reactions. This enzyme's only function is to decompose hydrogen peroxide into water and oxygen, helping prevent accumulation of hydrogen peroxide that might damage cells.

Each enzyme must be able to "recognize" its specific substrate. This ability to identify a substrate depends upon the shape of an enzyme molecule. That is, each enzyme's polypeptide chain twists and coils into a unique three-dimensional conformation that fits the special shape of its substrate molecule.

RECONNECT TO CHAPTER 2, PROTEINS, PAGE 54.

> The action of the enzyme catalase is obvious when using hydrogen peroxide to cleanse a wound. Injured cells release catalase, and when hydrogen peroxide contacts them, bubbles of oxygen are set free. The resulting foam removes debris from inaccessible parts of the wound.

During an enzyme-catalyzed reaction, regions of the enzyme molecule called **active sites** temporarily combine with portions of the substrate, forming an enzyme-substrate complex. This interaction strains chemical bonds in the substrate in a way that makes a particular chemical reaction more likely to occur. When it does, the enzyme is released in its original form, able to bind another substrate molecule (fig. 4.4). Note that many

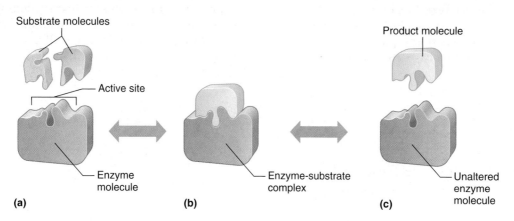

FIGURE 4.4

An enzyme catalyzed reaction. (Many enzyme-catalyzed reactions, as depicted here, are reversible.) In the forward reaction, (*a*) the shape of the substrate molecules fit the shape of the enzyme's active site; (*b*) when the substrate molecules temporarily combine with the enzyme, a chemical reaction occurs; the result is (*c*) a product molecule and an unaltered enzyme.

enzyme reactions are reversible and in some cases the same enzyme catalyzes both directions.

Enzyme catalysis can be summarized as follows:

$$\text{Substrate} + \text{Enzyme} \rightarrow \begin{array}{c}\text{Enzyme-}\\ \text{substrate}\\ \text{complex}\end{array} \rightarrow \begin{array}{c}\text{Product}\\ \text{(changed}\\ \text{substrate)}\end{array} + \begin{array}{c}\text{Enzyme}\\ \text{(unchanged)}\end{array}$$

The speed of an enzyme-catalyzed reaction depends partly on the number of enzyme and substrate molecules in the cell. The reaction occurs more rapidly if the concentration of the enzyme or the concentration of the substrate increases. Also, the efficiency of different kinds of enzymes varies greatly. Thus, some enzymes can process only a few substrate molecules per second, whereas others can handle thousands or nearly a million substrate molecules per second.

Cellular metabolism includes hundreds of different chemical reactions, each controlled by a specific kind of enzyme. Often sequences of enzyme-controlled reactions, called **metabolic pathways,** lead to synthesis or breakdown of particular biochemicals (fig. 4.5). Thus, hundreds of different kinds of enzymes are present in every cell.

Enzyme names are often derived from the names of their substrates, with the suffix *-ase* added. For example, a lipid-splitting enzyme is called a *lipase,* a protein-splitting enzyme is a *protease,* and a starch-(amylum) splitting enzyme is an *amylase.* Similarly, *sucrase* is an enzyme that splits the sugar sucrose, *maltase* splits the sugar maltose, and *lactase* splits the sugar lactose.

Cofactors and Coenzymes

Often an enzyme is inactive until it combines with a non-protein component that either helps the active site attain its appropriate shape or helps bind the enzyme to its substrate. Such a substance, called a **cofactor,** may be an ion of an element, such as copper, iron, or zinc, or it may be a small organic molecule, called a **coenzyme** (ko-en′zīm). Coenzymes are often composed of vitamin molecules or incorporate altered forms of vitamin molecules into their structures. An example of a coenzyme is coenzyme A, involved in cellular respiration, discussed in the next section.

Vitamins are essential organic substances that human cells cannot synthesize (or may not synthesize in sufficient quantities) and therefore must come from the diet. Since vitamins provide coenzymes that can, like enzymes, function again and again, cells require very small quantities of vitamins. Chapter 18 (pp. 705–712) discusses vitamins further.

Factors That Alter Enzymes

Almost all enzymes are proteins, and like other proteins, they can be denatured by exposure to excessive heat, radiation, electricity, certain chemicals, or fluids with extreme pH values. For example, many enzymes become inactive at 45°C, and nearly all of them are denatured at 55°C. Some poisons are chemicals that denature enzymes. Cyanide, for instance, can interfere with respiratory enzymes and damage cells by halting their energy-obtaining processes.

Certain microorganisms, colorfully called "extremophiles," live in conditions of extremely high or low heat, salinity, or pH. Their enzymes have evolved to function under these conditions and are useful in industrial processes that are too harsh to use other enzymes.

1 What is an enzyme?

2 How can an enzyme control the rate of a metabolic reaction?

3 How does an enzyme "recognize" its substrate?

4 What is the role of a cofactor?

5 What factors can denature enzymes?

Energy for Metabolic Reactions

Energy is the capacity to change something; it is the ability to do work. Therefore, we recognize energy by what it can do. Common forms of energy include heat, light, sound, electrical energy, mechanical energy, and chemical energy.

Although energy cannot be created or destroyed, it can be changed from one form to another. An ordinary incandescent light bulb changes electrical energy to heat and light, and an automobile engine changes the chemical energy in gasoline to heat and mechanical energy.

Changes occur in the human body as a characteristic of life—whenever this happens, energy is being transferred. Thus, all metabolic reactions involve energy in some form.

Release of Chemical Energy

Most metabolic processes depend on chemical energy. This form of energy is held in the chemical bonds that link atoms into molecules and is released when these

FIGURE 4.5

A metabolic pathway consists of a series of enzyme-controlled reactions leading to formation of a product. Each new substrate is the product of the previous reaction.

bonds break. Burning a marshmallow over a campfire releases the chemical energy held within the bonds of substances in the marshmallow as heat and light. Similarly, when a marshmallow is eaten, digested, and absorbed, cells "burn" glucose molecules from that marshmallow in a process called **oxidation** (ok"si-da'shun). The energy released by oxidation of glucose is used to promote cellular metabolism. There are obviously some important differences between the oxidation of substances inside cells and the burning of substances outside them.

Burning in nonliving systems (such as starting a fire in a fireplace) usually requires a relatively large amount of energy to begin, and most of the energy released escapes as heat or light. In cells, enzymes initiate oxidation by decreasing the activation energy. Also, by transferring energy to special energy-carrying molecules, cells are able to capture almost half of the energy released in the form of chemical energy. The rest escapes as heat, which helps maintain body temperature.

Cellular respiration is the process that releases energy from molecules such as glucose and makes it available for cellular use. The chemical reactions of cellular respiration must occur in a particular sequence, each one controlled by a different enzyme. Some of these enzymes are in the cell's cytosol, whereas others are in the mitochondria. Such precision of activity suggests that the enzymes are physically positioned in the exact sequence as that of the reactions they control. Indeed, the enzymes responsible for some of the reactions of cellular respiration are located in tiny, stalked particles on the membranes (cristae) within the mitochondria (see chapter 3, p. 73).

1 What is energy?

2 How does cellular oxidation differ from burning?

3 Define cellular respiration.

Cellular Respiration

Cellular respiration occurs in three distinct, yet interconnected, series of reactions: **glycolysis** (gli-kol'ĭ-sis), the **citric acid cycle,** and the **electron transport chain** (oxidative phosphorylation) (fig. 4.6). The products of these reactions include carbon dioxide, water, and energy. Although most of the energy is lost as heat, almost half is captured in a form that the cell can use through the synthesis of **ATP (adenosine triphosphate),** an energy-rich molecule.

Cellular respiration includes aerobic reactions (require oxygen) and anaerobic reactions (do not require oxygen). For each glucose molecule that is decomposed completely by cellular respiration, up to thirty-eight molecules of ATP can be produced. All but two ATP molecules are formed by the aerobic reactions.

ATP Molecules

Each ATP molecule consists of three main parts—an adenine, a ribose, and three phosphates in a chain (fig. 4.7). The third phosphate of ATP is attached by a high-energy bond, and the chemical energy stored in that bond may be quickly transferred to another molecule in a metabolic process. When such an energy transfer occurs, the terminal, high-energy bond of the ATP molecule breaks, releasing its energy. Energy from the breakdown of ATP powers cellular work such as skeletal muscle contraction, active transport across cell membranes, or secretion.

An ATP molecule that loses its terminal phosphate becomes an ADP (adenosine diphosphate) molecule, which has only two phosphates. However, ATP can be resynthesized from an ADP by using energy released from cellular respiration to reattach a phosphate, a process known as **phosphorylation** (fos"fōr-ĭ-la'shun). Thus, as shown in figure 4.8, ATP and ADP molecules shuttle back and forth between the energy-releasing reactions of cellular respiration and the energy-utilizing reactions of the cell.

ATP is not the only kind of energy-carrying molecule within a cell, but it is the primary one. Without enough ATP, cells quickly die.

1 What is meant by anaerobic reactions? Aerobic reactions?

2 What happens to the energy that cellular respiration releases?

3 What are the final products of cellular respiration?

4 What is the function of ATP molecules?

Glycolysis

Both aerobic and anaerobic pathways begin with glycolysis. Literally "the breaking of glucose," glycolysis is a series of ten enzyme-catalyzed reactions that break down the 6-carbon glucose molecule into two 3-carbon pyruvic acid molecules. Glycolysis occurs in the cytosol (see fig. 4.6), and because it does not itself require oxygen, it is sometimes referred to as the *anaerobic phase of cellular respiration.*

Glycolysis can be summarized by three main events (fig. 4.9):

1. First, glucose is phosphorylated by the addition of two phosphate groups, one at each end of the molecule. Although this step requires ATP, it "primes" the molecule for some of the energy-releasing reactions that occur later on.

2. Second, the 6-carbon glucose molecule is split into two 3-carbon molecules.

3. Third, the electron carrier NADH is produced, ATP is synthesized, and two 3-carbon pyruvic acid molecules result. Note that some of the reactions of glycolysis release hydrogen atoms. The electrons of

Glycolysis

1 The 6-carbon sugar glucose is broken down into two 3-carbon pyruvic acid molecules with a net gain of 2 ATP and the release of high energy electrons.

Citric Acid Cycle

2 The 3-carbon pyruvic acids generated by glycolysis enter the mitochondria. Each loses a carbon (generating CO_2) and is combined with a coenzyme to form a 2-carbon acetyl coenzyme A (acetyl CoA). More high energy electrons are released.

3 Each acetyl CoA combines with a 4-carbon oxaloacetic acid to form the 6-carbon citric acid, for which the cycle is named. For each citric acid a series of reactions removes 2 carbons (generating 2 CO_2's), synthesizes 1 ATP and releases more high energy electrons. The figure shows 2 ATP, resulting directly from 2 turns of the cycle per glucose molecule that enters glycolysis.

Electron Transport Chain

4 The high energy electrons still contain most of the chemical energy of the original glucose molecule. Special carrier molecules bring the high energy electrons to a series of enzymes that convert much of the remaining energy to more ATP molecules. The other products are heat and water. The requirement of oxygen in this last step is why the overall process is called aerobic respiration.

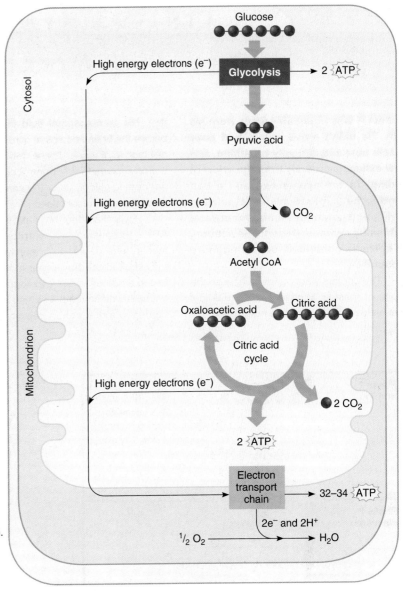

FIGURE 4.6
Glycolysis occurs in the cytosol and does not require oxygen. The aerobic reactions of cellular respiration occur in the mitochondria and only in the presence of oxygen. The products include ATP, heat, CO_2, and water.

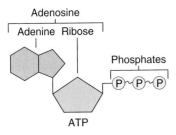

FIGURE 4.7
An ATP molecule consists of an adenine, a ribose, and three phosphates. The wavy lines connecting the last two phosphates represent high-energy chemical bonds. When broken, these bonds provide energy, which the cell uses for metabolic processes.

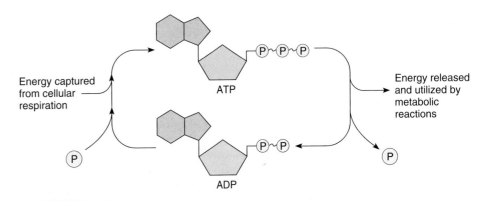

FIGURE 4.8
ATP provides energy for cellular reactions. Cellular respiration generates ATP.

OVERRIDING A BLOCK IN GLYCOLYSIS

Michael P. was noticeably weak from his birth. He didn't move much, had poor muscle tone and difficulty breathing, and grew exhausted merely from the effort of feeding. At the age of two and a half months, he suffered his first seizure, staring and jerking his limbs for several frightening minutes. Despite medication, his seizures continued, occurring more frequently.

The doctors were puzzled because the results of most of Michael's many medical tests were normal—with one notable excep-tion. His cerebrospinal fluid (the fluid that bathes the brain and spinal cord) was unusu-ally high in glucose. These deficiencies told the physicians that Michael's cells were not performing glycolysis or the anaerobic reac-tions of cellular respiration.

Hypothesizing that a profound lack of ATP was causing the symptoms, medical researchers decided to intervene beyond the block in the boy's metabolic pathway, taking a detour to energy production. When Michael was seven and a half months old, he began a diet rich in certain fatty acids.

Within four days, he appeared to be healthy for the very first time! The diet had resumed cellular respiration at the point of acetyl coenzyme A formation by supplying an alternative to glucose. Other children with similar symptoms have since enjoyed spec-tacular recoveries similar to Michael's thanks to the dietary intervention, but doc-tors do not yet know the long-term effects of the therapy. This medical success story, however, illustrates the importance of the energy pathways—and how valuable our understanding of them can be. ∎

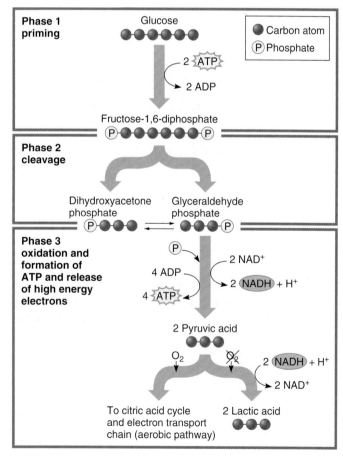

FIGURE 4.9

Glycolysis breaks down glucose in three stages: (1) phosphoryla-tion, (2) splitting, and (3) production of NADH and ATP. Each glu-cose molecule broken down by glycolysis yields a net gain of 2 ATP.

these hydrogen atoms contain much of the energy associated with the chemical bonds of the original glucose molecule. To keep this energy in a form the cell can use, these hydrogen atoms are passed in pairs to molecules of the hydrogen carrier NAD$^+$ (nicotinamide adenine dinucleotide). In this reaction, two of the electrons and one hydrogen nucleus bind to NAD$^+$ to form NADH. The remaining hydrogen nucleus (a hydrogen ion) is released as follows:

$$NAD^+ + 2H \rightarrow NADH + H^+$$

NADH delivers these high-energy electrons to the electron transport chain elsewhere in the mitochondria, where most of the ATP will be synthesized.

ATP is also synthesized directly in glycolysis. After subtracting the two ATP used in the priming step, this gives a net yield of two ATP per molecule of glucose.

Disruptions in glycolysis or the reactions that follow it can devastate health. Clinical Application 4.1 illus-trates how medical sleuths traced one boy's unusual com-bination of symptoms to a block in glycolysis.

Anaerobic Reactions

For glycolysis to continue, NADH + H$^+$ must be able to deliver electrons to the electron transport chain, thus replenishing the cellular supply of NAD$^+$. In the presence of oxygen, this is exactly what happens. Oxygen acts as the final electron acceptor at the end of the electron transport chain, enabling the chain to continue processing electrons and recycling NAD^+.

Under anaerobic conditions, however, the electron transport chain has nowhere to unload its electrons, and it can no longer accept new electrons from NADH. As an alternative, $NADH + H^+$ can give its electrons and hydrogens back to pyruvic acid in a reaction that forms **lactic acid.** Although this regenerates NAD^+, the buildup of lactic acid eventually inhibits glycolysis, and ATP production declines. The lactic acid diffuses into the blood, and when oxygen levels return to normal, the liver converts the lactic acid back into pyruvic acid, which can finally enter the aerobic pathway.

Human muscle cells that are working so strenuously that their production of pyruvic acid exceeds the oxygen supply begin to produce lactic acid. In this condition of "oxygen debt," the muscle cells are forced to utilize solely the anaerobic pathway, which provides fewer ATPs per glucose molecule than does the aerobic reactions of cellular respiration. The accumulation of lactic acid contributes to the feeling of muscle fatigue and cramps. Walking after cramping at the end of a race can make a runner feel better by hastening the depletion of lactic acid.

Lactic acid formation occurs in an interesting variety of circumstances. Coaches measure lactic acid levels in swimmers' and sprinters' blood to assess their physical condition. Lactic acid accumulates to triple the normal levels in the bloodstreams of children who vigorously cry when they are being prepared for surgery, but not in children who are calm and not crying. This suggests that lactic acid formation accompanies stress.

Aerobic Reactions

If enough oxygen is available, the pyruvic acid generated by glycolysis can continue through the aerobic pathways (see fig. 4.6). These reactions include the synthesis of **acetyl coenzyme A** (as'ĕ-til ko-en'zīm A) or acetyl CoA, the citric acid cycle, and the electron transport chain. In addition to carbon dioxide and water, the aerobic reactions themselves yield up to thirty-six ATP molecules per glucose.

This sequence of reactions begins with pyruvic acid produced by glycolysis moving from the cytosol into the mitochondria (fig. 4.10). From each pyruvic acid, enzymes inside the mitochondria remove two hydrogen atoms, a carbon atom, and two oxygen atoms, generating NADH and a CO_2 and leaving a 2-carbon acetic acid. The acetic acid then combines with a molecule of coenzyme A

(derived from the vitamin pantothenic acid) to form acetyl CoA. CoA "carries" the acetic acid into the citric acid cycle.

Citric Acid Cycle

The citric acid cycle begins when a 2-carbon acetyl CoA molecule combines with a 4-carbon oxaloacetic acid molecule to form the 6-carbon citric acid and CoA (fig. 4.10). The CoA can be used again to combine with acetic acid to form acetyl CoA. The citric acid is changed through a series of reactions back into oxaloacetic acid. The cycle repeats as long as oxygen and pyruvic acid are supplied to the mitochondrion.

The citric acid cycle has three important consequences:

1. One ATP is produced directly for each citric acid molecule that goes through the cycle.

2. For each citric acid molecule, eight hydrogen atoms with high-energy electrons are transferred to the hydrogen carriers NAD^+ and the related FAD (flavine adenine dinucleotide):

$$NAD^+ + 2H \rightarrow NADH + H^+$$

$$FAD + 2H \rightarrow FADH_2$$

3. As the 6-carbon citric acid reacts to form the 4-carbon oxaloacetic acid, two carbon dioxide molecules are produced.

The carbon dioxide produced by the formation of acetyl CoA and in the citric acid cycle dissolves in the cytoplasm, diffuses from the cell, and enters the bloodstream. Eventually, the respiratory system excretes the carbon dioxide.

Electron Transport Chain

The hydrogen and high-energy electron carriers (NADH and $FADH_2$) generated by glycolysis and the citric acid cycle now hold most of the energy contained in the original glucose molecule. In order to couple this energy to ATP synthesis, the high-energy electrons are handed off to the electron transport chain, which is a series of enzyme complexes that carry and pass electrons along from one to another. These complexes dot the folds of the inner mitochondrial membranes (see chapter 3, p. 73), which, if stretched out, may be forty-five times as long as the cell membrane in some cells. The electron transport chain passes each electron along, gradually lowering the electron's energy level and transferring that energy to ATP synthase, an enzyme complex that uses this energy to phosphorylate ADP to form ATP (fig. 4.11). These reactions, known as oxidation/reduction reactions, are described further in Appendix C, pages 975–976.

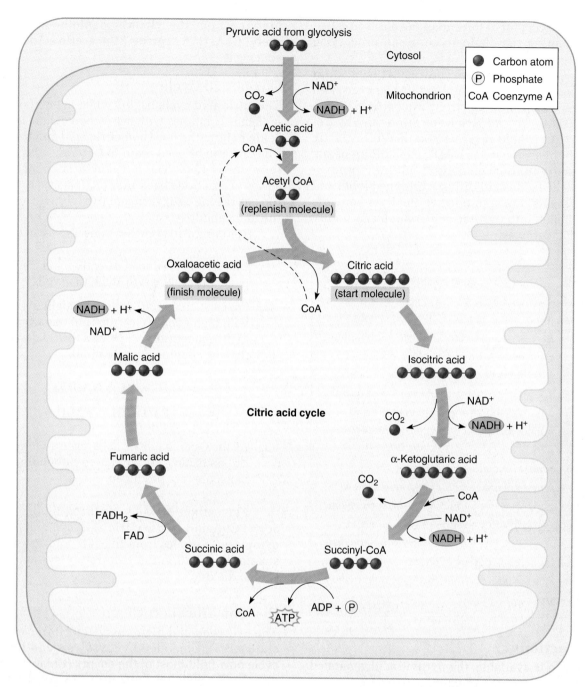

FIGURE 4.10

For each turn of the citric acid cycle (two "turns" or citric acids per glucose), one ATP is produced directly, eight hydrogens with high-energy electrons are released, and two CO_2 molecules are produced.

Neither glycolysis nor the citric acid cycle uses oxygen directly, although they are part of the aerobic metabolism of glucose. Instead, the final enzyme of the electron transport chain gives up a pair of electrons that combine with two hydrogen ions (provided by the hydrogen carriers) and an atom of oxygen to form a water molecule:

$$2e^- + 2H^+ + 1/2\ O_2 \rightarrow H_2O$$

Thus, oxygen is the final electron "carrier." In the absence of oxygen, electrons cannot continue to pass through the electron transport chain, and the aerobic reactions of cellular respiration grind to a halt.

Figure 4.12 summarizes the steps in glucose metabolism. More detailed descriptions of glycolysis and the citric acid cycle may be found in Appendix C, A Closer Look at Cellular Respiration, pages 974–977.

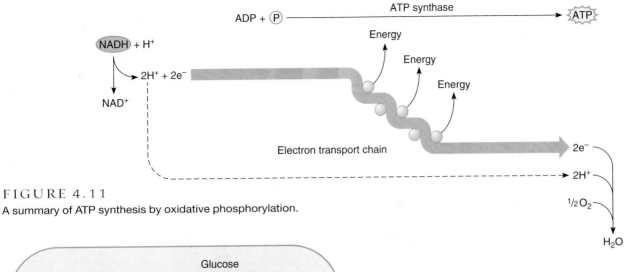

FIGURE 4.11
A summary of ATP synthesis by oxidative phosphorylation.

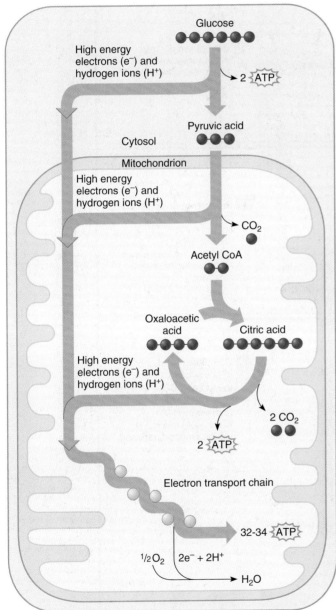

FIGURE 4.12
An overview of aerobic respiration, including the net yield of ATP at each step per molecule of glucose.

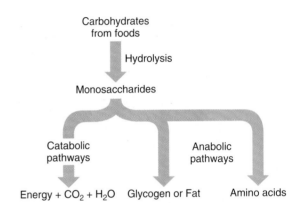

FIGURE 4.13
Hydrolysis breaks down carbohydrates from foods into monosaccharides. The resulting molecules may enter catabolic pathways and be used as energy sources, or they may enter anabolic pathways and be stored as glycogen or fat or converted to amino acids.

Carbohydrate Storage

Metabolic pathways are usually interconnected in ways that enable certain molecules to enter more than one pathway. For example, carbohydrate molecules from foods may enter catabolic pathways and be used to supply energy, or they may enter anabolic pathways and be stored or be converted to some of the twenty different amino acids (fig. 4.13).

Excess glucose in cells may enter anabolic carbohydrate pathways and be linked into storage forms such as glycogen. Most cells can produce glycogen, but liver and muscle cells store the greatest amounts. Following a meal, when blood glucose concentration is relatively high, liver cells obtain glucose from the blood and synthesize glycogen. Between meals, when blood glucose concentration is lower, the reaction is reversed, and glucose is released into the blood. This mechanism ensures that cells throughout the body have a continual supply of glucose to support cellular respiration.

Glucose can also react to form fat molecules, which are later deposited in adipose tissues. This happens when a person takes in more carbohydrates than can be stored as glycogen or are required for normal activities. Because the body has an almost unlimited capacity to perform this type of anabolism, overeating, even if mostly carbohydrates, can cause weight gain (overweight).

This section has considered the metabolism of glucose, although lipids and proteins can also be broken down to release energy for ATP synthesis. In all three cases, the final process is aerobic respiration, and the most common entry point is into the citric acid cycle as acetyl CoA (fig. 4.14). These pathways are described in detail in chapter 18 (pp. 697–699).

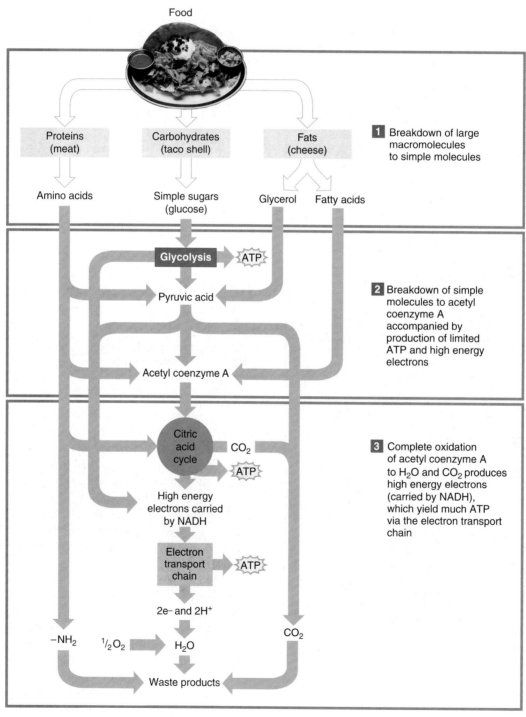

FIGURE 4.14

A summary of the breakdown (catabolism) of proteins, carbohydrates, and fats.

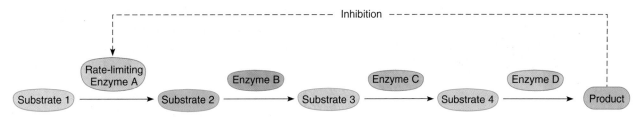

FIGURE 4.15

A negative feedback mechanism may control a rate-limiting enzyme in a metabolic pathway. The product of the pathway inhibits the enzyme.

Regulation of Metabolic Pathways

The rate at which a metabolic pathway functions is often determined by a regulatory enzyme that catalyzes one of its steps. The number of molecules of such a regulatory enzyme is limited. Consequently, the enzymes can become saturated when the substrate concentration exceeds a certain level. Once this happens, increasing the substrate concentration no longer affects the reaction rate.

As a rule, such a **rate-limiting enzyme** is the first enzyme in a series. This position is important because some intermediate substance of the pathway might accumulate if an enzyme occupying another location in the sequence were rate limiting.

Often the product of a metabolic pathway inhibits the rate-limiting regulatory enzyme. This type of control is an example of negative feedback. Accumulating product inhibits the pathway, and synthesis of the product falls. When the concentration of product decreases, the inhibition lifts, and more product is synthesized. In this way, a single enzyme can control a whole pathway, stabilizing the rate of production (fig. 4.15).

1   What is a rate-limiting enzyme?

2   How can negative feedback control a metabolic pathway?

Nucleic Acids and Protein Synthesis

Because enzymes control the metabolic pathways that enable cells to survive, cells must have information for producing these specialized proteins. Many other proteins are important in physiology as well, such as blood proteins, the proteins that form muscle and connective tissues, and the antibodies that protect against infection. The information that instructs a cell to synthesize a particular protein is held in the sequence of building blocks of **deoxyribonucleic acid (DNA),** the genetic material. As we will see later in this chapter, the correspondence between a unit of DNA information and a particular amino acid constitutes the **genetic code** (je-net′ik kōd).

Genetic Information

Children resemble their parents because of inherited traits, but what actually passes from parents to a child is genetic information, in the form of DNA molecules from the parents' sex cells. Chromosomes are cell structures that carry the DNA. As an offspring develops, mitosis passes the information on to new cells. Genetic information "tells" cells how to construct a great variety of protein molecules, each with a specific function. The portion of a DNA molecule that contains the genetic information for making a particular protein is called a **gene** (jēn). Because enzymes control synthesis reactions, all four groups of organic molecules—proteins, carbohydrates, lipids, and nucleic acids—require genetic instructions.

The complete set of genetic instructions in a cell constitutes the **genome.** Viral genomes have been sequenced since the 1970s. Researchers began to decipher genome sequences of bacteria in 1995, working up through more complex organisms to humans in 2000. Not all of the human genome encodes protein—functions of many DNA sequences are not known. Chapter 24 (p. 921) discusses the human genome.

Recall from chapter 2 (p. 54) that nucleotides are the building blocks of nucleic acids. A nucleotide consists of a 5-carbon sugar (ribose or deoxyribose), a phosphate group, and one of several organic, nitrogen-containing (nitrogenous) bases (fig. 4.16). DNA and RNA nucleotides form long strands (polynucleotide chains) by alternately joining their sugar and phosphate portions, which provides a "backbone" structure (fig. 4.17).

A DNA molecule consists of two polynucleotide chains. The nitrogenous bases project from the sugar-phosphate backbone of one strand and bind, or pair, by hydrogen bonds to the nitrogenous bases of the second strand (fig. 4.18). The resulting structure is somewhat like a ladder, in which the rails represent the sugar and phosphate backbones of the two strands and the rungs represent the paired nitrogenous bases. Notice that the sugars forming the two backbones point in opposite directions. For this reason, the two strands are called *antiparallel.*

A DNA molecule is sleek and symmetrical because the bases pair in only two combinations, maintaining a

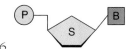

FIGURE 4.16

Each nucleotide of a nucleic acid consists of a 5-carbon sugar (S); a phosphate group (P); and an organic, nitrogenous base (B).

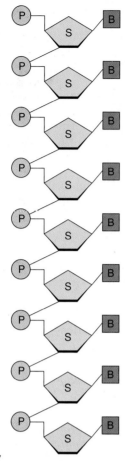

FIGURE 4.17

A polynucleotide chain consists of nucleotides connected by a sugar-phosphate backbone.

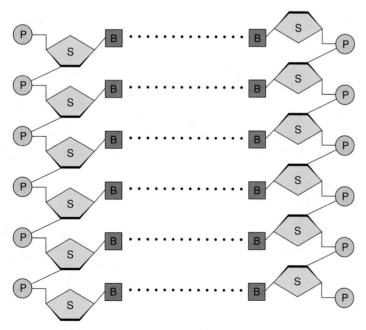

FIGURE 4.18

DNA consists of two polynucleotide chains. Hydrogen bonds (dotted lines) hold the nitrogenous bases of one strand to the nitrogenous bases of the second strand. Note that the sugars point in opposite directions—that is, the strands are antiparallel.

constant width of the overall structure. In a DNA nucleotide, the base may be one of four types: adenine, thymine, cytosine, or guanine. Adenine (A), a two-ring structure, binds to thymine (T), a single-ring structure. Guanine (G), a two-ring structure, binds to cytosine (C), a single-ring structure. These pairs—A with T, and G with C—are called **complementary base pairs** (fig. 4.19a). Because of this phenomenon, the sequence of one DNA strand can always be derived from the other by following the "base-pairing rules." For example, if the sequence of one strand of the DNA molecule is G, A, C, T, then the complementary strand's sequence is C, T, G, A. (The sequence of bases in one of these strands encodes the instructions for making a protein.)

The double-stranded DNA molecule twists to form a double helix, resembling a spiral staircase (fig. 4.19b). An individual DNA molecule may be several million base pairs long. In the nucleus, DNA is wound around complexes of proteins called histones to form chromatin (fig. 4.19b). Investigators can use DNA sequences to identify individuals (From Science to Technology 4.1). More detailed structures of DNA and its nucleotides are shown in Appendix D, pages 978–979.

Genetic Code

Genetic information specifies the correct sequence of the amino acids in a polypeptide chain. Each of the twenty different types of amino acids is represented in a DNA molecule by a triplet code, consisting of sequences of three nucleotides. That is, the sequence C, G, T in a DNA strand represents one kind of amino acid; the sequence G, C, A represents another kind; and T, T, A still another kind. Other sequences encode instructions for beginning or ending the synthesis of a protein molecule.

The sequence of nucleotides in a DNA molecule dictates the sequence of amino acids of a particular protein molecule and indicates how to start or stop the protein's synthesis. This method of storing information for protein synthesis is the genetic code. However, because DNA molecules are located in the nucleus and protein synthesis occurs in the cytoplasm, and because the cell must keep a permanent copy of the genetic instructions, the genetic information must somehow get from the nucleus into the cytoplasm for the cell to use it. RNA molecules accomplish this transfer of information.

The genetic code is said to be universal because all species on earth use the same DNA base triplets to specify the same amino acids. Researchers deciphered the code in the 1960s. When the media mentions an individual's genetic code or that scientists are currently breaking the code, what they really are referring to is the sequence of DNA bases comprising a certain gene or genome—not the genetic code (the correspondence between DNA triplet and amino acid).

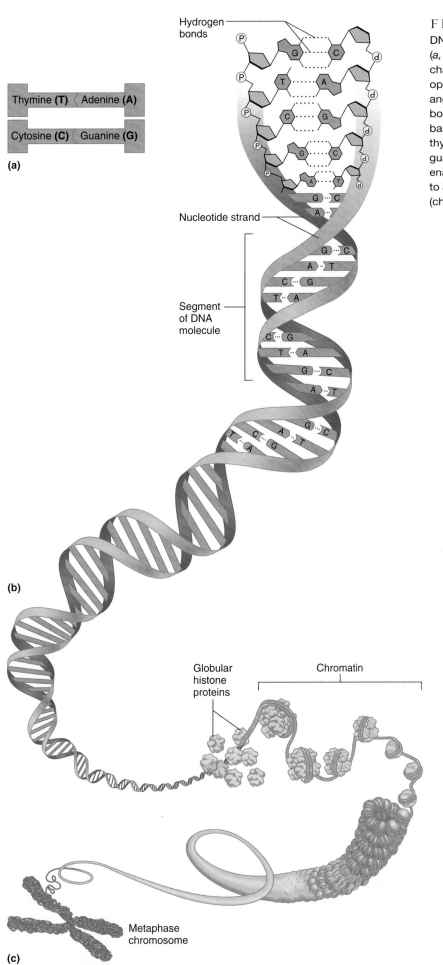

(a)

Thymine **(T)** — Adenine **(A)**

Cytosine **(C)** — Guanine **(G)**

Hydrogen bonds

Nucleotide strand

Segment of DNA molecule

(b)

Globular histone proteins

Chromatin

Metaphase chromosome

(c)

FIGURE 4.19

DNA and chromosome structure. (*a, b*) The two polynucleotide chains of a DNA molecule point in opposite directions (antiparallel) and are held together by hydrogen bonds between complementary base pairs—adenine (A) bonds to thymine (T); cytosine (C) bonds to guanine (G). (*c*) Histone proteins enable the long double helix to assume a compact form (chromosome).

DNA MAKES HISTORY

In July 1918, the last tsar of Russia, Nicholas II, and his family, the Romanovs, met gruesome deaths at the hands of Bolsheviks in a town in the Ural Mountains of central Russia. Captors led the tsar, tsarina, four daughters and one son, plus the family physician and three servants, to a cellar and shot them, bayoneting those who did not die quickly. The executioners stripped the bodies and loaded them onto a truck, which would take them to a mine shaft where they would be left. But the truck broke down, and the bodies were instead placed in a shallow grave, then damaged with sulfuric acid so that they could not be identified.

In July 1991, two Russian amateur historians found the grave, and based on its location, alerted the government that the long-sought bodies of the Romanov family might have been found. An official forensic examination soon determined that the skeletons were from nine individuals. The sizes of the skeletons indicated that three were children. The porcelain, platinum, and gold in the teeth of some of the skeletons suggested that they were royalty. The facial bones were so decomposed from the acid that conventional forensic tests were not possible. But one very valuable type of evidence remained—DNA. Forensic scientists extracted DNA from bone cells and mass-produced it for study using a technique called the polymerase chain reaction (PCR) described in Clinical Application 4.2.

By identifying DNA sequences specific to the Y chromosome, which is found only in males, the DNA detectives could tell which of the skeletons were from males. Then they delved into the DNA in mitochondria. Because these organelles pass primarily from mother to offspring, identifying a mitochondrial DNA pattern in a woman and children would establish her as their mother. This was indeed so for one of the women (with impressive dental work) and the children.

But a mother, her children, and some companions does not a royal family make. The researchers had to connect the skeletons to the royal family. Again they turned to DNA. Genetic material from one of the male skeletons shared certain rare DNA sequences with DNA from living descendants of the Romanovs. This man also had aristocratic dental work and shared DNA sequences with the children! The mystery of the fate of the Romanovs was apparently solved, thanks to the help of DNA.

DNA profiling is a general term for several techniques that are increasingly being used to compare the genetic material of individuals, to confirm or rule out relationships—such as blood relatedness, presence at a crime scene, or to identify accident victims. Recent applications of DNA profiling have exonerated several jailed innocent people, and identified Thomas Jefferson as a possible father of a son of his slave Sally Hemings. DNA profiling applications aren't confined to humans. It was used, for example, to identify the two strains of cultivated grapes that can be bred to yield sixteen popular varieties of wine grapes. ■

RNA Molecules

RNA (ribonucleic acid) molecules differ from DNA molecules in several ways. RNA molecules are single-stranded, and their nucleotides contain ribose rather than deoxyribose sugar. Like DNA, RNA nucleotides each contain one of four organic bases, but whereas adenine, cytosine, and guanine nucleotides occur in both DNA and RNA, thymine nucleotides are found only in DNA. In place of thymine nucleotides, RNA molecules contain uracil (U) nucleotides (fig. 4.20 and Appendix D, p. 979).

The first step in the delivery of information from the nucleus to the cytoplasm is the synthesis of a type of RNA called **messenger RNA** (mRNA). In messenger RNA synthesis, RNA nucleotides form complementary base pairs with a section of a strand of DNA that encodes a particular protein. However, just as the words in a sentence must be read in the correct order to make sense, the base sequence of a strand of DNA must be "read" in the correct direction. Furthermore, only one of the two antiparallel strands of DNA contains the genetic message. An enzyme called RNA polymerase determines the correct DNA strand and the right direction for RNA synthesis (fig. 4.21).

In mRNA synthesis, RNA polymerase binds to a promoter, which is a DNA base sequence that begins a gene. As a result of RNA polymerase binding, a section of the double-stranded DNA molecule unwinds and pulls apart, exposing a portion of the gene. RNA polymerase then moves along the strand, exposing other portions of the gene. At the same time, a molecule of mRNA forms as RNA nucleotides complementary to those along the DNA strand are strung together. For example, if the sequence of DNA bases is TACCCGAGG, the complementary bases in the developing mRNA molecule will be AUGGGCUCC, as figure 4.22 shows. (The other strand of DNA is not used in this process, but it is important in DNA replication, discussed later in the chapter.)

RNA polymerase continues to move along the DNA strand, exposing portions of the gene, until it reaches a spe-

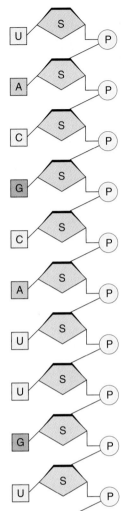

FIGURE 4.20

RNA differs from DNA in that it is single-stranded, contains ribose rather than deoxyribose, and has uracil (U) rather than thymine (T) as one of its four bases.

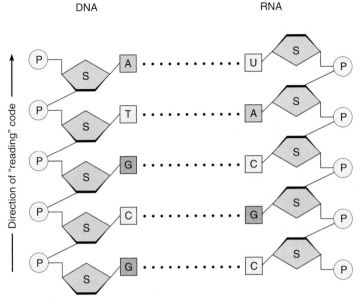

FIGURE 4.21

When an RNA molecule is synthesized beside a strand of DNA, complementary nucleotides bond as in a double-stranded DNA molecule, with one exception: RNA contains uracil nucleotides (U) in place of thymine nucleotides (T).

cial DNA base sequence (termination signal) that signals the end of the gene. At this point, the RNA polymerase releases the newly formed mRNA molecule and leaves the DNA. The DNA then rewinds and assumes its previous double helix structure. This process of copying DNA information into the structure of an mRNA molecule is called **transcription** (tranz-krip′-shun). Messenger RNA molecules can be hundreds or even thousands of nucleotides long.

Each amino acid in the protein to be synthesized was originally represented by a series of three bases in DNA. Those amino acids, in the proper order, are now represented by a series of three base sequences, called **codons,** (ko′donz) in mRNA. To complete the process of protein synthesis, mRNA must leave the nucleus and associate with a ribosome. There, the series of codons on mRNA are translated from the "language" of nucleic acids to the "language" of amino acids. This process is fittingly called **translation** (see fig. 4.22). Note that sixty-four possible DNA base triplets encode twenty different amino acids. This means that more than one codon can specify the same amino acid, a point we will return to soon. Table 4.1 compares DNA and RNA molecules.

TABLE 4.1	A Comparison of DNA and RNA Molecules	
	DNA	**RNA**
Main location	Part of chromosomes, in nucleus	Cytoplasm
5-carbon sugar	Deoxyribose	Ribose
Basic molecular structure	Double-stranded	Single-stranded
Organic bases included	Cytosine, guanine, adenine, thymine	Cytosine, guanine, adenine, uracil
Major functions	Contains genetic code for protein synthesis, replicates prior to mitosis	Messenger RNA carries transcribed DNA information to cytoplasm and acts as template for synthesis of protein molecules; transfer RNA carries amino acids to messenger RNA; ribosomal RNA provides structure and enzyme activity for ribosomes

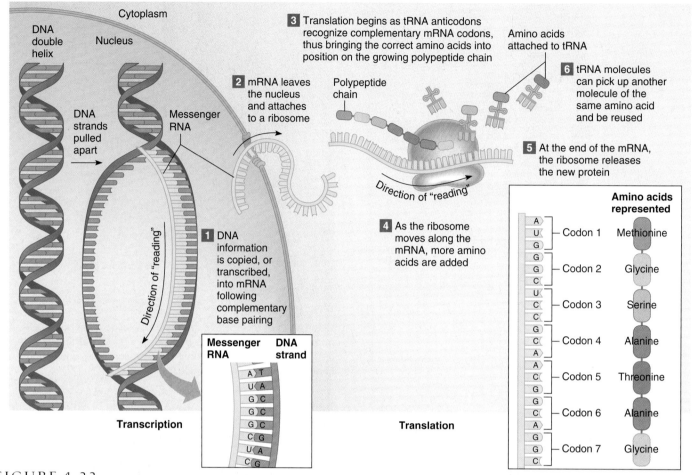

FIGURE 4.22

DNA information is transcribed into mRNA, which, in turn, is translated into a sequence of amino acids.

Until the early 1980s, all enzymes were thought to be proteins. Then, researchers found that a bit of RNA that they thought was contaminating a reaction in which RNA molecules are shortened actually contributed the enzymatic activity. The RNA enzymes were named "ribozymes." Because certain RNA molecules can carry information as well as function as enzymes—two biologically important properties—they may have been a bridge between chemicals and the earliest cell-like assemblages on earth long ago.

Protein Synthesis

Just as you have the opportunity to express yourself if you are called on in class, a gene that is transcribed and translated into a protein is said to be *expressed*. The proteins that result determine the function a cell performs in the body. Gene expression is the basis for cellular differentiation, described in chapter 3 (p. 92).

Synthesizing a protein molecule requires that the correct amino acid building blocks be present in the cytoplasm. Furthermore, these amino acids must align in the proper sequence along a strand of messenger RNA. A second kind of RNA molecule, synthesized in the nucleus and called **transfer RNA** (tRNA), aligns amino acids in a way that enables them to bond. A transfer RNA molecule consists of only seventy to eighty nucleotides and has a complex three-dimensional shape. The two ends of the tRNA molecule are most important for the "connector" function (fig. 4.22).

At one end, each transfer RNA molecule has a specific binding site for a particular amino acid. There is at least one type of transfer RNA molecule for each of the twenty amino acids. Before the transfer RNA can pick up its amino acid, the amino acid must be activated. Special enzymes catalyze this step. ATP provides the energy to form a bond between the amino acid and its transfer RNA molecule (fig. 4.23).

The other end of each transfer RNA molecule includes a region called the **anticodon** that contains three nucleotides in a particular sequence unique to that type of transfer RNA. These nucleotides bond only to a specific complementary mRNA codon. In this way, the appropriate transfer RNA carries its amino acid to the correct place in the sequence, as prescribed by the mRNA (fig. 4.23).

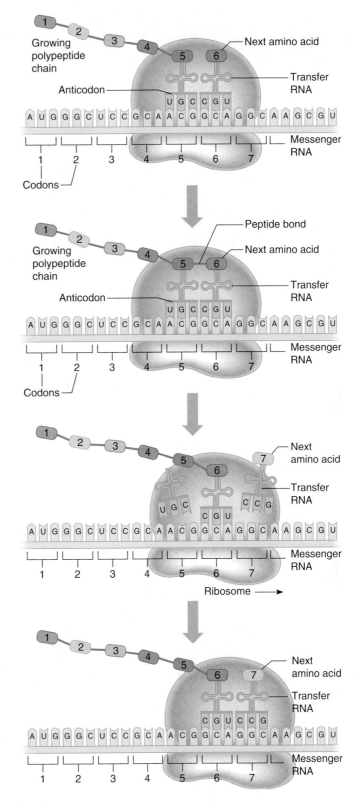

1 The transfer RNA molecule for the last amino acid added holds the growing polypeptide chain and is attached to its complementary codon on mRNA .

2 A second tRNA binds complementarily to the next codon, and in doing so brings the next amino acid into position on the ribosome. A peptide bond forms, linking the new amino acid to the growing polypeptide chain.

3 The tRNA molecule that brought the last amino acid to the ribosome is released to the cytoplasm, and will be used again. The ribosome moves to a new position at the next codon on mRNA.

4 A new tRNA complementary to the next codon on mRNA brings the next amino acid to be added to the growing polypeptide chain.

FIGURE 4.23
Protein synthesis occurs on ribosomes.

Although only twenty types of amino acids need be encoded, four bases can combine in triplets sixty-four different ways, so there are sixty-four different codons possible, and all of them occur in mRNA (table 4.2). Three of these codons do not have a corresponding transfer RNA.

They provide a "stop" signal, indicating the end of protein synthesis, much like the period at the end of this sentence. Sixty-one different transfer RNAs are specific for the remaining sixty-one codons, which means that more than one type of tRNA can correspond to the same amino

TABLE 4.2 Codons (mRNA Three Base Sequences)

First Letter	Second Letter: U		C		A		G		Third Letter
	U		**C**		**A**		**G**		
U	UUU	phenylalanine (phe)	UCU	serine (ser)	UAU	tyrosine (tyr)	UGU	cysteine (cys)	**U**
	UUC		UCC		UAC		UGC		**C**
	UUA	leucine (leu)	UCA		UAA	STOP	UGA	STOP	**A**
	UUG		UCG		UAG	STOP	UGG	tryptophan (trp)	**G**
C	CUU	leucine (leu)	CCU	proline (pro)	CAU	histidine (his)	CGU	arginine (arg)	**U**
	CUC		CCC		CAC		CGC		**C**
	CUA		CCA		CAA	glutamine (gln)	CGA		**A**
	CUG		CCG		CAG		CGG		**G**
A	AUU	isoleucine (ilu)	ACU	threonine (thr)	AAU	asparagine (asn)	AGU	serine (ser)	**U**
	AUC		ACC		AAC		AGC		**C**
	AUA		ACA		AAA	lysine (lys)	AGA	arginine (arg)	**A**
	AUG	START methionine (met)	ACG		AAG		AGG		**G**
G	GUU	valine (val)	GCU	alanine (ala)	GAU	aspartic acid (asp)	GGU	glycine (gly)	**U**
	GUC		GCC		GAC		GGC		**C**
	GUA		GCA		GAA	glutamic acid (glu)	GGA		**A**
	GUG		GCG		GAG		GGG		**G**

acid type. Because a given amino acid can be specified by more than one codon, the genetic code is said to be "degenerate." However, each type of tRNA can bind only its one particular amino acid, so the instructions are precise, and the corresponding codon will code only for that amino acid.

The binding of tRNA and mRNA occurs in close association with a ribosome. A ribosome is a tiny particle of two unequal-sized subunits composed of **ribosomal RNA** (rRNA) and protein. The smaller subunit of a ribosome binds to a molecule of messenger RNA near the codon at the beginning of the messenger strand. This action allows a transfer RNA molecule with the complementary anticodon to bring the amino acid it carries into position and temporarily join to the ribosome. A second transfer RNA molecule, complementary to the second codon on mRNA, then binds (with its activated amino acid) to an adjacent site on the ribosome. The first transfer RNA molecule then releases its amino acid, providing the energy for a peptide bond to form between the two amino acids (fig. 4.23). This process repeats again and again as the ribosome moves along the messenger RNA, adding amino acids one at a time to the developing polypeptide molecule. The enzymatic activity necessary for bonding of the amino acids comes from ribosomal proteins and some RNA molecules (ribozymes) in the larger subunit of the ribosome. This subunit also holds the growing chain of amino acids.

A molecule of messenger RNA usually associates with several ribosomes at the same time. Thus, several copies of that protein, each in a different stage of formation, may be present at any given moment.

As the polypeptide forms, proteins called *chaperones* fold it into its unique shape, and when the process is completed, the polypeptide is released as a separate functional molecule. The transfer RNA molecules, ribosomes, mRNA, and the enzymes can function repeatedly in protein synthesis.

ATP molecules provide the energy for protein synthesis. Because a protein may consist of many hundreds of amino acids and the energy from three ATP molecules is required to link each amino acid to the growing chain, a large fraction of a cell's energy supply supports protein synthesis. Table 4.3 summarizes protein synthesis.

The number of molecules of a particular protein that a cell synthesizes is generally proportional to the number of the corresponding messenger RNA molecules present. The rate at which messenger RNA is transcribed from DNA in the nucleus and the rate at which enzymes (ribonucleases) destroy the messenger RNA in the cytoplasm therefore control protein synthesis.

Proteins called *transcription factors* activate certain genes, thereby controlling which proteins a cell produces and how many copies form. A connective tissue cell might have many messenger RNAs representing genes that encode the protein collagen; a muscle cell would have abundant messenger RNAs encoding muscle proteins. Extracellular signals such as hormones and growth factors activate transcription factors.

1 What is the function of DNA?

2 How is information carried from the nucleus to the cytoplasm?

3 How are protein molecules synthesized?

TABLE 4.3 Protein Synthesis

Transcription (Within the Nucleus)

1. RNA polymerase binds to the base sequence of a gene.
2. This enzyme unwinds a portion of the DNA molecule, exposing part of the gene.
3. RNA polymerase moves along one strand of the exposed gene and catalyzes synthesis of an mRNA molecule, whose nucleotides are complementary to those of the strand of the gene.
4. When RNA polymerase reaches the end of the gene, the newly formed mRNA molecule is released.
5. The DNA molecule rewinds and closes the double helix.
6. The mRNA molecule passes through a pore in the nuclear envelope and enters the cytoplasm.

Translation (Within the Cytoplasm)

1. A ribosome binds to the mRNA molecule near the codon at the beginning of the messenger strand.
2. A tRNA molecule that has the complementary anticodon brings its amino acid to the ribosome.
3. A second tRNA brings the next amino acid to the ribosome.
4. A peptide bond forms between the two amino acids, and the first tRNA is released.
5. This process is repeated for each codon in the mRNA sequence as the ribosome moves along its length, forming a chain of amino acids.
6. As the chain of amino acids grows, it folds, with the help of chaperone proteins, into the unique conformation of a functional protein molecule.
7. The completed protein molecule (polypeptide) is released. The mRNA molecule, ribosome, and tRNA molecules are recycled.

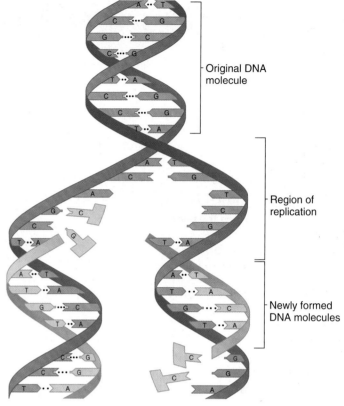

FIGURE 4.24

When a DNA molecule replicates, its original strands separate locally. A new strand of complementary nucleotides forms along each original strand.

Some antibiotic drugs fight infection by interfering with bacterial protein synthesis, RNA transcription, or DNA replication. Rifampin is a drug that blocks bacterial transcription by binding to RNA polymerase, preventing the gene's message from being transmitted. Streptomycin is an antibiotic that binds a bacterium's ribosomal subunits, braking protein synthesis to a halt. Quinolone blocks an enzyme that unwinds bacterial DNA, preventing both transcription and DNA replication. Humans have different ribosomal subunits and transcription and replication enzymes than bacteria, so the drugs do not affect these processes in us.

DNA Replication

When a cell divides, each newly formed cell must have a copy of the original cell's genetic information (DNA) so it will be able to synthesize the proteins necessary to build cellular parts and carry on metabolism. DNA **replication** (re"pli-ka'shun) is the process that creates an exact copy of a DNA molecule. It occurs during interphase of the cell cycle.

As replication begins, hydrogen bonds break between the complementary base pairs of the double strands comprising the DNA molecule. Then the double-stranded structure unwinds and pulls apart, exposing unpaired nucleotide bases. New nucleotides pair with the exposed bases, forming hydrogen bonds. An enzyme, DNA polymerase, catalyzes this base pairing. Enzymes then knit together the new sugar-phosphate backbone. In this way, a new strand of complementary nucleotides extends along each of the old (original) strands. Two complete DNA molecules result, each with one new and one original strand (fig. 4.24). During mitosis, the two DNA molecules that form the two chromatids of each of the chromosomes separate so that one of these DNA molecules passes to each of the new cells.

From Science to Technology 4.2 discusses the polymerase chain reaction (PCR), a method for mass-producing, or amplifying, genes. PCR has revolutionized biomedical science.

Changes in Genetic Information

The amount of genetic information held within a set of human chromosomes is enormous, equal to twenty sets of *Encyclopaedia Britannica.* Because each of the trillions of cells in an adult body results from mitosis

GENE AMPLIFICATION

The polymerase chain reaction (PCR) is a procedure that borrows a cell's machinery for DNA replication, allowing researchers to make many copies of a gene of interest. Starting materials are

- **two types of short DNA pieces known to bracket the gene of interest, called primers**
- **a large supply of DNA bases**
- **the enzymes that replicate DNA**

A simple test procedure rapidly builds up copies of the gene. Here's how it works.

In the first step of PCR, heat is used to separate the two strands of the target DNA—such as bacterial DNA in a body fluid sample from a person who has symptoms of an infection. Next, the temperature is lowered, and the two short DNA primers are added. The primers bind by complementary base pairing to the separated target strands. In the third step, DNA polymerase and bases are added. The DNA polymerase adds bases to the primers and builds a sequence complementary to the target sequence. The newly synthesized strands then act as templates in the next round of replication, which is immediately initiated by raising the temperature. All of this is done in an automated device called a thermal cycler that controls the key temperature changes.

The pieces of DNA accumulate geometrically. The number of amplified pieces of DNA equals 2^n where n equals the number of temperature cycles. After just twenty cycles, 1 million copies of the original sequence are in the test tube. Table 4A lists some diverse applications of PCR.

PCR's greatest strength is that it works on crude samples of rare and short DNA sequences, such as a bit of brain tissue on the bumper of a car, which in one criminal case led to identification of a missing person. PCR's greatest weakness, ironically, is its exquisite sensitivity. A blood sample submitted for diagnosis of an infection contaminated by leftover DNA from a previous run, or a stray eyelash dropped from the person running the reaction, can yield a false positive result. The technique is also limited in that a user must know the sequence to be amplified and that mutations can sometimes occur in the amplified DNA that are not present in the source DNA. ■

TABLE 4A	PCR Applications
PCR Has Been Used to Amplify:	

Genetic material from HIV in a human blood sample when infection has been so recent that antibodies are not yet detectable.

A bit of DNA in a preserved quagga (a relative of the zebra) and a marsupial wolf, which are recently extinct animals.

DNA in sperm cells found in the body of a rape victim so that specific sequences could be compared to those of a crime suspect.

Genes from microorganisms that cannot be grown or maintained in culture for study.

Mitochondrial DNA from various modern human populations. Comparisons of mitochondrial DNA sequences indicate that *Homo sapiens* originated in Africa, supporting fossil evidence.

DNA from the brain of a 7,000-year-old human mummy, which indicated that native Americans were not the only people to dwell in North America long ago.

DNA sequences unique to moose in hamburger meat, proving that illegal moose poaching had occurred.

DNA sequences in maggots in a decomposing human corpse, enabling forensic scientists to determine the time of death.

DNA in deteriorated road kills and carcasses washed ashore, to identify locally threatened species.

DNA in products illegally made from endangered species, such as powdered rhinoceros horn, sold as an aphrodisiac.

DNA sequences in animals that are unique to the bacteria that cause Lyme disease, providing clues to how the disease is transmitted.

DNA from genetically altered microbes that are released in field tests, to follow their dispersion.

DNA from a cell of an eight-celled human preembryo, to diagnose cystic fibrosis.

Y chromosome-specific DNA from a human egg fertilized in the laboratory to determine the sex.

A papilloma virus DNA sequence present in, and possibly causing, an eye cancer.

DNA from remains of journalist Daniel Pearl, who was beheaded in Pakistan.

(except for egg and sperm), genetic information had to be replicated many times and with a high degree of accuracy. DNA can peruse itself for errors and correct them, a process termed DNA repair. Still, occasionally a replication mistake occurs or DNA is damaged, altering the genetic information. Such a change in DNA is called a **mutation** (mu-ta′shun).

Some mutations can cause devastating medical conditions; occasionally, a mutation can confer an advantage. For example, up to 1% of the individuals of some populations have mutations that render their cells unable to become infected with HIV. These lucky people, thanks to their mutation, cannot contract AIDS. The vignette to chapter 3 (p. 62) describes how this mutation changes cells.

A type of genetic change that does not affect health is called a polymorphism. Researchers are currently identifying "single nucleotide polymorphisms"—called SNPs (pronounced "snips")—that are correlated to increased risk of developing certain disorders. SNP maps are helping researchers to extract meaningful medical information from human genome sequence data.

Nature of Mutations

Mutations can originate in a number of ways. In one common mechanism during DNA replication, a base may pair incorrectly with the newly forming strand, or extra bases may be added. Or, sections of DNA strands may be deleted, moved to other regions of the molecule, or even attached to other chromosomes. In any case, the consequences are similar—genetic information is changed. If a protein is constructed from this information, its molecular structure may be faulty and the function changed or absent. For example, the muscle weakness of Duchenne muscular dystrophy may result from a mutation in the gene encoding the protein dystrophin. The mutation may be a missing or changed nucleotide base or absence of the entire dystrophin gene. In each case, lack of dystrophin, which normally supports muscle cell membranes during contraction, causes the cells to collapse. The muscles weaken and atrophy. Figure 4.25 shows how the change of one base may cause another inherited illness, sickle cell disease.

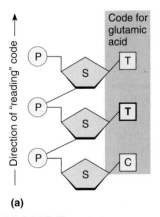
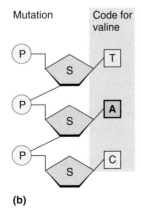

(a)   **(b)**

FIGURE 4.25

Mutation. (a) The DNA code for the amino acid glutamic acid is CTT. (b) If something happens to change the first thymine in this section of the molecule to adenine, the DNA code changes to CAT, which specifies the amino acid valine. The resulting mutation, when it occurs in the DNA that encodes the protein hemoglobin, causes sickle cell disease.

Fortunately, cells detect damage in their DNA molecules and use **repair enzymes** to clip out defective nucleotides in a single DNA strand and fill the resulting gap with nucleotides complementary to those on the remaining strand of DNA. This restores the original structure of the double-stranded DNA molecule.

If DNA is not repaired, illness may result. A class of disorders affects DNA repair. One such condition is xeroderma pigmentosum (XP). When other youngsters burst out of their homes on a sunny day to frolic outdoors, a child who has XP must cover up as completely as possible, wearing pants and long sleeves even in midsummer, and must apply sunscreen on every bit of exposed skin. Moderate sun exposure easily leads to skin sores or cancer. Even with all the precautions, the child's skin is a sea of freckles. Special camps and programs for children with XP allow them to play outdoors at night, when they are safe.

Effects of Mutations

The nature of the genetic code provides some protection against mutation. Sixty-one codons specify the twenty types of amino acids, and therefore, some amino acids correspond to more than one codon type. Usually, two or three codons specifying the same amino acid differ only in the third base of the codon. A mutation that changes the third codon base can encode the same amino acid. For example, the DNA triplets GGA and GGG each specify the amino acid proline. If a mutation changes the third position of GGA to a G, the amino acid for that position in the encoded protein does not change—it is still proline.

If a mutation alters a base in the second position, the substituted amino acid is very often similar in overall shape to the normal one, and the protein is not changed significantly enough to affect its function. This mutation, too, would go unnoticed. (An important exception is the mutation shown in fig. 4.25.) Yet another protection against mutation is that a person has two copies of each chromosome, and therefore of each gene. If one copy is mutated, the other may provide enough of the gene's normal function to maintain health. (This is more complicated for the sex chromosomes, X and Y, discussed in chapter 24, p. 932.) Finally, it also makes a difference whether a mutation occurs in the DNA of a body cell of an adult or in the DNA of a cell that is part of a developing embryo. In an adult, an altered cell might not be noticed because many normally functioning cells surround it. In the embryo, however, the abnormal cell might give rise to many cells forming the developing body. All the cells of a person's body could be defective if the mutation were present in the DNA of the fertilized egg.

TABLE 4.4	Commonly Encountered Mutagens	
Mutagen	**Source**	
Aflatoxin B	Fungi growing on peanuts and other foods	
2-amino 5-nitrophenol		
2,4-diaminoanisole		
2,5-diaminoanisole	Hair dye components	
2,4-diaminotoluene		
p-phenylenediamine		
Caffeine	Cola, tea, coffee	
Furylfuramide	Food additive	
Nitrosamines	Pesticides, herbicides, cigarette smoke	
Proflavine	Antiseptic in veterinary medicine	
Sodium nitrite	Smoked meats	
Tris (2,3-dibromopropyl phosphate)	Flame retardant in children's sleepwear	

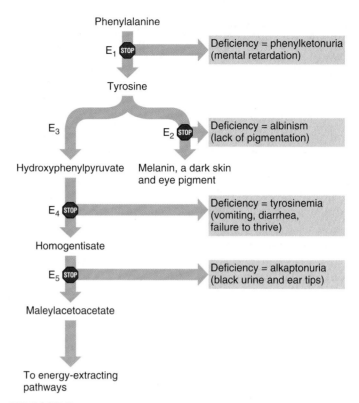

FIGURE 4.26

Four inborn errors of metabolism result from blocks affecting four enzymes in the pathway for the breakdown of phenylalanine, an amino acid. PKU results from a block of the first enzyme of the pathway (E_1). A block at E_2 leads to buildup of the amino acid tyrosine, and lack of its breakdown product, the pigment melanin, causes the pink eyes, white hair, and white skin of albinism. A block at E_4 can be deadly in infancy, and at E_5 leads to alkaptonuria, which causes severe joint pain and blackish deposits in the palate, ears, and eyes.

Mutations may occur spontaneously if a chemical quirk causes a base in an original DNA strand to be in an unstable form just as replication occurs there. Certain chemical substances, called **mutagens,** cause mutations. Researchers often use mutagens to intentionally alter gene function in order to learn how a gene normally acts. Table 4.4 lists some mutagens.

Ultraviolet radiation in sunlight is a familiar mutagen. It can cause an extra chemical bond to form between thymines that are adjacent on a DNA strand. This bond forms a kink, which can cause an incorrect base to be inserted during DNA replication. If sun damage is not extensive, repair enzymes remove the extra bonds, and replication proceeds. If damage is great, the cell dies. We experience this as a peeling sunburn. If a sun-damaged cell cannot be repaired or does not die, it often turns cancerous. This is why many years of sunburns can cause certain types of skin cancer.

A type of disorder called an "inborn error of metabolism" results from inheriting a mutation that alters an enzyme. Such an enzyme block in a biochemical pathway has two general effects: the substance that the enzyme normally acts on builds up, and the substance resulting from the enzyme's normal action becomes scarce. It is similar to blocking a garden hose: water pressure builds up behind the block, but no water comes out after it.

The biochemical excesses and deficiencies that an inborn error of metabolism triggers can drastically affect health. The specific symptoms depend upon which pathways and substances are affected. Figure 4.26 shows how blocks of different enzymes in one biochemical pathway lead to different sets of symptoms. Clinical Application 4.2 describes phenylketonuria (PKU), one of these inborn errors about which we know a great deal.

Because a gene consists of a sequence of hundreds of building blocks, mutation can alter a gene in many ways—just like a typographical error can occur on this page in many ways. Different mutations in the same gene can produce different severities of symptoms. The most common mutation in the gene that causes cystic fibrosis, for example, causes severe lung infection and obstruction and digestive difficulties, and affected individuals often die young. Other mutations are associated with less severe effects, such as frequent bronchitis or sinus infections. This second group generally lives longer than people with the more common mutation.

1 How are DNA molecules replicated?

2 What is a mutation?

3 How do mutations occur?

4 What kinds of mutations are of greatest concern?

5 How does the nature of the genetic code protect against mutation, to an extent?

PHENYLKETONURIA

In Oslo, Norway, in 1934, an observant mother of two mentally retarded children noticed that their soiled diapers had an odd, musty odor. She mentioned this to Ivar Folling, a relative who was a physician and biochemist. Folling was intrigued. Analyzing the children's urine, he found large amounts of the amino acid phenylalanine, which is usually present only in trace amounts because an enzyme catalyzes a chemical reaction that breaks it down. The children lacked this enzyme because they had inherited an inborn error of metabolism called phenylketonuria, or PKU. The buildup of phenylalanine causes mental retardation.

Researchers wondered if a diet very low in phenylalanine might prevent the mental retardation. The diet would include the other nineteen types of amino acids so that normal growth, which requires protein, could occur. The diet would theoretically alter the body's chemistry in a way that would counteract the overabundance of phenylalanine that the faulty genes caused.

In 1963, theory became reality when researchers devised a dietary treatment for this otherwise devastating illness (fig. 4A). The diet is very restrictive and difficult to follow, but it does prevent mental retardation. However, treated children may still have learning disabilities. We still do not know how long people with PKU should adhere to the diet, but it may be for their entire lives. ∎

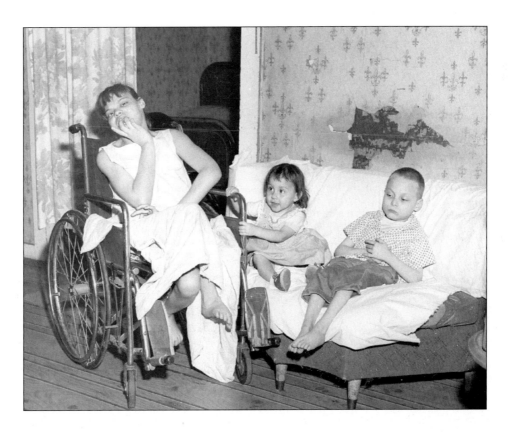

FIGURE 4A

These three siblings have each inherited PKU. The older two siblings—the girl in the wheelchair and the boy on the right—are mentally retarded because they were born before a diet that prevents symptoms became available. The child in the middle, although she also has inherited PKU, is of normal intelligence because she was lucky enough to have been born after the diet was invented.

Introduction (page 104)

A cell continuously carries on metabolic processes.

Metabolic Processes (page 104)

1. Anabolism
   a. Anabolism builds large molecules from smaller molecules.
   b. In dehydration synthesis, hydrogen atoms and hydroxyl groups are removed, water forms, and smaller molecules bind by sharing atoms.
   c. Complex carbohydrates are synthesized from monosaccharides, fats are synthesized from glycerol and fatty acids, and proteins are synthesized from amino acids.
2. Catabolism
   a. Catabolism breaks down larger molecules into smaller ones.
   b. In hydrolysis, a water molecule supplies a hydrogen atom to one portion of a molecule and a hydroxyl group to a second portion; the bond between these two portions breaks.
   c. Complex carbohydrates are decomposed into monosaccharides, fats are decomposed into glycerol and fatty acids, and proteins are decomposed into amino acids.

Control of Metabolic Reactions (page 106)

Enzymes control metabolic reactions.

1. Enzyme action
   a. Metabolic reactions require energy to start.
   b. Enzymes are proteins that increase the rate of specific metabolic reactions.
   c. An enzyme acts upon a molecule by temporarily combining with it and distorting its chemical structure.
   d. The shape of an enzyme molecule fits the shape of its substrate molecule.
   e. When an enzyme combines with its substrate, the substrate changes, enabling it to react, forming a product. The enzyme is released in its original form.
   f. The rate of enzyme-controlled reactions depends upon the numbers of enzyme and substrate molecules and the efficiency of the enzyme.
   g. Enzymes are usually named according to their substrates, with -ase at the end.
2. Cofactors and coenzymes
   a. Cofactors are additions to some enzymes that are necessary for their function.
   b. A cofactor may be an ion or a small organic molecule called a coenzyme.
   c. Vitamins, which are the sources of coenzymes, usually cannot be synthesized by human cells in adequate amounts.
3. Factors that alter enzymes
   a. Enzymes are proteins and can be denatured.
   b. Factors that may denature enzymes include excessive heat, radiation, electricity, certain chemicals, and extreme pH values.

Energy for Metabolic Reactions (page 107)

Energy is a capacity to produce change or to do work. Common forms of energy include heat, light, sound, electrical energy, mechanical energy, and chemical energy. Whenever changes take place, energy is being transferred.

1. Release of chemical energy
   a. Most metabolic processes utilize chemical energy that is released when molecular bonds are broken.
   b. The energy glucose releases during cellular respiration is used to promote metabolism.
   c. Enzymes in the cytoplasm and mitochondria control cellular respiration.

Cellular Respiration (page 108)

Metabolic processes usually have a number of steps that occur in a specific sequence. A sequence of enzyme-controlled reactions is called a metabolic pathway. Typically, metabolic pathways are interconnected.

1. ATP molecules
   a. Energy is captured in the bond of the terminal phosphate of each ATP molecule.
   b. Captured energy is released when the terminal phosphate bond of an ATP molecule is broken.
   c. An ATP molecule that loses its terminal phosphate becomes an ADP molecule.
   d. An ADP molecule can be converted to an ATP molecule by capturing energy and a phosphate.
   e. Thirty-eight molecules of ATP can be produced for each glucose molecule that is completely catabolized by cellular respiration.

2. Glycolysis
   a. Glycolysis, the first step of glucose catabolism, occurs in the cytosol and does not require oxygen.
   b. Glycolysis can be divided into three stages, in which some of the energy released is transferred to molecules of ATP.
   c. Some of the energy released in glycolysis is in the form of high-energy electrons attached to hydrogen carriers.

3. Anaerobic reactions
   a. Oxygen is the final electron acceptor in the aerobic reactions of cellular respiration.
   b. In the anaerobic reactions, NADH and H^+ instead donate electrons and hydrogens to pyruvic acid, generating lactic acid.
   c. Lactic acid builds up, eventually inhibiting glycolysis and ATP formation.
   d. When oxygen returns, liver cells convert lactic acid to pyruvic acid.

4. Aerobic reactions
   a. The second phase of glucose catabolism occurs in the mitochondria and requires oxygen.
   b. These reactions include the citric acid cycle and the electron transport chain.
   c. Considerably more energy is transferred to ATP molecules during the aerobic reactions of cellular respiration than during glycolysis.
   d. The products of the aerobic reactions of cellular respiration are heat, carbon dioxide, water, and energy.
   e. The citric acid cycle is a complex series of reactions that decompose molecules, release carbon dioxide, release hydrogen atoms that have high-energy electrons, and form ATP molecules.
   f. Hydrogen atoms from the citric acid cycle become hydrogen ions, which, in turn, combine with oxygen to form water molecules.
   g. High-energy electrons from hydrogen atoms enter an electron transport chain. Energy released from the chain is used to form ATP.

h. Each glucose molecule metabolized yields a maximum of thirty-eight ATP molecules.
i. Excess carbohydrates may enter anabolic pathways and be polymerized into and stored as glycogen or converted into fat.

5. Regulation of metabolic pathways
   a. A rate-limiting enzyme may regulate a metabolic pathway.
   b. A negative feedback mechanism in which the product of a pathway inhibits the regulatory enzyme may control the regulatory enzyme.
   c. The rate of product formation usually remains stable.

Nucleic Acids and Protein Synthesis (page 115)

DNA molecules contain information that tells a cell how to synthesize proteins, including enzymes.

1. Genetic information
   a. DNA information specifies inherited traits.
   b. A gene is a portion of a DNA molecule that contains the genetic information for making one kind of protein.
   c. The nucleotides of a DNA strand are in a particular sequence.
   d. The nucleotides pair with those of the second strand in a complementary fashion.

2. Genetic code
   a. The sequence of nucleotides in a DNA molecule represents the sequence of amino acids in a protein molecule.
   b. RNA molecules transfer genetic information from the nucleus to the cytoplasm.

3. RNA molecules
   a. RNA molecules are usually single-stranded, contain ribose instead of deoxyribose, and contain uracil nucleotides in place of thymine nucleotides.
   b. Messenger RNA molecules, which are synthesized in the nucleus, contain a nucleotide sequence that is complementary to that of an exposed strand of DNA.
   c. Messenger RNA molecules move into the cytoplasm, associate with ribosomes, and are templates for the synthesis of protein molecules.

4. Protein synthesis
   a. Molecules of transfer RNA position amino acids along a strand of messenger RNA.
   b. A ribosome binds to a messenger RNA molecule and allows a transfer RNA molecule to recognize its correct position on the messenger RNA.
   c. The ribosome contains enzymes required for the synthesis of the protein and holds the protein until it is completed.
   d. As the protein forms, it folds into a unique shape.
   e. ATP provides the energy for protein synthesis.

5. DNA replication
   a. Each new cell requires a copy of the original cell's genetic information.
   b. DNA molecules are replicated during interphase of the cell cycle.
   c. Each new DNA molecule contains one old strand and one new strand.

Changes in Genetic Information (page 123)

A DNA molecule contains a great amount of information. A change in the genetic information is a mutation. Not all changes to DNA are harmful.

1. Nature of mutations
   a. Mutations include several kinds of changes in DNA.
   b. A protein synthesized from an altered DNA sequence may function abnormally or not at all.
   c. Repair enzymes can correct some forms of DNA damage.

2. Effects of mutations
   a. The genetic code protects against some mutations.
   b. A mutation in a sex cell or fertilized egg or early embryo may have a more severe effect than a mutation in an adult because a greater proportion of the individual's cells are affected.

CRITICAL THINKING QUESTIONS

1. Because enzymes are proteins, they can denature. How does this explain the fact that changes in the pH of body fluids during illness may threaten life?

2. Some weight-reducing diets drastically limit intake of carbohydrates but allow many foods rich in fat and protein. What changes would such a diet cause in cellular metabolism? How would excretion of substances from the internal environment change?

3. Why are vitamins that function as coenzymes in cells required in extremely low concentrations?

4. What changes in concentrations of oxygen and carbon dioxide would you expect to find in the blood of a person who is forced to exercise on a treadmill beyond his or her normal capacity? How might these changes affect the pH of the person's blood?

5. How do the antibiotic actions of rifampin and streptomycin differ?

6. A student is accustomed to running 3 miles each afternoon at a slow, leisurely pace. One day, she runs a mile as fast as she can. Afterwards she is winded, with pains in her chest and leg muscles. She thought she was in great shape! What has she experienced, in terms of energy metabolism?

7. In fructose intolerance, a missing enzyme makes a person unable to utilize fructose, a simple sugar abundant in fruit. Infants with the condition have very low mental and motor function. Older children are very lethargic and mildly mentally retarded. By adulthood, the nervous system deteriorates, eventually causing mental illness and death. Molecules that are derived from fructose are intermediates in the first few reactions of glycolysis. The enzyme missing in people with fructose intolerance would normally catalyze these reactions. Considering this information about the whole-body and biochemical effects of fructose intolerance, suggest what might be happening on a cellular level to these people.

8. Write the sequence of the complementary strand of DNA to the sequence AGCGATTGCATGC. What is the sequence of mRNA that would be transcribed from the given sequence?

9. Explain why exposure to ultraviolet light in tanning booths may be dangerous.

REVIEW EXERCISES

1. Define *anabolism* and *catabolism*.
2. Distinguish between dehydration synthesis and hydrolysis.
3. Define *peptide bond*.
4. Define *enzyme*.
5. How does an enzyme interact with its substrate?
6. List three factors that increase the rates of enzyme-controlled reactions.
7. How are enzymes usually named?
8. Define *cofactor*.
9. Explain why humans require vitamins in their diets.
10. Explain how an enzyme may be denatured.
11. Define *energy*.
12. Explain how the oxidation of molecules inside cells differs from the burning of substances outside cells.
13. Define *cellular respiration*.
14. Distinguish between the anaerobic reactions and the aerobic reactions of cellular respiration.
15. Explain the importance of ATP to cellular processes.
16. Describe the relationship between ATP and ADP molecules.
17. Define *metabolic pathway*.
18. Describe the starting material and products of *glycolysis*.
19. State the products of the citric acid cycle.
20. How are carbohydrates stored?
21. Explain how one enzyme can regulate a metabolic pathway.
22. Describe how a negative feedback mechanism can help control a metabolic pathway.
23. Explain the chemical basis of genetic information.
24. Describe the chemical makeup of a gene.
25. Describe the general structure and components of a DNA molecule.
26. Distinguish between the functions of messenger RNA and transfer RNA.
27. Distinguish between transcription and translation.
28. Explain two functions of ribosomes in protein synthesis.
29. Distinguish between a codon and an anticodon.
30. Explain how a DNA molecule is replicated.
31. Define *mutation*, and explain how mutations may originate.
32. Define *repair enzyme*.
33. Explain how a mutation may affect an organism's cells—or not affect them.

WEB CONNECTIONS

Visit the Student Edition of the Online Learning Center at www.mhhe.com/shier10 **for answers to chapter questions, additional quizzes, interactive learning exercises, and other study tools.**

UNDERSTANDING WORDS

adip-, fat: *adip*ose tissue—tissue that stores fat.

chondr-, cartilage: *chondr*ocyte—cartilage cell.

-cyt, cell: osteo*cyt*e—bone cell.

epi-, upon, after, in addition: *epi*thelial tissue—tissue that covers all free body surfaces.

-glia, glue: neuro*glia*—cells that bind nervous tissue together.

hist-, web, tissue: *hist*ology—study of composition and function of tissues.

hyal-, resemblance to glass: *hyal*ine cartilage—flexible tissue containing chondrocytes.

inter-, among, between: *inter*calated disc—band of gap junctions between the ends of adjacent cardiac muscle cells.

macr-, large: *mac*rophage—large phagocytic cell.

neur-, nerve: *neur*on—nerve cell.

os-, bone: *os*seous tissue—bone tissue.

phag-, to eat: *phag*ocyte—cell that engulfs and destroys foreign particles.

pseud-, false: *pseud*ostratified epithelium—tissue with cells that appear to be in layers, but are not.

squam-, scale: *squam*ous epithelium—tissue with flattened or scalelike cells.

strat-, layer: *strat*ified epithelium—tissue whose cells are in layers.

stria-, groove: *stria*ted muscle—tissue whose cells have alternating light and dark cross-markings.

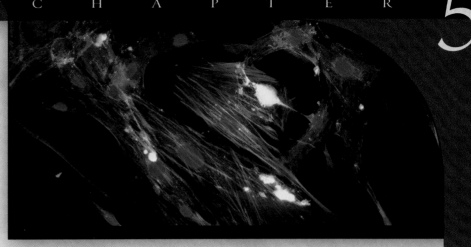

The fibroblast connective tissue cells shown here have been taken from fetal skin. Fibroblasts are responsible for forming connective tissue by secreting matrix material such as collagen. Immunofluorescent light micrograph (225×).

Tissues

CHAPTER OBJECTIVES

After you have studied this chapter, you should be able to

1. Describe the general characteristics and functions of epithelial tissue.

2. Name the types of epithelium and identify an organ in which each is found.

3. Explain how glands are classified.

4. Describe the general characteristics of connective tissue.

5. Describe the major cell types and fibers of connective tissue.

6. List the types of connective tissue within the body.

7. Describe the major functions of each type of connective tissue.

8. Distinguish among the three types of muscle tissue.

9. Describe the general characteristics and functions of nervous tissue.

10. Describe the four major types of membranes.

C ells aggregate to form tissues, and tissues interact to form organs. Dissecting a complex organ to observe how tissues comprise it is a commonly performed exercise; attempting to build an organ from its component cells and tissues is much more challenging.

The field of tissue engineering uses cells, synthetic materials, or combinations of them to fashion human body parts. Consider the task facing Nicolas L'Heureux, who recreated a small-diameter human blood vessel when a graduate student. Such a vessel has three layers: an innermost layer of tilelike endothelial cells that secrete anti-clotting agents, a middle layer of smooth muscle and elastic connective tissue, and an outer layer of fibroblasts and the collagen protein they secrete.

Previous attempts at producing a small blood vessel combined natural and synthetic ingredients in various ways, with mixed results. The goal is to keep the inner lining smooth enough to prevent blood clots from forming, but construct outer layers that are strong enough to keep the vessel open under the pressure of circulating blood. The trick, L'Heureux found, was to let the cells do the work—with a little help.

L'Heureux and his co-workers grew fibroblasts and smooth muscle cells in sheets. They then rolled the sheets around tubes through which nutrients circulated in, and cellular wastes circulated out. Then the researchers seeded endothelial cells onto the inner surface, where the cells knit a smooth inner lining. By allowing the fibroblasts to secrete the collagen, rather than supplying the protein directly, the vessels formed in a more natural way and persisted. Blood vessels engineered

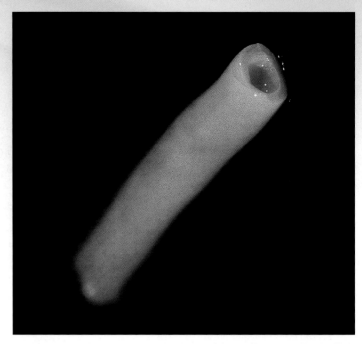

Recipe for a lab-built small-diameter blood vessel: Seed lining cells onto the inner surfaces of tubes of collagen-secreting cells and smooth muscle. These engineered blood vessels may someday replace damaged vessels in people's legs and hearts.

in this way may eventually be used to treat the thousands of people who need vascular grafts in their legs or new coronary arteries. ■

In all complex organisms, cells are organized into layers or groups called **tissues.** Although the cells of different tissues vary in size, shape, arrangement, and function, those within a tissue are quite similar.

∞ R E C O N N E C T   T O   C H A P T E R   3, I N T E R C E L L U L A R J U N C T I O N S ,   P A G E   6 7 .

Usually, tissue cells are separated by nonliving, intercellular materials that the cells produce. These inter-cellular materials vary in composition and amount from one tissue to another and may be solid, semisolid, or liquid. For example, a solid (mineral) separates bone tissue cells, whereas a liquid (plasma) separates blood tissue cells.

The tissues of the human body include four major types: *epithelial, connective, muscle,* and *nervous.* These tissues associate and interact to form organs that have specialized functions. Table 5.1 compares the four major tissue types.

TABLE 5.1	Tissues		
Type	**Function**	**Location**	**Distinguishing Characteristics**
Epithelial	Protection, secretion, absorption, excretion	Cover body surface, cover and line internal organs, compose glands	Lack blood vessels, cells readily divide, cells are tightly packed
Connective	Bind, support, protect, fill spaces, store fat, produce blood cells	Widely distributed throughout the body	Mostly have good blood supply, cells are farther apart than cells of epithelia, with matrix in between
Muscle	Movement	Attached to bones, in the walls of hollow internal organs, heart	Contractile
Nervous	Transmit impulses for coordination, regulation, integration, and sensory reception	Brain, spinal cord, nerves	Cells connect to each other and other body parts

This chapter examines in detail epithelial and connective tissues. Throughout this chapter, simplified line drawings (for example, fig. 5.1*a*) are included with each micrograph (for example, fig. 5.1*b*) to emphasize the distinguishing characteristics of the specific tissue. Chapter 9 discusses muscle tissue, and chapters 10 and 11 detail nervous tissue.

1 What is a tissue?

2 List the four major types of tissue.

Epithelial Tissues

General Characteristics

Epithelial tissues (ep"ĭ-the'le-al tish'ūz) are widespread throughout the body. Since epithelium covers organs, forms the inner lining of body cavities, and lines hollow organs, it always has a free surface—one that is exposed to the outside or to an open space internally. The underside of this tissue is anchored to connective tissue by a thin, nonliving layer called the **basement membrane.**

One of the ways that cancer cells spread is by secreting a substance that dissolves basement membranes. This enables cancer cells to invade adjacent tissue layers. Cancer cells also produce fewer adhesion proteins, or none at all, which allows them to invade surrounding tissue.

As a rule, epithelial tissues lack blood vessels. However, nutrients diffuse to epithelium from underlying connective tissues, which have abundant blood vessels.

Because epithelial cells readily divide, injuries heal rapidly as new cells replace lost or damaged ones. Skin cells and the cells that line the stomach and intestines are epithelial cells that are continually being damaged and replaced.

Epithelial cells are tightly packed, with little intercellular material. In many places, desmosomes attach one to another (see chapter 3, p. 67). Consequently, these cells form effective protective barriers in such structures as the outer layer of the skin and the inner lining of the mouth. Other epithelial functions include secretion, absorption, and excretion.

Epithelial tissues are classified according to the shape and number of layers of cells. Epithelial tissues that are composed of thin, flattened cells are *squamous;* those with cubelike cells are *cuboidal;* and those with elongated cells are *columnar;* those with single layers of cells are *simple;* those with two or more layers of cells are *stratified.* In the following descriptions, note that the free surfaces of epithelial cells are modified to reflect their specialized functions.

1 List the general characteristics of epithelial tissue.

2 Explain how epithelial tissues are classified.

Simple Squamous Epithelium

Simple squamous (skwa'mus) **epithelium** consists of a single layer of thin, flattened cells. These cells fit tightly together, somewhat like floor tiles, and their nuclei are usually broad and thin (fig. 5.1).

Substances pass rather easily through simple squamous epithelium, which is common at sites of diffusion and filtration. For instance, simple squamous epithelium lines the air sacs (alveoli) of the lungs where oxygen and carbon dioxide are exchanged. It also forms the walls of capillaries, lines the insides of blood and lymph vessels, and covers the membranes that line body cavities. However, because it is so thin and delicate, simple squamous epithelium is easily damaged.

Simple Cuboidal Epithelium

Simple cuboidal epithelium consists of a single layer of cube-shaped cells. These cells usually have centrally located, spherical nuclei (fig. 5.2).

Simple cuboidal epithelium covers the ovaries and lines the kidney tubules and ducts of certain glands, such as the salivary glands, pancreas, and liver. In the kidneys, it functions in tubular secretion and tubular reabsorption; in glands, it secretes glandular products.

Simple Columnar Epithelium

Simple columnar epithelium is composed of a single layer of elongated cells whose nuclei are usually at about the same level, near the basement membrane (fig. 5.3). The cells of this tissue can be ciliated or nonciliated. *Cilia,* which are 7–10 μm in length, extend from the free surfaces of the cells, and they move constantly. In the female reproductive tubes, cilia aid in moving egg cells through the oviducts to the uterus.

Nonciliated simple columnar epithelium lines the uterus and portions of the digestive tract, including the stomach and small and large intestines. Because its cells are elongated, this tissue is thick, which enables it to protect underlying tissues. The cells of simple columnar epithelium also secrete digestive fluids and absorb nutrients from digested foods.

Simple columnar cells, specialized for absorption, often have many tiny, cylindrical processes extending from their surfaces. These processes, called *microvilli,* are from 0.5 to 1.0 μm long. They increase the surface area of the cell membrane where it is exposed to substances being absorbed (fig. 5.4).

Typically, specialized, flask-shaped glandular cells are scattered among the cells of simple columnar epithelium. These cells, called *goblet cells,* secrete a protective fluid called *mucus* onto the free surface of the tissue (see fig. 5.3).

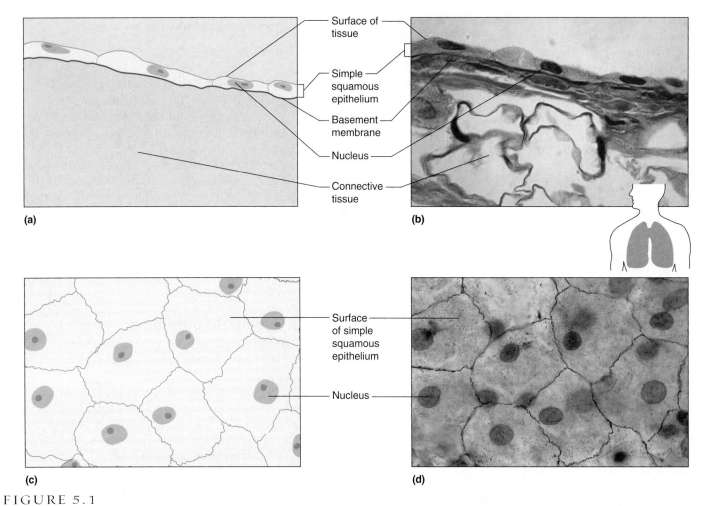

Labels for Figure 5.1:
- Surface of tissue
- Simple squamous epithelium
- Basement membrane
- Nucleus
- Connective tissue
- (a)
- (b)
- Surface of simple squamous epithelium
- Nucleus
- (c)
- (d)

FIGURE 5.1
Simple squamous epithelium consists of a single layer of tightly packed, flattened cells (670×). (a) and (b) side view, (c) and (d) surface view.

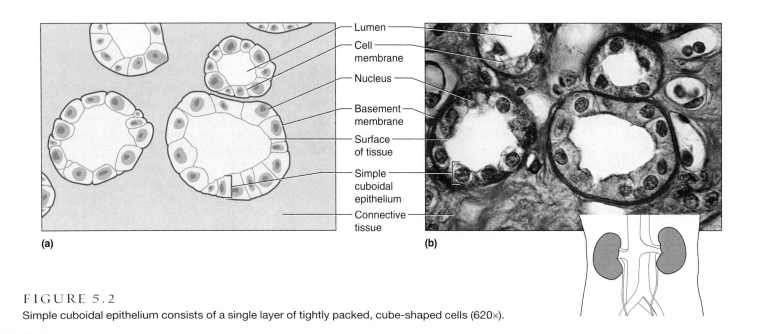

Labels for Figure 5.2:
- Lumen
- Cell membrane
- Nucleus
- Basement membrane
- Surface of tissue
- Simple cuboidal epithelium
- Connective tissue
- (a)
- (b)

FIGURE 5.2
Simple cuboidal epithelium consists of a single layer of tightly packed, cube-shaped cells (620×).

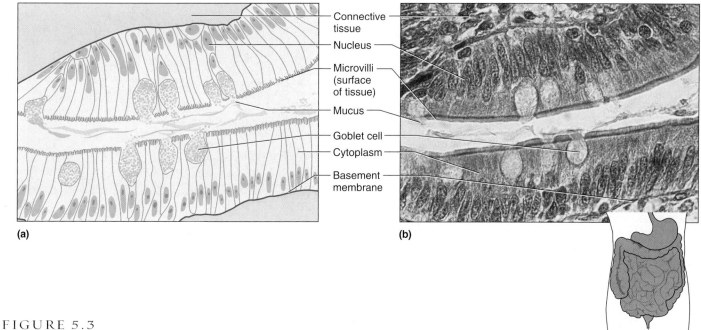

Connective tissue	
Nucleus	
Microvilli (surface of tissue)	
Mucus	
Goblet cell	
Cytoplasm	
Basement membrane	

(a)

(b)

FIGURE 5.3

Simple columnar epithelium consists of a single layer of elongated cells (400×).

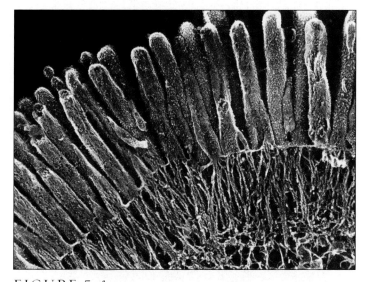

FIGURE 5.4

A scanning electron micrograph of microvilli, which fringe the exposed surfaces of some columnar epithelial cells (33,000×).

Pseudostratified Columnar Epithelium

The cells of **pseudostratified** (soo″do-strat′ĭ-fīd) **columnar epithelium** appear stratified or layered, but they are not. A layered effect occurs because the nuclei are at two or more levels in the row of aligned cells. However, the cells, which vary in shape, all reach the basement membrane, even though some of them may not contact the free surface.

The cells of pseudostratified columnar epithelium often are fringed with cilia. The cilia extend from the free surfaces of the cells. Goblet cells scattered throughout this tissue secrete mucus, which the cilia sweep away (fig. 5.5).

Pseudostratified columnar epithelium lines the passages of the respiratory system. Here, the mucous-covered linings are sticky and trap dust and microorganisms that enter with the air. The cilia move the mucus and its captured particles upward and out of the airways.

Stratified Squamous Epithelium

Stratified epithelium is named for the shape of the cells forming the outermost layers. **Stratified squamous epithelium** consists of many layers of cells, making this tissue relatively thick. Cells nearest the free surface are flattened the most, whereas those in the deeper layers, where cell division occurs, are cuboidal or columnar. As the newer cells grow, older ones are pushed farther and farther outward, where they flatten (fig. 5.6).

The outermost layer of the skin (epidermis) is stratified squamous epithelium. As the older cells are pushed outward, they accumulate a protein called *keratin,* then harden and die. This "keratinization" produces a covering of dry, tough, protective material that prevents water and other substances from escaping from underlying tissues and blocks chemicals and microorganisms from entering.

Stratified squamous epithelium also lines the oral cavity, throat, esophagus, vagina, and anal canal. In these parts, the tissue is not keratinized; it stays soft and moist, and the cells on its free surfaces remain alive.

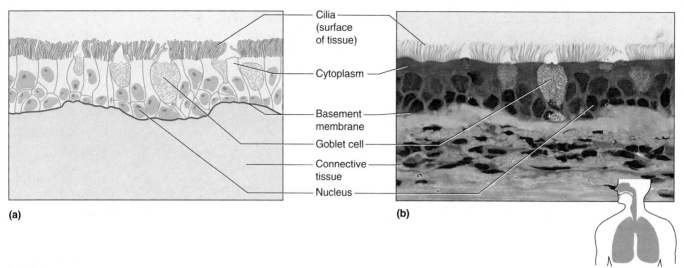

Cilia (surface of tissue)
Cytoplasm
Basement membrane
Goblet cell
Connective tissue
Nucleus

(a)

(b)

FIGURE 5.5

Pseudostratified columnar epithelium appears stratified because nuclei are located at different levels (255×).

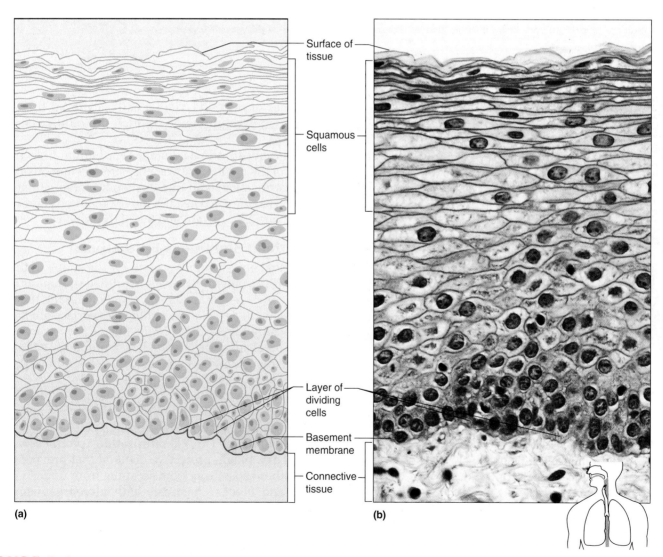

Surface of tissue
Squamous cells
Layer of dividing cells
Basement membrane
Connective tissue

(a)

(b)

FIGURE 5.6

Stratified squamous epithelium consists of many layers of cells (385×).

Stratified Cuboidal Epithelium

Stratified cuboidal epithelium consists of two or three layers of cuboidal cells that form the lining of a lumen (fig. 5.7). The layering of the cells provides more protection than the single layer affords.

Stratified cuboidal epithelium lines the larger ducts of the mammary glands, sweat glands, salivary glands, and pancreas. It also forms the lining of developing ovarian follicles and seminiferous tubules, which are parts of the female and male reproductive systems, respectively.

Stratified Columnar Epithelium

Stratified columnar epithelium consists of several layers of cells (fig. 5.8). The superficial cells are elongated, whereas the basal layers consist of cube-shaped cells.

Stratified columnar epithelium is in the vas deferens, part of the male urethra, and in parts of the pharynx.

Transitional Epithelium

Transitional epithelium (uroepithelium) is specialized to change in response to increased tension. It forms the inner lining of the urinary bladder and lines the ureters and part of the urethra. When the wall of one of these organs contracts, the tissue consists of several layers of cuboidal cells; however, when the organ is distended, the tissue stretches, and the physical relationships among the cells change. While distended, the tissue appears to contain only a few layers of cells (fig. 5.9). In addition to providing an expandable lining, transitional epithelium forms a barrier that helps prevent the contents of the urinary tract from diffusing back into the internal environment.

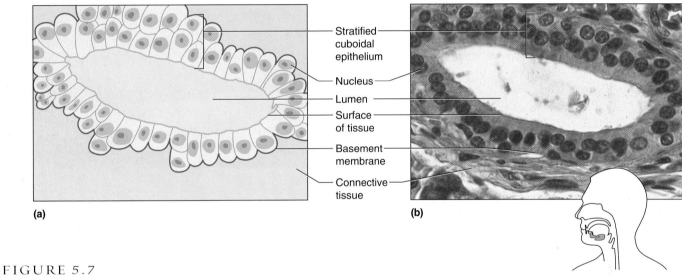

(a)   Stratified cuboidal epithelium — Nucleus — Lumen — Surface of tissue — Basement membrane — Connective tissue   (b)

FIGURE 5.7

Stratified cuboidal epithelium consists of two to three layers of cube-shaped cells surrounding a lumen (430×).

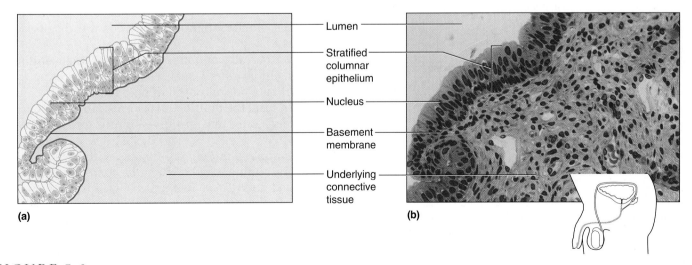

(a)   Lumen — Stratified columnar epithelium — Nucleus — Basement membrane — Underlying connective tissue   (b)

FIGURE 5.8

Stratified columnar epithelium consists of a superficial layer of columnar cells overlying several layers of cuboidal cells (220×).

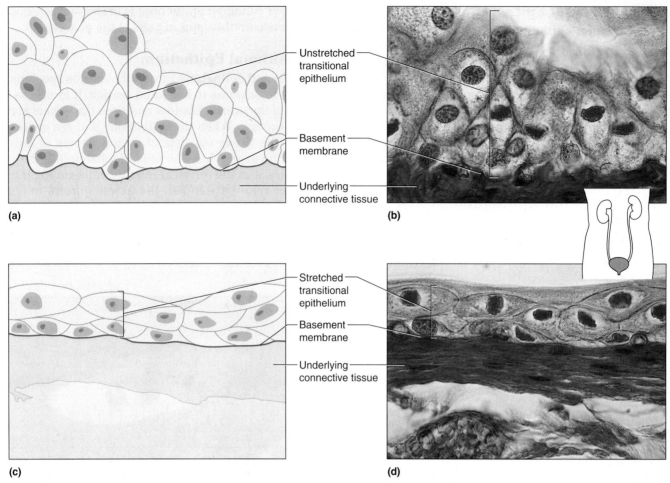

(a)

Unstretched
transitional
epithelium

Basement
membrane

Underlying
connective tissue

(b)

Stretched
transitional
epithelium

Basement
membrane

Underlying
connective tissue

(c)

(d)

FIGURE 5.9

Transitional epithelium. (*a* and *b*) When the organ wall contracts, transitional epithelium is unstretched and consists of many layers (675×).
(*c* and *d*) When the organ is distended, the tissue stretches and appears thinner (675×).

Up to 90% of all human cancers are *carcinomas,* which are growths that originate in epithelium. Most carcinomas begin on surfaces that contact the external environment, such as skin, linings of the airways in the respiratory tract, or linings of the stomach or intestines in the digestive tract. This observation suggests that the more common cancer-causing agents may not penetrate tissues very deeply.

1 Describe the structure of each type of epithelium.

2 Describe the special functions of each type of epithelium.

Glandular Epithelium

Glandular epithelium is composed of cells that are specialized to produce and secrete substances into ducts or into body fluids. Such cells are usually found within columnar or cuboidal epithelium, and one or more of these cells constitutes a *gland*. Glands that secrete their products into ducts that open onto some internal or external surface are called **exocrine glands.** Glands that secrete their products into tissue fluid or blood are called **endocrine glands.** (Endocrine glands are discussed in chapter 13.)

An exocrine gland may consist of a single epithelial cell (unicellular gland), such as a mucous-secreting goblet cell, or it may be composed of many cells (multicellular gland). In turn, the multicellular forms can be structurally subdivided into two groups—simple and compound glands.

A *simple gland* communicates with the surface by means of an unbranched duct, and a *compound gland* has a branched duct. These two types of glands can be further classified according to the shapes of their secretory portions. Glands that consist of epithelial-lined tubes are called *tubular glands;* those whose terminal portions form saclike dilations are called *alveolar glands* (acinar glands). Branching and coiling of the secretory portions may occur as well. Figure 5.10 illustrates several types of exocrine glands classified by structure. Table 5.2 summarizes the types of exocrine glands, lists their characteristics, and provides an example of each type.

Exocrine glands are also classified according to the ways these glands secrete their products. Glands that release fluid products by exocytosis are called **merocrine glands.** Glands that lose small portions of their glandular cell bodies during secretion are called **apocrine glands.** Glands that release entire cells are called **holocrine glands.** After release, the cells containing accumulated secretory products disintegrate, liberating their secretions

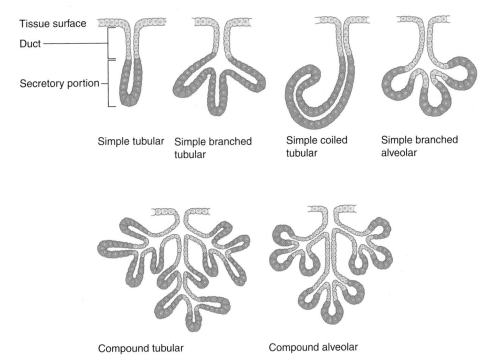

FIGURE 5.10
Structural types of exocrine glands.

TABLE 5.2	Types of Exocrine Glands	
Type	**Characteristics**	**Example**
Unicellular glands	A single secretory cell	Mucous-secreting goblet cell (see fig. 5.3)
Multicellular glands	Glands that consist of many cells	
Simple glands	Glands that communicate with surface by means of unbranched ducts	
1. Simple tubular gland	Straight tubelike gland that opens directly onto surface	Intestinal glands of small intestine (see fig. 17.3)
2. Simple coiled tubular gland	Long, coiled, tubelike gland; long duct	Eccrine (sweat) glands of skin (see fig. 6.9)
3. Simple branched tubular gland	Branched, tubelike gland; duct short or absent	Mucous glands in small intestine (see fig. 17.3)
4. Simple branched alveolar gland	Secretory portions of gland expand into saclike compartments along duct	Sebaceous gland of skin (see fig. 5.12)
Compound glands	Glands that communicate with surface by means of branched ducts	
1. Compound tubular gland	Secretory portions are coiled tubules, usually branched	Bulbourethral glands of male (see fig. 22.1)
2. Compound alveolar gland	Secretory portions are irregularly branched tubules with numerous saclike outgrowths	Salivary glands (see fig. 17.12)

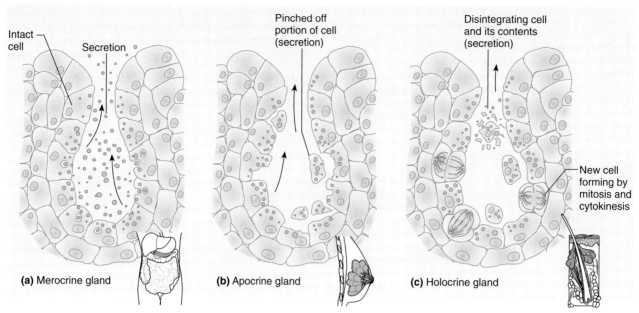

FIGURE 5.11

Glandular secretions. (*a*) Merocrine glands release secretions without losing cytoplasm. (*b*) Apocrine glands lose small portions of their cell bodies during secretion. (*c*) Holocrine glands release entire cells filled with secretory products.

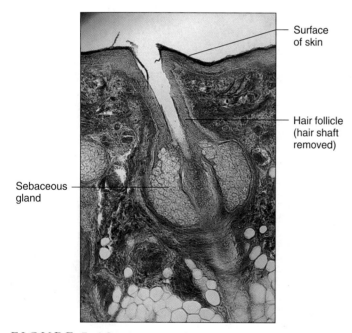

FIGURE 5.12

The sebaceous gland associated with a hair follicle is a simple branched alveolar gland that secretes entire cells (30×).

(figs. 5.11 and 5.12). Table 5.3 summarizes these glands and their secretions.

Most exocrine secretory cells are merocrine, and they can be further subdivided as either *serous cells* or *mucous cells.* The secretion of serous cells is typically watery, has a high concentration of enzymes, and is called

TABLE 5.3	Types of Glandular Secretions	
Type	**Description of Secretion**	**Example**
Merocrine glands	A fluid product released through the cell membrane by exocytosis	Salivary glands, pancreatic glands, sweat glands of the skin
Apocrine glands	Cellular product and portions of the free ends of glandular cells pinch off during secretion	Mammary glands, ceruminous glands lining the external ear canal
Holocrine glands	Entire cells laden with secretory products disintegrate	Sebaceous glands of the skin

serous fluid. Such cells are common in the linings of the body cavities. Mucous cells secrete a thicker fluid *mucus.* This substance is rich in the glycoprotein *mucin* and is abundantly secreted from the inner linings of the digestive and respiratory systems. Table 5.4 summarizes the characteristics of the different types of epithelial tissues.

1 Describe the structure of each type of epithelium.

2 Describe the special functions of each type of epithelium.

3 Distinguish between exocrine and endocrine glands.

4 Explain how exocrine glands are classified.

5 Distinguish between a serous cell and a mucous cell.

TABLE 5.4 Epithelial Tissues

Type	Description	Function	Location
Simple squamous epithelium	Single layer, flattened cells	Filtration, diffusion, osmosis, covers surface	Air sacs of lungs, walls of capillaries, linings of blood and lymph vessels
Simple cuboidal epithelium	Single layer, cube-shaped cells	Secretion, absorption	Surface of ovaries, linings of kidney tubules, and linings of ducts of certain glands
Simple columnar epithelium	Single layer, elongated cells	Protection, secretion, absorption	Linings of uterus, stomach, and intestines
Pseudostratified columnar epithelium	Single layer, elongated cells	Protection, secretion, movement of mucus and substances	Linings of respiratory passages
Stratified squamous epithelium	Many layers, top cells flattened	Protection	Outer layer of skin, linings of oral cavity, throat, vagina, and anal canal
Stratified cuboidal epithelium	2–3 layers, cube-shaped cells	Protection	Linings of larger ducts of mammary glands, sweat glands, salivary glands, and the pancreas
Stratified columnar epithelium	Top layer of elongated cells, lower layers of cube-shaped cells	Protection, secretion	Vas deferens, part of the male urethra, and parts of the pharynx
Transitional epithelium	Many layers of cube-shaped and elongated cells	Distensibility, protection	Inner lining of urinary bladder and linings of ureters and part of urethra
Glandular epithelium	Unicellular or multicellular	Secretion	Salivary glands, sweat glands, endocrine glands

Connective Tissues

General Characteristics

Connective tissues (kŏ-nek′tiv tish′ūz) comprise much of the body and are the most abundant type of tissue by weight. They bind structures, provide support and protection, serve as frameworks, fill spaces, store fat, produce blood cells, protect against infections, and help repair tissue damage.

Connective tissue cells are farther apart than epithelial cells, and they have an abundance of intercellular material, or **matrix** (ma′triks), between them. This matrix consists of fibers and a *ground substance* whose consistency varies from fluid to semisolid to solid. The ground substance binds, supports, and provides a medium through which substances may be transferred between the blood and cells within the tissue.

Connective tissue cells can usually divide. These tissues have varying degrees of vascularity, but in most cases, they have good blood supplies and are well nourished. Some connective tissues, such as bone and cartilage, are quite rigid. Loose connective tissue (areolar), adipose tissue, and dense connective tissue are more flexible.

Major Cell Types

Connective tissues contain a variety of cell types. Some of them are called *fixed cells* because they are usually present in stable numbers. These include fibroblasts and mast cells. Other cells, such as macrophages, are *wandering cells.* They temporarily appear in tissues, usually in response to an injury or infection.

The **fibroblast** (fi′bro-blast) is the most common kind of fixed cell in connective tissues. It is a large, star-shaped cell. Fibroblasts produce fibers by secreting protein into the matrix of connective tissues (fig. 5.13).

Macrophages (mak′ro-fājez), or histiocytes, originate as white blood cells (see chapter 14, p. 519) and are almost as numerous as fibroblasts in some connective tissues. They are usually attached to fibers but can detach and actively move about. Macrophages are specialized to carry on phagocytosis. Because they function as scavenger cells that can clear foreign particles from tissues, macrophages are an important defense against infection (fig. 5.14). They also play a role in immunity (see chapter 16, p. 621).

Mast cells are large and are widely distributed in connective tissues, where they are usually located near

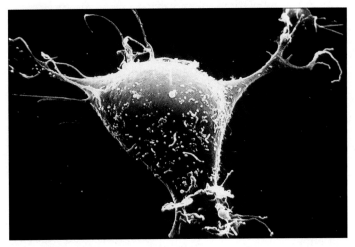

FIGURE 5.13
A scanning electron micrograph of a fibroblast (4,000×).

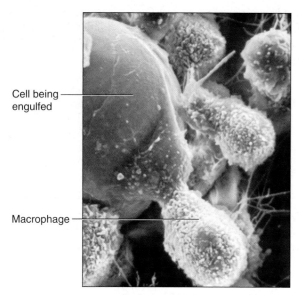

FIGURE 5.14

Cell being engulfed

Macrophage

FIGURE 5.14
Macrophages are scavenger cells common in connective tissues. This scanning electron micrograph shows a number of macrophages engulfing a larger cell (3,330×).

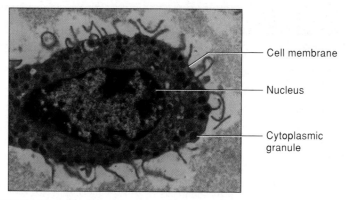

Cell membrane

Nucleus

Cytoplasmic granule

FIGURE 5.15
A transmission electron micrograph of a mast cell (5,000×).

blood vessels (fig. 5.15). They release *heparin,* a compound that prevents blood clotting. Mast cells also release *histamine,* a substance that promotes some of the reactions associated with inflammation and allergies, such as asthma and hay fever (see chapter 16, p. 630).

> Release of histamine stimulates inflammation by dilating the small arterioles that feed capillaries, the tiniest blood vessels. The resulting swelling and redness is inhospitable to infectious bacteria and viruses and also dilutes toxins. Inappropriate histamine release as part of an allergic response can be most uncomfortable. Allergy medications called antihistamines counter this misplaced inflammation.

Connective Tissue Fibers

Fibroblasts produce three types of connective tissue fibers: collagenous fibers, elastic fibers, and reticular fibers. Of these, collagenous and elastic fibers are the most abundant.

Collagenous (kol-laj′ĕ-nus) **fibers** are thick threads of the protein **collagen,** which is the major structural protein of the body. Collagenous fibers are grouped in long, parallel bundles, and they are flexible but only slightly elastic (fig. 5.16). More importantly, they have great tensile strength—that is, they can resist considerable pulling force. Thus, collagenous fibers are important components of body parts that hold structures together, such as *ligaments* (which connect bones to bones) and *tendons* (which connect muscles to bones).

Tissue containing abundant collagenous fibers is called *dense connective tissue.* Such tissue appears white,

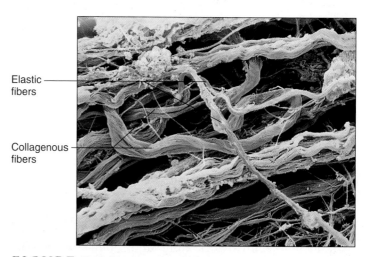

Elastic fibers

Collagenous fibers

FIGURE 5.16
Falsely colored scanning electron micrograph of collagenous fibers (yellow) and elastic fibers (blue) (4,100×).

and for this reason collagenous fibers of dense connective tissue are sometimes called white fibers. *Loose connective tissue,* on the other hand, has sparse collagenous fibers. Clinical Application 5.1, figure 5.17, and table 5.5 concern disorders that result from abnormal collagen.

> When skin is exposed to prolonged and intense sunlight, connective tissue fibers lose elasticity, and the skin stiffens and becomes leathery. In time, the skin may sag and wrinkle. Collagen injections may temporarily smooth out wrinkles. However, collagen applied as a cream to the skin does not combat wrinkles because collagen molecules are far too large to actually penetrate the skin.

Elastic fibers are composed of bundles of microfibrils embedded in a protein called **elastin.** These fibers branch, forming complex networks in various tissues. They are weaker than collagenous fibers but very elastic. That is,

ABNORMALITIES OF COLLAGEN

Much of the human body consists of the protein collagen. It accounts for more than 60% of the protein in bone and cartilage and provides 50–90% of the dry weight of skin, ligaments, tendons, and the dentine of teeth. Collagen is in the eyes, blood vessel linings, basement membranes, and connective tissue. It is not surprising, then, that defects in collagen cause a variety of medical problems.

Collagen abnormalities are devastating because this protein has an extremely precise structure that is easily disrupted, even by slight alterations that might exert little noticeable effect in other proteins. Collagen is sculpted from a precursor molecule called procollagen. Three procollagen chains coil and entwine to form a very regular triple helix.

Triple helices form as the procollagen is synthesized, but once secreted from the cell, the helices are trimmed. The collagen fibrils continue to associate outside the cell, building the networks that hold the body together. Collagen is rapidly synthesized and assembled into its rigid architecture. Many types of mutations can disrupt the protein's structure, including missing procollagen chains, kinks in the triple helix, failure to cut mature collagen, and defects in aggregation outside the cell.

Knowing which specific mutations cause disorders offers a way to identify the condition before symptoms arise. This can be helpful if early treatment can follow. A woman who has a high risk of developing hereditary osteoporosis, for example, might take calcium supplements before symptoms appear.

Aortic aneurysm is a more serious connective tissue disorder that can be presymptomatically detected if the underlying mutation is discovered. In aortic aneurysm, a weakened aorta (the largest blood vessel in the body, which emerges from the heart) bursts. Knowing that the mutant gene has not been inherited can ease worries—and knowing that it has been inherited can warn affected individuals to have frequent ultrasound exams so that aortic weakening can be detected early enough to correct with surgery. ■

TABLE 5.5 Collagen Disorders

Disorder	Molecular Defect	Symptoms
Chondrodysplasia	Collagen chains are too wide and asymmetric	Stunted growth; deformed joints
Dystrophic epidermolysis bullosa	Breakdown of collagen fibrils that attach skin layers to each other	Stretchy, easily scarred skin; lax joints
Hereditary osteoarthritis	Substituted amino acid in collagen chain alters shape	Painful joints
Osteogenesis imperfecta type I	Too few collagen triple helices	Easily broken bones; deafness; blue sclera (whites of the eyes)
Stickler syndrome	Short collagen chains	Joint pain; degeneration of retina and fluid around it

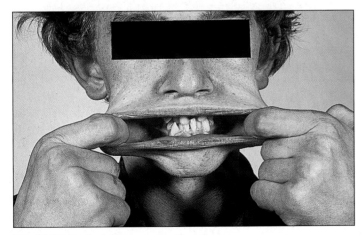

FIGURE 5.17
Abnormal collagen causes the stretchy skin of Ehlers-Danlos syndrome type I.

they are easily stretched or deformed and will resume their original lengths and shapes when the force acting upon them is removed. Elastic fibers are common in body parts that are normally subjected to stretching, such as the vocal cords and air passages of the respiratory system. Elastic fibers are sometimes called yellow fibers, because tissues amply supplied with them appear yellowish (see fig. 5.16).

Surgeons use elastin in foam, powder, or sheet form to prevent scar tissue adhesions from forming at the sites of tissue removal. Elastin is produced in bacteria that are genetically altered to contain human genes that instruct them to manufacture the human protein. This is cheaper than synthesizing elastin chemically and safer than obtaining it from cadavers.

TABLE 5.6	Components of Connective Tissue	
Component	**Characteristic**	**Function**
Fibroblasts	Widely distributed, large, star-shaped cells	Secrete proteins that become fibers
Macrophages	Motile cells sometimes attached to fibers	Clear foreign particles from tissues by phagocytosis
Mast cells	Large cells, usually located near blood vessels	Release substances that may help prevent blood clotting and promote inflammation
Collagenous fibers (white fibers)	Thick, threadlike fibers of collagen with great tensile strength	Hold structures together
Elastic fibers (yellow fibers)	Bundles of microfibrils embedded in elastin	Provide elastic quality to parts that stretch
Reticular fibers	Thin fibers of collagen	Form supportive networks within tissues

Reticular fibers are very thin collagenous fibers. They are highly branched and form delicate supporting networks in a variety of tissues. Table 5.6 summarizes the components of connective tissue.

1. What are the general characteristics of connective tissue?

2. What are the major types of fixed cells in connective tissue?

3. What is the primary function of fibroblasts?

4. What are the characteristics of collagen and elastin?

Categories of Connective Tissues

Connective tissue is broken down into two categories. *Connective tissue proper* includes loose connective tissue, adipose tissue, reticular connective tissue, dense connective tissue, and elastic connective tissue. The *specialized connective tissues* include cartilage, bone, and blood. Each type of connective tissue is described in the following sections.

Loose Connective Tissue

Loose connective tissue, or **areolar tissue** (ah-re′o-lar tish′u), forms delicate, thin membranes throughout the body. The cells of this tissue, mainly fibroblasts, are located some distance apart and are separated by a gel-like ground substance that contains many collagenous and elastic fibers that fibroblasts secrete (fig. 5.18).

Loose connective tissue binds the skin to the underlying organs and fills spaces between muscles. It lies beneath most layers of epithelium, where its many blood vessels nourish nearby epithelial cells.

Adipose Tissue

Adipose tissue (ad′ĭ-pōs tish′u), or fat, is another form of connective tissue. Certain cells within connective tissue (adipocytes) store fat in droplets within their cytoplasm. At first, these cells resemble fibroblasts, but as they accumulate fat, they enlarge, and their nuclei are pushed to one side (fig. 5.19). When adipocytes become so abundant that they crowd out other cell types, they form adipose tissue. This tissue lies beneath the skin, in spaces between muscles, around the kidneys, behind the eye-

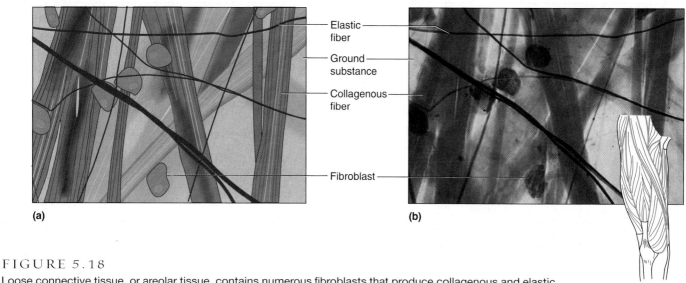

Elastic fiber

Ground substance

Collagenous fiber

Fibroblast

(a)   (b)

FIGURE 5.18

Loose connective tissue, or areolar tissue, contains numerous fibroblasts that produce collagenous and elastic fibers (700×).

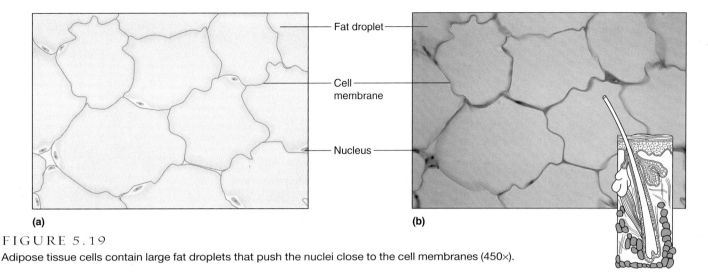

FIGURE 5.19
Adipose tissue cells contain large fat droplets that push the nuclei close to the cell membranes (450×).

balls, in certain abdominal membranes, on the surface of the heart, and around certain joints.

Adipose tissue cushions joints and some organs, such as the kidneys. It also insulates beneath the skin, and it stores energy in fat molecules.

A person is born with a certain number of fat cells. Because excess food calories are likely to be converted to fat and stored, the amount of adipose tissue in the body reflects diet or an endocrine disorder. During a period of fasting, adipose cells may lose their fat droplets, shrink, and become more like fibroblasts again.

> Infants and young children have a continuous layer of adipose tissue just beneath the skin, which gives their bodies a rounded appearance. In adults, this subcutaneous fat thins in some regions and remains thick in others. For example, in males, adipose tissue usually thickens in the upper back, arms, lower back, and buttocks; in females, it is more likely to develop in the breasts, buttocks, and thighs.

Reticular Connective Tissue

Reticular connective tissue is composed of thin, collagenous fibers in a three-dimensional network. It supports the walls of certain internal organs, such as the liver, spleen, and lymphatic organs (fig. 5.20).

Dense Connective Tissue

Dense connective tissue consists of many closely packed, thick, collagenous fibers, a fine network of elastic fibers, and a few cells, most of which are fibroblasts. Subclasses of this tissue are regular or irregular, according to how organized the fiber patterns are.

Collagenous fibers of regular dense connective tissue are very strong, enabling the tissue to withstand pulling forces (fig. 5.21). It often binds body parts together, as parts of tendons and ligaments. The blood supply to regular dense connective tissue is poor, slowing tissue repair. This is why a sprain, which damages tissues surrounding a joint, may take considerable time to heal.

Fibers of irregular dense connective tissue are thicker, interwoven, and more randomly organized. This

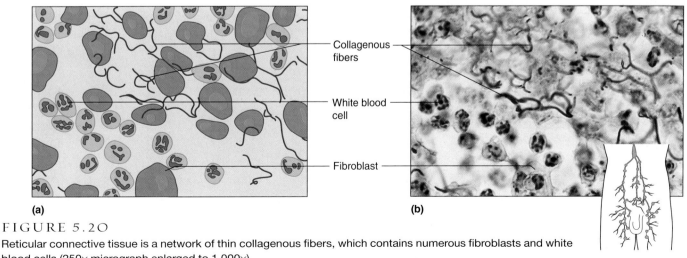

FIGURE 5.20
Reticular connective tissue is a network of thin collagenous fibers, which contains numerous fibroblasts and white blood cells (250× micrograph enlarged to 1,000×).

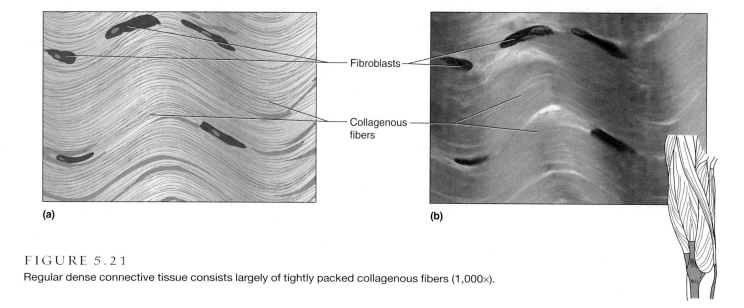

(a) Fibroblasts Collagenous fibers **(b)**

FIGURE 5.21

Regular dense connective tissue consists largely of tightly packed collagenous fibers (1,000×).

allows the tissue to sustain tension exerted from many different directions. Irregular dense connective tissue is found in the dermis, the inner skin layer.

Elastic Connective Tissue

Elastic connective tissue mainly consists of yellow, elastic fibers in parallel strands or in branching networks. Between these fibers are collagenous fibers and fibroblasts. This tissue is found in the attachments between vertebrae of the spinal column (ligamenta flava). It is also in the layers within the walls of certain hollow internal organs, including the larger arteries, some portions of the heart, and the larger airways, where it imparts an elastic quality (fig. 5.22).

1 Differentiate between loose connective tissue and dense connective tissue.

2 What are the functions of adipose tissue?

3 Distinguish between reticular and elastic connective tissues.

Cartilage

Cartilage (kar'ti-lij) is a rigid connective tissue. It provides support, frameworks, attachments, protects underlying tissues, and forms structural models for many developing bones.

Cartilage matrix is abundant and is largely composed of collagenous fibers embedded in a gel-like ground substance. This ground substance is rich in a protein-polysaccharide complex (chondromucoprotein) and contains a large amount of water. Cartilage cells, or **chondrocytes** (kon'dro-sītz), occupy small chambers called *lacunae* and thus lie completely within the matrix.

A cartilaginous structure is enclosed in a covering of connective tissue called *perichondrium*. Although cartilage tissue lacks a direct blood supply, blood vessels are in the surrounding perichondrium. Cartilage cells near the perichondrium obtain nutrients from these vessels by diffusion, which is aided by the water in the matrix. This lack of a direct blood supply is why torn

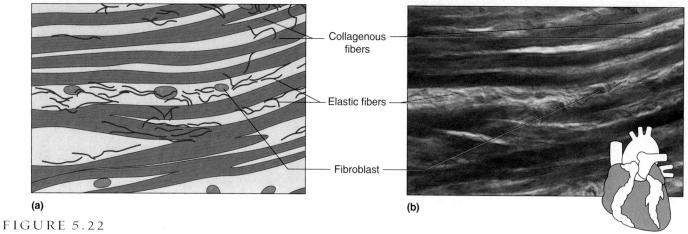

(a) Collagenous fibers Elastic fibers Fibroblast **(b)**

FIGURE 5.22

Elastic connective tissue contains many elastic fibers with collagenous fibers between them (170× micrograph enlarged to 680×).

cartilage heals slowly, and why chondrocytes do not divide frequently.

The three types of cartilage are distinguished by their different types of intercellular material. Hyaline cartilage has very fine collagenous fibers in its matrix, elastic cartilage contains a dense network of elastic fibers, and fibrocartilage has many large collagenous fibers.

Hyaline cartilage (fig. 5.23), the most common type, looks somewhat like white glass. It is found on the ends of bones in many joints, in the soft part of the nose, and in the supporting rings of the respiratory passages. Parts of an embryo's skeleton begin as hyaline cartilage "models" that bone gradually replaces. Hyaline cartilage is also important in the development and growth of most bones (see chapter 7, p. 187).

Elastic cartilage (fig. 5.24) is more flexible than hyaline cartilage because its matrix contains many elastic fibers. It provides the framework for the external ears and parts of the larynx.

Fibrocartilage (fig. 5.25), a very tough tissue, contains many collagenous fibers. It is a shock absorber for structures that are subjected to pressure. For example, fibrocartilage forms pads (intervertebral discs) between the individual bones (vertebrae) of the spinal column. It also cushions bones in the knees and in the pelvic girdle.

Bone

Bone (osseous tissue) is the most rigid connective tissue. Its hardness is largely due to mineral salts, such as calcium phosphate and calcium carbonate, in its matrix. This intercellular material also contains a great amount of collagen, whose fibers flexibly reinforce the mineral components of bone.

Bone internally supports body structures. It protects vital structures in the cranial and thoracic cavities and is an attachment for muscles. Bone also contains red marrow, which forms blood cells, and it stores and releases inorganic salts.

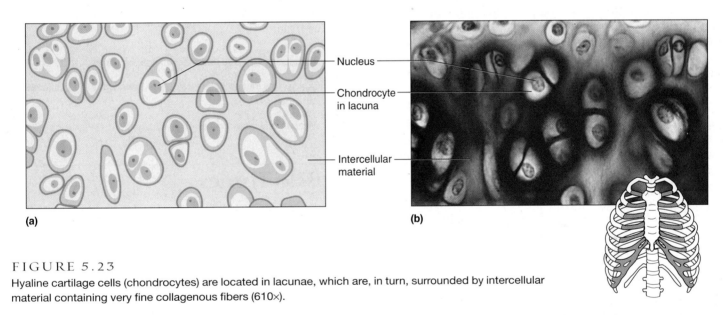

(a)   Nucleus   Chondrocyte in lacuna   Intercellular material   (b)

FIGURE 5.23
Hyaline cartilage cells (chondrocytes) are located in lacunae, which are, in turn, surrounded by intercellular material containing very fine collagenous fibers (610×).

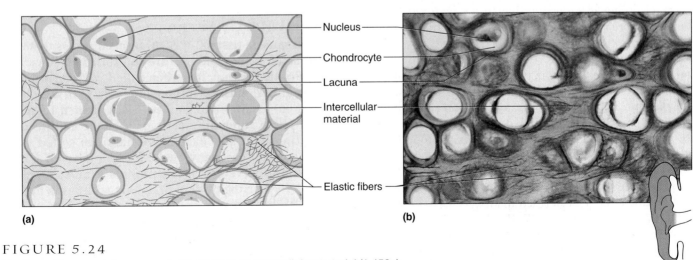

(a)   Nucleus   Chondrocyte   Lacuna   Intercellular material   Elastic fibers   (b)

FIGURE 5.24
Elastic cartilage contains many elastic fibers in its intercellular material (1,450×).

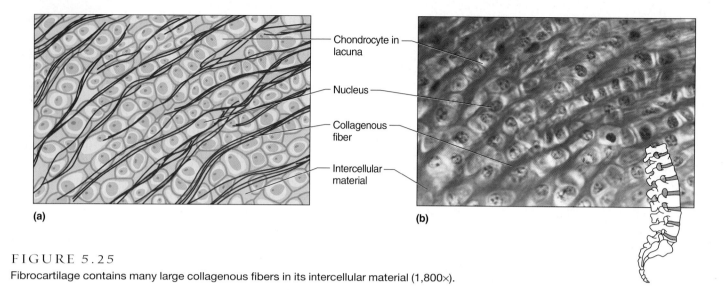

Chondrocyte in lacuna

Nucleus

Collagenous fiber

Intercellular material

(a)   (b)

FIGURE 5.25

Fibrocartilage contains many large collagenous fibers in its intercellular material (1,800×).

Bone matrix is deposited by bone cells, **osteocytes** (os′te-o-sītz), in thin layers called *lamellae,* which form concentric patterns around capillaries located within tiny longitudinal tubes called *central,* or *Haversian canals.* Osteocytes are located in lacunae that are rather evenly spaced between the lamellae. Consequently, osteocytes also form concentric circles (fig. 5.26).

In a bone, the osteocytes and layers of intercellular material, which are concentrically clustered around a central canal, form a cylinder-shaped unit called an **osteon** (os′te-on), or Haversian system. Many of these units cemented together form the substance of bone (see chapter 7, p. 184).

Each central canal contains a blood vessel, so every bone cell is fairly close to a nutrient supply. In addition, the bone cells have many cytoplasmic processes that extend outward and pass through minute tubes in the matrix called *canaliculi.* Gap junctions attach these cellular processes to the membranes of nearby cells (see chapter 3, p. 67). As a result, materials can move rapidly between blood vessels and bone cells. Thus, in spite of its inert appearance, bone is a very active tissue. Injured bone heals much more rapidly than does injured cartilage.

Blood

Blood, another type of connective tissue, is composed of cells that are suspended in a fluid intercellular matrix called *plasma.* These cells include *red blood cells, white blood cells,* and cellular fragments called *platelets* (fig. 5.27). Red blood cells transport gases; white blood cells fight infection; and platelets are involved in blood clotting. Most blood cells form in special tissues (hematopoietic tissues) in red marrow within the hollow parts of certain bones. Blood is described in chapter 14.

Of the blood cells, only the red cells function entirely within the blood vessels. White blood cells typically migrate from the blood through capillary walls. They enter connective tissues where they carry on their major activities, and they usually reside there until they die. Table 5.7 lists the characteristics of the types of connective tissue.

1 Describe the general characteristics of cartilage.

2 Explain why injured bone heals more rapidly than does injured cartilage.

3 What are the major components of blood?

Muscle Tissues

General Characteristics

Due to their elongated shape, the cells in **muscle tissues** (mus′el tish′uz) are sometimes called *muscle fibers.* Muscle tissues are contractile; they can shorten and thicken. As they contract, muscle cells pull at their attached ends, which moves body parts. The three types of muscle tissue (skeletal, smooth, and cardiac) are discussed in chapter 9.

Skeletal Muscle Tissue

Skeletal muscle tissue (fig. 5.28) forms muscles that usually attach to bones and that are controlled by conscious effort. For this reason, it is often called *voluntary* muscle tissue. Skeletal muscle cells are long—up to or more than 40 mm in length—and narrow—less than 0.1 mm in width. These threadlike cells of skeletal muscle have alternating light and dark cross-markings called *striations.* Each cell has many nuclei (multinucleate). A message from a nerve cell can stimulate a muscle cell to contract by causing protein filaments within the muscle cell to slide past one another. Then, the muscle cell relaxes. Skeletal muscles move the head, trunk, and limbs and enable us to make facial expressions, write, talk, and sing, as well as chew, swallow, and breathe.

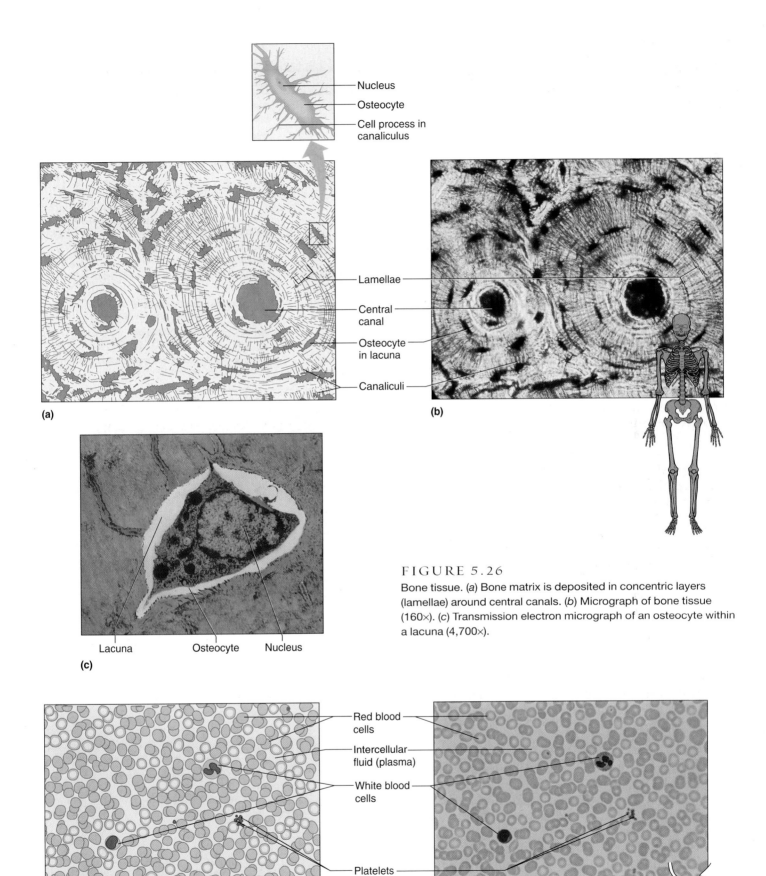

Nucleus
Osteocyte
Cell process in canaliculus

Lamellae
Central canal
Osteocyte in lacuna
Canaliculi

(a)

(b)

Lacuna   Osteocyte   Nucleus

(c)

FIGURE 5.26

Bone tissue. (*a*) Bone matrix is deposited in concentric layers (lamellae) around central canals. (*b*) Micrograph of bone tissue (160×). (*c*) Transmission electron micrograph of an osteocyte within a lacuna (4,700×).

Red blood cells
Intercellular fluid (plasma)
White blood cells
Platelets

(a)

(b)

FIGURE 5.27

Blood tissue consists of red blood cells, white blood cells, and platelets suspended in an intercellular fluid (425×).

TABLE 5.7   Connective Tissues

Type	Description	Function	Location
Loose connective tissue	Cells in fluid-gel matrix	Binds organs together, holds tissue fluids	Beneath the skin, between muscles, beneath epithelial tissues
Adipose tissue	Cells in fluid-gel matrix	Protects, insulates, and stores fat	Beneath the skin, around the kidneys, behind the eyeballs, on the surface of the heart
Reticular connective tissue	Cells in fluid-gel matrix	Supports	Walls of liver, spleen, and lymphatic organs
Dense connective tissue	Cells in fluid-gel matrix	Binds organs together	Tendons, ligaments, dermis
Elastic connective tissue	Cells in fluid-gel matrix	Provides elastic quality	Connecting parts of the spinal column, in walls of arteries and airways
Hyaline cartilage	Cells in solid-gel matrix	Supports, protects, provides framework	Ends of bones, nose, and rings in walls of respiratory passages
Elastic cartilage	Cells in solid-gel matrix	Supports, protects, provides flexible framework	Framework of external ear and part of larynx
Fibrocartilage	Cells in solid-gel matrix	Supports, protects, absorbs shock	Between bony parts of spinal column, parts of pelvic girdle, and knee
Bone	Cells in solid matrix	Supports, protects, provides framework	Bones of skeleton, middle ear
Blood	Cells and platelets in fluid matrix	Transports gases, defends against disease, clotting	Throughout the body within a closed system of blood vessels and heart chambers

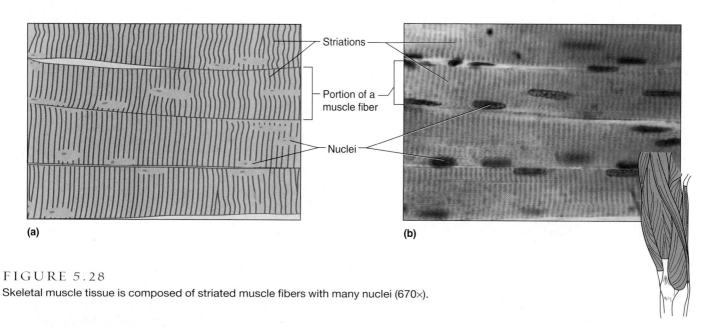

(a)   (b)

FIGURE 5.28
Skeletal muscle tissue is composed of striated muscle fibers with many nuclei (670×).

Smooth Muscle Tissue

Smooth muscle tissue (fig. 5.29) is called smooth because its cells lack striations. Smooth muscle cells are shorter than those of skeletal muscle and are spindle-shaped, each with a single, centrally located nucleus. This tissue comprises the walls of hollow internal organs, such as the stomach, intestines, urinary bladder, uterus, and blood vessels. Unlike skeletal muscle, smooth muscle usually cannot be stimulated to contract by conscious efforts. Thus, its actions are *involuntary.* For example, smooth muscle tissue moves food through the digestive tract, constricts blood vessels, and empties the urinary bladder.

Cardiac Muscle Tissue

Cardiac muscle tissue (fig. 5.30) is only in the heart. Its cells, which are striated, are joined end-to-end. The resulting muscle cells are branched and interconnected in complex networks. Each cardiac muscle cell has a single nucleus. Where one cell touches another cell is a specialized intercellular junction called an *intercalated disc,* seen only in cardiac tissue.

Cardiac muscle, like smooth muscle, is controlled involuntarily and, in fact, can continue to function without being stimulated by nerve impulses. This tissue makes up the bulk of the heart and pumps blood through the heart chambers and into blood vessels.

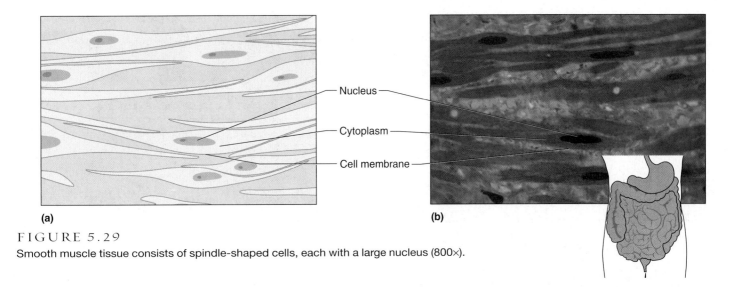

FIGURE 5.29
Smooth muscle tissue consists of spindle-shaped cells, each with a large nucleus (800×).

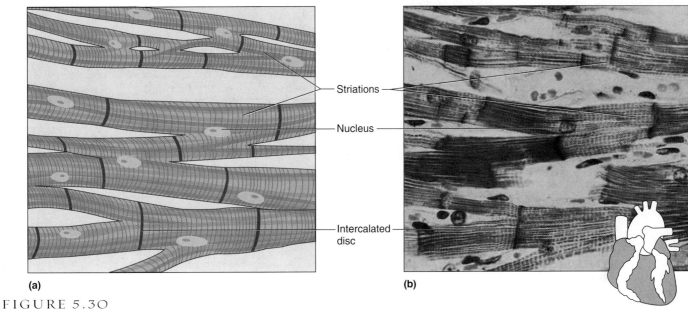

FIGURE 5.30
Cardiac muscle cells are branched and interconnected, with a single nucleus each (360×).

The cells of different tissues vary greatly in their abilities to divide. Cells that divide continuously include the epithelial cells of the skin and inner lining of the digestive tract and the connective tissue progenitor cells that form blood cells in red bone marrow. However, skeletal and cardiac muscle cells and nerve cells do not usually divide at all after differentiating.

Fibroblasts respond rapidly to injuries by increasing in number and fiber production. They are often the principal agents of repair in tissues that have limited abilities to regenerate. For instance, cardiac muscle tissue typically degenerates in the regions damaged by a heart attack. Fibroblasts then, over time, knit connective tissue that replaces the damaged cardiac muscle. A scar is formed.

1 List the general characteristics of muscle tissue.

2 Distinguish among skeletal, smooth, and cardiac muscle tissues.

Nervous Tissues

Nervous tissues (ner′vus tish′ūz) are found in the brain, spinal cord, and peripheral nerves. The basic cells are called *nerve cells,* or **neurons** (nu′ronz), and they are among the more highly specialized body cells. Neurons sense certain types of changes in their surroundings and respond by transmitting nerve impulses along cellular processes to other neurons or to muscles or glands (fig. 5.31). As a result of the extremely complex patterns by which neurons connect with each other and with muscle

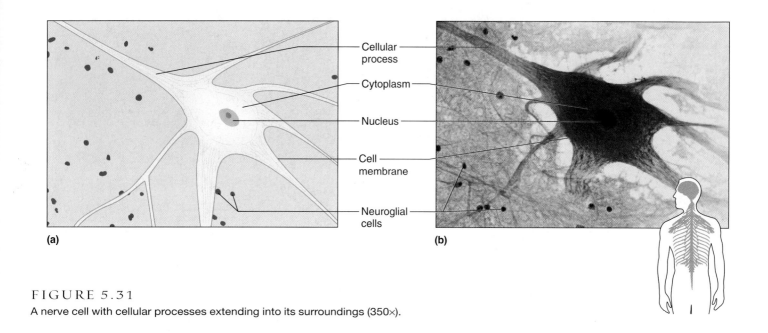

(a)

(b)

FIGURE 5.31
A nerve cell with cellular processes extending into its surroundings (350×).

TABLE 5.8	Muscle and Nervous Tissues						
Type	Description	Function	Location	Type	Description	Function	Location
Skeletal muscle tissue	Long, threadlike cells, striated, many nuclei	Voluntary movements of skeletal parts	Muscles usually attached to bones	Cardiac muscle tissue	Branched cells, striated, single nucleus	Heart movements	Heart muscle
Smooth muscle tissue	Shorter cells, single, central nucleus	Involuntary movements of internal organs	Walls of hollow internal organs	Nervous tissue	Cell with cytoplasmic extensions	Sensory reception and conduction of nerve impulses	Brain, spinal cord, and peripheral nerves

and gland cells, they can coordinate, regulate, and integrate many body functions.

In addition to neurons, nervous tissue includes **neuroglial cells** (nu-rog′le-ahl selz), or supporting cells, shown in figure 5.31. These cells support and bind the components of nervous tissue, carry on phagocytosis, and help supply nutrients to neurons by connecting them to blood vessels. They may also play a role in cell-to-cell communications. Nervous tissue is discussed in chapter 10.

Table 5.8 summarizes the general characteristics of muscle and nervous tissues. From Science to Technology 5.1 discusses bioengineered tissues.

1 Describe the general characteristics of nervous tissue.

2 Distinguish between neurons and neuroglial cells.

Types of Membranes

Two or more kinds of tissues grouped together and performing specialized functions constitute an **organ.** For example, **epithelial membranes,** which are thin, sheetlike structures that are usually composed of epithelial and underlying connective tissues and covering body surfaces and lining body cavities, are organs.

The three major types of epithelial membranes are *serous, mucous,* and *cutaneous.* Usually, these structures are thin. **Synovial membranes** (sĭ-no′ve-al mem′brānz), lining joints and discussed further in chapter 8 (pp. 257–258), are composed entirely of connective tissues.

Serous (se′rus) **membranes** line the body cavities that lack openings to the outside and reduce friction between the organs and cavity walls. They form the inner linings of the thorax and abdomen, and they cover the organs within these cavities (see fig. 1.11 and 1.12). A serous membrane consists of a layer of simple squamous epithelium (mesothelium) and a thin layer of loose connective tissue. Cells of a serous membrane secrete watery *serous fluid,* which helps lubricate the membrane surfaces.

Mucous (mu′kus) **membranes** line the cavities and tubes that open to the outside of the body. These include the oral and nasal cavities and the tubes of the digestive, respiratory, urinary, and reproductive systems. A mucous membrane consists of epithelium overlying a layer of loose connective tissue; however, the type of epithelium varies with the location of the membrane. For example,

TISSUE ENGINEERING

If an automobile or appliance part is damaged or malfunctions, replacing it is fairly simple. Not so for the human body. To replace a human body part, biomedical engineers must first learn how to replicate the combination of cells, biochemicals, and intercellular materials that comprise tissues and organs. Then physicians must dampen the immune response sufficiently for the body to accept the replacement. A solution to the challenge of replacing body parts is tissue engineering, which combines synthetic materials with cells.

The basic recipe for a bioengineered tissue is to place cells in or on a scaffolding sculpted from a synthetic material that is accepted in the body. The cells secrete substances as they normally would, or they may be genetically altered to overproduce their natural secreted products or supply entirely different ones with therapeutic benefit, such as growth factors that might make the implant more acceptable to the body.

New Skin and More

Burn patients can sometimes be helped by a bioengineered skin consisting of the patient's epidermal cells placed in sheets over dermal cells grown in culture. A nylon mesh framework supports both layers. This semisynthetic skin may also be useful for patients who have lost a great deal of skin in surgery to remove tattoos, cancers, or moles. Bioengineered skin is also used for *in vitro* toxicity testing. In many laboratories, it has replaced live animals in testing cosmetic ingredients. A replacement cartilage similar to the skin recipe, consisting of chondrocytes in collagen, may help replace joints destroyed by arthritis.

A scaled-down version of an engineered tissue, called a cell implant, offers a new route to drug delivery, placing cells that naturally manufacture vital substances precisely where a patient needs them. The cells are packaged so that they secrete without alerting the immune system. The cells of the implant are surrounded with a polymer membrane with holes small enough to allow nutrients in and the therapeutic biochemicals out, while excluding the larger molecules that trigger immune rejection.

Prime candidates for cell implants are pancreatic beta cells, which would secrete insulin to aid people with diabetes mellitus. Brain implants would secrete dopamine, providing the biochemical that is deficient in people who have Parkinson disease. ∎

---

stratified squamous epithelium lines the oral cavity, pseudostratified columnar epithelium lines part of the nasal cavity, and simple columnar epithelium lines the small intestine. Goblet cells within a mucous membrane secrete *mucus*.

The **cutaneous** (ku-ta′ne-us) **membrane** is an organ of the integumentary organ system and is more commonly called *skin*. It is described in detail in the next chapter.

 Name the four types of membranes, and explain how they differ.

CHAPTER SUMMARY

Introduction (page 132)

Cells are organized in layers or groups to form tissues. Intercellular materials, varying from solid to liquid, separate cells. The four major types of human tissue are epithelial tissues, connective tissues, muscle tissues, and nervous tissues.

Epithelial Tissues (page 133)

1. General characteristics
   a. Epithelial tissue covers all free body surfaces and is the major tissue of glands.
   b. A basement membrane anchors epithelium to connective tissue. Epithelial tissue lacks blood vessels, contains little intercellular material, and is continuously replaced.
   c. It functions in protection, secretion, absorption, and excretion.
2. Simple squamous epithelium
   a. This tissue consists of a single layer of thin, flattened cells through which substances pass easily.
   b. It functions in the exchange of gases in the lungs and lines blood vessels, lymph vessels, and membranes within the thorax and abdomen.
3. Simple cuboidal epithelium
   a. This tissue consists of a single layer of cube-shaped cells.
   b. It carries on secretion and absorption in the kidneys and various glands.
4. Simple columnar epithelium
   a. This tissue is composed of elongated cells whose nuclei are near the basement membrane.
   b. It lines the uterus and digestive tract, where it functions in protection, secretion, and absorption.
   c. Absorbing cells often possess microvilli.
   d. This tissue usually contains goblet cells that secrete mucus.
5. Pseudostratified columnar epithelium
   a. This tissue appears stratified because the nuclei are at two or more levels.

    b. Its cells may have cilia that move mucus over the surface of the tissue.

    c. It lines tubes of the respiratory system.

6. Stratified squamous epithelium
    a. This tissue is composed of many layers of cells, the topmost of which are flattened.
    b. It protects underlying cells from harmful environmental effects.
    c. It covers the skin and lines the oral cavity, throat, vagina, and anal canal.

7. Stratified cuboidal epithelium
    a. This tissue is composed of two or three layers of cube-shaped cells.
    b. It lines the larger ducts of the sweat glands, salivary glands, and pancreas.
    c. It functions in protection.

8. Stratified columnar epithelium
    a. The top layer of cells in this tissue contains elongated columns. Cube-shaped cells make up the bottom layers.
    b. It is in the vas deferens, part of the male urethra, and parts of the pharynx.
    c. This tissue functions in protection and secretion.

9. Transitional epithelium
    a. This tissue is specialized to become distended.
    b. It is in the walls of organs of the urinary tract.
    c. It helps prevent the contents of the urinary passageways from diffusing out.

10. Glandular epithelium
    a. Glandular epithelium is composed of cells that are specialized to secrete substances.
    b. A gland consists of one or more cells.
      (1) Exocrine glands secrete into ducts.
      (2) Endocrine glands secrete into tissue fluid or blood.
    c. Exocrine glands are classified according to the organization of their cells.
      (1) Simple glands have unbranched ducts.
      (2) Compound glands have branched ducts.
      (3) Tubular glands consist of simple epithelial-lined tubes.
      (4) Alveolar glands consist of saclike dilations connected to the surface by narrowed ducts.
    d. Exocrine glands are classified according to composition of their secretions.
      (1) Merocrine glands secrete watery fluids without loss of cytoplasm. Most secretory cells are merocrine.
        (a) Serous cells secrete watery fluid with a high enzyme content.
        (b) Mucous cells secrete mucus.
      (2) Apocrine glands lose portions of their cells during secretion.
      (3) Holocrine glands release cells filled with secretions.

Connective Tissues (page 141)

1. General characteristics
    a. Connective tissue connects, supports, protects, provides frameworks, fills spaces, stores fat, produces blood cells, protects against infection, and helps repair damaged tissues.
    b. Connective tissue cells usually have considerable intercellular material between them.
    c. This intercellular matrix consists of fibers and a ground substance.

2. Major cell types
    a. Fibroblasts produce collagenous and elastic fibers.
    b. Macrophages are phagocytes.
    c. Mast cells may release heparin and histamine and usually are near blood vessels.

3. Connective tissue fibers
    a. Collagenous fibers are composed of collagen and have great tensile strength.
    b. Elastic fibers are composed of microfibrils embedded in elastin and are very elastic.
    c. Reticular fibers are very fine collagenous fibers.

4. Categories of connective tissue
    a. Connective tissue proper includes loose connective tissue, adipose tissue, reticular connective tissue, dense connective tissue, and elastic connective tissue.
    b. Specialized connective tissues include cartilage, bone, and blood.

5. Loose connective tissue
    a. This tissue forms thin membranes between organs and binds them.
    b. It is beneath the skin and between muscles.
    c. Its intercellular spaces contain tissue fluid.

6. Adipose tissue
    a. Adipose tissue is a specialized form of connective tissue that stores fat, cushions, and insulates.
    b. It is found beneath the skin, in certain abdominal membranes, and around the kidneys, heart, and various joints.

7. Reticular connective tissue
    a. This tissue largely consists of thin, branched collagenous fibers.
    b. It supports the walls of the liver, spleen, and lymphatic organs.

8. Dense connective tissue
    a. This tissue is largely composed of strong, collagenous fibers that bind structures.
    b. Regular dense connective tissue is found in tendons and ligaments, whereas irregular tissue is found in the dermis.

9. Elastic connective tissue
    a. This tissue is mainly composed of elastic fibers.
    b. It imparts an elastic quality to the walls of certain hollow internal organs such as the lungs and blood vessels.

10. Cartilage
    a. Cartilage provides a supportive framework for various structures.
    b. Its intercellular material is composed of fibers and a gel-like ground substance.
    c. It lacks a direct blood supply and is slow to heal.
    d. Cartilaginous structures are enclosed in a perichondrium, which contains blood vessels.
    e. Major types are hyaline cartilage, elastic cartilage, and fibrocartilage.
    f. Cartilage is at the ends of various bones, in the ear, in the larynx, and in the pads between the bones of the spinal column, pelvic girdle, and knees.

11. Bone
    a. The intercellular matrix of bone contains mineral salts and collagen.
    b. Its cells usually form concentric circles around osteonic canals. Canaliculi connect the cells.
    c. It is an active tissue that heals rapidly.

12. Blood
    a. Blood is composed of cells suspended in fluid.
    b. Blood cells are formed by special tissue in the hollow parts of certain bones.

Muscle Tissues (page 148)

1. General characteristics
   a. Muscle tissue contracts, moving structures that are attached to it.
   b. Three types are skeletal, smooth, and cardiac muscle tissues.
2. Skeletal muscle tissue
   a. Muscles containing this tissue usually attach to bones and are controlled by conscious effort.
   b. Muscle cells are long and threadlike, containing several nuclei, with alternating light and dark cross-markings.
   c. Muscle cells contract when stimulated by nerve impulses, then immediately relax.
3. Smooth muscle tissue
   a. This tissue of spindle-shaped cells, each with one nucleus, is in the walls of hollow internal organs.
   b. Usually it is involuntarily controlled.
4. Cardiac muscle tissue
   a. This tissue is found only in the heart.
   b. Cells, each with a single nucleus, are joined by intercalated discs and form branched networks.
   c. Cardiac muscle tissue is involuntarily controlled.

Nervous Tissues (page 151)

1. Nervous tissue is in the brain, spinal cord, and peripheral nerves.

2. Neurons
   a. Neurons sense changes and respond by transmitting nerve impulses to other neurons or to muscles or glands.
   b. They coordinate, regulate, and integrate body activities.
3. Neuroglial cells
   a. Some of these cells bind and support nervous tissue.
   b. Others carry on phagocytosis.
   c. Still others connect neurons to blood vessels.
   d. Some are involved in cell-to-cell communication.

Types of Membranes (page 152)

1. Epithelial membranes
   a. Serous membranes
      (1) Serous membranes are organs that line body cavities lacking openings to the outside.
      (2) They are composed of epithelium and loose connective tissue.
      (3) Cells of serous membranes secrete watery serous fluid that lubricates membrane surfaces.
   b. Mucous membranes
      (1) Mucous membranes are organs that line cavities and tubes opening to the outside of the body.
      (2) They are composed of epithelium and loose connective tissue.
      (3) Cells of mucous membranes secrete mucus.
   c. The cutaneous membrane is the external body covering commonly called the skin.
2. Synovial membranes are organs that line joints.

CRITICAL THINKING QUESTIONS

1. Joints such as the elbow, shoulder, and knee contain considerable amounts of cartilage and loose connective tissue. How does this explain the fact that joint injuries are often very slow to heal?
2. Disorders of collagen are characterized by deterioration of connective tissues. Why would you expect such diseases to produce widely varying symptoms?
3. Sometimes, in response to irritants, mucous cells secrete excess mucus. What symptoms might this produce if it occurred in (a) the respiratory passageways or (b) the digestive tract?
4. Tissue engineering combines living cells with synthetic materials to create functional substitutes for human tissues. What components would you use to engineer replacement (a) skin, (b) blood, (c) bone, and (d) muscle?
5. Collagen and elastin are added to many beauty products. What type of tissue are they normally part of?
6. In the lungs of smokers, a process called metaplasia occurs where the normal lining cells of the lung are replaced by squamous metaplastic cells (many layers of squamous epithelial cells). Functionally, why is this an undesirable body reaction to tobacco smoke?
7. Cancer-causing agents (carcinogens) usually act on cells that are dividing. Which of the four tissues would carcinogens most influence? Least influence?

REVIEW EXERCISES

1. Define *tissue*.
2. Name the four major types of tissue found in the human body.
3. Describe the general characteristics of epithelial tissues.
4. Distinguish between simple epithelium and stratified epithelium.
5. Explain how the structure of simple squamous epithelium provides its function.
6. Name an organ that includes each of the following tissues, and give the function of the tissue:
   a. Simple squamous epithelium
   b. Simple cuboidal epithelium
   c. Simple columnar epithelium
   d. Pseudostratified columnar epithelium
   e. Stratified squamous epithelium
   f. Stratified cuboidal epithelium
   g. Stratified columnar epithelium
   h. Transitional epithelium
7. Define *gland*.
8. Distinguish between an exocrine gland and an endocrine gland.
9. Explain how glands are classified according to the structure of their ducts and the organization of their cells.
10. Explain how glands are classified according to the nature of their secretions.
11. Distinguish between a serous cell and a mucous cell.

12. Describe the general characteristics of connective tissue.
13. Define *matrix* and *ground substance.*
14. Describe three major types of connective tissue cells.
15. Distinguish between collagen and elastin.
16. Explain the difference between loose connective tissue and dense connective tissue.
17. Explain how the quantity of adipose tissue in the body reflects diet.
18. Distinguish between regular and irregular dense connective tissues.
19. Distinguish between elastic and reticular connective tissues.
20. Explain why injured dense connective tissue and cartilage are usually slow to heal.
21. Name the major types of cartilage, and describe their differences and similarities.
22. Describe how bone cells are organized in bone tissue.
23. Explain how bone cells receive nutrients.
24. Describe the composition of blood.
25. Describe the general characteristics of muscle tissues.
26. Distinguish among skeletal, smooth, and cardiac muscle tissues.
27. Describe the general characteristics of nervous tissue.
28. Distinguish between neurons and neuroglial cells.
29. Explain why a membrane is an organ.
30. Identify locations in the body of the four types of membranes.

WEB CONNECTIONS

Visit the Student Edition of the Online Learning Center at www.mhhe.com/shier10 **for answers to chapter questions, additional quizzes, interactive learning exercises, and other study tools.**

alb-, white: *alb*inism—condition characterized by a lack of pigment.

cut-, skin: sub*cut*aneous—beneath the skin.

derm-, skin: *derm*is—inner layer of the skin.

epi-, upon, after, in addition: *epi*dermis—outer layer of the skin.

follic-, small bag: hair *follic*le—tubelike depression in which a hair develops.

hol-, entire, whole: *hol*ocrine gland—gland discharges the entire cell containing the secretion.

kerat-, horn: *kerat*in—protein produced as epidermal cells die and harden.

melan-, black: *melan*in—dark pigment produced by certain cells.

por-, passage, channel: *por*e—opening by which a sweat gland communicates to the skin's surface.

seb-, grease: *seb*aceous gland—gland that secretes an oily substance.

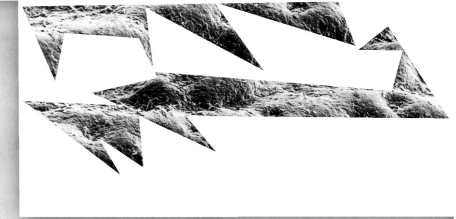

Falsely colored scanning electron micrograph of skin from the palm, with sweat pores resembling miniature craters. (40×).

Skin and the Integumentary System

CHAPTER OBJECTIVES

After you have studied this chapter, you should be able to

1. Describe the structure of the layers of the skin.

2. List the general functions of each layer of the skin.

3. Describe the accessory organs associated with the skin.

4. Explain the functions of each accessory organ.

5. Explain how the skin helps regulate body temperature.

6. Summarize the factors that determine skin color.

7. Describe the events that are part of wound healing.

8. Describe life-span changes in the integumentary system.

S mallpox is a viral infection that hideously scars the skin. It can be deadly, and in the past has decimated populations. From the years 1519 to 1521, the disease killed half of the Aztec population of Mexico, some 10 million people. In the 1600s, explorers in what is now the southeast United States found native American settlements where everyone had died from smallpox.

Smallpox is the first infectious disease for which a vaccine was developed, in 1796. Still, by 1966, when the World Health Organization began a smallpox eradication campaign, two million people died of the disease each year. Thanks to widespread vaccination, the last natural case occurred in 1977, and by 1982, nations no longer vaccinated people. Samples of the virus were maintained in the U.S. and the former Soviet Union, and at least 16 nations reportedly also kept the virus for possible use as a bioweapon.

Because of the cessation of vaccination, people born since 1972 in the U.S. are not immune to smallpox, and younger physicians have never seen a case of the disease. Limited vaccination is now resuming in many countries because of the fear of use of the virus as a bioweapon. Smallpox has re-entered the medical school curriculum.

Poxviruses are characteristically brick-shaped, and infect humans, mice, and rabbits. The smallpox virus enters the human respiratory tract and settles into the mucous membranes, then travels to nearby lymph nodes. During a 4 to 14 day latent period before symptoms begin, the virus enters the blood and then rapidly divides in giant wandering cells called macrophages. It also reproduces in the liver and spleen.

Symptoms begin dramatically. Over the course of 2 to 3 days, the person suffers from an abrupt headache, backache, and fever. Mucous membranes in the mouth and throat become inflamed, and the capillar-

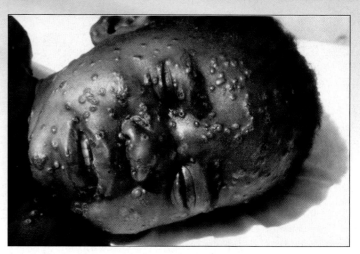

Smallpox causes disfiguring vesicles in the skin. The last natural case occurred in 1977.

ies of the dermis layer of the skin erupt into lesions. The skin becomes covered in small, red, flat sores called macules. These rise, swell and form liquid-filled vesicles, and finally crust over. The early lesions are packed with virus. During the fourth through seventh day of the rash, larger skin lesions may suddenly appear, and fever return. The face and extremities are the most severely affected areas of the skin. The death rate is about thirty percent; of the survivors, 65 to 80 percent are left with permanent scars. There is no treatment for smallpox.

In the U.S. today, people over the age of 30 still bear the marks on their upper arms from smallpox vaccination. Now that safer vaccines are being developed, a resumption of global vaccination should prevent a return of this pathogen. ■

Skin and Its Tissues

The skin, composed of several kinds of tissues, is one of the larger and more versatile organs of the body; it is vital in maintaining homeostasis (fig. 6.1). It is a protective covering that prevents many harmful substances, as well as microorganisms, from entering the body. Skin also retards water loss by diffusion from deeper tissues and helps regulate body temperature. It houses sensory receptors; contains immune system cells; synthesizes various chemicals, including vitamin D; and excretes small quantities of waste.

The skin includes two distinct tissue layers. The outer layer, called the **epidermis** (ep″i-der′mis), is composed of stratified squamous epithelium. The inner layer, or **dermis** (der′mis), is thicker than the epidermis, and is made up of connective tissue containing collagen and elastic fibers, epithelial tissue, smooth muscle tissue, nervous tissue, and blood. A *basement membrane* that is

anchored to the dermis by short fibrils separates the two skin layers.

A group of inherited conditions collectively called *epidermolysis bullosa* (EB) destroy the vital integrity of the skin's layered organization. Symptoms include very easy blistering and scarring. Different types of EB reflect the specific proteins affected. In EB simplex, blisters form only on the hands and feet and usually only during warm weather. EB simplex is an abnormality in the protein keratin in epidermal cells. In the severe dystrophic form, collagen fibers that anchor the dermis to the epidermis are abnormal, causing the layers to separate, forming many blisters. The basement membrane form of EB is so severe that it causes death in infancy. It is a defect in epiligrin, a protein that anchors the epidermis to the basement membrane.

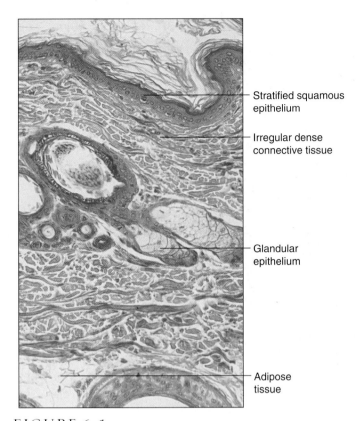

Stratified squamous
epithelium

Irregular dense
connective tissue

Glandular
epithelium

Adipose
tissue

FIGURE 6.1

An organ, such as the skin, is composed of several kinds of tissues
(30×).

Beneath the dermis, masses of loose connective and adipose tissues bind the skin to underlying organs. These tissues are not part of the skin. They form the **subcutaneous layer** (sub″ku-ta′ne-us la′er), or hypodermis (fig. 6.2).

1 List the general functions of the skin.

2 Name the tissue in the outer layer of the skin.

3 Name the tissues in the inner layer of the skin.

Epidermis

Since the epidermis is composed of stratified squamous epithelium, it lacks blood vessels. However, the deepest layer of epidermal cells, called the *stratum basale,* is close to the dermis and is nourished by dermal blood vessels. Cells of the stratum basale can divide and grow because they are well nourished. As new cells enlarge, they push the older epidermal cells away from the dermis toward the surface of the skin. The farther the cells travel, the poorer their nutrient supply becomes, and, in time, they die.

The cell membranes of older skin cells (keratinocytes) thicken and develop many desmosomes that fasten them to each other (see chapter 3, p. 67). At the

Because blood vessels in the dermis supply nutrients to the epidermis, interference with blood flow may kill epidermal cells. For example, when a person lies in one position for a prolonged period, the weight of the body pressing against the bed blocks the skin's blood supply. If cells die, the tissues begin to break down (necrosis), and a pressure ulcer (also called a decubitus ulcer or bedsore) may appear.

Pressure ulcers usually occur in the skin overlying bony projections, such as on the hip, heel, elbow, or shoulder. Frequently changing body position or massaging the skin to stimulate blood flow in regions associated with bony prominences can prevent pressure ulcers. In the case of a paralyzed person who cannot feel pressure or respond to it by shifting position, caregivers must turn the body often to prevent pressure ulcers.

same time, the cells begin to harden, a process called **keratinization** (ker″ah-tin″ĭ-za′shun), when strands of tough, fibrous, waterproof keratin proteins are synthesized and stored within the cell. As a result, many layers of tough, tightly packed dead cells accumulate in the epidermis, forming an outermost layer called the *stratum corneum.* The dead cells that compose it are eventually shed. This happens, for example, when the skin is rubbed briskly with a towel.

The structural organization of the epidermis varies from region to region. It is thickest on the palms of the hands and the soles of the feet, where it may be 0.8–1.4 mm thick. In most areas, only four layers can be distinguished. They are the *stratum basale* (stratum germinativum, or basal cell layer), which is the deepest layer; the *stratum spinosum,* a thick layer; the *stratum granulosum,* a granular layer; and the *stratum corneum,* a fully keratinized layer (horny layer). An additional layer, the *stratum lucidum* (between the stratum granulosum and the stratum corneum) is in the thickened skin of the palms and soles. The cells of these layers change shape as they are pushed toward the surface (fig. 6.3).

In body regions other than the palms and soles, the epidermis is usually very thin, averaging 0.07–0.12 mm. The stratum lucidum may be missing where the epidermis is thin. Table 6.1 describes the characteristics of each layer of the epidermis.

In healthy skin, production of epidermal cells is closely balanced with loss of dead cells from the stratum corneum, so that skin does not wear away completely. In fact, the rate of cell division increases where the skin is rubbed or pressed regularly, causing the growth of thickened areas called *calluses* on the palms and soles and keratinized conical masses on the toes called *corns.* Other changes in the skin include the common rashes described in table 6.2.

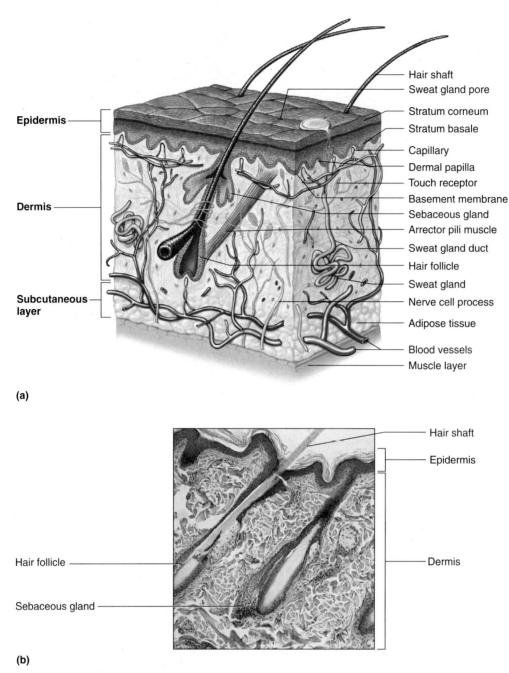

Hair shaft
Sweat gland pore
Stratum corneum
Stratum basale
Capillary
Dermal papilla
Touch receptor
Basement membrane
Sebaceous gland
Arrector pili muscle
Sweat gland duct
Hair follicle
Sweat gland
Nerve cell process
Adipose tissue
Blood vessels
Muscle layer

Epidermis

Dermis

Subcutaneous layer

(a)

Hair shaft
Epidermis
Dermis

Hair follicle
Sebaceous gland

(b)

FIGURE 6.2

Skin. (*a*) A section of skin. (*b*) A light micrograph depicting the layered structure of the skin (75×).

In psoriasis, a chronic skin disease, cells in the epidermis divide seven times more frequently than normal. Excess cells accumulate, forming bright red patches covered with silvery scales, which are keratinized cells. Medications used to treat cancer, such as methotrexate, are used to treat severe cases of psoriasis. Immune suppressing medications, such as topical corticosteroids, are used for chronic treatment of psoriasis. Five million people in the United States and 2% of all people worldwide have psoriasis.

Specialized cells in the epidermis called **melanocytes** produce the dark pigment **melanin** (mel'ah-nin) that provides skin color, discussed further on page 172 (fig. 6.4*a*). Melanin absorbs ultraviolet radiation in sunlight, preventing mutations in the DNA of skin cells and other damaging effects.

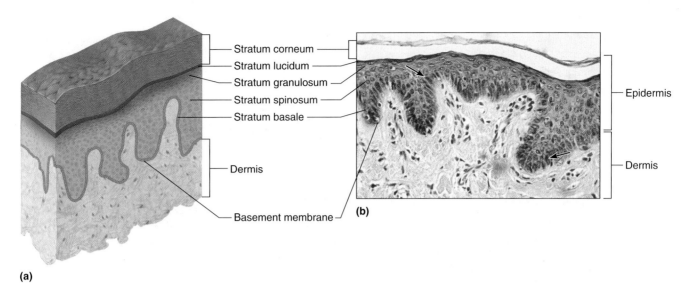

FIGURE 6.3

Epidermis. (a) The layers of the epidermis are distinguished by changes in cells as they are pushed toward the surface of the skin. (b) Melanocytes (arrows) that are mainly in the stratum basale, the deepest layer of the epidermis, produce the pigment melanin (240×).

TABLE 6.1	Layers of the Epidermis	
Layer	**Location**	**Characteristics**
Stratum corneum	Outermost layer	Many layers of keratinized, dead epithelial cells that are flattened and nonnucleated
Stratum lucidum	Between stratum corneum and stratum granulosum on soles and palms	Cells appear clear; nuclei, organelles, and cell membranes are no longer visible
Stratum granulosum	Beneath the stratum corneum	Three to five layers of flattened granular cells that contain shrunken fibers of keratin and shriveled nuclei
Stratum spinosum	Beneath the stratum granulosum	Many layers of cells with centrally located, large, oval nuclei and developing fibers of keratin; cells becoming flattened
Stratum basale (basal cell layer)	Deepest layer	A single row of cuboidal or columnar cells that divide and grow; this layer also includes melanocytes

TABLE 6.2	Rashes	
Illness	**Description of Rash**	**Cause**
Chicken pox	Tiny pustules start on back, chest, or scalp and spread for three to four days. Pustules form blisters, then crust, then fall away.	*Herpes varicella*
Fifth disease	Beginning with "slapped cheek" appearance, then red spots suddenly cover entire body, lasting up to two days.	*Human parvovirus B19*
Impetigo	Thin-walled blisters and thick, crusted lesions appear.	*Staphylococcus aureus, Streptococcus pyogenes*
Lyme disease	Large rash resembling a bull's-eye usually appears on thighs or trunk.	*Borrelia burgdorferi*
Rosacea	Flushing leads to sunburned appearance in center of face. Red pimples and then wavy red lines develop.	Unknown, but may be a microscopic mite living in hair follicles
Roseola infantum	Following high fever, red spots suddenly cover entire body, lasting up to two days.	*Herpesvirus 6*
Scarlet fever	Rash resembling sunburn with goose bumps begins below ears, on chest and underarms, and spreads to abdomen, limbs, and face. Skin may peel.	*Group A Streptococcus*
Shingles	Small, clear blisters appear on inflamed skin. Blisters enlarge, become cloudy, crust, then fall off.	The virus that causes chicken pox stays in peripheral nerves, affecting the area where the nerve endings reach the skin.

Acute sunburn (solar erythema) is an inflammatory reaction of the skin to excessive exposure to ultraviolet radiation in sunlight. The skin becomes very red, swollen, and painful, with discomfort peaking between 6 and 48 hours after exposure. Within a few days the skin may peel, as surface cell die and are shed. Peeling, an example of apoptosis (programmed cell death), prevents cancer from developing by ridding the body of susceptible cells. Microscopic skin changes begin within a half hour of intense sun exposure, including damage to cells in the upper, epidermal layer of the skin, and swelling of blood vessels in the deeper, dermal layer.

Continued sun exposure leads to tanning, as specialized skin cells produce more melanin pigment. At the same time, the stratum corneum, the outermost epidermal layer, thickens. Over time, sun overexposure hastens wrinkling and may produce a leathery feel as the skin loses elasticity. Frequent, severe sunburns, especially early in life, raise the risk of developing skin cancer.

Treatment for acute sunburn includes frequent cool baths, perhaps with oatmeal or baking soda added to soothe. Do not wash the area with a harsh soap, or use products with benzocaine, which can cause allergic reactions. Apply aloe for the first two days, but do not use petroleum jelly, ointments, or butters—these lock in the heat. Seek medical care if fever, blistering, dizziness, or visual disturbances develop, which are signs of sun poisoning.

To avoid sunburn, stay out of the sun between the hours of 10 AM and 3 PM, and when exposed, apply sunblock with and SPF factor of at least 15—even on a cloudy day. Certain medications can hasten or intensify the skin's reaction to sun. Tanning lotions, reflectors, sunlamps, or tanning booths may pose a risk for sunburn.

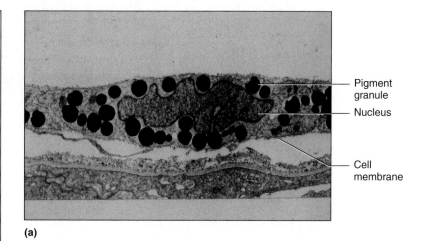

(a)

Pigment granule
Nucleus
Cell membrane

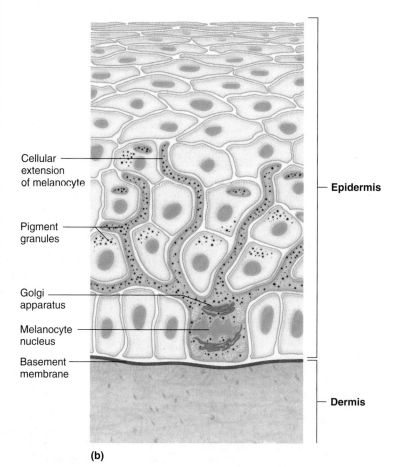

Cellular extension of melanocyte

Pigment granules

Golgi apparatus

Melanocyte nucleus

Basement membrane

Epidermis

Dermis

(b)

FIGURE 6.4

Melanocyte. (a) Transmission electron micrograph of a melanocyte with pigment-containing granules (10,600×). (b) A melanocyte may have pigment-containing extensions that pass between epidermal cells and transfer pigment into them.

Melanocytes lie in the deepest portion of the epidermis and in the underlying connective tissue of the dermis. Although melanocytes are the only cells that can produce melanin, the pigment also may be present in nearby epidermal cells. This happens because melanocytes have long, pigment-containing cellular extensions that pass upward between neighboring epidermal cells, and the extensions can transfer granules of melanin into these other cells by a process called *cytocrine secretion*. Nearby epidermal cells may contain more melanin than the melanocytes (fig. 6.4b). Clinical Application 6.1 discusses one consequence of excessive sun exposure—skin cancer.

1   Explain how the epidermis is formed.

2   Distinguish between the stratum basale and the stratum corneum.

3   What is the function of melanin?

SKIN CANCER

Like cigarette smoking, a deep, dark tan was once very desirable. A generation ago, a teenager might have spent hours on a beach, skin glistening with oil, maybe even using a reflecting device to concentrate sun exposure on the face. Today, as they lather on sunblock, many of these people realize that the tans of yesterday may cause cancer tomorrow. However, today many people increase their risk of developing skin cancer by spending time in tanning booths.

Cancer begins when ultraviolet radiation causes mutation in the DNA of a skin cell. People who inherit the very rare condition xeroderma pigmentosum are very prone to developing skin cancer because they lack DNA repair enzymes. They must be completely covered by clothing and sunblock when in the sun to avoid developing skin cancers (fig. 6A).

Skin cancer usually arises from non-pigmented epithelial cells within the deep layer of the epidermis or from pigmented melanocytes. Skin cancers originating from epithelial cells are called *cutaneous carcinomas* (basal cell carcinoma or squamous cell carcinoma); those arising from melanocytes are *cutaneous melanomas* (melanocarcinomas or malignant melanomas) (fig. 6B).

Cutaneous carcinomas are the most common type of skin cancer. They occur most frequently in light-skinned people over forty years of age. These cancers usually appear in persons who are regularly exposed to sunlight, such as farmers, sailors, athletes, and sunbathers.

A cutaneous carcinoma often develops from a hard, dry, scaly growth with a reddish base. The lesion may be flat or raised and usually firmly adheres to the skin, appearing most often on the neck, face, or scalp. Fortunately, cutaneous carcinomas are typically slow growing and can usually be cured completely by surgical removal or radiation treatment.

A cutaneous melanoma is pigmented with melanin, often with a variety of colored areas—variegated brown, black, gray, or blue. A melanoma usually has irregular rather than smooth outlines and may feel bumpy (fig. 6B).

People of any age may develop a cutaneous melanoma. These cancers seem to be caused by short, intermittent exposure to high-intensity sunlight. Thus, risk of melanoma increases in persons who stay indoors but occasionally sustain blistering sunburns.

Light-skinned people who burn rather than tan are at higher risk of developing a cutaneous melanoma. The cancer usually appears in the skin of the trunk, especially the back, or the limbs, arising from normal-appearing skin or from a mole (nevus). The lesion spreads horizontally through the skin, but eventually may thicken and grow downward into the skin, invading deeper tissues. Surgical removal during the horizontal growth phase can arrest the cancer. But once the lesion thickens and spreads into deeper tissues, it becomes more difficult to treat, and the survival rate is very low. A type of gene therapy called a "cancer vaccine" attempts to stimulate a person's immune system to locate and destroy melanoma cells that have spread.

The incidence of melanoma has been increasing rapidly for the past twenty years. To reduce the chances of occurrence, avoid exposure to high-intensity sunlight, use sunscreens and sunblocks, and examine the skin regularly. Report any unusual lesions—particularly those that change in color, shape, or surface texture—to a physician. ■

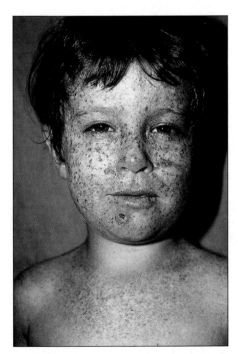

FIGURE 6A

This child has xeroderma pigmentosum. Sun exposure causes extreme freckling, and skin cancer is likely to develop because he lacks DNA repair enzymes. The large lesion on his chin is a skin cancer.

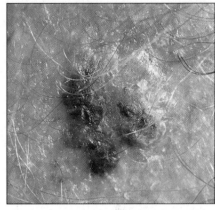

(a)

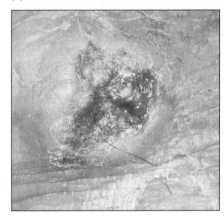

(b)

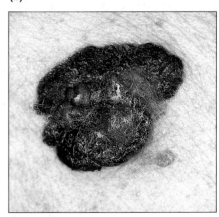

(c)

FIGURE 6B

Skin cancer. (*a*) Squamous cell carcinoma. (*b*) Basal cell carcinoma. (*c*) Malignant melanoma.

Dermis

The boundary between the epidermis and dermis is usually uneven. This is because the epidermis has ridges projecting inward and the dermis has conical *dermal papillae* passing into the spaces between the ridges (see fig. 6.2).

Fingerprints form from these undulations of the skin at the distal end of the palmar surface of a finger. Fingerprints are used for purposes of identification because they are individually unique. The pattern of a fingerprint is genetically determined, and the prints form during fetal existence. However, during a certain time early in development, fetal movements can change the print pattern. Because no two fetuses move exactly alike, even the fingerprints of identical twins are slightly different.

The dermis binds the epidermis to the underlying tissues. It is largely composed of irregular dense connective tissue that includes tough collagenous fibers and elastic fibers in a gel-like ground substance. Networks of these fibers give the skin toughness and elasticity. On the average, the dermis is 1.0–2.0 mm thick; however, it may be as thin as 0.5 mm or less on the eyelids or as thick as 3.0 mm on the soles of the feet.

The dermis also contains muscle fibers. Some regions, such as the skin that encloses the testes (scrotum), contain many smooth muscle cells that can wrinkle the skin when they contract. Other smooth muscles in the dermis are associated with accessory organs such as hair follicles and glands. Many skeletal muscle fibers are anchored to the dermis in the skin of the face. They help produce the voluntary movements associated with facial expressions.

Nerve cell processes are scattered throughout the dermis. Motor processes carry impulses to dermal muscles and glands, and sensory processes carry impulses away from specialized sensory receptors, such as touch receptors (see fig. 6.2).

One type of dermal sensory receptor, Pacinian corpuscles, is stimulated by heavy pressure, whereas another type, Meissner's corpuscles, senses light touch. Still other receptors respond to temperature changes or to factors that can damage tissues. Sensory receptors are discussed in chapter 12 (p. 424). The dermis also contains blood vessels, hair follicles, sebaceous glands, and sweat glands, which are discussed later in the chapter.

Subcutaneous Layer

The subcutaneous layer (hypodermis) beneath the dermis consists of loose connective and adipose tissues (see fig. 6.2). The collagenous and elastic fibers of this layer are continuous with those of the dermis. Most of these fibers run parallel to the surface of the skin, extending in all directions. As a result, no sharp boundary separates the dermis and the subcutaneous layer.

The adipose tissue of the subcutaneous layer insulates, helping to conserve body heat and impeding the entrance of heat from the outside. The amount of adipose tissue varies greatly with each individual's nutritional condition. It also varies in thickness from one region to another. For example, adipose tissue is usually thick over the abdomen, but absent in the eyelids.

The subcutaneous layer contains the major blood vessels that supply the skin. Branches of these vessels form a network (rete cutaneum) between the dermis and the subcutaneous layer. They, in turn, give off smaller vessels that supply the dermis above and the underlying adipose tissue.

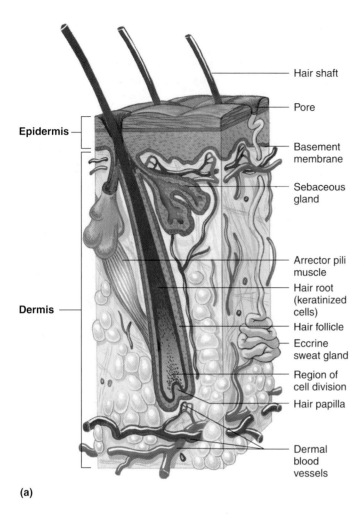

Hair shaft

Pore

Epidermis

Basement membrane

Sebaceous gland

Arrector pili muscle

Hair root (keratinized cells)

Hair follicle

Eccrine sweat gland

Dermis

Region of cell division

Hair papilla

Dermal blood vessels

(a)

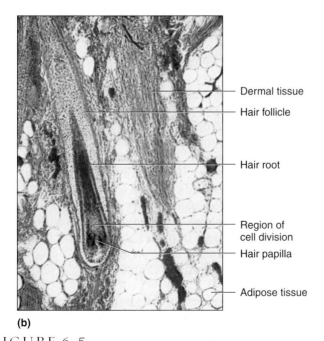

Dermal tissue

Hair follicle

Hair root

Region of cell division

Hair papilla

Adipose tissue

(b)

FIGURE 6.5

Hair follicle. (*a*) A hair grows from the base of a hair follicle when epidermal cells divide and older cells move outward and become keratinized. (*b*) Light micrograph of a hair follicle (160×).

1 What kinds of tissues make up the dermis?

2 What are the functions of these tissues?

3 What are the functions of the subcutaneous layer?

Accessory Organs of the Skin

Accessory organs of the skin extend downward from the epidermis and include hair follicles, nails, and skin glands. As long as accessory organs remain intact, severely burned or injured dermis can regenerate.

Hair Follicles

Hair is present on all skin surfaces except the palms, soles, lips, nipples, and parts of the external reproductive organs; however, it is not always well developed. For example, hair on the forehead is usually very fine.

Each hair develops from a group of epidermal cells at the base of a tubelike depression called a **hair follicle** (hār fol'i-kl). This follicle extends from the surface into the dermis and contains the hair *root,* the portion of hair embedded in the skin. The epidermal cells at its base are nourished from dermal blood vessels in a projection of connective tissue (hair papilla) at the deep end of the follicle. As these epidermal cells divide and grow, older cells are pushed toward the surface. The cells that move upward and away from the nutrient supply become keratinized and die. Their remains constitute the structure of a developing *hair shaft* that extends away from the skin surface. In other words, a hair is composed of dead epidermal cells (figs. 6.5 and 6.6). Both hair and epidermal cells develop from the same types of stem cells.

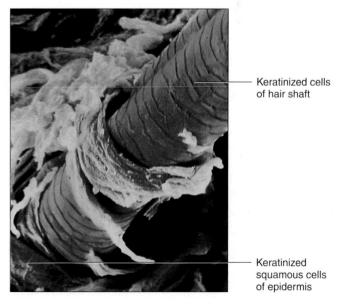

Keratinized cells of hair shaft

Keratinized squamous cells of epidermis

FIGURE 6.6

Scanning electron micrograph of a hair emerging from the epidermis (875×).

HAIR LOSS

A healthy person loses from twenty to 100 hairs a day as part of the normal growth cycle of hair. A hair typically grows for two to six years, rests for two to three months, then falls out. A new hair grows in its place. At any time, 90% of hair is in the growth phase.

In the United States, about 57.5 million people have some degree of baldness. Pattern baldness, in which the top of the head loses hair, affects 35 million men and 20 million women. The women tend to be past menopause, when lowered amounts of the hormone estrogen contribute to hair loss, which is more even on the scalp than it is in men. Pattern baldness is called *androgenic alopecia* because it is associated with testosterone, an androgenic (male) hormone. About 2.5 million people have an inherited condition called *alopecia areata,* in which the body manufactures antibodies that attack the hair follicles. This results in oval bald spots in mild cases but complete loss of scalp and body hair in severe cases.

Various conditions can cause temporary hair loss. Lowered estrogen levels shortly before and after giving birth may cause a woman's hair to fall out in clumps. Taking birth control pills, cough medications, certain antibiotics, vitamin A derivatives, antidepressants, and many other medications can also cause temporary hair loss. A sustained high fever may prompt hair loss six weeks to three months later.

Many people losing their hair seek treatment (fig. 6C). One treatment is minoxidil (Rogaine), a drug originally used to lower high blood pressure. Rogaine causes new hair to grow in 10 to 14% of cases, but in 90% of people, it slows hair loss. However, when a person stops taking it, any new hair falls out.

Hair transplants move hair follicles from a hairy part of a person's body to a bald part, and they are successful. Several other approaches, however, are potentially damaging to the scalp. Suturing on hair pieces often leads to scarring and infection. The Food and Drug Administration banned hair implants of high-density artificial fibers because they, too, become easily infected. Products called "thinning hair supplements" are conditioners, often found in ordinary shampoo, that merely make hair feel thicker.

FIGURE 6C
Being bald can be beautiful, but many people with hair loss seek ways to grow hair.

They are generally concoctions of herbs and the carbohydrate polysorbate. Labels claim the product "releases hairs trapped in the scalp." ∎

> *Folliculitis* is an inflammation of the hair follicles in response to bacterial infection. Folliculitis is more common in males and can occur during everyday shaving. One woman got a severe case by repeatedly using a loofah sponge containing bacteria.

Usually a hair grows for a time and then rests while it remains anchored in its follicle. Later, a new hair begins to grow from the base of the follicle, and the old hair is pushed outward and drops off. Sometimes, however, the hairs are not replaced. When this occurs in the scalp, the result is baldness, described in Clinical Application 6.2.

Genes determine hair color by directing the type and amount of pigment that epidermal melanocytes produce. For example, dark hair has much more melanin than blond hair. The white hair of a person with the inherited condition *albinism* lacks melanin altogether. Bright red hair contains an iron pigment (trichosiderin) that is not in hair of any other color. A mixture of pigmented and unpigmented hair usually appears gray.

A bundle of smooth muscle cells, forming the *arrector pili muscle* (see figs. 6.2 and 6.5a), attaches to each hair follicle. This muscle is positioned so that a short hair within the follicle stands on end when the muscle contracts. If a person is upset or very cold, nerve impulses may stimulate the arrector pili muscles to contract, raising gooseflesh, or goose bumps. Each hair follicle also has associated with it one or more sebaceous (oil-producing) glands, discussed later in the chapter.

> Some interesting hair characteristics are inherited. The direction of a cowlick is inherited, with a clockwise whorl being more common than a counterclockwise whorl. A white forelock, and hairy ears, elbows, nose tip, or palms are also inherited.

Nails

Nails are protective coverings on the ends of the fingers and toes. Each nail consists of a *nail plate* that overlies a surface of skin called the *nail bed.* Specialized epithelial cells that are continuous with the epithelium of the skin produce the nail bed. The whitish, thickened, half-moon–shaped region (lunula) at the base of a nail plate is the most active growing region. The epithelial cells here divide, and the newly formed cells are keratinized. This gives rise to tiny, keratinized scales that become part of the nail plate, pushing it forward over the nail bed. In time, the plate extends beyond the end of the nail bed and with normal use gradually wears away (fig. 6.7).

Nail appearance mirrors health. Bluish nail beds may reflect a circulation problem. A white nail bed or oval depressions in a nail can indicate anemia. A pigmented spot under a nail that isn't caused by an injury may be a melanoma. Horizontal furrows may result from a period of serious illness or indicate malnutrition. Certain disorders of the lungs, heart, or liver may cause extreme curvature of the nails. Red streaks in noninjured nails may be traced to rheumatoid arthritis, ulcers, or hypertension.

Skin Glands

Sebaceous glands (se-ba'shus glandz) (see fig. 6.2) contain groups of specialized epithelial cells and are usually associated with hair follicles. They are holocrine glands (see chapter 5, p. 139), and their cells produce globules of a fatty material that accumulate, swelling and bursting the cells. The resulting mixture of fatty material and cellular debris is called *sebum.*

Sebum is secreted into hair follicles through short ducts and helps keep the hairs and the skin soft, pliable, and waterproof (fig. 6.8). Acne results from excess sebum secretion (Clinical Application 6.3).

Sebaceous glands are scattered throughout the skin but are not on the palms and soles. In some regions, such as the lips, the corners of the mouth, and parts of the external reproductive organs, sebaceous glands open directly to the surface of the skin rather than being connected to hair follicles.

Sweat (swet) **glands,** or sudoriferous glands, are widespread in the skin. Each gland consists of a tiny tube that originates as a ball-shaped coil in the deeper dermis or superficial subcutaneous layer. The coiled portion of the gland is closed at its deep end and is lined with sweat-secreting epithelial cells. The most numerous sweat glands, called **eccrine** (ek'rin) **glands,** respond throughout life to body temperature elevated by environmental heat or physical exercise (fig. 6.9). These glands are common on the forehead, neck, and back, where they produce profuse sweat on hot days or during intense physical activity. They also cause the moisture that appears on the palms and soles when a person is emotionally stressed.

The fluid the eccrine sweat glands secrete is carried by a tube (duct) that opens at the surface as a *pore* (fig. 6.10). Sweat is mostly water, but it also contains small quantities of salts and wastes, such as urea and uric acid. Thus, sweating is also an excretory function.

The secretions of certain sweat glands, called **apocrine** (ap'o-krin) **glands,** develop a scent as they are metabolized by skin bacteria (see fig. 6.9). (Although they are

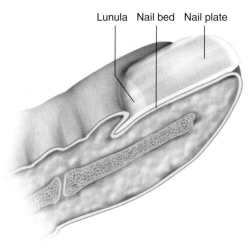

Lunula Nail bed Nail plate

FIGURE 6.7
Nails grow from epithelial cells that divide and become keratinized as the rest of the nail.

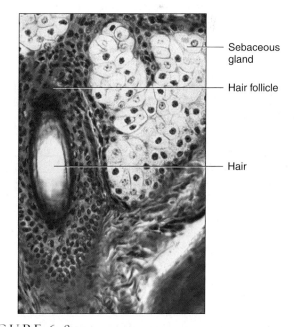

Sebaceous gland

Hair follicle

Hair

FIGURE 6.8
A sebaceous gland secretes sebum into a hair follicle, shown here in oblique section (300×).

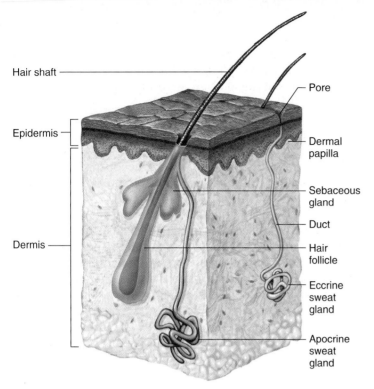

Hair shaft

Epidermis

Dermis

Pore

Dermal papilla

Sebaceous gland

Duct

Hair follicle

Eccrine sweat gland

Apocrine sweat gland

FIGURE 6.9

Note the difference in location of the ducts of the eccrine and apocrine sweat glands.

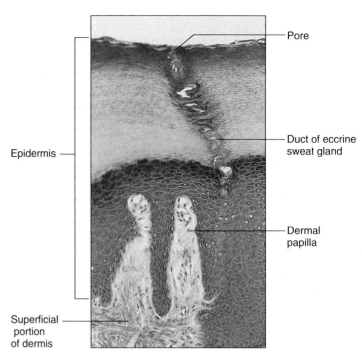

Pore

Epidermis

Duct of eccrine sweat gland

Dermal papilla

Superficial portion of dermis

FIGURE 6.10

Light micrograph of the epidermis showing the duct of an eccrine sweat gland (30×).

currently called apocrine, these glands secrete by the same mechanism as eccrine glands—see merocrine glands described in chapter 5, p. 139.) Apocrine sweat glands become active at puberty and can wet certain areas of the skin when a person is emotionally upset, frightened, or in pain. Apocrine sweat glands are also active during sexual arousal. In adults, the apocrine glands are most numerous in axillary regions, the groin, and the area around the nipples. Ducts of these glands open into hair follicles.

Other sweat glands are structurally and functionally modified to secrete specific fluids, such as the ceruminous glands of the external ear canal that secrete ear wax

and the female mammary glands that secrete milk (see chapter 23, p. 904). Table 6.3 summarizes skin glands.

1 Explain how a hair forms.

2 What causes gooseflesh?

3 How does the composition of a fingernail differ from that of a hair?

4 What is the function of the sebaceous glands?

5 Describe the locations of the sweat glands.

6 How do the functions of eccrine sweat glands and apocrine sweat glands differ?

TABLE 6.3	Skin Glands		
Type	**Description**	**Function**	**Location**
Sebaceous glands	Groups of specialized epithelial cells	Keep hair soft, pliable, waterproof	Near or connected to hair follicles, everywhere but on palms and soles
Eccrine sweat glands	Abundant sweat glands with odorless secretion	Lower body temperature	Originate in deep dermis or subcutaneous layer and open to surface on forehead, neck, and back
Apocrine sweat glands	Less numerous sweat glands with secretions that develop odors	Wet skin during pain, fear, emotional upset, and sexual arousal	Near hair follicles in armpit, groin, around nipples
Ceruminous glands	Modified sweat glands	Secrete earwax	External ear canal
Mammary glands	Modified sweat glands	Secrete milk	Breasts

ACNE

Many young people are all too familiar with *acne vulgaris,* a disorder of the sebaceous glands. Excess sebum and squamous epithelial cells clog the glands, producing blackheads and whiteheads (comedones). The blackness is not dirt but results from the accumulated cells blocking light. In addition, the clogged sebaceous gland provides an attractive environment for anaerobic bacteria that signals the immune system to trigger inflammation. The inflamed, raised area is a pimple (pustule).

A Hormonal Problem

Acne is the most common skin disease, affecting 80% of people at some time between the ages of eleven and thirty. It is largely hormonally induced. Just before puberty, the adrenal glands increase production of androgens, which stimulate increased secretion of sebum. At puberty, sebum production surges again. Acne usually develops because the sebaceous glands are extra responsive to androgens, but in some cases, androgens may be produced in excess.

Acne can cause skin blemishes far more serious than the perfect models in acne medication ads depict (fig. 6D). Scarring from acne can lead to emotional problems. Fortunately, several highly effective treatments are available.

What to Do—And Not Do

Acne is not caused by uncleanliness or eating too much chocolate or greasy food. Although cleansing products containing soaps, detergents, or astringents can remove surface sebum, they do not stop the flow of oil that contributes to acne. Abrasive products are actually harmful because they irritate the skin and increase inflammation.

Several acne treatments are available, but most take weeks to months to work. Women with acne are sometimes prescribed birth control pills because the estrogens counter androgen excess. Other drugs with estrogenic effects are available in Europe but not in the United States or Canada. Isotretinoin is a derivative of vitamin A that helps nearly all people achieve relief or even permanent cures, but it has several side effects and causes birth defects. Systemic antibiotics can treat acne by clearing bacteria from sebaceous glands. Topical treatments include tretinoin (another vitamin A derivative), salicylic acid (an aspirin solution), and benzoyl peroxide.

Treatment for severe acne requires a doctor's care. Drug combinations are tailored to the severity of the condition (table 6A). ■

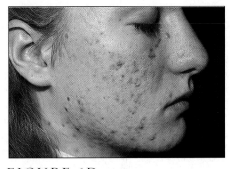

FIGURE 6D

Acne is a common skin condition usually associated with a surge of androgen activity—not eating chocolate, as was once believed.

TABLE 6A	Acne Treatments (By Increasing Severity)
Condition	**Treatment**
Noninflammatory comedonal acne (blackheads and whiteheads)	Topical tretinoin or salicylic acid
Papular inflammatory acne	Topical antibiotic
Widespread blackheads and pustules	Topical tretinoin and systemic antibiotic
Severe cysts	Systemic isotretinoin
Explosive acne (ulcerated lesions, fever, joint pain)	Systemic corticosteroids

Regulation of Body Temperature

The regulation of body temperature is vitally important because even slight shifts can disrupt the rates of metabolic reactions. Normally, the temperature of deeper body parts remains close to a set point of 37°C (98.6°F). The maintenance of a stable temperature requires that the amount of heat the body loses be balanced by the amount it produces. The skin plays a key role in the homeostatic mechanism that regulates body temperature.

RECONNECT WITH CHAPTER 1, HOMEOSTASIS, PAGE 10.

Heat Production and Loss

Heat is a product of cellular metabolism; thus, the more active cells of the body are the major heat producers. These cells include skeletal and cardiac muscle cells and the cells of certain glands, such as the liver.

When body temperature rises above the set point, nerve impulses stimulate structures in the skin and other organs to release heat. For example, during physical exercise, active muscles release heat, which the blood carries away. The warmed blood reaches the part of the brain (the hypothalamus) that controls the body's temperature set point, which signals muscles in the walls of specialized dermal blood vessels to relax. As the vessels dilate (vasodilation), more blood enters them, and some of the heat the blood carries escapes to the outside. At the same time, deeper blood vessels contract (vasoconstriction), diverting blood to the surface, and the skin reddens. The heart is stimulated to beat faster, moving more blood out of the deeper regions.

The primary means of body heat loss is **radiation** (ra-de-a'shun), by which infrared heat rays escape from warmer surfaces to cooler surroundings. These rays radiate in all directions, much like those from the bulb of a heat lamp.

Conduction and convection release less heat. In **conduction** (kon-duk'shun), heat moves from the body directly into the molecules of cooler objects in contact with its surface. For example, heat is lost by conduction into the seat of a chair when a person sits down. The heat loss continues as long as the chair is cooler than the body surface touching it.

Heat is also lost by conduction to the air molecules that contact the body. As air becomes heated, it moves away from the body, carrying heat with it, and is replaced by cooler air moving toward the body. This type of continuous circulation of air over a warm surface is **convection** (kon-vek'shun).

Still another means of body heat loss is **evaporation** (e-vap"o-ra'shun). When the body temperature rises above normal, the nervous system stimulates eccrine sweat glands to release sweat onto the surface of the skin. As this fluid evaporates (changes from a liquid to a gas), it carries heat away from the surface, cooling the skin.

When body temperature drops below the set point, the brain triggers a different set of responses in skin structures. Muscles in the walls of dermal blood vessels are stimulated to contract; this decreases the flow of heat-carrying blood through the skin, which tends to lose color, and helps reduce heat loss by radiation, conduction, and convection. At the same time, sweat glands remain inactive, decreasing heat loss by evaporation. If the body temperature continues to drop, the nervous system may stimulate muscle cells in the skeletal muscles throughout the body to contract slightly. This action requires an increase in the rate of cellular respiration and releases heat as a by-product. If this response does not raise the body temperature to normal, small groups of muscles may contract rhythmically with still greater force, causing the person to shiver, thus generating more heat. Figure 6.11 summarizes the body's temperature-regulating mechanism, and Clinical Application 6.4 examines two causes of elevated body temperature.

Problems in Temperature Regulation

The body's temperature-regulating mechanism does not always operate satisfactorily, and the consequences may

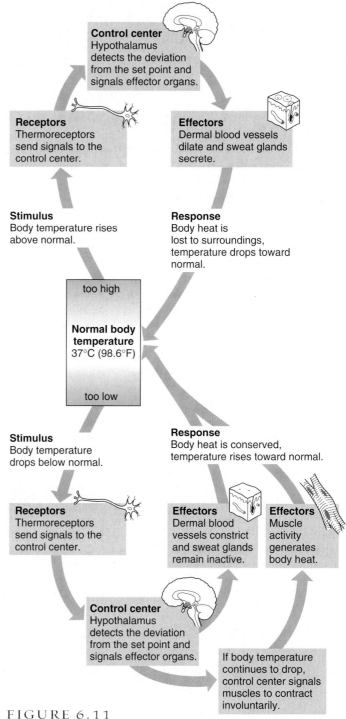

FIGURE 6.11

Body temperature regulation is an example of homeostasis.

be dangerous. For example, air can hold only a limited amount of water vapor, so on a hot, humid day, the air may become nearly saturated with water. At such times, the sweat glands may be activated, but the sweat cannot quickly evaporate. The skin becomes wet, but the person remains hot and uncomfortable. Body temperature may rise, a condition called hyperthermia. In addition, if the air temperature is high, heat loss by radiation is less effective. In fact, if the air temperature exceeds body tempera-

ELEVATED BODY TEMPERATURE

It was a warm June morning when the harried and hurried father strapped his five-month-old son Bryan into the backseat of his car and headed for work. Tragically, the father forgot to drop his son off at the babysitter's. When his wife called him at work late that afternoon to inquire why the child was not at the sitter's, the shocked father realized his mistake and hurried down to his parked car. But it was too late—Bryan had died. Left for ten hours in the car in the sun, all windows shut, the baby's temperature had quickly soared. Two hours after he was discovered, the child's temperature still exceeded 41°C (106°F).

Sarah L.'s case of elevated body temperature was more typical. She awoke with a fever of 40°C (104°F) and a terribly painful sore throat. At the doctor's office, a test revealed that Sarah had a *Streptococcus* infection. The fever was her body's attempt to fight the infection.

The true cases of young Bryan and Sarah illustrate two reasons why body temperature may rise—inability of the temperature homeostatic mechanism to handle an extreme environment and an immune system response to infection. In Bryan's case, sustained exposure to very high heat overwhelmed the temperature-regulating mechanism, resulting in hyperthermia. Body heat built up faster than it could dissipate, and body temperature rose, even though the set point of the thermostat was normal. His blood vessels dilated so greatly in an attempt to dissipate the excess heat that after a few hours, his cardiovascular system collapsed.

Fever is a special case of hyperthermia, in which temperature is elevated due to an elevated set point. In fever, molecules on the surfaces of the infectious agents (usually bacteria or viruses) stimulate phagocytes to release a substance called interleukin-1 (also called endogenous pyrogen, meaning "fire maker from within"). The bloodstream carries interleukin-1 to the hypothalamus, where it raises the set point controlling temperature. In response, the brain signals skeletal muscles to increase heat production, blood flow to the skin to decrease, and sweat glands to decrease secretion. As a result, body temperature rises to the new set point, and fever develops. The increased body temperature helps the immune system kill the pathogens.

Rising body temperature requires different treatments, depending on the degree of elevation. Hyperthermia in response to exposure to intense, sustained heat should be rapidly treated by administering liquids to replace lost body fluids and electrolytes, sponging the skin with water to increase cooling by evaporation, and covering the person with a refrigerated blanket. Fever can be lowered with ibuprofen or acetaminophen, or aspirin in adults. Some health professionals believe that a slightly elevated temperature should not be reduced (with medication or cold baths) because it may be part of a normal immune response. A high or prolonged fever, however, requires medical attention. ■

ture, the person may gain heat from the surroundings, elevating body temperature even higher.

> Because body temperature regulation depends on evaporation of sweat from the skin's surface and because high humidity hinders evaporation, athletes are advised to slow down their activities on hot, humid days. They should also stay out of the sunlight whenever possible and drink enough fluids to avoid dehydration. Such precautions can prevent the symptoms of heat exhaustion, which include fatigue, dizziness, headache, muscle cramps, and nausea.

Hypothermia, or lowered body temperature, can result from prolonged exposure to cold or as part of an illness. It can be extremely dangerous. Hypothermia begins with shivering and a feeling of coldness, but if not treated, progresses to mental confusion, lethargy, loss of reflexes and consciousness, and, eventually, a shutting down of major organs. If the temperature in the body's core drops just a few degrees, fatal respiratory failure or heart arrhythmia may result. However, the extremities can withstand drops of 20 to 30°F below normal.

Certain people are at higher risk for developing hypothermia. These include the very young and the very old, very thin individuals, and the homeless. Hypothermia can be prevented by dressing appropriately and staying active in the cold. A person suffering from hypothermia must be warmed gradually so that respiratory and cardiovascular functioning remain stable.

> Hypothermia is intentionally induced during certain surgical procedures involving the heart or central nervous system (brain or spinal cord). In heart surgery, for example, body temperature may be lowered to between 78°F (26°C) and 89°F (32°C). The cooling lowers the body's metabolic rate so that less oxygen is required. Hypothermia for surgery is accomplished by packing the patient in ice or by removing blood, cooling it, and returning it to the body.

1 Why is regulation of body temperature so important?

2 How is body heat produced?

3 How does the body lose excess heat?

4 How does the skin help regulate body temperature?

5 What are the dangers of hypothermia?

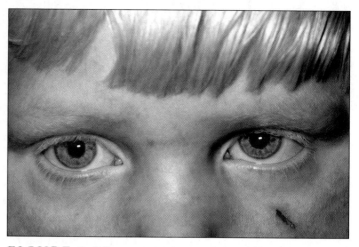

FIGURE 6.12

The pale eyes, skin, and hair of a person with albinism reflect lack of melanin. Albinism is inherited.

Skin Color

Heredity and the environment determine skin color.

Genetic Factors

Regardless of racial origin, all people have about the same number of melanocytes in their skin. Differences in skin color result from differences in the amount of melanin these cells produce, which is controlled by several genes. The more melanin, the darker the skin. The distribution and the size of pigment granules within melanocytes also influence skin color. The granules in very dark skin are single and large; those in lighter skin occur in clusters of two to four granules and are smaller. People who inherit mutant melanin genes have nonpigmented skin, part of *albinism*. It affects people of all races and also many other species (fig. 6.12).

Among the general U.S. population, only 1 in 20,000 people has albinism. Among the native Hopi people in Arizona, however, the incidence is 1 in 200. The reason for this is as much sociological as it is biological. Men with albinism help the women rather than risk severe sunburn in the fields with the other men. They disproportionately contribute to the next generation because they have more sexual contact with women.

Environmental Factors

Environmental factors such as sunlight, ultraviolet light from sunlamps, and X rays affect skin color. These factors rapidly darken existing melanin, and they stimulate melanocytes to produce more pigment and transfer it to nearby epidermal cells within a few days. This is why sunbathing tans skin. Unless exposure to sunlight continues, however, the tan fades as pigmented epidermal cells become keratinized and wear away.

Physiological Factors

Blood in the dermal vessels adds color to the skin. For example, when blood is well oxygenated, the blood pigment hemoglobin is bright red, making the skins of light-complexioned people appear pinkish. On the other hand, when the blood oxygen concentration is low, hemoglobin is dark red, and the skin appears bluish—a condition called *cyanosis*.

The state of the blood vessels also affects skin color. If the vessels are dilated, more blood enters the dermis, and the skin of a light-complexioned person reddens. This may happen when a person is overheated, embarrassed, or under the influence of alcohol. Conversely, conditions that constrict blood vessels cause the skin to lose this reddish color. Thus, if body temperature drops abnormally or if a person is frightened, the skin may appear pale.

A yellow-orange plant pigment called *carotene,* which is especially common in yellow vegetables, may give skin a yellowish cast if a person consumes too much of it. This results from accumulation of carotene in the adipose tissue of the subcutaneous layer. Illnesses may also affect skin color. A yellowish skin tone can also indicate *jaundice,* a consequence of liver malfunction.

Some newborns develop the yellowish skin of jaundice shortly after birth. A blood incompatibility or an immature liver can cause jaundice. An observant British hospital nurse discovered a treatment for newborn jaundice in 1958. She liked to take her tiny charges out in the sun, and she noticed that a child whose skin had a yellow pallor developed normal pigmentation when he lay in sunlight. However, the part of the child's body covered by a diaper and therefore not exposed to the sun remained yellow. Further investigation showed that sunlight enables the body to break down bilirubin, the liver substance that accumulates in the skin. Today, newborns who develop persistently yellowish skin may have to lie under artificial "bili lights" for a few days, clad only in protective goggles.

1 What factors influence skin color?

2 Which of these factors are genetic? Which are environmental?

Healing of Wounds and Burns

Inflammation is a normal response to injury or stress. Blood vessels in affected tissues dilate and become more permeable, allowing fluids to leak into the damaged tissues. Inflamed skin may become reddened, swollen, warm, and painful to touch. However, the dilated blood vessels provide the tissues with more nutrients and oxygen, which aids healing. The specific events in the healing process depend on the nature and extent of the injury.

Cuts

If a break in the skin is shallow, epithelial cells along its margin are stimulated to divide more rapidly than usual. The newly formed cells fill the gap.

If the injury extends into the dermis or subcutaneous layer, blood vessels break, and the escaping blood forms a clot in the wound. A clot consists mainly of a fibrous protein (fibrin) that forms from another protein in the plasma, blood cells, and platelets that become entrapped in the protein fibers. Tissue fluids seep into the area and dry. The blood clot and the dried fluids form a *scab* that covers and protects underlying tissues. Before long, fibroblasts migrate into the injured region and begin forming new collagenous fibers that bind the edges of the wound together. Suturing or otherwise closing a large break in the skin speeds this process. In addition, the connective tissue matrix releases *growth factors* that stimulate certain cells to divide and regenerate the damaged tissue.

As healing continues, blood vessels extend into the area beneath the scab. Phagocytic cells remove dead cells and other debris. Eventually, the damaged tissues are replaced, and the scab sloughs off. If the wound is extensive, the newly formed connective tissue may appear on the surface as a *scar.*

In large, open wounds, healing may be accompanied by formation of small, rounded masses called *granulations* that develop in the exposed tissues. A granulation consists of a new branch of a blood vessel and a cluster of collagen-secreting fibroblasts that the vessel nourishes. In time, some of the blood vessels are resorbed, and the fibroblasts move away, leaving a scar that is largely composed of collagenous fibers. Figure 6.13 shows the stages in the healing of a wound.

Burns

Slightly burned skin, such as from a minor sunburn, may become warm and reddened (erythema) as dermal blood vessels dilate. This response may be accompanied by mild edema, and, in time, the surface layer of skin may be shed. A burn injuring only the epidermis is called a *superficial partial-thickness* (first degree) *burn.* Healing usually occurs within a few days to two weeks, with no scarring.

A burn that destroys some epidermis as well as some underlying dermis is a *deep partial-thickness* (second degree) *burn.* Fluid escapes from damaged dermal capillaries, and as it accumulates beneath the outer layer of epidermal cells, blisters appear. The injured region becomes moist and firm and may vary in color from dark red to waxy white. Such a burn most commonly occurs as a result of exposure to hot objects, hot liquids, flames, or burning clothing.

The healing of a deep partial-thickness burn depends upon accessory organs of the skin that survive the injury because they are located deep in the dermis. These organs, which include hair follicles, sweat glands, and sebaceous glands, contain epithelial cells. During healing, these cells grow out onto the surface of the dermis, spread over it, and form a new epidermis. In time, the skin usually completely recovers, and scar tissue does not develop unless an infection occurs.

A burn that destroys the epidermis, dermis, and the accessory organs of the skin is called a *full-thickness* (third degree) *burn.* The injured skin becomes dry and leathery, and it may vary in color from red to black to white.

A full-thickness burn usually occurs as a result of immersion in hot liquids or prolonged exposure to hot objects, flames, or corrosive chemicals. Since most of the epithelial cells in the affected region are likely to be destroyed, spontaneous healing can occur only by growth of epithelial cells inward from the margin of the burn. If the injury is extensive, treatment may involve removing a thin layer of skin from an unburned region of the body and transplanting it to the injured area. This procedure is called an *autograft.*

If the burn is too extensive to replace with skin from other parts of the body, cadaveric skin from a skin bank may be used to cover the injury. In this case, the transplant, a *homograft,* is a temporary covering that decreases the size of the wound, helps prevent infection, and helps preserve deeper tissues. In time, after healing has begun, the temporary covering may be removed and replaced with an autograft, as skin becomes available in areas that have healed. However, skin grafts can leave extensive scars.

Various skin substitutes also may be used to temporarily cover extensive burns. These include amniotic membrane that surrounded a human fetus, and artificial membranes composed of silicone, polyurethane, or nylon together with a network of collagenous fibers. Another type of skin substitute comes from cultured human epithelial cells. In a laboratory, a bit of human skin the size of a postage stamp can grow to the size of a bathmat in about three weeks. Skin substitutes are a major focus of tissue engineering, discussed in From Science to Technology 5.2 (p. 153).

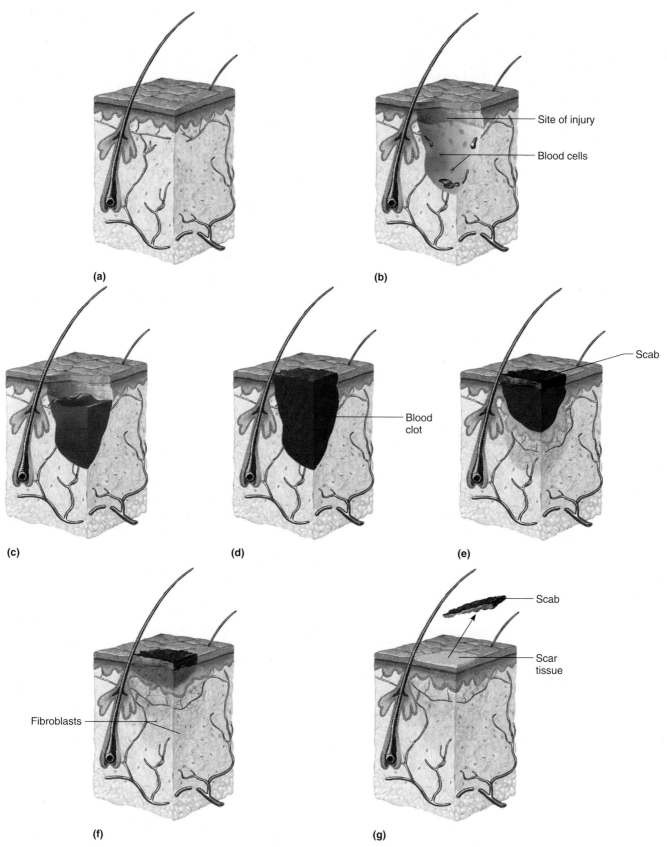

FIGURE 6.13

Healing of a wound. (*a*) If normal skin is (*b*) injured deeply, (*c*) blood escapes from dermal blood vessels, and (*d*) a blood clot soon forms. (*e*) The blood clot and dried tissue fluid form a scab that protects the damaged region. (*f*) Later, blood vessels send out branches, and fibroblasts migrate into the area. (*g*) The fibroblasts produce new connective tissue fibers, and when the skin is mostly repaired, the scab sloughs off.

The treatment of a burn patient requires estimating the extent of the body's surface that is affected. Physicians use the "rule of nines," subdividing the skin's surface into regions, each accounting for 9% (or some multiple of 9%) of the total surface area (fig. 6.14). This estimate is important in planning to replace body fluids and electrolytes lost from injured tissues and for covering the burned area with skin or skin substitutes.

1 What is the tissue response to inflammation?

2 What occurs within a healing wound to cause the sloughing of the scab?

3 Which type of burn is most likely to leave a scar? Why?

Life-Span Changes

We are more aware of aging-related changes in skin than in other organ systems, simply because we can easily see them. Aging skin affects appearance, temperature regulation, and vitamin D activation.

The epidermis maintains its thickness as the decades pass, but as the cell cycle slows, cells tend to grow larger and more irregular in shape. Skin may appear scaly because, at the microscopic level, more sulfur–sulfur bonds form within keratin molecules. Patches of pigment commonly called "age spots" or "liver spots" appear and grow (fig. 6.15). These are sites of oxidation of

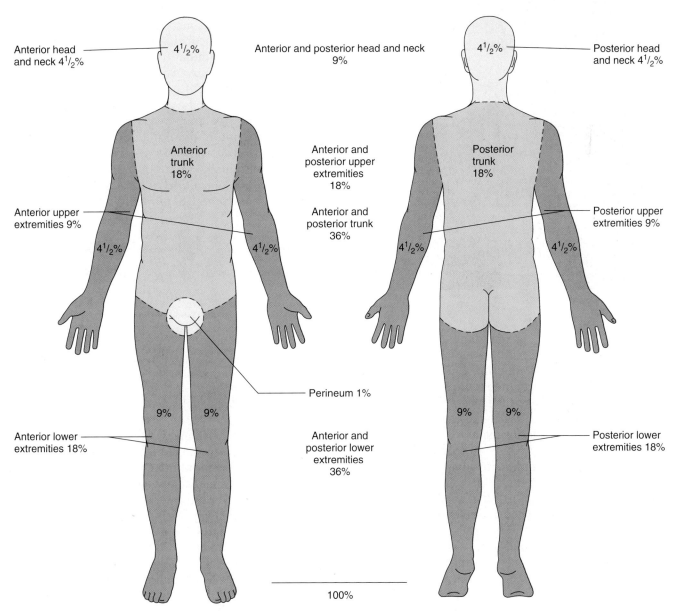

Anterior head and neck 4$\frac{1}{2}$%

Anterior and posterior head and neck 9%

Posterior head and neck 4$\frac{1}{2}$%

Anterior trunk 18%

Anterior and posterior upper extremities 18%

Posterior trunk 18%

Anterior upper extremities 9%

Anterior and posterior trunk 36%

Posterior upper extremities 9%

Perineum 1%

Anterior lower extremities 18%

Anterior and posterior lower extremities 36%

Posterior lower extremities 18%

100%

FIGURE 6.14

As an aid for estimating the extent of damage burns cause, the body is subdivided into regions, each representing 9% (or some multiple of 9%) of the total skin surface area.

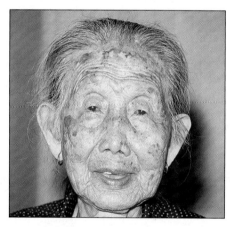

FIGURE 6.15

Aging-associated changes are very obvious in the skin.

Sensitivity to pain and pressure diminishes with age as the number of receptors falls. A ninety-year-old's skin has only one-third the number of such receptors as the skin of a young adult.

The ability to control temperature falters as the number of sweat glands in the skin falls, as the capillary beds that surround sweat glands and hair follicles shrink, and as the ability to shiver declines. In addition, the number of blood vessels in the deeper layers decreases, as does the ability to shunt blood towards the body's interior to conserve heat. As a result, an older person is less able to tolerate the cold and cannot regulate heat. An older person might set the thermostat ten to fifteen degrees higher than a younger person in the winter. Fewer blood vessels in and underlying the skin account for the pale complexions of some older individuals. Changes in the distribution of blood vessels also contribute to development of pressure sores in a bedridden person whose skin does not receive adequate stimulation.

Aging of the skin is also related to skeletal health. The skin is the site of activation of vitamin D, which requires exposure to the sun. Vitamin D is necessary for the absorption of calcium by bone tissue. Many older people do not get outdoors much, and the wavelengths of light that are important for vitamin D activation do not readily penetrate glass windows. In addition, older skin has a diminished ability to activate the vitamin. Therefore, homebound seniors often take vitamin D supplements to help maintain bone structure.

1 What changes occur in the epidermis and dermis with age?

2 How do the skin's accessory structures change over time?

3 Why do older people have more difficulty controlling body temperature than do younger people?

fats in the secretory cells of apocrine and eccrine glands and reflect formation of oxygen free radicals.

The dermis becomes reduced as synthesis of the connective tissue proteins collagen and elastin slows. The combination of a shrinking dermis and loss of some fat from the subcutaneous layer results in wrinkling and sagging of the skin. Fewer lymphocytes delay wound healing. Some of the changes in the skin's appearance result from specific deficits. Less oil from sebaceous glands means that the skin becomes considerably drier.

The skin's accessory structures also show signs of aging. Slowed melanin production causes hair to become gray or white as the follicle becomes increasingly transparent. Hair growth slows, the hairs thin, and the number of follicles decreases. Males may develop pattern baldness, which is hereditary but not often expressed in females. A diminished blood supply to the nail beds impairs their growth, dulling and hardening them.

Common Skin Disorders

athlete's foot (ath'-lētz foot) Fungus infection (*Tinea pedis*) usually in the skin of the toes and soles.

birthmark (berth'mark) Congenital blemish or spot on the skin, visible at birth or soon after.

boil (boil) Bacterial infection (furuncle) of the skin, produced when bacteria enter a hair follicle.

carbuncle (kar'bung-kl) Bacterial infection, similar to a boil, that spreads into the subcutaneous tissues.

cyst (sist) Liquid-filled sac or capsule.

eczema (ek'zĕ-mah) Noncontagious skin rash, often accompanied by itching, blistering, and scaling.

erythema (er″ĭ-the'mah) Reddening of the skin due to dilation of dermal blood vessels in response to injury or inflammation.

herpes (her'pēz) Infectious disease of the skin usually caused by the herpes simplex virus and

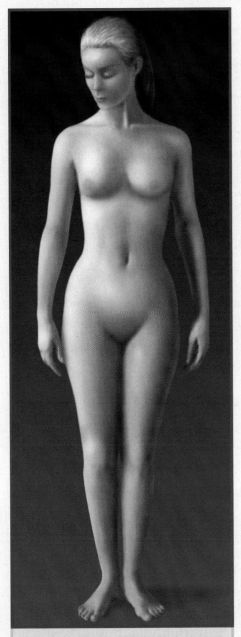

INTEGUMENTARY SYSTEM

The skin provides protection, contains sensory organs, and helps control body temperature.

Skeletal System

Vitamin D activated by the skin helps provide calcium for bone matrix.

Lymphatic System

The skin, acting as a barrier, provides an important first line of defense for the immune system.

Muscular System

Involuntary muscle contractions (shivering) work with the skin to control body temperature. Muscles act on facial skin to create expressions.

Digestive System

Excess calories may be stored as subcutaneous fat. Vitamin D activated by the skin stimulates dietary calcium absorption.

Nervous System

Sensory receptors provide information about the outside world to the nervous system. Nerves control the activity of sweat glands.

Respiratory System

Stimulation of skin receptors may alter respiratory rate.

Endocrine System

Hormones help to increase skin blood flow during exercise. Other hormones stimulate either the synthesis or the decomposition of subcutaneous fat.

Urinary System

The kidneys help compensate for water and electrolytes lost in sweat.

Cardiovascular System

Skin blood vessels play a role in regulating body temperature.

Reproductive System

Sensory receptors play an important role in sexual activity and in the suckling reflex.

characterized by recurring formations of small clusters of vesicles.

keloid (ke′loid) Elevated, enlarging fibrous scar usually initiated by an injury.

mole (mōl) Fleshy skin tumor (nevus) that is usually pigmented; colors range from brown to black.

pediculosis (pĕ-dik″u-lo′sis) Disease produced by an infestation of lice.

pruritus (proo-ri′tus) Itching of the skin.

pustule (pus′tūl) Elevated, pus-filled area.

scabies (ska′bēz) Disease resulting from an infestation of mites.

seborrhea (seb″o-re′ah) Hyperactivity of the sebaceous glands, accompanied by greasy skin and dandruff.

ulcer (ul′ser) Open sore.

urticaria (ur″tĭ-ka′re-ah) Allergic reaction of the skin that produces reddish, elevated patches (hives).

wart (wort) Flesh-colored, raised area caused by a viral infection.

CHAPTER SUMMARY

Skin and Its Tissues (page 158)

Skin is a protective covering, helps regulate body temperature, houses sensory receptors, synthesizes chemicals, and excretes wastes. It is composed of an epidermis and a dermis separated by a basement membrane. A subcutaneous layer, not part of the skin, lies beneath the dermis.

1. Epidermis
   a. The epidermis is a layer of stratified squamous epithelium that lacks blood vessels.
   b. The deepest layer, called stratum basale, contains cells that divide and grow.
   c. Epidermal cells undergo keratinization as they are pushed toward the surface.
   d. The outermost layer, called stratum corneum, is composed of dead epidermal cells.
   e. Production of epidermal cells balances the rate at which they are lost at the surface.
   f. Epidermis protects underlying tissues against water loss, mechanical injury, and the effects of harmful chemicals.
   g. Melanin protects underlying cells from the effects of ultraviolet light.
   h. Melanocytes transfer melanin to nearby epidermal cells.

2. Dermis
   a. The dermis is a layer composed of irregular dense connective tissue that binds the epidermis to underlying tissues.
   b. It also contains muscle cells, blood vessels, and nerve cell processes.
   c. Dermal blood vessels supply nutrients to all skin cells and help regulate body temperature.
   d. Nervous tissue is scattered through the dermis.
      (1) Some dermal nerve cell processes carry impulses to muscles and glands of the skin.
      (2) Other dermal nerve cell processes are associated with sensory receptors in the skin.

3. Subcutaneous layer
   a. The subcutaneous layer is composed of loose connective tissue and adipose tissue.
   b. Adipose tissue helps conserve body heat.
   c. This layer contains blood vessels that supply the skin.

Accessory Organs of the Skin (page 165)

1. Hair follicles
   a. Hair occurs in nearly all regions of the skin.
   b. Each hair develops from epidermal cells at the base of a tubelike hair follicle.
   c. As newly formed cells develop and grow, older cells are pushed toward the surface and undergo keratinization.
   d. A hair usually grows for a while, rests, and then is replaced by a new hair.
   e. Hair color is determined by genes that direct the type and amount of pigment in its cells.
   f. A bundle of smooth muscle cells and one or more sebaceous glands are attached to each hair follicle.

2. Nails
   a. Nails are protective covers on the ends of fingers and toes.
   b. They are produced by epidermal cells that undergo keratinization.

3. Skin glands
   a. Sebaceous glands secrete sebum, which helps keep skin and hair soft and waterproof.
   b. Sebaceous glands are usually associated with hair follicles.
   c. Sweat glands are located in nearly all regions of the skin.
   d. Each sweat gland consists of a coiled tube.
   e. Eccrine sweat glands, located on the forehead, neck, back, palms, and soles, respond to elevated body temperature or emotional stress.
   f. Sweat is primarily water but also contains salts and waste products.
   g. Apocrine sweat glands, located in the axillary regions, groin, and around the nipples, moisten the skin when a person is emotionally upset, scared, in pain, or sexually aroused.

Regulation of Body Temperature (page 169)

Regulation of body temperature is vital because heat affects the rates of metabolic reactions. Normal temperature of deeper body parts is close to a set point of 37°C (98.6°F).

1. Heat production and loss
   a. Heat is a by-product of cellular respiration.
   b. When body temperature rises above normal, more blood enters dermal blood vessels, and the skin reddens.
   c. Heat is lost to the outside by radiation, conduction, convection, and evaporation.
   d. Sweat gland activity increases heat loss by evaporation.
   e. If the body temperature drops below normal, dermal blood vessels constrict, causing the skin to lose color, and sweat glands become inactive.
   f. When heat is lost excessively, skeletal muscles involuntarily contract; this increases cellular respiration and produces additional heat.

2. Problems in temperature regulation
   a. Air can hold a limited amount of water vapor.
   b. When the air is saturated with water, sweat may fail to evaporate, and body temperature may remain elevated.
   c. Hypothermia is lowered body temperature. It causes shivering, mental confusion, lethargy, loss of reflexes and consciousness, and eventually major organ failure.

Skin Color (page 172)

All humans have about the same concentration of melanocytes. Skin color is largely due to the amount of melanin in the epidermis.

1. Genetic factors
   a. Each person inherits genes for melanin production.
   b. Dark skin is due to genes that cause large amounts of melanin to be produced; lighter skin is due to genes that cause lesser amounts of melanin to form.
   c. Mutant genes may cause a lack of melanin in the skin.
2. Environmental factors
   a. Environmental factors that influence skin color include sunlight, ultraviolet light, and X rays.
   b. These factors darken existing melanin and stimulate additional melanin production.
3. Physiological factors
   a. The oxygen content of the blood in dermal vessels may cause the skin of light-complexioned persons to appear pinkish or bluish.

b. Carotene in the subcutaneous layer may cause the skin to appear yellowish.
c. Disease may affect skin color.

Healing of Wounds and Burns (page 173)

Skin injuries trigger inflammation. The affected area becomes red, warm, swollen, and tender.

1. A cut in the epidermis is filled in by dividing epithelial cells. Clots close deeper cuts, sometimes leaving a scar where connective tissue fills in. Granulations form as part of the healing process.
2. A superficial partial-thickness burn heals quickly with no scarring. The area is warm and red. A burn penetrating to the dermis is a deep partial-thickness burn. It blisters. Deeper skin structures help heal this more serious type of burn. A full-thickness burn is the most severe and may require a skin graft.

Life-Span Changes (page 175)

1. Aging skin affects appearance as "age spots" or "liver spots" appear and grow, along with wrinkling and sagging.
2. Due to changes in the number of sweat glands and shrinking capillary beds in the skin, elderly people are less able to tolerate the cold and cannot regulate heat.
3. Older skin has a diminished ability to activate vitamin D necessary for skeletal health.

CRITICAL THINKING QUESTIONS

1. What special problems would result from the loss of 50% of a person's functional skin surface? How might this person's environment be modified to compensate partially for such a loss?
2. A premature infant typically lacks subcutaneous adipose tissue. Also, the surface area of an infant's body is relatively large compared to its volume. How do you think these factors affect the ability of an infant to regulate its body temperature?
3. As a rule, a superficial partial-thickness burn is more painful than one involving deeper tissues. How would you explain this observation?
4. Which of the following would result in the more rapid absorption of a drug: a subcutaneous injection or an intradermal injection? Why?

5. What methods might be used to cool the skin of a child experiencing a high fever? For each method you list, identify the means by which it promotes heat loss—radiation, conduction, convection, or evaporation.
6. How would you explain to an athlete the importance of keeping the body hydrated when exercising in warm weather?
7. Everyone's skin contains about the same number of melanocytes even though people come in many different colors. How is this possible?
8. How is skin peeling after a severe sunburn protective? How might a fever be protective?
9. Why would collagen and elastin added to skin creams be unlikely to penetrate the skin—as some advertisements imply they do?

REVIEW EXERCISES

1. Define *integumentary system*.
2. List six functions of skin.
3. Distinguish between the epidermis and the dermis.
4. Describe the subcutaneous layer.
5. Explain what happens to epidermal cells as they undergo keratinization.
6. List the layers of the epidermis.
7. Describe the function of melanocytes.
8. Describe the structure of the dermis.
9. Review the functions of dermal nervous tissue.
10. Explain the functions of the subcutaneous layer.
11. Distinguish between a hair and a hair follicle.
12. Review how hair color is determined.
13. Describe how nails are formed.
14. Explain the function of sebaceous glands.
15. Distinguish between eccrine and apocrine sweat glands.
16. Explain the importance of body temperature regulation.
17. Describe the role of the skin in promoting the loss of excess body heat.
18. Explain how body heat is lost by radiation.

19. Distinguish between conduction and convection.
20. Describe the body's responses to decreasing body temperature.
21. Review how air saturated with water vapor may interfere with body temperature regulation.
22. Explain how environmental factors affect skin color.
23. Describe three physiological factors that affect skin color.
24. Distinguish between the healing of shallow and deeper breaks in the skin.
25. Distinguish among first-, second-, and third-degree burns.
26. Describe possible treatments for a third-degree burn.
27. List three effects of aging on skin.

W E B   C O N N E C T I O N S

Visit the Student Edition of the Online Learning Center at www.mhhe.com/shier10 **for answers to chapter questions, additional quizzes, interactive learning exercises, and other study tools.**

UNDERSTANDING WORDS

ax-, axis: *ax*ial skeleton—upright portion of the skeleton that supports the head, neck, and trunk.

-blast, bud, a growing organism in early stages: osteo*blast*—cell that will form bone tissue.

canal-, channel: *canal*iculus—tubular passage.

carp-, wrist: *carp*als—wrist bones.

-clast, break: osteo*clast*—cell that breaks down bone tissue.

clav-, bar: *clav*icle—bone that articulates with the sternum and scapula.

condyl-, knob, knuckle: *condyl*e—rounded, bony process.

corac-, a crow's beak: *corac*oid process—beaklike process of the scapula.

cribr-, sieve: *cribr*iform plate—portion of the ethmoid bone with many small openings.

crist-, crest: *crist*a galli—bony ridge that projects upward into the cranial cavity.

fov-, pit: *fov*ea capitis—pit in the head of a femur.

glen-, joint socket: *glen*oid cavity—depression in the scapula that articulates with the head of a humerus.

inter-, among, between: *inter*vertebral disc—structure located between adjacent vertebrae.

intra-, inside: *intra*membranous bone—bone that forms within sheetlike masses of connective tissue.

lamell-, thin plate: *lamell*a—thin bony plate.

meat-, passage: auditory *meat*us—canal of the temporal bone that leads inward to parts of the ear.

odont-, tooth: *odont*oid process—toothlike process of the second cervical vertebra.

poie-, make, produce: hemato*poie*sis—process by which blood cells are formed.

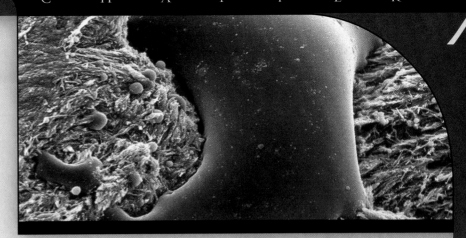

An osteoclast (blue) is a type of bone cell that removes old bone tissue, thereby playing a vital role in the remodeling of this tissue. Falsely colored scanning electron micrograph (800×).

Skeletal System

CHAPTER OBJECTIVES

After you have studied this chapter, you should be able to

1. Classify bones according to their shapes and name an example from each group.

2. Describe the general structure of a bone and list the functions of its parts.

3. Distinguish between intramembranous and endochondral bones and explain how such bones grow and develop.

4. Describe the effects of sunlight, nutrition, hormonal secretions, and exercise on bone development.

5. Discuss the major functions of bones.

6. Distinguish between the axial and appendicular skeletons and name the major parts of each.

7. Locate and identify the bones and the major features of the bones that comprise the skull, vertebral column, thoracic cage, pectoral girdle, upper limb, pelvic girdle, and lower limb.

8. Describe life-span changes in the skeletal system.

A s the hardest and therefore most enduring of human tissues, bone has persisted over time to provide clues to early humans and their forebears. Some glimpses into the past, courtesy of skeletal remains or fossils, include:

7300–6220 B.C. Skulls with circular holes are the earliest evidence of trepanation, a technique used to relieve pressure following a skull fracture or as a spiritual treatment for headache, tumors, or mental illness. A few of the people treated with trepanation were lucky—they survived, as evidenced by new bony growth over the holes made in their skulls. However, most trepanated skulls have gaping, drilled holes, indicating that the "treatment" was lethal.

2.8–2.6 million years ago "Mr. Ples" is the name anthropologists have given to the face and left side of a skull from Sterkfontein, South Africa, which once belonged to a member of *Australopithecus africanus* (see photo), a type of primate that preceded *Homo sapiens.* Using computer modeling to fashion a "virtual endocast" of the entire skull contents, researchers have estimated the cranial capacity of *A. africanus* at 515 cubic centimeters (cc). By comparison, a chimp's cranial capacity averages 370 cc, and a modern human's, 1,350. Expanded cranial capacity correlates to increase in intelligence.

3.5 million years ago Not all evidence of a skeletal system is in the form of preserved bone. On the Serengeti Plain are clues to our ancestors who first began to walk upright, a stance that freed their hands, perhaps making possible the development of tools. This evidence consists of shallow footprints where an animal called *Australopithecus afarensis* once lived. The prints reveal that it had long big toes and arched feet. ■

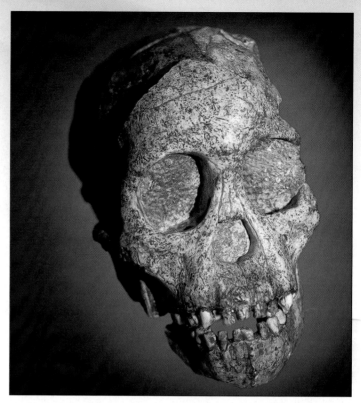

Australopithecus africanus lived from 2.8 to 2.6 million years ago. Our knowledge of this and other human ancestors comes from skeletal evidence.

Nonliving material in the matrix of bone tissue makes the whole organ appear to be inert. A bone also contains very active, living tissues. An individual bone is composed of a variety of tissues: bone tissue, cartilage, dense connective tissue, blood, and nervous tissue.

Bone Structure

The bones of the skeletal system differ greatly in size and shape, but they are similar in their structure, development, and functions.

Bone Classification

Bones are classified according to their shapes—long, short, flat, or irregular (fig. 7.1).

Long bones have long longitudinal axes and expanded ends. Examples are the forearm and thigh bones.

Short bones are somewhat cubelike, with their lengths and widths roughly equal. The bones of the wrists and ankles are examples of this type.

Flat bones are platelike structures with broad surfaces, such as the ribs, scapulae, and some bones of the skull.

Irregular bones have a variety of shapes and are usually connected to several other bones. Irregular bones include the vertebrae that comprise the backbone and many facial bones.

In addition to these four groups of bones, some authorities recognize a fifth group called **sesamoid bones,** or **round bones** (see fig. 7.47c). These bones are usually small and nodular and are embedded within tendons adjacent to joints, where the tendons are compressed. The kneecap (patella) is an example of a sesamoid bone.

Parts of a Long Bone

The femur, a long bone in the thigh, illustrates the structure of bone (fig. 7.2). At each end of such a bone is an expanded portion called an **epiphysis** (e-pif′ĭ-sis) (pl., *epiphyses*), which articulates (or forms a joint) with another bone. On its outer surface, the articulating portion of the epiphysis is coated with a layer of hyaline car-

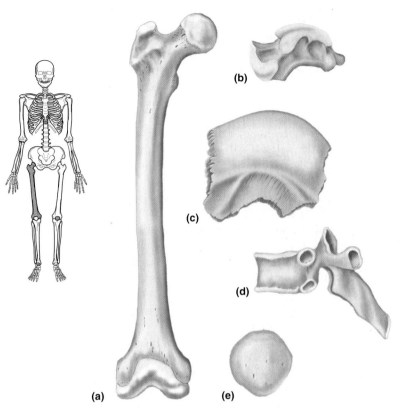

FIGURE 7.1

Bones are classified by shape. (*a*) The femur of the thigh is a long bone, (*b*) a tarsal bone of the ankle is a short bone, (*c*) a parietal bone of the skull is a flat bone, (*d*) a vertebra of the backbone is an irregular bone, and (*e*) the patella of the knee is a sesamoid bone. The whole-skeleton location icon highlights the bones used as examples for classification.

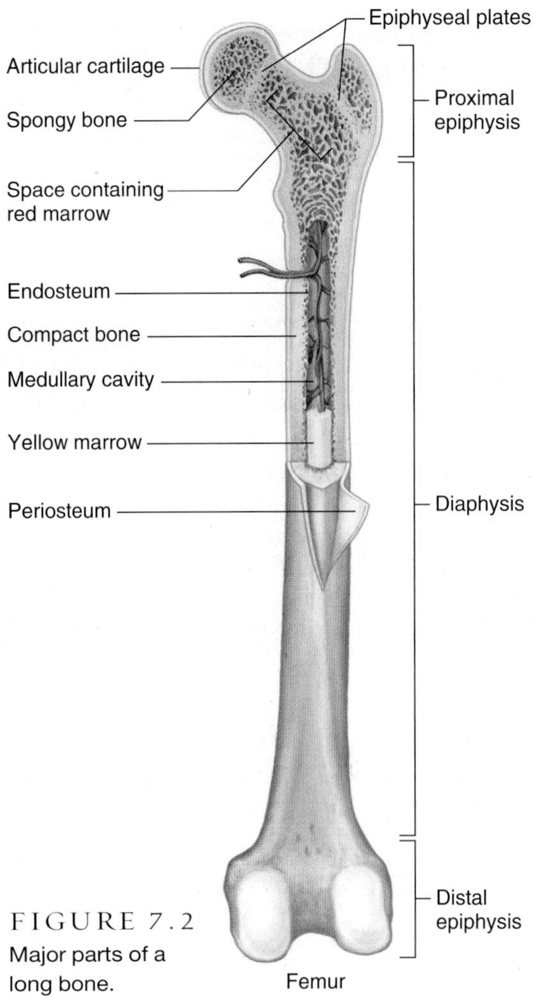

Epiphyseal plates

Articular cartilage

Spongy bone

Space containing red marrow

Proximal epiphysis

Endosteum

Compact bone

Medullary cavity

Yellow marrow

Periosteum

Diaphysis

Distal epiphysis

FIGURE 7.2

Major parts of a long bone.

Femur

tilage called **articular cartilage** (ar-tik′u-lar kar′tĭ-lij). The shaft of the bone, which is located between the epiphyses, is called the **diaphysis** (di-af′ĭ-sis).

Except for the articular cartilage on its ends, the bone is completely enclosed by a tough, vascular covering of fibrous tissue called the **periosteum** (per″e-os′te-um). This membrane is firmly attached to the bone, and the periosteal fibers are continuous with ligaments and tendons that are connected to the membrane. The periosteum also functions in the formation and repair of bone tissue.

A bone's shape makes possible its functions. Bony projections called *processes,* for example, provide sites for attachment of ligaments and tendons; grooves and openings are passageways for blood vessels and nerves; a depression of one bone might articulate with a process of another.

The wall of the diaphysis is mainly composed of tightly packed tissue called **compact bone** (kom′pakt bōn), or cortical bone. This type of bone has a continuous matrix with no gaps (fig. 7.3*a*).

The epiphyses, on the other hand, are largely composed of **spongy bone** (spun′je bōn), or cancellous bone,

with thin layers of compact bone on their surfaces (fig. 7.3*b*). Spongy bone consists of many branching bony plates called **trabeculae** (trah-bek′u-le). Irregular connecting spaces between these plates help reduce the bone's weight. The bony plates are most highly developed in the regions of the epiphyses that are subjected to compressive forces. Both compact and spongy bone are strong and resist bending.

A bone usually has both compact and spongy bone tissues. Short, flat, and irregular bones typically consist of a mass of spongy bone that is either covered by a layer of compact bone or sandwiched between plates of compact bone (fig. 7.3*c*).

Compact bone in the diaphysis of a long bone forms a semirigid tube with a hollow chamber called the **medullary cavity** (med′u-lār″e kav′ĭ te) that is continuous with the spaces of the spongy bone. A thin membrane, containing bone-forming cells, called **endosteum** (en-dos′te-um) lines these areas; and a specialized type of soft connective tissue called **marrow** (mar′o) fills them. Marrow exists in two forms, red marrow and yellow marrow, described later in the chapter (see also fig. 7.2).

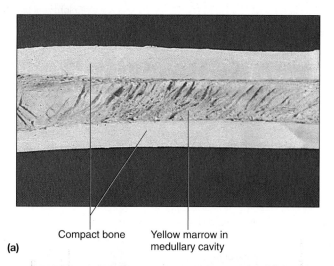

(a)

Compact bone   Yellow marrow in medullary cavity

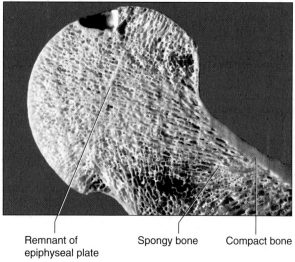

(b)

Remnant of epiphyseal plate   Spongy bone   Compact bone

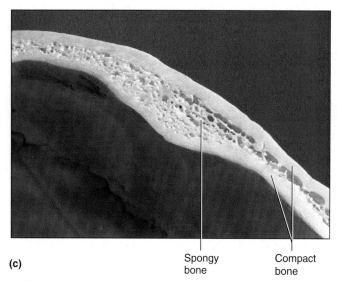

(c)

Spongy bone   Compact bone

FIGURE 7.3

Compact bone and spongy bone. (*a*) In a femur, the wall of the diaphysis consists of compact bone. (*b*) The epiphyses of the femur contain spongy bone enclosed by a thin layer of compact bone. (*c*) This skull bone contains a layer of spongy bone sandwiched between plates of compact bone.

Microscopic Structure

Recall from chapter 5 (p. 148) that bone cells called **osteocytes** (os'te-o-sītz) are located in tiny, bony chambers called lacunae, which form concentric circles around central canals (Haversian canals). Osteocytes transport nutrients and wastes to and from nearby cells by means of cellular processes passing through canaliculi. The intercellular material of bone tissue is largely collagen and inorganic salts. Collagen gives bone its strength and resilience, and inorganic salts make it hard and resistant to crushing.

Compact Bone

In compact bone, the osteocytes and layers of intercellular material concentrically clustered around a central canal form a cylinder-shaped unit called an **osteon** (os'te-on), sometimes called an Haversian system (fig. 7.4). Many of these units cemented together form the substance of compact bone (fig. 7.5). The orientation of the osteons resists compressive forces.

Each central canal contains blood vessels and nerve fibers surrounded by loose connective tissue. Blood in these vessels nourishes bone cells associated with the central canal via gap junctions between osteocytes.

Central canals pervade bone tissue longitudinally. Transverse *perforating canals* (Volkmann's canals) interconnect them. These canals contain larger blood vessels and nerves that allow the smaller blood vessels and nerve cell processes in the osteonic canals to communicate with the surface of the bone and the medullary cavity (see fig. 7.4).

Spongy Bone

Spongy bone is also composed of osteocytes and intercellular material, but the bone cells do not aggregate around central canals. Instead, the cells lie within the trabeculae and get nutrients from substances diffusing into the canaliculi that lead to the surface of these thin, bony plates.

Severe bone pain is a symptom of sickle cell disease, which is inherited. Under low oxygen conditions, abnormal hemoglobin (an oxygen-carrying protein) bends the red blood cells that contain it into a sickle shape, which obstructs circulation. X rays can reveal blocked arterial blood flow in bones of sickle cell disease patients.

1 Explain how bones are classified.

2 List five major parts of a long bone.

3 How do compact and spongy bone differ in structure?

4 Describe the microscopic structure of compact bone.

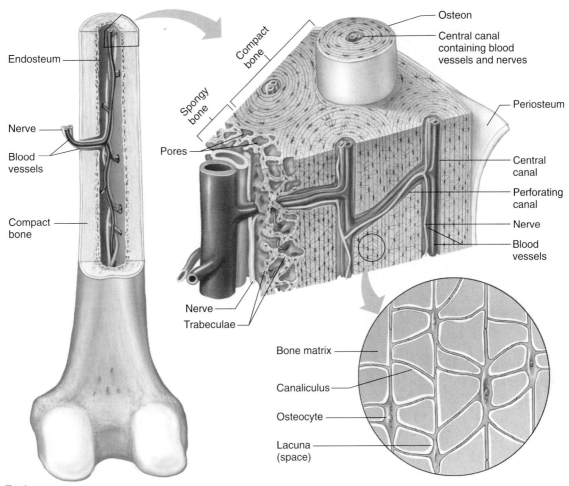

FIGURE 7.4

Compact bone is composed of osteons cemented together by bone matrix. Drawing is not to scale.

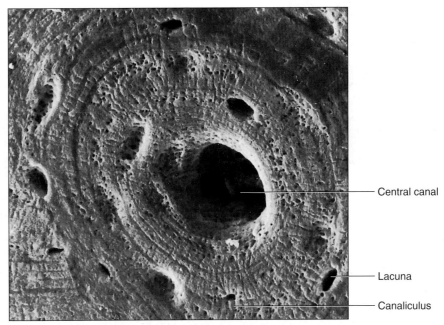

FIGURE 7.5

Scanning electron micrograph of a single osteon in compact bone (575×). *Tissues and Organs: A Text-Atlas of Scanning Electron Microscopy*, by R. G. Kessel and R. H. Kardon. © 1979 W. H. Freeman and Company.

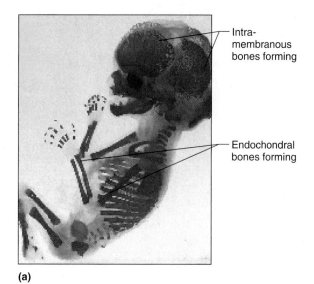

(a)

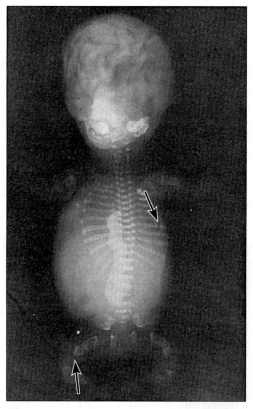

(b)

FIGURE 7.6

Fetal skeleton. (*a*) Note the stained bones of this fourteen-week fetus. (*b*) Bones can fracture even before birth. This fetus has numerous broken bones (arrows) because of an inherited defect in collagen called osteogenesis imperfecta which causes brittle bones. Often parents of such children are unfairly accused of child abuse when their children frequently break bones.

Bone Development and Growth

Parts of the skeletal system begin to form during the first few weeks of prenatal development, and bony structures continue to grow and develop into adulthood. Bones form by replacing existing connective tissue in one of two ways. Some bones originate within sheetlike layers of connective tissues; they are called *intramembranous bones*. Others begin as masses of cartilage that are later replaced by bone tissue; they are called *endochondral bones* (fig. 7.6).

Intramembranous Bones

The broad, flat bones of the skull are **intramembranous bones** (in"trah-mem'brah-nus bōnz). During their development (osteogenesis), membranelike layers of unspecialized, or primitive, connective tissues appear at the sites of the future bones. Dense networks of blood vessels supply these connective tissue layers, which may form around the vessels. These primitive cells enlarge and differentiate into bone-forming cells called **osteoblasts** (os'te-o-blasts), which, in turn, deposit bony matrix around themselves. As a result, spongy bone forms in all directions along blood vessels within the layers of primitive connective tissues. Later, some spongy bone may become compact bone as spaces fill with bone matrix.

As development continues, the osteoblasts may become completely surrounded by matrix, and in this manner, they become secluded within lacunae. At the same time, matrix enclosing the cellular processes of the osteoblasts gives rise to canaliculi. Once isolated in lacunae, these cells are called *osteocytes* (fig. 7.7).

Cells of the primitive connective tissue that persist outside the developing bone give rise to the periosteum. Osteoblasts on the inside of the periosteum form a layer of compact bone over the surface of the newly formed spongy bone.

FIGURE 7.7

Transmission electron micrograph (falsely colored) of an osteocyte isolated within a lacuna (4,700×).

This process of replacing connective tissue to form an intramembranous bone is called *intramembranous ossification*. Table 7.1 lists the major steps of the process.

Endochondral Bones

Most of the bones of the skeleton are **endochondral bones** (en'do-kon'dral bōnz). They develop from masses of hyaline cartilage shaped like future bony structures. These cartilaginous models grow rapidly for a time and then begin to change extensively. For example, cartilage cells enlarge and their lacunae grow. The surrounding matrix breaks down, and soon the cartilage cells die and degenerate.

As the cartilage decomposes, a periosteum forms from connective tissue that encircles the developing structure. Blood vessels and undifferentiated connective tissue cells invade the disintegrating tissue. Some of the invading cells differentiate into osteoblasts and begin to form spongy bone in the spaces previously housing the cartilage. Once completely surrounded by the bony matrix, osteoblasts are called osteocytes. As ossification continues, osteoblasts beneath the periosteum deposit compact bone around the spongy bone.

The process of forming an endochondral bone by the replacement of hyaline cartilage is called *endochondral ossification*. Its major steps are listed in table 7.1 and illustrated in figure 7.8.

In a long bone, bony tissue begins to replace hyaline cartilage in the center of the diaphysis. This region is

TABLE 7.1	Major Steps in Bone Development
Intramembranous Ossification	**Endochondral Ossification**
1. Sheets of primitive connective tissue appear at sites of future bones.	1. Masses of hyaline cartilage form models of future bones.
2. Primitive connective tissue cells collect around blood vessels in these layers.	2. Cartilage tissue breaks down. Periosteum develops.
3. Connective tissue cells differentiate into osteoblasts, which deposit spongy bone.	3. Blood vessels and differentiating osteoblasts from the periosteum invade the disintegrating tissue.
4. Osteoblasts become osteocytes when bony matrix completely surrounds them.	4. Osteoblasts form spongy bone in the space occupied by cartilage.
5. Connective tissue on the surface of each developing structure forms a periosteum.	5. Osteoblasts become osteocytes when bony matrix completely surrounds them.
6. Osteoblasts on the inside of the periosteum deposit compact bone over the spongy bone.	6. Osteoblasts beneath the periosteum deposit compact bone around spongy bone.

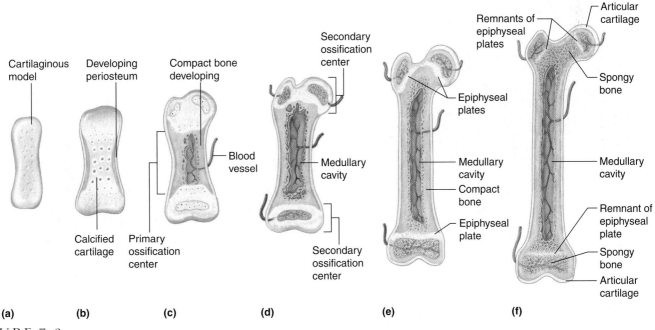

(a) (b) (c) (d) (e) (f)

FIGURE 7.8

Major stages (*a–f*) in the development of an endochondral bone. (Relative bone sizes are not to scale.)

called the *primary ossification center,* and bone develops from it toward the ends of the cartilaginous structure. Meanwhile, osteoblasts from the periosteum deposit a thin layer of compact bone around the primary ossification center. The epiphyses of the developing bone remain cartilaginous and continue to grow. Later, *secondary ossification centers* appear in the epiphyses, and spongy bone forms in all directions from them. As spongy bone is deposited in the diaphysis and in the epiphysis, a band of cartilage, called the **epiphyseal plate** (ep″ĭ-fiz′e-al plāt), or metaphysis, remains between the two ossification centers (see figs. 7.2, 7.3*b,* and 7.8).

Growth at the Epiphyseal Plate

In a long bone, the diaphysis is separated from the epiphysis by an epiphyseal plate. The cartilaginous cells of the epiphyseal plate occur in four layers, each of which may be several cells thick, as shown in figure 7.9. The first layer, closest to the end of the epiphysis, is composed of resting cells that do not actively participate in growth. This layer anchors the epiphyseal plate to the bony tissue of the epiphysis.

The second layer of the epiphyseal plate contains rows of many young cells undergoing mitosis. As new cells appear and as intercellular material forms around them, the cartilaginous plate thickens.

The rows of older cells, which are left behind when new cells appear, form the third layer, enlarging and thickening the epiphyseal plate still more. Consequently, the entire bone lengthens. At the same time, invading osteoblasts, which secrete calcium salts, accumulate in the intercellular matrix adjacent to the oldest cartilaginous cells, and as the matrix calcifies, the cells begin to die.

The fourth layer of the epiphyseal plate is quite thin. It is composed of dead cells and calcified intercellular substance.

In time, large, multinucleated cells called **osteoclasts** (os′te-o-klasts) break down the calcified matrix. These large cells originate by the fusion of single-nucleated white blood cells called monocytes (see chapter 14, p. 519). Osteoclasts secrete an acid that dissolves the inorganic component of the calcified matrix, and their lysosomal enzymes digest the organic components. Osteoclasts also phagocytize components of the bony matrix. After osteoclasts remove the matrix, bone-building osteoblasts invade the region and deposit bone tissue in place of the calcified cartilage.

A long bone continues to lengthen while the cartilaginous cells of the epiphyseal plates are active. How-

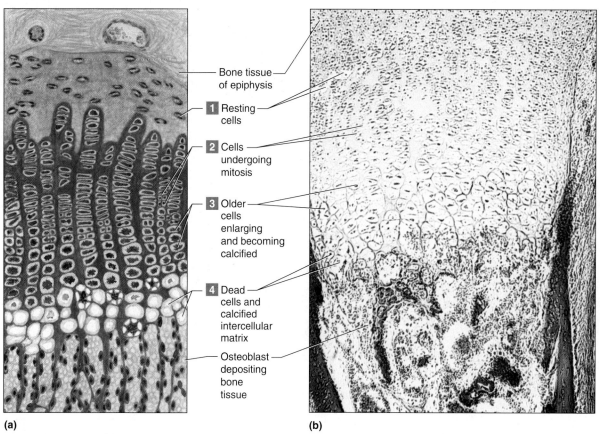

Bone tissue of epiphysis

1 Resting cells

2 Cells undergoing mitosis

3 Older cells enlarging and becoming calcified

4 Dead cells and calcified intercellular matrix

Osteoblast depositing bone tissue

(a)　　　　　　　　　**(b)**

FIGURE 7.9

Epiphyseal plate. (*a*) The cartilaginous cells of an epiphyseal plate lie in four layers, each of which may be several cells thick. (*b*) A micrograph of an epiphyseal plate (100×).

ever, once the ossification centers of the diaphysis and epiphyses meet and the epiphyseal plates ossify, lengthening is no longer possible in that end of the bone.

A developing bone thickens as compact bone is deposited on the outside, just beneath the periosteum. As this compact bone forms on the surface, osteoclasts erode other bone tissue on the inside (fig. 7.10). The resulting space becomes the medullary cavity of the diaphysis, which later fills with marrow.

The bone in the central regions of the epiphyses and diaphysis remains spongy, and hyaline cartilage on the ends of the epiphyses persists throughout life as articular cartilage. Table 7.2 lists the ages at which various bones ossify.

Homeostasis of Bone Tissue

After the intramembranous and endochondral bones form, the actions of osteoclasts and osteoblasts continually remodel them. Thus, throughout life, osteoclasts resorb bone tissue, and osteoblasts replace the bone. These opposing processes of *resorption* and *deposition* are well regulated so that the total mass of bone tissue within an adult skeleton normally remains nearly constant, even though 3% to 5% of bone calcium is exchanged each year.

A child's long bones are still growing if a radiograph shows epiphyseal plates (fig. 7.11). If a plate is damaged as a result of a fracture before it ossifies, elongation of that long bone may prematurely cease, or if growth continues, it may be uneven. For this reason, injuries to the epiphyses of a young person's bones are of special concern. On the other hand, an epiphysis is sometimes altered surgically in order to equalize growth of bones that are developing at very different rates.

In bone cancers, abnormally active osteoclasts destroy bone tissue. Interestingly, cancer of the prostate gland can have the opposite effect. If such cancer cells reach the bone marrow, as they do in most cases of advanced prostatic cancer, they stimulate osteoblast activity. This promotes formation of new bone on the surfaces of the bony trabeculae.

FIGURE 7.10
Micrograph of a bone-resorbing osteoclast (800×).

Developing medullary cavity

Osteoclast

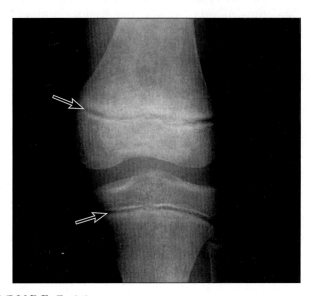

FIGURE 7.11
Radiograph showing the presence of epiphyseal plates (arrows) in a child's bones indicates that the bones are still lengthening.

TABLE 7.2	Ossification Timetable		
Age	**Occurrence**	**Age**	**Occurrence**
Third month of prenatal development	Ossification in long bones begins.	15 to 18 years (females) 17 to 20 years (males)	Bones of the upper limbs and scapulae completely ossify.
Fourth month of prenatal development	Most primary ossification centers have appeared in the diaphyses of bones.	16 to 21 years (females) 18 to 23 years (males)	Bones of the lower limbs and coxal bones completely ossify.
Birth to 5 years	Secondary ossification centers appear in the epiphyses.	21 to 23 years (females) 23 to 25 years (males)	Bones of the sternum, clavicles, and vertebrae completely ossify.
5 to 12 years in females, or 5 to 14 years in males	Ossification rapidly spreads from the ossification centers, and certain bones are ossifying.	By 23 years (females) By 25 years (males)	Nearly all bones completely ossify.

Factors Affecting Bone Development, Growth, and Repair

A number of factors influence bone development, growth, and repair. These include nutrition, exposure to sunlight, hormonal secretions, and physical exercise. For example, vitamin D is necessary for proper absorption of calcium in the small intestine. In the absence of this vitamin, calcium is poorly absorbed, and the inorganic salt portion of bone matrix lacks calcium, softening and thereby deforming bones. In children, this condition is called *rickets,* and in adults, it is called *osteomalacia.*

Vitamin D is relatively uncommon in natural foods, except for eggs. But it is readily available in milk and other dairy products fortified with vitamin D. Vitamin D also forms from a substance (dehydrocholesterol) produced by cells in the digestive tract or obtained in the diet. Dehydrocholesterol is carried by the blood to the skin, and when exposed to ultraviolet light from the sun, it is converted to a compound that becomes vitamin D.

Vitamins A and C are also required for normal bone development and growth. Vitamin A is necessary for osteoblast and osteoclast activity during normal development. Thus, deficiency of vitamin A may retard bone development. Vitamin C is required for collagen synthesis, so its lack also may inhibit bone development. In this case, osteoblasts produce less collagen in the intercellular material of the bone tissue, and the resulting bones are abnormally slender and fragile.

Hormones secreted by the pituitary gland, thyroid gland, parathyroid glands, and ovaries or testes affect bone growth and development. The pituitary gland, for instance, secretes **growth hormone,** which stimulates division of cartilage cells in the epiphyseal plates. In the absence of this hormone, the long bones of the limbs fail to develop normally, and the child has *pituitary dwarfism.* Such a person is very short but has normal body proportions. If excess growth hormone is released before the epiphyseal plates ossify, height may exceed 8 feet—a condition called **pituitary gigantism.** In an adult, secretion of excess growth hormone causes a condition called *acromegaly,* in which the hands, feet, and jaw enlarge (see chapter 13, page 481).

> Pituitary dwarfism is treated with human growth hormone (HGH). Today, HGH is plentiful and pure, thanks to recombinant DNA technology. Bacteria given the human gene for HGH secrete the hormone. Previously, HGH was pooled from donors or cadavers. This introduced the risk of transmitting infection. A controversial use of HGH is to give it to children who are of short stature, but not abnormally so, or to use it to enhance height with the goal of improving athletic ability.

Thyroid hormone stimulates replacement of cartilage in the epiphyseal plates of long bones with bone tissue. Thyroid hormone can halt bone growth by causing premature ossification of the epiphyseal plates. Deficiency of thyroid hormone also may stunt growth, because without its stimulation, the pituitary gland does not secrete enough growth hormone. In contrast to the bone-forming activity of thyroid hormone, parathyroid hormone stimulates an increase in the number and activity of osteoclasts.

Both male and female sex hormones (called androgens and estrogens, respectively) from the testes, ovaries, and adrenal glands promote formation of bone tissue. Beginning at puberty, these hormones are abundant, causing the long bones to grow considerably. However, sex hormones also stimulate ossification of the epiphyseal plates, and consequently they stop bone lengthening at a relatively early age. The effect of estrogens on the epiphyseal plates is somewhat stronger than that of androgens. For this reason, females typically reach their maximum heights earlier than males.

Physical stress also stimulates bone growth. For example, when skeletal muscles contract, they pull at their attachments on bones, and the resulting stress stimulates the bone tissue to thicken and strengthen (hypertrophy). Conversely, with lack of exercise, the same bone tissue wastes, becoming thinner and weaker (atrophy). This is why the bones of athletes are usually stronger and heavier than those of nonathletes (fig. 7.12). It is also why fractured bones immobilized in casts may shorten. Clinical Application 7.1 describes what happens when a bone breaks.

> Astronauts experience a one percent loss of bone mass per month in space. Under microgravity conditions, osteoblast activity diminishes and osteoclast activity increases, with greater loss in spongy compared to compact bone. Researchers predict that a 50 percent bone loss could occur on a several-year-long space flight, such as a mission to Mars.

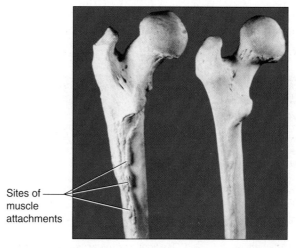

Sites of muscle attachments

FIGURE 7.12
Note the increased amount of bone at the sites of muscle attachments in the femur on the left. The thickened bone is better able to withstand the force resulting from muscle contraction.

1   Describe the development of an intramembranous bone.

2   Explain how an endochondral bone develops.

3   List the steps in the growth of a long bone.

4   Explain how nutritional factors affect bone development.

5   What effects do hormones have on bone growth?

6   How does physical exercise affect bone structure?

Bone Function

Bones shape, support, and protect body structures. They also act as levers that aid body movements, house tissues that produce blood cells, and store various inorganic salts.

Support and Protection

Bones give shape to structures such as the head, face, thorax, and limbs. They also provide support and protection. For example, the bones of the lower limbs, pelvis, and vertebral column support the body's weight. The bones of the skull protect the eyes, ears, and brain. Those of the rib cage and shoulder girdle protect the heart and lungs, whereas bones of the pelvic girdle protect the lower abdominal and internal reproductive organs.

Body Movement

Whenever limbs or other body parts move, bones and muscles interact as simple mechanical devices called **levers** (lev′erz). A lever has four basic components: (1) a rigid bar or rod, (2) a pivot or fulcrum on which the bar turns, (3) an object that is moved against resistance, and (4) a force that supplies energy for the movement of the bar.

A pair of scissors is a lever. The handle and blade form a rigid bar that rocks on a pivot near the center (the screw). The material to be cut by the blades represents the resistance, while the person on the handle end supplies the force needed for cutting the material.

Figure 7.13 shows the three types of levers, which differ in their arrangements. A first-class lever's parts are

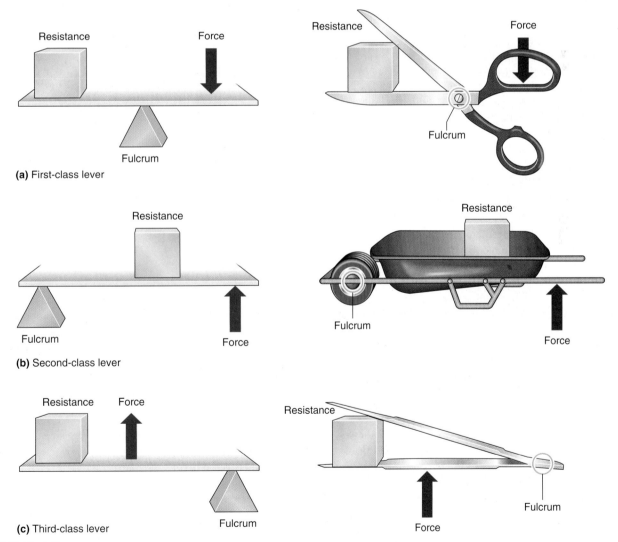

(a) First-class lever

(b) Second-class lever

(c) Third-class lever

FIGURE 7.13

Three types of levers. (a) A first-class lever is used in a pair of scissors, (b) a second-class lever is used in a wheelbarrow, and (c) a third-class lever is used in a pair of forceps.

FRACTURES

When seven-year-old Jacob fell from the tree limb he had been hanging from and held out his arm at an odd angle, it was obvious that he had broken a bone. An X ray at the hospital emergency room confirmed this, and Jacob spent the next six weeks with his broken arm immobilized in a cast.

Many of us have experienced fractured, or broken, bones. A fracture is classified by its cause and the nature of the break. For example, a break due to injury is a *traumatic fracture,* whereas one resulting from disease is a *spontaneous,* or *pathologic, fracture.*

A broken bone exposed to the outside by an opening in the skin is termed a *compound (open) fracture.* It has the added danger of infection, because microorganisms enter through the broken skin. A break protected by uninjured skin is a *closed fracture.* Figure 7A shows several types of traumatic fractures.

Repair of a Fracture

Whenever a bone breaks, blood vessels within it and its periosteum rupture, and the periosteum is likely to tear. Blood escaping from the broken vessels spreads through the damaged area and soon forms a blood clot, or *hematoma.* Vessels in surrounding tissues dilate, swelling and inflaming tissues.

Within days or weeks, developing blood vessels and large numbers of osteoblasts originating from the periosteum invade the hematoma. The osteoblasts rapidly divide in the regions close to the new blood vessels, building spongy bone

A *greenstick* fracture is incomplete, and the break occurs on the convex surface of the bend in the bone.

A *fissured* fracture involves an incomplete longitudinal break.

A *comminuted* fracture is complete and fragments the bone.

A *transverse* fracture is complete, and the break occurs at a right angle to the axis of the bone.

An *oblique* fracture occurs at an angle other than a right angle to the axis of the bone.

A *spiral* fracture is caused by twisting a bone excessively.

FIGURE 7A
Various types of fractures.

nearby. Granulation tissue develops, and in regions farther from a blood supply, fibroblasts produce masses of fibrocartilage.

Meanwhile, phagocytic cells begin to remove the blood clot as well as any dead or damaged cells in the affected area. Osteoclasts also appear and resorb bone fragments, aiding in "cleaning up" debris.

In time, fibrocartilage fills the gap between the ends of the broken bone. This

like those of a pair of scissors. Its pivot is located between the resistance and the force, making the sequence of components resistance–pivot–force. Other examples of first-class levers are seesaws and hemostats (devices used to clamp blood vessels).

The parts of a second-class lever are in the sequence pivot–resistance–force, as in a wheelbarrow. The parts of a third-class lever are in the sequence resistance–force–pivot.

Eyebrow tweezers or forceps used to grasp an object illustrate this type of lever.

The actions of bending and straightening the upper limb at the elbow illustrate bones and muscles functioning as levers (fig. 7.14). When the upper limb bends, the forearm bones represent the rigid bar; the elbow joint is the pivot; the hand is moved against the resistance provided by its weight; and the force is supplied by muscles

mass, termed a cartilaginous *callus,* is later replaced by bone tissue in much the same way that the hyaline cartilage of a developing endochondral bone is replaced. That is, the cartilaginous callus breaks down, blood vessels and osteoblasts invade the area, and a bony callus fills the space.

Typically, more bone is produced at the site of a healing fracture than is necessary to replace the damaged tissues. Osteoclasts remove the excess, and the final result is a bone shaped very much like the original. Figure 7B shows the steps in the healing of a fracture.

The rate of fracture repair depends upon several factors. For instance, if the ends of the broken bone are close together, healing is more rapid than if they are far apart. Physicians can help the bone-healing process. The first casts to immobilize fractured bones were introduced in Philadelphia in 1876, and soon after, doctors began using screws and plates internally to align healing bone parts. Today, orthopedic surgeons also use rods, wires, and nails. These devices have become lighter and smaller; many are built of titanium. A new approach, called a hybrid fixator, treats a broken leg using metal pins internally to align bone pieces. The pins are anchored to a metal ring device worn outside the leg. Also, some bones naturally heal more rapidly than others. The long bones of the upper limbs, for example, may heal in half the time required by the long bones of the lower limbs, as Jacob was happy to discover. He also healed quickly because of his young age. ■

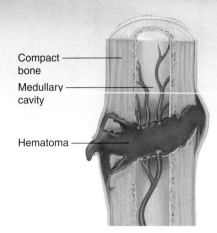

Compact bone

Medullary cavity

Hematoma

(a) Blood escapes from ruptured blood vessels and forms a hematoma.

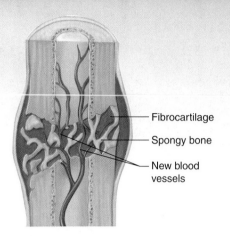

Fibrocartilage

Spongy bone

New blood vessels

(b) Spongy bone forms in regions close to developing blood vessels, and fibrocartilage forms in more distant regions.

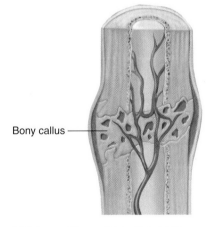

Bony callus

(c) A bony callus replaces fibrocartilage.

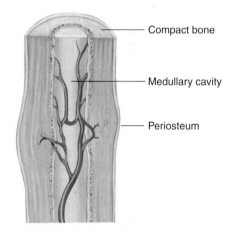

Compact bone

Medullary cavity

Periosteum

(d) Osteoclasts remove excess bony tissue, restoring new bone structure much like the original.

FIGURE 7B

Major steps (*a–d*) in the repair of a fracture.

on the anterior side of the arm. One of these muscles, the *biceps brachii,* is attached by a tendon to a projection (radial tuberosity) on the *radius* bone in the forearm, a short distance below the elbow. Since the parts of this lever are arranged in the sequence resistance–force–pivot, it is a third-class lever.

When the upper limb straightens at the elbow, the forearm bones again serve as the rigid bar, the hand as the resistance, and the elbow joint as the pivot. However, this time the *triceps brachii,* a muscle located on the posterior side of the arm, supplies the force. A tendon of this muscle is attached to a projection (olecranon process) of the ulna bone at the point of the elbow. Since the parts of the lever are arranged resistance–pivot–force, it is a first-class lever.

A second-class lever (pivot–resistance–force) is also demonstrated in the human body. The pivot is the

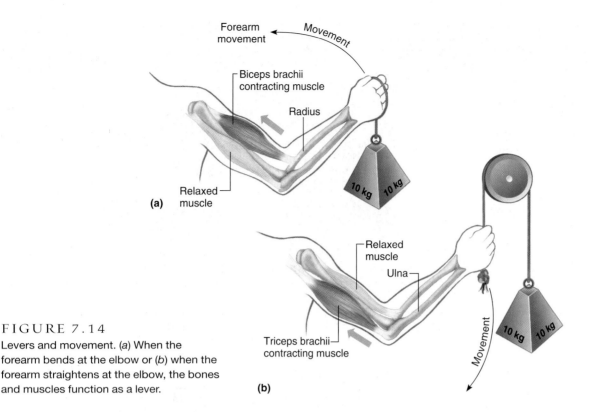

FIGURE 7.14

Levers and movement. (*a*) When the forearm bends at the elbow or (*b*) when the forearm straightens at the elbow, the bones and muscles function as a lever.

temporomandibular joint; and the resistance is supplied by muscles, attaching to a projection (coronoid process) and body of the mandible, that resist or oppose opening the mouth. The muscles attached to the chin area of the mandible provide the force that opens the mouth.

Levers provide a range of movements. Levers that move limbs, for example, are arranged in ways that produce rapid motions, whereas others, such as those that move the head, help maintain posture with minimal effort.

Blood Cell Formation

The process of blood cell formation, called **hematopoiesis** (he″mă-to-poi-e′sis), or hemopoiesis, begins in the yolk sac, which lies outside the embryo (see chapter 23, p. 892). Later in development, blood cells are manufactured in the liver and spleen, and still later, they form in bone marrow.

Marrow is a soft, netlike mass of connective tissue within the medullary cavities of long bones, in the irregular spaces of spongy bone, and in the larger central canals of compact bone tissue. There are two kinds of marrow—red marrow and yellow marrow. *Red marrow* functions in the formation of red blood cells (erythrocytes), white blood cells (leukocytes), and blood platelets. It is red because of the red, oxygen-carrying pigment **hemoglobin** contained within the red blood cells.

Red marrow occupies the cavities of most bones in an infant. With increasing age, however, yellow marrow replaces much of it. *Yellow marrow* stores fat and is inactive in blood cell production.

In an adult, red marrow is primarily found in the spongy bone of the skull, ribs, sternum, clavicles, verte-brae, and pelvis. If the blood cell supply is deficient, some yellow marrow may change back into red marrow and produce blood cells. Chapter 14 (p. 511) discusses blood cell formation.

Inorganic Salt Storage

Recall that the intercellular matrix of bone tissue contains collagen and inorganic mineral salts. The salts account for about 70% of the matrix by weight and are mostly tiny crystals of a type of calcium phosphate called *hydroxyapatite*. Clinical Application 7.2 discusses osteoporosis, a condition that results from loss of bone mineral.

The human body requires calcium for a number of vital metabolic processes, including blood clot formation, nerve impulse conduction, and muscle cell contraction. When the blood is low in calcium, parathyroid hormone stimulates osteoclasts to break down bone tissue, releasing calcium salts from the intercellular matrix into the blood. On the other hand, very high blood calcium inhibits osteoclast activity, and calcitonin from the thyroid gland stimulates osteoblasts to form bone tissue, storing excess calcium in the matrix (fig. 7.15). This mechanism is particularly important in developing bone matrix in children. The details of this homeostatic mechanism are presented in chapter 13, page 487.

In addition to storing calcium and phosphorus (as calcium phosphate), bone tissue contains lesser amounts of magnesium, sodium, potassium, and carbonate ions. Bones also accumulate certain harmful metallic elements such as lead, radium, and strontium, which are not normally present in the body but are sometimes accidentally ingested.

OSTEOPOROSIS

It is an all-too-familiar scenario. The elderly woman pulls herself out of bed, reaches for the night table for support, and misses. She falls, landing on her hip. A younger woman would pull herself up and maybe ache for a few minutes and develop a black-and-blue mark by the next day. But the eighty-year-old, with weakened, brittle bones, suffers a broken hip.

In **osteoporosis,** the skeletal system loses bone volume and mineral content. This disorder is associated with aging. Within affected bones, trabeculae are lost, and the bones develop spaces and canals, which enlarge and fill with fibrous and fatty tissues. Such bones easily fracture and may spontaneously break because they are no longer able to support body weight. For example, a person with osteoporosis may suffer a spontaneous fracture of the thigh bone (femur) at the hip or the collapse of sections of the backbone (vertebrae). Similarly, the distal portion of a forearm bone (radius) near the wrist may be fractured as a result of a minor stress.

In the United States, 200,000 senior citizens fracture their hips each year. For white women, the lifetime risk of hip fracture is 1 in 6; for African-American and Asian women, the risk is lower. For men the risk is about half that of women until the women reach menopause and lose bone mass at a much faster rate than men.

Factors that increase the risk of osteoporosis include low intake of dietary calcium and lack of physical exercise (particularly during the early growing years). However, excessively strenuous exercise in adolescence can delay puberty, which raises the risk of developing osteoporosis later in life for both sexes.

In females, declining levels of the hormone estrogen contribute to development of osteoporosis. The ovaries produce estrogen until menopause. At this time, women experience a rapid acceleration in the rate of bone density loss. Evidence of the estrogen-osteoporosis link comes from studies on women who have declining estrogen levels and increased risk of osteoporosis. These include young women who have had their ovaries removed, women who have anorexia nervosa (self-starvation) that stopped their menstrual cycles, and women past menopause. Drinking alcohol, smoking cigarettes, and inheriting certain genes may also increase a person's risk of developing osteoporosis.

Osteoporosis may be prevented if steps are taken early enough. Bone mass usually peaks at about age thirty-five. Thereafter, bone loss may exceed bone formation in both males and females. To reduce such loss, doctors suggest that people in their mid-twenties and older should take in 1,000–1,500 milligrams of calcium daily. An 8-ounce glass of nonfat milk, for example, contains about 275 milligrams of calcium. It is also recommended that people engage in regular exercise, especially walking or jogging, in which the bones support body weight. As a rule, women have about 30% less bone mass than men; after menopause, women typically lose bone mass twice as fast as men do. Also, people with osteoporosis can slow progress of the disease by taking a drug that is a form of the hormone calcitonin, if they can tolerate the side effect of throat irritation.

Confirming osteoporosis is sometimes difficult. A radiograph may not reveal a decrease in bone density until 20% to 30% of the bone tissue is lost. Noninvasive diagnostic techniques, however, can detect rapid changes in bone mass. These include a *densitometer scanner,* which measures the density of wrist bones, and *quantitative computed tomography,* which can visualize the density of other bones.

Alternatively, a physician may take a bone sample, usually from a hipbone, to directly assess the condition of the tissue. Such a biopsy may also be used to judge the effectiveness of treatment for bone disease. ■

Biomineralization—the combining of minerals with organic molecules, as occurs in bones—is seen in many animal species. Ancient Mayan human skulls have teeth composed of nacre, also known as "mother-of-pearl" and found on mollusk shells, but tooth roots of human bone. The Mayan dentists knew that somehow the human body recognizes a biomineral used in another species. Today, nacre is used to fill in bone lost in the upper jaw. The nacre not only does not evoke rejection by the immune system, but it also stimulates the person's osteoblasts to produce new bone tissue.

1 Name three major functions of bones.

2 Explain how parts of the upper limb form a first-class lever and a third-class lever.

3 Distinguish between the functions of red marrow and yellow marrow.

4 Explain regulation of the concentration of blood calcium.

5 List the substances normally stored in bone tissue.

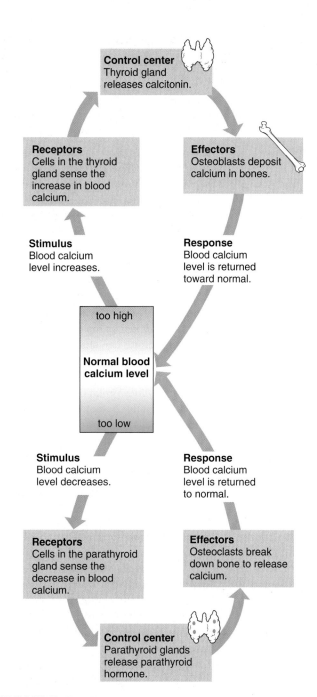

FIGURE 7.15
Hormonal regulation of bone calcium resorption and deposition.

TABLE 7.3	Bones of the Adult Skeleton	
1. Axial Skeleton		
a. Skull		22 bones
8 cranial bones		
frontal 1		
parietal 2		
occipital 1		
temporal 2		
sphenoid 1		
ethmoid 1		
14 facial bones		
maxilla 2		
palatine 2		
zygomatic 2		
lacrimal 2		
nasal 2		
vomer 1		
inferior nasal concha 2		
mandible 1		
b. Middle ear bones		6 bones
malleus 2		
incus 2		
stapes 2		
c. Hyoid		1 bone
hyoid bone 1		
d. Vertebral column		26 bones
cervical vertebra 7		
thoracic vertebra 12		
lumbar vertebra 5		
sacrum 1		
coccyx 1		
e. Thoracic cage		25 bones
rib 24		
sternum 1		
2. Appendicular Skeleton		
a. Pectoral girdle		4 bones
scapula 2		
clavicle 2		
b. Upper limbs		60 bones
humerus 2		
radius 2		
ulna 2		
carpal 16		
metacarpal 10		
phalanx 28		
c. Pelvic girdle		2 bones
coxa 2		
d. Lower limbs		60 bones
femur 2		
tibia 2		
fibula 2		
patella 2		
tarsal 14		
metatarsal 10		
phalanx 28		
	Total	206 bones

Skeletal Organization

Number of Bones

The number of bones in a human skeleton is often reported to be 206 (table 7.3), but the actual number varies from person to person. People may lack certain bones or have extra ones. For example, the flat bones of the skull usually grow together and tightly join along irregular lines called **sutures.** Occasionally, extra bones called sutural bones (wormian bones) develop in these sutures (fig. 7.16). Extra small, round sesamoid bones may develop in tendons, where they reduce friction in places where tendons pass over bony prominences.

Divisions of the Skeleton

For purposes of study, it is convenient to divide the skeleton into two major portions—an axial skeleton and an appendicular skeleton (fig. 7.17). The **axial skeleton** consists of the bony and cartilaginous parts that support and protect the organs of the head, neck, and trunk. These parts include the following:

1. **Skull.** The skull is composed of the *cranium* (brain case) and the *facial bones.*

2. **Hyoid bone.** The hyoid (hi′oid) bone is located in the neck between the lower jaw and the larynx (fig. 7.18). It does not articulate with any other bones but is fixed in position by muscles and ligaments. The hyoid bone supports the tongue and is an attachment for certain muscles that help move the tongue during swallowing. It can be felt approximately a finger's width above the anterior prominence of the larynx.

3. **Vertebral column.** The vertebral column, or spinal column, consists of many vertebrae separated by cartilaginous *intervertebral discs.* This column forms the central axis of the skeleton. Near its distal end, several vertebrae fuse to form the **sacrum** (sa′krum), which is part of the pelvis. A small,

rudimentary tailbone called the **coccyx** (kok′siks), is attached to the end of the sacrum.

4. **Thoracic cage.** The thoracic cage protects the organs of the thoracic cavity and the upper abdominal cavity. It is composed of twelve pairs of **ribs,** which articulate posteriorly with thoracic vertebrae. It also includes the **sternum** (ster′num), or breastbone, to which most of the ribs are attached anteriorly.

The **appendicular skeleton** consists of the bones of the upper and lower limbs and the bones that anchor the limbs to the axial skeleton. It includes the following:

1. **Pectoral girdle.** The pectoral girdle is formed by a **scapula** (scap′u-lah), or shoulder blade, and a **clavicle** (klav′ĭ-k′l), or collarbone, on both sides of the body. The pectoral girdle connects the bones of the upper limbs to the axial skeleton and aids in upper limb movements.

2. **Upper limbs.** Each upper limb consists of a **humerus** (hu′mer-us), or arm bone; two forearm bones—a **radius** (ra′de-us) and an **ulna** (ul′nah)—and a wrist and hand. The humerus, radius, and ulna articulate with each other at the elbow joint. At the distal end of the radius and ulna is the wrist. There are eight **carpals** (kar′palz), or wrist bones. The five bones of the palm are called **metacarpals** (met″ah-kar′palz), and the fourteen finger bones are called **phalanges** (fah-lan′jēz).

3. **Pelvic girdle.** The pelvic girdle is formed by two **coxae** (kok′se), or hipbones, which are attached to each other anteriorly and to the sacrum posteriorly. They connect the bones of the lower limbs to the axial skeleton and, with the sacrum and coccyx, form the **pelvis,** which protects the lower abdominal and internal reproductive organs.

4. **Lower limbs.** Each lower limb consists of a **femur** (fe′mur), or thigh bone; two leg bones—a large **tibia** (tib′e-ah), or shin bone, and a slender **fibula** (fib′u-lah), or calf bone—and an ankle and a foot. The femur and tibia articulate with each other at the knee joint, where the **patella** (pah-tel′ah), or kneecap, covers the anterior surface. At the distal ends of the tibia and fibula is the ankle. There are seven **tarsals** (tahr′salz), or ankle bones. The five bones of the instep are called **metatarsals** (met″ah-tar′salz), and the fourteen bones of the toes (like the fingers) are called **phalanges.** Table 7.4 defines some terms used to describe skeletal structures.

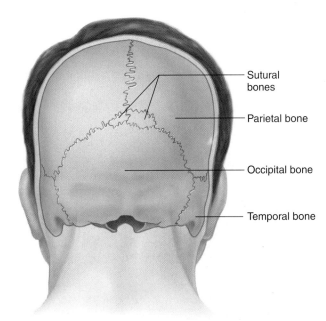

FIGURE 7.16

Sutural bones are extra bones that sometimes develop in sutures between the flat bones of the skull.

- Sutural bones
- Parietal bone
- Occipital bone
- Temporal bone

1 Distinguish between the axial and appendicular skeletons.

2 List the bones of the axial skeleton and of the appendicular skeleton.

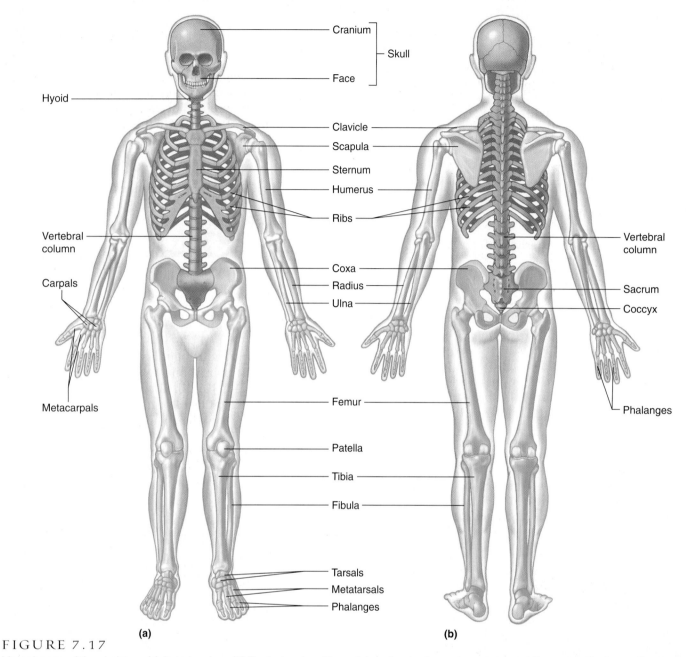

FIGURE 7.17

Major bones of the skeleton. (*a*) Anterior view. (*b*) Posterior view. The axial portion is shown in orange, and the appendicular portions are shown in green.

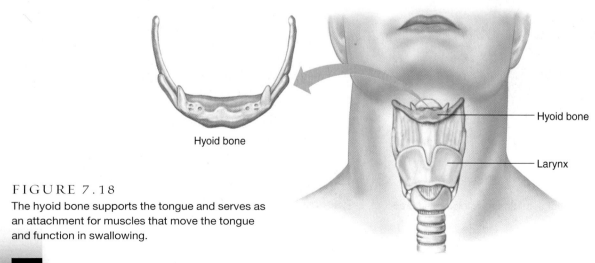

FIGURE 7.18

The hyoid bone supports the tongue and serves as an attachment for muscles that move the tongue and function in swallowing.

TABLE 7.4 Terms Used to Describe Skeletal Structures

Term	Definition	Example
condyle (kon′dīl)	A rounded process that usually articulates with another bone	Occipital condyle of the occipital bone (fig. 7.22)
crest (krest)	A narrow, ridgelike projection	Iliac crest of the ilium (fig. 7.50)
epicondyle (ep″ĭ-kon′dīl)	A projection situated above a condyle	Medial epicondyle of the humerus (fig. 7.45)
facet (fas′et)	A small, nearly flat surface	Facet of a thoracic vertebra (fig. 7.38)
fissure (fish′ūr)	A cleft or groove	Inferior orbital fissure in the orbit of the eye (fig. 7.20)
fontanel (fon″tah-nel′)	A soft spot in the skull where membranes cover the space between bones	Anterior fontanel between the frontal and parietal bones (fig. 7.33)
foramen (fo-ra′men)	An opening through a bone that usually serves as a passageway for blood vessels, nerves, or ligaments	Foramen magnum of the occipital bone (fig. 7.22)
fossa (fos′ah)	A relatively deep pit or depression	Olecranon fossa of the humerus (fig. 7.45)
fovea (fo′ve-ah)	A tiny pit or depression	Fovea capitis of the femur (fig. 7.53)
head (hed)	An enlargement on the end of a bone	Head of the humerus (fig. 7.45)
linea (lin′e-ah)	A narrow ridge	Linea aspera of the femur (fig. 7.53)
meatus (me-a′tus)	A tubelike passageway within a bone	External acoustic meatus of the ear (fig. 7.21)
process (pros′es)	A prominent projection on a bone	Mastoid process of the temporal bone (fig. 7.21)
ramus (ra′mus)	A branch or similar extension	Ramus of the mandible (fig. 7.31)
sinus (si′nus)	A cavity within a bone	Frontal sinus of the frontal bone (fig. 7.27)
spine (spīn)	A thornlike projection	Spine of the scapula (fig. 7.43)
suture (soo′cher)	An interlocking line of union between bones	Lambdoidal suture between the occipital and parietal bones (fig. 7.21)
trochanter (tro-kan′ter)	A relatively large process	Greater trochanter of the femur (fig. 7.53)
tubercle (tu′ber-kl)	A small, knoblike process	Tubercle of a rib (fig. 7.41)
tuberosity (tu″bĕ-ros′ĭ-te)	A knoblike process usually larger than a tubercle	Radial tuberosity of the radius (fig. 7.46)

Skull

A human skull usually consists of twenty-two bones that, except for the lower jaw, are firmly interlocked along sutures. Eight of these interlocked bones make up the cranium, and fourteen form the facial skeleton. The **mandible** (man′dĭ-b′l), or lower jawbone, is a movable bone held to the cranium by ligaments (figs. 7.19 and 7.21). Some facial and cranial bones together form the orbit of the eye (fig. 7.20). Plates 8–36 on pages 238–252 show a set of photographs of the human skull and its parts.

Cranium

The **cranium** (kra′ne-um) encloses and protects the brain, and its surface provides attachments for muscles that make chewing and head movements possible. Some of the cranial bones contain air-filled cavities called *sinuses,* which are lined with mucous membranes and connect by passageways to the nasal cavity. Sinuses reduce the weight of the skull and increase the intensity of the voice by serving as resonant sound chambers.

The eight bones of the cranium (table 7.5) are as follows:

1. **Frontal bone.** The frontal (frun′tal) bone forms the anterior portion of the skull above the eyes, including the forehead, the roof of the nasal cavity, and the roofs of the orbits (bony sockets) of the eyes. On the upper margin of each orbit, the frontal bone is marked by a *supraorbital foramen* (or *supraorbital notch* in some skulls) through which blood vessels and nerves pass to the tissues of the forehead. Within the frontal bone are two *frontal sinuses,* one above each eye near the midline. The frontal bone is a single bone in adults, but it develops in two parts (see fig. 7.33). These halves grow together and usually completely fuse by the fifth or sixth year of life.

2. **Parietal bones.** One parietal (pah-ri′ĕ-tal) bone is located on each side of the skull just behind the frontal bone. Each is shaped like a curved plate and has four borders. Together, the parietal bones form the bulging sides and roof of the cranium. They are fused at the midline along the *sagittal suture,* and they meet the frontal bone along the *coronal suture.*

3. **Occipital bone.** The occipital (ok-sip′ĭ-tal) bone joins the parietal bones along the *lambdoidal* (lam′doid-al) *suture.* It forms the back of the skull and the base of the cranium. A large opening on its lower surface, the *foramen magnum,* houses nerve fibers from the brain that pass through and enter the

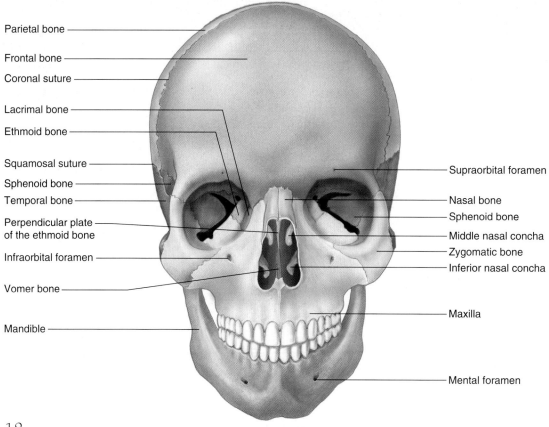

FIGURE 7.19
Anterior view of the skull.

Parietal bone

Frontal bone

Coronal suture

Lacrimal bone

Ethmoid bone

Squamosal suture

Sphenoid bone

Temporal bone

Perpendicular plate
of the ethmoid bone

Infraorbital foramen

Vomer bone

Mandible

Supraorbital foramen

Nasal bone

Sphenoid bone

Middle nasal concha

Zygomatic bone

Inferior nasal concha

Maxilla

Mental foramen

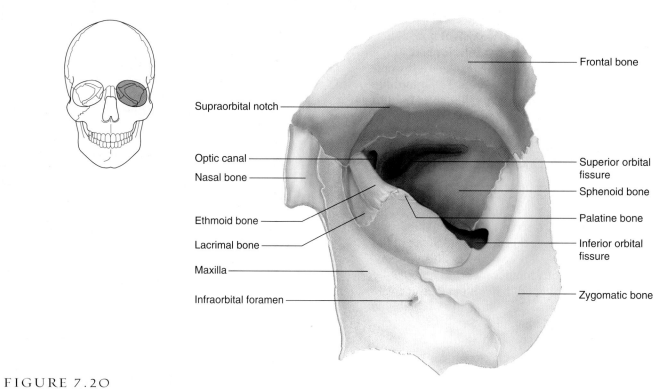

FIGURE 7.20
The orbit of the eye includes both cranial and facial bones.

Supraorbital notch

Optic canal

Nasal bone

Ethmoid bone

Lacrimal bone

Maxilla

Infraorbital foramen

Frontal bone

Superior orbital
fissure

Sphenoid bone

Palatine bone

Inferior orbital
fissure

Zygomatic bone

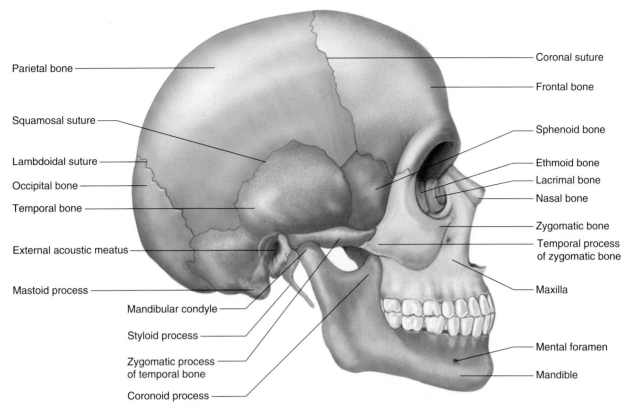

FIGURE 7.21
Right lateral view of the skull.

TABLE 7.5	Cranial Bones		
Name and Number	**Description**		**Special Features**
Frontal (1)	Forms forehead, roof of nasal cavity, and roofs of orbits		Supraorbital foramen, frontal sinuses
Parietal (2)	Form side walls and roof of cranium		Fused at midline along sagittal suture
Occipital (1)	Forms back of skull and base of cranium		Foramen magnum, occipital condyles
Temporal (2)	Form side walls and floor of cranium		External acoustic meatus, mandibular fossa, mastoid process, styloid process, zygomatic process
Sphenoid (1)	Forms parts of base of cranium, sides of skull, and floors and sides of orbits		Sella turcica, sphenoidal sinuses
Ethmoid (1)	Forms parts of roof and walls of nasal cavity, floor of cranium, and walls of orbits		Cribriform plates, perpendicular plate, superior and middle nasal conchae, ethmoidal sinuses, crista galli

vertebral canal to become part of the spinal cord. Rounded processes called *occipital condyles,* located on each side of the foramen magnum, articulate with the first vertebra (atlas) of the vertebral column.

4. **Temporal bones.** A temporal (tem′por-al) bone on each side of the skull joins the parietal bone along a *squamosal* (skwa-mo′sal) *suture.* The temporal bones form parts of the sides and the base of the cranium. Located near the inferior margin is an opening, the *external acoustic* (auditory) *meatus,*

which leads inward to parts of the ear. The temporal bones also house the internal ear structures and have depressions called the *mandibular fossae* (glenoid fossae) that articulate with condyles of the mandible. Below each external auditory meatus are two projections—a rounded *mastoid process* and a long, pointed *styloid process.* The mastoid process provides an attachment for certain muscles of the neck, whereas the styloid process anchors muscles associated with the tongue and pharynx. An opening near the mastoid process, the *carotid canal,* transmits the internal carotid artery. An opening between the temporal and occipital bones, the *jugular foramen,* accommodates the internal jugular vein (fig. 7.22).

The mastoid process may become infected. The tissues in this region of the temporal bone contain a number of interconnected air cells lined with mucous membranes that communicate with the middle ear. These spaces sometimes become inflamed when microorganisms spread into them from an infected middle ear (*otitis media*). The resulting mastoid infection, called *mastoiditis,* is of particular concern because nearby membranes that surround the brain may become infected.

A *zygomatic process* projects anteriorly from the temporal bone in the region of the external auditory meatus. It joins the *zygomatic bone* and helps form the prominence of the cheek.

5. **Sphenoid bone.** The sphenoid (sfe'noid) bone (fig. 7.23) is wedged between several other bones in the anterior portion of the cranium. It consists of a central part and two winglike structures that extend laterally toward each side of the skull. This bone helps form the base of the cranium, the sides of the skull, and the floors and sides of the orbits. Along the midline within the cranial cavity, a portion of the sphenoid bone indents to form the saddle-shaped *sella turcica* (sel'ah tur'si-ka). In this depression lies the pituitary gland, which hangs from the base of the brain by a stalk.

The sphenoid bone also contains two *sphenoidal sinuses.* These lie side by side and are separated by a bony septum that projects downward into the nasal cavity.

6. **Ethmoid bone.** The ethmoid (eth'moid) bone (fig. 7.24) is located in front of the sphenoid bone. It consists of two masses, one on each side of the nasal cavity, which are joined horizontally by thin *cribriform* (krib'rĭ-form) *plates.* These plates form part of the roof of the nasal cavity, and nerves associated with the sense of smell pass through tiny openings (*olfactory foramina*) in them. Portions of the ethmoid bone also form sections of the cranial floor, orbital walls, and nasal cavity walls. A *perpendicular plate* projects downward in the midline from the cribriform plates to form most of the nasal septum.

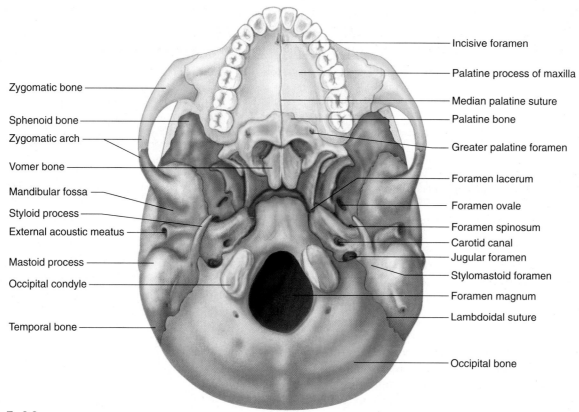

FIGURE 7.22
Inferior view of the skull.

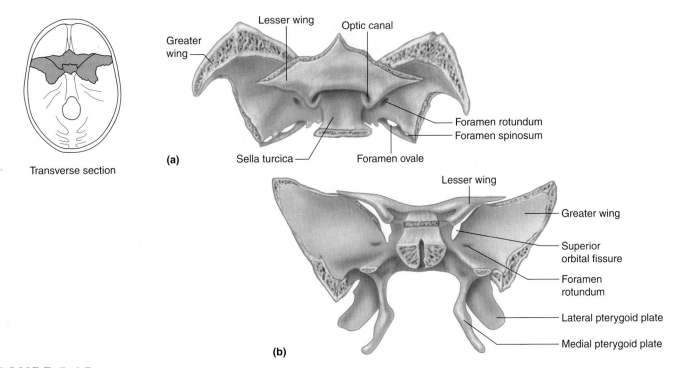

Lesser wing

Optic canal

Greater wing

Foramen rotundum

Foramen spinosum

(a)

Sella turcica

Foramen ovale

Lesser wing

Greater wing

Superior orbital fissure

Foramen rotundum

Lateral pterygoid plate

Medial pterygoid plate

(b)

FIGURE 7.23

The sphenoid bone. (*a*) Superior view. (*b*) Posterior view. (The sphenoidal sinuses are within the bone and are not visible in this representation.)

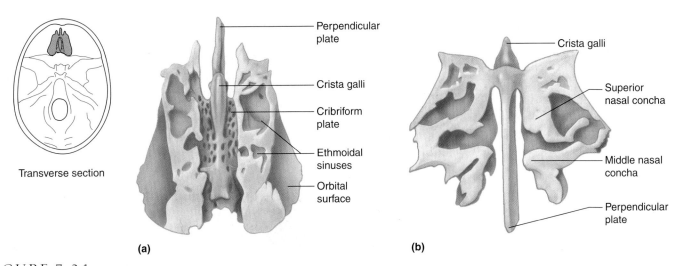

Transverse section

Perpendicular plate

Crista galli

Cribriform plate

Ethmoidal sinuses

Orbital surface

(a)

Crista galli

Superior nasal concha

Middle nasal concha

Perpendicular plate

(b)

FIGURE 7.24

The ethmoid bone (*a*) superior view and (*b*) posterior view.

Delicate, scroll-shaped plates called the *superior nasal concha* (kong′kah) and the *middle nasal concha* project inward from the lateral portions of the ethmoid bone toward the perpendicular plate. These bony plates support mucous membranes that line the nasal cavity. The mucous membranes, in turn, begin moistening, warming, and filtering air as it enters the respiratory tract. The lateral portions of the ethmoid bone contain many small air spaces, the

ethmoidal sinuses. Figure 7.25 shows various structures in the nasal cavity.

Projecting upward into the cranial cavity between the cribriform plates is a triangular process of the ethmoid bone called the *crista galli* (kris′tă gal′li; cock's comb). Membranes that enclose the brain attach to this process. Figure 7.26 shows a view of the cranial cavity.

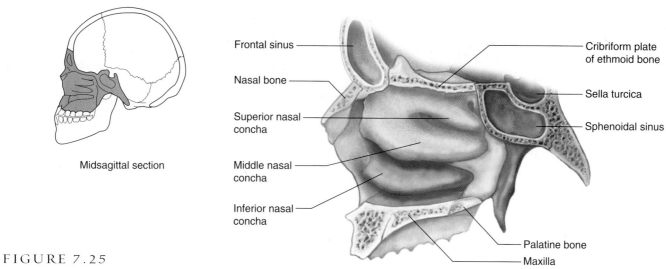

FIGURE 7.25
Lateral wall of the nasal cavity.

Frontal sinus
Nasal bone
Superior nasal concha
Middle nasal concha
Inferior nasal concha
Cribriform plate of ethmoid bone
Sella turcica
Sphenoidal sinus
Palatine bone
Maxilla
Midsagittal section

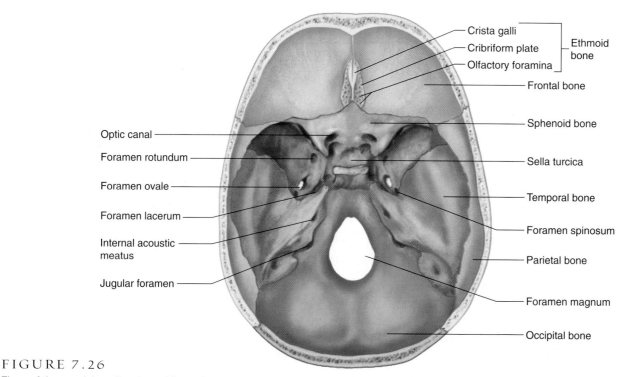

FIGURE 7.26
Floor of the cranial cavity, viewed from above.

Crista galli
Cribriform plate
Olfactory foramina
Ethmoid bone
Frontal bone
Sphenoid bone
Sella turcica
Temporal bone
Foramen spinosum
Parietal bone
Foramen magnum
Occipital bone
Optic canal
Foramen rotundum
Foramen ovale
Foramen lacerum
Internal acoustic meatus
Jugular foramen

Facial Skeleton

The **facial skeleton** consists of thirteen immovable bones and a movable lower jawbone. In addition to forming the basic shape of the face, these bones provide attachments for muscles that move the jaw and control facial expressions.

The bones of the facial skeleton are as follows:

1. **Maxillary bones.** The maxillary (mak′sĭ-ler″e) bones (pl., *maxillae,* mak-sĭl′e) form the upper jaw;

together they form the keystone of the face, since all the other immovable facial bones articulate with them.

Portions of these bones comprise the anterior roof of the mouth (*hard palate*), the floors of the orbits, and the sides and floor of the nasal cavity. They also contain the sockets of the upper teeth. Inside the maxillae, lateral to the nasal cavity, are *maxillary sinuses.* These spaces are the largest of the sinuses, and they extend from the floor of the orbits to the

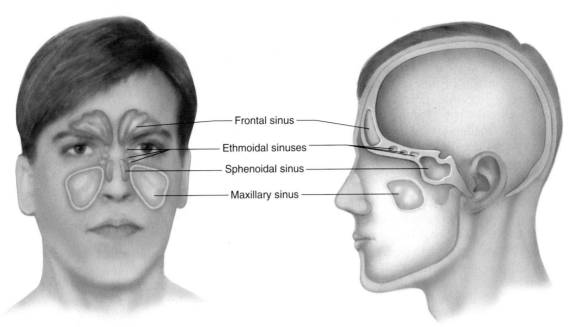

FIGURE 7.27
Locations of the sinuses.

roots of the upper teeth. Figure 7.27 shows the locations of the maxillary and other sinuses. Table 7.6 displays a summary of the sinuses.

During development, portions of the maxillary bones called *palatine processes* grow together and fuse along the midline, or median palatine suture. This forms the anterior section of the hard palate (see fig. 7.22).

The inferior border of each maxillary bone projects downward, forming an *alveolar* (al-ve′o-lar) *process.* Together these processes form a horseshoe-shaped *alveolar arch* (dental arch). Teeth occupy cavities in this arch (dental alveoli). Dense connective tissue binds teeth to the bony sockets (see chapter 17, p. 650).

Sometimes, fusion of the palatine processes of the maxillae is incomplete at birth; the result is a *cleft palate.* Infants with a cleft palate may have trouble suckling because of the opening between the oral and nasal cavities. A temporary prosthetic device (artificial palate) may be inserted within the mouth, or a special type of nipple can be placed on bottles until surgery can be performed to correct the condition.

2. **Palatine bones.** The L-shaped palatine (pal′ah-tīn) bones (fig. 7.28) are located behind the maxillae. The horizontal portions form the posterior section of the hard palate and the floor of the nasal cavity. The

TABLE 7.6	Sinuses of the Cranial and Facial Bones	
Sinuses	**Number**	**Location**
Frontal sinuses	2	Frontal bone above each eye and near the midline
Sphenoidal sinuses	2	Sphenoid bone above the posterior portion of the nasal cavity
Ethmoidal sinuses	2 groups of small spaces	Ethmoid bone on either side of the upper portion of the nasal cavity
Maxillary sinuses	2	Maxillary bones lateral to the nasal cavity and extending from the floor of the orbits to the roots of the upper teeth

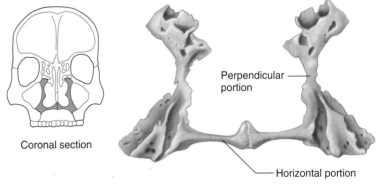

Coronal section

Perpendicular portion

Horizontal portion

FIGURE 7.28
The horizontal portions of the palatine bones form the posterior section of the hard palate, and the perpendicular portions help form the lateral walls of the nasal cavity.

perpendicular portions help form the lateral walls of the nasal cavity.

3. **Zygomatic bones.** The zygomatic (zi″go-mat′ik) bones are responsible for the prominences of the cheeks below and to the sides of the eyes. These bones also help form the lateral walls and the floors of the orbits. Each bone has a *temporal process,* which extends posteriorly to join the zygomatic process of a temporal bone. Together these processes form a *zygomatic arch* (see figs. 7.21 and 7.22).

4. **Lacrimal bones.** A lacrimal (lak′rĭ-mal) bone is a thin, scalelike structure located in the medial wall of each orbit between the ethmoid bone and the maxilla (see fig. 7.21). A groove in its anterior portion leads from the orbit to the nasal cavity, providing a pathway for a channel that carries tears from the eye to the nasal cavity.

5. **Nasal bones.** The nasal (na′zal) bones are long, thin, and nearly rectangular (see fig. 7.19). They lie side by side and are fused at the midline, where they form the bridge of the nose. These bones are attachments for the cartilaginous tissues that form the shape of the nose.

6. **Vomer bone.** The thin, flat vomer (vo′mer) bone is located along the midline within the nasal cavity. Posteriorly, it joins the perpendicular plate of the ethmoid bone, and together they form the nasal septum (figs. 7.29 and 7.30).

7. **Inferior nasal conchae.** The inferior nasal conchae (kong′ke) are fragile, scroll-shaped bones attached to the lateral walls of the nasal cavity. They are the largest of the conchae and are positioned below the superior and middle nasal conchae of the ethmoid bone (see figs. 7.19 and 7.25). Like the ethmoidal conchae, the inferior conchae support mucous membranes within the nasal cavity.

8. **Mandible.** The mandible (man′dĭ-b′l), or lower jawbone, is a horizontal, horseshoe-shaped body with a flat *ramus* projecting upward at each end. The rami are divided into a posterior *mandibular condyle* and an anterior *coronoid process* (fig. 7.31). The mandibular condyles articulate with the mandibular fossae of the temporal bones, whereas the coronoid processes provide attachments for muscles used in chewing. Other large chewing muscles are inserted on the lateral surfaces of the rami. A curved bar of

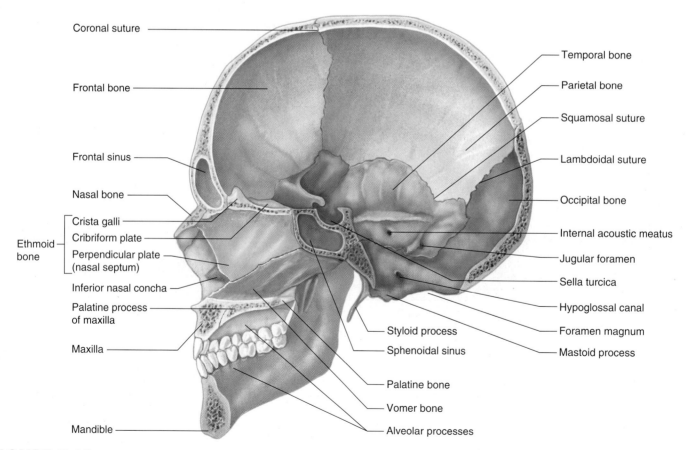

FIGURE 7.29
Sagittal section of the skull.

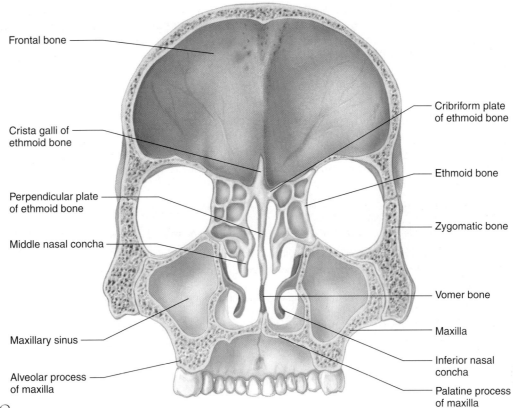

FIGURE 7.30
Coronal section of the skull (posterior view).

Frontal bone

Crista galli of ethmoid bone

Perpendicular plate of ethmoid bone

Middle nasal concha

Maxillary sinus

Alveolar process of maxilla

Cribriform plate of ethmoid bone

Ethmoid bone

Zygomatic bone

Vomer bone

Maxilla

Inferior nasal concha

Palatine process of maxilla

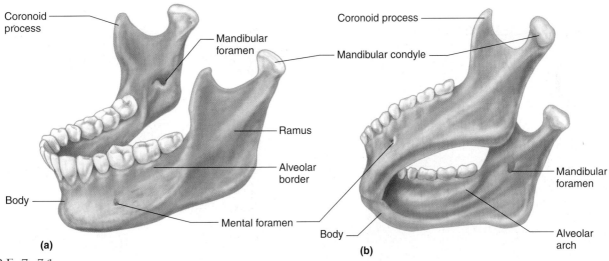

Coronoid process

Mandibular foramen

Coronoid process

Mandibular condyle

Ramus

Alveolar border

Body

Mental foramen

Body

Mandibular foramen

Alveolar arch

(a)

(b)

FIGURE 7.31
Mandible. (*a*) Left lateral view. (*b*) Inferior view.

bone on the superior border of the mandible, the *alveolar border,* contains the hollow sockets (dental alveoli) that bear the lower teeth.

On the medial side of the mandible, near the center of each ramus, is a *mandibular foramen.* This opening admits blood vessels and a nerve, which supply the roots of the lower teeth. Dentists inject anesthetic into the tissues near this foramen to temporarily block nerve impulse conduction and desensitize teeth on that side of the jaw. Branches of the blood vessels and the nerve emerge from the mandible through the *mental foramen,* which opens on the outside near the point of the jaw. They supply the tissues of the chin and lower lip.

Table 7.7 describes the fourteen facial bones. Figure 7.32 shows features of these bones on radiographs. Table 7.8 lists the major openings (*foramina*) and passageways through bones of the skull, as well as their general locations and the structures they transmit.

Infantile Skull

At birth, the skull is incompletely developed, with fibrous membranes connecting the cranial bones. These membranous areas are called **fontanels** (fon"tah-nel′z), or, more commonly, soft spots. They permit some movement between the bones so that the developing skull is partially compressible and can slightly change shape. This action, called *molding,* enables an infant's skull to more easily pass through the birth canal. Eventually, the fontanels close as the cranial bones grow together. The posterior fontanel usually closes about two months after birth; the sphenoid fontanel closes at about three months; the mastoid fontanel closes near the end of the first year; and the anterior fontanel may not close until the middle or end of the second year.

Other characteristics of an infantile skull (fig. 7.33) include a relatively small face with a prominent forehead

TABLE 7.7	Bones of the Facial Skeleton	
Name and Number	**Description**	**Special Features**
Maxillary (2)	Form upper jaw, anterior roof of mouth, floors of orbits, and sides and floor of nasal cavity	Alveolar processes, maxillary sinuses, palatine process
Palatine (2)	Form posterior roof of mouth, and floor and lateral walls of nasal cavity	
Zygomatic (2)	Form prominences of cheeks, and lateral walls and floors of orbits	Temporal process
Lacrimal (2)	Form part of medial walls of orbits	Groove that leads from orbit to nasal cavity
Nasal (2)	Form bridge of nose	
Vomer (1)	Forms inferior portion of nasal septum	
Inferior nasal conchae (2)	Extend into nasal cavity from its lateral walls	
Mandible (1)	Forms lower jaw	Body, ramus, mandibular condyle, coronoid process, alveolar process, mandibular foramen, mental foramen

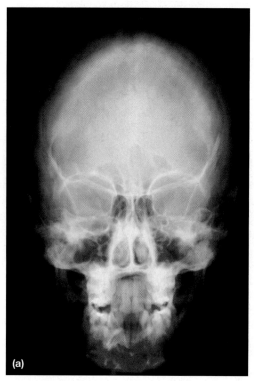
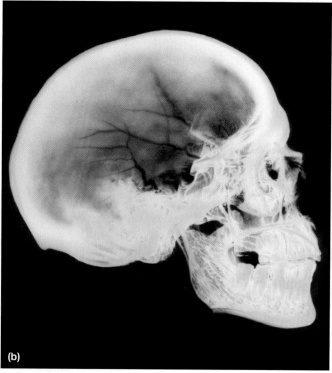

FIGURE 7.32

Falsely colored radiographs of the skull. (*a*) Anterior view and (*b*) right lateral view.

TABLE 7.8   Passageways Through Bones of the Skull

Passageway	Location	Major Stuctures Transmitted
Carotid canal (fig. 7.22)	Inferior surface of the temporal bone	Internal carotid artery, veins, and nerves
Foramen lacerum (fig. 7.22)	Floor of cranial cavity between temporal and sphenoid bones	Branch of pharyngeal artery (in life, opening is largely covered by fibrocartilage)
Foramen magnum (fig. 7.26)	Base of skull in occipital bone	Nerve fibers passing between the brain and spinal cord as it exits from the base of the brain, also certain arteries
Foramen ovale (fig. 7.22)	Floor of cranial cavity in sphenoid bone	Mandibular division of trigeminal nerve and veins
Foramen rotundum (fig. 7.26)	Floor of cranial cavity in sphenoid bone	Maxillary division of trigeminal nerve
Foramen spinosum (fig. 7.26)	Floor of cranial cavity in sphenoid bone	Middle meningeal blood vessels and branch of mandibular nerve
Greater palatine foramen (fig. 7.22)	Posterior portion of hard palate in palatine bone	Palatine blood vessels and nerves
Hypoglossal canal (fig. 7.29)	Near margin of foramen magnum in occipital bone	Hypoglossal nerve
Incisive foramen (fig. 7.22)	Incisive fossa in anterior portion of hard palate	Nasopalatine nerves, openings of vomeronasal organ
Inferior orbital fissure (fig. 7.20)	Floor of the orbit	Maxillary nerve and blood vessels
Infraorbital foramen (fig. 7.20)	Below the orbit in maxillary bone	Infraorbital blood vessels and nerves
Internal acoustic meatus (fig. 7.26)	Floor of cranial cavity in temporal bone	Branches of facial and vestibulocochlear nerves, and blood vessels
Jugular foramen (fig. 7.26)	Base of the skull between temporal and occipital bones	Glossopharyngeal, vagus and accessory nerves, and blood vessels
Mandibular foramen (fig. 7.31)	Inner surface of ramus of mandible	Inferior alveolar blood vessels and nerves
Mental foramen (fig. 7.31)	Near point of jaw in mandible	Mental nerve and blood vessels
Optic canal (fig. 7.20)	Posterior portion of orbit in sphenoid bone	Optic nerve and ophthalmic artery
Stylomastoid foramen (fig. 7.22)	Between styloid and mastoid processes	Facial nerve and blood vessels
Superior orbital fissure (fig. 7.20)	Lateral wall of orbit	Oculomotor, trochlear, and abducens nerves, and ophthalmic division of trigeminal nerve
Supraorbital foramen (fig. 7.19)	Upper margin or orbit in frontal bone	Supraorbital blood vessels and nerves

and large orbits. The jaw and nasal cavity are small, the sinuses are incompletely formed, and the frontal bone is in two parts. The skull bones are thin, but they are also somewhat flexible and thus are less easily fractured than adult bones.

In the infantile skull, a frontal suture (metopic suture) separates the two parts of the developing frontal bone in the midline. This suture usually closes before the sixth year; however, in a few adults, the frontal suture remains open.

1 Locate and name each of the bones of the cranium.

2 Locate and name each of the facial bones.

3 Explain how an adult skull differs from that of an infant.

Vertebral Column

The **vertebral column** extends from the skull to the pelvis and forms the vertical axis of the skeleton (fig. 7.34). It is composed of many bony parts called **vertebrae** (ver′tĕ-bre) that are separated by masses of fibrocartilage called *intervertebral discs* and are connected to one another by ligaments. The vertebral column supports the head and the trunk of the body, yet is flexible enough to permit movements, such as bending forward, backward, or to the side, and turning or rotating on the central axis. It also protects the spinal cord, which passes through a *vertebral canal* formed by openings in the vertebrae.

An infant has thirty-three separate bones in the vertebral column. Five of these bones eventually fuse to form the sacrum, and four others join to become the coccyx. As a result, an adult vertebral column has twenty-six bones.

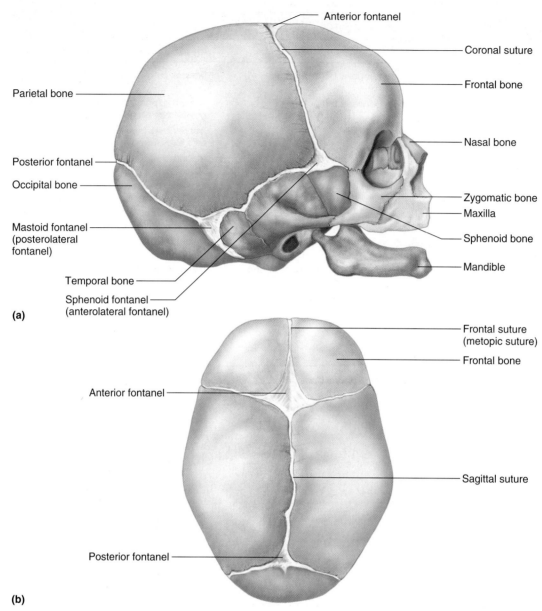

Parietal bone

Posterior fontanel

Occipital bone

Mastoid fontanel
(posterolateral
fontanel)

Temporal bone

Sphenoid fontanel
(anterolateral fontanel)

(a)

Anterior fontanel

Coronal suture

Frontal bone

Nasal bone

Zygomatic bone

Maxilla

Sphenoid bone

Mandible

Frontal suture
(metopic suture)

Frontal bone

Anterior fontanel

Sagittal suture

Posterior fontanel

(b)

FIGURE 7.33

Fontanels. (*a*) Right lateral view and (*b*) superior view of the infantile skull.

Normally, the vertebral column has four curvatures, which give it a degree of resiliency. The names of the curves correspond to the regions in which they occur, as shown in figure 7.34. The *thoracic* and *pelvic curvatures* are concave anteriorly and are called primary curves. The *cervical curvature* in the neck and the *lumbar curvature* in the lower back are convex anteriorly and are called secondary curves. The cervical curvature develops when a baby begins to hold up its head, and the lumbar curvature develops when the child begins to stand.

A Typical Vertebra

Although the vertebrae in different regions of the vertebral column have special characteristics, they also have features in common. A typical vertebra (fig. 7.35) has a drum-shaped *body*, which forms the thick, anterior portion of the bone. A longitudinal row of these vertebral bodies supports the weight of the head and trunk. The intervertebral discs, which separate adjacent vertebrae, are fastened to the roughened upper and lower surfaces of the vertebral bodies. These discs cushion and soften the forces caused by such movements as walking and jumping, which might otherwise fracture vertebrae or jar the brain. The bodies of adjacent vertebrae are joined on their anterior surfaces by *anterior longitudinal ligaments* and on their posterior surfaces by *posterior longitudinal ligaments.*

Projecting posteriorly from each vertebral body are two short stalks called *pedicles* (ped′ĭ-k′lz). They form the sides of the *vertebral foramen.* Two plates called *laminae*

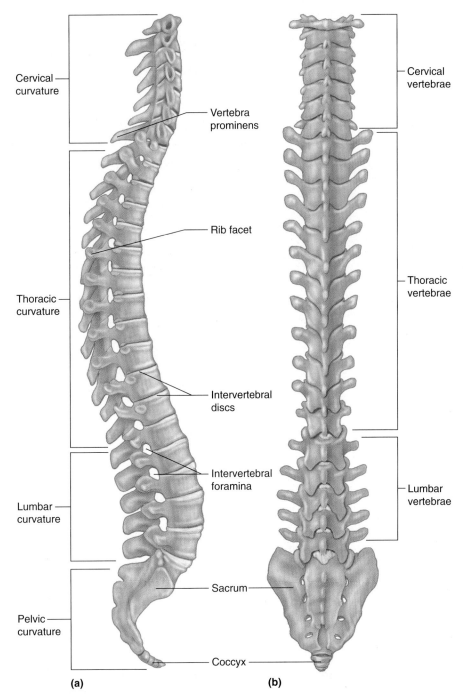

Cervical curvature

Vertebra prominens

Rib facet

Thoracic curvature

Intervertebral discs

Intervertebral foramina

Lumbar curvature

Pelvic curvature

Cervical vertebrae

Thoracic vertebrae

Lumbar vertebrae

Sacrum

Coccyx

(a) (b)

FIGURE 7.34

The curved vertebral column consists of many vertebrae separated by intervertebral discs. (a) Right lateral view. (b) Posterior view.

(lam'ĭ-ne) arise from the pedicles and fuse in the back to become a *spinous process.* The pedicles, laminae, and spinous process together complete a bony *vertebral arch* around the vertebral foramen, through which the spinal cord passes.

Between the pedicles and laminae of a typical vertebra is a *transverse process,* which projects laterally and posteriorly. Various ligaments and muscles are attached to the dorsal spinous process and the transverse processes.

Projecting upward and downward from each vertebral arch are *superior* and *inferior articulating processes.* These processes bear cartilage-covered facets by which each vertebra is joined to the one above and the one below it.

On the lower surfaces of the vertebral pedicles are notches that align to help form openings called *intervertebral foramina* (in"ter-ver'tĕ-bral fo-ram'ĭ-nah). These openings provide passageways for spinal nerves that proceed between adjacent vertebrae and connect to the spinal cord.

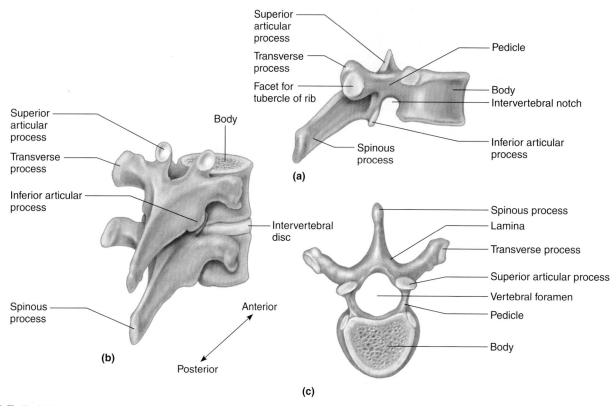

FIGURE 7.35
Typical thoracic vertebra. (*a*) Right lateral view. (*b*) Adjacent vertebrae join at their articular processes. (*c*) Superior view.

Gymnasts, high jumpers, pole vaulters, and other athletes who hyperextend and rotate their vertebral columns and stress them with impact sometimes fracture the portion of the vertebra between the superior and inferior articulating processes (the pars interarticularis). Such damage to the vertebra is called spondylolysis, and it is most common at L5.

Cervical Vertebrae

Seven **cervical vertebrae** comprise the bony axis of the neck. These are the smallest of the vertebrae, but their bone tissues are denser than those in any other region of the vertebral column.

The transverse processes of the cervical vertebrae are distinctive because they have *transverse foramina,* which are passageways for arteries leading to the brain. Also, the spinous processes of the second through the sixth cervical vertebrae are uniquely forked (bifid). These processes provide attachments for muscles.

The spinous process of the seventh vertebra is longer and protrudes beyond the other cervical spines. It is called the *vertebra prominens,* and because it can be felt through the skin, it is a useful landmark for locating other vertebral parts (see fig. 7.34).

Two of the cervical vertebrae, shown in figure 7.36, are of special interest. The first vertebra, or **atlas** (at′las), supports the head. It has practically no body or spine and appears as a bony ring with two transverse processes. On its superior surface, the atlas has two kidney-shaped *facets,* which articulate with the occipital condyles.

The second cervical vertebra, or **axis** (ak′sis), bears a toothlike *dens* (odontoid process) on its body. This process projects upward and lies in the ring of the atlas. As the head is turned from side to side, the atlas pivots around the dens (figs. 7.36 and 7.37).

Thoracic Vertebrae

The twelve **thoracic vertebrae** are larger than those in the cervical region. Each vertebra has a long, pointed spinous process, which slopes downward, and facets on the sides of its body, which articulate with a rib.

Beginning with the third thoracic vertebra and moving inferiorly, the bodies of these bones increase in size. Thus, they are adapted to bear increasing loads of body weight.

Lumbar Vertebrae

The five **lumbar vertebrae** in the small of the back (loin) support more weight than the superior vertebrae and have larger and stronger bodies. Compared to other types of

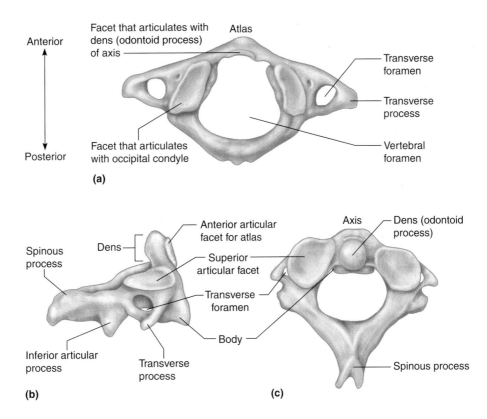

(a)

Anterior

Posterior

Facet that articulates with dens (odontoid process) of axis

Facet that articulates with occipital condyle

Atlas

Transverse foramen

Transverse process

Vertebral foramen

(b)

Spinous process

Dens

Anterior articular facet for atlas

Superior articular facet

Transverse foramen

Body

Inferior articular process

Transverse process

(c)

Axis

Dens (odontoid process)

Spinous process

FIGURE 7.36

Atlas and axis. (a) Superior view of the atlas. (b) Right lateral view and (c) superior view of the axis.

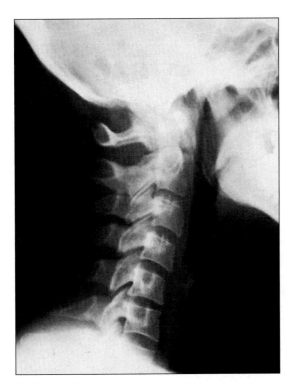

FIGURE 7.37

Radiograph of the cervical vertebrae.

vertebrae, the transverse processes of these vertebrae project posteriorly at sharp angles, whereas their short, thick spinous processes are nearly horizontal. Figure 7.38 compares the structures of the cervical, thoracic, and lumbar vertebrae.

> The painful condition of *spondylolisthesis* occurs when a vertebra slips out of place over the vertebra below it. Most commonly the fifth lumbar vertebra slides forward over the body of the sacrum. Persons with spondylolysis (see previous box) may be more likely to develop spondylolisthesis, as are gymnasts, football players, and others who flex or extend their vertebral columns excessively and forcefully.

Sacrum

The **sacrum** (sa′krum) is a triangular structure at the base of the vertebral column. It is composed of five vertebrae that develop separately but gradually fuse between ages eighteen and thirty. The spinous processes of these fused bones form a ridge of tubercles, the *median sacral crest.* Nerves and blood vessels pass through rows of openings,

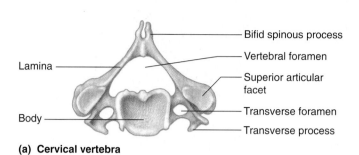

Lamina

Body

Bifid spinous process

Vertebral foramen

Superior articular facet

Transverse foramen

Transverse process

(a) Cervical vertebra

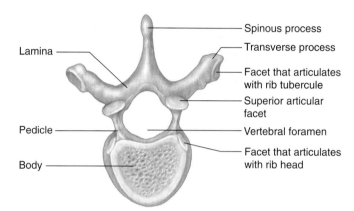

Lamina

Pedicle

Body

Spinous process

Transverse process

Facet that articulates with rib tubercule

Superior articular facet

Vertebral foramen

Facet that articulates with rib head

(b) Thoracic vertebra

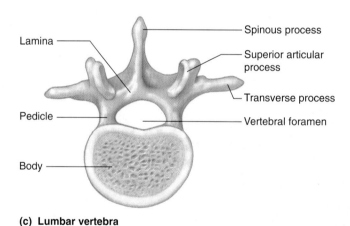

Lamina

Pedicle

Body

Spinous process

Superior articular process

Transverse process

Vertebral foramen

(c) Lumbar vertebra

FIGURE 7.38
Superior view of (*a*) a cervical vertebra, (*b*) a thoracic vertebra, and (*c*) a lumbar vertebra.

called the *dorsal sacral foramina,* located to the sides of the tubercles (fig. 7.39).

The sacrum is wedged between the coxae of the pelvis and is united to them at its *auricular surfaces* by fibrocartilage of the *sacroiliac joints.* The pelvic girdle transmits the body's weight to the legs at these joints (see fig. 7.17).

The sacrum forms the posterior wall of the pelvic cavity. The upper anterior margin of the sacrum, which represents the body of the first sacral vertebra, is called the *sacral promontory* (sa'kral prom'on-to"re). During a vaginal examination, a physician can feel this projection and use it as a guide in determining the size of the pelvis. This measurement is helpful in estimating how easily an

infant may be able to pass through a woman's pelvic cavity during childbirth.

The vertebral foramina of the sacral vertebrae form the *sacral canal,* which continues through the sacrum to an opening of variable size at the tip, called the *sacral hiatus* (hi-a'tus). This foramen exists because the laminae of the last sacral vertebra are not fused. On the ventral surface of the sacrum, four pairs of *anterior sacral foramina* provide passageways for nerves and blood vessels.

Coccyx

The **coccyx** (kok'siks), or tailbone, is the lowest part of the vertebral column and is usually composed of four vertebrae that fuse by the twenty-fifth year. Ligaments attach it

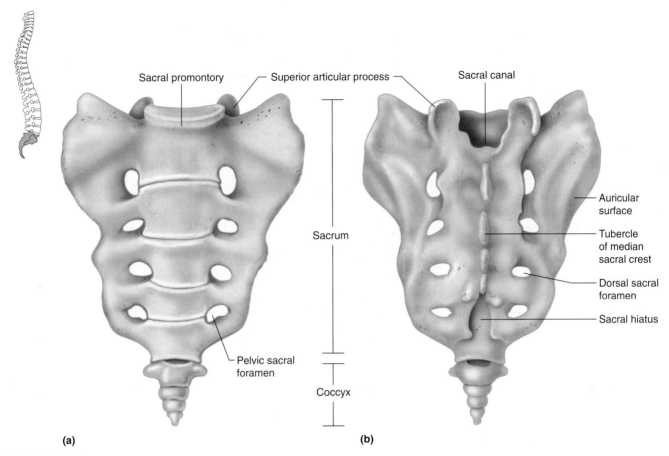

Sacral promontory — Superior articular process — Sacral canal

Sacrum

Pelvic sacral foramen

Coccyx

(a)

Auricular surface

Tubercle of median sacral crest

Dorsal sacral foramen

Sacral hiatus

(b)

FIGURE 7.39

Sacrum and coccyx. (a) Anterior view and (b) posterior view.

to the margins of the sacral hiatus (fig. 7.39). Sitting presses on the coccyx, and it moves forward, acting like a shock absorber. Sitting down with great force, as when slipping and falling on ice, can fracture or dislocate the coccyx. Table 7.9 summarizes the bones of the vertebral column, and Clinical Application 7.3 discusses disorders of the vertebral column.

1 Describe the structure of the vertebral column.

2 Explain the difference between the vertebral column of an adult and that of an infant.

3 Describe a typical vertebra.

4 How do the structures of cervical, thoracic, and lumbar vertebrae differ?

TABLE 7.9		Bones of the Vertebral Column			
Bones	**Number**	**Special Features**	**Bones**	**Number**	**Special Features**
Cervical vertebrae	7	Transverse foramina; facets of atlas articulate with occipital condyles of skull; dens of axis articulates with atlas; spinous processes of second through sixth vertebrae are bifid	Lumbar vertebrae	5	Large bodies; transverse processes that project posteriorly at sharp angles; short, thick spinous processes directed nearly horizontally
Thoracic vertebrae	12	Pointed spinous processes that slope downward; facets that articulate with ribs	Sacrum	5 vertebrae fused into 1 bone	Dorsal sacral foramina, auricular surfaces, sacral promontory, sacral canal, sacral hiatus, anterior sacral foramina
			Coccyx	4 vertebrae fused into 1 bone	Attached by ligaments to the margins of the sacral hiatus

DISORDERS OF THE VERTEBRAL COLUMN

Changes in the intervertebral discs may cause various problems. Each disc is composed of a tough, outer layer of fibrocartilage (annulus fibrosus) and an elastic central mass (nucleus pulposus). With age, these discs degenerate—the central masses lose firmness, and the outer layers thin and weaken, developing cracks. Extra pressure, as when a person falls or lifts a heavy object, can break the outer layers of the discs, squeezing out the central masses. Such a rupture may press on the spinal cord or on spinal nerves that branch from it. This condition, called a *ruptured,* or *herniated disc,* may cause back pain and numbness or loss of muscular function in the parts innervated by the affected spinal nerves.

A surgical procedure called a *laminectomy* may relieve the pain of a herniated disc by removing a portion of the posterior arch of a vertebra. This reduces the pressure on the affected nerve tissues. Alternatively, a protein-digesting enzyme (chymopapain) may be injected into the injured disc to shrink it.

Sometimes problems develop in the curvatures of the vertebral column because of poor posture, injury, or disease. An exaggerated thoracic curvature causes rounded shoulders and a hunchback. This condition, called *kyphosis,* occasionally develops in adolescents who undertake strenuous athletic activities. Unless corrected before bone growth completes, the condition can permanently deform the vertebral column.

Sometimes the vertebral column develops an abnormal lateral curvature, so that one hip or shoulder is lower than the other. This may displace or compress the thoracic and abdominal organs. With unknown cause, this condition, called *scoliosis,* is most common in adolescent females. It also may accompany such diseases as poliomyelitis, rickets, or tuberculosis. An accentuated lumbar curvature is called *lordosis,* or swayback.

As a person ages, the intervertebral discs tend to shrink and become more rigid, and compression is more likely to fracture the vertebral bodies. Consequently, height may decrease, and the thoracic curvature of the vertebral column may be accentuated, bowing the back. ■

Thoracic Cage

The **thoracic cage** includes the ribs, the thoracic vertebrae, the sternum, and the costal cartilages that attach the ribs to the sternum. These bones support the shoulder girdle and upper limbs, protect the viscera in the thoracic and upper abdominal cavities, and play a role in breathing (fig. 7.40).

Ribs

The usual number of **ribs** is twenty-four—one pair attached to each of the twelve thoracic vertebrae. Some individuals develop extra ribs associated with their cervical or lumbar vertebrae.

The first seven rib pairs, which are called the *true ribs* (vertebrosternal ribs), join the sternum directly by their costal cartilages. The remaining five pairs are called *false ribs* because their cartilages do not reach the sternum directly. Instead, the cartilages of the upper three false ribs (vertebrochondral ribs) join the cartilages of the seventh rib, whereas the last two rib pairs have no attachments to the sternum. These last two pairs (or sometimes the last three pairs) are called *floating ribs* (vertebral ribs).

A typical rib (fig. 7.41) has a long, slender shaft, which curves around the chest and slopes downward. On the posterior end is an enlarged *head* by which the rib articulates with a facet on the body of its own vertebra and with the body of the next higher vertebra. The neck of

the rib is flattened, lateral to the head, where ligaments attach. A *tubercle,* close to the head of the rib, articulates with the transverse process of the vertebra.

The costal cartilages are composed of hyaline cartilage. They are attached to the anterior ends of the ribs and continue in line with them toward the sternum.

Sternum

The **sternum** (ster'num), or breastbone, is located along the midline in the anterior portion of the thoracic cage. It is a flat, elongated bone that develops in three parts—an upper *manubrium* (mah-nu'bre-um), a middle *body,* and a lower *xiphoid* (zif'oid) *process* that projects downward (see fig. 7.40).

The sides of the manubrium and the body are notched where they articulate with costal cartilages. The manubrium also articulates with the clavicles by facets on its superior border. It usually remains as a separate bone until middle age or later, when it fuses to the body of the sternum.

> The manubrium and body of the sternum lie in different planes, so their line of union projects slightly forward. This projection, at the level of the second costal cartilage, is called the *sternal angle* (angle of Louis). It is commonly used as a clinical landmark to locate a particular rib (see fig. 7.40).

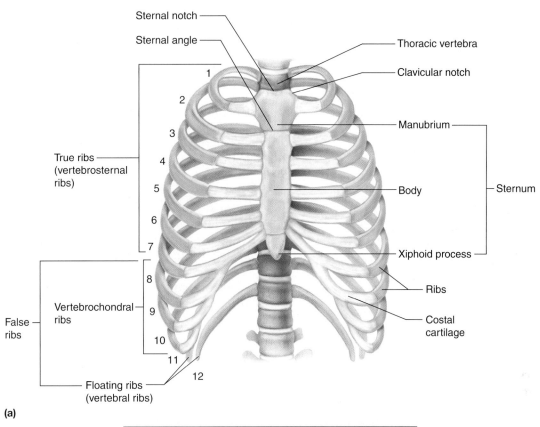

Sternal notch

Sternal angle

Thoracic vertebra

Clavicular notch

1

2

3

True ribs
(vertebrosternal
ribs)

4

5

6

7

Manubrium

Body

Sternum

Xiphoid process

8

Vertebrochondral
ribs

9

Ribs

False
ribs

10

Costal
cartilage

11

12

Floating ribs
(vertebral ribs)

(a)

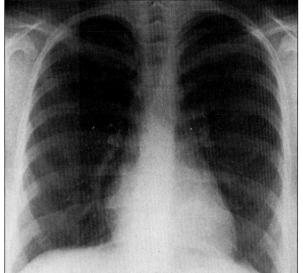

FIGURE 7.40

The thoracic cage includes (a) the thoracic vertebrae, the sternum, the ribs, and the costal cartilages that attach the ribs to the sternum. (b) Radiograph of the thoracic cage, anterior view. The light region behind the sternum and above the diaphragm is the heart.

(b)

The xiphoid process begins as a piece of cartilage. It slowly ossifies, and by middle age it usually fuses to the body of the sternum.

1 Which bones comprise the thoracic cage?

2 Describe a typical rib.

3 What are the differences among true, false, and floating ribs?

Red marrow within the spongy bone of the sternum produces blood cells into adulthood. Since the sternum has a thin covering of compact bone and is easy to reach, samples of its marrow may be removed to diagnose diseases. This procedure, a *sternal puncture,* suctions (aspirates) some marrow through a hollow needle. (Marrow may also be removed from the iliac crest of a coxal bone.)

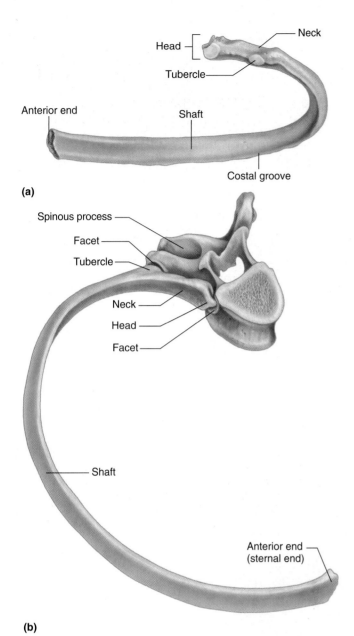

(a)

Head
Neck
Tubercle
Anterior end
Shaft
Costal groove

(b)

Spinous process
Facet
Tubercle
Neck
Head
Facet
Shaft
Anterior end (sternal end)

FIGURE 7.41
A typical rib. (*a*) Posterior view. (*b*) Articulations of a rib with a thoracic vertebra (superior view).

Pectoral Girdle

The **pectoral** (pek'to-ral) **girdle,** or shoulder girdle, is composed of four parts—two clavicles (collarbones) and two scapulae (shoulder blades). Although the word *girdle* suggests a ring-shaped structure, the pectoral girdle is an incomplete ring. It is open in the back between the scapulae, and the sternum separates its bones in front. The pectoral girdle supports the upper limbs and is an attachment for several muscles that move them (fig. 7.42).

Clavicles

The **clavicles** (klav'ĭ-k'lz) are slender, rodlike bones with elongated S-shapes (fig. 7.42). Located at the base of the

neck, they run horizontally between the sternum and the shoulders. The medial (or sternal) ends of the clavicles articulate with the manubrium, and the lateral (or acromial) ends join processes of the scapulae.

The clavicles brace the freely movable scapulae, helping to hold the shoulders in place. They also provide attachments for muscles of the upper limbs, chest, and back. Because of its elongated double curve, the clavicle is structurally weak. If compressed lengthwise due to abnormal pressure on the shoulder, it is likely to fracture.

Scapulae

The **scapulae** (skap'u-le) are broad, somewhat triangular bones located on either side of the upper back. They have flat bodies with concave anterior surfaces. The posterior surface of each scapula is divided into unequal portions by a *spine.* Above the spine is the *supraspinous fossa,* and below the spine is the *infraspinous fossa.* This spine leads to a *head,* which bears two processes—an *acromion* (ah-kro'me-on) *process* that forms the tip of the shoulder and a *coracoid* (kor'ah-koid) *process* that curves anteriorly and inferiorly to the clavicle (fig. 7.43). The acromion process articulates with the clavicle and provides attachments for muscles of the upper limb and chest. The coracoid process also provides attachments for upper limb and chest muscles. On the head of the scapula between the processes is a depression called the *glenoid cavity* (glenoid fossa of the scapula). It articulates with the head of the arm bone (humerus).

The scapula has three borders. The *superior border* is on the superior edge. The *axillary,* or *lateral border,* is directed toward the upper limb. The *vertebral,* or *medial border,* is closest to the vertebral column, about 5 cm away.

1 Which bones form the pectoral girdle?

2 What is the function of the pectoral girdle?

In the epic poem the *Iliad,* Homer describes a man whose "shoulders were bent and met over his chest." The man probably had a rare inherited condition, called cleidocranial dysplasia, in which certain bones do not grow. The skull consists of small fragments joined by connective tissue, rather than large, interlocking hard bony plates. The scapulae are stunted or missing.

Cleidocranial dysplasia was first reported in a child in the huge Arnold family, founded by a Chinese immigrant to South Africa. The child had been kicked by a horse, and X rays revealed that the fontanels atop the head had never closed. The condition became known as "Arnold head." In 1997, researchers traced the condition to a malfunctioning gene that normally instructs certain cells to specialize as bone. Mice missing both copies of this gene develop a skeleton that is completely cartilage—bone never replaces the original cartilage model.

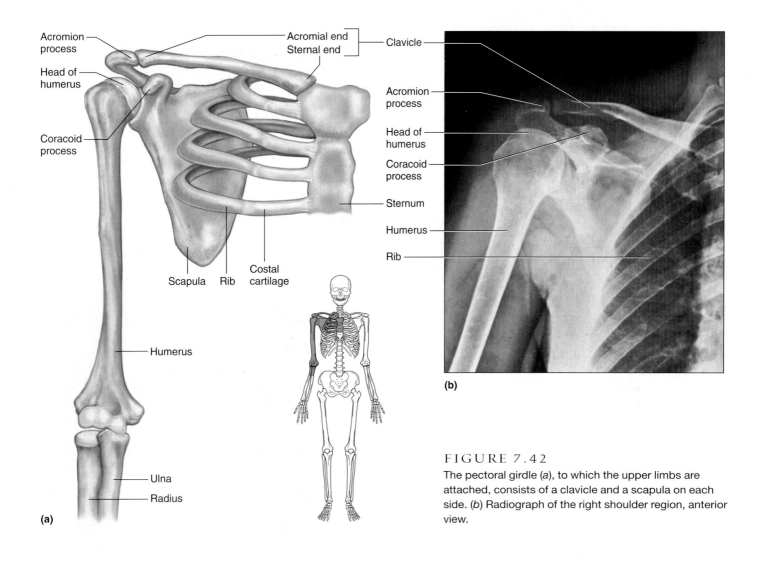

(a)

Acromion process
Head of humerus
Coracoid process
Humerus
Ulna
Radius
Acromial end
Sternal end
Clavicle
Scapula   Rib   Costal cartilage

(b)

Clavicle
Acromion process
Head of humerus
Coracoid process
Sternum
Humerus
Rib

FIGURE 7.42

The pectoral girdle (*a*), to which the upper limbs are attached, consists of a clavicle and a scapula on each side. (*b*) Radiograph of the right shoulder region, anterior view.

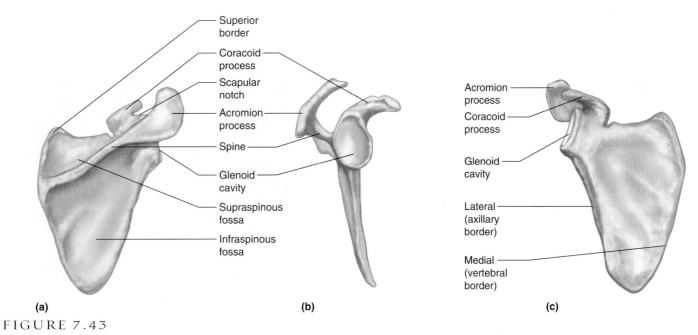

(a)

Superior border
Coracoid process
Scapular notch
Acromion process
Spine
Glenoid cavity
Supraspinous fossa
Infraspinous fossa

(b)

(c)

Acromion process
Coracoid process
Glenoid cavity
Lateral (axillary border)
Medial (vertebral border)

FIGURE 7.43

Scapula. (*a*) Posterior surface of the right scapula. (*b*) Lateral view showing the glenoid cavity that articulates with the head of the humerus. (*c*) Anterior surface.

Upper Limb

The bones of the upper limb form the framework of the arm, forearm, and hand. They also provide attachments for muscles, and they function as levers that move limb parts. These bones include a humerus, a radius, an ulna, carpals, metacarpals, and phalanges (fig. 7.44).

Humerus

The **humerus** (fig. 7.45) is a long bone that extends from the scapula to the elbow. At its upper end is a smooth, rounded *head* that fits into the glenoid cavity of the scapula. Just below the head are two processes—a *greater tubercle* on the lateral side and a *lesser tubercle* on the anterior side. These tubercles provide attachments for muscles that move the upper limb at the shoulder.

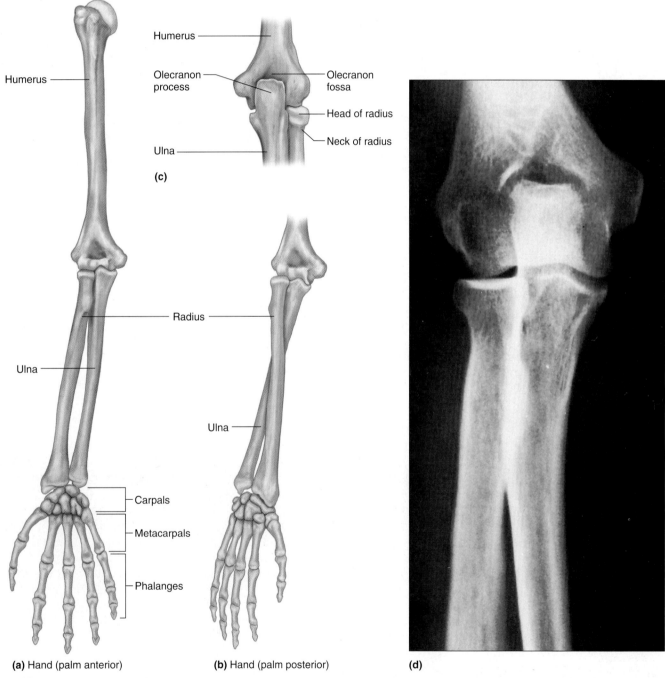

(a) Hand (palm anterior)   **(b)** Hand (palm posterior)   **(d)**

FIGURE 7.44

Right upper limb. (*a*) Anterior view with the hand, palm anterior and (*b*) with the hand, palm posterior. (*c*) Posterior view of the right elbow.
(*d*) Radiograph of the right elbow and forearm, anterior view.

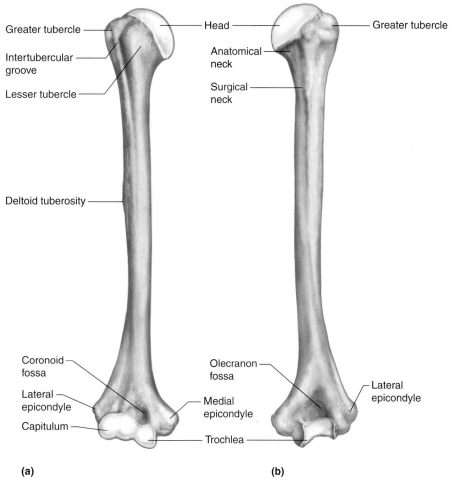

Greater tubercle

Intertubercular groove

Lesser tubercle

Deltoid tuberosity

Coronoid fossa

Lateral epicondyle

Capitulum

Head

Anatomical neck

Surgical neck

Olecranon fossa

Medial epicondyle

Trochlea

Greater tubercle

Lateral epicondyle

(a)

(b)

FIGURE 7.45

Humerus. (a) Anterior surface and (b) posterior surface of the right humerus.

Between them is a narrow furrow, the *intertubercular groove,* through which a tendon passes from a muscle in the arm (biceps brachii) to the shoulder.

The narrow depression along the lower margin of the head that separates it from the tubercles is called the *anatomical neck.* Just below the head and the tubercles of the humerus is a tapering region called the *surgical neck,* so named because fractures commonly occur there. Near the middle of the bony shaft on the lateral side is a rough V-shaped area called the *deltoid tuberosity.* It provides an attachment for the muscle (deltoid) that raises the upper limb horizontally to the side.

At the lower end of the humerus are two smooth *condyles*—a knoblike *capitulum* (kah-pit'u-lum) on the lateral side and a pulley-shaped *trochlea* (trok'le-ah) on the medial side. The capitulum articulates with the radius at the elbow, whereas the trochlea joins the ulna.

Above the condyles on either side are *epicondyles,* which provide attachments for muscles and ligaments of the elbow. Between the epicondyles anteriorly is a depression, the *coronoid* (kor'o-noid) *fossa,* that receives a process of the ulna (coronoid process) when the elbow bends. Another depression on the posterior surface, the

olecranon (o"lek'ra-non) *fossa,* receives an olecranon process when the upper limb straightens at the elbow.

Radius

The **radius,** located on the thumb side of the forearm, is somewhat shorter than its companion, the ulna (fig. 7.46). The radius extends from the elbow to the wrist and crosses over the ulna when the hand is turned so that the palm faces backward.

A thick, disclike *head* at the upper end of the radius articulates with the capitulum of the humerus and a notch of the ulna (radial notch). This arrangement allows the radius to rotate freely.

On the radial shaft just below the head is a process called the *radial tuberosity.* It is an attachment for a muscle (biceps brachii) that bends the upper limb at the elbow. At the distal end of the radius, a lateral *styloid* (sti'loid) *process* provides attachments for ligaments of the wrist.

Ulna

The **ulna** is longer than the radius and overlaps the end of the humerus posteriorly. At its proximal end, the ulna has

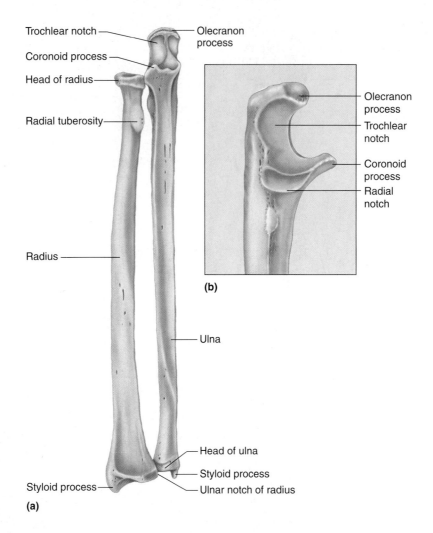

Trochlear notch
Coronoid process
Head of radius
Radial tuberosity
Radius
Ulna
Head of ulna
Styloid process
Ulnar notch of radius
Styloid process
Olecranon process

Olecranon process
Trochlear notch
Coronoid process
Radial notch

(a)

(b)

FIGURE 7.46

Radius and ulna. (*a*) The head of the right radius articulates with the radial notch of the ulna, and the head of the ulna articulates with the ulnar notch of the radius. (*b*) Lateral view of the proximal end of the right ulna.

a wrenchlike opening, the *trochlear notch* (semilunar notch), that articulates with the trochlea of the humerus. A process lies on either side of this notch. The *olecranon process,* located above the trochlear notch, provides an attachment for the muscle (triceps brachii) that straightens the upper limb at the elbow. During this movement, the olecranon process of the ulna fits into the olecranon fossa of the humerus. Similarly, the *coronoid process,* just below the trochlear notch, fits into the coronoid fossa of the humerus when the elbow bends.

At the distal end of the ulna, its knoblike *head* articulates laterally with a notch of the radius (ulnar notch) and with a disc of fibrocartilage inferiorly (fig. 7.46). This disc, in turn, joins a wrist bone (triquetrum). A medial *styloid process* at the distal end of the ulna provides attachments for ligaments of the wrist.

Many a thirtyish parent of a young little leaguer or softball player becomes tempted to join in. But if he or she has not pitched in many years, sudden activity may break the forearm. Forearm pain while pitching is a signal that a fracture could happen. Medical specialists advise returning to the pitching mound gradually. Start with twenty pitches, five days a week, for two to three months before regular games begin. By the season's start, 120 pitches per daily practice session should be painless.

Hand

The hand is made up of the wrist, palm, and fingers. The skeleton of the wrist consists of eight small **carpal bones** that are firmly bound in two rows of four bones each. The resulting compact mass is called a *carpus* (kar′pus). The carpus is rounded on its proximal surface, where it articulates with the radius and with the fibrocartilaginous disc on the ulnar side. The carpus is concave anteriorly, forming a canal through which tendons and nerves extend to the palm. Its distal surface articulates with the metacarpal bones. Figure 7.47 names the individual bones of the carpus.

Five **metacarpal bones,** one in line with each finger, form the framework of the palm. These bones are cylindrical, with rounded distal ends that form the knuckles of a clenched fist. The metacarpals articulate proximally with the carpals and distally with the phalanges. The metacarpal on the lateral side is the most freely movable; it permits the thumb to oppose the fingers when grasping something. These bones are numbered 1 to 5, beginning with the metacarpal of the thumb (fig. 7.47).

The **phalanges** are the finger bones. There are three in each finger—a proximal, a middle, and a distal phalanx—and two in the thumb. (The thumb lacks a middle phalanx.) Thus, each hand has fourteen finger bones. Table 7.10 summarizes the bones of the pectoral girdle and upper limbs.

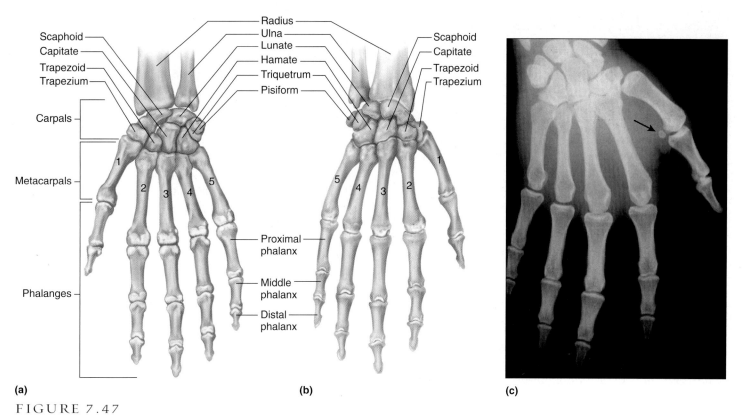

FIGURE 7.47

Hand. (a) Anterior view and (b) posterior view of the right hand. (c) Radiograph of the right hand. Note the small sesamoid bone associated with the joint at the base of the thumb (arrow).

TABLE 7.10	Bones of the Pectoral Girdle and Upper Limbs	
Name and Number	**Location**	**Special Features**
Clavicle (2)	Base of neck between sternum and scapula	Sternal end, acromial end
Scapula (2)	Upper back, forming part of shoulder	Body, spine, head, acromion process, coracoid process, glenoid cavity
Humerus (2)	Arm, between scapula and elbow	Head, greater tubercle, lesser tubercle, intertubercular groove, surgical neck, deltoid tuberosity, capitulum, trochlea, medial epicondyle, lateral epicondyle, coronoid fossa, olecranon fossa
Radius (2)	Lateral side of forearm, between elbow and wrist	Head, radial tuberosity, styloid process, ulnar notch
Ulna (2)	Medial side of forearm, between elbow and wrist	Trochlear notch, olecranon process, head, styloid process, radial notch
Carpal (16)	Wrist	Arranged in two rows of four bones each
Metacarpal (10)	Palm	One in line with each finger and thumb
Phalanx (28)	Finger	Three in each finger; two in each thumb

1   Locate and name each of the bones of the upper limb.

2   Explain how the bones of the upper limb articulate with one
     another.

It is not uncommon for a baby to be born with an extra finger
or toe, but since the extra digit is usually surgically removed
early in life, hands like the ones in figure 7.48 are rare.
Polydactyly ("many digits") is an inherited trait. It is common
in cats. A lone but popular male cat brought the trait from
England to colonial Boston. Polydactyly is also common
among the Amish people.

Pelvic Girdle

The **pelvic girdle** consists of the two coxae, hipbones or
innominate bones, which articulate with each other anteriorly and with the sacrum posteriorly (fig. 7.49). The
sacrum, coccyx, and pelvic girdle together form the bowl-
shaped *pelvis.* The pelvic girdle supports the trunk of the
body, provides attachments for the lower limbs, and protects the urinary bladder, the distal end of the large intestine, and the internal reproductive organs. The body's
weight is transmitted through the pelvic girdle to the
lower limbs and then onto the ground.

Coxae

Each **coxa,** or hipbone, develops from three parts—an
ilium, an ischium, and a pubis. These parts fuse in the
region of a cup-shaped cavity called the *acetabulum* (as"ĕ-
tab'u-lum). This depression, on the lateral surface of the

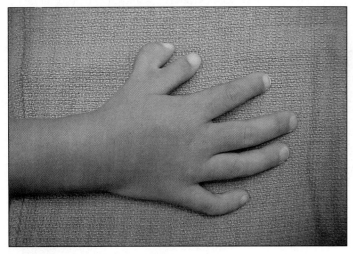

FIGURE 7.48
A person with polydactyly has extra digits.

hipbone, receives the rounded head of the femur or thigh-
bone (fig. 7.50).

The **ilium** (il'e-um), which is the largest and most
superior portion of the coxa, flares outward, forming the
prominence of the hip. The margin of this prominence is
called the *iliac crest.* The smooth, concave surface on the
anterior aspect of the ilium is the *iliac fossa.*

Posteriorly, the ilium joins the sacrum at the *sacroiliac* (sa"kro-il'e-ak) *joint.* Anteriorly, a projection of the
ilium, the *anterior superior iliac spine,* can be felt lateral
to the groin. This spine provides attachments for ligaments and muscles and is an important surgical landmark.

A common injury in contact sports such as football is bruising
the soft tissues and bone associated with the anterior superior iliac spine. Wearing protective padding can prevent this
painful injury, called a *hip pointer.*

On the posterior border of the ilium is a *posterior
superior iliac spine.* Below this spine is a deep indenta-
tion, the *greater sciatic notch,* through which a number of
nerves and blood vessels pass.

The **ischium** (is'ke-um), which forms the lowest portion of the coxa, is L-shaped, with its angle, the *ischial
tuberosity,* pointing posteriorly and downward. This
tuberosity has a rough surface that provides attachments
for ligaments and lower limb muscles. It also supports the
weight of the body during sitting. Above the ischial
tuberosity, near the junction of the ilium and ischium, is a
sharp projection called the *ischial spine.* Like the sacral
promontory, this spine, which can be felt during a vaginal
examination, is used as a guide for determining pelvis
size. The distance between the ischial spines is the shortest diameter of the pelvic outlet.

The **pubis** (pu'bis) constitutes the anterior portion of
the coxa. The two pubic bones come together at the midline to form a joint called the *symphysis pubis* (sim'fĭ-sis
pu'bis). The angle these bones form below the symphysis
is the *pubic arch.*

A portion of each pubis passes posteriorly and
downward to join an ischium. Between the bodies of
these bones on either side is a large opening, the *obturator
foramen,* which is the largest foramen in the skeleton. An
obturator membrane covers and nearly closes this foramen (see figs. 7.49 and 7.50).

Greater and Lesser Pelves

If a line were drawn along each side of the pelvis from the
sacral promontory downward and anteriorly to the upper
margin of the symphysis pubis, it would mark the *pelvic
brim* (linea terminalis). This margin separates the lower,
or lesser (true), pelvis from the upper, or greater (false),
pelvis (fig. 7.51).

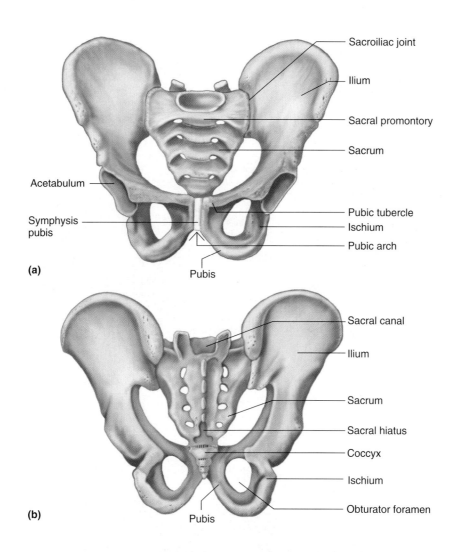

(a)

Sacroiliac joint

Ilium

Sacral promontory

Sacrum

Pubic tubercle

Ischium

Pubic arch

Acetabulum

Symphysis pubis

Pubis

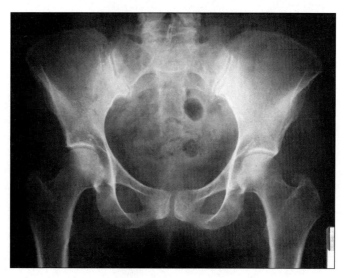

(b)

Sacral canal

Ilium

Sacrum

Sacral hiatus

Coccyx

Ischium

Obturator foramen

Pubis

(c)

FIGURE 7.49

Pelvic girdle. (*a*) Anterior view. (*b*) Posterior view. This girdle provides an attachment for the lower limbs and together with the sacrum and coccyx forms the pelvis. (*c*) Radiograph of the pelvic girdle.

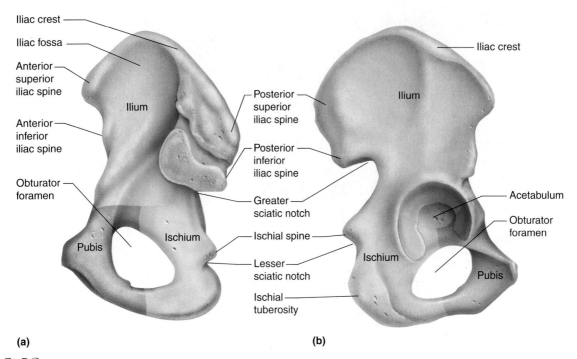

FIGURE 7.50

Coxa. (*a*) Medial surface of the right coxa. (*b*) Right lateral view.

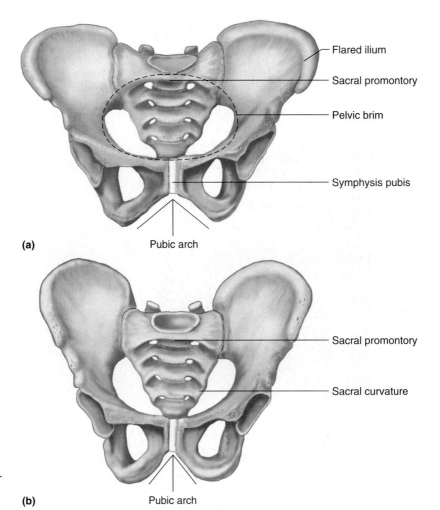

FIGURE 7.51

The female pelvis is usually wider in all diameters and roomier than that of the male. (*a*) Female pelvis. (*b*) Male pelvis.

The *greater pelvis* is bounded posteriorly by the lumbar vertebrae, laterally by the flared parts of the iliac bones, and anteriorly by the abdominal wall. The false pelvis helps support the abdominal organs.

The *lesser pelvis* is bounded posteriorly by the sacrum and coccyx and laterally and anteriorly by the lower ilium, ischium, and pubis bones. This portion of the pelvis surrounds a short, canal-like cavity that has an upper inlet and a lower outlet. An infant passes through this cavity during childbirth.

Differences between Male and Female Pelves

Some basic structural differences distinguish the male and the female pelves, even though it may be difficult to find all of the "typical" characteristics in any one individual. These differences arise from the function of the female pelvis as a birth canal. Usually, the female iliac bones are more flared than those of the male, and consequently, the female hips are usually broader than the male's. The angle of the female pubic arch may be greater, there may be more distance between the ischial spines and the ischial tuberosities, and the sacral curvature may be shorter and flatter. Thus, the female pelvic cavity is usually wider in all diameters than that of the male. Also, the bones of the female pelvis are usually lighter, more delicate, and show less evidence of muscle attachments (fig. 7.51). Table 7.11 summarizes some of the differences between the male and female skeletons.

1 Locate and name each bone that forms the pelvis.

2 Name the bones that fuse to form a coxa.

3 Distinguish between the greater pelvis and the lesser pelvis.

4 How are male and female pelves different?

Lower Limb

The bones of the lower limb form the frameworks of the thigh, leg, and foot. They include a femur, a tibia, a fibula, tarsals, metatarsals, and phalanges (fig. 7.52).

Femur

The **femur,** or thigh bone, is the longest bone in the body and extends from the hip to the knee. A large, rounded *head* at its proximal end projects medially into the acetabulum of the coxa. On the head, a pit called the *fovea capitis* marks the attachment of a ligament. Just below the head are a constriction, or *neck,* and two large processes—a superior, lateral *greater trochanter* and an inferior, medial *lesser trochanter.* These processes provide attachments for muscles of the lower limbs and buttocks. On the posterior surface in the middle third of the shaft is a longitudinal crest called the *linea aspera.* This rough strip is an attachment for several muscles (fig. 7.53).

At the distal end of the femur, two rounded processes, the *lateral* and *medial condyles,* articulate with the tibia of the leg. A patella also articulates with the femur on its distal anterior surface.

On the medial surface at its distal end is a prominent *medial epicondyle,* and on the lateral surface is a *lateral epicondyle.* These projections provide attachments for muscles and ligaments.

Patella

The **patella,** or kneecap, is a flat sesamoid bone located in a tendon that passes anteriorly over the knee (see fig. 7.52). Because of its position, the patella controls the angle at which this tendon continues toward the tibia, so it functions in lever actions associated with lower limb movements.

As a result of a blow to the knee or a forceful unnatural movement of the leg, the patella sometimes slips to one side. This painful condition is called a *patellar dislocation.* Doing exercises that strengthen muscles associated with the knee and wearing protective padding can prevent knee displacement. Unfortunately, once the soft tissues that hold the patella in place are stretched, patellar dislocation tends to recur.

TABLE 7.11	Differences Between the Male and Female Skeletons
Part	**Differences**
Skull	Male skull is larger and heavier, with more conspicuous muscular attachments. Male forehead is shorter, facial area is less round, jaw larger, and mastoid processes and supraorbital ridges more prominent than those of a female.
Pelvis	Male pelvic bones are heavier, thicker, and have more obvious muscular attachments. The obturator foramina and the acetabula are larger and closer together than those of a female.
Pelvic cavity	Male pelvic cavity is narrower in all diameters and is longer, less roomy, and more funnel-shaped. The distances between the ischial spines and between the ischial tuberosities are less than in a female.
Sacrum	Male sacrum is narrower, sacral promontory projects forward to a greater degree, and sacral curvature is bent less sharply posteriorly than in a female.
Coccyx	Male coccyx is less movable than that of a female.

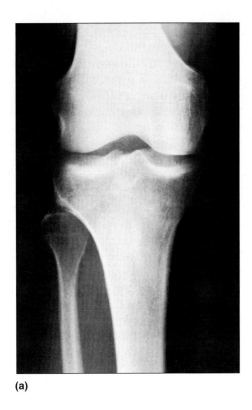

(a)

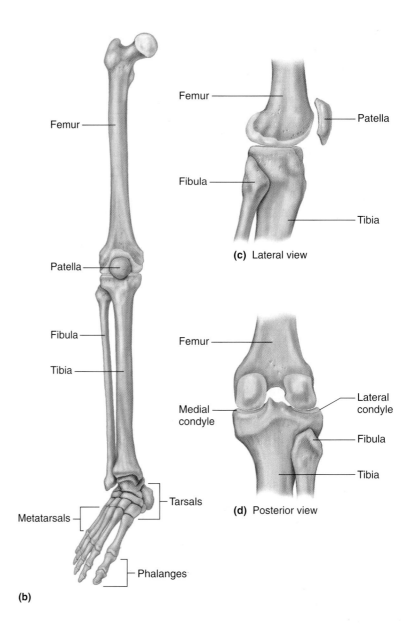

Femur

Patella

Fibula

Tibia

(c) Lateral view

Femur

Medial condyle

Lateral condyle

Fibula

Tibia

(d) Posterior view

Femur

Patella

Fibula

Tibia

Tarsals

Metatarsals

Phalanges

(b)

FIGURE 7.52

Parts of the lower limb. (*a*) Radiograph of the right knee (anterior view), showing the ends of the femur, tibia, and fibula. Thinner areas of bone, such as part of the head of the fibula and the patella, barely show in this radiograph. (*b*) Anterior view of the right lower limb. (*c*) Lateral view of the right knee. (*d*) Posterior view of the right knee.

Tibia

The **tibia,** or shin bone, is the larger of the two leg bones and is located on the medial side. Its proximal end is expanded into *medial* and *lateral condyles,* which have concave surfaces and articulate with the condyles of the femur. Below the condyles, on the anterior surface, is a process called the *tibial tuberosity,* which provides an attachment for the *patellar ligament* (a continuation of the patella-bearing tendon). A prominent *anterior crest* extends downward from the tuberosity and attaches connective tissues in the leg.

At its distal end, the tibia expands to form a prominence on the inner ankle called the *medial malleolus* (mah-le′o-lus), which is an attachment for ligaments. On its lateral side is a depression that articulates with the fibula (fig. 7.54). The inferior surface of the tibia's distal end articulates with a large bone (the talus) in the ankle.

The skeleton is particularly vulnerable to injury during the turbulent teen years, when bones grow rapidly. Athletic teens sometimes develop Osgood-Schlatter disease, which is a painful swelling of a bony projection of the tibia below the knee. Overusing the thigh muscles to straighten the lower limb irritates the area, causing the swelling. Usually a few months of rest and no athletic activity allows the bone to heal on its own. Rarely, a cast must be used to immobilize the knee.

Fibula

The **fibula** is a long, slender bone located on the lateral side of the tibia. Its ends are slightly enlarged into a proximal *head* and a distal *lateral malleolus.* The head articu-

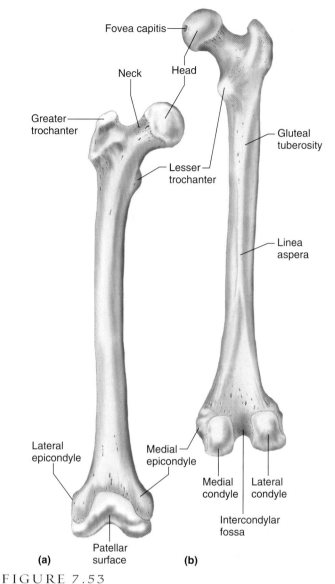

FIGURE 7.53

Femur. (a) Anterior surface and (b) posterior surface of the right femur.

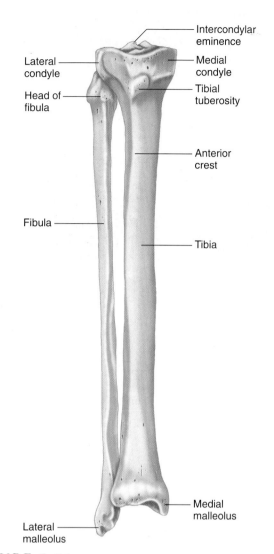

FIGURE 7.54

Bones of the right leg, anterior view.

lates with the tibia just below the lateral condyle; however, it does not enter into the knee joint and does not bear any body weight. The lateral malleolus articulates with the ankle and protrudes on the lateral side (fig. 7.54).

Foot

The foot is made up of the ankle, the instep, and the toes. The ankle or *tarsus* (tahr'sus) is composed of seven **tarsal bones.** These bones are arranged so that one of them, the **talus** (ta'lus), can move freely where it joins the tibia and fibula, thus forming the ankle. The remaining tarsal bones are bound firmly together, forming a mass supporting the talus. Figures 7.55 and 7.56 name the individual bones of the tarsus.

The largest of the tarsals, the **calcaneus** (kal-ka'ne-us), or heel bone, is located below the talus where it pro-

jects backward to form the base of the heel. The calcaneus helps support the weight of the body and provides an attachment for muscles that move the foot.

The instep or *metatarsus* (met"ah-tahr'sus) consists of five elongated **metatarsal bones,** which articulate with the tarsus. They are numbered 1 to 5, beginning on the medial side (fig. 7.56). The heads at the distal ends of these bones form the ball of the foot. The tarsals and metatarsals are arranged and bound by ligaments to form the arches of the foot. A longitudinal arch extends from the heel to the toe, and a transverse arch stretches across the foot. These arches provide a stable, springy base for the body. Sometimes, however, the tissues that bind the metatarsals weaken, producing fallen arches, or flat feet.

The **phalanges** of the toes are shorter, but otherwise similar to those of the fingers, and align and articulate

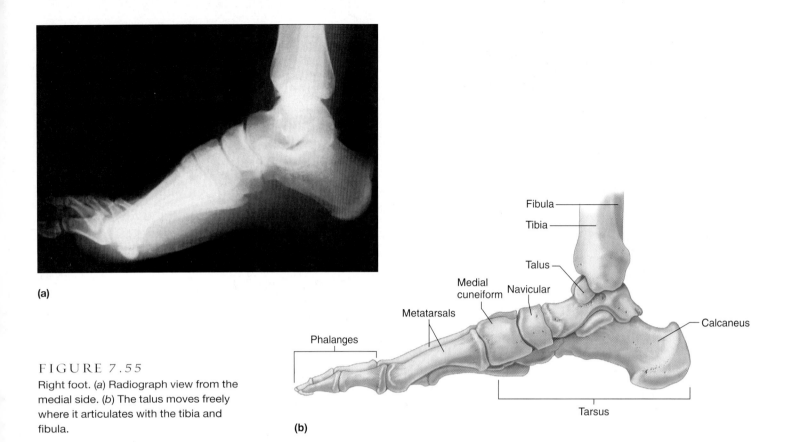

FIGURE 7.55

Right foot. (*a*) Radiograph view from the medial side. (*b*) The talus moves freely where it articulates with the tibia and fibula.

(a)

Fibula

Tibia

Talus

Medial cuneiform

Navicular

Metatarsals

Calcaneus

Phalanges

Tarsus

(b)

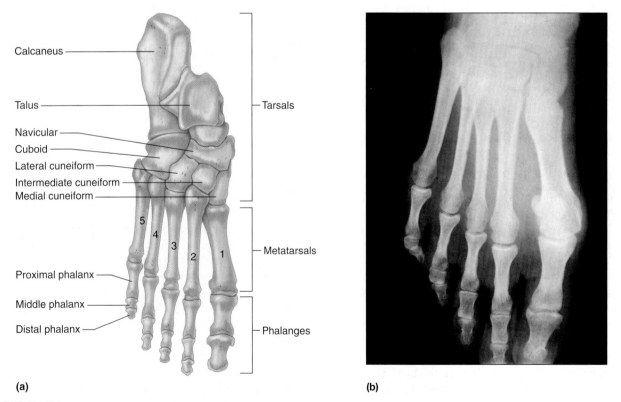

Calcaneus

Talus

Navicular

Cuboid

Lateral cuneiform

Intermediate cuneiform

Medial cuneiform

Tarsals

5 4 3 2 1

Metatarsals

Proximal phalanx

Middle phalanx

Distal phalanx

Phalanges

(a)

(b)

FIGURE 7.56

Right foot. (*a*) Viewed superiorly. (*b*) Radiograph of the right foot viewed superiorly.

TABLE 7.12   Bones of the Pelvic Girdle and Lower Limbs

Name and Number	Location	Special Features
Coxa (2)	Hip, articulating with the other coxa anteriorly and with the sacrum posteriorly	Ilium, iliac crest, anterior superior iliac spine, ischium, ischial tuberosity, ischial spine, obturator foramen, acetabulum, pubis
Femur (2)	Thigh, between hip and knee	Head, fovea capitis, neck, greater trochanter, lesser trochanter, linea aspera, lateral condyle, medial condyle, gluteal tuberosity, intercondylar fossa
Patella (2)	Anterior surface of knee	A flat sesamoid bone located within a tendon
Tibia (2)	Medial side of leg, between knee and ankle	Medial condyle, lateral condyle, tibial tuberosity, anterior crest, medial malleolus, intercondylar eminence
Fibula (2)	Lateral side of leg, between knee and ankle	Head, lateral malleolus
Tarsal (14)	Ankle	Freely movable talus that articulates with leg bones; calcaneus that forms the base of the heel; five other tarsal bones bound firmly together
Metatarsal (10)	Instep	One in line with each toe, arranged and bound by ligaments to form arches
Phalanx (28)	Toe	Three in each toe, two in great toe

with the metatarsals. Each toe has three phalanges—a proximal, a middle, and a distal phalanx—except the great toe, which has only two because it lacks the middle phalanx (fig. 7.56). Table 7.12 summarizes the bones of the pelvic girdle and lower limbs.

> An infant with two casts on her feet is probably being treated for clubfoot, a very common birth defect in which the foot twists out of its normal position, turning in, out, up, down, or some combination of these directions. Clubfoot probably results from arrested development during fetal existence, but the precise cause is not known. Clubfoot can almost always be corrected with special shoes, or surgery, followed by several months in casts to hold the feet in the correct position.

1 Locate and name each of the bones of the lower limb.

2 Explain how the bones of the lower limb articulate with one another.

3 Describe how the foot is adapted to support the body.

Life-Span Changes

Aging-associated changes in the skeletal system are apparent at the cellular and whole-body levels. Most obvious is the incremental decrease in height that begins at about age thirty, with a loss of about 1/16 of an inch a year. In the later years, compression fractures in the vertebrae may contribute significantly to loss of height (fig. 7.57). Overall, as calcium levels fall and bone material gradually vanishes, the skeleton loses strength, and the

FIGURE 7.57
The bones change to different degrees and at different rates over a lifetime.

bones become brittle and increasingly prone to fracture. However, the continued ability of fractures to heal reveals that the bone tissue is still alive and functional.

Components of the skeletal system and individual bones change to different degrees and at different rates over a lifetime. Gradually, osteoclasts come to outnumber osteoblasts, which means that bone is eaten away in the remodeling process at a faster rate than it is replaced—resulting in more spaces in bones. The bone thins, its strength waning. Bone matrix changes, with the ratio of mineral to protein increasing, making bones more brittle and prone to fracture. Beginning in the third decade of life, bone matrix is removed faster than it is laid down. By age thirty-five, all of us start to lose bone mass.

Trabecular bone, due to its spongy, less compact nature, shows the changes of aging first, as they thin, increasing in porosity and weakening the overall structure. The vertebrae consist mostly of trabecular bone. It is also found in the upper part of the femur, whereas the shaft is more compact bone. The fact that trabecular bone weakens sooner than compact bone destabilizes the femur, which is why it is a commonly broken bone among the elderly.

Compact bone loss begins at around age forty and continues at about half the rate of loss of trabecular bone. As remodeling continues throughout life, older osteons disappear as new ones are built next to them. With age, the osteons may coalesce, further weakening the overall structures as gaps form.

Bone loss is slow and steady in men, but in women, it is clearly linked to changing hormone levels. In the first decade following menopause, 15 to 20% of trabecular bone is lost, which is two to three times the rate of loss in men and premenopausal women. During the same time, compact bone loss is 10 to 15%, which is three to four times the rate of loss in men and premenopausal women. By about age seventy, both sexes are losing bone at about the same rate. By very old age, a woman may have only half the trabecular and compact bone mass as she did in her twenties, whereas a very elderly man may have one-third less bone mass.

Falls among the elderly are common and have many causes (see table 7.13). The most common fractures, after vertebral compression and hip fracture, are of the wrist, leg, and pelvis. Aging-related increased risk of fracture usually begins at about age fifty. Because healing is slowed, pain from a broken bone may persist for months. Preserving skeletal health may involve avoiding falls, taking calcium supplements, getting enough vitamin D, avoiding carbonated beverages (phosphates deplete bone), and getting regular exercise.

1 Why is bone lost faster, with aging, than bone replacement?

2 In which bones do fractures most commonly occur in the elderly?

TABLE 7.13	Reasons for Falls Among the Elderly
Overall frailty	
Decreased muscle strength	
Decreased coordination	
Side effects of medication	
Slowed reaction time due to stiffening joints	
Poor vision and/or hearing	
Disease (cancer, infection, arthritis)	

Clinical Terms Related to the Skeletal System

achondroplasia (a-kon″dro-pla′ze-ah) Inherited condition that retards formation of cartilaginous bone. The result is a type of dwarfism.

acromegaly (ak″ro-meg′ah-le) Abnormal enlargement of facial features, hands, and feet in adults as a result of overproduction of growth hormone.

Colles fracture (kol′ēz-frak′tūre) Fracture at the distal end of the radius that displaces the smaller fragment posteriorly.

epiphysiolysis (ep″ĭ-fiz″e-ol′ĭ-sis) Separation or loosening of the epiphysis from the diaphysis of a bone.

laminectomy (lam″ĭ-nek′to-me) Surgical removal of the posterior arch of a vertebra, usually to relieve symptoms of a ruptured intervertebral disc.

lumbago (lum-ba′go) Dull ache in the lumbar region of the back.

orthopedics (or″tho-pe′diks) Medical specialty that prevents, diagnoses, and treats diseases and abnormalities of the skeletal and muscular systems.

ostealgia (os″te-al′je-ah) Pain in a bone.

ostectomy (os-tek′to-me) Surgical removal of a bone.

osteitis (os″te-i′tis) Inflammation of bone tissue.

osteochondritis (os″te-o-kon-dri′tis) Inflammation of bone and cartilage tissues.

osteogenesis (os″te-o-jen′ĕ-sis) Bone development.

osteogenesis imperfecta (os″te-o-jen′ĕ-sis im-per-fek′ta) Inherited condition of deformed and abnormally brittle bones.

osteoma (os″te-o′mah) Tumor composed of bone tissue.

osteomalacia (os″te-o-mah-la′she-ah) Softening of adult bone due to a disorder in calcium and phosphorus metabolism, usually caused by vitamin D deficiency.

osteomyelitis (os″te-o-mi″ĕ-li′tis) Bone inflammation caused by the body's reaction to bacterial or fungal infection.

osteonecrosis (os″te-o-ne-kro′sis) Death of bone tissue. This condition most commonly occurs in the femur head in elderly persons and may be due to obstructed arteries supplying the bone.

osteopathology (os″te-o-pah-thol′o-je) Study of bone diseases.

osteopenia (os″te-o-pe′ni-ah) Decrease in bone mass due to reduction in rate of bone tissue formation.

osteoporosis (os″te-o-po-ro′sis) Decreased bone mineral content.

osteotomy (os″te-ot′o-me) Cutting a bone.

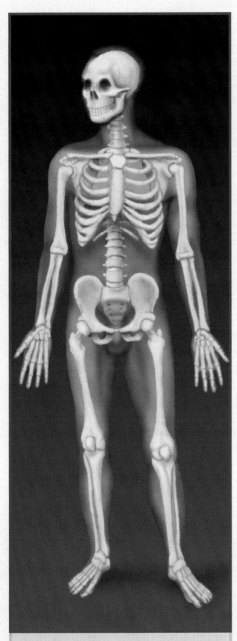

SKELETAL SYSTEM

Bones provide support, protection, and movement and also play a role in calcium balance.

Integumentary System

Vitamin D, activated in the skin, plays a role in calcium availability for bone matrix.

Lymphatic System

Cells of the immune system originate in the bone marrow.

Muscular System

Muscles pull on bones to cause movement.

Digestive System

Absorption of dietary calcium provides material for bone matrix.

Nervous System

Proprioceptors sense the position of body parts. Pain receptors warn of trauma to bone. Bones protect the brain and spinal cord.

Respiratory System

Ribs and muscles work together in breathing.

Endocrine System

Some hormones act on bone to help regulate blood calcium levels.

Urinary System

The kidneys and bones work together to help regulate blood calcium levels.

Cardiovascular System

Blood transports nutrients to bone cells. Bone helps regulate plasma calcium levels, important to heart function.

Reproductive System

The pelvis helps support the uterus during pregnancy. Bone may provide a source of calcium during lactation.

Introduction (page 182)

Individual bones are the organs of the skeletal system. A bone contains very active tissues.

Bone Structure (page 182)

Bone structure reflects its function.

1. Bone classification
   Bones are grouped according to their shapes—long, short, flat, irregular, or round (sesamoid).
2. Parts of a long bone
   a. Epiphyses at each end are covered with articular cartilage and articulate with other bones.
   b. The shaft of a bone is called the diaphysis.
   c. Except for the articular cartilage, a bone is covered by a periosteum.
   d. Compact bone has a continuous matrix with no gaps.
   e. Spongy bone has irregular interconnecting spaces between bony plates.
   f. Both compact and spongy bone are strong and resist bending.
   g. The diaphysis contains a medullary cavity filled with marrow.
3. Microscopic structure
   a. Compact bone contains osteons cemented together.
   b. Central canals contain blood vessels that nourish the cells of osteons.
   c. Perforating canals connect central canals transversely and communicate with the bone's surface and the medullary cavity.
   d. Diffusion from the surface of thin bony plates nourishes cells of spongy bones.

Bone Development and Growth (page 186)

1. Intramembranous bones
   a. Certain flat bones of the skull are intramembranous bones.
   b. They develop from layers of connective tissues.
   c. Osteoblasts within the membranous layers form bone tissue.
   d. Mature bone cells are called osteocytes.
   e. Primitive connective tissue gives rise to the periosteum.
2. Endochondral bones
   a. Most of the bones of the skeleton are endochondral.
   b. They develop as hyaline cartilage that is later replaced by bone tissue.
   c. Primary ossification center appears in the diaphysis, whereas secondary ossification centers appear in the epiphyses.
   d. An epiphyseal plate remains between the primary and secondary ossification centers.
3. Growth at the epiphyseal plate
   a. An epiphyseal plate consists of layers of cells: resting cells, young dividing cells, older enlarging cells, and dying cells.
   b. The epiphyseal plates are responsible for lengthening.
   c. Long bones continue to lengthen until the epiphyseal plates are ossified.
   d. Growth in thickness is due to ossification beneath the periosteum.
   e. The action of osteoclasts forms the medullary cavity.
4. Homeostasis of bone tissue
   a. Osteoclasts and osteoblasts continually remodel bone.
   b. The total mass of bone remains nearly constant.
5. Factors affecting bone development, growth, and repair
   a. Deficiencies of vitamin A, C, or D result in abnormal development.
   b. Insufficient secretion of pituitary growth hormone may result in dwarfism; excessive secretion may result in gigantism, or acromegaly.
   c. Deficiency of thyroid hormone delays bone growth.
   d. Male and female sex hormones promote bone formation and stimulate ossification of the epiphyseal plates.

Bone Function (page 191)

1. Support and protection
   a. Bones shape and form body structures.
   b. Bones support and protect softer, underlying tissues.
2. Body movement
   a. Bones and muscles function together as levers.
   b. A lever consists of a rod, a pivot (fulcrum), a resistance, and a force that supplies energy.
   c. Parts of a first-class lever are arranged resistance–pivot–force; of a second-class lever, pivot–resistance–force; of a third-class lever, resistance–force–pivot.
3. Blood cell formation
   a. At different ages, hematopoiesis occurs in the yolk sac, the liver, the spleen, and the red bone marrow.
   b. Red marrow houses developing red blood cells, white blood cells, and blood platelets.
4. Inorganic salt storage
   a. The intercellular material of bone tissue contains large quantities of calcium phosphate in the form of hydroxyapatite.
   b. When blood calcium ion concentration is low, osteoclasts resorb bone, releasing calcium salts.
   c. When blood calcium ion concentration is high, osteoblasts are stimulated to form bone tissue and store calcium salts.
   d. Bone stores small amounts of sodium, magnesium, potassium, and carbonate ions.
   e. Bone tissues may accumulate lead, radium, or strontium.

Skeletal Organization (page 196)

1. Number of bones
   a. Usually a human skeleton has 206 bones, but the number may vary.
   b. Extra bones in sutures are called sutural bones.
2. Divisions of the skeleton
   a. The skeleton can be divided into axial and appendicular portions.
   b. The axial skeleton consists of the skull, hyoid bone, vertebral column, and thoracic cage.
   c. The appendicular skeleton consists of the pectoral girdle, upper limbs, pelvic girdle, and lower limbs.

Skull (page 199)

The skull consists of twenty-two bones, which include eight cranial bones and fourteen facial bones.

1. Cranium
   a. The cranium encloses and protects the brain and provides attachments for muscles.
   b. Some cranial bones contain air-filled sinuses that help reduce the weight of the skull.
   c. Cranial bones include the frontal bone, parietal bones, occipital bone, temporal bones, sphenoid bone, and ethmoid bone.
2. Facial skeleton
   a. Facial bones form the basic shape of the face and provide attachments for muscles.
   b. Facial bones include the maxillary bones, palatine bones, zygomatic bones, lacrimal bones, nasal bones, vomer bone, inferior nasal conchae, and mandible.
3. Infantile skull
   a. Incompletely developed bones, connected by fontanels, enable the infantile skull to change shape slightly during childbirth.
   b. Infantile skull bones are thin, somewhat flexible, and less easily fractured.

Vertebral Column (page 209)

The vertebral column extends from the skull to the pelvis and protects the spinal cord. It is composed of vertebrae separated by intervertebral discs. An infant has thirty-three vertebral bones, and an adult has twenty-six. The vertebral column has four curvatures—cervical, thoracic, lumbar, and pelvic.

1. A typical vertebra
   a. A typical vertebra consists of a body, pedicles, laminae, spinous process, transverse processes, and superior and inferior articulating processes.
   b. Notches on the upper and lower surfaces of the pedicles on adjacent vertebrae form intervertebral foramina through which spinal nerves pass.
2. Cervical vertebrae
   a. Cervical vertebrae comprise the bones of the neck.
   b. Transverse processes have transverse foramina.
   c. The atlas (first vertebra) supports the head.
   d. The dens of the axis (second vertebra) provides a pivot for the atlas when the head is turned from side to side.
3. Thoracic vertebrae
   a. Thoracic vertebrae are larger than cervical vertebrae.
   b. Their long spinous processes slope downward, and facets on the sides of bodies articulate with the ribs.
4. Lumbar vertebrae
   a. Vertebral bodies of lumbar vertebrae are large and strong.
   b. Their transverse processes project posteriorly at sharp angles, and their spinous processes are directed horizontally.
5. Sacrum
   a. The sacrum, formed of five fused vertebrae, is a triangular structure that has rows of dorsal sacral foramina.
   b. It is united with the coxae at the sacroiliac joints.
   c. The sacral promontory provides a guide for determining the size of the pelvis.

6. Coccyx
   a. The coccyx, composed of four fused vertebrae, forms the lowest part of the vertebral column.
   b. It acts as a shock absorber when a person sits.

Thoracic Cage (page 216)

The thoracic cage includes the ribs, thoracic vertebrae, sternum, and costal cartilages. It supports the pectoral girdle and upper limbs, protects viscera, and functions in breathing.

1. Ribs
   a. Twelve pairs of ribs are attached to the twelve thoracic vertebrae.
   b. Costal cartilages of the true ribs join the sternum directly; those of the false ribs join indirectly or not at all.
   c. A typical rib has a shaft, head, and tubercles that articulate with the vertebrae.
2. Sternum
   a. The sternum consists of a manubrium, body, and xiphoid process.
   b. It articulates with costal cartilages and clavicles.

Pectoral Girdle (page 218)

The pectoral girdle is composed of two clavicles and two scapulae. It forms an incomplete ring that supports the upper limbs and provides attachments for muscles that move the upper limbs.

1. Clavicles
   a. Clavicles are rodlike bones that run horizontally between the sternum and shoulders.
   b. They hold the shoulders in place and provide attachments for muscles.
2. Scapulae
   a. The scapulae are broad, triangular bones with bodies, spines, heads, acromion processes, coracoid processes, glenoid cavities, supraspinous and infraspinous fossae, superior borders, axillary borders, and vertebral borders.
   b. They articulate with the humerus of each upper limb and provide attachments for muscles of the upper limbs and chest.

Upper Limb (page 220)

Limb bones provide the frameworks and attachments of muscles and function in levers that move the limb and its parts.

1. Humerus
   a. The humerus extends from the scapula to the elbow.
   b. It has a head, greater tubercle, lesser tubercle, intertubercular groove, anatomical neck, surgical neck, deltoid tuberosity, capitulum, trochlea, epicondyles, coronoid fossa, and olecranon fossa.
2. Radius
   a. The radius is located on the thumb side of the forearm between the elbow and wrist.
   b. It has a head, radial tuberosity, styloid process, and ulnar notch.
3. Ulna
   a. The ulna is longer than the radius and overlaps the humerus posteriorly.
   b. It has a trochlear notch, olecranon process, coronoid process, head, styloid process, and radial notch.
   c. It articulates with the radius laterally and with a disc of fibrocartilage inferiorly.

4. Hand
    a. The wrist includes eight carpals that form a carpus.
    b. The palm has five metacarpals.
    c. The five fingers have fourteen phalanges.

Pelvic Girdle (page 224)

The pelvic girdle consists of two coxae that articulate with each other anteriorly and with the sacrum posteriorly. The sacrum, coccyx, and pelvic girdle form the pelvis. The girdle provides support for body weight and attachments for muscles and protects visceral organs.

1. Coxae
   Each coxa consists of an ilium, ischium, and pubis, which are fused in the region of the acetabulum.
    a. Ilium
       (1) The ilium, the largest portion of the coxa, joins the sacrum at the sacroiliac joint.
       (2) It has an iliac crest with anterior and posterior superior iliac spines and iliac fossae.
    b. Ischium
       (1) The ischium is the lowest portion of the coxa.
       (2) It has an ischial tuberosity and ischial spine.
    c. Pubis
       (1) The pubis is the anterior portion of the coxa.
       (2) Pubis bones are fused anteriorly at the symphysis pubis.
2. Greater and lesser pelves
    a. The lesser pelvis is below the pelvic brim; the greater pelvis is above it.
    b. The lesser pelvis functions as a birth canal; the greater pelvis helps support abdominal organs.
3. Differences between male and female pelves
    a. Differences between male and female pelves are related to the function of the female pelvis as a birth canal.
    b. Usually the female pelvis is more flared; pubic arch is broader; distance between the ischial spines and the ischial tuberosities is greater; and sacral curvature is shorter.

Lower Limb (page 227)

Bones of the lower limb provide the frameworks of the thigh, leg, ankle, and foot.

1. Femur
    a. The femur extends from the hip to the knee.
    b. It has a head, fovea capitis, neck, greater trochanter, lesser trochanter, linea aspera, lateral condyle, and medial condyle.
2. Patella
    a. The patella is a flat, round, or sesamoid bone in the tendon that passes anteriorly over the knee.
    b. It controls the angle of this tendon and functions in lever actions associated with lower limb movements.
3. Tibia
    a. The tibia is located on the medial side of the leg.
    b. It has medial and lateral condyles, tibial tuberosity, anterior crest, and medial malleolus.
    c. It articulates with the talus of the ankle.
4. Fibula
    a. The fibula is located on the lateral side of the tibia.
    b. It has a head and lateral malleolus that articulates with the ankle but does not bear body weight.
5. Foot
    a. The foot consists of the tarsus, metatarsus, and five toes.
    b. It includes the talus that helps form the ankle, six other tarsals, five metatarsals, and fourteen phalanges.

Life-Span Changes (page 231)

Aging-associated changes in the skeleton are apparent at the cellular and whole-body levels.

1. Incremental decrease in height begins at about age thirty.
2. Gradually, bone loss exceeds bone replacement.
    a. In the first decade following menopause, bone loss occurs more rapidly in women than in men or premenopausal women. By age seventy, both sexes are losing bone at about the same rate.
    b. Aging increases risk of bone fractures.

CRITICAL THINKING QUESTIONS

1. What steps do you think should be taken to reduce the chances of bones accumulating abnormal metallic elements such as lead, radium, and strontium?

2. Why do you think incomplete, longitudinal fractures of bone shafts (greenstick fractures) are more common in children than in adults?

3. When a child's bone is fractured, growth may be stimulated at the epiphyseal plate. What problems might this extra growth cause in an upper or lower limb before the growth of the other limb compensates for the difference in length?

4. Why do elderly persons often develop bowed backs and appear shorter than they were in earlier years?

5. How might the condition of an infant's fontanels be used to evaluate skeletal development? How might the fontanels be used to estimate intracranial pressure?

6. Why are women more likely than men to develop osteoporosis? What steps can reduce the risk of developing this condition?

7. How does the structure of a bone make it strong yet lightweight?

8. Archeologists discover skeletal remains of humanlike animals in Ethiopia. Examination of the bones suggests that the remains represent four types of individuals. Two of the skeletons have bone densities that are 30% less than those of the other two skeletons. The skeletons with the lower bone mass also have broader front pelvic bones. Within the two groups defined by bone mass, smaller skeletons have bones with evidence of epiphyseal plates, but larger bones have only a thin line where the epiphyseal plates should be. Give the age group and gender of the individuals in this find.

REVIEW EXERCISES

Part A

1. List four groups of bones based upon their shapes, and name an example from each group.
2. Sketch a typical long bone, and label its epiphyses, diaphysis, medullary cavity, periosteum, and articular cartilages.
3. Distinguish between spongy and compact bone.
4. Explain how central canals and perforating canals are related.
5. Explain how the development of intramembranous bone differs from that of endochondral bone.
6. Distinguish between osteoblasts and osteocytes.
7. Explain the function of an epiphyseal plate.
8. Explain how a bone grows in thickness.
9. Define *osteoclast.*
10. Explain how osteoclasts and osteoblasts regulate bone mass.
11. Describe the effects of vitamin deficiencies on bone development.
12. Explain the causes of pituitary dwarfism and gigantism.
13. Describe the effects of thyroid and sex hormones on bone development.
14. Explain the effects of exercise on bone structure.
15. Provide several examples to illustrate how bones support and protect body parts.
16. Describe a lever, and explain how its parts may be arranged to form first- , second- , and third-class levers.
17. Explain how upper limb movements function as levers.
18. Describe the functions of red and yellow bone marrow.
19. Explain the mechanism that regulates the concentration of blood calcium ions.
20. List three substances that may be abnormally stored in bone.
21. Distinguish between the axial and appendicular skeletons.
22. Name the bones of the cranium and the facial skeleton.
23. Explain the importance of fontanels.
24. Describe a typical vertebra.
25. Explain the differences among cervical, thoracic, and lumbar vertebrae.
26. Describe the locations of the sacroiliac joint, the sacral promontory, and the sacral hiatus.
27. Name the bones that comprise the thoracic cage.
28. List the bones that form the pectoral and pelvic girdles.
29. Name the bones of the upper limb.
30. Name the bones that comprise a coxa.
31. List the major differences that may occur between the male and female pelves.
32. List the bones of the lower limb.
33. Describe changes in trabecular bone and compact bone with aging.
34. List factors that may preserve skeletal health.

Part B

Match the parts listed in column I with the bones listed in column II.

I		II	
1.	Coronoid process	A.	Ethmoid bone
2.	Cribriform plate	B.	Frontal bone
3.	Foramen magnum	C.	Mandible
4.	Mastoid process	D.	Maxillary bone
5.	Palatine process	E.	Occipital bone
6.	Sella turcica	F.	Temporal bone
7.	Supraorbital notch	G.	Sphenoid bone
8.	Temporal process	H.	Zygomatic bone
9.	Acromion process	I.	Femur
10.	Deltoid tuberosity	J.	Fibula
11.	Greater trochanter	K.	Humerus
12.	Lateral malleolus	L.	Radius
13.	Medial malleolus	M.	Scapula
14.	Olecranon process	N.	Sternum
15.	Radial tuberosity	O.	Tibia
16.	Xiphoid process	P.	Ulna

WEB CONNECTIONS

Visit the Student Edition of the Online Learning Center at www.mhhe.com/shier10 **for answers to chapter questions, additional quizzes, interactive learning exercises, and other study tools.**

Human Skull

The following set of reference plates is presented to help you locate some of the more prominent features of the human skull. As you study these photographs, it is important to remember that individual human skulls vary in every characteristic. Also, the photographs in this set depict bones from several different skulls.

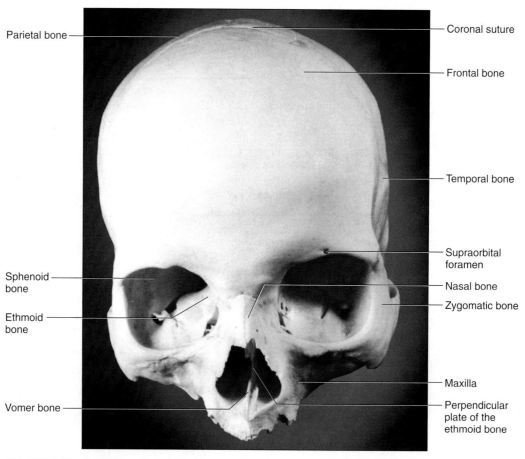

Parietal bone —

Coronal suture —

Frontal bone —

Temporal bone —

Supraorbital foramen —

Sphenoid bone —

Nasal bone —

Zygomatic bone —

Ethmoid bone —

Vomer bone —

Maxilla —

Perpendicular plate of the ethmoid bone —

PLATE EIGHT
The skull, frontal view.

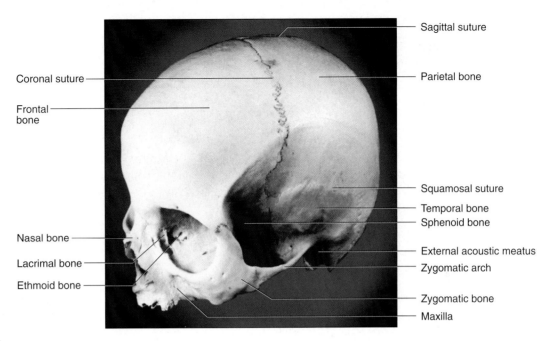

Sagittal suture

Coronal suture

Frontal bone

Parietal bone

Squamosal suture

Temporal bone

Sphenoid bone

Nasal bone

Lacrimal bone

Ethmoid bone

External acoustic meatus

Zygomatic arch

Zygomatic bone

Maxilla

PLATE NINE
The skull, left anterolateral view.

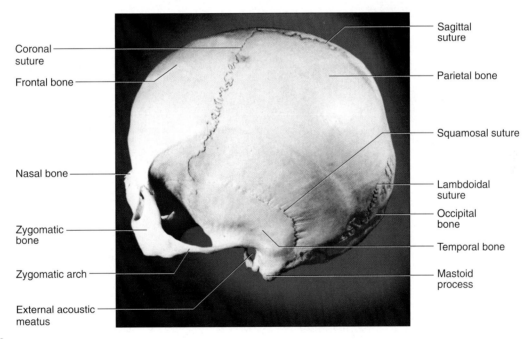

Coronal suture

Frontal bone

Nasal bone

Zygomatic bone

Zygomatic arch

External acoustic meatus

Sagittal suture

Parietal bone

Squamosal suture

Lambdoidal suture

Occipital bone

Temporal bone

Mastoid process

PLATE TEN
The skull, left posterolateral view.

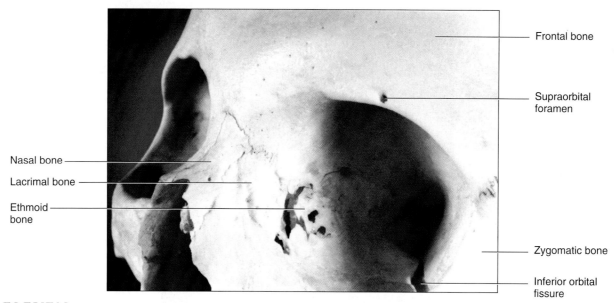

Frontal bone

Supraorbital foramen

Nasal bone

Lacrimal bone

Ethmoid bone

Zygomatic bone

Inferior orbital fissure

PLATE ELEVEN
Bones of the left orbital region.

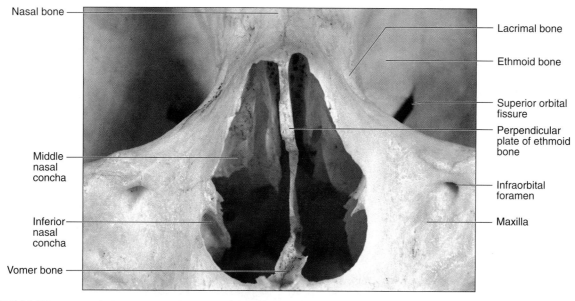

Nasal bone

Lacrimal bone

Ethmoid bone

Superior orbital fissure

Perpendicular plate of ethmoid bone

Infraorbital foramen

Maxilla

Middle nasal concha

Inferior nasal concha

Vomer bone

PLATE TWELVE
Bones of the anterior nasal region.

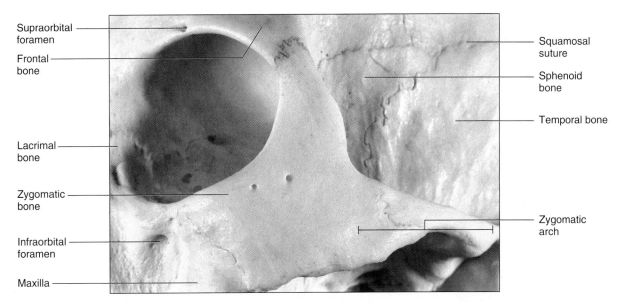

Supraorbital foramen

Frontal bone

Lacrimal bone

Zygomatic bone

Infraorbital foramen

Maxilla

Squamosal suture

Sphenoid bone

Temporal bone

Zygomatic arch

PLATE THIRTEEN
Bones of the left zygomatic region.

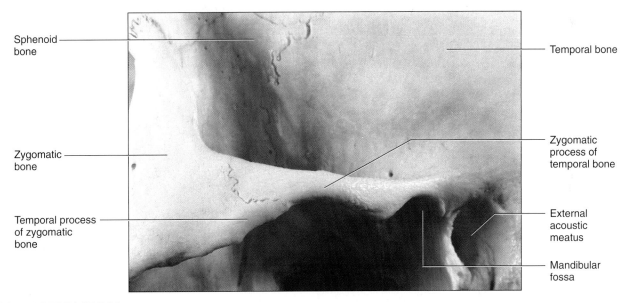

Sphenoid bone

Zygomatic bone

Temporal process of zygomatic bone

Temporal bone

Zygomatic process of temporal bone

External acoustic meatus

Mandibular fossa

PLATE FOURTEEN
Bones of the left temporal region.

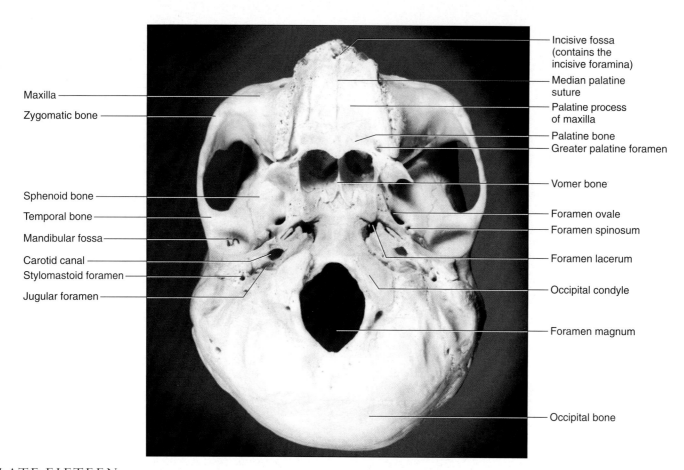

Maxilla

Zygomatic bone

Sphenoid bone

Temporal bone

Mandibular fossa

Carotid canal

Stylomastoid foramen

Jugular foramen

Incisive fossa
(contains the
incisive foramina)

Median palatine
suture

Palatine process
of maxilla

Palatine bone

Greater palatine foramen

Vomer bone

Foramen ovale

Foramen spinosum

Foramen lacerum

Occipital condyle

Foramen magnum

Occipital bone

PLATE FIFTEEN
The skull, inferior view.

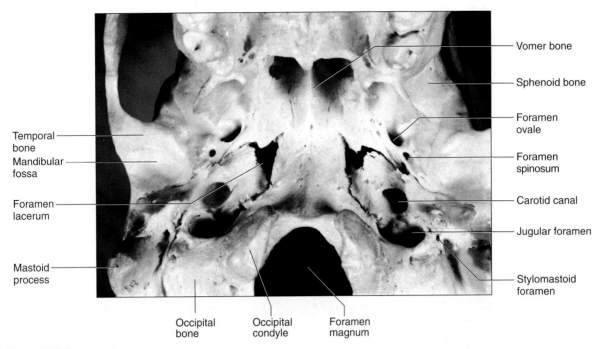

Temporal
bone

Mandibular
fossa

Foramen
lacerum

Mastoid
process

Vomer bone

Sphenoid bone

Foramen
ovale

Foramen
spinosum

Carotid canal

Jugular foramen

Stylomastoid
foramen

Occipital
bone

Occipital
condyle

Foramen
magnum

PLATE SIXTEEN
Base of the skull, sphenoid region.

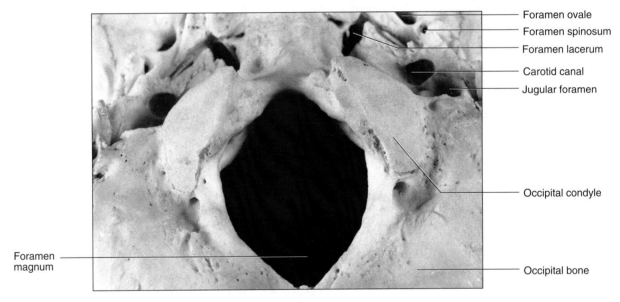

Foramen ovale

Foramen spinosum

Foramen lacerum

Carotid canal

Jugular foramen

Occipital condyle

Foramen
magnum

Occipital bone

PLATE SEVENTEEN
Base of the skull, occipital region.

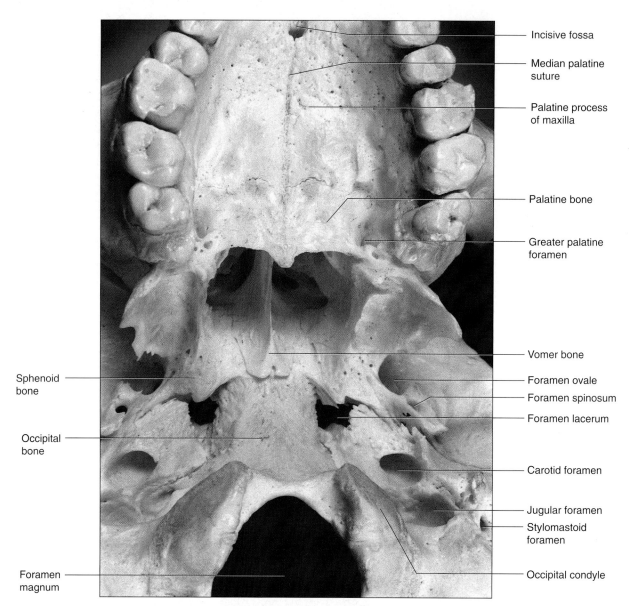

Incisive fossa

Median palatine suture

Palatine process of maxilla

Palatine bone

Greater palatine foramen

Vomer bone

Foramen ovale

Foramen spinosum

Foramen lacerum

Carotid foramen

Jugular foramen

Stylomastoid foramen

Occipital condyle

Sphenoid bone

Occipital bone

Foramen magnum

PLATE EIGHTEEN
Base of the skull, maxillary region.

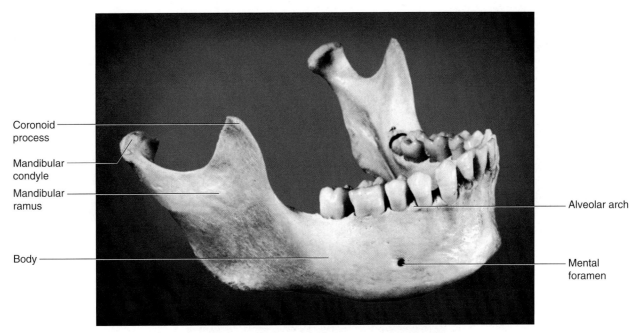

Coronoid process

Mandibular condyle

Mandibular ramus

Body

Alveolar arch

Mental foramen

PLATE NINETEEN
Mandible, lateral view.

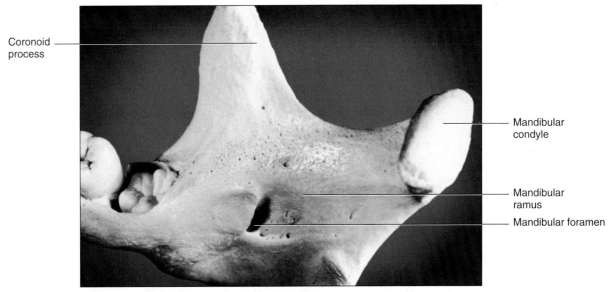

Coronoid process

Mandibular condyle

Mandibular ramus

Mandibular foramen

PLATE TWENTY
Mandible, medial surface of right ramus.

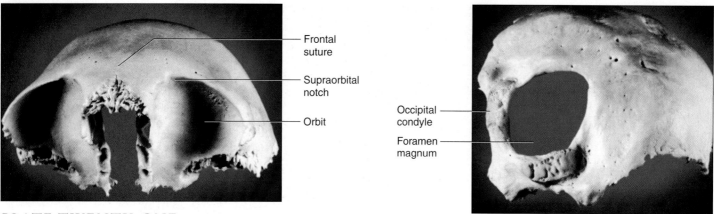

PLATE TWENTY-ONE
Frontal bone, anterior view.

Frontal suture
Supraorbital notch
Orbit

PLATE TWENTY-TWO
Occipital bone, inferior view.

Occipital condyle
Foramen magnum

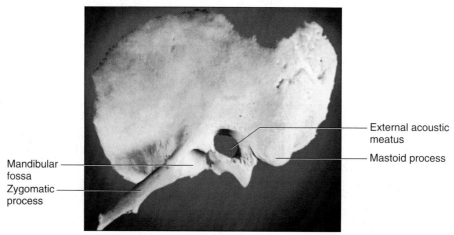

PLATE TWENTY-THREE
Temporal bone, left lateral view.

External acoustic meatus
Mastoid process
Mandibular fossa
Zygomatic process

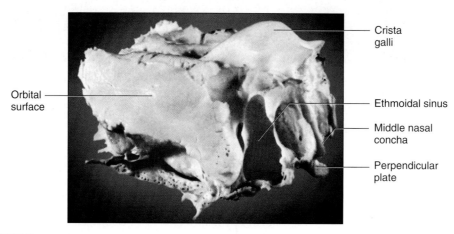

PLATE TWENTY-FOUR
Ethmoid bone, right lateral view.

Crista galli
Orbital surface
Ethmoidal sinus
Middle nasal concha
Perpendicular plate

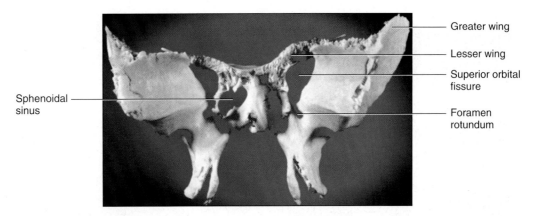

Greater wing

Lesser wing

Superior orbital
fissure

Sphenoidal
sinus

Foramen
rotundum

PLATE TWENTY-FIVE
Sphenoid bone, anterior view.

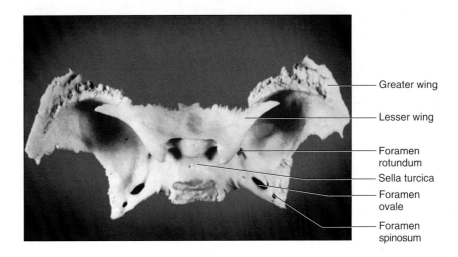

Greater wing

Lesser wing

Foramen
rotundum

Sella turcica

Foramen
ovale

Foramen
spinosum

PLATE TWENTY-SIX
Sphenoid bone, superior view.

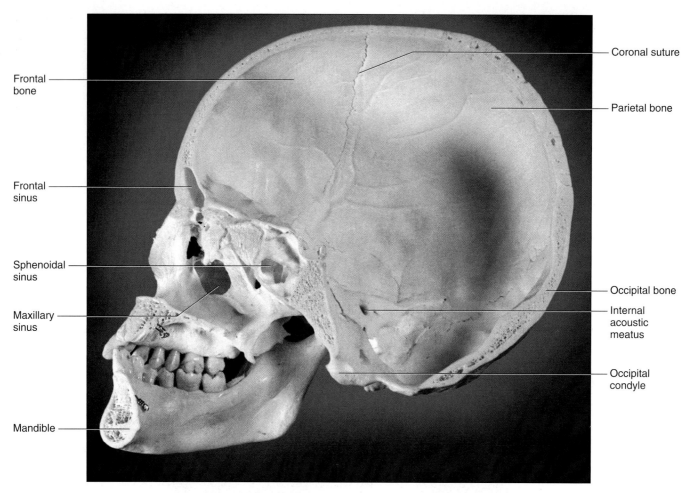

Frontal bone

Coronal suture

Frontal sinus

Parietal bone

Sphenoidal sinus

Maxillary sinus

Occipital bone

Internal acoustic meatus

Occipital condyle

Mandible

PLATE TWENTY-SEVEN

The skull, sagittal section.

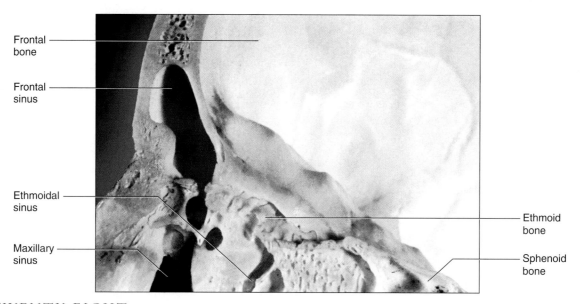

Frontal bone

Frontal sinus

Ethmoidal sinus

Maxillary sinus

Ethmoid bone

Sphenoid bone

PLATE TWENTY-EIGHT

Ethmoidal region, sagittal section.

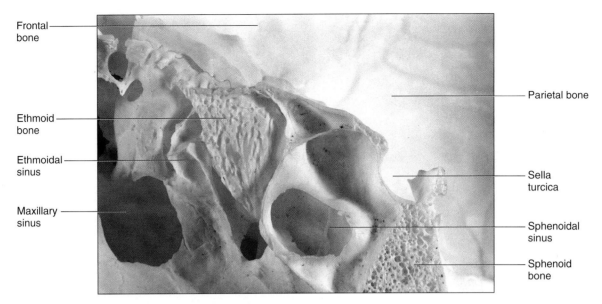

Frontal bone

Ethmoid bone

Ethmoidal sinus

Maxillary sinus

Parietal bone

Sella turcica

Sphenoidal sinus

Sphenoid bone

PLATE TWENTY-NINE
Sphenoidal region, sagittal section.

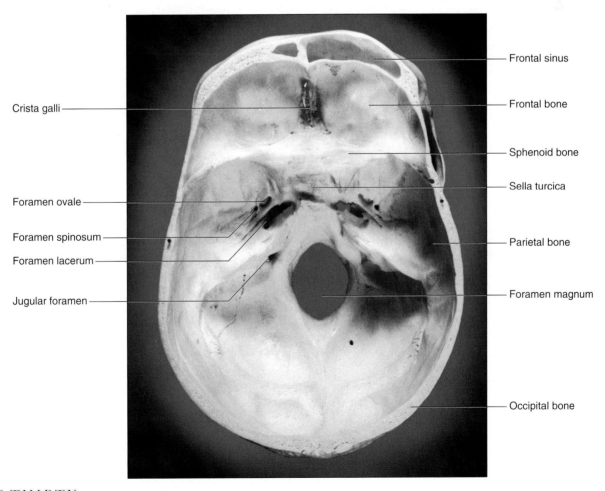

Crista galli

Foramen ovale

Foramen spinosum

Foramen lacerum

Jugular foramen

Frontal sinus

Frontal bone

Sphenoid bone

Sella turcica

Parietal bone

Foramen magnum

Occipital bone

PLATE THIRTY
The skull, floor of the cranial cavity.

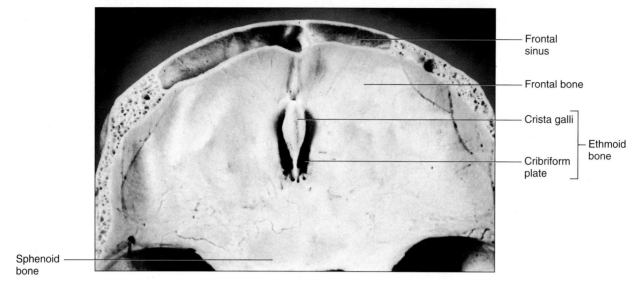

Frontal
sinus

Frontal bone

Crista galli

Ethmoid
bone

Cribriform
plate

Sphenoid
bone

PLATE THIRTY-ONE
Frontal region, transverse section.

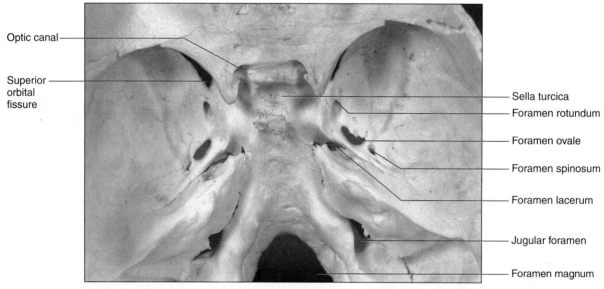

Optic canal

Superior
orbital
fissure

Sella turcica

Foramen rotundum

Foramen ovale

Foramen spinosum

Foramen lacerum

Jugular foramen

Foramen magnum

PLATE THIRTY-TWO
Sphenoidal region, floor of the cranial cavity.

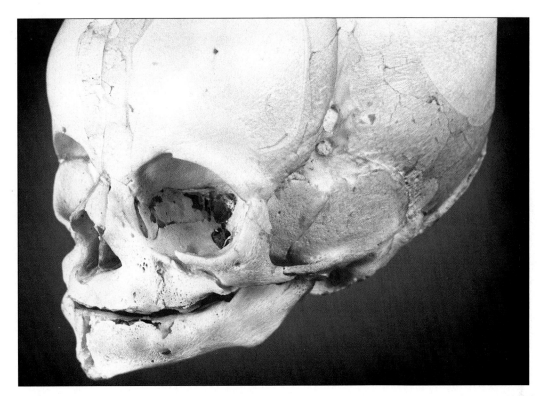

PLATE THIRTY-THREE
Skull of a fetus, left anterolateral view.

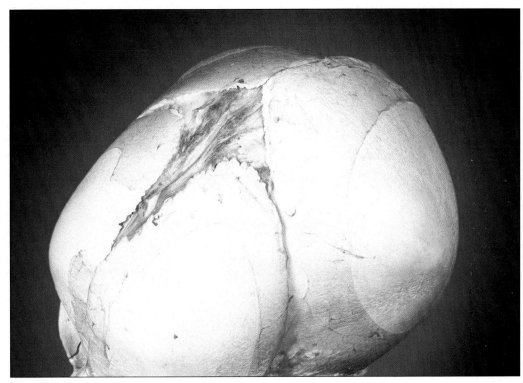

PLATE THIRTY-FOUR
Skull of a fetus, left superior view.

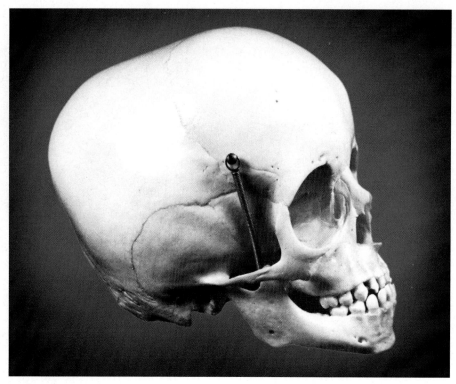

PLATE THIRTY-FIVE
Skull of a child, right lateral view.

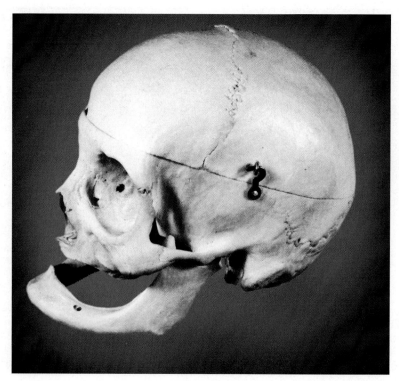

PLATE THIRTY-SIX
Skull of an aged person, left lateral view. (Note that this skull has been cut postmortem to allow the removal of the cranium.)

UNDERSTANDING WORDS

acetabul-, vinegar cup: *acetabul*um—
depression of the coxa that articulates
with the head of the femur.

annul-, ring: *annul*ar ligament—ring-shaped
band of connective tissue below the
elbow joint that encircles the head of
the radius.

arth-, joint: *arth*rology—study of joints and
ligaments.

burs-, bag, purse: prepatellar *burs*a—fluid-
filled sac between the skin and the
patella.

condyl-, knob: medial *condyl*e—rounded bony
process at the distal end of the femur.

fov-, pit: *fov*ea capitis—pit in the head of the
femur to which a ligament is attached.

glen-, joint socket: *glen*oid cavity—depression
in the scapula that articulates with the
head of the humerus.

labr-, lip: glenoidal *labr*um—rim of
fibrocartilage attached to the margin of
the glenoid cavity.

ov-, egglike: syn*ov*ial fluid—thick fluid within
a joint cavity that resembles egg white.

sutur-, sewing: *sutur*e—type of joint in which
flat bones are interlocked by a set of tiny
bony processes.

syndesm-, binding together: *syndesm*osis—
type of joint in which the bones are held
together by long fibers of connective
tissue.

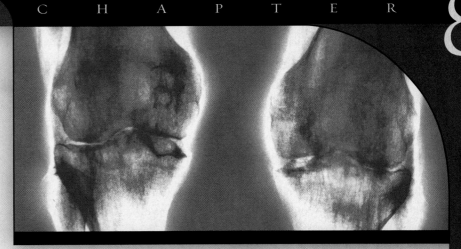

As shown in this falsely colored radiograph, rheumatoid arthritis has caused the
symmetrical inflammation and erosion of these knee joints. Drugs and, in severe
cases, joint replacements are used to treat this painful and debilitating condition.

Joints of the Skeletal System

CHAPTER OBJECTIVES

After you have studied this chapter, you should be able to

1. Explain how joints can be classified according to the type of tissue
that binds the bones together.

2. Describe how bones of fibrous joints are held together.

3. Describe how bones of cartilaginous joints are held together.

4. Describe the general structure of a synovial joint.

5. List six types of synovial joints and name an example of each type.

6. Explain how skeletal muscles produce movements at joints, and
identify several types of joint movements.

7. Describe the shoulder joint and explain how its articulating parts
are held together.

8. Describe the elbow, hip, and knee joints and explain how their
articulating parts are held together.

9. Describe the life-span changes in joints.

G out is a metabolic disorder in which lack of an enzyme blocks recycling of two of the four DNA nucleotides called purines. As a result, uric acid crystals accumulate in joints, causing great pain.

In humans, gout mostly affects the small joints in the foot, usually those of the great toes. For many years, gout was attributed solely to eating a great deal of red meat, which is rich in purines. Today, we know that while such a diet may exacerbate gout, a genetic abnormality causes the illness. Yet researchers recently discovered evidence that is consistent with the association of gout and eating red meat—signs of the condition in *Tyrannosaurus rex!*

An arthritis specialist and two paleontologists examined a cast of the right forearm of a dinosaur named Sue, a long-ago resident of the Hell Creek Formation in South Dakota, whose fossilized remains were found in 1990 jutting from the ground. Although telltale uric acid crystals had long since decomposed, X rays revealed patterns of bone erosion that could have resulted only from gout. The researchers examined only Sue's forearm, however, because she had been discovered on native American land and had been illegally traded by a fossil dealer. As a result of this dubious background, the Federal Bureau of Investigation had confiscated Sue. So the researchers examined bones from 83 other dinosaurs but found evidence of gout in only one other individual.

Sue had a hard life. Her facial bones and a lower limb bone were broken, and a tooth was found embedded in a rib, a legacy of an ancient battle. Whatever the reason for her injuries, Sue may have experienced the same kind of persistent pain that humans do. She is now on display at the Field Museum in Chicago. ■

Joints, or **articulations** (ar-tik′u-la″shunz), are functional junctions between bones. They bind parts of the skeletal system, make possible bone growth, permit parts of the skeleton to change shape during childbirth, and enable the body to move in response to skeletal muscle contractions.

Classification of Joints

Joints vary considerably in structure and function. However, they can be classified by the type of tissue that binds the bones at each junction. Three general groups are fibrous joints, cartilaginous joints, and synovial joints.

Joints can also be grouped according to the degree of movement possible at the bony junctions. In this scheme, joints are classified as immovable (synarthrotic), slightly movable (amphiarthrotic), and freely movable (diarthrotic). The structural and functional classification schemes overlap somewhat. Currently, structural classification is the one most commonly used.

Fibrous Joints

Fibrous (fi′brus) **joints** are so named because the dense connective tissue holding them together contains many collagenous fibers. They lie between bones that closely contact one another. The three types of fibrous joints are

1. **Syndesmosis** (sin″des-mo′sis). In this type of joint, the bones are bound by long fibers of connective tissue that form an *interosseous ligament.* Because this ligament is flexible and may be twisted, the joint may permit slight movement and thus is amphiarthrotic (am″fe-ar-thro′tik). A syndesmosis is at the distal ends of the tibia and fibula, where they join to form the tibiofibular articulation (fig. 8.1).

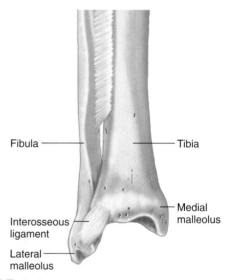

Fibula — Tibia

Medial malleolus

Interosseous ligament

Lateral malleolus

FIGURE 8.1
The articulation between the distal ends of the tibia and fibula is an example of a syndesmosis.

2. **Suture** (su′chur). Sutures are only between flat bones of the skull, where the broad margins of adjacent bones grow together and unite by a thin layer of dense connective tissue called a *sutural ligament.* Recall from chapter 7 (p. 208) that the infantile skull is incompletely developed, with several of the bones connected by membranous areas called *fontanels* (see fig. 7.33). These areas allow the skull to change shape slightly during childbirth, but as the bones continue to grow, the fontanels close, and sutures replace them. With time, some of the bones at sutures interlock by tiny bony processes. Such a suture is in the adult human skull where the parietal and occipital bones meet to form the lambdoidal suture. Because they are

immovable, sutures are synarthrotic (sin'ar-thro'tik) joints (figs. 8.2 and 8.3).

3. **Gomphosis** (gom-fo'sis). A gomphosis is a joint formed by the union of a cone-shaped bony process in a bony socket. The peglike root of a tooth fastened

to a jawbone by a *periodontal ligament* is such a joint. This ligament surrounds the root and firmly attaches it to the jaw with bundles of thick collagenous fibers. A gomphosis is a synarthrotic joint (fig. 8.4).

1 What is a joint?

2 How are joints classified?

3 Describe three types of fibrous joints.

4 What is the function of the fontanels?

Cartilaginous Joints

Hyaline cartilage or fibrocartilage connects the bones of **cartilaginous** (kar"tĭ-lah'jin-us) **joints.** The two types are

1. **Synchondrosis** (sin"kon-dro'sis). In a synchondrosis, bands of hyaline cartilage unite the bones. Many of these joints are temporary structures that disappear during growth. An example is an immature long bone where a band of hyaline cartilage (the epiphyseal plate) connects an epiphysis to a diaphysis. This cartilage band participates in bone lengthening and, in time, is replaced with bone. When ossification completes, usually before the age of twenty-five years, movement no longer occurs at the joint. Thus, the joint is synarthrotic (see fig. 7.11).

   Another synchondrosis occurs between the manubrium (sternum) and the first rib, which are directly united by costal cartilage (fig. 8.5). This joint is also synarthrotic, but permanent. The joints between the costal cartilages and the sternum of ribs 2 through 7 are usually synovial joints.

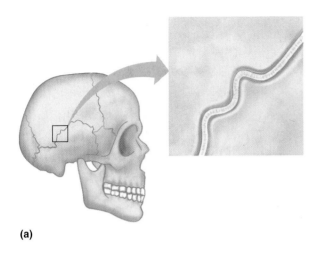

(a)

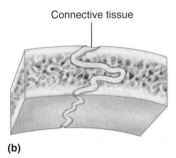

Connective tissue

(b)

FIGURE 8.2

Fibrous joints. (*a*) The fibrous joints between the bones of the skull are immovable and are called sutures. (*b*) A thin layer of connective tissue connects the bones at the suture.

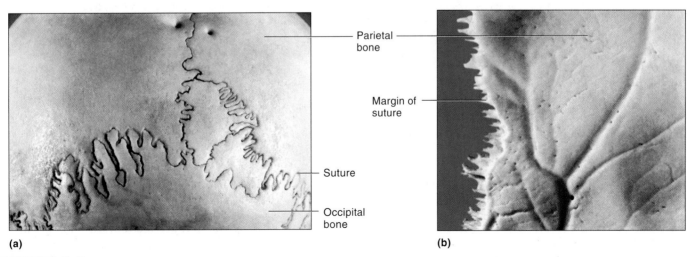

Parietal bone

Margin of suture

Suture

Occipital bone

(a)

(b)

FIGURE 8.3

Cranial sutures. (*a*) Sutures between the parietal and occipital bones of the skull. (*b*) The inner margin of a parietal suture. The grooves on the inside of this parietal bone mark the paths of blood vessels located near the brain's surface.

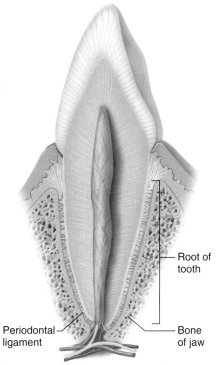

FIGURE 8.4

The articulation between the root of a tooth and the jawbone is a gomphosis.

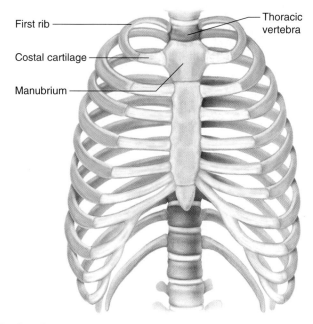

FIGURE 8.5

The articulation between the first rib and the manubrium is a synchondrosis.

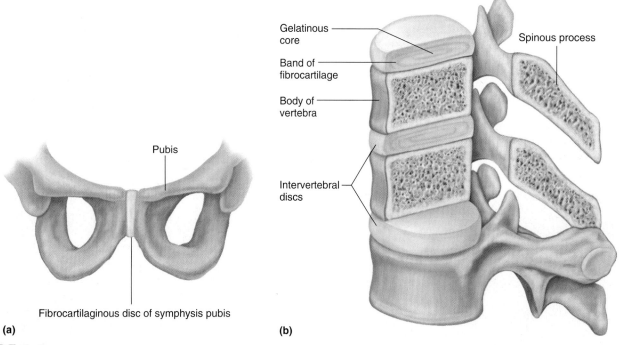

(a)

(b)

FIGURE 8.6

Fibrocartilage composes (a) the symphysis pubis that separates the pubic bones and (b) the intervertebral discs that separate vertebrae.

2. **Symphysis** (sim'fi-sis). The articular surfaces of the bones at a symphysis are covered by a thin layer of hyaline cartilage, and the cartilage, in turn, is attached to a pad of springy fibrocartilage. A limited amount of movement occurs at such a joint whenever forces compress or deform the cartilaginous pad. An example of this type of joint is the symphysis pubis in the pelvis, which allows maternal pelvic bones to shift as an infant passes through the birth canal (fig. 8.6a).

The joint formed by the bodies of two adjacent vertebrae separated by an intervertebral disc is also a symphysis (fig. 8.6b and reference plate 51). Each intervertebral disc is composed of a band of fibrocartilage (annulus fibrosus) that surrounds a gelatinous core (nucleus pulposus). The disc absorbs shocks and helps equalize pressure between the vertebrae when the body moves. Since each disc is slightly flexible, the combined movement of many of the joints in the vertebral column allows the back to bend forward or to the side or to twist. Because these joints allow slight movements, they are amphiarthrotic joints.

Synovial Joints

Most joints of the skeletal system are **synovial** (si-no've-al) **joints,** and because they allow free movement, they are diarthrotic (di"ar-thro'tik). These joints are more complex structurally than fibrous or cartilaginous joints. They consist of articular cartilage, a joint capsule, and a synovial membrane, which secretes synovial fluid.

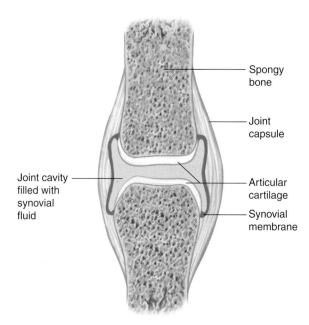

General Structure of a Synovial Joint

The articular ends of the bones in a synovial joint are covered with a thin layer of hyaline cartilage (fig. 8.7). This layer, which is called the **articular cartilage,** resists wear

FIGURE 8.7
The generalized structure of a synovial joint.

and minimizes friction when it is compressed as the joint moves.

Typically, the bone beneath articular cartilage (subchondral plate) contains cancellous bone, which is somewhat elastic (fig. 8.7). This plate absorbs shocks, helping protect the joint from stresses caused by the load of body weight and by forces produced by contracting muscles.

Excessive mechanical stress due to obesity or certain athletic activities may fracture a subchondral plate. Although such fractures usually heal, the bone that regenerates may be less elastic than the original, reducing its protective function.

A tubular **joint capsule** (articular capsule) that has two distinct layers holds together the bones of a synovial joint. The outer layer largely consists of dense connective tissue, whose fibers attach to the periosteum around the circumference of each bone of the joint near its articular end. Thus, the outer fibrous layer of the capsule completely encloses the other parts of the joint. It is, however, flexible enough to permit movement and strong enough to help prevent the articular surfaces from being pulled apart.

Bundles of strong, tough collagenous fibers called **ligaments** (lig'ah-mentz) reinforce the joint capsule and help bind the articular ends of the bones. Some ligaments appear as thickenings in the fibrous layer of the capsule, whereas others are *accessory structures* located outside the capsule. In either case, these structures help prevent excessive movement at the joint. That is, the ligament is relatively inelastic, and it tightens when the joint is stressed.

The inner layer of the joint capsule consists of a shiny, vascular lining of loose connective tissue called the

synovial membrane. This membrane, which is only a few cells thick, covers all of the surfaces within the joint capsule, except the areas the articular cartilage covers. The synovial membrane surrounds a closed sac called the *synovial cavity,* into which the membrane secretes a clear, viscous fluid called **synovial fluid.** In some regions, the surface of the synovial membrane has villi as well as larger folds and projections that extend into the cavity. Besides filling spaces and irregularities of the joint cavity, these extensions increase the surface area of the synovial membrane. The membrane may also store adipose tissue and form movable fatty pads within the joint. This multifunctional membrane also reabsorbs fluid, which is important when a joint cavity is injured or infected.

Synovial fluid has a consistency similar to uncooked egg white, and it moistens and lubricates the smooth cartilaginous surfaces within the joint. It also helps supply articular cartilage with nutrients that are obtained from blood vessels of the synovial membrane. The volume of synovial fluid in a joint cavity is usually just enough to cover the articulating surfaces with a thin film of fluid. The amount of synovial fluid in the cavity of the knee is 0.5 mL or less.

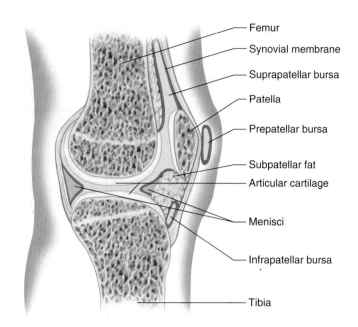

FIGURE 8.8
Menisci separate the articulating surfaces of the femur and tibia. Several bursae are associated with the knee joint.

A physician can determine the cause of joint inflammation or degeneration (arthritis) by aspirating a sample of synovial fluid from the affected joint using a procedure called arthrocentesis. Bloody fluid with lipid on top indicates a fracture extending into the joint. Clear fluid is found in osteoarthritis, a degeneration of collagen in the joint that is inherited or degenerative. Cloudy, yellowish fluid may indicate the autoimmune disorder rheumatoid arthritis, and crystals in the synovial fluid signal gout. If the fluid is cloudy but red-tinged and containing pus, a bacterial infection may be present that requires prompt treatment. Normal synovial fluid has 180 or fewer leukocytes (white blood cells) per cubic mm. If the fluid is infected, the leukocyte count exceeds 2,000.

Some synovial joints are partially or completely divided into two compartments by discs of fibrocartilage called **menisci** (me-nis′ke) (sing., *meniscus*) located between the articular surfaces. Each meniscus attaches to the fibrous layer of the joint capsule peripherally, and its free surface projects into the joint cavity. In the knee joint, crescent-shaped menisci cushion the articulating surfaces and help distribute body weight onto these surfaces (fig. 8.8).

Certain synovial joints are also associated with fluid-filled sacs called **bursae** (ber′se). Each bursa has an inner lining of synovial membrane, which may be continuous with the synovial membrane of a nearby joint cavity.

These sacs contain synovial fluid and are commonly located between the skin and underlying bony prominences, as in the case of the patella of the knee or the olecranon process of the elbow. Bursae cushion and aid the movement of tendons that glide over bony parts or over other tendons. The names of bursae indicate their locations. Figure 8.8 shows a *suprapatellar bursa,* a *prepatellar bursa,* and an *infrapatellar bursa.*

1 Describe two types of cartilaginous joints.

2 What is the function of an intervertebral disc?

3 Describe the structure of a synovial joint.

4 What is the function of the synovial fluid?

Articular cartilage, like other cartilaginous structures, lacks a direct blood supply (see chapter 5, p. 146). Surrounding synovial fluid supplies oxygen, nutrients, and other vital chemicals. Normal body movements force these substances into the joint cartilage. When a joint is immobilized or is not used for a long time, inactivity may cause degeneration of the articular cartilage. The degeneration may reverse when joint movements resume. However, it is important to avoid exercises that greatly compress the tissue during the period of regeneration. Otherwise, chondrocytes in the thinned cartilage may be injured, hindering repair.

Types of Synovial Joints

The articulating bones of synovial joints have a variety of shapes that allow different kinds of movement. Based upon their shapes and the movements they permit, these joints can be classified into six major types—ball-and-socket joints, condyloid joints, gliding joints, hinge joints, pivot joints, and saddle joints.

1. A **ball-and-socket joint** consists of a bone with a globular or slightly egg-shaped head that articulates with the cup-shaped cavity of another bone. Such a joint allows a wider range of motion than does any other kind, permitting movements in all planes, as well as rotational movement around a central axis. The hip and shoulder contain joints of this type (fig. 8.9a).

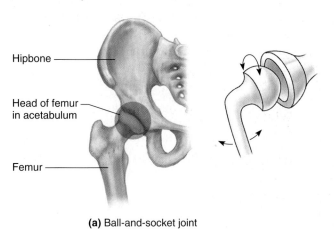

(a) Ball-and-socket joint

Hipbone
Head of femur in acetabulum
Femur

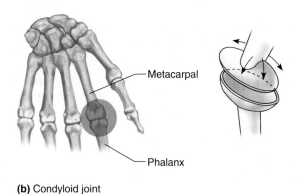

(b) Condyloid joint

Metacarpal
Phalanx

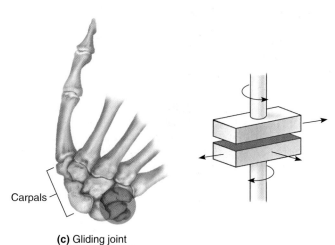

(c) Gliding joint

Carpals

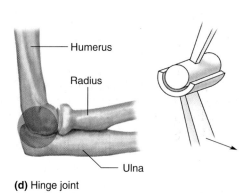

(d) Hinge joint

Humerus
Radius
Ulna

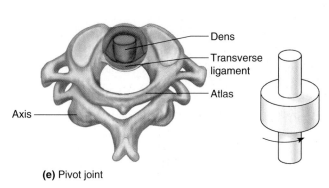

(e) Pivot joint

Dens
Transverse ligament
Atlas
Axis

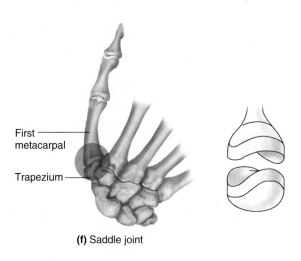

(f) Saddle joint

First metacarpal
Trapezium

FIGURE 8.9
Types and examples of synovial (freely movable) joints.

2. In a **condyloid joint,** the ovoid condyle of one bone fits into the elliptical cavity of another bone, as in the joints between the metacarpals and phalanges. This type of joint permits a variety of movements in different planes; rotational movement, however, is not possible (fig. 8.9*b*).

3. The articulating surfaces of **gliding joints** are nearly flat or slightly curved. These joints allow sliding or back-and-forth motion and twisting movements. Most of the joints within the wrist and ankle, as well as those between the articular processes of adjacent vertebrae, belong to this group (fig. 8.9*c*). The sacroiliac joints and the joints formed by ribs 2 through 7 connecting with the sternum are also gliding joints.

4. In a **hinge joint,** the convex surface of one bone fits into the concave surface of another, as in the elbow and the joints of the phalanges. Such a joint resembles the hinge of a door in that it permits movement in one plane only (fig. 8.9*d*).

5. In a **pivot joint,** the cylindrical surface of one bone rotates within a ring formed of bone and fibrous tissue of a ligament. Movement at such a joint is limited to rotation around a central axis. The joint between the proximal ends of the radius and the ulna, where the head of the radius rotates in a ring formed by the radial notch of the ulna and a ligament (annular ligament), is of this type. Similarly, a pivot joint functions in the neck as the head turns from side to side. In this case, the ring formed by a ligament (transverse ligament) and the anterior arch of the atlas rotates around the dens of the axis (fig. 8.9*e*).

6. A **saddle joint** forms between bones whose articulating surfaces have both concave and convex regions. The surface of one bone fits the complementary surface of the other. This physical relationship permits a variety of movements, mainly in two planes, as in the case of the joint between the carpal (trapezium) and the metacarpal of the thumb (fig. 8.9*f*).

Table 8.1 summarizes the types of joints.

TABLE 8.1	Types of Joints		
Type of Joint	**Description**	**Possible Movements**	**Example**
Fibrous	Articulating bones fastened together by thin layer of dense connective tissue containing many collagenous fibers		
1. *Syndesmosis* (amphiarthrotic)	Bones bound by interosseous ligament	Joint flexible and may be twisted	Tibiofibular articulation
2. *Suture* (synarthrotic)	Flat bones united by sutural ligament	None	Parietal bones articulate at sagittal suture of skull
3. *Gomphosis* (synarthrotic)	Cone-shaped process fastened in bony socket by periodontal ligament	None	Root of tooth united with mandible
Cartilaginous	Articulating bones connected by hyaline cartilage or fibrocartilage		
1. *Synchondrosis* (synarthrotic)	Bones united by bands of hyaline cartilage	Movement occurs during growth process until ossification occurs	Joint between epiphysis and diaphysis of a long bone
2. *Symphysis* (amphiarthrotic)	Articular surfaces separated by thin layers of hyaline cartilage attached to band of fibrocartilage	Limited movement, as when back is bent or twisted	Joints between bodies of vertebrae
Synovial (diarthrotic)	Articulating bones surrounded by a joint capsule of ligaments and synovial membranes; ends of articulating bones covered by hyaline cartilage and separated by synovial fluid		
1. *Ball-and-socket*	Ball-shaped head of one bone articulates with cup-shaped socket of another	Movements in all planes; rotation	Shoulder, hip
2. *Condyloid*	Oval-shaped condyle of one bone articulates with elliptical cavity of another	Variety of movements in different planes, but no rotation	Joints between metacarpals and phalanges
3. *Gliding*	Articulating surfaces are nearly flat or slightly curved	Sliding or twisting	Joints between various bones of wrist and ankle
4. *Hinge*	Convex surface of one bone articulates with concave surface of another	Flexion and extension	Elbow and joints of phalanges
5. *Pivot*	Cylindrical surface of one bone articulates with ring of bone and fibrous tissue	Rotation	Joint between proximal ends of radius and ulna
6. *Saddle*	Articulating surfaces have both concave and convex regions; surface of one bone fits the complementary surface of another	Variety of movements, mainly in two planes	Joint between carpal and metacarpal of thumb

Types of Joint Movements

Skeletal muscle action produces movements at synovial joints. Typically, one end of a muscle is attached to a relatively immovable or fixed part on one side of a joint, and the other end of the muscle is fastened to a movable part on the other side. When the muscle contracts, its fibers pull its movable end (*insertion*) toward its fixed end (*origin*), and a movement occurs at the joint.

The following terms describe movements at joints that occur in different directions and in different planes (figs. 8.10, 8.11, and 8.12):

flexion (flek′shun) Bending parts at a joint so that the angle between them decreases and the parts come closer together (bending the lower limb at the knee).

extension (ek-sten′shun) Straightening parts at a joint so that the angle between them increases and the parts move farther apart (straightening the lower limb at the knee).

hyperextension (hi″per-ek-sten′shun) Excess extension of the parts at a joint, beyond the anatomical position (bending the head back beyond the upright position).

dorsiflexion (dor″sĭ-flek′shun) Bending the foot at the ankle toward the shin (bending the foot upward).

plantar flexion (plan′tar flek′shun) Bending the foot at the ankle toward the sole (bending the foot downward).

abduction (ab-duk′shun) Moving a part away from the midline (lifting the upper limb horizontally to form a right angle with the side of the body).

adduction (ah-duk′shun) Moving a part toward the midline (returning the upper limb from the horizontal position to the side of the body).

rotation (ro-ta′shun) Moving a part around an axis (twisting the head from side to side). Medial rotation involves movement toward the midline, whereas lateral rotation involves movement in the opposite direction.

circumduction (ser″kum-duk′shun) Moving a part so that its end follows a circular path (moving the finger in a circular motion without moving the hand).

supination (soo″pĭ-na′shun) Turning the hand so the palm is upward or facing anteriorly (in anatomical position).

pronation (pro-na′shun) Turning the hand so the palm is downward or facing posteriorly (in anatomical position).

eversion (e-ver′zhun) Turning the foot so the sole faces laterally.

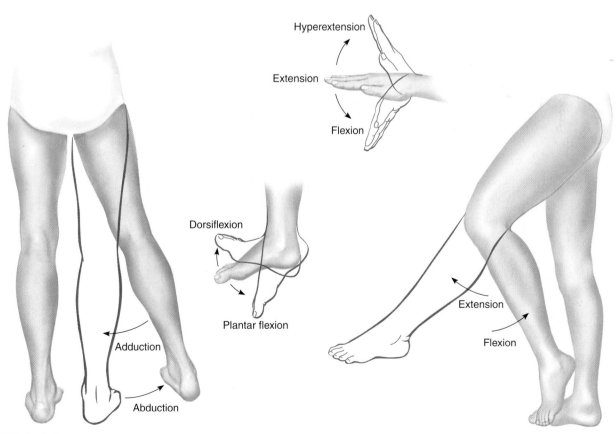

FIGURE 8.10

Joint movements illustrating adduction, abduction, dorsiflexion, plantar flexion, hyperextension, extension, and flexion.

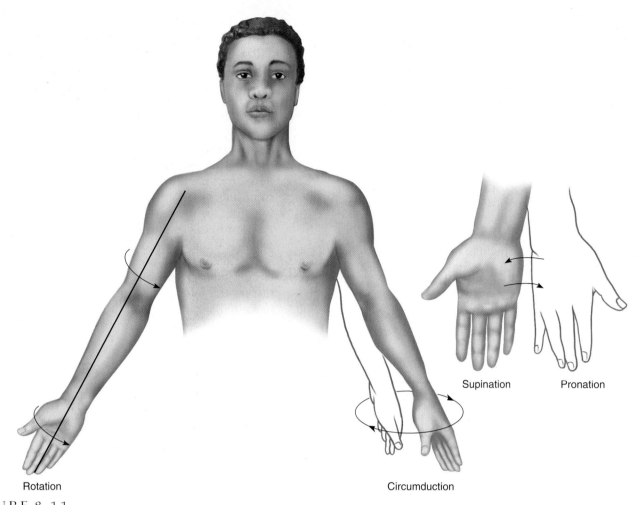

Supination   Pronation

Rotation   Circumduction

FIGURE 8.11

Joint movements illustrating rotation, circumduction, pronation, and supination.

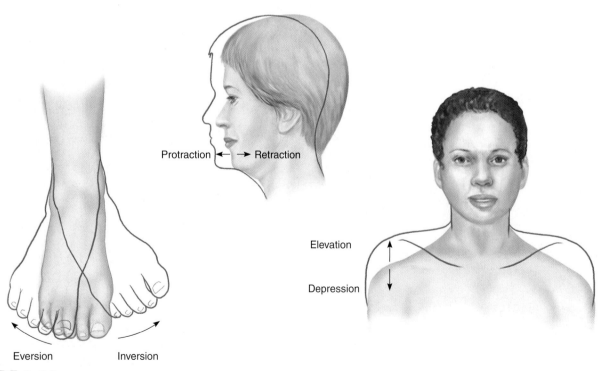

Protraction ← → Retraction

Elevation

Depression

Eversion   Inversion

FIGURE 8.12

Joint movements illustrating eversion, inversion, retraction, protraction, elevation, and depression.

inversion (in-ver′zhun) Turning the foot so the sole faces medially.

protraction (pro-trak′shun) Moving a part forward (thrusting the chin forward).

retraction (re-trak′shun) Moving a part backward (pulling the chin backward).

elevation (el″ĕ-va′shun) Raising a part (shrugging the shoulders).

depression (de-presh′un) Lowering a part (drooping the shoulders).

Where movements of body parts are part of the definition, we will simply describe movements of those parts, for example, adduction of the lower limb or rotation of the head. Special cases also fall herein, as with plantar flexion of the foot. Other movements are described by the change in geometry at a joint, such as the action of the biceps brachii, flexion at the elbow. Here we will go with the more descriptive "flexion of the forearm at the elbow." Table 8.2 lists information on several joints.

TABLE 8.2	Joints of the Body		
Joint	**Location**	**Type of Joint**	**Type of Movement**
Skull	Cranial and facial bones	Suture, fibrous	Immovable, synarthrotic
Temporomandibular	Temporal bone, mandible	Modified hinge, synovial	Elevation, depression, protraction, retraction, diarthrotic
Atlanto-occipital	Atlas, occipital bone	Condyloid, synovial	Flexion, extension, diarthrotic
Atlantoaxial	Atlas, axis	Pivot, synovial	Rotation
Intervertebral	Between vertebral bodies	Symphysis, cartilaginous	Slight movement, amphiarthrotic
Intervertebral	Between articular processes	Gliding, synovial	Flexion, extension, slight rotation, diarthrotic
Sacroiliac	Sacrum and hipbone	Gliding, synovial	Little to no movement, diarthrotic
Vertebrocostal	Vertebrae and ribs	Gliding, synovial	Slight movement during breathing, diarthrotic
Sternoclavicular	Sternum and clavicle	Gliding, synovial	Slight movement when shrugging shoulders, diarthrotic
Sternocostal	Sternum and rib 1	Synchondrosis, cartilaginous	Immovable, synarthrotic
Sternocostal	Sternum and ribs 2–7	Gliding, synovial	Slight movement during breathing, diarthrotic
Acromioclavicular	Scapula and clavicle	Gliding, synovial	Protraction, retraction, elevation, depression, diarthrotic
Shoulder (glenohumeral)	Humerus and scapula	Ball-and-socket, synovial	Flexion, extension, adduction, abduction, rotation, circumduction, diarthrotic
Elbow	Humerus and ulna	Hinge, synovial	Flexion, extension, diarthrotic
Proximal radioulnar	Radius and ulna	Pivot, synovial	Rotation, diarthrotic
Distal radioulnar	Radius and ulna	Syndesmosis, fibrous	Slight movement, amphiarthrotic
Wrist (radiocarpal)	Radius and carpals	Condyloid, synovial	Flexion, extension, adduction, abduction, circumduction, diarthrotic
Intercarpal	Adjacent carpals	Gliding, synovial	Slight movement, diarthrotic
Carpometacarpal	Carpal and metacarpal 1	Saddle, synovial	Flexion, extension, adduction, abduction, diarthrotic
Carpometacarpal	Carpals and metacarpals 2–5	Condyloid, synovial	Flexion, extension, adduction, abduction, diarthrotic
Metacarpophalangeal	Metacarpal and proximal phalanx	Condyloid, synovial	Flexion, extension, adduction, abduction, diarthrotic
Interphalangeal	Adjacent phalanges	Hinge, synovial	Flexion, extension, diarthrotic
Symphysis pubis	Pubic bones	Symphysis, cartilaginous	Slight movement, amphiarthrotic
Hip	Hipbone and femur	Ball-and-socket, synovial	Flexion, extension, adduction, abduction, rotation, circumduction, diarthrotic
Knee (tibiofemoral)	Femur and tibia	Modified hinge, synovial	Flexion, extension, slight rotation when flexed, diarthrotic
Knee (femoropatellar)	Femur and patella	Gliding, synovial	Slight movement, diarthrotic
Proximal tibiofibular	Tibia and fibula	Gliding, synovial	Slight movement, diarthrotic
Distal tibiofibular	Tibia and fibula	Syndesmosis, fibrous	Slight movement, amphiarthrotic
Ankle (talocrural)	Talus, tibia, and fibula	Hinge, synovial	Dorsiflexion, plantar flexion, slight circumduction, diarthrotic
Intertarsal	Adjacent tarsals	Gliding, synovial	Inversion, eversion, diarthrotic
Tarsometatarsal	Tarsals and metatarsals	Gliding, synovial	Slight movement, diarthrotic
Metatarsophalangeal	Metatarsal and proximal phalanx	Condyloid, synovial	Flexion, extension, adduction, abduction, diarthrotic

Examples of Synovial Joints

The shoulder, elbow, hip, and knee are large, freely movable joints. Although these joints have much in common, each has a unique structure that makes possible its specific function.

Shoulder Joint

The **shoulder joint** is a ball-and-socket joint that consists of the rounded head of the humerus and the shallow glenoid cavity of the scapula. The coracoid and acromion processes of the scapula protect these parts, and dense connective tissue and muscle hold them together.

The joint capsule of the shoulder is attached along the circumference of the glenoid cavity and the anatomical neck of the humerus. Although it completely envelopes the joint, the capsule is very loose, and by itself is unable to keep the bones of the joint in close contact. However, muscles and tendons surround and reinforce the capsule, keeping together the articulating parts of the shoulder (fig. 8.13).

The tendons of several muscles intimately blend with the fibrous layer of the shoulder joint capsule, forming the *rotator cuff*, which reinforces and supports the shoulder joint. Throwing a ball can create powerful decelerating forces that injure the rotator cuff.

The ligaments that help prevent displacement of the articulating surfaces of the shoulder joint include the following (fig. 8.14):

1. **Coracohumeral** (kor″ah-ko-hu′mer-al) **ligament.** This ligament is composed of a broad band of connective tissue that connects the coracoid process of the scapula to the greater tubercle of the humerus. It strengthens the superior portion of the joint capsule.

2. **Glenohumeral** (gle″no-hu′mer-al) **ligaments.** These include three bands of fibers that appear as thickenings in the ventral wall of the joint capsule. They extend from the edge of the glenoid cavity to the lesser tubercle and the anatomical neck of the humerus.

3. **Transverse humeral ligament.** This ligament consists of a narrow sheet of connective tissue fibers

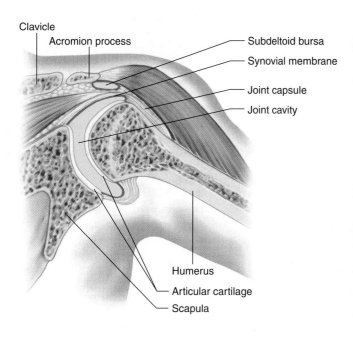

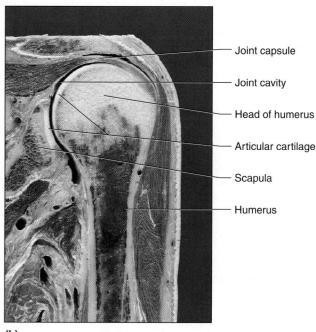

(a)

(b)

FIGURE 8.13

Shoulder joint. (*a*) The shoulder joint allows movements in all directions. Note that a bursa is associated with this joint. (*b*) Photograph of the shoulder joint (coronal section).

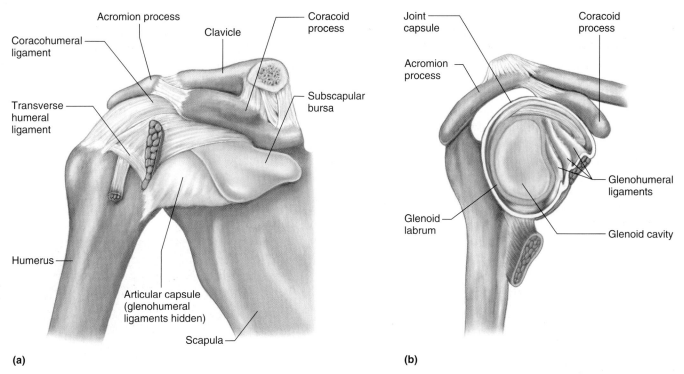

Acromion process

Clavicle

Coracoid process

Coracohumeral ligament

Transverse humeral ligament

Subscapular bursa

Humerus

Articular capsule (glenohumeral ligaments hidden)

Scapula

(a)

Joint capsule

Coracoid process

Acromion process

Glenohumeral ligaments

Glenoid labrum

Glenoid cavity

(b)

FIGURE 8.14

Ligaments associated with the shoulder joint. (*a*) Ligaments hold together the articulating surfaces of the shoulder. (*b*) The glenoid labrum is a ligament composed of fibrocartilage.

that runs between the lesser and the greater tubercles of the humerus. Together with the intertubercular groove of the humerus, the ligament forms a canal (retinaculum) through which the long head of the biceps brachii muscle passes.

4. **Glenoid labrum** (gle′noid la′brum). This structure is composed of fibrocartilage. It is attached along the margin of the glenoid cavity and forms a rim with a thin, free edge that deepens the cavity.

Several bursae are associated with the shoulder joint. The major ones include the *subscapular bursa* located between the joint capsule and the tendon of the subscapularis muscle, the *subdeltoid bursa* between the joint capsule and the deep surface of the deltoid muscle, the *subacromial bursa* between the joint capsule and the undersurface of the acromion process of the scapula, and the *subcoracoid bursa* between the joint capsule and the coracoid process of the scapula. Of these, the subscapular bursa is usually continuous with the synovial cavity of the joint cavity, and although the others do not communicate with the joint cavity, they may be connected to each other (see figs. 8.13 and 8.14).

The shoulder joint is capable of a very wide range of movement, due to the looseness of its attachments and the relatively large articular surface of the humerus compared to the shallow depth of the glenoid cavity. These movements include flexion, extension, abduction, adduction,

rotation, and circumduction. Motion occurring simultaneously in the joint formed between the scapula and the clavicle may also aid such movements.

Because the bones of the shoulder joint are mainly held together by supporting muscles rather than by bony structures and strong ligaments, the joint is somewhat weak. Consequently, the articulating surfaces may become displaced or dislocated easily. Such a *dislocation* most commonly occurs with a forceful impact during abduction, as when a person falls on an outstretched arm. This movement may press the head of the humerus against the lower part of the joint capsule where its wall is thin and poorly supported by ligaments. Dislocations commonly affect joints of the shoulders, knees, fingers, and jaw.

Elbow Joint

The **elbow joint** is a complex structure that includes two articulations—a hinge joint between the trochlea of the humerus and the trochlear notch of the ulna and a gliding joint between the capitulum of the humerus and a shallow depression (fovea) on the head of the radius. A joint capsule completely encloses and holds together these unions (fig. 8.15). Ulnar and radial collateral ligaments thicken the two joints, and fibers from a muscle (brachialis) in the arm reinforce its anterior surface.

The **ulnar collateral ligament,** which is a thick band of dense connective tissue, is located in the medial wall of the capsule. The anterior portion of this ligament connects the medial epicondyle of the humerus to the medial margin of the coronoid process of the ulna. Its posterior part is attached to the medial epicondyle of the humerus and to the olecranon process of the ulna (fig. 8.16a).

The **radial collateral ligament,** which strengthens the lateral wall of the joint capsule, is a fibrous band extending between the lateral epicondyle of the humerus and the *annular ligament of the radius.* The annular ligament, in turn, attaches to the margin of the trochlear notch of the ulna, and it encircles the head of the radius, keeping the head in contact with the radial notch of the ulna. The elbow joint capsule encloses the resulting radioulnar joint so that its function is closely associated with the elbow (fig. 8.16b).

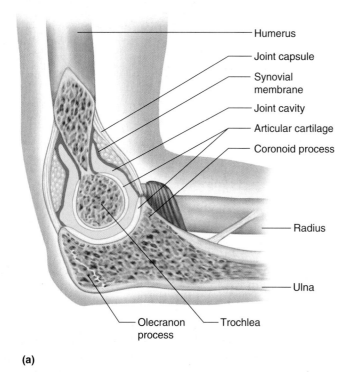

(a)

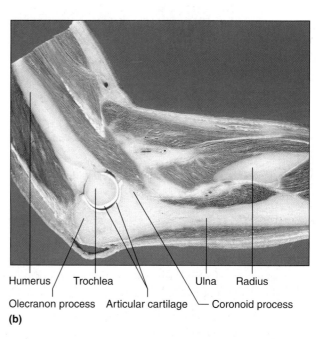

(b)

FIGURE 8.15

Elbow joint. (*a*) The elbow joint allows hinge movements, as well as pronation and supination of the hand. (*b*) Photograph of the elbow joint (sagittal section).

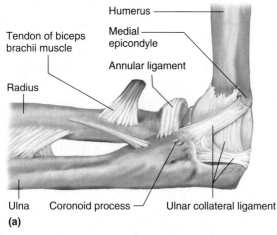

(a)

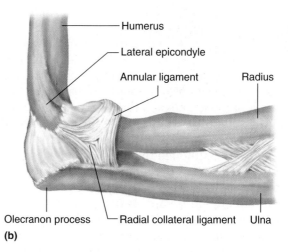

(b)

FIGURE 8.16

Ligaments associated with the elbow joint. (*a*) The ulnar collateral ligament, medial view, and (*b*) the radial collateral ligament strengthen the capsular wall of the elbow joint, lateral view.

The *synovial membrane* that forms the inner lining of the elbow capsule projects into the joint cavity between the radius and ulna and partially divides the joint into humerus–ulnar and humerus–radial portions. Also, varying amounts of adipose tissue form fatty pads between the synovial membrane and the fibrous layer of the joint capsule. These pads help protect nonarticular bony areas during joint movements.

The only movements that can occur at the elbow between the humerus and ulna are hinge-type movements—flexion and extension. The head of the radius, however, is free to rotate in the annular ligament. This movement allows pronation and supination of the hand.

1 Which parts help keep together the articulating surfaces of the shoulder joint?

2 What factors allow an especially wide range of motion in the shoulder?

3 Which structures form the hinge joint of the elbow?

4 Which parts of the elbow permit pronation and supination of the hand?

Arthroscopy enables a surgeon to visualize the interior of a joint and even perform diagnostic or therapeutic procedures, guided by the image on a video screen. An arthroscope is a thin, tubular instrument about 25 cm long containing optical fibers that transmit an image. The surgeon inserts the device through a small incision in the joint capsule. It is far less invasive than conventional surgery. Many runners have undergone uncomplicated arthroscopy and raced just weeks later.

Arthroscopy is combined with a genetic technique called the polymerase chain reaction (PCR) to rapidly diagnose infection. Guided by an arthroscope, the surgeon samples a small piece of the synovial membrane. PCR detects and amplifies specific DNA sequences, such as those of bacteria. For example, the technique can rapidly diagnose Lyme disease by detecting DNA from the causative bacterium *Borrelia burgdorferi.* This is valuable because a variety of bacteria can infect joints, and choosing the appropriate antibiotic, based on knowing the type of bacterium, is crucial for fast and complete recovery.

Hip Joint

The **hip joint** is a ball-and-socket joint that consists of the head of the femur and the cup-shaped acetabulum of the coxa. A ligament (ligamentum capitis) attaches to a pit (fovea capitis) on the head of the femur and to connective tissue within the acetabulum. This attachment, however, seems to have little importance in holding the articulating bones together, but rather carries blood vessels to the head of the femur (fig. 8.17).

(a)

(b)

FIGURE 8.17

Hip joint. (*a*) The acetabulum provides the socket for the head of the femur in the hip joint. (*b*) The pit (fovea capitis) in the femur's head marks attachment of a ligament that surrounds blood vessels and nerves.

A horseshoe-shaped ring of fibrocartilage (acetabular labrum) at the rim of the acetabulum deepens the cavity of the acetabulum. It encloses the head of the femur and helps hold it securely in place. In addition, a heavy, cylindrical joint capsule that is reinforced with still other ligaments surrounds the articulating structures and connects the neck of the femur to the margin of the acetabulum (fig. 8.18).

The major ligaments of the hip joint include the following (fig. 8.19):

1. **Iliofemoral** (il″e-o-fem′o-ral) **ligament**. This ligament consists of a Y-shaped band of very strong fibers that connects the anterior inferior iliac spine of the coxa to a bony line (intertrochanteric line) extending between the greater and lesser trochanters of the femur. The iliofemoral ligament is the strongest ligament in the body.

2. **Pubofemoral** (pu″bo-fem′o-ral) **ligament.** The pubofemoral ligament extends between the superior

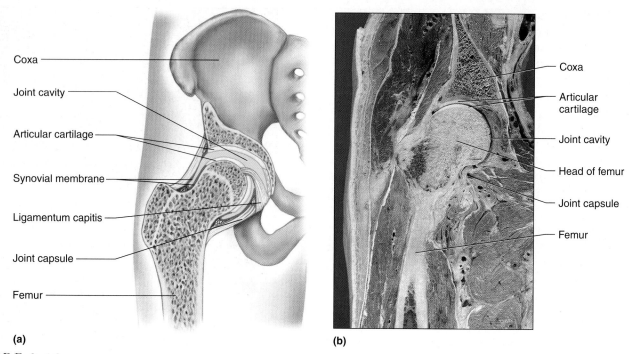

FIGURE 8.18

Hip joint. (*a*) A ring of cartilage in the acetabulum and a ligament-reinforced joint capsule hold together the hip joint. (*b*) Photograph of the hip joint (coronal section).

Labels in (a): Coxa, Joint cavity, Articular cartilage, Synovial membrane, Ligamentum capitis, Joint capsule, Femur

Labels in (b): Coxa, Articular cartilage, Joint cavity, Head of femur, Joint capsule, Femur

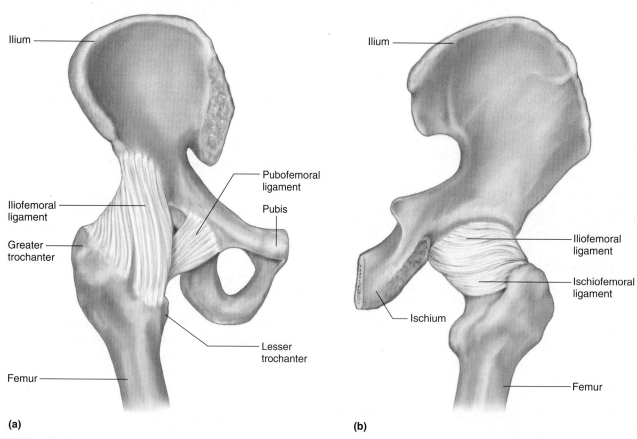

Labels in (a): Ilium, Iliofemoral ligament, Greater trochanter, Femur, Pubofemoral ligament, Pubis, Lesser trochanter

Labels in (b): Ilium, Ischium, Iliofemoral ligament, Ischiofemoral ligament, Femur

FIGURE 8.19

The major ligaments of the right hip joint. (*a*) Anterior view. (*b*) Posterior view.

REPLACING JOINTS

Surgeons use several synthetic materials to replace joints that are severely damaged by arthritis or injury. Metals such as cobalt, chromium, and titanium alloys are used to replace larger joints, whereas silicone polymers are more commonly used to replace smaller joints. Such artificial joints must be durable yet not provoke immune system rejection. They must also allow normal healing to occur and not move surrounding structures out of their normal positions.

Before the advent of joint replacements, surgeons removed damaged or diseased joint surfaces, hoping that scar tissue filling in the area would restore mobility. This type of surgery was rarely successful. In the 1950s, Alfred Swanson, an army surgeon in Grand Rapids, Michigan, invented the first joint implants using silicone polymers. By 1969, after much refinement, the first silicone-based joint implants became available. These devices provided flexible hinges for joints of the toes, fingers, and wrists. Since then, more than two dozen joint replacement models have been developed, and more than a million people have them, mostly in the hip.

A surgeon inserts a joint implant in a procedure called implant resection arthroplasty. The surgeon first removes the surface of the joint bones and excess cartilage. Next, the centers of the tips of abutting bones are hollowed out, and the stems of the implant are inserted here. The hinge part of the implant lies between the bones, aligning them yet allowing them to bend, as they would at a natural joint. Bone cement fixes the implant in place. Finally, the surgeon repairs the tendons, muscles, and ligaments. As the site of the implant heals, the patient must exercise the joint. A year of physical therapy may be necessary to fully benefit from replacement joints.

Newer joint replacements use materials that resemble natural body chemicals. Hip implants, for example, may bear a coat of hydroxyapatite, which interacts with natural bone. Instead of filling in spaces with bone cement, some investigators are testing a variety of porous coatings that allow bone tissue to grow into the implant area. ■

portion of the pubis and the iliofemoral ligament. Its fibers also blend with the fibers of the joint capsule.

3. **Ischiofemoral** (is″ke-o-fem′o-ral) **ligament.** This ligament consists of a band of strong fibers that originates on the ischium just posterior to the acetabulum and blends with the fibers of the joint capsule.

Muscles surround the joint capsule of the hip. The articulating parts of the hip are held more closely together than those of the shoulder, allowing considerably less freedom of movement. The structure of the hip joint, however, still permits a wide variety of movements, including extension, flexion, abduction, adduction, rotation, and circumduction. The hip is one of the joints most frequently replaced (Clinical Application 8.1).

Knee Joint

The **knee joint** is the largest and most complex of the synovial joints. It consists of the medial and lateral condyles at the distal end of the femur and the medial and lateral condyles at the proximal end of the tibia. In addition, the femur articulates anteriorly with the patella. Although the knee functions largely as a modified hinge joint (allowing flexion and extension), the articulations between the femur and tibia are condyloid (allowing some rotation when the knee is flexed), and the joint between the femur and patella is a gliding joint.

The *joint capsule* of the knee is relatively thin, but ligaments and the tendons of several muscles greatly strengthen it. For example, the fused tendons of several muscles in the thigh cover the capsule anteriorly. Fibers from these tendons descend to the patella, partially enclose it, and continue downward to the tibia. The capsule attaches to the margins of the femoral and tibial condyles as well as between these condyles (fig. 8.20).

The ligaments associated with the joint capsule that help keep the articulating surfaces of the knee joint in contact include the following (fig. 8.21):

1. **Patellar** (pah-tel′ar) **ligament.** This ligament is a continuation of a tendon from a large muscle group in the thigh (quadriceps femoris). It consists of a strong, flat band that extends from the margin of the patella to the tibial tuberosity.

2. **Oblique popliteal** (ŏ′blēk pop-lit′e-al) **ligament.** This ligament connects the lateral condyle of the femur to the margin of the head of the tibia.

3. **Arcuate** (ar′ku-āt) **popliteal ligament.** This ligament appears as a Y-shaped system of fibers that extends from the lateral condyle of the femur to the head of the fibula.

4. **Tibial collateral** (tib′e-al kŏ-lat′er-al) **ligament** (medial collateral ligament). This ligament is a broad, flat band of tissue that connects the medial condyle of the femur to the medial condyle of the tibia.

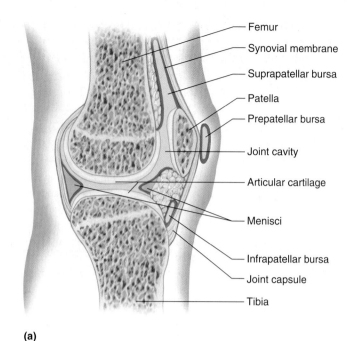

(a)

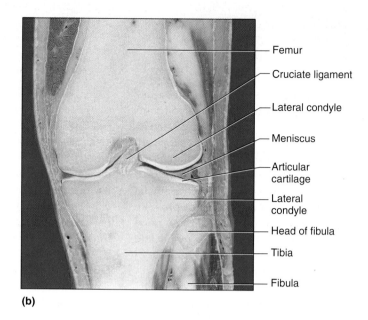

(b)

FIGURE 8.20

Knee joint. (a) The knee joint is the most complex of the synovial joints (sagittal section). (b) Photograph of the knee joint (coronal section).

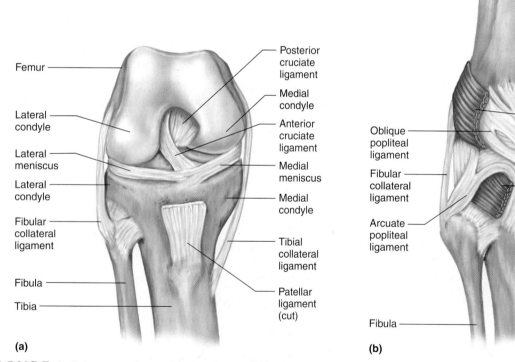

(a)

(b)

FIGURE 8.21

Ligaments within the knee joint help to strengthen it. (a) Anterior view of right knee (patella removed). (b) Posterior view of left knee.

5. **Fibular** (fib'u-lar) **collateral ligament** (lateral collateral ligament). This ligament consists of a strong, round cord located between the lateral condyle of the femur and the head of the fibula.

In addition to the ligaments that strengthen the joint capsule, two ligaments within the joint, called **cruciate** (kroo'she-āt) **ligaments,** help prevent displacement of the articulating surfaces. These strong bands of fibrous tissue stretch upward between the tibia and the femur, crossing each other on the way. They are named according to their positions of attachment to the tibia. Thus, the *anterior cruciate ligament* originates from the anterior intercondylar area of the tibia and extends to the lateral condyle of the femur. The *posterior cruciate ligament* connects the posterior intercondylar area of the tibia to the medial condyle of the femur.

The young soccer player, running at full speed, suddenly switches direction and is literally stopped in her tracks by a popping sound followed by a searing pain in her knee. Two hours after she veered toward the ball, her knee is swollen and painful, due to bleeding within the joint. She has torn the anterior cruciate ligament, a serious knee injury.

Two fibrocartilaginous *menisci* separate the articulating surfaces of the femur and tibia and help align them. Each meniscus is roughly C-shaped, with a thick rim and a thinner center, and attaches to the head of the tibia. The medial and lateral menisci form depressions that fit the corresponding condyles of the femur (fig. 8.21).

Several bursae are associated with the knee joint. These include a large extension of the knee joint cavity called the *suprapatellar bursa,* located between the anterior surface of the distal end of the femur and the muscle group (quadriceps femoris) above it; a large *prepatellar bursa* between the patella and the skin; and a smaller *infrapatellar bursa* between the proximal end of the tibia and the patellar ligament (see fig. 8.8).

As with a hinge joint, the basic structure of the knee joint permits flexion and extension. However, when the knee is flexed, rotation is also possible. Clinical Application 8.2 discusses some common joint disorders.

Tearing or displacing a meniscus is another common knee injury, usually resulting from forcefully twisting the knee when the leg is flexed. Since the meniscus is composed of fibrocartilage, this type of injury heals very slowly. Also, a torn and displaced portion of cartilage jammed between the articulating surfaces impedes movement of the joint. Following such a knee injury, the synovial membrane may become inflamed (acute synovitis) and secrete excess fluid, distending the joint capsule so that the knee swells above and on the sides of the patella.

1 Which structures help keep the articulating surfaces of the hip together?

2 What types of movement does the structure of the hip permit?

3 What types of joints are within the knee?

4 Which parts help hold together the articulating surfaces of the knee?

Life-Span Changes

Joint stiffness is an early sign of aging. By the fourth decade, a person may notice that the first steps each morning become difficult. Changes in collagen structure lie behind the increasing stiffness (fig. 8.22). However, the

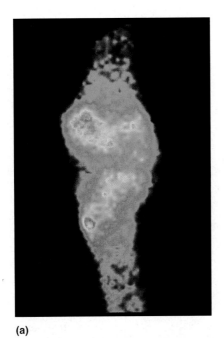
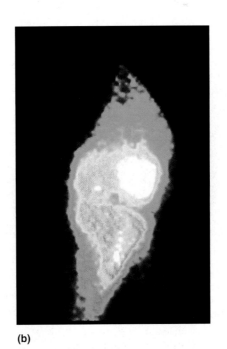

(a)   (b)

FIGURE 8.22

Nuclear scan of (a) good and (b) bad knees. The different colors in (b) indicate changes within the tissues associated with degeneration.

JOINT DISORDERS

Joints have a tough job. They must support weight, provide a great variety of body movements, and are used very frequently. In addition to this normal wear and tear, these structures are sometimes subjected to injury from overuse, infection, an immune system launching a misplaced attack, or degeneration. Here is a look at some common joint problems.

Sprains

Sprains result from overstretching or tearing the connective tissues, including cartilage, ligaments, and tendons associated with a joint, but they do not dislocate the articular bones (fig. 8A). Usually forceful wrenching or twisting sprains the wrist or ankles. For example, inverting an ankle too far can sprain it by stretching the ligaments on its lateral surface. Severe injuries may pull these tissues loose from their attachments.

A sprained joint is painful and swollen, restricting movement. Immediate treatment of a sprain is rest; more serious cases require medical attention. However, immobilization of a joint, even for a brief period, causes bone resorption and weakens ligaments. Consequently, exercise may help strengthen the joint.

Bursitis

Overuse of a joint or stress on a bursa may cause *bursitis,* an inflammation of a bursa. The bursa between the heel bone (calcaneus) and the Achilles tendon may become inflamed as a result of a sudden increase in physical activity using the feet. Similarly, a form of bursitis called tennis elbow affects the bursa between the olecranon process and the skin. Bursitis is treated with rest. Medical attention may be necessary.

Arthritis

Arthritis is a disease that causes inflamed, swollen, and painful joints. More than a hundred different types of arthritis affect 50 million people in the United States. Arthritis can also be part of other syndromes (table 8A). The most common causes of arthritis are discussed in the following section.

Rheumatoid Arthritis (RA)

Rheumatoid arthritis, an autoimmune disorder (a condition in which the immune system attacks the body's healthy tissues), is the most painful and debilitating form of arthritis. The synovial membrane of a joint becomes inflamed and thickens, forming a mass called a pannus. Then, the articular cartilage is damaged, and fibrous tissue infiltrates, interfering with joint movements. In time, the joint may ossify, fusing the articulating bones (bony ankylosis). Joints severely damaged by RA may be surgically replaced.

RA may affect many joints or only a few. It is usually a systemic illness, accompanied by fatigue, muscular atrophy, anemia, and osteoporosis, as well as changes in the skin, eyes, lungs, blood vessels, and heart. RA usually affects adults, but there is a juvenile form.

Osteoarthritis

Osteoarthritis, a degenerative disorder, is the most common type of arthritis (fig. 8B). It usually occurs with aging, but an inherited form may appear as early as one's thirties. A person may first become aware of osteoarthritis when a blow to the affected joint produces pain that is much more intense than normal. Gradually, the area of the affected joint deforms. For example, arthritic fingers take on a gnarled appearance, or a knee may bulge.

In osteoarthritis, articular cartilage softens and disintegrates gradually, roughening the articular surfaces. Joints become painful, with restricted movement. For example, arthritic fingers may lock into place while a person is playing the guitar or tying a shoelace. Osteoarthritis most often affects joints that are used the most over a lifetime, such as those of the fingers, hips, knees, and the lower parts of the vertebral column.

Nonsteroidal anti-inflammatory drugs (NSAIDs) have been used for many years to control osteoarthritis symptoms. Highly effective NSAIDs, called COX-2 inhibitors, relieve inflammation without the gastrointestinal side effects of older drugs. Celebrex (celecoxib) is used to treat osteoarthritis and rheumatoid arthritis. Vioxx (rofecoxib) is used to treat osteoarthritis. Exercise can keep stiff joints more flexible, and such simple measures as wearing gloves in the winter can alleviate symptoms.

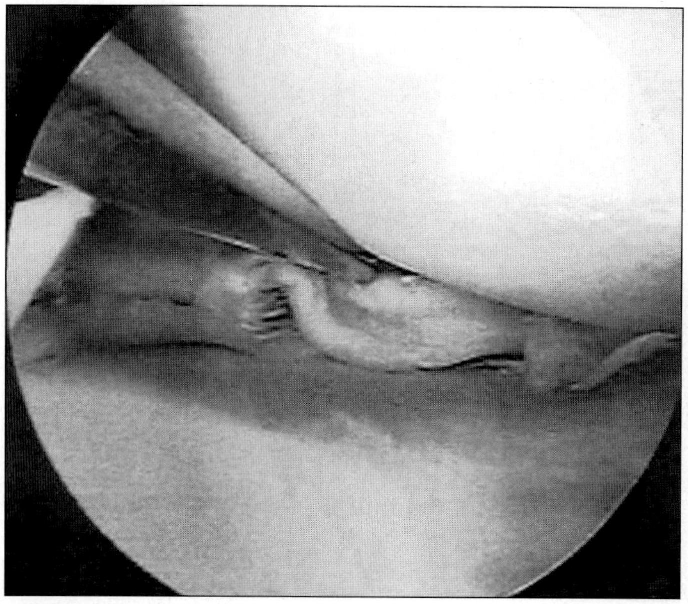

FIGURE 8A
Arthroscopic view of a torn meniscus in the knee and arthroscopic scissors. Because fibrocartilage does not heal well, in many cases of torn meniscus the only treatment option is to cut out the damaged portion.

TABLE 8A — Different Types of Arthritis

Some More-Common Forms of Arthritis

Type	Incidence in the United States
Osteoarthritis	20.7 million
Rheumatoid arthritis	2.1 million
Spondyloarthropathies	2.5 million

Some Less-Common Forms of Arthritis

Type	Incidence in the United States	Age of Onset	Symptoms
Gout	1.6 million (85% male)	>40	Sudden onset of extreme pain and swelling of a large joint
Juvenile rheumatoid arthritis	100,000	<18	Joint stiffness, often in knee
Scleroderma	300,000	30–50	Skin hardens and thickens
Systemic lupus erythematosus	500,000 (>90% female)	teens–50s	Fever, weakness, upper body rash, joint pain
Kawasaki disease	Hundreds of cases in local outbreaks	6 months–11 years	Fever, joint pain, red rash on palms and soles, heart complications
Strep A infection	100,000	any age	Confusion, body aches, shock, low blood pressure, dizziness, arthritis, pneumonia
Lyme disease	15,000	any age	Arthritis, malaise, neurologic and cardiac manifestations

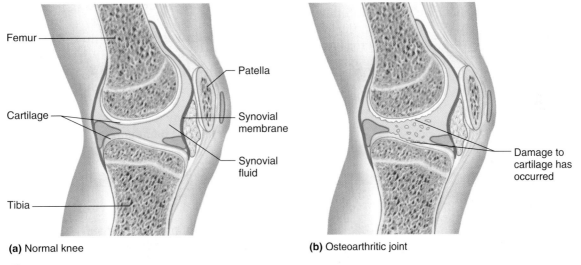

(a) Normal knee   **(b)** Osteoarthritic joint

FIGURE 8B

An inherited defect in collagen or prolonged wear and tear destroys joints in osteoarthritis.

Lyme Arthritis

Lyme disease is a bacterial infection passed in a tick bite that causes intermittent arthritis of several joints, usually weeks after the initial symptoms of rash, fatigue, and flu-like aches and pains. Lyme arthritis was first observed in Lyme, Connecticut, where an astute woman kept a journal after noticing that many of her young neighbors had what appeared to be the very rare juvenile form of rheumatoid arthritis. Her observations led Allen Steere, a Yale University rheumatologist, to trace the illness to a tick-borne bacterial infection. Antibiotic treatment beginning as soon as the early symptoms of Lyme disease are recognized can prevent development of the associated arthritis.

Other types of bacteria that cause arthritis include common *Staphylococcus* and *Streptococcus* species, *Neisseria gonorrhoeae* (which causes the sexually transmitted disease gonorrhea), and *Mycobacterium* (which causes tuberculosis). Arthritis may also be associated with AIDS, because the immune system breakdown raises the risk of infection by bacteria that can cause arthritis. ■

joints actually age slowly, and exercise can lessen or forestall stiffness.

The fibrous joints are the first to change, as the four types of fontanels close the bony plates of the skull at two, three, twelve, and eighteen to twenty-four months of age. Other fibrous joints may accumulate bone matrix over time, causing them to bring bones into closer apposition, even fusing. The fibrous joints actually strengthen over a lifetime.

Synchondroses that connect epiphyses to diaphyses in long bones disappear as the skeleton grows and develops. Another synchondrosis is the joint that links the first rib to the manubrium (sternum). As water content decreases and deposition of calcium salts increases, this cartilage stiffens. Ligaments lose their elasticity as the collagen fibers become more tightly cross-linked. Breathing may become labored, and movement more restrained.

Aging also affects symphysis joints, which consist of a pad of fibrocartilage sandwiched between thin layers of hyaline cartilage. In the intervertebral discs, less water diminishes the flexibility of the vertebral column and impairs the ability of the soft centers of the discs to absorb shocks. The discs may even collapse on themselves slightly, contributing to the loss of height in the elderly. The stiffening spine gradually restricts the range of motion.

Loss of function in synovial joints begins in the third decade of life, but progresses slowly. Fewer capillaries serving the synovial membrane slows the circulation of synovial fluid, and the membrane may become infiltrated with fibrous material and cartilage. As a result, the joint may lose elasticity, stiffening. More collagen cross-links shorten and stiffen ligaments, affecting the range of motion. This may, in turn, upset balance and retard the ability to respond in a protective way to falling, which may explain why older people are more likely to be injured in a fall than younger individuals.

1 Which type of joint is the first to show signs of aging?

2 Describe the loss of function in synovial joints as a progressive process.

Clinical Terms Related to Joints

ankylosis (ang″kĭ-lo′sis) Abnormal stiffness of a joint or fusion of bones at a joint, often due to damage of the joint membranes from chronic rheumatoid arthritis.

arthralgia (ar-thral′je-ah) Pain in a joint.

arthrocentesis (ar″thro-sen-te′sis) Puncture of and removal of fluid from a joint cavity.

arthrodesis (ar′thro-de′sis) Surgery to fuse the bones at a joint.

arthrogram (ar′thro-gram) Radiograph of a joint after an injection of radiopaque fluid into the joint cavity.

arthrology (ar-throl′o-je) Study of joints and diseases of them.

arthropathy (ar-throp′ah-the) Any joint disease.

arthroplasty (ar′thro-plas″te) Surgery to make a joint more movable.

arthrostomy (ar-thros′to-me) Surgical opening of a joint to allow fluid drainage.

arthrotomy (ar-throt′o-me) Surgical incision of a joint.

hemarthrosis (hem″ar-thro′sis) Blood in a joint cavity.

hydrarthrosis (hi″drar-thro′sis) Accumulation of fluid within a joint cavity.

luxation (luk-sa′shun) Dislocation of a joint.

subluxation (sub″luk-sa′shun) Partial dislocation of a joint.

synovectomy (sin″o-vek′to-me) Surgical removal of the synovial membrane of a joint.

C H A P T E R   S U M M A R Y

Introduction (page 254)

A joint forms wherever two or more bones meet. Joints are the functional junctions between bones.

Classification of Joints (page 254)

Joints are classified according to the type of tissue that binds the bones together.

1. Fibrous joints
   a. Bones at fibrous joints are fastened tightly together by a layer of dense connective tissue with many collagenous fibers.
   b. There are three types of fibrous joints.
      (1) A syndesmosis is characterized by bones bound by relatively long fibers of connective tissue.
      (2) A suture occurs where flat bones are united by a thin layer of connective tissue and become interlocked by a set of bony processes.
      (3) A gomphosis is formed by the union of a cone-shaped bony process to a bony socket.
2. Cartilaginous joints
   a. A layer of cartilage holds together bones of cartilaginous joints.
   b. There are two types of cartilaginous joints.
      (1) A synchondrosis is characterized by bones united by hyaline cartilage that may disappear as a result of growth.
      (2) A symphysis is a joint whose articular surfaces are covered by hyaline cartilage and attached to a pad of fibrocartilage.

3. Synovial joints
   a. Synovial joints have a more complex structure than other types of joints.
   b. These joints include articular cartilage, a joint capsule, and a synovial membrane.

General Structure of a Synovial Joint (page 257)

Articular cartilage covers articular ends of bones.

1. A joint capsule strengthened by ligaments holds bones together.
2. A synovial membrane that secretes synovial fluid lines the inner layer of a joint capsule.
3. Synovial fluid moistens, provides nutrients, and lubricates the articular surfaces.
4. Menisci divide some synovial joints into compartments.
5. Some synovial joints have fluid-filled bursae.
   a. Bursae are usually located between the skin and underlying bony prominences.
   b. Bursae cushion and aid movement of tendons over bony parts.
   c. Bursae are named according to their locations.

Types of Synovial Joints (page 259)

1. Ball-and-socket joints
   a. In a ball-and-socket joint, the globular head of a bone fits into the cup-shaped cavity of another bone.
   b. These joints permit a wide variety of movements.
   c. The hip and shoulder are ball-and-socket joints.
2. Condyloid joints
   a. A condyloid joint consists of an ovoid condyle of one bone fitting into an elliptical cavity of another bone.
   b. This joint permits a variety of movements.
   c. The joints between the metacarpals and phalanges are condyloid.
3. Gliding joints
   a. Articular surfaces of gliding joints are nearly flat.
   b. These joints permit the articular surfaces to slide back and forth.
   c. Most of the joints of the wrist and ankle are gliding joints.
4. Hinge joints
   a. In a hinge joint, the convex surface of one bone fits into the concave surface of another bone.
   b. This joint permits movement in one plane only.
   c. The elbow and the joints of the phalanges are the hinge type.
5. Pivot joints
   a. In a pivot joint, a cylindrical surface of one bone rotates within a ring of bone and fibrous tissue.
   b. This joint permits rotational movement.
   c. The articulation between the proximal ends of the radius and the ulna is a pivot joint.
6. Saddle joints
   a. A saddle joint forms between bones that have complementary surfaces with both concave and convex regions.
   b. This joint permits a variety of movements.
   c. The articulation between the carpal and metacarpal of the thumb is a saddle joint.

Types of Joint Movements (page 261)

1. Muscles acting at synovial joints produce movements in different directions and in different planes.

2. Joint movements include flexion, extension, hyperextension, dorsiflexion, plantar flexion, abduction, adduction, rotation, circumduction, supination, pronation, eversion, inversion, elevation, depression, protraction, and retraction.

Examples of Synovial Joints (page 264)

1. Shoulder joint
   a. The shoulder joint is a ball-and-socket joint that consists of the head of the humerus and the glenoid cavity of the scapula.
   b. A cylindrical joint capsule envelops the joint.
      (1) The capsule is loose and by itself cannot keep the articular surfaces together.
      (2) It is reinforced by surrounding muscles and tendons.
   c. Several ligaments help prevent displacement of the bones.
   d. Several bursae are associated with the shoulder joint.
   e. Because its parts are loosely attached, the shoulder joint permits a wide range of movements.
2. Elbow joint
   a. The elbow has a hinge joint between the humerus and the ulna and a gliding joint between the humerus and the radius.
   b. The joint capsule is reinforced by collateral ligaments.
   c. A synovial membrane partially divides the joint cavity into two portions.
   d. The joint between the humerus and the ulna permits flexion and extension only.
3. Hip joint
   a. The hip joint is a ball-and-socket joint between the femur and the coxa.
   b. A ring of fibrocartilage deepens the cavity of the acetabulum.
   c. The articular surfaces are held together by a heavy joint capsule that is reinforced by ligaments.
   d. The hip joint permits a wide variety of movements.
4. Knee joint
   a. The knee joint includes two condyloid joints between the femur and the tibia and a gliding joint between the femur and the patella.
   b. Ligaments and tendons strengthen the relatively thin joint capsule.
   c. Several ligaments, some of which are within the joint capsule, bind articular surfaces.
   d. Two menisci separate the articulating surfaces of the femur and the tibia.
   e. Several bursae are associated with the knee joint.
   f. The knee joint permits flexion and extension; when the lower limb is flexed at the knee, some rotation is possible.

Life-Span Changes (page 271)

1. Joint stiffness is often the earliest sign of aging.
   a. Collagen changes cause the feeling of stiffness.
   b. Regular exercise can lessen the effects.
2. Fibrous joints are the first to begin to change and strengthen over a lifetime.
3. Synchondroses of the long bones disappear with growth and development.
4. Changes in symphysis joints of the vertebral column diminish flexibility and decrease height.
5. Over time, synovial joints lose elasticity.

CRITICAL THINKING QUESTIONS

1. How would you explain to an athlete why damaged joint ligaments and cartilages are so slow to heal following an injury?

2. Compared to the shoulder and hip joints, in what way is the knee joint poorly protected and thus especially vulnerable to injuries?

3. Based upon your knowledge of joint structures, which do you think could be more satisfactorily replaced by a prosthetic device, a hip joint or a knee joint? Why?

4. If a patient's forearm and elbow were immobilized by a cast for several weeks, what changes would you expect to occur in the bones of the upper limb?

5. Why is it important to encourage an inactive patient to keep all joints mobile, even if it is necessary to have another person or a device move the joints (passive movement)?

6. How would you explain to a person with a dislocated shoulder that the shoulder is likely to become more easily dislocated in the future?

REVIEW EXERCISES

Part A

1. Define *joint*.
2. Explain how joints are classified.
3. Compare the structure of a fibrous joint with that of a cartilaginous joint.
4. Distinguish between a syndesmosis and a suture.
5. Describe a gomphosis, and name an example.
6. Compare the structures of a synchondrosis and a symphysis.
7. Explain how the joints between vertebrae permit movement.
8. Describe the general structure of a synovial joint.
9. Describe how a joint capsule may be reinforced.
10. Explain the function of the synovial membrane.
11. Explain the function of synovial fluid.
12. Define *meniscus*.
13. Define *bursa*.
14. List six types of synovial joints, and name an example of each type.
15. Describe the movements permitted by each type of synovial joint.
16. Name the parts that comprise the shoulder joint.
17. Name the major ligaments associated with the shoulder joint.
18. Explain why the shoulder joint permits a wide range of movements.
19. Name the parts that comprise the elbow joint.
20. Describe the major ligaments associated with the elbow joint.
21. Name the movements permitted by the elbow joint.
22. Name the parts that comprise the hip joint.
23. Describe how the articular surfaces of the hip joint are held together.
24. Explain why there is less freedom of movement in the hip joint than in the shoulder joint.
25. Name the parts that comprise the knee joint.
26. Describe the major ligaments associated with the knee joint.
27. Explain the function of the menisci of the knee.
28. Describe the locations of the bursae associated with the knee joint.
29. Describe the process of aging as it contributes to the stiffening of fibrous, cartilaginous, and synovial joints.

Part B

Match the movements in column I with the descriptions in column II.

	I		II
1.	Rotation	A.	Turning palm upward
2.	Supination	B.	Decreasing angle between parts
3.	Extension	C.	Moving part forward
4.	Eversion	D.	Moving part around an axis
5.	Protraction	E.	Turning the foot so the sole faces laterally
6.	Flexion	F.	Increasing angle between parts
7.	Pronation	G.	Lowering a part
8.	Abduction	H.	Turning palm downward
9.	Depression	I.	Moving part away from midline

WEB CONNECTIONS

Visit the Student Edition of the Online Learning Center at www.mhhe.com/shier10 for answers to chapter questions, additional quizzes, interactive learning exercises, and other study tools.

UNDERSTANDING WORDS

calat-, something inserted: inter*calat*ed disc—membranous band that connects cardiac muscle cells.

erg-, work: syn*erg*ist—muscle that works together with a prime mover to produce a movement.

fasc-, bundle: *fasc*iculus—bundle of muscle fibers.

-gram, something written: myo*gram*—recording of a muscular contraction.

hyper-, over, more: muscular *hyper*trophy—enlargement of muscle fibers.

inter-, between: *inter*calated disc—membranous band that connects cardiac muscle cells.

iso-, equal: *iso*tonic contraction—contraction during which the tension in a muscle remains unchanged.

laten-, hidden: *laten*t period—period between a stimulus and the beginning of a muscle contraction.

myo-, muscle: *myo*fibril—contractile fiber of a muscle cell.

reticul-, a net: sarcoplasmic *reticul*um—network of membranous channels within a muscle fiber.

sarco-, flesh: *sarco*plasm—substance (cytoplasm) within a muscle fiber.

syn-, together: *syn*ergist—muscle that works with a prime mover to produce a movement.

tetan-, stiff: *tetan*ic contraction—sustained muscular contraction.

-tonic, stretched: iso*tonic* contraction—contraction during which the tension of a muscle remains unchanged.

-troph, well fed: muscular hyper*troph*y—enlargement of muscle fibers.

voluntar-, of one's free will: *voluntar*y muscle—muscle that can be controlled by conscious effort.

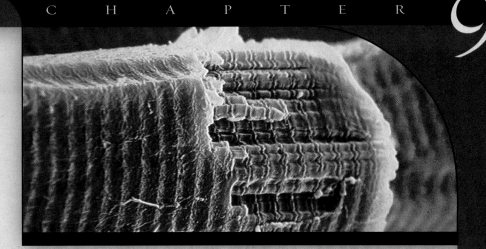

Falsely colored scanning electron micrograph (SEM) of normal human striated muscle fibers, revealing the characteristic banding pattern of the constituent myofibrils (1,900×).

Muscular System

CHAPTER OBJECTIVES

After you have studied this chapter, you should be able to

1. Describe how connective tissue is part of the structure of a skeletal muscle.

2. Name the major parts of a skeletal muscle fiber and describe the function of each part.

3. Explain the major events that occur during muscle fiber contraction.

4. Explain how energy is supplied to the muscle fiber contraction mechanism, how oxygen debt develops, and how a muscle may become fatigued.

5. Distinguish between fast and slow twitch muscle fibers.

6. Distinguish between a twitch and a sustained contraction.

7. Describe how exercise affects skeletal muscles.

8. Explain how various types of muscular contractions produce body movements and help maintain posture.

9. Distinguish between the structures and functions of a multiunit smooth muscle and a visceral smooth muscle.

10. Compare the contraction mechanisms of skeletal, smooth, and cardiac muscle fibers.

11. Explain how the locations of skeletal muscles help produce movements and how muscles interact.

12. Identify and locate the major skeletal muscles of each body region and describe the action of each muscle.

ike many things in life, individual muscles aren't appreciated until we see what happens when they do not work. For children with Moebius syndrome, absence of the sixth and seventh cranial nerves, which carry impulses from the brain to the muscles of the face, leads to an odd collection of symptoms.

The first signs of Moebius syndrome are typically difficulty sucking, excessive drooling, and sometimes crossed eyes. The child has difficulty swallowing and chokes easily, cannot move the tongue well, and is very sensitive to bright light because he or she cannot squint or blink or even avert the eyes. Special bottles and feeding tubes can help the child eat, and surgery can correct eye defects.

Children with Moebius syndrome are slow to reach developmental milestones but do finally walk. As they get older, if they are lucky, they are left with only one symptom, but it is a rather obvious one—inability to form facial expressions.

A young lady named Chelsey Thomas called attention to this very rare condition when she underwent two surgeries that would enable her

to smile. In 1995 and 1996, when she was seven years old, Chelsey had two transplants of nerve and muscle tissue from her legs to either side of her mouth, supplying the missing "smile apparatus." Gradually, she acquired the subtle, and not-so-subtle, muscular movements of the mouth that make the human face so expressive. Chelsey inspired several other youngsters to undergo "smile surgery." Publicity about her surgery informed many health care professionals about this extremely rare condition. ■

The three types of muscle tissues are skeletal, smooth, and cardiac, as described in chapter 5 (pages 148–150). This chapter focuses on the skeletal muscles, which are usually attached to bones and are under conscious control.

Structure of a Skeletal Muscle

A skeletal muscle is an organ of the muscular system. It is composed primarily of skeletal muscle tissue, nervous tissue, blood, and connective tissues.

Connective Tissue Coverings

An individual skeletal muscle is separated from adjacent muscles and held in position by layers of dense connective tissue called **fascia** (fash′e-ah). This connective tissue surrounds each muscle and may project beyond the end of its muscle fibers to form a cordlike **tendon.** Fibers in a tendon intertwine with those in the periosteum of a bone, attaching the muscle to the bone. In other cases, the connective tissues associated with a muscle form broad, fibrous sheets called **aponeuroses** (ap″o-nu-ro′sēz), which

may attach to bone, or, in some cases, the coverings of adjacent muscles (figs. 9.1 and 9.2).

The layer of connective tissue that closely surrounds a skeletal muscle is called the *epimysium.* Another layer of connective tissue, called the *perimysium,* extends inward from the epimysium and separates the muscle tissue into small sections. These sections contain bundles of skeletal muscle fibers called *fascicles* (fasciculi). Each muscle fiber within a fascicle (fasciculus) lies within a layer of connective tissue in the form of a thin covering called *endomysium* (figs. 9.2 and 9.3). Layers of connective tissue, therefore, enclose and separate all parts of a skeletal muscle. This arrangement allows the parts to move somewhat independently. Also, many blood vessels and nerves pass through these layers.

A tendon or the connective tissue sheath of a tendon (tenosynovium) may become painfully inflamed and swollen following an injury or the repeated stress of athletic activity. These conditions are called *tendinitis* and *tenosynovitis,* respectively. The tendons most commonly affected are those associated with the joint capsules of the shoulder, elbow, hip, and knee and those involved with moving the wrist, hand, thigh, and foot.

A *compartment* is the space that contains a particular group of muscles, blood vessels, and nerves, all tightly enclosed by fascia. The limbs have many such compartments. If an injury causes fluid, such as blood from an internal hemorrhage, to accumulate within a compartment, the pressure inside will rise. The increased pressure, in turn, may interfere with blood flow into the region, reducing the supply of oxygen and nutrients to the affected tissues. This condition, called *compartment syndrome,* often produces severe, unrelenting pain. Persistently elevated compartmental pressure may irreversibly damage the enclosed muscles and nerves. Treatment for compartment syndrome may require immediate intervention by a surgical incision through the fascia (fasciotomy) to relieve the pressure and restore circulation.

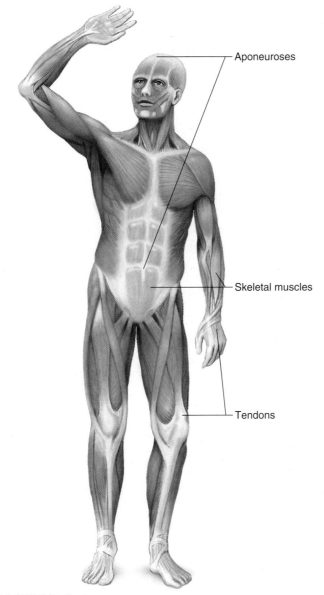

FIGURE 9.1

Tendons attach muscles to bones, whereas aponeuroses attach muscles to other muscles.

The fascia associated with each individual organ of the muscular system is part of a complex network of fasciae that extends throughout the body. The portion of the network that surrounds and penetrates the muscles is called *deep fascia.* It is continuous with the *subcutaneous fascia* that lies just beneath the skin, forming the subcutaneous layer described in chapter 6 (p. 164). The network is also continuous with the *subserous fascia* that forms the connective tissue layer of the serous membranes covering organs in various body cavities and lining those cavities (see chapter 5, p. 152).

Skeletal Muscle Fibers

Recall from chapter 5 (p. 148) that a skeletal muscle fiber is a single muscle cell (see fig. 5.28). Each fiber forms from many undifferentiated cells that fuse during development. Each resulting multinucleated muscle fiber is a thin, elongated cylinder with rounded ends that attach to the connective tissues associated with a muscle. Just beneath the muscle cell membrane (*sarcolemma*), the cytoplasm (*sarcoplasm*) of the fiber contains many small, oval nuclei and mitochondria. The sarcoplasm also has abundant, parallel, threadlike structures called **myofibrils** (mi″o-fi′-brilz) (fig. 9.4*a*)

The myofibrils play a fundamental role in the muscle contraction mechanism. They contain two kinds of protein filaments: Thick filaments composed of the protein **myosin** (mi′o-sin), and thin filaments composed primarily of the protein **actin** (ak′tin). (Two other thin filament proteins, troponin and tropomyosin, will be discussed later.) The organization of these filaments produces the alternating light and dark striations characteristic of skeletal muscle (and cardiac muscle) fibers. The striations form a repeating pattern of units called **sarcomeres** (sar′ko-mērz) along each muscle fiber. The myofibrils may be thought of as sarcomeres joined end to end. (fig. 9.4*a*).

The striation pattern of skeletal muscle has two main parts. The first, the *I bands* (the light bands), are composed of thin actin filaments held by direct attachments to structures called *Z lines,* which appear in the center of the I bands. The second part of the striation pattern consists of the *A bands* (the dark bands), which are composed of thick myosin filaments overlapping thin actin filaments (fig. 9.4*b*).

Note that the A band consists not only of a region where thick and thin filaments overlap, but also a slightly lighter central region (*H zone*) consisting only of thick filaments. The A band includes a thickening known as the *M line,* which consists of proteins that help hold the thick filaments in place (fig. 9.4*b*). The myosin filaments are also held in place by the Z lines but are attached to them by a large protein called **titin** (connectin) (fig. 9.5). A sarcomere extends from one Z line to the next.

Thick filaments are composed of many molecules of myosin. Each myosin molecule consists of two twisted protein strands with globular parts called *cross-bridges* (heads) that project outward along their lengths. Thin filaments consist of double strands of actin twisted into a helix. Actin molecules are globular, and each has a binding site to which the cross-bridges of a myosin molecule can attach (fig. 9.6).

Two other types of protein, **troponin** and **tropomyosin,** associate with actin filaments. Troponin molecules have three protein subunits and are attached to actin. Tropomyosin molecules are rod-shaped and occupy the longitudinal grooves of the actin helix. Each tropomyosin is held in place by a troponin molecule, forming a troponin-tropomyosin complex (fig. 9.6).

Within the sarcoplasm of a muscle fiber is a network of membranous channels that surrounds each myofibril and runs parallel to it. These membranes form the **sarcoplasmic reticulum,** which corresponds to the endoplasmic

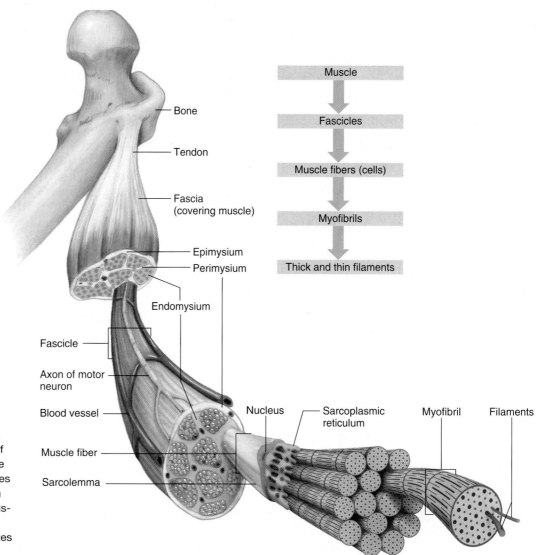

Muscle → Fascicles → Muscle fibers (cells) → Myofibrils → Thick and thin filaments

Bone

Tendon

Fascia (covering muscle)

Epimysium
Perimysium
Endomysium

Fascicle

Axon of motor neuron

Blood vessel

Muscle fiber

Sarcolemma

Nucleus

Sarcoplasmic reticulum

Myofibril

Filaments

FIGURE 9.2

A skeletal muscle is composed of a variety of tissues, including layers of connective tissue. Fascia covers the surface of the muscle, epimysium lies beneath the fascia, and perimysium extends into the structure of the muscle where it separates muscle cells into fascicles. Endomysium separates individual muscle fibers.

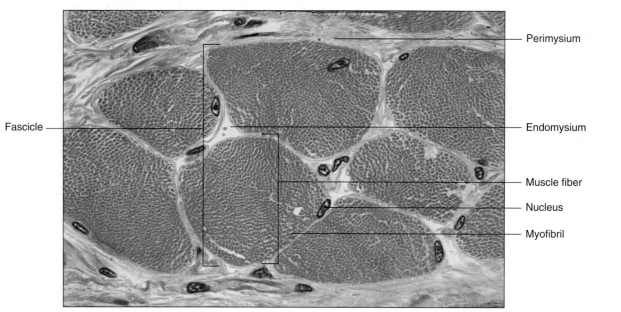

Perimysium

Fascicle

Endomysium

Muscle fiber

Nucleus

Myofibril

FIGURE 9.3

Scanning electron micrograph of a fascicle (fasciculus) surrounded by its connective tissue sheath, the perimysium. Muscle fibers within the fascicle are surrounded by endomysium (320×).

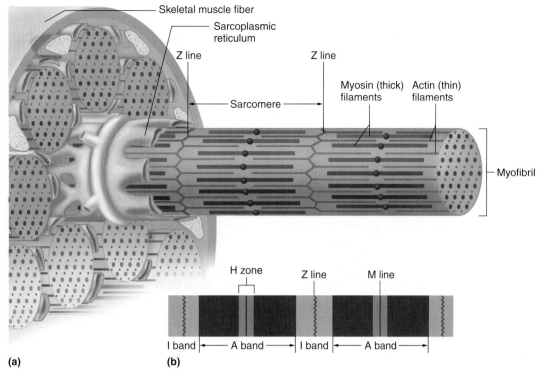

(a)   (b)

FIGURE 9.4

Skeletal muscle fiber. (*a*) A skeletal muscle fiber contains numerous myofibrils, each consisting of (*b*) repeating units called sarcomeres. The characteristic striations of a sarcomere reflect the organization of actin and myosin filaments.

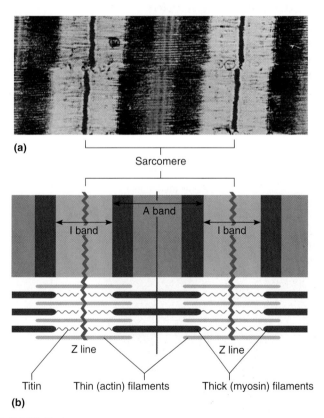

(a)

(b)

FIGURE 9.5

A sarcomere. (*a*) Micrograph (16,000×). (*b*) The relationship of thin and thick filaments in a sarcomere. The size of the H zone may change depending on the degree of filament overlap. Compare with the size of the H zone and filament overlap in fig. 9.4*a* and *b*.

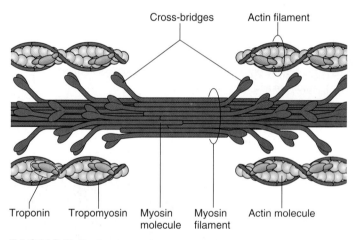

FIGURE 9.6

Thick filaments are composed of the protein myosin, and thin filaments are composed of the protein actin. Myosin molecules have cross-bridges that extend toward nearby actin filaments.

reticulum of other cells (see figs. 9.2 and 9.4). A set of membranous channels, the **transverse tubules** (T-tubules), extends into the sarcoplasm as invaginations continuous with the sarcolemma and contains extracellular fluid. Each transverse tubule lies between two enlarged portions of the sarcoplasmic reticulum called **cisternae,** and these three structures form a **triad** near the region where the actin and myosin filaments overlap (fig. 9.7).

Although muscle fibers and the connective tissues associated with them are flexible, they can tear if overstretched. This type of injury is common in athletes and is called a *muscle strain.* The seriousness of the injury depends on the degree of damage the tissues sustain. In a mild strain, only a few muscle fibers are injured, the fascia remains intact, and little function is lost. In a severe strain, many muscle fibers as well as fascia tear, and muscle function may be lost completely. A severe strain is very painful and is accompanied by discoloration and swelling of tissues due to ruptured blood vessels. Surgery may be required to reconnect the separated tissues.

---

1 Describe how connective tissue is associated with a skeletal muscle.

2 Describe the general structure of a skeletal muscle fiber.

3 Explain why skeletal muscle fibers appear striated.

4 Explain the physical relationship between the sarcoplasmic reticulum and the transverse tubules.

Skeletal Muscle Contraction

A muscle fiber contraction is a complex interaction of several cellular and chemical constituents. The final result is a movement within the myofibrils in which the filaments of actin and myosin slide past one another, shortening the sarcomeres. When this happens, the muscle fiber shortens and pulls on its attachments.

Actin, myosin, troponin, and tropomyosin are abundant in muscle cells. Scarcer proteins are also vital to muscle function. This is the case for a rod-shaped muscle protein called *dystrophin.* It accounts for only 0.002% of total muscle protein in skeletal muscle, but its absence causes the devastating inherited disorder Duchenne muscular dystrophy, a disease that usually affects boys. Dystrophin binds to the inside face of muscle cell membranes, supporting them against the powerful force of contraction. Without even these minute amounts of dystrophin, muscle cells burst and die. Other forms of muscular dystrophy result from abnormalities of other proteins to which dystrophin attaches.

Neuromuscular Junction

Each skeletal muscle fiber is connected to an extension (a nerve axon) of a **motor neuron** (mo'tor nu'ron) that passes outward from the brain or spinal cord. Normally a skeletal muscle fiber contracts only upon stimulation by a motor neuron.

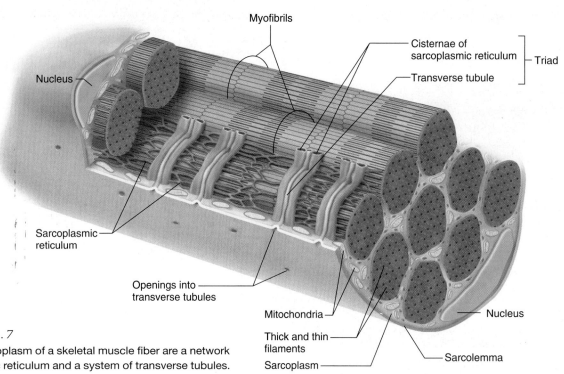

FIGURE 9.7
Within the sarcoplasm of a skeletal muscle fiber are a network of sarcoplasmic reticulum and a system of transverse tubules.

The site where the axon and muscle fiber meet is called a **neuromuscular junction** (myoneural junction). There, the muscle fiber membrane is specialized to form a **motor end plate,** where nuclei and mitochondria are abundant and the sarcolemma is extensively folded (fig. 9.8).

A muscle fiber usually has a single motor end plate. Motor neuron axons, however, are densely branched. By means of these branches, one motor neuron axon may connect to many muscle fibers. Together, a motor neuron and the muscle fibers it controls constitute a **motor unit** (mo'tor u'nit) (fig. 9.9).

In the summer months of the early 1950s, parents in the United States lived in terror of their children contracting poliomyelitis, a viral infection that attacks nerve cells that stimulate skeletal muscles to contract. In half of the millions of affected children, fever, headache, and nausea rapidly progressed to a stiffened back and neck, drowsiness, and then the feared paralysis, usually of the lower limbs or muscles that control breathing or swallowing. Today, many a middle-aged person with a limp owes this slight disability to polio.

Vaccines introduced in the middle 1950s ended the nightmare of polio in many nations, but the disease has not been globally eradicated because not everyone has been vaccinated. In addition, a third of the 1.6 million polio survivors in the United States are experiencing the fatigue, muscle weakness and atrophy, and difficulty breathing of post-polio syndrome. Researchers think that this condition begins 10 to 40 years after severe poliomyelitis, when surviving motor neurons that grew extra axon branches to compensate for lost neurons degenerate from years of overuse. West Nile virus infection causes a poliomyelitis-like syndrome.

A small gap called the **synaptic cleft** separates the membrane of the neuron and the membrane of the muscle fiber. The cytoplasm at the distal ends of the nerve fiber is rich in mitochondria and contains many tiny vesicles (synaptic vesicles) that store chemicals called **neurotransmitters** (nu″ro-trans′mit-erz).

Stimulus for Contraction

Acetylcholine (ACh) is the neurotransmitter that motor neurons use to control skeletal muscle. ACh is synthesized in the cytoplasm of the motor neuron and is stored in synaptic vesicles near the distal end of its axon. When a nerve impulse (or action potential, described in chapter 10, pp. 350–353) reaches the end of the axon, some of these vesicles release acetylcholine into the synaptic cleft (see fig. 9.8).

Acetylcholine diffuses rapidly across the synaptic cleft, combines with ACh receptors on the motor end-plate, and stimulates the muscle fiber. The response is a

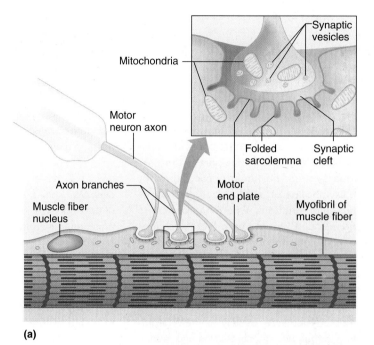

(a)

(b)

FIGURE 9.8
Neuromuscular junction. (*a*) A neuromuscular junction includes the end of a motor neuron and the motor end plate of a muscle fiber. (*b*) Micrograph of a neuromuscular junction (500×).

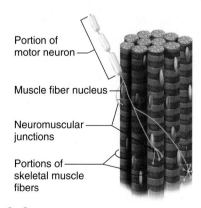

FIGURE 9.9
Muscle fibers within a motor unit may be distributed throughout the muscle.

MYASTHENIA GRAVIS

In an autoimmune disorder, the immune system attacks part of the body. In myasthenia gravis (MG), that part is the nervous system, particularly receptors for acetylcholine on muscle cells at neuromuscular junctions, where neuron meets muscle cell. People with MG have one-third the normal number of acetylcholine receptors at these junctions. On a whole-body level, this causes weak and easily fatigued muscles.

MG affect hundreds of thousands of people worldwide, usually women, beginning in their twenties or thirties and men in their sixties and seventies. The specific symptoms depend upon the site of attack. For 85% of patients, the disease causes generalized muscle weakness. Many people develop a characteristic flat smile and nasal voice and have difficulty chewing and swallowing due to affected facial and neck muscles. Many have limb weakness. About 15% of patients experience the illness only in the muscles surrounding their eyes. The disease reaches crisis level when respiratory muscles are affected, requiring a ventilator to support breathing. MG does not affect sensation or reflexes.

Until 1958, MG was a serious threat to health, with a third of patients dying, a third worsening, and only a third maintaining or improving their condition. Today, most people with MG can live near-normal lives, thanks to a combination of the following treatments:

- Drugs that inhibit acetylcholinesterase, which boosts availability of acetylcholine.
- Removing the thymus gland, which oversees much of the immune response.
- Immunosuppressant drugs.
- Intravenous antibodies to bind and inactivate the ones causing the damage.
- Plasma exchange, which rapidly removes the damaging antibodies from the circulation. This helps people in crisis. ■

muscle impulse, an electrical signal that is very much like a nerve impulse. A muscle impulse changes the muscle cell membrane in a way that transmits the impulse in all directions along and around the muscle cell, into the transverse tubules, into the sarcoplasm, and ultimately to the sarcoplasmic reticulum and the cisternae. Clinical Application 9.1 discusses myasthenia gravis, in which the immune system attacks certain neuromuscular junctions.

When the bacterium *Clostridium botulinum* grows in an anaerobic (oxygen-poor) environment, such as in a can of unrefrigerated food, it produces a toxin that prevents the release of acetylcholine from nerve terminals if ingested by a person. Symptoms include nausea, vomiting, and diarrhea; headache, dizziness, and blurred or double vision; and finally, weakness, hoarseness, and difficulty swallowing and, eventually, breathing. Physicians can administer an antitoxin substance that binds to and inactivates botulinum toxin in the bloodstream, stemming further symptoms, although not correcting damage already done.

Excitation Contraction Coupling

The sarcoplasmic reticulum has a high concentration of calcium ions compared to the cytosol. This is due to active transport of calcium ions (calcium pump) in the membrane of the sarcoplasmic reticulum. In response to a muscle impulse, the membranes of the cisternae become more permeable to these ions, and the calcium ions diffuse out of the cisternae into the cytosol of the muscle fiber (see fig. 9.7).

RECONNECT TO CHAPTER 3, ACTIVE TRANSPORT, PAGE 85

When a muscle fiber is at rest, the troponin-tropomyosin complexes block the binding sites on the actin molecules and thus prevent the formation of linkages with myosin cross-bridges (fig. 9.10 *1*). As the concentration of calcium ions in the cytosol rises, however, the calcium ions bind to the troponin, changing its shape (conformation) and altering the position of the tropomyosin. The movement of the tropomyosin molecules exposes the binding sites on the actin filaments, allowing linkages to form between myosin cross-bridges and actin (fig. 9.10 *2*).

RECONNECT TO CHAPTER 2, PROTEINS, PAGE 54

The Sliding Filament Theory

The sarcomere is considered the functional unit of skeletal muscles. This is because contraction of an entire skeletal muscle can be described in terms of the shortening of sarcomeres within it. According to the **sliding filament theory,**

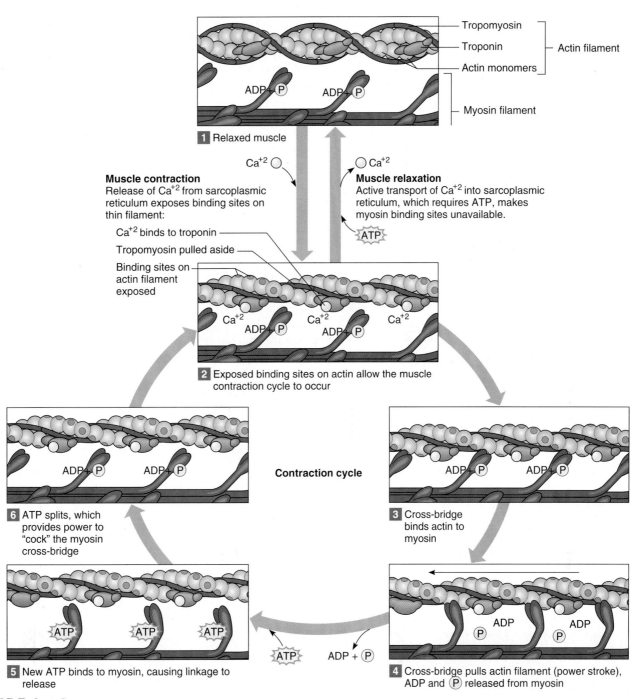

FIGURE 9.10

According to the sliding filament theory (1–3) when calcium ion concentration rises, binding sites on actin filaments open, and cross-bridges attach. (4) Upon binding to actin, cross-bridges spring from the cocked position and pull on actin filaments. (5) ATP binds to the cross-bridge (but is not yet broken down), causing it to release from the actin filament. (6) ATP breakdown provides energy to "cock" the unattached myosin cross-bridge. As long as ATP and calcium ions are present, the cycle continues. When calcium ion concentration is low, the muscle remains relaxed.

when sarcomeres shorten, the thick and thin filaments do not themselves change length. Rather, they just slide past one another, with the thin filaments moving toward the center of the sarcomere from both ends. As this occurs, the H zones and the I bands get narrower, the regions of overlap widen, and the Z lines move closer together, shortening the sarcomere (fig. 9.11).

Cross-bridge Cycling

The force that shortens the sarcomeres comes from cross-bridges pulling on the thin filaments. A myosin cross-bridge can attach to an actin binding site and bend slightly, pulling on the actin filament. Then the head can release, straighten, combine with another binding site further down the actin filament, and pull again (see fig. 9.10).

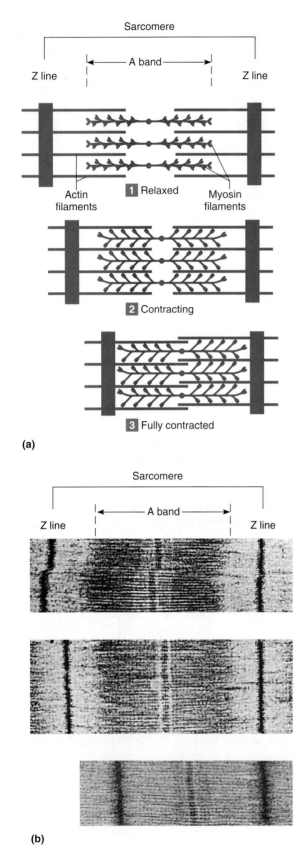

(a)

Sarcomere

A band

Z line   Z line

Actin filaments   **1** Relaxed   Myosin filaments

2 Contracting

3 Fully contracted

Sarcomere

A band

Z line   Z line

(b)

FIGURE 9.11

When a skeletal muscle contracts (a), individual sarcomeres shorten as thick and thin filaments slide past one another. (b) Transmission electron micrograph showing a sarcomere shortening during muscle contraction (23,000×).

Myosin cross-bridges contain the enzyme **ATPase,** which catalyzes the breakdown of ATP to ADP and phosphate. This reaction releases energy (see chapter 4, p. 108) that provides the force for muscle contraction. Breakdown of ATP puts the myosin cross-bridge in a "cocked" position (fig. 9.10 6). When a muscle is stimulated to contract, a cocked cross-bridge attaches to actin (9.10 3) and pulls the actin filament toward the center of the sarcomere, shortening the sarcomere and thus shortening the muscle (9.10 4). When another ATP binds, the cross-bridge is first released from the actin binding site (9.10 5), then breaks down the ATP to return to the cocked position (9.10 6). This cross-bridge cycle may repeat over and over, as long as ATP is present and nerve impulses cause ACh release at that neuromuscular junction.

Relaxation

When nerve impulses cease, two events relax the muscle fiber. First, the acetylcholine that remains in the synapse is rapidly decomposed by an enzyme called **acetylcholinesterase.** This enzyme is present in the synapse and on the membranes of the motor end plate. The action of acetylcholinesterase prevents a single nerve impulse from continuously stimulating a muscle fiber.

If acetylcholine receptors at the motor end plate are too few, or blocked, muscles cannot receive the signal to contract. This may occur as the result of a disease, such as myasthenia gravis, or exposure to a poison, such as nerve gas. A drug called pyridostigmine bromide is used to treat myasthenia gravis. The drug inhibits the enzyme (acetylcholinesterase) that normally breaks down acetylcholine, keeping the neurotransmitter around longer. It was given to veterans of the Persian Gulf War who reported muscle aches in the months following their military service. Health officials reasoned that the drug's effect on myasthenia gravis might also help restore muscle function if the veterans' symptoms arose from exposure to nerve gas during the war. Acetylcholinesterase inhibitors are also used as insecticides. The buildup of acetylcholine causes an insect to twitch violently, then die.

Second, when ACh is broken down, the stimulus to the sarcolemma and the membranes within the muscle fiber ceases. The calcium pump (which requires ATP) quickly moves calcium ions back into the sarcoplasmic reticulum, decreasing the calcium ion concentration of the cytosol. The cross-bridge linkages break (see figure 9.10 6—remember, this also requires ATP, although it is not broken down in this step), and tropomyosin rolls back into its groove, preventing any cross-bridge attachment (see fig. 9.10 1). Consequently, the muscle fiber relaxes. Table 9.1 summarizes the major events leading to muscle contraction and relaxation.

TABLE 9.1	Major Events of Muscle Contraction and Relaxation

Muscle Fiber Contraction	Muscle Fiber Relaxation
1. The distal end of a motor neuron releases acetylcholine.	1. Acetylcholinesterase decomposes acetylcholine, and the muscle fiber membrane is no longer stimulated.
2. Acetylcholine diffuses across the gap at the neuromuscular junction.	2. Calcium ions are actively transported into the sarcoplasmic reticulum.
3. The sarcolemma is stimulated, and a muscle impulse travels over the surface of the muscle fiber and deep into the fiber through the transverse tubules and reaches the sarcoplasmic reticulum.	3. ATP causes linkages between actin and myosin filaments to break without ATP breakdown.
4. Calcium ions diffuse from the sarcoplasmic reticulum into the sarcoplasm and bind to troponin molecules.	4. Cross-bridges recock.
5. Tropomyosin molecules move and expose specific sites on actin filaments.	5. Troponin and tropomyosin molecules inhibit the interaction between myosin and actin filaments.
6. Actin and myosin filaments form linkages.	6. Muscle fiber remains relaxed, yet ready until stimulated again.
7. Actin filaments are pulled inward by myosin cross-bridges.	
8. Muscle fiber shortens as a contraction occurs.	

It is important to remember that ATP is necessary for both muscle contraction and for muscle relaxation. The trigger for contraction is the increase in cytosolic calcium in response to stimulation by ACh from a motor neuron.

A few hours after death, the skeletal muscles partially contract, fixing the joints. This condition, called *rigor mortis,* may continue for seventy-two hours or more. It results from an increase in membrane permeability to calcium ions, which promotes cross-bridge attachment, and a decrease in availability of ATP in the muscle fibers, which prevents cross-bridge release from actin. Thus, the actin and myosin filaments of the muscle fibers remain linked until the muscles begin to decompose.

1 Describe a neuromuscular junction.

2 Define motor unit.

3 List four proteins associated with myofibrils, and explain their structural and functional relationships.

4 Explain how the filaments of a myofibril interact during muscle contraction.

5 Explain how a motor nerve impulse can trigger a muscle contraction.

Energy Sources for Contraction

The energy used to power the interaction between actin and myosin filaments during muscle fiber contraction comes from ATP molecules. However, a muscle fiber has only enough ATP to contract briefly. Therefore, when a fiber is active, ATP must be regenerated.

The initial source of energy available to regenerate ATP from ADP and phosphate is **creatine phosphate.** Like ATP, creatine phosphate contains a high-energy phosphate bond, and it is actually four to six times more abundant in muscle fibers than ATP. Creatine phosphate, however, cannot directly supply energy to a cell's energy-utilizing reactions. Instead, it stores excess energy released from mitochondria. Thus, whenever sufficient ATP is present, an enzyme in the mitochondria (creatine phosphokinase) promotes the synthesis of creatine phosphate, which stores the excess energy in its phosphate bond (fig. 9.12).

As ATP is decomposed to ADP, the energy from creatine phosphate molecules is transferred to these ADP molecules, quickly converting them back into ATP. The

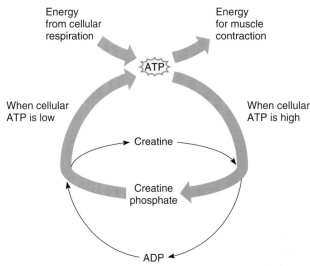

FIGURE 9.12
A muscle cell uses energy released in cellular respiration to synthesize ATP. ATP is then used to power muscle contraction or to synthesize creatine phosphate. Later, creatine phosphate may be used to synthesize ATP.

amount of ATP and creatine phosphate in a skeletal muscle, however, is usually not sufficient to support maximal muscle activity for more than about ten seconds during an intense contraction. As a result, the muscle fibers in an active muscle soon depend upon cellular respiration of glucose as a source of energy for synthesizing ATP. Typically, a muscle stores glucose in the form of glycogen.

Oxygen Supply and Cellular Respiration

Recall from chapter 4 (p. 108) that glycolysis, the early phase of cellular respiration, occurs in the cytoplasm and is *anaerobic,* not dependent on oxygen. This phase only partially breaks down energy-supplying glucose and releases only a few ATP molecules. The complete breakdown of glucose occurs in the mitochondria and is *aerobic,* requiring oxygen. This process, which includes the complex series of reactions of the *citric acid cycle* and *electron transport chain*, produces many ATP molecules.

Blood carries the oxygen necessary to support the aerobic reactions of cellular respiration from the lungs to body cells. Oxygen is transported within the red blood cells loosely bound to molecules of hemoglobin, the pigment responsible for the red color of blood. In regions of the body where the oxygen concentration is relatively low, oxygen is released from hemoglobin and becomes available for the aerobic reactions of cellular respiration.

Another pigment, **myoglobin,** is synthesized in muscle cells and imparts the reddish brown color of skeletal muscle tissue. Like hemoglobin, myoglobin can loosely combine with oxygen and, in fact, has a greater attraction for oxygen than does hemoglobin. Myoglobin can temporarily store oxygen in muscle tissue, which reduces a muscle's requirement for a continuous blood supply during contraction. This oxygen storage is important because blood flow may decrease during muscular contraction when contracting muscle fibers compress blood vessels (fig. 9.13).

Oxygen Debt

When a person is resting or moderately active, the respiratory and cardiovascular systems can usually supply sufficient oxygen to the skeletal muscles to support the aerobic reactions of cellular respiration. However, when skeletal muscles are used more strenuously, these systems may not be able to supply enough oxygen to sustain the aerobic reactions of cellular respiration. The muscle fibers must increasingly utilize the anaerobic reactions of cellular respiration for energy. This can lead to a rapid increase in blood levels of lactic acid, termed the **lactic acid threshold** (anaerobic threshold).

Chapter 4 (p. 108) discussed how under anaerobic conditions, glycolysis breaks glucose down to pyruvic acid and converts it to lactic acid, which diffuses out of

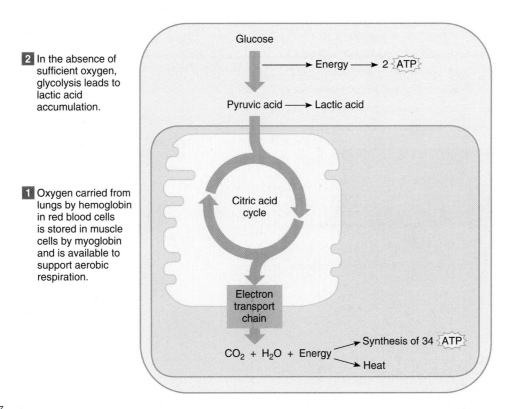

2 In the absence of sufficient oxygen, glycolysis leads to lactic acid accumulation.

1 Oxygen carried from lungs by hemoglobin in red blood cells is stored in muscle cells by myoglobin and is available to support aerobic respiration.

Glucose → Energy → 2 ATP

Pyruvic acid → Lactic acid

Citric acid cycle

Electron transport chain

$CO_2 + H_2O$ + Energy → Synthesis of 34 ATP → Heat

FIGURE 9.13

The oxygen required to support the aerobic reactions of cellular respiration is carried in the blood and stored in myoglobin. In the absence of sufficient oxygen, anaerobic reactions use pyruvic acid to produce lactic acid. The maximum number of ATPs generated per glucose molecule varies with cell type; in skeletal muscle, it is 36 (2 + 34).

the muscle fibers and is carried in the bloodstream to the liver. Liver cells can react lactic acid to form *glucose,* but this requires energy from ATP (fig. 9.14). During strenuous exercise, available oxygen is primarily used to synthesize ATP for muscle contraction rather than to make ATP for converting lactic acid into glucose. Consequently, as lactic acid accumulates, a person develops an **oxygen debt** that must be repaid at a later time. The amount of oxygen debt roughly equals the amount of oxygen liver cells require to convert the accumulated lactic acid into glucose, plus the amount the muscle cells require to resynthesize ATP and creatine phosphate, and restore their original concentrations. It also reflects the oxygen needed to restore blood and tissue oxygen levels to preexercise levels.

The metabolic capacity of a muscle may change with training. Thus, with high-intensity exercise that depends more on glycolysis for ATP, a muscle will synthesize more glycolytic enzymes, and its capacity for glycolysis will increase. With aerobic exercise, more capillaries and mitochondria will appear, and the muscles' capacity for the aerobic reactions of cellular respiration is greater.

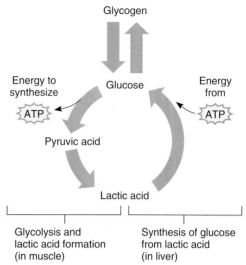

FIGURE 9.14
Liver cells can convert lactic acid, generated by muscles anaerobically, to glucose.

The runners are on the starting line, their muscles primed for a sprint. Glycogen will be broken down to release glucose, and creatine phosphate will supply high-energy phosphate groups to replenish ATP stores by phosphorylating ADP. The starting gun fires. Energy comes first from residual ATP, but almost instantaneously, creatine phosphate begins donating high-energy phosphates to ADP, regenerating ATP. Meanwhile, oxidation of glucose ultimately produces more ATP. But because the runner cannot take in enough oxygen to meet the high demand, most ATP is generated in glycolysis. Formation of lactic acid causes fatigue and possibly leg muscle cramps as the runner crosses the finish line. Already, her liver is actively converting lactic acid back to pyruvic acid and storing glycogen. In her muscles, creatine phosphate begins to build again.

Muscle Fatigue

A muscle exercised persistently for a prolonged period may lose its ability to contract, a condition called *fatigue.* This condition may result from a number of causes, including decreased blood flow, ion imbalances across the sarcolemma resulting from repeated stimulation, and psychological loss of the desire to continue the exercise. However, muscle fatigue is most likely to arise from accumulation of lactic acid in the muscle as a result of anaerobic ATP production. The lowered pH from the lactic acid prevents muscle fibers from responding to stimulation.

Occasionally a muscle fatigues and cramps at the same time. A cramp is a painful condition in which a muscle undergoes a sustained, involuntary contraction. Cramps are thought to occur when changes in the extra-

cellular fluid, particularly a decreased electrolyte concentration, surrounding the muscle fibers and their motor neurons somehow trigger uncontrolled stimulation of the muscle.

As muscle metabolism shifts from aerobic ATP production to anaerobic ATP production, lactic acid begins to accumulate in muscles and to appear in the bloodstream (lactic acid threshold). This leads to muscle fatigue. How quickly this happens varies from individual to individual, although people who regularly exercise aerobically produce less lactic acid than those who do not. Physically fit people make less lactic acid, because the strenuous exercise of aerobic training stimulates new capillaries to grow within the muscles, supplying more oxygen and nutrients to the muscle fibers. Such physical training also causes muscle fibers to produce additional mitochondria, increasing their ability to carry on the aerobic reactions of cellular respiration. Some muscle fibers may be more likely to accumulate lactic acid than others, as described in the section titled "Fast and Slow Twitch Muscle Fibers."

Heat Production

Heat is a by-product of cellular respiration; all active cells generate heat. Since muscle tissue represents such a large proportion of the total body mass, it is a major source of heat.

Less than half of the energy released in cellular respiration is available for use in metabolic processes; the rest becomes heat. Active muscles release a great deal of heat. Blood transports this heat throughout the body, which helps to maintain body temperature. Homeostatic mechanisms promote heat loss when the temperature of the internal environment begins to rise (see chapters 1 and 6, pp. 10–11 and 170, respectively).

1. What are the sources of energy used to regenerate ATP?

2. What are the sources of oxygen required for the aerobic reactions of cellular respiration?

3. How do lactic acid and oxygen debt relate to muscle fatigue?

4. What is the relationship between cellular respiration and heat production?

Muscular Responses

One way to observe muscle contraction is to remove a single muscle fiber from a skeletal muscle and connect it to a device that senses and records changes in the fiber's length. An electrical stimulator is usually used to promote muscle contraction.

Threshold Stimulus

When an isolated muscle fiber is exposed to a series of stimuli of increasing strength, the fiber remains unresponsive until a certain strength of stimulation is applied. This minimal strength required to cause contraction is called the **threshold stimulus** (thresh'old stim'u-lus). An impulse in a motor neuron normally releases enough ACh to bring the muscle fibers in its motor unit to threshold.

Recording a Muscle Contraction

The response of a single muscle fiber to the ACh released by a single action potential is called a **twitch.** A twitch involves a *period of contraction,* when the fiber pulls at its attachments, followed by a *period of relaxation,* during which pulling force declines. These events can be recorded, and the resulting pattern is called a **myogram** (fig. 9.15). Note that a twitch has a brief delay between the time of stimulation and the beginning of contraction. This is the **latent period,** which in human muscle may be less than 0.01 seconds.

If a muscle fiber is exposed to two stimuli (of threshold strength or greater) too quickly, it may respond with a twitch to the first stimulus but not to the second. This is because it takes an instant following a contraction for muscle fibers to become responsive to further stimulation. Thus, for a very brief moment following stimulation, a muscle fiber remains unresponsive. This time is called the **refractory period.**

A resting muscle fiber that is not brought to threshold will not contract. If a threshold strength or above stimulus is applied to a resting muscle fiber, enough calcium ions are released from the sarcoplasmic reticulum to fully activate cross-bridge binding. The actual force generated by that fiber depends on its length when stimulated, but at any given length, it will either contract or not. This has been termed the **all-or-none response.**

The length to which a muscle is stretched before stimulation affects the force it will develop when stimulated. If a muscle is stretched well beyond its normal resting length, the force will decrease. This is because sarcomeres become so long that myosin cross-bridges cannot reach binding sites on the thin filaments and cannot participate in contraction. Conversely, at very short muscle lengths, the sarcomere becomes compressed, and further shortening is not possible. During normal activities, muscles contract at their optimal lengths. Some activities, such as walking up stairs two steps at a time or lifting something from an awkward position, put muscles at a disadvantageous length and compromise performance.

To record how a whole muscle responds to stimulation, a skeletal muscle can be removed from a frog or other small animal and mounted in a special device. The muscle is then stimulated electrically, and when it contracts, it pulls on a lever. The lever's movement is recorded as a myogram. Because the myogram results from the combined twitches of muscle fibers taking part in the contraction, it looks essentially the same as the twitch contraction depicted in figure 9.15.

Twitch contractions are of little importance; rather, sustained contractions of whole muscles enable us to perform everyday activities. In the whole muscle, the degree of tension developed reflects (1) the frequency at which individual fibers are stimulated and (2) how many fibers take part in the overall contraction of the muscle.

Summation

The force that a muscle fiber can generate is not limited to the maximum force of a single twitch (fig. 9.16*a*). A muscle fiber exposed to a series of stimuli of increasing frequency reaches a point when it is unable to completely relax before the next stimulus in the series arrives. When this happens, the individual twitches begin to combine, and the muscle contraction becomes sustained. In such a

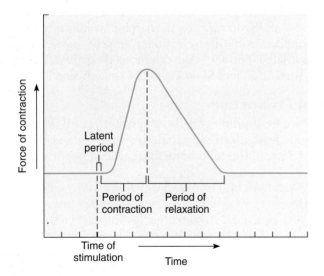

FIGURE 9.15
A myogram of a single muscle twitch.

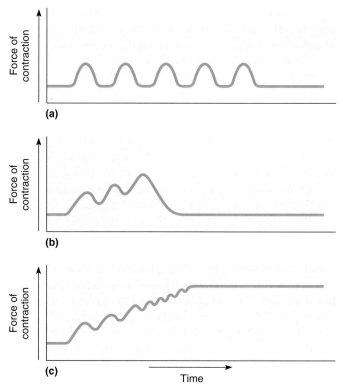

FIGURE 9.16
Myograms of (a) a series of twitches, (b) summation, and (c) a tetanic contraction. Note that stimulation frequency increases from one myogram to the next.

sustained contraction, the force of individual twitches combines by the process of **summation** (fig. 9.16b). When the resulting forceful, sustained contraction lacks even partial relaxation, it is called a **tetanic** (te-tan-ik) **contraction** (tetanus) (fig. 9.16c).

Recruitment of Motor Units

The number of muscle fibers in a motor unit varies considerably. The fewer muscle fibers in the motor units, however, the more precise the movements that can be produced in a particular muscle. For example, the motor units of the muscles that move the eyes may contain fewer than ten muscle fibers per motor unit and can produce very slight movements. Conversely, the motor units of the large muscles in the back may include a hundred or more muscle fibers. When these motor units are stimulated, the movements that result are less gradual compared to those of the eye.

Since the muscle fibers within a muscle are organized into motor units and each motor unit is controlled by a single motor neuron, all the muscle fibers in a motor unit are stimulated at the same time. Therefore, a motor unit also responds in an all-or-none manner. A whole muscle, however, does not behave like this, because it is composed of many motor units controlled by different motor neurons, some of which are more easily stimulated than others. Thus, if only the more easily stimulated motor neurons are involved, few motor units contract. At higher intensities of stimulation, other motor neurons respond, and more motor units are activated. Such an increase in the number of activated motor units is called *multiple motor unit summation,* or **recruitment** (re-kroot' ment). As the intensity of stimulation increases, recruitment of motor units continues until finally all possible motor units are activated in that muscle.

Sustained Contractions

During sustained contractions, smaller motor units, which have smaller diameter axons, tend to be recruited earlier. The larger motor units, which contain larger diameter axons, respond later and more forcefully. The product is a sustained contraction of increasing strength.

Typically, many action potentials are triggered in a motor neuron when it is called into action, thus individual twitches do not normally occur. Tetanic contractions of muscle fibers are common. On the whole-muscle level, contractions are smooth rather than irregular or jerky because the spinal cord stimulates contractions in different sets of motor units at different moments.

Tetanic contractions occur frequently in skeletal muscles during everyday activities. In many cases, the condition occurs in only a portion of a muscle. For example, when a person lifts a weight or walks, sustained contractions are maintained in the upper limb or lower limb muscles for varying lengths of time. These contractions are responses to a rapid series of stimuli transmitted from the brain and spinal cord on motor neurons.

Even when a muscle appears to be at rest, a certain amount of sustained contraction is occurring in its fibers. This is called **muscle tone** (tonus), and it is a response to nerve impulses originating repeatedly in the spinal cord and traveling to a few muscle fibers. The result is a continuous state of partial contraction.

Muscle tone is particularly important in maintaining posture. Tautness in the muscles of the neck, trunk, and lower limbs enables a person to hold the head upright, stand, or sit. If tone is suddenly lost, such as when a person loses consciousness, the body will collapse. Muscle tone is maintained in health but is lost if motor nerve axons are cut or if diseases interfere with conduction of nerve impulses.

> When skeletal muscles are contracted very forcefully, they may generate up to 50 pounds of pull for each square inch of muscle cross section. Consequently, large muscles such as those in the thigh can pull with several hundred pounds of force. Occasionally, this force is so great that the tendons of muscles tear away from their attachments to the bones.

Types of Contractions

Sometimes muscles shorten when they contract. For example, if a person lifts an object, the muscles remain taut, their attached ends pull closer together, and the object is moved. This type of contraction is termed **isotonic** (equal force—change in length), and because shortening occurs, it is called **concentric.**

Another type of isotonic contraction, called a lengthening or an **eccentric contraction,** occurs when the force a muscle generates is less than that required to move or lift an object, as in laying a book down on a table. Even in such a contraction, cross-bridges are working but not generating enough force to shorten the muscle.

At other times, a skeletal muscle contracts, but the parts to which it is attached do not move. This happens, for instance, when a person pushes against a wall. Tension within the muscles increases, but the wall does not move, and the muscles remain the same length. Contractions of this type are called **isometric** (equal length—change in force). Isometric contractions occur continuously in postural muscles that stabilize skeletal parts and hold the body upright. Figure 9.17 illustrates isotonic and isometric contractions.

Most body actions involve both isotonic and isometric contraction. In walking, for instance, certain leg and thigh muscles contract isometrically and keep the limb stiff as it touches the ground, while other muscles contract isotonically, bending the limb and lifting it. Similarly, walking down stairs involves eccentric contraction of certain thigh muscles.

Fast and Slow Twitch Muscle Fibers

Muscle fibers vary in contraction speed (slow twitch or fast twitch) and in whether they produce ATP oxidatively or glycolytically. Three combinations of these characteristics are found in humans. Slow-twitch fibers (type I) are always oxidative and are therefore resistant to fatigue. Fast-twitch fibers (type II) may be primarily glycolytic (fatigable) or primarily oxidative (fatigue resistant).

Slow-twitch (type I) fibers, such as those found in the long muscles of the back, are often called *red fibers* because they contain the red, oxygen-storing pigment myoglobin. These fibers are well supplied with oxygen-carrying blood. In addition, red fibers contain many mitochondria, an adaptation for the aerobic reactions of cellular respiration. These fibers have a high respiratory capacity and can generate ATP fast enough to keep up with the ATP breakdown that occurs when they contract. For this reason, these fibers can contract for long periods without fatiguing.

Fast-twitch glycolytic fibers (type IIa) are often called *white fibers* because they contain less myoglobin and have a poorer blood supply than red fibers. They include fibers found in certain hand muscles as well as in muscles that move the eye. These fibers have fewer mitochondria and thus have a reduced respiratory capacity. However, they have a more extensive sarcoplasmic reticulum to store and reabsorb calcium ions, and their ATPase is faster than that of red fibers. Because of these factors, white muscle fibers can contract rapidly, although they tend to fatigue as lactic acid accumulates and as the ATP and the biochemicals to regenerate ATP are depleted.

A third kind of fiber, the fast-twitch fatigue-resistant fibers (type IIb), are sometimes called *intermediate fibers.* These fibers have the fast-twitch speed associated with white fibers combined with a substantial oxidative capacity more characteristic of red fibers.

While some muscles may have mostly one fiber type or another, all muscles contain a combination of fiber types. The speed of contraction and aerobic capacities of the fibers present reflect the specialized functions of the muscle. For example, muscles that move the eyes contract about ten times faster than those that maintain posture,

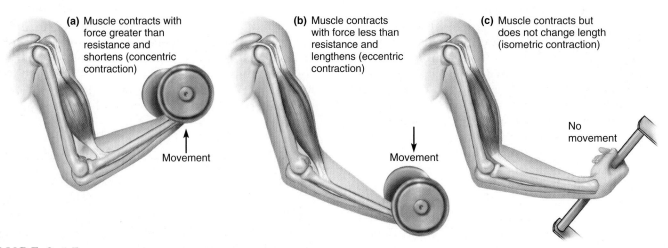

(a) Muscle contracts with force greater than resistance and shortens (concentric contraction)

Movement

(b) Muscle contracts with force less than resistance and lengthens (eccentric contraction)

Movement

(c) Muscle contracts but does not change length (isometric contraction)

No movement

FIGURE 9.17

Types of muscle contractions. (*a* and *b*) Isotonic contractions include concentric and eccentric contractions. (*c*) Isometric contractions occur when a muscle contracts but does not shorten.

USE AND DISUSE OF SKELETAL MUSCLES

Skeletal muscles are very responsive to use and disuse. Those that are forcefully exercised tend to enlarge. This phenomenon is called *muscular hypertrophy.* Conversely, a muscle that is not used *atrophies*—it decreases in size and strength.

The way a muscle responds to use also depends on the type of exercise. For instance, when a muscle contracts weakly, as during swimming and running, its slow, fatigue-resistant red fibers are most likely to be activated. As a result, these fibers develop more mitochondria and more extensive capillary networks. Such changes increase the fibers' abilities to resist fatigue during prolonged exercise, although their sizes and strengths may remain unchanged.

Forceful exercise, such as weightlifting, in which a muscle exerts more than 75% of its maximum tension, uses the muscle's fast, fatigable white fibers. In response, existing muscle fibers develop new filaments of actin and myosin, and as their diameters increase, the entire muscle enlarges. However, no new muscle fibers are produced during hypertrophy.

Since the strength of a contraction is directly proportional to the diameter of the muscle fibers, an enlarged muscle can contract more strongly than before. However, such a change does not increase the muscle's ability to resist fatigue during activities such as running or swimming.

If regular exercise stops, capillary networks shrink, and the number of mitochondria within the muscle fibers falls. Actin and myosin filaments diminish, and the entire muscle atrophies. Injured limbs immobilized in casts, or accidents or diseases that interfere with motor nerve impulses, commonly cause muscle atrophy. A muscle that cannot be exercised may shrink to less than one-half its usual size within a few months.

Muscle fibers whose motor neurons are severed not only shrink but also may fragment and, in time, be replaced by fat or fibrous tissue. However, reinnervation of such a muscle within the first few months following an injury can restore function. ■

and the muscles that move the limbs contract at intermediate rates. Clinical Application 9.2 discusses very noticeable effects of muscle use and disuse.

Birds that migrate long distances have abundant dark, slow-twitch muscles—this is why their meat is dark. In contrast, chickens that can only flap around the barnyard have abundant fast-twitch muscles, and mostly white meat.

World-class distance runners are the human equivalent of the migrating bird. Their muscles may contain over 90% slow-twitch fibers! In some European nations, athletic coaches measure slow-twitch to fast-twitch muscle fiber ratios to predict who will excel at long-distance events and who will fare better in sprints.

1 Define *threshold stimulus.*

2 What is an all-or-none response?

3 Distinguish between a twitch and a sustained contraction.

4 Define *muscle tone.*

5 Explain the differences between isometric and isotonic contractions.

6 Distinguish between fast-contracting and slow-contracting muscles fibers.

Smooth Muscles

The contractile mechanisms of smooth and cardiac muscles are essentially the same as those of skeletal muscles. However, the cells of these tissues have important structural and functional differences.

Smooth Muscle Fibers

As discussed in chapter 5 (p. 150), smooth muscle cells are shorter than the fibers of skeletal muscle, and they have single, centrally located nuclei. Smooth muscle cells are elongated with tapering ends and contain filaments of actin and myosin in myofibrils that extend throughout their lengths. However, the filaments are very thin and more randomly organized than those in skeletal muscle fibers. As a result, smooth muscle cells lack striations. They also lack transverse tubules, and their sarcoplasmic reticula are not well developed.

The two major types of smooth muscles are multiunit and visceral. In **multiunit smooth muscle,** the muscle fibers are less well organized and function as separate units, independent of neighboring cells. Smooth muscle of this type is found in the irises of the eyes and in the walls of blood vessels. Typically, multiunit smooth muscle contracts only after stimulation by motor nerve impulses or certain hormones.

Visceral smooth muscle (single-unit smooth muscle) is composed of sheets of spindle-shaped cells held in close contact by gap junctions. The thick portion of each cell lies next to the thin parts of adjacent cells. Fibers of visceral smooth muscle respond as a single unit. When one fiber is stimulated, the impulse moving over its surface may excite adjacent fibers that, in turn, stimulate others. Some visceral smooth muscle cells also display *rhythmicity*—a pattern of spontaneous repeated contractions.

These two features of visceral smooth muscle—transmission of impulses from cell to cell and rhythmicity—are largely responsible for the wavelike motion called **peristalsis** that occurs in certain tubular organs (see chapter 17, p. 648). Peristalsis consists of alternate contractions and relaxations of the longitudinal and circular muscles. These movements help force the contents of a tube along its length. In the intestines, for example, peristaltic waves move masses of partially digested food and help to mix them with digestive fluids. Peristalsis in the ureters moves urine from the kidneys to the urinary bladder.

Visceral smooth muscle is the more common type of smooth muscle and is found in the walls of hollow organs, such as the stomach, intestines, urinary bladder, and uterus. Usually there are two thickness of smooth muscle in the walls of these organs. The fibers of the outer coats are directed longitudinally, whereas those of the inner coats are arranged circularly. These muscular layers change the sizes and shapes of these organs as they function.

Smooth Muscle Contraction

Smooth muscle contraction resembles skeletal muscle contraction in a number of ways. Both mechanisms reflect reactions of actin and myosin; both are triggered by membrane impulses and release of calcium ions; and both use energy from ATP molecules. There are, however, significant differences between smooth and skeletal muscle action. For example, smooth muscle fibers lack troponin, the protein that binds to calcium ions in skeletal muscle. Instead, smooth muscle uses a protein called *calmodulin,* which binds to calcium ions released when its fibers are stimulated, thus activating the actin-myosin contraction mechanism. In addition, much of the calcium necessary for smooth muscle contraction diffuses into the cell from the extracellular fluid.

Acetylcholine, the neurotransmitter in skeletal muscle, as well as *norepinephrine,* affect smooth muscle. Each of these neurotransmitters stimulates contractions in some smooth muscles and inhibits contractions in others. The discussion of the autonomic nervous system in chapter 11 (p. 408) describes these actions in greater detail.

Hormones affect smooth muscles by stimulating or inhibiting contraction in some cases and altering the degree of response to neurotransmitters in others. For example, during the later stages of childbirth, the hormone oxytocin stimulates smooth muscles in the wall of the uterus to contract (see chapter 23, pp. 901–902).

Stretching of smooth muscle fibers can also trigger contractions. This response is particularly important to the function of visceral smooth muscle in the walls of certain hollow organs, such as the urinary bladder and the intestines. For example, when partially digested food stretches the wall of the intestine, automatic contractions move the contents away.

Smooth muscle is slower to contract and slower to relax than skeletal muscle. On the other hand, smooth muscle can forcefully contract longer with the same amount of ATP. Unlike skeletal muscle, smooth muscle fibers can change length without changing tautness; because of this, smooth muscles in the stomach and intestinal walls can stretch as these organs fill, holding the pressure inside the organs constant.

1 Describe the two major types of smooth muscle.

2 What special characteristics of visceral smooth muscle make peristalsis possible?

3 How is smooth muscle contraction similar to skeletal muscle contraction?

4 How do the contraction mechanisms of smooth and skeletal muscles differ?

Cardiac Muscle

Cardiac muscle appears only in the heart. It is composed of striated cells joined end to end, forming fibers that are interconnected in branching, three-dimensional networks. Each cell contains a single nucleus and many filaments of actin and myosin similar to those in skeletal muscle. A cardiac muscle cell also has a well-developed sarcoplasmic reticulum, a system of transverse tubules, and many mitochondria. However, the cisternae of the sarcoplasmic reticulum of a cardiac muscle fiber are less developed and store less calcium than those of a skeletal muscle fiber. On the other hand, the transverse tubules of cardiac muscle fibers are larger than those in skeletal muscle, and they release many calcium ions into the sarcoplasm in response to a single muscle impulse.

The calcium ions in transverse tubules come from the fluid outside the muscle fiber. Thus, extracellular calcium partially controls the strength of cardiac muscle contraction and enables cardiac muscle fibers to contract longer than skeletal muscle fibers can.

Drugs called calcium channel blockers are used to treat irregular heart rhythms. They do this by blocking ion channels that admit extracellular calcium into cardiac muscle cells.

The opposing ends of cardiac muscle cells are connected by cross-bands called *intercalated discs.* These bands are actually complex membrane junctions. Not only do they help join cells and transmit the force of contraction from cell to cell, but the intercellular junctions of the fused membranes of intercalated discs allow ions to diffuse between the cells. This allows muscle impulses to travel rapidly from cell to cell (see figs. 5.30 and 9.18).

When one portion of the cardiac muscle network is stimulated, the impulse passes to other fibers of the network, and the whole structure contracts as a unit (a *syncytium*); that is, the network responds to stimulation in an all-or-none manner. Cardiac muscle is also self-exciting and rhythmic. Consequently, a pattern of contraction and relaxation repeats again and again, causing the rhythmic contraction of the heart. Also, the refractory period of cardiac muscle is longer than in skeletal muscle and lasts until the contraction ends. Thus, sustained or tetanic contractions do not occur in the heart muscle. Table 9.2 summarizes characteristics of the three types of muscles.

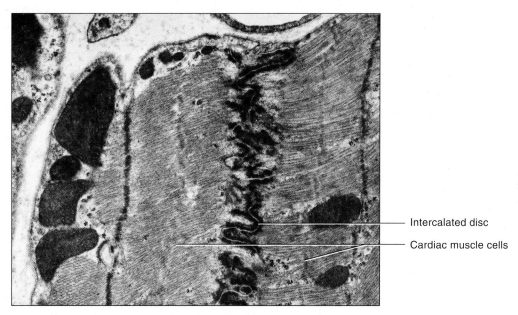

Intercalated disc

Cardiac muscle cells

FIGURE 9.18

The intercalated discs of cardiac muscle, shown in this transmission electron micrograph, bind adjacent cells and allow ions to move between cells (12,500×).

TABLE 9.2	Characteristics of Muscle Tissues		
	Skeletal	**Smooth**	**Cardiac**
Dimensions			
Length	Up to 30 cm	30–200 µm	50–100 µm
Diameter	10–100 µm	3–6 µm	14 µm
Major location	Skeletal muscles	Walls of hollow organs	Wall of the heart
Major function	Movement of bones at joints; maintenance of posture	Movement of walls of hollow organs; peristalsis; vasoconstriction	Pumping action of the heart
Cellular characteristics			
Striations	Present	Absent	Present
Nucleus	Multiple nuclei	Single nucleus	Single nucleus
Special features	Transverse tubule system is well developed	Lacks transverse tubules	Transverse tubule system is well developed; intercalated discs separate cells
Mode of control	Voluntary	Involuntary	Involuntary
Contraction characteristics	Contracts and relaxes relatively rapidly	Contracts and relaxes relatively slowly; some types self-exciting; rhythmic	Network of fibers contracts as a unit; self-exciting; rhythmic; remains refractory until contraction ends

1. How is cardiac muscle similar to skeletal muscle?

2. How does cardiac muscle differ from skeletal muscle?

3. What is the function of intercalated discs?

4. What characteristic of cardiac muscle causes the heart to contract as a unit?

Skeletal Muscle Actions

Skeletal muscles generate a great variety of body movements. The action of each muscle mostly depends upon the kind of joint it is associated with and the way the muscle is attached on either side of that joint.

Origin and Insertion

Recall from chapter 8 (p. 261) that one end of a skeletal muscle is usually fastened to a relatively immovable or fixed part, and the other end is connected to a movable part on the other side of a joint. The immovable end is called the **origin** of the muscle, and the movable end is called its **insertion.** When a muscle contracts, its insertion is pulled toward its origin (fig. 9.19). The head of a muscle is the part nearest its origin.

Some muscles have more than one origin or insertion. The *biceps brachii* in the arm, for example, has two origins. This is reflected in its name *biceps,* meaning "two heads." As figure 9.19 shows, one head of the muscle is attached to the coracoid process of the scapula, and the other head arises from a tubercle above the glenoid cavity of the scapula. The muscle extends along the anterior surface of the humerus and is inserted by a single tendon on the radial tuberosity of the radius. When the biceps brachii contracts, its insertion is pulled toward its origin, and the elbow bends.

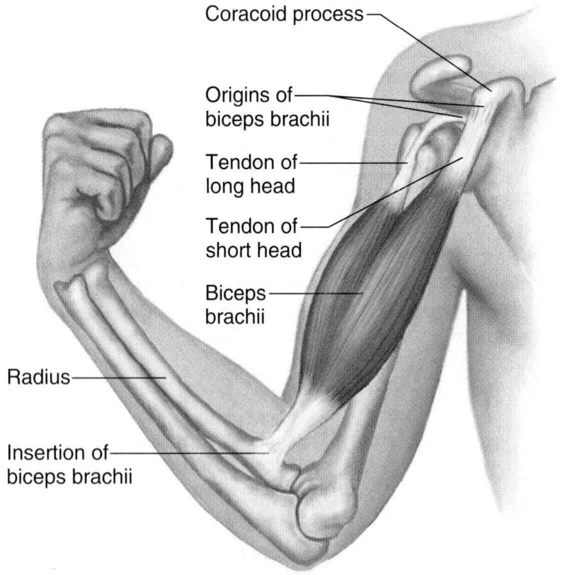

Interaction of Skeletal Muscles

Skeletal muscles almost always function in groups. As a result, when a particular body part moves, a person must do more than contract a single muscle; instead, after learning to make a particular movement, the person wills the movement to occur, and the nervous system stimulates the appropriate group of muscles.

By carefully observing body movements, it is possible to determine the roles of particular muscles. For instance, abduction of the arm requires contracting the *deltoid* muscle, which is said to be the **prime mover** or **agonist.** A prime mover is the muscle primarily responsible for producing an action. However, while a prime mover is acting, certain nearby muscles also contract. When a deltoid muscle contracts, nearby muscles help hold the shoulder steady and in this way make the action of the prime mover more effective. Muscles that contract and assist a prime mover are called **synergists** (sin′er-jists).

Still other muscles act as **antagonists** (an-tag′o-nists) to prime movers. These muscles can resist a prime mover's action and cause movement in the opposite direction—the antagonist of the prime mover that raises the upper limb can lower the upper limb, or the antagonist of the prime mover that bends the upper limb can straighten it. If both a prime mover and its antagonist contract simultaneously, the structure they act upon remains rigid. Similarly, smooth body movements depend upon the antagonists' relaxing and giving way to the prime movers whenever the prime movers contract. Once again, the nervous system controls these complex actions, as described in chapter 11 (p. 395).

FIGURE 9.19
The biceps brachii has two heads that originate on the scapula. A tendon inserts this muscle on the radius.

Labels: Coracoid process, Origins of biceps brachii, Tendon of long head, Tendon of short head, Biceps brachii, Radius, Insertion of biceps brachii

The movements termed "flexion" and "extension" describe changes in the angle between bones that meet at a joint. For example, flexion of the elbow joint refers to a movement of the forearm that decreases the angle at the elbow joint. Alternatively, one could say that flexion at the elbow results from the action of the biceps brachii on the radius of the forearm.

Since students often find it helpful to think of movements in terms of the specific actions of the muscles involved, we may also describe flexion and extension in these terms. Thus, the action of the biceps brachii may be described as "flexion of the forearm at the elbow" and the action of the quadriceps group as "extension of the leg at the knee." We believe that this occasional departure from strict anatomical terminology eases understanding and learning.

1 Distinguish between the origin and the insertion of a muscle.

2 Define *prime mover.*

3 What is the function of a synergist? An antagonist?

Major Skeletal Muscles

This section concerns the locations, actions, origins, and insertions of some of the major skeletal muscles. The tables that summarize the information concerning groups of these muscles also include the names of nerves that supply the individual muscles within each group. Chapter 11 (pp. 400–408) presents the origins and pathways of these nerves.

Figures 9.20 and 9.21 show the locations of superficial skeletal muscles—that is, those near the surface. Notice that the names of muscles often describe them. A name may indicate a muscle's size, shape, location, action, number of attachments, or the direction of its fibers, as in the following examples:

pectoralis major A muscle of large size (*major*) located in the pectoral region (chest).
deltoid Shaped like a delta or triangle.
extensor digitorum Extends the digits (fingers or toes).
biceps brachii A muscle with two heads (*biceps*), or points of origin, located in the brachium or arm.

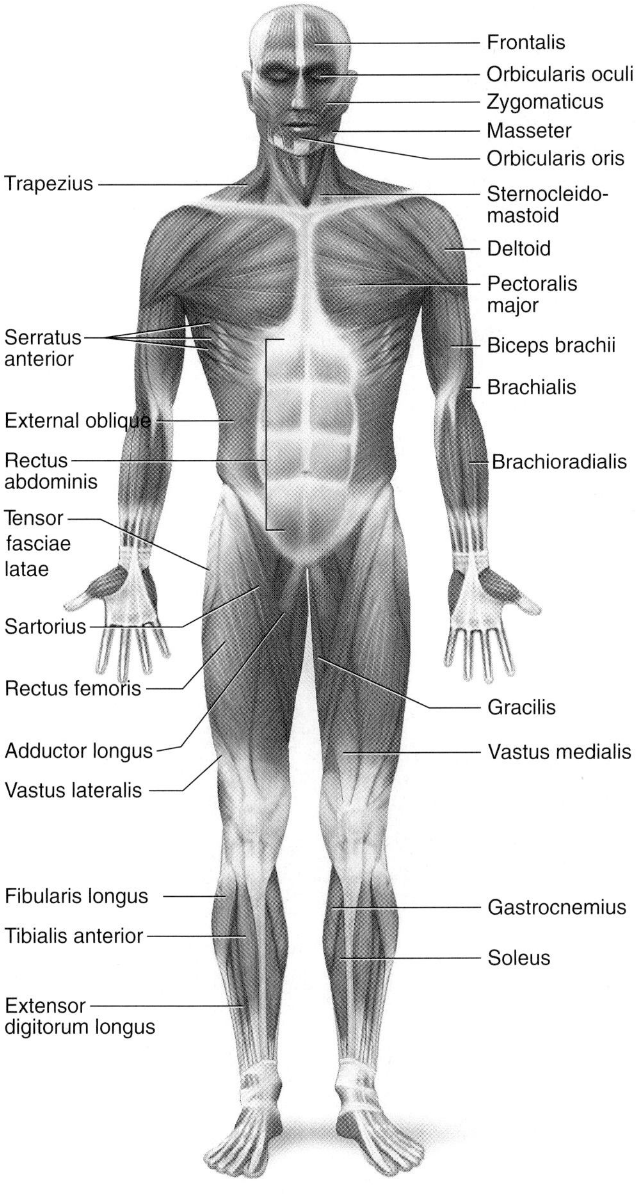

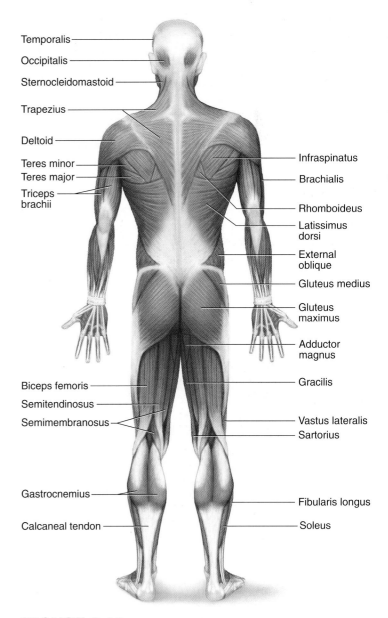

FIGURE 9.20
Anterior view of superficial skeletal muscles.

FIGURE 9.21
Posterior view of superficial skeletal muscles.

sternocleidomastoid Attached to the sternum, clavicle, and mastoid process.

external oblique Located near the outside, with fibers that run obliquely or in a slanting direction.

Muscles of Facial Expression

A number of small muscles beneath the skin of the face and scalp enable us to communicate feelings through facial expression. Many of these muscles are located around the eyes and mouth, and they make possible such expressions as surprise, sadness, anger, fear, disgust, and pain. As a group, the muscles of facial expression connect the bones of the skull to connective tissue in regions of the overlying skin. Figure 9.22 and reference plate 61 show these muscles, and table 9.3 lists them. The muscles of facial expression include the following:

Epicranius	Buccinator
Orbicularis oculi	Zygomaticus
Orbicularis oris	Platysma

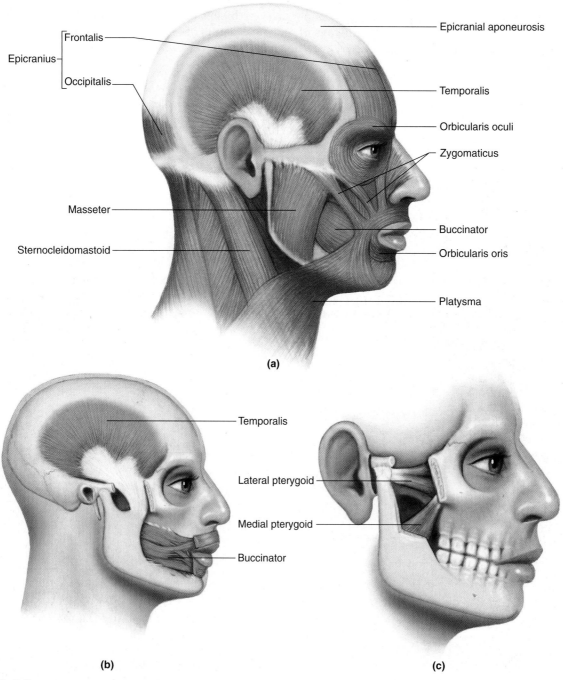

(a)

(b)

(c)

FIGURE 9.22

Muscles of head and face. (*a*) Muscles of facial expression and mastication; isolated views of (*b*) the temporalis and buccinator muscles and (*c*) the lateral and medial pterygoid muscles.

TABLE 9.3 — Muscles of Facial Expression

Muscle	Origin	Insertion	Action	Nerve Supply
Epicranius	Occipital bone	Skin and muscles around eye	Raises eyebrow as when surprised	Facial n.
Orbicularis oculi	Maxillary and frontal bones	Skin around eye	Closes eye as in blinking	Facial n.
Orbicularis oris	Muscles near the mouth	Skin of central lip	Closes lips, protrudes lips as for kissing	Facial n.
Buccinator	Outer surfaces of maxilla and mandible	Orbicularis oris	Compresses cheeks inward as when blowing air	Facial n.
Zygomaticus	Zygomatic bone	Orbicularis oris	Raises corner of mouth as when smiling	Facial n.
Platysma	Fascia in upper chest	Lower border of mandible	Draws angle of mouth downward as when pouting	Facial n.

The **epicranius** (ep″ĭ-kra′ne-us) covers the upper part of the cranium and consists of two muscular parts—the *frontalis* (frun-ta′lis), which lies over the frontal bone, and the *occipitalis* (ok-sip″ĭ-ta′lis), which lies over the occipital bone. These muscles are united by a broad, tendinous membrane called the *epicranial aponeurosis,* which covers the cranium like a cap. Contraction of the epicranius raises the eyebrows and horizontally wrinkles the skin of the forehead, as when a person expresses surprise. Headaches often result from sustained contraction of this muscle.

The **orbicularis oculi** (or-bik′u-la-rus ok′u-li) is a ringlike band of muscle, called a *sphincter muscle,* that surrounds the eye. It lies in the subcutaneous tissue of the eyelid and closes or blinks the eye. At the same time, it compresses the nearby tear gland, or *lacrimal gland,* aiding the flow of tears over the surface of the eye. Contraction of the orbicularis oculi also causes the folds, or crow's feet, that radiate laterally from the corner of the eye.

The **orbicularis oris** (or-bik′u-la-rus o′ris) is a sphincter muscle that encircles the mouth. It lies between the skin and the mucous membranes of the lips, extending upward to the nose and downward to the region between the lower lip and chin. The orbicularis oris is sometimes called the kissing muscle because it closes and puckers the lips.

The **buccinator** (buk′sĭ-na″tor) is located in the wall of the cheek. Its fibers are directed forward from the bones of the jaws to the angle of the mouth, and when they contract, the cheek is compressed inward. This action helps hold food in contact with the teeth when a person is chewing. The buccinator also aids in blowing air out of the mouth, and for this reason, it is sometimes called the trumpeter muscle.

The **zygomaticus** (zi″go-mat′ik-us) extends from the zygomatic arch downward to the corner of the mouth. When it contracts, the corner of the mouth is drawn upward, as in smiling or laughing.

The **platysma** (plah-tiz′mah) is a thin, sheetlike muscle whose fibers extend from the chest upward over the neck to the face. It pulls the angle of the mouth downward, as in pouting. The platysma also helps lower the mandible. The muscles that move the eye are described in chapter 12 (p. 448).

Muscles of Mastication

Four pairs of muscles attached to the mandible produce chewing movements. Three pairs of these muscles close the lower jaw, as in biting; the fourth pair can lower the jaw, cause side-to-side grinding motions of the mandible, and pull the mandible forward, causing it to protrude. The muscles of mastication are shown in figure 9.22 and reference plate 61 and are listed in table 9.4. They include the following:

Masseter	Medial pterygoid
Temporalis	Lateral pterygoid

The **masseter** (mas-se′ter) is a thick, flattened muscle that can be felt just in front of the ear when the teeth are clenched. Its fibers extend downward from the zygomatic arch to the mandible. The masseter raises the jaw, but it can also control the rate at which the jaw falls open in response to gravity (fig. 9.22a).

The **temporalis** (tem-po-ra′lis) is a fan-shaped muscle located on the side of the skull above and in front of the ear. Its fibers, which also raise the jaw, pass downward beneath the zygomatic arch to the mandible (fig. 9.22a and b). Tensing this muscle is associated with temporomandibular joint syndrome, discussed in Clinical Application 9.3.

The **medial pterygoid** (ter′ĭ-goid) extends back and downward from the sphenoid, palatine, and maxillary bones to the ramus of the mandible. It closes the jaw (fig. 9.22c) and moves it from side to side.

The fibers of the **lateral pterygoid** extend forward from the region just below the mandibular condyle to the sphenoid bone. This muscle can open the mouth, pull the

TMJ SYNDROME

Facial pain, headache, ringing in the ears, a clicking jaw, insomnia, teeth sensitive to heat or cold, backache, dizziness, and pain in front of the ears are aches and pains that may all result from temporomandibular joint (TMJ) syndrome. This condition is caused by a misaligned jaw or simply by a habit of grinding or clenching the teeth. These conditions may stress the temporomandibular joint, the articulation between the mandibular condyle of the mandible and the mandibular fossa of the temporal bone. Loss of coordination of these structures affects the nerves that pass through the neck and jaw region, causing the symptoms. In TMJ syndrome, tensing a muscle in the forehead can cause a headache, or a spasm in the muscle that normally opens the auditory tubes during swallowing can impair ability to clear the ears.

Doctors diagnose TMJ syndrome using an electromyograph, in which electrodes record muscle activity in four pairs of head and neck muscle groups. A form of treatment is transcutaneous electrical nerve stimulation (TENS), which stimulates the facial muscles for up to an hour. Another treatment is an orthotic device fitted by a dentist. Worn for three to six months, the device fine-tunes the action of jaw muscles to form a more comfortable bite. Finally, once the correct bite is determined, a dentist can use bonding materials to alter shapes of certain teeth to provide a more permanent treatment for TMJ syndrome. ■

TABLE 9.4 Muscles of Mastication

Muscle	Origin	Insertion	Action	Nerve Supply
Masseter	Lower border of zygomatic arch	Lateral surface of mandible	Elevates mandible	Trigeminal n.
Temporalis	Temporal bone	Coronoid process and anterior ramus of mandible	Elevates mandible	Trigeminal n.
Medial pterygoid	Sphenoid, palatine, and maxillary bones	Medial surface of mandible	Elevates mandible and moves it from side to side	Trigeminal n.
Lateral pterygoid	Sphenoid bone	Anterior surface of mandibular condyle	Depresses and protracts mandible and moves it from side to side	Trigeminal n.

mandible forward to make it protrude, and move the mandible from side to side (fig. 9.22c).

When, in 1995, two dentists examined an eyeless cadaver's skull from an unusual perspective, they discovered an apparently newly seen muscle in the head. Named the sphenomandibularis, the muscle extends about an inch and a half from behind the eyes to the inside of the jawbone and may assist chewing movements. In traditional dissection from the side, the new muscle's origin and insertion are not visible, so it may have appeared to be part of the larger and overlying temporalis muscle. Although the sphenomandibularis inserts on the inner side of the jawbone, as does the temporalis, it originates differently, on the sphenoid bone. The dentists then identified the sphenomandibularis in twenty-five other cadavers, and other researchers found it in live patients undergoing MRI scans.

Muscles That Move the Head and Vertebral Column

Paired muscles in the neck and back flex, extend, and rotate the head and hold the torso erect (figs. 9.23 and 9.25 and table 9.5). They include the following:

Sternocleidomastoid   Semispinalis capitis
Splenius capitis   Erector spinae

The **sternocleidomastoid** (ster″no-kli″do-mas′toid) is a long muscle in the side of the neck that extends upward from the thorax to the base of the skull behind the ear. When the sternocleidomastoid on one side contracts, the face turns to the opposite side. When both muscles contract, the head bends toward the chest. If other muscles fix the head in position, the sternocleidomastoids can raise the sternum, aiding forceful inhalation (fig. 9.25 and table 9.5).

The **splenius capitis** (sple′ne-us kap′ĭ-tis) is a broad, straplike muscle in the back of the neck. It connects the base of the skull to the vertebrae in the neck and upper

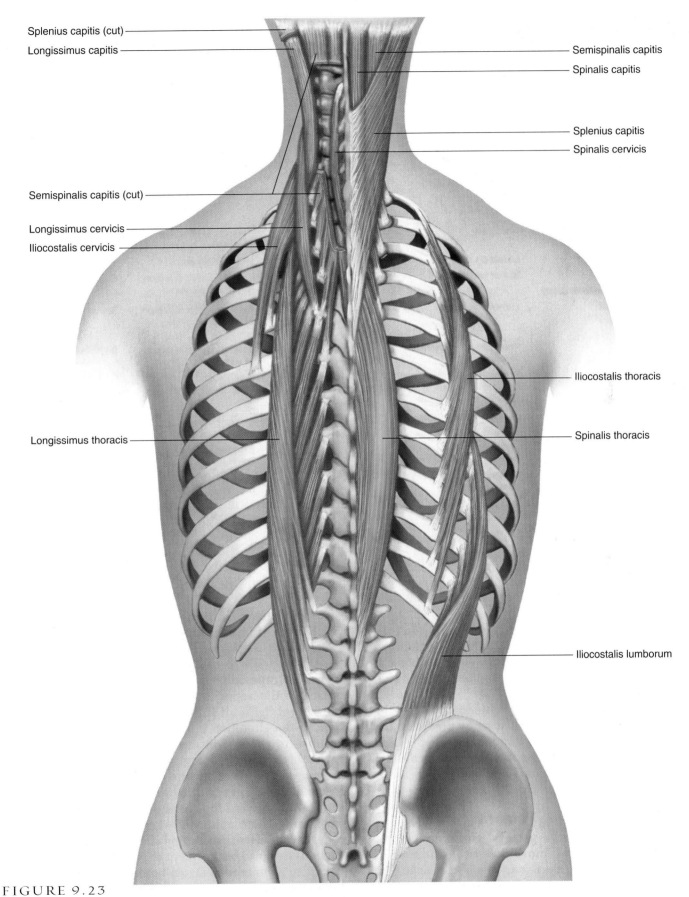

Splenius capitis (cut)
Longissimus capitis
Semispinalis capitis
Spinalis capitis
Splenius capitis
Spinalis cervicis
Semispinalis capitis (cut)
Longissimus cervicis
Iliocostalis cervicis
Iliocostalis thoracis
Longissimus thoracis
Spinalis thoracis
Iliocostalis lumborum

FIGURE 9.23

Deep muscles of the back and the neck help move the head (posterior view) and hold the torso erect. The splenius capitis and semispinalis capitis are removed on the left to show underlying muscles.

CHAPTER NINE *Muscular System*

TABLE 9.5 Muscles That Move the Head and Vertebral Column

Muscle	Origin	Insertion	Action	Nerve Supply
Sternocleidomastoid	Anterior surface of sternum and upper surface of clavicle	Mastoid process of temporal bone	Pulls head to one side, flexes neck or elevates sternum	Accessory, C2 and C3 cervical nerves
Splenius capitis	Spinous processes of lower cervical and upper thoracic vertebrae	Occipital bone	Rotates head, bends head to one side, or extends neck	Cervical nerves
Semispinalis capitis	Processes of lower cervical and upper thoracic vertebrae	Occipital bone	Extends head, bends head to one side, or rotates head	Cervical and thoracic spinal nerves
Erector spinae				
Iliocostalis (lateral) group				
Iliocostalis lumborum	Iliac crest	Lower six ribs	Extends lumbar region of vertebral column	Lumbar spinal nerves
Iliocostalis thoracis	Lower six ribs	Upper six ribs	Holds spine erect	Thoracic spinal nerves
Iliocostalis cervicis	Upper six ribs	Fourth through sixth cervical vertebrae	Extends cervical region of vertebral column	Cervical spinal nerves
Longissimus (intermediate) group				
Longissimus thoracis	Lumbar vertebrae	Thoracic and upper lumbar vertebrae and ribs 9 and 10	Extends thoracic region of vertebral column	Spinal nerves
Longissimus cervicis	Fourth and fifth thoracic vertebrae	Second through sixth cervical vertebrae	Extends cervical region of vertebral column	Spinal nerves
Longissimus capitis	Upper thoracic and lower cervical vertebrae	Mastoid process of temporal bone	Extends and rotates head	Cervical spinal nerves
Spinalis (medial) group				
Spinalis thoracis	Upper lumbar and lower thoracic vertebrae	Upper thoracic vertebrae	Extends vertebral column	Spinal nerves
Spinalis cervicis	Ligamentum nuchae and seventh cervical vertebra	Axis	Extends vertebral column	Spinal nerves
Spinalis capitis	Upper thoracic and lower cervical vertebrae	Occipital bone	Extends vertebral column	Spinal nerves

thorax. A splenius capitis acting singly rotates the head and bends it toward one side. Acting together, these muscles bring the head into an upright position (fig. 9.23 and table 9.5).

The **semispinalis capitis** (sem″e-spi-na′lis kap′ĭ-tis) is a broad, sheetlike muscle extending upward from the vertebrae in the neck and thorax to the occipital bone. It extends the head, bends it to one side, or rotates it (fig. 9.23 and table 9.5).

Erector spinae muscles run longitudinally along the back, with origins and insertions at many places on the axial skeleton. These muscles extend and rotate the head and maintain the erect position of the vertebral column. Erector spinae can be subdivided into lateral, intermediate, and medial groups (table 9.5).

Muscles That Move the Pectoral Girdle

The muscles that move the pectoral girdle are closely associated with those that move the arm. A number of these chest and shoulder muscles connect the scapula to nearby bones and move the scapula upward, downward, forward, and backward (figs. 9.24, 9.25, 9.26; reference plates 63, 64; table 9.6). Muscles that move the pectoral girdle include the following:

Trapezius Serratus anterior
Rhomboideus major Pectoralis minor
Levator scapulae

The **trapezius** (trah-pe′ze-us) is a large, triangular muscle in the upper back that extends horizontally from the base of the skull and the cervical and thoracic vertebrae to the shoulder. Its fibers are arranged into three groups—upper, middle, and lower. Together these fibers rotate the scapula. The upper fibers acting alone raise the scapula and shoulder, as when the shoulders are shrugged to express a feeling of indifference. The middle fibers pull the scapula toward the vertebral column, and the lower fibers draw the scapula and shoulder downward. When other muscles fix the shoulder in position, the trapezius can pull the head backward or to one side (fig. 9.24).

The **rhomboideus** (rom-boid′-ē-us) **major** connects the upper thoracic vertebrae to the scapula. It raises the scapula and adducts it (fig. 9.24).

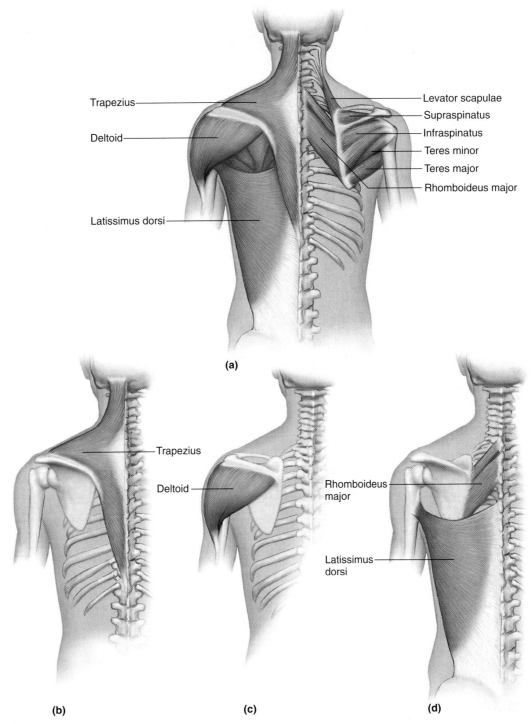

Trapezius

Deltoid

Latissimus dorsi

Levator scapulae

Supraspinatus

Infraspinatus

Teres minor

Teres major

Rhomboideus major

(a)

Trapezius

Deltoid

Rhomboideus major

Latissimus dorsi

(b)   **(c)**   **(d)**

FIGURE 9.24

Muscles of the shoulder and back. (*a*) Muscles of the posterior shoulder. The right trapezius is removed to show underlying muscles. Isolated views of (*b*) trapezius, (*c*) deltoid, and (*d*) rhomboideus and latissimus dorsi muscles.

The **levator scapulae** (le-va′tor scap′u-lē) is a strap-like muscle that runs almost vertically through the neck, connecting the cervical vertebrae to the scapula. It elevates the scapula (figs. 9.24 and 9.26).

The **serratus anterior** (ser-ra′tus an-te′re-or) is a broad, curved muscle located on the side of the chest. It arises as fleshy, narrow strips on the upper ribs and

extends along the medial wall of the axilla to the ventral surface of the scapula. It pulls the scapula downward and anteriorly and is used to thrust the shoulder forward, as when pushing something (fig. 9.25).

The **pectoralis** (pek″to-ra′lis) **minor** is a thin, flat muscle that lies beneath the larger pectoralis major. It extends laterally and upward from the ribs to the scapula and pulls

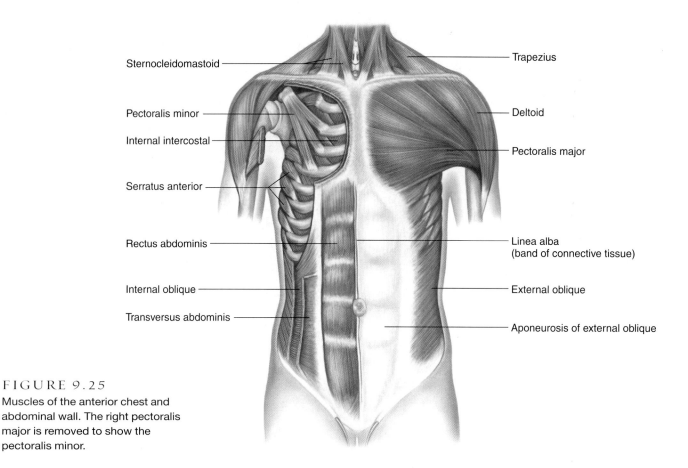

FIGURE 9.25
Muscles of the anterior chest and abdominal wall. The right pectoralis major is removed to show the pectoralis minor.

TABLE 9.6	Muscles That Move the Pectoral Girdle			
Muscle	**Origin**	**Insertion**	**Action**	**Nerve Supply**
Trapezius	Occipital bone and spines of cervical and thoracic vertebrae	Clavicle, spine, and acromion process of scapula	Rotates scapula; various fibers raise scapula, pull scapula medially, or pull scapula and shoulder downward	Accessory n.
Rhomboideus major	Spines of upper thoracic vertebrae	Medial border of scapula	Raises and adducts scapula	Dorsal scapular n.
Levator scapulae	Transverse processes of cervical vertebrae	Medial margin of scapula	Elevates scapula	Dorsal scapular and cervical nerves
Serratus anterior	Outer surfaces of upper ribs	Ventral surface of scapula	Pulls scapula anteriorly and downward	Long thoracic n.
Pectoralis minor	Sternal ends of upper ribs	Coracoid process of scapula	Pulls scapula forward and downward or raises ribs	Pectoral n.

A small, triangular region, called the *triangle of auscultation,* is located in the back where the trapezius overlaps the superior border of the latissimus dorsi and the underlying rhomboideus major. This area, which is near the medial border of the scapula, enlarges when a person bends forward with the arms folded across the chest. By placing the bell of a stethoscope within the triangle of auscultation, a physician can usually clearly hear the sounds of the respiratory organs.

the scapula forward and downward. When other muscles fix the scapula in position, the pectoralis minor can raise the ribs and thus aid forceful inhalation (fig. 9.25).

Muscles That Move the Arm

The arm is one of the more freely movable parts of the body because muscles connect the humerus to regions of the pectoral girdle, ribs, and vertebral column. These muscles can be grouped according to their primary actions—flexion, extension, abduction, and rotation (figs.

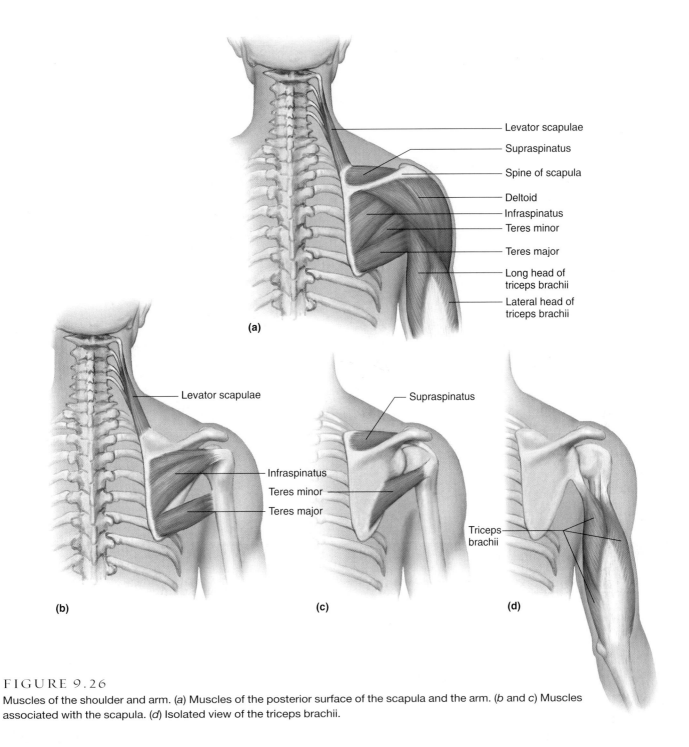

Levator scapulae
Supraspinatus
Spine of scapula
Deltoid
Infraspinatus
Teres minor
Teres major
Long head of triceps brachii
Lateral head of triceps brachii

(a)

Levator scapulae

Infraspinatus
Teres minor
Teres major

(b)

Supraspinatus

Infraspinatus
Teres minor

(c)

Triceps brachii

(d)

FIGURE 9.26

Muscles of the shoulder and arm. (*a*) Muscles of the posterior surface of the scapula and the arm. (*b* and *c*) Muscles associated with the scapula. (*d*) Isolated view of the triceps brachii.

9.26, 9.27, 9.28; reference plates 62, 63, 64; table 9.7). Muscles that move the arm include the following:

Flexors	**Abductors**
Coracobrachialis	Supraspinatus
Pectoralis major	Deltoid

Extensors	**Rotators**
Teres major	Subscapularis
Latissimus dorsi	Infraspinatus
	Teres minor

Flexors

The **coracobrachialis** (kor″ah-ko-bra′ke-al-is) extends from the scapula to the middle of the humerus along its medial surface. It flexes and adducts the arm (figs. 9.27 and 9.28).

The **pectoralis major** is a thick, fan-shaped muscle located in the upper chest. Its fibers extend from the center of the thorax through the armpit to the humerus. This muscle primarily pulls the arm forward and across the chest. It can also rotate the humerus medially and adduct the arm from a raised position (fig. 9.25).

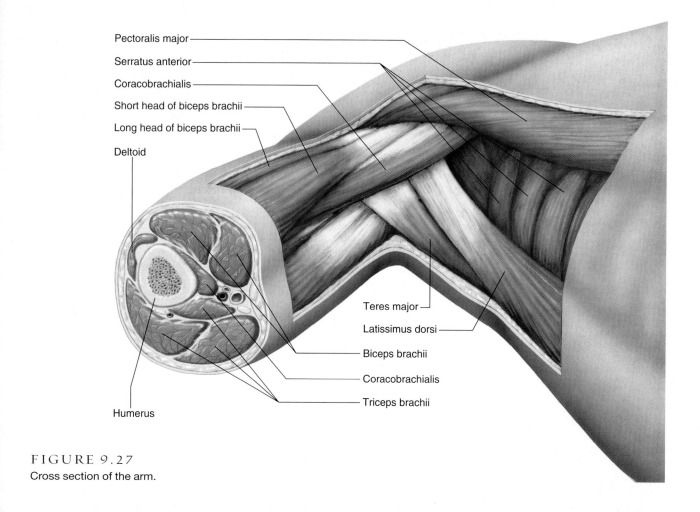

Pectoralis major
Serratus anterior
Coracobrachialis
Short head of biceps brachii
Long head of biceps brachii
Deltoid

Teres major
Latissimus dorsi
Biceps brachii
Coracobrachialis
Triceps brachii

Humerus

FIGURE 9.27
Cross section of the arm.

TABLE 9.7	Muscles That Move the Arm			
Muscle	**Origin**	**Insertion**	**Action**	**Nerve Supply**
Coracobrachialis	Coracoid process of scapula	Shaft of humerus	Flexes and adducts the arm	Musculocutaneus n.
Pectoralis major	Clavicle, sternum, and costal cartilages of upper ribs	Intertubercular groove of humerus	Flexes, adducts, and rotates arm medially	Pectoral n.
Teres major	Lateral border of scapula	Intertubercular groove of humerus	Extends, adducts, and rotates arm medially	Lower subscapular n.
Latissimus dorsi	Spines of sacral, lumbar, and lower thoracic vertebrae, iliac crest, and lower ribs	Intertubercular groove of humerus	Extends, adducts, and rotates the arm medially, or pulls the shoulder downward and back	Thoracodorsal n.
Supraspinatus	Posterior surface of scapula above spine	Greater tubercle of humerus	Abducts the arm	Suprascapular n.
Deltoid	Acromion process, spine of the scapula, and the clavicle	Deltoid tuberosity of humerus	Abducts, extends, and flexes arm	Axillary n.
Subscapularis	Anterior surface of scapula	Lesser tubercle of humerus	Rotates arm medially	Subscapular n.
Infraspinatus	Posterior surface of scapula below spine	Greater tubercle of humerus	Rotates arm laterally	Suprascapular n.
Teres minor	Lateral border of scapula	Greater tubercle of humerus	Rotates arm laterally	Axillary n.

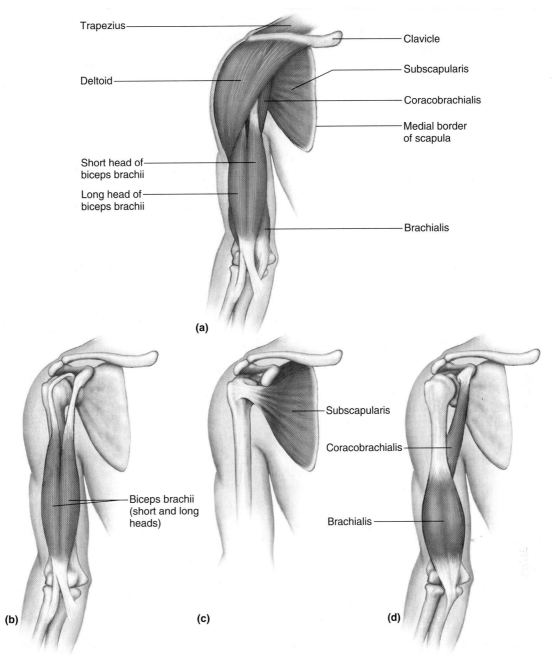

FIGURE 9.28
Muscles of the shoulder and arm. (*a*) Muscles of the anterior shoulder and the arm, with the rib cage removed. (*b, c,* and *d*) Isolated views of muscles associated with the arm.

Extensors

The **teres** (te′rēz) **major** connects the scapula to the humerus. It extends the humerus and can also adduct and rotate the arm medially (figs. 9.24 and 9.26).

The **latissimus dorsi** (lah-tis′ĭ-mus dor′si) is a wide, triangular muscle that curves upward from the lower back, around the side, and to the armpit. It can extend and adduct the arm and rotate the humerus medially. It also pulls the shoulder downward and back. This muscle is used to pull the arm back in swimming, climbing, and rowing (figs. 9.24 and 9.27).

Abductors

The **supraspinatus** (su″prah-spi′na-tus) is located in the depression above the spine of the scapula on its posterior surface. It connects the scapula to the greater tubercle of the humerus and abducts the arm (figs. 9.24 and 9.26).

The **deltoid** (del′toid) is a thick, triangular muscle that covers the shoulder joint. It connects the clavicle and scapula to the lateral side of the humerus and abducts the arm. The deltoid's posterior fibers can extend the humerus, and its anterior fibers can flex the humerus (fig. 9.24).

Rotators

The **subscapularis** (sub-scap′u-lar-is) is a large, triangular muscle that covers the anterior surface of the scapula. It connects the scapula to the humerus and rotates the arm medially (fig. 9.28).

The **infraspinatus** (in″frah-spi′na-tus) occupies the depression below the spine of the scapula on its posterior surface. The fibers of this muscle attach the scapula to the humerus and rotate the arm laterally (fig. 9.26).

The **teres minor** is a small muscle connecting the scapula to the humerus. It rotates the arm laterally (figs. 9.24 and 9.26).

Muscles That Move the Forearm

Most forearm movements are produced by muscles that connect the radius or ulna to the humerus or pectoral girdle. A group of muscles located along the anterior surface of the humerus flexes the forearm at the elbow, whereas a single posterior muscle extends this joint. Other muscles cause movements at the radioulnar joint and rotate the forearm.

The muscles that move the forearm are shown in figures 9.28, 9.29, 9.30, 9.31, in reference plates 63, 65, and

are listed in table 9.8, grouped according to their primary actions. They include the following:

Flexors	Extensor	Rotators
Biceps brachii	Triceps brachii	Supinator
Brachialis		Pronator teres
Brachioradialis		Pronator quadratus

Flexors

The **biceps brachii** (bi′seps bra′ke-i) is a fleshy muscle that forms a long, rounded mass on the anterior side of the arm. It connects the scapula to the radius and flexes the forearm at the elbow and rotates the hand laterally (supination), as when a person turns a doorknob or screwdriver (fig. 9.28).

The **brachialis** (bra′ke-al-is) is a large muscle beneath the biceps brachii. It connects the shaft of the humerus to the ulna and is the strongest flexor of the elbow (fig. 9.28).

The **brachioradialis** (bra″ke-o-ra″de-a′lis) connects the humerus to the radius. It aids in flexing the elbow (fig. 9.29).

Extensor

The **triceps brachii** (tri′seps bra′ke-i) has three heads and is the only muscle on the back of the arm. It connects the humerus and scapula to the ulna and is the primary extensor of the elbow (figs. 9.26 and 9.27).

Rotators

The **supinator** (su′pĭ-na-tor) is a short muscle whose fibers run from the ulna and the lateral end of the humerus to the radius. It assists the biceps brachii in rotating the forearm laterally (supination) (fig. 9.29).

The **pronator teres** (pro-na′tor te′rēz) is a short muscle connecting the ends of the humerus and ulna to

TABLE 9.8	Muscles That Move the Forearm			
Muscle	**Origin**	**Insertion**	**Action**	**Nerve Supply**
Biceps brachii	Coracoid process and tubercle above glenoid cavity of scapula	Radial tuberosity of radius	Flexes forearm at elbow and rotates hand laterally	Musculocutaneous n.
Brachialis	Anterior shaft of humerus	Coronoid process of ulna	Flexes forearm at elbow	Musculocutaneous, median, and radial nerves
Brachioradialis	Distal lateral end of humerus	Lateral surface of radius above styloid process	Flexes forearm at elbow	Radial n.
Triceps brachii	Tubercle below glenoid cavity and lateral and medial surfaces of humerus	Olecranon process of ulna	Extends forearm at elbow	Radial n.
Supinator	Lateral epicondyle of humerus and crest of ulna	Lateral surface of radius	Rotates forearm laterally	Radial n.
Pronator teres	Medial epicondyle of humerus and coronoid process of ulna	Lateral surface of radius	Rotates forearm medially	Median n.
Pronator quadratus	Anterior distal end of ulna	Anterior distal end of radius	Rotates forearm medially	Median n.

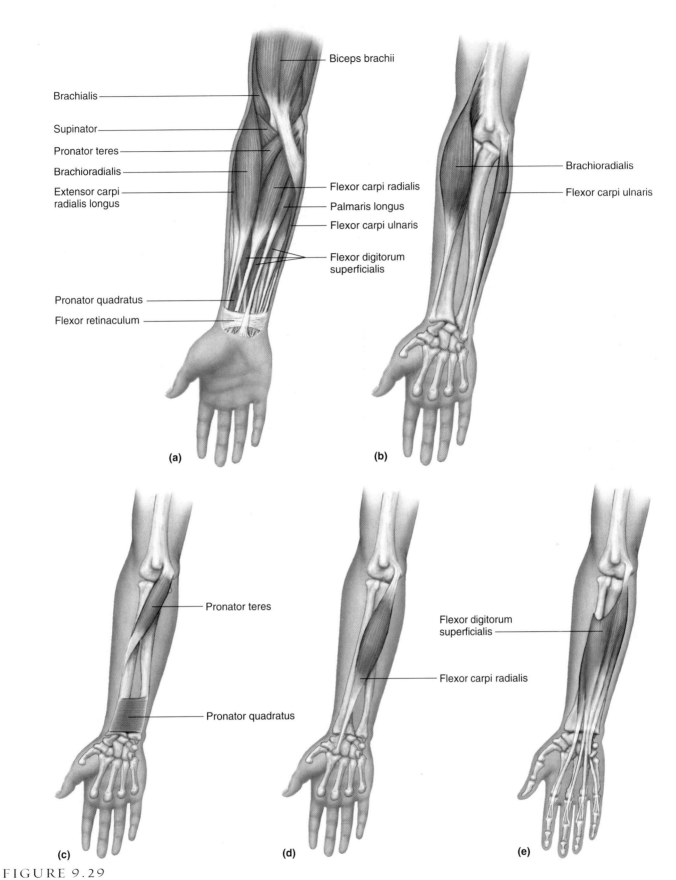

FIGURE 9.29

Muscles of the arm and forearm. (*a*) Muscles of the anterior forearm. (*b–e*) Isolated views of muscles associated with the anterior forearm.

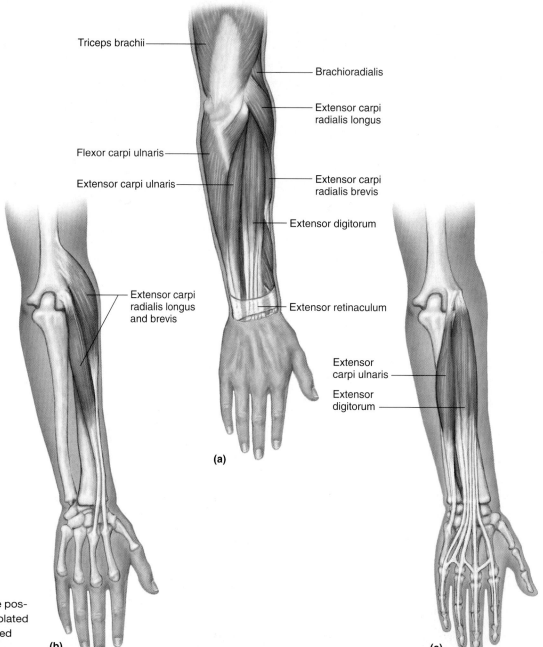

Triceps brachii
Brachioradialis
Extensor carpi radialis longus
Flexor carpi ulnaris
Extensor carpi ulnaris
Extensor carpi radialis brevis
Extensor digitorum
Extensor carpi radialis longus and brevis
Extensor retinaculum
Extensor carpi ulnaris
Extensor digitorum

(a)

(b)

(c)

FIGURE 9.30
Muscles of the arm and forearm. (a) Muscles of the posterior forearm. (b and c) Isolated views of muscles associated with the posterior forearm.

the radius. It rotates the arm medially, as when the hand is turned so the palm is facing downward (pronation) (fig. 9.29).

The **pronator quadratus** (pro-na'tor kwod-ra'tus) runs from the distal end of the ulna to the distal end of the radius. It assists the pronator teres in rotating the arm medially (fig. 9.29).

Muscles That Move the Hand

Movements of the hand include movements of the wrist and fingers. Many of the involved muscles originate from the distal end of the humerus and from the radius and ulna. The two major groups of these muscles are flexors

on the anterior side of the forearm and extensors on the posterior side. Figures 9.29, 9.30, 9.31, reference plate 65, and table 9.9 concern these muscles. The muscles that move the hand include the following:

Flexors	Extensors
Flexor carpi radialis longus	Extensor carpi radialis
Flexor carpi ulnaris	Extensor carpi radialis brevis
Palmaris longus	
Flexor digitorum profundus	Extensor carpi ulnaris
Flexor digitorum superficialis	Extensor digitorum

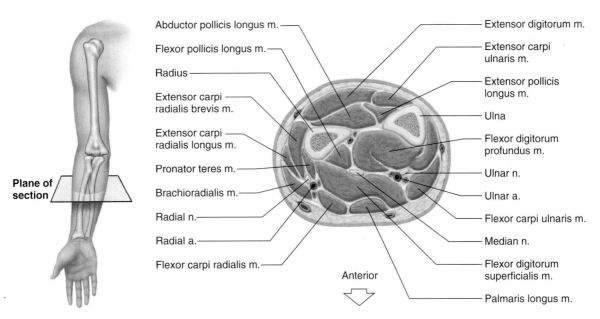

FIGURE 9.31

A cross section of the forearm (superior view) (*a.* stands for artery, *m.* stands for muscle, and *n.* stands for nerve.)

TABLE 9.9	Muscles That Move the Hand			
Muscle	**Origin**	**Insertion**	**Action**	**Nerve Supply**
Flexor carpi radialis	Medial epicondyle of humerus	Base of second and third metacarpals	Flexes wrist and abducts hand	Median n.
Flexor carpi ulnaris	Medial epicondyle of humerus and olecranon process	Carpal and metacarpal bones	Flexes wrist and adducts hand	Ulnar n.
Palmaris longus	Medial epicondyle of humerus	Fascia of palm	Flexes the wrist	Median n.
Flexor digitorum profundus	Anterior surface of ulna	Bases of distal phalanges in fingers 2–5	Flexes distal joints of fingers	Median and ulnar nerves
Flexor digitorum superficialis	Medial epicondyle of humerus, coronoid process of ulna, and radius	Tendons of fingers	Flexes fingers and wrist	Median n.
Extensor carpi radialis longus	Distal end of humerus	Base of second metacarpal	Extends wrist and abducts hand	Radial n.
Extensor carpi radialis brevis	Lateral epicondyle of humerus	Base of second and third metacarpals	Extends wrist and abducts hand	Radial n.
Extensor carpi ulnaris	Lateral epicondyle of humerus	Base of fifth metacarpal	Extends wrist and adducts hand	Radial n.
Extensor digitorum	Lateral epicondyle of humerus	Posterior surface of phalanges in fingers 2–5	Extends fingers	Radial n.

Flexors

The **flexor carpi radialis** (flek′sor kar-pi′ra″de-a′lis) is a fleshy muscle that runs medially on the anterior side of the forearm. It extends from the distal end of the humerus into the hand, where it is attached to metacarpal bones. The flexor carpi radialis flexes the wrist and abducts the hand (fig. 9.29).

The **flexor carpi ulnaris** (flek′sor kar-pi′ ul-na′ris) is located along the medial border of the forearm. It con-nects the distal end of the humerus and the proximal end of the ulna to carpal and metacarpal bones. It flexes the wrist and adducts the hand (fig. 9.29).

The **palmaris longus** (pal-ma′ris long′gus) is a slen-der muscle located on the medial side of the forearm between the flexor carpi radialis and the flexor carpi ulnaris. It connects the distal end of the humerus to fascia of the palm and flexes the wrist (fig. 9.29).

The **flexor digitorum profundus** (flek'sor dij"ĭ-to'rum pro-fun'dus) is a large muscle that connects the ulna to the distal phalanges. It flexes the distal joints of the fingers, as when a fist is made (fig. 9.31).

The **flexor digitorum superficialis** (flek'sor dij"ĭ-to'rum su"per-fish"e-a'lis) is a large muscle located beneath the flexor carpi ulnaris. It arises by three heads—one from the medial epicondyle of the humerus, one from the medial side of the ulna, and one from the radius. It is inserted in the tendons of the fingers and flexes the fingers and, by a combined action, flexes the wrist (fig. 9.29).

Extensors

The **extensor carpi radialis longus** (eks-ten'sor kar-pi' ra"de-a'lis long'gus) runs along the lateral side of the forearm, connecting the humerus to the hand. It extends the wrist and assists in abducting the hand (figs. 9.30 and 9.31).

The **extensor carpi radialis brevis** (eks-ten'sor kar-pi' ra"de-a'lis brev'ĭs) is a companion of the extensor carpi radialis longus and is located medially to it. This muscle runs from the humerus to metacarpal bones and extends the wrist. It also assists in abducting the hand (figs. 9.30 and 9.31).

The **extensor carpi ulnaris** (eks-ten'sor kar-pi' ul-na'ris) is located along the posterior surface of the ulna and connects the humerus to the hand. It extends the wrist and assists in adducting the hand (figs. 9.30 and 9.31).

The **extensor digitorum** (eks-ten'sor dij"ĭ-to rum) runs medially along the back of the forearm. It connects the humerus to the posterior surface of the phalanges and extends the fingers (figs. 9.30 and 9.31).

A structure called the *extensor retinaculum* consists of a group of heavy connective tissue fibers in the fascia of the wrist (fig. 9.30). It connects the lateral margin of the radius with the medial border of the styloid process of the ulna and certain bones of the wrist. The retinaculum gives off branches of connective tissue to the underlying wrist bones, creating a series of sheathlike compartments through which the tendons of the extensor muscles pass to the wrist and fingers.

Muscles of the Abdominal Wall

The walls of the chest and pelvic regions are supported directly by bone, but those of the abdomen are not. Instead, the anterior and lateral walls of the abdomen are composed of layers of broad, flattened muscles. These muscles connect the rib cage and vertebral column to the pelvic girdle. A band of tough connective tissue, called the **linea alba** (lin'e-ah al'bah), extends from the xiphoid process of the sternum to the symphysis pubis. It is an attachment for some of the abdominal wall muscles.

Contraction of these muscles decreases the volume of the abdominal cavity and increases the pressure inside. This action helps force air out of the lungs during forceful exhalation and also aids in defecation, urination, vomiting, and childbirth.

The abdominal wall muscles are shown in figure 9.32, reference plate 62, and are listed in table 9.10. They include the following:

External oblique   Transversus abdominis
Internal oblique   Rectus abdominis

TABLE 9.10	Muscles of the Abdominal Wall			
Muscle	**Origin**	**Insertion**	**Action**	**Nerve Supply**
External oblique	Outer surfaces of lower ribs	Outer lip of iliac crest and linea alba	Tenses abdominal wall and compresses abdominal contents	Intercostal nerves 7–12
Internal oblique	Crest of ilium and inguinal ligament	Cartilages of lower ribs, linea alba, and crest of pubis	Same as above	Intercostal nerves 7–12
Transversus abdominis	Costal cartilages of lower ribs, processes of lumbar vertebrae, lip of iliac crest, and inguinal ligament	Linea alba and crest of pubis	Same as above	Intercostal nerves 7–12
Rectus abdominis	Crest of pubis and symphysis pubis	Xiphoid process of sternum and costal cartilages	Same as above; also flexes vertebral column	Intercostal nerves 7–12

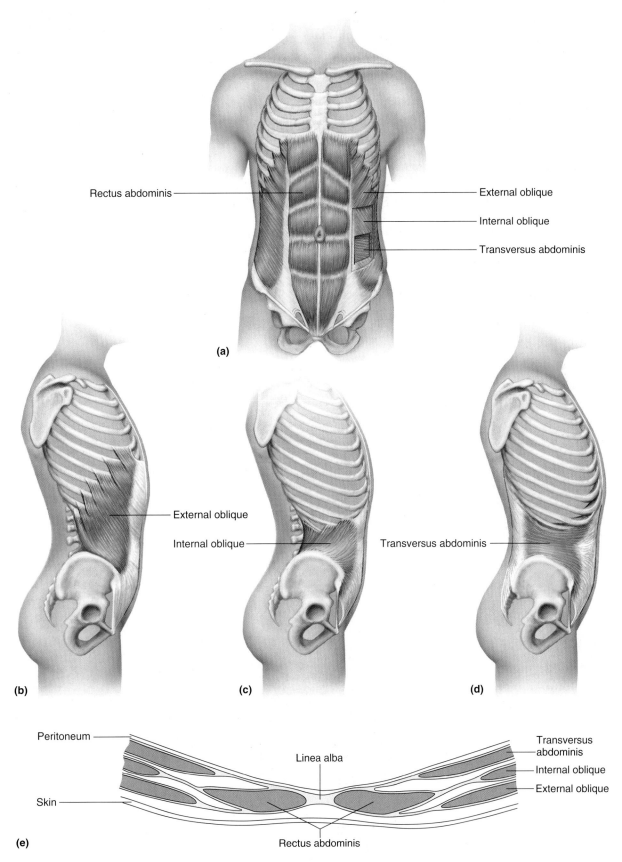

Rectus abdominis

External oblique

Internal oblique

Transversus abdominis

(a)

External oblique

Internal oblique

(b)

(c)

Transversus abdominis

(d)

Peritoneum

Transversus abdominis

Internal oblique

External oblique

Linea alba

Skin

Rectus abdominis

(e)

FIGURE 9.32

Muscles of the abdominal wall. (*a–d*) Isolated muscles of the abdominal wall. (*e*) Transverse section through the abdominal wall.

The **external oblique** (eks-ter′nal ŏ-blēk) is a broad, thin sheet of muscle whose fibers slant downward from the lower ribs to the pelvic girdle and the linea alba. When this muscle contracts, it tenses the abdominal wall and compresses the contents of the abdominal cavity.

Similarly, the **internal oblique** (in-ter′nal ŏ-blēk) is a broad, thin sheet of muscle located beneath the external oblique. Its fibers run up and forward from the pelvic girdle to the lower ribs. Its function is similar to that of the external oblique.

The **transversus abdominis** (trans-ver′sus ab-dom′ĭ-nis) forms a third layer of muscle beneath the external and internal obliques. Its fibers run horizontally from the lower ribs, lumbar vertebrae, and ilium to the linea alba and pubic bones. It functions in the same manner as the external and internal obliques.

The **rectus abdominis** (rek′tus ab-dom′ĭ-nis) is a long, straplike muscle that connects the pubic bones to the ribs and sternum. Three or more fibrous bands cross the muscle transversely, giving it a segmented appearance. The muscle functions with other abdominal wall muscles to compress the contents of the abdominal cavity, and it also helps to flex the vertebral column.

Muscles of the Pelvic Outlet

Two muscular sheets span the outlet of the pelvis—a deeper **pelvic diaphragm** and a more superficial **urogenital diaphragm.** The pelvic diaphragm forms the floor of the pelvic cavity, and the urogenital diaphragm fills the space within the pubic arch. Figure 9.33 and table 9.11 show the muscles of the male and female pelvic outlets. They include the following:

Pelvic Diaphragm	Urogenital Diaphragm
Levator ani	Superficial transversus perinei
Coccygeus	Bulbospongiosus
	Ischiocavernosus
	Sphincter urethrae

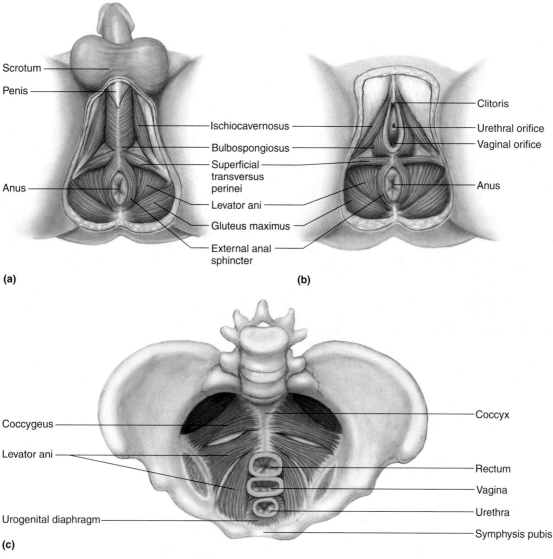

(a)
(b)
(c)

FIGURE 9.33

External view of muscles of (a) the male pelvic outlet and (b) the female pelvic outlet. (c) Internal view of female pelvic and urogenital diaphragms.

TABLE 9.11   Muscles of the Pelvic Outlet

Muscle	Origin	Insertion	Action	Nerve Supply
Levator ani	Pubic bone and ischial spine	Coccyx	Supports pelvic viscera and provides sphincterlike action in anal canal and vagina	Pudendal n.
Coccygeus	Ischial spine	Sacrum and coccyx	Same as above	S4 and S5 nerves
Superficial transversus perinei	Ischial tuberosity	Central tendon	Supports pelvic viscera	Pudendal n.
Bulbospongiosus	Central tendon	Males: Urogenital diaphragm and fascia of penis	Males: Assists emptying of urethra	Pudendal n.
		Females: Pubic arch and root of clitoris	Females: Constricts vagina	
Ischiocavernosus	Ischial tuberosity	Pubic arch	Assists function of bulbospongiosus	Pudendal n.
Sphincter urethrae	Margins of pubis and ischium	Fibers of each unite with those from other side	Opens and closes urethra	Pudendal n.

Pelvic Diaphragm

The **levator ani** (le-va′tor ah-ni′) muscles form a thin sheet across the pelvic outlet. They are connected at the midline posteriorly by a ligament that extends from the tip of the coccyx to the anal canal. Anteriorly, they are separated in the male by the urethra and the anal canal, and in the female by the urethra, vagina, and anal canal. These muscles help support the pelvic viscera and provide sphincterlike action in the anal canal and vagina.

An *external anal sphincter* that is under voluntary control and an *internal anal sphincter* that is formed of involuntary muscle fibers of the intestine encircle the anal canal and keep it closed.

The **coccygeus** (kok-sij′e-us) is a fan-shaped muscle that extends from the ischial spine to the coccyx and sacrum. It aids the levator ani.

Urogenital Diaphragm

The **superficial transversus perinei** (su″per-fish′al transver′sus per″ĭ-ne′i) consists of a small bundle of muscle fibers that passes medially from the ischial tuberosity along the posterior border of the urogenital diaphragm. It assists other muscles in supporting the pelvic viscera.

In males, the **bulbospongiosus** (bul″bo-spon″je-o′sus) muscles are united surrounding the base of the penis. They assist in emptying the urethra. In females, these muscles are separated medially by the vagina and constrict the vaginal opening. They can also retard the flow of blood in veins, which helps maintain an erection in the penis of the male and in the clitoris of the female.

The **ischiocavernosus** (is″ke-o-kav″er-no′sus) muscle is a tendinous structure that extends from the ischial tuberosity to the margin of the pubic arch. It assists the bulbospongiosus muscle.

The **sphincter urethrae** (sfingk′ter u-re′thrē) are muscles that arise from the margins of the pubic and ischial bones. Each arches around the urethra and unites with the one on the other side. Together they act as a sphincter that closes the urethra by compression and opens it by relaxation, thus helping control the flow of urine.

Muscles That Move the Thigh

The muscles that move the thigh are attached to the femur and to some part of the pelvic girdle. (An important exception is the sartorius, described later.) They can be separated into anterior and posterior groups. The muscles of the anterior group primarily flex the thigh; those of the posterior group extend, abduct, or rotate it. The muscles in these groups are shown in figures 9.34, 9.35, 9.36, 9.37, in reference plates 66 and 67, and are listed in table 9.12. Muscles that move the thigh include the following:

Anterior Group	**Posterior Group**
Psoas major	Gluteus maximus
Iliacus	Gluteus medius
	Gluteus minimus
	Tensor fasciae latae

Still another group of muscles, attached to the femur and pelvic girdle, adducts the thigh. This group includes the following:

Pectineus	Adductor magnus
Adductor longus	Gracilis

Anterior Group

The **psoas** (so′as) **major** is a long, thick muscle that connects the lumbar vertebrae to the femur. It flexes the thigh (fig. 9.34).

The **iliacus** (il′e-ak-us), a large, fan-shaped muscle, lies along the lateral side of the psoas major. The iliacus and the psoas major are the primary flexors of the thigh, and they advance the lower limb in walking movements (fig. 9.34).

Posterior Group

The **gluteus maximus** (gloo′te-us mak′si-mus) is the largest muscle in the body and covers a large part of each buttock. It connects the ilium, sacrum, and coccyx to the

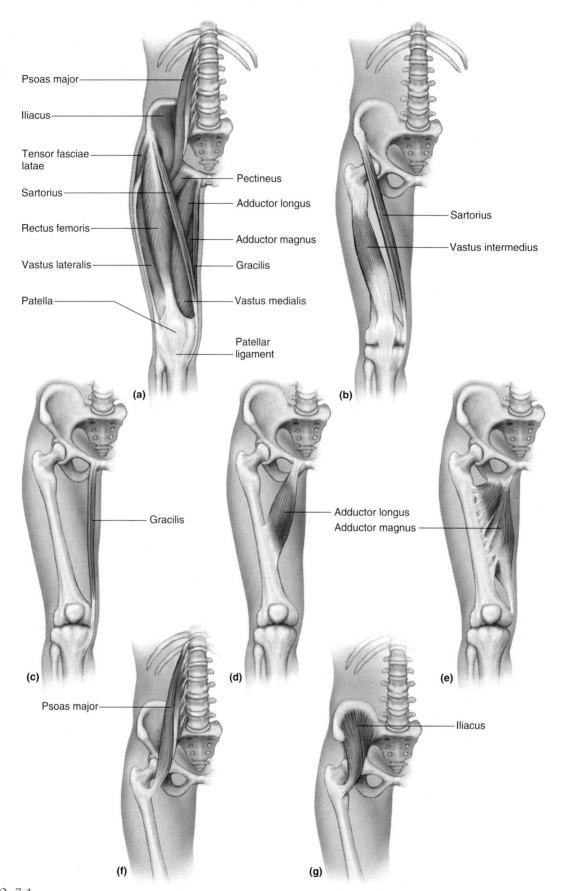

Psoas major

Iliacus

Tensor fasciae latae

Sartorius

Rectus femoris

Vastus lateralis

Patella

Pectineus

Adductor longus

Adductor magnus

Gracilis

Vastus medialis

Patellar ligament

(a)

Sartorius

Vastus intermedius

(b)

Gracilis

(c)

(d)

Adductor longus

Adductor magnus

(e)

Psoas major

(f)

Iliacus

(g)

FIGURE 9.34

Muscles of the thigh and leg. (a) Muscles of the anterior right thigh. Isolated views of (b) the vastus intermedius, (c–e) adductors of the thigh, (f–g) flexors of the thigh.

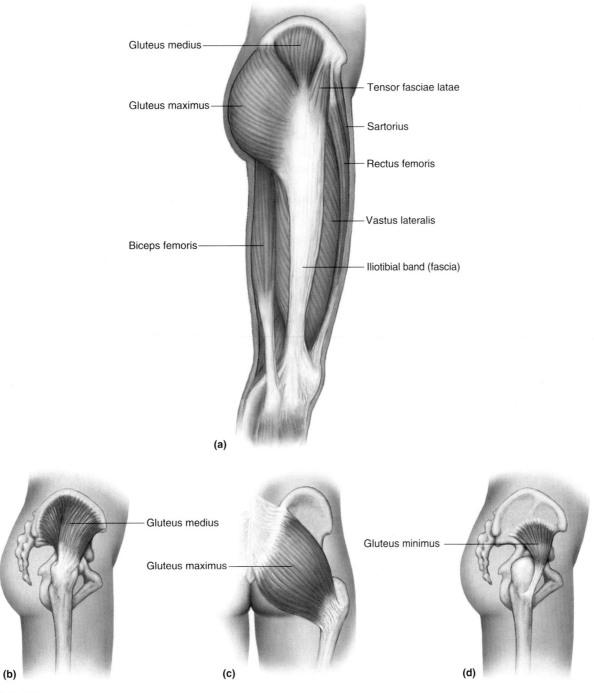

Gluteus medius

Gluteus maximus

Tensor fasciae latae

Sartorius

Rectus femoris

Vastus lateralis

Biceps femoris

Iliotibial band (fascia)

(a)

Gluteus medius

Gluteus maximus

Gluteus minimus

(b)   **(c)**   **(d)**

FIGURE 9.35
Muscles of the thigh and leg. (*a*) Muscles of the lateral right thigh. (*b–d*) Isolated views of the gluteal muscles.

femur by fascia of the thigh and extends the thigh. The gluteus maximus helps to straighten the lower limb at the hip when a person walks, runs, or climbs. It is also used to raise the body from a sitting position (fig. 9.35).

The **gluteus medius** (gloo′te-us me′de-us) is partly covered by the gluteus maximus. Its fibers extend from the ilium to the femur, and they abduct the thigh and rotate it medially (fig. 9.35).

The **gluteus minimus** (gloo′te-us min′ĭ-mus) lies beneath the gluteus medius and is its companion in attachments and functions (fig. 9.35).

The **tensor fasciae latae** (ten′sor fash′e-e lah-tē) connects the ilium to the iliotibial band (fascia of the thigh), which continues downward to the tibia. This muscle abducts and flexes the thigh and rotates it medially (fig. 9.35).

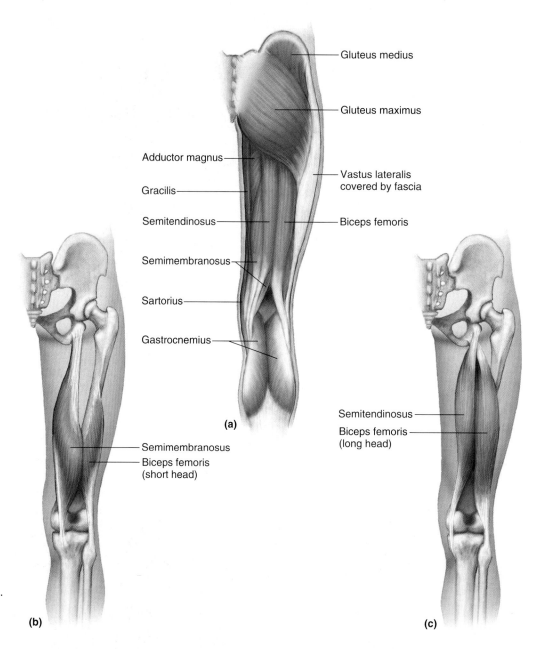

FIGURE 9.36
Muscles of the thigh and leg.
(*a*) Muscles of the posterior right thigh.
(*b* and *c*) Isolated views of muscles
that flex the leg at the knee.

The gluteus medius and gluteus minimus help support and maintain the normal position of the pelvis. If these muscles are paralyzed as a result of injury or disease, the pelvis tends to drop to one side whenever the foot on that side is raised. Consequently, the person walks with a waddling limp called the *gluteal gait.*

Thigh Adductors

The **pectineus** (pek-tin′e-us) muscle runs from the spine of the pubis to the femur. It adducts and flexes the thigh (fig. 9.34).

The **adductor longus** (ah-duk′tor long′gus) is a long, triangular muscle that runs from the pubic bone to the femur. It adducts the thigh and assists in flexing and rotating it laterally (fig. 9.34).

The **adductor magnus** (ah-duk′tor mag′nus) is the largest adductor of the thigh. It is a triangular muscle that connects the ischium to the femur. It adducts the thigh and assists in extending and rotating it laterally (fig. 9.34).

The **gracilis** (gras′il-is) is a long, straplike muscle that passes from the pubic bone to the tibia. It adducts the thigh and flexes the leg at the knee (fig. 9.34).

Muscles That Move the Leg

The muscles that move the leg connect the tibia or fibula to the femur or to the pelvic girdle. They fall into two major groups—those that flex the knee and those that extend it. The muscles of these groups are shown in figures 9.34, 9.35, 9.36, 9.37, in reference plates 66 and 67,

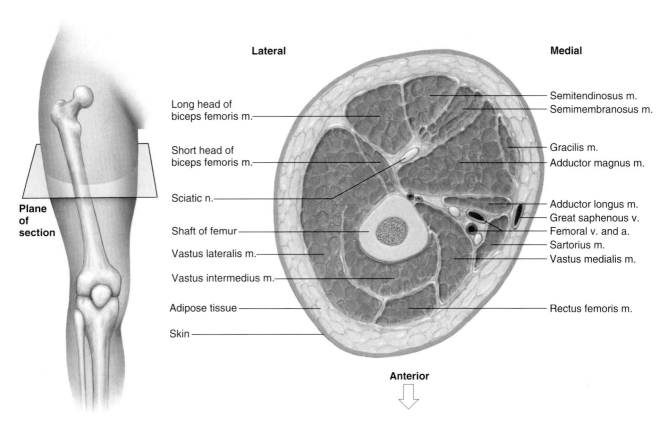

Lateral

Medial

Long head of biceps femoris m.

Short head of biceps femoris m.

Sciatic n.

Shaft of femur

Vastus lateralis m.

Vastus intermedius m.

Adipose tissue

Skin

Semitendinosus m.
Semimembranosus m.

Gracilis m.
Adductor magnus m.

Adductor longus m.
Great saphenous v.
Femoral v. and a.
Sartorius m.
Vastus medialis m.

Rectus femoris m.

Plane of section

Anterior

FIGURE 9.37

A cross section of the thigh (superior view). (*a.* stands for artery, *v.* stands for vein, *m.* stands for muscle, and *n.* stands for nerve.)

TABLE 9.12	Muscles That Move the Thigh			
Muscle	**Origin**	**Insertion**	**Action**	**Nerve Supply**
Psoas major	Lumbar intervertebral discs; bodies and transverse processes of lumbar vertebrae	Lesser trochanter of femur	Flexes thigh	Branches of L1-3 nerves
Iliacus	Iliac fossa of ilium	Lesser trochanter of femur	Flexes thigh	Femoral n.
Gluteus maximus	Sacrum, coccyx, and posterior surface of ilium	Posterior surface of femur and fascia of thigh	Extends thigh at hip	Inferior gluteal n.
Gluteus medius	Lateral surface of ilium	Greater trochanter of femur	Abducts and rotates thigh medially	Superior gluteal n.
Gluteus minimus	Lateral surface of ilium	Greater trochanter of femur	Same as gluteus medius	Superior gluteal n.
Tensor fasciae latae	Anterior iliac crest	Iliotibial band (fascia of thigh)	Abducts, flexes, and rotates thigh medially	Superior gluteal n.
Pectineus	Spine of pubis	Femur distal to lesser trochanter	Adducts and flexes thigh	Obturator and femoral nerves
Adductor longus	Pubic bone near symphysis pubis	Posterior surface of femur	Adducts, flexes, and rotates thigh laterally	Obturator n.
Adductor magnus	Ischial tuberosity	Posterior surface of femur	Adducts, extends, and rotates thigh laterally	Obturator and branch of sciatic n.
Gracilis	Lower edge of symphysis pubis	Medial surface of tibia	Adducts thigh and flexes leg at the knee	Obturator n.

TABLE 9.13   Muscles That Move the Leg

Muscle	Origin	Insertion	Action	Nerve Supply
Hamstring Group				
Biceps femoris	Ischial tuberosity and linea aspera of femur	Head of fibula and lateral condyle of tibia	Flexes and rotates leg laterally and extends thigh	Tibial n.
Semitendinosus	Ischial tuberosity	Medial surface of tibia	Flexes and rotates leg medially and extends thigh	Tibial n.
Semimembranosus	Ischial tuberosity	Medial condyle of tibia	Flexes and rotates leg medially and extends thigh	Tibial n.
Sartorius	Anterior superior iliac spine	Medial surface of tibia	Flexes leg and thigh, abducts and rotates thigh laterally	Femoral n.
Quadriceps Femoris Group				
Rectus femoris	Spine of ilium and margin of acetabulum			
Vastus lateralis	Greater trochanter and posterior surface of femur	Patella by common tendon, which continues as patellar ligament to tibial tuberosity	Extends leg at knee	Femoral n.
Vastus medialis	Medial surface of femur			
Vastus intermedius	Anterior and lateral surfaces of femur			

and are listed in table 9.13. Muscles that move the leg include the following:

Flexors	**Extensor**
Biceps femoris	Quadriceps femoris group
Semitendinosus	
Semimembranosus	
Sartorius	

Flexors

As its name implies, the **biceps femoris** (bi'seps fem'or-is) has two heads, one attached to the ischium and the other attached to the femur. This muscle passes along the back of the thigh on the lateral side and connects to the proximal ends of the fibula and tibia. The biceps femoris is one of the hamstring muscles, and its tendon (hamstring) can be felt as a lateral ridge behind the knee. This muscle flexes and rotates the leg laterally and extends the thigh (figs. 9.35 and 9.36).

The **semitendinosus** (sem"e-ten'dĭ-no-sus) is another hamstring muscle. It is a long, bandlike muscle on the back of the thigh toward the medial side, connecting the ischium to the proximal end of the tibia. The semitendinosus is so named because it becomes tendinous in the middle of the thigh, continuing to its insertion as a long, cordlike tendon. It flexes and rotates the leg medially and extends the thigh (fig. 9.36).

The **semimembranosus** (sem"e-mem'brah-no-sus) is the third hamstring muscle and is the most medially located muscle in the back of the thigh. It connects the ischium to the tibia and flexes and rotates the leg medially and extends the thigh (fig. 9.36).

The **sartorius** (sar-to're-us) is an elongated, straplike muscle that passes obliquely across the front of the thigh and then descends over the medial side of the knee. It connects the ilium to the tibia and flexes the leg and the thigh. It can also abduct the thigh and rotate it laterally (figs. 9.34 and 9.35).

The tendinous attachments of the hamstring muscles to the ischial tuberosity are sometimes torn as a result of strenuous running or kicking motions. This painful injury is commonly called "pulled hamstrings" and is usually accompanied by internal bleeding from damaged blood vessels that supply the muscles.

Extensor

The large, fleshy muscle group called the **quadriceps femoris** (kwod'rĭ-spes fem'or-is) occupies the front and sides of the thigh and is the primary extensor of the knee. It is composed of four parts—*rectus femoris, vastus lateralis, vastus medialis,* and *vastus intermedius* (figs. 9.34 and 9.37). These parts connect the ilium and femur to a common *patellar tendon,* which passes over the front of the knee and attaches to the patella. This tendon then continues as the *patellar ligament* to the tibia.

Occasionally, as a result of traumatic injury in which muscle, such as the quadriceps femoris, is compressed against an underlying bone, new bone tissue may begin to develop within the damaged muscle. This condition is called *myositis ossificans*. When the bone tissue matures several months following the injury, surgery can remove the newly formed bone.

Muscles That Move the Foot

Movements of the foot include movements of the ankle and toes. A number of muscles that move the foot are located in the leg. They attach the femur, tibia, and fibula to bones of the foot and are responsible for moving the foot upward (dorsiflexion) or downward (plantar flexion) and turning the foot so the toes are inward (inversion) or outward (eversion). These muscles are shown in figures 9.38, 9.39, 9.40, 9.41, in reference plates 68, 69, 70, and are listed in table 9.14. Muscles that move the foot include the following:

Dorsal Flexors
Tibialis anterior
Fibularis tertius
Extensor digitorum longus

Invertor
Tibialis posterior

Plantar Flexors
Gastrocnemius
Soleus
Flexor digitorum longus

Evertor
Fibularis longus

Dorsal Flexors

The **tibialis anterior** (tib"e-a'lis an-te're-or) is an elongated, spindle-shaped muscle located on the front of the leg. It arises from the surface of the tibia, passes medially over the distal end of the tibia, and attaches to bones of the foot. Contraction of the tibialis anterior causes dorsiflexion and inversion of the foot (fig. 9.38).

The **fibularis (peroneus) tertius** (fib"u-la'ris ter'shus) is a muscle of variable size that connects the fibula to the lateral side of the foot. It functions in dorsiflexion and eversion of the foot (fig. 9.38).

The **extensor digitorum longus** (eks-ten'sor dij"ĭ-to'rum long'gus) is situated along the lateral side of the leg just behind the tibialis anterior. It arises from the proximal end of the tibia and the shaft of the fibula. Its tendon divides into four parts as it passes over the front of the ankle. These parts continue over the surface of the foot and attach to the four lateral toes. The actions of the extensor digitorum longus include dorsiflexion of the foot, eversion of the foot, and extension of the toes (figs. 9.38 and 9.39).

Plantar Flexors

The **gastrocnemius** (gas"trok-ne'me-us) on the back of the leg forms part of the calf. It arises by two heads from the femur. The distal end of this muscle joins the strong *calcaneal tendon* (Achilles tendon), which descends to the heel and attaches to the calcaneus. The gastrocnemius is a powerful plantar flexor of the foot that aids in pushing the body forward when a person walks or runs. It also flexes the leg at the knee (figs. 9.39 and 9.40).

TABLE 9.14	Muscles That Move the Foot			
Muscle	**Origin**	**Insertion**	**Action**	**Nerve Supply**
Tibialis anterior	Lateral condyle and lateral surface of tibia	Tarsal bone (cuneiform) and first metatarsal	Dorsiflexion and inversion of foot	Deep fibular n.
Fibularis tertius	Anterior surface of fibula	Dorsal surface of fifth metatarsal	Dorsiflexion and eversion of foot	Deep fibular n.
Extensor digitorum longus	Lateral condyle of tibia and anterior surface of fibula	Dorsal surfaces of second and third phalanges of four lateral toes	Dorsiflexion and eversion of foot and extension of toes	Deep fibular n.
Gastrocnemius	Lateral and medial condyles of femur	Posterior surface of calcaneus	Plantar flexion of foot and flexion of leg at knee	Tibial n.
Soleus	Head and shaft of fibula and posterior surface of tibia	Posterior surface of calcaneus	Plantar flexion of foot	Tibial n.
Flexor digitorum longus	Posterior surface of tibia	Distal phalanges of four lateral toes	Plantar flexion and inversion of foot and flexion of four lateral toes	Tibial n.
Tibialis posterior	Lateral condyle and posterior surface of tibia and posterior surface of fibula	Tarsal and metatarsal bones	Plantar flexion and inversion of foot	Tibial n.
Fibularis longus	Lateral condyle of tibia and head and shaft of fibula	Tarsal and metatarsal bones	Plantar flexion and eversion of foot; also supports arch	Superficial fibular n.

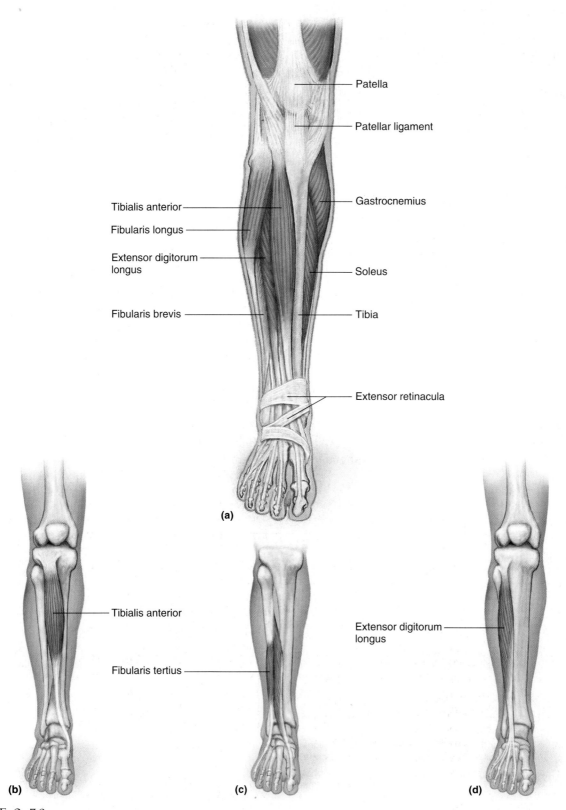

FIGURE 9.38
Muscles of the leg. (*a*) Muscles of the anterior right leg. (*b–d*) Isolated views of muscles associated with the anterior right leg.

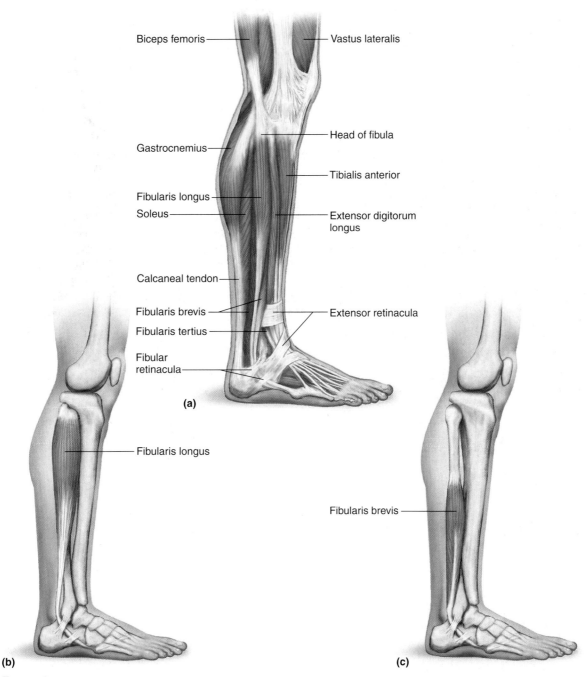

Biceps femoris
Vastus lateralis
Gastrocnemius
Head of fibula
Tibialis anterior
Fibularis longus
Soleus
Extensor digitorum longus
Calcaneal tendon
Fibularis brevis
Fibularis tertius
Extensor retinacula
Fibular retinacula

(a)

Fibularis longus

(b)

Fibularis brevis

(c)

FIGURE 9.39

Muscles of the leg. (*a*) Muscles of the lateral right leg. Isolated views of (*b*) fibularis longus and (*c*) fibularis brevis.

Strenuous athletic activity may partially or completely tear the calcaneal (Achilles) tendon. This injury occurs most frequently in middle-aged athletes who run or play sports that involve quick movements and directional changes. A torn calcaneal tendon usually requires surgical treatment.

The **soleus** (so′le-us) is a thick, flat muscle located beneath the gastrocnemius, and together these two muscles form the calf of the leg. The soleus arises from the tibia and fibula, and it extends to the heel by way of the calcaneal tendon. It acts with the gastrocnemius to cause plantar flexion of the foot (figs. 9.39 and 9.40).

The **flexor digitorum longus** (flek′sor dij″ĭ-to′rum long′gus) extends from the posterior surface of the tibia to the foot. Its tendon passes along the plantar surface of the foot. There the muscle divides into four parts that attach to the terminal bones of the four lateral toes. This muscle assists in plantar flexion of the foot, flexion of the four lateral toes, and inversion of the foot (fig. 9.40).

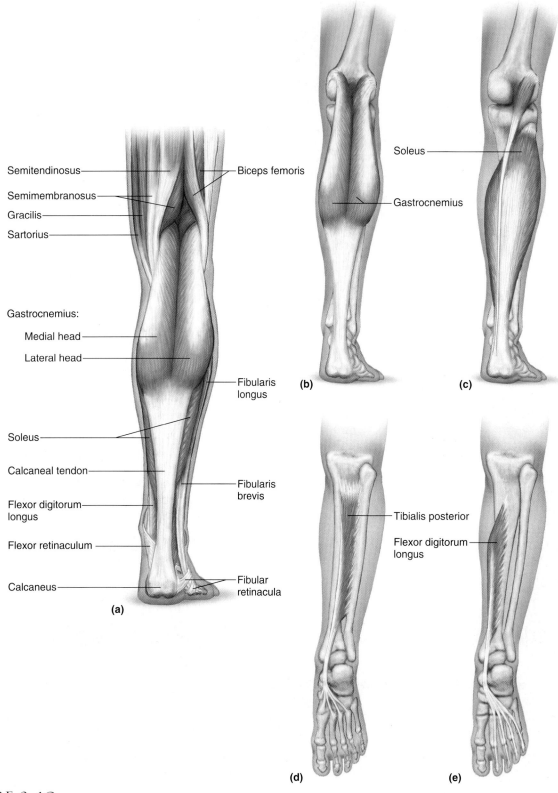

FIGURE 9.40

Muscles of the leg. (*a*) Muscles of the posterior right leg. (*b–e*) Isolated views of muscles associated with the posterior right leg.

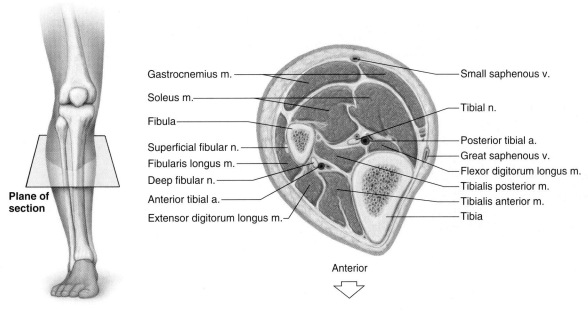

FIGURE 9.41

A cross section of the leg (superior view). (*a.* stands for artery, *v.* stands for vein, *m.* stands for muscle, and *n.* stands for nerve.)

Labels on figure:
Gastrocnemius m.
Soleus m.
Fibula
Superficial fibular n.
Fibularis longus m.
Deep fibular n.
Anterior tibial a.
Extensor digitorum longus m.
Small saphenous v.
Tibial n.
Posterior tibial a.
Great saphenous v.
Flexor digitorum longus m.
Tibialis posterior m.
Tibialis anterior m.
Tibia
Plane of section
Anterior

Invertor

The **tibialis posterior** (tib″e-a′lis pos-tēr′e-or) is the deepest of the muscles on the back of the leg. It connects the fibula and tibia to the ankle bones by means of a tendon that curves under the medial malleolus. This muscle assists in inversion and plantar flexion of the foot (fig. 9.40).

Evertor

The **fibularis** (peroneus) **longus** (fib″u-la′ris long′gus) is a long, straplike muscle located on the lateral side of the leg. It connects the tibia and the fibula to the foot by means of a stout tendon that passes behind the lateral malleolus. It everts the foot, assists in plantar flexion, and helps support the arch of the foot (figs. 9.39 and 9.41).

As in the wrist, fascia in various regions of the ankle thicken to form retinacula. Anteriorly, for example, *extensor retinacula* connect the tibia and fibula as well as the calcaneus and fascia of the sole. These retinacula form sheaths for tendons crossing the front of the ankle (fig. 9.39).

Posteriorly, on the inside, a *flexor retinaculum* runs between the medial malleolus and the calcaneus and forms sheaths for tendons passing beneath the foot (fig. 9.40). *Fibular retinacula* connect the lateral malleolus and the calcaneus, providing sheaths for tendons on the lateral side of the ankle (fig. 9.39).

Life-Span Changes

Signs of aging in the muscular system begin to appear in one's forties, although a person can still be very active. At a microscopic level, supplies of the molecules that enable muscles to function—myoglobin, ATP, and creatine phosphate—decline. The diameters of some muscle fibers may shrink, as the muscle layers in the walls of veins thicken, making the vessels more rigid and less elastic. Very gradually, the muscles become smaller, drier, and capable of less forceful contraction. Connective tissue and adipose cells begin to replace some muscle tissue. By age eighty, nearly half the muscle mass has atrophied, due to a decline in motor neuron activity. Diminishing muscular strength slows reflexes.

Exercise can help maintain a healthy muscular system, countering the less effective oxygen delivery that results from the decreased muscle mass that accompanies aging. Exercise also maintains the flexibility of blood vessels, which can decrease the likelihood of hypertension developing. A physician should be consulted before starting any exercise program.

According to the National Institute on Aging, exercise should include strength training and aerobics, with stretching before and after. Strength training consists of weight lifting or using a machine that works specific muscles against a resistance, performed so that the same muscle is not exercised on consecutive days. Strength training increases muscle mass, and the resulting stronger muscles can alleviate pressure on the joints, which may lessen arthritis pain. Aerobic exercise improves oxygen utilization by muscles and increases endurance. Stretching increases flexibility and decreases muscle strain, while improving blood flow to all muscles. A side benefit of regular exercise, especially among older individuals, is fewer bouts of depression.

1 What changes are associated with an aging muscular system?

2 Describe two types of recommended exercise.

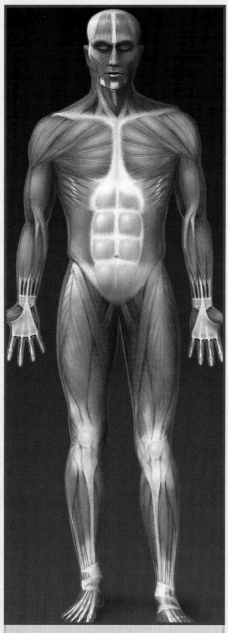

MUSCULAR SYSTEM

Muscles provide the force for moving body parts.

Integumentary System

The skin increases heat loss during skeletal muscle activity. Sensory receptors function in the reflex control of skeletal muscles.

Lymphatic System

Muscle action pumps lymph through lymphatic vessels.

Skeletal System

Bones provide attachments that allow skeletal muscles to cause movement.

Digestive System

Skeletal muscles are important in swallowing. The digestive system absorbs needed nutrients.

Nervous System

Neurons control muscle contractions.

Respiratory System

Breathing depends on skeletal muscles. The lungs provide oxygen for muscle cells and excrete carbon dioxide.

Endocrine System

Hormones help increase blood flow to exercising skeletal muscles.

Urinary System

Skeletal muscles help control expulsion of urine from the urinary bladder.

Cardiovascular System

Blood flow delivers oxygen and nutrients and removes wastes. Cardiac muscle pumps blood, smooth muscle in vessel walls enables vasoconstriction, vasodilation.

Reproductive System

Skeletal muscles are important in sexual activity.

Clinical Terms Related to the Muscular System

contracture (kon-trak'tūr) Condition in which there is great resistance to the stretching of a muscle.

convulsion (kun-vul'shun) Series of involuntary contractions of various voluntary muscles.

electromyography (e-lek″tro-mi-og'rah-fe) Technique for recording the electrical changes that occur in muscle tissues.

fibrillation (fi″bri-la'shun) Spontaneous contractions of individual muscle fibers, producing rapid and uncoordinated activity within a muscle.

fibrosis (fi-bro'sis) Degenerative disease in which connective tissue with many fibers replaces skeletal muscle tissue.

fibrositis (fi″bro-si'tis) Inflammation of connective tissues with many fibers, especially in the muscle fascia. This disease is also called muscular rheumatism.

muscular dystrophy (mus'ku-lar dis'tro-fe) Progressive muscle weakness and atrophy caused by deficient dystrophin protein.

myalgia (mi-al'je-ah) Pain resulting from any muscular disease or disorder.

myasthenia gravis (mi″as-the'ne-ah grav'is) Chronic disease characterized by muscles that are weak and easily fatigued. It results from the immune system's attack on neuromuscular junctions so that stimuli are not transmitted from motor neurons to muscle fibers.

myokymia (mi″o-ki'me-ah) Persistent quivering of a muscle.

myology (mi-ol'o-je) Study of muscles.

myoma (mi-o'mah) Tumor composed of muscle tissue.

myopathy (mi-op'ah-the) Any muscular disease.

myositis (mi″o-si'tis) Inflammation of skeletal muscle tissue.

myotomy (mi-ot'o-me) Cutting of muscle tissue.

myotonia (mi″o-to'ne-ah) Prolonged muscular spasm.

paralysis (pah-ral'ĭ-sis) Loss of ability to move a body part.

paresis (pah-re'sis) Partial or slight paralysis of the muscles.

shin splints (shin' splints) Soreness on the front of the leg due to straining the anterior leg muscles, often as a result of walking up and down hills.

torticollis (tor″tĭ-kol'is) Condition in which the neck muscles, such as the sternocleidomastoids, contract involuntarily. It is more commonly called wryneck.

CHAPTER SUMMARY

Introduction (page 278)

The three types of muscle tissue are skeletal, smooth, and cardiac.

Structure of a Skeletal Muscle (page 278)

Skeletal muscles are composed of nervous, vascular, and various connective tissues, as well as skeletal muscle tissue.

1. Connective tissue coverings
   a. Fascia covers each skeletal muscle.
   b. Other connective tissues surround cells and groups of cells within the muscle's structure.
   c. Fascia is part of a complex network of connective tissue that extends throughout the body.

2. Skeletal muscle fibers
   a. Each skeletal muscle fiber is a single muscle cell, which is the unit of contraction.
   b. Muscle fibers are cylindrical cells with many nuclei.
   c. The cytoplasm contains mitochondria, sarcoplasmic reticulum, and myofibrils of actin and myosin.
   d. The arrangement of the actin and myosin filaments causes striations. (I bands, Z lines, A bands, H zone and M line.)
   e. Cross-bridges of myosin filaments form linkages with actin filaments. The reaction between actin and myosin filaments provides the basis for contraction.
   f. When a fiber is at rest, troponin and tropomyosin molecules interfere with linkage formation. Calcium ions remove the inhibition.
   g. Transverse tubules extend from the cell membrane into the cytoplasm and are associated with the cisternae of the sarcoplasmic reticulum.

Skeletal Muscle Contraction (page 282)

Muscle fiber contraction results from a sliding movement of actin and myosin filaments that shortens the muscle fiber.

1. Neuromuscular junction
   a. Motor neurons stimulate muscle fibers to contract.
   b. The motor end plate of a muscle fiber lies on one side of a neuromuscular junction.
   c. One motor neuron and the muscle fibers associated with it constitute a motor unit.
   d. In response to a nerve impulse, the end of a motor nerve fiber secretes a neurotransmitter, which diffuses across the junction and stimulates the muscle fiber.

2. Stimulus for contraction
   a. Muscle fiber is usually stimulated by acetylcholine released from the end of a motor nerve fiber.
   b. Acetylcholinesterase decomposes acetylcholine to prevent continuous stimulation.
   c. Stimulation causes a muscle fiber to conduct an impulse that travels over the surface of the sarcolemma and reaches the deep parts of the fiber by means of the transverse tubules.

3. Excitation contraction coupling
   a. A muscle impulse signals the sarcoplasmic reticulum to release calcium ions.
   b. Linkages form between myosin and actin, and the actin filaments move inward, shortening the sarcomere.

4. The Sliding Filament Theory
   a. The sarcomere, defined by striations, is the functional unit of skeletal muscle.
   b. When thick and thin myofilaments slide past one another, the sarcomeres shorten. The muscle contracts.
5. Cross-bridge cycling
   a. A myosin cross-bridge can attach to an actin binding site and pull on the actin filament. The myosin head can then release the actin and combine with another active binding site farther down the actin filament, and pull again.
   b. The breakdown of ATP releases energy that provides the repetition of the cross-bridge cycle.
6. Relaxation
   a. Acetylcholine remaining in the synapse is rapidly decomposed by acetylcholinesterase, preventing continuous stimulation of a muscle fiber.
   b. The muscle fiber relaxes when calcium ions are transported back into the sarcoplasmic reticulum.
   c. Cross-bridge linkages break and do not re-form—the muscle fiber relaxes.
7. Energy sources for contraction
   a. ATP supplies the energy for muscle fiber contraction.
   b. Creatine phosphate stores energy that can be used to synthesize ATP as it is decomposed.
   c. Active muscles depend upon cellular respiration for energy.
8. Oxygen supply and cellular respiration
   a. Anaerobic reactions of cellular respiration yield few ATP molecules, whereas aerobic reactions of cellular respiration provide many ATP molecules.
   b. Hemoglobin in red blood cells carries oxygen from the lungs to body cells.
   c. Myoglobin in muscle cells stores some oxygen temporarily.
9. Oxygen debt
   a. During rest or moderate exercise, oxygen is sufficient to support the aerobic reactions of cellular respiration.
   b. During strenuous exercise, oxygen deficiency may develop, and lactic acid may accumulate as a result of the anaerobic reactions of cellular respiration.
   c. The amount of oxygen needed to convert accumulated lactic acid to glucose and to restore supplies of ATP and creatine phosphate is called oxygen debt.
10. Muscle fatigue
   a. A fatigued muscle loses its ability to contract.
   b. Muscle fatigue is usually due to the effects of accumulation of lactic acid.
   c. Athletes usually produce less lactic acid than nonathletes because of their increased ability to supply oxygen and nutrients to muscles.
11. Heat production
   a. Muscles represent an important source of body heat.
   b. Most of the energy released by cellular respiration is lost as heat.

Muscular Responses (page 290)

1. Threshold stimulus is the minimal stimulus needed to elicit a muscular contraction.
2. Recording a muscle contraction
   a. A twitch is a single, short contraction of a muscle fiber.
   b. A myogram is a recording of an electrically stimulated isolated muscle pulling a lever.
   c. The latent period is the time between stimulus and responding muscle contraction.

   d. During the refractory period immediately following contraction, a muscle cannot respond.
   e. If a muscle fiber contracts at all, it will contract completely. This has been termed the all-or-none response.
   f. The length to which a muscle is stretched before stimulation affects the force it will develop.
      (1) Normal activities occur at optimal length.
      (2) Too long or too short decreases force.
   g. Sustained contractions are more important than twitch contractions in everyday activities.
3. Summation
   a. A rapid series of stimuli may produce summation of twitches and sustained contraction.
   b. Forceful, sustained contraction without relaxation is a tetanic contraction.
4. Recruitment of motor units
   a. Muscles whose motor units contain small numbers of muscle fibers produce finer movements.
   b. Motor units respond in an all-or-none manner.
   c. At low intensity of stimulation, relatively small numbers of motor units contract.
   d. At increasing intensities of stimulation, other motor units are recruited until the muscle contracts with maximal tension.
5. Sustained contractions
   a. When contractions fuse, the strength of contraction may increase due to recruitment of fibers.
   b. Even when a muscle is at rest, its fibers usually maintain tone—that is, remain partially contracted.
6. Types of contractions
   a. One type of isotonic contraction occurs when a muscle contracts and its ends are pulled closer together. Because shortening occurs, it is called a concentric contraction.
   b. Another type of isotonic contraction occurs when the force a muscle generates is less than that required to move or lift an object. This lengthening contraction is called an eccentric contraction.
   c. When a muscle contracts but its attachments do not move, the contraction is called isometric.
   d. Most body movements involve both isometric and isotonic contractions.
7. Fast and slow twitch muscle fibers
   a. The speed of contraction is related to a muscle's specific function.
   b. Slow-contracting, or red, muscles can generate ATP fast enough to keep up with ATP breakdown and can contract for long periods.
   c. Fast-contracting, or white, muscles have reduced ability to carry on the aerobic reactions of cellular respiration and tend to fatigue relatively rapidly.

Smooth Muscles (page 293)

The contractile mechanisms of smooth and cardiac muscles are similar to those of skeletal muscle.

1. Smooth muscle fibers
   a. Smooth muscle cells contain filaments of myosin and actin.
   b. They lack transverse tubules, and the sarcoplasmic reticula are not well developed.
   c. Types include multiunit smooth muscle and visceral smooth muscle.
   d. Visceral smooth muscle displays rhythmicity.
   e. Peristalsis aids movement of material through hollow organs.

2. Smooth muscle contraction
   a. In smooth muscles, calmodulin binds to calcium ions and activates the contraction mechanism.
   b. Both acetylcholine and norepinephrine are neurotransmitters for smooth muscles.
   c. Hormones and stretching affect smooth muscle contractions.
   d. With a given amount of energy, smooth muscle can maintain a contraction for a longer time than can skeletal muscle.
   e. Smooth muscles can change length without changing tautness.

Cardiac Muscle (page 294)

1. Cardiac muscle contracts for a longer time than skeletal muscle because transverse tubules supply extra calcium ions.
2. Intercalated discs connect the ends of adjacent cardiac muscle cells and hold the cells together.
3. A network of fibers contracts as a unit and responds to stimulation in an all-or-none manner.
4. Cardiac muscle is self-exciting, rhythmic, and remains refractory until a contraction is completed.

Skeletal Muscle Actions (page 296)

1. Origin and insertion
   a. The movable end of attachment of a skeletal muscle to a bone is its insertion, and the immovable end is its origin.
   b. Some muscles have more than one origin or insertion.
2. Interaction of skeletal muscles
   a. Skeletal muscles function in groups.
   b. A prime mover is responsible for most of a movement; synergists aid prime movers; antagonists can resist the movement of a prime mover.
   c. Smooth movements depend upon antagonists giving way to the actions of prime movers.

Major Skeletal Muscles (page 297)

Muscle names often describe sizes, shapes, locations, actions, number of attachments, or direction of fibers.

1. Muscles of facial expression
   a. These muscles lie beneath the skin of the face and scalp and are used to communicate feelings through facial expression.
   b. They include the epicranius, orbicularis oculi, orbicularis oris, buccinator, zygomaticus, and platysma.
2. Muscles of mastication
   a. These muscles are attached to the mandible and are used in chewing.
   b. They include the masseter, temporalis, medial pterygoid, and lateral pterygoid.
3. Muscles that move the head and vertebral column
   a. Muscles in the neck and back move the head.
   b. They include the sternocleidomastoid, splenius capitis, semispinalis capitis, and erector spinae.
4. Muscles that move the pectoral girdle
   a. Most of these muscles connect the scapula to nearby bones and are closely associated with muscles that move the arm.
   b. They include the trapezius, rhomboideus major, levator scapulae, serratus anterior, and pectoralis minor.

5. Muscles that move the arm
   a. These muscles connect the humerus to various regions of the pectoral girdle, ribs, and vertebral column.
   b. They include the coracobrachialis, pectoralis major, teres major, latissimus dorsi, supraspinatus, deltoid, subscapularis, infraspinatus, and teres minor.
6. Muscles that move the forearm
   a. These muscles connect the radius and ulna to the humerus and pectoral girdle.
   b. They include the biceps brachii, brachialis, brachioradialis, triceps brachii, supinator, pronator teres, and pronator quadratus.
7. Muscles that move the hand
   a. These muscles arise from the distal end of the humerus and from the radius and ulna.
   b. They include the flexor carpi radialis, flexor carpi ulnaris, palmaris longus, flexor digitorum profundus, flexor digitorum superficialis, extensor carpi radialis longus, extensor carpi radialis brevis, extensor carpi ulnaris, and extensor digitorum.
   c. An extensor retinaculum forms sheaths for tendons of the extensor muscles.
8. Muscles of the abdominal wall
   a. These muscles connect the rib cage and vertebral column to the pelvic girdle.
   b. They include the external oblique, internal oblique, transversus abdominis, and rectus abdominis.
9. Muscles of the pelvic outlet
   a. These muscles form the floor of the pelvic cavity and fill the space of the pubic arch.
   b. They include the levator ani, coccygeus, superficial transversus perinei, bulbospongiosus, ischiocavernosus, and sphincter urethrae.
10. Muscles that move the thigh
    a. These muscles are attached to the femur and to some part of the pelvic girdle.
    b. They include the psoas major, iliacus, gluteus maximus, gluteus medius, gluteus minimus, tensor fasciae latae, pectineus, adductor longus, adductor magnus, and gracilis.
11. Muscles that move the leg
    a. These muscles connect the tibia or fibula to the femur or pelvic girdle.
    b. They include the biceps femoris, semitendinosus, semimembranosus, sartorius, and the quadriceps femoris group.
12. Muscles that move the foot
    a. These muscles attach the femur, tibia, and fibula to various bones of the foot.
    b. They include the tibialis anterior, fibularis tertius, extensor digitorum longus, gastrocnemius, soleus, flexor digitorum longus, tibialis posterior, and fibularis longus.
    c. Retinacula form sheaths for tendons passing to the foot.

Life-Span Changes (page 325)

1. Beginning in one's forties, supplies of ATP, myoglobin, and creatine phosphate begin to decline.
2. By age eighty, muscle mass may be halved. Reflexes slow. Adipose cells and connective tissue replace some muscle tissue.
3. Exercise is very beneficial in maintaining muscle function.

CRITICAL THINKING QUESTIONS

1. Why do you think athletes generally perform better if they warm up by exercising lightly before a competitive event?

2. Following childbirth, a woman may lose urinary control (incontinence) when sneezing or coughing. Which muscles of the pelvic floor should be strengthened by exercise to help control this problem?

3. What steps might be taken to minimize atrophy of skeletal muscles in patients who are confined to bed for prolonged times?

4. As lactic acid and other substances accumulate in an active muscle, they stimulate pain receptors, and the muscle may feel sore. How might the application of heat or substances that dilate blood vessels help relieve such soreness?

5. Several important nerves and blood vessels course through the muscles of the gluteal region. In order to avoid the possibility of damaging such parts, intramuscular injections are usually made into the lateral, superior portion of the gluteus medius. What landmarks would help you locate this muscle in a patient?

6. Following an injury to a nerve, the muscles it supplies with motor nerve fibers may become paralyzed. How would you explain to a patient the importance of moving the disabled muscles passively or contracting them with electrical stimulation?

REVIEW EXERCISES

Part A

1. List the three types of muscle tissue.
2. Distinguish between a tendon and an aponeurosis.
3. Describe the connective tissue coverings of a skeletal muscle.
4. Distinguish among deep fascia, subcutaneous fascia, and subserous fascia.
5. List the major parts of a skeletal muscle fiber, and describe the function of each part.
6. Describe a neuromuscular junction.
7. Define *motor unit,* and explain how the number of fibers within a unit affects muscular contractions.
8. Explain the function of a neurotransmitter substance.
9. Describe the major events that occur when a muscle fiber contracts.
10. Explain how ATP and creatine phosphate function in muscle contraction.
11. Describe how oxygen is supplied to skeletal muscles.
12. Describe how an oxygen debt may develop.
13. Explain how muscles may become fatigued and how a person's physical condition may affect tolerance to fatigue.
14. Explain how the actions of skeletal muscles affect maintenance of body temperature.
15. Define *threshold stimulus.*
16. Explain *all-or-none response.*
17. Describe the staircase effect.
18. Explain *recruitment.*
19. Explain how a skeletal muscle can be stimulated to produce a sustained contraction.
20. Distinguish between a tetanic contraction and muscle tone.
21. Distinguish between concentric and eccentric contractions, and explain how each is used in body movements.
22. Distinguish between fast-contracting and slow-contracting muscles.
23. Compare the structures of smooth and skeletal muscle fibers.
24. Distinguish between multiunit and visceral smooth muscles.
25. Define *peristalsis,* and explain its function.
26. Compare the characteristics of smooth and skeletal muscle contractions.
27. Compare the structures of cardiac and skeletal muscle fibers.
28. Compare the characteristics of cardiac and skeletal muscle contractions.
29. Distinguish between a muscle's origin and its insertion.
30. Define *prime mover, synergist,* and *antagonist.*

Part B

Match the muscles in column I with the descriptions and functions in column II.

I

1. Buccinator
2. Epicranius
3. Lateral pterygoid
4. Platysma
5. Rhomboideus major
6. Splenius capitis
7. Temporalis
8. Zygomaticus
9. Biceps brachii
10. Brachialis
11. Deltoid
12. Latissimus dorsi
13. Pectoralis major
14. Pronator teres
15. Teres minor
16. Triceps brachii
17. Biceps femoris
18. External oblique
19. Gastrocnemius
20. Gluteus maximus
21. Gluteus medius
22. Gracilis
23. Rectus femoris
24. Tibialis anterior

II

A. Inserted on the coronoid process of the mandible
B. Draws the corner of the mouth upward
C. Can raise and adduct the scapula
D. Can pull the head into an upright position
E. Consists of two parts—the frontalis and the occipitalis
F. Compresses the cheeks
G. Extends over the neck from the chest to the face
H. Pulls the jaw from side to side
I. Primary extensor of the elbow
J. Pulls the shoulder back and downward
K. Abducts the arm
L. Rotates the arm laterally
M. Pulls the arm forward and across the chest
N. Rotates the arm medially
O. Strongest flexor of the elbow
P. Strongest supinator of the forearm
Q. Inverts the foot
R. A member of the quadriceps femoris group
S. A plantar flexor of the foot
T. Compresses the contents of the abdominal cavity
U. Largest muscle in the body
V. A hamstring muscle
W. Adducts the thigh
X. Abducts the thigh

Part C

Which muscles can you identify in the bodies of these models whose muscles are enlarged by exercise?

WEB CONNECTIONS

Visit the Student Edition of the Online Learning Center at www.mhhe.com/shier10 **for answers to chapter questions, additional quizzes, interactive learning exercises, and other study tools.**

Surface Anatomy

The following set of reference plates is presented to help you locate some of the more prominent surface features in various regions of the body. For the most part, the labeled structures are easily seen or palpated through the skin. As a review, you may want to locate as many of these features as possible on your own body.

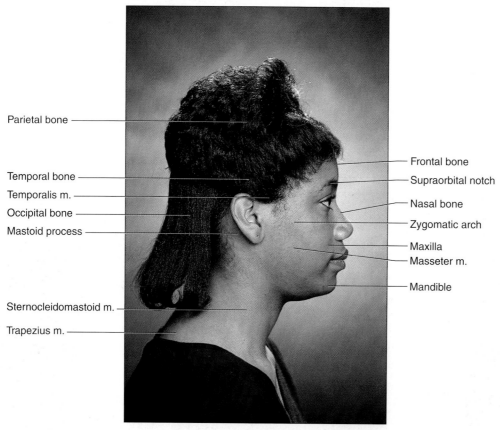

Parietal bone

Temporal bone

Temporalis m.

Occipital bone

Mastoid process

Sternocleidomastoid m.

Trapezius m.

Frontal bone

Supraorbital notch

Nasal bone

Zygomatic arch

Maxilla

Masseter m.

Mandible

PLATE THIRTY-SEVEN
Surface anatomy of head and neck, lateral view. (*m.* stands for muscle.)

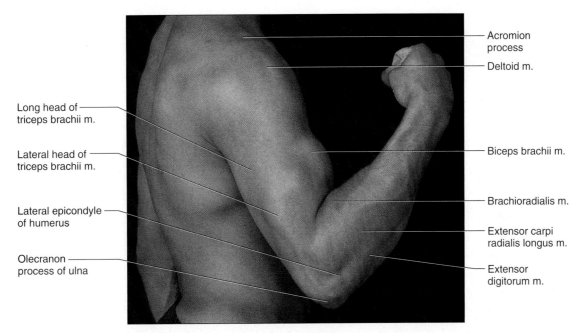

Acromion
process

Deltoid m.

Long head of
triceps brachii m.

Lateral head of
triceps brachii m.

Biceps brachii m.

Brachioradialis m.

Lateral epicondyle
of humerus

Extensor carpi
radialis longus m.

Olecranon
process of ulna

Extensor
digitorum m.

PLATE THIRTY-EIGHT
Surface anatomy of upper limb and thorax, lateral view. (*m.* stands for muscle.)

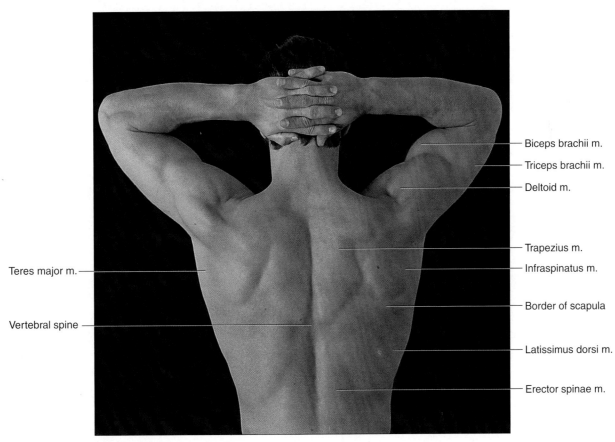

Biceps brachii m.

Triceps brachii m.

Deltoid m.

Trapezius m.

Teres major m.

Infraspinatus m.

Border of scapula

Vertebral spine

Latissimus dorsi m.

Erector spinae m.

PLATE THIRTY-NINE
Surface anatomy of back and upper limbs, posterior view. (*m.* stands for muscle.)

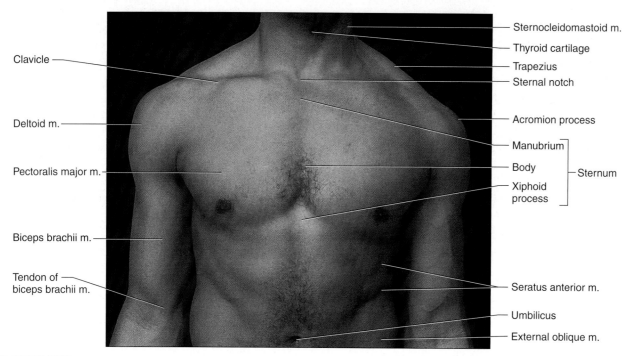

Clavicle

Deltoid m.

Pectoralis major m.

Biceps brachii m.

Tendon of
biceps brachii m.

Sternocleidomastoid m.

Thyroid cartilage

Trapezius

Sternal notch

Acromion process

Manubrium

Body

Xiphoid
process

Sternum

Seratus anterior m.

Umbilicus

External oblique m.

PLATE FORTY

Surface anatomy of torso and arms, anterior view. (*m.* stands for muscle.)

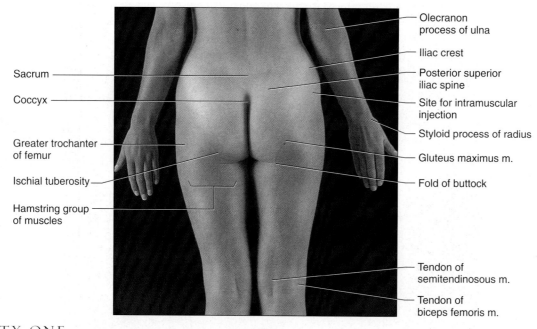

Sacrum

Coccyx

Greater trochanter
of femur

Ischial tuberosity

Hamstring group
of muscles

Olecranon
process of ulna

Iliac crest

Posterior superior
iliac spine

Site for intramuscular
injection

Styloid process of radius

Gluteus maximus m.

Fold of buttock

Tendon of
semitendinosous m.

Tendon of
biceps femoris m.

PLATE FORTY-ONE

Surface anatomy of torso and thighs, posterior view. (*m.* stands for muscle.)

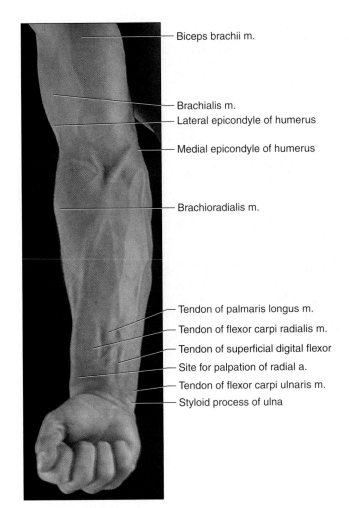

- Biceps brachii m.
- Brachialis m.
- Lateral epicondyle of humerus
- Medial epicondyle of humerus
- Brachioradialis m.
- Tendon of palmaris longus m.
- Tendon of flexor carpi radialis m.
- Tendon of superficial digital flexor
- Site for palpation of radial a.
- Tendon of flexor carpi ulnaris m.
- Styloid process of ulna

PLATE FORTY-TWO

Surface anatomy of forearm, anterior view. (*m.* stands for muscle and *a.* stands for artery.)

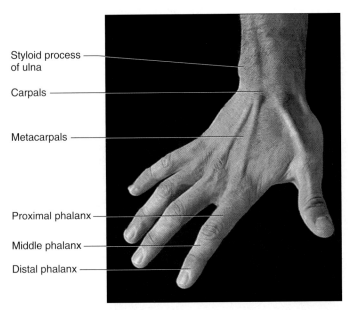

- Styloid process of ulna
- Carpals
- Metacarpals
- Proximal phalanx
- Middle phalanx
- Distal phalanx

PLATE FORTY-THREE

Surface anatomy of the hand.

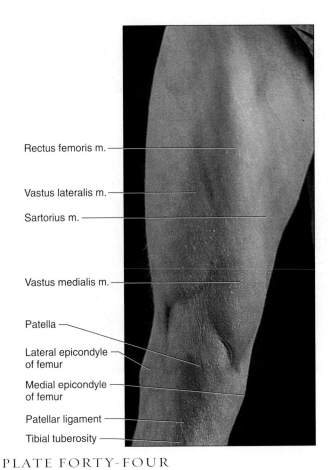

- Rectus femoris m.
- Vastus lateralis m.
- Sartorius m.
- Vastus medialis m.
- Patella
- Lateral epicondyle of femur
- Medial epicondyle of femur
- Patellar ligament
- Tibial tuberosity

PLATE FORTY-FOUR

Surface anatomy of knee and surrounding area, anterior view. (*m.* stands for muscle.)

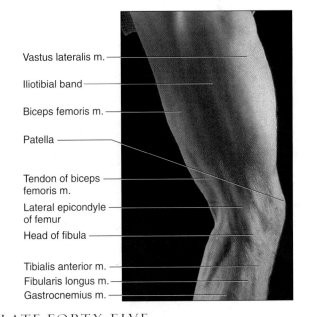

- Vastus lateralis m.
- Iliotibial band
- Biceps femoris m.
- Patella
- Tendon of biceps femoris m.
- Lateral epicondyle of femur
- Head of fibula
- Tibialis anterior m.
- Fibularis longus m.
- Gastrocnemius m.

PLATE FORTY-FIVE

Surface anatomy of knee and surrounding area, lateral view. (*m.* stands for muscle.)

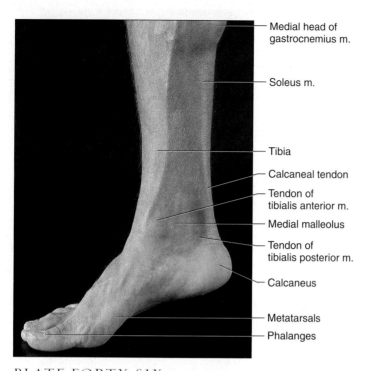

Medial head of
gastrocnemius m.

Soleus m.

Tibia

Calcaneal tendon

Tendon of
tibialis anterior m.

Medial malleolus

Tendon of
tibialis posterior m.

Calcaneus

Metatarsals

Phalanges

PLATE FORTY-SIX
Surface anatomy of ankle and leg, medial view. (*m.* stands for
muscle.)

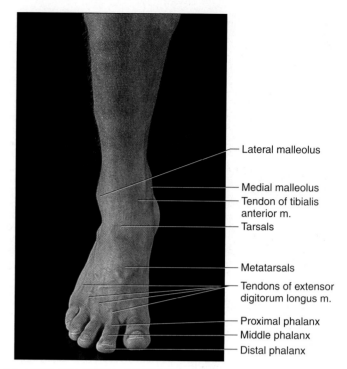

Lateral malleolus

Medial malleolus

Tendon of tibialis
anterior m.

Tarsals

Metatarsals

Tendons of extensor
digitorum longus m.

Proximal phalanx

Middle phalanx

Distal phalanx

PLATE FORTY-SEVEN
Surface anatomy of ankle and foot. (*m.* stands for muscle.)

10

UNDERSTANDING WORDS

astr-, starlike: *astr*ocyte—star-shaped neuroglial cell.

ax-, axle: *ax*on—cylindrical nerve fiber that carries impulses away from a neuron cell body.

dendr-, tree: *dendr*ite—branched nerve fiber that serves as the receptor surface of a neuron.

ependym-, tunic: *ependym*a—neuroglial cells that line spaces within the brain and spinal cord.

-lemm, rind or peel: neuri*lemm*a—sheath that surrounds the myelin of a nerve fiber.

moto-, moving: *moto*r neuron—neuron that stimulates a muscle to contract or a gland to release a secretion.

multi-, many: *multi*polar neuron—neuron with many fibers extending from the cell body.

oligo-, few: *oligo*dendrocyte—small neuroglial cell with few cellular processes.

peri-, all around: *peri*pheral nervous system—portion of the nervous system that consists of the nerves branching from the brain and spinal cord.

saltator-, a dancer: *saltator*y conduction—nerve impulse conduction in which the impulse seems to jump from node to node along the nerve fiber.

sens-, feeling: *sens*ory neuron—neuron that can be stimulated by a sensory receptor and conducts impulses into the brain or spinal cord.

syn-, together: *syn*apse—junction between two neurons.

uni-, one: *uni*polar—neuron with only one fiber extending from the cell body.

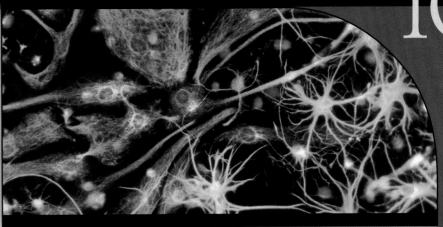

These progenitor cells, cultured in the laboratory, can specialize further to become mature neuroglial cells. The star-shaped astrocyte progenitors (light green) will supply neurons with nutrients. They are the most common type of neuroglial cell. Cell nuclei are stained blue. Immunofluorescent light micrograph (750×).

Nervous System I

Basic Structure and Function

CHAPTER OBJECTIVES

After you have studied this chapter, you should be able to

1. Explain the general functions of the nervous system.

2. Describe the general structure of a neuron.

3. Explain how neurons are classified.

4. Name four types of neuroglial cells and describe the functions of each.

5. Explain how an injured nerve fiber may regenerate.

6. Explain how a membrane becomes polarized.

7. Describe the events that lead to the conduction of a nerve impulse.

8. Explain how a nerve impulse is transmitted from one neuron to another.

9. Distinguish between excitatory and inhibitory postsynaptic potentials.

10. Explain two ways impulses are processed in neuronal pools.

For many years, brain neurons were thought to be incapable of dividing. Instead, people acquired new skills because their brains had more than enough neurons, with a nearly infinite number of connections possible. The human brain, however, does harbor very small numbers of neural stem cells, which can differentiate into neurons or into the neuroglial cells that support them. Recognition of what one researcher calls "brain marrow" took many years.

Researchers had shown as long ago as 1912 that in rats, some neurons in the hippocampus, the memory center, can divide. Still, the idea that neurons could no longer undergo mitosis persisted. Then in the mid-1980s, researchers identified dividing neurons that accompany learning of birdsong in chickadees and canaries. The cell division rate peaked when young birds needed to learn their songs to survive. In experiments that placed the birds' food farther away than usual, the division rate of neurons in the hippocampus rose as the birds had to sing longer to communicate the food location.

Identifying neural stem cells in humans proved more challenging than doing so in rats and birds, simply because the material is hard to obtain. In the late 1990s, researchers applied a chemical, bromodeoxyuridine (BrdU), to slices of brain tissue from tree shrews and marmosets, as model organisms closer to humans in an evolutionary sense than rats or birds. BrdU is preferentially taken up by dividing cells, and so the fact that marked cells showed up in the brain slices confirmed that cell division indeed occurs in cells in the mammalian brain.

Researchers at the Salk Institute in La Jolla, California, extended the BrdU staining approach to identifying neural stem cells in humans by asking several patients being treated with BrdU for cancers of the tongue or larynx to donate their brains upon their deaths. The brains revealed actively dividing neural stem cells in a region of the hippocampus called the dentate gyrus. These cells divide to generate more stem cells, and also give rise to cells that migrate to other areas of the brain,

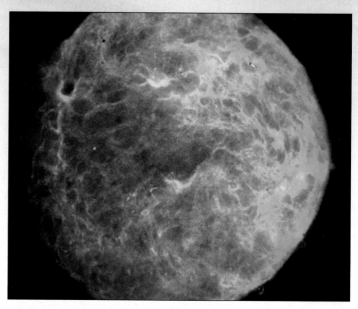

Neurospheres cultured in the laboratory consist of neural stem cells. These cells can divide and differentiate to give rise to neurons and neuroglial cells.

where they differentiate as either neurons or neuroglial cells. Further experiments identified neural stem cells near spaces in the brain called ventricles, and in the olfactory bulb, where the sense of smell originates.

The fact that the human brain contains reservoirs of cells that are capable of division and differentiation has clinical implications, if researchers can learn how these cells migrate and specialize. Neural stem cells can be grown in the laboratory, obtained from cadavers, or from the individual who requires treatment of a brain-related disorder. That is, it might be possible to treat neurodegenerative conditions, such as Parkinson disease or multiple sclerosis, from within, by coaxing a person's own neural stem cells to heal the damage. ■

General Functions of the Nervous System

The nervous system is composed predominantly of neural tissue but also includes some blood vessels and connective tissue. Neural tissue consists of two cell types: nerve cells, or **neurons** (nu′ronz), and **neuroglial** (nu-rog′le-ahl) **cells** (or neuroglia). Neurons are specialized to react to physical and chemical changes in their surroundings. Small cellular processes called **dendrites** (den′drītz) receive the input, and a longer process called an **axon** (ak′son), or nerve fiber, carries the information away from the cell in the form of bioelectric signals called **nerve impulses** (fig. 10.1). **Nerves** are bundles of axons. Neuroglial cells were once thought only to fill spaces and surround or support neurons. Today, we know that they have many other functions, including nourishing neurons and perhaps even sending and receiving messages.

An important part of the nervous system at the cellular level is not a cell at all, but the small spaces between neurons, called **synapses** (sin′aps-ez). Much of the effort of the nervous system centers on sending and receiving electrochemical messages from neuron to neuron at synapses. The actual carriers of this information are biological messenger molecules called **neurotransmitters** (nu″ro-trans-mit′erz).

The organs of the nervous system can be divided into two groups. One group, consisting of the brain and spinal cord, forms the **central nervous system** (sen′tral ner′vus sis′tem), or **CNS,** and the other, composed of the nerves (cranial and spinal nerves) that connect the central nervous system to other body parts, is called the **peripheral nervous system** (pĕ-rif′er-al ner′vus sis′tem), or **PNS** (fig. 10.2). Together these systems provide three general functions—sensory, integrative, and motor.

Structures called **sensory receptors** at the ends of peripheral neurons provide the sensory function of the

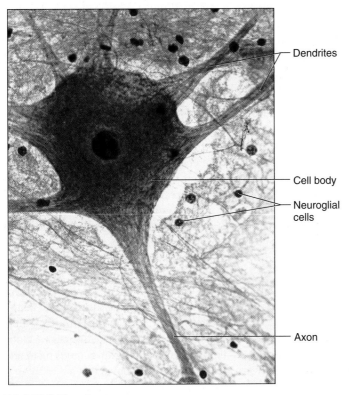

FIGURE 10.1

Neurons are the structural and functional units of the nervous system (600×). Neuroglial cells surround the neuron, appearing as dark dots. Note the location of the neuron processes (dendrites and a single axon).

Dendrites

Cell body

Neuroglial cells

Axon

nervous system (see chapter 11, p. 373). These receptors gather information by detecting changes inside and outside the body. They monitor external environmental factors such as light and sound intensities as well as the temperature, oxygen concentration, and other conditions of the body's internal environment.

Sensory receptors convert their information into nerve impulses, which are then transmitted over peripheral nerves to the central nervous system. There the signals are integrated—that is, they are brought together, creating sensations, adding to memory, or helping produce thoughts. Following integration, conscious or subconscious decisions are made and then acted upon by means of motor functions.

The motor functions of the nervous system employ neurons that carry impulses from the central nervous system to responsive structures called *effectors*. These effectors are outside the nervous system and include muscles that contract in response to nerve impulse stimulation, and glands that secrete when stimulated. The motor portion of the peripheral nervous system can be subdivided into the somatic and the autonomic nervous systems. Generally the **somatic nervous system** is involved in conscious (voluntary) activities, such as skeletal muscle contraction. The **autonomic nervous system** controls viscera, such as the heart and various glands, and thus controls subconscious (involuntary) actions.

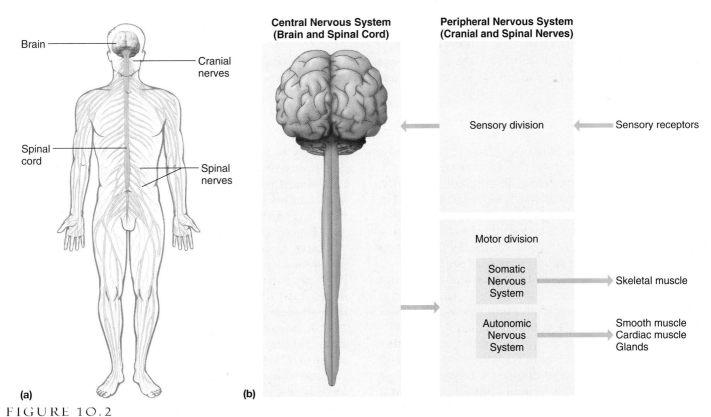

(a)

(b)

FIGURE 10.2

Nervous system. (*a*) The nervous system includes the central nervous system (brain and spinal cord) and the peripheral nervous system (cranial nerves and spinal nerves). (*b*) The nervous system receives information from sensory receptors and initiates responses through effector organs (muscles and glands).

MIGRAINE

The signs of a migraine are unmistakable—a pounding head, waves of nausea, sometimes shimmering images in the peripheral visual field, and extreme sensitivity to light or sound. Inherited susceptibilities and environmental factors probably cause migraines. Environmental triggers include sudden exposure to bright light, eating a particular food (chocolate, red wine, nuts, and processed meats top the list), lack of sleep, stress, high altitude, stormy weather, and excessive caffeine or alcohol intake. Because 70% of the millions of people who suffer from migraine worldwide are women, hormonal influences may also be involved.

Although it is considered a headache, a migraine attack is actually a response to changes in the diameters of blood vessels in the face, head, and neck. Constriction followed by dilation of these vessels causes head pain (usually on one side), nausea and perhaps vomiting, and sensitivity to light.

Migraine Types

The two major variants of migraine are called "classic" and "common." Ten to 15% of sufferers experience classic migraine, which lasts four to six hours and begins with an "aura" of light in the peripheral vision. Common migraine usually lacks an aura and may last for three to four days. A third, very rare type, familial hemiplegic migraine, may lead neurologists to finally understand precisely how all migraines occur.

Familial hemiplegic migraine runs in families. In addition to severe head pain, it paralyzes one side of the body for a few hours to a few days and may cause loss of consciousness. This form of migraine results from a single amino acid change in a neuronal protein calcium channel in three brain regions (cerebellum, brain stem, and hippocampus). Interestingly, two other types of mutations in the responsible gene cause two different nervous system disorders. A shortened protein causes episodic ataxia, a movement disorder that makes a person periodically walk as if intoxicated. Extra copies of a particular amino acid in the protein cause spinocerebellar ataxia type 6, which causes chronic lack of coordination.

Treatments

Learning what goes awry in the rare, inherited form of migraine may shed light on how more common forms of the disease begin and progress. The researchers who discovered the gene for familial hemiplegic migraine are already using the clues in the calcium channel protein to develop a drug that can prevent migraine attacks.

Current drug treatments, although very effective, must be taken as soon as symptoms begin. The first drug to directly stop a migraine in its tracks was Imitrex (sumatriptan), which became available in 1996. Imitrex mimics the action of the neurotransmitter serotonin, levels of which fluctuate during an attack. The drug constricts blood vessels in the brain, decreasing blood flow to certain areas. Newer drugs more precisely target the neurons that are affected in a migraine attack—specifically, those in an area called the trigeminal nucleus. These neurons control cerebral blood vessel dilation. Imitrex can cause cardiac side effects because it also binds to serotonin receptors on blood vessels in the heart.

Drugs can help about 85% of migraine sufferers. With several new drugs in development and a new understanding of how this painful condition develops, the future is bright for vanquishing migraine. ■

The nervous system can detect changes in the body, make decisions on the basis of the information received, and stimulate muscles or glands to respond. Typically, these responses counteract the effects of the changes, and in this way, the nervous system helps maintain homeostasis. Clinical Application 10.1 discusses migraine headaches, a common medical problem attributed to the nervous system that may involve its blood supply as well as neurons.

Neurons vary considerably in size and shape, but they share certain features. For example, every neuron has a **cell body,** dendrites, and an axon. Figure 10.3 shows some of the other structures common to neurons.

A neuron's cell body (soma or perikaryon) contains granular cytoplasm, mitochondria, lysosomes, a Golgi apparatus, and many microtubules. A network of fine threads called **neurofibrils** extends into the axons and supports them. Scattered throughout the cytoplasm are many membranous packets of **chromatophilic substance** (Nissl bodies), which consist of rough endoplasmic reticulum. Cytoplasmic inclusions in neurons contain glycogen, lipids, or pigments such as melanin.

Near the center of the neuron cell body is a large, spherical nucleus with a conspicuous nucleolus. Mature neurons generally do not divide; neural stem cells do.

Dendrites are usually highly branched, providing receptive surfaces to which processes from other neurons communicate. (In some kinds of neurons, the cell body itself provides such a receptive surface.) Often the dendrites have tiny, thornlike spines (dendritic spines) on their surfaces, which are contact points for other neurons.

A neuron may have many dendrites, but only one axon. The axon, which often arises from a slight elevation of the cell body (axonal hillock), is a slender, cylindrical process with a nearly smooth surface and uniform diameter. It is specialized to conduct nerve impulses away from the cell body. The cytoplasm of the axon includes many mitochondria, microtubules, and neurofibrils (ribosomes are found only in the cell body). The axon may give off

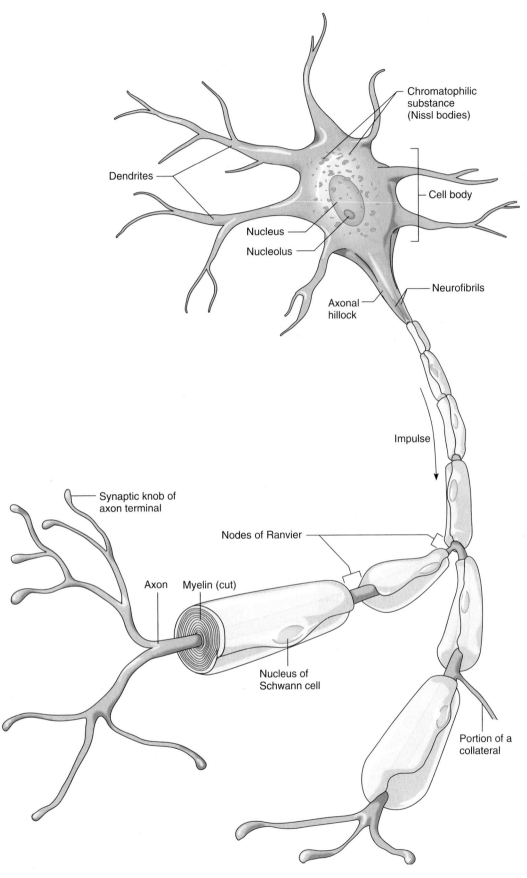

Chromatophilic
substance
(Nissl bodies)

Dendrites

Cell body

Nucleus

Nucleolus

Neurofibrils

Axonal
hillock

Impulse

Synaptic knob of
axon terminal

Nodes of Ranvier

Axon   Myelin (cut)

Nucleus of
Schwann cell

Portion of a
collateral

FIGURE 10.3
A common neuron.

branches, called *collaterals.* Near its end, an axon may have many fine extensions, each with a specialized ending called an *axon terminal.* This ends as a *synaptic knob* very close to the receptive surface of another cell, separated only by a space called the **synaptic cleft.**

In addition to conducting nerve impulses, an axon conveys biochemicals that are produced in the neuron cell body, which can be quite a task in these very long cells. This process, called *axonal transport,* involves vesicles, mitochondria, ions, nutrients, and neurotransmitters that move from the cell body to the ends of the axon.

The larger axons of peripheral neurons are encased in lipid-rich sheaths formed by layers of cell membranes of neuroglial cells called **Schwann cells,** which wind tightly, somewhat like a bandage wrapped around a finger. The layers are composed of **myelin** (mi'ĕ-lin), which has a higher proportion of lipid than other surface membranes. This coating is called a *myelin sheath.* The portions of the Schwann cells that contain most of the cytoplasm and the nuclei remain outside the myelin sheath and comprise a **neurilemma** (nu″ri-lem′mah), or *neurilemmal sheath,* which surrounds the myelin sheath (fig. 10.4). Narrow

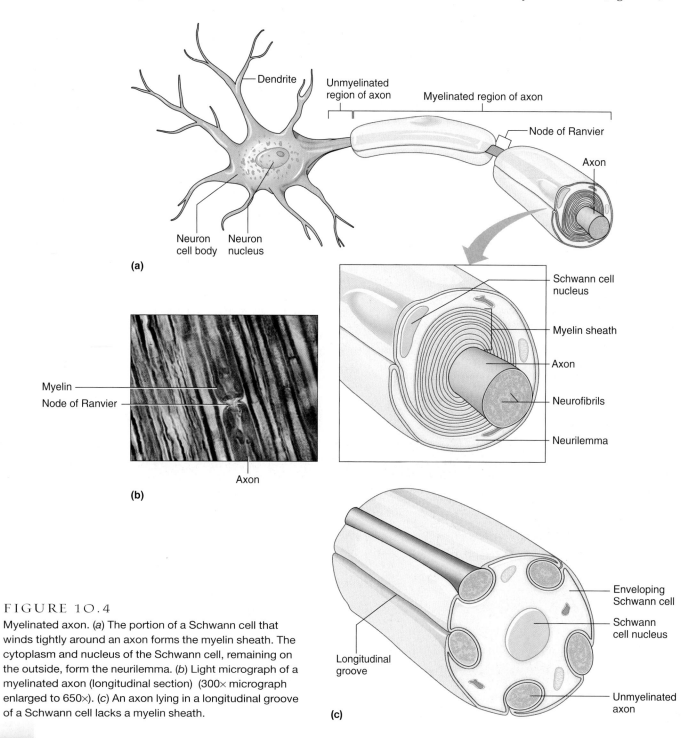

FIGURE 10.4

Myelinated axon. (*a*) The portion of a Schwann cell that winds tightly around an axon forms the myelin sheath. The cytoplasm and nucleus of the Schwann cell, remaining on the outside, form the neurilemma. (*b*) Light micrograph of a myelinated axon (longitudinal section) (300× micrograph enlarged to 650×). (*c*) An axon lying in a longitudinal groove of a Schwann cell lacks a myelin sheath.

MULTIPLE SCLEROSIS

On the television program The West Wing, two days before the state of the union address, President Jeb Bartlett fainted. His wife, a physician, became alarmed at her husband's high fever, because she alone knew that he had relapsing-remitting multiple sclerosis (MS). Symptoms come and go, but a fever or stress can trigger an episode of the neurological disorder. Dr. Bartlett had been secretly treating her husband with injections of beta interferon, an immune system biochemical that helps some people with MS.

President Bartlett is fictional, but unfortunately, MS is not. In North America, 300,000 people are affected. Diagnosis is based on symptoms and repeated magnetic resonance imaging (MRI) scans, which can track development of lesions. The first symptoms are often blurred vision and numb legs or arms, but because these are intermittent, diagnosis may take awhile. About 70 percent of affected individuals first notice symptoms between the ages of 20 and 40; the earliest known age of onset is 3 years, and the latest, 67 years. Some affected individuals eventually become permanently paralyzed. Women are twice as likely to develop MS as men, and Caucasians are more often affected than people of other races.

In MS, the myelin coating in various sites through the brain and spinal cord becomes inflamed and is eventually destroyed, leaving hard scars, called scleroses, that block the underlying neurons from transmitting messages. Muscles that no longer receive input from motor neurons stop contracting, and eventually, they atrophy. Symptoms reflect the specific neurons affected. Shortcircuiting in one part of the brain may affect fine coordination in one hand; if another brain part is affected, vision may be altered.

What might destroy myelin in MS? A virus may cause the body's immune system to attack the cells producing myelin. This would happen if viruses lay latent in nerve cells, then emerged years later bearing proteins also found on nerve cells. The immune system, interpreting the proteins as foreign, would attack the viruses as well as the neurons (an example of an autoimmune response).

A virus is suspected for a few reasons: viral infections can strip neurons of their myelin sheaths; viral infections can cause repeated bouts of symptoms; and most compelling, MS is much more common in some geographical regions (the temperate zones of Europe, South America, and North America) than others, suggesting a pattern of infection. However, studies in rats indicate that MS destroys oligodendrocyte progenitor cells, which suggests that once myelin is gone, the body cannot replace it.

Current clinical trials are exploring new treatments for MS. Stem cells taken from a person's bone marrow or separated from the bloodstream may be coaxed to replace myelin-producing progenitor cells and implanted where they are needed. Alpha interferon and other immune system biochemicals are being tested, and beta interferon is being tested with various other drugs. Hormones (testosterone, estriol, and methyl prednisolone), marijuana extracts, bee venom, ginkgo biloba and a low fat diet with fatty acid supplements are other approaches under investigation. ■

gaps in the myelin sheath between Schwann cells are called **nodes of Ranvier** (fig. 10.4).

Schwann cells also enclose, but do not wind around, the smallest axons of peripheral neurons. Consequently, these axons lack myelin sheaths. Instead, the axon or a group of axons may lie partially or completely in a longitudinal groove of Schwann cells.

Axons that have myelin sheaths are called *myelinated* (medullated) axons, and those that lack these sheaths are *unmyelinated axons* (fig. 10.5). Groups of myelinated axons appear white. Masses of such axons impart color to the *white matter* in the brain and spinal cord, but here another kind of neuroglial cell called an oligodendrocyte produces myelin. In the brain and spinal cord, myelinated axons lack neurilemma.

Unmyelinated nerve tissue appears gray. Thus, the *gray matter* within the brain and spinal cord contains many unmyelinated axons and neuron cell bodies. Clinical Application 10.2 discusses multiple sclerosis, in which neurons in the brain and spinal cord lose their myelin.

1 List the general functions of the nervous system.

2 Describe a neuron.

3 Explain how an axon in the peripheral nervous system becomes myelinated.

Classification of Neurons and Neuroglial Cells

Neurons vary in size and shape. They may differ in the length and size of their axons and dendrites and in the number of processes by which they communicate with other neurons.

Neurons also vary in function. Some carry impulses into the brain or spinal cord; others carry impulses out from the brain or spinal cord; and still others conduct impulses from neuron to neuron within the brain or spinal cord.

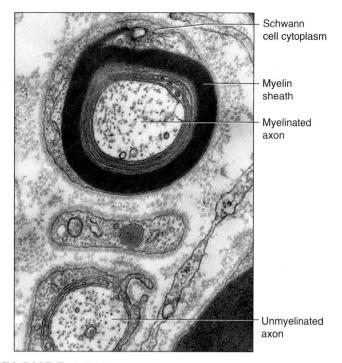

FIGURE 10.5

A transmission electron micrograph of myelinated and unmyelinated axons in cross section (30,000×).

Schwann cell cytoplasm

Myelin sheath

Myelinated axon

Unmyelinated axon

Classification of Neurons

On the basis of *structural differences,* neurons can be classified into three major groups, as figure 10.6 shows. Each type of neuron is specialized to send a nerve impulse in one direction, originating at a sensitive region of the axon called the **trigger zone.**

1. **Bipolar neurons.** The cell body of a bipolar neuron has only two processes, one arising from either end. Although these processes are similar in structure, one is an axon and the other is a dendrite. Such neurons are found within specialized parts of the eyes, nose, and ears.

2. **Unipolar neurons.** Each unipolar neuron has a single process extending from its cell body. A short distance from the cell body, this process divides into two branches, which really function as a single axon: One branch (peripheral process) is associated with dendrites near a peripheral body part. The other branch (central process) enters the brain or spinal cord. The cell bodies of some unipolar neurons aggregate in specialized masses of nerve tissue called *ganglia,* which are located outside the brain and spinal cord.

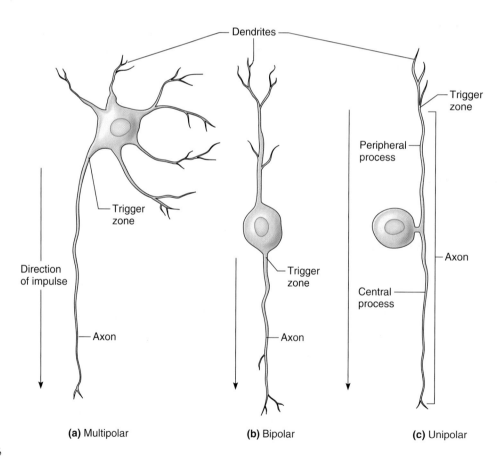

Dendrites

Trigger zone

Direction of impulse

Trigger zone

Axon

(a) Multipolar

Trigger zone

Axon

(b) Bipolar

Trigger zone

Peripheral process

Axon

Central process

(c) Unipolar

FIGURE 10.6

Structural types of neurons include (*a*) the multipolar neuron, (*b*) the bipolar neuron, and (*c*) the unipolar neuron. Note in each case the "trigger zone" at the initial portion of the axon.

3. **Multipolar neurons.** Multipolar neurons have many processes arising from their cell bodies. Only one is an axon; the rest are dendrites. Most neurons whose cell bodies lie within the brain or spinal cord are of this type. The neuron illustrated in figure 10.3 is multipolar.

Neurons can also be classified by *functional differences* into the following groups, depending on whether they carry information into the central nervous system (CNS), completely within the CNS, or out of the CNS (fig. 10.7).

1. **Sensory neurons** (afferent neurons) carry nerve impulses from peripheral body parts into the brain or spinal cord. These neurons have specialized *receptor ends* at the tips of their dendrites, or they have dendrites that are near *receptor cells* in the skin or in certain sensory organs.

   Changes that occur inside or outside the body are likely to stimulate receptor ends or receptor cells, triggering sensory nerve impulses. The impulses travel on sensory neuron axons into the brain or spinal cord. Most sensory neurons are unipolar, as shown in figure 10.7, although some are bipolar.

2. **Interneurons** (also called association or internuncial neurons) lie within the brain or spinal cord. They are multipolar and form links between other neurons. Interneurons transmit impulses from one part of the brain or spinal cord to another. That is, they may direct incoming sensory impulses to appropriate regions for processing and interpreting. Other incoming impulses are transferred to motor neurons.

3. **Motor neurons** (efferent neurons) are multipolar and carry nerve impulses out of the brain or spinal cord to effectors—structures that respond, such as muscles or glands. For example, when motor impulses reach muscles, they contract; when motor impulses reach glands, they release secretions.

   Two specialized groups of motor neurons, accelerator and inhibitory neurons, innervate smooth and cardiac muscles. *Accelerator neurons* increase muscular activities, whereas *inhibitory neurons* decrease such actions.

   Motor neurons that control skeletal muscle are under voluntary (conscious) control. Other motor neurons, such as those that control glands and smooth and cardiac muscle, are largely under involuntary control.

Table 10.1 summarizes the classification of neurons.

Classification of Neuroglial Cells

Neurons and neuroglial cells are intimately related, arising from the same neural stem cells and remaining associated throughout their existence. Neuroglial cells were once thought to be mere bystanders to neural function, providing scaffolding and controlling the sites at which neurons contact one another (figs. 10.8 and 10.9). These important cells have additional functions. In the embryo, neuroglial cells guide neurons to their positions and may stimulate them to specialize. Neuroglial cells also produce the growth factors that nourish neurons and remove ions and neurotransmitters that accumulate between neurons, enabling them to continue transmitting information. In cell culture experiments, certain types of neuroglial cells (astrocytes) signal neurons to form and maintain synapses.

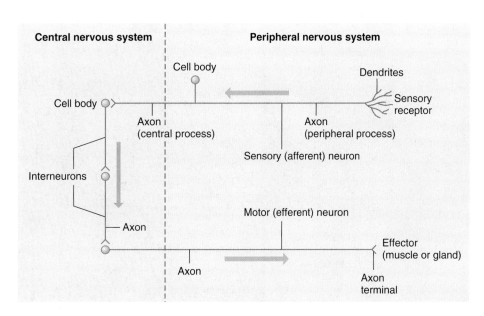

FIGURE 10.7

Sensory (afferent) neurons carry information into the central nervous system (CNS), interneurons are completely within the CNS, and motor (efferent) neurons carry instructions to the peripheral nervous system (PNS).

TABLE 10.1

Types of Neurons

A. Classified by Structure

Type	Structural Characteristics	Location
1. Bipolar neuron	Cell body with a process, arising from each end, one axon and one dendrite	In specialized parts of the eyes, nose, and ears
2. Unipolar neuron	Cell body with a single process that divides into two branches and functions as an axon	Cell body in ganglion outside the brain or spinal cord
3. Multipolar neuron	Cell body with many processes, one of which is an axon, the rest dendrites	Most common type of neuron in the brain and spinal cord

B. Classified by Function

Type	Functional Characteristics	Structural Characteristics
1. Sensory neuron	Conducts nerve impulses from receptors in peripheral body parts into the brain or spinal cord	Most unipolar; some bipolar
2. Interneuron	Transmits nerve impulses between neurons within the brain and spinal cord	Multipolar
3. Motor neuron	Conducts nerve impulses from the brain or spinal cord out to effectors—muscles or glands	Multipolar

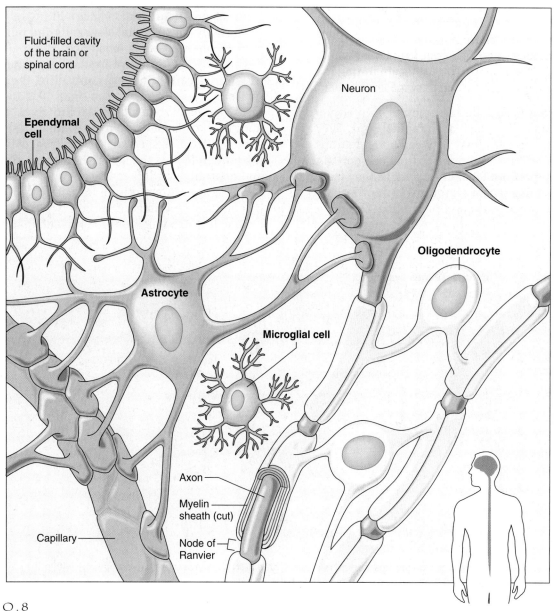

FIGURE 10.8

Types of neuroglial cells in the central nervous system include the microglial cell, oligodendrocyte, astrocyte, and ependymal cell. Cilia are on most ependymal cells during development and early childhood, but in the adult are mostly on ependymal cells in the ventricles of the brain.

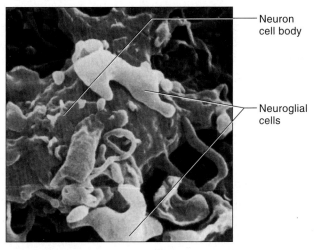

Neuron cell body

Neuroglial cells

FIGURE 10.9

A scanning electron micrograph of a neuron cell body and some of the neuroglial cells associated with it (1,000×). (Tissues and Organs: *A Text-Atlas of Scanning Electron Microscopy,* by R. G. Kessel and R. H. Kardon, © 1979 W. H. Freeman and Company.)

Schwann cells are the neuroglia of the peripheral nervous system. The central nervous system contains the following types of neuroglial cells:

1. **Astrocytes.** As their name implies, astrocytes are star-shaped cells. They are commonly found between neurons and blood vessels, where they provide support and hold structures together by means of abundant cellular processes. Astrocytes aid metabolism of certain substances, such as glucose, and they may help regulate the concentrations of important ions, such as potassium ions, within the interstitial space of nervous tissue. Astrocytes also respond to injury of brain tissue and form a special type of scar tissue, which fills spaces and closes gaps in the CNS. These multifunctional cells may also have a nutritive function, regulating movement of substances from blood vessels to neurons and bathing nearby neurons in growth factors. Astrocytes also play an important role in the blood-brain barrier,

which restricts movement of substances between the blood and the CNS (see Clinical Application 3.2, p. 69). Gap junctions link astrocytes to one another, forming protein-lined channels through which calcium ions travel, possibly stimulating neurons.

2. **Oligodendrocytes.** Oligodendrocytes resemble astrocytes but are smaller and have fewer processes. They commonly occur in rows along myelinated axons, and they form myelin in the brain and spinal cord.

   Unlike the Schwann cells of the peripheral nervous system, oligodendrocytes can send out a number of processes, each of which forms a myelin sheath around a nearby axon. In this way, a single oligodendrocyte may provide myelin for many axons. However, these cells do not form neurilemma.

3. **Microglia.** Microglial cells are small and have fewer processes than other types of neuroglial cells. These cells are scattered throughout the central nervous system, where they help support neurons and phagocytize bacterial cells and cellular debris. They usually increase in number whenever the brain or spinal cord is inflamed because of injury or disease.

4. **Ependyma.** Ependymal cells are cuboidal or columnar in shape and may have cilia. They form the inner lining of the *central canal* that extends downward through the spinal cord. Ependymal cells also form a one-cell-thick epithelial-like membrane that covers the inside of spaces within the brain called *ventricles* (see chapter 11, p. 369). Throughout the ventricles, gap junctions join ependymal cells to one another. They form a porous layer through which substances diffuse freely between the interstitial fluid of the brain tissues and the fluid (cerebrospinal fluid) within the ventricles.

   Ependymal cells also cover the specialized capillaries called *choroid plexuses* that are associated with the ventricles of the brain. Here they help regulate the composition of the cerebrospinal fluid.

Neuroglial cells form more than half of the volume of the brain. Table 10.2 summarizes their characteristics.

TABLE 10.2	Types of Neuroglial Cells of the Central Nervous System	
Type	**Characteristics**	**Functions**
Astrocytes	Star-shaped cells between neurons and blood vessels	Structural support, formation of scar tissue, transport of substances between blood vessels and neurons, communicate with one another and with neurons, mop up excess ions and neurotransmitters, induce synapse formation
Oligodendrocytes	Shaped like astrocytes, but with fewer cellular processes, occur in rows along axons	Form myelin sheaths within the brain and spinal cord, produce nerve growth factors
Microglia	Small cells with few cellular processes and found throughout the CNS	Structural support and phagocytosis (immune protection)
Ependyma	Cuboidal and columnar cells in the inner lining of the ventricles of the brain and the central canal of the spinal cord	Form a porous layer through which substances diffuse between the interstitial fluid of the brain and spinal cord and the cerebrospinal fluid

Regeneration of Nerve Axons

Injury to the cell body usually kills the neuron, and because mature neurons do not divide, it is not replaced, unless neural stem cells become stimulated to proliferate. However, a damaged peripheral axon may regenerate. For example, if injury or disease separates an axon in a peripheral nerve from its cell body, the distal portion of the axon and its myelin sheath deteriorate within a few weeks. Macrophages remove the fragments of myelin and other cellular debris. The proximal end of the injured axon develops sprouts shortly after the injury. Influenced by nerve growth factors that nearby neuroglial cells secrete, one of these sprouts may grow into a tube formed by remaining basement membrane and connective tissue. At the same time, any remaining Schwann cells proliferate along the length of the degenerating portion and form new myelin around the growing axon.

Myelin begins to form on axons during the fourteenth week of prenatal development. By the time of birth, many axons are not completely myelinated. All myelinated axons have begun to develop sheaths by the time a child starts to walk, and myelination continues into adolescence.

Excess myelin seriously impairs nervous system functioning. In Tay-Sachs disease, an inherited defect in a lysosomal enzyme causes myelin to accumulate, burying neurons in fat. The affected child begins to show symptoms by six months of age, gradually losing sight, hearing, and muscle function until death occurs by age four. Thanks to genetic screening among people of eastern European descent who are most likely to carry this gene, fewer than ten children are born in the United States with this condition each year.

Growth of a regenerating axon is slow (3 to 4 millimeters per day), but eventually the new axon may reestablish the former connection (fig. 10.10). Nerve growth factors, secreted by neuroglial cells, may help direct the growing axon. However, the regenerating axon may still end up in the wrong place, so full function often does not return.

If an axon of a neuron within the central nervous system is separated from its cell body, the distal portion of the axon will degenerate, but more slowly than a separated axon in the peripheral nervous system. However, axons within the central nervous system lack a neurilemma, and the myelin-producing oligodendrocytes fail to proliferate following an injury. Consequently, if the proximal end of a damaged axon begins to grow, there is no tube of sheath cells to guide it. Therefore, regeneration is unlikely.

If a peripheral nerve is severed, it is very important that the two cut ends be connected as soon as possible so that the regenerating sprouts of the axons can more easily reach the tubes formed by the basement membranes and connective tissues on the distal side of the gap. When the gap exceeds 3 millimeters, the regenerating axons may form a tangled mass called a *neuroma*. It is composed of sensory axons and is painfully sensitive to pressure. Neuromas sometimes complicate a patient's recovery following limb amputation.

1 What is a neuroglial cell?

2 Name and describe four types of neuroglial cells.

3 What are some functions of neuroglial cells?

4 Explain how an injured peripheral axon might regenerate.

5 Explain why functionally significant regeneration is unlikely in the central nervous system.

Cell Membrane Potential

A cell membrane is usually electrically charged, or *polarized,* so that the inside is negatively charged with respect to the outside. This polarization is due to an unequal distribution of positive and negative ions on either side of the membrane, and it is particularly important in the conduction of muscle and nerve impulses.

Distribution of Ions

Potassium ions (K^+) are the major intracellular positive ion (cation), and sodium ions (Na^+) are the major extracellular cation. The distribution is created largely by the sodium–potassium pump (Na^+/K^+ pump), which actively transports sodium ions out of the cell and potassium ions into the cell. It is also in part due to channels in the cell membrane that determine membrane permeability to these ions. These channels, formed by membrane proteins, can

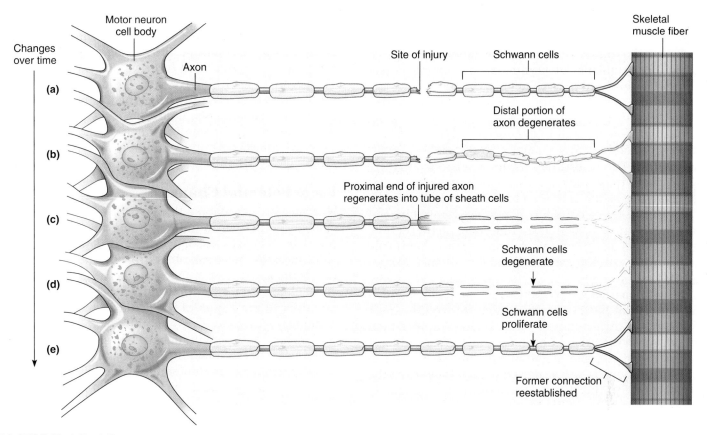

FIGURE 10.10

If a myelinated axon is injured, the following events may occur over several weeks to months: (*a*) The proximal portion of the axon may survive, but (*b*) the portion distal to the injury degenerates. (*c* and *d*) In time, the proximal portion may develop extensions that grow into the tube of basement membrane and connective tissue cells that the axon previously occupied and (*e*) possibly reestablish the former connection. Nerve growth factors that neuroglial cells secrete assist in the regeneration process.

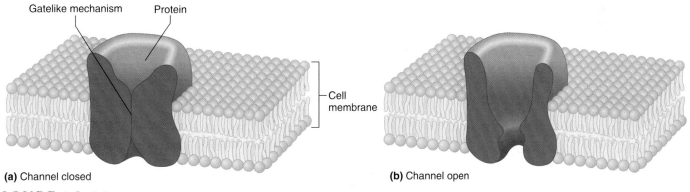

(a) Channel closed   **(b)** Channel open

FIGURE 10.11

A gatelike mechanism can (*a*) close or (*b*) open some of the channels in cell membranes through which ions pass.

be quite selective; that is, a particular channel may allow only one kind of ion to pass through and exclude all other ions of different size and charge. Thus, even though concentration gradients are present for sodium and potassium, the ability of these ions to diffuse across the cell membrane depends on the presence of channels.

✂ RECONNECT TO CHAPTER 3, CELL MEMBRANE, PAGE 66

Some channels are always open, whereas others may be either open or closed, somewhat like a gate. Both chemical and electrical factors can affect the opening and closing of these *gated channels* (fig. 10.11).

Resting Potential

A resting nerve cell is one that is not being stimulated to send a nerve impulse. Under resting conditions, nongated (always open) channels determine the membrane permeability to sodium and potassium ions.

Sodium and potassium ions follow the laws of diffusion described in chapter 3 (p. 80) and show a net movement from areas of high concentration to areas of low concentration across a membrane as their permeabilities permit. The resting cell membrane is only slightly permeable to these ions, but the membrane is more permeable to potassium ions than to sodium ions. Also, the cytoplasm of these cells has many negatively charged ions, called anions, which include phosphate (PO_4^{-2}), sulfate (SO_4^{-2}), and proteins, that are synthesized inside the cell and cannot diffuse through cell membranes (fig. 10.12a).

If we consider an imaginary neuron, before a membrane potential has been established, we would expect potassium to diffuse out of the cell more rapidly than sodium could diffuse in. This means that every millisecond (as the membrane potential is being established in our imaginary cell), a few more positive ions leave the cell than enter it (fig. 10.12a). As a result, the outside of the membrane gains a slight surplus of positive charges, and the inside reflects a surplus of the impermeable negatively charged ions. This creates a separation of positive and negative electrical charges between the inside and outside surfaces of the cell membrane (fig. 10.12b). All this time, the cell continues to expend metabolic energy in the form of ATP to actively transport sodium and potassium ions in opposite directions, thus maintaining the concentration gradients for those ions responsible for their diffusion in the first place.

The difference in electrical charge between two points is measured in units called volts. It is called a potential difference because it represents stored electrical energy that can be used to do work at some future time. The potential difference across the cell membrane is called the **membrane potential** (transmembrane potential) and is measured in millivolts.

> With the resting membrane potential established, a few sodium ions and potassium ions continue to diffuse across the cell membrane. The negative membrane potential helps sodium ions enter the cell despite sodium's low permeability, but it hinders potassium ions from leaving the cell despite potassium's higher permeability. The net effect is that three sodium ions "leak" into the cell for every two potassium ions that "leak" out. The Na^+/K^+ pump exactly balances these leaks by pumping three sodium ions out for every two potassium ions it pumps in.

In the case of a resting neuron, one that is not sending impulses or being affected by other neurons, the membrane potential is termed the **resting potential** (resting membrane potential) and has a value of −70 millivolts. The negative sign is relative to the inside of the cell and is due to the excess negative charges on the inside of the cell membrane. To understand how the resting potential provides the energy for sending a nerve impulse down the axon, we must first understand how neurons respond to signals called stimuli.

Local Potential Changes

Neurons are excitable; that is, they can respond to changes in their surroundings. Some neurons, for example, detect changes in temperature, light, or pressure outside the body, whereas others respond to signals from inside the body, often from other neurons. In either case, such changes or stimuli usually affect the membrane potential in the region of the membrane exposed to the stimulus.

Typically, the environmental change affects the membrane potential by opening a gated ion channel. If, as a result, the membrane potential becomes more negative than the resting potential, the membrane is *hyperpolarized.* If the membrane becomes less negative (more positive) than the resting potential, the membrane is *depolarized.*

Local potential changes are graded. This means that the amount of change in the membrane potential is directly proportional to the intensity of the stimulation. For example, if the membrane is being depolarized, the greater the stimulus, the greater the depolarization. If neurons are depolarized sufficiently, the membrane potential reaches a level called the *threshold* (thresh'old) *potential,* so-called because events are set into motion that result in an **action potential,** the basis for the nerve impulse.

In many cases, a single depolarizing stimulus is not sufficient to bring the membrane potential to threshold. However, if another stimulus of the same type arrives before the effect of the first one subsides, the local potential change is greater. This additive phenomenon is called **summation** (sum-ma'shun). Through summation, several subthreshold potential changes may combine to reach threshold. At threshold, an action potential is produced in an axon.

Action Potentials

The first part of the axon, known as the *initial segment,* is often referred to as the *trigger zone* (see fig. 10.6) because it contains a great number of voltage-gated sodium channels. At the resting membrane potential, these sodium channels remain closed, but when threshold is reached, they open for an instant, briefly increasing sodium permeability. Sodium ions diffuse inward across that part of the cell membrane and down their concentration gradient, aided by the attraction of the sodium ions to the negative electrical condition on the inside of the membrane.

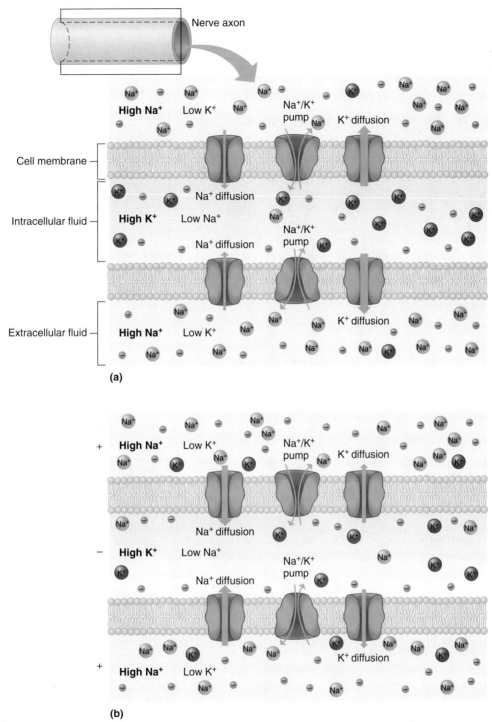

FIGURE 10.12

Development of the resting membrane potential. (*a*) Active transport creates a concentration gradient across the cell membrane for sodium ions (Na^+) and potassium ions (K^+). K^+ diffuses out of the cell rather slowly, but nonetheless faster than Na^+ can diffuse in. (*b*) This unequal diffusion results in a net loss of positive charge and a resultant excess of negative charge inside the membrane. (Count the ions. Note the loss of positive charges from inside the membrane.)

As the sodium ions rush inward, the membrane potential changes from its resting value (fig. 10.13*a*) and momentarily becomes positive on the inside (this is still considered depolarization). At the peak of the action potential, membrane potential may reach $^+$30mV (fig. 10.13*b*).

The voltage-gated sodium channels close quickly, but at almost the same time, slower voltage-gated potassium channels open and briefly increase potassium permeability. As potassium ions diffuse outward across that part of the membrane, the inside of the membrane

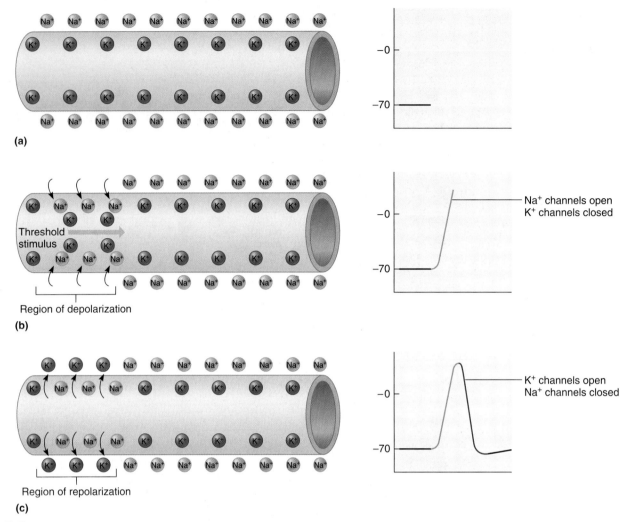

(a)

(b)

Threshold stimulus

Region of depolarization

Na⁺ channels open
K⁺ channels closed

(c)

Region of repolarization

K⁺ channels open
Na⁺ channels closed

FIGURE 10.13

At rest (a), the membrane potential is about –70 millivolts. When the membrane reaches threshold (b), voltage-sensitive sodium channels open, some Na⁺ diffuses inward, and the membrane is depolarized. Soon afterward (c), voltage-sensitive potassium channels open, K⁺ diffuses out, and the membrane is repolarized. (Negative ions not shown.)

becomes negatively charged once more. The membrane is thus repolarized (note in fig. 10.13c that it hyperpolarizes for an instant). The voltage-gated potassium channels then close as well. In this way, the resting potential is quickly reestablished, and it remains in the resting state until it is stimulated again (fig. 10.14). The active transport mechanism in the membrane works to maintain the original concentrations of sodium and potassium ions.

Axons are capable of action potentials, but the cell body and dendrites are not. An action potential at the trigger zone causes an electric current to flow a short distance down the axon, which stimulates the adjacent membrane to its threshold level, triggering another action potential. The second action potential causes another electric current to flow farther down the axon. This sequence of events results in a series of action potentials occurring sequentially all the way to the end of the axon without

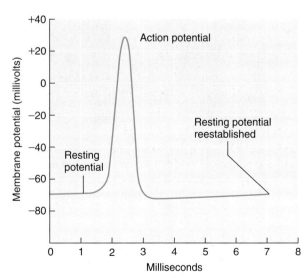

FIGURE 10.14

An oscilloscope records an action potential.

decreasing in amplitude, even if branches occur. The propagation of action potentials along an axon is the nerve impulse (fig. 10.15).

A nerve impulse is similar to the muscle impulse mentioned in chapter 9, pages 283–284. In the muscle fiber, stimulation at the motor end plate triggers an impulse to travel over the surface of the fiber and down into its transverse tubules. See table 10.3 for a summary of the events leading to the conduction of a nerve impulse.

All-or-None Response

Nerve impulse conduction is an all-or-none response. In other words, if a neuron responds at all, it responds completely. Thus, a nerve impulse is conducted whenever a stimulus of threshold intensity or above is applied to an axon and all impulses carried on that axon are the same strength. A greater intensity of stimulation produces more impulses per second, not a stronger impulse.

Refractory Period

For a very short time following passage of a nerve impulse, a threshold stimulus will not trigger another impulse on an axon. This brief period, called the **refractory period,** has two parts. During the *absolute refractory period,* which lasts about 1/2,500 of a second, the axon's membrane is changing in sodium permeability and cannot be stimulated. This is followed by a *relative refractory period,* during which the membrane is reestablishing its resting potential. While the membrane is in the relative refractory period, even though repolarization is incomplete, a threshold stimulus of high intensity may trigger an impulse.

As time passes, the intensity of stimulation required to trigger an impulse decreases until the axon's original excitability is restored. This return to the resting state usually takes from 10 to 30 milliseconds.

The refractory period limits how many action potentials may be generated in a neuron in a given amount of

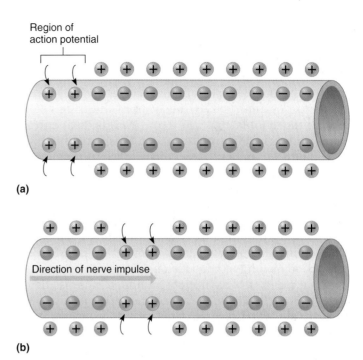

(a)

(b) Direction of nerve impulse

(c)

FIGURE 10.15

Nerve impulse. (a) An action potential in one region stimulates the adjacent region, and (b and c) a wave of action potentials (a nerve impulse) moves along the axon.

TABLE 10.3	Events Leading to Nerve Impulse Conduction

1. Nerve cell membrane maintains resting potential by diffusion of Na^+ and K^+ down their concentration gradients as the cell pumps them up the gradients.
2. Neurons receive stimulation, causing local potentials, which may sum to reach threshold.
3. Sodium channels in a local region of the membrane open.
4. Sodium ions diffuse inward, depolarizing the membrane.
5. Potassium channels in the membrane open.
6. Potassium ions diffuse outward, repolarizing the membrane.
7. The resulting action potential causes an electric current that stimulates adjacent portions of the membrane.
8. Series of action potentials occurs sequentially along the length of the axon as a nerve impulse.

time. Remembering that the action potential itself takes about a millisecond, and adding the time of the absolute refractory period to this, the maximum theoretical frequency of impulses in a neuron is about 700 per second. In the body, this limit is rarely achieved—frequencies of about 100 impulses per second are common.

Impulse Conduction

An unmyelinated axon conducts an impulse over its entire surface. A myelinated axon functions differently. Myelin contains a high proportion of lipid that excludes water and water-soluble substances. Thus, myelin serves as an insulator and prevents almost all flow of ions through the membrane that it encloses.

It might seem that the myelin sheath would prevent conduction of a nerve impulse, and this would be true if the sheath were continuous. However, nodes of Ranvier between Schwann cells or oligodendrocytes interrupt the sheath (see fig. 10.3). At these nodes, the axon membrane contains channels for sodium and potassium ions that open during a threshold depolarization.

When a myelinated axon is stimulated to threshold, an action potential occurs at the trigger zone. This causes an electric current to flow away from the trigger zone through the cytoplasm of the axon. As this local current reaches the first node, it stimulates the membrane to its threshold level, and an action potential occurs there, sending an electric current to the next node. Consequently, a nerve impulse traveling along a myelinated axon involves action potentials occurring only at the nodes. Because the action potentials appear to jump from node to node, this type of impulse conduction is called **saltatory conduction.** Conduction on myelinated axons is many times faster than conduction on unmyelinated axons (fig. 10.16).

The speed of nerve impulse conduction is also determined by the diameter of the axon—the greater the diameter, the faster the impulse. For example, an impulse on a thick, myelinated axon, such as that of a motor neuron associated with a skeletal muscle, might travel 120 meters per second, whereas an impulse on a thin, unmyelinated axon, such as that of a sensory neuron associated with the skin, might move only 0.5 meter per second. Clinical Application 10.3 discusses factors that influence nerve impulse conduction.

1 Summarize how a resting potential is achieved.

2 Explain how a polarized axon responds to stimulation.

3 List the major events that occur during an action potential.

4 Define *refractory period.*

5 Explain how impulse conduction differs in myelinated and unmyelinated axons.

The Synapse

Nerve impulses pass from neuron to neuron at synapses (fig. 10.17). A **presynaptic neuron** brings the impulse to the synapse and, as a result, stimulates or inhibits a **postsynaptic neuron.** A narrow space or synaptic cleft, or gap, separates the two neurons (fig. 10.18). The two cells are connected functionally, not physically. The process by which the impulse in the presynaptic neuron signals the postsynaptic neuron is called **synaptic transmission.**

Synaptic Transmission

A nerve impulse travels along the axon to the axon terminal. Axons usually have several rounded synaptic knobs at their terminals, which dendrites lack. These knobs contain arrays of membranous sacs, called synaptic vesicles, that contain neurotransmitter molecules. When a nerve impulse reaches a synaptic knob, voltage-sensitive calcium channels open, and calcium diffuses inward from the extracellular fluid. The increased calcium concentration inside the cell initiates a series of events that causes the synaptic vesicles to fuse with the cell membrane, releasing their neurotransmitter by exocytosis.

Synapses provide the informational potential of the nervous system, as billions of neurons make many trillions of connections. The human brain at birth contains 60 to 100 billion neurons. If that number is equated to the number of trees in the Amazon rain forest, then the number of synapses can be compared to the number of leaves on those 60 to 100 billion trees.

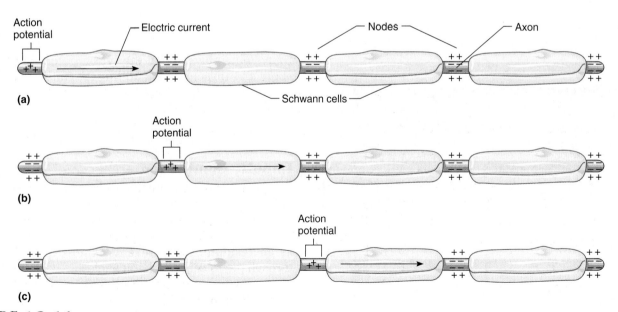

(a)

(b)

(c)

FIGURE 10.16
On a myelinated axon, a nerve impulse appears to jump from node to node.

FACTORS AFFECTING IMPULSE CONDUCTION

A number of substances alter axon membrane permeability to ions. For example, calcium ions are required to close sodium channels in axon membranes during an action potential. Consequently, if calcium is deficient, sodium channels remain open, and sodium ions diffuse through the membrane again and again so that impulses are transmitted repeatedly. If these spontaneous impulses travel along axons to skeletal muscle fibers, the muscles continuously spasm (tetanus or tetany). This can occur in women during pregnancy as the devel-

oping fetus uses maternal calcium. Tetanic contraction may also occur when the diet lacks calcium or vitamin D or when prolonged diarrhea depletes the body of calcium.

A small increase in the concentration of extracellular potassium ions causes the resting potential of nerve fibers to be less negative (partially depolarized). As a result, the threshold potential is reached with a less intense stimulus than usual. The affected fibers are very excitable, and the person may experience convulsions.

If the extracellular potassium ion concentration is greatly decreased, the resting potentials of the nerve fibers may become so negative that action potentials cannot occur. In this case, impulses are not triggered, and muscles become paralyzed.

Certain anesthetic drugs, such as procaine, decrease membrane permeability to sodium ions. In the tissue fluids surrounding an axon, these drugs prevent impulses from passing through the affected region. Consequently, the drugs keep impulses from reaching the brain, preventing perception of touch and pain. ■

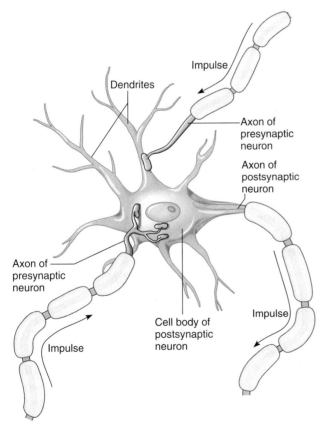

FIGURE 10.17
For an impulse to continue from one neuron to another, it must cross the synaptic cleft at a synapse. A synapse usually occurs between an axon and a dendrite or between an axon and a cell body.

Released neurotransmitter molecules diffuse across the synaptic cleft and react with specific receptor molecules in or on the postsynaptic neuron membrane. Effects of neurotransmitters may vary. Some open ion channels, and others close them. Because these ion channels respond to neurotransmitter molecules, they are called *chemically gated,* in contrast to the voltage-gated ion channels involved in action potentials. Changes in chemically gated ion channels create local potentials, called **synaptic potentials,** which enable one neuron to influence another.

Synaptic Potentials

Synaptic potentials are graded and can depolarize or hyperpolarize the receiving cell membrane. For example, if a neurotransmitter binds to a postsynaptic receptor and opens sodium ion channels, the ions diffuse inward, depolarizing the membrane, possibly triggering an action potential. This type of membrane change is called an **excitatory postsynaptic potential** (EPSP), and it lasts for about 15 milliseconds.

If a different neurotransmitter binds other receptors and increases membrane permeability to potassium ions, these ions diffuse outward, hyperpolarizing the membrane. Since an action potential is now less likely to occur, this change is called an **inhibitory postsynaptic potential** (IPSP). Some inhibitory neurotransmitters open chloride ion channels. In this case, if sodium ions enter the cell, negative chloride ions are free to follow, opposing the depolarization.

Within the brain and spinal cord, each neuron may receive the synaptic knobs of a thousand or more axons

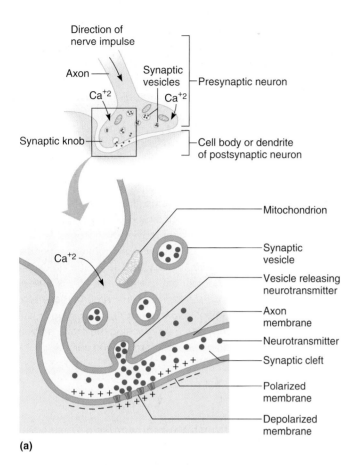

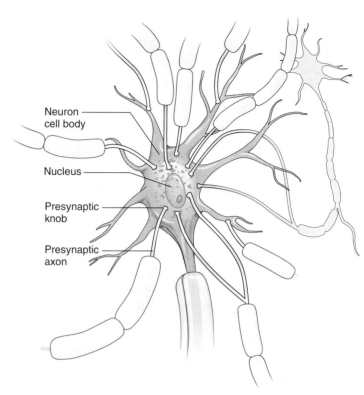

FIGURE 10.19
The synaptic knobs of many axons may communicate with the cell body of a neuron.

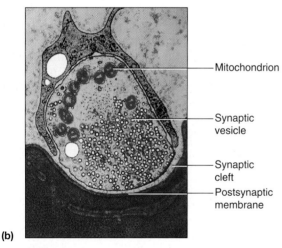

FIGURE 10.18
The synapse. (*a*) When a nerve impulse reaches the synaptic knob at the end of an axon, synaptic vesicles release a neurotransmitter that diffuses across the synaptic cleft. (*b*) A transmission electron micrograph of a synaptic knob filled with synaptic vesicles (37,500×).

on its dendrites and cell body. Furthermore, at any moment, some of the postsynaptic potentials are excitatory on a particular neuron, while others are inhibitory (fig. 10.19).

The integrated sum of the EPSPs and IPSPs determines whether an action potential results. If the net effect is more excitatory than inhibitory, threshold may be reached, and an action potential triggered. Conversely, if the net effect is inhibitory, no impulse is transmitted.

Summation of the excitatory and inhibitory effects of the postsynaptic potentials commonly takes place at the trigger zone, usually in a proximal region of the axon, but found in some dendrites as well (see fig. 10.6). This region has an especially low threshold for triggering an action potential; thus, it serves as a decision-making part of the neuron.

1 Describe a synapse.

2 Explain the function of a neurotransmitter.

3 Distinguish between an EPSP and an IPSP.

4 Describe the net effects of EPSPs and IPSPs.

Neurotransmitters

The nervous system produces at least thirty different kinds of neurotransmitters. Some neurons release only one type; others produce two or three kinds. Neurotransmitters include *acetylcholine,* which stimulates skeletal muscle contractions (see chapter 9, p. 283); a group of compounds called *monoamines* (such as epinephrine, norepinephrine, dopamine, and serotonin), which are formed by modifying amino acid molecules; a group of unmodified *amino acids* (such as glycine, glutamic acid, aspartic acid, and gamma-aminobutyric acid—GABA); and a large group of *peptides* (such as enkephalins and substance P), which are short chains of amino acids.

Most types of neurotransmitters are synthesized in the cytoplasm of the synaptic knobs and stored in synaptic vesicles. When an action potential passes along the membrane of a synaptic knob, it increases the membrane's permeability to calcium ions by opening its calcium ion channels. Calcium ions diffuse inward, and in response, some of the synaptic vesicles fuse with the presynaptic membrane and release their contents by exocytosis into the synaptic cleft. The more calcium that enters the synaptic knob, the more vesicles release neurotransmitter. Table 10.4 lists the major neurotransmitters and their actions. Tables 10.5 and 10.6 list disorders and drugs that alter neurotransmitter levels, respectively.

TABLE 10.4 Some Neurotransmitters and Representative Actions		
Neurotransmitter	**Location**	**Major Actions**
Acetylcholine	CNS	Involved in control of skeletal muscle actions
	PNS	Stimulates skeletal muscle contraction at neuromuscular junctions. May excite or inhibit at autonomic nervous system synapses
Biogenic amines		
Norepinephrine	CNS	Creates a sense of feeling good; low levels may lead to depression
	PNS	May excite or inhibit autonomic nervous system actions, depending on receptors
Dopamine	CNS	Creates a sense of feeling good; deficiency in some brain areas associated with Parkinson disease
	PNS	Limited actions in autonomic nervous system; may excite or inhibit, depending on receptors
Serotonin	CNS	Primarily inhibitory; leads to sleepiness; action is blocked by LSD, enhanced by selective serotonin reuptake inhibitor drugs
Histamine	CNS	Release in hypothalamus promotes alertness
Amino acids		
GABA	CNS	Generally inhibitory
Glutamate	CNS	Generally excitatory
Neuropeptides		
Substance P	PNS	Excitatory; pain perception
Endorphins, enkephalins	CNS	Generally inhibitory; reduce pain by inhibiting substance P release
Gases		
Nitric oxide	CNS	May play a role in memory
	PNS	Vasodilation

TABLE 10.5 Disorders Associated with Neurotransmitter Imbalances		
Condition	**Symptoms**	**Imbalance of Neurotransmitter in Brain**
Alzheimer disease	Memory loss, depression, disorientation, dementia, hallucinations, death	Deficient acetylcholine
Clinical depression	Debilitating, inexplicable sadness	Deficient norepinephrine and/or serotonin
Epilepsy	Seizures, loss of consciousness	Excess GABA leads to excess norepinephrine and dopamine
Huntington disease	Personality changes, loss of coordination, uncontrollable dancelike movements, death	Deficient GABA
Hypersomnia	Excessive sleeping	Excess serotonin
Insomnia	Inability to sleep	Deficient serotonin
Mania	Elation, irritability, overtalkativeness, increased movements	Excess norepinephrine
Myasthenia gravis	Progressive muscular weakness	Deficient acetylcholine receptors at neuromuscular junctions
Parkinson disease	Tremors of hands, slowed movements, muscle rigidity	Deficient dopamine
Schizophrenia	Inappropriate emotional responses, hallucinations	Deficient GABA leads to excess dopamine
Sudden infant death syndrome ("crib death")	Baby stops breathing, dies if unassisted	Excess dopamine
Tardive dyskinesia	Uncontrollable movements of facial muscles	Deficient dopamine

TABLE 10.6 **Drugs That Alter Neurotransmitter Levels**

Drug	Neurotransmitter Affected*	Mechanism of Action	Effect
Tryptophan	Serotonin	Stimulates neurotransmitter synthesis	Sleepiness
Reserpine	Norepinephrine	Decreases packaging of neurotransmitter into vesicles	Decreases blood pressure
Curare	Acetylcholine	Blocks receptor binding	Muscle paralysis
Valium	GABA	Enhances receptor binding	Decreases anxiety
Nicotine	Acetylcholine Dopamine	Activates receptors Elevates levels	Increases alertness Sense of pleasure
Cocaine	Dopamine	Blocks reuptake	Euphoria
Tricyclic antidepressants	Norepinephrine Serotonin	Blocks reuptake Blocks reuptake	Mood elevation Mood elevation
Monoamine oxidase inhibitors	Norepinephrine	Blocks enzymatic degradation of neurotransmitter in presynaptic cell	Mood elevation
Selective serotonin reuptake inhibitors	Serotonin	Blocks reuptake	Mood elevation, anti-anxiety agent

Others may be affected as well.

TABLE 10.7 **Events Leading to Neurotransmitter Release**

1. Action potential passes along an axon and over the surface of its synaptic knob.
2. Synaptic knob membrane becomes more permeable to calcium ions, and they diffuse inward.
3. In the presence of calcium ions, synaptic vesicles fuse to synaptic knob membrane.
4. Synaptic vesicles release their neurotransmitter by exocytosis into the synaptic cleft.
5. Synaptic vesicles become part of the membrane.
6. The added membrane provides material for endocytotic vesicles.

RECONNECT TO CHAPTER 3, EXOCYTOSIS, PAGE 87

After a vesicle releases its neurotransmitter, it becomes part of the cell membrane. Endocytosis eventually returns it to the cytoplasm, where it can provide material to form new secretory vesicles. Table 10.7 summarizes this vesicle trafficking.

In order to keep signal duration short, enzymes in synaptic clefts and on postsynaptic membranes rapidly decompose some neurotransmitters. The enzyme **acetylcholinesterase,** for example, decomposes acetylcholine on postsynaptic membranes. Other neurotransmitters are transported back into the synaptic knob of the presynaptic neuron or into nearby neurons or neuroglial cells, a process called *reuptake.* The enzyme **monoamine oxidase** inactivates the monoamine neurotransmitters epinephrine and norepinephrine after reuptake. This enzyme is found in mitochondria in the synaptic knob. Destruction or removal of neurotransmitter prevents continuous stimulation of the postsynaptic neuron.

Neuropeptides

Neurons in the brain or spinal cord synthesize **neuropeptides.** These peptides act as neurotransmitters or as *neuromodulators*—substances that alter a neuron's response to a neurotransmitter or block the release of a neurotransmitter.

Among the neuropeptides are the *enkephalins* that occur throughout the brain and spinal cord. Each enkephalin molecule is a chain of five amino acids. Synthesis of enkephalins increases during periods of painful stress, and they bind to the same receptors in the brain (opiate receptors) as the narcotic morphine. Enkephalins relieve pain sensations and probably have other functions as well.

Another morphinelike peptide, called *beta endorphin,* is found in the brain and cerebrospinal fluid. It acts longer than enkephalins and is a much more potent pain reliever (Clinical Application 10.4).

Substance P is a neuropeptide that consists of eleven amino acids and is widely distributed throughout the nervous system. It functions as a neurotransmitter (or perhaps as a neuromodulator) in the neurons that transmit pain impulses into the spinal cord and on to the brain. Enkephalins and endorphins may relieve pain by inhibiting the release of substance P from pain-transmitting neurons.

Impulse Processing

The way the nervous system processes nerve impulses and acts upon them reflects, in part, the organization of neurons and their axons within the brain and spinal cord.

Neuronal Pools

Interneurons, the neurons completely within the central nervous system, are organized into **neuronal pools.** These are groups of neurons that make synaptic connections with each other and work together to perform a common

OPIATES IN THE HUMAN BODY

Opiate drugs, such as morphine, heroin, codeine, and opium, are potent pain-killers derived from the poppy plant. These drugs alter pain perception, making it easier to tolerate, and elevate mood.

The human body produces its own opiates, called endorphins (for "endogenous morphine"), that are peptides. Like the poppy-derived opiates that they structurally resemble, endorphins influence mood and perception of pain.

The discovery of endorphins began in 1971 in research laboratories at Stanford University and the Johns Hopkins School of Medicine, where researchers exposed pieces of brain tissue from experimental mammals to morphine. The morphine was radioactively labeled (some of the atoms were radioactive isotopes) so researchers could follow its destination in the brain.

The morphine indeed bound to receptors on the membranes of certain nerve cells, particularly in the neurons that transmit pain. Why, the investigators wondered, would an animal's brain contain receptors for a chemical made by a poppy? Could a mammal's body manufacture its own opiates? The opiate receptor, then, would normally bind the body's own opiates (the endorphins) but would also be able to bind the chemically similar compounds made by the poppy. Over the next few years, researchers identified several types of endorphins in the human brain and associated their release with situations involving pain relief, such as acupuncture and analgesia to mother and child during childbirth.

The existence of endorphins explains why some people who are addicted to opiate drugs such as heroin experience withdrawal pain when they stop taking the drug. Initially, the body interprets the frequent binding of heroin to its endorphin receptors as an excess of endorphins. To bring the level down, the body slows its own production of endorphins. Then, when the addict stops taking the heroin, the body is caught short of opiates (heroin and endorphins). The result is pain.

Opiate drugs can be powerfully addicting when abused—that is, taken repeatedly by a person who is not in pain. These same drugs, however, are extremely useful in dulling severe pain, particularly in terminal illnesses. ∎

function, even though their cell bodies are often in different parts of the central nervous system. Each pool receives input from neurons (which may be part of other pools), and each pool generates output. Neuronal pools may have excitatory or inhibitory effects on other pools or on peripheral effectors.

As a result of incoming impulses and neurotransmitter release, a particular neuron of a neuronal pool is likely to receive a combination of excitation by some presynaptic neurons and inhibition by others. If the net effect is excitatory, threshold may be reached, and an outgoing impulse triggered. If the net effect is excitatory but subthreshold, an impulse will not be triggered, but because the neuron is close to threshold, it will be much more responsive to any further excitatory stimulation, a condition called **facilitation** (fah-sil″ĭ-ta′shun).

Convergence

Any single neuron in a neuronal pool may receive impulses from two or more other neurons. Axons originating from different parts of the nervous system leading to the same neuron exhibit **convergence** (kon-ver′jens).

Incoming impulses often represent information from various sensory receptors that detect changes. Convergence allows the nervous system to collect, process, and respond to information.

Convergence makes it possible for a neuron to sum impulses from different sources. For example, if a neuron receives subthreshold stimulation from one input neuron, it may reach threshold if it receives additional stimulation from a second input neuron. Thus, an output impulse triggered from this neuron reflects summation of impulses from two different sources. Such an output impulse may travel to a particular effector and evoke a response (fig. 10.20a).

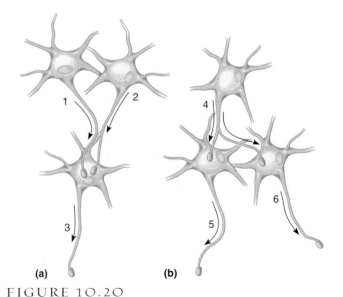

(a)   **(b)**

FIGURE 10.20

Impulse processing in neuronal pools. (a) Axons of neurons 1 and 2 converge to the cell body of neuron 3. (b) The axon of neuron 4 diverges to the cell bodies of neurons 5 and 6.

DRUG ADDICTION

Drug abuse and addiction are ancient as well as contemporary problems. A 3,500-year-old Egyptian document decries reliance on opium. In the 1600s, a smokable form of opium enslaved many Chinese, and the Japanese and Europeans discovered the addictive nature of nicotine. During the American Civil War, morphine was a widely used painkiller; cocaine was introduced a short time later to relieve veterans addicted to morphine. Today, abuse of drugs intended for medical use continues. LSD was originally used in psychotherapy but was abused in the 1960s as a hallucinogen. PCP was an anesthetic before being abused in the 1980s.

Why do people become addicted to certain drugs? Answers lie in the complex interactions of neurons, drugs, and individual behaviors.

The Role of Receptors

Eating hot fudge sundaes is highly enjoyable, but we usually don't feel driven to consume them repeatedly. Why do certain drugs compel a person to repeatedly use them, even when knowing that doing so can be dangerous—the definition of addiction? The biology of neurotransmission helps to explain how we, and other animals, become addicted to certain drugs.

Understanding how neurotransmitters fit receptors can explain the actions of certain drugs. When a drug alters the activity of a neurotransmitter on a postsynaptic neuron, it either halts or enhances synaptic transmission. A drug that binds to a receptor, blocking a neurotransmitter from bind-

ing there, is called an antagonist. A drug that activates the receptor, triggering an action potential, or that helps a neurotransmitter to bind, is called an agonist. The effect of a drug depends upon whether it is an antagonist or an agonist; on the particular behaviors the affected neurotransmitter normally regulates; and in which parts of the brain drugs affect neurotransmitters and their binding to receptors. Many addictive substances bind to receptors for the neurotransmitter dopamine, in a brain region called the nucleus accumbens.

With repeated use of an addictive substance, the number of receptors it targets can decline. This means that the person must use more of the drug to feel the same effect. Neural pathways that use the neurotransmitter norepinephrine control arousal, dreaming, and mood. Amphetamine enhances norepinephrine activity, thereby heightening alertness and mood. Amphetamine's structure is so similar to that of norepinephrine that it binds to norepinephrine receptors and triggers the same changes in the postsynaptic membrane.

Cocaine has a complex mechanism of action, both blocking reuptake of norepinephrine and binding to molecules that transport dopamine to postsynaptic cells. Cocaine's rapid and short-lived "high" reflects its short stay in the brain—its uptake takes just four to six minutes, and within twenty minutes, the drug loses half its activity.

GABA is an inhibitory neurotransmitter used in a third of the brain's synapses. The drug valium causes relaxation and inhibits seizures and anxiety by helping GABA bind

to receptors on postsynaptic neurons. Valium is therefore a GABA agonist.

Nicotine Addiction

Many medical professionals agree that cigarette smoking is highly addictive (figs. 10A and 10B). According to the Diagnostic & Statistical Manual of Mental Disorders, a person addicted to tobacco

1. must smoke more to attain the same effects (tolerance) over time;
2. experiences withdrawal symptoms when smoking stops, including weight gain, difficulty concentrating, insomnia, restlessness, anxiety, depression, slowed metabolism, and lowered heart rate;
3. smokes more often and for longer than intended;
4. spends considerable time obtaining cigarettes;
5. devotes less time to other activities;
6. continues to smoke despite knowing it is unhealthy;
7. wants to stop, but cannot easily do so.

Nicotine causes addiction, and the addiction supplies enough of the other chemicals in cigarette smoke to destroy health. The site of nicotine's activity is the neuron.

An activated form of nicotine binds protein receptors, called nicotinic receptors, that are parts of cell membranes of certain brain neurons. These receptors normally receive the neurotransmitter acetylcholine. When sufficient nicotine binds, a channel within the receptor opens, allowing positive ions to enter the neuron (fig. 10C). When a certain number of positive ions enter, the neu-

Divergence

Although a neuron has a single axon, axons may branch many times. Thus, impulses leaving a neuron of a neuronal pool may exhibit **divergence** (di-ver′jens) by reaching several other neurons. For example, one neuron may stimulate two others; each of these, in turn, may stimulate several others, and so forth. Such a pattern of diverging axons can

amplify an impulse—that is, spread it to increasing numbers of neurons within the pool (fig. 10.20*b*).

As a result of divergence, an impulse originating from a single neuron in the central nervous system may be amplified so that enough impulses reach the motor units within a skeletal muscle to cause forceful contraction. Similarly, an impulse originating from a sensory receptor may diverge and reach

FIGURE 10A

Celebrities helped glamorize smoking in the past, and some still do.

SURGEON GENERAL'S WARNING: Smoking Causes Lung Cancer, Heart Disease, Emphysema, And May Complicate Pregnancy.

FIGURE 10B

This photograph, published by the American Medical Association in 1944, shows a man using a prosthetic hand to light a cigarette—indicating that even the medical community then accepted smoking as a routine part of life. Today, physicians feel quite differently, as the Surgeon General's warning on cigarette packages and advertisements indicates.

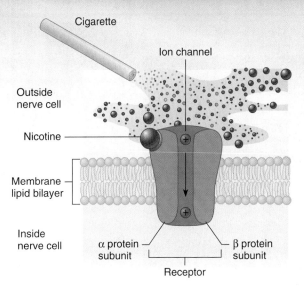

FIGURE 10C

The seat of tobacco addiction lies in nicotine's binding to nerve cell surface receptors that normally bind the neurotransmitter acetylcholine. Not only does nicotine alter the receptor so that positive ions enter the cell, triggering dopamine release, but the chemical's repeated presence in a heavy smoker stimulates excess receptors to accumulate—although they soon become nonfunctional. Nicotine's effects are complex.

ron is stimulated to release (by exocytosis) the neurotransmitter dopamine from its other end. The dopamine provides the pleasurable feelings associated with smoking. Addiction stems from two sources, researchers hypothesize—seeking the good feelings of sending off all that dopamine and avoiding painful withdrawal symptoms.

Binding nicotinic receptors isn't the only effect of nicotine on the brain. When a smoker increases the number of cigarettes smoked, the number of nicotinic receptors on the brain cells increases. This happens because of the way that the nicotine binding impairs the recycling of receptor proteins by endocytosis, so receptors are produced faster than they are taken apart. However, after a period of steady nicotine exposure,

many of the receptors malfunction and no longer admit the positive ions that trigger the nerve impulse. This may be why as time goes on it takes more nicotine to produce the same effects.

Many questions remain concerning the biological effects of tobacco smoking. Why don't lab animals experience withdrawal? Why do people who have successfully stopped smoking often start again six months later, even though withdrawal eases within two weeks of quitting? Why do some people become addicted easily, yet others smoke only a few cigarettes a day and can stop anytime? While scientists try to answer these questions, society must deal with questions of rights and responsibilities that cigarette smoking causes. ∎

several different regions of the central nervous system, where the resulting impulses can be processed and acted upon.

The nervous system enables us to experience the world and to think and feel emotion. This organ system is also very sensitive to outside influences. Clinical Application 10.5 discusses one way that an outside influence can affect the nervous system—drug addiction.

1 Define *neuropeptide.*

2 What is a neuronal pool?

3 Define *facilitation.*

4 What is convergence?

5 What is the relationship between divergence and amplification?

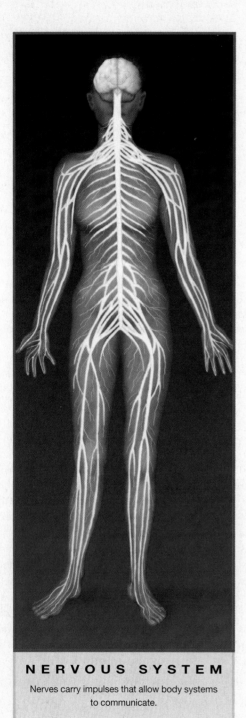

NERVOUS SYSTEM

Nerves carry impulses that allow body systems to communicate.

Integumentary System

Sensory receptors provide the nervous system with information about the outside world.

Lymphatic System

Stress may impair the immune response.

Skeletal System

Bones protect the brain and spinal cord and help maintain plasma calcium, which is important to neuron function.

Digestive System

The nervous system can influence digestive function.

Muscular System

Nerve impulses control movement and carry information about the position of body parts.

Respiratory System

The nervous system alters respiratory activity to control oxygen levels and blood pH.

Endocrine System

The hypothalamus controls secretion of many hormones.

Urinary System

Nerve impulses affect urine production and elimination.

Cardiovascular System

Nerve impulses help control blood flow and blood pressure.

Reproductive System

The nervous system plays a role in egg and sperm formation, sexual pleasure, childbirth, and nursing.

CHAPTER SUMMARY

General Functions of the Nervous System (page 338)

1. The nervous system is composed of neural tissue, including neurons and neuroglial cells, blood vessels and connective tissue.
2. Organs of the nervous system are divided into the central and peripheral nervous systems.
3. Sensory receptors detect changes in internal and external body conditions.
4. Integrative functions bring sensory information together and make decisions that motor functions act upon.
5. Motor impulses stimulate effectors to respond.
   a. The motor portion of the peripheral nervous system involved in conscious (voluntary) activities is the somatic nervous system.
   b. The motor portion of the peripheral nervous system involved in subconscious (involuntary) activities is the autonomic nervous system.
6. Neuron structure
   a. A neuron includes a cell body, cell processes, and the organelles usually found in cells.
   b. Dendrites and the cell body provide receptive surfaces.
   c. A single axon arises from the cell body and may be enclosed in a myelin sheath and a neurilemma.

Classification of Neurons and Neuroglial Cells (page 343)

Neurons differ in structure and function.

1. Classification of neurons
   a. On the basis of structure, neurons are classified as bipolar, unipolar, or multipolar.
   b. On the basis of function, neurons are classified as sensory neurons, interneurons, or motor neurons.
2. Classification of neuroglial cells
   a. Neuroglial cells make up a large portion of the nervous system and have several functions.
   b. They fill spaces, support neurons, hold nervous tissue together, play a role in the metabolism of glucose, help regulate potassium ion concentration, produce myelin, carry on phagocytosis, rid synapses of excess ions and neurotransmitters, nourish neurons, and stimulate synapse formation.
   c. They include Schwann cells in the peripheral nervous system and astrocytes, oligodendrocytes, microglia, and ependymal cells in the central nervous system.
3. Regeneration of nerve fibers
   a. If a neuron cell body is injured, the neuron is likely to die unless neural stem cells proliferate and produce replacements.
   b. If a peripheral axon is severed, its distal portion will die, but under the influence of nerve growth factors, the proximal portion may regenerate and reestablish its former connections, provided a tube of connective tissue guides it.
   c. Significant regeneration is not likely in the central nervous system.

Cell Membrane Potential (page 348)

A cell membrane is usually polarized as a result of an unequal distribution of ions on either side. Channels in membranes that allow passage of some ions but not others control ion distribution.

1. Distribution of ions
   a. Membrane ion channels, formed by proteins, may be always open or sometimes open and sometimes closed.
   b. Potassium ions pass more readily through resting neuron cell membranes than do sodium and calcium ions.
2. Resting potential
   a. A high concentration of sodium ions is on the outside of the membrane, and a high concentration of potassium ions is on the inside.
   b. Large numbers of negatively charged ions, which cannot diffuse through the cell membrane, are inside the cell.
   c. In a resting cell, more positive ions leave the cell than enter it, so the inside of the cell membrane develops a negative charge with respect to the outside.
3. Local potential changes
   a. Stimulation of a membrane affects its resting potential in a local region.
   b. The membrane is depolarized if it becomes less negative; it is hyperpolarized if it becomes more negative.
   c. Local potential changes are graded and subject to summation.
   d. Reaching threshold potential triggers an action potential.
4. Action potentials
   a. At threshold, sodium channels open and sodium ions diffuse inward, depolarizing the membrane.
   b. About the same time, potassium channels open and potassium ions diffuse outward, repolarizing the membrane.
   c. This rapid change in potential is an action potential.
   d. Many action potentials can occur before active transport reestablishes the original resting potential.
   e. The propagation of action potentials along a nerve fiber is an impulse.
5. All-or-none response
   a. A nerve impulse is an all-or-none response to a stimulus of threshold intensity applied to an axon.
   b. All the impulses conducted on an axon are the same.
6. Refractory period
   a. The refractory period is a brief time following passage of a nerve impulse when the membrane is unresponsive to an ordinary stimulus.
   b. During the absolute refractory period, the membrane cannot be stimulated; during the relative refractory period, the membrane can be stimulated with a high-intensity stimulus.
7. Impulse conduction
   a. Unmyelinated axons conduct impulses that travel over their entire surfaces.
   b. Myelinated axons conduct impulses that travel from node to node.
   c. Impulse conduction is more rapid on myelinated axons with large diameters.

The Synapse (page 354)

A synapse is a junction between two cells. A synaptic cleft is the gap between parts of two cells at a synapse. Synapses can occur between two neurons, a receptor cell and a neuron, or a neuron and an effector.

1. Synaptic transmission
   a. Impulses usually travel from dendrite or cell body, then along the axon to a synapse.
   b. Axons have synaptic knobs at their distal ends that secrete neurotransmitters.

c. The neurotransmitter is released when a nerve impulse reaches the end of an axon, and the neurotransmitter diffuses across the synaptic cleft.

d. A neurotransmitter reaching the dendrite or cell body on the distal side of the cleft triggers a nerve impulse.

2. Synaptic potentials
   a. Some neurotransmitters can depolarize postsynaptic membranes, triggering an action potential. This is an excitatory postsynaptic potential (EPSP).
   b. Others hyperpolarize the membranes, inhibiting action potentials. This is an inhibitory postsynaptic potential (IPSP).
   c. EPSPs and IPSPs are summed in a trigger zone of the neuron.

3. Neurotransmitters
   a. The nervous system produces at least thirty types of neurotransmitters.
   b. Calcium ions diffuse into synaptic knobs in response to action potentials, releasing neurotransmitters.
   c. Neurotransmitters are quickly decomposed or removed from synaptic clefts.

4. Neuropeptides
   a. Neuropeptides are chains of amino acids.
   b. Some neuropeptides are neurotransmitters or neuromodulators.

c. They include enkephalins, endorphins, and substance P.

Impulse Processing (page 358)

The way impulses are processed reflects the organization of neurons in the brain and spinal cord.

1. Neuronal pools
   a. Neurons are organized into pools within the central nervous system.
   b. Each pool receives, processes, and may conduct impulses away.
   c. Each neuron in a pool may receive excitatory and inhibitory stimuli.
   d. A neuron is facilitated when it receives subthreshold stimuli and becomes more excitable.

2. Convergence
   a. Impulses from two or more incoming axons may converge on a single neuron.
   b. Convergence enables a neuron to sum impulses from different sources.

3. Divergence
   a. Impulses leaving a pool may diverge by passing onto several output axons.
   b. Divergence amplifies impulses.

CRITICAL THINKING QUESTIONS

1. A drug called tacrine slows breakdown of acetylcholine in synaptic clefts. Which illness discussed in the chapter might tacrine theoretically treat?

2. Is Imitrex, a drug used to treat migraine, an agonist or an antagonist?

3. How would you explain the following observations?
   a. When motor nerve fibers in the leg are severed, the muscles they innervate become paralyzed; however, in time, control over the muscles often returns.
   b. When motor nerve fibers in the spinal cord are severed, the muscles they control become permanently paralyzed.

4. People who inherit familial periodic paralysis often develop very low blood potassium concentrations. How would you explain the fact that the paralysis may disappear quickly when potassium ions are administered intravenously?

5. What might be deficient in the diet of a pregnant woman who is complaining of leg muscle cramping? How would you explain this to her?

6. Why are rapidly growing cancers that originate in nervous tissue more likely to be composed of neuroglial cells than of neurons?

7. How are multiple sclerosis and Tay-Sachs disease opposite one another?

REVIEW EXERCISES

1. Distinguish between neurons and neuroglial cells.
2. Explain the relationship between the central nervous system and the peripheral nervous system.
3. List three general functions of the nervous system.
4. Describe the generalized structure of a neuron.
5. Define *myelin*.
6. Distinguish between myelinated and unmyelinated axons.
7. Explain how neurons are classified on the basis of their structure.
8. Explain how neurons are classified on the basis of their function.
9. Discuss the functions of each type of neuroglial cell.
10. Describe how an injured axon may regenerate.
11. Explain how a membrane may become polarized.
12. Define *resting potential*.
13. Distinguish between depolarizing and hyperpolarizing.
14. List the changes that occur during an action potential.

15. Distinguish between action potentials and nerve impulses.
16. Define *refractory period*.
17. Define *saltatory conduction*.
18. Define *synapse*.
19. Explain how a nerve impulse is transmitted from one neuron to another.
20. Explain the role of calcium in the release of neurotransmitters.
21. Define *neuropeptide*.
22. Distinguish between excitatory and inhibitory postsynaptic potentials.
23. Describe the "trigger zone" of a neuron.
24. Describe the relationship between an input neuron and its neuronal pool.
25. Define *facilitation*.
26. Distinguish between convergence and divergence.
27. Explain how nerve impulses are amplified.

WEB CONNECTIONS

Visit the Student Edition of the Online Learning Center at www.mhhe.com/shier10 **for answers to chapter questions, additional quizzes, interactive learning exercises, and other study tools.**

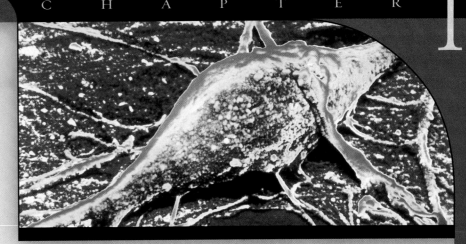

11

UNDERSTANDING WORDS

cephal-, head: en*cephal*itis—inflammation of the brain.

chiasm-, cross: optic *chiasm*a—X-shaped structure produced by the crossing over of optic nerve fibers.

flacc-, flabby: *flacc*id paralysis—paralysis characterized by loss of tone in muscles innervated by damaged axons.

funi-, small cord or fiber: *funi*culus—major nerve tract or bundle of myelinated axons within the spinal cord.

gangli-, swelling: *gangli*on—mass of neuron cell bodies.

mening-, membrane: *mening*es—membranous coverings of the brain and spinal cord.

plex-, interweaving: choroid *plex*us—mass of specialized capillaries associated with spaces in the brain.

Falsely colored scanning electron micrograph (SEM) of a single neuron of the human cerebral cortex—the outer gray matter of the brain. (4,600×)

Nervous System II

Divisions of the Nervous System

CHAPTER OBJECTIVES

After you have studied this chapter, you should be able to

1. Describe the coverings of the brain and spinal cord.
2. Describe the formation and function of cerebrospinal fluid.
3. Describe the structure of the spinal cord and its major functions.
4. Describe a reflex arc.
5. Define reflex behavior.
6. Name the major parts of the brain and describe the functions of each.
7. Distinguish among motor, sensory, and association areas of the cerebral cortex.
8. Explain hemisphere dominance.
9. Explain the stages in memory storage.
10. Explain the functions of the limbic system and the reticular formation.
11. List the major parts of the peripheral nervous system.
12. Describe the structure of a peripheral nerve and how its fibers are classified.
13. Name the cranial nerves and list their major functions.
14. Explain how spinal nerves are named and their functions.
15. Describe the general characteristics of the autonomic nervous system.
16. Distinguish between the sympathetic and the parasympathetic divisions of the autonomic nervous system.
17. Describe a sympathetic and a parasympathetic nerve pathway.
18. Explain how the autonomic neurotransmitters differently affect visceral effectors.

S eptember 13, 1848, was a momentous day for Phineas Gage, a young man who worked in Vermont evening out terrain for railroad tracks. To blast away rock, he would drill a hole, fill it with gunpowder, cover that with sand, insert a fuse, and then press down with an iron rod called a tamping iron. The explosion would go down into the rock.

But on that fateful September day, Gage began pounding on the tamping iron before his co-worker had put down the sand. The gunpowder exploded outward, slamming the inch-thick, 40-inch-long iron rod straight through Gage's skull. It pierced his brain like an arrow propelled through a soft melon, shooting out the other side of his head. Remarkably, Gage stood up just a few moments later, fully conscious and apparently unharmed by the hole just blasted through his head.

As it turned out, Gage was harmed in the freak accident, but in ways so subtle that they were not at first evident. His friends reported that "Gage was no longer Gage." Although retaining his intellect and abilities to move, speak, learn, and remember, Gage's personality dramatically changed. Once a trusted, honest, and dedicated worker, the 25-year-old became irresponsible, shirking work, cursing, and pursuing what his doctor termed "animal propensities."

Researchers as long ago as 1868 hypothesized that the tamping iron had ripped out a part of Gage's brain controlling personality. In 1994, computer analysis more precisely pinpointed the damage to the famous brain of Phineas Gage, which, along with the tamping iron, wound up in a museum at Harvard University. Researchers reconstructed the trajectory of the tamping iron, localizing two small areas in the front of the brain that control rational decision making and processing of emotion.

More than a hundred years after Gage's accident, in 1975, twenty-one-year-old Karen Ann Quinlan drank an alcoholic beverage after taking a prescription sedative, and her heart and lungs stopped functioning. When she was found, Quinlan had no pulse, was not breathing, had dilated pupils, and was unresponsive. Cardiopulmonary resuscitation restored her pulse, but once at the hospital, she was placed on a ventilator. Within twelve hours, some functions returned—her pupils constricted, she moved, gagged, grimaced, and even opened her eyes. Within a few months, she could even breathe unaided for short periods.

Because Quinlan's responses were random and not purposeful, and she was apparently unaware of herself and her environment, she was said to be in a *persistent vegetative state.* Her basic life functions were intact, but she had to be fed and given water intravenously. Fourteen months after Quinlan took the fateful pills and alcohol, her parents made a request that was to launch the right-to-die movement. They asked that Quinlan be taken off of life support. Doctors removed Quinlan's ventilator, and she lived for nine more years in a nursing home before dying of infection. She never regained awareness.

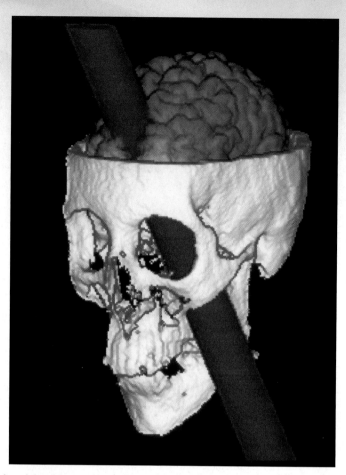

A rod that blasted through the head of a young railway worker has taught us much about the biology of personality.

Throughout Quinlan's and her family's ordeal, researchers tried to fathom what had happened to her. A CAT scan performed five years after the accident showed atrophy in two major brain regions, the cerebrum and the cerebellum. But when researchers analyzed Karen Ann Quinlan's brain in 1993, they were surprised. The most severely damaged part of her brain was the thalamus, an area thought to function merely as a relay station to higher brain structures. Quinlan's tragic case revealed that the thalamus is also important in processing thoughts, in providing the awareness and responsiveness that makes a person a conscious being.

The cases of Phineas Gage and Karen Ann Quinlan dramatically illustrate the function of the human brain by revealing what can happen when it is damaged. Nearly every aspect of our existence depends upon the brain and other parts of the nervous system, from thinking and feeling; to sensing, perceiving, and responding to the environment; to carrying out vital functions such as breathing and heartbeat. This chapter describes how the billions of neurons and neuroglial cells comprising the nervous system interact to enable us to survive and to enjoy the world around us. ■

The central nervous system (CNS) consists of the brain and the spinal cord. The brain is the largest and most complex part of the nervous system. It includes the two cerebral hemispheres, the diencephalon, the brainstem, and the cerebellum, which is described in detail in the section titled "Brain." The brain includes about one hundred billion (10^{11}) multipolar neurons and countless branches of the axons by which these neurons communicate with each other and with neurons elsewhere in the nervous system.

The brainstem connects the brain and spinal cord and allows two-way communication between them. The spinal cord, in turn, provides two-way communication between the central nervous system (CNS) and the peripheral nervous system (PNS).

Bones, membranes, and fluid surround the organs of the central nervous system. More specifically, the brain lies within the cranial cavity of the skull, whereas the spinal cord occupies the vertebral canal within the vertebral column. Beneath these bony coverings, membranes called **meninges,** located between the bone and the soft tissues of the nervous system, protect the brain and spinal cord (fig. 11.1*a*).

Meninges

The meninges (sing., *meninx*) have three layers—dura mater, arachnoid mater, and pia mater (fig. 11.1*b*). The **dura mater** is the outermost layer. It is primarily composed of tough, white, dense connective tissue and contains many blood vessels and nerves. It attaches to the inside of the cranial cavity and forms the internal periosteum of the surrounding skull bones (see reference plate 53).

In some regions, the dura mater extends inward between lobes of the brain and forms supportive and protective partitions (table 11.1). In other areas, the dura mater splits into two layers, forming channels called *dural sinuses,* shown in figure 11.1*b*. Venous blood flows through these channels as it returns from the brain to vessels leading to the heart.

The dura mater continues into the vertebral canal as a strong, tubular sheath that surrounds the spinal cord. It is attached to the cord at regular intervals by a band of pia mater (denticulate ligaments) that extends the length of the spinal cord on either side. The dural sheath terminates as a blind sac at the level of the second sacral vertebra, below the end of the spinal cord. The sheath around the spinal cord is not attached directly to the vertebrae but is separated by an *epidural space,* which lies between the dural sheath and the bony walls (fig. 11.2). This space contains blood vessels, loose connective tissue, and adipose tissue that provide a protective pad around the spinal cord.

TABLE 11.1	Partitions of the Dura Mater
Partition	**Location**
Falx cerebelli	Separates the right and left cerebellar hemispheres
Falx cerebri	Extends downward into the longitudinal fissure, and separates the right and left cerebral hemispheres (fig. 11.1*b*)
Tentorium cerebelli	Separates the occipital lobes of the cerebrum from the cerebellum (fig. 11.1*a*)

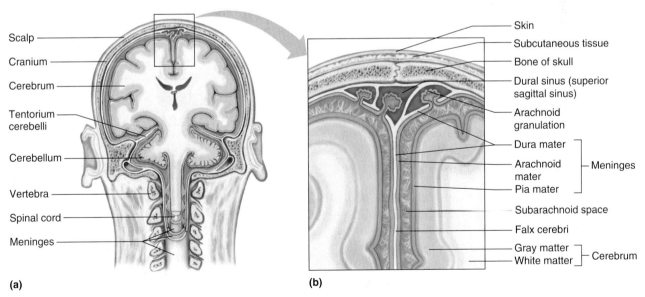

(a)

(b)

FIGURE 11.1

Meninges. (*a*) Membranes called meninges enclose the brain and spinal cord. (*b*) The meninges include three layers: dura mater, arachnoid mater, and pia mater.

CHAPTER ELEVEN   *Nervous System II*

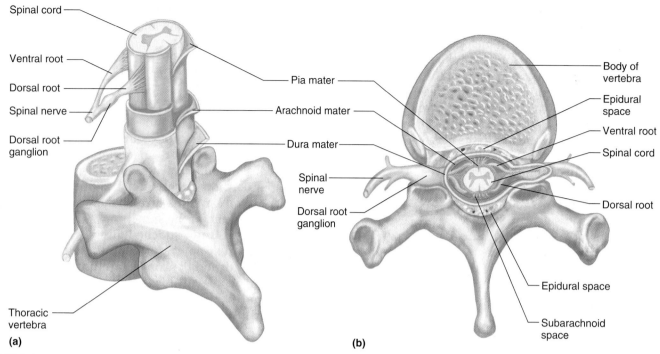

FIGURE 11.2

Meninges of the spinal cord. (*a*) The dura mater ensheaths the spinal cord. (*b*) Tissues forming a protective pad around the cord fill the epidural space between the dural sheath and the bone of the vertebra.

A blow to the head may rupture some blood vessels associated with the brain, and the escaping blood may collect in the space beneath the dura mater. This condition, called *subdural hematoma,* can increase pressure between the rigid bones of the skull and the soft tissues of the brain. Unless the accumulating blood is promptly evacuated, compression of the brain may lead to functional losses or even death.

Meningitis is an inflammation of the meninges. Bacteria or viruses that invade the cerebrospinal fluid are the usual causes of this condition. Meningitis may affect the dura mater, but it is more commonly limited to the arachnoid and pia maters.

Meningitis occurs most often in infants and children and is considered a serious childhood infection. Possible complications of this disease include loss of vision or hearing, paralysis, mental retardation, and death.

The **arachnoid mater** is a thin, weblike membrane that lacks blood vessels and is located between the dura and pia maters. It spreads over the brain and spinal cord but generally does not dip into the grooves and depressions on their surfaces. Many thin strands extend from its undersurface and are attached to the pia mater. Between the arachnoid and pia maters is a *subarachnoid space,* which contains the clear, watery **cerebrospinal fluid** (ser″ĕ-bro-spi′nal floo′id), or CSF.

The **pia mater** is very thin and contains many nerves, as well as blood vessels that nourish the underlying cells of the brain and spinal cord. The pia mater is attached to the surfaces of these organs and follows their irregular contours, passing over the high areas and dipping into the depressions.

1 Describe the meninges.

2 Name the layers of the meninges.

3 Explain where cerebrospinal fluid is located.

Ventricles and Cerebrospinal Fluid

Interconnected cavities called **ventricles** (ven′trĭ-klz) are located within the cerebral hemispheres and brainstem (fig. 11.3 and reference plates 53 and 54). These spaces are continuous with the central canal of the spinal cord and are filled with cerebrospinal fluid.

The largest ventricles are the *lateral ventricles,* which are the first and second ventricles (the first ventricle in the left cerebral hemisphere and the second ventricle in the right cerebral hemisphere). They extend anteriorly and posteriorly into the cerebral hemispheres.

A narrow space that constitutes the *third ventricle* is located in the midline of the brain beneath the corpus callosum, which is a bridge of axons that links the two cerebral hemispheres. This ventricle communicates with the

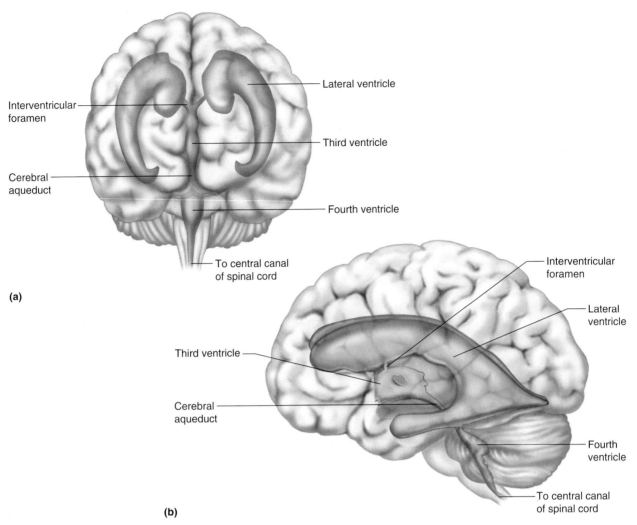

FIGURE 11.3

Ventricles in the brain. (*a*) Anterior view of the ventricles within the cerebral hemispheres and brainstem. (*b*) Lateral view.

lateral ventricles through openings (*interventricular foramina*) in its anterior end.

The *fourth ventricle* is located in the brainstem just in front of the cerebellum. A narrow canal, the *cerebral aqueduct* (aqueduct of Sylvius), connects it to the third ventricle and passes lengthwise through the brainstem. This ventricle is continuous with the central canal of the spinal cord and has openings in its roof that lead into the subarachnoid space of the meninges.

Tiny, reddish cauliflowerlike masses of specialized capillaries from the pia mater, called **choroid plexuses** (ko'roid plek'sus-ez), secrete cerebrospinal fluid. These structures project into the cavities of the ventricles (fig. 11.4). A single layer of specialized ependymal cells (see chapter 10, p. 347) joined closely by tight junctions covers the choroid plexuses. In much the same way that astrocytes provide a barrier between the blood and the brain interstitial fluid (blood-brain barrier), these cells block passage of water-soluble substances between the blood and the cerebrospinal fluid (blood-CSF barrier). At the same time, the cells selectively transfer certain substances from the blood into the cerebrospinal fluid by facilitated diffusion and transfer other substances by active transport (see chapter 3, pp. 81 and 85), thus regulating the composition of the cerebrospinal fluid.

Most of the cerebrospinal fluid arises in the lateral ventricles, from where it slowly circulates into the third and fourth ventricles and into the central canal of the spinal cord. It also enters the subarachnoid space of the meninges by passing through the wall of the fourth ventricle near the cerebellum.

Humans secrete nearly 500 milliliters of cerebrospinal fluid daily. However, only about 140 milliliters are in the nervous system at any time, because cerebrospinal fluid is continuously reabsorbed into the blood. The CSF is reabsorbed through tiny, fingerlike structures called *arachnoid granulations* that project from the subarachnoid space into the blood-filled dural sinuses (fig. 11.4).

Cerebrospinal fluid is a clear, somewhat viscid liquid that differs in composition from the fluid that

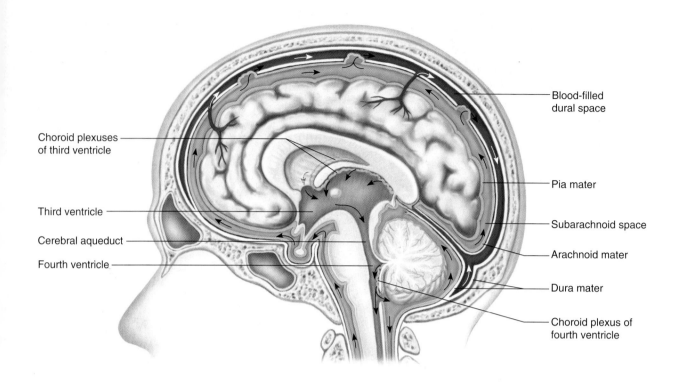

Choroid plexuses
of third ventricle

Third ventricle

Cerebral aqueduct

Fourth ventricle

Blood-filled
dural space

Pia mater

Subarachnoid space

Arachnoid mater

Dura mater

Choroid plexus of
fourth ventricle

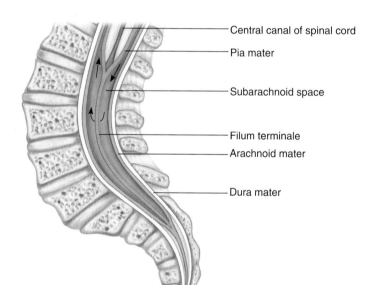

Central canal of spinal cord

Pia mater

Subarachnoid space

Filum terminale

Arachnoid mater

Dura mater

FIGURE 11.4

Choroid plexuses in ventricle walls secrete
cerebrospinal fluid. The fluid circulates
through the ventricles and central canal,
enters the subarachnoid space, and is
reabsorbed into the blood of the dural
sinuses through arachnoid granulations.
(Spinal nerves are not shown.)

leaves the capillaries in other parts of the body. Specifically, it contains a greater concentration of sodium and lesser concentrations of glucose and potassium than do other extracellular fluids. Its function is nutritive as well as protective. Cerebrospinal fluid helps maintain a stable ionic concentration in the central nervous system and provides a pathway to the blood for waste. The cerebrospinal fluid may also supply information about the internal environment to autonomic centers in the hypothalamus and brainstem, because the fluid forms from blood plasma and therefore its composition reflects changes in body fluids. Clinical Application

11.1 discusses the pressure that cerebrospinal fluid generates.

Because cerebrospinal fluid occupies the subarachnoid space of the meninges, it completely surrounds the brain and spinal cord. In effect, these organs float in the fluid. The CSF protects them by absorbing forces that might otherwise jar and damage their delicate tissues.

1 Where are the ventricles of the brain located?

2 How does cerebrospinal fluid form?

3 Describe the pattern of cerebrospinal fluid circulation.

CEREBROSPINAL FLUID PRESSURE

Because cerebrospinal fluid (CSF) is secreted and reabsorbed continuously, the fluid pressure in the ventricles remains relatively constant. However, infection, a tumor, or a blood clot can interfere with the fluid's circulation, increasing pressure within the ventricles (intracranial pressure). This can collapse cerebral blood vessels, retarding blood flow. Brain tissues may be injured by being forced against the skull.

A *lumbar puncture* (spinal tap) measures CSF pressure. A physician inserts a fine, hollow needle into the subarachnoid space between the third and fourth or between the fourth and fifth lumbar verte-

brae—below the end of the spinal cord (fig. 11A). An instrument called a *manometer* measures the pressure of the fluid, which is usually about 130 millimeters of water (10 millimeters of mercury). At the same time, samples of CSF may be withdrawn and tested for the presence of abnormal constituents. Red blood cells in the CSF, for example, may indicate a hemorrhage in the central nervous system.

A temporary drain inserted into the subarachnoid space between the fourth and fifth lumbar vertebrae can relieve pressure. In a fetus or infant whose cranial sutures

have not yet united, increasing intracranial pressure (ICP) may cause an enlargement of the cranium called *hydrocephalus,* or "water on the brain." A shunt to relieve hydrocephalus drains fluid away from the cranial cavity and into the digestive tract, where it is either reabsorbed into the blood or excreted (fig. 11B). ■

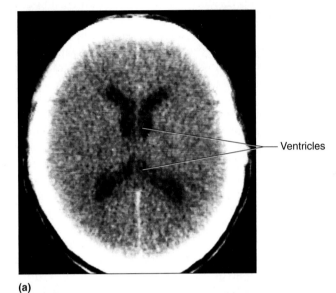

Ventricles

(a)

Ventricles

(b)

FIGURE 11B

CT scans of the human brain. (*a*) Normal ventricles. (*b*) Ventricles enlarged by accumulated fluid.

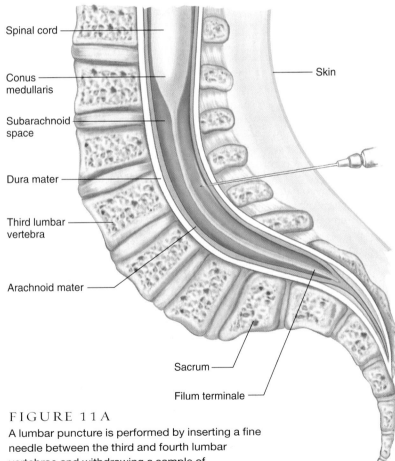

Spinal cord

Conus medullaris

Subarachnoid space

Dura mater

Third lumbar vertebra

Arachnoid mater

Skin

Sacrum

Filum terminale

Coccyx

FIGURE 11A

A lumbar puncture is performed by inserting a fine needle between the third and fourth lumbar vertebrae and withdrawing a sample of cerebrospinal fluid from the subarachnoid space. (For clarity, spinal nerves are not shown.)

Spinal Cord

The **spinal cord** is a slender column of nervous tissue that is continuous with the brain and extends downward through the vertebral canal. The spinal cord begins where nervous tissue leaves the cranial cavity at the level of the foramen magnum (see reference plate 55). The cord tapers to a point and terminates near the intervertebral disc that separates the first and second lumbar vertebrae (fig. 11.5*a*).

Structure of the Spinal Cord

The spinal cord consists of thirty-one segments, each of which gives rise to a pair of **spinal nerves.** These nerves branch to various body parts and connect them with the central nervous system.

In the neck region, a thickening in the spinal cord, called the *cervical enlargement,* supplies nerves to the upper limbs. A similar thickening in the lower back, the *lumbar enlargement,* gives off nerves to the lower limbs. Just inferior to the lumbar enlargement, the spinal cord tapers to a structure called the *conus medullaris.* From this tip, nervous tissue, including axons of both motor and sensory neurons, extends downward to become spinal nerves at the remaining lumbar and sacral levels. Originating from among them, a thin cord of connective tissue descends to the upper surface of the coccyx. This cord is called the *filum terminale* (fig. 11.5*b*). The filum terminale and the spinal nerves below the conus medullaris form a structure that resembles a horse's tail, the *cauda equina.*

Two grooves, a deep *anterior median fissure* and a shallow *posterior median sulcus,* extend the length of the spinal cord, dividing it into right and left halves. A cross section of the cord (fig. 11.6) reveals that it consists of white matter surrounding a core of gray matter. The pattern the gray matter produces roughly resembles a butterfly with its wings outspread. The upper and lower wings of gray matter are called the *posterior horns* and the *anterior horns,* respectively. Between them on either side is a protrusion of gray matter called the *lateral horn.* Motor neurons with relatively large cell bodies in the anterior horns (anterior horn cells) give rise to axons that pass out through spinal nerves to various skeletal muscles. However, the majority of neurons in the gray matter are interneurons (see chapter 10, p. 345).

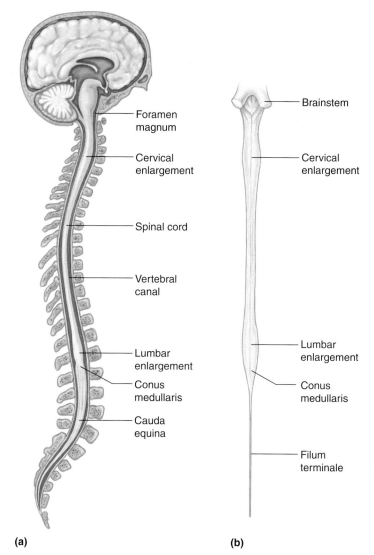

(a)   **(b)**

FIGURE 11.5

Spinal cord. (*a*) The spinal cord begins at the level of the foramen magnum. (*b*) Posterior view of the spinal cord with the spinal nerves removed.

A horizontal bar of gray matter in the middle of the spinal cord, the *gray commissure,* connects the wings of the gray matter on the right and left sides. This bar surrounds the **central canal,** which is continuous with the ventricles of the brain and contains cerebrospinal fluid. The central canal is prominent during embryonic development, but it becomes almost microscopic in an adult.

The gray matter divides the white matter of the spinal cord into three regions on each side—the *anterior, lateral,* and *posterior columns* (or *funiculi*). Each column consists of longitudinal bundles of myelinated nerve fibers that comprise major nerve pathways called **nerve tracts.**

Functions of the Spinal Cord

The spinal cord has two main functions. First, it is a center for spinal reflexes. Second, it is a conduit for nerve impulses to and from the brain.

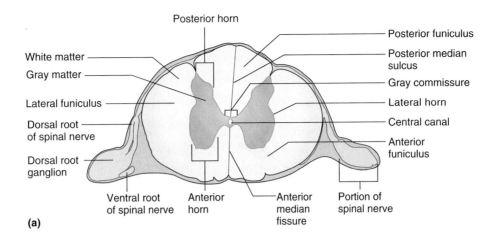

Posterior horn

White matter

Gray matter

Lateral funiculus

Dorsal root
of spinal nerve

Dorsal root
ganglion

Ventral root
of spinal nerve

Anterior
horn

Anterior
median
fissure

Portion of
spinal nerve

Posterior funiculus

Posterior median
sulcus

Gray commissure

Lateral horn

Central canal

Anterior
funiculus

(a)

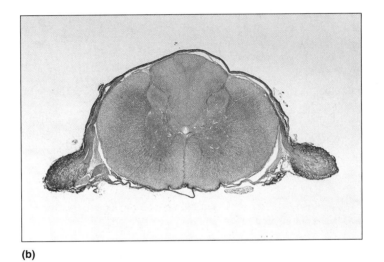

(b)

FIGURE 11.6
Spinal cord. (*a*) A cross section
of the spinal cord. (*b*) Identify
the parts of the spinal cord in
this micrograph (7.5×).

Reflex Arcs

Nerve impulses follow nerve pathways as they travel through the nervous system. The simplest of these pathways, including only a few neurons, constitutes a **reflex** (re′fleks) **arc.** Reflex arcs carry out the simplest responses—**reflexes.**

A reflex arc begins with a **receptor** (re-sep′tor) at the end of a sensory neuron. This neuron usually leads to several interneurons within the central nervous system, which serve as a processing center, or *reflex center.* Fibers from these interneurons may connect with interneurons in other parts of the nervous system. They also communicate with motor neurons, whose fibers pass outward from the central nervous system to **effectors** (e-fek′torz). Reflexes whose arcs pass through the spinal cord are called **spinal reflexes** (fig. 11.7).

Reflex Behavior

Reflexes are automatic, subconscious responses to changes (stimuli) within or outside the body. They help maintain homeostasis by controlling many involuntary processes such as heart rate, breathing rate, blood pressure, and digestion. Reflexes also carry out the automatic actions of swallowing, sneezing, coughing, and vomiting.

The *knee-jerk reflex* (patellar tendon reflex) is an example of a simple monosynaptic reflex so-called because it uses only two neurons—a sensory neuron communicating directly to a motor neuron. Striking the patellar ligament just below the patella initiates this reflex. The quadriceps femoris muscle group, which is attached to the patella by a tendon, is pulled slightly, stimulating stretch receptors within the muscle group. These receptors, in turn, trigger impulses that pass along the peripheral process (see fig. 10.7) of the axon of a unipolar sensory neuron axon into the lumbar region of the spinal cord. Within the spinal cord, the sensory axon synapses with a motor neuron. The impulse then continues along the axon of the motor neuron and travels back to the quadriceps femoris. The muscles respond by contracting, and the reflex is completed as the leg extends (fig. 11.8).

The knee-jerk reflex helps maintain an upright posture. For example, if a person is standing still and the knee begins to bend in response to gravity, the quadriceps

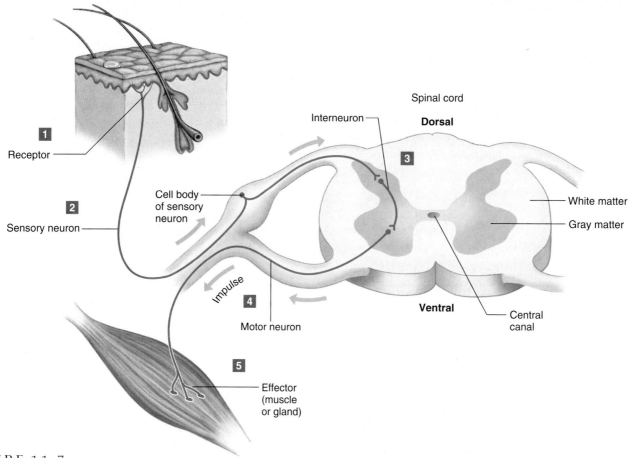

FIGURE 11.7

A reflex arc usually includes a receptor (1), a sensory neuron (2), integration center (3), a motor neuron (4), and an effector (5). In this example of a spinal reflex, the integration center is in the spinal cord.

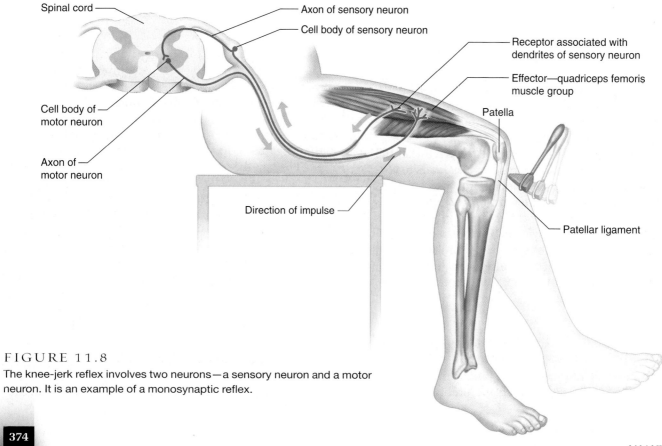

FIGURE 11.8

The knee-jerk reflex involves two neurons—a sensory neuron and a motor neuron. It is an example of a monosynaptic reflex.

femoris is stretched, the reflex is triggered, and the leg straightens again. Adjustments within the stretch receptors themselves keep the reflex responsive at different muscle lengths.

Another type of reflex, called a *withdrawal reflex* (fig. 11.9), occurs when a person touches something painful, as in stepping on a tack, activating skin receptors and sending sensory impulses to the spinal cord. There the impulses pass on to interneurons of a reflex center and are directed to motor neurons. The motor neurons transmit signals to the flexor muscles of the leg and thigh, which contract in response, pulling the foot away from the painful stimulus. At the same time, some of the incoming impulses stimulate interneurons that inhibit the action of the antagonistic extensor muscles (reciprocal innervation). This inhibition allows the flexor muscles to effectively withdraw the affected part.

While flexor muscles on the affected side (ipsilateral side) contract, the flexor muscles of the other limb (contralateral side) are inhibited. Furthermore, the extensor muscles on the contralateral side contract, helping to support the body weight that has been shifted to that side. This phenomenon, called a *crossed extensor reflex,* is due to interneuron pathways within the reflex center of the spinal cord that allow sensory impulses arriving on one side of the cord to pass across to the other side and produce an opposite effect (fig. 11.10).

Concurrent with the withdrawal reflex, other interneurons in the spinal cord carry sensory impulses upward to the brain. The person becomes aware of the experience and may feel pain.

A withdrawal reflex protects because it prevents or limits tissue damage when a body part touches something potentially harmful. Table 11.2 summarizes the

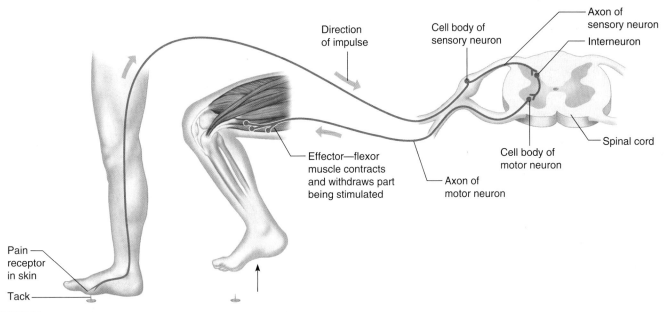

FIGURE 11.9

A withdrawal reflex involves a sensory neuron, an interneuron, and a motor neuron.

TABLE 11.2	Parts of a Reflex Arc	
Part	**Description**	**Function**
Receptor	The receptor end of a dendrite or a specialized receptor cell in a sensory organ	Sensitive to a specific type of internal or external change
Sensory neuron	Dendrite, cell body, and axon of a sensory neuron	Transmits nerve impulse from the receptor into the brain or spinal cord
Interneuron	Dendrite, cell body, and axon of a neuron within the brain or spinal cord	Serves as processing center; conducts nerve impulse from the sensory neuron to a motor neuron
Motor neuron	Dendrite, cell body, and axon of a motor neuron	Transmits nerve impulse from the brain or spinal cord out to an effector
Effector	A muscle or gland	Responds to stimulation by the motor neuron and produces the reflex or behavioral action

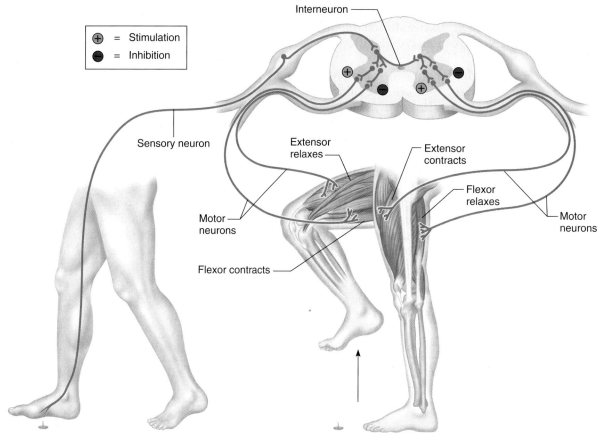

FIGURE 11.10
When the flexor muscle on one side is stimulated to contract in a withdrawal reflex, the extensor muscle on the opposite side also contracts. This helps to maintain balance.

Key:
⊕ = Stimulation
⊖ = Inhibition

Labels: Interneuron, Sensory neuron, Extensor relaxes, Extensor contracts, Flexor relaxes, Motor neurons, Motor neurons, Flexor contracts

components of a reflex arc. Clinical Application 11.2 discusses some familiar reflexes.

1 What is a nerve pathway?

2 Describe a reflex arc.

3 Define *reflex.*

4 Describe the actions that occur during a withdrawal reflex.

Ascending and Descending Tracts

The nerve tracts of the spinal cord together with the spinal nerves provide a two-way communication system between the brain and body parts outside the nervous system. The tracts that conduct sensory impulses to the brain are called **ascending tracts;** those that conduct motor impulses from the brain to motor neurons reaching muscles and glands are called **descending tracts.**

These tracts are comprised of axons. Typically, all the axons within a given tract originate from neuron cell bodies located in the same part of the nervous system and end together in some other part. The names that identify nerve tracts often reflect these common origins and termi-

nations. For example, a *spinothalamic tract* begins in the spinal cord and carries sensory impulses associated with the sensations of pain and touch to the thalamus of the brain (part of the diencephalon). A *corticospinal tract* originates in the cortex of the brain (the superficial portion of the cerebrum) and carries motor impulses downward through the spinal cord and spinal nerves. These impulses control skeletal muscle movements.

Ascending Tracts   Among the major ascending tracts of the spinal cord are the following (fig. 11.11):

1. **Fasciculus gracilis** (fah-sik′u-lus gras′il-is) and **fasciculus cuneatus** (ku′ne-at-us). These tracts are located in the posterior funiculi of the spinal cord. Their fibers conduct sensory impulses from the skin, muscles, tendons, and joints to the brain, where they are interpreted as sensations of touch, pressure, and body movement.

   At the base of the brain in an area called the medulla oblongata most of the fasciculus gracilis and fasciculus cuneatus fibers cross over (decussate) from one side to the other—that is, those ascending

USES OF REFLEXES

Since normal reflexes depend on normal neuron functions, reflexes are commonly used to obtain information concerning the condition of the nervous system. An anesthesiologist, for instance, may try to initiate a reflex in a patient who is being anesthetized in order to determine how the anesthetic drug is affecting nerve functions. Also, in the case of injury to some part of the nervous system, observing reflexes may reveal the location and extent of damage.

Injury to any component of a reflex arc alters its function. For example, a *plantar reflex* is normally initiated by stroking the sole of the foot, and the usual response is flexion of the foot and toes. However, damage to certain nerve pathways (corticospinal tract) may trigger an abnormal response called the *Babinski reflex,* which is a dorsiflexion, extending the great toe upward and fanning apart the smaller toes. If the injury is minor, the response may consist of plantar flexion with failure of the great toe to flex, or plantar flexion followed by dorsiflexion. The Babinski reflex is normally present in infants up to the age of twelve months and is thought to reflect immaturity in their corticospinal tracts.

Other reflexes that may be tested during a neurological examination include the following:

1. *Biceps-jerk reflex.* Extending a person's forearm at the elbow elicits this reflex. The examiner's finger is placed on the inside of the extended elbow over the tendon of the biceps muscle, and the finger is tapped. The biceps contracts in response, and the forearm flexes at the elbow.
2. *Triceps-jerk reflex.* Flexing a person's forearm at the elbow and tapping the short tendon of the triceps muscle close to its insertion near the tip of the elbow elicits this reflex. The muscle contracts in response, and the forearm extends slightly.
3. *Abdominal reflexes.* These reflexes occur when the examiner strokes the skin of the abdomen. For example, a dull pin drawn from the sides of the abdomen upward toward the midline and above the umbilicus causes the abdominal muscles underlying the skin to contract, and the umbilicus moves toward the stimulated region.
4. *Ankle-jerk reflex.* Tapping the calcaneal tendon just above its insertion on the calcaneus elicits this reflex. The response is plantar flexion, produced by contraction of the gastrocnemius and soleus muscles.
5. *Cremasteric reflex.* This reflex is elicited in males by stroking the upper inside of the thigh. In response, the testis on the same side is elevated by contracting muscles. ■

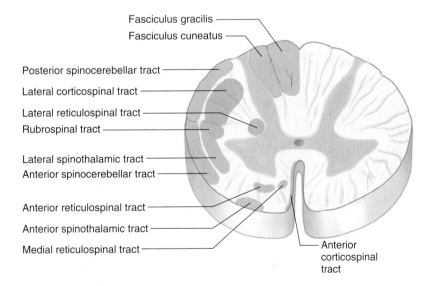

Fasciculus gracilis
Fasciculus cuneatus
Posterior spinocerebellar tract
Lateral corticospinal tract
Lateral reticulospinal tract
Rubrospinal tract
Lateral spinothalamic tract
Anterior spinocerebellar tract
Anterior reticulospinal tract
Anterior spinothalamic tract
Medial reticulospinal tract
Anterior corticospinal tract

FIGURE 11.11

Major ascending and descending tracts within a cross section of the spinal cord. Ascending tracts are in orange, descending tracts in tan. (Tracts are shown only on one side.)

on the left side of the spinal cord pass across to the right side, and vice versa. As a result, the impulses originating from sensory receptors on the left side of the body reach the right side of the brain, and those originating on the right side of the body reach the left side of the brain (fig. 11.12).

2. **Spinothalamic** (spi″no-thah-lam′ik) **tracts.** The *lateral* and *anterior spinothalamic tracts* are located in the lateral and anterior funiculi, respectively. Impulses in these tracts cross over in the spinal cord. The lateral tracts conduct impulses from various body regions to the brain and give rise to sensations of pain and temperature (fig. 11.12). Impulses carried on fibers of the anterior tracts are interpreted as touch and pressure.

3. **Spinocerebellar** (spi″no-ser″ĕ-bel′ar) **tracts.** The *posterior* and *anterior spinocerebellar tracts* lie near the surface in the lateral funiculi of the spinal cord. Fibers in the posterior tracts remain uncrossed, whereas those in the anterior tracts cross over in the medulla. Impulses conducted on their fibers originate in the muscles of the lower limbs and trunk and then travel to the cerebellum of the brain. These impulses coordinate muscular movements.

Descending Tracts   The major descending tracts of the spinal cord are shown in figure 11.11. They include the following:

1. **Corticospinal** (kor″tĭ-ko-spi′nal) **tracts.** The *lateral* and *anterior corticospinal tracts* occupy the lateral and anterior funiculi, respectively. Most of the fibers of the lateral tracts cross over in the lower portion of the medulla oblongata. Some fibers of the anterior tracts cross over at various levels of the spinal cord. The corticospinal tracts conduct motor impulses from the brain to spinal nerves and outward to various skeletal muscles. Thus, they help control voluntary movements (fig. 11.13).

   The corticospinal tracts are sometimes called *pyramidal tracts* after the pyramid-shaped regions in the medulla oblongata through which they pass. Other descending tracts are called *extrapyramidal tracts,* and they include the reticulospinal and rubrospinal tracts.

2. **Reticulospinal** (rĕ-tik″u-lo-spi′nal) **tracts.** The *lateral reticulospinal tracts* are located in the lateral funiculi, whereas the *anterior* and *medial reticulospinal tracts* are in the anterior funiculi. Some fibers in the lateral tracts cross over, whereas others remain uncrossed. Those of the anterior and medial tracts remain uncrossed. Motor impulses transmitted on the reticulospinal tracts originate in

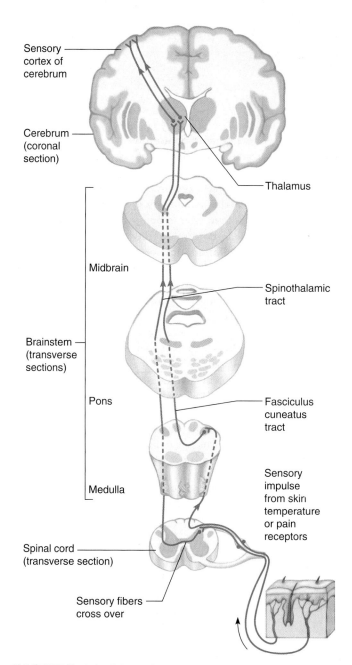

FIGURE 11.12

Sensory impulses originating in skin touch receptors ascend in the fasciculus cuneatus tract and cross over in the medulla of the brain. Pain and temperature information ascends in the lateral spinothalamic tract, which crosses over in the spinal cord.

the brain and control muscular tone and activity of sweat glands.

3. **Rubrospinal** (roo″bro-spi′nal) **tracts.** The fibers of the rubrospinal tracts cross over in the brain and pass through the lateral funiculi. They carry motor impulses from the brain to skeletal muscles, and they coordinate muscles and control posture.

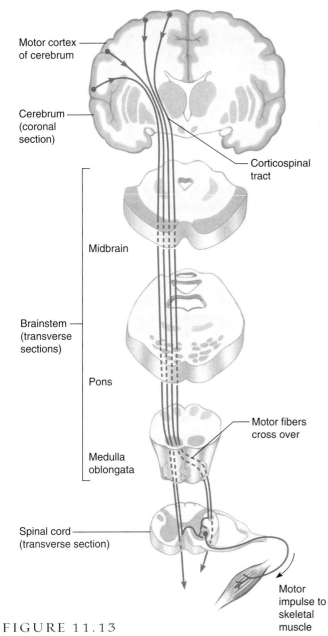

Motor cortex
of cerebrum

Cerebrum
(coronal
section)

Corticospinal
tract

Midbrain

Brainstem
(transverse
sections)

Pons

Motor fibers
cross over

Medulla
oblongata

Spinal cord
(transverse section)

Motor
impulse to
skeletal
muscle

FIGURE 11.13

Most motor fibers of the corticospinal tract begin in the cerebral cortex, cross over in the medulla, and descend in the spinal cord, where they synapse with neurons whose fibers lead to spinal nerves supplying skeletal muscles. Some fibers cross over in the spinal cord.

A hemi-lesion of the spinal cord (severed on only one side) affecting the corticospinal and spinothalamic tracts can cause Brown-Séquard syndrome. Because ascending tracts cross over at different levels, the injured side of the body becomes paralyzed and loses touch sensation. The other side of the body retains movement but loses sensations of pain and temperature.

TABLE 11.3	Nerve Tracts of the Spinal Cord	
Tract	**Location**	**Function**
Ascending Tracts		
1. Fasciculus gracilis and fasciculus cuneatus	Posterior funiculi	Conduct sensory impulses associated with the senses of touch, pressure, and body movement from skin, muscles, tendons, and joints to the brain
2. Spinothalamic tracts (lateral and anterior)	Lateral and anterior funiculi	Conduct sensory impulses associated with the senses of pain, temperature, touch, and pressure from various body regions to the brain
3. Spinocerebellar tracts (posterior and anterior)	Lateral funiculi	Conduct sensory impulses required for the coordination of muscle movements from muscles of the lower limbs and trunk to the cerebellum
Descending Tracts		
1. Corticospinal tracts (lateral and anterior)	Lateral and anterior funiculi	Conduct motor impulses associated with voluntary movements from the brain to skeletal muscles
2. Reticulospinal tracts (lateral, anterior, and medial)	Lateral and anterior funiculi	Conduct motor impulses associated with the maintenance of muscle tone and the activity of sweat glands from the brain
3. Rubrospinal tracts	Lateral funiculi	Conduct motor impulses associated with muscular coordination and the maintenance of posture from the brain

Table 11.3 summarizes the nerve tracts of the spinal cord. Clinical Application 11.3 describes injuries to the spinal cord.

Amyotrophic lateral sclerosis (ALS), also known as Lou Gehrig's disease, begins with slight stiffening and weakening of the upper and lower limbs, loss of finger dexterity, wasting hand muscles, severe muscle cramps, and difficulty swallowing. Muscle function declines throughout the body, and usually the person dies within five years from respiratory muscle paralysis. Some people, however, live many years with ALS, such as noted astronomer and author Stephen Hawking.

In ALS, motor neurons degenerate within the spinal cord, brainstem, and cerebral cortex. Fibrous tissue replaces them. By studying some of the 10% of ALS patients who inherit the disorder, researchers traced a cause to an abnormal form of an enzyme, superoxide dismutase, which normally dismantles oxygen free radicals, which are toxic by-products of metabolism.

SPINAL CORD INJURIES

On a bright May morning in 1995, actor Christopher Reeve sustained a devastating spinal cord injury when the horse that he was riding in a competition failed to clear a hurdle. Reeve hurled forward, striking his head on the fence. He landed on the grass—unconscious, not moving or breathing.

Reeve had broken the first and second cervical vertebrae, between the neck and the brainstem. Someone performed mouth-to-mouth resuscitation until paramedics inserted a breathing tube and then stabilized him on a board. At a nearby hospital, Reeve received methylprednisolone, a drug that can save a fifth of the damaged neurons by reducing inflammation. Reeve was then flown to a larger medical center for further treatment.

Reeve's rehabilitation has been slow, yet inspiring. Despite discouraging words from physicians, he has persisted in trying to exercise. Suspended from a harness, he moves his feet over a treadmill. He moves other muscles in a swimming pool and rides a special recumbent bicycle, with electrical stimulation to his legs enabling him to pedal an hour a day. Five years after the accident, Reeve gradually started to move his fingers, and then his hips and legs, although he still requires a wheelchair and a respirator. More than 500 other people with spinal cord injuries have improved with exercise, too. Reeve's motto has given hope to many: "nothing is impossible."

Thousands of people sustain spinal cord injuries each year. During the first few days the vertebrae are compressed and may break, which sets off action potentials in neurons, killing many of them. Dying neurons release calcium ions, which activate tissue-degrading enzymes. Then white blood cells arrive and produce inflammation that can destroy healthy as well as damaged neurons. Axons tear, myelin coatings are stripped off, and vital connections between nerves and muscles are cut. The tissue cannot regenerate.

The severity of a spinal cord injury depends on the extent and location of damage. Normal spinal reflexes require two-way communication between the spinal cord and the brain. Injuring nerve pathways depresses the cord's reflex activities in sites below the injury. At the same time, sensations and muscular tone in the parts the affected fibers innervate lessen. This condition, *spinal shock,* may last for days or weeks, although normal reflex activity may eventually return. However, if nerve fibers are severed, some of the cord's functions may be permanently lost.

Less severe injuries to the spinal cord, as from a blow to the head, whiplash, or rupture of an intervertebral disc, compress or distort the cord. Pain, weakness, and muscular atrophy in the regions the damaged nerve fibers supply may occur.

The most common cause of severe direct injury to the spinal cord is vehicular accidents (fig. 11C). Regardless of the cause, if nerve fibers in ascending tracts are cut, sensations arising from receptors below the level of the injury are lost. Damage to descending tracts results in loss of motor functions. For example, if the right lateral corticospinal tract is severed in the neck near the first cervical vertebra, control of the voluntary muscles in the right upper and lower limbs is lost, paralyzing them (hemiplegia). Problems of this type in fibers of the descending tracts produce *upper motor neuron syndrome,* characterized by *spastic paralysis* in which muscle tone increases, with very little atrophy of the muscles. However, uncoordinated reflex activity (hyperreflexia) usually occurs, when the flexor and extensor muscles of affected limbs alternately spasm.

Injury to motor neurons or their fibers in the horns of the spinal cord results in *lower motor neuron syndrome.* It is characterized by *flaccid paralysis,* a total loss of muscle tone and reflex activity, and the muscles atrophy.

Several new treatments are on the horizon for spinal cord injuries. They work in three ways:

1. *Limiting damage during the acute phase.* An experimental drug called GM1 ganglioside is a carbohydrate that is normally found on neuron cell membranes. It blocks the actions of amino acids that function as excitatory neurotransmitters, which cuts the deadly calcium ion influx into cells. It also blocks apoptosis (programmed cell death) and stimulates synthesis of nerve growth factor.

2. *Restoring or compensating for function.* A new drug called 4-aminopyridine blocks potassium channels on neurons. This boosts electrical transmission and compensates for the myelin-stripping effects of the injury. Being developed for patients injured at least eighteen months previously, this drug can restore some sexual, bowel, and bladder function.

3. *Regeneration.* Paralyzed rats given implants of human neural stem cells have regained the ability to walk. It may be possible one day to culture stem cells, taken from a person's bone marrow or skin, to become neural stem cells, which can then be used to "patch" a severed spinal cord. ■

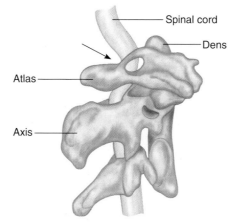

FIGURE 11C

A dislocation of the atlas may cause a compression injury to the spinal cord.

Labels: Spinal cord; Dens; Atlas; Axis

1 Describe the structure of the spinal cord.

2 What are ascending and descending tracts?

3 What is the consequence of fibers crossing over?

4 Name the major tracts of the spinal cord, and list the kinds of impulses each conducts.

Brain

The **brain** contains nerve centers associated with sensory functions and is responsible for sensations and perceptions. It issues motor commands to skeletal muscles and carries on higher mental functions, such as memory and reasoning. It also contains centers that coordinate muscular movements, as well as centers and nerve pathways that regulate visceral activities. In addition to overseeing the function of the entire body, the brain also provides characteristics such as personality.

Brain Development

The basic structure of the brain reflects the way it forms during early (embryonic) development. It begins as the neural tube that gives rise to the central nervous system. The portion that becomes the brain has three major cavities, or vesicles, at one end—the *forebrain* (prosencephalon), *midbrain* (mesencephalon), and *hindbrain* (rhombencephalon) (fig. 11.14). Later, the forebrain divides into anterior and posterior portions (telencephalon and diencephalon, respectively), and the hindbrain partially divides into two parts (metencephalon and myelencephalon). The resulting five cavities persist in the mature brain as the fluid-filled *ventricles* and the tubes that connect them. The tissue surrounding the spaces differentiates into the structural and functional regions of the brain.

The wall of the anterior portion of the forebrain gives rise to the *cerebrum* and *basal nuclei,* whereas the posterior portion forms a section of the brain called the *diencephalon.* The region the midbrain produces continues to be called the *midbrain* in the adult structure, and the hindbrain gives rise to the *cerebellum, pons,* and *medulla oblongata* (fig. 11.15 and table 11.4). Together, the midbrain, pons, and medulla oblongata comprise the **brainstem** (brān'stem), which attaches the brain to the spinal cord.

On a cellular level, the brain develops as specific neurons attract others by secreting growth hormones.

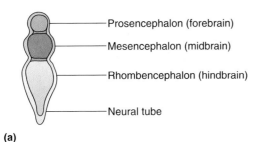

(a)

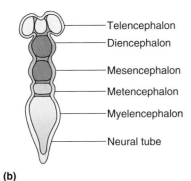

(b)

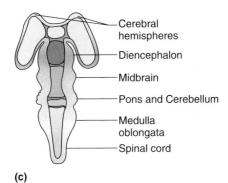

(c)

FIGURE 11.14

Brain development. (*a*) The brain develops from a tubular structure with three cavities. (*b*) The cavities persist as the ventricles and their interconnections. (*c*) The wall of the tube gives rise to various regions of the brain, brainstem, and spinal cord.

TABLE 11.4	Structural Development of the Brain	
Embryonic Vesicle	**Spaces Produced**	**Regions of the Brain Produced**
Forebrain (prosencephalon)		
Anterior portion (telencephalon)	Lateral ventricles	Cerebrum Basal ganglia
Posterior portion (diencephalon)	Third ventricle	Thalamus Hypothalamus Posterior pituitary gland Pineal gland
Midbrain (mesencephalon)	Cerebral aqueduct	Midbrain
Hindbrain (rhombencephalon)		
Anterior portion (metencephalon)	Fourth ventricle	Cerebellum Pons
Posterior portion (myelencephalon)	Fourth ventricle	Medulla oblongata

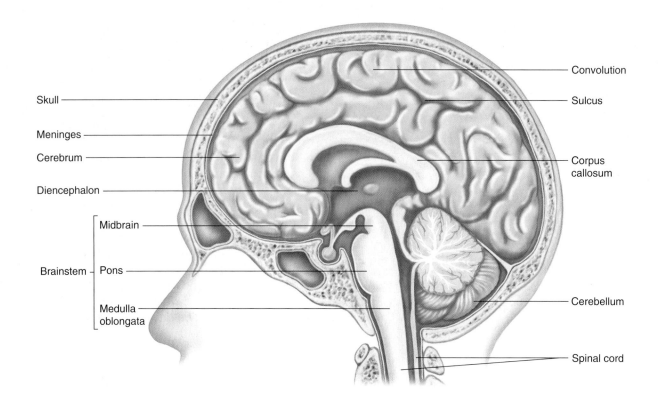

Convolution

Skull

Sulcus

Meninges

Cerebrum

Corpus callosum

Diencephalon

Midbrain

Brainstem — Pons

Medulla oblongata

Cerebellum

Spinal cord

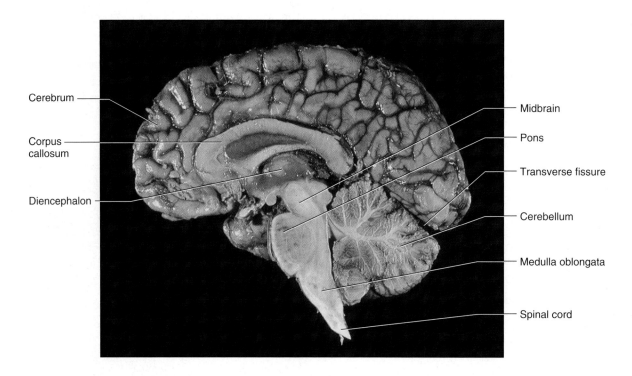

Cerebrum

Midbrain

Corpus callosum

Pons

Transverse fissure

Diencephalon

Cerebellum

Medulla oblongata

Spinal cord

FIGURE 11.15

The major portions of the brain include the cerebrum, the diencephalon, the cerebellum, and the brainstem. (See also reference plate 76.)

Apoptosis (programmed cell death) destroys excess neural connections.

Structure of the Cerebrum

The **cerebrum** (ser'ē-brum), which develops from the anterior portion of the forebrain, is the largest part of the mature brain. It consists of two large masses, or **cerebral hemispheres** (ser'ĕ-bral hem'i-sfĕrz), which are essentially mirror images of each other (fig. 11.16 and reference plate 49). A deep bridge of nerve fibers called the **corpus callosum** connects the cerebral hemispheres. A layer of dura mater called the *falx cerebri* separates them (see fig. 11.1).

Many ridges called **convolutions,** or *gyri* (sing., *gyrus*), separated by grooves, mark the cerebrum's surface. Generally, a shallow to somewhat deep groove is called a **sulcus,** and a very deep groove is called a **fissure.** The pattern of these elevations and depressions is complex, and it is distinct in all normal brains. For example, a *longitudinal fissure* separates the right and left cerebral hemispheres; a *transverse fissure* separates the cerebrum from the cerebellum; and sulci divide each hemisphere into lobes (see figs. 11.15 and 11.16).

A fetus or newborn with *anencephaly* has a face and lower brain structures but lacks most higher brain structures. A newborn with this anomaly survives only a day or two, and sometimes the parents donate the organs.

Anencephaly is a type of neural tube defect (NTD). It occurs at about the twenty-eighth day of prenatal development, when a sheet of tissue that normally folds to form a neural tube, which develops into the central nervous system, remains open at the top. A less-serious NTD is *spina bifida,* in which an opening farther down the neural tube causes a lesion in the spine. The most serious form of this condition results in paralysis from that point downward. Sometimes spina bifida can be improved or even corrected with surgery.

The precise cause of neural tube defects is not known, but it involves folic acid; taking supplements of this vitamin sharply cuts the recurrence risk among women who have had an affected child. Most pregnant women take a blood test at the fifteenth week of pregnancy to detect fluid leaking from an NTD.

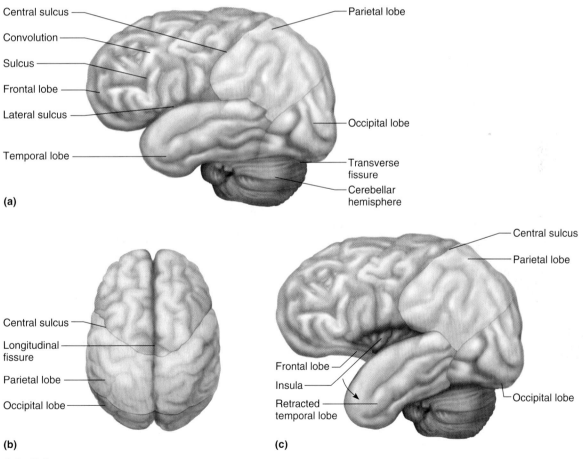

(a)

(b)

(c)

FIGURE 11.16

Colors in this figure distinguish the lobes of the cerebral hemispheres. (*a*) Lateral view of the right hemisphere. (*b*) Hemispheres viewed from above. (*c*) Lateral view of the right hemisphere with the insula exposed.

In a disorder called *lissencephaly* ("smooth brain"), a newborn has a smooth cerebral cortex, completely lacking the characteristic convolutions. Absence of a protein early in prenatal development prevents certain neurons from migrating within the brain, which blocks formation of convolutions. The child is profoundly mentally retarded, with frequent seizures and other neurological problems.

The lobes of the cerebral hemispheres (fig. 11.16) are named after the skull bones that they underlie. They include the following:

1. **Frontal lobe.** The frontal lobe forms the anterior portion of each cerebral hemisphere. It is bordered posteriorly by a *central sulcus* (fissure of Rolando), which passes out from the longitudinal fissure at a right angle, and inferiorly by a *lateral sulcus* (fissure of Sylvius), which exits the undersurface of the brain along its sides.

2. **Parietal lobe.** The parietal lobe is posterior to the frontal lobe and is separated from it by the central sulcus.

3. **Temporal lobe.** The temporal lobe lies inferior to the frontal and parietal lobes and is separated from them by the lateral sulcus.

4. **Occipital lobe.** The occipital lobe forms the posterior portion of each cerebral hemisphere and is separated from the cerebellum by a shelflike extension of dura mater called the *tentorium cerebelli.* The occipital lobe and the parietal and temporal lobes have no distinct boundary.

5. **Insula.** The insula (island of Reil) is a lobe located deep within the lateral sulcus and is so named because it is covered by parts of the frontal, parietal, and temporal lobes. A *circular sulcus* separates it from them.

A thin layer of gray matter (2 to 5 millimeters thick) called the **cerebral cortex** (ser'ĕ-bral kor'teks) constitutes the outermost portion of the cerebrum. It covers the convolutions, dipping into the sulci and fissures. The cerebral cortex contains nearly 75% of all the neuron cell bodies in the nervous system.

Just beneath the cerebral cortex is a mass of white matter that makes up the bulk of the cerebrum. This mass contains bundles of myelinated nerve fibers that connect neuron cell bodies of the cortex with other parts of the nervous system. Some of these fibers pass from one cerebral hemisphere to the other by way of the corpus callosum, and others carry sensory or motor impulses from the cortex to nerve centers in the brain or spinal cord.

Functions of the Cerebrum

The cerebrum provides higher brain functions: interpreting impulses from sense organs, initiating voluntary muscular movements, storing information as memory, and retrieving this information in reasoning. The cerebrum is also the seat of intelligence and personality.

Functional Regions of the Cortex

The regions of the cerebral cortex that perform specific functions have been located using a variety of techniques. Persons who have suffered brain disease or injury, such as Karen Ann Quinlan and Phineas Gage, or have had portions of their brains removed surgically, have provided clues to the functions of the impaired brain regions.

In other studies, areas of cortices have been exposed surgically and stimulated mechanically or electrically, with researchers observing the responses in certain muscles or the specific sensations that result. As a result of such investigations, researchers have divided the cerebral cortex into sections known as motor, sensory, and association areas. They overlap somewhat.

Motor Areas

The *primary motor areas* of the cerebral cortex lie in the precentral gyri of the frontal lobes just in front of the central sulcus and in the anterior wall of this sulcus (fig. 11.17). The nervous tissue in these regions contains many large *pyramidal cells,* named for their pyramid-shaped cell bodies.

Impulses from these pyramidal cells travel downward through the brainstem and into the spinal cord on the *corticospinal tracts.* Most of the nerve fibers in these tracts cross over from one side of the brain to the other within the brainstem and descend as lateral corticospinal tracts. Other fibers, in the anterior corticospinal tracts, cross over at various levels of the spinal cord (see fig. 11.13).

Within the spinal cord, the corticospinal fibers synapse with motor neurons in the gray matter of the anterior horns. Axons of the motor neurons lead outward through peripheral nerves to voluntary muscles. Impulses transmitted on these pathways in special patterns and frequencies are responsible for fine movements in skeletal muscles. More specifically, as figure 11.18 shows, cells in the upper portions of the motor areas send impulses to muscles in the thighs and legs; those in the middle portions control muscles in the arms and forearms; and those in lower portions activate muscles of the head, face, and tongue.

The *reticulospinal* and *rubrospinal tracts* coordinate and control motor functions that maintain balance and posture. Many of these fibers pass into the basal ganglia on the way to the spinal cord. Some of the impulses conducted on these pathways normally inhibit muscular actions.

In addition to the primary motor areas, certain other regions of the frontal lobe control motor functions. For example, a region called *Broca's area* is just anterior to the

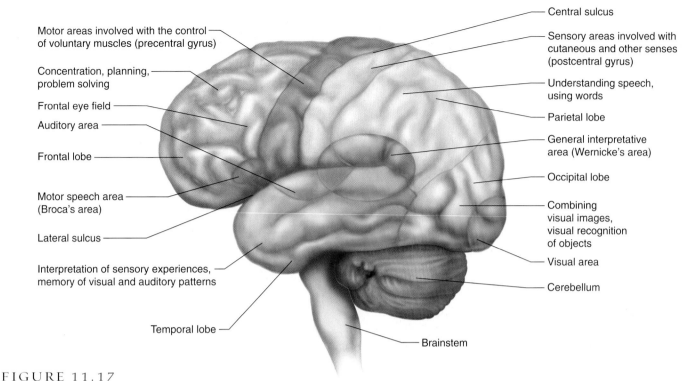

FIGURE 11.17

Some motor, sensory, and association areas of the left cerebral cortex.

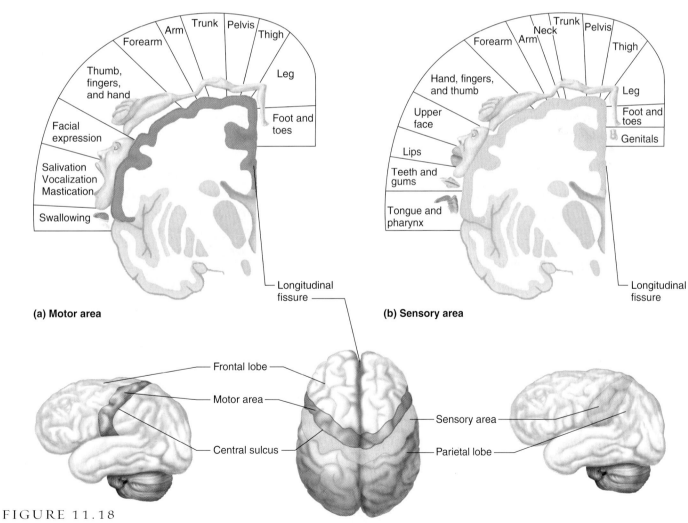

FIGURE 11.18

Functional regions of the cerebral cortex. (*a*) Motor areas that control voluntary muscles (only left hemisphere shown). (*b*) Sensory areas involved with cutaneous and other senses (only left hemisphere shown).

primary motor cortex and superior to the lateral sulcus (see fig. 11.17), usually in the left cerebral hemisphere. It coordinates the complex muscular actions of the mouth, tongue, and larynx, which make speech possible. A person with an injury to this area may be able to understand spoken words but may be unable to speak.

Above Broca's area is a region called the *frontal eye field*. The motor cortex in this area controls voluntary movements of the eyes and eyelids. Nearby is the cortex responsible for movements of the head that direct the eyes. Another region just in front of the primary motor area controls the muscular movements of the hands and fingers that make such skills as writing possible (see fig. 11.17).

An injury to the motor system may impair the ability to produce purposeful muscular movements. Such a condition that affects use of the upper and lower limbs, head, or eyes is called *apraxia*. When apraxia affects the speech muscles, disrupting speaking ability, it is called *aphasia*.

Sensory Areas

Sensory areas in several lobes of the cerebrum interpret impulses from sensory receptors, producing feelings or sensations. For example, the sensations of temperature, touch, pressure, and pain in the skin arise in the postcentral gyri of the anterior portions of the parietal lobes along the central sulcus and in the posterior wall of this sulcus (see fig. 11.17). The posterior parts of the occipital lobes provide vision, whereas the superior posterior portions of the temporal lobes contain the centers for hearing. The sensory areas for taste are near the bases of the central sulci along the lateral sulci, and the sense of smell arises from centers deep within the cerebrum.

Like motor fibers, sensory fibers, such as those in the *fasciculus cuneatus tract,* cross over in the spinal cord or the brainstem (see fig. 11.12). Thus, the centers in the right central hemisphere interpret impulses originating from the left side of the body, and vice versa. However, the sensory areas concerned with vision receive impulses from both eyes, and those concerned with hearing receive impulses from both ears.

Association Areas

Association areas are regions of the cerebral cortex that are not primarily sensory or motor in function, and interconnect with each other and with other brain structures. These areas occupy the anterior portions of the frontal lobes and are widespread in the lateral portions of the parietal, temporal, and occipital lobes. They analyze and interpret sensory experiences and help provide memory, reasoning, verbalizing, judgment, and emotions (see fig. 11.17).

The association areas of the frontal lobes provide higher intellectual processes, such as concentrating, planning, and complex problem solving. The anterior and inferior portions of these lobes (prefrontal areas) control emotional behavior and produce awareness of the possible consequences of behavior.

The parietal lobes have association areas that help interpret sensory information and aid in understanding speech and choosing words to express thoughts and feelings. Awareness of the form of objects, including one's own body parts, stems from the posterior regions of these lobes.

The association areas of the temporal lobes and the regions at the posterior ends of the lateral sulci interpret complex sensory experiences, such as those needed to understand speech and to read. These regions also store memories of visual scenes, music, and other complex sensory patterns.

The occipital lobes have association areas adjacent to the visual centers. These are important in analyzing visual patterns and combining visual images with other sensory experiences—as when one recognizes another person.

Of particular importance is the region where the parietal, temporal, and occipital association areas join near the posterior end of the lateral sulcus. This region, called the *general interpretative area* (Wernicke's area), plays the primary role in complex thought processing. It receives input from multiple sensory areas and consolidates the information. This is communicated to other brain areas that respond appropriately. The general interpretive area makes it possible for a person to recognize words and arrange them to express a thought, and to read and understand ideas presented in writing. Table 11.5 summarizes the functions of the cerebral lobes.

A person with *dyslexia* sees letters separately and must be taught to read in a different way than people whose nervous systems allow them to group letters into words. Three to 10% of people have dyslexia. The condition probably has several causes, with inborn visual and perceptual skills interacting with the way the child learns to read. Dyslexia has nothing to do with intelligence—many brilliant thinkers were "slow" in school because educators had not yet learned how to help them.

---

1 How does the brain form during early development?

2 Describe the cerebrum.

3 List the general functions of the cerebrum.

4 Where in the brain are the primary motor and sensory regions located?

5 Explain the functions of association areas.

TABLE 11.5	Functions of the Cerebral Lobes			
Lobe	**Functions**		**Lobe**	**Functions**
Frontal lobes	Motor areas control movements of voluntary skeletal muscles.		Temporal lobes	Sensory areas are responsible for hearing.
	Association areas carry on higher intellectual processes for concentrating, planning, complex problem solving, and judging the consequences of behavior.			Association areas interpret sensory experiences and remember visual scenes, music, and other complex sensory patterns.
Parietal lobes	Sensory areas provide sensations of temperature, touch, pressure, and pain involving the skin.		Occipital lobes	Sensory areas are responsible for vision.
	Association areas function in understanding speech and in using words to express thoughts and feelings.			Association areas combine visual images with other sensory experiences.

Hemisphere Dominance

Both cerebral hemispheres participate in basic functions, such as receiving and analyzing sensory impulses, controlling skeletal muscles on opposite sides of the body, and storing memory. However, in most persons, one side acts as a *dominant hemisphere* for certain other functions.

In over 90% of the population, for example, the left hemisphere is dominant for the language-related activities of speech, writing, and reading. It is also dominant for complex intellectual functions requiring verbal, analytical, and computational skills. In other persons, the right hemisphere is dominant, and in some, the hemispheres are equally dominant.

Tests indicate that the left hemisphere is dominant in 90% of right-handed adults and in 64% of left-handed ones. The right hemisphere is dominant in 10% of right-handed adults and in 20% of left-handed ones. The hemispheres are equally dominant in the remaining 16% of left-handed persons. As a consequence of hemisphere dominance, Broca's area on one side almost completely controls the motor activities associated with speech. For this reason, over 90% of patients with language impairment stemming from problems in the cerebrum have disorders in the left hemisphere.

In addition to carrying on basic functions, the nondominant hemisphere specializes in nonverbal functions, such as motor tasks that require orientation of the body in space, understanding and interpreting musical patterns, and visual experiences. It also provides emotional and intuitive thought processes. For example, although the region in the nondominant hemisphere that corresponds to Broca's area does not control speech, it influences the emotional aspects of spoken language.

Nerve fibers of the *corpus callosum,* which connect the cerebral hemispheres, enable the dominant hemisphere to control the motor cortex of the nondominant hemisphere. These fibers also transfer sensory information reaching the nondominant hemisphere to the general interpretative area of the dominant one, where the information can be used in decision making.

Memory

Memory, one of the most astonishing capabilities of the brain, is the consequence of learning. Whereas learning is the acquisition of new knowledge, memory is the persistence of that learning, with the ability to access it at a later time. Two types of memory, short term and long term, have been recognized for many years, and researchers are now beginning to realize that they differ in characteristics other than duration.

Short-term, or "working," memories are thought to be electrical in nature. Neurons may be connected in a circuit so that the last in the series stimulates the first. As long as the pattern of stimulation continues, the thought is remembered. When the electrical events cease, so does the memory—unless it enters long-term memory.

Long-term memory probably changes the structure or function of neurons in ways that enhance synaptic transmission, perhaps by establishing certain patterns of synaptic connections. Synaptic patterns fulfill two requirements of long-term memory. First, there are enough synapses to encode an almost limitless number of memories—each of the 10 billion neurons in the cortex can make tens of thousands of synaptic connections to other neurons, forming 60 trillion synapses. Second, a certain pattern of synapses can remain unchanged for years.

Understanding how neurons in different parts of the brain encode memories and how short-term memories are converted to long-term memories (a process called **memory consolidation**) is at the forefront of research into the functioning of the human brain. According to one theory called **long-term synaptic potentiation,** primarily in an area of the cerebral cortex called the **hippocampus,** frequent, nearly simultaneous, and repeated stimulation of the same neurons strengthens their synaptic connections.

CEREBRAL INJURIES AND ABNORMALITIES

The specific symptoms associated with a cerebral injury or abnormality depend upon the areas and extent of damage. A person with damage to the association areas of the frontal lobes may have difficulty concentrating on complex mental tasks, appearing disorganized and easily distracted.

If the general interpretative area of the dominant hemisphere is injured, the person may be unable to interpret sounds as words or to understand written ideas. However, the dominance of one hemisphere usually does not become established until after five or six years of age. Consequently, if the general interpretative area is destroyed in a child, the corresponding region of the other side of the brain may be able to take over the functions, and the child's language abilities may develop normally. If such an injury occurs in an adult, the nondominant hemisphere may develop only limited interpretative functions, producing a severe intellectual dis-

ability. Following are three common cerebral abnormalities.

- In a *concussion,* the brain is jarred against the cranium, usually as a result of a blow to the head, causing loss of consciousness. Short-term memory loss, mental cloudiness, difficulty concentrating and remembering, and a fierce headache may occur in the days after a concussion, but recovery is usually complete.
- *Cerebral palsy* (CP) is motor impairment at birth, often stemming from a brain anomaly occurring during prenatal development. In the past, most cases of CP were blamed on "birth trauma," but recently, researchers determined that the most common cause is a blocked cerebral blood vessel, which leads to atrophy of the brain region deprived of its blood supply. Birth trauma and brain infection cause some cases.

CP affects about 1 in every 1,000 births and is especially prevalent among premature babies. One-half to two-thirds of affected babies improve and can even outgrow the condition by age seven. Sometimes seizures or learning disabilities are present. Clinicians classify CP by the number of limbs and the types of neurons affected.

- In a "stroke," or *cerebrovascular accident* (CVA), a sudden interruption in blood flow in a vessel supplying brain tissues damages the cerebrum. The affected blood vessel may rupture, bleeding into the brain, or be blocked by a clot. In either case, brain tissues downstream from the vascular accident die or permanently lose function. Temporary interruption in cerebral blood flow, perhaps by a clot that quickly breaks apart, produces a much less serious *transient ischemic attack* (TIA). ∎

This strengthening results in more frequent action potentials triggered in postsynaptic cells in response to the repeated stimuli. Clinical Application 11.4 discusses some common causes of damage to the cerebrum.

Medical researchers have gained insight into the role of the hippocampus by observing the unusual behaviors and skills of people in whom these structures have been damaged. In 1953, a surgeon removed parts of the hippocampus and another area called the amygdala of a young man called H. M., thinking this drastic action might relieve his severe epilepsy. His seizures indeed became less frequent, but H. M. suffered a profound loss in the ability to consolidate short-term memories into long-term ones. As a result, events in H. M.'s life fade from memory as quickly as they occur. He is unable to recall any events that took place since surgery, living today as if it was the 1950s. He can read the same magazine article repeatedly with renewed interest each time. With practice, he improves skills that require procedural memory, such as puzzle solving. But, since factual memory is impossible, he insists that he has never seen the puzzle before!

Basal Nuclei

The **basal nuclei** (basal ganglia) are masses of gray matter located deep within the cerebral hemispheres. They are called the *caudate nucleus,* the *putamen,* and the *globus pallidus,* and they develop from the anterior portion of the forebrain (fig. 11.19). The neuron cell bodies that the basal nuclei contain relay motor impulses originating in the cerebral cortex and passing into the brainstem and spinal cord. The basal nuclei produce most of the inhibitory neurotransmitter *dopamine.* Impulses from the basal nuclei normally inhibit motor functions, thus controlling certain muscular activities. Clinical Application 11.5 discusses Parkinson disease, in which neurons in the basal nuclei degenerate.

1. What is hemisphere dominance?

2. What are the functions of the nondominant hemisphere?

3. Distinguish between short-term and long-term memory.

4. What is the function of the basal nuclei?

Diencephalon

The **diencephalon** (di″en-sef′ah-lon) develops from the posterior forebrain and is located between the cerebral hemispheres and above the brainstem (see figs. 11.15 and 11.20).

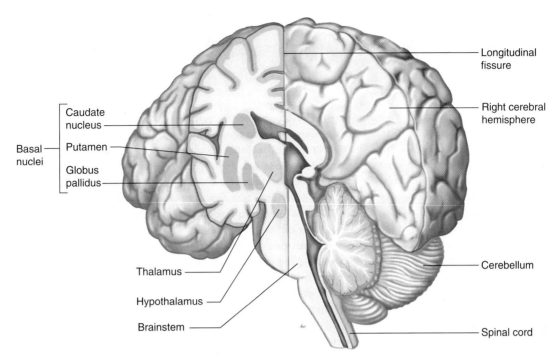

FIGURE 11.19
A coronal section of the left cerebral hemisphere reveals some of the basal nuclei.

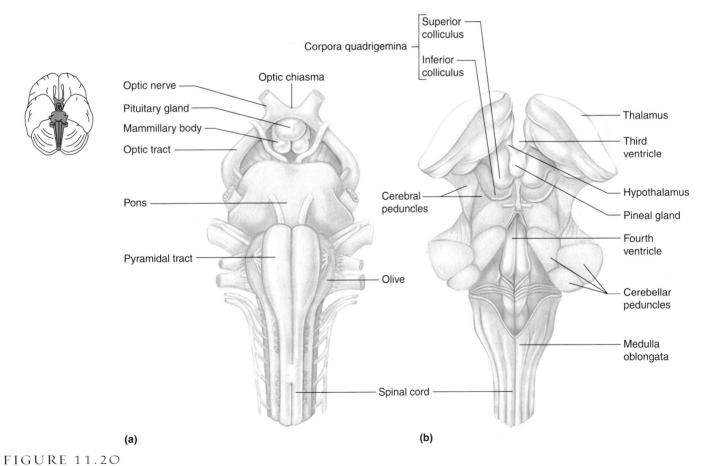

(a) **(b)**

FIGURE 11.20
Brainstem. (*a*) Ventral view of the brainstem. (*b*) Dorsal view of the brainstem with the cerebellum removed, exposing the fourth ventricle.

PARKINSON DISEASE

Actor Michael J. Fox was only 29 years old when he consulted a physician about a mysterious twitch in a finger. Fox, veteran of two hit television comedies and a long list of films, was shocked to receive a diagnosis of Parkinson disease, a condition that usually begins much later in life. He was one of the ten percent of the millions of people with Parkinson disease worldwide to experience symptoms before age forty.

Fox kept his diagnosis to himself, but by the late 1990s, his co-workers began to notice symptoms that emerged when medication wore off—rigidity, a shuffling and off-balance gait, and poor small motor control. It was difficult to ignore Fox's expressionless, mask-like face, a characteristic of Parkinson disease called hypomimia. Most frustrating to Fox was his worsening ability to communicate. His voice became so weakened that it took a huge effort to speak, a symptom called hypophia. When he could speak, he found that even though his brain could string thoughts into coherent sentences, the muscles of his jaw, lips, and tongue could not utter them. Oddest of all was micrographia, the tendency of his handwriting to become extremely small.

Parkinson disease also causes the sensation of not being able to stay in one spot.

By 1998, chased by relentless tabloid reporters eager to expose his illness, Fox publicly disclosed his condition. He continued to act on his television program Spin City, but in 2000, he quit, and began the Michael J. Fox Foundation for Parkinson's Research, which provides rapid funding for research into curing the disease. Today he continues to testify before Congress about the potential benefits of stem cell research and therapy, which he calls "the gateway" to curing Parkinson and other neurodegenerative disorders.

In Parkinson disease, neurons in the basal nuclei that synthesize the neurotransmitter dopamine degenerate. The resulting deficiency of dopamine in the striatum causes the motor symptoms. In some patients, non-motor symptoms develop too, including depression, dementia, constipation, incontinence, sleep problems, and orthostatic hypotension (dizziness upon standing).

No treatments can cure or slow the course of Parkinson disease, but replacing or enhancing utilization of dopamine can temporarily alleviate symptoms. The stan-

dard treatment for many years has been levodopa, which is a precursor to dopamine that can cross the blood-brain barrier. Once in the brain, levodopa is converted to dopamine. Fox takes levodopa so that he can do interviews without jumping around constantly. The medication allows him to live normally for short periods—riding bikes with his wife, or fishing with his son. But always, too soon, he begins to feel the telltale twitches, rigidity, and vibrations, and knows that the symptoms have returned.

Drug treatment for Parkinson disease is temporary, and becomes less effective over time. The brain becomes dependent on the external supply of dopamine and decreases its own production further, so that eventually higher doses of levodopa are needed to achieve the effect. Unfortunately, taking too much levodopa leads to another condition, called tardive dyskinesia, that produces uncontrollable facial tics and spastic extensions of the limbs. Tardive dyskinesia may result from effects of excess dopamine in areas other than those affected in Parkinson disease.

Surgery can alleviate Parkinson's symptoms. Fox underwent thalamotomy, in which an electrode caused a lesion in his

---

It surrounds the third ventricle and is largely composed of gray matter. Within the diencephalon, a dense mass, called the **thalamus** (thal'ah-mus), bulges into the third ventricle from each side. Another region of the diencephalon that includes many nuclei is the **hypothalamus** (hi"po-thal'ah-mus). It lies below the thalamic nuclei and forms the lower walls and floor of the third ventricle (see reference plates 49 and 53).

Other parts of the diencephalon include (1) the **optic tracts** and the **optic chiasma** that is formed by the optic nerve fibers crossing over; (2) the **infundibulum,** a conical process behind the optic chiasma to which the pituitary gland is attached; (3) the **posterior pituitary gland,** which hangs from the floor of the hypothalamus; (4) the **mammillary** (mam'ĭ-ler"e) **bodies,** which are two rounded structures behind the infundibulum; and (5) the **pineal gland,** which forms as a cone-shaped evagination from the roof of the diencephalon (see chapter 13, pp. 498–499).

The thalamus is a selective gateway for sensory impulses ascending from other parts of the nervous system to the cerebral cortex. It receives all sensory impulses (except those associated with the sense of smell) and channels them to appropriate regions of the cortex for interpretation. In addition, all regions of the cerebral cortex can communicate with the thalamus by means of descending fibers.

The thalamus seems to transmit sensory information by synchronizing action potentials. Consider vision. An image on the retina stimulates the *lateral geniculate nucleus* (LGN) region of the thalamus, which then sends action potentials to a part of the visual cortex. Researchers have observed that those action potentials are synchronized—that is, fired simultaneously—by the LGN's neurons only if the stimuli come from a single object, such as a bar. If the stimulus is two black dots, the resulting thalamic action potentials are not synchronized. The syn-

thalamus. The procedure calmed a violent shaking in his left arm. Pallidotomy causes lesions in the globus pallidus internus, a part of the basal ganglia, and the approach is also used on an area posterior to the thalamus. Deep brain implants of electrodes may also control some symptoms.

Implants of dopamine-producing cells has had limited success in some patients. Fox and many others are most excited about the possibility of using neural stem cells—

from cadavers or patients themselves—to replace degenerating dopamine-producing neurons.

The causes of Parkinson disease—and whether it really is just one syndrome or many—aren't known. Parkinson's-like symptoms have been attributed to use of certain designer drugs and exposure to pesticides. The severe Parkinson disease that afflicts former heavyweight champ Mohammed Ali may have been caused by

his frequent and violent blows to the head.

Several genes may increase risk of developing Parkinson disease, but in most cases, it is not inherited. However, in one large family with several affected members, Parkinson disease is caused by a mutation in the gene that encodes a protein called alpha-synuclein, which is found in the basal nuclei. When abnormal, the protein folds improperly, forming deposits in the brain. Although Parkinson disease is rarely inherited, understanding how the condition occurs in the rare familial variants may provide clues to helping the many others who have this debilitating illness. ■

FIGURE 11D
Professional boxers are at higher risk of developing Parkinson disease from repeated blows to the head. Muhammed Ali has Parkinson disease from many years of head injuries. Actor Michael J. Fox first experienced symptoms of Parkinson disease at age 29.

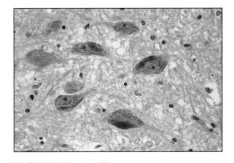

FIGURE 11E
The chemical composition of Lewy bodies, which are characteristic of the brains of people with Parkinson disease, may provide clues to the cause of the condition.

chronicity of action potentials, therefore, may be a way that the thalamus selects which stimuli to relay to higher brain structures. Therefore, the thalamus is not only a messenger but also an editor.

Nerve fibers connect the hypothalamus to the cerebral cortex, thalamus, and parts of the brainstem so that it can receive impulses from them and send impulses to them. The hypothalamus maintains homeostasis by regulating a variety of visceral activities and by linking the nervous and endocrine systems.

The hypothalamus regulates

1. Heart rate and arterial blood pressure.

2. Body temperature.

3. Water and electrolyte balance.

4. Control of hunger and body weight.

5. Control of movements and glandular secretions of the stomach and intestines.

6. Production of neurosecretory substances that stimulate the pituitary gland to release hormones that help regulate growth, control various glands, and influence reproductive physiology.

7. Sleep and wakefulness.

Structures in the region of the diencephalon also are important in controlling emotional responses. For example, portions of the cerebral cortex in the medial parts of the frontal and temporal lobes connect with the hypothalamus, thalamus, basal nuclei, and other deep nuclei. Together, these structures comprise a complex called the **limbic system.**

The limbic system controls emotional experience and expression and can modify the way a person acts. It produces such feelings as fear, anger, pleasure, and sorrow. The limbic system seems to recognize upsets in a person's physical or psychological condition that might threaten life. By causing pleasant or unpleasant feelings about experiences, the limbic system guides a person into behavior that may increase the chance of survival. In addition, portions of the limbic system interpret sensory impulses from the receptors associated with the sense of smell (olfactory receptors).

Brainstem

The **brainstem** connects the brain to the spinal cord. It consists of the midbrain, pons, and medulla oblongata. These structures include many tracts of nerve fibers and masses of gray matter called *nuclei* (see figs. 11.15 and 11.20).

Midbrain

The **midbrain** (mesencephalon) is a short section of the brainstem between the diencephalon and the pons. It contains bundles of myelinated nerve fibers that join lower parts of the brainstem and spinal cord with higher parts of the brain. The midbrain includes several masses of gray matter that serve as reflex centers. It also contains the *cerebral aqueduct* that connects the third and fourth ventricles (fig. 11.21).

Two prominent bundles of nerve fibers on the underside of the midbrain comprise the *cerebral peduncles.* These fibers include the corticospinal tracts and are the main motor pathways between the cerebrum and lower parts of the nervous system (see fig. 11.20). Beneath the cerebral peduncles are some large bundles of sensory fibers that carry impulses upward to the thalamus.

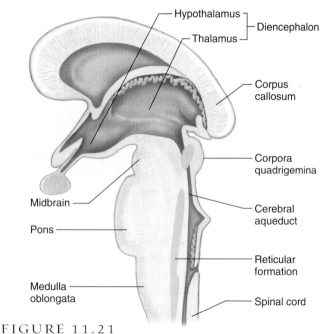

FIGURE 11.21

The reticular formation (shown in green) extends from the superior portion of the spinal cord into the diencephalon.

Two pairs of rounded knobs on the superior surface of the midbrain mark the location of four nuclei, known collectively as *corpora quadrigemina.* The upper masses (superior colliculi) contain the centers for certain visual reflexes, such as those responsible for moving the eyes to view something as the head turns. The lower ones (inferior colliculi) contain the auditory reflex centers that operate when it is necessary to move the head to hear sounds more distinctly (see fig. 11.20).

Near the center of the midbrain is a mass of gray matter called the *red nucleus.* This nucleus communicates with the cerebellum and with centers of the spinal cord, and it provides reflexes that maintain posture. It appears red because it is richly supplied with blood vessels.

Pons

The **pons** appears as a rounded bulge on the underside of the brainstem where it separates the midbrain from the medulla oblongata (see fig. 11.20). The dorsal portion of the pons largely consists of longitudinal nerve fibers, which relay impulses to and from the medulla oblongata and the cerebrum. Its ventral portion contains large bundles of transverse nerve fibers, which transmit impulses from the cerebrum to centers within the cerebellum.

Several nuclei of the pons relay sensory impulses from peripheral nerves to higher brain centers. Other nuclei function with centers of the medulla oblongata to regulate the rate and depth of breathing.

Medulla Oblongata

The **medulla oblongata** (mĕ-dul'ah ob"long-ga'tah) is an enlarged continuation of the spinal cord, extending from the level of the foramen magnum to the pons (see fig. 11.20). Its dorsal surface flattens to form the floor of the fourth ventricle, and its ventral surface is marked by the corticospinal tracts, most of whose fibers cross over at this level. On each side of the medulla oblongata is an oval swelling called the *olive,* from which a large bundle of nerve fibers arises and passes to the cerebellum.

Because of the medulla oblongata's location, all the ascending and descending nerve fibers connecting the brain and spinal cord must pass through it. As in the spinal cord, the white matter of the medulla surrounds a central mass of gray matter. Here, however, the gray matter breaks up into nuclei that are separated by nerve fibers. Some of these nuclei relay ascending impulses to the other side of the brainstem and then on to higher brain centers. The *nucleus gracilis* and the *nucleus cuneatus,* for example, receive sensory impulses from fibers of the fasciculus gracilis and the fasciculus cuneatus and pass them on to the thalamus or the cerebellum.

Other nuclei within the medulla oblongata control vital visceral activities. These centers include the following:

1. **Cardiac center.** Peripheral nerves transmit impulses originating in the cardiac center to the heart, where they increase or decrease heart rate.

2. **Vasomotor center.** Certain cells of the vasomotor center initiate impulses that travel to smooth muscles in the walls of blood vessels and stimulate them to contract, constricting the vessels (vasoconstriction) and thereby increasing blood pressure. A decrease in the activity of these cells can produce the opposite effect—dilation of the blood vessels (vasodilation) and a consequent drop in blood pressure.

3. **Respiratory center.** The respiratory center acts with centers in the pons to regulate the rate, rhythm, and depth of breathing.

Some nuclei within the medulla oblongata are centers for certain nonvital reflexes, such as those associated with coughing, sneezing, swallowing, and vomiting. However, since the medulla also contains vital reflex centers, injuries to this part of the brainstem are often fatal.

Reticular Formation

Scattered throughout the medulla oblongata, pons, and midbrain is a complex network of nerve fibers associated with tiny islands of gray matter. This network, the **reticular formation** (rĕ-tik′u-lar fōr-ma′shun), or reticular activating system, extends from the superior portion of the spinal cord into the diencephalon (fig. 11.21). Its intricate system of nerve fibers connects centers of the hypothalamus, basal nuclei, cerebellum, and cerebrum with fibers in all the major ascending and descending tracts.

When sensory impulses reach the reticular formation, it responds by activating the cerebral cortex into a state of wakefulness. Without this arousal, the cortex remains unaware of stimulation and cannot interpret sensory information or carry on thought processes. Thus, decreased activity in the reticular formation results in sleep. If the reticular formation is injured and ceases to function, the person remains unconscious, even with strong stimulation. This is called a comatose state.

A person in a persistent vegetative state is occasionally awake, but not aware; a person in a coma is not awake or aware. Sometimes following a severe injury, a person will become comatose and then gradually enter a persistent vegetative state. Coma and persistent vegetative state are also seen in the end stage of neurodegenerative disorders such as Alzheimer disease; when there is an untreatable mass in the brain, such as a blood clot or tumor; or in anencephaly, when a newborn lacks higher brain structures.

The reticular formation also filters incoming sensory impulses. Impulses judged to be important, such as those originating in pain receptors, are passed on to the cerebral cortex, while others are disregarded. This selective action of the reticular formation frees the cortex from what would otherwise be a continual bombardment of sensory stimulation and allows it to concentrate on more significant information. The cerebral cortex can also activate the reticular system, so intense cerebral activity tends to keep a person awake. In addition, the reticular formation regulates motor activities so that various skeletal muscles move together evenly, and it inhibits or enhances certain spinal reflexes.

Types of Sleep

The two types of normal sleep are *slow wave* and *rapid eye movement* (REM). Slow-wave sleep (also called non-REM sleep) occurs when a person is very tired, and it reflects decreasing activity of the reticular formation. It is restful, dreamless, and accompanied by reduced blood pressure and respiratory rate. Slow-wave sleep may range from light to heavy and is usually described in four stages. It may last from seventy to ninety minutes. Slow-wave and REM sleep alternate.

REM sleep is also called "paradoxical sleep" because some areas of the brain are active. As its name implies, the eyes can be seen rapidly moving beneath the eyelids. Cats and dogs in REM sleep sometimes twitch their limbs. In humans, REM sleep usually lasts from five to fifteen minutes. This "dream sleep" is apparently very important. If a person lacks REM sleep for just one night, sleep on the next night makes up for it. During REM sleep, heart and respiratory rates are irregular. Certain drugs, such as marijuana and alcohol, interfere with REM sleep. Table 11.6 describes several disorders of sleep.

The person to go the longest without sleep was a seventeen-year-old student who stayed awake, in a sleep laboratory under medical supervision, for 264 hours. Fortunately, he suffered no ill effects, but during his ordeal, he was irritable and had blurred vision, slurred speech, and memory lapses. Toward the end, he seemed confused about his identity.

Rats experimentally deprived of sleep do not fare as well. In a study conducted at the University of Chicago, rats kept awake by a moving floor developed skin sores and hormonal and metabolic changes, dying within eleven to thirty-two days. Control rats allowed to nap survived.

1 What are the major functions of the thalamus? Of the hypothalamus?

2 How may the limbic system influence a person's behavior?

3 Which vital reflex centers are located in the brainstem?

4 What is the function of the reticular formation?

5 Describe two types of sleep.

CHAPTER ELEVEN *Nervous System II*

TABLE 11.6   Sleep Disorders

Disorder	Symptoms	Percent of Population
Fatal familial insomnia	Inability to sleep, emotional instability, hallucinations, stupor, coma, death within thirteen months of onset around age fifty, both slow-wave and REM sleep abolished.	Very rare
Insomnia	Inability to fall or remain asleep.	10%
Narcolepsy	Abnormal REM sleep causes extreme daytime sleepiness, begins between ages of fifteen and twenty-five.	0.02–0.06%
Obstructive sleep apnea syndrome	Upper airway collapses repeatedly during sleep, blocking breathing. Snoring and daytime sleepiness.	4–5%
Parasomnias	Sleepwalking, sleeptalking, and night terrors outgrown.	<5% of children
REM-sleep behavior disorder	Excessive motor activity during REM sleep, which disturbs continuous sleep.	Very rare
Restless-leg syndrome	Brief, repetitive leg jerks during sleep. Leg pain forces person to get up several times a night.	Very rare
Sleep paralysis	Inability to move for up to a few minutes after awakening or when falling asleep.	Very rare

Cerebellum

The **cerebellum** (ser″ĕ-bel′um) is a large mass of tissue located inferior to the occipital lobes of the cerebrum and posterior to the pons and medulla oblongata (see fig. 11.15). It consists of two lateral hemispheres partially separated by a layer of dura mater called the *falx cerebelli.* A structure called the *vermis* connects the cerebellar hemispheres at the midline.

Like the cerebrum, the cerebellum is primarily composed of white matter with a thin layer of gray matter, the **cerebellar cortex,** on its surface. This cortex dou-

bles over on itself in a series of complex folds that have myelinated nerve fibers branching into them. A cut into the cerebellum reveals a treelike pattern of white matter, called the *arbor vitae,* that is surrounded by gray matter. A number of nuclei lie deep within each cerebellar hemisphere. The largest and most important is the *dentate nucleus.*

The cerebellum communicates with other parts of the central nervous system by means of three pairs of nerve tracts called **cerebellar peduncles** (ser″ĕ-bel′ar pe-dung′k′ls) (fig. 11.22). One pair, the *inferior peduncles,*

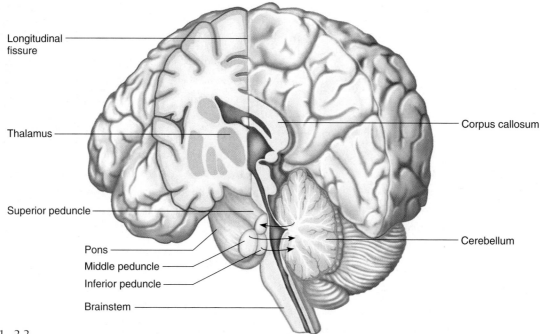

FIGURE 11.22

The cerebellum, which is located below the occipital lobes of the cerebrum, communicates with other parts of the nervous system by means of the cerebellar peduncles.

brings sensory information concerning the actual position of body parts such as limbs and joints to the cerebellum via the spinal cord and medulla oblongata. The *middle peduncles* transmit information from the cerebral cortex about the desired position of these body parts. After integrating and analyzing the information from these two sources, the cerebellum sends correcting impulses from the dentate nucleus via the *superior peduncles* to the midbrain (see figs. 11.21 and 11.22). These corrections are incorporated into motor impulses that travel downward through the pons, medulla oblongata, and spinal cord in the appropriate patterns to move the body in the desired way.

Overall, the cerebellum integrates sensory information concerning the position of body parts and coordinates skeletal muscle activity and maintains posture. It receives sensory impulses from receptors in muscles, tendons, and joints (proprioceptors) and from special sense organs, such as the eyes and ears (see chapter 12, p. 445). For example, the cerebellum uses sensory information from the semicircular canals of the inner ears concerning the motion and position of the head to help maintain equilibrium. Damage to the cerebellum is likely to result in tremors, inaccurate movements of voluntary muscles, loss of muscle tone, a reeling walk, and loss of equilibrium.

Table 11.7 summarizes the characteristics and functions of the major parts of the brain. Clinical Application 11.6 discusses how brain waves reflect brain activity.

1 Where is the cerebellum located?

2 What are the major functions of the cerebellum?

3 What kinds of receptors provide information to the cerebellum?

Peripheral Nervous System

The **peripheral nervous system** (PNS) consists of the nerves that branch from the central nervous system (CNS), connecting it to other body parts. The PNS includes the *cranial nerves* that arise from the brain and the *spinal nerves* that arise from the spinal cord.

The peripheral nervous system can also be subdivided into somatic and autonomic nervous systems. Generally, the **somatic nervous system** consists of the cranial and spinal nerve fibers that connect the CNS to the skin and skeletal muscles, so it oversees conscious activities. The **autonomic nervous system** (aw″to nom′ik ner′vus sis′tem) includes fibers that connect the CNS to viscera such as the heart, stomach, intestines, and various glands. Thus, the autonomic nervous system controls

TABLE 11.7	Major Parts of the Brain	
Part	**Characteristics**	**Functions**
1. Cerebrum	Largest part of the brain; two hemispheres connected by the corpus callosum	Controls higher brain functions, including interpreting sensory impulses, initiating muscular movements, storing memory, reasoning, and determining intelligence
2. Basal nuclei (ganglia)	Masses of gray matter deep within the cerebral hemispheres	Relay stations for motor impulses originating in the cerebral cortex and passing into the brainstem and spinal cord
3. Diencephalon	Includes masses of gray matter (thalamus and hypothalamus)	The thalamus is a relay station for sensory impulses ascending from other parts of the nervous system to the cerebral cortex; the hypothalamus helps maintain homeostasis by regulating visceral activities and by linking the nervous and endocrine systems
4. Brainstem	Connects the cerebrum to the spinal cord	
a. Midbrain	Contains masses of gray matter and bundles of nerve fibers that join the spinal cord to higher regions of the brain	Contains reflex centers that move the eyes and head, and maintains posture
b. Pons	A bulge on the underside of the brainstem that contains masses of gray matter and nerve fibers	Relays nerve impulses to and from the medulla oblongata and cerebrum; helps regulate rate and depth of breathing
c. Medulla oblongata	An enlarged continuation of the spinal cord that extends from the foramen magnum to the pons and contains masses of gray matter and nerve fibers	Conducts ascending and descending impulses between the brain and spinal cord; contains cardiac, vasomotor, and respiratory control centers and various nonvital reflex control centers
5. Cerebellum	A large mass of tissue located below the cerebrum and posterior to the brainstem; includes two lateral hemispheres connected by the vermis	Communicates with other parts of the CNS by nerve tracts; integrates sensory information concerning the position of body parts; and coordinates muscle activities and maintains posture

BRAIN WAVES

Brain waves are recordings of fluctuating electrical changes in the brain. To obtain such a recording, electrodes are positioned on the surface of a surgically exposed brain (an electrocorticogram, ECoG) or on the outer surface of the head (an electroencephalogram, EEG). These electrodes detect electrical changes in the extracellular fluid of the brain in response to changes in potential among large groups of neurons. The resulting signals from the electrodes are amplified and recorded. Brain waves originate from the cerebral cortex but also reflect activities in other parts of the brain that influence the cortex, such as the reticular formation. Because the intensity of electrical changes is proportional to the degree of neuronal activity, brain waves vary markedly in amplitude and frequency between sleep and wakefulness.

Brain waves are classified as alpha, beta, theta, and delta waves. *Alpha waves* are recorded most easily from the posterior regions of the head and have a frequency of 8–13 cycles per second. They occur when a person is awake but resting, with the eyes closed. These waves disappear during sleep, and if the wakeful person's eyes open, higher-frequency beta waves replace the alpha waves.

Beta waves have a frequency of more than 13 cycles per second and are usually recorded in the anterior region of the head. They occur when a person is actively engaged in mental activity or is under tension.

Theta waves have a frequency of 4–7 cycles per second and occur mainly in the parietal and temporal regions of the cerebrum. They are normal in children but do not usually occur in adults. However, some adults produce theta waves in early stages of sleep or at time of emotional stress.

Delta waves have a frequency below 4 cycles per second and occur during sleep. They originate from the cerebral cortex when it is not being activated by the reticular formation (fig. 11F).

Brain wave patterns can be useful for diagnosing disease conditions, such as distinguishing types of seizure disorders (epilepsy) and locating brain tumors. Brain waves are also used to determine when *brain death* has occurred. Brain death, characterized by the cessation of neuronal activity, may be verified by an EEG that lacks waves (isoelectric EEG). However, drugs that greatly depress brain functions must be excluded as the cause of the flat EEG pattern before brain death can be confirmed. ■

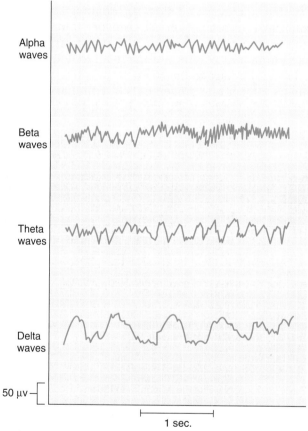

Alpha waves

Beta waves

Theta waves

Delta waves

50 µv

1 sec.

FIGURE 11F

Brain waves record fluctuating electrical changes in the brain.

---

subconscious actions. Table 11.8 outlines the subdivisions of the nervous system.

Structure of Peripheral Nerves

A peripheral nerve consists of connective tissue surrounding bundles of nerve fibers. The outermost layer of the connective tissue, called the *epineurium,* is dense and includes many collagenous fibers. Each bundle of nerve fibers (fascicle) is, in turn, enclosed in a sleeve of looser connective tissue called the *perineurium.* A small amount of loose connective tissue called *endoneurium* surrounds individual nerve fibers (figs. 11.23 and 11.24). Blood ves-

TABLE 11.8	Subdivisions of the Nervous System

1. **Central nervous system (CNS)**
   a. Brain
   b. Spinal cord
2. **Peripheral nervous system (PNS)**
   a. Cranial nerves arising from the brain
      (1) Somatic fibers connecting to the skin and skeletal muscles
      (2) Autonomic fibers connecting to viscera
   b. Spinal nerves arising from the spinal cord
      (1) Somatic fibers connecting to the skin and skeletal muscles
      (2) Autonomic fibers connecting to viscera

sels in the epineurium and perineurium give rise to a network of capillaries in the endoneurium that provides oxygen and nutrients to the neurons.

The term "muscle fiber" refers to a muscle cell, whereas the term "nerve fiber" refers to a cellular process, or part of a cell. The terminology for the connective tissue holding them together, however, is quite similar. In both cases, for example, fibers are bundled into fascicles, whereas epineurium in nerves corresponds to epimysium in muscles, and so forth (fig. 11.23).

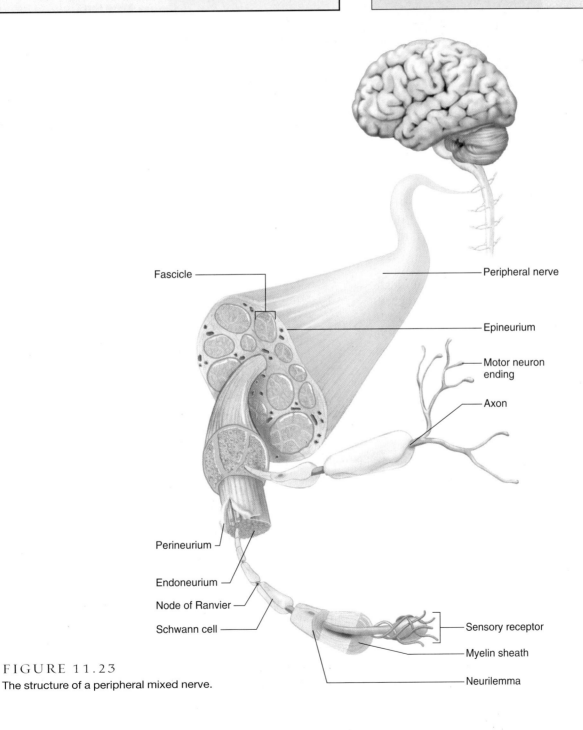

FIGURE 11.23

The structure of a peripheral mixed nerve.

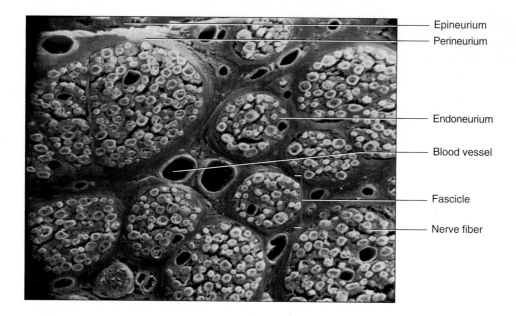

Epineurium
Perineurium
Endoneurium
Blood vessel
Fascicle
Nerve fiber

FIGURE 11.24
Scanning electron micrograph of a peripheral nerve in cross section (350×). Note the bundles or fascicles of nerve fibers. Fibers include axons of motor neurons as well as peripheral processes of sensory neurons. Copyright by R.G. Kessel and R.H. Kardon, *Tissues and Organs: A Text-Atlas of Scanning Electron Microscopy.* 1979 (W.H. Freeman & Co.).

Nerve Fiber Classification

Like nerve fibers, nerves that conduct impulses into the brain or spinal cord are called **sensory nerves,** and those that carry impulses to muscles or glands are termed **motor nerves.** Most nerves, however, include both sensory and motor fibers, and they are called **mixed nerves.**

Nerves originating from the brain that communicate with other body parts are called **cranial nerves,** whereas those originating from the spinal cord that communicate with other body parts are called **spinal nerves.** The nerve fibers within these structures can be subdivided further into four groups as follows:

1. **General somatic efferent fibers** carry motor impulses outward from the brain or spinal cord to skeletal muscles and stimulate them to contract.

2. **General visceral efferent fibers** carry motor impulses outward from the brain or spinal cord to various smooth muscles and glands associated with internal organs, causing certain muscles to contract or glands to secrete.

3. **General somatic afferent fibers** carry sensory impulses inward to the brain or spinal cord from receptors in the skin and skeletal muscles.

4. **General visceral afferent fibers** carry sensory impulses to the central nervous system from blood vessels and internal organs.

The term *general* in each of these categories indicates that the fibers are associated with general structures such as the skin, skeletal muscles, glands, and viscera. Three other groups of fibers, found only in cranial nerves, are associated with more specialized, or *special,* structures:

1. **Special somatic efferent fibers** carry motor impulses outward from the brain to the muscles used in chewing, swallowing, speaking, and forming facial expressions.

2. **Special visceral afferent fibers** carry sensory impulses inward to the brain from the olfactory and taste receptors.

3. **Special somatic afferent fibers** carry sensory impulses inward to the brain from the receptors of sight, hearing, and equilibrium.

Cranial Nerves

Twelve pairs of **cranial nerves** arise from the underside of the brain. Except for the first pair, which begins within the cerebrum, these nerves originate from the brainstem. They pass from their sites of origin through foramina of the skull and lead to areas of the head, neck, and trunk.

Although most cranial nerves are mixed nerves, some of those associated with special senses, such as smell and vision, contain only sensory fibers. Others that innervate muscles and glands are primarily composed of motor fibers and have only limited sensory functions. These are neurons associated with certain receptors (*proprioceptors*) that respond to the rate or degree of contraction of skeletal muscles. Because these fibers contribute directly to motor control, cranial nerves whose only sensory component is from such proprioceptors are usually considered motor nerves. This pertains to cranial nerves III, IV, VI, XI, and XII.

Neuron cell bodies to which the sensory fibers in the cranial nerves attach are located outside the brain and are usually in groups called *ganglia* (sing., *ganglion*). On the other hand, motor neuron cell bodies are typically located within the gray matter of the brain.

Cranial nerves are designated by numbers or name. The numbers indicate the order in which the nerves arise from the brain, from anterior to posterior. The names describe primary functions or the general distribution of their fibers (fig. 11.25).

The first pair of cranial nerves, the *olfactory nerves (I),* are associated with the sense of smell and contain only sensory neurons. These neurons synapse with bipolar neurons, located in the lining of the upper nasal cavity, that serve as *olfactory receptor cells.* Axons from these receptors pass upward through the cribriform plates of the ethmoid bone. The synapses occur in the *olfactory bulbs,* which are extensions of the cerebral cortex, located just beneath the frontal lobes. Sensory impulses travel from the olfactory bulbs along *olfactory tracts* to cerebral centers where they are interpreted. The result of this interpretation is the sensation of smell.

The second pair, the **optic nerves** (II), are sensory and lead from the eyes to the brain and are associated with vision. The cell bodies of these neurons form ganglion cell layers within the eyes, and their axons pass through the *optic foramina* of the orbits and continue into the visual nerve pathways of the brain (see chapter 12, p. 461).

The third pair, the **oculomotor** (ok"u-lo-mo'tor) **nerves** (III), arise from the midbrain and pass into the orbits of the eyes. One component of each nerve connects to a number of voluntary muscles, including those that raise the eyelids and four of the six muscles that move the eye.

A second portion of each oculomotor nerve is part of the autonomic nervous system, supplying involuntary muscles inside the eyes. These muscles help adjust the amount of light that enters the eyes and help focus the lenses of the eyes. This nerve is considered motor, with some proprioceptive fibers.

The fourth pair, the **trochlear** (trok'le-ar) **nerves** (IV), are the smallest cranial nerves. They arise from the midbrain and carry motor impulses to a fifth pair of external eye muscles, the *superior oblique muscles,* which are not supplied by the oculomotor nerves. This nerve is considered motor, with some proprioceptive fibers.

The fifth pair, the **trigeminal** (tri-jem'i-nal) **nerves** (V), are the largest of the cranial nerves and arise from the pons. They are mixed nerves, with the sensory portions more extensive than the motor portions. Each sensory component includes three large branches, called the ophthalmic, maxillary, and mandibular divisions (fig. 11.26).

The *ophthalmic division* consists of sensory fibers that bring impulses to the brain from the surface of the eye, the tear gland, and the skin of the anterior scalp, forehead, and upper eyelid. The fibers of the *maxillary division* carry sensory impulses from the upper teeth, upper gum, and upper lip, as well as from the mucous lining of the palate and facial skin. The *mandibular division* includes both motor and sensory fibers. The sensory branches transmit

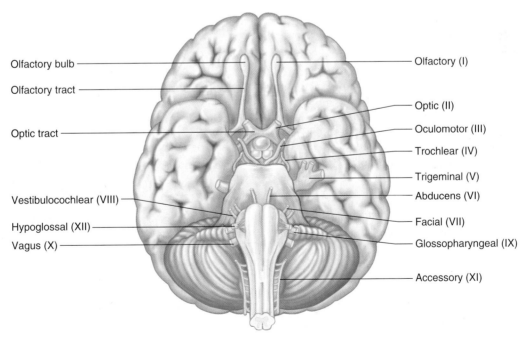

FIGURE 11.25

The cranial nerves, except for the first pair, arise from the brainstem. They are identified either by numbers indicating their order, their function, or the general distribution of their fibers.

CHAPTER ELEVEN *Nervous System II*

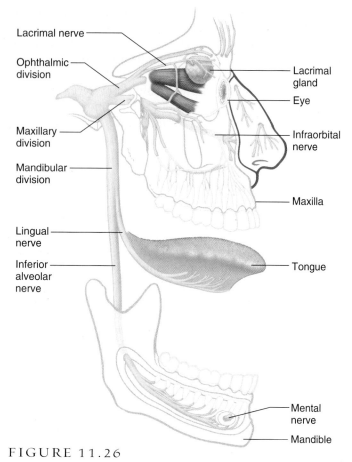

Lacrimal nerve

Ophthalmic division

Lacrimal gland

Eye

Maxillary division

Infraorbital nerve

Mandibular division

Maxilla

Lingual nerve

Inferior alveolar nerve

Tongue

Mental nerve

Mandible

FIGURE 11.26

Each trigeminal nerve has three large branches that supply various regions of the head and face: the ophthalmic division, the maxillary division, and the mandibular division.

impulses from the scalp behind the ears, the skin of the jaw, the lower teeth, the lower gum, and the lower lip. The motor branches supply the muscles of mastication and certain muscles in the floor of the mouth.

The sixth pair, the **abducens** (ab-du'senz) **nerves** (VI), are quite small and originate from the pons near the medulla oblongata. They enter the orbits of the eyes and supply motor impulses to the remaining pair of external eye muscles, the *lateral rectus muscles.* This nerve is considered motor, with some proprioceptive fibers.

A disorder of the trigeminal nerve called *trigeminal neuralgia* (tic douloureux) causes severe recurring pain in the face and forehead on the affected side. If drugs cannot control the pain, surgery may be used to sever the sensory portion of the nerve. However, the patient loses sensations in other body regions that the sensory branch supplies. Consequently, after such surgery, care must be taken when eating or drinking hot foods or liquids, and the mouth must be inspected daily for food particles or damage to the cheeks from biting.

The seventh pair of cranial nerves, the **facial** (fa'shal) **nerves** (VII), are mixed nerves that arise from the lower part of the pons and emerge on the sides of the face. Their sensory branches are associated with taste receptors on the anterior two-thirds of the tongue, and some of their motor fibers transmit impulses to muscles of facial expression. Still other motor fibers of these nerves function in the autonomic nervous system by stimulating secretions from tear glands and certain salivary glands (submandibular and sublingual glands) (fig. 11.27).

The eighth pair, the **vestibulocochlear** (ves-tib"u-lo-kok'le-ar) **nerves** (VIII, acoustic, or auditory, nerves), are sensory nerves that arise from the medulla oblongata. Each of these nerves has two distinct parts—a vestibular branch and a cochlear branch.

The neuron cell bodies of the *vestibular branch* fibers are located in ganglia near the vestibule and semicircular canals of the inner ear. These structures contain receptors that sense changes in the position of the head and, in response, initiate and send impulses to the cerebellum, where they are used in reflexes that maintain equilibrium.

The neuron cell bodies of the *cochlear branch* fibers are located in a ganglion of the cochlea, a part of the inner ear that houses the hearing receptors. Impulses from this branch pass through the medulla oblongata and midbrain on their way to the temporal lobe, where they are interpreted.

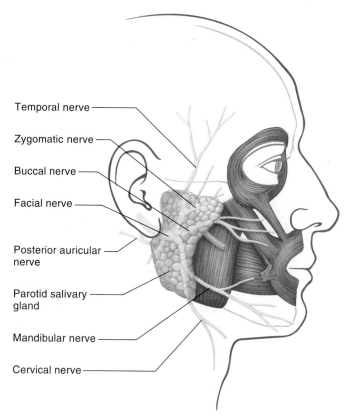

Temporal nerve

Zygomatic nerve

Buccal nerve

Facial nerve

Posterior auricular nerve

Parotid salivary gland

Mandibular nerve

Cervical nerve

FIGURE 11.27

The facial nerves are associated with taste receptors on the tongue and with muscles of facial expression.

The ninth pair, the **glossopharyngeal** (glos"o-fah-rin'je-al) **nerves** (IX), are associated with the tongue and pharynx. These nerves arise from the medulla oblongata, and, although they are mixed nerves, their predominant fibers are sensory. These fibers carry impulses from the lining of the pharynx, tonsils, and posterior third of the tongue to the brain. Fibers in the motor component of the glossopharyngeal nerves innervate constrictor muscles in the wall of the pharynx that function in swallowing.

The tenth pair, the **vagus** (va'gus) **nerves** (X), originate in the medulla oblongata and extend downward through the neck into the chest and abdomen. These nerves are mixed, containing both somatic and autonomic branches, with the autonomic fibers predominant.

Among the somatic components of the vagus nerves are motor fibers that carry impulses to muscles of the larynx. These fibers are associated with speech and swallowing reflexes that employ muscles in the soft palate and pharynx. Vagal sensory fibers carry impulses from the lin-

ings of the pharynx, larynx, and esophagus and from the viscera of the thorax and abdomen to the brain. Autonomic motor fibers of the vagus nerves supply the heart and many smooth muscles and glands in the viscera of the thorax and abdomen (fig. 11.28).

The eleventh pair, the **accessory** (ak-ses'o-re) **nerves** (XI, spinal accessory), originate in the medulla oblongata and the spinal cord; thus, they have both cranial and spinal branches. Each *cranial branch* of an accessory nerve joins a vagus nerve and carries impulses to muscles of the soft palate, pharynx, and larynx. The *spinal branch* descends into the neck and supplies motor fibers to the trapezius and sternocleidomastoid muscles. This nerve is considered motor, with some proprioceptive fibers.

The twelfth pair, the **hypoglossal** (hi"po-glos'al) **nerves** (XII), arise from the medulla oblongata and pass into the tongue. They primarily consist of fibers that carry impulses to muscles that move the tongue in speaking, chewing, and swallowing. This nerve is considered

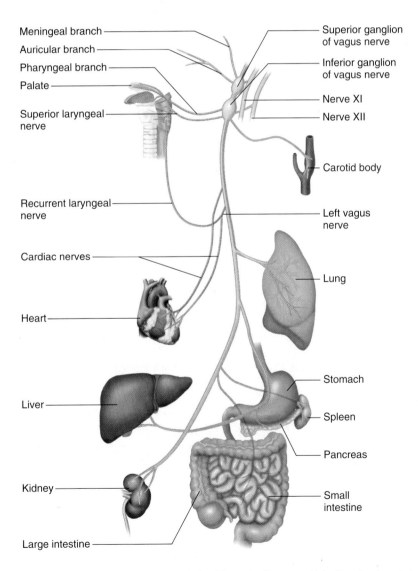

FIGURE 11.28

The vagus nerves (only the left vagus is shown) extend from the medulla oblongata downward into the chest and abdomen to supply many organs.

TABLE 11.9 Functions of Cranial Nerves

Nerve	Type	Function
I Olfactory	Sensory	Sensory fibers transmit impulses associated with the sense of smell.
II Optic	Sensory	Sensory fibers transmit impulses associated with the sense of vision.
III Oculomotor	Primarily motor	Motor fibers transmit impulses to muscles that raise the eyelids, move the eyes, adjust the amount of light entering the eyes, and focus the lenses.
		Some sensory fibers transmit impulses associated with proprioceptors.
IV Trochlear	Primarily motor	Motor fibers transmit impulses to muscles that move the eyes.
		Some sensory fibers transmit impulses associated with proprioceptors.
V Trigeminal	Mixed	
Ophthalmic division		Sensory fibers transmit impulses from the surface of the eyes, tear glands, scalp, forehead, and upper eyelids.
Maxillary division		Sensory fibers transmit impulses from the upper teeth, upper gum, upper lip, lining of the palate, and skin of the face.
Mandibular division		Sensory fibers transmit impulses from the scalp, skin of the jaw, lower teeth, lower gum, and lower lip.
		Motor fibers transmit impulses to muscles of mastication and to muscles in the floor of the mouth.
VI Abducens	Primarily motor	Motor fibers transmit impulses to muscles that move the eyes.
		Some sensory fibers transmit impulses associated with proprioceptors.
VII Facial	Mixed	Sensory fibers transmit impulses associated with taste receptors of the anterior tongue.
		Motor fibers transmit impulses to muscles of facial expression, tear glands, and salivary glands.
VIII Vestibulocochlear	Sensory	
Vestibular branch		Sensory fibers transmit impulses associated with the sense of equilibrium.
Cochlear branch		Sensory fibers transmit impulses associated with the sense of hearing.
IX Glossopharyngeal	Mixed	Sensory fibers transmit impulses from the pharynx, tonsils, posterior tongue, and carotid arteries.
		Motor fibers transmit impulses to salivary glands and to muscles of the pharynx used in swallowing.
X Vagus	Mixed	Somatic motor fibers transmit impulses to muscles associated with speech and swallowing; autonomic motor fibers transmit impulses to the viscera of the thorax and abdomen.
		Sensory fibers transmit impulses from the pharynx, larynx, esophagus, and viscera of the thorax and abdomen.
XI Accessory	Primarily motor	
Cranial branch		Motor fibers transmit impulses to muscles of the soft palate pharynx and larynx.
Spinal branch		Motor fibers transmit impulses to muscles of the neck and back; some proprioceptor input.
XII Hypoglossal	Primarily motor	Motor fibers transmit impulses to muscles that move the tongue; some proprioceptor input.

motor, with some proprioceptive fibers. Table 11.9 summarizes the functions of the cranial nerves.

1 Define *peripheral nervous system.*

2 Distinguish between somatic and autonomic nerve fibers.

3 Describe the structure of a peripheral nerve.

4 Distinguish among sensory, motor, and mixed nerves.

5 Name the cranial nerves, and list the major functions of each.

Spinal Nerves

Thirty-one pairs of spinal nerves originate from the spinal cord. They are mixed nerves, and they provide two-way communication between the spinal cord and parts of the upper and lower limbs, neck, and trunk.

Spinal nerves are not named individually but are grouped by the level from which they arise, with each nerve numbered in sequence (fig. 11.29). Thus, there are eight pairs of *cervical nerves* (numbered C1 to C8), twelve pairs of *thoracic nerves* (numbered T1 to T12), five pairs of *lumbar nerves* (numbered L1 to L5), five pairs of *sacral nerves* (numbered S1 to S5), and one pair of *coccygeal nerves* (Co).

The nerves arising from the superior part of the spinal cord pass outward almost horizontally, whereas those from the inferior portions of the spinal cord descend at sharp angles. This arrangement is a conse-

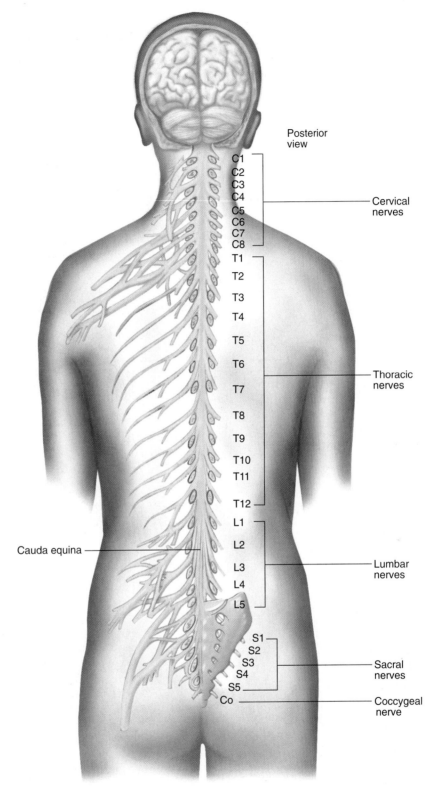

Posterior
view

C1
C2
C3
C4
C5 — Cervical
C6   nerves
C7
C8

T1
T2
T3
T4
T5
T6
T7 — Thoracic
       nerves
T8
T9
T10
T11
T12

L1
L2
L3 — Lumbar
L4   nerves
L5

Cauda equina

S1
S2
S3 — Sacral
S4   nerves
S5
Co — Coccygeal
       nerve

FIGURE 11.29
The thirty-one pairs of spinal nerves are grouped according to the level from which they arise and are numbered in sequence.

quence of growth. In early life, the spinal cord extends the entire length of the vertebral column, but with age, the column grows more rapidly than the cord. Thus, the adult spinal cord ends at the level between the first and second lumbar vertebrae, so the lumbar, sacral, and coccygeal nerves descend to their exits beyond the end of the cord. These descending nerves form a structure called the *cauda equina* (horse's tail) (fig. 11.29).

Each spinal nerve emerges from the cord by two short branches, or roots, which lie within the vertebral column. The **dorsal root** (posterior, or sensory, root) can be identified by an enlargement called the *dorsal root ganglion.* This ganglion contains the cell bodies of the sensory neurons whose dendrites conduct impulses inward from the peripheral body parts. The axons of these neurons extend through the dorsal root and into the spinal cord, where they form synapses with dendrites of other neurons.

An area of skin that the sensory nerve fibers of a particular spinal nerve innervate is called a *dermatome.* Der-matomes are highly organized, but they vary considerably in size and shape, as figure 11.30 indicates. A map of the dermatomes is often useful in localizing the sites of injuries to dorsal roots or the spinal cord.

The **ventral root** (anterior, or motor, root) of each spinal nerve consists of axons from the motor neurons whose cell bodies are located within the gray matter of the cord. A ventral root and a dorsal root unite to form a spinal nerve, which extends outward from the vertebral canal through an *intervertebral foramen.* Just beyond its foramen, each spinal nerve branches. One of these parts, the small *meningeal branch,* reenters the vertebral canal

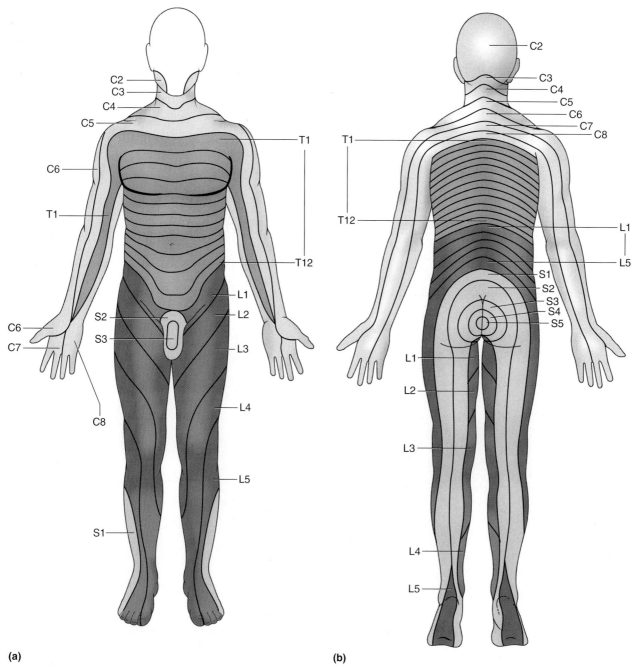

(a)   **(b)**

FIGURE 11.30

Dermatomes (*a*) on the anterior body surface and (*b*) on the posterior surface. Note that spinal nerve C1 does not supply any skin area.

through the intervertebral foramen and supplies the meninges and blood vessels of the cord, as well as the intervertebral ligaments and the vertebrae.

As figure 11.31 shows, a *posterior branch* (posterior ramus) of each spinal nerve turns posteriorly and innervates the muscles and skin of the back. The main portion of the nerve, the *anterior branch* (anterior ramus), continues forward to supply muscles and skin on the front and sides of the trunk and limbs.

The spinal nerves in the thoracic and lumbar regions have a fourth, or *visceral branch,* which is part of the autonomic nervous system. Except in the thoracic region, anterior branches of the spinal nerves combine to form complex networks called **plexuses** instead of continuing directly to the peripheral body parts. In a plexus, the fibers of various spinal nerves are sorted and recombined, so fibers associated with a particular peripheral body part reach it in the same nerve, even though the fibers originate from different spinal nerves (fig. 11.32).

Cervical Plexuses

The **cervical plexuses** lie deep in the neck on either side. They are formed by the anterior branches of the first four cervical nerves. Fibers from these plexuses supply the muscles and skin of the neck. In addition, fibers from the third, fourth, and fifth cervical nerves pass into the right and left **phrenic** (fren'ik) **nerves,** which conduct motor impulses to the muscle fibers of the diaphragm.

Brachial Plexuses

The anterior branches of the lower four cervical nerves and the first thoracic nerve give rise to **brachial plexuses.** These networks of nerve fibers are located deep within the shoulders between the neck and the axillae (armpits). The major branches emerging from the brachial plexuses include the following (fig. 11.33):

1. *Musculocutaneous nerves* supply muscles of the arms on the anterior sides and the skin of the forearms.

2. *Ulnar nerves* supply muscles of the forearms and hands and the skin of the hands.

3. *Median nerves* supply muscles of the forearms and muscles and skin of the hands.

4. *Radial nerves* supply muscles of the arms on the posterior sides and the skin of the forearms and hands.

5. *Axillary nerves* supply muscles and skin of the superior, lateral, and posterior regions of the arm.

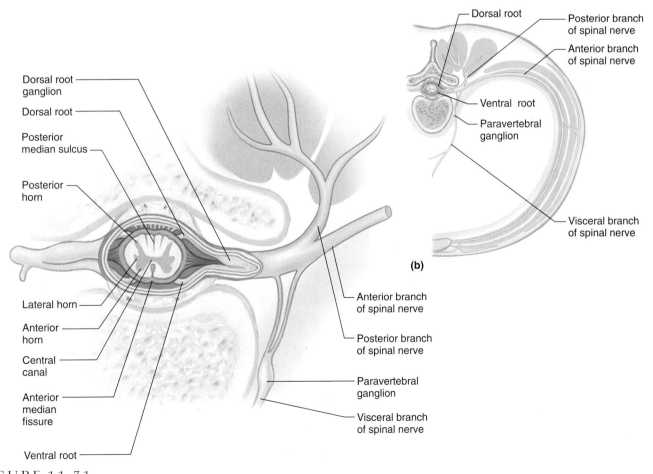

FIGURE 11.31

Spinal nerve. (*a*) Each spinal nerve has a posterior and an anterior branch. (*b*) The thoracic and lumbar spinal nerves also have a visceral branch.

CHAPTER ELEVEN   *Nervous System II*

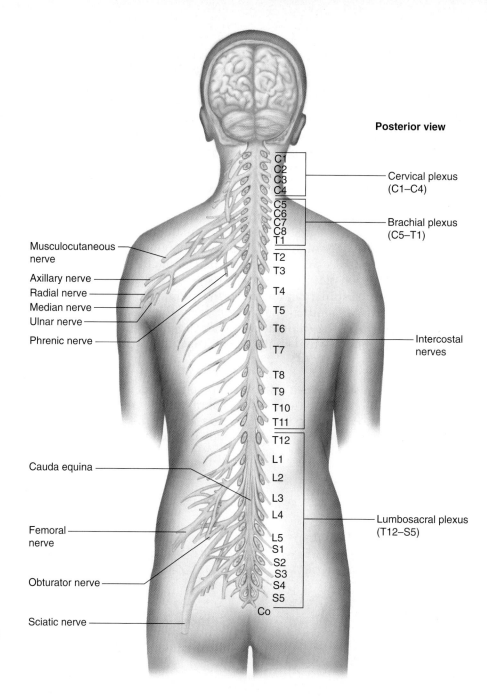

Cervical plexus (C1–C4)

Brachial plexus (C5–T1)

Musculocutaneous nerve

Axillary nerve

Radial nerve

Median nerve

Ulnar nerve

Phrenic nerve

Intercostal nerves

Cauda equina

Femoral nerve

Lumbosacral plexus (T12–S5)

Obturator nerve

Sciatic nerve

C1
C2
C3
C4
C5
C6
C7
C8
T1
T2
T3
T4
T5
T6
T7
T8
T9
T10
T11
T12
L1
L2
L3
L4
L5
S1
S2
S3
S4
S5
Co

FIGURE 11.32

The anterior branches of the spinal nerves in the thoracic region give rise to intercostal nerves. Those in other regions combine to form complex networks called plexuses.

Other nerves associated with the brachial plexus that innervate various skeletal muscles include the following:

1. The *lateral* and *medial pectoral nerves* supply the pectoralis major and pectoralis minor muscles.

2. The *dorsal scapular nerve* supplies the rhomboideus major and levator scapulae muscles.

3. The *lower subscapular nerve* supplies the subscapularis and teres major muscles.

4. The *thoracodorsal nerve* supplies the latissimus dorsi muscle.

5. The *suprascapular nerve* supplies the supraspinatus and infraspinatus muscles.

Lumbosacral Plexuses

The **lumbosacral** (lum″bo-sa′kral) **plexuses** are formed by the last thoracic nerve and the lumbar, sacral, and coccygeal nerves. These networks of nerve fibers extend from the lumbar region of the back into the pelvic cavity, giving rise to a number of motor and sensory fibers associated with the lower abdominal wall, external genitalia, buttocks, thighs, legs, and feet. The major branches of these plexuses include the following (fig. 11.34):

1. The *obturator nerves* supply the adductor muscles of the thighs.

2. The *femoral nerves* divide into many branches, supplying motor impulses to muscles of the thighs

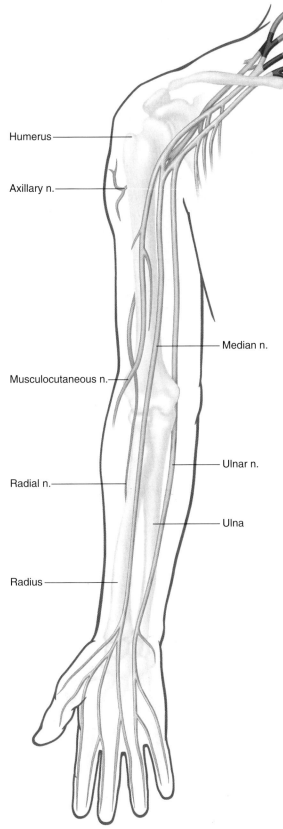

C5
C6
C7
C8
T1

Humerus

Axillary n.

Median n.

Musculocutaneous n.

Ulnar n.

Radial n.

Ulna

Radius

FIGURE 11.33
Nerves of the brachial plexus.

and legs and receiving sensory impulses from the skin of the thighs and legs.

3. The *sciatic nerves* are the largest and longest nerves in the body. They pass downward into the buttocks and descend into the thighs, where they divide into *tibial* and *common fibular nerves.* The many branches of these nerves supply muscles and skin in the thighs, legs, and feet.

Other nerves associated with the lumbosacral plexus that innervate various skeletal muscles include the following:

1. The *pudendal nerve* supplies the muscles of the perineum.

2. The *inferior* and *superior gluteal nerves* supply the gluteal muscles and the tensor fasciae latae muscle.

The anterior branches of the thoracic spinal nerves do not enter a plexus. Instead, they travel into spaces between the ribs and become **intercostal** (in″ter-kos′tal) **nerves.** These nerves supply motor impulses to the intercostal muscles and the upper abdominal wall muscles. They also receive sensory impulses from the skin of the thorax and abdomen. Clinical Application 11.7 discusses injuries to the spinal nerves.

1 How are spinal nerves grouped?

2 Describe how a spinal nerve joins the spinal cord.

3 Name and locate the major nerve plexuses.

Autonomic Nervous System

The autonomic nervous system is the part of the peripheral nervous system that functions independently (autonomously) and continuously, without conscious effort. This system controls visceral activities by regulating the actions of smooth muscles, cardiac muscles, and various glands. It oversees heart rate, blood pressure, breathing rate, body temperature, and other visceral activities that aid in maintaining homeostasis. Portions of the autonomic nervous system also respond during times of emotional stress and prepare the body to meet the demands of strenuous physical activity.

General Characteristics

Reflexes in which sensory signals originate from receptors within the viscera and the skin regulate autonomic activities. Afferent nerve fibers transmit these signals to nerve centers within the brain or spinal cord. In response, motor impulses travel out from these centers on efferent nerve fibers within cranial and spinal nerves.

Typically, these efferent fibers lead to ganglia outside the central nervous system. The impulses they carry

SPINAL NERVE INJURIES

Birth injuries, dislocations, vertebral fractures, stabs, gunshot wounds, and pressure from tumors can all injure spinal nerves. Suddenly bending the neck, called whiplash, can compress the nerves of the cervical plexuses, causing persistent headache and pain in the neck and skin, which the cervical nerves supply. If a broken or dislocated vertebra severs or damages the phrenic nerves associated with the cervical plexuses, partial or complete paralysis of the diaphragm may result.

Intermittent or constant pain in the neck, shoulder, or upper limb may result from prolonged abduction of the upper limb, as in painting or typing. This is due to too much pressure on the brachial plexus. This condition, called *thoracic outlet syndrome,* may also result from a congenital skeletal malformation that compresses the plexus during upper limb and shoulder movements.

Degenerative changes may compress an intervertebral disc in the lumbar region, producing *sciatica,* which causes pain in the lower back and gluteal region that can radiate to the thigh, calf, ankle, and foot. Sciatica is most common in middle-aged people, particularly distance runners. It usually compresses spinal nerve roots between L2 and S1, some of which contain fibers of the sciatic nerve. Rest, drugs, or surgery are used to treat sciatica.

In *carpal tunnel syndrome,* repeated hand movements, such as typing, inflame the tendons that pass through the carpal tunnel, which is a space between bones in the wrist. The swollen tendons compress the median nerve in the wrist, causing pain to shoot up the upper limb. Surgery or avoiding repetitive hand movements can relieve symptoms. ∎

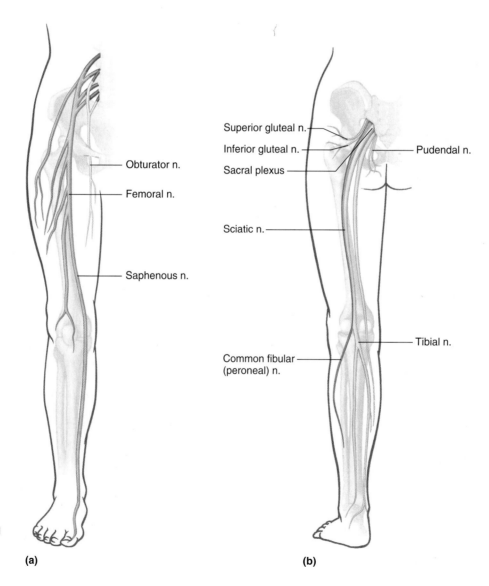

FIGURE 11.34
Nerves of the lumbosacral plexus. (*a*) Anterior view. (*b*) Posterior view.

(a)

(b)

are integrated within the ganglia and are relayed to various organs (muscles or glands) that respond by contracting, secreting, or being inhibited. The integrative function of the ganglia provides the autonomic nervous system with some degree of independence from the brain and spinal cord, and the visceral efferent nerve fibers associated with these ganglia comprise the autonomic nervous system.

The autonomic nervous system includes two divisions, called the **sympathetic** (sim″pah-thet′ik) and **parasympathetic** (par″ah-sim″pah-thet′ik) **divisions,** that interact. For example, many organs have nerve fibers from each of the divisions. Impulses on one set of fibers may activate an organ, whereas impulses on the other set inhibit it. Thus, the divisions may function antagonistically, regulating the actions of some organs by alternately activating or inhibiting them.

The functions of the autonomic divisions are varied; that is, each activates some organs and inhibits others. This reveals that the divisions have important functional differences. The sympathetic division primarily prepares the body for energy-expending, stressful, or emergency situations. Conversely, the parasympathetic division is most active under ordinary, restful conditions. It also counterbalances the effects of the sympathetic division and restores the body to a resting state following a stressful experience. For example, during an emergency, the sympathetic division increases heart and breathing rates; following the emergency, the parasympathetic division decreases these activities.

Autonomic Nerve Fibers

All of the nerve fibers of the autonomic nervous system are efferent, or motor, fibers. In the motor pathways of the somatic nervous system, a single neuron typically links the central nervous system and a skeletal muscle. In the autonomic system, motor pathways include two neurons, as figure 11.35 shows. The cell body of one neuron is located in the brain or spinal cord. Its axon, the **preganglionic** (pre″gang-gle-on′ik) **fiber,** leaves the CNS and synapses with one or more nerve fibers whose cell bodies are housed within an autonomic ganglion. The axon of

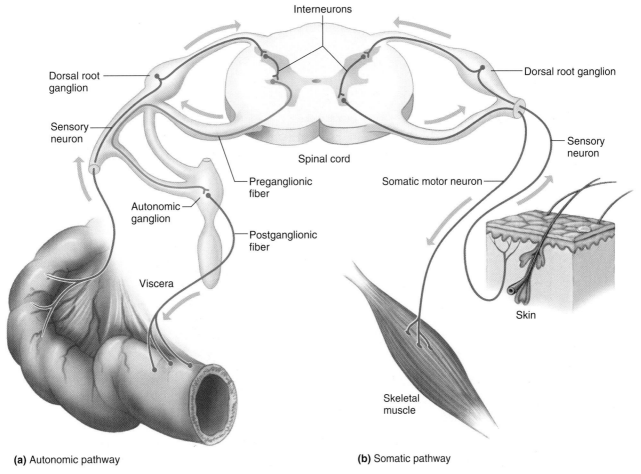

(a) Autonomic pathway

(b) Somatic pathway

FIGURE 11.35

Motor pathways. (a) Autonomic pathways include two neurons between the central nervous system and an effector. (b) Somatic pathways usually have a single neuron between the central nervous system and an effector.

such a second neuron is called a **postganglionic** (pōst″ gang-gle-on′ik) **fiber,** and it extends to a visceral effector.

Sympathetic Division

Within the sympathetic division (thoracolumbar division), the preganglionic fibers originate from neurons within the lateral horn of the spinal cord. These neurons are found in all of the thoracic segments and in the upper two or three lumbar segments of the cord. Their axons exit through the ventral roots of spinal nerves along with various somatic motor fibers.

After traveling a short distance, preganglionic fibers leave the spinal nerves through branches called *white rami* (sing., *ramus*) and enter sympathetic ganglia. Two groups of such ganglia, called **paravertebral ganglia,** are located in chains along the sides of the vertebral column. These ganglia, with the fibers that connect them, comprise the **sympathetic trunks** (fig. 11.36).

The paravertebral ganglia lie just beneath the parietal pleura in the thorax and beneath the parietal peritoneum in the abdomen (see chapter 1, p. 14). Although these ganglia are located some distance from the viscera they help control, other sympathetic ganglia are nearer to the viscera. The *collateral ganglia,* for example, are found within the abdomen, closely associated with certain large blood vessels (fig. 11.37).

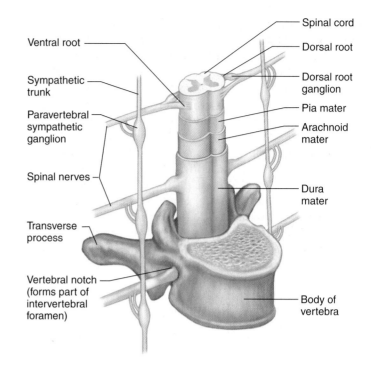

FIGURE 11.36
A chain of paravertebral ganglia extends along each side of the vertebral column.

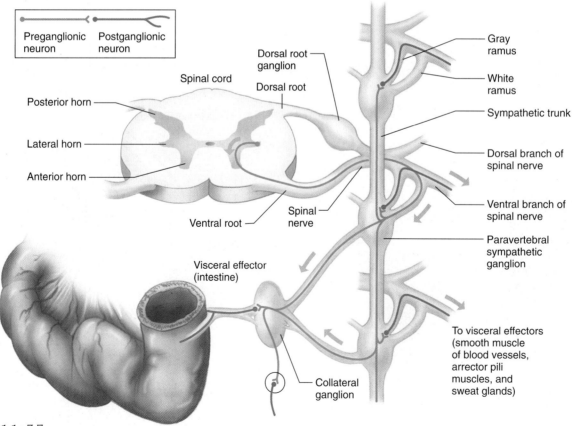

FIGURE 11.37
Sympathetic fibers leave the spinal cord in the ventral roots of spinal nerves, enter paravertebral ganglia, and synapse with other neurons that extend to visceral effectors.

Some of the preganglionic fibers that enter paravertebral ganglia synapse with neurons within these ganglia. Other fibers extend through the ganglia and pass up or down the sympathetic trunk and synapse with neurons in ganglia at higher or lower levels within the chain. Still other fibers pass through to collateral ganglia before they synapse. Typically, a preganglionic axon will synapse with several other neurons within a sympathetic ganglion.

The axons of the second neurons in sympathetic pathways, the postganglionic fibers, extend out from the sympathetic ganglia to visceral effectors. Those leaving paravertebral ganglia usually pass through branches called **gray rami** and return to a spinal nerve before proceeding to an effector (fig. 11.37). These branches appear gray because the postganglionic axons generally are unmyelinated, whereas the preganglionic axons in the white rami are nearly all myelinated.

An important exception to the usual arrangement of sympathetic fibers occurs in a set of preganglionic fibers that pass through the sympathetic ganglia and extend out to the medulla of each adrenal gland. These fibers terminate within the glands on special hormone-secreting cells that release **norepinephrine** (20%) and **epinephrine** (80%) when they are stimulated. Chapter 13 (p. 492) discusses the functions of the adrenal medulla and its hormones. Figure 11.38 shows the sympathetic division.

Parasympathetic Division

The preganglionic fibers of the parasympathetic division (craniosacral division) arise from neurons in the midbrain, pons, and medulla oblongata of the brainstem and from the sacral region of the spinal cord (fig. 11.39). From there, they lead outward on cranial or sacral nerves to ganglia located near or within various organs (terminal ganglia). The short postganglionic fibers continue from the ganglia to specific muscles or glands within these organs (fig. 11.40). Parasympathetic preganglionic axons are usually myelinated, and the parasympathetic postganglionic fibers are unmyelinated.

The parasympathetic preganglionic fibers associated with parts of the head are included in the oculomotor, facial, and glossopharyngeal nerves. Those fibers that innervate organs of the thorax and upper abdomen are parts of the vagus nerves. (The vagus nerves carry about 75% of all parasympathetic fibers.) Preganglionic fibers arising from the sacral region of the spinal cord lie within the branches of the second through the fourth sacral spinal nerves, and they carry impulses to viscera within the pelvic cavity (see fig. 11.39).

1 What is the general function of the autonomic nervous system?

2 How are the divisions of the autonomic system distinguished?

3 Describe a sympathetic nerve pathway and a parasympathetic nerve pathway.

Autonomic Neurotransmitters

The preganglionic fibers of the sympathetic and parasympathetic divisions all secrete acetylcholine, and for this reason they are called **cholinergic** (ko″lin-er′jik) **fibers.** The parasympathetic postganglionic fibers are also cholinergic fibers. Most sympathetic postganglionic fibers, however, secrete norepinephrine (noradrenalin) and are called **adrenergic** (ad″ren-er′jik) **fibers** (see fig. 11.40). Exceptions to this include the sympathetic postganglionic fibers that stimulate sweat glands and a few sympathetic fibers to blood vessels in skin (which cause vasodilation); these fibers secrete acetylcholine and therefore are cholinergic (adrenergic sympathetic fibers to blood vessels cause vasoconstriction).

The different postganglionic neurotransmitters (mediators) are responsible for the different effects that the sympathetic and parasympathetic divisions have on organs. Since each division can activate some effectors and inhibit others, it is not surprising that the divisions usually are antagonistic. For example, the sympathetic nervous system increases heart rate and dilates pupils, whereas parasympathetic stimulation decreases heart rate and constricts pupils. However, this is not always the case. For example, the diameters of most blood vessels lack parasympathetic innervation and are thus regulated by the sympathetic division. Smooth muscles in the walls of these vessels are continuously stimulated by sympathetic impulses; they are thereby maintained in a state of partial contraction called *sympathetic tone.* Decreasing sympathetic stimulation allows the muscular walls of such blood vessels to relax, increasing their diameters (vasodilation). Conversely, increasing sympathetic stimulation vasoconstricts vessels. Table 11.10 summarizes the effects of adrenergic and cholinergic fibers on various visceral effectors.

Actions of Autonomic Neurotransmitters

As in the case of stimulation at neuromuscular junctions (see chapter 9, p. 283) and synapses (see chapter 10, p. 354), the actions of autonomic neurotransmitters result from their binding to protein receptors in the membranes of effector cells. Receptor binding alters the membrane. For example, the membrane's permeability to certain ions may increase, and in smooth muscle cells, an action potential followed by muscular contraction may result. Similarly, a gland cell may respond to a change in its membrane by secreting a product.

Acetylcholine can combine with two types of cholinergic receptors, called *muscarinic receptors* and *nicotinic receptors.* These receptor names come from *muscarine,* a toxin from a fungus that can activate muscarinic receptors, and *nicotine,* the toxin of tobacco that can activate nicotinic receptors. The muscarinic receptors are located in the membranes of effector cells at the ends of all postganglionic parasympathetic nerve fibers and at the ends of the cholinergic sympathetic fibers. Responses from these

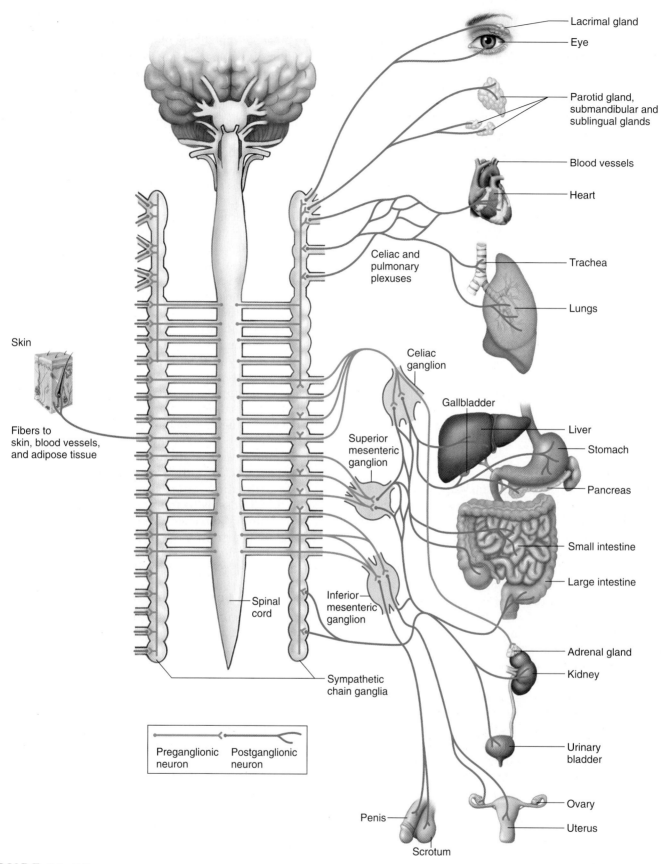

Lacrimal gland
Eye
Parotid gland, submandibular and sublingual glands
Blood vessels
Heart
Trachea
Lungs
Celiac ganglion
Gallbladder
Liver
Stomach
Pancreas
Small intestine
Large intestine
Adrenal gland
Kidney
Urinary bladder
Ovary
Uterus

Skin
Fibers to skin, blood vessels, and adipose tissue
Celiac and pulmonary plexuses
Superior mesenteric ganglion
Inferior mesenteric ganglion
Spinal cord
Sympathetic chain ganglia
Preganglionic neuron
Postganglionic neuron
Penis
Scrotum

FIGURE 11.38

The preganglionic fibers of the sympathetic division of the autonomic nervous system arise from the thoracic and lumbar regions of the spinal cord. Note that the adrenal medulla is innervated directly by a preganglionic fiber.

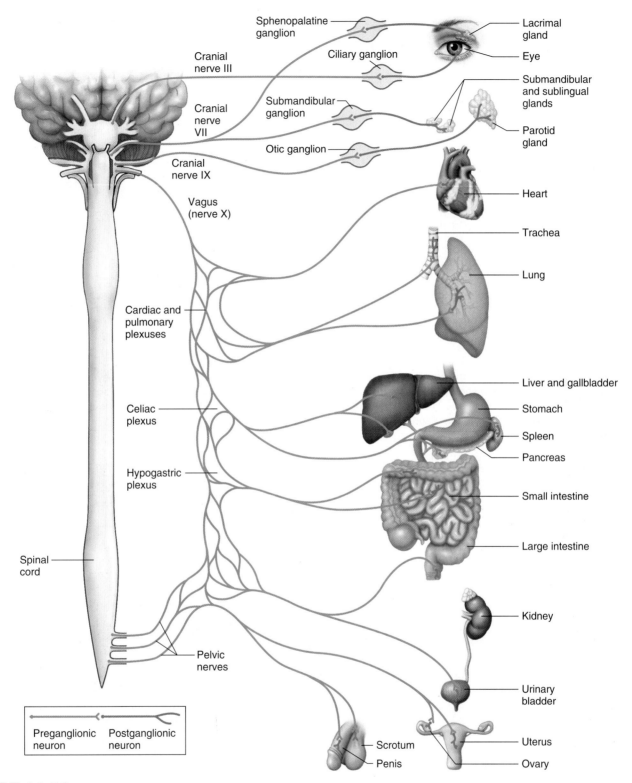

Sphenopalatine ganglion
Cranial nerve III
Ciliary ganglion
Lacrimal gland
Eye
Submandibular ganglion
Cranial nerve VII
Submandibular and sublingual glands
Parotid gland
Otic ganglion
Cranial nerve IX
Vagus (nerve X)
Heart
Trachea
Cardiac and pulmonary plexuses
Lung
Liver and gallbladder
Stomach
Celiac plexus
Spleen
Pancreas
Hypogastric plexus
Small intestine
Large intestine
Spinal cord
Kidney
Pelvic nerves
Urinary bladder
Preganglionic neuron   Postganglionic neuron
Scrotum
Penis
Uterus
Ovary

FIGURE 11.39

The preganglionic fibers of the parasympathetic division of the autonomic nervous system arise from the brain and sacral region of the spinal cord.

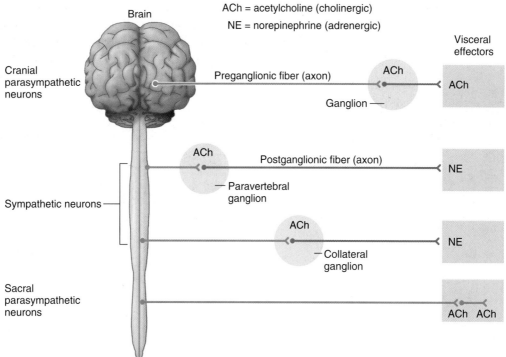

FIGURE 11.40

Most sympathetic fibers are adrenergic and secrete norepinephrine at the ends of the postganglionic fiber; parasympathetic fibers are cholinergic and secrete acetylcholine at the ends of the postganglionic fibers. Two arrangements of parasympathetic postganglionic fibers are seen in both cranial and sacral portions. Similarly, sympathetic paravertebral and collateral ganglia are seen in both the thoracic and lumbar portions of the nervous system. (Note, in this diagrammatic representation, dendrites are not shown.)

TABLE 11.10	Effects of Autonomic Stimulation on Various Effectors	
Effector Location	**Response to Sympathetic Stimulation**	**Response to Parasympathetic Stimulation**
Integumentary system		
Apocrine glands	Increased secretion	No action
Eccrine glands	Increased secretion (cholinergic effect)	No action
Special senses		
Iris of eye	Dilation	Constriction
Tear gland	Slightly increased secretion	Greatly increased secretion
Endocrine system		
Adrenal cortex	Increased secretion	No action
Adrenal medulla	Increased secretion	No action
Digestive system		
Muscle of gallbladder wall	Relaxation	Contraction
Muscle of intestinal wall	Decreased peristaltic action	Increased peristaltic action
Muscle of internal anal sphincter	Contraction	Relaxation
Pancreatic glands	Reduced secretion	Greatly increased secretion
Salivary glands	Reduced secretion	Greatly increased secretion
Respiratory system		
Muscles in walls of bronchioles	Dilation	Constriction
Cardiovascular system		
Blood vessels supplying muscles	Constriction (alpha adrenergic) Dilation (beta adrenergic)	No action
Blood vessels supplying skin	Constriction	No action
Blood vessels supplying heart (coronary arteries)	Dilation (beta adrenergic) Constriction (alpha adrenergic)	No action
Muscles in wall of heart	Increased contraction rate	Decreased contraction rate
Urinary system		
Muscle of bladder wall	Relaxation	Contraction
Muscle of internal urethral sphincter	Contraction	Relaxation
Reproductive systems		
Blood vessels to clitoris and penis	No action	Dilation leading to erection of clitoris and penis
Muscles associated with internal reproductive organs	Male ejaculation, female orgasm	

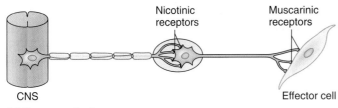

(a) Parasympathetic neurons

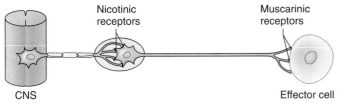

(b) Sympathetic neurons (cholinergic)

(c) Somatic motor neuron

FIGURE 11.41

Receptors. (*a, b,* and *c*) Muscarinic receptors occur in the membranes of effector cells at the axon terminals of autonomic cholinergic neurons. Nicotinic receptors are found in the membranes of postganglionic autonomic neurons and skeletal muscle fibers.

receptors are excitatory and occur relatively slowly. The nicotinic receptors are in the synapses between the preganglionic and postganglionic neurons of the parasympathetic and sympathetic pathways. They produce rapid, excitatory responses (fig. 11.41). (Receptors at neuromuscular junctions of skeletal muscles are nicotinic.)

Epinephrine and norepinephrine are the two chemical mediators of the sympathetic nervous system. The adrenal gland releases both as hormones, but only norepinephrine is released as a neurotransmitter by the sympathetic nervous system. These biochemicals can then combine with adrenergic receptors of effector cells.

The two major types of adrenergic receptors are termed *alpha* and *beta.* Exciting them elicits different responses in the effector organs. For example, stimulation of the alpha receptors in vascular smooth muscle causes vasoconstriction, whereas stimulation of the beta receptors in bronchial smooth muscle causes relaxation leading to bronchodilation. Furthermore, although norepinephrine has a somewhat stronger effect on alpha receptors, both of these mediators can stimulate both kinds of receptors. Consequently, the way each of these adrenergic substances influences effector cells depends on the relative numbers of alpha and beta receptors in the cell membranes.

The enzyme acetylcholinesterase rapidly decomposes the acetylcholine that cholinergic fibers release. (Recall that this decomposition also occurs at the neuromuscular junctions of skeletal muscle.) Thus, acetylcholine usually affects the postsynaptic membrane for only a fraction of a second.

Much of the norepinephrine released from adrenergic fibers is removed from the synapse by active transport back into the nerve endings. The enzyme monoamine oxidase found in mitochondria then inactivates norepinephrine. This may take a few seconds, during which some molecules may diffuse into nearby tissues or the bloodstream, where other enzymes decompose them. On the other hand, some norepinephrine molecules may escape decomposition and remain active for awhile. For these reasons, norepinephrine is likely to produce a more prolonged effect than acetylcholine. In fact, when the adrenal medulla releases norepinephrine and epinephrine into the blood in response to sympathetic stimulation, these substances may trigger sympathetic responses in organs throughout the body that last up to thirty seconds.

Many drugs influence autonomic functions. Some, like ephedrine, enhance sympathetic effects by stimulating release of norepinephrine from postganglionic sympathetic nerve endings. Others, like reserpine, inhibit sympathetic activity by preventing norepinephrine synthesis. Another group of drugs, which includes pilocarpine, produces parasympathetic effects, and some, like atropine, block the action of acetylcholine on visceral effectors.

Control of Autonomic Activity

Although the autonomic nervous system has some independence resulting from impulse integration within its ganglia, it is largely controlled by the brain and spinal cord. For example, recall the control centers in the medulla oblongata for cardiac, vasomotor, and respiratory activities. These reflex centers receive sensory impulses from viscera by means of vagus nerve fibers and use autonomic nerve pathways to stimulate motor responses in various muscles and glands. Thus, they control the autonomic nervous system. Similarly, the hypothalamus helps regulate body temperature, hunger, thirst, and water and electrolyte balance by influencing autonomic pathways.

Still higher levels within the brain, including the limbic system and the cerebral cortex, control the autonomic nervous system during emotional stress. In this way, the autonomic pathways can affect emotional expression and behavior. Subsequent chapters that deal with individual organs and organ systems discuss regulation of particular organs.

Life-Span Changes

The redundancies and overlap of function built into our nervous systems ensure that we can perceive and interact with the environment for many decades. In a sense, aging of this organ system actually begins before birth, as apoptosis, a form of programmed cell death, occurs in the brain, essentially carving out the structures that will remain. This normal dying off of neurons continues throughout life. When brain apoptosis fails, disease results. The brains of individuals who die of schizophrenia as young adults contain the same numbers of neurons as do newborns. These extra neurons produce the extra dopamine that can lead to the hallucinations that are the hallmark of this illness.

By age thirty, the die-off of neurons accelerates somewhat, although pockets of neural stem cells lining the ventricles retain the capacity to differentiate new neurons and neuroglial cells. Over an average lifetime, the brain shrinks by about 10%, with more loss among the gray matter than the white. Neuron loss is uneven, with cell death in the temporal lobe greatest, but very little death among certain neuron clusters in the brainstem. By age ninety, the frontal cortex has lost about half its neurons—but this deficit doesn't necessarily translate into loss of function.

The nervous system changes over time in several ways. The number of dendritic branches in the cerebral cortex falls. Signs of slowing neurotransmission include decreasing levels of neurotransmitters, the enzymes necessary to synthesize them, and the numbers of postsynaptic receptors. The rate of action potential propagation may decrease by 5 to 10%. Nervous system disorders that may begin to cause symptoms in older adulthood include stroke, depression, Alzheimer disease, Parkinson disease, and multi-infarct dementia.

Noticeable signs of a normally aging nervous system include fading memory and slowed responses and reflexes. Decline in function of the sympathetic nervous system may cause transient drops in blood pressure, which, in turn, may cause fainting. By the seventh decade, waning ability of nerves in the ankle to respond to vibrations from walking may affect balance, raising the risk of falling. Poor eyesight, anemia, inner ear malfunction, and effects of drugs also contribute to poor balance in the later years. Because of these factors, nearly a third of individuals over age sixty-five have at least one serious fall a year.

Changes in sleep patterns accompany aging, reflecting the functioning of the reticular activating system. Older individuals generally sleep fewer hours per night than they once did, experiencing transient difficulty in getting to sleep and staying asleep, with more frequent movements when they are sleeping. Many have bouts with insomnia, sometimes not sleeping more than an hour or two a night. Changing electroencephalogram patterns indicate that stage IV slow-wave sleep as well as REM sleep diminishes. All of these changes may result in daytime sleepiness.

Clinical Terms Related to the Nervous System

analgesia (an"al-je'ze-ah) Loss or reduction in the ability to sense pain, but without loss of consciousness.

analgesic (an"al-je'sik) Pain-relieving drug.

anesthesia (an"es-the'ze-ah) Loss of feeling.

aphasia (ah-fa'ze-ah) Disturbance or loss in the ability to use words or to understand them, usually due to damage to cerebral association areas.

apraxia (ah-prak'se-ah) Impairment in a person's ability to make correct use of objects.

ataxia (ah-tak'se-ah) Partial or complete inability to coordinate voluntary movements.

cordotomy (kor-dot'o-me) Surgical procedure severing a nerve tract within the spinal cord, usually to relieve intractable pain.

craniotomy (kra"ne-ot'o-me) Surgical procedure opening a part of the skull.

encephalitis (en"sef-ah-li'tis) Inflammation of the brain and meninges characterized by drowsiness and apathy.

epilepsy (ep'ĭ-lep"se) Temporary disturbances in normal brain impulses that may be accompanied by convulsive seizures and loss of consciousness.

Huntington disease (hunt'ing-tun di-zez') Hereditary disorder of the brain producing progressively worsening, uncontrollable dancelike movements and personality changes.

laminectomy (lam"i-nek'to-me) Surgical removal of the posterior arch of a vertebra, usually to relieve symptoms of a ruptured intervertebral disc that is pressing on a spinal nerve.

neuralgia (nu-ral'je-ah) Sharp, recurring pain associated with a nerve, usually caused by inflammation or injury.

neuritis (nu-ri'tis) Inflammation of a nerve.
vagotomy (va-got'o-me) Surgical severing of a vagus nerve.

CHAPTER SUMMARY

Introduction (page 367)
Bone and protective membranes called meninges surround the brain and spinal cord.

Meninges (page 367)
1. The meninges consist of a dura mater, arachnoid mater, and pia mater.
2. Cerebrospinal fluid occupies the space between the arachnoid and pia maters.

Ventricles and Cerebrospinal Fluid (page 368)
1. Ventricles are connected cavities within the cerebral hemispheres and brainstem.
2. Cerebrospinal fluid fills the ventricles.
3. Choroid plexuses in the walls of the ventricles secrete cerebrospinal fluid.
4. Ependymal cells of the choroid plexus regulate the composition of cerebrospinal fluid.
5. Cerebrospinal fluid circulates through the ventricles and is reabsorbed into the blood of the dural sinuses.

Spinal Cord (page 372)
The spinal cord is a nerve column that extends from the brain into the vertebral canal. It terminates at the level between the first and second lumbar vertebrae.

1. Structure of the spinal cord
   a. The spinal cord is composed of thirty-one segments, each of which gives rise to a pair of spinal nerves.
   b. It is characterized by a cervical enlargement, a lumbar enlargement, and two deep longitudinal grooves that divide it into right and left halves.
   c. White matter surrounds a central core of gray matter.
   d. The white matter is composed of bundles of myelinated nerve fibers.
2. Functions of the spinal cord
   a. The spinal cord is the center for spinal reflexes.
      (1) Reflexes are automatic, subconscious responses to changes.
      (2) They help maintain homeostasis.
      (3) The knee-jerk reflex employs only two neurons.
      (4) Withdrawal reflexes are protective actions.
   b. The cord provides a two-way communication system between the brain and structures outside the nervous system.
      (1) Ascending tracts carry sensory impulses to the brain; descending tracts carry motor impulses to muscles and glands.
      (2) Many of the fibers in the ascending and descending tracts cross over in the spinal cord or brain.

Brain (page 381)
The brain is the largest and most complex part of the nervous system. It contains nerve centers that are associated with sensations. The brain issues motor commands and carries on higher mental functions.

1. Brain development
   a. Brain structure reflects the way it forms.
   b. The brain develops from a neural tube with three cavities—the forebrain, midbrain, and hindbrain.
   c. The cavities persist as ventricles, and the walls give rise to structural and functional regions.
2. Structure of the cerebrum
   a. The cerebrum consists of two cerebral hemispheres connected by the corpus callosum.
   b. Its surface is marked by ridges and grooves; sulci divide each hemisphere into lobes.
   c. The cerebral cortex is a thin layer of gray matter near the surface.
   d. White matter consists of myelinated nerve fibers that interconnect neurons within the nervous system and communicate with other body parts.
3. Functions of the cerebrum
   a. The cerebrum is concerned with higher brain functions, such as thought, reasoning, interpretation of sensory impulses, control of voluntary muscles, and memory storage.
   b. The cerebral cortex has sensory, motor, and association areas.
   c. The primary motor regions lie in the frontal lobes near the central sulcus. Other areas of the frontal lobes control special motor functions.
   d. Areas that interpret sensory impulses from the skin are located in the parietal lobes near the central sulcus; other specialized sensory areas are in the temporal and occipital lobes.
   e. Association areas analyze and interpret sensory impulses and provide memory, reasoning, verbalizing, judgment, and emotions.
   f. One cerebral hemisphere usually dominates for certain intellectual functions.
   g. Short-term memory is probably electrical. Long-term memory is thought to be encoded in patterns of synaptic connections.
4. Basal nuclei
   a. Basal nuclei are masses of gray matter located deep within the cerebral hemispheres.
   b. They relay motor impulses originating in the cerebral cortex and aid in controlling motor activities.
5. Diencephalon
   a. The diencephalon contains the thalamus and hypothalamus.
   b. The thalamus selects incoming sensory impulses and relays them to the cerebral cortex.
   c. The hypothalamus is important in maintaining homeostasis.
   d. The limbic system produces emotional feelings and modifies behavior.
6. Brainstem
   a. The brainstem extends from the base of the brain to the spinal cord.
   b. The brainstem consists of the midbrain, pons, and medulla oblongata.

c. The midbrain contains reflex centers associated with eye and head movements.

d. The pons transmits impulses between the cerebrum and other parts of the nervous system and contains centers that help regulate rate and depth of breathing.

e. The medulla oblongata transmits all ascending and descending impulses and contains several vital and nonvital reflex centers.

f. The reticular formation filters incoming sensory impulses, arousing the cerebral cortex into wakefulness in response to meaningful impulses.

g. Normal sleep results from decreasing activity of the reticular formation, and paradoxical sleep occurs when activating impulses are received by some parts of the brain, but not by others.

7. Cerebellum

a. The cerebellum consists of two hemispheres connected by the vermis.

b. A thin cortex of gray matter surrounds the white matter of the cerebellum.

c. The cerebellum functions primarily as a reflex center, coordinating skeletal muscle movements and maintaining equilibrium.

Peripheral Nervous System (page 395)

The peripheral nervous system consists of cranial and spinal nerves that branch out from the brain and spinal cord to all body parts. It can be subdivided into somatic and autonomic portions.

1. Structure of peripheral nerves

a. A nerve consists of a bundle of nerve fibers surrounded by connective tissues.

b. The connective tissues form an outer epineurium, a perineurium enclosing bundles of nerve fibers, and an endoneurium surrounding each fiber.

2. Nerve fiber classification

a. Nerves are cordlike bundles of nerve fibers. Nerves can be classified as sensory nerves, motor nerves, or mixed nerves, depending on which type of fibers they contain.

b. Nerve fibers within the central nervous system can be subdivided into groups with general and special functions.

3. Cranial nerves

a. Twelve pairs of cranial nerves connect the brain to parts in the head, neck, and trunk.

b. Although most cranial nerves are mixed, some are pure sensory, and others are primarily motor.

c. The names of cranial nerves indicate their primary functions or the general distributions of their fibers.

d. Some cranial nerve fibers are somatic, and others are autonomic.

4. Spinal nerves

a. Thirty-one pairs of spinal nerves originate from the spinal cord.

b. These mixed nerves provide a two-way communication system between the spinal cord and the upper limbs, lower limbs, neck, and trunk.

c. Spinal nerves are grouped according to the levels from which they arise, and they are numbered sequentially.

d. Each nerve emerges by a dorsal and a ventral root.

(1) A dorsal root contains sensory fibers and has a dorsal root ganglion.

(2) A ventral root contains motor fibers.

e. Just beyond its foramen, each spinal nerve divides into several branches.

f. Most spinal nerves combine to form plexuses that direct nerve fibers to a particular body part.

Autonomic Nervous System (page 407)

The autonomic nervous system functions without conscious effort. It is concerned primarily with regulating visceral activities that maintain homeostasis.

1. General characteristics

a. Autonomic functions are reflexes controlled from centers in the hypothalamus, brainstem, and spinal cord.

b. Autonomic nerve fibers are associated with ganglia where impulses are integrated before distribution to effectors.

c. The integrative function of the ganglia provides a degree of independence from the central nervous system.

d. The autonomic nervous system consists of the visceral efferent fibers associated with these ganglia.

e. The autonomic nervous system is subdivided into two divisions—sympathetic and parasympathetic.

f. The sympathetic division prepares the body for stressful and emergency conditions.

g. The parasympathetic division is most active under ordinary conditions.

2. Autonomic nerve fibers

a. The autonomic fibers are efferent, or motor.

b. Sympathetic fibers leave the spinal cord and synapse in ganglia.

(1) Preganglionic fibers pass through white rami to reach paravertebral ganglia.

(2) Paravertebral ganglia and interconnecting fibers comprise the sympathetic trunks.

(3) Preganglionic fibers synapse within paravertebral or collateral ganglia.

(4) Postganglionic fibers usually pass through gray rami to reach spinal nerves before passing to effectors.

(5) A special set of sympathetic preganglionic fibers passes through ganglia and extends to the adrenal medulla.

c. Parasympathetic fibers begin in the brainstem and sacral region of the spinal cord and synapse in ganglia near various organs or in the organs themselves.

3. Autonomic neurotransmitters

a. Sympathetic and parasympathetic preganglionic fibers secrete acetylcholine.

b. Most sympathetic postganglionic fibers secrete norepinephrine and are adrenergic; postganglionic parasympathetic fibers secrete acetylcholine and are cholinergic.

c. The different effects of the autonomic divisions are due to the different neurotransmitters the postganglionic fibers release.

4. Actions of autonomic neurotransmitters

a. Neurotransmitters combine with receptors and alter cell membranes.

b. There are two types of cholinergic receptors and two types of adrenergic receptors.

c. How cells respond to neurotransmitters depends upon the number and type of receptors in their membranes.

d. Acetylcholine acts very briefly; norepinephrine and epinephrine may have more prolonged effects.

5. Control of autonomic activity

a. The central nervous system largely controls the autonomic nervous system.

b. The medulla oblongata uses autonomic fibers to regulate cardiac, vasomotor, and respiratory activities.

c. The hypothalamus uses autonomic fibers in regulating visceral functions.

d. The limbic system and cerebral cortex control emotional responses through the autonomic nervous system.

Life-Span Changes (page 416)

Aging of the nervous system is a gradual elimination of cells and, eventually, slowed functioning.

1. Apoptosis of brain neurons begins before birth.
2. Neuron loss among brain regions is uneven.
3. In adulthood, numbers of dendrites in the cerebral cortex fall, as more generally neurotransmission slows.
4. Nervous system changes in older persons increase the risk of falling.
5. Sleep problems are common in the later years.

CRITICAL THINKING QUESTIONS

1. If a physician plans to obtain a sample of spinal fluid from a patient, what anatomical site can be safely used, and how should the patient be positioned to facilitate this procedure?
2. What functional losses would you expect to observe in a patient who has suffered injury to the right occipital lobe of the cerebral cortex? To the right temporal lobe?
3. The Brown-Séguard syndrome is due to an injury on one side of the spinal cord. It is characterized by paralysis below the injury and on the same side as the injury, and by loss of sensations of temperature and pain on the opposite side. How would you explain these symptoms?
4. The biceps-jerk reflex employs motor neurons that exit from the spinal cord in the 5th spinal nerve (C5), that is, fifth from the top of the cord. The triceps-jerk reflex involves motor neurons in the 7th spinal nerve (C7). How might these reflexes be used to help locate the site of damage in a patient with a neck injury?
5. Substances used by intravenous drug abusers are sometimes obtained in tablet form and are crushed and dissolved before they are injected. Such tablets may contain fillers, such as talc or cornstarch, that may obstruct tiny blood vessels in the cerebrum. What problems might these obstructions create?
6. In planning treatment for a patient who has had a cerebrovascular accident (CVA), why would it be important to know whether the CVA was caused by a ruptured or obstructed blood vessel?
7. What symptoms might the sympathetic division of the autonomic nervous system produce in a patient who is experiencing stress?
8. How would you distinguish between a patient in a coma and one in a persistent vegetative state?

REVIEW EXERCISES

1. Name the layers of the meninges, and explain their functions.
2. Describe the location of cerebrospinal fluid within the meninges.
3. Describe the location of the ventricles of the brain.
4. Explain how cerebrospinal fluid is produced and how it functions.
5. Describe the structure of the spinal cord.
6. Describe a reflex arc.
7. Define *reflex*.
8. Describe a withdrawal reflex.
9. Name the major ascending and descending tracts of the spinal cord, and list the functions of each.
10. Explain the consequences of nerve fibers crossing over.
11. Describe how the brain develops.
12. Describe the structure of the cerebrum.
13. Define *cerebral cortex*.
14. Describe the location and function of the primary motor areas of the cortex.
15. Describe the location and function of Broca's area.
16. Describe the location and function of the sensory areas of the cortex.
17. Explain the function of the association areas of the lobes of the cerebrum.
18. Define *hemisphere dominance*.
19. Explain the function of the corpus callosum.
20. Distinguish between short-term and long-term memory.
21. Describe the location and function of the basal nuclei.
22. Name the parts of the diencephalon, and describe the general functions of each.
23. Define the limbic system, and explain its functions.
24. Name the parts of the midbrain, and describe the general functions of each.
25. Describe the pons and its functions.
26. Describe the medulla oblongata and its functions.
27. Describe the location and function of the reticular formation.
28. Distinguish between normal and paradoxical sleep.
29. Describe the functions of the cerebellum.
30. Distinguish between the somatic and autonomic nervous systems.
31. Describe the structure of a peripheral nerve.
32. Distinguish among sensory, motor, and mixed nerves.
33. List four general types of nerve fibers.
34. Name, locate, and describe the major functions of each pair of cranial nerves.
35. Explain how the spinal nerves are grouped and numbered.
36. Define *cauda equina*.
37. Describe the structure of a spinal nerve.
38. Define *plexus*, and locate the major plexuses of the spinal nerves.
39. Distinguish between the sympathetic and parasympathetic divisions of the autonomic nervous system.
40. Explain how autonomic ganglia provide a degree of independence from the central nervous system.
41. Distinguish between a preganglionic fiber and a postganglionic fiber.
42. Define *paravertebral ganglion*.
43. Trace a sympathetic nerve pathway through a ganglion to an effector.

44. Explain why the effects of the sympathetic and parasympathetic autonomic divisions differ.

45. Distinguish between cholinergic and adrenergic nerve fibers.

46. Define *sympathetic tone.*

47. Explain how autonomic neurotransmitters influence the actions of effector cells.

48. Distinguish between alpha adrenergic and beta adrenergic receptors.

49. Describe three examples in which the central nervous system employs autonomic nerve pathways.

W E B   C O N N E C T I O N S

Visit the Student Edition of the Online Learning Center at www.mhhe.com/shier10 **for answers to chapter questions, additional quizzes, interactive learning exercises, and other study tools.**

UNDERSTANDING WORDS

aud-, to hear: *aud*itory—pertaining to hearing.

choroid, skinlike: *choroid* coat—middle, vascular layer of the eye.

cochlea, snail: *cochlea*—coiled tube within the inner ear.

corn-, horn: *corn*ea—transparent outer layer in the anterior portion of the eye.

iris, rainbow: *iris*—colored, muscular part of the eye.

labyrinth, maze: *labyrinth*—complex system of connecting chambers and tubes of the inner ear.

lacri-, tears: *lacri*mal gland—tear gland.

lut-, yellow: macula *lut*ea—yellowish spot on the retina.

macula, spot: *macula* lutea—yellowish spot on the retina.

malle-, hammer: *malle*us—one of the three bones in the middle ear.

oculi-, eye: orbicularis *oculi*—muscle associated with the eyelid.

olfact-, to smell: *olfact*ory—pertaining to the sense of smell.

palpebra, eyelid: levetor *palpebra*e superioris—muscle associated with the eyelid.

photo-, light: *photo*receptor—specialized structures in the eye responsive to light.

scler-, hard: *scler*a—tough, outer protective layer of the eye.

therm-, heat: *therm*oreceptor—receptor sensitive to changes in temperature.

tympan-, drum: *tympan*ic membrane—eardrum.

vitre-, glass: *vitre*ous humor—clear, jellylike substance within the eye.

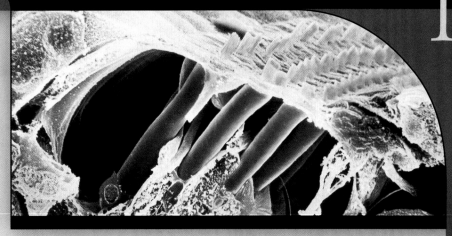

The organ of Corti, in the inner ear, has rows of hair cells. Each hair cell bears up to 100 hairs, which capture and transduce mechanical energy from sound into neural messages that travel to the brain (1,700×).

Somatic and Special Senses

CHAPTER OBJECTIVES

After you have studied this chapter, you should be able to

1. Name five kinds of receptors and explain the function of each.

2. Explain how receptors stimulate sensory impulses.

3. Explain how a sensation is produced.

4. Distinguish between somatic and special senses.

5. Describe the receptors associated with the senses of touch and pressure, temperature, and pain.

6. Describe how the sense of pain is produced.

7. Explain the importance of stretch receptors in muscles and tendons.

8. Explain the relationship between the senses of smell and taste.

9. Name the parts of the ear and explain the function of each part.

10. Distinguish between static and dynamic equilibrium.

11. Name the parts of the eye and explain the function of each part.

12. Explain how the eye refracts light.

13. Explain how the brain perceives depth and distance.

14. Describe the visual nerve pathway.

J ohn Dalton, a famous English chemist, saw things differently than most people. In a 1794 lecture, he described his visual world. Sealing wax that appeared red to other people was as green as a leaf to Dalton and his brother. Pink wildflowers were blue, and Dalton perceived the cranesbill plant as "sky blue" in daylight, but "very near yellow, but with a tincture of red" in candlelight. He concluded, " . . . that part of the image which others call red, appears to me little more than a shade, or defect of light." The Dalton brothers, like 7% of males and 0.4% of females today, had the inherited trait of color blindness.

Dalton was very curious about the cause of his color blindness, so he made arrangements with his personal physician, Joseph Ransome, to dissect his eyes after he died. Ransome snipped off the back of one eye, removing the retina, where the cone cells that provide color vision are nestled among the more abundant rod cells that impart black-and-white vision. Because Ransome could see red and green normally when he peered through the back of his friend's eyeball, he concluded that it was not an abnormal filter in front of the eye that altered color vision.

Fortunately, Ransome stored the eyes in dry air, where they remained relatively undamaged. In 1994, Dalton's eyes underwent DNA analysis at London's Institute of Ophthalmology. The research showed that Dalton's remaining retina lacked one of three types of pigments, called photopigments, that enable cone cells to capture certain incoming wavelengths of light.

Although people have studied color blindness for centuries, we are still learning more about it. Recently, researchers investigated why color blind men lacking cones that capture green light are affected to different degrees. They discovered that color blind men who can discern a few shades of green have red cone cells that can detect some wavelengths of light that fall within the green region of the spectrum. Color vision may be more complex than we had thought. ■

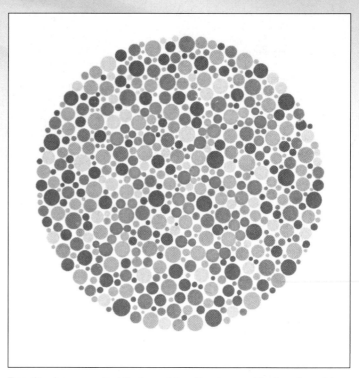

People who are color blind must function in a multicolored world. To help them overcome the disadvantage of not seeing important color differences, researchers have developed computer algorithms that convert colored video pictures into shades they can see. This circle of dots is a test to determine whether someone is color blind. Affected individuals cannot see a different color in certain of the circles in such a drawing. As a result, their brains cannot perceive the embedded pattern that forms the number 16 that others can see.

The above has been reproduced from *Ishihara's Tests for Colour Blindness published by Kanehara & Co. Ltd. Tokyo, Japan, but tests for colour blindness cannot be conducted with this material. For accurate testing, the original plates should be used.*

Our senses make our lives meaningful, connecting us to the sights and sounds, smells and textures of the world. Sensory receptors are the portals that link our nervous systems to the outside environment. Some senses are considered so distinctive that they are termed special, their receptors part of large sensory organs in the head.

All senses work in basically the same way. Sensory receptors are specialized cells or multicellular structures that collect information from the environment and stimulate neurons to send impulses along sensory fibers to the brain. There the cerebral cortex forms a perception, a person's particular view of the stimulus. Table 12.1 outlines the pathways from sensation to perception that describe an apple.

Recall from chapter 11 (p. 372) that the terms "axon" and nerve fiber are used synonymously. Also recall that unipolar neurons, which include most sensory neurons, have an unusual structure in which the portion of the neuron associated with the dendrites, called a peripheral process, is considered to behave like an axon. Because of this, and for simplicity, the neuron processes which bring sensory information into the CNS will be called sensory fibers or afferent fibers, no matter what type of neuron is involved.

Receptors and Sensations

Sensory receptors are diverse but share certain features. Each type of receptor is particularly sensitive to a distinct kind of environmental change and is much less sensitive to other forms of stimulation.

Information Flow	Smell	Taste	Sight	Hearing
Sensory receptors	Olfactory cells in nose	Taste bud receptor cells	Rods and cones in retina	Hair cells in cochlea
↓	↓	↓	↓	↓
Impulse in sensory fibers	Olfactory nerve fibers	Sensory fibers in various cranial nerves	Optic nerve fibers	Auditory nerve fibers
↓	↓	↓	↓	↓
Impulse reaches CNS	Cerebral cortex	Cerebral cortex	Midbrain and cerebral cortex	Midbrain and cerebral cortex
↓	↓	↓	↓	↓
Sensation (new experience, recalled memory)	A pleasant smell	A sweet taste	A small, round, red object	A crunching sound
↓	↓	↓	↓	↓
Perception	The smell of an apple	The taste of an apple	The sight of an apple	Biting into an apple

Receptor Types

Five general groups of sensory receptors are known, based on their sensitivities to changes in one of the following factors:

1. *Chemical concentration.* Receptors that respond to changes in the concentration of chemical substances are called **chemoreceptors** (ke″mo-re-sep′torz). Receptors associated with the senses of smell and taste are of this type. Chemoreceptors in internal organs detect changes in the blood concentrations of substances, including oxygen, hydrogen ions, and glucose.

2. *Tissue damage.* Tissue damage stimulates **pain receptors** (nociceptors). Triggering factors include exposure to excess mechanical, electrical, thermal, or chemical energy.

3. *Temperature change.* Receptors sensitive to temperature change are called **thermoreceptors** (ther″mo-re-sep′torz).

4. *Mechanical forces.* A number of sensory receptors sense mechanical forces, such as changes in the pressure or movement of fluids. As a rule, these **mechanoreceptors** (mek″ah-no re-sep′torz) detect changes that deform the receptors. For example, **proprioceptors** (pro″pre-o-sep′torz) sense changes in the tensions of muscles and tendons, whereas **baroreceptors** (pressoreceptors) in certain blood vessels can detect changes in blood pressure. Similarly, **stretch receptors** in the lungs sense degree of inflation.

5. *Light intensity.* Light receptors, or **photoreceptors** (fo″to-re-sep′torz), are only in the eyes. They respond to light energy of sufficient intensity.

Sensory Impulses

Sensory receptors can be ends of neurons or other kinds of cells located close to them. In either case, stimulation causes local changes in their membrane potentials (receptor potentials), generating a graded electric current that reflects the intensity of stimulation (see chapter 10, p. 350).

If a receptor is a neuron and the change in membrane potential reaches threshold, an action potential is generated, and a sensory impulse results on the afferent fiber. However, if the receptor is another type of cell, its receptor potential must be transferred to a neuron to trigger an action potential. Peripheral nerves transmit sensory impulses to the central nervous system, where they are analyzed and interpreted within the brain.

Sensations

A **sensation** occurs when the brain becomes aware of sensory impulses. A perception occurs when the brain interprets those sensory impulses. Because all the nerve impulses that travel away from sensory receptors into the central nervous system are alike, the resulting sensation depends on which region of the cerebral cortex receives the impulse. For example, impulses reaching one region are always interpreted as sounds, and those reaching another portion are always sensed as touch.

It makes little difference how receptors are stimulated. Pain receptors, for example, can be stimulated by heat, cold, or pressure, but the sensation is always the same because, in each case, the same part of the brain interprets the resulting nerve impulses as pain. Similarly, factors other than light, such as a sharp blow to the head, may trigger nerve impulses in visual receptors. When this happens, the person may "see" lights, even though no light is entering the eye, since any impulses reaching the visual cortex are interpreted as light. Normally receptors only respond to specific stimuli, so the brain creates the correct sensation for that particular stimulus.

At the same time that a sensation forms, the cerebral cortex interprets it to seem to come from the receptors being stimulated. This process is called **projection** because the brain projects the sensation back to its apparent source. Projection allows a person to pinpoint the region of stimulation. Thus, we are aware that the eyes see an apple, the nose smells it, and the ears hear the teeth crunch into it.

Sensory Adaptation

When sensory receptors are continuously stimulated, many of them undergo an adjustment called **sensory adaptation**. Somehow the receptor membrane becomes less responsive to the stimulus, and the resulting receptor potential generated is less likely to bring the neuron to threshold. As the receptors adapt, impulses leave them at decreasing rates, until finally these receptors may completely fail to send signals. Once receptors adapt, impulses are triggered only if the strength of the stimulus changes.

Imagine a person walking into a fish market. At first, the fishy odor seems intense, but it becomes less and less noticeable as the smell (olfactory) receptors adapt. Soon, the once offensive odor is barely noticeable, but if the person leaves and reenters, he or she again smells the stench.

1   List the five general types of sensory receptors.

2   What do all types of receptors have in common?

3   Explain how a sensation occurs.

4   What is sensory adaptation?

Somatic Senses

Somatic senses are those whose sensory receptors are associated with the skin, muscles, joints, and viscera. These senses can be divided into three groups:

1. Senses associated with changes at the body surface (exteroceptive senses), which include the senses of touch, pressure, temperature, and pain.

2. Senses associated with changes in muscles and tendons and in body position (proprioceptive senses).

3. Senses associated with changes in viscera (visceroceptive senses). Except for visceral pain, these senses are discussed in subsequent chapters.

Touch and Pressure Senses

The senses of touch and pressure derive from three kinds of receptors (fig. 12.1). As a group, these receptors sense mechanical forces that deform or displace tissues. The touch and pressure receptors include the following:

1. **Free nerve endings.** These receptors are common in epithelial tissues, where they lie between epithelial cells. They are associated with sensations of touch and pressure (fig. 12.1a).

2. **Meissner's corpuscles.** These are small, oval masses of flattened connective tissue cells in connective tissue sheaths. Two or more sensory nerve fibers branch into each corpuscle and end within it as tiny knobs.

   Meissner's corpuscles are abundant in the hairless portions of the skin, such as the lips, fingertips, palms, soles, nipples, and external genital organs. They sense motion of objects that barely contact the skin, interpreting impulses from them as the sensation of light touch. They are also used when a person touches something to judge its texture (fig. 12.1b).

3. **Pacinian corpuscles.** These sensory bodies are relatively large, ellipsoidal structures composed of connective tissue fibers and cells. They are common in the deeper subcutaneous tissues of the hands, feet, penis, clitoris, urethra, and breasts and also in tendons of muscles and ligaments of joints (fig. 12.1c). Heavy pressure stimulates Pacinian corpuscles. They may also detect vibrations in tissues.

Temperature Senses

Temperature receptors (thermoreceptors) include two groups of free nerve endings located in the skin. Those that respond to warmer temperatures are called *warm receptors,* and those that respond to colder temperatures are called *cold receptors.* The warm receptors are most sensitive to temperatures above 25°C (77°F) and become unresponsive at temperatures above 45°C (113°F). As 45°C is approached, pain receptors are also triggered, producing a burning sensation.

Cold receptors are most sensitive to temperatures between 10°C (50°F) and 20°C (68°F). If the temperature drops below 10°C, pain receptors are stimulated, and the person feels a freezing sensation.

At intermediate temperatures, the brain interprets sensory input from different combinations of these receptors as a particular temperature sensation. Both warm and cold receptors rapidly adapt, so within about a minute of continuous stimulation, the sensation of warm or cold begins to fade. This is why we quickly become comfortable after jumping into a cold swimming pool or submerging into a steaming hot tub.

Sense of Pain

Receptors that consist of free nerve endings sense pain. These receptors are widely distributed throughout the skin and internal tissues, except in the nervous tissue of the brain, which lacks pain receptors.

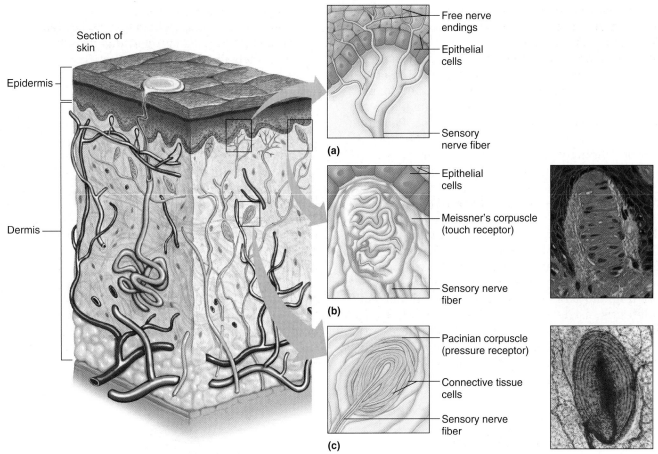

Section of skin

Epidermis

Dermis

- Free nerve endings
- Epithelial cells
- Sensory nerve fiber

(a)

- Epithelial cells
- Meissner's corpuscle (touch receptor)
- Sensory nerve fiber

(b)

- Pacinian corpuscle (pressure receptor)
- Connective tissue cells
- Sensory nerve fiber

(c)

FIGURE 12.1

Touch and pressure receptors include (*a*) free ends of sensory nerve fibers, (*b*) Meissner's corpuscle (with 225× micrograph), and (*c*) Pacinian corpuscle (with 50× micrograph).

Pain Receptors

The pain receptors are protective in that they are stimulated when tissues are damaged. Pain sensation is usually perceived as unpleasant, signaling that action be taken to remove the source of the stimulation.

Most pain receptors can be stimulated by more than one type of change. However, some pain receptors are most sensitive to mechanical damage. Others are particularly sensitive to extremes in temperature. Some pain receptors are most responsive to chemicals, such as hydrogen ions, potassium ions, or specific breakdown products of proteins, histamine, and acetylcholine. A deficiency of blood (ischemia) and thus a deficiency of oxygen (hypoxia) in a tissue, or stimulation of mechanical-sensitive receptors, also triggers pain sensation. For example, pain elicited during a muscle cramp results from interruption of blood flow that occurs as the sustained contraction squeezes capillaries, as well as from the stimulation of mechanical-sensitive pain receptors. Also, when blood flow is interrupted, pain-stimulating chemicals accumulate. Increasing blood flow through the sore tissue may relieve the resulting pain, and this is why heat is sometimes applied to reduce muscle soreness. The

heat dilates blood vessels and thus promotes blood flow, which helps reduce the concentration of the pain-stimulating substances. In some conditions, accumulating chemicals lower the thresholds of pain receptors, making inflamed tissues more sensitive to heat or pressure than before.

Pain receptors adapt very little, if at all. Once such a receptor is activated, even by a single stimulus, it may continue to send impulses into the central nervous system for some time.

Visceral Pain

As a rule, pain receptors are the only receptors in viscera whose stimulation produces sensations. Pain receptors in these organs respond differently to stimulation than those associated with surface tissues. For example, localized damage to intestinal tissue during surgical procedures may not elicit any pain sensations, even in a conscious person. However, when visceral tissues are subjected to more widespread stimulation, as when intestinal tissues are stretched or when the smooth muscles in the intestinal walls undergo spasms, a strong pain sensation may follow. Once again, the resulting pain results from stimulation of mechanoreceptors and from decreased blood flow accompanied by lower tissue oxygen levels and accumulation of pain-stimulating chemicals.

Visceral pain may feel as if it is coming from some part of the body other than the part being stimulated—a phenomenon called **referred pain**. For example, pain originating in the heart may be referred to the left shoulder or the medial surface of the left upper limb. Pain from the lower esophagus, stomach, or small intestine may seem to be coming from the upper central (epigastric) region of the abdomen. Pain from the urogenital tract may be referred to the lower central (hypogastric) region of the abdomen or to the sides between the ribs and the hip (fig. 12.2).

Referred pain may derive from *common nerve pathways* that sensory impulses coming both from skin areas and from internal organs use. Pain impulses from the heart seem to be conducted over the same nerve pathways as those from the skin of the left shoulder and the inside of the left arm, as shown in figure 12.3. During a heart attack, the cerebral cortex may incorrectly interpret the source of the impulses as the shoulder and the medial surface of the left upper limb, rather than the heart.

Pain originating in the parietal layers of thoracic and abdominal membranes—parietal pleura, parietal pericardium, or parietal peritoneum—is usually not referred; instead, such pain is felt directly over the area being stimulated.

Pain Nerve Pathways

The nerve fibers that conduct impulses away from pain receptors are of two main types: acute pain fibers and chronic pain fibers.

The *acute pain fibers* (also known as A-delta fibers) are thin, myelinated nerve fibers. They conduct nerve impulses rapidly, at velocities up to 30 meters per second. These impulses are associated with the sensation of sharp pain, which typically seems to originate in a local area of skin. This type of pain seldom continues after the pain-producing stimulus stops.

The *chronic pain fibers* (C fibers) are thin, unmyelinated nerve fibers. They conduct impulses more slowly than acute pain fibers, at velocities up to 2 meters per second. These impulses cause the dull, aching pain sensation that may be widespread and difficult to pinpoint. Such pain may continue for some time after the original

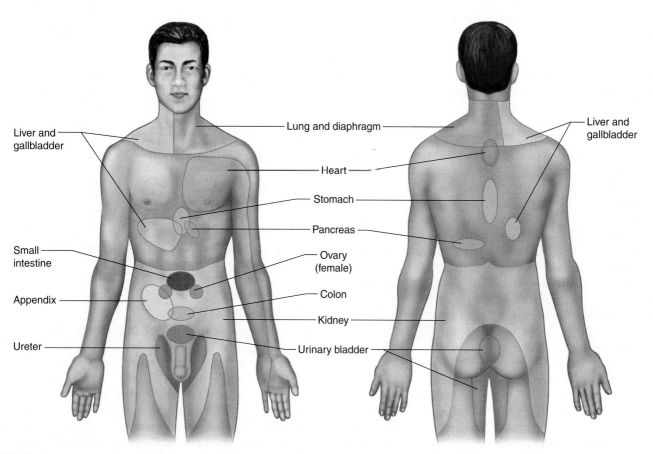

FIGURE 12.2
Surface regions to which visceral pain may be referred.

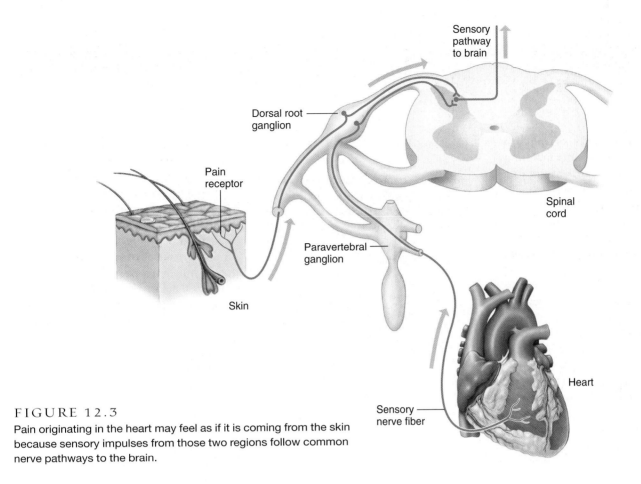

FIGURE 12.3

Pain originating in the heart may feel as if it is coming from the skin because sensory impulses from those two regions follow common nerve pathways to the brain.

Labels in figure:
Sensory pathway to brain
Dorsal root ganglion
Pain receptor
Spinal cord
Paravertebral ganglion
Skin
Heart
Sensory nerve fiber

stimulus ceases. Although acute pain is usually sensed as coming from the surface, chronic pain is likely to be felt in deeper tissues as well as in the skin. Visceral pain impulses are usually carried on C fibers.

Commonly, an event that stimulates pain receptors will trigger impulses on both types of pain fibers. This causes a dual sensation—a sharp, pricking pain, then a dull, aching one. The aching pain is usually more intense and may worsen over time. Chronic pain that resists relief and control can be debilitating.

Pain impulses that originate from tissues of the head reach the brain on sensory fibers of the fifth, seventh, ninth, and tenth cranial nerves. All other pain impulses travel on sensory fibers of spinal nerves, and they pass into the spinal cord by way of the dorsal roots of these spinal nerves.

Upon reaching the spinal cord, pain impulses enter the gray matter of the posterior horn, where they are processed (see chapter 11, p. 372). The fast-conducting fibers synapse with long nerve fibers that cross over to the opposite side of the spinal cord and ascend in the lateral spinothalamic tracts. The impulses carried on the slow-conducting fibers pass through one or more interneurons before reaching the long fibers that cross over and ascend to the brain.

Within the brain, most of the pain fibers terminate in the reticular formation (see chapter 11, p. 393), and from there are conducted on fibers of still other neurons to the thalamus, hypothalamus, and cerebral cortex.

Regulation of Pain Impulses

Awareness of pain occurs when pain impulses reach the level of the thalamus—that is, even before they reach the cerebral cortex. However, the cerebral cortex must judge the intensity of pain and locate its source. The cerebral cortex is also responsible for emotional and motor responses to pain.

∞ RECONNECT TO CHAPTER 10, POSTSYNAPTIC POTENTIALS, PAGE 355

Still other parts of the brain, including areas of gray matter in the midbrain, pons, and medulla oblongata, regulate the flow of pain impulses from the spinal cord (see chapter 11, pp. 392–393). Impulses from special neurons in these areas descend in the lateral funiculus to various levels within the spinal cord. The impulses stimulate the ends of certain nerve fibers to release biochemicals that can block pain signals by inhibiting presynaptic nerve fibers in the posterior horn of the spinal cord.

CANCER PAIN AND CHRONIC PAIN

One of the shortcomings of modern medicine is that many patients suffering from the pain of advanced, untreatable cancer do not receive adequate medication to ease their final days. Studies show that 50% of people with cancer are in pain at the time of their diagnosis, and 90% of those in advanced stages are in pain.

Doctors are beginning to advocate pain relief for all people with cancer who require it, not only for humanitarian reasons but for clinical ones. Studies on rats with lung cancer show that tumors spread more rapidly in animals not given pain medication. Plus, the widespread belief that giving narcotics would addict cancer patients has been shown to be unfounded. Narcotics are much more likely to be addicting when they are taken to provide euphoria than when they are taken to relieve severe pain.

Drugs used to help cancer patients include

- Anti-inflammatory agents such as aspirin and ibuprofen.
- Weak narcotics such as hydrocodone.
- Strong narcotics such as morphine.
- Opiates delivered directly to the spine via an implanted reservoir.

Many patients are offered the option of patient controlled analgesia, in which they determine their own dosage schedule.

Chronic pain is of three types: lower back pain, migraine, and myofascial syndrome (inflammation of muscles and their fascia). A variety of approaches treat chronic pain:

- Biofeedback. A mechanical device detects and amplifies an autonomic body function, such as blood pressure or heart rate. A person made aware of the measurement can concentrate to attempt to alter it to within a normal range.
- Anti-inflammatory drugs.
- Stretching exercises.
- Trigger point injections of local anesthetic drugs into cramping muscles.
- Some antidepressant drugs raise serotonin levels in the central nervous system, relieving some chronic pain.
- Transcutaneous electrical nerve stimulation (TENS) places electrodes on nerves causing pain. The patient feels a tingling sensation, then pain relief.
- A nerve block is invasive and interrupts a pain signal by freezing (cryotherapy) or introducing an anesthetic drug (neurolysis). Side effects include numbness or paralysis.
- A dorsal column stimulator consists of electrodes implanted near the spinal cord. ∎

Among the inhibiting substances released in the posterior horn are neuropeptides called *enkephalins* and the monoamine *serotonin* (see chapter 10, p. 358). Enkephalins can suppress both acute and chronic pain impulses; thus, they can relieve strong pain sensations, much as morphine and other opiate drugs do. In fact, enkephalins were discovered because they bind to the same receptor sites on neuron membranes as does morphine. Serotonin stimulates other neurons to release enkephalins.

Cannabinoids are substances in the plant *Cannibus sativa,* the source of marijuana, that may relieve pain. Anecdotal evidence for such an effect dates to A.D. 315. Neurons in areas of the brain, brainstem, and peripheral nervous system have receptors for cannabinoids. In some states, marijuana is legal for use to improve appetite in people who have AIDS or cancer.

Another group of neuropeptides with pain-suppressing, morphinelike actions are the *endorphins.* They are found in the pituitary gland and in regions of the nervous system, such as the hypothalamus, that transmit pain impulses. Enkephalins and endorphins are released in response to extreme pain impulses, providing natural pain control. Clinical Application 12.1 discusses treatments for severe pain.

1 Describe three types of touch and pressure receptors.

2 Describe thermoreceptors.

3 What types of stimuli excite pain receptors?

4 What is referred pain?

5 Explain how neuropeptides control pain.

Stretch Receptors

Stretch receptors are proprioceptors that send information to the spinal cord and brain concerning the lengths and tensions of muscles. The two main kinds of stretch receptors are muscle spindles and Golgi tendon organs; however, no sensation results when they are stimulated.

Muscle spindles are located in skeletal muscles near their junctions with tendons. Each spindle consists of one or more small, modified skeletal muscle fibers (intrafusal fibers) enclosed in a connective tissue sheath. Near its center, each intrafusal fiber has a specialized nonstriated region with the end of a sensory nerve fiber wrapped around it (fig. 12.4a).

The striated portions of the intrafusal fiber contract to keep the spindle taut at different muscle lengths. Thus, if the whole muscle is stretched, the muscle spindle is also stretched, triggering sensory nerve impulses on its nerve fiber. Such sensory impulses travel into the spinal cord and onto motor fibers leading back to the same muscle, contracting it. This action, called a **stretch reflex,** opposes the lengthening of the muscle and helps maintain the desired position of a limb in spite of gravitational or other forces tending to move it (see chapter 11, p. 373).

Golgi tendon organs are found in tendons close to their attachments to muscles. Each is connected to a set of skeletal muscle fibers and is innervated by a sensory neuron (fig. 12.4b). These receptors have high thresholds and are stimulated by increased tension. Sensory impulses from them produce a reflex that inhibits contraction of the muscle whose tendon they occupy. Thus, the Golgi tendon organs stimulate a reflex with an effect opposite that of a stretch reflex. The Golgi tendon reflex also helps maintain posture, and it protects muscle attachments from being pulled away from their insertions by excessive tension. Table 12.2 summarizes the somatic receptors and their functions.

1. Describe a muscle spindle.

2. Explain how muscle spindles help maintain posture.

3. Where are Golgi tendon organs located?

4. What is the function of Golgi tendon organs?

TABLE 12.2	Somatic Receptors	
Type	**Function**	**Sensation**
Free nerve endings (mechanoreceptors)	Detect changes in pressure	Touch, pressure
Meissner's corpuscles (mechanoreceptors)	Detect objects moving over the skin	Touch, texture
Pacinian corpuscles (mechanoreceptors)	Detect changes in pressure	Deep pressure, vibrations
Free nerve endings (thermoreceptors)	Detect changes in temperature	Heat, cold
Free nerve endings (pain receptors)	Detect tissue damage	Pain
Muscle spindles (mechanoreceptors)	Detect changes in muscle length	None
Golgi tendon organs (mechanoreceptors)	Detect changes in muscle tension	None

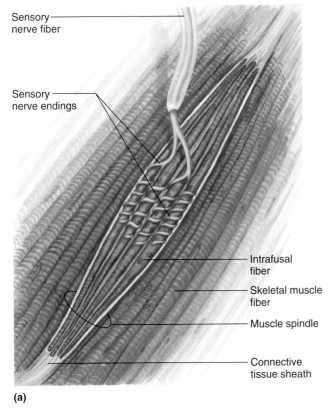

Sensory nerve fiber

Sensory nerve endings

Intrafusal fiber

Skeletal muscle fiber

Muscle spindle

Connective tissue sheath

(a)

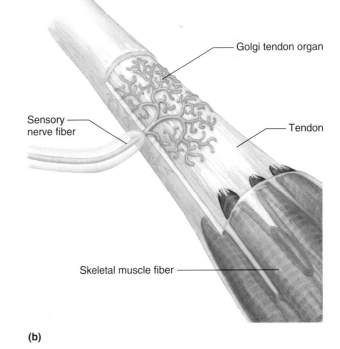

Golgi tendon organ

Sensory nerve fiber

Tendon

Skeletal muscle fiber

(b)

FIGURE 12.4

Stretch receptors maintain posture. (a) Increased muscle length stimulates muscle spindles, which stimulate muscle contraction. (b) Golgi tendon organs occupy tendons, where they inhibit muscle contraction.

"The song was full of glittering orange
diamonds."

"The paint smelled blue."

"The sunset was salty."

"The pickle tasted like a rectangle."

One in 2,000 people has a condition called synesthesia, in which the brain interprets a stimulus to one sense as coming from another. Most commonly, letters, numbers, or periods of time evoke specific colors. These associations are involuntary, are very specific, and persist over a lifetime. For example, a person might report

that 3 is always mustard yellow, or Thursday brown.

Synesthesia seems to be inherited and is more common in women. One of the authors of this book (R. L.) has it—to her days of the week and months are specific colors. Synesthesia has been attributed to an immature nervous system that cannot sort out sensory stimuli or to altered brain circuitry that routes stimuli to the wrong part of the cerebral cortex. It may be an inability that is common in infants, but tends to be lost in most people.

PET (positron emission tomography) scanning reveals a physical basis to synes-

thesia. Brain scans of six nonsynesthetes were compared with those of six synesthetes who reported associating words with colors. Cortical blood flow was monitored while a list of words was read aloud to both groups. Interestingly, cortical blood flow was greatly elevated in the synesthetes compared with the nonsynesthetes. Furthermore, while blood flow was increased in word-processing areas for both groups, the scans revealed that areas important in vision and color processing were also lit up in those with synesthesia. ■

Special Senses

Special senses are those whose sensory receptors are within large, complex sensory organs in the head. These senses and their respective organs include the following:

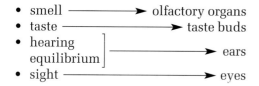

- smell ——————→ olfactory organs
- taste ——————→ taste buds
- hearing
  equilibrium]——————→ ears
- sight ——————→ eyes

Sense of Smell

The ability to detect the strong scent of a fish market, the antiseptic odor of a hospital, the aroma of a ripe melon—and thousands of other smells—is possible thanks to a yellowish patch of tissue the size of a quarter high up in the nasal cavity. This fabric of sensation is actually a layer of 12 million specialized cells.

Olfactory Receptors

Olfactory (ol-fak'to-re) *receptors,* used to sense smells, are similar to those for taste in that they are chemoreceptors sensitive to chemicals dissolved in liquids. These two chemical senses function closely together and aid in food selection, because we smell food at the same time we taste it. In fact, it is often difficult to tell what part of a food sensation is due to smell and what part is due to taste. For this reason, an onion tastes quite different when sampled with the nostrils closed, because much of the usual onion sensation is due to odor. Similarly, if excessive mucous

secretions from an upper respiratory infection (such as a cold) cover the olfactory receptors, food may seem to lose its taste. About 75 to 80% of flavor actually derives from the sense of smell. Clinical Application 12.2 discusses an unusual type of sensory abnormality.

Olfactory Organs

The **olfactory organs,** which contain the olfactory receptors, also include epithelial supporting cells. These organs appear as yellowish brown masses surrounded by pinkish mucous membrane. They cover the upper parts of the nasal cavity, the superior nasal conchae, and a portion of the nasal septum (fig. 12.5).

The **olfactory receptor cells** are bipolar neurons surrounded by columnar epithelial cells. These neurons have knobs at the distal ends of their dendrites covered with hairlike cilia. The cilia project into the nasal cavity and are the sensitive portions of the receptors (fig. 12.6). A person's 12 million olfactory receptor cells each have ten to twenty cilia.

Chemicals that stimulate olfactory receptors, called odorant molecules, enter the nasal cavity as gases, where they must dissolve at least partially in the watery fluids that surround the cilia before they can be detected by bonding to receptor proteins on the cilia. Odorant molecules bind to about 500 different types of olfactory receptors that are part of the cell membranes of the olfactory receptor cells, depolarizing them and thereby generating nerve impulses. In addition, signaling proteins inside the receptor cell translate the chemical signal (binding of the odorant molecule to the receptor protein) into the electrochemical language of the nervous system.

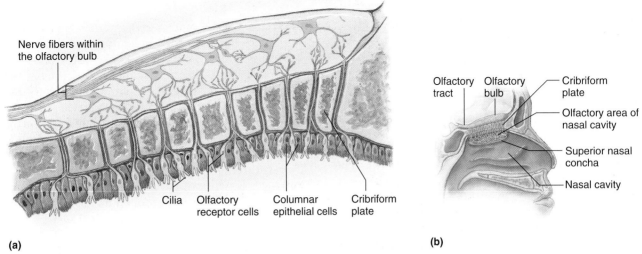

(a)

(b)

FIGURE 12.5

Olfactory receptors. (*a*) Columnar epithelial cells support olfactory receptor cells, which have cilia at their distal ends. (*b*) The olfactory area is associated with the superior nasal concha.

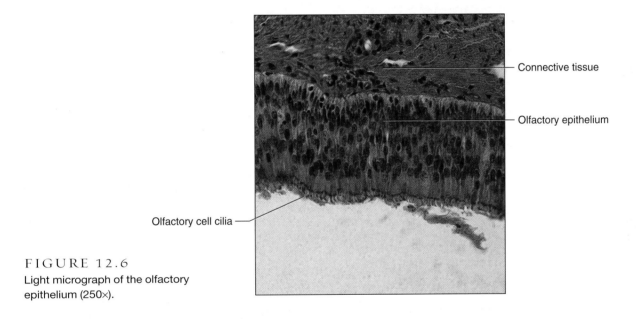

FIGURE 12.6

Light micrograph of the olfactory epithelium (250×).

The expert nose of the bloodhound is due to its 4 billion olfactory cells. A bloodhound's olfactory epithelium spread out covers 59 square inches, compared to 1 1/2 square inches in a human. Still, the human sense of smell is nothing to sneeze at—people can detect one molecule of green pepper smell in a gaseous sea of 3 trillion other molecules. Our 12 million olfactory receptor cells and their 500 receptor types allow us to discern some 10,000 scents. But, without air, we cannot smell, as early astronauts could attest; eating through tubes, early astronauts could not smell their food. Currently, astronauts eat in pressurized cabins, unaffected by the vacuum surrounding their spacecraft.

Sensory receptors are not the same as membrane receptors. Sensory receptors may be as small as individual cells or as large as complex organs such as the eye or ear. They respond to sensory stimuli. Membrane receptors are molecules such as proteins and glycoproteins located on the cell membranes. They allow cells, such as neurons and olfactory receptor cells, to respond to specific molecules. Thus, the olfactory receptors are cells that respond to chemical stimuli, but they depend on cell membrane receptors to do so.

Olfactory Nerve Pathways

Once olfactory receptors are stimulated, nerve impulses travel along their axons through tiny openings in the cribriform plates of the ethmoid bone. These fibers (which form the first cranial nerves) synapse with neurons located in the enlargements of the **olfactory bulbs,** structures that lie on either side of the crista galli of the ethmoid bone (see figs. 7.26 and 12.5).

Within the olfactory bulbs, the sensory impulses are analyzed, and as a result, additional impulses travel along the **olfactory tracts** to portions of the limbic system (see chapter 11, p. 392), a brain center for memory and emotions. This is why we may become nostalgic over a scent from the past. A whiff of the perfume that grandma used to wear may bring back a flood of memories. The input to the limbic system also explains why odors can alter mood so easily. For example, the scent of new-mown hay or rain on a summer's morning generally makes us feel good. The main interpreting areas for the olfactory impulses (olfactory cortex) are located deep within the temporal lobes and at the bases of the frontal lobes, anterior to the hypothalamus.

Olfactory Stimulation

Biologists are not certain how stimulated receptors encode specific smells, but a leading hypothesis is that each odor likely stimulates a distinct set of receptor cells that in turn have distinct sets of receptor proteins. The brain then recognizes the particular combination as an *olfactory code.* For example, perhaps there are ten types of odor receptors. Banana might stimulate receptors 2, 4, and 7; garlic, receptors 1, 5, and 9. Some investigators have proposed seven primary odors, but others hypothesize that the number is much higher and may reflect the functioning of hundreds of genes.

Because the olfactory organs are high in the nasal cavity above the usual pathway of inhaled air, sniffing and forcing air over the receptor areas may be necessary to smell a faint odor. Olfactory receptors undergo sensory adaptation rather rapidly, so the intensity of a smell drops about 50% within a second following the stimulation. Within a minute, the receptors may become almost insensitive to a given odor, but even though they have adapted to one scent, their sensitivity to other odors persists.

The olfactory receptor neurons are the only nerve cells in direct contact with the outside environment. Because of their exposed positions, these neurons are subject to damage; they are the only example of damaged neurons that are regularly replaced from stem cells.

1 Where are the olfactory receptors located?

2 Trace the pathway of an olfactory impulse from a receptor to the cerebrum.

Sense of Taste

Taste buds are the special organs of taste. They resemble orange sections and are primarily located on the surface of the tongue where they are associated with tiny elevations called **papillae** (figs. 12.7 and 12.8). Taste buds are also scattered in the roof of the mouth, the linings of the cheeks, and the walls of the pharynx.

Taste Receptors

Each taste bud includes a group of modified epithelial cells, which are the **taste cells** (gustatory cells) that function as receptors. Each of our 10,000 taste buds houses 50 to 150 taste cells. The taste bud also includes epithelial supporting cells. The entire structure is somewhat spherical, with an opening, the **taste pore,** on its free surface. Tiny projections (microvilli), called **taste hairs,** protrude from the outer ends of the taste cells and jut out through the taste pore. These taste hairs are the sensitive parts of the receptor cells.

Interwoven among and wrapped around the taste cells is a network of nerve fibers. The ends of these fibers closely contact the receptor cell membranes. A stimulated receptor cell triggers an impulse on a nearby nerve fiber, which travels into the brain.

A chemical to be tasted must dissolve in the watery fluid surrounding the taste buds. The salivary glands supply this fluid called saliva. To demonstrate the importance of saliva, blot your tongue and try to taste some dry food; then repeat the test after moistening your tongue with saliva.

As is the case for smell, the mechanism of tasting probably involves combinations of chemicals binding specific receptors on taste hair surfaces, altering membrane polarization, and thereby generating sensory impulses on nearby nerve fibers. The degree of change is directly proportional to the concentration of the stimulating substance.

Taste Sensations

The four *primary taste sensations* are *sweet, sour, salty,* and *bitter.* Some investigators recognize other taste sensations, including *alkaline, metallic,* and most recently, *umami,* which is a response to a salt of the amino acid glutamate (see box that follows). Each of the many flavors we experience daily results from one of the primary sensations or from a combination of two or more of them. The way we experience flavors may also reflect the concentration of chemicals as well as the sensations of smell, texture (touch), and temperature. Furthermore, chemicals in some foods—such as capsaisin in chili peppers—may stimulate pain receptors that cause a burning sensation.

The taste sensation called "umami" (oo-MOM'-ee) has long been recognized in Japan but has only recently come to the attention of western taste researchers. Apparently difficult to describe, umami has been defined as "savory," "pungent," "meaty," or simply "delicious" or "perfect" and seems to encompass all of the other taste sensations. Umami arises from taste receptors that are specific for monosodium glutamate (MSG), which is present naturally in many foods, and added to others to enhance flavor. MSG is a salt of the amino acid glutamate.

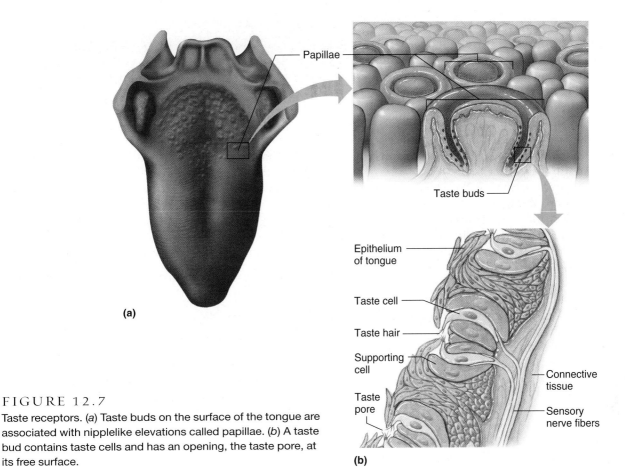

Papillae

Taste buds

Epithelium
of tongue

Taste cell

Taste hair

Supporting
cell

Connective
tissue

Taste
pore

Sensory
nerve fibers

(a)

(b)

FIGURE 12.7

Taste receptors. (*a*) Taste buds on the surface of the tongue are associated with nipplelike elevations called papillae. (*b*) A taste bud contains taste cells and has an opening, the taste pore, at its free surface.

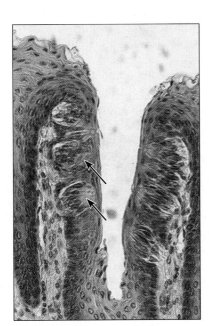

FIGURE 12.8

A light micrograph of some taste buds (arrows) (225×).

Current evidence suggests that all taste cells respond to more than one taste sensation, although for a given taste cell, one taste sensation is likely to predominate. Due to the distribution of taste cells, responsiveness to particular sensations may vary from one region of the tongue to another. For example, sensitivity to a sweet stimulus may peak at the tip of the tongue, whereas responsiveness to sour is greatest at the margins of the tongue, and to bitter at the back of the tongue. Receptors that are particularly responsive to salt are quite widely distributed. In practice, perhaps due to the fact that all taste receptor cells are sensitive to multiple taste stimuli to some degree, individual responses vary widely. More importantly, it may be the pattern of these responses from differentially sensitive receptor cells that provides the brain with the information necessary to create what we call taste.

Sweet receptors are usually stimulated by carbohydrates, but a few inorganic substances, including some salts of lead and beryllium, also elicit sweet sensations. Acids stimulate *sour receptors.* The intensity of a sour sensation is roughly proportional to the concentration of the hydrogen ions in the substance being tasted. Ionized inorganic salts mainly stimulate *salt receptors.* The quality of the sensation that each salt produces depends upon the kind of positively charged ion, such as Na$^+$ from table salt, that it releases into solution. A variety of chemicals stimulates *bitter receptors,* including many organic compounds. Inorganic salts of magnesium and calcium produce bitter sensations, too. Extreme sensitivity to bitter tastes is inherited—this is why diet colas taste sweet to some people but are very bitter to others.

One group of bitter compounds of particular interest are the *alkaloids,* which include a number of poisons such as strychnine, nicotine, and morphine. Spitting out bitter substances may be a protective mechanism to avoid ingesting poisonous alkaloids in foods.

Taste receptors, like olfactory receptors, rapidly undergo sensory adaptation. One can avoid the resulting loss of taste by moving bits of food over the surface of the tongue to stimulate different receptors at different moments.

Although taste cells are very close to the surface of the tongue and are therefore exposed to environmental wear and tear, the sense of taste is not as likely to diminish with age as is the sense of smell. This is because taste cells are modified epithelial cells and divide continually. A taste cell functions for only about three days before it is replaced.

The sense of taste reflects what happens to food as it is chewed. Most foods are chemically complex, so they stimulate different receptors. In an experiment to track the actual act of tasting, chemists collected samples of air from participants' nostrils as they bit into juicy red tomatoes. An analytical technique called mass spectrometry revealed that chewing activates a sequence of chemical reactions in the tomato as its tissues are torn, releasing first aromatic hydrocarbons, then after a thirty-second delay, products of fatty acid breakdown, and, finally, several alcohols. This gradual release of stimulating molecules is why we experience a series of flavors as we savor a food.

Taste Nerve Pathways

Sensory impulses from taste receptors located in the anterior two-thirds of the tongue travel on fibers of the facial nerve (VII); impulses from receptors in the posterior one-third of the tongue and the back of the mouth pass along the glossopharyngeal nerve (IX); and impulses from receptors at the base of the tongue and the pharynx travel on the vagus nerve (X) (see chapter 11, p. 401). These cranial nerves conduct the impulses into the medulla oblongata. From there, the impulses ascend to the thalamus and are directed to the gustatory cortex of the cerebrum, located in the parietal lobe along a deep portion of the lateral sulcus. Clinical Application 12.3 and table 12.3 discuss disorders of smell and taste.

TABLE 12.3	Types of Smell and Taste Disorders	
Doctors use these terms when discussing taste and smell disorders:		
	Smell	**Taste**
Loss of sensation	Anosmia	Ageusia
Diminished sensation	Hyposmia	Hypogeusia
Heightened sensation	Hyperosmia	Hypergeusia
Distorted sensation	Dysosmia	Dysgeusia

1 Why is saliva necessary to taste?

2 Name the four primary taste sensations.

3 What characteristic of taste receptors helps maintain a sense of taste with age?

4 Trace a sensory impulse from a taste receptor to the cerebral cortex.

Sense of Hearing

The organ of hearing, the *ear,* has external, middle, and inner sections. In addition to making hearing possible, the ear provides the sense of equilibrium.

External Ear

The external ear consists of all of the structures that face the outside. These include an outer, funnel-like structure called the **auricle** (pinna) and an S-shaped tube, the *external acoustic* (ah-ko͞os′tik) *meatus* (external auditory canal) that leads inward for about 2.5 centimeters (fig. 12.9). The meatus terminates with the **tympanic membrane** (eardrum).

The external acoustic meatus passes into the temporal bone. Near this opening, hairs guard the tube. The opening and tube are lined with skin that contains many modified sweat glands called *ceruminous glands,* which secrete wax (cerumen). The hairs and wax help keep large foreign objects, such as insects, out of the ear.

Vibrations are transmitted through matter as sound waves. Just as the sounds of some musical instruments are produced by vibrating strings or reeds, the sounds of the human voice are caused by vibrating vocal folds in the larynx. The auricle of the ear helps collect sound waves traveling through air and directs them into the external acoustic meatus.

After entering the meatus, the sound waves pass to the end of the tube and alter the pressure on the tympanic membrane. The tympanic membrane is a semitransparent membrane covered by a thin layer of skin on its outer surface and by mucous membrane on the inside. It has an oval margin and is cone-shaped, with the apex of the cone directed inward. The tympanic membrane moves back and forth in response to sound waves, reproducing the vibrations of the sound-wave source.

SMELL AND TASTE DISORDERS

Imagine a spicy slice of pizza, or freshly brewed coffee, and your mouth waters in anticipation. But for millions of people, the senses of smell and taste are dulled, distorted, or gone altogether. Many more of us get some idea of their plight when a cold temporarily stifles these senses.

Compared to the loss of hearing or sight, being unable to taste or smell normally may seem more an oddity than an illness. But those with such ailments would probably disagree. In some situations, a poor or absent sense of smell can even be dangerous. One person died in a house fire because he did not smell the smoke in time to escape.

The direct connection between the outside environment and the brain makes the sense of smell very vulnerable to damage. Smell and taste disorders can be trig-

gered by colds and flu, allergies, nasal polyps, swollen mucous membranes inside the nose, a head injury, chemical exposure, a nutritional or metabolic problem, or a disease. In many cases, a cause cannot be identified.

Drugs can alter taste and smell in many ways, affecting cell turnover, the neural conduction system, the status of receptors, and changes in nutritional status. Consider what happened to twelve hikers touring Peru and Bolivia. A day before a long hike, three of them had begun taking acetazolamide (Diamox), a drug that prevents acute mountain sickness. The night after the climb, the group went out for beer. To three of the people, the brew tasted unbearably bitter, and a drink of cola to wash away the taste was equally offensive. At fault: acetazolamide.

Drugs containing sulfur atoms squelch taste. They include the anti-inflammatory drug penicillamine, the antihypertensive drug captopril (Capoten), and transdermal (patch) nitroglycerin to treat chest pain. The antibiotic tetracycline and the antiprotozoan metronidazole (Flagyl) cause a metallic taste. Cancer chemotherapy and radiation treatment often alter taste and smell.

Exposure to toxic chemicals can affect taste and smell, too. A forty-five year-old woman from Altoona, Pennsylvania, suddenly found that once-pleasant smells had become offensive. Her doctor traced her problem to inhaling a paint stripper. Hydrocarbon solvents in the product—toluene, methanol, and methylene chloride—were responsible for her *cacosmia*, the association of an odor of decay with normally inoffensive stimuli. ■

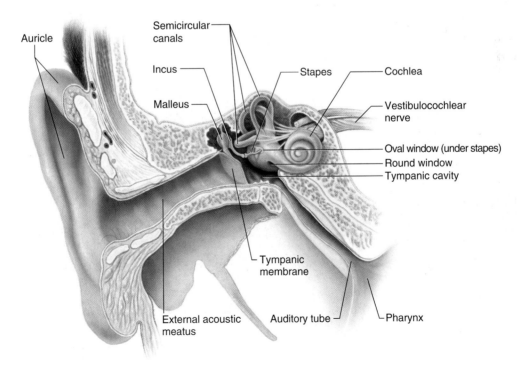

FIGURE 12.9
Major parts of the ear.

Middle Ear

The **middle ear,** or the **tympanic cavity** is an air-filled space in the temporal bone that separates the external and internal ears. It is bounded by the tympanic membrane laterally and the inner ear medially and contains three small bones called **auditory ossicles.**

The three **auditory ossicles,** called the *malleus,* the *incus,* and the *stapes,* are attached to the wall of the tympanic cavity by tiny ligaments and are covered by mucous membrane. These bones bridge the tympanic membrane and the inner ear, transmitting vibrations between these parts. Specifically, the malleus is attached to the tympanic membrane, helping to maintain its conical shape. When the tympanic membrane vibrates, the malleus vibrates in unison with it. The malleus vibrates the incus, and the incus passes the movement on to the stapes. Ligaments hold the stapes to an opening in the wall of the tympanic cavity called the **oval window.** Vibration of the stapes, which acts like a piston at the oval window, moves a fluid within the inner ear. These vibrations of the fluid stimulate the hearing receptors (fig. 12.9).

In addition to transmitting vibrations, the auditory ossicles form a lever system that helps increase (amplify) the force of the vibrations as they pass from the tympanic membrane to the oval window. Also, because the ossicles transmit vibrations from the large surface of the tympanic membrane to a much smaller area at the oval window, the vibrational force concentrates as it travels from the external to the inner ear. As a result of these two factors, the pressure (per square millimeter) that the stapes applies at the oval window is about twenty-two times greater than that which sound waves exert on the tympanic membrane.

The middle ear also contains two small skeletal muscles that are attached to the auditory ossicles and are controlled involuntarily. One of them, the *tensor tympani,* is inserted on the medial surface of the malleus and is anchored to the cartilaginous wall of the auditory tube. When it contracts, it pulls the malleus inward. The other muscle, the *stapedius,* is attached to the posterior side of the stapes and the inner wall of the tympanic cavity. It pulls the stapes outward (fig. 12.10). These muscles are

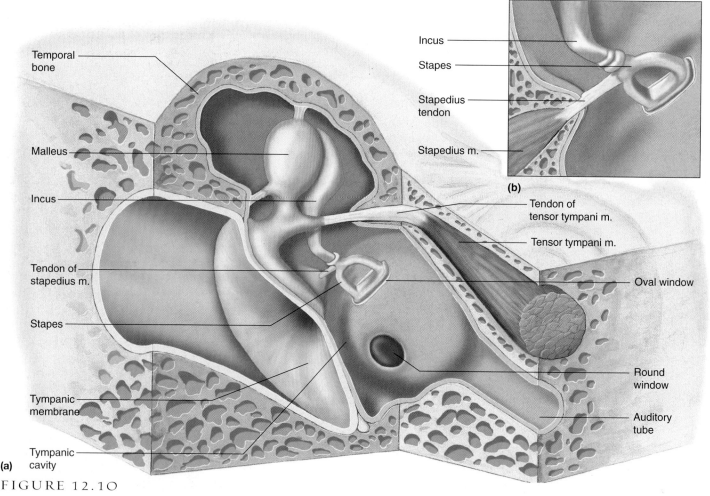

FIGURE 12.10

Two small muscles attached to the (*a*) malleus and (*b*) stapes, the tensor tympani and the stapedius, are effectors in the tympanic reflex. Figure 12.9 does not show these muscles.

the effectors in the **tympanic reflex,** which is elicited in about one-tenth second following a loud, external sound. When the reflex occurs, the muscles contract, and the malleus and stapes move. As a result, the bridge of ossicles in the middle ear becomes more rigid, reducing its effectiveness in transmitting vibrations to the inner ear.

The tympanic reflex reduces pressure from loud sounds that might otherwise damage the hearing receptors. The tympanic reflex is also elicited by ordinary vocal sounds, as when a person speaks or sings, and this action muffles the lower frequencies of such sounds, improving the hearing of higher frequencies, which are common in human vocal sounds. In addition, the tensor tympani muscle also steadily pulls on the tympanic membrane. This is important because a loose tympanic membrane would not be able to transmit vibrations effectively to the auditory ossicles.

The muscles of the middle ear take 100 to 200 milliseconds to contract. For this reason, the tympanic reflex cannot protect the hearing receptors from the effects of loud sounds that occur very rapidly, such as those from an explosion or a gunshot. On the other hand, this protective mechanism can reduce the effects of intense sounds that arise slowly, such as the roar of thunder.

Auditory Tube

An **auditory tube** (eustachian tube) connects each middle ear to the throat. This tube allows air to pass between the tympanic cavity and the outside of the body by way of the throat (nasopharynx) and mouth. It helps maintain equal air pressure on both sides of the tympanic membrane, which is necessary for normal hearing (see fig. 12.10).

The function of the auditory tube becomes noticeable during rapid change in altitude. As a person moves from a high altitude to a lower one, the air pressure on the outside of the tympanic membrane steadily increases. As a result, the tympanic membrane may be pushed inward, out of its normal position, impairing hearing.

When the air pressure difference is great enough, some air may force its way up through the auditory tube into the middle ear. This equalizes the pressure on both sides of the tympanic membrane, which moves back into its regular position, causing a popping sound as normal hearing returns. A reverse movement of air ordinarily occurs when a person moves from a low altitude to a higher one.

The auditory tube is usually closed by valvelike flaps in the throat, which may inhibit air movements into the middle ear. Swallowing, yawning, or chewing aid in opening the valves and can hasten equalization of air pressure.

Signs of an ear infection in an infant or toddler are hard to miss—irritability, screaming incessantly for no apparent reason, perhaps fever or tugging on the affected ear. A doctor viewing the painful ear with an instrument called an otoscope sees a red and bulging tympanic membrane. The diagnosis: *otitis media,* or a middle ear infection.

Ear infections occur because the mucous membranes that line the auditory tubes are continuous with the linings of the middle ears, creating a conduit for bacteria infecting the throat or nasal passages to travel to the ear. This route to infection is greater in young children because their auditory tubes are shorter than they are in adults. Half of all children in the United States have an ear infection by the first birthday, and 90% have one by age six.

Physicians treat acute otitis media with antibiotics. Because recurrent infections may cause hearing loss and interfere with learning, children with recurrent otitis media are often fitted with tympanostomy tubes, which are inserted into affected ears during a brief surgical procedure. The tubes form a small tunnel through the tympanic membrane so the ears can drain. By the time the tubes fall out, the child has usually outgrown the susceptibility to ear infections.

Inner Ear

The inner ear is a complex system of intercommunicating chambers and tubes called a **labyrinth** (lab'i-rinth). Each ear has two such regions—the osseous labyrinth and the membranous labyrinth.

The *osseous labyrinth* is a bony canal in the temporal bone; the *membranous labyrinth* is a tube that lies within the osseous labyrinth and has a similar shape (fig. 12.11*a*). Between the osseous and membranous labyrinths is a fluid called *perilymph,* which cells in the wall of the bony canal secrete. Within the membranous labyrinth is a slightly different fluid called *endolymph.*

The parts of the labyrinths include a **cochlea** (kok'le-ah) that functions in hearing and three **semicircular canals** that provide a sense of equilibrium. A bony chamber called the **vestibule,** located between the cochlea and the semicircular canals, houses membranous structures that serve both hearing and equilibrium.

The cochlea is shaped like a snail shell, coiled around a bony core (modiolus) with a thin, bony shelf (spiral lamina) that wraps around the core like a spiral staircase (fig. 12.11*b*). The shelf divides the bony labyrinth of the cochlea into upper and lower compartments. The upper compartment, called the *scala vestibuli,* leads from the oval window to the apex of the spiral. The lower compartment, the *scala tympani,* extends from the apex of the cochlea to a membrane-covered opening in the wall

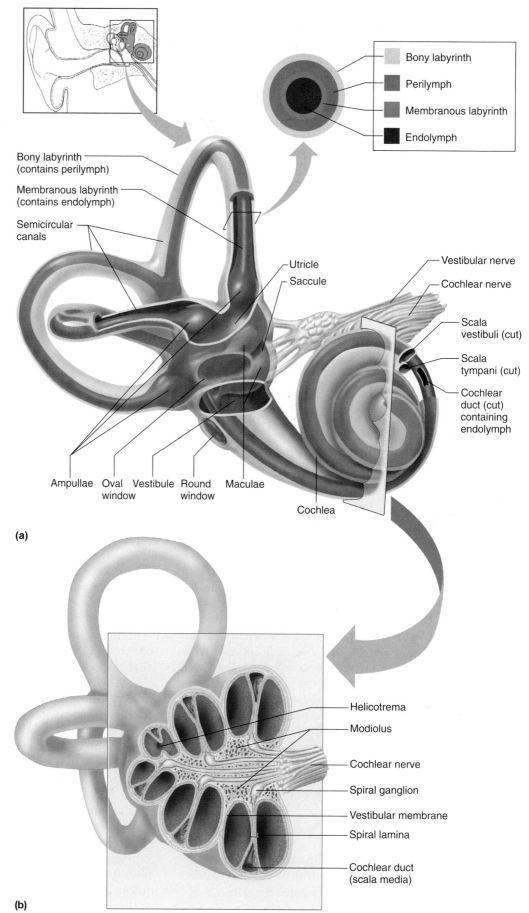

Bony labyrinth
Perilymph
Membranous labyrinth
Endolymph

Bony labyrinth
(contains perilymph)

Membranous labyrinth
(contains endolymph)

Semicircular
canals

Utricle

Saccule

Vestibular nerve

Cochlear nerve

Scala
vestibuli (cut)

Scala
tympani (cut)

Cochlear
duct (cut)
containing
endolymph

Ampullae Oval Vestibule Round Maculae
window window

Cochlea

(a)

Helicotrema

Modiolus

Cochlear nerve

Spiral ganglion

Vestibular membrane

Spiral lamina

Cochlear duct
(scala media)

(b)

F I G U R E 1 2 . 1 1

Within the inner ear (a) perilymph
separates the osseous labyrinth of
the inner ear from the membranous
labyrinth, which contains endolymph.
(b) The spiral lamina coils around a
bony core, the modiolus.

of the inner ear called the **round window.** These compartments constitute the bony labyrinth of the cochlea, and they are filled with perilymph. At the apex of the cochlea, the fluids in the chambers are connected by a small opening (helicotrema) (figs. 12.11*b* and 12.12).

A portion of the membranous labyrinth within the cochlea, called the *cochlear duct* (scala media), lies between the two bony compartments and is filled with endolymph. The cochlear duct ends as a closed sac at the apex of the cochlea. The duct is separated from the scala vestibuli by a *vestibular membrane* (Reissner's membrane) and from the scala tympani by a *basilar membrane* (fig. 12.12).

The basilar membrane extends from the bony shelf of the cochlea and forms the floor of the cochlear duct. It contains many thousands of stiff, elastic fibers whose lengths vary, becoming progressively longer from the base of the cochlea to its apex. Vibrations entering the perilymph at the oval window travel along the scala vestibuli and pass through the vestibular membrane to enter the endolymph of the cochlear duct, where they move the basilar membrane. After passing through the basilar membrane, the vibrations enter the perilymph of the scala tympani, and their forces are dissipated into the air in the tympanic cavity by movement of the membrane covering the round window.

The **organ of Corti,** which contains about 16,000 hearing receptor cells, is located on the upper surface of the basilar membrane and stretches from the apex to the base of the cochlea. The receptor cells, called **hair cells,** are in four parallel rows, with many hairlike processes (stereocilia) that extend into the endolymph of the cochlear duct. Above these hair cells is a *tectorial membrane,* which is attached to the bony shelf of the cochlea

and passes like a roof over the receptor cells, contacting the tips of their hairs (figs. 12.13 and 12.14).

Different frequencies of vibration move different parts of the basilar membrane. A particular sound frequency causes the hairs of a specific group of receptor cells to bend against the tectorial membrane. Other frequencies deflect other sets of receptor cells.

Hearing receptor cells are epithelial cells, but they respond to stimuli somewhat like neurons (see chapter 10, pp. 350–353). For example, when a receptor cell is at rest, its membrane is polarized. When its hairs bend selective ion channels open, and its cell membrane depolarizes. The membrane then becomes more permeable, specifically to calcium ions. The receptor cell has no axon or dendrites, but it does have neurotransmitter-containing vesicles in the cytoplasm near its base. In the presence of calcium ions, some of these vesicles fuse with the cell membrane and release neurotransmitter to the outside. The neurotransmitter stimulates the ends of nearby sensory nerve fibers, and in response, they transmit nerve impulses along the cochlear branch of the vestibulocochlear nerve (cranial nerve VIII) to the brain.

The ear of a young person with normal hearing can detect sound waves with frequencies varying from about 20 to 20,000 or more vibrations per second. The range of greatest sensitivity is between 2,000 and 3,000 vibrations per second (fig. 12.15).

Auditory Nerve Pathways

The cochlear branches of the vestibulocochlear nerves enter the auditory nerve pathways that extend into the medulla oblongata and proceed through the midbrain to the thalamus. From there they pass into the auditory cortices of the temporal lobes of the cerebrum, where they

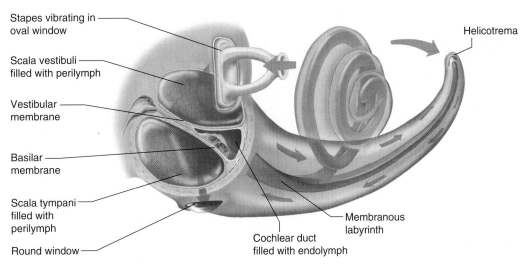

Stapes vibrating in oval window

Scala vestibuli filled with perilymph

Vestibular membrane

Basilar membrane

Scala tympani filled with perilymph

Round window

Cochlear duct filled with endolymph

Membranous labyrinth

Helicotrema

FIGURE 12.12

The cochlea is a coiled, bony canal with a membranous tube (labyrinth) inside. If the cochlea could be unwound, the membranous labyrinth would be seen ending as a closed sac at the apex where the bony canal makes a u-turn.

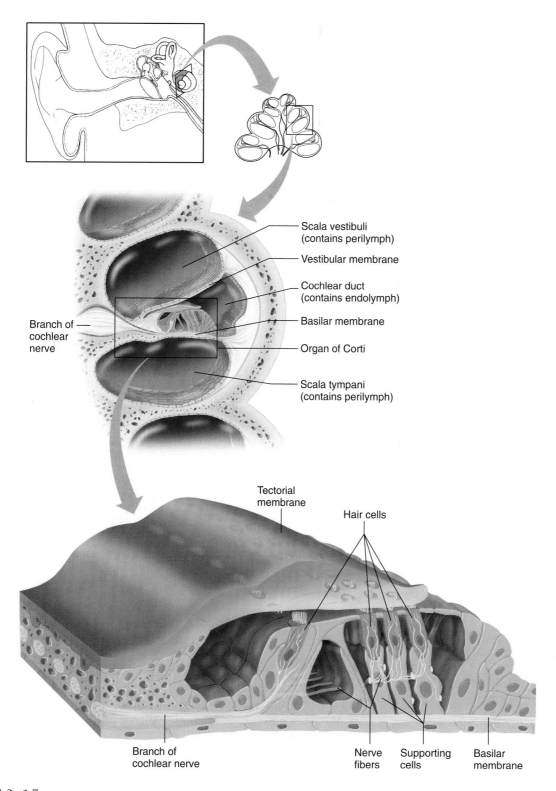

Scala vestibuli
(contains perilymph)

Vestibular membrane

Cochlear duct
(contains endolymph)

Basilar membrane

Organ of Corti

Scala tympani
(contains perilymph)

Branch of
cochlear
nerve

Tectorial
membrane

Hair cells

Branch of
cochlear nerve

Nerve
fibers

Supporting
cells

Basilar
membrane

FIGURE 12.13

Cochlea. (*a*) Cross section of the cochlea. (*b*) Organ of Corti and the tectorial membrane.

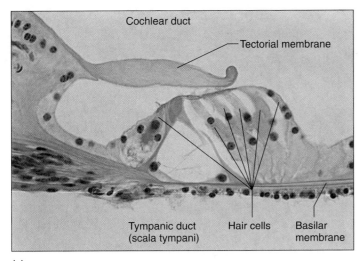

(a)

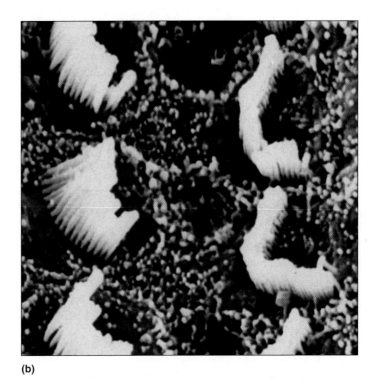

(b)

FIGURE 12.14

Organ of Corti. (*a*) A micrograph of the organ of Corti and the tectorial membrane (300×). (*b*) A scanning electron micrograph of hair cells in the organ of Corti, looking down on the "hairs" (bright yellow) (6,700×).

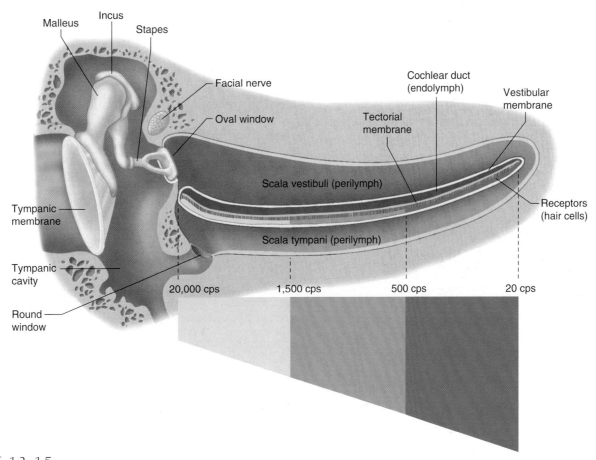

FIGURE 12.15

Receptors in regions of the cochlear duct sense different frequencies of vibration, expressed in cycles per second (cps).

are interpreted. On the way, some of these fibers cross over, so that impulses arising from each ear are interpreted on both sides of the brain. Consequently, damage to a temporal lobe on one side of the brain is not necessarily accompanied by complete hearing loss in the ear on that side (fig. 12.16).

Table 12.4 summarizes the pathway of vibrations through the parts of the middle and inner ears. Clinical Application 12.4 examines types of hearing loss.

Units called *decibels* (dB) measure sound intensity. The decibel scale begins at 0 dB, which is the intensity of the sound that is least perceptible by a normal human ear. The decibel scale is logarithmic, so a sound of 10 dB is 10 times as intense as the least perceptible sound; a sound of 20 dB is 100 times as intense; and a sound of 30 dB is 1,000 times as intense.

On the decibel scale, a whisper has an intensity of about 40 dB, normal conversation measures 60–70 dB, and heavy traffic produces about 80 dB. A sound of 120 dB, such as that commonly produced by the amplified sound at a rock concert, produces discomfort, and a sound of 140 dB, such as that emitted by a jet plane at takeoff, causes pain. Frequent or prolonged exposure to sounds with intensities above 90 dB can cause permanent hearing loss.

1 Describe the external, middle, and inner ears.

2 Explain how sound waves are transmitted through the parts of the ear.

3 Describe the tympanic reflex.

4 Distinguish between the osseous and membranous labyrinths.

5 Explain the function of the organ of Corti.

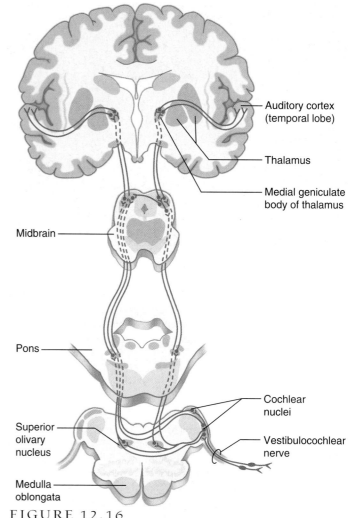

FIGURE 12.16

The auditory nerve pathway extends into the medulla oblongata, proceeds through the midbrain to the thalamus, and passes into the auditory cortex of the cerebrum.

TABLE 12.4	Steps in the Generation of Sensory Impulses from the Ear
1. Sound waves enter the external acoustic meatus.	7. A receptor cell becomes depolarized; its membrane becomes more permeable to calcium ions.
2. Waves of changing pressures cause the tympanic membrane to reproduce the vibrations coming from the sound-wave source.	8. In the presence of calcium ions, vesicles at the base of the receptor cell release neurotransmitter.
3. Auditory ossicles amplify and transmit vibrations to the end of the stapes.	9. Neurotransmitter stimulates the ends of nearby sensory neurons.
4. Movement of the stapes at the oval window transmits vibrations to the perilymph in the scala vestibuli.	10. Sensory impulses are triggered on fibers of the cochlear branch of the vestibulocochlear nerve.
5. Vibrations pass through the vestibular membrane and enter the endolymph of the cochlear duct.	11. The auditory cortex of the temporal lobe interprets the sensory impulses.
6. Different frequencies of vibration in endolymph move specific regions of the basilar membrane, thus stimulating specific sets of receptor cells.	

HEARING LOSS

Several factors can impair hearing, including interference with transmission of vibrations to the inner ear (*conductive deafness*) or damage to the cochlea or the auditory nerve and its pathways (*sensorineural deafness*). Disease, injury, and heredity all can impair hearing. There are more than 100 forms of inherited deafness, many of which are part of syndromes. About 8% of people have hearing loss.

About 95% of cases of hearing loss are conductive. One cause is accumulated dry wax or a foreign object in the ear, which plugs the acoustic meatus. Changes in the tympanic membrane or auditory ossicles can also block hearing. The tympanic membrane may harden as a result of disease, becoming less responsive to sound waves, or an injury may tear or perforate it.

A common disorder of the auditory ossicles is *otosclerosis,* in which new bone is deposited abnormally around the base of the stapes. This interferes with the ossicles' movement, which is necessary to transmit vibrations to the inner ear. Surgery often can restore some hearing to a person with otosclerosis by chipping away the bone that holds the stapes in position, or replacing the stapes with a wire or plastic substitute.

Two tests used to diagnose conductive deafness are the Weber test and the Rinne test. In the Weber test, the handle of a vibrating tuning fork is pressed against the forehead. A person with normal hearing perceives the sound coming from directly in front, whereas a person with sound conduction blockage in one middle ear hears the sound coming from the impaired side.

In the Rinne test, a vibrating tuning fork is held against the bone behind the ear. After the sound is no longer heard by conduction through the bones of the skull, the fork is moved to just in front of the external acoustic meatus. In middle ear conductive deafness, the vibrating fork can no longer be heard, but a normal ear will continue to hear its tone.

Very loud sounds can cause sensorineural deafness. If exposure is brief, hearing loss may be temporary, but when exposure is repeated and prolonged, such as occurs in foundries, near jackhammers, or on a firing range, impairment may be permanent. Many rock musicians, and their fans, have hearing loss from years of performing or hearing loud concerts. Such hearing loss begins as the hair cells develop blisterlike bulges that eventually pop. The tissue beneath the hair cells swells and softens until the hair cells and sometimes the neurons leaving the cochlea become blanketed with scar tissue and degenerate. Other causes of sensorineural deafness include tumors in the central nervous system, brain damage as a result of vascular accidents, and the use of certain drugs.

Because hearing loss and other ear problems can begin gradually, it is important to be aware of their signs, which may include the following:

- difficulty hearing people talking softly
- inability to understand speech when there is background noise
- ringing in the ears
- dizziness
- loss of balance

New parents should notice whether their infant responds to sounds in a way that indicates normal hearing. Before 1993, 50% of hearing-impaired infants were not diagnosed until age two. Since then, the federal government has advised hearing exams as part of a well-baby visit to a doctor. If the baby's responses indicate a possible problem, the next step is to see an audiologist, who identifies and measures hearing loss.

Often a hearing aid can help people with conductive hearing loss. A hearing aid has a tiny microphone that picks up sound waves and converts them to electrical signals, which are then amplified so that the person can hear them. An ear mold holds the device in place, either behind the outer ear, in the outer ear, or in the ear canal.

A cochlear implant enables people with sensorineural hearing loss to detect some sounds, although it usually remains difficult to discern distinct words. The device converts sound waves to electrical signals, which stimulate neurons in the cochlea. ∎

Sense of Equilibrium

The feeling of equilibrium derives from two senses—*static equilibrium* (stat′ik e″kwĭ-lib′re-um) and *dynamic equilibrium* (di-nam′ik e″kwĭ-lib′re-um). Different sensory organs provide these two components of equilibrium. The organs associated with static equilibrium sense the position of the head, maintaining stability and posture when the head and body are still. When the head and body suddenly move or rotate, the organs of dynamic equilibrium detect such motion and aid in maintaining balance.

Static Equilibrium

The organs of **static equilibrium** are located within the vestibule, a bony chamber between the semicircular canals and the cochlea. More specifically, the membranous labyrinth inside the vestibule consists of two expanded chambers—a **utricle** and a **saccule.** The larger utricle communicates with the saccule and the membranous portions of the semicircular canals; the saccule, in turn, communicates with the cochlear duct (fig. 12.17).

The utricle and saccule each has a small patch of hair cells and supporting cells called a **macula** (mak′u-lah) on

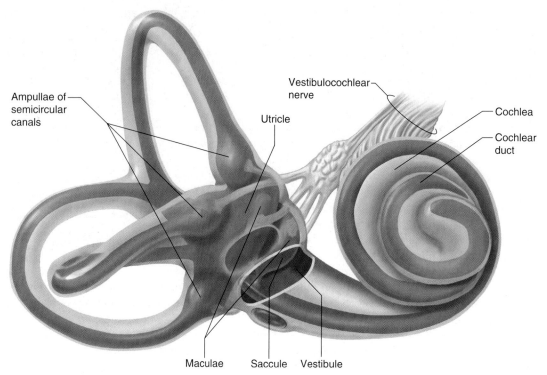

FIGURE 12.17

The saccule and utricle, which are expanded portions of the membranous labyrinth, are located within the bony chamber of the vestibule. (Compare with figure 12.11.)

its wall. When the head is upright, the hairs of the macula in the utricle project vertically, while those in the saccule project horizontally. In each case, the hairs contact a sheet of gelatinous material (otolithic membrane) that has crystals of calcium carbonate (otoliths) embedded on its surface. These particles add weight to the gelatinous sheet, making it more responsive to changes in position. The hair cells, which are sensory receptors, have nerve fibers wrapped around their bases. These fibers are associated with the vestibular portion of the vestibulocochlear nerve.

Gravity stimulates hair cells to respond. This usually occurs when the head bends forward, backward, or to one side. Such movements tilt the gelatinous mass of one or more maculae, and as the gelatinous material sags in response to gravity, the hairs projecting into it bend. This action stimulates the hair cells, and they signal their associated nerve fibers (figs. 12.18 and 12.19). The resulting nerve impulses travel into the central nervous system by means of the vestibular branch of the vestibulocochlear nerve, informing the brain of the head's position. The brain responds to this information by sending motor impulses to skeletal muscles, and they may contract or relax appropriately to maintain balance.

The maculae also participate in the sense of dynamic equilibrium. For example, if the head or body is thrust forward or backward abruptly, the gelatinous mass of the maculae lags slightly behind, and the hair cells are stimulated. In this way, the maculae aid the brain in detecting movements such as falling and in maintaining posture while walking.

Dynamic Equilibrium

Each semicircular canal follows a circular path about 6 millimeters in diameter. The three bony semicircular canals lie at right angles to each other and occupy three different planes in space. Two of them, the *superior canal* and the *posterior canal*, stand vertically, whereas the third, the *lateral canal*, is horizontal. Their orientations closely approximate the three body planes (see chapter 1, pp. 21–23).

Suspended in the perilymph of each bony canal is a membranous semicircular canal that ends in a swelling called an **ampulla** (am-pul′lah). The ampullae communicate with the utricle of the vestibule.

An ampulla contains a septum that crosses the tube and houses a sensory organ. Each of these organs, called a **crista ampullaris,** has a number of sensory hair cells and supporting cells. As in the maculae, the hairs of the hair cells extend upward into a dome-shaped gelatinous mass called the *cupula.* Also, the hair cells are connected at their bases to nerve fibers that make up part of the vestibular branch of the vestibulocochlear nerve (fig. 12.20).

Rapid turns of the head or body stimulate the hair cells of the crista. At such times, the semicircular canals

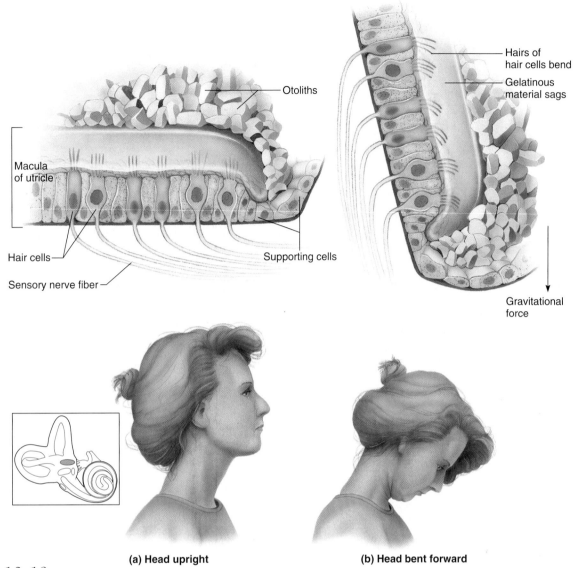

Otoliths

Hairs of
hair cells bend

Gelatinous
material sags

Macula
of utricle

Hair cells

Sensory nerve fiber

Supporting cells

Gravitational
force

(a) Head upright

(b) Head bent forward

FIGURE 12.18

The maculae respond to changes in head position. (*a*) Macula of the utricle with the head in an upright position. (*b*) Macula of the utricle with the head bent forward.

FIGURE 12.19

Scanning electron micrograph of hairs of hair cells, such as those found in the utricle and saccule (8,000×).

move with the head or body, but the fluid inside the membranous canals tends to remain stationary because of inertia. This bends the cupula in one or more of the canals in a direction opposite that of the head or body movement, and the hairs embedded in it also bend. This bending of the hairs stimulates the hair cells to signal their associated nerve fibers, and, as a result, impulses travel to the brain (fig. 12.21).

Parts of the cerebellum are particularly important in interpreting impulses from the semicircular canals. Analysis of such information allows the brain to predict the consequences of rapid body movements, and by modifying signals to appropriate skeletal muscles, the cerebellum can maintain balance.

Other sensory structures aid in maintaining equilibrium. Various proprioceptors, particularly those associated with the joints of the neck, inform the brain about the position of body parts. The eyes detect changes in posture

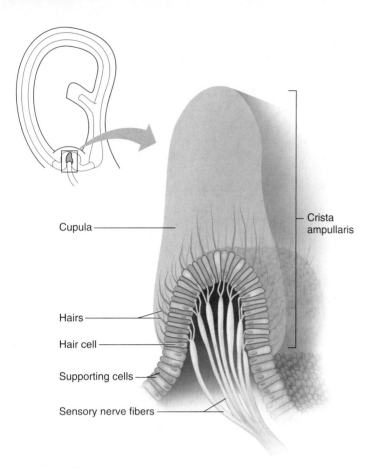

Cupula

Crista ampullaris

Hairs

Hair cell

Supporting cells

Sensory nerve fibers

FIGURE 12.20

A crista ampullaris is located within the ampulla of each semicircular canal.

(a) Head in still position

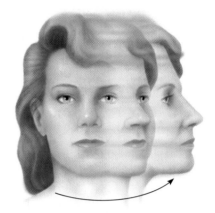

(b) Head rotating

that result from body movements. Such visual information is so important that even if the organs of equilibrium are damaged, keeping the eyes open and moving slowly is sufficient to maintain normal balance.

Motion sickness is a disturbance of the inner ear's sensation of balance. Nine out of ten people have experienced this nausea and vomiting, usually when riding in a car or on a boat. Astronauts began reporting a form of motion sickness called space adaptation syndrome in 1968, when spacecraft were made roomy enough for astronauts to move about while in flight.

Although the cause of motion sickness is not known, one theory is that it results when visual information contradicts the inner ear's sensation that one is motionless. Consider a woman riding in a car. Her inner ears tell her that she is not moving, but the passing scenery tells her eyes that she is moving. The problem is compounded if she tries to read. The brain reacts to these seemingly contradictory sensations by signaling a "vomiting center" in the medulla oblongata.

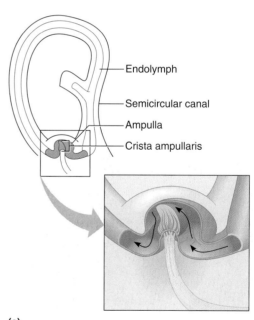

Endolymph

Semicircular canal

Ampulla

Crista ampullaris

(c)

FIGURE 12.21

Equilibrium. (a) When the head is stationary, the cupula of the crista ampullaris remains upright. (b) When the head is moving rapidly, (c) the cupula bends opposite the motion of the head, stimulating sensory receptors.

1 Distinguish between the senses of static and dynamic equilibrium.

2 Which structures provide the sense of static equilibrium? Of dynamic equilibrium?

3 How does sensory information from other receptors help maintain equilibrium?

Sense of Sight

A number of **accessory organs** assist the visual receptors, which are in the eyes. These include the eyelids and lacrimal apparatus that help protect the eyes and a set of extrinsic muscles that move them.

Visual Accessory Organs

Each eye, lacrimal gland, and associated extrinsic muscles are housed within the pear-shaped orbital cavity of the skull. The orbit, which is lined with the periosteums of various bones, also contains fat, blood vessels, nerves, and connective tissues.

Each *eyelid* (palpebra) is composed of four layers—skin, muscle, connective tissue, and conjunctiva. The skin of the eyelid, which is the thinnest skin of the body, covers the lid's outer surface and fuses with its inner lining near the margin of the lid (fig. 12.22).

The muscles that move the eyelids include the *orbicularis oculi* and the *levator palpebrae superioris*. Fibers of the orbicularis oculi encircle the opening between the lids and spread out onto the cheek and forehead. This muscle acts as a sphincter that closes the lids when it contracts.

Fibers of the levator palpebrae superioris muscle arise from the roof of the orbit and are inserted in the connective tissue of the upper lid. When these fibers contract, the upper lids are raised, and the eye opens.

The connective tissue layer of the eyelid, which helps give it form, contains many modified sebaceous

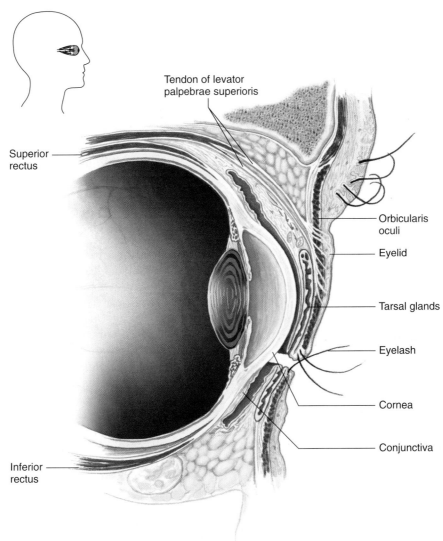

FIGURE 12.22
Sagittal section of the closed eyelids and the anterior portion of the eye.

glands (tarsal glands). Ducts carry the oily secretions of these glands to openings along the borders of the lids. This secretion helps keep the lids from sticking together.

The **conjunctiva** is a mucous membrane that lines the inner surfaces of the eyelids and folds back to cover the anterior surface of the eyeball, except for its central portion (cornea). Although the tissue that lines the eyelids is relatively thick, the conjunctiva that covers the eyeball is very thin. It is also freely movable and quite transparent, so that blood vessels are clearly visible beneath it.

> A child in school with "pinkeye" is usually sent straight home. Bacteria cause this highly contagious form of inflammation of the conjunctiva, or *conjunctivitis.* Viral conjunctivitis is not usually contagious. Allergy or exposure to an irritating chemical may also cause conjunctivitis.

The *lacrimal apparatus* consists of the **lacrimal gland,** which secretes tears, and a series of *ducts,* which carry the tears into the nasal cavity (fig. 12.23). The gland is located in the orbit, superior and lateral to the eye. It secretes tears continuously, and they pass out through tiny tubules and flow downward and medially across the eye.

Two small ducts (superior and inferior canaliculi) collect tears, and their openings (puncta) can be seen on the medial borders of the eyelids. From these ducts, the fluid moves into the *lacrimal sac,* which lies in a deep

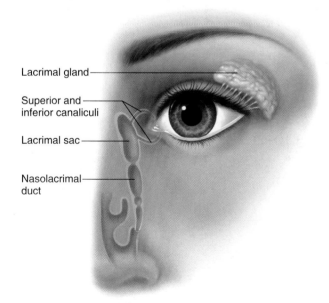

Lacrimal gland

Superior and inferior canaliculi

Lacrimal sac

Nasolacrimal duct

FIGURE 12.23

The lacrimal apparatus consists of a tear-secreting gland and a series of ducts.

groove of the lacrimal bone, and then into the *nasolacrimal duct,* which empties into the nasal cavity.

Glandular cells of the conjunctiva also secrete a tear-like liquid that, together with the secretion of the lacrimal gland, moistens and lubricates the surface of the eye and the lining of the lids. Tears contain an enzyme, called *lysozyme,* that has antibacterial properties, reducing the chance of eye infections.

Tear glands secrete excessively when a person is upset or when the conjunctiva is irritated. Tears spill over the edges of the eyelids, and the nose fills with fluid. When a person cries, parasympathetic nerve fibers carry motor impulses to the lacrimal glands.

The **extrinsic muscles** of the eye arise from the bones of the orbit and are inserted by broad tendons on the eye's tough outer surface. Six such muscles move the eye in various directions (fig. 12.24). Although any given eye movement may use more than one of them, each muscle is associated with one primary action, as follows:

1. **Superior rectus**—rotates the eye upward and toward the midline.

2. **Inferior rectus**—rotates the eye downward and toward the midline.

3. **Medial rectus**—rotates the eye toward the midline.

4. **Lateral rectus**—rotates the eye away from the midline.

5. **Superior oblique**—rotates the eye downward and away from the midline.

6. **Inferior oblique**—rotates the eye upward and away from the midline.

The motor units of the extrinsic eye muscles contain the fewest muscle fibers (five to ten) of any muscles in the body, so they can move the eyes with great precision. Also, the eyes move together so that they align when looking at something. Such alignment is the result of complex motor adjustments that contract certain eye muscles while relaxing their antagonists. For example, when the eyes move to the right, the lateral rectus of the right eye and the medial rectus of the left eye must contract. At the same time, the medial rectus of the right eye and the lateral rectus of the left eye must relax. A person whose eyes are not coordinated well enough to align has *strabismus.* Table 12.5 summarizes the muscles associated with the eyelids and eye.

1 Explain how the eyelid is moved.

2 Describe the conjunctiva.

3 What is the function of the lacrimal apparatus?

4 Describe the function of each extrinsic eye muscle.

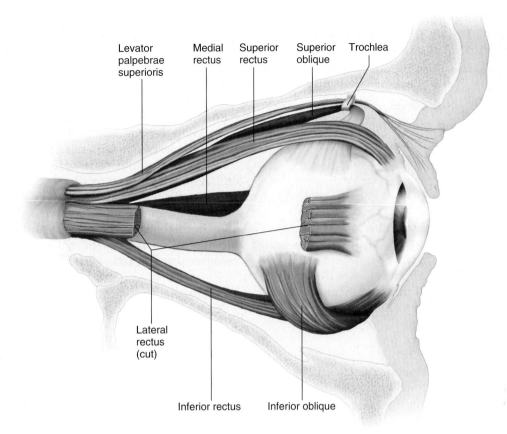

Levator palpebrae superioris | Medial rectus | Superior rectus | Superior oblique | Trochlea

Lateral rectus (cut)

Inferior rectus | Inferior oblique

FIGURE 12.24

The extrinsic muscles of the right eye (lateral view). Note the pulleylike trochlea that the superior oblique passes through.

TABLE 12.5	Muscles Associated with the Eyelids and Eyes					
Skeletal Muscles **Name**	**Innervation**	**Function**	**Smooth Muscles** **Name**	**Innervation**	**Function**
Muscles of the eyelids Orbicularis oculi	Facial nerve (VII)	Closes eye	Ciliary muscles	Oculomotor nerve (III) parasympathetic fibers	Relax suspensory ligaments
Levator palpebrae superioris	Oculomotor nerve (III)	Opens eye	Iris, circular muscles	Oculomotor nerve (III) parasympathetic fibers	Constrict pupil
Extrinsic muscles of the eyes Superior rectus	Oculomotor nerve (III)	Rotates eye upward and toward midline	Iris, radial muscles	Sympathetic fibers	Dilate pupil
Inferior rectus	Oculomotor nerve (III)	Rotates eye downward and toward midline			
Medial rectus	Oculomotor nerve (III)	Rotates eye toward midline			
Lateral rectus	Abducens nerve (VI)	Rotates eye away from midline			
Superior oblique	Trochlear nerve (IV)	Rotates eye downward and away from midline			
Inferior oblique	Oculomotor nerve (III)	Rotates eye upward and away from midline			

Structure of the Eye

The eye is a hollow, spherical structure about 2.5 centimeters in diameter. Its wall has three distinct layers—an outer *fibrous tunic,* a middle *vascular tunic,* and an inner *nervous tunic.* The spaces within the eye are filled with fluids that support its wall and internal structures and help maintain its shape. Figure 12.25 shows the major parts of the eye.

The Outer Tunic

The anterior sixth of the outer tunic bulges forward as the transparent **cornea** (kor'ne-ah), which is the window of the eye and helps focus entering light rays. It is largely composed of connective tissue with a thin surface layer of epithelium. The cornea is transparent because it contains few cells and no blood vessels. The cells and collagenous fibers form unusually regular patterns.

In contrast, the cornea is well supplied with nerve fibers that enter its margin and radiate toward its center. These fibers are associated with many pain receptors that have very low thresholds. Cold receptors are also abundant in the cornea, but heat and touch receptors are not.

Along its circumference, the cornea is continuous with the **sclera** (skle'rah), the white portion of the eye. The sclera makes up the posterior five-sixths of the outer tunic and is opaque due to many large, haphazardly arranged collagenous and elastic fibers. The sclera protects the eye and is an attachment for the extrinsic muscles.

In the back of the eye, the **optic** (op'tik) **nerve** and blood vessels pierce the sclera. The dura mater that encloses these structures is continuous with the sclera.

The Middle Tunic

The middle, or vascular, tunic of the eyeball (uveal layer) includes the choroid coat, the ciliary body, and the iris. The **choroid coat,** in the posterior five-sixths of the globe of the eye, loosely joins the sclera. Blood vessels pervade the choroid coat and nourish surrounding tissues. The choroid coat also contains numerous pigment-producing melanocytes that give it a brownish black appearance. The melanin of these cells absorbs excess light and helps keep the inside of the eye dark.

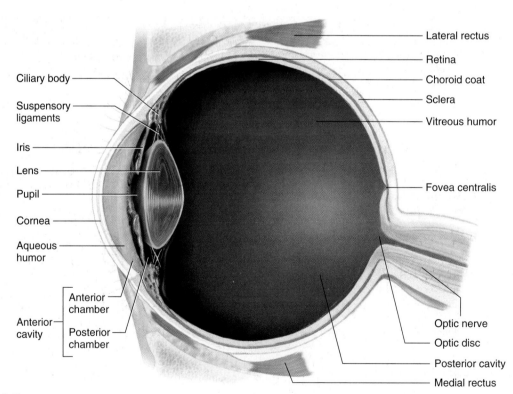

FIGURE 12.25

Transverse section of the right eye (superior view).

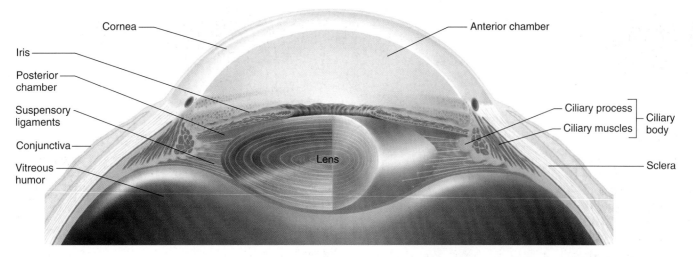

Cornea

Iris

Posterior chamber

Suspensory ligaments

Conjunctiva

Vitreous humor

Anterior chamber

Ciliary process

Ciliary muscles

Ciliary body

Lens

Sclera

FIGURE 12.26
Anterior portion of the eye.

In 1905, doctors transplanted the cornea of an eleven-year-old boy who lost his eye in an accident into a man whose cornea had been destroyed by a splash of a caustic chemical, marking one of the first successful human organ transplants. Today, corneal transplants are commonly used to treat corneal disease, the most common cause of blindness worldwide. In this procedure, called a *penetrating keratoplasty*, a piece of donor cornea replaces the central two-thirds of the defective cornea. These transplants are highly successful because the cornea lacks blood vessels, and therefore, the immune system does not have direct access to the new, "foreign" tissue. Unfortunately, as is the case for many transplantable body parts, donor tissue is in short supply.

The **ciliary body,** which is the thickest part of the middle tunic, extends forward from the choroid coat and forms an internal ring around the front of the eye. Within the ciliary body are many radiating folds called *ciliary processes* and two distinct groups of muscle fibers that constitute the *ciliary muscles.* Figure 12.26 shows these structures.

Many strong but delicate fibers, called *suspensory ligaments* (zonular fibers), extend inward from the ciliary processes and hold the transparent **lens** in position. The distal ends of these fibers are attached along the margin of a thin capsule that surrounds the lens. The body of the lens, which lacks blood vessels, lies directly behind the iris and pupil and is composed of specialized epithelial cells.

The cells of the lens originate from a single layer of epithelium beneath the anterior portion of the lens capsule. The cells divide, and the new cells on the surface of the lens capsule differentiate into columnar cells called

lens fibers, which constitute the substance of the lens. Lens fiber production continues slowly throughout life, thickening the lens from front to back. Simultaneously, the deeper lens fibers are compressed toward the center of the structure (fig. 12.27).

The lens capsule is a clear, membranelike structure largely composed of intercellular material. It is quite elastic, a quality that keeps it under constant tension. As a

FIGURE 12.27
A scanning electron micrograph of the long, flattened lens fibers (2,650×). Note the fingerlike junctions where one fiber joins another.

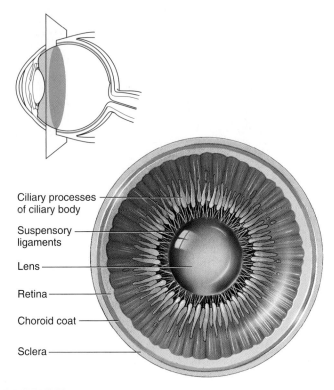

Ciliary processes
of ciliary body

Suspensory
ligaments

Lens

Retina

Choroid coat

Sclera

FIGURE 12.28
Lens and ciliary body viewed from behind.

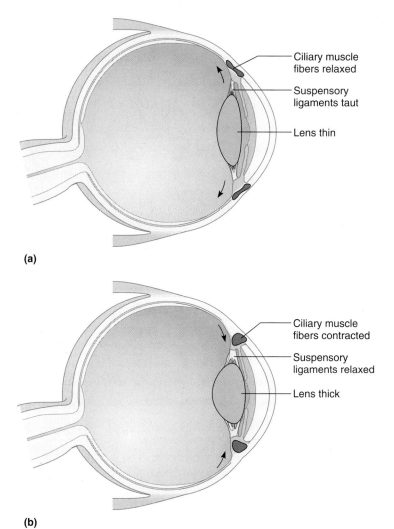

Ciliary muscle
fibers relaxed

Suspensory
ligaments taut

Lens thin

(a)

Ciliary muscle
fibers contracted

Suspensory
ligaments relaxed

Lens thick

(b)

FIGURE 12.29
In accommodation, (a) the lens thins as ciliary muscle fibers relax.
(b) The lens thickens as the ciliary muscle fibers contract.

result, the lens can assume a globular shape. However, the suspensory ligaments attached to the margin of the capsule are also under tension, and they pull outward, flattening the capsule and the lens (fig. 12.28).

If the tension on the suspensory ligaments relaxes, the elastic capsule rebounds, and the lens surface becomes more convex. This change occurs in the lens when the eye focuses to view a close object and is called **accommodation** (ah-kom″o-da′shun).

The ciliary muscles relax the suspensory ligaments during accommodation. One set of these muscle fibers forms a circular sphincterlike structure around the ciliary processes. The fibers of the other set extend back from fixed points in the sclera to the choroid coat. When the circular muscle fibers contract, the diameter of the ring formed by the ciliary processes decreases; when the other fibers contract, the choroid coat is pulled forward, and the ciliary body shortens. Both of these actions relax the suspensory ligaments, thickening the lens. In this thickened state, the lens is focused for viewing closer objects than before (fig. 12.29).

To focus on a distant object, the ciliary muscles relax, increasing tension on the suspensory ligaments. The lens thins again.

1 Describe the outer and middle tunics of the eye.

2 What factors contribute to the transparency of the cornea?

3 How does the shape of the lens change during accommodation?

The **iris** is a thin diaphragm mostly composed of connective tissue and smooth muscle fibers. Seen from the outside, it is the colored portion of the eye. The iris extends forward from the periphery of the ciliary body and lies between the cornea and the lens. It divides the space separating these parts, which is called the *anterior cavity,* into an *anterior chamber* (between the cornea and the iris) and a *posterior chamber* (between the iris and the vitreous humor, occupied by the lens).

The epithelium on the inner surface of the ciliary body continuously secretes a watery fluid called **aqueous humor** into the posterior chamber. The fluid circulates from this chamber through the **pupil,** a circular opening in the center of the iris, and into the anterior chamber (fig. 12.30). Aqueous humor fills the space between the cornea and the lens, providing nutrients and maintaining the shape of the front of the eye. It subsequently leaves the anterior chamber through veins and a special drainage canal, the scleral venous sinus (canal of Schlemm),

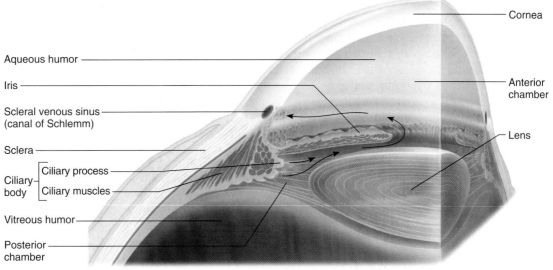

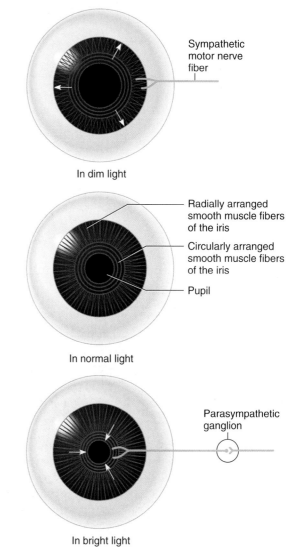

FIGURE 12.30

Aqueous humor (blue arrows), which is secreted into the posterior chamber, circulates into the anterior chamber and leaves it through the scleral venous sinus (canal of Schlemm).

located in its wall at the junction of the cornea and the sclera.

The smooth muscle fibers of the iris form two groups, a *circular set* and a *radial set*. These muscles control the size of the pupil, through which light passes. The circular set of muscle fibers acts as a sphincter, and when it contracts, the pupil gets smaller (constricts), and the intensity of the light entering decreases. When the radial muscle fibers contract, the diameter of the pupil increases (dilates), and the intensity of the light entering increases.

The sizes of the pupils change constantly in response to pupillary reflexes triggered by such factors as light intensity, gaze, accommodation, and variations in emotional state. For example, bright light elicits a reflex, and impulses travel along parasympathetic nerve fibers to the *circular muscles* of the irises. The pupils constrict in response. Conversely, in dim light, impulses travel on sympathetic nerve fibers to the *radial muscles* of the irises, and the pupils dilate (fig. 12.31).

The amount and distribution of melanin in the irises and the density of the tissue within the body of the iris determine eye color. If melanin is present only in the epithelial cells on the iris's posterior surface, the iris scatters more wavelengths of light, causing it to look blue or even green. When the same distribution of melanin occurs with denser-than-usual tissue within the body of the iris, it appears gray. When melanin is within the body of the iris as well as in the epithelial covering, the iris appears brown.

The Inner Tunic The inner tunic of the eye consists of the **retina** (ret'i-nah), which contains the visual receptor cells (photoreceptors). This nearly transparent sheet of tissue is continuous with the optic nerve in the back of the eye and extends forward as the inner lining of the eyeball. It ends just behind the margin of the ciliary body.

FIGURE 12.31

Dim light stimulates the radial muscles of the iris to contract, and the pupil dilates. Bright light stimulates the circular muscles of the iris to contract, and the pupil constricts.

The retina is thin and delicate, but its structure is quite complex. It has distinct layers, including pigmented epithelium, neurons, nerve fibers, and limiting membranes (figs. 12.32 and 12.33).

There are five major groups of retinal neurons. The nerve fibers of three of these groups—the *receptor cells, bipolar neurons,* and *ganglion cells*—provide a direct pathway for impulses triggered in the receptors to the optic nerve and brain. The nerve fibers of the other two groups of retinal cells, called *horizontal cells* and *amacrine cells,* pass laterally between retinal cells (see fig. 12.32). The horizontal and amacrine cells modify the impulses transmitted on the fibers of the direct pathway.

In the central region of the retina is a yellowish spot called the **macula lutea** that occupies about 1 square millimeter. A depression in its center, called the **fovea cen-**

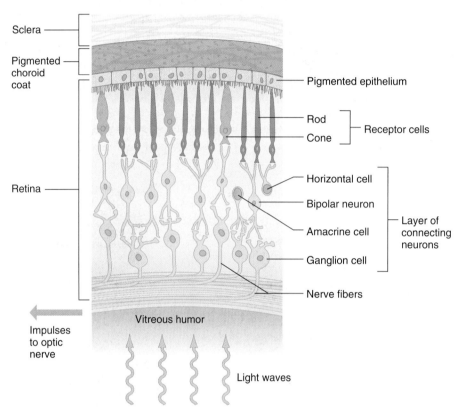

FIGURE 12.32
The retina consists of several cell layers.

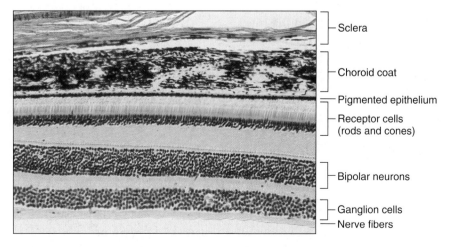

FIGURE 12.33
Note the layers of cells and nerve fibers in this light micrograph of the retina (75×).

tralis, is in the region of the retina that produces the sharpest vision.

Just medial to the fovea centralis is an area called the **optic disc** (fig. 12.34). Here the nerve fibers from the retina leave the eye and become parts of the optic nerve. A central artery and vein also pass through at the optic disc. These vessels are continuous with capillary networks of the retina, and together with vessels in the underlying choroid coat, they supply blood to the cells of the inner tunic. Because the optic disc lacks receptor cells, it is commonly referred to as the *blind spot* of the eye.

The space enclosed by the lens, ciliary body, and retina is the largest compartment of the eye and is called the *posterior cavity.* It is filled with a transparent, jellylike fluid called **vitreous humor,** which together with some collagenous fibers comprise the **vitreous body.** The vitreous body supports the internal structures of the eye and helps maintain its shape.

In summary, light waves entering the eye must pass through the cornea, aqueous humor, lens, vitreous humor, and several layers of the retina before they reach the photoreceptors (see fig. 12.32). Table 12.6 summarizes the layers of the eye.

1 Explain the origin of aqueous humor and trace its path through the eye.

2 How is the size of the pupil regulated?

3 Describe the structure of the retina.

Light Refraction

When a person sees something, either the object is giving off light, or light waves are reflected from it. These light waves enter the eye, and an image of what is seen focuses upon the retina. The light rays must bend to be focused, a phenomenon called **refraction** (re-frak′shun).

Refraction occurs when light waves pass at an oblique angle from a medium of one optical density into a medium of a different optical density. For example, as

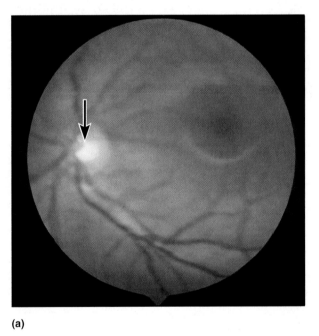

(a)

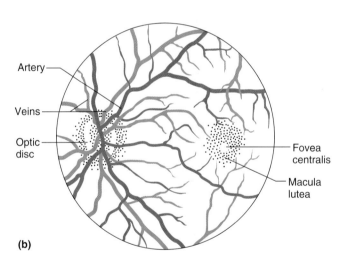

Artery —
Veins —
Optic disc —
— Fovea centralis
— Macula lutea

(b)

FIGURE 12.34
Retina. (*a*) Nerve fibers leave the eye in the area of the optic disc (arrow) to form the optic nerve (53×). (*b*) Major features of the retina.

TABLE 12.6	Layers of the Eye			
Layer/Tunic	**Posterior Portion**	**Function**	**Anterior Portion**	**Function**
Outer layer	Sclera	Protection	Cornea	Light transmission and refraction
Middle layer	Choroid coat	Blood supply, pigment prevents reflection	Ciliary body, iris	Accommodation; controls light intensity
Inner layer	Retina	Photoreception, impulse transmission	None	

figure 12.35 shows, when light passes obliquely from a less-dense medium such as air into a denser medium such as glass, or from air into the cornea of the eye, the light is bent toward a line perpendicular to the surface between these substances. When the surface between such refracting media is curved, a lens is formed. A lens with a *convex* surface causes light waves to converge,

and a lens with a *concave* surface causes light waves to diverge (fig. 12.36). Clinical Application 12.5 discusses some familiar problems with refraction.

The convex surface of the cornea refracts light waves from objects outside the eye, providing about 75% of the total refractive power of the eye. The light is refracted again by the convex surface of the lens and to a lesser extent by the surfaces of the fluids within the eye chambers.

If the shape of the eye is normal, light waves are focused sharply upon the retina, much as a motion-picture image is focused on a screen for viewing. Unlike the motion-picture image, however, the one formed on the retina is upside down and reversed from left to right (fig. 12.37). When the visual cortex of the cerebrum interprets such an image, it corrects this, and objects are seen in their real positions.

Light waves coming from objects more than 20 feet away are traveling in nearly parallel lines, and the cornea and the lens in its more flattened or "at rest" condition focuses the light waves on the retina. Light waves arriving from objects less than 20 feet away, however, reach the eye along more divergent lines—in fact, the closer the object, the more divergent the lines.

Divergent light waves tend to focus behind the retina unless something increases the refracting power of the eye. Accommodation accomplishes this increase, thickening

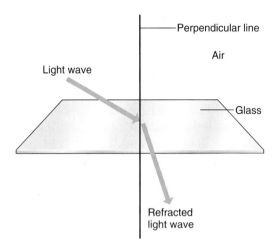

FIGURE 12.35
When light passes at an oblique angle from air into glass, the light waves bend toward a line perpendicular to the surface of the glass.

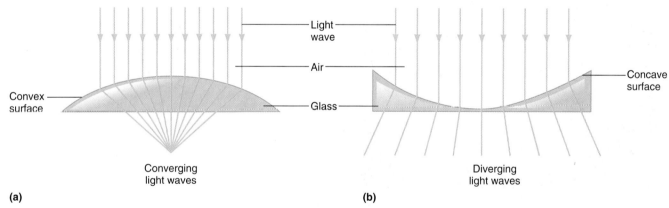

FIGURE 12.36
Light waves passing through a lens. (*a*) A lens with a convex surface causes light waves to converge. (*b*) A lens with a concave surface causes them to diverge.

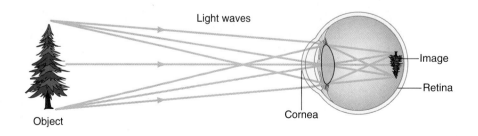

FIGURE 12.37
The image of an object forms upside down on the retina.

REFRACTION DISORDERS

The elastic quality of the lens capsule lessens with time. People over forty-five years of age are often unable to accommodate sufficiently to read the fine print in books and newspapers or on medicine bottles. Their eyes remain focused for distant vision. This condition is termed *presbyopia,* or farsightedness of age. Eyeglasses or contact lenses can usually make up for the eye's loss of refracting power.

Other visual problems result from eyeballs that are too short or too long for sharp focusing. If an eye is too short, light waves are not focused sharply on the retina because their point of focus lies behind it. A person with this condition may be able to bring the image of distant objects into focus by accommodation, but this requires contraction of the ciliary muscles at times when these muscles are at rest in a normal eye. Still more accommodation is necessary to view closer objects, and the person may suffer from ciliary muscle fatigue, pain, and headache when doing close work.

People with short eyeballs are usually unable to accommodate enough to focus on the very close objects. They are *farsighted.* Eyeglasses or contact lenses with *convex* surfaces can remedy this condition (hyperopia) by focusing images closer to the front of the eye.

If an eyeball is too long, light waves are focused in front of the retina, blurring the image. In other words, the refracting power of the eye, even when the lens is flattened, is too great. Although a person with this problem may be able to focus on close objects by accommodation, distance vision is invariably poor. For this reason, the person is said to be *nearsighted.* Eyeglasses or contact lenses with *concave* surfaces that focus images farther from the front of the eye treat nearsightedness (myopia) (figs. 12A and 12B).

Another refraction problem, *astigmatism,* reflects a defect in the curvature of the cornea or the lens. The normal cornea has a spherical curvature, like the inside of a ball; an astigmatic cornea usually has an elliptical curvature, like the bowl of a spoon. As a result, some portions of an image are in focus on the retina, but other portions are blurred, and vision is distorted.

Without corrective lenses, astigmatic eyes tend to accommodate back and forth reflexly in an attempt to sharpen focus. The consequence of this continual action is often ciliary muscle fatigue and headache. ■

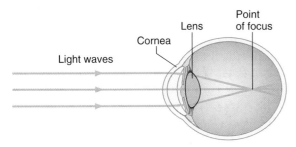

(a) Eye too long (myopia)

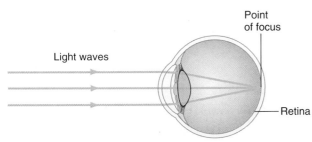

(b) Normal eye

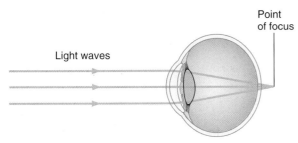

(c) Eye too short (hyperopia)

FIGURE 12A

Point of focus. (*a*) If an eye is too long, the focus point of images lies in front of the retina. (*b*) In a normal eye, the focus point is on the retina. (*c*) If an eye is too short, the focus point lies behind the retina.

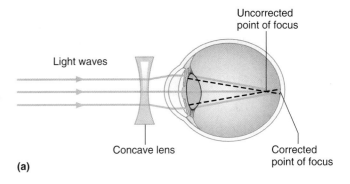

(a)

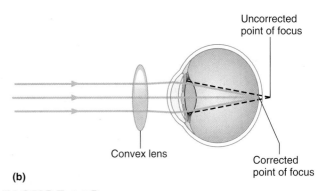

(b)

FIGURE 12B

Corrective lenses. (*a*) A concave lens corrects nearsightedness. (*b*) A convex lens corrects farsightedness.

the lens. As the lens thickens, light waves converge more strongly so that diverging light waves coming from close objects focus on the retina.

1. What is refraction?

2. What parts of the eye provide refracting surfaces?

3. Why is it necessary to accommodate for viewing close objects?

Visual Receptors

The photoreceptors of the eye are modified neurons of two distinct kinds. One group of receptor cells, called **rods,** have long, thin projections at their terminal ends. The cells of the other group, called **cones,** have short, blunt projections. The retina contains about 100 million rods and 3 million cones.

Rods and cones are found in a deep layer of the retina, closely associated with a layer of pigmented epithelium (see figs. 12.32 and 12.33). The projections from the receptors extend into the pigmented layer and contain light-sensitive visual pigments.

The epithelial pigment of the retina absorbs light waves that the receptor cells do not absorb, and together with the pigment of the choroid coat, the epithelial pigment keeps light from reflecting off the surfaces inside the eye. The pigment layer also stores vitamin A, which the receptor cells use to synthesize visual pigments.

> Researchers can grow retinal epithelial cells in laboratory cultures, and the cells retain their pigment. This means that someday scientists may be able to grow tissue that can be implanted into a person's eye to treat some forms of blindness.

The visual receptors are stimulated only when light reaches them. Thus, when a light image is focused on an area of the retina, some receptors are stimulated and send impulses to the brain. However, the impulse leaving each activated receptor provides only a small portion of the information required for the brain to interpret a total scene.

> *Albinism* is an inherited condition in which an enzyme necessary to produce pigment is missing, causing very pale, highly sun-sensitive skin. More severe forms of albinism also affect the eyes, making vision blurry and intolerant to light. A person may squint even in very faint light. The extrasensitivity is due to the fact that light reflects inside the lenses, over-stimulating visual receptors. The eyes of many people with albinism also dart about uncontrollably, a condition called *nystagmus.*

Rods and cones function differently. Rods are hundreds of times more sensitive to light than are cones, and as a result, rods provide vision in dim light. In addition, rods produce colorless vision, whereas cones can detect colors.

Cones provide sharp images, whereas rods produce more general outlines of objects. This difference is due to the fact that nerve fibers from many rods may converge, and their impulses may be transmitted to the brain on the same nerve fiber (see chapter 10, p. 359). Thus, if light stimulates a rod, the brain cannot tell which one of many receptors has actually been stimulated. Such a convergence of impulses occurs to a much lesser degree among cones, so when a cone is stimulated, the brain is able to pinpoint the stimulation more accurately (fig. 12.38).

The area of sharpest vision, the fovea centralis in the macula lutea, lacks rods but contains densely packed cones with few or no converging fibers. Also, the overlying layers of the retina, as well as the retinal blood vessels, are displaced to the sides in the fovea, which more fully exposes the receptors to incoming light. Consequently, to view something in detail, a person moves the eyes so that the important part of an image falls upon the fovea centralis.

The concentration of cones decreases in areas farther away from the macula lutea, whereas the concentration of rods increases in these areas. Also, the degree of convergence among the rods and cones increases toward the periphery of the retina. As a result, the visual sensations from images focused on the sides of the retina tend to be blurred compared with those focused on the central portion of the retina.

> The bony orbit usually protects the eye, but a forceful blow can displace structures within and around the eye. The suspensory ligaments may tear and the lens become dislocated into the posterior cavity, or the retina may pull away from the underlying vascular choroid coat. Once the retina is detached, photoreceptor cells may die because of lack of oxygen and nutrients. Unless such a *detached retina* is repaired surgically, this injury may cause visual loss or blindness.

Visual Pigments

Both rods and cones contain light-sensitive pigments that decompose when they absorb light energy. The light-sensitive pigment in rods is called **rhodopsin** (ro-dop′sin), or visual purple, and it is embedded in membranous discs that are stacked within these receptor cells (fig. 12.39). A single rod cell may have 2,000 interconnected discs, derived from the cell membrane. In the presence of light, rhodopsin molecules break down into molecules of a colorless protein called *opsin* and a yellowish organic

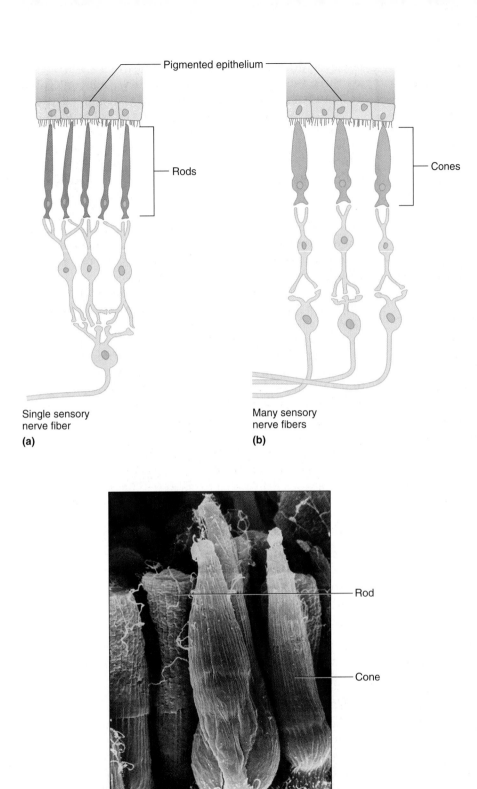

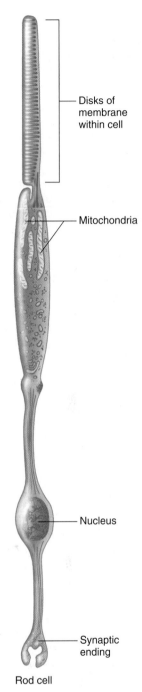

Disks of membrane within cell

Mitochondria

Nucleus

Synaptic ending

Rod cell

FIGURE 12.39
Rhodopsin is embedded in discs of membrane that are stacked within the rod cells.

Pigmented epithelium

Rods

Cones

Single sensory nerve fiber

(a)

Many sensory nerve fibers

(b)

Rod

Cone

(c)

FIGURE 12.38
Rods and cones. (*a*) A single sensory nerve fiber transmits impulses from several rods to the brain. (*b*) Separate sensory nerve fibers transmit impulses from cones to the brain. (*c*) Scanning electron micrograph of rods and cones (1,350×).

molecule called *retinal* (retinene) that is synthesized from vitamin A.

In darkness, sodium channels in portions of the receptor cell membranes are kept open by a nucleotide called *cyclic guanosine monophosphate* (cGMP). When rhodopsin molecules absorb light, they change shape and release opsin, in mere trillionths of a second. The released opsin then becomes an active enzyme. This enzyme activates a second enzyme (transducin), which, in turn, activates still another enzyme (phosphodiesterase). The third enzyme of this series breaks down cGMP, and as the concentration of cGMP decreases, sodium channels close, and the receptor cell membrane hyperpolarizes (see chapter 10, p. 350). The degree of hyperpolarization is directly proportional to the intensity of the light stimulating the receptor cells.

The hyperpolarization reaches the synaptic end of the cell, inhibiting release of neurotransmitter. Through a complex mechanism, decreased release of neurotransmitter by photoreceptor cells either stimulates or inhibits nerve impulses (action potentials) in nearby retinal neurons. Consequently, complex patterns of nerve impulses travel away from the retina, through the optic nerve, and into the brain, where they are interpreted as vision.

In bright light, nearly all of the rhodopsin in the rods decomposes, sharply reducing the sensitivity of these receptors. The cones continue to function, however, and in bright light, we therefore see in color. In dim light, rhodopsin can be regenerated from opsin and retinal faster than it is broken down. This regeneration requires cellular energy, which ATP provides (see chapter 4, p. 108). Under these conditions, the rods continue to function and the cones remain unstimulated. Hence, we see only shades of gray in dim light.

The light sensitivity of an eye whose rods have converted the available opsin and retinal to rhodopsin increases about 100,000 times, and the eye is said to be *dark adapted.* A person needs a dark-adapted eye to see in dim light. For example, when going from daylight into a darkened theater, it may be difficult to see well enough to locate a seat, but soon the eyes adapt to the dim light, and vision improves. Later, leaving the theater and entering the sunlight may cause discomfort or even pain. This occurs at the moment that most of the rhodopsin decomposes in response to the bright light. At the same time, the light sensitivity of the eyes decreases greatly, and they become *light adapted.*

Too little vitamin A in the diet reduces the quantity of retinal, impairing rhodopsin production and sensitivity of the rods. The result is poor vision in dim light, called night blindness.

The light-sensitive pigments of cones, called *iodopsins,* are similar to rhodopsin in that they are composed of retinal combined with a protein; the protein, however, differs from the protein in the rods. In fact, there are three sets of cones within the retina, each containing an abundance of one of three different visual pigments.

The wavelength of a particular kind of light determines the color perceived from it. For example, the shortest wavelengths of visible light are perceived as violet, whereas the longest wavelengths of visible light are seen as red. One type of cone pigment (erythrolabe) is most sensitive to red light waves, another (chlorolabe) to green light waves, and a third (cyanolabe) to blue light waves. The sensitivities of these pigments do overlap somewhat. For example, both red and green light pigments are sensitive to orange light waves. On the other hand, red pigment absorbs orange light waves more effectively.

The color perceived depends upon which sets of cones the light in a given image stimulates. If all three types of sets of cones are stimulated, the light is perceived as white, and if none are stimulated, it is seen as black.

Examination of the retinas of different people reveals that individuals have unique patterns of cone types, all apparently able to provide color vision. Some parts of the retina are even normally devoid of one particular type, yet the brain integrates information from all over to "fill in the gaps," creating a continuous overall image. People who lack a cone type, though, due to a mutation, are color blind.

As primates, we humans enjoy a more multicolored world than many other mammals. This is because the visual systems of nonprimate mammals funnel input from groups of photoreceptor cells into the CNS. That is, several photoreceptors signal the same bipolar neurons, which, in turn, pool their input to ganglion cells. Primates are the only mammals to have three types of cones (others have two), and it appears that primates excel in color vision because the rods and cones connect individually to neural pathways to the brain.

Stereoscopic Vision

Stereoscopic vision (stereopsis) simultaneously perceives distance, depth, height, and width of objects. Such vision is possible because the pupils are 6–7 centimeters apart. Consequently, objects that are close (less than 20 feet away) produce slightly different retinal images. That is, the right eye sees a little more of one side of an object, while the left eye sees a little more of the other side. These two images are somehow superimposed and interpreted by the visual cortex of the brain. The result is the perception of a single object in three dimensions (fig. 12.40).

Because stereoscopic vision depends on vision with two eyes (binocular vision), it follows that a one-eyed person is less able to judge distance and depth accurately. To compensate, a person with one eye can use the relative sizes and positions of familiar objects as visual clues.

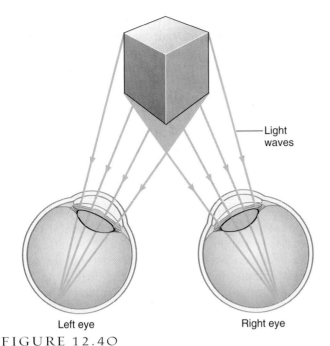

FIGURE 12.40
Stereoscopic vision results from formation of two slightly different retinal images.

Left eye   Right eye

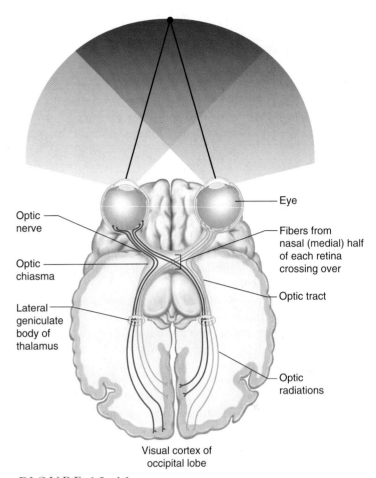

FIGURE 12.41
The visual pathway includes the optic nerve, optic chiasma, optic tract, and optic radiations.

> A woman had a stroke that damaged part of her visual cortex, so that she could no longer integrate images to perceive motion. She saw movement as a series of separate, static images. Her deficit had profound effects on her life. She could not pour a drink, because she could not tell when the cup would overflow. She could not cross a street because she could not detect cars moving toward her.

optic radiations, and the pathways lead to the visual cortex of the occipital lobes (fig. 12.41).

> Since each visual cortex receives impulses from each eye, a person may develop partial blindness in both eyes if either visual cortex is injured. For example, if the right visual cortex (or the right optic tract) is injured, sight may be lost in the temporal side of the right eye and the nasal side of the left eye. Similarly, damage to the central portion of the optic chiasma, where fibers from the nasal sides of the eyes cross over, blinds the nasal sides of both eyes.

Visual Nerve Pathways

As mentioned in chapter 11 (p. 399), the axons of the ganglion cells in the retina leave the eyes to form the *optic nerves.* Just anterior to the pituitary gland, these nerves give rise to the X-shaped *optic chiasma,* and within the chiasma, some of the fibers cross over. More specifically, the fibers from the nasal (medial) half of each retina cross over, whereas those from the temporal (lateral) sides do not. Thus, fibers from the nasal half of the left eye and the temporal half of the right eye form the right *optic tract;* fibers from the nasal half of the right eye and the temporal half of the left eye form the left optic tract.

The nerve fibers continue in the optic tracts, and just before they reach the thalamus, a few of them leave to enter nuclei that function in various visual reflexes. Most of the fibers, however, enter the thalamus and synapse in its posterior portion (lateral geniculate body). From this region, the visual impulses enter nerve pathways called

Fibers not leading to the thalamus conduct visual impulses downward into the brainstem. These impulses are important for controlling head and eye movements associated with tracking an object visually, for controlling the simultaneous movements of both eyes, and for controlling certain visual reflexes, such as those that move the iris muscles.

1 Distinguish between the rods and the cones of the retina.

2 Explain the roles of visual pigments.

3 What factors make stereoscopic vision possible?

4 Trace the pathway of visual impulses from the retina to the occipital cortex.

Life-Span Changes

We often first become aware of aging-associated changes through diminished senses. By age forty, a book may need to be held farther away from the eyes so that the person can focus on what seems to be print that is smaller than it used to be. By the fifties, the senses of smell and taste may begin to diminish, which usually reflects anosmia, a loss of olfactory receptors.

By age sixty, a quarter of the population experiences noticeable hearing loss, and from ages sixty-five to seventy-four, the percentage reaches a third. Half of all people over age eighty-five cannot hear adequately. Age-related hearing loss may be the result of decades of cumulative damage to the sensitive hair cells of the organ of Corti in the inner ear. It becomes more difficult to hear high pitches, as well as particular sounds, such as *f, g, s, t, sh, th, z* and *ch.* Hearing loss may also be due to a degeneration or failure of nerve pathways to the brain. This condition, called presbycusis, may affect the ability to understand speech. It usually worsens very gradually. Tinnitus is also more common among older adults than other age groups. This is a ringing or roaring in the ears, which may be persistent or intermittent. Hearing aids can often restore some hearing. It is important for those who live with people suffering from hearing loss to recognize the problem. A person to whom the ordinary sounds of life are hopelessly garbled may show, quite understandably, symptoms of paranoia, depression, or social withdrawal.

Vision may decline with age for several reasons. "Dry eyes" are common. Too few tears, or poor quality tears, lead to itching and burning eyes, and diminished vision. In some cases, too many tears result from oversensitivity to environmental effects, such as wind, intense light, or a change in temperature.

With age, tiny dense clumps of gel or crystal-like deposits form in the vitreous humor. When these clumps cast shadows on the retina, the person sees small moving specks in the field of vision. These specks, or *floaters,* are most apparent when looking at a plain background, such as the sky or a blank wall. Also with age, the vitreous humor may shrink and pull away from the retina. This may mechanically stimulate receptor cells of the retina, and the person may see flashes of light.

The inability to read small print up close, called presbyopia, results from a loss of elasticity in the lens, preventing it from changing shape easily. After age seventy, the iris cannot dilate as well as it once did, cutting the amount of light that can enter the eye by 50%. Brighter lights can counter this effect.

A disorder called *glaucoma* sometimes develops in the eyes as a person ages when the rate of aqueous humor formation exceeds the rate of its removal. Fluid accumulates in the anterior chamber of the eye, and the fluid pressure rises. As this pressure is transmitted to all parts of the eye, in time, the blood vessels that supply the receptor cells of the retina may squeeze shut, cutting off the nutrient and oxygen supply. The result may eventually be permanent blindness.

Drugs, or traditional or laser surgery to promote the outflow of aqueous humor can treat glaucoma if it is diagnosed early. However, since glaucoma in its early stages typically produces no symptoms, discovery of the condition usually depends on measuring the intraocular pressure using an instrument called a *tonometer.*

A common eye disorder particularly in older people is *cataract.* The lens or its capsule slowly becomes cloudy, opaque, and discolored, adding a yellowish tinge to a person's view of the world. Clear images cannot focus on the retina, and in time, the person may become blind. Cataract is often treated by removing the lens with a laser and replacing it with an artificial implant. Afterwards, patients report that their surroundings are no longer yellow!

Several conditions affect the retinas of an older person. Age-related macular degeneration is an impairment of the macula, the most sensitive part of the retina, which blurs images. Retinal detachment becomes more common. People with diabetes are at high risk of developing diabetic retinopathy, which is an interference with retinal function because of damaged blood vessels or growth of new ones that block vision.

Despite these various problems, many older individuals continue to enjoy sharp, functional senses well into the upper decades of life.

1 Why do smell and taste diminish with age?

2 What are some causes of age-related hearing loss?

3 Describe visual problems that are likely to arise with age.

Clinical Terms Related to the Senses

amblyopia (am"ble-o'pe-ah) Dim vision due to a cause other than a refractive disorder or lesion.
anopia (an-o'pe-ah) Absence of an eye.
audiometry (aw"de-om'ĕ-tre) Measurement of auditory acuity for various frequencies of sound waves.
blepharitis (blef"ah-ri'tis) Inflammation of the margins of the eyelids.

causalgia (kaw-zal′je-ah) Persistent, burning pain usually associated with injury to a limb.

conjunctivitis (kon-junk″tĭ-vi′tis) Inflammation of the conjunctiva.

diplopia (di-plo′pe-ah) Double vision, or the sensation of seeing two objects when only one is viewed.

emmetropia (em″ĕ-tro′pe-ah) Normal condition of the eyes; eyes with no refractive defects.

enucleation (e-nu″kle-a′shun) Removal of the eyeball.

exophthalmos (ek″sof-thal′mos) Condition in which the eyes protrude abnormally.

hemianopsia (hem″e-an-op′se-ah) Defective vision affecting half of the visual field.

hyperalgesia (hi″per-al-je′ze-ah) Abnormally increased sensitivity to pain.

iridectomy (ir″ĭ-dek′to-me) Surgical removal of part of the iris.

iritis (i-ri′tis) Inflammation of the iris.

keratitis (ker″ah-ti′tis) Inflammation of the cornea.

labyrinthectomy (lab″i-rin-thek′to-me) Surgical removal of the labyrinth.

labyrinthitis (lab″i-rin-thi′tis) Inflammation of the labyrinth.

Ménière's disease (men″e-ārz′ dĭ-zez) Inner ear disorder characterized by ringing in the ears, increased sensitivity to sounds, dizziness, and loss of hearing.

neuralgia (nu-ral′je-ah) Pain resulting from inflammation of a nerve or a group of nerves.

neuritis (nu-ri′tis) Inflammation of a nerve.

nystagmus (nis-tag′mus) Involuntary oscillation of the eyes.

otitis media (o-ti′tis me′de-ah) Inflammation of the middle ear.

otosclerosis (o″to-skle-ro′sis) Formation of spongy bone in the inner ear, which often causes deafness by fixing the stapes to the oval window.

pterygium (tĕ-rij′e-um) Abnormally thickened patch of conjunctiva that extends over part of the cornea.

retinitis pigmentosa (ret″ĭ-ni′tis pig″men-to′sa) Inherited progressive retinal sclerosis that deposits pigment in the retina and causes retinal atrophy.

retinoblastoma (ret″i-no-blas-to′mah) Inherited, highly malignant tumor arising from immature retinal cells.

tinnitus (tĭ-ni′tus) Ringing or buzzing in the ears.

tonometry (to-nom′e-tre) Measurement of fluid pressure within the eyeball.

trachoma (trah-ko′mah) Bacterial disease of the eye, characterized by conjunctivitis, which may lead to blindness.

tympanoplasty (tim″pah-no-plas′te) Surgical reconstruction of the middle ear bones and establishment of continuity from the tympanic membrane to the oval window.

uveitis (u″ve-i′tis) Inflammation of the uvea, the region of the eye that includes the iris, the ciliary body, and the choroid coat.

vertigo (ver′tĭ-go) Sensation of dizziness.

CHAPTER SUMMARY

Introduction (page 422)

Sensory receptors are sensitive to environmental changes and initiate impulses to the brain and spinal cord.

Receptors and Sensations (page 422)

1. Receptor types
   a. Each type of receptor is sensitive to a distinct type of stimulus.
   b. The major types of receptors include the following:
      (1) Chemoreceptors, sensitive to changes in chemical concentration.
      (2) Pain receptors, sensitive to tissue damage.
      (3) Thermoreceptors, sensitive to temperature changes.
      (4) Mechanoreceptors, sensitive to mechanical forces.
      (5) Photoreceptors, sensitive to light.
2. Sensory impulses
   a. When receptors are stimulated, changes occur in their membrane potentials.
   b. Receptor potentials are transferred to nerve fibers, triggering action potentials.
3. Sensations
   a. Sensations are feelings resulting from sensory stimulation.
   b. A particular part of the sensory cortex interprets every impulse that reaches it in the same way.
   c. The cerebral cortex projects a sensation back to the region of stimulation.
4. Sensory adaptations are adjustments of sensory receptors to continuous stimulation. Impulses are triggered at slower and slower rates.

Somatic Senses (page 424)

Somatic senses receive information from receptors in skin, muscles, joints, and viscera. They can be grouped as exteroceptive, proprioceptive, and visceroceptive senses.

1. Touch and pressure senses
   a. Free ends of sensory nerve fibers are the receptors for the sensations of touch and pressure.
   b. Meissner's corpuscles are the receptors for the sensations of light touch.
   c. Pacinian corpuscles are the receptors for the sensations of heavy pressure and vibrations.

2. Thermoreceptors include two sets of free nerve endings that are heat and cold receptors.
3. Sense of pain
   a. Pain receptors
      (1) Pain receptors are free nerve endings that tissue damage stimulates.
      (2) Pain receptors provide protection, do not adapt rapidly, and can be stimulated by changes in temperature, mechanical force, and chemical concentration.
   b. The only receptors in viscera that provide sensations are pain receptors. These receptors are most sensitive to certain chemicals and lack of blood flow. The sensations they produce feel as if they come from some other part of the body (referred pain).
   c. Pain nerve pathways
      (1) The two main types of pain fibers are acute pain fibers and chronic pain fibers.
      (2) Acute pain fibers are fast conducting; chronic pain fibers are slower conducting.
      (3) Pain impulses are processed in the dorsal horn of the spinal cord, and they ascend in the spinothalamic tracts.
      (4) Within the brain, pain impulses pass through the reticular formation before being conducted to the cerebral cortex.
   d. Regulation of pain impulses
      (1) Awareness of pain occurs when impulses reach the thalamus.
      (2) The cerebral cortex judges the intensity of pain and locates its source.
      (3) Impulses descending from the brain cause neurons to release pain-relieving substances, such as enkephalins and serotonin.
      (4) Endorphin is a pain-relieving biochemical produced in the brain.
   e. Certain neuropeptides synthesized in the brain and spinal cord inhibit pain impulses.
4. Stretch receptors
   a. Stretch receptors provide information about the condition of muscles and tendons.
   b. Muscle spindles are stimulated when a muscle is relaxed, and they initiate a reflex that contracts the muscle.
   c. Golgi tendon organs are stimulated when muscle tension increases, and they initiate a reflex that relaxes the muscle.

Special Senses (page 430)

Special senses are those whose receptors occur in relatively large, complex sensory organs of the head.

1. Sense of smell
   a. Olfactory receptors
      (1) Olfactory receptors are chemoreceptors that chemicals dissolved in liquid stimulate.
      (2) Olfactory receptors function together with taste receptors and aid in food selection.
   b. Olfactory organs
      (1) The olfactory organs consist of receptors and supporting cells in the nasal cavity.
      (2) Olfactory receptors are neurons with cilia that sense lipid-soluble chemicals.
   c. Olfactory nerve pathways.
      (1) Nerve impulses travel from the olfactory receptors through the olfactory nerves, olfactory bulbs, and olfactory tracts.
      (2) They go to interpreting centers in the limbic system of the brain.
   d. Olfactory stimulation
      (1) Olfactory impulses may result when various gaseous molecules combine with specific sites on the cilia of the receptor cells.
      (2) Olfactory receptors adapt rapidly.
      (3) Olfactory receptors are often damaged by environmental factors and are replaced from a pool of stem cells.
2. Sense of taste
   a. Taste receptors
      (1) Taste buds consist of receptor cells and supporting cells.
      (2) Taste cells have taste hairs that are sensitive to particular chemicals dissolved in water.
      (3) Taste hair surfaces have receptor sites to which chemicals combine and trigger impulses to the brain.
   b. Taste sensations
      (1) The four primary taste sensations are sweet, sour, salty, and bitter.
      (2) Various taste sensations result from the stimulation of one or more sets of taste receptors.
      (3) Each of the four primary kinds of taste cells is particularly sensitive to a certain group of chemicals.
   c. Taste nerve pathways
      (1) Sensory impulses from taste receptors travel on fibers of the facial, glossopharyngeal, and vagus nerves.
      (2) These impulses are carried to the medulla and ascend to the thalamus and then to the gustatory cortex in the parietal lobes.
3. Sense of hearing
   a. The external ear includes the auricle, the external acoustic meatus, and the tympanic membrane. It collects sound waves created by vibrating objects.
   b. Middle ear
      (1) Auditory ossicles of the middle ear conduct sound waves from the tympanic membrane to the oval window of the inner ear. They also increase the force of these waves.
      (2) Skeletal muscles attached to the auditory ossicles provide the tympanic reflex, which protects the inner ear from the effects of loud sounds.
   c. Auditory tubes connect the middle ears to the throat and help maintain equal air pressure on both sides of the tympanic membranes.
   d. Inner ear
      (1) The inner ear consists of a complex system of connected tubes and chambers—the osseous and membranous labyrinths. It includes the cochlea, which houses the organ of Corti.
      (2) The organ of Corti contains the hearing receptors that vibrations in the fluids of the inner ear stimulate.
      (3) Different frequencies of vibrations stimulate different sets of receptor cells; the human ear can detect sound frequencies from about 20 to 20,000 vibrations per second.

e. Auditory nerve pathways
   (1) The nerve fibers from hearing receptors travel in the cochlear branch of the vestibulocochlear nerves.
   (2) Auditory impulses travel into the medulla oblongata, midbrain, and thalamus and are interpreted in the temporal lobes of the cerebrum.
4. Sense of equilibrium
   a. Static equilibrium maintains the stability of the head and body when they are motionless. The organs of static equilibrium are located in the vestibule.
   b. Dynamic equilibrium balances the head and body when they are moved or rotated suddenly. The organs of this sense are located in the ampullae of the semicircular canals.
   c. Other structures that help maintain equilibrium include the eyes and the proprioceptors associated with certain joints.
5. Sense of sight
   a. Visual accessory organs include the eyelids and lacrimal apparatus that protect the eye and the extrinsic muscles that move the eye.
   b. Structure of the eye
      (1) The wall of the eye has an outer, a middle, and an inner tunic that function as follows:
         (a) The outer layer (sclera) is protective, and its transparent anterior portion (cornea) refracts light entering the eye.
         (b) The middle layer (choroid coat) is vascular and contains pigments that help keep the inside of the eye dark.
         (c) The inner layer (retina) contains the visual receptor cells.
      (2) The lens is a transparent, elastic structure. The ciliary muscles control its shape.
      (3) The iris is a muscular diaphragm that controls the amount of light entering the eye; the pupil is an opening in the iris.
      (4) Spaces within the eye are filled with fluids (aqueous and vitreous humors) that help maintain its shape.
   c. Light refraction
      (1) Light waves are primarily refracted by the cornea and lens to focus an image on the retina.
      (2) The lens must thicken to focus on close objects.

d. Visual receptors
   (1) The visual receptors are rods and cones.
   (2) Rods are responsible for colorless vision in relatively dim light, and cones provide color vision.
e. Visual pigments
   (1) A light-sensitive pigment in rods (rhodopsin) decomposes in the presence of light and triggers a complex series of reactions that initiate nerve impulses on the optic nerve.
   (2) Three sets of cones provide color vision. Each set contains a different light-sensitive pigment, and each set is sensitive to a different wavelength of light; the color perceived depends on which set or sets of cones are stimulated.
f. Stereoscopic vision
   (1) Stereoscopic vision provides perception of distance and depth.
   (2) Stereoscopic vision occurs because of the formation of two slightly different retinal images that the brain superimposes and interprets as one image in three dimensions.
   (3) A one-eyed person uses relative sizes and positions of familiar objects to judge distance and depth.
g. Visual nerve pathways
   (1) Nerve fibers from the retina form the optic nerves.
   (2) Some fibers cross over in the optic chiasma.
   (3) Most of the fibers enter the thalamus and synapse with others that continue to the visual cortex of the occipital lobes.
   (4) Other impulses pass into the brainstem and function in various visual reflexes.

Life-Span Changes (page 462)

Diminished senses are often one of the first noticeable signs of aging.

1. Age-related hearing loss may reflect damage to hair cells of the organ of Corti, degeneration of nerve pathways to the brain, or tinnitus.
2. Age-related visual problems include dry eyes, floaters and light flashes, presbyopia, glaucoma, cataracts, retinal detachment, and macular degeneration.

CRITICAL THINKING QUESTIONS

1. How would you interpret the following observation? A person enters a tub of water and reports that it is too warm, yet a few moments later says the water feels comfortable, even though the water temperature is unchanged.

2. Why are some serious injuries, such as those produced by a bullet entering the abdomen, relatively painless, whereas others, such as those that crush the skin, are quite painful?

3. Labyrinthitis is an inflammation of the tissues of the inner ear. What symptoms would you expect to observe in a patient with this disorder?

4. Sometimes, as a result of an injury to the eye, the retina detaches from its pigmented epithelium. Assuming that the retinal tissues remain functional, what is likely to happen to the person's vision if the retina moves unevenly toward the interior of the eye?

5. The auditory tubes of a child are shorter and directed more horizontally than those of an adult. How might this explain the greater prevalence of middle ear infections in children compared to adults?

6. A patient with heart disease experiences pain at the base of the neck and in the left shoulder and arm after exercise. How would you explain to the patient the origin of this pain?

REVIEW EXERCISES

1. List five groups of sensory receptors, and name the kind of change to which each is sensitive.
2. Explain how sensory receptors stimulate sensory impulses.
3. Define *sensation*.
4. Explain the projection of a sensation.
5. Define *sensory adaptation*.
6. Explain how somatic senses can be grouped.
7. Describe the functions of free nerve endings, Meissner's corpuscles, and Pacinian corpuscles.
8. Explain how thermoreceptors function.
9. Compare pain receptors with other types of somatic receptors.
10. List the factors that are likely to stimulate visceral pain receptors.
11. Define *referred pain*.
12. Explain how neuropeptides relieve pain.
13. Distinguish between muscle spindles and Golgi tendon organs.
14. Explain how the senses of smell and taste function together to create the perception of the flavors of foods.
15. Describe the olfactory organ and its function.
16. Trace a nerve impulse from the olfactory receptor to the interpreting centers of the brain.
17. Explain how an olfactory code distinguishes odor stimuli.
18. Explain how the salivary glands aid the taste receptors.
19. Name the four primary taste sensations, and indicate a stimulus for each.
20. Explain why taste sensation is less likely to diminish with age than olfactory sensation.
21. Trace the pathway of a taste impulse from the receptor to the cerebral cortex.
22. Distinguish among the external, middle, and inner ear.
23. Trace the path of a sound vibration from the tympanic membrane to the hearing receptors.
24. Describe the functions of the auditory ossicles.
25. Describe the tympanic reflex, and explain its importance.
26. Explain the function of the auditory tube.
27. Distinguish between the osseous and the membranous labyrinths.
28. Describe the cochlea and its function.
29. Describe a hearing receptor.
30. Explain how a hearing receptor stimulates a sensory neuron.
31. Trace a nerve impulse from the organ of Corti to the interpreting centers of the cerebrum.
32. Describe the organs of static and dynamic equilibrium and their functions.
33. Explain how the sense of vision helps maintain equilibrium.
34. List the accessory organs that aid in maintaining equilibrium, and describe the functions of each.
35. Name the three layers of the eye wall, and describe the functions of each.
36. Describe how accommodation is accomplished.
37. Explain how the iris functions.
38. Distinguish between aqueous humor and vitreous humor.
39. Distinguish between the macula lutea and the optic disc.
40. Explain how light waves focus on the retina.
41. Distinguish between rods and cones.
42. Explain why cone vision is generally more acute than rod vision.
43. Describe the function of rhodopsin.
44. Explain how the eye adapts to light and dark.
45. Describe the relationship between light wavelengths and color vision.
46. Define *stereoscopic vision*.
47. Explain why a person with binocular vision is able to judge distance and depth of close objects more accurately than a one-eyed person.
48. Trace a nerve impulse from the retina to the visual cortex.

WEB CONNECTIONS

Visit the Student Edition of the Online Learning Center at www.mhhe.com/shier10 **for answers to chapter questions, additional quizzes, interactive learning exercises, and other study tools.**

13

UNDERSTANDING WORDS

cort-, bark, rind: adrenal *cort*ex—outer portion of an adrenal gland.

-crin, to secrete: endo*crine*—pertaining to internal secretions.

diuret-, to pass urine: *diuret*ic—substance that promotes the production of urine.

endo-, within: *endo*crine gland—gland that releases its secretion internally into a body fluid.

exo-, outside: *exo*crine gland—gland that releases its secretion to the outside through a duct.

horm-, impetus, impulse: *horm*one—substance that a cell secretes that affects another cell.

hyper-, above: *hyper*thyroidism—condition resulting from an above-normal secretion of thyroid hormone.

hypo-, below: *hypo*thyroidism—condition resulting from a below-normal secretion of thyroid hormone.

lact-, milk: pro*lact*in—hormone that promotes milk production.

med-, middle: adrenal *med*ulla—middle section of an adrenal gland.

para-, beside: *para*thyroid glands—set of glands located near the surface of the thyroid gland.

toc-, birth: oxy*toc*in—hormone that stimulates the uterine muscles to contract during childbirth.

-tropic, influencing: adrenocortico*tropic* hormone—a hormone secreted by the anterior pituitary gland that stimulates the adrenal cortex.

vas-, vessel: *vas*opressin—substance that causes blood vessel walls to contract.

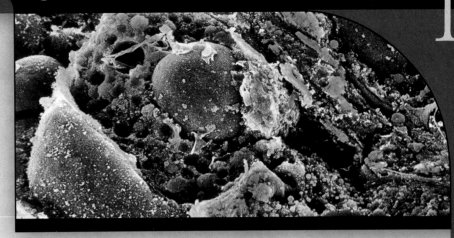

These cells in the adrenal cortex secrete glucocorticoid hormones, which have several effects on metabolism (9,300×).

Endocrine System

CHAPTER OBJECTIVES

After you have studied this chapter, you should be able to

1. Distinguish between endocrine and exocrine glands.

2. Describe how hormones can be classified according to their chemical composition.

3. Explain how steroid and nonsteroid hormones affect target cells.

4. Discuss how negative feedback mechanisms regulate hormonal secretion.

5. Explain how the nervous system controls hormonal secretion.

6. Name and describe the locations of the major endocrine glands and list the hormones they secrete.

7. Describe the general functions of the various hormones.

8. Explain how the secretion of each hormone is regulated.

9. Distinguish between physical and psychological stress.

10. Describe the general stress response.

11. Describe some of the changes associated with aging of the endocrine system.

he sweet-smelling urine that is the hallmark of type I (insulin-dependent) diabetes mellitus was noted as far back as an Egyptian papyrus from 1500 B.C. In A.D. 96 in Greece, Aretaeus of Cappadocia described the condition as a "melting down of limbs and flesh into urine." One of the first to receive as a drug insulin, a hormone that his body could not produce, was a three-year-old boy. In December 1922, before treatment, he weighed only 15 pounds. The boy rapidly improved after beginning insulin treatment, doubling his weight in just two months.

Insulin and the gland that produces it—the pancreas—are familiar components of the endocrine system. Understanding type I diabetes mellitus provides a fascinating glimpse into the evolution of medical technology that continues today.

In 1921, Canadian physiologists Sir Frederick Grant Banting and Charles Herbert Best discovered the link between lack of insulin and diabetes. They induced diabetes symptoms in a dog by removing its pancreas, then cured it by administering insulin from another dog's healthy pancreas. Just a year later, people with diabetes—such as the starving three-year-old—began to receive insulin extracted from pigs or cattle.

So it went until 1982, when pure human insulin became available by genetically altering bacteria to produce the human protein. Human insulin helped those who were allergic to the product from pigs or cows. Today, people receive insulin in a variety of ways, discussed in Clinical Application 13.4. Although a person with type I diabetes mellitus today is considerably healthier than the boy on the brink of the discovery of insulin, implants, injections, and aerosols to deliver insulin cannot exactly duplicate the function of the pancreas. Better understanding of the endocrine system will lead to better treatment of this and other hormonal disorders. ◼

General Characteristics of the Endocrine System

The **endocrine system** is so named because the cells, tissues, and organs that comprise it, collectively called endocrine glands, secrete substances into the internal environment. The secreted substances, called **hormones,** diffuse from the interstitial fluid into the bloodstream and eventually act on cells, called **target cells,** some distance away.

Other glands secrete substances into the internal environment, and although they are not hormones by the traditional definition, they function in similar fashion as messenger molecules and are sometimes referred to as "local hormones." These include **paracrine** secretions, which enter the interstitial fluid but affect only neighboring cells, and **autocrine** secretions, which affect only the secreting cell itself.

Another category of substances, secreted by **exocrine glands,** enter tubes or ducts that lead to body surfaces. In contrast to endocrine secretions, exocrine secretions are released externally. Two examples are stomach acid reaching the lumen of the digestive tract and sweat being released at the skin's surface (fig. 13.1).

The interrelationships of the glands of the endocrine system, although not well understood, are vividly obvious in families that have an inherited cancer syndrome called *multiple endocrine neoplasia* (MEN). Different glands are affected in different individuals within a family, although the genetic cause is the same. One family member might have a tumor of the adrenal glands called pheochromocytoma; another might have thyroid cancer; yet a third relative might have parathyroid hyperplasia, a precancerous condition.

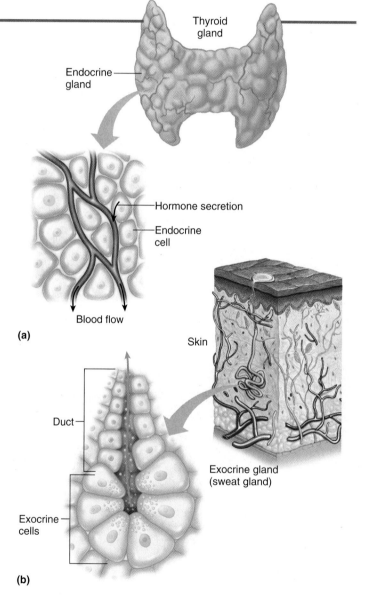

FIGURE 13.1

Types of glands. (*a*) Endocrine glands release hormones into the internal environment (body fluids). (*b*) Exocrine glands secrete to the outside environment, through ducts that lead to body surfaces.

TABLE 13.1	A Comparison Between the Nervous System and the Endocrine System	
	Nervous System	**Endocrine System**
Cells	Neurons	Glandular epithelium
Chemical signal	Neurotransmitter	Hormone
Specificity of action	Receptors on postsynaptic cell	Receptors on target cell
Speed of onset	Seconds	Seconds to hours
Duration of action	Very brief unless neuronal activity continues	May be brief or may last for days even if secretion ceases

Cells of the endocrine system and the nervous system communicate using chemical signals that bind to receptor molecules. Table 13.1 summarizes some similarities and differences between the two systems. In contrast to the nervous system, which releases neurotransmitter molecules into synapses, the endocrine system releases hormones into the bloodstream, which carries these messenger molecules everywhere. However, the endocrine system is also precise, because only target cells can respond to a hormone (fig. 13.2). A hormone's target cells have specific receptors that other cells lack. These receptors are proteins or glycoproteins with binding sites for a specific hormone.

Endocrine glands and their hormones help regulate metabolic processes. They control the rates of certain chemical reactions, aid in transporting substances through membranes, and help regulate water balance, electrolyte balance, and blood pressure. Endocrine hormones also play vital roles in reproduction, development, and growth.

Small groups of specialized cells produce some hormones. However, the larger endocrine glands—the pituitary gland, thyroid gland, parathyroid glands, adrenal glands, and pancreas—are the subject of this chapter (fig. 13.3). Subsequent chapters discuss several other hormone-secreting glands and tissues.

Hormone Action

Hormones are released into the extracellular spaces surrounding endocrine cells, from where they diffuse into the bloodstream and are carried to all parts of the body. Although some hormones may have widespread effects,

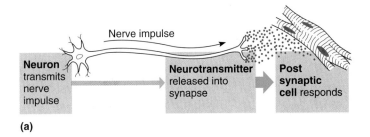

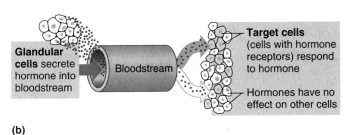

(a)

(b)

FIGURE 13.2

Chemical communication. (*a*) Neurons release neurotransmitters into a synapse, affecting postsynaptic cells. (*b*) Glands release hormones into the bloodstream. Blood carries hormone molecules throughout the body, but only target cells respond.

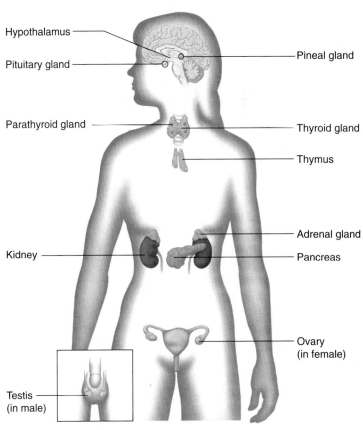

FIGURE 13.3

Locations of major endocrine glands.

in all cases, hormones affect only their target cells. This is because only target cells have receptors for any particular hormone. The other chemical messengers, paracrine and autocrine substances, also bind to specific receptors, and some examples of these are included in the chapter.

Chemistry of Hormones

Chemically, most hormones are either steroids (or steroid-like substances) that are synthesized from cholesterol (see chapter 2, p. 52), or they are nonsteroids (including amines, peptides, proteins, or glycoproteins that are synthesized from amino acids). Hormones are organic compounds. They can stimulate changes in target cells even if present in extremely low concentrations.

Steroid Hormones

Steroids (ste'roidz) are lipids that include complex rings of carbon and hydrogen atoms (fig. 13.4a). Steroids differ by the types and numbers of atoms attached to these rings and the ways they are joined (see fig. 2.16). All steroid

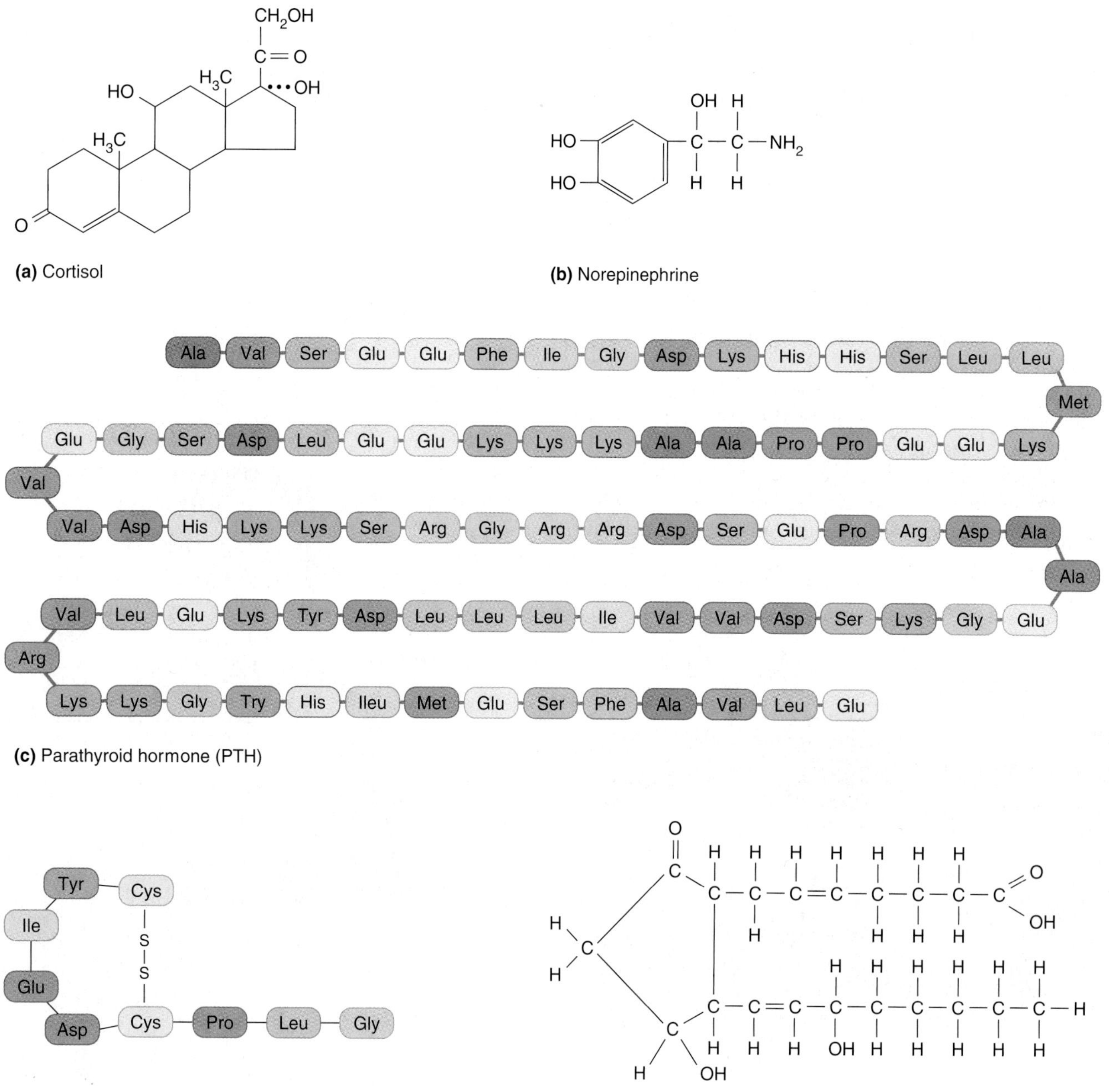

(a) Cortisol

(b) Norepinephrine

(c) Parathyroid hormone (PTH)

(d) Oxytocin

(e) Prostaglandin PGE$_2$

FIGURE 13.4

Structural formulas of (*a*) a steroid hormone (cortisol) and (*b*) an amine hormone (norepinephrine). Amino acid sequences of (*c*) a protein hormone (PTH) and (*d*) a peptide hormone (oxytocin). Structural formula of (*e*) a prostaglandin (PGE$_2$).

hormones are derived from cholesterol, including sex hormones (such as testosterone and the estrogens) and those the adrenal cortex secretes (including aldosterone and cortisol). Vitamin D is a modified steroid and can be converted into a hormone, as is discussed later in this chapter, in the section titled "Parathyroid Hormone" (see also chapter 18, p. 707).

Nonsteroid Hormones

Hormones called *amines,* including norepinephrine and epinephrine, are derived from amino acids. These hormones are also synthesized in the adrenal medulla (the inner portion of the adrenal gland) from the amino acid tyrosine (fig. 13.4*b*).

Protein hormones are composed of long chains of amino acids, linked to form intricate molecular structures (see chapter 2, pp. 52–54 and fig. 13.4*c*). They include the hormone secreted by the parathyroid gland and some of those secreted by the anterior pituitary gland. Certain other hormones secreted from the anterior pituitary gland are *glycoproteins,* which consist of proteins joined to carbohydrates.

The *peptide* hormones are short chains of amino acids (fig. 13.4*d*). This group includes hormones associated with the posterior pituitary gland and some produced in the hypothalamus.

Another group of compounds, called *prostaglandins* (pros"tah-glan'dinz), are paracrine substances. They regulate neighboring cells. Prostaglandins are lipids (20-carbon fatty acids that include 5-carbon rings) and are synthesized from a type of fatty acid (arachidonic acid) in cell membranes (fig. 13.4*e*). Prostaglandins are produced in a wide variety of cells, including those of the liver, kidneys, heart, lungs, thymus gland, pancreas, brain, and reproductive organs.

Table 13.2 lists the names and abbreviations of some of the hormones discussed in this chapter. Table 13.3 and figure 13.4 summarize the chemical composition of hormones. Other hormones related to specific organ systems are discussed in their appropriate chapters.

TABLE 13.2	Hormone Names and Abbreviations		
Source	**Name**	**Abbreviation**	**Synonym**
Hypothalamus	Corticotropin-releasing hormone	CRH	
	Gonadotropin-releasing hormone	GnRH	Luteinizing hormone releasing hormone (LHRH)
	Somatostatin	SS	Growth hormone release-inhibiting hormone (GIH)
	Growth hormone-releasing hormone	GHRH	
	Prolactin release-inhibiting hormone	PIH	Dopamine
	Prolactin-releasing factor*	PRF*	
	Thyrotropin-releasing hormone	TRH	
Anterior pituitary gland	Adrenocorticotropic hormone	ACTH	Corticotropin
	Follicle-stimulating hormone	FSH	Follitropin
	Growth hormone	GH	Somatotropin (STH)
	Luteinizing hormone	LH	Lutropin, interstitial cell-stimulating hormone (ICSH)
	Prolactin	PRL	
	Thyroid-stimulating hormone	TSH	Thyrotropin
Posterior pituitary gland	Antidiuretic hormone	ADH	Vasopressin
	Oxytocin	OT	
Thyroid gland	Calcitonin		
	Thyroxine	T_4	
	Triiodothyronine	T_3	
Parathyroid gland	Parathyroid hormone	PTH	Parathormone
Adrenal medulla	Epinephrine	EPI	Adrenalin
	Norepinephrine	NE	Noradrenalin
Adrenal cortex	Aldosterone		
	Cortisol		Hydrocortisone
Pancreas	Glucagon		
	Insulin		
	Somatostatin	SS	

*"Factor" is used because prolactin-releasing hormone has been hypothesized, but not yet identified.

TABLE 13.3	Types of Hormones	
Type of Compound	**Formed from**	**Examples**
Amines	Amino acids	Norepinephrine, epinephrine
Peptides	Amino acids	ADH, OT, TRH, SS, GnRH
Proteins	Amino acids	PTH, GH, PRL
Glycoproteins	Protein and carbohydrate	FSH, LH, TSH
Steroids	Cholesterol	Estrogens, testosterone, aldosterone, cortisol

TABLE 13.4	Sequence of Steroid Hormone Action

1. Endocrine gland secretes steroid hormone.
2. Steroid hormone diffuses through target cell membrane and enters nucleus.
3. Hormone combines with a receptor molecule.
4. Steroid hormone-receptor complex binds to DNA and promotes transcription of messenger RNA.
5. Messenger RNA enters the cytoplasm and directs protein synthesis.
6. Newly synthesized proteins produce hormone's specific effects.

1 What is a hormone?

2 How do endocrine glands and exocrine glands differ?

3 How are hormones chemically classified?

Actions of Hormones

Hormones exert their effects by altering metabolic processes. For example, a hormone might change the activity of an enzyme necessary for synthesizing a particular substance or alter the rate at which particular chemicals are transported through cell membranes.

Hormones may reach all cells but affect only those that have appropriate receptors. Each hormone receptor is a protein or glycoprotein molecule that has a *binding site* for a specific hormone. A hormone delivers its message to a cell by uniting with the binding site of its receptor. The more receptors the hormone binds on its target cells, the greater the response.

Steroid Hormones

Steroid hormones are insoluble in water. They are carried in the bloodstream weakly bound to plasma proteins in such a way that allows them to be released in sufficient quantity to affect their target cells. However, unlike amine, peptide, and protein hormones, steroid hormones are soluble in the lipids that make up the bulk of cell membranes. For this reason, steroid hormones can diffuse into cells relatively easily and may enter any cell in the body. Once inside a target cell, steroid hormones combine (usually within the nucleus) with specific protein receptors. The resulting *hormone-receptor complex* binds within the nucleus to a particular region of the DNA and either activates or inhibits specific genes. The activated genes, in turn, are transcribed into messenger RNA (mRNA), which enters the cytoplasm where it directs synthesis of specific proteins. The newly synthesized proteins, which may be enzymes, transport proteins, or even hormone receptors, bring about the cellular changes associated with the particular hormone (fig.

13.5, table 13.4, and Clinical Application 13.1). An example is the steroid hormone **aldosterone** (al'do-ster-ōn", al-dos'ter-ōn), from the adrenal gland, whose action is to stimulate sodium retention by the kidneys. In response to aldosterone, cells that form tubules within the kidney begin to synthesize more Na^+/K^+ pumps, the proteins that actively transport these ions across the cell membrane, retaining sodium.

In some cases, steroid hormones may inhibit a particular gene, so transcription does not occur. In this case, the cellular response results from decreased levels of a particular protein.

Nonsteroid Hormones

A nonsteroid hormone, such as an amine, peptide, or protein, usually combines with specific receptor molecules on the target cell membrane. Each receptor molecule is a protein that has a *binding site* and an *activity site.* The hormone combines with the binding site of the receptor. This causes the receptor's activity site to interact with other membrane proteins. Receptor binding may alter the function of enzymes or membrane transport mechanisms, changing the concentrations of still other cellular components. The hormone that triggers this cascade of biochemical activity is considered a *first messenger.* The biochemicals in the cell that induce the changes that are recognized as responses to the hormone are called *second messengers.*

Many hormones use **cyclic adenosine monophosphate** (cyclic AMP, or cAMP) as a second messenger. In this mechanism, a hormone binds to its receptor, and the resulting hormone-receptor complex activates a protein called a **G protein,** which then activates an enzyme called **adenylate cyclase** (ah-den'ĭ-lat si'klās), an integral membrane protein with its active site facing the inside of the cell. The activated enzyme removes two phosphates from ATP and circularizes it, forming *cyclic AMP* (fig. 13.6). Cyclic AMP, in turn, activates another set of enzymes called **protein kinases** (ki'nās-ez). Protein kinases transfer phosphate groups from ATP molecules to protein substrate

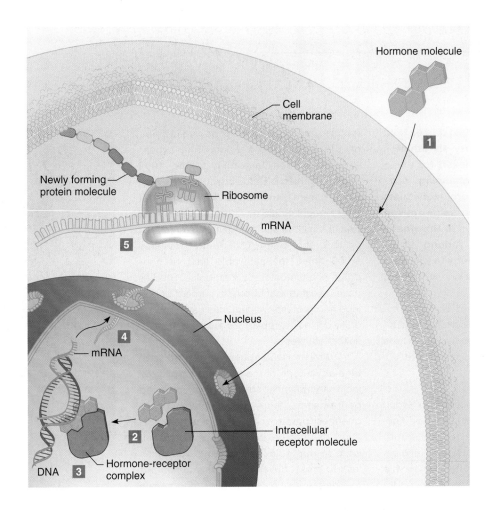

FIGURE 13.5

Steroid hormones. (1) A steroid hormone crosses a cell membrane and (2) combines with a protein receptor, usually in the nucleus. (3) The hormone-receptor complex activates synthesis of specific messenger RNA (mRNA) molecules. (4) The mRNA molecules leave the nucleus and enter the cytoplasm (5) where they guide synthesis of the encoded proteins. Note: In the bloodstream, most molecules of a particular steroid are bound to proteins. Only the few that are not bound are free to enter cells, as shown here.

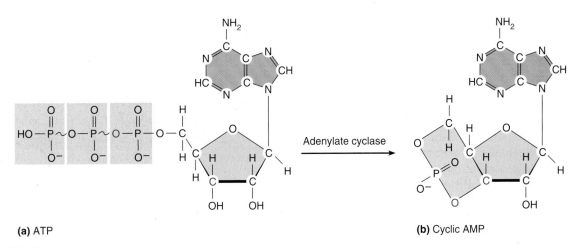

(a) ATP

(b) Cyclic AMP

FIGURE 13.6

Adenylate cyclase catalyzes conversion of (a) ATP molecules into cyclic AMP (b). The atoms forming the new bond are shown in red.

USING HORMONES TO IMPROVE ATHLETIC PERFORMANCE

In the 2000 summer Olympic games held in Sydney, Australia, thirty-six athletes and coaches were dismissed for using banned performance-enhancing substances—after many more individuals had been ejected for the same reason following drug tests given in the weeks preceding the games. Among those seeking to go beyond biology to win were runners, weight lifters, wrestlers, cyclists, and rowers.

The general focus of performance enhancement is misuse of certain powerful hormones of the endocrine system. Three types of approaches are described here.

Steroids

In the 1988 summer Olympics held in Seoul, South Korea, Canadian Ben Johnson flew past his competitors in the 100-meter run. But seventy-two hours later, officials rescinded the gold medal he won for his record-smashing time of 9.79 seconds, after a urine test revealed traces of the drug stanozolol, a synthetic stand-in for the steroid hormone testosterone (fig. 13A). Johnson's natural testosterone level was only 15% of normal—evidence of negative

feedback acting because of an outside supply of the hormone. Yet Johnson's experience was soon forgotten. In the 1992 summer games in Barcelona, Spain, several athletes were dismissed for using drugs that they thought would have steroidlike effects. And in the 2000 summer games, a urine test on U.S. shot-putter C.J. Hunter revealed 1,000 times the allowable limit of nandrolone, a testosterone metabolite. Studies have revealed that about 30 percent of college and professional athletes use anabolic steroids, as do up to 20 percent of high school athletes.

Athletes who abuse steroids do so to take advantage of the hormone's ability to increase muscular strength. But improved performance today may have consequences tomorrow. Steroids hasten adulthood, stunting height and causing early hair loss. In males, excess steroid hormones lead to breast development, and in females to a deepened voice, hairiness, and a male physique. The kidneys, liver, and heart may be damaged, and atherosclerosis may develop because steroids raise LDL and lower HDL—the opposite of a healthy cholesterol profile. In males, the body mistakes

FIGURE 13A

Canadian track star Ben Johnson ran away with the gold medal in the 100-meter race at the 1988 Summer Olympics—but then had to return the award when traces of a steroid drug showed up in his urine. Drug abuse continues to be a problem, among amateur as well as professional athletes.

molecules. This phosphorylation alters the shapes of the substrate molecules and converts some of them from inactive forms into active ones.

The activated proteins then alter various cellular processes to bring about the effect of that particular hormone (fig. 13.7). The response of any particular cell to such a hormone is determined not only by the type of membrane receptors present, but also by the kinds of protein substrate molecules in the cell. Table 13.5 summarizes these actions. Cellular responses to second messenger activation include altering membrane permeabilities, activating enzymes, promoting synthesis of certain proteins, stimulating or inhibiting specific metabolic pathways, promoting cellular movements, and initiating secretion of hormones and other substances. A specific example is the action of epinephrine to raise blood sugar during periods of physical stress. Epinephrine acts through the second messenger cAMP to

increase the activity of the enzyme that breaks down liver glycogen, leading to increased glucose that can diffuse out of liver cells and enter the bloodstream.

Because many hormones utilize the cAMP-mediated second messenger system to exert their effects, an abnormality in this signaling system can lead to symptoms from many endocrine glands. In McCune-Albright syndrome, for example, a defect in the G protein that activates adenylate cyclase results in the conversion of ATP to cAMP even without hormonal stimulation. As a result, cells in the pituitary, thyroid, gonads, and adrenal glands secrete hormones in excess. These conditions are inherited because the components of the second messenger system are proteins.

the synthetic steroids for the natural hormone and lowers its own production of testosterone—as Ben Johnson found out. Infertility may result. Steroids can also cause psychiatric symptoms, including delusions, depression, and violence.

Growth Hormone

Some athletes take human growth hormone (HGH) preparations to supplement the effects of steroids, because HGH enlarges muscles, as steroids strengthen them. HGH has been available thanks to recombinant DNA technology (a type of genetic modification) since 1985, and it is prescribed to treat children with certain forms of inherited dwarfism. However, HGH is available from other nations and can be obtained illegally to enhance athletic performance. Unlike steroids, HGH has a half-life of only seventeen to forty-five minutes, which means that it becomes undetectable in body fluids within an hour. There was no urine or blood test for it at the 2000 Olympics.

Boosting the Blood's Oxygen-Carrying Capacity

Red blood cells carry oxygen to muscles. Therefore, increasing the number of red blood cells can, theoretically, increase oxygen delivery to muscles and thereby enhance endurance. Swedish athletes introduced a practice called "blood doping" in 1972. The athletes would have blood removed a month or more prior to performance, then have the blood reinfused, boosting the red blood cell supply. Today's high-tech version of blood doping is to take erythropoietin (EPO), which is a hormone secreted from the kidneys that signals the bone marrow to produce more red blood cells. Like human growth hormone, EPO is manufactured using recombinant DNA technology. It is used legitimately to treat certain forms of anemia.

Using EPO to improve athletic performance is ill advised. In 1987, it led to heart attacks and death in more than two dozen cyclists from the Netherlands. Runners and swimmers also use EPO. In the 2000 summer Olympic games, some athletes tried a different way to boost their red blood cell supplies—an experimental preparation of cow hemoglobin.

Almost as interesting as the ways in which some athletes seek to manipulate the endocrine system is the ways in which they seek to escape detection. For example, in the 2000 Olympics, some athletes used a water-soluble steroid compound that is metabolized much faster than testosterone,

so it doesn't leave a trace. But officials were quick to respond, instituting a test to determine the ratio of testosterone to the impostor, epitestosterone.

An athlete can time drug use based on understanding how the body metabolizes a particular substance. For example, Irish swimmer Michelle Smith won four medals in the 1996 Atlanta Olympics and passed all drug tests. But some months later, a random urine test for another competition showed a superlethal level of alcohol—which she had presumably added to the urine sample to disguise banned drugs. She was banned from the 2000 games for sample tampering. Similarly, East German athletes took home 216 medals in the 1976 and 1980 Olympics combined, but in 1978, between games, they admitted to steroid use in all sports except sailing!

Some athletes caught in the act offer creative explanations for their altered personal biochemistries. A Latvian rower claimed he'd taken an herbal supplement, and not steroids, as did a U.S. shot-putter. A German runner claimed a competitor had spiked his toothpaste with steroids. And a track coach from Uzbekistan said that his athletes used growth hormone to treat hair loss. ∎

TABLE 13.5	Sequence of Actions of Nonsteroid Hormone Using Cyclic AMP

1. Endocrine gland secretes nonsteroid hormone.
2. Body fluid carries hormone to its target cell.
3. Hormone combines with receptor site on membrane of its target cell, activating G protein.
4. Adenylate cyclase molecules are activated within target cell's membrane.
5. Adenylate cyclase converts ATP into cyclic AMP.
6. Cyclic AMP activates protein kinases.
7. These enzymes activate protein substrates in the cell that change metabolic processes.
8. Cellular changes produce the hormone's effects.

Another enzyme, phosphodiesterase, quickly inactivates cAMP, so its action is short-lived. For this reason, a continuing response in a target cell depends upon a continuing signal from hormone molecules combining with receptors in the target cell membrane. Since enzymes continuously degrade hormones, a sustained response depends on continued hormone secretion.

Hormones whose actions depend upon cyclic AMP include releasing hormones from the hypothalamus; thyroid-stimulating hormone TSH, adrenocorticotropic hormone ACTH, follicle-stimulating hormone FSH, and luteinizing hormone LH from the anterior pituitary gland; antidiuretic hormone ADH from the posterior pituitary gland; parathyroid hormone PTH from the parathyroid glands; norepinephrine and epinephrine from the adrenal glands; calcitonin from the thyroid gland; and glucagon from the pancreas.

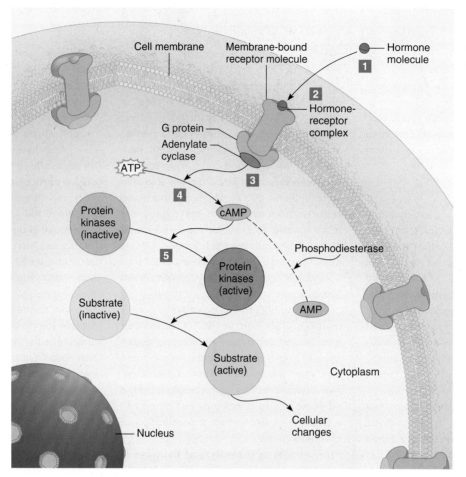

FIGURE 13.7

Nonsteroid hormones. (*1*) Body fluids carry nonsteroid hormone molecules to the target cell, where they combine (*2*) with receptor sites on the cell membrane. (*3*) This may activate molecules of adenylate cyclase, which (*4*) catalyze conversion of ATP into cyclic AMP. (*5*) Cyclic AMP promotes a series of reactions leading to the cellular changes associated with the hormone's action. The dashed arrow indicates breakdown of cAMP, which turns off the cellular response to the hormone.

Certain nonsteroid hormones employ second messengers other than cAMP. For example, a second messenger called diacylglycerol (DAG), like cAMP, activates a protein kinase leading to a cellular response.

In another mechanism, a hormone binding to its receptor increases calcium ion concentration within the target cell. Such a hormone may stimulate transport of calcium ions inward through the cell membrane or induce release of calcium ions from cellular storage sites via a second messenger called inositol triphosphate (IP3). The calcium ions combine with the protein *calmodulin* (see chapter 9, p. 294), altering its molecular structure in a way that activates the molecule. Activated calmodulin can then interact with enzymes, altering their activities and thus eliciting diverse responses.

Still another hormonal mechanism uses *cyclic guanosine monophosphate* (cyclic GMP, or cGMP). Like cAMP, cGMP is a nucleotide derivative and functions in much the same manner as a second messenger.

Two of the amine hormones, collectively referred to as thyroid hormone, can enter cells and exert their effects much the way steroids do. That is, they form a hormone-receptor complex that activates transcription of specific genes. The newly synthesized protein then brings about the cellular response.

Cellular response to a steroid hormone (and thyroid hormone) is directly proportional to the number of hormone-receptor complexes that form. In contrast, response to a hormone operating through a second messenger is greatly amplified. This is possible because many second messenger molecules can be activated in response to just a few hormone-receptor complexes. Because of such amplification, cells are highly sensitive to changes in the concentrations of nonsteroid hormones.

1 How does a steroid hormone act on its target cells?

2 How does a nonsteroid hormone act on its target cells?

3 What is a second messenger?

Prostaglandins

Prostaglandins are paracrine substances, acting locally, that are very potent and present in very small quantities.

They are not stored in cells but are synthesized just before they are released. They are rapidly inactivated.

Some prostaglandins regulate cellular responses to hormones. For example, different prostaglandins can either activate or inactivate adenylate cyclase in cell membranes, thereby controlling production of cAMP and altering the cell's response to a hormone.

Prostaglandins produce a variety of effects. Some prostaglandins can relax smooth muscle in the airways of the lungs and in the blood vessels, dilating these passageways. Yet other prostaglandins can contract smooth muscle in the walls of the uterus, causing menstrual cramps and labor contractions. They stimulate secretion of hormones from the adrenal cortex and inhibit secretion of hydrochloric acid from the wall of the stomach. Prostaglandins also influence movements of sodium ions and water in the kidneys, help regulate blood pressure, and have powerful effects on both male and female reproductive physiology. When tissues are injured, prostaglandins promote inflammation (see chapter 16, p. 618).

Understanding prostaglandin function has medical applications. Drugs such as aspirin and certain steroids that relieve the joint pain of rheumatoid arthritis inhibit production of prostaglandins in the synovial fluid of affected joints. Daily doses of aspirin may reduce the risk of heart attack by altering prostaglandin activity. Prostaglandins may be used as drugs to dilate constricted blood vessels to relieve hypertension.

1 What are prostaglandins?

2 Describe one possible function of prostaglandins.

3 What kinds of effects do prostaglandins produce?

Control of Hormonal Secretions

Hormones are continually excreted in the urine and broken down by various enzymes, primarily in the liver. Therefore, increasing or decreasing blood levels of a hormone requires increased or decreased secretion. Hormone secretion is precisely regulated.

Control Sources

Hormone secretion must be precisely controlled so that these biochemicals can effectively maintain the internal environment. Hormone secretion is controlled in three ways, all of which employ **negative feedback** (see chapter 1, p. 10). In each case, an endocrine gland or the system controlling it is sensitive to the concentration of the hormone the gland secretes, a process the hormone controls, or an action the hormone has on the internal environment (fig. 13.8).

1. The hypothalamus controls the anterior pituitary gland's release of **tropic hormones,** which are hormones that stimulate other endocrine glands to release hormones (fig. 13.8a). The hypothalamus constantly receives information about the internal environment from neural connections and cerebrospinal fluid, made possible by its location near the thalamus and the third ventricle (fig. 13.9).

2. The nervous system stimulates some glands directly. The adrenal medulla, for example, secretes its hormones (epinephrine and norepinephrine) in response to preganglionic sympathetic nerve impulses. The secretory cells replace the postganglionic sympathetic neurons, which would normally secrete norepinephrine alone as a neurotransmitter (see fig. 13.8b).

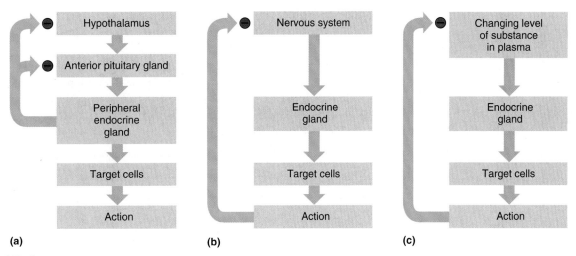

FIGURE 13.8

Control of the endocrine system occurs in three ways: (a) the hypothalamus and anterior pituitary, (b) the nervous system directly, and (c) glands that respond directly to changes in the internal environment. Negative feedback inhibition is indicated by ●.

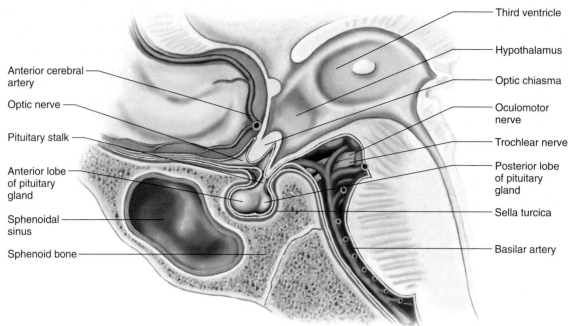

FIGURE 13.9

The pituitary gland is attached to the hypothalamus and lies in the sella turcica of the sphenoid bone.

3. Another group of glands responds directly to changes in the composition of the internal environment. For example, when the blood glucose level rises, the pancreas secretes insulin, and when the blood glucose level falls, it secretes glucagon, as we shall see later in the chapter in the section titled "Hormones of the Pancreatic Islets" (see fig. 13.8c).

In each of the above cases, as hormone levels rise in the blood and the hormone exerts its effects, negative feedback inhibits the system, and hormone secretion decreases. Then, as hormone levels in the blood decrease and the hormone's effects are no longer taking place, inhibition of the system ceases, and secretion of that hormone increases again (fig. 13.10). As a result of negative feedback, hormone levels in the bloodstream remain relatively stable, tending to fluctuate slightly above and below an average value (fig. 13.11).

1 How does the nervous system help regulate hormonal secretions?

2 How does a negative feedback system control hormonal secretion?

Pituitary Gland

The **pituitary** (pǐ-tu´ǐ-tar″e) **gland** (hypophysis) is about 1 centimeter in diameter and is located at the base of the brain. It is attached to the hypothalamus by the pituitary

stalk, or *infundibulum,* and lies in the sella turcica of the sphenoid bone, as figure 13.9 shows.

The pituitary gland consists of two distinct portions: an *anterior lobe* (adenohypophysis) and a *posterior lobe* (neurohypophysis). The anterior lobe secretes a number of hormones, including growth hormone (GH), thyroid-stimulating hormone (TSH), adrenocorticotropic hormone (ACTH), follicle-stimulating hormone (FSH), luteinizing hormone (LH), and prolactin (PRL). Although the cells of the posterior lobe (pituicytes) do not synthesize any hormones, two important hormones, antidiuretic hormone (ADH) and oxytocin (OT), are secreted by special neurons called **neurosecretory cells** whose nerve endings secrete into the bloodstream within the posterior lobe. The cell bodies of these neurosecretory cells are in the hypothalamus.

During fetal development, a narrow region appears between the anterior and posterior lobes of the pituitary gland. Called the *intermediate lobe* (pars intermedia), this region produces melanocyte-stimulating hormone (MSH), which regulates the formation of melanin—the pigment in the skin and in portions of the eyes and brain. The region atrophies during prenatal development and appears only as a vestige in adults.

The brain controls most of the pituitary gland's activities (fig. 13.12). The pituitary gland's posterior lobe releases hormones into the bloodstream in response to

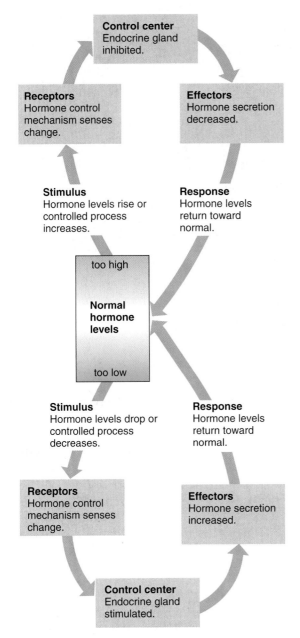

FIGURE 13.10
Hormone secretion is under the control of negative feedback.

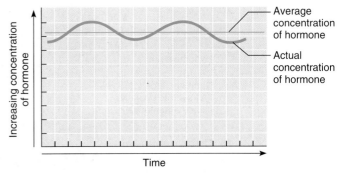

FIGURE 13.11
As a result of negative feedback, hormone concentrations remain relatively stable, although they may fluctuate slightly above and below average concentrations.

nerve impulses from the hypothalamus. A different mechanism controls the anterior lobe. Here releasing hormones from the hypothalamus primarily control secretions. These releasing hormones are carried in the blood via a capillary bed associated with the hypothalamus. The vessels merge to form the **hypophyseal** (hi"po-fiz'e-al) **portal veins** that pass downward along the pituitary stalk and give rise to a capillary bed in the anterior lobe. Thus, substances released into the blood from the hypothalamus are carried directly to the anterior lobe. The hypothalamus, therefore, is an endocrine gland itself, yet it also controls other endocrine glands. This is also true of the anterior pituitary.

> The arrangement of two capillaries in series is quite unusual and is called a *portal system*. It occurs in only three places in the body: the hepatic portal vein connects intestinal capillaries to special liver capillaries called sinusoids; the efferent arteriole of kidney nephrons connects two sets of capillaries; and the hypophyseal portal vein gives rise to a capillary net in the anterior lobe of the pituitary gland.

Upon reaching the anterior lobe of the pituitary, each of the hypothalamic releasing hormones acts on a specific population of cells. Some of the resulting actions are inhibitory (prolactin release-inhibiting hormone and somatostatin), but most stimulate the anterior pituitary to release hormones that stimulate the secretions of peripheral endocrine glands. In many of these cases, important negative feedback relationships regulate hormone levels in the bloodstream. Figure 13.13 shows this general relationship.

1 Where is the pituitary gland located?

2 List the hormones that the anterior and posterior lobes of the pituitary gland secrete.

3 Explain how the hypothalamus controls the actions of the pituitary gland.

Anterior Pituitary Hormones

The anterior lobe of the pituitary gland is enclosed in a dense capsule of collagenous connective tissue and largely consists of epithelial tissue organized in blocks around many thin-walled blood vessels. Within the

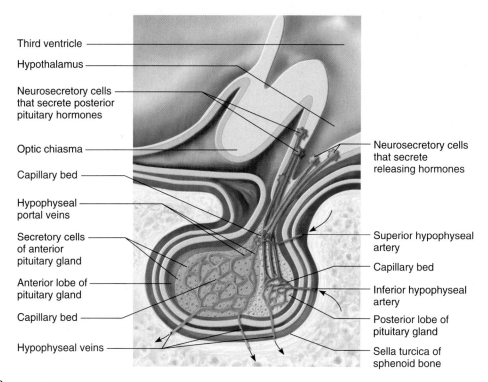

FIGURE 13.12

Hypothalamic releasing hormones stimulate cells of the anterior lobe to secrete hormones. Nerve impulses originating in the hypothalamus stimulate nerve endings in the posterior lobe of the pituitary gland to release hormones.

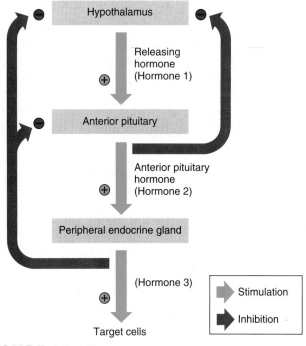

FIGURE 13.13

Hypothalamic control of the peripheral endocrine glands may utilize as many as three types of hormones, with multiple negative feedback controls. (⊕ = stimulation; ⊖ = inhibition)

epithelial tissue are five types of secretory cells. They are *somatotropes* that secrete GH, *mammatropes* that secrete PRL, *thyrotropes* that secrete TSH, *corticotropes* that secrete ACTH, and *gonadotropes* that secrete FSH and LH (figs. 13.14 and 13.15). In males, LH (luteinizing hormone) is known as ICSH (interstitial cell-stimulating hormone) because it affects the interstitial cells of the testes (see chapter 22, p. 845).

Growth hormone, which is also called *somatotropin* (STH), is a protein that stimulates cells to enlarge and more rapidly divide. It enhances the movement of amino acids through the cell membranes and increases the rate of protein synthesis. GH also decreases the rate at which cells utilize carbohydrates and increases the rate at which they use fats.

Growth hormone is secreted in pulses, especially during sleep. Two biochemicals from the hypothalamus control its secretion. They are released alternately, exerting opposite effects. *Growth hormone-releasing hormone* (GHRH) stimulates secretion of GH, and *somatostatin* (SS) inhibits secretion.

Nutritional state seems to play a role in control of GH. For example, more GH is released during periods of protein deficiency and abnormally low blood glucose concentration. Conversely, when blood protein and glucose concentrations increase, growth hormone secretion

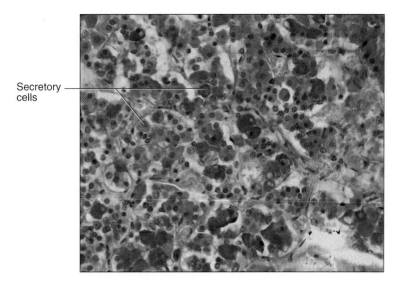

Secretory cells

FIGURE 13.14

Light micrograph of the anterior pituitary gland (240×).

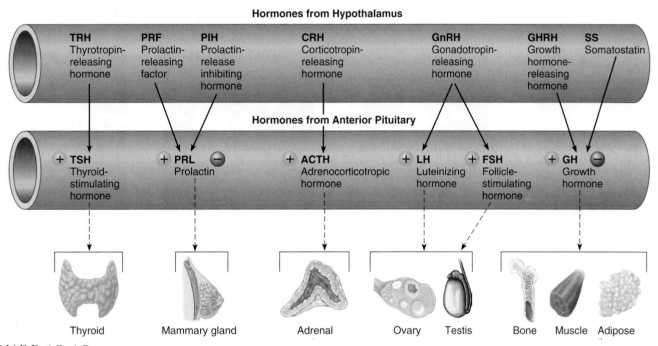

Hormones from Hypothalamus

TRH	PRF	PIH	CRH	GnRH	GHRH	SS
Thyrotropin-releasing hormone	Prolactin-releasing factor	Prolactin-release inhibiting hormone	Corticotropin-releasing hormone	Gonadotropin-releasing hormone	Growth hormone-releasing hormone	Somatostatin

Hormones from Anterior Pituitary

+ TSH	+ PRL −	+ ACTH	+ LH	+ FSH	+ GH −
Thyroid-stimulating hormone	Prolactin	Adrenocorticotropic hormone	Luteinizing hormone	Follicle-stimulating hormone	Growth hormone

Thyroid   Mammary gland   Adrenal   Ovary   Testis   Bone   Muscle   Adipose

FIGURE 13.15

Hormones released from the hypothalamus, the corresponding hormones released from the anterior lobe of the pituitary gland, and their target organs.

decreases. Apparently, the hypothalamus is able to sense changes in the concentrations of certain blood nutrients, and it releases (GHRH) in response to some of them.

Growth hormone can stimulate elongation of bone tissue directly, but its effect on cartilage requires a mediator substance. The liver releases a biochemical called insulin-like growth factor-I (IGF-I), a somatomedin, in response to GH. IGF I promotes growth of cartilage. Clini-

cal Application 13.2 discusses some clinical uses of growth hormone.

Prolactin is a protein, and as its name suggests, it promotes breast-milk production. In males, excess PRL decreases secretion of luteinizing hormone (LH). Because LH is necessary for production of male sex hormones (androgens), excess prolactin secretion may cause a male to produce too few sex hormones and become impotent.

GROWTH HORMONE UPS AND DOWNS

Insufficient secretion of growth hormone during childhood produces *hypopituitary dwarfism.* Body proportions and mental development are normal, but because secretion of other anterior pituitary hormones is also below normal, additional hormone deficiency symptoms may appear. For example, a child with this condition often fails to develop adult sexual features unless he or she receives hormone therapy.

Human growth hormone, manufactured using recombinant DNA technology, is a valuable drug in treating hypopituitary dwarfism. Treatment must begin before the bones completely ossify. The hormone also has some controversial uses. Some people want to use it to increase height in children who are short, but not abnormally so. A few years ago, growth hormone was given experimentally to older individuals to see if it would slow aging-related changes. Although muscle tone improved and the participants reported feeling well, side effects arose, and the value of such treatment was not confirmed. In another application, bovine growth hormone is given to dairy cows to increase their milk production.

Oversecretion of growth hormone in childhood may result in *gigantism,* in which height may eventually exceed 8 feet. Gigantism is usually caused by a tumor of the pituitary gland, which secretes excess pituitary hormones and GH. As a result, a giant often suffers from other metabolic disturbances.

If growth hormone is oversecreted in an adult after the epiphyses of the long bones have ossified, the person does not grow taller. The soft tissues, however, continue to enlarge and the bones thicken, producing a large tongue, nose, hands and feet, and a protruding jaw. This condition, *acromegaly,* is also often associated with a pituitary tumor (fig. 13B). ■

(a) (b) (c) (d)

FIGURE 13B

Oversecretion of growth hormone in adulthood causes acromegaly. Note the changes in this woman's facial features at ages (*a*) nine, (*b*) sixteen, (*c*) thirty-three, and (*d*) fifty-two.

Two biochemicals from the hypothalamus may regulate prolactin secretion. One of these, *prolactin release-inhibiting hormone* (PIH), restrains secretion of prolactin. The other, *prolactin-releasing factor* (PRF), is thought to stimulate prolactin secretion, but PRF has not yet been identified.

Thyroid-stimulating hormone, also called *thyrotropin,* is a glycoprotein. It controls secretion of certain hormones from the thyroid gland. TSH can also stimulate growth of the gland, and abnormally high TSH levels may lead to an enlarged thyroid gland, or *goiter.*

The hypothalamus partially regulates TSH secretion by producing thyrotropin-releasing hormone (TRH). Circulating thyroid hormones help regulate TSH secretion by inhibiting release of TRH and TSH; therefore, as the blood concentration of thyroid hormones increases, secretions of TRH and TSH decline (fig. 13.16).

External factors influence release of TRH and TSH. These include exposure to extreme cold, which is accompanied by increased hormonal secretion, and emotional stress, which sometimes increases hormonal secretion and other times decreases secretion.

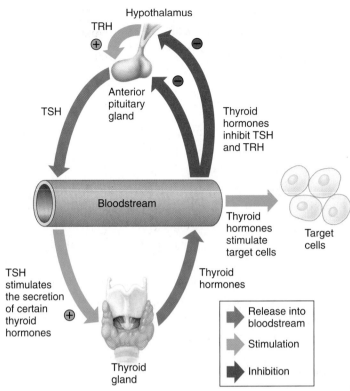

FIGURE 13.16

Thyrotropin-releasing hormone (TRH) from the hypothalamus stimulates the anterior pituitary gland to release thyroid-stimulating hormone (TSH), which stimulates the thyroid gland to release hormones. These thyroid hormones reduce the secretion of TSH and TRH. (⊕ = stimulation; ⊖ = inhibition)

1 How does growth hormone affect the cellular metabolism of carbohydrates, fats, and proteins?

2 What are the functions of prolactin in females? In males?

3 How is TSH secretion regulated?

Adrenocorticotropic (ad-re″no-kor″te-ko-trop′ik) **hormone** is a peptide that controls the manufacture and secretion of certain hormones from the outer layer (cortex) of the adrenal gland. The secretion of ACTH is regulated in part by *corticotropin-releasing hormone* (CRH), which is released from the hypothalamus in response to decreased concentrations of adrenal cortical hormones. Stress can increase secretion of ACTH by stimulating release of CRH.

Both **follicle-stimulating hormone** and **luteinizing** (loo′te-in-iz′eng) **hormone** are glycoproteins and are called *gonadotropins,* which means they exert their actions on the gonads or reproductive organs. FSH, for example, is responsible for growth and development of follicles that house egg cells in the ovaries. It also stimulates the follicular cells to secrete a group of female sex hormones, collectively called *estrogen* (or estrogens).

In males, FSH stimulates the production of sperm cells in the testes. LH promotes secretion of sex hormones in both males and females and is essential for release of egg

cells from the ovaries. Other functions of the gonadotropins and their interactions are discussed in chapter 22.

The mechanism that regulates secretion of gonadotropins is not well understood. However, it is known that the hypothalamus secretes a *gonadotropin-releasing hormone* (GnRH). The hypothalamus does not secrete this hormone in significant amounts until puberty. Gonadotropins are virtually absent in the body fluids of infants and children.

1 What is the function of ACTH?

2 Describe the functions of FSH and LH in a male and in a female.

3 What is a gonadotropin?

Posterior Pituitary Hormones

Unlike the anterior lobe of the pituitary gland, which is primarily composed of glandular epithelial cells, the posterior lobe largely consists of nerve fibers and neuroglial cells (*pituicytes*). The neuroglial cells support the nerve fibers that originate in the hypothalamus. The hypothalamic cells that give rise to these fibers are called neurosecretory cells because their secretions function not as neurotransmitters but as hormones.

Specialized neurons in the hypothalamus produce the two hormones associated with the posterior pituitary—**antidiuretic** (an″ti-di″u-ret′ik) **hormone** and **oxytocin** (ok″se-to′sin) (see fig. 13.12). These hormones travel down axons through the pituitary stalk to the posterior pituitary and are stored in vesicles (secretory granules) near the ends of the axons. The hormones are released into the blood in response to nerve impulses coming from the hypothalamus.

Antidiuretic hormone and oxytocin are short polypeptides with similar sequences (fig. 13.17). A *diuretic* is a

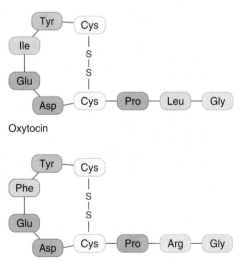

FIGURE 13.17

The structure of oxytocin differs from that of ADH by only two amino acids, yet they function differently.

substance that increases urine production. An *antidiuretic,* then, is a chemical that decreases urine formation. ADH produces its antidiuretic effect by causing the kidneys to reduce the volume of water they excrete. In this way, ADH plays an important role in regulating the concentration of body fluids (see chapter 20, p. 793).

Drinking alcohol is often followed by frequent and copious urination. This is because alcohol (ethyl alcohol) inhibits ADH secretion. A person must replace the lost body fluid to maintain normal water balance. Although it seems counterintuitive, drinking too much beer can lead to dehydration because the body loses more fluid than it replaces.

ADH present in sufficient concentrations contracts certain smooth muscles, including those in the walls of blood vessels. As a result, vascular resistance and blood pressure may increase. For this reason, ADH is also called *vasopressin.* Although ADH is seldom present in quantities sufficient to cause high blood pressure, its secretion increases following severe blood loss. Blood pressure may drop as a consequence of profuse bleeding. In this situation, ADH's vasopressor effect may help to minimize the drop and return blood pressure toward normal.

ADH's two effects—vasoconstriction and water retention—are possible because the hormone binds two different receptors on target cells. The binding of ADH to V1 receptors increases the concentration of the second messenger inositol triphosphate, which increases intracellular calcium ion concentration, leading to vasoconstriction. The second receptor, V2, is found on parts of the kidneys' microscopic tubules called collecting ducts. ADH binding there activates the cAMP second messenger system, which ultimately causes collecting duct cells to reabsorb water that would otherwise be excreted as urine.

The hypothalamus regulates secretion of ADH. Certain neurons in this part of the brain, called *osmoreceptors,* sense changes in the concentration of body fluids. For example, if a person is dehydrating due to a lack of water intake, the solutes in blood become more concentrated. The osmoreceptors, sensing the resulting increase in osmotic pressure, signal the posterior pituitary to release ADH, which causes the kidneys to retain water.

On the other hand, if a person drinks a large volume of water, body fluids become more dilute, which inhibits the release of ADH. Consequently, the kidneys excrete more dilute urine until the concentration of body fluids returns to normal.

Blood volume also affects ADH secretion. Increased blood volume stretches the walls of certain blood vessels, stimulating volume receptors that signal the hypothalamus to inhibit release of ADH. However, if hemorrhage decreases blood volume, these receptors are stretched less and therefore send fewer inhibiting impulses. As a result, ADH secretion increases, and as before, ADH causes the kidneys to conserve water. This helps prevent further volume loss.

A baby first displayed symptoms at five months of age—he drank huge volumes of water. By thirteen months, he had become severely dehydrated, although he drank nearly continuously. His parents were constantly changing his wet diapers. Doctors finally diagnosed a form of the condition *diabetes insipidus,* which impairs ADH regulation of water balance. The boy was drinking sufficient fluids, but his kidneys could not retain the water. ADH V2 receptors on the kidney collecting ducts were defective. The hormone could bind, but the receptor failed to trigger cAMP formation. The boy's ADH was still able to constrict blood vessels because the V1 receptors were unaffected. A high-calorie diet and providing lots of water preserved the boy's mental abilities, but he remained small for his age. Tumors and injury affecting the hypothalamus and posterior pituitary can also cause diabetes insipidus.

Oxytocin also has an antidiuretic action, but less so than ADH. In addition, oxytocin can contract smooth muscles in the uterine wall, playing a role in the later stages of childbirth. The uterus becomes more sensitive to oxytocin's effects during pregnancy. Stretching of uterine and vaginal tissues late in pregnancy, caused by the growing fetus, initiates nerve impulses to the hypothalamus, which then signals the posterior pituitary to release oxytocin, which, in turn, stimulates the uterine contractions of labor.

In the breasts, oxytocin contracts certain cells near the milk-producing glands and their ducts. In lactating breasts, this action forces liquid from the milk glands into the milk ducts and ejects the milk.

The mechanical stimulation of suckling initiates nerve impulses that travel to the mother's hypothalamus, which responds by signaling the posterior pituitary to release oxytocin, which, in turn, stimulates milk release. Thus, milk is normally not ejected from the milk glands and ducts until the baby suckles. The fact that milk is ejected from both breasts in response to suckling is a reminder that all target cells respond to a hormone.

Oxytocin has no established function in males, although it is present in the male posterior pituitary. There is evidence that it may stimulate the movement of certain fluids in the male reproductive tract during sexual activity. Table 13.6 reviews the hormones of the pituitary gland.

Anterior Lobe Hormone	Action	Source of Control
Growth hormone (GH)	Stimulates increase in size and rate of division of body cells; enhances movement of amino acids through membranes; promotes growth of long bones	Secretion inhibited by somatostatin (SS) and stimulated by growth hormone-releasing hormone (GHRH) from the hypothalamus
Prolactin (PRL)	Sustains milk production after birth; amplifies effect of LH in males	Secretion inhibited by prolactin release-inhibiting hormone (PIH) and may be stimulated by yet to be identified prolactin-releasing factor (PRF) from the hypothalamus
Thyroid-stimulating hormone (TSH)	Controls secretion of hormones from the thyroid gland	Thyrotropin-releasing hormone (TRH) from the hypothalamus
Adrenocorticotropic hormone (ACTH)	Controls secretion of certain hormones from the adrenal cortex	Corticotropin-releasing hormone (CRH) from the hypothalamus
Follicle-stimulating hormone (FSH)	Development of egg-containing follicles in ovaries; stimulates follicular cells to secrete estrogen; in males, stimulates production of sperm cells	Gonadotropin-releasing hormone (GnRH) from the hypothalamus
Luteinizing hormone (LH or ICSH in males)	Promotes secretion of sex hormones; releases egg cell in females	Gonadotropin-releasing hormone (GnRH) from the hypothalamus

Posterior Lobe Hormone	Action	Source of Control
Antidiuretic hormone (ADH)	Causes kidneys to reduce water excretion; in high concentration, raises blood pressure	Hypothalamus in response to changes in blood water concentration and blood volume
Oxytocin (OT)	Contracts muscles in uterine wall and those associated with milk-secreting glands	Hypothalamus in response to stretch in uterine and vaginal walls and stimulation of breasts

If the uterus is not sufficiently contracting to expel a fully developed fetus, oxytocin is sometimes given intravenously to stimulate uterine contractions, thus inducing labor. Oxytocin is also administered to the mother following childbirth to ensure that the uterine muscles contract enough to squeeze broken blood vessels closed, minimizing the danger of hemorrhage.

1. What is the function of ADH?
2. How is the secretion of ADH controlled?
3. What effects does oxytocin produce in females?

Thyroid Gland

The **thyroid gland** (thi′roid gland), as figure 13.18 shows, is a very vascular structure that consists of two large lateral lobes connected by a broad isthmus. It is located just below the larynx on either side and anterior to the trachea. It has a special ability to remove iodine from the blood.

Structure of the Gland

A capsule of connective tissue covers the thyroid gland, which is made up of many secretory parts called *follicles.* The cavities within these follicles are lined with a single layer of cuboidal epithelial cells and are filled with a clear viscous *colloid,* which consists primarily of a glycoprotein called *thyroglobulin.* The follicular cells produce and secrete hormones that may either be stored in the colloid or released into nearby capillaries (fig. 13.19). Other hormone-secreting cells, called *extrafollicular cells* (C cells), lie outside the follicles.

Thyroid Hormones

The thyroid gland produces three important hormones. The follicular cells synthesize two of these, which have marked effects on the metabolic rates of body cells. The extrafollicular cells produce the third type of hormone, which influences blood concentrations of calcium and phosphate ions.

The two important thyroid hormones that affect cellular metabolic rates are **thyroxine** (thi-rok′sin), or tetraiodothyronine (also called T_4 because it includes four atoms of iodine), and **triiodothyronine** (tri″i-o″do-thi′ro-nēn), also called T_3 because it includes three atoms of iodine. These hormones help regulate the metabolism of carbohydrates,

TABLE 13.6   Hormones of the Pituitary Gland

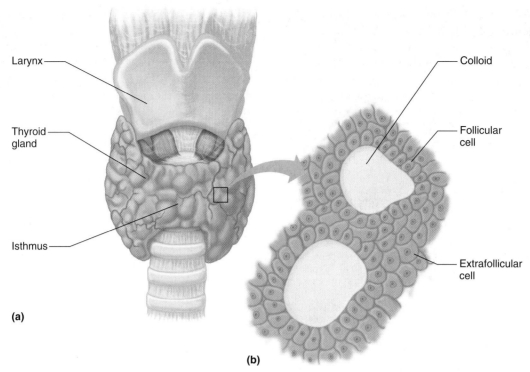

FIGURE 13.18
Thyroid gland. (*a*) The thyroid gland consists of two lobes connected anteriorly by an isthmus. (*b*) Follicular cells secrete thyroid hormones.

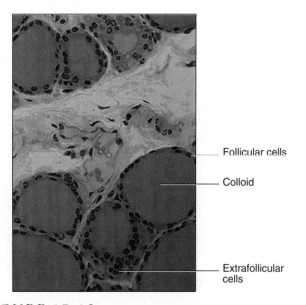

FIGURE 13.19
A light micrograph of thyroid gland tissue (240×). The open spaces that follicular cells surround are filled with colloid.

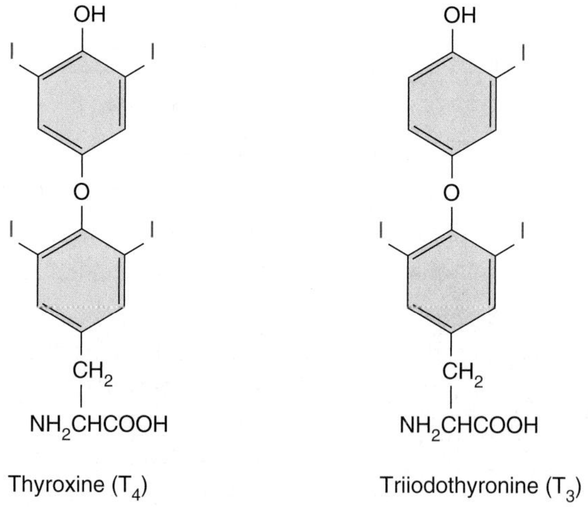

FIGURE 13.20
The hormones thyroxine and triiodothyronine have very similar molecular structures.

Thyroxine (T_4)

Triiodothyronine (T_3)

lipids, and proteins. Specifically, thyroxine and triiodothyronine increase the rate at which cells release energy from carbohydrates, enhance the rate of protein synthesis, and stimulate breakdown and mobilization of lipids. These hormones are the major factors determining how many calories the body must consume at rest in order to maintain life, the *basal metabolic rate* (BMR). They are essential for normal growth and development and for maturation of the nervous system (fig. 13.20). TSH from the anterior pituitary gland controls levels of thyroid hormones.

Follicular cells require iodine salts (iodides) to produce thyroxine and triiodothyronine. Such salts are normally obtained from foods, and after they have been absorbed from the intestine, the blood carries some of them in the form of iodide (I^-) to the thyroid gland. An efficient active transport protein called the *iodide pump* moves the iodides into the follicular cells, where they are converted to iodine and concentrated. The iodine, with the amino acid tyrosine, is used to synthesize these thyroid hormones.

Follicular cells synthesize *thyroglobulin,* whose protein portion is rich in molecules of the amino acid tyrosine, many of which have already had iodine attached by an enzymatic reaction. As the thyroglobulin protein twists and coils into its tertiary structure, bonds form between some of the tyrosine molecules, creating potential thyroid hormones waiting to be released. The follicular cells take up molecules of thyroglobulin by endocytosis, break down the protein, and release the individual thyroid hormones into the bloodstream. When the thyroid hormone levels in the bloodstream drop below a certain level, this process occurs more rapidly, returning thyroid hormone levels to normal.

Once in the blood, thyroid hormones combine with blood proteins (alpha globulins) and are transported to body cells. Triiodothyronine is nearly five times more potent, but thyroxine accounts for at least 95% of circulating thyroid hormones.

The thyroid gland produces **calcitonin,** which is usually not referred to as a thyroid hormone because it is synthesized by the C cells, which are distinct from the gland's follicles. Calcitonin plays a role in the control of blood calcium and phosphate ion concentrations. It helps lower concentrations of calcium and phosphate ions by decreasing the rate at which they leave the bones and enter extracellular fluids by inhibiting the bone-destroying activity of osteoclasts (see chapter 7, p. 194). At the same time, calcitonin increases the rate at which calcium and phosphate ions are deposited in bone matrix by stimulating activity of osteoblasts. Calcitonin also increases the excretion of calcium ions and phosphate ions by the kidneys.

Calcitonin secretion is stimulated by a high blood calcium ion concentration, as may occur following absorption of calcium ions from a recent meal. Certain hormones also prompt its secretion, such as gastrin, which is released from active digestive organs. Calcitonin helps prevent prolonged elevation of blood calcium ion concentration after eating.

Research suggests that calcitonin may be most important during physiological stress. For example, its actions help protect bones from resorption during pregnancy and lactation, when calcium is needed for growth of the fetus and synthesis of breast milk. In the young, calcitonin stimulates the increase in bone deposition associated with growth.

Table 13.7 summarizes the actions and sources of control of the thyroid hormones. Clinical Application 2.1, table 13.8, and figures 13.21, 13.22, and 13.23 discuss disorders of the thyroid gland.

TABLE 13.7	Hormones of the Thyroid Gland	
Hormone	**Action**	**Source of Control**
Thyroxine (T_4)	Increases rate of energy release from carbohydrates; increases rate of protein synthesis; accelerates growth; stimulates activity in the nervous system	TSH from the anterior pituitary gland
Triiodothyronine (T_3)	Same as above, but five times more potent than thyroxine	Same as above
Calcitonin	Lowers blood calcium and phosphate ion concentrations by inhibiting release of calcium and phosphate ions from bones and by increasing the rate at which calcium and phosphate ions are deposited in bones	Elevated blood calcium ion concentration, digestive hormones

TABLE 13.8	Disorders of the Thyroid Gland
Condition	**Mechanism/Symptoms**
Hyperthyroid	
Graves disease	Autoantibodies (against self) bind TSH receptors on thyroid cell membranes, mimicking action of TSH, overstimulating gland (hyperthyroidism); this is an exothalmic goiter
Hyperthyroidism	High metabolic rate, sensitivity to heat, restlessness, hyperactivity, weight loss, protruding eyes, goiter
Hypothyroid	
Hashimoto disease	Autoantibodies (against self) attack thyroid cells, producing hypothyroidism
Hypothyroidism (infantile)	Cretinism—stunted growth, abnormal bone formation, mental retardation, low body temperature, sluggishness
Hypothyroidism (adult)	Myxedema—low metabolic rate, sensitivity to cold, sluggishness, poor appetite, swollen tissues, mental dullness
Simple goiter	Deficiency of thyroid hormones due to iodine deficiency; because no thyroid hormones inhibit pituitary release of TSH, thyroid is overstimulated and enlarges but functions below normal (hypothyroidism)

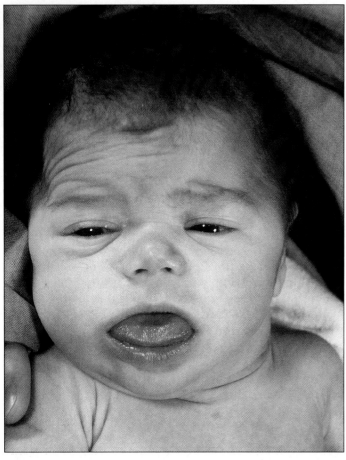

FIGURE 13.21
Cretinism is due to an underactive thyroid gland during infancy and childhood.

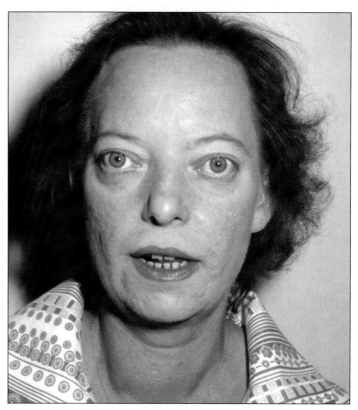

FIGURE 13.22
Hyperthyroidism may cause the eyes to protrude.

1. Where is the thyroid gland located?

2. Which hormones of the thyroid gland affect carbohydrate metabolism, the mobilization of lipids, and protein synthesis?

3. What substance is essential for the production of thyroxine and triiodothyronine?

4. How does calcitonin influence the concentrations of blood calcium and phosphate ions?

On a spring morning in 1991, then U.S. President George Bush set out on his daily run. After only a few hundred yards, he stopped, winded and experiencing an irregular heartbeat. Over the preceding two months, his wife, Barbara, had inexplicably lost 18 pounds, and her eyesight had become blurry and her eyes dry. Both Bushes were feeling unusually fidgety. Blood tests found that they had hyperthyroidism—and so did their dog, Millie. Because of the three cases, the White House environment was carefully studied, but no cause was ever identified.

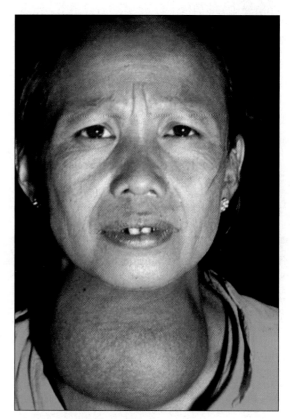

FIGURE 13.23
An iodine deficiency causes simple (endemic) goiter and results in high levels of TSH.

Parathyroid Glands

The **parathyroid glands** (par"ah-thi'roid glandz) are located on the posterior surface of the thyroid gland, as figure 13.24 shows. Usually there are four of them—a superior and an inferior gland associated with each of the thyroid's lateral lobes. The parathyroid glands secrete a hormone that regulates the concentrations of calcium and phosphate ions in the blood.

Structure of the Glands

Each parathyroid gland is a small, yellowish brown structure covered by a thin capsule of connective tissue. The body of the gland consists of many tightly packed secretory cells that are closely associated with capillary networks (fig. 13.25).

Parathyroid Hormone

The parathyroid glands secrete a protein, **parathyroid hormone** (PTH), or *parathormone* (see fig. 13.4). This hormone increases blood calcium ion concentration and decreases blood phosphate ion concentration through actions in the bones, kidneys, and intestines.

The intercellular matrix of bone tissue contains a considerable amount of calcium phosphate and calcium carbonate. PTH stimulates bone resorption by osteoclasts and inhibits the activity of osteoblasts (see chapter 7, p. 194). As bone resorption increases, calcium and phosphate ions are released into the blood. At the same time, PTH causes the kidneys to conserve blood calcium ions and to excrete more phosphate ions in the urine. PTH also indirectly stimulates absorption of calcium ions from food in the intestine by influencing metabolism of vitamin D.

Vitamin D (cholecalciferol) is synthesized from dietary cholesterol, which intestinal enzymes convert into provitamin D (7-dehydrocholesterol). This provitamin is largely stored in the skin, and exposure to the ultraviolet wavelengths of sunlight changes it to vitamin D. Some vitamin D also comes from foods.

Cancer cells in nonendocrine tissues sometimes secrete hormones, producing symptoms of an *endocrine paraneoplastic syndrome*. A patient with this unusual syndrome in 1994 led to discovery of a new hormone. In a middle-aged woman with oncogenic osteomalacia, a small tumor in the soft tissue of her thigh disrupted phosphate homeostasis, causing her bones to demineralize, leading to her bone pain. To the surprise of medical researchers, the tumor was not producing parathyroid hormone but a new hormone, tentatively named phosphaturic factor, which increases excretion of phosphate in urine, weakening the bones. The woman recovered when her tumor was removed, but it grew back and symptoms returned.

The liver changes vitamin D to hydroxycholecalciferol, which is carried in the bloodstream or is stored in tissues. When PTH is present, hydroxycholecalciferol can be changed in the kidneys into an active form of vitamin

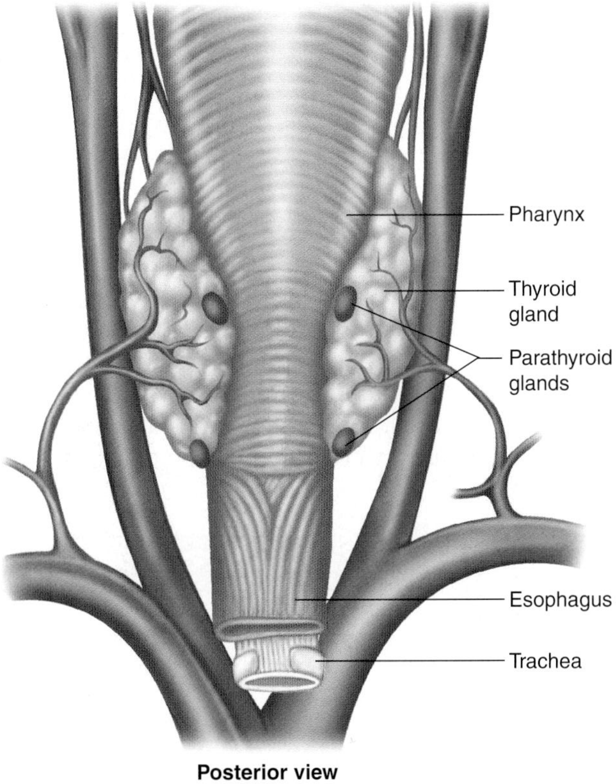

![Parathyroid glands illustration]

Pharynx

Thyroid gland

Parathyroid glands

Esophagus

Trachea

Posterior view

FIGURE 13.24

The parathyroid glands are embedded in the posterior surface of the thyroid gland.

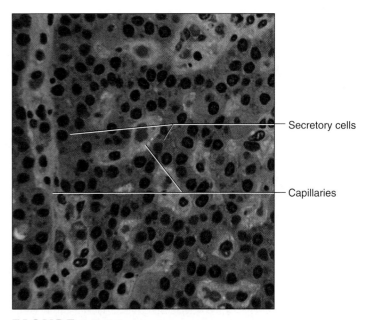

Secretory cells

Capillaries

FIGURE 13.25

Light micrograph of the parathyroid gland (540×).

D (dihydroxycholecalciferol), which controls absorption of calcium ions from the intestine (fig. 13.26).

A negative feedback mechanism operating between the parathyroid glands and the blood calcium ion concentration regulates secretion of PTH (fig. 13.27). As the concentration of blood calcium ions rises, less PTH is secreted; as the concentration of blood calcium ions drops, more PTH is released.

The opposite effects of calcitonin and PTH maintain calcium ion homeostasis. This is important in a number of physiological processes. For example, as the blood calcium ion concentration drops (hypocalcemia), the nervous system becomes abnormally excitable, and impulses may be triggered spontaneously. As a result, muscles, including the respiratory muscles, may undergo tetanic contractions, and the person may die due to a failure to breathe. In contrast, an abnormally high concentration of blood calcium ions (hypercalcemia) depresses the nervous system. Consequently, muscle contractions are weak, and reflexes are sluggish. Table 13.9 lists parathyroid disorders.

1 Where are the parathyroid glands located?

2 How does parathyroid hormone help regulate the concentrations of blood calcium and phosphate ions?

3 How does the negative feedback system of the parathyroid glands differ from that of the thyroid gland?

Adrenal Glands

The **adrenal glands** (suprarenal glands) are closely associated with the kidneys. A gland sits atop each kidney like a cap and is embedded in the mass of adipose tissue that encloses the kidney.

Structure of the Glands

The adrenal glands are shaped like pyramids. Each adrenal gland is very vascular and consists of two parts (fig. 13.28). The central portion is the adrenal medulla, and the outer part is the adrenal cortex. These regions are not sharply divided, but they are distinct glands that secrete different hormones.

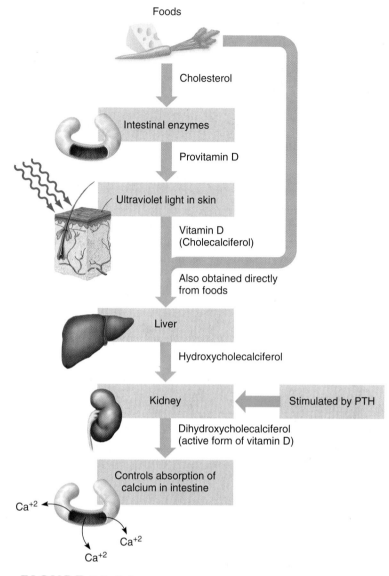

FIGURE 13.26
Mechanism by which PTH promotes calcium absorption in the intestine.

The **adrenal medulla** (ah-dre'nal me-dul'ah) consists of irregularly shaped cells grouped around blood vessels. These cells are intimately connected with the sympathetic division of the autonomic nervous system. In fact, these adrenal medullary cells are modified postganglionic

TABLE 13.9	Disorders of the Parathyroid Glands		
Condition	**Symptoms/Mechanism**	**Cause**	**Treatment**
Hyperparathyroidism	Fatigue, muscular weakness, painful joints, altered mental functions, depression, weight loss, bone weakening. Increased PTH secretion overstimulates osteoclasts.	Tumor	Remove tumor, correct bone deformities
Hypoparathyroidism	Muscle cramps and seizures. Decreased PTH secretion reduces osteoclast activity, diminishing blood calcium ion concentration.	Inadvertent surgical removal; injury	Calcium salt injections, massive doses of vitamin D

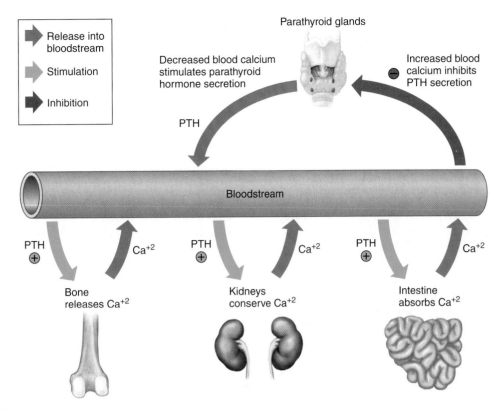

FIGURE 13.27

Parathyroid hormone (PTH) stimulates bone to release calcium (Ca^{+2}) and the kidneys to conserve calcium. It indirectly stimulates the intestine to absorb calcium. The resulting increase in blood calcium concentration inhibits secretion of PTH. (⊕ = stimulation; ⊖ = inhibition.)

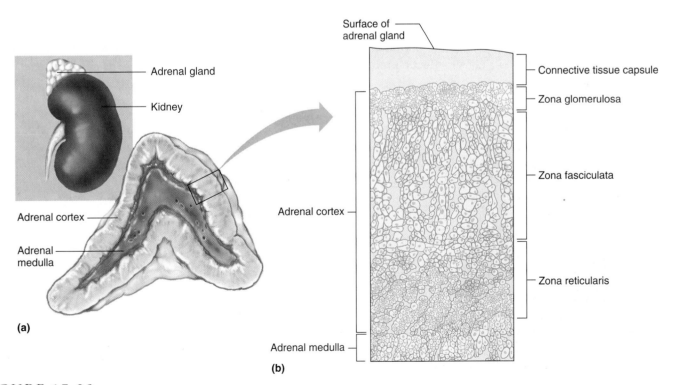

FIGURE 13.28

Adrenal glands. (*a*) An adrenal gland consists of an outer cortex and an inner medulla. (*b*) The cortex consists of three layers, or zones, of cells.

neurons, and preganglionic autonomic nerve fibers lead to them directly from the central nervous system (see chapter 11, pp. 409–410).

The **adrenal cortex** (ah-dre'nal kor'teks) makes up the bulk of the adrenal gland. It is composed of closely packed masses of epithelial layers that form an outer, a middle, and an inner zone of the cortex—the zona glomerulosa, the zona fasciculata, and the zona reticularis, respectively (fig. 13.29).

Hormones of the Adrenal Medulla

The cells of the adrenal medulla (chromaffin cells) produce, store, and secrete two closely related hormones, **epinephrine** (ep"ĭ-nef'rin), also called adrenalin and **norepinephrine** (nor"ep-ĭ-nef'rin), also called noradrenalin. Both of these substances are a type of amine called a *catecholamine,* and they have similar molecular structures and physiological functions (fig. 13.30). In fact, epinephrine is synthesized from norepinephrine.

The synthesis of these hormones begins with the amino acid tyrosine. In the first step of the process, an enzyme (tyrosine hydroxylase) in the secretory cells converts tyrosine into a substance called *dopa.* A second enzyme (dopa decarboxylase) changes dopa to dopamine, and a third enzyme (dopamine betahydroxylase) converts dopamine to norepinephrine. Still another enzyme (phenylethanolamine N-methyltransferase) converts the norepinephrine to epinephrine. About 15% of the norepinephrine is stored unchanged. The hormones are kept in tiny vesicles (chromaffin granules), much like neurotransmitters are stored in vesicles in neurons.

The effects of the adrenal medullary hormones generally resemble those that result when sympathetic nerve fibers stimulate their effectors. These effects include increased heart rate and increased force of cardiac muscle contraction, elevated blood pressure, increased breathing rate, and decreased activity in the digestive system (see table 11.10). The hormonal effects last up to ten times longer than the neurotransmitter effects because the hormones are removed from the tissues relatively slowly.

The ratio of the two hormones in the adrenal medullary secretion varies with different physiological conditions, but usually it is about 80% epinephrine and 20% norepinephrine. Although these hormones' effects are generally similar, certain effectors respond differently to them. These differences are due to the relative numbers of alpha and beta receptors in the membranes of the effector cells. Both hormones can stimulate both classes of receptors, although norepinephrine has a greater effect on alpha receptors. Table 13.10 compares some of the differences in the effects of these hormones.

Impulses arriving on sympathetic nerve fibers stimulate the adrenal medulla to release its hormones at the same time sympathetic impulses stimulate other effectors. As a rule, these impulses originate in the hypothalamus in response to stress. Thus, the medullary secretions function together with the sympathetic division of the autonomic nervous system in preparing the body for energy-expending action—"fight or flight."

∞ R E C O N N E C T   T O   C H A P T E R
1 1 ,   S Y M P A T H E T I C
D I V I S I O N ,   P A G E S   4 1 0 – 4 1 1

1 Describe the location and structure of the adrenal glands.

2 Name the hormones the adrenal medulla secretes.

3 What general effects do hormones secreted by the adrenal medulla produce?

4 What usually stimulates release of hormones from the adrenal medulla?

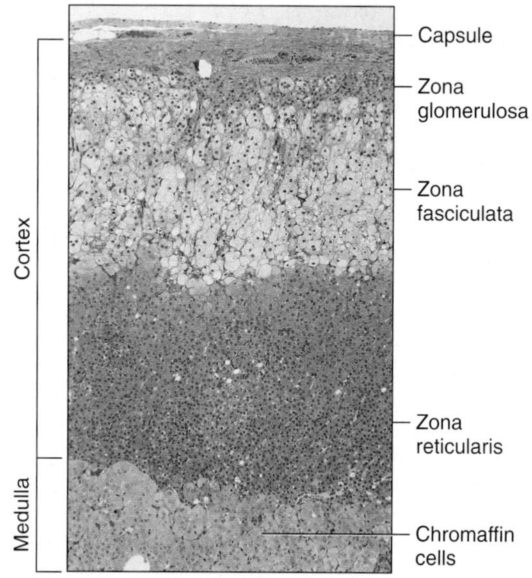

FIGURE 13.29
Light micrograph of the adrenal medulla and the adrenal cortex (75×).

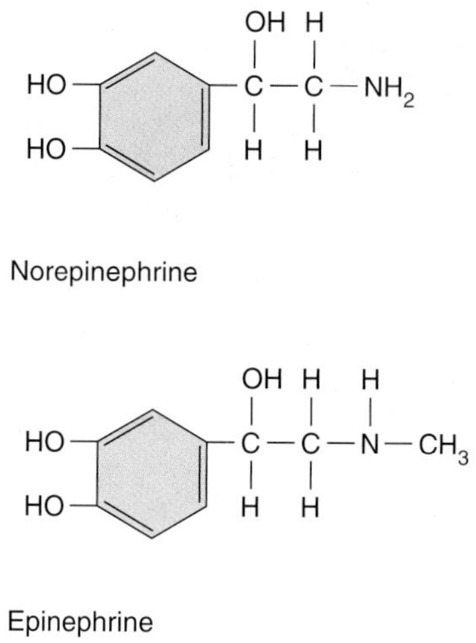

Norepinephrine

Epinephrine

FIGURE 13.30
Epinephrine and norepinephrine have similar molecular structures and similar functions.

TABLE 13.10

Structure or Function Affected	Epinephrine	Norepinephrine
Heart	Rate increases	Rate increases
	Force of contraction increases	Force of contraction increases
Blood vessels	Vessels in skeletal muscle vasodilate, decreasing resistance to blood flow	Blood flow to skeletal muscles increases, resulting from constriction of blood vessels in skin and viscera
Systemic blood pressure	Some increase due to increased cardiac output	Great increase due to vasoconstriction, counteracted in muscle blood vessels during exercise
Airways	Dilated	Some dilation
Reticular formation of brain	Activated	Little effect
Liver	Promotes breakdown of glycogen to glucose, increasing blood sugar level	Little effect on blood sugar level
Metabolic rate	Increases	Increases

Comparative Effects of Epinephrine and Norepinephrine (TABLE 13.10)

Hormones of the Adrenal Cortex

The cells of the adrenal cortex produce more than thirty different steroids, including several hormones (corticosteroids). Unlike the adrenal medullary hormones, without which a person can survive, some of those released by the cortex are vital. In fact, in the absence of adrenal cortical secretions, a person usually dies within a week if he or she does not receive extensive electrolyte therapy. The most important adrenal cortical hormones are aldosterone, cortisol, and certain sex hormones.

Aldosterone

Cells in the outer zone (zona glomerulosa) of the adrenal cortex synthesize **aldosterone.** This hormone is called a *mineralocorticoid* because it helps regulate the concentration of mineral electrolytes, such as sodium and potassium ions. More specifically, aldosterone causes the kidney to conserve sodium ions and to excrete potassium ions. The cells that secrete aldosterone respond directly to changes in the composition of blood plasma. However, whereas an increase in plasma potassium strongly stimulates these cells, a decrease in plasma sodium only slightly stimulates them. Control of aldosterone secretion is indirectly linked to plasma sodium level by the **renin-angiotensin system.**

Groups of specialized kidney cells (juxtaglomerular cells) are able to respond to changes in blood pressure and the plasma sodium ion concentration. If the level of either of these factors decreases, the cells release an enzyme called **renin.** Renin decomposes a blood protein called **angiotensinogen** (an″je-o-ten′sin-o-jen), which releases a peptide called **angiotensin I.** Another enzyme (angiotensin-converting enzyme, or ACE) in the lungs converts angiotensin I into another form, **angiotensin II,** which is carried in the bloodstream (fig. 13.31). When angiotensin II reaches the adrenal cortex, it stimulates the release of aldosterone. ACTH is necessary for aldosterone secretion to respond to this and other stimuli.

Aldosterone, in conserving sodium ions, indirectly retains water by osmosis. This helps maintain blood sodium ion concentration and blood volume (fig. 13.31). Angiotensin II is also a powerful vasoconstrictor and helps maintain systemic blood pressure by constricting blood vessels.

ACE inhibitors are a class of drugs used to treat some forms of high blood pressure (hypertension). They work by competing with angiotensin-converting enzyme, blocking formation of angiotensin II and preventing inactivation of a substance called bradykinin, a vasodilator. Both effects lead to dilation of blood vessels, lowering blood pressure.

Cortisol

Cortisol (hydrocortisone) is a *glucocorticoid,* which means it affects glucose metabolism. It is produced in the middle zone (zona fasciculata) of the adrenal cortex and has a molecular structure similar to aldosterone (fig. 13.32). In addition to affecting glucose, cortisol influences protein and fat metabolism. Among the more important actions of cortisol are the following:

1. It inhibits the synthesis of protein in various tissues, increasing blood concentration of amino acids.

2. It promotes the release of fatty acids from adipose tissue, increasing the use of fatty acids as an energy source and decreasing the use of glucose as an energy source.

3. It stimulates liver cells to synthesize glucose from noncarbohydrates (gluconeogenesis), such as circulating amino acids and glycerol, thus increasing blood glucose concentration.

Cortisol's actions help keep the blood glucose concentration within the normal range between meals. These actions are important because just a few hours without food can exhaust liver glycogen, another major source of glucose.

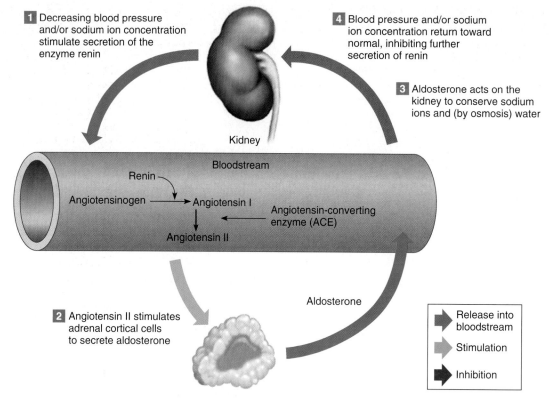

FIGURE 13.31

Aldosterone increases blood volume and pressure by promoting conservation of sodium ions and water (steps *1–4*).

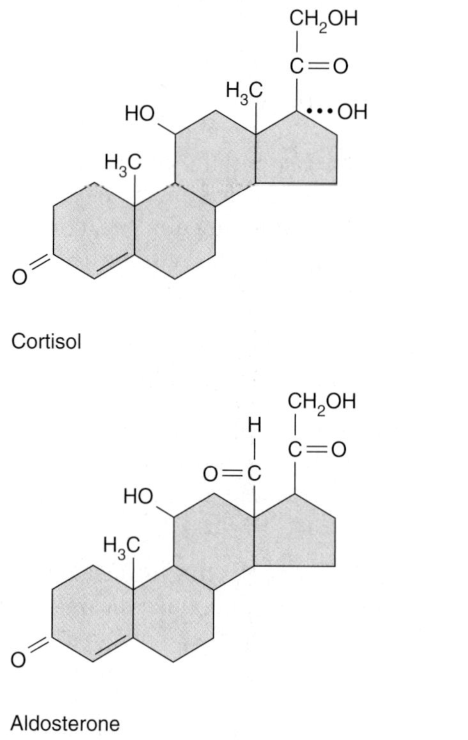

Cortisol

Aldosterone

FIGURE 13.32

Cortisol and aldosterone are steroids with similar molecular structures.

A negative feedback mechanism much like that controlling the thyroid hormones T_3 and T_4 regulates cortisol release. It involves the hypothalamus, anterior pituitary gland, and adrenal cortex. The hypothalamus secretes CRH (corticotropin-releasing hormone) into the hypophyseal portal veins, which carry the CRH to the anterior pituitary gland, stimulating it to secrete ACTH. In turn, ACTH stimulates the adrenal cortex to release cortisol. Cortisol inhibits release of both CRH and ACTH. As concentration of these substances falls, cortisol production drops.

Cortisol and related compounds are often used as drugs to reduce inflammation. They relieve pain by

- decreasing permeability of capillaries, preventing leakage of fluids that swell surrounding tissues
- stabilizing lysosomal membranes, preventing release of their enzymes, which destroy tissue
- inhibiting prostaglandin synthesis

Unfortunately, the concentration of cortisol compounds necessary to stifle inflammation is toxic, so these drugs can be used for only a short time. They are used to treat autoimmune disorders, allergies, asthma, and patients who have received organ transplants or tissue grafts.

The set point of the feedback loop controlling cortisol secretion changes from time to time, altering hormone output to meet the demands of changing conditions. For example, under stress—injury, disease, extreme temperature, or emotional upset—nerve impulses send the brain information concerning the stressful condition. In response, brain centers signal the hypothalamus to release more CRH, leading to a higher concentration of cortisol until the stress subsides (fig. 13.33).

Sex Hormones

Cells in the inner zone (zona reticularis) of the adrenal cortex produce sex hormones. These hormones are male (adrenal androgens), but some of them are converted into female hormones (estrogens) by the skin, liver, and adipose tissues. These hormones may supplement the supply of sex hormones from the gonads and stimulate early development of the reproductive organs. Also, adrenal androgens may play a role in controlling the female sex drive. Table 13.11 summarizes the actions of the cortical hormones. Clinical Application 13.3 discusses some of the effects of a malfunctioning adrenal gland on health.

1 Name the important hormones of the adrenal cortex.

2 What is the function of aldosterone?

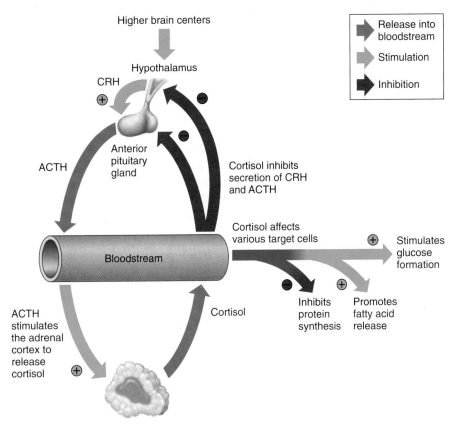

FIGURE 13.33

Negative feedback regulates cortisol secretion, similar to the regulation of thyroid hormone secretion (see fig. 13.16). (⊕ = stimulation; ⊖ = inhibition.)

TABLE 13.11	Hormones of the Adrenal Cortex	
Hormone	**Action**	**Factors Regulating Secretion**
Aldosterone	Helps regulate the concentration of extracellular electrolytes by conserving sodium ions and excreting potassium ions	Electrolyte concentrations in body fluids and renin-angiotensin mechanism
Cortisol	Decreases protein synthesis, increases fatty acid release, and stimulates glucose synthesis from noncarbohydrates	CRH from the hypothalamus and ACTH from the anterior pituitary gland
Adrenal androgens	Supplement sex hormones from the gonads; may be converted into estrogens	

DISORDERS OF THE ADRENAL CORTEX

John F. Kennedy's beautiful bronze complexion may have resulted not from sunbathing, but from a disorder of the adrenal glands. When he ran for the presidency in 1960, Kennedy knew he had *Addison disease,* **but his staff kept his secret, for fear it would affect his career. Kennedy had almost no adrenal tissue but was able to function by receiving mineralocorticoids and glucocorticoids, the standard treatment.**

In Addison disease, the adrenal cortex does not secrete hormones sufficiently due to immune system attack (autoimmunity) or an infection such as tuberculosis. Signs and symptoms include decreased blood sodium, increased blood potassium, low blood glucose level (hypoglycemia), dehydration, low blood pressure, frequent infections, fatigue, nausea and vomiting, loss of appetite, and increased skin pigmentation. Some sufferers experience salt cravings—one woman reported eating bowls and bowls of salty chicken noodle soup, with pickles and briny pickle juice added! Without treatment, death comes within days from severe disturbances in electrolyte balance.

An adrenal tumor or oversecretion of ACTH by the anterior pituitary causes hypersecretion of glucocorticoids (primarily cortisol), resulting in *Cushing syndrome.* It may also result from taking corticosteroid drugs for many years, such as to treat asthma or rheumatoid arthritis. Tissue protein level plummets, due to muscle wasting and loss of bone tissue. Blood glucose level remains elevated, and excess sodium is retained. As a result, tissue fluid increases, blood pressure rises, and the skin appears puffy. The skin may appear thin due to inhibition of collagen synthesis by the excess cortisol. Adipose tissue deposited in the face and back produce a characteristic "moon face" and "buffalo hump." Increase in adrenal sex hormone secretion may masculinize a female, causing growth of facial hair and a deepening voice. Other symptoms include extreme fatigue, sleep disturbances, skin rashes, headache, and leg muscle cramps.

Treatment of Cushing syndrome attempts to reduce ACTH secretion. This may entail removing a tumor in the pituitary gland or partially or completely removing the adrenal glands.

Both Addison disease and Cushing syndrome are rare, and for this reason, they are often misdiagnosed, or, in early stages, the patient's report of symptoms is not taken seriously. Addison disease affects thirty-nine to sixty people of every million, and Cushing syndrome affects five to twenty-five people per million. ∎

---

3   What does cortisol do?

4   How are blood concentrations of aldosterone and cortisol regulated?

Pancreas

The **pancreas** (pan′kre-as) contains two major types of secretory tissues. This organization of cell types reflects its dual function as an exocrine gland that secretes digestive juice through a duct and an endocrine gland that releases hormones into body fluids.

Structure of the Gland

The pancreas is an elongated, somewhat flattened organ that is posterior to the stomach and behind the parietal peritoneum (fig. 13.34). It is attached to the first section of the small intestine (duodenum) by a duct, which transports its digestive juice into the intestine. The digestive functions of the pancreas are discussed in chapter 17 (p. 665).

The endocrine portion of the pancreas consists of cells grouped around blood vessels. These groups, called *pancreatic islets (islets of Langerhans),* include three distinct types of hormone-secreting cells—*alpha cells,* which secrete glucagon; *beta cells,* which secrete insulin; and *delta cells,* which secrete somatostatin (fig. 13.35).

Hormones of the Pancreatic Islets

Glucagon is a protein that stimulates the liver to break down glycogen into glucose (glycogenolysis) and to convert noncarbohydrates, such as amino acids, into glucose (gluconeogenesis). Glucagon also stimulates breakdown of fats into fatty acids and glycerol.

In a negative feedback system, a low concentration of blood glucose stimulates release of glucagon from the alpha cells. When blood glucose concentration returns toward normal, glucagon secretion decreases. This mechanism prevents hypoglycemia from occurring at times when glucose concentration is relatively low, such as between meals, or when glucose is being used rapidly—during periods of exercise, for example.

The hormone **insulin** is also a protein, and its main effect is exactly opposite that of glucagon. Insulin stimulates the liver to form glycogen from glucose and inhibits conversion of noncarbohydrates into glucose. Insulin also has the special effect of promoting the facilitated diffusion (see chapter 3, p. 82) of glucose through the mem-

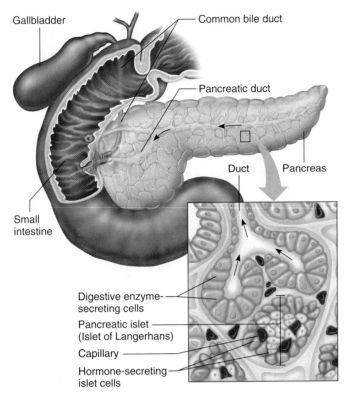

Gallbladder
Common bile duct
Pancreatic duct
Duct
Pancreas
Small intestine
Digestive enzyme-secreting cells
Pancreatic islet (Islet of Langerhans)
Capillary
Hormone-secreting islet cells

FIGURE 13.34

The hormone-secreting cells of the pancreas are grouped in clusters, or islets, that are closely associated with blood vessels. Other pancreatic cells secrete digestive enzymes into ducts.

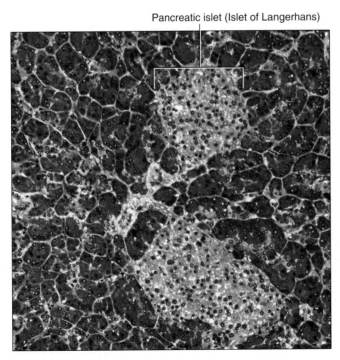

Pancreatic islet (Islet of Langerhans)

FIGURE 13.35

Light micrograph of a pancreatic islet in the pancreas (200×).

branes of cells bearing insulin receptors. These cells include those of cardiac muscle, adipose tissues, and resting skeletal muscles (glucose uptake by exercising skeletal muscles is not dependent on insulin). Insulin action decreases the concentration of blood glucose, promotes transport of amino acids into cells, and increases protein synthesis. It also stimulates adipose cells to synthesize and store fat.

An enzyme called glucokinase enables beta cells to "sense" glucose level, important information in determining rates of synthesis of glucagon and insulin. One form of a rare type of diabetes mellitus, maturity-onset diabetes of the young (MODY), is caused by a mutation in a gene encoding glucokinase—the beta cells cannot accurately assess when they must produce insulin. Other genetic mutations that cause MODY alter insulin's structure, secretion, or cell surface receptors, or the ability of liver cells to form glycogen in response to insulin. MODY is treated with drugs or dietary modification.

A negative feedback system sensitive to the concentration of blood glucose regulates insulin secretion. When glucose concentration is relatively high, as may occur following a meal, the beta cells release insulin. By promoting formation of glycogen in the liver and entrance of glucose into adipose and muscle cells, insulin helps prevent excessive rise in the blood glucose concentration (hyperglycemia). Then, when the glucose concentration falls, between meals or during the night, insulin secretion decreases (fig. 13.36).

As insulin concentration falls, less glucose enters the adipose and muscle cells, and the glucose remaining in the blood is available for cells that lack insulin receptors to use, such as nerve cells. Neurons readily tap the energy in a continuous supply of glucose to produce ATP.

Nerve cells, including those of the brain, obtain glucose by a facilitated diffusion mechanism that is not dependent on insulin, but rather only on the blood glucose concentration. For this reason, nerve cells are particularly sensitive to changes in the blood glucose concentration. Conditions that cause such changes—excess insulin secretion, for example—are likely to affect brain functions.

At the same time that insulin concentration is decreasing, glucagon secretion is increasing. Therefore, these hormones function together to maintain a relatively constant blood glucose concentration, despite great variations in the amounts of ingested carbohydrates.

Somatostatin (similar to the hypothalamic hormone), which the delta cells release, helps regulate glucose metabolism by inhibiting secretion of glucagon and

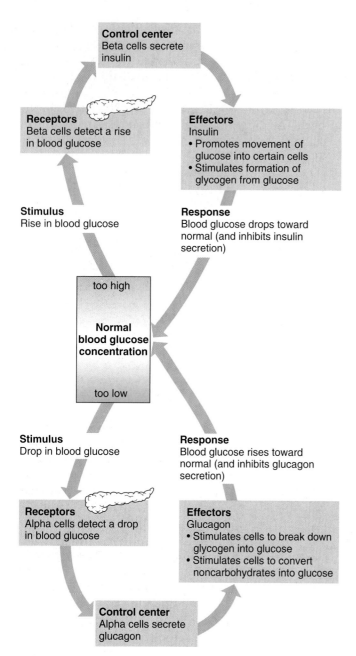

Control center
Beta cells secrete insulin

Receptors
Beta cells detect a rise in blood glucose

Effectors
Insulin
• Promotes movement of glucose into certain cells
• Stimulates formation of glycogen from glucose

Stimulus
Rise in blood glucose

Response
Blood glucose drops toward normal (and inhibits insulin secretion)

too high

Normal blood glucose concentration

too low

Stimulus
Drop in blood glucose

Response
Blood glucose rises toward normal (and inhibits glucagon secretion)

Receptors
Alpha cells detect a drop in blood glucose

Effectors
Glucagon
• Stimulates cells to break down glycogen into glucose
• Stimulates cells to convert noncarbohydrates into glucose

Control center
Alpha cells secrete glucagon

FIGURE 13.36

Insulin and glucagon function together to stabilize blood glucose concentration. Negative feedback responding to blood glucose concentration controls the levels of both hormones.

insulin. Table 13.12 summarizes the hormones of the pancreatic islets, and Clinical Application 13.4 discusses diabetes mellitus, a derangement of the control of glucose metabolism.

1 What is the name of the endocrine portion of the pancreas?

2 What is the function of glucagon?

3 What is the function of insulin?

4 How are the secretions of glucagon and insulin controlled?

5 Why are nerve cells particularly sensitive to changes in blood glucose concentration?

Hypoglycemia, or low blood glucose level due to excess insulin in the bloodstream, causes episodes of shakiness, weakness, and anxiety. Following a diet of frequent, small meals that are low in carbohydrates and high in protein can often control symptoms by preventing the surges of insulin that lower the blood glucose level. Hypoglycemia is most often seen when a person with diabetes injects too much insulin, but it can also reflect a tumor of the insulin-producing cells of the pancreas, or it may occur transiently following very strenuous exercise.

Other Endocrine Glands

Additional organs that produce hormones and, therefore, are parts of the endocrine system include the pineal gland, thymus gland, reproductive glands, and certain glands of the digestive tract, the heart, and the kidneys.

Pineal Gland

The **pineal gland** (pin'e-al gland) is a small, oval structure located deep between the cerebral hemispheres, where it attaches to the upper portion of the thalamus near the roof of the third ventricle. It largely consists of specialized *pineal cells* and supportive *neuroglial cells* (see fig. 11.20b).

TABLE 13.12	Hormones of the Pancreatic Islets	
Hormone	**Action**	**Source of Control**
Glucagon	Stimulates the liver to break down glycogen and convert noncarbohydrates into glucose; stimulates breakdown of fats	Blood glucose concentration
Insulin	Promotes formation of glycogen from glucose, inhibits conversion of noncarbohydrates into glucose, and enhances movement of glucose through adipose and muscle cell membranes, decreasing blood glucose concentration; promotes transport of amino acids into cells; enhances synthesis of proteins and fats	Blood glucose concentration
Somatostatin	Helps regulate carbohydrates	Not determined

DIABETES MELLITUS

Life for a person with type I (insulin-dependent) *diabetes mellitus* **(IDDM) means constant awareness of the illness—usually several insulin injections a day, frequent finger punctures to monitor blood glucose level, a restrictive diet, and concern over complications, which include loss of vision, leg ulcers, and kidney damage. The many symptoms of this form of diabetes mellitus reflect disturbances in carbohydrate, protein, and fat metabolism.**

In Latin, *diabetes* means "increased urine output," and *mellitus* means "honey," referring to urine's sugar content. Lack of insulin decreases movement of glucose into skeletal muscle and adipose cells, inhibiting glycogen formation. As a result, blood glucose concentration rises (hyperglycemia). When blood glucose reaches a certain level, the kidneys begin to excrete the excess, and glucose appears in the urine (glycosuria). Water follows the glucose by osmosis, causing excess urinary water loss. The person becomes dehydrated and very thirsty.

Untreated diabetes mellitus decreases protein synthesis, causing tissues to waste away as glucose-starved cells use protein as an energy source. Weight falls, and wounds cannot heal. Fatty acids accumulate in the blood as a result of decreased fat synthesis and storage. Ketone bodies, a by-product of fat metabolism, also build up in the blood. They are excreted in the urine as sodium salts, and a large volume of water follows by osmosis, intensifying dehydration and lowering sodium ion concentration in the blood. Accumulation of ketones and loss of sodium ions lead to metabolic acidosis, a condition that lowers the pH of body fluids. Acidosis and dehydration adversely affect brain neurons. Without treatment (insulin replacement), the person becomes disoriented and may enter a diabetic coma and die.

Type I diabetes is also called juvenile-onset diabetes, because it usually appears before age twenty. About 15% of people with diabetes mellitus have this form. It is an autoimmune disorder in which the immune system attacks pancreatic beta cells, ultimately destroying them and halting insulin secretion. Treatment for type I diabetes is administering enough insulin to control carbohydrate metabolism. Insulin is typically injected two or three times a day, or provided continually by an implanted insulin pump. Insulin aerosols in development may replace one or two of the daily injections, but a person still needs one shot of long-acting insulin a day. Transplant of islet cells from a donor has been in development since 1971. It is currently in clinical trials. It is technically challenging because only 2 percent of a pancreas consists of the needed cells. The cells often die in culture, so the transplant must take place soon after an organ becomes available. In the future, stem cell infusions may be able to supplement beta cells.

A milder form of diabetes, type II or noninsulin-dependent diabetes mellitus (NIDDM), begins gradually, usually in people over forty. Cells lose insulin receptors and cannot respond to insulin. Heredity and a lifestyle of overeating and underexercising are risk factors for developing type II diabetes. Treatment includes careful control of diet to avoid foods that stimulate insulin production, exercising, drugs, and maintaining desirable body weight.

The glucose-tolerance test is used to diagnose both major types of diabetes mellitus. The patient ingests a known quantity of glucose, and blood glucose concentration is measured at intervals to assess glucose utilization. If the person has diabetes, blood glucose concentration rises greatly and remains elevated for several hours. In a healthy person, glucose rise is less dramatic, and the level returns to normal in about one-and-a-half hours. ∎

The pineal gland secretes a hormone, **melatonin,** which is synthesized from serotonin. Varying patterns of light and dark outside the body control the gland's activities with environmental information arriving by means of nerve impulses. In the presence of light, nerve impulses originating in the retinas of the eyes travel to the hypothalamus. From the hypothalamus, they enter the reticular formation and then pass downward into the spinal cord. In the spinal cord, the impulses travel along sympathetic nerve fibers back into the brain, and finally they reach the pineal gland. In response to impulses that light triggers, melatonin secretion from the pineal gland decreases.

In the absence of light, nerve impulses from the eyes are decreased, and secretion of melatonin increases. Melatonin secretion is part of the regulation of **circadian rhythms,** which are patterns of repeated activity associated with the environmental cycles of day and night. Circadian rhythms responding to light and dark include sleep/wake rhythms and seasonal cycles of fertility in many mammals. Melatonin binds to two types of receptors on brain neurons, one that is very abundant and one that is relatively uncommon. The major receptors are found on cells of the suprachiasmatic nucleus, a region that regulates the circadian clock. Binding to the second, less abundant receptors, however, induces sleepiness.

Melatonin inhibits secretion of gonadotropins from the anterior pituitary gland and helps regulate the female reproductive cycle. It may also control onset of puberty (see chapter 22, p. 858).

Thymus Gland

The **thymus gland** (thi′mus gland), which lies in the mediastinum posterior to the sternum and between the lungs, is large in young children but diminishes in size with age. This gland secretes a group of hormones, called **thymosins,** that affect production and differentiation of certain white blood cells (T lymphocytes). The thymus gland plays an important role in immunity and is discussed in chapter 16 (p. 614).

Reproductive Glands

The reproductive organs that secrete important hormones include the **ovaries,** which produce estrogens and progesterone; the **placenta,** which produces estrogens, progesterone, and a gonadotropin; and the **testes,** which produce testosterone. These glands and their secretions are discussed in chapter 22.

Digestive Glands

The digestive glands that secrete hormones are generally associated with the linings of the stomach and small intestine. These structures and their secretions are described in chapter 17 (pp. 660 and 665).

Other Hormone-Producing Organs

Other organs that produce hormones include the heart, which secretes *atrial natriuretic peptide* (chapter 15, p. 546), and the kidneys, which secrete a hormone that stimulates red blood cell production called *erythropoietin* (chapter 14, p. 514).

1   Where is the pineal gland located?

2   What is the function of the pineal gland?

3   Where is the thymus gland located?

Stress and Its Effects

Because survival depends upon maintaining homeostasis, factors that change the body's internal environment are potentially life threatening. When such dangers are sensed, nerve impulses are directed to the hypothalamus, triggering physiological responses that resist a loss of homeostasis. These responses often include increased activity in the sympathetic division of the autonomic nervous system and increased secretion of adrenal hormones. A factor capable of stimulating such a response is called a **stressor,** and the condition it produces in the body is called **stress.**

Types of Stress

Stressors may be physical, psychological, or a combination of both.

Physical stress threatens tissues. This includes extreme heat or cold, decreased oxygen concentration, infections, injuries, prolonged heavy exercise, and loud sounds. Often physical stress is accompanied by unpleasant or painful sensations.

Psychological stress results from thoughts about real or imagined dangers, personal losses, unpleasant social interactions (or lack of social interactions), or any factors that threaten a person. Feelings of anger, fear, grief, anxiety, depression, and guilt cause psychological stress. In other instances, psychological stress may stem from pleasant stimuli, such as friendly social contact, feelings of joy or happiness, or sexual arousal. The factors that produce psychological stress vary greatly from person to person. A situation that is stressful to one person may not affect another, and what is stressful at one time may not be at another time.

Responses to Stress

The hypothalamus controls response to stress, termed the *general stress* (or *general adaptation*) *syndrome.* This response, evoked to stress of any kind, maintains homeostasis.

Recall that the hypothalamus receives information from nearly all body parts, including visceral receptors, the cerebral cortex, the reticular formation, and limbic system. At times of stress, the hypothalamus responds to incoming impulses by activating the "fight or flight" response. More specifically, sympathetic impulses from the hypothalamus raise blood glucose concentration, the level of blood glycerol and fatty acids, heart rate, blood pressure and breathing rate, and dilate the air passages. The response also shunts blood from the skin and digestive organs into the skeletal muscles and increases secretion of epinephrine from the adrenal medulla. The epinephrine, in turn, intensifies these sympathetic responses and prolongs their effects (fig. 13.37).

At the same time that sympathetic activity increases, the hypothalamus's release of corticotropin-releasing hormone (CRH) stimulates the anterior pituitary gland to secrete ACTH, which increases the adrenal cortex's secretion of cortisol. Cortisol supplies cells with amino acids and extra energy sources and diverts glucose from skeletal muscles to brain tissue (fig. 13.37). Stress can also trigger release of glucagon from the pancreas, growth hormone (GH) from the anterior pituitary, and antidiuretic hormone (ADH) from the posterior pituitary gland. Secretion of renin from the kidney may also be stimulated.

Glucagon and growth hormone help mobilize energy sources, such as glucose, glycerol, and fatty acids, and stimulate cells to take up amino acids, facilitating repair of injured tissues. ADH stimulates the kidneys to retain water. This action decreases urine output and helps to maintain blood volume—particularly important if a person is bleeding or sweating heavily. Renin, by increasing angiotensin II levels, helps stimulate the kidneys to retain sodium (through aldosterone), and through the vasoconstrictor action of angiotension II contributes to maintaining blood pressure. Table 13.13 summarizes the body's reactions to stress.

TABLE 13.13	Major Events in the General Stress Syndrome

1. As a result of stress, nerve impulses are transmitted to the hypothalamus.

2. Sympathetic impulses arising from the hypothalamus increase blood glucose concentration, blood glycerol concentration, blood fatty acid concentration, heart rate, blood pressure, and breathing rate. They dilate air passages, shunt blood into skeletal muscles, and increase secretion of epinephrine from the adrenal medulla.

3. Epinephrine intensifies and prolongs sympathetic actions.

4. The hypothalamus secretes CRH, which stimulates secretion of ACTH by the anterior pituitary gland.

5. ACTH stimulates release of cortisol by the adrenal cortex.

6. Cortisol increases the concentration of blood amino acids, releases fatty acids, and forms glucose from noncarbohydrate sources.

7. Secretion of glucagon from the pancreas and growth hormone from the anterior pituitary increase.

8. Glucagon and growth hormone aid mobilization of energy sources and stimulate uptake of amino acids by cells.

9. Secretion of ADH from the posterior pituitary increases.

10. ADH promotes the retention of water by the kidneys, which increases blood volume.

11. Renin increases blood levels of angiotensin II, which acts as a vasoconstrictor and also stimulates aldosterone secretion by the adrenal cortex.

12. Aldosterone stimulates sodium retention by the kidneys.

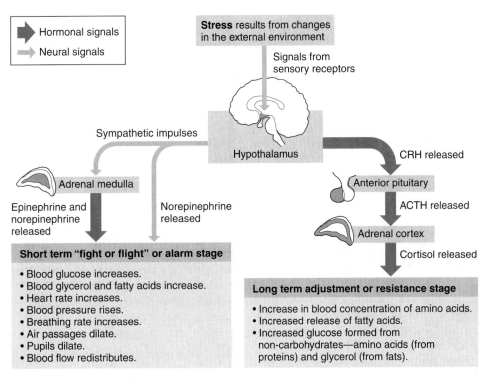

FIGURE 13.37

During stress, the hypothalamus helps prepare the body for "fight or flight" by triggering sympathetic impulses to various organs. It also stimulates epinephrine release, intensifying the sympathetic responses. The hypothalamus also stimulates the adrenal cortex to release cortisol, which promotes longer-term responses that resist the effects of stress.

The seventeen-year-old was visiting her family physician for the third time. She had recurrent stomach pains and seemed to have a constant respiratory infection. Unlike her previous visits, this time the woman seemed noticeably upset. Suspecting that her physical symptoms stemmed from a struggle to deal with a stressful situation, the doctor looked for signs—increased heart and respiratory rate, elevated blood pressure, and excessive sweating. He took a blood sample to measure levels of epinephrine and cortisol, while prodding her to talk about whatever was bothering her.

The young woman's cortisol was indeed elevated, which could account for her gastrointestinal pain and high blood pressure, as well as her impaired immunity. On the doctor's advice, she began seeing a psychologist. Her symptoms began to abate when she was able to discuss the source of her stress—a family member's illness and anxiety about beginning college.

1 What is stress?

2 Distinguish between physical stress and psychological stress.

3 Describe the general stress syndrome.

Life-Span Changes

With age, general changes in the glands of the endocrine system are a decrease in size and increase in the proportion of each gland that is fibrous in nature. At the cellular level, lipofuscin pigment accumulates as the glands age. Functionally, hormone levels may change with advancing years. Treatments for endocrine disorders associated with aging are usually straightforward—supplementing deficient hormones, or removing part of an overactive gland or using drugs to block the action of an overabundant hormone.

Aging affects different hormones in characteristic ways. For growth hormone, with age, the typical surge in secretion that occurs at night lessens somewhat. Lower levels of GH are associated with declining strength in the skeleton and muscles with advancing age. However, supplementing older people with GH in an attempt to duplicate the effects of exercise is dangerous because excess hormone raises blood pressure and blood glucose levels and may enlarge the spleen, liver, and kidneys.

Levels of antidiuretic hormone increase with age, but this is due to slowed breakdown in the liver and kidneys, rather than increased synthesis. As a result, the kidneys are stimulated to reabsorb more water.

The thyroid gland shrinks with age, as individual follicles shrink and become separated by increasing amounts of fibrous connective tissue. Nodules, which may be benign or cancerous, become more common with age. Upon autopsy, many individuals are found to have thyroid nodules that were never detected. Although blood levels of T_3 and T_4 may diminish with age, in general, the thyroid gland's control over the metabolism of various cell types is maintained throughout life. Calcitonin levels decline with age, which raises the risk of osteoporosis.

Parathyroid function differs between the sexes with age. Secretion peaks in males at about age fifty, whereas in women, the level of parathyroid hormone decreases until about age forty, after which it rises and contributes to osteoporosis risk. Fat accumulates between the cells of the parathyroid glands.

The adrenal glands illustrate the common theme of aging-related physical changes, yet continued function. Fibrous connective tissue, lipofuscin pigment, and increased numbers of abnormal cells characterize the aging adrenal glands. However, thanks to the fine-tuning of negative feedback systems, blood levels of glucocorticoids and mineralocorticoids usually remain within the normal range, although the ability to maintain homeostasis of osmotic pressure, blood pressure, acid/base balance and sodium and potassium ion distributions may falter with age.

The most obvious changes in endocrine function that occur with age involve blood glucose regulation. The pancreas may be able to maintain secretion of insulin and glucagon, but lifestyle changes, such as increase in fat intake and less exercise, may lead to an increase in blood insulin level. The development of insulin resistance—the decreased ability of muscle, liver, and fat cells to take in glucose even in the presence of insulin—reflects impaired ability of these target cells to respond to the hormone, rather than compromised pancreatic function. Blood glucose buildup may signal the pancreas to secrete more insulin, setting the stage for type II diabetes mellitus.

The daily fall and rise of melatonin levels may even out somewhat with age, which may alter control of the sleep/wake cycle. Changes to the tempo of the body clock may, in turn, affect secretion of other hormones.

The thymus gland begins to noticeably shrink before age twenty, with accompanying declining levels of thymosins. By age sixty, thymosin secretion is nil. The result is a slowing of the maturation of B and T cells, which accounts for increased susceptibility to infections with age.

1 What general types of changes occur in the glands of the endocrine system with aging?

2 How do the structures and functions of particular endocrine glands change over a lifetime?

Integumentary System

Melanocytes produce skin pigment in response to hormonal stimulation.

Lymphatic System

Hormones stimulate lymphocyte production.

Skeletal System

Hormones act on bones to control calcium balance.

Digestive System

Hormones help control digestive system activity.

Muscular System

Hormones help increase blood flow to exercising muscles.

Respiratory System

Decreased oxygen causes hormonal stimulation of red blood cell production; red blood cells transport oxygen and carbon dioxide.

Nervous System

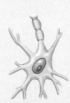

Neurons control the secretions of the anterior and posterior pituitary glands and the adrenal medulla.

Urinary System

Hormones act on the kidneys to help control water and electrolyte balance.

Cardiovascular System

Hormones are carried in the bloodstream; some have direct actions on the heart and blood vessels.

Reproductive System

Sex hormones play a major role in development of secondary sex characteristics, egg, and sperm.

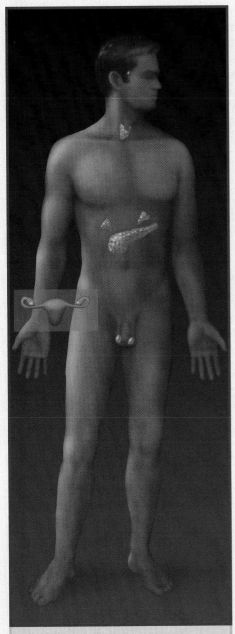

ENDOCRINE SYSTEM

Glands secrete hormones that have a variety of effects on cells, tissues, organs, and organ systems.

Clinical Terms Related to the Endocrine System

adrenalectomy (ah-dre″nah-lek′to-me) Surgical removal of the adrenal glands.

adrenogenital syndrome (ah-dre″no-jen′ĭ-tal sin′drōm) Group of symptoms associated with changes in sexual characteristics as a result of increased secretion of adrenal androgens.

exophthalmos (ek″sof-thal′mos) Abnormal protrusion of the eyes.

hirsutism (her′sut-izm) Excess hair growth, especially in women.

hypercalcemia (hi″per-kal-se′me-ah) Excess blood calcium.

hyperglycemia (hi″per-gli-se′me-ah) Excess blood glucose.

hypocalcemia (hi″po-kal-se′me-ah) Deficient blood calcium.

hypophysectomy (hi-pof″ĭ-sek′to-me) Surgical removal of the pituitary gland.

parathyroidectomy (par″ah-thi″roi-dek′to-me) Surgical removal of the parathyroid glands.

pheochromocytoma (fe-o-kro″mo-si-to′mah) Type of tumor found in the adrenal medulla and usually accompanied by high blood pressure.

polyphagia (pol″e-fa′je-ah) Excessive eating.

thymectomy (thi-mek′to-me) Surgical removal of the thymus gland.

thyroidectomy (thi″roi-dek′to-me) Surgical removal of the thyroid gland.

thyroiditis (thi″roi-di′tis) Inflammation of the thyroid gland.

CHAPTER SUMMARY

General Characteristics of the Endocrine System (page 468)

Endocrine glands secrete their products into body fluids (the internal environment); exocrine glands secrete their products into ducts that lead to the outside of the body. As a group, endocrine glands regulate metabolic processes.

Hormone Action (page 469)

Endocrine glands secrete hormones that affect target cells possessing specific receptors.

1. Chemistry of hormones
   a. Steroid hormones
   b. Nonsteroid hormones
2. Actions of hormones
   a. Steroid hormones
      (1) Steroid hormones enter target cells and combine with receptors to form complexes.
      (2) These complexes activate specific genes in the nucleus, which direct synthesis of specific proteins.
      (3) The degree of cellular response is proportional to the number of hormone-receptor complexes formed.
   b. Nonsteroid hormones
      (1) Nonsteroid hormones combine with receptors in the target cell membrane.
      (2) A hormone-receptor complex stimulates membrane proteins, such as adenylate cyclase, to induce the formation of second messenger molecules.
      (3) A second messenger, such as cAMP, activates protein kinases.
      (4) Protein kinases activate certain protein substrate molecules, which, in turn, change cellular processes.
      (5) The cellular response to a nonsteroid hormone is amplified because the enzymes induced by a small number of hormone-receptor complexes can catalyze formation of a large number of second messenger molecules.
3. Prostaglandins
   a. Prostaglandins are paracrine substances present in small quantities that have powerful hormonelike effects.
   b. Prostaglandins modulate hormones that regulate formation of cyclic AMP.

Control of Hormonal Secretions (page 477)

The concentration of each hormone in the body fluids is precisely regulated.

1. Control sources
   a. Other glands secrete hormones in response to releasing hormones the hypothalamus secretes.
   b. Some endocrine glands secrete in response to nerve impulses.
   c. Some glands secrete in response to changes in the plasma concentration of a substance.
2. Negative feedback systems
   a. In a negative feedback system, a gland is sensitive to the concentration of a substance it regulates.
   b. When the concentration of the regulated substance reaches a certain concentration, it inhibits the gland.
   c. As the gland secretes less hormone, the controlled substance also decreases.

Pituitary Gland (page 478)

The pituitary gland, which is attached to the base of the brain, has an anterior lobe and a posterior lobe. Releasing hormones from the hypothalamus control most pituitary secretions.

1. Anterior pituitary hormones
   a. The anterior pituitary consists largely of epithelial cells, and it secretes GH, PRL, TSH, ACTH, FSH, and LH.
   b. Growth hormone (GH)
      (1) Growth hormone stimulates body cells to grow and divide.

(2) Growth hormone-releasing hormone and somatostatin from the hypothalamus control GH secretion.
  c. Prolactin (PRL)
    (1) PRL promotes breast development and stimulates milk production.
    (2) In males, prolactin decreases secretion of LH (ICSH).
    (3) Prolactin release-inhibiting hormone from the hypothalamus restrains secretion of prolactin, whereas the yet to be identified prolactin-releasing factor is thought to promote its secretion.
  d. Thyroid-stimulating hormone (TSH)
    (1) TSH controls secretion of hormones from the thyroid gland.
    (2) The hypothalamus, by secreting thyrotropin-releasing hormone, regulates TSH secretion.
  e. Adrenocorticotropic hormone (ACTH)
    (1) ACTH controls the secretion of certain hormones from the adrenal cortex.
    (2) The hypothalamus, by secreting corticotropin-releasing hormone, regulates ACTH secretion.
  f. Follicle-stimulating hormone (FSH) and luteinizing hormone (LH) are gonadotropins that affect the reproductive organs.
2. Posterior pituitary hormones
  a. The posterior lobe of the pituitary gland largely consists of neuroglial cells and nerve fibers that originate in the hypothalamus.
  b. The two hormones of the posterior pituitary are produced in the hypothalamus.
  c. Antidiuretic hormone (ADH)
    (1) ADH causes the kidneys to excrete less water.
    (2) In high concentration, ADH constricts blood vessel walls, raising blood pressure.
    (3) The hypothalamus regulates ADH secretion.
  d. Oxytocin (OT)
    (1) Oxytocin has an antidiuretic effect and can contract muscles in the uterine wall.
    (2) OT also contracts certain cells associated with production and ejection of milk from the milk glands of the breasts.

Thyroid Gland (page 485)

The thyroid gland is located in the neck and consists of two lateral lobes.

1. Structure of the gland
  a. The thyroid gland consists of many hollow secretory parts called follicles.
  b. The follicles are fluid filled and store the hormones the follicle cells secrete.
  c. Extrafollicular cells secrete calcitonin.
2. Thyroid hormones
  a. Thyroxine and triiodothyronine
    (1) These hormones increase the rate of metabolism, enhance protein synthesis, and stimulate lipid breakdown.
    (2) These hormones are needed for normal growth and development and for maturation of the nervous system.
  b. Calcitonin
    (1) Calcitonin lowers blood calcium and phosphate ion concentrations.
    (2) This hormone prevents prolonged elevation of calcium after a meal.

Parathyroid Glands (page 489)

The parathyroid glands are located on the posterior surface of the thyroid.

1. Structure of the glands
  a. Each gland is small and yellow-brown, within a thin connective tissue capsule.
  b. Each gland consists of secretory cells that are well supplied with capillaries.
2. Parathyroid hormone (PTH)
  a. PTH increases blood calcium ion concentration and decreases blood phosphate ion concentration.
  b. PTH stimulates resorption of bone tissue, causes the kidneys to conserve calcium ions and excrete phosphate ions, and indirectly stimulates absorption of calcium ions from the intestine.
  c. A negative feedback mechanism operating between the parathyroid glands and the blood regulates these glands.

Adrenal Glands (page 490)

The adrenal glands are located atop the kidneys.

1. Structure of the glands
  a. Each adrenal gland consists of a medulla and a cortex.
  b. The adrenal medulla and adrenal cortex are distinct glands that secrete different hormones.
2. Hormones of the adrenal medulla
  a. The adrenal medulla secretes epinephrine and norepinephrine.
  b. These hormones are synthesized from tyrosine and are similar chemically.
  c. These hormones produce effects similar to those of the sympathetic nervous system.
  d. Sympathetic impulses originating from the hypothalamus stimulate secretion of these hormones.
3. Hormones of the adrenal cortex
  a. The cortex produces several types of steroids that include hormones.
  b. Aldosterone
    (1) It causes the kidneys to conserve sodium ions and water and to excrete potassium ions.
    (2) It is secreted in response to increased potassium ion concentration or presence of angiotensin II.
    (3) By conserving sodium ions and water, it helps maintain blood volume and pressure.
  c. Cortisol
    (1) It inhibits protein synthesis, releases fatty acids, and stimulates glucose formation from noncarbohydrates.
    (2) A negative feedback mechanism involving secretion of CRH from the hypothalamus and ACTH from the anterior pituitary gland controls its level.
  d. Adrenal sex hormones
    (1) These hormones are of the male type although some can be converted into female hormones.
    (2) They supplement the sex hormones produced by the gonads.

Pancreas (page 496)

The pancreas secretes digestive juices as well as hormones.

1. Structure of the gland
  a. The pancreas is posterior to the stomach and is attached to the small intestine.

b. The endocrine portion, which is called the pancreatic islets (islets of Langerhans), secretes glucagon, insulin, and somatostatin.
2. Hormones of the pancreatic islets
   a. Glucagon stimulates the liver to produce glucose, increasing concentration of blood glucose. It also breaks down fat.
   b. Insulin moves glucose through cell membranes, stimulates its storage, promotes protein synthesis, and stimulates fat storage.
   c. Nerve cells lack insulin receptors and depend upon diffusion for a glucose supply.
   d. Somatostatin inhibits insulin and glucagon release.

Other Endocrine Glands (page 498)

1. Pineal gland
   a. The pineal gland is attached to the thalamus near the roof of the third ventricle.
   b. Postganglionic sympathetic nerve fibers innervate it.
   c. It secretes melatonin, which inhibits secretion of gonadotropins from the anterior pituitary gland.
   d. It may help regulate the female reproductive cycle.
2. Thymus gland
   a. The thymus gland lies posterior to the sternum and between the lungs.
   b. It shrinks with age.
   c. It secretes thymosin, which affects the production of certain lymphocytes that, in turn, provide immunity.
3. Reproductive glands
   a. The ovaries secrete estrogens and progesterone.
   b. The placenta secretes estrogens, progesterone, and a gonadotropin.
   c. The testes secrete testosterone.
4. The digestive glands include certain glands of the stomach and small intestine that secrete hormones.

5. Other hormone-producing organs include the heart and kidneys.

Stress and Its Effects (page 500)

Stress occurs when the body responds to stressors that threaten the maintenance of homeostasis. Stress responses include increased activity of the sympathetic nervous system and increased secretion of adrenal hormones.

1. Types of stress
   a. Physical stress results from environmental factors that are harmful or potentially harmful to tissues.
   b. Psychological stress results from thoughts about real or imagined dangers.
   c. Factors that produce psychological stress vary with the individual and the situation.
2. Responses to stress
   a. Responses to stress maintain homeostasis.
   b. The hypothalamus controls a general stress syndrome.

Life-Span Changes (page 502)

Endocrine glands tend to shrink and accumulate fibrous connective tissue, fat, and lipofuscin, but hormonal activities usually remain within the normal range.

1. GH levels even out, as muscular strength declines.
2. ADH levels increase due to slowed breakdown.
3. The thyroid shrinks but control of metabolism continues.
4. Decreasing levels of calcitonin and parathyroid hormone increase osteoporosis risk.
5. The adrenal glands show aging-related changes, but negative feedback maintains functions.
6. Muscle, liver, and fat cells may develop insulin resistance.
7. Changes in melatonin secretion affect the body clock.
8. Thymosin production declines, hampering infectious disease resistance.

CRITICAL THINKING QUESTIONS

1. Based on your understanding of the actions of glucagon and insulin, would a person with diabetes mellitus be likely to require more insulin or more sugar following strenuous exercise? Why?
2. What problems might result from the prolonged administration of cortisol to a person with a severe inflammatory disease?
3. How might the environment of a patient with hyperthyroidism be modified to minimize the drain on body energy resources?
4. Which hormones should be administered to an adult whose anterior pituitary gland has been removed? Why?
5. A patient who has lost a large volume of blood will secrete excess aldosterone from the adrenal cortex. What effect will this increased secretion have on the patient's blood concentrations of sodium and potassium ions?
6. Both growth hormone and growth hormone-releasing hormone have been successfully used to promote growth in children with short statures. What is the difference in the ways these hormones produce their effects?

REVIEW EXERCISES

1. What is an endocrine gland?
2. Define *hormone* and *target cell*.
3. Explain how hormones can be grouped on the basis of their chemical composition.
4. Explain how steroid hormones influence cells.
5. Distinguish between the binding site and the activity site of a receptor molecule.
6. Explain how nonsteroid hormones may function through the formation of cAMP.
7. Explain how nonsteroid hormones may function through an increase in intracellular calcium ion concentration.
8. Explain how the cellular response to a hormone operating through a second messenger is amplified.
9. Define *prostaglandins*, and explain their general function.
10. Describe a negative feedback system.
11. Define *releasing hormone*, and provide an example of one.
12. Describe the location and structure of the pituitary gland.
13. List the hormones the anterior pituitary gland secretes.
14. Explain how the brain controls pituitary gland activity.
15. Explain how growth hormone produces its effects.
16. List the major factors that affect growth hormone secretion.
17. Summarize the functions of prolactin.
18. Describe regulation of concentrations of circulating thyroid hormones.
19. Explain the control of secretion of ACTH.
20. List the major gonadotropins, and explain the general functions of each.
21. Compare the cellular structures of the anterior and posterior lobes of the pituitary gland.
22. Name the hormones associated with the posterior pituitary, and explain their functions.
23. Explain how the release of ADH is regulated.
24. Describe the location and structure of the thyroid gland.
25. Name the hormones the thyroid gland secretes, and list the general functions of each.
26. Define *iodine pump*.
27. Describe the location and structure of the parathyroid glands.
28. Explain the general functions of parathyroid hormone.
29. Describe mechanisms that regulate the secretion of parathyroid hormone.
30. Distinguish between the adrenal medulla and the adrenal cortex.
31. List the hormones produced by the adrenal medulla, and describe their general functions.
32. List the steps in the synthesis of adrenal medullary hormones.
33. Name the most important hormones of the adrenal cortex, and describe the general functions of each.
34. Describe the regulation of the secretion of aldosterone.
35. Describe control of cortisol secretion.
36. Describe the location and structure of the pancreas.
37. List the hormones the pancreatic islets secrete, and describe the general functions of each.
38. Summarize how the secretion of hormones from the pancreas is regulated.
39. Describe the location and general function of the pineal gland.
40. Describe the location and general function of the thymus gland.
41. Distinguish between a stressor and stress.
42. List several factors that cause physical and psychological stress.
43. Describe the general stress syndrome.
44. Which components of the endocrine system change the most as a person ages?

WEB CONNECTIONS

Visit the Student Edition of the Online Learning Center at www.mhhe.com/shier10 **for answers to chapter questions, additional quizzes, interactive learning exercises, and other study tools.**

UNDERSTANDING WORDS

agglutin-, to glue together: *agglutin*ation— clumping together of red blood cells.

bil-, bile: *bil*irubin—pigment excreted in the bile.

-crit, to separate: hemato*crit*—percentage by volume of cells in a blood sample, determined by separating the cells from the plasma.

embol-, stopper: *embol*ism—obstruction of a blood vessel.

erythr-, red: *erythr*ocyte—red blood cell.

hem-, blood: *hem*oglobin—red pigment responsible for the color of blood.

hepar-, liver: *hepar*in—anticoagulant secreted by liver cells.

leuko-, white: *leuko*cyte—white blood cell.

-lys, to break up: fibrino*lys*in—protein-splitting enzyme that can digest fibrin.

macro-, large: *macro*phage—large phagocytic cell.

-osis, abnormal condition: leukocyt*osis*— condition in which white blood cells are overproduced.

-poie, make, produce: erythro*poie*tin— hormone that stimulates the production of red blood cells.

poly-, many: *poly*cythemia—condition in which red blood cells are overproduced.

-sta, halt, make stand: hemo*sta*sis—arrest of bleeding from damaged blood vessels.

thromb-, clot: *thromb*ocyte—blood platelet involved in the formation of a blood clot.

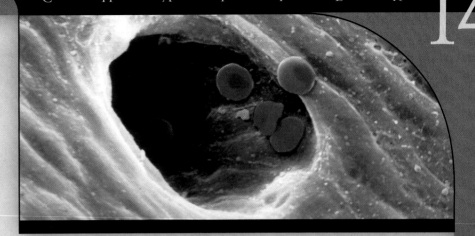

Red blood cells move from a large blood vessel into a much smaller capillary. A red blood cell travels about 900 miles during its four month existence. Falsely colored scanning electron micrograph (1,400×).

Blood

CHAPTER OBJECTIVES

After you have studied this chapter, you should be able to

1. Describe the general characteristics of blood and discuss its major functions.

2. Distinguish among the formed elements of the blood.

3. Explain the significance of red blood cell counts and how they are used to diagnose disease.

4. Discuss the life cycle of a red blood cell.

5. Explain the control of red blood cell production.

6. Distinguish among the five types of white blood cells and give the function(s) of each type.

7. List the major components of plasma and describe the functions of each.

8. Define *hemostasis* and explain the mechanisms that help to achieve it.

9. Describe the major steps in hemostasis.

10. Explain how to prevent coagulation.

11. Explain blood typing and how it is used to avoid adverse reactions following blood transfusions.

12. Describe how blood reactions may occur between fetal and maternal tissues.

F inding a substitute for human blood is a tall order, especially trying to mimic the major functions of this connective tissue. Blood carries oxygen (red blood cells), provides protection against infection (white blood cells), promotes clotting (platelets), and carries other vital substances. Efforts to replace blood in the past have sought to fill in the missing fluid volume or to reproduce blood's oxygen-carrying role. Red blood cell substitutes used in times past have included wine, ale, milk, plant resins, urine, and opium! The search for blood substitutes intensified after each of the two world wars because injured soldiers desperately needed transfusions, and again when the AIDS pandemic made transfusions dangerous unless blood is properly screened.

A red blood cell substitute must meet several requirements: It must carry oxygen and give it up to tissues, be nontoxic, be storeable, function until the body can take over, and not provoke an immune response. The nine red blood cell substitutes currently in clinical trials are of two basic types. Perfluorocarbons are synthetic chemicals that carry dissolved oxygen. These were actually developed in the 1960s, and a famous photo shows a mouse apparently drowning in a beaker of the chemical-but still breathing even though submerged.

The second type of red blood cell substitute dismantles red blood cells and isolates the oxygen-carrying hemoglobin molecules, which are then linked in various ways. The starting material is usually cow's blood or old stored human blood.

In the future, red blood cell replacements may come from stem cells. In one experiment, researchers coaxed stem cells from human umbilical cord blood to divide about 100 times in a laboratory dish, in the presence of growth factors that steered development towards forming mature red blood cells. The human cells developed in the dish until just before the final stage, when they eject their nuclei. Because the cells would not do this in the dish, the researchers infused the cells into mice that lack immune systems. Sure enough, development continued in the mice, and they soon had mature human red blood cells. Not only did the

At this biotechnology company, hemoglobin is purified in efforts to develop a red blood cell substitute.

human cells lack nuclei, indicating that they had matured, but the hemoglobin they produced was of the adult variety, not the fetal form of hemoglobin that was in the original umbilical cord blood. Translated to humans, the researchers estimate that a million cord blood stem cells can be cultured in 10 days to yield 6 to 10 billion cells, which, once infused into a person, would continue proliferating to yield 600 billion to a trillion cells-that's three to five times the body's daily red blood cell output. ■

Blood signifies life, and for good reason—it has many vital functions. This complex mixture of cells, cell fragments, and dissolved biochemicals transports nutrients, oxygen, wastes, and hormones; helps maintain the stability of the interstitial fluid; and distributes heat. The blood, heart, and blood vessels form the cardiovascular system and link the body's internal and external environments.

Blood is a type of connective tissue whose cells are suspended in a liquid material. Blood is vital in transporting substances between body cells and the external environment, thereby promoting homeostasis.

Blood and Blood Cells

Whole blood is slightly heavier and three to four times more viscous than water. Its cells, which form mostly in

red bone marrow, include red blood cells and white blood cells. Blood also contains cellular fragments called blood platelets (fig. 14.1). The cells and platelets are termed "formed elements" of the blood, in contrast to the liquid portion.

Blood Volume and Composition

Blood volume varies with body size, changes in fluid and electrolyte concentrations, and the amount of adipose tissue. Blood volume is typically about 8% of body weight. An average-sized adult has a blood volume of about 5 liters.

If a blood sample stands in a tube for awhile and is prevented from clotting, the cells separate from the liquid portion of the blood and settle to the bottom. Centrifuging the sample quickly packs the cells into the lower part of the centrifuge tube, as figure 14.2 shows.

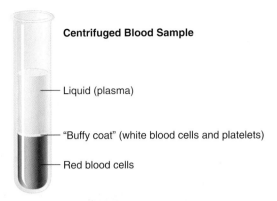

Centrifuged Blood Sample

Liquid (plasma)

"Buffy coat" (white blood cells and platelets)

Red blood cells

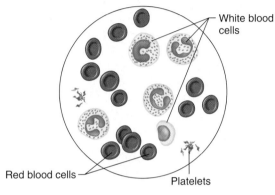

Peripheral Blood Smear

White blood cells

Red blood cells

Platelets

FIGURE 14.1

Blood consists of a liquid portion called plasma and a solid portion that includes red blood cells, white blood cells, and platelets. (Note: When blood components are separated, the white blood cells and platelets form a thin layer, called the "buffy coat," between the plasma and the red blood cells.)

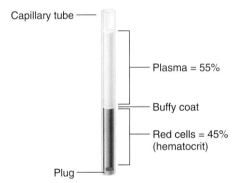

Capillary tube

Plasma = 55%

Buffy coat

Red cells = 45% (hematocrit)

Plug

FIGURE 14.2

If a blood-filled capillary tube is centrifuged, the red cells pack in the lower portion, and the percentage of red cells (hematocrit) can be determined. Values shown are within the normal range for healthy humans.

The percentage of cells and liquid in the blood sample can then be calculated.

A blood sample is usually about 45% cells by volume. This percentage is called the **hematocrit** (HCT), or **packed cell volume** (PCV). The remaining 55% of a blood sample is clear, straw-colored **plasma** (plaz′mah). Appendix B, Laboratory Tests of Clinical Significance, (pp. 970–972) lists val-

ues for the hematocrit and other common blood tests in healthy individuals.

In addition to red blood cells, which comprise more than 99% of the blood cells, the formed elements of the blood include white blood cells and platelets. Plasma is a complex mixture that includes water, amino acids, proteins, carbohydrates, lipids, vitamins, hormones, electrolytes, and cellular wastes.

The Origin of Blood Cells

Blood cells originate in red bone marrow from **hematopoietic** (he″mat-o-poi-et′ik) **stem cells** (fig. 14.3). A stem cell can divide to give rise to specialized cells (more differentiated) as well as more stem cells. As hematopoietic stem cells divide, the new cells (progenitor cells) respond to different secreted growth factors, called **colony-stimulating factors,** that turn on some genes and turn off others. This exposure to growth factors ultimately sculpts the distinctive formed elements of blood, including the cellular components of the immune system. A protein called *thrombopoietin* (TPO) stimulates large cells called **megakaryocytes** to proliferate. These cells eventually come apart to yield platelets.

R E C O N N E C T   W I T H
C H A P T E R   3 ,   S T E M   A N D
P R O G E N I T O R   C E L L S ,
P A G E S   9 4 – 9 6 .

1 What are the major components of blood?

2 What factors affect blood volume?

3 How is hematocrit determined?

4 How do blood cells form?

Characteristics of Red Blood Cells

Red blood cells, or **erythrocytes** (ĕ-rith′ro-sītz), are tiny, approximately 7.5 μm in diameter. They are biconcave discs, which means that they are thin near their centers and thicker around their rims (fig. 14.4). This special shape is an adaptation for the red blood cell's function of transporting gases; it increases the surface area through which gases can diffuse. The shape also places the cell membrane closer to oxygen-carrying *hemoglobin* molecules within the cell. Because of its shape, a red blood cell can readily squeeze through the narrow passages of capillaries.

Each red blood cell is about one-third hemoglobin by volume. This protein is responsible for the color of the blood. The rest of the cell mainly consists of membrane, water, electrolytes, and enzymes. When the hemoglobin combines with oxygen, the resulting *oxyhemoglobin* is bright red, and when the oxygen is released, the resulting *deoxyhemoglobin* is darker. Blood rich in deoxyhemoglobin may appear bluish when it is viewed through blood vessel walls.

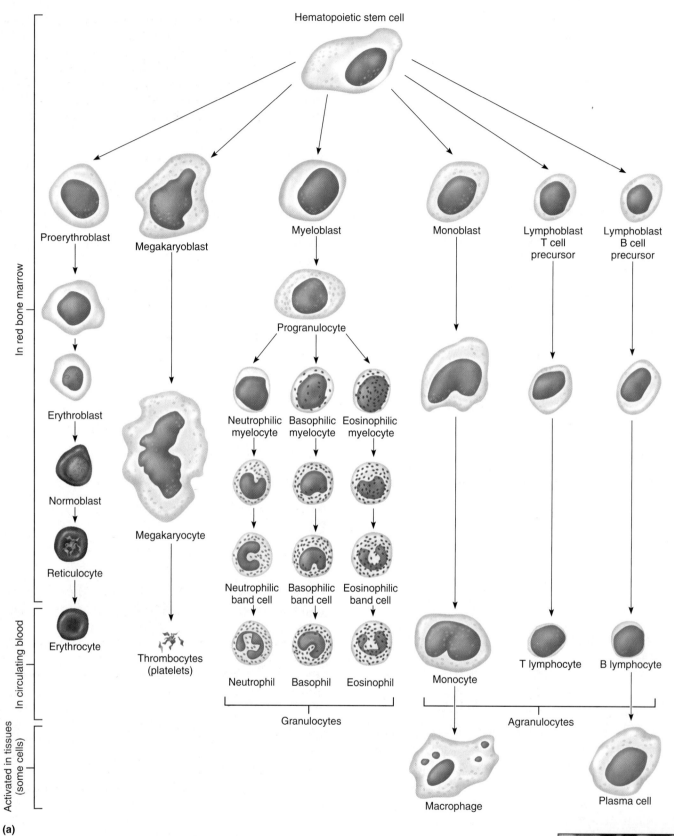

Hematopoietic stem cell

In red bone marrow

Proerythroblast

Megakaryoblast

Myeloblast

Monoblast

Lymphoblast
T cell
precursor

Lymphoblast
B cell
precursor

Erythroblast

Progranulocyte

Normoblast

Neutrophilic
myelocyte

Basophilic
myelocyte

Eosinophilic
myelocyte

Reticulocyte

Megakaryocyte

Neutrophilic
band cell

Basophilic
band cell

Eosinophilic
band cell

In circulating blood

Erythrocyte

Thrombocytes
(platelets)

Neutrophil

Basophil

Eosinophil

Monocyte

T lymphocyte

B lymphocyte

Granulocytes

Agranulocytes

Activated in tissues
(some cells)

Macrophage

Plasma cell

(a)

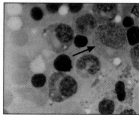

(b)

FIGURE 14.3

Blood cells. (a) Development of blood cells from hematopoietic stem cells (HSCs) in bone marrow. (b) Light micrograph of a HSC (arrow) in red bone marrow (500×).

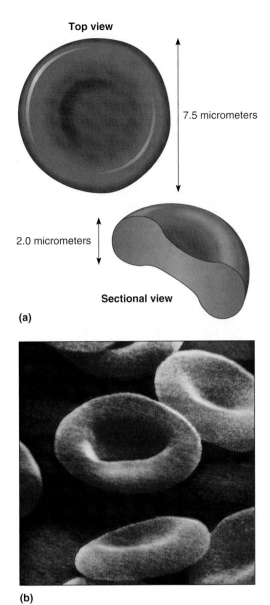

Top view

7.5 micrometers

2.0 micrometers

Sectional view

(a)

(b)

FIGURE 14.4

Red blood cells. (*a*) The biconcave shape of a red blood cell makes possible its function. (*b*) Falsely colored scanning electron micrograph of human red blood cells (5,000×).

A person experiencing prolonged oxygen deficiency (hypoxia) may become *cyanotic.* The skin and mucous membranes appear bluish due to an abnormally high blood concentration of deoxyhemoglobin. Exposure to low temperature may also result in cyanosis. Such exposure constricts superficial blood vessels, which slows blood flow and removes more oxygen than usual from blood flowing through the vessels.

Red blood cells have nuclei during their early stages of development but extrude them as the cells mature, which provides more space for hemoglobin. Since they lack nuclei, red blood cells cannot synthesize messenger RNA or divide. Because they also lack mitochondria, red blood cells produce ATP through glycolysis only and use none of the oxygen they carry. As long as cytoplasmic enzymes function, these can carry on vital energy-releasing processes. With time, however, red blood cells become less and less active. Typically, this leads to more rigid red blood cells that are more likely to be damaged or worn and eventually removed by the spleen and liver.

In *sickle cell disease,* a single DNA base change causes an incorrect amino acid to be incorporated into the protein portion of hemoglobin, causing hemoglobin to crystallize in a low oxygen environment. This bends the red blood cells containing the hemoglobin into a sickle shape, which blocks circulation in small vessels, causing excruciating joint pain and damaging many organs. As the spleen works harder to recycle the abnormally short-lived red blood cells, infection becomes likely.

Most children with sickle cell disease are diagnosed at birth and receive antibiotics daily for years to prevent infection. Hospitalization for blood transfusions may be necessary if the person experiences painful sickling "crises" of blocked circulation.

A bone marrow transplant can completely cure sickle cell disease but has a 15% risk of causing death. Another treatment is a drug used to treat cancer called hydroxyurea. It activates dormant genes that produce a slightly different form of hemoglobin in the fetus. Because of the presence of the functional fetal hemoglobin, the sickle hemoglobin cannot crystallize as quickly as it otherwise would. Sickling is delayed, which enables red blood cells carrying sickled hemoglobin to reach the lungs faster—where fresh oxygen restores the cells' normal shapes.

1 Describe a red blood cell.

2 How does the biconcave shape of a red blood cell make possible its function?

3 What is the function of hemoglobin?

Red Blood Cell Counts

The number of red blood cells in a cubic millimeter (mm^3) of blood is called the *red blood cell count* (RBCC or RCC). Although this number varies from time to time even in healthy individuals, the typical range for adult males is 4,600,000–6,200,000 cells per mm^3, and that for adult females is 4,200,000–5,400,000 cells per mm^3. For children, the average range is 4,500,000–5,100,000 cells per mm^3. These values may vary slightly with the hospital, physician, and type of equipment used to make blood cell counts. The number of red blood cells generally increases after several days following strenuous exercise, or an increase in altitude, due to the body cells' increased oxygen demand.

> The equivalent units of microliters (µL) are sometimes used in place of mm³ in describing blood cell counts. For example, 4,600,000 cells per mm³ is equivalent to 4,600,000 cells per µL.

Since an increasing number of circulating red blood cells increases the blood's *oxygen-carrying capacity,* changes in this number may affect health. For this reason, red blood cell counts are routinely consulted to help diagnose and evaluate the courses of various diseases.

Red Blood Cell Production and Its Control

Red blood cell formation (erythropoiesis) initially occurs in the yolk sac, liver, and spleen. After an infant is born, these cells are produced almost exclusively by tissue lining the spaces in bones, filled with red bone marrow.

∞ RECONNECT WITH CHAPTER 7, BLOOD CELL FORMATION, PAGE 194.

Within the red bone marrow, hematopoietic stem cells give rise to **erythroblasts** (ĕ-rith′ro-blastz) that can synthesize hemoglobin molecules at the rate of 2 million to 3 million per second. The erythroblasts also divide and give rise to many new cells. The nuclei of these newly formed cells soon shrink and are extruded by being pinched off in thin coverings of cytoplasm and cell membrane. The resulting cells are erythrocytes. Some of these young red cells may contain a netlike structure (reticulum) for a day or two. This network is the remainder of the endoplasmic reticulum, and such cells are called **reticulocytes** (rĕ-tik′u-lo-sitz). This is the stage that exits the bone marrow to enter the blood. When the reticulum degenerates, the cells are fully mature.

The average life span of a red blood cell is 120 days. During that time, a red blood cell travels through the body about 75,000 times. Many of these cells are removed from the circulation each day, yet the number of cells in the circulating blood remains relatively stable. This observation suggests a *homeostatic* control of the rate of red blood cell production.

A *negative feedback mechanism* utilizing the hormone **erythropoietin** (e-rith″ro-poi′ĕ-tin) controls the rate of red blood cell formation. In response to prolonged oxygen deficiency, erythropoietin is released, primarily from the kidneys and to a lesser extent from the liver. (In a fetus, the liver is the main site of erythropoietin production.) At high altitudes, for example, although the percentage of oxygen in the air remains the same, the atmospheric pressure

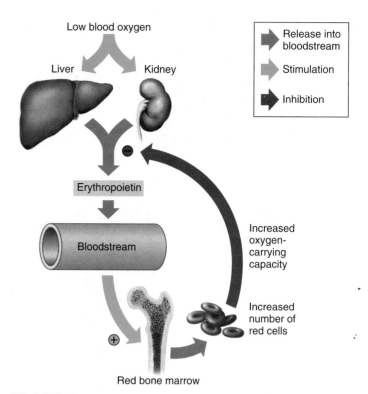

FIGURE 14.5

Low blood oxygen causes the kidneys and liver to release erythropoietin, which stimulates the production of red blood cells that carry oxygen to tissues.

decreases and availability of oxygen is reduced. The amount of oxygen delivered to the tissues initially decreases. As figure 14.5 shows, this drop in oxygen triggers release of erythropoietin, which travels via the blood to the red bone marrow and stimulates increased erythrocyte production.

After a few days, many new red blood cells begin to appear in the circulating blood. The increased rate of production continues until the number of erythrocytes in the circulation supplies sufficient oxygen to tissues. When the oxygen level in the air returns to normal, erythropoietin release decreases, and the rate of red blood cell production returns to normal as well.

> Several members of a large Danish family have inherited a condition called erythrocytosis. Their reticulocytes are extra sensitive to erythropoietin, and as a result, these individuals produce about 25% more red blood cells than normal. The result: great physical endurance. One man from this family has won three Olympic gold medals and two world championships in cross-country skiing.

Other conditions that can lower oxygen levels and stimulate erythropoietin release include loss of blood, which decreases the oxygen-carrying capacity of the car-

TABLE 14.1	Major Events in the Hormonal Control of Red Blood Cell Production

1. The kidney and liver tissues experience an oxygen deficiency.
2. These tissues release the hormone erythropoietin.
3. Erythropoietin travels to the red bone marrow and stimulates an increase in production of red blood cells.
4. As increasing numbers of red blood cells are released into the circulation, the oxygen-carrying capacity of the blood rises.
5. The oxygen concentration in the kidney and liver tissues increases, and the release of erythropoietin decreases.

diovascular system, and chronic lung diseases, which decrease the respiratory surface area available for gas exchange. Table 14.1 summarizes the steps in the regulation of red blood cell production.

1 What is the typical red blood cell count for an adult male? For an adult female?

2 Where are red blood cells produced?

3 How does a red blood cell change as it matures?

4 How is red blood cell production controlled?

Dietary Factors Affecting Red Blood Cell Production

The availability of two B-complex vitamins—vitamin B_{12} and folic acid—significantly influences red blood cell production. These vitamins are required for DNA synthesis, so they are necessary for the growth and division of all cells. Since cell division occurs at a particularly high rate in hematopoietic tissue, this tissue is especially vulnerable to deficiency of either of these vitamins. Lack of vitamin B_{12} is usually due to a disorder in the stomach lining rather than to a dietary deficiency, because certain cells in the stomach secrete a substance called *intrinsic factor* that is needed for absorption of vitamin B_{12}.

Iron is required for hemoglobin synthesis. Although much of the iron released during the decomposition of hemoglobin is available for reuse, some iron is lost each day and must be replaced. Only a small fraction of ingested iron is absorbed. Iron absorption is slow, although the rate varies with the total amount of iron in the body. When iron stores are low, absorption rate increases, and when the tissues are becoming saturated with iron, the rate greatly decreases. Figure 14.6 summarizes the life cycle of a red blood cell. The dietary factors that affect red blood cell production are summarized in table 14.2.

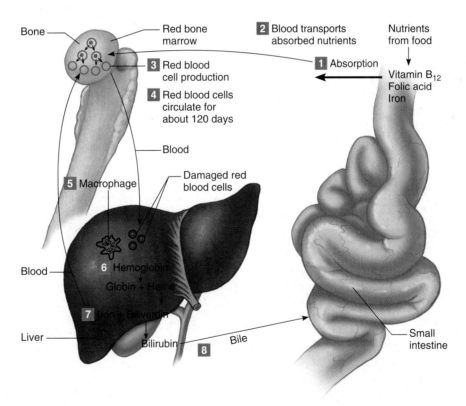

FIGURE 14.6

Life cycle of a red blood cell. (*1*) The small intestine absorbs essential nutrients. (*2*) Blood transports nutrients to red bone marrow. (*3*) In the marrow, red blood cells arise from the division of less-specialized progenitor cells. (*4*) Mature red blood cells are released into the bloodstream, where they circulate for about 120 days. (*5*) Macrophages destroy damaged red blood cells in the spleen and liver. (*6*) Hemoglobin liberated from red blood cells is broken down into heme and globin. (*7*) Iron from heme returns to red bone marrow and is reused. (*8*) Biliverdin and bilirubin are excreted in the bile.

TABLE 14.2	Dietary Factors Affecting Red Blood Cell Production	
Substance	**Source**	**Function**
Vitamin B$_{12}$ (requires intrinsic factor for absorption via small intestine)	Absorbed from small intestine	DNA synthesis
Iron	Absorbed from small intestine; conserved during red blood cell destruction and made available for reuse	Hemoglobin synthesis
Folic acid	Absorbed from small intestine	DNA synthesis

Vitamin C increases absorption of iron in the digestive tract. Drinking orange juice with a meal is a good way to boost iron intake.

TABLE 14.3	Types of Anemia	
Type	**Cause**	**Defect**
Aplastic anemia	Toxic chemicals, radiation	Damaged bone marrow
Hemolytic anemia	Toxic chemicals	Red blood cells destroyed
Iron deficiency anemia	Dietary lack of iron	Hemoglobin deficient
Pernicious anemia	Inability to absorb vitamin B$_{12}$	Excess of immature cells
Sickle cell disease	Defective gene	Red blood cells abnormally shaped
Thalassemia	Defective gene	Hemoglobin deficient; red blood cells short-lived

A deficiency of red blood cells or a reduction in the amount of the hemoglobin they contain results in a condition called **anemia.** This reduces the oxygen-carrying capacity of the blood, and the affected person may appear pale and lack energy (fig. 14.7 and table 14.3). A pregnant woman may become anemic if she doesn't eat iron-rich foods, because her blood volume increases due to fluid retention to accommodate the requirements of the fetus.

This increased blood volume decreases the hematocrit. Clinical Application 14.1 discusses another disorder of red blood cells that may have had an impact on American history.

1 Which vitamins are necessary for red blood cell production?

2 Why is iron required for the development of red blood cells?

In the absence of intrinsic factor, vitamin B$_{12}$ absorption decreases, causing the red bone marrow to form abnormally large, irregularly shaped, thin-membraned fragile cells. This condition, called *pernicious anemia,* can cause permanent brain damage if not treated promptly with vitamin B$_{12}$ injections. Taking excess folic acid—as pregnant women do to prevent neural tube defects in the fetus—can mask a vitamin B$_{12}$ deficiency. These women must be careful to follow a balanced diet and get sufficient vitamin B$_{12}$.

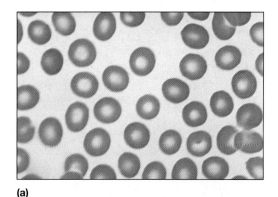

(a)

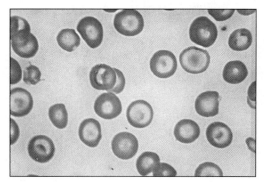

(b)

FIGURE 14.7

Red blood cells, normal and abnormal. (a) Light micrograph of normal human erythrocytes (1,200×). (b) Light micrograph of erythrocytes from a person with hypochromic anemia (1,000×).

Destruction of Red Blood Cells

Red blood cells are quite elastic and flexible, and they readily bend as they pass through small blood vessels. With age, however, these cells become more fragile, and they are frequently damaged simply by passing through capillaries, particularly those in active muscles.

KING GEORGE III AND PORPHYRIA VARIEGATA

Some historians blame the American Revolution on a blood disease—an abnormality of hemoglobin that afflicted King George III, who ruled England at the time. So puzzling were his symptoms that not until this century did medical researchers discover the underlying disorder, called *porphyria variegata.*

At age fifty, the king first experienced abdominal pain and constipation, followed by weak limbs, fever, a rapid pulse, hoarseness, and dark red urine. Next, nervous system symptoms began, including insomnia, headaches, visual problems, restlessness, delirium, convulsions, and stupor. His confused and racing thoughts, combined with his ripping off his wig and running about

naked while at the peak of a fever, convinced court observers that the king was mad. Just as Parliament was debating his ability to rule, he mysteriously recovered.

But George III's plight was far from over. He suffered a relapse thirteen years later, then again three years after that. Always the symptoms appeared in the same order, beginning with abdominal pain, fever, and weakness and progressing to the nervous system symptoms. Finally, an attack in 1811 placed him in an apparently permanent stupor, and he was dethroned by the Prince of Wales. He lived for several more years, experiencing further episodes of his odd affliction.

In George III's time, physicians were permitted to do very little to the royal body,

basing diagnoses on what the patient told them. Twentieth-century researchers found that George III's red urine was caused by an inborn error of metabolism. In porphyria variegata, because of the absence of an enzyme, part of the blood pigment hemoglobin, called a porphyrin ring, is routed into the urine instead of being broken down and metabolized by cells. Porphyrin builds up and attacks the nervous system, causing many of the other symptoms. Examination of the medical records of King George III's descendants reveals several of them also had symptoms of porphyria variegata. The underlying defect in red blood cell recycling had appeared in its various guises as different problems. ■

In a condition called *hereditary hemochromatosis,* the small intestine absorbs too much iron from food. The metal accumulates in the liver, pancreas, heart, and endocrine glands and makes the skin appear bronze. A symptom is increased susceptibility to bacterial infection, because bacteria thrive in the iron-rich tissues. Hereditary hemochromatosis often goes undiagnosed for many years. Usually women do not develop symptoms until after menopause, when their menstrual periods no longer relieve the condition each month. Treatment is simple—periodically removing blood. Without treatment, premature death results from chronic liver disease or heart failure. Hereditary hemochromatosis affects 0.3% to 0.5% of many populations, and up to 10% of individuals are carriers.

Damaged or worn red blood cells rupture as they pass through the spleen or liver. In these organs, macrophages (see chapter 5, p. 141) phagocytize and destroy damaged red blood cells and their contents. Hemoglobin molecules liberated from the red blood cells break down into their four component polypeptide "globin" chains, each surrounding a heme group. Heme further decomposes into iron and a greenish pigment called **biliverdin.** The iron, combined with a protein called *transferrin,* may be carried by the blood to the hematopoi-

etic (red blood cell–forming) tissue in the red bone marrow and reused in synthesizing new hemoglobin. About 80% of the iron is stored in the liver cells in the form of an iron-protein complex called *ferritin.* In time, the biliverdin is converted to an orange pigment called **bilirubin.** Biliverdin and bilirubin are excreted in the bile as bile pigments (fig. 14.8). Table 14.4 summarizes the process of red blood cell destruction.

1 What happens to damaged red blood cells?

2 What are the products of hemoglobin breakdown?

TABLE 14.4	Major Events in Red Blood Cell Destruction

1. Squeezing through the capillaries of active tissues damages red blood cells.
2. Macrophages in the liver and spleen phagocytize damaged red blood cells.
3. Hemoglobin from the red blood cells is decomposed into heme and globin.
4. Heme is decomposed into iron and biliverdin.
5. Iron is made available for reuse in the synthesis of new hemoglobin or is stored in the liver as ferritin.
6. Some biliverdin is converted into bilirubin.
7. Biliverdin and bilirubin are excreted in bile as bile pigments.

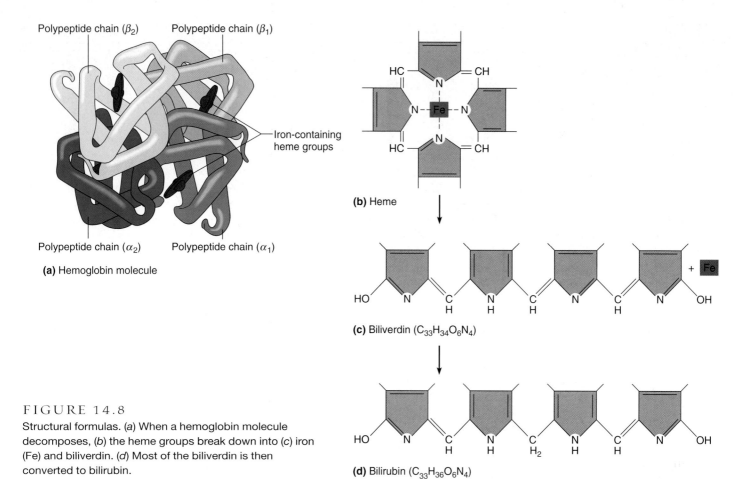

FIGURE 14.8

Structural formulas. (*a*) When a hemoglobin molecule decomposes, (*b*) the heme groups break down into (*c*) iron (Fe) and biliverdin. (*d*) Most of the biliverdin is then converted to bilirubin.

Labels in figure:
Polypeptide chain (β_2) Polypeptide chain (β_1)
Iron-containing heme groups
Polypeptide chain (α_2) Polypeptide chain (α_1)
(a) Hemoglobin molecule
(b) Heme
(c) Biliverdin ($C_{33}H_{34}O_6N_4$)
(d) Bilirubin ($C_{33}H_{36}O_6N_4$)

Types of White Blood Cells

White blood cells, or **leukocytes** (lu′ko-sītz), protect against disease. Leukocytes develop from hematopoietic stem cells in response to hormones. These hormones fall into two groups—**interleukins** (in″ter-loo′kinz) and **colony-stimulating factors (CSFs)**. Interleukins are numbered, while most colony-stimulating factors are named for the cell population they stimulate. Blood transports white blood cells to sites of infection. White blood cells may then leave the bloodstream, as described later in this chapter.

Normally, five types of white cells are in circulating blood, distinguished by size, the nature of the cytoplasm, the shape of the nucleus, and staining characteristics. Some types of leukocytes have granular cytoplasm and make up a group called **granulocytes** (gran′u-lo-sītz″), whereas others lack cytoplasmic granules and are called **agranulocytes** (ah-gran′u-lo-sītz).

A typical granulocyte is about twice the size of a red blood cell. The members of this group include neutrophils, eosinophils, and basophils. These cells develop in the red bone marrow in much the same manner as red blood cells. However, they have a relatively short life span, averaging about twelve hours.

A defect in the ability of granulocytes to destroy pathogenic bacteria can make infection deadly. In *chronic granulomatous disease,* granulocytes engulf bacteria and bring them into the cell, but once there, cannot produce superoxide, a chemical that kills the bacteria. As a result, certain infections rage out of control.

Neutrophils (nu′tro-filz) have fine cytoplasmic granules that appear light purple with a combination of acid and base stains. The nucleus of an older neutrophil is lobed and consists of two to five sections (segments, so these cells are sometimes called *segs*) connected by thin strands of chromatin. For this reason, they are also called *polymorphonuclear* leukocytes. Younger neutrophils are also called *bands* because their nuclei are C-shaped. Neutrophils are the first white blood cells to arrive at an infection site. These cells phagocytize bacteria, fungi, and some viruses. Neutrophils account for 54% to 62% of the leukocytes in a typical blood sample from an adult (fig. 14.9).

Eosinophils (e″o-sin′o-filz) contain coarse, uniformly sized cytoplasmic granules that stain deep red in acid stain. The nucleus usually has only two lobes

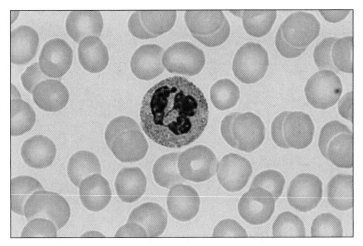

FIGURE 14.9
A neutrophil has a lobed nucleus with two to five components (1,050×).

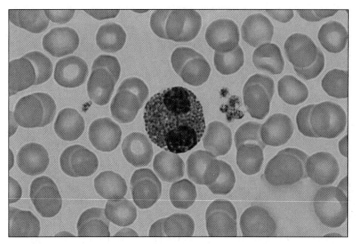

FIGURE 14.10
An eosinophil has red-staining cytoplasmic granules (1,050×).

(bilobed). Eosinophils moderate allergic reactions and defend against parasitic worm infestation. These cells make up 1% to 3% of the total number of circulating leukocytes (fig. 14.10).

Basophils (ba'so-filz) are similar to eosinophils in size and in the shape of their nuclei. However, they have fewer, more irregularly shaped cytoplasmic granules than eosinophils, and these granules appear deep blue in basic stain. A basophil's granules can obscure a view of the nucleus. Basophils migrate to damaged tissues where they release *histamine,* which promotes inflammation, and *heparin,* which inhibits blood clotting, thus increasing blood flow to injured tissues. This type of leukocyte usually accounts for less than 1% of the leukocytes (fig. 14.11).

The leukocytes of the agranulocyte group include monocytes and lymphocytes. Monocytes generally arise from red bone marrow. Lymphocytes are formed in the organs of the lymphatic system as well as in the red bone marrow.

Monocytes (mon'o-sītz) are the largest blood cells, two to three times greater in diameter than red blood cells. Their nuclei are spherical, kidney-shaped, oval, or lobed. Monocytes leave the bloodstream and become *macrophages* that phagocytize bacteria, dead cells, and other debris in the tissues. They usually make up 3% to 9% of the leukocytes in a blood sample and live for several weeks or even months (fig. 14.12).

Lymphocytes (lim'fo-sītz) are usually only slightly larger than erythrocytes. A typical lymphocyte contains a large, spherical nucleus surrounded by a thin rim of cytoplasm. The major types of lymphocytes are *T cells* and *B cells,* both important in *immunity.* T cells directly attack microorganisms, tumor cells, and transplanted cells. B cells produce **antibodies** (an'ti-bod″ez) (see chapter 16, p. 621), which are proteins that attack foreign molecules.

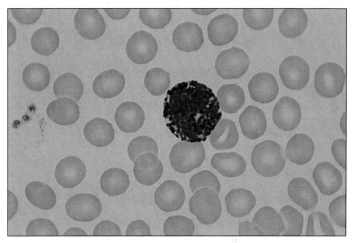

FIGURE 14.11
A basophil has cytoplasmic granules that stain deep blue (1,050×).

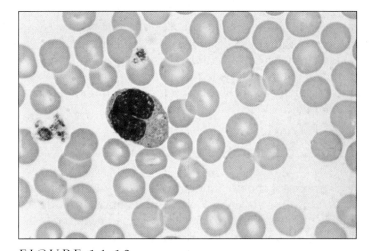

FIGURE 14.12
A monocyte may leave the bloodstream and become a macrophage (1,050×).

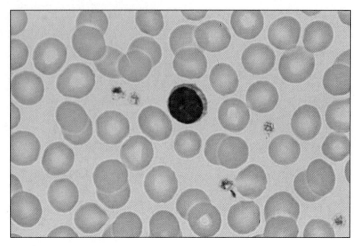

FIGURE 14.13
The lymphocyte contains a large, spherical nucleus (1,050×).

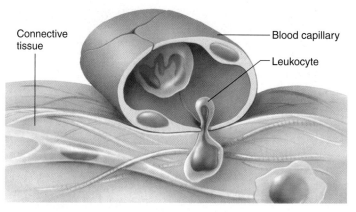

FIGURE 14.14
In a type of movement called diapedesis, leukocytes squeeze between the cells of a capillary wall and enter the tissue space outside the blood vessel.

Lymphocytes account for 25% to 33% of the circulating leukocytes. They may live for years (fig. 14.13).

1. Which hormones are necessary for the development of white blood cells in red bone marrow?

2. Distinguish between granulocytes and agranulocytes.

3. List five types of white blood cells, and explain how they differ from one another.

4. Describe the function of each type of white blood cell.

Functions of White Blood Cells

Leukocytes protect against infection in various ways. Some leukocytes phagocytize bacterial cells in the body, and others produce proteins (*antibodies*) that destroy or disable foreign particles.

Leukocytes can squeeze between the cells that form the walls of the smallest blood vessels. This movement, called **diapedesis** (di″ah-pĕ-de′sis), allows the white blood cells to leave the circulation (fig. 14.14). A series of proteins called cellular adhesion molecules help guide leukocytes to the site of injury, a process called leukocyte trafficking. Once outside the blood, they move through interstitial spaces using a form of self-propulsion called *ameboid motion.*

RECONNECT TO CHAPTER 3, CELLULAR ADHESION MOLECULES, PAGE 67.

The most mobile and active phagocytic leukocytes are neutrophils and monocytes. Although neutrophils are unable to ingest particles much larger than bacterial cells, monocytes can engulf larger structures. Monocytes contain numerous lysosomes, which are filled with digestive enzymes that break down organic molecules in captured bacteria. Neutrophils and monocytes often become so engorged with digestive products and bacterial toxins that they also die.

When microorganisms invade human tissues, basophils respond by releasing biochemicals that dilate local blood vessels. For example, histamine dilates smaller blood vessels and makes the smallest vessels leaky. As more blood flows through the smallest vessels, the tissues redden and copious fluids leak into the interstitial spaces. The swelling that this inflammatory reaction produces delays the spread of invading microorganisms into other regions (see chapter 5, p. 142). At the same time, damaged cells release chemicals that attract leukocytes. This phenomenon is called **positive chemotaxis** (poz′ĭ-tiv ke″mo-tak′sis) and, when combined with diapedesis, brings many white blood cells into inflamed areas quickly (fig. 14.15).

As bacteria, leukocytes, and damaged cells accumulate in an inflamed area, a thick fluid called *pus* often forms and remains while the invading microorganisms are active. If the pus cannot escape to the outside of the body or into a body cavity, it may remain trapped in the tissues for some time. Eventually, surrounding cells absorb it.

1. How do white blood cells fight infection?

2. Which white blood cells are the most active phagocytes?

3. How do white blood cells reach microorganisms that are outside blood vessels?

White Blood Cell Counts

The procedure used to count white blood cells is similar to that used for counting red blood cells. However, before a *white blood cell count* (WBCC or WCC) is made, the red blood cells in the blood sample are destroyed so they will not be mistaken for white blood cells. Normally, a cubic

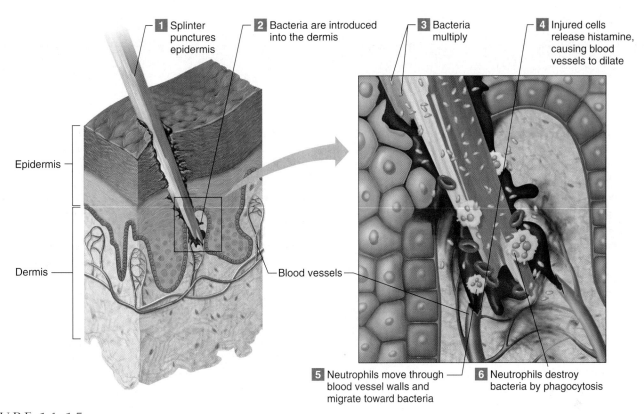

1 Splinter punctures epidermis

2 Bacteria are introduced into the dermis

3 Bacteria multiply

4 Injured cells release histamine, causing blood vessels to dilate

Epidermis

Dermis

Blood vessels

5 Neutrophils move through blood vessel walls and migrate toward bacteria

6 Neutrophils destroy bacteria by phagocytosis

FIGURE 14.15

When bacteria invade the tissues, leukocytes migrate into the region and destroy the microbes by phagocytosis.

millimeter of blood includes 5,000 to 10,000 white blood cells.

The total number and percentages of different white blood cell types are of clinical interest. A rise in the number of circulating white blood cells may indicate infection. A total number of white blood cells exceeding 10,000 per mm^3 of blood constitutes **leukocytosis** (loo″ko-si-to′sis), indicating acute infection, such as appendicitis. Leukocytosis may also follow vigorous exercise, emotional disturbances, or great loss of body fluids.

A total white blood cell count below 5,000 per mm^3 of blood is called **leukopenia** (loo″ko-pe′ne-ah). Such a deficiency may accompany typhoid fever, influenza, measles, mumps, chicken pox, AIDS, or poliomyelitis. Leukopenia may also result from anemia or from lead, arsenic, or mercury poisoning.

A *differential white blood cell count* (DIFF) lists percentages of the types of leukocytes in a blood sample. This test is useful because the relative proportions of white blood cells may change in particular diseases. The number of neutrophils, for instance, usually increases during bacterial infections, and eosinophils may become more abundant during certain parasitic infections and allergic reactions. In HIV infection and AIDS, the numbers of a type of lymphocyte called helper T cells plummet.

Table 14.5 lists some disorders that alter the numbers of particular types of white blood cells. Clinical

TABLE 14.5	White Blood Cell Alterations
White Blood Cell Population Change	**Illness**
Elevated lymphocytes	Hairy cell leukemia, whooping cough, mononucleosis
Elevated eosinophils	Tapeworm infestation, hookworm infestation, allergic reactions
Elevated monocytes	Typhoid fever, malaria, tuberculosis
Elevated neutrophils	Bacterial infections
Too few helper T cells (lymphocytes)	AIDS

Application 14.2 examines leukemia, cancer of white blood cells.

1 What is the normal human white blood cell count?

2 Distinguish between leukocytosis and leukopenia.

3 What is a differential white blood cell count?

Blood Platelets

Platelets (plāt′letz), or **thrombocytes** (throm′bo-sĭtz), are not complete cells. They arise from very large cells in the red bone marrow, called *megakaryocytes* (meg″ah-kar′e-o-sĭtz),

LEUKEMIA

The young woman had noticed symptoms for several months before she finally went to the doctor. At first it was just fatigue and headaches, which she attributed to studying for final exams. She had frequent colds and bouts of fever, chills, and sweats that she thought were just minor infections. When she developed several bruises and bone pain and noticed that her blood did not clot very quickly after cuts and scrapes, she consulted her physician, who examined her and took a blood sample. One glance at a blood smear under a microscope alarmed the doctor—there were far too few red blood cells and platelets and too many white blood cells. She sent the sample to a laboratory to diagnose the type of *leukemia*, or cancer of the white blood cells, that was causing her patient's symptoms.

The young woman had *myeloid leukemia.* Her red bone marrow was producing too many granulocytes, but they were immature cells, unable to fight infection (fig. 14A). This explained the frequent illnesses. The leukemic cells were crowding out red blood cells and their precursors in the red marrow, causing her anemia and resulting fatigue. Platelet deficiency (thrombocytopenia) led to an increased tendency to bleed. Finally, spread of the cancer cells outside the marrow painfully weakened the surrounding bone. Eventually, if she wasn't treated, the cancer cells would spread outside the cardiovascular system, causing other tissues that would normally not produce white blood cells to do so.

A second type of leukemia, distinguished by the source of the cancer cells, is *lymphoid leukemia.* These cancer cells are lymphocytes, produced in lymph nodes. Many of the symptoms are similar to those of myeloid leukemia. Sometimes a person has no leukemia symptoms at all, and a routine blood test detects the condition.

Leukemia is also classified as acute or chronic. An acute condition appears suddenly, symptoms progress rapidly, and death occurs in a few months without treatment. Chronic forms begin more slowly and may remain undetected for months or even years or, in rare cases, decades. Without treatment, life expectancy after symptoms develop is about three years. With treatment, 50% to 80% of patients enter remission, a period of stability that may become a cure. Chemotherapy may be necessary for a year or two to increase the chances of long remission.

Leukemia treatment includes correcting symptoms with blood transfusions, treating infections, and using drugs that kill cancer cells. Several drugs in use for many years have led to increases in cure rates, particularly for acute lymphoid leukemia in children. A new drug called Gleevec has had spectacular success in treating a type of chronic leukemia. If other treatments fail, a bone marrow transplant can cure leukemia, but the procedure is very risky. Stem cell transplants, using cells from donated umbilical cord blood, can also cure leukemia. Therefore, people with leukemia often have many treatment options. ■

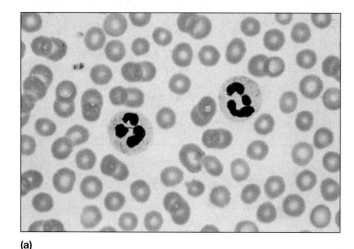

(a)

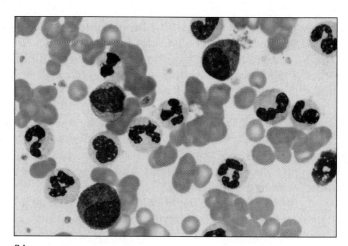

(b)

FIGURE 14A

Leukemia and blood cells. (*a*) Normal blood cells (700×). (*b*) Blood cells from a person with granulocytic leukemia, a type of myeloid leukemia (700×). Note the increased number of leukocytes.

that fragment a little like a shattered plate, releasing small sections of cytoplasm—platelets—into the circulation. The larger fragments of the megakaryocytes shrink and become platelets as they pass through the blood vessels of the lungs.

Each platelet lacks a nucleus and is less than half the size of a red blood cell. It is capable of ameboid movement and may live for about ten days. In normal blood, the *platelet count* varies from 130,000 to 360,000 platelets per mm^3.

Platelets help repair damaged blood vessels by sticking to broken surfaces. They release **serotonin,** which contracts smooth muscles in the vessel walls, reducing blood flow. Table 14.6 summarizes the characteristics of blood cells and platelets.

1   What is the normal human blood platelet count?

2   What is the function of blood platelets?

Blood Plasma

Plasma is the clear, straw-colored, liquid portion of the blood in which the cells and platelets are suspended. It is approximately 92% water and contains a complex mixture of organic and inorganic biochemicals. Functions of plasma constituents include transporting nutrients, gases, and vitamins; helping to regulate fluid and electrolyte balance; and maintaining a favorable pH. Figure 14.16 shows the chemical makeup of plasma.

Plasma Proteins

By weight, **plasma proteins** are the most abundant dissolved substances (solutes) in plasma. These proteins remain in the blood and interstitial fluids and ordinarily are not used as energy sources. The three main plasma protein groups are albumins, globulins, and fibrinogen. The groups differ in chemical composition and physiological function.

Albumins (al-bu'minz) are the smallest of the plasma proteins, yet account for 60% of these proteins by weight. They are synthesized in the liver, and because they are so plentiful, albumins are an important determinant of the *osmotic pressure* of the plasma.

Recall from chapter 3 (p. 83) that the presence of an impermeant solute on one side of a selectively permeable membrane creates an osmotic pressure and that water always diffuses toward a greater osmotic pressure. Because plasma proteins are too large to pass through the capillary walls, they are impermeant, and they create an osmotic pressure that tends to hold water in the capillaries despite the fact that blood pressure tends to force water out of capillaries by filtration (see chapter 3, p. 84). The term *colloid osmotic pressure* is often used to describe this osmotic effect due to the plasma proteins.

By maintaining the colloid osmotic pressure of plasma, albumins and other plasma proteins help regulate water movement between the blood and the tissues. In doing so, they help control blood volume, which, in turn, directly affects blood pressure (see chapter 15, p.

TABLE 14.6	Cellular Components of Blood		
Component	**Description**	**Number Present**	**Function**
Red blood cell (erythrocyte)	Biconcave disk without a nucleus, about one-third hemoglobin	4,200,000 to 6,200,000 per mm^3	Transports oxygen and carbon dioxide
White blood cell (leukocyte)		5,000 to 10,000 per mm^3	Destroys pathogenic microorganisms and parasites and removes worn cells
Granulocytes	About twice the size of red blood cells; cytoplasmic granules are present		
1. Neutrophil	Nucleus with two to five lobes; cytoplasmic granules stain light purple in combined acid and base stains	54%–62% of white blood cells present	Phagocytizes small particles
2. Eosinophil	Nucleus bilobed; cytoplasmic granules stain red in acid stain	1%–3% of white blood cells present	Kills parasites and helps control inflammation and allergic reactions
3. Basophil	Nucleus lobed; cytoplasmic granules stain blue in basic stain	Less than 1% of white blood cells present	Releases heparin and histamine
Agranulocytes	Cytoplasmic granules are absent		
1. Monocyte	Two to three times larger than a red blood cell; nuclear shape varies from spherical to lobed	3%–9% of white blood cells present	Phagocytizes large particles
2. Lymphocyte	Only slightly larger than a red blood cell; its nucleus nearly fills cell	25%–33% of white blood cells present	Provides immunity
Platelet (thrombocyte)	Cytoplasmic fragment	130,000 to 360,000 per mm^3	Helps control blood loss from broken vessels

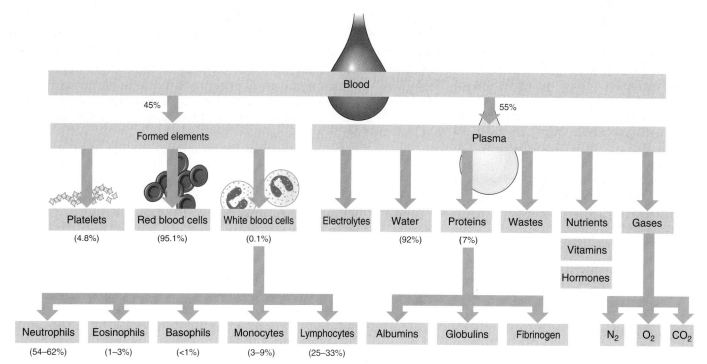

FIGURE 14.16
Blood composition.

573). For this reason, it is important that the concentration of plasma proteins remains relatively stable. Albumins also bind and transport certain molecules, such as bilirubin, free fatty acids, many hormones, and certain drugs.

> If the concentration of plasma proteins falls, tissues swell, a condition called *edema*. This may result from starvation or a protein-deficient diet, either of which requires the body to use protein for energy, or from an impaired liver that cannot synthesize plasma proteins. As concentration of plasma proteins drops, so does the colloid osmotic pressure, sending fluids into the intercellular spaces.

Globulins (glob'u-linz), which make up about 36% of the plasma proteins, can be further subdivided into *alpha, beta,* and *gamma globulins.* The liver synthesizes alpha and beta globulins. They have a variety of functions, including transport of lipids and fat-soluble vitamins. Lymphatic tissues produce the gamma globulins, which are a type of antibody (see chapter 16, p. 625).

Fibrinogen (fi-brin'o-jen), which constitutes about 4% of the plasma protein, plays a primary role in blood coagulation. Synthesized in the liver, it is the largest of the plasma proteins. The function of fibrinogen is discussed later in this chapter under the section titled

TABLE 14.7	Plasma Proteins		
Protein	**Percentage of Total**	**Origin**	**Function**
Albumin	60%	Liver	Helps maintain colloid osmotic pressure
Globulin	36%		
Alpha globulins		Liver	Transport lipids and fat-soluble vitamins
Beta globulins		Liver	Transport lipids and fat-soluble vitamins
Gamma globulins		Lymphatic tissues	Constitute the antibodies of immunity
Fibrinogen	4%	Liver	Plays a key role in blood coagulation

"Blood Coagulation." Table 14.7 summarizes the characteristics of the plasma proteins.

1 List three types of plasma proteins.

2 How do albumins help maintain water balance between the blood and the tissues?

3 Which of the globulins functions in immunity?

4 What is the role of fibrinogen?

Gases and Nutrients

The most important *blood gases* are oxygen and carbon dioxide. Plasma also contains a considerable amount of dissolved nitrogen, which ordinarily has no physiological function. Chapter 19 (pp. 760–763) discusses blood gases and their transport.

As a rule, blood gases are evaluated using a fresh sample of whole blood obtained from an artery. This blood is cooled to decrease the rates of metabolic reactions, and an anticoagulant is added to prevent clotting. In the laboratory, the levels of oxygen and carbon dioxide of the blood are determined, the blood pH is measured, and the plasma bicarbonate concentration is calculated. Such information is used to diagnose and treat disorders of circulation, respiration, and electrolyte balance. Appendix B (pp. 970–972) lists average values for these laboratory tests.

The *plasma nutrients* include amino acids, simple sugars, nucleotides, and lipids absorbed from the digestive tract. For example, plasma transports glucose from the small intestine to the liver, where it may be stored as glycogen or converted to fat. If blood glucose concentration drops below the normal range, glycogen may be broken down into glucose, as described in chapter 13 (p. 496).

Recently absorbed amino acids are also carried to the liver. Here they may be used to manufacture proteins, or deaminated and used as an energy source (see chapter 18, p. 699).

Plasma lipids include fats (triglycerides), phospholipids, and cholesterol. Because lipids are not water soluble and plasma is almost 92% water, these lipids are carried in the plasma by joining with proteins, forming **lipoprotein** (lip″o-pro′te-in) complexes. These lipoproteins are relatively large and consist of a surface layer of phospholipid, cholesterol, and protein surrounding a triglyceride core. The protein constituents of lipoproteins in the outer layer, called *apoproteins* or apolipoproteins, can combine with receptors on the membranes of specific target cells. Lipoprotein molecules vary in the proportions of the lipids they contain.

Because lipids are less dense than proteins, as the proportion of lipids in a lipoprotein increases, the density of the particle decreases. Conversely, as the proportion of lipids decreases, the density increases. Lipoproteins are classified on the basis of their densities, which reflect their composition. *Chylomicrons* mainly consist of triglycerides absorbed from the small intestine. *Very low-density lipoproteins* (VLDL) have a relatively high concentration of triglycerides. *Low-density lipoproteins* (LDL) have a relatively high concentration of cholesterol

and are the major cholesterol-carrying lipoproteins. *High-density lipoproteins* (HDL) have a relatively high concentration of protein and a lower concentration of lipids.

Many residents of Tangier Island in the Chesapeake Bay have inherited a deficiency in a particular blood lipoprotein from the original settlers, who arrived in 1686. A low blood level of alpha lipoprotein, which normally transports cholesterol in the blood, leads to low blood cholesterol. However, excess cholesterol accumulates elsewhere, such as in the thymus gland and in scavenging macrophages.

Chylomicrons, discussed further in chapter 17 (p. 678), transport dietary fats to muscle and adipose cells. Similarly, very low-density lipoproteins, produced in the liver, transport triglycerides synthesized from excess dietary carbohydrates. After VLDL molecules deliver their loads of triglycerides to adipose cells, an enzyme, lipoprotein lipase, converts their remnants to low-density lipoproteins. Because most of the triglycerides have been removed, LDL molecules have a higher cholesterol content than do the original VLDL molecules. Various cells, including liver cells, have surface receptors that combine with apoproteins associated with LDL molecules. These cells slowly remove LDL from plasma by receptor-mediated endocytosis, supplying cells with cholesterol (see chapter 3, p. 87).

After chylomicrons deliver their triglycerides to cells, their remnants are transferred to high-density lipoproteins. These HDL molecules, which form in the liver and small intestine, transport chylomicron remnants to the liver, where they rapidly enter cells by receptor-mediated endocytosis. The liver disposes of the cholesterol it obtains in this manner by secreting it into bile or by using it to synthesize bile salts. Table 14.8 summarizes the characteristics and functions of these lipoproteins.

Much of the cholesterol and bile salts in bile are later reabsorbed by the small intestine and transported back to the liver, and the secretion-reabsorption cycle repeats. During each cycle, some of the cholesterol and bile salts escape reabsorption, reach the large intestine, and are eliminated with the feces.

1 Which gases are in plasma?

2 Which nutrients are in blood plasma?

3 How are triglycerides transported in plasma?

4 How is cholesterol eliminated from the liver?

Nonprotein Nitrogenous Substances

Molecules that contain nitrogen atoms but are not proteins comprise a group called **nonprotein nitrogenous substances** (NPN). In plasma, this group includes amino

TABLE 14.8	Plasma Lipoproteins	
Lipoprotein	**Characteristics**	**Functions**
Chylomicron	High concentration of triglycerides	Transports dietary fats to muscle and adipose cells
Very low-density lipoprotein (VLDL)	Relatively high concentration of triglycerides; produced in the liver	Transports triglycerides from the liver to adipose cells
Low-density lipoprotein (LDL)	Relatively high concentration of cholesterol; formed from remnants of VLDL molecules that have given up their triglycerides	Delivers cholesterol to various cells, including liver cells
High-density lipoprotein (HDL)	Relatively high concentration of protein and low concentration of lipids	Transports to the liver remnants of chylomicrons that have given up their triglycerides

acids, urea, uric acid, creatine (kre′ah-tin), and creatinine (kre-at′ĭ-nin). Amino acids come from protein digestion and amino acid absorption. Urea and uric acid are products of protein and nucleic acid catabolism, respectively, and creatinine results from the metabolism of creatine. As discussed in chapter 9 (p. 287), creatine occurs as **creatine phosphate** in muscle and brain tissues as well as in the blood, where it stores high-energy phosphate bonds, much like those of ATP molecules.

Normally, the concentration of nonprotein nitrogenous substances remains relatively stable because protein intake and utilization are balanced with excretion of nitrogenous wastes. Because about half of the NPN substances is urea, which the kidneys ordinarily excrete, a rise in the blood urea nitrogen (BUN) may suggest a kidney disorder. Excess protein catabolism or infection may also elevate BUN.

Plasma Electrolytes

Recall that electrolytes release ions when dissolved in water. Blood plasma contains a variety of these ions, often themselves called *electrolytes.* Plasma electrolytes are absorbed from the intestine or released as by-products of cellular metabolism. They include sodium, potassium, calcium, magnesium, chloride, bicarbonate, phosphate, and sulfate ions. Of these, sodium and chloride ions are the most abundant. Bicarbonate ions are important in maintaining the osmotic pressure and the pH of plasma, and like other plasma constituents, they are regulated so that their blood concentrations remain relatively stable. These electrolytes are discussed in chapter 21 (p. 813) in connection with water and electrolyte balance.

1 What is a nonprotein nitrogenous substance?

2 Why does kidney disease increase the blood concentration of these substances?

3 What are the sources of plasma electrolytes?

Hemostasis

Hemostasis (he″mo-sta′sis) refers to the stoppage of bleeding, which is vitally important when blood vessels are damaged. Following an injury to the blood vessels, several actions may help to limit or prevent blood loss, including blood vessel spasm, platelet plug formation, and blood coagulation. These mechanisms are most effective in minimizing blood losses from small vessels. Injury to a larger vessel may result in a severe hemorrhage that requires special treatment.

Blood Vessel Spasm

Cutting or breaking a smaller blood vessel stimulates the smooth muscles in its wall to contract, an event called *vasospasm.* Blood loss lessens almost immediately, and the ends of the severed vessel may close completely. This effect results from direct stimulation of the vessel wall as well as from reflexes elicited by pain receptors in the injured tissues.

Although the reflex response may last only a few minutes, the effect of the direct stimulation usually continues for about thirty minutes. By then, a blockage called a *platelet plug* has formed, and blood is coagulating. Also, platelets release serotonin, which contracts smooth muscles in the blood vessel walls. This vasoconstriction further helps to reduce blood loss.

Platelet Plug Formation

Platelets adhere to exposed ends of injured blood vessels. They adhere to any rough surface, particularly to the collagen in connective tissue underlying the endothelial lining of blood vessels.

When platelets contact collagen, their shapes change drastically, and numerous spiny processes begin to protrude from their membranes. At the same time, platelets adhere to each other, forming a platelet plug in the vascular break. A plug may control blood loss from a small break, but a larger one may require a blood clot to halt bleeding. Figure 14.17 shows the steps in platelet plug formation.

1 What is hemostasis?

2 How does a blood vessel spasm help control bleeding?

3 Describe the formation of a platelet plug.

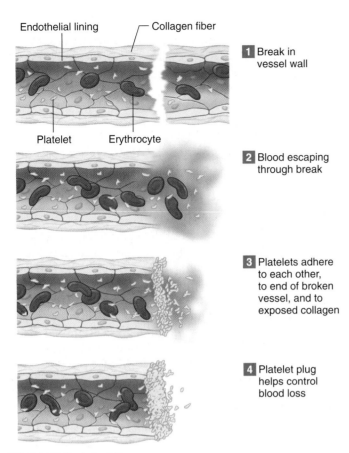

Endothelial lining — Collagen fiber

1 Break in vessel wall

Platelet   Erythrocyte

2 Blood escaping through break

3 Platelets adhere to each other, to end of broken vessel, and to exposed collagen

4 Platelet plug helps control blood loss

FIGURE 14.17
Steps in platelet plug formation.

TABLE 14.9	Hemostatic Mechanisms	
Mechanism	**Stimulus**	**Effect**
Blood vessel spasm	Direct stimulus to vessel wall or to pain receptors; platelets release serotonin	Smooth muscles in vessel wall contract reflexly; vasoconstriction helps maintain prolonged vessel spasm
Platelet plug formation	Exposure of platelets to rough surfaces or to collagen of connective tissue	Platelets adhere to rough surfaces and to each other, forming a plug
Blood coagulation	Cellular damage and blood contact with foreign surfaces result in activation of factors that favor coagulation	Blood clot forms as a result of a series of reactions, terminating in the conversion of fibrinogen into fibrin

Blood Coagulation

Coagulation (ko-ag″u-la′shun), the most effective hemostatic mechanism, causes formation of a *blood clot* by a series of reactions, each one activating the next in a chain reaction, or *cascade.* Coagulation may occur extrinsically or intrinsically. Release of biochemicals from broken blood vessels or damaged tissues triggers the **extrinsic clotting mechanism.** Blood contact with foreign surfaces in the absence of tissue damage stimulates the **intrinsic clotting mechanism.** These responses are described in the next sections.

Blood coagulation is very complex and utilizes many biochemicals called *clotting factors.* They are designated by Roman numerals indicating the order of their discovery. Vitamin K is necessary for some clotting factors to function. Whether or not the blood coagulates depends on the balance between factors that promote coagulation (procoagulants) and others that inhibit it (anticoagulants). Normally, the anticoagulants prevail, and the blood does not clot. However, as a result of injury (trauma), biochemicals that favor coagulation may increase in concentration, and the blood may coagulate.

The major event in blood clot formation is conversion of the soluble plasma protein *fibrinogen* (factor I) into insoluble threads of the protein **fibrin.** Activation of certain plasma proteins by still other protein factors triggers conversion of fibrinogen to fibrin. Table 14.9 summarizes the three primary hemostatic mechanisms.

Extrinsic Clotting Mechanism

The extrinsic clotting mechanism is triggered when blood contacts damaged blood vessel walls or tissues outside blood vessels. Such damaged tissues release a complex of substances called *tissue thromboplastin* (factor III) that is associated with disrupted cell membranes. Tissue thromboplastin activates factor VII, which combines with and activates factor X. Further, factor X combines with and activates factor V. These reactions, which also require calcium ions (factor IV), lead to production and release of *prothrombin activator* by the platelets.

Prothrombin (factor II) is an alpha globulin that the liver continually produces and is thus a normal constituent of plasma. In the presence of calcium ions, prothrombin activator converts prothrombin into **thrombin** (factor IIa). Thrombin, in turn, catalyzes a reaction that fragments fibrinogen (factor I). The fibrinogen fragments join, forming long threads of fibrin. Fibrinogen is a soluble plasma protein, but fibrin is insoluble. Thrombin also activates factor XIII, which strengthens and stabilizes fibrin threads.

Once fibrin threads form, they stick to exposed surfaces of damaged blood vessels, creating a meshwork that entraps blood cells and platelets (fig. 14.18). The resulting mass is a blood clot, which may block a vascular break and prevent further blood loss.

The amount of prothrombin activator that appears in the blood is directly proportional to the degree of tissue damage. Once a blood clot begins to form, it promotes still more clotting, because thrombin also acts

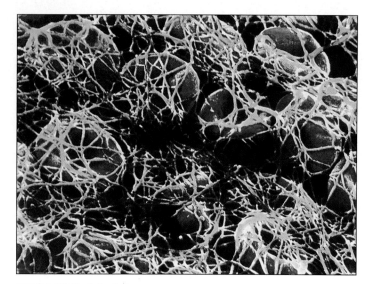

FIGURE 14.18

A scanning electron micrograph of fibrin threads (2,800×).

directly on blood clotting factors other than fibrinogen, causing prothrombin to form still more thrombin. This type of self-initiating action is an example of a **positive feedback system,** in which the original action stimulates more of the same type of action. Such a mechanism produces unstable conditions and can operate for only a short time in a living system, because life depends on the maintenance of a stable internal environment (see chapter 1, p. 9).

Normally, blood flow throughout the body prevents formation of a massive clot within the cardiovascular system by rapidly carrying excess thrombin away and keeping its concentration too low to enhance further clotting. Also, a substance called *antithrombin,* present in the blood and on the surfaces of endothelial cells that line blood vessels, limits thrombin formation. Consequently, blood coagulation is usually limited to blood that is standing still or moving slowly, and clotting ceases where a clot contacts circulating blood.

Intrinsic Clotting Mechanism

Unlike extrinsic clotting, all of the components necessary for intrinsic clotting are in the blood. Activation of a substance called the *Hageman factor* (factor XII) initiates intrinsic clotting. This happens when blood is exposed to a foreign surface such as collagen in connective tissue instead of the smooth endothelial lining of intact blood vessels or when blood is stored in a glass container. Activated factor XII activates factor XI, which activates factor IX. Factor IX then joins with factor VIII and platelet phospholipids to activate factor X. These reactions, which also depend on the presence of calcium ions, lead to the production of prothrombin activator. The subsequent steps of blood clot formation are the same as those described for the extrinsic mechanism (fig. 14.19). Table 14.10 compares extrinsic and intrinsic clotting mechanisms. Table 14.11 lists the clotting factors, their sources, and clotting mechanisms.

TABLE 14.10	Blood Coagulation	
Steps	**Extrinsic Clotting Mechanism**	**Intrinsic Clotting Mechanism**
Trigger	Damage to vessel or tissue	Blood contacts foreign surface
Initiation	Tissue thromboplastin	Hageman factor
Series of reactions involving several clotting factors and calcium ions (Ca^{+2}) lead to the production of:	Prothrombin activator	Prothrombin activator
Prothrombin activator and calcium ions cause the conversion of:	Prothrombin to thrombin	Prothrombin to thrombin
Thrombin causes fragmentation, then joining of:	Fibrinogen to fibrin	Fibrinogen to fibrin

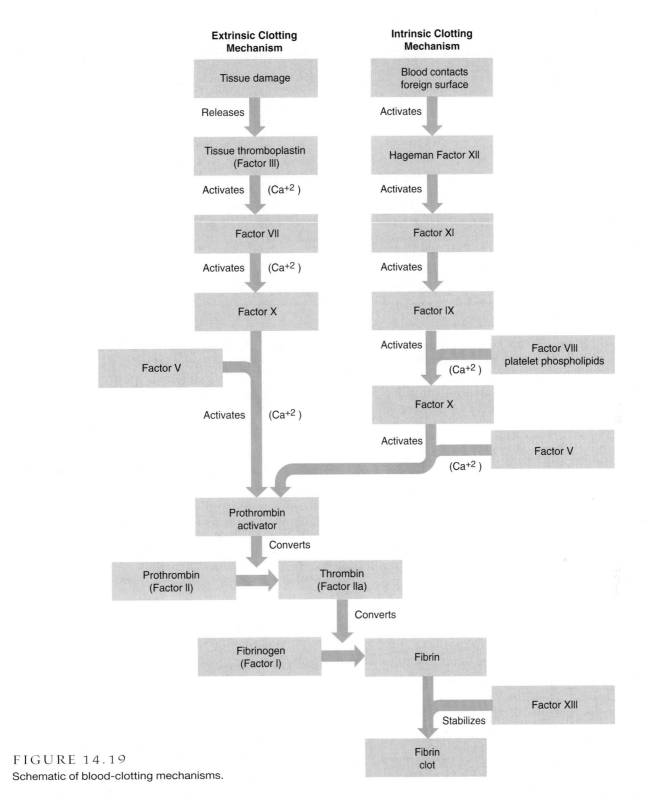

FIGURE 14.19
Schematic of blood-clotting mechanisms.

Fate of Blood Clots

After a blood clot forms, it soon begins to retract as the tiny processes extending from the platelet membranes adhere to strands of fibrin within the clot and contract. The blood clot shrinks, pulling the edges of the broken vessel closer together and squeezing a fluid called **serum** from the clot. Serum is essentially plasma minus all of its fibrinogen and most of the other clotting factors. Platelets associated with a blood clot also release *platelet-derived growth factor* (PDGF), which stimulates smooth muscle cells and fibroblasts to repair damaged blood vessel walls.

Fibroblasts invade blood clots that form in ruptured vessels, producing connective tissue with numerous fibers throughout the clots, which helps strengthen and

TABLE 14.11 | Clotting Factors

Clotting Factor	Source	Mechanism(s)
I (fibrinogen)	Synthesized in liver	Extrinsic and intrinsic
II (prothrombin)	Synthesized in liver, requires vitamin K	Extrinsic and intrinsic
III (tissue thromboplastin)	Damaged tissue	Extrinsic
IV (calcium ions)	Diet, bone	Extrinsic and intrinsic
V (proaccelerin)	Synthesized in liver, released by platelets	Extrinsic and intrinsic
VII (serum prothrombin conversion accelerator)	Synthesized in liver, requires vitamin K	Extrinsic
VIII (antihemophilic factor)	Released by platelets and endothelial cells	Intrinsic
IX (plasma thromboplastin component)	Synthesized in liver, requires vitamin K	Intrinsic
X (Stuart-Prower factor)	Synthesized in liver, requires vitamin K	Extrinsic and intrinsic
XI (plasma thromboplastin antecedent)	Synthesized in liver	Intrinsic
XII (Hageman factor)	Synthesized in liver	Intrinsic
XIII (fibrin-stabilizing factor)	Synthesized in liver, released by platelets	Extrinsic and intrinsic

seal vascular breaks. Many clots, including those that form in tissues as a result of blood leakage (hematomas), disappear in time. In clot dissolution, fibrin threads absorb a plasma protein called *plasminogen* (profibrinolysin). Then a substance called plasminogen activator released from the lysosomes of damaged tissue cells converts plasminogen to *plasmin.* Plasmin is a protein-splitting enzyme that can digest fibrin threads and other proteins associated with blood clots. Plasmin formation may dissolve a whole clot; however, clots that fill large blood vessels are seldom removed naturally.

A blood clot abnormally forming in a vessel is a **thrombus** (throm'bus). If the clot dislodges or if a fragment of it breaks loose and is carried away by the blood flow, it is called an **embolus** (em'-bo-lus). Generally, emboli continue to move until they reach narrow places in vessels where they may lodge and block blood flow, causing an **embolism.**

A blood clot forming in a vessel that supplies a vital organ, such as the heart (coronary thrombosis) or the brain (cerebral thrombosis), blocks blood flow and kills tissues the vessel serves (*infarction*) and may be fatal. A blood clot that travels and then blocks a vessel that supplies a vital organ, such as the lungs (pulmonary embolism), affects the portion of the organ that the blocked blood vessel supplies.

Drugs based on "clot-busting" biochemicals can be lifesavers. *Tissue plasminogen activator* (tPA) may restore blocked coronary or cerebral circulation if given within four hours of a heart attack or stroke. A drug derived from bacteria called *streptokinase* may also be successful, for a fraction of the cost. Another plasminogen activator used as a drug is *urokinase,* an enzyme produced in certain kidney cells. Heparin and coumadin are drugs that interfere with clot formation.

Abnormal clot formations are often associated with conditions that change the endothelial linings of vessels. For example, in *atherosclerosis* (ath"er-o"skle-ro'sis), accumulations of fatty deposits change arterial linings, sometimes initiating inappropriate clotting. This is the most common cause of thrombosis in medium-sized arteries (fig. 14.20).

Coagulation may also occur in blood that is flowing too slowly. The concentration of clot-promoting substances may increase to a critical level instead of being carried away by more rapidly moving blood, and a clot may form. This event is the usual cause of thrombosis in veins.

1 Distinguish between extrinsic and intrinsic clotting mechanisms.

2 What is the basic event in blood clot formation?

3 What factors initiate the formation of fibrin?

4 What prevents the formation of massive clots throughout the cardiovascular system?

5 Distinguish between a thrombus and an embolus.

6 How might atherosclerosis promote the formation of blood clots?

Prevention of Coagulation

In a healthy cardiovascular system, the endothelium of the blood vessels partly prevents spontaneous blood clot formation. This smooth lining discourages the accumulation of platelets and clotting factors. Endothelial cells also produce a prostaglandin (see chapter 13, p. 476) called *prostacyclin* (PGI_2), which inhibits the adherence of platelets to the inner surface of healthy blood vessel walls.

When a clot is forming, fibrin threads latch onto or *adsorb* thrombin, thus helping prevent the spread of the clotting reaction. A plasma alpha globulin, *antithrombin,* inactivates additional thrombin by binding to it and blocking its action on fibrinogen. In addition, basophils and

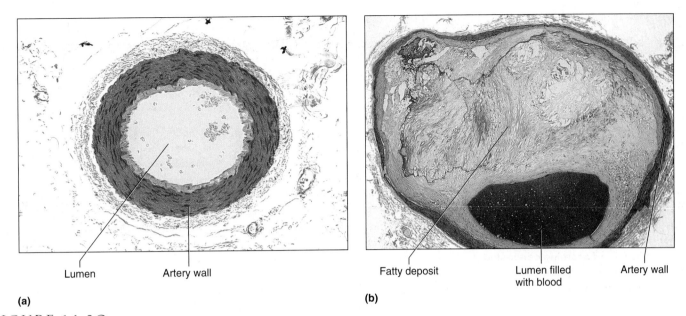

(a)   Lumen   Artery wall   **(b)**   Fatty deposit   Lumen filled with blood   Artery wall

FIGURE 14.20

Artery cross sections. (*a*) Light micrograph of a normal artery (90×). (*b*) The inner wall of an artery changed as a result of atherosclerosis (55×).

mast cells in the connective tissue surrounding capillaries secrete the anticoagulant *heparin.* This substance interferes with formation of prothrombin activator, prevents the action of thrombin on fibrinogen, and promotes removal of thrombin by antithrombin and fibrin adsorption.

Thrombocytopenia (throm"bo-si"to-pe'ne-ah) is a tendency to bleed because of a platelet count that drops below 100,000 platelets per cubic millimeter of blood. Symptoms include bleeding easily; capillary hemorrhages throughout the body; and small, bruiselike spots on the skin. Thrombocytopenia is a common side effect of cancer chemotherapy and radiation treatments and can also develop as a complication of pregnancy, leukemia, bone marrow transplantation, infectious disease, cardiac surgery, or anemia.

Transfusion of platelets is the conventional treatment for thrombocytopenia. Another treatment is thrombopoietin (TPO), which stimulates formation and maturation of megakaryocytes and thereby boosts platelet levels.

Heparin-secreting cells are particularly abundant in the liver and lungs, where capillaries trap small blood clots that commonly form in the slow-moving blood of veins. These cells secrete heparin continually, preventing additional clotting in the cardiovascular system. Table 14.12 summarizes clot-inhibiting factors. Clinical Application 14.3 discusses an ancient anticlotting treatment that is in use again—biochemicals in the saliva of leeches. Clinical Application 14.4 discusses clotting disorders.

1 How does the lining of a blood vessel help prevent blood clot formation?

2 What is the function of antithrombin?

3 How does heparin help prevent blood clot formation?

Blood Groups and Transfusions

Lamb blood was used in early blood transfusion experiments, which date from the late 1600s. By the 1800s, human blood was being used. Results were unpredictable—some recipients were cured, but some were

TABLE 14.12	Factors That Inhibit Blood Clot Formation		
Factor	**Action**	**Factor**	**Action**
Smooth lining of blood vessel	Prevents activation of intrinsic blood clotting mechanism	Antithrombin in plasma	Interferes with the action of thrombin
Prostacyclin	Inhibits adherence of platelets to blood vessel wall	Heparin from mast cells and basophils	Interferes with the formation of prothrombin activator
Fibrin threads	Adsorbs thrombin		

THE RETURN OF THE MEDICINAL LEECH

It had taken surgeon Joseph Upton ten hours to sew the five-year-old's ear back on, after a dog had bitten it off. At first the operation appeared to be a success, but after four days, trouble began. Blood flow in the ear was blocked. Close examination showed that the arteries that the surgeon had repaired were fine, but the smaller veins were becoming congested. So Dr. Upton tried an experimental technique—he applied twenty-four leeches to the wound area.

The leeches latched on for up to an hour each, drinking the boy's blood. Leech saliva contains several biochemicals, one of which is a potent anticoagulant called hirudin, in honor of its source, the medicinal leech *Hirudo medicinalis*. Unlike conventional anticlotting agents such as heparin, which are short-acting, hirudin works for up to twenty-four hours after the leech has drunk its fill and dropped off. Hirudin specifically blocks thrombin in veins. The long-acting leech biochemical gave the boy's ear time to heal.

Leeches have long been part of medical practice, with references hailing back to the ancient Egyptians 2,500 years ago (fig. 14B). The leech's popularity peaked in Europe in the nineteenth century, when French physicians alone used more than a billion of them a year, to drain "bad humours" from the body to cure nearly every ill. Use of leeches fell in the latter half of the nineteenth century. They were rediscovered by Yugoslav plastic surgeons in 1960 and by French microsurgeons in the early 1980s. In 1985, Dr. Upton made headlines and brought leeches into the limelight by saving the boy's ear at Children's Hospital in Boston.

A leech's bite does not hurt, patients say. But for those unwilling to have one or more 3-inch long, slimy green-gray invertebrates picnicking on a wound, hirudin is also available as a drug called hirulog, produced by recombinant DNA technology (fig. 14C). ■

FIGURE 14B
For centuries, bloodletting with leeches was believed to cure many ills. This woman in seventeenth-century Belgium applies a medicinal leech to her arm.

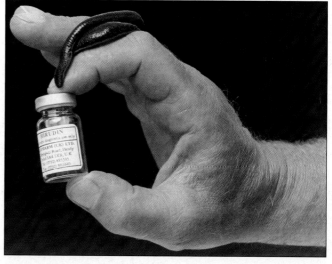

FIGURE 14C
Microsurgeons sometimes use leeches to help maintain blood flow through veins in patients after reattaching severed ears or digits. An anticoagulant in the leech's saliva keeps the blood thin enough to flow. © Biopharm (USA) Limited 1994.

killed when their kidneys failed under the strain of handling clumping red blood cells when blood types were incompatible. So poor was the success rate that, by the late 1800s, many nations banned transfusions.

Around this time, Austrian physician Karl Landsteiner began investigating why transfusions sometimes worked and sometimes did not. In 1900, he determined that blood was of differing types and that only certain combinations of them were compatible. In 1910, identification of the ABO blood antigen gene explained the observed blood type incompatibilities. Today, twenty different genes are known to contribute to the surface fea-

LIVING WITH HEMOPHILIA

Don Miller was born in 1949 and is semi-retired from running the math library at the University of Pittsburgh. Today, he has a sheep farm. On June 1, 1999, he was the first hemophilia patient to receive a disabled virus that delivered a functional gene for clotting factor VIII to his bloodstream. Within weeks, he began to experience results. Miller is one of the first of a new breed of patients—people helped by gene therapy. Here he describes his life with hemophilia.

"The hemophilia was discovered when I was circumcised, and I almost bled to death, but the doctors weren't really sure until I was about eighteen months old. No one where I was born was familiar with it.

"When I was three, I fell out of my crib and I was black and blue from my waist to the top of my head. The only treatment then was whole blood replacement. So I learned not to play sports. A minor sprain would take a week or two to heal. One time I fell at my grandmother's house and had a one-inch-long cut on the back of my leg. It took five weeks to stop bleeding, just leaking real slowly. I didn't need whole blood replacement, but if I moved a little the wrong way, it would open and bleed again.

"I had transfusions as seldom as I could. The doctors always tried not to infuse me until it was necessary. Of course there was no AIDS then, but there were problems with transmitting hepatitis through blood transfusions, and other blood-borne diseases. All that whole blood can kill you from kidney failure. When I was nine or ten I went to the hospital for intestinal polyps. I was operated on and they told me I'd have a 10 percent chance of pulling through. I met other kids there with hemophilia who died from kidney failure due to the amount of fluid from all the transfusions. Once a year I went to the hospital for blood tests. Some years I went more often than that. Most of the time I would just lay there and bleed. My joints don't work from all the bleeding.

"By the time I got married, treatment had progressed to gamma globulin from plasma. I married at twenty, the day of my graduation from college. By then I was receiving gamma globulin from donated plasma and small volumes of cryoprecipitate, which is the factor VIII clotting protein that my body cannot produce, pooled from many donors. We decided not to have children because that would end the hemophilia in the family.

"I'm one of the oldest patients at the Pittsburgh Hemophilia Center, so I knew people. I was HIV negative, and over age twenty-five, which is what they want. By that age a lot of people with hemophilia are HIV positive, because they lived through the time period when we had no choice but to use pooled cryoprecipitate before recombinant factor VIII was available. I lucked out. I took so little cryoprecipitate that I wasn't exposed to very much. And, I had the time. The gene therapy protocol involves showing up three times a week.

"The treatment is three infusions, one a day for three days, on an outpatient basis. So far there have been no side effects. Once the gene therapy is perfected, it will be a three-day treatment. A dosage study will follow this one, which is just for safety. Animal studies showed it's best given over three days. I go in once a week to be sure there is no adverse reaction. They hope it will be a one-time treatment. The virus will lodge in the liver and keep replicating.

"In the eight weeks before the infusion, I used eight doses of factor. In the fourteen weeks since then, I've used three. Incidents that used to require treatment no longer do. As long as I don't let myself feel stressed, I don't have spontaneous bleeding. I've had two nosebleeds that stopped within minutes without treatment, with only a trace of blood on the handkerchief, as opposed to hours of dripping.

"I'm somewhat more active, but fifty years of wear and tear won't be healed by this gene therapy. Two of the treatments I required started from overdoing activity, so now I'm trying to find the middle ground." ∎

---

tures of red blood cells, which determine compatibility between blood types.

Antigens and Antibodies

The clumping of red blood cells when testing blood compatibility or resulting from a transfusion reaction is called **agglutination** (ah-gloo″tĭna′shun). This phenomenon is due to a reaction between red blood cell surface molecules called **antigens** (an′-ti-jenz), formerly called *agglutinogens,* and protein **antibodies** (an′ti-bod″ez), formerly called *agglutinins,* carried in the plasma. Although many different antigens are associated with human erythrocytes, only a few of them are likely to produce serious transfusion reactions. These include the antigens of the ABO group and those of the Rh group. Avoiding the mixture of certain kinds of antigens and antibodies prevents adverse transfusion reactions.

A mismatched blood transfusion quickly produces telltale signs of agglutination—anxiety; breathing difficulty; facial flushing; headache; and severe pain in the neck, chest, and lumbar area. Red blood cells burst, releasing free hemoglobin. Macrophages phagocytize the hemoglobin, converting it to bilirubin, which may sufficiently accumulate to cause the yellow skin of jaundice. Free hemoglobin in the kidneys may ultimately cause them to fail.

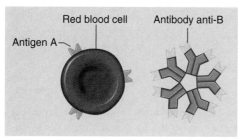
Type A blood

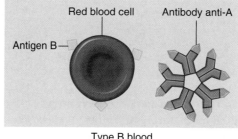
Type B blood

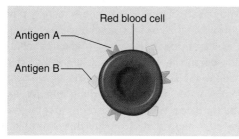
Type AB blood

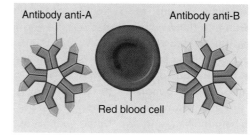
Type O blood

FIGURE 14.21

Different combinations of antigens and antibodies distinguish blood types. Cells and antibodies not drawn to scale.

ABO Blood Group

The *ABO blood group* is based on the presence (or absence) of two major antigens on red blood cell membranes—*antigen A* and *antigen B*. A person's erythrocytes contain one of four antigen combinations: only A, only B, both A and B, or neither A nor B. An individual with only antigen A has *type A blood;* a person with only antigen B has *type B blood;* one with both antigen A and antigen B has *type AB blood;* and one with neither antigen A nor antigen B has *type O blood.*

Certain antibodies are synthesized in the plasma about two to eight months following birth. Specifically, whenever antigen A is absent in the red blood cells, an antibody called *anti-A* is produced, and whenever antigen B is absent, an antibody called *anti-B* is manufactured. Therefore, persons with type A blood also have antibody anti-B in their plasma; those with type B blood have antibody anti-A; those with type AB blood have neither antibody; and those with type O blood have both anti-A and anti-B (fig. 14.21 and table 14.13). The antibodies anti-A and anti-B are large and do not cross the placenta. Thus, a pregnant woman and her fetus may be of different ABO blood types, and agglutination in the fetus will not occur.

The percentage of blood types in human populations reflects history and migration patterns. Type B blood is found in 5% to 10% of the English and Irish, but gradually increases eastward, reaching 25% to 30% in the former Soviet Union. Blood type frequencies can also reveal who were conquerors and who were conquered. For example, the frequencies of ABO blood types are very similar in northern Africa, the Near East, and Southern Spain—exactly the places where Arabs ruled until 1492.

The major concern in blood transfusion procedures is that the cells in the donated blood not clump due to antibodies present in the recipient's plasma. For example, a person with type A (anti-B) blood must not receive blood of type B or AB, either of which would clump in the presence of anti-B in the recipient's type A blood. Likewise, a person with type B (anti-A) blood must not be given type A or AB blood, and a person with type O (anti-A and anti-B) blood must not be given type A, B, or AB blood (fig. 14.22).

Because type AB blood lacks both anti-A and anti-B antibodies, an AB person can receive a transfusion of blood of any other type. For this reason, type AB persons are sometimes called *universal recipients.* However, type A (anti-B) blood, type B (anti-A) blood, and type O (anti-A and anti-B) blood still contain antibodies (either anti-A

TABLE 14.13	Antigens and Antibodies of the ABO Blood Group	
Blood Type	**Antigen**	**Antibody**
A	A	anti-B
B	B	anti-A
AB	A and B	Neither anti-A nor anti-B
O	Neither A nor B	Both anti-A and anti-B

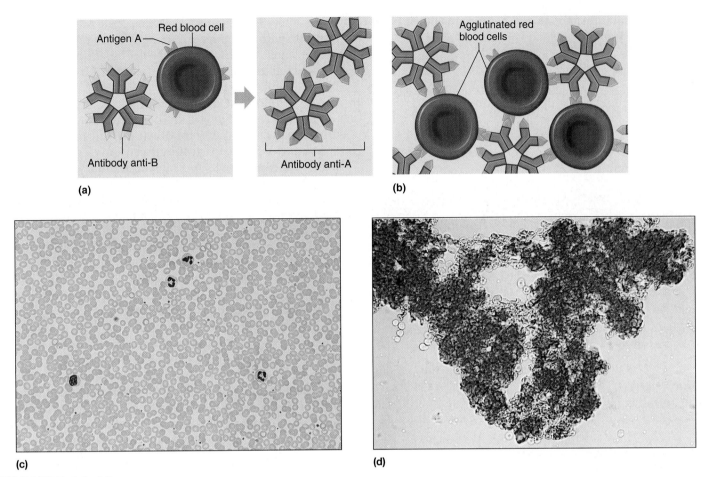

FIGURE 14.22

Agglutination. (*a*) If red blood cells with antigen A are added to blood containing antibody anti-A, (*b*) the antibodies react with the antigens, causing clumping (agglutination). (*c*) Nonagglutinated blood (210×). (*d*) Agglutinated blood (220×). Cells and antibodies in *a* and *b* not drawn to scale.

and/or anti-B) that could agglutinate type AB cells. Consequently, even for AB individuals, it is always best to use donor blood of the same type as the recipient blood. If the matching type is not available and type A, B, or O is used, it should be transfused slowly and in limited amounts so that the donor blood is well diluted by the recipient's larger blood volume. This precaution usually avoids serious reactions between the donor's antibodies and the recipient's antigens.

Type O blood lacks antigens A and B. Therefore, this type theoretically could be transfused into persons with blood of any other type. Individuals with type O blood are sometimes called *universal donors*. Type O blood, however, does contain both anti-A and anti-B antibodies, and if it is given to a person with blood type A, B, or AB, it too should be transfused slowly and in limited amounts to minimize the chance of an adverse reaction. When type O blood is given to blood types A, B, or AB, it is generally transfused as "packed cells," meaning the plasma has been removed. This also minimizes adverse reactions due to the anti-A and anti-B antibodies found in the plasma of type O blood. Table 14.14 summarizes preferred blood

TABLE 14.14	Preferred and Permissible Blood Types for Transfusions	
Blood Type of Recipient	**Preferred Blood Type of Donor**	**Permissible Blood Type of Donor (In an Extreme Emergency)**
A	A	A, O
B	B	B, O
AB	AB	AB, A, B, O
O	O	O

types for normal transfusions and permissible blood types for emergency transfusions.

1 Distinguish between antigens and antibodies.

2 What is the main concern when blood is transfused from one individual to another?

3 Why is a type AB person called a universal recipient?

4 Why is a type O person called a universal donor?

When is type O blood not really type O blood? Blood typing involves adding known antibodies to a blood sample to see if the blood will clump, demonstrating the presence of corresponding antigens on the red blood cell membranes. A person with a rare genetic condition called the Bombay phenotype lacks an enzyme that inserts a particular sugar onto red blood cell surfaces. Without that sugar, the A and B antigens cannot bind. The result is blood that tests as O (because it lacks A and B antigens) but can genetically be of any ABO type—A, B, AB, or O. Although the Bombay phenotype does not affect health, it can sometimes explain a child's ABO type that cannot be derived from those of the parents.

Rh Blood Group

The *Rh blood group* was named after the rhesus monkey in which it was first studied. In humans, this group includes several Rh antigens (factors). The most impor-

tant of these is *antigen D:* however, if any of the antigen D and other Rh antigens are present on the red blood cell membranes, the blood is said to be *Rh-positive.* Conversely, if the red blood cells lack the Rh antigens, the blood is called *Rh-negative.*

As in the case of antigens A and B, the presence (or absence) of Rh antigens is an inherited trait. Unlike anti-A and anti-B, antibodies for Rh (*anti-Rh*) do not appear spontaneously. Instead, they form only in Rh-negative persons in response to special stimulation.

If an Rh-negative person receives a transfusion of Rh-positive blood, the recipient's antibody-producing cells are stimulated by the presence of the Rh antigens and will begin producing *anti-Rh antibodies.* Generally, no serious consequences result from this initial transfusion, but if the Rh-negative person—who is now sensitized to Rh-positive blood—receives another transfusion of Rh-positive blood some months later, the donated red blood cells are likely to agglutinate.

A related condition may occur when an Rh-negative woman is pregnant with an Rh-positive fetus for the first time. Such a pregnancy may be uneventful; however, at the time of this infant's birth (or if a miscarriage occurs), the placental membranes that separated the maternal blood from the fetal blood during the pregnancy tear, and some of the infant's Rh-positive blood cells may enter the maternal circulation. These Rh-positive cells may then stimulate the maternal tissues to begin producing anti-Rh antibodies (fig. 14.23).

If a woman who has already developed anti-Rh antibodies becomes pregnant with a second Rh-positive fetus,

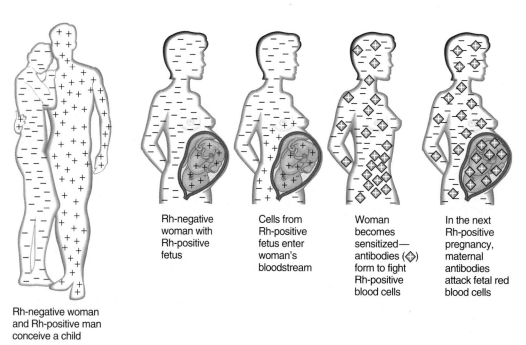

Rh-negative woman and Rh-positive man conceive a child

Rh-negative woman with Rh-positive fetus

Cells from Rh-positive fetus enter woman's bloodstream

Woman becomes sensitized— antibodies (◇) form to fight Rh-positive blood cells

In the next Rh-positive pregnancy, maternal antibodies attack fetal red blood cells

FIGURE 14.23

If a man who is Rh-positive and a woman who is Rh-negative conceive a child who is Rh-positive, the woman's body may manufacture antibodies that attack future Rh-positive offspring.

these anti-Rh antibodies, called hemolysins, cross the placental membrane and destroy the fetal red cells. The fetus then develops a condition called *erythroblastosis fetalis* (ĕrith″ro-blas-to′sis fe′tal-iz), or hemolytic disease of the newborn.

1 What is the Rh blood group?

2 What are two ways that Rh incompatibility can arise?

Erythroblastosis fetalis is extremely rare today because obstetricians carefully track Rh status. An Rh⁻ woman who might carry an Rh⁺ fetus is given an injection of a drug called RhoGAM. This is actually anti-Rh antibodies, which bind to and shield any Rh⁺ fetal cells that might contact the woman's cells, sensitizing her immune system. RhoGAM must be given within seventy-two hours of possible contact with Rh⁺ cells—such situations include giving birth, terminating a pregnancy, miscarrying, or undergoing amniocentesis (a prenatal test in which a needle is inserted into the uterus).

Clinical Terms Related to the Blood

anisocytosis (an-i″so-si-to′sis) Abnormal variation in the size of erythrocytes.

antihemophilic plasma (an″ti-he″mo-fil′ik plaz′mah) Normal blood plasma that has been processed to preserve an antihemophilic factor.

citrated whole blood (sit′rāt-ed hōl blud) Normal blood to which a solution of acid citrate has been added to prevent coagulation.

dried plasma (drīd plaz′mah) Normal blood plasma that has been vacuum dried to prevent the growth of microorganisms.

hemorrhagic telangiectasia (hem″o-raj′ik tel-an″je-ek-ta′ze-ah) Inherited tendency to bleed from localized lesions of the capillaries.

heparinized whole blood (hep′er-i-nīzed″ hōl blud) Normal blood to which a solution of heparin has been added to prevent coagulation.

macrocytosis (mak″ro-si-to′sis) Abnormally large erythrocytes.

microcytosis (mi″kro-si-to′sis) Abnormally small erythrocytes.

neutrophilia (nu″tro-fil′e-ah) Increase in the number of circulating neutrophils.

packed red cells Concentrated suspension of red blood cells from which the plasma has been removed.

pancytopenia (pan″si-to-pe′ne-ah) Abnormal depression of all the cellular components of blood.

poikilocytosis (poi″ki-lo-si-to′sis) Irregularly shaped erythrocytes.

purpura (per′pu-rah) Spontaneous bleeding into the tissues and through the mucous membranes.

septicemia (sep″ti-se′me-ah) Reproduction of disease-causing microorganisms in the blood.

spherocytosis (sfēr″o-si-to′sis) Hemolytic anemia caused by defective proteins supporting the cell membranes of red blood cells. The cells are abnormally spherical.

thalassemia (thal″ah-se′me-ah) Group of hereditary hemolytic anemias resulting from very thin, fragile erythrocytes. Globin chains are missing.

CHAPTER SUMMARY

Introduction (page 510)

Blood is often considered a type of connective tissue whose cells are suspended in a liquid intercellular material. It transports substances between the body cells and the external environment and helps maintain a stable internal environment.

Blood and Blood Cells (page 510)

Blood contains red blood cells, white blood cells, platelets, and plasma.

1. Blood volume and composition
   a. Blood volume varies with body size, fluid and electrolyte balance, and adipose tissue content.
   b. Blood can be separated into formed elements and liquid portions.
      (1) The formed elements portion is mostly red blood cells.
      (2) The liquid plasma includes water, nutrients, hormones, electrolytes, and cellular wastes.

2. The origin of blood cells
   a. Blood cells develop from hematopoietic stem cells in red bone marrow.
   b. Cells descended from stem cells respond to colony-stimulating factors to specialize.
   c. Thrombopoietin stimulates megakaryocytes to give rise to platelets, much as erythropoietin stimulates formation of red blood cells.

3. Characteristics of red blood cells
   a. Red blood cells are biconcave discs with shapes that provide increased surface area and place their cell membranes close to internal structures.
   b. They contain hemoglobin, which combines loosely with oxygen.
   c. The mature form lacks nuclei but contains enzymes needed for energy-releasing processes.

4. Red blood cell counts
   a. The red blood cell count equals the number of cells per mm³ of blood.

b. The average count may range from approximately 4,000,000 to 6,000,000 cells per mm^3.
c. Red blood cell count is related to the oxygen-carrying capacity of the blood and is used in diagnosing and evaluating the courses of diseases.

5. Red blood cell production and its control
   a. During fetal development, red blood cells form in the yolk sac, liver, and spleen; later, red blood cells are produced by the red bone marrow.
   b. The number of red blood cells remains relatively stable.
   c. A negative feedback mechanism involving erythropoietin from the kidneys and liver controls rate of red blood cell production.
      (1) Erythropoietin is released in response to low oxygen levels.
      (2) High altitude, loss of blood, or chronic lung disease can lower oxygen concentration.

6. Dietary factors affecting red blood cell production
   a. The availability of vitamin B$_{12}$, iron, and folic acid affects red blood cell production.
   b. The rate of iron absorption varies with the amount of iron in the body.

7. Destruction of red blood cells
   a. Red blood cells are fragile and are damaged while moving through capillaries.
   b. Macrophages in the liver and spleen phagocytize damaged red blood cells.
   c. Hemoglobin molecules are decomposed, and the iron they contain is recycled.
   d. Biliverdin and bilirubin are pigments released from hemoglobin.

8. Types of white blood cells
   a. White blood cells control infection.
   b. Granulocytes include neutrophils, eosinophils, and basophils.
   c. Agranulocytes include monocytes and lymphocytes.

9. White blood cells fight infection
   a. Neutrophils and monocytes phagocytize foreign particles.
   b. Chemicals released by damaged cells attract and stimulate leukocytes.
   c. Eosinophils kill parasites and help control inflammation and allergic reactions.
   d. Basophils release heparin, which inhibits blood clotting, and histamine to increase blood flow to injured tissues.
   e. Lymphocytes are involved in immunity and produce antibodies that attack specific foreign antigens.

10. White blood cell counts
    a. Normal total white blood cell counts vary from 5,000 to 10,000 cells per mm^3 of blood.
    b. The number of white blood cells may change in abnormal conditions such as infections, emotional disturbances, or excessive loss of body fluids.
    c. A differential white blood cell count indicates the percentages of various types of leukocytes present.

11. Blood platelets
    a. Blood platelets are fragments of megakaryocytes that detach and enter the circulation.
    b. The normal count varies from 130,000 to 360,000 platelets per mm^3.
    c. Platelets help close breaks in blood vessels.

Blood Plasma (page 523)

Plasma is the liquid part of the blood that is composed of water and a mixture of organic and inorganic substances. It transports nutrients and gases, helps regulate fluid and electrolyte balance, and helps maintain stable pH.

1. Plasma proteins
   a. Plasma proteins remain in blood and interstitial fluids and are not normally used as energy sources.
   b. Three major groups exist.
      (1) Albumins help maintain the osmotic pressure of plasma.
      (2) Globulins include antibodies. They provide immunity and transport lipids and fat-soluble vitamins.
      (3) Fibrinogen functions in blood clotting.

2. Gases and nutrients
   a. Gases in plasma include oxygen, carbon dioxide, and nitrogen.
   b. Plasma nutrients include amino acids, simple sugars, and lipids.
      (1) Glucose is stored in the liver as glycogen and is released whenever the blood glucose concentration falls.
      (2) Amino acids are used to synthesize proteins and are deaminated for use as energy sources.
      (3) Lipoproteins function in the transport of lipids.

3. Nonprotein nitrogenous substances
   a. Nonprotein nitrogenous substances are composed of molecules that contain nitrogen atoms but are not proteins.
   b. They include amino acids, urea, uric acid, creatine, and creatinine.
      (1) Urea and uric acid are products of catabolism.
      (2) Creatinine results from the metabolism of creatine.
   c. Levels of these substances usually remain stable; an increase may indicate a kidney disorder.

4. Plasma electrolytes
   a. Plasma electrolytes are absorbed from the intestines and are released as by-products of cellular metabolism.
   b. They include ions of sodium, potassium, calcium, magnesium, chloride, bicarbonate, phosphate, and sulfate.
   c. They are important in the maintenance of osmotic pressure and pH.

Hemostasis (page 526)

Hemostasis refers to the stoppage of bleeding. Hemostatic mechanisms are most effective in controlling blood loss from small vessels.

1. Blood vessel spasm (vasospasm)
   a. Smooth muscles in walls of smaller blood vessels reflexly contract following injury.
   b. Platelets release serotonin that stimulates vasoconstriction and helps maintain vessel spasm.

2. Platelet plug formation
   a. Platelets adhere to rough surfaces and exposed collagen.
   b. Platelets adhere together at the sites of injuries and form platelet plugs in broken vessels.

3. Blood coagulation
   a. Blood clotting, the most effective means of hemostasis, involves a series of reactions wherein each reaction stimulates the next reaction (cascade), which may be initiated by extrinsic or intrinsic mechanisms.

b. The extrinsic clotting mechanism is triggered when blood contacts damaged tissue.

c. The intrinsic clotting mechanism is triggered when blood contacts a foreign surface.

d. Clot formation depends on the balance between clotting factors that promote clotting and those that inhibit clotting.

e. The basic event of coagulation is the conversion of soluble fibrinogen into insoluble fibrin.

f. After forming, the clot retracts and pulls the edges of a broken vessel closer together.

g. A thrombus is an abnormal blood clot in a vessel; an embolus is a clot or fragment of a clot that moves in a vessel.

h. Fibroblasts invade a clot, forming connective tissue throughout.

i. Protein-splitting enzymes may eventually destroy a clot.

4. Prevention of coagulation
   a. The smooth lining of blood vessels discourages the accumulation of platelets.
   b. As a clot forms, fibrin adsorbs thrombin and prevents the reaction from spreading.
   c. Antithrombin interferes with the action of excess thrombin.
   d. Some cells secrete heparin, an anticoagulant.

Blood Groups and Transfusions (page 531)

Blood can be typed on the basis of the surface structures of its cells.

1. Antigens and antibodies
   a. Red blood cell membranes may contain specific antigens, and blood plasma may contain antibodies against certain of these antigens.

b. Blood typing uses known antibodies to identify antigens on red blood cell membranes.

2. ABO blood group
   a. Blood can be grouped according to the presence or absence of antigens A and B.
   b. Whenever antigen A is absent, antibody anti-A is present; whenever antigen B is absent, antibody anti-B is present.
   c. Adverse transfusion reactions are avoided by preventing the mixing of red blood cells that contain an antigen with plasma that contains the corresponding antibody.
   d. Adverse reactions are due to agglutination (clumping) of the red blood cells.

3. Rh blood group
   a. Rh antigens are present on the red blood cell membranes of Rh-positive blood; they are absent in Rh-negative blood.
   b. If an Rh-negative person is exposed to Rh-positive blood, anti-Rh antibodies are produced in response.
   c. Mixing Rh-positive red cells with plasma that contains anti-Rh antibodies agglutinates the positive cells.
   d. If an Rh-negative female is pregnant with an Rh-positive fetus, some of the positive cells may enter the maternal blood at the time of birth and stimulate the maternal tissues to produce anti-Rh antibodies.
   e. Anti-Rh antibodies in maternal blood may pass through the placental tissues and react with the red blood cells of an Rh-positive fetus.

CRITICAL THINKING QUESTIONS

1. What change would you expect to occur in the hematocrit of a person who is dehydrated? Why?

2. Erythropoietin is available as a drug. Why would athletes abuse it?

3. If a patient with inoperable cancer is treated using a drug that reduces the rate of cell division, what changes might occur in the blood cell count? How might the patient's environment be modified to compensate for the effects of these changes?

4. Hypochromic (iron-deficiency) anemia is common among aging persons who are admitted to hospitals for other conditions. What environmental and sociological factors might promote this form of anemia?

5. How would you explain to a patient with leukemia, who has a greatly elevated white blood cell count, the importance of avoiding bacterial infections?

6. If a woman whose blood is Rh-negative and contains anti-Rh antibodies is carrying a fetus with Rh-negative blood, will the fetus be in danger of developing erythroblastosis fetalis? Why or why not?

7. In the United States, between 1977 and 1985, more than 10,000 men contracted the human immunodeficiency virus (HIV) from contaminated factor VIII that they received to treat hemophilia. What are two abnormalities in the blood of these men?

8. Why do patients with liver diseases commonly develop blood clotting disorders?

9. Why can a person receive platelets donated by anyone but must receive a particular type of whole blood?

REVIEW EXERCISES

1. List the major components of blood.
2. Define *hematocrit*, and explain how it is determined.
3. Describe a red blood cell.
4. Distinguish between oxyhemoglobin and deoxyhemoglobin.
5. Explain what is meant by a *red blood cell count*.
6. Describe the life cycle of a red blood cell.
7. Define *erythropoietin*, and explain its function.
8. Explain how vitamin B_{12} and folic acid deficiencies affect red blood cell production.
9. List two sources of iron that can be used for the synthesis of hemoglobin.
10. Distinguish between biliverdin and bilirubin.
11. Distinguish between granulocytes and agranulocytes.
12. Name five types of leukocytes, and list the major functions of each type.
13. Explain the significance of white blood cell counts as aids to diagnosing diseases.
14. Describe a blood platelet, and explain its functions.
15. Name three types of plasma proteins, and list the major functions of each type.
16. Define *lipoprotein*.
17. Define *apoprotein*.
18. Distinguish between low-density lipoprotein and high-density lipoprotein.
19. Name the sources of VLDL, LDL, HDL, and chylomicrons.
20. Describe how lipoproteins are removed from plasma.
21. Explain how cholesterol is eliminated from plasma and from the body.
22. Define *nonprotein nitrogenous substances*, and name those commonly present in plasma.
23. Name several plasma electrolytes.
24. Define *hemostasis*.
25. Explain how blood vessel spasms are stimulated following an injury.
26. Explain how a platelet plug forms.
27. List the major steps leading to the formation of a blood clot.
28. Indicate the trigger and outline the steps for extrinsic clotting and for intrinsic clotting.
29. Distinguish between fibrinogen and fibrin.
30. Describe a positive feedback system that operates during blood clotting.
31. Define *serum*.
32. Distinguish between a thrombus and an embolus.
33. Explain how a blood clot may be removed naturally from a blood vessel.
34. Describe how blood coagulation may be prevented.
35. Name a vitamin required for blood clotting.
36. Distinguish between an antigen and an antibody.
37. Explain the basis of ABO blood types.
38. Explain why a person with blood type AB is sometimes called a universal recipient.
39. Explain why a person with blood type O is sometimes called a universal donor.
40. Distinguish between Rh-positive and Rh-negative blood.
41. Describe how a person may become sensitized to Rh-positive blood.
42. Describe *erythroblastosis fetalis*, and explain how this condition may develop.

WEB CONNECTIONS

Visit the Student Edition of the Online Learning Center at www.mhhe.com/shier10 **for answers to chapter questions, additional quizzes, interactive learning exercises, and other study tools.**

UNDERSTANDING WORDS

angio-, vessel: *angio*tensin—substance that constricts blood vessels.

ather-, porridge: *ather*osclerosis—deposits of plaque in arteries.

brady-, slow: *brady*cardia—abnormally slow heartbeat.

diastol-, dilation: *diastol*ic pressure—blood pressure when the ventricle of the heart is relaxed.

edem-, swelling: *edem*a—condition in which fluids accumulate in the tissues and cause them to swell.

-gram, something written: electrocardio*gram*—recording of the electrical changes in the myocardium during a cardiac cycle.

lun-, moon: semi*lun*ar valve—valve with crescent-shaped flaps.

my-, muscle: *my*ocardium—muscle tissue within the wall of the heart.

papill-, nipple: *papill*ary muscle—small mound of muscle within a ventricle of the heart.

phleb-, vein: *phleb*itis—inflammation of a vein.

scler-, hard: arterio*scler*osis—loss of elasticity and hardening of a blood vessel wall.

syn-, together: *syn*cytium—mass of merging cells that act together.

systol-, contraction: *systol*ic pressure—blood pressure resulting from a ventricular contraction.

tachy-, rapid: *tachy*cardia—abnormally fast heartbeat.

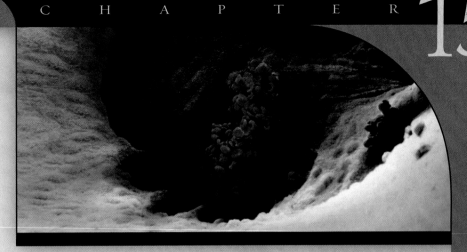

A thrombus (blood clot) partially blocks this artery entering the heart (300×). This situation causes myocardial infarction (heart attack).

Cardiovascular System

CHAPTER OBJECTIVES

After you have studied this chapter, you should be able to

1. Name the organs of the cardiovascular system and discuss their functions.

2. Identify and locate the major parts of the heart and discuss the function of each part.

3. Trace the pathway of the blood through the heart and the vessels of the coronary circulation.

4. Discuss the cardiac cycle and explain how it is controlled.

5. Identify the parts of a normal ECG pattern and discuss the significance of this pattern.

6. Compare the structures and functions of the major types of blood vessels.

7. Describe how substances are exchanged between blood in capillaries and the tissue fluid surrounding body cells.

8. Explain how blood pressure is produced and controlled.

9. Describe the mechanisms that aid in returning venous blood to the heart.

10. Compare the pulmonary and systemic circuits of the cardiovascular system.

11. Identify and locate the major arteries and veins of the pulmonary and systemic circuits.

12. Describe life-span changes in the cardiovascular system.

A man rushing to catch a flight at a busy airport stops suddenly, looks about in confusion, and collapses. People congregate around him, as a woman runs to a device mounted on a nearby wall. It is an external defibrillator, and looks like a laptop computer. The woman learned how to use it in a cardiopulmonary resuscitation class. She brings it over to the man, opens it, and places electrode pads over the man's chest, as indicated in a drawing on the inner cover of the defibrillator. Then the device speaks, "Analyzing heart rhythm," it declares as a computer assesses the heart rhythm. After a short pause, the device says, "charging, stand clear," and then "push button." The woman does so, and the device delivers a shock to the man's chest. It assesses the heart rhythm again, and instructs the woman to deliver a second shock. Soon, the man recovers, just as emergency technicians arrive.

The external defibrillators found in airports, malls, and other public places can save the life of a person suffering sudden cardiac arrest. One study conducted at Chicago's O'Hare and Midland airports found that over a ten month period, external defibrillators saved 64 percent of the people they shocked. Without defibrillation, only 5 to 7 percent of people survive sudden cardiac arrest. Each minute, the odds of survival shrink by 10 percent, and after six minutes, brain damage is irreversible. Sudden cardiac arrest can result from an accelerated heartbeat (ventricular tachycardia) or a chaotic and irregular heartbeat (ventricular fibrillation). The electrical malfunction that usually causes these conditions may result from an artery blocked with plaque, or from build up of scar tissue from a previous myocardial infarction (heart attack.)

For people who know that they have an inherited disorder that causes sudden cardiac arrest (by having suffered an event and then had genetic tests), a device called an implantable cardioverter defibrillator (ICD) can be placed under the skin of the chest in a one-hour procedure. Like the external defibrillator, the ICD monitors heart rhythm, and when the telltale deviations of ventricular tachycardia or ventricular fibrillation begin, it delivers a shock.

ICDs have been so successful in preventing subsequent cardiac arrests that they may soon be offered to people at high risk for the condition. The two major risk factors are having had a previous myocardial infarction and a low ejection fraction, which is the volume of blood

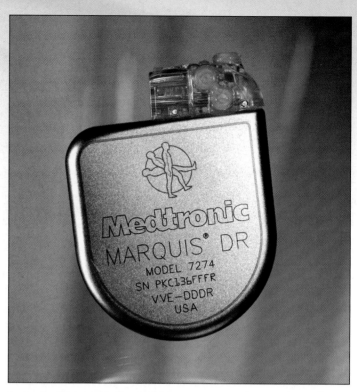

An implantable defibrillator delivers a shock to a heart whose ventricles are wildly contracting, restoring a normal heartbeat.

pumped with each heartbeat. Normal ejection fraction is 50 to 60 percent; low is below 30 to 40 percent. Scarring lowers the ejection fraction. An echocardiogram, which is an ultrasound scan of the heart, can reveal the ejection fraction.

In a four-year study at 76 medical centers involving 1,232 patients with the two risk factors, use of an ICD decreased death rate from sudden cardiac arrest by 31 percent. One person who already has an ICD to prevent sudden cardiac arrest is vice president Dick Cheney, who received his device on June 30, 2001. He had not suffered cardiac arrest, but has the classic risk factors. Many nations already use the device as a preventative. In the United States, wider use of ICDs could save thousands of the 300,000 who die each year of sudden cardiac arrest. ∎

The **cardiovascular system** includes the heart and the blood vessels. A functional cardiovascular system is vital for survival, because without blood circulation, the tissues lack oxygen and nutrients, and wastes accumulate. Under such conditions, the cells soon begin irreversible change, which quickly leads to death. Figure 15.1 shows the general pattern of blood transport in the cardiovascular system.

Structure of the Heart

The heart is a hollow, cone-shaped, muscular pump located within the mediastinum of the thorax and resting upon the diaphragm.

Size and Location of the Heart

Heart size varies with body size. However, an average adult's heart is generally about 14 centimeters long and 9 centimeters wide (fig. 15.2).

The heart is bordered laterally by the lungs, posteriorly by the vertebral column, and anteriorly by the sternum (fig. 15.3 and reference plates 50, 56, 71, and 72). Its *base,* which is attached to several large blood vessels, lies beneath the second rib. Its distal end extends downward and to the left, terminating as a bluntly pointed *apex* at the level of the fifth intercostal space. For this reason, it is possible to sense the *apical heartbeat* by feeling or listening to the chest wall between the fifth and sixth ribs, about 7.5 centimeters to the left of the midline.

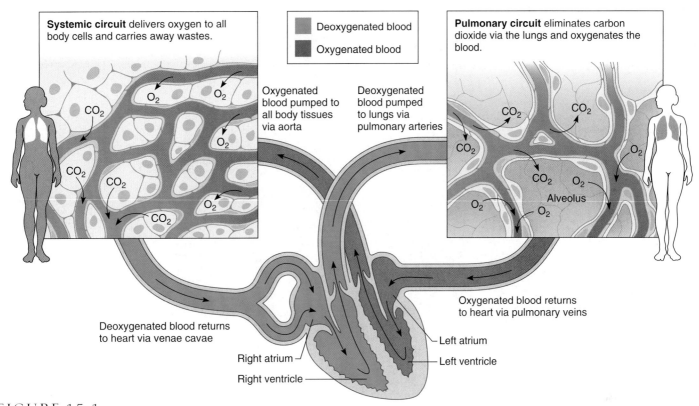

Systemic circuit delivers oxygen to all body cells and carries away wastes.

Deoxygenated blood

Oxygenated blood

Pulmonary circuit eliminates carbon dioxide via the lungs and oxygenates the blood.

O_2 O_2

CO_2

CO_2

O_2

CO_2 CO_2

O_2

CO_2

CO_2 CO_2

CO_2

CO_2 O_2

Alveolus

O_2 O_2

O_2

Oxygenated blood pumped to all body tissues via aorta

Deoxygenated blood pumped to lungs via pulmonary arteries

Oxygenated blood returns to heart via pulmonary veins

Deoxygenated blood returns to heart via venae cavae

Left atrium

Left ventricle

Right atrium

Right ventricle

FIGURE 15.1

The cardiovascular system transports blood throughout the body, delivering oxygen to cells and removing carbon dioxide.

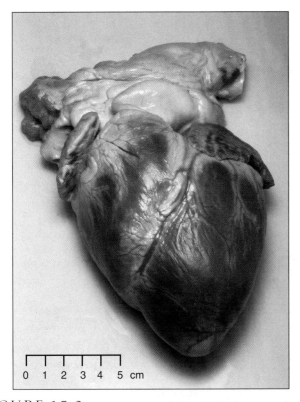

0 1 2 3 4 5 cm

FIGURE 15.2

Anterior view of a human heart. (Since this photo is not life size, a proportionately reduced rule has been included to help the student grasp the true size of the organ.)

Coverings of the Heart

The **pericardium** (per″i-kar′de-um), or pericardial sac, is a covering that encloses the heart and the proximal ends of the large blood vessels to which it attaches. The pericardium consists of an outer fibrous bag, the *fibrous pericardium,* that surrounds a more delicate, double-layered serous membrane. The inner layer of this serous membrane, the *visceral pericardium* (epicardium), covers the heart. At the base of the heart, the visceral pericardium turns back upon itself to become the *parietal pericardium,* which forms the inner lining of the fibrous pericardium.

The fibrous pericardium is a tough, protective sac largely composed of dense connective tissue. It is attached to the central portion of the diaphragm, the posterior of the sternum, the vertebral column, and the large blood vessels emerging from the heart (see figs. 1.9*b*, 15.4 and reference plates 56 and 57). Between the parietal and visceral layers of the pericardium is a space, the *pericardial cavity,* that contains a small volume of serous fluid that the pericardial membranes secrete. This fluid reduces friction between the pericardial membranes as the heart moves within them.

In *pericarditis,* inflammation of the pericardium due to viral or bacterial infection produces adhesions that attach the layers of the pericardium. This condition is very painful and interferes with heart movements.

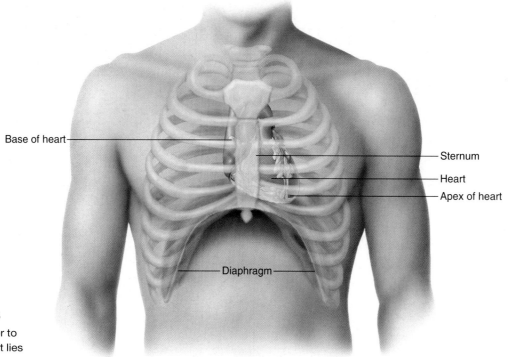

FIGURE 15.3

The heart is posterior to the sternum, where it lies upon the diaphragm.

Base of heart

Sternum

Heart

Apex of heart

Diaphragm

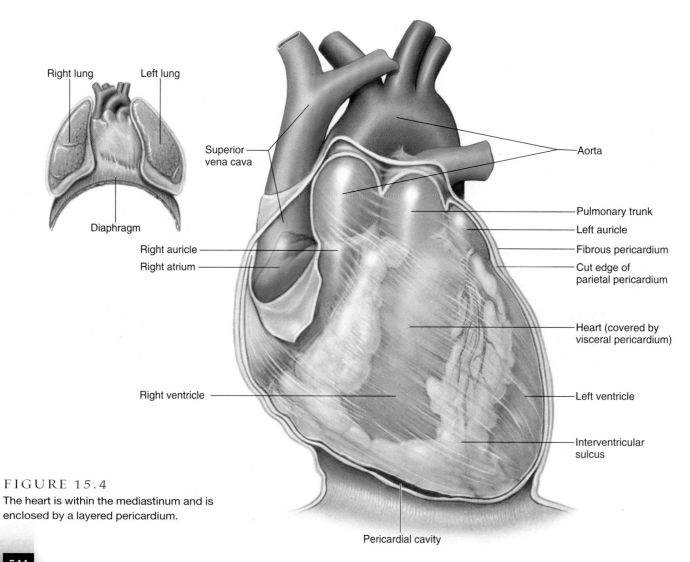

FIGURE 15.4

The heart is within the mediastinum and is enclosed by a layered pericardium.

Right lung

Left lung

Diaphragm

Superior vena cava

Aorta

Pulmonary trunk

Left auricle

Fibrous pericardium

Cut edge of parietal pericardium

Right auricle

Right atrium

Heart (covered by visceral pericardium)

Right ventricle

Left ventricle

Interventricular sulcus

Pericardial cavity

1 Where is the heart located?

2 Where would you listen to hear the apical heartbeat?

3 Distinguish between the visceral pericardium and the parietal pericardium.

4 What is the function of the fluid in the pericardial cavity?

Wall of the Heart

The wall of the heart is composed of three distinct layers: an outer epicardium, a middle myocardium, and an inner endocardium (fig. 15.5).

The **epicardium** (ep"i-kar'de-um), which corresponds to the visceral pericardium, is protective. It is a serous membrane that consists of connective tissue covered by epithelium, and it includes blood capillaries, lymph capillaries, and nerve fibers. The deeper portion of the epicardium often contains fat, particularly along the paths of coronary arteries and cardiac veins that provide blood flow through the myocardium.

The middle layer, or **myocardium** (mi"o-kar'de-um), is thick and largely consists of the cardiac muscle tissue that pumps blood out of the heart chambers. The muscle fibers are arranged in planes, separated by connective tissues that are richly supplied with blood capillaries, lymph capillaries, and nerve fibers.

The inner layer, or **endocardium** (en"do-kar'de-um), consists of epithelium and underlying connective tissue. The endocardium also contains blood vessels and some specialized cardiac muscle fibers called *Purkinje fibers,* described later in this chapter in the section titled "Cardiac Conduction System."

The endocardium lines all of the heart chambers and covers the structures, such as the heart valves, that project into them. This inner lining is also continuous with the inner lining of the blood vessels (endothelium) attached to the heart. Table 15.1 summarizes the characteristics of the three layers of the heart wall.

Heart Chambers and Valves

Internally, the heart is divided into four hollow chambers, two on the left and two on the right. The upper chambers, called **atria** (a'tre-ah) (sing., *atrium*), have thin walls and receive blood returning to the heart. Small, earlike projections, called **auricles** (aw'rĕ-klz), extend anteriorly from the atria (see fig. 15.4). The lower chambers, the **ventricles** (ven'tri-klz), force the blood out of the heart into arteries.

A structure called the *interatrial septum* separates the right from the left atrium. An *interventricular septum* separates the two ventricles. The atrium on each side communicates with its corresponding ventricle through an opening called the **atrioventricular orifice** (a"tre-o-ven-trik'u-lar or'ĭ-fis), guarded by an *atrioventricular valve (A-V valve).*

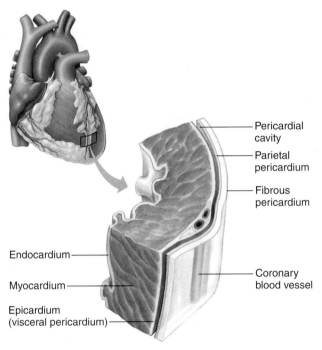

Pericardial cavity

Parietal pericardium

Fibrous pericardium

Endocardium

Myocardium

Epicardium (visceral pericardium)

Coronary blood vessel

FIGURE 15.5

The heart wall has three layers: an endocardium, a myocardium, and an epicardium.

TABLE 15.1	Wall of the Heart	
Layer	**Composition**	**Function**
Epicardium (visceral pericardium)	Serous membrane of connective tissue covered with epithelium and including blood capillaries, lymph capillaries, and nerve fibers	Forms a protective outer covering; secretes serous fluid
Myocardium	Cardiac muscle tissue separated by connective tissues and including blood capillaries, lymph capillaries, and nerve fibers	Contracts to pump blood from the heart chambers
Endocardium	Membrane of epithelium and underlying connective tissue, including blood vessels and specialized muscle fibers	Forms a protective inner lining of the chambers and valves

Grooves on the surface of the heart mark the divisions between its chambers, and they also contain major blood vessels that supply the heart tissues. The deepest of these grooves is the **atrioventricular** (coronary) **sulcus,** which encircles the heart between the atria and ventricles. Two **interventricular** (anterior and posterior) **sulci** mark the septum that separates the right and left ventricles (see fig. 15.4).

When increasing blood volume stretches muscle cells associated with the atria, the cells secrete a peptide hormone called *atrial natriuretic peptide* (ANP). ANP inhibits release of renin from the kidneys and of aldosterone from the adrenal cortex. The overall result is increased excretion of sodium ions and water from the kidneys and lowered blood volume and blood pressure. Researchers are investigating use of ANP to treat high blood pressure.

---

1 Describe the layers of the heart wall.

2 Name and locate the four chambers of the heart.

3 Name the orifices between the upper and the lower chambers of the heart.

The right atrium receives blood from two large veins: the *superior vena cava* and the *inferior vena cava.* These veins return blood that is low in oxygen from tissues. A smaller vein, the *coronary sinus,* also drains blood into the right atrium from the myocardium of the heart itself.

A large **tricuspid valve** (right atrioventricular valve) guards the atrioventricular orifice between the right atrium and the right ventricle. It is composed of three leaflets, or cusps, as its name implies. This valve permits the blood to move from the right atrium into the right ventricle and prevents it from moving in the opposite direction. The cusps fold passively out of the way against the ventricular wall when the blood pressure is greater on the atrial side, and they close passively when the pressure is greater on the ventricular side (figs. 15.6, 15.7, 15.8, and 15.9).

Strong, fibrous strings, called *chordae tendineae* (kor′de ten′dĭ-ne), are attached to the cusps on the ventricular side. These strings originate from small mounds of cardiac muscle tissue, the **papillary muscles** (pap′ĭ-ler″e mus′elz), that project inward from the walls of the ventricle. The papillary muscles contract when the ventricle contracts. As the tricuspid valve closes, these muscles pull on the chordae tendineae and prevent the cusps from swinging backwards into the atrium.

The right ventricle has a thinner muscular wall than the left ventricle. This right chamber pumps the blood a fairly short distance to the lungs against a relatively low resistance to blood flow. The left ventricle, on the other hand, must force the blood to all the other parts of the body against a much greater resistance to flow.

When the muscular wall of the right ventricle contracts, the blood inside its chamber is put under increasing pressure, and the tricuspid valve closes passively. As a result, the only exit is through the *pulmonary trunk,* which divides to form the left and right *pulmonary arteries.* At the base of this trunk is a **pulmonary valve** (pulmonary semilunar valve), which consists of three cusps (see fig. 15.8). This valve opens as the right ventricle contracts. However, when the ventricular muscles relax, the blood begins to back up in the pulmonary trunk. This closes the pulmonary valve, preventing a return flow into the ventricle. Unlike the tricuspid valve, the pulmonary valve does not have chordae tendineae or papillary muscles attached to its cusps.

The left atrium receives the blood from the lungs through four *pulmonary veins*—two from the right lung and two from the left lung. The blood passes from the left atrium into the left ventricle through the atrioventricular orifice, which a valve guards. This valve consists of two leaflets and is named the **mitral valve** (shaped like a miter, a turbanlike headdress) or left atrioventricular valve or bicuspid valve. It prevents the blood from flowing back into the left atrium from the ventricle when the ventricle contracts. As with the tricuspid valve, the papillary muscles and the chordae tendineae prevent the cusps of the mitral valve from swinging backwards into the atrium.

When the left ventricle contracts, the mitral valve closes passively, and the only exit is through a large artery called the *aorta.* Its branches distribute blood to all parts of the body.

Mitral valve prolapse (MVP) is common, affecting up to 6% of the U.S. population. In this condition, one (or both) of the cusps of the mitral valve stretches and bulges into the left atrium during ventricular contraction. The valve usually continues to function adequately, but sometimes, blood regurgitates into the left atrium. Through a stethoscope, a regurgitating MVP sounds like a click at the end of ventricular contraction, then a murmur as blood goes back through the valve into the left atrium. Symptoms of MVP include chest pain, palpitations, fatigue, and anxiety.

The mitral valve can be damaged by certain species of *Streptococcus* bacteria. Endocarditis, an inflammation of the endocardium due to infection, appears as a plantlike growth on the valve. People with MVP are particularly susceptible to endocarditis. Individuals with MVP must take antibiotics before undergoing dental work to prevent *Streptococcus* bacteria in the mouth from migrating through the blood to the heart and causing infection.

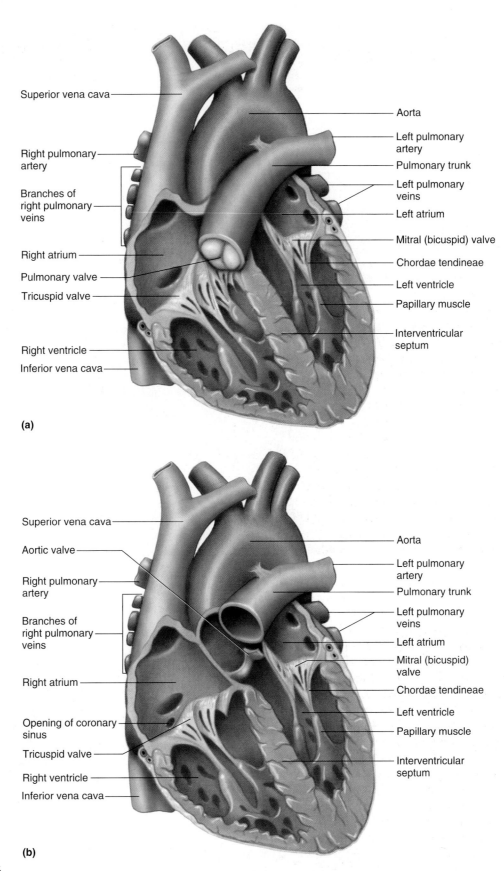

(a)

- Superior vena cava
- Right pulmonary artery
- Branches of right pulmonary veins
- Right atrium
- Pulmonary valve
- Tricuspid valve
- Right ventricle
- Inferior vena cava
- Aorta
- Left pulmonary artery
- Pulmonary trunk
- Left pulmonary veins
- Left atrium
- Mitral (bicuspid) valve
- Chordae tendineae
- Left ventricle
- Papillary muscle
- Interventricular septum

(b)

- Superior vena cava
- Aortic valve
- Right pulmonary artery
- Branches of right pulmonary veins
- Right atrium
- Opening of coronary sinus
- Tricuspid valve
- Right ventricle
- Inferior vena cava
- Aorta
- Left pulmonary artery
- Pulmonary trunk
- Left pulmonary veins
- Left atrium
- Mitral (bicuspid) valve
- Chordae tendineae
- Left ventricle
- Papillary muscle
- Interventricular septum

FIGURE 15.6

Coronal sections of the heart (*a*) showing the connection between the right ventricle and the pulmonary trunk and (*b*) showing the connection between the left ventricle and the aorta, as well as the four hollow chambers.

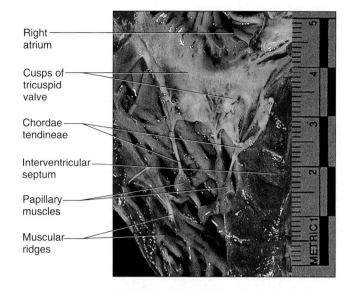

Right atrium

Cusps of tricuspid valve

Chordae tendineae

Interventricular septum

Papillary muscles

Muscular ridges

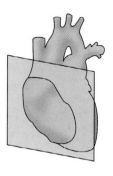

FIGURE 15.7
Photograph of a human tricuspid valve.

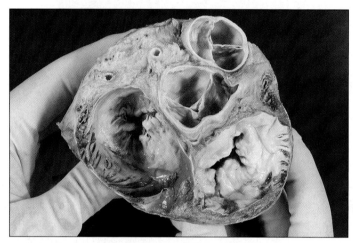

FIGURE 15.8
Photograph of the pulmonary and aortic valves of the heart (superior view). Figure 15.9 labels the valves as seen in this photograph.

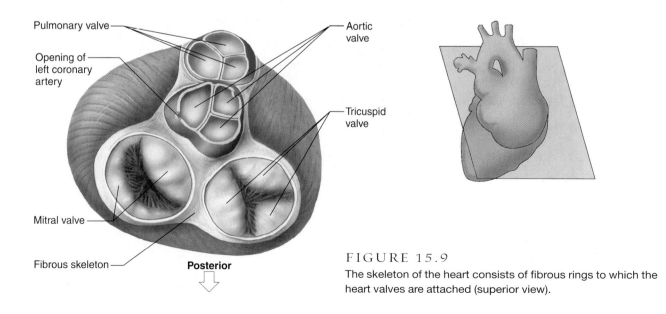

Pulmonary valve

Aortic valve

Opening of left coronary artery

Tricuspid valve

Mitral valve

Fibrous skeleton

Posterior

FIGURE 15.9
The skeleton of the heart consists of fibrous rings to which the heart valves are attached (superior view).

TABLE 15.2 | Valves of the Heart

Valve	Location	Function
Tricuspid valve	Right atrioventricular orifice	Prevents blood from moving from right ventricle into right atrium during ventricular contraction
Pulmonary valve	Entrance to pulmonary trunk	Prevents blood from moving from pulmonary trunk into right ventricle during ventricular relaxation
Mitral valve	Left atrioventricular orifice	Prevents blood from moving from left ventricle into left atrium during ventricular contraction
Aortic valve	Entrance to aorta	Prevents blood from moving from aorta into left ventricle during ventricular relaxation

At the base of the aorta is an **aortic valve** (aortic semilunar valve) that consists of three cusps (see fig. 15.8). It opens and allows blood to leave the left ventricle as it contracts. When the ventricular muscles relax, this valve closes and prevents blood from backing up into the ventricle.

The mitral and tricuspid valves are also called atrioventricular valves because they are between atria and ventricles. The pulmonary and aortic valves are also called semilunar because of the half-moon shapes of their cusps. Table 15.2 summarizes the locations and functions of the heart valves.

1 Which blood vessels carry blood into the right atrium?

2 Where does blood go after it leaves the right ventricle?

3 Which blood vessels carry blood into the left atrium?

4 What prevents blood from flowing back into the ventricles when they relax?

Skeleton of the Heart

Rings of dense connective tissue surround the pulmonary trunk and aorta at their proximal ends. These rings are continuous with others that encircle the atrioventricular orifices. They provide firm attachments for the heart valves and for muscle fibers and prevent the outlets of the atria and ventricles from dilating during contraction. The fibrous rings, together with other masses of dense connective tissue in the portion of the septum between the ventricles (interventricular septum), constitute the *skeleton of the heart* (see fig. 15.9).

Path of Blood Through the Heart

Blood that is low in oxygen and high in carbon dioxide enters the right atrium through the venae cavae and the coronary sinus. As the right atrial wall contracts, the blood passes through the right atrioventricular orifice and enters the chamber of the right ventricle (fig. 15.10).

When the right ventricular wall contracts, the tricuspid valve closes the right atrioventricular orifice, and the blood moves through the pulmonary valve into the pulmonary trunk and its branches (pulmonary arteries). From these vessels, blood enters the capillaries associated with the alveoli (microscopic air sacs) of the lungs. Gas exchange occurs between the blood in the capillaries and the air in the alveoli. The freshly oxygenated blood, which is now relatively low in carbon dioxide, returns to the heart through the pulmonary veins that lead to the left atrium.

The left atrial wall contracts, and the blood moves through the left atrioventricular orifice and into the chamber of the left ventricle. When the left ventricular wall contracts, the mitral valve closes the left atrioventricular orifice, and the blood passes through the aortic valve into the aorta and its branches. Figure 15.11 summarizes the path the blood takes as it moves through the heart to the alveolar capillaries and systemic capillaries, then back to the heart.

Magnetic resonance imaging (MRI) can image coronary arteries. Blood flow appears as a bright signal, and areas of diminished or absent blood flow, or blood turbulence, appear as blank areas. This approach is less invasive than the standard procedure of *coronary angiography,* in which a catheter is snaked through a blood vessel into the heart and a contrast agent is used to show heart structure.

Blood Supply to the Heart

The first two branches of the aorta, called the right and left **coronary arteries,** supply blood to the tissues of the heart. Their openings lie just beyond the aortic valve (fig. 15.12).

One branch of the left coronary artery, the *circumflex artery,* follows the atrioventricular sulcus between the left atrium and the left ventricle. Its branches supply blood to the walls of the left atrium and the left ventricle. Another branch of the left coronary artery, the *anterior interventricular artery* (or *left anterior descending artery*), travels in the anterior interventricular sulcus, and its branches supply the walls of both ventricles.

The right coronary artery passes along the atrioventricular sulcus between the right atrium and the right ventricle. It gives off two major branches—a *posterior*

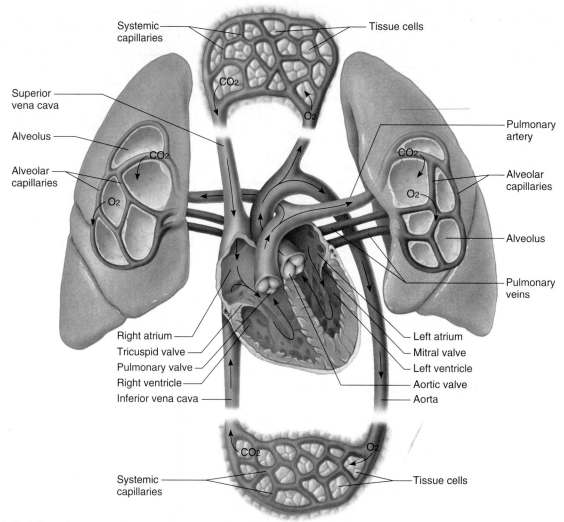

Systemic capillaries

Tissue cells

Superior vena cava

Pulmonary artery

Alveolus

Alveolar capillaries

Alveolar capillaries

CO_2

O_2

Alveolus

Pulmonary veins

Right atrium

Left atrium

Tricuspid valve

Mitral valve

Pulmonary valve

Left ventricle

Right ventricle

Aortic valve

Inferior vena cava

Aorta

CO_2

O_2

Systemic capillaries

Tissue cells

FIGURE 15.1O

The right ventricle forces blood to the lungs, whereas the left ventricle forces blood to all other body parts. (Structures are not drawn to scale.)

interventricular artery, which travels along the posterior interventricular sulcus and supplies the walls of both ventricles, and a *marginal artery,* which passes along the lower border of the heart. Branches of the marginal artery supply the walls of the right atrium and the right ventricle (figs. 15.13 and 15.14).

Because the heart must beat continually to supply blood to the tissues, myocardial cells require a constant supply of freshly oxygenated blood. The myocardium contains many capillaries fed by branches of the coronary arteries. The smaller branches of these arteries usually have connections (anastomoses) between vessels that provide alternate pathways for blood, called collateral circulation.

In most body parts, blood flow in arteries peaks during ventricular contraction. However, blood flow in the vessels of the myocardium is poorest during ventricular contraction. This is because the muscle fibers of the myocardium compress nearby vessels as they contract, interfering with blood flow. Also, the openings into the coronary arteries are partially blocked as the flaps of the

aortic valve open. Conversely, during ventricular relaxation, the myocardial vessels are no longer compressed, and the orifices of the coronary arteries are not blocked by the aortic valve. This increases blood flow into the myocardium.

Blood that has passed through the capillaries of the myocardium is drained by branches of the **cardiac veins,** whose paths roughly parallel those of the coronary arteries. As figure 15.13*b* shows, these veins join the **coronary sinus,** an enlarged vein on the posterior surface of the heart in the atrioventricular sulcus. The coronary sinus empties into the right atrium. Figure 15.15 summarizes the path of blood that supplies the tissues of the heart.

1 Which structures make up the skeleton of the heart?

2 Review the path of blood through the heart.

3 How does blood composition differ in the right and left ventricle?

4 Which vessels supply blood to the myocardium?

5 How does blood return from the cardiac tissues to the right atrium?

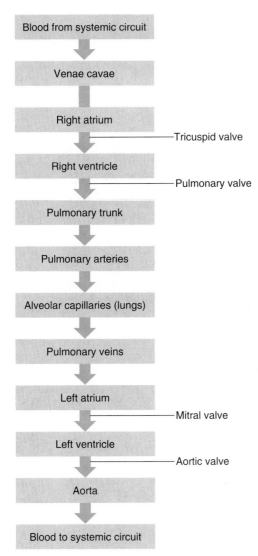

FIGURE 15.11
Path of blood through the heart and pulmonary circuit.

A thrombus or embolus that blocks or narrows a coronary artery branch deprives myocardial cells of oxygen, producing ischemia and painful *angina pectoris*. The pain usually occurs during physical activity, when oxygen requirements exceed supply. Emotional stress may also trigger angina pectoris.

Angina pectoris may cause a heavy pressure, tightening, or squeezing sensation in the chest. The pain is usually felt behind the sternum or in the anterior portion of the upper thorax, but may radiate to the neck, jaw, throat, shoulder, upper limb, back, or upper abdomen. Symptoms include profuse perspiration (diaphoresis), difficulty breathing (dyspnea), nausea, or vomiting.

A blood clot that completely obstructs a coronary artery or one of its branches (coronary thrombosis) kills part of the heart. This is a *myocardial infarction* (MI), more commonly known as a heart attack.

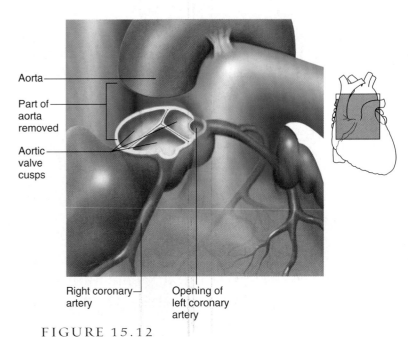

FIGURE 15.12
The openings of the coronary arteries lie just beyond the aortic valve.

Heart Actions

The heart chambers function in coordinated fashion. Their actions are regulated so that atria contract, called atrial **systole** (sis'to-le), while ventricles relax, called ventricular **diastole** (di-as'to-le); then ventricles contract (ventricular systole) while atria relax (atrial diastole). Then the atria and ventricles both relax for a brief interval. This series of events constitutes a complete heartbeat, or **cardiac cycle** (kar'de-ak si'kl).

Cardiac Cycle

During a cardiac cycle, the pressure within the heart chambers rises and falls. Pressure in the ventricles is low early in diastole, and the pressure difference between atria and ventricles causes the A-V valves to open and the ventricles to fill. About 70% of the returning blood enters the ventricles prior to contraction, and ventricular pressure gradually increases. When the atria contract, the remaining 30% of returning blood is pushed into the ventricles, and ventricular pressure increases a bit more. Then, as the ventricles contract, ventricular pressure rises sharply, and as soon as the ventricular pressure exceeds atrial pressure, the A-V valves close. At the same time, the papillary muscles contract, and by pulling on the chordae tendineae, they prevent the cusps of the A-V valves from bulging too far into the atria.

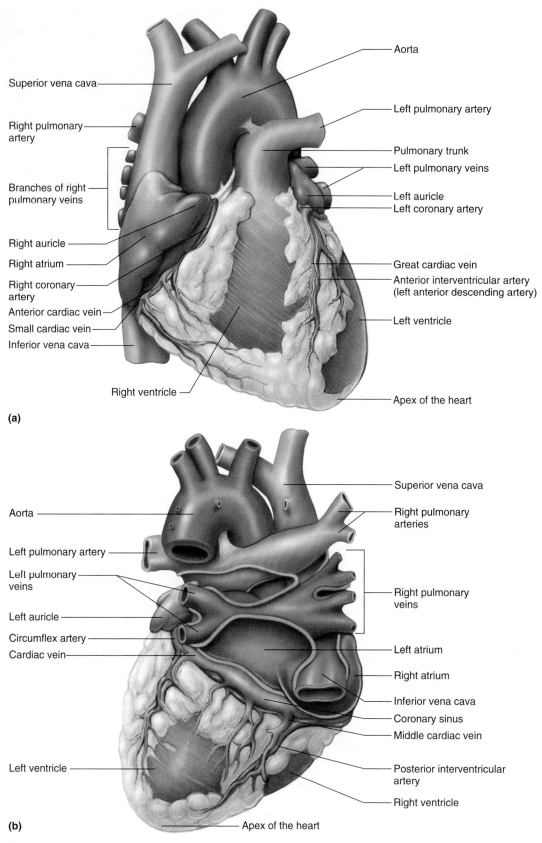

(a)

Superior vena cava

Right pulmonary artery

Branches of right pulmonary veins

Right auricle

Right atrium

Right coronary artery

Anterior cardiac vein

Small cardiac vein

Inferior vena cava

Right ventricle

Aorta

Left pulmonary artery

Pulmonary trunk

Left pulmonary veins

Left auricle

Left coronary artery

Great cardiac vein

Anterior interventricular artery (left anterior descending artery)

Left ventricle

Apex of the heart

(b)

Aorta

Left pulmonary artery

Left pulmonary veins

Left auricle

Circumflex artery

Cardiac vein

Left ventricle

Superior vena cava

Right pulmonary arteries

Right pulmonary veins

Left atrium

Right atrium

Inferior vena cava

Coronary sinus

Middle cardiac vein

Posterior interventricular artery

Right ventricle

Apex of the heart

FIGURE 15.13

Blood vessels associated with the surface of the heart. (*a*) Anterior view. (*b*) Posterior view.

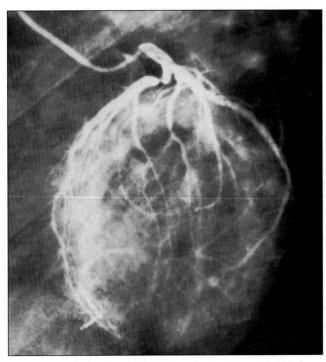

FIGURE 15.14

An angiogram (radiograph) of the coronary arteries is a diagnostic procedure used to examine specific blood vessels.

During ventricular contraction, the A-V valves remain closed. The atria are now relaxed, and pressure in the atria is quite low, even lower than venous pressure. As a result, blood flows into the atria from the large, attached veins. That is, as the ventricles are contracting, the atria are filling, already preparing for the next cardiac cycle (fig. 15.16).

As ventricular systole progresses, ventricular pressure continues to increase until it exceeds the pressure in the pulmonary trunk (right side) and aorta (left side). At this point, the pressure differences across the semilunar valves cause the pulmonary and aortic valves to open, and blood is ejected from each valve's respective ventricle into these arteries.

As blood flows out of the ventricles, ventricular pressure begins to drop, and it drops even further as the ventricles begin to relax. When ventricular pressure is lower than the blood pressure in the aorta and pulmonary trunk, the pressure difference is reversed, and the semilunar valves close. The ventricles continue to relax, and as soon as ventricular pressure is less than atrial pressure, the A-V valves open, and the ventricles begin to fill once more. Atria and ventricles are both relaxed for a brief interval. The graph in figure 15.17 summarizes some of the changes that occur during a cardiac cycle.

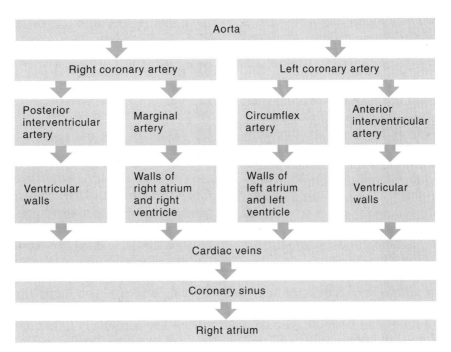

FIGURE 15.15

Path of blood through the coronary circulation.

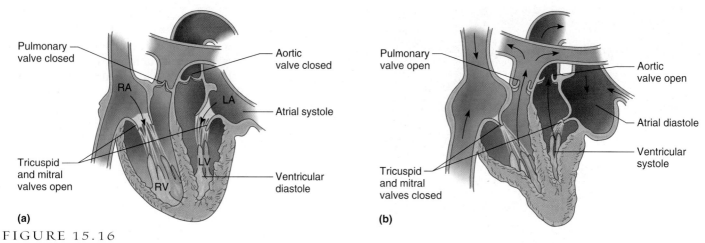

FIGURE 15.16
The atria (*a*) empty during atrial systole and (*b*) fill with blood during atrial diastole.

Heart Sounds

A heartbeat heard through a stethoscope sounds like "lubb-dupp." These sounds are due to vibrations in the heart tissues produced as the blood flow is suddenly speeded or slowed with the contraction and relaxation of the heart chambers, and with the opening and closing of the valves.

The first part of a heart sound (*lubb*) occurs during the ventricular contraction, when the A-V valves are closing. The second part (*dupp*) occurs during ventricular relaxation, when the pulmonary and aortic valves are closing (fig. 15.17).

Sometimes during inspiration, the interval between the closure of the pulmonary and the aortic valves is long enough that a sound coming from each of these events can be heard. In this case, the second heart sound is said to be *split*.

Heart sounds are of particular interest because they can indicate the condition of the heart valves. For example, inflammation of the endocardium (endocarditis) may change the shapes of the valvular cusps. Then, the cusps may close incompletely, and some blood may leak back through the valve. This produces an abnormal sound called a *murmur*. The seriousness of a murmur depends on the extent of valvular damage. Many heart murmurs are harmless. Fortunately for those who have serious problems, it is often possible to repair the damaged valves or to replace them. From Science to Technology 15.1 describes treatments for a failing heart.

Using a stethoscope, it is possible to hear sounds associated with the aortic and pulmonary valves by listening from the second intercostal space on either side of the sternum. The *aortic sound* comes from the right, and the *pulmonic sound* from the left. The sound associated with the mitral valve can be heard from the fifth intercostal space at the nipple line on the left. The sound of the tricuspid valve can be heard at the fifth intercostal space just to the right of the sternum (fig. 15.18).

Cardiac Muscle Fibers

Recall that cardiac muscle fibers function like those of skeletal muscles, but the fibers connect in branching networks (chapter 9, p. 294). Stimulation to any part of the network sends impulses throughout the heart, which contracts as a unit.

A mass of merging cells that act as a unit is called a **functional syncytium** (funk'shun-al sin-sish'e-um). Two such structures are in the heart—in the atrial walls and in the ventricular walls. These masses of cardiac muscle fibers are separated from each other by portions of the heart's fibrous skeleton, except for a small area in the right atrial floor. In this region, the *atrial syncytium* and the *ventricular syncytium* are connected by fibers of the cardiac conduction system.

1 Describe the pressure changes that occur in the atria and ventricles during a cardiac cycle.

2 What causes heart sounds?

3 What is a functional syncytium?

4 Where are the functional syncytia of the heart located?

Cardiac Conduction System

Throughout the heart are clumps and strands of specialized cardiac muscle tissue whose fibers contain only a few myofibrils. Instead of contracting, these areas initiate and distribute impulses (cardiac impulses) throughout the myocardium. They comprise the **cardiac conduction system,** which coordinates the events of the cardiac cycle.

A key portion of this conduction system is the **sinoatrial** (si"no-a'tre-al) **node,** or S-A node, a small, elongated mass of specialized cardiac muscle tissue just beneath the epicardium. It is located in the right atrium near the opening of the superior vena cava, and its fibers are continuous with those of the atrial syncytium.

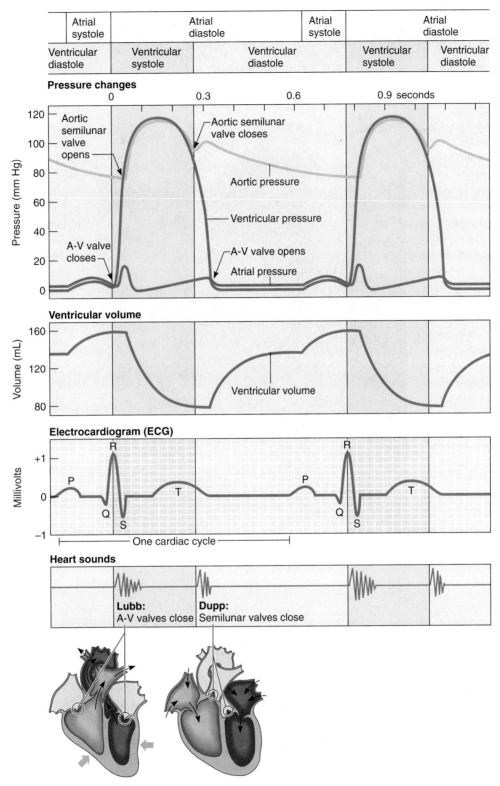

	Atrial systole	Atrial diastole		Atrial systole	Atrial diastole	
Ventricular diastole	Ventricular systole	Ventricular diastole			Ventricular systole	Ventricular diastole

Pressure changes

0   0.3   0.6   0.9 seconds

Pressure (mm Hg)

120
100 — Aortic semilunar valve opens
80
60 — Ventricular pressure
40
20 — A-V valve closes
0

Aortic semilunar valve closes

Aortic pressure

A-V valve opens

Atrial pressure

Ventricular volume

Volume (mL)

160

120

80

Ventricular volume

Electrocardiogram (ECG)

Millivolts

+1

R   P   R   P   T

0

Q   Q

S   S

−1

One cardiac cycle

Heart sounds

Lubb: A-V valves close

Dupp: Semilunar valves close

FIGURE 15.17

A graph of some of the changes that occur in the heart during a cardiac cycle.

The cells of the S-A node can reach threshold on their own, and their membranes contact one another. Without stimulation from nerve fibers or any other outside agents, the nodal cells initiate impulses that spread into the surrounding myocardium and stimulate the cardiac muscle fibers to contract.

S-A node activity is rhythmic. The S-A node initiates one impulse after another, seventy to eighty times

REPLACING THE HEART

When Tina Orbacz was pregnant with her second child, she attributed her increasing fatigue to her pregnant state. But a month after her son's birth, she was even more exhausted. Plus, she had lost weight during her pregnancy. Finally, cardiologists discovered that she was suffering from heart failure due to a birth defect, called an *atrial septal defect,* that weakened the tissue between the atria. Because her heart was failing, and a complication of this disorder is lung failure, Tina needed a heart-lung transplant. After a three year wait, she received the needed organs from a teenager who had died in an accident. Because Tina was young and her cell surface molecules closely matched those of the donated heart and lungs, she has done extraordinarily well. Today, she exercises regularly and leads a normal life.

In a *heart transplant,* the recipient's failing heart is removed, except for the posterior walls of the right and left atria and their connections to the venae cavae and pulmonary veins. The donor heart is similarly prepared and is attached to the atrial cuffs remaining in the recipient's thorax. Finally, the recipient's aorta and pulmonary arteries are connected to those of the donor heart.

Because of the shortage of donor hearts and the severity of the illnesses that lead to heart failure, many people die waiting for a transplant. Fortunately there are alternatives.

A mechanical half-heart, called a *left ventricular assist device* (LVAD), can often maintain cardiac function long enough for a donor heart to become available. A LVAD helped Mike Dorsey, a forty-one-year-old father of six. Although he had to stay in the

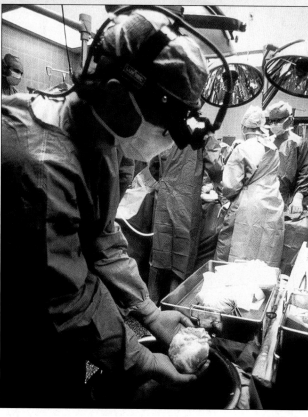

A heart transplant can save a life. A heart that might have died with its donor years ago can provide a new lease on life for a recipient, thanks to our understanding of the immune system—and a well-trained medical team!

hospital to wear his LVAD, the device worked so well that he was able to exercise and help with office work while awaiting his transplant. The LVAD enabled him to increase his physical fitness, which contributed to the success of his eventual heart transplant six months later. A few patients in England too ill to receive transplants are surviving with permanently implanted LVADs.

A newer treatment is an implantable artificial heart, which replaces the ventricles. On July 2, 2001, 50-year-old Bob Tools became the first recipient. Only weeks from death at the time of the seven hour surgery,

the device enabled him to live five more months. The two-pound, titanium and plastic cardiac stand-in consists of an internal motor-driven hydraulic pump, battery and electronics package; and an external battery pack. The electronics component manages the rate and force of the pump's actions, tailoring them to the patient's condition. Several other patients have survived an average of two months with the implantable artificial heart.

In the future, treatment for heart failure may entail coaxing stem cells already there to divide and produce daughter cells that differentiate into exactly what is needed to heal the damaged tissue. Evidence that this happens naturally comes from hearts transplanted from women to men. In one study, at varying times after the transplant, and after the recipients had died (of a variety of causes), researchers detected cells in the donor hearts that had the telltale Y chromosome of males. This meant that the recipient's cells had migrated to and mingled with donor heart cells. Many of the recipient's cells had already specialized into connective tissue, cardiac muscle tissue, and epithelium—precisely what was required to accept the new part. These experiments showed that a recipient's cells infiltrate a transplanted heart and eventually provide new specialized cells. The stem cells may come from the bit of recipient tissue to which the new organ is stitched, or migrate from the bone marrow and then differentiate. Researchers hypothesize that the smaller, healthy female hearts, stressed in their new surroundings in larger, unhealthy bodies, produce growth factors and cell signaling molecules that direct the healing process. ■

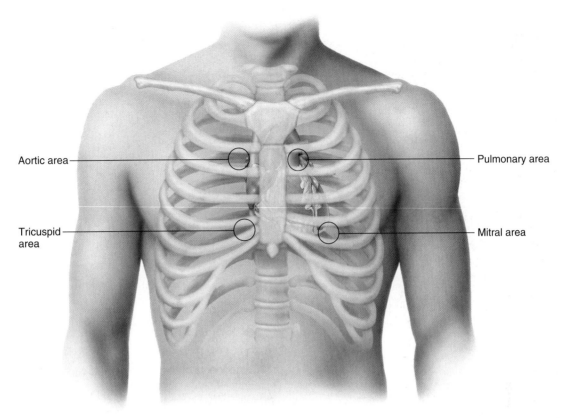

Aortic area

Pulmonary area

Tricuspid area

Mitral area

FIGURE 15.18
Thoracic regions where the sounds of each heart valve are most easily heard.

a minute in an adult. Because it generates the heart's rhythmic contractions, the S-A node is often called the **pacemaker.** From the S-A node, bundles of atrial muscle, called *internodal atrial muscle,* preferentially conduct impulses along tracts to specific regions of the heart.

As a cardiac impulse travels from the S-A node into the atrial syncytium, it goes from cell to cell via gap junctions. The right and left atria contract almost simultaneously. Instead of passing directly into the ventricular syncytium, which is separated from the atrial syncytium by the fibrous skeleton of the heart, the cardiac impulse passes along fibers of the conduction system that are continuous with atrial muscle fibers. These conducting fibers lead to a mass of specialized cardiac muscle tissue called the **atrioventricular node,** or A-V node. This node is located in the inferior portion of the septum that separates the atria (interatrial septum) and just beneath the endocardium. It provides the only normal conduction pathway between the atrial and ventricular syncytia, because the fibrous skeleton does not conduct the impulse.

The fibers that conduct the cardiac impulse into the A-V node (junctional fibers) have very small diameters, and because small fibers conduct impulses slowly, they delay transmission of the impulse. The impulse is delayed further as it moves through the A-V node, allowing time for the atria to contract completely so they empty all their blood into the ventricles prior to ventricular contraction.

Once the cardiac impulse reaches the distal side of the A-V node, it passes into a group of large fibers that make up the **A-V bundle** (atrioventricular bundle or bundle of His), and the impulse moves rapidly through them. The A-V bundle enters the upper part of the interventricular septum and divides into right and left bundle branches that lie just beneath the endocardium. About halfway down the septum, the branches give rise to enlarged **Purkinje** (poor-kin′je) **fibers.** These larger fibers carry the impulse to distant regions of the ventricular myocardium much faster than cell-to-cell conduction could. Thus, the massive ventricular myocardium contracts as a functioning unit.

The base of the aorta, which contains the aortic valves, is enlarged and protrudes somewhat into the interatrial septum close to the A-V bundle. Consequently, inflammatory conditions, such as bacterial endocarditis affecting the aortic valves (aortic valvulitis), may also affect the A-V bundle.

If a portion of the bundle is damaged, it may no longer conduct impulses normally. As a result, cardiac impulses may reach the two ventricles at different times so that they fail to contract together. This condition is called a *bundle branch block.*

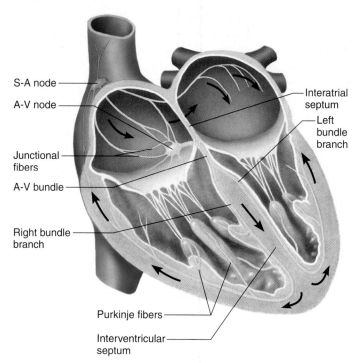

FIGURE 15.19
The cardiac conduction system.

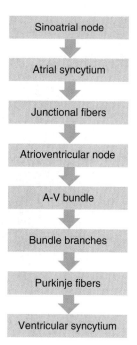

FIGURE 15.20
Components of the cardiac conduction system.

The Purkinje fibers spread from the interventricular septum into the papillary muscles, which project inward from the ventricular walls, and then continue downward to the apex of the heart. There they curve around the tips of the ventricles and pass upward over the lateral walls of these chambers. Along the way, the Purkinje fibers give off many small branches, which become continuous with cardiac muscle fibers. These parts of the conduction system are shown in figure 15.19 and are summarized in figure 15.20.

The muscle fibers in the ventricular walls form irregular whorls. When impulses on the Purkinje fibers stimulate these muscle fibers, the ventricular walls contract with a twisting motion (fig. 15.21). This action squeezes blood out of the ventricular chambers and forces it into the aorta and pulmonary trunk.

Another property of the conduction system is that the Purkinje fibers transmit the impulse to the apex of the heart first. As a result, contraction begins at the apex and pushes the blood superiorly toward the aortic and pulmonary semilunar valves, rather than having the impulse begin superiorly and push blood toward the apex, as it would if the impulse traveled from cell to cell.

1 What is the function of the cardiac conduction system?

2 What kinds of tissues make up the cardiac conduction system?

3 How is a cardiac impulse initiated?

4 How is a cardiac impulse transmitted from the right atrium to the other heart chambers?

A significant percentage of cases of heart failure in adults of African descent may be due to an inherited condition called familial amyloidosis. A protein called amyloid forms deposits in the heart, causing angina (chest pain), failure of cardiac muscle function (cardiomyopathy), blockage of conduction of electrical impulses, and disturbed heart rhythm (arrhythmia). Echocardiography can detect the amyloid deposits that thicken the ventricular walls. It is important to distinguish amyloidosis from other forms of arrhythmias, because drug treatments are different.

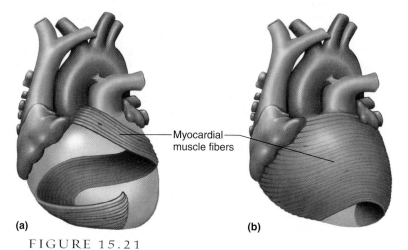

(a)   (b)

FIGURE 15.21
The muscle fibers within the ventricular walls are arranged in patterns of whorls. The fibers of groups (a) and (b) surround both ventricles in these anterior views of the heart.

Electrocardiogram

An **electrocardiogram** (e-lek"tro-kar'de-o-gram") (ECG) is a recording of the electrical changes that occur in the myocardium during a cardiac cycle. (This pattern occurs as action potentials stimulate cardiac muscle fibers to con-

tract, but it is not the same as individual action potentials.) Because body fluids can conduct electrical currents, such changes can be detected on the surface of the body.

To record an ECG, electrodes are placed on the skin and connected by wires to an instrument that responds to very weak electrical changes by moving a pen or stylus on a moving strip of paper. Up-and-down movements of the pen correspond to electrical changes in the myocardium. Because the paper moves past the pen at a known rate, the distance between pen deflections indicates time elapsing between phases of the cardiac cycle.

As figure 15.22*a* illustrates, a normal ECG pattern includes several deflections, or *waves,* during each cardiac cycle. Between cycles, the muscle fibers remain polarized, with no detectable electrical changes. Consequently, the

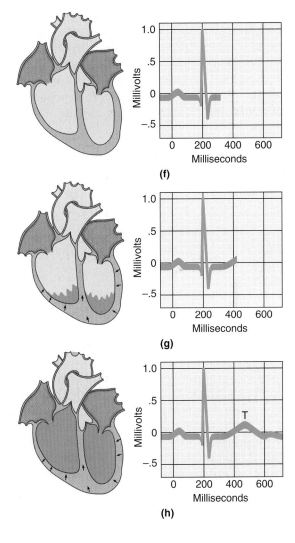

FIGURE 15.22

ECG pattern. (*a*) A normal ECG. In this set of drawings (*b–h*), the yellow areas of the hearts indicate where depolarization is occurring, and the green areas indicate where tissues are repolarizing; the portion of the ECG pattern produced at each step is shown by the continuation of the line on the graph paper.

pen does not move and simply marks along the baseline. When the S-A node triggers a cardiac impulse, the atrial fibers depolarize, producing an electrical change. The pen moves, and at the end of the electrical change, returns to the base position. This first pen movement produces a *P wave,* corresponding to depolarization of the atrial fibers that will lead to contraction of the atria (fig. 15.22*b–d*).

When the cardiac impulse reaches the ventricular fibers, they rapidly depolarize. Because the ventricular walls are thicker than those of the atria, the electrical change is greater, and the pen deflects more. When the electrical change ends, the pen returns to the baseline, leaving a mark called the *QRS complex,* which usually consists of a *Q wave,* an *R wave,* and an *S wave.* This complex appears due to depolarization of the ventricular fibers just prior to the contraction of the ventricular walls (fig. 15.22*e* and *f*).

The electrical changes occurring as the ventricular muscle fibers repolarize slowly produce a *T wave* as the pen deflects again, ending the ECG pattern (fig. 15.22*g* and *h*). The record of the atrial repolarization seems to be missing from the pattern because the atrial fibers repolarize at the same time that the ventricular fibers depolarize. Thus, the QRS complex obscures the recording of the atrial repolarization.

> In addition to the waves that comprise the classic electrocardiogram are repeating subpatterns of other waves that occur at different timescales and in an irregular pattern. Although it may seem counterintuitive, this complex, varying backdrop to the cardiac cycle seems to be necessary for health. It is disrupted in congestive heart failure.

Physicians use ECG patterns to assess the heart's ability to conduct impulses. For example, the time period between the beginning of a P wave and the beginning of a QRS complex (*P-Q interval,* or if the initial portion of the QRS wave is upright, *P-R interval*) indicates the time for the cardiac impulse to travel from the S-A node through the A-V node. Ischemia or other problems affecting the fibers of the A-V conduction pathways can increase this P-Q interval. Similarly, injury to the A-V bundle can extend the QRS complex, because it may take longer for an impulse to spread throughout the ventricular walls (fig. 15.23).

1 What is an electrocardiogram?

2 Which cardiac events do the P wave, QRS complex, and T wave represent?

Regulation of the Cardiac Cycle

The volume of blood pumped changes to accommodate cellular requirements. For example, during strenuous

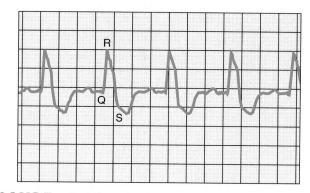

FIGURE 15.23
A prolonged QRS complex may result from damage to the A-V bundle fibers.

exercise, skeletal muscles require more blood, and heart rate increases in response. Since the S-A node normally controls heart rate, changes in this rate often involve factors that affect the pacemaker, such as the motor impulses carried on the parasympathetic and sympathetic nerve fibers (see figs. 11.38, 11.39, 15.24, 15.38, and 15.39).

The parasympathetic fibers that innervate the heart arise from neurons in the medulla oblongata and make up parts of the *vagus nerves.* Most of these fibers branch to the S-A and A-V nodes. When the nerve impulses reach nerve fiber endings, they secrete acetylcholine, which decreases S-A and A-V nodal activity. As a result, heart rate decreases.

The vagus nerves continually carry impulses to the S-A and A-V nodes, braking heart action. Consequently, parasympathetic activity can change heart rate in either direction. An increase in the impulses slows the heart, and a decrease in the impulses releases the parasympathetic "brake" and increases heart rate.

Sympathetic fibers reach the heart by means of the *accelerator nerves,* whose branches join the S-A and A-V nodes as well as other areas of the atrial and ventricular myocardium. The endings of these fibers secrete norepinephrine in response to nerve impulses, which increases the rate and force of myocardial contractions.

The *cardiac control center* of the medulla oblongata maintains balance between the inhibitory effects of the parasympathetic fibers and the excitatory effects of the sympathetic fibers. In this region of the brain, masses of neurons function as *cardioinhibitor* and *cardioaccelerator reflex centers.* These centers receive sensory impulses from throughout the cardiovascular system and relay motor impulses to the heart in response. For example, receptors that are sensitive to stretch are located in certain regions of the aorta (aortic arch) and in the carotid arteries (carotid sinuses). These receptors, called *baroreceptors* (pressoreceptors), can detect changes in blood pressure. Rising pressure stretches the receptors, and they signal the cardioinhibitor center in the medulla. In response, the medulla sends parasympathetic motor impulses to the

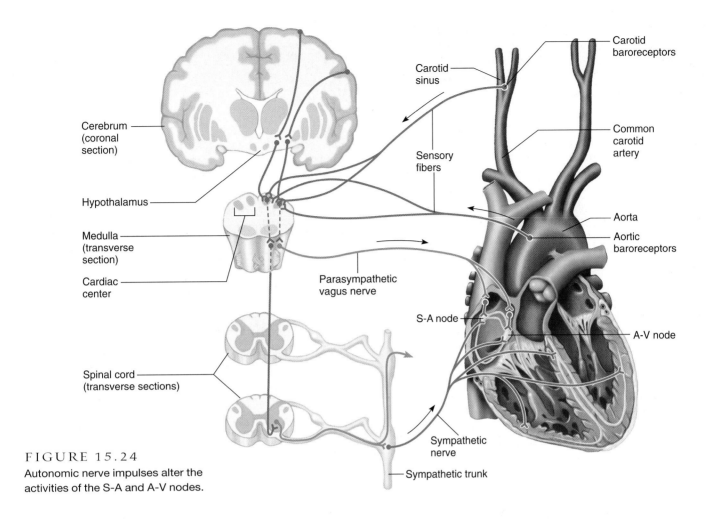

FIGURE 15.24
Autonomic nerve impulses alter the activities of the S-A and A-V nodes.

heart, decreasing the heart rate. This action helps lower blood pressure toward normal (fig. 15.24).

Another regulatory reflex involves stretch receptors in the venae cavae near the entrances to the right atrium. If venous blood pressure increases abnormally in these vessels, the receptors signal the cardioaccelerator center, and sympathetic impulses flow to the heart. As a result, heart rate and force of contraction increase, and the venous pressure is reduced.

Impulses from the cerebrum or hypothalamus also influence the cardiac control center. Such impulses may decrease heart rate, as occurs when a person faints following an emotional upset, or they may increase heart rate during a period of anxiety.

Two other factors that influence heart rate are temperature change and certain ions. Rising body temperature increases heart action, which is why heart rate usually increases during fever. On the other hand, abnormally low body temperature decreases heart action.

Of the ions that influence heart action, the most important are potassium (K^+) and calcium (Ca^{+2}). Potassium affects the electrical potential of the cell membrane, altering its ability to reach the threshold for conducting an impulse (see chapter 10, p. 350). The sarcoplasmic reticula of cardiac muscle fibers have less calcium than do the sarcoplasmic reticula of skeletal muscle fibers. Therefore, cardiac muscle depends more on extracellular (blood-borne) calcium. Although homeostatic mechanisms normally maintain the concentrations of these ions within narrow ranges, these mechanisms sometimes fail, and the consequences can be serious or even fatal. Clinical Application 15.1 examines abnormal heart rhythms.

Excess potassium ions (hyperkalemia) alters the usual polarized state of the cardiac muscle fibers, decreasing the rate and force of contractions. Very high potassium ion concentration may block conduction of cardiac impulses, and heart action may suddenly stop (cardiac arrest). Conversely, if the potassium concentration drops below normal (hypokalemia), the heart may develop a potentially life-threatening abnormal rhythm (arrhythmia).

Excess calcium ions (hypercalcemia) increase heart action, introducing danger that the heart will undergo a prolonged contraction. Conversely, low calcium ion concentration (hypocalcemia) depresses heart action because these ions help initiate muscle contraction.

ARRHYTHMIAS

Each year, thousands of people die from a fast or irregular heartbeat. These are types of altered heart rhythm called *arrhythmia.*

In *fibrillation,* small areas of the myocardium contract in an uncoordinated, chaotic fashion (fig. 15A). As a result, the myocardium fails to contract as a whole, and blood is no longer pumped. Atrial fibrillation is not life threatening because the ventricles still pump blood, but ventricular fibrillation is often deadly. Ventricular fibrillation can be caused by an obstructed coronary artery, toxic drug exposure, electric shock, or traumatic injury to the heart or chest wall. A defibrillator device can deliver a shock to restore a normal heartbeat, as the chapter opening vignette describes.

An abnormally fast heartbeat, usually more than 100 beats per minute, is called *tachycardia.* Increase in body temperature, nodal stimulation by sympathetic fibers, certain drugs or hormones, heart disease, excitement, exercise, anemia, or shock can all cause tachycardia. Figure 15B shows the ECG of a tachycardic heart.

Bradycardia means a slow heart rate, usually fewer than sixty beats per minute. Decreased body temperature, nodal stimulation by parasympathetic impulses, or certain drugs may cause bradycardia. It also may occur during sleep. Figure 15C shows the ECG of a bradycardic heart. Athletes sometimes have unusually slow heartbeats

because their hearts pump a greater-than-normal volume of blood with each beat. The slowest heartbeat recorded in a healthy athlete was twenty-five beats per minute!

A *premature beat* occurs before it is expected in a normal series of cardiac cycles. Cardiac impulses originating from unusual (ectopic) regions of the heart prob-

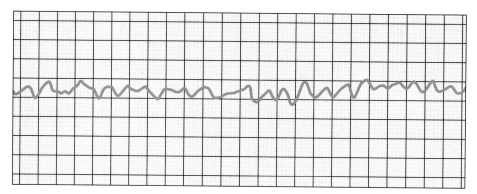

FIGURE 15A
Ventricular fibrillation is rapid, uncoordinated depolarization of the ventricles.

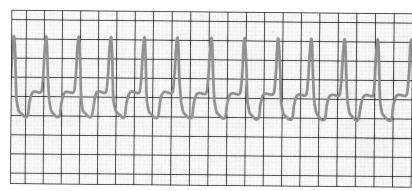

FIGURE 15B
Tachycardia is a rapid heartbeat.

---

1 Which nerves supply parasympathetic fibers to the heart? Which nerves supply sympathetic fibers?

2 How do parasympathetic and sympathetic impulses help control heart rate?

3 How do changes in body temperature affect heart rate?

Blood Vessels

The blood vessels are organs of the cardiovascular system, and they form a closed circuit of tubes that carries blood from the heart to cells and back again. These vessels include arteries, arterioles, capillaries, venules, and veins. The arteries and arterioles conduct blood away from the ventricles of the heart and lead to the capillaries. The capillaries are sites of exchange of substances between blood and the body cells, and the venules and veins return blood from the capillaries to the atria. From Science to Technology 15.2 describes angiogenesis, the formation of new blood vessels in the body.

Arteries and Arterioles

Arteries (ar-te′rēz) are strong, elastic vessels that are adapted for carrying the blood away from the heart under high pressure. These vessels subdivide into progressively

ably cause a premature beat. That is, the impulse originates from a site other than the S-A node. Cardiac impulses may arise from ischemic tissues or from muscle fibers irritated by disease or drugs.

A heart chamber *flutters* when it contracts regularly, but very rapidly, such as 250–350 times per minute. Although normal hearts may flutter occasionally, this condition is more likely to be due to damage to the myocardium (fig. 15D).

Any interference or block in cardiac impulse conduction may cause arrhythmia, the type varying with the location and extent of the block. Such arrhythmias arise because certain cardiac tissues other than the S-A node can function as pacemakers.

The S-A node usually initiates 70 to 80 heartbeats per minute, called a sinus rhythm. If the S-A node is damaged, impulses originating in the A-V node may travel upward into the atrial myocardium and downward into the ventricular walls, stimulating them to contract. Under the influence of the A-V node acting as a *secondary pacemaker,* the heart may continue to pump blood, but at a rate of forty to sixty beats per minute, called a nodal rhythm. Similarly, the Purkinje fibers can initiate cardiac impulses, contracting the heart fifteen to forty times per minute.

An *artificial pacemaker* can treat a disorder of the cardiac conduction system. This device includes an electrical pulse generator and a lead wire that communicates with a portion of the myocardium. The pulse generator contains a permanent battery that provides energy and a microprocessor that can sense the cardiac rhythm and signal the heart to alter its contraction rate.

An artificial pacemaker is surgically implanted beneath the patient's skin in the shoulder. An external programmer adjusts its functions from the outside. The first pacemakers, made in 1958, were crude. Today, thanks to telecommunications advances, a physician can check a patient's pacemaker over the phone! A device called a pacemaker-cardioverter-defibrillator can correct both abnormal heart rhythm and cardiac arrest. ■

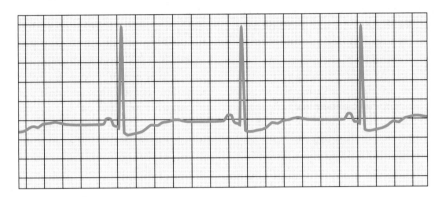

FIGURE 15C
Bradycardia is a slow heartbeat.

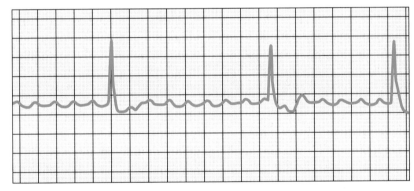

FIGURE 15D
Atrial flutter is an abnormally rapid rate of atrial depolarization.

thinner tubes and eventually give rise to the finer branched **arterioles** (ar-te′re-olz).

The wall of an artery consists of three distinct layers, or *tunics,* shown in figure 15.25a. The innermost layer, tunica interna (intima), is composed of a layer of simple squamous epithelium, called *endothelium,* that rests on a connective tissue membrane that is rich in elastic and collagenous fibers.

The endothelial lining of an artery provides a smooth surface that allows blood cells and platelets to flow through without being damaged. Additionally, endothelium helps prevent blood clotting by secreting biochemicals that inhibit platelet aggregation (see chapter 14, p. 530). Endothelium also may help regulate local blood flow by secreting substances that either dilate or constrict blood vessels.

The middle layer, tunica media, makes up the bulk of the arterial wall. It includes smooth muscle fibers, which encircle the tube, and a thick layer of elastic connective tissue. The connective tissue gives the vessel a tough elasticity that enables it to withstand the force of blood pressure and, at the same time, to stretch and accommodate the sudden increase in blood volume that accompanies ventricular contraction.

The outer layer, tunica externa (adventitia), is thin and chiefly consists of connective tissue with irregularly

CHAPTER FIFTEEN   *Cardiovascular System*

ALTERING ANGIOGENESIS

Angiogenesis is the formation of new blood vessels. Under the influence of specific growth factors, endothelial cells divide and assemble into the tubules that form capillaries as well as the innermost linings of larger blood vessels. In normal development, angiogenesis is crucial to build a blood supply to serve a growing body. New blood vessels are needed to deliver nutrients, hormones, and growth factors to tissues and to remove wastes. Angiogenesis is also essential for healing. After a heart attack, for example, new vessels form in the remaining healthy cardiac muscle to supply blood.

As with most biological processes, angiogenesis must be highly controlled. Excess, deficient, or inappropriate angiogenesis can cause, or worsen, a variety of illnesses. By understanding how angiogenesis proceeds, medical researchers are developing ways to direct new blood vessel formation, with two specific applications in mind—healing hearts and starving cancerous tumors.

Heart Attacks: Promoting Angiogenesis

An errant clot blocks a coronary artery. Within seconds, the localized lack of oxygen stimulates muscle cells to release hypoxia-inducible factor (HIF-1). This is a transcription factor, a protein that activates several genes. Activated HIF-1 restores homeostasis by stimulating glycolysis (anaerobic respiration); signalling the kidneys to produce erythropoietin, which boosts the red blood cell supply; and triggering angiogenesis by turning on production of vascular endothelial growth factor (VEGF). The growth factor stimulates certain cells to proliferate and aggregate to form capillaries, which, eventually, restore some blood flow to the blocked cardiac muscle. Fibroblast growth factor also assists in angiogenesis.

When natural angiogenesis isn't sufficient, part of the heart dies. Coronary bypass surgery and angioplasty are treatments that restore blood flow, but for patients who cannot undergo these procedures or whose blockages are in vessels too narrow or difficult to reach, harnessing and targeting angiogenesis may help to save starved heart parts. One approach is to package growth factors in time-release capsules that are implanted near small vessels while large ones are being surgically bypassed. In one clinical trial, this technique increased blood flow to the area and halted chest pain. Another strategy is gene therapy, which delivers the genes that encode the growth factors to oxygen-starved areas of the heart.

Cancer Treatment: Preventing Angiogenesis

A tumor surrounds itself with blood vessels. Once it reaches the size of a pinhead, a tumor secretes growth factors that stimulate nearby capillaries to sprout new branches that extend toward it. Endothelial cells within the tumor assemble into sheets, roll into tubules, and, eventually, snake out of the tumor as new capillaries. Other cancer cells wrap around the capillaries, spreading out on this scaffolding into nearby tissues. Some cancer cells enter blood vessels and travel to other parts of the body. For a time, maybe even years, these secondary tumors stay small, adhering to the outsides of the blood vessels that delivered them. But when the primary tumor is removed, angiogenesis-promoting growth factors wash over the tumors, and they grow.

In the 1970s, researchers began to study the antiangiogenesis factors that keep secondary tumors small, to develop them as cancer treatments. Two candidates that have had spectacular results in mice are angiostatin and endostatin. So far clinical trials in humans have shown endostatin to be safe even at high doses; trials for efficacy have not yet been completed. (Angiostatin is derived from plasminogen and endostatin from collagen.) Unlike other cancer treatments, angiostatin and endostatin can be used over and over without the targeted tissue becoming resistant, as is the case with conventional drugs. In addition to these two drug candidates, at least two dozen others are in clinical trials, including an old drug thalidomide; alpha interferon, an immune system biochemical; and antibodies that disable the VEGF that is vital for new blood vessel formation. By combining different antiangiogenesis factors, it may be possible to make tumors disappear entirely—as happens experimentally in mice. ∎

---

organized elastic and collagenous fibers. This layer attaches the artery to the surrounding tissues. It also contains minute vessels (vasa vasorum) that give rise to capillaries and provide blood to the more external cells of the artery wall. (The opening vignette for chapter 5 includes a photo of a semisynthetic blood vessel substitute.)

The sympathetic branches of the autonomic nervous system innervate smooth muscle in artery and arteriole walls. *Vasomotor fibers* stimulate the smooth muscle cells to contract, reducing the diameter of the vessel. This is called **vasoconstriction** (vas″o-kon-strik′-shun). If vasomotor impulses are inhibited, the muscle fibers relax, and the diameter of the vessel increases. This is called **vasodilation** (vas″o-di-la′shun). Changes in the diameters of arteries and arterioles greatly influence blood flow and pressure.

Although the walls of the larger arterioles have three layers similar to those of arteries, the middle and outer layers thin as the arterioles approach the capillaries. The wall of a very small arteriole consists only of an endothe-

Artery

Vein

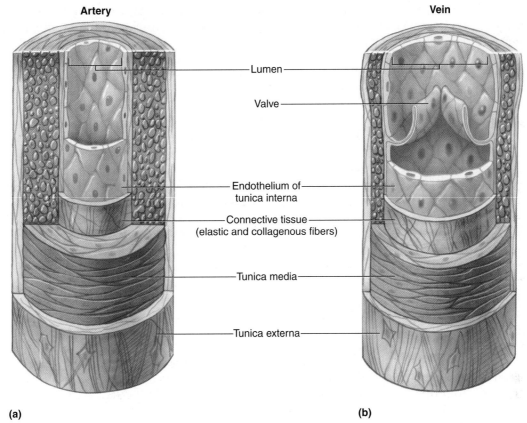

Lumen

Valve

Endothelium of
tunica interna

Connective tissue
(elastic and collagenous fibers)

Tunica media

Tunica externa

(a)

(b)

FIGURE 15.25

Blood vessels. (*a*) The wall of an artery. (*b*) The wall of a vein.

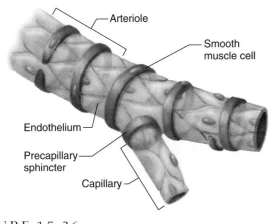

Arteriole

Smooth
muscle cell

Endothelium

Precapillary
sphincter

Capillary

FIGURE 15.26

The smallest arterioles have only a few smooth muscle fibers in
their walls. Capillaries lack these fibers.

lial lining and some smooth muscle fibers, surrounded by
a small amount of connective tissue (figs. 15.26 and
15.27). Arterioles, which are microscopic continuations
of arteries, give off branches called *metarterioles* that, in
turn, join capillaries.

The arteriole and metarteriole walls are adapted for
vasoconstriction and vasodilation in that their muscle
fibers respond to impulses from the autonomic nervous

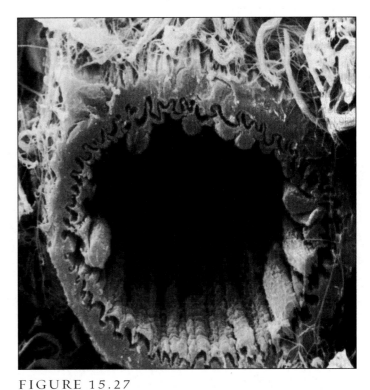

FIGURE 15.27

Scanning electron micrograph of an arteriole cross section (2,200×).
© R.G. Kessel and R.H. Kardon. *Tissues and Organs: A Text-Atlas
of Scanning Electron Microscopy,* 1979.

system by contracting or relaxing. Thus, these vessels help control the flow of blood into the capillaries.

Sometimes metarterioles connect directly to venules, and blood entering them can bypass the capillaries. These connections between arteriole and venous pathways, shown in figure 15.28, are called *arteriovenous shunts.*

1 Describe the wall of an artery.

2 What is the function of the smooth muscle in the arterial wall?

3 How is the structure of an arteriole different from that of an artery?

Capillaries

Capillaries (kap'i-lar″ez) are the smallest diameter blood vessels. They connect the smallest arterioles and the smallest venules. Capillaries are extensions of the inner linings of arterioles in that their walls are endothelium—a single layer of squamous epithelial cells (fig. 15.29a). These thin walls form the semipermeable layer through which substances in the blood are exchanged for substances in the tissue fluid surrounding body cells.

Capillary Permeability

The openings or intercellular channels in the capillary walls are thin slits where endothelial cells overlap. The sizes of these openings, and consequently the permeability of the capillary wall, vary from tissue to tissue. For example, the openings are relatively small in the capillaries of smooth, skeletal, and cardiac muscle, whereas those in capillaries associated with endocrine glands, the kidneys, and the lining of the small intestine are larger.

Capillaries with the largest openings include those of the liver, spleen, and red bone marrow. These capillaries are discontinuous, and the distance between their cells appears as little cavities (sinusoids) in the organ. Discontinuous capillaries allow large proteins and even intact cells to pass through as they enter or leave the circulation (fig. 15.29b and c). Clinical Application 3.2 discusses the blood-brain barrier, the protective tight capillaries in the brain. The barrier is not present in the pituitary and pineal glands and parts of the hypothalamus.

Capillary Arrangement

The higher a tissue's rate of metabolism, the denser its capillary networks. Muscle and nerve tissues, which use abundant oxygen and nutrients, are richly supplied with capillaries; cartilaginous tissues, the epidermis, and the cornea, where metabolism is slow, lack capillaries.

> If the capillaries of an adult were unwound and spread end to end, they would cover from 25,000 to 60,000 miles.

The patterns of capillary arrangement also differ in various body parts. For example, some capillaries pass directly from arterioles to venules, but others lead to

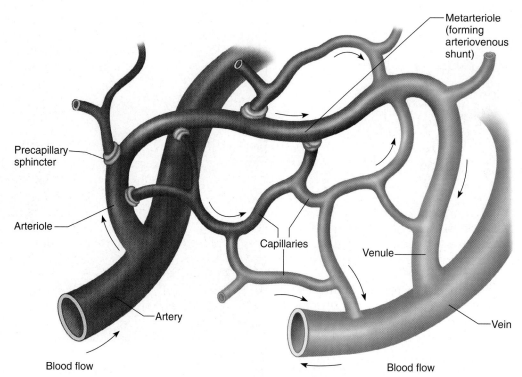

Metarteriole (forming arteriovenous shunt)

Precapillary sphincter

Arteriole

Capillaries

Venule

Artery

Vein

Blood flow

Blood flow

FIGURE 15.28

Some metarterioles provide arteriovenous shunts by connecting arterioles directly to venules.

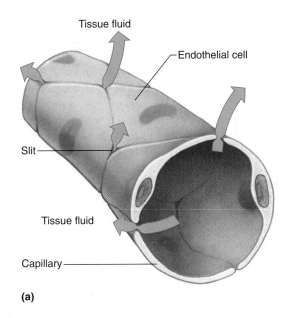

(a)

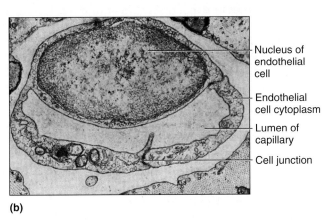

Nucleus of
endothelial
cell

Endothelial
cell cytoplasm

Lumen of
capillary

Cell junction

(b)

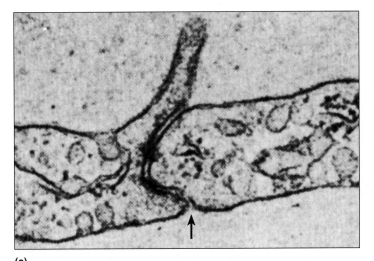

(c)

FIGURE 15.29

Capillary structure. (a) Substances are exchanged between the blood and tissue fluid through openings (slits) separating endothelial cells. (b) Transmission electron micrograph of a capillary cross section (11,500×). (c) Note the narrow slitlike openings at the cell junctions (arrow) (micrograph b enlarged to 62,500×).

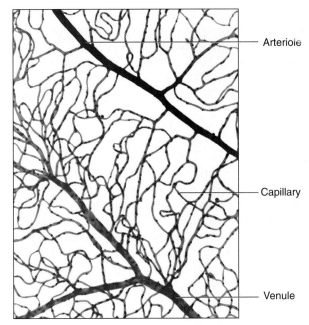

Arteriole

Capillary

Venule

FIGURE 15.30

Light micrograph of a capillary network (100×).

highly branched networks (fig. 15.30). Such physical arrangements make it possible for the blood to follow different pathways through a tissue that are attuned to cellular requirements.

Blood flow can vary among tissues as well. During exercise, for example, blood is directed into the capillary networks of the skeletal muscles, where the cells require more oxygen and nutrients. At the same time, the blood bypasses some of the capillary networks in the tissues of the digestive tract, where demand for blood is less critical. Conversely, when a person is relaxing after a meal, blood can be shunted from the inactive skeletal muscles into the capillary networks of the digestive organs.

Regulation of Capillary Blood Flow

The distribution of blood in the various capillary pathways is mainly regulated by the smooth muscles that encircle the capillary entrances. As figure 15.28 shows, these muscles form *precapillary sphincters,* which may close a capillary by contracting or open it by relaxing. The precapillary sphincters respond to the needs of the cells. When the cells have low concentrations of oxygen and nutrients, the precapillary sphincters relax, and blood flow increases; when cellular requirements have been met, the precapillary sphincters may contract again.

1 Describe a capillary wall.

2 What is the function of a capillary?

3 What controls blood flow into capillaries?

Exchanges in the Capillaries

The vital function of exchanging gases, nutrients, and metabolic by-products between the blood and the tissue fluid surrounding body cells occurs in the capillaries. The biochemicals exchanged move through the capillary walls by diffusion, filtration, and osmosis.

 R E C O N N E C T   T O   C H A P T E R 3 ,   M O V E M E N T S   I N T O   A N D O U T   O F   T H E   C E L L ,   P A G E S 8 0 – 8 4 .

Diffusion is the most important means of transfer. Because blood entering certain capillaries carries high concentrations of oxygen and nutrients, these substances diffuse through the capillary walls and enter the tissue fluid. Conversely, the concentrations of carbon dioxide and other wastes are generally greater in the tissues, and such wastes tend to diffuse into the capillary blood.

The paths these substances follow depend primarily on their solubilities in lipids. Substances that are soluble in lipid, such as oxygen, carbon dioxide, and fatty acids, can diffuse through most areas of the cell membranes that make up the capillary wall because the membranes are largely lipid. Lipid-insoluble substances, such as water, sodium ions, and chloride ions, diffuse through pores in the cell membranes and through the slitlike openings between the endothelial cells that form the capillary wall (see fig. 15.29). Plasma proteins generally remain in the blood because they are not soluble in the lipid portions of the endothelial cell membranes, and they are too large to diffuse through the membrane pores or slitlike openings between the endothelial cells of most capillaries.

In *filtration, hydrostatic pressure* forces molecules through a membrane. In the capillaries, the blood pres-sure generated when ventricle walls contract provides the force for filtration.

Blood pressure also moves blood through the arter-ies and arterioles. This pressure decreases as the distance from the heart increases because of friction (peripheral resistance) between the blood and the vessel walls. For this reason, blood pressure is greater in the arteries than in the arterioles and greater in the arterioles than in the capillaries. It is similarly greater at the arteriolar end of a capillary than at the venular end.

The walls of arteries and arterioles are too thick to allow blood components to pass through. However, the hydrostatic pressure of the blood pushes small molecules through capillary walls by filtration. This effect primarily occurs at the arteriolar ends of capillaries, whereas diffu-sion takes place along their entire lengths.

The presence of an impermeable solute on one side of a cell membrane creates an osmotic pressure. Because plasma proteins are trapped within the capillaries, they cre-ate an osmotic pressure that draws water into the capillaries. The term *colloid osmotic pressure* is often used to describe this osmotic effect due solely to the plasma proteins.

The effect of capillary blood pressure, favoring fil-tration, and the plasma colloid osmotic pressure, favor-ing reabsorption, have opposite actions. At the arteriolar end of capillaries, the blood pressure is higher (41.3 mm Hg outward) than the colloid osmotic pressure (28 mm Hg inward), so at the arteriolar end of the capillary, filtra-tion predominates. At the venular end, the colloid osmotic pressure is essentially unchanged (28 mm Hg inward), but the blood pressure has decreased due to resistance through the capillary (21.3 mm Hg outward). Thus, at the venular end, reabsorption predominates (fig. 15.31). (The interstitial fluid also has hydrostatic pres-

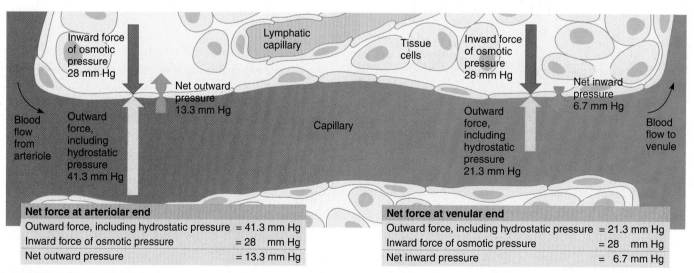

Net force at arteriolar end	
Outward force, including hydrostatic pressure	= 41.3 mm Hg
Inward force of osmotic pressure	= 28 mm Hg
Net outward pressure	= 13.3 mm Hg

Net force at venular end	
Outward force, including hydrostatic pressure	= 21.3 mm Hg
Inward force of osmotic pressure	= 28 mm Hg
Net inward pressure	= 6.7 mm Hg

FIGURE 15.31

Water and other substances leave capillaries because of a net outward pressure at the capillaries' arteriolar ends. Water enters at the capillar-ies' venular ends because of a net inward pressure. Substances move in and out along the length of the capillaries according to their respec-tive concentration gradients.

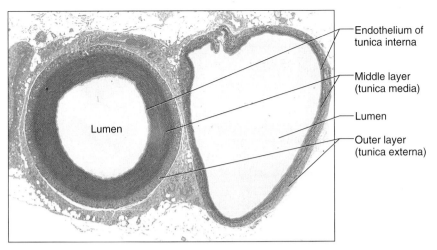

FIGURE 15.32

Note the structural differences in these cross sections of an artery and a vein (90×). Micrograph is falsely colored.

Labels: Endothelium of tunica interna; Middle layer (tunica media); Lumen; Outer layer (tunica externa); Lumen

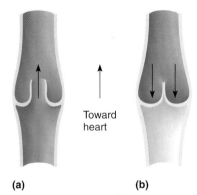

FIGURE 15.33

Venous valves. (a) allow blood to move toward the heart, but (b) prevent blood from moving backward away from the heart.

(a) (b)

Toward heart

sure and osmotic pressure, but the values are quite low and tend to cancel each other out; as such, they can be omitted from this discussion.)

Normally, more fluid leaves the capillaries than returns to them. *Lymphatic vessels* collect the excess fluid and return it to the venous circulation. This mechanism is discussed in chapter 16 (p. 611).

Sometimes unusual events increase blood flow to capillaries, and excess fluid enters spaces between tissue cells (interstitial spaces). This may occur, for instance, in response to certain chemicals such as *histamine* that vasodilate the metarterioles and increase capillary permeability. Enough fluid may leak out of the capillaries to overwhelm lymphatic drainage, and affected tissues become swollen (edematous) and painful.

1 What forces are responsible for the exchange of substances between blood and the tissue fluid?

2 Why is the fluid movement out of a capillary greater at its arteriolar end than at its venular end?

3 Since more fluid leaves the capillary than returns to it, how is the remainder returned to the vascular system?

If the right ventricle of the heart is unable to pump blood out as rapidly as it enters, other parts of the body may develop edema because the blood backs up into the veins, venules, and capillaries, increasing blood pressure in these vessels. As a result of this increased *back pressure,* osmotic pressure of the blood in the venular ends of the capillaries is less effective in attracting water from tissue fluid, and the tissues swell. This is true particularly in the lower extremities if the person is upright, or in the back if the person is supine. In the terminal stages of heart failure, edema is widespread, and fluid accumulates in the peritoneal cavity of the abdomen. This condition is called *ascites.*

Venules and Veins

Venules (ven'ūlz) are the microscopic vessels that continue from the capillaries and merge to form **veins** (vānz). The veins, which carry blood back to the atria, follow pathways that roughly parallel those of the arteries.

The walls of veins are similar to those of arteries in that they are composed of three distinct layers. However, the middle layer of the venous wall is poorly developed. Consequently, veins have thinner walls that contain less smooth muscle and less elastic tissue than those of comparable arteries, but their lumens have a greater diameter (figs. 15.25 and 15.32).

Many veins, particularly those in the upper and lower limbs, contain flaplike *valves* (called semilunar valves), which project inward from their linings. Valves, shown in figure 15.33, are usually composed of two leaflets that are pushed closed if the blood begins to back up in a vein. These valves aid in returning blood to the heart because they are open as long as the flow is toward the heart but close if it is in the opposite direction.

Veins also function as *blood reservoirs,* useful in times of blood loss. For example, in hemorrhage accompanied by a drop in arterial blood pressure, sympathetic nerve impulses reflexly stimulate the muscular walls of the veins. The resulting venous constrictions help maintain blood pressure by returning more blood to the heart. This mechanism ensures a nearly normal blood flow even when as much as 25% of the blood volume is lost. Figure 15.34 illustrates the relative volumes of blood in the veins and other blood vessels.

Table 15.3 summarizes the characteristics of blood vessels. Clinical Application 15.2 examines disorders of blood vessels.

1 How does the structure of a vein differ from that of an artery?

2 What are the functions of veins and venules?

3 How does venous circulation help to maintain blood pressure when hemorrhaging causes blood loss?

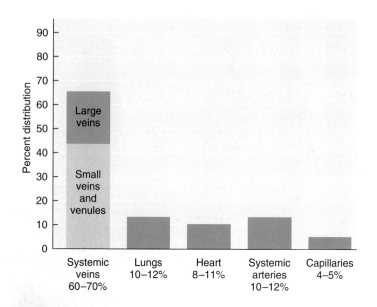

FIGURE 15.34

Most of the blood volume is contained within the veins and venules.

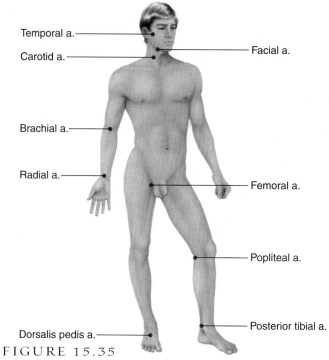

FIGURE 15.35

Sites where an arterial pulse is most easily detected (*a.* stands for artery).

Blood Pressure

Blood pressure is the force the blood exerts against the inner walls of the blood vessels. Although this force occurs throughout the vascular system, the term *blood pressure* most commonly refers to pressure in arteries supplied by branches of the aorta (systemic arteries).

Arterial Blood Pressure

The arterial blood pressure rises and falls in a pattern corresponding to the phases of the cardiac cycle. That is, when the ventricles contract (ventricular systole), their walls squeeze the blood inside their chambers and force it into the pulmonary trunk and aorta. As a result, the pressures in these arteries increase sharply. The maximum pressure achieved during ventricular contraction is called the **systolic pressure.** When the ventricles relax (ventricular diastole), the arterial pressure drops, and the lowest pressure that remains in the arteries before the next ventricular contraction is termed the **diastolic pressure.**

The surge of blood entering the arterial system during a ventricular contraction distends the elastic walls of the arteries, but the pressure begins to drop almost immediately as the contraction ends, and the arterial walls recoil. This alternate expanding and recoiling of the arterial wall can be felt as a *pulse* in an artery that runs close to the surface. Figure 15.35 shows several sites where a

TABLE 15.3	Characteristics of Blood Vessels	
Vessel	**Type of Wall**	**Function**
Artery	Thick, strong wall with three layers—an endothelial lining, a middle layer of smooth muscle and elastic tissue, and an outer layer of connective tissue	Carries blood under relatively high pressure from the heart to arterioles
Arteriole	Thinner wall than an artery but with three layers; smaller arterioles have an endothelial lining, some smooth muscle tissue, and a small amount of connective tissue	Connects an artery to a capillary, helps control the blood flow into a capillary by vasoconstricting or vasodilating
Capillary	Single layer of squamous epithelium	Provides a membrane through which nutrients, gases, and wastes are exchanged between the blood and tissue fluid; connects an arteriole to a venule
Venule	Thinner wall, less smooth muscle and elastic tissue than in an arteriole	Connects a capillary to a vein
Vein	Thinner wall than an artery but with similar layers; the middle layer is more poorly developed; some have flaplike valves	Carries blood under relatively low pressure from a venule to the heart; valves prevent a backflow of blood; serves as blood reservoir

CLINICAL APPLICATION

BLOOD VESSEL DISORDERS

In the arterial disease *atherosclerosis,* deposits of fatty materials, particularly cholesterol, form within the intima and inner lining of the arterial walls. Such deposits, called *plaque,* protrude into the lumens of the vessels and interfere with blood flow (fig. 15E). Furthermore, plaque often forms a surface texture that can initiate formation of a blood clot, increasing the risk of developing thrombi or emboli that cause blood deficiency (*ischemia*) or tissue death (*necrosis*) downstream from the obstruction.

The walls of affected arteries may degenerate, losing their elasticity and becoming hardened or *sclerotic.* In this stage of the disease, called *arteriosclerosis,* a sclerotic vessel may rupture under the force of blood pressure.

Risk factors for developing atherosclerosis include a fatty diet, elevated blood pressure, tobacco smoking, obesity, and lack of physical exercise (see chapter 18, pp. 697–698). Emotional and genetic factors may also increase susceptibility to atherosclerosis.

If atherosclerosis so weakens the wall of an artery that blood pressure dilates a region of it, a pulsating sac called an *aneurysm* may form. Aneurysms tend to grow. If the resulting sac develops by a longitudinal splitting of the middle layer of the arterial wall, it is called a *dissecting aneurysm.* An aneurysm may cause symptoms by pressing on nearby organs, or it may rupture and produce a great loss of blood.

Aneurysms may also result from trauma, high blood pressure, infections, inherited disorders such as Marfan syndrome, or congenital defects in blood vessels. Common sites of aneurysms include the thoracic and abdominal aorta and an arterial circle at the base of the brain (circle of Willis).

Phlebitis, or inflammation of a vein, is relatively common. It may occur in association with an injury or infection or after surgery, or it may develop for no apparent reason.

If inflammation is restricted to a superficial vein, such as the greater or lesser saphenous veins, blood flow may be rechanneled through other vessels. But if it occurs in a deep vein, such as the tibial, peroneal, popliteal, or femoral veins, the consequences can be quite serious, particularly if the blood within the affected vessel clots and blocks normal circulation. This condition, called *thrombophlebitis,* introduces a risk that a blood clot within a vein will detach, move with the venous blood, pass through the heart, and lodge in the pulmonary arterial system within a lung. Such an obstruction is called a *pulmonary embolism.*

Varicose veins are abnormal and irregular dilations in superficial veins, particularly in the legs. This condition is usually associated with prolonged, increased back pressure within the affected vessels due to gravity, as occurs when a person stands. Crossing the legs or sitting in a chair so that its edge presses against the area behind the knee can obstruct venous blood flow and aggravate varicose veins.

Increased venous back pressure stretches and widens the veins. Because the valves within these vessels do not change size, they soon lose their abilities to block the backward flow of blood, and blood tends to accumulate in the enlarged regions.

Increased venous pressure is also accompanied by rising pressure within the venules and capillaries that supply the veins. Consequently, tissues in affected regions typically become edematous and painful.

Heredity, pregnancy, obesity, and standing for long periods raise the risk of developing varicose veins. Elevating the legs above the level of the heart or putting on support hosiery before arising in the morning can relieve discomfort. Intravenous injection of a substance that destroys veins (a sclerosing agent) or surgical removal of the affected veins may be necessary. ■

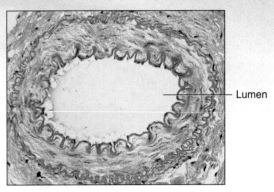

(a)

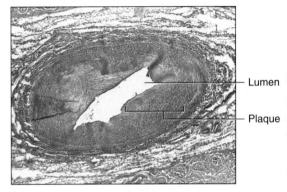

(b)

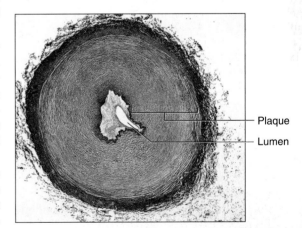

(c)

FIGURE 15E

Development of atherosclerosis. (*a*) Normal arteriole. (*b, c*) Accumulation of plaque on the inner wall of the arteriole.

MEASUREMENT OF ARTERIAL BLOOD PRESSURE

Systemic arterial blood pressure usually is measured using an instrument called a sphygmomanometer (sfig″mo-mah-nom′ĕ-ter) (fig. 15F). This device consists of an inflatable cuff connected by tubing to a compressible bulb and a pressure gauge. The bulb is used to pump air into the cuff, and a rise in pressure is indicated on the pressure gauge. The pressure in the cuff is expressed in millimeters of mercury (mm Hg) based on older equipment that used a glass tube containing a column of mercury in place of a pressure gauge. (These devices are being discontinued because of the danger of mercury.)

To measure arterial blood pressure, the cuff of the sphygmomanometer is usually wrapped around the arm so that it surrounds the brachial artery. Air is pumped into the cuff until the cuff pressure exceeds the pressure in that artery. As a result, the vessel is squeezed closed and its blood flow stopped. At this moment, if the diaphragm of a stethoscope is placed over the brachial artery at the distal border of the cuff, no sounds can be heard from the vessel because the blood flow is interrupted. As air is slowly released from the cuff, the air pressure inside it decreases. When the cuff

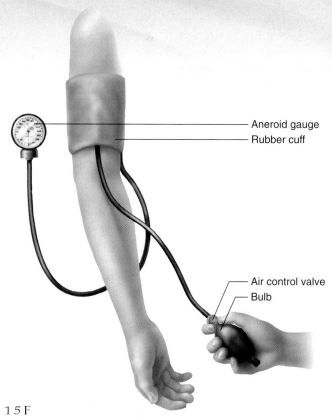

Aneroid gauge
Rubber cuff

Air control valve
Bulb

FIGURE 15F

A sphygmomanometer is used to measure arterial blood pressure. The use of the column of mercury is the most accurate measurement, but due to environmental concerns, it is being replaced by alternative gauges and digital readouts.

pulse can be detected. The radial artery, for example, courses near the surface at the wrist and is commonly used to sense a person's radial pulse.

The radial pulse rate is equal to the rate at which the left ventricle contracts, and for this reason, it can be used to determine heart rate. A pulse can also reveal something about blood pressure, because an elevated pressure produces a pulse that feels strong and full, whereas a low pressure produces a pulse that is weak and easily compressed. Clinical Application 15.3 describes how to measure arterial blood pressure.

1 Distinguish between systolic and diastolic blood pressure.

2 Which cardiac event causes systolic pressure? Diastolic pressure?

3 What causes a pulse in an artery?

Factors That Influence Arterial Blood Pressure

Arterial pressure depends on a variety of factors. These include heart action, blood volume, resistance to flow, and blood viscosity (fig. 15.36).

Heart Action

In addition to producing blood pressure by forcing blood into the arteries, heart action determines how much blood enters the arterial system with each ventricular contraction. The volume of blood discharged from the ventricle with each contraction is called the **stroke volume** and equals about 70 milliliters in an average-weight male at rest. The volume discharged from the ventricle per minute is called the **cardiac output.** It is calculated by multiplying the stroke volume by the heart rate in beats per minute. (Cardiac output = stroke volume × heart rate.) Thus, if the stroke volume is 70 milliliters and the heart

pressure is approximately equal to the systolic blood pressure within the brachial artery, the artery opens enough for a small amount of blood to spurt through. This movement produces a sharp sound (Korotkoff's sound) that can be heard through the stethoscope. The pressure indicated on the pressure gauge when this first tapping sound is heard represents the *arterial systolic pressure* (SP).

As the cuff pressure continues to drop, a series of increasingly louder sounds can be heard. Then, when the cuff pressure is approximately equal to that within the fully opened artery, the sounds become abruptly muffled and disappear. The pressure indicated on the pressure gauge when this happens represents the *arterial diastolic pressure* (DP). The sound results from turbulence that occurs when the artery narrows.

The results of a blood pressure measurement are reported as a fraction, such as 120/80. In this notation, the upper number indicates the systolic pressure in mm Hg (SP), and the lower number indicates the diastolic pressure in mm Hg (DP). Figure 15G shows how these pressures decrease as distance from the left ventricle increases.

The difference between the systolic and diastolic pressures (SP-DP), which is called the *pulse pressure* (PP), is generally about 40 mm Hg.

The average pressure in the arterial system is also of interest because it represents the force that is effective throughout the cardiac cycle for driving blood to the tissues. This force, called the *mean arterial pressure,* is approximated by adding the diastolic pressure and one-third of the pulse pressure (DP + 1/3PP). ■

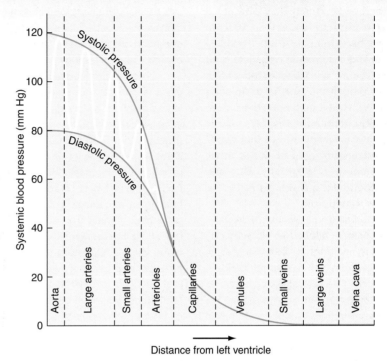

FIGURE 15G

Blood pressure decreases as the distance from the left ventricle increases. Systolic pressure occurs during maximal ventricular contraction. Diastolic pressure occurs when the ventricles relax.

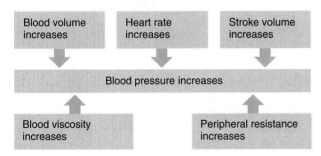

FIGURE 15.36

Some of the factors that influence arterial blood pressure.

rate is 72 beats per minute, the cardiac output is 5,040 milliliters per minute.

Blood pressure varies with the cardiac output. If either the stroke volume or the heart rate increases, so does the cardiac output, and, as a result, blood pressure initially rises. Conversely, if the stroke volume or the heart rate decreases, the cardiac output decreases, and blood pressure also initially decreases.

Blood Volume

Blood volume equals the sum of the formed elements and plasma volumes in the vascular system. Although the blood volume varies somewhat with age, body size, and sex, it is usually about 5 liters for adults or 8% of body weight in kilograms (1 kilogram of water equals 1 liter).

Blood volume can be determined by injecting a known volume of an indicator, such as radioactive iodine, into the blood. After a time that allows for thorough mixing, a blood sample is withdrawn, and the concentration of the indicator measured. The total blood volume is calculated using this formula: blood volume = amount of indicator injected/concentration of indicator in blood sample.

SPACE MEDICINE

When the rescue team approached the space shuttle *Atlantis* just after it landed on September 26, 1996, they brought a stretcher, expecting to carry off Mission Specialist Shannon Lucid, Ph.D. The fifty-three-year-old biochemist had just spent 188 days aboard the Russian *Mir* space station, as part of the NASA/*Mir* Science Program (fig. 15H). Lucid had spent more time in space than any other U.S. astronaut, and it was widely known that about 70% of astronauts cannot stand at all upon reencountering gravity. But she walked, albeit a little wobbly, the 25 feet to the crew transporter.

The human body evolved under conditions of constant gravity. So when a body is exposed to microgravity (very low gravity) or weightlessness for extended periods, changes occur. The field of space medicine examines anatomic and physiologic responses to conditions in space. Shannon Lucid was expected to require the stretcher because of decreased muscle mass, mineral-depleted bones, and low blood volume. The 400 hours that she logged on the *Mir*'s treadmill and stationary bicycle may have helped her stay in shape.

Lucid was poked and prodded, monitored and tested, as medical researchers attempted to learn how six months in space affects cardiovascular functioning, respiratory capacity, mood, blood chemistry, circadian rhythms, muscular strength, body fluid composition, and many other aspects of anatomy and physiology.

Feeling unsteady upon returning to earth is one of the better-studied physiologic responses to low-gravity conditions. It is called orthostatic intolerance. Normally, gravity helps blood circulate in the lower limbs. In microgravity or no gravity, blood pools in blood vessels in the center of the body, registering on receptors there. The body interprets this as excess blood, and in response, signals the kidneys to excrete more fluid. But there really isn't an increased blood volume. On return to earth, the body actually has a pint to a quart less blood than it should, up to a 10 to 20% decrease in total blood volume. If blood vessels cannot constrict sufficiently to counter the plummeting blood pressure, orthostatic intolerance results. To minimize the effect, astronauts wear lower-body suction suits, which apply a vacuum force that helps draw blood into the blood vessels of the lower limbs. Maintaining fluid intake helps prevent dehydration. ■

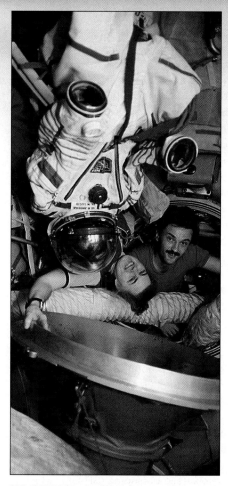

FIGURE 15H

Shannon Lucid's 188-day stay in space revealed to researchers much about the body's responses to microgravity conditions. While aboard the space station *Mir,* Lucid conducted experiments on quail embryos and growth of protein crystals.

Blood pressure is normally directly proportional to the volume of the blood within the cardiovascular system. Thus, any changes in the blood volume can initially alter the blood pressure. For example, if a hemorrhage reduces blood volume, blood pressure initially drops. If a transfusion restores normal blood volume, normal pressure may be reestablished. Blood volume can also fall if the fluid balance is upset, as happens in dehydration. Fluid replacement can reestablish normal blood volume and pressure. Clinical Application 15.4 describes how the unusual conditions of microgravity in outer space affect the distribution of blood volume and control of blood pressure.

Peripheral Resistance

Friction between blood and the walls of the blood vessels produces a force called **peripheral resistance** (pe-rif′er-al re-zis′tans), which hinders blood flow. Blood pressure must overcome this force if the blood is to continue flowing. Therefore, factors that alter the peripheral resistance change blood pressure. For example, contraction of smooth muscles in the walls of contracting arterioles increases the peripheral resistance by constricting these vessels. Blood tends to back up into the arteries supplying the arterioles, and the arterial pressure rises. Dilation of the arterioles has the opposite effect—peripheral resistance lessens, and the arterial blood pressure drops in response (fig. 15.37).

Because arterial walls are quite elastic, when the ventricles discharge a surge of blood, arteries swell. Almost immediately, the elastic tissues recoil, and the vessel walls press against the blood inside. This action helps force the blood onward against the peripheral resistance in arterioles and capillaries. It is this recoiling of the arteries that maintains blood pressure during diastole. If there were no elasticity in the arterial walls, blood pressure would fall to zero between ventricular contractions. Elastic recoil also converts the intermittent flow of blood, which is characteristic of the arterial system, into a more continuous movement through the capillaries.

Viscosity

The **viscosity** (vis-kos′ĭ-te) of a fluid is a physical property that derives from the ease with which its molecules flow past one another. The greater the viscosity, the greater the resistance to flow.

Blood cells and some plasma proteins increase blood viscosity. Since the greater the blood's resistance to flowing, the greater the force needed to move it through the vascular system, it is not surprising that blood pressure rises as blood viscosity increases and drops as blood viscosity decreases.

Although the viscosity of blood normally remains stable, any condition that alters the concentrations of blood cells or specific plasma proteins may alter blood viscosity. For example, anemia may decrease viscosity and consequently lower blood pressure. Excess red blood cells increase viscosity and blood pressure.

1 How is cardiac output calculated?

2 How are cardiac output and blood pressure related?

3 How does blood volume affect blood pressure?

4 What is the relationship between peripheral resistance and blood pressure? Between blood viscosity and blood pressure?

Control of Blood Pressure

Blood pressure (BP) is determined by cardiac output (CO) and peripheral resistance (PR) according to this relationship: BP = CO × PR. Maintenance of normal blood pressure therefore requires regulation of these two factors (fig. 15.38).

Cardiac output depends on the stroke volume and heart rate. Stroke volume, the amount of blood pumped in a single beat, is reflected by the difference between **end-diastolic volume (EDV)**, the volume of blood in each ventricle at the end of ventricular diastole, and **end-systolic volume (ESV)**, the volume of blood in each ventricle at the end of the ventricular systole. Mechanical, neural, and chemical factors affect stroke volume and heart rate.

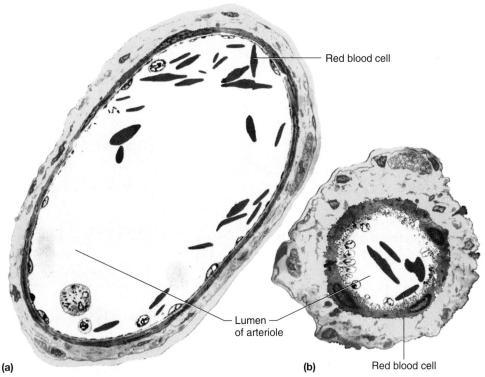

Red blood cell

Lumen of arteriole

Red blood cell

(a)

(b)

FIGURE 15.37

Vasodilation and vasoconstriction. (*a*) Relaxation of smooth muscle in the arteriole wall produces dilation, whereas (*b*) contraction of the smooth muscle causes constriction (*a* and *b* 1,500×).

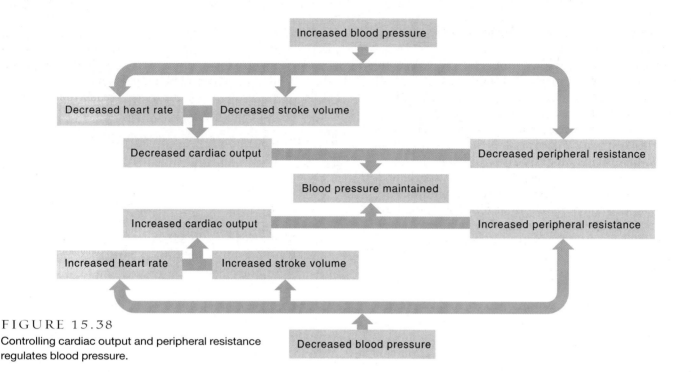

FIGURE 15.38

Controlling cardiac output and peripheral resistance regulates blood pressure.

Cardiac output is limited by the amount of blood returning to the ventricles, called the *venous return.* Usually, however, stroke volume can be increased by sympathetic stimulation, which increases the force of ventricular contraction. Since only about sixty percent of the end-diastolic volume is pumped out in a normal contraction, increasing the force of ventricular contraction may increase that fraction and help maintain stroke volume if venous return should decrease.

Another mechanism, called the Frank-Starling law of the heart (Starling's Law of the heart) increases stroke volume independently of sympathetic stimulation. As blood enters the ventricles, myocardial fibers are mechanically stretched. This constitutes the **preload.** The greater the EDV, the greater the preload. Within limits, the greater the length of these fibers, the greater the force with which they contract. This relationship between fiber length and force of contraction is called the **Frank-Starling law of the heart,** or Starling's law of the heart. This becomes important, for example, during exercise, when venous return increases. The more blood that enters the heart from the veins, the greater the ventricular distension, the stronger the contraction, the greater the stroke volume, and the greater the cardiac output.

∞ R E C O N N E C T   T O   C H A P T E R   9 ,   R E C O R D I N G   A   M U S C L E   C O N T R A C T I O N ,   P A G E   2 9 0 .

The Frank-Starling law of the heart helps ensure that the cardiac output is equal to the venous return. This further ensures that the volume of blood pumped by both ventricles is, on the average, equal.

Recall that baroreceptors in the walls of the aortic arch and carotid sinuses sense changes in blood pressure.

If arterial pressure increases, nerve impulses travel from the receptors to the *cardiac center* of the medulla oblongata. This center relays parasympathetic impulses to the S-A node in the heart, and heart rate decreases in response. As a result of this *cardioinhibitor reflex,* cardiac output falls, and blood pressure decreases toward the normal level. Figure 15.39 summarizes this mechanism.

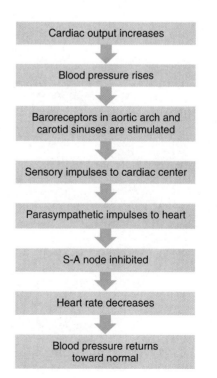

FIGURE 15.39

If blood pressure rises, baroreceptors initiate the cardioinhibitor reflex, which lowers the blood pressure.

Conversely, decreasing arterial blood pressure initiates the *cardioaccelerator reflex,* which involves sympathetic impulses to the S-A node. As a result, the heart beats faster. This response increases cardiac output, increasing arterial pressure.

Recall that epinephrine increases heart rate (chapter 13, p. 493) and consequently alters cardiac output and blood pressure. Other factors that increase heart rate and blood pressure include emotional responses, such as fear and anger; physical exercise; and a rise in body temperature.

Changes in arteriole diameters regulate peripheral resistance. Because blood vessels with smaller diameters offer a greater resistance to blood flow, factors that cause arteriole vasoconstriction increase peripheral resistance, and factors causing vasodilation decrease resistance.

The *vasomotor center* of the medulla oblongata continually sends sympathetic impulses to the smooth muscles in the arteriole walls, keeping them in a state of tonic contraction, which helps maintain the peripheral resistance associated with normal blood pressure. Because the vasomotor center responds to changes in blood pressure, it can increase peripheral resistance by increasing its outflow of sympathetic impulses, or it can decrease such resistance by decreasing its sympathetic outflow. In the latter case, the vessels undergo vasodilation as sympathetic stimulation decreases.

Whenever arterial blood pressure suddenly increases, baroreceptors in the aortic arch and carotid sinuses signal the vasomotor center, and the sympathetic outflow to the arteriole walls falls (fig. 15.40). The resulting vasodilation decreases peripheral resistance, and blood pressure decreases toward the normal level.

Similarly, if blood pressure drops, as following a hemorrhage, the vasomotor center increases sympathetic outflow. The resulting release of epinephrine and norepinephrine vasoconstricts most systemic vessels, increasing peripheral resistance. This helps return blood pressure toward normal.

The vasomotor center's control of vasoconstriction and vasodilation is especially important in the arterioles of the *abdominal viscera* (splanchnic region). These vessels, if fully dilated, could accept nearly all the blood of the body and send the arterial pressure toward zero. Thus, control of their diameters is essential in regulating normal peripheral resistance.

Certain chemicals, including carbon dioxide, oxygen, and hydrogen ions, also influence peripheral resistance by affecting precapillary sphincters and smooth muscles in arteriole and metarteriole walls. For example, increasing blood carbon dioxide, decreasing blood oxygen, and lowering of the blood's pH relaxes these muscles in the systemic circulation. This increases local blood flow to tissues with high metabolic rates, such as exercising skeletal muscles.

Other chemicals also influence peripheral resistance and thus blood pressure. Nitric oxide, produced by endo-

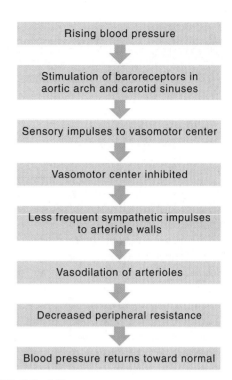

FIGURE 15.40
Dilating arterioles helps regulate blood pressure.

thelial cells, and bradykinin, formed in the blood, are both vasodilators. The hormone angiotensin plays a role in vasoconstriction; and endothelin, released by cells of the endothelium, is a powerful vasoconstrictor. Clinical Application 15.5 discusses high blood pressure.

1 What factors affect cardiac output?

2 Explain the Frank-Starling law of the heart.

3 What is the function of the baroreceptors in the walls of the aortic arch and carotid sinuses?

4 How does the vasomotor center control peripheral resistance?

Venous Blood Flow

Blood pressure decreases as the blood moves through the arterial system and into the capillary networks, so little pressure remains at the venular ends of capillaries (see fig. 15G). Instead, blood flow through the venous system is only partly the direct result of heart action and depends on other factors, such as skeletal muscle contraction, breathing movements, and vasoconstriction of veins. For example, contracting skeletal muscles press on veins, moving blood from one valve section to another. This massaging action of contracting skeletal muscles helps push the blood through the venous system toward the heart (fig. 15.41).

Respiratory movements also move venous blood. During inspiration, the pressure within the thoracic cavity is reduced as the diaphragm contracts and the rib cage

HYPERTENSION

Hypertension, or high blood pressure, is persistently elevated arterial pressure. It is one of the more common diseases of the cardiovascular system in industrialized nations.

High blood pressure with unknown cause is called *essential* (also primary or idiopathic) *hypertension.* Elevated blood pressure that is a consequence of another problem, such as arteriosclerosis or kidney disease, is called *secondary hypertension.*

Arteriosclerosis is accompanied by decreased elasticity of the arterial walls and narrowed vessel lumens, which raise blood pressure. Kidney diseases often produce changes that interfere with blood flow to kidney cells. In response, the affected tissues may release an enzyme called *renin* that leads to the production of *angiotensin II,* a powerful vasoconstrictor that increases peripheral resistance in the arterial system, raising arterial pressure (fig. 15I). Angiotensin II also stimulates the adrenal cortex to release *aldosterone,* which stimulates the kidneys to retain sodium ions and water. The resulting increase in blood volume contributes to increased blood pressure. Normally, this mechanism ensures that a decrease in blood flow to the kidneys

is followed by an increase in arterial pressure, which, in turn, restores blood flow to the kidneys. If the decreased blood flow is the result of disease, such as atherosclerosis, the mechanism may cause high blood pressure and promote further deterioration of the arterial system.

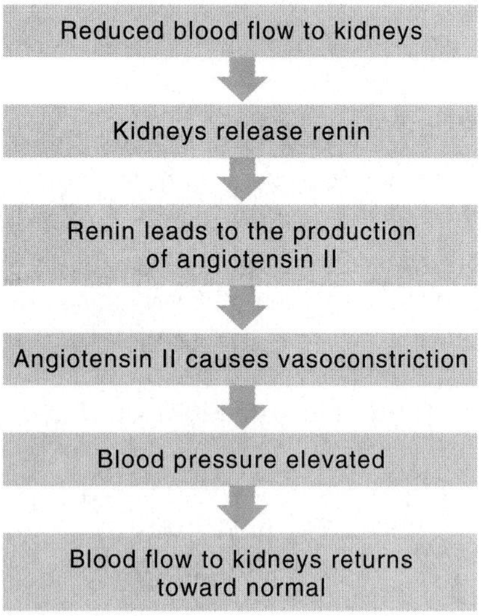

In some individuals, high sodium intake leads to vasoconstriction, raising blood pressure. Obesity also is a risk factor for hypertension because it tends to increase peripheral resistance. Psychological stress, which activates sympathetic nerve impulses that cause generalized vasoconstriction, may also lead to hypertension. Yet another cause of hypertension may be an inability of endothelium to respond to a relaxing factor, leading to vasoconstriction.

Hypertension is called a "silent killer" because it may not have direct symptoms, yet can set the stage for serious cardiovascular complications. For example, as the left ventricle works harder to pump blood at a higher pressure, the myocardium thickens, enlarging the heart. If the coronary blood vessels cannot support this overgrowth, parts of the heart muscle die and become replaced with fibrous tissue. Eventually, the enlarged and weakened heart dies.

Hypertension also contributes to the development of atherosclerosis. As arteries accumulate plaque, a *coronary thrombosis* or a *coronary embolism* may occur. Similar changes in the arteries of the brain increase the chances of a *cerebral vascular accident* (CVA), or stroke, due to a cerebral thrombosis, embolism, or hemorrhage.

When an embolus or hemorrhage causes a stroke, paralysis and other functional losses appear suddenly. A thrombus-caused stroke is slower. It may begin with clumsiness, progress to partial visual loss, then affect speech. One arm becomes paralyzed, then a day later, perhaps an entire side of the body is affected. Table 15A lists risk factors for a stroke.

A *transient ischemic attack* (TIA, or "ministroke") is a temporary block in a small artery. Symptoms include difficulty in speaking or understanding speech; numbness or

weakness in the face, upper limb, lower limb, or one side; dizziness; falling; an unsteady gait; blurred vision; or blindness. These symptoms typically resolve within twenty-four hours with no lasting effects, but may be a warning of an impending, more serious stroke.

Treatment of hypertension varies and may include exercising regularly, controlling weight, reducing stress, and limiting the diet to foods that are low in sodium. Drugs, such as diuretics and/or inhibitors of sympathetic nerve activity, may help control blood pressure. Diuretics increase urinary excretion of sodium and water, reducing the volume of body fluids. Sympathetic inhibitors block the synthesis of neurotransmitters, such as norepinephrine, or block receptor sites of effector cells. Table 15B describes how drugs that treat hypertension work. ■

FIGURE 15I
Renin stimulates production of angiotensin II, which elevates blood pressure.

| Reduced blood flow to kidneys |
| Kidneys release renin |
| Renin leads to the production of angiotensin II |
| Angiotensin II causes vasoconstriction |
| Blood pressure elevated |
| Blood flow to kidneys returns toward normal |

TABLE 15A Risk Factors for Stroke

Alcohol consumption
Diabetes
Elevated serum cholesterol
Family history of cardiovascular disease
Hypertension
Smoking
Transient ischemic attacks

TABLE 15B Drugs to Treat Hypertension

Type of Drug	Mechanism of Action
Angiotensin-converting enzyme (ACE) inhibitors	Block formation of angiotensin II, preventing vasoconstriction
Beta blockers	Lower heart rate
Calcium channel blockers	Dilate blood vessels by keeping calcium ions out of muscle cells in vessel walls
Diuretics	Increase urine output, lowering blood volume

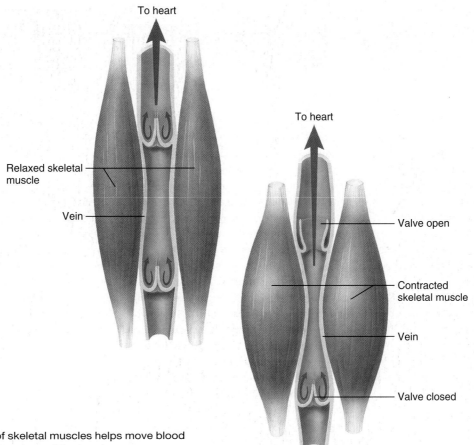

FIGURE 15.41
The massaging action of skeletal muscles helps move blood through the venous system toward the heart.

Labels in figure:
To heart
Relaxed skeletal muscle
Vein
To heart
Valve open
Contracted skeletal muscle
Vein
Valve closed

moves upward and outward. At the same time, the pressure within the abdominal cavity is increased as the diaphragm presses downward on the abdominal viscera. Consequently, blood is squeezed out of the abdominal veins and forced into thoracic veins. During exercise, these respiratory movements act with skeletal muscle contractions to increase return of venous blood to the heart.

The veins also provide a blood reservoir that can adapt its capacity to changes in blood volume (see fig. 15.34). If some blood is lost and blood pressure falls, vein constriction (*venoconstriction*) can force blood out of this reservoir, returning venous blood to the heart. By maintaining venous return, venoconstriction helps to maintain blood pressure.

Central Venous Pressure

Because all the veins, except those returning to the heart from the lungs, drain into the right atrium, the pressure within this heart chamber is called *central venous pressure*. This pressure is of special interest because it affects the pressure within the peripheral veins. For example, if the heart is beating weakly, the central venous pressure increases, and blood backs up in the venous network, raising its pressure too. However, if the heart is beating forcefully, the central venous pressure and the pressure within the venous network decrease.

As a result of disease or injury, blood or tissue fluid may accumulate in the pericardial cavity, increasing pressure. This condition, called *acute cardiac tamponade*, can be life threatening. As the pressure around the heart increases, it may compress the heart, interfere with the flow of blood into its chambers, and prevent pumping action. An early symptom of acute cardiac tamponade may be increased central venous pressure, with visible engorgement of the veins in the neck.

Other factors that increase the flow of blood into the right atrium, and thus elevate the central venous pressure, include increase in blood volume or widespread venoconstriction. An increase in central venous pressure can lead to peripheral edema because the resulting higher capillary hydrostatic pressure favors movement of fluid into the tissues. Clinical Application 15.6 discusses the effects of exercise on the heart and blood vessels.

1 What is the function of the venous valves?

2 How do skeletal muscles affect venous blood flow?

3 How do respiratory movements affect venous blood flow?

4 What factors stimulate venoconstriction?

EXERCISE AND THE CARDIOVASCULAR SYSTEM

We all know that exercise is good for the heart. Yet each year, a few individuals die of sudden cardiac arrest while shoveling snow, running, or engaging in some other strenuous activity. The explanation for this apparent paradox is that exercise *is* good for the heart—but only if it is a regular part of life.

Physiological responses to intense aerobic exercise generally increase blood flow, and therefore oxygen delivery, to active muscles. In muscles, vasodilation opens more capillaries. At the same time, vasoconstriction diminishes blood flow where it is not immediately needed, such as to the digestive tract. Blood flow, however, is maintained in the brain and kidneys, which require a steady stream of oxygen and nutrients to function. Respiratory movements and skeletal muscle activity increase venous return to the heart. As venous return to the heart increases, ventricular walls stretch, stimulating them to contract with greater force. Heart rate increases as well.

The cardiovascular system adapts to exercise as a way of life. The conditioned athlete experiences increases in heart pumping efficiency, blood volume, blood hemoglobin concentration, and the number of mitochondria in muscle fibers. All of these adaptations improve oxygen delivery to, and utilization by, muscle tissue.

An athlete's heart typically changes in response to these increased demands and may enlarge 40% or more. Myocardial mass increases, the ventricular cavities expand, and the ventricle walls thicken. Stroke volume increases, and heart rate decreases, as does blood pressure. To a physician unfamiliar with a conditioned cardiovascular system, a trained athlete may appear to be abnormal!

The cardiovascular system responds beautifully to a slow, steady buildup in exercise frequency and intensity. It does not react well to sudden demands—such as a person who never exercises suddenly shoveling snow or running 3 miles. Although sedentary people have a two- to sixfold increased risk of cardiac arrest while exercising than when not, people in shape have little or no excess risk while exercising.

How much exercise is enough to benefit the cardiovascular system? To achieve the benefits of exercise, the heart rate must be elevated to 70% to 85% of its "theoretical maximum" for at least half an hour three times a week. You can calculate your theoretical maximum by subtracting your age from 220. If you are eighteen years old, your theoretical maximum is 202 beats per minute. Seventy to 85% of this value is 141 to 172 beats per minute. Some good activities for raising the heart rate are tennis, skating, skiing, handball, vigorous dancing, hockey, basketball, biking, and fast walking.

It is wise to consult a physician before starting an exercise program. People over the age of thirty are advised to have a stress test, which is an electrocardiogram taken while exercising. (The standard electrocardiogram is taken at rest.) An arrhythmia that appears only during exercise may indicate heart disease that has not yet produced symptoms. ■

Paths of Circulation

The blood vessels can be divided into two major pathways. The **pulmonary circuit** consists of vessels that carry blood from the heart to the lungs and back to the heart. The **systemic circuit** carries blood from the heart to all parts of the body, except the lungs, and back again (fig. 15.42). The systemic circuit includes the coronary circulation, which supplies the heart itself and has already been described.

The pathways described in the following sections are those of an adult. Chapter 23 (pp. 898–901) describes the somewhat different fetal pathways.

Pulmonary Circuit

Blood enters the pulmonary circuit as it leaves the right ventricle through the pulmonary trunk. The pulmonary trunk extends upward and posteriorly from the heart, and about 5 centimeters above its origin, it divides into the right and left pulmonary arteries. These branches penetrate the right and left lungs, respectively. Within the lungs, they diverge into *lobar branches* (three on the right side and two on the left) that accompany the main divisions of the bronchi (airways) into the lobes of the lungs. After repeated divisions, the lobar branches give rise to arterioles that continue into the capillary networks associated with the walls of the alveoli (air sacs) (fig. 15.42).

The blood in the arteries and arterioles of the pulmonary circuit is low in oxygen and high in carbon dioxide. Gases are exchanged between the blood and the air as the blood moves through the *alveolar capillaries,* discussed in chapter 19 (p. 758).

Because the right ventricle contracts with less force than the left ventricle, the arterial pressure in the pulmonary circuit is less than that in the systemic circuit. Therefore, the alveolar capillary pressure is low.

The force that moves fluid out of an alveolar capillary is 23 mm Hg; the force pulling fluid into it is 22 mm

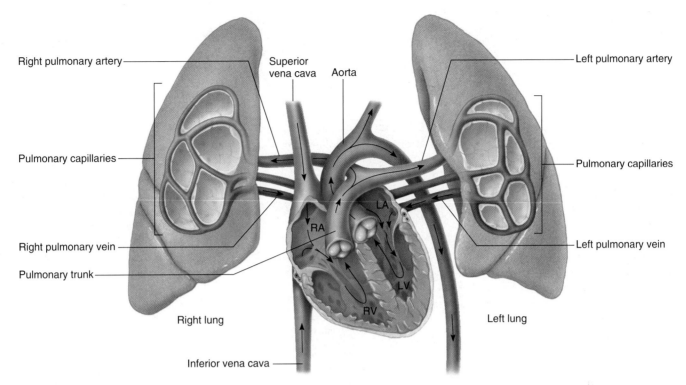

Right pulmonary artery

Superior vena cava

Aorta

Left pulmonary artery

Pulmonary capillaries

Pulmonary capillaries

Right pulmonary vein

LA

Left pulmonary vein

Pulmonary trunk

RA

LV

RV

Right lung

Left lung

Inferior vena cava

FIGURE 15.42

Blood reaches the lungs through branches of the pulmonary arteries, and it returns to the heart through pulmonary veins.

Hg. Thus, such a capillary has a net filtration pressure of 1 mm Hg. This pressure causes a slight, continuous flow of fluid into the narrow interstitial space between the alveolar capillary and the alveolus.

The epithelial cells of the alveoli are so tightly joined that sodium, chloride, and potassium ions, as well as glucose and urea, enter the interstitial space but usually fail to enter the alveoli. This helps maintain a high osmotic pressure in the interstitial fluid. Consequently, osmosis rapidly moves any water that gets into the alveoli back into the interstitial space. Although the alveolar surface must be moist to allow diffusion of oxygen and carbon dioxide, this mechanism prevents excess water from entering the alveoli and helps keep the alveoli from filling with fluid (fig. 15.43).

Fluid in the interstitial space may be drawn back into the alveolar capillaries by the somewhat higher osmotic pressure of the blood. Alternatively, lymphatic vessels (see chapter 16, pp. 609–611) may return it to the circulation.

As a result of the gas exchanges between the blood and the alveolar air, blood entering the venules of the pulmonary circuit is rich in oxygen and low in carbon dioxide. These venules merge to form small veins, and they, in turn, converge to form larger veins. Four *pulmonary veins,* two from each lung, return blood to the left atrium, and this completes the vascular loop of the pulmonary circuit.

Pulmonary edema, in which lungs fill with fluid, can accompany a failing left ventricle or a damaged mitral valve. A weak left ventricle may be unable to move the normal volume of blood into the systemic circuit. Blood backing up into the pulmonary circuit increases pressure in the alveolar capillaries, flooding the interstitial spaces with fluid. Increasing pressure in the interstitial fluid may rupture the alveolar membranes, and fluid may enter the alveoli more rapidly than it can be removed. This reduces the alveolar surface available for gas exchange, and the person may suffocate.

Systemic Circuit

Freshly oxygenated blood moves from the left atrium into the left ventricle. Contraction of the left ventricle forces this blood into the systemic circuit. This circuit includes the aorta and its branches that lead to all of the body tissues, as well as the companion system of veins that returns blood to the right atrium.

1 Distinguish between the pulmonary and systemic circuits of the cardiovascular system.

2 Trace the path of blood through the pulmonary circuit from the right ventricle.

3 Explain why the alveoli normally do not fill with fluid.

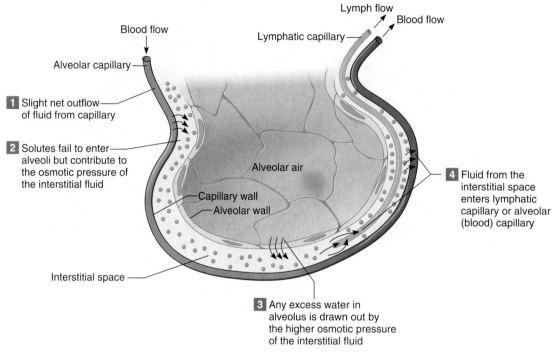

Blood flow

Alveolar capillary

1 Slight net outflow of fluid from capillary

2 Solutes fail to enter alveoli but contribute to the osmotic pressure of the interstitial fluid

Capillary wall
Alveolar wall

Interstitial space

Lymph flow
Blood flow
Lymphatic capillary

Alveolar air

4 Fluid from the interstitial space enters lymphatic capillary or alveolar (blood) capillary

3 Any excess water in alveolus is drawn out by the higher osmotic pressure of the interstitial fluid

FIGURE 15.43
Cells of the alveolar wall are tightly joined. The relatively high osmotic pressure of the interstitial fluid draws water out of them.

Arterial System

The **aorta** is the largest diameter artery in the body. It extends upward from the left ventricle, arches over the heart to the left, and descends just anterior and to the left of the vertebral column.

Principal Branches of the Aorta

The first portion of the aorta is called the *ascending aorta.* Located at its base are the three cusps of the aortic valve, and opposite each cusp is a swelling in the aortic wall called an **aortic sinus.** The right and left *coronary arteries* arise from two of these sinuses. Blood flow into these arteries is intermittent and is driven by the elastic recoil of the aortic wall following contraction of the left ventricle.

Several small structures called **aortic bodies** are located within the epithelial lining of the aortic sinuses. These bodies contain chemoreceptors that sense blood concentrations of oxygen and carbon dioxide.

Three major arteries originate from the *arch of the aorta* (aortic arch). They are the brachiocephalic (innominate) artery, the left common carotid artery, and the left subclavian artery. The aortic arch contains baroreceptors that detect changes in blood pressure.

The **brachiocephalic** (brak″e-o-sĕ-fal′ik) **artery** supplies blood to the tissues of the upper limb and head, as its name suggests. It is the first branch from the aortic arch

and rises upward through the mediastinum to a point near the junction of the sternum and the right clavicle. There it divides, giving rise to the right **common carotid** (kah-rot′id) **artery,** which carries blood to the right side of the neck and head, and the right **subclavian** (sub-kla′ve-an) **artery,** which leads into the right arm. Branches of the subclavian artery also supply blood to parts of the shoulder, neck, and head.

The left *common carotid artery* and the left *subclavian artery* are respectively the second and third branches of the aortic arch. They supply blood to regions on the left side of the body corresponding to those supplied by their counterparts on the right (fig. 15.44 and reference plates 71, 72, and 73).

Although the upper part of the *descending aorta* is positioned to the left of the midline, it gradually moves medially and finally lies directly in front of the vertebral column at the level of the twelfth thoracic vertebra. The portion of the descending aorta above the diaphragm is the **thoracic aorta** (tho-ras′ik a-or′tah), and it gives off numerous small branches to the thoracic wall and the thoracic viscera. These branches, the *bronchial, pericardial,* and *esophageal arteries,* supply blood to the structures for which they were named. Other branches become *mediastinal arteries,* supplying various tissues within the mediastinum, and *posterior intercostal arteries,* which pass into the thoracic wall.

Below the diaphragm, the descending aorta becomes the **abdominal aorta,** and it gives off branches to the

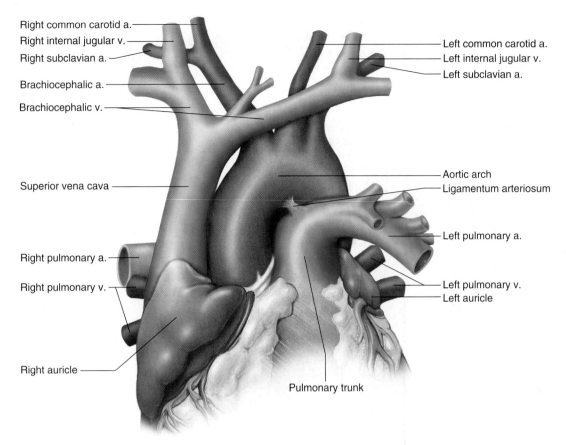

Right common carotid a.
Right internal jugular v.
Right subclavian a.
Brachiocephalic a.
Brachiocephalic v.
Superior vena cava
Right pulmonary a.
Right pulmonary v.
Right auricle
Left common carotid a.
Left internal jugular v.
Left subclavian a.
Aortic arch
Ligamentum arteriosum
Left pulmonary a.
Left pulmonary v.
Left auricle
Pulmonary trunk

FIGURE 15.44

The major blood vessels associated with the heart. (*a.* stands for artery, *v.* stands for vein.)

abdominal wall and various abdominal organs. These branches include the following:

1. **Celiac** (se′le-ak) **artery.** This single vessel gives rise to the left *gastric, splenic,* and *hepatic arteries,* which supply upper portions of the digestive tract, the spleen, and the liver, respectively. (*Note:* The hepatic artery supplies the liver with about one-third of its blood flow, and this blood is oxygen-rich. The remaining two-thirds of the liver's blood flow arrives by means of the hepatic portal vein and is oxygen-poor.)

2. **Phrenic** (fren′ik) **arteries.** These paired arteries supply blood to the diaphragm.

3. **Superior mesenteric** (mes″en-ter′ik) **artery.** The superior mesenteric is a large, unpaired artery that branches to many parts of the intestinal tract, including the jejunum, ileum, cecum, ascending colon, and transverse colon.

4. **Suprarenal** (soo″prah-re′nal) **arteries.** This pair of vessels supplies blood to the adrenal glands.

5. **Renal** (re′nal) **arteries.** The renal arteries pass laterally from the aorta into the kidneys. Each artery

then divides into several lobar branches within the kidney tissues.

6. **Gonadal** (go′nad-al) **arteries.** In a female, paired ovarian arteries arise from the aorta and pass into the pelvis to supply the ovaries. In a male, *spermatic arteries* originate in similar locations. They course downward and pass through the body wall by way of the *inguinal canal* to supply the testes.

7. **Inferior mesenteric artery.** Branches of this single artery lead to the descending colon, the sigmoid colon, and the rectum.

8. **Lumbar arteries.** Three or four pairs of lumbar arteries arise from the posterior surface of the aorta in the region of the lumbar vertebrae. These arteries supply muscles of the skin and the posterior abdominal wall.

9. **Middle sacral artery.** This small, single vessel descends medially from the aorta along the anterior surfaces of the lower lumbar vertebrae. It carries blood to the sacrum and coccyx.

The abdominal aorta terminates near the brim of the pelvis, where it divides into right and left *common iliac arteries.* These vessels supply blood to lower regions of the abdominal wall, the pelvic organs, and the lower extremities (fig. 15.45). Table 15.4 summarizes the main branches of the aorta.

Arteries to the Neck, Head, and Brain

Branches of the subclavian and common carotid arteries supply blood to structures within the neck, head, and brain (figs. 15.46 and 15.47). The main divisions of the

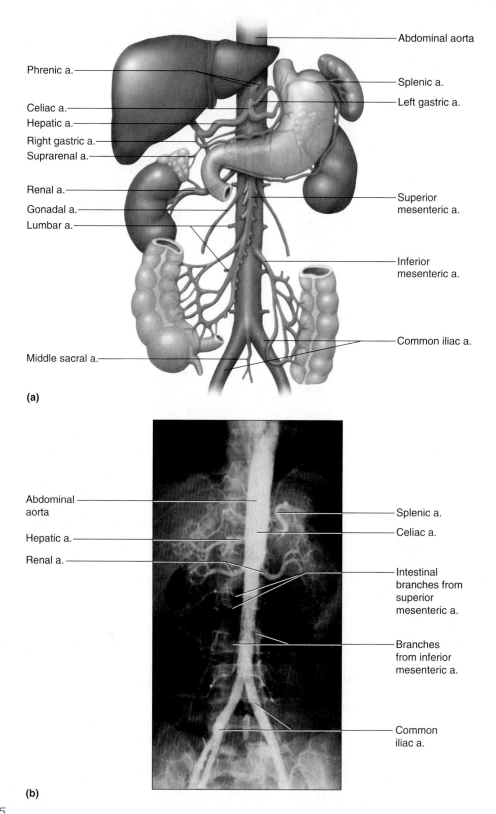

(a)

(b)

FIGURE 15.45

Abdominal aorta. (*a*) Its major branches. (*b*) Angiogram (radiograph). (*a.* stands for artery.)

TABLE 15.4   The Aorta and Its Principal Branches   585

Portion of Aorta	Major Branch	General Regions or Organs Supplied	Portion of Aorta	Major Branch	General Regions or Organs Supplied
Ascending aorta	Right and left coronary arteries	Heart	Abdominal aorta	Celiac artery	Organs of upper digestive tract
Arch of aorta	Brachiocephalic artery	Right upper limb, right side of head		Phrenic artery	Diaphragm
	Left common carotid artery	Left side of head		Superior mesenteric artery	Portions of small and large intestines
	Left subclavian artery	Left upper limb		Suprarenal artery	Adrenal gland
				Renal artery	Kidney
				Gonadal artery	Ovary or testis
Descending aorta				Inferior mesenteric artery	Lower portions of large intestine
Thoracic aorta	Bronchial artery	Bronchi		Lumbar artery	Posterior abdominal wall
	Pericardial artery	Pericardium		Middle sacral artery	Sacrum and coccyx
	Esophageal artery	Esophagus		Common iliac artery	Lower abdominal wall, pelvic organs, and lower limb
	Mediastinal artery	Mediastinum			
	Posterior intercostal artery	Thoracic wall			

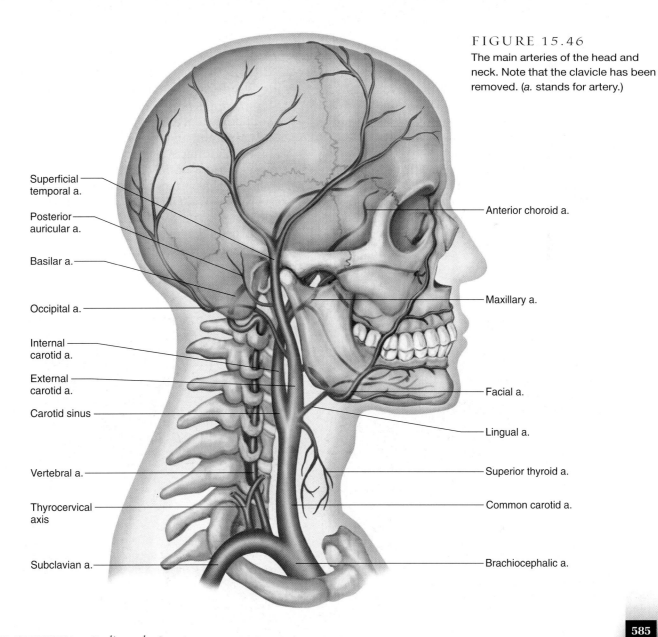

FIGURE 15.46
The main arteries of the head and neck. Note that the clavicle has been removed. (*a.* stands for artery.)

Superficial temporal a.
Posterior auricular a.
Basilar a.
Occipital a.
Internal carotid a.
External carotid a.
Carotid sinus
Vertebral a.
Thyrocervical axis
Subclavian a.

Anterior choroid a.
Maxillary a.
Facial a.
Lingual a.
Superior thyroid a.
Common carotid a.
Brachiocephalic a.

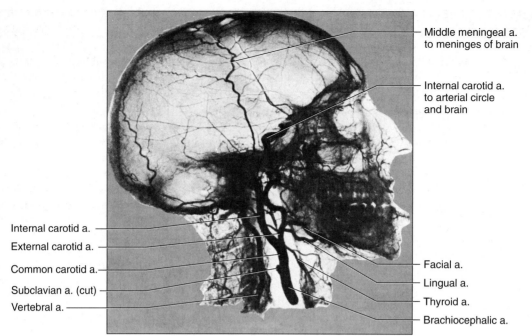

FIGURE 15.47

An angiogram of the arteries associated with the head. (*a.* stands for artery.)

subclavian artery to these regions are the vertebral, thyrocervical, and costocervical arteries. The common carotid artery communicates with these regions by means of the internal and external carotid arteries.

The **vertebral arteries** arise from the subclavian arteries in the base of the neck near the tips of the lungs. They pass upward through the foramina of the transverse processes of the cervical vertebrae and enter the skull by way of the foramen magnum. Along their paths, these vessels supply blood to vertebrae and to their associated ligaments and muscles.

Within the cranial cavity, the vertebral arteries unite to form a single *basilar artery.* This vessel passes along the ventral brainstem and gives rise to branches leading to the pons, midbrain, and cerebellum. The basilar artery terminates by dividing into two *posterior cerebral arteries* that supply portions of the occipital and temporal lobes of the cerebrum. The posterior cerebral arteries also help form the **cerebral arterial circle** (*circle of Willis*) at the base of the brain, which connects the vertebral artery and internal carotid artery systems (fig. 15.48). The union of these systems provides alternate pathways through which blood can

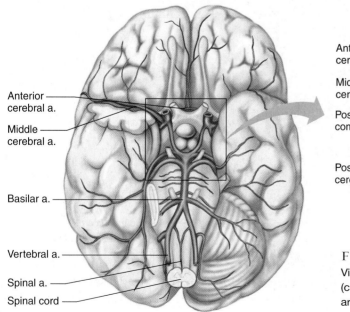

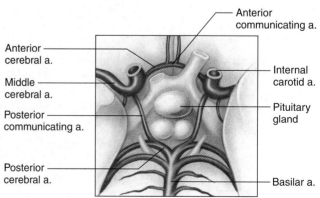

FIGURE 15.48

View of inferior surface of the brain. The cerebral arterial circle (circle of Willis) is formed by the anterior and posterior cerebral arteries, which join the internal carotid arteries. (*a.* stands for artery.)

reach brain tissues in the event of an arterial occlusion. It also equalizes blood pressure in the brain's blood supply.

The **thyrocervical** (thi″ro-ser′vĭ-kal) **arteries** are short vessels that give off branches at the thyrocervical axis to the thyroid gland, parathyroid glands, larynx, trachea, esophagus, and pharynx, as well as to various muscles in the neck, shoulder, and back. The **costocervical** (kos″to-ser′vĭ-kal) **arteries,** which are the third vessels to branch from the subclavians, carry blood to muscles in the neck, back, and thoracic wall.

The left and right *common carotid arteries* ascend deeply within the neck on either side. At the level of the upper laryngeal border, they divide to form the internal and external carotid arteries.

The **external carotid artery** courses upward on the side of the head, giving off branches to structures in the neck, face, jaw, scalp, and base of the skull. The main vessels that originate from this artery include the following:

1. *Superior thyroid artery* to the hyoid bone, larynx, and thyroid gland.

2. *Lingual artery* to the tongue, muscles of the tongue, and salivary glands beneath the tongue.

3. *Facial artery* to the pharynx, palate, chin, lips, and nose.

4. *Occipital artery* to the scalp on the back of the skull, the meninges, the mastoid process, and various muscles in the neck.

5. *Posterior auricular artery* to the ear and the scalp over the ear.

The external carotid artery terminates by dividing into *maxillary* and *superficial temporal arteries.* The maxillary artery supplies blood to the teeth, gums, jaws, cheek, nasal cavity, eyelids, and meninges. The temporal artery extends to the parotid salivary gland and to various surface regions of the face and scalp.

The **internal carotid artery** follows a deep course upward along the pharynx to the base of the skull. Entering the cranial cavity, it provides the major blood supply to the brain. The major branches of the internal carotid artery include the following:

1. *Ophthalmic artery* to the eyeball and to various muscles and accessory organs within the orbit.

2. *Posterior communicating artery* that forms part of the circle of Willis.

3. *Anterior choroid artery* to the choroid plexus within the lateral ventricle of the brain and to nerve structures within the brain.

The internal carotid artery terminates by dividing into *anterior* and *middle cerebral arteries.* The middle cerebral artery passes through the lateral tissue and supplies the lateral surface of the cerebrum, including the primary motor and sensory areas of the face and upper limbs, the optic radiations, and the speech area (see chapter 11, pp. 384–386). The anterior cerebral artery extends anteriorly between the cerebral hemispheres and supplies the medial surface of the brain.

Near the base of each internal carotid artery is an enlargement called a **carotid sinus.** Like the aortic sinuses, these structures contain baroreceptors that control blood pressure. A number of small epithelial masses, called **carotid bodies,** also occur in the wall of the carotid sinus. These bodies are very vascular and contain chemoreceptors that act with those of the aortic bodies to regulate circulation and respiration.

Arteries to the Shoulder and Upper Limb

The subclavian artery, after giving off branches to the neck, continues into the arm (fig. 15.49). It passes between the clavicle and the first rib and becomes the axillary artery.

The **axillary artery** supplies branches to structures in the axilla and the chest wall, including the skin of the shoulder, part of the mammary gland, the upper end of the humerus, the shoulder joint, and muscles in the back, shoulder, and chest. As this vessel leaves the axilla, it becomes the brachial artery.

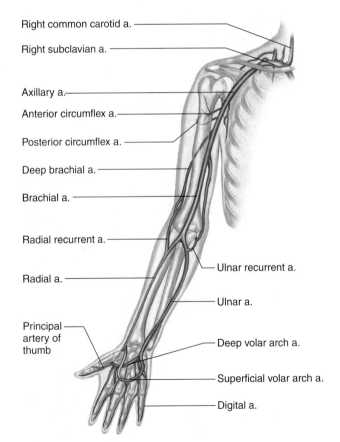

FIGURE 15.49

The main arteries to the shoulder and upper limb. (a. stands for artery.)

The **brachial artery** courses along the humerus to the elbow. It gives rise to a *deep brachial artery* that curves posteriorly around the humerus and supplies the triceps muscle. Shorter branches pass into the muscles on the anterior side of the arm, whereas others descend on each side to the elbow and connect with arteries in the forearm. The resulting arterial network allows blood to reach the forearm even if a portion of the distal brachial artery becomes obstructed.

Within the elbow, the brachial artery divides into an ulnar artery and a radial artery. The **ulnar artery** leads downward on the ulnar side of the forearm to the wrist. Some of its branches join the anastomosis around the elbow joint, whereas others supply blood to flexor and extensor muscles in the forearm.

The **radial artery,** a continuation of the brachial artery, travels along the radial side of the forearm to the wrist. As it nears the wrist, it comes close to the surface and provides a convenient vessel for taking the pulse (radial pulse). Branches of the radial artery join the anastomosis of the elbow and supply the lateral muscles of the forearm.

At the wrist, the branches of the ulnar and radial arteries join to form a network of vessels. Arteries arising from this network supply blood to structures in the wrist, hand, and fingers.

Arteries to the Thoracic and Abdominal Walls

Blood reaches the thoracic wall through several vessels. These include branches from the subclavian artery and the thoracic aorta (fig. 15.50).

The subclavian artery contributes to this supply through a branch called the **internal thoracic artery.** This vessel originates in the base of the neck and passes downward on the pleura and behind the cartilages of the upper six ribs. It gives off two *anterior intercostal arteries* to each of the upper six intercostal spaces; these two arteries supply the intercostal muscles, other intercostal tissues, and the mammary glands.

The *posterior intercostal arteries* arise from the thoracic aorta and enter the intercostal spaces between the third through the eleventh ribs. These arteries give off branches that supply the intercostal muscles, the vertebrae, the spinal cord, and deep muscles of the back.

Branches of the *internal thoracic* and *external iliac* arteries provide blood to the anterior abdominal wall. Paired vessels originating from the abdominal aorta, including the *phrenic* and *lumbar arteries,* supply blood to structures in the posterior and lateral abdominal wall.

Arteries to the Pelvis and Lower Limb

The abdominal aorta divides to form the **common iliac** (il'e-ak) **arteries** at the level of the pelvic brim. These vessels provide blood to the pelvic organs, gluteal region, and lower limbs.

Each common iliac artery descends a short distance and divides into an internal (hypogastric) branch and an external branch. The **internal iliac artery** gives off numerous branches to various pelvic muscles and visceral structures, as well as to the gluteal muscles and the external genitalia. Parts of figure 15.51 show important branches of this vessel, including the following:

1. *Iliolumbar artery* to the ilium and muscles of the back.

2. *Superior and inferior gluteal arteries* to the gluteal muscles, pelvic muscles, and skin of the buttocks.

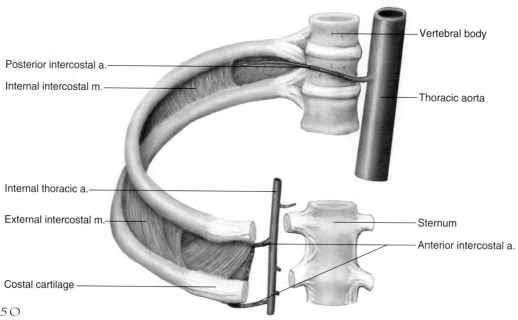

Posterior intercostal a.
Internal intercostal m.
Internal thoracic a.
External intercostal m.
Costal cartilage
Vertebral body
Thoracic aorta
Sternum
Anterior intercostal a.

FIGURE 15.50

Arteries that supply the thoracic wall. (*a.* stands for artery, *m.* stands for muscle.)

588

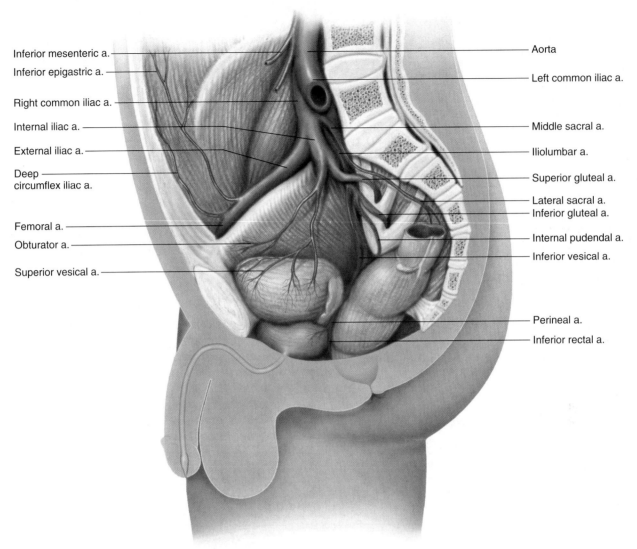

Inferior mesenteric a.
Inferior epigastric a.
Right common iliac a.
Internal iliac a.
External iliac a.
Deep circumflex iliac a.
Femoral a.
Obturator a.
Superior vesical a.

Aorta
Left common iliac a.
Middle sacral a.
Iliolumbar a.
Superior gluteal a.
Lateral sacral a.
Inferior gluteal a.
Internal pudendal a.
Inferior vesical a.
Perineal a.
Inferior rectal a.

FIGURE 15.51
Arteries that supply the pelvic region. (*a.* stands for artery.)

3. *Internal pudendal artery* to muscles in the distal portion of the alimentary canal, the external genitalia, and the hip joint.

4. *Superior* and *inferior vesical arteries* to the urinary bladder. In males, these vessels also supply the seminal vesicles and the prostate gland.

5. *Middle rectal artery* to the rectum.

6. *Uterine artery* to the uterus and vagina.

The **external iliac artery** provides the main blood supply to the lower limbs (fig. 15.52). It passes downward along the brim of the pelvis and gives off two large branches—an *inferior epigastric artery* and a *deep circumflex iliac artery.* These vessels supply the muscles and skin in the lower abdominal wall. Midway between the symphysis pubis and the anterior superior iliac spine of the ilium, the external iliac artery becomes the femoral artery.

The **femoral** (fem′or-al) **artery,** which passes fairly close to the anterior surface of the upper thigh, gives off

many branches to muscles and superficial tissues of the thigh. These branches also supply the skin of the groin and the lower abdominal wall. Important subdivisions of the femoral artery include the following:

1. *Superficial circumflex iliac artery* to the lymph nodes and skin of the groin.

2. *Superficial epigastric artery* to the skin of the lower abdominal wall.

3. *Superficial* and *deep external pudendal arteries* to the skin of the lower abdomen and external genitalia.

4. *Deep femoral artery* (the largest branch of the femoral artery) to the hip joint and muscles of the thigh.

5. *Deep genicular artery* to distal ends of thigh muscles and to an anastomosis around the knee joint.

As the femoral artery reaches the proximal border of the space behind the knee (popliteal fossa), it becomes the

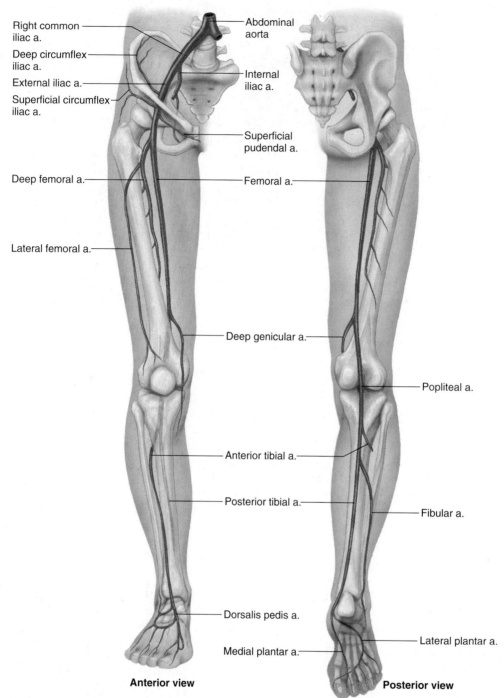

Right common iliac a.

Deep circumflex iliac a.

External iliac a.

Superficial circumflex iliac a.

Abdominal aorta

Internal iliac a.

Superficial pudendal a.

Deep femoral a.

Femoral a.

Lateral femoral a.

Deep genicular a.

Popliteal a.

Anterior tibial a.

Posterior tibial a.

Fibular a.

Dorsalis pedis a.

Medial plantar a.

Lateral plantar a.

Anterior view

Posterior view

FIGURE 15.52

Main branches of the external iliac artery. (a. stands for artery.)

popliteal (pop″lĭ-te′al) **artery.** Branches of this artery supply blood to the knee joint and to certain muscles in the thigh and calf. Also, many of its branches join the anastomosis of the knee and help provide alternate pathways for blood in the case of arterial obstructions. At the lower border of the popliteal fossa, the popliteal artery divides into the anterior and posterior tibial arteries.

The **anterior tibial** (tib′e-al) **artery** passes downward between the tibia and the fibula, giving off branches to the

skin and muscles in anterior and lateral regions of the leg. It also communicates with the anastomosis of the knee and with a network of arteries around the ankle. This vessel continues into the foot as the *dorsalis pedis artery,* which supplies blood to the foot and toes.

The **posterior tibial artery,** the larger of the two popliteal branches, descends beneath the calf muscles, giving off branches to the skin, muscles, and other tissues of the leg along the way. Some of these vessels join the anastomoses of the knee and ankle. As it passes between the medial malleolus and the heel, the posterior tibial artery divides into the *medial* and *lateral plantar arteries.* Branches from these arteries supply blood to tissues of the heel, foot, and toes.

The largest branch of the posterior tibial artery is the *fibular artery,* which travels downward along the fibula and contributes to the anastomosis of the ankle. Figure 15.53 shows the major vessels of the arterial system.

1 Describe the structure of the aorta.

2 Name the vessels that arise from the aortic arch.

3 Name the branches of the thoracic and abdominal aorta.

4 Which vessels supply blood to the head? To the upper limb? To the abdominal wall? To the lower limb?

Venous System

Venous circulation returns blood to the heart after gases, nutrients, and wastes are exchanged between the blood and body cells.

Characteristics of Venous Pathways

The vessels of the venous system begin with the merging of capillaries into venules, venules into small veins, and small veins into larger ones. Unlike the arterial pathways, however, those of the venous system are difficult to follow. This is because the vessels commonly connect in irregular networks, so many unnamed tributaries may join to form a relatively large vein.

On the other hand, the larger veins typically parallel the courses of named arteries, and these veins often have the same names as their counterparts in the arterial system. For example, the renal vein parallels the renal artery, and the common iliac vein accompanies the common iliac artery.

The veins that carry the blood from the lungs and myocardium back to the heart have already been described. The veins from all the other parts of the body converge into two major pathways, the *superior* and *inferior venae cavae,* which lead to the right atrium.

Veins from the Head, Neck, and Brain

The **external jugular** (jug'u-lar) **veins** drain blood from the face, scalp, and superficial regions of the neck. These vessels descend on either side of the neck, passing over the sternocleidomastoid muscles and beneath the platysma. They empty into the *right* and *left subclavian veins* in the base of the neck (fig. 15.54).

The **internal jugular veins,** which are somewhat larger than the external jugular veins, arise from numerous veins and venous sinuses of the brain and from deep veins in various parts of the face and neck. They pass downward through the neck beside the common carotid

arteries and also join the subclavian veins. These unions of the internal jugular and subclavian veins form large **brachiocephalic (innominate) veins** on each side. These vessels then merge in the mediastinum and give rise to the **superior vena cava,** which enters the right atrium.

A lung cancer, enlarged lymph node, or an aortic aneurysm can compress the superior vena cava, interfering with return of blood from the upper body to the heart. This produces pain, shortness of breath, distension of veins draining into the superior vena cava, and swelling of tissues in the face, head, and lower limbs. Restriction of blood flow to the brain may threaten life.

Veins from the Upper Limb and Shoulder

A set of deep veins and a set of superficial ones drain the upper limb. The deep veins generally parallel the arteries in each region and are given similar names, such as the *radial vein, ulnar vein, brachial vein,* and *axillary vein.* The superficial veins connect in complex networks just beneath the skin. They also communicate with the deep vessels of the upper limb, providing many alternate pathways through which the blood can leave the tissues (fig. 15.55).

The main vessels of the superficial network are the basilic and cephalic veins. They arise from anastomoses in the hand and wrist on the ulnar and radial sides, respectively.

The **basilic** (bah-sil'ik) **vein** passes along the back of the forearm on the ulnar side for a distance and then curves forward to the anterior surface below the elbow. It continues ascending on the medial side until it reaches the middle of the arm. There it penetrates the tissues deeply and joins the *brachial vein.* As the basilic and brachial veins merge, they form the *axillary vein.*

The **cephalic** (sĕ-fal'ik) **vein** courses upward on the lateral side of the upper limb from the hand to the shoulder. In the shoulder, it pierces the tissues and empties into the axillary vein. Beyond the axilla, the axillary vein becomes the *subclavian vein.*

In the bend of the elbow, a *median cubital vein* ascends from the cephalic vein on the lateral side of the forearm to the basilic vein on the medial side. This large vein is usually visible. It is often used as a site for *venipuncture,* when it is necessary to remove a sample of blood for examination or to add fluids to the blood.

Veins from the Abdominal and Thoracic Walls

Tributaries of the brachiocephalic and azygos veins drain the abdominal and thoracic walls. For example, the *brachiocephalic vein* receives blood from the *internal thoracic vein,* which generally drains the tissues the internal

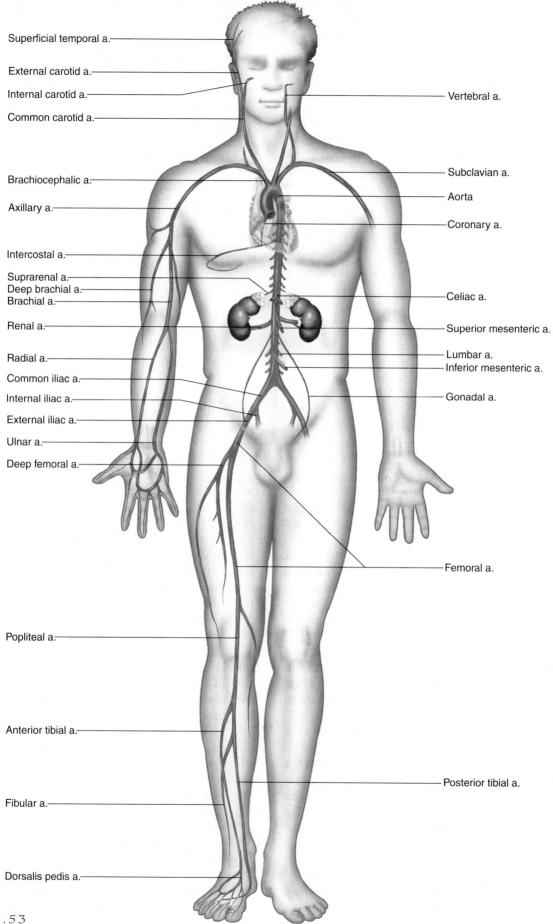

Superficial temporal a.

External carotid a.

Internal carotid a.

Common carotid a.

Vertebral a.

Brachiocephalic a.

Subclavian a.

Axillary a.

Aorta

Coronary a.

Intercostal a.

Suprarenal a.

Deep brachial a.

Brachial a.

Celiac a.

Renal a.

Superior mesenteric a.

Radial a.

Lumbar a.

Common iliac a.

Inferior mesenteric a.

Internal iliac a.

Gonadal a.

External iliac a.

Ulnar a.

Deep femoral a.

Femoral a.

Popliteal a.

Anterior tibial a.

Posterior tibial a.

Fibular a.

Dorsalis pedis a.

FIGURE 15.53

Major vessels of the arterial system (*a.* stands for artery.)

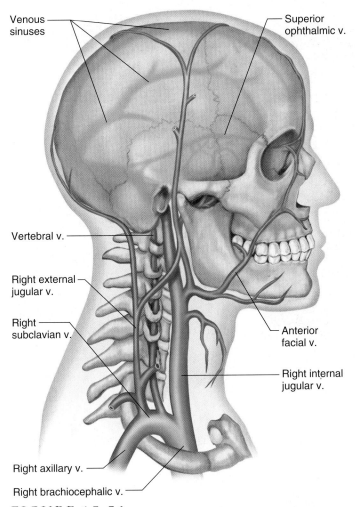

FIGURE 15.54
The major veins of the brain, head, and neck. Note that the clavicle has been removed. (*v.* stands for vein.)

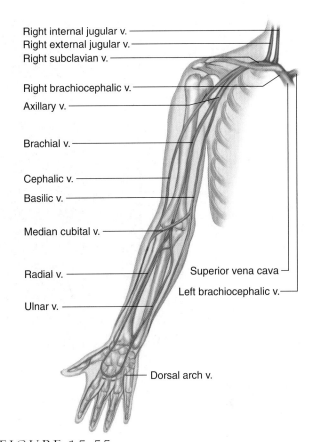

FIGURE 15.55
The main veins of the upper limb and shoulder. (*v.* stands for vein.)

thoracic artery supplies. Some *intercostal veins* also empty into the brachiocephalic vein (fig. 15.56).

The **azygos** (az'ĭ-gos) **vein** originates in the dorsal abdominal wall and ascends through the mediastinum on the right side of the vertebral column to join the superior vena cava. It drains most of the muscular tissue in the abdominal and thoracic walls.

Tributaries of the azygos vein include the *posterior intercostal veins* on the right side, which drain the intercostal spaces, and the *superior* and *inferior hemiazygos veins,* which receive blood from the posterior intercostal veins on the left. The right and left *ascending lumbar veins,* with tributaries that include vessels from the lumbar and sacral regions, also connect to the azygos system.

Veins from the Abdominal Viscera

Veins carry blood directly to the atria of the heart, except those that drain the abdominal viscera (fig. 15.57). They originate in the capillary networks of the stomach, intestines, pancreas, and spleen and carry blood from

these organs through a **hepatic portal** (por'tal) **vein** to the liver (fig. 15.58). There the blood enters capillary-like **hepatic sinusoids** (hĕ-pat'ik si'nŭ-soidz). This unique venous pathway is called the **hepatic portal system.**

The tributaries of the hepatic portal vein include the following vessels:

1. Right and left *gastric veins* from the stomach.

2. *Superior mesenteric vein* from the small intestine, ascending colon, and transverse colon.

3. *Splenic vein* from a convergence of several veins draining the spleen, the pancreas, and a portion of the stomach. Its largest tributary, the *inferior mesenteric vein,* brings blood upward from the descending colon, sigmoid colon, and rectum.

About 80% of the blood flowing to the liver in the hepatic portal system comes from the capillaries in the stomach and intestines and is oxygen-poor, but rich in nutrients. As discussed in chapter 17 (pp. 666–669), the liver handles these nutrients in a variety of ways. It regulates blood glucose concentration by polymerizing excess glucose into glycogen for storage or by breaking down glycogen into glucose when blood glucose concentration drops below normal.

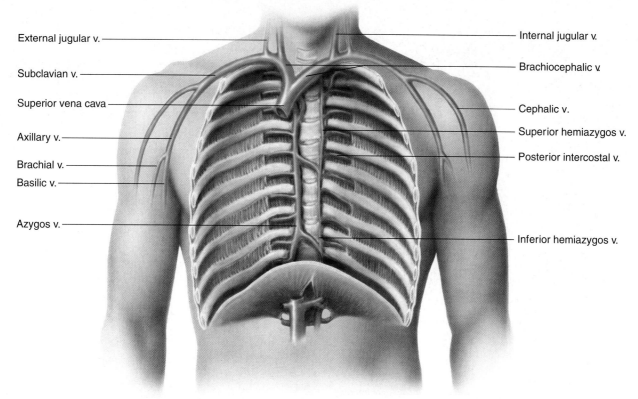

FIGURE 15.56

Veins that drain the thoracic wall. (*v.* stands for vein.)

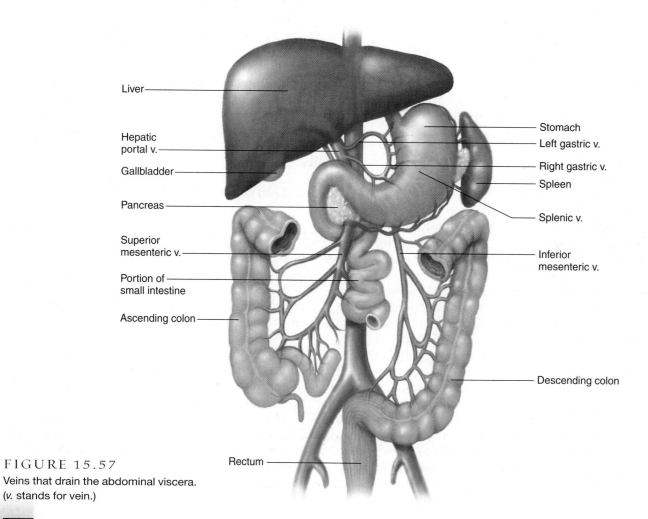

FIGURE 15.57

Veins that drain the abdominal viscera.
(*v.* stands for vein.)

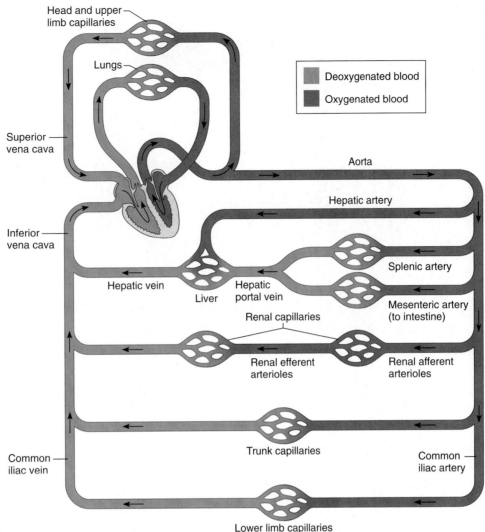

Head and upper limb capillaries

Lungs

Superior vena cava

Aorta

Deoxygenated blood
Oxygenated blood

Hepatic artery

Inferior vena cava

Hepatic vein

Liver

Hepatic portal vein

Splenic artery

Mesenteric artery (to intestine)

Renal capillaries

Renal efferent arterioles

Renal afferent arterioles

Trunk capillaries

Common iliac vein

Common iliac artery

Lower limb capillaries

FIGURE 15.58

In this schematic drawing of the cardiovascular system, note how the hepatic portal vein drains one set of the capillaries and leads to another set. A similar relationship exists in the kidneys.

The liver helps regulate blood concentrations of recently absorbed amino acids and lipids by modifying them into forms cells can use, by oxidizing them, or by changing them into storage forms. The liver also stores certain vitamins and detoxifies harmful substances.

Blood in the hepatic portal vein nearly always contains bacteria that have entered through intestinal capillaries. Large *Kupffer cells* lining the hepatic sinusoids phagocytize these microorganisms, removing them from the portal blood before it leaves the liver.

After passing through the hepatic sinusoids of the liver, the blood in the hepatic portal system travels through a series of merging vessels into **hepatic veins.** These veins empty into the *inferior vena cava,* returning the blood to the general circulation.

Other veins empty into the inferior vena cava as it ascends through the abdomen. They include the *lumbar, gonadal, renal, suprarenal,* and *phrenic veins.* These vessels drain regions that arteries with corresponding names supply.

Veins from the Lower Limb and Pelvis

As in the upper limb, veins that drain the blood from the lower limb can be divided into deep and superficial groups (fig. 15.59). The deep veins of the leg, such as the *anterior* and *posterior tibial veins,* have names that correspond to the arteries they accompany. At the level of the knee, these vessels form a single trunk, the **popliteal vein.** This vein continues upward through the thigh as the **femoral vein,** which, in turn, becomes the **external iliac vein.**

The superficial veins of the foot, leg, and thigh connect to form a complex network beneath the skin. These vessels drain into two major trunks: the small and great saphenous veins.

The **small saphenous** (sah-fe'nus) **vein** begins in the lateral portion of the foot and passes upward behind the lateral malleolus. It ascends along the back of the calf, enters the popliteal fossa, and joins the popliteal vein.

The **great saphenous vein,** which is the longest vein in the body, originates on the medial side of the foot. It ascends in front of the medial malleolus and extends upward along

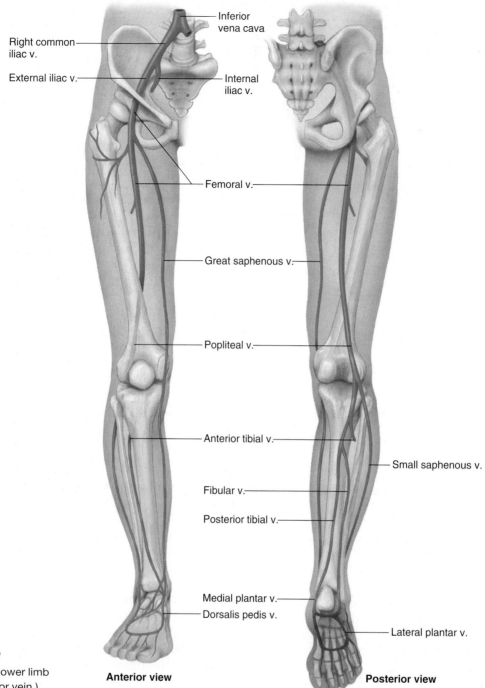

Inferior
vena cava

Right common
iliac v.

External iliac v.

Internal
iliac v.

Femoral v.

Great saphenous v.

Popliteal v.

Anterior tibial v.

Small saphenous v.

Fibular v.

Posterior tibial v.

Medial plantar v.
Dorsalis pedis v.

Lateral plantar v.

Anterior view

Posterior view

FIGURE 15.59

The main veins of the lower limb
and pelvis. (*v.* stands for vein.)

the medial side of the leg and thigh. In the thigh just below
the inguinal ligament, it penetrates deeply and joins the
femoral vein. Near its termination, the great saphenous vein
receives tributaries from a number of vessels that drain the
upper thigh, groin, and lower abdominal wall.

In addition to communicating freely with each other,
the saphenous veins communicate extensively with the
deep veins of the leg and thigh. Blood can thus return to
the heart from the lower extremities by several routes.

In the pelvic region, vessels leading to the **internal
iliac vein** carry blood away from organs of the reproductive, urinary, and digestive systems. This vein is formed

by tributaries corresponding to the branches of the internal iliac artery, such as the *gluteal, pudendal, vesical, rectal, uterine,* and *vaginal veins.* Typically, these veins have
many connections and form complex networks (plexuses)
in the regions of the rectum, urinary bladder, and prostate
gland (in the male) or uterus and vagina (in the female).

The internal iliac veins originate deep within the
pelvis and ascend to the pelvic brim. There they unite with
the right and left external iliac veins to form the **common
iliac veins.** These vessels, in turn, merge to produce the
inferior vena cava at the level of the fifth lumbar vertebra.
Figure 15.60 shows the major vessels of the venous system.

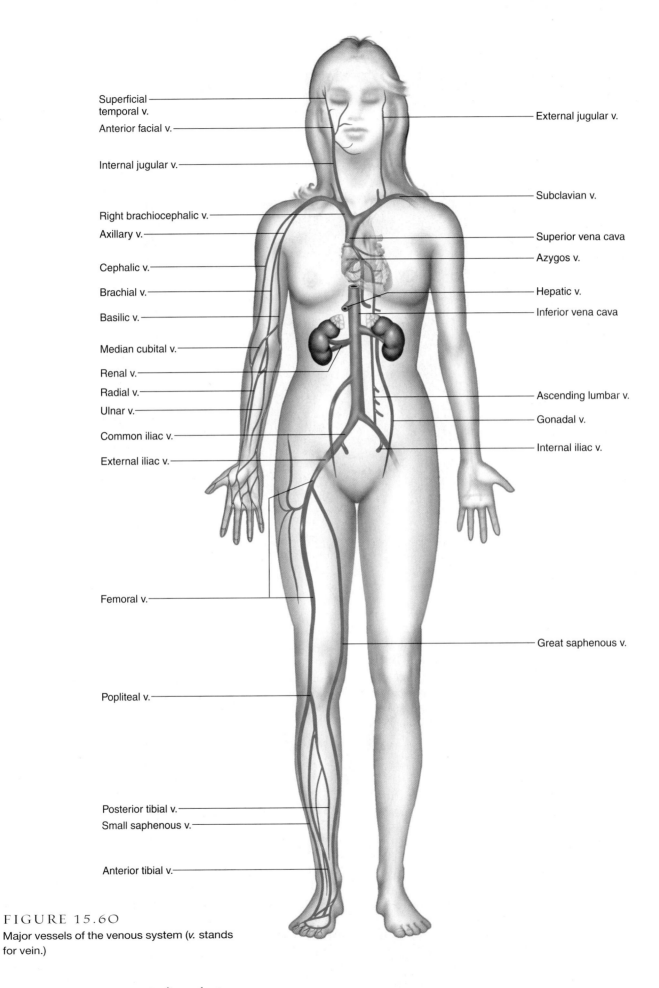

Superficial temporal v.

Anterior facial v.

Internal jugular v.

Right brachiocephalic v.

Axillary v.

Cephalic v.

Brachial v.

Basilic v.

Median cubital v.

Renal v.

Radial v.

Ulnar v.

Common iliac v.

External iliac v.

Femoral v.

Popliteal v.

Posterior tibial v.

Small saphenous v.

Anterior tibial v.

External jugular v.

Subclavian v.

Superior vena cava

Azygos v.

Hepatic v.

Inferior vena cava

Ascending lumbar v.

Gonadal v.

Internal iliac v.

Great saphenous v.

FIGURE 15.60

Major vessels of the venous system (*v.* stands for vein.)

MOLECULAR CAUSES OF CARDIOVASCULAR DISEASE

A variety of inherited and environmental factors contribute to causing cardiovascular disease. Many cases are probably due to a fatty diet and sedentary lifestyle, against a backdrop of genetic predisposition. Disorders of the heart and blood vessels caused by single genes are very rare, but understanding how they arise can provide insights that are useful in developing treatments for more prevalent forms of disease. For example, widely used cholesterol-lowering drugs called statins were developed based on analyzing the molecular malfunction behind the one-in-a-million inherited condition familial hypercholesterolemia.

A Connective Tissue Defect

In January 1986, volleyball champion Flo Hyman left the court during a game in Japan, collapsed, and suddenly died. Her aorta had burst, and death was instant. Hyman had *Marfan syndrome,* an inherited condition that also caused the characteristics that led her to excel in her sport—her great height and long fingers (fig. 15J).

In Marfan syndrome, an abnormal form of a connective tissue protein called fibrillin weakens the aorta wall. After Flo Hyman died, her siblings were examined, and her brother Michael was found to have a weakened aorta. By surgically repairing his aorta and giving him drugs to control his blood pressure and heart rate, physicians enabled him to avoid the sudden death that claimed his famous sister. Testing for the

(a)

(b)

FIGURE 15J

Aneurysm. (*a*) Two years after she led the 1984 U.S. women's volleyball team to a silver medal in the Olympics, Flo Hyman died suddenly when her aorta burst, a symptom of Marfan syndrome. (*b*) Note the swelling (aneurysm) of the aorta in the heart on the right. A burst aneurysm here is fatal.

The chapter concludes with Clinical Application 15.7, which looks at molecular explanations of certain cardiovascular disorders, and Clinical Application 15.8, which discusses coronary artery disease.

1 Name the veins that return blood to the right atrium.

2 Which major veins drain blood from the head? From the upper limbs? From the abdominal viscera? From the lower limbs?

Life-Span Changes

Tracking the changes that are part of normal aging of the cardiovascular system is difficult because of the high incidence of disease affecting the heart and blood vessels, which increases exponentially with age. For example, 60% of men over age sixty have at least one narrowed coronary artery; the same is true for women over age eighty. Signs of cardiovascular disease may appear long

causative gene can alert physicians to affected individuals before the dangerous swelling in the aorta begins.

A Myosin Defect

Each year, one or two seemingly healthy young people die suddenly during a sports event, usually basketball. The cause of death is often *familial hypertrophic cardiomyopathy,* an inherited overgrowth of the heart muscle. The defect in this disorder is different than that behind Marfan syndrome. It is an abnormality in one of the myosin chains that comprise cardiac muscle. Again, detecting the responsible gene can alert affected individuals to their increased risk of sudden death. They can adjust the type of exercise they do to avoid stressing the cardiovascular system.

A Metabolic Glitch

Sometimes inherited heart disease strikes very early in life. Jim D. died at four days of age, two days after suffering cardiac arrest. Two years later, his parents had another son. Like Jim, Kerry seemed normal at birth, but at thirty-six hours of age, his heart rate plummeted, he had a seizure, and he stopped breathing. He was resuscitated. A blood test revealed excess long-chain fatty acids, indicating a metabolic disorder, an inability to utilize fatty acids. Lack of food triggered the symptoms because the boys could not use fatty acids for energy, as healthy people do. Kerry was able to survive for three years by following a diet low in fatty acids and eating frequently. Once he became comatose because he missed a meal. Eventually, he died of respiratory failure.

Kerry and Jim had inherited a deficiency of a mitochondrial enzyme that processes long-chain fatty acids. Because this is a primary energy source for cardiac muscle, their tiny hearts failed.

Controlling Cholesterol

Low-density lipoprotein (LDL) receptors on liver cells admit cholesterol into the cells, keeping the lipid from building up in the bloodstream and occluding arteries. When LDL receptors bind cholesterol, they activate a negative feedback system that temporarily halts the cell's production of cholesterol. In the severe form of *familial hypercholesterolemia,* a person inherits two defective copies of the gene encoding the LDL receptors. Yellowish lumps of cholesterol can be seen behind the knees and elbows, and heart failure usually causes death in childhood. People who inherit one defective gene have a milder form of the illness. They tend to develop coronary artery disease in young adulthood, but can delay symptoms by following a heart-healthy diet and regularly exercising. These people have half the normal number of LDL receptors.

In Niemann-Pick type C disease, a defective protein disturbs the fate of cholesterol inside cells. Normally, the protein escorts cholesterol out of a cell's lysosomes (a type of organelle), which triggers the negative feedback mechanism that shuts off cholesterol synthesis. When the protein is absent or malfunctions, the cell keeps producing cholesterol and LDL receptors. Coronary artery disease develops, which is typically fatal in childhood.

Homing in on Homocysteine

Homocysteine is an amino acid that forms when a different amino acid, methionine, loses a methyl (CH_3) group. Normally, enzymes tack the methyl back onto homocysteine, regenerating methionine, or metabolize homocysteine to yield yet a third type of amino acid, cysteine. If an enzyme deficiency called *homocystinuria* causes homocysteine to build up in the blood, changes in arterial linings develop that encourage cholesterol plaque deposition, and the risks of heart attack and stroke rise dramatically. The complex biochemical pathways that recycle homocysteine to methionine, or break it down to cysteine, require three vitamins—folic acid, B_6, and B_{12}. The details of these pathways were deciphered in the 1960s, based on study of a handful of children with the rare homocystinuria. Their artery linings were like those of a much older person with severe atherosclerosis.

In the 1970s, a pathologist, Kilmer McCully, hypothesized that if extremely high levels of homocysteine in the blood of these children caused their severe heart disease, then perhaps more moderately elevated levels in others are responsible for more common forms of cardiovascular disease. Since then, many clinical trials have confirmed the correlation between elevated blood homocysteine and increased risk of cardiovascular disease. We do not yet know whether lowering homocysteine level improves heart and blood vessel health. If this is the case, the treatment for homocysteine-related cardiovascular disease is straightforward—obtaining sufficient folic acid, B_6, and B_{12}. ■

before symptoms arise. Autopsies of soldiers killed in the Korean and Vietnam wars, for example, reveal significant plaque buildup in the arterial walls of otherwise healthy young men. Some degree of cholesterol deposition in blood vessels may be part of normal aging in many types of animals, but the accumulation is great enough to lead to overt disease in humans. This is because we can extend our lives in ways that other animals cannot, and hidden disease conditions can become problems if one lives long enough!

Assessing cardiac output over a lifetime vividly illustrates how cardiovascular disease prevalence can interfere with studying the changes associated with normal aging. Recall that cardiac output is the ability of the heart to meet the body's oxygen requirements and is calculated as the heart rate in beats per minute multiplied by the stroke volume in milliliters per beat. For many years, studies indicated that cardiac output declines with age, but when researchers began to screen participants for hidden heart disease with treadmill stress tests, then evaluated only

CORONARY ARTERY DISEASE

Dave R., a fifty-two-year-old overweight accountant, had been having occasional chest pains for several months. The mild pain occurred during his usual weekend tennis match, and he attributed it to indigestion. The discomfort almost always diminished after the game, but recently, the pain seemed more severe and prolonged. Dave asked his physician about the problem.

The physician explained that Dave was probably experiencing *angina pectoris,* a symptom of *coronary artery disease* (CAD), and suggested that he undergo an *exercise stress test.* Dave walked on a treadmill, increasing speed and incline while he exercised, and an ECG was recorded, and his blood pressure monitored. Near the end of the test, when Dave's heart reached the desired rate, a small quantity of radioactive thallium-201 was injected into a vein. A *scintillation counter* scanned Dave's heart to determine if branches of his coronary arteries carried the blood marked with the thallium uniformly throughout the myocardium (see fig. 15.13).

The test revealed that Dave was developing CAD. In addition, he had hypertension and high blood cholesterol. The physician advised Dave to stop smoking; to reduce his intake of foods high in saturated fats, cholesterol, and sodium; and to exercise regularly. He was given medications to lower his blood pressure and to relieve the pain of angina. The doctor also cautioned Dave to avoid stressful situations and to lose weight.

Six months later, in spite of following medical advice, Dave suffered a heart attack—a sign that blood flow to part of his myocardium had been obstructed, producing oxygen deficiency (ischemia). The attack began as severe, crushing chest pain, shortness of breath, and sweating. Paramedics stabilized Dave's condition and transported him to a hospital. There, a cardiologist concluded from an ECG that Dave's heart attack was caused by a blood clot obstructing a coronary artery (occlusive coronary thrombosis). The cardiologist administered a thrombolytic ("clot-busting") drug, tissue plasminogen activator (tPA) intravenously.

After some time, the ECG showed that the blood vessel remained partially obstructed, so the cardiologist ordered a *coronary angiogram.* In this X-ray procedure, which was conducted in the cardiac catheterization laboratory, a thin plastic catheter was passed through a guiding sheath inserted into the femoral artery of Dave's right inguinal area. From there, the catheter was pushed into the aorta until it reached the region of the opening to the left coronary artery, and then near the opening to the right coronary artery.

X-ray fluoroscopy monitored the progress of the catheter. Each time the catheter was in proper position, a radiopaque dye (contrast medium) was released from its distal end into the blood. X-ray images that revealed the path of the dye as it entered a coronary artery and its branches were recorded on videotape and on motion-picture film, which were later analyzed frame by frame. A single severe narrowing was discovered near the origin of Dave's left anterior descending artery. The cardiologist decided to perform *percutaneous transluminal coronary angioplasty* (PTCA) in order to enlarge the opening (lumen) of that vessel.

The PTCA was performed by passing another plastic catheter through the guiding sheath used for the angiogram. This second tube had a tiny deflated balloon at its tip, and when the balloon was located in the region of the arterial narrowing, it was inflated for a short time with relatively high pressure. The inflating balloon compressed the atherosclerotic plaque (atheroma) that was obstructing the arterial wall. The expanding balloon also stretched the blood vessel wall, thus widening its lumen (recanalization). Blood flow to the myocardial tissue downstream from the obstruction improved immediately.

About 50% of the time, a vessel opened with PTCA becomes occluded again, because the underlying disease persists. To prevent this *restenosis,* the doctor inserted a *coronary stent,* which is an expandable tube or coil that holds the vessel wall open. The cardiologist had two other options that have a slightly higher risk of causing damage. She might have vaporized the plaque obstructing the vessel with an excimer laser pulse delivered along optical fibers threaded through the catheter. Or, she could have performed atherectomy, in which a cutting device attached to the balloon inserted into the catheter spins, removing plaque by withdrawing it on the catheter tip.

Should the coronary stent fail, or an obstruction block another heart vessel, Dave might benefit from *coronary bypass surgery.* A portion of his internal mammary artery inside his chest wall or his great saphenous vein would be removed and stitched (with the vein reversed to allow blood flow through the valves) between the aorta and the blocked coronary artery at a point beyond the obstruction, restoring circulation through the heart. ■

---

individuals with completely healthy cardiovascular systems, they discovered that cardiac output at rest is maintained as a person ages. It does decline during exercise for some people, however.

The heart normally remains the same size or may even shrink a little with age, but cardiovascular disease may cause enlargement. It is normal for the proportion of the heart that is cardiac muscle to decline with age, because cardiac muscle cells do not divide—that is, they are not replaced (unless stem cells are activated by injury). The number and size of cardiac muscle cells fall, and lipofuscin pigments become especially prominent in this cell type. Fibrous connective tissue and adipose tissue fill in the spaces left by the waning population of cardiac muscle cells, thickening the endocardium. Adipose cells may also accumulate in the ventricle walls and the septum between them. As a result, the left ventricular wall may be up to 25% thicker at age eighty than it was at age thirty.

The heart valves tend to thicken and become more rigid after age sixty, but these changes may actually begin as early as the third decade. The valves may calcify— some calcification of the aortic valve in particular after the seventh decade is very common and may be considered a normal part of aging.

Just as the heart need not falter with age, so does the cardiac conduction system remain functional despite change. The sinoatrial and atrioventricular nodes and the atrioventricular bundle become more elastic. However, these changes may alter the ECG pattern.

Systolic blood pressure increases with age; a blood pressure reading of 140/90 is not considered abnormal in an older person. In about 40% of the elderly, the systolic pressure exceeds 160. The increase seems to be due to the decreasing diameters and elasticity of arteries, an effect that is dampened somewhat by regular exercise. Resting heart rate declines with age, from 145 or more beats per minute in a fetus, to 140 beats per minute in a newborn, then levels out in an adult to about 70 (range of 60–99) beats per minute.

In the vascular system, changes that are part of aging are most apparent in the arteries. The tunica interna thickens. Dividing smooth muscle cells in the tunica media may push up the endothelium in places, and over time, the lumens of the larger arteries narrow. Rigidity increases as collagen, calcium, and fat are deposited as elastin production declines. Arterial elasticity at age seventy is only about half of what it was at age twenty. The arterioles have diminished ability to contract in response to cold temperatures and to dilate in response to heat, contributing to the loss of temperature control that is common among the elderly. The extent of change in arteries may reflect how much stress they are under—that is, not all arteries "age" at the same rate.

Veins may accumulate collagen and calcify but, in general, do not change as much with age as do arteries. Thickened patches may appear in the inner layer, and fibers in the valves, but venous diameters are large enough that these changes have little impact on function. The venous supply to many areas is so redundant that alternate vessels can often take over for damaged ones. The number of capillaries declines with age. The once-sleek endothelium changes as the cells become less uniform in size and shape. The endothelial inner linings of blood vessels are important to health because these cells release nitric oxide, which signals the vessels to dilate to increase blood flow, which counters atherosclerosis and thrombosis.

At least one study demonstrates that exercise can help maintain a "young" vascular system. The study compared the vascular endothelial linings of athletic and sedentary individuals of various ages and found that the status of the vessels of the exercising elderly were very similar to those of either athletic or sedentary people in their twenties. This finding is consistent with results of the Honolulu Heart Program, which found that walking 1.5 miles each day correlates to lowered heart disease risk in older people.

Overall, aging-related changes affect many components of the cardiovascular system. But in the absence of disease, the system is so fine-tuned and redundant that effective oxygen delivery can continue well into the later decades of life.

1 Explain why the heart may enlarge with age.

2 Describe what happens to resting heart rate with age.

Clinical Terms Related to the Cardiovascular System

anastomosis (ah-nas″to-mo′sis) Connection between two blood vessels, sometimes produced surgically.

angiospasm (an′je-o-spazm″) Muscular spasm in the wall of a blood vessel.

arteriography (ar″te-re-og′rah-fe) Injection of radiopaque solution into the vascular system for an X-ray examination of arteries.

asystole (a-sis′to-le) Condition in which the myocardium fails to contract.

cardiac tamponade (kar′de-ak tam″po-nād′) Compression of the heart by fluid accumulating within the pericardial cavity.

congestive heart failure (kon-jes′tiv hart fāl′yer) Inability of the left ventricle to pump adequate blood to cells.

cor pulmonale (kor pul-mo-na′le) Heart-lung disorder of pulmonary hypertension and hypertrophy of the right ventricle.

embolectomy (em″bo-lek′to-me) Removal of an embolus through an incision in a blood vessel.

endarterectomy (en″dar-ter-ek′to-me) Removal of the inner wall of an artery to reduce an arterial occlusion.

palpitation (pal″pi-ta′shun) Awareness of a heartbeat that is unusually rapid, strong, or irregular.

pericardiectomy (per″i-kar″de-ek′to-me) Excision of the pericardium.

phlebitis (flĕ-bi′tis) Inflammation of a vein, usually in the lower limbs.

phlebotomy (flĕ-bot′o-me) Incision or puncture of a vein to withdraw blood.

sinus rhythm (si′nus rithm) The normal cardiac rhythm regulated by the S-A node.

thrombophlebitis (throm″bo-fle-bi′tis) Formation of a blood clot in a vein in response to inflammation of the venous wall.

valvotomy (val-vot′o-me) Incision of a valve.

venography (ve-nog′rah-fe) Injection of radiopaque solution into the vascular system for X-ray examination of veins.

INNERCONNECTIONS
CARDIOVASCULAR SYSTEM

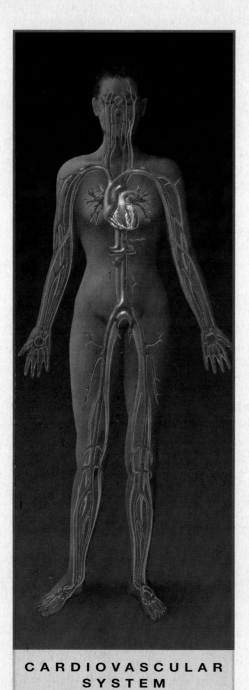

CARDIOVASCULAR SYSTEM

The heart pumps blood through as many as 60,000 miles of blood vessels, delivering nutrients to, and removing wastes from, all body cells.

Integumentary System

Changes in skin blood flow are important in temperature control.

Lymphatic System

The lymphatic system returns tissue fluids to the bloodstream.

Skeletal System

Bones help control plasma calcium levels.

Digestive System

The digestive system breaks down nutrients into forms readily absorbed by the bloodstream.

Muscular System

Blood flow increases to exercising skeletal muscle, delivering oxygen and nutrients and removing wastes. Muscle actions help the blood circulate.

Respiratory System

The respiratory system oxygenates the blood and removes carbon dioxide. Respiratory movements help the blood circulate.

Nervous System

The brain depends on blood flow for survival. The nervous system helps control blood flow and blood pressure.

Urinary System

The kidneys clear the blood of wastes and substances present in the body. The kidneys help control blood pressure and blood volume.

Endocrine System

Hormones are carried in the bloodstream. Some hormones directly affect the heart and blood vessels.

Reproductive System

Blood pressure is important in normal function of the sex organs.

602

UNIT FOUR

CHAPTER SUMMARY

Introduction (page 542)

The cardiovascular system provides oxygen and nutrients to tissues and removes wastes.

Structure of the Heart (page 542)

1. Size and location of the heart
   a. The heart is about 14 centimeters long and 9 centimeters wide.
   b. It is located within the mediastinum and rests on the diaphragm.
2. Coverings of the heart
   a. A layered pericardium encloses the heart.
   b. The pericardial cavity is a space between the visceral and parietal layers of the pericardium.
3. Wall of the heart
   a. The wall of the heart has three layers.
   b. These layers include an epicardium, a myocardium, and an endocardium.
4. Heart chambers and valves
   a. The heart is divided into four chambers—two atria and two ventricles—that communicate through atrioventricular orifices on each side.
   b. Right chambers and valves
      (1) The right atrium receives blood from the venae cavae and coronary sinus.
      (2) The tricuspid valve guards the right atrioventricular orifice.
      (3) The right ventricle pumps blood into the pulmonary trunk.
      (4) A pulmonary valve guards the base of the pulmonary trunk.
   c. Left chambers and valves
      (1) The left atrium receives blood from the pulmonary veins.
      (2) The mitral valve guards the left atrioventricular orifice.
      (3) The left ventricle pumps blood into the aorta.
      (4) An aortic valve guards the base of the aorta.
5. Skeleton of the heart
   a. The skeleton of the heart consists of fibrous rings that enclose the bases of the pulmonary artery, aorta, and atrioventricular orifices.
   b. The fibrous rings provide attachments for valves and muscle fibers and prevent the orifices from excessively dilating during ventricular contractions.
6. Path of blood through the heart
   a. Blood that is relatively low in oxygen and high in carbon dioxide enters the right side of the heart from the venae cavae and coronary sinus and then is pumped into the pulmonary circulation.
   b. After the blood is oxygenated in the lungs and some of its carbon dioxide is removed, it returns to the left side of the heart through the pulmonary veins.
   c. From the left ventricle, it moves into the aorta.
7. Blood supply to the heart
   a. The coronary arteries supply blood to the myocardium.
   b. It is returned to the right atrium through the cardiac veins and coronary sinus.

Heart Actions (page 551)

1. Cardiac cycle
   a. The atria contract (atrial systole) while the ventricles relax (ventricular diastole); the ventricles contract (ventricular systole) while the atria relax (atrial diastole).
   b. Pressure within the chambers rises and falls in cycles.
2. Heart sounds
   a. Heart sounds can be described as *lubb-dupp*.
   b. Heart sounds are due to the vibrations that the valve movements produce.
   c. The first part of the sound occurs as A-V valves close, and the second part is associated with the closing of pulmonary and aortic valves.
3. Cardiac muscle fibers
   a. Cardiac muscle fibers connect to form a functional syncytium.
   b. If any part of the syncytium is stimulated, the whole structure contracts as a unit.
   c. Except for a small region in the floor of the right atrium, the fibrous skeleton separates the atrial syncytium from the ventricular syncytium.
4. Cardiac conduction system
   a. This system, composed of specialized cardiac muscle tissue, initiates and conducts depolarization waves through the myocardium.
   b. Impulses from the S-A node pass slowly to the A-V node; impulses travel rapidly along the A-V bundle and Purkinje fibers.
   c. Muscle fibers in the ventricular walls form whorls that squeeze blood out of the contracting ventricles.
5. Electrocardiogram (ECG)
   a. An ECG records electrical changes in the myocardium during a cardiac cycle.
   b. The pattern contains several waves.
      (1) The P wave represents atrial depolarization.
      (2) The QRS complex represents ventricular depolarization.
      (3) The T wave represents ventricular repolarization.
6. Regulation of the cardiac cycle
   a. Physical exercise, body temperature, and concentration of various ions affect heartbeat.
   b. Branches of sympathetic and parasympathetic nerve fibers innervate the S-A and A-V nodes.
      (1) Parasympathetic impulses decrease heart action; sympathetic impulses increase heart action.
      (2) The cardiac center in the medulla oblongata regulates autonomic impulses to the heart.

Blood Vessels (page 562)

The blood vessels form a closed circuit of tubes that transport blood between the heart and body cells. The tubes include arteries, arterioles, capillaries, venules, and veins.

1. Arteries and arterioles
   a. The arteries are adapted to carry relatively high pressure blood away from the heart.
   b. The arterioles are branches of arteries.
   c. The walls of arteries and arterioles consist of layers of endothelium, smooth muscle, and connective tissue.
   d. Autonomic fibers that can stimulate vasoconstriction or vasodilation innervate smooth muscles in vessel walls.

2. Capillaries

Capillaries connect arterioles and venules. The capillary wall is a single layer of cells that forms a semipermeable membrane.

a. Capillary permeability
   (1) Openings in the capillary walls are thin slits between endothelial cells.
   (2) The sizes of the openings vary from tissue to tissue.
   (3) Endothelial cells of brain capillaries are tightly fused, forming a blood-brain barrier through which substances move by facilitated diffusion.

b. Capillary arrangement
   Capillary density varies directly with tissue metabolic rates.

c. Regulation of capillary blood flow
   (1) Precapillary sphincters regulate capillary blood flow.
   (2) Precapillary sphincters open when cells are low in oxygen and nutrients and close when cellular needs are met.

3. Exchanges in the capillaries
   a. Gases, nutrients, and metabolic by-products are exchanged between the capillary blood and the tissue fluid.
   b. Diffusion provides the most important means of transport.
   c. Diffusion pathways depend on lipid solubilities.
   d. Plasma proteins generally remain in the blood.
   e. Filtration, which is due to the hydrostatic pressure of blood, causes a net outward movement of fluid at the arteriolar end of a capillary.
   f. Osmosis due to colloid osmotic pressure causes a net inward movement of fluid at the venular end of a capillary.
   g. Some factors cause fluids to accumulate in the tissues.

4. Venules and veins
   a. Venules continue from capillaries and merge to form veins.
   b. Veins carry blood to the heart.
   c. Venous walls are similar to arterial walls but are thinner and contain less muscle and elastic tissue.

Blood Pressure (page 570)

Blood pressure is the force blood exerts against the insides of blood vessels.

1. Arterial blood pressure
   a. The arterial blood pressure is produced primarily by heart action; it rises and falls with phases of the cardiac cycle.
   b. Systolic pressure occurs when the ventricle contracts; diastolic pressure occurs when the ventricle relaxes.

2. Factors that influence arterial blood pressure
   a. Heart action, blood volume, resistance to flow, and blood viscosity influence arterial blood pressure.
   b. Arterial pressure increases as cardiac output, blood volume, peripheral resistance, or blood viscosity increases.

3. Control of blood pressure
   a. Blood pressure is controlled in part by the mechanisms that regulate cardiac output and peripheral resistance.
   b. Cardiac output depends on the volume of blood discharged from the ventricle with each beat (stroke volume) and on the heart rate.
      (1) The more blood that enters the heart, the stronger the ventricular contraction, the greater the stroke volume, and the greater the cardiac output.

      (2) The cardiac center of the medulla oblongata regulates heart rate.
   c. Changes in the diameter of arterioles, controlled by the vasomotor center of the medulla oblongata, regulate peripheral resistance.

4. Venous blood flow
   a. Venous blood flow is not a direct result of heart action; it depends on skeletal muscle contraction, breathing movements, and venoconstriction.
   b. Many veins contain flaplike valves that prevent blood from backing up.
   c. Venous constriction can increase venous pressure and blood flow.

5. Central venous pressure
   a. Central venous pressure is the pressure in the right atrium.
   b. Factors that influence it alter the flow of blood into the right atrium.
   c. It affects pressure within the peripheral veins.

Paths of Circulation (page 580)

1. Pulmonary circuit
   a. The pulmonary circuit consists of vessels that carry blood from the right ventricle to the lungs, alveolar capillaries, and vessels that lead back to the left atrium.
   b. Alveolar capillaries exert less pressure than those of the systemic circuit.
   c. Tightly joined epithelial cells of alveoli walls prevent most substances from entering the alveoli.
   d. Osmotic pressure rapidly draws water out of alveoli into the interstitial fluid, so alveoli do not fill with fluid.

2. Systemic circuit
   a. The systemic circuit is composed of vessels that lead from the heart to all body parts (including vessels supplying the heart itself), except the lungs, and back to the heart.
   b. It includes the aorta and its branches as well as the system of veins that return blood to the right atrium.

Arterial System (page 582)

1. Principal branches of the aorta
   a. The branches of the ascending aorta include the right and left coronary arteries.
   b. The branches of the aortic arch include the brachiocephalic, left common carotid, and left subclavian arteries.
   c. The branches of the descending aorta include the thoracic and abdominal groups.
   d. The abdominal aorta terminates by dividing into right and left common iliac arteries.

2. Arteries to the neck, head, and brain include branches of the subclavian and common carotid arteries.

3. Arteries to the shoulder and upper limb
   a. The subclavian artery passes into the arm, and in various regions, it is called the axillary and brachial artery.
   b. Branches of the brachial artery include the ulnar and radial arteries.

4. Arteries to the thoracic and abdominal walls
   a. Branches of the subclavian artery and thoracic aorta supply the thoracic wall.
   b. Branches of the abdominal aorta and other arteries supply the abdominal wall.

5. Arteries to the pelvis and lower limb
   The common iliac artery supplies the pelvic organs, gluteal region, and lower limb.

Venous System (page 591)

1. Characteristics of venous pathways
   a. The veins return blood to the heart.
   b. Larger veins usually parallel the paths of major arteries.
2. Veins from the head, neck, and brain
   a. The jugular veins drain these regions.
   b. Jugular veins unite with subclavian veins to form the brachiocephalic veins.
3. Veins from the upper limb and shoulder
   a. Sets of superficial and deep veins drain the upper limb.
   b. The major superficial veins are the basilic and cephalic veins.
   c. The median cubital vein in the bend of the elbow is often used as a site for venipuncture.
4. Veins from the abdominal and thoracic walls include tributaries of the brachiocephalic, and azygos veins drain these walls.
5. Veins from the abdominal viscera
   a. The blood from the abdominal viscera generally enters the hepatic portal system and is carried to the liver.
   b. The blood in the hepatic portal system is rich in nutrients.
   c. The liver helps regulate the blood concentrations of glucose, amino acids, and lipids.
   d. Phagocytic cells in the liver remove bacteria from the portal blood.
   e. From the liver, hepatic veins carry blood to the inferior vena cava.
6. Veins from the lower limb and pelvis
   a. Sets of deep and superficial veins drain these regions.
   b. The deep veins include the tibial veins, and the superficial veins include the saphenous veins.

Life-Span Changes (page 598)

1. Some degree of cholesterol deposition in blood vessels may be a normal part of aging, but accumulation is great enough to lead to overt disease.
2. Fibrous connective tissue and adipose tissue enlarge the heart by filling in when the number and size of cardiac muscle cells fall.
3. Blood pressure increases with age, while resting heart rate decreases with age.
4. Moderate exercise correlates to lowered risk of heart disease in older people.

CRITICAL THINKING QUESTIONS

1. Given the way capillary blood flow is regulated, do you think it is wiser to rest or to exercise following a heavy meal? Explain.
2. If a patient develops a blood clot in the femoral vein of the left lower limb and a portion of the clot breaks loose, where is the blood flow likely to carry the embolus? What symptoms are likely?
3. When a person strains to lift a heavy object, intrathoracic pressure increases. What do you think will happen to the rate of venous blood returning to the heart during such lifting? Why?
4. Why is a ventricular fibrillation more likely to be life threatening than an atrial fibrillation?
5. Cirrhosis of the liver, a disease commonly associated with alcoholism, obstructs blood flow through the hepatic blood vessels. As a result, the blood backs up, and the capillary pressure greatly increases in the organs drained by the hepatic portal system. What effects might this increasing capillary pressure produce, and which organs would it affect?
6. If a cardiologist inserted a catheter into a patient's right femoral artery, which arteries would the tube have to pass through in order to reach the entrance of the left coronary artery?
7. How might the results of a cardiovascular exam differ for an athlete in top condition and a sedentary, overweight individual?
8. Cigarette smoke contains thousands of chemicals, including nicotine and carbon monoxide. Nicotine constricts blood vessels. Carbon monoxide prevents oxygen from binding to hemoglobin. How do these two components of smoke affect the cardiovascular system?
9. What structures and properties should an artificial heart have?

REVIEW EXERCISES

1. Describe the general structure, function, and location of the heart.
2. Describe the pericardium.
3. Compare the layers of the cardiac wall.
4. Identify and describe the locations of the chambers and the valves of the heart.
5. Describe the skeleton of the heart, and explain its function.
6. Trace the path of blood through the heart.
7. Trace the path of blood through the coronary circulation.
8. Describe a cardiac cycle.
9. Describe the pressure changes that occur in the atria and ventricles during a cardiac cycle.
10. Explain the origin of heart sounds.
11. Describe the arrangement of the cardiac muscle fibers.
12. Distinguish between the roles of the S-A node and the A-V node.
13. Explain how the cardiac conduction system controls the cardiac cycle.
14. Describe and explain the normal ECG pattern.
15. Discuss how the nervous system regulates the cardiac cycle.

16. Describe two factors other than the nervous system that affect the cardiac cycle.
17. Distinguish between an artery and an arteriole.
18. Explain control of vasoconstriction and vasodilation.
19. Describe the structure and function of a capillary.
20. Describe the function of the blood-brain barrier.
21. Explain control of blood flow through a capillary.
22. Explain how diffusion functions in the exchange of substances between blood plasma and tissue fluid.
23. Explain why water and dissolved substances leave the arteriolar end of a capillary and enter the venular end.
24. Describe the effect of histamine on a capillary.
25. Distinguish between a venule and a vein.
26. Explain how veins function as blood reservoirs.
27. Distinguish between systolic and diastolic blood pressures.
28. Name several factors that influence the blood pressure, and explain how each produces its effect.
29. Describe the control of blood pressure.
30. List the major factors that promote the flow of venous blood.
31. Define *central venous pressure.*
32. Distinguish between the pulmonary and systemic circuits of the cardiovascular system.
33. Trace the path of blood through the pulmonary circuit.
34. Explain why the alveoli normally do not fill with fluid.
35. Describe the aorta, and name its principal branches.
36. Describe the relationship between the major venous pathways and the major arterial pathways.
37. List and describe the changes occurring in the cardiovascular system as a result of aging.

W E B   C O N N E C T I O N S

Visit the Student Edition of the Online Learning Center at www.mhhe.com/shier10 **for answers to chapter questions, additional quizzes, interactive learning exercises, and other study tools.**

UNDERSTANDING WORDS

auto-, self: *auto*immune disease—condition in which the immune system attacks the body's own tissues.

-gen, become, be produced: aller*gen*—substance that stimulates an allergic response.

humor-, moisture, fluid: *humor*al immunity—immunity resulting from antibodies in body fluids.

immun-, free, exempt: *immun*ity—resistance to (freedom from) a specific disease.

inflamm-, to set on fire: *inflamm*ation—localized redness, heat, swelling, and pain in the tissues.

nod-, knot: *nod*ule—small mass of lympho-cytes surrounded by connective tissue.

path-, disease, sickness: *path*ogen—disease-causing agent.

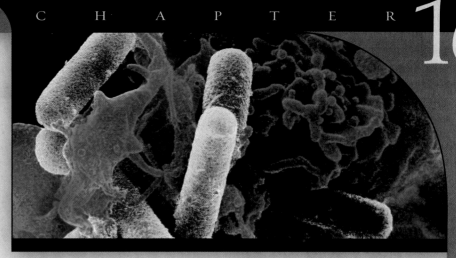

A falsely colored leukocyte (white blood cell) engulfs *Bacillus cereus* bacteria, and will use enzymes to dismantle it (13,000×).

Lymphatic System and Immunity

CHAPTER OBJECTIVES

After you have studied this chapter, you should be able to

1. Describe the general functions of the lymphatic system.

2. Identify the locations of the major lymphatic pathways.

3. Describe how tissue fluid and lymph form and explain the function of lymph.

4. Explain how lymphatic circulation is maintained and describe the consequence of lymphatic obstruction.

5. Describe a lymph node and its major functions.

6. Describe the location of the major chains of lymph nodes.

7. Discuss the functions of the thymus and spleen.

8. Distinguish between innate (nonspecific) and adaptive (specific) defenses and provide examples of each.

9. Explain how two major types of lymphocytes are formed, activated, and how they function in immune mechanisms.

10. Name the major types of immunoglobulins and discuss their origins and actions.

11. Distinguish between primary and secondary immune responses.

12. Distinguish between active and passive immunity.

13. Explain how allergic reactions, tissue rejection reactions, and autoimmunity arise from immune mechanisms.

14. Describe life-span changes in immunity.

The young woman was admitted to the emergency department for sudden onset of difficulty breathing. Although she had a history of asthma, this was something different, for she was also flushed and had vomited. An astute medical student taking a quick history from the woman's roommates discovered that she had just eaten cookies from a vending machine in their dorm. Suspecting that the cookies may have contained peanuts, the student alerted the attending physician, who treated the woman for suspected peanut allergy—giving oxygen, an antihistamine, a steroid drug, and epinephrine. She recovered.

Peanut allergy is common and on the rise, but only in certain westernized countries. In the United States, 6 to 8 percent of children under the age of four and 2 percent of the population over 10 years of age are allergic to peanuts. About 30,000 people react each year, and about 200 die.

Peculiarities of peanuts and our fondness for them may explain why allergy is apparently increasing. Three glycoproteins in peanuts are allergens, causing the misdirected immune response that constitutes an allergy. These glycoproteins are highly concentrated in the peanut, and when ingested, they disturb the intestinal lining in such a way that they enter the circulation rapidly, without being digested. That is, many allergens gain quick and easy access to the welcoming cells of the immune system that congregate beneath the intestinal lining.

Compounding the rapidity with which peanut allergens flood the bloodstream is the fact that people in the United States are bombarded with peanuts. One large study found that virtually everyone has eaten a peanut by two years of age, usually in peanut butter. This is sufficient exposure to set the stage for later allergy in genetically predisposed individuals. The fact that the average age of first allergic reaction to peanuts is 14 months suggests that the initial exposure—necessary to "prime" the immune system for future response—happens either through breast milk or in the uterus, an hypothesis consistent with the popularity of peanuts among the pregnant and nursing. Countries where peanuts are rarely eaten, such as Denmark and Norway, have very low incidence of peanut allergy.

The method of peanut preparation in the United States—dry roasting—may make the three glycoproteins that evoke the allergic response more active. In China, where peanuts are equally popular but are boiled or fried, allergy is rare. However, children of Chinese immigrants in the United States have the same incidence of peanut allergy as other children in the country, supporting the idea that method of preparation contributes to allergenicity.

It is obviously important for people who know that they have a peanut allergy to avoid peanuts. Airlines stopped serving peanuts—once a staple—because of the prevalence of allergic reactions. Parents of children with allergies should read all food labels, because peanuts are ingredients of many candies and sauces. Parents should become familiar with the signs of the allergy attack and if they know a child is allergic to peanuts carry an injectible "pen" of epinephrine. ■

The lymphatic system is closely associated with the cardiovascular system because it includes a network of vessels that assist in circulating body fluids. Lymphatic vessels transport excess fluid away from interstitial spaces in most tissues and return it to the bloodstream (fig. 16.1). Special lymphatic capillaries (*lacteals*) located in the lining of the small intestine absorb digested fats, then transport the fats to venous circulation. The organs of the lymphatic system also help defend the body against infection by disease-causing agents.

Lymphatic Pathways

The **lymphatic pathways** (lim-fat′ik path′wāz) begin as lymphatic capillaries that merge to form larger lymphatic vessels. These, in turn, lead to larger vessels that unite with the veins in the thorax.

Lymphatic Capillaries

Lymphatic capillaries are microscopic, closed-ended tubes. They extend into the interstitial spaces, forming complex networks that parallel the networks of the blood capillaries (fig. 16.2). The walls of the lymphatic capillaries are similar to those of the blood capillaries. Each consists of a single layer of squamous epithelial cells called endothelium. These thin walls make it possible for tissue fluid (interstitial fluid) from the interstitial space to enter the lymphatic capillaries. Fluid inside a lymphatic capillary is called **lymph** (limf).

Lymphatic Vessels

The walls of **lymphatic vessels** are similar to those of veins, but thinner. Each is composed of three layers: an endothelial lining, a middle layer of smooth muscle and elastic fibers, and an outer layer of connective tissue. Also like those veins below the heart, the lymphatic vessels have semilunar valves, which help prevent backflow of lymph. Figure 16.3 shows one of these valves.

The larger lymphatic vessels lead to specialized organs called **lymph nodes** (limf nōdz). After leaving the nodes, the vessels merge to form still larger lymphatic trunks.

Lymphatic Trunks and Collecting Ducts

The **lymphatic trunks,** which drain lymph from the lymphatic vessels, are named for the regions they serve. For example, the *lumbar trunk* drains lymph from the lower limbs, lower abdominal wall, and pelvic organs; the *intestinal trunk* drains the abdominal viscera; the *inter-*

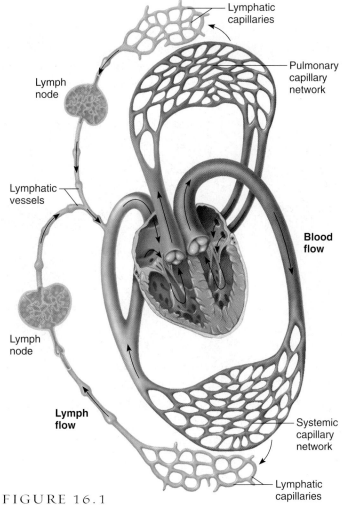

FIGURE 16.1

Schematic representation of lymphatic vessels transporting fluid from interstitial spaces to the bloodstream.

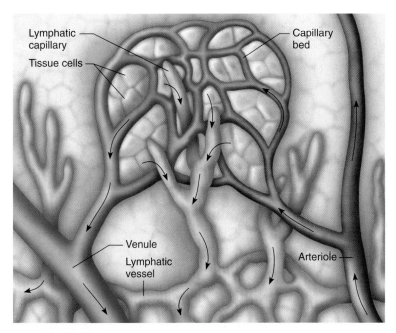

FIGURE 16.2

Lymphatic capillaries are microscopic, closed-ended tubes that begin in the interstitial spaces of most tissues.

costal and *bronchomediastinal trunks* drain lymph from portions of the thorax; the *subclavian trunk* drains the upper limb; and the *jugular trunk* drains portions of the neck and head. These lymphatic trunks then join one of two **collecting ducts**—the thoracic duct or the right lymphatic duct. Figure 16.4 shows the location of the major lymphatic trunks and collecting ducts, and figure 16.5 shows a lymphangiogram, or radiograph, of some lymphatic vessels and lymph nodes.

The **thoracic duct** is the larger and longer of the two collecting ducts. It begins in the abdomen, passes upward through the diaphragm beside the aorta, ascends anterior to the vertebral column through the mediastinum, and empties into the left subclavian vein near the junction of the left jugular vein. This duct drains lymph from the intestinal, lumbar, and intercostal trunks, as well as from the left subclavian, left jugular, and left bronchomediastinal trunks.

The **right lymphatic duct** originates in the right thorax at the union of the right jugular, right subclavian, and right bronchomediastinal trunks. It empties into the right subclavian vein near the junction of the right jugular vein.

After leaving the two collecting ducts, lymph enters the venous system and becomes part of the plasma prior to the blood's return to the right atrium. Thus, lymph from the lower body regions, the left upper limb, and the left side of the head and neck enters the thoracic duct; lymph from the right side of the head and neck, the right upper limb, and the right thorax enters the right lymphatic duct (fig. 16.6). Figure 16.7 summarizes the lymphatic pathway.

The skin is richly supplied with lymphatic capillaries. Consequently, if the skin is broken, or if something is injected into it (such as venom from a stinging insect), foreign substances rapidly enter the lymphatic system.

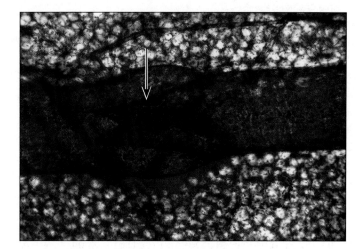

FIGURE 16.3

Light micrograph of the flaplike valve (arrow) within a lymphatic vessel (25×).

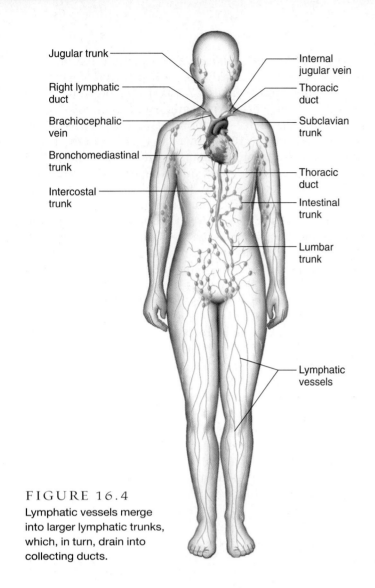

FIGURE 16.4
Lymphatic vessels merge into larger lymphatic trunks, which, in turn, drain into collecting ducts.

Jugular trunk
Right lymphatic duct
Brachiocephalic vein
Bronchomediastinal trunk
Intercostal trunk
Internal jugular vein
Thoracic duct
Subclavian trunk
Thoracic duct
Intestinal trunk
Lumbar trunk
Lymphatic vessels

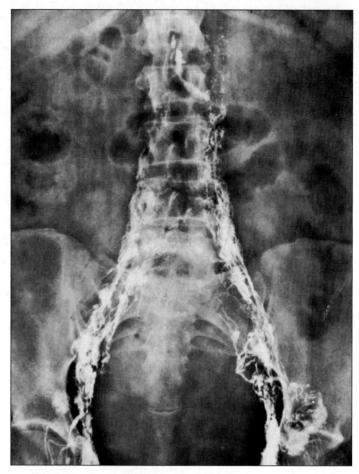

FIGURE 16.5
A lymphangiogram (radiograph) of the lymphatic vessels and lymph nodes of the pelvic region.

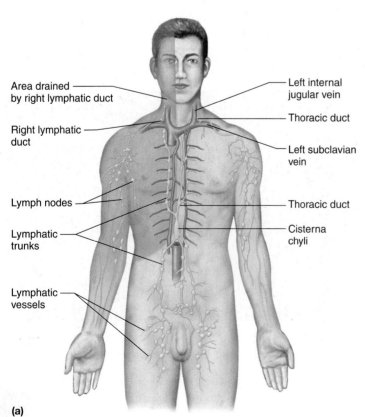

Area drained by right lymphatic duct
Right lymphatic duct
Lymph nodes
Lymphatic trunks
Lymphatic vessels
Left internal jugular vein
Thoracic duct
Left subclavian vein
Thoracic duct
Cisterna chyli

(a)

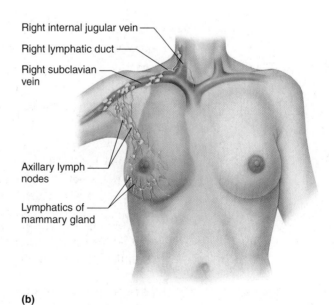

Right internal jugular vein
Right lymphatic duct
Right subclavian vein
Axillary lymph nodes
Lymphatics of mammary gland

(b)

FIGURE 16.6
Lymphatic pathways. (*a*) The right lymphatic duct drains lymph from the upper right side of the body, whereas the thoracic duct drains lymph from the rest of the body. (*b*) Lymph drainage of the right breast illustrates a localized function of the lymphatic system. Surgery to remove a cancerous breast can disrupt this drainage, causing painful swelling.

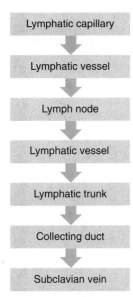

FIGURE 16.7
The lymphatic pathway.

1   What are the general functions of the lymphatic system?

2   Through which lymphatic structures would lymph pass in traveling from a lower limb back to the bloodstream?

Tissue Fluid and Lymph

Lymph is essentially tissue fluid that has entered a lymphatic capillary. Thus, lymph formation depends upon tissue fluid formation.

Tissue Fluid Formation

Capillary blood pressure filters water and small molecules from the plasma. The resulting fluid has much the same composition as the plasma (including nutrients, gases, and hormones), with the important exception of the plasma proteins, which are generally too large to pass through the capillary walls. The osmotic effect of these (called the *plasma colloid osmotic pressure*) helps to draw fluid back into the capillaries by osmosis.

∞ RECONNECT TO CHAPTER 15, EXCHANGES IN THE CAPILLARIES, PAGE 568.

Lymph Formation

Filtration from the plasma normally exceeds reabsorption, leading to the net formation of tissue fluid (interstitial fluid). This increases the interstitial fluid hydrostatic pressure somewhat, favoring movement of tissue fluid into lymphatic capillaries, forming lymph. Thus, lymph formation prevents the accumulation of excess tissue fluid, or *edema* (ĕ'de-mah).

Lymph Function

Lymphatic vessels in the small intestine play a major role in the absorption of dietary fats (chapter 17, p. 678). Lymph also returns to the bloodstream any very small proteins that the blood capillaries filtered. At the same time, lymph transports foreign particles, such as bacteria or viruses, to lymph nodes.

Although these proteins and foreign particles cannot easily enter blood capillaries, the lymphatic capillaries are adapted to receive them. Specifically, the epithelial cells that form the walls of lymphatic vessels overlap each other but are not attached. This configuration, shown in figure 16.8, creates flaplike valves in the lymphatic capillary wall, which are pushed inward when the pressure is greater on the outside of the capillary but close when the pressure is greater on the inside.

The epithelial cells of the lymphatic capillary wall are also attached to surrounding connective tissue cells by thin protein filaments. As a result, the lumen of a lymphatic capillary remains open even when the outside pressure is greater than the pressure within the lymph capillary.

1   What is the relationship between tissue fluid and lymph?

2   How do proteins in tissue fluid affect lymph formation?

3   What are the major functions of lymph?

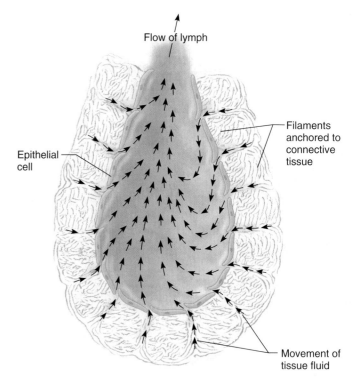

FIGURE 16.8
Tissue fluid enters lymphatic capillaries through flaplike valves between epithelial cells.

Lymph Movement

The hydrostatic pressure of tissue fluid drives lymph into lymphatic capillaries. However, muscular activity largely influences the movement of lymph through the lymphatic vessels.

Lymph Flow

Lymph, like venous blood, is under relatively low hydrostatic pressure. As a result, it may not flow readily through the lymphatic vessels without help from contracting skeletal muscles in the limbs, pressure changes from the action of skeletal muscles used in breathing, and contraction of smooth muscles in the walls of the larger lymphatic trunks. Lymph flow peaks during physical exercise, due to the actions of skeletal muscles and pressure changes associated with breathing.

Contracting skeletal muscles compress lymphatic vessels. This squeezing action moves the lymph inside a vessel, but because the lymphatic vessels contain valves that prevent backflow, the lymph can move only toward a collecting duct. Additionally, the smooth muscles in the walls of the larger lymphatic trunks may contract and compress the lymph inside, forcing the fluid onward.

Breathing aids lymph circulation by creating a relatively low pressure in the thorax during inhalation. At the same time, the contracting diaphragm increases the pressure in the abdominal cavity. Consequently, lymph is squeezed out of the abdominal vessels and forced into the thoracic vessels. Once again, valves within the lymphatic vessels prevent lymph backflow.

Obstruction of Lymph Movement

The continuous movement of fluid from interstitial spaces into blood capillaries and lymphatic capillaries stabilizes the volume of fluid in these spaces. Conditions that interfere with lymph movement cause tissue fluid to accumulate within the interstitial spaces, producing edema. For example, a surgeon removing a cancerous breast tumor also usually removes nearby axillary lymph nodes to prevent associated lymphatic vessels from transporting cancer cells to other sites (metastasis). Removing the lymphatic tissue can obstruct drainage from the upper limb, causing edema.

1 What factors promote lymph flow?

2 What is the consequence of lymphatic obstruction?

Lymph Nodes

Lymph nodes (lymph glands) are located along the lymphatic pathways. They contain large numbers of *lymphocytes* and *macrophages,* which fight invading microorganisms.

Structure of a Lymph Node

Lymph nodes vary in size and shape but are usually less than 2.5 centimeters long and are somewhat bean-shaped. Figure 16.9 illustrates a section of a typical lymph node.

The indented region of a bean-shaped node is called the **hilum,** and it is the portion through which the blood vessels and nerves connect with the structure. The lymphatic vessels leading to a node (afferent vessels) enter separately at various points on its convex surface, but the

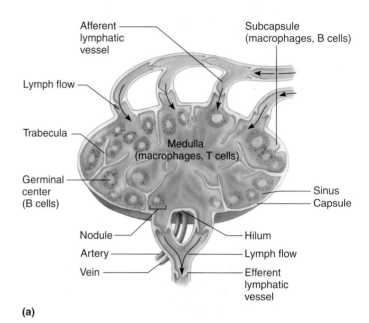

(a)

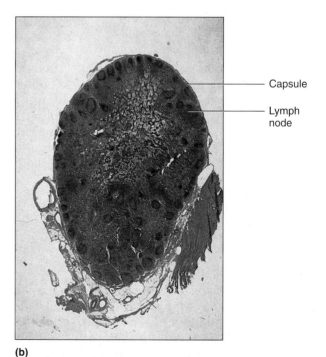

(b)

FIGURE 16.9

Lymph node. (*a*) A section of a lymph node. (*b*) Light micrograph of a lymph node (5×).

lymphatic vessels leaving the node (efferent vessels) exit from the hilum.

A capsule of connective tissue with numerous fibers encloses each lymph node. The capsule extends into the node and partially subdivides it into compartments called **lymph nodules,** with lighter-staining germinal centers that contain dense masses of actively dividing lymphocytes and macrophages. These nodules are the structural units of the lymph node.

Nodules also occur singly or in groups associated with the mucous membranes of the respiratory and digestive tracts. The *tonsils,* described in chapter 17 (pp. 649–650), are composed of partially encapsulated lymph nodules. Also, aggregates of nodules called *Peyer's patches* are scattered throughout the mucosal lining of the distal portion of the small intestine. Within the Peyer's patches are scattered cells, called M cells, through which certain ingested molecules pass by transcytosis. After the molecules pass through the M cells, they face lymphocytes and other immune system cells that then may initiate an immune response.

The spaces within the node, called **lymph sinuses,** provide a complex network of chambers and channels through which lymph circulates. Lymph enters a lymph node through *afferent lymphatic vessels,* moves slowly through the lymph sinuses, and leaves through *efferent lymphatic vessels* (fig. 16.10).

Superficial lymphatic vessels inflamed by bacterial infection appear as red streaks beneath the skin, a condition called *lymphangitis.* Inflammation of the lymph nodes, called *lymphadenitis,* often follows. Affected nodes enlarge and may be quite painful.

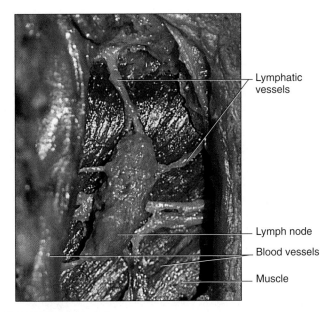

FIGURE 16.10
Lymph enters and leaves a lymph node through lymphatic vessels.

Locations of Lymph Nodes

The lymph nodes generally occur in groups or chains along the paths of the larger lymphatic vessels throughout the body but are absent in the central nervous system. The major locations of the lymph nodes, shown in figure 16.11, are as follows:

1. **Cervical region.** Nodes in the cervical region follow the lower border of the mandible, anterior to and posterior to the ears, and deep within the neck along the paths of the larger blood vessels. These nodes are associated with the lymphatic vessels that drain the skin of the scalp and face, as well as the tissues of the nasal cavity and pharynx.

2. **Axillary region.** Nodes in the underarm region receive lymph from vessels that drain the upper limbs, the wall of the thorax, the mammary glands (breasts), and the upper wall of the abdomen.

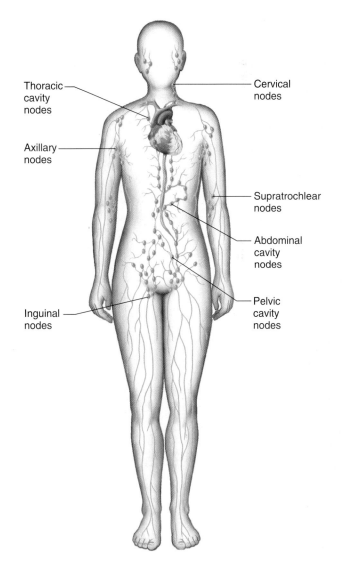

FIGURE 16.11
Major locations of lymph nodes.

3. **Inguinal region.** Nodes in the inguinal region receive lymph from the lower limbs, the external genitalia, and the lower abdominal wall.

4. **Pelvic cavity.** Within the pelvic cavity, nodes primarily occur along the paths of the iliac blood vessels. They receive lymph from the lymphatic vessels of the pelvic viscera.

5. **Abdominal cavity.** Nodes within the abdominal cavity form chains along the main branches of the mesenteric arteries and the abdominal aorta. These nodes receive lymph from the abdominal viscera.

6. **Thoracic cavity.** Nodes of the thoracic cavity occur within the mediastinum and along the trachea and bronchi. They receive lymph from the thoracic viscera and from the internal wall of the thorax.

7. **Supratrochlear region.** These nodes are located superficially on the medial side of the elbow. They often enlarge in children as a result of infections from cuts and scrapes on the hands.

> The illness described as "swollen glands" actually refers to enlarged cervical lymph nodes associated with throat or respiratory infection.

Functions of Lymph Nodes

Lymph nodes have two primary functions: filtering potentially harmful particles from lymph before returning it to the bloodstream and monitoring the body fluids (immune surveillance) provided by lymphocytes and macrophages. Along with the red bone marrow, the lymph nodes are centers for lymphocyte production. These cells attack invading viruses, bacteria, and other parasitic cells that lymphatic vessels bring to the nodes. Macrophages in the nodes engulf and destroy foreign substances, damaged cells, and cellular debris.

1 Distinguish between a lymph node and a lymph nodule.

2 What factors promote the flow of lymph through a node?

3 In what body regions are lymph nodes most abundant?

4 What are the major functions of lymph nodes?

Thymus and Spleen

Two other lymphatic organs, whose functions are similar to those of the lymph nodes, are the thymus and the spleen.

Thymus

The **thymus** (thi'mus) gland is a soft, bilobed structure enclosed in a connective tissue capsule (fig. 16.12a). It is located within the mediastinum, anterior to the aortic arch and posterior to the upper part of the body of the sternum, and extends from the root of the neck to the pericardium. The thymus varies in size from person to person, and it is usually larger during infancy and early childhood. After puberty, the thymus shrinks, and in an adult, it may be quite small (fig. 16.13). In elderly persons, adipose and connective tissues replace the normal lymphatic tissue.

Connective tissues extend inward from the thymus's surface, subdividing it into lobules (see fig. 16.12b). The lobules contain many lymphocytes that developed from progenitor cells in the bone marrow. Most of these cells (thymocytes) are inactivated; however, some mature into **T lymphocytes,** which leave the thymus and provide immunity. Epithelial cells within the thymus secrete protein hormones called *thymosins,* which stimulate maturation of T lymphocytes after they leave the thymus and migrate to other lymphatic tissues.

Spleen

The **spleen** (splēn) is the largest lymphatic organ. It is located in the upper left portion of the abdominal cavity, just inferior to the diaphragm, posterior and lateral to the stomach (see fig. 16.12a and reference plates 4, 5, and 6).

The spleen resembles a large lymph node in that it is enclosed in connective tissue that extends inward from the surface and partially subdivides the organ into chambers, or lobules. The organ also has a hilum on one surface through which blood vessels and nerves enter. However, unlike the sinuses of a lymph node, the spaces (venous sinuses) within the chambers of the spleen are filled with blood instead of lymph.

The tissues within the lobules of the spleen are of two types. The *white pulp* is distributed throughout the spleen in tiny islands. This tissue is composed of nodules (splenic nodules), which are similar to those in lymph nodes and contain many lymphocytes. The *red pulp,* which fills the remaining spaces of the lobules, surrounds the venous sinuses. This pulp contains numerous red blood cells, which impart its color, plus many lymphocytes and macrophages (fig. 16.14).

Blood capillaries within the red pulp are quite permeable. Red blood cells can squeeze through the pores in these capillary walls and enter the venous sinuses. The older, more fragile red blood cells may rupture as they make this passage, and the resulting cellular debris is removed by macrophages within the splenic sinuses.

> During fetal development, pulp cells of the spleen produce blood cells, much as red bone marrow cells do after birth. As the time of birth approaches, this splenic function ceases. However, in certain diseases, such as *erythroblastosis fetalis,* in which many red blood cells are destroyed, the splenic pulp cells may resume their hematopoietic activity.

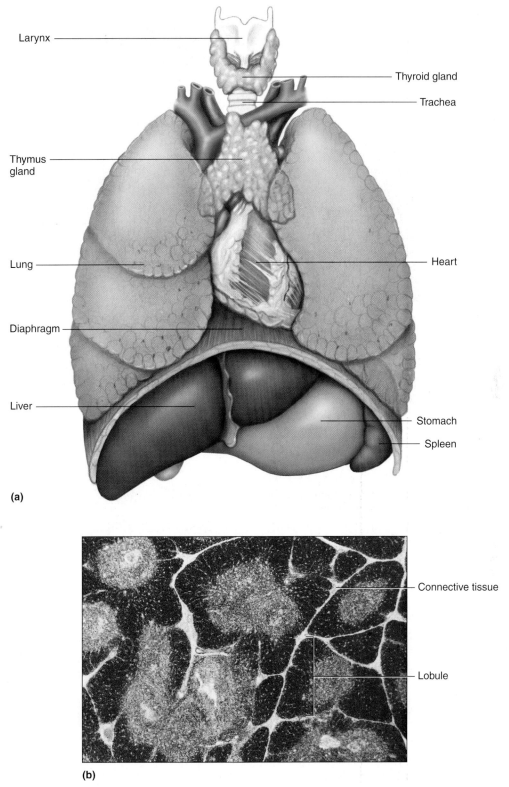

Larynx

Thyroid gland

Trachea

Thymus gland

Lung

Heart

Diaphragm

Liver

Stomach

Spleen

(a)

Connective tissue

Lobule

(b)

FIGURE 16.12

Thymus and spleen. (a) The thymus gland is bilobed and located between the lungs and superior to the heart. The spleen is located inferior to the diaphragm and posterior and lateral to the stomach. (b) A cross section of the thymus (20×). Note how the gland is subdivided into lobules.

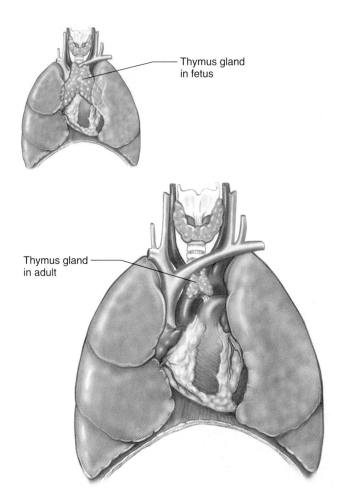

FIGURE 16.13

Compared to other thoracic organs, the thymus in the fetus is quite large, but in the adult is quite small. Figure is not to scale.

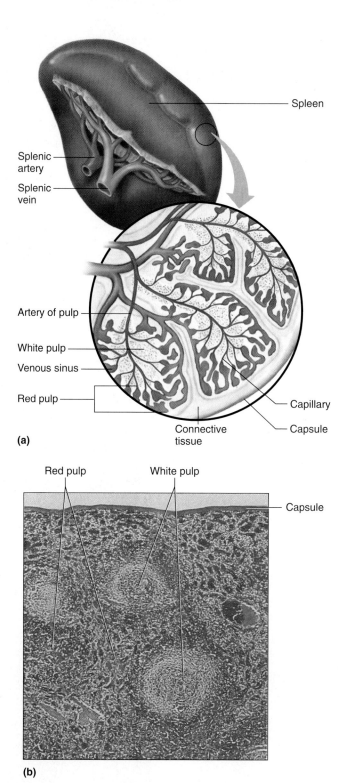

FIGURE 16.14

Spleen. (*a*) The spleen resembles a large lymph node. (*b*) Light micrograph of the spleen (15×).

The macrophages engulf and destroy foreign particles, such as bacteria, that may be carried in the blood as it flows through the sinuses. The lymphocytes of the spleen, like those of the thymus, lymph nodes, and nodules, also help defend the body against infections. Thus, the spleen filters blood much as the lymph nodes filter lymph. Table 16.1 summarizes the characteristics of the major organs of the lymphatic system.

1 Why are the thymus and spleen considered organs of the lymphatic system?

2 What are the major functions of the thymus and the spleen?

Body Defenses Against Infection

The presence and multiplication of a disease-causing agent, or **pathogen** (path′o-jen), causes an infection. Pathogens include simple microorganisms such as bacteria, complex microorganisms such as protozoa, and even spores of multicellular organisms, such as fungi. Viruses are pathogens, but they are not considered organisms because their structure is far simpler than that of a living cell and they must infect a living cell to reproduce. An infection may be present even though an individual feels well. People who are infected with the human immuno-

TABLE 16.1	Major Organs of the Lymphatic System	
Organ	**Location**	**Function**
Lymph nodes	In groups or chains along the paths of larger lymphatic vessels	Filter foreign particles and debris from lymph; produce and house lymphocytes that destroy foreign particles in lymph; house macrophages that engulf and destroy foreign particles and cellular debris carried in lymph
Thymus	Within the mediastinum posterior to the upper portion of the body of the sternum	Houses lymphocytes; differentiates thymocytes into T lymphocytes
Spleen	In upper left portion of the abdominal cavity, inferior to the diaphragm, posterior and lateral to the stomach	Houses macrophages that remove foreign particles, damaged red blood cells, and cellular debris from the blood; contains lymphocytes

In San Francisco in 1982, Simon Guzman became one of the very first recorded individuals to succumb to AIDS. He was the first person whose death was attributed to a parasitic infection previously seen only in sheep—*cryptosporidiosis.* This infection, which causes relentless diarrhea, illustrates a hallmark of AIDS: alteration of species resistance.

deficiency virus (HIV), which causes AIDS, often live for many years in good health before becoming ill.

The human body can prevent entry of pathogens or destroy them if they enter. Some mechanisms are quite general and protect against many types of pathogens, providing **innate** (*nonspecific*) **defense.** They function the same way regardless of the invader or the number of times a particular type of organism invades. These defenses include species resistance, mechanical barriers, enzyme action, interferon, fever, natural killer cells, inflammation, and phagocytosis. Other protective mechanisms are very precise, targeting certain pathogens with an **adaptive** (*specific*) **defense,** or immunity. These more-directed responses are carried out by specialized lymphocytes that recognize foreign molecules (nonself antigens) in the body and act against them.

Innate (Nonspecific) Defenses

Species Resistance

Species resistance refers to the fact that a given kind of organism, or species (such as the human species, *Homo sapiens*), develops diseases that are unique to it. A species may be resistant to diseases that affect other species because its tissues somehow fail to provide the temperature or chemical environment that a particular pathogen requires. For example, humans are subject to infections by the infectious agents that cause measles, mumps, gonorrhea, and syphilis, but other animal species are not. Similarly, humans are resistant to certain forms of malaria and tuberculosis that affect birds. However, new influenza strains that affect humans may come from birds, especially poultry.

Mechanical Barriers

The skin and mucous membranes lining the passageways of the respiratory, digestive, urinary, and reproductive systems create **mechanical barriers** that prevent the entrance of some infectious agents. As long as these barriers remain intact, many pathogens are unable to penetrate them. Another protection is that the epidermis sloughs off, removing superficial bacteria with it. In addition, the mucus-coated ciliated epithelium, described in chapter 19 (p. 733), that lines the respiratory passages entraps particles and sweeps them out of the airways and into the pharynx, where they are swallowed. Along with the hair that traps infectious agents associated with the skin and mucous membranes is the fluid (sweat and mucus) that rinses away microorganisms. These barriers provide a *first line of defense.* The rest of the nonspecific defenses discussed in this section are part of the *second line of defense.*

Chemical Barriers

Enzymes in body fluids provide a **chemical barrier** to pathogens. Gastric juice, for example, contains the protein-splitting enzyme pepsin and has a low pH due to the presence of hydrochloric acid. The combined effect of pepsin and hydrochloric acid kills many pathogens that enter the stomach. Similarly, tears contain the enzyme lysozyme, which acts against certain bacteria on the eyes. The accumulation of salt from perspiration also kills certain bacteria on the skin.

Interferons (in″ter-fēr′onz) are hormonelike peptides that certain cells produce, including lymphocytes and fibroblasts. Once released from a virus-infected cell, interferon binds to receptors on uninfected cells, stimulating them to synthesize proteins that block replication of a variety of viruses. Thus, interferon's effect is nonspecific. Interferons also stimulate phagocytosis and enhance the activity of certain other cells that help to resist infections and the growth of tumors.

Other antimicrobial biochemicals are defensins and collectins. **Defensins** are peptides produced by neutrophils and other types of granular white blood cells, and in the intestinal epithelium, the urogenital tract, kidneys, and the skin. Recognition of a nonself cell surface or viral particle triggers the expression of genes that encode defensins. Some defensins act by making holes in bacterial cell walls and membranes, which are sufficient to cripple the microbes. **Collectins** are proteins that provide broad protection against bacteria, yeasts, and some viruses. These proteins home in on slight differences in the structures and arrangements of sugars that protrude from the surfaces of pathogens. Collectins detect not only the sugar molecules, but the pattern in which they are clustered, grabbing on much like velcro clings to fabric, thus making the pathogen more easily phagocytized.

Fever

A **fever** is a nonspecific defense that offers powerful protection. A fever begins as a viral or bacterial infection stimulates lymphocytes to proliferate, producing cells that secrete a substance called *interleukin-1* (IL-1), which is also more colorfully known as endogenous pyrogen ("fire maker from within"). IL-1 raises the thermoregulatory set point in the brain's hypothalamus to maintain a higher body temperature. Fever indirectly counters microbial growth because higher body temperature causes the liver and spleen to sequester iron, which reduces the level of iron in the blood. Since bacteria and fungi require more iron as temperature rises, their growth and reproduction in a fever-ridden body slows and may cease. Phagocytic cells also attack more vigorously when the temperature rises.

Natural Killer (NK) Cells

Natural killer (NK) cells are a small population of lymphocytes, distinctly different from the lymphocytes involved in the adaptive (specific) defense mechanisms (discussed later in this chapter). NK cells defend the body against various viruses and cancer by secreting cytolytic substances called **perforins.** Perforins cause the cell membrane to disintegrate, destroying the infected cell. NK cells also secrete chemicals that enhance inflammation.

Inflammation

Inflammation produces localized redness, swelling, heat, and pain (see chapter 5, p. 142). The redness is a result of blood vessel dilation and the consequent increase of blood flow and volume within the affected tissues (hyperemia). This effect, coupled with an increase in permeability of nearby capillaries and subsequent leakage of protein-rich fluid into tissue spaces, swells tissues

(edema). The heat comes from the entry of blood from deeper body parts, which is generally warmer than that near the surface. Pain results from stimulation of nearby pain receptors. Most inflammation is a tissue response to pathogen invasion but alternately can be caused by physical factors (heat, ultraviolet light) or chemical factors (acids, bases).

White blood cells accumulate at the sites of inflammation, where some of them help control pathogens by phagocytosis. Neutrophils are the first to arrive at the site, followed by monocytes. Monocytes pass through capillary walls (diapedsis), becoming macrophages that remove pathogens from surrounding tissues. In bacterial infections, the resulting mass of white blood cells, bacterial cells, and damaged tissue may form a thick fluid called *pus.*

Tissue fluids (exudate) also collect in inflamed tissues. These fluids contain fibrinogen and other clotting factors that may stimulate a network of fibrin threads to form within the affected region. Later, fibroblasts may arrive and form fibers around the area until it is enclosed in a sac of connective tissue containing many fibers. This action helps to inhibit the spread of pathogens and toxins to adjacent tissues.

Once an infection is controlled, phagocytic cells remove dead cells and other debris from the site of inflammation. Cell division replaces lost cells. Table 16.2 summarizes the process of inflammation.

Phagocytosis

Phagocytosis (fag"o-si-to'sis) removes foreign particles from the lymph as it moves from the interstitial spaces to the bloodstream. Phagocytes in the blood vessels and in the tissues of the spleen, liver, or bone marrow usually remove particles that reach the blood. Recall from chapter 14 (pp. 518–519) that the most active phagocytic cells of the blood are *neutrophils* and *monocytes*. Chemicals released from injured tissues attract these cells (chemotaxis). Neutrophils engulf and digest smaller particles; monocytes phagocytize larger ones.

Monocytes are relatively nonmotile phagocytes that occupy lymph nodes, the spleen, liver, and lungs. Monocytes give rise to *macrophages* (histiocytes), which become fixed in various tissues and attach to the inner walls of

TABLE 16.2	Major Actions of an Inflammation Response	
Action	**Result**
Blood vessels dilate. Capillary permeability increases and fluid leaks into tissue spaces.	Tissues become red, swollen, warm, and painful.
White blood cells invade the region.	Pus may form as white blood cells, bacterial cells, and cellular debris accumulate.
Tissue fluids containing clotting factors seep into the area.	A clot containing threads of fibrin may form.
Fibroblasts arrive.	A connective tissue sac may form around the injured tissues.
Phagocytes are active.	Bacteria, dead cells, and other debris are removed.
Cells divide.	Newly formed cells replace injured ones.

TABLE 16.3	Types of Innate (Nonspecific) Defenses
Type	**Description**
Species resistance	A species is resistant to certain diseases to which other species are susceptible.
Mechanical barriers	Unbroken skin and mucous membranes prevent the entrance of some infectious agents.
Chemical barriers	Enzymes in various body fluids kill pathogens. pH extremes and high salt concentration also harm pathogens. Interferons induce production of other proteins that block reproduction of viruses, stimulate phagocytosis, and enhance the activity of cells such that they resist infection and the growth of tumors. Defensins damage bacterial cell walls and membranes. Collectins grab onto microbes.
Fever	Elevated body temperature inhibits microbial growth and increases phagocytic activity.
Natural killer cells	Distinct type of lymphocyte that secretes substances that cause lysis of virus-infected cells and cancer cells.
Inflammation	A tissue response to injury that helps prevent the spread of infectious agents into nearby tissues.
Phagocytosis	Neutrophils, monocytes, and macrophages engulf and destroy foreign particles and cells.

blood and lymphatic vessels. A macrophage can engulf up to 100 bacteria, compared to the twenty or so bacteria that a neutrophil can engulf. Monocytes, macrophages, and neutrophils constitute the **mononuclear phagocytic system** (reticuloendothelial system). Table 16.3 summarizes the types of innate (nonspecific) defenses.

1 What is an infection?

2 Explain seven innate (nonspecific) defense mechanisms.

Adaptive (Specific) Defenses or Immunity

Immunity (i-mu′ni-te) is resistance to particular pathogens or to their toxins or metabolic by-products. An immune response is based upon the ability to distinguish molecules that are part of the body ("self") from those that are not ("nonself," or foreign). Such molecules that can elicit an immune response are called **antigens** (an′ti-jenz). Lymphocytes and macrophages that recognize specific nonself antigens carry out immune responses.

Antigens

Before birth, cells inventory the proteins and other large molecules in the body, learning to identify these as "self." The lymphatic system responds to nonself, or foreign, antigens, but not normally to self antigens. Receptors on lymphocyte surfaces enable the cells to recognize foreign antigens.

Antigens may be proteins, polysaccharides, glycoproteins, or glycolipids. The antigens that are most effective in eliciting an immune response are large and complex, with few repeating parts. Sometimes, a smaller molecule that cannot by itself stimulate an immune response combines with a larger one, which makes it able to do so (antigenic). Such a small molecule is called a **hapten** (hap′ten). Stimulated lymphocytes react either to the hapten or to the larger molecule of the combination. Haptens are found in certain drugs, such as penicillin; in household and industrial chemicals; in dust particles; and in products of animal skins (dander).

Lymphocyte Origins

During fetal development, red bone marrow releases undifferentiated lymphocytes into the circulation (fig. 16.15). About half of these cells reach the thymus, where they remain for a time. Here, these thymocytes specialize into **T lymphocytes,** or **T cells.** ("T" refers to *thymus-derived* lymphocytes.) Later, the blood transports T cells, where they comprise 70% to 80% of the circulating lymphocytes. T cells reside in lymphatic organs and are particularly abundant in the lymph nodes, the thoracic duct, and the white pulp of the spleen.

Other lymphocytes are thought to remain in the red bone marrow until they differentiate into **B lymphocytes,** or **B cells.** (Historically, the "B" stands for *bursa of*

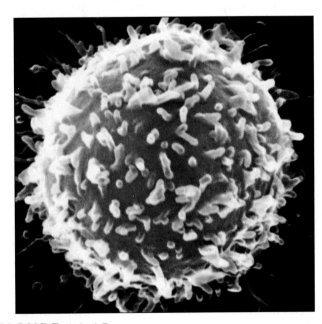

FIGURE 16.15
Falsely colored scanning electron micrograph of a human circulating lymphocyte (8,600×).

Fabricius, an organ in the chicken where these cells were first identified.) The blood distributes B cells, which constitute 20% to 30% of circulating lymphocytes. B cells settle in lymphatic organs along with T cells and are abundant in lymph nodes, the spleen, bone marrow, and intestinal lining (fig. 16.16).

1 What is immunity?

2 What is the difference between an antigen and a hapten?

3 How do T cells and B cells originate?

Lymphocyte Functions

T cells and B cells respond to antigens they recognize in different ways. T cells attach to foreign, antigen-bearing cells, such as bacterial cells, and interact directly—that is, by cell-to-cell contact. This is called the **cellular immune response,** or cell-mediated immunity.

T cells (and some macrophages) also synthesize and secrete polypeptides called **cytokines** (or more specifically, lymphokines) that enhance certain cellular responses to antigens. For example, *interleukin-1* and *interleukin-2* stimulate synthesis of several cytokines

from other T cells. In addition, interleukin-1 helps activate T cells, whereas interleukin-2 causes T cells to proliferate and activates a certain type of T cell (cytotoxic T cells). Other cytokines called *colony-stimulating factors* (CSFs) stimulate production of leukocytes in the red bone marrow, cause B cells to grow and mature, and activate macrophages. Certain cytokine combinations shut off the immune response. Table 16.4 summarizes several cytokine types.

T cells may also secrete toxins that kill their antigen-bearing target cells: growth-inhibiting factors that prevent target cell growth, or interferon that inhibits proliferation of viruses and tumor cells. Several types of T cells have distinct functions.

B cells attack foreign antigens in a different way. They differentiate into **plasma cells,** which produce and secrete large globular proteins called **antibodies** (an'tĭ-bod"ēz), or **immunoglobulins** (im"u-no-glob'u-linz). A plasma cell is an antibody factory, as evidenced by its characteristically huge Golgi apparatus. At the peak of an infection, a plasma cell may produce and secrete 2,000 antibody molecules a second! Body fluids carry antibodies, which then react in various ways to destroy specific

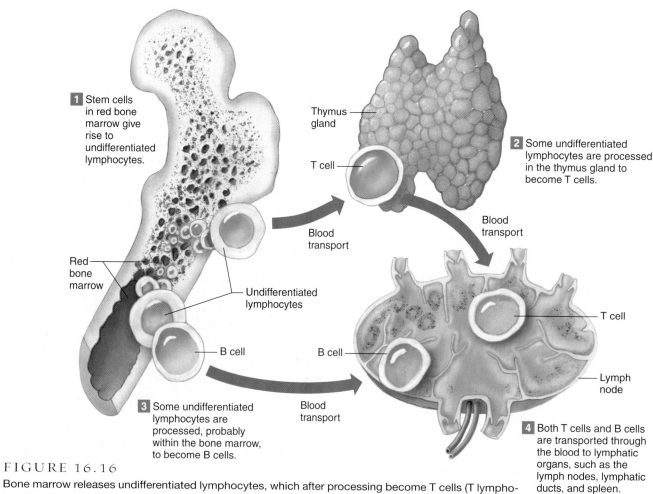

1 Stem cells in red bone marrow give rise to undifferentiated lymphocytes.

Red bone marrow

Undifferentiated lymphocytes

B cell

3 Some undifferentiated lymphocytes are processed, probably within the bone marrow, to become B cells.

Blood transport

Thymus gland

T cell

2 Some undifferentiated lymphocytes are processed in the thymus gland to become T cells.

Blood transport

B cell

T cell

Lymph node

4 Both T cells and B cells are transported through the blood to lymphatic organs, such as the lymph nodes, lymphatic ducts, and spleen.

FIGURE 16.16

Bone marrow releases undifferentiated lymphocytes, which after processing become T cells (T lymphocytes) or B cells (B lymphocytes). Note that in the fetus, the medullary cavity contains red marrow.

TABLE 16.4	Types of Cytokines
Cytokine	**Function**
Colony-stimulating factors	Stimulate bone marrow to produce lymphocytes
Interferons	Block viral replication, stimulate macrophages to engulf viruses, stimulate B cells to produce antibodies, attack cancer cells
Interleukins	Control lymphocyte differentiation and growth
Tumor necrosis factor	Stops tumor growth, releases growth factors, causes fever that accompanies bacterial infection, stimulates lymphocyte differentiation

antigens or antigen-bearing particles. This antibody-mediated immune response is called the **humoral immune response** ("humoral" refers to fluid).

Each person has millions of varieties of T and B cells. Because the members of each variety originate from a single early cell, they are all alike, forming a **clone** (klōn) of cells (identical cells originating from division of a single cell). The members of each variety have a particular type of antigen receptor on their cell membranes that can respond only to a specific antigen. Table 16.5 compares the characteristics of T cells and B cells.

T Cells and the Cellular Immune Response

A lymphocyte must be activated before it can respond to an antigen. T cell activation requires the presence of processed fragments of antigen attached to the surface of another kind of cell, called an **antigen-presenting cell** (accessory cell). Macrophages, B cells, and several other cell types can be antigen-presenting cells.

T cell activation begins when a macrophage phagocytizes a bacterium, digesting it in its lysosomes. Some bacterial antigens exit the lysosomes and move to the macrophage's surface. Here, they are displayed on the cell membrane near certain protein molecules that are part of a group of proteins called the **major histocompatibility complex (MHC)** or *human leukocyte associated (HLA) antigens* because they were first identified on white blood cells. MHC antigens help T cells recognize that an antigen is foreign, not self. Class I MHC antigens are within cell membranes of all body cells except red blood cells. Class II MHC antigens occur on the surfaces of antigen-presenting cells, thymus cells, and activated T cells.

A specialized type of T cell, called a **helper T cell,** becomes activated when its antigen receptors combine with displayed foreign antigens (fig. 16.17). The activated helper T cell stimulates the B cell to produce antibodies that are specific for the displayed antigen.

A type of helper T cell called a CD4 cell is the prime target of HIV, the virus that causes AIDS. (CD4 stands for the "cluster-of-differentiation" antigen it bears that enables it to recognize a macrophage displaying a foreign antigen.) Considering the role of CD4 helper T cells as key players in establishing immunity—they stimulate B cells and secrete cytokines—it is no wonder that harming them destroys immunity.

Another type of T cell is a **cytotoxic T cell,** which recognizes nonself antigens that cancerous cells or virally infected cells display on their surfaces near certain MHC proteins. A cytotoxic T cell becomes activated when it combines with an antigen that fits its receptors. Then, the T cell proliferates, enlarging its clone of cells. Cytotoxic T cells then bind to the surfaces of antigen-bearing cells, where they release a protein called *perforin* that cuts pore-like openings, destroying these cells. Cytotoxic T cells continually monitor the body's cells, recognizing and eliminating tumor cells and cells infected with viruses.

Memory T cells are among the many T cells produced upon initial exposure to an antigen, but they include only those cells not responding to the antigen at that time. These cells provide for a no-delay response to any future exposure to the same antigen, with immediate differentiation into cytotoxic T cells. This response generally vanquishes the invading pathogen before it can cause the body to produce signs and symptoms of disease.

1 What are the functions of T cells and B cells?

2 How do T cells become activated?

3 What is the function of cytokines?

4 How do cytotoxic T cells destroy antigen-bearing cells?

TABLE 16.5	A Comparison of T Cells and B Cells	
Characteristic	**T Cells**	**B Cells**
Origin of undifferentiated cell	Red bone marrow	Red bone marrow
Site of differentiation	Thymus	Probably the red bone marrow
Primary locations	Lymphatic tissues, 70%-80% of the circulating lymphocytes	Lymphatic tissues, 20%-30% of the circulating lymphocytes
Primary functions	Provide cellular immune response in which T cells interact directly with the antigens or antigen-bearing agents to destroy them	Provide humoral immune response in which B cells interact indirectly, producing antibodies that destroy the antigens or antigen-bearing agents

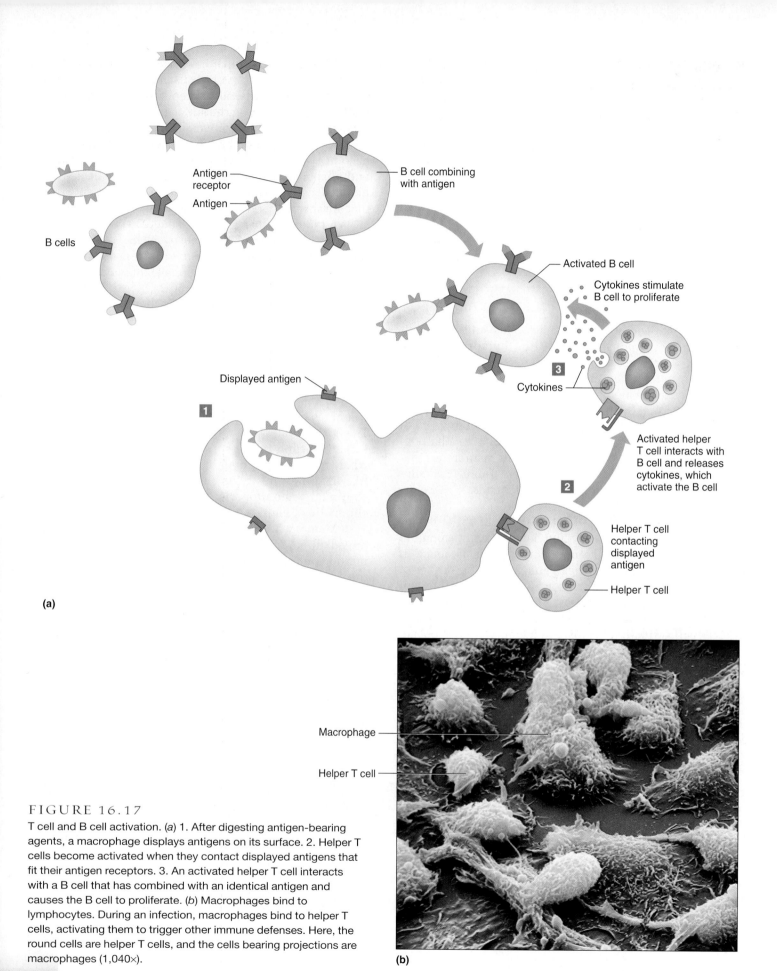

B cells

Antigen receptor

Antigen

B cell combining with antigen

Activated B cell

Cytokines stimulate B cell to proliferate

Cytokines

3

Displayed antigen

1

2

Activated helper T cell interacts with B cell and releases cytokines, which activate the B cell

Helper T cell contacting displayed antigen

Helper T cell

(a)

Macrophage

Helper T cell

(b)

FIGURE 16.17

T cell and B cell activation. (*a*) 1. After digesting antigen-bearing agents, a macrophage displays antigens on its surface. 2. Helper T cells become activated when they contact displayed antigens that fit their antigen receptors. 3. An activated helper T cell interacts with a B cell that has combined with an identical antigen and causes the B cell to proliferate. (*b*) Macrophages bind to lymphocytes. During an infection, macrophages bind to helper T cells, activating them to trigger other immune defenses. Here, the round cells are helper T cells, and the cells bearing projections are macrophages (1,040×).

B Cells and the Humoral Immune Response

A B cell may become activated when it encounters an antigen whose molecular shape fits the shape of the B cell's antigen receptors. In response to the receptor-antigen combination, the B cell divides repeatedly, expanding its clone. However, most antigens require T cell "help" to activate B cells.

When an activated helper T cell encounters a B cell that has already combined with an identical foreign antigen, the helper cell releases certain cytokines. These cytokines stimulate the B cell to proliferate, thus enlarging its clone of antibody-producing cells (fig. 16.18). The cytokines also attract macrophages and leukocytes into inflamed tissues and help keep them there. T cells can also suppress antibody formation by releasing cytokines that inhibit B cell function.

Some members of the activated B cell's clone differentiate further into *memory cells* (fig. 16.19). Like memory T cells, these memory B cells respond rapidly to subsequent exposure to a specific antigen.

Other members of the activated B cell's clone differentiate further into plasma cells, which secrete antibodies similar in structure to the antigen-receptor molecules on the original B cell's surface. These antibodies can combine with the antigen-bearing agent that has invaded the body, and react against it. Table 16.6 summarizes the steps leading to antibody production as a result of B and T cell activities.

A single type of B cell carries information to produce a single type of antibody. However, different B cells respond to different antigens on a pathogen's surface. Therefore, an immune response may include several types of antibodies manufactured against a single microbe

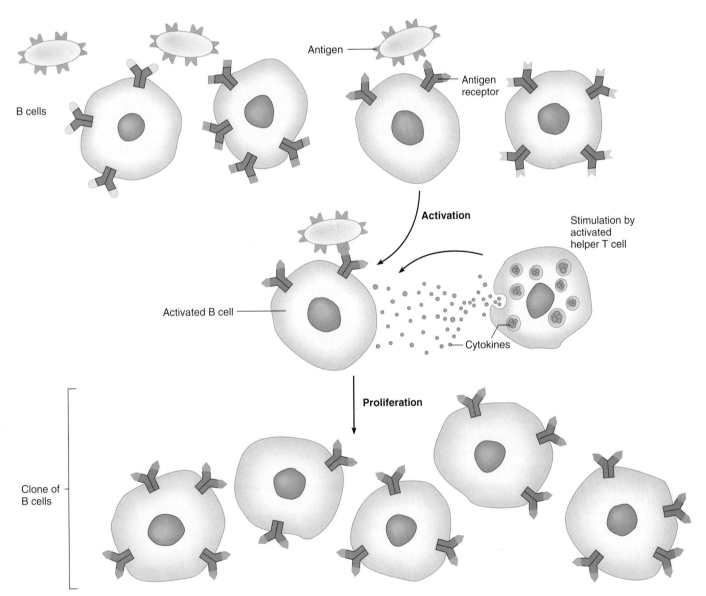

FIGURE 16.18

When a B cell encounters an antigen that fits its antigen receptor, it becomes activated and proliferates, thus enlarging its clone. Note that all cells in the clone have the same antigen receptor.

IMMUNOTHERAPY

At the turn of the last century, German bacteriologist Paul Ehrlich developed the concept of the "magic bullet"—a substance that could enter the body and destroy diseased cells, yet spare the healthy ones. The biochemicals and cells of the immune system, with their great specificity for attacking foreign tissue, would be ideal magic bullets. *Immunotherapy* **uses immune system components to fight disease—both the humoral immune response (antibodies) and the cellular immune response (cytokines).**

Monoclonal Antibodies— Targeting Immunity

Tapping the specificity of a single B cell and using its single type, or *monoclonal,* antibody to target a specific antigen (such as on a cancer or bacterial cell) awaited finding a way to entice the normally short-lived mature B cells into persisting in culture. In 1975, British researchers Cesar Milstein and Georges Köhler devised *monoclonal antibody* (MAb) technology, an ingenious way to capture the antibody-making capacity of a single B cell.

Milstein and Köhler injected a mouse with antigen-laden red blood cells from a sheep. They then isolated a single B cell from the mouse's spleen and fused it with a cancerous white blood cell from a mouse. The result was a fused cell, or *hybridoma,* with a valuable pair of talents: Like the B cell, it produces large amounts of a single antibody type; like the cancer cell, it divides continuously (fig. 16A).

MAbs are used in basic research, veterinary and human health care, and agriculture. Cell biologists use pure antibodies to localize and isolate proteins. Diagnostic MAb "kits" detect tiny amounts of a single molecule. Most kits consist of a paper strip impregnated with a MAb, to which the user adds a body fluid. For example, a woman who suspects she is pregnant places drops of her urine onto the paper. A color change ensues if the MAb binds to human chorionic gonadotropin (see chapter 23, p. 884), indicating pregnancy.

MAbs can highlight cancer before it can be detected by other means. The MAb is attached to a radioactive chemical, which is then detected when the MAb binds an antigen unique to the cancer cell surfaces.

Detecting a cancer's recurrence with a MAb requires only an injection followed by a painless imaging procedure.

MAbs can ferry conventional cancer treatments to where they are needed and limit their toxicity by sparing healthy tissue. Drugs or radioactive chemicals are attached to MAbs that deliver them to antigens on cancer cells. When injected into a patient, the MAb and its cargo are engulfed by the cancer cells, which are destroyed. Although MAbs were originally derived from mice, human versions that prevent allergic reactions are now available.

Cytokines

Immunotherapy experiments were difficult to do in the late 1960s because cytokines and antibodies could be obtained only in small amounts from cadavers. In the 1970s, recombinant DNA and monoclonal antibody technologies made it possible to make unlimited amounts of pure proteins—just as the AIDS epidemic was making it essential to find a purer source of biochemicals than cadavers.

Interferon was the first cytokine to be tested on a large scale. Although it did not

TABLE 16.6	Steps in Antibody Production

B Cell Activities

1. Antigen-bearing agents enter tissues.
2. B cell becomes activated when it encounters an antigen that fits its antigen receptors.
3. Either alone or more often in conjunction with helper T cells the activated B cell proliferates, enlarging its clone.
4. Some of the newly formed B cells differentiate further to become plasma cells.
5. Plasma cells synthesize and secrete antibodies whose molecular structure is similar to the activated B cell's antigen receptors.
6. Antibodies combine with antigen-bearing agents, helping to destroy them.

T Cell Activities

1. Antigen-bearing agents enter tissues.
2. Accessory cell, such as a macrophage, phagocytizes antigen-bearing agent, and the macrophage's lysosomes digest the agent.
3. Antigens from the digested antigen-bearing agents are displayed on the surface membrane of the accessory cell.
4. Helper T cell becomes activated when it encounters a displayed antigen that fits its antigen receptors.
5. Activated helper T cell releases cytokines when it encounters a B cell that has previously combined with an identical antigen-bearing agent.
6. Cytokines stimulate the B cell to proliferate.
7. Some of the newly formed B cells differentiate into antibody-secreting plasma cells.
8. Antibodies combine with antigen-bearing agents, helping to destroy them.

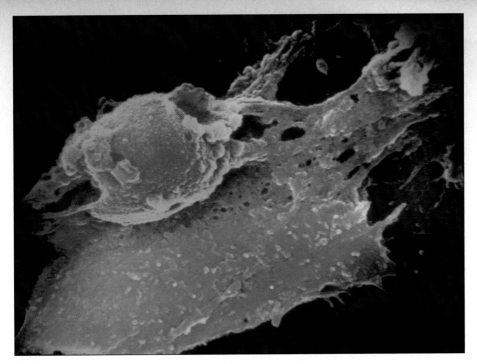

or virus. This is called a **polyclonal response.** From Science to Technology 16.1 discusses how researchers use clones of single B cells to produce single, or monoclonal, antibodies.

Antibody Molecules

Antibodies (or immunoglobulins) are soluble, globular proteins that constitute the *gamma globulin* fraction of plasma proteins (see chapter 14, p. 524). Each immunoglobulin molecule begins with four chains of amino acids that are linked together by pairs of sulfur atoms that are attached to each other by disulfide bonds. The four chains form a Y-shaped structure (fig. 16.20). Two of these amino acid chains are identical **light chains** (L-chains), and two are identical **heavy chains** (H-chains). The heavy chains contain about twice as many amino acids as the light chains. The five major types of immunoglobulin molecules are distinguished by a particular kind of heavy chain. Most of the types of immunoglobulin molecules consist of a single Y-shaped structure, but some immunoglobulins contain as many as five (see fig. 14.21).

As with other proteins, the sequences of amino acids of the heavy and light chains confer the unique, three-dimensional structure (conformation) of each immunoglobulin. This special conformation, in turn, imparts the physiological properties of the molecule. For example, one end of each of the heavy and light chains consists of variable sequences of amino acids (variable regions). These regions are specialized to react with the shape of a specific antigen molecule.

Antibodies can bind to certain antigens because of the conformation of the variable regions. The antibody contorts to form a pocket around the antigen. These specialized ends of the antibody molecule are called **antigen-binding sites,** and the particular parts that actually bind the antigen are called **idiotypes** (id′e-o-tīpz′).

The remaining portions of the chains are termed constant regions because their amino acid sequences are very similar from molecule to molecule. Constant regions impart other properties of the immunoglobulin molecule, such as its ability to bond to cellular structures or to combine with certain chemicals (see fig. 16.20).

Types of Immunoglobulins

Of the five major types of immunoglobulins, three constitute the bulk of the circulating antibodies. They are immunoglobulin G, which accounts for about 80% of the antibodies; immunoglobulin A, which makes up about 13%; and immunoglobulin M, which is responsible for about 6%. Immunoglobulin D and immunoglobulin E account for the remainder of the antibodies.

Immunoglobulin G (IgG) is in plasma and tissue fluids and is particularly effective against bacteria, viruses, and toxins. IgG also activates a group of enzymes called

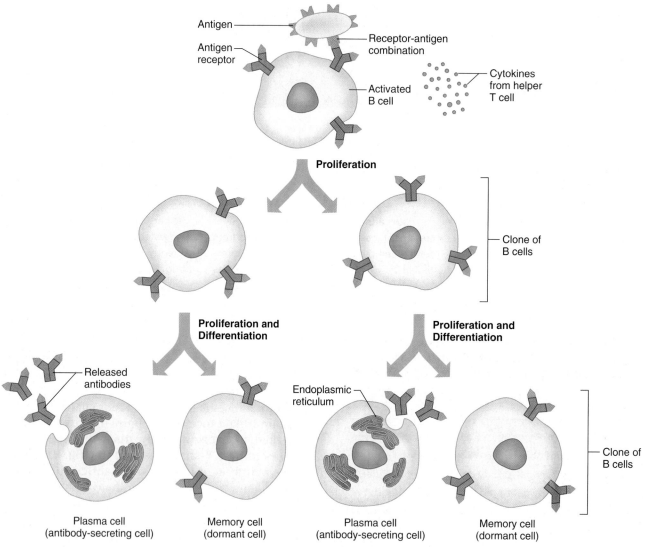

FIGURE 16.19

An activated B cell proliferates after stimulation by cytokines released by helper T cells. The B cell's clone enlarges. Some cells of the clone give rise to antibody-secreting plasma cells and others to dormant memory cells.

626

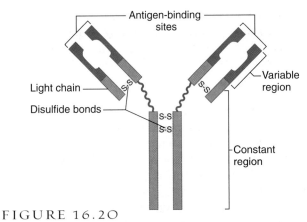

FIGURE 16.20
An immunoglobulin molecule consists basically of two identical light chains of amino acids and two identical heavy chains of amino acids.

Immunoglobulin D (IgD) is found on the surfaces of most B cells, especially those of infants. IgD acts as an antigen receptor and is important in activating B cells (see fig. 16.18).

Immunoglobulin E (IgE) appears in exocrine secretions along with IgA. It is associated with allergic reactions that are described later in this chapter in the section titled "Allergic Reactions." Table 16.7 summarizes the major immunoglobulins and their functions.

1 How are B cells activated?

2 How does the antibody response protect against diverse infections?

3 What is an immunoglobulin?

4 Describe the structure of an immunoglobulin molecule, and name the five major types of immunoglobulins.

Antibody Actions

In general, antibodies react to antigens in three ways. Antibodies directly attack antigens, activate a set of enzymes (*complement*) that attack the antigens, or stimulate localized changes that help prevent spread of the antigens.

In a direct attack, antibodies combine with antigens and cause them to clump (agglutinate) or to form insoluble substances (precipitation). Such actions make it easier for phagocytic cells to engulf the antigen-bearing agents and eliminate them. In other instances, antibodies cover the toxic portions of antigen molecules and neutralize their effects (neutralization). However, under normal conditions, complement activation is more important in protecting against infection than is direct antibody attack.

Complement (kom'plě-ment) is a group of proteins (complement system) in plasma and other body fluids. When certain IgG or IgM antibodies combine with antigens, they expose reactive sites on the antibody constant regions. This triggers a series of reactions leading to activation of the complement proteins, which, in turn, produce a variety of effects, including coating the antigen-antibody complexes (opsonization), making the complexes more susceptible to phagocytosis; attracting macrophages and neutrophils into the region (chemotaxis); clumping

complement, which is described in the following section titled "Antibody Actions."

Immunoglobulin A (IgA) is commonly found in exocrine gland secretions. It is in breast milk, tears, nasal fluid, gastric juice, intestinal juice, bile, and urine.

A newborn does not yet have its own antibodies but does retain IgG that passed through the placenta from the mother. These maternal antibodies protect the infant against some illnesses to which the mother is immune. The maternal antibody supply begins to fall just about when the infant begins to manufacture its own antibodies. The newborn also receives IgA from colostrum, a substance secreted from the mother's breasts for the first few days after birth. Antibodies in colostrum protect against certain digestive and respiratory infections.

Immunoglobulin M (IgM) is a type of antibody that develops in the plasma in response to contact with certain antigens in foods or bacteria. The antibodies anti-A and anti-B, described in chapter 14 (p. 534), are examples of IgM. IgM also activates complement.

TABLE 16.7	Characteristics of Major Immunoglobulins	
Type	**Occurrence**	**Major Function**
IgG	Tissue fluid and plasma	Defends against bacterial cells, viruses, and toxins; activates complement
IgA	Exocrine gland secretions	Defends against bacterial cells and viruses
IgM	Plasma	Reacts with antigens occurring naturally on some red blood cell membranes following certain blood transfusions; activates complement
IgD	Surface of most B lymphocytes	B cell activation
IgE	Exocrine gland secretions	Promotes inflammation and allergic reactions

antigen-bearing cells; rupturing membranes of foreign cells (lysis); and altering the molecular structure of viruses, making them harmless. Other proteins promote inflammation, which helps prevent the spread of infectious agents.

Immunoglobulin E promotes inflammation that may be so intense that it damages tissues. This antibody is usually attached to the membranes of widely distributed *mast cells* (see chapter 5, p. 141). When antigens combine with the antibodies, the resulting antigen-antibody complexes stimulate mast cells to release biochemicals, such as histamine, that cause the changes associated with inflammation, such as vasodilation and edema. Table 16.8 summarizes the actions of antibodies.

1 In what general ways do antibodies function?

2 What is the function of complement?

3 How is complement activated?

Immune Responses

When B cells or T cells become activated after first encountering the antigens for which they are specialized to react, their actions constitute a **primary immune response.** During such a response, plasma cells release antibodies (IgM, followed by IgG) into the lymph. The antibodies are transported to the blood and then throughout the body, where they help destroy antigen-bearing agents. Production and release of antibodies continues for several weeks.

Following a primary immune response, some of the B cells produced during proliferation of the clone remain dormant and serve as *memory cells* (see fig. 16.19). If the identical antigen is encountered in the future, the clones of these memory cells enlarge, and they can respond rapidly with IgG to the antigen to which they were previously sensitized. These memory B

cells along with the previously discussed memory T cells produce a **secondary immune response.** Cells in lymph nodes called *follicular dendritic cells* may help memory by harboring and slowly releasing viral antigens after an initial infection. This constantly stimulates memory B cells, which present the antigens to T cells, maintaining immunity.

As a result of a primary immune response, detectable concentrations of antibodies usually appear in the body fluids within five to ten days following an exposure to antigens. If the identical antigen is encountered some time later, a secondary immune response may produce additional antibodies within a day or two (fig. 16.21). Although newly formed antibodies may persist in the body for only a few months, or perhaps a few years, memory cells live much longer. A secondary immune response may be very long-lasting.

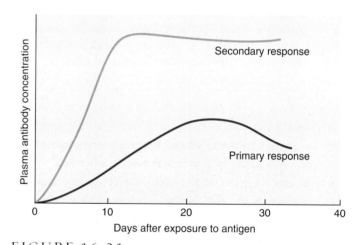

FIGURE 16.21

A primary immune response produces a lesser concentration of antibodies than does a secondary immune response.

TABLE 16.8	Actions of Antibodies		
General Action		**Type of Effect**	**Description**
Direct Attack			
		Agglutination	Antigens clump
		Precipitation	Antigens become insoluble
		Neutralization	Antigens lose toxic properties
Activation of Complement (Presence of antibodies combined with antigens)		Opsonization	Alters antigen cell membranes so cells are more susceptible to phagocytosis
		Chemotaxis	Attracts macrophages and neutrophils into the region
		Agglutination	Clumping together of antigen-bearing cells
		Lysis	Allows rapid movement of water and ions into the foreign cell causing osmotic rupture of the foreign cell
		Neutralization	Altering the molecular structure of viruses making them harmless
Localized Changes			
		Inflammation	Helps prevent the spread of antigens

Practical Classification of Immunity

A generation ago, it was common at certain times of the year for grade-school classrooms to be nearly empty, due to several infectious "childhood diseases," including measles, mumps, rubella, and chicken pox. However, each child usually suffered each illness only once, thanks to *naturally acquired active immunity*. This form of immunity develops after a primary immune response and is a response to exposure to a live pathogen and development of symptoms.

Today, most children in developed countries do not contract measles, mumps, rubella, or chicken pox because they develop another type of active immunity, produced in response to receiving a preparation called a **vaccine** (vak'sēn). A vaccine contains an antigen that can stimulate a primary immune response against a particular disease-causing agent but does not produce the severe symptoms of that disease.

A vaccine might contain bacteria or viruses that have been killed or attenuated (weakened) so that they cannot cause a serious infection, or it may contain a toxoid, which is a toxin from an infectious organism that has been chemically altered to destroy its dangerous effects. A vaccine may even consist of a single glycoprotein or similar large molecule from the pathogen's surface, which may be sufficient evidence of a foreign antigen to alert the immune system. A vaccine causes a person to develop *artificially acquired active immunity.*

Viruses whose genetic material rapidly mutates present a great challenge to vaccine development because their surfaces, which serve as antigens, change. It is a little like fighting an enemy who is constantly changing disguises. For this reason, pharmaceutical companies must develop a new vaccine against influenza each year. HIV is particularly changeable, which has severely hampered efforts to produce a vaccine.

Sometimes a person who has been exposed to infection requires protection against a disease-causing microorganism but lacks the time necessary to develop active immunity from receiving a vaccine. This happens with hepatitis A, a viral infection of the liver. In such a case, it may be possible to inject the person with antiserum (ready-made antibodies). These antibodies may be obtained from gamma globulin separated from the blood plasma of persons who have already developed immunity against the disease.

An injection of antibodies or antitoxin (antibodies against a toxin) provides *artificially acquired passive immunity.* This type of immunity is called passive because the recipient's cells do not produce the antibodies. Such immunity is short-term, seldom lasting more than a few weeks. Furthermore, because the recipient's lymphocytes might not have time to react to the pathogens for which protection was needed, susceptibility to infection may persist.

During pregnancy, certain antibodies (IgG) pass from the maternal blood into the fetal bloodstream. Receptor-mediated endocytosis (see chapter 3, p. 87) utilizing receptor sites on cells of the fetal yolk sac accomplishes the transfer (see chapter 23, p. 892). These receptor sites bind to a region common to the structure of IgG molecules. After entering the fetal cells, the antibodies are secreted into the fetal blood. As a result, the fetus acquires limited immunity against the pathogens for which the pregnant woman has developed active immunities. Thus, the fetus has *naturally acquired passive immunity,* which may persist for six months to a year after birth. Table 16.9 summarizes the types of immunity.

1 Distinguish between a primary and a secondary immune response.

2 Explain the difference between active and passive immunities.

TABLE 16.9 — Practical Classification of Immunity

Type	Mechanism	Result
Naturally acquired active immunity	Exposure to live pathogens	Stimulation of an immune response with symptoms of a disease
Artificially acquired active immunity	Exposure to a vaccine containing weakened or dead pathogens or their components	Stimulation of an immune response without the severe symptoms of a disease
Artificially acquired passive immunity	Injection of gamma globulin containing antibodies or antitoxin	Short-term immunity without stimulating an immune response
Naturally acquired passive immunity	Antibodies passed to fetus from pregnant woman with active immunity	Short-term immunity for newborn without stimulating an immune response

Allergic Reactions

Both allergic reactions and immune responses entail sensitizing of lymphocytes or combining of antigens with antibodies. An allergic reaction, however, is an immune attack against a nonharmful substance and can damage tissues. An allergy is also called a hypersensitivity reaction. One form of allergic reaction can occur in almost anyone, but another form affects only people who have inherited an ability to produce exaggerated immune responses. The antigens that trigger allergic responses are called **allergens** (al′er-jenz).

An *immediate-reaction* (type I or anaphylactic) *allergy* occurs within minutes after contact with an allergen. Persons with this type of allergy have an inherited tendency to overproduce IgE antibodies in response to certain antigens. IgE normally comprises a tiny fraction of plasma proteins.

An immediate-reaction allergy activates B cells, which become sensitized when the allergen is first encountered. Subsequent exposures to the allergen trigger allergic reactions. In the initial exposure, IgE attaches to the membranes of widely distributed mast cells and basophils. When a subsequent allergen-antibody reaction occurs, these cells release allergy mediators such as *histamine, prostaglandin D₂,* and *leukotrienes* (fig. 16.22). These substances cause a variety of physiological effects, including dilation of blood vessels, increased vascular permeability that swells tissues, contraction of bronchial and intestinal smooth muscles, and increased mucous production. The result is a severe inflammation reaction that is responsible for the symptoms of the allergy, such as hives, hay fever, asthma, eczema, or gastric disturbances.

Anaphylactic shock is a severe form of immediate-reaction allergy, in which mast cells release allergy mediators throughout the body. The person may at first feel an inexplicable apprehension, and then suddenly, the entire body itches and breaks out in red hives. Vomiting and diarrhea may follow. The face, tongue, and larynx begin to swell, and breathing becomes difficult. Unless the person receives an injection of epinephrine (adrenaline) and sometimes a tracheotomy (an incision into the windpipe to restore breathing), he or she will lose consciousness and may die within five minutes. Anaphylactic shock most often results from an allergy to penicillin or insect stings. Fortunately, thanks to prompt medical attention and avoidance of allergens by people who know they have allergies, only a few hundred people a year die of anaphylactic shock. The peanut allergy described in the chapter opening essay causes many of the symptoms of anaphylactic shock, but usually not the sensation of the throat closing.

One theory of the origin of allergies, particularly anaphylactic shock, is that they evolved at a time when insect bites and the natural substances from which antibiotics such as penicillin are made threatened human survival. Today, that once-protective response is an overreaction. The observation that IgE protects against roundworm and flatworm infections, in addition to taking part in allergic reactions, supports the idea that this antibody class is a holdover from times past, when challenges to the immune system differed from what they are today.

Hypersensitivities that take one to three hours to develop include *antibody-dependent cytotoxic reactions* (type II) and *immune complex reactions* (type III). In an antibody-dependent cytotoxic reaction, an antigen binds to a specific cell, stimulating phagocytosis and complement-mediated lysis of the antigen. A transfusion reaction to mismatched blood is a type II hypersensitivity reaction. In an immune complex reaction, widespread antigen-antibody complexes cannot be cleared from the circulation by phagocytosis and lysis. As a result, the complexes may block small vessels, which damages the tissues that they reach. **Autoimmunity,** the loss of the ability to tolerate self-antigens, illustrates this type of hypersensitivity reaction. It is discussed later in the chapter in the section titled "Autoimmunity."

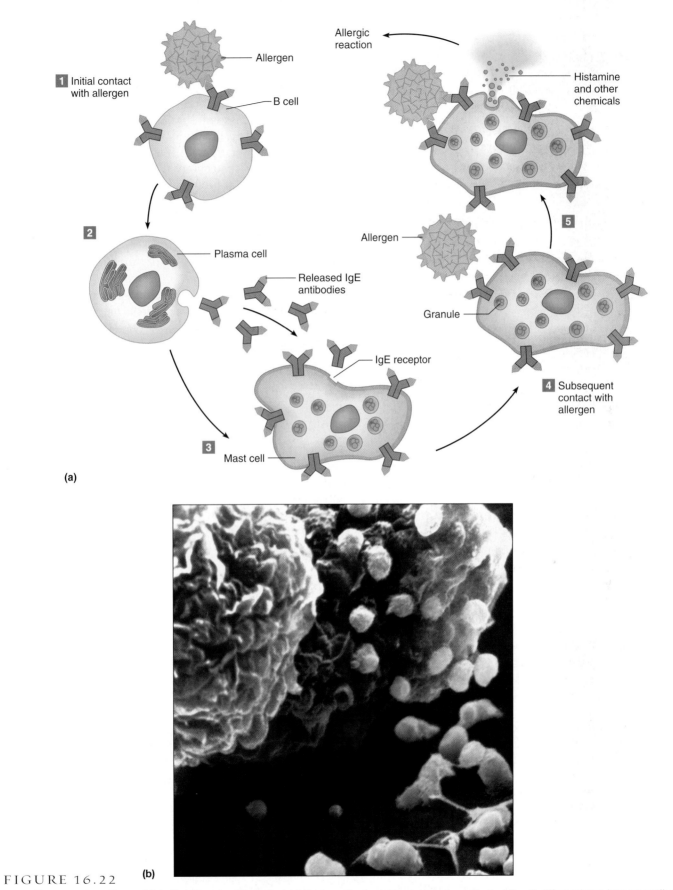

(a)

1 Initial contact with allergen

Allergen

B cell

2

Plasma cell

Released IgE antibodies

IgE receptor

3 Mast cell

Allergic reaction

Histamine and other chemicals

5

Allergen

Granule

4 Subsequent contact with allergen

(b)

FIGURE 16.22

Immediate-reaction allergy. (*a*) 1. B cells are activated when they contact an allergen. 2. An activated B cell differentiates into an antibody-secreting plasma cell. 3. Antibodies attach to mast cells. 4. When allergens are encountered, they combine with the antibodies on the mast cells. 5. The mast cells release allergy mediators, which cause the symptoms of the allergy attack. (*b*) A mast cell releases histamine granules (3,000×).

A *delayed-reaction allergy* (type IV) may affect anyone. It results from repeated exposure of the skin to certain chemicals—commonly, household or industrial chemicals or some cosmetics. After repeated contacts, the presence of the foreign substance activates T cells, many of which collect in the skin. The T cells and the macrophages they attract release chemical factors, which, in turn, cause eruptions and inflammation of the skin (dermatitis). This reaction is called *delayed* because it usually takes about forty-eight hours to occur.

Transplantation and Tissue Rejection

When a car breaks down, replacing the damaged or malfunctioning part often fixes the trouble. The same is sometimes true for the human body. Transplanted tissues and organs include corneas, kidneys, lungs, pancreases, bone marrow, pieces of skin, livers, and hearts. The danger the immune system poses to transplanted tissue is that the recipient's cells may recognize the donor's tissues as foreign and attempt to destroy the transplanted tissue. Such a response is called a **tissue rejection reaction.** The transplanted tissue may also produce substances that harm the recipient's tissue, a response called graft-versus-host disease (GVHD).

Tissue rejection resembles the cellular immune response against a foreign antigen. The greater the antigenic difference between the cell surface molecules (MHC antigens discussed earlier in this chapter on page 621) of the recipient tissues and the donor tissues, the more rapid and severe the rejection reaction. Matching the cell surface molecules of donor and recipient tissues can minimize the rejection reaction. This means locating a donor whose tissues are antigenically similar to those of the person needing a transplant—a procedure much like matching the blood of a donor with that of a recipient before giving a blood transfusion.

The four major varieties of grafts (transplant tissue) include

- *Isograft.* Tissue is taken from a genetically identical twin.
- *Autograft.* Tissue is taken from elsewhere in a person's body. (Technically, this is not a transplant because it occurs within an individual.)
- *Allograft.* Tissue comes from an individual who is not genetically identical to the recipient, but of the same species.
- *Xenograft.* Tissue comes from a different species, such as pigs and baboons.

Table 16.10 presents examples of transplants.

Immunosuppressive drugs are used to reduce rejection of transplanted tissues. These drugs interfere with the recipient's immune response by suppressing formation of antibodies or production of T cells, thereby dampening the humoral and cellular immune responses. Unfortunately, the use of immunosuppressive drugs can leave a recipient more prone to infections. It is not uncommon for a patient to survive a transplant but die of infection because of a weakened immune system. The first immunosuppressive drug, cyclosporin, was discovered in a soil sample from Switzerland in the early 1980s. New drugs are more effective at selectively suppressing only those parts of the immune response that target transplanted tissue. Drugs that target different parts of the organ rejection immune response are often teamed. For example, a drug called Rapamune (sirolimus) is administered with cyclosporine to kidney recipients.

TABLE 16.10		Transplant Types
Type	**Donor**	**Example**
Isograft	Identical twin	Bone marrow transplant from a healthy twin to a twin who has leukemia
Autograft	Self	Skin graft from one part of body to replace burned skin
Allograft	Same species	Kidney transplant from relative or closely matched donor
Xenograft	Different species	Heart valves from a pig

Less drastic than an organ transplant is a cell implant, which consists of small pieces of tissue. Implants of liver cells may treat cirrhosis; pancreas cells may treat diabetes; skeletal muscle cells may replace heart muscle damaged in a heart attack or treat muscular dystrophy; adrenal gland cells may treat Parkinson disease; and brain cell implants may treat Alzheimer and Huntington diseases. All are still experimental or in the animal testing process.

Autoimmunity

Sometimes the immune system fails to distinguish self from nonself, producing antibodies, called **autoantibodies,** and cytotoxic T cells that attack and damage tissues and organs of the body. This attack against self is called *autoimmunity.* The signs and symptoms of autoimmune disorders reflect the cell types under attack. For example, in autoimmune hemolytic anemia, autoantibodies destroy red blood cells. In autoimmune ulcerative colitis, colon cells are the target, and severe abdominal pain results. Table 16.11 lists some autoimmune disorders.

Why might the immune system attack body tissues? Perhaps a virus, while replicating within a human cell, "borrows" proteins from the host cell's surface and incorporates them onto its own surface. When the immune system "learns" the surface of the virus to destroy it, it also learns to attack the human cells that normally bear the particular protein. Another explanation of autoimmunity is that somehow T cells never learn in the thymus to distinguish self from nonself.

A third possible route of autoimmunity is when a nonself antigen coincidentally resembles a self antigen.

This may explain type I diabetes, which is a deficiency of insulin, the hormone required to transport glucose from the blood into cells that use it (see Clinical Application 13.4). Type I diabetes affects 1 in 500 people under the age of twenty—they must inject insulin several times a day.

Part of a protein on insulin-producing cells matches part of bovine serum albumin (BSA), which is a protein in cow's milk. Perhaps children with an allergy to cow's milk develop antibodies against BSA, which later attack the similar-appearing pancreas cells, causing type I diabetes. An ongoing study is tracking the health of children with family histories of type I diabetes who have avoided drinking cow's milk, compared to similarly high-risk children who have drunk cow's milk.

Some disorders traditionally thought to be autoimmune in origin may in fact have a more bizarre cause—fetal cells persisting in a woman's circulation, even decades after the fetus has grown up! In response to an as yet unknown trigger, the fetal cells, perhaps "hiding" in a tissue such as skin, emerge, stimulating antibody production. If we didn't know the fetal cells were there, the resulting antibodies and symptoms would appear to be an autoimmune disorder. This mechanism, called microchimerism ("small mosaic"), may explain the higher prevalence of autoimmune disorders among women. It was discovered in a disorder called scleroderma, which means "hard skin" (figure 16.23).

Patients describe scleroderma, which typically begins between ages forty-five and fifty-five, as "the body turning to stone." Symptoms include fatigue, swollen joints, stiff fingers, and a masklike face. The hardening may affect blood vessels, the lungs, and the esophagus, too. Clues that scleroderma is a delayed

TABLE 16.11	Autoimmune Disorders	
Disorder	**Symptoms**	**Antibodies Against**
Glomerulonephritis	Lower back pain	Kidney cell antigens that resemble streptococcal bacteria antigens
Graves disease	Restlessness, weight loss, irritability, increased heart rate and blood pressure	Thyroid gland antigens near thyroid-stimulating hormone receptor, causing overactivity
Type I diabetes mellitus	Thirst, hunger, weakness, emaciation	Pancreatic beta cells
Hemolytic anemia	Fatigue and weakness	Red blood cells
Multiple sclerosis	Weakness, incoordination, speech disturbances, visual complaints	Myelin in the white matter of the central nervous system
Myasthenia gravis	Muscle weakness	Receptors for neurotransmitters on skeletal muscle
Pernicious anemia	Fatigue and weakness	Binding site for vitamin B on cells lining stomach
Rheumatic fever	Weakness, shortness of breath	Heart valve cell antigens that resemble streptococcal bacteria antigens
Rheumatoid arthritis	Joint pain and deformity	Cells lining joints
Systemic lupus erythematosus	Red rash on face, prolonged fever, weakness, kidney damage	DNA, neurons, blood cells
Ulcerative colitis	Lower abdominal pain	Colon cells

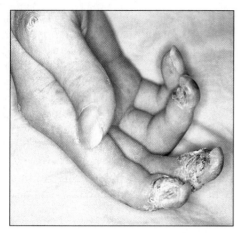

1 How are allergic reactions and immune reactions similar yet different?

2 How does a tissue rejection reaction involve an immune response?

3 How is autoimmunity an abnormal functioning of the immune response?

FIGURE 16.23

Scleroderma hardens the skin. Some cases appear to be caused by a long-delayed reaction of the immune system to cells retained from a fetus—even decades earlier.

response to persisting fetal cells include the following observations:

- It is much more common among women.
- Symptoms resemble those of graft-versus-host disease (GVHD), in which transplanted tissue produces chemicals that destroy the body. Antigens on cells in scleroderma lesions match those involved in GVHD.
- Mothers who have scleroderma and their sons have cell surfaces that are more similar than those of unaffected mothers and their sons. Perhaps the similarity of cell surfaces enabled the fetal cells to escape surveillance and destruction by the woman's immune system. Female fetal cells probably have the same effect, but this is more difficult to demonstrate because these cells cannot be distinguished from maternal cells by the presence of a Y chromosome.

It's possible that other disorders traditionally considered to be autoimmune may actually reflect an immune system response to lingering fetal cells.

Chronic fatigue syndrome is a poorly understood immune system imbalance. The condition begins suddenly, producing fatigue so great that getting out of bed is an effort. Chills, fever, sore throat, swollen glands, muscle and joint pain, and headaches are also symptoms. The various disabling aches and pains reflect an overactive immune system. Affected people have up to forty times the normal amount of interleukin-2 and too many cytotoxic T cells, yet too little interferon. It is as if the immune system mounts a defense and then doesn't know when to shut it off. The cause of chronic fatigue syndrome is not known.

Life-Span Changes

In a sense, aging of the immune system actually begins before birth, when nonself T cells are selected for destruction, via programmed cell death (apoptosis), in the thymus. The immune system begins to decline early in life. The thymus gland reaches its maximal size in adolescence and then slowly shrinks. By age seventy, the thymus is one-tenth the size it was at the age of ten, and the immune system is only 25% as powerful.

The declining strength of the immune response is why elderly people have a higher risk of developing cancer and succumb more easily to infections that they easily fought off at an earlier age, such as influenza, tuberculosis, and pneumonia. Encephalitis due to infection by the West Nile virus, a newly described illness, may cause very minor symptoms in young people, but kill the elderly. HIV infection progresses to AIDS faster in people older than forty. AIDS is more difficult to diagnose in older people, sometimes because physicians do not initially suspect the condition, instead attributing the fatigue, confusion, loss of appetite, and swollen glands to other causes. However, 11% of new cases of AIDS occur in those over age fifty.

Interestingly, numbers of T cells diminish only slightly with increasing age, and numbers of B cells not at all. However, activity levels change for both types of lymphocytes. Because T cell function controls production of B cells, effects on B cells are secondary. The antibody response to antigens is slower, and as a result, vaccines that would ordinarily be effective in one dose may require an extra dose. The proportions of the different antibody classes shift, with IgA and IgG increasing, and IgM and IgE decreasing. A person may produce more autoantibodies than at a younger age, increasing the risk of developing an autoimmune disorder.

Because of the declining function of the immune system, elderly people may not be candidates for certain medical treatments that suppress immunity, such as cancer chemotherapy and steroids to treat inflammatory disorders. Overall, the immune system makes it possible for us to survive in a world that is also home to many microorganisms. Clinical Application 16.1 looks at the devastation of immunity that is AIDS.

1 When is maximum size of the thymus reached?

2 Explain the decline in strength of the immune response in elderly people.

IMMUNITY BREAKDOWN: AIDS

Natural History of a Modern Plague

In late 1981 and early 1982, physicians from large cities in the United States began reporting to the Centers for Disease Control and Prevention cases of formerly rare infections in otherwise healthy young men. Some of the infections were prevalent in the general population, such as herpes simplex and cytomegalovirus, but in these young men were unusually severe. Some infections were caused by organisms known to infect only nonhuman animals. Other infections, particularly pneumonia caused by the microorganism *Pneumocystis carinii,* and a cancer, Kaposi sarcoma, were known only in individuals whose immune systems were suppressed (fig. 16B). The bodies of the sick young men had become nesting places for all types of infectious agents, including viruses, bacteria, protozoans, and fungi. The infections were *opportunistic,* which means that they took advantage of a weakened immune system.

As the infections spread, a portrait of a lethal disease emerged. *Acquired immune deficiency syndrome,* or AIDS, starts with recurrent fever, weakness, and weight loss. Then usually after a relatively healthy period, infections begin. The human immunodeficiency virus (HIV) that causes AIDS can be present for a decade or longer before a person feels ill. Five percent of infected people have remained healthy for more than fifteen years.

HIV infection gradually shuts down the immune system. First, HIV enters macrophages, impairing this first line of defense. In these cells and later in helper T cells, the virus adheres with its surface protein, called gp120, to two coreceptors on the host cell surface, called CD4 and CCR5. Once the virus enters the cell, a viral enzyme reverse transcriptase, catalyzes the construction of a DNA strand comple-

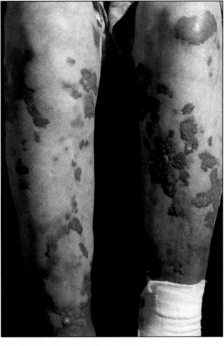

FIGURE 16B

Prior to the appearance of AIDS, Kaposi sarcoma was a rare cancer seen only in elderly Jewish and Italian men and in people with suppressed immune systems. In these groups, it produces purplish patches on the lower limbs, but in AIDS patients, Kaposi sarcoma patches appear all over the body and sometimes internally too. These lower limbs display characteristic lesions.

mentary to the viral RNA sequence (the virus has RNA as its genetic material). The initial viral DNA strand replicates to form a DNA double helix, which enters the cell's nucleus and inserts into a chromosome. The viral DNA sequences are then transcribed and translated. The cell fills with viral pieces, which are assembled into complete new viral particles that eventually burst from the cell.

Once helper T cells begin to die at a high rate, bacterial infections begin, because B cells aren't activated to produce antibodies. Much later in infection, HIV variants arise

that bind to receptors called CXCR4 on cytotoxic T cells, killing them. Loss of these cells renders the body very vulnerable to other infections and to cancers.

HIV replicates quickly, mutates quickly, and can hide. The virus is especially prone to mutation, because it cannot repair replication errors and because those errors occur frequently—because of the "sloppiness" of reverse transcriptase. The immune system simply cannot keep up; antibodies against one viral variant are useless against another. For several years, the bone marrow produces 2 billion new T and B cells a day to counter the million to a billion new HIV particles that burst daily from shattered cells.

So genetically diverse is the population of HIV in a human host that within days of initial infection, variants can arise that resist the drugs used to treat AIDS. HIV's changeable nature has important clinical implications. Combining drugs that act in different ways provides the greatest chance of slowing the disease process. Several classes of drugs target HIV infection at various stages. The first drugs developed, such as AZT, ddI, ddC, and 3TC, block viral replication. Drugs called protease inhibitors prevent HIV from processing its proteins to a functional size, crippling the virus. A third class of drugs, called fusion inhibitors, block the binding of HIV to T cell surfaces. Combining drugs can keep viral load low and delay symptom onset and progression, although viral variants emerge that resist the drugs. The goal is to enable infected people to live normal life spans in relatively good health. More than 200 drugs are also available to treat AIDS-associated opportunistic infections and cancers.

Better understanding of the biology of HIV infection, plus new drug weapons and clues from survivors, are providing what has long been lacking in the global fight against AIDS—hope. Slowly, HIV infection is becoming a chronic illness, rather than a death sentence. ■

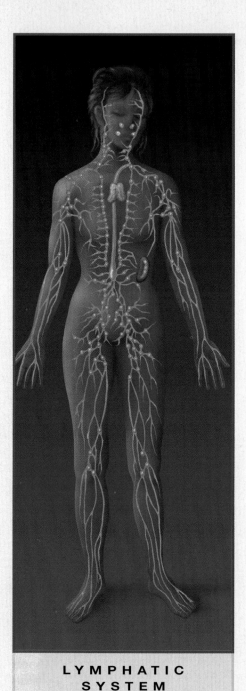

LYMPHATIC SYSTEM

The lymphatic system is an important link between the interstitial fluid and the plasma; it also plays a major role in the response to infection.

Integumentary System

The skin is a first line of defense against infection.

Skeletal System

Cells of the immune system originate in the bone marrow.

Muscular System

Muscle action helps pump lymph through the lymphatic vessels.

Nervous System

Stress may impair the immune response.

Endocrine System

Hormones stimulate lymphocyte production.

Cardiovascular System
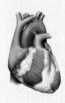
The lymphatic system returns tissue fluid to the bloodstream. Lymph originates as interstitial fluid, formed by the action of blood pressure.

Digestive System

Lymph plays a major role in the absorption of fats.

Respiratory System

Cells of the immune system patrol the respiratory system to defend against infection.

Urinary System

The kidneys control the volume of extracellular fluid, including lymph.

Reproductive System

Special mechanisms inhibit the female immune system in its attack of sperm as foreign invaders.

Clinical Terms Related to the Lymphatic System and Immunity

asplenia (ah-sple′ne-ah) Absence of a spleen.

immunocompetence (im″u-no-kom′pe-tens) Ability to produce an immune response to the presence of antigens.

immunodeficiency (im″u-no-de-fish′en-se) Inability to produce an immune response.

lymphadenectomy (lim-fad″ĕ-nek′to-me) Surgical removal of lymph nodes.

lymphadenopathy (lim-fad″ĕ-nop′ah-the) Enlargement of the lymph nodes.

lymphadenotomy (lim-fad″ĕ-not′o-me) Incision of a lymph node.

lymphocytopenia (lim″fo-si″to-pe′ne-ah) Too few lymphocytes in the blood.

lymphocytosis (lim″fo-si″to′sis) Too many lymphocytes in the blood.

lymphoma (lim-fo′mah) Tumor composed of lymphatic tissue.

lymphosarcoma (lim″fo-sar-ko′mah) Cancer within the lymphatic tissue.

splenectomy (sple-nek′to-me) Surgical removal of the spleen.

splenitis (sple-ni′tis) Inflammation of the spleen.

splenomegaly (sple″no-meg′ah-le) Abnormal enlargement of the spleen.

splenotomy (sple-not′o-me) Incision of the spleen.

thymectomy (thi-mek′to-me) Surgical removal of the thymus.

thymitis (thi-mi′tis) Inflammation of the thymus.

C H A P T E R   S U M M A R Y

Introduction (page 608)

The lymphatic system is closely associated with the cardiovascular system. It transports excess fluid to the bloodstream, absorbs fats, and helps defend the body against disease-causing agents.

Lymphatic Pathways (page 608)

1. Lymphatic capillaries
   a. Lymphatic capillaries are microscopic, closed-ended tubes that extend into interstitial spaces.
   b. They receive lymph through their thin walls.
   c. Lacteals are lymphatic capillaries in the villi of the small intestine.
2. Lymphatic vessels
   a. Lymphatic vessels are formed by the merging of lymphatic capillaries.
   b. They have walls similar to veins, only thinner, and possess valves that prevent backflow of lymph.
   c. Larger lymphatic vessels lead to lymph nodes and then merge into lymphatic trunks.
3. Lymphatic trunks and collecting ducts
   a. Lymphatic trunks drain lymph from large body regions.
   b. Trunks lead to two collecting ducts—the thoracic duct and the right lymphatic duct.
   c. Collecting ducts join the subclavian veins.

Tissue Fluid and Lymph (page 611)

1. Tissue fluid formation
   a. Tissue fluid originates from plasma and includes water and dissolved substances that have passed through the capillary wall.
   b. It generally lacks large proteins, but some smaller proteins leak into interstitial spaces.
   c. As the protein concentration of tissue fluid increases, colloid osmotic pressure increases.
2. Lymph formation
   a. Increasing hydrostatic pressure within interstitial spaces forces some tissue fluid into lymphatic capillaries, and this fluid becomes lymph.
   b. Lymph formation prevents accumulation of excess tissue fluid (edema).
3. Lymph function
   a. Lymph returns the smaller protein molecules and fluid to the bloodstream.
   b. It transports foreign particles to the lymph nodes.

Lymph Movement (page 612)

1. Lymph flow
   a. Lymph is under low pressure and may not flow readily without external aid.
   b. Lymph is moved by the contraction of skeletal muscles and low pressure in the thorax created by breathing movements.
2. Obstruction of lymph movement
   a. Any condition that interferes with the flow of lymph results in edema.
   b. Obstruction of lymphatic vessels due to surgical removal of lymph nodes causes edema in the affected area.

Lymph Nodes (page 612)

1. Structure of a lymph node
   a. Lymph nodes are usually bean-shaped, with blood vessels, nerves, and efferent lymphatic vessels attached to the indented region; afferent lymphatic vessels enter at points on the convex surface.
   b. Lymph nodes are enclosed in connective tissue that extends into the nodes and subdivides them into nodules.
   c. Nodules contain masses of lymphocytes and macrophages, as well as spaces through which lymph flows.
2. Locations of lymph nodes
   a. Lymph nodes aggregate in groups or chains along the paths of larger lymphatic vessels.
   b. They primarily occur in cervical, axillary, and inguinal regions and within the pelvic, abdominal, and thoracic cavities.
3. Functions of lymph nodes
   a. Lymph nodes filter potentially harmful foreign particles from the lymph before it is returned to the bloodstream.

b. Lymph nodes are centers for the production of lymphocytes that act against foreign particles.

c. They contain macrophages that remove foreign particles from lymph.

Thymus and Spleen (page 614)

1. Thymus
   a. The thymus is a soft, bilobed organ located within the mediastinum.
   b. It slowly shrinks after puberty.
   c. It is composed of lymphatic tissue subdivided into lobules.
   d. Lobules contain lymphocytes, most of which are inactive, that develop from precursor cells in the red bone marrow.
   e. Some lymphocytes leave the thymus and provide immunity.
   f. The thymus secretes thymosin, which stimulates lymphocytes that have migrated to other lymphatic tissues.

2. Spleen
   a. The spleen is located in the upper left portion of the abdominal cavity.
   b. It resembles a large lymph node that is encapsulated and subdivided into lobules by connective tissue.
   c. Spaces within splenic lobules are filled with blood.
   d. The spleen contains many macrophages and lymphocytes, which filter foreign particles and damaged red blood cells from the blood.

Body Defenses Against Infection (page 616)

The presence and reproduction of pathogens cause infection. Pathogens include bacteria, complex single-celled organisms, fungi, and viruses. An infection may be present without immediately causing symptoms. The body has innate (nonspecific) and adaptive (specific) defenses against infection.

Innate (Nonspecific) Defenses (page 617)

1. Species resistance
   Each species is resistant to certain diseases that may affect other species but is susceptible to diseases other species may resist.

2. Mechanical barriers
   a. Mechanical barriers include skin and mucous membranes.
   b. Intact mechanical barriers prevent entrance of some pathogens.

3. Chemical barriers
   a. Enzymes in gastric juice and tears kill some pathogens.
   b. Low pH in the stomach prevents growth of some bacteria.
   c. High salt concentration in perspiration kills some bacteria.
   d. Interferons stimulate uninfected cells to synthesize antiviral proteins that block proliferation of viruses; stimulate phagocytosis; and enhance activity of cells that help resist infections and stifle tumor growth.
   e. Defensins make holes in bacterial cell walls and membranes.
   f. Collectins provide broad protection against a wide variety of microbes by grabbing onto them.

4. Fever
   a. Viral or bacterial infection stimulates certain lymphocytes to secrete IL-1, which temporarily raises body temperature.
   b. Physical factors, such as heat or ultraviolet light, or chemical factors, such as acids or bases, can cause fever.
   c. Elevated body temperature and the resulting decrease in blood iron level and increased phagocytic activity hamper infection.

5. Natural killer cells
   Natural killer cells defend the body against a variety of viruses and cancer by secreting cytolytic perforins to destroy infected cells.

6. Inflammation
   a. Inflammation is a tissue response to damage, injury, or infection.
   b. The response includes localized redness, swelling, heat, and pain.
   c. Chemicals released by damaged tissues attract white blood cells to the site.
   d. Clotting may occur in body fluids that accumulate in affected tissues.
   e. Connective tissue containing many fibers may form a sac around the injured tissue and thus aid in preventing the spread of pathogens.

7. Phagocytosis
   a. The most active phagocytes in blood are neutrophils and monocytes; monocytes give rise to macrophages, which remain fixed in tissues.
   b. Phagocytic cells associated with the linings of blood vessels in the red bone marrow, liver, spleen, and lymph nodes constitute the mononuclear phagocytic system.
   c. Phagocytes remove foreign particles from tissues and body fluids.

Adaptive (Specific) Defenses or Immunity (page 619)

1. Antigens
   a. Before birth, body cells inventory "self" proteins and other large molecules.
   b. After inventory, lymphocytes develop receptors that allow them to differentiate between nonself (foreign) and self antigens.
   c. Nonself antigens combine with T cell and B cell surface receptors and stimulate these cells to cause an immune reaction.
   d. Haptens are small molecules that can combine with larger ones, becoming antigenic.

2. Lymphocyte origins
   a. Lymphocytes originate in red bone marrow and are released into the blood before they differentiate.
   b. Some reach the thymus where they mature into T cells.
   c. Others, the B cells, mature in the red bone marrow.
   d. Both T cells and B cells reside in lymphatic tissues and organs.

3. Lymphocyte functions
   a. T cells respond to antigens by cell-to-cell contact (cellular immune response).
   b. T cells secrete cytokines, such as interleukins, that enhance cellular responses to antigens.
   c. T cells may also secrete substances that are toxic to their target cells.
   d. B cells interact with antigen-bearing agents indirectly, providing the humoral immune response.
   e. Varieties of T cells and B cells number in the millions.
   f. The members of each variety respond only to a specific antigen.
   g. As a group, the members of each variety form a clone.

4. T cells and the cellular immune response
   a. T cells are activated when an antigen-presenting cell displays a foreign antigen.
   b. When a macrophage acts as an accessory cell, it phagocytizes an antigen-bearing agent, digests the agent, and displays the antigens on its cell membrane in association with certain MHC proteins.
   c. A helper T cell becomes activated when it encounters displayed antigens for which it is specialized to react.
   d. Once activated, helper T cells stimulate B cells to produce antibodies.
   e. CD4 helper T cells stimulate humoral and cellular immunity. HIV cripples these cells.
   f. Cytotoxic T cells recognize foreign antigens on tumor cells and cells whose surfaces indicate that they are infected by viruses. Stimulated cytotoxic T cells secrete perforin to destroy its target cells.
   g. Memory T cells allow for immediate response to second and subsequent exposure to the same antigen.
5. B cells and humoral immunity
   a. B cell activation
      (1) A B cell is activated when it encounters an antigen that fits its antigen receptors.
      (2) An activated B cell proliferates (especially when stimulated by a T cell), enlarging its clone.
      (3) Some activated B cells specialize into antibody-producing plasma cells.
      (4) Antibodies react against the antigen-bearing agent that stimulated their production.
      (5) An individual's diverse B cells defend against a very large number of pathogens.
   b. Antibody molecules
      (1) Antibodies are soluble proteins called immunoglobulins.
      (2) They constitute the gamma globulin fraction of plasma.
      (3) Each immunoglobulin molecule consists of four chains of amino acids linked together.
      (4) Variable regions at the ends of these chains are specialized into antigen binding sites to react with antigens.
   c. Types of immunoglobulins
      (1) The five major types of immunoglobulins are IgG, IgA, IgM, IgD, and IgE.
      (2) IgG, IgA, and IgM make up most of the circulating antibodies.
   d. Antibody actions
      (1) Antibodies directly attach to antigens, activate complement, or stimulate local tissue changes that are unfavorable to antigen-bearing agents.
      (2) Direct attachment results in agglutination, precipitation, or neutralization.
      (3) Activated proteins of complement attract phagocytes, alter infected cells so they become more susceptible to phagocytosis, and lyse foreign cell membranes.
6. Immune responses
   a. B cells or T cells first encountering an antigen for which they are specialized to react constitutes a primary immune response.
      (1) During this response, antibodies are produced for several weeks.
      (2) Some B cells remain dormant as memory cells, aided by follicular dendritic cells.
   b. A secondary immune response occurs rapidly as a result of memory cell response if the same antigen is encountered later.

7. Practical classification of immunity
   a. A person who encounters a pathogen and has a primary immune response develops naturally acquired active immunity.
   b. A person who receives a vaccine containing a dead or weakened pathogen, or part of it, develops artificially acquired active immunity.
   c. A person who receives an injection of antibodies or antitoxin has artificially acquired passive immunity.
   d. When antibodies pass through a placental membrane from a pregnant woman to her fetus, the fetus develops naturally acquired passive immunity.
   e. Active immunity lasts much longer than passive immunity.
8. Allergic reactions
   a. Allergic or hypersensitivity reactions are excessive misdirected immune responses that may damage tissues.
   b. Immediate-reaction allergy is an inborn ability to overproduce IgE.
      (1) Allergic reactions result from mast cells bursting and releasing allergy mediators such as histamine and serotonin.
      (2) The released chemicals cause allergy symptoms such as hives, hay fever, asthma, eczema, or gastric disturbances.
      (3) In anaphylactic shock, allergy mediators flood the body, causing severe symptoms, including decreased blood pressure and difficulty breathing.
   c. Antibody-dependent cytotoxic allergic reactions occur when blood transfusions are mismatched.
   d. Immune complex allergic reactions involve autoimmunity, which is an immune reaction against self antigens.
   e. Delayed-reaction allergy, which can occur in anyone and inflame the skin, results from repeated exposure to allergens.
9. Transplantation and tissue rejection
   a. A transplant recipient's immune system may react against the donated tissue in a tissue rejection reaction.
   b. Matching cell surface molecules (MHC antigens) of donor and recipient tissues and using immunosuppressive drugs can minimize tissue rejection.
   c. Immunosuppressive drugs may increase susceptibility to infection.
   d. Transplants may take place between genetically identical twins, from one body part to another, between unrelated individuals of the same species, or between individuals of different species.
10. Autoimmunity
   a. In autoimmune disorders, the immune system manufactures autoantibodies that attack one's own body tissues.
   b. Autoimmune disorders may result from a previous viral infection, faulty T cell development, or reaction to a nonself antigen that resembles a self antigen.
   c. Retained fetal cells can cause a condition that resembles an autoimmune disorder.

Life-Span Changes (page 634)

1. The immune system begins to decline early in life, in part due to the decreasing size of the thymus.
2. Numbers of T cells and B cells do not change significantly, but activity levels do.
3 Proportions of the different antibody classes shift.

CRITICAL THINKING QUESTIONS

1. How can removal of enlarged lymph nodes for microscopic examination aid in diagnosing certain conditions?

2. Why is injecting a substance into the skin like injecting it into the lymphatic system?

3. Why does vaccination provide long-lasting protection against a disease, whereas gamma globulin (IgG) provides only short-term protection?

4. When a breast is surgically removed to treat breast cancer, the lymph nodes in the nearby axillary region are sometimes excised also. Why is this procedure likely to cause swelling of the upper limb on the treated side?

5. What functions of the lymphatic system would be affected by being born without a thymus?

6. More people need transplants than there are organs available. Discuss the pros and cons of the following proposed rationing systems for determining who should receive transplants: (a) first come, first served; (b) people with the best tissue and blood-type match; (c) patients whose need for an organ is caused by infection or disease, as opposed to those whose need for an organ was preventable, such as a lung destroyed by smoking; (d) the youngest people; (e) the wealthiest people; (f) the most important people.

7. Why is a transplant consisting of fetal tissue less likely to provoke an immune rejection response than tissue from an adult?

8. T cells "learn" to recognize self from nonself during prenatal development. How could this learning process be altered to prevent allergies? To enable a person to accept a transplant?

9. Some parents keep their preschoolers away from other children to prevent them from catching illnesses. How might these well-meaning parents actually be harming their children?

10. A xenograft is tissue from a nonhuman animal used to replace a body part in a human. For example, pigs are being bred to provide cardiovascular spare parts because their hearts and blood vessels are similar to ours. To increase the likelihood of such a xenotransplant working, researchers genetically engineer pigs to produce human antigens on their cell surfaces. How can this improve the chances of a human body not rejecting such a transplant?

REVIEW EXERCISES

1. Explain how the lymphatic system is related to the cardiovascular system.

2. Trace the general pathway of lymph from the interstitial spaces to the bloodstream.

3. Identify and describe the locations of the major lymphatic trunks and collecting ducts.

4. Distinguish between tissue fluid and lymph.

5. Describe the primary functions of lymph.

6. Explain why physical exercise promotes lymphatic circulation.

7. Explain how a lymphatic obstruction leads to edema.

8. Describe the structure of a lymph node, and list its major functions.

9. Locate the major body regions occupied by lymph nodes.

10. Describe the structure and functions of the thymus.

11. Describe the structure and functions of the spleen.

12. Distinguish between innate (nonspecific) and adaptive (specific) body defenses against infection.

13. Explain *species resistance.*

14. Name two mechanical barriers to infection.

15. Describe how enzymatic actions function as defense mechanisms against pathogens.

16. Distinguish among the chemical barriers (interferons, defensins, and collectins), and give examples of their different actions.

17. List possible causes of fever, and explain the benefits of fever.

18. Describe natural killer cells and their action.

19. List the major effects of inflammation, and explain why each occurs.

20. Identify the major phagocytic cells in the blood and other tissues.

21. Distinguish between an antigen and a hapten.

22. Review the origin of T cells and B cells.

23. Explain the cellular immune response.

24. Define *cytokine.*

25. List three types of T cells and describe the function of each in the immune response.

26. Define *clone of lymphocytes.*

27. Explain the humoral immune response.

28. Explain how a B cell is activated.

29. Explain the function of *plasma cells.*

30. Describe an immunoglobulin molecule.

31. Distinguish between the variable region and the constant region of an immunoglobulin molecule.

32. List the major types of immunoglobulins, and describe their main functions.

33. Describe three ways in which antibody direct attack on an antigen helps in the removal of that antigen.

34. Explain the functions of complement.

35. Distinguish between a primary and a secondary immune response.

36. Distinguish between active and passive immunity.

37. Define *vaccine.*

38. Explain how a vaccine produces its effect.

39. Describe how a fetus may obtain antibodies from the maternal blood.

40. Explain the relationship between an allergic reaction and an immune response.

41. Distinguish between an antigen and an allergen.

42. Describe how an immediate-reaction allergic response may occur.

43. List the major events leading to a delayed-reaction allergic response.

44. Explain the relationship between tissue rejection and an immune response.

45. Describe two methods used to reduce the severity of a tissue rejection reaction.

46. How do immunosuppressive drugs increase the likelihood of success of a transplant, yet place a patient at a higher risk of contracting infections?

47. Explain the relationship between autoimmunity and an immune response.

48. Describe the causes for a decline in the strength of the immune response in the elderly.

W E B   C O N N E C T I O N S

Visit the Student Edition of the Online Learning Center at www.mhhe.com/shier10 **for answers to chapter questions, additional quizzes, interactive learning exercises, and other study tools.**

UNDERSTANDING WORDS

aliment-, food: *aliment*ary canal—tubelike portion of the digestive system.

cari-, decay: dental *cari*es—tooth decay.

cec-, blindness: *cec*um—blind-ended sac at the beginning of the large intestine.

chym-, juice: *chym*e—semifluid paste of food particles and gastric juice formed in the stomach.

decidu-, falling off: *decidu*ous teeth—teeth that are shed during childhood.

frenul-, bridle, restraint: *frenul*um—membranous fold that anchors the tongue to the floor of the mouth.

gastr-, stomach: *gastr*ic gland—portion of the stomach that secretes gastric juice.

hepat-, liver: *hepat*ic duct—duct that carries bile from the liver to the common bile duct.

hiat-, opening: esophageal *hiat*us—opening through which the esophagus penetrates the diaphragm.

lingu-, tongue: *lingu*al tonsil—mass of lymphatic tissue at the root of the tongue.

peri-, around: *peri*stalsis—wavelike ring of contraction that moves material along the alimentary canal.

pyl-, gatekeeper, door: *pyl*oric sphincter—muscle that serves as a valve between the stomach and small intestine.

rect-, straight: *rect*um—distal portion of the large intestine.

sorpt-, to soak up: ab*sorpt*ion—uptake of substances.

vill-, hairy: *vill*i—tiny projections of mucous membrane in the small intestine.

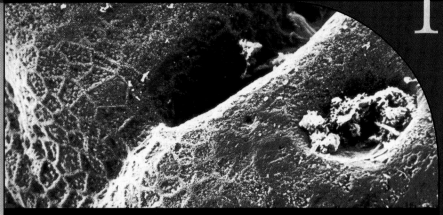

The gastric pit at the upper right of this falsely colored scanning electron micrograph of a section of mucosa of the stomach contains gastric glands that produce and secrete hydrochloric acid and the digestive enzyme pepsin (950×).

Digestive System

CHAPTER OBJECTIVES

After you have studied this chapter, you should be able to

1. Name and describe the locations and major organs of the digestive system.

2. Describe the general functions of each digestive organ.

3. Describe the structure of the wall of the alimentary canal.

4. Explain how the contents of the alimentary canal are mixed and moved.

5. List the enzymes the digestive organs and glands secrete and describe the function of each.

6. Describe how digestive secretions are regulated.

7. Explain how digestive reflexes control movement of material through the alimentary canal.

8. Describe the mechanisms of swallowing, vomiting, and defecating.

9. Explain how the products of digestion are absorbed.

10. Describe aging-related changes in the digestive system.

Fear of constipation is ancient, based on the idea that the large intestine contains poison. An Egyptian papyrus from the sixteenth century B.C. traces the origins of many diseases to various decomposing foods in the lower digestive tract. The discovery in the late eighteenth century that bacteria are normal residents of a human's intestines added to the concept of "intestinal autointoxication," a belief that bacteria mixed with leftovers from digestion could poison us from within. Many societies have attributed a variety of ills to constipation, which was thought to be a consequence of an increasingly urban lifestyle accompanied by less exercise and an unhealthy diet.

In the 1920s and 1930s, people widely feared that constipation would cause "sewer-like blood." Parents forced children to defecate daily, preferably in the morning so they wouldn't need to fret over a missed movement all day long. People discovered that eating bran helped them meet this daily requirement, and brands of bran flourished, one even called DinaMite. People tried all sorts of gadgets, ate various "cleansing" foods, from yeast to yogurt; and many people even had parts of their large intestines removed to lessen the likelihood that the foul contents would kill them.

In the second half of the twentieth century, after antibiotics had helped control many infectious diseases, attention turned to cancer. The dietary connection to constipation extended to cancer, and the idea that certain foods can either cause or prevent cancers of the large intestine or rectum (colorectal cancer) arose, based largely on studies of populations. People whose diets were low in meat and fat and high in fruits and vegetables tended to have a lower incidence of colorectal cancer than populations whose diets were fatty. In the 1970s, the "fiber hypothesis" gained favor, echoing the earlier popularity of bran cereal. However, two studies published in 2000 showed that, in more than 3,500 individuals, low-fat, high-fiber diets had no effect on the recurrence of intestinal polyps, which are growths that often precede development of cancer.

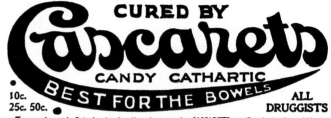

In the first half of the twentieth century, an industry revolved around the perceived necessity of a daily bowel movement.

These results were confusing, because epidemiological studies continue to show associations between high-fiber diets and lower incidence of colorectal cancer. Further studies are needed. It is possible that some other aspect of these cultures—such as exercise habits or the way meat is prepared—prevents colorectal cancer. Meanwhile, bran, fruits, and vegetables remain healthful foods that can prevent constipation. ■

Digestion (di-jest′yun) is the mechanical and chemical breakdown of foods into forms that cell membranes can absorb. Mechanical digestion breaks large pieces into smaller ones without altering their chemical composition. Chemical digestion breaks food into simpler chemicals. The organs of the **digestive system** carry out these processes, as well as ingestion, propulsion, absorption, and defecation.

The digestive system consists of the **alimentary canal** (al″i-men′tar-e kah-nal′), extending from the mouth to the anus, and several accessory organs, which release secretions into the canal. The alimentary canal includes the mouth, pharynx, esophagus, stomach, small intestine, large intestine, and anal canal. The accessory organs include the salivary glands, liver, gallbladder, and pan-

creas. Figure 17.1 and reference plates 4, 5, and 6 show the major organs of the digestive system.

The digestive system originates from the inner layer (endoderm) of the embryo, which folds to form the tube of the alimentary canal. The accessory organs develop as buds from the tube.

General Characteristics of the Alimentary Canal

The alimentary canal is a muscular tube about 8 meters long that passes through the body's ventral cavity. The structure of its wall, how it moves food, and its innervation are similar throughout its length (fig. 17.2).

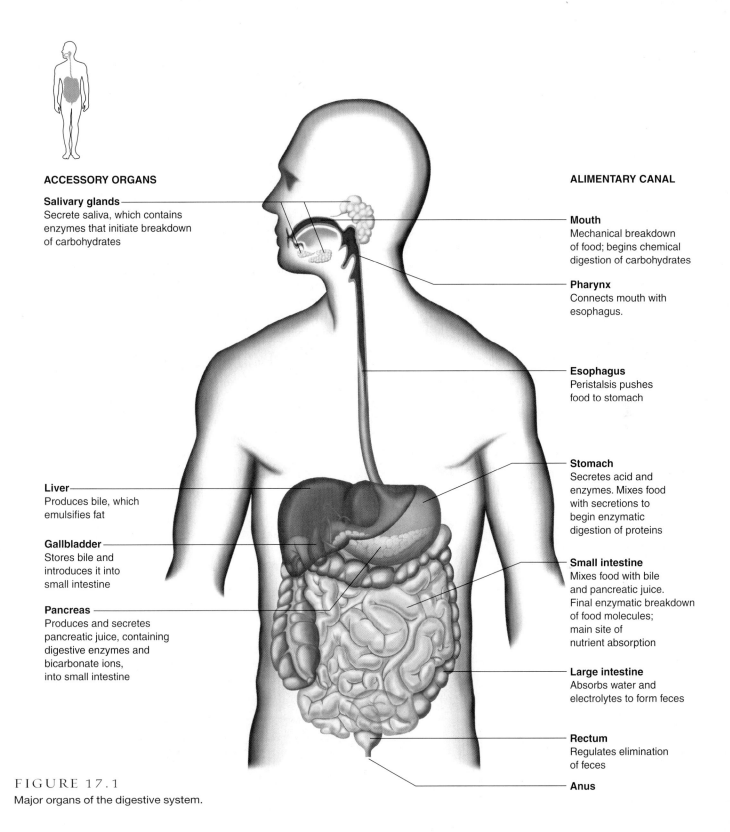

ACCESSORY ORGANS

Salivary glands
Secrete saliva, which contains
enzymes that initiate breakdown
of carbohydrates

Liver
Produces bile, which
emulsifies fat

Gallbladder
Stores bile and
introduces it into
small intestine

Pancreas
Produces and secretes
pancreatic juice, containing
digestive enzymes and
bicarbonate ions,
into small intestine

ALIMENTARY CANAL

Mouth
Mechanical breakdown
of food; begins chemical
digestion of carbohydrates

Pharynx
Connects mouth with
esophagus.

Esophagus
Peristalsis pushes
food to stomach

Stomach
Secretes acid and
enzymes. Mixes food
with secretions to
begin enzymatic
digestion of proteins

Small intestine
Mixes food with bile
and pancreatic juice.
Final enzymatic breakdown
of food molecules;
main site of
nutrient absorption

Large intestine
Absorbs water and
electrolytes to form feces

Rectum
Regulates elimination
of feces

Anus

FIGURE 17.1
Major organs of the digestive system.

Structure of the Wall

The wall of the alimentary canal consists of four distinct layers that are developed to different degrees from region to region. Although the four-layered structure persists throughout the alimentary canal, certain regions are specialized for particular functions. Beginning with the innermost tissues, these layers, shown in figure 17.3, include the following:

1. **Mucosa** (mu-ko'sah), or **mucous membrane.** This layer is formed of surface epithelium, underlying connective tissue (lamina propria), and a small

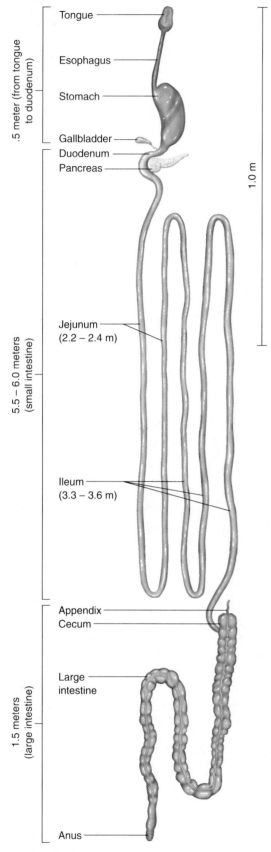

.5 meter (from tongue to duodenum)

Tongue
Esophagus
Stomach
Gallbladder
Duodenum
Pancreas

1.0 m

Jejunum
(2.2 – 2.4 m)

5.5 – 6.0 meters (small intestine)

Ileum
(3.3 – 3.6 m)

Appendix
Cecum

Large intestine

1.5 meters (large intestine)

Anus

FIGURE 17.2

The alimentary canal is a muscular tube about 8 meters long.

amount of smooth muscle (muscularis mucosae). In some regions, it develops folds and tiny projections, which extend into the lumen of the digestive tube and increase its absorptive surface area. It may also contain glands that are tubular invaginations into which the lining cells secrete mucus and digestive enzymes. The mucosa protects the tissues beneath it and carries on secretion and absorption.

2. **Submucosa** (sub″mu-ko′sah). The submucosa contains considerable loose connective tissue as well as glands, blood vessels, lymphatic vessels, and nerves. Its vessels nourish the surrounding tissues and carry away absorbed materials.

3. **Muscular layer.** This layer, which provides movements of the tube, consists of two coats of smooth muscle tissue. The fibers of the inner coat encircle the tube. When these *circular fibers* (they are actually closed spirals) contract, the diameter of the tube decreases. The fibers of the outer muscular coat run lengthwise. When these *longitudinal fibers* (open spirals) contract, the tube shortens.

4. **Serosa** (se-ro′sah), or **serous layer.** The serous layer, or outer covering of the tube, is composed of the *visceral peritoneum,* which is formed of epithelium on the outside and connective tissue beneath. The cells of the serosa protect underlying tissues and secrete serous fluid, which moistens and lubricates the tube's outer surface so that the organs (lined with the parietal peritoneum) slide freely within the abdominal cavity and against one another.

Table 17.1 summarizes the characteristics of these layers.

Movements of the Tube

The motor functions of the alimentary canal are of two basic types—*mixing movements* and *propelling movements* (fig. 17.4). Mixing occurs when smooth muscles in small segments of the tube contract rhythmically. For example, when the stomach is full, waves of muscular

TABLE 17.1	Layers of the Wall of the Alimentary Canal	
Layer	**Composition**	**Function**
Mucosa	Epithelium, connective tissue, smooth muscle	Protection, secretion, absorption
Submucosa	Loose connective tissue, blood vessels, lymphatic vessels, nerves	Nourishes surrounding tissues, transports absorbed materials
Muscular layer	Smooth muscle fibers arranged in circular and longitudinal groups	Movements of the tube and its contents
Serosa	Epithelium, connective tissue	Protection, lubrication

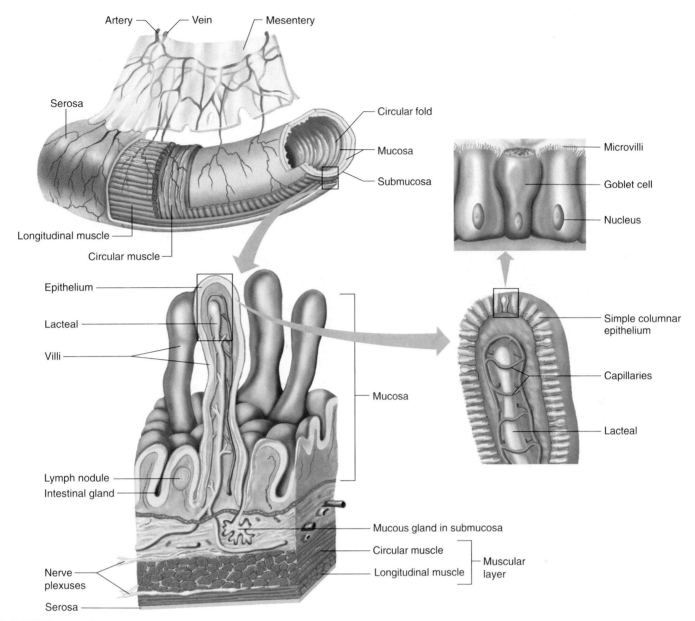

FIGURE 17.3

The wall of the small intestine, as in other portions of the alimentary canal, includes four layers: an inner mucosa, a submucosa, a muscular layer, and an outer serosa.

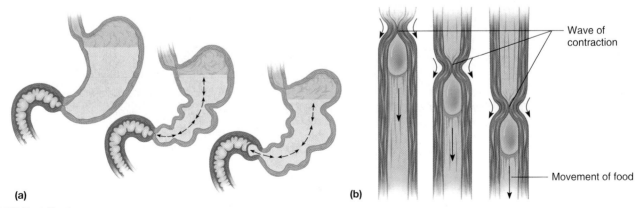

(a)

(b)

FIGURE 17.4

Movements through the alimentary canal. (a) Mixing movements occur when small segments of the muscular wall of the alimentary canal rhythmically contract. (b) Peristaltic waves move the contents along the canal.

contractions move along its wall from one end to the other. These waves occur every twenty seconds or so, and they mix foods with the digestive juices that the mucosa secretes.

Propelling movements include a wavelike motion called **peristalsis** (per″i-stal′sis). When peristalsis occurs, a ring of contraction appears in the wall of the tube. At the same time, the muscular wall just ahead of the ring relaxes—a phenomenon called *receptive relaxation.* As the wave moves along, it pushes the tubular contents ahead of it. Peristalsis begins when food expands the tube. It causes the sounds that can be heard through a stethoscope applied to the abdominal wall.

A device the size of a medicine capsule can image the alimentary canal, revealing blockages and sites of bleeding. The patient swallows the capsule, which contains a camera, a light source, radio transmitter, and batteries. About six hours later, it transmits images from digestive headquarters—the small intestine—to a device worn on the physician's belt. The information goes to a computer, and still or video images are downloaded. The device, which is disposable, leaves the body in the feces within a day or two. The "GI camera" is based on a device called a Heidelberg capsule used to monitor stomach acid. Soon to come is a capsule with longer-lasting batteries and better light to image the large intestine.

Innervation of the Tube

Branches of the sympathetic and parasympathetic divisions of the autonomic nervous system extensively innervate the alimentary canal. These nerve fibers, mainly associated with the tube's muscular layer, maintain muscle tone and regulate the strength, rate, and velocity of muscular contractions. Many of the postganglionic fibers are organized into a network or plexus of neurons within the wall of the canal (see fig. 17.3). The *submucosal plexus* is important in controlling secretions by the gastrointestinal tract. The *myenteric plexus* of the muscular layer controls gastrointestinal motility. The plexuses also contain sensory neurons.

Parasympathetic impulses generally increase the activities of the digestive system. Some of these impulses originate in the brain and are conducted through branches of the vagus nerves to the esophagus, stomach, pancreas, gallbladder, small intestine, and proximal half of the large intestine. Other parasympathetic impulses arise in the sacral region of the spinal cord and supply the distal half of the large intestine.

Sympathetic nerve impulses' effects on digestive actions usually are opposite those of the parasympathetic division. That is, sympathetic impulses inhibit certain digestive actions. For example, such impulses inhibit

mixing and propelling movements, but contract sphincter muscles in the wall of the alimentary canal, blocking movement of materials through the tube.

So extensive are the nerve plexuses of the gastrointestinal tract that it is sometimes said to have a "second brain." The small intestine, for example, has at least 100 million neurons, many glial cells, and abundant and diverse neurotransmitters, neuropeptides, and growth factors.

1 What are the general functions of the digestive system?

2 Which organs constitute the digestive system?

3 Describe the wall of the alimentary canal.

4 Name the two types of movements that occur in the alimentary canal.

5 How do parasympathetic nerve impulses affect digestive actions? What effect do sympathetic nerve impulses have?

Mouth

The **mouth,** which is the first portion of the alimentary canal, receives food and begins digestion by mechanically breaking up the solid particles into smaller pieces and mixing them with saliva. This action is called mastication. The mouth also functions as an organ of speech and sensory reception. It is surrounded by the lips, cheeks, tongue, and palate and includes a chamber between the palate and tongue called the *oral cavity,* as well as a narrow space between the teeth, cheeks, and lips called the *vestibule* (fig. 17.5 and reference plate 49).

Cheeks and Lips

The **cheeks** form the lateral walls of the mouth. They consist of outer layers of skin, pads of subcutaneous fat, muscles associated with expression and chewing, and inner linings of moist, stratified squamous epithelium.

Because cheek cells are easily removed, they are a practical source of DNA for genetic tests. "Cheekbrush tests" identify carriers of certain inherited disorders. The patient swishes a brush on the inside of the cheek; then the doctor sends the brush to a laboratory. Here, cheek cells are removed from the brush, and the DNA is extracted and analyzed for the presence of gene variants, such as those that cause cystic fibrosis. Cheekbrush tests are also used in forensics to obtain DNA to be used in DNA profiling to identify individuals.

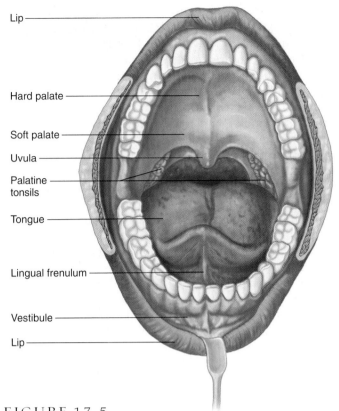

Lip

Hard palate

Soft palate

Uvula

Palatine tonsils

Tongue

Lingual frenulum

Vestibule

Lip

FIGURE 17.5

The mouth is adapted for ingesting food and preparing it for digestion.

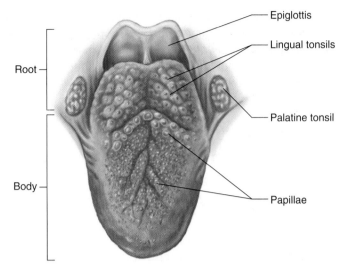

Epiglottis

Lingual tonsils

Root

Palatine tonsil

Body

Papillae

FIGURE 17.6

The surface of the tongue, superior view.

The **lips** are highly mobile structures that surround the mouth opening. They contain skeletal muscles and sensory receptors useful in judging the temperature and texture of foods. Their normal reddish color is due to the many blood vessels near their surfaces. The external borders of the lips mark the boundaries between the skin of the face and the mucous membrane that lines the alimentary canal.

Tongue

The **tongue** (tung) is a thick, muscular organ that occupies the floor of the mouth and nearly fills the oral cavity when the mouth is closed. Mucous membrane covers the tongue, which is connected in the midline to the floor of the mouth by a membranous fold called the **lingual frenulum** (ling' gwahl fren'u-lum).

The *body* of the tongue is largely composed of skeletal muscle fibers that run in several directions. These muscles mix food particles with saliva during chewing and move food toward the pharynx during swallowing. The surface of the tongue has rough projections, called **papillae** (pah-pil'a). Some of these provide friction, which helps handle food. Other papillae contain most of the taste buds (fig. 17.6). Some taste buds are scattered elsewhere in the mouth, particularly in children.

The posterior region, or *root,* of the tongue is anchored to the hyoid bone. It is covered with rounded masses of lymphatic tissue called **lingual tonsils** (ton'silz).

Palate

The **palate** (pal'at) forms the roof of the oral cavity and consists of a hard anterior part and a soft posterior part. The *hard palate* is formed by the palatine processes of the maxillary bones in front and the horizontal portions of the palatine bones in back. The *soft palate* forms a muscular arch, which extends posteriorly and downward as a cone-shaped projection called the **uvula** (u'vu-lah).

During swallowing, muscles draw the soft palate and the uvula upward. This action closes the opening between the nasal cavity and the pharynx, preventing food from entering the nasal cavity.

In the back of the mouth, on either side of the tongue and closely associated with the palate, are masses of lymphatic tissue called **palatine** (pal'ah-tīn) **tonsils.** These structures lie beneath the epithelial lining of the mouth and, like other lymphatic tissues, help protect the body against infections (see chapter 16, p. 613).

The palatine tonsils are common sites of infection and when inflamed, produce *tonsillitis.* Infected tonsils may swell so greatly that they block the passageways of the pharynx and interfere with breathing and swallowing. Because the mucous membranes of the pharynx, auditory tubes, and middle ears are continuous, such an infection can spread from the throat into the middle ears (otitis media).

When tonsillitis occurs repeatedly and does not respond to antibiotic treatment, the tonsils are sometimes surgically removed. Such tonsillectomies are done less often today than they were a generation ago because the tonsils' role in immunity is now recognized.

Other masses of lymphatic tissue, called **pharyngeal** (fah-rin′je-al) **tonsils,** or *adenoids,* are on the posterior wall of the pharynx, above the border of the soft palate. If the adenoids enlarge and block the passage between the pharynx and the nasal cavity, they also may be surgically removed (fig. 17.7).

1 What are the functions of the mouth?

2 How does the tongue function as part of the digestive system?

3 What is the role of the soft palate in swallowing?

4 Where are the tonsils located?

Teeth

The **teeth** are the hardest structures in the body. They are not considered part of the skeletal system because they contain at least two types of proteins that are not also found in bone, and their structure is different.

Teeth develop in sockets within the alveolar processes of the mandibular and maxillary bones. Teeth are unique structures in that two sets form during development (fig. 17.8). The members of the first set, the *primary teeth* (deciduous teeth), usually erupt through the gums (gingiva) at regular intervals between the ages of six months and two to four years. The ten primary teeth in each jaw are located from the midline toward the sides in the following sequence: central incisor, lateral incisor, cuspid (canine), first molar, and second molar.

The primary teeth are usually shed in the same order they appeared, after their roots are resorbed. Then, the *secondary (permanent) teeth* push the primary teeth out of their sockets. This secondary set consists of thirty-two teeth—sixteen in each jaw—and they are arranged from the midline as follows: central incisor, lateral incisor, cuspid (canine), first bicuspid (premolar), second bicuspid (premolar), first molar, second molar, and third molar (fig. 17.9). Table 17.2 summarizes the types and numbers of primary and secondary teeth.

The permanent teeth usually begin to erupt at six years, but the set may not be completed until the third molars appear between seventeen and twenty-five years.

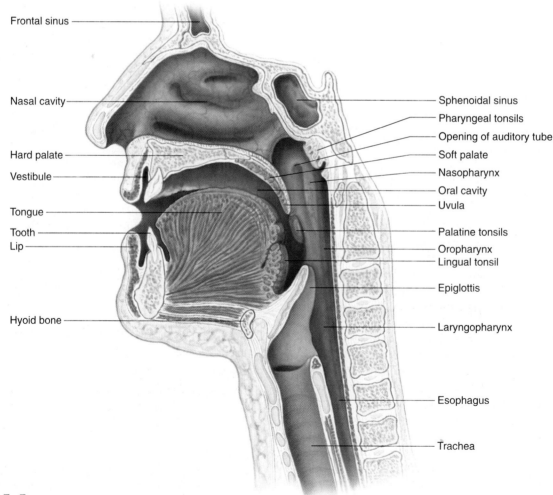

Frontal sinus

Nasal cavity

Hard palate

Vestibule

Tongue

Tooth

Lip

Hyoid bone

Sphenoidal sinus

Pharyngeal tonsils

Opening of auditory tube

Soft palate

Nasopharynx

Oral cavity

Uvula

Palatine tonsils

Oropharynx

Lingual tonsil

Epiglottis

Laryngopharynx

Esophagus

Trachea

FIGURE 17.7

Sagittal section of the mouth, nasal cavity, and pharynx.

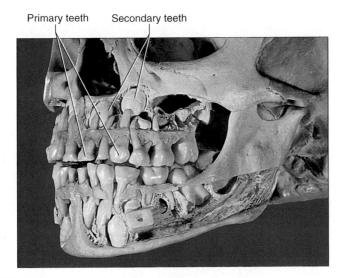

Primary teeth   Secondary teeth

FIGURE 17.8

This partially dissected child's skull reveals primary and developing secondary teeth in the maxilla and mandible.

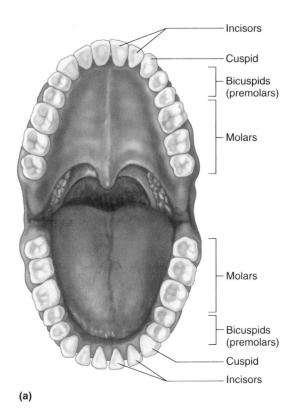

Incisors
Cuspid
Bicuspids (premolars)
Molars
Molars
Bicuspids (premolars)
Cuspid
Incisors

(a)

TABLE 17.2		Primary and Secondary Teeth		
Primary Teeth (Deciduous)		**Secondary Teeth (Permanent)**		
Type	*Number*	*Type*		*Number*
Incisor		Incisor		
Central	4	Central		4
Lateral	4	Lateral		4
Cuspid	4	Cuspid		4
		Bicuspid		
		First		4
		Second		4
Molar		Molar		
First	4	First		4
Second	4	Second		4
		Third		4
Total	20	**Total**		32

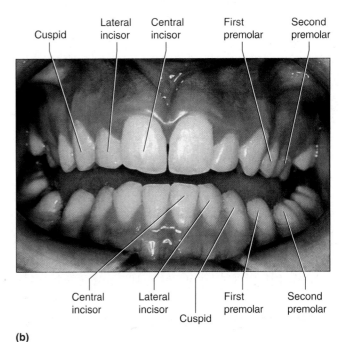

Cuspid   Lateral incisor   Central incisor   First premolar   Second premolar

Central incisor   Lateral incisor   Cuspid   First premolar   Second premolar

(b)

FIGURE 17.9

Permanent teeth. (a) The secondary teeth of the upper and lower jaws. (b) Anterior view of the secondary teeth.

Sometimes these third molars, which are also called wisdom teeth, become wedged in abnormal positions within the jaws and fail to erupt. Such teeth are said to be *impacted* and must be removed to alleviate pain.

The teeth break food into smaller pieces, which begins mechanical digestion. Chewing increases the surface area of the food particles, enabling digestive enzymes to interact more effectively with nutrient molecules.

Different teeth are adapted to handle food in different ways. The *incisors* are chisel-shaped, and their sharp edges bite off large pieces of food. The *cuspids* are cone-shaped, and they grasp and tear food. The *bicuspids* and *molars* have somewhat flattened surfaces and are specialized for grinding food particles.

Each tooth consists of two main portions—the *crown,* which projects beyond the gum, and the *root,* which is anchored to the alveolar process of the jaw. The region where these portions meet is called the *neck* of the tooth. Glossy, white *enamel* covers the crown. Enamel mainly consists of calcium salts and is the hardest substance in the body. Unfortunately, if abrasive action or injury damages enamel, it is not replaced. Enamel also tends to wear away with age.

The bulk of a tooth beneath the enamel is composed of a living cellular tissue called *dentin,* a substance much like bone, but somewhat harder. Dentin, in turn, surrounds the tooth's central cavity (pulp cavity), which contains a combination of blood vessels, nerves, and connective tissue called pulp. Blood vessels and nerves reach this cavity through tubular *root canals,* which extend upward into the root. Tooth loss is most often associated with diseases of the gums (gingivitis) and the dental pulp (endodontitis).

The root is enclosed by a thin layer of bonelike material called *cementum,* which is surrounded by a *periodontal ligament* (periodontal membrane). This ligament contains bundles of thick collagenous fibers, which pass between the cementum and the bone of the alveolar process, firmly attaching the tooth to the jaw. The ligament also contains blood vessels and nerves near the surface of the cementum-covered root (fig. 17.10).

The mouth parts and their functions are summarized in table 17.3. Clinical Application 17.1 describes the effect of bacteria on teeth.

1. How do primary teeth differ from secondary teeth?

2. How are types of teeth adapted to provide specialized functions?

3. Describe the structure of a tooth.

4. Explain how a tooth is attached to the bone of the jaw.

Salivary Glands

The **salivary** (sal′ĭ-ver-e) **glands** secrete saliva. This fluid moistens food particles, helps bind them, and begins the chemical digestion of carbohydrates. Saliva is also a solvent, dissolving foods so that they can be tasted, and it helps cleanse the mouth and teeth. Bicarbonate ions (HCO_3^-) in saliva help buffer the acid concentration so that the pH of saliva usually remains near neutral,

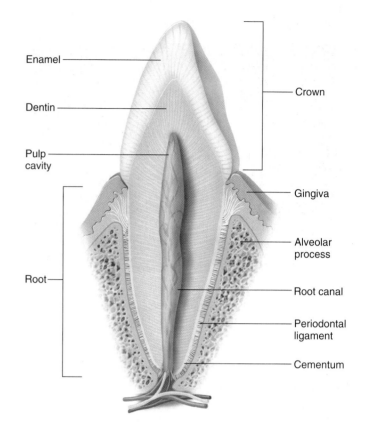

FIGURE 17.10
A section of a cuspid tooth.

between 6.5 and 7.5. This is a favorable range for the action of the salivary enzyme and protects the teeth from exposure to acids in foods.

Many minor salivary glands are scattered throughout the mucosa of the tongue, palate, and cheeks. They continuously secrete fluid so that the lining of the mouth remains moist. There are also three pairs of major salivary glands: the parotid glands, the submandibular glands, and the sublingual glands.

Salivary Secretions

Within a salivary gland are two types of secretory cells, *serous cells* and *mucous cells,* that occur in varying pro-

TABLE 17.3	Mouth Parts and Their Functions in Digestion				
Part	**Location**	**Function**	**Part**	**Location**	**Function**
Cheeks	Form lateral walls of mouth	Hold food in mouth; muscles function in chewing	Tongue	Occupies floor of mouth	Mixes food with saliva; moves food toward pharynx; contains taste receptors
Lips	Surround mouth opening	Contain sensory receptors used to judge characteristics of foods	Palate	Forms roof of mouth	Holds food in mouth; directs food to pharynx
			Teeth	In sockets of mandibular and maxillary bones	Break food particles into smaller pieces; help mix food with saliva during chewing

DENTAL CARIES

Sticky foods, such as caramel, lodge between the teeth and in the crevices of molars, feeding bacteria such as *Actinomyces*, *Streptococcus mutans*, and *Lactobacillus*. These microbes metabolize carbohydrates in the food, producing acid by-products that destroy tooth enamel and dentin (fig. 17A). The bacteria also produce sticky substances that hold them in place.

If a person eats candy bars but does not brush the teeth soon afterward, the actions of the acid-forming bacteria will produce decay, called *dental caries.* Unless a dentist cleans and fills the resulting cavity that forms where enamel is destroyed, the damage will spread to the underlying dentin. The tooth becomes very sensitive.

Dental caries can be prevented in several ways:

1. Brush and floss teeth regularly.
2. Have regular dental exams and cleanings.
3. Drink fluoridated water or receive a fluoride treatment. Fluoride is actually incorporated into the enamel's chemical structure, strengthening it.
4. Apply a sealant to children and adolescents' teeth that have crevices that might hold onto decay-causing bacteria. The sealant is a coating that keeps acids from eating away at tooth enamel.

One dental researcher took an unconventional approach to preventing dental caries that, understandably, was never commercialized. He invented a mouthwash consisting of mutant bacteria that would replace *Streptococcus mutans* but would not decay enamel. Consumer acceptance of a mutant bacterial brew was an obstacle! ■

FIGURE 17A

Actinomyces bacteria (falsely colored) clinging to teeth release acids that decay tooth enamel (1,250×).

portions within different glands. Serous cells produce a watery fluid that contains a digestive enzyme called **amylase** (am′ĭ-lās). This enzyme splits starch and glycogen molecules into disaccharides—the first step in the chemical digestion of carbohydrates. Mucous cells secrete a thick liquid called **mucus,** which binds food particles and acts as a lubricant during swallowing.

Like other digestive structures, the salivary glands are innervated by branches of both sympathetic and parasympathetic nerves. Impulses arriving on sympathetic fibers stimulate the gland cells to secrete a small quantity of viscous saliva. Parasympathetic impulses, on the other hand, elicit the secretion of a large volume of watery saliva. Such parasympathetic impulses are activated reflexly when a person sees, smells, tastes, or even thinks about pleasant foods. Conversely, if food looks, smells, or tastes unpleasant, parasympathetic activity is inhibited, so less saliva is produced, and swallowing may become difficult.

Major Salivary Glands

The **parotid** (pah-rot′id) **glands** are the largest of the major salivary glands. Each gland lies anterior to and somewhat inferior to each ear, between the skin of the cheek and the masseter muscle. A *parotid duct* (Stensen's duct) passes from the gland inward through the buccinator muscle, entering the mouth just opposite the upper second molar on either side of the jaw. The parotid glands secrete a clear, watery fluid that is rich in amylase (figs. 17.11 and 17.12*a*).

The **submandibular** (sub"man-dib'u-lar) **glands** are located in the floor of the mouth on the inside surface of the lower jaw. The secretory cells of these glands are predominantly serous, with a few mucous cells. The submandibular glands secrete a more viscous fluid than the parotid glands (figs. 17.11 and 17.12*b*). The ducts of the submandibular glands (Wharton's ducts) open inferior to the tongue, near the lingual frenulum.

The **sublingual** (sub-ling'gwal) **glands** are the smallest of the major salivary glands. They are found on the floor of the mouth inferior to the tongue. Because their cells are primarily the mucous type, their secretions, which enter the mouth through many separate ducts (Rivinus's ducts), tend to be thick and stringy (figs. 17.11 and 17.12*c*). Table 17.4 summarizes the characteristics of the major salivary glands.

1 What is the function of saliva?

2 What stimulates the salivary glands to secrete saliva?

3 Where are the major salivary glands located?

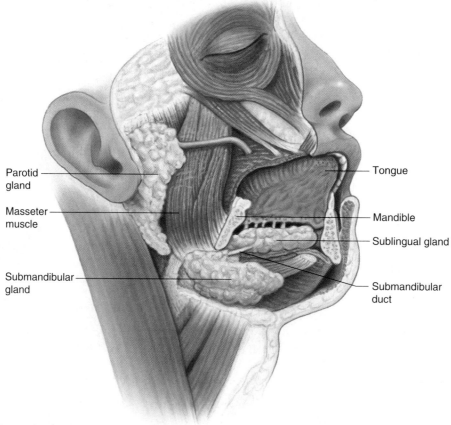

FIGURE 17.11
Locations of the major salivary glands.

TABLE 17.4	The Major Salivary Glands		
Gland	**Location**	**Duct**	**Type of Secretion**
Parotid glands	Anterior to and somewhat inferior to the ears between the skin of the cheeks and the masseter muscles	Parotid ducts pass through the buccinator muscles and enter the mouth opposite the upper second molars	Clear, watery serous fluid, rich in amylase
Submandibular glands	In the floor of the mouth on the inside surface of the mandible	Ducts open inferior to the tongue near the frenulum	Primarily serous fluid but with some mucus; more viscous than parotid secretion
Sublingual glands	In the floor of the mouth inferior to the tongue	Many separate ducts	Primarily thick, stringy mucus

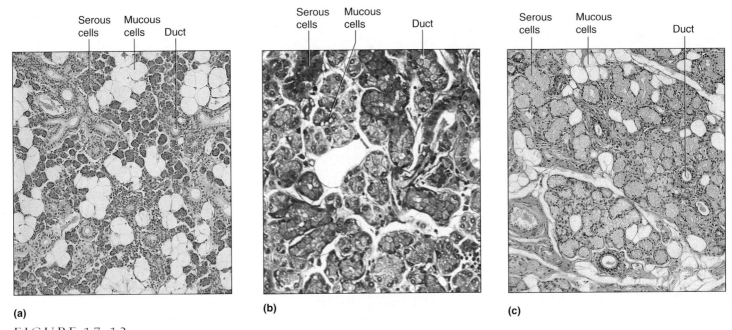

(a)

(b)

(c)

FIGURE 17.12
Light micrographs of (a) the parotid salivary gland (75×), (b) the submandibular salivary gland (180×), and (c) the sublingual salivary gland (80×).

Pharynx and Esophagus

The pharynx is a cavity posterior to the mouth from which the tubular esophagus leads to the stomach. The pharynx and the esophagus do not digest food, but both are important passageways, and their muscular walls function in swallowing.

Structure of the Pharynx

The **pharynx** (far'inks) connects the nasal and oral cavities with the larynx and esophagus (see fig. 17.7). It can be divided into the following parts:

1. The **nasopharynx** (na″zo-far'inks) is located superior to the soft palate. It communicates with the nasal cavity and provides a passageway for air during breathing. The auditory tubes, which connect the pharynx with the middle ears, open through the walls of the nasopharynx (see chapter 12, p. 437).

2. The **oropharynx** (o″ro-far'inks) is posterior to the mouth. It opens posterior to the soft palate into the nasopharynx and projects downward to the upper border of the epiglottis. This portion is a passageway for food moving downward from the mouth and for air moving to and from the nasal cavity.

3. The **laryngopharynx** (lah-ring″go-far'inks) is located just inferior to the oropharynx. It extends from the upper border of the epiglottis downward to the lower border of the cricoid cartilage of the larynx and is a passageway to the esophagus.

The muscles in the walls of the pharynx form inner circular and outer longitudinal groups (fig. 17.13). The circular muscles, called *constrictor muscles,* pull the walls inward during swallowing. The *superior constrictor muscles,* which are attached to bony processes of the skull and mandible, curve around the upper part of the pharynx. The *middle constrictor muscles* arise from projections on the hyoid bone and fan around the middle of the pharynx. The *inferior constrictor muscles* originate from cartilage of the larynx and pass around the lower portion of the pharyngeal cavity. Some of the lower inferior constrictor muscle fibers contract most of the time, which prevents air from entering the esophagus during breathing.

Although the pharyngeal muscles are skeletal muscles, they are under voluntary control only in the sense that swallowing (deglutition) can be voluntarily initiated. Complex reflexes control the precise actions of these muscles during swallowing.

Swallowing Mechanism

Swallowing reflexes can be divided into three stages. In the first stage, which is voluntary, food is chewed and mixed with saliva. Then, the tongue rolls this mixture into a mass (bolus) and forces it into the pharynx. The second stage of swallowing begins as food reaches the pharynx and stimulates sensory receptors around the pharyngeal opening. This triggers the swallowing

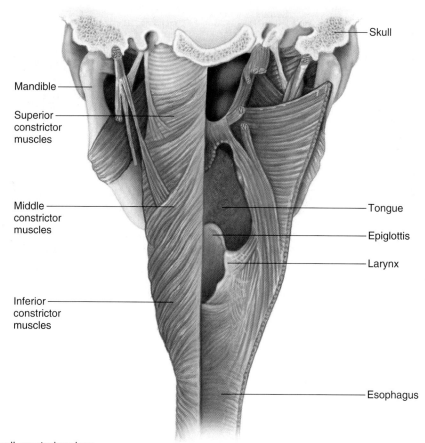

Mandible

Superior
constrictor
muscles

Middle
constrictor
muscles

Inferior
constrictor
muscles

Skull

Tongue

Epiglottis

Larynx

Esophagus

FIGURE 17.13
Muscles of the pharyngeal wall, posterior view.

reflexes, illustrated in figure 17.14, which includes the following actions:

1. The soft palate (including the uvula) raises, preventing food from entering the nasal cavity.

2. The hyoid bone and the larynx are elevated; the epiglottis closes off the top of the trachea so that food is less likely to enter.

3. The tongue is pressed against the soft palate and uvula, sealing off the oral cavity from the pharynx.

4. The longitudinal muscles in the pharyngeal wall contract, pulling the pharynx upward toward the food.

5. The lower portion of the inferior constrictor muscles relaxes, opening the esophagus.

6. The superior constrictor muscles contract, stimulating a peristaltic wave to begin in other pharyngeal muscles. This wave forces the food into the esophagus.

As the swallowing reflexes occur, breathing is momentarily inhibited. Then, during the third stage of swallowing, peristalsis transports the food in the esophagus to the stomach.

Esophagus

The **esophagus** (ĕ-sof'ah-gus) is a straight, collapsible tube about 25 centimeters long. It provides a passageway for food, and its muscular wall propels food from the pharynx to the stomach. The esophagus descends through the thorax posterior to the trachea, passing through the mediastinum. It penetrates the diaphragm through an opening, the *esophageal hiatus* (ĕ-sof"ah-je'al hi-a'tus), and is continuous with the stomach on the abdominal side of the diaphragm (figs. 17.15, 17.16 and reference plates 57, 73).

Mucous glands are scattered throughout the submucosa of the esophagus. Their secretions moisten and lubricate the inner lining of the tube.

In a *hiatal hernia,* a portion of the stomach protrudes through a weakened area of the diaphragm, through the esophageal hiatus and into the thorax. As a result of a hiatal hernia, regurgitation (reflux) of gastric juice into the esophagus may inflame the esophageal mucosa, causing heartburn, difficulty in swallowing, or ulceration and blood loss. In response to the destructive action of gastric juice, columnar epithelium may replace the squamous epithelium that normally lines the esophagus (see chapter 5, page 135). This condition, called *Barrett's esophagus,* increases the risk of developing esophageal cancer.

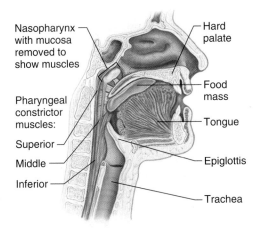

(a) The tongue forces food into the pharynx.

Nasopharynx with mucosa removed to show muscles
Hard palate
Food mass
Pharyngeal constrictor muscles:
Superior
Middle
Inferior
Tongue
Epiglottis
Trachea

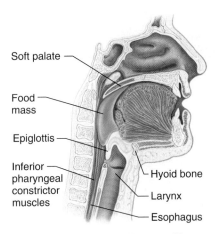

(b) The soft palate, hyoid bone, and larynx are raised, the tongue is pressed against the palate, the epiglottis closes, and the inferior constrictor muscles relax so that the esophagus opens.

Soft palate
Food mass
Epiglottis
Inferior pharyngeal constrictor muscles
Hyoid bone
Larynx
Esophagus

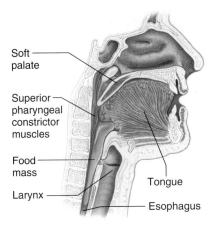

(c) Superior constrictor muscles contract and force food into the esophagus.

Soft palate
Superior pharyngeal constrictor muscles
Food mass
Larynx
Tongue
Esophagus

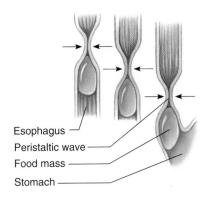

(d) Peristaltic waves move food through the esophagus to the stomach.

Esophagus
Peristaltic wave
Food mass
Stomach

FIGURE 17.14

Steps in the swallowing reflex.

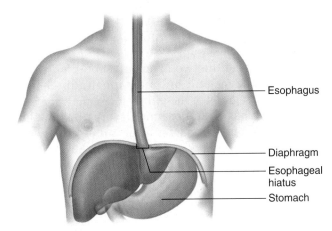

FIGURE 17.15

The esophagus functions as a passageway between the pharynx and the stomach.

Esophagus
Diaphragm
Esophageal hiatus
Stomach

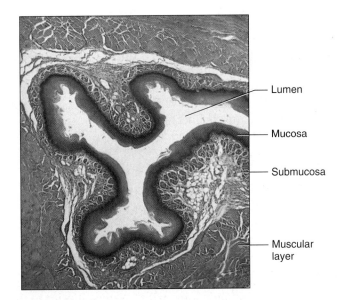

FIGURE 17.16

This cross section of the esophagus shows its muscular wall (10×).

Lumen
Mucosa
Submucosa
Muscular layer

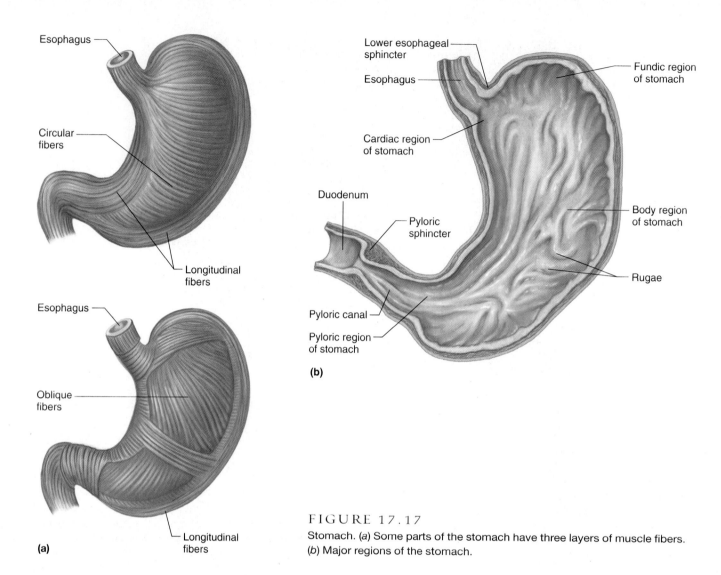

Esophagus

Circular fibers

Longitudinal fibers

Esophagus

Oblique fibers

Longitudinal fibers

(a)

Lower esophageal sphincter

Esophagus

Cardiac region of stomach

Duodenum

Pyloric sphincter

Pyloric canal

Pyloric region of stomach

Fundic region of stomach

Body region of stomach

Rugae

(b)

FIGURE 17.17

Stomach. (*a*) Some parts of the stomach have three layers of muscle fibers. (*b*) Major regions of the stomach.

Just superior to the point where the esophagus joins the stomach, some of the circular muscle fibers have increased sympathetic muscle tone, forming the **lower esophageal sphincter** (lo′er ĕ-sof″ah-je′al sfingk′ter), or cardiac sphincter (fig. 17.17). These fibers usually remain contracted, and they close the entrance to the stomach. In this way, they help prevent regurgitation of the stomach contents into the esophagus. When peristaltic waves reach the stomach, the muscle fibers that guard its entrance relax briefly to allow the swallowed food to enter.

1 Describe the regions of the pharynx.

2 List the major events that occur during swallowing.

3 What is the function of the esophagus?

Stomach

The **stomach** (stum′ak) is a J-shaped, pouchlike organ, about 25–30 centimeters long, which hangs inferior to the diaphragm in the upper left portion of the abdominal cav-

ity (see figs. 17.1 and 17.15; reference plates 4 and 5). It has a capacity of about one liter or more, and its inner lining is marked by thick folds (rugae) of the mucosal and submucosal layers that disappear when its wall is distended. The stomach receives food from the esophagus, mixes it with gastric juice, initiates the digestion of proteins, carries on limited absorption, and moves food into the small intestine.

In addition to the two layers of smooth muscle—an inner circular layer and an outer longitudinal layer—found in other regions of the alimentary canal, some parts of the stomach have another inner layer of oblique fibers. This third innermost muscular layer is most highly developed near the opening of the esophagus and in the body of the stomach (fig. 17.17).

Parts of the Stomach

The stomach, shown in figures 17.17 and 17.18 and reference plate 51, can be divided into the cardiac, fundic, body, and pyloric regions. The *cardiac region* is a small area near the esophageal opening (cardia). The *fundic*

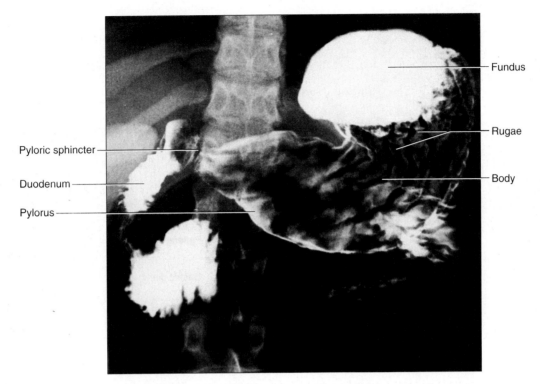

FIGURE 17.18
Radiograph of a stomach. (*Note:* A radiopaque compound the patient swallowed appears white in the radiograph.)

region, which balloons superior to the cardiac portion, is a temporary storage area and sometimes fills with swallowed air. This produces a gastric air bubble, which may be used as a landmark on a radiograph of the abdomen. The dilated *body region,* which is the main part of the stomach, is located between the fundic and pyloric portions. The *pyloric region* (antrum) is a funnel-shaped portion that narrows and becomes the *pyloric canal* as it approaches the small intestine.

At the end of the pyloric canal, the circular layer of fibers in its muscular wall thickens, forming a powerful muscle, the **pyloric sphincter.** This muscle is a valve that controls gastric emptying.

> *Hypertrophic pyloric stenosis* is a birth defect in which muscle overgrowth blocks the pyloric canal. The newborn vomits, with increasing force. To diagnose the condition, a radiograph is taken of the area after the infant drinks formula containing a radiopaque barium compound. Surgical splitting of the muscle blocking the passageway from stomach to small intestine is necessary to enable the infant to eat normally. Pyloric stenosis can occur later in life as a result of ulcers or cancer.

Gastric Secretions

The mucous membrane that forms the inner lining of the stomach is thick, its surface studded with many small openings. These openings, called *gastric pits,* are located at the ends of tubular **gastric glands** (oxyntic glands) (fig. 17.19). Although their structure and the composition of their secretions vary in different parts of the stomach, gastric glands generally contain three types of secretory cells. One type, the *mucous cell* (goblet cell), occurs in the necks of the glands near the openings of the gastric pits. The other types, *chief cells* (peptic cells) and *parietal cells* (oxyntic cells), are found in the deeper parts of the glands (fig. 17.19). The chief cells secrete digestive enzymes, and the parietal cells release a solution containing hydrochloric acid. The products of the mucous cells, chief cells, and parietal cells together form **gastric juice.**

Of the several digestive enzymes in gastric juice, **pepsin** is by far the most important. The chief cells secrete it as an inactive, nonerosive enzyme precursor called **pepsinogen.** When pepsinogen contacts the hydrochloric acid from the parietal cells, however, it breaks down rapidly into pepsin. Pepsin, in turn, can also break down pepsinogen into more pepsin.

Pepsin begins the digestion of nearly all types of dietary protein. This enzyme is most active in an acid environment, which the hydrochloric acid in gastric juice provides.

Gastric juice contains small quantities of a fat-splitting enzyme, *gastric lipase.* However, its action is weak due in part to the low pH of gastric juice. Gastric lipase acts mainly on butterfat.

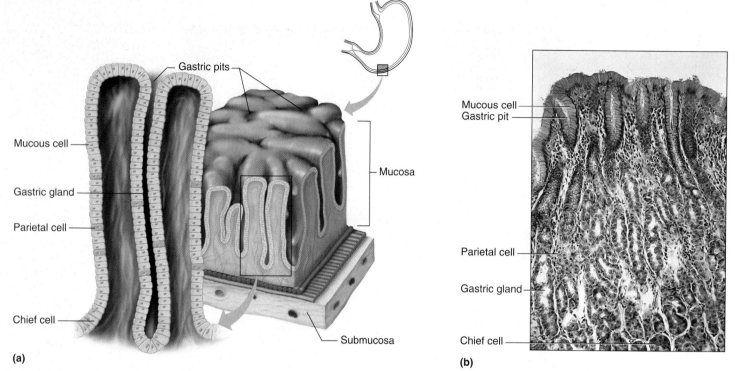

FIGURE 17.19

Lining of the stomach. (*a*) Gastric glands include mucous cells, parietal cells, and chief cells. The mucosa of the stomach is studded with gastric pits that are the openings of the gastric glands. (*b*) A light micrograph of cells associated with the gastric glands (50×).

Much of what we know about the stomach's functioning comes from a French-Canadian explorer, Alexis St. Martin, who in 1822 accidentally shot himself in the abdomen. His extensive injuries eventually healed, but a hole, called a fistula, was left, allowing observers to look at his stomach in action. A U.S. Army surgeon, William Beaumont, spent eight years watching food digesting in the stomach, noting how the stomach lining changed in response to stress.

In 1984, another chapter unfolded in our knowledge of digestive function when medical resident Barry Marshall at Royal Perth Hospital in western Australia performed a daring experiment. His mentor, J. Robin Warren, had hypothesized that a bacterial infection causes gastritis (inflammation of the stomach lining) and peptic ulcers (sores in the lining of the esophagus, stomach, or small intestine). At the time, the prevailing view was that a poor diet and stressful lifestyle caused these conditions. Marshall concocted what he called "swamp water"—and drank billions of bacteria. He developed gastritis, which, fortunately, cleared up. A colleague who repeated the experiment developed an ulcer and required antibiotics. After a decade of debate, the medical community finally concurred that the bacterium *Helicobacter pylori*, which thrives under acidic conditions, indeed causes many cases of gastritis and peptic ulcers. A short course of antibiotics and acid-lowering drugs has replaced lifelong treatments.

The mucous cells of the gastric glands secrete copious thin mucus. In addition, the cells of the mucous membrane, associated with the inner lining of the stomach and between the gastric glands, release a more viscous and alkaline secretion, which coats the inside of the stomach wall. This coating is especially important because pepsin can digest the proteins of stomach tissues, as well as those in foods. Thus, the coating normally prevents the stomach from digesting itself.

Still another component of gastric juice is **intrinsic factor** (in-trin'sik fak'tor). The parietal cells of the gastric glands secrete intrinsic factor, which is required for vitamin B_{12} absorption from the small intestine. Table 17.5 summarizes the components of gastric juice.

1 Where is the stomach located?

2 What are the secretions of the chief cells and parietal cells?

3 Which is the most important digestive enzyme in gastric juice?

4 How does the stomach keep from digesting itself?

Regulation of Gastric Secretions

Gastric juice is produced continuously, but the rate varies considerably and is controlled both neurally and hormonally. Within the gastric glands, specialized cells closely associated with the parietal cells secrete the hormone *somatostatin,* which inhibits acid secretion. However, acetylcholine (ACh) released from nerve endings in response to parasympathetic impulses arriving on the

TABLE 17.5	Major Components of Gastric Juice	
Component	**Source**	**Function**
Pepsinogen	Chief cells of the gastric glands	Inactive form of pepsin
Pepsin	Formed from pepsinogen in the presence of hydrochloric acid	A protein-splitting enzyme that digests nearly all types of dietary protein
Hydrochloric acid	Parietal cells of the gastric glands	Provides the acid environment needed for the conversion of pepsinogen into pepsin and for the action of pepsin
Mucus	Goblet cells and mucous glands	Provides a viscous, alkaline protective layer on the stomach wall
Intrinsic factor	Parietal cells of the gastric glands	Required for absorption of vitamin B_{12}

vagus nerves suppresses the secretion of somatostatin and stimulates the gastric glands to secrete large amounts of gastric juice, which is rich in hydrochloric acid and pepsinogen. These parasympathetic impulses also stimulate certain stomach cells, mainly in the pyloric region, to release a peptide hormone called **gastrin,** which increases the secretory activity of gastric glands (fig. 17.20). Furthermore, parasympathetic impulses and gastrin promote release of *histamine* from gastric mucosal cells, which, in turn, stimulates additional gastric secretion.

Histamine effectively promotes secretion of gastric acid. Drugs that block the histamine receptors of gastric mucosal cells (H_2-blockers) are used to inhibit excess gastric acid secretion.

Gastric secretion occurs in three stages—the cephalic, gastric, and intestinal phases. The *cephalic phase* begins before food reaches the stomach and possibly even before eating. In this stage, parasympathetic reflexes operating through the vagus nerves stimulate gastric secretion at the taste, smell, sight, or thought of food. The hungrier the person is, the greater the gastric secretion. The cephalic phase is responsible for 30% to 50% of the secretory response to a meal.

The *gastric phase* of gastric secretion, which accounts for 40% to 50% of the secretory activity, starts when food enters the stomach. The presence of food and the distension of the stomach wall trigger the stomach to release gastrin, which stimulates production of still more gastric juice.

As food enters the stomach and mixes with gastric juice, the pH of the contents rises, which enhances gastrin secretion. Consequently, the pH of the stomach contents drops. As the pH approaches 3.0, secretion of gastrin is inhibited. When the pH reaches 1.5, gastrin secretion ceases.

Gastrin stimulates cell growth in the mucosa of the stomach and intestines, except where gastrin is produced. This effect helps replace mucosal cells damaged by normal stomach function, disease, or medical treatments.

For the stomach to secrete hydrochloric acid, hydrogen ions are removed from the blood, and an equivalent number of alkaline bicarbonate ions are released into the blood. Following a meal, the blood concentration of bicarbonate ions increases, and the urine excretes excess bicarbonate ions. This phenomenon is called the *alkaline tide.*

The *intestinal phase* of gastric secretion, which accounts for about 5% of the total secretory response to a meal, begins when food leaves the stomach and enters the small intestine. When food first contacts the intestinal wall, it stimulates intestinal cells to release a hormone, *intestinal gastrin,* that again enhances gastric gland secretion.

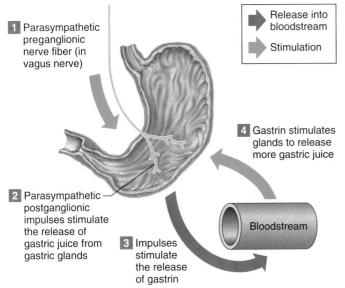

1 Parasympathetic preganglionic nerve fiber (in vagus nerve)

Release into bloodstream
Stimulation

4 Gastrin stimulates glands to release more gastric juice

2 Parasympathetic postganglionic impulses stimulate the release of gastric juice from gastric glands

3 Impulses stimulate the release of gastrin

Bloodstream

FIGURE 17.20
The secretion of gastric juice is regulated in part by parasympathetic nerve impulses that stimulate the release of gastric juice and gastrin.

As more food moves into the small intestine, a sympathetic reflex triggered by acid in the upper part of the small intestine inhibits secretion of gastric juice from the stomach wall. Also, proteins and fats in this region of the intestine stimulate release of the peptide hormone **cholecystokinin** (ko"le-sis"to-ki'nin) from the intestinal wall, which decreases gastric motility. Similarly, fats in the small intestine stimulate intestinal cells to release *intestinal somatostatin,* which inhibits release of gastric juice. Overall, these actions decrease gastric secretion and motility as the small intestine fills with food. Table 17.6 summarizes the phases of gastric secretion.

1 What controls gastric juice secretion?

2 Distinguish among the cephalic, gastric, and intestinal phases of gastric secretion.

3 What is the function of cholecystokinin?

Gastric Absorption

Gastric enzymes begin breaking down proteins, but the stomach wall is not well-adapted to absorb digestive products. The stomach absorbs only some water and certain salts, as well as certain lipid-soluble drugs. Most nutrients are absorbed in the small intestine. Alcohol, which is not a nutrient, is absorbed both in the small intestine and stomach. This is why the intoxicating effects of alcohol are felt soon after consuming alcoholic beverages.

Mixing and Emptying Actions

Food entering the stomach stretches the smooth muscles in its wall. The stomach may enlarge, but its muscles maintain their tone, and internal pressure of the stomach normally is unchanged. A person may eat more than the

TABLE 17.6	Phases of Gastric Secretion
Phase	**Action**
Cephalic phase	The sight, taste, smell, or thought of food triggers parasympathetic reflexes. Gastric juice is secreted in response.
Gastric phase	Food in stomach chemically and mechanically stimulates release of gastrin, which, in turn, stimulates secretion of gastric juice; reflex responses also stimulate gastric juice secretion.
Intestinal phase	As food enters the small intestine, it stimulates intestinal cells to release intestinal gastrin, which, in turn, promotes the secretion of gastric juice from the stomach wall.

stomach can comfortably hold, and when this happens, the internal pressure may rise enough to stimulate pain receptors. The result is a stomachache. Clinical Application 17.2 discusses this common problem along with its associated indigestion.

Following a meal, the mixing movements of the stomach wall aid in producing a semifluid paste of food particles and gastric juice called **chyme** (kīm). Peristaltic waves push the chyme toward the pyloric region of the stomach, and as chyme accumulates near the pyloric sphincter, this muscle begins to relax. Stomach contractions push chyme a little (5–15 milliliters) at a time into the small intestine. These contraction waves push most of the chyme backward into the stomach, mixing it further. The lower esophageal sphincter prevents reflux of stomach contents into the esophagus. Figure 17.21 illustrates this process.

The rate at which the stomach empties depends on the fluidity of the chyme and the type of food. Liquids usually pass through the stomach quite rapidly, but solids

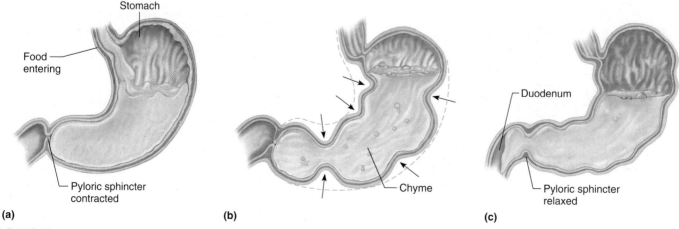

(a)   **(b)**   **(c)**

FIGURE 17.21

Stomach movements. (*a*) As the stomach fills, its muscular wall stretches, but the pyloric sphincter remains closed. (*b*) Mixing movements combine food and gastric juice, creating chyme. (*c*) Peristaltic waves move the chyme toward the pyloric sphincter, which relaxes and admits some chyme into the duodenum.

OH, MY ACHING STOMACH!

At the barbecue, Perry W. consumed two burgers, three hot dogs, beans in a spicy sauce, loads of chips, several beers, and ice cream. Later, a feeling of fullness became abdominal pain, then heartburn, as his stomach contents backed up into his esophagus.

Perry found temporary relief with an over-the-counter antacid product, which quickly raised the pH of the stomach. These products usually include a compound containing either sodium, calcium, magnesium, or aluminum. Another ingredient in some products is simethicone, which breaks up gas bubbles in the digestive tract. If antacids do not help within a few minutes or they are used for longer than two weeks, a doctor should be consulted. The problem may be more serious than overeating.

Avoiding acid indigestion and heartburn is a more healthful approach than gorging and then reaching for medication—or even taking products that lower acid production before a large or spicy meal. Some tips:

- Avoid large meals. The more food, the more acid the stomach produces.
- Eat slowly so that stomach acid secretion is more gradual.
- Do not lie down immediately after eating. Being upright enables gravity to help food move along the alimentary canal.
- If prone to indigestion or heartburn, avoid caffeine, which increases stomach acid secretion.
- Cigarettes and alcohol irritate the stomach lining and relax the sphincter at the junction between the stomach and the esophagus. This makes it easier for food to return to the esophagus, causing heartburn.
- Do not eat acidic foods, such as citrus fruits and tomatoes, unless it is at least three hours before bedtime.
- Use a pillow that elevates the head six to eight inches above the stomach. ■

remain until they are well mixed with gastric juice. Fatty foods may remain in the stomach three to six hours; foods high in proteins move through more quickly; carbohydrates usually pass through more rapidly than either fats or proteins.

As chyme fills the duodenum, internal pressure on the organ increases, stretching the intestinal wall. These actions stimulate sensory receptors in the wall, triggering an **enterogastric reflex** (en″ter-o-gas′trik re′fleks). The name of this reflex, like those of other digestive reflexes, describes the origin and termination of reflex impulses. That is, the enterogastric reflex begins in the small intestine (*entero*) and ends in the stomach (*gastro*). As a result of the enterogastric reflex, fewer parasympathetic impulses arrive at the stomach, inhibiting peristalsis, and intestinal filling slows (fig. 17.22). If chyme entering the intestine is fatty, the intestinal wall releases the hormone cholecystokinin, which further inhibits peristalsis.

Vomiting results from a complex reflex that empties the stomach in the reverse of the normal direction. Irritation or distension in the stomach or intestines can trigger vomiting. Sensory impulses travel from the site of stimulation to the *vomiting center* in the medulla oblongata, and motor responses follow. These include taking a deep breath, raising the soft palate and thus closing the nasal cavity, closing the opening to the trachea (glottis), relaxing the circular muscle fibers at the base of the esophagus, contracting the diaphragm so it presses downward over the stomach, and contracting the abdominal wall muscles to increase pressure inside the abdominal cavity. As a

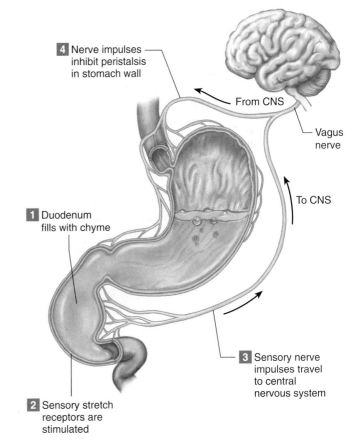

4 Nerve impulses inhibit peristalsis in stomach wall

From CNS

Vagus nerve

To CNS

1 Duodenum fills with chyme

3 Sensory nerve impulses travel to central nervous system

2 Sensory stretch receptors are stimulated

FIGURE 17.22

The enterogastric reflex partially regulates the rate at which chyme leaves the stomach.

result, the stomach is squeezed from all sides, forcing its contents upward and out through the esophagus, pharynx, and mouth.

Activity in the vomiting center can be stimulated by drugs (emetics), by toxins in contaminated foods, and sometimes by rapid changes in body motion. In this last situation, sensory impulses from the labyrinths of the inner ears reach the vomiting center and can produce motion sickness. The vomiting center can also be activated by stimulation of higher brain centers through sights, sounds, odors, tastes, emotions, or mechanical stimulation of the back of the pharynx.

Nausea emanates from activity in the vomiting center or in nerve centers near it. During nausea, stomach movements usually are diminished or absent, and duodenal contents may move back into the stomach.

1 How is chyme produced?

2 What factors influence how quickly chyme leaves the stomach?

3 Describe the enterogastric reflex.

4 Describe the vomiting reflex.

5 What factors may stimulate the vomiting reflex?

Pancreas

The **pancreas,** discussed as an endocrine gland in chapter 13 (p. 496), also has an exocrine function—secretion of a digestive juice called **pancreatic juice** (pan″kre-at′ik jōōs).

Structure of the Pancreas

The pancreas is closely associated with the small intestine and is located posterior to the parietal peritoneum. It extends horizontally across the posterior abdominal wall, with its head in the C-shaped curve of the duodenum (portion of the small intestine) and its tail against the spleen (fig. 17.23 and reference plate 59).

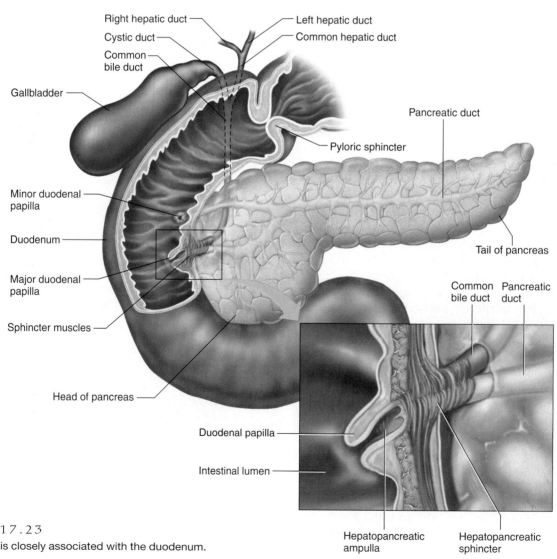

FIGURE 17.23

The pancreas is closely associated with the duodenum.

The cells that produce pancreatic juice, called *pancreatic acinar cells,* make up the bulk of the pancreas. These cells form clusters called *acini* (acinus, singular) around tiny tubes into which they release their secretions. The smaller tubes unite to form larger ones, which, in turn, give rise to a *pancreatic duct* extending the length of the pancreas and transporting pancreatic juice to the small intestine. The pancreatic duct usually connects with the duodenum at the same place where the bile duct from the liver and gallbladder joins the duodenum, although other connections may be present (see figs. 13.34 and 17.23).

The pancreatic and bile ducts join at a short, dilated tube called the *hepatopancreatic ampulla* (ampulla of Vater). A band of smooth muscle, called the *hepatopancreatic sphincter* (sphincter of Oddi), surrounds this ampulla.

Pancreatic Juice

Pancreatic juice contains enzymes that digest carbohydrates, fats, proteins, and nucleic acids. The carbohydrate-digesting enzyme, **pancreatic amylase,** splits molecules of starch or glycogen into disaccharides. The fat-digesting enzyme, **pancreatic lipase,** breaks triglyceride molecules into fatty acids and monoglycerides. (A monoglyceride molecule consists of one fatty acid bound to glycerol.)

The protein-splitting (proteolytic) enzymes are **trypsin, chymotrypsin,** and **carboxypeptidase.** Each of these enzymes splits the bonds between particular combinations of amino acids in proteins. Because no single enzyme can split all possible combinations of amino acids, several enzymes are necessary to completely digest protein molecules.

The proteolytic enzymes are stored in inactive forms within tiny cellular structures called *zymogen granules.* These enzymes, like gastric pepsin, are secreted in inactive forms and must be activated by other enzymes after they reach the small intestine. For example, the pancreatic cells release inactive **trypsinogen,** which is activated to trypsin when it contacts the enzyme *enterokinase,* which the mucosa of the small intestine secretes. Chymotrypsin and carboxypeptidase are activated, in turn, by trypsin. This mechanism prevents enzymatic digestion of proteins within the secreting cells and the pancreatic ducts.

A painful condition called *acute pancreatitis* results from a blockage in the release of pancreatic juice. Trypsinogen, activated as pancreatic juice builds up, digests parts of the pancreas. Alcoholism, gallstones, certain infections, traumatic injuries, or the side effects of some drugs can cause pancreatitis.

Pancreatic juice contains two types of **nucleases,** which are enzymes that break down nucleic acid molecules into nucleotides. It also has a high concentration of bicarbonate ions that makes the juice alkaline. This alkalinity provides a favorable environment for the actions of the digestive enzymes and helps neutralize the acidic chyme as it arrives from the stomach. At the same time, the alkaline condition in the small intestine blocks the action of pepsin, which might otherwise damage the duodenal wall.

Regulation of Pancreatic Secretion

The nervous and endocrine systems regulate release of pancreatic juice, much as they regulate gastric and small intestinal secretions. For example, during the cephalic and gastric phases of gastric secretion, parasympathetic impulses stimulate the pancreas to release digestive enzymes. A peptide hormone, **secretin,** stimulates the pancreas to secrete a large quantity of fluid when acidic chyme enters the duodenum. Secretin is released into the blood from the duodenal mucous membrane in response to the acid in chyme. The pancreatic juice secreted at this time contains few, if any, digestive enzymes but has a high concentration of bicarbonate ions that neutralize the acid in chyme (fig. 17.24).

In cystic fibrosis, abnormal chloride channels in certain cells entrap ions inside, drawing water into the cells from interstitial spaces. This dries out secretions in the lungs and pancreas, leaving a very sticky mucus that impairs the functioning of these organs and encourages infection by certain types of bacteria. When the pancreas is plugged with mucus, its secretions, containing digestive enzymes, cannot reach the duodenum. Individuals with cystic fibrosis must take digestive enzyme supplements—in capsule or powder form that can be mixed with a soft food such as applesauce—to prevent malnutrition.

Proteins and fats in chyme in the duodenum also stimulate the release of the hormone cholecystokinin from the intestinal wall. As in the case of secretin, cholecystokinin reaches the pancreas by way of the bloodstream. Pancreatic juice secreted in response to cholecystokinin has a high concentration of digestive enzymes.

1 Where is the pancreas located?

2 List the enzymes in pancreatic juice.

3 What are the functions of the enzymes in pancreatic juice?

4 What regulates secretion of pancreatic juice?

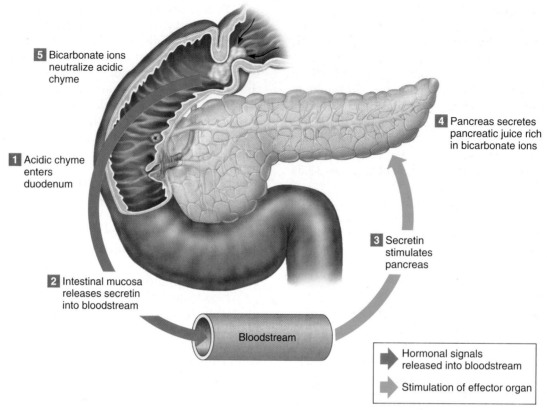

FIGURE 17.24
Acidic chyme entering the duodenum from the stomach stimulates the release of secretin, which, in turn, stimulates the release of pancreatic juice.

Liver

The **liver,** the largest internal organ, is located in the upper right quadrant of the abdominal cavity, just inferior to the diaphragm. It is partially surrounded by the ribs and extends from the level of the fifth intercostal space to the lower margin of the ribs. It is reddish brown in color and well supplied with blood vessels (figs. 17.25, 17.26, 17.27 and reference plates 51, 57, 74).

Liver Structure

A fibrous capsule encloses the liver, and connective tissue divides the organ into a large *right lobe* and a smaller *left lobe.* The *falciform ligament* is a fold of visceral peritoneum that separates the lobes and fastens the liver to the abdominal wall anteriorly. The liver also has two minor lobes, the *quadrate lobe,* near the gallbladder, and the *caudate lobe,* close to the vena cava (fig. 17.26). The area where the four lobes meet and blood vessels and ducts enter or exit the liver is the *porta hepatis.*

A fold of visceral peritoneum called the *coronary ligament* attaches the liver to the diaphragm on its superior surface. Each lobe is separated into many tiny **hepatic lobules,** which are the liver's functional units (fig. 17.27). A lobule consists of many *hepatic cells* radiating outward from a *central vein.* Vascular channels called **hepatic sinusoids** separate platelike groups of these cells from each other. Blood from the digestive tract, which is carried in the *hepatic portal vein* (see chapter 15, p. 593), brings newly absorbed nutrients into the sinusoids (fig. 17.28). At the same time, oxygenated blood from the hepatic artery mixes freely with the blood containing nutrients, then flows through the liver sinusoids and nourishes the hepatic cells.

Often blood in the portal veins contains some bacteria that have entered through the intestinal wall. However, large **Kupffer cells,** which are fixed to the inner lining (endothelium) of the hepatic sinusoids, remove most of the bacteria from the blood by phagocytosis. Then the blood passes into the *central veins* of the hepatic lobules and moves out of the liver via the hepatic vein.

Within the liver lobules are many fine *bile canals,* which receive secretions from the hepatic cells. The canals of neighboring lobules unite to form larger ducts and then converge to become the **hepatic ducts.** These ducts merge, in turn, to form the *common hepatic duct.*

Liver Functions

The liver carries on many important metabolic activities. Recall from chapter 13 (p. 496) that the liver plays a key role in carbohydrate metabolism by helping maintain the

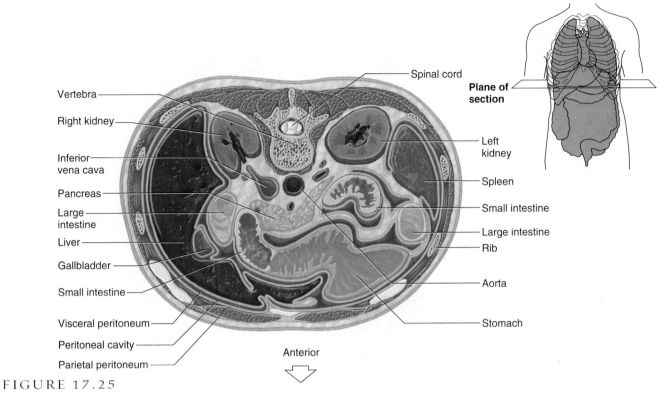

FIGURE 17.25

This transverse section of the abdomen reveals the liver and other organs within the upper portion of the abdominal cavity.

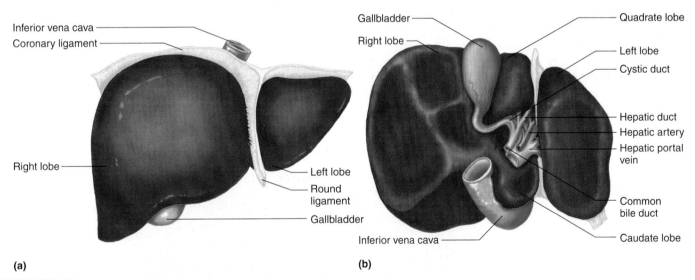

(a)

(b)

FIGURE 17.26

Lobes of the liver, viewed (*a*) anteriorly and (*b*) inferiorly.

normal concentration of blood glucose. Liver cells that respond to hormones such as insulin and glucagon lower the blood glucose level by polymerizing glucose to glycogen and raise the blood glucose level by breaking down glycogen to glucose or by converting noncarbohydrates into glucose.

The liver's effects on lipid metabolism include oxidizing fatty acids at an especially high rate (see chapter 18, p. 698); synthesizing lipoproteins, phospholipids, and cholesterol; and converting portions of carbohydrate and protein molecules into fat molecules. The blood transports fats synthesized in the liver to adipose tissue for storage.

The most vital liver functions are probably those related to protein metabolism. They include deaminating amino acids; forming urea (see chapter 18, p. 699); synthesizing plasma proteins, such as clotting factors (see

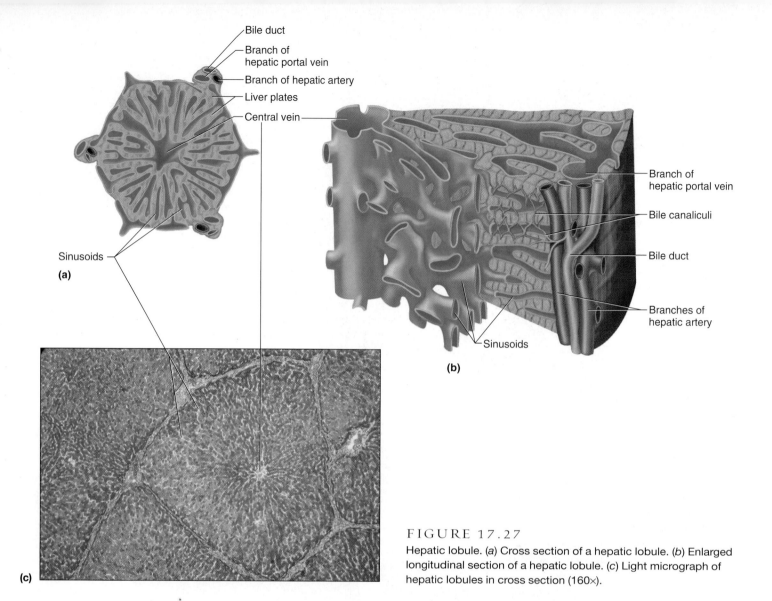

Bile duct
Branch of hepatic portal vein
Branch of hepatic artery
Liver plates
Central vein

Sinusoids

(a)

Branch of hepatic portal vein

Bile canaliculi

Bile duct

Branches of hepatic artery

Sinusoids

(b)

(c)

FIGURE 17.27

Hepatic lobule. (*a*) Cross section of a hepatic lobule. (*b*) Enlarged longitudinal section of a hepatic lobule. (*c*) Light micrograph of hepatic lobules in cross section (160×).

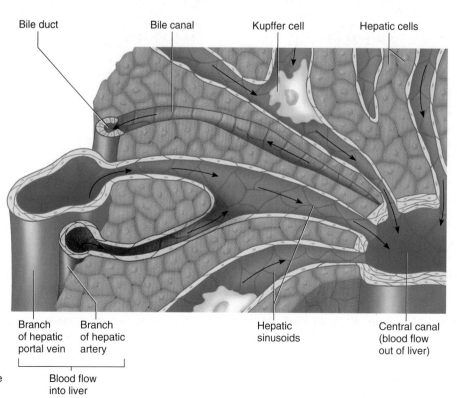

Bile duct   Bile canal   Kupffer cell   Hepatic cells

Branch of hepatic portal vein   Branch of hepatic artery   Hepatic sinusoids   Central canal (blood flow out of liver)

Blood flow into liver

FIGURE 17.28

The paths of blood and bile within a hepatic lobule.

REPLACING THE LIVER

Liver transplants have been performed since 1964, and today are often combined to treat systemic illnesses with other organs, such as pancreases, kidneys, and sections of intestine. An 11-month-old transplanted in 1982 is still alive and well, attesting to the success of the procedure. The publicity over that case led to passage of the National Organ Transplant Act in 1984, which in turn led to establishment of the National Organ Procurement and Transplant Network. Today, people can donate portions of their livers to help relatives with liver disease, although in at least one case, the donor died in his attempt to save his brother.

Continuing problems with liver transplants are the scarcity of donor organs, and the rapidity with which liver failure kills. Each year in the United States, only about 4,500 of the 12,000 or so individuals requiring livers survive long enough to undergo a transplant. A person can survive only a few days once the liver stops functioning. In fulminant hepatic failure, for example, an otherwise healthy, young person suddenly experiences liver failure, caused by expo-

sure to a toxin, reaction to a drug, or a viral infection. Jaundice and fatigue progress rapidly to coma and death. This is what might have happened to Sheryl Orlyk, a 34-year-old who entered a hospital in Chicago with the condition on April 2, 2002. She became the first person to receive the "extracorporeal liver assist device" (ELAD), a liver stand-in that cleansed her blood until a cadaver organ became available on May 4. ELAD performed so well that she was able to return home in just 10 days.

ELAD is the first type of bioartificial liver to undergo clinical trials. It is called "bioartificial" because it has synthetic as well as biological components. The device consists of two chambers that are filled with hollow fibers that house millions of continuously dividing human liver cells (hepatocytes). ELAD functions like an artificial kidney (dialysis machine). A patient's plasma is separated from the blood and passed through the device, where the liver cells remove toxins, as they would as part of a natural organ. The plasma is then filtered, the formed elements added back, and the blood reinfused into the patient.

In the past, bioartificial livers used cells from pigs, which could provoke an allergic reaction or introduce viruses. These devices could only be used for 6 to 8 hours a day, for a few days. In contrast, ELAD can be used continuously for up to 10 days. So far, it works. In the first clinical trial, 12 of 15 (80 percent) of patients who used ELAD survived to be successfully transplanted, compared to 5 of 9 (56 percent) of patients who did not use the device. Larger trials are underway.

Further down the clinical road is "therapeutic liver repopulation," in which implants of hepatocytes from donors will replace and eventually restore damaged or diseased liver tissue. Stem cell therapy is also being developed. Researchers have discovered that certain stem cells in the bone marrow can travel to the liver, where they differentiate into hepatocyte progenitor cells, which can then mature into functional hepatocytes. This means that someday liver disease may be treatable with a bone marrow transplant, or even with an infusion of stem cells into the bloodstream. Only five to ten percent of the liver need be replaced to restore function. These approaches are in the animal testing stage of development. ∎

---

chapter 14, p. 527); and converting certain amino acids to other amino acids.

> Bacteria in the intestine produce ammonia, which is carried in the blood to the liver, where it is converted to urea. When this liver function fails, concentration of blood ammonia sharply rises, causing *hepatic coma,* a condition that can lead to death.

The liver also stores many substances, including glycogen, iron, and vitamins A, D, and B_{12}. Extra iron from the blood combines with a protein (apoferritin) in liver cells, forming *ferritin.* The iron is stored in this form until blood iron concentration falls, when some of the iron is released. Thus, the liver is important in iron homeostasis.

Liver cells help destroy damaged red blood cells and phagocytize foreign antigens. The liver removes toxic substances such as alcohol from the blood (detoxification). The liver can also serve as a blood reservoir, storing 200 to 400 milliliters of blood. The liver's role in digestion is to secrete bile. Table 17.7 summarizes the major functions of the liver. Clinical Application 17.3 discusses hepatitis, an inflammation of the liver.

1 Describe the location of the liver.

2 Describe a hepatic lobule.

3 Review liver functions.

4 Which liver function participates in digestion?

Composition of Bile

Bile (bīl) is a yellowish green liquid that hepatic cells continuously secrete. In addition to water, it contains *bile*

HEPATITIS

Hepatitis is an inflammation of the liver caused by viral infection, or, more rarely, from reaction to a drug, alcoholism, or autoimmunity. There are several types of hepatitis.

Liver Inflammation Causes Distinct Symptoms

Hepatitis A is one of the least severe forms of this common illness. For the first few days, symptoms include mild headache, low fever, fatigue, lack of appetite, nausea and vomiting, and sometimes stiff joints. By the end of the first week, more distinctive symptoms arise, including a rash, pain in the upper right quadrant of the abdomen, dark and foamy urine, and pale feces. The skin and sclera of the eyes begin to turn yellow from accumulating bile pigments (jaundice). Great fatigue may continue for two or three weeks, and then gradually, the person begins to feel better.

At the other end of the hepatitis spectrum is fulminant hepatitis, which is rare and can be caused by any of several viruses. Symptoms start suddenly and severely, and behavior and personality may change. Without medical attention, the condition progresses to kidney and liver failure, or coma.

Hepatitis B produces chronic symptoms that persist for more than six months. Perhaps as many as 300 million people worldwide are carriers of hepatitis B. They do not have symptoms but can infect others. Five percent of such carriers eventually develop liver cancer.

An Alphabet of Viral Causes

Several types of viruses can cause hepatitis. At the beginning of the twentieth century, before investigators knew how to distinguish viruses by their nucleic acid sequences, two types of hepatitis were defined epidemiologically. So-called "infectious hepatitis," which was transmitted from person to person, was later attributed to the hepatitis A virus. "Serum hepatitis" was transmitted by blood and was later found to be caused by the hepatitis B virus. Hepatitis A often arose from food handlers who did not properly wash after using the bathroom, establishing a fecal-oral route of transmission. Hepatitis B was more often passed sexually.

By the mid-1970s, technology enabled physicians to identify either hepatitis A or hepatitis B virus. But then a problem arose: many cases of what appeared to be hepatitis were not caused by either of the known viral types. These were called "non-A non-B" hepatitis—which just meant that researchers did not know what caused them. Then in the 1980s, the "non-A and non-B" viruses began to be identified.

Hepatitis C affects 3.9 million people in the United States and 275,000 in Canada. Before blood supplies began being screened for hepatitis C, most cases of this form of the illness arose from transfusions. In the 1980s, 240,000 new cases were identified in the United States each year. Today only about 34,000 new cases occur each year, and most were caused by intravenous drug use. Only 20 percent of people with hepatitis C experience the typical symptoms of fatigue, jaundice, abdominal pain, nausea and dark urine. Some people find relief with the immune system biochemical interferon, or the antiviral drug ribavirin.

Hepatitis D is bloodborne, usually associated with intravenous drug use, and occurs in people already infected with hepatitis B. It kills about 20% of the people it infects. Hepatitis E infection is more common in developing nations, where it often severely affects pregnant women.

Since 1994, two new hepatitis viruses have been identified. Very little is known about the hepatitis F virus, but it can pass from human feces to infect other primates. Hepatitis G is very rare but seems to account for a significant percentage of cases of fulminant hepatitis. However, in people with healthy immune systems, it produces symptoms so mild that they may not even be noticed. ■

TABLE 17.7	Major Functions of the Liver		
General Function	**Specific Function**	**General Function**	**Specific Function**
Carbohydrate metabolism	Polymerizes glucose to glycogen, breaks down glycogen to glucose, and converts noncarbohydrates to glucose	Protein metabolism	Deaminates amino acids; forms urea; synthesizes plasma proteins; converts certain amino acids to other amino acids
Lipid metabolism	Oxidizes fatty acids; synthesizes lipoproteins, phospholipids, and cholesterol; converts portions of carbohydrate and protein molecules into fats	Storage	Stores glycogen, vitamins A, D, and B_{12}, iron, and blood
		Blood filtering	Removes damaged red blood cells and foreign substances by phagocytosis
		Detoxification	Removes toxins from the blood
		Secretion	Secretes bile

salts, bile pigments, cholesterol, and electrolytes. Of these, bile salts are the most abundant. They are the only bile substances that have a digestive function.

Hepatic cells use cholesterol to produce bile salts, and in secreting these salts, they release some cholesterol into the bile. Cholesterol by itself has no special function in bile or in the alimentary canal.

Bile pigments (bilirubin and biliverdin) are breakdown products of hemoglobin from red blood cells. These pigments are normally excreted in the bile (see chapter 14, p. 517). The yellowish skin, sclerae, and mucous membranes of jaundice result from excess deposition of bile pigments.

> Jaundice can have several causes. In *obstructive jaundice,* bile ducts are blocked. In *hepatocellular jaundice,* the liver is diseased. In *hemolytic jaundice,* red blood cells are destroyed too rapidly.

Gallbladder

The **gallbladder** is a pear-shaped sac located in a depression on the inferior surface of the liver. It is connected to the **cystic duct,** which, in turn, joins the hepatic duct (see fig. 17.26 and reference plate 59). The gallbladder has a capacity of 30–50 milliliters, is lined with columnar epithelial cells, and has a strong muscular layer in its wall. It stores bile between meals, concentrates bile by reabsorbing water, and releases bile into the duodenum when stimulated by cholecystokinin from the small intestine.

The **common bile duct** is formed by the union of the common hepatic and cystic ducts. It leads to the duodenum, where the hepatopancreatic sphincter muscle guards its exit (see fig. 17.23). This sphincter normally remains contracted so that bile collects in the common bile duct and backs up into the cystic duct. When this happens, the bile flows into the gallbladder, where it is stored.

Bile salts, bile pigments, and cholesterol become increasingly concentrated as the gallbladder lining reabsorbs some water and electrolytes. Although the cholesterol normally remains in solution, under certain conditions it may precipitate and form solid crystals. If cholesterol continues to come out of solution, the crystals enlarge, forming

> A famous photograph of former U.S. President Lyndon Johnson depicts him displaying the 8-inch scar on his abdomen from where his gallbladder was removed to treat gallstones. Today, many such procedures—called *cholecystectomies*— are performed using a laser, which leaves four tiny cuts, each a quarter to a half inch long. Recovery from the laser procedure is much faster than from traditional surgery.

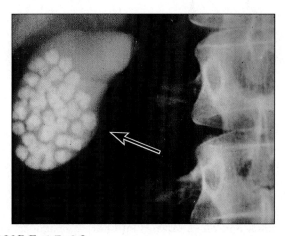

FIGURE 17.29
Falsely colored radiograph of a gallbladder that contains gallstones (arrow).

gallstones (fig. 17.29). This may happen if the bile is too concentrated, hepatic cells secrete too much cholesterol, or the gallbladder is inflamed (cholecystitis). Gallstones in the bile duct may block the flow of bile, causing obstructive jaundice and considerable pain. Clinical Application 17.4 discusses disorders of the gallbladder.

Regulation of Bile Release

Normally, bile does not enter the duodenum until cholecystokinin stimulates the gallbladder to contract. The intestinal mucosa releases this hormone in response to proteins and fats in the small intestine. (Recall from the discussion earlier in this chapter, the action of cholecystokinin to stimulate pancreatic enzyme secretion.) The hepatopancreatic sphincter usually remains contracted until a peristaltic wave in the duodenal wall approaches it. Then, just before the wave reaches it, the sphincter relaxes, and bile squirts into the duodenum (fig. 17.30). Table 17.8 summarizes the hormones that help control digestion.

Functions of Bile Salts

Bile salts aid digestive enzymes and enhance absorption of fatty acids and certain fat-soluble vitamins. Molecules of fats clump into *fat globules.* Bile salts affect fat globules much like a soap or detergent would, reducing surface tension and breaking fat globules into smaller droplets. This action is called **emulsification.** Monoglycerides that form from the action of pancreatic lipase on triglyceride molecules aid emulsification. This emulsification greatly increases the total surface area of the fatty substance, and the tiny droplets mix with water. Lipases can then digest the fat molecules more effectively.

Bile salts aid in absorption of fatty acids and cholesterol by forming complexes (micelles) that are very soluble in chyme and that epithelial cells can more easily absorb. The fat-soluble vitamins A, D, E, and K are also absorbed. Lack of bile salts results in poor lipid absorption and vitamin deficiencies.

GALLBLADDER DISEASE

Molly G., an overweight, forty-seven-year-old college administrator and mother of four, had been feeling healthy until recently. Then she regularly began to feel pain in the upper right quadrant of her abdomen (see fig. 1.23*b*). Sometimes the discomfort seemed to radiate around to her back and move upward into her right shoulder. Most commonly, she felt this pain after her evening meal; occasionally it also occurred during the night, awakening her. After an episode of particularly severe pain accompanied by sweating (diaphoresis) and nausea, Molly approached her physician.

During an examination of Molly's abdomen, the physician discovered tenderness in the epigastric region (see fig. 1.23*a*). She decided that Molly might be experiencing the symptoms of *acute cholecystitis*—an inflammation of the gallbladder. The physician recommended that Molly have a *cholecystogram*—an X ray of the gallbladder.

Molly took tablets containing a contrast medium the night before the X-ray procedure. This schedule allowed time for the small intestine to absorb the substance, which was carried to the liver and excreted into the bile. Later, the bile and contrast medium would be stored and concentrated in the gallbladder and would make the contents of the gallbladder opaque to X rays.

Molly's cholecystogram (see fig. 17.29) revealed several stones (calculi) in her gallbladder, a condition called *cholelithiasis*. Because Molly's symptoms were worsening, her physician recommended that she consult an abdominal surgeon about undergoing a *cholecystectomy*—surgical removal of the gallbladder.

During the surgical procedure, an incision was made in Molly's right subcostal region. Her gallbladder was excised from the liver. Then the surgeon explored the cystic duct (see fig. 17.26) and hepatic ducts for stones but found none.

Unfortunately, following her recovery from surgery, Molly's symptoms persisted. Her surgeon ordered a *cholangiogram*—an X-ray series of the bile ducts. This study showed a residual stone at the distal end of Molly's common bile duct (see fig. 17.23).

The surgeon extracted the residual stone using a *fiber-optic endoscope,* a long, flexible tube that can be passed through the patient's esophagus and stomach and into the duodenum. This instrument enables a surgeon to observe features of the gastrointestinal tract by viewing them directly through the eyepiece of the endoscope or by watching a monitor. A surgeon can also perform manipulations using specialized tools that are passed through the endoscope to its distal end.

In Molly's case, the surgeon performed an *endoscopic papillotomy*—an incision of the hepatopancreatic sphincter (see fig. 17.23) by applying an electric current to a wire extending from the end of the endoscope. She then removed the exposed stone by manipulating a tiny basket at the tip of the endoscope. ■

TABLE 17.8	Hormones of the Digestive Tract	
Hormone	**Source**	**Function**
Gastrin	Gastric cells, in response to food	Increases secretory activity of gastric glands
Intestinal gastrin	Cells of small intestine, in response to chyme	Increases secretory activity of gastric glands
Somatostatin	Gastric cells	Inhibits secretion of acid by parietal cells
Intestinal somatostatin	Intestinal wall cells, in response to fats	Inhibits secretion of acid by parietal cells
Cholecystokinin	Intestinal wall cells, in response to proteins and fats in the small intestine	Decreases secretory activity of gastric glands and inhibits gastric motility; stimulates pancreas to secrete fluid with a high digestive enzyme concentration; stimulates gallbladder to contract and release bile
Secretin	Cells in the duodenal wall, in response to acidic chyme entering the small intestine	Stimulates pancreas to secrete fluid with a high bicarbonate ion concentration

The mucous membrane of the small intestine reabsorbs nearly all of the bile salts, along with fatty acids. The bile salts are carried in the blood to the liver, where hepatic cells resecrete them into the bile ducts. Liver cells synthesize bile salts, which replace the small quantities that are lost in the feces.

1 Explain how bile originates.

2 Describe the function of the gallbladder.

3 How is secretion of bile regulated?

4 How do bile salts function in digestion?

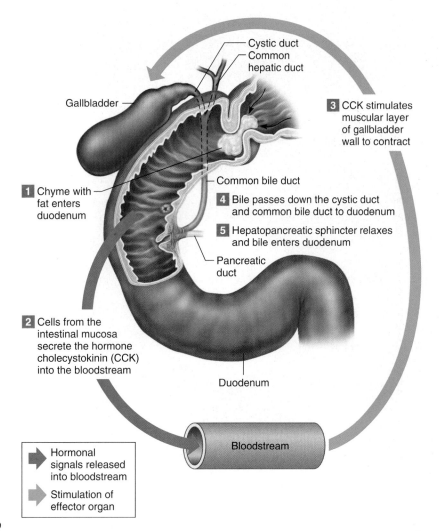

- Cystic duct
- Common hepatic duct
- Gallbladder
- **3** CCK stimulates muscular layer of gallbladder wall to contract
- **1** Chyme with fat enters duodenum
- Common bile duct
- **4** Bile passes down the cystic duct and common bile duct to duodenum
- **5** Hepatopancreatic sphincter relaxes and bile enters duodenum
- Pancreatic duct
- **2** Cells from the intestinal mucosa secrete the hormone cholecystokinin (CCK) into the bloodstream
- Duodenum
- Bloodstream

→ Hormonal signals released into bloodstream

→ Stimulation of effector organ

FIGURE 17.30

Fatty chyme entering the duodenum stimulates the gallbladder to release bile.

Small Intestine

The **small intestine** is a tubular organ that extends from the pyloric sphincter to the beginning of the large intestine. With its many loops and coils, it fills much of the abdominal cavity (see fig. 17.1 and reference plates 4 and 5). Although the small intestine is 5.5–6.0 meters (18–20 feet) long in a cadaver when the muscular wall lacks tone, it may be only half this long in a living person.

The small intestine receives secretions from the pancreas and liver. It also completes digestion of the nutrients in chyme, absorbs the products of digestion, and transports the remaining residues to the large intestine.

Parts of the Small Intestine

The small intestine, shown in figures 17.31 and 17.32 and in reference plates 51, 58, 74, and 75, consists of three portions: the duodenum, the jejunum, and the ileum.

The **duodenum** (du″o-de′num), which is about 25 centimeters long and 5 centimeters in diameter, lies posterior

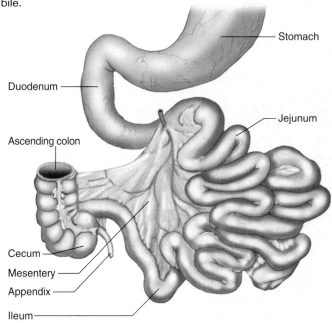

- Stomach
- Duodenum
- Jejunum
- Ascending colon
- Cecum
- Mesentery
- Appendix
- Ileum

FIGURE 17.31

The three parts of the small intestine are the duodenum, the jejunum, and the ileum.

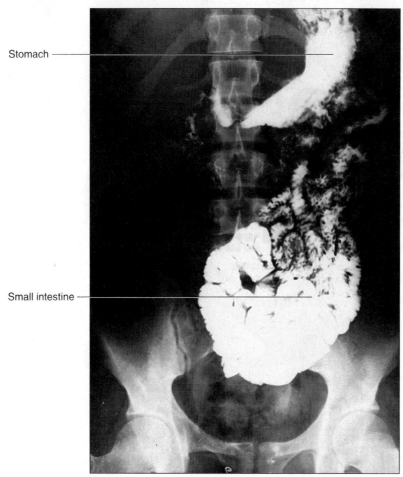

Stomach

Small intestine

FIGURE 17.32
Radiograph showing a normal small intestine containing a radiopaque substance that the patient ingested.

to the parietal peritoneum (retroperitoneal). It is the shortest and most fixed portion of the small intestine. The duodenum follows a C-shaped path as it passes anterior to the right kidney and the upper three lumbar vertebrae.

The remainder of the small intestine is mobile and lies free in the peritoneal cavity. The proximal two-fifths of this portion is the **jejunum** (jĕ-joo'num), and the remainder is the **ileum** (il'e-um). There is no distinct separation between the jejunum and ileum, but the diameter of the jejunum is usually greater, and its wall is thicker, more vascular, and more active than that of the ileum.

The jejunum and ileum are suspended from the posterior abdominal wall by a double-layered fold of peritoneum called **mesentery** (mes'en-ter"e) (fig. 17.33). The mesentery supports the blood vessels, nerves, and lymphatic vessels that supply the intestinal wall.

A filmy, double fold of peritoneal membrane called the *greater omentum* drapes like an apron from the stomach over the transverse colon and the folds of the small intestine. If infections occur in the wall of the alimentary canal, cells from the omentum may adhere to the inflamed region and help wall it off so that the infection is less likely to enter the peritoneal cavity (fig. 17.34).

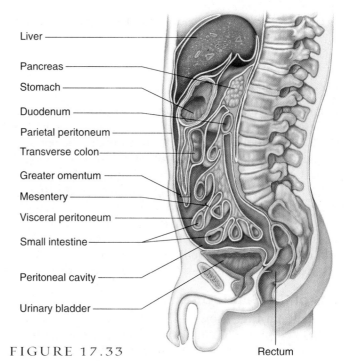

Liver

Pancreas

Stomach

Duodenum

Parietal peritoneum

Transverse colon

Greater omentum

Mesentery

Visceral peritoneum

Small intestine

Peritoneal cavity

Urinary bladder

Rectum

FIGURE 17.33
Mesentery formed by folds of the peritoneal membrane suspends portions of the small intestine from the posterior abdominal wall.

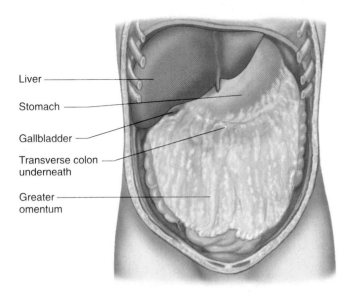

FIGURE 17.34
The greater omentum hangs like an apron over the abdominal organs.

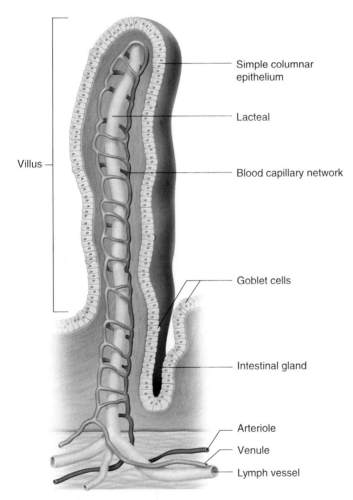

FIGURE 17.35
Structure of a single intestinal villus.

Structure of the Small Intestinal Wall

Throughout its length, the inner wall of the small intestine has a velvety appearance. This is due to many tiny projections of mucous membrane called **intestinal villi** (figs. 17.35 and 17.36; see fig. 17.3). These structures are most numerous in the duodenum and the proximal portion of the jejunum. They project into the passageway, or **lumen,** of the alimentary canal, contacting the intestinal contents. Villi greatly increase the surface area of the intestinal lining, aiding absorption of digestive products.

Each villus consists of a layer of simple columnar epithelium and a core of connective tissue containing blood capillaries, a lymphatic capillary called a **lacteal,** and nerve fibers. At their free surfaces, the epithelial cells have many fine extensions called *microvilli* that form a brushlike border and greatly increase the surface area of the intestinal cells, enhancing absorption further (see figs. 17.3 and 17.37). The blood and lymph capillaries carry away absorbed nutrients, and impulses transmitted by the nerve fibers can stimulate or inhibit activities of the villus.

Between the bases of adjacent villi are tubular **intestinal glands** (crypts of Lieberkühn) that extend downward into the mucous membrane. The deeper layers of the small intestinal wall are much like those of other parts of the alimentary canal in that they include a submucosa, a muscular layer, and a serosa.

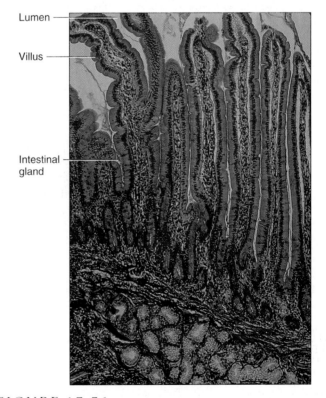

FIGURE 17.36
Light micrograph of intestinal villi from the wall of the duodenum (50×).

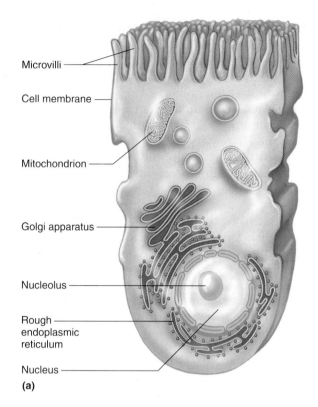

Microvilli

Cell membrane

Mitochondrion

Golgi apparatus

Nucleolus

Rough endoplasmic reticulum

Nucleus

(a)

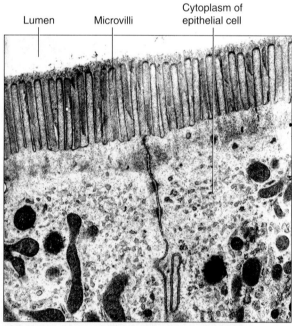

Lumen   Microvilli   Cytoplasm of epithelial cell

(b)

FIGURE 17.37

Intestinal epithelium. (*a*) Microvilli increase the surface area of intestinal epithelial cells. (*b*) Transmission electron micrograph of microvilli (16,000×).

The lining of the small intestine has many circular folds of mucosa, called *plicae circulares,* that are especially well developed in the lower duodenum and upper jejunum (fig. 17.38). Together with the villi and micro-

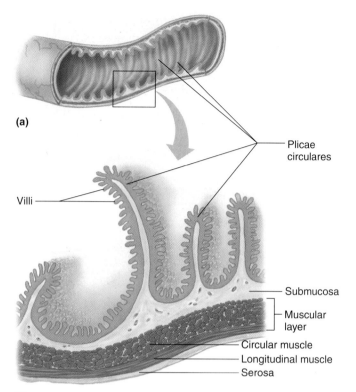

(a)

Plicae circulares

Villi

Submucosa

Muscular layer

Circular muscle

Longitudinal muscle

Serosa

(b)

FIGURE 17.38

Section of small intestine. (*a*) The inner lining of the small intestine contains many circular folds, the plicae circulares. (*b*) A longitudinal section through some of these folds.

villi, these folds help increase the surface area of the intestinal lining.

Secretions of the Small Intestine

In addition to the mucous-secreting goblet cells, which are abundant throughout the mucosa of the small intestine, many specialized *mucous-secreting glands* (Brunner's glands) are in the submucosa within the proximal portion of the duodenum. These glands secrete large quantities of viscid, alkaline mucus in response to certain stimuli.

> The epithelial cells that form the lining of the small intestine are continually replaced. New cells form within the intestinal glands by mitosis and migrate outward onto the villus surface. When the migrating cells reach the tip of the villus, they are shed. This *cellular turnover* renews the small intestine's epithelial lining every three to six days.

The intestinal glands at the bases of the villi secrete abundant watery fluid (see fig. 17.35). The villi rapidly reabsorb this fluid, which carries digestive products into the villi. The fluid the intestinal glands secrete has a pH that is nearly neutral (6.5–7.5), and it lacks digestive enzymes. However, the epithelial cells of the intestinal mucosa have digestive enzymes embedded in the membranes of the microvilli on their luminal surfaces. These enzymes break down food molecules just before absorption takes place. They include **peptidases,** which split peptides into their constituent amino acids; **sucrase, maltase,** and **lactase,** which split the disaccharides sucrose, maltose, and lactose into monosaccharides glucose, fructose, and galactose; and **intestinal lipase,** which splits fats into fatty acids and glycerol. Table 17.9 summarizes the sources and actions of the major digestive enzymes.

Many adults do not produce sufficient lactase to adequately digest lactose, or milk sugar. In this *lactose intolerance,* lactose remains undigested, increasing osmotic pressure of the intestinal contents and drawing water into the intestines. At the same time, intestinal bacteria metabolize undigested sugar, producing organic acids and gases. The overall result of lactose intolerance is bloating, intestinal cramps, and diarrhea. To avoid these symptoms, people with lactose intolerance can take lactase in pill form before eating dairy products. Infants with lactose intolerance can drink formula based on soybeans rather than milk. Genetic evidence suggests that lactose intolerance may be the "normal" condition, with the ability to digest lactose the result of a mutation that occurred recently in our evolutionary past. Because lactose intolerance usually affects adults, and is not seen in other primates, it may not have adversely affected prehistoric humans, who did not live long enough to experience symptoms.

Regulation of Small Intestinal Secretions

Since mucus protects the intestinal wall in the same way it protects the stomach lining, it is not surprising that mucous secretion increases in response to mechanical stimulation and the presence of irritants, such as gastric juice. Consequently, stomach contents entering the small intestine stimulate the duodenal mucous glands to release large quantities of mucus.

Goblet cells and intestinal glands secrete their products when they are stimulated by direct contact with chyme, which provides both chemical and mechanical stimulation. Distension of the intestinal wall activates the nerve plexuses within the wall and stimulates parasympathetic reflexes that also trigger release of small intestine secretions.

1. Describe the parts of the small intestine.

2. What is the function of an intestinal villus?

3. Distinguish between intestinal villi and microvilli.

4. How is surface area maximized in the small intestine?

5. What is the function of the intestinal glands?

6. List intestinal digestive enzymes.

Absorption in the Small Intestine

Because villi greatly increase the surface area of the intestinal mucosa, the small intestine is the most important absorbing organ of the alimentary canal. In fact, the small intestine is so effective in absorbing digestive products, water, and electrolytes, that very little absorbable material reaches its distal end.

TABLE 17.9	Summary of the Major Digestive Enzymes		
Enzyme	**Source**	**Digestive Action**	
Salivary Enzyme			
Amylase	Salivary glands	Begins carbohydrate digestion by breaking down starch and glycogen to disaccharides	
Gastric Enzymes			
Pepsin	Gastric glands	Begins protein digestion	
Lipase	Gastric glands	Begins butterfat digestion	
Pancreatic Enzymes			
Amylase	Pancreas	Breaks down starch and glycogen into disaccharides	
Lipase	Pancreas	Breaks down fats into fatty acids and glycerol	
Trypsin, chymotrypsin	Pancreas	Breaks down proteins or partially digested proteins into peptides	
Carboxypeptidase	Pancreas	Breaks down peptides into amino acids	
Nucleases	Pancreas	Breaks down nucleic acids into nucleotides	
Intestinal Enzymes			
Peptidase	Mucosal cells	Breaks down peptides into amino acids	
Sucrase, maltase, lactase	Mucosal cells	Breaks down disaccharides into monosaccharides	
Lipase	Mucosal cells	Breaks down fats into fatty acids and glycerol	
Enterokinase	Mucosal cells	Shortens trypsinogen into trypsin	

Carbohydrate digestion begins in the mouth with the activity of salivary amylase and is completed in the small intestine by enzymes from the intestinal mucosa and pancreas. The resulting monosaccharides are absorbed by the villi and enter blood capillaries (fig. 17.39). Monosaccharides are absorbed by facilitated diffusion or active transport (see chapter 3, pp. 81 and 85).

Protein digestion begins in the stomach as a result of pepsin activity and is completed in the small intestine by enzymes from the intestinal mucosa and the pancreas. During this process, large protein molecules are broken down into amino acids, which are then absorbed into the villi by active transport and are carried away by the blood (fig. 17.40).

Fat molecules are digested almost entirely by enzymes from the intestinal mucosa and pancreas (fig. 17.41). The resulting fatty acid molecules are absorbed in the following steps: (1) The fatty acid molecules dissolve in the epithelial cell membranes of the villi and diffuse through them. (2) The endoplasmic reticula of the cells use the fatty acids to resynthesize fat molecules similar to those previously digested. (3) These fats collect in clusters that become encased in protein. (4) The resulting large molecules of lipoprotein are called *chylomicrons,* and they make their way to the lacteal of the villus. (5) Periodic contractions of smooth muscles in the villus help empty the lacteal into the cysterna chyli (see fig. 16.6*a*), an expansion of the thoracic duct. The

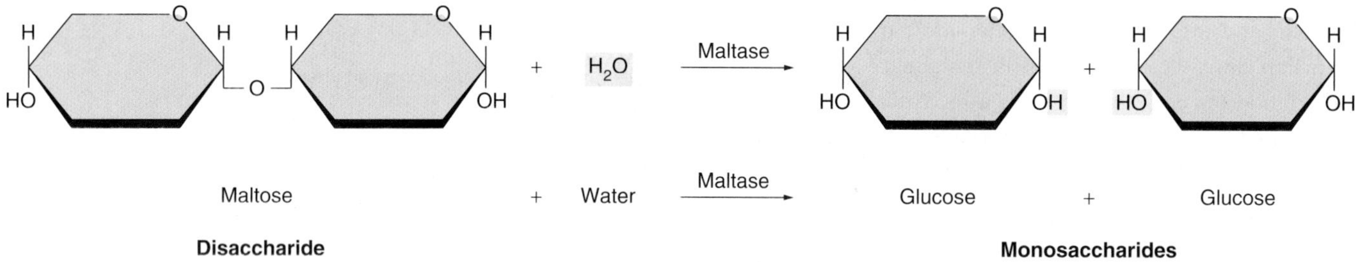

FIGURE 17.39

Digestion breaks down complex carbohydrates into disaccharides, which are then broken down into monosaccharides, which are small enough for intestinal villi to absorb. The monosaccharides then enter the bloodstream.

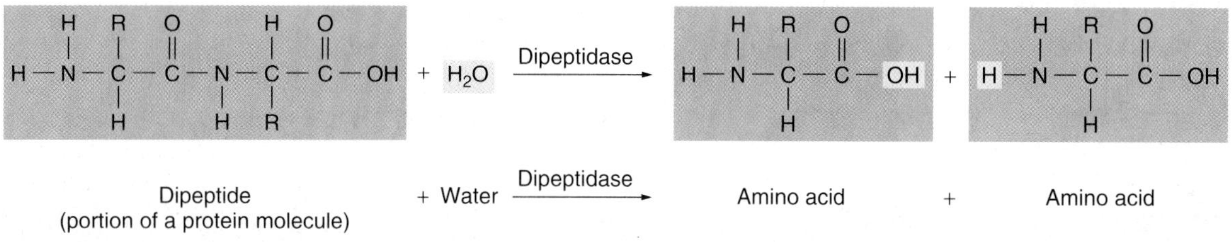

FIGURE 17.40

The amino acids that result from dipeptide digestion are absorbed by intestinal villi and enter the blood.

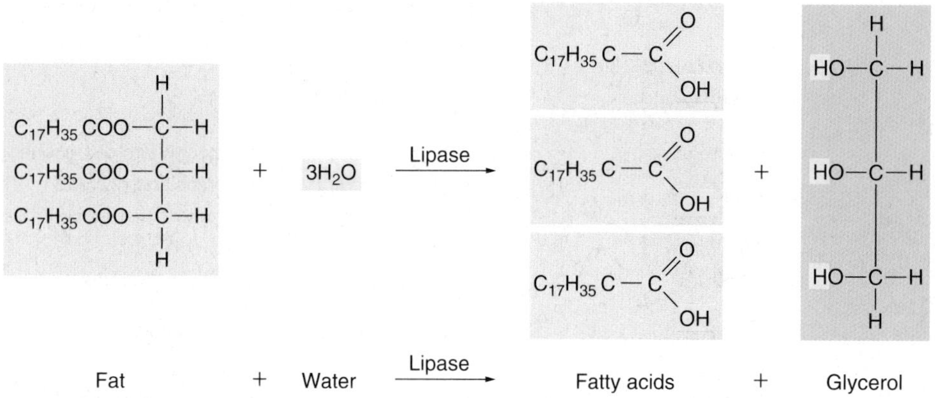

FIGURE 17.41

Fatty acids and glycerol result from fat digestion. Intestinal villi absorb them, and most are resynthesized into fat molecules before they enter the lacteals.

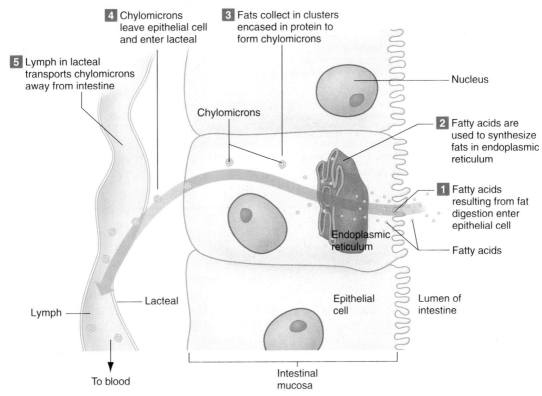

4 Chylomicrons leave epithelial cell and enter lacteal

3 Fats collect in clusters encased in protein to form chylomicrons

5 Lymph in lacteal transports chylomicrons away from intestine

Chylomicrons

Nucleus

2 Fatty acids are used to synthesize fats in endoplasmic reticulum

1 Fatty acids resulting from fat digestion enter epithelial cell

Endoplasmic reticulum

Fatty acids

Lymph

Lacteal

Epithelial cell

Lumen of intestine

To blood

Intestinal mucosa

FIGURE 17.42
Fat absorption has several steps.

lymph carries the chylomicrons to the bloodstream (fig. 17.42).

The blood transports chylomicrons to capillaries of muscle and adipose tissues. A specific type of protein (apoprotein) associated with the surface of chylomicrons activates an enzyme (lipoprotein lipase) that is attached to the inner lining of such capillaries. As a result of the enzyme's action, fatty acids and monoglycerides are released from the chylomicrons, and they enter muscle or adipose cells for use as energy sources or to be stored. The remnants of the chylomicrons travel in the blood to the liver, where they bind to receptors on the surface of liver cells. The remnants quickly enter liver cells by receptor-mediated endocytosis (see chapter 3, p. 87), and lysosomes degrade them. Some fatty acids with short carbon chains may be absorbed directly into the blood capillary of the villus without being converted back into fat.

In addition to absorbing the products of carbohydrate, protein, and fat digestion, the intestinal villi absorb electrolytes and water. Certain ions, such as those of sodium, potassium, chloride, nitrate, and bicarbonate, are readily absorbed; but others, including ions of calcium, magnesium, and sulfate, are poorly absorbed.

Electrolytes are usually absorbed by active transport, and water by osmosis. Thus, even though the intestinal contents may be hypertonic to the epithelial cells at first, as nutrients and electrolytes are absorbed, they become

---

Be sure to read the labels on cakes claiming to be fat free. Often these products contain diglycerides, rather than the triglycerides found in dietary fats. Once diglycerides are broken down to their constituent fatty acids during digestion and enter the small intestine epithelium, they are reformed into fat. So, yes, these cakes are technically no-fat, but the result after digestion may be the same as eating the real thing.

In *malabsorption,* the small intestine digests, but does not absorb, some nutrients. Causes of malabsorption include surgical removal of a portion of the small intestine, obstruction of lymphatic vessels due to a tumor, or interference with the production and release of bile as a result of liver disease.

Another cause of malabsorption is a reaction to *gluten,* found in certain grains, especially wheat and rye. This condition is called *celiac disease.* Microvilli are damaged, and in severe cases, villi may be destroyed. Both of these effects reduce the absorptive surface of the small intestine, preventing absorption of some nutrients. Symptoms of malabsorption include diarrhea, weight loss, weakness, vitamin deficiencies, anemia, and bone demineralization.

TABLE 17.10	Intestinal Absorption of Nutrients	
Nutrient	**Absorption Mechanism**	**Means of Transport**
Monosaccharides	Facilitated diffusion and active transport	Blood in capillaries
Amino acids	Active transport	Blood in capillaries
Fatty acids and glycerol	Facilitated diffusion of glycerol; diffusion of fatty acids into cells	
	(a) Most fatty acids are resynthesized back into fats and incorporated in chylomicrons for transport.	Lymph in lacteals
	(b) Some fatty acids with relatively short carbon chains are transported without being changed back into fats.	Blood in capillaries
Electrolytes	Diffusion and active transport	Blood in capillaries
Water	Osmosis	Blood in capillaries

slightly hypotonic to the cells. Then, water follows the nutrients and electrolytes into the villi by osmosis. Table 17.10 summarizes the absorption process.

1 Which substances resulting from digestion of carbohydrate, protein, and fat molecules does the small intestine absorb?

2 Which ions does the small intestine absorb?

3 What transport mechanisms do intestinal villi use?

4 Describe how fatty acids are absorbed and transported.

Movements of the Small Intestine

Like the stomach, the small intestine carries on mixing movements and peristalsis. The major mixing movement is called *segmentation,* in which small, ringlike contractions occur periodically, cutting the chyme into segments and moving it back and forth. Segmentation also slows the movement of chyme through the small intestine.

Peristaltic waves propel chyme through the small intestine. These waves are usually weak, and they stop after pushing the chyme a short distance. Consequently, chyme moves slowly through the small intestine, taking from three to ten hours to travel its length.

As might be expected, parasympathetic impulses enhance both mixing and peristaltic movements, and sympathetic impulses inhibit them. Reflexes involving parasympathetic impulses to the small intestine sometimes originate in the stomach. For example, food filling the stomach distends its wall, triggering a reflex (gastroenteric reflex) that greatly increases peristaltic activity in the small intestine. Another reflex is initiated when the duodenum fills with chyme, stretching its wall. This reflex speeds movement through the small intestine.

If the small intestine wall becomes overdistended or irritated, a strong *peristaltic rush* may pass along the entire length of the organ, sweeping chyme into the large intestine so quickly that water, nutrients, and electrolytes that would normally be absorbed are not. The result is *diarrhea,* a condition in which defecation becomes more frequent and the stools become watery. Prolonged diarrhea causes imbalances in water and electrolyte concentrations.

The **ileocecal sphincter** joins the small intestine's ileum to the large intestine's cecum. Normally, this sphincter remains constricted, preventing the contents of the small intestine from entering the large intestine, and at the same time keeping the contents of the large intestine from backing up into the ileum. However, eating a meal elicits a gastroileal reflex that increases peristalsis in the ileum and relaxes the sphincter, forcing some of the contents of the small intestine into the cecum.

1 Describe the movements of the small intestine.

2 How are the movements of the small intestine initiated?

3 What is a peristaltic rush?

4 What stimulus relaxes the ileocecal sphincter?

Large Intestine

The **large intestine** is so named because its diameter is greater than that of the small intestine. This portion of the alimentary canal is about 1.5 meters long, and it begins in the lower right side of the abdominal cavity where the ileum joins the cecum. From there, the large intestine ascends on the right side, crosses obliquely to the left, and descends into the pelvis. At its distal end, it opens to the outside of the body as the anus.

The large intestine absorbs water and electrolytes from the chyme remaining in the alimentary canal. It also reabsorbs and recycles water and remnants of digestive secretions. The large intestine also forms and stores feces.

Parts of the Large Intestine

The large intestine consists of the cecum, the colon, the rectum, and the anal canal. Figures 17.43 and 17.44 and reference plates 51, 52, 58, and 75 depict the large intestine.

The **cecum,** at the beginning of the large intestine, is a dilated, pouchlike structure that hangs slightly inferior to the ileocecal opening. Projecting downward from it is a narrow tube with a closed end called the **vermiform** (wormlike) **appendix.** The human appendix has no

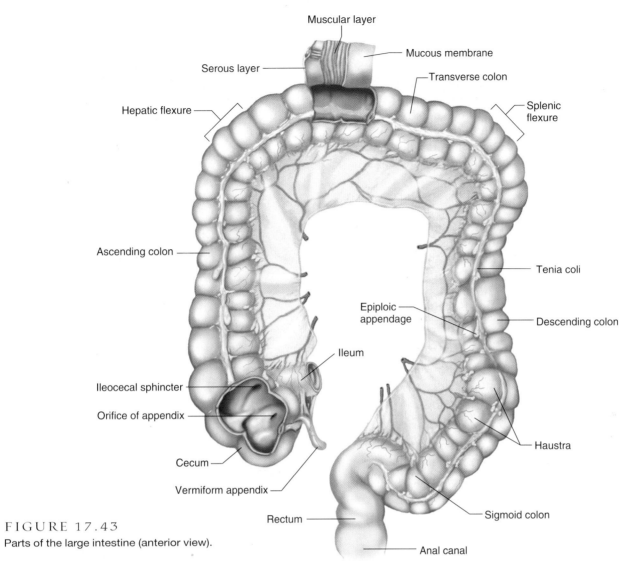

FIGURE 17.43
Parts of the large intestine (anterior view).

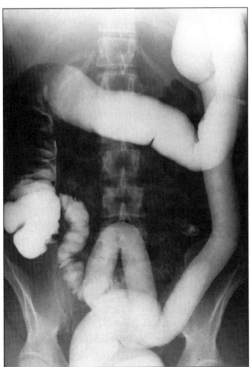

FIGURE 17.44
Radiograph of the large intestine containing a radiopaque substance that the patient ingested.

known digestive function. However, it contains lymphatic tissue.

In *appendicitis,* the appendix becomes inflamed and infected. Surgery is required to prevent the appendix from rupturing. If it breaks open, the contents of the large intestine may enter the abdominal cavity and cause a serious infection of the peritoneum called *peritonitis.*

The **colon** is divided into four portions—the ascending, transverse, descending, and sigmoid colons. The **ascending colon** begins at the cecum and extends upward against the posterior abdominal wall to a point just inferior to the liver. There it turns sharply to the left (as the right colic, or hepatic, flexure) and becomes the **transverse colon.** The transverse colon is the longest and most movable part of the large intestine. It is suspended by a fold of peritoneum and sags in the middle below the stomach. As the transverse colon approaches the spleen, it turns abruptly downward (as the left colic, or splenic, flexure) and becomes the **descending colon.** At the brim of the pelvis, the descending colon makes an S-shaped curve, called the **sigmoid colon,** and then becomes the rectum.

The **rectum** lies next to the sacrum and generally follows its curvature. It is firmly attached to the sacrum by the peritoneum, and it ends about 5 centimeters inferior to the tip of the coccyx, where it becomes the anal canal (fig. 17.45).

The **anal canal** is formed by the last 2.5 to 4.0 centimeters of the large intestine. The mucous membrane in the canal is folded into a series of six to eight longitudinal *anal columns.* At its distal end, the canal opens to the outside as the **anus.** Two sphincter muscles guard the anus—an *internal anal sphincter muscle,* composed of smooth muscle under involuntary control, and an *external anal sphincter muscle,* composed of skeletal muscle under voluntary control.

Hemorrhoids are, literally, a pain in the rear. Enlarged and inflamed branches of the rectal vein in the anal columns cause intense itching, sharp pain, and sometimes bright red bleeding. The hemorrhoids may be internal (which do not produce symptoms) or bulge out of the anus. Causes of hemorrhoids include anything that puts prolonged pressure on the delicate rectal tissue, including obesity, pregnancy, constipation, diarrhea, and liver disease.

Eating more fiber-rich foods and drinking lots of water can usually prevent or cure hemorrhoids. Warm soaks in the tub, cold packs, and careful wiping of painful areas also helps, as do external creams and ointments. Surgery—with a scalpel or a laser—can remove severe hemorrhoids.

---

1 What is the general function of the large intestine?

2 Describe the parts of the large intestine.

3 Distinguish between the internal sphincter muscle and the external sphincter muscle of the anus.

Structure of the Large Intestinal Wall

The wall of the large intestine includes the same types of tissues found in other parts of the alimentary canal but also has some unique features. The large intestinal wall lacks the villi and plicae circularis characteristic of the small intestine. Also, the layer of longitudinal muscle fibers does not uniformly cover its wall; the fibers are in three distinct bands (teniae coli) that extend the entire length of the colon. These bands exert tension lengthwise on the wall, creating a series of pouches (haustra). The large intestinal wall also has small collections of fat (epiploic appendages) in the serosa on its outer surface (fig. 17.46).

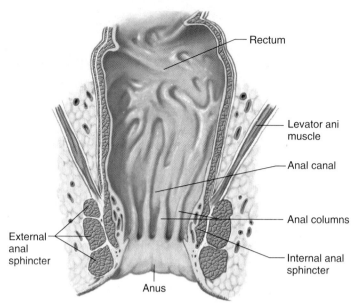

FIGURE 17.45
The rectum and the anal canal are at the distal end of the alimentary canal.

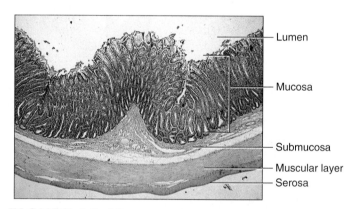

FIGURE 17.46
Light micrograph of the large intestinal wall (64×).

Functions of the Large Intestine

Unlike the small intestine, which secretes digestive enzymes and absorbs the products of digestion, the large intestine has little or no digestive function. However, the mucous membrane that forms the inner lining of the large intestine contains many tubular glands. Structurally, these glands are similar to those of the small intestine, but they are composed almost entirely of goblet cells. Consequently, mucus is the only significant secretion of this portion of the alimentary canal (fig. 17.47).

Mechanical stimulation from chyme and parasympathetic impulses control the rate of mucous secretion. In both cases, the goblet cells respond by increasing mucous production, which, in turn, protects the intestinal wall against the abrasive action of materials passing through it. Mucus also holds particles of fecal matter together, and, because it is alkaline, mucus helps control the pH of the

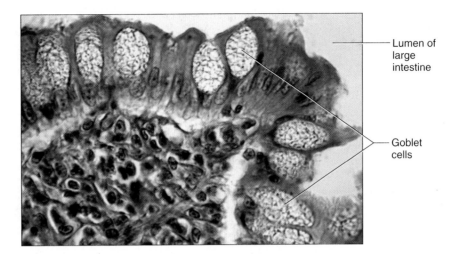

FIGURE 17.47

Light micrograph of the large intestinal mucosa (560×).

Labels: Lumen of large intestine; Goblet cells

large intestinal contents. This is important because acids are sometimes released from the feces as a result of bacterial activity.

Chyme entering the large intestine usually has few nutrients remaining in it and mostly consists of materials not digested or absorbed in the small intestine. It also contains water, electrolytes, mucus, and bacteria.

Intestinal gas is 99% nitrogen and oxygen taken in while breathing and eating, plus methane (CH_4), carbon dioxide (CO_2), and hydrogen contributed from the bacterial fermentation of undigested food. The characteristic odor comes from bacterial action on the nitrogen and sulfur in proteins, which yields pungent-smelling ammonia (NH_3) and foul hydrogen sulfide (H_2S). Most people release a half liter of intestinal gas a day. Foods rich in sulfur-containing amino acids make intestinal gas more foul. These include beans, broccoli, bran, brussels sprouts, cabbage, cauliflower, and onions.

Absorption in the large intestine is normally limited to water and electrolytes, and this usually occurs in the proximal half of the tube. Electrolytes such as sodium ions can be absorbed by active transport, while the water follows passively, entering the mucosa by osmosis. As a result, about 90% of the water that enters the large intestine is absorbed, and little sodium or water is lost in the feces.

The many bacteria that normally inhabit the large intestine, called *intestinal flora,* break down some of the molecules that escape the actions of human digestive enzymes. For instance, cellulose, a complex carbohydrate in food of plant origin, passes through the alimentary canal almost unchanged, but colon bacteria can break down cellulose and use it as an energy source. These bacteria, in turn, synthesize certain vitamins, such as K, B_{12}, thiamine, and riboflavin, which the intestinal mucosa

absorbs. Bacterial actions in the large intestine may produce intestinal gas (flatus).

1 How does the structure of the large intestine differ from that of the small intestine?

2 What substances does the large intestine absorb?

3 What useful substances do bacteria inhabiting the large intestine produce?

Hirschsprung disease causes extreme, chronic constipation and abdominal distension. The part of the large intestine distal to the distension lacks innervation. As a result, the person does not feel the urge to defecate. The problem begins in the embryo, when an abnormal gene prevents neurons from migrating to this portion of the gastrointestinal tract. Surgery may be used to treat Hirschsprung disease, which was once lethal.

Movements of the Large Intestine

The movements of the large intestine—mixing and peristalsis—are similar to those of the small intestine, although usually slower. The mixing movements break the fecal matter into segments and turn it so that all portions are exposed to the intestinal mucosa. This helps absorb water and electrolytes.

The peristaltic waves of the large intestine are different from those of the small intestine. Instead of occurring frequently, they happen only two or three times each day. These waves produce *mass movements* in which a large section of the intestinal wall constricts vigorously, forcing the intestinal contents to move toward the rectum. Typically, mass movements follow a meal, as a result of the gastrocolic reflex. Irritation of the intestinal mucosa can also trigger such movements. For instance, a person suffering from an inflamed colon (colitis) may experience frequent mass movements.

DISORDERS OF THE LARGE INTESTINE

The large intestine (colon) is the source of familiar digestive discomforts as well as more serious disorders.

Diverticulosis and Inflammatory Bowel Disease

In diverticulosis, parts of the intestinal wall weaken, and the inner mucous membrane protrudes through (fig. 17B). If chyme accumulates in the outpouching and becomes infected (diverticulitis), antibiotics or surgical removal of the area may become necessary. Lack of dietary fiber may set the stage for diverticulosis. The condition does not occur in populations that eat high fiber diets, and began to appear in the United States only after refined foods were introduced in the middle of the twentieth century.

Inflammatory bowel disease includes *ulcerative colitis* and *Crohn disease*. Ulcerative colitis is an extensive inflammation and ulceration of the large intestine, affecting primarily the mucosa and submucosa. The resulting bouts of bloody diarrhea and cramps may last for days or weeks and may recur frequently or only very rarely. The severe diarrhea leads to weight loss and electrolyte imbalances and may develop into colon cancer or affect other organs, including the skin, eyes, or liver. Crohn disease is similar to ulcerative colitis in that it produces diarrhea and abdominal cramps, but the diarrhea may not be bloody, and complications such as cancer are atypical. Surgery is often used to treat inflammatory bowel disease. The cause or causes of inflammatory bowel disease aren't known with certainty but may involve autoimmunity, a genetic predisposition, or a bacterial infection.

Colorectal Cancer

Cancer of the large intestine or rectum, known as *colorectal cancer,* is the fourth most prevalent cancer in the United States and the second most common cause of cancer death. More than 30,000 new cases are diagnosed each year, and more than 56,000 people die of the condition. Put another way, an individual faces a one in eighteen lifetime risk of developing colorectal cancer. However, the condition does tend to run in families.

Symptoms of colorectal cancer include

- a change in frequency or consistency of bowel movements
- blood in the feces
- a narrowing of feces
- abdominal discomfort or pain
- weight loss
- fatigue
- unexplained vomiting

Several diagnostic tests can detect colorectal cancer, described in figure 17C and table 17A. These tests are of two general types—looking for clues in feces or exploring the large intestine with an endoscopic device. The key to saving more lives is to know whom to test. For example, the more invasive procedures are usually reserved for individuals with strong family histories of colorectal cancer. But an experiment that performed colonoscopy on 3,121 healthy volunteers without strong family histories of

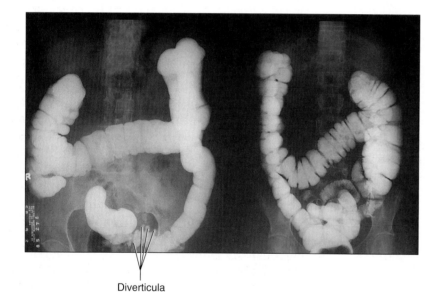

Diverticula

FIGURE 17B

Diverticulosis of the large intestine. The large intestine on the right is healthy. The organ on the left shows diverticula on the left side.

When it is appropriate to defecate, a person usually can initiate a *defecation reflex* by holding a deep breath and contracting the abdominal wall muscles. This action increases the internal abdominal pressure and forces feces into the rectum. As the rectum fills, its wall is distended and the defecation reflex is triggered, stimulating peristaltic waves in the descending colon, and the internal anal sphincter relaxes. At the same time, other reflexes involving the sacral region of the spinal cord strengthen the peristaltic waves, lower the diaphragm, close the glottis, and contract the abdominal wall muscles. These actions additionally increase the internal abdominal pressure and squeeze the rectum. The external anal sphincter is signaled to relax, and the feces are

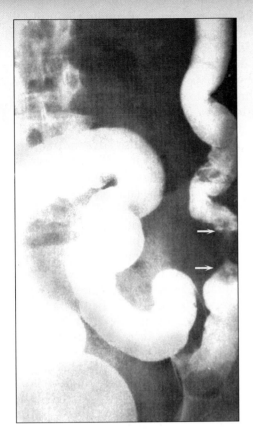

FIGURE 17C

One diagnostic test for colon cancer is a double-contrast barium enema. The barium highlights the lower digestive tract, revealing in this radiograph an obstruction caused by a tumor.

TABLE 17A	Diagnostic Tests for Colorectal Cancer
Diagnostic Test	**Description**
Digital rectal exam	Physician palpates large intestine and rectum.
Double-contrast barium enema	X-ray exam following ingestion of contrast agent highlights blockages in large intestine
Fecal occult blood test	Blood detected in feces sample
Colorectal cancer gene test (experimental)	Mutations associated with colorectal cancer detected in DNA of cells shed with feces
Sigmoidoscopy	Endoscope views rectum and lower colon
Colonoscopy	Endoscope views rectum and entire colon

the illness revealed that 1% of them had advanced colorectal cancer! More than a third had some abnormal growths in their colons. Therefore, existing screening protocols may miss some early cases.

An alternative to invasively screening the general population is to develop a fecal test that is more accurate than the standard *fecal occult blood test,* which has a 10% false positive rate. A more accurate fecal test screens the DNA from cells in feces for mutations that are associated with colorectal cancer. (A strictly inherited form of colorectal cancer entails mutations present in the sperm or egg; more common cancers arise from somatic mutations, which originate in cells of the affected tissues.) The fecal gene test for colorectal cancer will not replace colonoscopy—still the most accurate test—but will help physicians select patients for it.

Colorectal cancer develops gradually. First, a cell lining the large intestine begins to divide more frequently than others, and the accumulating cells enter a precancerous state. Next, the growth forms a polyp, which is benign. Months or even years later, the polyp becomes cancerous. Still other genetic changes control the cancer's spread.

Treatment for colorectal cancer is to remove the affected tissue. If a large portion of the intestine is removed, surgery is used to construct a new opening for feces to exit the body. The free end of the intestine is attached to an opening created through the skin of the abdomen, and a bag is attached to the opening to collect the fecal matter. This procedure is called a colostomy.

The future is bright for early detection and even prevention of colorectal cancer. Some people may be more likely than others to develop colorectal cancer because they inherit a susceptibility gene. If a susceptibility test can be developed, individuals can lower their risk by eating high-fiber foods and exercising.

A new class of nonsteroidal anti-inflammatory drugs (NSAIDs) called *Cox-2 inhibitors* may help to prevent colorectal cancer. These drugs block the enzyme cyclooxygenase-2, which is necessary to convert a substance in cell membranes (arachidonic acid) into prostaglandins, which cause inflammation. The drugs were initially approved to treat arthritis. Researchers discovered that people who take older NSAIDs, such as daily aspirin, have half the normal incidence of intestinal polyps. Cox-2 inhibitors were tested in people known to have genes that predispose them to developing colorectal cancer. (The older drugs also block a second enzyme which causes gastrointestinal cramping.) Cox-2 inhibitors are now routinely used to treat an inherited form of colon cancer (familial adenomatous polyposis). Clinical trials are underway to determine whether these drugs can prevent, or at least delay onset of, other forms of colorectal and other cancers. ■

forced to the outside. A person can voluntarily inhibit defecation by contracting the external anal sphincter.

Feces

Feces (fe′sēz) are composed of materials that were not digested or absorbed, along with water, electrolytes, mucus, and bacteria. Usually the feces are about 75% water, and their color derives from bile pigments altered by bacterial action.

The pungent odor of the feces results from a variety of compounds that bacteria produce. These compounds include phenol, hydrogen sulfide, indole, skatole, and ammonia. Clinical Application 17.5 examines conditions affecting the large intestine.

1 How does peristalsis in the large intestine differ from peristalsis in the small intestine?

2 List the major events that occur during defecation.

3 Describe the composition of feces.

Life-Span Changes

Changes to the digestive system that are associated with the passing years are slow and slight, so most people can enjoy eating a variety of foods as they grow older. Maintaining healthy teeth, of course, is vital to obtaining adequate nutrition. This requires frequent dental checkups, cleanings and plaque removal, plus care of the gums. Tooth loss due to periodontal disease becomes more likely after age thirty-five.

Despite regular dental care, some signs of aging may affect the teeth. The enamel often thins from years of brushing, teeth grinding, and eating acidic foods. Thinning enamel may make the teeth more sensitive to hot and cold foods. At the same time, the cementum may thicken. The dentin heals more slowly and enlarges as the pulp shrinks. Loss of neurons in the pulp may make it more difficult to be aware of tooth decay. The gums recede, creating more pockets to harbor the bacteria whose activity contributes to periodontal disease. The teeth may loosen as the bones of the jaw weaken. On a functional level, older people sometimes do not chew their food thoroughly, swallowing larger chunks of food that may present a choking hazard.

A common complaint of older individuals is "dry mouth," or xerostomia. This condition is not a normal part of aging—studies have shown that the oldest healthy people make just as much saliva as healthy younger people. Dry mouth is common, however, because it is a side effect of more than 400 medications, many of which are more likely to be taken by older persons. These include antidepressants, antihistamines, and drugs that treat cancer or hypertension. In addition, radiation and chemotherapy used to treat cancer can cause mouth sores and tooth decay. It is a good idea for cancer patients to coordinate dental visits with other aspects of their care.

Once past the mouth, food travels through a gastrointestinal tract that declines gradually in efficiency with age. A slowing of peristalsis may result in frequent heartburn as food backs up into the esophagus. The stomach lining thins with age, and secretion of hydrochloric acid, pepsin, and intrinsic factor decline. Exit of chyme from the stomach slows. Overall, these changes may affect the rate at which certain medications are absorbed.

Because the small intestine is the site of absorption of nutrients, it is here that noticeable signs of aging on digestion arise. Subtle shifts in the microbial species that inhabit the small intestine alter the rates of absorption of particular nutrients. With age, the small intestine becomes less efficient at absorbing vitamins A, D, and K and the mineral zinc. This raises the risk of deficiency symptoms—effects on skin and vision due to a lack of vitamin A; weakened bones from inadequate vitamin D; impaired blood clotting seen in vitamin K deficiency; and slowed healing, decreased immunity, and altered taste evidenced in zinc deficiency.

Many people who have inherited lactose intolerance begin to notice the telltale cramping after eating dairy foods in the middle years. They must be careful that by avoiding dairy products, they do not also lower their calcium intake. Less hydrochloric acid also adversely affects the absorption of calcium, as well as iron. Too little intrinsic factor may lead to vitamin B_{12} deficiency anemia.

The lining of the large intestine changes too, thinning and containing less smooth muscle and mucus. A dampening of the responsiveness of the smooth muscle to neural stimulation slows peristalsis, ultimately causing constipation. Compounding this common problem is a loss of elasticity in the walls of the rectum and declining strength and responsiveness of the internal and external sphincters.

The accessory organs to digestion age too, but not necessarily in ways that affect health. Both the pancreas and the liver are large organs with cells to spare, so a decline in their secretion abilities does not usually hamper digestion. Only 10% of the pancreas and 20% of the liver are required to digest foods. However, the liver may not be able to detoxify certain medications as quickly as it once did. The gallbladder becomes less sensitive to cholecystokinin, but in a classic feedback response, cells of the intestinal mucosa secrete more of it into the bloodstream, and the gallbladder continues to be able to contract. The bile ducts widen in some areas, but the end of the common bile duct narrows as it approaches the small intestine. As long as gallstones do not become entrapped in the ducts, the gallbladder generally functions well into the later years.

1 Describe the effects of aging on the teeth.

2 What conditions might be caused by the slowing of peristalsis in the digestive tract that occurs with aging?

Clinical Terms Related to the Digestive System

achalasia (ak″ah-la′ze-ah) Failure of the smooth muscle to relax at some junction in the digestive tube, such as that between the esophagus and stomach.
achlorhydria (ah″klōr-hi′dre-ah) Lack of hydrochloric acid in gastric secretions.
aphagia (ah-fa′je-ah) Inability to swallow.
cholecystitis (ko″le-sis-ti′tis) Inflammation of the gallbladder.

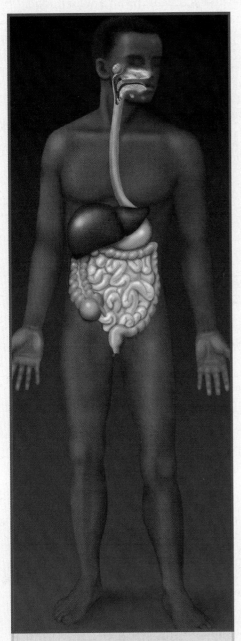

DIGESTIVE SYSTEM

The digestive system ingests, digests, and absorbs nutrients for use by all body cells.

Integumentary System

Vitamin D activated in the skin plays a role in absorption of calcium from the digestive tract.

Cardiovascular System

The bloodstream carries absorbed nutrients to all body cells.

Skeletal System

Bones are important in mastication. Calcium absorption is necessary to maintain bone matrix.

Lymphatic System

The lymphatic system plays a major role in the absorption of fats.

Muscular System

Muscles are important in mastication, swallowing, and the mixing and moving of digestion products through the gastrointestinal tract.

Respiratory System

The digestive system and the respiratory system share common anatomical structures.

Nervous System

The nervous system can influence digestive system activity.

Urinary System

The kidneys and liver work together to activate vitamin D.

Endocrine System

Hormones can influence digestive system activity.

Reproductive System

In a woman, nutrition is essential for conception and normal development of an embryo and fetus.

cholelithiasis (ko″le-li-thi′ah-sis) Stones in the gallbladder.

cholestasis (ko″le-sta′sis) Blockage in bile flow from the gallbladder.

cirrhosis (si-ro′sis) Liver condition in which the hepatic cells degenerate and the surrounding connective tissues thicken.

diverticulitis (di″ver-tik″u-li′tis) Inflammation of small pouches (diverticula) that form in the lining and wall of the colon.

dumping syndrome (dum′ping sin′drōm) Symptoms, including diarrhea, that often occur following a gastrectomy.

dysentery (dis′en-ter″e) Intestinal infection, caused by viruses, bacteria, or protozoans, that causes diarrhea and cramps.

dyspepsia (dis-pep′se-ah) Indigestion; difficulty in digesting a meal.

dysphagia (dis-fa′je-ah) Difficulty in swallowing.

enteritis (en″tĕ-ri′tis) Inflammation of the intestine.

esophagitis (e-sof″ah-ji′tis) Inflammation of the esophagus.

gastrectomy (gas-trek′to-me) Partial or complete removal of the stomach.

gastrostomy (gas-tros′to-me) The creation of an opening in the stomach wall through which food and liquids may be administered when swallowing is not possible.

glossitis (glŏ-si′tis) Inflammation of the tongue.

ileitis (il″e-i′tis) Inflammation of the ileum.

ileus (il′e-us) Obstruction of the intestine due to an inhibition of motility or a mechanical cause.

pharyngitis (far″in-ji′tis) Inflammation of the pharynx.

pylorospasm (pi-lor′o-spazm) Spasm of the pyloric portion of the stomach or of the pyloric sphincter.

pyorrhea (pi″o-re′ah) Inflammation of the dental periosteum with pus formation.

stomatitis (sto″mah-ti′tis) Inflammation of the lining of the mouth.

C H A P T E R   S U M M A R Y

Introduction (page 644)

Digestion is the process of mechanically and chemically breaking down foods so that they can be absorbed. The digestive system consists of an alimentary canal and several accessory organs that carry out the processes of ingestion, propulsion, digestion, absorption, and defecation.

General Characteristics of the Alimentary Canal (page 644)

Regions of the alimentary canal perform specific functions.

1. Structure of the wall
   a. The wall consists of four layers.
   b. These layers include the mucosa, submucosa, muscular layer, and serosa.
2. Movements of the tube
   a. Motor functions include mixing and propelling movements.
   b. Peristalsis is responsible for propelling movements.
   c. The wall of the tube undergoes receptive relaxation just ahead of a peristaltic wave.
3. Innervation of the tube
   a. The tube is innervated by branches of the sympathetic and parasympathetic divisions of the autonomic nervous system.
   b. Parasympathetic impulses generally increase digestive activities; sympathetic impulses generally inhibit digestive activities.
   c. Sympathetic impulses contract certain sphincter muscles, controlling movement through the alimentary canal.

Mouth (page 648)

The mouth is adapted to receive food and begin preparing it for digestion. It also serves as an organ of speech and sensory perception.

1. Cheeks and lips
   a. Cheeks form the lateral walls of the mouth.
   b. Lips are highly mobile and possess a variety of sensory receptors useful in judging the characteristics of food.
2. Tongue
   a. The tongue is a thick, muscular organ that mixes food with saliva and moves it toward the pharynx.
   b. The rough surface of the tongue handles food and contains taste buds.
   c. Lingual tonsils are located on the root of the tongue.
3. Palate
   a. The palate comprises the roof of the mouth and includes hard and soft portions.
   b. The soft palate, including the uvula, closes the opening to the nasal cavity during swallowing.
   c. Palatine tonsils are located on either side of the tongue in the back of the mouth.
   d. Tonsils consist of lymphatic tissues.
4. Teeth
   a. Two sets of teeth develop in sockets of the mandibular and maxillary bones.
   b. There are twenty primary and thirty-two secondary teeth.
   c. Teeth mechanically break food into smaller pieces, increasing the surface area exposed to digestive actions.
   d. Different kinds of teeth are adapted to handle foods in different ways, such as biting, grasping, or grinding.
   e. Each tooth consists of a crown and root and is composed of enamel, dentin, pulp, nerves, and blood vessels.
   f. A tooth is attached to the alveolar process by collagenous fibers of the periodontal ligament.

Salivary Glands (page 652)

Salivary glands secrete saliva, which moistens food, helps bind food particles, begins chemical digestion of carbohydrates, makes taste possible, helps cleanse the mouth, and regulates pH in the mouth.

1. Salivary secretions
   a. Salivary glands include serous cells that secrete digestive enzymes and mucous cells that secrete mucus.
   b. Parasympathetic impulses stimulate the secretion of serous fluid.
2. Major salivary glands
   a. The parotid glands are the largest, and they secrete saliva rich in amylase.
   b. The submandibular glands in the floor of the mouth produce viscous saliva.
   c. The sublingual glands in the floor of the mouth primarily secrete mucus.

Pharynx and Esophagus (page 655)

The pharynx and esophagus serve as passageways.

1. Structure of the pharynx
   a. The pharynx is divided into a nasopharynx, oropharynx, and laryngopharynx.
   b. The muscular walls of the pharynx contain fibers in circular and longitudinal groups.
2. Swallowing mechanism
   a. Swallowing occurs in three stages.
      (1) Food is mixed with saliva and forced into the pharynx.
      (2) Involuntary reflex actions move the food into the esophagus.
      (3) Peristalsis transports food to the stomach.
   b. Swallowing reflexes momentarily inhibit breathing.
3. Esophagus
   a. The esophagus passes through the mediastinum and penetrates the diaphragm.
   b. Circular muscle fibers at the distal end of the esophagus help prevent regurgitation of food from the stomach.

Stomach (page 658)

The stomach receives food, mixes it with gastric juice, carries on a limited amount of absorption, and moves food into the small intestine.

1. Parts of the stomach
   a. The stomach is divided into cardiac, fundic, body, and pyloric regions.
   b. The lower esophageal sphincter serves as a valve between the esophagus and the stomach.
   c. The pyloric sphincter serves as a valve between the stomach and the small intestine.
2. Gastric secretions
   a. Gastric glands secrete gastric juice.
   b. Gastric juice contains pepsin, hydrochloric acid, lipase, and intrinsic factor.
3. Regulation of gastric secretions
   a. Parasympathetic impulses and the hormone gastrin enhance gastric secretion.
   b. The three stages of gastric secretion are the cephalic, gastric, and intestinal phases.
   c. The presence of food in the small intestine reflexly inhibits gastric secretions.
4. Gastric absorption
   a. The stomach is not well-adapted for absorption.
   b. A few substances such as water and other small molecules are absorbed through the stomach wall.
5. Mixing and emptying actions
   a. As the stomach fills, its wall stretches, but its internal pressure remains unchanged.
   b. Mixing movements aid in producing chyme; peristaltic waves move chyme into the pyloric region.
   c. The muscular wall of the pyloric region regulates chyme movement into the small intestine.
   d. The rate of emptying depends on the fluidity of the chyme and the type of food present.
   e. The upper part of the small intestine fills, and an enterogastric reflex inhibits peristalsis in the stomach.
   f. Vomiting results from a complex reflex that has many stimuli.

Pancreas (page 664)

The pancreas is closely associated with the duodenum.

1. Structure of the pancreas
   a. It produces pancreatic juice that is secreted into a pancreatic duct.
   b. The pancreatic duct leads to the duodenum.
2. Pancreatic juice
   a. Pancreatic juice contains enzymes that can split carbohydrates, proteins, fats, and nucleic acids.
   b. Pancreatic juice has a high bicarbonate ion concentration that helps neutralize chyme and causes the intestinal contents to be alkaline.
3. Regulation of pancreatic secretion
   a. Secretin from the duodenum stimulates the release of pancreatic juice that contains few digestive enzymes but has a high bicarbonate ion concentration.
   b. Cholecystokinin from the intestinal wall stimulates the release of pancreatic juice that has a high concentration of digestive enzymes.

Liver (page 666)

The liver is located in the upper right quadrant of the abdominal cavity.

1. Liver structure
   a. The liver is a highly vascular organ, enclosed in a fibrous capsule, and divided into lobes.
   b. Each lobe consists of hepatic lobules, the functional units of the liver.
   c. Bile from the lobules is carried by bile canals to hepatic ducts that unite to form the common bile duct.
2. Liver functions
   a. The liver has many functions. It metabolizes carbohydrates, lipids, and proteins; stores some substances; filters blood; destroys toxins; and secretes bile.
   b. Bile is the only liver secretion that directly affects digestion.
3. Composition of bile
   a. Bile contains bile salts, bile pigments, cholesterol, and electrolytes.
   b. Only the bile salts have digestive functions.
   c. Bile pigments are products of red blood cell breakdown.
4. Gallbladder
   a. The gallbladder stores bile between meals.
   b. A sphincter muscle controls release of bile from the common bile duct.
   c. Gallstones may form within the gallbladder.
5. Regulation of bile release
   a. Cholecystokinin from the small intestine stimulates bile release.
   b. The sphincter muscle at the base of the common bile duct relaxes as a peristaltic wave in the duodenal wall approaches.

6. Functions of bile salts
    a. Bile salts emulsify fats and aid in the absorption of fatty acids, cholesterol, and certain vitamins.
    b. Bile salts are reabsorbed in the small intestine.

Small Intestine (page 673)

The small intestine extends from the pyloric sphincter to the large intestine. It receives secretions from the pancreas and liver, completes digestion of nutrients, absorbs the products of digestion, and transports the residues to the large intestine.

1. Parts of the small intestine
    a. The small intestine consists of the duodenum, jejunum, and ileum.
    b. The small intestine is suspended from the posterior abdominal wall by mesentery.
2. Structure of the small intestinal wall
    a. The wall is lined with villi that greatly increase the surface area and aid in mixing and absorption.
    b. Microvilli on the free ends of epithelial cells increase the surface area even more.
    c. Intestinal glands are located between the villi.
    d. Circular folds in the lining of the intestinal wall also increase its surface area.
3. Secretions of the small intestine
    a. Intestinal glands secrete a watery fluid that lacks digestive enzymes but provides a vehicle for moving chyme to the villi.
    b. Digestive enzymes embedded in the surfaces of microvilli split molecules of sugars, proteins, and fats.
4. Regulation of small intestinal secretions
    a. Secretion is stimulated by gastric juice, chyme, and reflexes stimulated by distension of the small intestinal wall.
5. Absorption in the small intestine
    a. Blood capillaries in the villi absorb monosaccharides, amino acids, fatty acids, and glycerol.
    b. Blood capillaries in the villi also absorb water and electrolytes.
    c. Fat molecules with longer chains of carbon atoms enter the lacteals of the villi; fatty acids with short carbon chains enter the blood capillaries of the villi.
6. Movements of the small intestine
    a. Movements include mixing by segmentation and peristalsis.
    b. Overdistension or irritation may stimulate a peristaltic rush and result in diarrhea.

c. The ileocecal sphincter controls movement of the intestinal contents from the small intestine into the large intestine.

Large Intestine (page 680)

The large intestine absorbs water and electrolytes and forms and stores feces.

1. Parts of the large intestine
    a. The large intestine consists of the cecum, colon, rectum, and anal canal.
    b. The colon is divided into ascending, transverse, descending, and sigmoid portions.
2. Structure of the large intestinal wall
    a. The large intestine wall resembles the wall in other parts of the alimentary canal.
    b. The large intestine wall has a unique layer of longitudinal muscle fibers arranged in distinct bands.
3. Functions of the large intestine
    a. The large intestine has little or no digestive function, although it secretes mucus.
    b. Mechanical stimulation and parasympathetic impulses control the rate of mucous secretion.
    c. The large intestine absorbs water and electrolytes.
    d. Many bacteria inhabit the large intestine, where they break down certain undigestible substances and synthesize certain vitamins.
4. Movements of the large intestine
    a. Movements are similar to those in the small intestine.
    b. Mass movements occur two to three times each day.
    c. A reflex stimulates defecation.
5. Feces
    a. The large intestine forms and stores feces.
    b. Feces consist of water, undigested material, mucus, and bacteria.
    c. The color of feces is due to bile pigments that have been altered by bacterial action.

Life-Span Changes (page 686)

1. Older people sometimes do not chew food thoroughly because thinning enamel makes teeth more sensitive to hot and cold foods, gums recede, and teeth may loosen.
2. Slowing peristalsis in the digestive tract may cause heartburn and constipation.
3. Aging affects nutrient absorption in the small intestine.
4. Accessory organs to digestion also age, but not necessarily in ways that affect health.

CRITICAL THINKING QUESTIONS

1. How would removal of 95% of the stomach (subtotal gastrectomy) to treat severe ulcers or cancer affect the digestion and absorption of foods? How would the patients have to alter eating habits? Why?
2. Why may a person with inflammation of the gallbladder (cholecystitis) also develop an inflammation of the pancreas (pancreatitis)?
3. What effect is a before-dinner alcoholic cocktail likely to have on digestion? Why are such beverages inadvisable for persons with ulcers?
4. What type of acid-alkaline disorder is likely to develop if the stomach contents are repeatedly lost by vomiting over a prolonged period? What acid-alkaline disorder may develop as a result of prolonged diarrhea?
5. Several years ago, an extract from kidney beans was sold in health-food stores as a "starch blocker." Advertisements claimed that one could eat a plate of spaghetti, yet absorb none of it, because starch-digesting enzyme function would be blocked. The kidney bean product indeed kept salivary amylase from functioning. However, people who took the starch blocker developed abdominal pain, bloating, and gas. Suggest a reason for the ill effects of the supposed starch blocker.

1. List and describe the locations of the major parts of the alimentary canal.
2. List and describe the locations of the accessory organs of the digestive system.
3. Name the four layers of the wall of the alimentary canal.
4. Distinguish between mixing movements and propelling movements.
5. Define *peristalsis*.
6. Explain the relationship between peristalsis and receptive relaxation.
7. Describe the general effects of parasympathetic and sympathetic impulses on the alimentary canal.
8. Discuss the functions of the mouth and its parts.
9. Distinguish among lingual, palatine, and pharyngeal tonsils.
10. Compare the primary and secondary teeth.
11. Explain how the various types of teeth are adapted to perform specialized functions.
12. Describe the structure of a tooth.
13. Explain how a tooth is anchored in its socket.
14. List and describe the locations of the major salivary glands.
15. Explain how the secretions of the salivary glands differ.
16. Discuss the digestive functions of saliva.
17. Name and locate the three major regions of the pharynx.
18. Describe the mechanism of swallowing.
19. Explain the function of the esophagus.
20. Describe the structure of the stomach.
21. List the enzymes in gastric juice, and explain the function of each enzyme.
22. Explain how gastric secretions are regulated.
23. Describe the mechanism that controls the emptying of the stomach.

24. Describe the enterogastric reflex.
25. Explain the mechanism of vomiting.
26. Describe the locations of the pancreas and the pancreatic duct.
27. List the enzymes in pancreatic juice, and explain the function of each enzyme.
28. Explain how pancreatic secretions are regulated.
29. Describe the structure of the liver.
30. List the major functions of the liver.
31. Describe the composition of bile.
32. Trace the path of bile from a bile canal to the small intestine.
33. Explain how gallstones form.
34. Define *cholecystokinin*.
35. Explain the functions of bile salts.
36. List and describe the locations of the parts of the small intestine.
37. Name the enzymes of the intestinal mucosa, and explain the function of each enzyme.
38. Explain the regulation of the secretions of the small intestine.
39. Describe the functions of the intestinal villi.
40. Summarize how each major type of digestive product is absorbed.
41. Explain control of the movement of the intestinal contents.
42. List and describe the locations of the parts of the large intestine.
43. Explain the general functions of the large intestine.
44. Describe the defecation reflex.
45. What are the effects of altered rates of absorption, due to aging, in the small intestine?
46. How does digestive function change with age?

Visit the Student Edition of the Online Learning Center at www.mhhe.com/shier10 **for answers to chapter questions, additional quizzes, interactive learning exercises, and other study tools.**

bas-, base: *bas*al metabolic rate—metabolic rate of body under resting (basal) conditions.

calor-, heat: *calor*ie—unit used to measure heat or energy content of foods.

carot-, carrot: *carot*ene—yellowish plant pigment responsible for the color of carrots and other yellowish plant parts.

lip-, fat: *lip*ids—fat or fatlike substance insoluble in water.

mal-, bad, abnormal: *mal*nutrition—poor nutrition resulting from lack of food or failure to use available foods to best advantage.

-meter, measure: calori*meter*—instrument used to measure the caloric content of food.

nutri-, nourish: *nutri*ent—substance needed to nourish cells.

obes-, fat: *obes*ity—condition in which the body contains excess fat.

pell-, skin: *pell*agra—vitamin deficiency condition characterized by inflammation of the skin and other symptoms.

Polarized light micrograph of crystals of vitamin C (120×).

Nutrition and Metabolism

CHAPTER OBJECTIVES

After you have studied this chapter, you should be able to

1. Define *nutrition, nutrients,* and *essential nutrients.*

2. List the major sources of carbohydrates, lipids, and proteins.

3. Describe how cells utilize carbohydrates, lipids, and amino acids.

4. Define *nitrogen balance.*

5. Explain how the energy values of foods are determined.

6. Explain the factors that affect an individual's energy requirements.

7. Define *energy balance.*

8. Explain what is meant by desirable weight.

9. List the fat-soluble and water-soluble vitamins and summarize the general functions of each vitamin.

10. Distinguish between a vitamin and a mineral.

11. List the major minerals and trace elements and summarize the general functions of each.

12. Describe an adequate diet.

13. Distinguish between primary and secondary malnutrition.

14. List factors that may lead to inadequate nutrition later in life.

When gymnast Christy Henrich was buried on a Friday morning in July 1994, she weighed 61 pounds. Three weeks earlier, she had weighed an unbelievable 47 pounds. Between those two dates, she celebrated her twenty-second birthday.

Christy suffered from *anorexia nervosa,* a psychological disorder common among professional athletes in which the patient perceives herself as fat—even though she is obviously not. A year before her death, Christy told a newspaper reporter, "My life is a horrifying nightmare. It feels like there's a beast inside me, like a monster."

Christy's decline into self-starvation began in 1988, when a judge at a gymnastics competition told her she'd have to lose weight if she wanted to make the U.S. Olympic team. Christy weighed 90 pounds. From then on, with only a few respites, she ate nearly nothing, exercised many hours each day, and used laxatives to drop her weight lower and lower. Finally, it dropped so low that her vital organs could not function.

Eating disorders such as anorexia nervosa were once recognized almost exclusively among girls and young women. But the male of the species can have distorted body image too—and act on it. One survey of eight-year-old boys revealed that more than a third of them had attempted weight loss. A variation on the eating disorder theme increasingly being seen among boys and young men is called muscle dysmorphia or, more commonly, "bigorexia." It is the drive to appear muscular, and may entail taking amino acid food supplements. Without a doctor's guidance, altering nutrition so drastically to achieve a popular body form is dangerous. ■

World-class gymnast Christy Henrich died of complications of the self-starvation eating disorder anorexia nervosa in July 1994. In this photo, taken 11 months before her death, she weighed under 60 pounds. Concern over weight gain propelled her down the path of this deadly nutritional illness.

The human body requires fuel as well as materials to develop, grow, and heal. Nutrients from food fulfill these requirements. However, like many physiological processes, nutrition is very much a matter of balance. Too few nutrients, and disorders associated with malnutrition result. Too many nutrients, and obesity is the consequence.

Why We Eat

Nutrients (nu′tre-ents) are chemical substances supplied from the environment that an organism requires for survival. The **macronutrients,** needed in bulk, include carbohydrates, proteins, and fats. **Micronutrients** are essential in small daily doses and include vitamins and minerals. The body also requires water.

In countries with adequate food supplies, most healthy individuals can obtain nourishment by eating a variety of foods and limiting fat intake. People who do not eat meat products can also receive adequate nutrition but must pay more attention to food choices to avoid developing nutrient deficiencies. For example, eliminating red meat also means eliminating an excellent source of iron,

copper, zinc, and vitamin B_{12}. The fiber that often makes up much of a vegetarian's diet, although very healthful in many ways, also decreases absorption of iron. Therefore, a vegetarian must be careful to obtain sufficient iron from nonmeat sources. This is easily done, providing proper **nutrition** (adequate nutrients) when sources, actions, and interactions of nutrients are considered. Fortified foods, green leafy vegetables, and especially whole grains provide many of the nutrients also present in meat. Table 18.1 lists the different types of vegetarian diets.

TABLE 18.1	Types of Vegetarian Diets
Type	**Food Restrictions**
Vegan	No animal foods
Ovo-vegetarian	Eggs allowed; no dairy or meat
Lacto-vegetarian	Dairy allowed; no eggs or meat
Lacto-ovo-vegetarian	Dairy and eggs allowed; no meat
Pesco-vegetarian	Dairy, eggs, and fish allowed; no other meat
Semivegetarian	Dairy, eggs, chicken, and fish allowed; no other meat

Digestion breaks down nutrients to sizes that can be absorbed and transported in the bloodstream. **Metabolism** refers to the ways that nutrients are altered chemically and used in anabolism (building up or synthesis) and catabolism (breaking down) of chemical compounds to support the activities of life. (Chapter 4, pp. 113–114, introduced metabolism of carbohydrates.) Nutrients that human cells cannot synthesize, such as certain amino acids, are particularly important and are therefore called **essential nutrients.**

We eat to obtain the nutrients that power the activities of life. Eating is a complex, finely tuned homeostatic mechanism that balances nutrient intake with nutrient utilization. Several types of interacting hormones control appetite by affecting parts of the hypothalamus called the arcuate and paraventricular nuclei. These hormones can be classed by how quickly they exert their effects (Table 18.2).

Insulin, secreted from the pancreas, and **leptin,** secreted from adipocytes throughout the body, regulate fat stores in the long term. Insulin stimulates adipocytes and certain other cells to take up glucose, and promotes glucose molecules to link to form glycogen, a storage carbohydrate. Eating stimulates adipocytes to secrete leptin, which signals the hypothalamus to suppress appetite and increase metabolic rate—a negative feedback response to a meal. Bloodstream levels of leptin parallel increasing fat stores. Low leptin levels signal low fat stores and possible starvation. Metabolism slows to conserve energy, and appetite increases. Inherited leptin deficiency causes extreme obesity, because the body cannot sense its fat stores. Another appetite control molecule, melanocortin, is produced by neurons in the hypothalamus and suppresses appetite when weight is gained.

Cholecystokinin, secreted from the small intestine and **ghrelin,** produced in the stomach, work in the short term. Cholecystokinin signals satiety after eating, whereas ghrelin stimulates appetite, its level diminishing with satiety. Only recently discovered, ghrelin increases appetite and leads to obesity when infused into rodents, and causes short term hunger when given to people. Therefore, a compound that blocks ghrelin production or activity might help people to lose weight. The success of gastric bypass surgery in leading to weight loss may be due in part to decreasing ghrelin secretion that results from removing stomach tissue.

Endocrine cells in the small and large intestines secrete neuropeptide Y proteins, which act over an intermediate time period, peaking in the bloodstream between meals. Neuropeptide Y proteins integrate incoming information from leptin, insulin, cholecystokinin, and glucocorticoids, and may delay eating for up to 12 hours. Altogether, these hormones maintain homeostasis of lipid levels in the blood. Drug developers are focusing on these weight-control proteins in the never-ending search for obesity treatments.

Carbohydrates

Carbohydrates are organic compounds and include the sugars and starches. The energy held in their chemical bonds is used to power cellular processes.

Carbohydrate Sources

Carbohydrates are ingested in a variety of forms. Complex carbohydrates include the *polysaccharides,* which include starch from plant foods and glycogen from meats. Foods containing starch and glycogen usually have many other nutrients, including valuable vitamins and minerals. The simple carbohydrates include *disaccharides* from milk sugar, cane sugar, beet sugar, and molasses and *monosaccharides* from honey and fruits. Digestion breaks complex carbohydrates down to monosaccharides, which are small enough to be absorbed.

Cellulose is a complex carbohydrate that is abundant in our food—it provides the crunch to celery and the crispness to lettuce. We cannot digest cellulose, and most of it passes through the alimentary canal largely unchanged. However, cellulose provides bulk (also called fiber or roughage) against which the muscular wall of the digestive system can push, facilitating the movement of food. *Hemicellulose, pectin,* and *lignin* are other plant carbohydrates that provide fiber.

Carbohydrate Utilization

The monosaccharides that are absorbed from the digestive tract include *fructose, galactose,* and *glucose.* Liver enzymes convert fructose and galactose into glucose

TABLE 18.2	Substances That Control Appetite	
Substance	**Site of Secretion**	**Function**
Long term		
Insulin	Pancreas	Stimulates adipocytes to admit glucose and store fat; glycogen synthesis
Leptin	Adipocytes	Suppresses appetite and increases metabolic rate after eating
Melanocortin	Hypothalamic neurons	Senses high leptin levels that indicate weight gain; suppresses appetite
Intermediate		
Neuropeptide Y proteins	Small and large intestines	Integrates signals to time next meal
Short term		
Cholecystokinin	Small intestine	Decreases gastric activity as small intestine fills with food
Ghrelin	Stomach	Stimulates appetite

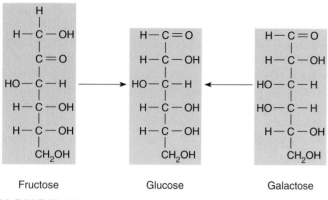

FIGURE 18.1

Liver enzymes catalyze reactions that convert the monosaccharides fructose and galactose into glucose.

Fructose   Glucose   Galactose

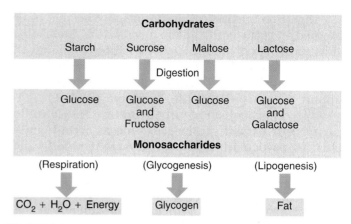

FIGURE 18.2

Monosaccharides from foods are used for energy, stored as glycogen, or reacted to produce fat.

(fig. 18.1). Recall that glucose is the form of carbohydrate that is most commonly oxidized in glycolysis for cellular fuel.

∞ R E C O N N E C T   T O   C H A P T E R   4 ,
C E L L U L A R   R E S P I R A T I O N ,
P A G E S   1 0 8 – 1 1 3 .

> Sugar substitutes provide concentrated sweetness, so fewer calories are needed to sweeten a food. Aspartame, which is a dipeptide (two joined amino acids), is 200 times as sweet as table sugar (sucrose). A tiny amount of it equals the sweetness of a teaspoonful or more of sugar. Saccharin, another sugar substitute, is 300 times as sweet as sugar.

Some excess glucose is polymerized to form *glycogen* (glycogenesis) and stored as a glucose reserve in the liver and muscles. Glucose can be rapidly mobilized from glycogen (glycogenolysis) when it is required to supply energy. However, only a certain amount of glycogen can be stored. Excess glucose beyond what is stored as glycogen is usually converted into fat and stored in adipose tissue (fig. 18.2).

Cells also use carbohydrates as starting materials for such vital biochemicals as the 5-carbon sugars *ribose* and *deoxyribose*. These sugars are required for the production of the nucleic acids RNA and DNA and the disaccharide *lactose* (milk sugar), which is synthesized when the breasts are actively secreting milk.

Many cells can also obtain energy by oxidizing fatty acids. However, some cells, such as neurons, normally depend on a continuous supply of glucose for survival. (Under some conditions, such as prolonged starvation, other fuel sources may become available for neurons.) Even a temporary decrease in the glucose supply may seriously impair nervous system function. Consequently, the body requires a minimum amount of carbohydrate. If

adequate carbohydrates are not supplied in foods, the liver may convert some noncarbohydrates, such as amino acids from proteins or glycerol from fats, into glucose—a process called *gluconeogenesis*. Thus, the requirement for glucose has physiological priority over the need to synthesize certain other substances, such as proteins, from available amino acids.

1 List several common sources of carbohydrates.

2 In what form are carbohydrates utilized as a cellular fuel?

3 Explain what happens to excess glucose in the body.

4 Name two uses of carbohydrates other than supplying energy.

5 How does the body obtain glucose when its food supply of carbohydrates is insufficient?

> An adult's liver stores about 100 grams of glycogen, and muscle tissue stores another 200 grams, providing enough reserve to meet energy demands for about twelve hours when the person is resting. Whether these stores are filled depends on diet. People live on widely varying amounts of carbohydrates, often reflecting economic conditions. In the United States, a typical adult's diet supplies about 50% of total body energy from carbohydrates. In Asian countries where rice comprises much of the diet, carbohydrates contribute even more to the diet.

Carbohydrate Requirements

Because carbohydrates provide the primary source of fuel for cellular processes, the need for carbohydrates varies with individual energy requirements. Therefore, physically active individuals require more carbohydrates than those who are sedentary. The minimal requirement for carbohydrates in the human diet is unknown. It is estimated, however, that an intake of at least 125–175 grams daily is necessary to spare protein (that is, to avoid protein breakdown) and to avoid metabolic disorders resulting from excess fat utilization. An average diet includes 200–300 grams of carbohydrates daily.

1 Why do the daily requirements for carbohydrates vary from person to person?

2 What is the daily minimum requirement for carbohydrates?

Lipids

Lipids are organic compounds that include fats, oils, and fatlike substances such as phospholipids and cholesterol (see chapter 2, pp. 50–52). They supply energy for cellular processes and help build structures, such as cell membranes. The most common dietary lipids are the fats called *triglycerides* (tri-glis′er-īdz) (see fig. 2.14).

Lipid Sources

Triglycerides are found in plant- and animal-based foods. Saturated fats (which should comprise no more than 10% of the diet) are mainly found in foods of animal origin, such as meat, eggs, milk, and lard, as well as in palm and coconut oil. Unsaturated fats are contained in seeds, nuts, and plant oils.

Cholesterol is abundant in liver and egg yolk and is present in smaller amounts in whole milk, butter, cheese, and meats. Foods of plant origin do not contain cholesterol. This is why a label on a plant-based food claiming that it is "cholesterol-free" states the obvious.

> Be wary of claims that a food product is "99% fat-free." This usually refers to percentage by weight—not calories, which is what counts. A creamy concoction that is 99% fat-free may be largely air and water, and therefore in that form, fat comprises very little of it. But when the air is compressed and the water absorbed, as happens in the stomach, the fat percentage may skyrocket.

Lipid Utilization

Foods contain lipids in the form of phospholipids, cholesterol, or, most commonly, fats (triglycerides). A triglyceride consists of a glycerol portion and three fatty acids.

Lipids provide a variety of physiological functions; however, fats are mainly used to supply energy. Gram for gram, fats contain more than twice as much chemical energy as carbohydrates or proteins. This is why people trying to lose weight are advised to minimize fats in their diets.

Before a triglyceride molecule can release energy, it must undergo hydrolysis. Digestion breaks triglycerides down into fatty acids and glycerol. After being absorbed, these products are transported in the lymph to the blood, then on to tissues. As figure 18.3 shows, some of the resulting fatty acid portions can then react to form molecules of acetyl coenzyme A (acetyl CoA) by a series of reactions called **beta oxidation,** which occurs in the mitochondria.

In the first phase of beta-oxidation, fatty acids are activated. This change requires energy from ATP and a special group of enzymes called thiokinases. Each of these enzymes can act upon a fatty acid that has a particular carbon chain length.

Once fatty acid molecules have been activated, other enzymes called **fatty acid oxidases** that are located within mitochondria break them down. This phase of the reactions removes successive two-carbon segments of fatty acid chains. The liver converts some of these segments into acetyl coenzyme A molecules. Excess acetyl CoA is converted into compounds called **ketone bodies,** such as acetone, which later may be changed back to acetyl coenzyme A. In either case, the citric acid cycle can oxidize

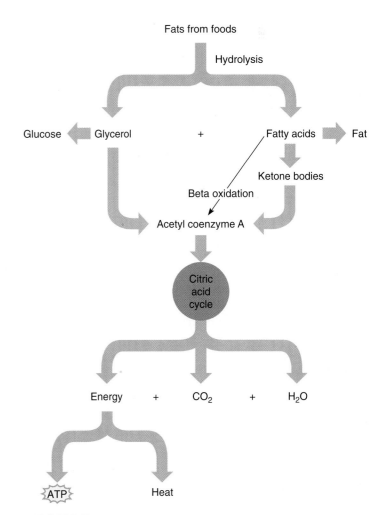

FIGURE 18.3

The body digests fat from foods into glycerol and fatty acids, which may enter catabolic pathways and provide energy.

the acetyl coenzyme A molecules. The glycerol portions of the triglyceride molecules can also enter metabolic pathways leading to the citric acid cycle, or they can be used to synthesize glucose.

> When ketone bodies form faster than they can be decomposed, some of them are eliminated through the lungs and kidneys. When this happens, the ketone acetone may impart a fruity odor to the breath and urine. This can occur when a person fasts, forcing body cells to metabolize a large amount of fat, and in persons suffering from diabetes mellitus who develop a serious imbalance in pH called acidosis, which results when acetone and other acidic ketones accumulate.

Glycerol and fatty acid molecules resulting from the hydrolysis of fats can also react to form fat molecules by anabolic processes and be stored in fat tissue. Additional fat molecules can be synthesized from excess glucose or amino acids.

The liver can convert fatty acids from one form to another. However, the liver cannot synthesize certain fatty acids, which are called **essential fatty acids.** *Linoleic acid,* for example, is an essential fatty acid required to synthesize phospholipids, which, in turn, are necessary for constructing cell membranes and myelin sheaths, and for transporting circulating lipids. Linoleic acid can be converted to another essential fatty acid, *arachidonic acid,* used in the synthesis of prostaglandins. Good sources of linoleic acid include corn oil, cottonseed oil, and soy oil. *Linolenic acid* is also an essential fatty acid.

The liver uses free fatty acids to synthesize triglycerides, phospholipids, and lipoproteins that may then be released into the blood (fig. 18.4). Thus, the liver regulates circulating lipids. It also controls the total amount of cholesterol in the body by synthesizing cholesterol and releasing it into the blood or by removing cholesterol from the blood and excreting it into the bile. The liver uses cholesterol to produce bile salts. Cholesterol is not used as an energy source, but it does provide structural material for cell and organelle membranes, and it furnishes starting materials for the synthesis of certain sex hormones and hormones produced by the adrenal cortex.

Adipose tissue stores excess triglycerides. If the blood lipid concentration drops (in response to fasting, for example), some of these stored triglycerides are hydrolyzed into free fatty acids and glycerol and then released into the bloodstream.

Lipid Requirements

The lipid content of human diets varies widely. A person who eats mostly burgers, fries, and shakes may consume 50% or more of total daily calories from fat. For a vegetarian, the percentage may be far lower. The American Heart Association advises that the diet not exceed 30% of total daily calories from fat.

The amounts and types of fats required for health are unknown. Since linoleic acid is an essential fatty acid, to prevent deficiency conditions from developing, nutritionists recommend that infants fed formula receive 3% of the energy intake in the form of linoleic acid. A typical adult diet consisting of a variety of foods usually provides adequate fats. Dietary fats must also supply the required amounts of fat-soluble vitamins.

1 Which foods commonly supply lipids?

2 Which fatty acids are essential nutrients?

3 What is the role of the liver in the utilization of lipids?

4 What is the function of cholesterol?

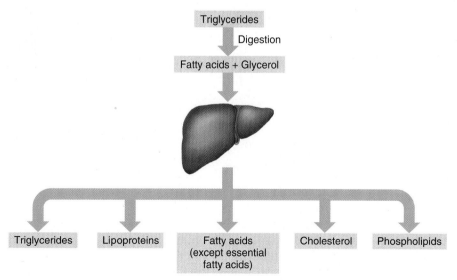

FIGURE 18.4

The liver uses fatty acids to synthesize a variety of lipids.

Proteins

Proteins are polymers of amino acids. They have a wide variety of functions. When dietary proteins are digested, the resulting amino acids are absorbed and transported by the blood to cells. Many of these amino acids are used to form new protein molecules, as specified by DNA base sequences. These new proteins include enzymes that control the rates of metabolic reactions; clotting factors; the keratins of skin and hair; elastin and collagen of connective tissue; plasma proteins that regulate water balance; the muscle components actin and myosin; certain hormones; and antibodies, which protect against infection (fig. 18.5).

Protein molecules may also supply energy. To do this, they must first be broken down into amino acids. The amino acids then undergo **deamination,** a process that occurs in the liver that removes the nitrogen-containing portions (—NH_2 groups) from the amino acids. These —NH_2 groups subsequently react to form a waste called **urea** (u-re′ah). The liver therefore produces urea from amino groups formed by deamination of amino acids. The blood carries urea to the kidneys, where it is excreted in urine.

Depending upon the particular amino acids involved, the remaining deaminated portions are decomposed by several pathways. Some of these pathways lead to formation of acetyl coenzyme A, and others more directly lead to steps of the citric acid cycle. As energy is released from the cycle, some of it is captured in molecules of ATP (fig. 18.6). If energy is not required immediately, the deaminated portions of the amino acids may react to form glucose or fat molecules in other metabolic pathways (see fig. 18.5).

A few hours after a meal, protein catabolism, through the process of gluconeogenesis (see chapter 13, p. 496), becomes a major source of blood glucose. However, metabolism in most tissues soon shifts away from glucose and toward fat catabolism as a source of ATP. Thus, energy needs are met in a way that spares proteins for tissue building and repair, rather than being broken down and reassembled into carbohydrates to supply energy. Using structural proteins to generate energy causes the tissue-wasting characteristic of starvation.

Protein Sources

Foods rich in proteins include meats, fish, poultry, cheese, nuts, milk, eggs, and cereals. Legumes, including beans and peas, contain lesser amounts of protein.

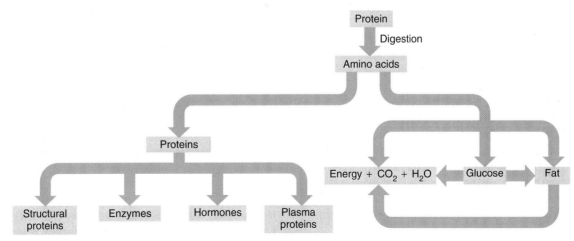

FIGURE 18.5

Proteins are digested to their constituent amino acids. These amino acids are then linked, following genetic instructions, to build new proteins. Free amino acids are also used to supply energy under certain conditions.

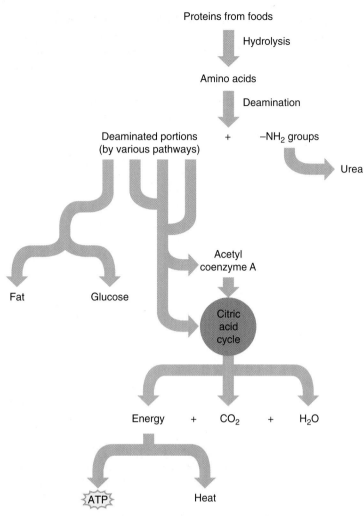

FIGURE 18.6

The body digests proteins from foods into amino acids, but must deaminate these smaller molecules before they can be used as energy sources.

TABLE 18.3	Amino Acids in Foods	
Alanine	Glycine	Proline
Arginine (ch)	Histidine (ch)	Serine
Asparagine	Isoleucine (e)	Threonine (e)
Aspartic acid	Leucine (e)	Tryptophan (e)
Cysteine	Lysine (e)	Tyrosine
Glutamic acid	Methionine (e)	Valine (e)
Glutamine	Phenylalanine (e)	

Eight essential amino acids (e) cannot be synthesized by human cells and must be provided in the diet. Two additional amino acids (ch) are essential in growing children.

meats, and eggs, contain adequate amounts of the essential amino acids to maintain human body tissues and promote normal growth and development. **Incomplete proteins,** such as *zein* in corn, which has too little of the essential amino acids tryptophan and lysine, are unable by themselves to maintain human tissues or to support normal growth and development.

A protein in wheat called *gliadin* is an example of a *partially complete protein.* Although it does not contain enough lysine to promote human growth, it contains enough to maintain life.

Individual plant proteins typically do not provide enough of one or more essential amino acids for adequate nutrition of a person. Combining appropriate plant foods can provide a diversity of dietary amino acids. For example, beans are low in methionine but have enough lysine. Rice lacks lysine but has enough methionine. A meal of beans and rice offers enough of both types of amino acids.

Plants can be genetically modified to make their protein more "complete." For example, genetic instructions for producing the amino acid tryptophan inserted into corn cells can compensate for the low levels of this nutrient normally found in corn.

1 How do cells utilize proteins?

2 Which foods are rich sources of protein?

3 Why are some amino acids called essential?

4 Distinguish between a complete protein and an incomplete protein.

Nitrogen Balance

In a healthy adult, proteins are continuously built up and broken down. This occurs at different rates in different tissues, but the overall gain of body proteins equals the loss, producing a state of **dynamic equilibrium** (di-nam'ik e"kwĭ-lib're-um). Since proteins contain a high percent-

The human body can synthesize many amino acids (nonessential amino acids). However, eight amino acids the adult body needs (ten required for growing children) cannot be synthesized sufficiently or at all, and they are called **essential amino acids.** This term refers only to dietary intake since all amino acids are required for normal protein synthesis. Table 18.3 lists the amino acids found in foods and indicates those that are essential.

All twenty types of amino acids must be in the body at the same time for growth and tissue repair to occur. In other words, if the diet lacks one essential amino acid, the cells cannot synthesize protein. Since essential amino acids are not stored, those present but not used in protein synthesis are oxidized as energy sources or are converted into carbohydrates or fats.

Proteins are classified as complete or incomplete on the basis of the amino acid types they provide. The **complete proteins,** which include those available in milk,

age of nitrogen, dynamic equilibrium also brings **nitrogen balance** (ni'tro-jen bal'ans)—a condition in which the amount of nitrogen taken in is equal to the amount excreted.

A person who is starving has a *negative nitrogen balance* because the amount of nitrogen excreted as a result of amino acid oxidation exceeds the amount the diet replaces. Conversely, a growing child, a pregnant woman, or an athlete in training is likely to have a *positive nitrogen balance* because more protein is being built into new tissue and less is being used for energy or excreted.

Protein Requirements

In addition to supplying essential amino acids, proteins provide nitrogen and other elements for the synthesis of nonessential amino acids and certain nonprotein nitrogenous substances. The amount of protein individuals require varies according to body size, metabolic rate, and nitrogen balance condition.

For an average adult, nutritionists recommend a daily protein intake of about 0.8 gram per kilogram (0.4 gram per pound) of body weight or 10% of a person's diet.

For a pregnant woman, who needs to maintain a positive nitrogen balance, the recommendation adds 30 grams of protein per day. Similarly, a nursing mother requires an additional 20 grams of protein per day to maintain a high level of milk production.

In addition to tissue wasting, protein deficiency may also decrease the level of plasma proteins, which decreases the colloid osmotic pressure of the plasma. As a result, fluids collect in the tissues, producing a condition called *nutritional edema.* Table 18.4 summarizes the sources, uses, and requirements for carbohydrate, lipid, and protein nutrients.

 RECONNECT TO CHAPTER 15, EXCHANGES IN THE CAPILLARIES, PAGES 568–569.

1 What are the physiological functions of proteins?

2 What is a negative nitrogen balance? A positive nitrogen balance?

3 How can inadequate nutrition cause edema?

	TABLE 18.4	**Carbohydrate, Lipid, and Protein Nutrients**				

Nutrient	**Sources and RDA* for Adults**	**Calories per Gram**	**Utilization**	**Conditions Associated with**		
				Excesses		**Deficiencies**
Carbohydrate	Primarily from starch and sugars in foods of plant origin and from glycogen in meats 125–175 g	4.1	Oxidized for energy; used in production of ribose, deoxyribose, and lactose; stored in liver and muscles as glycogen; converted to fats and stored in adipose tissue	Obesity, dental caries, nutritional deficits		Metabolic acidosis
Lipid	Meats, eggs, milk, lard, plant oils 80–100 g	9.5	Oxidized for energy; production of triglycerides, phospholipids, lipoproteins, and cholesterol, stored in adipose tissue; glycerol portions of fat molecules may be used to synthesize glucose	Obesity, increased serum cholesterol, increased risk of heart disease		Weight loss, skin lesions
Protein	Meats, cheese, nuts, milk, eggs, cereals, legumes 0.8 g/kg body weight	4.1	Production of protein molecules used to build cell structure and to function as enzymes or hormones; used in the transport of oxygen, regulation of water balance, control of pH, formation of antibodies; amino acids may be broken down and oxidized for energy or converted to carbohydrates or fats for storage	Obesity		Extreme weight loss, wasting, anemia, growth retardation

RDA = recommended dietary allowance.

Energy Expenditures

Carbohydrates, fats, and proteins supply energy. Since energy is required for all metabolic processes, energy availability is of prime importance to cell survival. If the diet is deficient in energy-supplying nutrients, structural molecules may gradually be consumed, leading to death. On the other hand, excess intake of energy supplying nutrients may lead to obesity.

Energy Values of Foods

The amount of potential energy a food contains can be expressed as **calories** (kal′o-rēz), which are units of heat. Although a calorie is defined as the amount of heat required to raise the temperature of a gram of water by 1 degree Celsius (°C), the calorie used to measure food energy is 1,000 times greater. This *large calorie* (Cal.) equals the amount of heat required to raise the temperature of a kilogram (1,000 grams) of water by 1°C (actually from 14.5°C to 15.5°C) and is also equal to 4.184 joules, with a joule being the international unit of heat and energy. A large calorie is sometimes called a *kilocalorie,* but it is customary in nutritional studies to refer to it simply as a calorie.

Figure 18.7 shows a bomb calorimeter, which is used to measure the caloric contents of foods. It consists of a metal chamber submerged in a known volume of water. A food sample is dried, weighed, and placed in a nonreactive dish inside the chamber. The chamber is filled with oxygen gas and submerged in the water. Then, the food is ignited and allowed to oxidize completely. Heat released from the food raises the temperature of the surrounding water, and the change in temperature is measured. Since the volume of the water is known, the amount of heat released from the food can be calculated in calories.

Caloric values determined in a bomb calorimeter are somewhat higher than the amount of energy that metabolic oxidation actually releases because nutrients generally are not completely absorbed from the digestive tract. Also, the body does not completely oxidize amino acids but excretes parts of them in urea or uses them to synthesize other nitrogenous substances. When such losses are considered, cellular oxidation yields on the average about 4.1 calories from 1 gram of carbohydrate, 4.1 calories from 1 gram of protein, and 9.5 calories from 1 gram of fat. The fact that more than twice as much energy is derived from equal amounts by weight of fats as from either proteins or carbohydrates is one reason why avoiding fatty foods helps weight loss. Fats also encourage weight gain because they add flavor to food, which can cause overeating.

1 What term designates the potential energy in a food?

2 How can the energy value of a food be determined?

3 What is the energy value of a gram of carbohydrate? A gram of protein? A gram of fat?

Energy Requirements

The amount of energy required to support metabolic activities for twenty-four hours varies from person to person. The factors that influence energy needs include the individual's basal metabolic rate, degree of muscular activity, body temperature, and rate of growth.

The **basal metabolic rate** (ba′sal met″ah-bol′ik rāt), or BMR, measures the rate at which the body expends energy under *basal conditions*—when a person is awake and at rest; after an overnight fast; and in a comfortable, controlled environment. Tests of thyroid function can be used to estimate a person's BMR.

The amount of oxygen the body consumes is directly proportional to the amount of energy cellular respiration releases. The BMR, therefore, reveals the total amount of energy expended in a given time period to support the activities of such organs as the brain, heart, lungs, liver, and kidneys.

The average adult BMR indicates a requirement for approximately 1 calorie of energy per hour for each kilogram of body weight. However, this requirement varies with such factors as sex, body size, body temperature, and level of endocrine gland activity. For example, since heat loss is directly proportional to the body surface area, and a smaller person has a greater surface area relative to body mass, he or she will have a higher BMR. Males tend to have higher metabolic rates than females. As body temperature increases, BMR increases, and as the blood level of thyroxine or epinephrine increases, so does the BMR. The BMR can also increase when the level of physical activity increases during the day.

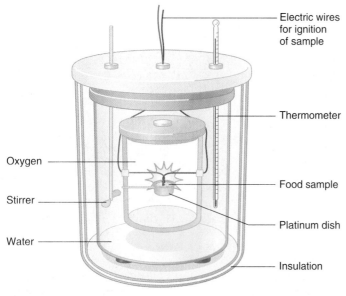

FIGURE 18.7
A bomb calorimeter measures the caloric content of a food sample.

The labels on the figure read:
- Electric wires for ignition of sample
- Thermometer
- Oxygen
- Food sample
- Stirrer
- Platinum dish
- Water
- Insulation

Maintaining the basal metabolic rate usually requires the body's greatest expenditure of energy. The energy required to support voluntary muscular activity comes next, though this amount varies greatly with the type of activity (table 18.5). For example, the energy to maintain posture while sitting at a desk might require 100 calories per hour above the basal need, whereas running or swimming might require 500–600 calories per hour.

Maintenance of body temperature may require additional energy expenditure, particularly in cold weather. In this case, extra energy may be expended by involuntary muscular contractions, such as shivering, or through voluntary muscular actions, such as walking. Growing children and pregnant women, because their bodies are actively producing new tissues, also require more calories.

Energy Balance

A state of **energy balance** exists when caloric intake in the form of foods equals caloric output from the basal metabolic rate and muscular activities. Under these conditions, body weight remains constant, except perhaps for slight variations due to changes in water content.

If, on the other hand, caloric intake exceeds the output, a *positive energy balance* occurs, and tissues store excess nutrients. This increases body weight, since an excess of 3,500 calories can be stored as a pound of fat. Conversely, if caloric output exceeds input, the energy balance is negative, and stored materials are mobilized from the tissues for oxidation, causing weight loss.

1 What is basal metabolic rate?

2 What factors influence the BMR?

3 What is *energy balance*?

Desirable Weight

The most obvious and common nutritional disorders reflect calorie imbalances, which may result from societal and geographic factors. Obesity is a common problem in nations where food is plentiful and diverse. The

tendency to become obese may be a holdover from thousands of years ago, when the ability to store energy in the form of fat was a survival advantage when food supplies were scarce or erratic. Today in many African nations, natural famines combined with political unrest cause mass starvation. Starvation is considered later in the chapter.

It is difficult to determine a desirable body weight. In the past, weight standards were based on average weights and heights within a certain population, and the degrees of underweight and overweight were expressed as percentage deviations from these averages. These standards reflected the gradual gain in weight that usually occurs with age. Later, medical researchers recognized that such an increase in weight after the age of twenty-five to thirty years is not necessary and may not be healthy. Consequently, standards of *desirable weights* were prepared. More recent height–weight guidelines are based upon the characteristics of people who live the longest. These weights are somewhat more lenient than those in the desirable weight charts.

Overweight is defined as exceeding desirable weight by 10% to 20%. A person who is more than 20% above the desired weight is considered to be *obese*, although **obesity** (o-bēs'ĭ-te) is more correctly defined as excess adipose tissue. Therefore, overweight and obesity are not the same. For example, as figure 18.8 shows, an athlete or a person whose work involves heavy muscular activity may be overweight, but not obese. Clinical Application 18.1 discusses obesity.

(a)   **(b)**

FIGURE 18.8

Weight. (*a*) An obese person is overweight and has excess adipose tissue. (*b*) An athlete may be overweight due to muscle overgrowth but is not considered obese. In fact, many athletes have very low percentages of body fat.

TABLE 18.5	Calories Used During Various Activities
Activity	**Calories (per Hour)**
Walking up stairs	1,100
Running (jogging)	570
Swimming	500
Vigorous exercise	450
Slow walking	200
Dressing and undressing	118
Sitting at rest	100

OBESITY

The U.S. National Institutes of Health considers obesity a "killer disease," and for good reason. A person who is obese—defined as 20% above "ideal" weight based on population statistics considering age, sex, and build—is at higher risk for diabetes, digestive disorders, heart disease, kidney failure, hypertension, stroke, and cancers of the female reproductive organs and the gallbladder. The body is enormously strained to support the extra weight—miles of blood vessels are needed to nourish the additional pounds. More than 60% of adults in the United States are overweight or obese. Obesity is the second leading cause of preventable death, following cigarette smoking. People in the United States are overweight because of overeating. The average person today consumes 3,700 calories daily, compared to 3,100 in the 1960s.

Obesity refers to extra pounds of fat. The proportion of fat in a human body ranges from 5% to more than 50%, with "normal" for males falling between 12% and 23% and for females between 16% and 28%. An elite athlete may have a body fat level as low as 4%. Fat distribution also affects health. Excess poundage above the waist is linked to increased risk of heart disease, diabetes, hypertension, and lipid disorders. Figure 18A shows how to estimate obesity using a measurement called body mass index.

Both heredity and the environment contribute to obesity. Dozens of genes interact to control energy balance and therefore body weight. The observation that identical twins reared in different households can grow into adults of vastly different weights

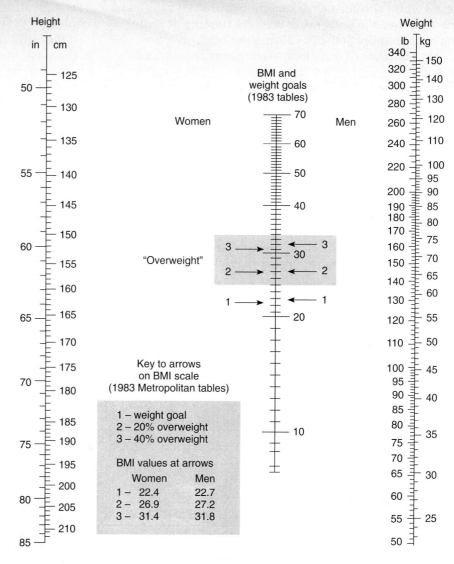

FIGURE 18A

Body mass index (BMI) is a measurement that estimates obesity. BMI equals weight in kilograms divided by height in meters, squared (BMI = weight [kg]/height [m]^2). To calculate your BMI, draw a line from your height on the left side of the figure to your weight on the right side. The point where the line crosses the scale in the middle is your BMI. A value above 27.2 for men and 26.9 for women indicates obesity. A value above 30 is cause for concern; a value above 35 may be life-threatening.

When a person needs to gain weight, diet can be altered to include more calories and to emphasize particular macronutrients. For example, a person recovering from a debilitating illness might consume more carbohydrates, whereas a bodybuilder might eat extra protein to hasten muscle development. An infant also needs to gain weight rapidly, which is best accomplished by drinking human milk. The high fat content of human milk is important for the rapid growth of the infant's brain.

1   What is *desirable weight*?

2   Distinguish between overweight and obesity.

3   Under what conditions is weight gain desirable?

indicates that environment influences weight too. Even the environment before birth can affect body weight later. Individuals who were born at full term, but undernourished as fetuses, are at high risk of later obesity. Physiological changes that countered starvation in the uterus cause obesity when they persist.

Certain genes encode proteins that connect sensations in the gastrointestinal tract with centers in the hypothalamus that control hunger and satiety. It is how we satisfy those signals—what we eat—that provides the environmental component to body weight. A certain set of genes may have led to a trim figure in a human many thousand years ago, when food had to be hunted or gathered—and meat was leaner. Today, with the myriad of mouth-watering food choices, those same genes do not foster slimness in a person who takes in many more calories than he or she expends.

Treatment for Obesity
Diet and Exercise

A safe goal for weight loss using dietary restriction and exercise is 1 pound of fat per week. A pound of fat contains 3,500 calories of energy, so that pound can be shed by an appropriate combination of calorie cutting and exercise. This might mean eating 500 calories less per day or exercising off 500 calories each day. Actually more than a pound of weight will drop because water is lost as well as fat.

Dieting should apply to the energy-providing nutrients (carbohydrates, proteins, and fats) but never to the vitamins and minerals. A rule of thumb is to leave the proportion of protein calories about the same or slightly increased, cut fat calories in half, and cut carbohydrates by a third. Choose foods that you like and distribute them into three or four balanced meals of 250 to 500 calories each. Restricting unhealthful fats is one effective way to lose weight, because a gram of fat contains 9 calories, compared to 4 calories for a gram of protein or carbohydrate. Excessive protein intake that accompanies eating such foods as beef can damage the kidneys.

Ideally, weight loss can be accomplished by changing diet and exercise habits. However, realistically, two-thirds of those who lose weight regain it within five years. Physicians are increasingly regarding obesity as a chronic illness that for some may require more drastic measures than dieting and exercising.

Drug Therapy

Some physicians recommend drug therapy if the BMI exceeds 30 or if it exceeds 27 and the person also has hypertension, diabetes mellitus, or hyperlipidemia. Several types of "diet drugs" are no longer in use, including amphetamines, which carried the risk of addiction, and the combination of fenfluramine and phentermine, which shed weight but damaged heart valves.

Newer antiobesity drugs target fat in diverse ways. Tetrahydrolipostatin, marketed as Orlistat and Xenical, inhibits the function of pancreatic lipase, preventing the digestion and absorption of about a third of dietary fat. The fat is eliminated, causing loose feces. This effect is not disruptive as long as the person follows a low-fat diet. The unpleasantness of the loose feces tends to keep users on the diet!

Another approach to treating obesity is developing drugs that block the proteins (adipocyte transcription factors) that enable fibroblasts to specialize into fat cells (adipocytes). Researchers are investigating ways to manipulate appetite-control hormones, such as ghrelin and leptin, in ways that promote weight loss.

Surgery

Surgery is recommended for people whose BMI exceeds 40 or if it exceeds 35 and medical problems are present. Such procedures are called "bariatric surgery," from the Greek for "weight treatment." Today, the most common type of bariatric surgery staples off a portion of the stomach so that it can hold only 2 tablespoons of food at a time, and bypassing part of the small intestine. This gastric bypass is of two types—2.5 to 5 feet of duodenum or 12 to 15 feet, which encompasses part of the jejunum too. Because of the shortened alimentary canal, the individual must drastically lower food intake. Overeating leads to "dumping syndrome," a very unpleasant combination of weakness, nausea, sweating, and faintness.

Bariatric surgery targeting the stomach and the small intestine evolved from separate procedures that were performed in the 1980s, with only limited success. Patients undergoing early stomach stapling tended to eat too much and burst the staples! More extensive, earlier intestinal bypasses led to malnutrition and liver failure. The combination surgery, however, performed on more than 40,000 people a year in the United States, has led to weight losses of more than 100 pounds, and disappearance of several obesity-associated illnesses in some patients, including hypertension, arthritis, back pain, varicose veins, sleep apnea, and type II diabetes mellitus. Singer Carnie Wilson and meteorologist Al Roker have publicized their remarkable success with the procedure. ∎

Vitamins

Vitamins (vi′tah-minz) are organic compounds (other than carbohydrates, lipids, and proteins) required in small amounts for normal metabolic processes but that body cells cannot synthesize in adequate amounts. Vitamins are essential nutrients that must come from foods or from **provitamins,** which are precursor substances.

Vitamins can be classified on the basis of solubility, since some are soluble in fats (or fat solvents) and others are soluble in water. Those that are *fat-soluble* include vitamins A, D, E, and K; the *water-soluble* group includes

TABLE 18.6	Vitamin Fallacies and Facts
Fallacy	**Fact**
The more vitamins, the better	Too much of a water-soluble vitamin results in excretion of the vitamin through urination; too much of a fat-soluble vitamin can harm health
A varied diet provides all needed vitamins	Many people do need vitamin supplements, particularly pregnant and breast-feeding women
Vitamins provide energy	Vitamins do not directly supply energy; they aid in the release of energy from carbohydrates, fats, and proteins

the B vitamins and vitamin C. Table 18.6 lists, and corrects, some common misconceptions about vitamins.

Different species have different vitamin requirements. For example, ascorbic acid is a required vitamin (C) in humans, guinea pigs, and Indian fruit bats, but not in other animals, which can manufacture their own ascorbic acid.

Fat-Soluble Vitamins

Because fat-soluble vitamins dissolve in fats, they associate with lipids and are influenced by the same factors that affect lipid absorption. For example, the presence of bile salts in the intestine promotes absorption of these vitamins. As a group, the fat-soluble vitamins are stored in moderate quantities within various tissues, and because they are fairly resistant to heating, cooking or food processing does not usually destroy them.

1   What are vitamins?

2   How are vitamins classified?

3   How do bile salts affect the absorption of fat-soluble vitamins?

Vitamin A occurs in several forms, including retinol and retinal (retinene). Body cells synthesize this vitamin from a group of yellowish plant pigments, which are provitamins called *carotenes* (fig. 18.9). Excess vitamin A or its precursors are mainly stored in the liver, which regulates their concentration in the body. An adult's liver stores enough vitamin A to supply body requirements for a year. Infants and children usually lack such reserves and are therefore more likely to develop vitamin A deficiencies if their diets are inadequate.

Vitamin A is relatively stable to the effects of heat, acids, and bases. However, it is readily destroyed by oxidation and is unstable in light.

Vitamin A is important in vision. Retinal is used to synthesize *rhodopsin* (visual purple) in the rods of the retina and may be required for production of light-sensitive pigments in the cones as well. The vitamin also functions in the synthesis of mucoproteins and mucopolysaccharides, in development of normal bones and teeth, and in maintenance of epithelial cells in skin and mucous membranes. Vitamin A and beta carotenes also act as **antioxidants** (an"te-ok'sĭ-dantz) by readily combining with oxygen and certain oxygen-containing molecules that have unshared electrons, which makes them highly reactive and damaging to cellular structures. These unstable molecules are called oxygen free radicals, and they accumulate in certain diseases and with age.

RECONNECT TO CHAPTER 2, BONDING OF ATOMS, PAGE 43.

Only foods of animal origin such as liver, fish, whole milk, butter, and eggs are sources of vitamin A. However, the vitamin's precursor, carotene, is widespread in leafy green vegetables and in yellow or orange vegetables and fruits.

Excess vitamin A produces peeling skin, hair loss, nausea, headache, and dizziness, a condition called *hypervitaminosis A*. Chronic overdoses of the vitamin may inhibit growth and cause the bones and joints to

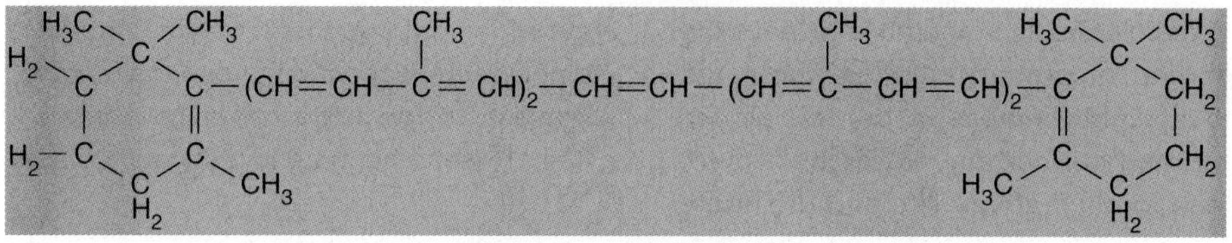

FIGURE 18.9

A molecule of beta carotene can react to form two molecules of retinal, which, in turn, can react to form retinol.

Beta carotene

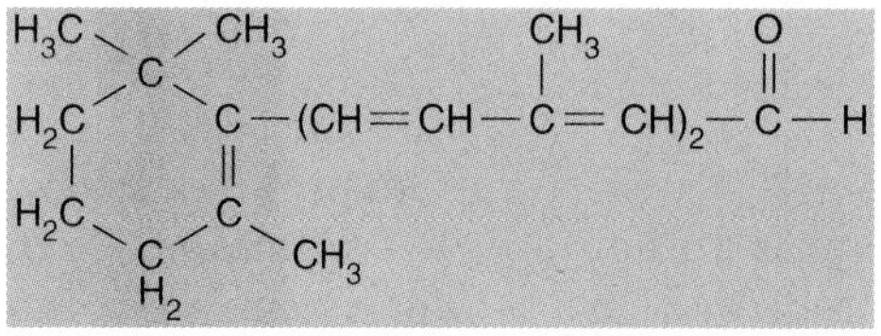

Retinal (retinene)

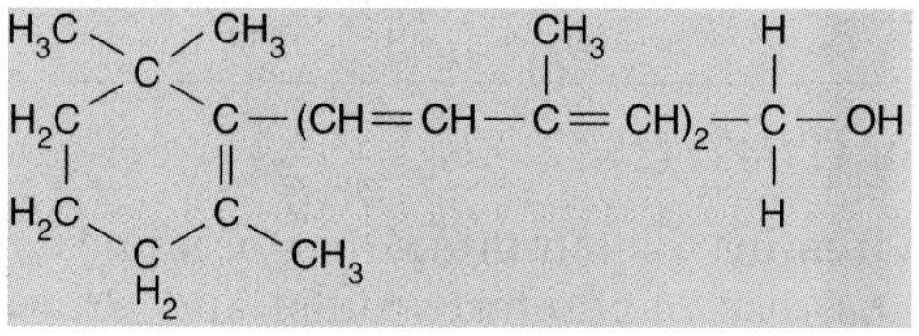

Retinol

degenerate. "Megadosing" on fat-soluble vitamins is particularly dangerous during pregnancy. Some forms of vitamin A, in excess, can cause birth defects.

"Yellow rice" is a genetically modified variety that, thanks to genes from a bacterium and petunia, manufactures beta carotene, a vitamin A precursor. In addition, alteration of a rice gene enables the plant to double its iron content. Once these traits are bred into a commercial strain of rice, the new crop may help prevent vitamin A and iron deficiencies in malnourished people living in developing nations. It will be available to them free of charge.

A deficiency of vitamin A causes *night blindness,* in which a person cannot see normally in dim light. Xeropthalmia, a dryness of the conjunctiva and cornea, is due to vitamin A deficiency. Vitamin A deficiency also causes degenerative changes in certain epithelial tissues, and the body becomes more susceptible to infection.

1 What chemical in the body is the precursor to vitamin A?

2 What conditions destroy vitamin A?

3 Which foods are good sources of vitamin A?

Vitamin D is a group of steroids that have similar properties. One of these substances, vitamin D_3 (cholecalciferol), is found in foods such as milk, egg yolk, and fish liver oils. Vitamin D_2 (ergocalciferol) is produced commercially by exposing a steroid obtained from yeasts (ergosterol) to ultraviolet light. Vitamin D can also be synthesized from dietary cholesterol that has been metabolized to provitamin D by intestinal enzymes, then stored in the skin and exposed to ultraviolet light (see chapter 13, pp. 489–490).

Like other fat-soluble vitamins, vitamin D resists the effects of heat, oxidation, acids, and bases. It is primarily stored in the liver and is less abundant in the skin, brain, spleen, and bones.

As it is needed, vitamin D stored in the form of hydroxycholecalciferol is released into the blood. When parathyroid hormone is present, this form of vitamin D is converted in the kidneys into an active form of the vitamin (dihydroxycholecalciferol). This substance, in turn, is carried as a hormone in the blood to the intestines where it stimulates production of calcium-binding protein. Here, it promotes absorption of calcium and phosphorus, ensuring that adequate amounts of these minerals are available in the blood for tooth and bone formation and metabolic processes.

Because natural foods are often poor sources of vitamin D, it is often added to food during processing. For example, homogenized, nonfat, and evaporated milk are typically fortified with vitamin D. *Fortified* means essential nutrients have been added to a food where they originally were absent or scarce. *Enriched* means essential nutrients have been partially replaced in a food that has lost nutrients during processing.

Excess vitamin D, or *hypervitaminosis D,* produces diarrhea, nausea, and weight loss. Over time, it may also cause calcification of certain soft tissues and irreversible kidney damage.

In children, vitamin D deficiency results in *rickets,* in which the bones and teeth fail to develop normally (fig. 18.10). In adults or in the elderly who have little exposure to sunlight, such a deficiency may lead to *osteomalacia,* in which the bones decalcify and weaken due to disturbances in calcium and phosphorus metabolism. Risk of developing vitamin deficiency increases in people who stay out of the sun or liberally use sun block to prevent skin cancer. However, just 5 minutes of sun exposure two to three times a week can maintain skeletal health without elevating skin cancer risk.

Because older people tend to be outdoors less than younger individuals, the Institute of Medicine suggests

FIGURE 18.10

Vitamin D deficiency causes rickets, in which the bones and teeth do not develop normally.

TABLE 18.7	Vitamin D Requirements with Age
Age Range (Years)	**International Units of Vitamin D**
<50	200
50–70	400
70+	600

that daily vitamin D intake be increased with age (see table 18.7).

1 Where is vitamin D stored?

2 What are the functions of vitamin D?

3 Which foods are good sources of vitamin D?

4 What are symptoms of vitamin D excess and deficiency?

Vitamin E includes a group of compounds, the most active of which is *alpha-tocopherol.* This vitamin is resistant to the effects of heat, acids, and visible light but is unstable in bases and in the presence of ultraviolet light or oxygen. Vitamin E is a strong antioxidant.

Vitamin E is found in all tissues but is primarily stored in the muscles and adipose tissue. It is also highly concentrated in the pituitary and adrenal glands.

The precise functions of vitamin E are unknown, but it is thought to act as an antioxidant by inhibiting the breakdown of polyunsaturated fatty acids in the tissues. It may also help maintain the stability of cell membranes.

Vitamin E is widely distributed among foods. Its richest sources are oils from cereal seeds such as wheat germ. Other good sources are salad oils, margarine, shortenings, fruits, nuts, and vegetables. Since this vitamin is so easily obtained, deficiency conditions are rare.

1 Where is vitamin E stored?

2 What are the functions of vitamin E?

3 Which foods are good sources of vitamin E?

Vitamin K, like the other fat-soluble vitamins, occurs in several chemical forms. One of these, vitamin K_1 (phylloquinone), is found in foods, whereas another, vitamin K_2, is produced by bacteria (*Escherichia coli*) that normally inhabit the human intestinal tract. These vitamins resist the effects of heat but are destroyed by oxidation or by exposure to acids, bases, or light. The liver stores them to a limited degree.

Vitamin K primarily functions in the liver, where it is necessary for the formation of several proteins needed for blood clotting, including *prothrombin* (see chapter 14, p. 527). Consequently, deficiency of vitamin K causes prolonged blood clotting time and a tendency to hemorrhage.

The richest sources of vitamin K are leafy green vegetables. Other good sources are egg yolk, pork liver, soy oil, tomatoes, and cauliflower. Table 18.8 summarizes the fat-soluble vitamins and their properties.

TABLE 18.8	Fat-Soluble Vitamins				
Vitamin	**Characteristics**	**Functions**	**Sources and RDA* for Adults**	**Conditions Associated with**	
				Excesses	**Deficiencies**
Vitamin A	Occurs in several forms; synthesized from carotenes; stored in liver; stable in heat, acids, and bases; unstable in light	An antioxidant necessary for synthesis of visual pigments, mucoproteins, and mucopolysaccharides; for normal development of bones and teeth; and for maintenance of epithelial cells	Liver, fish, whole milk, butter, eggs, leafy green vegetables, yellow and orange vegetables and fruits 4,000–5,000 IU**	Nausea, headache, dizziness, hair loss, birth defects	Night blindness, degeneration of epithelial tissues
Vitamin D	A group of steroids; resistant to heat, oxidation, acids, and bases; stored in liver, skin, brain, spleen, and bones	Promotes absorption of calcium and phosphorus; promotes development of teeth and bones	Produced in skin exposed to ultraviolet light; in milk, egg yolk, fish liver oils, fortified foods 400 IU	Diarrhea, calcification of soft tissues, renal damage	Rickets, bone decalcification and weakening
Vitamin E	A group of compounds; resistant to heat and visible light; unstable in presence of oxygen and ultraviolet light; stored in muscles and adipose tissue	An antioxidant; prevents oxidation of vitamin A and polyunsaturated fatty acids; may help maintain stability of cell membranes	Oils from cereal seeds, salad oils, margarine, shortenings, fruits, nuts, and vegetables 30 IU	Hypertension	Rare, uncertain effects
Vitamin K	Occurs in several forms; resistant to heat but destroyed by acids, bases, and light; stored in liver	Required for synthesis of prothrombin, which functions in blood clotting	Leafy green vegetables, egg yolk, pork liver, soy oil, tomatoes, cauliflower 55–70 µg	None known	Easy bruising and bleeding

*RDA = recommended dietary allowance.
**IU = international unit.

1 Where in the body is vitamin K synthesized?

2 What is the function of vitamin K?

3 Which foods are good sources of vitamin K?

Water-Soluble Vitamins

The water-soluble vitamins include the B vitamins and vitamin C. The **B vitamins** are several compounds that are essential for normal cellular metabolism. They help oxidize (remove electrons from) carbohydrates, lipids, and proteins during cellular respiration. Since the B vitamins often occur together in foods, they are usually referred to as the *vitamin B complex*. Members of this group differ chemically and functionally.

The B-complex vitamins include the following:

1. **Thiamine,** or **vitamin B_1**. In its pure form, thiamine is a crystalline compound called thiamine hydrochloride. It is destroyed by exposure to heat and oxygen, especially in alkaline environments. (See fig. 18.16 for its molecular structure.)

   Thiamine is part of a coenzyme called *cocarboxylase,* which oxidizes carbohydrates. More specifically, thiamine is required for pyruvic acid to enter the citric acid cycle (see chapter 4, p. 111); in the absence of this vitamin, pyruvic acid accumulates in the blood. Thiamine also functions as a coenzyme in the synthesis of the sugar ribose, which is part of the nucleic acid RNA.

   Thiamine is primarily absorbed through the wall of the duodenum and is transported by the blood to body cells. Only small amounts are stored in the tissues, and excess is excreted in the urine.

   Since vitamin B_1 oxidizes carbohydrates, the cellular requirements vary with caloric intake. It is recommended that an adult diet contain 0.5 milligram (mg) of thiamine for every 1,000 calories ingested daily.

   Good sources of thiamine are lean meats, liver, eggs, whole-grain cereals, leafy green vegetables, and legumes.

A mild deficiency of thiamine produces loss of appetite, fatigue, and nausea. Prolonged deficiency leads to a disease called *beriberi,* which causes gastrointestinal disturbances, mental confusion, muscular weakness and paralysis, and heart enlargement. In severe cases, the heart may fail.

2. **Riboflavin,** or **vitamin B_2**. Riboflavin is a yellowish brown crystalline substance that is relatively stable to the effects of heat, acids, and oxidation but is destroyed by exposure to bases and ultraviolet light. This vitamin is part of several enzymes and coenzymes that are known as *flavoproteins.* One such coenzyme, FAD, is an electron carrier in the citric acid cycle and electron transport chain of aerobic respiration. Flavoproteins are essential for the oxidation of glucose and fatty acids and for cellular growth. The absorption of riboflavin is regulated by an active transport system that controls the amount entering the intestinal mucosa. Riboflavin is carried in the blood combined with blood proteins called *albumins.* Excess riboflavin in the blood is excreted in the urine, and any that remains unabsorbed in the intestine is lost in the feces.

   The amount of riboflavin the body requires varies with caloric intake. About 0.6 mg of riboflavin per 1,000 calories is sufficient to meet daily cellular requirements.

   Riboflavin is widely distributed in foods, and rich sources include meats and dairy products. Leafy green vegetables, whole-grain cereals, and enriched cereals provide lesser amounts. Vitamin B_2 deficiency produces dermatitis and blurred vision.

3. **Niacin.** Niacin, which is also known as *nicotinic acid,* occurs in plant tissues and is stable in the presence of heat, acids, and bases. After ingestion, it is converted to a physiologically active form called *niacinamide* (fig. 18.11). Niacinamide is the form of niacin that is present in foods of animal origin.

   Niacin functions as part of two coenzymes (coenzyme I, also called NAD [fig. 18.12], and coenzyme II, called NADP) that play essential roles in the oxidation of glucose, acting as electron carriers in glycolysis, the citric acid cycle, and the electron transport chain, as well as in the synthesis of proteins and fats. These coenzymes are also

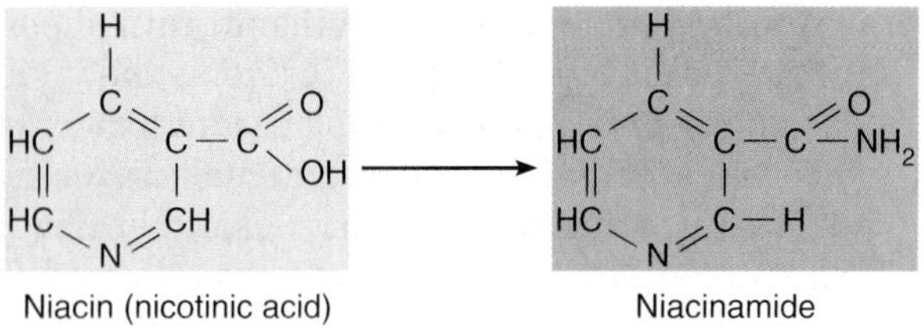

FIGURE 18.11

Enzymes catalyze reactions that convert niacin from foods into physiologically active niacinamide.

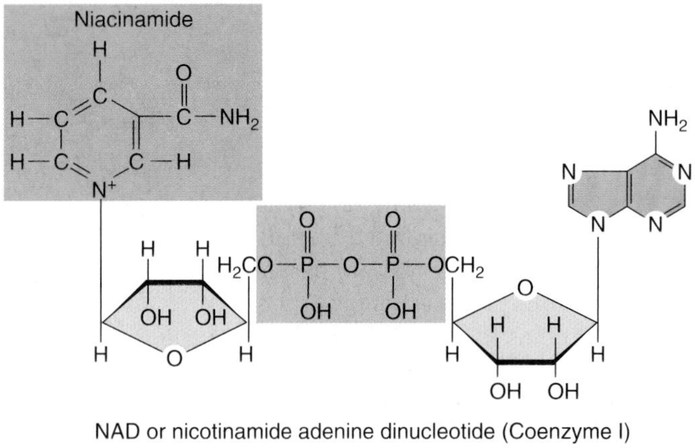

(Niacin (nicotinic acid) → Niacinamide)

FIGURE 18.12

Niacinamide is incorporated into molecules of coenzyme I.

NAD or nicotinamide adenine dinucleotide (Coenzyme I)

required for the synthesis of the sugars (ribose and deoxyribose) that are part of nucleic acids.

Niacin is readily absorbed from foods, and human cells synthesize it from the essential amino acid *tryptophan*. Consequently, the daily requirement for niacin varies with tryptophan intake. Nutritionists recommend a daily niacin (or niacin equivalent) intake of 6.6 mg per 1,000 calories.

Rich sources of niacin (and tryptophan) include liver, lean meats, peanut butter, and legumes. Milk is a poor source of niacin but a good source of tryptophan.

Historically, niacin deficiencies have been associated with diets largely consisting of corn and corn products, which are very low in niacin and lack adequate tryptophan. Such a deficiency causes a disease called *pellagra* that produces dermatitis, inflammation of the digestive tract, diarrhea, and mental disorders.

Although pellagra is relatively rare today, it was a serious problem in the rural South of the United States in the early 1900s. Pellagra is less common in cultures that extensively use corn treated with lime ($CaCO_3$) to release niacin from protein binding in corn. It sometimes occurs in people with chronic alcoholism who have substituted alcohol for foods.

4. **Pantothenic acid,** or **vitamin B_5.** Pantothenic acid is a yellowish oil that is destroyed by heat, acids, and bases. It functions as part of a complex molecule called *coenzyme A,* which, in turn, reacts with intermediate products of carbohydrate and fat metabolism to become *acetyl coenzyme A,* which enters the citric acid cycle. Pantothenic acid is therefore essential to cellular energy release.

A daily adult intake of 4–7 mg of pantothenic acid is adequate. Most diets provide sufficient amounts, since deficiencies are rare, and no clearly defined set of deficiency symptoms is known. Good sources of pantothenic acid include meats, whole-grain cereals, legumes, milk, fruits, and vegetables.

5. **Vitamin B_6.** Vitamin B_6 is a group of three compounds that are chemically similar, as figure 18.13 shows. They are called *pyridoxine, pyridoxal,* and *pyridoxamine.* These compounds have similar actions and are fairly stable in the presence of heat and acids. Oxidation or exposure to bases or ultraviolet light destroys them. The vitamin B_6 compounds function as coenzymes that are essential in a wide variety of metabolic pathways, including those that synthesize proteins, amino acids, antibodies, and nucleic acids, as well as the conversion of tryptophan to niacin.

Since vitamin B_6 functions in the metabolism of nitrogen-containing substances, the requirement for this vitamin varies with the protein content of the diet rather than with caloric intake. The recommended daily allowance of vitamin B_6 is 2.0

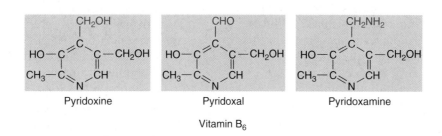

Pyridoxine

Pyridoxal

Pyridoxamine

Vitamin B_6

FIGURE 18.13

Vitamin B_6 includes three similar chemical compounds.

710

mg, but because it is so widespread in foods, deficiency conditions are quite rare. Good sources of vitamin B_6 include liver, meats, bananas, avocados, beans, peanuts, whole-grain cereals, and egg yolk.

6. **Cyanocobalamin,** or **vitamin B_{12}.** Cyanocobalamin has a complex molecular structure that contains a single atom of the element *cobalt* (fig. 18.14). In its pure form, this vitamin is red. It is stable to the effects of heat but is inactivated by exposure to light or strong acids or bases.

   Secretion of *intrinsic factor* from the parietal cells of the gastric glands regulates cyanocobalamin absorption. Intrinsic factor combines with cyanocobalamin and facilitates its transfer through the epithelial lining of the small intestine and into the blood. Calcium ions must be present for the process to take place.

   Various tissues store cyanocobalamin, particularly those of the liver. An average adult has a reserve sufficient to supply cells for three to five years. This vitamin is essential for the functions of all cells. It is part of coenzymes required for the synthesis of nucleic acids and the metabolism of carbohydrates and fats. Vitamin B_{12} is important to

erythrocyte production. Cyanocobalamin also help form myelin in the central nervous system.

Only foods of animal origin contain cyanocobalamin. Good sources include liver, meats, milk, cheese, and eggs. In most countries, dietary lack of this vitamin is rare, although strict vegetarians may develop a deficiency.

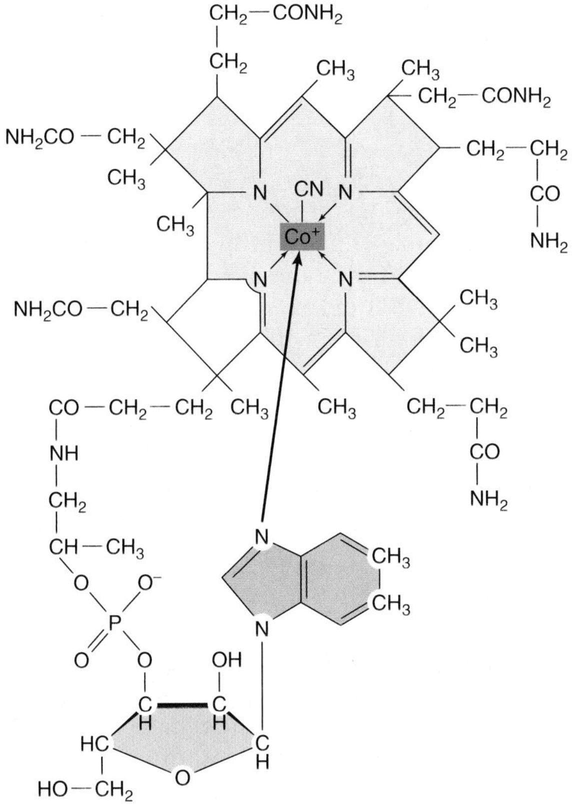

> When the gastric glands of some individuals fail to secrete adequate amounts of intrinsic factor, cyanocobalamin is poorly absorbed. This leads to *pernicious anemia,* in which abnormally large red blood cells called macrocytes are produced when bone marrow cells do not divide properly because of defective DNA synthesis.

7. **Folacin,** or **folic acid.** Folacin is a yellow crystalline compound that exists in several forms. It is easily oxidized in an acid environment and is destroyed by heat in alkaline solutions; consequently, this vitamin may be lost in foods that are stored or cooked.

   Folacin is readily absorbed from the digestive tract and is stored in the liver, where it is converted to a physiologically active substance called *folinic acid.* Folinic acid functions as a coenzyme that is necessary for the metabolism of certain amino acids and for the synthesis of DNA. It also acts with cyanocobalamin in promoting production of normal red blood cells.

   Good sources of folacin include liver, leafy green vegetables, whole-grain cereals, and legumes. A deficiency of folacin leads to *megaloblastic anemia,* which is characterized by a reduced number of normal red blood cells and the presence of large, nucleated red cells. Folacin deficiency has been linked to neural tube defects, in which the tube that becomes the central nervous system in a fetus fails to close entirely. Neural tube defects include spina bifida and anencephaly. Taking synthetic folic acid supplements just before and during pregnancy can greatly reduce the risk of a neural tube defect.

Vitamin B_{12} (cyanocobalamin)

FIGURE 18.14

Vitamin B_{12}, which has the most complex molecular structure of the vitamins, contains cobalt (Co).

> Naturally occurring folate is a mixture of compounds that have the same activity as synthetic folic acid, also called pteroylmonoglutamic acid. However, synthetic folic acid is much more stable and enters the bloodstream much more readily. This difference has led to confusion. For example, 200 micrograms of synthetic folic acid is prescribed to treat anemia, but the same effect requires 400 micrograms of folate from foods. Synthetic folic acid used to enrich grain foods has a greater effect on health than folate.

8. **Biotin.** Biotin is a simple compound that is stable to the effects of heat, acids, and light but may be destroyed by oxidation or bases. (See fig. 18.16 for the molecular structure of biotin.)

Biotin is a coenzyme in metabolic pathways for amino acids and fatty acids. It also plays a role in the synthesis of the purine nitrogenous bases of nucleic acids.

Metabolically active organs such as the brain, liver, and kidneys store small quantities of biotin. Bacteria that inhabit the intestinal tract synthesize biotin. The vitamin is widely distributed in foods, and dietary deficiencies are rare. Good sources include liver, egg yolk, nuts, legumes, and mushrooms.

1 Which biochemicals comprise the vitamin B complex?

2 Which foods are good sources of vitamin B complex?

3 Which of the B-complex vitamins can be synthesized from tryptophan?

4 What is the general function of each member of the B complex?

Vitamin C, or ascorbic acid. Ascorbic acid is a crystalline compound that contains six carbon atoms. Chemically, it is similar to the monosaccharides (fig. 18.15). Vitamin C is one of the least stable of the vitamins in that oxidation, heat, light, or bases destroy it. However, vitamin C is fairly stable in acids.

Ascorbic acid is necessary for the production of the connective tissue protein *collagen,* for conversion of folacin to folinic acid, and in the metabolism of certain amino acids. It also promotes iron absorption and synthesis of certain hormones from cholesterol.

Although vitamin C is not stored in any great amount, tissues of the adrenal cortex, pituitary gland, and intestinal glands contain high concentrations. Excess vitamin C is excreted in the urine or oxidized.

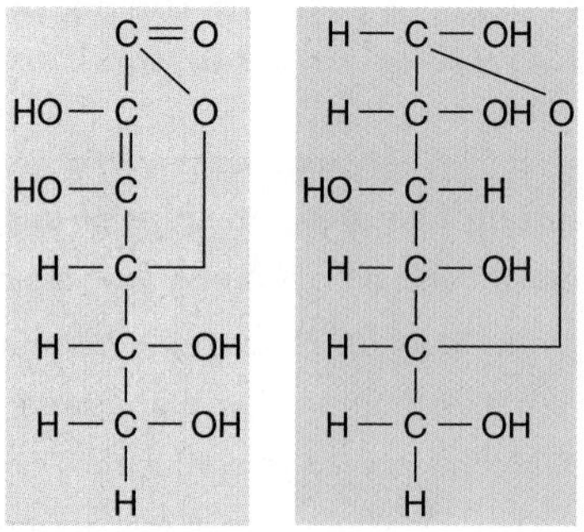

Individual requirements for ascorbic acid may vary. Ten mg per day is sufficient to prevent deficiency symptoms, and 80 mg per day saturate the tissues within a few weeks. Many nutritionists recommend a daily adult intake of 60 mg, which is enough to replenish normal losses and to provide a satisfactory level for cellular requirements.

Ascorbic acid is fairly widespread in plant foods; particularly high concentrations are found in citrus fruits and tomatoes. Leafy green vegetables are also good sources.

Prolonged deficiency of ascorbic acid leads to *scurvy,* which occurs more frequently in infants and children. Scurvy produces abnormal bone development and swollen, painful joints. Because of a tendency for cells to pull apart in scurvy, the gums may swell and bleed easily, resistance to infection is lowered, and wounds heal slowly. If a woman takes large doses of ascorbic acid during pregnancy, the newborn may develop symptoms of scurvy when the daily dose of the vitamin drops after birth. Table 18.9 summarizes the water-soluble vitamins and their characteristics.

1 What factors destroy vitamin C?

2 What is the function of vitamin C?

3 Which foods are good sources of vitamin C?

4 What are symptoms of vitamin C deficiency?

Minerals

Carbohydrates, lipids, proteins, and vitamins are all organic compounds. Dietary **minerals** are inorganic elements that are essential in human metabolism. These elements are usually extracted from the soil by plants, and humans obtain them from plant foods or from animals that have eaten plants.

Characteristics of Minerals

Minerals are responsible for about 4% of body weight and are most concentrated in the bones and teeth. The minerals *calcium* and *phosphorus* are very abundant in these tissues.

Minerals are usually incorporated into organic molecules. For example, phosphorus is found in phospholipids, iron in hemoglobin, and iodine in thyroxine. However, some minerals are part of inorganic compounds, such as the calcium phosphate of bone. Other minerals are free ions, such as the sodium, chloride, and calcium ions in the blood.

Minerals comprise parts of the structural materials of all cells. They also assist enzymes, contribute to the osmotic pressure of body fluids, and help conduct nerve impulses, contract muscle fibers, coagulate blood, and

FIGURE 18.15
Vitamin C is chemically similar to some 6-carbon monosaccharides.

| TABLE 18.9 | Water-Soluble Vitamins | 713 |

Vitamin	Characteristics	Functions	Sources and RDA* for Adults	Conditions Associated with	
				Excesses	*Deficiencies*
Thiamine (Vitamin B₁)	Destroyed by heat and oxygen, especially in alkaline environment	Part of coenzyme required for oxidation of carbohydrates; coenzyme required in ribose synthesis	Lean meats, liver, eggs, whole-grain cereals, leafy green vegetables, legumes 1.5 mg	None known	Beriberi, muscular weakness, heart enlarges
Riboflavin (Vitamin B₂)	Stable to heat, acids, and oxidation; destroyed by bases and ultraviolet light	Part of enzymes and coenzymes, such as FAD, required for oxidation of glucose and fatty acids and for cellular growth	Meats, dairy products, leafy green vegetables, whole-grain cereals 1.7 mg	None known	Dermatitis, blurred vision
Niacin (Nicotinic acid)	Stable to heat, acids, and bases; converted to niacinamide by cells; synthesized from tryptophan	Part of coenzymes NAD and NADP required for oxidation of glucose and synthesis of proteins, fats, and nucleic acids	Liver, lean meats, peanuts, legumes 20 mg	Hyperglycemia, vasodilation, gout	Pellagra, photosensitive dermatitis, diarrhea, mental disorders
Pantothenic acid	Destroyed by heat, acids, and bases	Part of coenzyme A required for oxidation of carbohydrates and fats	Meats, whole-grain cereals, legumes, milk, fruits, vegetables 10 mg	None known	Rare, loss of appetite, mental depression, muscle spasms
Vitamin B₆	Group of three compounds; stable to heat and acids; destroyed by oxidation, bases, and ultraviolet light	Coenzyme required for synthesis of proteins and various amino acids, for conversion of tryptophan to niacin, for production of antibodies, and for nucleic acid synthesis	Liver, meats, bananas, avocados, beans, peanuts, whole-grain cereals, egg yolk 2 mg	Numbness	Rare, convulsions, vomiting, seborrhea lesions
Cyanocobalamin (Vitamin B₁₂)	Complex, cobalt-containing compound; stable to heat; inactivated by light, strong acids, and strong bases; absorption regulated by intrinsic factor from gastric glands; stored in liver	Part of coenzyme required for synthesis of nucleic acids and for metabolism of carbohydrates; plays role in myelin synthesis required for normal red blood cell production	Liver, meats, milk, cheese, eggs 3–6 µg	None known	Pernicious anemia
Folacin (Folic acid)	Occurs in several forms; destroyed by oxidation in acid environment or by heat in alkaline environment; stored in liver where it is converted into folinic acid	Coenzyme required for metabolism of certain amino acids and for DNA synthesis; promotes production of normal red blood cells	Liver, leafy green vegetables, whole-grain cereals, legumes 0.4 mg	None known	Megaloblastic anemia
Biotin	Stable to heat, acids, and light; destroyed by oxidation and bases	Coenzyme required for metabolism of amino acids and fatty acids and for nucleic acid synthesis	Liver, egg yolk, nuts, legumes, mushrooms 0.3 mg	None known	Rare, elevated blood cholesterol, nausea, fatigue, anorexia
Ascorbic acid (Vitamin C)	Chemically similar to monosaccharides; stable in acids but destroyed by oxidation, heat, light, and bases	Required for collagen production, conversion of folacin to folinic acid, and metabolism of certain amino acids; promotes absorption of iron and synthesis of hormones from cholesterol	Citrus fruits, tomatoes, potatoes, leafy green vegetables 60 mg	Exacerbates gout and kidney stone formation	Scurvy, lowered resistance to infection, wounds heal slowly

*RDA = recommended dietary allowance.

maintain pH. The physiologically active form of minerals is the ionized form, such as Ca^{+2}.

Homeostatic mechanisms regulate the concentrations of minerals in body fluids. This ensures that excretion of minerals matches intake.

1 How do minerals differ from other nutrients?

2 What are the major functions of minerals?

3 Which are the most abundant minerals in the body?

Major Minerals

Calcium and phosphorus account for nearly 75% by weight of the mineral elements in the body; thus, they are **major minerals** (macrominerals). Other major minerals, each of which accounts for 0.05% or more of body weight, include potassium, sulfur, sodium, chlorine, and magnesium. Descriptions of the major minerals follow:

1. **Calcium.** Calcium (Ca) is widely distributed in cells and body fluids, even though 99% of the body's supply is in the inorganic salts of the bones and teeth. It is essential for nerve impulse conduction, muscle fiber contraction, and blood coagulation. Calcium also decreases the permeability of cell membranes and activates certain enzymes.

    The amount of calcium absorbed varies with a number of factors. For example, the proportion of calcium absorbed increases as the body's need for calcium increases. Vitamin D and high protein intake promote calcium absorption; increased motility of the digestive tract or an excess intake of fats decreases absorption. Consequently, the amount of dietary calcium needed to supply cells may vary. However, a daily intake of 800 mg is sufficient to cover adult requirements in spite of variations in absorption.

    Only a few foods contain significant amounts of calcium. Milk and milk products and fish with bones, such as salmon or sardines, are the richest sources. Leafy green vegetables, such as mustard greens, turnip greens, and kale, are good sources, but because one must consume large amounts of these vegetables to obtain sufficient minerals, most people must regularly consume milk or milk products to maintain an adequate intake of calcium.

    Calcium deficiency in children causes stunted growth, misshapen bones, and enlarged wrists and ankles. In adults, such a deficiency may remove calcium from the bones, thinning them and raising risk of fracture. Because calcium is required for normal closing of the sodium channels in nerve cell membranes, too little calcium (hypocalcemia) can cause tetany. Extra calcium demands in pregnancy can cause cramps.

2. **Phosphorus.** Phosphorus (P) is responsible for about 1% of total body weight, and most of it is incorporated in the calcium phosphate of bones and teeth. The remainder serves as structural components and plays important roles in nearly all metabolic reactions. Phosphorus is a constituent of nucleic acids, many proteins, some enzymes, and some vitamins. It is also found in the phospholipids of cell membranes, in the energy-carrying molecule ATP, and in the phosphates of body fluids that regulate pH. (Review the molecular structure of ATP in fig. 4.7.)

    The recommended daily adult intake of phosphorus is 800 mg, and since this mineral is abundant in protein foods, diets adequate in proteins are also adequate in phosphorus. Phosphorus-rich foods include meats, cheese, nuts, whole-grain cereals, milk, and legumes.

1 What are the functions of calcium?

2 What are the functions of phosphorus?

3 Which foods are good sources of calcium and phosphorus?

3. **Potassium.** Potassium (K) is widely distributed throughout the body and is concentrated inside cells rather than in extracellular fluids. On the other hand, sodium, which has similar chemical properties, is concentrated outside cells. The ratio of potassium to sodium within a cell is 10:1, whereas the ratio outside the cell is 1:28.

    Potassium helps maintain intracellular osmotic pressure and pH. It promotes reactions of carbohydrate and protein metabolism and plays a vital role in establishing the membrane potential that occurs in nerve impulse conduction and muscle fiber contraction.

    Nutritionists recommend a daily adult intake of 2.5 grams (2,500 mg) of potassium. Because this mineral is widely distributed in foods, a typical adult diet provides between 2 and 6 grams each day. Dietary potassium deficiency is rare, but it may occur for other reasons. For example, when a person has diarrhea, the intestinal contents may pass through the digestive tract so rapidly that potassium absorption is greatly reduced. Vomiting or using diuretic drugs may also deplete potassium. The consequences of such losses may include muscular weakness, cardiac abnormalities, and edema.

    Foods rich in potassium are avocados, dried apricots, meats, milk, peanut butter, potatoes, and bananas. Citrus fruits, apples, carrots, and tomatoes provide lesser amounts.

1 How is potassium distributed in the body?

2 What is the function of potassium?

3 Which foods are good sources of potassium?

4. **Sulfur.** Sulfur (S) is responsible for about 0.25% of body weight and is widely distributed through tissues. It is particularly abundant in skin, hair, and nails. Most sulfur is part of the amino acids *methionine* and *cysteine.* Other sulfur-containing compounds include thiamine, insulin, and biotin (fig. 18.16). In addition, sulfur is a constituent of mucopolysaccharides found in cartilage, tendons, and bones and of sulfolipids that are in the liver, kidneys, salivary glands, and brain.

No daily requirement for sulfur has been established. It is thought, however, that a diet providing adequate amounts of protein will also meet the body's sulfur requirement. Good food sources of this mineral include meats, milk, eggs, and legumes.

5. **Sodium.** About 0.15% of adult body weight is sodium (Na), which is widely distributed throughout the body. Only about 10% of this mineral is inside the cells, and about 40% is within the extracellular fluids. The remainder is bound to the inorganic salts of bones.

Sodium is readily absorbed from foods by active transport. The kidneys regulate the blood concentration of sodium under the influence of the adrenal cortical hormone *aldosterone,* which causes the kidneys to reabsorb sodium while expelling potassium.

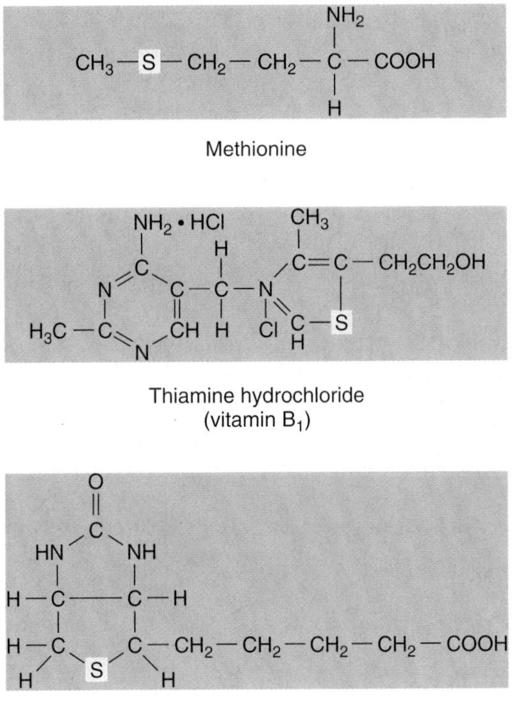

Methionine

Thiamine hydrochloride
(vitamin B$_1$)

Biotin

FIGURE 18.16
Three examples of essential sulfur-containing nutrients.

Sodium makes a major contribution to the solute concentration of extracellular fluids and thus helps regulate water movement between cells and their surroundings. It is necessary for nerve impulse conduction and muscle fiber contraction and helps to move substances, such as chloride ions, through cell membranes (see chapter 21, p. 817).

The usual human diet probably provides more than enough sodium to meet the body's requirements. Sodium may be lost as a result of diarrhea, vomiting, kidney disorders, sweating, or using diuretics. Sodium loss may cause a variety of symptoms, including nausea, muscular cramps, and convulsions.

The amount of sodium naturally present in foods varies greatly, and it is commonly added to foods in the form of table salt (sodium chloride). In some geographic regions, drinking water contains significant concentrations of sodium. Foods high in sodium include cured ham, sauerkraut, cheese, and graham crackers.

1 In which compounds and tissues of the body is sulfur found?

2 Which hormone regulates the blood concentration of sodium?

3 What are the functions of sodium?

6. **Chlorine.** Chlorine (Cl) in the form of chloride ions is widely distributed throughout the body, although it is most highly concentrated in cerebrospinal fluid and in gastric juice. Together with sodium, chlorine helps to maintain the solute concentration of extracellular fluids, regulate pH, and maintain electrolyte balance. It is also essential for the formation of hydrochloric acid in gastric juice, and it functions in the transport of carbon dioxide by red blood cells.

Chlorine and sodium are usually obtained together in the form of table salt (sodium chloride), and as in the case of sodium, an ordinary diet usually provides considerably more chlorine than the body requires. Vomiting, diarrhea, kidney disorders, sweating, or using diuretics can deplete chlorine in the body.

7. **Magnesium.** Magnesium (Mg) is responsible for about 0.05% of body weight and is found in all cells. It is particularly abundant in bones in the form of phosphates and carbonates.

Magnesium is important in ATP-forming reactions in mitochondria, as well as in breaking down ATP to ADP. Therefore, it is important in providing energy for cellular processes.

Magnesium absorption in the intestinal tract adapts to dietary intake of the mineral. When the intake of magnesium is high, a smaller percentage is absorbed from the intestinal tract, and when the intake is low, a larger percentage is absorbed.

Absorption increases as protein intake increases, and decreases as calcium and vitamin D intake increase. Bone tissue stores a reserve supply of magnesium, and excess is excreted in the urine.

The recommended daily allowance of magnesium is 300 mg for females and 350 mg for males. A typical diet usually provides only about 120 mg of magnesium for every 1,000 calories, barely meeting the body's needs. Good sources of magnesium include milk and dairy products (except butter), legumes, nuts, and leafy green vegetables. Table 18.10 summarizes the major minerals.

TABLE 18.10	Major Minerals				
Mineral	**Distribution**	**Functions**	**Sources and RDA* for Adults**	**Conditions Associated with**	
				Excesses	**Deficiencies**
Calcium (Ca)	Mostly in the inorganic salts of bones and teeth	Structure of bones and teeth; essential for nerve impulse conduction, muscle fiber contraction, and blood coagulation; increases permeability of cell membranes; activates certain enzymes	Milk, milk products, leafy green vegetables 800 mg	Kidney stones	Stunted growth, misshapen bones, fragile bones
Phosphorus (P)	Mostly in the inorganic salts of bones and teeth	Structure of bones and teeth; component in nearly all metabolic reactions; constituent of nucleic acids, many proteins, some enzymes, and some vitamins; occurs in cell membrane, ATP, and phosphates of body fluids	Meats, cheese, nuts, whole-grain cereals, milk, legumes 800 mg	None known	Stunted growth
Potassium (K)	Widely distributed; tends to be concentrated inside cells	Helps maintain intracellular osmotic pressure and regulate pH; promotes metabolism; required for nerve impulse conduction and muscle fiber contraction	Avocados, dried apricots, meats, nuts, potatoes, bananas 2,500 mg	None known	Muscular weakness, cardiac abnormalities, edema
Sulfur (S)	Widely distributed; abundant in skin, hair, and nails	Essential part of various amino acids, thiamine, insulin, biotin, and mucopolysaccharides	Meats, milk, eggs, legumes None established	None known	None known
Sodium (Na)	Widely distributed; large proportion occurs in extracellular fluids and bound to inorganic salts of bone	Helps maintain osmotic pressure of extracellular fluids and regulate water movement; required for conduction of nerve impulses and contraction of muscle fibers; aids in regulation of pH and in transport of substances across cell membranes	Table salt, cured ham, sauerkraut, cheese, graham crackers 2,500 mg	Hypertension, edema	Nausea, muscle cramps, convulsions
Chlorine (Cl)	Closely associated with sodium; most highly concentrated in cerebrospinal fluid and gastric juice	Helps maintain osmotic pressure of extracellular fluids, regulate pH, and maintain electrolyte balance; essential in formation of hydrochloric acid; aids transport of carbon dioxide by red blood cells	Same as for sodium None established	Vomiting	Muscle cramps
Magnesium (Mg)	Abundant in bones	Required in metabolic reactions in mitochondria associated with ATP production; helps breakdown of ATP to ADP	Milk, dairy products, legumes, nuts, leafy green vegetables 300–350 mg	Diarrhea	Neuromuscular disturbances

*RDA = recommended dietary allowance.

1 Where are chloride ions most highly concentrated in the body?

2 Where is magnesium stored?

3 What factors influence the absorption of magnesium from the intestinal tract?

Trace Elements

Trace elements (microminerals) are essential minerals found in minute amounts, each making up less than 0.005% of adult body weight. They include iron, manganese, copper, iodine, cobalt, zinc, fluorine, selenium, and chromium.

Iron (Fe) is most abundant in the blood; is stored in the liver, spleen, and bone marrow; and is found to some extent in all cells. Iron enables *hemoglobin* molecules in red blood cells to carry oxygen (fig. 18.17). Iron is also part of *myoglobin,* which stores oxygen in muscle cells. In addition, iron assists in vitamin A synthesis, is incorporated into a number of enzymes, and is included in the cytochrome molecules that participate in ATP-generating reactions.

An adult male requires from 0.7 to 1 mg of iron daily, and a female needs 1.2 to 2 mg. A typical diet supplies about 10 to 18 mg of iron each day, but only 2% to 10% of the iron is absorbed. For some people, this may not be enough iron. Eating foods rich in vitamin C along with iron-containing foods can increase absorption of this important mineral.

Pregnant women require extra iron to support the formation of a placenta and the growth and development of a fetus. Iron is also required for the synthesis of hemoglobin in a fetus as well as in a pregnant woman, whose blood volume increases by a third.

Liver is the only really rich source of dietary iron, and since liver is not a very popular food, iron is one of the more difficult nutrients to obtain from natural sources in adequate amounts. Foods that contain some iron include lean meats; dried apricots, raisins, and prunes; enriched whole-grain cereals; legumes; and molasses.

Manganese (Mn) is most concentrated in the liver, kidneys, and pancreas. It is necessary for normal growth and development of skeletal structures and other connective tissues. Manganese is part of enzymes that are essential for the synthesis of fatty acids and cholesterol, for urea formation, and for the normal functions of the nervous system.

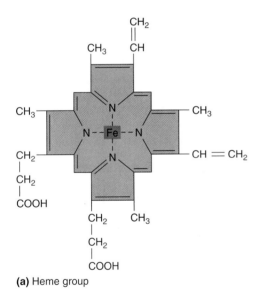

(a) Heme group

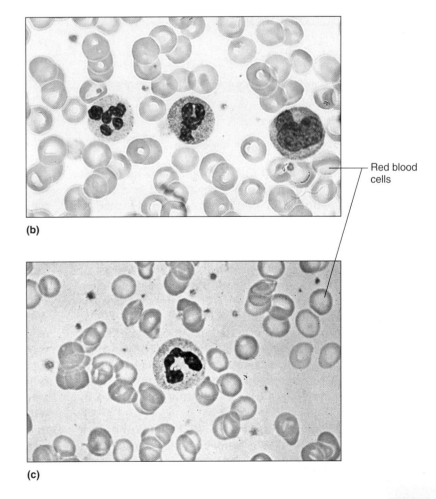

(b)

(c)

Red blood cells

FIGURE 18.17

Iron in hemoglobin. (a) A hemoglobin molecule contains four heme groups, each of which houses a single iron atom (Fe) that can combine with oxygen. Iron deficiency anemia can result from a diet poor in iron-containing foods. The red blood cells in (b) are normal (400×), but many of those in (c) are small and pale (280×). They contain too little hemoglobin, because iron is lacking in the diet. Vegetarians must be especially careful to consume sufficient iron.

The daily requirement for manganese is 2.5–5 mg. The richest sources include nuts, legumes, and whole-grain cereals; leafy green vegetables and fruits are good sources.

A compulsive disorder that may result from mineral deficiency is *pica,* in which people consume huge amounts of nondietary substances such as ice chips, soil, sand, laundry starch, clay and plaster, and even such strange things as hair, toilet paper, matchheads, inner tubes, mothballs, and charcoal. The condition is named for the magpie bird, *Pica pica,* which eats a range of odd things.

Pica affects people of all cultures and was noted as early as 40 B.C. The connection to dietary deficiency stems from the observation that slaves suffering from pica in colonial America recovered when their diets improved, particularly when they were given iron supplements. Another clue comes from a variation on pica called geophagy—"eating dirt"—that affects many types of animals, including humans. Researchers discovered that when parrots eat a certain claylike soil in their native Peru, soil particles bind alkaloid toxins in their seed food and carry the toxins out of the body. Perhaps pica in humans is protective in some way, too.

1 What is the primary function of iron?

2 Why does the usual diet provide only a narrow margin of safety in supplying iron?

3 How is manganese utilized?

4 Which foods are good sources of manganese?

Copper (Cu) is found in all body tissues but is most highly concentrated in the liver, heart, and brain. It is essential for hemoglobin synthesis, bone development, melanin production, and formation of myelin within the nervous system.

A daily intake of 2 mg of copper is sufficient to supply cells. Because a typical adult diet has about 2–5 mg of this mineral, adults seldom develop copper deficiencies. Foods rich in copper include liver, oysters, crabmeat, nuts, whole-grain cereals, and legumes.

Iodine (I) is found in minute quantities in all tissues but is highly concentrated within the thyroid gland. Its only known function is as an essential component of thyroid hormones. (The molecular structures of two of these hormones, thyroxine and triiodothyronine, are shown in fig. 13.20.)

A daily intake of 1 microgram (0.001 mg) of iodine per kilogram of body weight is adequate for most adults. Since the iodine content of foods varies with the iodine content of soils in different geographic regions, many people use *iodized* table salt to season foods to prevent deficiencies.

Cobalt (Co) is widely distributed in the body because it is an essential part of cyanocobalamin (vitamin B_{12}). It is also necessary for the synthesis of several important enzymes.

The amount of cobalt required in the daily diet is unknown. This mineral is found in a great variety of foods, and the quantity in the average diet is apparently sufficient. Good sources of cobalt include liver, lean meats, and milk.

Zinc (Zn) is most concentrated in the liver, kidneys, and brain. It is part of many enzymes involved in digestion, respiration, and bone and liver metabolism. It is also necessary for normal wound healing and for maintaining the integrity of the skin.

The daily requirement for zinc is about 15 mg, and most diets provide 10–15 mg. Since only a portion of this amount may be absorbed, zinc deficiencies may occur. The richest sources of zinc are meats; cereals, legumes, nuts, and vegetables provide lesser amounts.

Fluorine (F), as part of the compound fluoroapatite, replaces hydroxyapatite in teeth, strengthening the enamel and preventing dental caries. **Selenium** (Se) is stored in the liver and kidneys. It is a constituent of certain enzymes and participates in heart function. This mineral is found in lean meats, whole-grain cereals, and onions. **Chromium** (Cr) is widely distributed throughout the body and regulates glucose utilization. It is found in liver, lean meats, yeast, and pork kidneys. Table 18.11 summarizes the characteristics of trace elements.

The term "dietary supplement" is used to refer to minerals, vitamins, carbohydrates, proteins, and fats—the micronutrients and macronutrients. Clinical Application 18.2 discusses the current broadened meaning of "dietary supplement."

1 How is copper used?

2 What is the function of iodine?

3 Why might zinc deficiencies be common?

Healthy Eating

An adequate diet provides sufficient energy (calories), essential fatty acids, essential amino acids, vitamins, and minerals to support optimal growth and to maintain and repair body tissues. However, because individual nutrient requirements vary greatly with age, sex, growth rate, amount of physical activity, and level of stress, as well as with genetic and environmental factors, it is not possible to design a diet that is adequate for everyone. However, nutrients are so widely distributed in foods that satisfactory amounts and combinations can usually be obtained in spite of individual food preferences.

Trace Element	Distribution	Functions	Sources and RDA* for Adults	Conditions Associated with	
				Excesses	**Deficiencies**
Iron (Fe)	Primarily in blood; stored in liver, spleen, and bone marrow	Part of hemoglobin molecule; catalyzes formation of vitamin A; incorporated into a number of enzymes	Liver, lean meats, dried apricots, raisins, enriched whole-grain cereals, legumes, molasses 10–18 mg	Liver damage	Anemia
Manganese (Mn)	Most concentrated in liver, kidneys, and pancreas	Occurs in enzymes required for fatty acids and cholesterol synthesis, formation of urea, and normal functioning of the nervous system	Nuts, legumes, whole-grain cereals, leafy green vegetables, fruits 2.5–5 mg	None known	None known
Copper (Cu)	Most highly concentrated in liver, heart, and brain	Essential for hemoglobin synthesis, bone development, melanin production, and myelin formation	Liver, oysters, crabmeat, nuts, whole-grain cereals, legumes 2–3 mg	Rare	Rare
Iodine (I)	Concentrated in thyroid gland	Essential component for synthesis of thyroid hormones	Food content varies with soil content in different geographic regions; iodized table salt 0.15 mg	Decreased synthesis of thyroid hormones	Goiter
Cobalt (Co)	Widely distributed	Component of cyanocobalamin; required for synthesis of several enzymes	Liver, lean meats, milk None established	Heart disease	Pernicious anemia
Zinc (Zn)	Most concentrated in liver, kidneys, and brain	Constituent of several enzymes involved in digestion, respiration, bone metabolism, liver metabolism; necessary for normal wound healing and maintaining integrity of the skin	Meats, cereals, legumes, nuts, vegetables 15 mg	Slurred speech, problems walking	Depressed immunity, loss of taste and smell, learning difficulties
Fluorine (F)	Primarily in bones and teeth	Component of tooth structure	Fluoridated water 1.5–4 mg	Mottled teeth	None known
Selenium (Se)	Concentrated in liver and kidneys	Occurs in enzymes	Lean meats, fish, cereals 0.05–2 mg	Vomiting, fatigue	None known
Chromium (Cr)	Widely distributed	Essential for use of carbohydrates	Liver, lean meats, wine 0.05–2 mg	None known	None known

*RDA = recommended dietary allowance.

It is very difficult to keep track of the different nutrients in a diet and be certain that an adequate amount of each is consumed daily. Nutritionists have devised several ways to help consumers make healthy food choices. Most familiar is the RDA guideline that has appeared on several charts in the chapter. *RDA* stands for United States Recommended Daily Allowance. An RDA is actually the upper limit of another measurement, called the Recommended Dietary Allowance, which lists optimal calorie intake for each sex at various ages and the amounts of vitamins and minerals needed to avoid deficiency or excess conditions. The RDA values on food packages are set high, ensuring that most people who fol-

low them receive sufficient amounts of each nutrient. Government panels meet every five years to evaluate the RDAs in light of new data.

Placing foods into groups is a simpler way to follow a healthy diet. Diagrams called **food pyramids** such as the one in figure 18.18, show at a glance that complex carbohydrates should comprise the bulk of the diet. A four-food-group plan from the 1950s depicted meat as equal to grains, dairy products, and fruits and vegetables in number of servings. A plan from the 1940s had eight categories, including separate groups for butter and margarine, and for eggs—foods now associated with the development of heart disease. In the 1920s, an entire food

DIETARY SUPPLEMENTS—PROCEED WITH CAUTION

Displayed prominently among the standard vitamin and mineral preparations in the pharmacy or health food store is a dizzying collection of products (fig. 18B). Some obviously come from organisms, such as bee pollen and shark cartilage; others have chemical names, such as glucosamine with chondroitin. Still others have a mystical aura, such as St. John's Wort. These "dietary supplements" are neither food nor drug, although they do contain active compounds that may function as pharmaceuticals in the human body.

The Dietary Supplements Health and Education Act (DSHEA) of 1994 amended earlier regulations in the United States, in response to consumer demand to have more control over dietary approaches to maintaining health. While the act loosens safety requirements for these products, it also calls for further research into how they work.

Past definitions of "dietary supplement" meant only essential nutrients—carbohydrates, proteins, fats, vitamins, or minerals. The 1994 act expanded the definition to

"a product (other than tobacco) that is intended to supplement the diet that bears or contains one or more of the following dietary ingredients: a vitamin, a mineral, an herb or other botanical, an amino acid, a dietary substance for use by man to supplement the diet by increasing the total daily intake, or a concentrate, metabolite, constituent, extract, or combinations of these ingredients."

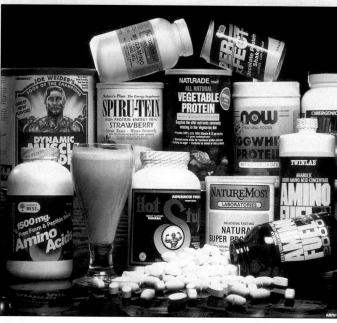

FIGURE 18B

Food supplements often contain or consist of natural components that function as drugs in the human body. Proceed with caution!

Labels cannot claim that a dietary supplement diagnoses, prevents, mitigates, treats, or cures any specific disease. The language is very positive. For example, Valerian root "promotes restful sleep," St. John's Wort "may help enhance mood," echinacea and goldenseal "may help support the immune system," and glucosamine with chondroitin "improves skeletal function." Curative or preventative claims must be firmly backed up with evidence. Examples include the ability of folic acid to reduce the incidence of neural tube defects, and of calcium to reduce the risk of osteoporosis. Many label claims, however, are unclear. One product promoted as "supporting the healthy functioning of the heart muscle," for example, consists of sheep spleen, pig intestine, various unspecified cow parts, mushrooms, pea extract, grains and soy. There isn't a hint as to how this complex concoction affects cardiac muscle.

Because many dietary supplements contain pharmaceutical agents, a physician should be consulted before using these products, particularly if a person has a serious illness or is taking medication, because the active ingredients in supplements may interact with other drugs. For example, the active ingredient in St. John's Wort, hypericin, lowers blood levels of nearly half of all prescription drugs, by interfering with liver enzymes that metabolize many drugs. Some patients have experienced intracranial hemorrhage after taking ginkgo biloba, a tree extract reported to enhance memory.

Certain dietary supplements are of dubious value. For example, the marketing of shark cartilage followed initial studies that suggested sharks do not get cancer. Since sharks have cartilaginous skeletons, the idea arose that their cartilage somehow protects against cancer. It turned out that sharks indeed get cancer, and shark cartilage has no magical properties. Similarly, anyone who understands the basics of cellular respiration realizes why supplements of pyruvic acid or ATP are not necessary to boost energy levels. Some health-food stores sell DNA, which is merely very expensive brewer's yeast, and totally unnecessary, since any food consisting of cells is packed with DNA. The list is quite long of supplements with little scientific evidence of value. Yet dietary supplements are a multibillion dollar industry. ∎

group was devoted to sweets! A new food pyramid will distinguish healthful fats (mono- and polyunsaturated fats) from dangerous (saturated) fats, and indicate how quickly certain complex carbohydrates are broken down to release glucose into the bloodstream. Whole grains

(brown) do this slowly, delaying hunger, compared to bleached grains (white).

When making individual food choices, it helps to read and understand food labels. Disregard claims such as "light" and "low fat" and skip right to the calories of

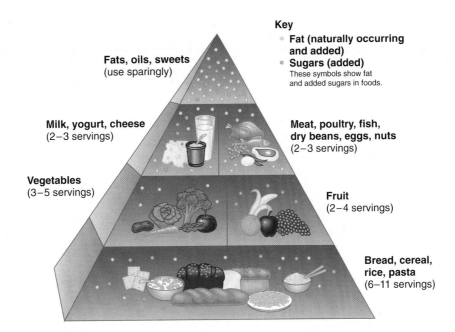

Key
- Fat (naturally occurring and added)
- Sugars (added)
These symbols show fat and added sugars in foods.

Fats, oils, sweets
(use sparingly)

Milk, yogurt, cheese
(2–3 servings)

Meat, poultry, fish,
dry beans, eggs, nuts
(2–3 servings)

Vegetables
(3–5 servings)

Fruit
(2–4 servings)

Bread, cereal,
rice, pasta
(6–11 servings)

FIGURE 18.18

The U.S. Department of Agriculture introduced this food pyramid in 1992 as a guideline to healthy eating. In contrast to former food-group plans, the pyramid gives an instant idea of which foods should make up the bulk of the diet—whole grains, fruits, and vegetables. A new food pyramid will distinguish among types of fats and how quickly particular complex carbohydrates are broken down to release glucose into the bloodstream (the glycemic index).

different ingredients. "Light" can mean many things: that the nutritionally altered form contains one-third fewer calories or half the fat of the reference food, that the sodium content has been reduced by 50%, or it may describe the texture and color of the food. "Low fat" indicates 3 grams of fat or less per serving. Many times when fat is removed, sugar is added, so the calories may be more, not less. Ideally, fat percentage should be below 30, and that of saturated fats below 10. Ingredients are listed in descending order by weight. Clinical Application 18.3 discusses some ways that understanding nutrition can help athletic performance.

1 What is an adequate diet?

2 What factors influence individual needs for nutrients?

3 Describe the various ways consumers can make wise food choices.

Malnutrition

Malnutrition (mal″nu-trish′un) is poor nutrition that results from a lack of essential nutrients or a failure to utilize them. It may result from *undernutrition* and produce the symptoms of deficiency diseases, or it may be due to *overnutrition* arising from excess nutrient intake.

The factors leading to malnutrition vary. For example, a deficiency condition may stem from lack of availability or poor quality of food. On the other hand, malnutrition may result from overeating or taking too many vitamin supplements. Malnutrition from diet alone is called *primary malnutrition.*

The Food and Drug Administration allows the following specific food and health claims:

- Dietary calcium decreases risk of osteoporosis (a bone-thinning condition).
- A low-fat diet lowers risk of some cancers.
- A diet low in saturated fat and cholesterol lowers risk of coronary heart disease.
- Fiber, fruits and vegetables, and whole grains reduce the risk of some cancers and coronary heart disease.
- Lowering sodium intake lowers blood pressure.
- Folic acid lowers the risk of neural tube defects.

Take any other health claims with a grain of salt!

Secondary malnutrition occurs when an individual's characteristics make a normally adequate diet insufficient. For example, a person who secretes very little bile salts may develop a deficiency of fat-soluble vitamins because bile salts promote absorption of fats. Likewise, severe and prolonged emotional stress may lead to secondary malnutrition, because stress can change hormonal concentrations, and such changes may result in amino acid breakdown or excretion of nutrients.

Starvation

A healthy human can stay alive for fifty to seventy days without food. In prehistoric times, this margin allowed survival during seasonal famines. In some areas of Africa today, famine is not a seasonal event but a constant condition, and millions of people have starved to death.

NUTRITION AND THE ATHLETE

Can a marathoner, cross-country skier, weight lifter, or competitive swimmer eat to win? A diet of 60% or more carbohydrate, 18% protein, and 22% fat should be adequate to support frequent, strenuous activity.

Macronutrients

As the source of immediate energy, carbohydrates are the athlete's best friend. Athletes should get the bulk of their carbohydrates from vegetables and grains in frequent meals, because the muscles can store only 1,800 calories worth of glycogen.

Many people erroneously believe that an athlete needs protein or amino acid supplements. Excess dietary protein, however, can strain the kidneys in ridding the body of the excess nitrogen, dehydrating the athlete as more water is used in urine. The only evidence that clearly supports a benefit is that consuming amino acids soon after exercise stimulates muscle cells to take up amino acids, thereby perhaps accelerating manufacture of muscle proteins.

The American Dietetic Association suggests that athletes eat 1 gram of protein per kilogram of weight per day, compared to 0.8 gram for nonathletes. Athletes should not rely solely on meat for protein, because these foods can be high in fat. Supplements are necessary for only young athletes at the start of training, under a doctor's supervision. Too little protein in an athlete is linked to "sports anemia," in which hemoglobin levels decline and blood may appear in the urine.

Water

A sedentary person loses a quart of water a day as sweat; an athlete may lose 2 to 4 quarts of water an hour! To stay hydrated, athletes should drink 3 cups of cold water two hours before an event, then 2 more cups fifteen minutes before the event, and small amounts every fifteen minutes during the event. They should drink afterward too. Another way to determine water needs is to weigh in before and after training. For each pound lost, athletes should drink a pint of water. They should also avoid sugary fluids, which slow water's trip through the digestive system, and alcohol, which increases fluid loss.

Vitamins and Minerals

If an athlete eats an adequate, balanced diet, vitamin supplements are not needed. Supplements of sodium and potassium are usually not needed either, because the active body naturally conserves these nutrients. To be certain of enough sodium, athletes may want to salt their food; to get enough potassium, they can eat bananas, dates, apricots, oranges, or raisins.

A healthy pregame meal should be eaten two to five hours before the game, provide 500 to 1,500 calories, and include 4 or 5 cups of fluid. The pregame meal should also be high in carbohydrates, which taste good, provide energy, and are easy to digest.

Creatine

Athletes should be wary of dietary supplements that may be harmful. Consider creatine. Among professional football players, 25 to 75% use creatine, and surveys of high school athletes show that about 8% of them take the supplements. Many athletes think that creatine is a safe alternative to steroids for bulking up muscles, but taken in excess, it actually destroys muscles.

Cells normally synthesize creatine from the amino acids arginine, glycine, and methionine. Creatine provides energy to muscle cells by phosphorylating ADP to generate ATP, and is then metabolized to creatinine. Taking too much creatine can disturb the distribution of water in tissues, as a 24-year-old weight lifter painfully learned. One morning, he awoke with pain in his thighs so severe that he went to an emergency department, where he passed blood and protein in his urine, his breathing became labored, and he rapidly developed an enlarged heart and lungs. Water was rushing into his skeletal muscle cells, swelling vital organs.

The weight lifter's enlarged muscles were not stronger, as the creatine product label had promised, but breaking down, from a condition called rhabdomyolysis. He had greatly exceeded the recommended dose schedule, thinking that more is better. Although creatine supplementation had never been monitored for more than 10 weeks, this young athlete took large doses for more than a year! Many months of physical therapy enabled him to recover.

The Food and Drug Administration has received many adverse event reports of muscle cramps, seizures, diarrhea, loss of appetite, muscle strains, and dehydration associated with creatine use. In 1997, three college wrestlers died from dehydration associated with using this dietary supplement. ∎

Starvation is also seen in hunger strikers, in prisoners of concentration camps, and in sufferers of psychological eating disorders such as *anorexia nervosa* and *bulimia.*

Whatever the cause, the starving human body begins to digest itself. After only one day without eating, the body's reserves of sugar and starch are gone. Next, the body extracts energy from fat and then from muscle protein. By the third day, hunger ceases as the body uses energy from fat reserves. Gradually, metabolism slows to conserve energy, blood pressure drops, the pulse slows, and chills set in. Skin becomes dry and hair falls out as the proteins in these structures are broken down to release amino acids that are used for the more vital functioning of the brain, heart, and lungs. When the immune system's antibody proteins are dismantled for their amino acids, protection against infection declines. Mouth sores and anemia develop, the heart beats irregularly, and bone begins to degenerate. After several weeks without food, coordination is gradually lost. Near the end, the starving human is blind, deaf, and emaciated.

Marasmus and Kwashiorkor

Lack of nutrients is called **marasmus** (mah-raz'mus), and it causes people to resemble living skeletons (fig. 18.19*a*). Children under the age of two with marasmus often die of measles or other infections, their immune systems too weakened to fight off normally mild viral illnesses.

Some starving children do not look skeletal but have protruding bellies. These youngsters suffer from a form of protein starvation called **kwashiorkor** (kwash-e-or'kor), which in the language of Ghana means "the evil spirit which infects the first child when the second child is born." (fig. 18.19*b*) Kwashiorkor typically appears in a child who has recently been weaned from the breast, usually because of the birth of a sibling. The switch from protein-rich breast milk to the protein-poor gruel that is the staple of many developing nations is the source of this protein deficiency. The children's bellies swell with fluid, which is filtered from capillaries in greater than normal quantities due to a lack of plasma proteins. This condition is called **ascites** (ah-si'tēz). Their skin may develop lesions. Infections overwhelm the body as the immune system becomes depleted of its protective antibodies.

Anorexia Nervosa

Anorexia nervosa (an"o-rek'se-ah ner'vo-sah) is self-imposed starvation. The condition is reported to affect 1 out of 250 adolescents, and 95% of them are female, although the true number among males is not known and may be higher than has been thought. The sufferer, typically a well-behaved adolescent girl from an affluent family, perceives herself to be overweight and eats barely enough to survive. She is terrified of gaining weight and usually loses 25% of her original body weight. In addition to eating only small amounts of low-calorie foods, she further loses weight by vomiting, by taking laxatives and diuretics, or by exercising intensely. Her eating behavior is often ritualized. She may meticulously arrange her meager meal on her plate or consume only a few foods. She develops low blood pressure, a slowed or irregular heartbeat, constipation, and constant chilliness. She stops menstruating as her body fat level plunges. Like any starving person, the hair becomes brittle, and the skin dries out. To conserve body heat, she may develop soft, pale, fine body hair called lanugo, normally seen only on a developing fetus.

When the person with anorexia reaches an obviously emaciated state, her parents usually have her hospitalized, where she is fed intravenously so that she does not starve to death or die suddenly of heart failure due to a mineral imbalance. She also receives psychotherapy

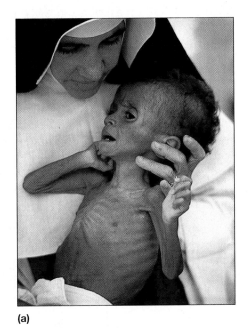
(a)

(b)

FIGURE 18.19
Two types of starvation in the young. (*a*) This child, suffering from marasmus, did not have adequate nutrition as an infant. (*b*) These children suffer from kwashiorkor. Although they may have received adequate nourishment from breast milk early in life, they became malnourished when their diet switched to a watery, white extract from cassava that looks like milk but has very little protein. The lack of protein in the diet causes edema and the ascites that swells their bellies.

and nutritional counseling. Despite these efforts, 15% to 21% of people with anorexia die.

Anorexia nervosa has no known physical cause. One hypothesis is that the person is rebelling against approaching womanhood. Indeed, her body is astonishingly child-like, and she has often ceased to menstruate. She typically has low self esteem and believes that others, particularly her parents, are controlling her life. Her weight is something that she can control. Anorexia can be a one-time, short-term experience or a lifelong obsession.

Bulimia

A person suffering from **bulimia** (bu-lim'e-ah) is often of normal weight. She eats whatever she wants, often in huge amounts, but she then rids her body of the thousands of extra calories by vomiting, taking laxatives, or exercising frantically. For an estimated one in five college students, the majority of them female, "bingeing and purging" appears to be a way of coping with stress.

Sometimes a dentist is the first to spot bulimia by observing a patient with teeth decayed from frequent vomiting. The backs of her hands may bear telltale scratches from efforts to induce vomiting. Her throat is raw and her esophageal lining ulcerated from the stomach acid forced forward by vomiting. The binge and purge cycle is very hard to break, even with psychotherapy and nutritional counseling.

A person with bulimia tends to eat soft foods that can be consumed in large amounts quickly with minimal chewing. Reveals one athletic young woman, "For me a binge consists of a pound of cottage cheese, a head of lettuce, a steak, a loaf of Italian bread, a 10-ounce serving of broccoli, spinach or a head of cabbage, a cake, an 18-ounce pie, with a quart or half gallon of ice cream. When my disease is at its worst, I eat raw oatmeal with butter, laden with mounds of sugar, or a loaf of white bread with butter and syrup poured over it." She follows her "typical" 20,000-calorie binge with hours of bicycle riding, running, and swimming.

1 What is primary malnutrition? Secondary malnutrition?

2 What happens to the body during starvation?

3 How do marasmus and kwashiorkor differ?

4 How do anorexia nervosa and bulimia differ?

Life-Span Changes

Dietary requirements remain generally the same throughout life, but the ability to acquire those nutrients may change drastically. Changing nutrition with age often reflects effects of medical conditions and social and economic circumstances. Medications can dampen appetite directly through side effects such as nausea or altered taste perception or alter a person's mood in a way that prevents eating. Poverty may take a greater nutritional toll on older people who either cannot get out to obtain food or who give whatever is available to younger people.

Medical conditions that affect the ability to obtain adequate nutrition include depression, tooth decay and periodontal disease, diabetes mellitus, lactose intolerance, and alcoholism. These conditions may lead to deficiencies that are not immediately obvious. Vitamin A deficiency, for example, may take months or years to become noticeable because the liver stores this fat-soluble vitamin. Calcium depletion may not produce symptoms, even as the mineral is taken from bones. The earliest symptom of malnutrition, fatigue, may easily be attributed to other conditions or ignored.

Evidence for vitamin D deficiency related to sun avoidance has a long history. The link between lack of sunlight and development of rickets was noted in 1822, and a century later, researchers realized that sun exposure helps reverse the disease in children. Other evidence comes from diverse sources, such as women who wear veils and naval personnel serving three-month tours of duty on submarines.

The basal metabolic rate (BMR) changes with age. It rises from birth to about age five and then declines until adolescence, when it peaks again. During adulthood, the BMR drops in parallel to decreasing activity levels and shrinking muscle mass. In women, it may spike during pregnancy and breastfeeding, when caloric requirements likewise increase. Table 18.12 shows changes in the BMR for adults who are healthy and engage in regular, light exercise.

For all ages, weight gain occurs when energy in exceeds energy out, and weight loss happens when energy out exceeds energy in. Age fifty seems to be a key point in energy balance. For most people, energy balance is positive, and weight is maintained before this age, but afterwards, weight may creep up. However, being aware of a decrease in activity, and curbing food consumption accordingly, enables many people over the age of fifty to maintain their weight.

TABLE 18.12	Energy Requirements Decline with Age	
	cal/day	
Age	**Female**	**Male**
23–50	2,000	2,700
51–74	1,800	2,400
75+	1,600	2,050

It is important to obtain a good balance of nutrients and enough energy throughout life. Numerous studies have linked caloric restriction to increased longevity—in such species as mice and fruit flies. However, these observations cannot be extrapolated to humans because the experimental laboratory animals, although given very little food, were otherwise kept extremely healthy. Human starvation is usually the consequence of many other problems and is more likely to lead to malnutrition than increased longevity.

1 List factors that affect nutrient acquisition.

2 Describe changes in BMR throughout life.

Clinical Terms Related to Nutrition and Metabolism

cachexia (kah-kek′se-ah) State of chronic malnutrition and physical wasting.

casein (ka′se-in) Primary protein found in milk.

celiac disease (se′le-ak dĭ-zēz′) Inability to digest or use fats and carbohydrates.

emaciation (e-ma″se-a′shun) Extreme leanness due to tissue wasting.

hyperalimentation (hi″per-al″ĭ-men-ta′shun) Long-term intravenous nutrition.

hypercalcemia (hi″per-kal-se′me-ah) Excess calcium in the blood.

hypercalciuria (hi″per-kal-se-u′re-ah) Excess excretion of calcium in the urine.

hyperglycemia (hi″per-gli-se′me-ah) Excess glucose in the blood.

hyperkalemia (hi″per-kah-le′me-ah) Excess potassium in the blood.

hypernatremia (hi″per-nah-tre′me-ah) Excess sodium in the blood.

hypoalbuminemia (hi″po-al-bu″mĭ-ne′me-ah) Low level of albumin in the blood.

hypoglycemia (hi″po-gli-se′me-ah) Low level of glucose in the blood.

hypokalemia (hi″po-kah-le′me-ah) Low level of potassium in the blood.

hyponatremia (hi″po-nah-tre′me-ah) Low level of sodium in the blood.

isocaloric (i″so-kah-lo′rik) Containing equal amounts of heat energy.

lipogenesis (lip″o-jen′e-sis) The formation of fat.

nyctalopia (nik″tah-lo′pe-ah) Night blindness.

polyphagia (pol″e-fa′je-ah) Overeating.

proteinuria (pro″te-i-nu′re-ah) Protein in the urine.

CHAPTER SUMMARY

Why We Eat (page 694)

Nutrients include carbohydrates, lipids, proteins, vitamins, and minerals. The ways nutrients are used to support life processes constitute metabolism. Essential nutrients are required for health and body cells cannot synthesize them. Macronutrients include carbohydrates, lipids, and proteins. Micronutrients are vitamins and minerals. Water is also essential. Hormones communicate from the gastrointestinal tract to the hypothalamus to control appetite, and monitor fat stores.

Carbohydrates (page 695)

Carbohydrates are organic compounds that are primarily used to supply cellular energy.

1. Carbohydrate sources
   a. Carbohydrates are ingested in a variety of forms.
   b. Polysaccharides, disaccharides, and monosaccharides are carbohydrates.
   c. Cellulose is a polysaccharide that human enzymes cannot digest, but it provides bulk that facilitates movement of intestinal contents.
2. Carbohydrate utilization
   a. Carbohydrates are absorbed as monosaccharides.
   b. Enzymes in the liver catalyze reactions that convert fructose and galactose into glucose.
   c. Oxidation releases energy from glucose.
   d. Excess glucose is stored as glycogen or reacted to produce fat.
3. Carbohydrate requirements
   a. Most carbohydrates supply energy; some are used to produce sugars.

b. Some cells require a continuous supply of glucose to survive.
   c. If glucose is scarce, amino acids may react to produce glucose.
   d. Humans survive with a wide range of carbohydrate intakes.
   e. Poor nutritional status is usually related to low intake of nutrients other than carbohydrates.

Lipids (page 697)

Lipids are organic compounds that supply energy and are used to build cell structures. They include fats, phospholipids, and cholesterol.

1. Lipid sources
   a. Triglycerides are obtained from foods of plant and animal origins.
   b. Cholesterol is mostly obtained in foods of animal origin.
2. Lipid utilization
   a. Before fats can be used as an energy source, they must be broken down into glycerol and fatty acids.
   b. Beta oxidation decomposes fatty acids.
      (1) Beta oxidation activates fatty acids and breaks them down into segments of two carbon atoms each.
      (2) Fatty acid segments are converted into acetyl coenzyme A, which can then be oxidized in the citric acid cycle.
   c. The liver and adipose tissue control triglyceride metabolism.

d. Liver enzymes can alter the molecular structures of fatty acids.
e. Linoleic acid, linolenic acid, and arachidonic acid are essential fatty acids.
f. The liver regulates cholesterol level by synthesizing or excreting it.

3. Lipid requirements
a. Humans survive with a wide range of lipid intakes.
b. The amounts and types of lipids needed for health are unknown.
c. Fat intake must be sufficient to carry fat-soluble vitamins.

Proteins (page 699)

Proteins are organic compounds that serve as structural materials, act as enzymes, and provide energy. Amino acids are incorporated into various structural and functional proteins, including enzymes. During starvation, tissue proteins may be used as energy sources; thus, the tissues waste away.

1. Protein sources
a. Proteins are mainly obtained from meats, dairy products, cereals, and legumes.
b. During digestion, proteins are broken down into amino acids.
c. The resulting amino acids can be used as building materials, to form enzymes, or as energy sources.
d. Before amino acids can be used as energy sources, they must be deaminated.
e. The deaminated portions of amino acids can be broken down into carbon dioxide and water or used to produce glucose or fat.
f. Eight amino acids are essential for adults, whereas ten are essential for growing children.
g. All essential amino acids must be present at the same time in order for growth and repair of tissues to take place.
h. Complete proteins contain adequate amounts of all the essential amino acids needed to maintain the tissues and promote growth.
i. Incomplete proteins lack adequate amounts of one or more essential amino acids.

2. Nitrogen balance
a. In healthy adults, the gain of protein equals the loss of protein, and a nitrogen balance exists.
b. A starving person has a negative nitrogen balance; a growing child, a pregnant woman, or an athlete in training usually has a positive nitrogen balance.

3. Protein requirements
a. Proteins and amino acids are needed to supply essential amino acids and nitrogen for the synthesis of nitrogen-containing molecules.
b. The consequences of protein deficiencies are particularly severe among growing children.

Energy Expenditures (page 702)

Energy is of prime importance to survival and may be obtained from carbohydrates, fats, or proteins.

1. Energy values of foods
a. The potential energy values of foods are expressed in calories.
b. When energy losses due to incomplete absorption and incomplete oxidation are taken into account, 1 gram of carbohydrate or 1 gram of protein yields about 4 calories, whereas 1 gram of fat yields about 9 calories.

2. Energy requirements
a. The amount of energy required varies from person to person.
b. Factors that influence energy requirements include basal metabolic rate, muscular activity, body temperature, and nitrogen balance.

3. Energy balance
a. Energy balance exists when caloric intake equals caloric output.
b. If energy balance is positive, body weight increases; if energy balance is negative, body weight decreases.

4. Desirable weight
a. The most common nutritional disorders involve caloric imbalances.
b. Average weights of persons 25–30 years of age are desirable for older persons as well.
c. Height–weight guidelines are based on longevity.
d. A person who exceeds the desirable weight by 10% to 20% is called overweight.
e. A person whose body contains excess fatty tissue is obese.

Vitamins (page 705)

Vitamins are organic compounds (other than carbohydrates, lipids, and proteins) that are essential for normal metabolic processes and cannot be synthesized by body cells in adequate amounts.

1. Fat-soluble vitamins
a. General characteristics
(1) Fat-soluble vitamins are carried in lipids and are influenced by the same factors that affect lipid absorption.
(2) They are fairly resistant to the effects of heat; thus, they are not destroyed by cooking or food processing.
b. Vitamin A
(1) Vitamin A occurs in several forms, is synthesized from carotenes, and is stored in the liver.
(2) It is an antioxidant required for production of visual pigments.
c. Vitamin D
(1) Vitamin D is a group of related steroids.
(2) It is found in certain foods and is produced commercially; it can also be synthesized in the skin.
(3) When needed, vitamin D is converted by the kidneys to an active form that functions as a hormone and promotes the intestine's absorption of calcium and phosphorus.
d. Vitamin E
(1) Vitamin E is also an antioxidant.
(2) It is stored in muscles and adipose tissue.
(3) It likely prevents breakdown of polyunsaturated fatty acids and stabilizes cell membranes.
e. Vitamin K
(1) Vitamin K_1 occurs in foods; vitamin K_2 is produced by certain bacteria that normally inhabit the intestinal tract.
(2) Some is stored in the liver.
(3) It is used to produce prothrombin, which is required for blood clotting.

2. Water-soluble vitamins
a. General characteristics
(1) Water-soluble vitamins include the B vitamins and vitamin C.
(2) B vitamins make up a group called the vitamin B complex and oxidize carbohydrates, lipids, and proteins.

b. Vitamin B complex
   (1) Thiamine (vitamin B_1)
       (a) Thiamine functions as part of coenzymes that oxidize carbohydrates and synthesize essential sugars.
       (b) Small amounts are stored in the tissues; excess is excreted in the urine.
       (c) Quantities needed vary with caloric intake.
   (2) Riboflavin (vitamin B_2)
       (a) Riboflavin functions as part of several enzymes and coenzymes that are essential to the oxidation of glucose and fatty acids.
       (b) Its absorption is regulated by an active transport system; excess is excreted in the urine.
       (c) Quantities required vary with caloric intake.
   (3) Niacin (nicotinic acid)
       (a) Niacin functions as part of coenzymes required for the oxidation of glucose and for the synthesis of proteins and fats.
       (b) It can be synthesized from tryptophan; daily requirement varies with the tryptophan intake.
   (4) Pantothenic acid
       (a) Pantothenic acid functions as part of coenzyme A; thus, it is essential for energy-releasing mechanisms.
       (b) Most diets provide sufficient amounts; deficiencies are rare.
   (5) Vitamin B_6
       (a) Vitamin B_6 is a group of compounds that function as coenzymes in metabolic pathways that synthesize proteins, certain amino acids, antibodies, and nucleic acids.
       (b) Its requirement varies with protein intake.
   (6) Cyanocobalamin (vitamin B_{12})
       (a) The cyanocobalamin molecule contains cobalt.
       (b) Its absorption is regulated by the secretion of intrinsic factor from the gastric glands.
       (c) It functions as part of coenzymes needed for nucleic acid synthesis and for the metabolism of carbohydrates and fats.
   (7) Folacin (folic acid)
       (a) Liver enzymes catalyze reactions that convert folacin to physiologically active folinic acid.
       (b) It is a coenzyme needed for the metabolism of certain amino acids, DNA synthesis, and the normal production of red blood cells.
   (8) Biotin
       (a) Biotin is a coenzyme required for the metabolism of amino acids and fatty acids, and for nucleic acid synthesis.
       (b) It is stored in metabolically active organs.
c. Ascorbic acid (vitamin C)
   (1) Vitamin C is closely related chemically to monosaccharides.
   (2) It is required for collagen production, the metabolism of certain amino acids, and iron absorption.
   (3) It is not stored in large amounts; excess is excreted in the urine.

Minerals (page 712)

1. Characteristics of minerals
   a. Minerals are responsible for about 4% of body weight.
   b. About 75% by weight of the minerals are found in bones and teeth as calcium and phosphorus.
   c. Minerals are usually incorporated into organic molecules, although some are in inorganic compounds or are free ions.
   d. They comprise structural materials, function in enzymes, and play vital roles in various metabolic processes.
   e. Homeostatic mechanisms regulate mineral concentrations.
   f. The physiologically active form of minerals is the ionized form.
2. Major minerals
   a. Calcium
      (1) Calcium is essential for forming bones and teeth, conducting nerve impulses, contracting muscle fibers, clotting blood, and activating various enzymes.
      (2) Existing calcium concentration, vitamin D, protein intake, and motility of the digestive tract affect calcium absorption.
   b. Phosphorus
      (1) Phosphorus is incorporated into the salts of bones and teeth.
      (2) It participates in nearly all metabolic reactions as a constituent of nucleic acids, proteins, and some vitamins.
      (3) It also is in the phospholipids of cell membranes, in ATP, and in phosphates of body fluids.
   c. Potassium
      (1) Potassium is concentrated inside cells.
      (2) It maintains osmotic pressure, regulates pH, metabolizes carbohydrates and proteins, conducts nerve impulses, and contracts muscle fibers.
   d. Sulfur
      (1) Sulfur is incorporated into two of the twenty amino acids.
      (2) It is also in thiamine, insulin, biotin, and mucopolysaccharides.
   e. Sodium
      (1) Most sodium is in extracellular fluids or is bound to the inorganic salts of bone.
      (2) The kidneys, under the influence of aldosterone, regulate the blood concentration of sodium.
      (3) Sodium helps maintain solute concentration and regulates water balance.
      (4) It is essential for conducting nerve impulses, contracting muscle fibers, and moving substances through cell membranes.
   f. Chlorine
      (1) Chlorine is closely associated with sodium as chloride ions.
      (2) It acts with sodium to help maintain osmotic pressure, regulate pH, and maintain electrolyte balance.
      (3) It is essential for hydrochloric acid formation and for carbon dioxide transport by red blood cells.
   g. Magnesium
      (1) Magnesium is abundant in the bones as phosphates and carbonates.
      (2) It functions in ATP production and in the breakdown of ATP to ADP.
      (3) A reserve supply of magnesium is stored in the bones; excesses are excreted in the urine.
3. Trace elements
   a. Iron
      (1) Iron is part of hemoglobin in red blood cells and myoglobin in muscles.

    (2) A reserve supply of iron is stored in the liver, spleen, and bone marrow.

    (3) It is required to catalyze vitamin A formation; it is also incorporated into various enzymes and the cytochrome molecules.

b. Manganese
    (1) Most manganese is concentrated in the liver, kidneys, and pancreas.
    (2) It is necessary for normal growth and development of skeletal structures and other connective tissues; it is essential for the synthesis of fatty acids, cholesterol, and urea.

c. Copper
    (1) Most copper is concentrated in the liver, heart, and brain.
    (2) It is required for hemoglobin synthesis, bone development, melanin production, and myelin formation.

d. Iodine
    (1) Iodine is most highly concentrated in the thyroid gland.
    (2) It is an essential component of thyroid hormones.
    (3) It is often added to foods as iodized table salt.

e. Cobalt
    (1) Cobalt is widely distributed throughout the body.
    (2) It is an essential part of cyanocobalamin and is required for the synthesis of several enzymes.

f. Zinc
    (1) Zinc is most concentrated in the liver, kidneys, and brain.
    (2) It is a constituent of several enzymes that take part in digestion, respiration, and metabolism.

g. Fluorine
    (1) The teeth concentrate fluorine.
    (2) It is incorporated into enamel and prevents dental caries.

h. Selenium
    (1) The liver and kidneys store selenium.
    (2) It is a constituent of certain enzymes.

i. Chromium
    (1) Chromium is widely distributed throughout the body.
    (2) It regulates glucose utilization.

Healthy Eating (page 718)

1. An adequate diet provides sufficient energy and essential nutrients to support optimal growth, as well as maintenance and repair, of tissues.
2. Individual needs vary so greatly that it is not possible to design a diet that is adequate for everyone.
3. Devices to help consumers make healthy food choices include Recommended Daily Allowances, Recommended Dietary Allowances, food group plans, food pyramids, and food labels.
4. Malnutrition
    a. Poor nutrition is due to lack of foods or failure to wisely use available foods.
    b. Primary malnutrition is due to poor diet.
    c. Secondary malnutrition is due to an individual characteristic that makes a normal diet inadequate.
5. Starvation
    a. A person can survive fifty to seventy days without food.
    b. A starving body digests itself, starting with carbohydrates, then fats, then proteins.
    c. Symptoms include low blood pressure, slow pulse, chills, dry skin, hair loss, and poor immunity. Finally, vital organs cease to function.
    d. Marasmus is lack of all nutrients.
    e. Kwashiorkor is protein starvation.
    f. Anorexia nervosa is a self-starvation eating disorder.
    g. Bulimia is an eating disorder characterized by bingeing and purging.

Life-Span Changes (page 724)

1. Changing nutrition with age reflects medical conditions and social and economic circumstances.
2. Basal metabolic rate rises in early childhood, declines, then peaks again in adolescence, with decreasing activity during adulthood.
3. Weight gain, at any age, occurs when energy in exceeds energy out, and weight loss occurs when energy out exceeds energy in.

CRITICAL THINKING QUESTIONS

1. For each of the following diets, indicate how the diet is nutritionally unsound (if it is) and why it would be easy or difficult to follow.
    a. For the first ten days of the Beverly Hills diet, only fruit is eaten. On day 10, you can eat a bagel and butter, and then only fruit until day 19, when you can eat steak or lobster. The cycle repeats, adding more meat. This diet is based on "conscious combining"—the idea that eating certain combinations of foods leads to weight loss.
    b. The Weight Loss Clinic diet consists of 800 calories per day, with 46.1% protein, 35.2% carbohydrate, and 18.7% fat.
    c. The macrobiotic diet includes 10% to 20% protein, 70% carbohydrate, and 10% fat, with a half hour of walking each day. Most familiar foods are forbidden, but you can eat many unusual foods—such as rice cakes, seaweed, barley stew, pumpkin soup, rice gruel, kasha and onions, millet balls, wheat berries, and parsnip chips.
    d. The No Aging diet maintains that eating foods rich in nucleic acids (RNA and DNA) can prolong life, since these are the genetic materials. Recommended foods include sardines, salmon, calves' liver, lentils, and beets.

2. Why does the blood sugar concentration of a person whose diet is low in carbohydrates remain stable?

3. A young man takes several vitamin supplements each day, claiming that they give him energy. Is he correct? Why or why not?

4. A soccer coach advises his players to eat a hamburger and French fried potatoes about two hours before a game. Suggest a more sensible pregame meal.

5. Anorexia nervosa is a form of starvation. If it is a nutritional problem, then why should treatment include psychotherapy?

6. Why do starving children often die of infections that are usually mild in well-nourished children?

7. Using nutrient tables, calculate the number of grams of carbohydrate, lipid, and protein that you eat in a typical day, and the total calories in these foods. Suggest ways to improve your diet.

8. Examine the label information on the packages of a variety of dry breakfast cereals. Which types of cereals provide the best sources of vitamins and minerals? Which major nutrients are lacking in these cereals?

9. If a person decided to avoid eating meat and other animal products, such as milk, cheese, and eggs, what foods might be included in the diet to provide essential amino acids?

10. How might a diet be modified in order to limit the intake of cholesterol?

11. How do you think the nutritional requirements of a healthy twelve-year-old boy, a twenty-four-year-old pregnant woman, and a healthy sixty-year-old man differ?

REVIEW EXERCISES

1. Define *essential nutrient*.
2. List some common sources of carbohydrates.
3. Summarize the importance of cellulose in the diet.
4. Explain what happens to excess glucose in the body.
5. Explain why a temporary drop in the blood glucose concentration may impair nervous system functioning.
6. List some of the factors that affect an individual's need for carbohydrates.
7. Define *triglyceride*.
8. List some common sources of lipids.
9. Define *beta oxidation*.
10. Explain how fats may provide energy.
11. Describe the liver's role in fat metabolism.
12. Discuss the functions of cholesterol.
13. Define *deamination,* and explain its importance.
14. List some common sources of protein.
15. Distinguish between essential and nonessential amino acids.
16. Explain why all of the essential amino acids must be present before growth can occur.
17. Distinguish between complete and incomplete proteins.
18. Review the major functions of amino acids.
19. Define *nitrogen balance*.
20. Explain why a protein deficiency may be accompanied by edema.
21. Define *calorie*.
22. Explain how the caloric values of foods are determined.
23. Define *basal metabolic rate*.
24. List some of the factors that affect the BMR.
25. Define *energy balance*.
26. Explain what is meant by *desirable weight*.
27. Distinguish between overweight and obesity.
28. Discuss the general characteristics of fat-soluble vitamins.
29. List the fat-soluble vitamins, and describe the major functions of each vitamin.
30. List some good sources for each of the fat-soluble vitamins.
31. Explain what is meant by the *vitamin B complex.*
32. List the water-soluble vitamins, and describe the major functions of each vitamin.
33. List some good sources for each of the water-soluble vitamins.
34. Discuss the general characteristics of the mineral nutrients.
35. List the major minerals, and describe the major functions of each mineral.
36. List some good sources for each of the major minerals.
37. Distinguish between a major mineral and a trace element.
38. List the trace elements, and describe the major functions of each trace element.
39. List some good sources of each of the trace elements.
40. Define *adequate diet*.
41. Explain various methods to eat an adequate diet.
42. Define *malnutrition*.
43. Distinguish between primary and secondary malnutrition.
44. Discuss bodily changes during starvation.
45. Distinguish among marasmus, kwashiorkor, anorexia nervosa, and bulimia.
46. Describe some medical conditions that affect the ability to obtain adequate nutrition as a person ages.

WEB CONNECTIONS

Visit the Student Edition of the Online Learning Center at www.mhhe.com/shier10 **for answers to chapter questions, additional quizzes, interactive learning exercises, and other study tools.**

Respiratory System

UNDERSTANDING WORDS

alveol-, small cavity: *alveol*us—microscopic air sac within a lung.

bronch-, windpipe: *bronch*us—primary branch of the trachea.

carcin-, spreading sore: *carcin*oma—type of cancer.

carin-, keel-like: *carin*a—ridge of cartilage between the right and left bronchi.

cric-, ring: *cric*oid cartilage—ring-shaped mass of cartilage at the base of the larynx.

epi-, upon: *epi*glottis—flaplike structure that partially covers the opening into the larynx during swallowing.

hem-, blood: *hem*oglobin—pigment in red blood cells.

inhal-, to breathe in: *inhal*ation—to take air into the lungs.

phren-, mind, diaphragm: *phren*ic nerve—nerve associated with the cervical plexuses that stimulate the muscle fibers of the diaphragm to contract.

tuber-, swelling: *tuber*culosis—disease characterized by the formation of fibrous masses within the lungs.

Falsely colored electron micrograph of the lining of the trachea (windpipe) consisting of mucous-secreting goblet cells (brown) amid ciliated epithelium (red) (2,400×). The mucus entraps inhaled particles, while the cilia sweep them up and out of the respiratory tract.

CHAPTER OBJECTIVES

After you have studied this chapter, you should be able to

1. List the general functions of the respiratory system.

2. Name and describe the locations of the organs of the respiratory system.

3. Describe the functions of each organ of the respiratory system.

4. Explain how inspiration and expiration are accomplished.

5. Name and define each of the respiratory air volumes and capacities.

6. Explain how the alveolar ventilation rate is calculated.

7. List several nonrespiratory air movements and explain how each occurs.

8. Locate the respiratory center and explain how it controls normal breathing.

9. Discuss how various factors affect the respiratory center.

10. Describe the structure and function of the respiratory membrane.

11. Explain how the blood transports oxygen and carbon dioxide.

12. Describe the effects of aging on the respiratory system.

F or about six weeks following September 11, 2001, when terrorists crashed two jet planes into the World Trade Center, an invisible shower of small particles rained down on lower Manhattan. When residents returned to their apartments in Battery Park City downwind of the site, they found a fine film coating everything. By early November, with fires at the site still burning, recovery workers continued to use masks, while people who had returned to nearby places of business reported coughs and irritated eyes. Health officials did their best with limited information, advising that no one be exposed to the tainted air without a mask for more than two hours. Said Thomas Cahill, an authority at the University of California, Davis, on the composition and transport of airborne particles, "The air from Ground Zero was laden with extremely high amounts of very small particles, probably associated with high temperatures in the underground debris pile. No one has ever reported a situation like the one we see in the World Trade Center samples."

Environmental scientists from the U.S. Department of Energy collected daily air samples from October 23 through mid-December and sent the samples to investigators at Davis, who classified the particulates by size, shape, and composition. The researchers knew what to look for by what had burned—concrete and glass released silicon; jet fuel oil contributed sulfur, nickel, vanadium, and other metals; computers and electrical devices released lead. They also analyzed breakdown products of wood, plastics, and carpets.

The size of inhaled particles is as important as their composition, in terms of health effects. Coarse particles, which range from 5 to 12 micrometers in diameter, are usually ejected from the respiratory system in coughs or sneezes but can worsen certain allergies and asthma. The researchers found that the level of coarse particles did not fall after it rained, as expected, and the particles were covered in soot, indicating that the ongoing fires were continuously spewing out new particles.

Very fine particles, which range from 0.09 to 0.24 micrometers in diameter, pass the body's initial barriers and may be deposited deeper in

Lower Manhattan, September 11, 2001.

the respiratory system, or even reach other organ systems. These particles include abundant sulfur compounds, tiny bits of silicon, and metals, including vanadium and nickel from fuel oil, titanium from concrete, and iron, copper, and zinc. Because many of these metals had never been seen before at these levels, safety guidelines do not exist—analysis of health effects is ongoing.

During the post-9/11 period, returning residents and workers were advised to install air filters and to use wet mops and rags instead of vacuum cleaners and brooms, which can resuspend settled particles. By the time the results of the initial air analysis were released—February 11, 2002—the air quality in lower Manhattan had almost returned to near-normal. But longer-term effects on the functioning of the respiratory system continue to be monitored. ■

The respiratory system consists of passages that filter incoming air and transport it into the body, into the lungs, and to the many microscopic air sacs where gases are exchanged. The entire process of exchanging gases between the atmosphere and body cells is called **respiration** (res″pĭ-ra′shun). It consists of several events:

- Movement of air in and out of the lungs, commonly called breathing, or *ventilation.*
- Exchange of gases between the air in the lungs and the blood, sometimes called *external respiration.*
- Transport of gases by the blood between the lungs and body cells.

- Exchange of gases between the blood and the body cells, sometimes called *internal respiration.*
- Oxygen utilization and production of carbon dioxide by body cells is part of the process of *cellular respiration.*

Why We Breathe

Respiration occurs on a macroscopic level—a function provided by an organ system. However, the reason that body cells must exchange gases—that is, take up oxygen and rid themselves of carbon dioxide—is apparent at the cellular and molecular levels.

∞ RECONNECT TO CHAPTER 4,
AEROBIC REACTIONS,
PAGE 111

Cellular respiration enables cells to harness energy held in the chemical bonds of nutrient molecules. In aerobic reactions, cells liberate energy from these molecules by removing electrons and channeling them through a series of carriers called the electron transport chain. At the end of this chain, electrons are combined with oxygen atoms and hydrogen ions to make water. Without oxygen, these reactions cease.

Besides producing ATP, the aerobic reactions produce carbon dioxide (CO_2), a metabolic waste which combines with body water to form carbonic acid, helping to maintain blood pH. Too much CO_2, however, will lower the blood pH, threatening homeostasis. The role of the respiratory system, therefore, is both to provide oxygen for aerobic reactions, and to eliminate CO_2 rapidly enough to maintain the pH of the internal environment.

Organs of the Respiratory System

The organs of the respiratory system can be divided into two groups, or tracts. Those in the *upper respiratory tract* include the nose, nasal cavity, sinuses, and pharynx. Those in the *lower respiratory tract* include the larynx, trachea, bronchial tree, and lungs (fig. 19.1).

Nose

The nose is covered with skin and is supported internally by muscle, bone, and cartilage. Its two *nostrils* (external nares) provide openings through which air can enter and leave the nasal cavity. Many internal hairs guard these openings, preventing entry of large particles carried in the air.

Nasal Cavity

The **nasal cavity,** a hollow space behind the nose, is divided medially into right and left portions by the **nasal septum.** This cavity is separated from the cranial cavity by the cribriform plate of the ethmoid bone and from the oral cavity by the hard palate.

> The nasal septum may bend during birth or shortly before adolescence. Such a *deviated septum* may obstruct the nasal cavity, making breathing difficult.

As figure 19.2 shows, **nasal conchae** (turbinate bones) curl out from the lateral walls of the nasal cavity on each side, dividing the cavity into passageways called the *superior, middle,* and *inferior meatuses* (see chapter 7,

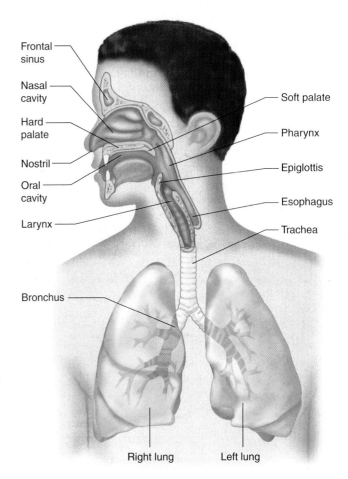

FIGURE 19.1
Organs of the respiratory system.

p. 206). They also support the mucous membrane that lines the nasal cavity and help increase its surface area.

The upper posterior portion of the nasal cavity, below the cribriform plate, is slitlike, and its lining contains the olfactory receptors that provide the sense of smell. The remainder of the cavity conducts air to and from the nasopharynx.

The mucous membrane lining the nasal cavity contains pseudostratified ciliated epithelium that is rich in mucous-secreting goblet cells (see chapter 5, p. 135). It also includes an extensive network of blood vessels and normally appears pinkish. As air passes over the membrane, heat radiates from the blood and warms the air, adjusting its temperature to that of the body. In addition, evaporation of water from the mucous lining moistens the air. The sticky mucus the mucous membrane secretes entraps dust and other small particles entering with the air.

As the cilia of the epithelial cells move, a thin layer of mucus and any entrapped particles are pushed toward the pharynx (fig. 19.3). When the mucus reaches the pharynx, it is swallowed. In the stomach, gastric juice is likely

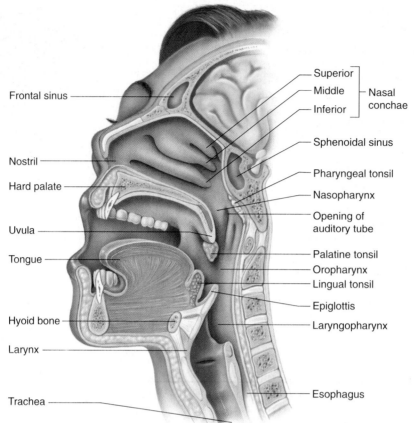

FIGURE 19.2

Major features of the upper respiratory tract.

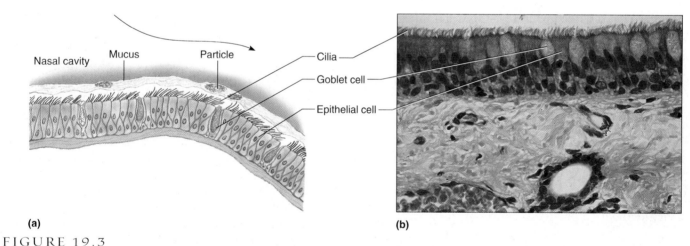

(a)   **(b)**

FIGURE 19.3

Mucus movement in the respiratory tract. (*a*) Cilia move mucus and trapped particles from the nasal cavity to the pharynx. (*b*) Micrograph of ciliated epithelium in respiratory tract (275×)

to destroy microorganisms in the mucus, including disease-causing forms. Thus, the filtering that the mucous membrane provides not only prevents particles from reaching the lower air passages, but also helps prevent respiratory infections. Clinical Application 19.1 discusses how cigarette smoking impairs the respiratory system, beginning with the cleansing mucus and cilia.

1 What is respiration?

2 Which organs constitute the respiratory system?

3 What is the function of the mucous membrane that lines the nasal cavity?

4 What is the function of the cilia on the cells that line the nasal cavity?

THE EFFECTS OF CIGARETTE SMOKING ON THE RESPIRATORY SYSTEM

Damage to the respiratory system from cigarette smoking is slow, progressive, and deadly. A healthy respiratory system is continuously cleansed. The mucus produced by the respiratory tubules traps dirt and disease-causing organisms, which cilia sweep toward the mouth, where they can be eliminated. Smoking greatly impairs this housekeeping. With the very first inhalation of smoke, the beating of the cilia slows. With time, the cilia become paralyzed and, eventually, disappear altogether. The loss of cilia leads to the development of smoker's cough. The cilia no longer effectively remove mucus, so the individual must cough it up. Coughing is usually worse in the morning because mucus has accumulated during sleep.

To make matters worse, excess mucus is produced and accumulates, clogging the air passageways. Pathogenic organisms that are normally removed now have easier access to the respiratory surfaces, and the resulting lung congestion favors their growth. This is why smokers are sick more often than nonsmokers. In addition, a lethal chain reaction begins. Smoker's cough leads to chronic bronchitis, caused by destroyed respiratory cilia. Mucus production increases and the lining of the bronchioles thickens, making breathing difficult. The bronchioles lose elasticity and are no longer able to absorb the pressure changes accompanying coughing. As a result, a cough can increase the air pressure within the alveoli (microscopic air sacs) enough to rupture the delicate alveolar walls; this condition is the hallmark of smoking-induced *emphysema.* The burst alveoli cause worsening of the cough, fatigue, wheezing, and impaired breathing. Emphysema is fifteen times more common among individuals who smoke a pack of cigarettes a day than among nonsmokers.

Simultaneous with the structural changes progressing to emphysema may be cellular changes leading to lung cancer. First, cells in the outer border of the bronchial lin-

ing begin to divide more rapidly than usual. Eventually, these cells displace the ciliated cells. Their nuclei begin to resemble those of cancerous cells—large and distorted with abnormal numbers of chromosomes. Up to this point, the damage can be repaired if smoking ceases. If smoking continues, these cells may eventually break through the basement membrane and begin dividing within the lung tissue, forming a tumor with the potential of spreading throughout lung tissue (figs. 19A and 19B) and beyond, such as to the brain or bones. Eighty percent of lung cancer cases are due to cigarette smoking. Only 13% of lung cancer patients live as long as five years after the initial diagnosis.

It pays to quit. Much of the damage to the respiratory system can be repaired. Cilia are restored, and the thickening of alveolar walls due to emphysema can be reversed. But ruptured alveoli are gone forever. The nicotine in tobacco smoke causes a powerful dependency by binding to certain receptors on brain cells. ■

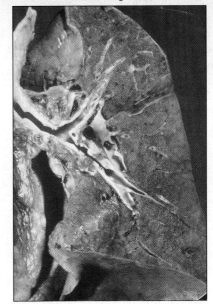
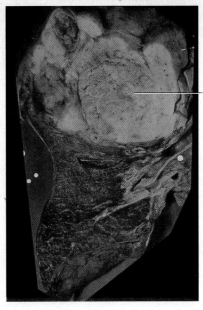

Normal lung tissue · Cancerous lung tissue · Tumor

FIGURE 19A

The lung on the left is healthy. A cancerous tumor has invaded the lung on the right, taking up nearly half of the lung space.

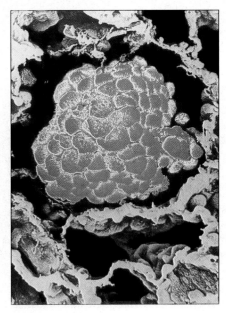

FIGURE 19B

Lung cancer may begin as a tiny tumor growing in an alveolus, a falsely colored microscopic air sac (125×).

Sinuses

Recall from chapter 7 (p. 205) that the *sinuses* (paranasal sinuses) are air-filled spaces located within the *maxillary, frontal, ethmoid,* and *sphenoid bones* of the skull (fig. 19.4). These spaces open into the nasal cavity and are lined with mucous membranes that are continuous with the lining of the nasal cavity. Consequently, mucus secretions drain from the sinuses into the nasal cavity. Membranes that are inflamed and swollen because of nasal infections or allergic reactions (sinusitis) may block this drainage, increasing pressure within a sinus and causing headache.

The sinuses reduce the weight of the skull. They also serve as resonant chambers that affect the quality of the voice.

It is possible to illuminate a person's frontal sinus in a darkened room by holding a small flashlight just beneath the eyebrow. Similarly, holding the flashlight in the mouth illuminates the maxillary sinuses.

1 Where are the sinuses located?

2 What are the functions of the sinuses?

Pharynx

The **pharynx** (throat) is located posterior to the oral cavity and between the nasal cavity and the larynx. It is a passageway for food moving from the oral cavity to the esophagus and for air passing between the nasal cavity and the larynx (see fig. 19.2). It also aids in producing the sounds of speech. The subdivisions of the pharynx—the nasopharynx, oropharynx, and laryngopharynx—are described in chapter 17 (p. 655).

Larynx

The **larynx** is an enlargement in the airway superior to the trachea and inferior to the pharynx. It is a passageway for air moving in and out of the trachea and prevents foreign objects from entering the trachea. The larynx also houses the *vocal cords* (see reference plates 49 and 71).

The larynx is composed of a framework of muscles and cartilages bound by elastic tissue. The largest of the cartilages are the thyroid, cricoid, and epiglottic cartilages (fig. 19.5). These structures are single. The other laryngeal cartilages—the arytenoid, corniculate, and cuneiform cartilages—are paired.

The **thyroid cartilage** was named for the thyroid gland that covers its lower area. This cartilage is the shieldlike structure that protrudes in the front of the neck and is sometimes called the Adam's apple. The protrusion typically is more prominent in males than in females because of an effect of male sex hormones on the development of the larynx.

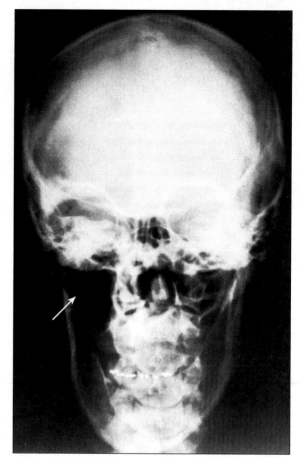

(a)

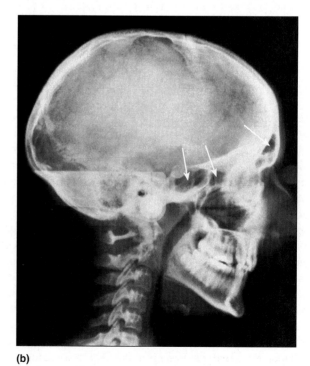

(b)

FIGURE 19.4

Radiograph of a skull (*a*) from the anterior view and (*b*) from the lateral view, showing air-filled sinuses (arrows) within the bones.

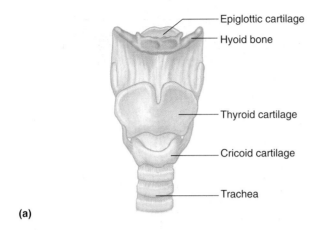

Epiglottic cartilage
Hyoid bone
Thyroid cartilage
Cricoid cartilage
Trachea

(a)

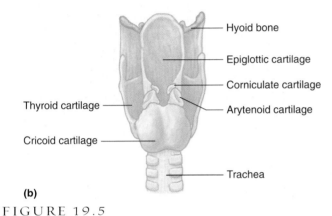

Hyoid bone
Epiglottic cartilage
Corniculate cartilage
Arytenoid cartilage
Thyroid cartilage
Cricoid cartilage
Trachea

(b)

FIGURE 19.5
Larynx. (*a*) Anterior and (*b*) posterior views of the larynx.

The **cricoid cartilage** lies inferior to the thyroid cartilage. It marks the lowermost portion of the larynx.

The **epiglottic cartilage,** the only one of the laryngeal cartilages that is elastic, not hyaline, cartilage, is attached to the upper border of the thyroid cartilage and supports a flaplike structure called the **epiglottis.** The epiglottis usually stands upright and allows air to enter the larynx. During swallowing, however, muscular contractions raise the larynx, and the base of the tongue presses the epiglottis downward. As a result, the epiglottis partially covers the opening into the larynx, helping prevent foods and liquids from entering the air passages.

The pyramid-shaped **arytenoid cartilages** are located superior to and on either side of the cricoid cartilage. Attached to the tips of the arytenoid cartilages are the tiny, conelike **corniculate cartilages.** These cartilages are attachments for muscles that help regulate tension on the vocal cords during speech and aid in closing the larynx during swallowing.

The **cuneiform cartilages** are small, cylindrical structures in the mucous membrane between the epiglottic and the arytenoid cartilages. They stiffen the soft tissues in this region.

Inside the larynx, two pairs of horizontal folds composed of muscle tissue and connective tissue with a covering of mucous membrane extend inward from the lateral walls. The upper folds (vestibular folds) are called *false vocal cords* because they do not produce sounds. Muscle fibers within these folds help close the larynx during swallowing.

The lower folds are the *true vocal cords.* They contain elastic fibers and are responsible for vocal sounds, which are created when air is forced between these folds, causing them to vibrate from side to side. This action generates sound waves, which can be formed into words by changing the shapes of the pharynx and oral cavity and by using the tongue and lips. Figure 19.6 shows both pairs of folds.

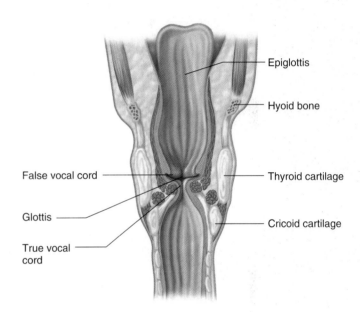

Epiglottis
Hyoid bone
False vocal cord
Thyroid cartilage
Glottis
Cricoid cartilage
True vocal cord

(a)

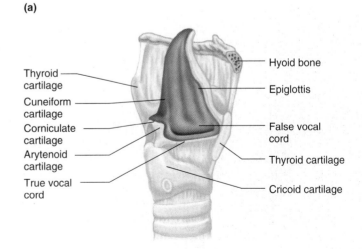

Thyroid cartilage
Cuneiform cartilage
Corniculate cartilage
Arytenoid cartilage
True vocal cord
Hyoid bone
Epiglottis
False vocal cord
Thyroid cartilage
Cricoid cartilage

(b)

FIGURE 19.6
Larynx. (*a*) Coronal section and (*b*) sagittal section of the larynx.

Changing tension on the vocal cords, by contracting or relaxing laryngeal muscles, controls *pitch* (musical tone) of the voice. Increasing the tension produces a higher pitch, and decreasing the tension creates a lower pitch.

The *intensity* (loudness) of a vocal sound depends upon the force of the air passing over the vocal cords. Stronger blasts of air result in greater vibration of the vocal cords and louder sound.

During normal breathing, the vocal cords remain relaxed, and the opening between them, called the **glottis** (glot′is), is a triangular slit. However, when food or liquid is swallowed, muscles close the glottis within the false vocal folds. Along with closing of the epiglottis, this action helps prevent food or liquid from entering the trachea (fig. 19.7). The mucous membrane that lines the lar-

ynx continues to filter incoming air by entrapping particles and moving them toward the pharynx by ciliary action.

1 What part of the respiratory tract is shared with the alimentary canal?

2 Describe the structure of the larynx.

3 How do the vocal cords produce sounds?

4 What is the function of the glottis? Of the epiglottis?

Trachea

The **trachea** (windpipe) is a flexible cylindrical tube about 2.5 centimeters in diameter and 12.5 centimeters in length. It extends downward anterior to the esophagus and into the thoracic cavity, where it splits into right and left bronchi (fig. 19.8 and reference plate 50).

The inner wall of the trachea is lined with a ciliated mucous membrane that contains many goblet cells. This membrane continues to filter the incoming air and to move entrapped particles upward into the pharynx where the mucous can be swallowed.

Within the tracheal wall are about twenty C-shaped pieces of hyaline cartilage, one above the other. The open

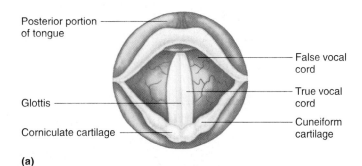

(a)

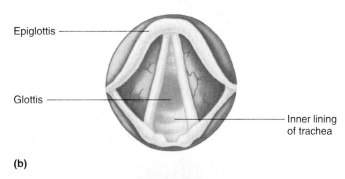

(b)

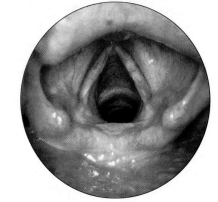

(c)

FIGURE 19.7

The vocal cords as viewed from above with the glottis (a) closed and (b) open. (c) Photograph of the glottis and vocal folds.

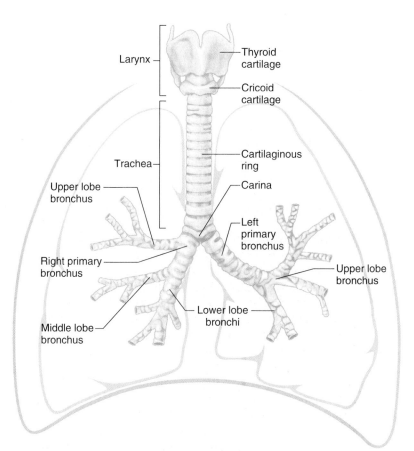

FIGURE 19.8

The trachea transports air between the larynx and the bronchi.

ends of these incomplete rings are directed posteriorly, and the gaps between their ends are filled with smooth muscle and connective tissues (figs. 19.9 and 19.10). These cartilaginous rings prevent the trachea from collapsing and blocking the airway. At the same time, the soft tissues that complete the rings in the back allow the nearby esophagus to expand as food moves through it on the way to the stomach.

A blocked trachea can cause asphyxiation in minutes. If swollen tissues, excess secretions, or a foreign object obstruct the trachea, making a temporary, external opening in the tube so that air can bypass the obstruction is lifesaving. This procedure, shown in figure 19.11, is called a *tracheostomy.*

On December 13, 1799, George Washington spent the day walking on his estate in a freezing rain. The next day, he had trouble breathing and swallowing. Several doctors were called in. One suggested a tracheostomy, cutting a hole in the throat so that the president could breathe. He was voted down. The other physicians suggested bleeding the patient, plastering his throat with bran and honey, and placing beetles on his legs to produce blisters. No treatment was provided, and within a few hours, Washington's voice became muffled, breathing was more labored, and he was restless. For a short time he seemed euphoric, and then he died.

George Washington had *epiglottitis,* an inflammation that swells the epiglottis to ten times its normal size. A tracheostomy might have saved his life.

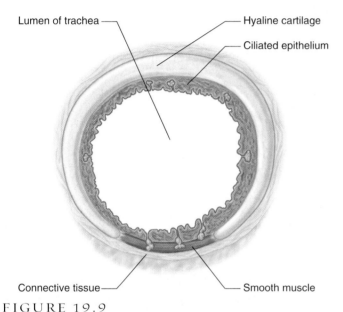

FIGURE 19.9
Cross section of the trachea. Note the C-shaped rings of hyaline cartilage in the wall.

Bronchial Tree

The **bronchial tree** (brong′ke-al trē) consists of branched airways leading from the trachea to the microscopic air sacs in the lungs. Its branches begin with the right and left **primary bronchi,** which arise from the trachea at the level of the fifth thoracic vertebrae. The openings of the primary bronchi are separated by a ridge of cartilage called the *carina* (see fig. 19.8). Each bronchus, accompanied by large blood vessels, enters its respective lung.

Branches of the Bronchial Tree

A short distance from its origin, each primary bronchus divides into **secondary,** or **lobar, bronchi** (two on the left

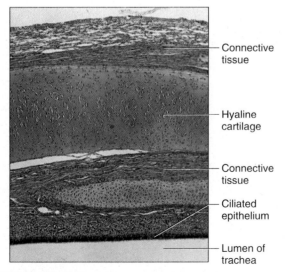

FIGURE 19.10
Light micrograph of a section of the tracheal wall (63×).

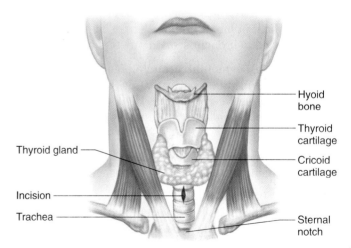

FIGURE 19.11
A tracheostomy may be performed to allow air to bypass an obstruction within the larynx.

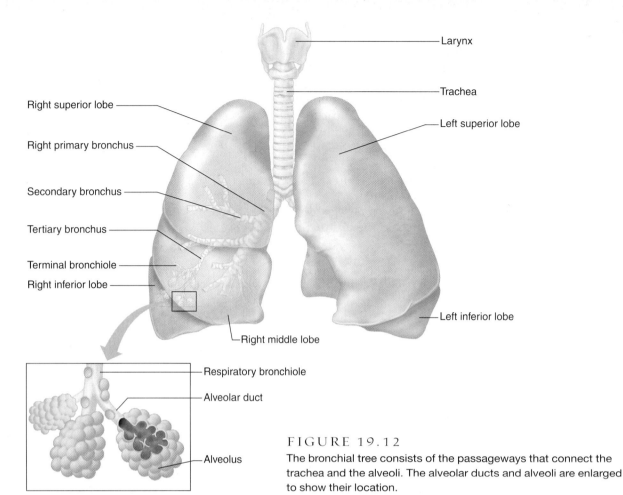

Right superior lobe

Right primary bronchus

Secondary bronchus

Tertiary bronchus

Terminal bronchiole

Right inferior lobe

Larynx

Trachea

Left superior lobe

Left inferior lobe

Right middle lobe

Respiratory bronchiole

Alveolar duct

Alveolus

FIGURE 19.12

The bronchial tree consists of the passageways that connect the trachea and the alveoli. The alveolar ducts and alveoli are enlarged to show their location.

and three on the right) that, in turn, branch repeatedly into finer and finer tubes (figs. 19.12 and 19.13). When stripped of their associated blood vessels and tissues, the airways appear as an upside down tree. The successive divisions of these branches from the trachea to the microscopic air sacs follow:

1. **Right** and **left primary bronchi.**

2. **Secondary,** or **lobar, bronchi.** Three branch from the right primary bronchus, and two branch from the left.

3. **Tertiary,** or **segmental, bronchi.** Each of these branches supplies a portion of the lung called a *bronchopulmonary segment.* Usually there are ten such segments in the right lung and eight in the left lung.

4. **Intralobular bronchioles.** These small branches of the segmental bronchi enter the basic units of the lung—the *lobules.*

5. **Terminal bronchioles.** These tubes branch from a bronchiole. Fifty to eighty terminal bronchioles occupy a lobule of the lung.

6. **Respiratory bronchioles.** Two or more respiratory bronchioles branch from each terminal bronchiole. Short and about 0.5 millimeter in diameter, these

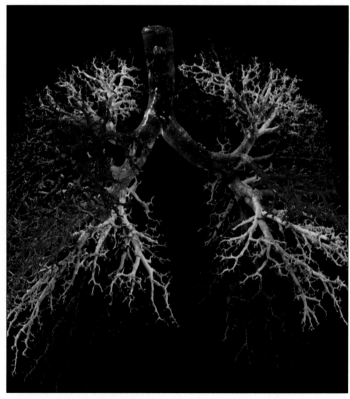

FIGURE 19.13

A plastic cast of the bronchial tree.

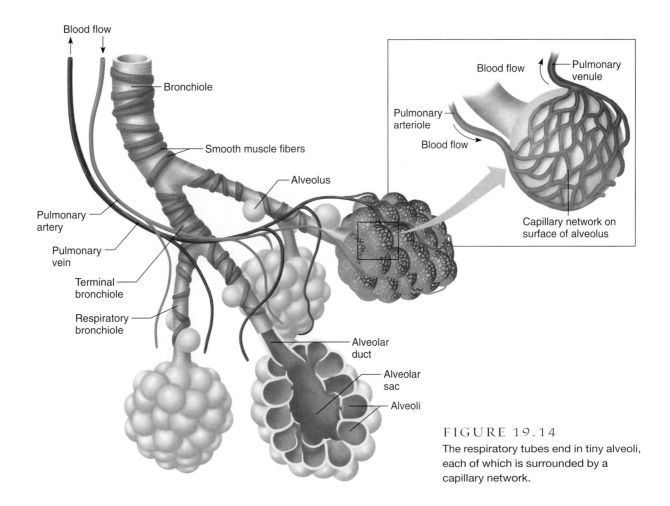

FIGURE 19.14
The respiratory tubes end in tiny alveoli, each of which is surrounded by a capillary network.

structures are called "respiratory" because a few air sacs bud from their sides, making them able to take part in gas exchange.

7. **Alveolar ducts.** Alveolar ducts branch from each respiratory bronchiole (fig. 19.14).

8. **Alveolar sacs.** Alveolar sacs are thin-walled, closely packed outpouchings of the alveolar ducts.

9. **Alveoli.** Alveoli are thin-walled, microscopic air sacs that open to an alveolar sac. Thus, air can diffuse freely from the alveolar ducts, through the alveolar sacs, and into the alveoli (fig. 19.15).

Dust particles, asbestos fibers, and other pollutants travel at speeds of 200 centimeters per second in the trachea but slow to 1 centimeter per second when deep in the lungs. Gravity deposits such particles, particularly at branchpoints in the respiratory tree. It is a little like traffic backing up at an exit to a highway.

Structure of the Respiratory Tubes

The structure of a bronchus is similar to that of the trachea, but the C-shaped cartilaginous rings are replaced with cartilaginous plates where the bronchus enters the lung. These plates are irregularly shaped and completely surround the tube. However, as finer and finer branch tubes

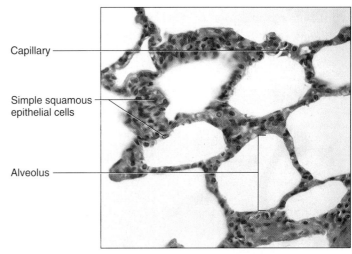

FIGURE 19.15
Light micrograph of alveoli (250×).

appear, the amount of cartilage decreases, and it finally disappears in the bronchioles, which have diameters of about 1 millimeter. Meanwhile, a layer of smooth muscle that surrounds the tube just beneath the mucosa becomes more prominent. This muscular layer remains in the wall to the ends of the respiratory bronchioles, and only a few muscle fibers are in the walls of the alveolar ducts.

Elastic fibers are scattered among the smooth muscle cells and are abundant in the connective tissue that surrounds the respiratory tubes. These fibers play an important role in breathing, as is explained later in this chapter.

As the tubes become smaller in diameter, the type of cells that line them changes. The lining of the larger tubes consists of pseudostratified, ciliated columnar epithelium and mucous-secreting goblet cells. However, along the way, the number of goblet cells and the height of the other epithelial cells decline, and cilia become scarcer. In the finer tubes, beginning with the respiratory bronchioles, the lining is cuboidal epithelium; in the alveoli, it is simple squamous epithelium closely associated with a dense network of capillaries. The mucous lining gradually thins, until none appears in the alveoli.

A flexible optical instrument called a *fiberoptic bronchoscope* is used to examine the trachea and bronchial tree. This procedure (bronchoscopy) is used in diagnosing tumors or other pulmonary diseases, and to locate and remove aspirated foreign bodies in the air passages.

Functions of the Respiratory Tubes and Alveoli

The branches of the bronchial tree are air passages, which continue to filter incoming air and distribute it to the alveoli in all parts of the lungs. The alveoli, in turn, provide a large surface area of thin epithelial cells through which gas exchanges can occur (fig. 19.16). If the 300 million alveoli in the human lung were spread out, they would cover an area of between 70 and 80 square meters—nearly half the size of a tennis court.

During gas exchange, oxygen diffuses through the alveolar walls and enters the blood in nearby capillaries. Carbon dioxide diffuses from the blood through these walls and enters the alveoli (figs. 19.17 and 19.18).

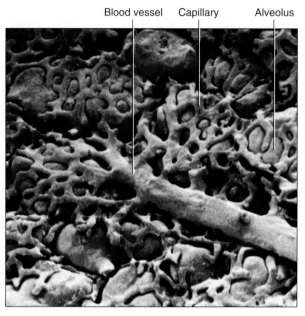

FIGURE 19.17

Falsely colored scanning electron micrograph of casts of alveoli and associated capillary networks. These casts were prepared by filling the alveoli and blood vessels with resin and later removing the soft tissues by digestion, leaving only the resin casts (420×).
Tissues and Organs:
A Text-Atlas of Scanning Electron Microscopy, by Richard D. Kessel and Randy Kardon. © 1979 W. H. Freeman and Company.

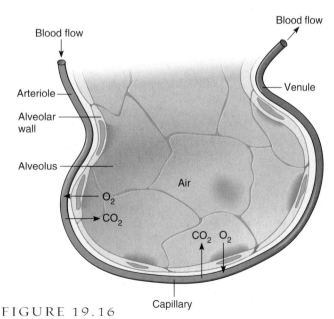

FIGURE 19.16

Oxygen (O_2) diffuses from the air within the alveolus into the capillary, while carbon dioxide (CO_2) diffuses from the blood within the capillary into the alveolus.

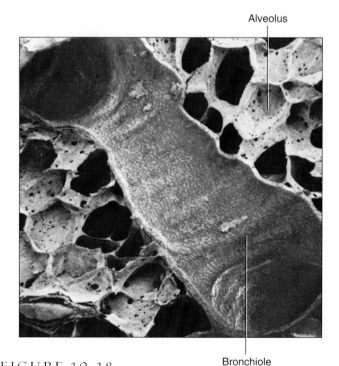

FIGURE 19.18

Falsely colored scanning electron micrograph of lung alveoli and a bronchiole (70×).

In severe cases of the inherited illness *cystic fibrosis,* airways become clogged with thick, sticky mucus, which attracts bacteria. As damaged white blood cells accumulate at the infection site, their DNA may leak out and further clog the area. A treatment that moderately eases breathing is deoxyribonuclease (DNase), an enzyme normally found in the body that degrades the accumulating extracellular DNA.

1 What is the function of the cartilaginous rings in the tracheal wall?

2 How do the right and left bronchi differ in structure?

3 List the branches of the bronchial tree.

4 Describe the changes in structure that occur in the respiratory tubes as their diameters decrease.

5 How are gases exchanged in the alveoli?

Several techniques enable a person who has stopped breathing to survive. In *artificial respiration,* a person blows into the mouth of a person who has stopped breathing. The oxygen in the rescuer's exhaled breath can keep the victim alive.

In *extracorporeal membrane oxygenation,* blood is pumped out of the body and across a gas-permeable membrane that adds oxygen and removes carbon dioxide, simulating lung function. Such a device can keep a person alive until he or she recovers from other problems, but is too costly and cumbersome to maintain life indefinitely.

A lung assist device, called an *intravascular oxygenator,* consists of hundreds of tiny porous hair-thin fibers surgically implanted in the inferior vena cava. Here, deoxygenated blood returning to the heart receives oxygen and is rid of carbon dioxide—but only at about 30% the capacity of a healthy respiratory system.

Lungs

The lungs are soft, spongy, cone-shaped organs located in the thoracic cavity. The right and left lungs are separated medially by the heart and the mediastinum, and they are enclosed by the diaphragm and the thoracic cage (see figs. 1.9, 19.19 and reference plates 56, 57, and 71).

Each lung occupies most of the thoracic space on its side and is suspended in the cavity by a bronchus and some large blood vessels. These tubular structures enter the lung on its medial surface through a region called the **hilum.** A layer of serous membrane, the *visceral pleura,* is

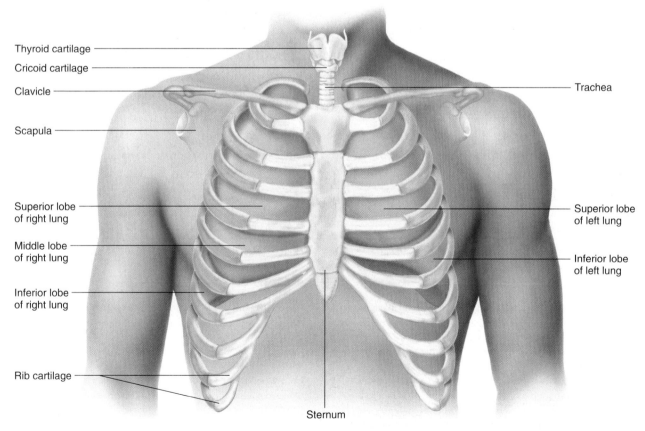

Thyroid cartilage

Cricoid cartilage

Clavicle

Scapula

Superior lobe of right lung

Middle lobe of right lung

Inferior lobe of right lung

Rib cartilage

Trachea

Superior lobe of left lung

Inferior lobe of left lung

Sternum

FIGURE 19.19

Locations of the lungs within the thoracic cavity.

firmly attached to the surface of each lung, and this membrane folds back at the hilus to become the *parietal pleura*. The parietal pleura, in turn, forms part of the mediastinum and lines the inner wall of the thoracic cavity (fig. 19.20).

There is no significant space between the visceral and parietal pleurae, since they are essentially in contact with each other. The potential space between them, the **pleural cavity,** contains only a thin film of serous fluid that lubricates the adjacent pleural surfaces, reducing friction as they move against one another during breathing. This fluid also helps hold the pleural membranes together.

The right lung is larger than the left lung, and it is divided by fissures into three parts, called the superior, middle, and inferior lobes. The left lung is similarly divided and consists of two parts, a superior and an inferior lobe.

A lobar bronchus of the bronchial tree supplies each lobe. A lobe also has connections to blood and lymphatic vessels and is enclosed by connective tissues. Connective tissue further subdivides a lobe into **lobules,** each of which contains terminal bronchioles together with their alveolar ducts, alveolar sacs, alveoli, nerves, and associated blood and lymphatic vessels.

Table 19.1 summarizes the characteristics of the major parts of the respiratory system. Clinical Application 19.2 considers substances that irritate the lungs.

1 Where are the lungs located?

2 What is the function of the serous fluid within the pleural cavity?

3 How does the structure of the right lung differ from that of the left lung?

4 What kinds of structures make up a lung?

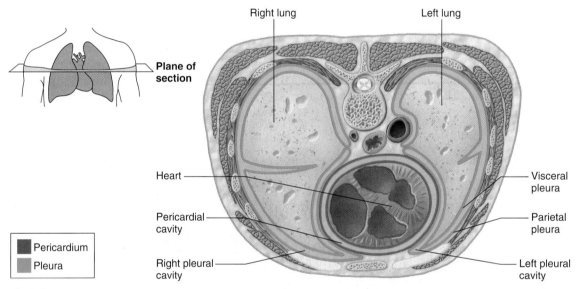

FIGURE 19.20
The potential spaces between the pleural membranes called the left and right pleural cavities are shown here as actual spaces.

TABLE 19.1	Parts of the Respiratory System	
Part	**Description**	**Function**
Nose	Part of face centered above the mouth and inferior to the space between the eyes	Nostrils provide entrance to nasal cavity; internal hairs begin to filter incoming air
Nasal cavity	Hollow space behind nose	Conducts air to pharynx; mucous lining filters, warms, and moistens incoming air
Sinuses	Hollow spaces in various bones of the skull	Reduce weight of the skull; serve as resonant chambers
Pharynx	Chamber posterior to the oral cavity and between the nasal cavity and larynx	Passageway for air moving from nasal cavity to larynx and for food moving from oral cavity to esophagus
Larynx	Enlargement at the top of the trachea	Passageway for air; prevents foreign objects from entering trachea; houses vocal cords
Trachea	Flexible tube that connects larynx with bronchial tree	Passageway for air; mucous lining continues to filter air
Bronchial tree	Branched tubes that lead from the trachea to the alveoli	Conducts air to the alveoli; mucous lining continues to filter incoming air
Lungs	Soft, cone-shaped organs that occupy a large portion of the thoracic cavity	Contain the air passages, alveoli, blood vessels, connective tissues, lymphatic vessels, and nerves of the lower respiratory tract

LUNG IRRITANTS

The lungs are exquisitely sensitive to the presence of inhaled particles. Such exposures can cause a variety of symptoms, both acute and chronic, that range from a persistent cough to cancer.

Asbestos

Asbestos, a naturally occurring mineral, was once widely used in buildings and on various products because it resists burning and chemical damage. Asbestos easily crumbles into fibers, which, when airborne, can enter human respiratory passages. Asbestos-related problems include

- asbestosis (shortness of breath resulting from scars in lungs)
- lung cancer
- mesothelioma (a rare cancer of the pleural membrane)

Asbestos fibers that are longer than 5 micrometers (0.0002 inch) and thinner than 2 micrometers (0.00008 inch) can cause illness when inhaled. Table 19A indicates how risk of becoming ill rises with duration of exposure.

Although asbestos clearly causes respiratory illness, it does so only if it is disturbed so that fibers break free and become airborne. Experts must determine whether it is safer to encapsulate asbestos in a building and leave it in place or remove it. Often, removing asbestos is actually more dangerous because this releases fibers. Today, synthetic fiberglass or plastics are used instead of asbestos.

Berylliosis

Beryllium is an element used in fluorescent powders, metal alloys, and in the nuclear power industry. A small percentage of workers exposed to beryllium dust or vapor develop an immune response, which damages the lungs. Symptoms include cough, shortness of breath, fatigue, loss of appetite, and weight loss. Fevers and night sweats indicate the role of the immune system. Radiographs show granuloma scars in the lungs. Pulmonary function tests and simply listening to breath sounds with a stethoscope reveal impaired breathing.

Symptoms of berylliosis typically begin about a decade after the first exposure, but this response time can range from several months to as long as forty years. Many people who worked with the element in the Rocky Flats plant in Colorado have developed berylliosis, and they are being monitored at the National Jewish Medical and Research Center in Denver. Affected individuals include those who directly contacted the element frequently, as well as support staff such as secretaries who probably inhaled beryllium.

Berylliosis can be distinguished from other lung ailments with a blood test that detects antibodies to beryllium. Affected individuals and those who do not have symptoms but know that they were exposed to beryllium are advised to have periodic blood tests and chest radiographs to detect the condition early. The steroid drug prednisone is used to control symptoms.

A Disorder with Many Names

Repeatedly inhaling dust of organic origin can cause a lung irritation called extrinsic allergic alveolitis. An acute form of this reaction impairs breathing and causes a fever a few hours after encountering dust. In the chronic form, lung changes occur gradually over several years. The condition is associated with several occupations and has a variety of colorful names:

Bathtub refinisher's lung
Bird breeder disease
Cheese worker's lung
Enzyme detergent sensitivity
Farmer lung
Laboratory technician's lung
Maltworker lung
Maple bark stripper disease
Mushroom picker disease
Plastic worker's lung
Popcorn worker's lung
Poultry raiser disease
Snuff taker's lung
Wheat weevil disease ■

TABLE 19A	Asbestos-Related Respiratory Illness	
Situation	**Level of Exposure (fibers/cubic centimeter)**	**Cancer Cases per Million Exposed People**
Asbestos workers with twenty years' exposure	10 fiber/cc	200,000
Permissible upper limit in buildings today	0.1 fiber/cc	2,000
Child in school with asbestos six hours/day	0.0005 fiber/cc	6
Most modern buildings	0.0002 fiber/cc	4

Breathing Mechanism

Breathing, which is also called ventilation, is the movement of air from outside the body into the bronchial tree and alveoli, followed by a reversal of this air movement. The actions responsible for these air movements are termed **inspiration** (in″spi-ra′shun), or inhalation, and **expiration** (ek″spi-ra′shun), or exhalation.

Inspiration

Atmospheric pressure due to the weight of the air is the force that moves air into the lungs. At sea level, this pressure is sufficient to support a column of mercury about 760 millimeters high in a tube. Thus, normal air pressure equals 760 millimeters (mm) of mercury (Hg). (Other units are in common usage: 760 mm Hg = 760 Torr = 1 atmosphere.)

Air pressure is exerted on all surfaces in contact with the air, and because people breathe air, the inside surfaces of their lungs are also subjected to pressure. In other words, when the respiratory muscles are at rest, the pressures on the inside of the lungs and alveoli and on the outside of the thoracic wall are about the same (fig. 19.21).

Pressure and volume are related in an opposite, or inverse, way. For example, pulling back on the plunger of a syringe increases the volume inside the barrel, lowering the air pressure inside. Atmospheric pressure then pushes outside air into the syringe (fig. 19.22*a*). In contrast, pushing on the plunger of a syringe reduces the volume inside the syringe, but the pressure inside increases, forcing air out into the atmosphere (fig. 19.22*b*). The movement of air into and out of the lungs occurs in much the same way.

If the pressure inside the lungs and alveoli (intra-alveolar pressure) decreases, outside air will then be pushed into the airways by atmospheric pressure. This is what happens during normal inspiration, and it involves the action of muscle fibers within the dome-shaped *diaphragm.*

The diaphragm is located just inferior to the lungs. It consists of an anterior group of skeletal muscle fibers (costal fibers), which originate from the ribs and sternum, and a posterior group (crural fibers), which originate from the vertebrae. Both groups of muscle fibers are inserted on a tendinous central portion of the diaphragm (reference plate 71).

The muscle fibers of the diaphragm are stimulated to contract by impulses carried by the *phrenic nerves,* which are associated with the cervical plexuses. When this occurs, the diaphragm moves downward, the thoracic cavity enlarges, and the intra-alveolar pressure falls about 2 mm Hg below that of atmospheric pressure. In response to this decreased pressure, air is forced into the airways by atmospheric pressure (fig. 19.23).

While the diaphragm is contracting and moving downward, the *external (inspiratory) intercostal muscles* and certain thoracic muscles may be stimulated to contract. This action raises the ribs and elevates the sternum, increasing the size of the thoracic cavity even more. As a result, the intra-alveolar pressure falls farther, and atmospheric pressure forces more air into the airways.

Lung expansion in response to movements of the diaphragm and chest wall depends on movements of the pleural membranes. Any separation of the pleural membranes decreases pressure in the intrapleural space, holding these membranes together. In addition, only a thin film of serous fluid separates the parietal pleura on the inner wall of the thoracic cavity from the visceral pleura attached to the surface of the lungs. The water molecules in this fluid greatly attract the pleural membranes and each other, helping to hold the moist surfaces of the pleural membranes tightly together, much as a wet coverslip sticks to a microscope slide. As a result of these factors, when the intercostal muscles move the thoracic wall

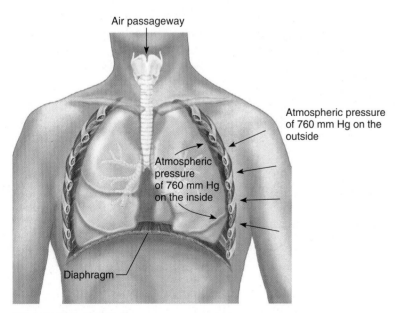

FIGURE 19.21
When the lungs are at rest, the pressure on the inside of the lungs is equal to the pressure on the outside of the thorax.

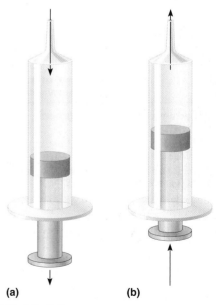

FIGURE 19.22
Moving the plunger of a syringe causes air to move (*a*) in or (*b*) out of the syringe. Air movements in and out of the lungs occur in much the same way.

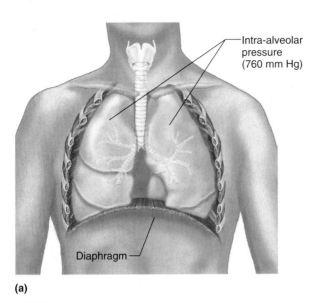

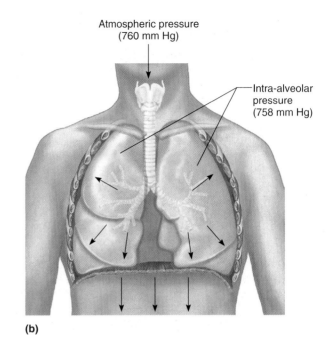

FIGURE 19.23

Normal inspiration. (*a*) Prior to inspiration, the intra-alveolar pressure is 760 mm Hg. (*b*) The intra-alveolar pressure decreases to about 758 mm Hg as the thoracic cavity enlarges, and atmospheric pressure forces air into the airways.

upward and outward, the parietal pleura moves too, and the visceral pleura follows it. This helps expand the lung in all directions.

Although the moist pleural membranes help expand the lungs, the moist inner surfaces of the alveoli have the opposite effect. Here the attraction of water molecules to each other creates a force called **surface tension** that makes it difficult to inflate the alveoli and may actually collapse them. Certain alveolar cells, however, synthesize a mixture of lipoproteins called **surfactant,** which is secreted continuously into alveolar air spaces. Surfactant reduces the alveoli's tendency to collapse, especially when lung volumes are low, and makes it easier for inspiratory efforts to inflate the alveoli. Table 19.2 summarizes the steps of inspiration.

Born two months early, Benjamin McClatchey weighed only 2 pounds, 13 ounces. Like many of the 380,000 "preemies" born each year in the United States, Benjamin had *respiratory distress syndrome* (RDS). His lungs were too immature to produce sufficient surfactant, and as a result, they could not overcome the force of surface tension enough to inflate.

A decade ago, Benjamin might not have survived RDS. But with the help of a synthetic surfactant sprayed or dripped into his lungs through an endotracheal tube and a ventilator machine designed to assist breathing in premature infants, he survived. Unlike conventional ventilators, which force air into the lungs at pressures that could damage delicate newborn lungs, the high-frequency ventilator used on preemies delivers the lifesaving oxygen in tiny, gentle puffs.

TABLE 19.2	Major Events in Inspiration

1. Nerve impulses travel on phrenic nerves to muscle fibers in the diaphragm, contracting them.
2. As the dome-shaped diaphragm moves downward, the thoracic cavity expands.
3. At the same time, the external intercostal muscles may contract, raising the ribs and expanding the thoracic cavity further.
4. The intra-alveolar pressure decreases.
5. Atmospheric pressure, which is greater on the outside, forces air into the respiratory tract through the air passages.
6. The lungs fill with air.

If a person needs to take a deeper than normal breath, the diaphragm and external intercostal muscles contract more forcefully. Additional muscles, such as the pectoralis minors and sternocleidomastoids, can also be used to pull the thoracic cage further upward and outward, enlarging the thoracic cavity, and decreasing intra-alveolar pressure even more (fig. 19.24).

The ease with which the lungs can expand as a result of pressure changes occurring during breathing is called *compliance* (distensibility). In a normal lung, compliance decreases as lung volume increases, because an inflated lung is more difficult to expand than a lung at

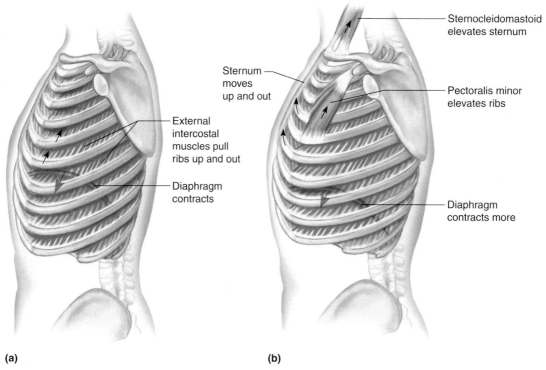

(a)   (b)

FIGURE 19.24

Maximal inspiration. (*a*) Shape of the thorax at the end of normal inspiration. (*b*) Shape of the thorax at the end of maximal inspiration, aided by contraction of the sternocleidomastoid and pectoralis minor muscles.

rest. Conditions that obstruct air passages, destroy lung tissue, or impede lung expansion in other ways also decrease compliance.

Expiration

The forces responsible for normal resting expiration come from *elastic recoil* of lung tissues and from surface tension. The lungs contain a considerable amount of elastic tissue, which stretches as the lungs expand during inspiration. When the diaphragm lowers, the abdominal organs inferior to it are compressed. As the diaphragm and the external intercostal muscles relax following inspiration, the elastic tissues cause the lungs to recoil, and they return to their original shapes. Similarly, elastic tissues within the abdominal organs cause them to spring back into their previous shapes, pushing the diaphragm upward. At the same time, surface tension that develops between the moist surfaces of the alveolar linings shrinks alveoli. Each of these factors increases the intra-alveolar pressure about 1 mm Hg above atmospheric pressure, so the air inside the lungs is forced out through the respiratory passages. Because normal resting expiration occurs without the contraction of muscles, it is considered a passive process.

The recoil of the elastic fibers within the lung tissues reduces pressure in the pleural cavity. Consequently, the pressure between the pleural membranes (intrapleural pressure) is usually about 4 mm Hg less than atmospheric pressure.

Because of the low intrapleural pressure, the visceral and parietal pleural membranes are held closely together, and no significant space normally separates them in the pleural cavity. However, if the thoracic wall is punctured, atmospheric air may enter the pleural cavity and create a substantial space between the membranes. This condition is called *pneumothorax*, and when it occurs, the lung on the affected side may collapse because of its elasticity.

Pneumothorax may be treated by covering the chest wound with an impermeable bandage, passing a tube (chest tube) through the thoracic wall into the pleural cavity, and applying suction to the tube. The suction reestablishes negative pressure within the cavity, and the collapsed lung expands.

If a person needs to exhale more air than normal, the posterior *internal* (*expiratory*) *intercostal muscles* can be contracted. These muscles pull the ribs and sternum downward and inward, increasing the pressure in the lungs. Also, the *abdominal wall muscles,* including the external and internal obliques, the transversus abdominis, and the rectus abdominis, can be used to squeeze the abdominal organs inward. Thus, the abdominal wall muscles can increase pressure in the abdominal cavity

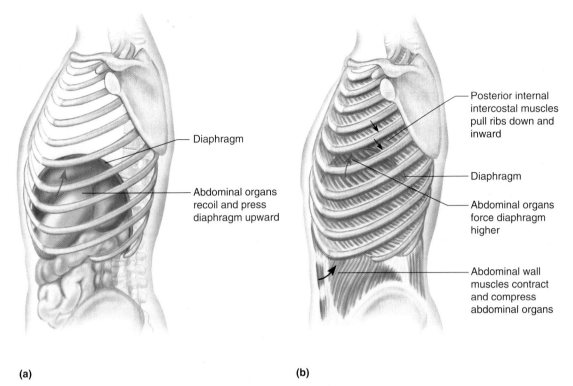

(a)   **(b)**

FIGURE 19.25

Expiration. (*a*) Normal resting expiration is due to elastic recoil of the lung tissues and the abdominal organs. (*b*) Contraction of the abdominal wall muscles and posterior internal intercostal muscles aids maximal expiration.

and force the diaphragm still higher against the lungs, squeezing additional air out of the lungs (fig. 19.25). Table 19.3 summarizes the steps in expiration.

1 Describe the events in inspiration.

2 How does surface tension aid in expanding the lungs during inspiration?

3 What forces are responsible for normal expiration?

Respiratory Volumes and Capacities

Different degrees of effort in breathing move different volumes of air in or out of the lungs. The measurement of such air volumes is called *spirometry,* and it describes four distinct **respiratory volumes.**

One inspiration plus the following expiration is called a **respiratory cycle.** The amount of air that enters or leaves during a respiratory cycle is termed the **tidal volume.** About 500 milliliters (mL) of air enter during a normal, resting inspiration. On the average, the same amount leaves during a normal, resting expiration. Thus, the **resting tidal volume** is about 500 mL (fig. 19.26).

During forced maximal inspiration, a quantity of air in addition to the resting tidal volume enters the lungs. This additional volume is called the **inspiratory reserve volume** (complemental air), and it equals about 3,000 mL.

TABLE 19.3	Major Events in Expiration

1. The diaphragm and external respiratory muscles relax.
2. Elastic tissues of the lungs and thoracic cage, which were stretched during inspiration, suddenly recoil, and surface tension collapses alveolar walls.
3. Tissues recoiling around the lungs increase the intra-alveolar pressure.
4. Air is squeezed out of the lungs.

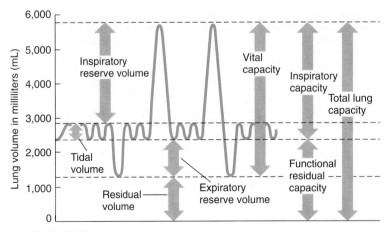

FIGURE 19.26

Respiratory volumes and capacities.

During a maximal forced expiration, about 1,100 mL of air in addition to the resting tidal volume can be expelled from the lungs. This quantity is called the **expiratory reserve volume** (supplemental air). However, even after the most forceful expiration, about 1,200 mL of air remains in the lungs. This is called the **residual volume.**

Residual air remains in the lungs at all times, and consequently, newly inhaled air always mixes with air already in the lungs. This prevents the oxygen and carbon dioxide concentrations in the lungs from fluctuating greatly with each breath.

Once the respiratory volumes are known, four *respiratory capacities* can be calculated by combining two or more of the volumes. If the inspiratory reserve volume (3,000 mL) is combined with the tidal volume (500 mL) and the expiratory reserve volume (1,100 mL), the total is termed the **vital capacity** (4,600 mL). This capacity is the maximum amount of air a person can exhale after taking the deepest breath possible.

The tidal volume (500 mL) plus the inspiratory reserve volume (3,000 mL) gives the **inspiratory capacity** (3,500 mL), which is the maximum volume of air a person can inhale following a resting expiration. Similarly, the expiratory reserve volume (1,100 mL) plus the residual volume (1,200 mL) equals the **functional residual capacity** (2,300 mL), which is the volume of air that remains in the lungs following a resting expiration.

The vital capacity plus the residual volume equals the **total lung capacity** (about 5,800 mL) (fig. 19.26). This total varies with age, sex, and body size.

Some of the air that enters the respiratory tract during breathing fails to reach the alveoli. This volume (about 150 mL) remains in the passageways of the trachea, bronchi, and bronchioles. Since gas exchanges do not occur through the walls of these passages, this air is said to occupy *anatomic dead space.*

Occasionally, air sacs in some regions of the lungs are nonfunctional due to poor blood flow in the adjacent cap-illaries. This creates *alveolar dead space.* The anatomic and alveolar dead space volumes combined equal *physiologic dead space.* In a normal lung, the anatomic and physiologic dead spaces are essentially the same (about 150 mL).

A spirometer (fig. 19.27) is used to measure respiratory air volumes (except the residual volume). These measurements are used to evaluate the course of respiratory illnesses, such as emphysema, pneumonia, lung cancer, and bronchial asthma. Table 19.4 summarizes the respiratory air volumes and capacities.

1. What is tidal volume?

2. Distinguish between inspiratory and expiratory reserve volumes.

3. How is vital capacity measured?

4. How is the total lung capacity calculated?

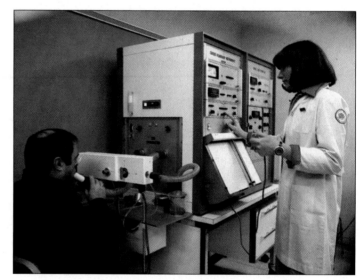

FIGURE 19.27
A spirometer can be used to measure respiratory air volumes.

TABLE 19.4	Respiratory Air Volumes and Capacities	
Name	**Volume***	**Description**
Tidal volume (TV)	500 mL	Volume moved in or out of the lungs during a respiratory cycle
Inspiratory reserve volume (IRV)	3,000 mL	Volume that can be inhaled during forced breathing in addition to resting tidal volume
Expiratory reserve volume (ERV)	1,100 mL	Volume that can be exhaled during forced breathing in addition to resting tidal volume
Residual volume (RV)	1,200 mL	Volume that remains in the lungs at all times
Inspiratory capacity (IC)	3,500 mL	Maximum volume of air that can be inhaled following exhalation of resting tidal volume: IC = TV + IRV
Functional residual capacity (FRC)	2,300 mL	Volume of air that remains in the lungs following exhalation of resting tidal volume: FRC = ERV + RV
Vital capacity (VC)	4,600 mL	Maximum volume of air that can be exhaled after taking the deepest breath possible: VC = TV + IRV + ERV
Total lung capacity (TLC)	5,800 mL	Total volume of air that the lungs can hold: TLC = VC + RV

*Values are typical for a tall, young adult.

Alveolar Ventilation

The amount of new atmospheric air that is moved into the respiratory passages each minute is called the *minute ventilation*. It equals the tidal volume multiplied by the breathing rate. Thus, if the tidal volume is 500 mL and the breathing rate is 12 breaths per minute, the minute ventilation is 500 mL × 12, or 6,000 mL per minute. However, much of the new air remains in the physiologic dead space.

The volume of new air that does reach the alveoli and is available for gas exchange is calculated by subtracting the physiologic dead space (150 mL) from the tidal volume (500 mL). The resulting volume (350 mL) multiplied by the breathing rate (12 breaths per minute) is the *alveolar ventilation rate* (4,200 mL per minute). The alveolar ventilation rate is a major factor affecting the concentrations of oxygen and carbon dioxide in the alveoli.

> In the microgravity environment of space, the gas exchange capacity of the alveoli increases by 28%. This is because blood flow in the pulmonary capillaries and ventilation of the alveoli are more uniform than on earth in the presence of gravity. Eight astronauts aboard two space shuttle missions provided this information by performing various pulmonary function tests. They worked in Spacelab, a small, cylindrical pressurized laboratory that is taken into space aboard the shuttle.

Nonrespiratory Air Movements

Air movements that occur in addition to breathing are called *nonrespiratory movements.* They are used to clear air passages, as in coughing and sneezing, or to express emotions, as in laughing and crying.

Nonrespiratory movements usually result from *reflexes,* although sometimes they are initiated voluntarily. A cough, for example, can be produced through conscious effort or may be triggered by a foreign object in an air passage.

Coughing involves taking a deep breath, closing the glottis, and forcing air upward from the lungs against the closure. Then the glottis is suddenly opened, and a blast of air is forced upward from the lower respiratory tract. Usually this rapid rush of air removes the substance that triggered the reflex.

> The most sensitive areas of the air passages are in the larynx, the carina, and in regions near the branches of the major bronchi. The distal portions of the bronchioles (respiratory bronchioles), alveolar ducts, and alveoli lack a nerve supply. Consequently, before any material in these parts can trigger a cough reflex, it must be moved into the larger passages of the respiratory tract.

A *sneeze* is much like a cough, but it clears the upper respiratory passages rather than the lower ones. This reflex is usually initiated by a mild irritation in the lining of the nasal cavity, and, in response, a blast of air is forced up through the glottis. This time, the air is directed into the nasal passages by depressing the uvula, thus closing the opening between the pharynx and the oral cavity.

Laughing involves taking a breath and releasing it in a series of short expirations. *Crying* consists of very similar movements, and sometimes it is necessary to note a person's facial expression in order to distinguish laughing from crying.

A *hiccup* is caused by sudden inspiration due to a spasmodic contraction of the diaphragm while the glottis is closed. Air striking the vocal folds causes the sound of the hiccup. We do not know the function, if any, of hiccups.

Yawning may aid respiration by providing an occasional deep breath. During normal, quiet breathing, not all of the alveoli are ventilated, and some blood may pass through the lungs without becoming well oxygenated. This low blood oxygen concentration somehow triggers the yawn reflex, prompting a very deep breath that ventilates more alveoli.

Table 19.5 summarizes the characteristics of nonrespiratory air movements. Clinical Application 19.3 discusses respiratory problems that affect ventilation.

TABLE 19.5	Nonrespiratory Air Movements	
Air Movement	**Mechanism**	**Function**
Coughing	Deep breath is taken, glottis is closed, and air is forced against the closure; suddenly the glottis is opened, and a blast of air passes upward	Clears lower respiratory passages
Sneezing	Same as coughing, except air moving upward is directed into the nasal cavity by depressing the uvula	Clears upper respiratory passages
Laughing	Deep breath is released in a series of short expirations	Expresses happiness
Crying	Same as laughing	Expresses sadness
Hiccuping	Diaphragm contracts spasmodically while glottis is closed	No useful function known
Yawning	Deep breath is taken	Ventilates a larger proportion of the alveoli and aids oxygenation of the blood
Speech	Air is forced through the larynx, causing vocal cords to vibrate; words are formed by lips, tongue, and soft palate	Vocal communication

RESPIRATORY DISORDERS THAT DECREASE VENTILATION: BRONCHIAL ASTHMA AND EMPHYSEMA

Injuries to the respiratory center or to spinal nerve tracts that transmit motor impulses may paralyze breathing muscles. Paralysis may also be due to a disease, such as *poliomyelitis*, that affects parts of the central nervous system and injures motor neurons. Sometimes, other muscles, by increasing their responses, can compensate for functional losses of a paralyzed muscle. Otherwise, mechanical ventilation is necessary. More common disorders that decrease ventilation are bronchial asthma and emphysema.

Bronchial asthma is usually an allergic reaction to foreign antigens in the respiratory tract, such as inhaled pollen. Cells of the larger airways secrete abundant mucus, which traps allergens. Ciliated columnar epithelial cells move the mucus up and out of the bronchi, then up and out of the trachea. Thus, upper respiratory structures remain relatively clear. However, in the lower respiratory areas, mucus drainage plus edematous secretions accumulate because fewer cells are ciliated. The allergens and secretions irritate smooth muscles, stimulating bronchioconstrictions.

Breathing becomes increasingly difficult, and inhalation produces a characteristic wheezing sound as air moves through narrowed passages.

A person with asthma usually finds it harder to force air out of the lungs than to bring it in. This is because inspiration utilizes powerful breathing muscles, and, as they contract, the lungs expand, opening the air passages. Expiration, on the other hand, is a passive process due to elastic recoil of stretched tissues. Expiration also compresses the tissues and constricts the bronchioles, further impairing air movement through the narrowed air passages.

Increase in the prevalence of asthma in the United States in recent years may be due to a too-clean environment, especially for children. Many studies have shown that children who are with others and contract minor respiratory infections, as well as children raised with cats or dogs, are less likely to develop asthma than are children who do not have these exposures. This association of a primed immune system with lower risk of developing asthma is called the hygiene hypothesis.

Emphysema is a progressive, degenerative disease that destroys many alveolar walls. As a result, clusters of small air sacs merge into larger chambers, which decreases the total surface area of the alveolar walls. At the same time, the alveolar walls lose their elasticity, and the capillary networks associated with the alveoli diminish (fig. 19C).

Because of the loss of tissue elasticity, a person with emphysema finds it increasingly difficult to force air out of the lungs. Abnormal muscular efforts are required to compensate for the lack of elastic recoil that normally contributes to expiration. About 3% of the 2 million people in the United States who have emphysema inherit the condition; the majority of the other cases are due to smoking or other respiratory irritants.

An experimental treatment for severe emphysema is lung volume reduction surgery. As its name suggests, the procedure reduces lung volume, which opens collapsed airways and eases breathing. So far, it seems to noticeably improve lung function (as measured by distance walked in six minutes) and quality of life. ■

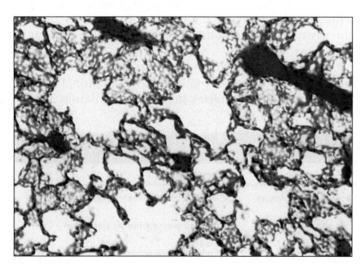 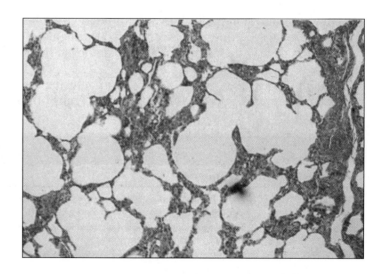

FIGURE 19C
Comparison of lung tissues. (*a*) Normal lung tissue (100×). (*b*) As emphysema develops, alveoli merge, forming larger chambers (100×).

1. How is the minute ventilation calculated? The alveolar ventilation rate?

2. Which nonrespiratory air movements help clear the air passages?

3. Which nonrespiratory air movements are used to express emotions?

4. What seems to be the function of a yawn?

Control of Breathing

Normal breathing is a rhythmic, involuntary act that continues when a person is unconscious.

Respiratory Center

Groups of neurons in the brainstem comprise the **respiratory center,** which controls breathing. This center periodically initiates impulses that travel on cranial and spinal nerves to breathing muscles, causing inspiration and expiration. The respiratory center also adjusts the rate and depth of breathing to meet cellular needs for supply of oxygen and removal of carbon dioxide, even during strenuous physical exercise.

The components of the respiratory center are widely scattered throughout the pons and medulla oblongata. However, two areas of the respiratory center are of special interest. They are the rhythmicity area of the medulla and the pneumotaxic area of the pons (fig. 19.28).

The **medullary rhythmicity area** includes two groups of neurons that extend throughout the length of the medulla oblongata. They are called the dorsal respiratory group and the ventral respiratory group.

The *dorsal respiratory group* is responsible for the basic rhythm of breathing. The neurons of this group emit bursts of impulses that signal the diaphragm and other inspiratory muscles to contract. The impulses of each burst begin weakly, strengthen for about two seconds, and cease abruptly. The breathing muscles that contract in response to the impulses cause the volume of air entering the lungs to increase steadily. The neurons remain inactive while expiration occurs passively, and then they emit another burst of inspiratory impulses so that the inspiration–expiration cycle repeats.

The *ventral respiratory group* is quiescent during normal breathing. However, when more forceful breathing is necessary, the neurons in this group generate impulses that increase inspiratory movement. Other neurons in the group activate the muscles associated with forceful expiration.

A condition called *sleep apnea* is responsible for some cases of sudden infant death and for snoring. Babies who have difficulty breathing just after birth are often sent home with monitoring devices, which sound an alarm when the child stops breathing, alerting parents to resuscitate the infant. The position in which the baby sleeps seems to affect the risk of sleep apnea—sleeping on the back or side is safest during the first year of life.

Adults with sleep apnea may cease breathing for ten to twenty seconds, hundreds of times a night. Bedmates may be aware of the problem because the frequent cessation in breathing causes snoring. The greatest danger of adult sleep apnea is the fatigue, headache, depression, and drowsiness that follows during waking hours.

Sleep apnea is diagnosed in a sleep lab, which monitors breathing during slumber. One treatment for sleep apnea is *nasal continuous positive airway pressure.* A device is strapped onto the nose at night that regulates the amount of air sent into and out of the respiratory system. Much simpler and more comfortable are inexpensive tape devices that hold the nostrils open.

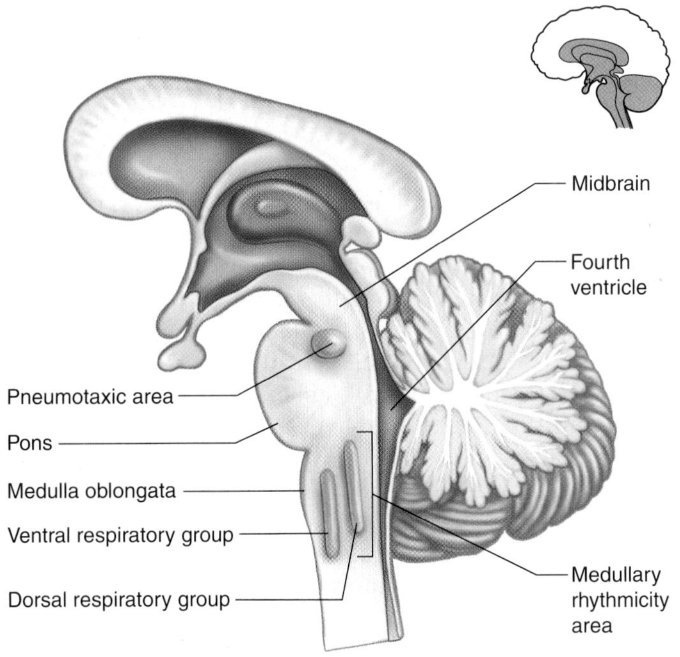

FIGURE 19.28
The respiratory center is located in the pons and the medulla oblongata.

The neurons in the **pneumotaxic area** continuously transmit impulses to the dorsal respiratory group and regulate the duration of inspiratory bursts originating from the dorsal group. In this way the pneumotaxic neurons control the rate of breathing. More specifically, when the pneumotaxic signals are strong, the inspiratory bursts are shorter, and the rate of breathing increases (tachypnea); when the pneumotaxic signals are weak, the inspiratory

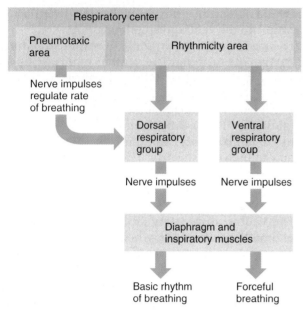

FIGURE 19.29
The medullary rhythmicity area and pneumotaxic area of the pons within the respiratory center control breathing.

bursts are longer, and the rate of breathing decreases (bradypnea) (fig. 19.29).

1 Where is the respiratory center located?

2 Describe how the respiratory center maintains a normal breathing pattern.

3 Explain how the breathing pattern may be changed.

Factors Affecting Breathing

In a mixture of gases such as air, each gas accounts for a portion of the total pressure the mixture produces. The amount of pressure each gas contributes is called the **partial pressure** of that gas and is proportional to its concentration. For example, because air is 21% oxygen, oxygen accounts for 21% of the atmospheric pressure (21% of 760 Hg), or 160 mm Hg (.21 × 760 = 160). Thus, the partial pressure of oxygen, symbolized P_{O_2}, in atmospheric air is 160 mm Hg. Similarly, the partial pressure of carbon dioxide (P_{CO_2}) in air is 0.3 mm Hg.

Gas molecules from the air may enter, or dissolve, in liquid. This is what happens when carbon dioxide is added to a carbonated beverage or when inspired gases dissolve in the blood in the alveolar capillaries. Although the calculation of the concentration of a dissolved gas is a bit complicated, it turns out that the partial pressure of a gas dissolved in a liquid is by definition equal to the partial pressure of that gas in the air the liquid has equilibrated with. Thus, instead of concentrations of oxygen and carbon dioxide in the body fluids, we will refer to P_{O_2} and P_{CO_2}.

A number of factors influence breathing rate and depth. These include P_{O_2} and P_{CO_2} in body fluids, the degree to which lung tissues are stretched, and emotional state. For example, **central chemoreceptors** are found in chemosensitive areas within the respiratory center, located in the ventral portion of the medulla oblongata near the origins of the vagus nerves. These chemoreceptors are very sensitive to changes in P_{CO_2} and blood pH. If P_{CO_2} or the hydrogen ion concentration rises, the central chemoreceptors signal the respiratory center, and breathing rate increases.

The similarity of the effects of carbon dioxide and hydrogen ions is a consequence of the fact that carbon dioxide combines with water in the cerebrospinal fluid to form carbonic acid (H_2CO_3):

$$CO_2 + H_2O \rightarrow H_2CO_3$$

The carbonic acid thus formed soon ionizes, releasing hydrogen ions (H^+) and bicarbonate ions (HCO_3^-):

$$H_2CO_3 \rightarrow H^+ + HCO_3^-$$

It is the presence of hydrogen ions rather than the carbon dioxide that influences the central chemoreceptors. In any event, breathing rate and tidal volume increase when a person inhales air rich in carbon dioxide or when body cells produce excess carbon dioxide or hydrogen ions. These changes increase alveolar ventilation. As a result, more carbon dioxide is exhaled, and the blood P_{CO_2} and hydrogen ion concentration return toward normal.

> Adding carbon dioxide to air can stimulate the rate and depth of breathing. Ordinary air is about 0.04% carbon dioxide. If a patient inhales air containing 4% carbon dioxide, breathing rate usually doubles.

Low blood P_{O_2} has little direct effect on the central chemoreceptors associated with the respiratory center. Instead, changes in the blood P_{O_2} are primarily sensed by **peripheral chemoreceptors** in specialized structures called the *carotid bodies* and *aortic bodies,* which are located in the walls of the carotid sinuses and aortic arch (fig. 19.30). When decreased P_{O_2} stimulates these peripheral receptors, impulses are transmitted to the respiratory center, and the breathing rate and tidal volume increase, thus increasing alveolar ventilation. This mechanism is usually not triggered until the P_{O_2} reaches a very low level; thus, oxygen seems to play only a minor role in the control of normal respiration.

The peripheral chemoreceptors of the carotid and aortic bodies are also stimulated by changes in the blood P_{CO_2} and pH. However, CO_2 and hydrogen ions have a much greater effect on the central chemoreceptors of the respiratory center than they do on the carotid and aortic bodies.

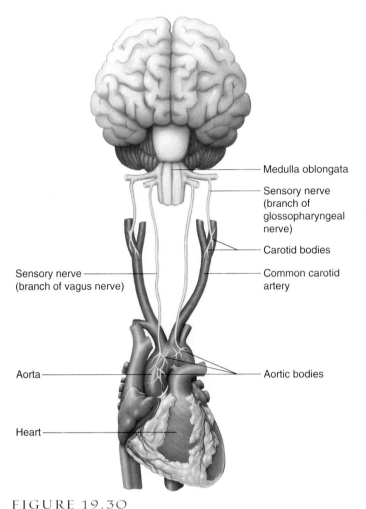

FIGURE 19.30

Decreased blood oxygen concentration stimulates peripheral chemoreceptors in the carotid and aortic bodies.

Medulla oblongata

Sensory nerve (branch of glossopharyngeal nerve)

Carotid bodies

Common carotid artery

Sensory nerve (branch of vagus nerve)

Aorta

Aortic bodies

Heart

An exception to normal respiratory control may occur in patients who have chronic obstructive pulmonary diseases (COPD), such as asthma, bronchitis, and emphysema. These patients gradually adapt to high concentrations of carbon dioxide, and for them, low oxygen concentrations may serve as a necessary respiratory stimulus. When such a patient is placed on 100% oxygen, the low arterial P_{O_2} may be corrected, the stimulus removed, and breathing may stop.

An *inflation reflex* (Hering-Breuer reflex) helps regulate the depth of breathing. This reflex occurs when stretch receptors in the visceral pleura, bronchioles, and alveoli are stimulated as lung tissues are stretched. The sensory impulses of the reflex travel via the vagus nerves to the pneumotaxic area of the respiratory center and shorten the duration of inspiratory movements. This action prevents overinflation of the lungs during forceful breathing (fig. 19.31).

Emotional upset or strong sensory stimulation may alter the normal breathing pattern. Gasping and rapid breathing are familiar responses to fear, anger, shock, excitement, horror, surprise, sexual stimulation, or even the chill of stepping into a cold shower. Because control of the respiratory muscles is voluntary, we can alter breathing pattern consciously, or even stop it altogether for a short time. During childbirth, for example, women often concentrate on controlling their breathing, which distracts them from the pain.

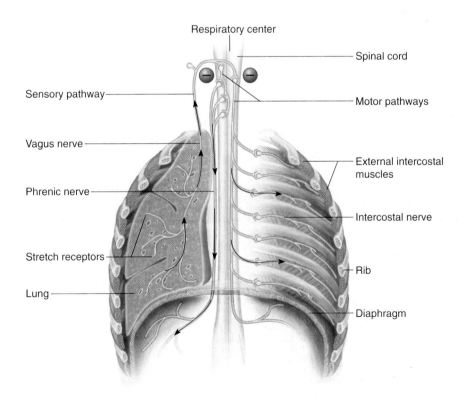

Respiratory center

Spinal cord

Sensory pathway

Motor pathways

Vagus nerve

External intercostal muscles

Phrenic nerve

Intercostal nerve

Stretch receptors

Rib

Lung

Diaphragm

FIGURE 19.31

In the process of inspiration, motor impulses travel from the respiratory center to the diaphragm and external intercostal muscles, which contract and cause the lungs to expand. This expansion stimulates stretch receptors in the lungs to send inhibiting impulses to the respiratory center, preventing overinflation.

EXERCISE AND BREATHING

Moderate to heavy exercise greatly increases the amount of oxygen skeletal muscles use. A young man at rest utilizes about 250 milliliters of oxygen per minute but may require 3,600 milliliters per minute during maximal exercise. While oxygen utilization is increasing, carbon dioxide production increases also. Since decreased blood oxygen and increased blood carbon dioxide concentration stimulate the respiratory center, it is not surprising that exercise is accompanied by increased breathing rate. However, studies reveal that blood oxygen and carbon dioxide concentrations usually do not change during exercise—this reflects the respiratory system's effectiveness in obtaining oxygen and releasing carbon dioxide to the outside.

The cerebral cortex and the proprioceptors associated with muscles and joints are also implicated in the increased breathing rate associated with exercise (see chapter 12, p. 423). The cortex transmits stimulating impulses to the respiratory center whenever it signals the skeletal muscles to contract. At the same time, muscular movements stimulate the proprioceptors, triggering a *joint reflex*. In this reflex, sensory impulses are transmitted from the proprioceptors to the respiratory center, and breathing accelerates.

The increase in breathing rate during exercise requires increased blood flow to skeletal muscles. Thus, exercise increases demand on both the respiratory and the cardiovascular systems. If either of these systems fails to keep pace with cellular demands, the person will begin to feel out of breath. This sensation, however, is usually due to the inability of the cardiovascular system to move enough blood between the lungs and the cells, rather than to the inability of the respiratory system to provide enough air. ■

If a person decides to stop breathing, the blood concentrations of carbon dioxide and hydrogen ions begin to rise, and the concentration of oxygen falls. These changes (primarily the increased CO_2) stimulate the respiratory center, and soon the need to inhale overpowers the desire to hold the breath—much to the relief of parents when young children threaten to hold their breaths until they turn blue! On the other hand, a person can increase the breath-holding time by breathing rapidly and deeply in advance. (This could be dangerous, see box that follows.) This action, termed **hyperventilation** (hi″per-ven″tĭ-la′shun), lowers the blood carbon dioxide concentration below normal. Following hyperventilation, it takes longer than usual for the carbon dioxide concentration to reach the level needed to override the conscious effort of breath holding.

Table 19.6 discusses factors affecting breathing. Clinical Application 19.4 focuses on one influence on breathing—exercise.

Sometimes a person who is emotionally upset may hyperventilate, become dizzy, and lose consciousness. This is due to a lowered carbon dioxide concentration followed by a rise in pH (respiratory alkalosis) and a localized vasoconstriction of cerebral arterioles, decreasing blood flow to nearby brain cells. Hampered oxygen supply to the brain causes fainting. A person should never hyperventilate to help hold the breath while swimming, because the person may lose consciousness under water and drown.

TABLE 19.6	Factors Affecting Breathing		
Factors	**Receptors Stimulated**	**Response**	**Effect**
Stretch of tissues	Stretch receptors in visceral pleura, bronchioles, and alveoli	Inhibits inspiration	Prevents overinflation of lungs during forceful breathing
Low blood P_{O_2}	Chemoreceptors in carotid and aortic bodies	Increases alveolar ventilation	Increases blood oxygen concentration
High blood P_{CO_2}	Chemosensitive areas of the respiratory center	Increases alveolar ventilation	Decreases blood carbon dioxide concentration and cerebrospinal fluid hydrogen ion concentration
High cerebrospinal fluid hydrogen ion concentration	Chemosensitive areas of the respiratory center	Increases alveolar ventilation	Decreases blood carbon dioxide concentration and cerebrospinal fluid hydrogen ion concentration

1. Describe the inflation reflex.

2. Which chemical factors affect breathing?

3. How does hyperventilation decrease respiratory rate?

Alveolar Gas Exchanges

The tubelike parts of the respiratory system move air in and out of the air passages. The alveoli are the sites of the vital process of gas exchange between the air and the blood.

Alveoli

Alveoli (al-ve'o-li) are microscopic air sacs clustered at the distal ends of the finest respiratory tubes—the alveolar ducts. Each alveolus consists of a tiny space surrounded by a thin wall that separates it from adjacent alveoli. Tiny openings, called **alveolar pores,** in the walls of some alveoli may permit air to pass from one alveolus to another (fig. 19.32). This arrangement provides alternate air pathways if the passages in some portions of the lung become obstructed.

Phagocytic cells called *alveolar macrophages* are in alveoli and in the pores connecting the air sacs. These macrophages phagocytize airborne agents, including bacteria, thereby cleaning the alveoli (fig. 19.33).

Respiratory Membrane

Part of the wall of an alveolus is made up of cells that secrete pulmonary surfactant, described earlier (type II cells), but the bulk of the wall of an alveolus consists of an inner lining of simple squamous epithelium (type I cells) and a dense network of capillaries, which are also lined with simple squamous epithelial cells (fig. 19.33). Thin basement membranes separate the layers of these flat-

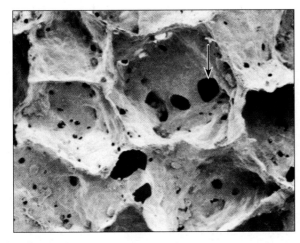

FIGURE 19.32
Alveolar pores (arrow) allow air to pass from one alveolus to another (300×).

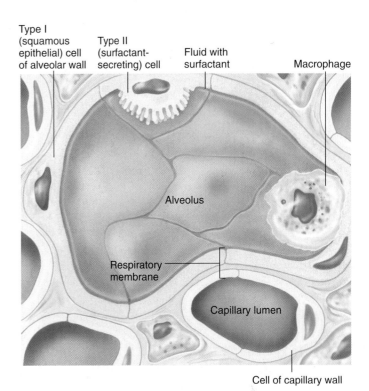

FIGURE 19.33
The respiratory membrane consists of the walls of the alveolus and the capillary.

tened cells, and in the spaces between them are elastic and collagenous fibers that help support the alveolar wall. Thus, two thicknesses of epithelial cells and basement membranes separate the air in an alveolus and the blood in a capillary. These layers make up the **respiratory membrane** (alveolar-capillary membrane), through which gas exchange occurs between the alveolar air and the blood (figs. 19.34 and 19.35).

Diffusion Through the Respiratory Membrane

Molecules diffuse from regions where they are in higher concentration toward regions where they are in lower concentration. Thus, in determining the direction of diffusion of a solute, we must know the concentration gradient. In the case of gases, it is more convenient to think in terms of a partial pressure gradient, such that a gas will diffuse from an area of higher partial pressure to an area of lower partial pressure.

RECONNECT TO CHAPTER 3, DIFFUSION, PAGE 81.

When a mixture of gases dissolves in blood, the resulting concentration of each dissolved gas is proportional to its partial pressure. Each gas diffuses between blood and its surroundings from areas of higher partial pressure to areas of lower partial pressure until the partial pressures in the two regions reach equilibrium. For

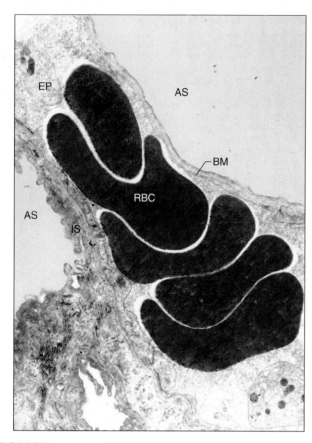

FIGURE 19.34

Falsely colored electron micrograph of a capillary located between alveoli (7,000×). (*AS,* alveolar space; *RBC,* red blood cell; *BM,* basement membrane; *IS,* interstitial connective tissue; *EP,* epithelial nucleus.)

example, the P_{CO_2} of blood entering the pulmonary capillaries is 45 mm Hg, but the P_{CO_2} in alveolar air is 40 mm Hg. Because of the difference in these partial pressures, carbon dioxide diffuses from blood, where its partial pressure is higher, across the respiratory membrane and

into alveolar air. When blood leaves the lungs, its P_{CO_2} is 40 mm Hg, which is the same as the P_{CO_2} of alveolar air. Similarly, the P_{O_2} of blood entering the pulmonary capillaries is 40 mm Hg, but reaches 104 mm Hg as oxygen diffuses from alveolar air into the blood. Thus, since equilibrium is reached, blood leaves the alveolar capillaries with a P_{O_2} of 104 mm Hg. (Some venous blood draining the bronchi and bronchioles mixes with this blood before returning to the heart, so the P_{O_2} of systemic arterial blood is 95 mm Hg.) Because of the large volume of air always in the lungs, as long as breathing continues, alveolar P_{O_2} stays relatively constant at 104 mm Hg. Clinical Application 19.5 examines illnesses that result from impaired gas exchange.

The respiratory membrane is normally quite thin (about 1 micrometer thick), and gas exchange is rapid. However, a number of factors may affect diffusion across the respiratory membrane. More surface area, shorter distance, greater solubility of gases, and a steeper partial pressure gradient all favor increased diffusion. Diseases that harm the respiratory membrane, such as pneumonia, or reduce the surface area for diffusion, such as emphysema, impair gas exchange. These conditions may require increased P_{O_2} for treatment.

The respiratory membrane is normally so thin that certain soluble chemicals other than carbon dioxide may diffuse into alveolar air and be exhaled. This is why breath analysis can reveal alcohol in the blood or acetone can be smelled on the breath of a person who has untreated diabetes mellitus. Breath analysis may also detect substances associated with kidney failure, certain digestive disturbances, and liver disease.

1 Describe the structure of the respiratory membrane.

2 What is partial pressure?

3 What causes oxygen and carbon dioxide to move across the respiratory membrane?

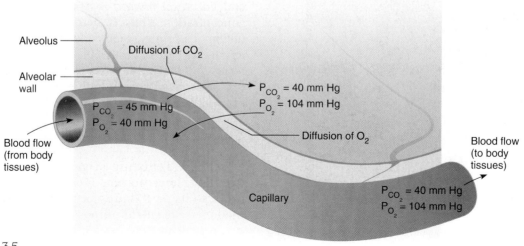

FIGURE 19.35

Gases are exchanged between alveolar air and capillary blood because of differences in partial pressures.

DISORDERS THAT IMPAIR GAS EXCHANGE: PNEUMONIA, TUBERCULOSIS, AND ADULT RESPIRATORY DISTRESS SYNDROME

Five-year-old Carly had what her parents at first thought was just a "bug" that was passing through the family. But after twelve hours of flulike symptoms, Carly's temperature shot up to 105°F, her chest began to hurt, and her breathing became rapid and shallow. Later that day, a chest radiograph confirmed what the doctor suspected—Carly had *pneumonia.* **Apparently, the bacteria that had caused a mild upper respiratory infection in her parents and sisters had taken a detour in her body, infecting her lower respiratory structures instead.**

Antibiotics successfully treated Carly's bacterial pneumonia. A viral infection, or, as is often the case in people with AIDS, *Pneumocystis carinii* infection, can also cause pneumonia. For all types of pneumonia, the events within the infected lung are similar: alveolar linings swell with edema and become abnormally permeable, allowing fluids and white blood cells to accumulate in the air sacs. As the alveoli fill, the surface area available for gas exchange diminishes. Breathing becomes difficult. Untreated, pneumonia can kill.

Tuberculosis is a different type of lung infection, caused by the bacterium *Myco-bacterium tuberculosis* (fig. 19D). Fibrous connective tissue develops around the sites of infection, forming structures called *tubercles.* By walling off the bacteria, the tubercles help stop their spread. Sometimes this protective mechanism fails, and the bacteria flourish throughout the lungs and may even spread to other organs. In the later stages of infection, other types of bacteria may cause secondary infections. As lung tissue is destroyed, the surface area for gas exchange decreases. In addition, the widespread fibrous tissue thickens the respiratory membrane, further restricting gas exchange. A variety of drugs are used to treat tuberculosis, but in recent years, strains resistant to drugs have arisen, and these can be swiftly deadly.

Another type of condition that impairs gas exchange is *atelectasis.* This is the collapse of a lung, or some part of it, together with the collapse of the blood vessels that supply the affected region. Obstruction of a respiratory tube, such as by an inhaled foreign object or excess mucus secretion, may cause atelectasis. The air in the alveoli beyond the obstruction is absorbed, and as the air pressure in the alveoli decreases, their elastic walls collapse, and they can no longer function. Fortunately, after a portion of a lung collapses, the functional regions that remain are often able to carry on enough gas exchange to sustain the body cells.

Adult respiratory distress syndrome (ARDS) is a special form of atelectasis in which alveoli collapse. It has a variety of causes, all of which damage lung tissues. These include pneumonia and other infections, near drowning, aspiration of stomach acid into the respiratory system, or physical trauma to the lungs from an injury or surgical procedure. Anesthetics can suppress surfactant production, causing postsurgical difficulty breathing for twenty-four to forty-eight hours or until surfactant production returns to normal. This damage disrupts the respiratory membrane that separates the air in the alveoli from the blood in the pulmonary capillaries, allowing protein-rich fluid to escape from the capillaries and flood the alveoli. They collapse in response, and surfactant is nonfunctional. Blood vessels and airways narrow, greatly elevating blood pressure in the lungs. Delivery of oxygen to tissues is seriously impaired. ARDS is fatal about 60% of the time. ■

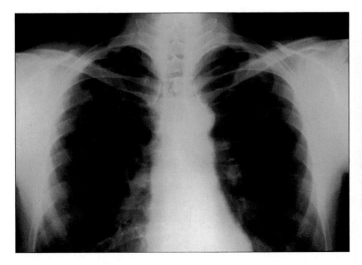

Healthy lungs

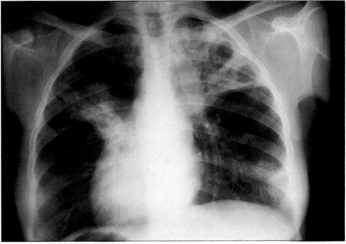

Tuberculosis

FIGURE 19D

Healthy lungs appear dark and clear on a radiograph. Lungs with tuberculosis have cloudy areas where fibrous tissue grows, walling off infected areas.

CLINICAL APPLICATION

EFFECTS OF HIGH ALTITUDE

Each year, about 100,000 mountain climbers experience varying degrees of altitude sickness, because at high elevations, the proportion of oxygen in air remains the same (about 21%), but the P_{O_2} decreases. When a person ascends rapidly, oxygen diffuses more slowly from the alveoli into the blood, and the hemoglobin becomes less saturated with oxygen. In some individuals, the body's efforts to get more oxygen—increased breathing and heart rate and enhanced red blood cell and hemoglobin production—cannot keep pace with the plummeting oxygen supply.

Severe altitude sickness includes a condition called high altitude pulmonary edema (HAPE). Symptoms include sudden severe headache, nausea and vomiting, rapid heart rate and breathing, and a cyanotic (blue) cast to the skin.

Much of what we know about HAPE was learned at a "mountain hut" on Punta Gnifetti on the Italian-Swiss border. Italian physiologist Angelo Mosso established this facility at the turn of the nineteenth century, where he monitored signs and symptoms in soldiers who climbed the mountain. Mosso described HAPE as an "inflammation of the lungs," but more recent investigations suggest that the inflammation is a consequence of the reaction to hypoxia, not a cause of it.

Another recent study indicates why certain individuals have a higher risk of developing HAPE than others. Researchers compared the ability of nasal epithelium to transport sodium ions in mountain climbers who experienced HAPE to mountain climbers who did not. The affected individuals had a third lower ability to transport sodium. Because lowered oxygen (hypoxia) suppresses synthesis of the protein subunits that form sodium ion channels, it is possible that high risk individuals inherit impaired ability to transport sodium ions that worsens sufficiently under low oxygen conditions to cause symptoms.

Ascending slowly, not overdoing exercise level, and perhaps taking a drug called nifedipine a few hours before a hike can lower the risk of developing HAPE. It is treated by giving oxygen and coming down from the mountain. But too often, this condition is fatal. ■

Exposure to high oxygen concentration (hyperoxia) for a prolonged time may damage lung tissues, particularly capillary walls. Excess fluid may escape the capillaries and flood the alveolar air spaces, interfering with gas exchange, which can be lethal. Similarly, hyperoxia can damage the retinal capillaries of premature infants, causing *retrolental fibroplasia* (RLF), a condition that may lead to blindness.

Gas Transport

The blood transports oxygen and carbon dioxide between the lungs and the body cells. As these gases enter the blood, they dissolve in the liquid portion, the plasma, or combine chemically with other atoms or molecules.

Oxygen Transport

Almost all the oxygen (over 98%) is carried in the blood bound to the protein **hemoglobin** (he″mo-glo′bin) in red blood cells. The iron in hemoglobin provides the color of these blood cells. The remainder of the oxygen is dissolved in the blood plasma.

Hemoglobin consists of two types of components called *heme* and *globin* (see chapter 14, p. 517). Globin is a protein that contains 574 amino acids in four polypeptide chains. Each chain is associated with a heme group, and each heme group contains an atom of iron. Each iron atom can loosely combine with an oxygen molecule. As oxygen dissolves in blood, it rapidly combines with

hemoglobin, forming a new compound called **oxyhemoglobin** (ok″se-he′mo-glo′bin). Each hemoglobin molecule can combine with a maximum of four oxygen molecules.

The P_{O_2} determines the amount of oxygen that combines with hemoglobin. The greater the P_{O_2}, the more oxygen will combine with hemoglobin, until the hemoglobin molecules are saturated with oxygen (fig. 19.36). At normal arterial P_{O_2} (95 mm Hg), hemoglobin is essentially completely saturated.

The chemical bonds that form between oxygen and hemoglobin molecules are relatively unstable, and as the

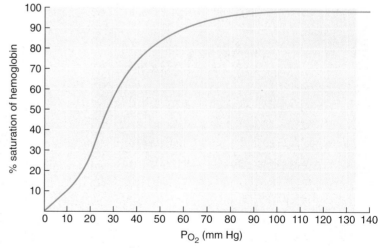

Oxyhemoglobin dissociation at 38°C

FIGURE 19.36
Hemoglobin is completely saturated at normal systemic arterial P_{O_2} but readily releases oxygen at the P_{O_2} of the body tissues.

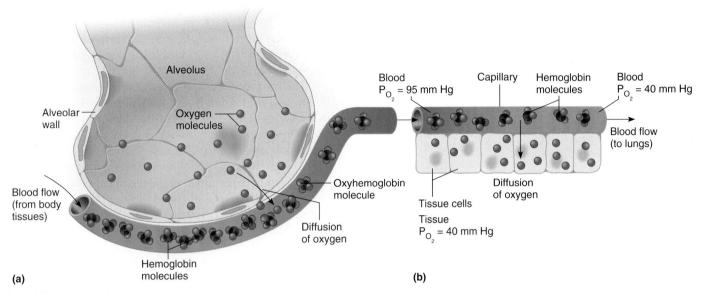

FIGURE 19.37
Blood transports oxygen. (*a*) Oxygen molecules, entering the blood from the alveolus, bond to hemoglobin, forming oxyhemoglobin. (*b*) In the regions of the body cells, oxyhemoglobin releases oxygen. Note that much oxygen is still bound to hemoglobin at the P_{O_2} of systemic venous blood.

P_{O_2} decreases, oxygen is released from oxyhemoglobin molecules (fig. 19.36). This happens in tissues, where cells have used oxygen in respiration. The free oxygen diffuses from the blood into nearby cells, as figure 19.37 shows.

Increasing blood concentration of carbon dioxide (P_{CO_2}), acidity, and temperature all increase the amount of oxygen that oxyhemoglobin releases (figs. 19.38, 19.39, and 19.40). These influences explain why more oxygen is released from the blood to the skeletal muscles during periods of exercise. The increased muscular activity accompanied by increased oxygen use increases the P_{CO_2}, decreases the pH, and raises the local temperature. At the same time, less-active cells receive less oxygen.

Carbon monoxide (CO) is a toxic gas produced in gasoline engines and some stoves as a result of incomplete combustion of fuels. It is also a component of tobacco smoke. Carbon monoxide is toxic because it combines with hemoglobin many times more effectively than does oxygen and therefore does not readily dissociate from hemoglobin. Thus, when a person breathes carbon monoxide, less hemoglobin is available for oxygen transport, and cells are deprived of oxygen. The effects of carbon monoxide on hemoglobin may cause the lower average birth weights of infants born to women who smoked while pregnant.

Treatment for carbon monoxide poisoning is to administer oxygen in high concentration to replace some of the carbon monoxide bound to hemoglobin molecules. Carbon dioxide is usually given simultaneously to stimulate the respiratory center, which, in turn, increases breathing rate. Rapid breathing helps reduce the concentration of carbon monoxide in the alveoli.

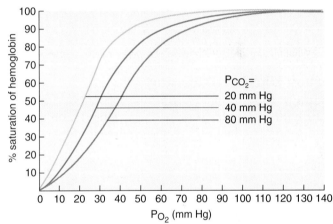

Oxyhemoglobin dissociation at 38°C

FIGURE 19.38
The amount of oxygen released from oxyhemoglobin increases as the P_{CO_2} increases.

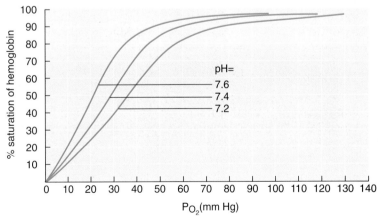

Oxyhemoglobin dissociation at 38°C

FIGURE 19.39
The amount of oxygen released from oxyhemoglobin increases as the blood pH decreases.

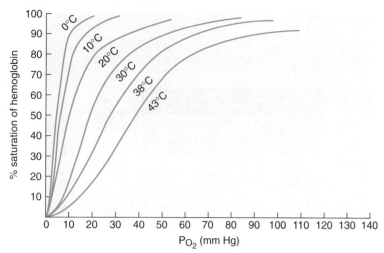

Oxyhemoglobin dissociation at various temperatures

FIGURE 19.40
The amount of oxygen released from oxyhemoglobin increases as the blood temperature increases.

1 How is oxygen transported from the lungs to body cells?

2 What factors affect the release of oxygen from oxyhemoglobin?

Carbon Dioxide Transport

Blood flowing through capillaries gains carbon dioxide because the tissues have a high P_{CO_2}. This carbon dioxide is transported to the lungs in one of three forms: as carbon dioxide dissolved in plasma, as part of a compound formed by bonding to hemoglobin, or as part of a bicarbonate ion (fig. 19.41).

The amount of carbon dioxide that dissolves in plasma is determined by its partial pressure. The higher the P_{CO_2} of the tissues, the more carbon dioxide will go into solution. However, only about 7% of the carbon dioxide is transported in this form.

Unlike oxygen, which combines with the iron atoms of hemoglobin molecules, carbon dioxide bonds with the amino groups (—NH_2) of these molecules. Consequently, oxygen and carbon dioxide do not directly compete for binding sites—a hemoglobin molecule can transport both gases at the same time.

Carbon dioxide combining with hemoglobin forms a loosely bound compound called **carbaminohemoglobin** (kar-bam″ĭ-no-he″mo-glo″bin). This molecule readily decomposes in regions where the P_{CO_2} is low, releasing its carbon dioxide. Although this method of transporting carbon dioxide is theoretically quite effective, carbaminohemoglobin forms relatively slowly. Only about 15% to 25% of the total carbon dioxide is carried this way.

The most important carbon dioxide transport mechanism involves the formation of **bicarbonate ions** (HCO_3^-). Recall that carbon dioxide reacts with water to form carbonic acid (H_2CO_3). This reaction occurs slowly in the blood plasma, but much of the carbon dioxide diffuses into the red blood cells. These cells contain an enzyme, **carbonic anhydrase** (kar-bon′ik an-hi′drās),

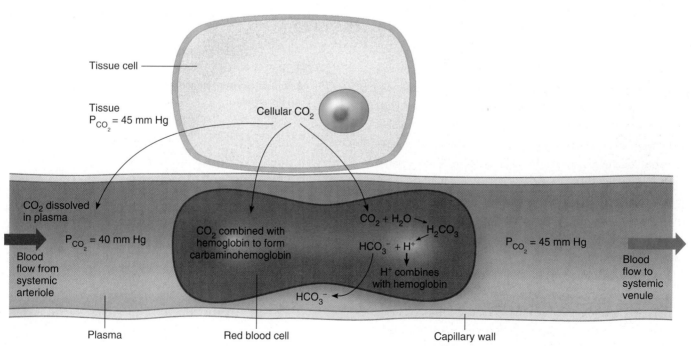

FIGURE 19.41
Carbon dioxide produced by tissue cells is transported in the blood plasma in a dissolved state, combined with hemoglobin, or in the form of bicarbonate ions (HCO_3^-).

which speeds the reaction between carbon dioxide and water.

The resulting carbonic acid dissociates almost immediately, releasing hydrogen ions (H^+) and bicarbonate ions (HCO_3^-):

$$H_2CO_3 \rightarrow H^+ + HCO_3^-$$

These new hydrogen ions might be expected to lower blood pH, but this reaction occurs in the systemic capillaries, where deoxyhemoglobin is generated. Deoxyhemoglobin is an excellent buffer because hydrogen ions combine readily with it. The bicarbonate ions diffuse out of the red blood cells and enter the blood plasma. As much as 70% of the carbon dioxide transported in the blood is carried in this form.

As the bicarbonate ions leave the red blood cells and enter the plasma, *chloride ions,* which also have negative charges, are repelled electrically, and they move from the plasma into the red blood cells. This exchange in position of the two negatively charged ions, shown in figure 19.42, maintains the ionic balance between the red blood cells and the plasma. It is termed the **chloride shift.**

As blood passes through the capillaries of the lungs, the dissolved carbon dioxide diffuses into the alveoli, in response to the relatively low P_{CO_2} of the alveolar air. As the plasma P_{CO_2} drops, hydrogen ions and bicarbonate ions in the red blood cells recombine to form carbonic acid, and under the influence of carbonic anhydrase, the carbonic acid quickly yields new carbon dioxide and water:

$$H^+ + HCO_3^- \rightarrow H_2CO_3 \rightarrow CO_2 + H_2O$$

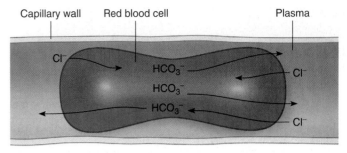

FIGURE 19.42

As bicarbonate ions (HCO_3^-) diffuse out of the red blood cell, chloride ions (Cl^-) from the plasma diffuse into the cell, maintaining the electrical balance between ions. This exchange of ions is called the chloride shift.

Carbaminohemoglobin also releases its carbon dioxide, and both of these events contribute to the P_{CO_2} of the alveolar capillary blood. Carbon dioxide diffuses out of the blood until an equilibrium is established between the P_{CO_2} of the blood and the P_{CO_2} of the alveolar air. Figure 19.43 summarizes this process, and table 19.7 summarizes the transport of blood gases.

1 Describe three ways carbon dioxide can be transported from cells to the lungs.

2 How can hemoglobin carry oxygen and carbon dioxide at the same time?

3 How do bicarbonate ions help buffer the blood (maintain its pH)?

4 What is the chloride shift?

5 How is carbon dioxide released from the blood into the lungs?

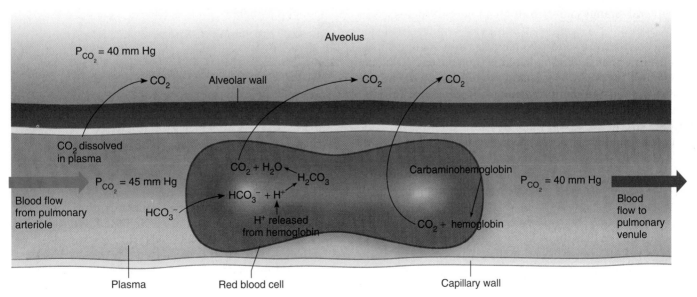

FIGURE 19.43

In the lungs, carbon dioxide diffuses from the blood into the alveoli.

TABLE 19.7	Gases Transported in Blood	
Gas	**Reaction**	**Substance Transported**
Oxygen	Combines with iron atoms of hemoglobin molecules	Oxyhemoglobin
Carbon dioxide	About 7% dissolves in plasma	Carbon dioxide
	About 23% combines with the amino groups of hemoglobin molecules	Carbaminohemoglobin
	About 70% reacts with water to form carbonic acid; the carbonic acid then dissociates to release hydrogen ions and bicarbonate ions	Bicarbonate ions

Life-Span Changes

Changes in the respiratory system over a lifetime reflect both the accumulation of environmental influences and the effects of aging in other organ systems. The lungs and respiratory passageways of a person who has breathed only clean air are pinker and can exchange gases much more efficiently as the years pass than can the respiratory system of a person who has breathed polluted air and smoked for many years. Those who have been exposed to foul air are more likely to develop chronic bronchitis, emphysema, and/or lung cancer. Long-term exposure to particulates in the workplace can also raise the risk of developing these conditions. Still, many age-associated changes in the respiratory system are unavoidable.

With age, protection of the lungs and airways falters, as ciliated epithelial cells become fewer, and their cilia less active or gone. At the same time, mucus thickens; the swallowing, gagging, and coughing reflexes slow; and macrophages lose their efficiency in phagocytizing bacteria. These changes combine to slow the clearance of pathogens from the lungs and respiratory passages, which increases susceptibility to and severity of respiratory infections.

Several changes contribute to an overall increase in effort required to breathe that accompanies aging. Cartilage between the sternum and ribs calcifies and stiffens, and skeletal shifts change the shape of the thoracic cavity into a "barrel chest" as posture too changes with age. In the bronchioles, fibrous connective tissue replaces some smooth muscle, decreasing contractility. As muscles lose strength, breathing comes to depend more upon the diaphragm. The vital capacity, which reaches a maximum by age forty, may drop by a third by the age of seventy years.

Keeping fresh air in the lungs becomes more difficult with age. As the farthest reaches of the bronchiole walls thin, perhaps in response to years of gravity, they do not stay as open as they once did, trapping residual air in the lower portions of the lungs. Widening of the bronchi and alveolar ducts increases dead space. The lungs can still handle the same volume of air, but a greater proportion of that air is "stale," reflecting lessened ability to move air in and out. The maximum minute ventilation drops by 50% from age twenty to age eighty.

Aging-associated changes occur at the microscopic level too. The number of alveoli is about 24 million at birth, peaking at 300 million by age eight years. The number remains constant throughout life, but the alveoli expand. Alveolar walls thin and may coalesce, and the depth of alveoli begins to diminish by age forty, decreasing the surface area available for gas exchange—about three square feet per year. In addition, an increase in the proportion of collagen to elastin and a tendency of the collagen to cross-link impair the ability of alveoli to expand fully. As a result, oxygen transport from the alveoli to the blood, as well as oxygen loading onto hemoglobin in red blood cells, becomes less efficient. Diffusion of CO_2 out of the blood and through the alveolar walls slows too.

As with other organ systems, the respiratory system undergoes specific changes, but these may be unnoticeable at the whole-body level. A person who is sedentary or engages only in light activity would probably not be aware of the slowing of air flow in and out of the respiratory system. Unaccustomed exercise, however, would quickly reveal how difficult breathing has become compared to in years past.

1 How does the environment influence the effects of aging on the respiratory system?

2 What aging-related changes raise the risk of respiratory infection?

3 How do alveoli change with age?

Clinical Terms Related to the Respiratory System

anoxia (ah-nok'se-ah) Absence or a deficiency of oxygen within tissues.
asphyxia (as-fik'se-ah) Deficiency of oxygen and excess carbon dioxide in the blood and tissues.
atelectasis (at"e-lek'tah-sis) Collapse of a lung or some portion of it.
bradypnea (brad"e-ne'ah) Abnormally slow breathing.
bronchitis (brong-ki'tis) Inflammation of the bronchial lining.

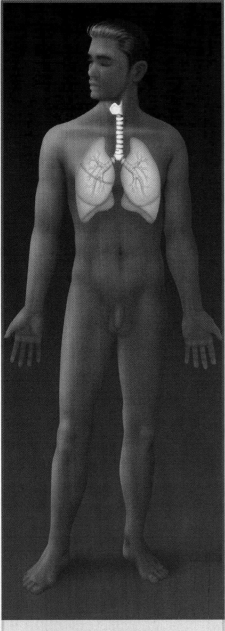

RESPIRATORY SYSTEM

The respiratory system provides oxygen for the internal environment and excretes carbon dioxide.

Integumentary System

Stimulation of skin receptors may alter respiratory rate.

Cardiovascular System

As the heart pumps blood through the lungs, the lungs oxygenate the blood and excrete carbon dioxide.

Skeletal System

Bones provide attachments for muscles involved in breathing.

Lymphatic System

Cells of the immune system patrol the lungs and defend against infection.

Muscular System

The respiratory system eliminates carbon dioxide produced by exercising muscles.

Digestive System

The digestive system and respiratory system share openings to the outside.

Nervous System

The brain controls the respiratory system. The respiratory system helps control pH of the internal environment.

Urinary System

The kidneys and the respiratory system work together to maintain blood pH. The kidneys compensate for water lost through breathing.

Endocrine System

Hormonelike substances control the production of red blood cells that transport oxygen and carbon dioxide.

Reproductive System

Respiration increases during sexual activity. Fetal "respiration" begins before birth.

Cheyne-Stokes respiration (chān stōks res"pi-ra'shun) Irregular breathing pattern of a series of shallow breaths that increase in depth and rate, followed by breaths that decrease in depth and rate.

dyspnea (disp'ne-ah) Difficulty in breathing.

eupnea (up-ne'ah) Normal breathing.

hemothorax (he"mo-tho'raks) Blood in the pleural cavity.

hypercapnia (hi"per-kap'ne-ah) Excess carbon dioxide in the blood.

hyperoxia (hi"per-ok'se-ah) Excess oxygenation of the blood.

hyperpnea (hi"perp-ne'ah) Increase in the depth and rate of breathing.

hyperventilation (hi"per-ven"ti-la'shun) Prolonged, rapid, and deep breathing.

hypoxemia (hi"pok-se'me-ah) Deficiency in the oxygenation of the blood.

hypoxia (hi-pok'se-ah) Diminished availability of oxygen in the tissues.

lobar pneumonia (lo'ber nu-mo'ne-ah) Pneumonia that affects an entire lobe of a lung.

pleurisy (ploo'rĭ se) Inflammation of the pleural membranes.

pneumoconiosis (nu"mo-ko"ne-o'sis) Accumulation of particles from the environment in the lungs and the reaction of the tissues.

pneumothorax (nu"mo-tho'raks) Entrance of air into the space between the pleural membranes, followed by collapse of the lung.

rhinitis (ri-ni'tis) Inflammation of the nasal cavity lining.

sinusitis (si"nu-si'tis) Inflammation of the sinus cavity lining.

tachypnea (tak"ip-ne'ah) Rapid, shallow breathing.

tracheotomy (tra"ke-ot'o-me) Incision in the trachea for exploration or the removal of a foreign object.

CHAPTER SUMMARY

Introduction (page 732)

The respiratory system includes the passages that transport air to and from the lungs and the air sacs in which gas exchanges occur. Respiration is the entire process by which gases are exchanged between the atmosphere and the body cells.

Why We Breathe (page 732)

Respiration is necessary because of cellular respiration. Cells require oxygen to extract maximal energy from nutrient molecules and to rid themselves of carbon dioxide, a metabolic waste.

Organs of the Respiratory System (page 733)

The respiratory system includes the nose, nasal cavity, sinuses, pharynx, larynx, trachea, bronchial tree, and lungs. The upper respiratory tract includes the nose, nasal cavity, sinuses, and pharynx; the lower respiratory tract includes the larynx, trachea, bronchial tree, and lungs.

1. Nose
   a. Bone and cartilage support the nose.
   b. Nostrils provide entrances for air.
2. Nasal cavity
   a. The nasal cavity is a space posterior to the nose.
   b. The nasal septum divides it medially.
   c. Nasal conchae divide the cavity into passageways and help increase the surface area of the mucous membrane.
   d. Mucous membrane filters, warms, and moistens incoming air.
   e. Particles trapped in the mucus are carried to the pharynx by ciliary action and are swallowed.
3. Sinuses
   a. Sinuses are spaces in the bones of the skull that open into the nasal cavity.
   b. They are lined with mucous membrane that is continuous with the lining of the nasal cavity.

4. Pharynx
   a. The pharynx is posterior to the mouth, between the nasal cavity and the larynx.
   b. It provides a common passage for air and food.
   c. It aids in creating vocal sounds.
5. Larynx
   a. The larynx is an enlargement at the top of the trachea.
   b. It is a passageway for air and helps prevent foreign objects from entering the trachea.
   c. It is composed of muscles and cartilages; some of these cartilages are single, whereas others are paired.
   d. It contains the vocal cords, which produce sounds by vibrating as air passes over them.
      (1) The pitch of a sound is related to the tension on the cords.
      (2) The intensity of a sound is related to the force of the air passing over the cords.
   e. The glottis and epiglottis help prevent food and liquid from entering the trachea.
6. Trachea
   a. The trachea extends into the thoracic cavity in front of the esophagus.
   b. It divides into the right and left bronchi.
   c. The mucous lining continues to filter incoming air.
   d. Incomplete cartilaginous rings support the wall.
7. Bronchial tree
   a. The bronchial tree consists of branched air passages that connect the trachea to the air sacs.
   b. The branches of the bronchial tree include primary bronchi, lobar bronchi, segmental bronchi, intralobular bronchioles, terminal bronchioles, respiratory bronchioles, alveolar ducts, alveolar sacs, and alveoli.
   c. Structure of the respiratory tubes
      (1) As tubes branch, the amount of cartilage in the walls decreases, and the muscular layer becomes more prominent.
      (2) Elastic fibers in the walls aid breathing.
      (3) The epithelial lining changes from pseudostratified and ciliated to cuboidal and

simple squamous as the tubes become progressively smaller.
  d. Functions of the respiratory tubes include distribution of air and exchange of gases between the alveolar air and the blood.
8. Lungs
  a. The left and right lungs are separated by the mediastinum and are enclosed by the diaphragm and the thoracic cage.
  b. The visceral pleura is attached to the surface of the lungs; parietal pleura lines the thoracic cavity.
  c. The right lung has three lobes, and the left lung has two.
  d. Each lobe is composed of lobules that contain alveolar ducts, alveolar sacs, alveoli, nerves, blood vessels, lymphatic vessels, and connective tissues.

Breathing Mechanism (page 745)

Inspiration and expiration movements are accompanied by changes in the size of the thoracic cavity.

1. Inspiration
  a. Atmospheric pressure forces air into the lungs.
  b. Inspiration occurs when the intra-alveolar pressure is reduced.
  c. The intra-alveolar pressure is reduced when the diaphragm moves downward and the thoracic cage moves upward and outward.
  d. Surface tension holding the pleural membranes together aids lung expansion.
  e. Surfactant reduces surface tension within the alveoli.
2. Expiration
  a. The forces of expiration come from the elastic recoil of tissues and from surface tension within the alveoli.
  b. Expiration can be aided by thoracic and abdominal wall muscles that pull the thoracic cage downward and inward and compress the abdominal organs inward and upward.
3. Respiratory volumes and capacities
  a. One inspiration followed by one expiration is called a respiratory cycle.
  b. The amount of air that moves in or out during a respiratory cycle is the tidal volume.
  c. Additional air that can be inhaled is the inspiratory reserve volume; additional air that can be exhaled is the expiratory reserve volume.
  d. Residual air remains in the lungs and is mixed with newly inhaled air.
  e. The inspiratory capacity is the maximum volume of air a person can inhale following exhalation of the tidal volume.
  f. The functional residual capacity is the volume of air that remains in the lungs following the exhalation of the tidal volume.
  g. The vital capacity is the maximum amount of air a person can exhale after taking the deepest breath possible.
  h. The total lung capacity is equal to the vital capacity plus the residual air volume.
  i. Air in the anatomic and alveolar dead spaces is not available for gas exchange.
4. Alveolar ventilation
  a. Minute ventilation is calculated by multiplying the tidal volume by the breathing rate.
  b. Alveolar ventilation rate is calculated by subtracting the physiologic dead space from the tidal volume and multiplying the result by the breathing rate.

  c. The alveolar ventilation rate is a major factor affecting the exchange of gases between the alveolar air and the blood.
5. Nonrespiratory air movements
  a. Nonrespiratory air movements are air movements other than breathing.
  b. They include coughing, sneezing, laughing, crying, hiccuping, and yawning.

Control of Breathing (page 753)

Normal breathing is rhythmic and involuntary, although the respiratory muscles can be controlled voluntarily.

1. Respiratory center
  a. The respiratory center is located in the brainstem and includes parts of the medulla oblongata and pons.
  b. The medullary rhythmicity area includes two groups of neurons.
    (1) The dorsal respiratory group is responsible for the basic rhythm of breathing.
    (2) The ventral respiratory group increases inspiratory and expiratory movements during forceful breathing.
  c. The pneumotaxic area regulates the rate of breathing.
2. Factors affecting breathing
  a. The partial pressure of a gas is determined by the concentration of that gas in a mixture of gases or the concentration of gas dissolved in a liquid.
  b. Chemicals, lung tissue stretching, and emotional state affect breathing.
  c. Chemosensitive areas (central chemoreceptors) are associated with the respiratory center.
    (1) Carbon dioxide combines with water to form carbonic acid, which, in turn, releases hydrogen ions in the CSF.
    (2) Stimulation of these areas increases alveolar ventilation.
  d. Peripheral chemoreceptors are in the carotid bodies and aortic bodies of certain arteries.
    (1) These chemoreceptors sense low oxygen concentration.
    (2) When oxygen concentration is low, alveolar ventilation increases.
  e. Stretching the lung tissues triggers an inflation reflex.
    (1) This reflex reduces the duration of inspiratory movements.
    (2) This prevents overinflation of the lungs during forceful breathing.
  f. Hyperventilation decreases carbon dioxide concentration, but *this is very dangerous when associated with breath holding during underwater swimming.*

Alveolar Gas Exchanges (page 757)

Gas exchanges between the air and the blood occur within the alveoli.

1. Alveoli
  a. The alveoli are tiny sacs clustered at the distal ends of the alveolar ducts.
  b. Some alveoli open into adjacent air sacs that provide alternate pathways for air when passages are obstructed.
2. Respiratory membrane
  a. The respiratory membrane consists of the alveolar and capillary walls.
  b. Gas exchanges take place through these walls.

3. Diffusion through the respiratory membrane
   a. Gases diffuse from regions of higher partial pressure toward regions of lower partial pressure.
   b. Oxygen diffuses from the alveolar air into the blood; carbon dioxide diffuses from the blood into the alveolar air.

Gas Transport (page 760)

Blood transports gases between the lungs and the body cells.

1. Oxygen transport
   a. Oxygen is mainly transported in combination with hemoglobin molecules.
   b. The resulting oxyhemoglobin is relatively unstable and releases its oxygen in regions where the P_{O_2} is low.
   c. More oxygen is released as the blood concentration of carbon dioxide increases, as the blood becomes more acidic, and as the blood temperature increases.
2. Carbon dioxide transport
   a. Carbon dioxide may be carried in solution, either as dissolved CO_2, CO_2 bound to hemoglobin, or as a bicarbonate ion.
   b. Most carbon dioxide is transported in the form of bicarbonate ions.

   c. Carbonic anhydrase, an enzyme, speeds the reaction between carbon dioxide and water to form carbonic acid.
   d. Carbonic acid dissociates to release hydrogen ions and bicarbonate ions.

Life-Span Changes (page 764)

The lungs, respiratory passageways, and alveoli undergo aging-associated changes that are exacerbated by exposure to polluted air. However, the increased work required to breathe with age is typically not noticeable unless one engages in vigorous exercise.

1. Exposure to pollutants, smoke, and other particulates raises risk of developing diseases of the respiratory system.
2. Loss of cilia, thickening of mucus, and impaired macrophages raise the risk of infection.
3. Calcified cartilage, skeletal changes, altered posture, and replacement of smooth muscle with fibrous connective tissue in bronchioles make breathing more difficult. Vital capacity diminishes.
4. The lungs contain a greater proportion of "stale" air.
5. Alveoli coalesce and become shallower, slowing gas exchange.

CRITICAL THINKING QUESTIONS

1. If the upper respiratory passages are bypassed with a tracheostomy, how might the air entering the trachea be different from air normally passing through this canal? What problems might this cause for the patient?
2. Certain respiratory disorders, such as emphysema, reduce the capacity of the lungs to recoil elastically. Which respiratory air volumes will this condition affect? Why?
3. What changes would you expect to occur in the relative concentrations of blood oxygen and carbon dioxide in a patient who breathes rapidly and deeply for a prolonged time? Why?

4. If a person has stopped breathing and is receiving pulmonary resuscitation, would it be better to administer pure oxygen or a mixture of oxygen and carbon dioxide? Why?
5. The air pressure within the passenger compartment of a commercial aircraft may be equivalent to an altitude of 8,000 feet. What problem might this create for a person with a serious respiratory disorder?
6. Patients experiencing asthma attacks are often advised to breathe through pursed (puckered) lips. How might this help reduce the symptoms of the asthma?

REVIEW EXERCISES

1. Describe the general functions of the respiratory system.
2. Distinguish between the upper and lower respiratory tracts.
3. Explain how the nose and nasal cavity filter incoming air.
4. Name and describe the locations of the major sinuses, and explain how a sinus headache may occur.
5. Distinguish between the pharynx and the larynx.
6. Name and describe the locations and functions of the cartilages of the larynx.
7. Distinguish between the false vocal cords and the true vocal cords.
8. Compare the structure of the trachea with the structure of the branches of the bronchial tree.
9. List the successive branches of the bronchial tree, from the primary bronchi to the alveoli.
10. Describe how the structure of the respiratory tube changes as the branches become finer.
11. Explain the functions of the respiratory tubes.
12. Distinguish between visceral pleura and parietal pleura.

13. Name and describe the locations of the lobes of the lungs.
14. Explain how normal inspiration and forced inspiration are accomplished.
15. Define surface tension, and explain how it aids the breathing mechanism.
16. Define surfactant, and explain its function.
17. Define compliance.
18. Explain how normal expiration and forced expiration are accomplished.
19. Distinguish between vital capacity and total lung capacity.
20. Distinguish among anatomic, alveolar, and physiologic dead spaces.
21. Distinguish between minute respiratory volume and alveolar ventilation rate.
22. Compare the mechanisms of coughing and sneezing, and explain the function of each.
23. Explain the function of yawning.
24. Describe the location of the respiratory center, and name its major components.

25. Describe how the basic rhythm of breathing is controlled.
26. Explain the function of the pneumotaxic area of the respiratory center.
27. Explain why increasing blood concentrations of carbon dioxide and hydrogen ions have similar effects on the respiratory center.
28. Describe the function of the peripheral chemoreceptors in the carotid and aortic bodies of certain arteries.
29. Describe the inflation reflex.
30. Discuss the effects of emotions on breathing.
31. Define *hyperventilation*, and explain how it affects the respiratory center.
32. Define *respiratory membrane*, and explain its function.

33. Explain the relationship between the partial pressure of a gas and its rate of diffusion.
34. Summarize the gas exchanges that occur through the respiratory membrane.
35. Describe how oxygen is transported in blood.
36. List three factors that increase release of oxygen from the blood.
37. Explain why carbon monoxide is toxic.
38. List three ways that carbon dioxide is transported in blood.
39. Explain the function of carbonic anhydrase.
40. Define *chloride shift.*
41 Describe the changes that make it harder to breathe with advancing years.

W E B   C O N N E C T I O N S

Visit the Student Edition of the Online Learning Center at www.mhhe.com/shier10 **for answers to chapter questions, additional quizzes, interactive learning exercises, and other study tools.**

UNDERSTANDING WORDS

af-, to: *af*ferent arteriole—arteriole that leads to a nephron.

calyc-, small cup: major *calyc*es—cuplike subdivisions of the renal pelvis.

cort-, covering: renal *cort*ex—shell of tissue surrounding the inner region of a kidney.

cyst-, bladder: *cyst*itis—inflammation of the bladder.

detrus-, to force away: *detrus*or muscle—muscle within the bladder wall that causes urine to be expelled.

glom-, little ball: *glom*erulus—cluster of capillaries within a renal corpuscle.

juxta-, near to: *juxta*medullary nephron—nephron located near the renal medulla.

mict-, to pass urine: *mict*urition—process of expelling urine from the bladder.

nephr-, pertaining to the kidney: *nephr*on—functional unit of a kidney.

papill-, nipple: renal *papill*ae—small elevations that project into a renal calyx.

prox-, nearest: *prox*imal tubule—coiled portion of the renal tubule leading from the glomerular capsule.

ren-, kidney: *ren*al cortex—outer region of a kidney.

trigon-, triangular shape: *trigon*e—triangular area on the internal floor of the bladder.

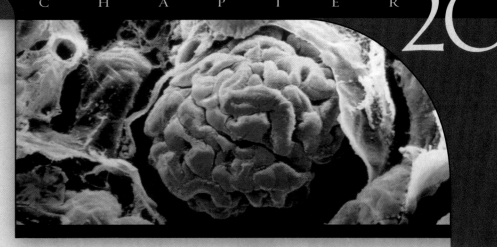

A renal corpuscle includes the glomerulus (red), which is a tangle of capillaries, and a glomerular capsule, which is an expansion at the receiving end of the renal tubule (520×). It is here that filtration of plasma, and urine formation, begins.

Urinary System

CHAPTER OBJECTIVES

After you have studied this chapter, you should be able to

1. Name the organs of the urinary system and list their general functions.

2. Describe the locations of the kidneys and the structure of a kidney.

3. List the functions of the kidneys.

4. Trace the pathway of blood through the major vessels within a kidney.

5. Describe a nephron and explain the functions of its major parts.

6. Explain how glomerular filtrate is produced and describe its composition.

7. Explain how various factors affect the rate of glomerular filtration and how this rate is regulated.

8. Discuss the role of tubular reabsorption in urine formation.

9. Explain why the osmotic concentration of the glomerular filtrate changes as it passes through a renal tubule.

10. Describe a countercurrent mechanism and explain how it helps concentrate urine.

11. Define tubular secretion and explain its role in urine formation.

12. Describe the structure of the ureters, urinary bladder, and urethra.

13. Discuss the process of micturition and explain how it is controlled.

14. Describe how the components of the urinary system change with age.

elicia had looked forward to summer camp all year, especially the overnight hikes. A three-day expedition in July was wonderful, but five days after returning to camp, Felicia developed severe abdominal cramps. So did seventeen other campers and two counselors, some of whom had bloody diarrhea, too. Several of the stricken campers were hospitalized, Felicia among them. Although the others improved in a few days and were released, Felicia suffered from a complication, called hemolytic uremic syndrome (HUS). Her urine had turned bloody, and she also had blood abnormalities—severe anemia and lack of platelets.

Camp personnel reported the outbreak to public health officials, who quickly recognized the signs of food poisoning and traced the illness to hamburgers cooked outdoors on the trip. The burgers were served rare, the red meat not hot enough to kill a strain of *Escherichia coli* bacteria that releases a poison called shigatoxin.

Most people who eat meat tainted with *E. coli* toxin become ill, but the damage usually is restricted to the digestive tract, producing cramps and diarrhea for several days. In about 6% of affected people, mostly children, HUS develops because the bloodstream transports the toxin to the kidneys. Here, the toxin destroys cells of the microscopic capillaries that normally filter proteins and blood cells from forming urine. With the filtration compromised, proteins and blood cells, as well as damaged kidney cells, appear in the urine.

HUS is a leading cause of acute renal (kidney) failure, killing 3% to 5% of affected children. Felicia was in the lucky majority. Blood clotted around the sites of her damaged kidney cells, and over a few weeks, new cells formed. Three weeks after the bloody urine began, her urine was once again clear, and she was healthy. ■

Cook that burger! Hemolytic uremic syndrome is a complication of infection by a strain of *E. coli* bacteria that produces shigatoxin. Destruction of the filtering capillaries in the kidney allows proteins and blood cells to enter urine.

The urinary system consists of a pair of glandular kidneys, which remove substances from the blood, form urine, and help regulate certain metabolic processes; a pair of tubular ureters, which transport urine from the kidneys; a saclike urinary bladder, which collects urine from the ureters and serves as a urine reservoir; and a tubular urethra, which conveys urine to the outside of the body. Figures 20.1 and 20.2 show these organs.

Kidneys

A **kidney** is a reddish brown, bean-shaped organ with a smooth surface. It is about 12 centimeters long, 6 centimeters wide, and 3 centimeters thick in an adult, and it is enclosed in a tough, fibrous capsule (tunic fibrosa).

Location of the Kidneys

The kidneys lie on either side of the vertebral column in a depression high on the posterior wall of the abdominal cavity. The upper and lower borders of the kidneys are generally at the levels of the twelfth thoracic and third lumbar vertebrae, respectively, although the positions of the kidneys may vary slightly with changes in posture and with breathing movements. The left kidney is usually about 1.5 to 2 centimeters higher than the right one.

The kidneys are positioned *retroperitoneally* (re″tro-per″ĭ-to-ne′al-le), which means they are behind the parietal peritoneum and against the deep muscles of the back. Connective tissue (renal fascia) and masses of adipose tissue (renal fat) surrounding the kidneys hold them in place (fig. 20.3 and reference plates 58, 59).

Kidney Structure

The lateral surface of each kidney is convex, but its medial side is deeply concave. The resulting medial depression leads into a hollow chamber called the **renal sinus.** Through the entrance to this sinus, termed the *hilum,* pass blood vessels, nerves, lymphatic vessels, and the ureter (see fig. 20.1).

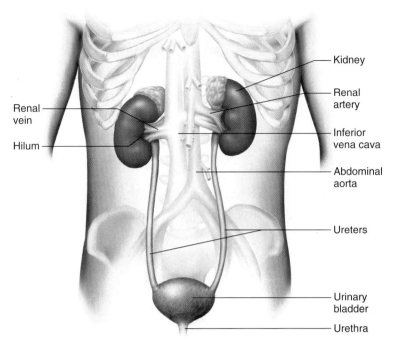

FIGURE 20.1

The urinary system includes the kidneys, ureters, urinary bladder, and urethra. Notice the relationship of these structures to the major blood vessels.

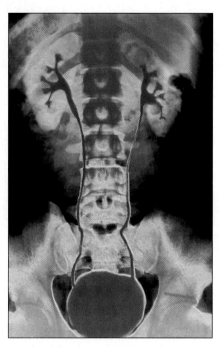

FIGURE 20.2

Structures of the urinary system are visible in this falsely colored radiograph.

The superior end of the ureter expands to form a funnel-shaped sac called the **renal pelvis,** which is located inside the renal sinus. The pelvis is subdivided into two or three tubes, called *major calyces* (sing., *calyx*), and they, in turn, are subdivided into eight to fourteen *minor calyces* (fig. 20.4). A series of small projections called *renal papillae* project into each minor calyx.

The kidney includes two distinct regions: an inner medulla and an outer cortex. The **renal medulla** (re'nal mĕ-dul'ah) is composed of conical masses of tissue called *renal pyramids,* the bases of which are directed toward the convex surface of the kidney, and the apexes of which form the renal papillae. The tissue of the medulla appears striated because it consists of microscopic tubules that lead from the cortex to the renal papillae.

Two common inherited kidney abnormalities are *polycystic kidney disease,* which affects adults, and *Wilms tumor,* which affects young children. In polycystic kidney disease, cysts present in the kidneys since childhood or adolescence begin to produce symptoms in the thirties, including abdominal pain, bloody urine, and elevated blood pressure. The condition can lead to renal failure. In Wilms tumor, pockets of cells within a child's kidney remain as they were in the embryo—they are unspecialized and divide rapidly, forming a cancerous tumor. Loss of a tumor suppressor gene causes Wilms tumor.

The **renal cortex** (re'nal kor'teks), which appears somewhat granular, forms a shell around the medulla. Its tissue dips into the medulla between the renal pyramids, forming *renal columns.* The cortex itself is surrounded by the **renal capsule,** a fibrous membrane that helps maintain the shape of the kidney and provides some protection (figs. 20.4 and 20.5).

Functions of the Kidneys

The main function of the kidneys is to regulate the volume, composition, and pH of body fluids. In the process, the kidneys remove metabolic wastes from the blood and excrete them to the outside. These wastes include nitrogenous and sulfur-containing products of protein metabolism. The kidneys also help control the rate of red blood cell formation by secreting the hormone *erythropoietin* (see chapter 14, p. 514), regulate blood pressure by secreting the enzyme *renin* (see chapter 13, p. 493), and regulate absorption of calcium ions by activating *vitamin D* (see chapter 13, pp. 489–490).

Medical technology can take over the role of a kidney. In *hemodialysis,* a person's blood is rerouted across an artificial membrane that "cleanses" it, removing substances that would normally be excreted in the urine. A patient usually must use this artificial kidney three times a week, for several hours each time. Clinical Application 20.1 further discusses hemodialysis.

1   Where are the kidneys located?

2   Describe the structure of a kidney.

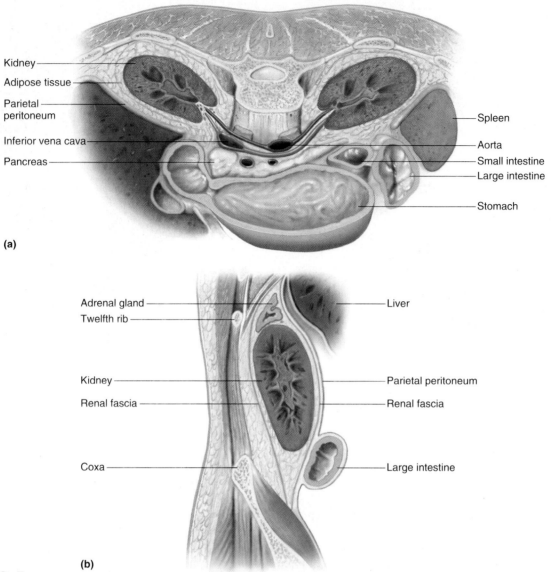

(a)

Kidney
Adipose tissue
Parietal peritoneum
Inferior vena cava
Pancreas

Spleen
Aorta
Small intestine
Large intestine
Stomach

Adrenal gland
Twelfth rib

Liver

Kidney
Renal fascia

Parietal peritoneum
Renal fascia

Coxa

Large intestine

(b)

FIGURE 20.3

The kidneys are located retroperitoneally. (*a*) Transverse section through the posterior abdominal cavity including the kidneys, which are located behind the parietal peritoneum. Adipose and other connective tissues surround and support the kidneys. (*b*) Sagittal section through the posterior abdominal cavity showing the kidney.

3 Name the functional unit of the kidney.

4 What are the general functions of the kidneys?

About two in every ten patients with renal failure can use a procedure that can be done at home called *continuous ambulatory peritoneal dialysis* instead of hemodialysis. The patient infuses a solution into the abdominal cavity through a permanently implanted tube. The solution stays in for four to eight hours, while it takes up substances that would normally be excreted into urine. Then the patient drains the waste-laden solution out of the tube, replacing it with clean fluid. Infection is a risk associated with this procedure.

Renal Blood Vessels

The **renal arteries,** which arise from the abdominal aorta, supply blood to the kidneys (see fig. 20.5). These arteries transport a large volume of blood. When a person is at rest, the renal arteries usually carry from 15% to 30% of the total cardiac output into the kidneys, although the kidneys account for only 1% of body weight.

A renal artery enters a kidney through the hilum and gives off several branches, called the *interlobar arteries,* which pass between the renal pyramids. At the junction between the medulla and the cortex, the interlobar arteries branch to form a series of incomplete arches, the *arcuate arteries* (arciform arteries), which, in turn, give rise to *interlobular arteries.* The final branches of the interlobular arteries, called **afferent arterioles** (af′er-ent ar-te′re-ōlz), lead to the nephrons, the functional units of the kidneys.

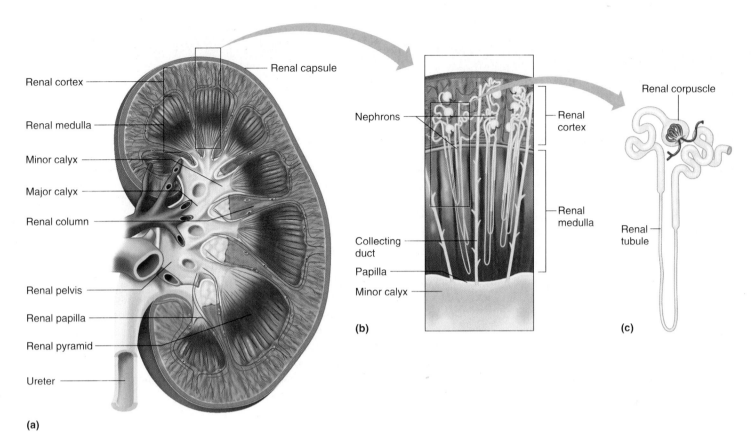

(a)

FIGURE 20.4

The kidney. (*a*) Longitudinal section of a kidney. (*b*) Part of a renal pyramid containing nephrons. (*c*) A single nephron.

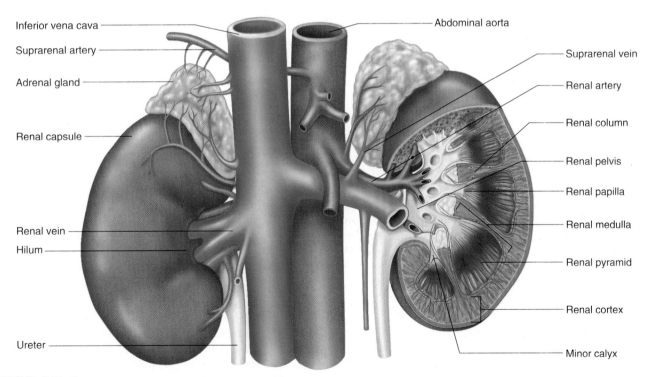

FIGURE 20.5

Blood vessels associated with the kidneys and adrenal glands. Note their relationship with the renal pelvis and ureters.

CHRONIC KIDNEY FAILURE

Charles B., a forty-three-year-old muscular construction worker, had been feeling unusually tired for several weeks, with occasional dizziness and difficulty sleeping. More recently he had noticed a burning pain in his lower back, just below his rib cage, and his urine had darkened. In addition, his feet, ankles, and face seemed swollen. His wife suggested that he consult their family physician about these symptoms.

The physician found that Charles had elevated blood pressure (hypertension) and that the regions of his kidneys were sensitive to pressure. A urinalysis revealed excess protein (proteinuria) and blood (hematuria). Blood tests indicated elevated blood urea nitrogen (BUN), elevated serum creatinine, and decreased serum protein (hypoproteinemia) concentrations.

The physician told Charles that he probably had *chronic glomerulonephritis,* an inflammation of the capillaries within the glomeruli of the renal nephrons, and that this was an untreatable progressive degenerative disease. Microscopic examination of a small sample of kidney tissue (biopsy) later confirmed the diagnosis.

In spite of medical treatment and careful attention to his diet, Charles's condi-

tion deteriorated rapidly. When it appeared that most of his kidney function had been lost (end-stage renal disease, or ESRD), he was offered artificial kidney treatments (hemodialysis).

To prepare Charles for hemodialysis, a vascular surgeon created a fistula in his left forearm by surgically connecting an artery to a vein. The greater pressure of the blood in the artery that now flowed directly into the vein swelled the vein, making it more accessible.

During hemodialysis treatment, a hollow needle was inserted into the vein of the fistula near its arterial connection. This allowed the blood to flow, with the aid of a blood pump, through a tube leading to the blood compartment of a dialysis machine. Within this compartment, the blood passed over a selectively permeable membrane. On the opposite side of the membrane was a dialysate solution with a controlled composition. Negative pressure on the dialysate side of the membrane, created by a vacuum pump, increased the movement of fluid through the membrane. At the same time, waste and excess electrolytes diffused from the blood through the membrane and entered the dialysate solution. The blood was then returned through a tube to the vein of the fistula.

In order to maintain favorable blood concentrations of waste, electrolytes, and water, Charles had to undergo hemodialysis three times per week, with each treatment lasting three to four hours. During the treatments, he was given an anticoagulant to prevent blood clotting, an antibiotic drug to control infections, and an antihypertensive drug to reduce his blood pressure.

Charles was advised to carefully control his intake of water, sodium, potassium, proteins, and total calories between treatments. He was also asked to consider another option for the treatment of ESRD—a kidney transplant—which could free him from the time-consuming dependence on hemodialysis.

In a transplant, a kidney from a living donor or a cadaver, whose tissues are antigenically similar (histocompatible) to those of the recipient, is placed in the depression on the medial surface of the right or left ilium (iliac fossa). The renal artery and vein of the donor kidney are connected to the recipient's iliac artery and vein, respectively, and the kidney's ureter is attached to the dome of the recipient's urinary bladder. The patient must then remain on immunosuppressant drugs to prevent rejection of the transplant. ■

Venous blood is returned through a series of vessels that generally correspond to the arterial pathways. For example, the venous blood passes through interlobular, arcuate, interlobar, and renal veins. The **renal vein** then joins the inferior vena cava as it courses through the abdominal cavity. (Figs. 20.6 and 20.7 show branches of the renal arteries and veins.)

Nephrons

Structure of a Nephron

Each kidney contains about 1 million functional units called **nephrons** (nefronz). Each nephron consists of a **renal corpuscle** (re′nal kor′pusl) and a **renal tubule** (re′nal tu′būl) (see fig. 20.4).

A renal corpuscle consists of a filtering unit composed of a tangled cluster of blood capillaries called a **glomerulus** (glo-mer′u-lus) and a surrounding thin-walled, saclike structure called a **glomerular** (Bowman's) **capsule.** Afferent arterioles give rise to these capillaries, which lead to **efferent arterioles** (ef′er-ent ar-te′re-ōlz) (fig. 20.7). Filtration of fluid from the glomerular capillaries is the first step in urine formation.

The glomerular capsule is an expansion at the end of a renal tubule that receives the fluid filtered at the glomerulus. It is composed of two layers of squamous epithelial cells: a visceral layer that closely covers the glomerulus and an outer parietal layer that is continuous with the visceral layer and with the wall of the renal tubule (fig. 20.8).

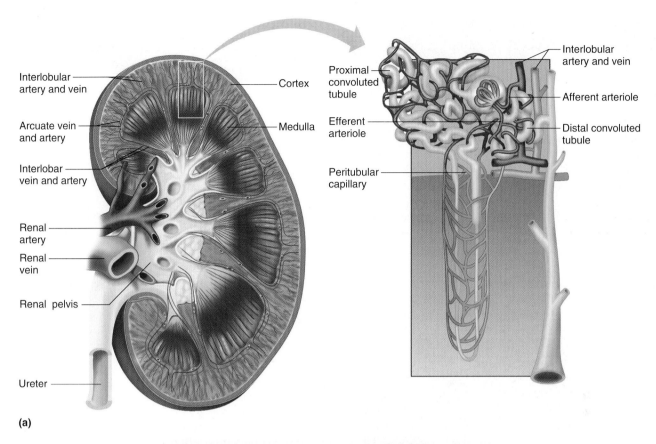

Interlobular artery and vein

Cortex

Arcuate vein and artery

Medulla

Interlobar vein and artery

Renal artery

Renal vein

Renal pelvis

Ureter

(a)

Proximal convoluted tubule

Interlobular artery and vein

Efferent arteriole

Afferent arteriole

Distal convoluted tubule

Peritubular capillary

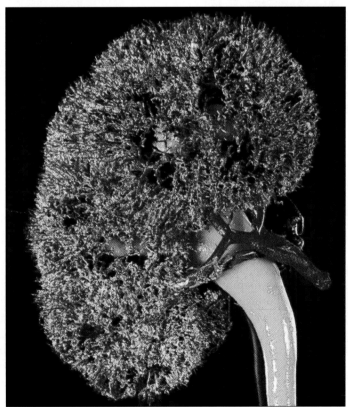

(b)

FIGURE 20.6

Renal blood vessels. (*a*) Main branches of the renal artery and vein. (*b*) Corrosion cast of the renal arterial system. Not all blood vessels associated with the nephron are shown.

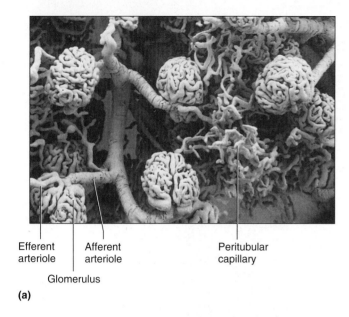

Efferent arteriole
Afferent arteriole
Glomerulus
Peritubular capillary

(a)

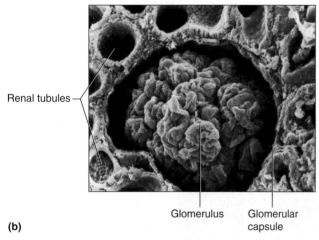

Renal tubules

Glomerulus
Glomerular capsule

(b)

FIGURE 20.7

Blood vessels associated wtih nephrons. (*a*) A scanning electron micrograph of a cast of the renal blood vessels associated with the glomeruli (200×). (*b*) Scanning electron micrograph of a glomerular capsule surrounding a glomerulus (480×). *Tissues and Organs: A Text-Atlas of Scanning Electron Microscopy,* by R. G. Kessel and R. H. Kardon. ©1979 W. H. Freeman and Company.

The cells of the parietal layer are typical squamous epithelial cells; however, those of the visceral layer are highly modified epithelial cells called *podocytes.* Each podocyte has several primary processes extending from its cell body, and these processes, in turn, bear numerous secondary processes, or *pedicels.* The pedicels of each cell interdigitate with those of adjacent podocytes, and the clefts between them form a complicated system of *slit pores* (figs. 20.8 and 20.9).

The renal tubule leads away from the glomerular capsule and becomes highly coiled. This coiled portion of the tubule is the *proximal convoluted tubule.* Following

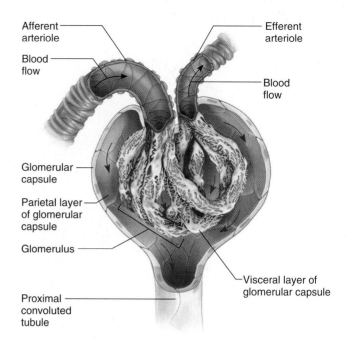

Afferent arteriole
Blood flow
Efferent arteriole
Blood flow
Glomerular capsule
Parietal layer of glomerular capsule
Glomerulus
Visceral layer of glomerular capsule
Proximal convoluted tubule

FIGURE 20.8

The glomerular capsule has both a visceral layer and a parietal layer.

the proximal convoluted tubule is the **nephron loop** (loop of Henle). The proximal convoluted tubule dips toward the renal pelvis to become the *descending limb* of the nephron loop. The tubule then curves back toward its renal corpuscle and forms the *ascending limb* of the nephron loop. The ascending limb returns to the region of the renal corpuscle, where it becomes highly coiled again and is called the *distal convoluted tubule.* This distal portion is shorter and straighter than the proximal tubule.

Several distal convoluted tubules merge in the renal cortex to form a *collecting duct* (collecting tubule), which is technically not part of the nephron. The collecting duct passes into the renal medulla, widening as it joins other collecting ducts. The resulting tube empties into a minor calyx through an opening in a renal papilla. Figures 20.10 and 20.11 show the parts of a nephron. Clinical Application 20.2 examines glomerulonephritis, an inflammation of the glomeruli.

Juxtaglomerular Apparatus

Near its origin, the distal convoluted tubule passes between the afferent and efferent arterioles and contacts them. At the point of contact, the epithelial cells of the distal convoluted tubule are quite tall and densely packed. These cells comprise a structure called the *macula densa.*

Close by, in the wall of the afferent arteriole near its attachment to the glomerulus, are large, vascular smooth muscle cells called *juxtaglomerular cells.* Together with the cells of the macula densa, they constitute the **juxtaglomerular apparatus** (juks"tah-glo-mer'u-lar ap"ah-ra'tus)

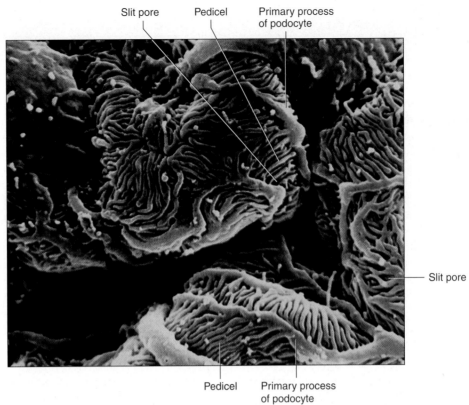

FIGURE 20.9

Scanning electron micrograph of a portion of a glomerulus (8,000×). Note the slit pores between the pedicels.

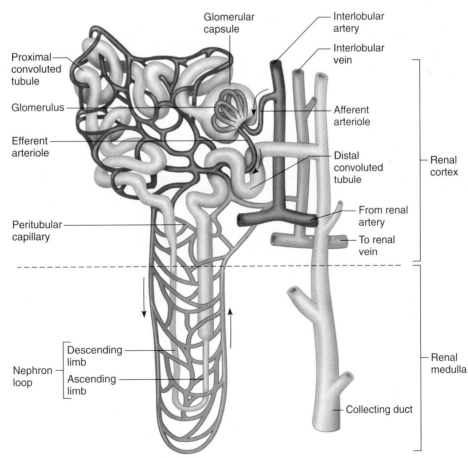

FIGURE 20.10

Structure of a nephron and the blood vessels associated with it.

GLOMERULONEPHRITIS

Nephritis is an inflammation of the kidney. Glomerulonephritis is an inflammation of the glomeruli, and it may be acute or chronic and can lead to renal failure.

Acute glomerulonephritis (AGN) usually results from an abnormal immune reaction that develops one to three weeks following bacterial infection by beta-hemolytic *Streptococcus*. As a rule, the infection occurs in some other part of the body and does not affect the kidneys directly. Instead, bacterial antigens trigger an immune reaction. Antibodies against these antigens form insoluble immune complexes (see chapter 16, p. 630)

that travel in the bloodstream to the kidneys. The antigen-antibody complexes are deposited in and block the glomerular capillaries, which become further obstructed as the inflammatory response sends many white blood cells to the region. Those capillaries remaining open may become abnormally permeable, sending plasma proteins and red blood cells into the urine.

Most glomerulonephritis patients eventually regain normal kidney function; however, in severe cases, renal functions may fail completely, and without treatment, the person is likely to die within a week or so.

Chronic glomerulonephritis is a progressive disease in which increasing numbers of nephrons are slowly damaged until finally the kidneys are unable to function. This condition is usually associated with certain diseases other than streptococcal infections, and it also involves formation of antigen-antibody complexes that precipitate and accumulate in the glomeruli. The resulting inflammation is prolonged, and it is accompanied by fibrous tissue replacing glomerular membranes. As this happens, the functions of the nephrons are permanently lost, and eventually the kidneys fail. ■

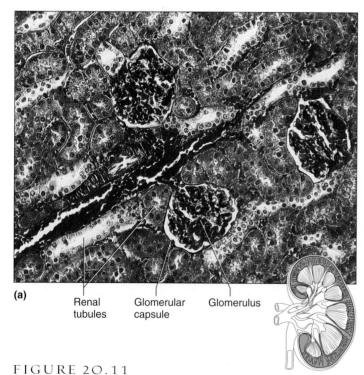

(a)

Renal tubules   Glomerular capsule   Glomerulus

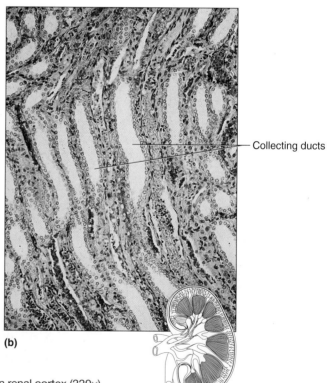

Collecting ducts

(b)

FIGURE 20.11

Renal cortex and renal medulla. (a) Light micrograph of a section of the human renal cortex (220×). (b) Light micrograph of the renal medulla (200×).

(complex). This structure is important in regulating the secretion of renin (see chapter 13, p. 493) (fig. 20.12).

Cortical and Juxtamedullary Nephrons

Most nephrons have corpuscles located in the renal cortex near the surface of the kidney. These *cortical nephrons* have relatively short nephron loops that usually do not reach the renal medulla.

Another group, called *juxtamedullary nephrons,* have corpuscles close to the renal medulla, and their nephron loops extend deep into the medulla. Although these nephrons represent only about 20% of the total,

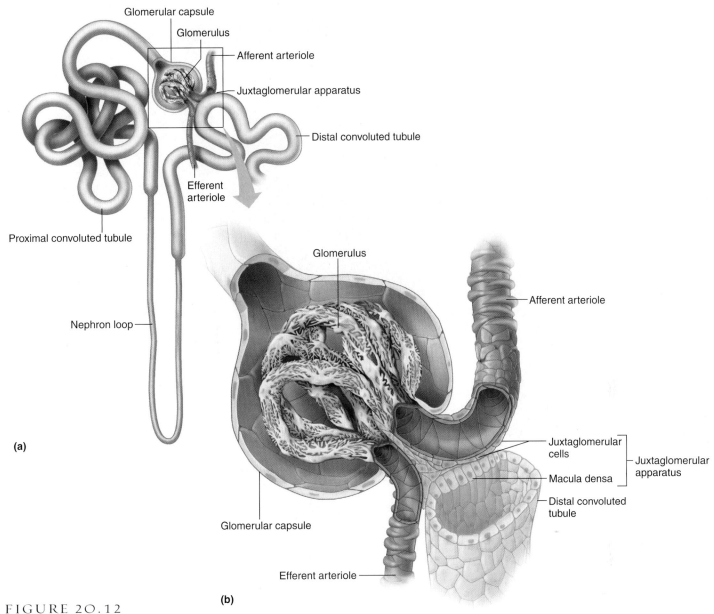

FIGURE 20.12

Juxtaglomerular apparatus. (a) Location of the juxtaglomerular apparatus. (b) Enlargement of a section of the juxtaglomerular apparatus, which consists of the macula densa and the juxtaglomerular cells.

they are important in regulating water balance (fig. 20.13).

Blood Supply of a Nephron

The cluster of capillaries that forms a glomerulus arises from an afferent arteriole whose diameter is greater than that of other arterioles. Blood passes through the capillaries of the glomerulus, then (minus any filtered fluid) enters an efferent arteriole (rather than a venule), whose diameter is smaller than that of the afferent arteriole. The greater resistance to blood flow of the efferent arteriole causes blood to back up into the glomerulus. This results in a relatively high pressure in the glomerular capillaries compared to capillaries elsewhere.

The efferent arteriole branches into a complex network of capillaries that surrounds the renal tubule called the **peritubular capillary** (per″i-tū′bu-lar kap′i-ler″e) **system.** Blood in the system has passed through two arterioles and is under relatively low pressure (see fig. 20.10). Branches of this system that primarily receive blood from the efferent arterioles of the juxtamedullary nephrons form capillary loops called *vasa recta.* These loops dip into the renal medulla and are closely associated with the loops of the juxtamedullary nephrons

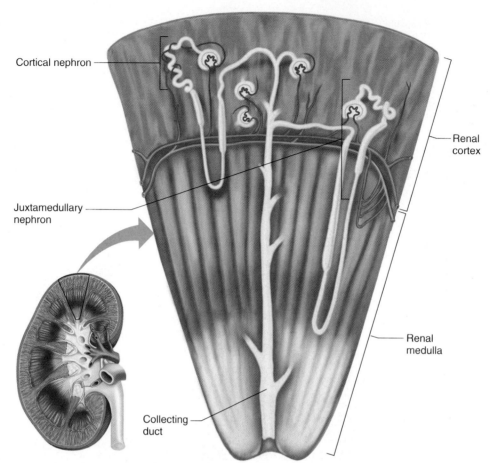

FIGURE 20.13

Cortical nephrons are close to the surface of a kidney; juxtamedullary nephrons are near the renal medulla.

(fig. 20.14). After flowing through the vasa recta, blood returns to the renal cortex, where it joins blood from other branches of the peritubular capillary system and enters the venous system of the kidney. Figure 20.15 summarizes the pathway that blood follows as it passes through the blood vessels of the kidney and nephron.

1 Describe the system of vessels that supplies blood to the kidney.

2 Name the parts of a nephron.

3 Which structures comprise the juxtaglomerular apparatus?

4 Distinguish between a cortical nephron and a juxtamedullary nephron.

5 Describe the blood supply of a nephron.

Urine Formation

The main function of the nephrons is to control the composition of body fluids and remove wastes from the blood. The product is **urine,** which is excreted from the body. It contains wastes and excess water and electrolytes.

Urine formation begins with filtration of plasma by the glomerular capillaries, a process called **glomerular** **filtration** (glo-mer′u-lar fil-tra′shun). However, glomerular filtration produces 180 liters of fluid more than four times total body water, every twenty-four hours. Glomerular filtration could not continue for very long unless most of this filtered fluid were returned to the internal environment. The kidney accomplishes this by the process of **tubular reabsorption,** selectively reclaiming just the right amounts of substances, such as water, electrolytes, and glucose, that the body requires. Waste products and excess substances are allowed out of the body. Some substances that the body must eliminate, such as hydrogen ions and certain toxins, are removed even faster than through filtration alone by the process of **tubular secretion.** In other words, the following relationship determines the volume of substances excreted in the urine:

$$\text{urinary excretion} = \text{glomerular filtration}$$
$$+ \text{ tubular secretion}$$
$$- \text{ tubular reabsorption}$$

The final product of these processes is **urine.** The kidneys contribute to homeostasis by maintaining the composition of the internal environment.

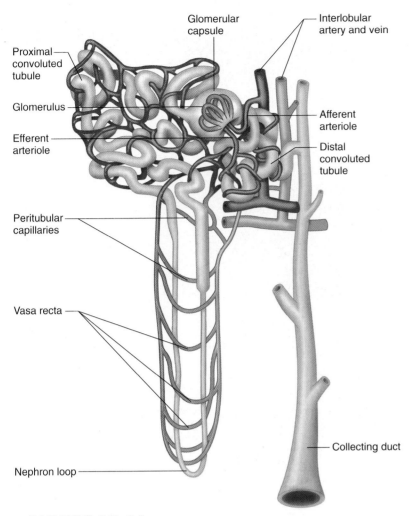

FIGURE 20.14
The capillary loop of the vasa recta is closely associated with the nephron loop of a juxtamedullary nephron.

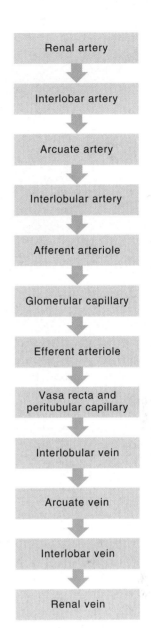

FIGURE 20.15
Pathway of blood through the blood vessels of the kidney and nephron.

Glomerular Filtration

Urine formation begins when water and other small dissolved molecules and ions are filtered out of the glomerular capillary plasma and into the glomerular capsules by the process of *glomerular filtration*. Large molecules, such as proteins, are restricted primarily because of their size. The filtration of these materials through the capillary walls is much like the filtration that occurs at the arteriole ends of other capillaries throughout the body. The glomerular capillaries, however, are many times more permeable to small molecules than the capillaries in other tissues, due to the many tiny openings (fenestrae) in their walls (fig. 20.16).

⧖ RECONNECT TO CHAPTER 15, EXCHANGES IN THE CAPILLARIES, PAGES 568–569

The glomerular capsule receives the resulting **glomerular filtrate,** which has about the same composition as the filtrate that becomes tissue fluid elsewhere in

the body. That is, glomerular filtrate is mostly water and the same solutes as in blood plasma, except for the larger protein molecules. More specifically, glomerular filtrate contains water, glucose, amino acids, urea, uric acid, creatine, creatinine, and sodium, chloride, potassium, calcium, bicarbonate, phosphate, and sulfate ions. Table 20.1 shows the relative concentrations of some of the substances in the blood plasma, glomerular filtrate, and urine.

Filtration Pressure

The main force that moves substances by filtration through the glomerular capillary wall is the hydrostatic pressure of the blood inside, as in the case for other capillaries. (Recall that glomerular capillary pressure is high

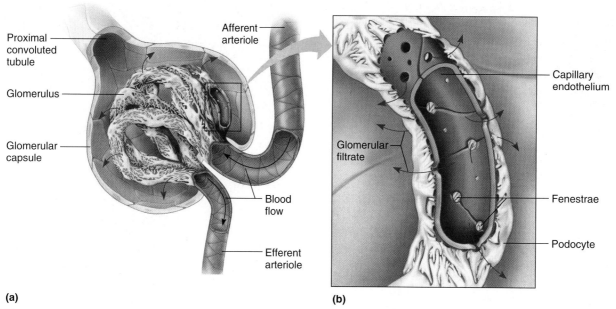

Proximal convoluted tubule

Glomerulus

Glomerular capsule

Afferent arteriole

Blood flow

Efferent arteriole

Capillary endothelium

Glomerular filtrate

Fenestrae

Podocyte

(a)

(b)

FIGURE 20.16

Glomerular filtration. (a) The first step in urine formation is filtration of substances out of glomerular capillaries and into the glomerular capsule. (b) Glomerular filtrate passes through the fenestrae of the capillary endothelium.

TABLE 20.1	Relative Concentrations of Plasma, Glomerular Filtrate, and Urine Components		
	Concentrations (mEq/L)		
Substance	**Plasma**	**Glomerular Filtrate**	**Urine**
Sodium (Na$^+$)	142	142	128
Potassium (K$^+$)	5	5	60
Calcium (Ca^{+2})	4	4	5
Magnesium (Mg^{+2})	3	3	15
Chloride (Cl$^-$)	103	103	134
Bicarbonate (HCO$_3^-$)	27	27	14
Sulfate (SO$_4^{-2}$)	1	1	33
Phosphate (PO$_4^{-3}$)	2	2	40
(mEq/L (milliequivalents per liter) is a commonly used measure of concentration based on how many charges an ion carries. For a substance with a charge of 1, such as Cl$^-$, a mEq is equal to a millimole.)			
	Concentrations (mg/100 mL)		
Substance	**Plasma**	**Glomerular Filtrate**	**Urine**
Glucose	100	100	0
Urea	26	26	1,820
Uric acid	4	4	53
Creatinine	1	1	196

compared to other capillaries.) The osmotic pressure of the blood plasma in the glomerulus and the hydrostatic pressure inside the glomerular capsule also influence glomerular filtration.

The colloid osmotic pressure of the plasma caused by plasma proteins is always higher than that of the glomerular filtrate (except in some kinds of kidney disease). This draws water back into the glomerular capillaries, opposing filtration. Any increase in glomerular capsule hydrostatic pressure also opposes filtration (fig. 20.17).

The net effect of all of these forces is called **net filtration pressure,** and it is normally positive, favoring filtra-

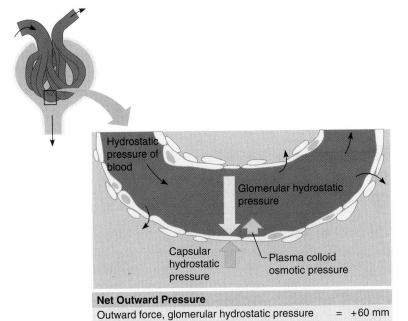

Hydrostatic pressure of blood

Glomerular hydrostatic pressure

Capsular hydrostatic pressure

Plasma colloid osmotic pressure

Net Outward Pressure	
Outward force, glomerular hydrostatic pressure	= +60 mm
Inward force of plasma colloid osmotic pressure	= −32 mm
Inward force of capsular hydrostatic pressure	= −18 mm
Net outward pressure	= +10 mm

FIGURE 20.17
The hydrostatic and osmotic pressure of the plasma and the hydrostatic pressure of the fluid in the glomerular capsule affect the rate of glomerular filtration.

tion at the glomerulus. Net filtration pressure is calculated as follows:

**Net filtration pressure =
force favoring filtration − forces opposing filtration**
(glomerular capillary hydrostatic pressure)   (capsular hydrostatic pressure and glomerular capillary osmotic pressure)

The concentrations of certain components of the blood plasma can be used to evaluate kidney functions. For example, if the kidneys are functioning inadequately, the plasma concentrations of urea (as indicated by a blood urea nitrogen test) and of creatinine may increase as much as tenfold above normal.

Filtration Rate

The glomerular filtration rate (GFR) is directly proportional to the net filtration pressure. Consequently, the factors that affect the glomerular hydrostatic pressure, glomerular plasma osmotic pressure, or hydrostatic pressure in the glomerular capsule will also affect the rate of filtration (fig. 20.17).

Normally, glomerular hydrostatic pressure is the most important factor determining net filtration pressure

and GFR. Since each glomerular capillary is located between two arterioles—the afferent and efferent arterioles—any change in the diameters of these vessels is likely to change glomerular hydrostatic pressure, changing glomerular filtration rate. The afferent arteriole, through which the blood enters the glomerulus, may vasoconstrict in response to stimulation by sympathetic nerve impulses. If this occurs, net filtration pressure in that glomerulus decreases, and filtration rate drops. If, on the other hand, the efferent arteriole (through which the blood leaves the glomerulus) vasoconstricts, blood backs up into the glomerulus, net filtration pressure increases, and filtration rate rises. Vasodilation of these vessels produces opposite effects.

If arterial blood pressure drops drastically, as may occur during *shock,* the glomerular hydrostatic pressure may fall below the level required for filtration, leading to acute renal failure. At the same time, the epithelial cells of the renal tubules may fail to receive sufficient nutrients to maintain their high rates of metabolism. As a result, cells may die (tubular necrosis), and renal functions may be lost permanently, resulting in chronic renal failure.

The colloid osmotic pressure of the glomerular plasma also influences net filtration pressure and the rate of filtration. In other systemic capillaries, filtration occurs at the beginning of the capillary, but the osmotic effect of the plasma proteins predominates at the capillary, and most filtered fluid is thus reabsorbed. The small excess remaining eventually becomes lymph.

Because of the relatively high hydrostatic pressure in the glomerular capillaries, much more fluid is filtered than by capillaries elsewhere. In fact, as filtration occurs through the capillary wall, proteins remaining in the plasma raise the colloid osmotic pressure within the glomerular capillaries. Despite this, the glomerular capillary hydrostatic pressure is sufficiently great that the net filtration pressure is normally positive. That is, the forces favoring filtration in the glomerular capillaries always predominate. Of course, conditions that lower plasma colloid osmotic pressure, such as a decrease in plasma protein concentration, would increase filtration rate.

The hydrostatic pressure in the glomerular capsule is another factor that may affect net filtration pressure and rate. This capsular pressure sometimes changes as a result of an obstruction, such as a stone in a ureter or an enlarged prostate gland pressing on the urethra. If this occurs, fluids back up into the renal tubules and raise the hydrostatic pressure in the glomerular capsules. Because any increase in capsular pressure opposes glomerular filtration, filtration rate may decrease significantly.

At rest, the kidneys receive approximately 25% of the cardiac output, and about 20% of the blood plasma is filtered as it flows through the glomerular capillaries. This means that in an average adult, the glomerular filtration rate for the nephrons of both kidneys is about 125 milliliters per minute, or 180,000 milliliters (180 liters) in twenty-four hours. Assuming that the blood plasma volume is about 3 liters, the production of 180 liters of filtrate in twenty-four hours means that all of the plasma must be filtered through the glomeruli about sixty times each day (fig. 20.18). Since this twenty-four-hour volume is nearly 45 gallons, it is obvious that not all of it is excreted as urine. Instead, most of the fluid that passes through the renal tubules is reabsorbed and reenters the plasma.

The volume of plasma the kidneys filter also depends on the *surface area* of the glomerular capillaries. This surface area is estimated to be about 2 square meters—approximately equal to the surface area of an adult's skin.

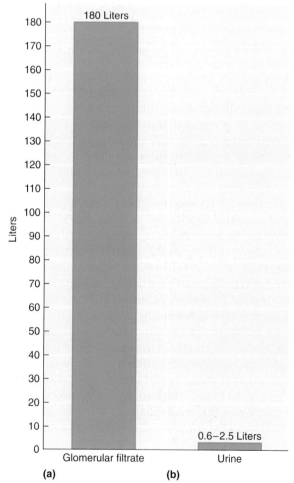

FIGURE 20.18

Relative amounts of (a) glomerular filtrate and (b) urine formed in twenty-four hours.

An injury to a kidney can be more dangerous than an injury to another organ. An injured kidney produces a protein called *transforming growth factor beta,* which causes scars to form. The scars further damage the kidney, impairing its function.

1 What processes occur in urine formation?

2 How is filtration pressure calculated?

3 What factors influence the rate of glomerular filtration?

Control of Filtration Rate

In general, glomerular filtration rate remains relatively constant through a process called **autoregulation**. However, certain conditions override autoregulation. GFR may increase, for example, when body fluids are in excess and decrease when the body must conserve fluid.

Recall from chapter 11 (p. 414) that sympathetic nervous system fibers synapse with the vascular smooth muscle of arterioles. Reflexes responding to changes in blood pressure and volume control the activity of these sympathetic fibers. If blood pressure and volume drop, vasoconstriction of the afferent arterioles results, decreasing filtration pressure and thus GFR. The result is an appropriate decrease in the rate of urine formation when the body must conserve water. If receptors detect excess body fluids, vasodilation of the afferent arteriole results, increasing filtration pressure and GFR.

A second control of GFR is the hormonelike **renin-angiotensin system.** The juxtaglomerular cells of the afferent arterioles secrete an enzyme, **renin,** in response to stimulation from sympathetic nerves and pressure-sensitive cells called **renal baroreceptors** that are in the afferent arteriole. These factors stimulate renin secretion if blood pressure drops. The macula densa also controls renin secretion. Cells of the macula densa sense the concentrations of sodium, potassium, and chloride ions in the distal renal tubule. Decreasing levels of these ions stimulate renin secretion.

Once in the bloodstream, *renin* reacts with the plasma protein *angiotensinogen* to form *angiotensin I.* An enzyme, *angiotensin-converting enzyme* (ACE), present on capillary endothelial cells (particularly in the lungs), rapidly converts angiotensin I to *angiotensin II.*

Angiotensin II has a number of renal effects that help maintain sodium balance, water balance, and blood pressure (fig. 20.19). As a vasoconstrictor, it affects both the afferent and efferent arterioles. Although afferent arteriolar constriction decreases GFR, efferent arteriolar constriction minimizes the decrease, thus contributing to autoregulation of GFR. Angiotensin II has a major effect on the kidneys through the adrenal cortical hormone aldosterone, which stimulates sodium reabsorption in the distal convoluted tubule. By stimulating aldosterone

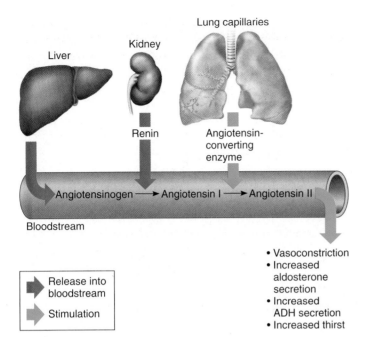

FIGURE 20.19

The formation of angiotensin II in the bloodstream involves several organs and includes multiple actions that conserve sodium and water.

secretion, angiotensin II helps to reduce the amount of sodium excreted in the urine.

The hormone **atrial natriuretic peptide** (ANP) also affects sodium excretion. ANP secretion increases when the atria of the heart stretch due to increased blood volume. ANP stimulates sodium excretion through a number of mechanisms, including increasing GFR.

Tubular Reabsorption

If the composition of the glomerular filtrate entering the renal tubule is compared with that of the urine leaving the tubule, it is obvious that the fluid changes as it passes through the tubule (see table 20.1). For example, glucose is present in the filtrate but absent in the urine. In contrast, urea and uric acid are considerably more concentrated in urine than they are in the glomerular filtrate. Such changes in fluid composition are largely the result of *tubular reabsorption* (tu'bu-lar re-ab-sorp'shun), the process by which substances are transported out of the tubular fluid, through the epithelium of the renal tubule, and into the interstitial fluid. These substances then diffuse into the peritubular capillaries (fig. 20.20).

Tubular reabsorption returns substances to the internal environment. The term *tubular* is used because this process is controlled by the epithelial cells that make up the renal tubules. In tubular reabsorption, substances must first cross the cell membrane facing the inside of the

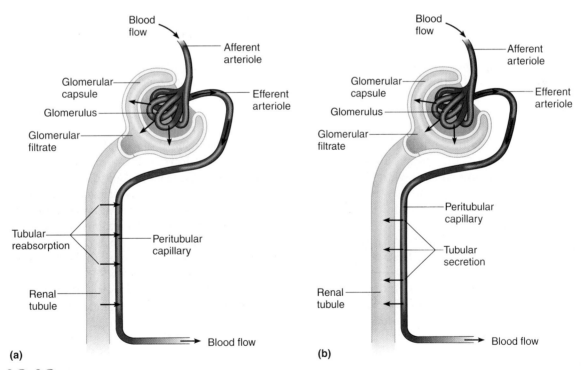

FIGURE 20.20

Two processes in addition to glomerular filtration contribute to urine formation. (*a*) Reabsorption transports substances from the glomerular filtrate into the blood within the peritubular capillary. (*b*) Secretion transports substances from the blood within the peritubular capillary into the renal tubule.

CHAPTER TWENTY *Urinary System*

tubule (mucosal surface) and then the cell membrane facing the interstitial fluid (serosal surface).

The basic rules for movements across cell membranes apply to tubular reabsorption. Substances moving down a concentration gradient must either be lipid soluble or there must be a carrier or channel for that substance. Active transport, requiring ATP, may move substances uphill against a concentration gradient. If active transport is involved at any step of the way, the process is considered active tubular reabsorption. In all other cases, the process is considered passive.

∞ R E C O N N E C T   T O   C H A P T E R   3 ,
M O V E M E N T S   I N T O   A N D
O U T   O F   T H E   C E L L ,   P A G E S
8 0 – 8 6

Peritubular capillary blood is under relatively low pressure because it has already passed through two arterioles. Also, the wall of the peritubular capillary is more permeable than that of other capillaries. Finally, the relatively high rate of glomerular filtration has increased the protein concentration and, thus, the colloid osmotic pressure of the peritubular capillary plasma. All of these factors enhance the rate of fluid reabsorption from the renal tubule.

Tubular reabsorption occurs throughout the renal tubule. However, most of it occurs in the proximal convoluted portion. The epithelial cells in this portion have many *microvilli* that form a "brush border" on their free surfaces facing the tubular lumen. These tiny extensions greatly increase the surface area exposed to the glomerular filtrate and enhance reabsorption.

Segments of the renal tubule are adapted to reabsorb specific substances, using particular modes of transport. Glucose reabsorption, for example, occurs through the walls of the proximal convoluted tubule by active transport. Water also is rapidly reabsorbed through the epithelium of the proximal convoluted tubule by osmosis; however, portions of the distal convoluted tubule and collecting duct are almost impermeable to water. This characteristic of the distal convoluted tubule is important in the regulation of urine concentration and volume, as described in a subsequent section.

Recall that active transport requires carrier proteins in a cell membrane. The molecule to be transported binds to the carrier; the carrier changes shape, releases the transported molecule on the other side of the cell membrane, and then returns to its original position and repeats the process. Such a mechanism has a *limited transport capacity;* that is, it can transport only a certain number of molecules in a given length of time because the number of carriers is limited.

Usually all of the glucose in the glomerular filtrate is reabsorbed because there are enough carrier molecules to transport it. When the plasma glucose concentration increases to a critical level, called the **renal plasma threshold,** more glucose molecules are in the filtrate than the active transport mechanism can handle. As a result, some glucose remains in the filtrate and is excreted in the urine. This explains why the elevated blood glucose of diabetes mellitus results in glucose in the urine.

Any increase in urine volume is called *diuresis.* Nonreabsorbed glucose in the tubular fluid draws water into the renal tubule by osmosis, thus increasing urine volume. Such an increase is called an *osmotic diuresis.*

Glucose in the urine is called *glucosuria.* It may follow intravenous administration of glucose, or eating a candy bar, or it may occur in a person with diabetes mellitus. In type I diabetes, blood glucose concentration rises because of insufficient insulin secretion from the pancreas.

One in three people who have diabetes mellitus sustains kidney damage, as evidenced by protein in their urine. Following a low-protein diet can slow the loss of kidney function. High protein diets, popular for weight loss, can damage the kidneys.

Amino acids enter the glomerular filtrate and are reabsorbed in the proximal convoluted tubule. Three different active transport mechanisms reabsorb different groups of amino acids, whose members have similar structures. As a result, normally only a trace of amino acids remains in the urine.

The glomerular filtrate is nearly free of protein, but a number of smaller protein molecules, such as albumin, squeeze through the glomerular capillaries. These proteins are taken up by *endocytosis* through the brush border of epithelial cells lining the proximal convoluted tubule. Once they are inside an epithelial cell, the proteins are degraded to amino acids and moved into the blood of the peritubular capillary.

The epithelium of the proximal convoluted tubule also reabsorbs creatine; lactic, citric, uric, and ascorbic (vitamin C) acids; and phosphate, sulfate, calcium, potassium, and sodium ions. Active transport mechanisms with limited transport capacities reabsorb all of these chemicals. Such substances begin to appear in the urine when their concentrations in the glomerular filtrate exceed their respective renal plasma thresholds. Clinical Application 20.3 discusses how the nephrotic syndrome causes plasma proteins to appear in the urine.

Sodium and Water Reabsorption

Water reabsorption occurs passively by osmosis, primarily in the proximal convoluted tubule, and is closely associated with the active reabsorption of sodium ions. In the proximal convoluted tubule, if sodium reabsorption increases, water reabsorption increases; if sodium reabsorption decreases, water reabsorption decreases also.

THE NEPHROTIC SYNDROME

The *nephrotic syndrome* is a set of symptoms that often appears in patients with renal diseases. It causes considerable loss of plasma proteins into the urine (proteinuria), widespread edema, and increased susceptibility to infections.

Plasma proteins are lost into the urine because of increased permeability of the glomerular membranes, which accompanies renal disorders such as glomerulonephritis. As a consequence of a decreasing plasma protein concentration (hypoproteinemia), the plasma osmotic pressure falls, increasing filtration pressure in capillaries throughout the body. This may lead to widespread, severe edema as a large volume of fluid accumulates in the interstitial spaces within the tissues and in body spaces such as the abdominal cavity, pleural cavity, pericardial cavity, and joint cavities.

Also, as edema develops, blood volume decreases and blood pressure drops. These changes may activate the renin-angiotensin system, leading to the release of aldosterone from the adrenal cortex (see chapter 13, p. 493), which, in turn, stimulates the kidneys to conserve sodium ions and water. This action reduces the urine output and may aggravate the edema.

The nephrotic syndrome sometimes appears in young children who have *lipoid nephrosis.* The cause of this condition is unknown, but it alters the epithelial cells of the glomeruli so that the glomerular membranes enlarge and distort, allowing proteins through. ■

Much of the sodium ion reabsorption occurs in the proximal segment of the renal tubule by active transport (sodium pump mechanism). When the positively charged sodium ions (Na^+) are moved through the tubular wall, negatively charged ions, including chloride ions (Cl^-), phosphate ions (PO_4^{-3}), and bicarbonate ions (HCO_3^-), accompany them. This movement of negatively charged ions is due to the electrochemical attraction between particles of opposite electrical charge. Although this movement of negatively charged ions depends on active transport of sodium, it is considered a passive process because it does not require a direct expenditure of cellular energy. Active transport reabsorbs some of these ions, such as HCO_3^- and PO_4^{-3}.

As more sodium ions are reabsorbed into the peritubular capillary along with negatively charged ions, the concentration of solutes within the peritubular blood might be expected to increase. However, since water diffuses through cell membranes from regions of lesser solute concentration (hypotonic) toward regions of greater solute concentration (hypertonic), water moves by osmosis, following the ions from the renal tubule into the peritubular capillary.

The proximal convoluted tubule reabsorbs about 70% of the filtered sodium, other ions, and water. By the end of the proximal convoluted tubule, osmotic equilibrium is reached, and the remaining tubular fluid is isotonic (fig. 20.21).

Active transport continues to reabsorb sodium ions as the tubular fluid moves through the nephron loop, the distal convoluted tubule, and the collecting duct. Consequently, almost all of the sodium and water (97% to 99%) that enters the renal tubules as part of the glomerular filtrate may be reabsorbed before the urine is excreted.

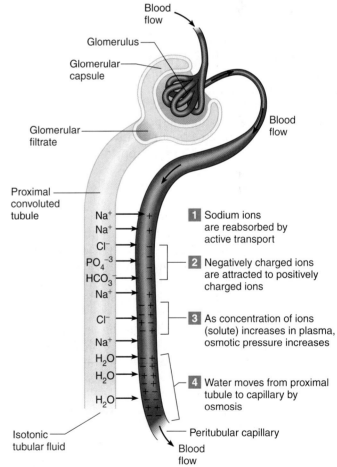

FIGURE 20.21

In the proximal portion of the renal tubule, osmosis reabsorbs water in response to active transport reabsorbing sodium and other solutes.

However, aldosterone controls sodium reabsorption, and antidiuretic hormone controls water reabsorption. Under the influence of these hormones, reabsorption of sodium and water can change to keep conditions in the body fluids constant. Chapter 21 (pp. 812–816) discusses the specific effects of these hormones.

Recall that the kidneys filter an extremely large volume of fluid (180 liters) each day. If 99% of the glomerular filtrate is reabsorbed, the remaining 1% excreted includes a relatively large amount of sodium and water (table 20.2). On the other hand, if sodium and water reabsorption decrease to 97% of the amount filtered, the amount excreted triples! Therefore, small changes in the tubular reabsorption of sodium and water result in large changes in urinary excretion of these substances.

1 How is the peritubular capillary adapted for reabsorption?

2 Which substances present in the glomerular filtrate are not normally present in urine?

3 Which mechanisms reabsorb solutes from glomerular filtrate?

4 Define *renal plasma threshold*.

5 Describe the role of passive transport in urine formation.

Tubular Secretion

In *tubular secretion* (tu′bu-lar se-kre′shun), certain substances move from the plasma of the peritubular capillary into the fluid of the renal tubule. As a result, the amount of a particular chemical excreted in the urine may exceed the amount filtered from the plasma in the glomerulus (see fig. 20.20). As in the case of tubular reabsorption, the term *tubular* is used because the epithelial cells of the renal tubules control the process.

Active transport mechanisms similar to those that function in reabsorption secrete some substances. However, the secretory mechanisms transport substances in the opposite direction. For example, the epithelium of the proximal convoluted tubules actively secretes certain organic compounds, including penicillin and histamine, into the tubular fluid.

Hydrogen ions are also actively secreted throughout the entire renal tubule. As a result, urine is usually acidic by the time it is excreted, although the urinary pH can vary considerably. The secretion of hydrogen ions is important in regulating the pH of body fluids, as chapter 21 (p. 821) explains.

Surgery is the primary treatment for cancer of the kidneys. However, in half of all cases, cancer returns, usually in the lungs. A treatment for this metastatic kidney cancer is the immune system cytokine interleukin-2. It is administered intravenously in cycles in a hospital setting because of sometimes severe side effects. Interleukin-2 stimulates the immune system to attack tumor cells. In about 15% of patients on the therapy, tumors shrink.

Most of the potassium ions in the glomerular filtrate are actively reabsorbed in the proximal convoluted tubule, but some may be secreted in the distal convoluted tubule and collecting duct. During this process, the active reabsorption of sodium ions out of the tubular fluid under the influence of aldosterone produces a negative electrical charge within the tube. Because positively charged potassium ions (K^+) are attracted to regions that are negatively charged, these ions move passively through the tubular epithelium and enter the tubular fluid. Potassium ions are also secreted actively (fig. 20.22).

To summarize, urine forms as a result of the following:

- Glomerular filtration of materials from blood plasma.
- Reabsorption of substances, including glucose; water; urea; proteins; creatine; amino, lactic, citric, and uric acids; and phosphate, sulfate, calcium, potassium, and sodium ions.
- Secretion of substances, including penicillin, histamine, phenobarbital, hydrogen ions, ammonia, and potassium ions.

1 Define *tubular secretion*.

2 Which substances are actively secreted? Passively secreted?

3 How does the reabsorption of sodium affect the secretion of potassium?

Regulation of Urine Concentration and Volume

Hormones such as aldosterone and ANP affect the solute concentration of urine, particularly sodium. However, the ability of the kidneys to maintain the internal environment rests in large part on their ability to concentrate urine by reabsorbing large volumes of water.

TABLE 20.2	Average Values for Sodium and Water Filtration, Reabsorption, and Excretion		
	Amount Filtered per Day	**Amount Reabsorbed per Day (%)**	**Amount Excreted per Day**
Water (L)	180	178.2 (99%)	1.8 (1%)
Na$^+$ (g)	630	626.8 (99.5%)	3.2 (0.5%)

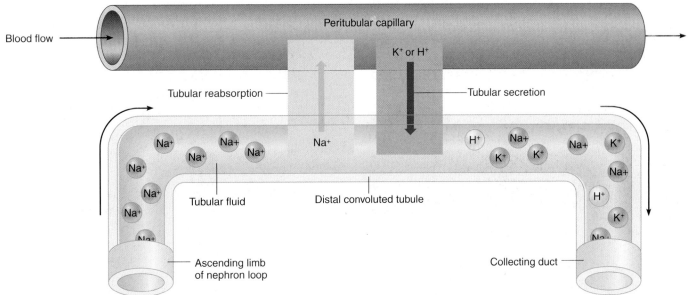

FIGURE 20.22

In the distal convoluted tubule, potassium ions (or hydrogen ions) may be passively secreted in response to the active reabsorption of sodium ions.

In contrast to conditions in the proximal convoluted tubule, the tubular fluid reaching the distal convoluted tubule is hypotonic because of changes that occur through the loop segment of each nephron. The cells lining the distal convoluted tubule and the collecting duct that follows continue to reabsorb sodium ions (chloride ions follow passively) under the influence of aldosterone, which the adrenal cortex secretes (see chapter 13, p. 493). In addition, the interstitial fluid surrounding the collecting ducts is hypertonic, particularly in the medulla. These might seem to be ideal conditions for water reabsorption as well. However, the cells lining the later portion of the distal convoluted tubule and the collecting duct are impermeable to water unless antidiuretic hormone (ADH) is present. Thus, water inside the tubule may be excreted, forming dilute urine.

As discussed in chapter 13 (pp. 483–484), neurosecretory cells in the hypothalamus produce ADH. The posterior lobe of the pituitary gland releases ADH in response to decreasing concentration of water in the body fluids or to decreasing blood volume and blood pressure. When ADH reaches the kidney, it stimulates these cells to insert water channels (aquaporins) into their membranes. This greatly increases permeability to water of the epithelial cell linings of the distal convoluted tubule and the collecting duct; consequently, water rapidly moves out of these segments by osmosis, especially where the collecting duct passes through the extremely hypertonic medulla. The urine becomes more concentrated, and water is retained in the internal environment (fig. 20.23).

A **countercurrent mechanism** involving the nephron loops, particularly of the juxtamedullary nephrons, ensures that the medullary interstitial fluid becomes hypertonic. This mechanism is possible because the descending and ascending limbs of the nephron loops lie parallel and very close to one another. The mechanism is named partly for the fact that fluid moving down the descending limb creates a current that is counter to that of the fluid moving up in the ascending limb.

The different parts of the nephron loop have important functional differences. For example, the epithelial lining in the thick upper portion of the ascending limb (thick segment) is relatively impermeable to water. However, the epithelium does actively reabsorb sodium and chloride ions (some potassium is actively reabsorbed as well). As these solutes accumulate in the interstitial fluid outside the ascending limb, it becomes hypertonic, while the tubular fluid inside becomes hypotonic because it is losing its solute.

In contrast to the ascending limb, the epithelium of the descending limb (thin segment) is quite permeable to water, but relatively impermeable to solutes. Because this segment is surrounded by hypertonic fluid created by the ascending limb, water tends to leave the descending limb by osmosis. The contents of the descending limb become more concentrated, or hypertonic (fig. 20.24).

The very concentrated tubular fluid now moves into the ascending limb, and sodium chloride (NaCl) is again actively reabsorbed into the medullary interstitial fluid, raising the interstitial NaCl concentration further. With the increased interstitial fluid solute concentration, even more water diffuses out of the descending limb, further increasing the salt concentration of the tubular fluid. Each time this circuit is completed, the concentration of NaCl increases, or multiplies. For this reason, the mechanism is called a *countercurrent multiplier*. In humans, this strategy creates a tubular fluid solute concentration near the tip of the loop that is more than four times the solute concentration of plasma (fig. 20.24).

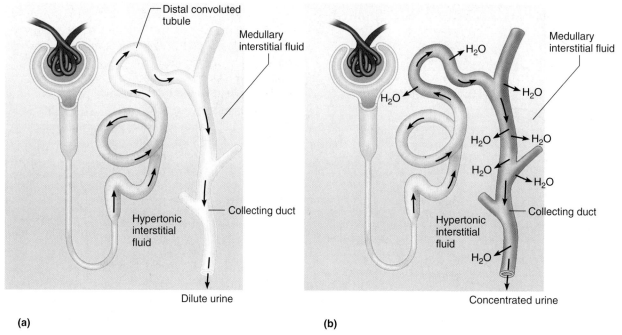

FIGURE 20.23

Urine concentrating mechanism. (*a*) The distal convoluted tubule and collecting duct are impermeable to water, so water may be excreted as dilute urine. (*b*) If ADH is present, however, these segments become permeable, and water is reabsorbed by osmosis into the hypertonic medullary interstitial fluid.

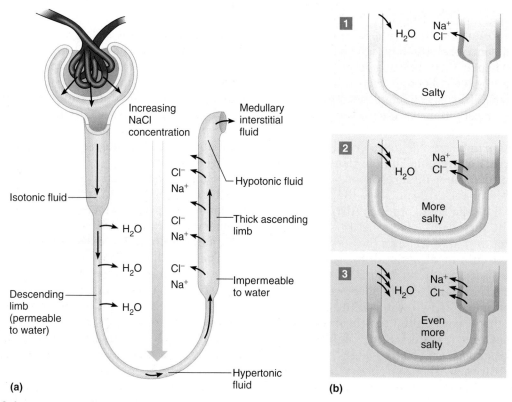

FIGURE 20.24

The countercurrent multiplier. (*a*) The solute concentration of interstitial fluid in the medulla equilibrates with tubular fluid, which loses water in the descending limb, and thus becomes hypertonic by the tip of the nephron loop. The ascending limb of the loop actively reabsorbs solute. (*b*) Active solute reabsorption from the ascending limb of the loop causes even more water loss from the descending limb as tubular fluid continues to flow. The countercurrent multiplier progressively increases the solute concentration of the interstitial fluid, up to a maximum near the tip of the loop more than four times that of plasma.

The solute concentration of the tubular fluid progressively decreases toward the renal cortex. Since the descending limb of the loop is permeable to water, the interstitial fluid at any level of the loop is essentially in equilibrium with the fluid in the tubule. Thus, the concentration gradient in the loop is also found in the interstitial fluid (fig. 20.24).

The vasa recta is another countercurrent mechanism that maintains the NaCl concentration gradient in the renal medulla. Blood flows slowly down the descending portion of the vasa recta, and NaCl enters it by diffusion. Then, as the blood moves back up toward the renal cortex, most of the NaCl diffuses from the blood and reenters the medullary interstitial fluid. Consequently, the bloodstream carries little NaCl away from the renal medulla, preserving the gradient (fig. 20.25).

To summarize, the countercurrent multiplier creates a large concentration gradient for water reabsorption in the interstitial fluid surrounding the distal convoluted tubules and the collecting ducts of the nephron. The epithelial lining of these structures is impermeable to water, unless ADH is present. The higher the blood levels of ADH, the more permeable the epithelial lining becomes, leading to increased water reabsorption and the production of concentrated urine. In this way, soluble wastes and other substances can be excreted in a minimum of water, thus minimizing the loss of body water when dehydration is a threat. If the body fluids contain excess water, ADH secretion decreases, and the epithelial linings of the distal convoluted tubule and the collecting duct become less permeable to water. Less water is reabsorbed, and the urine is more dilute. Table 20.3 summarizes the role of ADH in urine production. Table 20.4 summarizes the functions of different parts of the nephron.

Urea and Uric Acid Excretion

Urea is a by-product of amino acid catabolism in the liver, and its plasma concentration reflects the amount of protein in the diet. Urea enters the renal tubule by filtration. About 50% of it is reabsorbed (passively) by diffusion, but the remainder is excreted in the urine.

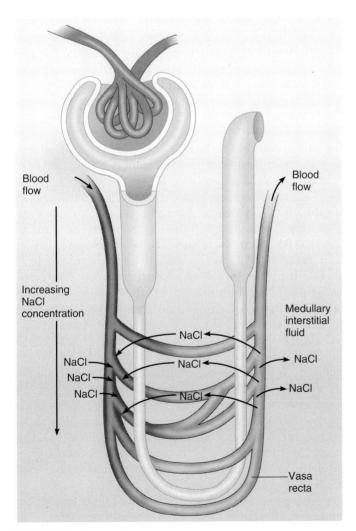

Blood flow

Increasing NaCl concentration

NaCl
NaCl
NaCl
NaCl

NaCl
NaCl
NaCl

Blood flow

Medullary interstitial fluid

NaCl
NaCl
NaCl

Vasa recta

FIGURE 20.25

A countercurrent mechanism in the vasa recta helps maintain the NaCl concentration gradient in the medullary interstitial fluid (see fig. 20.14).

In the inborn error of metabolism gout, uric acid crystals are deposited in certain joints, particularly of the great toe, causing severe pain. Treatments include taking drugs that increase the kidneys' excretion of uric acid and block an enzyme in the biosynthetic pathway for uric acid; limiting intake of foods that are sources of uric acid, including organ meats, anchovies, and sardines; maintaining a healthy weight; and drinking more fluids to dilute the urine, which enhances uric acid excretion.

Gout is an illness with a long history in medicine. Hippocrates mentioned it, and in 1793, English physician Alfred Baring Garrod isolated and implicated uric acid from the blood of a patient with gout and noted that affected individuals often had relatives suffering from it too. At that time, gout was thought to be the result of being a lazy glutton!

TABLE 20.3	Role of ADH in Regulating Urine Concentration and Volume

1. Concentration of water in the blood decreases.
2. Increase in the osmotic pressure of body fluids stimulates osmoreceptors in the hypothalamus.
3. Hypothalamus signals the posterior pituitary gland to release ADH.
4. Blood carries ADH to the kidneys.
5. ADH causes the distal convoluted tubules and collecting ducts to increase water reabsorption by osmosis.
6. Urine becomes more concentrated, and urine volume decreases.

TABLE 20.4	Functions of Nephron Components
Part	**Function**
Renal Corpuscle	
Glomerulus	Filtration of water and dissolved substances from the plasma
Glomerular capsule	Receives the glomerular filtrate
Renal Tubule	
Proximal convoluted tubule	Reabsorption of glucose; amino acids; creatine; lactic, citric, uric, and ascorbic acids; phosphate, sulfate, calcium, potassium, and sodium ions by active transport
	Reabsorption of proteins by endocytosis
	Reabsorption of water by osmosis
	Reabsorption of chloride ions and other negatively charged ions by electro-chemical attraction
	Active secretion of substances such as penicillin, histamine, creatinine, and hydrogen ions
Descending limb of nephron loop	Reabsorption of water by osmosis
Ascending limb of nephron loop	Reabsorption of sodium, potassium, and chloride ions by active transport
Distal convoluted tubule	Reabsorption of sodium ions by active transport
	Reabsorption of water by osmosis
	Active secretion of hydrogen ions
	Secretion of potassium ions both actively and by electrochemical attraction
Collecting Duct	Reabsorption of water by osmosis

(Note: Although the collecting duct is not anatomically part of the nephron, it is functionally connected.)

Urea moves within the kidney in other ways, including the countercurrent multiplier mechanism that concentrates the medullary interstitial fluid. As a result, urea contributes to the reabsorption of water from the collecting duct.

Uric acid is a product of the metabolism of certain nucleic acid bases (the purines, adenine and guanine). Active transport completely reabsorbs the uric acid that is filtered. The fact that some uric acid (equal to approximately 10% of the amount filtered) is excreted in the urine reflects uric acid secretion into the renal tubule.

1 Describe a countercurrent mechanism.

2 How does the hypothalamus regulate urine concentration and volume?

3 Explain how urea and uric acid are excreted.

Urine Composition

Urine composition reflects the amounts of water and solutes that the kidneys must eliminate from the body or retain in the internal environment to maintain homeostasis. It varies considerably from time to time because of differences in dietary intake and physical activity. Urine

is about 95% water and usually also consists of urea from the catabolism of amino acids, uric acid from the catabolism of nucleic acids, and creatinine from metabolism of creatine. Urine may also contain a trace of amino acids, as well as electrolytes whose concentrations reflect the amounts included in the diet (see table 20.1). Appendix B (pp. 972–973) lists the normal concentrations of urine components.

Abnormal constituents of urine may not indicate illness. Glucose in the urine may result from a sugary meal or may occur toward the end of pregnancy; protein may appear in the urine following vigorous physical exercise; ketones appear in the urine during a prolonged fast or when a person follows a very low calorie or low carbohydrate diet.

The volume of urine produced usually varies between 0.6 and 2.5 liters per day. Such factors as fluid intake, environmental temperature, relative humidity of the surrounding air, and a person's emotional condition, respiratory rate, and body temperature influence the exact urine volume. An output of 50–60 milliliters of urine per hour is considered normal, and an output of less than 30 milliliters per hour may indicate kidney failure.

Renal Clearance

The rate at which a particular chemical is removed from the plasma indicates kidney efficiency. It is called *renal clearance.*

Tests of renal clearance detect glomerular damage or monitor the progression of renal disease. One such test, the *inulin clearance test,* uses *inulin* (not to be confused with insulin), a complex polysaccharide found in certain plant roots. In the test, a known amount of inulin is infused into the blood at a constant rate. The inulin passes freely through the glomerular membranes, so its concentration in the glomerular filtrate equals that in the plasma. In the renal tubule, inulin is not reabsorbed to any significant degree, nor is it secreted. Consequently, the rate at which it appears in the urine can be used to calculate the rate of glomerular filtration.

Similarly, the kidneys remove creatinine from the blood. Creatinine is produced at a constant rate during muscle metabolism. Like inulin, creatinine is filtered, but neither reabsorbed nor secreted by the kidneys. The *creatinine clearance test,* which compares a patient's blood and urine creatinine concentrations, can also be used to calculate the GFR. A significant advantage is that the bloodstream normally has a constant level of creatinine. Therefore, a single measurement of plasma creatinine levels provides a rough index of kidney function. For example, significantly elevated plasma creatinine levels suggest that GFR is greatly reduced. Because nearly all of the creatinine the kidneys filter normally appears in the urine, a change in the rate of creatinine excretion may reflect renal failure.

Another plasma clearance test uses *para-aminohippuric acid* (PAH), a substance that filters freely through the glomerular membranes. However, unlike inulin, any PAH remaining in the peritubular capillary plasma after filtration is secreted into the proximal convoluted tubules. Therefore, essentially all PAH passing through the kidneys appears in the urine. For this reason, the rate of PAH clearance can be used to calculate the rate of plasma flow through the kidneys. Then, if the hematocrit is known (see chapter 14, p. 511), the rate of total blood flow through the kidneys can also be calculated.

Parents of infants may be startled when a physician hospitalizes their child for an illness that in an adult might be considered mild—a day or two of vomiting and diarrhea. Because the kidneys of infants and young children are unable to concentrate urine and conserve water as effectively as those of adults, they can lose water rapidly, which may lead to dehydration. A 20-pound infant can lose a pound in just a day of an acute viral illness, and this is a sufficiently significant proportion of body weight to warrant hospitalization, where intravenous fluids are given to restore water and electrolyte balance (see chapter 21, p. 808).

1 List the normal constituents of urine.

2 What is the normal hourly output of urine? The minimal hourly output?

Elimination of Urine

After forming along the nephrons, urine passes from the collecting ducts through openings in the renal papillae and enters the major and minor calyces of the kidney. From there it passes through the renal pelvis, into a ureter, and into the urinary bladder. The urethra delivers urine to the outside.

Ureters

Each **ureter** is a tubular organ about 25 centimeters long, which begins as the funnel-shaped renal pelvis. It extends downward posterior to the parietal peritoneum and parallel to the vertebral column. Within the pelvic cavity, it courses forward and medially to join the urinary bladder from underneath.

The wall of a ureter is composed of three layers. The inner layer, or *mucous coat,* includes several thicknesses of transitional epithelial cells and is continuous with the linings of the renal tubules and the urinary bladder. The middle layer, or *muscular coat,* largely consists of smooth muscle fibers arranged in circular and longitudinal bun-

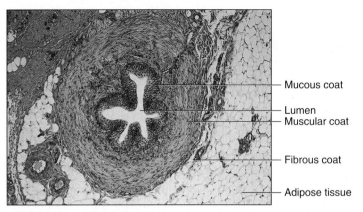

FIGURE 20.26
Cross section of a ureter (75×).

dles. The outer layer, or *fibrous coat,* is composed of connective tissue (fig. 20.26).

Muscular peristaltic waves, originating in the renal pelvis, help move the urine along the length of the ureter. The presence of urine in the renal pelvis initiates these waves, whose frequency keeps pace with the rate of urine formation. If this rate is high, a peristaltic wave may occur every few seconds; if the rate is low, a wave may occur every few minutes.

Because the linings of the ureters and the urinary bladder are continuous, bacteria may ascend from the bladder into the ureters, causing infection. An inflammation of the bladder, called *cystitis,* is more common in women than in men because the female urethral pathway is shorter. Inflammation of the ureter is called *ureteritis.*

When a peristaltic wave reaches the urinary bladder, it spurts urine into the bladder. A flaplike fold of mucous membrane covers the opening where the urine enters. This fold acts as a valve, allowing urine to enter the bladder from the ureter but preventing it from backing up from the bladder into the ureter.

If a ureter becomes obstructed, such as by a small kidney stone (renal calculus) in its lumen, strong peristaltic waves are initiated in the proximal portion of the tube, which may help move the stone into the bladder. The presence of a stone usually also stimulates a sympathetic reflex (ureterorenal reflex) that constricts the renal arterioles and reduces urine production in the affected kidney.

1 Describe the structure of a ureter.

2 How is urine moved from the renal pelvis to the urinary bladder?

3 What prevents urine from backing up from the urinary bladder into the ureters?

4 How does an obstruction in a ureter affect urine production?

Urinary Bladder

The **urinary bladder** is a hollow, distensible, muscular organ. It is located within the pelvic cavity, posterior to the symphysis pubis and inferior to the parietal peritoneum (fig. 20.27 and reference plate 52). In a male, the bladder lies posteriorly against the rectum, and in a female, it contacts the anterior walls of the uterus and vagina.

The pressure of surrounding organs alters the spherical shape of the bladder. When the bladder is empty, its inner wall forms many folds, but as it fills with urine, the wall becomes smoother. At the same time, the superior surface of the bladder expands upward into a dome.

When greatly distended, the bladder pushes above the pubic crest and into the region between the abdominal wall and the parietal peritoneum. The dome can reach the level of the umbilicus and press against the coils of the small intestine.

The internal floor of the bladder includes a triangular area called the *trigone,* which has an opening at each of its three angles (fig. 20.28). Posteriorly, at the base of the trigone, the openings are those of the ureters. Anteriorly, at the apex of the trigone, is a short, funnel-shaped extension called the *neck* of the bladder, which contains the opening into the urethra. The trigone generally remains in a fixed position, even though the rest of the bladder distends and contracts.

The wall of the urinary bladder consists of four layers. The inner layer, or *mucous coat,* includes several thicknesses of transitional epithelial cells, similar to those lining the ureters and the upper portion of the urethra. The thickness of this tissue changes as the bladder expands and contracts. During distension, the tissue appears to be only two or three cells thick, but during contraction, it appears to be five or six cells thick (see fig. 5.9).

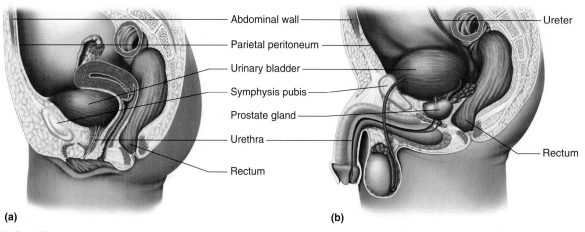

(a)   **(b)**

Abdominal wall — Parietal peritoneum — Urinary bladder — Symphysis pubis — Prostate gland — Urethra — Rectum — Ureter — Rectum

FIGURE 20.27

The urinary bladder is located within the pelvic cavity and behind the symphysis pubis. (*a*) In a female, it lies in contact with the uterus and vagina. (*b*) In a male, it lies against the rectum.

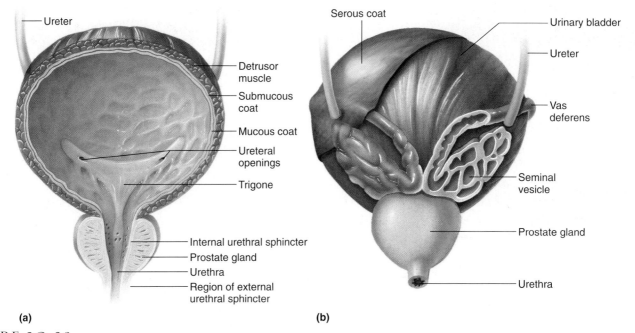

(a)

- Ureter
- Detrusor muscle
- Submucous coat
- Mucous coat
- Ureteral openings
- Trigone
- Internal urethral sphincter
- Prostate gland
- Urethra
- Region of external urethral sphincter

(b)

- Serous coat
- Urinary bladder
- Ureter
- Vas deferens
- Seminal vesicle
- Prostate gland
- Urethra

FIGURE 2O.28

A male urinary bladder. (*a*) Longitudinal section. (*b*) Posterior view.

The second layer of the bladder wall is the *submucous coat.* It consists of connective tissue and contains many elastic fibers.

The third layer of the bladder wall, the *muscular coat,* is primarily composed of coarse bundles of smooth muscle fibers. These bundles are interlaced in all directions and at all depths, and together they comprise the **detrusor muscle** (de-truz′or mus′l). The portion of the detrusor muscle that surrounds the neck of the bladder forms an *internal urethral sphincter.* Sustained contraction of this sphincter muscle prevents the bladder from emptying until the pressure within it increases to a certain level. The detrusor muscle has parasympathetic nerve fibers that function in the reflex that passes urine.

The outer layer of the wall, the *serous coat,* consists of the parietal peritoneum. It is found only on the upper surface of the bladder. Elsewhere, the outer coat is composed of fibrous connective tissue (fig. 20.29).

1   Describe the trigone of the urinary bladder.

2   Describe the structure of the bladder wall.

3   What kind of nerve fibers supply the detrusor muscle?

Urethra

The **urethra** is a tube that conveys urine from the urinary bladder to the outside of the body. Its wall is lined with mucous membrane and contains a thick layer of longitudinal smooth muscle fibers. The urethral wall also contains many mucous glands, called *urethral glands,* which secrete mucus into the urethral canal (fig. 20.30).

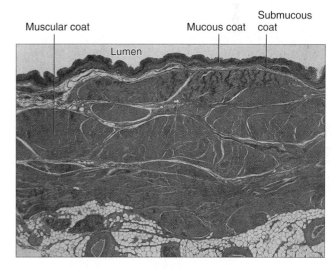

- Muscular coat
- Mucous coat
- Submucous coat
- Lumen

FIGURE 2O.29

Light micrograph of the human urinary bladder wall (6×).

In a female, the urethra is about 4 centimeters long. It passes forward from the bladder, courses below the symphysis pubis, and empties between the labia minora. Its opening, the *external urethral orifice* (urinary meatus), is located anterior to the vaginal opening and about 2.5 centimeters posterior to the clitoris (fig. 20.31*a*).

In a male, the urethra, which functions both as a urinary canal and a passageway for cells and secretions from the reproductive organs, can be divided into three sections: the prostatic urethra, the membranous urethra, and the penile urethra (see fig. 20.31*b* and reference plate 60).

The **prostatic urethra** is about 2.5 centimeters long and passes from the urinary bladder through the *prostate*

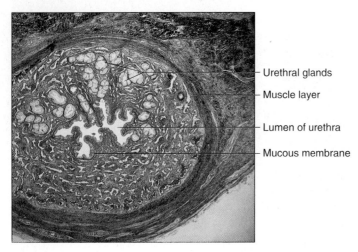

FIGURE 20.30
Cross section of the urethra (10×).

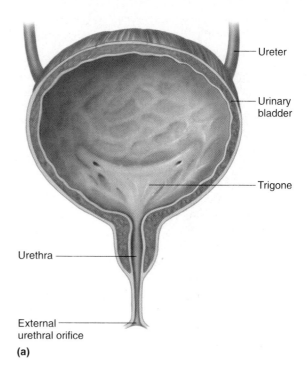

(a)

gland, which is located just below the bladder. Ducts from reproductive structures join the urethra in this region.

The **membranous urethra** is about 2 centimeters long. It begins just distal to the prostate gland, passes through the urogenital diaphragm, and is surrounded by the fibers of the external urethral sphincter muscle.

The **penile urethra** is about 15 centimeters long and passes through the corpus spongiosum of the penis, where erectile tissue surrounds it. This portion of the urethra terminates with the *external urethral* orifice at the tip of the penis.

1   Describe the structure of the urethra.

2   How does the urethra of a male differ from that of a female?

Micturition

Urine leaves the urinary bladder by the **micturition** (mik″tu-rish′un) or urination reflex. The detrusor muscle contracts, and contractions of muscles in the abdominal wall and pelvic floor may help, as well as fixation of the thoracic wall and diaphragm. In micturition, the *external urethral sphincter* also relaxes. This muscle, which is part of the urogenital diaphragm (see chapter 9, p. 314), surrounds the urethra about 3 centimeters from the bladder and is composed of voluntary skeletal muscle tissue.

Distension of the bladder wall as it fills with urine stimulates the urge to urinate. The wall expands, stimulating stretch receptors, which triggers the micturition reflex.

The *micturition reflex center* is located in the sacral portion of the spinal cord. When sensory impulses from the stretch receptors signal the reflex center, parasympathetic motor impulses travel out to the detrusor muscle, which contracts rhythmically in response. A sensation of urgency accompanies this action.

The urinary bladder may hold as much as 600 milliliters of urine. The desire to urinate usually appears when it contains about 150 milliliters. Then, as urine volume

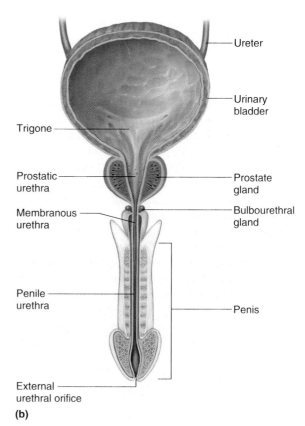

(b)

FIGURE 20.31
Urinary bladder and urethra (*a*) of the female (longitudinal section) and (*b*) of the male (longitudinal section).

20.4

CLINICAL APPLICATION

URINALYSIS: CLUES TO HEALTH

Urine has long fascinated medical minds. As a folk remedy, urine has been used as a mouthwash, toothache treatment, and a cure for sore eyes. Hippocrates (460–377 B.C.) was the first to observe that the condition of the urine can reflect health, noting that frothy urine denoted kidney disease. During the Middle Ages, health practitioners consulted charts that matched certain urine colors to certain diseases. In the seventeenth century, British physicians diagnosed diabetes by having their medical students taste sugar in patients' urine. Today, urine composition is still used as a window on health and also to check for illicit drug use.

Certain inherited disorders can alter urine quite noticeably. The name *maple syrup urine disease* vividly describes what this inborn error of metabolism does to the urine. This condition, which causes mental retardation, results from a block in the breakdown pathways for certain amino acids. In *alkaptonuria,* one of the first inborn errors to be described, urine turns black when it is left to stand. This condition also produces painful arthritis and blackened ear tips. People with *Wilson disease* have an inherited inability to excrete copper. If they are properly diagnosed and given the drug penicillamine, they excrete a copper-colored urine.

Other genetic conditions alter urine without causing health problems. People with *beeturia* excrete dark pink urine after they eat beets. The problem for people with *urinary excretion of odoriferous component of asparagus* is obvious. Parents of newborns who have inherited *blue diaper syndrome* are in for a shock when they change their child's first diaper. Due to a defect in transport of the amino acid tryptophan in the small intestine, bacteria degrade the partially digested tryptophan, producing a compound that turns blue on contact with oxygen. ■

---

increases to 300 milliliters or more, the sensation of fullness becomes increasingly uncomfortable.

As the bladder fills with urine and its internal pressure increases, contractions of its wall intensify. When these contractions become strong enough to force the internal urethral sphincter open, another reflex signals the external urethral sphincter to relax, and the bladder may empty. However, because the external urethral sphincter is composed of skeletal muscle, it is under conscious control, and therefore usually remains contracted until a person decides to urinate. Nerve centers in the brainstem and cerebral cortex that inhibit the micturition reflex aid this control. When a person decides to urinate, the external urethral sphincter relaxes, and inhibition of the micturition reflex lifts. Nerve centers within the pons and the hypothalamus heighten the micturition reflex. The detrusor muscle contracts, and urine is excreted through the urethra. Within a few moments, the

neurons of the micturition reflex tire, the detrusor muscle relaxes, and the bladder begins to fill with urine again. Table 20.5 outlines the micturition process, and Clinical Application 20.4 discusses urinalysis and health.

> Damage to the spinal cord above the sacral region may abolish voluntary control of urination. However, if the micturition reflex center and its sensory and motor fibers are uninjured, micturition may continue to occur reflexively. In this case, the bladder collects urine until its walls stretch enough to trigger a micturition reflex, and the detrusor muscle contracts in response. This condition is called an *automatic bladder.*

1 Describe micturition.

2 How is it possible to consciously inhibit the micturition reflex?

TABLE 20.5	Major Events of Micturition

1. Urinary bladder distends as it fills with urine.
2. Stretch receptors in the bladder wall are stimulated, and they signal the micturition center in the sacral spinal cord.
3. Parasympathetic nerve impulses travel to the detrusor muscle, which responds by contracting rhythmically.
4. The need to urinate is sensed as urgent.
5. Voluntary contraction of the external urethral sphincter and inhibition of the micturition reflex by impulses from the brainstem and the cerebral cortex prevent urination.
6. Following the decision to urinate, the external urethral sphincter is relaxed, and impulses from the pons and the hypothalamus facilitate the micturition reflex.
7. The detrusor muscle contracts, and urine is expelled through the urethra.
8. Neurons of the micturition reflex center fatigue, the detrusor muscle relaxes, and the bladder begins to fill with urine again.

Incontinence is the loss of control of micturition. Stress incontinence, caused by pressure on the bladder, is particularly common among women who have had children, especially if they have gained weight. An effective treatment is at least two months of doing Kegel exercises, in which a woman contracts the muscles that support the bladder, several times daily. Treatments for severe cases include a tamponlike cone inserted into the vagina to raise the pelvic floor; a small foam pad placed over the urethra to catch small amounts of urine; collagen injections around the urethra to tighten it; and surgery. Many people use absorbent pads.

Nighttime bedwetting was noted as long ago as 1500 B.C. Treatments have ranged from drinking the broth from boiled hens' combs, to blocking the urethra at night, to punishment and ridicule. In many cases, this *nocturnal enuresis* is inherited. Drug treatment and pads to absorb urine help to manage affected children, who usually outgrow the condition.

Life-Span Changes

As with other organ systems, the urinary system is sufficiently redundant, in both structure and function, to mask aging-related changes. However, overall, the kidneys are slower to remove nitrogenous wastes and toxins and to compensate for changes to maintain homeostasis.

From the outside, the kidneys change with age, appearing scarred and grainy as arterioles serving the cortex constrict and fibrous connective tissue accumulates around the capsules. On the inside, kidney cells begin to die as early as age twenty years, but the gradual shrinkage is not generally noticeable until after age forty. By eighty years, the kidneys have lost about a third of their mass.

Kidney shrinkage is largely due to the gradual loss of glomeruli—they may atrophy, cease functioning, become blocked with fibrous connective tissue, or untwist. About 5% of glomeruli are abnormal by age forty; 37% are abnormal by age ninety. The progressive shut down of glomeruli decreases the surface area available for filtration, and as a result, glomerular filtration rate (GFR) begins to drop in the fourth decade of life. By age seventy-five, GFR is about half that in a young adult, falling from about 125 milliliters/minute to about 60. With this decline in function, proteins are more likely to get into the urine. About a third of the elderly have proteinuria.

Further along the nephron, the renal tubules thicken, accumulating coats of fat. They may shorten, forming small outpouches as cell death disrupts their sleek symmetry. Urine may become more dilute as reabsorption of sodium and glucose and other molecules becomes less efficient. The renal tubules also slow in their processing of certain drugs, which therefore remain in the circulation longer. It becomes harder to clear non-steroidal anti-inflammatory drugs such as aspirin, as well as opiates, antibiotics, urea, uric acid, creatinine, and various toxins. Therefore, a person's age should be taken into account when prescribing drugs. The pharmaceutical industry is beginning to test new drugs on people of a range of ages.

Cardiovascular changes slow the journey of blood through the kidneys. A college student's kidneys may process about a fourth of the cardiac output, or about 1,200 milliliters, per minute. Her eighty-year-old grandfather's kidneys can handle about half that volume. Starting at about age twenty, renal blood flow rate diminishes by about 1% per year. The blood vessels that serve the kidneys become slower to dilate or constrict in response to body conditions. At the same time, the kidneys' release of renin declines, hampering control of osmotic pressure, blood pressure, and sodium and potassium ion concentrations in the blood. The kidneys are also less able to activate vitamin D, which may contribute to the higher prevalence of osteoporosis among the elderly.

The bladder, ureters, and urethra change with the years too. These muscular organs lose elasticity and recoil with age, so in the later years, the bladder holds less than half of what it did in young adulthood, and may retain more urine after urination. In the elderly, the urge to urinate may become delayed, so when it does happen, it is sudden. Older individuals have to urinate at night more than younger people.

Controlling bladder function is a challenge at the beginning of life and much later too. A child usually learns to control urination by about age two or three years. Loss of bladder control, or incontinence, becomes more common in advanced years, although it is not considered a normal part of aging. It results from loss of muscle tone in the bladder, urethra and ureters. Incontinence affects 15% to 20% of women over sixty-five and half of all men. In women, incontinence reflects the stresses of childbirth and the effects of less estrogen during menopause. Bladder sphincter muscles atrophy, muscles in the pelvic floor weaken, and poor muscle tone develops in the smooth muscle of the urethra. In males, incontinence is usually a consequence of an enlarged prostate gland that presses on the bladder.

1 How do the kidneys change in appearance with advancing years?

2 What happens to glomeruli as a person ages?

3 How does kidney function change with age?

4 How do aging-related changes in the cardiovascular system affect the kidneys?

5 How do the ureters, bladder, and urethra change with age?

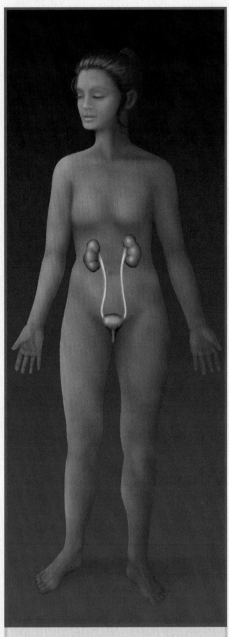

URINARY SYSTEM

The urinary system controls the composition of the internal environment.

Integumentary System

The urinary system compensates for water loss due to sweating. The kidneys and skin both play a role in vitamin D production.

Cardiovascular System

The urinary system controls blood volume. Blood volume and blood pressure play a role in determining water and solute excretion.

Skeletal System

The kidneys and bone tissue work together to control plasma calcium levels.

Lymphatic System

The kidneys control extracellular fluid volume and composition (including lymph).

Muscular System

Muscle tissue controls urine elimination from the bladder.

Digestive System

The kidneys compensate for fluids lost by the digestive system.

Nervous System

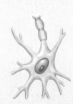

The nervous system influences urine production and elimination.

Respiratory System

The kidneys and the lungs work together to control the pH of the internal environment.

Endocrine System

The endocrine system influences urine production.

Reproductive System

The urinary system in males shares organs with the reproductive system. The kidneys compensate for fluids lost from the male and female reproductive systems.

Clinical Terms Related to the Urinary System

anuria (ah-nu're-ah) Absence of urine due to failure of kidney function or to an obstruction in a urinary pathway.

bacteriuria (bak-te"re-u're-ah) Bacteria in the urine.

cystectomy (sis-tek'to-me) Surgical removal of the urinary bladder.

cystitis (sis-ti'tis) Inflammation of the urinary bladder.

cystoscope (sis'to-skop) Instrument to visually examine the interior of the urinary bladder.

cystotomy (sis-tot'o-me) Incision of the wall of the urinary bladder.

diuresis (di"u-re'sis) Increased production of urine.

diuretic (di"u-ret'ik) Substance that increases urine production.

dysuria (dis-u're-ah) Painful or difficult urination.

enuresis (en"u-re'sis) Uncontrolled urination.

hematuria (hem"ah-tu're-ah) Blood in the urine.

incontinence (in-kon'ti-nens) Inability to control urination and/or defecation reflexes.

nephrectomy (ne-frek'to-me) Surgical removal of a kidney.

nephrolithiasis (nef"ro-li-thi'ah-sis) Kidney stones.

nephroptosis (nef'rop-to'sis) Movable or displaced kidney.

oliguria (ol"i-gu're-ah) Scanty output of urine.

polyuria (pol"e-u're-ah) Excess urine.

pyelolithotomy (pi"ĕ-lo-li-thot'o-me) Removal of a stone from the renal pelvis.

pyelonephritis (pi"ĕ-lo-ne-fri'tis) Inflammation of the renal pelvis.

pyelotomy (pi"ĕ-lot'o-me) Incision into the renal pelvis.

pyuria (pi-u're-ah) Pus (excess white blood cells) in the urine.

uremia (u-re'me-ah) Condition in which substances ordinarily excreted in the urine accumulate in the blood.

ureteritis (u-re"ter-i'tis) Inflammation of the ureter.

urethritis (u"re-thri'tis) Inflammation of the urethra.

CHAPTER SUMMARY

Introduction (page 772)

The urinary system consists of the kidneys, ureters, urinary bladder, and urethra.

Kidneys (page 772)

1. Location of the kidneys
   a. The kidneys are bean-shaped organs on either side of the vertebral column, high on the posterior wall of the abdominal cavity.
   b. They are positioned posterior to the parietal peritoneum and held in place by adipose and connective tissue.
2. Kidney structure
   a. A kidney contains a hollow renal sinus.
   b. The ureter expands into the renal pelvis, which, in turn, is divided into major and minor calyces.
   c. Renal papillae project into the renal sinus.
   d. Kidney tissue is divided into a medulla and a cortex.
3. Functions of the kidneys
   a. The kidneys remove metabolic wastes from the blood and excrete them to the outside.
   b. They also help regulate red blood cell production, blood pressure, calcium ion absorption, and the volume, composition, and pH of the blood.
4. Renal blood vessels
   a. Arterial blood flows through the renal artery, interlobar arteries, arcuate arteries, interlobular arteries, afferent arterioles, glomerular capillaries, efferent arterioles, and peritubular capillaries.
   b. Venous blood returns through a series of vessels that correspond to those of the arterial pathways.
5. Nephrons
   a. Structure of a nephron
      (1) A nephron is the functional unit of the kidney.
      (2) It consists of a renal corpuscle and a renal tubule.
         (a) The corpuscle consists of a glomerulus and a glomerular capsule.
         (b) Portions of the renal tubule include the proximal convoluted tubule, the nephron loop (ascending and descending limbs), and the distal convoluted tubule.
      (3) The nephron joins a collecting duct, which empties into the minor calyx of the renal pelvis.
   b. Juxtaglomerular apparatus
      (1) The juxtaglomerular apparatus is located at the point of contact between the distal convoluted tubule and the afferent and efferent arterioles.
      (2) It consists of the macula densa and the juxtaglomerular cells.
   c. Cortical and juxtamedullary nephrons
      (1) Cortical nephrons are the most numerous and have corpuscles near the surface of the kidney.
      (2) Juxtamedullary nephrons have corpuscles near the medulla.
   d. Blood supply of a nephron
      (1) The glomerular capillary receives blood from the afferent arteriole and passes it to the efferent arteriole.
      (2) The efferent arteriole gives rise to the peritubular capillary system, which surrounds the renal tubule.
      (3) Capillary loops, called vasa recta, dip down into the medulla.

Urine Formation (page 782)

Nephrons remove wastes from the blood and regulate water and electrolyte concentrations. Urine is the product of these functions.

1. Glomerular filtration
   a. Urine formation begins when water and dissolved materials are filtered out of the glomerular capillary.
   b. The glomerular capillaries are much more permeable than the capillaries in other tissues.
2. Filtration pressure
   a. Filtration is mainly due to hydrostatic pressure inside the glomerular capillaries.
   b. The osmotic pressure of the blood plasma and hydrostatic pressure in the glomerular capsule also affect filtration.
   c. Filtration pressure is the net force acting to move material out of the glomerulus and into the glomerular capsule.
   d. The composition of the filtrate is similar to that of tissue fluid.
3. Filtration rate
   a. The rate of filtration varies with the filtration pressure.
   b. Filtration pressure changes with the diameters of the afferent and efferent arterioles.
   c. As the osmotic pressure in the glomerulus increases, filtration decreases.
   d. As the hydrostatic pressure in a glomerular capsule increases, the filtration rate decreases.
   e. The kidneys produce about 125 milliliters of glomerular fluid per minute, most of which is reabsorbed.
   f. The volume of filtrate varies with the surface area of the glomerular capillary.
4. Control of filtration rate
   a. Glomerular filtration rate (GFR) remains relatively constant but may be increased or decreased when the need arises. Increased sympathetic nerve activity can decrease GFR.
   b. When tubular fluid NaCl concentration decreases, the macula densa causes the juxtaglomerular cells to release renin. This triggers a series of changes leading to vasoconstriction, which may affect GFR, and secretion of aldosterone, which stimulates tubular sodium reabsorption.
   c. Autoregulation is the ability of an organ or tissue to maintain a constant blood flow under certain conditions when the arterial blood pressure is changing.
5. Tubular reabsorption
   a. Substances are selectively reabsorbed from the glomerular filtrate.
   b. The peritubular capillary is adapted for reabsorption.
      (1) It carries low-pressure blood.
      (2) It is very permeable.
   c. Most reabsorption occurs in the proximal tubule, where the epithelial cells possess microvilli.
   d. Different modes of transport reabsorb various substances in particular segments of the renal tubule.
      (1) Glucose and amino acids are reabsorbed by active transport.
      (2) Water is reabsorbed by osmosis.
      (3) Proteins are reabsorbed by endocytosis.
   e. Active transport mechanisms have limited transport capacities.
   f. If the concentration of a substance in the filtrate exceeds its renal plasma threshold, the excess is excreted in the urine.
   g. Substances that remain in the filtrate are concentrated as water is reabsorbed.
   h. Sodium ions are reabsorbed by active transport.
      (1) Negatively charged ions accompany positively charged sodium ions out of the filtrate.
      (2) Water is passively reabsorbed by osmosis as sodium ions are actively reabsorbed.
6. Tubular secretion
   a. Tubular secretion transports certain substances from the plasma to the tubular fluid.
   b. Some substances are actively secreted.
      (1) These include various organic compounds and hydrogen ions.
      (2) The proximal and distal convoluted tubules secrete hydrogen ions.
   c. Potassium ions are secreted both actively and passively in the distal convoluted tubule and collecting duct.
7. Regulation of urine concentration and volume
   a. Most of the sodium ions are reabsorbed before the urine is excreted.
   b. Sodium ions are concentrated in the renal medulla by the countercurrent mechanism.
      (1) Chloride ions are actively reabsorbed in the ascending limb, and sodium ions follow them passively.
      (2) Tubular fluid in the ascending limb becomes hypotonic as it loses solutes.
      (3) Water leaves the descending limb by osmosis, and NaCl enters this limb by diffusion.
      (4) Tubular fluid in the descending limb becomes hypertonic as it loses water and gains NaCl.
      (5) As NaCl repeats this circuit, its concentration in the medulla increases.
   c. The vasa recta countercurrent mechanism helps maintain the NaCl concentration in the medulla.
   d. The distal convoluted tubule and collecting duct are impermeable to water, which therefore is excreted in urine.
   e. ADH from the posterior pituitary gland increases the permeability of the distal convoluted tubule and collecting duct, promoting water reabsorption.
8. Urea and uric acid excretion
   a. Urea is a by-product of amino acid metabolism.
      (1) It is passively reabsorbed by diffusion.
      (2) About 50% of the urea is excreted in urine.
      (3) A countercurrent mechanism involving urea helps in the reabsorption of water.
   b. Uric acid results from the metabolism of nucleic acids.
      (1) Most is reabsorbed by active transport.
      (2) Some is secreted into the renal tubule.
9. Urine composition
   a. Urine is about 95% water, and it usually contains urea, uric acid, and creatinine.
   b. It may contain a trace of amino acids and varying amounts of electrolytes, depending upon dietary intake.
   c. The volume of urine varies with the fluid intake and with certain environmental factors.
10. Renal clearance
    a. Renal clearance is the rate at which a chemical is removed from the plasma.
    b. The inulin clearance test, creatinine clearance test, and para-aminohippuric acid test can be used to calculate GFR.

Elimination of Urine (page 795)

1. Ureters
   a. The ureter is a tubular organ that extends from each kidney to the urinary bladder.
   b. Its wall has mucous, muscular, and fibrous layers.
   c. Peristaltic waves in the ureter force urine to the urinary bladder.
   d. Obstruction in the ureter stimulates strong peristaltic waves and a reflex that decreases urine production.
2. Urinary bladder
   a. The urinary bladder is a distensible organ that stores urine and forces it into the urethra.
   b. The openings for the ureters and urethra are located at the three angles of the trigone in the floor of the urinary bladder.
   c. Muscle fibers in the wall form the detrusor muscle.
   d. A portion of the detrusor muscle forms an internal urethral sphincter.
3. Urethra
   a. The urethra conveys urine from the urinary bladder to the outside.
   b. In females, it empties between the labia minora.
   c. In males, it conveys products of reproductive organs as well as urine.
      (1) Three portions of the male urethra are prostatic, membranous, and penile.
      (2) The urethra empties at the tip of the penis.
4. Micturition
   a. Micturition is the process of expelling urine.
   b. In micturition, the detrusor muscle contracts and the external urethral sphincter relaxes.
   c. Micturition reflex
      (1) Distension stimulates stretch receptors in the urinary bladder wall.
      (2) The micturition reflex center in the sacral portion of the spinal cord sends parasympathetic motor impulses to the detrusor muscle.
      (3) As the urinary bladder fills, its internal pressure increases, forcing the internal urethral sphincter open.
      (4) A second reflex relaxes the external urethral sphincter, unless its contraction is voluntarily controlled.
      (5) Nerve centers in the brainstem and cerebral cortex aid control of urination.

Life-Span Changes (page 800)

Distinctive changes occur in the kidneys, ureters, and urethra with age, but nephrons are so numerous that a healthy person is usually unaware of kidney shrinkage and slowed cleansing of the blood.

1. With age, the kidneys appear grainy and scarred.
2. GFR drops significantly with age as glomeruli atrophy, fill with connective tissue, or unwind.
3. Renal tubules accumulate fat on their outsides and become asymmetric. Reabsorption and secretion may slow or become impaired. Drugs remain longer in the circulation as a person ages.
4. Changes in the cardiovascular system slow the rate of processing through the urinary system. The kidneys slow in their response to changes, and are less efficient at activating vitamin D.
5. The urinary bladder, ureters, and urethra lose elasticity, with effects on the urge and timing of urination.

CRITICAL THINKING QUESTIONS

1. If an infant is born with narrowed renal arteries, what effect would this condition have on the volume of urine produced? Explain your answer.
2. Why are people with nephrotic syndrome, in which plasma proteins are lost into the urine, more susceptible to infections?
3. If a patient who has had major abdominal surgery receives intravenous fluids equal to the volume of blood lost during surgery, would you expect the volume of urine produced to be greater or less than normal? Why?
4. If a physician prescribed oral penicillin therapy for a patient with an infection of the urinary bladder, how would you describe for the patient the route by which the drug would reach the bladder?
5. If the blood pressure of a patient who is in shock as a result of a severe injury decreases greatly, how would you expect the volume of urine to change? Why?
6. Inflammation of the urinary bladder is more common in women than in men. How might this observation be related to the anatomy of the female and male urethras?

REVIEW EXERCISES

1. Name the organs of the urinary system, and list their general functions.
2. Describe the external and internal structure of a kidney.
3. List the functions of the kidneys.
4. Name the vessels the blood passes through as it travels from the renal artery to the renal vein.
5. Distinguish between a renal corpuscle and a renal tubule.
6. Name the structures fluid passes through as it travels from the glomerulus to the collecting duct.
7. Describe the location and structure of the juxtaglomerular apparatus.
8. Distinguish between cortical and juxtamedullary nephrons.
9. Distinguish among filtration, reabsorption, and secretion as they relate to urine formation.
10. Define *filtration pressure.*
11. Compare the composition of the glomerular filtrate with that of the blood plasma.
12. Explain how the diameters of the afferent and efferent arterioles affect the rate of glomerular filtration.
13. Explain how changes in the osmotic pressure of the blood plasma may affect the rate of glomerular filtration.

14. Explain how the hydrostatic pressure of a glomerular capsule affects the rate of glomerular filtration.
15. Describe two mechanisms by which the body regulates the filtration rate.
16. Define *autoregulation*.
17. Discuss how tubular reabsorption is a selective process.
18. Explain how the peritubular capillary is adapted for reabsorption.
19. Explain how the epithelial cells of the proximal convoluted tubule are adapted for reabsorption.
20. Explain why active transport mechanisms have limited transport capacities.
21. Define *renal plasma threshold*, and explain its significance in tubular reabsorption.
22. Explain how amino acids and proteins are reabsorbed.
23. Describe the effect of sodium reabsorption on the reabsorption of negatively charged ions.
24. Explain how sodium ion reabsorption affects water reabsorption.
25. Explain how hypotonic tubular fluid is produced in the ascending limb of the nephron loop.
26. Explain why fluid in the descending limb of the nephron loop is hypertonic.
27. Describe the function of ADH.
28. Explain how the renal tubule is adapted to secrete hydrogen ions.
29. Explain how potassium ions may be secreted passively.
30. Explain how urine may become concentrated as it moves through the collecting duct.
31. Compare the processes by which urea and uric acid are reabsorbed.
32. List the more common substances found in urine and their sources.
33. List some of the factors that affect the volume of urine produced each day.
34. Describe the structure and function of a ureter.
35. Explain how the muscular wall of the ureter aids in moving urine.
36. Discuss what happens if a ureter becomes obstructed.
37. Describe the structure and location of the urinary bladder.
38. Define *detrusor muscle*.
39. Distinguish between the internal and external urethral sphincters.
40. Compare the urethra of a female with that of a male.
41. Describe the micturition reflex.
42. Explain how the micturition reflex can be voluntarily controlled.
43. Describe the changes that occur in the urinary system with age.

WEB CONNECTIONS

Visit the Student Edition of the Online Learning Center at www.mhhe.com/shier10 **for answers to chapter questions, additional quizzes, interactive learning exercises, and other study tools.**

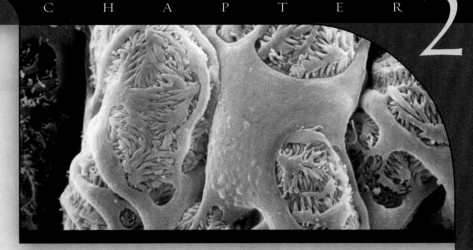

The kidneys play a major role in maintaining electrolyte balance in body fluids. This falsely colored scanning electron micrograph shows podocytes, which are part of the glomeruli in the kidneys (3,650×).

Water, Electrolyte, and Acid-Base Balance

CHAPTER OBJECTIVES

After you have studied this chapter, you should be able to

1. Explain water and electrolyte balance and discuss the importance of this balance.

2. Describe how the body fluids are distributed within compartments, how fluid composition differs between compartments, and how fluids move from one compartment to another.

3. List the routes by which water enters and leaves the body and explain how water input and output are regulated.

4. Explain how electrolytes enter and leave the body and how the input and output of electrolytes are regulated.

5. Explain acid-base balance.

6. Describe how hydrogen ion concentrations are expressed mathematically.

7. List the major sources of hydrogen ions in the body.

8. Distinguish between strong and weak acids and bases.

9. Explain how chemical buffer systems, the respiratory center, and the kidneys minimize changing pH values of the body fluids.

A ugust 2, 2001, was another 90° high-humidity day at training camp for the Minnesota Vikings in Mankato. The day before, offensive tackle Korey Stringer hadn't been able to participate in afternoon practice, citing exhaustion—but he vowed to make it the next morning. He did, but did not feel well. After vomiting three times, he walked over to an air-conditioned shelter, dizzy and breathing heavily. Trainers recognized the signs of heat exhaustion and took Stringer to a nearby medical facility, but it was too late. On arrival, Stringer's body temperature was a life-threatening 108°F, and he soon lost consciousness. To the shock and dismay of his teammates, he died at 1:50 the next morning.

Korey Stringer died of heatstroke, which occurs rapidly when the body is exposed to a heat index (heat considering humidity) of more than 105°F and body temperature rises to above 106°F. On that August day, the heat index was 110°F. Under these conditions, the body stops sweating, so heat can no longer be dissipated, and the organs fail. The situation is worse if the individual is heavy or if the body is covered. Stringer weighed 335 pounds and was exercising in full football gear.

During the heat wave of 2001, several athletes in their teens also succumbed to heatstroke in the weeks following Stringer's death. According to the Centers for Disease Control and Prevention, more than 300 people die in the United States each year from this preventable condition, most of them either elderly people or infants, who may have poor temperature control. Despite knowing the symptoms, heatstroke is unpredictable because people have different limits. In the wake of Stringer's death, many players remembered feeling dizzy or experiencing chills when the weather was hot, but continuing to exercise anyway. Athletic trainers typically weigh players twice a day and are alerted to possible heatstroke if an athlete suddenly loses 6 to 8 pounds. After Stringer's death, sports medicine specialists advised the National Football League to shorten or change the time of practices when heat and humidity become dangerous, to enforce water breaks, and to allow players at least a week to adjust to a different climate before wearing full

Korey Stringer was an offensive tackle for the Minnesota Vikings who died of heatstroke.

gear. Stringer's experience may save others by calling attention to the danger of heatstroke. Following is a list of the symptoms of heatstroke:

Headache

Dizziness

Exhaustion

Profuse sweating, which then stops

Dry, hot, and red skin

Pulse elevated as high as 180 beats per minute

Increased respiratory rate

Disorientation

Losing consciousness or having a seizure

Rapid rise in body temperature ■

The term *balance* suggests a state of equilibrium, and in the case of water and electrolytes, it means that the quantities entering the body equal the quantities leaving it. Maintaining such a balance requires mechanisms to ensure that lost water and electrolytes are replaced and that any excesses are excreted. As a result, the levels of water and electrolytes in the body remain relatively stable at all times.

It is important to remember that water balance and electrolyte balance are interdependent because electrolytes are dissolved in the water of body fluids. Consequently, anything that alters the concentrations of the electrolytes will alter the concentration of the water by adding solutes to it or by removing solutes from it. Like-

wise, anything that changes the concentration of the water will change the concentrations of the electrolytes by concentrating or diluting them.

Distribution of Body Fluids

Body fluids are not uniformly distributed. Instead, they occur in regions, or *compartments,* of different volumes that contain fluids of varying compositions. The movement of water and electrolytes between these compartments is regulated to stabilize their distribution and the composition of body fluids.

Fluid Compartments

The body of an average adult female is about 52% water by weight, and that of an average male is about 63% water. This difference between the sexes is due to the fact that females generally have more adipose tissue, which has lit-tle water. Males have more muscle tissue, which contains a great deal of water. Water in the body (about 40 liters), together with its dissolved electrolytes, is distributed into two major compartments: an intracellular fluid compart-ment and an extracellular fluid compartment (fig. 21.1).

The **intracellular** (in"trah-sel'u-lar) **fluid compart-ment** includes all the water and electrolytes that cell membranes enclose. In other words, intracellular fluid is the fluid within the cells, and, in an adult, it represents about 63% by volume of the total body water.

∞ R E C O N N E C T   T O   C H A P T E R   1 ,
H O M E O S T A S I S ,   P A G E   9

The **extracellular** (ek"strah-sel'u-lar) **fluid compart-ment** includes all the fluid outside the cells—within the tissue spaces (interstitial fluid), the blood vessels (plasma), and the lymphatic vessels (lymph). Epithelial layers sepa-rate a specialized fraction of the extracellular fluid from other extracellular fluids. This *transcellular (trans-sel'ular) fluid* includes cerebrospinal fluid of the central nervous system, aqueous and vitreous humors of the eyes, synovial fluid of the joints, serous fluid within the body cavities, and fluid secretions of the exocrine glands. The fluids of the extracellular compartment constitute about 37% by volume of the total body water (fig. 21.2).

Body Fluid Composition

Extracellular fluids generally have similar compositions, including high concentrations of sodium, chloride, cal-cium, and bicarbonate ions and lesser concentrations of potassium, magnesium, phosphate, and sulfate ions. The blood plasma fraction of extracellular fluid contains con-siderably more protein than do either interstitial fluid or lymph.

Intracellular fluid has high concentrations of potas-sium, phosphate, and magnesium ions. It includes a greater concentration of sulfate ions and lesser concentra-tions of sodium, chloride, and bicarbonate ions than does extracellular fluid. Intracellular fluid also has a greater

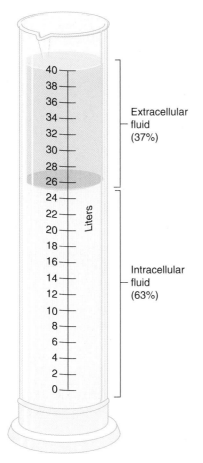

FIGURE 21.1

Of the 40 liters of water in the body of an average adult male, about two-thirds is intracellular, and one-third is extracellular.

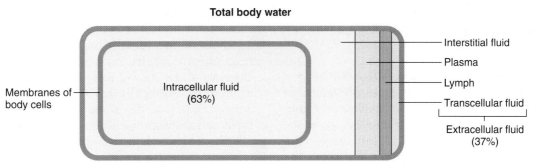

FIGURE 21.2

Cell membranes separate fluid in the intracellular compartment from fluid in the extracellular compartment. Approximately two-thirds of the water in the body is inside cells.

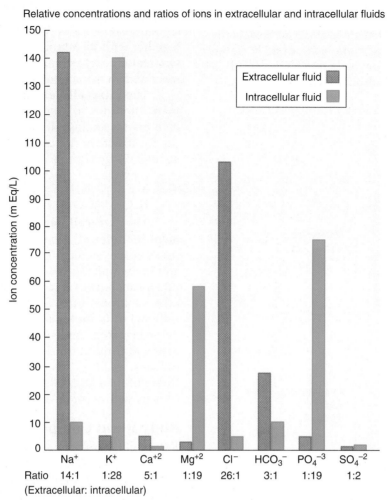

Relative concentrations and ratios of ions in extracellular and intracellular fluids

Ratio	Na⁺	K⁺	Ca⁺²	Mg⁺²	Cl⁻	HCO₃⁻	PO₄⁻³	SO₄⁻²
(Extracellular: intracellular)	14:1	1:28	5:1	1:19	26:1	3:1	1:19	1:2

FIGURE 21.3

Extracellular fluids have relatively high concentrations of sodium (Na^+), calcium (Ca^{+2}), chloride (Cl^-), and bicarbonate (HCO_3^-) ions. Intracellular fluid has relatively high concentrations of potassium (K^+), magnesium (Mg^{+2}), phosphate (PO_4^{-3}), and sulfate (SO_4^{-2}) ions.

concentration of protein than plasma. Figure 21.3 shows these relative concentrations.

1 How are fluid balance and electrolyte balance interdependent?

2 Describe the normal distribution of water within the body.

3 Which electrolytes are in higher concentrations in extracellular fluids? In intracellular fluid?

4 How does the concentration of protein vary in the various body fluids?

Movement of Fluid Between Compartments

Two major factors regulate the movement of water and electrolytes from one fluid compartment to another: hydrostatic pressure and osmotic pressure. For example, as explained in chapter 15 (p. 568), fluid leaves the plasma at the arteriolar ends of capillaries and enters the interstitial spaces because of the net outward force of *hydrostatic pressure* (blood pressure). Fluid returns to the plasma from the interstitial spaces at the venular ends of capillaries because of the net inward force of *colloid*

osmotic pressure. Likewise, as mentioned in chapter 16 (p. 611), fluid leaves the interstitial spaces and enters the lymph capillaries due to the hydrostatic pressure of the interstitial fluid. As a result of the circulation of lymph, interstitial fluid returns to the plasma.

Because hydrostatic pressure within the cells and surrounding interstitial fluid is ordinarily equal and remains stable, any net fluid movement that occurs is likely to be the result of changes in osmotic pressure (fig. 21.4). Recall that osmotic pressure is due to the presence of impermeant solutes on one side of a cell membrane. Because of the Na^+/K^+ pump, sodium (extracellular) and potassium (intracellular) ions behave as impermeant solutes and create an osmotic pressure. For example, since most cell membranes in the body are freely permeable to water, a decrease in extracellular sodium ion concentration causes a net movement of water from the extracellular compartment into the intracellular compartment by osmosis. The cell swells. Conversely, if the extracellular sodium ion concentration increases, cells shrink as they lose water. Although the solute composition of body fluids varies

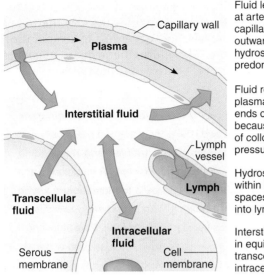

FIGURE 21.4
Net movements of fluids between compartments result from differences in hydrostatic and osmotic pressures.

Fluid leaves plasma at arteriolar end of capillaries because outward force of hydrostatic pressure predominates

Fluid returns to plasma at venular ends of capillaries because inward force of colloid osmotic pressure predominates

Hydrostatic pressure within interstitial spaces forces fluid into lymph capillaries

Interstitial fluid is in equilibrium with transcellular and intracellular fluids

between intracellular and extracellular compartments, water will "follow salt" and distribute by osmosis such that the water concentration (and total solute concentration) is essentially equal inside and outside cells.

1 Which factors control the movement of water and electrolytes from one fluid compartment to another?

2 How does the sodium ion concentration within body fluids affect the net movement of water between the compartments?

Different substances may be distributed to different compartments. For example, an infusion of 1 liter of isotonic sodium chloride solution is restricted largely to the extracellular fluid because of the active transport sodium pumps in cell membranes. In contrast, a liter of isotonic glucose solution may be given intravenously without damaging red blood cells, but as the glucose is metabolized aerobically, it reacts to release carbon dioxide and water. Thus, the liter of isotonic glucose yields a liter of water that can be distributed throughout intracellular and extracellular compartments.

Water Balance

Water balance exists when water intake equals water output. Homeostasis requires control of both water intake and water output. Ultimately, maintenance of the internal environment depends on thirst centers in the brain to vary water intake and on the kidneys' ability to vary water output.

Water Intake

The volume of water gained each day varies among individuals. An average adult living in a moderate environment takes in about 2,500 milliliters. Probably 60% is obtained from drinking water or beverages, and another 30% comes from moist foods. The remaining 10% is a by-product of the oxidative metabolism of nutrients, which is called **water of metabolism** (fig. 21.5a).

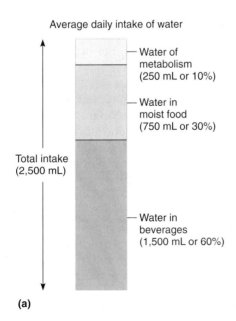

Average daily intake of water

Water of metabolism (250 mL or 10%)

Water in moist food (750 mL or 30%)

Total intake (2,500 mL)

Water in beverages (1,500 mL or 60%)

(a)

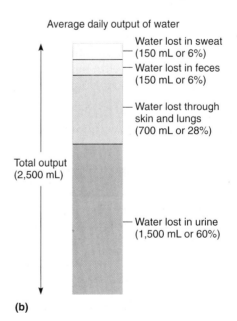

Average daily output of water

Water lost in sweat (150 mL or 6%)

Water lost in feces (150 mL or 6%)

Water lost through skin and lungs (700 mL or 28%)

Total output (2,500 mL)

Water lost in urine (1,500 mL or 60%)

(b)

FIGURE 21.5
Water balance. (a) Major sources of body water. (b) Routes by which the body loses water. Urine production is most important in the regulation of water balance.

Regulation of Water Intake

The primary regulator of water intake is thirst. The intense feeling of thirst derives from the osmotic pressure of extracellular fluids and a *thirst center* in the hypothalamus of the brain.

As the body loses water, the osmotic pressure of the extracellular fluids increases. Such a change stimulates *osmoreceptors* in the thirst center, and in response, the hypothalamus causes the person to feel thirsty and to seek water. A thirsty person usually has a dry mouth, caused by loss of extracellular water and resulting decreased flow of saliva.

The thirst mechanism is normally triggered whenever the total body water decreases by as little as 1%. The act of drinking and the resulting distension of the stomach wall trigger nerve impulses that inhibit the thirst mechanism. Thus, drinking stops long before the swallowed water is absorbed. This inhibition helps prevent the person from drinking more than is required to replace the volume lost, avoiding development of an imbalance. Table 21.1 summarizes the steps in this mechanism.

1 What is water balance?

2 Where is the thirst center located?

3 What stimulates fluid intake? What inhibits it?

Water Output

Water normally enters the body only through the mouth, but it can be lost by a variety of routes. These include obvious losses in urine, feces, and sweat (sensible perspiration), as well as evaporation of water from the skin (insensible perspiration) and from the lungs during breathing.

If an average adult takes in 2,500 milliliters of water each day, then 2,500 milliliters must be eliminated to maintain water balance. Of this volume, perhaps 60% will be lost in urine, 6% in feces, and 6% in sweat. About 28% will be lost by evaporation from the skin and lungs (fig. 21.5*b*). These percentages vary with such environmental factors as temperature and relative humidity and with physical exercise.

If insufficient water is taken in, water output must be reduced to maintain balance. Water lost by sweating is a necessary part of the body's temperature control mechanism; water lost in feces accompanies the elimination of undigested food materials; and water lost by evaporation is largely unavoidable. Therefore, the primary means of regulating water output is control of urine production.

Proteins called *aquaporins* form water-selective membrane channels that enable body cells, including red blood cells and cells in the proximal convoluted tubules and descending limbs of the nephron loops, to admit water. A mutation in one aquaporin gene (which instructs cells to manufacture a type of aquaporin protein) causes a form of *diabetes insipidus,* in which the renal tubules fail to reabsorb water. Rare individuals have been identified who lack certain other aquaporin genes, and they apparently have no symptoms. This suggests that cells have more than one way to admit water.

Regulation of Water Output

The distal convoluted tubules and collecting ducts of the nephrons regulate the volume of water excreted in the urine. The epithelial linings of these segments of the renal tubule remain relatively impermeable to water unless antidiuretic hormone (ADH) is present.

Recall from chapter 13 (p. 484) that osmoreceptors in the hypothalamus help control release of ADH. If the blood plasma becomes more concentrated because of excessive water loss, the osmoreceptors lose water by osmosis and shrink. This change triggers impulses that signal the posterior pituitary gland to release ADH. The ADH released into the bloodstream reaches the kidneys, where it increases the permeability of the distal convoluted tubules and collecting ducts. Consequently, water reabsorption increases, conserving water. This action resists further osmotic change in the plasma. In fact, the *osmoreceptor* (oz"mo-re-sep'tor)-*ADH mechanism* can reduce a normal urine production of 1,500 milliliters per day to about 500 milliliters per day when the body is dehydrated.

If a person drinks too much water, the plasma becomes less concentrated, and the osmoreceptors swell as they receive extra water by osmosis. In this instance, ADH release is inhibited, and the distal tubules and collecting ducts remain impermeable to water. Consequently, less water is reabsorbed and more urine produced. Table 21.2 summarizes the steps in this mechanism. Clinical Application 21.1 discusses disorders resulting from water imbalance.

1 By what routes does the body lose water?

2 What is the primary regulator of water loss?

3 What types of water loss are unavoidable?

4 How does the hypothalamus regulate water balance?

TABLE 21.1	Regulation of Water Intake

1. The body loses as little as 1% of its water.
2. An increase in the osmotic pressure of extracellular fluid due to water loss stimulates osmoreceptors in the thirst center.
3. Activity in the hypothalamus causes the person to feel thirsty and to seek water.
4. Drinking and the resulting distension of the stomach by water stimulate nerve impulses that inhibit the thirst center.
5. Water is absorbed through the walls of the stomach and small intestine.
6. The osmotic pressure of extracellular fluid returns to normal.

| TABLE 21.2 | Events in Regulation of Water Output | | 813 |

Dehydration	Excess Water Intake
1. Extracellular fluid becomes osmotically more concentrated.	1. Extracellular fluid becomes osmotically less concentrated.
2. Osmoreceptors in the hypothalamus are stimulated by the increase in the osmotic pressure of body fluids.	2. This change stimulates osmoreceptors in the hypothalamus.
3. The hypothalamus signals the posterior pituitary gland to release ADH into the blood.	3. The posterior pituitary gland decreases ADH release.
4. Blood carries ADH to the kidneys.	4. Renal tubules decrease water reabsorption.
5. ADH causes the distal convoluted tubules and collecting ducts to increase water reabsorption.	5. Urine output increases, and excess water is excreted.
6. Urine output decreases, and further water loss is minimized.	

Diuretics are chemicals that promote urine production. They produce their effects in different ways. Some, such as alcohol and certain narcotic drugs, promote urine formation by inhibiting ADH release. Certain other substances, such as caffeine, inhibit the reabsorption of sodium ions or other solutes in portions of the renal tubules. As a consequence, the osmotic pressure of the tubular fluid increases, reducing osmotic reabsorption of water and increasing urine volume.

Electrolyte Balance

An **electrolyte balance** (e-lek'tro-līt bal'ans) exists when the quantities of electrolytes (molecules that release ions in water) the body gains equal those lost (fig. 21.6).

Electrolyte Intake

The electrolytes of greatest importance to cellular functions release sodium, potassium, calcium, magnesium, chloride, sulfate, phosphate, bicarbonate, and hydrogen ions. These electrolytes are primarily obtained from foods, but they may also be found in drinking water and other beverages. In addition, some electrolytes are byproducts of metabolic reactions.

Regulation of Electrolyte Intake

Ordinarily, a person obtains sufficient electrolytes by responding to hunger and thirst. However, a severe electrolyte deficiency may cause *salt craving*, which is a strong desire to eat salty foods.

Electrolyte Output

The body loses some electrolytes by perspiring (sweat has about half the solute concentration of plasma). The quantities of electrolytes leaving vary with the amount of perspiration. More electrolytes are lost in sweat on warmer days and during strenuous exercise. Varying amounts of electrolytes are lost in the feces. The greatest electrolyte output occurs as a result of kidney function and urine production. The kidneys alter renal electrolyte losses to maintain the proper composition of body fluids.

Recall from chapter 2 (p. 44) that water molecules are polar, and molecules that have polar regions within them (such as carbohydrates and proteins) dissolve in water but remain intact, whereas molecules that are held together by ionic bonds (such as the electrolytes) dissociate in water to release ions.

The total solute concentration of a body fluid determines its *osmolarity*. One molecule of glucose yields one dissolved particle, and one molecule of sodium chloride yields two, a sodium ion and a chloride ion. Since the osmolarity of body solutions is determined by the total number of dissolved particles, irrespective of the source, the term *osmoles* is used. Thus, one mole of glucose yields one osmole of dissolved particles, and one mole of sodium chloride yields two osmoles. The total number of osmoles per liter gives the osmolarity of the solution.

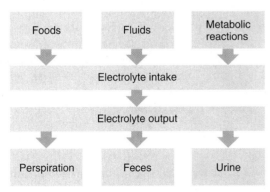

FIGURE 21.6
Electrolyte balance exists when the intake of electrolytes from all sources equals the output of electrolytes.

1. Which electrolytes are most important to cellular functions?

2. Which mechanisms ordinarily regulate electrolyte intake?

3. By what routes does the body lose electrolytes?

WATER BALANCE DISORDERS

Among the more common disorders involving an imbalance in the water of body fluids are dehydration, water intoxication, and edema.

Dehydration

In 1994, thousands of starving people died in the African nation of Rwanda. It wasn't lack of food that killed most of these people, but cholera, a bacterial infection that cripples the ability of intestinal mucosal cells to reabsorb water. The severe diarrhea that develops can kill in days, sometimes even hours. Dehydration is deadly.

Dehydration is a deficiency condition that occurs when output of water exceeds intake. It is a great problem for athletes, military personnel, and certain industrial workers. This condition may develop following excessive sweating or as a result of prolonged water deprivation accompanied by continued water output. In either case, as water is lost, the extracellular fluid becomes increasingly more concentrated, and water tends to leave cells by osmosis (fig. 21A). Dehydration may also accompany illnesses in which prolonged vomiting or diarrhea depletes body fluids.

During dehydration, the skin and mucous membranes of the mouth feel dry, and body weight drops. Severe hyperthermia may develop as the body temperature regulating mechanism falters due to lack of water for sweat. In severe dehydration, as waste products accumulate in the extracellular fluid, symptoms of cerebral disturbances, including mental confusion, delirium, and coma, may develop.

Because the kidneys of infants are less able to conserve water than are those of adults, infants are more likely to become dehydrated. Elderly people are also especially susceptible to developing water imbalances because the sensitivity of their thirst mechanisms decreases with age, and physical disabilities may make it difficult for them to obtain adequate fluids.

The treatment for dehydration is to replace the lost water and electrolytes. If only water is replaced, the extracellular fluid will become more dilute than normal. This may produce a condition called water intoxication.

Water Intoxication

Babies rushed to emergency rooms because they are having seizures sometimes are suffering from drinking too much water, a rare condition called *water intoxication.* This can occur when a baby under six months of age is given several bottles of water a day or very dilute infant formula. The hungry infant gobbles down the water, and soon its tissues swell with the excess fluid. When the serum sodium level drops, the eyes begin to flutter, and a seizure occurs. As extracellular fluid becomes hypotonic, water enters the cells rapidly by osmosis (fig. 21B). Coma resulting from swelling brain tissues may follow unless water intake is restricted and hypertonic salt solutions are given to draw water back into the extracellular fluid. Usually, recovery is complete within a few days.

Drinking too much water occurs in other ways. A medical journal case report describes a woman who drank two extra quarts of water a day while on a very low calorie diet, under the advice of a diet counselor. When the dieter was hospitalized after suffering severe lethargy, nausea, and weakness, tests revealed the extremely low levels of sodium in her body fluids, and she was diagnosed with "crash diet potomania." A similar situation is seen among people who consume too much beer.

Edema

Edema is an abnormal accumulation of extracellular fluid within the interstitial spaces. A variety of factors can cause it, including decrease in the plasma protein concentration (hypoproteinemia), obstructions in lymphatic vessels, increased venous pressure, and increased capillary permeability.

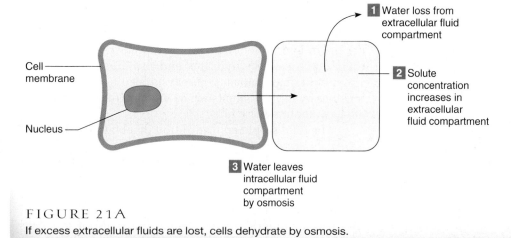

1 Water loss from extracellular fluid compartment

2 Solute concentration increases in extracellular fluid compartment

3 Water leaves intracellular fluid compartment by osmosis

Cell membrane

Nucleus

FIGURE 21A
If excess extracellular fluids are lost, cells dehydrate by osmosis.

Regulation of Electrolyte Output

The concentrations of positively charged ions, such as sodium (Na^+), potassium (K^+), and calcium (Ca^{+2}), are particularly important. For example, certain concentrations of these ions are vital for nerve impulse conduction, muscle fiber contraction, and maintenance of cell membrane permeability. Potassium is especially important in maintaining the resting potential of nerve and cardiac muscle cells, and abnormal potassium levels may cause these cells to function abnormally.

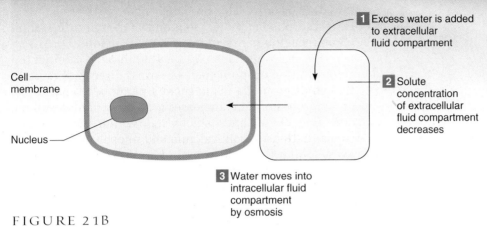

1 Excess water is added to extracellular fluid compartment

2 Solute concentration of extracellular fluid compartment decreases

3 Water moves into intracellular fluid compartment by osmosis

Cell membrane

Nucleus

FIGURE 21B

If excess water is added to the extracellular fluid compartment, cells gain water by osmosis.

Hypoproteinemia may result from failure of the liver to synthesize plasma proteins; kidney disease (glomerulonephritis) that damages glomerular capillaries, allowing proteins to enter the urine; or starvation, in which amino acid intake is insufficient to support synthesis of plasma proteins. In each of these instances, the plasma protein concentration is decreased, which decreases plasma osmotic pressure, reducing the normal return of tissue fluid to the venular ends of capillaries. Consequently, tissue fluid accumulates in the interstitial spaces.

As discussed in chapter 16 (p. 612), *lymphatic obstructions* may result from surgery or from parasitic infections of lymphatic vessels. Back pressure develops in the lymphatic vessels, which interferes with the normal movement of tissue fluid into them. At the same time, proteins that the lymphatic circulation ordinarily removes accumulate in the interstitial spaces, raising osmotic pressure of the interstitial fluid. This effect attracts still more fluid into the interstitial spaces.

If the outflow of blood from the liver into the inferior vena cava is blocked, the venous pressure within the liver and portal blood vessels increases greatly. This, in turn, raises pressure in liver sinusoids and intestinal capillaries. As a result, fluid with a high concentration of protein is exuded from the surfaces of the liver and intestine into the peritoneal cavity. This elevates the osmotic pressure of the abdominal fluid, which, in turn, attracts more water into the peritoneal cavity by osmosis. This condition, called *ascites,* distends the abdomen. It is quite painful.

Edema may also result from increased capillary permeability accompanying *inflammation*. Recall that inflammation is a response to tissue damage and usually releases chemicals such as histamine from damaged cells. Histamine causes vasodilation and increased capillary permeability, so excess fluid leaks out of the capillary and enters the interstitial spaces. Table 21A summarizes the factors that result in edema. ■

TABLE 21A	Factors Associated with Edema	
Factor	**Cause**	**Effect**
Low plasma protein concentration	Liver disease and failure to synthesize proteins; kidney disease and loss of proteins in urine; lack of proteins in diet due to starvation	Plasma osmotic pressure decreases; less fluid enters venular ends of capillaries by osmosis
Obstruction of lymph vessels	Surgical removal of portions of lymphatic pathways; certain parasitic infections	Back pressure in lymph vessels interferes with movement of fluid from interstitial spaces into lymph capillaries
Increased venous pressure	Venous obstructions or faulty venous valves	Back pressure in veins increases capillary filtration and interferes with return of fluid from interstitial spaces into venular ends of capillaries
Inflammation	Tissue damage	Capillaries become abnormally permeable; fluid leaks from plasma into interstitial spaces

Sodium ions account for nearly 90% of the positively charged ions in extracellular fluids. The kidneys and the hormone aldosterone provide the primary mechanism regulating these ions. Aldosterone, which the adrenal cortex secretes, increases sodium ion reabsorption in the distal convoluted tubules and collecting ducts of the nephrons. A decrease in sodium ion concentration in the extracellular fluid stimulates aldosterone secretion via the renin-angiotensin system, as described in chapter 20 (p. 786 and fig. 20.19).

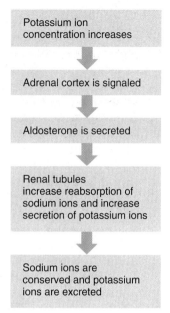

FIGURE 21.7

If the potassium ion concentration increases, the kidneys conserve sodium ions and excrete potassium ions.

Aldosterone also regulates *potassium ions.* An important stimulus for aldosterone secretion is a rising potassium ion concentration, which directly stimulates cells of the adrenal cortex. This hormone enhances the renal tubular reabsorption of sodium ions and, at the same time, stimulates renal tubular secretion of potassium ions (fig. 21.7).

Recall from chapter 13 (p. 489) that the calcium ion concentration dropping below normal directly stimulates the parathyroid glands to secrete parathyroid hormone. Parathyroid hormone increases activity in bone-resorbing cells (osteocytes and osteoclasts), which increases the concentrations of both calcium and phosphate ions in the extracellular fluids. Parathyroid hormone also indirectly stimulates calcium absorption from the intestine. Concurrently, this hormone causes the kidneys to conserve calcium ions (through increased tubular reabsorption) and increases the urinary excretion of phosphate ions. The increased phosphate excretion offsets the increased plasma phosphate. Thus, the net effect of the hormone is to return the *calcium ion* concentration of the extracellular fluid to normal levels but to maintain a normal *phosphate ion* concentration (fig. 21.8).

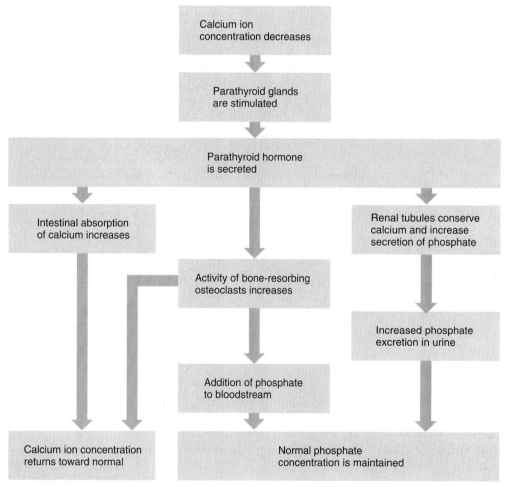

FIGURE 21.8

If the concentration of calcium ions decreases, parathyroid hormone increases calcium ion concentration. Increased urinary phosphate excretion offsets bone resorption (which increases blood phosphate levels) to maintain a normal concentration of phosphate ions.

Abnormal increases in blood calcium (hypercalcemia) sometimes result from hyperparathyroidism, in which excess secretion of PTH increases bone resorption. Hypercalcemia may also be caused by cancers, particularly those originating in the bone marrow, breasts, lungs, or prostate gland. Usually the increase in calcium occurs when cancer causes bone tissue to release ions. In other cases, however, the blood calcium concentration increases when cancer cells produce biochemicals that have physiological effects similar to parathyroid hormone. This most often occurs in lung cancer. Symptoms of cancer-induced hypercalcemia include weakness and fatigue, impaired mental function, headache, nausea, increased urine volume (polyuria), and increased thirst (polydipsia).

Abnormal decreases in blood calcium (hypocalcemia) may result from reduced availability of PTH following removal of the parathyroid glands, or from vitamin D deficiency, which may result from decreased absorption following gastrointestinal surgery or excess excretion due to kidney disease. Hypocalcemia may be life threatening because it may produce muscle spasms within the airways and cardiac arrhythmias. Administering calcium salts and high doses of vitamin D to promote calcium absorption can correct this condition.

Generally, the regulatory mechanisms that control positively charged ions secondarily control the concentrations of negatively charged ions. For example, chloride ions (Cl⁻), the most abundant negatively charged ions in the extracellular fluids, are passively reabsorbed from the renal tubules in response to the active reabsorption of sodium ions. That is, the negatively charged chloride ions are electrically attracted to the positively charged sodium ions and accompany them as they are reabsorbed.

Some negatively charged ions, such as phosphate ions (PO_4^{-3}) and sulfate ions (SO_4^{-2}), also are partially regulated by active transport mechanisms that have limited transport capacities. Thus, if the extracellular phosphate ion concentration is low, the phosphate ions in the renal tubules are conserved. On the other hand, if the renal plasma threshold is exceeded, the excess phosphate will be excreted in the urine. Clinical Application 21.2 discusses symptoms associated with sodium and potassium imbalances.

1 How does aldosterone regulate sodium and potassium ion concentration?

2 How is calcium regulated?

3 What mechanism regulates the concentrations of most negatively charged ions?

Acid-Base Balance

As discussed in chapter 2 (p. 46), electrolytes that ionize in water and release hydrogen ions are **acids.** Substances that combine with hydrogen ions are **bases.** Acid-base balance entails regulation of the hydrogen ion concentration of body fluids. Regulation of hydrogen ions is very important because slight changes in hydrogen ion concentrations can alter the rates of enzyme-controlled metabolic reactions, shift the distribution of other ions, or modify hormone actions. Recall that the internal environment is normally maintained between pH 7.35 and 7.45.

Sources of Hydrogen Ions

Most of the hydrogen ions in body fluids originate as by-products of metabolic processes, although the digestive tract may directly absorb small quantities. The major metabolic sources of hydrogen ions include the following. (All of these are reversible reactions but, for clarity, are presented as the net reaction only. Remember, it is the concentration of H⁺ at equilibrium that determines the pH.)

1. **Aerobic respiration of glucose.** This process produces carbon dioxide and water. Carbon dioxide diffuses out of the cells and reacts with water in the extracellular fluids to form *carbonic acid:*

$$CO_2 + H_2O \rightarrow H_2CO_3$$

The resulting carbonic acid then ionizes to release hydrogen ions and bicarbonate ions:

$$H_2CO_3 \rightarrow H^+ + HCO_3^-$$

2. **Anaerobic respiration of glucose.** Glucose metabolized anaerobically produces *lactic acid,* which adds hydrogen ions to body fluids.

3. **Incomplete oxidation of fatty acids.** The incomplete oxidation of fatty acids produces *acidic ketone bodies,* which increase hydrogen ion concentration.

4. **Oxidation of amino acids containing sulfur.** The oxidation of sulfur-containing amino acids yields *sulfuric acid* (H_2SO_4), which ionizes to release hydrogen ions.

5. **Breakdown (hydrolysis) of phosphoproteins and nucleic acids.** Phosphoproteins and nucleic acids contain phosphorus. Their oxidation produces *phosphoric acid* (H_3PO_4), which ionizes to release hydrogen ions.

The acids resulting from metabolism vary in strength. Thus, their effects on the hydrogen ion concentration of body fluids vary (fig. 21.9).

SODIUM AND POTASSIUM IMBALANCES

Extracellular fluids usually have high sodium ion concentrations, and intracellular fluid usually has high potassium ion concentration. The renal regulation of sodium is closely related to that of potassium because active reabsorption of sodium (under the influence of aldosterone) is accompanied by secretion (and excretion) of potassium. Thus, it is not surprising that conditions that alter sodium ion balance also affect potassium ion balance.

Such disorders can be summarized as follows:

1. *Low sodium concentration (hyponatremia).* Possible causes of hyponatremia include prolonged sweating, vomiting, or diarrhea; renal disease in which sodium is inadequately reabsorbed; adrenal cortex disorders in which aldosterone secretion is insufficient to promote the reabsorption of sodium (Addison disease); and drinking too much water.

    Possible effects of hyponatremia include the development of extracellular fluid that is hypotonic and promotes the movement of water into the cells by osmosis. This is accompanied by the symptoms of water intoxication described in Clinical Application 21.1.

2. *High sodium concentration (hypernatremia).* Possible causes of hypernatremia include excessive water loss by evaporation and diffusion, as may occur during high fever, or increased water loss accompanying diabetes insipidus, in one form of which ADH secretion is insufficient to maintain water conservation by the renal tubules and collecting ducts. Possible effects of hypernatremia include disturbances of the central nervous system, such as confusion, stupor, and coma.

3. *Low potassium concentration (hypokalemia).* Possible causes of hypokalemia include excessive release of aldosterone by the adrenal cortex (Cushing syndrome), which increases renal excretion of potassium; use of diuretic drugs that promote potassium excretion; kidney disease; and prolonged vomiting or diarrhea. Possible effects of hypokalemia include muscular weakness or paralysis, respiratory difficulty, and severe cardiac disturbances, such as atrial or ventricular arrhythmias.

4. *High potassium concentration (hyperkalemia).* Possible causes of hyperkalemia include renal disease, which decreases potassium excretion; use of drugs that promote renal conservation of potassium; insufficient secretion of aldosterone by the adrenal cortex (Addison disease); or a shift of potassium from the intracellular fluid to the extracellular fluid, a change that accompanies an increase in plasma hydrogen ion concentration (acidosis). Possible effects of hyperkalemia include paralysis of the skeletal muscles and severe cardiac disturbances, such as cardiac arrest. ■

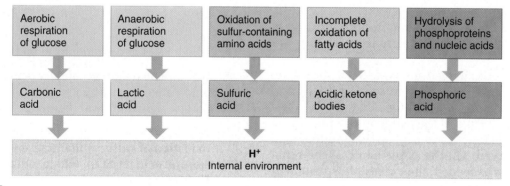

FIGURE 21.9
Some of the metabolic processes that provide hydrogen ions.

Strengths of Acids and Bases

Acids that ionize more completely (release more H⁺) are strong acids, and those that ionize less completely are weak acids. For example, the hydrochloric acid (HCl) of gastric juice is a strong acid and dissociates completely to release a lot of H⁺, but the carbonic acid (H_2CO_3) produced when carbon dioxide reacts with water is weak and dissociates less completely to release less H⁺.

Bases release ions, such as hydroxide ions (OH⁻), which can combine with hydrogen ions, thereby lowering their concentration. Thus, sodium hydroxide (NaOH), which releases hydroxide ions, and sodium bicarbonate (NaHCO$_3$), which releases bicarbonate ions (HCO$_3^-$), are bases. Strong bases dissociate to release more OH$^-$ or its equivalent than do weak bases. Often, the negative ions themselves are called bases. For example, HCO$_3^-$ acting as a base combines with H$^+$ from the strong acid HCl to form the weak acid carbonic acid (H$_2$CO$_3$).

Regulation of Hydrogen Ion Concentration

Either an acid shift or an alkaline (basic) shift in the body fluids could threaten the internal environment. However, normal metabolic reactions generally produce more acid than base. These reactions include cellular metabolism of glucose, fatty acids, and amino acids. Consequently, the maintenance of acid-base balance usually entails elimination of acid. This is accomplished in three ways:

1. Acid-base buffer systems

2. Respiratory excretion of carbon dioxide

3. Renal excretion of hydrogen ions

1 Explain why the regulation of hydrogen ion concentration is so important.

2 What are the major sources of hydrogen ions in the body?

Acid-Base Buffer Systems

Acid-base buffer systems occur in all of the body fluids and involve chemicals that combine with acids or bases when they are in excess. Buffers are substances that stabilize the pH of a solution, despite the addition of an acid or a base. More specifically, the chemical components of a buffer system can combine with strong acids to convert them into weak acids. Likewise, these buffers can combine with strong bases to convert them into weak bases. Such activity helps minimize pH changes in the body fluids. The three most important buffer systems in body fluids are the bicarbonate buffer system, the phosphate buffer system, and the protein buffer system.

In the following discussion, associated anions and cations have been omitted for clarity. For example, the weak base sodium bicarbonate (NaHCO$_3$) is represented by bicarbonate (HCO$_3^-$). Sodium is also the cation associated with the phosphate ions.

1. **Bicarbonate buffer system.** In the bicarbonate buffer system, which is present in both intracellular and extracellular fluids, the bicarbonate ion (HCO$_3^-$) acts as a weak base, and carbonic acid (H$_2$CO$_3$) acts as a weak acid. In the presence of excess hydrogen ions, bicarbonate ions combine with hydrogen ions to form carbonic acid, thus minimizing any increase in the hydrogen ion concentration of body fluids:

$$H^+ + HCO_3^- \rightarrow H_2CO_3$$

On the other hand, if conditions are basic or alkaline, carbonic acid dissociates to release bicarbonate ion and hydrogen ion:

$$H_2CO_3 \rightarrow H^+ + HCO_3^-$$

It is important to remember that even though this reaction releases bicarbonate ion, it is the increase of free hydrogen ions at equilibrium that is important in minimizing the shift toward a more alkaline pH.

2. **Phosphate buffer system.** The phosphate buffer system is also present in both intracellular and extracellular fluids. However, it is particularly important in the control of hydrogen ion concentration in the intracellular fluid and in renal tubular fluid and urine. This buffer system consists of two phosphate ions, dihydrogen phosphate (H$_2$PO$_4^-$) and monohydrogen phosphate (HPO$_4^{2-}$).

In the presence of excess hydrogen ions, monohydrogen phosphate ions act as a weak base, combining with hydrogen ions to form dihydrogen phosphate, thus minimizing any increase in the hydrogen ion concentration of the body fluids.

$$H^+ + HPO_4^{-2} \rightarrow H_2PO_4^-$$

On the other hand, if conditions are basic or alkaline, dihydrogen phosphate, acting as a weak acid, dissociates to release hydrogen ion:

$$H_2PO_4^- \rightarrow H^+ + HPO_4^{-2}$$

3. **Protein buffer system.** The protein acid-base buffer system consists of the plasma proteins, such as albumins, and certain proteins within the cells, including the hemoglobin of red blood cells.

As described in chapter 2 (p. 52), proteins are chains of amino acids. Some of these amino acids have freely exposed groups of atoms, called carboxyl groups. If the H$^+$ concentration drops, a carboxyl group (—COOH) can become ionized, releasing a hydrogen ion, thus resisting the pH change:

$$-COOH \leftrightarrow -COO^- + H^+$$

Notice that this is a reversible reaction. In the presence of excess hydrogen ions, the —COO$^-$ portions of the protein molecules accept hydrogen ions and become —COOH groups again. This action decreases the number of free hydrogen ions in the body fluids and again minimizes the pH change.

Some of the amino acids within a protein molecule also contain freely exposed amino groups (—NH$_2$). If the H$^+$ concentration rises, these amino groups can accept hydrogen ions in another reversible reaction:

$$—NH_2 + H^+ \leftrightarrow —NH_3^+$$

In the presence of excess hydroxyl ions (OH^-), the $—NH_3^+$ groups within protein molecules give up hydrogen ions and become $—NH_2$ groups again. These hydrogen ions then combine with hydroxyl ions to form water molecules. Once again, pH change is minimized. Thus, protein molecules can function as acids by releasing hydrogen ions under alkaline conditions or as bases by accepting hydrogen ions under acid conditions. This special property allows protein molecules to operate as an acid-base buffer system.

Hemoglobin is an especially important protein that buffers hydrogen ions. As explained in chapter 19 (p. 762), carbon dioxide, produced by cellular oxidation of glucose, diffuses through the capillary wall and enters the plasma and then the red blood cells. The red cells contain an enzyme, *carbonic anhydrase,* that speeds the reaction between carbon dioxide and water, producing carbonic acid:

$$CO_2 + H_2O \rightarrow H_2CO_3$$

The carbonic acid quickly dissociates, releasing hydrogen ions and bicarbonate ions:

$$H_2CO_3 \rightarrow H^+ + HCO_3^-$$

In the peripheral tissues, where CO_2 is generated, oxygen is used in the metabolism of glucose. As a result, hemoglobin has given up much of its oxygen and is in the form of deoxyhemoglobin. In this form, hemoglobin can bind hydrogen ions, thus acting as a buffer to minimize the pH change that would otherwise occur.

The above two reactions can be written as a single reversible reaction:

$$CO_2 + H_2O \rightarrow H_2CO_3 \leftrightarrow H^+ + HCO_3^-$$

Thus, in the peripheral tissues, where CO_2 levels are high, the reaction equilibrium shifts to the right, generating H^+, which is buffered by hemoglobin, and HCO_3^-, which becomes a plasma electrolyte. In the lungs, where oxygen levels are high, hemoglobin is no longer a good buffer, and it releases its H^+. However, the released H^+ combines with plasma HCO_3, shifting the reaction equilibrium to the left, generating carbonic acid, which quickly dissociates to form CO_2 and water. The water is added to the body fluids, and the CO_2 is exhaled. Because of this relationship to CO_2, carbonic acid is sometimes called a *volatile acid.*

Individual amino acids in body fluids can also function as acid-base buffers by accepting or releasing hydrogen ions. This is possible because every amino acid has an amino group ($—NH_2$) and a carboxyl group ($—COOH$).

To summarize, acid-base buffer systems take up hydrogen ions when body fluids are becoming more acidic and give up hydrogen ions when the fluids are becoming more basic (alkaline). Buffer systems convert stronger acids into weaker acids or convert stronger bases into weaker bases, as table 21.3 summarizes.

In addition to minimizing pH fluctuations, acid-base buffer systems in body fluids buffer each other. Consequently, whenever the hydrogen ion concentration begins to change, the chemical balances within all of the buffer systems change too, resisting the drift in pH.

Neurons are particularly sensitive to changes in the pH of body fluids. For example, if the interstitial fluid becomes more alkaline than normal (alkalosis), neurons become more excitable, and seizures may result. Conversely, acidic conditions (acidosis) depress neuron activity, and level of consciousness may decrease.

1 What is the difference between a strong acid or base and a weak acid or base?

2 How does a chemical buffer system help regulate pH of body fluids?

3 List the major buffer systems of the body.

Chemical buffer systems only temporarily solve the problem of acid-base balance. Ultimately, the body must

TABLE 21.3	Chemical Acid-Base Buffer Systems	
Buffer System	**Constituents**	**Actions**
Bicarbonate system	Bicarbonate ion (HCO_3^-)	Converts a strong acid into a weak acid
	Carbonic acid (H_2CO_3)	Converts a strong base into a weak base
Phosphate system	Monohydrogen phosphate ion (HPO_4^{-2})	Converts a strong acid into a weak acid
	Dihydrogen phosphate ($H_2PO_4^-$)	Converts a strong base into a weak base
Protein system (and amino acids)	$—NH_3^+$ group of an amino acid or protein	Releases hydrogen ions in the presence of excess base
	$—COO^-$ group of an amino acid or protein	Accepts hydrogen ions in the presence of excess acid

eliminate excess acid or base. The lungs (controlled by the respiratory center) and the kidneys accomplish this task.

Respiratory Excretion of Carbon Dioxide

The **respiratory center** in the brainstem helps regulate hydrogen ion concentrations in the body fluids by controlling the rate and depth of breathing. Specifically, if body cells increase their production of carbon dioxide, as occurs during periods of physical exercise, carbonic acid production increases. As the carbonic acid dissociates, the concentration of hydrogen ions increases, and the pH of the internal environment drops (see chapter 19, p. 754). Such an increasing concentration of carbon dioxide in the central nervous system and the subsequent increase in hydrogen ion concentration in the cerebrospinal fluid stimulate chemosensitive areas within the respiratory center.

In response, the respiratory center increases the depth and rate of breathing so that the lungs excrete more carbon dioxide. As a result, hydrogen ion concentration in body fluids returns toward normal, because the released carbon dioxide is in equilibrium with carbonic acid and CO_2 (fig. 21.10):

$$CO_2 + H_2O \leftrightarrow H_2CO_3 \leftrightarrow H^+ + HCO_3^-$$

Conversely, if body cells are less active, concentrations of carbon dioxide and hydrogen ions in body fluids remain relatively low. As a result, breathing rate and depth fall. This increases the carbon dioxide level in the body fluids, returning pH to normal. If the pH drops below normal, the respiratory center is stimulated to increase the rate and depth of breathing.

Thus, the activity of the respiratory center changes in response to shifts in the pH of the body fluids, reducing these shifts to a minimum. Because most of the hydrogen ions in the body fluids originate from carbonic acid produced when carbon dioxide reacts with water, the respiratory regulation of hydrogen ion concentration is important.

Renal Excretion of Hydrogen Ions

Nephrons help regulate the hydrogen ion concentration of the body fluids by excreting hydrogen ions in the urine. Recall from chapter 20 (p. 790) that the epithelial cells lining the proximal and distal convoluted tubules and the collecting ducts secrete these ions into the tubular fluid. The tubular secretion of hydrogen ions is linked to tubular reabsorption of bicarbonate ions. In this way, the kidneys also regulate the concentration of bicarbonate ions in body fluids. These mechanisms also help balance the quantities of sulfuric acid, phosphoric acid, and various organic acids that appear in body fluids as by-products of metabolic processes.

The metabolism of certain amino acids, for example, produces sulfuric and phosphoric acids. Consequently, a diet high in proteins may trigger excess acid formation. The kidneys compensate for such gains in acids by altering the rate of hydrogen ion secretion, thus resisting a shift in the pH of body fluids (fig. 21.11).

Once hydrogen ions are secreted, phosphates that were filtered into the fluid of the renal tubule buffer them. Ammonia aids in this buffering action.

Through deamination of certain amino acids, the cells of the renal tubules produce ammonia (NH_3), which diffuses readily through cell membranes and enters the renal tubule. When increase in the hydrogen ion concentration of body fluids is prolonged, the renal tubules increase ammonia production. Because ammonia is a weak base, it can accept hydrogen ions to form *ammonium ions* (NH_4^+):

$$H^+ + NH_3 \rightarrow NH_4^+$$

Cell membranes are quite impermeable to ammonium ions, which are trapped in the renal tubule as they form and are excreted with the urine. This mechanism helps to transport excess hydrogen ions to the outside and helps prevent the urine from becoming too acidic.

The various regulators of hydrogen ion concentration operate at different rates. Acid-base buffers function rapidly and can convert strong acids or bases into weak acids or bases almost immediately. For this reason, these chemical buffer systems are sometimes called the body's *first line of defense* against shifts in pH.

Physiological buffer systems, such as the respiratory and renal mechanisms, function more slowly and constitute *second line of defense*. The respiratory mechanism may require several minutes to begin resisting a change in pH, and the renal mechanism may require one to three days to regulate a changing hydrogen ion concentration (fig. 21.12). Clinical Application 21.3 examines the effects of acid-base imbalances.

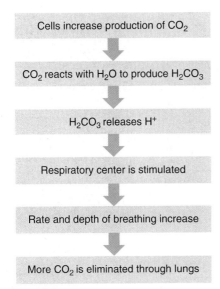

FIGURE 21.10

An increase in carbon dioxide elimination follows an increase in carbon dioxide production.

CHAPTER TWENTY-ONE *Water, Electrolyte, and Acid-Base Balance*

ACID-BASE IMBALANCES

Ordinarily, chemical and physiological buffer systems maintain the hydrogen ion concentration of body fluids within very narrow pH ranges. Abnormal conditions may disturb the acid-base balance. For example, the pH of arterial blood is normally 7.35–7.45. A value below 7.35 produces *acidosis* (as″i-do′sis). A pH above 7.45 produces *alkalosis* (al″kah-lo′sis). Such shifts in the pH of body fluids may be life threatening. In fact, a person usually cannot survive if the pH drops to 6.8 or rises to 8.0 for more than a few hours (fig. 21C).

Acidosis results from an accumulation of acids or a loss of bases, both of which cause abnormal increases in the hydrogen ion concentrations of body fluids. Conversely, alkalosis results from a loss of acids or an accumulation of bases accompanied by a decrease in hydrogen ion concentrations (fig. 21D).

The two major types of acidosis are *respiratory acidosis* and *metabolic acidosis*.

Factors that increase carbon dioxide levels, also increasing the concentration of carbonic acid (the respiratory acid), cause respiratory acidosis. Metabolic acidosis is due to an abnormal accumulation of any other acids in the body fluids or to a loss of bases, including bicarbonate ions. Similarly, the two major types of alkalosis are *respiratory alkalosis* and *metabolic alkalosis*. Excessive loss of carbon dioxide and consequent loss of carbonic acid cause respiratory alkalosis. Metabolic alkalosis is due to excessive loss of hydrogen ions or gain of bases.

Since in respiratory acidosis carbon dioxide accumulates, this can result from factors that hinder pulmonary ventilation (fig. 21E). These include the following:

1. Injury to the respiratory center of the brainstem, decreasing rate and depth of breathing.
2. Obstructions in air passages that interfere with air movement into the alveoli.
3. Diseases that decrease gas exchanges, such as pneumonia, or those that reduce surface area of the respiratory membrane, such as emphysema.

Any of these conditions can increase the level of carbonic acid and hydrogen ions in body fluids, lowering pH. Chemical buffers, such as hemoglobin, may resist this shift in pH. At the same time, increasing levels of carbon dioxide and hydrogen ions stimulate the respiratory center, increasing breathing rate and depth and thereby lowering carbon dioxide levels. Also, the kidneys may begin to excrete more hydrogen ions.

Eventually, thanks to these chemical and physiological buffers, the pH of the body fluids may return to normal. When this happens, the acidosis is said to be *compensated*.

The symptoms of respiratory acidosis result from depression of central nervous system function and include drowsiness, disorientation, and stupor. Evidence of respiratory insufficiency, such as labored breathing and cyanosis, is usually also evident. In *uncompensated acidosis,* the person may become comatose and die.

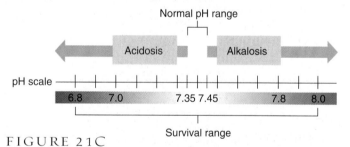

pH of arterial blood

FIGURE 21C

If the pH of arterial blood drops to 6.8 or rises to 8.0 for more than a few hours, the person usually cannot survive.

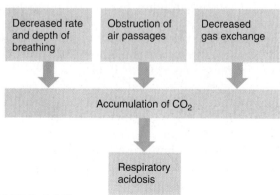

FIGURE 21E

Some of the factors that lead to respiratory acidosis.

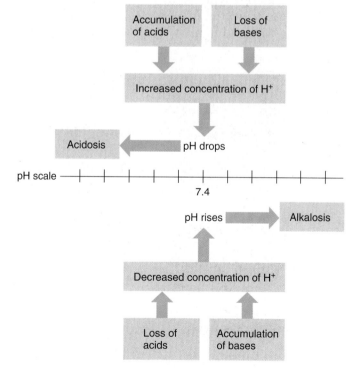

FIGURE 21D

Acidosis results from accumulation of acids or loss of bases. Alkalosis results from loss of acids or accumulation of bases.

Metabolic acidosis is due to either accumulation of nonrespiratory acids or loss of bases (fig. 21F). Factors that may lead to this condition include

1. Kidney disease that reduces glomerular filtration and fails to excrete the acids produced in metabolism (uremic acidosis).
2. Prolonged vomiting that loses the alkaline contents of the upper intestine and stomach contents. (Losing only the stomach contents produces metabolic alkalosis.)
3. Prolonged diarrhea, in which excess alkaline intestinal secretions are lost (especially in infants).
4. Diabetes mellitus, in which some fatty acids react to produce ketone bodies, such as *acetoacetic acid, beta-hydroxybutyric acid,* and *acetone.* Normally, these molecules are scarce, and cells oxidize them as energy sources. However, if fats are being utilized at an abnormally high rate, as may occur in diabetes mellitus, ketone bodies may accumulate faster than they can be oxidized, and spill over into the urine (ketonuria); in addition, the lungs may release acetone, which is volatile and imparts a fruity odor to the breath.

More seriously, the accumulation of acetoacetic acid and beta-hydroxybutyric acid may lower pH (ketonemic acidosis).

These acids may also combine with bicarbonate ions in the urine. As a result, excess bicarbonate ions are excreted, interfering with the function of the bicarbonate acid-base buffer system.

Whatever the cause, metabolic acidosis shifts pH downward. However, the following factors resist this shift: chemical buffer systems, which accept excess hydrogen ions; the respiratory center, which increases breathing rate and depth; and the kidneys, which excrete more hydrogen ions.

Respiratory alkalosis develops as a result of *hyperventilation,* described in chapter 19 (p. 756). Hyperventilation is accompanied by too great a loss of carbon dioxide and consequent decreases in carbonic acid and hydrogen ion concentrations (fig. 21G).

Hyperventilation may occur during periods of anxiety, although it may also accompany fever or poisoning from salicylates, such as aspirin. At high altitudes, hyperventilation may be a response to low oxygen pressure. Also, musicians, such as bass tuba players, who must provide a large volume of air when playing sustained passages, sometimes hyperventilate. In each case, rapid, deep breathing depletes carbon dioxide, and the pH of body fluids increases.

Chemical buffers, such as hemoglobin, that release hydrogen ions resist this pH change. Also, the lower levels of carbon dioxide and hydrogen ions stimulate the respiratory center to a lesser degree. This inhibits hyperventilation, thus reducing further carbon dioxide loss. At the same time, the kidneys decrease their secretion of hydrogen ions, and the urine becomes alkaline as bases are excreted.

The symptoms of respiratory alkalosis include lightheadedness, agitation, dizziness, and tingling sensations. In severe cases, impulses may be triggered spontaneously on peripheral nerves, and muscles may respond with tetanic contractions (see chapter 9, p. 291).

Metabolic alkalosis results from a great loss of hydrogen ions or from a gain in bases, both of which are accompanied by a rise in the pH of the blood (alkalemia) (fig. 21H). This condition may occur following gastric drainage (lavage), prolonged vomiting in which only the stomach contents are lost, or the use of certain diuretic drugs. Because gastric juice is very acidic, its loss leaves the body fluids with a net increase of basic substances and a pH shift toward alkaline values. Metabolic alkalosis may also develop as a result of ingesting too much antacid, such as sodium bicarbonate, to relieve the symptoms of indigestion. The symptoms of metabolic alkalosis include a decrease in the breathing rate and depth, which, in turn, results in an increased concentration of carbon dioxide in the blood. ■

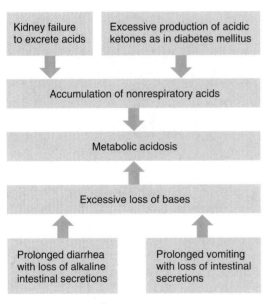

FIGURE 21F
Some of the factors that lead to metabolic acidosis.

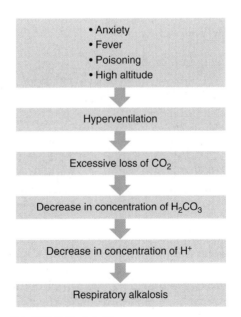

FIGURE 21G
Some of the factors that lead to respiratory alkalosis.

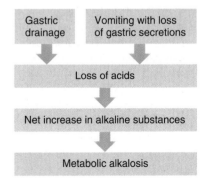

FIGURE 21H
Some of the factors that lead to metabolic alkalosis.

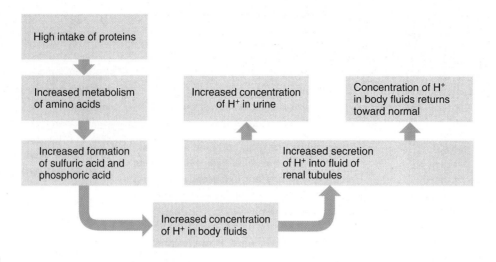

FIGURE 21.11

If the concentration of hydrogen ions in body fluids increases, the renal tubules increase their secretion of hydrogen ions into the urine.

1. How does the respiratory system help regulate acid-base balance?

2. How do the kidneys respond to excess hydrogen ions?

3. How do the rates at which chemical and physiological buffer systems act differ?

Clinical Terms Related to Water, Electrolyte, and Acid-Base Balance

acetonemia (as″e-to-ne′me-ah) Abnormal amounts of acetone in the blood.

acetonuria (as″e-to-nu′re-ah) Abnormal amounts of acetone in the urine.

albuminuria (al-bu′mĭ-nu′re-ah) Albumin in the urine.

anasarca (an″ah-sar′kah) Widespread accumulation of tissue fluid.

antacid (ant-as′id) Substance that neutralizes an acid.

anuria (ah-nu′re-ah) Absence of urine excretion.

azotemia (az″o-te′me-ah) Accumulation of nitrogenous wastes in the blood.

diuresis (di″u-re′sis) Increased production of urine.

glucosuria (glu″ko-su′re-ah) Excess sugar in the urine.

hyperglycemia (hi″per-gli-se′me-ah) Abnormally high blood sugar level.

hyperkalemia (hi″per-kah-le′me-ah) Excess potassium in the blood.

hypernatremia (hi″per-na-tre′me-ah) Excess sodium in the blood.

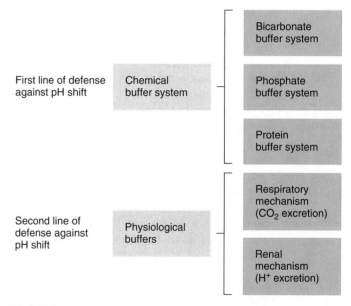

FIGURE 21.12

Chemical buffers act rapidly, while physiological buffers may require several minutes to several days to begin resisting a change in pH.

hyperuricemia (hi″per-u″rĭ-se′me-ah) Excess uric acid in the blood.

hypoglycemia (hi″po-gli-se′me-ah) Abnormally low blood sugar level.

ketonuria (ke″to-nu′re-ah) Ketone bodies in the urine.

ketosis (ke″to′sis) Acidosis due to excess ketone bodies in the body fluids.

proteinuria (pro″te-ĭ-nu′-e-ah) Protein in the urine.

uremia (u-re′me-ah) Toxic condition resulting from nitrogenous wastes in the blood.

CHAPTER SUMMARY

Introduction (page 808)

The maintenance of water and electrolyte balance requires that the quantities of these substances entering the body equal the quantities leaving it. Altering the water balance necessarily affects the electrolyte balance.

Distribution of Body Fluids (page 808)

1. Fluid compartments
   a. The intracellular fluid compartment includes the fluids and electrolytes cell membranes enclose.
   b. The extracellular fluid compartment includes all fluids and electrolytes outside cell membranes.
      (1) Interstitial fluid within tissue spaces
      (2) Plasma within blood
      (3) Lymph within lymphatic vessels
      (4) Transcellular fluid within body cavities
2. Body fluid composition
   a. Extracellular fluids
      (1) Extracellular fluids have high concentrations of sodium, chloride, calcium, and bicarbonate ions, with less potassium, calcium, magnesium, phosphate, and sulfate ions.
      (2) Plasma contains more protein than does either interstitial fluid or lymph.
   b. Intracellular fluid contains relatively high concentrations of potassium, magnesium, and phosphate ions; it also contains a greater concentration of sulfate ions and lesser concentrations of sodium, chloride, calcium, and bicarbonate ions than does extracellular fluid.
3. Movement of fluid between compartments
   a. Hydrostatic and osmotic pressure regulate fluid movements.
      (1) Fluid leaves plasma because of hydrostatic pressure and returns to plasma because of osmotic pressure.
      (2) Hydrostatic pressure drives fluid into lymph vessels.
      (3) Osmotic pressure regulates fluid movement in and out of cells.
   b. Sodium ion concentrations are especially important in fluid movement regulation.

Water Balance (page 811)

1. Water intake
   a. The volume of water taken in varies from person to person.
   b. Most water comes from consuming liquid or moist foods.
   c. Oxidative metabolism produces some water.
2. Regulation of water intake
   a. The thirst mechanism is the primary regulator of water intake.
   b. Drinking and the resulting stomach distension inhibit the thirst mechanism.
3. Water output
   a. Water is lost in a variety of ways.
      (1) It is excreted in the urine, feces, and sweat.
      (2) Insensible loss occurs through evaporation from the skin and lungs.
   b. Urine production regulates water output.

4. Regulation of water output
   a. The distal convoluted tubules and collecting ducts of the nephrons regulate water output.
      (1) ADH from the hypothalamus and posterior pituitary gland stimulates water reabsorption in these segments.
      (2) The mechanism involving ADH can reduce normal output of 1,500 milliliters to 500 milliliters per day.
   b. If excess water is taken in, the ADH mechanism is inhibited.

Electrolyte Balance (page 813)

1. Electrolyte intake
   a. The most important electrolytes in the body fluids are those that release ions of sodium, potassium, calcium, magnesium, chloride, sulfate, phosphate, and bicarbonate.
   b. These ions are obtained in foods and beverages or as by-products of metabolic processes.
2. Regulation of electrolyte intake
   a. Electrolytes are usually obtained in sufficient quantities in response to hunger and thirst mechanisms.
   b. In a severe electrolyte deficiency, a person may experience a salt craving.
3. Electrolyte output
   a. Electrolytes are lost through perspiration, feces, and urine.
   b. Quantities lost vary with temperature and physical exercise.
   c. The greatest electrolyte loss occurs as a result of kidney functions.
4. Regulation of electrolyte output
   a. Concentrations of sodium, potassium, and calcium ions in the body fluids are particularly important.
   b. The regulation of sodium ions involves the secretion of aldosterone from the adrenal glands.
   c. The regulation of potassium ions also involves aldosterone.
   d. Calcitonin from the thyroid gland and parathyroid hormone from the parathyroid glands regulate calcium ion concentration.
   e. The mechanisms that control positively charged ions secondarily regulate negatively charged ions.
      (1) Chloride ions are passively reabsorbed in renal tubules as sodium ions are actively reabsorbed.
      (2) Some negatively charged ions, such as phosphate ions, are reabsorbed partially by limited-capacity active transport mechanisms.

Acid-Base Balance (page 817)

1. Acids are electrolytes that release hydrogen ions. Bases combine with hydrogen ions.
   a. Aerobic respiration of glucose
      (1) Aerobic respiration of glucose produces carbon dioxide, which reacts with water to form carbonic acid.
      (2) Carbonic acid dissociates to release hydrogen and bicarbonate ions.
   b. Anaerobic respiration of glucose produces lactic acid.

c. Incomplete oxidation of fatty acids releases acidic ketone bodies.

d. Oxidation of sulfur-containing amino acids produces sulfuric acid.

e. Hydrolysis of phosphoproteins and nucleic acids gives rise to phosphoric acid.

2. Strengths of acids and bases
   a. Acids vary in the extent to which they ionize.
      (1) Strong acids, such as hydrochloric acid, ionize more completely.
      (2) Weak acids, such as carbonic acid, ionize less completely.
   b. Bases vary in strength also.
      (1) Strong bases, such as hydroxide ions, combine readily with hydrogen ions.
      (2) Weak bases, such as bicarbonate ions, combine with hydrogen ions less readily.

3. Regulation of hydrogen ion concentration
   a. Acid-base buffer systems
      (1) Buffer systems are composed of sets of two or more chemicals.
      (2) They convert strong acids into weaker acids or strong bases into weaker bases.
         (3) They include the bicarbonate buffer system, phosphate buffer system, and protein buffer system.
         (4) Buffer systems minimize pH changes.
   b. Respiratory excretion of carbon dioxide
      (1) The respiratory center is located in the brainstem.
      (2) It helps regulate pH by controlling the rate and depth of breathing.
      (3) Increasing carbon dioxide and hydrogen ion concentrations stimulates chemoreceptors associated with the respiratory center; breathing rate and depth increase, and carbon dioxide concentration decreases.
      (4) If the carbon dioxide and hydrogen ion concentrations are low, the respiratory center inhibits breathing.
   c. Renal excretion of hydrogen ions
      (1) Nephrons secrete hydrogen ions to regulate pH.
      (2) Phosphates buffer hydrogen ions in urine.
      (3) Ammonia produced by renal cells helps transport hydrogen ions to the outside of the body.
   d. Chemical buffers act rapidly; physiological buffers act more slowly.

CRITICAL THINKING QUESTIONS

1. An elderly, semiconscious patient is tentatively diagnosed as having acidosis. What components of the arterial blood will be most valuable in determining if the acidosis is of respiratory origin?

2. Some time ago, several newborns died due to an error in which sodium chloride was substituted for sugar in their formula. What symptoms would this produce? Why do you think newborns are more prone to the hazard of excess salt intake than adults?

3. Explain the threat to fluid and electrolyte balance in the following situation: A patient is being nutritionally maintained on concentrated solutions of hydrolyzed protein that are administered through a gastrostomy tube.

4. Describe what might happen to the plasma pH of a patient as a result of
   a. prolonged diarrhea
   b. suction of the gastric contents
   c. hyperventilation
   d. hypoventilation

5. Radiation therapy may damage the mucosa of the stomach and intestines. What effect might this have on the patient's electrolyte balance?

6. If the right ventricle of a patient's heart is failing, increasing the venous pressure, what changes might occur in the patient's extracellular fluid compartments?

REVIEW EXERCISES

1. Explain how water balance and electrolyte balance are interdependent.

2. Name the body fluid compartments, and describe their locations.

3. Explain how the fluids within these compartments differ in composition.

4. Describe how fluid movements between the compartments are controlled.

5. Prepare a list of sources of normal water gain and loss to illustrate how the input of water equals the output of water.

6. Define *water of metabolism.*

7. Explain how water intake is regulated.

8. Explain how the nephrons function in the regulation of water output.

9. List the most important electrolytes in the body fluids.

10. Explain how electrolyte intake is regulated.

11. List the routes by which electrolytes leave the body.

12. Explain how the adrenal cortex functions in the regulation of electrolyte output.

13. Describe the role of the parathyroid glands in regulating electrolyte balance.

14. Describe the role of the renal tubule in regulating electrolyte balance.

15. Distinguish between an acid and a base.

16. List five sources of hydrogen ions in the body fluids, and name an acid that originates from each source.

17. Distinguish between a strong acid and a weak acid, and name an example of each.

18. Distinguish between a strong base and a weak base, and name an example of each.

19. Explain how an acid-base buffer system functions.

20. Describe how the bicarbonate buffer system resists changes in pH.

21. Explain why a protein has acidic as well as basic properties.

22. Describe how a protein functions as a buffer system.
23. Describe the function of hemoglobin as a buffer of carbonic acid.
24. Explain how the respiratory center functions in the regulation of the acid-base balance.

25. Explain how the kidneys function in the regulation of the acid-base balance.
26. Describe the role of ammonia in the transport of hydrogen ions to the outside of the body.
27. Distinguish between a chemical buffer system and a physiological buffer system.

WEB CONNECTIONS

Visit the Student Edition of the Online Learning Center at www.mhhe.com/shier10 **for answers to chapter questions, additional quizzes, interactive learning exercises, and other study tools.**

UNDERSTANDING WORDS

andr-, man: *andr*ogens—male sex hormones.

contra-, against, counter: *contra*ception—prevention of fertilization.

crur-, lower part: *crur*a—diverging parts at the base of the penis by which it is attached to the pelvic arch.

ejacul-, to shoot forth: *ejacul*ation—process by which semen is expelled from the male reproductive tract.

fimb-, fringe: *fimb*riae—irregular extensions on the margin of the infundibulum of the uterine tube.

follic-, small bag: *follic*le—ovarian structure that contains an egg cell.

genesis, origin: spermato*genesis*—process by which sperm cells are formed.

gubern-, to steer, to guide: *gubern*aculum—fibromuscular cord that guides the descent of a testis.

labi-, lip: *labi*a minora—flattened, longitudinal folds that extend along the margins of the vestibule.

mamm-, breast: *mamm*ary gland—female accessory gland that secretes milk.

mast-, breast: *mast*itis—inflammation of the mammary gland.

mens-, month: *mens*es —monthly flow of blood from the female reproductive tract.

mons, an eminence: *mons* pubis—rounded elevation of fatty tissue overlying the symphysis pubis in a female.

oo-, egg: *oo*genesis—formation of an egg.

prim-, first: *prim*ordial follicle—ovarian follicle composed of an egg surrounded by a single layer of cells.

puber-, adult: *puber*ty—time when a person becomes able to reproduce.

zon-, belt: *zon*a pellucida—transparent layer surrounding an oocyte.

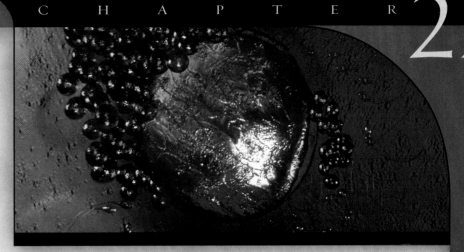

A few hundred sperm approach an egg, the winners of a race that several hundred million sperm began (970×). Only one sperm cell can fertilize the egg.

Reproductive Systems

CHAPTER OBJECTIVES

After you have studied this chapter, you should be able to

1. State the general functions of the male reproductive system.

2. Name the parts of the male reproductive system and describe the general functions of each part.

3. Outline the process of meiosis and explain how it mixes up parental genes.

4. Outline the process of spermatogenesis.

5. Trace the path sperm cells follow from their site of formation to the outside.

6. Describe the structure of the penis and explain how its parts produce an erection.

7. Explain how hormones control the activities of the male reproductive organs and the development of male secondary sex characteristics.

8. State the general functions of the female reproductive system.

9. Name the parts of the female reproductive system and describe the general functions of each part.

10. Outline the process of oogenesis.

11. Explain how hormones control the activities of the female reproductive organs and the development of female secondary sex characteristics.

12. Describe the major events that occur during a reproductive cycle.

13. List several methods of birth control and describe the relative effectiveness of each method.

14. List general symptoms of sexually transmitted diseases.

rectile dysfunction (impotence), in which the penis cannot become erect or sustain an erection, was until recently not often talked about. Then in the spring of 1998, along came Viagra® (sildenafil), a drug that enables about half of all men who take it to produce and maintain erections. The drug was originally developed to treat chest pain. Its effects on the penis were noted when participants in the clinical trials reported improved sex lives and refused to return extra pills! In most cases, Viagra appears to be safe and effective for men.

Erectile dysfunction has many causes, including underlying diseases such as diabetes mellitus; paralysis; treatments such as prostate surgery and many types of drugs, such as certain antidepressants; and lifestyle factors such as excess smoking or drinking alcohol. Side effects of Viagra include headache, facial flushing, and gastrointestinal upset. Men taking nitrate drugs to treat angina should not take Viagra, because the combination can cause life-threatening drops in blood pressure.

Viagra is based on sound science. The process of erection depends upon a very small molecule, nitric oxide (NO), that until recently was most widely known as a constituent of smog, cigarette smoke, and acid rain (NO should not be confused with the anesthetic nitrous oxide).

The penis consists of two chambers of spongy tissue that surround blood vessels. When the vessels fill with blood, as they do following sexual stimulation, the organ engorges and stiffens. The stimulation causes neurons as well as the endothelial cells that line the interiors of the blood vessels to release NO. The NO then enters muscle cells that form the middle layers of the blood vessels, relaxing them by activating a series of other chemicals. The vessels dilate and fill with blood, and the penis becomes erect. One such chemical, cGMP, must stay around for awhile for an erection to persist. Viagra works by blocking the enzyme that normally breaks down cGMP, thereby sustaining the erection.

Viagra is only one of several new treatments for erectile dysfunction. Evaluating them is challenging, because of a powerful placebo effect. Other therapies are being developed and tested in rats by measuring pressure in the penis. In more people-oriented investigations,

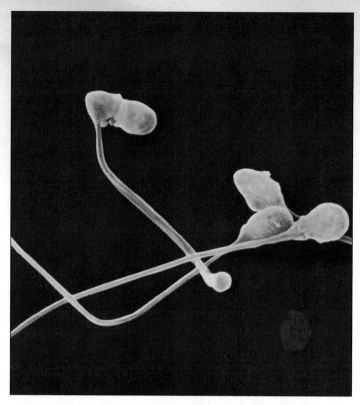

Even with normal penile function, having too many abnormally shaped sperm cells can impair a man's fertility (3,900×).

participants keep diaries that record "daily erectile activity" while taking a new drug versus taking a placebo, or they are asked to answer detailed questions about their sex lives. Some studies use more invasive techniques. In "penile plethysmography," a volunteer watches an erotic film after taking a drug or placebo, while a device measures engorgement at the base of the penis. In an "audiovisual stimulation penogram," the penis is attached to a mercury-based strain gauge, which assesses engorgement as the man watches a movie. ■

Organs of the Male Reproductive System

The organs of the male reproductive system are specialized to produce and maintain the male sex cells, or *sperm cells;* to transport these cells, together with supporting fluids, to the female reproductive tract; and to secrete male sex hormones.

The *primary sex organs* (gonads) of this system are the two testes in which the sperm cells (spermatozoa) and the male sex hormones are formed. The other structures of the male reproductive system are termed *accessory sex organs* (secondary sex organs). They include the internal reproductive organs and the external reproductive organs (fig. 22.1).

Testes

The **testes** (sing., *testis*) are ovoid structures about 5 centimeters in length and 3 centimeters in diameter. Both testes, each suspended by a spermatic cord, are contained within the cavity of the saclike *scrotum* (see fig. 22.1 and reference plate 52).

Descent of the Testes

In a male fetus, the testes originate from masses of tissue posterior to the parietal peritoneum, near the developing kidneys. Usually a month or two before birth, these organs descend to the lower abdominal cavity and pass through the abdominal wall into the scrotum.

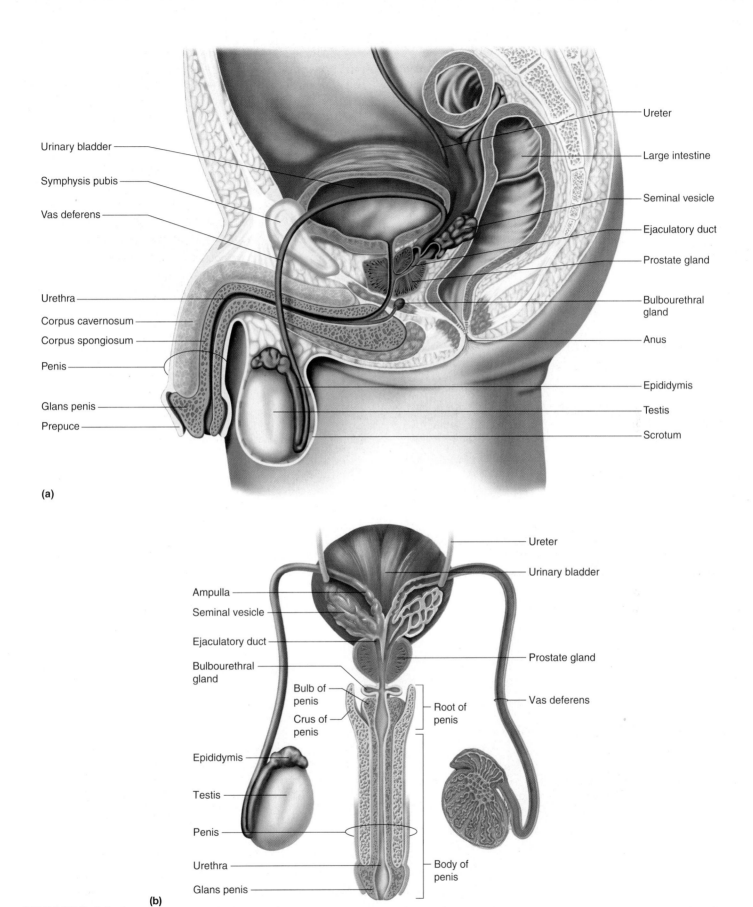

FIGURE 22.1

Male reproductive organs. (*a*) Sagittal view and (*b*) posterior view. The paired testes are the primary sex organs, and the other structures, both internal and external, are accessory sex organs.

The male sex hormone *testosterone,* which the developing testes secrete, stimulates the testes to descend. A fibromuscular cord called the **gubernaculum** (gu"ber-nak′u-lum) aids movement of the testes. This cord is attached to each developing testis and extends into the inguinal region of the abdominal cavity. It passes through the abdominal wall and is fastened to the skin on the outside of the scrotum. The testis descends, guided by the gubernaculum, passing through the **inguinal canal** (ing′ gwĭ-nal kah-nal′) of the abdominal wall and entering the scrotum, where it remains anchored by the gubernaculum. Each testis carries a developing *vas deferens,* blood vessels, and nerves. These structures later form parts of the **spermatic cord** by which the testis is suspended in the scrotum (fig. 22.2).

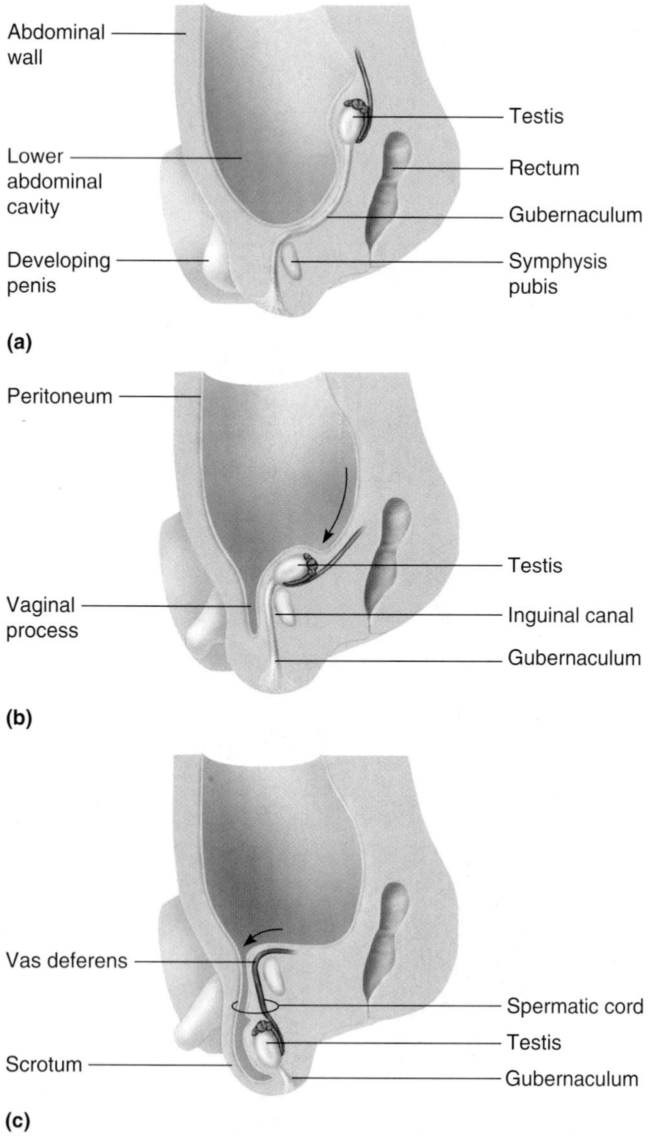

(a)

Abdominal wall
Lower abdominal cavity
Developing penis
Testis
Rectum
Gubernaculum
Symphysis pubis

(b)

Peritoneum
Vaginal process
Testis
Inguinal canal
Gubernaculum

(c)

Vas deferens
Scrotum
Spermatic cord
Testis
Gubernaculum

FIGURE 22.2
During fetal development, each testis descends through an inguinal canal and enters the scrotum (*a–c*).

If the testes fail to descend into the scrotum, they will not produce sperm cells because the temperature in the abdominal cavity is too high. If this condition, called *cryptorchidism,* is left untreated, the cells that normally produce sperm cells degenerate, and the male is infertile.

During the descent of a testis, a pouch of peritoneum, called the *vaginal process,* moves through the inguinal canal and into the scrotum. In about one-quarter of males, this pouch remains open, providing a potential passageway through which a loop of intestine may be forced by great abdominal pressure, producing an *indirect inguinal hernia.* If the protruding intestinal loop is so tightly constricted within the inguinal canal that its blood supply stops, the condition is called a *strangulated hernia.* Without prompt treatment, the strangulated tissues may die.

1   What are the primary sex organs of the male reproductive system?

2   Describe the descent of the testes.

3   What is the function of the gubernaculum, both during and after the descent of the testes?

4   What happens if the testes fail to descend into the scrotum?

Structure of the Testes

A tough, white, fibrous capsule called the *tunica albuginea* encloses each testis. Along its posterior border, the connective tissue thickens and extends into the organ, forming a mass called the *mediastinum testis.* From this structure, thin layers of connective tissue, called *septa,* pass into the testis and subdivide it into about 250 *lobules.*

A lobule contains one to four highly coiled, convoluted **seminiferous tubules** (se″-mĭ-nif′er-us too′būlz), each of which is approximately 70 centimeters long when uncoiled. These tubules course posteriorly and unite to form a complex network of channels called the *rete testis* (re′te tes′tis). The rete testis is located within the mediastinum testis and gives rise to several ducts that join a tube called the *epididymis.* The epididymis, in turn, is coiled on the outer surface of the testis.

The epithelial cells of the seminiferous tubules can give rise to *testicular cancer,* a common cancer in young men. In most cases, the first sign is a painless testis enlargement or a scrotal mass that attaches to a testis.

If a biopsy (tissue sample) reveals cancer cells, surgery is performed to remove the affected testis (orchiectomy). Radiation and/or chemotherapy often prevent the cancer from recurring and spreading.

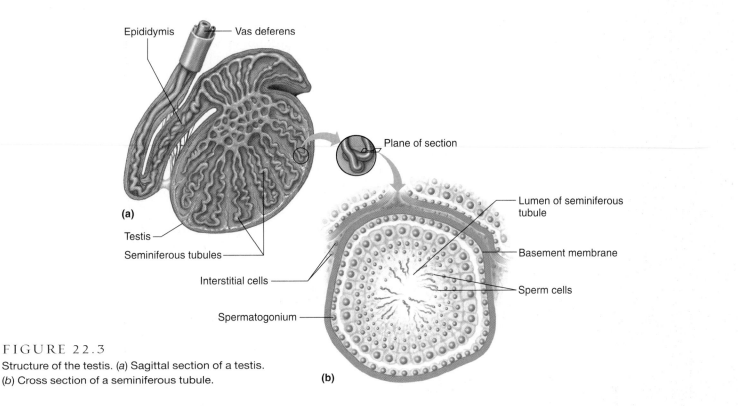

FIGURE 22.3
Structure of the testis. (*a*) Sagittal section of a testis.
(*b*) Cross section of a seminiferous tubule.

The seminiferous tubules are lined with a specialized stratified epithelium, which includes the spermatogenic cells that give rise to the sperm cells. Other specialized cells, called **interstitial cells** (in″ter-stish′al) (cells of Leydig), lie in the spaces between the seminiferous tubules. Interstitial cells produce and secrete male sex hormones (figs. 22.3 and 22.4).

1   Describe the structure of a testis.

2   Where are the sperm cells produced within the testes?

3   Which cells produce male sex hormones?

Formation of Sperm Cells

The epithelium of the seminiferous tubules consists of supporting cells (*sustentacular cells,* or Sertoli cells) and spermatogenic cells. The sustentacular cells are tall, columnar cells that extend the full thickness of the epithelium from its base to the lumen of the seminiferous tubule. The sustentacular cells support, nourish, and regulate the *spermatogenic cells,* which give rise to sperm cells (spermatozoa).

In the male embryo, the spermatogenic cells are undifferentiated and are called *spermatogonia.* Each of these cells contains 46 chromosomes (23 pairs) in its nucleus, the usual number for human cells (fig. 22.5).

During embryonic development, hormones stimulate the spermatogonia to become active. Some of the cells undergo mitosis. Each cell division gives rise to two new cells, one of which maintains the supply of undifferentiated cells, the other of which enlarges to become a *pri-*

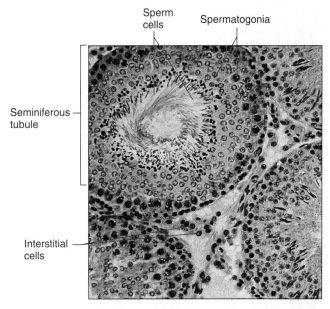

FIGURE 22.4
Light micrograph of seminiferous tubules (200×).

mary spermatocyte. Sperm production (spermatogenesis) (fig. 22.5) is arrested at this stage. At puberty, mitosis resumes giving rise to new spermatogonia. Testosterone secretion rises, and the primary spermatocytes then reproduce by a special type of cell division called **meiosis** (mi-o′sis).

Meiosis includes two successive divisions, called the *first* and *second meiotic divisions.* The first meiotic

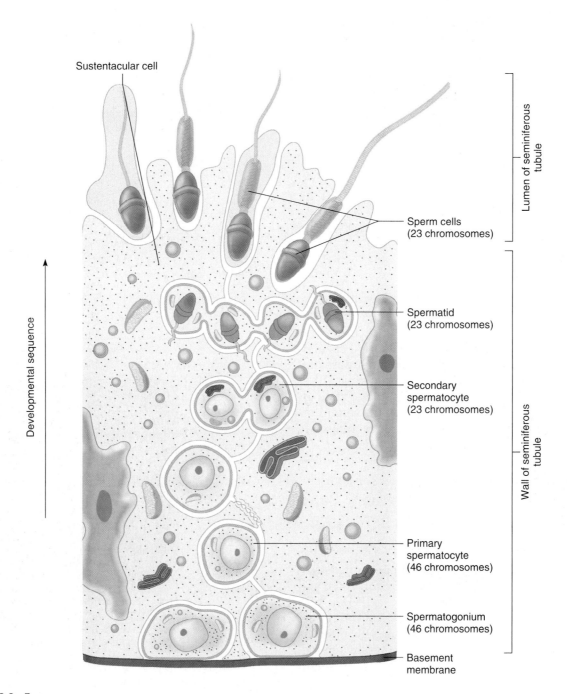

Sustentacular cell

Lumen of seminiferous
tubule

Sperm cells
(23 chromosomes)

Spermatid
(23 chromosomes)

Secondary
spermatocyte
(23 chromosomes)

Wall of seminiferous
tubule

Primary
spermatocyte
(46 chromosomes)

Spermatogonium
(46 chromosomes)

Basement
membrane

Developmental sequence

FIGURE 22.5

Spermatogonia give rise to primary spermatocytes by mitosis; the spermatocytes, in turn, give rise to sperm cells by meiosis.

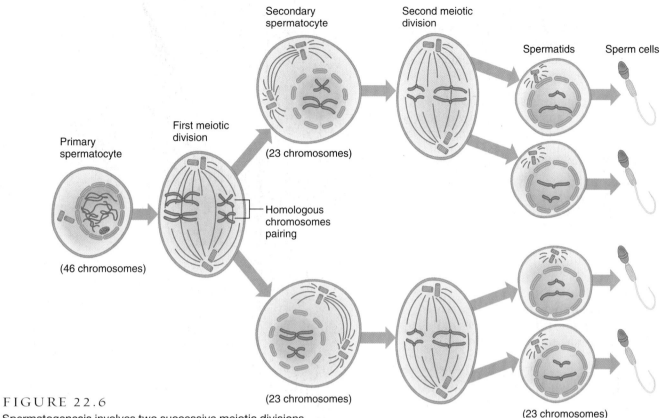

Secondary
spermatocyte

Second meiotic
division

Spermatids   Sperm cells

First meiotic
division

Primary
spermatocyte

(23 chromosomes)

Homologous
chromosomes
pairing

(46 chromosomes)

(23 chromosomes)

FIGURE 22.6
Spermatogenesis involves two successive meiotic divisions.

(23 chromosomes)

division (meiosis I) separates homologous chromosome pairs. Homologous pairs are the same, gene for gene. They may not be identical, however, because a gene may have variants, and the chromosome that comes from the person's mother may carry a different variant for the corresponding gene from the father's homologous chromosome. Before meiosis I, each homologous chromosome is replicated, so it consists of two complete DNA strands called *chromatids.* The chromatids of a replicated chromosome attach at regions called *centromeres.*

Each of the cells that undergo the second meiotic division (*meiosis II*) emerges with one member of each homologous pair, a condition termed **haploid.** That is, a haploid cell has one set of chromosomes. This second division separates the chromatids, producing cells that are still haploid, but whose chromosomes are no longer in the replicated form. After meiosis II, each of the chromatids has become an independent chromosome.

The steps of meiosis are clearer when considered in a time sequence. However, keep in mind that, like mitosis, meiosis is a continuous process. Considering it in steps simply makes it easier to follow. Refer to figure 22.6 throughout the following discussion.

1. **Prophase I.** Individual chromosomes appear as thin threads within the nucleus, then shorten and thicken. Nucleoli disappear, the nuclear membrane temporarily disassembles, and microtubules begin

to build the spindle that will separate the chromosomes. The DNA of the chromosomes has already been replicated.

As prophase I continues, homologous chromosomes pair up side by side and tightly intertwine. During this pairing, called *synapsis,* the chromatids of the homologous chromosomes contact one another at various points along their lengths. Often, the chromatids break in one or more places and exchange parts, forming chromatids with new combinations of genetic information (fig. 22.7).

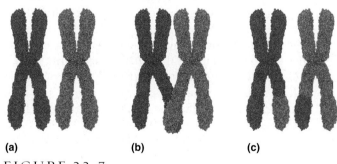

(a)   (b)   (c)

FIGURE 22.7
Crossing over mixes up traits. (*a*) Pairing of homologous chromosomes, (*b*) chromatids crossing over, (*c*) results of crossing over. The different colors represent the fact that one homologous chromosome comes from the individual's father and one from the mother.

Since one chromosome of a homologous pair is from a person's mother and the other is from the father, an exchange, or **cross-over,** between homologous chromosomes produces chromatids that contain genetic information from both parents.

2. **Metaphase I.** During the first metaphase, chromosome pairs line up about midway between the poles of the developing spindle, and they are held under great tension, like two groups of people playing tug-of-war. Each chromosome pair consists of two chromosomes, which equals four chromatids. Each chromosome attaches to spindle fibers from one pole. The chromosome alignment is random with respect to maternal and paternal origin of the chromosomes. That is, each of the 23 chromosomes contributed from the mother may be on the left or the right, and the same is true for the paternal chromosomes—it is similar to the number of ways that 23 pairs of children could line up, while maintaining the pairs. Chromosomes can line up with respect to each other in many, many combinations.

3. **Anaphase I.** Homologous chromosome pairs separate, and each replicated member moves to one end of the spindle. Thus, each new cell receives only one replicated member of a homologous pair of chromosomes, overall halving the chromosome number.

4. **Telophase I.** The original cell divides in two. Nuclear membranes form around the chromosomes, nucleoli reappear, and the spindle fibers disassemble into their constituent microtubules. Then the second meiotic division begins.

   Meiosis II (see fig. 22.6) is very similar to a mitotic division. During *prophase II,* chromosomes condense and reappear, still replicated. They move into positions midway between the poles of the developing spindle. In *metaphase II,* the replicated chromosomes attach to spindle fibers. In *anaphase II,* centromeres separate, freeing the chromatids to move to opposite poles of the spindles. The former chromatids are now considered to be chromosomes. In *telophase II,* each of the two cells resulting from meiosis I divides to form two cells. Therefore, each cell undergoing meiosis has the potential to produce four gametes. In males, the gametes mature into four sperm cells. In females, three of the products of meiosis are "cast aside" as polar bodies, and one cell becomes the egg.

Meiosis generates astounding genetic variety. Any one of a person's more than 8 million possible combinations of 23 chromosomes can combine with any one of the more than 6 million combinations of his or her mate, rais-

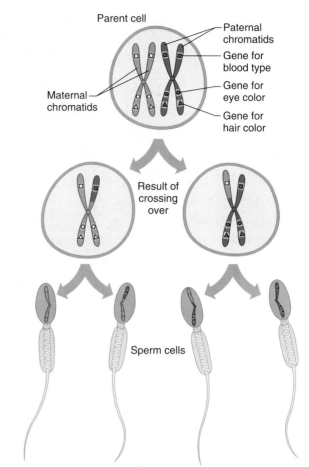

FIGURE 22.8
As a result of crossing over, the genetic information in sperm cells and egg cells varies from cell to cell. Colors represent parent of origin.

ing the potential variability to more than 70 trillion genetically unique individuals! Cross-over contributes even more genetic variability. Figure 22.8 illustrates in a simplified manner how maternal and paternal traits reassort during meiosis.

During **spermatogenesis** (sper″mah-to-jen′ĕ-sis), the primary spermatocytes each divides to form two *secondary spermatocytes.* Each of these cells, in turn, divides to form two *spermatids,* which mature into sperm cells. Meiosis reduces the number of chromosomes in each cell by one-half. Consequently, for each primary spermatocyte that undergoes meiosis, four sperm cells with 23 chromosomes in each of their nuclei are formed. Because the chromosome number is halved, when a sperm and egg join in a process called **fertilization** (fer″ti-li-za′shun), the new individual has a complete set of 23 pairs of chromosomes.

The spermatogonia are located near the wall of the seminiferous tubule. As spermatogenesis occurs, cells in more advanced stages are pushed along the sides of sustentacular cells toward the lumen of the seminiferous tubule.

Near the base of the epithelium, membranous processes from adjacent sustentacular cells fuse by specialized junctions (occluding junctions) into complexes that divide the tissue into two layers (see fig. 23.5). The spermatogonia are on one side of this barrier, and the cells in more advanced stages are on the other side. This membranous complex helps maintain a favorable environment for development of sperm cells by preventing certain large molecules from moving from the interstitial fluid of the basal epithelium into the region of the differentiating cells.

Sperm have fascinated biologists for centuries. Anton van Leeuwenhoek was the first to view human sperm under a microscope in 1678, concluding that they were parasites in semen. By 1685, he had modified his view, writing that sperm contain a preformed human being and are seeds requiring nurturing in a female to start a new life.

Spermatogenesis occurs continually in a male, starting at puberty. The resulting sperm cells collect in the lumen of each seminiferous tubule, then pass through the rete testis to the epididymis, where they accumulate and mature.

Structure of a Sperm Cell

A mature sperm cell is a tiny, tadpole-shaped structure about 0.06 millimeter long. It consists of a flattened head, a cylindrical body, and an elongated tail.

The oval *head* of a sperm cell is primarily composed of a nucleus and contains highly compacted chromatin consisting of 23 chromosomes. A small protrusion at its anterior end, called the *acrosome,* contains enzymes, including hyaluronidase, that aid the sperm cell in penetrating an egg cell (fig. 22.9).

The midpiece of a sperm contains a central, filamentous core and many mitochondria organized in a spiral. The

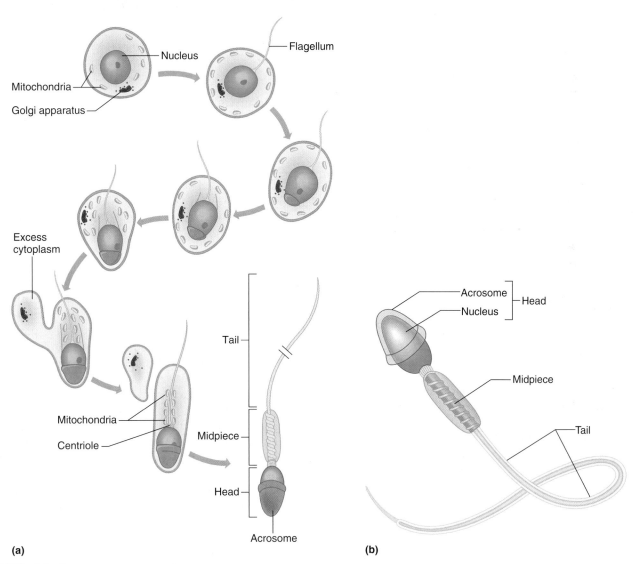

FIGURE 22.9
Sperm cell maturation. (*a*) The head of the sperm develops largely from the nucleus of the formative cell. (*b*) Parts of a mature sperm cell.

FIGURE 22.10
Falsely colored scanning electron micrograph of human sperm cells (1,400×).

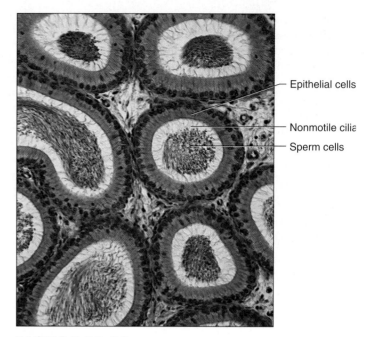

Epithelial cells

Nonmotile cilia

Sperm cells

FIGURE 22.11
Cross section of a human epididymis (145×).

tail (flagellum) consists of several microtubules enclosed in an extension of the cell membrane. The mitochondria provide ATP for the lashing movement of the tail that propels the sperm cell through fluid. The scanning electron micrograph in figure 22.10 shows a few mature sperm cells.

Many toxic chemicals that affect sperm hamper their ability to swim, so the cells cannot transmit the toxin to an egg. One notable exception is cocaine, which attaches to thousands of binding sites on human sperm cells, without apparently harming the cells or impeding their movements. Sperm can ferry cocaine to an egg, but it is not known what harm, if any, the drug causes. We do know that fetuses exposed to cocaine in the uterus may suffer a stroke, or, as infants, be unable to react normally to their surroundings.

1 Explain the function of the sustentacular cells in the seminiferous tubules.

2 Describe the major events that occur during meiosis.

3 How does meiosis provide genetic variability?

4 Describe spermatogenesis.

5 Describe the structure of a sperm cell.

Male Internal Accessory Organs

The internal accessory organs of the male reproductive system include the epididymides, vasa deferentia, ejaculatory ducts, and urethra, as well as the seminal vesicles, prostate gland, and bulbourethral glands.

Epididymis

Each **epididymis** (ep″ĭ-did′ĭ-mis) (pl., *epididymides*) is a tightly coiled, threadlike tube about 6 meters long (see figs. 22.1, 22.11, and reference plate 52). The epididymis is connected to ducts within a testis. It emerges from the top of the testis, descends along its posterior surface, and then courses upward to become the vas deferens.

The inner lining of the epididymis is composed of pseudostratified columnar cells that bear nonmotile cilia. These cells secrete glycogen and other substances that support stored sperm cells and promote their maturation.

When immature sperm cells reach the epididymis, they are nonmotile. However, as they travel through the epididymis as a result of rhythmic peristaltic contractions, they mature. Following this aging process, the sperm cells can move independently and fertilize egg cells (ova). However, they usually do not "swim" until after ejaculation.

Vas Deferens

Each **vas deferens** (vas def′er-enz) (pl., *vasa deferentia*), also called ductus deferens, is a muscular tube about 45 centimeters long that is lined with pseudostratified columnar epithelium (fig. 22.12). It begins at the lower end of the epididymis and passes upward along the medial side of a testis to become part of the spermatic cord. Each vas deferens passes through the inguinal canal, enters the abdominal cavity outside the parietal peritoneum, and courses over the pelvic brim. From there, it extends backward and medially into the pelvic cavity, where it ends behind the urinary bladder.

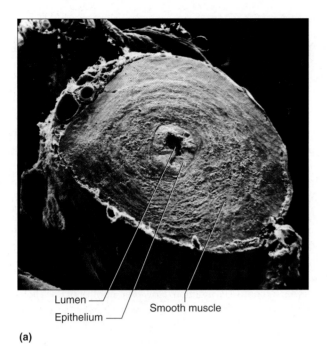

Lumen
Epithelium
Smooth muscle

(a)

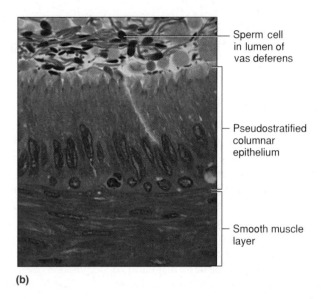

Sperm cell in lumen of vas deferens

Pseudostratified columnar epithelium

Smooth muscle layer

(b)

FIGURE 22.12

Vas deferens. (*a*) Falsely colored scanning electron micrograph of a cross section of the vas deferens (85×). (*b*) Light micrograph of the wall of the vas deferens (700×). *Tissues and Organs: A Text-Atlas of Scanning Electron Microscopy,* by R. G. Kessel and R. H. Kardon. © 1979 W. H. Freeman and Company.

Near its termination, the vas deferens dilates into a portion called the *ampulla.* Just outside the prostate gland, the tube becomes slender again and unites with the duct of a seminal vesicle. The fusion of these two ducts forms an **ejaculatory duct,** which passes through the prostate gland and empties into the urethra through a slit-like opening (see fig. 22.1).

Seminal Vesicle

A **seminal vesicle** (see fig. 22.1) is a convoluted, saclike structure about 5 centimeters long that is attached to the vas deferens near the base of the urinary bladder. The glandular tissue lining the inner wall of the seminal vesicle secretes a slightly alkaline fluid. This fluid helps regulate the pH of the tubular contents as sperm cells travel to the outside. The secretion of the seminal vesicle also contains *fructose,* a monosaccharide that provides energy to the sperm cells, and *prostaglandins,* which stimulate muscular contractions within the female reproductive organs, aiding the movement of sperm cells toward the egg cell.

At emission, the contents of the seminal vesicles empty into the ejaculatory ducts. This greatly increases the volume of the fluid discharged from the vas deferens.

1 Describe the structure of the epididymis.

2 Trace the path of the vas deferens.

3 What is the function of a seminal vesicle?

Prostate Gland

The **prostate** (pros'tāt) **gland** (see figs. 22.1 and 22.13) is a chestnut-shaped structure about 4 centimeters across and 3 centimeters thick that surrounds the beginning of the urethra, just inferior to the urinary bladder. It is composed of many branched tubular glands enclosed in connective tissue. Septa of connective tissue and smooth muscle

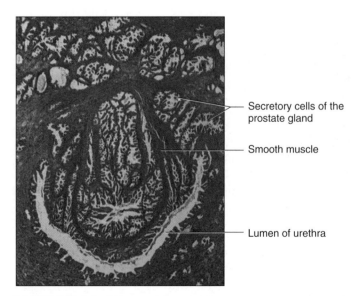

Secretory cells of the prostate gland

Smooth muscle

Lumen of urethra

FIGURE 22.13

Light micrograph of the prostate gland (10×).

PROSTATE ENLARGEMENT

The prostate gland is small in boys, begins to grow in early adolescence, and reaches adult size several years later. An adult's prostate gland is about the size of a walnut. Usually, the gland does not grow again until age fifty, when in half of all men, it enlarges enough to press on the urethra. This condition is called benign prostatic hypertrophy (BPH). As many as 90% of men over age seventy may have BPH. It produces a feeling of pressure on the bladder because it cannot empty completely, and the man feels the urge to urinate frequently. An early sign may be dribbling after urination. Retained urine can lead to infection and inflammation, bladder stones, or kidney disease.

Medical researchers do not know what causes prostate enlargement. Risk factors include a fatty diet, having had a vasectomy, and possibly occupational exposure to batteries or the metal cadmium. The enlargement may be benign or cancerous. Because prostate cancer is highly treatable if detected early, men should have their prostates examined regularly. Four out of five men who have prostate cancer are over age sixty-five.

Diagnostic tests for prostate cancer include a rectal exam; visualization of the prostate, urethra, and bladder with a device that is inserted through the penis, called a cytoscope; as well as a blood test to detect elevated prostate specific antigen (PSA), a cell surface protein normally found on prostate cells. Elevated PSA levels indicate an enlarged prostate, possibly from a benign or cancerous growth. Ultrasound may provide further information.

Table 22A summarizes treatments for an enlarged prostate. The components of treatment vary greatly from individual to individual. In some men, the recommended course is "watchful waiting," continuing to have frequent checkups to monitor the enlargement, but not taking action until symptoms arise. Surgery to treat prostate cancer is highly effective. It once commonly left a man incontinent and with erectile dysfunction. However, control of urination often returns within a few weeks, and newer surgical methods preserve the nerves that are necessary for erection to occur. ∎

TABLE 22A	Medical Treatments for an Enlarged Prostate Gland
Surgical removal of some prostate tissue or entire gland	
Radiation	
Drug (Proscar, or finasteride) to block testosterone's growth-stimulating effect on the prostate	
Alpha blocker drugs, which relax muscles near the prostate, relieving pressure	
Microwave energy delivered through a probe inserted into the urethra or rectum	
Balloon inserted into the urethra and inflated with liquid	
Liquid nitrogen delivered by a probe through the skin to freeze the tumor	
Device (stent) inserted between lobes of prostate to relieve pressure on the urethra	

extend inward from the capsule, separating the tubular glands. The ducts of these glands open into the urethra.

The prostate gland secretes a thin, milky fluid. This alkaline secretion neutralizes the sperm cell-containing fluid, which is acidic from accumulation of metabolic wastes from stored sperm cells. Prostatic fluid also enhances the motility of sperm cells, which remain relatively nonmotile in the acidic contents of the epididymis. In addition, the prostatic fluid helps neutralize the acidic secretions of the vagina, thus helping to sustain sperm cells that enter the female reproductive tract.

The prostate gland releases its secretions into the urethra as smooth muscles contract in its capsular wall. As this release occurs, the contents of the vas deferens and the seminal vesicles enter the urethra, which increases the volume of the fluid. Clinical Application 22.1 discusses the effects of prostate enlargement.

Bulbourethral Glands

The **bulbourethral** (bul″bo-u-re′thral) **glands** (Cowper's glands) are two small structures, each about a centimeter in diameter. They are located inferior to the prostate gland lateral to the membranous urethra and are enclosed by fibers of the external urethral sphincter muscle (see fig. 22.1).

The bulbourethral glands are composed of many tubes whose epithelial linings secrete a mucouslike fluid. This fluid is released in response to sexual stimulation and lubricates the end of the penis in preparation for sexual intercourse (coitus). However, females secrete most of the lubricating fluid for intercourse.

Semen

The fluid the urethra conveys to the outside during ejaculation is called **semen** (se′men). It consists of sperm cells from the testes and secretions of the seminal vesicles, prostate gland, and bulbourethral glands. Semen is slightly alkaline (pH about 7.5), and it includes prostaglandins and nutrients.

The volume of semen released at one time varies from 2 to 5 milliliters. The average number of sperm cells in the fluid is about 120 million per milliliter.

Sperm cells remain nonmotile while they are in the ducts of the testis and epididymis but begin to swim as they mix with the secretions of accessory glands. However, sperm cells cannot fertilize an egg cell until they enter the female reproductive tract. Development of this ability, called *capacitation,* entails changes that weaken the acrosomal membranes of the sperm cells. When sperm cells are placed with egg cells in a laboratory dish to achieve fertilization—a technique called *in vitro* fertilization, discussed in Clinical Application 23.1—chemicals are added to simulate capacitation.

Although sperm cells can live for many weeks in the ducts of the male reproductive tract, they usually survive only a day or two after being expelled to the outside, even when they are maintained at body temperature. On the other hand, sperm cells can be stored and kept viable for years if they are frozen at a temperature below −100°C. Clinical Application 22.2 describes some causes of male infertility.

1   Where is the prostate gland located?

2   What is the function of the prostate gland's secretion?

3   What is the function of the bulbourethral glands?

4   What are the components of semen?

Male External Reproductive Organs

The male external reproductive organs are the scrotum, which encloses the testes, and the penis. The urethra passes through the penis.

Scrotum

The **scrotum** is a pouch of skin and subcutaneous tissue that hangs from the lower abdominal region posterior to the penis. The subcutaneous tissue of the scrotal wall lacks fat but contains a layer of smooth muscle fibers that constitute the *dartos muscle.* Exposure to cold stimulates these muscles to contract, the scrotal skin to wrinkle, and the testes to move closer to the pelvic cavity, where they can absorb heat. Exposure to warmth stimulates the fibers to relax and the scrotum to hang loosely and provides an environment 3°C (about 5°F) below body temperature, which is more conducive to sperm production and survival.

A medial septum divides the scrotum into two chambers, each of which encloses a testis. Each chamber also contains a serous membrane, which covers the front and sides of the testis and the epididymis, helping to ensure that the testis and epididymis move smoothly within the scrotum (see fig. 22.1).

Penis

The **penis** is a cylindrical organ that conveys urine and semen through the urethra to the outside. It is also specialized to enlarge and stiffen by a process called *erection,* which enables it to be inserted into the vagina during sexual intercourse.

The *body,* or shaft, of the penis is composed of three columns of erectile tissue, which include a pair of dorsally located *corpora cavernosa* and a single, ventral *corpus spongiosum.* A tough capsule of white dense connective tissue called a *tunica albuginea* surrounds each column. Skin, a thin layer of subcutaneous tissue, and a layer of connective tissue enclose the penis (fig. 22.14).

The corpus spongiosum, through which the urethra extends, enlarges at its distal end to form a sensitive, cone-shaped **glans penis.** This structure covers the ends of the corpora cavernosa and bears the urethral opening—the *external urethral orifice.* The skin of the glans is very thin, hairless, and contains sensory receptors for sexual stimulation. A loose fold of skin called the *prepuce* (foreskin) begins just posterior to the glans and extends anteriorly to

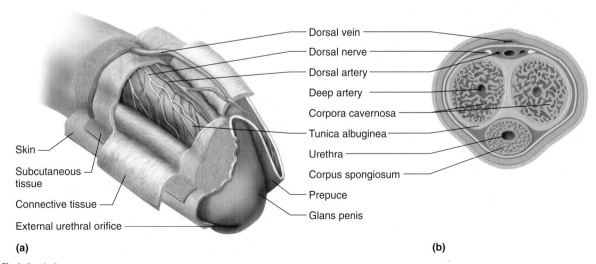

Skin

Subcutaneous tissue

Connective tissue

External urethral orifice

(a)

Dorsal vein

Dorsal nerve

Dorsal artery

Deep artery

Corpora cavernosa

Tunica albuginea

Urethra

Corpus spongiosum

Prepuce

Glans penis

(b)

FIGURE 22.14

Structure of the penis. (*a*) Interior and (*b*) cross section of the penis.

MALE INFERTILITY

Male infertility—the inability of sperm cells to fertilize an egg cell—has several causes. If during fetal development the testes do not descend into the scrotum, the higher temperature of the abdominal cavity or inguinal canal causes the developing sperm cells in the seminiferous tubules to degenerate. Certain diseases, such as mumps, may inflame the testes (orchitis), impairing fertility by destroying cells in the seminiferous tubules.

Both the quality and quantity of sperm cells are essential factors in the ability of a man to father a child. If a sperm head is misshapen, if a sperm cannot swim, or if there are simply too few sperm cells, completing the arduous journey to the well-protected egg may be impossible. Sometimes even a sperm cell that enters an egg is unsuccessful because it lacks the microtubules necessary to attract and merge the nuclei of the two cells.

In the past, sperm analysis was based on microscopic examination. Today, computer-aided sperm analysis (CASA) is standardizing and expanding criteria for normalcy in human male seminal fluid and the sperm cells it contains.

To analyze sperm, a man abstains from intercourse for two to three days, then provides a sperm sample, which must be examined within the hour. The man must also provide information about his reproductive history and possible exposure to toxins. The sperm sample is placed on a slide under a microscope, and then a video camera sends an image to a videocassette recorder, which projects a live or digitized image. The camera also sends the image to a computer, which traces sperm trajectories and displays them on a monitor or prints a hard copy (fig. 22A). Figure 22B shows a CASA of normal sperm, depicting how the swimming pattern alters as they travel.

CASA systems are also helpful in studies that use sperm as "biomarkers" of exposure to toxins. For example, the sperm of

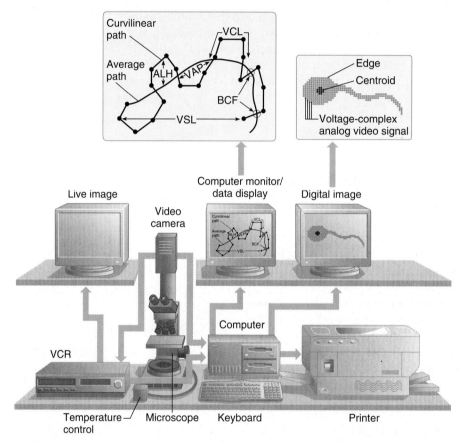

FIGURE 22A

Computer analysis improves the consistency and accuracy of describing sperm motility, morphology, and abundance.

cover it as a sheath. A surgical procedure called *circumcision* is used to remove the prepuce.

At the *root* of the penis, the columns of erectile tissue separate. The corpora cavernosa diverge laterally in the perineum and are firmly attached to the inferior surface of the pubic arch by connective tissue. These diverging parts form the *crura* (sing., *crus*) of the penis. The single corpus spongiosum is enlarged between the crura as the *bulb* of the penis, which is attached to membranes of the perineum.

1 Describe the structure of the penis.

2 What is circumcision?

3 How is the penis attached to the perineum?

Erection, Orgasm, and Ejaculation

During sexual stimulation, parasympathetic nerve impulses from the sacral portion of the spinal cord release the vasodilator nitric oxide, which dilates the

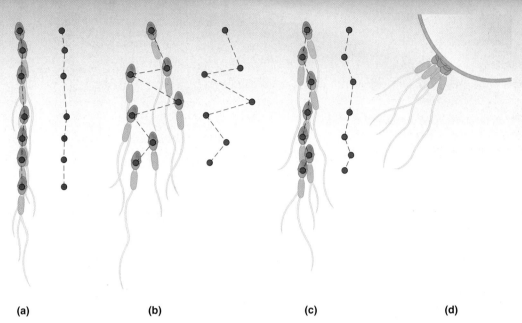

FIGURE 22B

A computer tracks sperm cell movements. In semen, sperm cells swim in a straight line (*a*), but as they are activated by biochemicals normally found in the woman's body, their trajectories widen (*b*). The sperm cells in (*c*) are in the mucus of a woman's cervix, and the sperm cells in (*d*) are attempting to digest through the structures surrounding an egg cell.

men who work in the dry-cleaning industry and are exposed to the solvent perchloroethylene (believed to damage sperm) were compared with sperm from men who work in the laundry industry and are exposed to many of the same chemicals except this one. CASA showed a difference in sperm motility that was directly related to level of exposure, as measured by exhalation of the chemical. This result supported the reproductive evidence: Although the men in both groups had the same numbers of children, the dry cleaners' partners took much longer to conceive than did the launderers' partners. Table 22B lists the components of a sperm analysis. ■

TABLE 22B	Semen Analysis
Characteristic	**Normal Value**
Volume	2–5 milliliters/ejaculate
Sperm cell density	60–150 million cells/milliliter
Percent motile	> 40%
Motile sperm cell density	> 24 million/milliliter
Average velocity	> 20 micrometers/second
Motility	> 8 micrometers/second
Percent normal morphology	> 80%
White blood cells	Occasional or absent

arteries leading into the penis, increasing blood flow into erectile tissues. At the same time, the increasing pressure of arterial blood entering the vascular spaces of the erectile tissue compresses the veins of the penis, reducing flow of venous blood away from the penis. Consequently, blood accumulates in the erectile tissues, and the penis swells and elongates, producing an **erection** (fig. 22.15).

The culmination of sexual stimulation is **orgasm** (or'gazm), a pleasurable feeling of physiological and psy-

chological release. Orgasm in the male is accompanied by emission and ejaculation.

Emission (e-mish'un) is the movement of sperm cells from the testes and secretions from the prostate gland and seminal vesicles into the urethra, where they mix to form semen. Emission occurs in response to sympathetic nerve impulses from the spinal cord, which stimulate peristaltic contractions in smooth muscles within the walls of the testicular ducts, epididymides, vasa deferentia, and ejaculatory ducts. At the same time, other

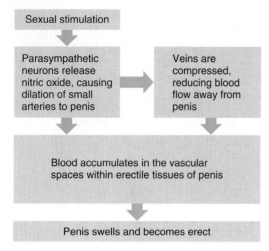

FIGURE 22.15
Mechanism of penile erection.

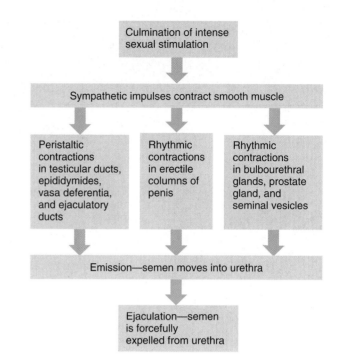

FIGURE 22.16
Mechanism of emission and ejaculation.

sympathetic impulses stimulate rhythmic contractions of the seminal vesicles and prostate gland.

As the urethra fills with semen, sensory impulses are stimulated and pass into the sacral portion of the spinal cord. In response, motor impulses are transmitted from the spinal cord to certain skeletal muscles at the base of the erectile columns of the penis, rhythmically contracting them. This increases the pressure within the erectile tissues and aids in forcing the semen through the urethra to the outside—a process called **ejaculation** (e-jak″u-la′shun).

The sequence of events during emission and ejaculation is coordinated so that the fluid from the bulbourethral glands is expelled first. This is followed by the release of fluid from the prostate gland, the passage of the sperm cells, and finally, the ejection of fluid from the seminal vesicles (fig. 22.16).

Immediately after ejaculation, sympathetic impulses constrict the arteries that supply the erectile tissue, reducing the inflow of blood. Smooth muscles within the walls of the vascular spaces partially contract again, and the veins of the penis carry the excess blood out of these spaces. The penis gradually returns to its flaccid state, and usually another erection and ejaculation cannot be triggered for a period of ten to thirty minutes or longer. Table 22.1 summarizes the functions of the male reproductive organs.

Spontaneous emission and ejaculation commonly occur in adolescent males during sleep. Changes in hormonal concentrations that accompany adolescent development and sexual maturation cause these *nocturnal emissions*.

1 What controls blood flow into the erectile tissues of the penis?

2 Distinguish among orgasm, emission, and ejaculation.

3 Review the events associated with emission and ejaculation.

TABLE 22.1	Functions of the Male Reproductive Organs
Organ	**Function**
Testis	
Seminiferous tubules	Produce sperm cells
Interstitial cells	Produce and secrete male sex hormones
Epididymis	Stores sperm cells undergoing maturation; conveys sperm cells to vas deferens
Vas deferens	Conveys sperm cells to ejaculatory duct
Seminal vesicle	Secretes an alkaline fluid containing nutrients and prostaglandins that helps neutralize the acidic components of semen
Prostate gland	Secretes an alkaline fluid that helps neutralize the acidic components of semen and enhances motility of sperm cells
Bulbourethral gland	Secretes fluid that lubricates end of the penis
Scrotum	Encloses, protects, and regulates temperature of testes
Penis	Conveys urine and semen to outside of body; inserted into the vagina during sexual intercourse; the glans penis is richly supplied with sensory nerve endings associated with feelings of pleasure during sexual stimulation

Hormonal Control of Male Reproductive Functions

Hormones secreted by the *hypothalamus,* the *anterior pituitary gland,* and the testes control male reproductive functions. These hormones initiate and maintain sperm cell production and oversee the development and maintenance of male sex characteristics.

Hypothalamic and Pituitary Hormones

Prior to ten years of age, a boy's body is reproductively immature. During this period, the body is childlike, and the spermatogenic cells of the testes are undifferentiated. Then a series of changes leads to development of a reproductively functional adult. The hypothalamus controls many of these changes.

Recall from chapter 13 (p. 483) that the hypothalamus secretes gonadotropin-releasing hormone (GnRH), which enters the blood vessels leading to the anterior pituitary gland. In response, the anterior pituitary gland secretes the **gonadotropins** (go-nad"o-trōp'inz) called *luteinizing hormone* (LH) and *follicle-stimulating hormone* (FSH). LH, which in males is called interstitial cell-stimulating hormone (ICSH), promotes development of the interstitial cells (cells of Leydig) of the testes, and

they, in turn, secrete male sex hormones. FSH stimulates the sustentacular cells of the seminiferous tubules to proliferate, grow, mature, and respond to the effects of the male sex hormone testosterone. Then, in the presence of FSH and testosterone, these cells stimulate the spermatogenic cells to undergo spermatogenesis, giving rise to sperm cells (fig. 22.17). The sustentacular cells also secrete a hormone called *inhibin,* which inhibits the anterior pituitary gland by negative feedback and thus prevents oversecretion of FSH.

Male Sex Hormones

Male sex hormones are termed **androgens** (an'dro-jenz). The interstitial cells of the testes produce most of them, but small amounts are synthesized in the adrenal cortex (see chapter 13, p. 495).

The hormone **testosterone** (tes-tos'te-rōn) is the most abundant androgen. It is secreted and transported in the blood, loosely attached to plasma proteins. Like other steroid hormones, testosterone combines with receptor molecules usually in the nuclei of its target cells (see chapter 13, p. 472). However, in many target cells, such as those in the prostate gland, seminal vesicles, and male external accessory organs, testosterone is first converted to another androgen called **dihydrotestosterone** (di-hi"dro-tes-tos'ter-ōn), which stimulates the cells of these

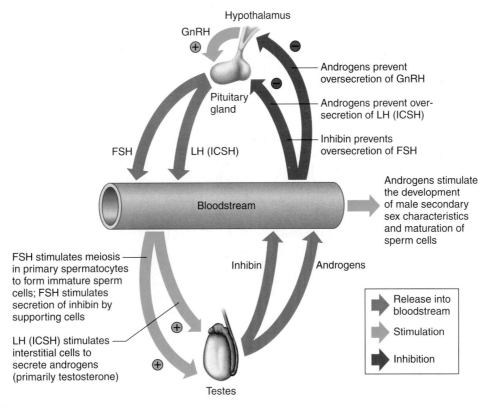

FIGURE 22.17

The hypothalamus controls maturation of sperm cells and development of male secondary sex characteristics. A negative feedback mechanism operating between the hypothalamus, the anterior lobe of the pituitary gland, and the testes controls the concentration of testosterone.

organs. Androgen molecules that do not reach receptors in target cells are usually changed by the liver into forms that can be excreted in bile or urine.

Testosterone secretion begins during fetal development and continues for a few weeks following birth, then nearly ceases during childhood. Between the ages of thirteen and fifteen, a young man's androgen production usually accelerates. This phase in development, when an individual becomes reproductively functional, is **puberty** (pu'ber-te). After puberty, testosterone secretion continues throughout the life of a male.

In a group of disorders called male pseudohermaphroditism, testes are usually present, but a block in testosterone synthesis prevents the fetus from developing male structures, and as a result, later, the child appears to be a girl. But at puberty, the adrenal glands begin to produce testosterone, as they normally do in any male. This leads to masculinization: The voice deepens, and muscles build up into a masculine physique; breasts do not develop, nor does menstruation occur. The clitoris may enlarge so greatly under the adrenal testosterone surge that it looks like a penis. Individuals with a form of this condition that is prevalent in the Dominican Republic are called *guevedoces,* which means "penis at age twelve."

Actions of Testosterone

Cells of the embryonic testes first produce testosterone after about eight weeks of development. This hormone stimulates the formation of the male reproductive organs, including the penis, scrotum, prostate gland, seminal vesicles, and ducts. Later in development, testosterone causes the testes to descend into the scrotum.

During puberty, testosterone stimulates enlargement of the testes (the primary male sex characteristic) and accessory organs of the reproductive system, as well as development of male *secondary sex characteristics,* which are special features associated with the adult male body. Secondary sex characteristics in the male include

1. Increased growth of body hair, particularly on the face, chest, axillary region, and pubic region. Sometimes growth of hair on the scalp decreases.

2. Enlargement of the larynx and thickening of the vocal folds, with lowering of the pitch of the voice.

3. Thickening of the skin.

4. Increased muscular growth, broadening shoulders, and narrowing of the waist.

5. Thickening and strengthening of the bones.

Testosterone also increases the rate of cellular metabolism and production of red blood cells by stimulating release of erythropoietin. For this reason, the average number of red blood cells in a cubic millimeter of blood is usually greater in males than in females. Testosterone stimulates sexual activity by affecting certain portions of the brain.

RECONNECT TO CHAPTER 14, RED BLOOD CELL PRODUCTION AND ITS CONTROL, PAGE 514.

Regulation of Male Sex Hormones

The extent to which male secondary sex characteristics develop is directly related to the amount of testosterone that the interstitial cells secrete. The hypothalamus regulates testosterone output through negative feedback (fig. 22.17).

As the concentration of testosterone in the blood increases, the hypothalamus becomes inhibited, decreasing its stimulation of the anterior pituitary gland by GnRH. As the pituitary's secretion of LH (ICSH) falls in response, the amount of testosterone the interstitial cells release decreases.

As the blood concentration of testosterone drops, the hypothalamus becomes less inhibited, and it once again stimulates the anterior pituitary gland to release LH. The increasing secretion of LH causes the interstitial cells to release more testosterone, and its blood concentration increases. Testosterone level decreases somewhat during and after the *male climacteric,* a decline in sexual function that occurs with aging. At any given age, the testosterone concentration in the male is regulated to remain relatively constant.

1. Which hormone initiates the changes associated with male sexual maturity?

2. Describe several male secondary sex characteristics.

3. List the functions of testosterone.

4. Explain how the secretion of male sex hormones is regulated.

Organs of the Female Reproductive System

The organs of the female reproductive system are specialized to produce and maintain the female sex cells, or *egg cells;* to transport these cells to the site of fertilization; to provide a favorable environment for a developing offspring; to move the offspring to the outside; and to produce female sex hormones.

The *primary sex organs* (gonads) of this system are the ovaries, which produce the female sex cells and sex hormones. The *accessory sex organs* of the female reproductive system are the internal and external reproductive organs (fig. 22.18; reference plates 5 and 6).

846

UNIT SIX

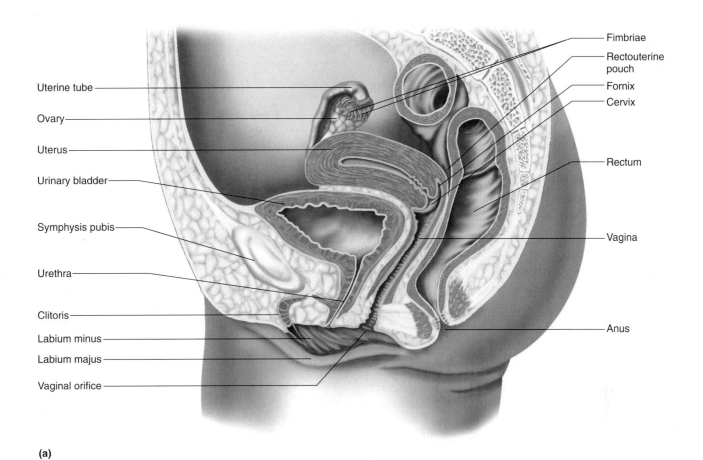

(a)

Fimbriae

Rectouterine pouch

Fornix

Cervix

Uterine tube

Ovary

Uterus

Rectum

Urinary bladder

Symphysis pubis

Urethra

Vagina

Clitoris

Labium minus

Labium majus

Vaginal orifice

Anus

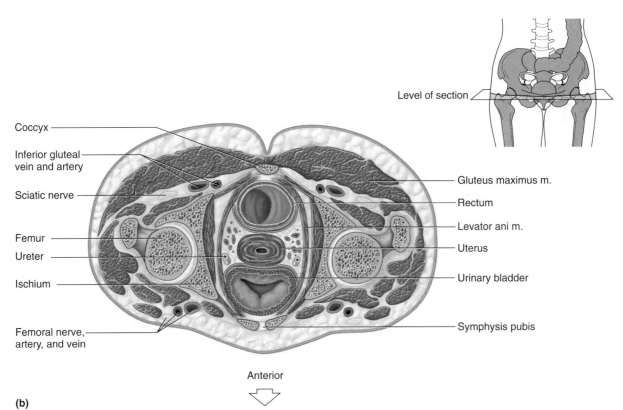

Level of section

Coccyx

Inferior gluteal vein and artery

Sciatic nerve

Femur

Ureter

Ischium

Femoral nerve, artery, and vein

Gluteus maximus m.

Rectum

Levator ani m.

Uterus

Urinary bladder

Symphysis pubis

Anterior

(b)

FIGURE 22.18

The ovaries are the primary female sex organs, and the other structures, both internal and external, are accessory sex organs. (*a*) Sagittal view. (*b*) Transverse section of the female pelvic cavity. (*m*. stands for muscle.)

Ovaries

The **ovaries** are solid, ovoid structures measuring about 3.5 centimeters in length, 2 centimeters in width, and 1 centimeter in thickness. An individual ovary is located in a shallow depression (ovarian fossa) on each side in the lateral wall of the pelvic cavity (fig. 22.18).

Ovary Attachments

Several ligaments help hold each ovary in position. The largest of these, formed by a fold of peritoneum, is called the *broad ligament.* It is also attached to the uterine tubes and the uterus.

A small fold of peritoneum, called the *suspensory ligament,* holds the ovary at its upper end. This ligament also contains the ovarian blood vessels and nerves. At its lower end, the ovary is attached to the uterus by a rounded, cordlike thickening of the broad ligament called the *ovarian ligament* (fig. 22.19).

Ovary Descent

Like the testes in a male fetus, the ovaries in a female fetus originate from masses of tissue posterior to the parietal peritoneum, near the developing kidneys. During development, these structures descend to locations just inferior to the pelvic brim, where they remain attached to the lateral pelvic wall.

Ovary Structure

The tissues of an ovary can be subdivided into two rather indistinct regions, an inner *medulla* and an outer *cortex.*

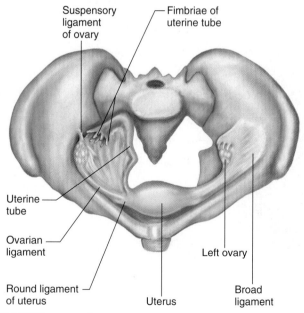

FIGURE 22.19

The ovaries are located on each side against the lateral walls of the pelvic cavity. The right uterine tube is retracted to reveal the ovarian ligament.

The ovarian medulla is mostly composed of loose connective tissue and contains many blood vessels, lymphatic vessels, and nerve fibers. The ovarian cortex consists of more compact tissue and has a granular appearance due to tiny masses of cells called *ovarian follicles.*

A layer of cuboidal epithelial cells (germinal epithelium) covers the free surface of the ovary. Just beneath this epithelium is a layer of dense connective tissue called the *tunica albuginea* (too'nĭ-kah al″bu-jin′e-ah).

1 What are the primary sex organs of the female?

2 Describe the descent of the ovary.

3 Describe the structure of an ovary.

Primordial Follicles

During prenatal development (before birth), small groups of cells in the outer region of the ovarian cortex form several million **primordial follicles.** Each of these structures consists of a single, large cell called a *primary oocyte,* which is closely surrounded by a layer of flattened epithelial cells called *follicular cells.*

Early in development, the primary oocytes begin to undergo meiosis, but the process soon halts and does not continue until puberty. Once the primordial follicles appear, no new ones form. Instead, the number of oocytes in the ovary steadily declines, as many of the oocytes degenerate. Of the several million oocytes formed originally, only a million or so remain at the time of birth, and perhaps 400,000 are present at puberty. Of these, probably fewer than 400 or 500 will be released from the ovary during the reproductive life of a female. Probably fewer than ten will go on to form a new individual!

A possible explanation for the increased incidence of chromosome defects in children of older mothers is that the eggs, having been present for several decades, had time to be extensively exposed to damaging agents, such as radiation, viruses, and toxins.

Oogenesis

Beginning at puberty, some primary oocytes are stimulated to continue meiosis. As in the case of sperm cells, the resulting cells have one-half as many chromosomes (23) in their nuclei as their parent cells—that is, one chromosome set. Egg formation is called **oogenesis** (o″o-jen′ĕ-sis).

When a primary oocyte divides, the distribution of the cytoplasm is unequal. One of the resulting cells, called a *secondary oocyte,* is large, and the other, called the *first polar body,* is very small.

The large secondary oocyte represents a future *egg cell* (ovum) that can be fertilized by uniting with a sperm cell. If this happens, the oocyte divides unequally to produce a tiny *second polar body* and a large fertilized egg cell, or **zygote** (zi′gōt), that can divide and develop into an **embryo** (em′bree-o) (fig. 22.20). An embryo is the stage of prenatal development when the rudiments of all organs form. The polar bodies have no further function, and they soon degenerate.

Formation of polar bodies may appear wasteful, but it has an important biological function. It allows for pro-

duction of an egg cell that has the massive amounts of cytoplasm and abundant organelles required to carry a zygote through the first few cell divisions, yet the right number of chromosomes.

1 Describe the major events of oogenesis.

2 How does polar body formation benefit an egg?

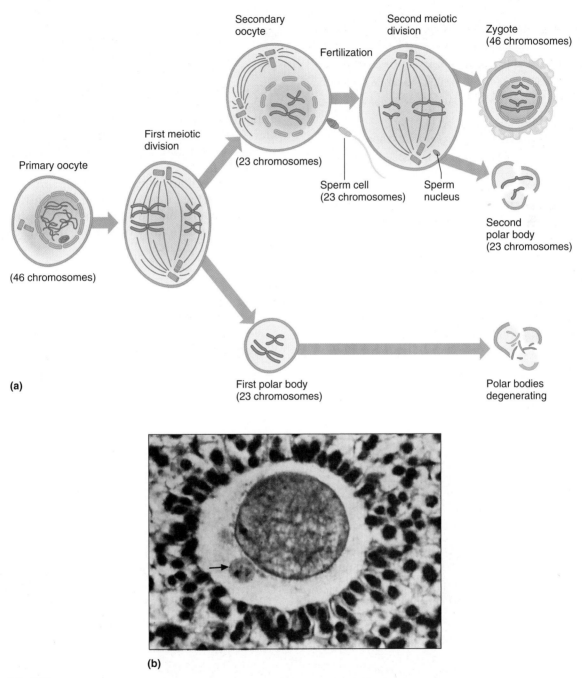

(a)

(b)

FIGURE 22.20

During oogenesis, (a) a single egg cell (secondary oocyte) results from meiosis in a primary oocyte. If the egg cell is fertilized, it generates a second polar body and becomes a zygote. (Note: The second meiotic division does not occur in the egg cell if it is not fertilized.) (b) Light micrograph of a secondary oocyte and a polar body (arrow) (700×).

Follicle Maturation

At puberty, the anterior pituitary gland secretes increased amounts of FSH, and the ovaries enlarge in response. At the same time, some of the primordial follicles mature (fig. 22.21). Within them, the oocytes enlarge, and the surrounding follicular cells divide mitotically, giving rise to a stratified epithelium composed of *granulosa cells.* A layer of glycoprotein, called the **zona pellucida** (zo′-nah pel-lu′-sid-ah), gradually separates the oocyte from the granulosa cells; at this stage, the structure is called a *primary follicle.*

Meanwhile, the ovarian cells outside the follicle become organized into two layers. The *inner vascular layer* (theca interna) is largely composed of steroid-secreting cells, plus some loose connective tissue and blood vessels. The *outer fibrous layer* (theca externa) consists of tightly packed connective tissue cells.

The follicular cells continue to proliferate, and when there are six to twelve layers of cells, irregular, fluid-filled spaces appear among them. These spaces soon join to form a single cavity (antrum), and the oocyte is pressed to one side of the follicle. At this stage, the follicle is about 0.2 millimeter in diameter and is called a *secondary follicle.*

Maturation of the follicle takes ten to fourteen days. The *mature follicle* (preovulatory, or Graafian, follicle) is about 10 millimeters or more in diameter, and its fluid-filled cavity bulges outward on the surface of the ovary, like a blister. The oocyte within the mature follicle is a large, spherical cell, surrounded by a thick zona pellucida, attached to a mantle of follicular cells (corona radiata). Processes from these follicular cells extend through the zona pellucida and supply nutrients to the oocyte (fig. 22.22).

Although as many as twenty primary follicles may begin maturing at any one time, one follicle (dominant follicle) usually outgrows the others. Typically, only the dominant follicle fully develops, and the other follicles degenerate (fig. 22.23).

Certain drugs used to treat female infertility, such as Clomid (clomiphene), may cause a woman to "superovulate." More than one follicle grows, more than one oocyte is released, and if all these oocytes are fertilized, the result is multiples. In 1997, an Iowa couple had septuplets after using a fertility drug.

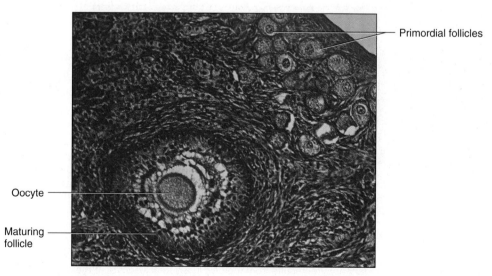

Primordial follicles

Oocyte

Maturing follicle

FIGURE 22.21
Light micrograph of the surface of a mammalian ovary (200×).

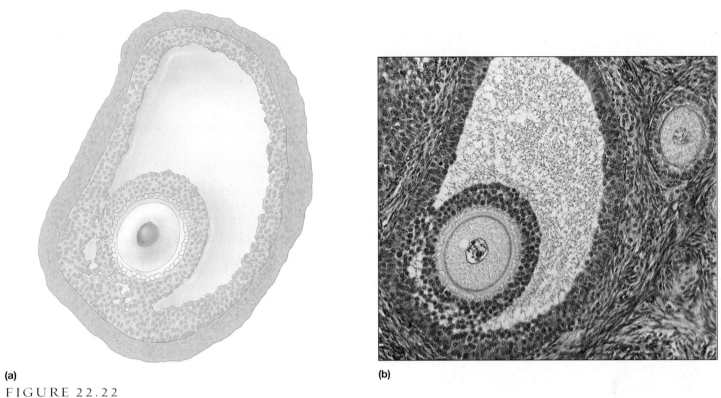

(a)

(b)

FIGURE 22.22
Ovarian follicle. (*a*) Structure of a mature (Graafian) follicle. (*b*) Light micrograph of a mature follicle (250×).

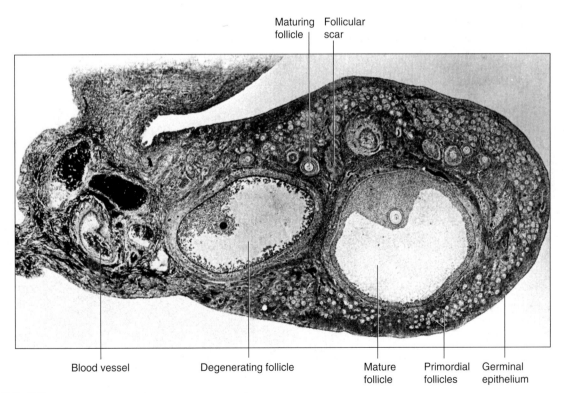

Maturing follicle | Follicular scar

Blood vessel | Degenerating follicle | Mature follicle | Primordial follicles | Germinal epithelium

FIGURE 22.23
Light micrograph of a mammalian (monkey) ovary (30×). If ovulation does not occur, the follicle degenerates.

Ovulation

As a follicle matures, its primary oocyte undergoes meiosis I, giving rise to a secondary oocyte and a first polar body. A process called **ovulation** (o"vu-la'shun) releases these cells from the follicle.

Hormonal stimulation (LH) from the anterior pituitary gland triggers ovulation, causing the mature follicle to swell rapidly and its wall to weaken. Eventually the wall ruptures, and the follicular fluid, accompanied by the oocyte, oozes outward from the surface of the ovary. Figure 22.24 shows expulsion of a mammalian oocyte.

After ovulation, the oocyte and one or two layers of follicular cells surrounding it are usually propelled to the opening of a nearby uterine tube. If the oocyte is not fertilized within a short time, it degenerates. Figure 22.25 illustrates maturation of a follicle and the release of an oocyte.

1 What changes occur in a follicle and its oocyte during maturation?

2 What causes ovulation?

3 What happens to an oocyte following ovulation?

Female Internal Accessory Organs

The internal accessory organs of the female reproductive system include a pair of uterine tubes, a uterus, and a vagina.

Uterine Tubes

The **uterine tubes** (fallopian tubes, or oviducts) are suspended by portions of the broad ligament and open near the ovaries. Each tube, which is about 10 centimeters long and 0.7 centimeters in diameter, passes medially to the uterus, penetrates its wall, and opens into the uterine cavity.

Near each ovary, a uterine tube expands to form a funnel-shaped **infundibulum** (in"fun-dib'u-lum), which

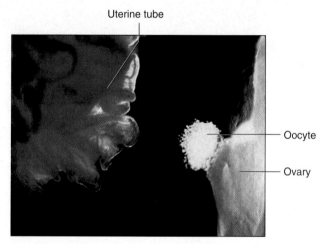

FIGURE 22.24

Light micrograph of a follicle during ovulation (75×).

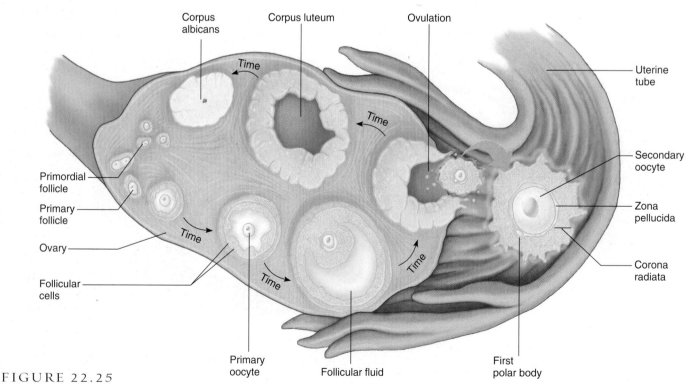

FIGURE 22.25

Within an ovary, as a follicle matures, a developing oocyte enlarges and becomes surrounded by follicular cells and fluid. Eventually, the mature follicle ruptures, releasing the secondary oocyte and layers of surrounding follicular cells.

partially encircles the ovary medially. On its margin, the infundibulum bears a number of irregular, branched extensions called **fimbriae** (fim′bre) (fig. 22.26). Although the infundibulum generally does not touch the ovary, one of the larger extensions (ovarian fimbria) connects directly to the ovary.

The wall of a uterine tube consists of an inner mucosal layer, a middle muscular layer, and an outer covering of peritoneum. The mucosal layer is drawn into many longitudinal folds and is lined with simple columnar epithelial cells, some of which are *ciliated* (fig. 22.27). The epithelium secretes mucus, and the cilia beat toward

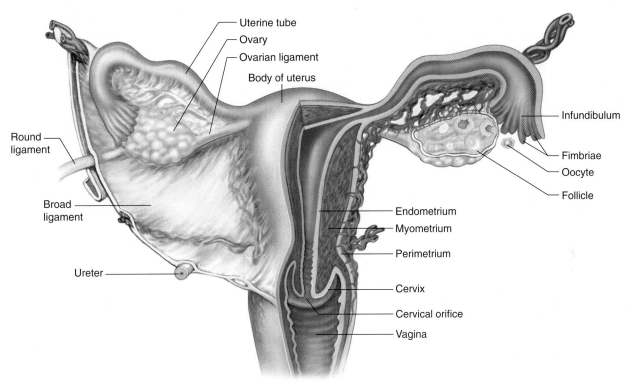

FIGURE 22.26
The funnel-shaped infundibulum of the uterine tube partially encircles the ovary (posterior view).

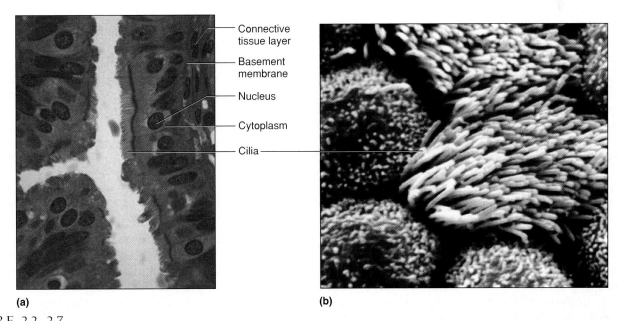

(a) (b)

FIGURE 22.27
Uterine tube. (*a*) Light micrograph of a uterine tube (800×). (*b*) Falsely colored scanning electron micrograph of ciliated cells that line the uterine tube (4,000×).

the uterus. These actions help attract the egg cell and expelled follicular fluid into the infundibulum following ovulation. Ciliary action and peristaltic contractions of the tube's muscular layer aid transport of the egg down the uterine tube.

Uterus

The **uterus** receives the embryo that develops from an egg cell that has been fertilized in the uterine tube, and sustains its development. It is a hollow, muscular organ, shaped somewhat like an inverted pear.

The *broad ligament,* which also attaches to the ovaries and uterine tubes, extends from the lateral walls of the uterus to the pelvic walls and floor, creating a drape across the top of the pelvic cavity (see fig. 22.26). A flattened band of tissue within the broad ligament, called the *round ligament,* connects the upper end of the uterus to the anterior pelvic wall (see figs. 22.19 and 22.26).

The size of the uterus changes greatly during pregnancy. In its nonpregnant, adult state, it is about 7 centimeters long, 5 centimeters wide (at its broadest point), and 2.5 centimeters in diameter. The uterus is located medially within the anterior portion of the pelvic cavity, superior to the vagina, and is usually bent forward over the urinary bladder.

The upper two-thirds, or *body,* of the uterus has a dome-shaped top, called the *fundus,* and is joined by the uterine tubes, which enter its wall at its broadest part. The lower one-third, or neck, of the uterus is called the **cervix.** This tubular part extends downward into the upper portion of the vagina. The cervix surrounds the opening called the *cervical orifice* (ostium uteri), through which the uterus opens to the vagina.

The uterine wall is thick and composed of three layers (fig. 22.28). The **endometrium** is the inner mucosal layer lining the uterine cavity. It is covered with columnar epithelium and contains abundant tubular glands. The **myometrium,** a very thick, muscular layer, largely consists of bundles of smooth muscle fibers in longitudinal, circular, and spiral patterns and is interlaced with connective tissues. During the monthly female reproductive cycles and during pregnancy, the endometrium and myometrium extensively change. The **perimetrium** consists of an outer serosal layer, which covers the body of the uterus and part of the cervix.

Vagina

The **vagina** is a fibromuscular tube, about 9 centimeters long, that extends from the uterus to the outside. It conveys uterine secretions, receives the erect penis during sexual intercourse, and provides the open channel for the offspring during birth.

The vagina extends upward and back into the pelvic cavity. It is posterior to the urinary bladder and urethra, anterior to the rectum, and attached to these structures by connective tissues. The upper one-fourth of the vagina is

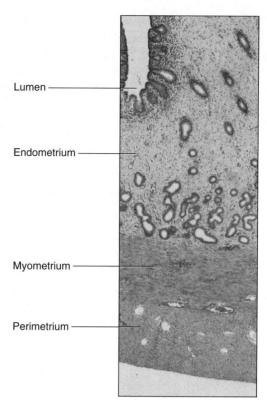

FIGURE 22.28
Light micrograph of the uterine wall (10.5×).

separated from the rectum by a pouch (rectouterine pouch). The tubular vagina also surrounds the end of the cervix, and the recesses between the vaginal wall and the cervix are termed *fornices* (sing., *fornix*). The fornices are clinically important because they are thin-walled and allow the physician to palpate the internal abdominal organs during a physical examination. Also, the posterior fornix, which is somewhat longer than the others, provides a surgical access to the peritoneal cavity through the vagina.

The *vaginal orifice* is partially closed by a thin membrane of connective tissue and stratified squamous epithelium called the **hymen.** A central opening of varying size allows uterine and vaginal secretions to pass to the outside.

The vaginal wall consists of three layers. The inner *mucosal layer* is stratified squamous epithelium and is drawn into many longitudinal and transverse ridges (vaginal rugae). This layer lacks mucous glands; the mucus found in the lumen of the vagina comes from the glands of the cervix and the vestibular glands at the mouth of the vagina.

The middle *muscular layer* of the vagina mainly consists of smooth muscle fibers in longitudinal and circular patterns. At the lower end of the vagina is a thin band of striated muscle. This band helps close the vaginal opening; however, a voluntary muscle (bulbospongiosus) is primarily responsible for closing this orifice.

The outer *fibrous layer* consists of dense connective tissue interlaced with elastic fibers. It attaches the vagina to surrounding organs.

1 How does an egg cell move into the infundibulum following ovulation?

2 How is a secondary oocyte moved along a uterine tube?

3 Describe the structure of the uterus.

4 What is the function of the uterus?

5 Describe the structure of the vagina.

Female External Reproductive Organs

The *external accessory organs* of the female reproductive system include the labia majora, the labia minora, the clitoris, and the vestibular glands. These structures that surround the openings of the urethra and vagina compose the **vulva** (fig. 22.29).

Labia Majora

The **labia majora** (sing., *labium majus*) enclose and protect the other external reproductive organs. They correspond to the scrotum of the male and are composed of rounded folds of adipose tissue and a thin layer of smooth muscle, covered by skin. On the outside, this skin includes hairs, sweat glands, and sebaceous glands, whereas on the inside, it is thinner and hairless.

The labia majora lie close together and are separated longitudinally by a cleft (pudendal cleft), which includes the urethral and vaginal openings. At their anterior ends,

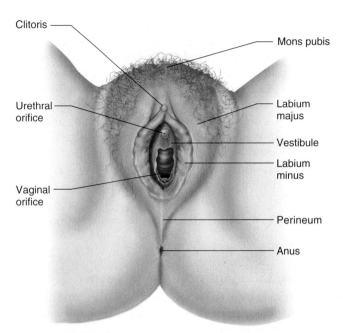

FIGURE 22.29
Female external reproductive organs and associated structures.

the labia merge to form a medial, rounded elevation of adipose tissue called the *mons pubis,* which overlies the symphysis pubis. At their posterior ends, the labia taper and merge into the perineum near the anus.

Labia Minora

The **labia minora** (sing., *labium minus*) are flattened longitudinal folds between the labia majora, and they extend along either side of a space called the **vestibule.** They are composed of connective tissue richly supplied with blood vessels, causing a pinkish appearance. Stratified squamous epithelium covers this tissue. Posteriorly, the labia minora merge with the labia majora, whereas anteriorly, they converge to form a hoodlike covering around the clitoris.

Clitoris

The **clitoris** (klit'o-ris) is a small projection at the anterior end of the vulva between the labia minora. It is usually about 2 centimeters long and 0.5 centimeter in diameter, including a portion embedded in surrounding tissues. The clitoris corresponds to the penis and has a similar structure. It is composed of two columns of erectile tissue called *corpora cavernosa.* A septum separates these columns, which are covered with dense connective tissue.

At the root of the clitoris, the corpora cavernosa diverge to form *crura,* which, in turn, attach to the sides of the pubic arch. At its anterior end, a small mass of erectile tissue forms a **glans,** which is richly supplied with sensory nerve fibers.

Vestibule

The labia minora enclose the vestibule. The vagina opens into the posterior portion of the vestibule, and the urethra opens in the midline, just anterior to the vagina and about 2.5 centimeters posterior to the glans of the clitoris.

A pair of **vestibular glands** (Bartholin's glands), corresponding to the bulbourethral glands in the male, lie on either side of the vaginal opening. Their ducts open into the vestibule near the lateral margins of the vaginal orifice.

Beneath the mucosa of the vestibule on either side is a mass of vascular erectile tissue. These structures are called the *vestibular bulbs.* They are separated from each other by the vagina and the urethra, and they extend forward from the level of the vaginal opening to the clitoris.

1   What is the male counterpart of the labia majora? Of the clitoris?

2   Which structures are within the vestibule?

Erection, Lubrication, and Orgasm

Erectile tissues located in the clitoris and around the vaginal entrance respond to sexual stimulation. Following such stimulation, parasympathetic nerve impulses from the sacral portion of the spinal cord inhibit sympathetic control of the arteries associated with the erectile tissues, causing them to dilate. As a result, inflow of blood increases, tissues swell, and the vagina expands and elongates.

If sexual stimulation is sufficiently intense, parasympathetic impulses stimulate the vestibular glands to secrete mucus into the vestibule. This secretion moistens and lubricates the tissues surrounding the vestibule and the lower end of the vagina, facilitating insertion of the penis into the vagina. Mucus secretion continuing during sexual intercourse helps prevent irritation of tissues that might occur if the vagina remained dry.

The clitoris is abundantly supplied with sensory nerve fibers, which are especially sensitive to local stimulation. The culmination of such stimulation is orgasm, the pleasurable sensation of physiological and psychological release.

Just prior to orgasm, the tissues of the outer third of the vagina engorge with blood and swell. This action increases the friction on the penis during intercourse.

Orgasm initiates a series of reflexes involving the sacral and lumbar portions of the spinal cord. In response to these reflexes, the muscles of the perineum and the walls of the uterus and uterine tubes contract rhythmically. These contractions help transport sperm cells through the female reproductive tract toward the upper ends of the uterine tubes (fig. 22.30).

Following orgasm, the flow of blood into the erectile tissues slackens, and the muscles of the perineum and reproductive tract relax. Consequently, the organs return to a state similar to that prior to sexual stimulation. Table 22.2 summarizes the functions of the female reproductive organs.

TABLE 22.2	Functions of the Female Reproductive Organs
Organ	**Function**
Ovary	Produces egg cells and female sex hormones
Uterine tube	Conveys egg cell toward uterus; site of fertilization; conveys developing embryo to uterus
Uterus	Protects and sustains embryo during pregnancy
Vagina	Conveys uterine secretions to outside of body; receives erect penis during sexual intercourse; provides open channel for offspring during birth process
Labia majora	Enclose and protect other external reproductive organs
Labia minora	Form margins of vestibule; protect openings of vagina and urethra
Clitoris	Glans is richly supplied with sensory nerve endings associated with feelings of pleasure during sexual stimulation
Vestibule	Space between labia minora that includes vaginal and urethral openings
Vestibular glands	Secrete fluid that moistens and lubricates vestibule

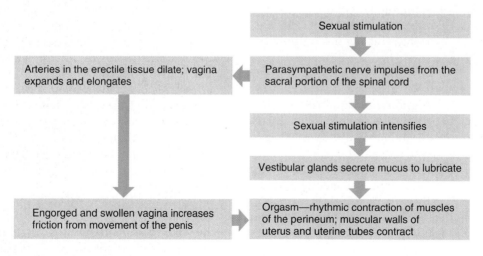

FIGURE 22.30
Mechanism of erection, lubrication, and orgasm in the human female.

1 What events result from parasympathetic stimulation of the female reproductive organs?

2 What changes occur in the vagina just prior to and during female orgasm?

3 How do the uterus and the uterine tubes respond to orgasm?

Hormonal Control of Female Reproductive Functions

The *hypothalamus,* the *anterior pituitary gland,* and the *ovaries* secrete hormones that control development and maintenance of female secondary sex characteristics, maturation of female sex cells, and changes that occur during the monthly reproductive cycle.

Female Sex Hormones

A girl's body is reproductively immature until about ten years of age. Then, the hypothalamus begins to secrete increasing amounts of GnRH, which, in turn, stimulate the anterior pituitary gland to release the gonadotropins FSH and LH. These hormones play primary roles in controlling female sex cell maturation and in producing female sex hormones.

Several tissues, including the ovaries, the adrenal cortices, and the placenta (during pregnancy), secrete female sex hormones. These hormones include the group of **estrogens** (es′tro-jenz) and **progesterone** (pro-jes′tĭ-rōn). *Estradiol* is the most abundant of the estrogens, which also include *estrone* and *estriol.*

The primary source of estrogens (in a nonpregnant female) is the ovaries, although some estrogens are also synthesized in adipose tissue from adrenal androgens. At puberty, under the influence of the anterior pituitary gland, the ovaries secrete increasing amounts of estrogens. Estrogens stimulate enlargement of accessory organs, including the vagina, uterus, uterine tubes, and ovaries, as well as the external structures, and is also responsible for the development and maintenance of female *secondary sex characteristics.* These are listed in figure 22.31 and include the following:

1. Development of the breasts and the ductile system of the mammary glands within the breasts.

2. Increased deposition of adipose tissue in the subcutaneous layer generally and in the breasts, thighs, and buttocks particularly.

3. Increased vascularization of the skin.

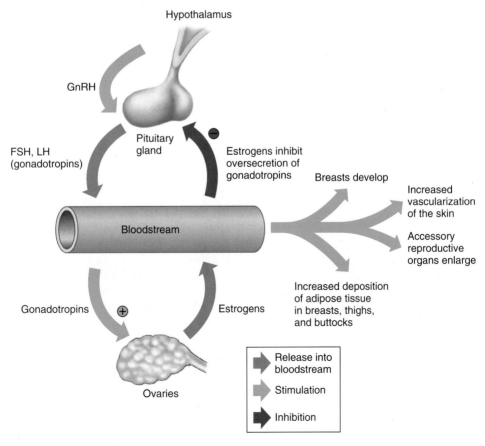

FIGURE 22.31
Control of female secondary sex development. Estrogens inhibit LH and FSH during most of the reproductive cycle.

The ovaries are also the primary source of progesterone (in a nonpregnant female). This hormone promotes changes that occur in the uterus during the female reproductive cycle, affects the mammary glands, and helps regulate secretion of gonadotropins from the anterior pituitary gland.

Certain other changes that occur in females at puberty are related to *androgen* (male sex hormone) concentrations. For example, increased growth of hair in the pubic and axillary regions is due to androgen secreted by the adrenal cortices. Conversely, development of the female skeletal configuration, which includes narrow shoulders and broad hips, is a response to a low concentration of androgen.

Female athletes who train for endurance events, such as the marathon, typically maintain about 6% body fat. Male endurance athletes usually have about 4% body fat. This difference of 50% in proportion of body fat reflects the actions of sex hormones in males and females. *Testosterone,* the male hormone, promotes deposition of protein throughout the body and especially in skeletal muscles. Estrogens, the female hormones, deposit adipose tissue in the breasts, thighs, buttocks, and the subcutaneous layer of the skin.

1 What stimulates sexual maturation in a female?

2 Name the major female sex hormones.

3 What is the function of estrogens?

4 What is the function of androgen in a female?

Female Reproductive Cycle

The female reproductive cycle is characterized by regular, recurring changes in the endometrium, which culminate in menstrual bleeding (menses). Such cycles usually begin near the thirteenth year of life and continue into middle age, then cease.

Women athletes may have disturbed reproductive cycles, ranging from diminished menstrual flow (oligomenorrhea) to complete stoppage (amenorrhea). The more active an athlete, the more likely are menstrual problems. This effect results from a loss of adipose tissue and a consequent decline in estrogens, which adipose tissue synthesizes from adrenal androgens.

A female's first reproductive cycle (menarche) occurs after the ovaries and other organs of the female reproductive control system mature and respond to certain hormones. Then, the hypothalamic secretion of gonadotropin-releasing hormone (GnRH) stimulates the anterior pituitary gland to release threshold levels of FSH (follicle-stimulating hormone) and LH (luteinizing hormone). As its name implies, FSH stimulates maturation of an ovarian *follicle.* The granulosa cells of the follicle produce increasing amounts of estrogens and some progesterone. LH also stimulates certain ovarian cells (theca interna) to secrete precursor molecules (testosterone) used to produce estrogens.

In a young female, estrogens stimulate development of various secondary sex characteristics. Estrogens secreted during subsequent reproductive cycles continue development of these traits and maintain them. Table 22.3 summarizes the hormonal control of female secondary sex characteristics.

Increasing concentration of estrogens during the first week or so of a reproductive cycle changes the uterine lining, thickening the glandular endometrium (proliferative phase). Meanwhile, the developing follicle completes maturation, and by the fourteenth day of the cycle, the follicle appears on the surface of the ovary as a blisterlike bulge. Within the follicle, the granulosa cells, which surround the oocyte and connect it to the inner wall, loosen. Follicular fluid accumulates rapidly.

While the follicle matures, estrogens that it secretes inhibit the release of LH from the anterior pituitary gland but allow LH to be stored in the gland. Estrogens also make the anterior pituitary cells more sensitive to the action of GnRH, which is released from the hypothalamus in rhythmic pulses about ninety minutes apart.

Near the fourteenth day of follicular development, the anterior pituitary cells finally respond to the pulses of GnRH and release stored LH. The resulting surge in LH concentration, which lasts for about thirty-six hours, weakens and ruptures the bulging follicular wall. At the same time, the oocyte and follicular fluid escape from the ovary (ovulation).

Following ovulation, the remnants of the follicle and the theca interna within the ovary change rapidly. The

TABLE 22.3	Hormonal Control of Female Secondary Sex Characteristics

1. The hypothalamus releases GnRH, which stimulates the anterior pituitary gland.
2. The anterior pituitary gland secretes FSH and LH.
3. FSH stimulates the maturation of a follicle.
4. Granulosa cells of the follicle produce and secrete estrogens; LH stimulates certain cells to secrete estrogen precursor molecules.
5. Estrogens are responsible for the development and maintenance of most of the female secondary sex characteristics.
6. Concentrations of androgen affect other secondary sex characteristics, including skeletal growth and growth of hair.
7. Progesterone, secreted by the ovaries, affects cyclical changes in the uterus and mammary glands.

space containing the follicular fluid fills with blood, which soon clots, and under the influence of LH, the follicular and thecal cells expand to form a temporary glandular structure within the ovary, called a **corpus luteum** ("yellow body") (see fig. 22.25).

Follicular cells secrete some progesterone during the first part of the reproductive cycle. However, corpus luteum cells secrete abundant progesterone and estrogens during the second half of the cycle. Consequently, as a corpus luteum is established, the blood concentration of progesterone increases sharply.

Progesterone causes the endometrium to become more vascular and glandular. It also stimulates the uterine glands to secrete more glycogen and lipids (secretory phase). As a result, the endometrial tissues fill with fluids containing nutrients and electrolytes, which provide a favorable environment for the development of an embryo.

High levels of estrogens and progesterone inhibit the release of LH and FSH from the anterior pituitary gland. Consequently, no other follicles are stimulated to develop when the corpus luteum is active. However, if the oocyte released at ovulation is not fertilized, the corpus luteum begins to degenerate (regress) about the twenty-fourth day of the cycle. Eventually, connective tissue replaces it. The remnant of such a corpus luteum is called a *corpus albicans* (see fig. 22.25).

When the corpus luteum ceases to function, concentrations of estrogens and progesterone decline rapidly, and in response, blood vessels in the endometrium constrict. This reduces the supply of oxygen and nutrients to the thickened endometrium, and these lining tissues (decidua) soon disintegrate and slough off. At the same time, blood escapes from damaged capillaries, creating a flow of blood and cellular debris, which passes through the vagina as the *menstrual flow* (menses). This flow usually begins about the twenty-eighth day of the cycle and continues for three to five days, while the concentrations of estrogens are relatively low.

The beginning of the menstrual flow marks the end of a reproductive cycle and the beginning of a new cycle. This cycle is summarized in table 22.4 and diagrammed in figure 22.32.

Because the blood concentrations of estrogens and progesterone are low at the beginning of the reproductive cycle, the hypothalamus and anterior pituitary gland are no longer inhibited. Consequently, the concentrations of FSH and LH soon increase, and a new follicle is stimulated to mature. As this follicle secretes estrogens, the uterine lining undergoes repair, and the endometrium begins to thicken again. Clinical Application 22.3 addresses some causes of infertility in the female.

Menopause

After puberty, reproductive cycles continue at regular intervals into the late forties or early fifties, when they usually become increasingly irregular. Then, in a few months

TABLE 22.4	Major Events in a Reproductive Cycle

1. The anterior pituitary gland secretes FSH and LH.
2. FSH stimulates maturation of a follicle.
3. Granulosa cells of the follicle produce and secrete estrogens.
   a. Estrogens maintain secondary sex traits.
   b. Estrogens cause the uterine lining to thicken.
4. The anterior pituitary gland releases a surge of LH, which stimulates ovulation.
5. Follicular and thecal cells become corpus luteum cells, which secrete estrogens and progesterone.
   a. Estrogens continue to stimulate uterine wall development.
   b. Progesterone stimulates the uterine lining to become more glandular and vascular.
   c. Estrogens and progesterone inhibit secretion of FSH and LH from the anterior pituitary gland.
6. If the egg cell is not fertilized, the corpus luteum degenerates and no longer secretes estrogens and progesterone.
7. As the concentrations of luteal hormones decline, blood vessels in the uterine lining constrict.
8. The uterine lining disintegrates and sloughs off, producing a menstrual flow.
9. The anterior pituitary gland is no longer inhibited and again secretes FSH and LH.
10. The reproductive cycle repeats.

or years, the cycles cease altogether. This period in life is called **menopause** (men'o-pawz), or female climacteric.

The cause of menopause is aging of the ovaries. After about thirty-five years of cycling, few primary follicles remain to respond to pituitary gonadotropins. Consequently, the follicles no longer mature, ovulation does not occur, and the blood concentration of estrogens plummets, although many women continue to synthesize some estrogens from adrenal androgens.

As a result of reduced concentrations of estrogens and lack of progesterone, the female secondary sex characteristics may change. The breasts, vagina, uterus, and uterine tubes may shrink, and the pubic and axillary hair may thin. The epithelial linings associated with urinary and reproductive organs may thin. There may be increased loss of bone matrix (osteoporosis) and thinning of the skin. Because the pituitary secretions of FSH and LH are no longer inhibited, these hormones may be released continuously for some time.

About 50% of women reach menopause by age fifty, and 85% reach it by age fifty-two. Of these, perhaps 20% have no unusual health effects—they simply stop menstruating. However, about 50% of women experience unpleasant vasomotor signs during menopause, including sensations of heat in the face, neck, and upper body called "hot flashes." Such a sensation may last for thirty seconds to five minutes and may be accompanied by chills and sweating. Women may also experience headache, backache, and fatigue during menopause. These vasomotor symptoms may result from changes in

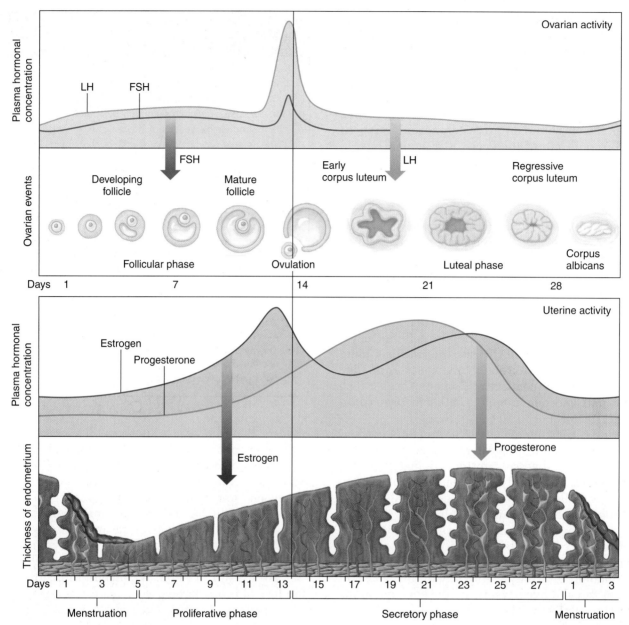

FIGURE 22.32
Major events in the female reproductive cycle.

the rhythmic secretion of GnRH by the hypothalamus in response to declining concentrations of sex hormones.

Before 2002, approximately 38 percent of women past menopause in the United States received estrogen replacement therapy (ERT), in the form of patches or oral medication, to alleviate signs of menopause and prevent osteoporosis. Long-term studies revealed, however, that the therapy may not only offer no benefits, but can increase the risk of breast cancer in some women.

1 Trace the events of the female reproductive cycle.

2 What effect does progesterone have on the endometrium?

3 What causes menstrual flow?

4 What are some changes that may occur at menopause?

Mammary Glands

The **mammary glands** are accessory organs of the female reproductive system that are specialized to secrete milk following pregnancy.

FEMALE INFERTILITY

For one out of six couples, trying for parenthood is a time of increasing concern, as pregnancy remains elusive. Physicians define infertility as the inability to conceive after a year of trying. A physical cause is found in 90% of cases, and 60% of the time, the abnormality lies in the female's reproductive system. Some specialists use the term subfertility to distinguish individuals and couples who can conceive unaided, but for whom this may take longer than is usual.

One of the more common causes of female infertility is hyposecretion of gonadotropic hormones from the anterior pituitary gland, followed by failure to ovulate (anovulation). This type of anovulatory cycle can sometimes be detected by testing the female's urine for *pregnanediol,* a product of progesterone metabolism. Since the concentration of progesterone normally rises following ovulation, no increase in pregnanediol in the urine during the latter part of the reproductive cycle suggests lack of ovulation.

Fertility specialists can treat absence of ovulation due to too little secretion of gonadotropic hormones by administering hCG (obtained from human placentas) or another ovulation-stimulating biochemical, human menopausal gonadotropin (hMG), which contains LH and FSH and is obtained from urine of women who are past menopause. However, either hCG or hMG may overstimulate the ovaries and cause many follicles to release egg cells simultaneously, resulting in multiple births if fertilization occurs.

Another cause of female infertility is *endometriosis,* in which tissue resembling the inner lining of the uterus (endometrium) grows in the abdominal cavity. This may happen if small pieces of the endometrium move up through the uterine tubes during menses and implant in the abdominal cavity. Here the tissue changes as it would in the uterine lining during the reproductive cycle. However, when the tissue begins to break down at the end of the cycle, it cannot be expelled to the outside. Instead, material remains in the abdominal cavity where it may irritate the lining (peritoneum) and cause considerable abdominal pain.

These breakdown products also stimulate formation of fibrous tissue (fibrosis), which may encase the ovary and prevent ovulation or obstruct the uterine tubes. Conception becomes impossible.

Some women become infertile as a result of infections, such as gonorrhea. Infections can inflame and obstruct the uterine tubes or stimulate production of viscous mucus that can plug the cervix and prevent entry of sperm.

The first step in finding the right treatment for a particular patient is to determine the cause of the infertility. Table 22C describes diagnostic tests that a woman who is having difficulty conceiving may undergo. ■

TABLE 22C	Tests to Assess Female Infertility
Test	**What It Checks**
Hormone levels	If ovulation occurs
Ultrasound	Placement and appearance of reproductive organs and structures
Postcoital test	Cervix examined soon after unprotected intercourse to see if mucus is thin enough to allow sperm through
Endometrial biopsy	Small piece of uterine lining sampled and viewed under microscope to see if it can support an embryo
Hysterosalpingogram	Dye injected into uterine tube and followed with scanner shows if tube is clear or blocked
Laparoscopy	Small, lit optical device inserted near navel to detect scar tissue blocking tubes, which ultrasound may miss
Laparotomy	Scar tissue in tubes removed through incision made for laparoscopy

Location of the Glands

The mammary glands are located in the subcutaneous tissue of the anterior thorax within the hemispherical elevations called *breasts.* The breasts overlie the *pectoralis major* muscles and extend from the second to the sixth ribs and from the sternum to the axillae.

A *nipple* is located near the tip of each breast at about the level of the fourth intercostal space. It is surrounded by a circular area of pigmented skin called the *areola* (fig. 22.33).

Structure of the Glands

A mammary gland is composed of fifteen to twenty irregularly shaped lobes. Each lobe contains glands (alveolar glands) and a duct (lactiferous duct) that leads to the nipple and opens to the outside. Dense connective and adipose tissues separate the lobes. These tissues also support the glands and attach them to the fascia of the underlying pectoral muscles. Other connective tissue, which forms dense strands called *suspensory ligaments,* extends inward from the dermis of the breast to the fascia, helping support

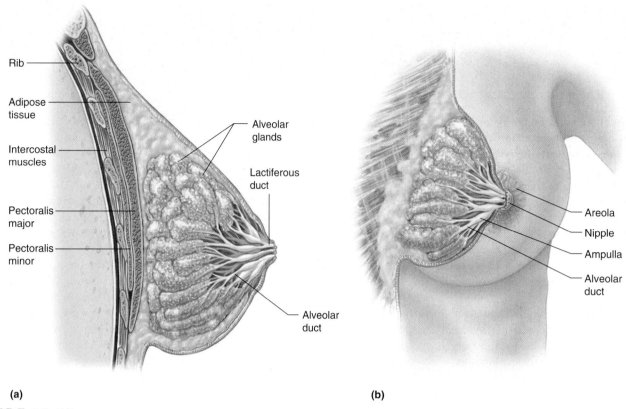

FIGURE 22.33
Structure of the female breast and mammary glands. (*a*) Sagittal section. (*b*) Anterior view.

the breast's weight. Clinical Application 22.4 discusses breast cancer.

Development of the Breasts

The mammary glands of boys and girls are similar. As children reach *puberty,* the glands in males do not develop, whereas ovarian hormones stimulate development of the glands in females. As a result, the alveolar glands and ducts enlarge, and fat is deposited so that each breast becomes surrounded by adipose tissue, except for the region of the areola. Chapter 23 (p. 904) describes the hormonal mechanism that stimulates mammary glands to produce and secrete milk.

1 Describe the structure of a mammary gland.

Birth Control

Birth control is the voluntary regulation of the number of offspring produced and the time they are conceived. This control requires a method of **contraception** (kon″trah-sep′shun) designed to avoid fertilization of an egg cell following sexual intercourse or to prevent implantation of a very early embryo. Table 22.5 describes several contraceptive approaches and devices and indicates their effectiveness.

Coitus Interruptus

Coitus interruptus is the practice of withdrawing the penis from the vagina before ejaculation, preventing entry of sperm cells into the female reproductive tract. This method of contraception often proves unsatisfactory and may result in pregnancy, since a male may find it difficult to withdraw just prior to ejaculation. Also, some semen containing sperm cells may reach the vagina before ejaculation occurs.

Rhythm Method

The *rhythm method* (also called timed coitus or natural family planning) requires abstinence from sexual intercourse a few days before and a few days after ovulation. The rhythm method results in a relatively high rate of pregnancy because accurately identifying infertile times to have intercourse is difficult. Another disadvantage of the rhythm method is that it requires adherence to a particular pattern of behavior and restricts spontaneity in sexual activity.

1 Why is coitus interruptus unreliable?

2 Describe the idea behind the rhythm method of contraception.

3 What factors make the rhythm method less reliable than some other methods of contraception?

| TABLE 22.5 | Birth Control Methods | | | 863 |

	Method	Mechanism	Advantages	Disadvantages	Pregnancies per Year per 100 Women*
	None				85
BARRIER AND SPERMICIDAL	Condom	Worn over penis or within vagina, keeps sperm out of vagina	Protection against sexually transmitted diseases (latex only)	Disrupts spontaneity, can break, reduces sensation in male	2–12
	Condom and spermicide	Worn over penis or within vagina, keeps sperm out of vagina, and kills sperm that escape	Protection against sexually transmitted diseases (latex only)	Disrupts spontaneity, reduces sensation in male	2–5
	Diaphragm and spermicide	Kills sperm and blocks uterus	Inexpensive	Disrupts spontaneity, messy, needs to be fitted by doctor	6–18
	Cervical cap and spermicide	Kills sperm and blocks uterus	Inexpensive, can be left in 24 hours	May slip out of place, messy, needs to be fitted by doctor	6–18
	Spermicidal foam or jelly	Kills sperm and blocks vagina	Inexpensive	Messy	3–21
	Spermicidal suppository	Kills sperm and blocks vagina	Easy to use and carry	Irritates 25% of users, male and female	3–15

HORMONAL	Combination estrogen and progestin
	  Lunelle™ injection
	  Nuvaring®
	  Ortho Evra® patch
	  Birth control pill
		Prevents ovulation and implantation, thickens cervical mucus	Does not interrupt spontaneity, lowers risk of some cancers, decreases menstrual flow	Raises risk of cardiovascular disease in some women, causes weight gain and breast tenderness	3

	Minipill	Thickens cervical mucus	Does not interrupt spontaneity	Menstrual changes	5
	Medroxyprogesterone acetate (Depo-Provera)	Prevents ovulation, alters uterine lining	Easy to use	Menstrual changes, weight gain	0.3
	Progesterone implant	Prevents ovulation, thickens cervical mucus	Easy to use	Menstrual changes	0.3
BEHAVIORAL	Rhythm method	No intercourse during fertile times	No cost	Difficult to do, hard to predict timing	20
	Withdrawal (coitus interruptus)	Removal of penis from vagina before ejaculation	No cost	Difficult to do	4–18
SURGICAL	Vasectomy	Sperm cells never reach penis	Permanent, does not interrupt spontaneity	Requires minor surgery	0.15
	Tubal ligation	Egg cells never reach uterus	Permanent, does not interrupt spontaneity	Requires surgery, entails some risk of infection	0.4
OTHER	Intrauterine device	Prevents implantation	Does not interrupt spontaneity	Severe menstrual cramps, increases risk of infection	3

*The lower figures apply when the contraceptive device is used correctly. The higher figures reflect human error in using birth control.

The effectiveness of the rhythm method can sometimes be increased by measuring and recording the woman's body temperature when she awakes each morning for several months. Body temperature typically rises about 0.6°F immediately following ovulation. However, this technique does not work for all women. More helpful may be an "ovulation predictor kit" that detects the surge in LH preceding ovulation.

Mechanical Barriers

Mechanical barriers prevent sperm cells from entering the female reproductive tract during sexual intercourse. One such device males use is a *condom*. It is a thin latex or natural membrane sheath placed over the erect penis before intercourse to prevent semen from entering the vagina (fig. 22.34a). A condom is inexpensive, and it may also help protect the user against contracting sexually transmitted diseases and prevent him from spreading them. However, men often feel that a condom decreases

TREATING BREAST CANCER

One in eight women will develop breast cancer at some point in her life (table 22D). Breast cancer is really several illnesses. As information on the human genome reveals the cellular and molecular characteristics that distinguish subtypes of the disease, treatments old and new are being increasingly tailored to individuals, at the time of diagnosis. This "rational" approach may delay progression of disease and increase survival rate for many women and enable them to avoid drug treatments that will not work.

Warning Signs

Changes that could signal breast cancer include a small area of thickened tissue, a dimple, a change in contour, or a flattened nipple or one that points in an unusual direction or produces a discharge. A woman can note these changes by performing a monthly "breast self-exam," in which she lies flat on her back with the arm raised behind her head and systematically feels all parts of each breast. But sometimes breast cancer gives no warning at all—early signs of fatigue and feeling ill may not occur until the disease has spread beyond the breast.

If a woman finds a lump in a breast, the next step is a physical exam, where a health-care provider palpates the breast and does a mammogram, which is an X-ray scan that can pinpoint the location and approximate extent of abnormal tissue (fig. 22C). An ultrasound scan can distinguish between a cyst (a fluid-filled sac of glandular tissue) and a tumor (a solid mass). If an area is suspicious, a thin needle is used to take a biopsy (sample) of the tissue, whose cells will be scrutinized for the telltale characteristics of cancer.

Eighty percent of the time, a breast lump is a sign of fibrocystic breast disease, which is benign (noncancerous). The lump

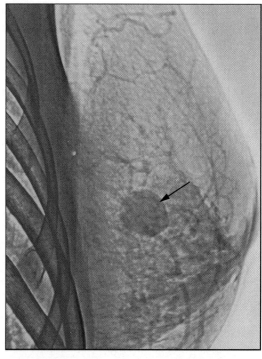

FIGURE 22C
Mammogram of a breast with a tumor (arrow).

TABLE 22D	Breast Cancer Risk		
By Age	**Odds**	**By Age**	**Odds**
25	1 in 19,608	60	1 in 24
30	1 in 2,525	65	1 in 17
35	1 in 622	70	1 in 14
40	1 in 217	75	1 in 11
45	1 in 93	80	1 in 10
50	1 in 50	85	1 in 9
55	1 in 33	95 or older	1 in 8

the sensitivity of the penis during intercourse. Also, its use interrupts the sex act.

A *female condom* resembles a small plastic bag. A woman inserts it into her vagina prior to intercourse. The device blocks sperm from reaching the cervix.

Another mechanical barrier is the *diaphragm.* It is a cup-shaped structure with a flexible ring forming the rim. The diaphragm is inserted into the vagina so that it covers the cervix, preventing entry of sperm cells into the uterus (fig. 22.34*b*).

To be effective, a diaphragm must be fitted for size by a physician, be inserted properly, and be used in conjunction with a chemical spermicide that is applied to the surface adjacent to the cervix and to the rim of the diaphragm. The device must be left in position for sev-eral hours following sexual intercourse. A diaphragm can be inserted into the vagina up to six hours before sexual contact.

Similar to but smaller than the diaphragm is the *cervical cap,* which adheres to the cervix by suction. A woman inserts it with her fingers before intercourse. Cervical caps have been used for centuries in different cultures and have been made of such varied substances as beeswax, lemon halves, paper, and opium poppy fibers.

Chemical Barriers

Chemical barrier contraceptives include creams, foams, and jellies that have spermicidal properties. Within the vagina, such chemicals create an environment that is unfavorable for sperm cells (fig. 22.34*c*).

may be a cyst or a solid, fibrous mass of connective tissue called a fibroadenoma. Treatment for fibrocystic breast disease includes taking vitamin E or synthetic androgens under a doctor's care, lowering caffeine intake, and examining unusual lumps further.

Surgery, Radiation, and Chemotherapies

If biopsied breast cells are cancerous, treatment usually begins with surgery. A lumpectomy removes a small tumor and some surrounding tissue; a simple mastectomy removes a breast; and a modified radical mastectomy removes the breast and surrounding lymph nodes but preserves the pectoral muscles. Radical mastectomies, which remove the muscles too, are rarely done anymore. In addition, a few lymph nodes are typically examined, which allows a physician to identify the ones that are affected and must be removed.

Most breast cancers are then treated with radiation and combinations of chemotherapeutic drugs, plus sometimes newer drugs that are targeted to certain types of breast cancer. Standard chemotherapies kill all rapidly dividing cells, and those used for breast cancer include fluorouracil, doxorubicin, cyclophosphamide, and methotrexate. A newer chemotherapeutic agent is paclitaxol, which was originally derived from the bark of yew trees. Drugs related to paclitaxol but that are less toxic are also used.

Many times physicians can minimize the side effects of chemotherapy with additional drugs and by using a regimen of lower but more frequent doses.

Drugs called selective estrogen receptor modulators (SERMs) are used for women whose cancer cells have receptors for estrogen. These drugs include tamoxifen and raloxifene. SERMs block the receptors so that estrogen cannot bind and trigger division of cancer cells. In contrast to standard chemotherapies, which are given for weeks or months, SERMs are taken for many years. Ongoing clinical trials are investigating whether tamoxifen can prevent cancer in certain women who are at very high risk for inherited forms of the illness.

Another new breast cancer drug, Herceptin, can help women whose cancer cells bear many receptors that bind a particular growth factor. Herceptin is a type of immune system biochemical called a monoclonal antibody. It prevents the growth factor from stimulating cell division. The type of breast cancer that it treats is extremely aggressive and tends to affect younger women. In a few cases, Herceptin has led to remarkable recoveries.

Prevention Strategies

Physicians advise women to have baseline mammograms by the age of forty, and yearly mammograms after that. Although a mammogram can detect a tumor up to two years before it can be felt, it can also miss

some tumors. Breast self-exam is also important in early detection.

Genetic tests are becoming available that can identify women who have inherited certain variants of genes—such as BRCA1, BRCA2, p53, and her-2/neu—that place them at very high risk for developing breast cancer. Women at high risk can be tested more frequently, and some have even had their breasts removed because they have inherited a gene variant that, in their families, predicts a very high risk of developing breast cancer. In one family, a genetic test told one woman whose two sisters and mother had inherited breast cancer that she had escaped their fate, and she canceled the surgery. Yet her young cousin, who thought she was free of the gene because it was inherited through her father, found by genetic testing that she would likely develop breast cancer. A subsequent mammogram revealed that the disease had already begun.

Only 5% to 10% of all breast cancers are inherited directly. This means that a person inherits one mutation that is present in all cells (a germinal mutation), and then the cancer starts when a second mutation occurs in a cell of the affected tissue (somatic mutation). Most cancers are caused by two mutations in a cell of the affected tissue. Much current research seeks to identify the environmental triggers that cause these somatic mutations. ■

Chemical barrier methods are fairly easy to use but have a high failure rate when used alone. They are more effective when used with a condom or diaphragm.

Combined Hormone Contraceptives

Combined hormone contraceptives employ estrogen and progestin in combination to prevent pregnancy. Various methods are used to administer the hormones, but all work on the same principle with about the same efficacy. A monthly injection of Lunelle™ is one such method. A small flexible chemical ring (Nuvaring®) may be inserted deep into the vagina once a month, remaining in place three out of four weeks. A plastic patch (Ortho Evra®) impregnated with the hormones may be applied to the skin on the buttocks, stomach, arm, or upper torso once a

week for three out of four weeks. The most commonly used method to deliver the hormones is orally, in pill form.

An *oral contraceptive,* or birth control pill, contains synthetic estrogen-like and progesterone-like substances. When a woman takes the pill daily, these drugs disrupt the normal pattern of gonadotropin secretion and prevent the surge in LH release that triggers ovulation. Oral contraceptives also interfere with buildup of the uterine lining that is necessary for implantation of a fertilized ovum (fig. 22.34*d*).

Oral contraceptives, if used correctly, prevent pregnancy nearly 100% of the time. However, they may cause nausea, retention of body fluids, increased pigmentation of the skin, and breast tenderness. Also, some women,

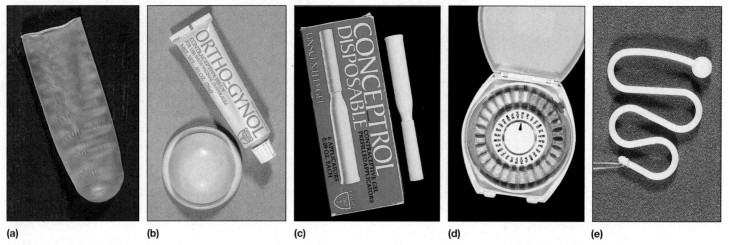

FIGURE 22.34

Devices and substances used for birth control include (*a*) male condom, (*b*) diaphragm, (*c*) spermicidal gel, (*d*) oral contraceptive, and (*e*) IUD.

(a) (b) (c) (d) (e)

particularly those over thirty-five years of age who smoke, may develop intravascular blood clots, liver disorders, or high blood pressure when using certain types of oral contraceptives.

Similar to, but different from the combined hormone birth control pill is the "minipill" which contains only progestin. The progestin thickens the cervical mucus so the sperm have difficulty reaching the egg. The minipill must be taken every day at approximately the same time for maximum effectiveness. It is still slightly less effective than combined hormone contraceptives.

Injectable Contraception

An intramuscular injection of Depo-Provera (medroxy-progesterone acetate) protects against pregnancy for three months by preventing maturation and release of a secondary oocyte. It also alters the uterine lining, making it less hospitable for a developing embryo. Because Depo-Provera is long-acting, it takes ten to eighteen months after the last injection for the effects to wear off.

Use of Depo-Provera requires a doctor's care, because of potential side effects and risks. The most common side effect is weight gain. Women with a history of breast cancer, depression, kidney disease, high blood pressure, migraine headaches, asthma, epilepsy, or diabetes, or strong family histories of these conditions, should probably not use this form of birth control.

Contraceptive Implants

A *contraceptive implant* is a set of small progesterone-containing capsules or rods, which are inserted surgically under the skin of a woman's arm or scapular region. The progesterone, which is released slowly from the implant, prevents ovulation in much the same way as do oral contraceptives. A contraceptive implant is effective for a period of up to five years, and its contraceptive action can be reversed by removing the device.

A large dose of high-potency estrogens can prevent implantation of a developing embryo in the uterus. Such a "morning-after pill," taken shortly after unprotected intercourse, promotes powerful contractions of smooth muscle in a woman's reproductive tract. This may dislodge and expel a fertilized egg or early embryo. However, if the embryo has already implanted, this treatment may harm it.

1 Describe two methods of contraception that use mechanical barriers.

2 What action can increase the effectiveness of chemical contraceptives?

3 What substances are contained in oral contraceptives?

4 Explain how oral contraceptives, injectable contraceptives, and contraceptive implants prevent pregnancy.

Intrauterine Devices

An *intrauterine device,* or *IUD,* is a small, solid object that a physician places within the uterine cavity. An IUD interferes with implantation, perhaps by inflaming the uterine tissues (fig. 22.34*e*).

An IUD may be spontaneously expelled from the uterus or produce abdominal pain or excessive menstrual bleeding. It may also harm the uterus or produce other serious health problems and should be checked at regular intervals by a physician. A few babies have been born with IUDs attached to them.

Surgical Methods

Surgical methods of contraception sterilize the male or female. In the male, a physician removes a small section of each vas deferens near the epididymis and ties the cut

866

ends of the ducts. This is a *vasectomy,* and it is a simple operation that produces few side effects, although it may cause some pain for a week or two.

After a vasectomy, sperm cells cannot leave the epididymis, thus they are excluded from the semen. However, sperm cells may already be present in portions of the ducts distal to the cuts. Consequently, the sperm count may not reach zero for several weeks.

The corresponding procedure in the female is called *tubal ligation.* The uterine tubes are cut and tied so that sperm cells cannot reach an egg cell.

Neither a vasectomy nor a tubal ligation changes hormonal concentrations or sex drives. These procedures, shown in figure 22.35, provide the most reliable forms of contraception. Reversing them requires microsurgery.

1 How does an IUD prevent pregnancy?

2 Describe the surgical methods of contraception for a male and for a female.

Sexually Transmitted Diseases

The twenty recognized **sexually transmitted diseases** (STDs) are often called "silent infections" because the early stages may not produce symptoms, especially in women (table 22.6). By the time symptoms appear, it is often too late to prevent complications or the spread of the infection to sexual partners. Because many STDs have similar symptoms, and some of the symptoms are also seen in diseases or allergies that are not sexually related, it is wise to consult a physician if one or a combination of these symptoms appears:

1. Burning sensation during urination

2. Pain in the lower abdomen

3. Fever or swollen glands in the neck

4. Discharge from the vagina or penis

5. Pain, itch, or inflammation in the genital or anal area

6. Pain during intercourse

7. Sores, blisters, bumps, or a rash anywhere on the body, particularly the mouth or genitals

8. Itchy, runny eyes

One possible complication of the STDs gonorrhea and chlamydia is **pelvic inflammatory disease,** in which bacteria enter the vagina and spread throughout the reproductive organs. The disease begins with intermittent cramps, followed by sudden fever, chills, weakness, and severe cramps. Hospitalization and intravenous antibiotics can stop the infection. The uterus and uterine tubes are often scarred, resulting in infertility and increased risk of ectopic pregnancy.

Acquired immune deficiency syndrome (AIDS) is a sexually transmitted disease. AIDS is a steady deterioration of the body's immune defenses and is caused by a virus. The body becomes overrun by infection and often cancer, diseases that the immune system usually conquers. The human immunodeficiency virus (HIV), that causes AIDS, is passed from one person to another in body fluids such as semen, blood, and milk. It is most frequently transmitted during unprotected intercourse or by using a needle containing contaminated blood.

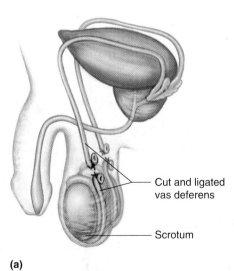

(a)

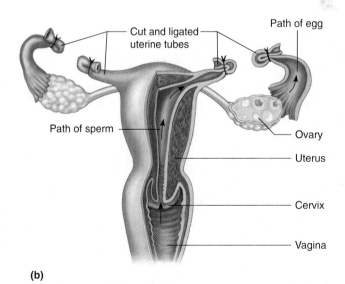

(b)

FIGURE 22.35

Surgical methods of birth control. (*a*) In a vasectomy, each vas deferens is cut and ligated. (*b*) In a tubal ligation, each uterine tube is cut and ligated.

TABLE 22.6		**Some Sexually Transmitted Diseases**				
Disease	**Cause**	**Symptoms**	**Number of Reported Cases (U.S.)**	**Effects on Fetus**	**Treatment**	**Complications**
Acquired immune deficiency syndrome	Human immunodeficiency virus	Fever, weakness, infections, cancer	> 14 million (infected)	Exposure to HIV and other infections	Drugs to treat or delay symptoms; no cure	Dementia
Chlamydia infection	*Chlamydia trachomatis* bacteria	Painful urination and intercourse, mucous discharge from penis or vagina	3–10 million	Premature birth, blindness, pneumonia	Antibiotics	Pelvic inflammatory disease, infertility, arthritis, ectopic pregnancy
Genital herpes	Herpes virus type II	Genital sores, fever	20 million	Brain damage, stillbirth	Antiviral drug (acyclovir)	Increased risk of cervical cancer
Genital warts	Human papilloma virus	Warts on genitals	1 million	None known	Chemical or surgical removal	Increased risk of cervical cancer
Gonorrhea	*Neisseria gonorrhoeae* bacteria	In women, usually none; in men, painful urination	2 million	Blindness, stillbirth	Antibiotics	Arthritis, rash, infertility, pelvic inflammatory disease
Syphilis	*Treponema pallidum* bacteria	Initial chancre sore usually on genitals or mouth; rash 6 months later; several years with no symptoms as infection spreads; finally damage to heart, liver, nerves, brain	90,000	Miscarriage, premature birth, birth defects, stillbirth	Antibiotics	Dementia

1 Why are sexually transmitted diseases often called "silent infections"?

2 Why are sexually transmitted diseases sometimes difficult to diagnose?

3 What are some common symptoms of sexually transmitted diseases?

Clinical Terms Related to the Reproductive Systems

amenorrhea (a-men″o-re′ah) Absence of menstrual flow, usually due to a disturbance in hormonal concentrations.

conization (ko″nĭ-za′shun) Surgical removal of a cone of tissue from the cervix for examination.

curettage (ku″rĕ-tahzh′) Surgical procedure in which the cervix is dilated and the endometrium of the uterus is scraped (commonly called D and C, for dilation and curettage).

dysmenorrhea (dis″men-o-re′ah) Painful menstruation.

endometritis (en″do-mĕ-tri′tis) Inflammation of the uterine lining.

epididymitis (ep″ĭ-did″i-mi′tis) Inflammation of the epididymis.

hematometra (hem″ah-to-me′trah) Accumulation of menstrual blood within the uterine cavity.

hydrocele (hi′dro-seal) Enlarged scrotum caused by accumulation of fluid along the spermatic cord.

hypospadias (hi″po-spay′dee-us) Male developmental anomaly in which the urethra opens on the underside of the penis.

hysterectomy (his″te-rek′to-me) Surgical removal of the uterus.

mastitis (mas″ti′tis) Inflammation of a mammary gland.

oophorectomy (o″of-o-rek′to-me) or **ovariectomy** (o″va-re-ek′to-me) Surgical removal of an ovary.

oophoritis (o″of-o-ri′tis) Inflammation of an ovary.

orchiectomy (or″ke-ek′to-me) Surgical removal of a testis.

orchitis (or-ki′tis) Inflammation of a testis.

prostatectomy (pros″tah-tek′to-me) Surgical removal of a portion or all of the prostate gland.

prostatitis (pros″tah-ti′tis) Inflammation of the prostate gland.

salpingectomy (sal″pin-jek′to-me) Surgical removal of a uterine tube.

salpingitis (sal″pin-ji′tis) Inflammation of the uterine tube.

vaginitis (vaj″ĭ-ni′tis) Inflammation of the vaginal lining.

varicocele (var′ĭ-ko-sēl″) Distension of the veins within the spermatic cord.

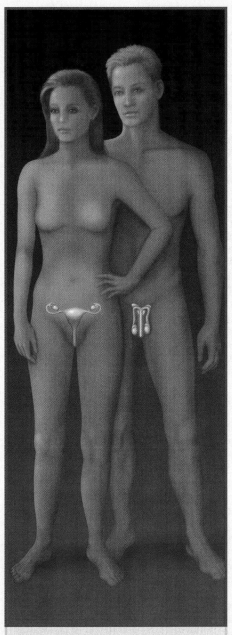

REPRODUCTIVE SYSTEMS

Gamete production, fertilization, fetal development, and childbirth are essential for survival of the species.

Integumentary System

Skin sensory receptors play a role in sexual pleasure.

Skeletal System

Bones can be a temporary source of calcium during lactation.

Muscular System

Skeletal, cardiac, and smooth muscles all play a role in reproductive processes and sexual activity.

Nervous System

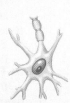

The nervous system plays a major role in sexual activity and sexual pleasure.

Endocrine System

Hormones control the production of ova in the female and sperm in the male.

Cardiovascular System

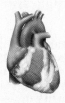

Blood pressure is necessary for the normal function of erectile tissue in the male and female.

Lymphatic System

Special mechanisms inhibit the female immune system from attacking sperm as foreign invaders.

Digestive System

Proper nutrition is essential for the formation of normal gametes and for normal fetal development during pregnancy.

Respiratory System

During pregnancy, the placenta provides oxygen to the fetus and removes carbon dioxide.

Urinary System

Male urinary and reproductive systems share common structures. Kidneys compensate for fluid loss from the reproductive systems. Pregnancy may cause fluid retention.

Introduction (page 830)

Various reproductive organs produce sex cells and sex hormones, sustain these cells, or transport them from place to place.

Organs of the Male Reproductive System (page 830)

The male reproductive organs produce and maintain sperm cells, transport these cells, and produce male sex hormones. The primary male sex organs are the testes, which produce sperm cells and male sex hormones. Accessory organs include the internal and external reproductive organs.

Testes (page 830)

1. Descent of the testes
   a. Testes originate posterior to the parietal peritoneum near the level of the developing kidneys.
   b. The gubernaculum guides the descent of the testes into the lower abdominal cavity and through the inguinal canal.
   c. Undescended testes fail to produce sperm cells because of the high abdominal temperature.
2. Structure of the testes
   a. The testes are composed of lobules separated by connective tissue and filled with the seminiferous tubules.
   b. The seminiferous tubules unite to form the rete testis that joins the epididymis.
   c. The seminiferous tubules are lined with epithelium, which produces sperm cells.
   d. The interstitial cells that produce male sex hormones occur between the seminiferous tubules.
3. Formation of sperm cells
   a. The epithelium lining the seminiferous tubules includes sustentacular cells and spermatogenic cells.
      (1) The sustentacular cells support and nourish the spermatogenic cells.
      (2) The spermatogenic cells give rise to spermatogonia.
   b. Meiosis consists of two divisions, each progressing through prophase, metaphase, anaphase, and telophase.
      (1) In the first meiotic division, homologous, replicated chromosomes (each consisting of two chromatids held together by a centromere) separate, and their number is halved.
      (2) In the second meiotic division, the chromatids part, producing four haploid cells from each diploid cell undergoing meiosis.
      (3) The meiotic products mature into sperm cells or oocyte and polar bodies.
      (4) Meiosis leads to genetic variability because of the random alignment of maternally and paternally derived chromosomes in metaphase I and crossing-over.
   c. The process of spermatogenesis produces sperm cells from spermatogonia.
      (1) Meiosis reduces the number of chromosomes in sperm cells by one-half (46 to 23).
      (2) Spermatogenesis produces four sperm cells from each primary spermatocyte.
   d. Membranous processes of adjacent sustentacular cells form a barrier within the epithelium.
      (1) The barrier separates early and advanced stages of spermatogenesis.
      (2) It helps provide a favorable environment for differentiating cells.
4. Structure of a sperm cell
   a. Sperm head contains a nucleus with 23 chromosomes.
   b. Sperm body contains many mitochondria.
   c. Sperm tail propels the cell.

Male Internal Accessory Organs (page 838)

1. Epididymis
   a. The epididymis is a tightly coiled tube on the outside of the testis that leads into the vas deferens.
   b. It stores and nourishes immature sperm cells and promotes their maturation.
2. Vas deferens
   a. The vas deferens is a muscular tube that forms part of the spermatic cord.
   b. It passes through the inguinal canal, enters the abdominal cavity, courses medially into the pelvic cavity, and ends behind the urinary bladder.
   c. It fuses with the duct from the seminal vesicle to form the ejaculatory duct.
3. Seminal vesicle
   a. The seminal vesicle is a saclike structure attached to the vas deferens.
   b. It secretes an alkaline fluid that contains nutrients, such as fructose, and prostaglandins.
   c. This secretion is added to sperm cells during emission.
4. Prostate gland
   a. This gland surrounds the urethra just below the urinary bladder.
   b. It secretes a thin, milky fluid, which enhances the motility of sperm cells and neutralizes the fluid containing the sperm cells as well as the acidic secretions of the vagina.
5. Bulbourethral glands
   a. These glands are two small structures inferior to the prostate gland.
   b. They secrete a fluid that lubricates the penis in preparation for sexual intercourse.
6. Semen
   a. Semen is composed of sperm cells and secretions of the seminal vesicles, prostate gland, and bulbourethral glands.
   b. This fluid is slightly alkaline and contains nutrients and prostaglandins.
   c. Sperm cells in semen begin to swim, but these sperm cells are unable to fertilize egg cells until they enter the female reproductive tract.

Male External Reproductive Organs (page 841)

1. Scrotum
   a. The scrotum is a pouch of skin and subcutaneous tissue that encloses the testes.
   b. The dartos muscle in the scrotal wall causes the skin of the scrotum to be held close to the testes or to hang loosely, thus regulating the temperature for sperm production and survival.

2. Penis
   a. The penis conveys urine and semen.
   b. It is specialized to become erect for insertion into the vagina during sexual intercourse.
   c. Its body is composed of three columns of erectile tissue surrounded by connective tissue.
   d. The root of the penis is attached to the pelvic arch and membranes of the perineum.
3. Erection, orgasm, and ejaculation
   a. During erection, the vascular spaces within the erectile tissue become engorged with blood as arteries dilate and veins are compressed.
   b. Orgasm is the culmination of sexual stimulation and is accompanied by emission and ejaculation.
   c. Semen moves along the reproductive tract as smooth muscle in the walls of the tubular structures contract, stimulated by a reflex.
   d. Following ejaculation, the penis becomes flaccid.

Hormonal Control of Male Reproductive Functions (page 845)

1. Hypothalamic and pituitary hormones
   The male body remains reproductively immature until the hypothalamus releases GnRH, which stimulates the anterior pituitary gland to release gonadotropins.
   a. FSH stimulates spermatogenesis.
   b. LH (ICSH) stimulates the interstitial cells to produce male sex hormones.
   c. Inhibin prevents oversecretion of FSH.
2. Male sex hormones
   a. Male sex hormones are called androgens.
   b. Testosterone is the most important androgen.
   c. Testosterone is converted into dihydrotestosterone in some organs.
   d. Androgens that fail to become fixed in tissues are metabolized in the liver and excreted.
   e. Androgen production increases rapidly at puberty.
3. Actions of testosterone
   a. Testosterone stimulates the development of the male reproductive organs and causes the testes to descend.
   b. It is responsible for the development and maintenance of male secondary sex characteristics.
4. Regulation of male sex hormones
   a. A negative feedback mechanism regulates testosterone concentration.
      (1) As the concentration of testosterone rises, the hypothalamus is inhibited, and the anterior pituitary secretion of gonadotropins is reduced.
      (2) As the concentration of testosterone falls, the hypothalamus signals the anterior pituitary to secrete gonadotropins.
   b. The concentration of testosterone remains relatively stable from day to day.

Organs of the Female Reproductive System (page 846)

The primary female sex organs are the ovaries, which produce female sex cells and sex hormones. Accessory organs are internal and external.

Ovaries (page 848)

1. Ovary attachments
   a. Several ligaments hold the ovaries in position.
   b. These ligaments include broad, suspensory, and ovarian ligaments.
2. Ovary descent
   a. The ovaries descend from posterior to the parietal peritoneum near the developing kidneys.
   b. They are attached to the pelvic wall just inferior to the pelvic brim.
3. Ovary structure
   a. The ovaries are subdivided into a medulla and a cortex.
   b. The medulla is composed of connective tissue, blood vessels, lymphatic vessels, and nerves.
   c. The cortex contains ovarian follicles and is covered by cuboidal epithelium.
4. Primordial follicles
   a. During prenatal development, groups of cells in the ovarian cortex form millions of primordial follicles.
   b. Each primordial follicle contains a primary oocyte and a layer of flattened epithelial cells.
   c. The primary oocyte begins to undergo meiosis, but the process soon halts and does not resume until puberty.
   d. The number of oocytes steadily declines throughout the life of a female.
5. Oogenesis
   a. Beginning at puberty, some oocytes are stimulated to continue meiosis.
   b. When a primary oocyte undergoes oogenesis, it gives rise to a secondary oocyte in which the original chromosome number is reduced by one-half (from 46 to 23).
   c. A secondary oocyte may be fertilized to produce a zygote.
6. Follicle maturation
   a. At puberty, FSH initiates follicle maturation.
   b. During maturation, the oocyte enlarges, the follicular cells proliferate, and a fluid-filled cavity appears and produces a secondary follicle.
   c. Ovarian cells surrounding the follicle form two layers.
   d. Usually only one follicle reaches full development.
7. Ovulation
   a. Ovulation is the release of an oocyte from an ovary.
   b. The oocyte is released when its follicle ruptures.
   c. After ovulation, the oocyte is drawn into the opening of the uterine tube.

Female Internal Accessory Organs (page 852)

1. Uterine tubes
   a. These tubes convey egg cells toward the uterus.
   b. The end of each uterine tube is expanded, and its margin bears irregular extensions.
   c. Ciliated cells that line the tube and peristaltic contractions in the wall of the tube move an egg cell into the tube's opening.
2. Uterus
   a. The uterus receives the embryo and sustains it during development.
   b. The uterine wall includes the endometrium, myometrium, and perimetrium.
3. Vagina
   a. The vagina connects the uterus to the vestibule.
   b. It receives the erect penis, conveys uterine secretions to the outside, and provides an open channel for the fetus during birth.
   c. The vaginal orifice is partially closed by a thin membrane, the hymen.
   d. Its wall consists of a mucosa, muscularis, and outer fibrous coat.

Female External Reproductive Organs (page 855)

1. Labia majora
   a. The labia majora are rounded folds of adipose tissue and skin that enclose and protect the other external reproductive parts.
   b. The upper ends form a rounded elevation over the symphysis pubis.
2. Labia minora
   a. The labia minora are flattened, longitudinal folds between the labia majora.
   b. They are well supplied with blood vessels.
3. Clitoris
   a. The clitoris is a small projection at the anterior end of the vulva; it corresponds to the male penis.
   b. It is composed of two columns of erectile tissue.
   c. Its root is attached to the sides of the pubic arch.
4. Vestibule
   a. The vestibule is the space between the labia minora that encloses the vaginal and urethral openings.
   b. The vestibular glands secrete mucus into the vestibule during sexual stimulation.
5. Erection, lubrication, and orgasm
   a. During periods of sexual stimulation, the erectile tissues of the clitoris and vestibular bulbs become engorged with blood and swollen.
   b. The vestibular glands secrete mucus into the vestibule and vagina.
   c. During orgasm, the muscles of the perineum, uterine wall, and uterine tubes contract rhythmically.

Hormonal Control of Female Reproductive Functions (page 857)

Hormones from the hypothalamus, anterior pituitary gland, and ovaries play important roles in the control of sex cell maturation and the development and maintenance of female secondary sex characteristics.

1. Female sex hormones
   a. A female body remains reproductively immature until about ten years of age when gonadotropin secretion increases.
   b. The most important female sex hormones are estrogens and progesterone.
      (1) Estrogens are responsible for the development and maintenance of most female secondary sex characteristics.
      (2) Progesterone causes changes in the uterus.
2. Female reproductive cycle
   a. The reproductive cycle is characterized by regularly recurring changes in the uterine lining culminating in menstrual flow.
   b. A reproductive cycle is initiated by FSH, which stimulates maturation of a follicle.
   c. Granulosa cells of a maturing follicle secrete estrogens, which are responsible for maintaining the secondary sex traits and thickening the uterine lining.
   d. Ovulation is triggered when the anterior pituitary gland releases a relatively large amount of LH.
   e. Following ovulation, the follicular cells and thecal cells give rise to the corpus luteum.
      (1) The corpus luteum secretes estrogens and progesterone, which cause the uterine lining to become more vascular and glandular.
      (2) If an oocyte is not fertilized, the corpus luteum begins to degenerate.
      (3) As the concentrations of estrogens and progesterone decline, the uterine lining disintegrates, causing menstrual flow.
   f. During this cycle, estrogens and progesterone inhibit the release of LH and FSH; as the concentrations of these hormones decline, the anterior pituitary secretes FSH and LH again, stimulating a new reproductive cycle.
3. Menopause
   a. Eventually the ovaries cease responding to FSH, and cycling ceases.
   b. Menopause is characterized by a low concentration of estrogens and a continuous secretion of FSH and LH.
   c. The female reproductive organs undergo varying degrees of regressive changes.

Mammary Glands (page 860)

1. Location of the glands
   a. The mammary glands are located in the subcutaneous tissue of the anterior thorax within the breasts.
   b. The breasts extend between the second and sixth ribs and from sternum to axillae.
2. Structure of the glands
   a. The mammary glands are composed of lobes that contain tubular glands.
   b. The lobes are separated by dense connective and adipose tissues.
   c. The mammary glands are connected to the nipple by ducts.
3. Development of the breasts
   a. Breasts of males remain nonfunctional.
   b. Estrogens stimulate breast development in females.
      (1) Alveolar glands and ducts enlarge.
      (2) Fat is deposited around and within the breasts.

Birth Control (page 862)

Voluntary regulation of the number of children produced and the time they are conceived is called birth control. This usually involves some method of contraception.

1. Coitus interruptus
   a. Coitus interruptus is withdrawal of the penis from the vagina before ejaculation.
   b. Some semen may be expelled from the penis before ejaculation.
2. Rhythm method
   a. Abstinence from sexual intercourse a few days before and after ovulation is the rhythm method.
   b. It is almost impossible to accurately predict the time of ovulation.
3. Mechanical barriers
   a. Males and females can use condoms.
   b. Females use diaphragms and cervical caps.
4. Chemical barriers
   a. Spermicidal creams, foams, and jellies are chemical barriers to conception.
   b. These provide an unfavorable environment in the vagina for sperm survival.
5. Combined hormone contraceptives
   a. A monthly injection, a flexible ring inserted deep into the vagina, or a plastic patch can deliver estrogen and progestin to prevent pregnancy.
   b. Tablets that contain synthetic estrogen-like and progesterone-like substances are taken by the woman.

c. They disrupt the normal pattern of gonadotropin secretion and prevent ovulation and the normal buildup of the uterine lining.
d. When used correctly, this method is almost 100% effective.
e. Some women develop undesirable side effects.
f. A minipill contains only progestin and must be taken at the same time daily.

6. Injectable contraceptives
a. Intramuscular injection with medroxyprogesterone acetate every three months.
b. High levels of hormone act similarly to oral contraceptives to prevent pregnancy.
c. Very effective if administered promptly at the end of the three months.
d. Women may experience side effects; in some women, use is contraindicated.

7. Contraceptive implants
a. A contraceptive implant consists of a set of progesterone-containing capsules or rods that are inserted under the skin.
b. Progesterone released from the implant prevents ovulation.

c. The implant is effective for years, and its action can be reversed by having it removed.

8. Intrauterine devices
a. An IUD is a solid object inserted in the uterine cavity.
b. It is thought to prevent pregnancy by interfering with implantation.
c. It may be expelled spontaneously or produce undesirable side effects.

9. Surgical methods
a. These are sterilization procedures.
(1) Vasectomy is performed in males.
(2) Tubal ligation is performed in females.
b. Surgical methods are the most reliable forms of contraception.

Sexually Transmitted Diseases (page 867)

1. Sexually transmitted diseases are passed during sexual contact and may go undetected for years.
2. Many of the sexually transmitted diseases share similar symptoms.

CRITICAL THINKING QUESTIONS

1. What changes, if any, might occur in the secondary sex characteristics of an adult male following removal of one testis? Following removal of both testes? Following removal of the prostate gland?
2. If a woman who is considering having a tubal ligation asks, "Will the operation cause me to go through my change of life early?", how would you answer?
3. What effect would it have on a woman's reproductive cycles if a single ovary were removed surgically? What effect would it have if both ovaries were removed?
4. As a male reaches adulthood, what will be the consequences if his testes have remained undescended since birth? Why?
5. What types of contraceptives provide the greatest protection against sexually transmitted diseases?
6. Some men are unable to become fathers because their spermatids do not mature into sperm. Injection of their

spermatids into their partner's secondary oocytes sometimes results in conception. A few men have fathered healthy babies this way. Why would this procedure work with spermatids but not with primary spermatocytes?

7. Understanding the causes of infertility can be valuable in developing new birth control methods. Cite a type of contraceptive based on each of the following causes of infertility: (a) failure to ovulate due to a hormonal imbalance; (b) a large fibroid tumor that disturbs the uterine lining; (c) endometrial tissue blocking uterine tubes; (d) low sperm count (too few sperm per ejaculate).

8. Sometimes, a sperm cell fertilizes a polar body rather than an oocyte. An embryo does not develop, and the fertilized polar body degenerates. Why is a polar body unable to support development of an embryo?

REVIEW EXERCISES

1. List the general functions of the male reproductive system.
2. Distinguish between the primary and accessory male reproductive organs.
3. Describe the descent of the testes.
4. Define *cryptorchidism.*
5. Describe the structure of a testis.
6. Explain the function of the sustentacular cells in the testis.
7. Outline the process of meiosis.
8. List two ways that meiosis provides genetic variability.
9. List the major steps in spermatogenesis.
10. Describe a sperm cell.
11. Describe the epididymis, and explain its function.

12. Trace the path of the vas deferens from the epididymis to the ejaculatory duct.
13. On a diagram, locate the seminal vesicles, and describe the composition of their secretion.
14. On a diagram, locate the prostate gland, and describe the composition of its secretion.
15. On a diagram, locate the bulbourethral glands, and explain the function of their secretion.
16. Describe the composition of semen.
17. Define *capacitation.*
18. Describe the structure of the scrotum.
19. Describe the structure of the penis.
20. Explain the mechanism that produces an erection of the penis.

21. Distinguish between emission and ejaculation.
22. Explain the mechanism of ejaculation.
23. Explain the role of GnRH in the control of male reproductive functions.
24. Distinguish between androgen and testosterone.
25. Define *puberty*.
26. Describe the actions of testosterone.
27. List several male secondary sex characteristics.
28. Explain the regulation of testosterone concentration.
29. List the general functions of the female reproductive system.
30. Distinguish between the primary and accessory female reproductive organs.
31. Describe how the ovaries are held in position.
32. Describe the descent of the ovaries.
33. Describe the structure of an ovary.
34. Define *primordial follicle*.
35. List the major steps in oogenesis.
36. Distinguish between a primary and a secondary follicle.
37. Describe how a follicle matures.
38. Define *ovulation*.
39. On a diagram, locate the uterine tubes, and explain their function.
40. Describe the structure of the uterus.
41. Describe the structure of the vagina.
42. Distinguish between the labia majora and the labia minora.
43. On a diagram, locate the clitoris, and describe its structure.
44. Define *vestibule*.
45. Describe the process of erection in the female reproductive organs.
46. Define *orgasm*.
47. Explain the role of GnRH in regulating female reproductive functions.
48. List several female secondary sex characteristics.
49. Define *reproductive cycle*.
50. Explain how a reproductive cycle is initiated.
51. Summarize the major events in a reproductive cycle.
52. Define *menopause*.
53. Describe how male and female sex cells are transported within the female reproductive tract.
54. Describe the structure of a mammary gland.
55. Define *contraception*.
56. List several methods of contraception, and explain how each prevents pregnancy.
57. List several sexually transmitted diseases.

W E B   C O N N E C T I O N S

Visit the Student Edition of the Online Learning Center at www.mhhe.com/shier10 **for answers to chapter questions, additional quizzes, interactive learning exercises, and other study tools.**

UNDERSTANDING WORDS

allant-, sausage: *allant*ois—tubelike structure that extends from the yolk sac into the connecting stalk of an embryo.

cleav-, to divide: *cleav*age—period of development characterized by a division of the zygote into smaller and smaller cells.

ect-, outside: *ect*oderm—outermost germ layer of embryo.

lacun-, pool: *lacun*a—space between the chorionic villi that fills with maternal blood.

lanug-, down: *lanug*o—fine hair covering the fetus.

mes-, middle: *mes*oderm—middle germ layer of embryo.

morul-, mulberry: *morul*a—embryonic structure consisting of a solid ball of about sixteen cells that resembles a mulberry.

nat-, to be born: pre*nat*al—period of development before birth.

ne-, new, young: *ne*onatal period—period of development including the first four weeks after birth.

post-, after: *post*natal period—period of development after birth.

pre-, before: *pre*natal period—period of development before birth.

sen-, old: *sen*escence—process of growing old.

troph-, nurture: *troph*oblast—cellular layer that surrounds the inner cell mass and helps nourish it.

umbil-, navel: *umbil*ical cord—structure attached to the fetal navel (umbilicus) that connects the fetus to the placenta.

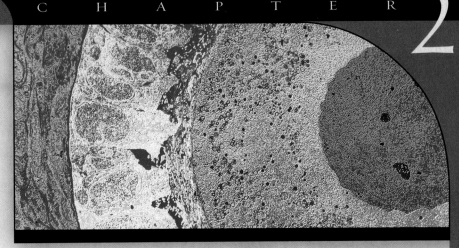

This primary oocyte will only complete meiosis if a sperm enters, and the nuclei of the female and male cell meet and merge (4,050×).

Pregnancy, Growth, and Development

CHAPTER OBJECTIVES

After you have studied this chapter, you should be able to

1. Distinguish between growth and development, prenatal and postnatal periods.

2. Define pregnancy; describe the process of fertilization and implantation.

3. Describe the major events of the period of cleavage.

4. Describe the hormonal changes and other changes in the maternal body during pregnancy.

5. Explain how the primary germ layers originate and list the structures each layer produces.

6. Describe the major events of the embryonic stage of development.

7. Describe the formation and function of the placenta.

8. Define *fetus* and describe the major events that occur during the fetal stage of development.

9. Trace the general path of blood through the fetal cardiovascular system.

10. Describe the birth process and explain the role of hormones in this process.

11. Describe the major cardiovascular and physiological adjustments that occur in the newborn.

12. Name the stages of development between the neonatal period and death and list the general characteristics of each stage.

A human uterus can best accommodate one fetus, and this is why most births are "singletons." About one in eighty pregnancies produces twins, and although these babies are often smaller and born earlier than singletons, most fare quite well. The picture isn't as bright as the number of fetuses increases. An Iowa couple became the parents of healthy septuplets in late 1997, and two of the children have lingering medical problems. A year later, a couple in Texas had octuplets, seven of whom survived. These families are two relative success stories.

After the McCaughey septuplets were born, Mario and Jane Simeone, of Tucson, Arizona, decided to tell the public that multiple pregnancies do not always have such happy endings. The Simeones learned this from their own tragedy. After six years of undergoing treatment for infertility, Jane delivered triplets on June 21, 1997, two girls and a boy, fifteen weeks premature. Within three weeks, both girls had died, and the boy, Mario Jr., remained hospitalized, gaining strength and weight. Mario Jr. came home by summer's end and has been healthy.

One in ten "multiples" does not survive to see a first birthday. Those that do are more likely to have seizures, blindness, cerebral palsy, and mental retardation than singletons. Many multiple conceptions and pregnancies end before survival is possible. ■

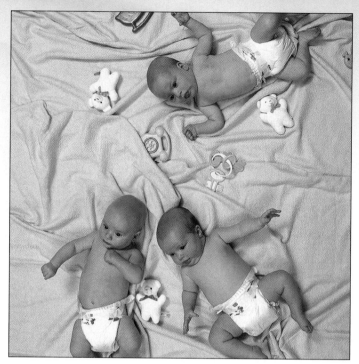

Multiples such as triplets and quadruplets are more likely to be born with health problems than "singletons."

A sperm cell and an egg cell unite, forming a zygote, and the journey of prenatal development begins. Following thirty-eight weeks of cell division, growth, and specialization into distinctive tissues and organs, a new human being enters the world.

Humans grow, develop, and age. **Growth** is an increase in size. It entails increase in cell numbers as a result of mitosis, followed by enlargement of the newly formed cells and of the body.

Development, which includes growth, is the continuous process by which an individual changes from one life phase to another. These life phases include a **prenatal period** (pre-na′tal pe′re-od), which begins with the fertilization of an egg cell and ends at birth, and a **postnatal period** (pōst-na′tal pe′re-od), which begins at birth and ends with death.

Pregnancy

Pregnancy (preg′nan-se) is the presence of a developing offspring in the uterus. It results from the union of the genetic packages of an egg cell and a sperm cell—an event called **fertilization** (fer″tǐ-lǐ za′shun).

Transport of Sex Cells

Ordinarily, before fertilization can occur, a secondary oocyte must be ovulated and enter a uterine tube. During sexual intercourse, the male deposits semen containing sperm cells in the vagina near the cervix. To reach the secondary oocyte, the sperm cells must then move upward through the uterus and uterine tube. Lashing of sperm tails and muscular contractions within the walls of the

Nausea and vomiting in pregnancy—more commonly known as morning sickness—may be a protective mechanism to shield a fetus from foods that might contain toxins or pathogens. The condition affects two in three pregnancies and coincides with the time in gestation when a woman's immune system is at its weakest. An analysis of more than 80,000 pregnant women found that they tend to have aversions to foods that spoil easily, such as eggs and meats, as well as to coffee and alcohol. Yet many pregnant women eat more fruits and vegetables than usual. In addition, in societies where the diet is mostly grains with little if any meat, incidence of morning sickness is much lower than in groups with more varied, and possibly dangerous, diets. Rates of morning sickness are highest in Japan, where raw fish is a dietary staple, and European countries, where undercooked meat is often eaten. Coincidence? Maybe. But more likely, evolution has selected for morning sickness where it correlates to, and possibly contributes to, better birth outcomes.

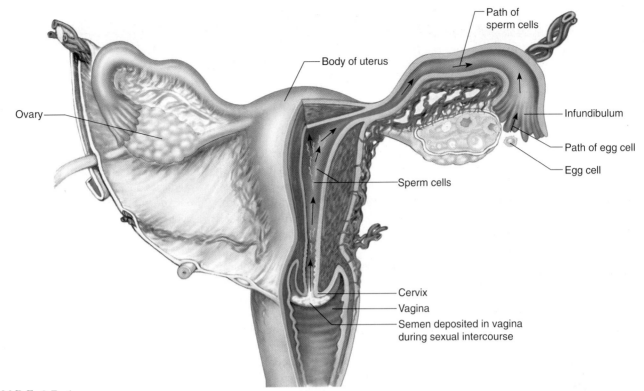

FIGURE 23.1

The paths of the egg and sperm cells through the female reproductive tract.

uterus and uterine tube, stimulated by prostaglandins in the semen, aid the sperm cells' journey. Also, under the influence of high concentrations of estrogens during the first part of the reproductive cycle, the uterus and cervix contain a thin, watery secretion that promotes sperm transport and survival. Conversely, during the latter part of the cycle, when the progesterone concentration is high, the female reproductive tract secretes a viscous fluid that hampers sperm transport and survival (fig. 23.1).

Sperm movement is inefficient. Even though as many as 200 million to 600 million sperm cells may be deposited in the vagina by a single ejaculation, only a few hundred ever reach a secondary oocyte. The journey to the upper portions of the uterine tube takes less than an hour following sexual intercourse. Although many sperm cells may reach a secondary oocyte, usually only one sperm cell actually fertilizes it (fig. 23.2). If a second sperm were to enter, the fertilized ovum would have three sets of chromosomes, and be very unlikely to develop very far. About one in a million births produces a severely deformed child who has inherited three sets of chromosomes.

A secondary oocyte may survive for only 12 to 24 hours following ovulation, whereas sperm cells may live up to 72 hours within the female reproductive tract. Consequently, sexual intercourse probably must occur earlier than 72 hours before ovulation or within 24 hours following ovulation if fertilization is to take place. From Science to Technology 23.1 describes assisted routes to conception.

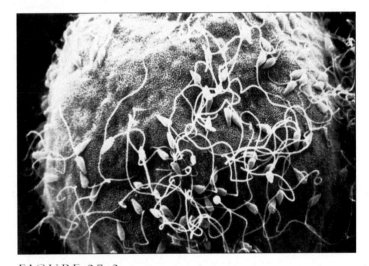

FIGURE 23.2

Scanning electron micrograph of sperm cells on the surface of an egg cell (1,200×).

Fertilization

When a sperm reaches a secondary oocyte, it invades the follicular cells that adhere to the oocyte's surface (corona radiata) and binds to the *zona pellucida* that surrounds the oocyte's cell membrane. The acrosome of a sperm cell attached to the zona pellucida releases an enzyme that helps the motile sperm penetrate the zona pellucida.

ASSISTED REPRODUCTIVE TECHNOLOGIES

Conception requires the meeting and merging of sperm and egg, which naturally occurs in the woman's reproductive tract. Abnormal gametes or blockages that impede this meeting of cells can result in infertility (inability to conceive). Assisted reproductive technologies can help couples conceive. The procedures usually involve a laboratory technique and sometimes participation of a third individual. These techniques are often costly and may take several attempts, and some have very low success rates. Most assisted reproductive technologies were developed in nonhuman animals. For example, the first artificial inseminations were performed in dogs in 1782, and the first successful *in vitro* fertilization was accomplished in 1959, in a rabbit. Here is a look at some procedures.

Donated Sperm—Artificial Insemination

In artificial insemination, a doctor places donated sperm in a woman's reproductive tract. A woman might seek artificial insemination if her partner is infertile or carries a gene for an inherited illness or if she desires to be a single parent. More than 250,000 babies have been born worldwide as a result of this procedure. The first human artificial inseminations by donor were done in the 1890s. For many years, physicians donated sperm, and this became a way for male medical students to earn a few extra dollars. By 1953, sperm could be frozen and stored for later use. Today, sperm banks freeze and store donated sperm and then provide it to physicians who perform artificial insemination. In 1983, the Sperm Bank of California became the first to ask donors if they wished to be contacted by their children years later. The first such meetings occurred in 2002, and were very successful.

A woman or couple choosing artificial insemination can select sperm from a catalog that lists the personal characteristics of the donors, including blood type; hair, skin, and eye color; build; and even educational level and interests. Of course, not all of these traits are inherited. Although artificial insemination has helped many people to become parents, it and other assisted reproductive technologies have led to occasional dilemmas (table 23A).

In Vitro Fertilization

In *in vitro* fertilization (IVF), which means "fertilization in glass," sperm meets egg outside the woman's body. The fertilized ovum divides two or three times and is then introduced into the egg donor's (or another woman's) uterus. If all goes well, a pregnancy begins.

A woman might undergo IVF if her ovaries and uterus work but her uterine tubes are blocked. To begin, she takes a hormone that hastens maturity of several oocytes. Using a laparoscope to view the ovaries and uterine tubes, a physician removes a few of the largest eggs and transfers them to a dish, then adds chemi-

TABLE 23A	Assisted Reproductive Dilemmas

1. A physician in California used his own sperm to artificially inseminate fifteen patients and told them that he had used sperm from anonymous donors.
2. A plane crash killed the wealthy parents of two early embryos stored at −320°F (−195°C) in a hospital in Melbourne, Australia. Adult children of the couple were asked to share their estate with two eight-celled balls.
3. Several couples in Chicago planning to marry discovered that they were half-siblings. Their mothers had been artificially inseminated with sperm from the same donor.
4. Two Rhode Island couples sued a fertility clinic for misplacing embryos.
5. A man sued his ex-wife for possession of their frozen embryos as part of the divorce settlement.
6. A man who donated sperm when he was healthy later developed a late-onset genetic disease, cerebellar ataxia. Each of the 18 children conceived using his sperm faces a 1 in 2 chance of having inherited the condition.

In "zona blasting," an experimental procedure to aid certain infertile men, an egg cell cultured in a laboratory dish is chemically stripped of its zona pellucida. The more vulnerable egg now presents less of a barrier to a sperm and is more easily fertilized.

The cell membranes of the sperm cell and the secondary oocyte fuse and sperm movement ceases. The sperm cell sheds its tail. At the same time, the oocyte cell membrane becomes unresponsive to other sperm cells. The union of the oocyte and sperm cell membranes also triggers some lysosome-like granules (cortical granules) just beneath the oocyte cell membrane to release enzymes that harden the zona pellucida. This reduces the chance that other sperm cells will penetrate, and it forms a protective layer around the newly formed fertilized egg cell.

The sperm nucleus enters the oocyte's cytoplasm and swells. The secondary oocyte then divides unequally to form a large cell and a tiny second polar body, which is later expelled. Therefore, female meiosis completes only after the sperm enters the egg. Next, the nuclei of the male and female cells unite. Their nuclear membranes disassemble, and their chromosomes mingle, completing the process of fertilization, diagrammed in figure 23.3.

cals similar to those in the female reproductive tract, and sperm.

If a sperm cannot penetrate the egg *in vitro,* it may be sucked up into a tiny syringe and injected using a tiny needle into the female cell (fig. 23A). This variant of IVF, called intracytoplasmic sperm injection (ICSI), is very successful, resulting in a 68% fertilization rate. It can help men with very low sperm counts, high numbers of abnormal sperm, or injuries or illnesses that prevent them from ejaculating. However, if ICSI is done to help a man become a father who has an inherited form of infertility, he may be passing the problem on to a son. In ICSI, minor surgery removes testicular tissue, from which viable sperm are isolated and injected into eggs. A day or so later, a physician transfers some of the resulting balls of eight or sixteen cells to the woman's uterus. The birth rate following IVF is about 17%, compared with 31% for natural conceptions (fig. 23B).

Gamete Intrafallopian Transfer

One reason that IVF rarely works is the artificial fertilization environment. A procedure called GIFT, which stands for gamete intrafallopian transfer, circumvents this problem by moving fertilization to the woman's body. A woman takes a superovulation drug for a week and then has several of her largest eggs removed. A man donates a sperm sample, and a physician separates the most active cells. The collected eggs and sperm are deposited together in the woman's uterine tube, at a site past any obstruction so that implantation can occur. GIFT is 26% successful.

In zygote intrafallopian transfer (ZIFT), a physician places an *in vitro* fertilized ovum in a woman's uterine tube. This is unlike IVF because the site of introduction is the uterine tube and unlike GIFT because fertilization occurs in the laboratory. Allowing the fertilized ovum to make its own way to the uterus seems to increase the chance that it will implant. ZIFT is 23% successful. ∎

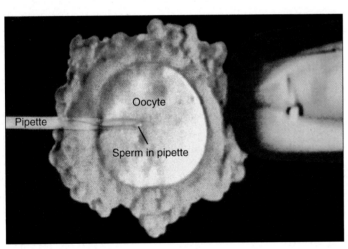

FIGURE 23A

Intracytoplasmic sperm injection (ICSI) enables some infertile men and men with spinal cord injuries and other illnesses to become fathers. A single sperm cell is injected into the cytoplasm of an egg.

FIGURE 23B

IVF worked well for Michele and Ray L'Esperance. Five fertilized ova implanted in Michele's uterus are now Erica, Alexandria, Veronica, Danielle, and Raymond.

A couple expecting a child can estimate the approximate time of conception (fertilization) by adding fourteen days to the date of the onset of the last menstrual period. They can predict the time of birth by adding 266 days to the fertilization date. Most babies are born within ten to fifteen days of this calculated time.

Obstetricians estimate the date of conception by scanning the embryo with ultrasound and comparing the crown-to-rump length to known values that are the average for each day of gestation. This approach is inaccurate if an embryo is smaller or larger than usual due to a medical problem.

Because the sperm cell and the egg cell each provide 23 chromosomes, the product of fertilization is a cell with 46 chromosomes—the usual number in a human cell. This cell, called a **zygote** (zi′gōt), is the first cell of the future offspring.

1 Distinguish between growth and development.

2 What factors aid the movements of the egg and sperm cells through the female reproductive tract?

3 Where in the female reproductive system does fertilization normally take place?

4 List the events of fertilization.

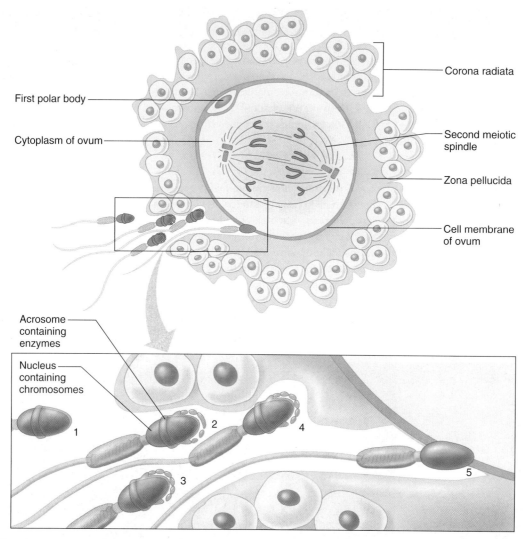

Labels on figure:
- First polar body
- Cytoplasm of ovum
- Corona radiata
- Second meiotic spindle
- Zona pellucida
- Cell membrane of ovum
- Acrosome containing enzymes
- Nucleus containing chromosomes
- 1, 2, 3, 4, 5

FIGURE 23.3

Steps in fertilization: (*1*) The sperm cell reaches the corona radiata surrounding the egg cell. (*2*) The acrosome of the sperm cell releases a protein-digesting enzyme. (*3* and *4*) The sperm cell penetrates the zona pellucida surrounding the egg cell. (*5*) The sperm cell's membrane fuses with the egg cell's membrane.

Prenatal Period

The prenatal period of development usually lasts for thirty-eight weeks from conception and can be divided into a period of cleavage, an embryonic stage, and a fetal stage.

Period of Cleavage

Conception occurs when the genetic packages of sperm and egg merge, forming a zygote. Thirty hours later, the zygote undergoes mitosis, giving rise to two new cells (fig. 23.4). These cells, in turn, divide to form four cells, which then divide to form eight cells, and so forth. These divisions take place rapidly with little time for the cells to grow. Thus, with each subsequent division, the resulting cells are smaller and smaller. This rapid cell division and distribution of the zygote's cytoplasm into progressively

smaller cells is called **cleavage** (klēv′ij), and the cells produced in this way are called *blastomeres.* The ball of cells that results from these initial cell divisions is also called a cleavage embryo. From Science to Technology 23.2 describes genetic tests of blastomeres.

The tiny mass of cells moves through the uterine tube to the uterine cavity, aided by the beating of cilia of the tubular epithelium and by weak peristaltic contractions of smooth muscles in the tubular wall. Secretions from the epithelial lining bring nutrients to the developing organism.

The trip to the uterus takes about three days, and by then, the structure consists of a solid ball, called a **morula** (mor′u-lah), of about sixteen cells (fig. 23.5). The morula remains free within the uterine cavity for about three days. Cell division continues, and the solid ball of cells

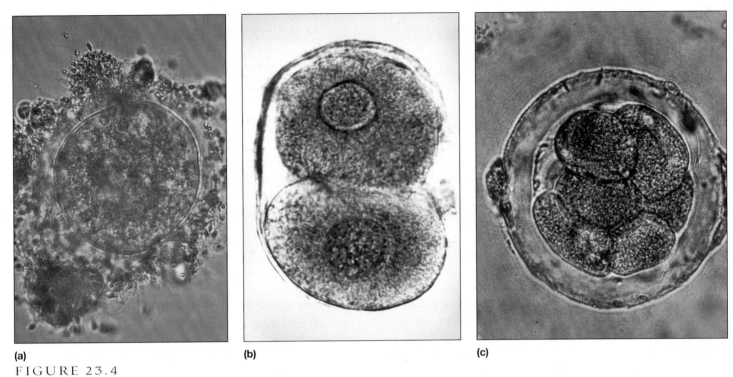

(a)   **(b)**   **(c)**

FIGURE 23.4

Light micrographs of (*a*) a human egg surrounded by follicular cells and sperm cells (250×), (*b*) the two-cell stage (600×), and (*c*) a morula (500×).

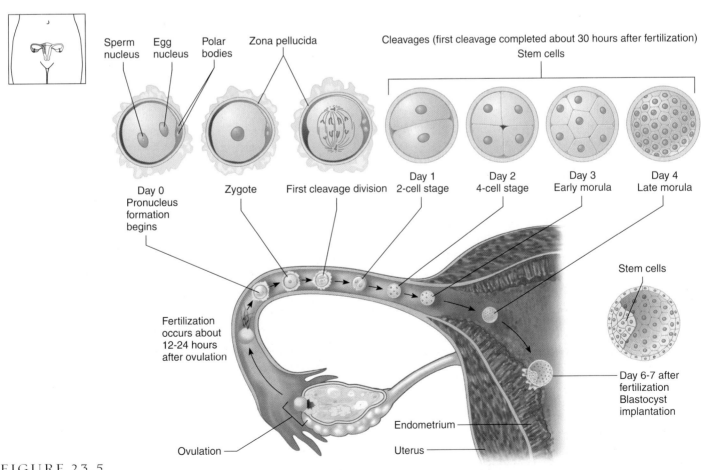

FIGURE 23.5

Early stages of human development.

CHAPTER TWENTY-THREE   *Pregnancy, Growth, and Development*

PREIMPLANTATION GENETIC DIAGNOSIS

Six-year-old Molly Nash would probably have died within a year or two of Fanconi anemia had she not received a very special gift from her baby brother Adam—his umbilical cord stem cells. Adam was not only free of the gene that causes the anemia, but his cell surfaces matched those of his sister, making a transplant very likely to succeed. But the parents didn't have to wait until Adam's birth in August 2000, to know that his cells could save Molly—they knew when he was a mere eight-celled cleavage embryo (fig. 23C).

When the Nashs learned that time was running out for Molly because they could not find a compatible bone marrow donor, they turned to preimplantation genetic diagnosis (PGD). Following *in vitro* fertilization, described in From Science to Technology 23.1, researchers at the Reproductive Genetics Institute at Illinois Masonic Medical Center removed a single cell from each of several eight-celled cleavage embryos and probed those cells for the disease-causing gene variant that ran in the family. They also scrutinized the HLA genes, which control rejection of a transplanted organ, and chose the ball of cells that would be Adam to implant into Lisa Nash's uterus. A month after the birth, physicians infused

the umbilical cord stem cells into his sister. Today, Molly is healthy.

Preimplantation genetic diagnosis works because of a feature of many animal embryos called indeterminate cleavage. That is, up until a certain point in early development, a cell or two can be removed, yet the remainder of the embryo can continue to develop normally if implanted into a uterus. Allen Handyside and colleagues at Hammersmith Hospital in London invented the technology in 1989. The first cases helped a few families to avoid devastating

inherited illnesses in their sons. Then, in 1992, Chloe O'Brien was born free of the cystic fibrosis that made her brother very ill, thanks to PGD. In 1994 came another milestone, when a girl was conceived and selected to provide umbilical cord stem cells that cured her teenage sister's leukemia.

So far, about 500 children have been born worldwide following PGD. In addition to enabling families to circumvent particular inherited conditions (table 23B), it enables couples who repeatedly lose early embryos

TABLE 23B	Some Genetic Diseases Detected with Preimplantation Genetic Diagnosis	
achondroplasia (dwarfism)		Huntington disease
adenosine deaminase deficiency (immune deficiency)		inborn errors of metabolism
		Gaucher disease
alpha-1-antitrypsin deficiency (emphysema)		ornithine transcarbamylase deficiency
		phenylketonuria
Alzheimer disease susceptibility		Tay-Sachs disease
beta thalassemia (anemia)		muscular dystrophies
cancer syndromes (p53 gene)		neurofibromatosis
cystic fibrosis		retinoblastoma
epidermolysis bullosa (skin disorder)		retinitis pigmentosa
Fanconi anemia		sickle cell disease
hemophilia A and B (clotting disorder)		spinal muscular atrophy

gradually hollows out. During this stage, the zona pellucida of the original egg cell degenerates, and the structure, now hollow and called a **blastocyst** (blas'to-sist), drops into one of the tubules in the endometrium. By the end of the first week of development, the blastocyst superficially implants in the endometrium (fig. 23.6*a*).

Within the blastocyst, cells in one region group to form an *inner cell mass* that eventually gives rise to the **embryo proper** (em'bre-o prop'er)—the body of the developing offspring. The cells that form the wall of the blastocyst make up the *trophoblast,* which develops into structures that assist the embryo.

About the sixth day, the blastocyst begins to attach to the uterine lining, aided by its secretion of proteolytic enzymes that digest a portion of the endometrium (fig. 23.6*b, c*). The blastocyst sinks slowly into the resulting

Sometimes two ovarian follicles release egg cells simultaneously, and if both are fertilized, the resulting zygotes can develop into fraternal (dizygotic) twins. Such twins are no more alike genetically than any brothers or sisters. Twins may also develop from a single fertilized egg (monozygotic twins). This may happen if two inner cell masses form within a blastocyst and each produces an embryo. Twins of this type usually share a single placenta, and they are identical genetically. Thus, they are always the same sex and are very similar in appearance.

depression, becoming completely buried in the uterine lining. At the same time, the uterine lining is stimulated to thicken below the implanting blastocyst, and cells of

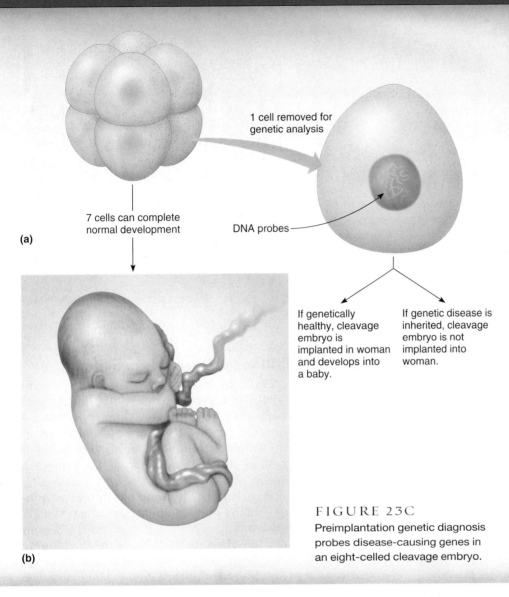

(a)

1 cell removed for genetic analysis

7 cells can complete normal development

DNA probes

If genetically healthy, cleavage embryo is implanted in woman and develops into a baby.

If genetic disease is inherited, cleavage embryo is not implanted into woman.

(b)

due to chromosome abnormalities—that is, they suffer repeat miscarriages—to select chromosomally normal embryos. Eventually, PGD may become a routine adjunct to IVF, because it ensures that only the healthiest embryos are implanted. This avoids multiple births and having to remove some embryos later in development so that the ones that are left have enough room to develop.

As with any technology, particularly one that affects conception of offspring, preimplantation genetic diagnosis has raised ethical concerns. At first in the 1990s, many people objected to the idea of intentionally conceiving a child to provide tissue to help an older sibling, but these outcries abated somewhat as the families involved showed great love for their younger children. A fear now is that human genome information will be used in conjunction with the technology to select children with less medically compelling characteristics—such as gender, inherited susceptibilities, athletic prowess, intelligence, personality traits, or appearance. ■

FIGURE 23C

Preimplantation genetic diagnosis probes disease-causing genes in an eight-celled cleavage embryo.

the trophoblast begin to produce tiny, fingerlike processes (microvilli) that grow into the endometrium. This process of the blastocyst nestling into the uterine lining is called **implantation** (im-plan-ta′shun), and it begins near the end of the first week and is completed during the second week of development (fig. 23.7).

The trophoblast secretes the hormone hCG, which maintains the corpus luteum during the early stages of pregnancy and keeps the immune system from rejecting the blastocyst. This hormone also stimulates synthesis of other hormones from the developing placenta. The **placenta** (plah-sen′tah) is a vascular structure, formed by the cells surrounding the embryo and cells of the endometrium, that attaches the embryo to the uterine wall and exchanges nutrients, gases, and wastes between the maternal blood and the embryo's blood.

Occasionally, the developing mass of cells implants in tissues outside the uterus, such as those of a uterine tube, an ovary, the cervix, or an organ in the abdominal cavity. The result is called an *ectopic pregnancy*. If a fertilized egg implants within the uterine tube, it is called a *tubal pregnancy*. A tubal pregnancy is dangerous to a pregnant woman and the developing offspring because the tube usually ruptures as the embryo enlarges and is accompanied by severe pain and heavy vaginal bleeding. Treatment is prompt surgical removal of the embryo and repair or removal of the damaged uterine tube.

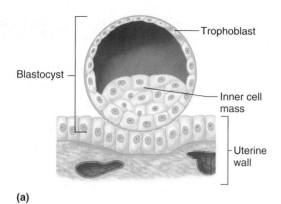

(a)

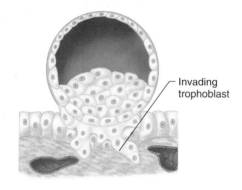

(b)

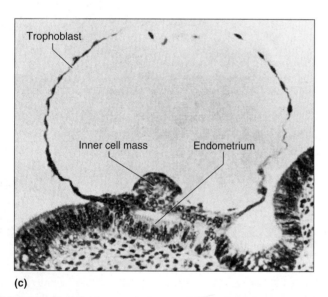

(c)

FIGURE 23.6

About the sixth day of development, the blastocyst (*a*) contacts the uterine wall and (*b*) begins to implant. The trophoblast, which will help form the placenta, secretes hCG, a hormone that maintains the pregnancy. (*c*) Light micrograph of a blastocyst from a monkey in contact with the endometrium of the uterine wall (150×).

1   What changes occur during cleavage?

2   How does a blastocyst attach to the endometrium?

3   In what ways does the endometrium respond to the activities of the blastocyst?

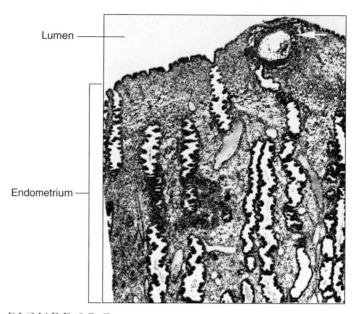

FIGURE 23.7

Light micrograph of a human cleavage embryo (arrow) implanting in the endometrium (18×).

Hormonal Changes During Pregnancy

During a typical reproductive cycle, the corpus luteum degenerates about two weeks after ovulation. Consequently, concentrations of estrogens and progesterone decline rapidly, the uterine lining is no longer maintained, and the endometrium sloughs off as menstrual flow. If this occurs following implantation, the embryo is lost (spontaneously aborted).

A hormone called **hCG (human chorionic gonadotropin)** normally helps prevent spontaneous abortion. A layer of cells, called a trophoblast, that secretes hCG and later helps form the placenta, surrounds the developing embryo. This hormone has properties similar to those of LH, and it maintains the corpus luteum, which continues secreting estrogens and progesterone. Thus, the uterine wall continues to grow and develop. At the same time, release of FSH and LH from the anterior pituitary gland is inhibited, so normal reproductive cycles cease (fig. 23.8).

Secretion of hCG continues at a high level for about two months, then declines to a low level by the end of four months. Although the corpus luteum persists throughout pregnancy, its function as a source of hormones becomes less important after the first three months (first trimester), when the placenta secretes sufficient estrogens and progesterone (fig. 23.9).

For the remainder of the pregnancy, *placental estrogens* and *placental progesterone* maintain the uterine wall. The placenta also secretes a hormone called **placental lactogen** that may stimulate breast development and prepare the mammary glands to secrete milk, with the aid of placental estrogens and progesterone. Placental progesterone and a polypeptide hormone called *relaxin* from the corpus luteum inhibit the smooth muscles in the

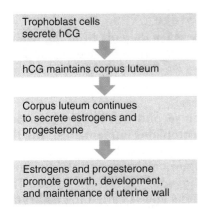

FIGURE 23.8

Mechanism that preserves the uterine lining during early pregnancy.

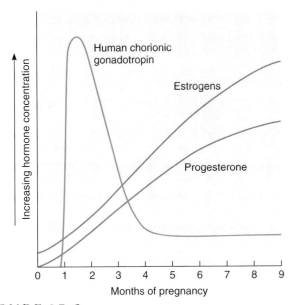

FIGURE 23.9

Relative concentrations of three hormones in maternal blood during pregnancy.

Detecting hCG in a woman's urine or blood is used to confirm pregnancy. The level of hCG in a pregnant woman's body fluids peaks at fifty to sixty days of gestation, then falls to a much lower level for the remainder of pregnancy. Later on, measuring hCG has other uses. If a woman miscarries but her blood still shows hCG, fetal tissue may remain in her uterus, and this material must be removed. At the fifteenth week of pregnancy, most women have a blood test that measures levels of three substances produced by the fetus—alpha fetoprotein, estriol (one of the estrogens), and hCG. If alpha fetoprotein and estriol are low but hCG is high, the fetus may have an extra chromosome, with severe effects on health.

myometrium, suppressing uterine contractions until the birth process begins.

The high concentration of placental estrogens during pregnancy enlarges the vagina and the external reproductive organs. Also, relaxin relaxes the ligaments holding the symphysis pubis and sacroiliac joints together. This action, which usually occurs during the last week of pregnancy, allows for greater movement at these joints, aiding passage of the fetus through the birth canal.

Other hormonal changes that occur during pregnancy include increased secretion of aldosterone from the adrenal cortex and of parathyroid hormone from the parathyroid glands. Aldosterone promotes renal reabsorption of sodium, leading to fluid retention. Parathyroid hormone helps to maintain a high concentration of maternal blood calcium, since fetal demand for calcium can cause hypocalcemia, which promotes cramps. Table 23.1 summarizes the hormonal changes of pregnancy.

RECONNECT TO CHAPTER 13, PARATHYROID GLANDS, PAGES 489–490.

1 What mechanism maintains the uterine wall during pregnancy?

2 What is the source of hCG during the first few months of pregnancy?

3 What is the source of the hormones that sustain the uterine wall during pregnancy?

4 What other hormonal changes occur during pregnancy?

TABLE 23.1	Hormonal Changes During Pregnancy

1. Following implantation, cells of the trophoblast begin to secrete hCG.
2. hCG maintains the corpus luteum, which continues secreting estrogens and progesterone.
3. As the placenta develops, it secretes large quantities of estrogens and progesterone.
4. Placental estrogens and progesterone
   a. stimulate the uterine lining to continue development.
   b. maintain the uterine lining.
   c. inhibit secretion of FSH and LH from the anterior pituitary gland.
   d. stimulate development of the mammary glands.
   e. inhibit uterine contractions (progesterone).
   f. enlarge the reproductive organs (estrogens).
5. Relaxin from the corpus luteum also inhibits uterine contractions and relaxes the pelvic ligaments.
6. The placenta secretes placental lactogen that stimulates breast development.
7. Aldosterone from the adrenal cortex promotes reabsorption of sodium.
8. Parathyroid hormone from the parathyroid glands helps maintain a high concentration of maternal blood calcium.

Other Changes During Pregnancy

Other changes in a woman's body respond to the increased demands of a growing fetus. As the fetus grows, the uterus enlarges greatly, and instead of being confined to its normal location in the pelvic cavity, it extends upward and may eventually reach the level of the ribs. The abdominal organs are displaced upward and compressed against the diaphragm. The enlarging uterus also presses on the urinary bladder.

> A pregnant woman is well aware of the effects of her expanding uterus. She can no longer eat large meals, develops heartburn often as stomach contents are pushed up into the esophagus, and frequently has to urinate as her uterus presses on her bladder.

As the placenta grows and develops, it requires more blood, and as the fetus enlarges, it needs more oxygen and produces more waste that must be excreted. The pregnant woman's blood volume, cardiac output, breathing rate, and urine production all increase to handle fetal growth.

The pregnant woman must eat more to obtain adequate nutrition for the fetus. Her intake must supply sufficient vitamins, minerals, and proteins for herself and the fetus. The fetal tissues have a greater capacity to capture available nutrients than do the maternal tissues. Consequently, if the pregnant woman's diet is inadequate, her body will usually show symptoms of a deficiency condition before fetal growth is adversely affected.

Embryonic Stage

The **embryonic stage** extends from the beginning of the second week through the eighth week of prenatal development. During this time, the placenta forms, the main internal organs develop, and the major external body structures appear.

During the second week of prenatal development, the blastocyst completes implantation, and the inner cell mass changes. A space, called the *amniotic cavity,* forms between the inner cell mass and the portion of the trophoblast that "invades" the endometrium. The inner cell mass then flattens and is called the **embryonic disc.** By the end of the second week, layers form.

The embryonic disc initially consists of two distinct layers: an outer *ectoderm* and an inner *endoderm.* A short time later, through a process called gastrulation, a third layer of cells, the *mesoderm,* forms between the ectoderm and endoderm. These three layers of cells are called the **primary germ layers** (pri′mer-e jerm la′erz) of the primordial embryo. All organs form from the primary germ layers. At this point, the embryo is termed a **gastrula** (gas′troo-lah). Also during this time, a structure called a connecting stalk appears. It attaches the embryo to the developing placenta (fig. 23.10). Table 23.2 summarizes the stages of early embryonic development.

Gastrulation is an important process in prenatal development because a cell's fate is determined by which layer it is in. *Ectodermal cells* give rise to the nervous system, portions of special sensory organs, the epidermis, hair, nails, glands of the skin, and linings of the mouth and anal canal. *Mesodermal cells* form all types of muscle tissue, bone tissue, bone marrow, blood, blood vessels,

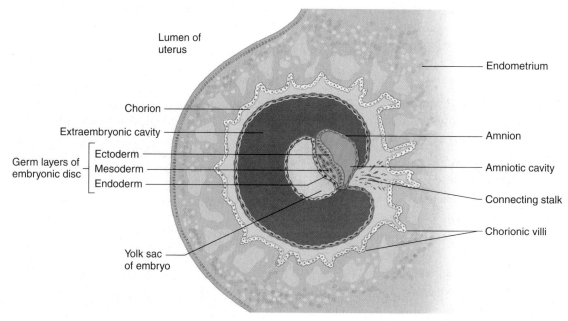

FIGURE 23.10

Early in the embryonic stage of development, the three primary germ layers form.

TABLE 23.2		Stages and Events of Early Human Prenatal Development
Stage	**Time Period**	**Principal Events**
Fertilized ovum	12-24 hours following ovulation	Oocyte fertilized; zygote has 23 pairs of chromosomes and is genetically distinct
Cleavage	30 hours to third day	Mitosis increases cell number
Morula	Third to fourth day	Solid ball of cells
Blastocyst	Fifth day through second week	Hollowed ball forms trophoblast (outside) and inner cell mass, which implants and flattens to form embryonic disc
Gastrula	End of second week	Primary germ layers form

lymphatic vessels, connective tissues, internal reproductive organs, kidneys, and the epithelial linings of the body cavities. *Endodermal cells* produce the epithelial linings of the digestive tract, respiratory tract, urinary bladder, and urethra (fig. 23.11). The primary germ layers also retain stem cells, a few of which persist in the adult.

As the embryo implants in the uterus, proteolytic enzymes from the trophoblast break down endometrial tissue, providing nutrients for the developing embryo. A second layer of cells begins to line the trophoblast, and together these two layers form a structure called the **chorion** (ko're-on). Soon, slender projections grow out from the trophoblast, including the new cell layer, eroding their way into the surrounding endometrium by continuing to secrete proteolytic enzymes. These projections become increasingly complex, and form the highly branched **chorionic villi,** which are well established by the end of the fourth week.

Continued secretion of proteolytic enzymes forms irregular spaces called **lacunae** in the endometrium around and between the chorionic villi. These spaces fill with maternal blood that escapes from endometrial blood vessels eroded by the enzyme action. At the same time, embryonic blood vessels carrying blood to and from the embryo extend through the connecting stalk and establish capillary networks in the developing chorionic villi. These embryonic vessels allow nutrient exchange with blood in the lacunae and provide for the increased nutrient needs of the growing embryo.

During the fourth week of development, the flat embryonic disc becomes cylindrical. By the end of week four, the head and jaws appear, the heart beats and forces blood through blood vessels, and tiny buds form, which will give rise to the upper and lower limbs (fig. 23.12).

During the fifth through the seventh weeks, as figure 23.13 shows, the head grows rapidly and becomes rounded and erect. The face, which is developing the eyes, nose, and mouth, appears more humanlike. The upper and lower limbs elongate, and fingers and toes form (fig. 23.14). By the end of the seventh week, all the main internal organs are established, and as these structures enlarge, the body takes on a humanlike appearance.

Until about the end of the eighth week, the chorionic villi cover the entire surface of the former trophoblast. However, as the embryo and the chorion surrounding it enlarge, only those villi that remain in contact with the endometrium endure. The others degenerate, and the portions of the chorion to which they were attached become smooth. Thus, the region of the chorion still in contact with the uterine wall is restricted to a disc-shaped area that becomes the placenta (fig. 23.15).

A thin membrane separates embryonic blood within the capillary of a chorionic villus from maternal blood in a lacuna. This membrane, called the **placental membrane,** is composed of the epithelium of the chorionic villus and the endothelium of the capillary inside the villus (fig. 23.16). Through this membrane, substances are exchanged between the maternal blood and the embryo's blood. Oxygen and nutrients diffuse from the maternal blood into the embryo's blood, and carbon dioxide and other wastes diffuse from the embryo's blood into the maternal blood. Active transport and pinocytosis also move various substances through the placental membrane.

1 Which major events occur during the embryonic stage of development?

2 Which tissues and structures develop from ectoderm? From mesoderm? From endoderm?

3 Describe the structure of a chorionic villus.

4 What is the function of the placental membrane?

5 How are substances exchanged between the embryo's blood and the maternal blood?

The embryonic portion of the placenta is composed of parts of the chorion and its villi; the maternal portion is composed of the area of the uterine wall (decidua basalis) to which the villi are attached. When it is fully formed, the placenta appears as a reddish brown disc, about 20 centimeters long and 2.5 centimeters thick. It usually weighs about 0.5 kilogram. Figure 23.17 shows the structure of the placenta.

While the placenta is forming from the chorion, a second membrane, called the **amnion** (am'ne-on), develops around the embryo. This membrane began to appear during the second week. Its margin is attached around the edge of the embryonic disc, and fluid called **amniotic fluid** fills the space between the amnion and the embryonic disc.

The developing placenta synthesizes progesterone from cholesterol in the maternal blood. Cells associated with the developing fetal adrenal glands use the placental progesterone to synthesize estrogens. The estrogens, in

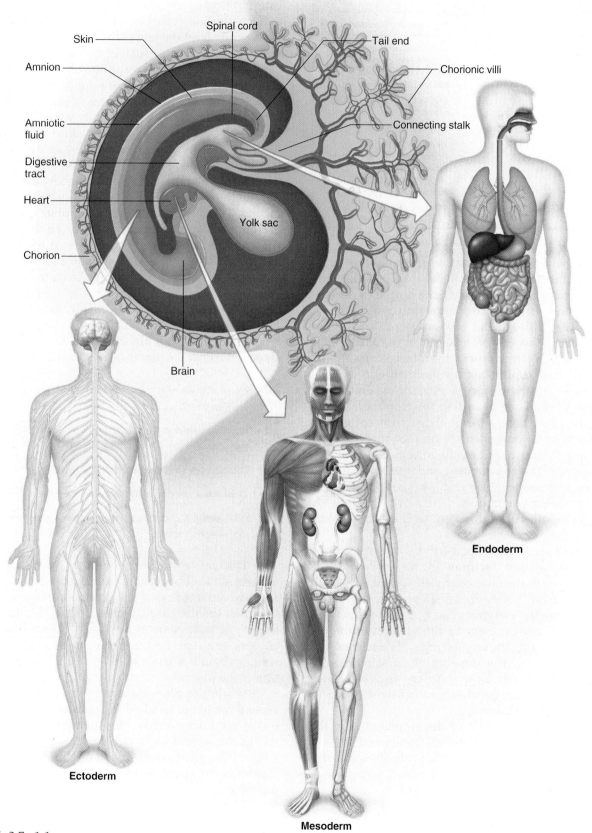

Skin

Spinal cord

Tail end

Amnion

Chorionic villi

Amniotic
fluid

Connecting stalk

Digestive
tract

Heart

Yolk sac

Chorion

Brain

Endoderm

Ectoderm

Mesoderm

FIGURE 23.11

Each of the primary germ layers forms a particular set of tissues and organs.

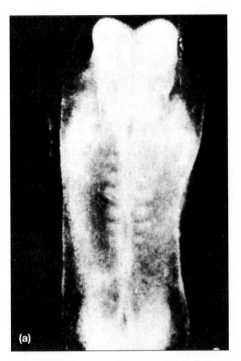 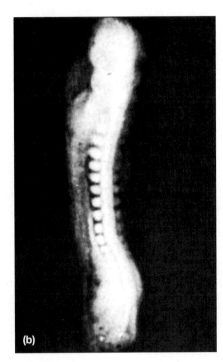 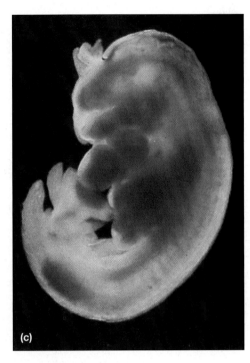

FIGURE 23.12

Embryo. (*a*) A human embryo at three weeks, posterior view; (*b*) at three and one-half weeks, lateral view; (*c*) at about four weeks, lateral view. (Figures are not to scale.)

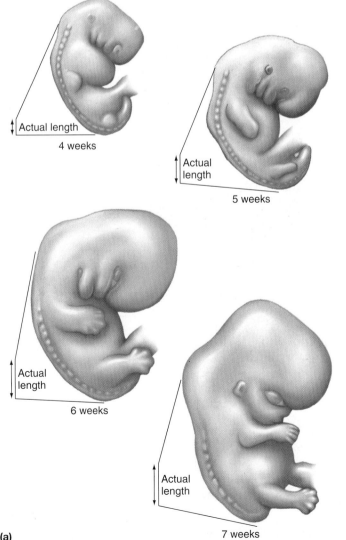

Actual length
4 weeks

Actual length
5 weeks

Actual length
6 weeks

Actual length
7 weeks

(a)

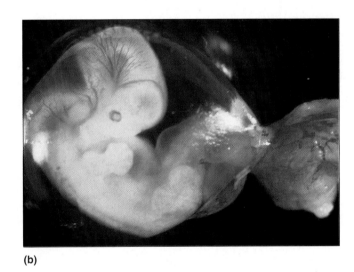

(b)

FIGURE 23.13

Development of an embryo. (*a*) In the fifth through the seventh weeks of gestation, the embryonic body and face develop a more humanlike appearance. (*b*) A human embryo after about six weeks of development.

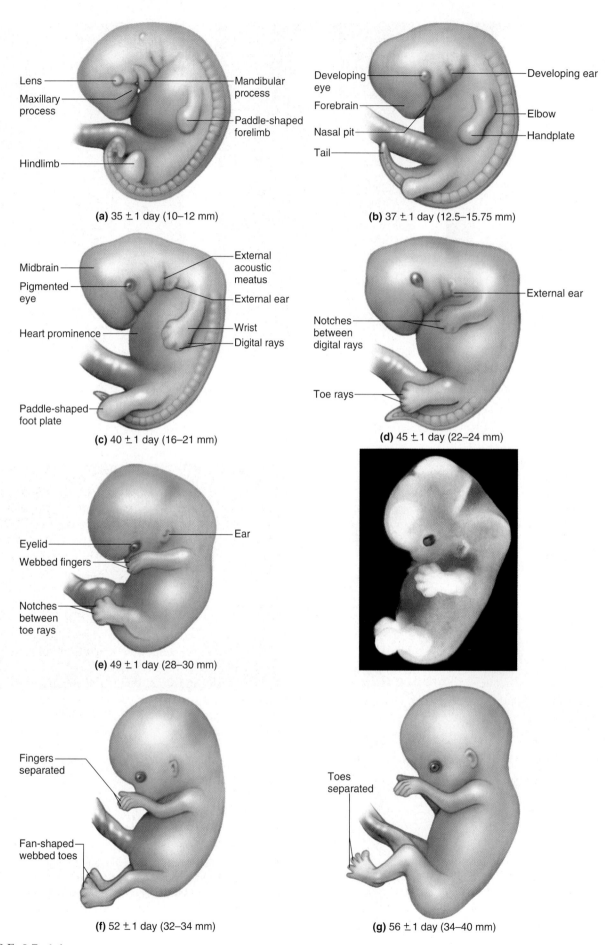

(a) 35 ± 1 day (10–12 mm)

Lens
Maxillary process
Hindlimb
Mandibular process
Paddle-shaped forelimb

(b) 37 ± 1 day (12.5–15.75 mm)

Developing eye
Forebrain
Nasal pit
Tail
Developing ear
Elbow
Handplate

(c) 40 ± 1 day (16–21 mm)

Midbrain
Pigmented eye
Heart prominence
Paddle-shaped foot plate
External acoustic meatus
External ear
Wrist
Digital rays

(d) 45 ± 1 day (22–24 mm)

Notches between digital rays
Toe rays
External ear

(e) 49 ± 1 day (28–30 mm)

Eyelid
Webbed fingers
Notches between toe rays
Ear

(f) 52 ± 1 day (32–34 mm)

Fingers separated
Fan-shaped webbed toes

(g) 56 ± 1 day (34–40 mm)

Toes separated

FIGURE 23.14

Changes occurring during the fifth (a–c), sixth (d), and seventh (e–g) weeks of development. The photo corresponds to forty-nine days of development.

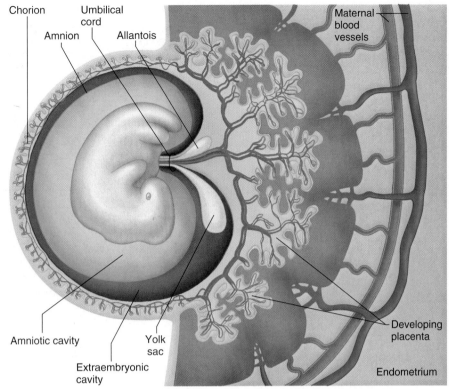

Chorion
Umbilical cord
Amnion
Allantois
Maternal blood vessels
Developing placenta
Endometrium
Amniotic cavity
Yolk sac
Extraembryonic cavity

FIGURE 23.15

As the amnion develops, it surrounds the embryo, and the umbilical cord begins to form from structures in the connecting stalk.

turn, promote changes in the maternal uterus and breasts and influence maternal metabolism and the development of various fetal organs.

As the embryo becomes more cylindrical, the margins of the amnion fold, enclosing the embryo in the amnion and amniotic fluid. The amnion envelopes the tissues on the underside of the embryo, particularly the connecting stalk, by which it is attached to the chorion and the developing placenta. In this manner, the **umbilical cord** (um-bil'ĭ-kal kord) forms (see fig. 23.15).

If a pregnant woman repeatedly ingests an addictive substance, her newborn may suffer from withdrawal symptoms when amounts of the chemical it is accustomed to receiving suddenly plummet. Newborn addiction occurs with certain addictive drugs of abuse, such as heroin; with certain prescription drugs used to treat anxiety; and even with very large doses of vitamin C. Although vitamin C is not addictive, if a fetus is accustomed to megadoses, after birth the sudden drop in vitamin C level may bring on symptoms of vitamin C deficiency.

The fully developed umbilical cord is about 1 centimeter in diameter and about 55 centimeters in length. It begins at the umbilicus of the embryo and inserts into the center of the placenta. The cord contains three blood vessels—two *umbilical arteries* and one *umbilical vein*—that transport blood between the embryo and the placenta (see fig. 23.17).

The umbilical cord also suspends the embryo in the *amniotic cavity.* The amniotic fluid provides a watery environment in which the embryo can grow freely without being compressed by surrounding tissues. The amniotic fluid also protects the embryo from being jarred by the movements of the woman's body.

In addition to the amnion and chorion, two other embryonic membranes form during development. They are the yolk sac and the allantois.

The **yolk sac** forms during the second week, and it is attached to the underside of the embryonic disc (see fig. 23.15). This structure forms blood cells in the early stages of development and gives rise to the cells that later become sex cells. The yolk sac also produces stem cells of the bone marrow, which are precursors to many cell types, but predominantly to blood cells. Portions of the yolk sac form the embryonic digestive tube as well. Part of the membrane derived from the yolk sac becomes incorporated into the umbilical cord, and the remainder lies in the cavity between the chorion and the amnion near the placenta.

The **allantois** (ah-lan'to-is) forms during the third week as a tube extending from the early yolk sac into the connecting stalk of the embryo. It, too, forms blood cells

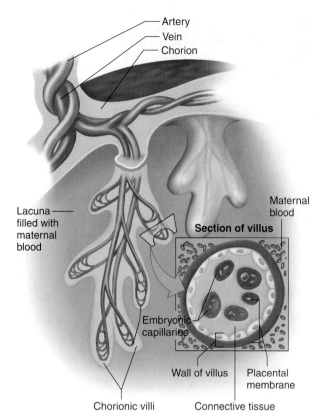

FIGURE 23.16

As illustrated in the section of villus (lower part of figure), the placental membrane consists of the epithelial wall of an embryonic capillary and the epithelial wall of a chorionic villus.

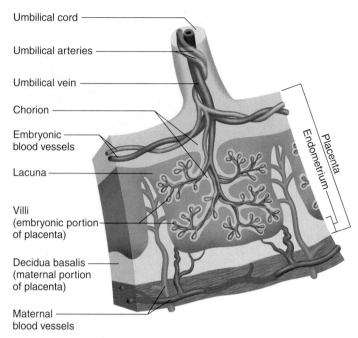

FIGURE 23.17

The placenta consists of an embryonic portion and a maternal portion.

and gives rise to the umbilical arteries and vein (see figs. 23.15 and 23.17).

Eventually, the amniotic cavity becomes so enlarged that the membrane of the amnion contacts the thicker chorion around it. The two membranes fuse into an *amniochorionic membrane* (see fig. 23.18).

By the beginning of the eighth week, the embryo is usually 30 millimeters long and weighs less than 5 grams. Although its body is quite unfinished, it looks human (fig. 23.19).

The *embryonic stage* concludes at the end of the eighth week. It is the most critical period of development, because during it, the embryo implants within the uterine wall, and all the essential external and internal body parts form. Disturbances to development during the embryonic stage can cause major malformations or malfunctions. This is why early prenatal care is very important.

Factors that cause congenital malformations by affecting an embryo during its period of rapid growth and development are called **teratogens.** Such agents include drugs, viruses, radiation, and even large amounts of otherwise healthful substances, such as fat-soluble vitamins. Each prenatal structure has a time in development, called its *critical period,* when it is sensitive to teratogens (fig. 23.20).

A critical period may extend over many months or be just a day or two. Neural tube defects, for example, are traced to day twenty-eight in development, when a sheet of ectoderm called the neural tube normally folds into a tube, which then develops into the central nervous system. When this process is disrupted, an opening remains in the spine (spina bifida) or in the brain (anencephaly).

In contrast, the critical period for the brain begins when the anterior neural tube begins to swell into a brain, and continues throughout gestation. This is why so many teratogens affect the brain. Clinical Application 23.1 discusses some teratogens and their effects.

RECONNECT TO CHAPTER 11, BRAIN DEVELOPMENT, PAGE 381.

1. Describe the development of the amnion.

2. Which blood vessels are in the umbilical cord?

3. What is the function of amniotic fluid?

4. What types of cells and other structures are derived from the yolk sac?

5. How do teratogens cause birth defects?

Fetal Stage

The **fetal stage** begins at the end of the eighth week of prenatal development and lasts until birth. During this period, growth is rapid, and body proportions change considerably. At the beginning of the fetal stage, the head is disproportionately large, and the lower limbs are

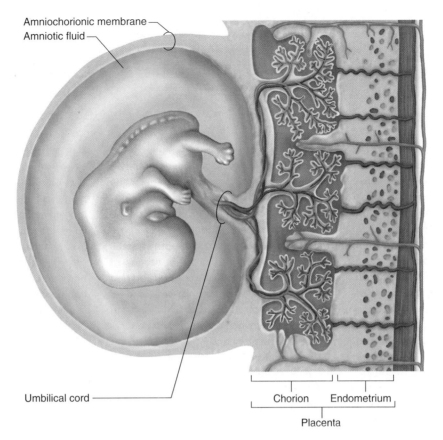

Amniochorionic membrane
Amniotic fluid
Umbilical cord
Chorion   Endometrium
Placenta

FIGURE 23.18

The developing placenta, composed of chorionic and endometrial capillaries, as it appears during the seventh week of development.

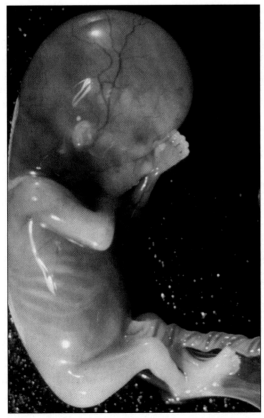

FIGURE 23.19

By the beginning of the eighth week of development, the embryonic body is recognizable as a human.

(a) When physical structures develop

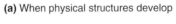
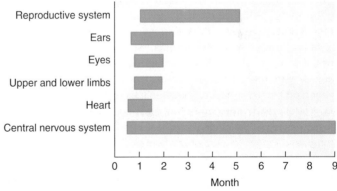

Reproductive system
Ears
Eyes
Upper and lower limbs
Heart
Central nervous system

0   1   2   3   4   5   6   7   8   9
Month

(b) When different teratogens disrupt development

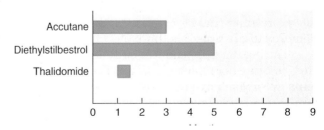

Accutane
Diethylstilbestrol
Thalidomide

0   1   2   3   4   5   6   7   8   9

FIGURE 23.20

Structures in the developing embryo and fetus are sensitive to specific teratogens at different times in gestation.

SOME CAUSES OF BIRTH DEFECTS

Thalidomide

The idea that the placenta protects the embryo and fetus from harmful substances was tragically disproven between 1957 and 1961, when 10,000 children in Europe were born with flippers in place of limbs. Doctors soon identified a mild tranquilizer, thalidomide, which all of the mothers of deformed infants had taken early in pregnancy, during the time of limb formation. Although some women in the United States did use thalidomide and had affected children, the United States was spared a thalidomide disaster because an astute government physician noted adverse effects of the drug on monkeys in experiments, and she halted use of the drug. However, thalidomide is used today to treat leprosy and certain blood disorders.

Rubella

The virus that causes rubella (German measles) is a powerful teratogen. Australian physicians first noted its effects in 1941, and a rubella epidemic in the United States in the early 1960s caused 20,000 birth defects and 30,000 stillbirths. Exposure in the first trimester leads to cataracts, deafness, and heart defects, and later exposure causes learning disabilities, speech and hearing problems, and type I diabetes mellitus. Widespread vaccination has slashed the incidence of this congenital rubella syndrome, and today it occurs only where people are not vaccinated.

Alcohol

A pregnant woman who has just one or two alcoholic drinks a day, or perhaps many drinks at a crucial time in prenatal development, risks *fetal alcohol syndrome* or the more prevalent *fetal alcohol effects* in her unborn child. Because the effects of small amounts of alcohol at different stages of pregnancy are not yet well understood and because each woman metabolizes alcohol slightly differently, it is best to avoid drinking alcohol entirely when pregnant or when trying to become pregnant.

A child with fetal alcohol syndrome has a characteristic small head, misshapen eyes, and a flat face and nose (fig. 23D). He or she grows slowly before and after birth. Intellect is impaired, ranging from minor learning disabilities to mental retardation. Teens and young adults with fetal alcohol syndrome are short and have small heads. Many individuals remain at early grade-school level. They often lack social and communication skills, such as understanding the consequences of actions, forming friendships, taking initiative, and interpreting social cues.

Problems in children of alcoholic mothers were noted by Aristotle more than twenty-three centuries ago. Today, fetal alcohol syndrome is the third most common cause of mental retardation in newborns. One to 3 in every 1,000 infants has the syndrome.

Cigarettes

Chemicals in cigarette smoke stress a fetus. Carbon monoxide crosses the placenta and plugs up the sites on the fetus's hemoglobin molecules that would normally bind oxygen. Other chemicals in smoke prevent nutrients from reaching the fetus. Studies comparing placentas of smokers and nonsmokers show that smoke-exposed placentas lack important growth factors. The result of all of these assaults is poor growth before and after birth. Cigarette smoking during pregnancy raises the risk of spontaneous abortion, stillbirth, prematurity, and low birth weight.

Nutrients and Malnutrition

Certain nutrients in large amounts, particularly vitamins, act in the body as drugs. The acne medication *isotretinoin* (Accutane) is a derivative of vitamin A that causes spontaneous abortions and defects of the heart, nervous system, and face. The tragic effects of this drug were noted exactly nine months after dermatologists began prescribing it to young women in the early 1980s. Today, the drug package bears prominent warnings, and it is never prescribed to pregnant women. A vitamin A-based drug used to treat psoriasis, as well as excesses of vitamin A itself, also cause birth defects. This is because some forms of vitamin A are stored in body fat for up to three years after ingestion.

Malnutrition during pregnancy causes intrauterine growth retardation (IUGR), which may have delayed health effects.

---

relatively short (fig. 23.21). Gradually, proportions come to more closely resemble those of a child.

During the third month, body lengthening accelerates, but growth of the head slows. The upper limbs of the **fetus** (fe′tus) achieve the relative length they will maintain throughout development, and ossification centers appear in most of the bones. By the twelfth week, the external reproductive organs are distinguishable as male or female. Figure 23.22 illustrates how these external

reproductive organs of the male and female differentiate from precursor structures.

In the fourth month, the body grows very rapidly and reaches a length of up to 20 centimeters and weighs about 170 grams. The lower limbs lengthen considerably, and the skeleton continues to ossify. The fetus has hair, nipples, and nails, and may even scratch itself.

In the fifth month, growth slows. The lower limbs achieve their final relative proportions. Skeletal muscles

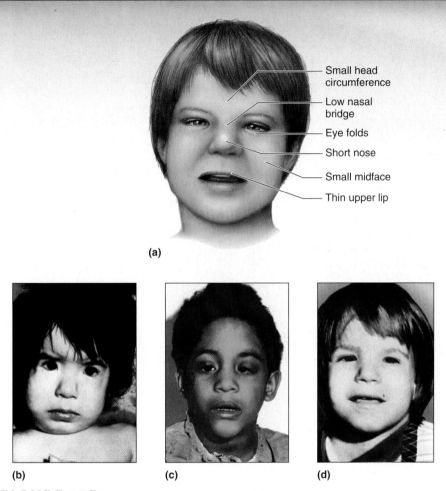

Small head circumference

Low nasal bridge

Eye folds

Short nose

Small midface

Thin upper lip

(a)

(b)   **(c)**   **(d)**

FIGURE 23D

Fetal alcohol syndrome. Some children whose mothers drank alcohol during pregnancy have characteristic flat faces (a) that are strikingly similar in children of different races (b–d). Women who drink excessively while pregnant have a 30% to 45% chance of having a child who is affected to some degree by prenatal exposure to alcohol. Two mixed drinks per day seems to be the level above which damage is likely to occur.

Fetal physiology adapts to starvation to best utilize available nutrients. Insulin resistance changes to compensate for lack of muscle tissue. Cardiovascular changes shunt blood to vital organs. Starvation also causes stress hormone levels to rise, arteries to stiffen, and too few kidney tubules to form. These changes set the stage for type II diabetes mellitus, hypertension, stroke, and coronary artery disease years later. Paradoxically, the infant is scrawny, but the older child tends to be obese, and difficulty losing weight may persist.

Many epidemiological investigations have linked IUGR to these conditions in the adult. One study of individuals who were fetuses during a seven-month famine in the Netherlands in 1943 documented increased incidence of spontaneous abortion, low birth weight, short stature, and delayed sexual development. Fifty years later, inability to maintain glucose homeostasis was common among these people. Other investigations on older individuals suggest that the conditions associated with what is being called "small baby syndrome" typically manifest after age sixty-five. Experiments on pregnant rats and sheep replicate the spectrum of disorders linked to IUGR and also indicate that somatostatin and glucocorticoid levels change with starvation in the uterus.

Occupational Hazards

Some teratogens are encountered in the workplace. Increased rates of spontaneous abortion and birth defects have been noted among women who work with textile dyes, lead, certain photographic chemicals, semiconductor materials, mercury, and cadmium. We do not know much about the role of the male in environmentally caused birth defects. Men whose jobs expose them to sustained heat, such as smelter workers, glass manufacturers, and bakers, may produce sperm that can fertilize an egg and possibly lead to spontaneous abortion or a birth defect. A virus or a toxic chemical carried in semen may also cause a birth defect. ■

contract, and the pregnant woman may feel fetal movements for the first time. Some hair grows on the fetal head, and fine, downy hair called lanugo covers the skin. A cheesy mixture of sebum from the sebaceous glands and dead epidermal cells (vernix caseosa) also coats the skin. The fetus, weighing about 450 grams and about 30 centimeters long, curls into the fetal position.

During the sixth month, the fetus gains a substantial amount of weight. Eyebrows and eyelashes appear. The skin is quite wrinkled and translucent. Blood vessels in the skin cause a reddish appearance.

In the seventh month, the skin becomes smoother as fat is deposited in the subcutaneous tissues. The eyelids, which fused during the third month, reopen. At the end of this month, a fetus is about 40 centimeters long.

In the final trimester, fetal brain cells rapidly form networks, as organs elaborate and grow. A layer of fat is laid down beneath the skin. The testes of males descend

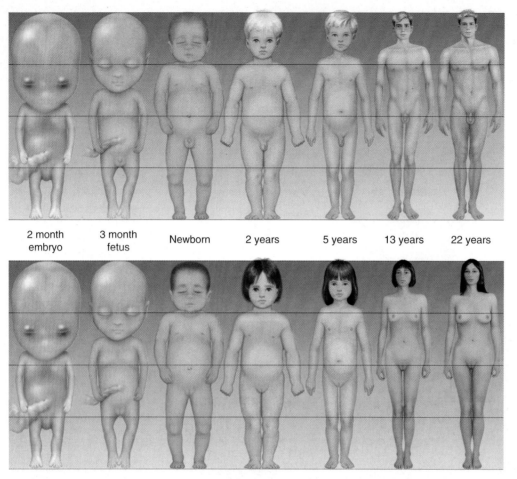

FIGURE 23.21

During development, body proportions change considerably.

from regions near the developing kidneys, through the inguinal canal, and into the scrotum (see chapter 22, p. 832). The digestive and respiratory systems mature last, which is why infants born prematurely often have difficulty digesting milk and breathing.

Premature infants' survival chances increase directly with age and weight. Survival is more likely if the lungs are sufficiently developed with the thin respiratory membranes necessary for rapid exchange of oxygen and carbon dioxide and if they produce enough surfactant to reduce alveolar surface tension (see chapter 19, p. 747). A fetus of less than twenty-four weeks or weighing less than 600 grams at birth seldom survives, even with intensive medical care. Neonatology is the medical field that cares for premature and ill newborns.

Approximately 266 days after a single sperm burrowed its way into an oocyte, a baby is ready to be born. It is about 50 centimeters long and weighs 2.7 to 3.6 kilograms. The skin has lost its downy hair but is still coated with sebum and dead epidermal cells. The scalp is usually covered with hair; the fingers and toes have well-developed nails; and the skull bones are largely ossified. As figure 23.23 shows, the fetus is usually positioned upside down with its head toward the cervix (*vertex position*).

The birth of a live, healthy baby is against the odds, considering human development from the beginning. Of every 100 secondary oocytes that are exposed to sperm, eighty-four are fertilized. Of these, sixty-nine implant in the uterus, forty-two survive one week or longer, thirty-seven survive six weeks or longer, and only thirty-one are born alive. Of those that do not survive to birth, about half have chromosomal abnormalities that are too severe to maintain life. Table 23.3 summarizes the stages of prenatal development.

1 What major changes occur during the fetal stage of development?

2 When can the sex of a fetus be determined?

3 How is a fetus usually positioned within the uterus at the end of pregnancy?

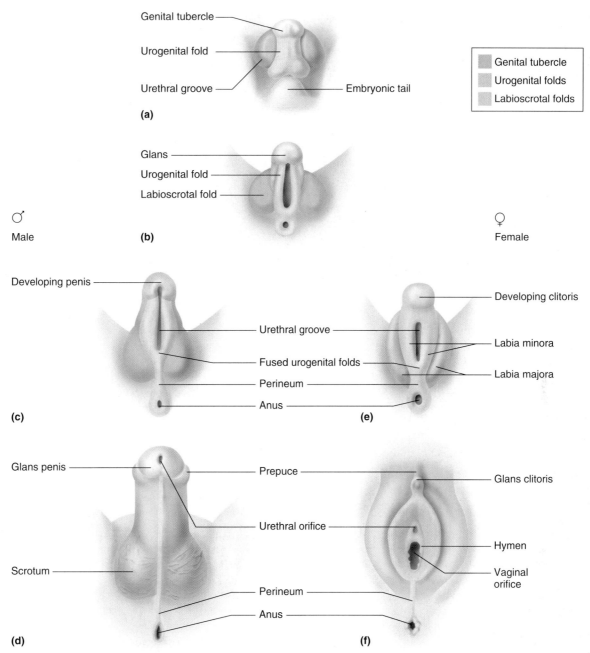

Genital tubercle

Urogenital fold

Urethral groove

Embryonic tail

(a)

Genital tubercle
Urogenital folds
Labioscrotal folds

Glans
Urogenital fold
Labioscrotal fold

♂
Male

(b)

♀
Female

Developing penis

Urethral groove

Fused urogenital folds

Perineum

Anus

(c)

Developing clitoris

Labia minora

Labia majora

(e)

Glans penis

Prepuce

Urethral orifice

Scrotum

Perineum

Anus

(d)

Glans clitoris

Hymen

Vaginal orifice

(f)

FIGURE 23.22

Formation of external reproductive organs. (*a* and *b*) The genital tubercle, urogenital fold, and labioscrotal folds that appear during the fourth week of development may differentiate into (*c* and *d*) male external reproductive organs or (*e* and *f*) female external reproductive organs.

Blake Schultz made medical history when he underwent major surgery seven weeks before birth. Ultrasound had revealed that his stomach, spleen, and intestines protruded through a hole in his diaphragm, the muscle sheet that separates the abdomen from the chest. This defect would have suffocated him shortly after birth were it not for pioneering surgery that exposed Blake's left side, gently tucked his organs in place, and patched the hole with a synthetic material.

Some prenatal medical problems can be treated by adminis-tering drugs to the pregnant woman or by altering her diet. An undersized fetus can receive a nutritional boost by putting the pregnant woman on a high-protein diet. It is also possible to treat prenatal medical problems directly: A tube inserted into the uterus can drain the dangerously swollen bladder of a fetus with a blocked urinary tract, providing relief until the problem can be surgically corrected at birth. A similar procedure can remove excess fluid from the brain of a fetus with hydrocephaly (a neural tube defect, also called "water on the brain").

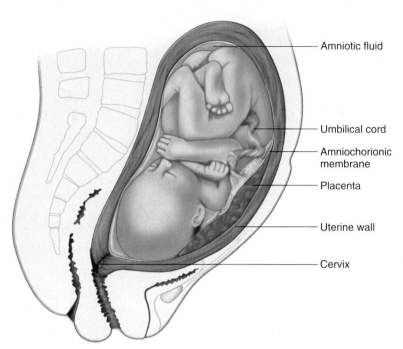

FIGURE 23.23

A full-term fetus is usually positioned with its head near the cervix.

TABLE 23.3	Stages of Prenatal Development		
Stage	**Time Period**	**Major Events**	
Period of cleavage	First week	Cells undergo mitosis, blastocyst forms; inner cell mass appears; blastocyst implants in uterine wall	
		Size: 1/4 inch (0.63 centimeter), weight: 1/120 ounce (0.21 gram)	
Embryonic stage	Second through eighth week	Inner cell mass becomes embryonic disc; primary germ layers form, embryo proper becomes cylindrical; main internal organs and external body structures appear; placenta and umbilical cord form, embryo proper is suspended in amniotic fluid	
		Size: 1 inch (2.5 centimeters), weight: 1/30 ounce (0.8 gram)	
Fetal stage	Ninth through twelfth week	Ossification centers appear in bones, sex organs differentiate, nerves and muscles coordinate so that the fetus can move its limbs	
		Size: 4 inches (10 centimeters), weight: 1 ounce (28 grams)	
	Thirteenth through sixteenth week	Body grows rapidly; ossification continues	
		Size: 8 inches (20 centimeters), weight: 6 ounces (170 grams)	
	Seventeenth through twentieth week	Muscle movements are stronger, and woman may be aware of slight flutterings; skin is covered with fine downy hair (lanugo) and coated with sebum mixed with dead epidermal cells (vernix caseosa)	
		Size: 12 inches (30.5 centimeters), weight: 1 pound (454 grams)	
	Twenty-first through thirty-eighth week	Body gains weight, subcutaneous fat deposited; eyebrows and lashes appear; eyelids reopen; testes descend	
		Size: 21 inches (53 centimeters), weight: 6 to 10 pounds (2.7 to 4.5 kilograms)	

Fetal Blood and Circulation

Throughout fetal development, the maternal blood supplies oxygen and nutrients and carries away wastes. These substances diffuse between the maternal and fetal blood through the placental membrane, and the umbilical blood vessels carry them to and from the fetus (fig. 23.24). Conse-quently, the fetal blood and vascular system are adapted to intrauterine existence. For example, the concentration of oxygen-carrying hemoglobin in the fetal blood is about 50% greater than in the maternal blood. Also, fetal hemoglobin has a greater affinity for oxygen than does an adult's hemoglobin. Thus, at the oxygen partial pressure of the

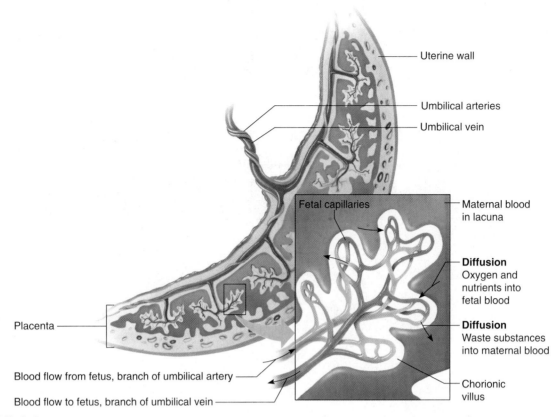

FIGURE 23.24

Oxygen and nutrients diffuse into the fetal blood from the maternal blood. Waste diffuses into the maternal blood from the fetal blood.

placental capillaries, fetal hemoglobin can carry 20% to 30% more oxygen than maternal hemoglobin.

In the fetal cardiovascular system, the *umbilical vein* transports blood rich in oxygen and nutrients from the placenta to the fetal body. This vein enters the body through the umbilical ring and travels along the anterior abdominal wall to the liver. About half the blood it carries passes into the liver, and the rest enters a vessel called the **ductus venosus** (duk′tus ve′no-sus), which bypasses the liver.

The ductus venosus extends a short distance and joins the inferior vena cava. There, oxygenated blood from the placenta mixes with deoxygenated blood from the lower parts of the fetal body. This mixture continues through the vena cava to the right atrium.

In an adult heart, the blood from the right atrium enters the right ventricle and is pumped through the pulmonary trunk and pulmonary arteries to the lungs. In the fetus, however, the lungs are nonfunctional, and the blood largely bypasses them. As blood from the inferior vena cava enters the fetal right atrium, much of it is shunted directly into the left atrium through an opening in the atrial septum. This opening is called the **foramen ovale** (fo-ra′men ova′le), and the blood passes through it because the blood pressure in the right atrium is somewhat greater than that in the left atrium. Furthermore, a small valve (septum primum) located on the left side of

the atrial septum overlies the foramen ovale and helps prevent blood from moving in the reverse direction.

The rest of the fetal blood entering the right atrium, including a large proportion of the deoxygenated blood entering from the superior vena cava, passes into the right ventricle and out through the pulmonary trunk. Only a small volume of blood enters the pulmonary circuit because the lungs are collapsed and their blood vessels have a high resistance to blood flow. However, enough blood reaches the lung tissues to sustain them.

Most of the blood in the pulmonary trunk bypasses the lungs by entering a fetal vessel called the **ductus arteriosus** (duk′tus ar-te″re-o′sus), which connects the pulmonary trunk to the descending portion of the aortic arch. As a result of this connection, the blood with a relatively low oxygen concentration, which is returning to the heart through the superior vena cava, bypasses the lungs and does not enter the portion of the aorta that branches to the heart and brain.

The more highly oxygenated blood that enters the left atrium through the foramen ovale mixes with a small amount of deoxygenated blood returning from the pulmonary veins. This mixture moves into the left ventricle and is pumped into the aorta. Some of it reaches the myocardium through the coronary arteries, and some reaches the brain tissues through the carotid arteries.

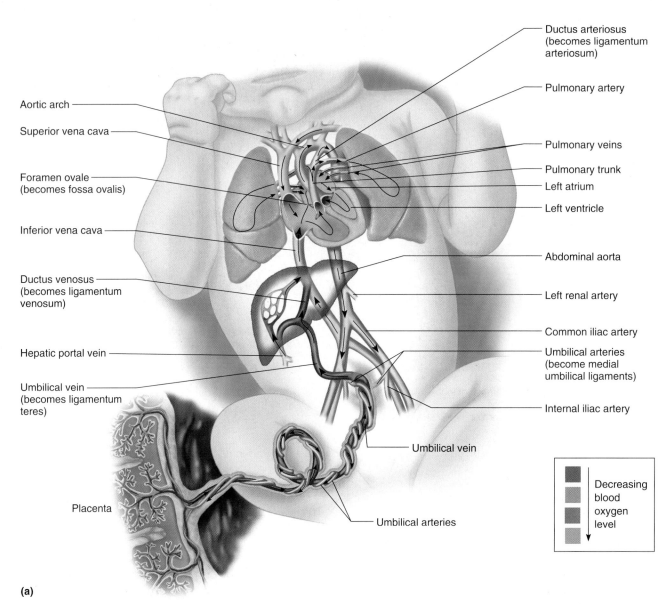

Aortic arch

Superior vena cava

Foramen ovale
(becomes fossa ovalis)

Inferior vena cava

Ductus venosus
(becomes ligamentum
venosum)

Hepatic portal vein

Umbilical vein
(becomes ligamentum
teres)

Placenta

Ductus arteriosus
(becomes ligamentum
arteriosum)

Pulmonary artery

Pulmonary veins

Pulmonary trunk

Left atrium

Left ventricle

Abdominal aorta

Left renal artery

Common iliac artery

Umbilical arteries
(become medial
umbilical ligaments)

Internal iliac artery

Umbilical vein

Umbilical arteries

Decreasing
blood
oxygen
level

(a)

FIGURE 23.25

The general pattern of fetal circulation is shown anatomically (a) and schematically (b).

The blood the descending aorta carries is partially oxygenated and partially deoxygenated. Some of it is carried into the branches of the aorta that lead to the lower regions of the body. The rest passes into the *umbilical arteries,* which branch from the internal iliac arteries and lead to the placenta. There the blood is reoxygenated (fig. 23.25).

The umbilical cord usually contains two arteries and one vein. Rarely, newborns have only one umbilical artery. This condition is often associated with other cardiovascular, urogenital, or gastrointestinal disorders. Because of the possibility of these conditions, the vessels within the severed cord are routinely counted following a birth.

Table 23.4 summarizes the major features of fetal circulation. At the time of birth, important adjustments must occur in the cardiovascular system when the placenta ceases to function and the newborn begins to breathe.

TABLE 23.4	Fetal Cardiovascular Adaptations
Adaptation	**Function**
Fetal blood	Has greater oxygen-carrying capacity than adult blood
Umbilical vein	Carries oxygenated blood from the placenta to the fetus
Ductus venosus	Conducts about half the blood from the umbilical vein directly to the inferior vena cava, thus bypassing the liver
Foramen ovale	Conveys a large proportion of the blood entering the right atrium from the inferior vena cava, through the atrial septum, and into the left atrium, thus bypassing the lungs
Ductus arteriosus	Conducts some blood from the pulmonary trunk to the aorta, thus bypassing the lungs
Umbilical arteries	Carry the blood from the internal iliac arteries to the placenta

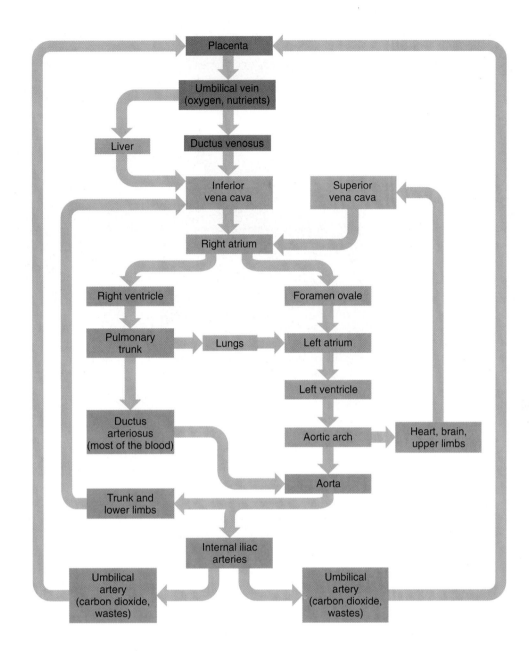

(b)

Clinical Application 23.2 describes a case in which fetal ultrasound revealed two hearts and bloodstreams, yet a single body.

1. How does the pattern of fetal circulation differ from that of an adult?

2. Which umbilical vessel carries oxygen-rich blood to the fetus?

3. What is the function of the ductus venosus?

4. What characteristic of fetal pulmonary circulation tends to shunt blood away from the lungs?

5. How does fetal circulation allow blood to bypass the lungs?

Birth Process

Pregnancy usually continues for forty weeks, or about nine calendar months, if it is measured from the begin-

ning of the last reproductive cycle. Pregnancy terminates with the *birth process* (parturition).

Birth is a complex, little-understood process. Progesterone plays a major role in its start. During pregnancy, this hormone suppresses uterine contractions. As the placenta ages, the progesterone concentration within the uterus declines, which may also stimulate synthesis of a prostaglandin that promotes uterine contractions. At the same time, the cervix begins to thin and then open. Changes in the cervix may begin a week or two before other signs of labor occur.

Stretching of the uterine and vaginal tissues late in pregnancy also stimulates the birth process. This may initiate nerve impulses to the hypothalamus, which, in turn, signals the posterior pituitary gland to release the hormone **oxytocin,** which stimulates powerful uterine contractions. Combined with the greater excitability of the

JOINED FOR LIFE

Patty Hensel's pregnancy in 1990 was uneventful. An ultrasound scan revealed an apparently normal fetus, although at one medical exam, Mike Hensel thought he heard two heartbeats.

A cesarean section was necessary because the baby was positioned bottom-first. To everyone's amazement, the baby had two heads and two necks, yet it appeared to share the rest of the body, with two legs and two arms in the correct places, and a third arm between the heads. The ultrasound had probably imaged the twins from an angle that superimposed one head on the other. Patty, dopey from medication, recalls hearing the word "Siamese" and thinking she had given birth to cats. She had delivered conjoined, or Siamese, twins.

The baby was actually two individuals, named Abigail and Brittany. Each twin had her own neck, head, heart, stomach, and gallbladder. Remarkably, each also had her own nervous system. The twins shared a large liver, a single bloodstream, and all organs below the navel, including the reproductive tract. They had three lungs and three kidneys.

Abby and Britty were strong and healthy. Doctors suggested surgery to separate the twins. Aware that only one child would likely survive surgery, Mike and Patty chose to let their daughters be. The girls are happy and active (fig. 23E). Abby and Britty have very distinctive personalities and attend school, swim, ride bikes, and play, like any other kids.

Conjoined twins occur in 1 in 50,000 births, and about 40% are stillborn. They

FIGURE 23E
Abby and Britty Hensel are conjoined twins, the result of incomplete twinning during the first two weeks of prenatal development.

can be attached in any of several ways. Abby and Britty Hensel were joined in a manner seen only four times before. They are the result of incomplete twinning, which probably occurred during the first two weeks of gestation. Because the girls have duplicated tissue derived from ectoderm, mesoderm, and endoderm, the partial twinning event must have occurred before the three germ layers were established, at day 14.

The term "Siamese twins" comes from Chang and Eng, who were born in Thailand, then called Siam, in 1811. They were joined by a ligament from the navel to the breastbone, which surgeons could easily correct today. Chang and Eng lived for sixty-three years, and each married.

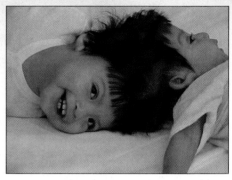

FIGURE 23F
Maria de Jesus and Maria Teresa were rare among the rare—conjoined twins attached at the head. Surgery successfully separated them.

Two percent of conjoined twins are attached at the head ("craniopagus"). This was the case for Maria de Jesus and Maria Teresa, born in Guatemala in 2001. The girls were attached in a way that they faced opposite each other, so they could not sit (fig. 23F). In a 22 hour operation on August 5, 2002, a team of surgeons separated them. Six weeks earlier, physicians had inserted an 8 inch silicone balloon under the shared scalps to stretch the skin. Meanwhile, the team worked with plastic models to find the best way to separate the twins and preserve facial structures. Fortunately, the girls had separate brains and cerebral arteries, but there was fear that they shared the sagittal sinus, the vein that carries blood from the brain to the heart. However, the surgery was a success. One girl received the sagittal sinus, and the other had sufficient collateral vessels to take over. They are doing well. ∎

myometrium due to the decline in progesterone secretion, oxytocin aids *labor* in its later stages.

During labor, muscular contractions force the fetus through the birth canal. Rhythmic contractions that begin at the top of the uterus and travel down its length force the contents of the uterus toward the cervix.

Since the fetus is usually positioned head downward, labor contractions force the head against the cervix. This action stretches the cervix, which elicits a reflex that stimulates still stronger labor contractions. Thus, a *positive feedback system* operates in which uterine contractions produce more intense uterine contractions until a maximum effort is achieved (fig. 23.26). At the same time, dilation of the cervix reflexly stimulates an increased release of oxytocin from the posterior pituitary gland.

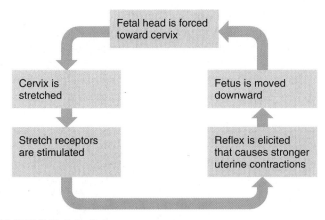

FIGURE 23.26
A positive feedback mechanism propels the birth process.

An infant passing through the birth canal can stretch and tear the tissues between the vulva and anus (perineum). Before the birth is complete, a physician may make an incision along the midline of the perineum from the vestibule to within 1.5 centimeters of the anus. This procedure, called an *epi-siotomy*, ensures that the perineal tissues are cut cleanly rather than torn, which aids healing.

As labor continues, abdominal wall muscles are stimulated to contract. These muscles also help move the fetus through the cervix and vagina to the outside. Table 23.5 summarizes some of the factors promoting labor. Figure 23.27 illustrates the steps of the birth process.

Following birth of the fetus, the placenta, which remains inside the uterus, separates from the uterine wall

TABLE 23.5	Factors Contributing to the Labor Process
1. As the time of birth approaches, secretion of progesterone declines, and its inhibiting effect on uterine contractions may lessen.
2. Decreasing progesterone concentration may stimulate synthesis of prostaglandins, which may initiate labor.
3. Stretching uterine tissues stimulates release of oxytocin from the posterior pituitary gland.
4. Oxytocin may stimulate uterine contractions and aid labor in its later stages.
5. As the fetal head stretches the cervix, a positive feedback mechanism results in stronger and stronger uterine contractions and a greater release of oxytocin.
6. Positive feedback stimulates abdominal wall muscles to contract with greater and greater force.
7. The fetus is forced through the birth canal to the outside.

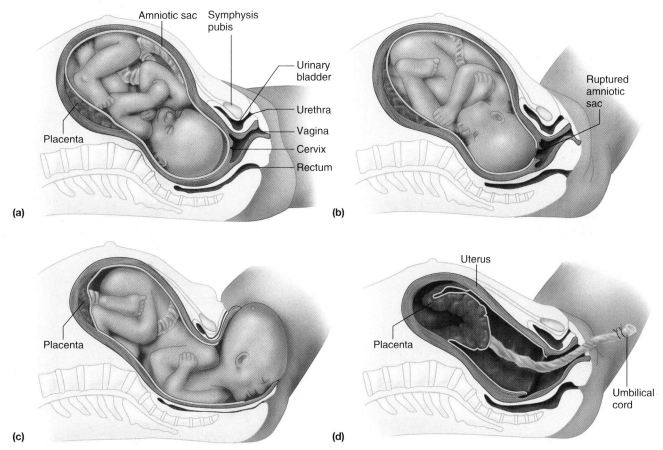

FIGURE 23.27
Stages in birth. (*a*) Fetal position before labor, (*b*) dilation of the cervix, (*c*) expulsion of the fetus, (*d*) expulsion of the placenta.

and is expelled by uterine contractions through the birth canal. This expulsion, termed the *afterbirth,* is accompanied by bleeding, because vascular tissues are damaged in the process. However, the loss of blood is usually minimized by continued contraction of the uterus that compresses the bleeding vessels. The action of oxytocin stimulates this contraction. Breast-feeding also contributes to returning the uterus to its original, prepregnancy size, as the suckling of the newborn stimulates the mother's posterior pituitary to release oxytocin.

For several weeks following childbirth, the uterus shrinks by a process called *involution.* Also, its endometrium sloughs off and is discharged through the vagina. The new mother passes a bloody and then yellowish discharge from the vagina for a few weeks. This is followed by the return of an epithelial lining characteristic of a nonpregnant female.

1 List some of the physiological changes that occur in a woman's body during pregnancy.

2 Describe the role of progesterone in initiating labor.

3 Explain how dilation of the cervix affects labor.

4 Explain how bleeding is controlled naturally after the placenta is expelled.

Milk Production and Secretion

During pregnancy, placental estrogens and progesterone stimulate further development of the mammary glands. Estrogens cause the ductile systems to grow and branch, and deposit abundant fat around them. Progesterone stimulates the development of the alveolar glands at the ends of the ducts. Placental lactogen also promotes these changes.

Because of hormonal activity, the breasts may double in size during pregnancy. At the same time, glandular tissue replaces the adipose tissue of the breasts. Beginning about the fifth week of pregnancy, the anterior pituitary gland releases increasing amounts of *prolactin.* Prolactin is synthesized from early pregnancy throughout gestation, peaking at the time of birth. However, milk secretion does not begin until after birth. This is because during pregnancy, placental progesterone inhibits milk production, and placental lactogen blocks the action of prolactin. Consequently, even though the mammary glands can secrete milk, none is produced. The micrographs in figure 23.28 compare the mammary gland tissues of a nonpregnant woman with those of a lactating woman.

Following childbirth and the expulsion of the placenta, the maternal blood concentrations of placental hormones decline rapidly. The action of prolactin is no longer inhibited. Prolactin stimulates the mammary glands to secrete large quantities of milk. This hormonal effect does not occur until two or three days following birth, and in the meantime, the glands secrete a thin, watery fluid called *colostrum.* Colostrum is rich in proteins, particularly protective antibodies, but has lower concentrations of carbohydrates and fats than milk.

Milk does not flow readily through the ductile system of the mammary gland but must be actively ejected as

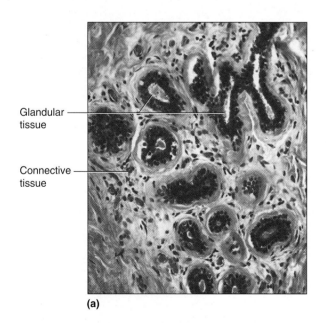

Glandular tissue

Connective tissue

(a)

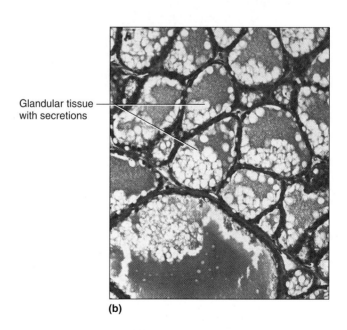

Glandular tissue with secretions

(b)

FIGURE 23.28

Mammary glands. (*a*) Light micrograph of a mammary gland in a nonpregnant woman (160×). (*b*) Light micrograph of an active (lactating) mammary gland (160×).

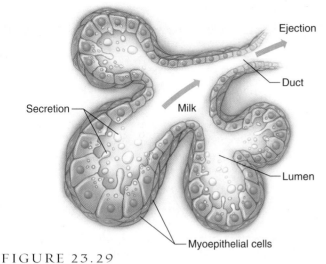

FIGURE 23.29

Myoepithelial cells contract to eject milk from an alveolar gland.

tinue secreting prolactin. Thus, prolactin is released as long as milk leaves the breasts. However, if milk is not removed regularly, the hypothalamus inhibits the secretion of prolactin, and within about one week, the mammary glands lose their capacity to produce milk.

A woman who is breast-feeding feels her milk "let down," or flood her breasts, when her infant suckles. If the baby nurses on a very regular schedule, the mother may feel the letdown shortly before the baby is due to nurse. The connection between mind and hormonal control of lactation is so strong that if a nursing mother simply hears a baby cry, her milk may flow. If this occurs in public, she can keep from wetting her shirt by pressing her arms strongly against her chest.

specialized *myoepithelial cells* surrounding the alveolar glands contract. A reflex action controls this process and is elicited when the breast is suckled or the nipple or areola is otherwise mechanically stimulated (fig. 23.29). Then, impulses from sensory receptors within the breasts travel to the hypothalamus, which signals the posterior pituitary gland to release oxytocin. The oxytocin reaches the breasts by means of the blood and stimulates the myoepithelial cells to contract (in both breasts). Within about thirty seconds, milk squirts into a suckling infant's mouth (fig. 23.30).

Sensory impulses triggered by mechanical stimulation of the nipples also signal the hypothalamus to con-

To wean a nursing child, it is best to stop breast-feeding gradually, by eliminating one feeding per day each week, for example. If a woman stops nursing abruptly, her breasts will become painfully engorged for several days.

A woman who is breast-feeding usually does not ovulate for several months. This may be because prolactin suppresses release of gonadotropins from the anterior pituitary gland. When a woman discontinues breast-feeding, the anterior pituitary no longer secretes prolactin. Then, FSH is released, and the reproductive cycle is activated. If a new mother does not wish to repeat her recent childbirth experience soon, she or her partner should be practicing contraception, because she will be fertile during the two weeks prior to the return of her menstrual period.

Table 23.6 summarizes the hormonal control of milk production, and table 23.7 lists some agents that adversely affect lactation or harm the child. Clinical Application 23.3 explains the benefits of breast-feeding.

1 How does pregnancy affect the mammary glands?

2 What stimulates the mammary glands to produce milk?

3 What causes milk to flow into the ductile system of a mammary gland?

4 What happens to milk production if milk is not regularly removed from the breast?

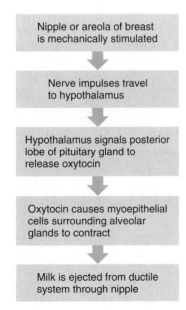

FIGURE 23.30

Mechanism that ejects milk from the breasts.

Postnatal Period

Following birth, both mother and newborn experience physiological and structural changes. The postnatal period of development lasts from birth until death. It can be divided into the neonatal period, infancy, childhood, adolescence, adulthood, and senescence.

TABLE 23.6	Hormonal Control of the Mammary Glands	
Before Pregnancy (Beginning of Puberty)		**Following Childbirth**
Ovarian hormones secreted during reproductive cycles stimulate alveolar glands and ducts of mammary glands to develop.		1. Placental hormonal concentrations decline, so the action of prolactin is no longer inhibited.
During Pregnancy		2. The breasts begin producing milk.
1. Estrogens cause the ductile system to grow and branch.		3. Mechanical stimulation of the breasts releases oxytocin from the posterior pituitary gland.
2. Progesterone stimulates development of alveolar glands.		4. Oxytocin stimulates ejection of milk from ducts.
3. Placental lactogen promotes development of the breasts.		5. As long as milk is removed, more prolactin is released; if milk is not removed, milk production ceases.
4. Prolactin is secreted throughout pregnancy, but placental progesterone inhibits milk production.		

TABLE 23.7	Agents Contraindicated During Breast-Feeding	
Agent	**Use**	**Effect on Lactation or Baby**
Doxorubicin, methotrexate	Cancer chemotherapy, psoriasis, rheumatoid arthritis	Immune suppression
Cyclosporine	Immune suppression in transplant patients	Immune suppression
Radioactive isotopes	Cancer diagnosis and therapy	Radioactivity in milk
Phenobarbitol	Anticonvulsant	Sedation, spasms on weaning
Oral contraceptives	Birth control	Decreased milk production
Caffeine (large amounts)	Food additive	Irritability, poor sleeping
Cocaine	Drug of abuse	Intoxication, seizures, vomiting, diarrhea
Ethanol (alcohol) (large amounts)	Drug of abuse	Weak, drowsy; infant decreases in length but gains weight; decreased milk ejection reflex
Heroin	Drug of abuse	Tremors, restlessness, vomiting, poor feeding
Nicotine	Drug of abuse	Diarrhea, shock, increased heart rate; lowered milk production
Phencyclidine	Drug of abuse	Hallucinations

Neonatal Period

The **neonatal** (ne″o-na′tal) **period,** which extends from birth to the end of the first four weeks, begins very abruptly at birth (fig. 23.31). At that moment, physiological adjustments must occur quickly because the newborn must suddenly do for itself what the mother's body had been doing for it. Thus, the newborn (neonate) must respire, obtain and digest nutrients, excrete wastes, and regulate body temperature. However, a newborn's most immediate need is to obtain oxygen and excrete carbon dioxide, so its first breath is critical.

The first breath must be particularly forceful because the newborn's lungs are collapsed and the airways are small, offering considerable resistance to air movement. Also, surface tension tends to hold the moist membranes of the lungs together. However, the lungs of a full-term fetus continuously secrete surfactant (see chapter 19, p. 747), which reduces surface tension. After the first powerful breath begins to expand the lungs, breathing eases.

A newborn's first breath is stimulated by increasing concentration of carbon dioxide, decreasing pH, low oxygen concentration, drop in body temperature, and mechanical stimulation during and after birth. Also, in response to the stress the fetus experiences during birth, blood concen-

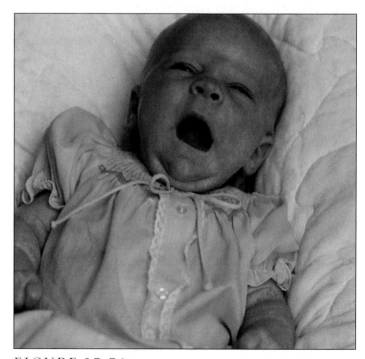

FIGURE 23.31

The neonatal period extends from birth to the end of the fourth week after birth.

HUMAN MILK—THE PERFECT FOOD FOR HUMAN BABIES

The female human body manufactures milk that is a perfect food for a human newborn in several ways. Human milk is rich in the lipids required for rapid brain growth, and it is low in protein. Cow's milk is the reverse, with three times as much protein as human milk. Much of cow milk protein is casein, which spurs a calf's rapid muscle growth, but forms hard-to-digest curds in a human baby's stomach. The protein in human milk has a balance of essential amino acids more suited to human growth and development than does the protein in cow's milk.

Human milk protects a newborn from many infections. For the first few days after giving birth, a new mother's breasts produce colostrum, which has less sugar and fat than mature milk but more protein, and is rich in antibodies. The antibodies protect the baby from such infections as *Salmonella* poisoning and polio. When the milk matures by a week to ten days, it has antibodies, enzymes, and white blood cells from the mother that continue infection protection. A milk protein called lactoferrin binds iron, making it unavailable to microorganisms that might use it to thrive in the newborn's digestive tract. Another biochemical in human milk, bifidus factor, encourages the growth of the bacteria *Lactobacillus bifidus,* which manufacture acids in the baby's digestive system that kill harmful bacteria.

A breast-fed baby typically nurses until he or she is full, not until a certain number of ounces have been drunk, which may explain why breast-fed babies are less likely to be obese than bottle-fed infants. Babies nurtured on human milk are also less likely to develop allergies to cow's milk. A nursing mother must eat about 500 calories per day more than usual to meet the energy requirements of milk production—but she also loses weight faster than a mother who bottle-feeds, because the fat reserves set aside during pregnancy are used to manufacture milk.

But breast-feeding is not the best choice for all women. It may be impossible to be present for each feeding or to provide milk. Also, many drugs a mother takes may enter breast milk and can affect the baby. Another disadvantage of breast-feeding is that the father cannot do it.

An alternative to breast-feeding is infant formula, which is usually cow's milk plus fats, proteins, carbohydrates, vitamins, and minerals added to make it as much like breast milk as possible. Although infant formula is nutritionally sound, the foul-smelling and bulkier bowel movements of the bottle-fed baby compared to the odorless, loose, more frequent and less abundant feces of a breast-fed baby indicate that breast milk is a more digestible first food than infant formula. ■

---

trations of epinephrine and norepinephrine rise significantly (see chapter 13, pp. 492–493). These hormones promote normal breathing by increasing the secretion of surfactant and dilating the airways.

For energy, the fetus primarily depends on glucose and fatty acids in the pregnant woman's blood. The newborn, on the other hand, is suddenly without an external source of nutrients. The mother will not produce milk for two to three days, by which time the infant's gastrointestinal tract will be able to digest it. However, the mother's breasts secrete *colostrum,* a fluid rich in nutrients and antibodies, until the milk comes in—an adaptation to the state of the newborn's digestive physiology. The newborn has a high metabolic rate, and its liver, which is not fully mature, may be unable to supply enough glucose to support metabolism. Consequently, the newborn utilizes stored fat for energy.

A newborn's kidneys are usually unable to produce concentrated urine, so they excrete a dilute fluid. For this reason, the newborn may become dehydrated and develop a water and electrolyte imbalance. Also, certain homeostatic control mechanisms may not function adequately. For example, during the first few days of life, body temperature may respond to slight stimuli by fluctuating above or below the normal level.

When the placenta ceases to function and breathing begins, changes occur in the newborn's cardiovascular system. Following birth, the umbilical vessels constrict. The arteries close first, and if the umbilical cord is not clamped or severed for a minute or so, blood continues to flow from the placenta to the newborn through the umbilical vein, adding to the newborn's blood volume.

The proximal portions of the umbilical arteries persist in the adult as the *superior vesical arteries* that supply blood to the urinary bladder. The more distal portions become solid cords (lateral umbilical ligaments). The umbilical vein becomes the cordlike *ligamentum teres* that extends from the umbilicus to the liver in an adult. The ductus venosus constricts shortly after birth and appears in the adult as a fibrous cord (ligamentum venosum) superficially embedded in the wall of the liver.

The foramen ovale closes as a result of blood pressure changes in the right and left atria. As blood ceases to flow from the umbilical vein into the inferior vena cava, the blood pressure in the right atrium falls. Also, as the lungs expand with the first breathing movements, resistance to blood flow through the pulmonary circuit decreases, more blood enters the left atrium through the pulmonary veins, and blood pressure in the left atrium increases.

As the blood pressure in the left atrium rises and that in the right atrium falls, the valve (septum primum) on the left side of the atrial septum closes the foramen ovale. In most individuals, this valve gradually fuses with the tissues along the margin of the foramen. In an adult, a depression called the *fossa ovalis* marks the site of the previous opening.

The ductus arteriosus, like other fetal vessels, constricts after birth. After this, blood can no longer bypass the lungs by moving from the pulmonary trunk directly into the aorta. In an adult, a cord called the *ligamentum arteriosum* represents the ductus arteriosus.

In *patent ductus arteriosus* (PDA), the ductus arteriosus fails to close completely. This condition is common in newborns whose mothers were infected with rubella virus (German measles) during the first three months of pregnancy.

After birth, the metabolic rate and oxygen consumption in neonatal tissues increase, in large part to maintain body temperature. If the ductus arteriosus remains open, the neonate's blood oxygen concentration may be too low to adequately supply body tissues, including the myocardium. If PDA is not corrected surgically, the heart may fail, even though the myocardium is normal.

Changes in the newborn's cardiovascular system are gradual. Although constriction of the ductus arteriosus may be functionally complete within fifteen minutes, the permanent closure of the foramen ovale may take up to a year. These cardiovascular changes are illustrated in figure 23.32 and summarized in table 23.8.

1 Define *neonatal period.*

2 What factors stimulate the first breath?

3 How do the kidneys of a newborn differ from those of an adult?

4 What is the fate of the foramen ovale? Of the ductus arteriosus?

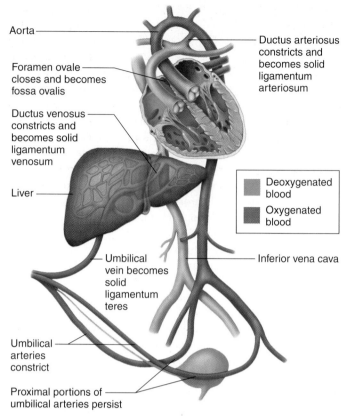

Aorta

Ductus arteriosus constricts and becomes solid ligamentum arteriosum

Foramen ovale closes and becomes fossa ovalis

Ductus venosus constricts and becomes solid ligamentum venosum

Liver

Deoxygenated blood

Oxygenated blood

Umbilical vein becomes solid ligamentum teres

Inferior vena cava

Umbilical arteries constrict

Proximal portions of umbilical arteries persist

FIGURE 23.32
Major changes occur in the newborn's cardiovascular system.

Infancy

The period of continual development extending from the end of the first four weeks to one year is called **infancy.** During this time, the infant grows rapidly and may triple its birth weight. Its teeth begin to erupt through the gums, and its muscular and nervous systems mature so that coordinated muscular activities become possible. The infant is soon able to follow objects visually; reach for and grasp objects; and sit, creep, and stand.

Infancy also brings the beginning of the ability to communicate. The infant learns to smile, laugh, and respond to

TABLE 23.8	Cardiovascular Adjustments in the Newborn		
Structure	**Adjustment**		**In the Adult**
Umbilical vein	Constricts		Becomes ligamentum teres that extends from the umbilicus to the liver
Ductus venosus	Constricts		Becomes ligamentum venosum that is superficially embedded in the wall of the liver
Foramen ovale	Closes by valvelike septum primum as blood pressure in right atrium decreases and blood pressure in left atrium increases		Valve fuses along margin of foramen ovale and is marked by a depression called the fossa ovalis
Ductus arteriosus	Constricts		Becomes ligamentum arteriosum that extends from the pulmonary trunk to the aorta
Umbilical arteries	Distal portions constrict		Distal portions become lateral umbilical ligaments; proximal portions function as superior vesical arteries

some sounds. By the end of the first year, the infant may be able to say two or three words. Often one of a child's first words is the name of a beloved pet.

Because infancy (as well as childhood) is a period of rapid growth, the infant has particular nutritional requirements. In addition to an energy source, the body requires proteins to provide the amino acids necessary to form new tissues; calcium and vitamin D to promote the development and ossification of skeletal structures; iron to support blood cell formation; and vitamin C for production of structural tissues such as cartilage and bone. By the time an infant is four months old, most of the circulating hemoglobin is the adult type.

Childhood

Childhood begins at the end of the first year and ends at puberty. During this period, growth continues at a high rate. The primary teeth appear, and then secondary teeth replace them. The child develops a high degree of voluntary muscular control and learns to walk, run, and climb. Bladder and bowel controls are established. The child learns to communicate effectively by speaking, and later, usually learns to read, write, and reason objectively. At the same time, the child is maturing emotionally.

1 Define *infancy.*

2 What developmental changes characterize infancy?

3 Define *childhood.*

4 What developmental changes characterize childhood?

Adolescence

Adolescence is the period of development between puberty and adulthood. It is a time of anatomical and physiological changes that result in reproductively functional individuals. Most of these changes are hormonally controlled, and they include the appearance of secondary sex characteristics as well as growth spurts in the muscular and skeletal systems.

Females usually experience these changes somewhat earlier than males, so early in adolescence, females may be taller and stronger than their male peers. On the other hand, females attain full growth at earlier ages, and in late adolescence, the average male is taller and stronger than the average female.

The periods of rapid growth in adolescence, which usually begin between the ages of eleven and thirteen in females and between thirteen and fifteen in males, cause increased demands for certain nutrients. It is not uncommon for a teenager to consume a huge plate of food, go back for more—and still remain thin. In addition to energy sources, foods must provide ample amounts of proteins, vitamins, and minerals to support growth of new tissues. Adolescence also brings increasing levels of motor skills, intellectual ability, and emotional maturity.

Adulthood

Adulthood (maturity) extends from adolescence to old age. As we age, we become gradually aware of certain declining functions—yet other abilities remain adequate throughout life. The "Life-Span Changes" sections in various previous chapters have described the effects of aging on particular organ systems. It is interesting to note the varying ages at which particular structures or functions peak.

By age eighteen, the human male is producing the highest level of the sex hormone testosterone that he will ever have in his lifetime, and sex drive is strong. In the twenties, muscle strength peaks in both sexes. Hair is its fullest, with each hair as thick as it will ever be. By the end of the third decade of life, obvious signs of aging may first appear as a loss in the elasticity of facial skin, producing small wrinkles around the mouth and eyes. Height is already starting to decrease, but not yet at a detectable level.

The age of thirty seems to be a developmental turning point. After this, hearing often becomes less acute. Heart muscle begins to thicken. The elasticity of the ligaments between the small bones in the back lessens, setting the stage for the slumping posture that becomes apparent in later years. Some researchers estimate that beginning roughly at age thirty, the human body becomes functionally less efficient by about 0.8% every year.

During their forties, many people weigh 10 to 20 pounds (4.5 to 9 kilograms) more than they did at the age of twenty, thanks to a slowing of metabolism and decrease in activity level. They may be 1/8 inch (0.3 centimeter) shorter, too. Hair may be graying as melanin-producing biochemical pathways lose efficiency, and some hair may fall out. Vision may become farsighted or nearsighted. The immune system is less efficient, making the body more prone to infection and cancer. Skeletal muscles tend to lose strength as more and more connective tissue appears within the muscles; the cardiovascular system is strained as the lumens of arterioles and arteries narrow due to accumulations of fatty deposits; skin loosens and wrinkles as elastic fibers in the dermis break down.

The early fifties bring further declines in the functioning of the human body. Nail growth slows, taste buds die, and the skin continues to lose its elasticity. For most people, the ability to see close objects becomes impaired, but for the nearsighted, vision improves. Women stop menstruating, although this does not necessarily mean an end to or loss of interest in sex. Delayed or reduced insulin release by the pancreas, in response to a glucose load, may lead to diabetes. By the decade's end, muscle mass and weight begin to decrease. A male produces less semen but is still sexually active. His voice may become higher as his vocal cords degenerate. A man has half the strength in his upper limb muscles and half the lung function as he did at age twenty-five. He is about 3/4 inch (2 centimeters) shorter.

The sixty-year-old may experience minor memory losses. A few million of the person's billions of brain cells have been lost over his or her lifetime, but for the most part, intellect remains quite sharp. By age seventy, height decreases a full inch (2.5 centimeters). Sagging skin and loss of connective tissue, combined with continued growth of cartilage, make the nose, ears, and eyes more prominent. Figure 23.33 outlines some of the anatomical and physiological changes that accompany aging.

Senescence

Senescence (se-nes′ens) is the process of growing old. It is a continuation of the degenerative changes that begin during adulthood. As a result, the body becomes less able to cope with the demands placed on it by the individual and by the environment.

Senescence is a result of the normal wear-and-tear of body parts over many years. For example, the cartilage covering the ends of bones at joints may wear away, leaving the joints stiff and painful. Other degenerative changes are caused by disease processes that interfere with vital functions, such as gas exchanges or blood circulation. Metabolic rate and distribution of body fluids may change. The rate of division of certain cell types declines, and immune responses weaken. As a result, the person becomes less able to repair damaged tissue and more susceptible to disease.

Decreasing efficiency of the central nervous system accompanies senescence. The person may lose some intellectual functions. Also, the physiological coordinating capacity of the nervous system may decrease, and homeostatic mechanisms may fail to operate effectively. Sensory functions decline with age also.

Death usually results, not from these degenerative changes, but from mechanical disturbances in the cardiovascular system, failure of the immune system, or disease processes that affect vital organs. However, the loss of function that often precedes death in the elderly is associated with inactivity, poor nutrition, and chronic disease. Table 23.9 summarizes the major phases of postnatal life and their characteristics, and table 23.10 lists some aging-related changes.

From 65% to 80% of all deaths in the United States take place in hospitals, often with painful and sometimes unwanted interventions to prolong life. One study found that about half of all conscious patients suffer severe pain prior to death. In Oregon, which has pioneered education on caring for the dying patient, a greater percentage of patients live out their last days at home, in nursing homes, or in hospices, which are facilities dedicated to providing comfort and support for the dying. The medical community is trying to remedy shortcomings in the treatment of the dying. Medical training is increasing emphasis on providing palliative care for the terminally ill. Such care seeks to make a patient comfortable, even if the treatment does not cure the disease or extend life.

Years

Decrease in height (about one inch); further decline in sense of taste; nose, ears, and eyes appear prominent as skin sags; facial fat decreases, but cartilage grows

Decrease in height (about 3/4 inch); decrease in lung capacity

Skin sags; decline in visual acuity; menopause; decline in sense of taste; decline in nail growth; increased risk of diabetes; decline in muscle mass, weight, metabolism, and memory

Back slumps; increase in weight; decrease in height; hair grays and thins; declining number of white blood cells; farsightedness

Peak hair thickness; skin less elastic; declines in hearing and height; peak of female sexuality; heart muscle thickens

Peak muscle strength

Peak of male sexuality

Sexual maturity

Thymus begins to shrink

Brain cells begin to die

FIGURE 23.33
Although many biological changes ensue as we grow older, photographs of actress Katharine Hepburn at various stages of her life indicate that we can age with great grace and beauty.

TABLE 23.9 — Stages in Postnatal Development

Stage	Time Period	Major Events
Neonatal period	Birth to end of fourth week	Newborn begins to carry on respiration, obtain nutrients, digest nutrients, excrete wastes, regulate body temperature, and make cardiovascular adjustments
Infancy	End of fourth week to one year	Growth rate is high; teeth begin to erupt; muscular and nervous systems mature so that coordinated activities are possible; communication begins
Childhood	One year to puberty	Growth rate is high; deciduous teeth erupt and are replaced by permanent teeth; high degree of muscular control is achieved; bladder and bowel controls are established; intellectual abilities mature
Adolescence	Puberty to adulthood	Person becomes reproductively functional and emotionally more mature; growth spurts occur in skeletal and muscular systems; high levels of motor skills are developed; intellectual abilities increase
Adulthood	Adolescence to old age	Person remains relatively unchanged anatomically and physiologically; degenerative changes begin to occur
Senescence	Old age to death	Degenerative changes continue; body becomes less and less able to cope with the demands placed upon it; death usually results from mechanical disturbances in the cardiovascular system or from disease processes that affect vital organs

TABLE 23.10 — Aging-Related Changes

Organ System	Aging-Related Changes
Integumentary system	Degenerative loss of collagenous and elastic fibers in dermis; decreased production of pigment in hair follicles; reduced activity of sweat and sebaceous glands
	Skin thins, wrinkles, and dries out; hair turns gray and then white
Skeletal system	Degenerative loss of bone matrix
	Bones become thinner, less dense, and more likely to fracture; stature may shorten due to compression of intervertebral discs and vertebrae
Muscular system	Loss of skeletal muscle fibers; degenerative changes in neuromuscular junctions
	Loss of muscular strength
Nervous system	Degenerative changes in neurons; loss of dendrites and synaptic connections; accumulation of lipofuscin in neurons; decreases in sensory sensitivities
	Decreasing efficiency in processing and recalling information; decreasing ability to communicate; diminished senses of smell and taste; loss of elasticity of lenses and consequent loss of ability to accommodate for close vision
Endocrine system	Reduced hormonal secretions
	Decreased metabolic rate; reduced ability to cope with stress; reduced ability to maintain homeostasis
Cardiovascular system	Degenerative changes in cardiac muscle; decrease in lumen diameters of arteries and arterioles
	Decreased cardiac output; increased resistance to blood flow; increased blood pressure
Lymphatic system	Decrease in efficiency of immune system
	Increased incidence of infections and neoplastic diseases; increased incidence of autoimmune diseases
Digestive system	Decreased motility in gastrointestinal tract; reduced secretion of digestive juices
	Reduced efficiency of digestion
Respiratory system	Degenerative loss of elastic fibers in lungs; reduced number of alveoli
	Reduced vital capacity; increase in dead air space; reduced ability to clear airways by coughing
Urinary system	Degenerative changes in kidneys; reduction in number of functional nephrons
	Reductions in filtration rate, tubular secretion, and reabsorption
Reproductive systems	
Male	Reduced secretion of sex hormones; enlargement of prostate gland; decrease in sexual energy
Female	Degenerative changes in ovaries; decrease in secretion of sex hormones
	Menopause; regression of secondary sex characteristics

1 How does the body change during adolescence?

2 Define *adulthood*.

3 What changes occur during adulthood?

4 What changes accompany senescence?

Aging

The aging process is difficult to analyze because of the intricate interactions of the body's organ systems. Breakdown of one structure ultimately affects the functioning of others. The medical field of gerontology examines the biological changes of aging at the molecular, cellular, organismal, and population levels. Aging is both passive and active.

Passive Aging

Aging as a passive process entails breakdown of structures and slowing of functions. At the molecular level, passive aging is seen in the degeneration of the elastin and collagen proteins of connective tissues, causing skin to sag and muscle to lose its firmness.

During a long lifetime, biochemical abnormalities accumulate. Mistakes occur throughout life when DNA replicates in dividing cells. Usually, repair enzymes correct this damage immediately. But over many years, exposure to chemicals, viruses, and radiation disrupts DNA repair mechanisms so that the error burden becomes too great to be fixed. The cell may die as a result of faulty genetic instructions.

Another sign of passive aging at the biochemical level is the breakdown of lipids. As aging membranes leak during lipid degeneration, a fatty, brown pigment called lipofuscin accumulates. Mitochondria also begin to break down in older cells, decreasing the supply of chemical energy to power the cell's functions.

The cellular degradation associated with aging may be set into action by highly reactive chemicals called **free radicals.** A molecule that is a free radical has an unpaired electron in its outermost valence shell. This causes the molecule to grab electrons from other molecules, destabilizing them, and a chain reaction of chemical instability begins that could kill the cell. Free radicals are a by-product of normal metabolism and also form by exposure to radiation or toxic chemicals. The bile pigment bilirubin protects against free radicals. Enzymes that usually inactivate free radicals diminish in number and activity in the later years. One such enzyme is *superoxide dismutase* (SOD).

Active Aging

Aging also entails new activities or the appearance of new substances. Lipofuscin granules, for example, may be considered an active sign of aging, but they result from the passive breakdown of lipids. Another example of active aging is autoimmunity, in which the immune sys-

tem turns against the body, attacking its cells as if they were invading organisms.

Active aging actually begins before birth, as certain cells die as part of the developmental program encoded in the genes. This process of programmed cell death, called **apoptosis** (ap"o-to'sis), occurs regularly in the embryo, degrading certain structures to pave the way for new ones. The number of neurons in the fetal brain, for example, is halved as those that make certain synaptic connections are spared from death. In the fetal thymus, T cells that do not recognize "self" cell surfaces die, thereby building the immune system. Throughout life, apoptosis enables organs to maintain their characteristic shapes.

Mitosis and apoptosis are opposite, but complementary, processes. That is, as organs grow, the number of cells in some regions increases, but in others, it decreases. Cell death is not a phenomenon only of the aged. It is a normal part of life. Clinical Application 23.4 discusses genetic disorders that greatly accelerate aging.

The Human Life Span

In the age-old quest for longer life, people have sampled everything from turtle soup to owl meat to human blood. A Russian-French microbiologist, Ilya Mechnikov, believed that a life span of 150 years could be achieved with the help of a steady diet of milk cultured with bacteria. He thought that the bacteria would live in the large intestine and somehow increase the human life span. (He died at age seventy-one.) Ironically, many people have died in pursuit of a literal "fountain of youth."

The human *life span*—the length of time that a human can theoretically live—is 120 years. Although most people succumb to disease or injury long before that point, in many countries the fastest growing age group is those over age eighty. These "oldest old," having passed the age when cancer and cardiovascular disease typically strike, are often quite healthy.

Life expectancy is a realistic projection of how long an individual will live, based on epidemiological information. In the United States, life expectancy is seventy-six and a half years. Yet in at least one African nation being decimated by the AIDS epidemic, life expectancy is only thirty-six years.

Life expectancy approaches life span as technology conquers diseases. Technology also alters the most prevalent killers. Development of antibiotic drugs removed some infectious diseases such as pneumonia and tuberculosis from the top of the "leading causes of death" list, a position that heart disease filled. Cancer is currently approaching heart disease as the most common cause of death in developed nations. Infections remain a major cause of death in less-developed countries. Table 23.11 lists the top causes of death in the United States, and Table 23.12 indicates how age is a factor in the nature of the most common causes of death.

Medical advances have greatly contributed to improved life expectancy. Antibiotics have tamed some

OLD BEFORE THEIR TIME

The progerias are inherited disorders that cause a person to live a lifetime in just a few years.

In *Hutchinson-Gilford syndrome,* which affects the Luciano brothers that appear in figure 23G, a child looks normal at birth. Within just a few years, however, the child acquires a shocking appearance of wrinkles, baldness, and the prominent facial features characteristic of advanced age. Arteries clog with fatty deposits. The child usually dies of a heart attack or a stroke by the age of twelve, although some patients live into their twenties. Only a few dozen cases of this syndrome have ever been reported.

An adult form of progeria called *Werner syndrome* becomes apparent before the twentieth birthday. The person usually dies before age fifty of disorders usually associated with advanced age, such as type II diabetes mellitus, atherosclerosis, and osteoporosis. Curiously, they usually do not develop hypertension or Alzheimer disease.

The cells of progeria patients show profound aging-related changes. Normal human cells in culture divide only fifty or so times, but cells from progeria patients divide only ten to thirty times, and die prematurely. Certain structures seen in normal cultured cells as they near the fifty-division limit (glycogen particles, lipofuscin granules, many lysosomes and vacuoles, and a few ribosomes) appear early in the cells of people with progeria. Understanding the mechanisms that cause these diseased cells to race through the aging process may help us to understand the biology of normal aging. ∎

FIGURE 23G

The Luciano brothers inherited progeria and appear far older than their years.

TABLE 23.11	Leading Causes of Death in the United States, 2000
Cause	**Percent of Total**
1. Heart disease	37.3
2. Cancer	29.1
3. Stroke	8.8
4. Chronic lower respiratory disease	6.4
5. Accidents	5.2
6. Diabetes mellitus	3.6
7. Pneumonia and influenza	3.4
8. Alzheimer disease	2.6
9. Nephritis, nephrotic syndrome, and nephrosis	2.0
10. Septicemia	1.6

Source: Centers for Disease Control and Prevention

TABLE 23.12	Leading Cause of Death in Different Age Groups in the United States, 1999
Age Group	**Cause**
1–34	Accidents
35–64	Cancer
65+	Heart disease

Source: Centers for Disease Control and Prevention

once-lethal infections, drugs enable many people with cancer to survive, and such advances as beta-blocking drugs and coronary bypass surgery have extended the lives of people with heart disease. However, a look at the history of two infectious diseases, pneumonia and tuberculosis, sounds a warning against complacency. The top two killers in 1900, pneumonia and tuberculosis, did not make the top five causes of death in the 1986 list. However, they reappear as numbers 4 and 5 in the 1993 list, in drug-resistant forms! The rise of new or renewed infectious diseases, such as AIDS and measles, also indicates that we cannot yet conquer all killers. Although we can alter our environment more than other species can, some forces of nature remain beyond our control.

1 Why is it difficult to sort out the causes of aging?

2 How is aging a passive process?

3 How is aging an active process?

4 Distinguish between life span and life expectancy.

Clinical Terms Related to Human Growth and Development

abortion (ah-bor'shun) Spontaneous or deliberate termination of pregnancy; a spontaneous abortion is commonly termed a miscarriage.

abruptio placentae (ab-rup'shē-o plah-cen'ta) Premature separation of the placenta from the uterine wall.

amniocentesis (am″ne-o-sen-te'sis) Technique in which a sample of amniotic fluid is withdrawn from the amniotic cavity by inserting a hollow needle through the pregnant woman's abdominal wall.

cesarean section (sĕ-sa're-an sek'shun) Delivery of a fetus through an abdominal incision.

dizygotic twins (di″zi-got'ik twinz) Twins resulting from the fertilization of two ova by two sperm cells. (fraternal twins)

eclampsia (ē-klamp'se-ah) Condition characterized by convulsions and coma that sometimes accompanies toxemia of pregnancy.

gestation (jes-ta'shun) Entire period of pregnancy.

hydatidiform mole (hi″dah-tid'ĭ-form mŏl) Abnormal pregnancy resulting from a pathologic ovum; a mass of cysts.

hydramnios (hi-dram'ne-os) Excess amniotic fluid.

hyperemesis gravidarum (hi″per-em'ĕ-sis grav'i-dar-um) Vomiting associated with pregnancy; morning sickness.

intrauterine transfusion (in″trah-u'ter-in trans-fu'zhun) Transfusion administered by injecting blood into the fetal peritoneal cavity before birth.

lochia (lo'ke-ah) Vaginal discharge following childbirth.

meconium (mĕ-ko'ne-um) Anal discharge from the digestive tract of a full-term fetus or a newborn.

monozygotic twins (mon″o-zi-got'ik twinz) Twins resulting from one sperm cell fertilizing one egg cell, which then splits. (identical twins)

perinatology (per″ĭ-na-tol'o-je) Branch of medicine concerned with the fetus after twenty-five weeks of development and with the newborn for the first four weeks after birth.

postpartum (pōst-par'tum) Occurring after birth.

teratology (ter″ah-tol'o-je) Study of substances that cause abnormal development and congenital malformations.

toxemia of pregnancy (tok-se'me-ah) Group of metabolic disturbances that may occur during pregnancy.

trimester (tri-mes'ter) Each third of the total period of pregnancy.

ultrasonography (ul'trah-son-og'rah-fe) Technique used to visualize the size and position of fetal structures from patterns of deflected ultrasonic waves.

CHAPTER SUMMARY

Introduction (page 876)

Growth refers to an increase in size; development is the process of changing from one phase of life to another.

Pregnancy (page 876)

Pregnancy is the presence of a developing offspring in the uterus.

1. Transport of sex cells
   a. Ciliary action aids movement of the egg cell to the uterine tube.
   b. A sperm cell moves by its tail lashing and muscular contraction in the female reproductive tract.
2. Fertilization
   a. With the aid of an enzyme, a sperm cell penetrates the zona pellucida.
   b. When a sperm cell penetrates an egg cell membrane, changes in the egg cell membrane and the zona pellucida prevent entry of additional sperm.

c. Fusion of the nuclei of a sperm and an egg cell complete fertilization.
d. The product of fertilization is a zygote with 46 chromosomes.

Prenatal Period (page 880)

1. Period of cleavage
   a. Fertilization occurs in a uterine tube and results in a zygote.
   b. The zygote undergoes mitosis, and the newly formed cells divide mitotically too.
   c. Each subsequent division produces smaller and smaller cells.
   d. A solid ball of cells (morula) forms, and it becomes a hollow ball called a blastocyst.
   e. The inner cell mass that gives rise to the embryo proper forms within the blastocyst.

f. The blastocyst implants in the uterine wall.
   (1) Enzymes digest the endometrium around the blastocyst.
   (2) Fingerlike processes from the blastocyst penetrate into the endometrium.
g. The period of cleavage lasts through the first week of development.
h. The trophoblast secretes hCG, which helps maintain the corpus luteum, helps protect the blastocyst against being rejected, and stimulates the developing placenta to secrete hormones.

2. Hormonal changes during pregnancy
   a. Embryonic cells produce hCG that maintains the corpus luteum, which continues to secrete estrogens and progesterone.
   b. Placental tissue produces high concentrations of estrogens and progesterone.
      (1) Estrogens and progesterone maintain the uterine wall and inhibit secretion of FSH and LH.
      (2) Progesterone and relaxin inhibit contractions of uterine muscles.
      (3) Estrogens cause enlargement of the vagina.
      (4) Relaxin helps relax the ligaments of the pelvic joints.
   c. The placenta secretes placental lactogen that stimulates the development of the breasts and mammary glands.
   d. During pregnancy, increasing secretion of aldosterone promotes retention of sodium and body fluid, and increasing secretion of parathyroid hormone helps maintain a high concentration of maternal blood calcium.

3. Other changes during pregnancy
   a. The uterus enlarges greatly.
   b. The woman's blood volume, cardiac output, breathing rate, and urine production increase.
   c. The woman's dietary needs increase, but if intake is inadequate, fetal tissues have priority for use of available nutrients.

4. Embryonic stage
   a. The embryonic stage extends from the second through the eighth weeks.
   b. It is characterized by the development of the placenta and the main internal and external body structures.
   c. The embryonic disc becomes cylindrical and is attached to the developing placenta by the connecting stalk.
   d. The cells of the inner cell mass fold inward, forming a gastrula that has two and then three primary germ layers.
      (1) Ectoderm gives rise to the nervous system, portions of the skin, the lining of the mouth, and the lining of the anal canal.
      (2) Mesoderm gives rise to muscles, bones, blood vessels, lymphatic vessels, reproductive organs, kidneys, and linings of body cavities.
      (3) Endoderm gives rise to linings of the digestive tract, respiratory tract, urinary bladder, and urethra.
   e. Chorionic villi develop and are surrounded by spaces filled with maternal blood.
   f. The embryo develops head, face, upper limbs, lower limbs, and mouth, and appears more humanlike.
   g. The placental membrane consists of the epithelium of the villi and the epithelium of the capillaries inside the villi.

   (1) Oxygen and nutrients diffuse from the maternal blood through the placental membrane and into the fetal blood.
   (2) Carbon dioxide and other wastes diffuse from the fetal blood through the placental membrane and into the maternal blood.
h. The placenta develops in the disc-shaped area where the chorion contacts the uterine wall.
   (1) The embryonic portion consists of the chorion and its villi.
   (2) The maternal portion consists of the uterine wall and attached villi.
i. A fluid-filled amnion develops around the embryo.
j. The umbilical cord is formed as the amnion envelopes the tissues attached to the underside of the embryo.
   (1) The umbilical cord includes two arteries and a vein.
   (2) It suspends the embryo in the amniotic cavity.
k. The chorion and amnion fuse.
l. The yolk sac forms on the underside of the embryonic disc.
   (1) It gives rise to blood cells and cells that later form sex cells.
   (2) It helps form the digestive tube.
m. The allantois extends from the yolk sac into the connecting stalk.
   (1) It forms blood cells.
   (2) It gives rise to the umbilical vessels.
n. By the beginning of the eighth week, the embryo is recognizable as a human.

5. Fetal stage
   a. This stage extends from the end of the eighth week and continues until birth.
   b. Existing structures grow and mature; only a few new parts appear.
   c. The body enlarges, upper and lower limbs reach final relative proportions, the skin is covered with sebum and dead epidermal cells, the skeleton continues to ossify, muscles contract, and fat is deposited in subcutaneous tissue.
   d. The fetus is full term at the end of the ninth month, which equals approximately 266 days.
      (1) It is about 50 centimeters long and weighs 6–8 pounds.
      (2) It is positioned with its head toward the cervix.

6. Fetal blood and circulation
   a. Umbilical vessels carry blood between the placenta and the fetus.
   b. Fetal blood carries a greater concentration of oxygen than does maternal blood.
   c. Blood enters the fetus through the umbilical vein and partially bypasses the liver by means of the ductus venosus.
   d. Blood enters the right atrium and partially bypasses the lungs by means of the foramen ovale.
   e. Blood entering the pulmonary trunk partially bypasses the lungs by means of the ductus arteriosus.
   f. Blood enters the umbilical arteries from the internal iliac arteries.

7. Birth
   a. Pregnancy usually lasts forty weeks from the beginning of the last reproductive cycle.
   b. During pregnancy, placental progesterone inhibits uterine contractions.

c. A variety of factors are involved with the birth process.
  (1) A decreasing concentration of progesterone and the release of prostaglandins may initiate the birth process.
  (2) The posterior pituitary gland releases oxytocin.
  (3) Uterine muscles are stimulated to contract, and labor begins.
  (4) A positive feedback mechanism causes stronger contractions and greater release of oxytocin.
d. Following the birth of the infant, placental tissues are expelled.

8. Milk production and secretion
   a. During pregnancy, the breasts change.
     (1) Estrogens cause the ductile system to grow.
     (2) Progesterone causes development of alveolar glands.
     (3) Prolactin is released during pregnancy, but progesterone inhibits milk production.
   b. Following childbirth, the concentrations of placental hormones decline.
     (1) The action of prolactin is no longer blocked.
     (2) The mammary glands begin to secrete milk.
   c. Reflex response to mechanical stimulation of the nipple causes the posterior pituitary to release oxytocin, which causes milk to be ejected from the alveolar ducts.
   d. As long as milk is removed from glands, more milk is produced; if milk is not removed, production ceases.
   e. During the period of milk production, the reproductive cycle is partially inhibited.

Postnatal Period (page 905)

1. Neonatal period
   a. This period extends from birth to the end of the fourth week.
   b. The newborn must begin to respire, obtain nutrients, excrete wastes, and regulate its body temperature.
   c. The first breath must be powerful in order to expand the lungs.
     (1) Surfactant reduces surface tension.
     (2) A variety of factors stimulate the first breath.
   d. The liver is immature and unable to supply sufficient glucose, so the newborn depends primarily on stored fat for energy.
   e. Immature kidneys cannot concentrate urine very well.
     (1) The newborn may become dehydrated.
     (2) Water and electrolyte imbalances may develop.
   f. Homeostatic mechanisms may function imperfectly, and body temperature may be unstable.
   g. The cardiovascular system changes when placental circulation ceases.
     (1) Umbilical vessels constrict.
     (2) The ductus venosus constricts.
     (3) The foramen ovale is closed by a valve as blood pressure in the right atrium falls and blood pressure in the left atrium rises.
     (4) The ductus arteriosus constricts.

2. Infancy
   a. Infancy extends from the end of the fourth week to one year of age.
   b. Infancy is a period of rapid growth.
     (1) The muscular and nervous systems mature, and coordinated activities become possible.
     (2) Communication begins.

c. Rapid growth depends on an adequate intake of proteins, vitamins, and minerals in addition to energy sources.

3. Childhood
   a. Childhood extends from the end of the first year to puberty.
   b. It is characterized by rapid growth, development of muscular control, and establishment of bladder and bowel control.

4. Adolescence
   a. Adolescence extends from puberty to adulthood.
   b. It is characterized by physiological and anatomical changes that result in a reproductively functional individual.
   c. Females may be taller and stronger than males in early adolescence, but the situation reverses in late adolescence.
   d. Adolescents develop high levels of motor skills, their intellectual abilities increase, and they continue to mature emotionally.

5. Adulthood
   a. Adulthood extends from adolescence to old age.
   b. The adult remains relatively unchanged physiologically and anatomically for many years.
   c. After age thirty, degenerative changes usually begin to occur.
     (1) Skeletal muscles lose strength.
     (2) The cardiovascular system becomes less efficient.
     (3) The skin loses its elasticity.
     (4) The capacity to produce sex cells declines.

6. Senescence
   a. Senescence is the process of growing old.
   b. Degenerative changes continue, and the body becomes less able to cope with demands placed upon it.
   c. Changes occur because of prolonged use, effects of disease, and cellular alterations.
   d. An aging person usually experiences losses in intellectual functions, sensory functions, and physiological coordinating capacities.
   e. Death usually results from mechanical disturbances in the cardiovascular system or from disease processes that affect vital organs.

Aging (page 912)

1. Passive aging
   a. Passive aging entails breakdown of structures and slowing or failure of functions.
   b. Connective tissue breaks down.
   c. DNA errors accumulate.
   d. Lipid breakdown in aging membranes releases lipofuscin.
   e. Free radical damage escalates.

2. Active aging
   a. In autoimmunity, the immune system attacks the body.
   b. Apoptosis is a form of programmed cell death. It occurs throughout life, shaping organs.

3. The human life span
   a. The theoretical maximum life span is 120 years.
   b. Life expectancy, based on real populations, is seventy-six and a half years in the United States, and may be quite lower in poorer nations and those ravaged by AIDS.
   c. Medical technology makes life expectancy more closely approach life span.

CRITICAL THINKING QUESTIONS

1. How would you explain the observation that twins resulting from a single fertilized egg cell can exchange body parts by tissue or organ transplant procedures without experiencing rejection reactions?

2. One of the more common congenital cardiac disorders is a ventricular septum defect in which an opening remains between the right and left ventricles. What problem would such a defect create as blood moves through the heart?

3. What symptoms may appear in a newborn if its ductus arteriosus fails to close?

4. What technology would enable a fetus born in the fourth month to survive in a laboratory setting? (This is not yet possible.)

5. Why is it important for a middle-aged adult who has neglected physical activity for many years to have a physical examination before beginning an exercise program?

6. If an aged relative came to live with you, what special provisions could you make in your household environment and routines that would demonstrate your understanding of the changes brought on by aging?

REVIEW EXERCISES

1. Define *growth* and *development.*
2. Describe the process of fertilization.
3. Describe the process of cleavage.
4. Distinguish between a morula and a blastocyst.
5. Describe the formation of the inner cell mass, and explain its significance.
6. Describe the process of implantation.
7. List three functions of hCG.
8. Explain the major hormonal changes that occur in the maternal body during pregnancy.
9. Describe the major nonhormonal changes that occur in the maternal body during pregnancy.
10. Explain how the primary germ layers form.
11. List the major body parts derived from ectoderm.
12. List the major body parts derived from mesoderm.
13. List the major body parts derived from endoderm.
14. Describe the formation of the placenta, and explain its functions.
15. Define *placental membrane.*
16. Distinguish between the chorion and the amnion.
17. Explain the function of amniotic fluid.
18. Describe the formation of the umbilical cord.
19. Explain how the yolk sac and the allantois are related, and list the functions of each.
20. Explain why the embryonic period of development is so critical.
21. Define *fetus.*
22. List the major changes that occur during the fetal stage of development.

23. Describe a full-term fetus.
24. Compare the properties of fetal hemoglobin with those of adult hemoglobin.
25. Explain how the fetal cardiovascular system is adapted for intrauterine life.
26. Trace the pathway of blood from the placenta to the fetus and back to the placenta.
27. Describe the role of progesterone in initiating the birth process.
28. Discuss the events that occur during the birth process.
29. Explain the roles of prolactin and oxytocin in milk production and secretion.
30. Distinguish between a newborn and an infant.
31. Explain why a newborn's first breath must be particularly forceful.
32. List some of the factors that stimulate the first breath.
33. Explain why newborns tend to develop water and electrolyte imbalances.
34. Describe the cardiovascular changes that occur in the newborn.
35. Describe the characteristics of an infant.
36. Distinguish between a child and an adolescent.
37. Define *adulthood.*
38. List some of the degenerative changes that begin during adulthood.
39. Define *senescence.*
40. List some of the factors that promote senescence.
41. Cite evidence of passive aging and active aging.

WEB CONNECTIONS

Visit the Student Edition of the Online Learning Center at www.mhhe.com/shier10 **for answers to chapter questions, additional quizzes, interactive learning exercises, and other study tools.**

UNDERSTANDING WORDS

chromo-, color: *chromo*some—a "colored body" in a cell's nucleus that includes the genes.

hetero-, other, different: *hetero*zygous— condition in which the members of a gene pair are different.

hom-, same, common: *hom*ologous chromosomes—pair of chromosomes that contain similar genetic information.

karyo-, nucleus: *karyo*type—a chart that displays chromosomes in size order.

mono-, one: *mono*somy—condition in which one kind of chromosome is present in only one copy.

phen-, show, be seen: *phen*otype—physical appearance that results from the way genes are expressed in an individual.

tri-, three: *tri*somy—condition in which one kind of chromosome is triply represented.

Color enhanced scanning tunneling electron micrograph of part of a DNA double helix (2,000,000×).

Genetics and Genomics

CHAPTER OBJECTIVES

After you have studied this chapter, you should be able to

1. Explain how gene discoveries are relevant to the study of anatomy and physiology and to health care.

2. Distinguish between genes and chromosomes.

3. Define genome.

4. Define the two types of chromosomes.

5. Explain how genes can have many alleles (variants), but a person can have only two alleles of a particular gene.

6. Distinguish among the modes of inheritance.

7. Explain how gene expression varies among individuals.

8. Describe how genes and the environment interact to produce complex traits.

9. Describe how traits are transmitted on the sex chromosomes and how gender affects gene expression.

10. Explain how deviations in chromosome number or arrangement can harm health and how these abnormalities are detected.

11. Explain how conditions caused by extra or missing chromosomes reflect a meiotic error.

12. Explain how gene therapy works.

T he year is 2006. With sequencing of the complete DNA instructions within a human cell—the genome—the focus of health care is becoming predictive. Devices called DNA microarrays, or simply DNA chips, can rapidly identify genes, providing profiles that indicate which diseases a person is at highest risk of developing and even how that person will likely react to particular drug treatments. Young people are encouraged to take such tests—if they want to—because they have time to try to prevent illnesses that have controllable environmental components. This is the choice that two college freshmen, Laurel and Peter, face. Each selects DNA microarray tests based on family background.

Laurel's brother, sister, and father smoke cigarettes, and her father's mother, also a smoker, died of lung cancer. Two relatives on her mother's side had colon cancer, and older relatives on both sides have Alzheimer disease. Laurel's DNA panel tests for gene variants that predispose her to developing addictions, such as genes that regulate her circadian (daily) rhythms and encode the receptor proteins on nerve cells that bind neurotransmitters; genes that cause colon or lung cancer; and genes associated with inherited forms of Alzheimer disease. Later in life, she may elect to have a prenatal DNA microarray test to detect inherited conditions in a fetus, or undergo a "toxicogenomics" screen to identify chemical sensitivities if she might encounter chemicals at her job that harm susceptible individuals.

Peter requests different tests, appropriate for his family history. Each winter he suffers from bronchitis and sometimes pneumonia, and his doctor once suggested that he might have mild cystic fibrosis (CF), especially since Peter's sister and mother also get bronchitis often. Unlike Laurel, Peter refuses a test for Alzheimer disease, even though his paternal grandfather died of it—he feels he could not bear knowing that the condition lay in his future. Because previous blood tests revealed elevated cholesterol and several relatives have suffered heart attacks or have hypertension, Peter takes a panel of tests to track cardiovascular disease risk, including gene variants that control blood clotting, blood pressure, homocysteine metabolism, and cholesterol synthesis, transport, and metabolism.

The DNA microarray tests are easy, for the patient. After completing a family history, each student provides a DNA sample by collecting cells from the inside of the cheek with a cotton swab. At a laboratory, DNA in the cells is extracted, cut into pieces, tagged with molecules that fluoresce under certain types of light, and finally the pieces are applied to a glass or nylon chip and the light applied. Once results are in—a pattern of colored dots on a square—a genetic counselor explains the findings.

Laurel learns that she is genetically predisposed to addictive behaviors and has a high risk of developing lung cancer—a dangerous

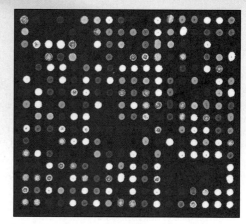

A DNA microarray, or "chip," identifies genes that are accessed in a particular cell to produce particular proteins. DNA microarrays will be used to confirm diagnoses based on signs and symptoms; predict future diseases that have a genetic basis; identify sensitivities to environmental agents; and predict which drugs will be effective to treat certain conditions in particular individuals.

combination. She knows that more than most people, she must avoid cigarettes and alcohol and other addictive drugs. Happily, she does not have genes that increase her chances of developing colon cancer or inherited Alzheimer disease.

Peter does have mild cystic fibrosis. He takes a different DNA microarray test that indicates which antibiotics will most effectively treat the frequent bronchitis and pneumonia. He might even be a candidate for gene therapy—periodically inhaling a preparation containing the normal version of his mutant gene delivered in a "disabled" virus that would otherwise cause a respiratory infection. Peter has several gene variants that elevate serum cholesterol level and blood pressure. By following a diet low in fat and high in fiber, exercising regularly, and having frequent cholesterol checks, he can help keep his cardiovascular system healthy. A third DNA microarray panel identifies the most effective cholesterol-lowering drug for him.

The genetic tests that Laurel and Peter take will become parts of their medical records, and they will add tests as their interests and health status change with age. But these medical records are confidential. Laws prevent employers and insurers from discrimination based on genetic information.

Although the scenario of Laurel and Peter is in the near future, each test described exists today. Human genome information promises to provide a wealth of new predictive and diagnostic tests and treatments for rare as well as common disorders. In the years to come, thanks to the avalanche of new genetic information, we will be learning what our own blueprints are and how they work to assemble a human body. ■

The Emerging Role of Genetics and Genomics in Medicine

Genetics (jĕ-net'iks), the study of inheritance of characteristics, concerns the transfer of information from generation to generation, which is termed heredity. That information is transmitted in the form of **genes** (jēnz), which consist of sequences of nucleotides of the nucleic acid DNA (see fig. 4.19). Genes are carried on rod-shaped structures called **chromosomes,** introduced in chapter 3 and revisited in figure 24.1. The transfer of genetic information between generations occurs through genes in the nuclei of eggs and sperm, via the process of meiosis discussed in chapter 22 (pp. 833–836).

A gene's nucleotide sequence tells a cell how to link a certain sequence of amino acids together to construct a specific protein molecule. Recall from chapter 4 (pp. 118–120) that the information in a DNA sequence is transcribed into a molecule of mRNA, which, in turn, is translated into a protein. The protein ultimately determines the trait associated with the gene, as figure 24.2 illustrates for cystic fibrosis (CF).

The complete set of genetic instructions in a human cell constitutes our **genome** (jĕ-nōme). The genome includes about 31,000 protein-encoding genes interspersed with other sequences, many of them highly repeated. Pieces of these genetic instructions, in the form of RNA, encode yet other proteins by combining segments. Therefore even though the human genome includes 31,000 or so genes, these specify 100,000 to 200,000 different proteins. In all cells except for the eggs and sperm, the DNA is distributed among 23 pairs of chromosomes, for a total of 46 chromosomes. These nonsex, or **somatic** cells, are said to be **diploid** because they have two sets of chromosomes. Therefore, a somatic cell contains two copies of the genome. Recall from chapter 22 (pp. 833 and 848) that sperm and eggs, which contain 23 individual chromosomes, are **haploid.** They have half the amount of genetic material of other cell types, or one copy of the genome.

Genetic information functions at several levels. It is encoded in chemicals, affects cells and tissues, and is expressed in the individual, yet is also passed to the next generation. At the population level, genetic change drives evolution.

Until recently, the field of genetics dealt mostly with single genes and the rare disorders that can be traced to the malfunction or absence of single genes. However, information from the human genome sequence is providing a new view of physiology as a complex interplay of gene function. Looking at the human body in terms of

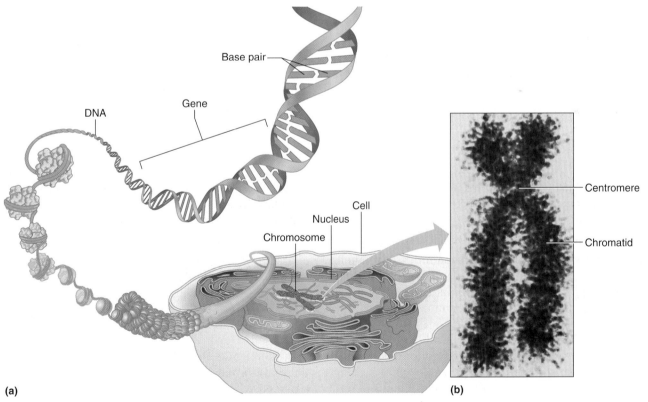

(a)

(b)

FIGURE 24.1

From DNA to gene to chromosome. (*a*) Chromosomes consist of a continuous DNA double helix and associated proteins and appear in the cell's nucleus just prior to cell division. (*b*) A transmission electron micrograph of a chromosome. Each half of the chromosome is a chromatid. Note the constriction, where the centromeres meet (25,000×).

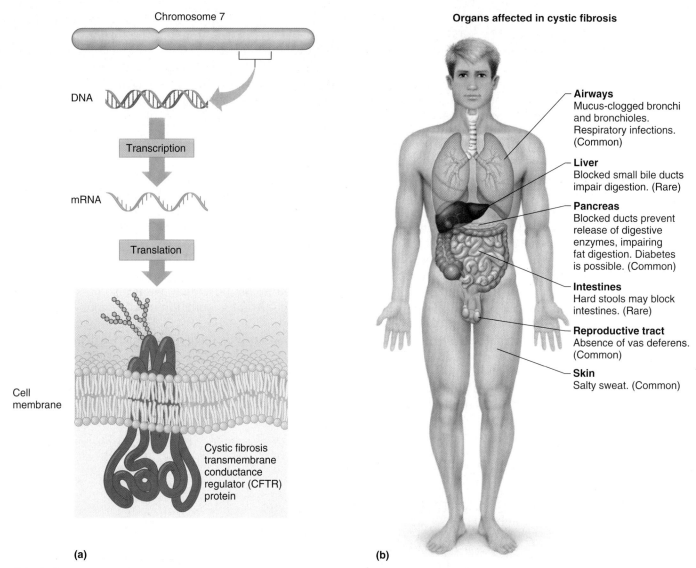

(a)

(b)

FIGURE 24.2

From gene to protein to person. (*a*) The gene encoding the CFTR protein, and causing cystic fibrosis when mutant, is on the seventh largest chromosome. CFTR protein folds into a channel that regulates the flow of chloride ions into and out of cells lining the respiratory tract, pancreas, intestines, and elsewhere. (*b*) In cystic fibrosis, the CFTR protein is abnormal, usually missing an amino acid. Its shape is altered, which entraps the chloride ions inside cells. Water entering these cells leaves behind very thick mucus and other secretions in the places highlighted in the illustration. The sticky secretions cause the symptoms of the illness. Source: Data from M.C. Iannuzi and F.S. Collins, "Reverse Genetics and Cystic Fibrosis" in *American Journal of Respiratory Cellular and Molecular Biology,* 2:309–316, 1990.

multiple, interacting genes is termed **genomics.** A related field, proteomics, focuses on the spectrum of proteins that specific cell types produce. A proteomics approach to studying breast cancer, for example, compares the thousands of types of proteins in a healthy milk duct lining cell to the proteins in the same type of cell that has become cancerous. Figure 24.3 views genomics at the whole body, cellular, and molecular levels.

The science of genetics has traditionally focused on disorders caused by single genes. However, the field now increasingly recognizes that genes provide our variability as well as illnesses, including eye, skin and hair color; height and body form; special talents; and hard-to-define

characteristics such as personality traits. Clinical Application 24.1 highlights a few interesting nonmedical traits rooted in the genes.

As important as genes are, they do not act alone. Often the environment influences how genes are expressed. The environment includes the chemical, physical, social, and biological factors surrounding an individual that influence his or her characteristics. For example, a person who inherits genes that confer susceptibility to smoking-induced lung cancer will probably not develop the illness if he or she never smokes and breathes clean air. Intelligence is a good example of a characteristic that has many genetic and environmental influences.

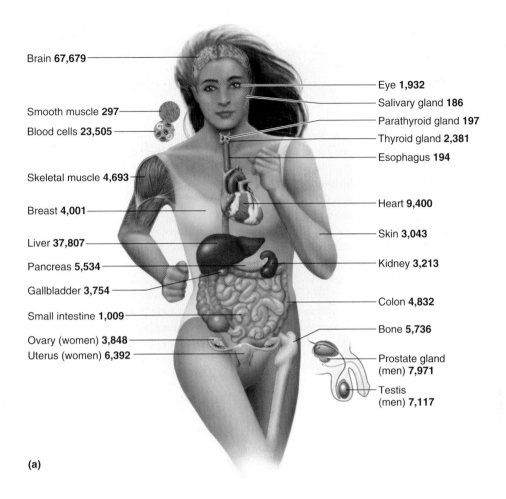

(a)

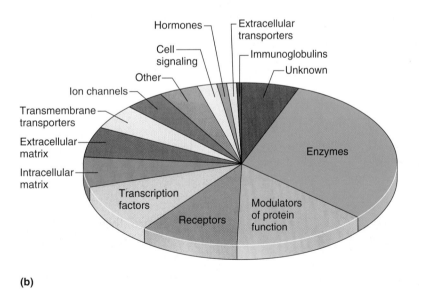

(b)

FIGURE 24.3

Representing the genome and proteome. One way to analyze genome data is to consider where genes function, both at the cellular and organ-system levels. In (*a*), the number of genes distributed by site of action exceeds the total gene number of about 31,000 because the same proteins may function in more than one organ. The proteome representation in (*b*) illustrates the relative contributions of different types of proteins to the functioning of a human, at the cellular level, from conception through old age.

IT'S ALL IN THE GENES

Do you have uncombable hair, misshapen toes or teeth, a pigmented tongue tip, or an inability to smell skunk? Do you lack teeth, eyebrows, eyelashes, nasal bones, thumbnails, or fingerprints? Can you roll your tongue or wiggle your ears? If so, you may find your unusual trait described in *Online Mendelian Inheritance in Man,* compiled by Johns Hopkins University geneticist Victor McKusick, which catalogs more than 10,000 known human genetic variants (www.ncbi.nlm.nih.gov/entrez/query.fcgi?db=OMIM). Most of the entries include family histories, clinical descriptions, molecular information, and how the trait is transmitted. Amidst the medical terminology can be found some fascinating inherited traits in humans, from top to toes.

Genes control whether hair is blond, brown, or black, whether or not it has red highlights, and whether it is straight, curly, or kinky. Widow's peaks, cowlicks, a whorl in the eyebrow, and white forelocks run in families, as do hairs with triangular cross sections. Some people have multicolored hairs like cats, and others have hair in odd places, such as on the elbows, nosetip, knuckles, palms of the hands, or soles of the feet. Teeth can be missing or extra, protuberant or fused, present at birth, or "shovel-shaped" or "snow-capped." A person can have a grooved tongue, duckbill lips, flared ears, egg-shaped pupils, three rows of eyelashes, spotted nails, or "broad thumbs and great toes." Extra breasts have been observed in humans and guinea pigs, and one family's claim to fame is a double nail on the littlest toes.

Unusual genetic variants can affect metabolism, sometimes with noticeable effects. Members of some families experience "urinary excretion of odoriferous component of asparagus" or "urinary excretion of beet pigment" after eating the implicated vegetables. In "blue diaper syndrome," an infant's inherited inability to break down an amino acid turns urine blue on contact with air.

One father and son could not open their mouths completely. "Dysmelodia" is the inability to carry a tune. Those with "odor blindness" cannot smell musk, skunk, cyanide, or freesia flowers. Motion sickness, migraine headaches, and stuttering may be inherited. Uncontrollable sneezing may be due to Achoo syndrome (an acronym for "autosomal dominant compelling helioophthalmic outburst" syndrome).

Figure 24A illustrates some more common genetic traits. ■

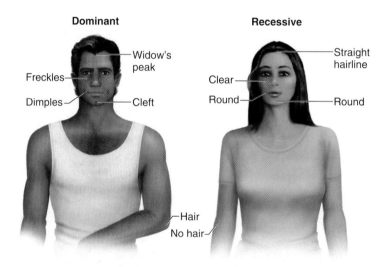

FIGURE 24A

Inheritance of some common traits: Freckles, dimples, widow's peak, hairy elbows, and a cleft chin.

---

1. What is genetics?

2. What are genes and chromosomes?

3. How has sequencing the human genome changed the field of genetics?

4. Distinguish among genetics, genomics, and proteomics.

Modes of Inheritance

The probability that a certain trait will occur in the offspring of two individuals can be determined by knowing how genes are distributed in meiosis and the combinations in which they can join at fertilization.

Chromosomes and Genes Come in Pairs

From the moment of conception, a human cell is diploid, containing two copies of each of the 23 different chromosomes. Chromosome charts called **karyotypes** are used to display the 23 chromosome pairs in size order (fig. 24.4). Pairs 1 through 22 are **autosomes** (aw'to-sōmz), which are chromosomes that do not carry genes that determine sex. The other two chromosomes, the X and the Y, determine sex and are called **sex chromosomes**. They are dis-

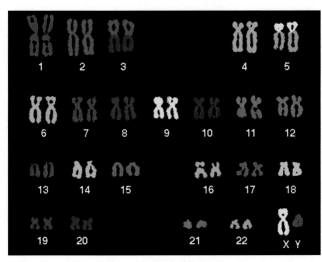

FIGURE 24.4

A normal human karyotype has the 22 pairs of autosomes aligned in size order, plus the sex chromosomes. In this karyotype, fluorescently tagged pieces of DNA are used as "probes" to bind to specific chromosomes, imparting vibrant colors. This technique is called FISH, which stands for fluorescence *in situ* hybridization.

cussed later in the chapter in the section titled "Matters of Sex."

Each chromosome includes hundreds or thousands of genes. Since we have two copies of each chromosome, we also have two copies of each gene, each located at the same position on the homologous chromosome pairs. Sometimes the members of a gene pair are alike, their DNA sequences specifying the same amino acid sequence of the protein product. However, because a gene consists of hundreds of nucleotide building blocks, it may exist in variant forms, called **alleles.** An individual who has two identical alleles of a gene is said to be **homozygous** (ho″mo-zi′gus) for that gene. A person with two different alleles for a gene is **heterozygous** (het′er o zi′gus) for it. A gene may have many alleles, but an individual person can have a maximum of two alleles for a particular gene.

> The allele that causes most cases of cystic fibrosis was discovered in 1989, and researchers immediately began developing a test to detect it. However, other alleles were soon discovered. Today, hundreds of mutations (changes) in the cystic fibrosis gene are known. Different allele combinations produce different combinations and severities of symptoms.

The particular combination of genes in a person's genome constitutes the **genotype** (je′no-tīp). The appearance or health condition of the individual that develops as a result of the ways the genes are expressed is termed the **phenotype** (fe′no-tīp). An allele is **wild type** if its asso-

ciated phenotype is either normal function or the most common expression in a particular population. Wild type is indicated with a + sign. An allele that is a change from wild type, perhaps producing an uncommon phenotype, is **mutant.** Disease-causing alleles are mutant.

Dominant and Recessive Inheritance

For many genes, in heterozygotes, one allele determines the phenotype. Such an allele whose action masks that of another allele is termed **dominant.** The allele whose expression is masked is **recessive.** Dominant alleles are usually indicated with a capital letter. A gene that causes a disease can be recessive or dominant. It may also be *autosomal* (carried on a nonsex chromosome) or *X-linked* (carried on the X chromosome) or *Y-linked* (carried on the Y chromosome). The more general and older term "sex-linked" refers to a gene on the X or Y chromosome.

Whether a trait is dominant or recessive, autosomal or carried on a sex chromosome, is called its *mode of inheritance.* This designation has important consequences in predicting the chance that offspring will inherit an illness or trait. The following rules emerge:

1. An autosomal condition is equally likely to affect either sex. X-linked characteristics affect males much more often than females, a point discussed later in the section "Sex Chromosomes and Their Genes."

2. A person most likely inherits a recessive condition from two parents who are each heterozygotes (carriers). The parents are usually healthy. For this reason, recessive conditions can "skip" generations.

3. A person who inherits a dominant condition has at least one affected parent. Therefore, dominant conditions do not skip generations. (An exception is if the dominant allele arises, as a new mutation, in the sperm or egg.) If, by chance, a dominant trait does not appear in a generation in a particular family, it does not reappear in subsequent generations, as a recessive trait might.

Cystic fibrosis is an example of an autosomal recessive disorder. The wild type allele for the CFTR gene, which is dominant over the disease-causing allele, specifies formation of chloride channels built of protein in the cell membrane of cells lining the pancreas, respiratory tract, intestine, testes, and other structures (see fig. 24.2). Certain recessive mutant alleles disrupt the structure and possibly the function of the chloride channels. An individual who inherits two such mutant alleles has cystic fibrosis and is homozygous recessive. A person inheriting only one recessive mutant allele plus a dominant wild type allele is a carrier and transmits the disease-causing allele in half of the gametes. A person who has two wild type alleles is homozygous dominant for the gene and does not have or carry CF. The three possible genotypes are associated with only two phenotypes, because both carriers and homozygous dominant individuals do not have the illness.

Using logic, understanding how chromosomes and genes are apportioned into gametes in meiosis, and knowing that mutant alleles that cause cystic fibrosis are autosomal recessive, we can predict genotypes and phenotypes of the next generation. Figure 24.5 illustrates two people who are heterozygous for a CF-causing allele. Half of the man's sperm contain the mutant allele, as do half of the woman's eggs. Because sperm and eggs combine at random, each offspring has a

- 25% chance of inheriting two wild type alleles (homozygous dominant, healthy, and not a carrier)
- 50% chance of inheriting a mutant allele from either parent (heterozygous and a carrier, but healthy)
- 25% chance of inheriting a mutant allele from each parent (homozygous recessive, has CF)

Genetic counselors use two tools to explain inheritance to families. A table called a **Punnett square** symbolizes the logic used to deduce the probabilities of particular genotypes occurring in offspring. The mother's alleles (for a particular gene) are listed atop the four boxes comprising the square, and the father's alleles are listed along the left side. Each box records an allele combination at fertilization.

A **pedigree** is a diagram that depicts family relationships and genotypes and phenotypes when they are known. Circles are females and squares are males; shaded-in symbols represent people who have a trait or condition; half-shaded symbols denote carriers. Roman numerals indicate generations. Figures 24.5 and 24.6 show Punnett squares and pedigrees.

In an autosomal recessive illness, an affected person's parents are usually carriers—they do not have the illness. Or, if the phenotype is mild, a parent might be homozygous recessive and affected. In an autosomal dominant condition, an affected person typically has an affected parent. He or she need inherit only one copy of the mutant allele to have the associated phenotype; in contrast, expression of an autosomal recessive condition requires two copies of the mutant allele.

An example of an autosomal dominant condition is Huntington disease (HD). Symptoms usually begin in the late thirties or early forties and include loss of coordination, uncontrollable dancelike movements, and personality changes, such as anger and irritability. Figure 24.6 shows the inheritance pattern for HD. If one parent has the mutant allele, half of his or her gametes will have it. Assuming the other parent does not have a mutant allele, each child conceived has a 1 in 2 chance of inheriting the gene and, eventually, developing the condition.

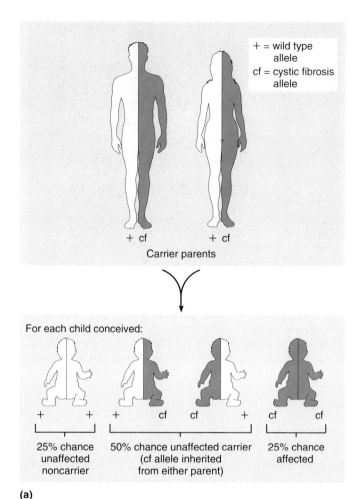

FIGURE 24.5

Inheritance of cystic fibrosis from carrier parents illustrates autosomal recessive inheritance. (*a*) Each child has a 25% chance of being unaffected and not a carrier, a 50% chance of being a carrier, and a 25% chance of being affected. Sexes are affected with equal frequency. A Punnett square (*b*) and a pedigree (*c*) are other ways of depicting this information. Symbols in the pedigree with both black and white indicate unaffected carriers. (Note that the pedigree illustrates the make-up of one possible family.)

Most of the 3,000 or so known human inherited disorders are autosomal recessive. These conditions tend to produce symptoms very early, sometimes before birth. Autosomal dominant conditions often have an adult onset. These conditions remain in populations because people have children before they know that

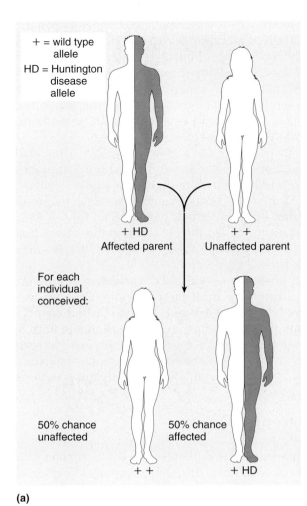

+ = wild type allele

HD = Huntington disease allele

+ HD
Affected parent

+ +
Unaffected parent

For each individual conceived:

50% chance unaffected
+ +

50% chance affected
+ HD

(a)

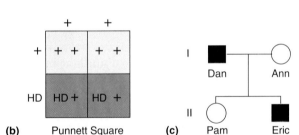

(b) Punnett Square

(c)

I Dan — Ann

II Pam — Eric

FIGURE 24.6

Inheritance of Huntington disease from a parent who will be affected in middle age illustrates autosomal dominant inheritance. (a) A person with just one HD allele develops the disease. A Punnett square (b) and pedigree (c) depict the inheritance of HD. The pedigree symbols for HD are completely filled in to indicate that the person is affected. Autosomal dominant conditions affect both sexes.

they have inherited the illness, knowledge that could cause some individuals to choose not to have children. Tests can detect certain genetic disorders before symptoms arise. The number of such tests will increase dramatically as researchers continue to analyze human genome information.

Certain recessive alleles that cause illness may remain in a population, even if they endanger health, because carrying them can protect against an infectious disease, a situation called *balanced polymorphism.* In sickle cell disease, for example, a mutation affecting just one DNA base causes the gene's product, the beta globin chain of hemoglobin, to aggregate under low oxygen conditions, which bends the red blood cell into a sickle shape that blocks blood flow. Carriers have only a few sickled cells, but these apparently are enough to make red blood cells inhospitable to malaria parasites. Therefore, carriers for sickle cell disease do not contract malaria, or they develop very mild cases.

Carriers of cystic fibrosis are resistant to diarrheal disorders, in which bacterial toxins open chloride channels in the small intestine. The carriers have some abnormal chloride channels, which renders these toxins ineffective.

Different Dominance Relationships

Most genes exhibit complete dominance or recessiveness. Interesting exceptions are incomplete dominance and codominance. In **incomplete dominance,** the heterozygous phenotype is intermediate between that of either homozygote. For example, in familial hypercholesterolemia (FH), a person with two disease-causing alleles completely lacks LDL (low-density lipoprotein) receptors on liver cells that take up cholesterol from the bloodstream (fig. 24.7). A person with one disease-causing allele (a heterozygote) has half the normal number of cholesterol receptors. Someone with two wild type alleles has the normal number of receptors. The associated phenotypes parallel the number of receptors—those with two mutant alleles die as children of heart attacks, individuals with one mutant allele may die in young or middle adulthood, and people with two wild type alleles do not develop this type of hereditary heart disease.

Different alleles that are both expressed in a heterozygote are **codominant.** For example, two of the three alleles of the *I* gene, which determines ABO blood type, are codominant (see fig. 14.21). People of blood type A have a molecule called antigen A on the surfaces of their red blood cells. Blood type B corresponds to red blood cells with antigen B. A person with type AB has red blood cells with both the A and B antigens, and the red cells of a person with type O blood have neither antigen.

The *I* gene encodes the enzymes that place the A and B antigens on red blood cell surfaces. The three alleles are I^A, I^B, and i. People with type A blood may be either genotype $I^A I^A$ or $I^A i$; type B corresponds to $I^B I^B$ or $I^B i$; type AB to $I^A I^B$; and type O to ii.

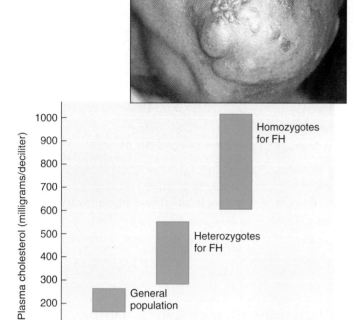

FIGURE 24.7

Incomplete dominance appears in the plasma cholesterol levels of heterozygotes and homozygotes for familial hypercholesterolemia [FH]. This condition is one of many that increase the cholesterol level in the blood, raising the risk of developing heart disease. The photograph shows cholesterol deposits on the elbow of a young man who is a homozygote for the disease-causing allele.

1   Distinguish between autosomes and sex chromosomes.

2   Distinguish between genotype and phenotype.

3   Distinguish between wild type and mutant alleles.

4   How do the modes of transmission of autosomal recessive and autosomal dominant inheritance differ?

5   Distinguish between incomplete dominance and codominance.

Gene Expression

The same allele combination can produce different degrees of the phenotype in different individuals, even siblings, because of influences such as nutrition, exposure to toxins, other illnesses, and the activities of other genes. A major goal of genomics will be to identify and understand these interactions.

Penetrance and Expressivity

Most disease-causing allele combinations are **completely penetrant,** which means that everyone who inherits it has some symptoms. A genotype is **incompletely penetrant** if some individuals do not express the phenotype. Polydactyly, having extra fingers or toes, is incompletely penetrant (see fig. 7.48). Some people who inherit the dominant allele have more than five digits on a hand or foot, yet others who are known to have the allele (because they have an affected parent and child) have the normal number of fingers and toes.

The penetrance of a gene is described numerically. If 80 of 100 people who have inherited the dominant polydactyly allele have extra digits, the allele is 80% penetrant. Inherited breast cancer is also incompletely penetrant—that is, not all women who inherit a mutant form of a gene called BRCA1 develop the cancer, but they do have a much greater risk of doing so than women who inherit the wild type allele.

A phenotype is **variably expressive** if the symptoms vary in intensity in different people. One person with polydactyly might have an extra digit on both hands and a foot; another might have two extra digits on both hands and both feet; a third person might have just one extra fingertip. Penetrance refers to the all-or-none expression of a genotype in an individual; expressivity refers to the severity of a phenotype.

Pleiotropy

A single genetic disorder can produce several symptoms, a phenomenon called **pleiotropy** (pleé-o-trope-ee). Family members who have different symptoms can appear to have different illnesses.

Pleiotropy is seen in genetic diseases that affect a single protein found in different parts of the body. This is the case for Marfan syndrome, an autosomal dominant defect in an elastic connective tissue protein called fibrillin. The fact that the protein is abundant in the lens of the eye, in the bones of the limbs, fingers, and ribs, and in the aorta explains the symptoms of lens dislocation, long limbs, spindly fingers, and a caved-in chest. The most serious symptom is a life-threatening weakening in the aorta wall, which sometimes causes the vessel to suddenly burst. If the weakening is found early, a synthetic graft can be used to patch that part of the vessel wall, saving the person's life. Clinical Application 14.1 discusses a pleiotropic disorder that left its mark on history, porphyria variegata.

Genetic Heterogeneity

The same phenotype may result from the actions of different genes, which is called **genetic heterogeneity** (jĕ-net'ik het"er-o-je-ne'ĭ-te). For example, the nearly 200 forms of hereditary deafness are each due to impaired actions of a different gene. The eleven types of clotting disorders reflect the many protein factors and enzymes that control this process. Any of several genes may also cause cleft palate, albinism, diabetes insipidus, colon cancer, and breast cancer. It is even likely that more than one gene can cause cystic fibrosis, when mutant. A study found that some people with clinically diagnosed mild cystic fibrosis have two wild type copies of the gene usually associated with the disorder.

1 Distinguish between penetrance and expressivity.

2 What is pleiotropy?

3 What is genetic heterogeneity?

Complex Traits

Most of the inherited disorders mentioned so far are **monogenic**—that is, they are determined by a single gene, and their expression is usually not greatly influenced by the environment. However, most if not all characteristics and disorders reflect input from the environment as well as genes.

Traits determined by more than one gene are termed **polygenic.** Usually, several genes each contribute to differing degrees toward molding the overall phenotype, which may vary greatly among individuals. A polygenic trait, with many degrees of expression because of the input of several genes, is said to be continuously varying. Height, skin color, and eye color are polygenic traits (figs. 24.8, 24.9 and 24.10).

Although the expression of a polygenic trait is continuous, we can categorize individuals into classes and calculate the frequencies of the classes. When we do this and plot the frequency for each phenotype class, a bell-shaped curve results. This curve indicating continuous variation of

(a)

(b)

FIGURE 24.8

Previous editions of this (and other) textbooks have used the photograph in (a) to illustrate the continuously varying nature of height. In the photo, taken around 1920, 175 cadets at the Connecticut Agricultural College lined up by height. In 1997, Professor Linda Strausbaugh asked her genetics students at the school, today the University of Connecticut at Storrs, to recreate the scene (b). They did, and confirmed the continuously varying nature of human height. But they also elegantly demonstrated how height has increased during the twentieth century. Improved nutrition has definitely played a role in expressing genetic potential for height. The tallest people in the old photograph (a) are 5'9" tall, whereas the tallest people in the more recent photograph (b) are 6'5" tall.

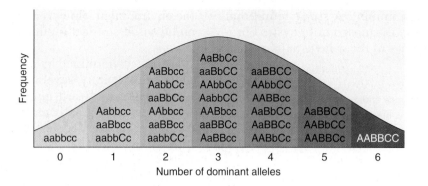

FIGURE 24.9

Variations in skin color. A model of three genes, with two alleles each, can explain some of the hues of human skin. In actuality, this trait likely involves many more than three genes.

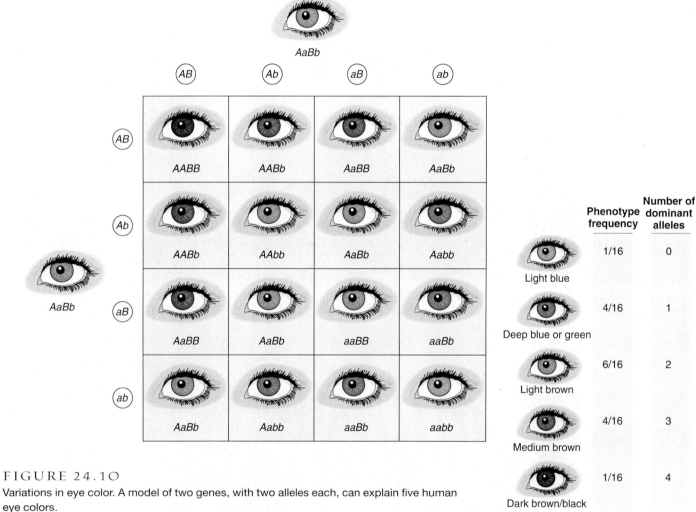

FIGURE 24.10

Variations in eye color. A model of two genes, with two alleles each, can explain five human eye colors.

a polygenic trait is strikingly similar for different characteristics, such as fingerprint patterns, height, eye color, and skin color. Even when different numbers of genes contribute to the phenotype, the curve is the same shape.

Eye color illustrates how interacting genes can mold a single trait. The colored part of the eye, the iris, darkens as melanocytes produce the pigment melanin. Blue eyes have just enough melanin to make the color opaque, and dark blue or green, brown or black eyes have increasingly more melanin in the iris. Unlike melanin in skin melanocytes, the pigment in the eye tends to stay in the cell that produces it.

Two genes, with two alleles each, can interact additively to account for five distinct eye colors—light blue, deep blue or green, light brown, medium brown, and dark brown/black. (It seems that manufacturers of mascara follow this two-gene scheme too!). If each dominant allele contributes a certain amount of pigment, then the greater the number of such alleles, the darker the eye color. If eye color is controlled by two genes *A* and *B,* each of which comes in two allelic forms *A* and *a* and *B* and *b,* then the lightest color would be genotype *aabb;* the darkest, *AABB.* The bell curve arises because there are more ways to inherit light brown eyes, with any two dominant alleles, than there are ways to inherit the other colors.

Traits molded by one or more genes plus the environment are termed **complex traits** (multifactorial traits). Height and skin color are multifactorial as well as polygenic, because environmental factors influence them: good nutrition enables a person to reach the height dictated by genes, and sun exposure affects skin color. Most of the more common illnesses, including heart disease, diabetes mellitus, hypertension, and cancers, are complex.

1 How does polygenic inheritance make possible many variations of a trait?

2 How may the environment influence gene expression?

3 How can two genes specify five phenotypes?

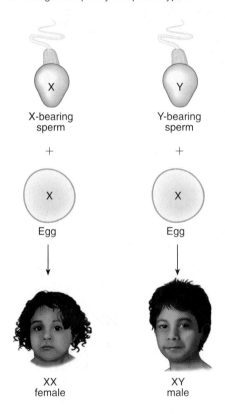

FIGURE 24.11

Sex determination. An egg contributes an X chromosome, and a sperm, either an X or a Y. If a Y-bearing sperm fertilizes an egg, the zygote is male (XY). If an X-bearing sperm fertilizes an egg, the zygote is female (XX). Sex is actually determined by a gene on the Y chromosome.

Matters of Sex

Human somatic (nonsex) cells include an X and a Y chromosome in males and two X chromosomes in females. All eggs carry a single X chromosome, and sperm carry either an X or a Y chromosome. Sex is determined at conception: a Y-bearing sperm fertilizing an egg conceives a male, and an X-bearing sperm conceives a female (fig. 24.11). The female is termed the homogametic sex because she has two of the same type of sex chromosome, and the human male is called the heterogametic sex because his two sex chromosomes are different. This is not the case for all types of animals. In birds, for example, the female is the heterogametic sex.

Sex Determination

The Y chromosome was first visualized with the use of a microscope in 1923, and its association with maleness was realized several years later. Researchers did not identify the gene responsible for being male until 1990. The SRY gene (*sex-determining region of the Y*) encodes a type of protein called a transcription factor, which switches on other genes that direct development of male structures in the embryo, while suppressing formation of female structures. Because a female lacks a Y chromosome, she also lacks an SRY gene, and the "default" option of female development ensues. Figure 24.12 shows the sex chromosomes.

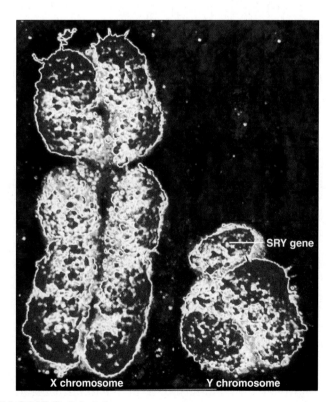

FIGURE 24.12

The X and Y chromosomes. The SRY gene, at one end of the short arm of the Y chromosome, starts the cascade of gene activity that directs development of a male (31,000×).

Sex Chromosomes and Their Genes

The X and Y chromosomes carry genes, but they are inherited in different patterns than are autosomal genes because of the different sex chromosome constitutions in males and females. Recall that traits transmitted on the X chromosome are X-linked, and on the Y, Y-linked. The X chromosome has more than 1,000 genes; the Y chromosome has only a few dozen genes.

Y-linked genes are considered in three groups, based on their similarity to X-linked genes. One group consists of genes at the tips of the Y chromosome that have counterparts on the X chromosome. These genes encode a variety of proteins that function in both sexes, participating in or controlling such activities as bone growth, signal transduction, the synthesis of hormones and receptors, and energy metabolism. The members of the second functional group of Y chromosome genes are very similar in DNA sequence to certain genes on the X chromosome, but they are not identical. These genes are expressed in nearly all tissues, including those found only in males. The third group of genes includes those unique to the Y chromosome. Many of them control male fertility, such as the SRY gene. Some cases of male infertility can be traced to tiny deletions of these parts of the Y chromosome. Other genes in this group encode proteins that participate in cell cycle control; proteins that regulate gene expression; enzymes; and protein receptors for immune system biochemicals.

Y-linked genes are transmitted only from fathers to sons, because only males have Y chromosomes. The differences in inheritance patterns of X-linked genes between females and males result from the fact that any gene on the X chromosome of a male is expressed in his phenotype, because he has no second allele on a second X chromosome to mask its expression. An allele on an X chromosome in a female may or may not be expressed, depending upon whether it is dominant or recessive and upon the nature of the allele on the second X chromosome. The human male is said to be **hemizygous** for X-linked traits because he has only one copy of each X chromosome gene, which is half what the female has. Red-green color blindness and the most common form of the clotting disorder hemophilia are recessive X-linked traits.

A male always inherits his Y chromosome from his father and his X chromosome from his mother. A female inherits one X chromosome from each parent. If a mother is heterozygous for a particular X-linked gene, her son has a 50% chance of inheriting either allele from her. X-linked genes are therefore passed from mother to son. Because a male does not receive an X chromosome from his father (he inherits the Y chromosome from his father), an X-linked trait is not passed from father to son.

Consider the inheritance of hemophilia A. It is passed from carrier mother to affected son with a risk of 50%, because he can inherit either her normal allele or the mutant one. A daughter has a 50% chance of inherit-ing the hemophilia allele and being a carrier like her mother and a 50% chance of not carrying the allele.

To remedy the seeming inequity of cells of a female having two X chromosomes compared to the male's one, female mammalian embryos shut off one X chromosome in each somatic cell. Which of a female's X chromosomes is silenced—the one she inherited from her mother or the one from her father—occurs randomly. Therefore, a female is a mosaic, with genes from her father's X chromosome expressed in some cells, and genes from her mother's in others. This X inactivation is detectable for some heterozygous, X-linked genes. A woman who is a carrier (a heterozygote) for Duchenne muscular dystrophy, for example, has a wild type allele for the dystrophin gene on one X chromosome and a disease-causing allele on the other. Cells in which the X chromosome bearing the wild type allele is inactivated do not produce the gene's protein product, dystrophin. However, cells in which the mutant allele is inactivated produce dystrophin. When a stain for dystrophin is applied to a sample of her muscle tissue, some cells turn blue, and others do not, revealing her carrier status. If by chance many of such a woman's wild type dystrophin alleles are turned off in her muscle cells, she may experience mild muscle weakness and is called a *manifesting heterozygote*.

A daughter can inherit an X-linked recessive disorder or trait if her father is affected and her mother is a carrier. She inherits one affected X chromosome from each parent. Without a biochemical test, though, a woman would not know that she is a carrier of an X-linked recessive trait unless she has an affected son.

For X-linked recessive traits that seriously impair health, affected males may not feel well enough to have children. Because a female affected by an X-linked trait must inherit the mutant allele from a carrier mother and an affected father, such traits that are nearly as common among females as males tend to be those associated with milder phenotypes. Color blindness is a mild X-linked trait—men who are color blind are as likely to have children as men with full color vision.

Dominant disease-causing alleles on the X chromosome are rarely seen. Males are usually much more severely affected than females, who have a second X to offer a protective effect. In a condition called incontinentia pigmenti, for example, an affected girl has swirls of pigment on her skin where melanin in the epidermis extends into the dermis. She may have abnormal teeth, sparse hair, visual problems, and seizures. However, males inheriting the dominant gene on their X chromosomes are so severely affected that they do not survive to be born.

Gender Effects on Phenotype

Certain autosomal traits are expressed differently in males and females, due to differences between the sexes.

A **sex-limited trait** affects a structure or function of the body that is present in only males or only females. Such a gene may be X-linked or autosomal. Beard growth and breast size are sex-limited traits. A woman cannot grow a beard because she does not manufacture sufficient hormones required for facial hair growth, but she can pass to her sons the genes that specify heavy beard growth. In animal breeding, milk yield and horn development are important sex-limited traits.

In **sex-influenced inheritance,** an allele is dominant in one sex but recessive in the other. Again, such a gene may be X-linked or autosomal. This difference in expression reflects hormonal differences between the sexes. For example, a gene for hair growth pattern has two alleles, one that produces hair all over the head and another that causes pattern baldness (fig. 24.13). The baldness allele is dominant in males but recessive in females, which is why

more men than women are bald. A heterozygous male is bald, but a heterozygous female is not. A bald woman would have two mutant alleles.

About 1% of human genes exhibit **genomic imprinting,** in which the the expression of a disorder differs depending upon which parent transmits the disease-causing gene or chromosome. The phenotype may differ in degree of severity, in age of onset, or even in the nature of the symptoms. The physical basis of genomic imprinting is that methyl (CH_3) groups are placed on the gene that is inherited from one parent, preventing it from being transcribed and translated.

1 Which chromosomes and genes determine sex?

2 What are the three classes of genes on the Y chromosome?

3 Why do X-linked recessive conditions appear most commonly in males?

4 How can gender affect gene expression?

(a)　　**(b)**　　**(c)**　　**(d)**

FIGURE 24.13

Pattern baldness is a sex-influenced trait and was a genetic trademark of the illustrious Adams family. (*a*) John Adams (1735–1826) was the second president of the United States. He was the father of (*b*) John Quincy Adams (1767–1848), who was the sixth president. John Quincy was the father of (*c*) Charles Francis Adams (1807–1886), who was a diplomat and the father of (*d*) Henry Adams (1838–1918), who was a historian.

Chromosome Disorders

Deviations from the normal human chromosome number of 46 produce syndromes because of the excess or deficit of genes. Rearrangement of chromosomes, such as an inversion of a section of a chromosome, or two nonhomologous chromosomes exchanging parts, may also cause symptoms. This may happen if the rearrangement disrupts a vital gene or if it results in "unbalanced" gametes that contain too little or too much genetic material. The following sections, "Polyploidy" and "Aneuploidy," take a closer look at specific types of chromosome aberrations.

Polyploidy

The most drastic upset in chromosome number is an entire extra set, a condition called **polyploidy.** This results from formation of a diploid (rather than a normal haploid) gamete. For example, if a haploid sperm fertilizes a diploid egg, the fertilized egg is *triploid,* with three copies of each chromosome. Most human polyploids cease developing as embryos or fetuses, but occasionally an infant survives for a few days, with many anomalies. Eight cases of tetraploidy (4 copies of each chromosome) have been reported. One such child, at age 26 months, had severe delayed growth and development, a small head with tiny features, and a heart defect.

Some organs normally have a few polyploid cells, with no adverse effects on health. Liver cells, for example, may be tetraploid or even octaploid (8 chromosome sets). Polyploidy is common in flowering plants but rare in animals.

Aneuploidy

Cells missing a chromosome or having an extra one are **aneuploid.** A normal chromosome number is termed **euploid.** Aneuploidy results from a meiotic error called **nondisjunction** (non"dis-jungk'shun) (fig. 24.14). In normal meiosis, pairs of homologous chromosomes separate, and each of the resulting gametes contains only one member of each pair. In nondisjunction, a chromosome pair fails to separate, either at the first or at the second meiotic division, producing a sperm or egg that has two copies of a particular chromosome or none, rather than the normal one copy. When such a gamete fuses with its mate at fertilization, the resulting zygote has either 47 or 45 chromosomes, instead of the normal 46.

Symptoms that result from aneuploidy depend upon which chromosome is missing or extra. Autosomal aneuploidy often results in mental retardation, possibly because so many genes affect brain function. Sex chromosome aneuploidy is less severe. Extra genetic material is apparently less dangerous than missing material, and this is why most children born with the wrong number of chromosomes have an extra one, called a **trisomy,** rather than a missing one, called a **monosomy.**

Aneuploid conditions have historically been named for the researchers or clinicians who identified them, but today, chromosome designations are preferred because they are more precise. Down syndrome, for example, refers to a distinct set of symptoms that are usually caused by trisomy 21. However, the syndrome may also arise from one copy of chromosome 21 exchanging parts with a different chromosome, which is a type of aberration called a *translocation.* Knowing whether a child with these symptoms has trisomy 21 or translocation Down syndrome is very important in a practical sense, because the probability of trisomy 21 recurring in a sibling is about 1 in 100, but the chance of translocation Down syndrome recurring is considerably greater. Clinical Application 24.2 takes a closer look at trisomy 21.

Trisomies 13 and 18 are the next most common autosomal aneuploids. Affected fetuses usually cease developing before birth. An infant with trisomy 13 has an underdeveloped face, extra and fused fingers and toes, heart defects, small adrenal glands, and a cleft lip or palate. An infant with trisomy 18 suffers many of the problems seen in trisomy 13, plus a peculiar positioning of the fingers and flaps of extra abdominal skin called a "prune belly."

Table 24.1 indicates the rarity of trisomies 13, 18, and 21 and that it is rarer still for an affected newborn to survive infancy. Trisomies of the other autosomes do not develop beyond the embryonic period.

Sex chromosome aneuploids are less severely affected than are autosomal aneuploids. XO syndrome (Turner syndrome) affects 1 in 2,000 newborn girls, but these represent only 1% of XO conceptions. Often the only symptom is a lag in sexual development, and with hormone supplements, life can be fairly normal, except for infertility.

About 1 in every 1,000 to 2,000 females has an extra X chromosome in each cell, a condition called triplo-X. Often the only associated characteristics are great height and menstrual irregularities. Males with an extra X chromosome have XXY syndrome (Klinefelter syndrome). Like XO females, many XXY males do not realize they have an unusual number of chromosomes until they encounter fertility problems. Associated characteristics

TABLE 24.1	Comparing and Contrasting Trisomies 13, 18, and 21	
Type of Trisomy	Incidence at Birth	Percent of Conceptions That Survive 1 Year After Birth
13 (Patau)	1/12,500–1/21,700	<5%
18 (Edward)	1/6,000–1/10,000	<5%
21 (Down)	1/800–1/826	85%

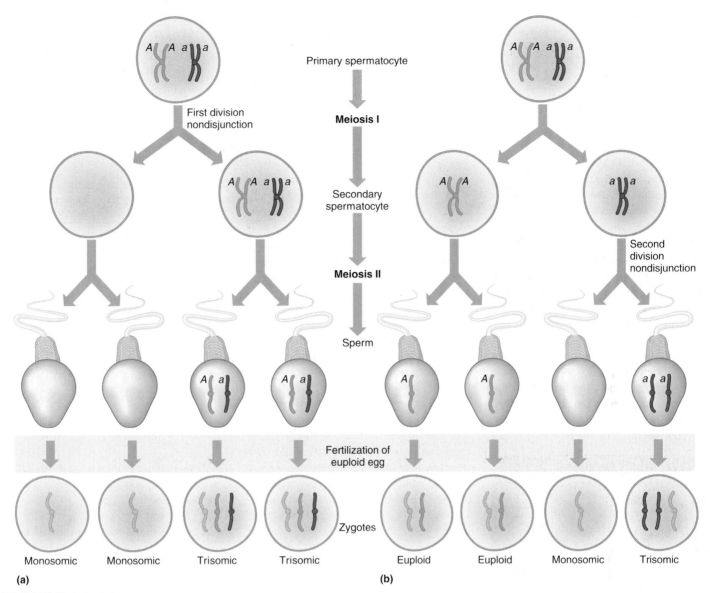

FIGURE 24.14

Extra or missing chromosomes constitute aneuploidy. Unequal division of chromosome pairs into sperm and egg cells can occur at either the first or the second meiotic division. (a) A single pair of chromosomes is unevenly partitioned into the two cells arising from the first division of meiosis in a male. The result: two sperm cells that have two copies of the chromosome and two sperm cells that have no copies of that chromosome. When a sperm cell with two copies of the chromosome fertilizes a normal egg cell, the zygote produced is trisomic for that chromosome; when a sperm cell lacking the chromosome fertilizes a normal egg cell, the zygote is monosomic for that chromosome. Symptoms depend upon which chromosome is involved. (b) This nondisjunction occurs at the second meiotic division. Because the two products of the first division are unaffected, two of the mature sperm are normal, and two are aneuploid. Egg cells can undergo nondisjunction as well, leading to zygotes with extra or missing chromosomes when they are fertilized by normal sperm cells.

are sexual underdevelopment (rudimentary testes and prostate glands and no pubic or facial hair), growth of breast tissue, long limbs, and large hands and feet. XXY syndrome affects 1 in every 500 to 2,000 male births.

One male in 1,000 has an extra Y chromosome, called XYY syndrome, or Jacobs syndrome. Until 1974, the extra chromosome was linked to criminal behavior, because the first studies to detect it were performed on inmates at a high-security mental facility. However, 96%

of men with XYY syndrome share only great height, acne, and speech and reading problems. It is possible that teachers, employers, parents, and others may expect more of these physically large boys and men than of their peers, and a small percentage of them cope with this stress by becoming aggressive.

A fertilized ovum that has one Y chromosome and no X chromosome has never been observed. Apparently, when a zygote lacks an X chromosome, so much genetic

DOWN SYNDROME

The most common autosomal aneuploid is *trisomy 21*, an extra chromosome 21. The characteristic slanted eyes and flat face of affected individuals prompted Sir John Langdon Haydon Down to coin the inaccurate term "mongolism" when he described the syndrome in 1886. As the medical superintendent of a facility for the profoundly mentally retarded, Down noted that about 10% of his patients resembled people of the Mongolian race. The resemblance is coincidental. Males and females of all races can have the syndrome.

A person with Down syndrome (either trisomy or translocation) is short and has straight, sparse hair and a tongue protruding through thick lips. The face has other telltale characteristics, including upward slanting eyes with "epicanthal" skin folds in

the inner corners and abnormally shaped ears. The hands have an abnormal pattern of creases, the joints are loose, and reflexes and muscle tone are poor. Developmental milestones (such as sitting, standing, and walking) are slow, and toilet training may take several years. Intelligence varies greatly, from profound mental retardation to being able to follow simple directions, read and use a computer. People with the syndrome have graduated from college.

Down syndrome (either type) is associated with many physical problems. Nearly 50% of affected people die before their first birthdays, often of heart or kidney defects, or of infections, which can be severe due to a suppressed immune system. Blockages in the digestive system are common and must be corrected surgically shortly after birth. An affected child is fifteen times more likely to develop leukemia than a healthy child, but this is still a low figure. Prenatal testing cannot reveal how severely affected the individual will be.

About 25% of people with either form of Down syndrome who live past age thirty-five develop the fibers and tangles of amyloid protein in their brains that are also seen in the brains of people who have died of Alzheimer disease. However, they may not have signs of dementia. (The risk among the general population of developing Alzheimer disease is 6%.) Both disorders seem to involve accelerated aging of part of the brain and accumulation of amyloid protein.

The likelihood of giving birth to a child with trisomy 21 Down syndrome increases dramatically with the age of the mother (table 24A). However, 80% of children with trisomy 21 are born to women under age thirty-five, because younger women are more likely to become pregnant and less

TABLE 24A	Risk of Trisomy 21 Increases with Maternal Age	
Maternal Age	Trisomy 21 Risk	Risk for Any Aneuploid
20	1/1,667	1/526
24	1/1,250	1/476
28	1/1,053	1/435
30	1/952	1/385
32	1/769	1/322
35	1/378	1/192
36	1/289	1/156
37	1/224	1/127
38	1/173	1/102
40	1/106	1/66
45	1/30	1/21
48	1/14	1/10

likely to undergo amniocentesis. About 5% of cases of trisomy 21 can be traced to nondisjunction in the sperm.

The age factor in Down syndrome may be due to the fact that meiosis in the female is completed after conception. The older a woman is, the longer her oocytes have been arrested on the brink of completing meiosis. During this time, the oocytes may have been exposed to chromosome-damaging chemicals or radiation. Other trisomies are more likely to occur among the offspring of older women, too. In the nineteenth century, when physicians noted that people with Down syndrome were often the youngest in their families, they attributed the condition to "maternal reproductive exhaustion."

Many of the medical problems that people with Down syndrome suffer are treatable, so life expectancy is now fifty-five years. In 1910, life expectancy was only to age nine! ■

Wendy Weisz has trisomy 21 Down syndrome. She enjoys art and has taken courses in it at Cuyahoga Community College.

material is missing that only a few, if any, cell divisions are possible.

Prenatal Tests

Several types of tests performed on pregnant women can identify anatomical or physiological features of fetuses that can indicate a chromosomal problem, or actually detect the abnormal chromosomes (fig. 24.15). An ultrasound scan, for example, can reveal the fusion of the eyes, cleft lip and/or palate, malformed nose, and extra fingers and toes that indicate trisomy 13 (fig. 24.16). A blood test performed on the woman during the fifteenth week of pregnancy detects levels of maternal serum markers (specifically, alpha fetoprotein, a form of estrogen, and human chorionic gonadotropin) that can indicate the underdeveloped liver that is a sign of trisomies 13, 18, and 21. Screening maternal serum markers is routine in the management of pregnancy.

After a maternal serum marker pattern indicates increased risk, the patient is offered **amniocentesis,** in which a needle is inserted into the amniotic sac and withdraws about 5 milliliters of fluid. Fetal fibroblasts in the sample are cultured and a karyotype constructed, which reveals extra, missing, or translocated chromosomes or smaller anomalies. However, amniocentesis does not reveal single gene defects unless DNA probes are applied to specific genes. Because amniocentesis has a risk of about 0.8% of being followed by miscarriage, it is typically performed on women whose risk of carrying an affected fetus is greater than this, which includes all women over age thirty-five and those with a family history of a chromosomal disorder on either parent's side.

Couples who have already had a child with a chromosome abnormality can elect to have **chorionic villus sampling** (CVS), which has the advantage of being performed as early as the tenth week from conception, but carries a higher risk of being followed by miscarriage than does amniocentesis. In CVS, a physician samples chorionic villus cells through the cervix. The basis of the test is that, theoretically, these cells are genetically identical to fetal cells because they too descend from the fertilized ovum. However, sometimes a mutation can occur in a villus cell only, or a fetal cell only, creating a false positive or false negative test result.

An experimental prenatal test, fetal cell sorting, is safer than amniocentesis or CVS because it samples only maternal blood, yet it provides the high accuracy of these tests. It is more accurate than measuring maternal serum markers. Fetal cell sorting separates out the rare fetal cells that normally cross the placenta and enter the woman's circulation; then a karyotype is constructed from the sampled cells. It can be performed early in pregnancy, but so far, it is too costly to be widely implemented.

Table 24.2 and figure 24.15 summarize the tests used to visualize fetal chromosomes as a window onto health. In the near future, DNA microarray analyses that screen for the activities of individual genes will likely replace these chromosome detection tests, as described in the chapter opener.

1 Why do deviations from the normal chromosome number of 46 affect health?

2 Distinguish between polyploidy and aneuploidy.

3 How do extra sets of chromosomes or extra individual chromosomes arise?

4 How are fetal chromosomes examined?

Gene Therapy

Understanding how an absent or malfunctioning gene causes disease can be a first step to preventing or treating the disease. **Gene therapy** alters, replaces, silences, or augments a gene's function to improve or prevent symptoms. As its name implies, gene therapy operates at the gene level, but treatment of some inherited disorders at the protein level is already standard medical practice. For example, a person with hemophilia receives the missing protein clotting factor, and someone with cystic fibrosis takes cow digestive enzymes to compensate for poor pancreatic function. Clinical Application 4.2 describes how a dietary regimen that restricts protein prevents the

TABLE 24.2	Prenatal Tests		
Procedure	**Time (Weeks)**	**Source**	**Information Provided**
Maternal serum markers	15–16	Maternal blood	Small liver may indicate increased risk of trisomy
Amniocentesis	14–16	Fetal skin, urinary bladder, digestive system cells in amniotic fluid	Karyotype of cell from fetus
CVS	10–12	Chorionic villi	Karyotype of cell from chorionic villus
Fetal cell sorting	Not yet established	Maternal blood	Karyotype of cell from fetus
Ultrasound	Any time	Applied externally or through vagina	Growth rate, head size, size and location of organs

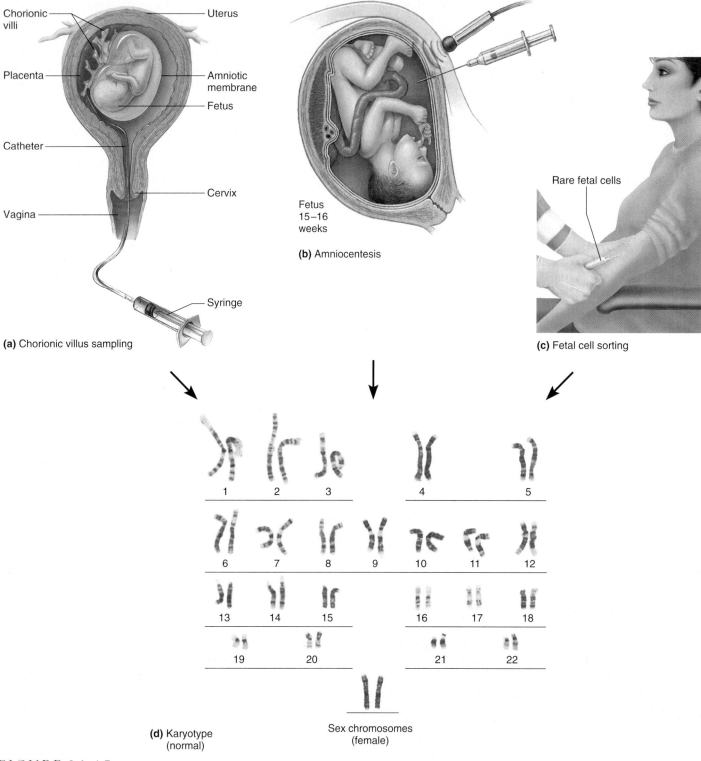

(a) Chorionic villus sampling

(b) Amniocentesis

Fetus
15–16
weeks

Rare fetal cells

(c) Fetal cell sorting

Chorionic villi

Uterus

Placenta

Amniotic membrane

Fetus

Catheter

Cervix

Vagina

Syringe

1 2 3 4 5

6 7 8 9 10 11 12

13 14 15 16 17 18

19 20 21 22

(d) Karyotype (normal)

Sex chromosomes (female)

FIGURE 24.15

Three ways to check a fetus's chromosomes. (*a*) Chorionic villus sampling (CVS) removes cells of the chorionic villi, whose chromosomes match those of the fetus because they all descend from the fertilized ovum. CVS is usually performed earlier than amniocentesis. (*b*) In amniocentesis, a needle is inserted into the uterus to collect a sample of amniotic fluid, which contains fetal cells. The cells are grown in the laboratory, and then dropped onto a microscope slide to spread the chromosomes. The chromosomes are then stained and arranged into a chromosome chart (karyotype). Amniocentesis is performed after the fifteenth week of gestation. (*c*) Fetal cell sorting separates fetal cells in the woman's circulation. (*d*) A genetic counselor interprets the results of these tests—a karyotype—for patients.

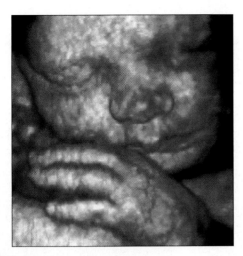

FIGURE 24.16

Ultrasound. In an ultrasound exam, sound waves are bounced off the embryo or fetus, and the pattern of deflected sound waves is converted into an image. By thirteen weeks, the face can be discerned. Ultrasound can now provide images that look three-dimensional.

mental retardation that is associated with an inborn error of metabolism, PKU.

Two Approaches to Gene Therapy

There are two basic types of gene therapy. **Heritable gene therapy** (germline gene therapy) introduces the genetic change into a sperm, egg, or fertilized egg, which corrects each cell of the resulting individual. The change is repeated in the person's gametes and can be passed to the next generation. Heritable gene therapy is not, and may never be, done in humans, but the resulting transgenic organisms of other species are useful. For example, transgenic mice that harbor human disease-causing genes are used to study the early stages of human diseases and to test treatments before they are tried on people. Transgenic cows and goats secrete human versions of therapeutic proteins, such as clotting factors and growth hormone. Transgenic crop plants have valuable traits, such as "golden rice," which manufactures beta carotene using genes from petunia and bacteria. Beta carotene is a precursor to vitamin A, and rice normally makes very little of it.

Nonheritable gene therapy (somatic gene therapy) targets only affected cells and therefore cannot be transmitted to the next generation. Nonheritable gene therapy for hemophilia, for example, provides genes that encode the needed clotting factors. A nonheritable gene therapy for cystic fibrosis is an aerosol containing a virus that has had its pathogenic genes removed and a functional human CFTR gene added. When the person inhales the aerosol, the needed gene enters airway epithelium, providing instructions to replace the nonfunctional ion channel protein that causes symptoms.

Experiments to develop gene therapies have been ongoing since 1990, with thousands of patients participating, to varying degrees of success. Progress has been slow. For example, gene therapy for hemophilia has been successful in a few patients, whereas gene therapy for CF so far has not provided a lasting cure. The age of gene therapy began with success, but has hit setbacks, discussed in Clinical Application 24.3.

Tools and Targets of Gene Therapy

Researchers use several methods to introduce therapeutic genes into cells. Healing DNA is linked to the genetic material of viruses that have had their disease-causing genes removed; in fatty bubbles called liposomes or complexed with other lipid molecules; "shot" along with metal particles into cells; and as "naked" preparations of DNA alone. The challenge in any nonheritable gene therapy is to target sufficient numbers of affected cells for a long enough time to exert a noticeable effect. A look at some specific gene therapy approaches provides a nice ending to our survey of human genetics. Figure 24.17 summarizes gene therapies.

Bone Marrow

Because bone marrow tissue includes the precursors of all mature blood cell types, it provides a route to treat blood disorders and immune deficiencies. Certain stem cells in bone marrow can also travel to other sites, such as muscle, liver, and brain, and differentiate there into, respectively, muscle, liver, or neural cells. Many new gene therapy targets might be reached via bone marrow.

Skin

Skin cells grow well in the laboratory. A person can donate a patch of skin the size of a letter on this page; after a genetic manipulation, the sample can grow to the size of a bathmat within just three weeks. The skin can then be grafted back onto the person. Skin grafts genetically modified to secrete therapeutic proteins, such as enzymes, clotting factors, or growth hormones, may provide a new drug delivery route.

Muscle

Muscle tissue is a good target for gene therapy for several reasons. It comprises about half of the body's mass, is easily accessible, and is near a blood supply. However, a challenge is to correct enough muscle cells to alleviate symptoms. For example, to treat Duchenne muscular dystrophy, researchers delivered the functional gene to immature muscle cells, but this approach worked only on small sections of muscle. An alternative strategy is to direct stem cells from bone marrow that can naturally migrate to muscle, where they differentiate and produce the needed protein.

Endothelium

Endothelium, a tissue which forms capillaries and lines the interiors of other blood vessels, can be altered to

GENE THERAPY SUCCESSES AND SETBACKS

Any new medical treatment or technology begins with creative minds and then brave volunteers. The first people to take new vaccines or to undergo new treatments know that they may give their lives in the process. Gene therapy, however, is unlike conventional drug therapy. It alters the genotype in a part of the body that has failed, and because the potentially therapeutic gene is usually delivered with other DNA, the body's reactions are unpredictable. Following are the stories of a few of the young people who have pioneered gene therapy.

ADA Deficiency—Early Success

Ashanti DaSilva was the first child to receive gene therapy. Shortly after noon on September 14, 1990, the four-year-old sat in a bed at the National Institutes of Health hospital in Bethesda, watching her own T cells, given copies of a missing gene, drip into her arm. Lack of the liver enzyme adenosine deaminase (ADA) caused an intermediate compound to accumulate that destroyed her T cells, toppling both her cellular and humoral immunity. Enzyme supplements had recently helped Ashanti avoid life-threatening infections, but gene therapy might offer a longer-lasting treatment. Eight-year-old Cynthia Cutshall joined the experiment a few months later. Each girl continued to receive the enzyme to prevent infection, as gradually, more T cells contained the healing gene.

Ashanti DaSilva rides her bike three years after she began receiving periodic gene therapy for ADA deficiency, an inherited lack of immunity. She is the first person to receive gene therapy—and it worked, although to a limited extent.

Ashanti and Cynthia's gene therapy had to be repeated often, because T cells die quickly. The next phase in the research was to "fix" umbilical cord stem cells in newborns diagnosed prenatally with the condition. These cells remain in the circulation longer. Three infants were treated, and as they've grown, gradually T cells capable of producing ADA have taken over. Their infections are easily controlled.

Jesse Gelsinger died at age eighteen of gene therapy to treat an inborn error of metabolism, ornithine transcarbamylase deficiency.

OTC Deficiency—A Setback

Jesse Gelsinger underwent gene therapy almost nine years to the day that Ashanti DaSilva received her treatment. But his experience sparked a reevaluation of the

secrete a substance directly into the bloodstream. For example, genetically modified endothelium might secrete insulin to treat diabetes mellitus or a clotting factor to treat hemophilia. Endothelium is implanted with collagen to provide support and angiogenesis factors to stimulate growth of capillaries.

Liver

The liver is a very important focus of gene therapy because it controls many bodily functions and because it can regenerate. A gene therapy that corrects just 5% of the 10 billion cells of the liver could produce an effect. For example, normal liver cells have low-density lipo-

protein (LDL) receptors on their surfaces, which bind cholesterol in the bloodstream and bring it into the cell. In familial hypercholesterolemia (FH) (see fig. 24.7), because liver cells lack LDL receptors, cholesterol accumulates on artery interiors. Liver cells genetically altered to have more LDL receptors can relieve the cholesterol buildup. One young woman who is heterozygous for FH had 15% of her liver removed, the cells isolated and given functional LDL receptor genes, and redelivered into the body through a major liver vein. Eighteen months later, the grafted liver cells bore more LDL receptors, and the woman's serum cholestrerol levels had improved.

technology in general. In September 1999, the eighteen-year-old died just days after receiving gene therapy, of an overwhelming and unanticipated immune system reaction.

Jesse had an inborn error of metabolism called *ornithine transcarbamylase (OTC) deficiency*. In this X-linked recessive disorder, the body lacks one of five enzymes required to break down amino acids liberated from dietary proteins. The nitrogen from the amino acids combines with hydrogen to form ammonia (NH_3), which rapidly accumulates in the bloodstream and travels to the brain. The condition usually causes irreversible coma within seventy-two hours of birth. Half of affected babies die within a month, and another quarter, by age five. The survivors can control their symptoms by eating a special low protein diet and taking drugs that bind ammonia.

Jesse wasn't diagnosed until he was two, because some of his cells could produce the enzyme, so his symptoms were mild. Still, when he went into a coma in December 1998 after a few days of not taking his medications, he and his father began to consider volunteering for a gene therapy trial they had read about. The next summer, four days after Jesse turned eighteen, he underwent testing at the University of Pennsylvania gene therapy center and was admitted to the trial. He knew he would not directly benefit soon, but he had wanted to try to help affected newborns.

The gene therapy consisted of an adenovirus, which causes the common cold, that carried a functional human OTC gene but had the genes that cause respiratory symptoms removed. The virus had already been used, usually safely, in about a quarter of the 330 gene therapy experiments done on more than 4,000 patients since 1990. Three groups of six patients were to receive three different doses, to identify the lowest dose that would fight the disease, but not cause dangerous side effects.

Jesse entered the hospital on Monday, September 13, after the seventeen others in the trial had already been treated and suffered nothing worse than a fever and aches and pains. Several billion modified viruses were sent into an artery leading into his liver. That night, Jesse developed a high fever. By morning, the whites of his eyes were yellow, indicating a high bilirubin level as his liver struggled to dismantle the hemoglobin bursting from shattered red blood cells. The ammonia level in his liver soared, as his blood clotting faltered. Jesse became disoriented, then comatose. By Wednesday, his lungs began to fail, and Jesse was placed on a ventilator. Thursday, other vital organs began to shut down, and by Friday, he was brain dead. His father turned off the life support.

It isn't entirely clear why the gene therapy killed Jesse Gelsinger. Perhaps a past infection with parvovirus led his immune system to attack the adenovirus, mistaking it for the cause of the former infection. Also, in the liver, the adenovirus had gone not to the targeted hepatocytes, but to a different cell type, the macrophages that trigger an immune response. The autopsy also revealed that the virus had spread to the spleen, lymph nodes, bone marrow, and elsewhere. In addition, Jesse's bone marrow lacked erythroid progenitor cells, indicating an underlying and undetected problem in hematopoiesis. Finally, examination of the adenovirus in Jesse's bloodstream revealed genetic changes—the gene therapy vector may have mutated.

Jesse Gelsinger wasn't the only gene therapy patient to become ill. In late 2002, two of eleven children in France who had received gene therapy to effectively treat inherited immune deficiency 3 years earlier developed leukemia. The virus that delivered the healing gene inserted into a different gene that apparently caused the cancer. These reports led to the discontinuation of some trials.

Efforts are underway to improve the safety of gene therapy. Viral vectors that can trigger new illness are being re-evaluated. Development of DNA microarray technology, which measures activities of many genes at once, will enable researchers to select patients based on more complete genetic information, including underlying conditions and how their immune systems would react to certain viruses. Many researchers compare gene therapy to organ transplantation, which also began slowly and with notable failures until the advent of immunosuppressant drugs and better ways to match donor and recipient transformed the technology into a standard medical practice. ■

Lungs

The respiratory tract is an excellent candidate for gene therapy because an aerosol can directly reach its lining cells, making it unnecessary to remove cells, alter them, and reimplant them. Lung epithelial cells take up inhaled genes and produce the proteins missing or abnormal in the inherited illness. For example, such gene therapy can provide alpha-l-antitrypsin, an enzyme whose absence causes a form of emphysema.

Nerve Tissue

Gene therapy of neurons is not feasible because these cells do not divide. Altering other cell types can circumvent this obstacle, such as neuroglial cells or fibroblasts that secrete nerve growth factor. Or, a therapeutic genetic change can be made in neural stem cells. Another route to nerve cell gene therapy is to send in a valuable gene attached to the herpes simplex virus, which remains in nerve cells after infection. Such a herpes gene carrier could alter a neuron's ability to secrete neurotransmitters.

Gene Therapy Against Cancer

Viruses may provide a treatment for a type of brain tumor called a *glioma,* which affects neuroglial cells. Cancerous neuroglial cells divide very rapidly, usually causing death

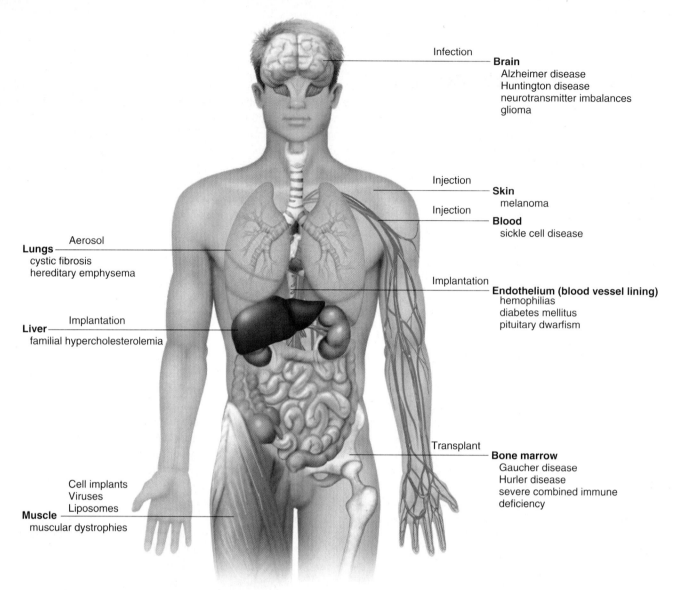

FIGURE 24.17

Sites of gene therapy and the methods used to introduce normal DNA.

within a year, even with aggressive treatment. A gene therapy approach infects fibroblasts with a virus bearing a gene from a herpes virus that makes the cells sensitive to a drug called ganciclovir. The altered fibroblasts are implanted near the tumor. There, the doctored virus infects nearby cancer cells, but not the healthy neurons, because they do not divide. When the patient takes ganciclovir, only the cells harboring the virus die—not healthy brain cells.

Another genetic approach to battling cancer is to enable tumor cells to produce immune system biochemicals, or to mark them so that the immune system more easily recognizes them. This approach is called a cancer vaccine. A treatment for the skin cancer melanoma, for example, alters tumor cells to display an antigen called HLA-B7, which stimulates the immune system to attack the cell.

1 What are the two basic types of gene therapy?

2 How does gene therapy work?

3 What are some of the ways that nonheritable gene therapy is being conducted?

Genomics and a New View of Anatomy and Physiology

Prior to sequencing the human genome, the field of human genetics focused almost exclusively on very rare inherited diseases. Today, human genetics and genomics have shifted focus to normal variations as well as conditions molded by interactions among genes and environmental factors. With this new way of looking at ourselves, physiology is not only being dissected at the cellular level, but at the level of the chemical signals that enable cells to interact to form tissues, and tissues to form organs. It is a new view of anatomy and physiology.

The Emerging Role of Genetics and Genomics in Medicine (page 921)

Genetics is the study of trait transmission through DNA passed in sperm and egg cells from generation to generation. Genes, which are parts of chromosomes, encode proteins. The human genome consists of about 31,000 protein-encoding genes plus many other sequences. Somatic cells are diploid; sex cells are haploid. Genomics considers many genes that interact with each other and the environment. Proteomics analyzes the spectrum of proteins that a particular cell type produces.

Modes of Inheritance (page 924)

1. Chromosomes and genes come in pairs
   a. Chromosome charts are called karyotypes.
   b. Chromosomes 1 through 22, numbered in decreasing size order, are autosomes. They do not determine sex.
   c. The X and Y chromosomes are sex chromosomes. They determine sex.
   d. Chromosomes and the genes they carry are paired.
   e. An allele is an alternate form of a gene. An individual can have two different alleles for a particular gene. The gene itself can have many alleles, because a gene consists of many building blocks, any of which may be altered.
   f. An individual with a pair of identical alleles for a particular gene is homozygous; if the alleles are different, the individual is heterozygous.
   g. The combination of genes present in an individual's cells constitutes a genotype; the appearance of the individual is its phenotype.
   h. A wild type allele provides normal or the most common function. A mutant allele causes disease or an unusual trait; it is a change from the wild type condition.
2. Dominant and recessive inheritance
   a. In the heterozygous condition, an allele that is expressed when the other is not is dominant. The masked allele is recessive.
   b. Recessive and dominant genes may be autosomal or X-linked or Y-linked.
   c. An autosomal recessive condition affects both sexes and may skip generations. The homozygous dominant and heterozygous individuals have normal phenotypes. The homozygous recessive individual has the condition. The heterozygote is a carrier. An affected individual inherits one mutant allele from each parent.
   d. An autosomal dominant condition affects both sexes and does not skip generations. A person inherits it from one parent, who is affected.
   e. Pedigrees and Punnett squares are used to depict modes of inheritance.
3. Different dominance relationships
   a. In incomplete dominance, a heterozygote has a phenotype intermediate between those of both homozygotes.
   b. In codominance, each of the alleles in the heterozygote is expressed.

Gene Expression (page 928)

1. Penetrance and expressivity
   a. A genotype is incompletely penetrant if not all individuals inheriting it express the phenotype.
   b. A genotype is variably expressive if it is expressed to different degrees in different individuals.

2. Pleiotropy
   a. A pleiotropic disorder has several symptoms, different subsets of which are expressed among individuals.
   b. Pleiotropy reflects a gene product that is part of more than one biochemical reaction or is found in several organs or structures.
3. Genetic heterogeneity
   a. Genetic heterogeneity refers to a phenotype that can be caused by alterations in more than one gene.
   b. The same symptoms may result from alterations in genes whose products are enzymes in the same biochemical pathway.

Complex Traits (page 929)

1. A trait caused by the action of a single gene is monogenic, and by the action of more than one gene, polygenic.
2. A trait caused by the action of one or more genes and the environment is complex.
3. Height, skin color, eye color, and many common illnesses are complex traits.
4. A frequency distribution for a polygenic trait forms a bell curve.

Matters of Sex (page 931)

A female has two X chromosomes; a male has one X and one Y chromosome. The X chromosome has many more genes than the Y.

1. Sex determination
   a. A male zygote forms when a Y-bearing sperm fertilizes an egg. A female zygote forms when an X-bearing sperm fertilizes an egg.
   b. A gene on the Y chromosome, called SRY, switches on genes in the embryo that promote development of male characteristics and suppresses genes that promote development of female characteristics.
2. Sex chromosomes and their genes
   a. Genes on the sex chromosomes are inherited differently than those on autosomes because the sexes differ in sex chromosome constitution.
   b. Y-linked genes are considered in three groups: those with counterparts on the X; those similar to genes on the X; and genes unique to the Y, many of which affect male fertility. Y-linked genes pass from fathers to sons.
   c. Males are hemizygous for X-linked traits; that is, they can have only one copy of an X-linked gene, because they have only one X chromosome.
   d. Females can be heterozygous or homozygous for genes on the X chromosome, because they have two copies of it.
   e. A male inherits an X-linked trait from a carrier mother. These traits are more common in males than in females.
   f. A female inherits an X-linked mutant gene from her carrier mother and/or from her father if the associated trait does not impair his ability to have children.
   g. Dominant X-linked traits are rarely seen because affected males typically die before birth.
3. Gender effects on phenotype
   a. Sex-limited traits affect structures or functions seen in only one sex and may be autosomal.
   b. Sex-influenced traits are dominant in one sex and recessive in the other.

c. In genomic imprinting, the severity, age of onset, or nature of symptoms varies according to which parent transmits the causative gene.

Chromosome Disorders (page 934)

Extra, missing, or rearranged chromosomes or parts of them can cause syndromes, because they either cause an imbalance of genetic material or disrupt a vital gene.

1. Polyploidy
   a. Polyploidy is an extra chromosome set.
   b. Polyploidy results from fertilization involving a diploid gamete.
   c. Human polyploids do not survive beyond a few days of birth.
2. Aneuploidy
   a. Cells with an extra or missing chromosome are aneuploid. Cells with the normal chromosome number are euploid.
   b. Aneuploidy results from nondisjunction, in which a chromosome pair does not separate, either in meiosis I or meiosis II, producing a gamete with a missing or extra chromosome. At fertilization, a monosomic or trisomic zygote results.
   c. A cell with an extra chromosome is trisomic. A cell with a missing chromosome is monosomic. Individuals with trisomies are more likely to survive to be born than those with monosomies.
   d. Autosomal aneuploids are more severe than sex chromosome aneuploids.

3. Prenatal tests
   a. Ultrasound can detect large-scale structural abnormalities and assess growth.
   b. Maternal serum marker tests indirectly detect a small fetal liver, which can indicate a trisomy.
   c. Amniocentesis samples and examines fetal chromosomes in amniotic fluid.
   d. Chorionic villus sampling obtains and examines chorionic villus cells, which descend from the fertilized egg and therefore are presumed to be genetically identical to fetal cells.
   e. Fetal cell sorting obtains and analyzes rare fetal cells in the maternal circulation.

Gene Therapy (page 937)

Gene therapy corrects the genetic defect causing symptoms.

1. Two approaches to gene therapy
   a. Heritable gene therapy alters all genes in an individual and therefore must be done on a gamete or fertilized egg. It is not done in humans but is useful in other species.
   b. Nonheritable gene therapy replaces or corrects defective genes in somatic cells, often those in which symptoms occur.
2. Tools and targets of gene therapy
   a. Healing genes are sent into cells in viruses, liposomes, blasted in, or as naked DNA.
   b. Gene therapies are being tested in various tissues and to treat cancer.
3. Genomics and a new view of anatomy and physiology

CRITICAL THINKING QUESTIONS

1. State possible advantages and disadvantages of DNA microarray tests performed shortly after birth to identify susceptibilities and inherited diseases that will likely affect the individual later in life.

2. A young couple is devastated when their second child is born and has PKU. Their older child is healthy, and no one else in the family has PKU. How is this possible?

3. A balding man undergoes a treatment that transfers some of the hair from the sides of his head, where it is still plentiful, to the top. Is he altering his phenotype or his genotype?

4. Bob and Joan know from a blood test that they are each heterozygous (carriers) for the autosomal recessive gene that causes sickle cell disease. If their first three children are healthy, what is the probability that their fourth child will have the disease?

5. A DNA microarray test includes several genes that cause cancer or increase sensitivity to substances that cause cancer. It also includes genes that confer high risk of addictive behaviors. The test is being developed to assess the risk that an individual who smokes will develop lung cancer. Do you think that such a test would be valuable or that it might be abused? Cite a reason for your answer.

6. In Hunter syndrome, lack of an enzyme leads to build up of sticky carbohydrates in the liver, spleen, and heart. The individual is also deaf and has unusual facial features. Hunter syndrome is inherited as an X-linked recessive condition. Intellect is usually unimpaired, and life span can be normal. A man who has mild Hunter syndrome has a child with a woman who is a carrier (heterozygote).

   a. What is the probability that a son inherits the syndrome?
   b. What is the chance that a daughter inherits it?
   c. What is the chance that a girl would be a carrier?

7. Amelogenesis imperfecta is an X-linked dominant condition that affects deposition of enamel onto teeth. Affected males have extremely thin enamel layers all over each tooth. Female carriers, however, have grooved teeth that result from uneven deposition of enamel. Explain the difference in phenotype between the sexes for this condition.

8. Why are medium-brown skin colors more common than very white or very black skin?

9. Why are there fewer Y-linked traits than there are Y-linked genes?

10. A woman aged forty receives genetic counseling before having an amniocentesis performed. She understands that her risk of carrying a fetus that has trisomy 21 Down syndrome is 1 in 106, but she is confused when the counselor explains that the risk of "any aneuploid" is 1 in 66. What does this mean?

11. Can a person who has been successfully treated for CF with an aerosol nongermline gene therapy still transmit the disease-causing allele to offspring? Cite a reason for your answer.

12. Parkinson disease is a movement disorder in which neurons in a part of the brain (the substantia nigra) can no longer produce the neurotransmitter dopamine, which is not a protein. Although Parkinson disease is not usually inherited, it may be treatable with gene therapy. What are two difficulties in developing gene therapy for Parkinson disease?

13. Cirrhosis of the liver, emphysema, and heart disease are all conditions that can be caused by a faulty gene or by a dangerous lifestyle habit (drinking alcohol, smoking, following a poor diet). When gene therapies become available for these conditions, should people with gene-caused disease be given priority in receiving the treatments? If not, what other criteria should be used for deciding who should receive a limited medical resource?

REVIEW EXERCISES

1. Discuss the relationship of DNA, genes, chromosomes, and the genome.
2. Discuss the origin of the 46 chromosomes in a human zygote.
3. Define *homologous chromosomes*.
4. Distinguish between
   - homozygote and heterozygote
   - autosome and sex chromosome
   - mutant and wild type
   - phenotype and genotype
   - incomplete dominance and codominance
   - haploid and diploid
   - penetrance and expressivity
   - germline and nongermline gene therapy
5. Explain how a gene can have many alleles.
6. Describe how cystic fibrosis is pleiotropic.
7. Explain why the frequency distributions of different complex traits give very similar bell curves.
8. Describe how the environment can influence gene expression.
9. Explain how genes and chromosomes determine gender.
10. Explain why Y-linked genes are passed only from fathers to sons.
11. Explain why the inheritance pattern of X-linked traits differs in males and females.
12. Explain why a male cannot inherit an X-linked trait from his father.
13. Explain why X-linked dominant traits are not seen in males.
14. Discuss how a sex-limited trait and a sex-influenced trait differ from an X-linked trait.
15. Explain how an individual with an extra set of chromosomes arises.
16. Explain how nondisjunction leads to aneuploidy.
17. Distinguish among four types of prenatal diagnostic tests.
18. Describe why heritable gene therapy is impractical in humans.
19. Explain how nonheritable gene therapy is being attempted in various human tissues.

WEB CONNECTIONS

Visit the Student Edition of the Online Learning Center at www.mhhe.com/shier10 **for answers to chapter questions, additional quizzes, interactive learning exercises, and other study tools.**

Human Cadavers

The following set of illustrations includes sagittal sections, transverse sections, and regional dissections of human cadavers. These photographs will help you visualize the spatial and proportional relationships between the major anatomic structures of actual specimens. The photographs can also serve as the basis for a review of the information you have gained from your study of the human organism.

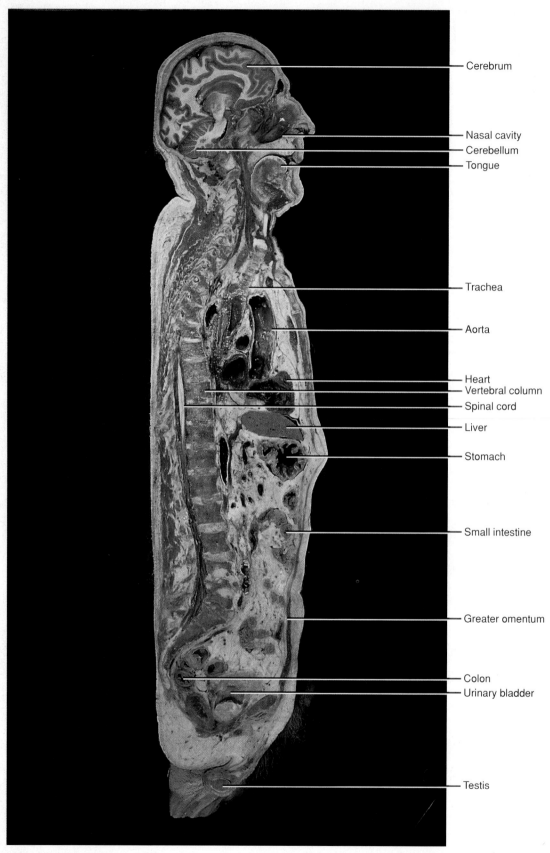

Cerebrum

Nasal cavity
Cerebellum
Tongue

Trachea

Aorta

Heart
Vertebral column
Spinal cord
Liver
Stomach

Small intestine

Greater omentum

Colon
Urinary bladder

Testis

PLATE FORTY-EIGHT
Saggital section of the head and trunk.

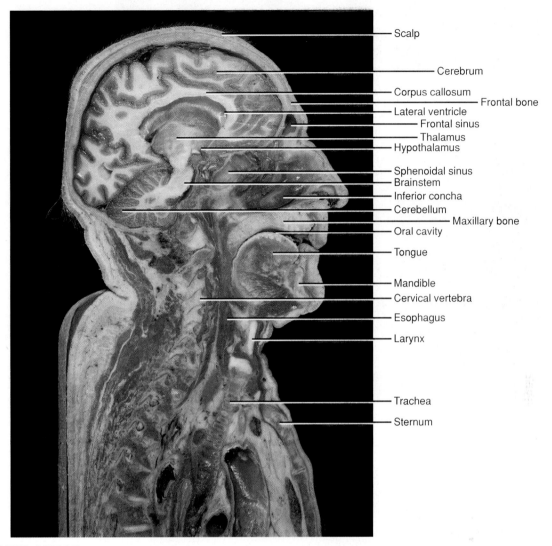

- Scalp
- Cerebrum
- Corpus callosum
- Frontal bone
- Lateral ventricle
- Frontal sinus
- Thalamus
- Hypothalamus
- Sphenoidal sinus
- Brainstem
- Inferior concha
- Cerebellum
- Maxillary bone
- Oral cavity
- Tongue
- Mandible
- Cervical vertebra
- Esophagus
- Larynx
- Trachea
- Sternum

PLATE FORTY-NINE
Saggital section of the head and neck.

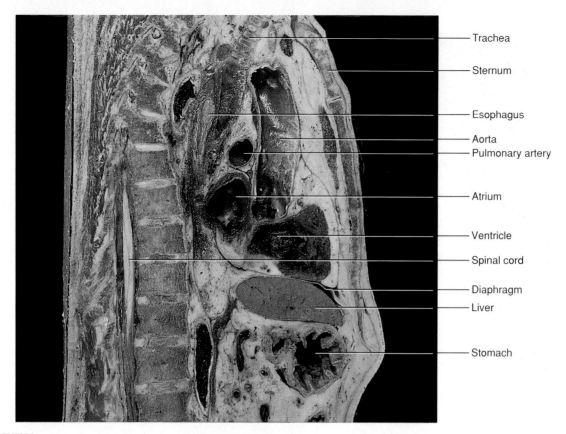

— Trachea

— Sternum

— Esophagus

— Aorta
— Pulmonary artery

— Atrium

— Ventricle

— Spinal cord

— Diaphragm
— Liver

— Stomach

PLATE FIFTY
Viscera of the thoracic cavity, sagittal section.

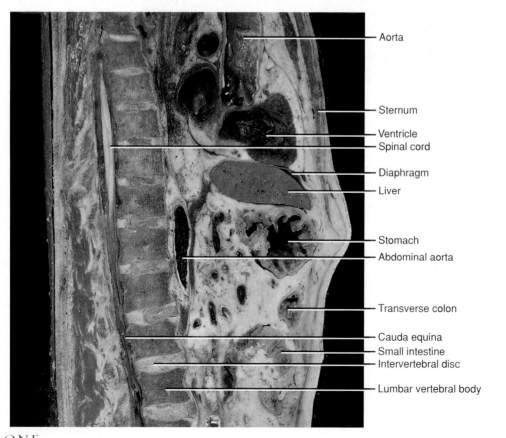

— Aorta

— Sternum

— Ventricle
— Spinal cord

— Diaphragm
— Liver

— Stomach
— Abdominal aorta

— Transverse colon

— Cauda equina
— Small intestine
— Intervertebral disc

— Lumbar vertebral body

PLATE FIFTY-ONE
Viscera of the abdominal cavity, sagittal section.

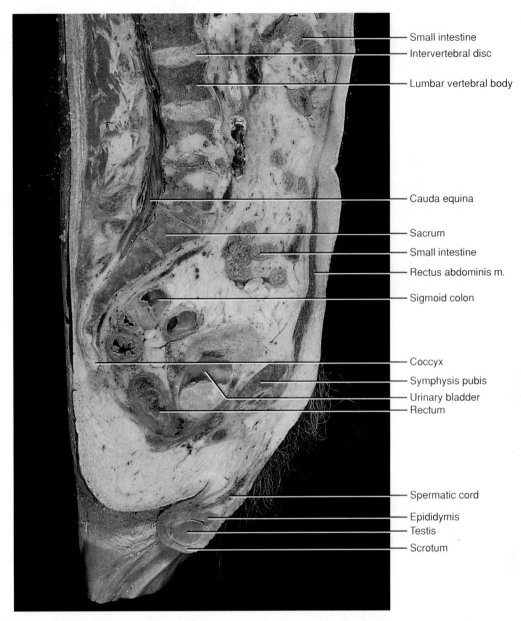

Small intestine
Intervertebral disc

Lumbar vertebral body

Cauda equina

Sacrum

Small intestine

Rectus abdominis m.

Sigmoid colon

Coccyx

Symphysis pubis

Urinary bladder

Rectum

Spermatic cord

Epididymis

Testis

Scrotum

PLATE FIFTY-TWO

Viscera of the pelvic cavity, sagittal section. (*m.* stands for muscle.)

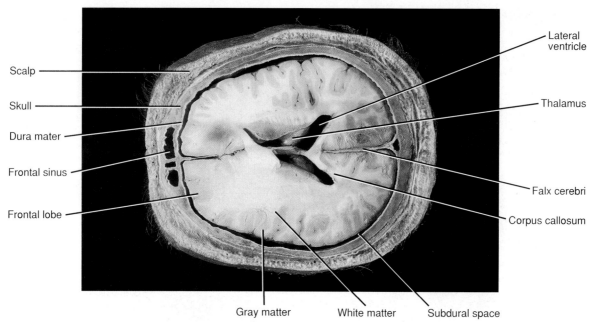

Scalp

Skull

Dura mater

Frontal sinus

Frontal lobe

Lateral ventricle

Thalamus

Falx cerebri

Corpus callosum

Gray matter   White matter   Subdural space

PLATE FIFTY-THREE

Transverse section of the head above the eyes, superior view.

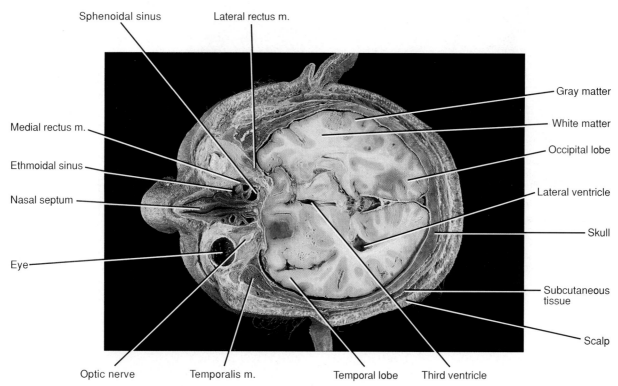

Sphenoidal sinus   Lateral rectus m.

Medial rectus m.

Ethmoidal sinus

Nasal septum

Eye

Gray matter

White matter

Occipital lobe

Lateral ventricle

Skull

Subcutaneous tissue

Scalp

Optic nerve   Temporalis m.   Temporal lobe   Third ventricle

PLATE FIFTY-FOUR

Transverse section of the head at the level of the eyes, superior view. (*m.* stands for muscle.)

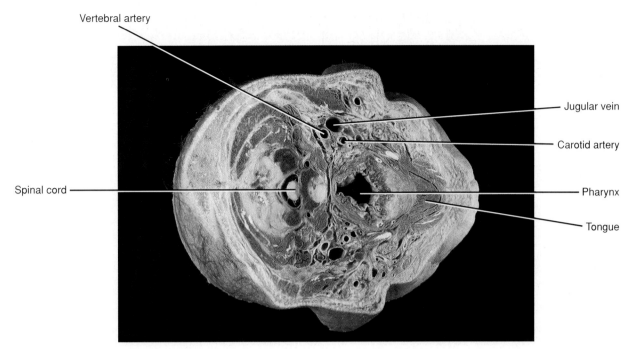

Vertebral artery

Jugular vein

Carotid artery

Spinal cord

Pharynx

Tongue

PLATE FIFTY-FIVE

Transverse section of the neck, inferior view.

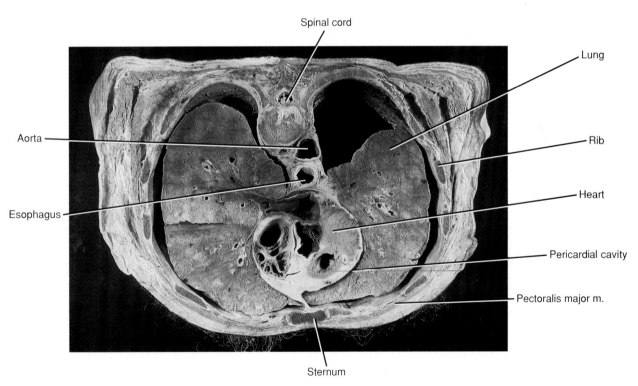

Spinal cord

Lung

Aorta

Rib

Heart

Esophagus

Pericardial cavity

Pectoralis major m.

Sternum

PLATE FIFTY-SIX

Transverse section of the thorax through the base of the heart, superior view. (*m.* stands for muscle.)

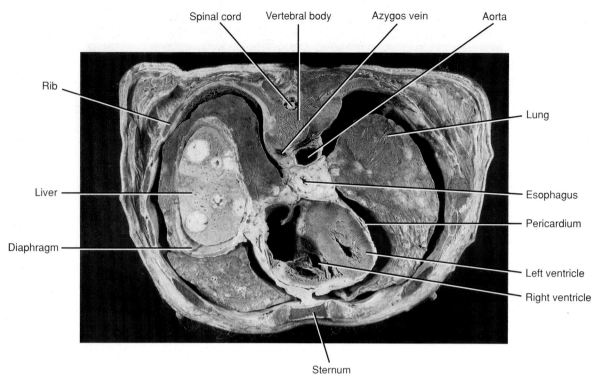

Spinal cord   Vertebral body   Azygos vein   Aorta

Rib

Lung

Liver

Esophagus

Pericardium

Diaphragm

Left ventricle

Right ventricle

Sternum

PLATE FIFTY-SEVEN
Transverse section of the thorax through the heart, superior view. (*m.* stands for muscle.)

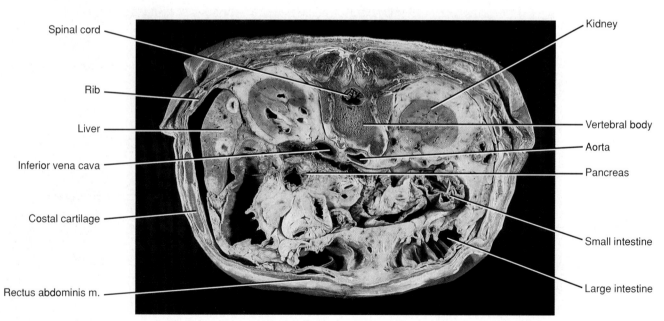

Spinal cord

Kidney

Rib

Liver

Vertebral body

Inferior vena cava

Aorta

Pancreas

Costal cartilage

Small intestine

Rectus abdominis m.

Large intestine

PLATE FIFTY-EIGHT
Transverse section of the abdomen through the kidneys, superior view. (*m.* stands for muscle.)

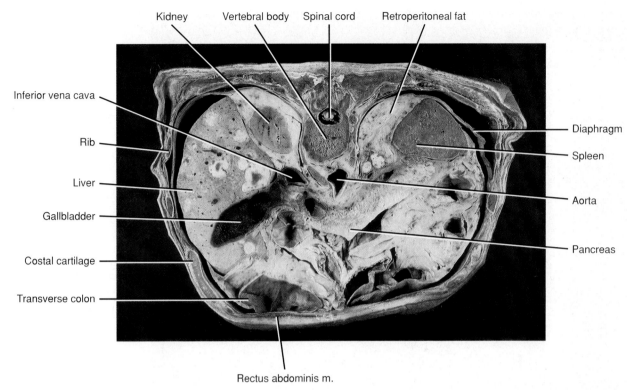

Kidney   Vertebral body   Spinal cord   Retroperitoneal fat

Inferior vena cava

Rib

Liver

Gallbladder

Costal cartilage

Transverse colon

Diaphragm

Spleen

Aorta

Pancreas

Rectus abdominis m.

PLATE FIFTY-NINE
Transverse section of the abdomen through the pancreas, superior view. (*m.* stands for muscle.)

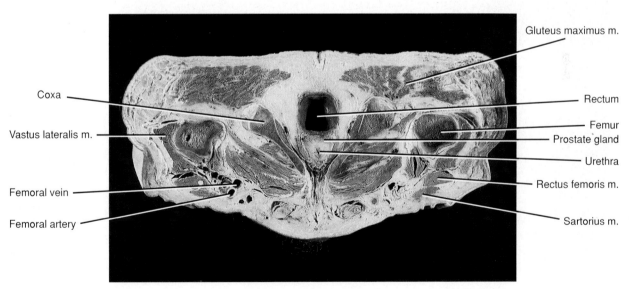

Gluteus maximus m.

Coxa

Vastus lateralis m.

Femoral vein

Femoral artery

Rectum

Femur

Prostate gland

Urethra

Rectus femoris m.

Sartorius m.

PLATE SIXTY
Transverse section of the male pelvic cavity, superior view. (*m.* stands for muscle.)

HUMAN CADAVERS

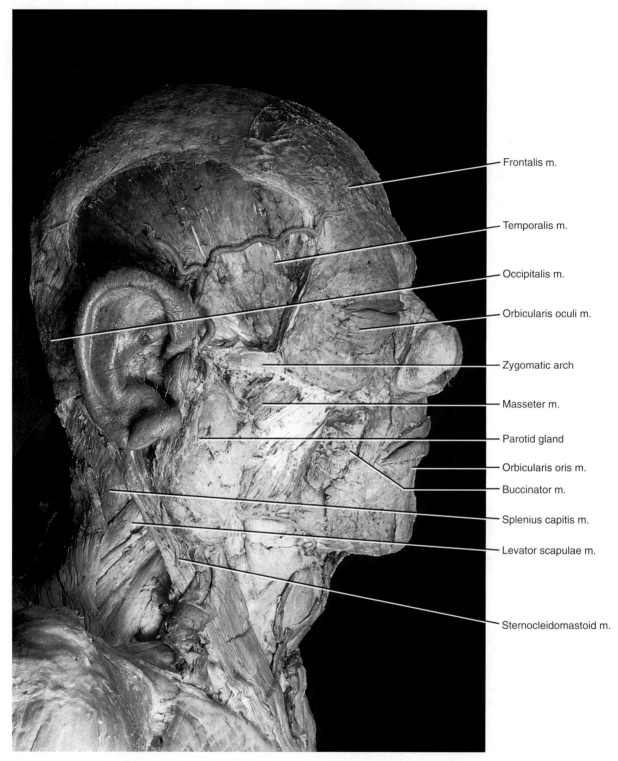

- Frontalis m.
- Temporalis m.
- Occipitalis m.
- Orbicularis oculi m.
- Zygomatic arch
- Masseter m.
- Parotid gland
- Orbicularis oris m.
- Buccinator m.
- Splenius capitis m.
- Levator scapulae m.
- Sternocleidomastoid m.

PLATE SIXTY-ONE
Lateral view of the head. (*m.* stands for muscle.)

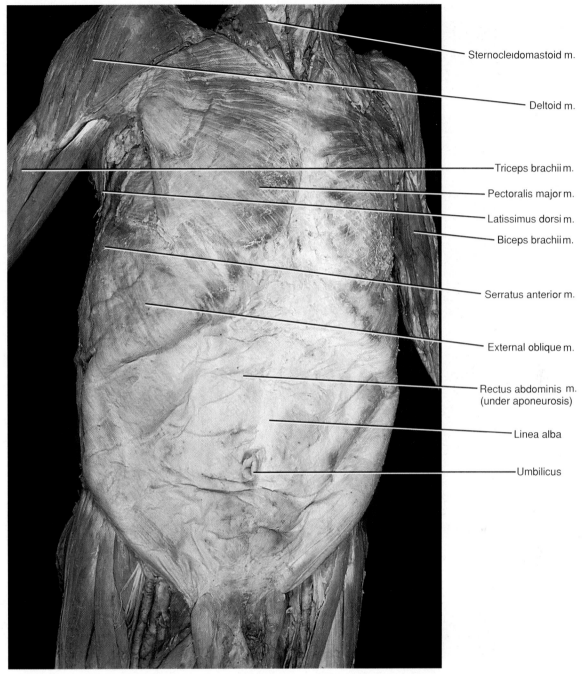

Sternocleidomastoid m.

Deltoid m.

Triceps brachii m.

Pectoralis major m.

Latissimus dorsi m.

Biceps brachii m.

Serratus anterior m.

External oblique m.

Rectus abdominis m.
(under aponeurosis)

Linea alba

Umbilicus

PLATE SIXTY-TWO
Anterior view of the trunk. (*m.* stands for muscle.)

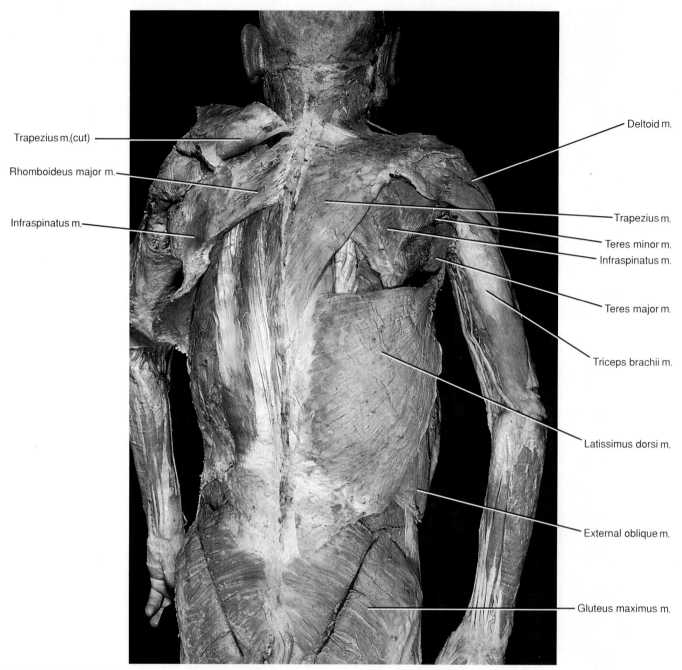

Trapezius m.(cut)

Rhomboideus major m.

Infraspinatus m.

Deltoid m.

Trapezius m.

Teres minor m.

Infraspinatus m.

Teres major m.

Triceps brachii m.

Latissimus dorsi m.

External oblique m.

Gluteus maximus m.

PLATE SIXTY-THREE

Posterior view of the trunk, with deep thoracic muscles exposed on the left. (*m.* stands for muscle.)

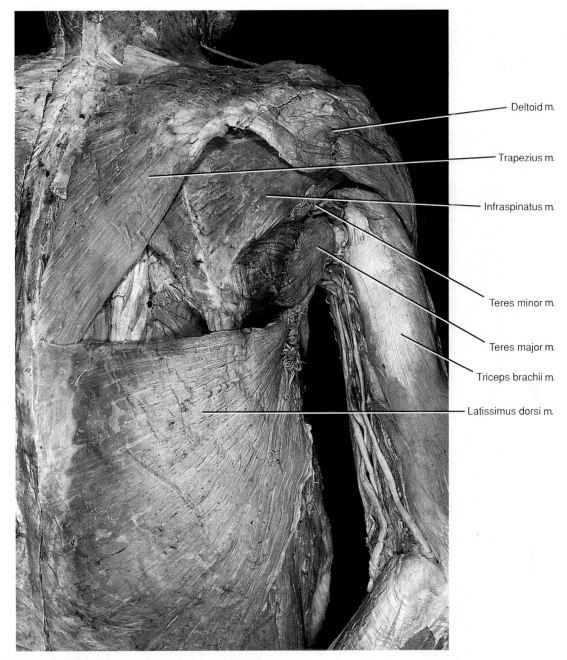

Deltoid m.

Trapezius m.

Infraspinatus m.

Teres minor m.

Teres major m.

Triceps brachii m.

Latissimus dorsi m.

PLATE SIXTY-FOUR
Posterior view of the right thorax and arm. (*m.* stands for muscle.)

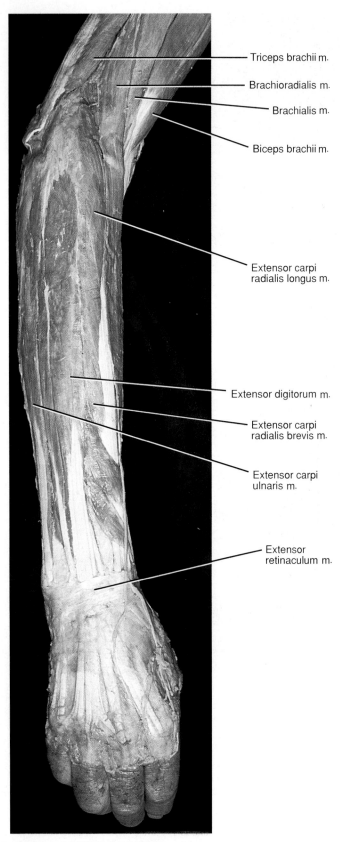

Triceps brachii m.

Brachioradialis m.

Brachialis m.

Biceps brachii m.

Extensor carpi radialis longus m.

Extensor digitorum m.

Extensor carpi radialis brevis m.

Extensor carpi ulnaris m.

Extensor retinaculum m.

PLATE SIXTY-FIVE

Posterior view of the right forearm and hand. (*m.* stands for muscle.)

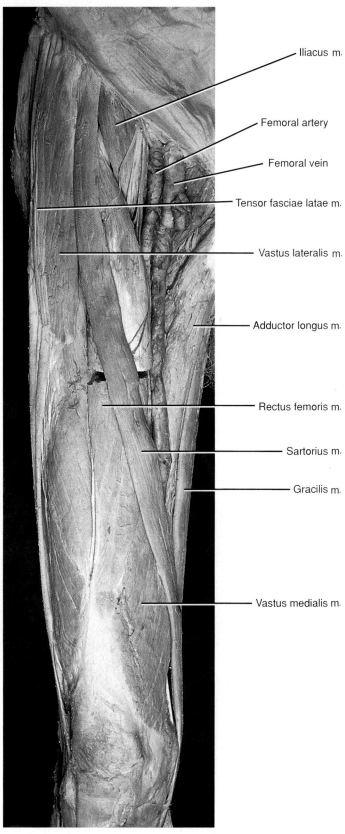

Iliacus m.

Femoral artery

Femoral vein

Tensor fasciae latae m.

Vastus lateralis m.

Adductor longus m.

Rectus femoris m.

Sartorius m.

Gracilis m.

Vastus medialis m.

PLATE SIXTY-SIX

Anterior view of the right thigh. (*m.* stands for muscle.)

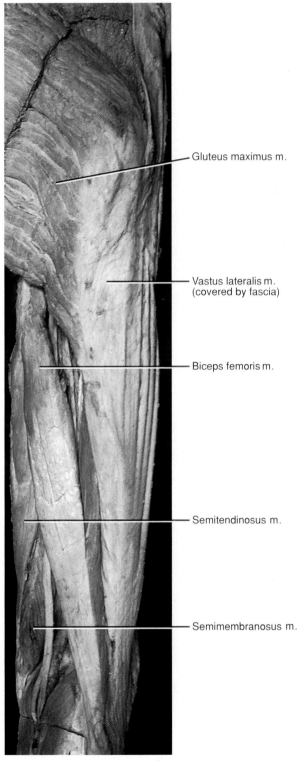

Gluteus maximus m.

Vastus lateralis m.
(covered by fascia)

Biceps femoris m.

Semitendinosus m.

Semimembranosus m.

PLATE SIXTY-SEVEN
Posterior view of the right thigh. (*m.* stands for muscle.)

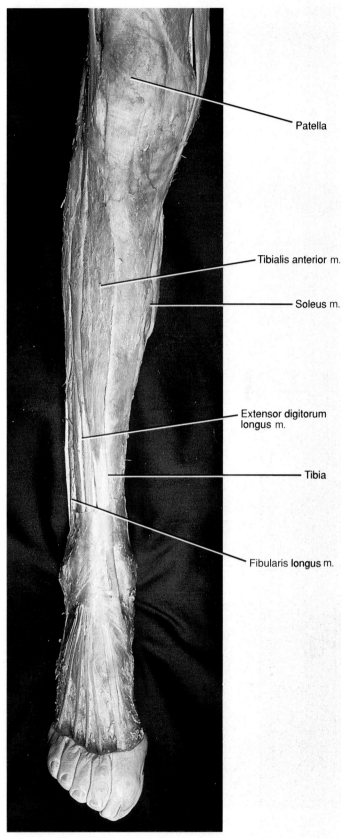

Patella

Tibialis anterior m.

Soleus m.

Extensor digitorum
longus m.

Tibia

Fibularis **longus** m.

PLATE SIXTY-EIGHT
Anterior view of the right leg. (*m.* stands for muscle.)

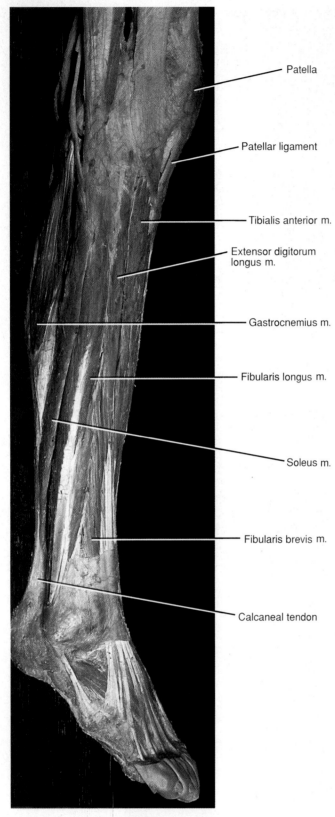

- Patella
- Patellar ligament
- Tibialis anterior m.
- Extensor digitorum longus m.
- Gastrocnemius m.
- Fibularis longus m.
- Soleus m.
- Fibularis brevis m.
- Calcaneal tendon

PLATE SIXTY-NINE
Lateral view of the right leg. (*m.* stands for muscle.)

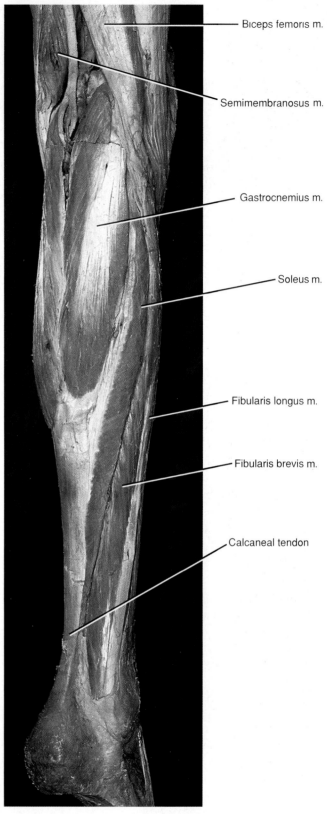

- Biceps femoris m.
- Semimembranosus m.
- Gastrocnemius m.
- Soleus m.
- Fibularis longus m.
- Fibularis brevis m.
- Calcaneal tendon

PLATE SEVENTY
Posterior view of the right leg. (*m.* stands for muscle.)

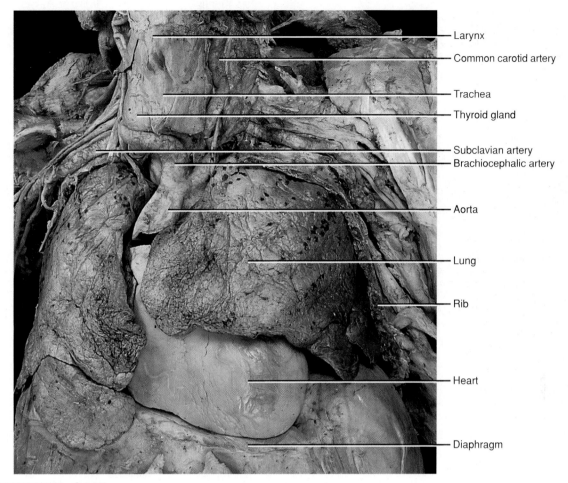

Larynx

Common carotid artery

Trachea

Thyroid gland

Subclavian artery

Brachiocephalic artery

Aorta

Lung

Rib

Heart

Diaphragm

PLATE SEVENTY-ONE

Thoracic viscera, anterior view. (Brachiocephalic veins have been removed to better expose the brachiocephalic artery and the aorta.)

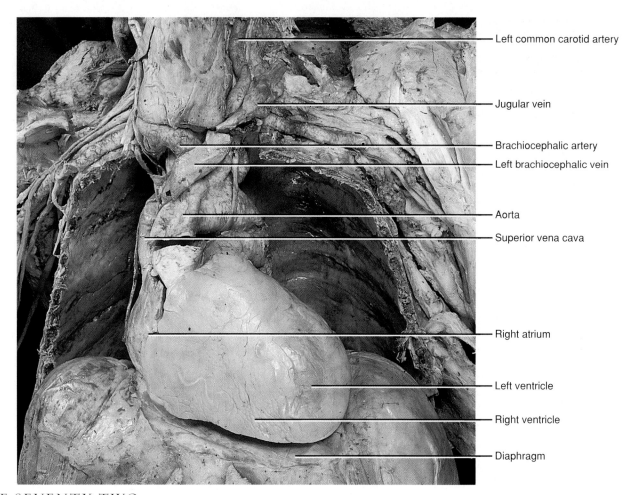

Left common carotid artery

Jugular vein

Brachiocephalic artery

Left brachiocephalic vein

Aorta

Superior vena cava

Right atrium

Left ventricle

Right ventricle

Diaphragm

PLATE SEVENTY-TWO
Thorax with the lungs removed, anterior view.

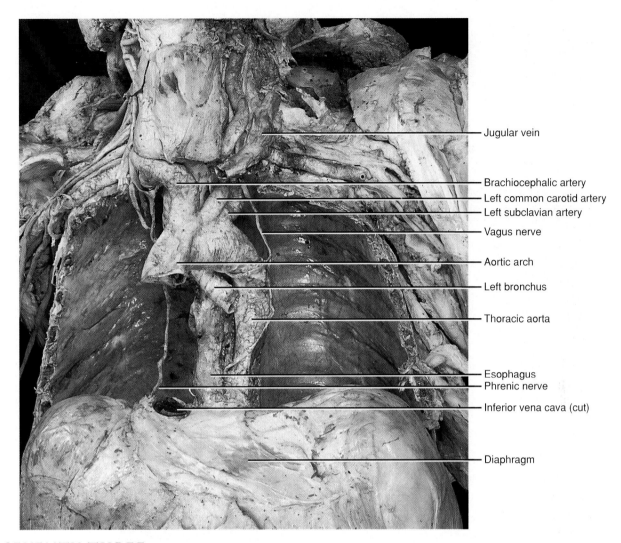

Jugular vein

Brachiocephalic artery
Left common carotid artery
Left subclavian artery
Vagus nerve

Aortic arch

Left bronchus

Thoracic aorta

Esophagus
Phrenic nerve

Inferior vena cava (cut)

Diaphragm

PLATE SEVENTY-THREE
Thorax with the heart and lungs removed, anterior view.

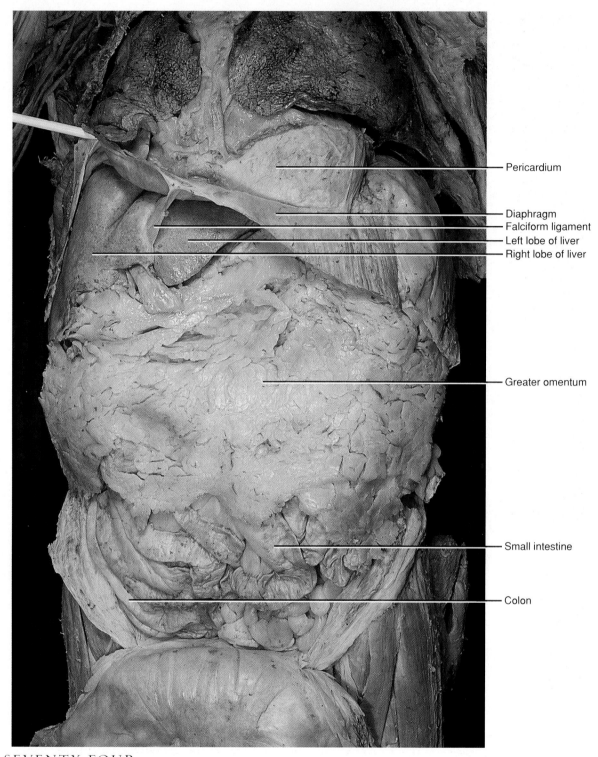

Pericardium

Diaphragm
Falciform ligament
Left lobe of liver
Right lobe of liver

Greater omentum

Small intestine

Colon

PLATE SEVENTY-FOUR
Abdominal viscera, anterior view.

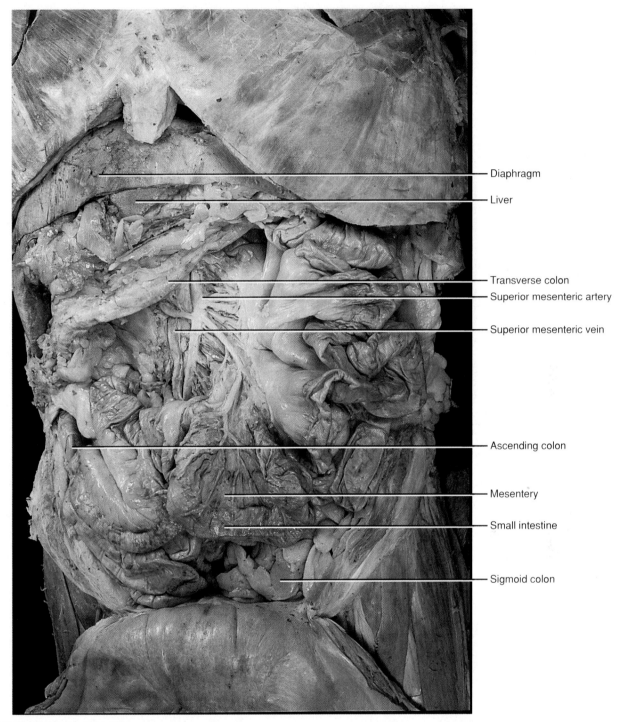

Diaphragm

Liver

Transverse colon

Superior mesenteric artery

Superior mesenteric vein

Ascending colon

Mesentery

Small intestine

Sigmoid colon

PLATE SEVENTY-FIVE
Abdominal viscera with the greater omentum removed, anterior view. (Small intestine has been displaced to the left.)

Anterior   **Posterior**

Corpus callosum ——————————

Frontal lobe of cerebrum ——————

Diencephalon/thalamus ——————

Diencephalon/hypothalamus ——————

Midbrain ————————————   —————— Occipital lobe
                                                                                                          of cerebrum

Pons ——————————————   —————— Cerebellum

Medulla oblongata ——————————   —————— Arbor vitae

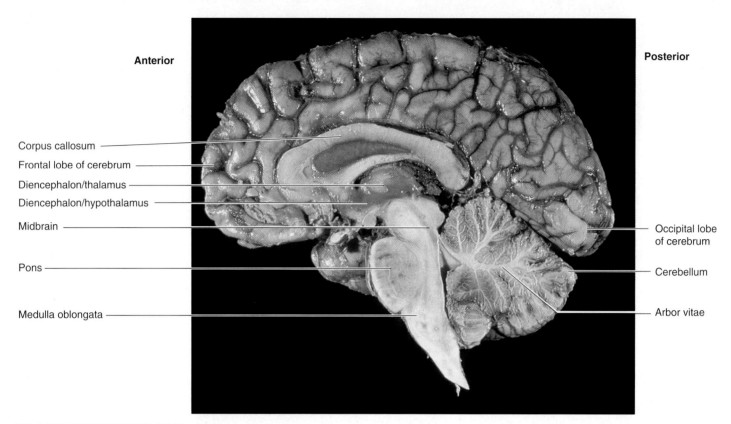

PLATE SEVENTY-SIX
Midsagittal section of the brain.

Periodic Table of Elements

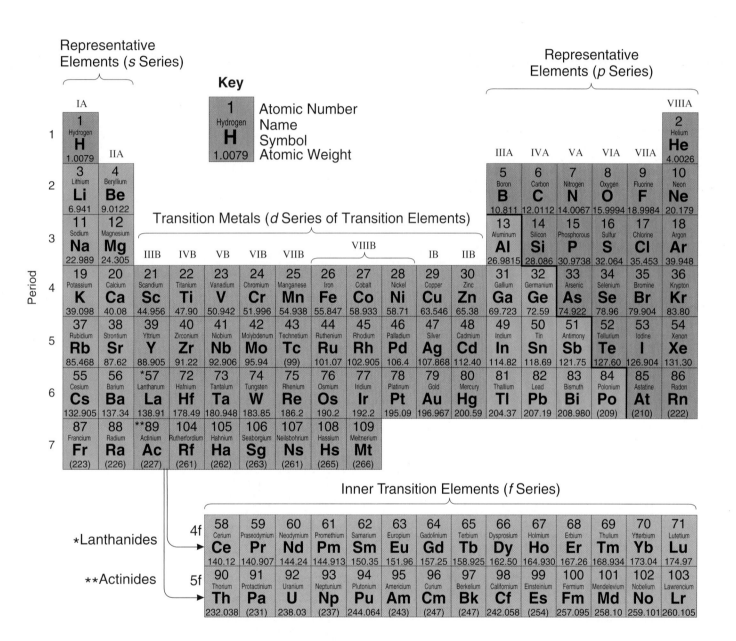

Elements 110 to 114 have been reported in experiments, but have not yet been confirmed.

Laboratory Tests of Clinical Importance

Common Tests Performed on Blood

TEST	NORMAL VALUES* (ADULT)	CLINICAL SIGNIFICANCE
Albumin (serum)	3.2–5.5 g/100 mL	Values increase in multiple myeloma and decrease with proteinuria and as a result of severe burns.
Albumin-globulin ratio, or A/G ratio (serum)	1.5:1 to 2.5:1	Ratio of albumin to globulin is lowered in kidney diseases and malnutrition.
Ammonia	12–55 µ mol/L	Values increase in severe liver disease, pneumonia, shock, and congestive heart failure.
Amylase (serum)	4–25 units/mL	Values increase in acute pancreatitis, intestinal obstructions, and mumps. They decrease in chronic pancreatitis, cirrhosis of the liver, and toxemia of pregnancy.
Bilirubin, total (serum)	0–1.0 mg/100 mL	Values increase in conditions causing red blood cell destruction or biliary obstruction.
Blood urea nitrogen, or BUN (plasma or serum)	8–25 mg/100 mL	Values increase in various kidney disorders and decrease in liver failure and during pregnancy.
Calcium (serum)	8.5–10.5 mg/100 mL	Values increase in hyperparathyroidism, hypervitaminosis D, and respiratory conditions that cause a rise in CO_2 concentration. They decrease in hypoparathyroidism, malnutrition, and severe diarrhea.
Carbon dioxide (serum)	24–30 mEq/L	Values increase in respiratory diseases, intestinal obstruction, and vomiting. They decrease in acidosis, nephritis, and diarrhea.
Chloride (serum)	100–106 mEq/L	Values increase in nephritis, Cushing syndrome, dehydration, and hyperventilation. They decrease in metabolic acidosis, Addison disease, diarrhea, and following severe burns.
Cholesterol, total (serum)	120–220 mg/100 mL (below 200 mg/100 mL recommended by the American Heart Association)	Values increase in diabetes mellitus and hypothyroidism. They decrease in pernicious anemia, hyperthyroidism, and acute infections.
Cholesterol, high-density lipoprotein (HDL)	Women: 30–80 mg/100 mL Men: 30–70 mg/100 mL	Values increase in liver disease. Decreased values are associated with an increased risk of atherosclerosis.
Cholesterol, low-density lipoprotein (LDL)	62–185 mg/100 mL	Increased values are associated with an increased risk of atherosclerosis.
Creatine (serum)	0.2–0.8 mg/100 mL	Values increase in muscular dystrophy, nephritis, severe damage to muscle tissue, and during pregnancy.
Creatinine (serum)	0.6–1.5 mg/100 mL	Values increase in various kidney diseases.
Ferritin (serum)	Men: 10–270 µg/100 mL Women: 5–280 µg/100 mL	Values correlate with total body iron store. They decrease with iron deficiency.
Globulin (serum)	2.3–3.5 g/100 mL	Values increase as a result of chronic infections.
Glucose (plasma)	70–110 mg/100 mL	Values increase in diabetes mellitus, liver diseases, nephritis, hyperthyroidism, and pregnancy. They decrease in hyperinsulinism, hypothyroidism, and Addison disease.
Hematocrit (whole blood)	Men: 40–54% Women: 37–47% Children: 35–49% (varies with age)	Values increase in polycythemia due to dehydration or shock. They decrease in anemia and following severe hemorrhage.

*These values may vary with hospital, physician, and type of equipment used to make measurements.

TEST	NORMAL VALUES* (ADULT)	CLINICAL SIGNIFICANCE
Hemoglobin (whole blood)	Men: 14–18 g/100 mL Women: 12–16 g/100 mL Children: 11.2–16.5 g/100 mL (varies with age)	Values increase in polycythemia, obstructive pulmonary diseases, congestive heart failure, and at high altitudes. They decrease in anemia, pregnancy, and as a result of severe hemorrhage or excessive fluid intake.
Iron (serum)	50–150 µg/100 mL	Values increase in various anemias and liver disease. They decrease in iron-deficiency anemia.
Iron-binding capacity (serum)	250–410 µg/100 mL	Values increase in iron-deficiency anemia and pregnancy. They decrease in pernicious anemia, liver disease, and chronic infections.
Lactic acid (whole blood)	0.6–1.8 mEq/L	Values increase with muscular activity and in congestive heart failure, severe hemorrhage, and shock.
Lactic dehydrogenase, or LDH (serum)	45–90 U/L	Values increase in pernicious anemia, myocardial infarction, liver disease, acute leukemia, and widespread carcinoma.
Lipids, total (serum)	450–850 mg/100 mL	Values increase in hypothyroidism, diabetes mellitus, and nephritis. They decrease in hyperthyroidism.
Magnesium	1.3–2.1 mEq/L	Values increase in renal failure, hypothyroidism, and Addison disease. They decrease in renal disease, liver disease, and pancreatitis.
Mean corpuscular hemoglobin (MCH)	26–32 pg/RBC	Values increase in macrocytic anemia. They decrease in microcytic anemia.
Mean corpuscular volume (MCV)	86–98 µ mm^3/RBC	Values increase in liver disease and pernicious anemia. They decrease in iron-deficiency anemia.
Osmolality	275–295 mOsm/kg	Values increase in dehydration, hypercalcemia, and diabetes mellitus. They decrease in hyponatremia, Addison disease, and water intoxication.
Oxygen saturation (whole blood)	Arterial: 96–100% Venous: 60–85%	Values increase in polycythemia and decrease in anemia and obstructive pulmonary diseases.
pH (whole blood)	7.35–7.45	Values increase due to mild vomiting, Cushing syndrome, and hyperventilation. They decrease as a result of hypoventilation, severe diarrhea, Addison disease, and diabetic acidosis.
Phosphatase acid (serum)	Women: 0.01–0.56 Sigma U/mL Men: 0.13–0.63 Sigma U/mL	Values increase in cancer of the prostate gland, hyperparathyroidism, certain liver diseases, myocardial infarction, and pulmonary embolism.
Phosphatase, alkaline (serum)	13–39 U/L	Values increase in hyperparathyroidism (and in other conditions that promote resorption of bone), liver diseases, and pregnancy.
Phosphorus (serum)	3.0–4.5 mg/100 mL	Values increase in kidney diseases, hypoparathyroidism, acromegaly, and hypervitaminosis D. They decrease in hyperparathyroidism.
Platelet count (whole blood)	150,000–350,000/mm^3	Values increase in polycythemia and certain anemias. They decrease in acute leukemia and aplastic anemia.
Potassium (serum)	3.5–5.0 mEq/L	Values increase in Addison disease, hypoventilation, and conditions that cause severe cellular destruction. They decrease in diarrhea, vomiting, diabetic acidosis, and chronic kidney disease.
Protein, total (serum)	6.0–8.4 g/100 mL	Values increase in severe dehydration and shock. They decrease in severe malnutrition and hemorrhage.
Prothrombin time (serum)	12–14 sec (one stage)	Values increase in certain hemorrhagic diseases, liver disease, vitamin K deficiency, and following the use of various drugs.
Red cell count (whole blood)	Men: 4,600,000–6,200,000/mm^3 Women: 4,200,000–5,400,000/mm^3 Children: 4,500,000–5,100,000/mm^3 (varies with age)	Values increase as a result of severe dehydration or diarrhea, and decrease in anemia, leukemia, and following severe hemorrhage.
Red cell distribution width (RDW)	8.5–11.5 microns	Variation in cell width changes with pernicious anemia.
Sedimentation rate, erythrocyte (whole blood)	Men: 1–13 mm/hr Women: 1–20 mm/hr	Values increase in infectious diseases, menstruation, pregnancy, and as a result of severe tissue damage.
Serum glutamic pyruvic transaminase (SGPT)	Women: 4–17 U/L Men: 6–24 U/L	Values increase in liver disease, pancreatitis, and acute myocardial infarction.

*These values may vary with hospital, physician, and type of equipment used to make measurements.

Common Tests Performed on Blood—*continued*

TEST	NORMAL VALUES* (ADULT)	CLINICAL SIGNIFICANCE
Sodium (serum)	135–145 mEq/L	Values increase in nephritis and severe dehydration. They decrease in Addison disease, myxedema, kidney disease, and diarrhea.
Thromboplastin time, partial (plasma)	35–45 sec	Values increase in deficiencies of blood factors VIII, IX, and X.
Thyroid-stimulating hormone (TSH)	0.5–5.0 µU/mL	Values increase in hypothyroidism and decrease in hyperthyroidism.
Thyroxine, or T_4 (serum)	4–12 µg/100 mL	Values increase in hyperthyroidism and pregnancy. They decrease in hypothyroidism.
Transaminases, or SGOT (serum)	7–27 units/mL	Values increase in myocardial infarction, liver disease, and diseases of skeletal muscles.
Triglycerides	40–150 mg/100 mL	Values increase in liver disease, nephrotic syndrome, hypothyroidism, and pancreatitis. They decrease in malnutrition and hyperthyroidism.
Triiodothyronine, or T_3 (serum)	75–195 ng/100 mL	Values increase in hyperthyroidism and decrease in hypothyroidism.
Uric acid (serum)	Men: 2.5–8.0 mg/100 mL Women: 1.5–6.0 mg/100 mL	Values increase in gout, leukemia, pneumonia, toxemia of pregnancy, and as a result of severe tissue damage.
White blood cell count, differential (whole blood)	Neutrophils 54–62% Eosinophils 1–3% Basophils <1% Lymphocytes 25–33% Monocytes 3–7%	Neutrophils increase in bacterial diseases; lymphocytes and monocytes increase in viral diseases; eosinophils increase in collagen diseases, allergies, and in the presence of intestinal parasites.
White blood cell count, total (whole blood)	5,000–10,000/mm³	Values increase in acute infections, acute leukemia, and following menstruation. They decrease in aplastic anemia and as a result of drug toxicity.

Common Tests Performed on Urine

TEST	NORMAL VALUES* (ADULT)	CLINICAL SIGNIFICANCE
Acetone and acetoacetate	0	Values increase in diabetic acidosis.
Albumin, qualitative	0 to trace	Values increase in kidney disease, hypertension, and heart failure.
Ammonia	20–70 mEq/L	Values increase in diabetes mellitus and liver diseases.
Bacterial count	Under 10,000/mL	Values increase in urinary tract infection.
Bile and bilirubin	0	Values increase in melanoma and biliary tract obstruction.
Calcium	Under 300 mg/24 hr	Values increase in hyperparathyroidism and decrease in hypoparathyroidism.
Creatinine (24 hours)	15–25 mg/kg body weight/day	Values increase in infections, and decrease in muscular atrophy, anemia, leukemia, and kidney diseases.
Creatinine clearance (24 hours)	100–140 mL/min	Values increase in renal diseases.
Glucose	0	Values increase in diabetes mellitus and various pituitary gland disorders.
Hemoglobin	0	Blood may occur in urine as a result of extensive burns, crushing injuries, hemolytic anemia, or blood transfusion reactions.
17-hydroxycorticosteroids	3–8 mg/24 hr	Values increase in Cushing syndrome and decrease in Addison disease.
Osmolality	850 mOsm/kg	Values increase in hepatic cirrhosis, congestive heart failure, and Addison disease. They decrease in hypokalemia, hypercalcemia, and diabetes insipidus.
pH	4.6–8.0	Values increase in urinary tract infections and chronic renal failure. They decrease in diabetes mellitus, emphysema, and starvation.
Phenylpyruvic acid	0	Values increase in phenylketonuria.

These values may vary with hospital, physician, and type of equipment used to make measurements.

Common Tests Performed on Urine—*continued*

TEST	NORMAL VALUES* (ADULT)	CLINICAL SIGNIFICANCE
Specific gravity (SG)	1.003–1.035	Values increase in diabetes mellitus, nephrosis, and dehydration. They decrease in diabetes insipidus, glomerulonephritis, and severe renal injury.
Urea	25–35 g/24 hr	Values increase as a result of excessive protein breakdown. They decrease as a result of impaired renal function.
Urea clearance	Over 40 mL blood cleared of urea/min	Values increase in renal diseases.
Uric acid	0.6–1.0 g/24 hr as urate	Values increase in gout and decrease in various kidney diseases.
Urobilinogen	0–4 mg/24 hr	Values increase in liver diseases and hemolytic anemia. They decrease in complete biliary obstruction and severe diarrhea.

These values may vary with hospital, physician, and type of equipment used to make measurements.

Glycolysis

Figure C.1 illustrates the chemical reactions of glycolysis. In the early steps of this metabolic pathway, the original glucose molecule is altered by the addition of phosphate groups (*phosphorylation*) and by the rearrangement of its atoms. ATP supplies the phosphate groups and the energy to drive these reactions. The result is a molecule of fructose bound to two phosphate groups (fructose-1,6-bisphosphate). This molecule is split through two separate reactions into two 3-carbon molecules (glyceraldehyde-3-phosphate). Since each of these is converted to pyruvic acid, the following reactions, 1 through 5, must be counted twice to account for breakdown of a single glucose molecule.

1. An inorganic phosphate group is added to glyceraldehyde-3-phosphate to form 1,3-bisphosphoglyceric acid, releasing two hydrogen atoms, to be used in ATP synthesis, described later.

2. 1,3-bisphosphoglyceric acid is changed to 3-phosphoglyceric acid. As this occurs, some energy in the form of a high-energy phosphate is transferred from the 1,3-bisphosphoglyceric acid to an ADP molecule, phosphorylating the ADP to ATP.

3. A slight alteration of 3-phosphoglyceric acid forms 2-phosphoglyceric acid.

4. A change in 2-phosphoglyceric acid converts it into phosphoenolpyruvic acid.

5. Finally, a high-energy phosphate is transferred from the phosphoenolpyruvic acid to an ADP molecule, phosphorylating it to ATP. A molecule of pyruvic acid remains.

Overall, one molecule of glucose is ultimately broken down to two molecules of pyruvic acid. Also, a total of four hydrogen atoms are released (step *a*), and four ATP molecules form (two in step *b* and two in step *e*). However, because two molecules of ATP are used early in glycolysis, there is a net gain of only two ATP molecules during this phase of cellular respiration.

In the presence of oxygen, each pyruvic acid molecule is oxidized to an acetyl group, which then combines with a molecule of coenzyme A (obtained from the vitamin pantothenic acid) to form acetyl coenzyme A. As this occurs, two more hydrogen atoms are released for each

molecule of acetyl coenzyme A formed. The acetyl coenzyme A is then broken down by means of the citric acid cycle, which figure C.2 illustrates.

Because obtaining energy for cellular metabolism is vital, disruptions in glycolysis or the reactions that follow it can devastate health. Clinical Application 4.1 tells how medical sleuths traced a boy's unusual combination of symptoms to a block in glycolysis.

Citric Acid Cycle

An acetyl coenzyme A molecule enters the citric acid cycle by combining with a molecule of oxaloacetic acid to form citric acid. As citric acid is produced, coenzyme A is released and thus can be used again to form acetyl coenzyme A from pyruvic acid. The citric acid is then changed by a series of reactions back into oxaloacetic acid, and the cycle may repeat.

Steps in the citric acid cycle release carbon dioxide and hydrogen atoms. More specifically, for each glucose molecule metabolized in the presence of oxygen, two molecules of acetyl coenzyme A enter the citric acid cycle. The cycle releases four carbon dioxide molecules and sixteen hydrogen atoms. At the same time, two more molecules of ATP form.

The released carbon dioxide dissolves in the cytoplasm and leaves the cell, eventually entering the bloodstream. Most of the hydrogen atoms released from the citric acid cycle, and those released during glycolysis and during the formation of acetyl coenzyme A, supply electrons used to produce ATP.

ATP Synthesis

Note that in figures C.1 and C.2 various metabolic reactions release hydrogen atoms. The electrons of these hydrogen atoms contain much of the energy associated with the chemical bonds of the original glucose molecule. To keep this energy in a usable form, these hydrogen atoms, with their high energy electrons, are passed in pairs to *hydrogen carriers*. One of these carriers is NAD^+ (nicotinamide adenine dinucleotide). When NAD^+ accepts a pair of hydrogen atoms, the two electrons and one hydrogen nucleus bind to NAD^+ to form NADH, and the remaining hydrogen nucleus (a hydrogen ion) is released as follows:

$$NAD^+ + 2H \rightarrow NADH + H^+$$

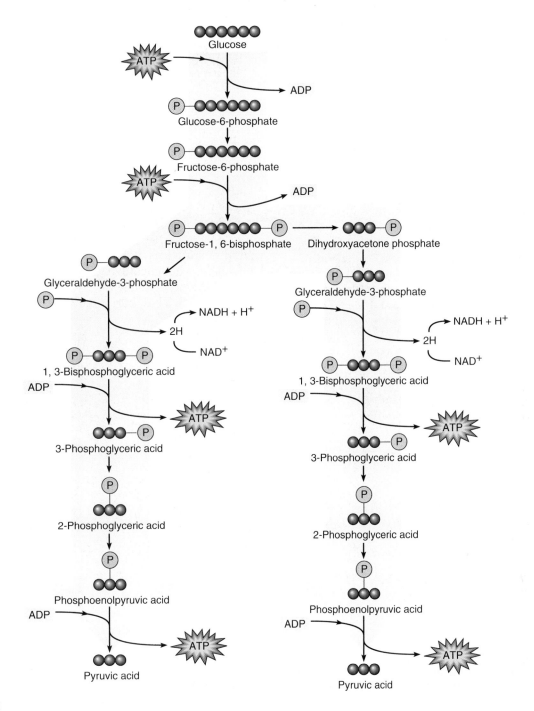

FIGURE C.1

Chemical reactions of glycolysis. There is a net production of 2 ATP molecules from each glucose molecule. The four hydrogen atoms released provide high-energy electrons that may be used to generate ATP in the electron transport chain, described later.

NAD⁺ is a coenzyme obtained from a vitamin (niacin), and when it combines with the energized electrons it is said to be *reduced.* Reduction results from the addition of electrons, often as part of hydrogen atoms. Another electron acceptor, FAD (flavine adenine dinucleotide), acts in a similar manner, combining with two electrons and two hydrogen nuclei to form $FADH_2$ (fig. C.2). In their reduced states, the hydrogen carriers NADH and $FADH_2$ now hold most of the energy once held in the bonds of the original glucose molecule.

Figure 4.11 shows that NADH can release the electrons and hydrogen nucleus. Since this reaction removes electrons, the resulting NAD⁺ is said to be *oxidized.* Oxidation results from the removal of electrons, often as

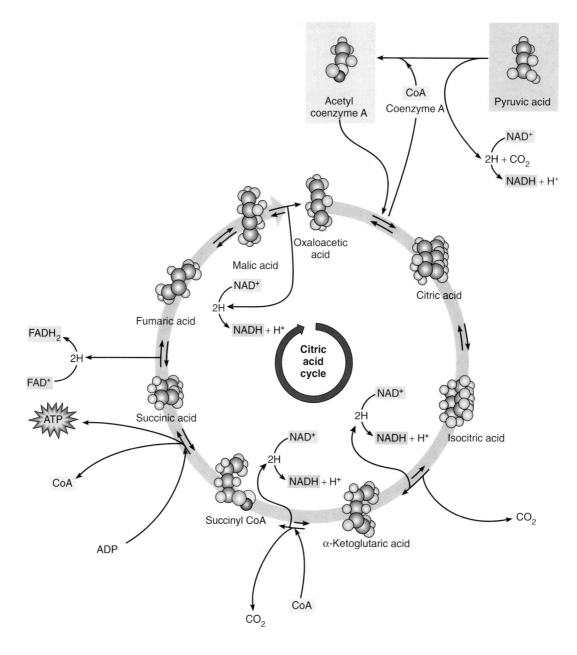

FIGURE C.2

Chemical reactions of the citric acid cycle. NADH and FADH$_2$ molecules carrying hydrogens are highlighted.

part of hydrogen atoms; it is the opposite of reduction. The two electrons this reaction releases pass to a series of electron carriers. The regenerated NAD$^+$ can once again accept electrons, and is recycled.

The molecules that act as electron carriers comprise an **electron transport chain** described in Chapter 4 (pp. 111–112). As electrons are passed from one carrier to another, the carriers are alternately reduced and oxidized as they accept or release electrons. The transported electrons gradually lose energy as they proceed down the chain.

Among the members of the electron transport chain are several proteins, including a set of iron-containing molecules called **cytochromes.** The chain is located in the inner membranes of the mitochondria (see chapter 3, p. 72). The folds of the inner mitochondrial membrane provide surface area on which the energy reactions take place. In a muscle cell, the inner mitochondrial membrane, if stretched out, may be as much as forty-five times as long as the cell membrane!

The final cytochrome of the electron transport chain (cytochrome oxidase) gives up a pair of electrons and causes two hydrogen ions (formed at the beginning of the sequence) to combine with an atom of oxygen. This process produces a water molecule:

$$2e^- + 2H^+ + 1/2\ O_2 \rightarrow H_2O$$

Thus, oxygen is the final electron acceptor. In the absence of oxygen, electrons cannot pass through the electron transport chain, NAD^+ cannot be regenerated, and aerobic respiration halts.

Note in figure 4.11 that as electrons pass through the electron transport chain, energy is released. Some of this energy is used by a mechanism involving the enzyme complex *ATP synthase* to combine phosphate and ADP by a high-energy bond (phosphorylation), forming ATP. Also note in figures C.1 and C.2 that twelve pairs of hydrogen atoms are released during the complete breakdown of one glucose molecule—two pairs from glycolysis, two pairs from the conversion of pyruvic acid to acetyl coenzyme A (one pair from each of two pyruvic acid molecules), and eight pairs from the citric acid cycle (four pairs for each of two acetyl coenzyme A molecules).

High-energy electrons from ten pairs of these hydrogen atoms produce thirty ATP molecules in the electron transport chain. Two pairs enter the chain differently and form four ATP molecules. Because this process of forming ATP involves both the oxidation of hydrogen atoms and the bonding of phosphate to ADP, it is called oxidative phosphorylation. Also, there is a net gain of two ATP molecules during glycolysis, and two ATP molecules form by direct enzyme action in two turns of the citric acid cycle. Thus, a maximum of thirty-eight ATP molecules form for each glucose molecule metabolized.

D APPENDIX

A Closer Look at DNA Structures

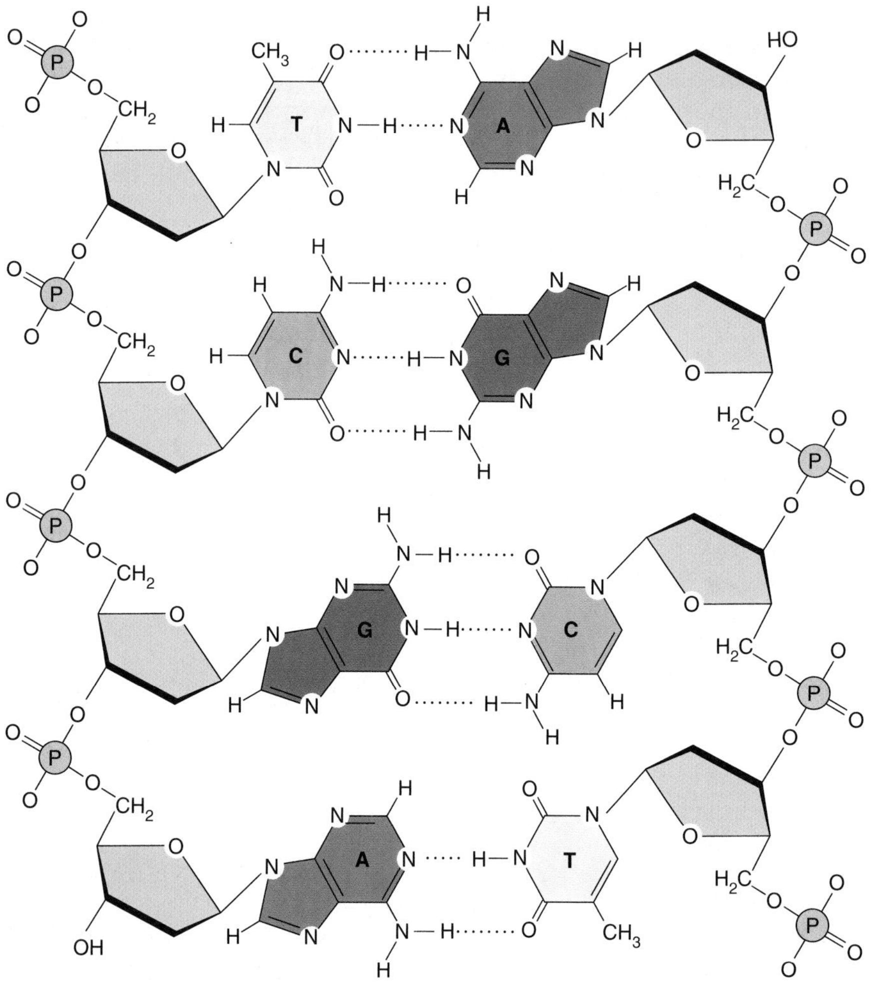

FIGURE D.1

The nucleotides of a double-stranded DNA molecule pair so that an adenine (A) of one strand hydrogen bonds to a thymine (T) of the other strand, and a guanine (G) of one strand hydrogen bonds to a cytosine (C) of the other. The dotted lines represent hydrogen bonds.

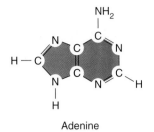

Adenine

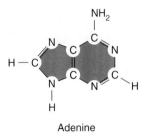

Thymine

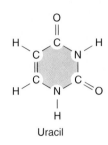

Adenine

Uracil

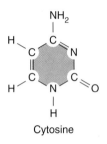

Cytosine

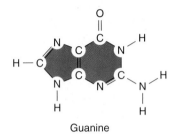

Guanine

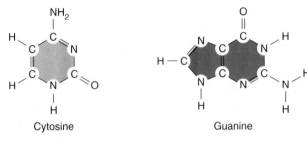

Cytosine

Guanine

FIGURE D.2

The deoxyribonucleotides contain adenine, thymine, cytosine, or guanine.

FIGURE D.3

The ribonucleotides contain adenine, uracil, cytosine, or guanine.

Glossary

Each word in this glossary is followed by a phonetic guide to pronunciation. In this guide, any unmarked vowel that ends a syllable or stands alone as a syllable has the long sound. Thus, *play* would be spelled *pla*. Any unmarked vowel that is followed by a consonant has the short sound. *Tough,* for instance, is spelled *tuf.* If a long vowel appears in the middle of a syllable (followed by a consonant), it is marked with the macron (¯), the sign for a long vowel. Thus, the word *plate* would be phonetically spelled plāt. Similarly, if a vowel stands alone or ends a syllable, but has the short sound, it is marked with a breve (˘).

A

abdominal (ab-dom′ĭ-nal)   Pertaining to the portion of the body between the diaphragm and the pelvis. p. 24

abdominal cavity (ab-dom′ĭ-nal kav′ĭ-tē)   The space between the diaphragm and the pelvic inlet that contains the abdominal viscera. p. 14

abdominopelvic cavity (ab-dom″ĭ-no-pel′vik kav′ĭ-tē)   The space between the diaphragm and the lower portion of the trunk of the body. p. 12

abduction (ab-duk′shun)   Movement of a body part away from the midline. p. 261

absorption (ab-sorp′shun)   The taking in of substances by cells or membranes. p. 8

accessory organ (ak-ses′o-re or′gan)   Organ that supplements the functions of other organs. p. 447

accommodation (ah-kom″o-da′shun)   Adjustment of the lens of the eye for close vision. p. 452

acetone (as′e-tōn)   One of the ketone bodies produced as a result of the oxidation of fats. p. 823

acetylcholine (as″ĕ-til-ko′lēn)   A type of neurotransmitter, which is a biochemical secreted at the axon ends of many neurons; transmits nerve impulses across synapses. p. 283

acetylcholinesterase (as″ĕ-til-ko″lin-es′ter-ās)   An enzyme that catalyzes breakdown of acetylcholine. p. 286

acetyl coenzyme A (as′ĕ-til ko-en′zīm)   An intermediate compound produced during the oxidation of carbohydrates and fats. p. 111

acid (as′id)   A substance that ionizes in water to release hydrogen ions. p. 46

acid-base buffer system (as′id-bās buf′er sis′tem)   A pair of chemicals, one a weak acid, the other a weak base, that resists pH changes. p. 819

acidosis (as″ĭ-do′sis)   A relative increase in the acidity of body fluids. p. 47

acromial (ah-kro′me-al)   Pertaining to the shoulder. p. 24

ACTH   Adrenocorticotropic hormone. p. 483

actin (ak′tin)   A protein in a muscle fiber that, together with myosin, is responsible for contraction and relaxation. p. 279

action potential (ak′shun po-ten′shal)   The sequence of electrical changes when a nerve cell membrane is exposed to a stimulus that exceeds its threshold. p. 350

activation energy (ak″tĭ-va′shun en′er-je)   Energy required to initiate a chemical reaction. p. 106

active site (ak′tiv sīt)   Region of an enzyme molecule that temporarily combines with a substrate. p. 106

active transport (ak′tiv trans′port)   Process that requires an expenditure of energy to move a substance across a cell membrane; moved against the concentration gradient. p. 85

acoustic (ah-kōōs′tik)   Pertaining to sound. p. 434

adduction (ah-duk′shun)   Movement of a body part toward the midline. p. 261

adenoids (ad′ĕ-noids)   The pharyngeal tonsils located in the nasopharynx. p. 650

adenosine diphosphate (ah-den′o-sēn di-fos′fāt)   Molecule produced when the terminal phosphate is lost from a molecule of adenosine triphosphate; ADP. p. 108

adenosine triphosphate (ah-den′o-sēn tri-fos′fāt)   Organic molecule that stores energy and releases energy for use in cellular processes; ATP. p. 73

adenylate cyclase (ah-den′ĭ-lāt si′klās)   An enzyme that is activated when certain hormones combine with receptors on cell membranes, causing ATP to release two phosphates and circularize, forming cyclic AMP. p. 472

ADH   Antidiuretic hormone. p. 483

adipose tissue (ad′ĭ-pōs tish′u)   Fat-storing tissue. p. 144

adolescence (ad″o-les′ens)   Period of life between puberty and adulthood. p. 909

ADP   Adenosine diphosphate. p. 108

adrenal cortex (ah-dre′nal kor′teks)   The outer portion of the adrenal gland. p. 491

adrenal gland (ah-dre′nal gland)   Endocrine gland located on the superior portion of each kidney. p. 490

adrenalin (ah-dren′ah-lin)   Epinephrine. A hormone produced by the adrenal glands. p. 492

adrenal medulla (ah-dre′nal me-dul′ah)   The inner portion of the adrenal gland. p. 490

adrenergic fiber (ad″ren-er′jik fi′ber)   An axon that secretes norepinephrine at the axon terminal. p. 411

adrenocorticotropic hormone (ah-dre″no-kor″te-ko-trōp′ik hor′mōn)   Hormone secreted by the anterior lobe of the pituitary gland that stimulates activity in the adrenal cortex; ACTH. p. 483

adulthood (ah-dult′hood)   Period of life between adolescence and senescence. p. 909

aerobic (a″er-ōb′ik)   Requiring molecular oxygen. p. 111

afferent (af′er-ent)   Conducting toward a center. For example, an afferent arteriole conveys blood to the glomerulus of a nephron within the kidney. p. 774

agglutination (ah-gloo″ti-na′shun) Clumping of blood cells in response to a reaction between an antibody and an antigen. p. 533

agonist (ag′o-nist) A prime mover. p. 296

agranulocyte (a-gran′u-lo-sīt) A nongranular leukocyte. p. 518

albumin (al-bu′min) A plasma protein that helps regulate the osmotic concentration of blood. p. 523

aldosterone (al-dos′ter-ōn) A hormone, secreted by the adrenal cortex, that regulates sodium and potassium ion concentrations and fluid volume. p. 472

alimentary canal (al″i-men′tar-e kah-nal′) The tubular portion of the digestive tract that leads from the mouth to the anus. p. 644

alkaline (al′kah-līn) Pertaining to or having the properties of a base or alkali; basic. p. 46

alkaline tide (al′kah-lin tīd) An increase in the blood concentration of bicarbonate ions following a meal. p. 661

alkaloid (al′kah-loid) A group of organic substances that are usually bitter in taste and have toxic effects. p. 434

alkalosis (al″kah-lo′sis) A relative increase in the alkalinity of body fluids. p. 47

allantois (ah-lan′to-is) An embryonic structure that forms the umbilical cord blood vessels. p. 891

allele (ah-lēl′) One of two or more different forms of a gene. p. 925

allergen (al′er-jen) A foreign substance capable of stimulating an allergic reaction. p. 630

all-or-none response (al′or-nun′ re-spons′) Phenomenon in which a muscle fiber or neuron completely responds when it is exposed to a stimulus of threshold strength. p. 290; 353

alpha receptor (al′fah re-sep′tor) Receptor on effector cell membrane that combines with epinephrine or norepinephrine. p. 415

alpha-tocopherol (al″fah-to-kof′er-ol) Vitamin E. p. 708

alveolar duct (al-ve′o-lar dukt) Fine tube that carries air to an air sac of the lungs. p. 741

alveolar pore (al-ve′o-lar pōr) Minute opening in the wall of an air sac which permits air to pass from one alveolus to another. p. 757

alveolar process (al-ve′o-lar pros′es) Projection on the border of the jaw in which the bony sockets of the teeth are located. p. 205

alveolus (al-ve′o-lus) An air sac of a lung; a saclike structure (pl., *alveoli*) p. 741

amacrine cell (am′ah-krin sel) A retinal neuron whose fibers pass laterally between other retinal cells. p. 654

amine (am′in) A type of nitrogen-containing organic compound, including the hormones secreted by the adrenal medulla. p. 471

amino acid (ah-me′no as′id) An organic compound of relatively small molecular size that contains an amino group ($-NH_2$) and a carboxyl group ($-COOH$); the structural unit of a protein molecule. p. 52

amniocentesis (am″ne-o-sen-te′sis) A procedure in which a sample of amniotic fluid is removed through the abdominal wall of a pregnant woman. Fetal cells in it are cultured and examined to check the chromosome complement. p. 937

amnion (am′ne-on) An extraembryonic membrane that encircles a developing fetus and contains amniotic fluid. p. 887

amniotic cavity (am″ne-ot′ik kav′ĭ-te) Fluid-filled space enclosed by the amnion. p. 891

amniotic fluid (am″ne-ot′ik floo′id) Fluid within the amniotic cavity that surrounds the developing fetus. p. 887

ampulla (am-pul′ah) An expansion at the end of each semicircular canal that contains a crista ampullaris. p. 444

amylase (am′ĭ-lās) An enzyme that hydrolyzes starch. p. 653

anabolism (ah-nab′o-liz″em) Metabolic process by which larger molecules are synthesized from smaller ones; anabolic metabolism. p. 104

anaerobic (an-a″er-ōb′ik) Absence of molecular oxygen. p. 110

anal canal (a′nal kah-nal′) The most distal two or three inches of the large intestine that open to the outside as the anus. p. 682

anaphase (an′ah-fāz) Stage in mitosis when replicated chromosomes separate and move to opposite poles of the cell. p. 92

anastomosis (ah-nas″to-mo′sis) A union of nerve fibers or blood vessels to form a network. p. 601

anatomical position (an″ah-tom′ĭ-kal po-zish′un) A body posture with the body erect, the face forward, the arms at the sides with the palms facing forward, and the toes pointing straight ahead. p. 21

anatomy (ah-nat′o-me) Branch of science dealing with the form and structure of body parts. p. 3

androgen (an′dro-jen) A male sex hormone such as testosterone. p. 845

anemia (ah-ne′me-ah) A condition of red blood cell or hemoglobin deficiency. p. 516

aneuploid (an′u-ploid) A cell with one or more extra or missing chromosomes. p. 934

aneurysm (an′u-rizm) A saclike expansion of a blood vessel wall. p. 143

angiotensin II (an″je-o-ten′sin too) A vasoconstricting biochemical that is produced when blood flow to the kidneys is reduced, elevating blood pressure. p. 493

angiotensinogen (an″je-o-ten-sin′o-jen) A serum globulin the liver secretes that renin converts to angiotensin. p. 493

anion (an′i-on) An atom carrying a negative charge due to one or more extra electrons. p. 350

anorexia nervosa (an″o-rek′se-ah ner-vo′sa) Disorder caused by the fear of becoming obese; includes loss of appetite and inability to maintain a normal minimum body weight. p. 723

anoxia (an-ok′se-ah) Abnormally low oxygen concentration of the tissues. p. 764

antagonist (an-tag′o-nist) A muscle that acts in opposition to a prime mover. p. 296

antebrachial (an″te-bra′ke-al) Pertaining to the forearm. p. 24

antecubital (an″te-ku′bi-tal) The region in front of the elbow joint. p. 24

anterior (an-te′re-or) Pertaining to the front. p. 21

anterior pituitary (an-te′re-or pi-tu′i-tār″e) The front lobe of the pituitary gland. p. 479

antibody (an′ti-bod″e) A protein that B cells of the immune system produce in response to the presence of a nonself antigen; it reacts with the antigen. p. 519

anticoagulant (an″tĭ-ko-ag′u-lant) A biochemical that inhibits blood clotting. p. 531

anticodon (an″ti-ko′don) Three contiguous nucleotides of a transfer RNA molecule that are complementary to a specific mRNA codon. p. 120

antidiuretic hormone (an″tĭ-di″u-ret′ik hor′mōn) Hormone released from the posterior lobe of the pituitary gland that enhances the conservation of water by the kidneys; ADH, vasopressin. p. 483

antigen (an′tĭ-jen) A chemical that stimulates cells to produce antibodies. p. 533

antigen-binding site (an′tĭ-jen-bīn′ding sīt) Specialized ends of antibodies that bind specific antigens. p. 626

antigen-presenting cell (an′tĭ-jen-pre-sen′ting cel) The cell that displays the antigen to the cells of the immune system so they can defend the body against that particular antigen. p. 621

antioxidant (an″tĭ-ok′sĭ-dant) A substance that inhibits oxidation of another substance. p. 706

antithrombin (an″tĭ-throm′bin) A substance that inhibits the action of thrombin and thus inhibits blood clotting. p. 530

anus (a′nus) Inferior outlet of the digestive tube. p. 682

aorta (a-or′tah) Major systemic artery that receives blood from the left ventricle. p. 582

aortic body (a-or′tik bod′e) A structure associated with the wall of the aorta that contains a group of chemoreceptors. p. 582

aortic sinus (a-or′tik si′nus) Swelling in the wall of the aorta that contains baroreceptors. p. 582

aortic valve (a-or′tik valv) Flaplike structures in the wall of the aorta near its origin that prevent blood from returning to the left ventricle of the heart. p. 549

apocrine gland (ap′o-krin gland) A type of gland whose secretions contain parts of secretory cells. p. 139

aponeurosis (ap″o-nu-ro′sis) A sheet of connective tissue by which certain muscles are attached to adjacent muscles. p. 278

apoptosis (ap″o-to′-sis) Programmed cell death. p. 94

appendicular (ap″en-dik′u-lar) Pertaining to the upper or lower limbs. p. 12

appendix (ah-pen′diks) A small, tubular appendage of lymphatic tissue that extends outward from the cecum of the large intestine; vermiform appendix. p. 452

aqueous humor (a′kwe-us hu′mor) Watery fluid that fills the anterior cavity of the eye. p. 452

arachnoid granulation (ah-rak′noid gran″u-la′shun) Fingerlike structure that projects from the subarachnoid space of the meninges into blood-filled dural sinuses and reabsorbs cerebrospinal fluid. p. 369

arachnoid mater (ah-rak′noid ma′ter) Delicate, weblike middle layer of the meninges. p. 368

arbor vitae (ar′bor vi′ta) Treelike pattern of white matter in a section of cerebellum. p. 394

areola (ah-re′o-lah) Pigmented region surrounding the nipple of the mammary gland or breast. p. 861

areolar tissue (ah-re′o-lar tish′u) Connective tissue composed mainly of fibers. p. 144

arrector pili muscle (ah-rek′tor pil′i mus′l) Smooth muscle in the skin

associated with a hair follicle. p. 166

arrhythmia (ah-rith′me-ah) Abnormal heart action characterized by a loss of rhythm. p. 561

arteriole (ar-te′re-ōl) A small branch of an artery that communicates with a capillary network. p. 563

arteriosclerosis (ar-te″re-o-sklĕ-ro′sis) Condition in which the walls of arteries thicken and lose their elasticity; hardening of the arteries. p. 571

artery (ar′ter-e) A vessel that transports blood away from the heart. p. 562

arthritis (ar-thri′tis) Joint inflammation. p. 272

articular cartilage (ar-tik′u-lar kar′tĭ-lij) Hyaline cartilage that covers the ends of bones in synovial joints. p. 183

articulation (ar-tik″u-la′shun) The joining of structures at a joint. p. 254

ascending colon (ah-send′ing ko′lon) Portion of the large intestine that passes upward on the right side of the abdomen from the cecum to the lower edge of the liver. p. 681

ascending tract (ah-send′ing trakt) Group of nerve fibers in the spinal cord that transmits sensory impulses upward to the brain. p. 376

ascites (ah-si′tez) Serous fluid accumulation in the abdominal cavity. p. 723

ascorbic acid (as-kor′bik as′id) One of the water-soluble vitamins; vitamin C. p. 712

assimilation (ah-sim″ĭ-la′shun) The action of chemically changing absorbed substances. p. 8

association area (ah-so″se-a′shun a′re-ah) Region of the cerebral cortex related to memory, reasoning, judgment, and emotional feelings. p. 386

astigmatism (ah-stig′mah-tizm) Visual defect due to errors in refraction caused by abnormal curvatures in the surface of the cornea or lens. p. 457

astrocyte (as′tro-sīt) A type of neuroglial cell that connects neurons to blood vessels. p. 347

atherosclerosis (ath″er-o-sklĕ-ro′sis) Condition in which fatty substances accumulate on the inner linings of arteries. p. 530

atmospheric pressure (at″mos-fer′ik presh′ur) Pressure exerted by the weight of the air; about 760 mm of mercury at sea level. p. 745

atom (at′om) Smallest particle of an element that has the properties of that element. p. 4

atomic number (ah-tom′ik num′ber) Number equal to the number of

protons in an atom of an element. p. 39

atomic weight (ah-tom′ik wāt) Number approximately equal to the number of protons plus the number of neutrons in an atom of an element. p. 39

ATP Adenosine triphosphate, the biological energy molecule. p. 73

ATPase Enzyme that causes ATP molecules to release the energy stored in their terminal phosphate bonds. p. 286

atrial natriuretic peptide (a′trē-al na″trē-u-ret′ik pep′tīd) A family of polypeptide hormones that increase sodium excretion. p. 787

atrioventricular bundle (a″tre-o-ven-trik′u-lar bun′dl) Group of specialized fibers that conduct impulses from the atrioventricular node to the Purkinje fibers in the ventricular muscle of the heart; A-V bundle; bundle of His. p. 557

atrioventricular node (a″tre-o-ven-trik′u-lar nōd) Specialized mass of cardiac muscle fibers located in the interatrial septum of the heart; functions in transmitting cardiac impulses from the sinoatrial node to the A-V bundle; A-V node. p. 557

atrioventricular orifice (a″tre-o-ven-trik′u-lar or′i-fis) Opening between the atrium and the ventricle on one side of the heart. p. 545

atrioventricular sulcus (a″tre-o-ven-trik′u-lar sul′kus) A groove on the surface of the heart that marks the division between an atrium and a ventricle. p. 546

atrioventricular valve (a″tre-o-ven-trik′u-lar valv) Cardiac valve located between an atrium and a ventricle. p. 546

atrium (a′tre-um) A chamber of the heart that receives blood from veins (pl., *atria*). p. 545

atrophy (at′ro-fe) A wasting away or decrease in size of an organ or tissue. p. 293

auditory (aw′di-to″re) Pertaining to the ear or to the sense of hearing. p. 437

auditory ossicle (aw′di-to″re os′i-kl) A bone of the middle ear. p. 436

auditory tube (aw′di-to″re toob) The tube that connects the middle ear cavity to the pharynx; eustachian tube. p. 437

auricle (aw′ri-kl) An earlike structure; the portion of the heart that forms the wall of an atrium. p. 434

autoantibody (aw″to-an′tĭ-bod″e) An antibody produced against oneself. p. 633

autocrine (aw′to-krin) A hormone that acts on the same cell that secreted it. p. 468

autoimmunity (aw″to-ĭ-mū′ni-tē)   An immune response against a person's own tissues; autoallergy. p. 633

autonomic nervous system (aw″to-nom′ik ner′vus sis′tem)   Portion of the nervous system that controls the actions of the viscera. p. 339

autoregulation (aw″to-reg″u-la′shun)   Ability of an organ or tissue to maintain a constant blood flow in spite of changing arterial blood pressure. p. 786

autosome (aw′to-sōm)   A chromosome other than a sex chromosome. p. 924

A-V bundle (bun′dl)   Atrioventricular bundle. p. 557

A-V node (nōd)   Atrioventricular node. p. 557

axial (ak′se-al)   Pertaining to the head, neck, and trunk. p. 12

axial skeleton (ak′se-al skel′ĕ-ton)   Portion of the skeleton that supports and protects the organs of the head, neck, and trunk. p. 197

axillary (ak′sĭ-ler″e)   Pertaining to the armpit. p. 24

axon (ak′son)   A nerve fiber; conducts a nerve impulse away from a neuron cell body. p. 338

axonal transport (ak′so-nal trans′port)   Transport of substances from the neuron cell body to an axon terminal. p. 342

B

ball-and-socket joint (bawl-and-sok′et joint)   A bone with a spherical mass on one end joined with a bone possessing a complementary hollow depression. p. 259

baroreceptor (bar″o-re-sep′tor)   Sensory receptor in the blood vessel wall stimulated by changes in pressure (pressoreceptor). p. 423

basal nuclei (bas′al nu′kle-i)   Mass of gray matter located deep within a cerebral hemisphere of the brain. p. 388

basal metabolic rate (ba′sal met″ah-bo′lic rāt)   Rate at which metabolic reactions occur when the body is at rest; BMR. p. 702

base (bās)   A substance that ionizes in water to release hydroxide ions (OH⁻) or other ions that combine with hydrogen ions. p. 46

basement membrane (bās′ment mem′brān)   A layer of nonliving material that anchors epithelial tissue to underlying connective tissue. p. 133

basophil (ba′so-fil)   White blood cell containing cytoplasmic granules that stain with basophilic dye. p. 519

beta oxidation (ba′tah ok″sĭ-da′shun)   Chemical process by which fatty acids are converted to molecules of acetyl which combine with coenzyme A to enter the citric acid cycle. p. 697

beta receptor (ba′tah re-sep′tor)   Receptor on an effector cell membrane that combines mainly with epinephrine and only slightly with norepinephrine. p. 415

bicarbonate buffer system (bi-kar′bo-nāt buf′er sis′tem)   A mixture of carbonic acid and sodium bicarbonate that weakens a strong base and a strong acid, respectively; resists a change in pH. p. 819

bicarbonate ion (bi-kar′bon-āt i′on) HCO_3^-. p. 819

bicuspid tooth (bi-kus′pid tooth)   A premolar that is specialized for grinding hard particles of food. p. 651

bicuspid valve (bi-kus′pid valv)   Mitral valve. p. 546

bile (bīl)   Fluid secreted by the liver and stored in the gallbladder. p. 669

bilirubin (bil″ĭ-roo′bin)   A bile pigment produced from hemoglobin breakdown. p. 517

biliverdin (bil″ĭ-ver′din)   A bile pigment produced from hemoglobin breakdown. p. 517

biochemistry (bi″o-kem′is-tre)   Branch of science dealing with the chemistry of living organisms. p. 38

biofeedback (bi″o-fēd′bak)   Procedure in which electronic equipment is used to help a person learn to consciously control certain visceral responses. p. 428

biotin (bi′o-tin)   A water-soluble vitamin; a member of the vitamin B complex. p. 712

bipolar neuron (bi-po′lar nu′ron)   A nerve cell whose cell body has only two processes, one an axon and the other a dendrite. p. 344

blastocyst (blas′to-sist)   An early stage of prenatal development that consists of a hollow ball of cells. p. 882

blood (blud)   Cells in a liquid matrix that circulate through the heart and vessels carrying oxygen and nutrients throughout the body. p. 148

B lymphocyte (B lim′fo-sīt)   Lymphocyte that reacts against foreign substances in the body by producing and secreting antibodies; B cell. p. 619

BMR   Basal metabolic rate. p. 702

bond (bond)   Connection between atoms in a compound. p. 40

bone (bōn)   Part of the skeleton composed of cells and inorganic, mineral matrix; also a connective tissue. p. 147

brachial (bra′ke-al)   Pertaining to the arm. p. 24

bradycardia (brad″e-kar′de-ah)   An abnormally slow heart rate or pulse rate. p. 562

brainstem (brān′stem)   Portion of the brain that includes the midbrain, pons, and medulla oblongata. p. 381

brain wave (brān wāv)   Recording of fluctuating electrical activity in the brain. p. 396

Broca's area (bro′kahz a′re-ah)   Region of the frontal lobe that coordinates complex muscular actions of the mouth, tongue, and larynx, making speech possible. p. 384

bronchial tree (brong′ke-al trē)   The bronchi and their branches that carry air from the trachea to the alveoli of the lungs. p. 739

bronchiole (brong′ke-ōl)   A small branch of a bronchus within the lung. p. 740

bronchus (brong′kus)   A branch of the trachea that leads to a lung (pl., *bronchi*). p. 739

buccal (buk′al)   Pertaining to the mouth and the inner lining of the cheeks. p. 24

buffer (buf′er)   A substance that can react with a strong acid or base to form a weaker acid or base, and thus resist a change in pH. p. 819

bulbourethral gland (bul″bo-u-re′thral gland)   Gland that secretes a viscous fluid into the male urethra at times of sexual excitement; Cowper's gland. p. 840

bulimia (bu-lim′e-ah)   A disorder of binge eating followed by purging. p. 724

bulk element (bulk el′ĕ-ment)   Basic chemical substance needed in large quantity. p. 38

bursa (bur′sah)   A saclike, fluid-filled structure, lined with synovial membrane, near a joint. p. 258

bursitis (bur-si′tis)   Inflammation of a bursa. p. 272

C

calcitonin (kal″sĭ-to′-nin)   Hormone secreted by the thyroid gland that helps regulate the level of blood calcium. p. 487

calorie (kal′o-re)   A unit used to measure heat energy and the energy contents of foods. p. 702

calorimeter (kal″o-rim′ĕ-ter)   A device used to measure the heat energy content of foods; bomb calorimeter. p. 702

canaliculus (kan″ah-lik′u-lus)   Microscopic canals that connect the lacunae of bone tissue. p. 148

capacitation (kah-pas″i-ta′shun)   Activation of a sperm cell to fertilize an egg cell. p. 841

capillary (kap′ĭ-ler″e)   A small blood vessel that connects an arteriole and a venule. p. 566

carbohydrate (kar″bo-hi′drāt)   An organic compound that contains carbon, hydrogen, and oxygen, with a 2:1 ratio of hydrogen to oxygen atoms. p. 48

carbonic anhydrase (kar-bon′ik an-hi′drās)   Enzyme that catalyzes the reaction between carbon dioxide and water to form carbonic acid. p. 762

carbon monoxide (kar′bon mon-ok′sīd)   A toxic gas that combines readily with hemoglobin to form a relatively stable compound; CO. p. 761

carbaminohemoglobin (kar-bam″ĭ-no-he″mo-glo′bin)   Compound formed by the union of carbon dioxide and hemoglobin. p. 762

carboxypeptidase (kar-bok″se-pep′ti-dās)   A protein-splitting enzyme in pancreatic juice. p. 665

cardiac center (kar′de-ak sen′ter)   A group of neurons in the medulla oblongata that controls heart rate. p. 392

cardiac conduction system (kar′de-ak kon-duk′shun sis′tem)   System of specialized cardiac muscle fibers that conducts cardiac impulses from the S-A node into the myocardium. p. 554

cardiac cycle (kar′de-ak si′kl)   A series of myocardial contractions that constitutes a complete heartbeat. p. 551

cardiac muscle (kar′de-ak mus′el)   Specialized type of muscle tissue found only in the heart. p. 150

cardiac output (kar′de-ak owt′poot)   A measurement calculated by multiplying the stroke volume in milliliters by the heart rate in beats per minute. p. 572

cardiac vein (kar′de-ak vān)   Blood vessel that returns blood from the venules of the myocardium to the coronary sinus. p. 550

cardiovascular (kar″de-o-vas′ku-lar)   Pertaining to the heart and blood vessels. p. 17

carina (kah-ri′nah)   A cartilaginous ridge located between the openings of the right and left bronchi. p. 739

carotene (kar′o-tēn)   A yellow, orange, or reddish pigment in plants and a precursor of vitamin A. p. 706

carotid bodies (kah-rot′id bod′ēz)   Masses of chemoreceptors located in the wall of the internal carotid artery near the carotid sinus. p. 587

carpal (kar′pal)   Bone of the wrist. p. 222

carpus (kar′pus)   The wrist; the wrist bones as a group. p. 222

cartilage (kar′tĭ-lij)   Type of connective tissue in which cells are located

within lacunae and are separated by a semisolid matrix. p. 146

cartilaginous joint (kar″tĭ-laj′ĭ-nus joint)   Two or more bones joined by cartilage. p. 255

catabolism (ka-tab′o-lizm)   Metabolic process that breaks down large molecules into smaller ones; catabolic metabolism. p. 104

catalase (kat′ah-lās)   An enzyme that catalyzes the decomposition of hydrogen peroxide. p. 76

catalyst (kat′ah-list)   A chemical that increases the rate of a chemical reaction, but is not permanently altered by the reaction. p. 45

cataract (kat′ah-rakt)   Loss of transparency of the lens of the eye. p. 462

catecholamine (kat″ĕ-kol′am-in)   A type of organic compound that includes epinephrine and norepinephrine. p. 492

cauda equina (kaw′da ek-wīn′a)   A group of spinal nerves that extends below the distal end of the spinal cord. p. 372

cecum (se′kum)   A pouchlike portion of the large intestine attached to the small intestine. p. 680

celiac (se′le-ak)   Pertaining to the abdomen. p. 24

cell (sel)   The structural and functional unit of an organism. p. 4

cell body (sel bod′e)   Portion of a nerve cell that includes a cytoplasmic mass and a nucleus, and from which the nerve fibers extend. p. 340

cell cycle (sel sī-kl)   The life cycle of a cell consisting of G_1 (growth), S (DNA synthesis), G_2 (growth), and mitosis (division). p. 90

cell membrane (sel mem′brān)   The selectively permeable outer boundary of a cell consisting of a phospholipid bilayer embedded with proteins. p. 62

cellular adhesion molecules (sel′u-lar ad-hee′zhon mol′ĕ-kūlz)   Proteins that guide cellular movement within the body; CAMs. p. 66

cellular immune response (sel′u-lar ĭ-mūn re-spons′)   The body's attack on foreign antigens carried out by T lymphocytes and their secreted products. p. 620

cellular respiration (sel′u-lar res″pĭ-ra′shun)   Process that releases energy from organic compounds in cells. p. 108

cellulose (sel′u-lōs)   A polysaccharide very abundant in plant tissues that human enzymes cannot break down. p. 695

cementum (se-men′tum)   Bonelike material that fastens the root of a tooth into its bony socket. p. 652

central canal (sen′tral kah-nal′)   Tube within the spinal cord that is continuous with the ventricles of the brain and contains cerebrospinal fluid; a tiny channel in bone tissue that contains a blood vessel; Haversian canal. p. 148

central nervous system (sen′tral ner′vus sis′tem)   Portion of the nervous system that consists of the brain and spinal cord; CNS. p. 338

centriole (sen′tre-ōl)   A cellular structure built of microtubules that organizes the mitotic spindle. p. 76

centromere (sen′tro-mēr)   Portion of a chromosome to which spindle fibers attach during mitosis. p. 90

centrosome (sen′tro-sōm)   Cellular organelle consisting of two centrioles. p. 76

cephalic (sĕ-fal′ik)   Pertaining to the head. p. 24

cerebellar cortex (ser″ĕ-bel′ar kor′teks)   The outer layer of the cerebellum. p. 394

cerebellar peduncles (ser″ĕ-bel′ar pe-dung′kl)   A bundle of nerve fibers connecting the cerebellum and the brainstem. p. 394

cerebellum (ser″ĕ-bel′um)   Portion of the brain that coordinates skeletal muscle movement. p. 394

cerebral aqueduct (ser′ĕ-bral ak′wĕ-dukt″)   Tube that connects the third and fourth ventricles of the brain. p. 369

cerebral cortex (ser′ĕ-bral kor′teks)   Outer layer of the cerebrum. p. 384

cerebral hemisphere (ser′ĕ-bral hem′ĭ-sfēr)   One of the large, paired structures that together constitute the cerebrum of the brain. p. 383

cerebrospinal fluid (ser″ĕ-bro-spi′nal floo′id)   Fluid occupying the ventricles of the brain, the subarachnoid space of the meninges, and the central canal of the spinal cord; CSF. p. 368

cerebrovascular accident (ser″ĕ-bro-vas′ku-lar ak′si-dent)   Sudden interruption of blood flow to the brain; a stroke. p. 388

cerebrum (ser′ĕ-brum)   Portion of the brain that occupies the upper part of the cranial cavity and provides higher mental functions. p. 383

cerumen (sĕ-roo′men)   Waxlike substance produced by cells that line the external ear canal. p. 434

cervical (ser′vĭ-kal)   Pertaining to the neck or to the cervix of the uterus. p. 24

cervix (ser′viks)   Narrow, inferior end of the uterus that leads into the vagina. p. 854

chemoreceptor (ke″mo-re-sep′tor)   A receptor that is stimulated by the binding of certain chemicals. p. 423

chemotaxis (ke″mo-tak′sis)   Attraction of leukocytes to chemicals released from damaged cells. p. 627

chief cell (chēf sel)   Cell of gastric gland that secretes various digestive enzymes, including pepsinogen. p. 659

childhood (chīld′hood)   Period of life between infancy and adolescence. p. 909

chloride shift (klo′rīd shift)   Movement of chloride ions from the blood plasma into red blood cells as bicarbonate ions diffuse out of the red blood cells into the plasma. p. 763

cholecystokinin (ko″le-sis″to-ki′nin)   Hormone the small intestine secretes that stimulates release of pancreatic juice from the pancreas and bile from the gallbladder. p. 662

cholesterol (ko-les′ter-ol)   A lipid produced by body cells used to synthesize steroid hormones and excreted into the bile. p. 698

cholinergic fiber (ko″lin-er′jik fi′ber)   A nerve fiber that secretes acetylcholine at the axon terminal. p. 411

chondrocyte (kon′dro-sīt)   A cartilage cell. p. 146

chorion (ko′re-on)   Extraembryonic membrane that forms the outermost covering around a fetus and contributes to formation of the placenta. p. 887

chorionic villi (ko″re-on′ik vil′i)   Projections that extend from the outer surface of the chorion and help attach an embryo to the uterine wall. p. 887

choroid coat (ko′roid kōt)   The vascular, pigmented middle layer of the wall of the eye. p. 450

choroid plexus (ko′roid plek′sus)   Mass of specialized capillaries from which cerebrospinal fluid is secreted into a ventricle of the brain. p. 369

chromatid (kro′mah-tid)   One half of a replicated chromosome or a single unreplicated chromosome. p. 90

chromatin (kro′mah-tin)   DNA and complexed protein that condenses to form chromosomes during mitosis. p. 79

chromatophilic substance (kro″mah-to-fil′ik sub′stans)   Membranous sacs within the cytoplasm of nerve cells that have ribosomes attached to their surfaces; Nissl bodies. p. 340

chromosome (kro′mo-sōm)   Rodlike structure that condenses from chromatin in a cell's nucleus during mitosis. p. 90

chylomicron (kil″o-mi′kron)   A microscopic droplet of fat in the blood following fat digestion. p. 678

chyme (kīm)   Semifluid mass of partially digested food that passes from the stomach to the small intestine. p. 662

chymotrypsin (ki″mo-trip′sin)   A protein-splitting enzyme in pancreatic juice. p. 665

cilia (sil′e-ah)   Microscopic, hairlike processes on the exposed surfaces of certain epithelial cells. p. 76

ciliary body (sil′e-er″e bod′e)   Structure associated with the choroid layer of the eye that secretes aqueous humor and contains the ciliary muscle. p. 451

circadian rhythm (ser″kah-de′an rithm)   A pattern of repeated behavior associated with the cycles of night and day. p. 499

circle of Willis (sir′kl uv wil′is)   An arterial ring located on the ventral surface of the brain. p. 586

circular muscles (ser′ku-lar mus′lz)   Muscles whose fibers are arranged in circular patterns, usually around an opening or in the wall of a tube; sphincter muscles. p. 453

circumduction (ser″kum-duk′shun)   Movement of a body part, such as a limb, so that the end follows a circular path. p. 261

cisternae (sis-ter′ne)   Enlarged portions of the sarcoplasmic reticulum near the actin and myosin filaments of a muscle fiber. p. 282

citric acid cycle (sit′rik as′id si′kl)   A series of chemical reactions that oxidizes certain molecules, releasing energy; Krebs cycle. p. 108

cleavage (klēv′ij)   The early successive divisions of the blastocyst cells into smaller and smaller cells. p. 880

clitoris (kli′to-ris)   Small erectile organ located in the anterior portion of the vulva; corresponds to the penis. p. 855

clone (klōn)   A group of cells that originate from a single cell and are therefore genetically identical. p. 621

CNS   Central nervous system. p. 338

coagulation (ko-ag″u-la′shun)   Blood clotting. p. 527

cocarboxylase (ko″kar-bok′sĭ-lās)   A coenzyme synthesized from thiamine that oxidizes carbohydrates. p. 709

cochlea (kok′le-ah)   Portion of the inner ear that contains the receptors of hearing. p. 437

codominant (ko-dom′ĭ-nant)   Both alleles fully expressed. p. 927

codon (ko′don)   A set of three nucleotides of a messenger RNA molecule corresponding to a particular amino acid. p. 119

coenzyme (ko-en′zīm)   A nonprotein substance that is necessary for the activity of a particular enzyme. p. 107

coenzyme A (ko-en′zīm)   Combines with acetyl to form acetyl coenzyme A which then enters the citric acid cycle. p. 710

cofactor (ko′fak-tor)   A small molecule or ion that must combine with an enzyme for activity. p. 107

collagen (kol′ah-jen)   Protein in the white fibers of connective tissues and in bone matrix. p. 142

collateral (ko-lat′er-al)   A branch of a nerve fiber or blood vessel. p. 342

collecting duct (ko-lek′ting dukt)   In the kidneys, a straight tubule that receives fluid from several nephrons. p. 778

collectins (ko-lek′tinz)   Proteins that provide broad protection against bacteria, yeasts, and some viruses. p. 617

colon (kolon)   Part of the large intestine. p. 681

colony-stimulating factor (ko′le-ne-stim′yu-lay″ting fak′tor)   A protein that stimulates differentiation and maturation of white blood cells. p. 511

color blindness (kul′er blīnd′nes)   An inherited inability to distinguish certain colors. p. 932

colostrum (ko-los′trum)   The first secretion of a woman's mammary glands after she gives birth. p. 904

common bile duct (kom′mon bīl dukt)   Tube that transports bile from the cystic duct to the duodenum. p. 671

compact bone (kom′pakt bōn)   Dense tissue in which cells are arranged in osteons with no apparent spaces. p. 183

compartment (com-part′ment)   A space occupied by a group of muscles, blood vessels, and nerves that is enclosed by fasciae. p. 808

complement (kom′plĕ-ment)   A group of proteins that are activated by the combination of antibody with antigen and enhance the reaction against foreign substances within the body. p. 627

complementary base pair (kom″plĕ-men′tă-re bās pār)   Hydrogen bond joins adenine and thymine or guanine and cytosine in DNA. Adenine bonds to uracil in RNA. p. 116

completely penetrant (kom-plēt′le pen′e-trent)   In genetics, indicates that the frequency of expression of a genotype is 100%. p. 928

complete protein (kom-plēt pro′tēn)   A protein that contains adequate amounts of the essential amino acids to maintain body tissues and to promote normal growth and development. p. 700

compound (kom′pownd)   A substance composed of two or more chemically bonded elements. p. 38

concussion (kon-kush′un)   Loss of consciousness due to a violent blow to the head. p. 388

condom (kon′dum)   A latex sheath used to cover the penis during sexual intercourse; used as a contraceptive and to minimize the risk of transmitting infection. p. 863

conduction (kon-duk′shun)   Movement of body heat into the molecules of cooler objects in contact with the body surface. p. 170

condyle (kon′dīl)   A rounded process of a bone, usually at the articular end. p. 199

condyloid joint (kon′dĭ-loid joint)   A bone with an ovoid projection at one end joined with a bone possessing a complementary elliptical cavity. p. 260

cone (kōn)   Color receptor in the retina of the eye. p. 458

conformation (kon-for-ma′shun)   The three-dimensional form of a protein, determined by its amino acid sequence and attractions and repulsions between amino acids; tertiary structure. p. 54

conjunctiva (kon″junk-ti′vah)   Membranous covering on the anterior surface of the eye. p. 448

connective tissue (kŏ-nek′tiv tish′u)   One of the basic types of tissue that includes bone, cartilage, blood, loose and fibrous connective tissue. p. 141

contraception (kon″trah-sep′shun)   A behavior or device that prevents fertilization. p. 862

contralateral (kon″trah-lat′er-al)   Positioned on the opposite side of something else. p. 21

convection (kon-vek′shun)   The transmission of heat from one substance to another through the circulation of heated air particles. p. 170

convergence (kon-ver′jens)   Nerve impulses arriving at the same neuron. p. 359

convolution (kon″vo-lu′shun)   An elevation on a structure's surface caused by infolding. p. 383

cornea (kor′ne-ah)   Transparent anterior portion of the outer layer of the eye wall. p. 450

coronal (ko-rō′nal)   A plane or section that divides a structure into front and back portions. p. 23

coronary artery (kor′o-na″re ar′ter-e)   An artery that supplies blood to the wall of the heart. p. 549

coronary sinus (kor′o-na″re si′nus)   A large vessel on the posterior surface of the heart into which the cardiac veins drain. p. 550

corpus callosum (kor′pus kah-lo′sum)   A mass of white matter within the brain, composed of nerve fibers connecting the right and left cerebral hemispheres. p. 383

corpus luteum (kor′pus lu′te-um)   Structure that forms from the tissues of a ruptured ovarian follicle and secretes female hormones. p. 859

cortex (kor′teks)   Outer layer of an organ such as the adrenal gland, cerebrum, or kidney. p. 492

cortical nephron (kor′tĭ-kl nef′ron)   A nephron with its corpuscle located in the renal cortex. p. 780

cortisol (kor′ti-sol)   A glucocorticoid secreted by the adrenal cortex. p. 493

costal (kos′tal)   Pertaining to the ribs. p. 24

countercurrent mechanism (kown′ter-kar″ent me′kĕ-nĭ″zm)   Part of the process by which the kidneys concentrate urine. p. 791

covalent bond (ko′va-lent bond)   Chemical bond formed by electron sharing between atoms. p. 43

coxal (kok′sel)   Pertaining to the hip. p. 24

cranial (kra′ne-al)   Pertaining to the cranium. p. 12

cranial cavity (kran′e-al kav′i-te)   A hollow space in the cranium containing the brain. p. 12

cranial nerve (kra′ne-al nerv)   Nerve that arises from the brain. p. 398

cranium (kra′ne-um)   Eight bones of the head. p. 199

creatine phosphate (kre′ah-tin fos′fāt)   A muscle biochemical that stores energy. p. 287

crest (krest)   A ridgelike projection of a bone. p. 199

cretinism (kre′tĭ-nizm)   A condition resulting from lack of thyroid secretion in an infant. p. 487

cricoid cartilage (kri′koid kar′tĭ-lij)   A ringlike cartilage that forms the lower end of the larynx. p. 737

crista ampullaris (kris′tah am-pul-lah′ris)   Sensory organ located within a semicircular canal that functions in the sense of dynamic equilibrium. p. 444

cross-over (kros o′ver)   The exchange of genetic material between homologous chromosomes during meiosis. p. 836

crural (krur′al)   Pertaining to the leg. p. 24

cubital (ku′bi-tal)   Pertaining to the elbow. p. 24

cuspid (kus′pid)   A canine tooth. p. 651

cutaneous (ku-ta′ne-us)   Pertaining to the skin. p. 153

cyanocobalamin (si″ah-no-ko-bal′ah-min)   Vitamin B_{12}. p. 711

cyanosis (si″ah-no′sis)   Bluish skin coloration due to decreased blood oxygen concentration. p. 513

cyclic adenosine monophosphate (sik′lik ah-den′o-sēn mon″o-fos′fāt)   A circularized derivative of ATP that responds to messages entering a cell and triggers the cell's response; cyclic AMP or cAMP. p. 472

cyclosporine (si″klo-spor′in)   A drug that suppresses the action of helper T cells, preventing rejection of transplanted tissue. p. 632

cystic duct (sis′tik dukt)   Tube that connects the gallbladder to the common bile duct. p. 671

cytochrome (si′to-krōm)   Protein within the inner mitochondrial membrane that is an electron carrier in cellular respiration (electron transport chain). p. 976

cytocrine secretion (si′to-krin se-kre′shun)   Transfer of granules of melanin from melanocytes into adjacent epithelial cells. p. 162

cytokine (si′to-kīn)   A type of protein secreted by a T lymphocyte that attacks viruses, virally infected cells, and cancer cells. p. 620

cytoplasm (si′to-plazm)   The contents of a cell excluding the nucleus and cell membrane. p. 70

cytoskeleton (si′to-skel″e-ton)   A system of protein tubules and filaments that reinforces a cell's three-dimensional form and provides scaffolding and transport tracts for organelles. p. 70

cytosol (si′to-sol)   The fluid matrix of the cytoplasm. p. 62

D

deamination (de-am″ĭ-na′shun)   Removing amino groups (NH_2) from amino acids. p. 699

deciduous teeth (de-sid′u-us tēth)   Teeth that are shed and replaced by permanent teeth; primary teeth. p. 650

decomposition (de-kom″po-zish′un)   The breakdown of molecules into simpler compounds. p. 44

deep (dēp)   Far beneath the surface. p. 21

defecation (def″ĕ-ka′shun)   The discharge of feces from the rectum through the anus. p. 684

defensin (di-fen′sin)   Antimicrobial peptide. p. 617

dehydration (de″hi-dra′shun)   Excessive water loss. p. 814

dehydration synthesis (de″hi-dra′shun sin′thĕ-sis)   Anabolic process that joins small molecules by releasing the equivalent of a water molecule; synthesis. p. 104

dendrite (den′drīt) Nerve fiber that transmits impulses toward a neuron cell body. p. 338

densitometer (den″si-tom′e-ter) An instrument used to measure the density of bone tissue. p. 195

dental caries (den′tal kar′ēz) Decalcification and decay of teeth. p. 653

dentin (den′tin) Bonelike substance that forms the bulk of a tooth. p. 652

deoxyhemoglobin (de-ok″sĭ-he′mo-glo′bin) Hemoglobin to which oxygen is not bound. p. 511

deoxyribonucleic acid (dē-ok′si-rī″bo-nu-klē′ik as′id) The genetic material; a double-stranded polymer of nucleotides, each containing a phosphate group, a nitrogenous base (adenine, thymine, guanine, or cytosine), and the sugar deoxyribose; DNA. p. 115

depolarization (de-po″lar-ĭ-za′shun) The loss of an electrical charge on the surface of a cell membrane. p. 350

deposition (dep″ō-zi′shun) The laying down of bony tissue. p. 189

depression (de-presh′un) Downward displacement. p. 263

dermatome (der′mah-tōm) An area of the body supplied by sensory nerve fibers associated with a particular dorsal root of a spinal nerve. p. 404

dermis (der′mis) The thick layer of the skin beneath the epidermis. p. 158

descending colon (de-send′ing ko′lon) Portion of the large intestine that passes downward along the left side of the abdominal cavity to the brim of the pelvis. p. 682

descending tract (de-send′ing trakt) Group of nerve fibers that carries nerve impulses downward from the brain through the spinal cord. p. 376

desmosome (des′mo-sōm) A specialized junction between cells, which serves as a ″spot weld.″ p. 67

detrusor muscle (de-trūz′or mus′l) Muscular wall of the urinary bladder. p. 797

dextrose (dek′strōs) Glucose. p. 48

diabetes insipidus (di″ah-be′tĕz in-sip′ĭ-dus) Extremely copious urine produced due to a deficiency of antidiuretic hormone or lack of ADH response. p. 484

diabetes mellitus (di″ah-be′tēz mel-li′tus) High blood glucose level and glucose in the urine due to a deficiency of insulin. p. 497

dialysis (di-al′ĭ-sis) Separation of smaller molecules from larger ones in a liquid. p. 774

diapedesis (di″ah-pĕ-de′sis) Squeezing of leukocytes between the cells of blood vessel walls. p. 520

diaphragm (di′ah-fram) A sheetlike structure largely composed of skeletal muscle and connective tissue that separates the thoracic and abdominal cavities; also a caplike contraceptive device inserted in the vagina. p. 12

diaphysis (di-af′ĭ-sis) The shaft of a long bone. p. 183

diastole (di-as′to-le) Phase of the cardiac cycle when a heart chamber wall relaxes. p. 551

diastolic pressure (di-a-stol′ik presh′ur) Arterial blood pressure during the diastolic phase of the cardiac cycle. p. 570

diencephalon (di″en-sef′ah-lon) Portion of the brain in the region of the third ventricle that includes the thalamus and hypothalamus. p. 388

differentiation (dif′er-en″she-a′shun) Cell specialization. p. 94

diffusion (dĭ-fu′zhun) Random movement of molecules from a region of higher concentration toward one of lower concentration. p. 80

digestion (di-jest′yun) Breaking down of large nutrient molecules into smaller molecules that can be absorbed; hydrolysis. p. 664

digital (di′ji-tal) Pertaining to the finger. p. 24

dihydrotestosterone (di-hi″dro-tes-tos′ter-ōn) Hormone produced from testosterone that stimulates certain cells of the male reproductive system. p. 845

dipeptide (di-pep′tīd) A molecule composed of two joined amino acids. p. 52

diploid (dip′loid) Body cell with two full sets of chromosomes, in humans 46. p. 921

disaccharide (di-sak′ah-rīd) A sugar produced by the union of two monosaccharides. p. 48

distal (dis′tal) Farther from the trunk or origin; opposite of proximal. p. 21

diuretic (di″u-ret′ik) A substance that increases urine production. p. 483

divergence (di-ver′jens) A spreading apart. p. 360

DNA Deoxyribonucleic acid. p. 115

dominant allele (dom′ĭ-nant ah-lēl′) The form of a gene that is expressed. p. 925

dorsal cavity (dor′sal kav′i-te) A hollow space in the posterior portion of the body containing the cranial cavity and vertebral canal. p. 12

dorsal root (dor′sal rōot) The sensory branch of a spinal nerve by which it joins the spinal cord. p. 404

dorsal root ganglion (dor′sal rōot gang′gle-on) Mass of sensory neuron cell bodies located in the dorsal root of a spinal nerve. p. 404

dorsiflexion (dor″si-flek′shun) Bending the foot upward. p. 261

dorsum (dor′sum) Pertaining to the back surface of a body part. p. 25

ductus arteriosus (duk′tus ar-te″re-o′sus) Blood vessel that connects the pulmonary artery and the aorta in a fetus. p. 899

ductus venosus (duk′tus ven-o′sus) Blood vessel that connects the umbilical vein and the inferior vena cava in a fetus. p. 899

duodenum (du″o-de′num) The first portion of the small intestine that leads from the stomach to the jejunum. p. 673

dural sinus (du′ral si′nus) Blood-filled channel formed by the splitting of the dura mater into two layers. p. 367

dura mater (du′rah ma′ter) Tough outer layer of the meninges. p. 367

dynamic equilibrium (di-nam′ik e″kwĭ-lib′re-um) Maintenance of balance when the head and body are suddenly moved or rotated. p. 444

dystrophin (dis′tre-fin) A protein comprising only 0.002% of the total protein in skeletal muscle that supports the cell membrane. Its absence causes muscular dystrophy. p. 282

E

eccentric contraction (ek-sen′trik kon-trak′shun) Force within a muscle less than that required to lift or move an object. p. 292

eccrine gland (ek′rin gland) Sweat gland that maintains body temperature. p. 167

ECG Electrocardiogram; EKG. p. 559

ectoderm (ek′to-derm) The outermost primary germ layer. p. 886

edema (ĕ-de′mah) Accumulation of fluid within the tissue spaces. p. 84

effector (ĕ-fek′tor) A muscle or gland that responds to stimulation. p. 10

efferent (ef′er-ent) Conducting away from the center. For example, an efferent arteriole conducts blood away from the glomerulus of a nephron. p. 776

ejaculation (e-jak″u-la′shun) Discharge of sperm-containing semen from the male urethra. p. 844

ejaculatory duct (e-jak′u-lah-to″re dukt) Tube, formed by the joining of the vas deferens and the tube from the seminal vesicle, that transports sperm to the urethra. p. 839

elastic cartilage (e-las′tik kar′tĭ-lij) Opaque, flexible connective tissue with branching yellow fibers throughout the matrix. p. 147

elastic fiber (e-las′tik fi′ber)   Yellow, stretchy, threadlike structure found in connective tissue. p. 142

elastin (e-las′tin)   Protein that comprises the yellow, elastic fibers of connective tissue. p. 142

electrocardiogram (e-lek″tro-kar′de-o-gram″)   A recording of the electrical activity associated with the heartbeat; ECG or EKG. p. 559

electrolyte (e-lek′tro-līt)   A substance that ionizes in a water solution. p. 46

electrolyte balance (e-lek′tro-līt bal′ans)   Condition when the quantities of electrolytes entering the body equal those leaving it. p. 813

electron (e-lek′tron)   A small, negatively charged particle that revolves around the nucleus of an atom. p. 38

electron transport chain (e-lek′tron trans′pohrt)   A series of oxidation-reduction reactions that takes high-energy electrons from glycolysis and the citric acid cycle to form water and ATP. p. 108

electrovalent bond (e-lek″tro-va′lent bond)   Chemical bond formed between two ions as a result of the transfer of electrons; ionic bond. p. 43

element (el′ĕ-ment)   A chemical substance with only one type of atom. p. 38

elevation (el-e-vā-shun)   Upward movement of a part of the body. p. 263

embolus (em′bo-lus)   A blood clot or gas bubble that is carried by the blood that may obstruct a blood vessel. p. 530

embryo (em′bre-o)   A prenatal stage of development after germ layers form but before the rudiments of all organs are present. p. 849

embryonic disc (em″brē-on′ik disk)   A flattened area in the cleavage embryo from which the embryo arises. p. 886

emission (e-mish′un)   The movement of sperm cells from the vas deferens into the ejaculatory duct and urethra. p. 843

emphysema (em″fĭ-se′mah)   Abnormal enlargement of the air sacs of the lungs. p. 752

emulsification (e-mul″sĭ-fĭ′ka′shun)   Breaking up of fat globules into smaller droplets by the action of bile salts. p. 671

enamel (e-nam′el)   Hard covering on the exposed surface of a tooth. p. 651

endocardium (en″do-kar′de-um)   Inner lining of the heart chambers. p. 545

endochondral bone (en′do-kon′dral bōn)   Bone that begins as hyaline

cartilage that is subsequently replaced by bone tissue. p. 187

endocrine gland (en″do-krīn gland)   A gland that secretes hormones directly into the blood or body fluids. p. 138

endocytosis (en″do-si-to′sis)   Physiological process by which a cell membrane envelopes a substance and draws it into the cell in a vesicle. p. 85

endoderm (en′do-derm)   The innermost primary germ layer. p. 886

endolymph (en′do-limf)   Fluid contained within the membranous labyrinth of the inner ear. p. 437

endometrium (en″do-me′tre-um)   The inner lining of the uterus. p. 854

endomysium (en″do-mis′e-um)   The sheath of connective tissue surrounding each skeletal muscle fiber. p. 278

endoneurium (en″do-nu′re-um)   Layer of loose connective tissue that surrounds individual nerve fibers. p. 396

endoplasmic reticulum (en-do-plaz′mic rē-tik′u-lum)   Organelle composed of a system of connected membranous tubules and vesicles. p. 70

endorphin (en-dor′fin)   A neuropeptide synthesized in the pituitary gland that suppresses pain. p. 358

endosteum (en-dos′tē-um)   Tissue lining the medullary cavity within the bone. p. 183

endothelium (en″do-the′le-um)   The layer of epithelial cells that forms the inner lining of blood vessels and heart chambers. p. 563

energy (en′er-je)   An ability to cause something to move and thus do work. p. 703

energy balance (en′er-je bal′ans)   When the caloric intake of the body equals its caloric output. p. 703

enkephalin (en-kef′ah-lin)   A neuropeptide that occurs in the brain and spinal cord; it inhibits pain impulses. p. 358

enterogastric reflex (en-ter-o-gas′trik re′fleks)   Inhibition of gastric (stomach) peristalsis and secretions when food enters the small intestine. p. 663

environment (en-vi′ron-ment)   Conditions and elements that make up the surroundings of the body. p. 9

enzyme (en′zīm)   A protein that catalyzes a specific biochemical reaction. p. 52

eosinophil (e″o-sin′o-fil)   White blood cell containing cytoplasmic granules that stain with acidic dye. p. 518

ependyma (ĕ-pen′dĭ-mah)   A membrane composed of neuroglial cells that lines the ventricles of the brain. p. 347

epicardium (ep″ĭ-kar′de-um)   The visceral portion of the pericardium on the surface of the heart. p. 545

epicondyle (ep″ĭ-kon′dīl)   A projection of a bone located above a condyle. p. 199

epidermis (ep″ĭ-der′mis)   Outer epithelial layer of the skin. p. 158

epididymis (ep″ĭ-did′ĭ-mis)   Highly coiled tubule that leads from the seminiferous tubules of the testis to the vas deferens. p. 838

epidural space (ep″ĭ-du′ral spās)   The space between the dural sheath of the spinal cord and the bone of the vertebral canal. p. 367

epigastric region (ep″ĭ-gas′trik re′jun)   The upper middle portion of the abdomen. p. 23

epiglottis (ep″ĭ-glot′is)   Flaplike cartilaginous structure located at the back of the tongue near the entrance to the trachea. p. 737

epimysium (ep″ĭ-mis′e-um)   The outer sheath of connective tissue surrounding a skeletal muscle. p. 278

epinephrine (ep″ĭ-nef′rin)   A hormone the adrenal medulla secretes during times of stress. p. 492

epineurium (ep″ĭ-nu′re-um)   Outermost layer of connective tissue surrounding a nerve. p. 396

epiphyseal plate (ep″ĭ-fiz′e-al plāt)   Cartilaginous layer within the long bone epiphysis that grows. p. 188

epiphysis (e-pif′ĭ-sis)   The end of a long bone. p. 182

epithelial membrane (ep″ĭ-the′le-al mem′brān)   Thin layer of tissue lining a cavity or covering a surface p. 152

epithelial tissue (ep″ĭ-the′le-al tish′u)   One of the basic types of tissue that covers all free body surfaces. p. 133

equilibrium (e″kwĭ-lib′re-um)   A state of balance between two opposing forces. p. 83

erection (ĕ-rek′shun)   The filling of penile tissues with blood, making the structure rigid and elevated. p. 843

erythroblast (ĕ-rith′ro-blast)   An immature red blood cell. p. 514

erythroblastosis fetalis (ĕ-rith″ro-blas-to′sis fe-tal′is)   A life-threatening condition of massive agglutination of the blood in the fetus or neonate due to the mother's anti-Rh antibodies reacting with the baby's Rh-positive red blood cells. p. 537

erythrocyte (ĕ-rith′ro-sīt)   A red blood cell. p. 511

erythropoietin (ĕ-rith″ro-poi′ĕ-tin) (EPo) A kidney hormone that promotes red blood cell formation. p. 514

esophageal hiatus (ĕ-sof″ah-je′al hi-a′tus) Opening in the diaphragm through which the esophagus passes. p. 656

esophagus (ĕ-sof′ah-gus) Tubular portion of the digestive tract that leads from the pharynx to the stomach. p. 656

essential amino acid (ĕ-sen′shal ah-me′no as′id) Amino acid required for health that body cells cannot synthesize in adequate amounts. p. 700

essential fatty acid (ĕ-sen′shal fat′e as′id) Fatty acid required for health that body cells cannot synthesize in adequate amounts. p. 698

essential nutrient (ĕ-sen′shal nu′trē-ent) A nutrient necessary for growth, normal functioning, and maintaining life that the diet must supply because the body cannot synthesize it. p. 695

estrogens (es′tro-jenz) A group of hormones (including estradiol, estrone, and estriol) that stimulate the development of female secondary sex characteristics and produce an environment suitable for fertilization, implantation, and growth of an embryo. p. 857

euploid (u′ploid) Having a balanced set of chromosomes. p. 934

evaporation (e″vap′o-ra-shun) Changing a liquid into a gas. p. 170

eversion (e-ver′zhun) Outward turning movement of the sole of the foot. p. 261

exchange reaction (eks-chānj re-ak′shun) A chemical reaction in which parts of two kinds of molecules trade positions. p. 44

excretion (ek-skre′shun) Elimination of metabolic wastes. p. 18

exocrine gland (ek′so-krin gland) A gland that secretes its products into a duct or onto a body surface. p. 138

exocytosis (eks″o-si-to′sis) Transport of a substance out of a cell in a vesicle. p. 85

expiration (ek″spĭ-ra′shun) Expulsion of air from the lungs. p. 745

extension (ek-sten′shun) Movement increasing the angle between parts at a joint. p. 261

extracellular fluid (ek″strah-sel′u-lar floo′id) Body fluids outside the individual cells. p. 809

extrapyramidal tract (ek″strah-pi-ram′i-dal trakt) Nerve tracts, other than the corticospinal tracts, that transmit impulses from the cerebral cortex to the spinal cord. p. 378

F

facet (fas′et) A small, flattened surface of a bone. p. 199

facilitated diffusion (fah-sil′ĭ-tāt′id dĭ-fu′zhun) Diffusion in which a carrier molecule transports a substance across a cell membrane from a region of higher concentration to a region of lower concentration. p. 81

facilitation (fah-sil″ĭ-tā′shun) Subthreshold stimulation of a neuron makes it more responsive to further stimulation. p. 359

fascia (fash′e-ah) A sheet of fibrous connective tissue that encloses a muscle. p. 278

fascicle (fas′ĭ-k′l) A small bundle of muscle fibers. p. 278

fat (fat) Adipose tissue; or an organic molecule containing glycerol and fatty acids. p. 50

fatty acid (fat′e as′id) A building block of a fat molecule. p. 50

fatty acid oxidase (fat′e as′id ok′si-days″) An enzyme that catalyzes the removal of hydrogen or electrons from a fatty acid molecule. p. 697

feces (fe′sēz) Material expelled from the digestive tract during defecation. p. 685

femoral (fem′or-al) Pertaining to the thigh. p. 25

ferritin (fer′ĭ-tin) An iron-protein complex that stores iron in liver cells. p. 517

fertilization (fer″tĭ-lĭ-za′shun) The union of an egg cell and a sperm cell. p. 836

fetus (fe′tus) A prenatal human after eight weeks of development. p. 894

fever (fe′ver) Elevation of body temperature above normal. p. 618

fibrillation (fi″brĭ-la′shun) Uncoordinated contraction of cardiac muscle fibers. p. 562

fibrin (fi′brin) Insoluble, fibrous protein formed from fibrinogen during blood coagulation. p. 527

fibrinogen (fi-brin′o-jen) Plasma protein converted into fibrin during blood coagulation. p. 524

fibroblast (fi′bro-blast) Cell that produces fibers and other intercellular materials in connective tissues. p. 141

fibrocartilage (fi″bro-kar′ti-lij) Strongest and most durable cartilage; made up of cartilage cells and many collagenous fibers. p. 147

fibrous joint (fi′brus joint) Two or more bones joined by connective tissue containing many fibers. p. 254

filtration (fil-tra′shun) Movement of material through a membrane as a result of hydrostatic pressure. p. 83

filtration pressure (fil-tra′shun presh′ur) Equal to the hydrostatic pressure of the blood entering the glomerulus minus the pressure of the opposing forces (the hydrostatic pressure within the glomerular capsule and the plasma osmotic pressure of the blood in the glomerulus). p. 783

fissure (fish′ūr) A narrow cleft separating parts, such as the lobes of the cerebrum. p. 199

flaccid paralysis (flak′sid pah-ral′ĭ-sis) Total loss of muscle tone when nerve fibers are damaged. p. 380

flagellum (flah-jel′um) Relatively long, motile process that extends from the surface of a cell. p. 76

flexion (flek′shun) Bending at a joint to decrease the angle between bones. p. 261

folacin (fōl′ah-sin) Vitamin that is part of the B complex group necessary for normal cellular synthetic processes; folic acid. p. 711

follicle (fol′ĭ-kl) A pouchlike depression or cavity. p. 858

follicle-stimulating hormone (fol′ĭ-kl stim′u-la″ting hor′mōn) A substance secreted by the anterior pituitary gland that stimulates development of an ovarian follicle in a female or production of sperm cells in a male; FSH. p. 483

follicular cells (fŏ-lik′u-lar selz) Ovarian cells that surround a developing egg cell and secrete female sex hormones. p. 859

fontanel (fon″tah-nel′) Membranous region between certain cranial bones in the skull of a fetus or infant. p. 208

food pyramid (food pēr′ah-mid) A triangular structure divided into sections representing different types and quantities of nutrients required by the body. p. 719

foramen (fo-ra′men) An opening, usually in a bone or membrane (pl., *foramina*). p. 199

foramen magnum (fo-ra′men mag′num) Opening in the occipital bone of the skull through which the spinal cord passes. p. 199

foramen ovale (fo-ra′men o-val′e) Opening in the interatrial septum of the fetal heart. p. 899

forebrain (fōr′brān) The anteriormost portion of the developing brain that gives rise to the cerebrum and basal ganglia. p. 381

formula (fōr′mu-lah) A group of symbols and numbers used to express the composition of a compound. p. 40

fossa (fos′ah) A depression in a bone or other part. p. 199

fovea (fo've-ah)   A tiny pit or depression. p. 199

fovea centralis (fo've-ah sen-tral'is)   Region of the retina, consisting of densely packed cones, that provides the greatest visual acuity. p. 454

fracture (frak'chur)   A break in a bone. p. 192

free radical (frē rad'eh-kel)   Highly reactive by-product of metabolism that can damage tissue. p. 912

frenulum (fren'u-lum)   A fold of tissue that anchors and limits movement of a body part. p. 649

frontal (frun'tal)   Pertaining to the forehead. p. 25

FSH   Follicle-stimulating hormone. p. 483

functional residual capacity (funk'shun-al re-zid'u-al kah-pas'i-te)   Amount of air remaining after a normal quiet respiration. p. 750

functional syncytium (funk'shun-al sin-sish'e-um)   Merging cells performing as a unit; those of the heart are joined electrically. A syncytium lacks cell boundaries, appearing as a multinucleated structure. p. 554

G

galactose (gah-lak'tōs)   A monosaccharide component of the disaccharide lactose. p. 48

gallbladder (gawl'blad-er)   Saclike organ associated with the liver that stores and concentrates bile. p. 671

gamete (gam'ēt)   A sex cell; either an egg cell or a sperm cell. p. 879

ganglion (gang'gle-on)   A mass of neuron cell bodies, usually outside the central nervous system (pl., *ganglia*). p. 399

gastric gland (gas'trik gland)   Gland within the stomach wall that secretes gastric juice. p. 659

gastric juice (gas'trik jo͞os)   Secretion of the gastric glands within the stomach. p. 659

gastric lipase (gas'trik lī'pās)   Fat-splitting enzyme in gastric juice. p. 659

gastrin (gas'trin)   Hormone secreted by the stomach lining that stimulates secretion of gastric juice. p. 661

gastrula (gas'troo-lah)   Embryonic stage following the blastula; cells differentiate into three layers: the endoderm, the mesoderm, and the ectoderm. p. 886

gene (jēn)   Portion of a DNA molecule that encodes the information to synthesize a protein, a control sequence, or tRNA or rRNA. The unit of inheritance. p. 115

genetic code (jĕ-net'ik kōd)   Information for synthesizing proteins that is encoded in the nucleotide sequence of DNA molecules. p. 115

genetic heterogeneity (jĕ-net'ik het″er-o-je-ne'ĭ-te)   Different genotypes that have identical phenotypes. p. 928

genetics (jĕ-net'iks)   The study of heredity. p. 921

genital (jen'i-tal)   Pertaining to the genitalia (internal and external organs of reproduction). p. 25

genome (jeh'nōm)   A complete set of genetic instructions for an organism. p. 115

genomics (je-nom'iks)   Study of genome. p. 921

genotype (je'no-tīp)   The combination of genes of an individual. p. 925

glans penis (glanz pe'nis)   Enlarged mass of corpus spongiosum at the end of the penis; may be covered by the foreskin. p. 841

gliding joint (glīd'eng joint)   Two bones with nearly flat surfaces joined together. p. 260

globin (glo'bin)   The protein portion of a hemoglobin molecule. p. 760

globulin (glob'u-lin)   A type of protein in blood plasma. p. 524

glomerular capsule (glo-mer'u-lar kap'sŭl)   Proximal portion of a renal tubule that encloses the glomerulus of a nephron; Bowman's capsule. p. 776

glomerular filtrate (glo-mer'u-lar fil'trāt)   Liquid that passes out of the glomerular capillaries in the kidney into the glomerular capsule. p. 783

glomerular filtration (glo-mer'u-lar fil-tra'shun)   Process in which blood pressure forces fluid through the glomerular capillaries in the kidney into the glomerular capsule. p. 782

glomerulus (glo-mer'u-lus)   A capillary tuft located within the glomerular capsule of a nephron. p. 776

glottis (glot'is)   Slitlike opening between the true vocal folds or vocal cords. p. 738

glucagon (gloo'kah-gon)   Hormone secreted by the pancreatic islets that releases glucose from glycogen. p. 496

glucocorticoid (gloo″ko-kor'tĭ-koid)   Any one of a group of hormones secreted by the adrenal cortex that influences carbohydrate, fat, and protein metabolism. p. 493

gluconeogenesis (gloo″ko-ne″o-jen'ĕ-sis)   Synthesis of glucose from noncarbohydrates such as amino acids. p. 696

glucose (gloo'kōs)   A monosaccharide in the blood that is the primary source of cellular energy. p. 48

gluteal (gloo'te-al)   Pertaining to the buttocks. p. 25

glycerol (glis'er-ol)   An organic compound that is a building block for fat molecules. p. 50

glycogen (gli'ko-jen)   A polysaccharide that stores glucose in the liver and muscles. p. 50

glycolysis (gli-kol'ĭ-sis)   The breakdown of glucose to pyruvic acid during cellular respiration. p. 108

glycoprotein (gli″ko-pro'te-in)   A compound composed of a carbohydrate combined with a protein. p. 72

glycosuria (gli″ko-sur'e-ah)   Presence of glucose in urine. p. 499

goblet cell (gob'let sel)   An epithelial cell that is specialized to secrete mucus. p. 133

goiter (goi'ter)   An enlarged thyroid gland. p. 482

Golgi apparatus (gol'jē ap″ah-ra'tus)   An organelle that prepares cellular products for secretion. p. 71

Golgi tendon organ (gol″jē ten'dun or'gan)   Sensory receptors in tendons close to muscle attachments that are involved in reflexes that help maintain posture. p. 429

gomphosis (gom-fo'sis)   Type of joint in which a cone-shaped process is fastened in a bony socket. p. 255

gonad (go'nad)   A sex cell-producing organ; an ovary or testis. p. 846

gonadotropin (go-nad″o-trōp'in)   A hormone that stimulates activity in the gonads. p. 845

G protein (g pro'tēn)   Organic compound which activates an enzyme that is bound to the inner surface of the cell membrane, eliciting a signal. p. 472

granulocyte (gran'u-lo-sīt)   A leukocyte that contains granules in its cytoplasm. p. 518

gray matter (grā mat'er)   Region of the central nervous system that generally lacks myelin and thus appears gray. p. 372

gray ramus (grā ra'mus)   A short nerve containing postganglionic axons returning to a spinal nerve. p. 411

groin (groin)   Region of the body between the abdomen and thighs. p. 25

growth (grōth)   Process by which a structure enlarges. p. 876

growth hormone (grōth hōr'mōn)   A hormone released by the anterior lobe of the pituitary gland that promotes the growth of the organism; GH or somatotropin. p. 190

gubernaculum (goo″ber'nak'u-lum)   A structure that guides another structure. p. 832

hair cell (hār sel)   Mechanoreceptor in the inner ear that lies between the basilar membrane and the tectorial membrane and triggers action potentials in fibers of the auditory nerve. p. 439

hair follicle (hār fol′ĭ-kl)   Tubelike depression in the skin in which a hair develops. p. 165

half-life (haf′līf)   The time it takes for one-half of the radioactivity of an amount of an isotope to decay. p. 41

haploid (hap′loid)   Sex cell with a single set of chromosomes, in humans 23. p. 835

hapten (hap′ten)   A small molecule that combines with a larger one; forms an antigen. p. 619

haustra (haws′trah)   Pouches in the wall of the large intestine. p. 682

hematocrit (he-mat′o-krit)   The volume percentage of red blood cells within a sample of whole blood. p. 511

hematoma (he″mah-to′mah)   A mass of coagulated blood within tissues or a body cavity. p. 530

hematopoiesis (hem″ah-to-poi-e′sis)   The production of blood and blood cells; hemopoiesis. p. 194

heme (hēm)   The iron-containing portion of a hemoglobin molecule. p. 760

hemizygous (hem″ĭ-zi′gus)   A gene carried on the Y chromosome in humans. p. 932

hemoglobin (he″mo-glo̱′bin)   Pigment of red blood cells that transports oxygen. p. 194

hemolysis (he-mol′ĭ-sis)   The rupture of red blood cells accompanied by the release of hemoglobin. p. 83

hemorrhage (hem′ŏ-rij)   Loss of blood from the cardiovascular system; bleeding. p. 709

hemostasis (he″mo-sta′sis)   The stoppage of bleeding. p. 526

heparin (hep′ah-rin)   A substance that interferes with the formation of a blood clot; an anticoagulant. p. 142

hepatic (hĕ-pat′ik)   Pertaining to the liver. p. 666

hepatic lobule (hĕ-pat′ik lob′ūl)   A functional unit of the liver. p. 666

hepatic sinusoid (hĕ-pat′ik si′nŭ-soid)   Vascular channel within the liver. p. 666

heredity (hĕ-red′ĭ-te)   The transmission of genetic information from parent to offspring. p. 921

heritable gene therapy (her′ĭ-tah-bl jēn ther′ah-pe)   Manipulation of genes in gametes or a fertilized egg to cure a medical condition. p. 939

heterozygous (het″er-o-zi′gus)   Different alleles in a gene pair. p. 925

hilum (hi′lum)   A depression where vessels, nerves, and other structures (bronchus, ureter, etc.) enter an organ. p. 612

hilus (hi′lus)   Hilum. p. 612

hindbrain (hīnd′brān)   Posteriormost portion of the developing brain that gives rise to the cerebellum, pons, and medulla oblongata. p. 381

hinge joint (hinj joint)   Two bones joined where the convex end of one bone fits into the complementary concave end of another. p. 260

hippocampus (hip″o-kam′pus)   A part of the cerebral cortex where memories form. p. 387

histamine (his′tah-min)   A substance released from stressed cells that promotes inflammation. p. 142

holocrine gland (ho′lo-krin gland)   A gland whose secretion contains entire secretory cells. p. 139

homeostasis (ho″me-o-sta′sis)   A state of equilibrium in which the internal environment of the body remains in the normal range. p. 9

homeostatic mechanism (ho″me-o-stat′ik mek′ah-nizm)   Process used to maintain normal internal environment within the body. p. 10

homozygous (ho″mo-zi′gus)   Identical alleles in a gene pair. p. 925

hormone (hor′mōn)   A substance that an endocrine gland secretes and that the blood or body fluids transport. p. 468

human chorionic gonadotropin (hu′man ko″re-on′ik gon″ah-do-tro′pin)   Hormone, secreted by an embryo, that helps form the placenta; hCG. p. 884

humoral immune response (hu′mor-al i-mūn′ ri-spons′)   Circulating antibodies' destruction of cells bearing foreign (nonself) antigens. p. 621

hyaline cartilage (hi′ah-līn kar′tĭ-lij)   Semitransparent, flexible connective tissue with an ultra-fine fiber matrix. p. 147

hydrogen bond (hi′ dro-jen bond)   A weak bond between a hydrogen atom and an atom of oxygen or nitrogen. p. 44

hydrolysis (hi-drol′ĭ-sis)   Enzymatically adding a water molecule to split a molecule into smaller portions. p. 105

hydrostatic pressure (hi″dro-stat′ik presh′ur)   Pressure exerted by fluids, such as blood pressure. p. 84

hydroxyapatite (hi-drok″se-ap′ah-tīt)   A type of crystalline calcium phosphate found in bone matrix. p. 194

hydroxide ion (hi-drok′sīd i′on)   OH⁻. p. 46

hymen (hi′men)   A membranous fold of tissue that partially covers the vaginal opening. p. 854

hyperextension (hi″per-ek-sten′shun)   Extreme extension; continuing extension beyond the anatomical position. p. 261

hyperglycemia (hi″per-gli-se′me-ah)   Excess blood glucose. p. 497

hyperkalemia (hi″per-kah-le′me-ah)   Elevated blood potassium. p. 818

hypernatremia (hi″per-nah-tre′me-ah)   Elevated blood sodium. p. 818

hyperparathyroidism (hi″per-par″ah-thi′roi-dizm)   Excess secretion of parathyroid hormone. p. 490

hyperpolarization (hi″per-po″lar-i-za′shun)   An increase in the negativity of the resting potential of a cell membrane. p. 350

hypertension (hi″per-ten′shun)   Elevated blood pressure. p. 578

hyperthyroidism (hi″per-thi′roi-dizm)   Oversecretion of thyroid hormones. p. 487

hypertonic (hi″per-ton′ik)   Condition in which a solution contains a greater osmotic pressure than the solution with which it is compared, usually the internal environment (body fluids). p. 83

hypertrophy (hi-per′tro-fe)   Enlargement of an organ or tissue. p. 293

hyperventilation (hi″per-ven″tĭ-la′shun)   Abnormally deep and prolonged breathing. p. 756

hypervitaminosis (hi″per-vi″tah-mĭ-no′sis)   Excessive intake of vitamins. p. 707

hypochondriac region (hi″po-kon′dre-ak re′jun)   The portion of the abdomen on either side of the epigastric region. p. 23

hypodermis (hi″po-der′mis)   Mainly composed of fat, this loose connective tissue layer is directly beneath the dermis; subcutaneous. p. 159

hypogastric region (hi″po-gas′trik re′jun)   The lower middle portion of the abdomen. p. 23

hypoglycemia (hi″po-gli-se′me-ah)   Abnormally low blood glucose. p. 496

hypokalemia (hi″po-kah-le′me-ah)   Low blood potassium. p. 818

hyponatremia (hi″po-nah-tre′me-ah)   Low blood sodium. p. 818

hypoparathyroidism (hi″po-par″ah-thi′roi-dizm)   Undersecretion of parathyroid hormone. p. 490

hypophysis (hi-pof′i-sis)   The pituitary gland. p. 478

hypoproteinemia (hi″po-pro″te-ĭ-ne′me-ah)   Low blood proteins. p. 815

hypothalamus (hi″po-thal′ah-mus)   A portion of the brain located below the thalamus and forming the floor of the third ventricle. p. 390

hypothyroidism (hi″po-thi′roi-dizm) Low secretion of thyroid hormones. p. 487

hypotonic (hi″po-ton′ik) Condition in which a solution contains a lesser osmotic pressure than the solution to which it is compared, usually the internal environment (body fluids). p. 83

hypoxia (hi-pok′se-ah) A deficiency of oxygen in the tissues. p. 513

I

idiotype (id′e-o-tīp′) The parts of an antibody's antigen binding site that are complementary in conformation to a particular antigen. p. 626

ileocecal sphincter (il″e-o-se′kal sfingk′ter) Ring of muscle fibers located at the distal end of the ileum where it joins the cecum. p. 680

ileum (il′e-um) Portion of the small intestine between the jejunum and the cecum. p. 674

iliac region (il′e-ak re′jun) Portion of the abdomen on either side of the lower middle, or hypogastric, region. p. 23

ilium (il′e-um) One of the bones of a coxa or hipbone. p. 224

immunity (ĭ-mu′nĭ-te) Resistance to the effects of specific disease-causing agents. p. 619

immunoglobulin (im″u-no-glob′u-lin) Globular plasma proteins that function as antibodies. p. 620

immunosuppressive drugs (im″u-no-sŭ-pres′iv drugz) Substances that suppress the immune response against transplanted tissue. p. 632

implantation (im″plan-ta′shun) The embedding of a cleavage embryo in the lining of the uterus. p. 883

impulse (im′puls) A wave of depolarization conducted along a nerve fiber or muscle fiber. p. 350

incisor (in-si′zor) One of the front teeth that is adapted for cutting food. p. 651

inclusion (in-kloo′zhun) A mass of lifeless chemical substance within the cytoplasm of a cell. p. 78

incomplete dominance (in″kom-plēt′ do′meh-nents) A heterozygote whose phenotype is intermediate between the phenotypes of the two homozygotes. p. 927

incompletely penetrant (in″kom-plēt′le pen′e-trent) When the frequency of genotype expression is less than 100%. p. 928

incomplete protein (in″kom-plēt′ pro′tēn) A protein that lacks adequate amounts of essential amino acids. p. 700

infancy (in′fan-se) The period of life from the fifth week after birth through the end of the first year. p. 908

infection (in-fek′shun) The invasion and multiplication of microorganisms in body tissues. p. 616

inferior (in-fēr′e-or) Situated below something else; pertaining to the lower surface of a part. p. 21

inflammation (in″flah-ma′shun) A tissue response to stress that is characterized by dilation of blood vessels and fluid accumulation in the affected region. p. 618

infrared ray (in″frah-red′ ra) A form of radiation energy, with wavelengths longer than visible light, by which heat moves from warmer surfaces to cooler surroundings. p. 170

infundibulum (in″fun-dib′u-lum) The stalk attaching the pituitary gland to the base of the brain. p. 390

ingestion (in-jes′chun) The taking of food or liquid into the body by way of the mouth. p. 7

inguinal (ing′gwĭ-nal) Pertaining to the groin region. p. 25

inguinal canal (ing′gwĭ-nal kah-nal′) Passage in the lower abdominal wall through which a testis descends into the scrotum. p. 832

inhibin (in′hib′in) A hormone secreted by cells of the testes and ovaries that inhibits the secretion of FSH from the anterior pituitary gland. p. 845

inorganic (in″or-gan′ik) Chemical substances that lack carbon and hydrogen. p. 47

insertion (in-ser′shun) The end of a muscle attached to a movable part. p. 296

inspiration (in″spĭ-ra′shun) Breathing in; inhalation. p. 745

inspiratory capacity (in-spi′rah-to″re kah-pas′i-te) Volume of air equal to the tidal volume plus the inspiratory reserve volume. p. 750

inspiratory reserve volume (in-spi′rah-to″re re-zerv′ vol′ūm) Maximum amount of air that can be inhaled. p. 749

insula (in′su-lah) A cerebral lobe deep within the lateral sulcus. p. 384

insulin (in′su-lin) A hormone secreted by the pancreatic islets that controls carbohydrate metabolism. p. 496

integumentary (in-teg-u-men′tar-e) Pertaining to the skin and its accessory organs. p. 16

intercalated disc (in-ter″kah-lāt′ed disk) Membranous boundary between adjacent cardiac muscle cells. p. 150

intercellular (in″ter-sel′u-lar) Between cells. p. 67

intercellular junction (in″ter-sel′u-lar junk′shun) Site of union between cells. p. 67

interferon (in″ter-fēr′on) A class of immune system chemicals (cytokines) that inhibit multiplication of viruses and growth of tumors. p. 617

interleukin (in″ter-lu′kin) A class of immune system chemicals (cytokines) with varied effects on the body. p. 518

internal environment (in-ter′nal en-vi′ron-ment) Conditions and elements that make up the inside of the body, surrounding the cells. p. 9

interneuron (in″ter-nu′ron) A neuron located between a sensory neuron and a motor neuron; intercalated; internuncial, or association neuron. p. 345

interphase (in′ter-fāz) Period between two cell divisions when a cell is carrying on its normal functions and prepares for division. p. 90

interstitial cell (in″ter-stish′al sel) A hormone-secreting cell between the seminiferous tubules of the testis. p. 833

interstitial fluid (in″ter-stish′al floo′id) Same as intercellular fluid. p. 611

intervertebral disc (in″ter-ver′tĕ-bral disk) A layer of fibrocartilage between the bodies of adjacent vertebrae. p. 209

intestinal gland (in-tes′tĭ-nal gland) Tubular gland at the base of a villus within the intestinal wall. p. 675

intestinal juice (in-tes′tĭ-nal jōōs) The secretion of the intestinal glands containing digestive enzymes. p. 676

intracellular (in″trah-sel′u-lar) Within cells. p. 809

intracellular fluid (in″trah-sel′u-lar floo′id) Fluid within cells. p. 809

intramembranous bone (in″trah-mem′brah-nus bōn) Bone that forms from membranelike layers of primitive connective tissue. p. 186

intrauterine device (in″trah-u′ter-in de-vīs) A solid object placed in the uterine cavity for purposes of contraception; IUD. p. 866

intrinsic factor (in-trin′sik fak′tor) A substance that gastric glands produce to promote absorption of vitamin B$_{12}$. p. 660

inversion (in-ver′zhun) Inward turning movement of the sole of the foot. p. 263

involuntary (in-vol′un-tār″e) Not consciously controlled; functions automatically. p. 150

iodopsin (i″o-dop′sin) A light-sensitive pigment within the cones of the retina. p. 460

ion (i′on) An atom or molecule with an electrical charge. p. 42

ionic bond (i-on′ik bond) A chemical bond formed between two ions by transfer of electrons. p. 43

ipsilateral (ip″sī-lat′er-al) On the same side. p. 21

iris (i′ris) Colored, muscular portion of the eye that surrounds the pupil and regulates its size. p. 452

ischemia (is-ke′me-ah) Deficiency of blood in a body part. p. 560

isometric contraction (i″so-met′rik kon-trak′shun) Muscular contraction in which the muscle length does not change. p. 292

isotonic contraction (i″so-ton′ik kon-trak′shun) Muscular contraction in which the muscle length changes. p. 292

isotonic (i″so-ton′ik) A solution with the same osmotic pressure as the solution with which it is compared, usually the internal environment (body fluids). p. 83

isotope (i′so-tōp) An atom that has the same number of protons as other atoms of an element but has a different number of neutrons in its nucleus. p. 39

IUD An intrauterine device. p. 866

J

jejunum (jĕ-joo′num) Portion of the small intestine located between the duodenum and the ileum. p. 674

joint (joint) The union of two or more bones; articulation. p. 254

joint capsule (joint kap′sul) An envelope, attached to the end of each bone at the joint, enclosing the cavity of a synovial joint. p. 257

juxtaglomerular apparatus (juks″tah-glo-mer′u-lār ap″ah-ra′tus) Structure located in the arteriolar walls near the glomerulus that regulates renal blood flow. p. 778

juxtamedullary nephron (juks″tah-med′u-lār-e nef′ron) A nephron with its corpuscle located near the renal medulla. p. 780

K

karyotype (kar′ē-o-tīp) A chart of the chromosomes arranged in homologous pairs. The human karyotype has 23 chromosome pairs. p. 924

keratin (ker′ah-tin) Protein in epidermis, hair, and nails. p. 159

keratinization (ker″ah-tin″ĭ-za′shun) The process by which cells form fibrils of keratin and harden. p. 159

ketone body (ke′tōn bod′e) Type of compound produced during fat catabolism, including acetone, acetoacetic acid, and betahydroxybutyric acid. p. 697

ketosis (ke″to′sis) An abnormal elevation of ketone bodies in body fluids. p. 824

kilocalorie (kil′o-kal″o-re) One thousand calories. p. 702

kilogram (kil′o-gram) A unit of weight equivalent to 1,000 grams. p. 702

kinase (ki′nas) An enzyme that activates a precursor form of another enzyme by adding a phosphate group. p. 472

Krebs cycle (krebz sī′kl) The citric acid cycle. p. 108

Kupffer cell (koop′fer sel) Large, fixed phagocyte in the liver that removes bacterial cells from the blood. p. 666

kwashiorkor (kwash″e-or′kor) Starvation resulting from a switch from breast milk to food deficient in nutrients. p. 723

L

labor (la′bor) The process of childbirth. p. 902

labyrinth (lab′ĭ-rinth) The system of connecting tubes within the inner ear, including the cochlea, vestibule, and semicircular canals. p. 437

lacrimal gland (lak′rĭ-mal gland) Tear-secreting gland. p. 448

lactase (lak′tās) Enzyme that catalyzes breakdown of lactose into glucose and galactose. p. 677

lactate (lak′tāt) Lactic acid. p. 111

lacteal (lak′te-al) A lymphatic capillary associated with a villus of the small intestine. p. 675

lactic acid (lak′tik as′id) An organic compound formed from pyruvic acid during the anaerobic reactions of cellular respiration. p. 111

lactose (lak′tōs) A disaccharide in milk; milk sugar. p. 48

lacuna (lah-ku′nah) A hollow cavity. p. 148

lamella (lah-mel′ah) A layer of matrix in bone tissue. p. 148

large intestine (lahrj in-tes′tin) The part of the gastrointestinal tract extending from the ileum to the anus; divided into the cecum, colon, rectum, and anal canal. p. 680

laryngopharynx (lah-ring″go-far′ingks) The lower portion of the pharynx near the opening to the larynx. p. 655

larynx (lar′ingks) Structure between the pharynx and trachea that houses the vocal cords. p. 736

latent period (la′tent pe′re-od) Time between the application of a stimulus and the beginning of a response in a muscle fiber. p. 290

lateral (lat′er-al) Pertaining to the side. p. 21

leukocyte (lu′ko-sīt) A white blood cell. p. 518

leukocytosis (lu″ko-si-to′sis) Too many white blood cells. p. 521

leukopenia (lu″ko-pe′ne-ah) Too few leukocytes in the blood. p. 521

lever (lev′er) A simple mechanical device consisting of a rod, fulcrum, weight, and a source of energy that is applied to some point on the rod. p. 191

ligament (lig′ah-ment) A cord or sheet of connective tissue binding two or more bones at a joint. p. 257

limbic system (lim′bik sis′tem) A group of connected structures within the brain that produces emotional feelings. p. 391

linea alba (lin′e-ah al′bah) A narrow band of tendinous connective tissue in the midline of the anterior abdominal wall. p. 312

lingual (ling′gwal) Pertaining to the tongue. p. 649

lipase (lī′pās) A fat-digesting enzyme. p. 677

lipid (lip′id) A fat, oil, or fatlike compound that usually has fatty acids in its molecular structure. p. 50

lipoprotein (lip″o-pro′te-in) A complex of lipid and protein. p. 525

liver (liv′er) A large, dark red organ in the upper part of the abdomen on the right side that detoxifies blood, stores glycogen and fat-soluble vitamins, and synthesizes proteins. p. 666

lobule (lob′ul) A small, well-defined portion of an organ. p. 744

lower esophageal sphincter (loh′er ĕ-sof″ah-je′al sfingk′ter) A ring of muscle at the distal end of the esophagus where it joins the stomach, that prevents food from re-entering the esophagus when the stomach contracts; cardiac sphincter. p. 658

lower limb (loh′er lim) Inferior appendage consisting of the thigh, leg, and foot. p. 227

lumbar (lum′bar) Pertaining to the region of the loins, part of back between the thorax and pelvis. p. 25

lumen (lu′men) Space within a tubular structure such as a blood vessel or intestine. p. 675

luteinizing hormone (lu′te-in-īz″ing hor′mōn) A hormone that the anterior pituitary secretes that controls formation of the corpus luteum in females and testosterone secretion in males; LH (ICSH in males). p. 483

lymph (limf)   Fluid that the lymphatic vessels carry. p. 611

lymph node (limf nōd)   A mass of lymphoid tissue located along the course of a lymphatic vessel. p. 612

lymphocyte (lim′fo-sīt)   A type of white blood cell that provides immunity; B cell or T cell. p. 519

lysosome (li′so-sōm)   Organelle that contains digestive enzymes. p. 74

M

macromineral (mak′ro-min″er-al)   An inorganic substance that is necessary for metabolism and is one of a group that accounts for 75% of the mineral elements within the body; major mineral. p. 714

macromolecule (mak′ro-mol′ĕ-kūl)   A very large molecule. p. 4

macrophage (mak′ro-fāj)   A large phagocytic cell. p. 141

macula (mak′u-lah)   A group of hair cells and supporting cells associated with an organ of static equilibrium. p. 443

macula lutea (mak′u-lah lu′te-ah)   A yellowish depression in the retina of the eye that is associated with acute vision. p. 454

major histocompatibility complex (ma′jŏr his″to-kom-pat″ĭ-bil′ĭ-te kom′pleks)   A cluster of genes that code for cell surface proteins; MHC. p. 621

malabsorption (mal″ab-sorp′shun)   Failure to absorb nutrients following digestion. p. 679

malignant (mah-lig′nant)   The power to threaten life; cancerous. p. 93

malnutrition (mal″nu-trish′un)   Physical symptoms resulting from lack of specific nutrients. p. 721

maltase (mawl′tās)   An enzyme that catalyzes breakdown of maltose into glucose. p. 677

maltose (mawl′tōs)   A disaccharide composed of two glucose molecules. p. 677

mammary (mam′ar-e)   Pertaining to the breast. p. 25

mammillary body (mam′ĭ-lar″e bod′e)   One of two small, rounded bodies posterior to the hypothalamus involved with reflexes associated with the sense of smell. p. 390

marasmus (mah-raz′mus)   Starvation due to profound nutrient deficiency. p. 723

marrow (mar′o)   Connective tissue that occupies the spaces within bones that includes stem cells. p. 183

mast cell (mast sel)   A cell to which antibodies, formed in response to allergens, attach, bursting the cell and releasing allergy mediators, which cause symptoms. p. 141

mastication (mast″ĭ-ka′shun)   Chewing movements. p. 299

matrix (ma′triks)   The intercellular substance of connective tissue. p. 141

matter (mat′er)   Anything that has weight and occupies space. p. 38

meatus (me-a′tus)   A passageway or channel, or the external opening of a passageway. p. 199

mechanoreceptor (mek″ah-no-re-sep′tor)   A sensory receptor that is sensitive to mechanical stimulation such as changes in pressure or tension. p. 423

medial (me′de-al)   Toward or near the midline. p. 21

mediastinum (me″de-ah-sti′num)   Tissues and organs of the thoracic cavity that form a septum between the lungs. p. 12

medulla (mĕ-dul′ah)   The inner portion of an organ. p. 490

medulla oblongata (mĕ-dul′ah ob″long-gah′tah)   Portion of the brain stem between the pons and the spinal cord. p. 392

medullary cavity (med′u-lār″e kav′ĭ-te)   Cavity within the diaphysis of a long bone containing marrow. p. 183

medullary rhythmicity area (med′u-lār″e rith-mi′si-te ār′e-a)   Area of the brainstem that controls the basic rhythm of inspiration and expiration. p. 753

megakaryocyte (meg″ah-kar′e-o-sīt)   A large bone marrow cell that shatters to yield blood platelets. p. 511

meiosis (mi-o′sis)   Cell division that halves the genetic material, resulting in egg and sperm cells (gametes). p. 90

Meissner's corpuscle (mīs′nerz kor′pus-l)   Sensory receptor close to the surface of the skin that is sensitive to light touch. p. 164

melanin (mel′ah-nin)   Dark pigment found in skin and hair. p. 160

melanocyte (mel′ah-no-sīt)   Melanin-producing cell. p. 160

melatonin (mel″ah-to′nin)   A hormone that the pineal gland secretes. p. 499

membrane potential (mem′brān po-ten′shal)   The unequal distribution of positive and negative ions on two sides of a membrane . p. 350

memory cell (mem′o-re sel)   B-lymphocyte or T-lymphocyte produced in a primary immune response that can be activated rapidly if the same antigen is encountered in the future. p. 621

menarche (mĕ-nar′ke)   The first menstrual period. p. 858

meninx (me′-ninks)   Membrane that covers the brain and spinal cord (*pl., meninges*). p. 367

meniscus (men-is′kus)   Fibrocartilage that separates the articulating surfaces of bones in the knee (pl., *menisci*). p. 258

menopause (men′o-pawz)   Termination of the female reproductive cycle. p. 859

menses (men′sēz)   Shedding of blood and tissue from the uterine lining at the end of a female reproductive cycle. p. 858

mental (men′tal)   Pertaining to the mind; pertaining to the chin body region. p. 25

merocrine gland (mer′o-krĭn gland)   Cells of structure remain intact while secreting products formed within the cell. p. 139

mesentery (mes′en-ter″e)   A fold of peritoneal membrane that attaches an abdominal organ to the abdominal wall. p. 674

mesoderm (mez′o-derm)   The middle primary germ layer. p. 886

messenger RNA (mes′in-jer RNA)   RNA that transmits information for a protein's amino acid sequence from the nucleus of a cell to the cytoplasm. p. 118

metabolic pathway (met″ah-bol′ik path′wa)   A series of linked enzymatically controlled chemical reactions. p. 107

metabolic rate (met″ah-bol′ic rāt)   The rate at which biochemicals are synthesized and broken down in cells. p. 702

metabolism (mĕ-tab′o-lizm)   All of the chemical reactions in cells that use or release energy. p. 7

metacarpal (met″ah-kar′pal)   Bone of the hand between the wrist and finger bones. p. 223

metaphase (met′ah-fāz)   Stage in mitosis when chromosomes align in the middle of the cell p. 92

metastasis (mĕ-tas′tah-sis)   The spread of disease from one body region to another; a characteristic of cancer. p. 93

metatarsal (met″ah-tar′sal)   Bone of the foot between the ankle and toe bones. p. 229

microfilament (mi″kro-fil′ah-ment)   A rod of the protein actin or myosin that provides structural support or movement in the cytoplasm. p. 77

microglial cell (mi-krog′le-al sel)   Neuroglial cell that supports neurons and phagocytizes. p. 347

micromineral (mi′kro-min″er-al)   An essential mineral present in a minute amount within the body; trace element. p. 715

microtubule (mi′kro-tu′būl) A hollow rod of the protein tubulin in the cytoplasm. p. 77

microvillus (mi″kro-vil′us) Cylindrical process that extends from some epithelial cell membranes and increases the membrane surface area. (*pl., microvilli*) p. 133

micturition (mik″tu-rish′un) Urination. p. 798

midbrain (mid′brān) A small region of the brainstem between the diencephalon and the pons. p. 392

mineral (min′er-al) Element not found in organic compounds that is essential in human metabolism. p. 712

mineralocorticoid (min″er-al-o-kor′tĭ-koid) A hormone the adrenal cortex secretes to influence electrolyte concentrations in body fluids. p. 493

mitochondrion (mi′to-kon′dre-on) Organelle housing enzymes that catalyze aerobic reactions of cellular respiration (pl., *mitochondria*). p. 72

mitosis (mi-to′sis) Division of a somatic cell to form two genetically identical cells. p. 90

mitral valve (mi′tral valv) Heart valve located between the left atrium and the left ventricle; bicuspid valve. p. 546

mixed nerve (mikst nerv) Nerve that includes both sensory and motor nerve fibers. p. 398

molar (mo′lar) A rear tooth with a flattened surface adapted for grinding food. p. 651

molecular formula (mo-lek′u-lar for′mu-lah) An abbreviation for the number of atoms of each element in a compound. p. 40

molecule (mol′ĕ-kūl) A particle composed of two or more joined atoms. p. 4

monoamine inhibitor (mon″o-am′in in-hib′i-tor) A substance that inhibits the action of the enzyme monoamine oxidase. p. 358

monocyte (mon′o-sīt) A type of white blood cell that is as a phagocyte. p. 519

monogenic (mon-o-jen′ik) Pertaining to a single gene. p. 929

monosaccharide (mon″o-sak′ah-rīd) A single sugar, such as glucose or fructose. p. 48

monosomy (mon′o-so″me) A cell missing one chromosome. p. 934

morula (mor′u-lah) An early stage in prenatal development; a solid ball of cells. p. 880

motor area (mo′tor a′re-ah) A region of the brain from which impulses to muscles or glands originate. p. 384

motor end plate (mo′tor end plāt) Specialized portion of a muscle fiber membrane at a neuromuscular junction. p. 283

motor nerve (mo′tor nerv) A nerve that consists of motor nerve fibers. p. 398

motor neuron (mo′tor nu′ron) A neuron that transmits impulses from the central nervous system to an effector. p. 282

motor unit (mo′tor unit) A motor neuron and its associated muscle fibers. p. 283

mucosa (mu-ko′sah) The membrane that lines tubes and body cavities that open to the outside of the body; mucous membrane. p. 645

mucous cell (mu′kus sel) Glandular cell that secretes mucus. p. 652

mucous membrane (mu′kus mem′brān) Mucosa. p. 152

mucus (mu′kus) Fluid secretion of the mucous cells. p. 153

multiple motor unit summation (mul′tĭ-pl mo′tor u′nit sum-mā′shun) A sustained muscle contraction of increasing strength in response to input from many motor units. p. 291

multipolar neuron (mul″tĭ-po′lar nu′ron) Nerve cell that has many processes arising from its cell body. p. 345

muscle impulse (mus′el im′puls) Impulse that travels along the sarcolemma and into the transverse tubules. p. 284

muscle spindle (mus′el spin′dul) Modified skeletal muscle fiber that can respond to changes in muscle length. p. 428

muscle tissue (mus′el tish′u) Contractile tissue consisting of filaments of actin and myosin, which slide past each other, shortening cells. p. 148

muscle tone (mus′el tōn) The contraction of some fibers in skeletal muscle at any given time. p. 291

mutagen (mu′tah-jen) An agent that can cause mutations. p. 126

mutant (mu′tant) An allele for a certain gene that has been altered from the "normal" condition. p. 925

mutation (mu-ta′shun) A change in a gene. p. 124

myelin (mi′ĕ-lin) Fatty material that forms a sheathlike covering around some nerve fibers. p. 342

myocardium (mi″o-kar′de-um) Muscle tissue of the heart. p. 545

myofibril (mi″o-fi′bril) Contractile fibers within muscle cells. p. 279

myoglobin (mi″o-glo′bin) A pigmented compound in muscle tissue that stores oxygen. p. 288

myogram (mi′o-gram) A recording of a muscular contraction. p. 290

myometrium (mi″o-me′tre-um) The layer of smooth muscle tissue within the uterine wall. p. 854

myoneural junction (mi″o-nu′ral jungk′shun) Site of union between a motor neuron axon and a muscle fiber. p. 283

myopia (mi-o′pe-ah) Nearsightedness. p. 457

myosin (mi′o-sin) A protein that, with actin, contracts and relaxes muscle fibers. p. 279

myxedema (mik″sĕ-de′mah) A deficiency of thyroid hormones in an adult. p. 487

N

nail (nāl) Horny plate at the distal end of a finger or toe. p. 167

nasal cavity (na′zal kav′ĭ-te) Space within the nose. p. 733

nasal concha (na′zal kong′kah) Shell-like bone extending outward from the wall of the nasal cavity; a turbinate bone. p. 733

nasal septum (na′zal sep′tum) A wall of bone and cartilage that separates the nasal cavity into two portions. p. 733

nasopharynx (na″zo-far″ingks) Portion of the pharynx associated with the nasal cavity. p. 655

natural killer cell (nat′u-ral kil′er sel) Lymphocyte that causes an infected or cancerous cell to burst. p. 618

negative feedback (neg′ah-tiv fēd′bak) A mechanism in which build up of a product causes suppression of its synthesis, also a component of physiological systems. p. 10

neonatal (ne″o-na′tal) The first four weeks of life. p. 906

nephron (nef′ron) The functional unit of a kidney, consisting of a renal corpuscle and a renal tubule. p. 776

nerve (nerv) A bundle of nerve fibers. p. 338

nerve fiber (nerv fi′ber) A thin process of a neuron (i.e., axon, dendrite). p. 372

nerve impulse (nerv im′puls) The electrochemical process of depolarization and repolarization along a nerve fiber. p. 338

nerve tract (nerv trakt) A long bundle of nerve fibers within the CNS having the same origin, function, and termination. p. 372

nervous tissue (ner′vus tish′u) Neurons and neuroglial cells composing the brain, spinal cord and nerves. p. 151

neurilemma (nur″ĭ-lem′ah) Sheath on the outside of some nerve fibers formed from Schwann cells. p. 342

neurofibril (nu″ro-fi′bril) Fine cytoplasmic thread that extends from the cell body into the process of a neuron. p. 340

neuroglial cell (nu-ro′gle-ahl sel) Specialized cell of the nervous system that produces myelin, communicates between cells, and maintains the ionic environment, as well as provides other functions. p. 152

neuromodulator (nu″ro-mod′u-lā-tor) A substance that alters a neuron's response to a neurotransmitter. p. 358

neuromuscular junction (nu″ro-mus′ku-lar jungk′shun) Point of contact between a nerve cell and muscle cell. p. 283

neuron (nu′ron) A nerve cell that consists of a cell body and its processes. p. 151

neuronal pool (nu′ro-nal po͞ol) An accumulation of nerve cells. p. 358

neuropeptide (nu″ro-pep′tīd) A peptide in the brain that functions as a neurotransmitter or neuromodulator. p. 358

neurosecretory cell (nu″ro-se-kre′to-re sel) Cell in the hypothalamus that functions as a neuron at one end but like an endocrine cell at the other, by receiving messages and secreting the hormones ADH and oxytocin. p. 478

neurotransmitter (nu″ro-trans-mit′er) Chemical that an axon end secretes to stimulate a muscle fiber to contract or a neuron to fire an impulse. p. 283

neutral (nu′tral) Neither acidic nor alkaline; pH 7. p. 46

neutron (nu′tron) An electrically neutral particle in an atomic nucleus. p. 38

neutrophil (nu′tro-fil) A type of phagocytic leukocyte. p. 518

niacin (ni′ah-sin) A vitamin of the B-complex group; nicotinic acid. p. 709

niacinamide (ni″ah-sin′ah-mīd) The physiologically active form of niacin. p. 709

nicotinic acid (nik″o-tin′ik as′id) Niacin. p. 709

nitrogen balance (ni′tro-jen bal′ans) Condition in which the quantity of nitrogen ingested equals the quantity excreted. p. 701

node of Ranvier (nōd of Ron′vee-ay) A short region of exposed (unmyelinated) axon between Schwann cells on neurons of the peripheral nervous system. p. 343

nodule (nod′ul) A small mass of tissue detected by touch. p. 613

nondisjunction (non″dis-jungk′shun) Failure of a pair of chromosomes to separate during meiosis. p. 934

nonheritable gene therapy (non-her′i-tah-bl jēn ther′ah-pe) Manipulation of genes in somatic cells to correct the effects of a mutation. p. 939

nonprotein nitrogenous substance (non-pro′tēn ni-troj′ĕ-nus sub′stans) A substance, such as urea or uric acid, that contains nitrogen but is not a protein. p. 525

norepinephrine (nor″ep-ĭ-nef′rin) A neurotransmitter released from the axons of some nerve fibers. p. 492

normal range (nor′mal rānj) Measurements or values obtained from a statistical sample of the healthy population for reference or comparison. p. 12

nuclear envelope (nu′kle-ar en′vĕ-lōp) Membrane surrounding the cell nucleus and separating it from the cytoplasm. p. 78

nuclear pore (nu′kle-ar pōr) A protein-lined channel in the nuclear envelope. p. 79

nuclease (nu′kle-ās) An enzyme that catalyzes decomposition of nucleic acids. p. 665

nucleic acid (nu-kle′ik as′id) A substance composed of bonded nucleotides; RNA or DNA. p. 54

nucleolus (nu-kle′o-lus) A small structure within the cell nucleus that contains RNA and proteins (pl., *nucleoli*). p. 79

nucleoplasm (nu′kle-o-plazm″) The contents of the cell nucleus. p. 79

nucleotide (nu′kle-o-tīd″) A building block of a nucleic acid molecule, consisting of a sugar, a nitrogenous base, and a phosphate group. p. 54

nucleus (nu′kle-us) A cellular organelle enclosed by a double-layered, porous membrane and containing DNA; the dense core of an atom that is composed of protons and neutrons (pl., *nuclei*). p. 38

nutrient (nu′tre-ent) A chemical that the body requires from the environment. p. 694

nutrition (nu-trish′un) The study of the sources, actions, and interactions of nutrients. p. 694

O

obesity (o-bēs′ĭ-te) Excess adipose tissue; exceeding desirable weight by more than 20%. p. 703

occipital (ok-sip′i-tal) Pertaining to the lower, back portion of the head. p. 25

olfactory (ol-fak′to-re) Pertaining to the sense of smell. p. 431

oligodendrocyte (ol″ĭ-go-den′dro-sīt) A type of neuroglial cell that forms myelin. p. 347

oncogene (ong′ko-jēn) A gene that normally controls cell division but when overexpressed leads to cancer. p. 94

oocyte (o′o-sīt) An immature egg cell. p. 848

oogenesis (o′o-jen′ĕ-sis) Differentiation of an egg cell. p. 848

ophthalmic (of-thal′mik) Pertaining to the eye. p. 587

optic (op′tik) Pertaining to the eye. p. 390

optic chiasma (op′tik ki-az′mah) X-shaped structure on the underside of the brain formed by a partial crossing over of optic nerve fibers. p. 390

optic disc (op′tik disk) Region in the retina of the eye where nerve fibers leave to become part of the optic nerve. p. 455

oral (o′ral) Pertaining to the mouth. p. 25

orbital (or′bĭ-tal) Pertaining to the body region of the eyeball; a region, in the atom, containing the electrons. p. 25

organ (or′gan) A structure consisting of a group of tissues with a specialized function. p. 5

organ of Corti (or′gan uv kor′te) An organ in the cochlear duct containing the receptors for hearing. It consists of hair cells and supporting cells. p. 439

organelle (or″gah-nel′) A part of a cell that performs a specialized function. p. 4

organic (or-gan′ik) Carbon-containing molecules. p. 47

organism (or′gah-nizm) An individual living thing. p. 5

organ system (or′gan sis′tem) Organs that function together. p. 5

orgasm (or′gaz-em) The culmination of sexual excitement. p. 843

orifice (or′ĭ-fis) An opening. p. 798

origin (or′ĭ-jin) End of a muscle that attaches to a relatively immovable part. p. 296

oropharynx (o″ro-far′ingks) Portion of the pharynx in the posterior part of the oral cavity. p. 655

osmoreceptor (oz″mo-re-sep′tor) Receptor that is sensitive to changes in the osmotic pressure of body fluids. p. 484

osmosis (oz-mo′sis) Diffusion of water through a selectively permeable membrane in response to a concentration gradient created by an impermeant solute. p. 82

osmotic pressure (oz-mot′ik presh′ur) The amount of pressure needed to stop osmosis; a solution's potential pressure caused by impermeant solute particles in the solution. p. 83

osseous tissue (os′e-us tish′u)   Bone tissue. p. 437

ossification (os″ĭ-fĭ-ka′shun)   The formation of bone tissue. p. 187

osteoblast (os′te-o-blast″)   A bone-forming cell. p. 186

osteoclast (os′te-o-klast″)   A cell that erodes bone. p. 188

osteocyte (os′te-o-sīt)   A mature bone cell. p. 148

osteon (os′te-on)   A cylinder-shaped unit containing bone cells that surround a central canal; Haversian system. p. 148

osteoporosis (os″te-o-po-ro′sis)   A condition in which bones break easily because calcium is removed from them faster than it is replaced. p. 194

otic (o′tik)   Pertaining to the ear. p. 25

otolith (o′to-lith)   A small particle of calcium carbonate associated with the receptors of equilibrium. p. 444

otosclerosis (o″to-sklĕ-ro′sis)   Abnormal formation of spongy bone within the ear that may interfere with the transmission of sound vibrations to hearing receptors. p. 443

oval window (o′val win′do)   Opening between the stapes and the inner ear. p. 436

ovarian (o-va′re-an)   Pertaining to the ovary. p. 848

ovary (o′var-e)   The primary reproductive organ of a female; an egg cell-producing organ. p. 848

ovulation (o″vu-la′shun)   The release of an egg cell from a mature ovarian follicle. p. 852

oxidase (ok′sĭ-dās)   An enzyme that promotes oxidation. p. 976

oxidation (ok″sĭ-da′shun)   Process by which oxygen is combined with another chemical; the removal of hydrogen or the loss of electrons; the opposite of reduction. p. 108

oxidative phosphorylation (ok′sĭ-da-tiv fos″fo-ri-la′shun)   The process of transferring electrons to form a high-energy phosphate bond by introducing a phosphate group to ADP and forming ATP. p. 977

oxygen debt (ok′sĭ-jen det)   The amount of oxygen that must be supplied following physical exercise to convert accumulated lactic acid to glucose. p. 289

oxyhemoglobin (ok″sĭ-he″mo-glo′bin)   Compound formed when oxygen combines with hemoglobin. p. 511

oxytocin (ok″sĭ-to′sin)   Hormone released by the posterior lobe of the pituitary gland that contracts smooth muscles in the uterus and mammary glands. p. 901

P

pacemaker (pās′māk-er)   Mass of specialized cardiac muscle tissue that controls the rhythm of the heartbeat; the sinoatrial node. p. 557

Pacinian corpuscle (pah-sin′e-an kor′pusl)   Nerve endings deep in the dermis providing perception of pressure. p. 164

packed cell volume (pakt sel vol′ūm)   The number of red cells in milliliters per 100 mL of centrifuged blood. p. 511

pain receptor (pān re″sep′tor)   Sensory nerve ending associated with the feeling of pain. p. 423

palate (pal′at)   The roof of the mouth. p. 649

palatine (pal′ah-tīn)   Pertaining to the palate. p. 649

palmar (pahl′mar)   Pertaining to the palm of the hand. p. 25

pancreas (pan′kre-as)   Glandular organ in the abdominal cavity that secretes hormones and digestive enzymes. p. 496

pancreatic (pan″kre-at′ik)   Pertaining to the pancreas. p. 496

pantothenic acid (pan″to-the′nik as′id)   A vitamin of the B-complex group; vitamin B_5. p. 710

papilla (pah-pil′ah)   Tiny, nipplelike projection. p. 432

papillary muscle (pap′ĭ-ler″e mus′el)   Muscle that extends inward from the ventricular walls of the heart and to which the chordae tendineae attach. p. 546

paracrine (par′ah-krin)   A type of endocrine secretion in which the hormone affects nearby cells. p. 468

paradoxical sleep (par″ah-dok′se-kal slēp)   Sleep in which some areas of the brain are active, producing dreams and rapid eye movements. p. 393

paralysis (pah-ral′i-sis)   Loss of ability to control voluntary muscular movements, usually due to a disorder of the nervous system. p. 380

paranasal sinus (par″ah-na′zal si-nus)   An air-filled cavity in a cranial bone; lined with mucous membrane and connected to the nasal cavity. p. 736

parasympathetic division (par″ah-sim″pah-thet′ik dĭ-vizh′un)   Portion of the autonomic nervous system that arises from the brain and sacral region of the spinal cord. p. 409

parathyroid gland (par″ah-thi′roid gland)   One of four small endocrine glands embedded in the posterior portion of the thyroid gland. p. 489

parathyroid hormone (par″ah-thi′roid hor′mōn)   Hormone secreted by the parathyroid glands that helps regulate the level of blood calcium and phosphate ions; PTH. p. 489

paravertebral ganglia (par″ah-ver′tĕ-bral gang′gle-ah)   Sympathetic ganglia that form chains along the sides of the vertebral column. p. 410

parietal (pah-ri′ĕ-tal)   Pertaining to the wall of an organ or cavity. p. 14

parietal cell (pah-ri′ĕ-tal sel)   Cell of a gastric gland that secretes hydrochloric acid and intrinsic factor. p. 659

parietal pleura (pah-ri′ĕ-tal ploo′rah)   Membrane that lines the inner wall of the thoracic cavity. p. 14

parotid glands (pah-rot′id glandz)   Large salivary glands located on the sides of the face just in front and below the ears. p. 653

partial pressure (par′shal presh′ur)   The pressure one gas produces in a mixture of gases. p. 754

parturition (par″tu-rish′un)   Childbirth. p. 901

patellar (pah-tel′ar)   Pertaining to the kneecap. p. 25

pathogen (path′o-jen)   A disease-causing agent. p. 616

pectoral (pek′tor-al)   Pertaining to the chest. p. 25

pectoral girdle (pek′tor-al ger′dl)   Portion of the skeleton that supports and attaches the upper limbs. p. 218

pedal (ped′al)   Pertaining to the foot. p. 25

pedigree (ped′ĭ-gre)   A chart showing the relationships of relatives and which ones have a particular trait. p. 926

pelvic (pel′vik)   Pertaining to the pelvis. p. 25

pelvic cavity (pel′vik kav′i-te)   Hollow place within the ring formed by the sacrum and coxae. p. 14

pelvic girdle (pel′vik ger′dl)   Portion of the skeleton to which the lower limbs attach. p. 224

pelvic inflammatory disease (pel′vik in-flam′ah-tore dĭ-zēz′)   An ascending infection of the upper female genital tract. p. 868

pelvis (pel′vis)   Bony ring formed by the sacrum and coxae. p. 224

penis (pe′nis)   External reproductive organ of the male through which the urethra passes. p. 841

pepsin (pep′sin)   Protein-splitting enzyme that the stomach gastric glands secrete. p. 659

pepsinogen (pep-sin′o-jen)   Inactive form of pepsin. p. 659

peptidase (pep′tĭ-dās)   An enzyme that catalyzes the breakdown of polypeptides. p. 677

peptide (pep′tīd)   Compound composed of two or more bonded amino acids. p. 52

peptide bond (pep′tīd bond)   Bond that forms between the carboxyl group of one amino acid and the amino group of another. p. 52

perception (per-sep′shun)   Mental awareness of sensory stimulation. p. 423

perforating canal (per′fo-rāt″eng kah-nal′)   A transverse channel that connects central canals within compact bone. p. 184

perforin (per′fo-rin)   Protein released by cytotoxic T cells that attaches to the antigen. p. 618

pericardial (per″ĭ-kar′de-al)   Pertaining to the pericardium. p. 14

pericardium (per″ĭ-kar′de-um)   Serous membrane that surrounds the heart. p. 543

perichondrium (per″ĭ-kon′dre-um)   Layer of fibrous connective tissue that encloses cartilaginous structures. p. 146

perilymph (per-ĭ-limf)   Fluid in the space between the membranous and osseous labyrinths of the inner ear. p. 437

perimetrium (per-ĭ-me′tre-um)   The outer serosal layer of the uterine wall. p. 854

perimysium (per″ĭ-mis′e-um)   Sheath of connective tissue that encloses a bundle of skeletal muscle fibers. p. 278

perineal (per″ĭ-ne′al)   Pertaining to the perineum. p. 25

perineum (per″ĭ-ne′um)   Body region between the scrotum or urethral opening and the anus. p. 842

perineurium (per″ĭ-nu-re-um)   Layer of connective tissue that encloses a bundle of nerve fibers within a nerve. p. 396

periodontal ligament (per″e-o-don′tal lig′ah-ment)   Fibrous membrane that surrounds a tooth and attaches it to the jawbone. p. 652

periosteum (per″e-os′te-um)   Fibrous connective tissue covering the surface of a bone. p. 183

peripheral (pĕ-rif′er-al)   Pertaining to parts near the surface or toward the outside. p. 21

peripheral nervous system (pĕ-rif′er-al ner′vus sis′tem)   The portions of the nervous system outside the central nervous system; PNS. p. 338

peripheral protein (pe-rif′er-al pro′tēn)   A globular protein associated with the inner surface of the cell membrane. p. 66

peripheral resistance (pĕ-rif′er-al re-zis′tans)   Resistance to blood flow due to friction between the blood and the walls of the blood vessels. p. 574

peristalsis (per″ĭ-stal′sis)   Rhythmic waves of muscular contraction in the walls of certain tubular organs. p. 294

peritoneal (per″ĭ-to-ne′al)   Pertaining to the peritoneum. p. 14

peritoneal cavity (per″ĭ-to-ne′al kav′ĭ-te)   The potential space between the parietal and visceral peritoneal membranes. p. 14

peritoneum (per″ĭ-to-ne′um)   A serous membrane that lines the abdominal cavity and encloses the abdominal viscera. p. 14

peritubular capillary (per′ĭ-tu′bu-lar kap′ĭ-ler″e)   Capillary that surrounds a renal tubule and functions in reabsorption and secretion during urine formation. p. 781

permeable (per′me-ah-bl)   Open to passage or penetration. p. 81

peroxisome (pĕ-roks′ĭ-sōm)   Membranous cytoplasmic vesicle that contains enzymes that catalyze reactions that produce and decompose hydrogen peroxide. p. 74

pH scale (pH skāl)   A shorthand notation for the hydrogen ion concentration used to indicate the acidic or alkaline condition of a solution; values range from 0 to 14. p. 46

phagocyte (fag′o-sīt)   Cell that ingests particulate matter. p. 85

phagocytosis (fag″o-si-to′sis)   Process by which a cell engulfs and digests solid substances. p. 85

phalanx (fa′langks)   A bone of a finger or toe (pl., *phalanges*). p. 223

pharynx (far′ingks)   Portion of the digestive tube between the mouth and the esophagus. p. 655

phenotype (fe′no-tīp)   The appearance of an individual due to the action of a particular set of genes. p. 925

phosphate buffer system (fos′fāt buf′er sis′tem)   Consists of a mix of sodium monohydrogen phosphate and sodium dihydrogen phosphate to weaken a strong base and a strong acid, respectively; to resist changes in pH. p. 819

phospholipid (fos″fo-lip′id)   A lipid that contains two fatty acid molecules and a phosphate group combined with a glycerol molecule. p. 50

phosphorylation (fos″for-ĭ-la′shun)   Metabolic process that adds a phosphate to an organic molecule. p. 108

photoreceptor (fo″to-re-sep′tor)   A nerve ending that is sensitive to light energy. p. 423

physiology (fiz″e-ol′o-je)   The branch of science that studies body functions. p. 3

pia mater (pi′ah ma′ter)   Inner layer of meninges that encloses the brain and spinal cord. p. 368

pineal gland (pin′e-al gland)   A small structure in the central part of the brain that secretes the hormone melatonin, which controls certain biological rhythms. p. 390

pinocytosis (pin″o-si-to′sis)   Process by which a cell engulfs droplets of fluid from its surroundings. p. 85

pituitary gland (pĭ-tu′ĭ-tār″e gland)   Endocrine gland attached to the base of the brain that consists of anterior and posterior lobes; the hypophysis. p. 478

pivot joint (piv′ut joint)   The end of a bone moving within a ring formed by another bone and connective tissue. p. 260

placenta (plah-sen′tah)   Structure that attaches the fetus to the uterine wall, providing for delivery of nutrients to and removal of wastes from the fetus. p. 891

placental lactogen (plah-sen′tahl lak′to-jen)   Hormone secreted by the placenta to inhibit maternal insulin activity during pregnancy. p. 884

plantar (plan′tar)   Pertaining to the sole of the foot. p. 25

plantar flexion (plan′tar flek′shun)   Bending the foot downward. p. 261

plasma (plaz′mah)   Fluid portion of circulating blood. p. 511

plasma cell (plaz′mah sel)   Antibody-producing cell that forms when activated B cells proliferate. p. 620

plasma protein (plaz′mah pro′tēn)   Protein dissolved in blood plasma. p. 523

plasmin (plaz′min)   A protein-splitting enzyme that can digest fibrin in a blood clot. p. 530

platelet (plāt′let)   Cytoplasmic fragment formed in the bone marrow that helps blood clot. p. 148

pleiotropy (plē′o-tro-pē)   A gene that has several expressions (phenotypes). p. 928

pleural (ploo′ral)   Pertaining to the pleura or membranes surrounding the lungs. p. 14

pleural cavity (ploo′ral kav′ĭ-te)   Potential space between pleural membranes. p. 14

pleural membranes (ploo′ral mem′brānz)   Serous membranes that enclose the lungs and line the chest wall. p. 14

plexus (plek′sus)   A network of interlaced nerves or blood vessels. p. 405

pluripotent (ploo-rip′o-tent)   A cell able to develop in any one of several possible ways. p. 94

pneumotaxic area (nu″mo-tax′ik ār′e-a) A portion of the respiratory control center in the pons of the brain. p. 753

PNS Peripheral nervous system p. 338

polar body (po′lar bod′e) Small, nonfunctional cell that is a product of meiosis in the female. p. 848

polar molecule (po′lar mol′ĕ-kūl) A combination of atoms in which the electrical charge is not distributed symmetrically. p. 44

polarization (po″lar-ĭ-za′shun) An electrical charge on a cell membrane surface due to an unequal distribution of positive and negative ions on either side of the membrane. p. 348

polydactyly (pol″e-dak′ti-le) Extra fingers or toes. p. 224

polygenic (pol″ĕ-jēn′ik) Pertaining to several different genes. p. 929

polymorphonuclear leukocyte (pol″e-mor″fo-nu′kle-ar lu′ko-sīt) A white blood cell with an irregularly lobed nucleus. p. 518

polynucleotide (pol″e-noo′-kle-o-tīd) A compound formed by the union of many nucleotides; a nucleic acid. p. 54

polypeptide (pol″e-pep′tīd) A compound formed by the union of many amino acid molecules. p. 52

polyploidy (pol′e-ploi″de) A condition in which a cell has one or more extra sets of chromosomes. p. 934

polysaccharide (pol″e-sak′ah-rīd) A carbohydrate composed of many joined monosaccharides. p. 48

pons (ponz) A portion of the brainstem above the medulla oblongata and below the midbrain. p. 392

popliteal (pop″lĭ-te′al) Pertaining to the region behind the knee. p. 25

positive chemotaxis (poz′ĭ-tiv ke″mo-tak′sis) Movement of a cell toward the greater concentration of a substance. p. 520

positive feedback (poz′ĭ-tiv fēd′bak) Process by which changes cause additional similar changes, producing unstable conditions. p. 528

posterior (pos-tēr′e-or) Toward the back; opposite of anterior. p. 21

postganglionic fiber (pōst″gang-gle-on′ik fi′ber) Autonomic nerve fiber located on the distal side of a ganglion. p. 410

postnatal (pōst-na′tal) After birth. p. 876

postsynaptic neuron (pōst″sĭ-nap′tik nu′ron) One of two adjacent neurons transmitting an impulse; cell situated after the synapse is crossed. p. 354

postsynaptic potential (pōst″sĭ-nap′tik po-ten′shal) Membrane polarization is increased (excitatory) or decreased (inhibitory) in the postsynaptic neuron with repeated stimulation over an excitatory or inhibitory pathway so that the neuron will either fire or have diminished responsiveness. p. 355

precursor (pre-ker′sor) Substance from which another substance forms. p. 705

preganglionic fiber (pre″gang-gle-on′ik fi′ber) Autonomic nerve fiber located on the proximal side of a ganglion. p. 409

pregnancy (preg′nan-se) The condition in which a female has a developing offspring in her uterus. p. 876

prenatal (pre-na′tal) Before birth. p. 876

presbycusis (pres″bĭ-ku′sis) Loss of hearing that accompanies old age. p. 462

presbyopia (pres″be-o′pe-ah) Loss of the eye's ability to accommodate due to declining elasticity in the lens; farsightedness of age. p. 457

presynaptic neuron (pre″sĭ-nap′tik nu′ron) One of two adjacent neurons transmitting an impulse; cell situated before the synapse is crossed. p. 354

primary germ layers (pri′ma-re jerm lā′erz) Three layers (endoderm, mesoderm, and ectoderm) of embryonic cells that develop into specific tissues and organs. p. 886

primary immune response (pri′ma-re ĭ-mūn′ re-spons′) The immune system's response to its first encounter with a foreign antigen. p. 628

primary reproductive organs (pri′ma-re re″pro-duk′tiv or′ganz) Sex cell-producing parts; testes in males and ovaries in females. p. 846

prime mover (prīm mōōv′er) Muscle responsible for a particular body movement. p. 296

product (prod′ukt) Something produced as the result of a chemical reaction. p. 44

profibrinolysin (pro″fi-brĭ-no-li′sin) The inactive form of fibrinolysin. p. 530

progenitor cell (pro-jen′ĭ-tor sel) Parent cell that gives rise to other cells. p. 94

progesterone (pro-jes′tĕ-rōn) A female hormone that the corpus luteum of the ovary and the placenta secrete. p. 857

projection (pro-jek′shun) Process by which the brain causes a sensation to seem to come from the region of the body being stimulated. p. 424

prolactin (pro-lak′tin) Hormone secreted by the anterior pituitary gland that stimulates the production of milk in the mammary glands; PRL. p. 481

pronation (pro-na′shun) Downward or backward rotation of the palm. p. 261

prophase (pro′fāz) Stage of mitosis when chromosomes become visible. p. 90

proprioceptor (pro″pre-o-sep′tor) A nerve ending that senses changes in muscle or tendon tension. p. 423

prostacyclin (pros″tah-si′klin) A substance released from endothelial cells that inhibits platelet adherence. p. 530

prostaglandins (pros′tah-glan′dins) A group of compounds that have powerful, hormonelike effects. p. 471

prostate gland (pros′tāt gland) Gland surrounding the male urethra below the urinary bladder that adds its secretion to semen just prior to ejaculation. p. 839

protein (pro′tēn) Nitrogen-containing organic compound composed of joined amino acid molecules. p. 52

protein buffer system (pro′tēn buf′er sis′tem) The amino acids of a protein accept or donate hydrogen ions to keep the concentration of hydrogen ions in solution constant; resist changes in pH. p. 819

prothrombin (pro-throm′bin) Plasma protein that functions in the formation of blood clots. p. 527

proton (pro′ton) A positively charged particle in an atomic nucleus. p. 38

protraction (pro-trak′shun) A forward movement of a body part. p. 263

proximal (prok′sĭ-mal) Closer to the trunk or origin; opposite of distal. p. 21

pseudostratified (soo″do-strat′ĭ-fīd) A single layer of cells appearing as more than one layer because the nuclei occupy different positions in the cells. p. 135

puberty (pu′ber-te) Stage of development in which the reproductive organs become functional. p. 846

pulmonary (pul′mo-ner″e) Pertaining to the lungs. p. 580

pulmonary circuit (pul′mo-ner″e ser′kit) System of blood vessels that carries blood between the heart and the lungs. p. 580

pulmonary valve (pul′mo-ner″e valv) Valve leading from the right ventricle to the pulmonary trunk (artery); pulmonary semilunar valve. p. 546

pulse (puls) The surge of blood felt through the walls of arteries due to the contraction of the heart ventricles. p. 570

Punnett square (pun′it skwair) Developed by an English geneticist, a grid that demonstrates possible progeny based on parental gametes. p. 926

pupil (pu′pil) Opening in the iris through which light enters the eye. p. 452

Purkinje fibers (pur-kin′je fi′berz) Specialized muscle fibers that conduct the cardiac impulse from the A-V bundle into the ventricular walls. p. 557

pyloric sphincter (pi-lor′ik sfingk′ter) A ring of muscle located between the stomach and the duodenum; controls food entry into the duodenum. p. 659

pyridoxine (pir′ĭ-dok′sēn) A vitamin of the B-complex group; vitamin B_6. p. 710

pyrogen (pi′ro-jen) A biochemical that increases body temperature. p. 618

pyruvic acid (pi-roo′vik as′id) An intermediate product of carbohydrate oxidation. p. 108

R

radiation (ra″de-a′shun) A form of energy that includes visible light, ultraviolet light, and X rays; the means by which body heat is lost in the form of infrared rays. p. 40; 170

radioactive (ra″de-o-ak′tiv) An atom that releases energy at a constant rate. p. 40

rate-limiting enzyme (rāt lim′ĭ-ting en′zīm) An enzyme, usually present in limited amount, that controls the rate of a metabolic pathway by regulating one of its steps. p. 115

reactant (re-ak′tant) Substance entering a chemical reaction. p. 44

receptor (re″sep′tor) A structure, usually protein, at the distal end of a sensory dendrite that can be stimulated. p. 10

receptor-mediated endocytosis (re″sep′tor-me′de-ā-tid en″do-si-to′sis) A selective uptake of molecules into a cell by binding to a specific receptor. p. 87

recessive allele (re-ses′iv ah-lēl) The form of a gene that is not expressed if the dominant form is present. p. 925

recruitment (re-krōōt′ment) Increase in number of motor units activated as intensity of stimulation increases. p. 291

rectum (rek′tum) The terminal end of the digestive tube between the sigmoid colon and the anus. p. 682

red blood cell (red blud sel) A disc-shaped cell, lacking a nucleus, that is packed with the oxygen-carrying molecule hemoglobin; erythrocyte. p. 148

red marrow (red mar′o) Blood cell-forming tissue in spaces within bones. p. 194

red muscle (red mus′el) Slow-contracting postural muscles that contain abundant myoglobin. p. 292

reduction (re-duk′shun) A chemical reaction in which electrons are gained; to properly realign a fractured bone. p. 111

referred pain (re-ferd′ pān) Pain that feels as if it is originating from a part other than the site being stimulated. p. 426

reflex (re′fleks) A rapid, automatic response to a stimulus. p. 373

reflex arc (re′fleks ark) A nerve pathway, consisting of a sensory neuron, interneuron, and motor neuron, that forms the structural and functional bases for a reflex. p. 373

refraction (re-frak′shun) A bending of light as it passes between media of different densities. p. 455

refractory period (re-frak′to-re pe′re-od) Time period following stimulation during which a neuron or muscle fiber will not respond to a stimulus. p. 290

relaxin (re-lak′sin) Hormone from the corpus luteum that inhibits uterine contractions during pregnancy. p. 884

releasing hormone (re-le′sing hor′mōn) A substance secreted by the hypothalamus whose target cells are in the anterior pituitary gland. p. 480

renal (re′nal) Pertaining to the kidney. p. 772

renal corpuscle (re′nal kor′pusl) Part of a nephron that consists of a glomerulus and a glomerular capsule; Malpighian corpuscle. p. 776

renal cortex (re′nal kor′teks) The outer portion of a kidney. p. 773

renal medulla (re′nal mĕ-dul′ah) The inner portion of a kidney. p. 773

renal pelvis (re′nal pel′vis) The cavity in a kidney that channels urine to the ureter. p. 773

renal plasma threshold (re′nal plaz′mah thresh′old) Concentration of a substance in blood at which it begins to be excreted in the urine. p. 788

renal tubule (re′nal tu′būl) Portion of a nephron that extends from the renal corpuscle to the collecting duct. p. 776

renin (re′nin) Enzyme that kidneys release to maintain blood pressure, plasma sodium, and blood volume. p. 493

repair enzyme (re-pār′ en′zīm) Protein that removes mismatched nucleotides from a section of DNA and replaces them with complementary nucleotides. p. 125

replication (rep″lĭ-ka′shun) Production of an exact copy of a DNA sequence. p. 123

repolarization (re-po″lar-ĭ-za′shun) Returning the cell membrane potential to resting potential. p. 352

reproduction (re″pro-duk′shun) Offspring formation. p. 19

reproductive cycle (re″pro-duk′tiv si′kl) Recurring changes in the uterine lining of a woman of reproductive age. p. 858

residual volume (re-zid′u-al vol′ūm) The amount of air remaining in the lungs after the most forceful expiration. p. 750

resorption (re-sorp′shun) Decomposition of a structure as a result of physiological activity. p. 189

respiration (res″pĭ-ra′shun) Cellular process that releases energy from nutrients; breathing. p. 732

respiratory capacity (re-spi′rah-to″re kah-pas′ĭ-te) Combination of any two or more respiratory volumes. p. 750

respiratory center (re-spi′rah-to″re sen′ter) Portion of the brainstem that controls the depth and rate of breathing. p. 393

respiratory cycle (re-spi′rah-to″re si′kl) An inspiration followed by an expiration. p. 749

respiratory membrane (re-spi′rah-to″re mem′brān) Membrane composed of a capillary wall, an alveolar wall, and their respective basement membranes through which blood and inspired air exchange gases. p. 757

respiratory volume (re-spi′rah-to″re vol′ūm) Any one of four distinct quantities of air within the lungs. p. 749

response (re-spons′) The action resulting from a stimulus. p. 375

resting potential (res′ting po-ten′shal) The difference in electrical charge between the inside and outside of an undisturbed nerve cell membrane. p. 350

resting tidal volume (res′ting tīd′al vol′ūm) Amount of air that enters and leaves the lungs during a normal, quiet inspiration. p. 749

reticular fiber (rĕ-tik′u-lar fi′ber) Threadlike structure within a network of like structures in connective tissue. p. 144

reticular formation (rĕ-tik′u-lar fōr-ma′shun) A complex network of nerve fibers within the brainstem that arouses the cerebrum. p. 393

reticulocyte (rĕ-tik'u-lo-sīt) An immature red blood cell that has a network of fibrils in its cytoplasm. p. 514

reticuloendothelial system (rĕ-tik″u-lo-en″do-the'le-al sis'tem) Tissue composed of widely scattered phagocytic cells. p. 619

retina (ret'ĭ-nah) Inner layer of the eye wall that contains the visual receptors. p. 453

retinal (ret'ĭ-nal) A form of vitamin A; retinene. p. 460

retinene (ret'ĭ-nēn) Chemical precursor of rhodopsin, a visual pigment. p. 460

retraction (rĕ-trak'shun) Movement of a part toward the back. p. 263

retroperitoneal (ret″ro-per″ĭ-to-ne'al) Located behind the peritoneum. p. 772

reversible reaction (re-ver'sĭ-b'l re-ak'shun) Chemical reaction in which the end products can change back into the reactants. p. 45

rhodopsin (ro-dop'sin) Light-sensitive pigment in the rods of the retina; visual purple. p. 458

rhythmicity area (rith-mis'ĭ-te a're-ah) A portion of the respiratory control center in the medulla. p. 753

riboflavin (ri″bo-fla'vin) A vitamin of the B-complex group; vitamin B$_2$. p. 709

ribonucleic acid (ri″bo-nu-kle'ik as'id) A single stranded polymer of nucleotides, each containing a phosphate group, a nitrogen base (adenine, uracil, cytosine, or guanine) and the sugar ribose; RNA. p. 118

ribose (ri'bōs) A 5-carbon sugar in RNA. p. 118

ribosomal RNA (ri-bo-sōm'al) A type of RNA that forms part of the ribosome. p. 122

ribosome (ri'bo-sōm) An organelle composed of RNA and protein that is a structural support for protein synthesis. p. 71

RNA Ribonucleic acid. p. 118

rod (rod) A type of light receptor that provides colorless vision. p. 458

rotation (ro-ta'shun) Movement turning a body part on its longitudinal axis. p. 261

round window (rownd win'do) A membrane-covered opening between the inner ear and the middle ear. p. 439

rugae (roo'je) Thick folds in the inner wall of the stomach that disappear when the stomach is distended. p. 658

S

saccule (sak'ūl) A saclike cavity that makes up part of the membranous labyrinth of the inner ear. p. 443

sacral (sa'kral) Pertaining to the five fused (pelvic) vertebrae at the distal end of the spinal column. p. 25

saddle joint (sad'l joint) Two bones joined each with a convex and concave surface that are complementary. p. 260

sagittal (saj'i-tal) A plane or section that divides a structure into right and left portions. p. 21

salivary gland (sal'ĭ-ver-e gland) A gland, associated with the mouth, that secretes saliva. p. 652

salt (sawlt) A compound produced by a reaction between an acid and a base. p. 46

saltatory conduction (sal'tah-tor-e kon-duk'shun) Nerve impulse conduction that seems to jump from one node to the next. p. 354

S-A node (nōd) Sinoatrial node. p. 554

sarcolemma (sar″ko-lem'ah) The cell membrane of a muscle fiber. p. 279

sarcomere (sar'ko-mēr) The structural and functional unit of a myofibril. p. 279

sarcoplasm (sar'ko-plazm) The cytoplasm within a muscle fiber. p. 279

sarcoplasmic reticulum (sar″ko-plaz'mik rĕ-tik'u-lum) Membranous network of channels and tubules within a muscle fiber, corresponding to the endoplasmic reticulum of other cells. p. 279

saturated fatty acid (sat'u-rāt″ed fat'e as'id) Fatty acid molecule that contains as many hydrogen atoms as possible, and therefore has no double bonds between its carbon atoms. p. 50

Schwann cell (shwahn sel) A type of neuroglial cell that surrounds a fiber of a peripheral nerve, forming the neurilemma and myelin. p. 342

sclera (skle'rah) White fibrous outer layer of the eyeball. p. 450

scrotum (skro'tum) A pouch of skin that encloses the testes. p. 841

sebaceous gland (sĕ-ba'shus gland) Gland of the skin that secretes sebum. p. 167

sebum (se'bum) Oily secretion of the sebaceous glands. p. 167

secondary immune response (sek'un-der″e i-mun' re-spons') The immune system's response to subsequent encounters with a foreign antigen. p. 628

secretin (se-kre'tin) Hormone that the small intestine secretes to stimulate the pancreas to release pancreatic juice. p. 665

secretion (se-kre'shun) Substance produced in and released from a gland cell. p. 139

selectively permeable (se-lĕk'tiv-le per'me-ah-b'l) A membrane that allows some molecules through but not others. p. 65

semen (se'men) Fluid containing sperm cells and secretions discharged from the male reproductive tract at ejaculation. p. 840

semicircular canal (sem'ĭ-ser'ku-lar kah-nal') Tubular structure within the inner ear that contains the receptors providing the sense of dynamic equilibrium. p. 437

seminal vesicle (sem″ĭ-nal ves'ĭ-kel) One of a pair of pouches that adds fructose and prostaglandins to sperm as semen forms. p. 839

seminiferous tubule (sem'ĭ-nif'er-us tu'būl) Tubule within the testes where sperm cells form. p. 832

senescence (sĕ-nes'ens) Aging. p. 910

sensation (sen-sa'shun) The brain's awareness of sensory nerve impulses. p. 423

sensory adaptation (sen'so-re ad″ap-ta'shun) Sensory receptors becoming less responsive after constant repeated stimulation. p. 424

sensory area (sen'so-re a're-ah) A portion of the cerebral cortex that receives and interprets sensory nerve impulses. p. 386

sensory nerve (sen'so-re nerv) A nerve composed of sensory nerve fibers. p. 398

sensory neuron (sen'so-re nu'ron) A neuron that transmits an impulse from a receptor to the central nervous system. p. 345

sensory receptor (sen'so-re re″sep'tor) A specialized dendrite that detects a particular stimulus and fires an action potential in response, which is transmitted to the central nervous system. p. 338

serosa (sēr-o'sah) Serous membrane composing the outer layer in the walls of the organs of digestion. p. 646

serotonin (se″ro-to'nin) A vasoconstrictor that blood platelets release when blood vessels break, controlling bleeding. Also a neurotransmitter. p. 523

serous cell (se'rus sel) A glandular cell that secretes a watery fluid with a high enzyme content. p. 652

serous fluid (se'rus floo'id) The secretion of a serous membrane. p. 152

serous membrane (se′rus mem′brān) Membrane that lines a cavity without an opening to the outside of the body. p. 152

serum (se′rum) The fluid portion of coagulated blood. p. 529

sesamoid bone (ses′ah-moid bōn) A round bone within tendons adjacent to joints. p. 182

set point (set point) Target value to be maintained by automatic control within the body. p. 10

sex chromosome (seks kro′mo-sōm) A chromosome that carries genes responsible for the development of characteristics associated with maleness or femaleness; an X or Y chromosome. p. 924

sex-influenced inheritance (seks-in′floo-enst in-her′ĭ-tens) Transmission of a trait that is dominant in one sex but recessive in the other. p. 933

sex-limited inheritance (seks′-lim′it-ed in-her′ĭ-tens) Transmission of a trait expressed in one sex only. p. 933

sex-linked inheritance (seks-linkt in-her′ĭ-tens) Transmission of a trait controlled by a gene on a sex chromosome (X or Y). p. 932

sexually transmitted disease (sek′shoo-ah-le trans-mi′ted dĭ-zēz′) Infection transmitted from one individual to another by direct contact during sexual activity; STD. p. 867

sigmoid colon (sig′moid ko′lon) S-shaped portion of the large intestine between the descending colon and the rectum. p. 682

signal transduction (sig′nahl trans-duk′shun) A series of biochemical reactions that allows cells to receive and respond to messages coming in through the cell membrane. p. 65

simple sugar (sim′pl shoog′ar) A monosaccharide. p. 48

sinoatrial node (si″no-a′tre-al nōd) Specialized tissue in the wall of the right atrium that initiates cardiac cycles; the pacemaker; S-A node. p. 554

sinus (si′nus) A cavity or space in a bone or other body part. p. 199

skeletal muscle (skel′ĕ-tal mus′l) Type of muscle tissue found in muscles attached to bones. p. 148

sliding filament theory (slī′ding fil′eh-ment the′o-re) Muscles contract when the thin (actin) and thick (myosin) filaments move past each other, shortening the skeletal muscle cells. p. 284

small intestine (smawl in-tes′tin) Part of the digestive tract extending from the stomach to the cecum; consisting of the duodenum, jejunum, and ileum. p. 673

smooth muscle (smōōth mus′el) Type of muscle tissue found in the walls of hollow viscera; visceral muscle. p. 150

sodium pump (so′de-um pump) The active transport mechanism that concentrates sodium ions on the outside of a cell membrane. p. 350

solute (sol′ūt) The substance dissolved in a solution. p. 82

solution (so-lu′shun) Homogenous mixture of substances (solutes) within a dissolving medium (solvent). p. 82

solvent (sol′vent) The liquid portion of a solution in which a solute is dissolved. p. 82

somatic (so-mat′ik) Pertaining to the body. p. 339

somatic cell (so-mat′ik sel) Any cell of the body other than the sex cells. p. 921

somatic nervous system (so-mat′-ik ner′vus sis′tem) Motor pathways of the peripheral nervous system that lead to the skin and skeletal muscles. p. 339

somatomedin (so″mah-to-me′din) A substance released from the liver in response to growth hormone that promotes the growth of cartilage. p. 481

somatostatin (so-mat′o-sta′tin) A hormone secreted by the pancreatic islets that inhibits the release of growth hormone. p. 480

somatotropin (so″mah-to-tro′pin) Growth hormone. p. 480

spastic paralysis (spas′tik pah-ral′ĭ-sis) A form of paralysis characterized by an increase in muscular tone without atrophy of the muscles involved. p. 380

special sense (spesh′al sens) Sense that involves receptors associated with specialized sensory organs, such as the eyes and ears. p. 430

species resistance (spe′sēz re-zis′tans) A natural ability of one type of organism to resist infection by pathogens that might cause disease in another type of organism. p. 617

spermatic cord (sper-mat′ik kord) A structure consisting of blood vessels, nerves, the vas deferens, and other vessels extending from the abdominal inguinal ring to the testis. p. 832

spermatid (sper′mah-tid) An intermediate stage in sperm cell formation. p. 836

spermatocyte (sper-mat′o-sīt) An early stage in sperm cell formation. p. 833

spermatogenesis (sper″mah-to-jen′ĕ-sis) Sperm cell production. p. 836

spermatogonium (sper″mah-to-go′ne-um) Undifferentiated spermatogenic cell in the outer portion of a seminiferous tubule. p. 833

sphincter (sfingk′ter) A circular muscle that closes an opening or the lumen of a tubular structure. p. 453

sphygmomanometer (sfĭg′mo-mah-nom′ĕ-ter) Instrument used for measuring blood pressure. p. 572

spinal (spi′nal) Pertaining to the spinal cord or to the vertebral canal. p. 372

spinal cord (spi′nal kord) Portion of the central nervous system extending from the brainstem through the vertebral canal. p. 372

spinal nerve (spi′nal nerv) Nerve that arises from the spinal cord. p. 372

spleen (splēn) A large organ in the upper left region of the abdomen. p. 614

spongy bone (spunj′e bōn) Bone that consists of bars and plates separated by irregular spaces; cancellous bone. p. 183

squamous (skwa′mus) Flat or platelike. p. 133

starch (starch) A polysaccharide that is common in foods of plant origin. p. 48

static equilibrium (stat′ik e′kwĭ-lib′re-um) The maintenance of balance when the head and body are motionless. p. 443

stem cell (stem sel) An unspecialized cell that divides to give rise to another stem cell and a cell that goes on to differentiate. p. 94

stereocilia (ste″re-o-sil′e-ah) The hairlike processes of the hair cells within the organ of Corti. p. 439

stereoscopic vision (ster″e-o-skop′ik vizh′un) Objects perceived as three-dimensional; depth perception. p. 460

sternal (ster′nal) Pertaining to the sternum. p. 25

steroid (ste′roid) A type of organic molecule including complex rings of carbon and hydrogen atoms. p. 470

stimulus (stim′u-lus) A change in the environment that triggers a response from an organism or cell (pl., *stimuli*). p. 283

stomach (stum′ak) Digestive organ between the esophagus and the small intestine. p. 658

strabismus (strah-biz′mus) Lack of visual coordination; crossed eyes. p. 448

stratified (strat′ĭ-fīd) Arranged in layers. p. 135

stratum basale (stra′tum ba′sal) The deepest layer of the epidermis in which the cells divide; stratum germinativum. p. 159

stratum corneum (stra′tum kor′ne-um) Outer horny layer of the epidermis. p. 159

stress (stres) Response to factors perceived as capable of threatening life. p. 500

stressor (stres′or) A factor capable of stimulating a stress response. p. 500

stretch receptor (strech re-sep′ter) Sensory nerve ending that responds to tension. p. 423

stretch reflex (strech re′fleks) Muscle contraction in response to stretching the muscle. p. 429

stroke volume (strōk vol′ūm) The amount of blood the ventricle discharges with each heartbeat. p. 572

structural formula (struk′cher-al fōr′mu-lah) A representation of the way atoms bond to form a molecule, using the symbols for each element and lines to indicate chemical bonds. p. 43

subarachnoid space (sub″ah-rak′noid spās) The space within the meninges between the arachnoid mater and the pia mater. p. 368

subcutaneous (sub″ku-ta′ne-us) Beneath the skin. p. 159

sublingual (sub-ling′gwal) Beneath the tongue. p. 654

submucosa (sub″mu-ko′sah) Layer of connective tissue underneath a mucous membrane. p. 646

substrate (sub′strāt) The substance upon which an enzyme acts. p. 106

sucrase (su′krās) Digestive enzyme that catalyzes the breakdown of sucrose. p. 677

sucrose (soo′krōs) A disaccharide; table sugar. p. 48

sugar (shoog′ar) Sweet carbohydrate. p. 48

sulcus (sul′kus) A shallow groove, such as that between convolutions on the surface of the brain (pl., *sulci*). p. 383

summation (sum-ma′shun) Increased force of contraction by a skeletal muscle fiber when twitches occur so rapidly that the next twitch occurs before the previous twitch relaxes. p. 291

superficial (soo′per-fish′al) Near the surface. p. 21

superior (su-pe′re-or) Structure higher than another structure. p. 21

supination (soo″pĭ-na′shun) Forearm rotation so that the palm faces upward when the arm is outstretched. p. 261

surface tension (ser′fas ten′shun) Force that holds moist membranes together due to the attraction of water molecules. p. 747

surfactant (ser-fak′tant) Substance the lungs produce that reduces the surface tension within the alveoli. p. 747

suture (soo′cher) An immovable joint, such as that between flat bones of the skull. p. 254

sweat gland (swet gland) Exocrine gland in skin that secretes a mixture of water, salt, urea, and other bodily wastes. p. 167

sympathetic nervous system (sim″pah-thet′ik ner′vus sis′tem) Portion of the autonomic nervous system that arises from the thoracic and lumbar regions of the spinal cord. p. 409

symphysis (sim′fĭ-sis) A slightly movable joint between bones separated by a pad of fibrocartilage. p. 256

synapse (sin′aps) The junction between the axon of one neuron and the dendrite or cell body of another neuron. p. 338

synaptic cleft (sĭ-nap′tik kleft) A narrow extracellular space between the presynaptic and postsynaptic neurons. p. 283

synaptic knob (sĭ-nap′tik nob) Tiny enlargement at the end of an axon that secretes a neurotransmitter. p. 342

synaptic potential (sĭ-nap′tik po-ten′shal) Electrical activity generated in the space between two neurons. p. 355

synaptic transmission (sĭ-nap′tik trans-mish′un) Communication of an impulse from one neuron to the next. p. 354

synchondrosis (sin″kon-dro′sis) Type of joint in which bones are united by bands of hyaline cartilage. p. 255

syndesmosis (sin″des-mo′sis) Type of joint in which the bones are united by relatively long fibers of connective tissue. p. 254

syndrome (sin′drōm) A group of symptoms that characterize a disease condition. p. 500

synergist (sin′er-jist) A muscle that assists the action of a prime mover. p. 296

synovial fluid (sĭno′ve-al floo′id) Fluid that the synovial membrane secretes. p. 258

synovial joint (sĭ-no′ve-al joint) A freely movable joint. p. 257

synovial membrane (sĭ-no′ve-al mem′brān) Membrane that forms the inner lining of the capsule of a freely movable joint. p. 258

synthesis (sin′thĕ-sis) Building large molecules from smaller ones that unite. p. 44

systemic circuit (sis-tem′ik ser′kit) The vessels that conduct blood between the heart and all body tissues except the lungs. p. 580

systole (sis′to-le) Phase of the cardiac cycle during which a heart chamber wall contracts. p. 551

systolic pressure (sis-tol′ik presh′ur) Arterial blood pressure during the systolic phase of the cardiac cycle. p. 570

T

tachycardia (tak″e-kar′de-ah) An abnormally rapid heartbeat. p. 562

target cell (tar′get sel) Specific cell on which a hormone exerts its effect. p. 468

tarsal (tahr′sal) Bone located in the area between the foot and leg. p. 229

tarsus (tar′sus) The ankle bones. p. 229

taste bud (tāst bud) Organ containing receptors associated with the sense of taste. p. 432

telophase (tel′o-fāz) Stage in mitosis when newly formed cells separate. p. 92

tendon (ten′don) A cordlike or bandlike mass of white fibrous connective tissue that connects a muscle to a bone. p. 278

teratogen (ter′ah-to-jen) A chemical or other environmental agent that causes a birth defect. p. 892

testis (tes′tis) Primary reproductive organ of a male; a sperm cell-producing organ (pl., *testes*). p. 830

testosterone (tes-tos′tĕ-rōn) Male sex hormone that the interstitial cells of the testes secrete. p. 845

tetanic contraction (tĕ-tan′ik kon-trak′shun) Continuous, forceful muscular contraction without relaxation. p. 291

thalamus (thal′ah-mus) A mass of gray matter at the base of the cerebrum in the wall of the third ventricle. p. 390

thermoreceptor (ther″mo-re-sep′tor) A sensory receptor sensitive to temperature changes; heat and cold receptors. p. 423

thiamine (thi′ah-min) Vitamin B_1. p. 709

thoracic (tho-ras′ik) Pertaining to the chest. p. 12

thoracic cavity (tho-ras′ik kav′i-te) Hollow place within the chest. p. 12

threshold potential (thresh′old po-ten′shal) Level of potential at which an action potential or nerve impulse is produced. p. 350

threshold stimulus (thresh′old stim′u-lus) Stimulation level that must be exceeded to elicit a nerve impulse or a muscle contraction. p. 290

thrombin (throm′bin) Blood-clotting enzyme that catalyzes formation of fibrin from fibrinogen. p. 527

thrombocyte (throm′bo-sīt) A blood platelet. p. 521

thrombocytopenia (throm″bo-si″to-pe′ne-ah) A low number of platelets in the circulating blood. p. 531

thrombus (throm′bus) A blood clot that remains at its site of formation in a blood vessel. p. 530

thymosins (thi′mo-sins) A group of peptides secreted by the thymus gland that increases production of certain types of white blood cells. p. 500

thymus (thi′mus) A glandular organ located in the mediastinum behind the sternum and between the lungs. p. 500

thyroglobulin (thi″ro-glob′u-lin) Biochemical secreted by cells of the thyroid gland that stores thyroid hormones. p. 487

thyroid gland (thi′roid gland) Endocrine gland just below the larynx and in front of the trachea that secretes thyroid hormones. p. 485

thyroid-stimulating hormone (thi′roid-stim″u-lāt′eng har′mōn) Chemical released from the anterior pituitary gland that acts to promote hormone production within the thyroid gland p. 482

thyrotropin (thi″ro-trōp′in) Hormone secreted by the anterior pituitary gland that stimulates the thyroid gland to secrete hormones; TSH. p. 482

thyroxine (thi-rok′sin) A hormone that the thyroid gland secretes; T₄. p. 485

tissue (tish′u) A group of similar cells that performs a specialized function. p. 5

titin (ti′tin) Protein that attaches myosin filaments to z lines. p. 279

T lymphocyte (T lim′fo-sīt) Lymphocyte that interacts directly with antigen-bearing cells and particles and secretes cytokines, producing the cellular immune response; T cell. p. 619

tonsils (ton′silz) Collections of lymphatic tissue in the throat. p. 649

total lung capacity (toh′tal lung kah-pas′i-te) Equal to the vital capacity plus the residual volume. p. 750

totipotent (to-tip′o-tent) Ability of a cell to differentiate into any type of cell. p. 94

trabecula (trah-bek′u-lah) Branching bony plate that separates irregular spaces within spongy bone. p. 183

trace element (trās el′ĕ-ment) Basic chemical substance needed in small quantity. p. 38

trachea (tra′ke-ah) Tubular organ that leads from the larynx to the bronchi. p. 738

transcellular fluid (trans″sel′u-lar floo′-id) A portion of the extracellular fluid, including the fluid within special body cavities. p. 809

transcription (trans-krip′shun) Manufacturing a complementary RNA from DNA. p. 119

transcytosis (trans″si-to′sis) Combination of receptor-mediated endocytosis and exocytosis used to move particles through a cell. p. 88

transferrin (trans′fer′rin) Blood plasma protein that transports iron. p. 517

transfer RNA (trans′fer RNA) RNA molecule that carries an amino acid to a ribosome in protein synthesis. p. 120

translation (trans-la′ shun) Assembly of an amino acid chain according to the sequence of base triplets in an mRNA molecule. p. 119

transverse (trans-vers′) At right angles to the long axis of a part; crosswise. p. 23

transverse colon (trans-vers′ ko′lon) Portion of the large intestine that extends across the abdomen from right to left below the stomach. p. 681

transverse tubule (trans-vers′ tu′būl) Membranous channel that extends inward from a muscle fiber membrane and passes through the fiber. p. 282

triad (tri′ad) Group of three things. p. 282

tricuspid valve (tri-kus′pid valv) Heart valve located between the right atrium and the right ventricle. p. 546

trigger zone (trig′ger zōn) Site on the neuron where an action potential is generated. p. 344

triglyceride (tri-glis′er-īd) A lipid composed of three fatty acids combined with a glycerol molecule. p. 50

triiodothyronine (tri″i-o″do-thi′ro-nēn) A type of thyroid hormone; T₃. p. 485

trisomy (tri′so-me) Condition in which a cell contains three chromosomes of a particular type instead of two. p. 934

trochanter (tro-kan′ter) A broad process on a bone. p. 199

trochlea (trok′le-ah) A pulley-shaped structure. p. 449

trophoblast (trof′o-blast) The outer cells of a blastocyst that help form the placenta and other extraembryonic membranes. p. 882

tropic hormone (trōp′ik hor′mōn) A hormone that has an endocrine gland as its target tissue. p. 477

tropomyosin (tro″po-mi′o-sin) Protein that blocks muscle contraction until calcium ions are present. p. 279

troponin (tro′po-nin) Protein that functions with tropomyosin to block muscle contraction until calcium ions are present. p. 279

trypsin (trip′sin) An enzyme in pancreatic juice that breaks down protein molecules. p. 665

trypsinogen (trip-sin′o-jen) Substance pancreatic cells secrete that is enzymatically cleaved to yield trypsin. p. 665

tubercle (tu′ber-kl) A small, rounded process on a bone. p. 199

tuberosity (tu″bĕ-ros′ĭ-te) An elevation or protuberance on a bone. p. 199

tubular reabsorption (too′bu-lar re″ab-sorp′shun) Process that transports substances out of the renal tubule into the interstitial fluid from which the substances diffuse into peritubular capillaries. p. 782

tubular secretion (too′bu-lar se-kre′shun) Process of substances moving out of the peritubular capillaries into the renal tubule for excretion in the urine. p. 782

tumor (too′mor) A tissue mass formed when cells lose division control. p. 93

tumor suppressor gene (too′mor sŭ-pres′or jēn) Section of DNA that codes for a protein that ordinarily inhibits cell division. p. 94

twitch (twich) A brief muscular contraction followed by relaxation. p. 290

tympanic membrane (tim-pan′ik mem′brān) A thin membrane that covers the auditory canal and separates the external ear from the middle ear; the eardrum. p. 434

U

ultratrace element (ul′trah-trās el′ĕ-ment) Basic chemical substance needed in very small quantity. p. 38

umbilical cord (um-bil′ĭ-kal kord) Cordlike structure that connects the fetus to the placenta. p. 891

umbilical region (um-bil′ĭ-kal re′jun) The central portion of the abdomen. p. 23

umbilicus (um-bil′ĭ-kus) Region to which the umbilical cord was attached; the navel. p. 891

unipolar neuron (un″ĭ-po′lar nu′ron) A neuron that has a single nerve fiber extending from its cell body. p. 344

unsaturated fatty acid (un-sat′u-rāt″ed fat′e as′id) Fatty acid molecule with one or more double bonds between the atoms of its carbon chain. p. 50

upper limb (uh′per lim) The superior appendage consisting of the arm, forearm, and hand. p. 220

urea (u-re′ah) A nonprotein nitrogenous substance produced as a result of protein metabolism. p. 699

ureter (u-re′ter) A muscular tube that carries urine from the kidney to the urinary bladder. p. 795

urethra (u-re′thrah)   Tube leading from the urinary bladder to the outside of the body. p. 797

uric acid (u′rik as′id)   Product of nucleic acid metabolism in the body. p. 794

urine (u′rin)   Wastes and excess water removed from the blood and excreted by the kidneys into the ureters to the urinary bladder and out of the body through the urethra. p. 782

uterine (u′ter-in)   Pertaining to the uterus. p. 852

uterine tube (u′ter-in tūb)   Tube that extends from the uterus on each side toward an ovary and transports sex cells; fallopian tube or oviduct. p. 852

uterus (u′ter-us)   Hollow muscular organ within the female pelvis in which a fetus develops. p. 854

utricle (u′trĭ-kl)   An enlarged portion of the membranous labyrinth of the inner ear. p. 443

uvula (u′vu-lah)   A fleshy portion of the soft palate that hangs down above the root of the tongue. p. 649

V

vaccine (vak′sēn)   A substance that contains antigens used to stimulate an immune response. p. 629

vacuole (vak′u-ōl)   A space or cavity within the cytoplasm of a cell. p. 77

vagina (vah-ji′nah)   Tubular organ that leads from the uterus to the vestibule of the female reproductive tract. p. 854

varicose veins (var′ĭ-kos vānz)   Abnormally swollen and enlarged veins, especially in the legs. p. 571

vasa recta (va′sah rek′tah)   A branch of the peritubular capillary that receives blood from the efferent arterioles of juxtamedullary nephrons. p. 781

vascular (vas′ku-lar)   Pertaining to blood vessels. p. 578

vas deferens (vas def′er-ens)   Tube that leads from the epididymis to the urethra of the male reproductive tract (pl., *vasa deferentia*). p. 838

vasoconstriction (vas″o-kon-strik′shun)   A decrease in the diameter of a blood vessel. p. 564

vasodilation (vas″o-di-la′shun)   An increase in the diameter of a blood vessel. p. 564

vasomotor center (vas″o-mo′tor sen′ter)   Neurons in the brainstem that control the diameter of the arteries. p. 393

vasopressin (vas″o-pres′in)   Antidiuretic hormone. p. 483

vein (vān)   A vessel that carries blood toward the heart. p. 569

vena cava (vēn′ah kāv′ah)   One of two large veins that convey deoxygenated blood to the right atrium of the heart. p. 591

ventral (ven′tral)   Pertaining to the front or anterior. p. 12

ventral cavity (ven′tral kav′i-te)   Hollow place within the body including the thoracic, abdominal, and pelvic cavities. p. 12

ventral root (ven′tral rōōt)   Motor branch of a spinal nerve by which it attaches to the spinal cord. p. 404

ventricle (ven′trĭ-kl)   A cavity, such as brain ventricles filled with cerebrospinal fluid, or heart ventricles that contain blood. p. 368

venule (ven′ūl)   A vessel that carries blood from capillaries to a vein. p. 569

vermiform appendix (ver′mĭ-form ah-pen′diks)   Appendix. p. 680

vertebral (ver′te-bral)   Pertaining to the bones of the spinal column. p. 25

vertebral canal (ver′te-bral kah-nal′)   Hollow place within the vertebrae containing the spinal cord. p. 12

vesicle (ves′ĭ-kal)   Membranous, cytoplasmic sac formed by an infolding of the cell membrane. p. 77

vestibule (ves′tĭ-būl)   A space at the opening to a canal. p. 437

villus (vil′us)   Tiny, fingerlike projection that extends outward from the inner lining of the small intestine (pl., *villi*). p. 675

viscera (vis′er-ah)   Organs within a body cavity. p. 12

visceral (vis′er-al)   Pertaining to the contents of a body cavity. p. 14

visceral peritoneum (vis′er-al per″ĭ-to-ne′um)   Membrane that covers organ surfaces within the abdominal cavity. p. 14

visceral pleura (vis′er-al ploo′rah)   Membrane that covers the surfaces of the lungs. p. 14

viscosity (vis-kos′ĭ-te)   The tendency for a fluid to resist flowing due to the internal friction of its molecules. p. 575

vital capacity (vi′tal kah-pas′i-te)   The maximum amount of air a person can exhale after taking the deepest breath possible. p. 750

vitamin (vi′tah-min)   An organic compound other than a carbohydrate, lipid, or protein that is needed for normal metabolism but that the body cannot synthesize in adequate amounts. p. 107

vitreous body (vit′re-us bod′e)   The collagenous fibers and fluid that occupy the posterior cavity of the eye. p. 455

vitreous humor (vit′re-us hu′mor)   The substance between the lens and the retina of the eye. p. 455

vocal cords (vo′kal kordz)   Folds of tissue within the larynx that produce sounds when they vibrate. p. 737

voluntary (vol′un-tār″e)   Capable of being consciously controlled. p. 148

vulva (vul′vah)   The external female reproductive parts that surround the vaginal opening. p. 855

W

water balance (wot′er bal′ans)   When the volume of water entering the body is equal to the volume leaving it. p. 811

water of metabolism (wot′er uv mĕ-tab′o-lizm)   Water produced as a by-product of metabolic processes. p. 811

white blood cell (whīt blud sel)   A cell that helps fight infection; leukocyte. p. 148

white muscle (whīt mus′el)   Fast-contracting skeletal muscle. p. 292

wild type (wīld tīp)   A phenotype or allele that is the most common for a certain gene in a population. p. 925

X

X-linked trait (eks-linkt′trāt)   Trait determined by a gene on an X chromosome. p. 932

x-ray (eks′ ray)   Used as a verb: to photograph using radiation. p. 672

X ray (eks ray)   Used as a noun: a photograph produced by radiation. May also be used as an adjective: X-ray. p. 192

Y

yellow marrow (yel′o mar′o)   Fat storage tissue found in the cavities within certain bones. p. 194

yolk sac (yōk sak)   An extraembryonic membrane connected to the embryo by a long, narrow tube. p. 891

Z

zonular fiber (zon′u-lar fi′ber)   A delicate fiber within the eye that extends from the ciliary process to the lens capsule. p. 451

zygote (zi′gōt)   Cell produced by the fusion of an egg and sperm; a fertilized egg cell. p. 849

zymogen granule (zi-mo′jen gran′-ūl)   A cellular structure that stores inactive forms of protein-splitting enzymes in a pancreatic cell. p. 665

Credits

Photos

Chapter 1

Opener: © Professor P. Motta/Dept. of Anatomy/University, "La Sapienza," Rome/Science Photo Library/Photo Researchers, Inc.; **Vignette:** L. Mulvehill/Photo Researchers, Inc.; **1.1:** Andreas Vesalius, 2nd book, *De Humani Corporis Fabrica,* 1543; **1A:** Alexander Tsiaras/Photo Researchers, Inc.; **1B:** Lutheran Hospital/Peter Arnold; **1C:** Science Photo Library/Photo Researchers, Inc.; **1.4:** © The McGraw-Hill Companies/Carol D. Jacobson, Ph.D., Dept. of Veterinary Anatomy, Iowa State University; **1.21a:** Patrick J. Lynch/Photo Researchers, Inc.; **1.21b:** Biophoto Associates/Photo Researchers, Inc.; **1.21c:** A. Glauberman/Photo Researchers, Inc.

Chapter 2

Opener: © Dr. Tim Evans/Science Photo Library/Photo Researchers, Inc.; **2A:** Mark Antman/The Image Works; **2Ba:** SIU/Peter Arnold; **2Da:** Science Photo Library/Photo Researchers, Inc.; **2Db:** CNRI/SPL/Photo Researchers, Inc.; **2E:** From M. I. Posner, 10/29/93, "Seeing the Mind," *Science* 262:673, Fig. 1, ©AAAS/Photo, Marcus Raichle; **2.23a–e:** Courtesy of John W. Hole, Jr.; **2.23f:** Courtesy of Ealing Corporation.

Chapter 3

Opener: © K. R. Porter/Photo Researchers, Inc.; **3.4:** © W. Omerod/Visuals Unlimited; **3.5a:** © Robert Becker/Custom Medical Stock Photo; **3.5b:** © R. Roseman/Custom Medical Stock Photo; **3.5c:** © Mohr/Custom Medical Stock Photo; **3.6a:** © Biophoto Associates/Photo Researchers, Inc.; **3.10a:** © Don W. Fawcett/Photo Researchers, Inc.; **3.11a:** © Gordon Leedale/Biophoto Associates; **3.13a:** © Bill Longcore/Photo Researchers, Inc.; **3.14:** © K. G. Murti/Visuals Unlimited; **3.15a:** © Don W. Fawcett/Visuals Unlimited; **3.16a:** © Oliver Meckes/Photo Researchers, Inc.; **3.17:** © Manfred Kage/Peter Arnold; **3.18:** © M. Schliwa/Visuals Unlimited; **3.19b:** © K. G. Murti/Visuals Unlimited; **3.20b:** © Stephen L. Wolfe; **3.26a, b, & c:** © David M. Phillips/Visuals Unlimited; **3.37:** © Ed Reschke; **3.38:** From *Scanning Electron Microscopy in Biology,* by R. G. Kessel and C. Y. Shih. ©1976 Springer-Verlag; **3.39:** © Dr. Tony Brain/SPL/Photo Researchers, Inc.; **3.45:** © Tony Brain/SPL/Photo Researchers, Inc.

Chapter 4

Opener: © Biophoto Associates/Science Photo Library/Photo Researchers, Inc.; **4A:** Courtesy of the March of Dimes Birth Defects Foundation.

Chapter 5

Opener: © Nancy Kedersha/Immunogen/Science Photo Library/Photo Researchers, Inc.; **Vignette:** © Federation of American Societies for Experimental Biology/Photo Courtesy of Nicolas L'Heureux; **5.1b & d, 5.2b:** © Ed Reschke; **5.3b:** © Victor Eroschenko; **5.4:** © Fawcett Hirokawa, Heuser/Photo Researchers, Inc.; **5.5b:** © Ed Reschke; **5.6b:** © G. W. Willis, M.D./Visuals Unlimited; **5.7b:** © Victor Eroschenko; **5.8b:** © Richard Kessel/Visuals Unlimited; **5.9b & d:** © Ed Reschke; **5.12:** © Ed Reschke/Peter Arnold; **5.13:** © David M. Phillips/Visuals Unlimited; **5.14:** © Manfred Kage/Peter Arnold; **5.15:** © Veronica Burmeister/Visuals Unlimited; **5.16 Bottom:** © P. Motta/ Science Photo Library/Photo Researchers, Inc.; **5.17:** © St. Mary's Hospital Medical School/Science Photo Library/Photo Researchers, Inc.; **5.18b, 5.19b:** © Ed Reschke; **5.20b:** © Ed Reschke/Peter Arnold; **5.21b:** © Ed Reschke; **5.22b:** © John D. Cunningham/Visuals Unlimited; **5.23b:** © Fred Hossler/Visuals Unlimited; **5.24b:** © Ed Reschke/Peter Arnold; **5.25b:** © R. Calentine/Visuals Unlimited; **5.26b:** © Victor B. Eichler, PhD; **5.26c:** © Biophoto Associates/Photo Researchers, Inc.; **5.27b:** © Ed Reschke/Peter Arnold; **5.28b, 5.29b:** © Ed Reschke; **5.30b:** © Manfred Kage/Peter Arnold; **5.31b:** © Ed Reschke.

Chapter 6

Opener: © Dr. Jeremy Burgess/Science Photo Library/Photo Researchers, Inc. **Vignette:** Center for Disease Control, U.S. Government; **6.1:** © The McGraw-Hill Companies/Carol D. Jacobson, Ph.D., Dept. of Veterinary Anatomy, Iowa State University; **6.2b:** © Victor Eroschenko; **6.3b:** © Victor B. Eichler, Ph.D.; **6.4a:** © M. Schliwa/Visuals Unlimited; **6A:** © Kenneth Greer/Visuals Unlimited; **6Ba, b, & c:** © Science Photo Library/Photo Researchers, Inc.; **6.5b:** © Victor B. Eichler, PhD.; **6.6:** © CNRI/ Science Photo Library/Photo Researchers, Inc.; **6C:** © Mark Richards/PhotoEdit; **6.8:** © Per H. Kjeldsen; **6.10:** © Ed Reschke; **6D:** © Kenneth Greer/Visuals Unlimited; **6.12:** © Mediscan, London; **6.15 Top** © Gunther/Explorer/Photo Researchers, Inc.; **6.15 Bottom:** © Kjell Sandvad/Visuals Unlimited.

Chapter 7

Opener: © David Scharf/Peter Arnold; **Vignette:** © John Reader/ Science Photo Library/Photo Researchers, Inc.; **7.3a:** © L. V. Bergman/The Bergman Collection; **7.3b & c:** Courtesy of John W. Hole, Jr.; **7.5:** *Tissues and Organs: A Text-Atlas of Scanning Electron Microscopy,* by R. G. Kessel and R. H. Kardon. ©1979 W. H. Freeman and Company; **7.6a:** © Biophoto Associates/Photo Researchers, Inc.; **7.6b:** Courtesy of Dr. Francis Collins; **7.7:** © Secchi, Lecaque, Roussel, Uclaf, CNRI/Science Photo Library/Photo Researchers, Inc.; **7.9b:** © Victor B. Eichler, Ph.D.; **7.10:** © Biophoto

Chapter 19

Opener: © Susumu Nishinaga/Science Photo Library/Photo Researchers, Inc.; **Vignette:** Courtesy of Wendy Josephs; **19.3b:** © Biophoto Associates/Photo Researchers, Inc.; **19A:** Used with permission of the American Cancer Society; **19B:** © Moredun Animal Health/SPL/Photo Researchers, Inc.; **19.4:** Courtesy of Eastman Kodak; **19.7c:** © CNRI/Phototake; **19.10:** © Ed Reschke: **19.13:** © imagingbody.com; **19.15:** © Dwight Kuhn; **19.17:** Copyright by R.G. Kessel and R.H. Kardon, *Tissues and Organs: A Text-Atlas of Scanning Electron Microscopy,* 1979 (W. H. Freeman & Co.); **19.18:** Courtesy of the American Lung Association: **19.27:** © Edward Lettau/Photo Researchers, Inc.; **19C a & b:** © Victor B. Eichler, PhD; **19.32:** Courtesy of the American Lung Association; **19.34:** © imagingbody.com; **19D a & b:** © CNRI/Phototake.

Chapter 20

Opener: © Prof. P. Motta/Dept. of Anatomy/University "La Sapienza" Rome/Science Photo Library/Photo Researchers, Inc., **Vignette:** © Tony Freeman/PhotoEdit; **20.2:** © CNRI/SPL/Photo Researchers, Inc.; **20.6b:** © L. V. Bergman/The Bergman Collection; **20.7a:** Copyright by R.G. Kessel and R.H. Kardon, *Tissues and Organs: A Text-Atlas of Scanning Electron Microscopy,* 1979 (W.H. Freeman & Co.); **20.7b:** Courtesy of R. B. Wilson MD, Eppley Institute for Research in Cancer, University of Nebraska Medical Center; **20.9:** © David M. Phillips/Visuals Unlimited; **20.11a:** © Biophoto Associates/Photo Researchers, Inc.; **20.11b:** © Manfred Kage/Peter Arnold; **20.26:** © Per H. Kjeldsen; **20.29:** © John D. Cunningham/Visuals Unlimited; **20.30:** © Ed Reschke.

Chapter 21

Opener: © Prof. P. M. Motta & M. Castellucci/Science Photo Library/Photo Researchers, Inc.; **Vignette:** © AP/PhotoFile/Tom Olmscheid.

Chapter 22

Opener: © BSIP, S & I/Photo Researchers, Inc.; **Vignette:** © Dr. Tony Brain/SPL/Photo Researchers, Inc.; **22.4:** Copyright by R.G. Kessel and R.H. Kardon, *Tissues and Organs: A Text-Atlas of Scanning Electron Microscopy,* 1979 (W.H. Freeman & Co.); **22.5a:** © Biophoto Associates/Photo Researchers, Inc.; **22.10:** © David M. Phillips/The Population Council/Photo Researchers, Inc.; **22.11:** © Ed Reschke; **22.12 a & b:** Copyright by R.G. Kessel and R.H. Kardon, *Tissues and Organs: A Text-Atlas of Scanning Electron Microscopy,* 1979 (W.H. Freeman & Co.); **22.13:** © Manfred Kage/Peter Arnold; **22.20b:** R. J. Blandau; **22.21:** © Ed Reschke; **22.22b:** © Ed Reschke/Peter Arnold; **22.23:** H. Mizoguchi, from W. Bloom & D. W. Fawcett, *A Textbook of Histology* 10th edition; **22.24:** © Dr. Landrum B. Shettles; **22.27a:** © Ed Reschke; **22.27b:** Mediscan, London; **22.28:** © The McGraw-Hill Companies, Inc./Carol D. Jacobson, PhD, Dept. of Veterinary Anatomy, Iowa State University; **22.34a & b:** © Biophoto Associates/Photo Researchers, Inc.; **22.34b:** © Biophoto Associates/Photo Researchers, Inc.; **22C:** Courtesy of Southern Illinois University School of Medicine/Biomedical; **22.37c, d & e:** © Bob Coyle.

Chapter 23

Opener: © Prof. P. Motta/Science Photo Library/Photo Researchers, Inc.; **23A:** Courtesy of Southern Illinois University School of Medicine/Biomedical; **23B:** © Taro Yamasaki/TimePix; **23.4a:** © A. Tsiara/Photo Researchers, Inc.; **23.4b:** © Omikron/Photo Researchers, Inc.; **23.4c:** © Petit Format/Nestle/Photo Researchers, Inc.; **23.6c, 23.7:** Courtesy of Ronan O'Rahilly, M.D. Carnegie Institute of Washington; **23.12a & b:** © Dr. Landrum B. Shettles; **23.12c:** © Petit Format/Nestle/Photo Researchers, Inc.; **23.13b:** © Carroll Weiss/Camera MD Studios; **23.14:** © Landrum B. Shettles; **23.19:** © Donald Yaeger/Camera M.D. Studios; **23Db, c, & d:** Streissguth, A.P. Landesman-Dwyer, S., Martin, J.C., & Smith, D. W. (1980) "Teratogenic effects of alcohol in humans and laboratory animals." *Science* 209 (18) 353–361; **23E:** © Steven Wewerka; **23F:** © Amy Waddell/UCLA Health Sciences; **23.28a–b:** © Biophoto Associates/Photo Researchers, Inc.; **23.29:** Donna Jernigan; **23.31 #1–5:** © UPI Bettmann/CORBIS; **23.31 #6:** © AP/Wide World; **23.31 #7 & 8:** © Bettmann/CORBIS; **23G:** © J. L. Bulcan/Liaison/Getty.

Chapter 24

Opener: © Lawrence Livermore Laboratory/Science Photo Library/Photo Researchers, Inc.; **Vignette:** Courtesy of Mergen, Ltd.; **24.1b:** From E. J. DuPraw, DNA and Chromosomes ©1970 Holt, Rhinehart and Winston, Inc.; **24.7:** From Jacques Genest Jr., Mark-Andre Lavole, "Images in Clinical Medicine," *New England Journal of Medicine,* P490:8-12-99. ©1999 Massachusetts Medical Society, All Rights Reserved; **24.8a:** Courtesy Library of Congress; **24.8b:** © Peter Morenus/University of Connecticut, at Storrs; **24.12:** © Biophoto Associates/Photo Researchers, Inc.; **24.13a, b, c & d:** © Bettmann/CORBIS; **page 936:** Courtesy of Coleen Weisz; **24.16:** © Bernard Bennot/Science Photo Library/Photo Researchers, Inc.; **page 940: (a)** © 1995 Jessica Boyatt, **(b)** Courtesy of Paul and Migdalia Gelsinger © 7/99 Migdalia Q. Gelsinger.

Reference Plates

8–36: Courtesy of John W. Hole, Jr.; **37, 40–41:** © The McGraw-Hill Companies, Inc./Joe De Grandis, photographer; **38–39, 42–47:** Kent M. Van De Graaff; **48–75:** © The McGraw-Hill Companies, Inc./Karl Rubin; **76:** © Martin Rotker/Photo Researchers.

Index

Note: Page numbers in *italics* refer to illustrations; page numbers in **boldface** refer to plates; page numbers followed by t refer to tables.

A

A band, 279, *281*
Abdomen
  imaging of, *56*
  quadrants of, 23, *23*
  regions of, 23, *23*
  transverse section of, *15*,
    954, 955
Abdominal aorta, 582–83, *584*,
  585t, *590, 773, 775, 900*, **950**
Abdominal cavity, *13*, 14, **36, 950**
Abdominal cavity lymph nodes,
  613, 614
Abdominal reflex, 377
Abdominal regions, 24, *24*
Abdominal viscera, *667*
  anterior view of, **966, 967**
  veins from, 593–95, *594*
Abdominal wall
  arteries to, 588
  muscles of, 312–14, 312t, 313t,
    748, *749*
  veins from, 591–92, *592*
Abdominopelvic cavity, 12–14, *13*
Abdominopelvic membranes, 14
Abducens nerve (VI), 399–400,
  399, 403t, 449t
Abduction, 261, *261*
Abductor pollicis longus
  muscle, *311*
ABO blood group, 532, 534–35,
  534, 534t, *535*, 536t, 927
Abortion, 914
  spontaneous, 884
Abruptio placentae, 914
Absorption
  as characteristic of life, 8t
  in large intestine, 683
  in small intestine, 677–80, *678*,
    679, 680t
Accelerator nerve, 560
Accelerator neuron, 345
Accessory nerve (XI), 302t, 304t,
  399, 401, 403t
  cranial branch of, 401
  spinal branch of, 401
Accessory organs
  of digestion, *645*
  of reproductive system
    female, 846, 852–55
    male, 830, 838–41
  of skin, 165–68
  visual, 447–49, *447, 448*

Accessory structures, of synovial
  joints, 257
Accommodation, 452, *452*, 456–57
Accutane, *893*, 894
Acetabulum, 224, *225, 226*, 267,
  267, 268
Acetaminophen, 171
Acetazolamide, 435
Acetic acid, *112*
Acetoacetate, 823, 972
Acetone, 697–98, 758, 823, 972
Acetonemia, 824
Acetonuria, 824
Acetylcholine, 283–84, 287t, 294,
  356, 357t, 358, 358t, 360,
  414–16, *414*, 560
Acetylcholine receptors, 283–84
Acetylcholinesterase, 286, 287t,
  358, 416
Acetylcholinesterase
  inhibitors, 286
Acetyl CoA, *109*, 111, *112, 113*,
  697, 697, 699, *700*, 710
Achalasia, 686
Achlorhydria, 686
Achondroplasia, 232, 882t
Achoo syndrome, 924
Acid(s), 45–46, 46t
  strengths of, 818–19
  volatile, 820
Acid-base balance, 817–24
  disorders of, 822–23, *822, 823*
Acid-base buffer systems, 47,
  819–21, 820t
Acidic solution, 47, *47*
Acid indigestion, 663
Acidosis, 47, 820, 822, *822*
  compensated, 822
  metabolic, 499, 701t,
    822–23, *823*
  respiratory, 822, *822*
  uncompensated, 822
Acinar cells, pancreatic, 665
Acini, 665
Acne, 167, 169, *169*
  treatment of, 169, 169t
Acquired immunodeficiency syn-
  drome. *See* AIDS
Acromegaly, 190, 232, 482, *482*
Acromial region, 24, *24*
Acromioclavicular joint, 263t
Acromion process, *219, 264, 265*,
  333, 334
Acrosome, 837, *837, 880*

Actin, 279, *281*, 282, *285*, 287t,
  293–94. *See also* Thin
  filaments
Actinomyces, 653, *653*
Action potential, 350–53, *352, 353*
Activation energy, 106, 108
Active aging, 912
Active immunity
  artificially acquired, 629, 630t
  naturally acquired, 629, 630t
Active site, 106–7, *106*
Active transport, 85, *85*, 89t
Activity site, on hormone
  receptors, 472
Acupuncture, 359
Acute leukemia, 522
Acute pain fibers, 426–27
Adam's apple, 736
Adams family, *933*
Adaptive (specific) defense, 617,
  619–34
Addiction, 920
Addison disease, 496, 818
Adduction, 261, *261*
Adductor brevis muscle, **35**
Adductor longus muscle, **33, 34**,
  35, 36, *297*, 315, *316*, 318,
  319, 319t, **960**
Adductor magnus muscle, **36**,
  297, 315, *316*, 318, *318*,
  319, 319t
Adenine, 116, *117*, 118, 119t, *979*
Adenoids. *See* Pharyngeal tonsils
Adenosine deaminase deficiency,
  882t, 940, *940*
Adenosis, 855
Adenylate cyclase, 472, *473*,
  475, 477
Adipocytes, *1*, 144
Adipocyte transcription
  factors, 705
Adipose cells, 62
Adipose tissue, 141, 144–45, *145*,
  150t, *159*, 698
  subcutaneous, 164
Adolescence, 909, 911t
ADP (adenosine diphosphate),
  108, *109*
Adrenal cortex, *467*, 471t, *491*,
  492, *492*
  disorders of, 496
  hormones of, 493–95, *494*,
    495, 495t
Adrenalectomy, 504

Adrenal glands, **35**, 411, *412, 469*,
  472, 490–96, *491*, 502,
  774, 775
Adrenal medulla, 471t, 477,
  490–92, *491, 492*
  hormones of, 492, *492, 493*
Adrenergic fibers, 414, *414*
Adrenergic receptors
  alpha, 415–16, 492
  beta, 415–16, 492
Adrenocorticotropic hormone,
  471t, 475, 478, 480, *481*,
  483, 485t, 494, *495*, 496,
  501, *501*
Adrenogenital syndrome, 504
Adrenoleukodystrophy, 75
Adulthood, 909–10, *910*, 911t
Adult respiratory distress
  syndrome, 759
Aerobic exercise, 325, 580
Aerobic reactions, of cellular respiration, 108, *109*, 111–12,
  112, 113, 288, *288*, 817
Afferent arterioles, 774, *777*,
  778, 779, 781, *783, 784*,
  785–86, *787*
Aflatoxin B, 127t
Afterbirth, 903–4, *903*
Age-related changes, 19–21,
  909–10, *910*
  in bone, 195, 231–32, *231*
  in cardiovascular system,
    598–601, 911t
  in digestive system, 686, 911t
  in endocrine system, 502, 911t
  in immune system, 634
  in integumentary system, 911t
  in joints, 271–72
  in lymphatic system, 911t
  in muscular system, 325, 911t
  in nervous system,
    416–17, 911t
  in nutrition and metabolism,
    724–25, 724t
  in reproductive system, 911t
  in respiratory system, 764, 911t
  in skeletal system, 216, 231–32,
    231, 911t
  in skin, 175–76, *176*
  in somatic and special
    senses, 462
  in urinary system, 800, 911t
Age spots, 175
Ageusia, 434t

Agglutination, 533, *535*, 627, 628t
Agglutinins, 533
Agglutinogens, 533
Aging, 912–14. *See also* Age-
    related changes
    active, 912
    passive, 912
Agonist (drug), 360
Agonist (muscle), 296
Agranulocytes, *512*, 518–19, 523t
Agricultural lifestyle, 3
AIDS (acquired immunodeficiency
    syndrome), 617, 635, *635,*
    867t, 868, 913t. *See also*
    Human immunodeficiency
    virus infection
    T cells in, 521, 521t
Air quality, 732
Albinism, *127*, 166, 172, *172*, 458
Albumin(s), 523–24, *524*, 524t, 709
    serum, 970
    urine, 972
Albumin-globulin ratio, 970
Albuminuria, 824
Alcohol, 906t
    consumption in pregnancy,
    894, *895*
Alcoholism, 665, 670, 709–10, 724
Aldosterone, 471–472, 471t, 493,
    494, 495t, 501, 578, 715,
    786–87, 789–91, 815–16,
    816, 885, 885t
Ali, Mohammed, 391, *391*
Alimentary canal, 644, *645*
    characteristics of, 644–48
    innervation of tube, 648
    length of, *646*
    movements of tube, 646–48, *647*
    wall of, 645–46
Alkalemia, 823
Alkaline taste, 432
Alkaline tide, 661
Alkaloids, 434
Alkalosis, 47, 820, 822, *822*
    metabolic, 822–23, *823*
    respiratory, 756, 822–23, *823*
Alkaptonuria, *127*, 799
Allantois, *891*, 892
Alleles, 925
Allergens, 164, 608, 630, *631*
Allergic contact dermatitis, 164
Allergy, 142, 494, 608, 630–32, *631*
    delayed-reaction, 632
    immediate-reaction, 630, *631*
    origin of, 630
Allograft, 632, 632t
All-or-none response
    of neurons, 353
    of skeletal muscle, 290
Aloe, 162
Alopecia areata, 166
Alpha-1–antitrypsin
    deficiency, 882t
Alpha blockers, 840
Alpha cells, 496
Alpha fetoprotein, 885, 937
Alpha helix, *53, 54*
Alpha radiation, 40, 45
Alpha tocopherol, 708

Alpha waves, 397, *397*
Altitude sickness, 760
Alveolar arch, 205, **245**
Alveolar capillaries, 580–81, *581*
Alveolar dead space, 750
Alveolar duct, *740*, 741, *741*
Alveolar gland, 138, *139, 140,*
    861, *862*
Alveolar macrophages, 757
Alveolar pore, 757, *757*
Alveolar process, 205, 650, *652*
Alveolar sacs, 741, *741*
Alveolar ventilation, 751, 756t
    in microgravity
    environment, 751
Alveolar ventilation rate, 751
Alveoli, 133, *550*, 581, *582*, *740*,
    741–42, *741, 742*
    age-related changes in, 764
    dental. *See* Dental alveoli
    gas exchange in, 757–58
Alzheimer disease, 21, 54, 56,
    69, 357t, 394, 417, 881t,
    936, *942*
Amacrine cells, 454, *454*
Amblyopia, 462
    suppression, 450
Ameboid motion, 520
Amenorrhea, 858, 868
Amino acids, 52
    absorption in small intestine,
    678, *678*, 680t
    blood, 525–26
    essential, 700
    excitatory, 356
    in foods, 700t
    metabolism of, 699, *699,*
    700, 817
    nonessential, 700
    protein synthesis, 120–22,
    120, 121
    structure of, 52, *52, 58*
    tubular reabsorption of, 788
Amino group, 52, *52*, 819–20
2–Amino 5–nitrophenol, 127t
4–Aminopyridine, 381
Ammonia, 669, 821, 941
    blood, 970
    urine, 972
Ammonium ions, 821
Amniocentesis, 914, 937, 937t, *938*
Amniochorionic membrane,
    892, *893*
Amnion, *886*, 887, *891*, 892
Amniotic cavity, 886, *886, 891*, 892
Amniotic fluid, 887, *893*
Amniotic sac, *903*
Amoeba, 9, *9*
Amphetamines, 360, 705
Ampulla
    of semicircular canals, *438,*
    444, *446*
    of vas deferens, 839
Amylase, 107
    pancreatic, 665, 677t
    salivary, 653, 677t, 678
    serum, 970
Amyloid, 54, 551, 936
Amyloidosis, familial, 551

Amyotropic lateral sclerosis,
    379–80
Anabolic steroids, 474
Anabolism, 104–5, *104, 105*
Anaerobic reactions, of cellular
    respiration, 108–11, *109,*
    288, *288*, 817
Anal canal, 680, *681*, 682, *682*
Anal column, 682, *682*
Analgesia, 417
Analgesic, 417
Anal sphincter
    external, *314*, 315, 682, *682,*
    684–85
    internal, 315, 682, *682*, 684
Anaphase
    meiosis I, 836
    meiosis II, 836
    mitotic, *91*, 92, 92t
Anaphylactic shock, 630
Anasarca, 824
Anastomosis, 601
Anatomical neck, of humerus,
    221, *221*
Anatomical position, 21
Anatomical terminology, 21–25
    for body regions, 23–25, *23, 24*
    for body sections, 21–23, *22*
    for relative position, 21
Anatomic dead space, 750
Anatomy, definition of, 3–4
Androgen(s), 169, 190
    adrenal, 495
    female, 858, 858t
    male, 845–46, *845*
Androgenic alopecia, 166
Anemia, 167, 516, *516*, 522,
    575, 701t
    types of, 516t
Anencephaly, 382, 394, 711, 892
Anesthesia, 417
Anesthetics, 355, 759
Aneuploidy, 934–37, 934t, *935*
Aneurysm, 571
    aortic, 591
Angina pectoris, 550, 600
Angiogenesis, 564
Angiogram, *584, 586*
Angioplasty, 564
Angiospasm, 601
Angiostatin, 564
Angiotensin, 577
Angiotensin I, 493, 786
Angiotensin II, 493, *494*, 501, 578,
    578, 786
Angiotensin-converting enzyme,
    493, 786, *787*
Angiotensin-converting enzyme
    inhibitors, 493, 578t
Angiotensinogen, 493, 786
Anisocytosis, 536
Ankle-jerk reflex, 377
Ankle joint, 263t
    bones of, 229–31, *230*
    surface anatomy of, **336**
Ankylosis, 274
Annular ligament of the radius,
    266, *266*
Annulus fibrosus, 216, 257

Anopia, 462
Anorexia nervosa, 195, 694,
    722–24
Anosmia, 434t
Anovulation, 861
Anoxia, 764
Antacids, 663, 823–24
Antagonist (drug), 360
Antagonist (muscle), 296
Antebrachial region, 24, *24*
Antecubital region, 24, *24*
Anterior, definition of, 21
Anterior branch, of spinal nerves,
    403, *406*
Anterior cavity, of eye, *450*, 452
Anterior chamber, of eye, *451,*
    452, *453*
Anterior column, 372
Anterior crest, of tibia, 228
Anterior fontanel, 208, *210*
Anterior horn, 372, *373*, 385,
    406, 411
Anterior regions, *24*
Anti-A antibody, 534–35, 534t, *535*
Anti-B antibody, 534–35, 534t, *535*
Antibiotics, interfering with bacte-
    rial protein synthesis, 123
Antibody, 52, 519, 524, 621, 623,
    625–26
    actions of, 627–28, *627*, 628t
    blood types, 533–34, *534*
    diversity among, 625
    maternal, 627, 629, 907
    monoclonal, 624, *625*, 865
    production of, 624t
    structure of, 625–26, *627*
Antibody-dependent cytotoxic
    reaction, 630
Antibody genes, 625
Anticoagulants, 527, 531–32
Anticodon, 120, *121, 120*
Anticonvulsants, 68t
Antidepressants, 428, 686
Antidiabetics, 68t
Antidiuretic, 484
Antidiuretic hormone, 471t, 475,
    478, 483–84, *483*, 485t,
    501–2, 790–91, 793, 793t,
    812–13, 813t
Antidiuretic hormone
    receptors, 484
    abnormal, 484
Antigen, 619
    blood types, 533–34, *534*
Antigen-binding site, 626
Antigen D, 536
Antigen-presenting cells, 621
Antihemophilic plasma, 536
Antihistamines, 69, 142, 686
Antihypertensives, 68t
Antiobesity drugs, 705
Antioxidants, 706, 708
Anti-Rh antibody, 536
Antiserum, 629
Antithrombin, 528, 530–31, 531t
Antrum, 850
Anuria, 802, 824
Anus, *645, 646,* 682, *831, 847*
Anxiety, 502

Aorta, *15, 543, 544,* 546, *547, 550, 552, 553,* 585t, 592, *595, 667, 755, 774,* **948, 950, 953, 954, 955, 963, 964**
  abdominal, **36,** 582–83, *584,* 585t, *590, 773, 775, 900,* **950**
  arch of, **33, 34, 35, 36,** 582, *583,* 585t, *900,* **965**
  ascending, 582, 585t
  branches of, 582–84, *582*
  descending, **35,** 582, 585t
  thoracic, 582, 585t, *588,* **965**
Aortic aneurysm, 143, 591
Aortic body, 582, 754, *755,* 756t
Aortic sinus, 582
Aortic sound, 554
Aortic valve, *547, 548,* 549, 549t, *550, 551,* 553–54, *554,* 601
Aortic valvulitis, 557
Apex, of heart, 542, *544, 552*
Aphagia, 686
Aphasia, 386, 417
Apical heartbeat, 542
Aplastic anemia, 516t
Apocrine gland, 139, *140,* 140t, 167–68, *168,* 168t
Aponeurosis, 278, *279*
Apoprotein(s), 525, 679
Apoprotein-B, 87
Apoptosis, 79, 94, 162, 381, 416, 912
Appendicitis, 681
Appendicular portion, 12
Appendicular skeleton, 197
Appendix, *646, 673,* 680–81, *681*
Appetite, 693, 695, 695t, 724
Apraxia, 386, 417
Aquaporins, 791, 812
Aqueous humor, *450,* 452, *453,* 462
Arachidonic acid, 698
Arachnoid granulations, *367, 369, 370*
Arachnoid mater, *367,* 368, *368, 370*
Arbor vitae, 395, **968**
Arch of the aorta, **33, 34, 35, 36,** 582, *583,* 585t, *900,* **965**
Arcuate artery, 774, *777, 783*
Arcuate popliteal ligament, 269, *270*
Arcuate vein, 776, *777, 783*
Areola, *30,* 861, *862*
Aretaeus of Cappadocia, 468
Arm
  muscles that move, 304–8, *305, 306,* 306t, *307*
  posterior view of, **959**
  surface anatomy of, **334**
"Arnold head," 218
Arousal, 393
Arrector pili muscle, *160, 165,* 166
Arrhythmia, 68, 551, 561–63, *562, 563,* 817–18
Arterial system, 582–91, *592*
Arteries, 562–66, *565, 566,* 570t.
  See also specific arteries
  wall of, 563, *565*
Arteriography, 601
Arterioles, 562–66, *565, 566, 567,* 570t, *577*

Arteriosclerosis, 571, 578
Arteriovenous shunt, 566, *566*
Arthralgia, 274
Arthritis, 3, 21, 70, 258, 272, 273t
Arthrocentesis, 258, 274
Arthrodesis, 274
Arthrogram, 274
Arthrology, 274
Arthropathy, 274
Arthroplasty, 269, 274
Arthroscopic scissors, *272*
Arthroscopy, 267
Arthrostomy, 274
Arthrotomy, 274
Articular cartilage, 183, *183, 187,* 257–58, *257, 258*
Articulating process
  inferior, 211, *212*
  superior, 211, *212*
Articulations. *See* Joint(s)
Artificial heart, 556
Artificial insemination, 878
Artificial kidney, 773
Artificially acquired active immunity, 629, 630t
Artificially acquired passive immunity, 629, 630t
Artificial pacemaker, 564
Artificial respiration, 743
Arytenoid cartilage, 736–37, *737*
Asbestos, 745, 745t
Asbestosis, 745
Ascending aorta, 582, 585t
Ascending colon, *673, 681, 681,* **967**
Ascending nerve tracts, 376–77, *377,* 379t
Ascites, 569, 723
Ascorbic acid. *See* Vitamin C
Aspartame, 696
Aspartic acid, 357
Asphyxia, 764
Asphyxiation, 739
Aspiration, 759
Aspirin, 171, 428, 477
Asplenia, 637
Assimilation, as characteristic of life, 8t
Assisted reproductive technologies, 878–79, *879*
  dilemmas in, 878t
Association area, of cerebral cortex, 386–87, 389
Aster, *91*
Asthma, 494, 630, 752, 755
Astigmatism, 457
Astrocytes, 69, *337, 346,* 347, 347t
Asystole, 601
Ataxia, 417
Atelectasis, 759, 764
Atherosclerosis, 21, 50, 530, *531,* 571, *571, 578,* 600
Athlete
  body fat, 858
  female, 858
    reproductive cycles in, 858
  heart of, 580
  heat-related illness in, 808
  nutrition and, 722
  water requirement of, 722

Athlete's foot, 176
Atlantoaxial joint, 263t
Atlanto-occipital joint, 263t
Atlas, 201, 212, *213*
  dislocation of, 380
Atmospheric pressure, 9, 746, *746, 747,* 748
Atom(s), 4, *5,* 5t, 38, 39t
  bonding of, 40–44
Atomic number, 39
Atomic radiation, 40
Atomic structure, 38–39, 40t
Atomic weight, 39
ATP (adenosine triphosphate), 73, 108, 714
  from citric acid cycle, *109,* 111, *112, 113, 974, 976*
  from electron transport chain, *109,* 111, *113*
  from fat metabolism, *697*
  from glycolysis, *109,* 110, *110, 974, 975*
  relationship to ADP, *109*
  structure of, 108, *109*
  synthesis of, 974–75
  use in muscle contraction, *285,* 286–89, *287,* 287t, 294
  use in protein synthesis, 122
  use in vision, 460
ATPase, 286
ATP synthase, *113, 977*
Atrial diastole, 551, 553, *554, 555*
Atrial fibrillation, 562
Atrial flutter, *563*
Atrial natriuretic peptide, 500, 546, 787
Atrial septal defect, 556
Atrial syncytium, 554
Atrial systole, 551, 553, *554, 555*
Atrioventricular bundle, 557, *558*
Atrioventricular node, 557, *558,* 560, *561*
Atrioventricular orifice, 545
Atrioventricular sulcus, 546
Atrioventricular valve, 545, 549, 553
Atrium, 545, **950**
  left, **34,** *543,* 546, *547,* 549, *550, 551, 552,* 581, 899, *900*
  right, *15,* **34,** *543, 544,* 546, *547,* 549, *550, 551, 552,* 579, 591, 899, **964**
Atrophy
  of bone, 190
  of muscles, 293
Atropine, 416
Audiometry, 462
Audiovisual stimulation penogram, 830
Auditory cortex, *442*
Auditory nerve pathways, 439–42, *442,* 442t
Auditory ossicles, 436
Auditory reflex center, 392
Auditory tube, *435, 436,* 437
Auricle of ear, 434, *435*
Auricle of heart, 545
  left, *544, 552, 583*
  right, *544, 552, 583*

Auricular artery, posterior, *585,* 587
Auricular nerve, posterior, *401*
Auricular surface, of sacrum, 214, *215*
Australopithecus afarensis, 182
Australopithecus africanus, 182, *182*
Autoantibodies, 633
Autocrine secretions, 468
Autograft, 173, 632, 632t
Autoimmune disorder, 284, 494, 499
Autoimmunity, 630, 633–34, 912
Automatic bladder, 800
Autonomic ganglion, 410
Autonomic nerve fibers, 410, *410*
Autonomic nervous system, 339, *339,* 396, 396t, 408, 415t
  characteristics of, 408–10
  control of, 416
  neurotransmitters of, 414–16, *414*
  parasympathetic division of, 413–14, *413,* 415t
  sympathetic division of, 410–11, *411, 412,* 415t
Autoregulation, 786
Autosomal aneuploidy, 934
Autosomal inheritance, 925
Autosome, 924–25, *925*
Awareness, 366
Axial portion, 12
Axial skeleton, 196t, 197
Axillary artery, **33,** 587, *587, 592*
Axillary lymph nodes, 613, *613*
Axillary nerve, 306t, 406, *407, 408*
Axillary region, 24, *24*
Axillary vein, **33,** 591, *593, 594, 597*
Axis, 212, *213*
Axon, 338, *339, 340, 341, 342,* 372.
  See also Nerve fiber
  diameter of, 354
  myelinated, *342, 343, 344,* 348, *354, 354*
  regeneration of, 348, *349*
  unmyelinated, *342, 343, 344*
Axonal hillock, 340, *341*
Axon terminal, 342
Azotemia, 824
AZT, 635
Azygos vein, *15,* 591, 593, *594, 597,* **954**

B

Babinski reflex, 377
Bacillus cereus, 104
Back, surface anatomy of, **333**
Background radiation, 45, 45t
Back pressure, 569
Bacteria
  of intestinal tract, 683, 708–9
  oral, 653, *653*
Bacterial count, urine, 972
Bacterial toxin, 104
Bacteriuria, 802
Balanced polymorphism, 927

Baldness, 166, *166*
Ball-and-socket joint, 259, *259, 260t, 263t*
Bands, *512,* 518
Banting, Sir Frederick Grant, 468
Bariatric surgery, 705
Barium enema, *685,* 685t
Baroreceptors, 423, 560, *561,* 576, *576,* 587, 786
Barrel chest, 764
Barrett's esophagus, 656
Basal body, *64,* 76
Basal cavity, *733*
Basal cell carcinoma, *163*
Basal conditions, 702
Basal metabolic rate, 486, 702–3, 724
Basal nuclei, 381, 382t, 388, *389,* 391, 396t
Base(s), 45–46, 46t
    strength of, 818–19
Base of heart, 542, *544*
Basement membrane, 133, 158, *160, 165*
Basic solution, 47, *47*
Basilar artery, *478, 585, 586,* 586
Basilar membrane, 439, *440, 441*
Basilic vein, 591, *593, 594,* 597
Basophils, *512,* 518–20, *519,* 523t, *524,* 530, 531t
B cells, 62, *512,* 519, 619–21, *620,* 621t, 628, *631*
    activation of, *622, 623*
    humoral immune response, 623–28, *623*
    memory, 623, *626,* 628
Beaumont, William, 660
Beeturia, 799, 924
Behavior, reflex, 373–76, *374, 375*
Benign joint hypermobility syndrome, 257
Benign prostatic hypertrophy, 840
Benign tumor, 93
Benzoyl peroxide, 169
Beriberi, 709, 713t
Berylliosis, 745
Best, Charles Herbert, 468
Beta blockers, 578t
Beta-carotene, 706–7, *706,* 939
Beta cells, 496–97
Beta-hydroxybutyric acid, 823
Beta-oxidation, 697, *697*
Beta-pleated sheet, *53,* 54
Beta radiation, 40, 45
Beta waves, 397, *397*
Bicarbonate, 754, 762–63, *762, 763,* 817, 819–20, 823
    blood, 525–26
    in cells, 48, *49*
    electrolyte balance, 813
    in extracellular fluid, *810*
    in pancreatic juice, 665, *666*
    in plasma, glomerular filtrate, and urine, 784t
Bicarbonate buffer system, 819, 820t
Biceps brachii muscle, **31,** 193, *194,* 221, *266,* 296–97, *296, 297, 306, 307,* 308, 308t, *309,* **333, 334, 335, 957, 960**

Biceps femoris muscle, *297, 317, 318, 319,* 320, 320t, *323, 324,* **334, 335, 961, 962**
Biceps-jerk reflex, 377
Bicuspid, 650–51, *651,* 651t
Bifidus factor, 907
Bigorexia, 694
Bile, *515*
    composition of, 669–71
    regulation of release of, 671, *673*
    urine, 972
Bile canaliculi, *668*
Bile duct, *668*
    common, *497, 664, 667,* 671–72, *673*
Bile pigments, 517, 517t, 671
Bile salts, 525, 669–71, 698
    functions of, 671–72
Bili lights, 172
Bilirubin, 172, *515,* 517, 517t, *518,* 671, 972
    serum, 970
Biliverdin, *515,* 517, 517t, *518,* 671
Binding sites, on hormone receptors, 472
Binge and purge cycle, 724
Binocular vision, 460
Bioartificial liver, 669
Biochemistry, 38
Biofeedback, for pain control, 428
Biogenic amines, 356, 357t, 471–72, 472t
Biomineralization, 195
Biopsy, 832, 864
Biotin, 712, 713t, 715, *715*
Bipolar neurons, 344, *344,* 346t, 454, *454*
Birth. *See* Childbirth
Birth control, 862–66, 863t, *866*
Birth control pills, 863t, 865, *866,* 906t
Birth defects, 894–95, *895*
Birthmark, 176
Bitter receptors, 434
Bitter taste, 432–33
Blackhead, 169
Bladder. *See* Urinary bladder
"Blast(s)," 96
Blastocyst, *881,* 882, *884,* 887t
    source of stem cells, 98
Blastomeres, 880
Blepharitis, 462
Blindness, 461
Blind spot, 455
Blister, 158, 164
Blood, 148, *149,* 150t
    coagulation of. *See* Coagulation
    composition of, 510–11, *511, 524*
    fetal, 898–901, *899, 900,* 900t, *901*
    gas transport in, 760–63, 764t
    oxygen-carrying capacity of, 511, 514, 516
    pH of, 47, 525–26, 822–23, *822,* 971
    site of gene therapy, *942*
    viscosity of, *573,* 575
Blood-brain barrier, 69
    malfunctioning, 69

Blood cells, 148, *149,* 523t. *See also specific cells*
    formation of, 194, 511, *511*
Blood-cerebrospinal fluid barrier, 369
Blood clot, 12, 192, *541,* 571. *See also* Coagulation
    fate of, 529–30
    formation of, 527–29, 527t
    in wound healing, 173, *174*
Blood doping, 475
Blood gases, *524,* 525
    evaluation of, 525
Blood groups, 531–36
    ABO, 532, 534–35, *534,* 534t, *535,* 536t, 927
    human history and migration patterns and, 534
    Rh, 536, *537*
Bloodhound, 431
Blood pressure, 84, *84,* 523, 568, 570–80, 786, *787,* 789, 791, 810
    age-related changes in, 601
    arterial, 570–75
    blood viscosity and, *573,* 575
    blood volume and, 573–74, *573*
    central venous pressure, 579
    heart action and, 572–73, *572*
    measurement of, 572–73, *572, 573*
    peripheral resistance and, *573,* 574–75
    regulation of, 11–12, 391, 477, 493, *494,* 546, 560–61, 575–77, *576, 577*
    venous blood flow and, 577–79, *579*
Blood reservoir
    liver, 669
    veins, 569, 579
Blood substitute, 510
Blood tests, 970–72
Blood transfusion. *See* Transfusion
Blood urea nitrogen, 526, 699, 776, 785, 970
Blood vessels, 562–69, 570t
    angiogenesis, 564
    disorders of, 571, *571,* 598–99
    embryonic, 887
    response to autonomic stimulation, 414
    site of gene therapy, *942*
    spasm of, 526, 527t
Blood volume, 484, 493, *494,* 501, 510–11, *511,* 523, 546, 573–74, *573,* 579, 601, *787,* 789, 791
    determination of, 574
    in pregnancy, 516
    within veins and venules, *570*
Blue diaper syndrome, 799, 924
Body
    of mandible, **245**
    of penis, *831,* 841
    of sternum, 216, *217*
    of uterus, 854
    of vertebrae, 210, *212, 214*
Body cavities, 12–14, *13, 14*

Body coverings, 16
Body fat, 703–4, 858
Body fluid(s)
    composition of, 809–10, *810*
    distribution of, 808–11
    movement of fluid between compartments, 810–11, *811*
    osmolarity of, 813
Body fluid compartments, 809, *809*
Body mass index, 704, *704*
Body parts, structure-function relationships in, 4, *4*
Body regions, 23–25, *23, 24*
Body sections, 21–23, *22*
Body temperature
    exercise and, 289
    heart rate and, 561
    maintenance of, 703
    following ovulation, 863
    regulation of, 10–11, *11,* 169–71, *171,* 176, 391
    problems in, 170–71
Body weight, 703–4, 724
    desirable, 703–4
    gaining weight, 704
    losing weight, 705
Boil, 176
Bolus, 655
Bombay phenotype, 536
Bomb calorimeter, 702, *702*
Bond, 40–44, *42, 43, 44. See also specific types of bonds*
Bone(s), 147–48, *149,* 150t
    age-related changes in, 195, 231–32, *231*
    blood cell formation in, 194
    body movement, 191–94
    classification of, 182, *183*
    compact, 183–84, *183, 184, 185,* 189, 193, 232
    development of, 186–90, 187t, 189t
    effect of exercise on, 190, *190,* 195, 232
    endochondral, 186, *186*
    flat, 182, *183*
    fracture of. *See* Fracture
    functions of, 191–96
    growth of, 186–90
    homeostasis of bone tissue, 190
    inorganic salt storage in, 194, *196*
    intramembranous, 186–87, *186*
    irregular, 182, *183*
    long, 182, *183,* 189
    microscopic structure of, 184, *185*
    number of, 196–97, 196t
    parts of, 182–83, *183*
    remodeling of, 189
    repair of, 190
    sesamoid, 182, *183*
    short, 182, *183*
    spongy, 183–84, *183, 184, 185,* 186–89, *193*
    structure of, 182–85
    trabecular, 232
Bone cancer, 189

Bone density, 195
Bone loss, 195, 232
  on space flight, 190
Bone marrow, 183, 194, 217
  site of gene therapy, 939, *942*
Bone marrow transplantation, 513,
  522, 669
Bone mass, 190, *190*, 195, 232
Bone pain, 184
Botulinum toxin, 284
Bovine serum albumin, 633
Brachial artery, **33**, *570*, 572, *587*,
  588, *592*
  deep, *587*, 588, *592*
Brachialis muscle, *297*, 307t, 308,
  308t, *309*, **335, 960**
Brachial plexus, **33**, 405–6,
  407, 408
Brachial region, 24, *24*
Brachial vein, 591, *593, 594, 597*
Brachiocephalic artery, **34, 36**, 582,
  583, 585, 585t, 586, 592,
  963, 964, 965
Brachiocephalic vein, **33, 35**, *583*,
  591, *593, 594, 597, 610* **964**
Brachioradialis muscle, *297*, 308,
  308t, *309*, 311, **333, 335, 960**
Bradycardia, 562, *563*
Bradykinin, 493, 577
Bradypnea, 754, 764
Brain, *339*, 367, 380–96
  age-related changes in, 417
  arteries of, 584–87, *585, 586*
  basal nuclei of, 388, *389*
  blood-brain barrier, 69
  cerebellum of. *See* Cerebellum
  cerebrum of. *See* Cerebrum
  development of, 380–81,
    382, 382t
  imaging of, 6–7, *7, 57, 371*, 430
  midsagittal section of, **968**
  parts of, 396t
  site of gene therapy, *942*
  synapses in, 354
  veins from, 591
Brain death, 397
Brainstem, 367, 381, *383, 385*,
  389, 392–94, *392, 393, 395*,
  396t, **949**
Brain tumor, 6–7, 69, 348, 394, 397
Brain waves, 397, *397*
Bran cereal, 644
BRCA1 gene, 928
Breast(s), **30**, 861, *862*
  development of, 862
Breast cancer, 98, 928
  inherited, 865
  risk for, 864t
  treatment of, 864–65, *864*
Breast-feeding, 12, 905, 907
  agents contraindicated
    during, 906t
Breast self-exam, 864–65
Breath analysis, 758
Breath-holding, 756
Breathing
  age-related changes in, 764
  control of, 753–56, 756t
  exercise and, 756

factors affecting, 754–56,
  755, 756t
first breath, 906
mechanism of, 745–53
Breathing rate, 756, 821
Broad ligament, 848, *848, 853*, 854
Broca's area, 385, *385*
Bronchi, *733, 738*, **965**
  primary, 739–40, *740*
  right, **35**
  secondary, 739–41, *740*
  tertiary, 740, *740*
Bronchial artery, 582, 585t
Bronchial asthma, 752
Bronchial tree, 739–43, 744t
  branches of, 739–41, *740, 741*
Bronchioles, 735, *741, 742*, 764
  intralobular, 740
  respiratory, 740–42, *740, 741*
  terminal, 740, *740, 741*
Bronchitis, 755, 764, 920
Bronchopulmonary segment, 740
Bronchoscopy, 742
Brown-Séquard syndrome, 379
Brush border, 788
Buccal nerve, *401*
Buccal region, 24, *24*
Buccinator muscle, 298–99, *298*,
  299t, **956**
Buffer systems, 47, 819–21, 820t
Buffy coat, *511*
Bulb, of penis, *831*, 842
Bulbospongiosus muscle, 314–15,
  314, 315t
Bulbourethral gland, 139t, *798*,
  831, 840, 844t
Bulimia, 722, 724
Bulk elements, 38
Bundle branch, *558*
Bundle branch block, 557
Burns, 153, 173–74
  deep partial-thickness (second
    degree), 173
  full-thickness (third degree), 173
  rule of nines, 175, *175*
  superficial partial-thickness
    (first degree), 173
Bursae, 258, *258*
Bursitis, 272
Bush, President George, 488
Buttock, **334**
B vitamins, 709–11, 713t

C

Cachexia, 725
Cacosmia, 435
Caffeine, 127t, 663, 813, 906t
Calcaneal tendon, *297*, 321, *323*,
  324, **336, 962**
  torn, 323
Calcaneus, 229, *230*, 324
Calcitonin, 194–95, *196*, 471t, 475,
  487, 487t, 490, 502
Calcium, 712, 714, 716t
  absorption of, 190, 489, *490*,
    491, 686, 707, 714
  blood, 194, *196*, 487, 489–90,
    491, 526, 784t, 817, 885

bone, 194, *196*, 487, 489, *491*
in cells, 48, *49*
in coagulation, 527, 528t,
  529, 530t
deficiency of, 714
dietary, 195, 714
electrolyte balance, 813,
  816, *816*
in extracellular fluid, *810*
in glomerular filtrate, 784t
in heart action, 561
homeostasis, *196*
in hormone action, 476
in human body, 39t
in muscle contraction, 284, *285*,
  286, 287t, 294
in nerve impulse transmission,
  354–55, 357
serum, 970
tubular reabsorption of, 788
urine, 784t, 972
Calcium carbonate, 489
Calcium channel(s), 68
  abnormal, 340
  drugs that affect, 68, 68t
Calcium channel blockers,
  294, 578t
Calcium phosphate, 194, 489,
  712, 714
Calcium pump, 284, 286
Callus
  bony, 193, *193*
  cartilaginous, 193
Callus (skin), 159
Calmodulin, 294, 476
Calorie, 702
Calorimeter, bomb, 702, *702*
Canaliculi, 148, *149, 185*, 186
Canal of Schlemm, 452, *453*
Cancer
  bladder, 796
  bone, 189
  breast, 98, 864–65, *864*, 928
  colorectal, 644, 684–85,
    685, 685t
  development of, 94, *95*
  esophageal, 656
  genetic changes in cancer
    cells, 796
  hypercalcemia in, 817
  invasion of adjacent tissue, 133
  kidney, 790
  lung, 591, 735, *735*, 745,
    764, 920
  metastasis of, 70, 93, *95*, 612
  pain of, 428
  prostate, 189, 840
  skin, 127, 162–63, *163*
  testicular, 832
  thyroid, 41, 468
  treatment of, 94
    angiogenesis inhibitors, 564
    cytokine-based, 625
    gene-therapy, 941–42
    monoclonal antibodies, 624
Cancer cells, 93
  characteristics of, 93–94,
    94, 94t
  invasion by, *95*

Cancer syndromes, 882t
Cancer vaccine, 163, 942
Cannabinoids, 428
Capacitation, 841
Capillary(ies), 566–69, 570t
  arrangement of, 566–67, *567*
  blood flow in, regulation of, 567
  exchanges in, 568–69, *568*
  filtration across, 84, *84*
  permeability of, 566
  structure of, 566, *567*
Capitate, *223*
Capitulum, 221, *221*
Captopril, 435
Carbaminohemoglobin, 762–63, *762*
Carbohydrates, 695
  calories per gram, 701t, 702
  in cells, 48–50, *49*, 55t
  dietary requirement for,
    696–97, 701t
  dietary sources of, 695, 701t
  digestion of, 678, *678*,
    695–96, *696*
  metabolism of, 105, *114*,
    695–96, *696*, 701t
  storage, 113–14, *113*
Carbon, in human body, 39t
Carbonate
  in cells, 48, *49*
  in human body, 194
Carbon dioxide
  blood, 525, 822, 970
    regulation of blood
      pressure, 577
  in cells, 48, 49t
  from citric acid cycle, 111, *112*
  diffusion across cell membrane,
    81, *82*
  diffusion from blood into
    alveoli, 742, *742*
  diffusion through respiratory
    membrane, 757–58, *758*
  exchange in capillaries, 568
  partial pressure of, 754, *755*,
    756, 756t, 758, *758*, 761,
    761, 763
  respiratory excretion of,
    821, *821*
  structure of, *58*
  transport in blood, 762–63, *762*,
    763, 764t
Carbonic acid, 48, 754, 763, 817,
  819–21
Carbonic anhydrase, 762–63, 820
Carbon monoxide, 48, 761
Carboxyl group, 52, *52*, 819
Carboxypeptidase, 665, 677t
Carbuncle, 176
Carcinoma, 138
Cardiac arrest, 542, 561, 580, 599
Cardiac center, 393, 576, *576*
Cardiac control center, 560–61, *561*
Cardiac cycle, 551–54, *554, 555*
  regulation of, 560–61, *561*
Cardiac muscle, 294–95, *296*
  rhymicity of, 295
Cardiac muscle fibers, 554
Cardiac muscle tissue, 150, *151*,
  152t, 295t

Cardiac output, 572, 575–77, *576,* 599–600
Cardiac plexus, *413*
Cardiac tamponade, 579, 601
Cardiac vein, 551, *553*
   anterior, *552*
   great, *552*
   middle, *552*
   small, *552*
Cardioaccelerator center, 560–61
Cardioaccelerator reflex, 577
Cardioinhibitor center, 560
Cardioinhibitor reflex, 576, *576*
Cardiology, 25
Cardiomyopathy, 551
Cardiovascular system, 8t, 17–18, *17,* 542, *543. See also* Blood vessels; Heart
   age-related changes in, 598–601, 911t
   effect of exercise on, 579–80, 601
   effect of weightlessness on, 574
   fetal, 898–901, *899, 900,* 900t, *901*
   molecular causes of disease, 598–99, *598*
   of newborn, 907–8, *908,* 908t
   system interconnections of, 602
Cardioverter defibrillator, implantable, 542, *542*
Carina, *738,* 739
Carotene, 172, 706
Carotid artery, *570,* **953**
   common, **31, 32, 34, 35, 36,** *582, 583, 585,* 585t, *586, 587, 587, 592, 755,* **963, 964, 965**
   external, *585, 586, 587, 592*
   internal, *585, 586, 587, 592*
Carotid body, 587, 754, *755,* 756t
Carotid canal, 201–2, *202,* 209t, **242, 243**
Carotid foramen, **244**
Carotid sinus, *585,* 587
Carpal(s), 196t, 197, *198, 220,* 222, *223,* 223t, **335**
Carpal region, 24, *24*
Carpal tunnel syndrome, 409
Carpometacarpal joint, 263t
Carpus, 222–23
Carrier (heterozygote), 925
Carrier protein, 82, *83, 85, 85*
Cartilage, 146–47, *147, 148*
   articular, 183, *183, 187,* 257–58, *257, 258*
   bioengineered, 153
   elastic, 147, *147,* 150t
   fibrocartilage. *See* Fibrocartilage
   hyaline, 147, *147,* 150t, 187, 189, 738–39, *739*
Cartilaginous joints, 255–57, *256,* 260t
Casein, 725
Cast (immobilization), 193
Catabolism, 104–5
Catalase, 76, 106
Catalyst, 45
Cataract, 462

Catecholamines, 492
Cauda equina, *372,* 402, *404, 407,* **950, 951**
Caudate nucleus, 388, *389*
Causalgia, 463
CCR5 receptors, 62, 635
CD4 receptors, 62, 621, 635
Cecum, **33, 34,** *646, 673, 680, 681*
Celebrex, 272
Celecoxib, 272
Celiac artery, **35,** *583, 584,* 585t, *592*
Celiac disease, 679, 725
Celiac ganglion, *412*
Celiac plexus, *413*
Celiac region, 24, *24*
Cell(s), 4–5, *5,* 5t, 9, 62. *See also specific organelles*
   age-related changes in, 19
   chemical constituents of, 47–55
   cytoplasm, 70–78
   death of, 79
   eukaryotic, 62
   inorganic substances in, 47–48, *49*
   movements into and out of, 80–89, 89t
   organic substances in, 48–55, 55t
   prokaryotic, 62
   shapes of, 62, *63*
   size of, 62, *63*
   structure of, *64*
   surface-to-volume ratio of, 92–93
Cell body, of neuron, 340, *341*
Cell cycle, 90–92, *90, 91*
   anaphase of, *91,* 92, 92t
   cytokinesis of, *90, 91,* 92, *93*
   interphase of, 90, *90, 91*
   metaphase of, *91,* 92, 92t
   mitosis of, 90–92, *90, 91*
   prophase of, 90–92, *90, 91,* 92t
   telophase of, *91,* 92, 92t
Cell division, control of, 92–94
Cell implants, 153, 348, 381, 633, *942*
Cell lineages, 97
Cell membrane, 62, 64–70, *64, 66, 67,* 80t
   active transport across, 85, *85,* 89t
   cellular adhesion molecules, 67–70, *70*
   characteristics of, 64–65, *65*
   diffusion across, 80–81, *81–82,* 89t
   endocytosis by, 85–87, *86, 87,* 89t
   exocytosis by, 85, *86, 87, 88,* 89t
   facilitated diffusion across, 81–82, *83,* 89t
   filtration across, 83–84, *84,* 89t
   intercellular junctions, 67, *69,* 70t
   membrane potential. *See* Membrane potential
   osmosis across, 82–83, *83,* 89t
   permeability of, 65–66, 81
   structure of, 65–67, *66, 67*
   transcytosis by, 88, *89,* 89t
Cell surface proteins, 67, 67t

Cellular adhesion molecules, 66–70, 67t, *70,* 520
Cellular immune response, 620–21, *622*
Cellular respiration, 108–15, 732–33, 974–77, *975, 976*
   aerobic reactions of, 108, *109,* 111–12, *112, 113,* 288, *288,* 817
   anaerobic reactions of, 108–11, *109,* 288, *288,* 817
   citric acid cycle. *See* Citric acid cycle
   definition of, 108
   electron transport chain. *See* Electron transport chain
   glycolysis. *See* Glycolysis
   in skeletal muscles, 288, *288*
Cellular turnover, in small intestine, 676
Cellulose, 48, 50, 695
Cementum, 652, *652*
Central canal, spinal cord, 148, *149,* 184, *185,* 347, 372, *373, 406*
Central chemoreceptors, 754
Central nervous system, 338, *339,* 367–96, 396t. *See also* Brain; Spinal cord
Central sulcus, 384, *384, 385, 386, 386*
Central vein, 666, *668*
Central venous pressure, 579
Centrioles, *64,* 76, *76,* 90, *91,* 92
Centromere, 90, *91,* 835, *921*
Centrosome, 76, *76,* 80t
Cephalic region, 24, *24*
Cephalic vein, 591, *593, 594, 597*
Cerebellar cortex, 395
Cerebellar hemisphere, *384*
Cerebellar peduncle, *392,* 395
   inferior, 395, *395*
   middle, 395, *395*
   superior, 395, *395*
Cerebellum, *367,* 381, *382,* 382t, *383, 385, 389,* 395, *395,* 396t, **948, 949, 968**
Cerebral aqueduct, 369, *370, 369,* 382t, 392, *393*
Cerebral arterial circle, 586, *586*
Cerebral artery
   anterior, *478, 586,* 587
   middle, *586,* 587
   posterior, 586
Cerebral cortex, 384, 427
   association area of, 386–87
   functional regions of, 385–88
Cerebral hemispheres, 381–84, *382*
   hemisphere dominance, 387–88
Cerebral palsy, 389
Cerebral peduncles, 392
Cerebral thrombosis, 530
Cerebral vascular accident. *See* Stroke
Cerebrospinal fluid, 368–70
   blood-cerebrospinal fluid barrier, 369
   composition of, 369–70
   production of, 369, *370*

Cerebrospinal fluid pressure, 371
Cerebrum, *367,* 381, 382t, *383,* 396t, **948, 949**
   functions of, 384–88, 387t
   injuries and abnormalities of, 389
   sensory areas of, 386, *386*
   structure of, 381–84
Ceruminous gland, 168, 168t, 434
Cervical cap, 863t, 864
Cervical curvature, 210, *211*
Cervical lymph nodes, 613, *613*
Cervical nerves, *401,* 402, *404*
Cervical orifice, *853,* 854
Cervical plexus, 405, *407,* 409
Cervical region, 24, *24*
Cervical vertebrae, 196t, *211,* 212, *213, 214,* 214t, 380, **949**
Cervix, *847, 853,* 854, *877,* 901
   dilation of, 902, *903*
Cesarean section, 914
CFTR gene, 925–26, *926*
Chaperone proteins, 122, 123t
Checkpoint, cell cycle, 90, *90, 91*
Cheek(s), 648–49, 652t
Cheekbrush tests, 648
Chemical barrier(s), to infection, 617–18, 619t
Chemical barrier contraception, 864–65
Chemical energy, release of, 107–8
Chemical reactions, 44–45
Chemoattractants, 70, *70*
Chemoreceptors, 423
   central, 754
   peripheral, 754
Chemotaxis, 520, 618, 627, *628*
Chemotherapy, 94
   for breast cancer, 865
Cheney, Vice President Dick, 542
Cherry red spot, 75
Chest pain, 600
Chewing. *See* Mastication
Cheyne-Stokes respiration, 766
Chicken pox, 161t
Chief cells, 659, 660, 661t
Child, skull of, **252**
Childbirth, 208, 214, 256, 294, 359, 755, 800, 901–4, *903*
Childhood, 909
Chlamydia infection, 867t
Chloride, 763
   blood, 526, 970
   in cells, 48, *49*
   electrolyte balance, 813, 817
   in extracellular fluid, *810*
   in plasma, glomerular filtrate, and urine, 784t
Chloride channels, 68
   abnormal, 68, 665, 922
   drugs that affect, 68, 68t
Chloride shift, 763, *763*
Chlorine, 39t, 715, 716t
Chlorolabe, 460
Cholangiogram, 672
Cholecalciferol, 707
Cholecystectomy, 671–72
Cholecystitis, 671–72, 686
Cholecystogram, 672

Cholecystokinin, 662, 665, 671, 672t, *673*, 686, 695, 695t
Cholelithiasis, 688
Cholera, 814
Cholestasis, 688
Cholesterol, 471, 489, 571
  bile, 525, 671
  blood, 87, 525, 599–600, 920, 940, 970
  dietary, 707
  entry into cells, 87
  membrane, *67*
  structure of, *51*
  synthesis of, 698, *698*
Cholinergic fibers, 414–15, *414*
Cholinergic receptors, *416*
Chondrocytes, 146–47, *147*
Chondrodysplasia, 143t
Chordae tendineae, 546, *547, 548*
Chorion, *886, 887, 891, 893*
Chorionic villi, *886, 887, 891, 899*
Chorionic villus sampling, 937, 937t, *938*
Choroid artery, anterior, *585, 587*
Choroid coat, 450, *450, 452, 454,* 455t
Choroid plexus, 347, 369, *370*
Chromaffin cells, 492, *492*
Chromaffin granules, 492
Chromatids, 90, *91,* 835–36, *835, 921*
Chromatin, *64,* 79, *79,* 80t, *91,* 117
Chromatophilic substance, 340, *341*
Chromium, 718, 719t
Chromosome(s), 76, 78, *103,* 115–16, *117,* 921, *921,* 924–25. *See also* Meiosis; Mitosis
Chromosome disorders, 934–37, 934t, *935*
  maternal age and, 848, 936t
Chronic bronchitis, 735, 764
Chronic fatigue syndrome, 634
Chronic granulomatous disease, 518
Chronic leukemia, 522
Chronic obstructive pulmonary disease, 755
Chronic pain fibers, 426–28
Chylomicrons, 525, 526t, 678–79, *679*
Chyme, 662
Chymopapain, 216
Chymotrypsin, 665, 677t
Cigarette smoking, 76, 906t
  effects on respiratory system, 735, *735*
  nicotine addiction, 360–61, *361*
  in pregnancy, 894
  smoker's cough, 735
  smoking cessation, 165, 735
Cilia, *64,* 76–77, *77,* 80t, 133, 135, *136*
Ciliary body, 450–51, *450, 451, 452, 453,* 455t
Ciliary ganglion, *413*
Ciliary muscles, 449t, 451–52, *452, 453,* 457

Ciliary plexus, *412*
Ciliary process, 451, *452, 453*
Ciliated epithelium, of respiratory tract, 733–34, *734*
Circadian rhythm, 499
Circular muscles, of iris, 453, *453*
Circular sulcus, 384
Circulation. *See also* Cardiovascular system
  as characteristic of life, 8t
  fetal, 898–901, *899, 900,* 900t, *901*
  paths of, 580–81
Circumcision, 842
Circumduction, 261, *262*
Circumflex artery, 549, *552, 553*
  anterior, *587*
  posterior, *587*
Cirrhosis, 688
Cisterna chyli, *610*
Cisternae
  of Golgi apparatus, 71–72, *72*
  of sarcoplasmic reticulum, 282, *282*
Citrated whole blood, 536
Citric acid, *109,* 111, *112, 113*
Citric acid cycle, 108, *109, 110,* 111, *112, 113, 114,* 288, *288,* 974, *976*
Clavicle, **30, 31,** 197, *198,* 218, *219,* 223t, *264, 265,* **334,** *743*
Cleavage, 880–84, *881, 884,* 887t
  indeterminate, 882
Cleavage embryo, 880
Cleavage furrow, *91*
Cleft palate, 205
Cleidocranial dysplasia, 218
Climacteric
  female. *See* Menopause
  male, 846
Clitoris, *847,* 855–56, *855, 856t, 897*
Clomid, 850
Clomiphene, 850
Clone, 621
Cloning, to produce therapeutic stem cells, 98, *99*
Clostridium botulinum, 284
Clot-busting drugs, 530, 600
Clotting. *See* Coagulation
Clotting factors, 2, 527–28, *529,* 530t, 533, 939
Coagulation, 527–31, 527t, 528t, *529*
  extrinsic clotting mechanism, 527–28, *528,* 528t
  fate of blood clots, 529–30
  intrinsic clotting mechanism, 527–28, 528t
  laboratory tests for, 528
  prevention of, 530–31, 531t
  vitamin K in, 708
Cobalt, 39t, 711, *711,* 718, 719t
Cobalt-60, 41
Cocaine, 358t, 360, 838, 906t
Cocarboxylase, 709
Coccygeal nerves, 402, *404*
Coccygeus muscle, 314–15, *314,* 315t

Coccyx, 196t, 197, *198,* 209, *211,* 214–15, 214t, *215,* 225, 227t, **334,** *847,* **951**
Cochlea, *435,* 437–49, *438, 440, 444*
Cochlear branch, of vestibulocochlear nerve, 401
Cochlear duct, *438,* 439, *439, 440, 441*
Cochlear implant, 443
Cochlear nerve, *438, 439, 442*
Codeine, 359, 428
Codominance, 927
Codon, 119–21, *121,* 122t, *120,* 123t
Coenzyme(s), 107
Coenzyme A, 710
Cofactors, 107
Cogwheel rigidity, 312
Coitus interruptus, 862, 863t
Cold receptors, 424
Coley, William, 625
Coley's toxins, 625
Colitis, 683
Collagen, *58,* 142, 147, 184, 712
  abnormal, 143, *143,* 143t, 158
  structure of, 142
Collagen injections, 142
Collagenous fibers, 142, *142, 144,* 144t, *145, 146, 147, 148*
Collateral(s) (axon branches), *341,* 342
Collateral circulation, 550
Collateral ganglia, 410
Collecting duct, 608–9, *611, 775, 778, 779,* 794t
Collectins, 618
Colles fracture, 232
Colloid, 485, *486*
Colloid osmotic pressure, 523–24, *568, 568,* 611, 784–85, 810
Colon, 680–81, **948, 966**
  ascending, **33, 34,** *673,* 681, *681,* **967**
  descending, **33, 35,** *681,* 682, *681,* **967**
  sigmoid, **34, 35,** *681,* 682, **951, 967**
  transverse, **33, 34,** *674,* 681, *681,* **950, 955, 967**
Colonoscopy, 684–85, 685t
Colony-stimulating factors, 511, 518, 620, 620t, 625
Color blindness, 422, 460, 932
  red-green, 932
Colorectal cancer, 644, 684–85, *685,* 685t
Color vision, 422, 460
Colostomy, 685
Colostrum, 627, 904, 907
Colposcope, 855
Coma, 393–94
  diabetic, 499
  hepatic, 669
Comedone, 169
Comminuted fracture, *192*
Common nerve pathways, 426, *427*
Communicating artery
  anterior, *586*
  posterior, *587*

Compact bone, 183–84, *183, 184, 185,* 189, *193,* 232
Compartment, 278
Compartment syndrome, 278
Complement, 626–27, 628t
Complementary base pairs, 116, 123, *123*
Complete penetrance, 928
Complete proteins, 700
Complex traits, 929–31, *929, 930*
Compliance, lung, 747–48
Compound(s), 38, 40
Compound fracture, 192
Compound gland, 138, *139,* 139t
Computed tomography scan, 2, 56–57, *56*
  of brain, *371*
  quantitative, 195
Computer-aided sperm analysis, 842–43, *842, 843*
Concave lens, 456, *456, 457*
Concentration gradient, 80–82, *81*
Concentric contraction, 292, *292*
Conception, date of, 879
Concussion, 389
Condom
  female, 863t, 864
  male, 863–64, 863t, *866*
Conduction (heat loss), 170
Conduction system, of heart, 554–58, *558,* 563, 601
Conductive deafness, 443
Condyles, 199t
  of femur, *229*
  of humerus, 221
  mandibular, *201,* 206, **245**
  occipital, 201, *202,* **242, 243, 244, 246, 248**
  of tibia, 228, *228, 229*
Condyloid joint, *259,* 260, 260t, 263t
Cones, 422, *454,* 458, *459,* 460
Confocal microscope, 65
Conformation, of proteins, *53,* 54
Congenital rubella syndrome, 894
Congestive heart failure, 601
Conization, 868
Conjoined twins, 902, *902*
Conjunctiva, *447,* 448
Conjunctivitis, 448, 463
Connecting stalk, 886, *886, 891,* 892
Connective tissue, *131,* 132, 132t, 141–48, 150t
  categories of, 144
  cell types in, 141–42, *141, 142,* 144t
  characteristics of, 141
  covering skeletal muscles, 278–79, *280*
  dense, 141–42, 145–46, *146,* 150t, *159*
  disorders of, 598, *598*
  elastic, 146, *146,* 150t
  fibers of, 142–44, *142, 144*
  loose, 141–42, 144, *144,* 150t
  reticular, 145, *145,* 150t
  specialized, 144
Connective tissue proper, 144

Consciousness, 366
Constipation, 644, 686
Constrictor muscles, of pharynx, 655, *656*
Contact dermatitis, 164
  allergic, 164
  irritant, 164
Contact inhibition, 93
Contact lens, 457
Continuous ambulatory peritoneal dialysis, 774
Contraception, 862
Contractile ring, 92
Contracture, 327
Contralateral, definition of, 21
Control center, 10, *10, 11*
Conus medullaris, 372, *372*
Convection, 170
Convergence, of nerve impulses, 359, *359*
Convex lens, 456, *456, 457*
Convolutions, cerebrum, 381–82, *383*
Convulsion, 327, 355
Copper, 39t, 718, 719t
Coracobrachialis muscle, **31, 32, 33,** 305, *306,* 306t, 307t
Coracohumeral ligament, 264, *265*
Coracoid process, 218, *219,* 265
Cordotomy, 417
Corn (skin), 159
Cornea, *447,* 450, *450, 451,* 455t, 456
Corneal transplant, 93, 451
Corniculate cartilage, 736–37, *737, 738*
Coronal suture, 199, *200, 201,* 206, *210,* **238, 239, 248**
Corona radiata, 850, *852, 877, 880*
Coronary angiography, 549, *553,* 600
Coronary artery, 548, 549, *551, 552, 553,* 582, 585t, *592*
Coronary artery disease, 599–600
Coronary bypass surgery, 564, 600
Coronary embolism, 578
Coronary ligament, 665, *667*
Coronary sinus, 546, *547,* 551, *552, 553*
Coronary stent, 600
Coronary thrombosis, 530, 550, 578, 600
Coronoid fossa, 221, *221*
Coronoid process, 194, *201,* 206, 222, *222,* **245, 266**
Corpora cavernosa, 841, *841,* 855
Corpora quadrigemina, 392, *392, 393*
Cor pulmonale, 601
Corpus albicans, *852, 859, 860*
Corpus callosum, 381, *383,* 388, *393, 395,* **949, 968**
Corpus cavernosum, *831*
Corpus luteum, *852, 859, 860,* 884
Corpus spongiosum, *831, 841, 841*
Cortex, of ovary, 848
Cortical granules, 878
Cortical nephron, 780–81, *782*

Corticospinal tract, 376–77, *379,* 381, 385
  anterior, *377, 378,* 379t
  lateral, *377, 378,* 379t
Corticosteroids, 160, 169t
Corticotropes, 480
Corticotropin-releasing hormone, 471t, *481,* 483, 494–95, *495,* 501, *501*
Cortisol, *470,* 471, 471t, 493–95, *494, 495,* 495t, 501, *501*
  elevated, 502
  to reduce inflammation, 494
Costal cartilage, *15,* 216, *217, 219,* 255, *256,* **954, 955**
Costal fibers, 746
Costal region, 24, *24*
Costocervical artery, 587
Coughing, 393, 751, 751t
Coumadin, 530
Countercurrent mechanism, 791, *792, 793*
Countercurrent multiplier, 791, 793
Covalent bond, 43, *43*
  double, 43, *43*
  single, 43
  triple, 43
Cowlick, 166
Coxae, 196t, 197, *198,* 224, *226,* 231t, *268,* **955**
Coxal region, 24, *24*
Cox-2 inhibitors, 685
Cranial branch, of accessory nerve, 401
Cranial cavity, 12, *13, 14,* 367
  floor of, *204,* **249**
Cranial nerves, *339,* 396–97, 396t, 399–401, *399, 400, 401, 402, 403*
  I (olfactory), *399, 400,* 403t
  II (optic), *392, 399, 400,* 403t, 450, *450,* 453, 455, *455,* 460–61, *461, 478,* **952**
  III (oculomotor), 399–400, *399,* 403t, *413, 414,* 449t, *478*
  IV (trochlear), 399–400, *399,* 403t, 449t, *478*
  V (trigeminal), 300t, *399, 400, 400,* 403t
  VI (abducens), 399–400, *399,* 403t, 449t
  VII (facial), 299t, *399,* 400–401, *401,* 403t, 414, 434, *441,* 449t
  VIII (vestibulocochlear), *399,* 401, 435, 439, *442,* 444, *444*
  IX (glossopharyngeal), *399,* 401, 403t, *413, 414,* 434
  X (vagus nerve), *399,* 401, *402,* 403t, *413, 414,* 434, 560, 661, 755, **965**
  XI (accessory), 302t, 304t, 399, *399,* 401, 403t
  XII (hypoglossal), 399, *399,* 401, 403t
Craniopagus twins, 902, *902*
Craniotomy, 417

Cranium, 197, *198,* 199–203, 201t, *203, 204, 205*
Crash diet potomania, 814
Creatine, 287, *287,* 526
  as dietary supplement, 722
  serum, 970
  tubular reabsorption of, 788
Creatine phosphate, 287, *287,* 289, 526
Creatine phosphokinase, 287
Creatinine, 526, 794
  in plasma and glomerular filtrate, 784t
  serum, 970
  urine, 776, 784t, 972
Creatinine clearance test, 794–95, 972
Cremasteric reflex, 377
Crest (bone), 199t
Cretinism, 488
Cribriform plate, 202, *203, 204, 206, 207,* **250,** 431
Cricoid cartilage, 736–37, *737, 738, 743*
Crista ampullaris, 444, *446*
Cristae, of mitochondria, 73, *74*
Crista galli, 203, *203, 204, 206, 207,* **246, 249, 250**
Critical period, 892, *893*
Crohn disease, 684
Cross-bridge(s), 279, *281, 284, 285*
Cross-bridge cycling, 285–86, *285, 286*
Crossed extensor reflex, 375
Crossing-over, *835, 836, 836*
Cross section, 23, *23*
Crown, of tooth, 651, *652*
Crow's feet, 299
Cruciate ligament, 270–71, *270*
  anterior, *270,* 271
    torn, 271
  posterior, *270,* 271
Crura, of penis, 842
Crural fibers, 746
Crural region, 24, *24*
Crying, 751, 751t
Cryoprecipitate, 533
Cryptorchidism, 832
Cryptosporidiosis, 617
Cubital region, 24, *24*
Cubital vein, median, 591, *593, 597*
Cuboid, *230*
Cuneiform bone
  lateral, *230*
  medial, *230*
Cuneiform cartilage, 736–37, *737, 738*
Cupula, 444–45, *446*
Curare, 358t
Curettage, 868
Cushing syndrome, 496
Cuspid, 650–51, *651,* 651t
Cutaneous carcinoma, 163, *163*
Cutaneous melanoma, 163, *163*
Cutaneous membrane, 153. *See also* Skin
Cuts, healing of skin, 173, *174*
Cutshall, Cynthia, 940

CXCR4 receptors, 635
Cyanide, 107
Cyanocobalamin, 599, 660, 669, 683, 711, *711,* 713t, 718
  in red cell production, 515–16, *515,* 516t
Cyanolabe, 460
Cyanosis, 172, 513
Cyclic adenosine monophosphate, 472, *473,* 474–75, *475*
Cyclic guanosine monophosphate, 460, 475, 830
Cyclin, 92
Cyclophosphamide, 865
Cyclosporine, 632, 906t
Cyst, 176
Cystectomy, 802
Cysteine, 715
Cysterna chyli, 678
Cystic duct, **34,** *664, 667, 671, 673*
Cystic fibrosis, 54, 68, 665, 709, 743, 882t, 920, *922*
  diarrheal disorders and, 927
  gene therapy in, 939, *942*
  inheritance of, 925–26, *926*
  mutation in, 127
Cystitis, 795, 802
Cystoscope, 802
Cystotomy, 802
Cytochrome(s), 976–77
Cytochrome oxidase, 977
Cytocrine secretion, 162
Cytokines, 620, 620t, 623–25, *626*
Cytokinesis, *91,* 92
Cytoplasm, 62
Cytoplasmic organelles, 62
Cytoscope, 840
Cytosine, 116, *117,* 118, 119t, *979*
Cytoskeleton, 70, *77, 78*
Cytosol, 62
Cytotoxic T cells, 620–21, 635

D

Dalton, John, 422
Dander, 619
Dark-adapted eye, 460
Dark meat, 293
Dartos muscle, 841
DaSilva, Ashanti, 940, *940*
ddC, 635
ddI, 635
Dead space
  alveolar, 750
  anatomic, 750
  physiologic, 750
Deafness
  conductive, 443
  sensorineural, 443
Deamination, 699, *700,* 821
Death, 910
  leading causes of, 912, 913t
Decibel, 442
Decidua basalis, 887, *891*
Decomposition, 44
Deep, definition of, 21
Defecation reflex, 684
Defensins, 617–18

Defibrillator, 562
 external, 542
De Humani Corporis Fabrica
 (Versalius), 3
Dehydration, 171, 484, 499,
 573–74, 722, 814, *814*
Dehydration synthesis, 104,
 104, 105
Dehydrocholesterol, 190, 489
Delayed-reaction allergy, 632
Delta cells, 496
Delta waves, 397, *397*
Deltoid muscle, **30, 31,** 296–97,
 297, 303, 304, 305, *305, 306,*
 307–8, *307,* **333, 334, 957,
 958, 959**
Deltoid process, **333**
Deltoid tuberosity, 221, *221*
Denaturation, of proteins, 54
Dendrites, 338, *339,* 340, *341,
 342,* 422
Dens, 212
Dense connective tissue, 141–42,
 145–46, *146,* 150t, *159*
 irregular, 145–46
 regular, 145, *146*
Densitometer scanner, 195
Dental alveoli, 205, 207
Dental caries, 653, *653,* 686, 701t,
 718, 724–25
Dentate gyrus, 338
Dentate nucleus, 395
Denticulate ligament, 367
Dentin, 652, *652*
Deoxyhemoglobin, 511
Deoxyribonuclease, 743
Deoxyribose, 54, *55,* 115, 119t, 696
Depolarized membrane, 350
Depo-Provera, 863t, 865–66
Depression (joint), *262,* 263
Depression (mental illness), 724
 clinical, 357t
Dermal papillae, *160,* 164
Dermatitis, contact, 164
Dermatology, 25
Dermatome, 402, *405*
Dermis, 158, *160, 161,* 164, *165*
Descending aorta, 582, 585t
Descending colon, **34,** *681,* 682
Descending nerve tracts, 376, *377,*
 378, 379t
Desirable weight, 703–4
Desmosomes, 67, *69,* 70t, 133, 159
Detached retina, 458, 462
Detoxification, 669, 686
Detrusor muscle, 797–99, *797*
Development, 96, 876. *See also*
 Postnatal period; Prenatal
 period
 of bone, 186–90, 187t, 189t
 of brain, 380–81, *382,* 382t
Deviated septum, 733
Diabetes insipidus, 484, 812
Diabetes mellitus, 153, 724, *942*
 ketone body production,
 698, 823
 maturity-onset diabetes of the
 young, 497
 retinopathy of, 462

type I (insulin-dependent), 468,
 499, 633, 633t, 788
type II (noninsulin-dependent),
 499, 502, 895
Diacylglycerol, 475
Diaminoanisole, 127t
Diamox, 435
Diapedesis, 520, *520,* 618
Diaphoresis, 550
Diaphragm (birth control), 863t,
 864, *866*
Diaphragm (muscle), 12, *13,* **32, 33,
 34, 35, 36,** *657,* 746–48, *748,
 755,* **950, 954, 955, 963, 964,
 965, 966, 967**
Diaphysis, 183, *183, 184,* 187, 255
Diarrhea, 680, 714–15, 772,
 814, 823
Diastole, 551
Diastolic pressure, 570, 573, *573*
Diencephalon, 381, *382, 383,*
 390–92, *393,* 396t, **968**
Diet
 for athletes, 722
 colorectal cancer and, 644
 effect on red cell production,
 515–16, 516t
 healthy eating, 718–24
 in pregnancy, 876, 886, 894–96
 weight loss, 705
Dietary supplements, 718,
 720–22, *720*
Diet drugs, 705
Diethylstilbestrol, 855, *893*
Differential white blood cell
 count, 521
Differentiated cells, 62, 96
Diffusion, 80–81, *81–82,* 89t
 facilitated, 81–82, *83,* 89t
 through respiratory membrane,
 757–58, *758*
Diffusional equilibrium, 80–81
Digestion, 644, 695
 as characteristic of life, 8t
Digestive enzymes, 677, 677t
Digestive enzyme supplements, 665
Digestive glands, 500
Digestive system, 8t, 18, *18,* 643–87,
 645. See also specific organs
 age-related changes in, 686, 911t
 system interconnections of, 687
Digital artery, *587*
Digital rectal exam, 685t, 840
Digital region, 24, *24*
Diglycerides, 679
Dihydrotestosterone, 845
Dihydroxyacetone phosphate, *110*
Dihydroxycholecalciferol, 707
Dimples, *924*
Dipeptidase, *678*
Dipeptide, 105
Diploid, 921
Diplopia, 450, 463
Disaccharides, 48, *49,* 695
 metabolism of, 104–5, *104*
Dishpan hands, 164
Dislocation
 of patella, 227
 of shoulder, 265

Dissecting aneurysm, 571
Disseminated intravascular
 clotting, 528
Distal, definition of, 21
Distal convoluted tubule, *777, 778,
 779, 781, 783,* 788, 791,
 791, 794t
Diuresis, 788, 802, 824
Diuretics, 68t, 483–84, 578, 578t,
 714–15, 802, 813, 823
Divergence, of nerve impulse, *359,*
 360–61
Diverticulitis, 688
Diverticulosis, 684
Dizygotic twins, 914
DNA (deoxyribonucleic acid), 54,
 115, *919,* 921. *See also*
 Chromosome(s); Gene
 antiparallel strands in, 115,
 116, 117
 double helix, 116, *978*
 genetic information, 115–16
 mitochondrial, 118
 mutations in. *See* Mutations
 repair of, 124–25, 127, 163, 912
 replication of, 123, *123,* 125
 structure of, *37, 55,* 115–16,
 116, 117, 119, 119t,
 978, 979
DNA chips. *See* DNA microarrays
DNA fingerprinting, 118
DNA microarrays, 920, *920,* 941
DNA polymerase, 123–24
DNA primer, 124
DNA probe, 850
DNA profiling, 648
DNase, 743
Dominant follicle, 850
Dominant inheritance, 925–27
 different dominance
 relationships, 927, *928*
 X-linked, 932
Dopa, 492
Dopa decarboxylase, 492
Dopamine, 153, 356, 357t, 358t,
 360–61, 388, 390–91,
 416, 492
Dopamine betahydroxylase, 492
Dopamine receptors, 360
Dorsal arch vein, *593*
Dorsal cavity, 12, *13*
Dorsal column stimulator, 428
Dorsalis pedis artery, *570,* 590,
 590, 592
Dorsalis pedis vein, *596*
Dorsal root, *368, 373,* 402,
 406, 411
Dorsal root ganglion, *368, 373,* 402,
 406, 410, 427
Dorsiflexion, 261, *261*
Dorsum, *24,* 25
Double-contrast barium enema,
 685, 685t
Double covalent bond, 43, *43*
Double helix, 116, *978*
Double vision, 450
Down syndrome, 934, 936, *936*
Doxorubicin, 865, 906t
Dried plasma, 536

Drug(s)
 affecting ion channels, 68, 68t
 causing hair loss, 166
 delivery by cell implant, 153
 delivery to brain, 69
 growth factors, 93
 routes of administration of, 164
 that alter autonomic
 functions, 416
 that alter neurotransmitter
 levels, 358t
 that alter taste and smell, 434
 in treatment of obesity, 705
Drug addiction, 359–61, *361*
Dry eyes, 462
Dry mouth, 686, 812
Ductus arteriosus, 899, *900,* 900t,
 908, *908,* 908t
Ductus venosus, 899, *900,* 900t,
 908, *908,* 908t
Dumping syndrome, 688, 705
Duodenum, *15,* **34, 35,** *658, 659,*
 663, 673–74, *673, 674*
Dural sheath, *368*
Dural sinus, 367, *367*
Dura mater, 367, *367, 368,*
 370, **952**
Dwarfism, hypopituitary, 190,
 482, *942*
Dynamic equilibrium, 443–46, 700
Dysentery, 688
Dysgeusia, 434t
Dyslexia, 387
Dysmelodia, 924
Dysmenorrhea, 868
Dysosmia, 434t
Dyspepsia, 688
Dysphagia, 688
Dyspnea, 550, 766
Dystrophic epidermolysis
 bullosa, 143
Dystrophin, 125, 282, 932
Dysuria, 802

E

Ear, 434–42
 external, 434
 inner, 437–39, *438,* 446
 middle, 436–37, *435, 436*
Ear infection, 437
Ear popping, 437
Ear wax, 168, 434, 443
Eating disorder, 694, 722–24
Eccentric contraction, 292, *292*
Eccrine glands, 139t, *165,* 167,
 168, 168t
Echinacea, 720
Echocardiography, 551
Eclampsia, 914
Ectoderm, 886, *886, 888*
Ectodermal cells, 886–87
Ectopic pregnancy, 883
Eczema, 176, 630
Edema, 84, 524, 569, 579, 611–12,
 618, 789, 814–15, 815t
 nutritional, 701
 pulmonary, 581
Edible vaccine, 629

Effector, 10, *10, 11,* 339, 373, *375,* 375t
Efferent arterioles, 479, 776, *777, 778, 779, 781, 783, 784, 785, 787*
Egg cells, 62, *63, 829,* 846, 921
transport of, 876–77, *877*
Ehlers-Danlos syndrome, *143*
Ehrlich, Paul, 69, 624
Ejaculation, 842–44, *844,* 877
Ejaculatory duct, *831,* 839
Elastic cartilage, 147, *147,* 150t
Elastic connective tissue, 146, *146,* 150t
Elastic fibers, 142–43, *142, 144,* 144t, *146, 147*
Elastic recoil, of lung tissue, 748
Elastin, 142
to prevent scar tissue adhesions, 143
Elbow joint, 263t, 265–67, *266*
movement at, 192–94, *194*
Elderly. *See also* Age-related changes
falls among, 232, 232t, 417
skull of, **252**
Electrocardiogram, *555,* 558–60, *559, 560, 562*
Electrocorticogram, 397
Electroencephalogram, 397
Electrolyte(s), 46–47
absorption in large intestine, 683
absorption in small intestine, 679, 680t
bile, 671
in cells, 48, *49*
plasma, 526
Electrolyte balance, 48, *807,* 808, 813–17, *813*
Electrolyte intake, 813
regulation of, 813
Electrolyte output, 813
regulation of, 814–17
Electromyography, 300, 327
Electron(s), 38–39, *39,* 39t
Electron configuration, 42
Electron microscope, 65, *65*
Electron shells, 40–41, *41*
Electron transport chain, 108, *109,* 110–12, *113, 114,* 288, *288,* 976–77
Elements, 38
in human body, 39t
Periodic Table of, 969
Elevation (joint), *262,* 263
Emaciation, 725
Embolectomy, 601
Embolism, 530
Embolus, 530, 550, 571
Embryo, 849
Embryonic disc, 886
Embryonic stage, 886–92, *886,* 887t, *888, 889, 890, 891,* 898t
Embryonic stem cells, 98, *99*
Embryo proper, 882
Emetics, 664
Emission, 839, 843–44, *844*
Emmetropia, 463
Emotion(s), 366, 387, 392

Emotional stress. *See* Psychological stress
Emphysema, 735, 752, *752,* 755, 758, 764, 822
hereditary, *942*
inherited, 882t
Emulsification, 671
Enamel, 651–52, *652,* 718
Encephalitis, 417
Endarterectomy, 601
End-diastolic volume, 575–76
Endocarditis, 546, 554, 557
Endocardium, 545, *545,* 545t
Endochondral bone, 186, *186*
Endochondral ossification, 187, *187,* 187t
Endocrine gland, 138, 468, *468*
Endocrine paraneoplastic syndrome, 489
Endocrine system, 8t, 17, *17,* 467–503. *See also specific glands and hormones*
age-related changes in, 502, 911t
characteristics of, 468–69, *468*
compared to nervous system, 469, *469,* 469t
system interconnections of, 503
Endocrinology, 25
Endocytosis, 85–87, *86, 87,* 89t
receptor-mediated, 85, 87, *87,* 89t, 525, 679
Endoderm, 886, *886, 888*
Endodermal cells, 887
Endodontitis, 652
Endogenous pyrogen, 171
Endolymph, 437, *446*
Endometriosis, 861
Endometritis, 868
Endometrium, *853,* 854, *854,* 858–59, *884*
Endomysium, 278, *280*
Endoneurium, 396, *398*
Endoplasmic reticulum, 70–71, *71,* 80t
rough, *64,* 71, *71*
smooth, *64,* 71, *71, 73*
Endorphins, 357t, 358–59, 428
beta endorphin, 358
Endoscopic papillotomy, 672
Endostatin, 564
Endosteum, 183, *183, 185*
Endosymbiont theory, 73
Endothelin, 577
Endothelium, 545, 563–66, *565*
site of gene therapy, 940
End-stage renal disease, 776
End-systolic volume, 575
Energy
activation, 106, 108
for metabolic reactions, 107–8
for muscle contraction, 287–88, *287, 288*
values of foods, 702, *702*
Energy balance, 703, 724
Energy expenditures, 702–4
Energy requirements, 702–3, 703t, 724, 724t
Energy shells, 40
Enkephalins, 357–58, 357t, 428

Enriched foods, 707
Enteritis, 688
Enterogastric reflex, 663, *663*
Enterokinase, 665, 677t
Enucleation, 463
Enuresis, 802
Environment, genetic-environmental interactions, 922
Enzyme(s), 52
action of, 106, *106*
active site of, 106–7, *106*
in cell membrane, 67t
coenzymes of, 107
cofactors of, 107
factors that alter, 107
names of, 107
rate-limiting, 115
RNA, 120
specificity of, 106
Enzyme-substrate complex, *106,* 107
Eosinophils, 518–19, *519,* 521, 521t, 523t, *524*
Ependymal cells, *346,* 347, 347t
Ephedrine, 416
Epicardium, 545, *545,* 545t
Epicondyles, 199t
of femur, 227, *229,* **335**
of humerus, 221, *221,* **333, 335**
Epicranial aponeurosis, *298,* 299
Epicranius muscle, 298–99, *298,* 299t
Epidemiology, 25
Epidermal growth factor, 93
Epidermis, 135, 158–62, *160, 161, 165*
Epidermolysis bullosa, 143t, 158, 882t
Epididymis, **33,** *831,* 832, *833,* 838, *838,* 844t, **951**
Epididymitis, 868
Epidural space, 367, *368*
Epigastric artery
inferior, 589, *589*
superficial, 589
Epigastric region, 23, *23*
Epiglottic cartilage, 736–37, *737*
Epiglottis, *649, 650,* 656, *656, 657, 733, 734,* 737
Epiglottitis, 739
Epilepsy, 56, 357t, 397, 417
Epiligrin, 158
Epimysium, 278, *280*
Epinephrine, 356, 358, 411, 415–16, 471, 471t, 474–75, 477, 491–92, *492,* 493t, 501, *501,* *576,* 577, 608, 630, 702
Epineurium, 396, *398*
Epiphyseal plate, *183, 187,* 188–90, *189,* 255
growth at, 188–89, *188*
Epiphysiolysis, 232
Epiphysis, 182, *183, 184*
Epiploic appendages, *681,* 682
Episiotomy, 903
Episodic ataxia, 340
Epitestosterone, 475
Epithelial cells, 62, *63*
Epithelial membranes, 152–53

Epithelial tissue, 132–40, 132t, 141t
Epithelium, 141t
columnar, 133
pseudostratified, 135, *136,* 141t
simple, 133, *135,* 141t
stratified, 137, *137,* 141t
cuboidal, 133
simple, 133, *134,* 141t
stratified, 137, *137,* 141t
glandular, 138–40, *139, 140,* 141t, *159*
simple, 133
squamous, 133
simple, 133, *134,* 141t
stratified, 135, *136,* 141t, *159*
stratified, 133
transitional, 137, *138,* 141t
Equilibrium, 443–46
dynamic, 443–46, 700
static, 443–44
Erectile dysfunction, 830, 840
Erection
female, 856, *856*
male, 830, 842–44, *844*
Erector spinae muscle, 302, 302t, **333**
Ergocalciferol, 707
Ergosterol, 707
Erythema, 176
Erythroblast(s), *512,* 514
Erythroblastosis fetalis, 536, 614
Erythrocytes. *See* Red blood cell(s)
Erythrocyte sedimentation rate, 971
Erythrocytosis, 514
Erythrolabe, 460
Erythropoiesis, 514
Erythropoietin, 475, 500, 514, *514,* 773, 846
Escherichia coli, 772, *772*
Esophageal artery, 582, 585t
Esophageal cancer, 656
Esophageal hiatus, 656, *657*
Esophagitis, 688
Esophagus, 15, **35, 36,** *645, 646, 650,* 656–58, *656, 657, 658, 733, 734,* **949, 950, 953, 954, 965**
Essential amino acids, 700
Essential fatty acids, 698
Essential hypertension, 578
Essential nutrients, 695
Estriol, 857, 885
Estrogen(s), 166, 169, 190, 195, 471, 483, 495, 500, 857–59, 858t, *860,* 884–85, *885,* 885t, 937
placental, 884–85, 887–88, 904
Estrogen replacement therapy, 860
Estrone, 857
Ethmoidal region, **248**
Ethmoidal sinus, 203, *203, 205,* 205t, **246, 248, 249,** 736, **952**
Ethmoid bone, 196t, *200, 201,* 201t, 202, *203, 204, 206, 207,* **238, 239, 240, 246, 248, 249, 250**
Eukaryotic cells, 62
Euploid, 934, *935*

Eupnea, 766
Evaporation, heat loss through, 170
Eversion, 261, *262*
Evolution, 921
 of mitochondria, 73
Exchange reaction, 44
Excitation contraction coupling, in skeletal muscle, 284
Excitatory postsynaptic potential, 355–56
Excretion, 18
 as characteristic of life, 8t
Exercise
 aerobic, 325, 580
 body temperature and, 289
 breathing and, 756
 calories used in various activities, 703t
 effect on bone, 190, *190,* 195, 232
 effect on cardiovascular system, 579–80, 601
 effect on skeletal muscles, 289, 293, 325
 heat production during, 170
 for weight loss, 705
Exercise stress test, 600
Exocrine gland, 138–39, *139,* 139t, 468, *468*
Exocytosis, 72, 85, *86, 87, 88,* 89t
Exophthalmos, 463, 504
Expiration, 745, 748–49, *749,* 749t, 752
Expiratory reserve volume, *749,* 750
Expressivity, 928
 variable, 928
Extension, 261, *261,* 296
Extensor carpi radialis brevis muscle, *310, 311,* 311t, 312, **960**
Extensor carpi radialis longus muscle, *309, 311,* 311t, 312, **333, 960**
Extensor carpi radialis muscle, *310*
Extensor carpi ulnaris muscle, *310, 311,* 311t, 312, **960**
Extensor digitorum longus muscle, *297,* 321, 321t, *322, 323, 325,* **336, 961, 962**
Extensor digitorum muscle, *297, 310, 311,* 311t, 312, **333, 960**
Extensor pollicis longus muscle, *311*
Extensor retinaculum, 312, *322, 323, 325,* **960**
External auditory meatus, 201, *201, 202,* **239, 241, 246,** 434, *435*
External ear, 434
External environment, *10*
External oblique muscle, **30, 31, 32,** *297,* 304, 312, 312t, *313, 314,* **334, 748, 957, 958**
External respiration, 732
Exteroceptive senses, 424
Extracellular fluid, 9, 809–10, *809, 810*
Extracorporeal liver assist device, 669

Extracorporeal membrane oxygenation, 743
Extracorporeal shock-wave lithotripsy, 796
Extrafollicular cells (C cells), thyroid, 485, 487
Extrapyramidal nerve tracts, 378
Extremophile, 107
Extrinsic allergic alveolitis, 745
Extrinsic clotting mechanism, 527–28, *528,* 528t
Extrinsic muscles, of eye, 448, *449,* 449t
Exudate, 618
Eye, 447–61, **952**
 age-related changes in, 457, 462
 color of, 453, 930–31, *930*
 dark-adapted, 460
 light-adapted, 460
 structure of, 450–55, *450, 451, 452, 453, 454, 455,* 455t
 visual accessory organs, 447–49, *447, 448*
Eyeglasses, 457, *457*
Eyelashes, *447*
Eyelid, 447, *447*
Eye movement, 461
Eye muscles, 400, 448, *449,* 449t

F

Facet, of vertebrae, 199t, 212
Facial artery, *570, 585, 586,* 587
Facial expression, 164, 278
 muscles of, 298–99, *298,* 299t
Facial nerve (VII), 299t, *399,* 400–401, *401,* 403t, 414, *434, 441,* 449t
Facial skeleton, 197, 204–8, *205,* 205t, *206, 207,* 208t
Facial vein, anterior, *593, 597*
Facilitated diffusion, 81–82, *83,* 89t
Facilitation, 359
FAD, 709, 975, *976*
$FADH_2$, 111, *112*
Fainting, 756
Falciform ligament, **32,** 666, **966**
Falls, among elderly, 232, 232t, 417
False ribs, 216, *217*
False vocal cords, 737, *737, 738*
Falx cerebelli, 367t, 395
Falx cerebri, *367,* 367t, 381, **952**
Familial adenomatous polyposis, 685
Fanconi anemia, 882, 882t
Farsightedness, 457, *457*
Fascia, 278–79, *280*
Fascicle, 28, *280,* 396, *398*
Fasciculus cuneatus tract, 376, *377, 378,* 379t, 386
Fasciculus gracilis tract, 376, *377,* 379t
Fasciotomy, 278
Fast-twitch fibers, 292–93
 fatigue-resistant, 292
 glycolytic, 292

Fat(s), 50. *See also* Fatty acid(s); Lipid(s)
 catabolism of, 105, *114*
 dietary, 50
 digestion of, 678, *678*
 synthesis of, 104–5, *105*
Fatal familial insomnia, 394t
Fat-free foods, 697
Fat globules, 671
Fat-soluble vitamins, 698, 705–8, 708t
Fat substitute, 699
Fatty acid(s), 50, *50, 51*
 abnormal metabolism of, 599
 absorption in small intestine, 678–79, *678, 679,* 680t
 absorption of, 671, 697
 beta-oxidation of, 697, *697*
 essential, 698
 incomplete oxidation of, 817
 saturated, 50, *50*
 structure of, *58*
 unsaturated, 50, *50*
Fatty acid oxidase, 697
Fecal occult blood test, 685, 685t
Feces, 685
 water output, 812
Feedback system
 negative feedback, 10, 115, *115,* 477–78, *479,* 494, 497, *498,* 514
 positive feedback, 12, 528, 902, *903*
Female climacteric, 859–60
Female infertility, 850, 861, 861t
Female reproductive cycle, 858–59, 859t, *860*
Female sex hormones, 857–58, *857*
Female torso, **30, 34, 35, 36**
Femoral artery, **31, 32, 34,** *319, 570,* 589–90, *589, 590, 592, 847,* **955, 960**
 deep, *529,* 589, *590*
 lateral, *590*
Femoral nerve, **31, 32, 33,** 319t, 320t, 406, *407, 409,* 847
Femoral region, 24, 25
Femoral vein, **30, 31, 32, 34,** *319,* 595, *596, 597, 847,* **955, 960**
Femur, **36,** *183, 184, 190,* 196t, 197, *198,* 227, *228, 229,* 231t, *258,* 267, *267, 268, 269, 270,* **335,** *847,* **955**
 fracture of, 232
Fenestrae, 783
Fenfluramine, 705
Ferritin, 517, 970
Fertilization, *829,* 836, *849,* 876–79, *877, 880, 881,* 887t
Fetal alcohol effects, 894
Fetal alcohol syndrome, 894, *895*
Fetal cells, in maternal circulation, 633–34, 937, 937t, *938*
Fetal cell sorting, 937, 937t, *938*
Fetal hemoglobin, 513, 898–99
Fetal stage, 892–901, *893, 896, 897, 898,* 898t

Fetus, 894
 blood of, 898–901, *899, 900,* 900t, *901*
 circulation in, 898–901, *899, 900,* 900t, *901*
 descent of ovaries, 848
 descent of testes, 830–32, *832, 842*
 maternal antibodies in, 629
 pituitary gland of, 478
 skeleton of, *186*
 skull of, **251**
 ultrasound examination of, 6, *6, 7,* 937t, *938*
Fever, 166, 171, 561, 618, 619t
 reduction of, 171
Fiber, dietary, 48, 682, 684, 694–95
Fiberoptic bronchoscope, 742
Fiberoptic endoscope, 672
Fibrillation, 327, 562
Fibrin, 173, 527, *528,* 528t, *529,* 531t
Fibrinogen, 524, *524,* 524t, 527, 528t, *529,* 530t, 531, 618
Fibroadenoma, 865
Fibroblast(s), *131,* 132, 141, *141, 144,* 144t, *145,* 151, 173, *174,* 529–30, 618
Fibroblast growth factor, 564
Fibrocartilage, 147, *148,* 150t, 192, *193, 256*
Fibrocystic breast disease, 864–65
Fibrosis, 327
Fibrositis, 327
Fibrous joint, 254–55, *255, 256,* 260t, 263t
Fibrous layer
 of ureter, 795, *795*
 of vagina, 854
Fibrous pericardium, 14, *15,* 543, *544, 545*
Fibrous proteins, 54
Fibrous tunic, 450
Fibula, 196t, 197, *198,* 228–29, *228, 229, 230,* 231t, *270,* **335**
Fibular artery, *590, 592*
Fibular collateral ligament, *270,* 272
Fibularis brevis muscle, *322, 324*
Fibularis longus muscle, *297,* 321, 321t, *322, 323, 324,* 325, *325,* **335**
Fibularis tertius muscle, 321, 321t, *323*
Fibular nerve
 deep, 321t, *325*
 superficial, 321t, *325*
Fibular retinaculum, *323, 324,* 325
Fifth disease, 161t
"Fight-or-flight" response, 492, 500–501
Filter paper, 84, *84*
Filtration
 across cell membrane, 83–84, *84,* 89t
 in capillaries, 568
 glomerular. *See* Glomerular filtration
Filum terminale, *370, 372*
Fimbriae, *847,* 853, *853*

Finasteride, 840
Fingerprints, 164
First breath, 906
First-class lever, 191–92, *191*
First meiotic division, 833–36
First messenger, 472
First trimester, 884
Fissure, 199t, 382
Fissured fracture, *192*
Fistula, 660, 776
Fixed cells, 141
Flaccid paralysis, 381
Flagella, *64*, 76–77, *77*, 80t
Flagyl, 435
Flat bone, 182, *183*
Flatus, 683
Flavoproteins, 709
Flexion, 261, *261*, 296
Flexor carpi radialis muscle, *309,*
    310, 311, *311,* 311t, *335*
Flexor carpi ulnaris muscle, *309,*
    310, 311, *311,* 311t, *335*
Flexor digitorum longus muscle,
    321, 321t, 323, *324, 325*
Flexor digitorum profundus
    muscle, *310,* 311, 311t, 312
Flexor digitorum superficialis
    muscle, *309,* 310, 311,
    311t, 312
Flexor pollicis longus muscle, *311*
Flexor retinaculum, *309,* 323,
    324, 325
Floaters, 462
Floating ribs, 216, *217*
Fluorine, 39t, 718, 719t
Fluorouracil, 865
Flutter, 563
Focus point, of eye, 457, *457*
Folic acid, 516, 599, 711, 713t
    deficiency of, 711
    during pregnancy, 382, 711
    in red cell production, 515–16,
        515, 516t
Folinic acid, 711
Follicle-stimulating hormone, 471t,
    475, 478, 480, *481,* 483, 485t,
    845, *845,* 857–59, *857,* 858t,
    860, 884–85, 885t, 905
Follicular cells, 848
Follicular dendritic cells, 628
Folliculitis, 166
Folling, Ivar, 126
Fontanels, 199t, 208, *210,* 254
Food
    amino acids in, 700t
    enriched, 707
    fat-free, 697
    fortified, 707
    low fat, 721
    as requirement of organisms,
        8, 9t
Food labels, 720–21
Food poisoning, 104, 772
Food pyramid, 719–20, *721*
Foot
    bones of, 229–31, *230*
    muscles that move, 321–25,
        321t, *322, 323, 324, 325*
    surface anatomy of, **336**

Footprints, prehistoric, 182
Foramen, 199t, 208. *See also*
    specific foramen
Foramen lacerum, *202, 204,* 209t,
    242, 243, 244, 249, 250
Foramen magnum, 199–200, *202,*
    204, 206, 209t, **242, 243,**
    244, 246, 249, 250, 372, *372*
Foramen ovale, *202, 203, 204,*
    209t, **242, 243, 244, 247,**
    249, 250, 899, *900,* 900t,
    907, *908,* 908t
Foramen rotundum, *203, 204,* 209t,
    247, 250
Foramen spinosum, *202, 203, 204,*
    209t, **242, 243, 244, 247,**
    249, 250
Forearm
    fracture of, 222
    muscles that move, 308–10,
        308t, *309, 310*
    posterior view of, **960**
    surface anatomy of, **335**
Forebrain, 380, *382,* 382t
Formed elements, in blood,
    510, *524*
Fornix, *847,* 854
Fortified foods, 707
Fossa, 199t
Fossa ovalis, 908, *908*
Fossil, 182
Fourth ventricle, 369, *370,*
    382t, *392*
Fovea, 199t
Fovea capitis, 227, *229,* 267, *267*
Fovea centralis, *450,* 454–55,
    455, 458
Fox, Michael J., 390, *391*
Fracture, 189, 192–93, *192, 193,* 195
    aging-related increased risk
        for, 232
    comminuted, *192*
    compound, 192
    of femur, 232
    fissured, *192*
    of forearm, 222
    greenstick, *192*
    of humerus, 308
    oblique, *192*
    repair of, 192–93, *193*
    spiral, *192*
    spontaneous (pathologic), 192
    transverse, *192*
    traumatic, 193
    treatment of, 193
    of vertebrae, 212, 381
Frank-Starling law of the heart,
    576, *576*
Fraternal twins, 882
Freckles, *924*
Freely movable joint, 254, 257
Free nerve endings, 424, *425,* 429t
Free radicals, 912
Frenulum, 649, *649*
Frontal bone, 196t, 199, *200, 201,*
    201t, *204, 206, 207,* 209,
    210, **238, 239, 240, 241, 246,**
    248, 249, 250, 332, 949
Frontal eye field, 385–86, *385*

Frontalis muscle, *297, 298,*
    299, **955**
Frontal lobe, 382–84, *384, 385,*
    386, 387, 387t, **952, 968**
Frontal region, *24, 25,* **250**
Frontal section, *22, 23*
Frontal sinus, *14,* 199, *204, 205,*
    205t, *206,* 248, *249, 250,*
    650, 733, 734, 736, **949, 952**
Frontal suture, 209, *210,* **246**
Fructose, 48, 695–96, *696,* 839
Fructose-1,6–diphosphate, *110*
Fulcrum, 191, *191*
Fumaric acid, *112*
Functional residual capacity, *749,*
    750, 750t
Functional syncytium, 554
Fundus, of uterus, 854
Furylfuramide, 127t
Fusion inhibitors, 635

G

GABA, 357, 357t, 358t, 360
Gage, Phineas, 366, *366*
Galactose, 48, 695–96, *696*
Gallbladder, *15,* **32, 33, 34,** 497,
    645, 646, 664, 667, 671, *671,*
    673, 686, **955**
    disorders of, 672
Gallium-67, 41
Gallstones, 665, 671–72, *671,* 686
Gamete intrafallopian transfer, 879
Gamma radiation, 40, 45
Ganciclovir, 942
Ganglia, *344,* 399, 408
Ganglion cells, 454, *454*
Gap junction, 67, *69,* 70t, 148, 557
Garrod, Alfred Baring, 793
Gas
    blood. *See* Blood gases
    intestinal, 683
Gas exchange. *See also*
        Capillary(ies); Lung(s)
    alveolar, 757–58
Gas transport, 760–63, 764t
Gastrectomy, 688
Gastric artery, 583, *584*
Gastric bypass surgery, 705
Gastric emptying, 659, 662–64,
    662, 663
Gastric glands, *643,* 659–61, *660*
Gastric juice, 617, 656, 659–60,
    661t, 818
Gastric pits, *643,* 659, *660*
Gastric vein, 593, *594*
Gastrin, 487, 661, 672t
Gastritis, 660
Gastrocnemius muscle, *270, 297,*
    318, 321, 321t, *322, 323,*
    324, 325, **335, 336, 962**
Gastrocolic reflex, 683
Gastroenteric reflex, 680
Gastroenterology, 25
Gastroileal reflex, 680
Gastrostomy, 688
Gastrula, 886, 887t
Gastrulation, 886
Gated channels, 349, *349*

Gaucher disease, 882t, *942*
Gelsinger, Jesse, *940,* 941
Gender, effects on phenotype,
    933, *933*
Gene, 115, 118, 921, *921,* 924–25
    genetic-environmental
        interactions, 922
Gene expression, 120, 928–29
General interpretative area, 385,
    387, 389
General stress syndrome, 500–501,
    501, 501t
Gene therapy, 348, 564, 920,
    937–42
    in cancer, 941–42
    in hemophilia, 533
    heritable, 939
    nonheritable, 939
    successes and setbacks,
        940–41, *940*
    tools and targets of, 939–42, *942*
Genetic code, 115–16, 122t
    degeneracy of, 122
    universality of, 116
Genetic counselor, 920, *938*
Genetic disease. *See also specific*
        diseases
    cheekbrush tests, 648
    preimplantation diagnosis of,
        882–83, 882t, *883*
Genetic heterogeneity, 928–29
Genetic information, 115–16, 921
    changes in, 123–27
Genetics, 921–24
Genetic tests, 920
Genicular artery, deep, 589, *590*
Genital herpes, 867t
Genital region, *24, 25*
Genital tubercle, *897*
Genital warts, 867t
Genome, 115, 920–21, *923*
Genomic imprinting, 933
Genomics, 921–24
Genotype, 925
Geophagy, 718
George III (King of England), 517
Geriatrics, 25
Germinal center, *612,* 613
Germinal epithelium, *833,* 836
Germinal mutation, 865
Gerontology, 25, 912
Gestation, 914
Ghrelin, 4, 695, 695t
GI camera, 648
Gigantism, 482
    pituitary, 190
Gingiva, *652*
Gingivitis, 652
Ginkgo biloba, 720
Gland(s), 138. *See also specific*
        glands
    apocrine, 139, *140,* 140t
    endocrine, 138
    exocrine, 138–39, *139,* 139t
    holocrine, 139, *140,* 140t
    merocrine, 139–40, *140,* 140t
    skin, 167–68, *167, 168,* 168t
Glandular epithelium, 138–40,
    139, 140, 141t, *159*

Glans, of clitoris, 855
Glans penis, *831*, 841, *841*, *897*
Glaucoma, 462
Gleevec, 522
Glenohumeral ligament, 264, *265*
Glenoid cavity, 218, *219*, *265*
Glenoid labrum, 265, *265*
Gliadin, 700
Gliding joint, *259*, 260, 260t, 263t
Glioma, 941–42, *942*
Globin, 760
Globular proteins, 54
Globulins, 524, *524*, 524t, 970
    alpha, 487, 524, 524t
    beta, 524, 524t
    gamma, 524, 524t, 625, 629
Globus pallidus, 388, *389*
Glomerular capsule, *771*, 776–78,
    778, *781*, *783*, *784*, *787*, 794t
Glomerular filtrate, 783, *784*, *787*
    composition of, 784t
    24–hour volume, 786, *786*
Glomerular filtration, 782–83, *784*
Glomerular filtration pressure,
    783–85, *785*
Glomerular filtration rate, 785–86,
    785, 800
    control of, 786–87, *787*
Glomerulonephritis, 633t, 815
    acute, 780
    chronic, 776, 780
Glomerulus, *771*, 776, *778*, *779*,
    781, *783*, *784*, *787*, 794t, 800
Glossitis, 688
Glossopharyngeal nerve (IX), *399*,
    401, 403t, *413*, 414, 434
Glottis, 738, *738*
Glucagon, 12, 471t, 475, 478,
    496–98, *498*, 498t,
    501–2, 667
Glucocorticoids, *467*, 493, 496, 502
Glucokinase, 497
Gluconeogenesis, 696, 699
Glucosamine with chondroitin, 720
Glucose, 48, 695–96, *696*
    blood, 12, 113, 474, 478,
        493–99, *498*, 525, 667,
        699, 784t, 970
    regulation of, 502
    cerebrospinal fluid, 110, 370
    glomerular filtrate, 784t
    movement across cell
        membranes, 82
    oxidation of, 108
    phosphorylation of, 108
    structure of, *49*, *58*
    tubular reabsorption of, 787–88
    urine, 499, 784t, 788, 794, 972
Glucose-tolerance test, 499
Glucosuria, 788, 824
Glutamic acid, 356–57, 357t
Gluteal artery
    inferior, 588, *589*, *847*
    superior, 588, *589*
Gluteal gait, 318
Gluteal nerve
    inferior, 319t, 407, *409*
    superior, 319t, 407, *409*
Gluteal region, *24*, 25

Gluteal tuberosity, *229*
Gluteal vein, 596, *847*
Gluten, 679
Gluteus maximus muscle, *297*,
    314, 315–17, *317*, *318*, 319t,
    847, **955**, **958**, **961**
Gluteus medius muscle, **36**, *297*,
    315, 317–18, *317*, *318*, 319t
Gluteus minimus muscle, 315,
    317–18, 319t
Glycerol, 50, *51*
Glycine, *58*, 356–57
Glycogen, 50, 113, *113*,
    695–96, *696*
Glycogenesis, 696, *696*
Glycogenolysis, 696, *696*
Glycolysis, 108–11, *109*, *110*, *114*,
    288–89, 974, *975*
    block in, 110
Glycoproteins, 52, 72, 471, 472t
Glycosuria, 499
GM1 ganglioside, 381
Goblet cells, 133, 135, *135*, *136*,
    138, 139t, 153, *647*, 659,
    661t, *675*, 677, 682, *683*,
    731, 733, *734*, 738
Goiter, 487t, *488*
Golden rice, 939
Goldenseal, 720
Golgi apparatus, *64*, 71–72, *72*,
    73, 80t
Golgi tendon organs, 428–29,
    429, 429t
Gomphosis, 255, *256*, 260t
Gonadal artery, 583, *584*, 585t, *592*
Gonadal vein, 595, *597*
Gonadotropes, 480
Gonadotropin(s), 483, 500, 845
    hyposecretion of, 861
Gonadotropin-releasing hormone,
    471t, *481*, 483, 845, *845*,
    857–58, *857*, 858t, 860, *860*
Gonorrhea, 861, 867t
Goose bumps, 166
Gout, 254, 258, 273t, 793
G_1 phase, 90, *90*
G_2 phase, 90, *90*
G proteins, 472, 475, *475*
Graafian follicle, *851*
Gracilis muscle, **33**, **35**, **36**, *297*,
    315, *316*, 318, *318*, *319*,
    319t, *324*, **960**
Graft-versus-host disease, 632, 634
Granulations, 173
Granulation tissue, *192*
Granulocytes, *512*, 518–19,
    522, 523t
Granulosa cells, 850, 858, 858t
Graves disease, 487t, 633t
Gray commissure, 372, *373*
Gray hair, 166, 176
Gray matter, 343, *365*, *367*, 372,
    373, **952**
Gray rami, 411, *411*
Greater omentum, **32**, 674, *674*,
    675, **948**, **966**
Greater pelvis, 224, 227
Greater trochanter, 227, *229*,
    268, **334**

Greater tubercle, 220, *221*
Great toe, 231, 254
Greenstick fracture, *192*
Ground substance, 141, 146
Growth, 876
    of bone, 186–90
    as characteristic of life, 8t
Growth factors, 93
    as drugs, 93
Growth hormone, 190, 471t, 478,
    480–81, *481*, 485t, 501–2
    abnormal amounts of, 482, *482*
    use by athletes, 475
Growth hormone-releasing
    hormone, 471t, 480–81, *481*
Guanine, 116, *117*, 118, 119t, *979*
Gubernaculum, 832
Guevedoces, 846
Gynecology, 25

H

Hageman factor, 528, 528t,
    529, 530t
Hair, 54
    color of, 166, 924
    growth of, 166
    loss of, 166, *166*
Hair cells, 439, *440*, *441*, 444, *445*
Hair dye, 127t
Hair follicles, *160*, 165–66,
    165, *167*
Hair papilla, 165, *165*
Hair root, 165, *165*
Hair shaft, *160*, 165, *165*
Hair transplant, 166
Hairy ears, 166
Hairy elbows, *924*
Half-life, 41
Hamate, *223*
Hamstring muscles, 320, 320t, **334**
    pulled hamstring, 320
Hand
    bones of, 222–23, *223*
    posterior view of, **960**
Handedness, 388
Haploid, 835, 921
Hapten, 619
Hard palate, 204, 649, *649*, *650*,
    652t, *657*, 733, *734*
Hashimoto disease, 487t
Haustra, *681*, 682
Hawking, Stephen, 379
Hay fever, 630
H_2-blockers, 661
Head
    arteries of, 584–87, *585*, *586*
    cavities within, 14, *14*
    imaging of, *56*
    lateral view of, **956**
    muscles that move, 300–302,
        301, 302t
    sagittal section of, **948**, **949**
    surface anatomy of, **332**
    transverse section of, **952**
    veins from, 591
Headache, 6–7, 299, 457
Head movement, 444–45, *445*,
    446, 461

Head of bone, 199t
    femur, 227
    fibula, 228, **335**
    humerus, *219*, *221*
    radius, *220*, 221, *222*
    ulna, *222*, *222*
Head of pancreas, *664*
Head of sperm, 837, *837*
Head position, 444, *445*
Head trauma, 368
Healthy eating, 718–24
Hearing, *421*, 434–42
    age-related changes in, 462
    loss of, 442–43
Hearing aid, 443, 462
Hearing receptor cells, 439
Heart, *15*, *33*, **948**, **953**, **963**. *See
    also* Cardiac *entries*
    actions of, 551–61
    age-related changes in, 600–601
    anterior view of, *543*
    artificial, 556
    blood supply to, 549–51, *551*,
        552, *553*
    chambers and valves of,
        545–49, *547*, *548*, 549t
    conduction system of, 554–58,
        558, *563*, 601
    coverings of, 543
    disorders of, 598–99
    magnetic resonance imaging
        of, 549
    path of blood through, 549,
        550, *551*
    size and location of, 542, *544*
    skeleton of, *548*, 549
    structure of, 542–51
    wall of, 545, *545*, 545t
Heart attack, 426, 530, *541*, 542,
    550, 564, 600
Heartbeat, apical, 542
Heartburn, 656, 663
Heart failure, 551, 556, 560, 569
Heart rate, 560–61, 572, *573*, *576*,
    577, 580
    regulation of, 391
    theoretical maximum, 580
Heart sounds, 554, *555*, *557*
    split, 554
Heart transplantation, 556, *556*
Heat, requirements of organisms,
    9, 9t
Heat exhaustion, 171, 808
Heat loss, 169–70
Heat production, 169–70
    by skeletal muscles, 289
Heatstroke, 808
Heavy chains, 625–26, *627*
Heidelberg capsule, 648
Height, 929, *929*, 931
Helicobacter pylori, 660
Helicotrema, *438*, 439
Helper T cells, 62, 621, *622*, 623,
    623, *626*, 635
Hemarthrosis, 274
Hematocrit, 511, *511*, 970
Hematology, 25
Hematoma, 2, 192, *193*, 530
    subdural, 368

Hematometra, 868
Hematopoiesis, 194
Hematopoietic stem cells, 94, 98
Hematuria, 776, 802
Heme, *515, 760*
Hemianopsia, 463
Hemiazygos vein
    inferior, 593, *594*
    superior, 593, *594*
Hemicellulose, 695
Hemiplegia, 381
Hemisphere dominance, 387–88
Hemizygote, 932
Hemochromatosis, 517
Hemodialysis, 773–74, 776
Hemoglobin, 54, 194, 288, *288,*
        511, 513, *515, 717, 717,*
        760–61, *760, 761, 762*
    breakdown of, 517, *518*
    buffer system, 820
    combined with carbon
        monoxide, 761
    development of red cell
        substitute, 510, *510*
    fetal, 513, 898–99
    structure of, *518, 760*
    synthesis of, 515
    urine, 972
    whole blood, 971
Hemolysis, 83, *84*
Hemolytic anemia, 516t, 633t
Hemolytic jaundice, 671
Hemolytic uremic syndrome,
    772, *772*
Hemophilia, 533, 939, *942*
    hemophilia A, 882t, 932
    hemophilia B, 882t
Hemopoietic stem cells, 511
Hemorrhage, 484, 573, 577
Hemorrhagic disease of the
    newborn, 709
Hemorrhagic telangiectasia, 537
Hemorrhoids, 682
Hemostasis, 526–31, 527t
Hemostat, 192
Hemothorax, 766
Henrich, Christy, 694, *694*
Heparin, 142, 519, 530–31, 531t
Heparinized whole blood, 537
Hepatic artery, 583, *584, 595,*
    667, 668
Hepatic cells, 666, *668*
Hepatic coma, 669
Hepatic duct, *664,* 666, *667,* 671
    common, *664,* 666, *673*
Hepatic flexure, *681*
Hepatic lobules, 666, *668*
Hepatic portal system, 593
Hepatic portal vein, 593, *594, 595,*
    666, *667, 668, 900*
Hepatic sinusoids, 593, 595, 666
Hepatic vein, 595, *595, 597,* 666
Hepatitis, 670
    fulminant, 670
Hepatocellular jaundice, 671
Hepatopancreatic ampulla,
    664, 665
Hepatopancreatic sphincter, *664,*
    665, 671, *673*

Hepburn, Katherine, *910*
Herceptin, 865
Heritable gene therapy, 939
Hernia
    hiatal, 656
    inguinal, 832
    strangulated, 832
Herniated disc, 216
Heroin, 359, 892, 906t
Herpes, 176, 867t
Heterogametic sex, 931
Heterozygote, 925–27
    manifesting, 932
Hiatal hernia, 656
Hiccup, 751, 751t
High altitude
    ear popping, 437
    hyperventilation, 823
High altitude pulmonary
    edema, 760
High-density lipoproteins, 525,
    526t, 970
Hilum
    of kidney, 772, *773, 775*
    of lymph node, 612, *612*
Hilus of lung, 743
Hindbrain, 380–81, *382,* 382t
Hinge joint, *259,* 260, 260t, 263t
Hip fracture, 195, 232
Hip joint, 263t, 267–69, *267, 268*
Hippocampus, 388
Hippocrates, 799
Hip pointer, 224
Hirschsprung disease, 683
Hirsutism, 504
Hirudin, 532
Hirulog, 532, *532*
Histamine, 142, 357t, 519–20, *521,*
    569, 628, 630, *631,* 661
Histiocytes, 618
Histology, 25
Histones, 79, 116, *117*
Hives, 630
HLA-B7 antigen, 942
HLA genes, 882
Holocrine gland, 139, *140,* 140t
Homeostasis, 2, 9–12, *10, 11,* 514
    body temperature regulation,
        170–71, *170*
    interactions of organ systems, *20*
Homocysteine, 599
Homocystinuria, 599
Homogametic sex, 931
Homograft, 173
Homologous chromosomes,
    835, *835*
Homozygote, 925
Hopi people, 172
Horizontal cells, 454, *454*
Hormone(s), 17, 52, 468. *See also*
    specific hormones
    action of, 472–76, *473, 476*
    amines, 471–72, 472t
    chemistry of, 470–71, *470*
    control of cell division by, 93
    control of milk production and
        secretion, 905, 906t
    control of secretion of,
        477–78, *477*

of digestive tract, 672t
    effect on smooth muscles, 294
    names and abbreviations
        for, 471t
    peptide, 471–72, 472t
    protein, 471–72, 472t
    steroid, 470–72, *470,* 472t,
        473, 476
    use by athletes, 474
Hormone receptor(s), 469, *469,*
    472, *473,* 476
Hormone-receptor complex,
    472, *473*
Hospice, 910
Hot flashes, 859
Human chorionic gonadotropin,
    861, 882, 884–85, *884, 885,*
    885t, 937
Human genome, 115
Human growth hormone,
    recombinant, 190
Human immunodeficiency virus
    (HIV), 62, 635, 867t
    transmission of, 868
    transport across lining of anus
        or vagina, 88, *89*
Human immunodeficiency virus
    (HIV) infection, 621
    in hemophiliacs, 533
    treatment of, 635
Human menopausal
    gonadotropin, 861
Human organism, 4
    organization of, 4–6, *5,* 5t,
        12–19
Human papilloma virus, 867t
Human remains, identification
    of, 118
Humeral ligament, transverse,
    264–65, *265*
Humerus, **33,** 196t, 197, *198, 219,*
    220–21, 220, 221, 223t, 264,
    264, 265, 266, **333, 335**
    fracture of, 308
Humoral immune response, 621,
    623–28, *623*
Humphrey, Vice President
    Hubert, 796
Hunger, 391, 705
Hunter, C.J., 474
Hunter-gatherer, 3
Huntington disease, 56, 357t, 417,
    882t, 926–27, *927, 942*
Hurler disease, *942*
Hutchinson-Gilford syndrome,
    913, *913*
Hyaline cartilage, 147, *147,* 150t,
    187, 189, 738–39, *739*
Hybrid fixator, 193
Hybridoma, 624, *625*
Hydatidiform mole, 914
Hydramnios, 914
Hydrarthrosis, 274
Hydrocele, 868
Hydrocephalus, 371, 896
Hydrochloric acid, 46, *643,* 661,
    661t, 818
Hydrogen bond, 44
    in DNA, *116, 117*

Hydrogen ions, 46–47, *47,* 47t
    electrolyte balance, 813
    in human body, 39t
    regulation of ion
        concentration, 819
    sources of, 817
    tubular excretion of, 790, *791,*
        821, *822*
Hydrogen peroxide, 76, 106
Hydrolysis, *104,* 105, *113*
Hydrophilic molecule, 65, *66, 67*
Hydrophobic molecule, 65, *66, 67*
Hydrostatic pressure, 9, 810
    of blood, 568, *568*
    glomerular, 785
Hydroxide ions, 818–19
Hydroxyapatite, 194
Hydroxycholecalciferol, 489–90,
    491, 707
17-Hydroxysteroids, 972
Hydroxyurea, 513
Hygiene hypothesis, of asthma, 752
Hymen, 854
Hyoid bone, 196t, 197, *198, 650,*
    656, *734, 737*
Hyperalgesia, 463
Hyperalimentation, 725
Hypercalcemia, 490, 504, 561,
    725, 817
    cancer-induced, 817
Hypercalciuria, 725
Hypercapnia, 766
Hypercholesterolemia, familial, 87,
    598–99, 927, *928,* 940, *942*
Hyperemesis gravidarum, 914
Hyperemia, 618
Hyperextension, 261, *261*
Hypergeusia, 434t
Hyperglycemia, 499, 504, 725, 824
Hypericin, 720
Hyperkalemia, 561, 725, 818, 824
Hyperkalemic periodic paralysis, 68
Hypernatremia, 725, 818, 824
Hyperopia, 457, *457*
Hyperosmia, 434t
Hyperoxia, 760, 766
Hyperparathyroidism, 490t, 817
Hyperpnea, 766
Hyperpolarized membrane, 350
Hyperreflexia, 381
Hypersomnia, 357t
Hypertension, 167, 477, 493, 578,
    578, 578t, 600, 776
    drug therapy in, 578t
    essential, 578
    secondary, 578
    treatment of, 165
Hyperthermia, 170–71, 814
Hyperthyroidism, 487t, 488, *488*
Hypertonic solution, 83, *84*
Hypertrophic cardiomyopathy,
    familial, 599
Hypertrophic pyloric stenosis, 659
Hypertrophy
    of bone, 190
    of muscles, 293
Hyperuricemia, 824
Hyperventilation, 756, 766, 823
Hypervitaminosis A, 706–7

Hypervitaminosis D, 707
Hypoalbuminemia, 725
Hypocalcemia, 490, 504, 561,
    714, 817
Hypochondriac region, 23, *23*
Hypodermic injection, 164
Hypogastric plexus, *413*
Hypogastric region, 23, *23*
Hypogeusia, 434t
Hypoglossal canal, *206*, 209t
Hypoglossal nerve (XII), 399, *399*,
    401, 403t
Hypoglycemia, 498, 725, 824
Hypokalemia, 561, 725, 818
Hypomimia, 390
Hyponatremia, 725, 818
Hypoparathyroidism, 490t
Hypophia, 390
Hypophyseal artery
    inferior, *480*
    superior, *480*
Hypophyseal portal vein, 479, *480*
Hypophysectomy, 504
Hypopituitary dwarfism, 482, *942*
Hypoproteinemia, 789,
    814–15, 815t
Hypoproteinuria, 776
Hyposmia, 434t
Hypospadias, 868
Hypothalamus, 382t, *389*, 390–91,
    392, 393, 469, 471t, 477,
    478, 479, *480, 481*, 482, *483*,
    949, 968
    hormonal control of female
        reproductive system, 857
    hormonal control of male
        reproductive system,
        845, *845*
Hypothermia, 171
    induced, 171
Hypothyroidism, 487t
Hypotonic solution, 83, *84*
Hypoxemia, 766
Hypoxia, 425, 513, 760, 766
Hypoxia-inducible factor, 564
Hysterectomy, 868
Hysterosalpingogram, 861t
H zone, 279, *281*, 285

I

I band, 279, *281, 285*
Ibuprofen, 171
Identical twins, 882
Idiotype, 626
Ileitis, 688
Ileocecal sphincter, 680, *681*
Ileum, **34**, 674, *681*
Ileus, 688
Iliac artery
    common, **34, 35**, 583–84, *584,
        585t, 589, 590*, 592, *900*
    deep circumflex, 589, *589, 590*
    external, 588–89, *589, 590, 592*
    internal, 588, *589, 590, 592, 900*
    superficial circumflex, 589, *590*
Iliac crest, **36**, 224, *226*, **334**
Iliac fossa, 224, *226*
Iliac region, 23, *23*

Iliac spine
    anterior inferior, *226*
    anterior superior, **30, 32, 36**,
        224, *226*, **334**
    posterior inferior, *226*
    posterior superior, 224,
        226, **334**
Iliacus muscle, **36**, 315, *316*,
    319t, **960**
Iliac vein
    common, 596, *596, 597*
    external, 595, *596, 597*
    internal, 596, *596, 597*
Iliocostalis cervicis muscle,
    301, 302t
Iliocostalis lumborum muscle,
    301, 302t
Iliocostalis thoracis muscle,
    301, 302t
Iliofemoral ligament, 267, *268*
Iliolumbar artery, 588, *589*
Iliotibial band, *317*
Iliotibial tract, **335**
Ilium, 224, *225, 226, 268*
Immediate-reaction allergy,
    630, *631*
Immovable joint, 254–55
Immune complex reaction, 630
Immune response
    cellular, 620–21, *622*
    humoral, 621, 623–28, *623*
    primary, 628, *628*
    secondary, 628, *628*
Immune surveillance, 614
Immune system, age-related
    changes in, 634
Immunity, 519, 619–34
    practical classification of,
        629, 630t
Immunocompetence, 637
Immunodeficiency, 637
Immunoglobulin(s), 621
    structure of, 625–26, *627*
    types of, 626–27
Immunoglobulin A (IgA),
    626–27, 627t
Immunoglobulin D (IgD), 627, 627t
Immunoglobulin E (IgE), 627–28,
    627t, 630, *631*
Immunoglobulin G (IgG),
    626–28, 627t
Immunoglobulin M (IgM),
    627–28, 627t
Immunology, 25
Immunosuppressant drugs, 632, 776
Immunotherapy, 624–25
Impacted tooth, 651
Impetigo, 161t
Implantable cardioverter
    defibrillator, 542, *542*
Implantation, 882–83, *884*, 886
Implant resection arthroplasty, 269
Inborn error of metabolism,
    126–27, *127*
Incisive foramen, *202*, 209t
Incisive fossa, **242, 244**
Incisor, 650–51, *651*, 651t
Inclusions, cellular, 78
Incomplete dominance, 927, *928*

Incomplete penetrance, 928
Incomplete proteins, 700
Incomplete twinning, 902
Incontinence, 800, 802, 840
    stress, 799
Incontinentia pigmenti, 932
Incus, 196t, *435*, 436, *436*, 441
Indeterminate cleavage, 882
Inert element, 42
Infant, 908–9, 911t. *See also*
    Newborn
    kidney of, 795, 814
    skull of, 208, *210*
Infant formula, 907
Infarction, 530
Infection, 616
    body defenses against, 616–17
Inferior, definition of, 21
Inferior canaliculi, 448, *448*
Inferior colliculus, 392, *392*
Inferior oblique muscle, 448,
    449, 449t
Inferior orbital fissure, **240**
Inferior rectus muscle, *447*, 448,
    449, 449t
Inferior vena cava, *15*, **35, 36**, 546,
    547, 550, 552, 581, 591,
    595–96, *595, 596, 597, 667*,
    773, 774, 775, 776, 900, **954,**
    955, 965
Infertility
    assisted reproductive
        technologies, 878–79, *879*
    female, 850, 861, 861t
    male, 842–43, *842, 843*, 932
Inflammation, 142, 173, 477, 494,
    500, 520, *521*, 618, 618t,
    619t, 627, 628t, 815, 815t
Inflammatory bowel disease, 684
Inflation reflex, 755
Infraorbital foramen, *200*, **240, 241**
Infraorbital nerve, *400*
Infrapatellar bursa, 258, *258,
    270, 271*
Infraspinatus muscle, *297, 303,
    305, 305*, 306t, 308, **333,**
    958, 959
Infraspinous fossa, 218, *219*
Infundibulum
    of diencephalon, 390, 478
    of uterine tube, 852–53, *853*
Inguinal canal, **32**, 583, 832
Inguinal hernia, 832
Inguinal lymph nodes, *613*, 614
Inguinal region, 23, *23, 24*, 25
Inheritance, modes of, 924–28
Inhibin, 845, *845*
Inhibitory neuron, 345
Inhibitory postsynaptic potential,
    355–56
Initial segment, of axon, 350
Injectable contraception, 865–66
Innate (nonspecific) defense,
    617–19, 619t
Inner cell mass, 882, *884*, 886
Inner ear, 437–39, *438*, 446
Inner membrane, of mitochondria,
    72–73, *74*, 111
Inner tunic, 453–55, 455t

Incomplete penetrance, 928
Inorganic salts
    in cells, 48
    storage in bone, 194, *196*
Inorganic substances, in cells,
    47–48, 49t
Inositol triphosphate, 476
Insect sting, 630
Insensible perspiration, 812
Insertion, of muscle, 261, 296, *296*
Insomnia, 357t, 394t, 417
Inspiration, 745–48, *746, 747*, 747t,
    748, 756t
Inspiratory capacity, *749*, 750, 750t
Inspiratory reserve volume, 749,
    749, 750t
Insula, 384, *384*
Insulin, 12, 82, 153, 468, 471t, 478,
    496–99, *498*, 498t, 502, 633,
    667, 695, 695t
Insulin-like growth factor-1, 481
Insulin receptors, 497, 499
Insulin resistance, 502, 895
Integral proteins, 66, 67t
Integrin, 70, *70*
Integumentary system, 8t, 16, *16*
    age-related changes in, 911t
Intensity of vocal sound, 738
Interatrial septum, 545, 557, *558*
Intercalated disc, 150, *151*,
    295, *295*
Intercarpal joint, 263t
Intercellular junctions, 67, *69*, 70t
Intercostal artery, *592*
    anterior, 588, *588*
    external, *588*
    internal, *588*
    posterior, 582, 585t, 588, *588*
Intercostal muscles
    external, **31, 32, 36**, 746–48,
        748, 755
    internal, **32, 36**, *304*, 748, *749*
Intercostal nerves, 312t, 407–8, *407*
Intercostal vein, 593
    inferior, *594*
    posterior, 593
Interferon(s), 617, 620t, 624–25, 670
Interferon alpha, 343, 564
Interferon beta, 343
Interleukin(s), 518, 620, 620t
Interleukin-1, 171, 618, 620
Interleukin-2, 620, 625, 790
Interlobar artery, 774, 777, 783
Interlobar vein, 776, *777, 783*
Interlobular artery, 774, *777,
    779, 783*
Interlobular vein, 776, *777, 779, 783*
Intermediate fibers, 292
Internal acoustic meatus, *204, 206*,
    209t, **248**
Internal environment, 9–10, *10*
Internal oblique muscle, **31, 32,**
    304, 312, 312t, *313*, 314, 748
Internal respiration, 732
Interneurons, 345, *345*, 346t, *374,
    375, 375*, 375t
Internodal atrial muscle, 557
Interosseous ligament, 254
Interphalangeal joint, 263t
Interphase, 90, *90*

1025

INDEX

Interstitial cells, *833, 834,* 844t, 845
Interstitial cell-stimulating
 hormone, 480, 845–46, *845.*
 See also Luteinizing
 hormone
Interstitial fluid, 608, 611, 810
Intertarsal joint, 263t
Intertubercular groove, 221, *221*
Interventricular artery
 anterior, 549, *552, 553*
 posterior, 549–50, *552, 553*
Interventricular foramen, 369, *369*
Interventricular septum, 545, *547,
 548, 558*
Interventricular sulcus, *544,* 546
Intervertebral disc, **36,** 147, 197,
 209–10, *211, 256,* 257, 274,
 409, **950, 951**
 changes in, 216
Intervertebral foramen, 211,
 211, 403
Intervertebral joint, 263t
Intestinal autointoxication, 644
Intestinal bypass surgery, 705
Intestinal gastrin, 661, 672t
Intestinal glands, 139t, *647,
 675–77, 675*
Intestinal lipase, 677, 677t
Intestinal somatostatin, 662, 672t
Intestinal villi, 675–76, *675,
 676, 678*
Intestine. *See* Large intestine;
 Small intestine
Intra-alveolar pressure, 746,
 746, 747
Intracellular fluid, 809–10, *809, 810*
Intracranial pressure, 371
Intracytoplasmic sperm injection,
 879, *879*
Intradermal injection, 164
Intrafusal fibers, 429, *429*
Intralobular bronchioles, 740
Intramembranous bone,
 186–87, *186*
Intramembranous ossification,
 187, 187t
Intramuscular injection, 164
Intrapleural pressure, 748
Intrauterine device, 863t, 866, *866*
Intrauterine growth retardation,
 894–95
Intrauterine transfusion, 914
Intravascular oxygenator, 743
Intrinsic clotting mechanism,
 527–28, 528t
Intrinsic factor, 516, 660, 661t, 711
Inulin, 794
Inulin clearance test, 794
Inversion, *262,* 263
In vitro fertilization, 841,
 878–79, *879*
Involuntary muscle, 150
Involution, 904
Iodide pump, 486–87
Iodine, 39t, 486–87, 718, 719t
 deficiency of, *488*
Iodine-131, 41
Iodized salt, 718
Iodopsins, 460

Ion(s), 39t, 42–43, *42,* 48, *49*
Ion channels
 abnormal, 68
 chemically gated, 355
 drugs that affect, 68, 68t
 gated, 349, *349*
 voltage-gated, 355
Ionic bond, *42,* 43–44
Ionizing radiation, 45, *45,* 45t
Ipsilateral, definition of, 21
Iridectomy, 463
Iris, 449t, 450, *450, 451,* 452–53,
 453, 455t
Iritis, 463
Iron, 669, 717, 719t
 absorption of, 516–17, 694
 cycle in body, 517, 517t, *518*
 in human body, 39t
 requirement in pregnancy, 717
 serum, 971
Iron-binding capacity, 971
Iron deficiency anemia, 516t, *717*
Irregular bone, 182, *183*
Ischemia, 425, 560, 571, 600
Ischial spine, 224, *226*
Ischial tuberosity, 224, *226,* **334**
Ischiocavernosus muscle, 314–15,
 314, 315t
Ischiofemoral ligament, *268,* 269
Ischium, 224, *225, 847*
Isocaloric foods, 725
Isocitric acid, *112*
Isograft, 632, 632t
Isometric contraction, 292, *292*
Isotonic contraction, 292, *292*
Isotonic solution, 83, *84*
Isotopes, 39–40
Isotretinoin, 169, 169t, *893,* 894

J

Jaundice, 172, 670–71
 hemolytic, 671
 hepatocellular, 671
 newborn, 172
 obstructive, 671
Jejunum, **34,** *673,* 674
Johnson, Ben, 474, *474*
Johnson, President Lyndon, 671
Joint(s). *See also specific joints*
 age-related changes in, 271–72
 ball-and-socket, 259, *259,*
 260t, 263t
 cartilaginous, 255–57, *256,* 260t
 classification of, 254–57
 condyloid, *259,* 260, 260t, 263t
 disorders of, 272–73
 fibrous, 254–55, *255, 256,*
 260t, 263t
 freely movable (diarthrotic),
 254, 257
 gliding, *259,* 260, 260t, 263t
 hinge, *259,* 260, 260t, 263t
 immovable (synarthrotic),
 254–55
 movements of, 261–63, *261, 262*
 pivot, *259,* 260, 260t, 263t
 replacement of, 269
 saddle, *259,* 260, 260t, 263t

 slightly movable
 (amphiarthrotic), 254
 synovial, 257–60, *257, 258, 259,
 260,* 260t, 263t, 264–71
Joint capsule, 257–58, *257*
Joint reflex, 756
Judgment, 387
Jugular foramen, 202, *202, 204,
 206,* 209t, **243, 244, 249, 250**
Jugular vein, **242, 953, 964, 965**
 external, **32,** *593, 594, 597*
 internal, **31, 32, 35,** *583,* 591,
 593, 594, 597, 610
Junctional fibers, 557, *558*
Juxtaglomerular apparatus,
 778–80, *781*
Juxtaglomerular cells, 493, 778,
 781, 786
Juxtamedullary nephrons, 780–81,
 782, 783

K

Kaposi sarcoma, 635, *635*
Karyokinesis, 90
Karyotype, 924–25, *925, 938*
Kawasaki disease, 273t
Kegel exercises, 799
Keloid, 178
Kennedy, John F., 496
Keratin, 159
 abnormal, 158
Keratinization, 135, 159
Keratinocytes, 159
Keratitis, 463
α-Ketoglutaric acid, *112*
Ketone bodies, 499, 697–98, *697,*
 794, 817, 823
Ketonemic acidosis, 823
Ketonuria, 823–24
Ketosis, 824
Kidney(s), **15, 35,** *667,* 772–82,
 954, 955. *See also* Renal
 entries
 age-related changes in, 800
 artificial, 773
 disorders of, 699
 functions of, 773
 of infant, 795, 814
 injured, 786
 location of, 772, *773*
 of newborn, 907
 renal blood vessels, 774–76, *775*
 structure of, 772–73, *773,
 774, 775*
Kidney cancer, 790
Kidney failure, 772
 acute, 785
 chronic, 776, 785
Kidney stones, 795–96
Kidney transplantation, 776
Kilocalorie, 702
Kinase, 92
Klinefelter syndrome, 934
Knee-jerk reflex, 373–75, *374*
Knee joint, *253, 258,* 263t, 269–71,
 270, 271
 injury to, 271
 surface anatomy of, **335**

Köhler, Georges, 624
Korotkoff's sound, 573
Kupffer cells, 595, 666, *668*
Kwashiorkor, 723, *723*
Kyphosis, 216

L

Labia majora, *847,* 855, *855,*
 856t, *897*
Labia minora, *847,* 855, *855,*
 856t, *897*
Labioscrotal folds, *897*
Labor (childbirth), 901–3, *903,* 903t
Laboratory tests, 970–73
Labyrinth, 437
Labyrinthectomy, 463
Labyrinthitis, 463
Lacrimal apparatus, 448, *448*
Lacrimal bone, 200, *201,* 206, 208t,
 239, 240, 241
Lacrimal duct, 448
Lacrimal glands, 299, *400,* 448, *448*
Lacrimal nerve, *400*
Lacrimal sac, 448, *448*
Lactase, 107, 677, 677t
Lactation. *See* Milk
Lacteals, 608, 675, *675, 678, 679*
Lactic acid, *110,* 817
 blood, 971
 cerebrospinal fluid, 110
 in muscle, 288–89, *288, 289*
 from pyruvic acid, 111, *112*
 tubular reabsorption of, 788
Lactic dehydrogenase, 971
Lactiferous duct, 861, *862*
Lactobacillus, 653
Lactoferrin, 907
Lacto-ovo-vegetarian, 694t
Lactose, 48, 696
Lactose intolerance, 677, 686, 724
Lacto-vegetarian, 694t
Lacunae
 of bone, 146, *147, 149,* 184, *185,
 186, 186*
 of endometrium, 887, *891*
Lambdoidal suture, 199, *201, 202,
 206,* **239**
Lamellae, of bone, 148, *149*
Laminae, of vertebrae, 210–11, *214*
Lamina propria, 645
Laminectomy, 216, 232, 417
Lan, Sang, 381
Landsteiner, Karl, 532
Lanugo, 723, 895
Laparoscopy, 861t
Laparotomy, 861t
Large calorie, 702
Large intestine, *15, 645, 646, 667,*
 680–86, *774,* **954**
 absorption in, 683
 disorders of, 684–85, *684, 685*
 functions of, 682–83
 movements of, 683–85
 parts of, 680–82, *681*
 structure of intestinal wall, 682,
 682, 683
Laryngopharynx, *650,* 655,
 734, 736

Larynx, **31, 34**, *198*, 656, *656*, *733*, *734*, 736–38, *737, 738*, 744t, **949, 963**
Latent period, in skeletal muscles, 290
Lateral, definition of, 21
Lateral column, 372
Lateral geniculate body, 461, *461*
Lateral geniculate nucleus, 391
Lateral horn, 372, *373, 406, 411*
Lateral rectus muscle, 400, 448, *449*, 449t, *450*, **952**
Lateral sulcus, 384, *384, 385*, 386
Lateral ventricle, 368–69, *369*, **949, 952**
Latissimus dorsi muscle, **31, 32**, *297, 303*, 305, *306*, 306t, 307, **333, 957, 958, 959**
Laughing, 751, 751t
Lead, 194
Leech, medicinal, 532, *532*
Leeuwenhoek, Anton van, 837
Left-handedness, 388
Left lower quadrant, of abdomen, 23, *23*
Left upper quadrant, of abdomen, 23, *23*
Left ventricular assist device, 556
Left ventricular failure, 581
Leg
    anterior view of, **961**
    lateral view of, **962**
    muscles that move, *317, 318–21, 318, 319*, 320t
    posterior view of, **962**
    surface anatomy of, **336**
Lens, *450*, 451–52, *451, 452, 453, 456*
    refraction disorders, 457, *457*
Lens fibers, 451, *451*
Leptin, 695, 695t
Lesser pelvis, 224, 227
Lesser trochanter, 227, *229, 268*
Lesser tubercle, 220, *221*
Leukemia, 522, *522*
    acute, 522
    chronic, 522
    lymphoid, 522
    myeloid, 522, *522*
    treatment of, 522
Leukocytes. *See* White blood cell(s)
Leukocytosis, 521
Leukopenia, 521
Leukotrienes, 630
Levator ani muscle, 314–15, *314*, 315t, *847*
Levator palpebrae superioris muscle, 447, *447, 449*, 449t
Levator scapulae muscle, 302–3, *303*, 304t, *305*, **956**
Levers, in body movement, 191–94, *191, 194*
Levodopa, 390
Lewy body, *391*
L'Heureux, Nicolas, 132
Life
    characteristics of, 6–7, 8t
    maintenance of, 7–12

Life expectancy, 912
Life span, 912–14
Life-span changes, 19–21
Ligament, 142, 183, 257. *See also specific ligaments*
Ligamenta flava, 146
Ligamentum arteriosum, *583, 900, 908, 908*
Ligamentum capitis, 267, *268*
Ligamentum teres, *900, 907*
Ligamentum venosum, *900, 907, 908*
Ligand, 68
Light-adapted eye, 460
Light chains, 625–26, *627*
Light microscope, 65
Lignin, 695
Limb buds, 887
Limbic system, 392, 432
Limp, 318
Linea, 199t
Linea alba, **31**, *304*, 312, **957**
Linea aspera, 227, *229*
Lingual artery, *585, 586*, 587
Lingual tonsils, 649, *650, 734*
Linoleic acid, 698
Linolenic acid, 698
Lipase, 107, 671, 677t, *678*
    gastric, 659
    intestinal, 677, 677t
    pancreatic, 665, 677t, 705
Lipid(s), 697
    blood, 525, 971
    calories per gram, 701t, 702
    in cells, 50–52, *50, 51*, 52t, 55t
    dietary requirements for, 698, 701t
    dietary sources of, 697, 701t
    digestion of, 697–98, *697, 698*
    membrane, 66
    metabolism of, 697–98, *697, 698*, 701t
    total serum, 971
Lipid bilayer, 65, *66, 67*
Lipofuscin, 502, 912
Lipogenesis, *696*, 725
Lipoid nephrosis, 789
Lipoprotein(s), 525, 667, 698, *698. See also specific types of lipoproteins*
    alpha lipoprotein, 525
Lipoprotein lipase, 679
Liposomes, *942*
Lips, 648–49, *649, 650*
Lissencephaly, 382
Liver, *15*, **32, 33, 34**, *645*, 666–71, *667, 668, 674, 774*, **948, 950, 954, 955, 966, 967**
    bioartificial, 669
    functions of, 666–69, 670t
    site of gene therapy, 940, *942*
    structure of, 666
    therapeutic liver repopulation, 669
Liver cells, 934
Liver failure, 669
Liver spots, 175
Liver transplantation, 669
Lobar pneumonia, 766

Lobes
    of liver, 666, **966**
    of lungs, *740, 743*, 744
Lobules
    of liver, 666, *668*
    of lungs, 740, 744
    of testes, 832
Local anesthetics, 68t
Lochia, 914
Long bone, 182–83, *183*, 189
Longissimus capitis muscle, *301*
Longissimus thoracis muscle, *301*
Longitudinal fissure, 382, *384, 389, 395*
Longitudinal ligament
    anterior, 210
    posterior, 210
Longitudinal section, 23, *23*
Long-QT syndrome, 68
Long-term memory, 388
Long-term synaptic potentiation, 388
Long thoracic nerve, 304t
Loose connective tissue, 141–42, 144, *144*, 150t
Lordosis, 216
"Lorenzo's Oil," 75
Low-density lipoprotein(s), 87, 525, 526t, 970
Low-density lipoprotein receptors, 87, 525, 599, 927, 940
Lower back pain, 428
Lower esophageal sphincter, 658, *658*
Lower limb
    anterior view of, **960, 961**
    arteries to, 588–91, *590*
    bones of, 196t, 197, 227–31, *228, 229, 230*, 231t
    lateral view of, **962**
    muscles that move, 315–25, *316, 317, 318, 319*, 319t, 320t, 321t, *322, 323, 324, 325*
    posterior view of, **961, 962**
    surface anatomy of, **334, 335, 336**
    veins from, 595–98, *596*
Lower motor neuron syndrome, 381
Lower respiratory tract, 733
Low fat foods, 721
LSD, 360
Lubrication, of female genitalia, 856
Lucid, Shannon, 574, *574*
Lumbago, 232
Lumbar artery, 583, *584*, 585t, 588, *592*
Lumbar curvature, 210, *211*
Lumbar nerves, 402, *404*
Lumbar puncture, 371, *371*
Lumbar region, 23, *23, 24*, 25
Lumbar vein, 593, 595
    ascending, *597*
Lumbar vertebrae, **36**, 196t, *211*, 212–13, *214*, 214t, **950, 951**
Lumbosacral plexus, 406–8, *407, 409*
Lumen, of alimentary canal, 675
Lumpectomy, 865

Lunate, *223*
Lung(s), *15*, **32, 33, 34**, 743–44, *743, 744*, 744t, **953, 954, 963**
    collapsed, 759
    irritants, 745, 745t
    of newborn, 906
    site of gene therapy, 941, *942*
Lung cancer, 591, 735, *735*, 745, 764, 920
Lung volume reduction surgery, 752
Lunula, 167, *167*
Luteinizing hormone, 471t, 475, 478, 480, *481*, 483, 485t, 845–46, 852, 857–59, *857*, 858t, *860*, 884–85, 885t
Luxation, 274
Lyme arthritis, 273
Lyme disease, 161t, 267, 273, 273t
Lymph, 608, *809*
    flow of, 612, 810
    formation of, 611
    function of, 611
Lymphadenectomy, 637
Lymphadenitis, 613
Lymphadenopathy, 637
Lymphadenotomy, 637
Lymphangiogram, 609, *610*
Lymphangitis, 613
Lymphatic capillaries, 608, *609, 611, 611*
Lymphatic duct, right, 609, *610*
Lymphatic pathways, 608–10, *609, 610, 611*
Lymphatic system, 8t, *17*, 18
    age-related changes in, 911t
    system interconnections of, 636
Lymphatic trunk, 608–9, *611*
    bronchomediastinal, 609, *610*
    intercostal, 608–9, *610*
    intestinal, 608, *610*
    jugular, 609, *610*
    lumbar, 608, *610*
    subclavian, 609
Lymphatic vessels, 569, 581, 608, *609, 610, 611*, 612, *613*
    afferent, *612*, 613
    efferent, *612*, 613
    obstructed, 612, 814–15, 815t
Lymph nodes, 608, *609, 610, 611*, 612–14, *613*, 617t
    abdominal cavity, *613*, 614
    axillary, 613, *613*
    cervical, 613, *613*
    functions of, 614
    inguinal, *613*, 614
    location of, 613–14, *613*
    structure of, 612–13, *612*
    supratrochlear, *613*, 614
    thoracic cavity, *613*, 614
Lymph nodules, 613
Lymphoblasts, *512*
Lymphocytes, 519–20, *520*, 523t, *524*, 612, 616, *619. See also* B cells; T cells
    elevated, 521
    functions of, 620–21
    origin of, 619–20, *620*
    production of, 614
Lymphocytopenia, 637

Lymphocytosis, 637
Lymphoid leukemia, 522
Lymphoma, 637
Lymphosarcoma, 637
Lymph sinuses, *612,* 613
Lysosomal storage disease, 75
Lysosomes, *64,* 74, *74,* 86–87
 abnormal, 74–75
Lysozyme, 448, 617

M

McCune-Albright syndrome, 475
McKusick, Victor, 924
Macrocytosis, 537
Macromolecules, 4, *5,* 5t
Macronutrients, 694, 722
Macrophages, 141, *142,* 144t, 158,
 512, 515, 519, 612, 616, 618,
 621, *622*
 alveolar, 757
Macula, *438,* 443–44, *444*
Macula densa, 778, *781,* 786
Macula lutea, 454, *455,* 458
Macular degeneration, 462
Mad cow disease, 54
Magnesium, 715–16, 716t, 971
 in cells, 48, *49*
 electrolyte balance, 813
 in human body, 39t, 194
 intracellular fluid, *810*
 in plasma, glomerular filtrate,
  and urine, 526, 784t
Magnetic resonance imaging, 6–7, *7*
 of heart, 549
Major calyx, 773, *775*
Major histocompatibility
  complex, 621
 class I antigens, 621
 class II antigens, 621
Malabsorption, 679
Malaria, sickle-cell disease and, 927
Male climacteric, 846
Male infertility, 842–43, *842,*
 843, 932
Male sex hormones, 845–46
 regulation of, 846
Male torso, **31, 32, 33**
Malic acid, *112*
Malignant tumor, 93
Malleolus
 lateral, 228, *229,* **336**
 medial, 228, *229,* **336**
Malleus, 196t, *435,* 436, *436, 441*
Malnutrition, 721, 724
 in pregnancy, 894–95
 primary, 721
 secondary, 721
Maltase, 107, 677, 677t, *678*
Maltose, *678*
Mammary artery, internal, 600
Mammary glands, 12, **30,** 168, 168t,
 860–62, 862, 904–5, *904, 905*
 location of, 861
 lymphatics of, *610*
 secretion of milk, *73*
 structure of, 861–62, *862*
Mammary region, *24,* 25
Mammatropes, 480

Mammillary body, 390, *392*
Mammogram, 864–65, *864*
Mandible, 194, 196t, 199, *200, 201,*
 206–7, *206, 207,* 208t, *210,*
 245, 248, 332, 400, 651, 949
Mandibular division, of trigeminal
  nerve, 400, *400*
Mandibular foramen, 207, 209t, **245**
Mandibular fossa, 201, *202,* **241,**
 242, 243, 246
Mandibular nerve, *401*
Mandibular ramus, **245**
Manganese, 39t, 717–18, 719t
Mania, 357t
Manifesting heterozygote, 932
Manometer, 371
Mantoux test, 632
Manubrium, 216, *217,* **334**
Maple syrup urine disease, 799
Marasmus, 723, *723*
Marfan syndrome, 571, 598,
 598, 928
Marginal artery, 550, *553*
Marijuana, 428
Marshall, Barry, 660
Masseter muscle, *297, 298,* 299,
 300t, **332,** *654,* **956**
Mass movements, 683
Mass spectrometry, 434
Mast cells, 141–42, *142,* 144t, 531,
 531t, 628, *631*
Mastectomy, modified radical, 865
Mastication, 434, 437, 648
 muscles of, *298,* 299–300, 300t
Mastitis, 868
Mastoid fontanel, 208, *210*
Mastoiditis, 202
Mastoid process, 201–2, *201, 202,*
 206, **239, 243, 246, 332**
Maternal serum marker, 937, 937t
Matrix, connective tissue, 141
Matter, structure of, 38–47, 39t
Maxilla, 196t, *200, 201, 202, 204,*
 206, 207, 210, **238, 239, 240,**
 241, 242, 332, *400, 651*
Maxillary artery, 585t, 587
Maxillary bone, 204–5, 208t, **949**
Maxillary division, of trigeminal
  nerve, 400, *400*
Maxillary region, **244**
Maxillary sinus, 204, *205,* 205t,
 207, **248, 249,** 736
M cells, 88, 613
Mean arterial pressure, 573
Mean corpuscular hemoglobin, 971
Mean corpuscular volume, 971
Measles, 629
Meatus, 199t
Mechanical barrier(s), to infection,
 617, 619t
Mechanical barrier contraception,
 863–64
Mechanical ventilation, 747, 752
Mechanoreceptors, 423, 426, 429t
Mechnikov, Ilya, 912
Meconium, 914
Medial, definition of, 21
Medial rectus muscle, 448, *449,*
 449t, *450,* **952**

Median nerve, 308t, *311,* 311t, 406,
 407, 408
Median section, 21, *22*
Mediastinal artery, 582, 585t
Mediastinum, 12, *13, 15*
Mediastinum testes, 832
Medicinal leech, 532, *532*
Medicinal science, history of, 3, *3*
Medulla oblongata, 376–77, 381,
 382t, *382, 383, 392,* 393,
 393, 396t, **968**
Medullary cavity, 183, *183*
Medullary rhythmicity area, 753,
 753, 754
Megakaryoblasts, *512*
Megakaryocytes, 511, *512,*
 521–23, 531
Megaloblastic anemia, 711, 713t
Meiosis, 90, 833–36, *835, 836,*
 848, *849*
 errors in, 934, *935*
Meiosis I, 835
Meiosis II, 835
Meissner's corpuscles, 164, 424,
 425, 429t
Melanin, *127,* 160, *161,* 162, *162,*
 166, 172, 453, 478, 930
Melanocortin-4 receptor, 695t
Melanocytes, 160, 162, *162,* 166,
 172, 930
Melanocyte-stimulating
  hormone, 478
Melanoma, *163,* 167, *942*
MELAS (mitochondrial
  encephalomyepathy, lactic
  acidosis, and strokelike
  episodes), 75
Melatonin, 499–500, 502
Membrane. *See specific*
 membranes; specific types
 of membranes
Membrane potential, 348–54. *See*
 also Action potential
 ion distribution, 348–49, *349*
 local potential changes, 350
 resting potential, 350, *351*
 threshold potential, 350, *352*
Membranous labyrinth, 437,
 438, 439
Membranous urethra, 798, *798*
Memory, 387–88, 417
 long-term, 388
 short-term, 388
Memory B cells, 623, *626,* 628
Memory consolidation, 388
Memory T cells, 621, 628
Menarche, 858
Ménière's disease, 463
Meningeal branch, of spinal
  nerves, 403
Meninges, 367–68, *367,* 367t,
 368, 383
Meningitis, 368
Meniscus, 258, *258, 270,* 271
 medial, *270*
 tearing or displacing, 271, *272*
Menopause, 195, 232, 517, 859–60
Menstrual cramps, 477
Menstrual cycle, 499

Menstrual flow, 858–59, *860*
Mental foramen, *200, 201,* 207,
 209t, **245**
Mental nerve, *400*
Mental region, *24,* 25
Merocrine gland, 139–40, *140,* 140t
Mesenteric artery, 595
 inferior, **35,** 583, *584,* 585t,
  589, 592
 superior, **35,** 583, *584,* 585t,
  592, **967**
Mesenteric ganglion
 inferior, *412*
 superior, *412*
Mesenteric vein
 inferior, 593, *594*
 superior, **35,** 593, *594,* **967**
Mesentery, **34,** *647, 673,* 674,
 674, **967**
Mesoderm, 886, *886, 888*
Mesodermal cells, 886–87
Mesothelioma, 745
Messenger RNA, 118–22, 119t,
 120, 121, 123t, 472, *473,* 921
Metabolic acidosis, 499, 701t,
 822–23, *823*
Metabolic alkalosis, 822–23, *823*
Metabolic pathway, 104, 107, *107*
 regulation of, 115, *115*
Metabolic rate, 485
Metabolism, 695
 age-related changes in,
  724–25, 724t
 control of reactions of, 106–7
 definition of, 7, 104
 energy for, 107–8
 inborn error of, 126–27, *127*
 metabolic processes, 104–6
 regulation of, 115, *115*
 water of, 811, *811*
Metacarpals, 196t, 197, *198, 220,*
 223, *223,* 223t, **335**
Metacarpophalangeal joint, 263t
Metallic taste, 432
Metaphase
 meiosis I, 836
 meiosis II, 836
 mitotic, *91,* 92, 92t
Metaphase plate, *91*
Metarterioles, 565
Metastasis, 70, 93, *95,* 612
Metatarsals, 196t, 197, *198,* 228,
 229, *230,* 231t, **336**
Metatarsophalangeal joint, 263t
Metatarsus, 229
Methionine, 715, *715*
Methotrexate, 160, 865
Methylprednisolone, 380
Metronidazole, 435
Micelles, 671
Microcytosis, 537
Microfilaments, 77–78, *78,* 80t
Microglial cells, *346,* 347, 347t
Micrograph, 65
Micrographia, 390
Microgravity, 574, *574*
Microliter, 514
Micrometer, 62
Micronutrients, 694

Microscope, 65, *65*
Microtubules, *64*, 76–78, *78*, 80t, 92
Microvilli, *64*, 133, *135*, *647*, 675, *676*, 677, 788
Micturition, 484, 798–99, 799t, 840
Micturition reflex, 798–99
Micturition reflex center, 798
Midbrain, 380–81, *382*, 382t, *383*, 392–93, *393*, 396t, **968**
Middle ear, 196t, 436–37, *435*, *436*
Middle ear cavity, 14, *14*
Middle tunic, 450–51, 455t
Midpiece, of sperm, 837–38, *837*
Migraine, 340, 428, 924
    classic, 340
    common, 340
    familial hemiplegic, 340
    treatment of, 340
Milk, 704, 860–62
    maternal antibodies in, 907
    perfect food for babies, 907
    production of, 904–5, *904*, *905*
    release of, 484
    secretion of, *73*, 904–5, *904*, *905*
Milk "let down," 905
Millivolt, 350
Milstein, Cesar, 624
Mineral(s), 712–18, 722
    characteristics of, 712–13
    major, 714–16, 716t
    trace elements, 717–18, 719t
Mineralocorticoids, 493, 496, 502
Minor calyx, 773, *775*
Minoxidil (Rogaine), 166
Mitochondria, *64*, 72–73, *74*, 111, 80t
    DNA of, 118
    evolution of, 73
Mitochondrial Eve, 73
Mitochondrial myopathy, 73
Mitosis, 90–92, *90*, *91*
Mitotic clock, 92
Mitral valve, 546, *547*, *548*, 549, 549t, *550*, *551*, 554, 581
Mitral valve prolapse, 546
Mixed nerve, 397, *398*
Mixing movements, of alimentary canal, 646–48, *647*
M line, 279, *281*
Modiolus, 437, *438*
Moebius syndrome, 278
Molar, 650–51, *651*, 651t
Molding, of infant's skull, 208
Mole (nevus), 163, 178
Molecular formula, 40, *43*
Molecules, 4, *5*, 5t, 39t, 40
Monoamine oxidase, 358, 416
Monoamine oxidase inhibitors, 358t
Monoblasts, *512*
Monoclonal antibodies, 624, *625*, 865
Monocytes, 188, *512*, 519–20, *519*, 523t, *524*, 618
    elevated, 521t
Monogenic traits, 929
Mononuclear phagocytic system, 619

Monosaccharides, 48, *49*, 695, *696*
    absorption in small intestine, 678, 680t
    metabolism of, 104–5, *104*
    structure of, *58*
Monosodium glutamate, 432
Monosomy, 934, *935*
Monosynaptic reflex, 373
Monounsaturated fatty acids, 50
Monozygotic twins, 914
Mons pubis, **30**, 855, *855*
Morning-after pill, 866
Morning sickness, 876
Morphine, 359–60, 428, 434
Morula, 880, *881*, 887t
Mosso, Angelo, 760
Mother-of-pearl, 195
Motion sickness, 165, 446, 924
Motor area, of cerebral cortex, 385–86, *385*, *386*
Motor end plate, 283, *283*, 286
Motor nerve, 397
Motor neuron, 282, *283*, 345, *345*, 346t, 373, *374*, 375, *375*, 375t
Motor unit, 283, *283*
    recruitment of, 291
Mouth, *645*, *646*, 648–52, *649*, *650*, 652t
Movement, 16, 191–94
    as characteristic of life, 8t
    of joints, 261–63, *261*, *262*
Mucin, 140
Mucopolysaccharides, 715
Mucosa
    of alimentary canal, 645, 646t
    of large intestine, 682, *682*, *683*
Mucosal layer
    of uterine tube, 853
    of vagina, 854
Mucous cells, 140, 659–60, *660*. *See also* Goblet cells
    of salivary glands, 652–53
Mucous coat
    of ureter, 795, *795*
    of urinary bladder, 796, *797*
Mucous gland, 139t
Mucous membranes, 152–53
    as barrier to infection, 617
    of respiratory tract, 733, *734*
Mucus, 133, 135, *135*, 140, 153, 653, 661t, 682, 731
Mucus-secreting glands, 676
Multicellular gland, 138, 139t
Multi-infarct dementia, 417
Multiple births, 876, *876*, 883
Multiple endocrine neoplasia, 468
Multiple sclerosis, 343
Multipolar neuron, *344*, 345, 346t
Multiunit smooth muscle, 293
Mumps, 629, 842
Murmur, 546, 554
Muscarine, 415
Muscarinic receptors, 415, *416*
Muscle(s). *See also specific muscles*
    involuntary, 150
    site of gene therapy, 939, *942*
    skeletal. *See* Skeletal muscle(s)
    voluntary, 148

Muscle cells, 62, *63*
    lactic acid in, 111
Muscle cramps, 111, 289, 425
Muscle dysmorphia, 694
Muscle fatigue, 111, 289
Muscle fiber, 148, *150*, 279–82, *280*, *281*, *282*
Muscle impulse, 284
Muscle relaxants, 68t
Muscle spindle, 428–29, *429*, 429t
Muscle strain, 282
Muscle tissue, 132, 132t, 148–51, 152t
    cardiac muscle, 150, *151*, 152t
    characteristics of, 148
    skeletal muscle, 148, *149*, 152t
    smooth muscle, 150, *151*, 152t
Muscle tone, 291
Muscular dystrophy, 327, 882t, *942*
    Duchenne, 125, *125*, 282, 932, 939
Muscularis mucosae, 646
Muscular layer
    of alimentary canal, 646, 646t, *647*
        circular fibers, 646, *647*
        longitudinal fibers, 646, *647*
    of large intestine, *681*, 682, *682*
    of small intestine, 675, *676*
    of stomach, 658, *658*
    of ureter, 795, *795*
    of urinary bladder, 797
    of uterine tube, 853
    of vagina, 854
Muscular system, 8t, 16, *16*
    age-related changes in, 325, 911t
    effect of weightlessness on, 574
    system interconnections of, 326
Musculocutaneous nerve, **33**, 306t, 308t, 405, *407*, *408*
Mutagens, 127, 127t
Mutations, 94, *95*, 124–25
    effects of, 125–27
    germinal, 865
    nature of, 125, *125*, 282, 932, 939
    protection against, 125
    somatic, 865
Myalgia, 327
Myasthenia gravis, 284, 286, 327, 357t, 633t
Myelin, 75, 342–43, 348
Myelinated axon, *342*, 343, *344*, 348, 354, *354*
Myelin sheath, 75, 342–43, *342*, *398*
Myeloblasts, *512*
Myelocytes, *512*
Myeloid leukemia, 522, *522*
Myenteric plexus, 648
Myoblasts, 96
Myocardial infarction. *See* Heart attack
Myocardium, 545, *545*, 545t, 550
Myoepithelial cells, 905, *905*
Myofascial syndrome, 428
Myofibrils, 77, *277*, 279, *280*, *281*, *282*
Myoglobin, 54, 288, *288*, 292, 717
Myogram, 290, *290*, *291*

Myokymia, 327
Myology, 327
Myoma, 327
Myometrium, *853*, 854, *854*
Myopathy, 327
    mitochondrial, 73
Myopia, 457, *457*
Myosin, 279, *281*, 282, 285, 287t, 293–94. *See also* Thick filaments
    abnormal, 599
Myositis, 327
Myositis ossificans, 321
Myotomy, 327
Myotonia, 327

N

Nacre, 195
NAD, 110, 709–10, *710*, 974–75, *975*, *976*
NADH
    from citric acid cycle, 111, *112*
    from conversion of pyruvic acid to lactic acid, 111, *112*
    from glycolysis, 108, *110*
    regeneration of NAD+ from, 111
NADP, 709–10
Nail(s), 167, *167*
    disorders affecting, 167
Nail bed, 167, *167*
Nail plate, 167, *167*
Nandrolone, 474
Narcolepsy, 394t
Nasal bone, 196t, *200*, *201*, *204*, 206, 208t, *210*, **238**, **239**, **240**, **332**
Nasal cavity, 14, *14*, 199, 203, *204*, 205, *205*, 430, *431*, 448, *650*, 744t, **948**
Nasal concha, 733
    inferior, 196t, *200*, *204*, 206, *206*, *207*, 208t, **240**, **734**, **949**
    middle, 203, *204*, *207*, **240**, **246**, **734**
    superior, 203, *203*, *204*, *431*, 734
Nasal continuous positive airway pressure, 753
Nasal region, 24, 25, **240**
Nasal septum, *206*, 733, **952**
    deviated septum, 733
Nasopharynx, 437, *650*, 655, *734*, 736
National Organ Procurement and Transplant Network, 669
Natural killer cells, 618
Naturally acquired active immunity, 629, 630t
Naturally acquired passive immunity, 629, 630t
Nausea, 664, 876
Navicular, *230*
Near drowning, 759
Nearsightedness, 457, *457*
Neck
    arteries of, 584–87, *585*, *586*
    sagittal section of, **949**
    surface anatomy of, **332**
    transverse section of, **953**
    veins from, 591

Neck of bone, radius, *220*
Neck of urinary bladder, 796
Necrosis, 79, 159, 571
Negative feedback, 10, 115, *115*, 477, *479*, 494, 497, *498*, 514
Negative nitrogen balance, 701
Neisseria gonorrhoeae, 867t
Neonatal period, 906–8, *906*, 911t
Neonatology, 25, 896
Nephrectomy, 802
Nephritis, 780
Nephrolithiasis, 802
Nephrology, 25
Nephron(s), *775*
  age-related changes in, 800
  blood supply of, 781–82
  cortical, 780–81, *782*
  juxtaglomerular apparatus, 778–80
  juxtamedullary, 780–81, *782, 783*
  structure of, *775*, 776–78
Nephron loop, 778, *781, 783*
  ascending limb of, 778, *779, 791, 792, 793*, 794t
  descending limb of, 778, *779, 791, 792, 793*, 794t
Nephroptosis, 802
Nephrotic syndrome, 788–89
Nerve, 338
Nerve block, 428
Nerve fiber, 372, 396, *398. See also* Axon
  classification of, 397–99
Nerve gas, 69, 286
Nerve growth factor, 348, *349*
Nerve impulse, 16, 338–39, 353, *353*
  conduction of, 353–54, *353*, 353t
  convergence of, 359, *359*
  divergence of, *359*, 360–61
  factors affecting conduction, 355
  processing of, 358–61
Nerve tracts, 372
Nervous system, 8t, 16–17, *17*
  age-related changes in, 416–17, 911t
  central. *See* Central nervous system
  compared to endocrine system, 469, *469*, 469t
  control of endocrine secretions, 477
  functions of, 338–43
  peripheral. *See* Peripheral nervous system
  system interconnections of, 362
Nervous tissue, 132, 132t, 151–52, 152t
  site of gene therapy, 941
Nervous tunic, 450
Net filtration pressure, 784–85
Neuralgia, 417, 463
Neural stem cells, 94, 98, 338, *338*, 381, 391
Neural tube, 380, *382*, 892
Neural tube defect, 382, 711, 892
Neurilemma, 342, *342, 398*
Neuritis, 417, 463
Neurofibrils, 340, *341*
Neurofibromatosis, 882t

Neuroglial cells, 152, *152, 337, 338, 339*, 483, 498
  abnormal, 348
  classification of, 345–47, *346, 347*, 347t
  functions of, 345
Neurology, 25
Neuroma, 348
Neuromodulators, 358
Neuromuscular junction, 282–83, *283*, 287t
Neurons, 151–52, *152, 338, 339, 341*, 343–48, *365*
  accelerator, 345
  action potential, 350–53, *352, 353*
  all-or-none response of, 353
  bipolar, 344, *344*, 346t, 454, *454*
  cell division in, 338
  classification of, 344–45, 346t
  depolarized membranes, 350
  hyperpolarized membrane, 350
  impulse conduction. *See* Nerve impulse
  inhibitory, 345
  membrane potential. *See* Membrane potential
  motor, 282, *283*, 345, *345*, 346t, 373, *374, 375, 375*, 375t
  multipolar, 344, 345, 346t
  pH sensitivity of, 820
  pools of, 358–59, *359*
  postsynaptic, 354, *355*
  presynaptic, 354, *355, 356*
  refractory period of, 353
  regeneration of, 348, *349*, 381
  sensory, 345, *345*, 346t, 373, *374, 375, 375*, 375t
  unipolar, 344, *344*, 346t, 422
Neuropeptides, 358, 428
Neuropeptide Y, 695, 695t
Neurosecretory cells, 478, *480*
Neurospheres, *338*
Neurotransmitters, 283, 294, 338, 354, 356–58, *356*, 357t, 469, *469*
  abnormal, *942*
  age-related changes in, 417
  of autonomic nervous system, 414–16, *414*
  disorders associated with imbalances, 357t
  drugs that alter levels of, 358t
  reuptake of, 358
  in vision, 460
Neutralization, 627, 628t
Neutral solution, 46
Neutrons, 38–39, *39*, 39t
Neutrophil(s), *512*, 518, *519*, 520–21, *521*, 523t, *524*, 618
  elevated, 521t
Neutrophilia, 537
Newborn
  cardiovascular system of, 907–8, *908*, 908t
  maternal antibodies in, 627
  premature, 747, 760, 896
  vitamin K deficiency in, 709
  withdrawal symptoms, 892

Newborn jaundice, 172
Niacin, 709–10, *710*, 713t
  deficiency of, 710
Niacinamide, 709, *710*
Nicholas II (Tsar of Russia), 118
Nicotine, 358t, 360, 415, 434, 735, 906t
  addiction to, 360–61, *361*
Nicotinic receptors, 360–61, 415, *416*
Niemann-Pick type C disease, 599
Nifedipine, 760
Night blindness, 460, 707, 708t
Nipple, **30**, 861, *862*
Nitrate drugs, 830
Nitric oxide, 48, 577, 830, 842–43
Nitrogen, in human body, 39t
Nitrogen balance, 700–701
  negative, 701
  positive, 701
Nitrogenous base, 115
Nitroglycerin, 435
Nitrosamines, 127t
Nocturnal emissions, 844
Nocturnal enuresis, 799
Nodal rhythm, 563
Nodes of Ranvier, *341, 342, 343, 398*
Noise pollution, 443
Nondisjunction, 934, *935*
Nonelectrolytes, 47
Nonessential amino acids, 700
Nonheritable gene therapy, 939
Nonprotein nitrogenous substances, in blood, 525–26
Nonrespiratory air movements, 751, 751t
"Nonself," 67, 619
Nonsteroidal anti-inflammatory drugs, 272, 685
Norepinephrine, 294, 356, 358, 358t, 360, 411, 414–16, *414, 470*, 471, 471t, 475, 477, 491–92, *492*, 493t, 577
Norepinephrine receptors, 360
Normal range, 12
Normal saline, 83
Normoblasts, *512*
Nose, 733, *733*, 744t
Nostrils, 733, *733, 734*
Nuclear envelope, 62, 78, *79*, 80t, *91, 92*
Nuclear pore, 79, *79*
Nuclear transfer, somatic cell, 98, *99*
Nuclear waste, 45
Nuclease, 665, 677t
Nucleic acids. *See also* DNA; RNA
  breakdown of, 817
  in cells, 54–55, *55*, 55t
  synthesis of, 115–20
Nucleolus, *64*, 79, *79*, 80t, 92
Nucleoplasm, 79
Nucleotides, 54, *54*, 115, *115*
Nucleus, atomic, 38–39, *39*, 39t
Nucleus, cell, 62, *64*, 78–79, *79*
Nucleus cuneatus, 393
Nucleus gracilis, 393

Nucleus pulposus, 216, 257
Nutrients, 694
  essential, 695
  exchange in capillaries, 568
  plasma, 525
Nutrition, 694
  age-related changes in, 724–25, 724t
  for athletes, 722
Nutritional edema, 701
Nyctalopia, 725
Nystagmus, 458, 463

O

Oberrothenback, Germany, 45
Obesity, 578, 701t, 703–5, *703, 704*
  treatment for, 705
Oblique fracture, *192*
Oblique popliteal ligament, 269, *270*
Oblique section, 22, *23*
Obsessive-compulsive disorder, 56–57
Obstetrics, 25
Obstructive jaundice, 671
Obstructive sleep apnea syndrome, 394t
Obturator artery, *589*
Obturator foramen, **36**, 224, *225, 226*
Obturator nerve, 319t, 406, *407, 409*
Occipital artery, 585, 587
Occipital bone, 196t, *197*, 199–200, 201, 201t, *202, 204, 206, 210*, **239, 242, 243, 244, 246, 248, 249, 255, 332**
Occipitalis muscle, *297, 298, 299*, **955**
Occipital lobe, 384, *384, 385*, 386–87, 387t, **952, 968**
Occipital region, 24, 25, **243**
Occluding junction, 837
Occupational teratogens, 895
Octet rule, 42
Oculomotor nerve (III), 399–400, *399*, 403t, *413, 414*, 449t, *478*
Odorant molecules, 430–31
Odor blindness, 924
Olecranon fossa, 221, *221*
Olecranon process, 193, *220, 222, 222, 266*, **333, 334**
Olestra, 699
Olfactory bulb, 338, *399*, 400, *431, 432*
Olfactory cortex, 432
Olfactory epithelium, *431*
  of bloodhound, 431
Olfactory foramen, 202, *204*
Olfactory nerve (I), *399*, 400, 403t
Olfactory nerve pathways, 431–32
Olfactory organs, 430–31, *431*
Olfactory receptor(s), 430–31, *431*, 733
Olfactory receptor cells, 400, 430, *431*
Olfactory tract, *399*, 400, *431, 432*

Oligodendrocytes, 343, *346, 347–48, 347t*
Oligomenorrhea, 858
Oliguria, 802
Olivary nucleus, superior, *442*
Olive, 393
Oncogene, 94, *95*
Oncology, 25
Online Mendelian Inheritance in Man, 924
Oocyte, 852, *853*
   primary, 848, *849, 852,* 875
   secondary, 848–49, *849, 852,* 877–78
Oogenesis, 848–49, *849*
Oophorectomy, 868
Oophoritis, 868
Ophthalmic artery, 587
Ophthalmic division, of trigeminal nerve, 400, *400*
Ophthalmic vein, superior, *593*
Ophthalmology, 25
Opiate(s), 359, 428
Opiate receptors, 359
Opium, 359–60
Opsin, 458–60
Opsonization, 627, *628,* 628t
Optic canal, *200, 203, 204,* 209t, **250**
Optic chiasma, 390, *392,* 461, *461, 478, 480*
Optic disc, *450, 455,* 455
Optic foramen, 400
Optic nerve (II), *392, 399,* 400, 403t, 450, *450,* 453, 455, *455,* 460–61, *461, 478,* **952**
Optic radiations, 461, *461*
Optic tract, 390, *392, 399,* 461, *461*
Oral cavity, 14, *14,* 648, *650, 733,* **949**
Oral contraceptives, 863t, 865, *866,* 906t
Oral region, *24,* 25
Orbicularis oculi muscle, *297, 298–99, 298,* 299t, 447, *447,* 449t, **956**
Orbicularis oris muscle, *297, 298–99, 298,* 299t, **956**
Orbit, 199, *200,* **246,** 458
Orbital cavity, 14, *14*
Orbital fissure
   inferior, *200,* 209t
   superior, *200, 203,* 209t
Orbital region, *24,* 25, **240**
Orchiectomy, 832, 868
Orchitis, 842, 868
Organ, 5, *5,* 5t
   age-related changes in, 21
Organelles, 4, *5,* 5t, 62, *64,* 80t. *See also specific organelles*
Organic substances, 47
   in cells, 48–55, 55t
Organism, 5, *5,* 5t
   requirements of, 8–9, 9t
Organization, levels of, 4–6, *5,* 5t
Organ of Corti, *421,* 439, *440*
Organ systems, 5, *5,* 5t, 8t, 16–19, *16–20*
   age-related changes in, 19
   interactions of organ systems, *20*

Orgasm
   female, 856, *856*
   male, 842–44, *844*
Origin, of muscle, 261, 296, *296*
Orlisat, 705
Ornithine transcarbamylase deficiency, 882t, 941
Oropharynx, *650,* 655, *734, 736*
Orthopedics, 25, 232
Orthostatic intolerance, 574
Osgood-Schlatter disease, 228
Osmolality
   blood, 971
   urine, 973
Osmolarity, of body fluids, 813
Osmoles, 813
Osmoreceptor(s), 484, 812
Osmoreceptor-ADH mechanism, 812
Osmosis, 82–83, *83,* 89t, 814, *814*
Osmotic diuresis, 788
Osmotic pressure, 83, 810
Osseous labyrinth, 437, *438*
Ossification
   endochondral, 187, *187,* 187t
   intramembranous, 187, 187t
Ossification center
   primary, *187,* 188, 189t
   secondary, *187,* 188, 189t
Ostealgia, 232
Ostectomy, 232
Osteitis, 232
Osteoarthritis, 143t, 258, 272, *273,* 273t
Osteoblasts, 96, 186–88, 187t, *188,* 190, 192, 194, 231, 489
Osteochondritis, 232
Osteoclasts, *181,* 188–90, *189,* 192–94, *193, 196,* 231
Osteocytes, 148, *149,* 184, *185,* 186, *186,* 187t
Osteogenesis, 186, 232
Osteogenesis imperfecta, 143t, *186,* 232
Osteoma, 232
Osteomalacia, 190, 232, 489, 707
Osteomyelitis, 232
Osteon, 148, 184, *185,* 232
Osteonecrosis, 232
Osteopathology, 232
Osteopenia, 232
Osteoporosis, 143, 195, 232, 502, 800, 859–60
Osteotomy, 232
Otic ganglion, *413*
Otic region, *24,* 25
Otolith(s), 444, *445*
Otolithic membrane, 444
Otosclerosis, 443, 463
Otoscope, 437
Outer membrane, of mitochondria, 72–73, *74*
Outer tunic, 450, 455t
Oval window, *435,* 436, *436, 438, 439, 441*
Ovarian follicle, 848
   maturation of, 850, *850, 851,* 858

Ovarian ligament, 848, *848, 853*
Ovariectomy, 868
Ovary, **34, 35,** *469,* 500, *847, 848–52, 852, 853,* 856t
   attachments of, 848
   descent of, 848
   hormones of, 857
   structure of, 848, *848*
Overnutrition, 721
Overweight, 114, 703, *703*
Ovo-vegetarian, 694t
Ovulation, 852, *852,* 861, 877
Ovulation predictor kit, 863
Oxaloacetic acid, *109,* 111, *112, 113*
Oxidation, 108, 975
Oxygen
   blood, 12, 525
   in cells, 48, 49t
   diffusion across cell membrane, 81, *82*
   diffusion from alveoli into capillaries, 742, *742*
   diffusion through respiratory membrane, 757–58, *758*
   as electron acceptor, 112, *113*
   exchange in capillaries, 568
   in human body, 39t
   partial pressure of, 754, *755,* 756, 756t, 758, *758*
   as requirement of organisms, 8, 9t
   supply for skeletal muscles, 288
   transport in blood, 511, 514, 516, 760–61, *760, 761, 762,* 764t
Oxygen consumption, 702
Oxygen debt, 111, 288–89, *289*
Oxygen saturation, 971
Oxyhemoglobin, 511, 760–61, *760, 761, 762*
Oxytocin, 294, *470,* 471t, 478, 483–84, *483,* 485t, 901–2, 904–5, *905,* 906t
   induction of labor with, 485

P

Pacemaker, 557
   artificial, 563
   secondary, 563
Pacinian corpuscles, 164, 424, *425,* 429t
Packed cell volume, 511
Packed red cells, 537
Paclitaxol, 865
Paganini, Niccolò, 257
Pain, 424–28
   acute, 426–27
   bone, 184
   cancer, 428
   chronic, 426–28
   pain nerve pathways, 426–27
   referred, 426, *426*
   regulation of pain impulses, 427–28
   visceral, 425–26, *426*
Pain receptors, 423–28, 429t
Palatine bone, 196t, *200,* 202, *204,* 205–6, *205,* 208t, **242, 244**

Palatine foramen, greater, *202,* 209t, **242, 244**
Palatine process, 205, **242, 244**
Palatine suture, median, *202,* **242, 244**
Palatine tonsils, 649, *649, 650, 734*
Palliative care, 910
Pallidotomy, 391
Palmaris longus muscle, *309, 310,* 311, *311,* 311t, **335**
Palmar region, *24,* 25
Palpitation, 601
Pancreas, 12, *15,* **35,** *645, 646, 667, 674, 774,* **954, 955**
   endocrine functions of, 96–98, 468, *469,* 471t, 478, *497, 498,* 498t
   exocrine functions of, 664–66, *664, 666*
   regulation of pancreatic secretions, 665, *666*
   structure of, 496, *497,* 664–65, *664*
Pancreatic amylase, 665, 677t
Pancreatic duct, *497,* 664, 665, 673
Pancreatic islets, 496, *497*
   hormones of, 496–98
   islet cell transplantation, 499
Pancreatic juice, 664–65
Pancreatic lipase, 665, 677t, 705
Pancreatitis
   acute, 665
   hereditary, 38
Pancytopenia, 537
Pannus, 272
Pantothenic acid, 710, 713t
Papillae, tongue, *433,* 649
Papillary muscles, 546, *547, 548*
Pap smear test, 855
Para-aminohippuric acid clearance test, 795
Paracrine secretions, 468
Paralysis, 327, 355, 409, 752
   flaccid, 381
   spastic, 381
Parasomnia, 394t
Parasympathetic division, 408–10, 413–14, *413,* 415t
Parasympathetic innervation
   of alimentary canal, 648
   of heart, 560, *561*
   of salivary glands, 653
   of small intestine, 680
Parathyroidectomy, 504
Parathyroid gland, *469,* 471t, 489–90, *489, 490,* 502
   disorders of, 490t
   hyperplasia of, 468
Parathyroid hormone, 190, 194, *196, 470,* 471t, 475, 489–90, *490, 491,* 707, 816, *816,* 885, 885t
Paravertebral ganglia, *406,* 410–11, *411, 427*
Paresis, 327
Parietal bone, *183,* 196t, *197,* 199, *200, 201,* 201t, *204, 206, 210,* **238, 239, 248, 249,** *255,* **332**

Parietal cells, 659, *660*, 661t
Parietal lobe, 384, *384, 385, 386,*
387, 387t
Parietal pericardium, 14, *15,* 543,
544, 545
Parietal peritoneum, 14, *15, 667,*
674, 774, 796
Parietal pleura, 14, *15,* 744, *744,*
746–47
Parkinson disease, 56, 69, 153,
312, 357t, 390–91, *391,* 417
Parotid duct, 654
Parotid gland, *401,* 653–54, *654,*
654t, *655,* **956**
Partially complete proteins, 700
Partial pressure, 754
Partial thromboplastin time,
528, 972
Parturition. *See* Childbirth
Passive aging, 912
Passive immunity
artificially acquired, 629, 630t
naturally acquired, 629, 630t
Patella, *183,* 196t, 197, *198,* 227,
228, 231t, *258,* 269, *322,*
335, 961, 962
dislocation of, 227
Patellar ligament, 228, 269, *316,*
320, *322,* **335,** 373, *374,* **962**
Patellar region, *24,* 25
Patellar tendon, 320
Patent ductus arteriosus, 908
Pathogen, 616
Pathology, 25
Patient-controlled analgesia, 428
Pattern baldness, 166, 176, *933*
PCP, 360
Peanut allergy, 608, 630
Pectin, 695
Pectineus muscle, 315, *316,*
318, 319t
Pectoral girdle, 197, 218–19, *219*
muscles that move, 302–4, *303,*
304, 304t
Pectoralis major muscle, **30, 31,**
297, *297,* 304, 305, *306,*
306t, **334,** 861, *862,* **953, 957**
Pectoralis minor muscle, **31,**
302–4, *304,* 304t, 747,
748, 862
Pectoral nerve, 304t, 306t
lateral, 406
medial, 406
Pectoral region, *24,* 25
Pedal region, *24,* 25
Pediatrics, 25
Pedicels, 778, *779*
Pedicle, 210, *212*
Pediculosis, 178
Pedigree, 926, *926, 927*
Pellagra, 710, 713t
Pelvic cavity, *13,* 14, 214
sagittal section of, **951**
transverse section of, **955**
Pelvic curvature, 210, *211*
Pelvic diaphragm, 314–15, *314*
Pelvic girdle, 196t, 197, 223t,
224–27, *225, 226,* 231t

Pelvic inflammatory disease, 868
Pelvic lymph nodes, *613,* 614
Pelvic outlet, muscles of, 314–15,
314, 315t
Pelvic region, *24,* 25
Pelvic sacral foramen, *36*
Pelvis, 197, 224
arteries to, 588–91
female, *226,* 227
greater, 224, 227
lesser, 224, 227
male, *226,* 227
veins from, 595–98, *596*
Penetrance, 928
complete, 928
incomplete, 928
Penetrating keratoplasty, 451
Penicillamine, 435
Penicillin, 106, 619, 630
Penile plethysmography, 830
Penile urethra, 798, *798*
Penis, **32, 33,** 798, *798,* 830, *831,*
841–42, *841,* 844t, *897*
Pepsin, 617, *643,* 659–60, 661t,
677t, 678
Pepsinogen, 659, 661, 661t
Peptic ulcers, 660
Peptidase, 677, 677t
Peptide bond, 52, *52,* 105, *105,*
121, 122, 123t
Peptide hormone, 471–72, 472t
Perchloroethylene, 843
Percutaneous transluminal
coronary angioplasty, 600
Perfluorocarbons, 510
Perforating canal, 184, *185*
Perforins, 618
Pericardial artery, 582, 585t
Pericardial cavity, *13,* 14, *15,* **35,**
544, 545, **953**
Pericardial membranes, *14*
Pericardial sac, **32**
Pericardiectomy, 601
Pericarditis, 543
Pericardium, 543, **954, 966**
fibrous, 543, *544, 545*
parietal, 543, *544, 545*
visceral, 543, *544*
Perichondrium, 146
Perilymph, 437, *438,* 439
Perimetrium, *853,* 854, *854*
Perimysium, 278, *280*
Perinatology, 914
Perineal artery, *589*
Perineal region, *24,* 25
Perineum, *855*
Perineurium, 396, *398*
Periodic Table, 969
Periodontal disease, 686, 724
Periodontal ligament, 255, *256,*
652, *652*
Periosteum, 183, *183, 185,* 186–87
Peripheral, definition of, 21
Peripheral blood smear, *511*
Peripheral chemoreceptors, 754
Peripheral nerves
injury to, 348
structure of, 396, *398*

Peripheral nervous system, 338,
339, 367, 396–417, 396t. *See*
also Autonomic nervous
system; Somatic nervous
system
cranial nerves. *See* Cranial
nerves
spinal nerves. *See* Spinal
nerves
Peripheral process, 422
Peripheral proteins, 66–67
Peripheral resistance, *573,* 574–75,
576, 577–78
Peristalsis, 294, *647,* 648, 656, *657,*
662, 680, 683, 686, 795
Peristaltic rush, 680
Peritoneal cavity, 14, *15, 667,* 674
Peritoneal membranes, 14
Peritoneum, 853
parietal, *667, 674, 774, 796*
visceral, 646, *667, 674*
Peritonitis, 681
Peritubular capillaries, *777, 779,*
781, *783, 787,* 788
Permanent wave, 54
Pernicious anemia, 516, 516t, *633,*
711, 713t
Peroneal artery, 591
Peroneal nerve
common, 407, *409*
tibial, 407
Peroneal vein, *596*
Peroneus brevis muscle, **962**
Peroneus longus muscle, **961, 962**
Peroxisomes, 74–76, 80t
abnormal, 75
Perpendicular plate, of ethmoid
bone, **238, 246**
Persistent vegetative state, 366, 394
Personality, 366
Perspiration, 11, 813
insensible, 812
sensible, 812
Pesco-vegetarian, 694t
Peyer's patches, 613
p21 gene, 21
pH, of blood, 47, 525–26,
822–23, *822*
Phagocytes, 85
Phagocytosis, 85–87, *86,* 89t, 141,
618–19, 619t
Phalanges
of fingers, 196t, 197, *198, 220,*
223, *223,* 223t, **335**
of toes, 196t, 197, *198, 228,*
229–31, *230,* 231t, **336**
Pharmacology, 25
Pharyngeal tonsils, 649–50,
650, 734
Pharyngitis, 688
Pharynx, *645, 646,* 655, *656, 733,*
736, 744t, **953**
Phencyclidine, 906t
Phenobarbital, 906t
Phenotype, 925
effects of gender on, 933, *933*
Phentermine, 705
Phenylalanine, 126

Phenylalaninediamine, 127t
Phenylethanolamine N-
methyltransferase, 492
Phenylketonuria, 126, *126,*
127, 882t
Phenylpyruvic acid, 972
Pheochromocytoma, 468, 504
Phlebitis, 571, 601
Phlebotomy, 601
Phosphatase, serum, 971
Phosphate
blood, 487, 489, 526, 784t
in cells, 48, *49*
electrolyte balance, 813,
816–17, *816*
in glomerular filtrate, 784t
intracellular fluid, *810*
tubular reabsorption of, 788
urine, 489, 784t
Phosphate buffer system, 819, 820t
Phosphaturic factor, 489
Phosphodiesterase, 460, 475
Phospholipids, 50, *51,* 52t, 698, *698*
membrane, 65–66, *66*
Phosphoproteins, 817
Phosphoric acid, 817, 821
Phosphorus, 712, 714, 716t
absorption of, 707
in human body, 39t
serum, 971
Phosphorylation, 108
Photoreceptors, 423, 453, 458
Phrenic artery, 583, *584,* 585t, 588
Phrenic nerve, 405, *407, 409,* 746,
755, **965**
Phrenic vein, 595
pH scale, 46–47, *47,* 47t
Phylloquinone, 708
Physical stress, 500
Physical training, 289
Physiological steady state, 81
Physiologic dead space, 750
Physiology, definition of, 3–4
Pia mater, 367–68, *367, 368, 370*
Pica, 718
Pig(s), as organ donors, 632
Pigmented epithelium, *454,*
458, *459*
Pilocarpine, 416
Pimple, 169
Pineal cells, 498
Pineal gland, 390, *392,* 469,
498–500
Pinkeye, 448
Pinocytosis, 85, *86,* 87, 89t, 788
Pisiform, *223*
Pitching, 222
Pitch of voice, 738
Pituicytes, 478, 483
Pituitary dwarfism, 190
Pituitary gigantism, 190
Pituitary gland, 202, *392,* 469,
478–85, *478*
anterior, 471t, 477–83, *478, 480,*
481, 483, 485t
fetal, 478
hormonal control of female
reproductive system, 857

hormonal control of male reproductive system, 845, *845*
intermediate lobe of, 478
posterior, 382t, 390, 471t, 478, *478*, 483–84, *483*, 485t
Pituitary stalk, *478*
Pivot joint, *259*, 260, 260t, 263t
Placenta, 500, 857, 882, 884, 886–87, *891, 893, 899, 900*
expulsion of, 903–4, *903*
Placental estrogens, 884–85, 887–88, 904
Placental lactogen, 884, 885t, 904
Placental membrane, 887
Placental progesterone, 884, 887, 904
Plantar artery
lateral, *590*, 591
medial, *590*, 591
Plantar flexion, 261, *261*
Plantar reflex, 377
Plantar region, *24*, 25
Plantar vein
lateral, *596*
medial, *596*
Plaque, atherosclerotic, 571, *571*
Plasma, *149*, 511, *511*, 523–26, *524, 809*
composition of, 784t
Plasma cells, *512*, 621, 623, *626, 631*
Plasma exchange, 284
Plasma proteins, 523–24, 524t, 568, 701, 789
Plasma thromboplastin, 530t
Plasmin, 530
Plasminogen, 530
Plasminogen activator, 530
Platelet(s), 148, *149*, 194, 510–11, *511, 512*, 521–23, 523t, *524*, 529
production of, *512*
Platelet count, 523, 531, 971
Platelet-derived growth factor, 529
Platelet plug, 526
formation of, 526, *527*, 527t
Platelet transfusion, 531
Platysma muscle, 298–99, *298*, 299t
Pleiotropy, 928
Pleura
parietal, 744, *744*, 746–47
visceral, 743, *744*
Pleural cavity, *13*, 14, *15*, **35**, 744, *744*
Pleural fluid, 746
Pleural membranes, 14, 746–47
Pleurisy, 766
Plexus, 403
Plicae circulares, 676, *676*
Pluripotent cells, 94
Pneumoconiosis, 766
Pneumocystis carinii pneumonia, 635, 759
Pneumonia, 758–59, 822, 913–14, 913t
Pneumotaxic area, 753–54, *753, 754*
Pneumothorax, 748, 766
Podiatry, 25

Podocytes, 778, *779, 784, 807*
Poikilocytosis, 537
Poison ivy, 164
Polar body, 836
first, 848–49, *849, 852, 880*
second, 849, *849, 878*
Polar body biopsy, 850
Polar molecule, 44, *44*
Poliomyelitis, 283, 752
Polyclonal response, 625
Polycystic kidney disease, 773
Polydactyly, 224, *224*
Polydipsia, 817
Polygenic traits, 929, *929*
Polymerase chain reaction, 118, 267
principles of reaction, 124
uses of, 124t
Polymorphism, 125
Polymorphonuclear leukocytes, 518
Polyp, colonic, 644, 685
Polypeptide, 105
Polyphagia, 504, 725
Polyploidy, 934
Polysaccharides, 48, *49*, 104, *104*, 695
Polyunsaturated fatty acids, 50
Polyuria, 802, 817
Pons, 381, *382*, 382t, *383, 393, 393*, 396t, **968**
Popliteal artery, *570, 590, 590*, 592
Popliteal region, *24*, 25
Popliteal vein, 595, *596, 597*
Popliteus muscle, *270*
Porphyria variegata, 517
Porphyrin ring, 517
Porta hepatis, 665
Portal system, 479
Positive chemotaxis, 520
Positive feedback, 12, 528, 902, *903*
Positive nitrogen balance, 701
Positron emission tomography, 56–57, *57*
of brain, 430
Postcoital test, 861, 861t
Posterior, definition of, 21
Posterior branch, of spinal nerves, 403, *406*
Posterior cavity, of eye, *450*, 455
Posterior chamber, of eye, *451*, 452, *453*
Posterior column, 372
Posterior fontanel, 208, *210*
Posterior horn, 372, *411*, 427–28
Posterior regions, *24*
Postganglionic fibers, 410, *410*
Postnatal period, 876, 905–11, 911t
Postpartum, 914
Post-polio syndrome, 283
Postsynaptic neuron, 354, *355*
Posture, 291, 445–46
Potassium, 714, 716t
in action potential, 351–52, *352, 353*
blood, 493, 526, 971
in cells, 48, *49*
cerebrospinal fluid, 370
dietary, 714

electrolyte balance, 813–14, *816*
in heart action, 561
in human body, 39t, 194
imbalances in, 818
intracellular fluid, *810*
in membrane potential, 348–50, *349, 351*
in nerve impulse conduction, 353–55, 353t
in plasma, glomerular filtrate, and urine, 784t
tubular reabsorption of, 788
tubular secretion of, 790, *791*
Potassium channels, 68
abnormal, 68
drugs that affect, 68, 68t
Poxvirus, 158
P-Q interval, 560
Precapillary sphincter, 567, 577
Precipitation, 627, 628t
Pregame meal, 722
Preganglionic fibers, 410, *410*
Pregnancy, 854, 876–905
alcohol consumption in, 894, *895*
birth defects, 894–95, *895*
blood volume in, 516
changes in woman's body during, 886
cigarette smoking in, 894
diet in, 876, 886, 894–96
ectopic, 883
embryonic stage of, 886–92, *886, 887*t, *888, 889*, 898t
fetal blood and circulation, 898–901, *899, 900*, 900t, *901*
fetal stage of, 892–901, *893, 896, 897, 898*, 898t
folic acid requirement in, 382, 711
hormonal changes during, 884–85, *885*, 885t
iron requirement in, 717
multiple, 876, *876*, 883
Rh incompatibility in, 536, *537*
tubal, 883
Pregnancy test, 624, 885
Pregnanediol, 861
Preimplantation genetic diagnosis, 882–83, 882t, *883*
Preload, 576
Premature beat, 562–63
Premature infant, 747, 760, 896
Prenatal period, 876–901
Prenatal surgery, 896
Prenatal tests, 937, 937t, *938*
Prepatellar bursa, 258, *258*, *270*, 271
Prepuce, *831*, 841–42, *841*
Presbycusis, 462
Presbyopia, 457, 462
Pressure, requirements of organisms, 9, 9t
Pressure receptors, 424, *425*, 429t
Pressure senses, 424, *425*
Pressure ulcer, 159, 176
Presynaptic neuron, 354, *355, 356*
Primary follicle, 850

Primary germ layers, 886–87, *886, 888*
Primary immune response, 628, *628*
Primary malnutrition, 721
Primary sex organs
female, 846
male, 830
Primary teeth, 650, 651t
Prime mover, 296
Primordial follicle, 848, *851, 852*
maturation of, 850, *850, 851*, 858
P-R interval, 560
Proaccelerin, 530t
Procaine, 355
Process, bony, 183, 199t
Procoagulants, 527
Procollagen, 143
Products, 44–45
Proerythroblasts, *512*
Proflavine, 127t
Progenitor cells, 94, *95, 96, 97*
Progeria, 913, *913*
Progesterone, 500, 857–59, 858t, *860*, 884–85, *885*, 885t, 901
placental, 884, 887, 904
Progranulocytes, *512*
Projection (sensory), 424
Prokaryotic cells, 62
Prolactin, 471t, 478, 481–82, *481*, 485t, 904–5, 906t
Prolactin release-inhibiting hormone, 471t, 479, *481*, 482
Prolactin-releasing factor, 471t, *481*, 482
Promoter, 118
Pronation, 261, *262*
Pronator quadratus muscle, 308–9, 308t
Pronator teres muscle, 308–10, 308t, *309, 311*
Propelling movements, of alimentary canal, 646–48, *647*
Prophase
meiosis I, 835
meiosis II, 836
mitotic, 90–92, *90, 91*, 92t
Proprioceptive senses, 424
Proprioceptors, 395, 399, 423, 428–29, *429*, 756
Proscar, 840
Prostacyclin, 530, 531t
Prostaglandins, *470*, 471, 476–77, 630, 685, 698, 839, 876, 901
Prostate cancer, 189, 840
Prostatectomy, 868
Prostate gland, *796*, 797–98, *797, 798, 831*, 839–40, *839*, 844t, **955**
enlarged, 800, 840, 840t
Prostate specific antigen, 840
Prostatic fluid, 840
Prostatic urethra, 797–98, *798*
Prostatitis, 868
Protease, 107
Protease inhibitors, 635

Protein(s), 699
  blood, 971
  calories per gram, 701t, 702
  in cells, 52–54, 55t
  complete, 700
  conformation of, *53, 54*
  denaturation of, 54
  dietary requirement for,
      701, 701t
  dietary sources of, 701t
  digestion of, 678, *678,* 699,
      699, 700
  incomplete, 700
  kwashiorkor, 723, *723*
  membrane, 66, *67,* 67t
  metabolism of, 105, *114,* 699,
      699, 700, 701t
  misfolding of, 54
  nitrogen balance, 700–701
  partially complete, 700
  primary structure of, 52–54, *53*
  quaternary structure of, *53, 54*
  secondary structure of, *53, 54*
  secretion of, 72
  structure of, *58*
  synthesis of, 105, *105,* 119–22,
      120, 121
  tertiary structure of, *53, 54*
  three-dimensional shape of, 52
  urine, 794
Protein buffer system,
    819–20, 820t
Protein-deficient diet, 524
Protein hormones, 471–72, 472t
Protein kinase, 472–75
Proteinuria, 725, 776, 789,
    800, 824
Proteome, *923*
Proteomics, 921–22
Prothrombin, 527, 528t, *529,*
    530t, 708
Prothrombin activator, 527–28,
    528t, *529,* 531
Prothrombin time, 528, 971
Protons, 38–39, *39,* 39t
Protraction, *262,* 263
Provitamins, 705
Proximal, definition of, 21
Proximal convoluted tubule, *777,*
    778, 778, 779, 781, 783, 784,
    788–89, 789, 794t
Pruritus, 178
Pseudohermaphroditism, 846
Pseudostratified columnar epithe-
    lium, 135, *136,* 141t
Psoas major muscle, 315,
    316, 319t
Psoriasis, 160
Psychiatry, 25
Psychological stress, 416, 500, 578,
    755–56
Pteroylmonoglutamic acid, 711
Pterygium, 463
Pterygoid muscle
  lateral, *298,* 299–300, 300t
  medial, *298,* 299, 300t
Pterygoid plate
  lateral, *203*
  medial, *203*

Puberty, 169, 190, 499
  female, 848, 862, 909
  male, 833, 846, 909
Pubic arch, 224, *225, 226*
Pubic tubercle, *225*
Pubis, 224
Pubofemoral ligament,
    267–69, *268*
Pudendal artery
  deep external, 589
  internal, 589, *589*
  superficial, 589, *590*
Pudendal cleft, 855
Pudendal nerve, 407, *409*
Pudendal vein, 596
Pulled hamstring, 320
Pulmonary artery, **34,** *543,* 546,
    549, *550, 551, 741, 900,* **950**
  left, *547, 552,* 580, *581, 583*
  lobar branches of, 580
  right, *547, 552,* 580, *581, 583*
Pulmonary circuit, *543,* 580–81,
    581, 899
Pulmonary edema, 581
  high altitude, 760
Pulmonary embolism, 530, 571
Pulmonary plexus, *412, 413*
Pulmonary trunk, **33, 34,** *544,* 546,
    547, 551, 552, 580, *581, 583,*
    899, *900*
Pulmonary valve, 546, *547, 548,*
    549t, *550, 551,* 553–54, *554*
Pulmonary vein, **34,** *543,* 546, *547,*
    550, 551, 552, 581, *581, 583,*
    741, 900
Pulmonic sound, 554
Pulp, of tooth, 652, *652*
Pulse, 570–71, *570*
Pulse pressure, 573
Punnett square, 926, *926, 927*
Pupil, *450,* 452–53
  size of, 453, *453*
Pupillary reflex, 453, *453*
Purified protein derivative, 632
Purine, 254
Purkinje fibers, 545, 557–58, *558*
Purpura, 537
Pus, 520
Pustule, 178
Putamen, 388, *389*
P wave, 560
Pyelolithotomy, 802
Pyelonephritis, 802
Pyelotomy, 802
Pyloric canal, *658,* 659
Pyloric sphincter, *658,* 659,
    659, 664
Pyloric stenosis, hypertrophic, 659
Pylorospasm, 688
Pyorrhea, 688
Pyramidal cells, 385
Pyramidal nerve tracts, 378
Pyridostigmine bromide, 286
Pyridoxal, 710–11, *710*
Pyridoxamine, 710–11, *710*
Pyridoxine, 710–11, *710*
Pyruvic acid, 108, *109, 110,* 111,
    112, 113
Pyuria, 802

Q

QRS complex, *559,* 560, *560*
Quadriceps femoris muscles, 271,
    320, 320t, 373, *374*
Quantitative computed
    tomography, 195
Quinlan, Karen Ann, 366
Quinolone, 123
Q wave, 560, *560*

R

Radial artery, *311,* 572, 588, *592*
Radial collateral ligament,
    266, *266*
Radial muscles, of iris, 453, *453*
Radial nerve, 308t, *311,* 311t, 406,
    407, 408
Radial notch, 221, *222*
Radial recurrent artery, *587*
Radial tuberosity, 193, 221, *222*
Radial vein, 591, *593, 597*
Radiation (heat loss), 170
Radioactive isotopes, 40, 906t
  detection of, *41*
  physiological studies using,
      41, *41*
Radioactive waste, 45
Radiographs, 56
Radiology, 25
Radiotherapy, for breast
    cancer, 865
Radioulnar joint
  distal, 263t
  proximal, 263t
Radium, 194
Radius, 193, 196t, 197, *198, 220,*
    221, *222, 223,* 223t, *266,*
    311, **334**
Raloxifene, 865
Ramus, 199t
  of mandible, 206
Ransome, Joseph, 422
Rapamune, 632
Rash, 161t
Rate-limiting enzyme, 115
Reactants, 44–45
Reasoning, 387
Receptive relaxation, 648
Receptor(s), 10, *10,* 66, 67t, 373,
    374, 375, 375t. *See also*
    specific types of receptors
Receptor cells, of retina, 454
Receptor-mediated endocytosis,
    85, 87, *87,* 89t, 525, 679
Recessive inheritance, 925–27
  X-linked, 932
Reciprocal innervation, 375
Recommended Daily
    Allowance, 719
Recruitment, of motor units, 291
Rectal artery
  inferior, *589*
  middle, 589
Rectal vein, 596, 682
Rectouterine pouch, *847,* 854
Rectum, **34, 36,** *645, 646,* 680, *681,*
    682, *682,* 796, *847,* **951, 955**

Rectus abdominis muscle, **30, 31,**
    297, 304, 312, 312t, *313,*
    314, 748, **951, 954, 955, 957**
Rectus femoris muscle, **31, 33, 34,**
    35, *297, 316, 317, 319,* 320,
    320t, **335, 955, 960**
Red blood cell(s), *63,* 148, *149,*
    194, *509,* 510–11, *511, 513,*
    516, 523t, *524,* 717, *717*
  blood doping, 475
  blood groups, 531–36
  characteristics of, 511–13
  destruction of, 516–17, 517t, 669
  hemolysis of, 83, *84*
  life cycle of, *515*
  life span of, 514
  production of, *512,* 514–15, *514*
    control of, 514, 515t
    factors affecting, 515–16,
        515, 516t
Red blood cell counts, 513–14, 971
Red blood cell distribution
    width, 971
Red blood cell substitute, 510
Red fibers, 292
Red-green color blindness, 932
Red hair, 166
Red marrow, 147–48, 183, 194,
    217, 514, *515,* 620
Red nucleus, 393
Red pulp, of spleen, 614, *616*
Reduction, 975
Reeve, Christopher, 380
Refecoxib, 272
Referred pain, 426, *426*
Reflex
  monosynaptic, 373
  spinal, 372–73
  uses of, 377
Reflex arc, 373, 375t
Reflex behavior, 373–76, *374, 375*
Reflex center, 373
Refraction, 455–58, *455*
  disorders of, 457, *457*
Refractory period
  absolute, 353
  of neurons, 353
  relative, 353
  in skeletal muscles, 290
Regeneration, of neurons, 348,
    349, 381
Relative position, 21
Relaxin, 884–85, 885t
REM sleep, 394
REM-sleep behavior disorder, 394t
Renal artery, 583, *584,* 585t, *592,*
    773, 774, 775, 777, 779,
    783, 900
Renal baroreceptors, 786
Renal capsule, 773, *775*
Renal clearance, 794–95
Renal column, 773, *775*
Renal corpuscle, *771, 775,*
    776, 794t
Renal cortex, 773, *775, 777,*
    779, 780
Renal failure. *See* Kidney failure
Renal fascia, 772, *774*
Renal fat, 772

Renal medulla, 773, *775, 777, 779, 780*
Renal papillae, 773, *775, 778*
Renal pelvis, *775, 796*
Renal plasma threshold, 788
Renal pyramid, 773, *775*
Renal sinus, 772
Renal tubule, *775, 776, 778, 787,* 794t
Renal vein, *597,* 773, *775, 776, 777, 779, 783*
Renin, 493, 501, 578, *578,* 786, *787*
Renin-angiotensin system, 493, 786, 789, 815
Repair enzymes, 125, 127, 163
Replication, 123, *123,* 125
Reproduction, 19
    as characteristic of life, 8t
Reproductive cycle, female, 858–59, 859t, *860*
Reproductive glands, 500
Reproductive system, 8t
    female, 19, *19. See also specific components*
        accessory sex organs, 846
        age-related changes in, 911t
        external reproductive organs, 855–56, *855*
        functions of organs of, 856
        hormonal control of, 857–60, *857*
        internal accessory organs, 852–55
        organs of, 846–55, *847*
        prenatal development of, 894, *897*
    male, 19, *19. See also specific components*
        accessory sex organs, 830, 838–41
        age-related changes in, 911t
        external reproductive organs, 841–44
        functions of organs of, 844t
        hormonal control of, 845–46, *845*
        organs of, 830–38, *831*
        prenatal development of, 894, *897*
    system interconnections of, 869
Reserpine, 358t, 416
Residual volume, *749,* 750, 750t
Respiration, 732–33
    artificial, 743
    cellular. *See* Cellular respiration
    as characteristic of life, 8t
    external, 732
    internal, 732
    venous blood flow and, 577–79
Respiratory acidosis, 822, *822*
Respiratory alkalosis, 756, 822–23, *823*
Respiratory bronchioles, 740–42, *740, 741*
Respiratory capacities, 749–50, *749, 750*
Respiratory center, 393, 753–54, *754,* 821, *821*

Respiratory cycle, 749
Respiratory distress syndrome, 747
Respiratory group
    dorsal, 753, *753, 754*
    ventral, 753, *753, 754*
Respiratory membrane, 757–58, *757, 758*
Respiratory system, 8t, 18, *18, 732*
    age-related changes in, 764, 911t
    disorders that decrease ventilation, 752
    disorders that impair gas exchange, 759
    effect of cigarette smoking on, 735, *735*
    organs of, 733–44, *733,* 744t
    system interconnections of, 765
Respiratory tubes
    functions of, *742, 742*
    structure of, 741–42
Respiratory volumes, 749–50, *749, 750*
Response, 10, *10, 11*
Responsiveness, as characteristic of life, 8t
Restenosis, 600
Resting potential, 350, *351*
Resting tidal volume, 749
Restless-leg syndrome, 394
Rete cutaneum, 164
Rete testis, 832
Reticular connective tissue, 145, *145,* 150t
Reticular fibers, 144, 144t
Reticular formation, 393–94, *393,* 427
Reticulocytes, *512,* 514
Reticulospinal tract, 385
    anterior, *377, 378,* 379t
    lateral, *377, 378,* 379t
    medial, *377, 378,* 379t
Retina, *450, 452,* 453–56, *454, 455,* 455t, *456, 458*
    culture of epithelial cells from, 458
    detached, 458, 462
    neurons of, 454, *454*
Retinal, 460, 706, *706*
Retinitis pigmentosa, 463, 882t
Retinoblastoma, 463, 882t
Retinol, 706, *706*
Retraction, *262,* 263
Retrolental fibroplasia, 760
Retroperitoneum, 674, 772
Reuptake, of neurotransmitters, 358
Reverse transcriptase, 635
Reversible reaction, 44–45
R group, 52, *52*
Rhabdomyolysis, 722
Rh blood group, 536, *537*
    incompatibilities in pregnancy, 536, *537*
Rheumatic fever, 633t
Rheumatoid arthritis, 167, *253,* 258, 272, 273t, 477, 633t
    juvenile, 273t
Rhinitis, 766
Rh-negative blood, 536
Rhodopsin, 458, *459,* 460, 706

RhoGAM, 536
Rhomboideus muscle, *297,* 302, *303,* 304t, **958**
Rhythmicity
    of cardiac muscle, 295
    of smooth muscle, 294
Rhythm method, 862–63, 863t
Rib(s), *15, 36,* 196t, 197, *198,* 216, *217, 218, 219,* 255, *256, 743,* **953, 954, 955, 963**
    false, 216, *217*
    floating, 216, *217*
    true, 216, *217*
Ribavirin, 670
Riboflavin, 683, 709, 713t
    deficiency of, 709
Ribonuclease, 122
Ribose, 54, *55,* 115, 119t, 696
Ribosomal proteins, 122
Ribosomal RNA, 119t, 122
Ribosomes, *64,* 71, 79, 80t, *120, 121,* 122, 123t
Ribozyme, 120, 122
Rickets, 190, 707, *707,* 708t, 724
Rifampin, 123
Right lower quadrant, of abdomen, 23, *23*
Right upper quadrant, of abdomen, 23, *23*
Rigor mortis, 287
Rinne test, 443
Rivinus's duct, 654
RNA (ribonucleic acid), 54, 118–19
    structure of, *55,* 115, 118–19, *119,* 119t
    synthesis of, 118, *119*
RNA polymerase, 118–19, 123, 123t
Rods, 422, *454, 458, 459,* 460, 706
Romanov family, 118
Root
    of penis, *831,* 842
    of tongue, 649, *649*
    of tooth, 651–52, *652*
Root canal, 652
Rosacea, 161t
Roseola infantum, 161t
Rotation, 261, *262*
Rotator cuff, 264
Rough endoplasmic reticulum, *64,* 71, *71*
Round ligament, *34, 667,* 853, 854
Round window, 435, 436, 438, 439, *439, 441*
Rubella, 629, 894, 908
Rubrospinal tract, *377, 378,* 379t, 385
Rugae, 658, *659*
Rule of nines, 175, *175*
Ruptured disc, 216
R wave, 560, *560*

S

Saccharin, 696
Saccule, *438,* 443, *444, 445*
Sacral artery
    lateral, *589*
    middle, 583, *584,* 585t, *589*
Sacral canal, 214, *215, 225*

Sacral crest, 213, *215*
Sacral foramen
    anterior, **36,** 214, *215*
    dorsal, 214, *215*
Sacral hiatus, 214, *215,* 225
Sacral nerves, 402, *404*
Sacral plexus, *409*
Sacral promontory, 214, *215, 225, 226*
Sacral region, *24,* 25
Sacroiliac joint, 214, 224, *225,* 263t
Sacrum, **36,** 196t, 197, *198,* 209, *211,* 213–14, 214t, *215,* 225, 227t, **334, 951**
Saddle joint, *259,* 260, 260t, 263t
Sagittal section, 21, *22*
Sagittal sinus, 902
Sagittal suture, 199, *210,* **239**
St. John's Wort, 720
St. Martin, Alexis, 660
Salicylic acid, 169, 169t
Saliva, 432, 653
Salivary amylase, 677t, 678
Salivary glands, 139t, *645,* 652–54, *654,* 654t, 655t
    major glands, 653–54, 653t
    secretions of, 653–54
Salpingectomy, 868
Salpingitis, 868
Salt(s), 45–46, 46t, 48
Saltatory conduction, 354, *354*
Salt receptors, 433–34
Salty taste, 432–33
Saphenous nerve, *409*
Saphenous vein, 596
    great, **30, 31, 34,** *319, 325,* 595–96, *596, 597,* 600
    small, *325,* 595, *596, 597*
Sarcolemma, 279, *280, 282*
Sarcomere, 279, *281,* 285
Sarcoplasm, 279
Sarcoplasmic reticulum, 279–82, *280, 281, 282*
Sartorius muscle, **30, 31, 32, 34, 35,** *297, 316, 317, 318, 319,* 320, 320t, *324,* **335, 955, 960**
Satiety, 695, 705
Saturated fatty acids, 50, *50*
Scab, 173, *174*
Scabies, 178
Scala tympani, 437, *438, 439,* 440
Scala vestibuli, 437, *438, 439,* 440
Scalp, **949, 952**
Scanning electron microscope, 65, *65*
Scanning probe microscope, 65
Scaphoid, *223*
Scapula, 197, *198,* 218, *219,* 223t, 264, *264, 265,* **333, 743**
Scapular nerve, dorsal, 304t, 406
Scar, 143, 151, 173, *174*
Scarlet fever, 161t
Schizophrenia, 357t, 416
Schwann cells, 342–43, *342, 398*
Sciatica, 409
Sciatic nerve, *319,* 319t, 407, *407*
Sciatic notch
    greater, 224, *226*
    lesser, *226*

Scintillation counter, *41*, 600
Sclera, 450, *450, 451, 452, 454*, 455t
Scleroderma, 273t, 633–34, *634*
Sclerosing agent, 571
Sclerosis, 343
Scoliosis, 216
Scrotum, **33**, 830, *831*, 832, 841, 844t, *897*, **951**
Scurvy, 712, 713t
Seasonal affective disorder, 500
Sebaceous glands, 139t, *140, 160, 165, 166–67, 167*, 168t, 169
Seborrhea, 178
Sebum, 167
Secondary follicle, 850
Secondary hypertension, 578
Secondary immune response, 628, *628*
Secondary malnutrition, 721
Secondary pacemaker, 563
Secondary sex characteristics
    female, 857–59
    male, 846
Secondary teeth, 650–51, *651*, 651t
Second-class lever, *191*, 192–94
Second meiotic division, 833–36
Second messenger, 472, 474–75, *475*
Secretin, 665, *666*, 672t
Secretion, of proteins, 72
Secretory vesicles, *64*, 72
Segmentation, 680
Selectin, 70, *70*
Selective estrogen receptor modulators, 865
Selectively permeable membrane, 65
Selective serotonin reuptake inhibitors, 358t
Selenium, 718, 719t
"Self," 67, 619
Sella turcica, 202, *203, 204, 206*, **247, 249, 250**, 478, *478*
Semen, 840–41, 843–44
Semen analysis, 843t
Semicircular canals, *435*, 437, *438*, 444–45, *444, 446*
    lateral canal, 444–46
    posterior canal, 444–46
    superior canal, 444–46
Semilunar valve, 549
Semimembranosus muscle, *297, 319*, 320, 320t, *324*, **961, 962**
Seminal vesicles, *831*, 839
Seminiferous tubules, 832, *833, 834*, 842, 844t
Semispinalis capitis muscle, *301*, 302, 302t
Semitendinosus muscle, *297, 318, 319*, 320, 320t, *324*, **334, 961**
Semivegetarian, 694t
Senescence, 910, 911t
Sensations, 423–24
Sensible perspiration, 812
Sensorineural deafness, 443
Sensory adaptation, 424
Sensory area, of cerebrum, 386, *386*

Sensory impulse, 423
Sensory nerve, 397
Sensory neuron, 345, *345*, 346t, 373, *374*, 375, *375*, 375t
Sensory receptors, 164, 338–39, *339*, 422–24. *See also specific types of receptors*
Septicemia, 537
Septum primum, 899, 908
Serosa
    of alimentary canal, 646, 646t
    of large intestine, *681*, 682, *682*
    of small intestine, 675, *676*
Serotonin, 340, 356, 357t, 358t, 428, 523
Serous cells, 140
    of salivary glands, 652–53
    of urinary bladder, 797, *797*
Serous fluid, 140, 152
Serous membranes, 14, 152
Serratus anterior muscle, **30, 31**, *297*, 302–3, *304*, 304t, *306*, **334, 957**
Serum glutamic pyruvic transaminase, 971
Sesamoid bone, 182, *183*
Set point, 10, *10, 11*, 12
Severe combined immune deficiency, 629, *942*
Sex chromosome(s), 924–25, *925*, 931–32, *931*
Sex chromosome aneuploidy, 934
Sex determination, 931, *931*
Sex hormones
    adrenal, 495, 495t
    female, 857–58, *857*
    male, 845–46
Sex-influenced traits, 933, *933*
Sex-limited traits, 933
Sexual intercourse, 840, 842–44, *844, 877, 877*
Sexually transmitted disease, 867–68, *867*, 867t
Shigatoxin, 772, *772*
Shingles, 161t
Shin splints, 327
Shivering, 11, 170
Short bone, 182, *183*
Short stature, 190, 482
Short-term memory, 388
Shoulder joint, 263t, 264–65, *264, 265*
    arteries to, 587–88
    dislocation of, 265
    veins from, 591, *592*
Sickle cell disease, 125, *125*, 184, 513, 516t, 882t, 927, *942*
Sight. *See* Vision
Sigmoid colon, *681*, 682, **951, 967**
Sigmoidoscopy, 685t
Signal transduction, 65
Sildenafil, 830
Simethicone, 663
Simple columnar epithelium, 133, *135*, 141t
Simple cuboidal epithelium, 133, *134*, 141t
Simple gland, 138, *139*, 139t

Simple squamous epithelium, 133, *134*, 141t
Single-cell organisms, 9, *9*
Single covalent bond, 43
Single nucleotide polymorphism, 125
Sinoatrial node, 554–57, *558, 560, 561*
Sinus, 199, 199t, 736, *736*, 744t
Sinusitis, 766
Sinusoids, 566
    hepatic, 593, 595, 666
Sinus rhythm, 563, 601
Sirolimus, 632
Sister chromatids, *91*
Skeletal muscle(s). *See also specific muscles*
    action of, 296, *296*
    age-related changes in, 325
    atrophy of, 293
    connective tissue coverings of, 278–79, *280*
    contraction of, 282–90, 287t
        all-or-none response, 290
        concentric, 292, *292*
        cross-bridge cycling, 285–86, *285, 286*
        eccentric, 292, *292*
        energy sources for, 287–88, *287, 288*
        excitation contraction coupling, 284
        force generated by, 291
        heat production, 289
        isometric, 292, *292*
        isotonic, 292, *292*
        latent period of, 290
        oxygen supply for, 288
        recording of, 290, *290*
        refractory period of, 290
        sliding filament theory of, 284–86, *285, 286*
        stimulus for, 283–84, *283*
        sustained, 291
        tetanic, 291, *291*, 355, 490
        types of, 292, *292*
        venous blood flow and, 577, *579*
    effect of exercise on, 289, 293, 325
    fast fibers, 292–93
    fibers of, 279–82, *280, 281, 282*
    hypertrophy of, 293
    insertion of, 261, 296, *296*
    interactions of muscles, 296
    lactic acid accumulation in, 288–89, *288, 289*
    in lymph flow, 612
    major muscles, 297–325
    neuromuscular junction. *See* Neuromuscular junction
    origin of, 261, 296, *296*
    oxygen debt, 288–89, *289*
    recruitment of motor units, 291
    relaxation of, *285*, 286–87, 287t
    slow fibers, 292–93
    structure of, 278–82

summation, 290–91, *291*
    threshold stimulus of, 290
    use and disuse of, 293
Skeletal muscle tissue, 148, *149*, 152t, 295t
Skeletal system, 8t, 16, *16*. *See also* Bone(s); Joint(s)
    age-related changes in, 216, 231–32, *231*, 911t
    system interconnections of, 233
    terminology used for, 199t
Skeleton
    appendicular, 197
    axial, 196t, 197
    divisions of, 197, *198*
    fetal, *186*
    male versus female, 227t
Skeleton of the heart, *548*, 549
Skin, 153, *157*, 158–76
    accessory organs of, 165–68
    age-related changes in, 175–76, *176*
    as barrier to infection, 617
    bioengineered, 153
    color of, 160, 172, 930–31, *930*
    dermis, 158, *160, 161*, 164, *165*
    epidermis, 158–62, *160, 161, 165*
    glands of, 167–68, *167, 168*, 168t
    regulation of body temperature, 169–71, *171*
    site of gene therapy, 939, *942*
    subcutaneous layer, 145, 159, *160*, 164
Skin cancer, 127, 162–63, *163*
Skin graft, 93, 173
Skin substitute, 173
Skull, *184*, 196t, 197, *198*, 199–208, 263t, **952**
    anterior view of, *200*
    anterolateral view of, **239**
        Australopithecus, 182, *182*
    base of, **243, 244**
    of child, **252**
    coronal section of, *207*
    of elderly, **252**
    fetal, **251**
    frontal view of, **238**
    infantile, 208, *210*
    inferior view of, *202*, **242**
    lateral view of, *201*
    male versus female, 227t
    passageways through, 209t
    posterolateral view of, **239**
    radiograph of, *208*
    sagittal section of, *206*, **248**
Sleep, 391, 393, 417
    REM, 394, 394t
    slow wave, 394
Sleep apnea, 753
Sleep deprivation, 394
Sleep disorders, 394t
Sleep paralysis, 394
Sleep/wake cycles, 499
Slightly movable joint, 254
Slit pore, 778, *779*
Slow-twitch fibers, 292–93

Slow wave sleep, 394
Small intestine, *15*, **32, 33**, *645, 646, 647, 667, 673–80, 674, 774,* **948, 950, 951, 954, 966, 967**
    absorption in, 677–80, *678, 679,* 680t
    cellular turnover in, 676
    movements of, 680
    parts of, 673–74, *673, 674*
    secretions of, 676–77
    structure of intestinal wall, 675–76, *675, 676*
Smallpox, 158, *158*, 629
Smell, 392, 400, 430–32, *431*
    disorders of, 434t, 435
"Smile surgery," 278
Smith, Michelle, 475
Smoker's cough, 735
Smoking. *See* Cigarette smoking
Smooth endoplasmic reticulum, *64*, 71, *71, 73*
Smooth muscle(s), 293–94
    contraction of, 294, 484
    multiunit, 293
    rhythmicity of, 294
    visceral, 293–94
Smooth muscle cells, 62, *63*
Smooth muscle fibers, 293–94
Smooth muscle tissue, 150, *151*, 152t, 295t
Sneezing, 393, 751t
Snoring, 753
Sodium, 715, 716t
    in action potential, 351–52, *352, 353*
    blood, 493, *494*, 526, 972
    in cells, 48, *49*
    cerebrospinal fluid, 370
    dietary, 578, 715
    electrolyte balance, 813, 815, *816*
    excretion of, 790t
    extracellular fluid, 810, *810*
    in human body, 39t, 194
    imbalances in, 818
    in membrane potential, 348–50, *349, 351*
    in nerve impulse conduction, 353–54, 353t
    in plasma, glomerular filtrate, and urine, 784t
    tubular reabsorption of, 788–91, *790*, 790t, *792*
Sodium balance, 786–87, *787*
Sodium channels, 68
    abnormal, 68
    drugs that affect, 68, 68t
Sodium chloride, *42*, 43, 45–46, *46*, 715
Sodium hydroxide, 46
Sodium nitrite, 127t
Sodium-potassium pump, 85, 348–50, *349, 351*, 472, 789, 810–11
Soft palate, 649, *649, 650*, 652t, 656, *657, 733*
Soleus muscle, *297*, 321, 321t, *322*, 323, *323, 324, 325*, **336, 961, 962**

Somatic afferent fibers
    general, 399
    special, 399
Somatic cell(s), 921
Somatic cell nuclear transfer, 98, *99*
Somatic efferent fibers
    general, 399
    special, 399
Somatic mutation, 865
Somatic nervous system, 339, *339*, 396, 396t
Somatic receptors, 429t
Somatic senses, 421–63
    age-related changes in, 462
Somatostatin, 471t, 479–80, *481*, 496–98, 498t, 660, 672t
    intestinal, 662, 672t
Somatotropes, 480
Sour receptors, 433
Sour taste, 432–33
Space adaptation syndrome, 446
Space medicine, 574, *574*, 751
Spastic paralysis, 381
Specialized connective tissue, 144
Special senses, 421–63
    age-related changes in, 462
Species resistance, 617, 619t
Specific gravity, of urine, 973
Speech, 401, 751t
Spermatic artery, 583
Spermatic cord, **32**, 832, **951**
Spermatids, *834, 835*, 836
Spermatocytes
    primary, 833, *834, 835*, 836
    secondary, *834, 835*, 836
Spermatogenesis, 833, 836
Spermatogenic cells, 833
Spermatogonia, 833, *834*, 836–37
Sperm bank, 878
Sperm cells, 77, *77*, 829, 830, 833, *834, 835*, 921
    abnormal, *830*
    formation of, 833–37, *834, 835, 836, 837*
    motility of, 841, 843
        toxic chemicals that affect, 838
    structure of, 837–38, *837, 838*
    transport of, 876–77, *877*
Spermicide, 863t, *866*
S phase, 90, *90*
Sphenoidal region, **243, 249, 250**
Sphenoidal sinus, *14*, 202, *204, 205*, 205t, **247, 248, 249,** *478, 650, 734, 736*, **949, 952**
Sphenoid bone, 196t, *200, 201*, 201t, 202, *202, 203, 204, 210*, **238, 239, 241, 242, 243, 244, 247, 248, 249, 250,** *478*
Sphenoid fontanel, 208, *210*
Sphenomandibularis muscle, 300
Sphenopalatine ganglion, *413*
Spherocytosis, 537
Sphincter muscle, 299
Sphincter urethrae muscle, 314–15, 315t
Sphygmomanometer, 572, *572*

Spina bifida, 382, 711, 892
Spinal artery, *586*
Spinal branch, of accessory nerve, 401
Spinal cord, *15*, 339, 367, 372–80, *383*, **948, 950, 953, 954, 955**
    anterior median fissure, 372
    ascending tracts of, 376–77, *377*, 379t
    cervical enlargement of, 372, *372*
    cross section of, *373*
    descending tracts of, 376, *377*, 378, 379t
    functions of, 372–79
    injury to, 98, 380–81, *380*, 381t, 800
        hemi-lesion, 379
        treatment of, 381
    lumbar enlargement of, 372, *372*
    posterior median sulcus of, 372
    structure of, 372, *372*
Spinalis capitis muscle, *301*, 302t
Spinalis cervicis muscle, *301*, 302t
Spinalis thoracis muscle, *301*, 302t
Spinal muscular atrophy, 882t
Spinal nerves, 302t, *339, 368*, 372, 396–97, 396t, 401–8
    anterior branch of, 403, *406*
    injury to, 409
    meningeal branch of, 403
    posterior branch of, 403, *406*
    visceral branch of, 403, *406*
Spinal reflex, 372–73
Spinal shock, 380
Spindle fibers, *91, 92*, 836
Spine of bone, 199t
    of scapula, 218
Spinocerebellar ataxia type 6, 340
Spinocerebellar tract
    anterior, *377, 378*, 379t
    posterior, *377, 378*, 379t
Spinothalamic tract, 376
    anterior, *377, 378*, 379t
    lateral, *377, 378*, *378*, 379t, 427
Spinous process, 211, *212, 214*
Spiral fracture, *192*
Spiral ganglion, *438*
Spiral lamina, 437, *438*
Spirometer, 750, *750*
Spleen, *15*, **33, 34, 35**, 614, *615, 616*, 617t, *774*, **955**
Splenectomy, 637
Splenic artery, *584, 595, 616*
Splenic flexure, *681*
Splenic vein, 593, *594*
Splenitis, 637
Splenius capitis muscle, 300–302, *301*, 302t, **956**
Splenotomy, 637
Spondyloarthropathy, 273t
Spondylolisthesis, 213
Spondylolysis, 212–13
Spongy bone, 183–84, *183, 184, 185*, 186–89, *193*
Spontaneous (pathologic) fracture, 192

Sprain, 272
Squamosal suture, *200*, 201, *201, 206*, **239, 241**
Squamous cell carcinoma, *163*
SRY gene, 931–32, *931*
Stanozolol, 474
Stapedius muscle, 436–37, *436*
Stapedius tendon, *436*
Stapes, 196t, *435*, 436, *436, 441*
Starch, 48, 50, 695
Starvation, 524, 701, 703, 721–24
Static equilibrium, 443–44
Statins, 87, 598
Stem cells, 94, *95, 96, 97*, 390, 510, 556, *620*, 881, 887, 939–40
    therapeutic, 98, *99*
Stereocilia, 439
Stereoscopic vision, 460, *461*
Sternal angle, 216, *217*
Sternal notch, *217*, **334**
Sternal puncture, 217
Sternal region, *24*, 25
Sternocleidomastoid muscle, **30, 31**, *297, 298*, 300, 302t, *304*, **332, 334**, 747, *748*, **956, 957**
Sternocostal joint, 263t
Sternum, *15*, **32**, 196t, 197, *198*, 216, *217, 219*, **334**, *544, 743*, **949, 950, 953, 954**
Steroid, 50–51, *51*, 52t
Steroid hormones, 470–72, *470*, 472t, *473*, 476
    use by athletes, 474
Stethoscope, 554
Stickler syndrome, 143t
Stomach, *15*, **32, 33, 34**, 645, 646, 657, 658–64, 667, *674, 774*, **948, 950**
    absorption in, 662
    body region of, *658, 659, 659*
    cardiac region of, 658–59, *658*
    fundic region of, *658, 659, 659*
    mixing and emptying actions of, 662–64, *662, 663*
    parts of, 658–59, *658*
    pyloric region of, *658, 659*
    secretions of, 659–60, *660*, 661t
        cephalic phase of, 661, 662t
        gastric phase of, 661, 662t
        intestinal phase of, 661, 662t
        regulation of, 660–62, *661*
Stomach stapling, 705
Stomatitis, 688
Stop codon, 121
Strabismus, 448
Strain, muscle, 282
Strangulated hernia, 832
Stratified columnar epithelium, 137, *137*, 141t
Stratified cuboidal epithelium, 137, *137*, 141t
Stratified squamous epithelium, 135, *136*, 141t, *159*
Stratum basale, 159, *160, 161*, 161t
Stratum corneum, 159, *160, 161*, 161t, 162
Stratum granulosum, 159, *161*, 161t
Stratum lucidum, 159, *161*, 161t

Stratum spinosum, 159, *161*, 161t
Strength training, 325
Streptococcal infection, 273t, 780
Streptococcus mutans, 653
Streptokinase, 530
Streptomycin, 123
Stress, 500–502
  physical, 500
  psychological. *See*
    Psychological stress
  responses to, 500–502, *501*, 501t
Stress incontinence, 799
Stressor, 500
Stress test, 580, 600
Stretch receptors, 423, 428–29,
  429, 756
Stretch reflex, 429
Striations, 148, *277*, 279
Stringer, Korey, 808
Stroke, 389, 417, 461, 530, 578
  risk factors for, 578t
Stroke volume, 572, *573, 576*, 580
Strontium, 194
Structural formula, 43, *43*
Strychnine, 434
Stuttering, 924
Styloid process, 201, *201, 202,*
  221–22, *222*, **334, 335**
Stylomastoid foramen, 209t, **242,**
  243, 244
Subacromial bursa, 265
Subarachnoid space, *367*, 368–69,
  368, 370
Subchondral plate, 257
Subclavian artery, **34**, *585, 586,*
  592, **963, 965**
  left, **35**, 582, *583*, 585t
  right, **36**, 582, *583*, 587
Subclavian vein, **32, 33**, 591,
  594, 597
  left, **34, 35**, 591, *610*
  right, 591, *593*
Subcoracoid bursa, 265
Subcutaneous fascia, 279
Subcutaneous fat, 145, 164
Subcutaneous injection, 164
Subcutaneous layer, 159, *160*, 164
Subdeltoid bursa, *264*
Subdural hematoma, 368
Subdural space, **952**
Subfertility, 861
Sublingual gland, 654, *654,*
  654t, *655*
Subluxation, 274
Submandibular duct, *654*
Submandibular ganglion, *413*
Submandibular gland, 654, *654,*
  654t, *655*
Submucosa
  of alimentary canal, 646, 646t
  of large intestine, 682, *682*
  of small intestine, 675, *676*
Submucosal plexus, 648
Submucous coat, of urinary
  bladder, 797, *797*
Subpatellar fat, *258*
Subscapular bursa, 265, *265*
Subscapularis muscle, **32**, 305,
  307, 307t, 308

Subscapular nerve, 306t, 406
Subserous fascia, 279
Substance P, 357–58, 357t
Substrate, 106
Succinic acid, *112*
Succinyl-CoA, *112*
Suckling, 12, 484, 905
Sucrase, 677, 677t
Sucrose, 48
Sudden cardiac arrest, 542, 580
Sudden infant death syndrome,
  357t, 753
Sugars, 48
Sugar substitute, 696
Sulcus, 382, *383, 384*
Sulfate
  blood, 526
  in cells, 48, *49*
  electrolyte balance, 813, 817
  intracellular fluid, *810*
  in plasma, glomerular filtrate,
    and urine, 784t
  tubular reabsorption of, 788
Sulfolipids, 715
Sulfur, 39t, 715, *715*, 716t
Sulfuric acid, 817, 821
Sumatriptan, 340
Summation, 290–91, *291*, 350
Sunburn, 162–63
Sun poisoning, 162
Suntan, 172
Superficial, definition of, 21
Superficial digital flexor, **335**
Superficial transversus perinei
  muscle, 314–15, *315*, 315t
Superior, definition of, 21
Superior canaliculi, 448, *448*
Superior colliculus, 392, *392*
Superior oblique muscle, 400, 448,
  449, 449t
Superior orbital fissure, **240,**
  247, 250
Superior rectus muscle, *447*, 448,
  449, 449t
Superior vena cava, **34, 35**, *544,*
  546, 547, 550, 552, 581, 583,
  591, *593, 594, 595, 597,*
  900, **964**
  compression of, 591
Superovulation, 850
Superoxide dismutase, 912
Supination, 261, *262*
Supinator muscle, 308, 308t, *309*
Suppression amblyopia, 450
Suprachiasmatic nucleus, 499
Supraorbital foramen, 199, *200,*
  209t, **238, 240, 241**
Supraorbital notch, *200*, **246, 332**
Suprapatellar bursa, 258, *258,*
  270, 271
Suprarenal artery, 583, *584*, 585t,
  592, 775
Suprarenal vein, 595, *775*
Suprascapular nerve, 306t, 406
Supraspinatus muscle, *303*, 305,
  305, 306t, 307
Supraspinous fossa, 218, *219*
Supratrochlear lymph nodes,
  613, 614

Surface area-to-volume
  relationship, 92–93
Surface tension, 747–48
Surfactant, 747, 759, 896, 906
  synthetic, 747
Surgery, prenatal, 896
Surgical methods of contraception,
  866, *867*
Surgical neck, of humerus,
  221, *221*
Suspensory ligament
  of eye, 450, 451–52, *451, 452*
  of mammary gland, 861–62
  of ovary, 848, *848*
Sustained contraction, 291
Sustentacular cells, 833, 845
Sutural bone, 197, *197*
Sutural ligament, 254
Suture, 196, *197*, 199t, 254–55,
  255, 260t, 263t
Suturing, 173
Swallowing, *198*, 393, 401, 437,
  649, 655–56, *657*, 737–38
S wave, 560, *560*
Sweat gland, *160*, 165, 167–68,
  168, 168t, 170
Sweat gland duct, *160*
Sweat gland pore, *157, 160,*
  167, *168*
Sweating, 170
Sweet receptors, 433
Sweet taste, 432–33
Swollen glands, 614
Sympathetic division, 408–10,
  412, 415t
  regulation of glomerular
    filtration rate, 786
Sympathetic innervation
  of alimentary canal, 648
  of blood vessels, 564
  of heart, 560, *561*
  of salivary glands, 653
  of small intestine, 680
Sympathetic tone, 414
Sympathetic trunk, 410, *411*
Symphysis, 256–57, *256,*
  260t, 263t
Symphysis pubis, **35, 36**, 224, *225,*
  226, 256, *256*, 263t, *796,*
  831, 847, **951**
Synapse, 338, 354–58, *355, 356*
Synapsis, 835
Synaptic cleft, 283, *283*, 342,
  355, 356
Synaptic knob, *341*, 342, 354,
  356, 357
Synaptic potential, 355–56
Synaptic transmission, 354–55, *355*
Synaptic vesicles, *283*, 354,
  356, 357
Synchondrosis, 255, *256*, 260t, 263t
Syncytium, 295
  functional, 554
Syndesmosis, 254, *254*, 260t, 263t
Synergists (muscles), 296
Synesthesia, 430
Synovectomy, 274
Synovial cavity, 258
Synovial fluid, 257–58, *257*

Synovial joint, 260, 260t, 263t. *See*
  also specific joints
  examples of, 264–71
  structure of, 257–58, *257, 258*
  types of, 259–60, *259*, 260t
Synovial membrane, 152, 257–58,
  257, 258, 267
Synovitis, acute, 271
Synthesis, 44
alpha-Synuclein, 391
Syphilis, 867t
Systemic circuit, *543*, 580–81
Systemic lupus erythematosus,
  273t, 633t
Systole, 551
Systolic pressure, 570, 573,
  573, 601

T

Tachycardia, 562, *562*
Tachypnea, 753, 766
Tail, of sperm, 837–38, *837*
Talus, 228–29, *230*
Tamoxifen, 865
Tanning, 162
Tanning booth, 163
Tardive dyskinesia, 357t, 390
Target cells, 17, 468–69, 472, *480*
Tarsal(s), *183*, 196t, 197, *198*, 228,
  229, *230*, 231t, **336**
Tarsal glands, *447*, 448
Tarsal region, *24*, 25
Tarsometatarsal joint, 263t
Tarsus, *230*
Taste, 432–34, *433*
  age-related changes in, 462
  disorders of, 434t, 435
Taste buds, 432, *433*
Taste cells, 432, *433*
Taste hairs, 432
Taste nerve pathways, 434
Taste pore, 432, *433*
Taste receptors, 432, *433*, 434
Taste sensations, 432–34
Tay-Sachs disease, 75, 348
3TC, 635
T cells, 500, *512*, 519, 614, 619–21,
  620, 621t, 628
  activation of, 621, *622*
  in AIDS, 521, 521t
  cellular immune response and,
    621, *622*
  cytotoxic, 620–21, 635
  helper, 62, 621, *622*, 623, *623,*
    626, 635
  memory, 621, 628
  site of gene therapy, 940
Tears, 448, 462, 617
Tectorial membrane, 439, *440, 441*
Teeth, *256*, 650–52, *650*, 652t. *See*
  also Dental caries
  dental care, 686
  impacted, 651
  loss of, 652
  primary, 650, 651t
  secondary, 650–51, *651*, 651t
  wisdom, 651
Telomere, 92

Telophase
  meiosis I, 836
  meiosis II, 836
  mitotic, *91*, 92, 92t
Temperature, 9
Temperature senses, 424
Temporal artery, *570*
  superficial, *585*, 587, *592*
Temporal bone, 196t, *197*, *200*,
    201, *201*, 201t, *202*, *204*,
    206, *210*, **238**, **239**, **241**, **242**,
    243, **246**, **332**, **436**
Temporalis muscle, *297*, *298*, 299,
    300t, **332**, **952**, **955**
Temporal lobe, 384, *384*, *385*,
    386–87, 387t, **952**
Temporal nerve, *401*
Temporal process, 206
Temporal region, **241**
Temporal vein, superficial, *597*
Temporomandibular joint, 194, 263t
Temporomandibular joint
    syndrome, 300
Tendinitis, 278
Tendon, 142, 183, 278, *279*, *280*
Teniae coli, *681*, 682
Tennis elbow, 272
Tenosynovitis, 278
Tenosynovium, 278
Tensor fasciae latae muscle, **31, 34,**
    35, *297*, 315, *316*, 317, *317*,
    319t, **960**
Tensor tympani muscle,
    436–37, *436*
Tentorium cerebelli, *367*, 367t, 384
Teratogens, 892, *893*, 894–95, *895*
Teratology, 914
Teres major muscle, **32**, *297*, *303*,
    305, *305*, *306*, 306t, 307,
    333, **958**, **959**
Teres minor muscle, *297*, *303*, 305,
    305, 306t, 308, **958**, **959**
Terminal bronchioles, 740,
    740, *741*
Terminally ill, 910
Termination signal, 119
Testes, **33**, *469*, 500, 830–38, *831*,
    841, 844t, **948**, **951**
  descent of, 830–32, *832*, 842
  structure of, 832–33, *833*, *834*
Testicular cancer, 832
Testosterone, 166, 471, 474, 500,
    832–33, 840, 845–46,
    858, 909
Tetanic contraction, 291, *291*,
    355, 490
Tetracycline, 435
Tetrahydrolipostatin, 705
Tetraploidy, 934
Thalamotomy, 390–91
Thalamus, 366, 382t, *389*, 390–91,
    392, *393*, 395, **949**, **952**, **968**
Thalassemia, 516t, 537, 882t
Thalidomide, *893*, 894
Thallium-201, 41, 600
Theca externa, 850
Theca interna, 850
Therapeutic liver repopulation, 669
Thermal cycler, 124

Thermoreceptors, 423–24, 429t
Thermostat, 10, *11*
Theta waves, 397, *397*
Thiamine, 683, 709, 713t, 715, *715*
  deficiency of, 709
Thick filaments, 279, *281*, *282*,
    285–86, *285*, *286*, 287t
Thigh
  anterior view of, **960**
  muscles that move, 315–18,
    316, *317*, *318*, *319*, 319t
  posterior view of, **961**
  surface anatomy of, **334**
Thin filaments, 279, *281*, *282*,
    285–86, *285*, *286*, 287t
"Thinning hair supplements," 166
Thiokinase, 697
Third-class lever, *191*, 192
Third ventricle, 368–69, *369*, *370*,
    382t, *392*, *478*, *480*, **952**
Thirst, 12, 812, 812t, 814
Thirst center, 812, 812t
Thomas, Chelsey, 278
Thoracic aorta, 582, 585t, *588*, **965**
Thoracic artery, internal, 588, *588*
Thoracic cage, 196t, 197, 216–17,
    217–18
Thoracic cavity, 12, *13*, **36**, **950**
Thoracic cavity lymph nodes,
    613, 614
Thoracic curvature, 210, *211*
Thoracic duct, 609, *610*, 678
Thoracic membranes, 14
Thoracic nerves, 402, *404*
Thoracic outlet syndrome, 409
Thoracic vein, internal, 591
Thoracic vertebrae, 196t, *211*, 212,
    212, *214*, 214t, *217*, *368*
Thoracic wall
  arteries to, 588
  veins from, 591–92, *592*, *594*
Thoracodorsal nerve, 306t, 406
Thorax
  anterior view of, **963, 964, 965**
  posterior view of, **959**
  surface anatomy of, **333**
  transverse section of, *15*,
    953, **954**
Thought processing, 387
Threshold potential, 350, *352*
Threshold stimulus, for skeletal
    muscles, 290
Thrombin, 527–28, 528t, *529*,
    530–31
Thrombocytopenia, 522, 531
Thrombophlebitis, 571, 601
Thromboplastin time, partial,
    528, 972
Thrombopoietin, 511, 531
Thrombosis, 530
Thrombus, 530, *541*, 550, 571, 578
Thymectomy, 504, 637
Thymine, 116, *117*, *119*, 119t, *979*
Thymitis, 637
Thymosins, 500, 502, 614
Thymus gland, 284, *469*, 500, 502,
    614, *615*, *616*, 617t, *620*,
    629, 634
Thyrocervical artery, *585*, 587

Thyroglobulin, 485, *486*, 487
Thyroid artery, *586*
  superior, *585*, 587
Thyroid cancer, 41, 468
Thyroid cartilage, **32, 33, 334**, 736,
    737, *743*
Thyroidectomy, 504
Thyroid follicular cells, 486–87
Thyroid gland, **31, 32, 33**, *469*, 471t,
    485–89, *489*, 502, 718, **963**
  disorders of, 487t
  iodine-131 scan of, 41, *41*
  structure of, 485, *486*
Thyroid hormones, 190, 476,
    485–88, *486*, 487t, *488*, 718
Thyroiditis, 504
Thyroid nodules, 502
Thyroid-stimulating hormone,
    471t, 475, 478, 480, *481*,
    482, *483*, 485t, *488*, *972*
Thyrotropes, 480
Thyrotropin-releasing hormone,
    471t, 481, 482, *483*
Thyroxine, 471t, 485–87, *486*,
    487t, 702, *972*
Tibia, 196t, 197, *198*, 228, *228*, *229*,
    230, 231t, *258*, 269, *270*,
    336, **961**
Tibial artery
  anterior, 325, 590, *590*, *592*
  posterior, 325, *570*, 590–91,
    590, *592*
Tibial collateral ligament, 269, *270*
Tibialis anterior muscle, *297*, 321,
    321t, *322*, *323*, 325, **335**,
    336, **961**, **962**
Tibialis posterior muscle, 321,
    321t, *324*, 325, *325*, **336**
Tibial nerve, 320t, 321t, *325*, 409
Tibial tuberosity, 228, *229*, **335**
Tibial vein
  anterior, 595, *596*, *597*
  posterior, 595, *596*, *597*
Tibiofibular articulation, 254, *254*
Tidal volume, 749, *749*, 750t
Tight junction, 67, 69, *69*, 70t
Tinnitus, 462–63
Tissue, 5, *5*, 5t, 132, 132t. *See also*
    specific tissues
  age-related changes in, 19
  cell division in, 151
Tissue engineering, 132, *132*, 153
Tissue fluid, 608, 810
  formation of, 611
Tissue plasminogen activator,
    530, 600
Tissue rejection reaction, 632
Tissue repair, 151
Tissue thromboplastin, 527, 528t,
    529, 530t
Titin, 279, *281*
Tongue, *400*, 433–34, *433*, 646, 649,
    649, 650, 652t, 654, 656, *656*,
    657, 734, **948**, **949**, **953**
Tonometry, 462–63
Tonsil(s), 613
  lingual, 649, *650*, 734
  palatine, 649, *649*, *650*, 734
  pharyngeal, 649–50, *650*, 734

Tonsillectomy, 649
Tonsillitis, 649
Tools, Bob, 556
Tooth. *See* Teeth
Torso, surface anatomy of, **334**
Torticollis, 327
Total body water, *809*
Total lung capacity, *749*, 750, 750t
Totipotent cells, 94, *99*
Touch, 424, *425*, 429t
Touch receptors, *160*, 164, 424,
    425, 429t
Toxemia of pregnancy, 914
Toxicity testing, *in vitro*, 153
Toxicogenomics, 920
Toxicology, 25
Toxin, bacterial, 104
Trabeculae, 183–84, *185*
Trabecular bone, 232
Trace elements, 38, 39t,
    717–19, 719t
Trachea, **32, 34, 35**, *650*, 731, *733*,
    734, *737*, 738–39, *738*, *739*,
    743, 744t, **948**, **949**, **950**, **963**
  obstruction of, 739
Tracheostomy, 739, *739*
Tracheotomy, 630, 766
Trachoma, 463
Transaminases, *972*
Transcellular fluid, *809*, *809*
Transcription, 119, *120*, 123t
Transcription factors, 122, 564
Transcutaneous electrical nerve
    stimulation, 300, 428
Transcytosis, 88, *89*, 89t
Transdermal patch, 165
Transducer, 6
Transducin, 460
Transferrin, 517
Transfer RNA, 119t, *120*, 120–22,
    121, 123t
Transforming growth factor
    beta, 786
Transfusion, 510, 531–36, 573
  intrauterine, 914
  of mismatched blood, 533
Transfusion reaction, 630
Transient ischemic attack,
    389, 578
Transitional epithelium, 137,
    138, 141t
Translation, 119–22, *120*, *121*, 123t
Translocation, 934
Transmissible spongiform
    encephalopathy, 54
Transmission electron microscope,
    65, *65*
Transplantation, 632, 632t. *See
    also specific organs*
Transport vesicles, 72
Transverse colon, *674*, *681*, *681*,
    950, **955**, **967**
Transverse fissure, 382, *383*, *384*
Transverse foramen, 212, *214*
Transverse fracture, *192*
Transverse process, of vertebrae,
    211, *212*
Transverse section, 22, *23*
Transverse tubules, 282, *282*, 287t

Transversus abdominis muscle, **31, 36**, *304*, 312, 312t, *313*, 314, 748
Trapezium, *223*
Trapezius muscle, **30**, *297*, 302, *303, 304*, 304t, *307*, **332, 333, 334, 958, 959**
Trapezoid bone, *223*
Traumatic fracture, 192
Tremor, 312
Trepanation, 182
Treponema pallidum, 867t
Tretinoin, 169, 169t
Triad, 282, *282*
Triangle of auscultation, 304
Triceps brachii muscle, 193, *194, 222, 297, 305, 306*, 308, 308t, *310*, **333, 957, 958, 959, 960**
Triceps-jerk reflex, 377
Trichosiderin, 166
Tricuspid valve, 546, *547, 548, 549*, 549t, *550, 551*, 554, *554*
Tricyclic antidepressants, 358t
Trigeminal nerve (V), 300t, *399, 400*, **400**, 403t
    mandibular division of, 400, *400*
    maxillary division of, 400, *400*
    ophthalmic division of, 400, *400*
Trigeminal neuralgia, 400
Trigger point injections, 428
Trigger zone, of axon, 344, *344*, 350
Triglycerides, 50, *51, 52*, 697, 972
    blood, 525
    digestion of, 697–98, *697, 698*
    metabolism of, 697–98, *697, 698*
Trigone, 796, *797, 798*
Triiodothyronine, 471t, 485–87, *486*, 487t, 972
Trimester, 914
Triple covalent bond, 43
Triploid, 934
Triplo-X, 934
Triquetum, 222, *223*
Tris(2,3-dibromopropyl phosphate), 127t
Trisomy, 934, *935*
Trisomy 13, 934, 934t
Trisomy 18, 934, 934t
Trisomy 21. *See* Down syndrome
Tristearin, 50
Trochanter, 199t
Trochlea, 221, *266*
Trochlear nerve (IV), 399–400, *399*, 403t, 449t, *478*
Trochlear notch, 222, *222*
Trophoblast, 882–84, *884*, 887
Tropic hormones, 477
Tropomyosin, 279, *281*, 282, 284, 287t
Troponin, 279, *281*, 282, 284, 287t
True ribs, 216, *217*
True vocal cords, 737, *737, 738*
Trunk
    anterior view of, **957**
    posterior view of, **958**
    sagittal section of, **948**

Trypsin, 38, 665, 677t
Trypsinogen, 665
Tryptophan, 358t, 710, 799
Tubal ligation, 863t, 866, *867*
Tubal pregnancy, 883
Tubercle (bone), 119t, 216
Tubercle (tuberculosis), 759
Tuberculin skin test, 632
Tuberculosis, 759, *759*, 913–14, 913t
Tuberosity, 199t
Tubular gland, 138, *139*
Tubular necrosis, 785
Tubular reabsorption, 782, 787–90, *787, 789*
Tubular secretion, 782, *787*, 790, *791*
Tubulin, 77, 92
Tumor, 93
Tumor necrosis factor, 620t, 625
Tumor suppressor gene, 94, *95*, 773
Tunic, 563
Tunica albuginea, 832, 841, *841*, 848
Tunica externa, 563–64, *565*
Tunica interna, *565*
Tunica media, 563, *565*
Tunic fibrosa, 772
Turner syndrome, 934
T wave, *559*, 560
Twins, 876
    conjoined, 902, *902*
    fraternal, 882
    identical, 882
Twitch, 290, *290, 291*
Tympanic membrane, 434, *435, 436*, 437, *441*
Tympanic reflex, 437
Tympanoplasty, 463
Tympanostomy tubes, 437
Tyrannosaurus rex, 254
Tyrosine, 492
Tyrosine hydroxylase, 492
Tyrosinemia, *127*

U

Ulcer, 167, 178
Ulcerative colitis, 633t, 684
Ulna, 196t, 197, *198, 220*, 221–22, *222, 223*, 223t, *266, 311*, **333, 335**
Ulnar artery, *311*, 587, *588, 592*
Ulnar collateral ligament, 266, *266*
Ulnar nerve, *311*, 311t, 406, *407, 408*
Ulnar notch, 222
Ulnar recurrent artery, *587*
Ulnar vein, 591, *593, 597*
Ultrasonography, 6–7, *6*, 914
    estimating date of conception, 879
    prenatal, 937t, *939*
Ultratrace elements, 38
Ultraviolet radiation, mutagenicity of, 127
Umami, 432
Umbilical artery, *891*, 892, 899–900, *900*, 900t, 907, *908*, 908t

Umbilical cord, *891*, 892, *893*, 900, *900, 903*
Umbilical cord blood, 98, 510, 522
Umbilical ligament, *900*, 907
Umbilical region, *24*, 25
Umbilical vein, *891*, 892, *899, 900*, 900t, 907, 908t
Umbilicus, **30, 334, 957**
Undernutrition, 721
Underweight, 703
Unicellular gland, 138, 139t
Unipolar neurons, 344, *344*, 346t, 422
Universal donor, 535
Universal recipient, 535
Unmyelinated axon, 342, 343, *344*
Unsaturated fatty acids, 50, *50*
Upper limb
    arteries to, 587–88
    bones of, 196t, 197, 220–23, *220, 221, 222, 223*, 223t
    muscles that move, 304–12, *305, 306*, 306t, *307*, 308t, *309, 310*, 311t, *312*
    posterior view of, **959, 960**
    surface anatomy of, **333, 334, 335**
    veins from, 591, *592*
Upper motor neuron syndrome, 381
Upper respiratory tract, 733
Uracil, 118, *119*, 119t, *979*
Urbanization, 3
Urea, 526, 669
    excretion of, 793–94
    in plasma, glomerular filtrate, and urine, 784t, 785
    tubular reabsorption of, 787
    urine, 973
Urea clearance, 973
Uremia, 802, 824
Uremic acidosis, 823
Ureter, **34, 35**, *773, 775, 777*, 795–96, *795, 796, 797, 798*, 800, *831, 847*
Ureteritis, 795, 802
Urethra, **36**, *773, 796*, 797–98, *797, 798*, 800, *831, 841, 847*, **955**
    membranous, 798, *798*
    penile, 798, *798*
    prostatic, 797–98, *798*
Urethral glands, 797
Urethral groove, *897*
Urethral orifice, external, 797–98, *798, 841, 841*
Urethral sphincter
    external, *797*, 798–99
    internal, 797, *797*, 799
Urethritis, 802
Uric acid, 254, 526
    excretion of, 793–94
    in plasma, glomerular filtrate, and urine, 784t
    serum, 972
    tubular reabsorption of, 787–88
    urine, 973
Urinalysis, 799, 972–73

Urinary bladder, **32, 33, 34, 35**, *674, 773*, 795–97, *796, 797*, 800, *831, 847*, **948, 951**
    automatic bladder, 800
    cancer of, 796
Urinary excretion of odoriferous component of asparagus, 799, 924
Urinary system, 8t, 18, *18*, 772, *773*. See also specific organs
    age-related changes in, 800
    system interconnections of, 801
Urination. *See* Micturition
Urine, 782
    composition of, 784t, 794
    concentration of, 790–93, *792, 793*
    elimination of, 795–99
    formation of, 782–95
    24-hour volume, 786, *786*
    pH of, 972
    volume of, 788, 790–94, *792, 793*, 798–99, 812
    water output, 812
Urobilinogen, 973
Urogenital diaphragm, 314–15, *314*
Urogenital fold, *897*
Urokinase, 530
Urology, 25
Urticaria, 178
Uterine artery, 589
Uterine tube, **34**, *847*, 852–54, *852, 853*, 856t, 877, 883
Uterine vein, 596
Uterus, **34, 35**, *847, 848, 853*, 854, *854*, 856t. *See also* Pregnancy
    contractions of, 12, 294, 477, 484–85, 901–2
Utricle, *438*, 443, *444, 445*
Uveitis, 463
Uvula, 649, *649, 650, 734*

V

Vaccine, 629
    cancer, 163, 942
    edible, 629
    polio, 283
    smallpox, 158
Vagina, **36**, *847, 853*, 854, 856t, 877, *903*
Vaginal orifice, *847*, 854, *855, 897*
Vaginal process, 832, *832*
Vaginal vein, 596
Vaginitis, 868
Vagotomy, 417
Vagus nerve (X), *399*, 401, *402*, 403t, *413*, 414, 434, 560, 661, 755, **965**
Valerian root, 720
Valium, 358t, 360
Valvotomy, 601
Variable expressivity, 928
Variable regions, of antibodies, 626, *627*
Varicocele, 868
Varicose veins, 571
Vasa recta, 781, *783*, 793

Vasa vasorum, 564
Vascular endothelial growth
    factor, 564
Vascular tunic, 450
Vas deferens, **33**, *797, 831,* 832,
    833, 838–39, *839,* 844t
Vasectomy, 840, 863t, 866, *867*
Vasoconstriction, 11, 170, 393,
    414, 484, 564, *575, 576,*
    577–78, 578, 580, 785
Vasodilation, 11, 170, 393, 414,
    564, *575,* 577, 580, 785
Vasomotor center, 393, *576,* 577
Vasomotor fibers, 564
Vasomotor symptoms, of
    menopause, 859–60
Vasospasm, 526, 527t
Vastus intermedius muscle, **35**,
    316, 318, 320, 320t
Vastus lateralis muscle, **33, 34, 35,**
    297, 316, 317, 318, 319, 320,
    320t, *323,* **335, 955, 960, 961**
Vastus medialis muscle, **33, 34,** *297,*
    316, 319, 320, 320t, **335, 960**
Vegan, 694t
Vegetarian diet, 694, 694t
Veins, 569, 570t. *See also specific*
    veins
    blood volume within, *570*
    valves of, 569, *569,* 571
    wall of, *565*
Venipuncture, 591
Venoconstriction, 579
Venography, 601
Venous blood flow, 577–79, *579*
Venous pathways, characteristics
    of, 591
Venous pressure, 571
    increased, 814–15, 815t
Venous return, 580
Venous sinus, *593*
Venous system, 591–98, *597*
Ventilation, 732. *See also*
    Breathing
    alveolar, 751, 756t
    respiratory disorders that
        decrease, 752
Ventral cavity, 12, *13*
Ventral root, *368, 373,* 402–3,
    406, 411
Ventricles of brain, 347, 368–70,
    369, 370, 381, 382t, **949, 952**
Ventricles of heart, 545
    left, *15,* **34,** *543, 544,* 546, *547,*
        549, *550, 551, 552,* 899,
        900, **954, 964**
    right, *15,* **34,** *543, 544,* 546, *547,*
        549, *550, 551, 552,* 569,
        580, 899, *950,* **954, 964**
Ventricular diastole, 551, 553,
    554, 555
Ventricular fibrillation, 542,
    562, *562*
Ventricular syncytium, 554
Ventricular systole, 551, 553–54,
    554, 555
Ventricular tachycardia, 542
Venules, *567,* 569, *570,* 570t
Verbalizing, 387

Vermiform appendix, **33, 34,**
    680–81, *681*
Vermis, 395
Vernix caseosa, 895
Vertebrae, *15, 183,* 197, 209–11
    cervical, 196t, *211,* 212, *213,*
        214, 214t, 380
    fracture of, 212, 381
    lumbar, 196t, *211,* 212–13,
        214, 214t
    thoracic, 196t, *211,* 212, *212,*
        214, 214t, *217, 368*
Vertebral arch, 211
Vertebral artery, *585,* 586, *586,*
    592, **953**
Vertebral body, **954, 955**
Vertebral canal, 209, 367, 372
Vertebral cavity, 12, *13*
Vertebral column, 196t, 197, *198,*
    209–15, **948**
    curvatures of, 210, *211*
    disorders of, 216
    muscles that move, 300–302,
        301, 302t
Vertebral foramen, 210, *214*
Vertebral region, *24, 25*
Vertebral spine, **333**
Vertebral vein, *593*
Vertebra prominens, 212
Vertebrocostal joint, 263t
Vertex position, 897, *898*
Vertigo, 463
Very low-density lipoproteins,
    525, 526t
Vesical artery
    inferior, 589, *589*
    superior, 589, *589,* 907
Vesical vein, 596
Vesicles, 77, 80t
Vesicle trafficking, 72
Vestibular branch, of
    vestibulocochlear nerve, 401
Vestibular bulb, 856
Vestibular glands, 856, 856t
Vestibular membrane, *438,* 439,
    439, 440, 441
Vestibular nerve, *438*
Vestibule (enclosed by labia
    minora), 855–56, 856t
Vestibule (inner ear), 437, *438,* 443
Vestibule (mouth), 648, *649, 650*
Vestibulocochlear nerve (VIII), *399,*
    401, *435,* 439, *442, 444, 444*
    cochlear branch of, 401
    vestibular branch of, 401
Vetter, David, 629
Viagra, 830
Villi, intestinal. *See* Intestinal villi
Vioxx, 272
Virchow, Rudolph, 75, 94
Viscera, 12
Visceral afferent fibers
    general, 399
    special, 399
Visceral branch, of spinal nerves,
    403, *406*
Visceral efferent fibers,
    general, 399

Visceral pericardium, 14, *15,*
    543, *544*
Visceral peritoneum, 14, *15,* 646,
    667, 674
Visceral pleura, 14, *15,* 743, *744*
Visceral smooth muscle, 293–94
Visceroceptive senses, 424
Vision, 400, 447–61
    age-related changes in, 462
    binocular, 460
    color, 422, 460
    double, 450
    stereoscopic, 460, *461*
Vision screening programs, 450
Visual cortex, 391, 461
Visual nerve pathways, 461, *461*
Visual pigments, 458–60
Visual receptors, 458
Vital capacity, *749,* 750, 750t, 764
Vital signs, 2
Vitamin(s), 705, 722
    coenzymes and, 107
    fallacies and facts about, 706t
    fat-soluble, 698, 705–8, 708t
    water-soluble, 705, 709–12, 713t
Vitamin A, 190, 458, 460, 669, 671,
    686, 706–7, *706,* 708t, 939
    deficiency of, 190, 460, 707
Vitamin B₁. *See* Thiamine
Vitamin B₂. *See* Riboflavin
Vitamin B₅. *See* Pantothenic acid
Vitamin B₆, 599, 710, 713t
Vitamin B₁₂. *See* Cyanocobalamin
Vitamin B complex, 709–11, 713t
Vitamin C, 190, 516, 706, 712, *712,*
    713t, 788, 892
    deficiency of, 712
Vitamin D, 176, 190, 471, 489–90,
    490, 669, 671, 686, 707–8,
    707, 708t, 714, 773, 800
    deficiency of, 190, 355, **707,**
    707, 724, 817
Vitamin E, 671, 708, 708t, 865
Vitamin K, 527, 671, 683, 686,
    708–9, 713t
Vitreous body, 455
Vitreous humor, *450, 451, 453,* 455
Vocal cords, 736–38, *737, 738*
    false, 737, *737, 738*
    true, 737, *737, 738*
Volar arch artery
    deep, *587*
    superficial, *587*
Volatile acid, 820
Voluntary muscles, 148
Vomer bone, 196t, *200, 202,* 206,
    207, 208t, **238, 240, 242,**
    243, 244
Vomiting, 393, 663, 714–15, 814,
    823, 876
Vomiting center, 446, 663–64

W

Wakefulness, 391, 393
Wandering cells, 141
Warm receptors, 424
Warren, J. Robin, 660
Wart, 178

Washington, George, 739
Waste products, exchange in
    capillaries, 568
Water
    absorption in intestine, 679,
        680t, 683
    in cells, 47–48, 49t
    content of human body, 809, *809*
    excretion of, 790t, 791
    osmosis, 82–83, *83*
    polarity of, 44, *44, 46*
    requirements of athletes, 722
    requirements of organisms, 8, 9t
    structure of, *42, 43, 44, 58*
    tubular reabsorption of,
        788–90, *790,* 790t
Water balance, 786–87, *787,* 808,
    811–13, *811*
    disorders of, 814–15, *814, 815*
Water intake, 811, *811*
    regulation of, 812, 812t
Water intoxication, 814
Water of metabolism, 811, *811*
Water output, *811,* 812
    regulation of, 812, 813t
Water retention, 484
Water-soluble vitamins, 705,
    709–12, 713t
Weaning, 905
Weber test, 443
Weight. *See* Body weight
Weightlessness, 574, *574*
Weisz, Wendy, *936*
Werner syndrome, 913
West Nile virus, 283
Wharton's duct, 654
Wheezing, 752
Whiplash, 409
White blood cell(s), *63,* 148, *149,*
    194, 510–11, *511,* 523t, *524,*
    607, 618
    alterations in, 521t
    functions of, 70, *70,* 520, *521*
    production of, *512*
    types of, 518–20, *519, 520*
White blood cell count, 520–21,
    521t, 972
    differential, 972
White fibers, 292
White forelock, 166, 924
Whitehead, 169
White matter, 343, *367,* 372, *373,* **952**
White meat, 293
White pulp, of spleen, 614, *616*
White rami, 410, *411*
Widow's peak, 924, *924*
Wild type, 925
Wilson disease, 799
Wisdom teeth, 651
Withdrawal reflex, 375, *375, 376*
Withdrawal symptoms, in
    newborn, 892
World Trade Center disaster, air
    quality after September 11,
    2001, 732
Wound healing, 93, 173, *174*
Wrinkles, 19, 142, 162, 164
Wrist joint, 263t
    bones of, 222–23, *223*

X

X chromosome, 924–25, *925,* 931–32, *931*
 inactivation of, 932
Xenical, 705
Xenograft, 632, 632t
Xenotransplantation, 632
Xeroderma pigmentosum, 125, 163, *163*
Xerophthalmia, 707
Xerostomia, 686
Xiphoid process, 216–17, *217,* **334**

X-linked trait, 925, 932
 dominant, 932
 recessive, 932
X-ray fluoroscopy, 600
X-ray imaging, 56
XYY syndrome, 935

Y

Yawning, 437, 751, 751t
Y chromosome, 924–25, *925,* 931–32, *931*
Yellow cake, 45
Yellow marrow, 183, *183,* 194

Yellow rice, 707
Y-linked inheritance, 925, 932
Yolk sac, *891,* 892
Yucca Mountain, Nevada, 45

Z

Zinc, 39t, 686, 718, 719t
Z line, 279, *281,* 285
Zona blasting, 878
Zona fasciculata, *491,* 492–93, *492*
Zona glomerulosa, *491,* 492–93, *492*
Zona pellucida, 850, *852,* 877–78, *880*

Zona reticularis, *491,* 492, *492,* 495
Zygomatic arch, *202,* 206, **239, 241, 332, 956**
Zygomatic bone, 196t, *200, 201, 202, 202,* 206, *207,* 208t, *210,* **238, 239, 240, 241, 242**
Zygomatic nerve, *401*
Zygomatic process, 202, **246**
Zygomatic region, **241**
Zygomaticus muscle, *297,* 298–99, *298,* 299t
Zygote, 849, *849,* 879–80, *881*
Zygote intrafallopian transfer, 879
Zymogen granules, 665